GRAM-POSITIVE PATHOGENS

GRAM-POSITIVE PATHOGENS

SECOND EDITION

Edited by

VINCENT A. FISCHETTI, RICHARD P. NOVICK,
JOSEPH J. FERRETTI, DANIEL A. PORTNOY, and JULIAN I. ROOD

ASM
PRESS

Washington, D.C.

Library of Congress Cataloging-in-Publication Data

Gram-positive pathogens / edited by Vincent A. Fischetti ... [et al.].—2nd ed.
    p. ; cm.
  Includes bibliographical references and index.
  ISBN-13: 978-1-55581-343-7 (hardcover : alk. paper)
  ISBN-10: 1-55581-343-7 (hardcover : alk. paper) 1. Gram-positive bacterial infections. 2. Gram-
positive bacteria.
  [DNLM: 1. Gram-Positive Bacteria. 2. Bacterial Vaccines. 3. Drug Resistance, Microbial.
4. Gram-Positive Bacterial Infections. QW 142 G745 2006] I. Fischetti, Vincent A.
II. American Society for Microbiology.

  QR201.G76G73 2006
  614.5′79—dc22
  2005026022

Address editorial correspondence to ASM Press, 1752 N St., N.W., Washington, DC 20036-2904, U.S.A.

Send orders to: ASM Press, P.O. Box 605, Herndon, VA 20172, U.S.A.
Phone: 800-546-2416; 703-661-1593
Fax: 703-661-1501
E-mail: books@asmusa.org
Online: http://estore.asm.org

Cover illustrations (clockwise from top left):
Spore-forming bacteria: Morphology of *Clostridium botulinum* type A cells viewed by phase-contrast microscopy. The spore-bearing cells of other serotypes of *C. botulinum* and *C. tetani* also have characterized morphologies typically with swelling of the rod-shaped vegetative cell. See chapter 55.
Streptococci: Tantalum and tungsten surface replica of critical-point-dried group A streptococci, revealing the distribution of the surface molecules and cell division in parallel with the growing chain. Magnification, ×96,000. Courtesy of V. A. Fischetti.
Staphylococci: Localization of *atl* gene products on the cell surface of *S. aureus* during the division cycle as determined by scanning electron microscopy. See chapter 36.
Listeria: The surface of a macrophage infected with *Listeria monocytogenes*. The single listeria bacillus is being propelled by a polymerizing actin tail. Reproduced from the *Journal of Cell Biology* (**109:**1597–1608, 1989), with permission of the publisher.

Contents

Contributors / ix
Preface / xiii

SECTION I
THE GRAM-POSITIVE CELL WALL / 1
SECTION EDITOR: Vincent A. Fischetti

1 Ultrastructure of Gram-Positive Cell Walls / 3
TERRY J. BEVERIDGE AND VALÉRIO R. F. MATIAS

2 Surface Proteins on Gram-Positive Bacteria / 12
VINCENT A. FISCHETTI

SECTION II
THE STREPTOCOCCUS / 27
SECTION EDITORS: Vincent A. Fischetti and Joseph J. Ferretti

A. GROUP A STREPTOCOCCI

3 Intracellular Invasion by *Streptococcus pyogenes*: Invasins, Host Receptors, and Relevance to Human Disease / 29
BEINAN WANG, DAVID CUE, AND P. PATRICK CLEARY

4 Capsular Polysaccharide of Group A Streptococci / 37
MICHAEL R. WESSELS

5 Toxins and Superantigens of Group A Streptococci / 47
JOHN K. McCORMICK, MARNIE L. PETERSON, AND PATRICK M. SCHLIEVERT

6 Genetics of Group A Streptococci / 59
KYU HONG CHO AND MICHAEL CAPARON

7 Cross-Reactive Antigens of Group A Streptococci / 74
MADELEINE W. CUNNINGHAM

8 Extracellular Matrix Interactions with Gram-Positive Pathogens / 89
GURSHARAN S. CHHATWAL AND KLAUS T. PREISSNER

9 Streptococcus-Mediated Host Cell Signaling / 100
VIJAY PANCHOLI

10 Vaccine Approaches To Protect against Group A Streptococcal Pharyngitis / 113
VINCENT A. FISCHETTI

11 The Bacteriophages of Group A Streptococci / 123
W. MICHAEL McSHAN

12 Molecular Epidemiology, Ecology, and Evolution of Group A Streptococci / 143
DEBRA E. BESSEN AND SUSAN K. HOLLINGSHEAD

B. GROUP B STREPTOCOCCI

13 Pathogenic Mechanisms and Virulence Factors of Group B Streptococci / 152
VICTOR NIZET AND CRAIG E. RUBENS

14 Surface Structures of Group B Streptococci Important in Human Immunity / 169
LAWRENCE C. MADOFF, LAWRENCE C. PAOLETTI, AND DENNIS L. KASPER

15 Epidemiology of Group B Streptococcal Infections / 186
ANNE SCHUCHAT AND SHARON BALTER

C. GROUP C AND G STREPTOCOCCI

16 Genetics and Pathogenicity Factors of Group C and G Streptococci / 196
HORST MALKE

17 Pathogenicity Factors in Group C and G Streptococci / 213
GURSHARAN S. CHHATWAL, DAVID J. McMILLAN, AND SUSANNE R. TALAY

18 Group C and Group G Streptococcal Infections: Epidemiologic and Clinical Aspects / 222
GIO J. BARACCO AND ALAN L. BISNO

D. *STREPTOCOCCUS PNEUMONIAE*

19 The Cell Wall of *Streptococcus pneumoniae* / 230
ALEXANDER TOMASZ AND WERNER FISCHER

20 *Streptococcus pneumoniae* Capsular Polysaccharide / 241
JAMES C. PATON AND JUDY K. MORONA

21 *Streptococcus pneumoniae*: Invasion and Inflammation / 253
CARLOS J. ORIHUELA AND ELAINE TUOMANEN

22 Phase Variation of *Streptococcus pneumoniae* / 268
JEFFREY N. WEISER

23 Genetics of *Streptococcus pneumoniae* / 275
JANET YOTHER AND SUSAN K. HOLLINGSHEAD

24 Pneumococcal Vaccines / 289
D. E. BRILES, J. C. PATON, E. SWIATLO, AND M. J. CRAIN

E. ENTEROCOCCI

25 Pathogenicity of Enterococci / 299
LYNN E. HANCOCK AND MICHAEL S. GILMORE

26 Enterococcal Genetics / 312
KEITH E. WEAVER

F. ORAL STREPTOCOCCI

27 Pathogenesis of Oral Streptococci / 332
R. R. B. RUSSELL

28 The Virulence Properties of *Streptococcus mutans* / 340
HOWARD K. KURAMITSU

29 Genetics of *sanguinis* Group Streptococci / 347
HOWARD F. JENKINSON AND M. MARGARET VICKERMAN

G. LACTOCOCCI

30 Genetics of Lactococci / 356
PHILIPPE GAUDU, YUJI YAMAMOTO, PETER RUHDAL JENSEN, KARIN HAMMER, AND ALEXANDRA GRUSS

SECTION III
THE STAPHYLOCOCCUS / 369
SECTION EDITOR: Richard P. Novick

31 Diagnostics, Typing, and Taxonomy / 371
WOLFGANG WITTE, BIRGIT STROMMENGER, AND GUIDO WERNER

32 The *Staphylococcus aureus* NCTC 8325 Genome / 381
ALLISON F. GILLASPY, VERONICA WORRELL, JOSHUA ORVIS, BRUCE A. ROE, DAVID W. DYER, AND JOHN J. IANDOLO

33 Genetics: Accessory Elements and Genetic Exchange / 413
NEVILLE FIRTH AND RONALD A. SKURRAY

34 Carbohydrate Catabolism: Pathways and Regulation / 427
REINHOLD BRÜCKNER AND RALF ROSENSTEIN

35 Respiration and Small-Colony Variants of *Staphylococcus aureus* / 434
RICHARD A. PROCTOR

36 The Staphylococcal Cell Wall / 443
ALEXANDER TOMASZ

37 Staphylococcal Capsule / 456
CHIA Y. LEE AND JEAN C. LEE

38 Staphylococcus aureus Exotoxins / 464
GREGORY A. BOHACH

39 Extracellular Enzymes / 478
STAFFAN ARVIDSON

40 Staphylococcal Sortases and Surface Proteins / 486
ANDREA C. DEDENT, LUCIANO A. MARRAFFINI, AND OLAF SCHNEEWIND

41 Staphylococcal Pathogenesis and Pathogenicity Factors: Genetics and Regulation / 496
RICHARD P. NOVICK

42 Staphylococcus aureus—Eukaryotic Cell Interactions / 517
CARLOS ARRECUBIETA AND FRANKLIN D. LOWY

43 The Epidemiology of Staphylococcus Infections / 526
FRED C. TENOVER AND RACHEL J. GORWITZ

44 Animal Models of Experimental Staphylococcus aureus Infection / 535
L. VINCENT COLLINS AND ANDRZEJ TARKOWSKI

45 Cellular and Extracellular Defenses against Staphylococcal Infections / 544
JERROLD WEISS, ARNOLD S. BAYER, AND MICHAEL YEAMAN

46 Biology and Pathogenicity of Staphylococcus epidermidis / 560
CHRISTINE HEILMANN AND GEORG PETERS

47 Biology and Pathogenicity of Staphylococci Other than Staphylococcus aureus and Staphylococcus epidermidis / 572
ANNE TRISTAN, GERARD LINA, JEROME ETIENNE, AND FRANÇOIS VANDENESCH

48 Antibiotic Resistance in the Staphylococci / 587
STEVEN J. PROJAN AND ALEXEY RUZIN

SECTION IV
THE LISTERIAE / 599
SECTION EDITOR: Daniel A. Portnoy

49 Epidemiology and Clinical Manifestations of Listeria monocytogenes Infection / 601
WALTER F. SCHLECH III

50 Listeria monocytogenes Infection of Mice: an Elegant Probe To Dissect Innate and T-Cell Immune Responses / 609
JODIE S. HARING AND JOHN T. HARTY

51 Genetic Tools for Use with Listeria monocytogenes / 620
DARREN E. HIGGINS, CARMEN BUCHRIESER, AND NANCY E. FREITAG

52 Regulation of Virulence Genes in Pathogenic Listeria spp. / 634
WERNER GOEBEL, STEFANIE MÜLLER-ALTROCK, AND JÜRGEN KREFT

53 Cell Biology of Invasion and Intracellular Growth by Listeria monocytogenes / 646
JAVIER PIZARRO-CERDÁ AND PASCALE COSSART

SECTION V
SPORE-FORMING PATHOGENS AND GRAM-POSITIVE MEMBERS OF THE ACTINOMYCETALES / 657
SECTION EDITOR: Julian I. Rood

54 Bacillus anthracis / 659
THERESA M. KOEHLER

55 Clostridial Genetics / 672
DENA LYRAS AND JULIAN I. ROOD

56 Neurotoxigenic Clostridia / 688
ERIC A. JOHNSON

57 Enterotoxic Clostridia: Clostridium perfringens Type A and Clostridium difficile / 703
BRUCE A. McCLANE, DAVID M. LYERLY, AND TRACY D. WILKINS

58 Histotoxic Clostridia / 715
DENNIS L. STEVENS AND JULIAN I. ROOD

ACTINOMYCETALES

59 Corynebacterium diphtheriae: Iron-Mediated Activation of DtxR and Regulation of Diphtheria Toxin Expression / 726
JOHN F. LOVE AND JOHN R. MURPHY

60 Actinomyces and Arcanobacterium spp.: Host-Microbe Interactions / 736
B. HELEN JOST AND STEPHEN J. BILLINGTON

61 The Pathogenesis of *Nocardia* / 750
BLAINE L. BEAMAN

SECTION VI
**ANTIBIOTIC RESISTANCE
MECHANISMS / 767**
SECTION EDITOR: Richard P. Novick

**62 Mechanisms of Resistance to β-Lactam
Antibiotics / 769**
DOUGLAS S. KERNODLE

**63 Resistance to Glycopeptides in Gram-
Positive Pathogens / 782**
HENRY S. FRAIMOW AND PATRICE COURVALIN

**64 Tetracycline Resistance Determinants in
Gram-Positive Bacteria / 801**
LAURA M. McMURRY AND STUART B. LEVY

**65 Mechanisms of Quinolone
Resistance / 821**
DAVID C. HOOPER

Index / 835

Contributors

CARLOS ARRECUBIETA
Division of Infectious Diseases, Department of Medicine,
Columbia University, College of Physicians and Surgeons,
New York, NY 10032

STAFFAN ARVIDSON
Microbiology and Tumor Biology Center, Karolinska
Institutet, S-171 77 Stockholm, Sweden

SHARON BALTER
Department of Public Health and Mental Hygiene,
New York City Department of Health and Mental Hygiene,
New York, NY 10013

GIO J. BARACCO
Department of Medicine Division of Infectious Diseases,
University of Miami School of Medicine, Miami, FL 33125

ARNOLD S. BAYER
Department of Medicine, David Geffen School of Medicine,
UCLA Box 466, Los Angeles, CA 90095

BLAINE L. BEAMAN
Department of Medical Microbiology, University of California
School of Medicine, Davis, CA 95616-5224

DEBRA E. BESSEN
Department of Microbiology and Immunology, New York
Medical College, Valhalla, NY 10595

TERRY J. BEVERIDGE
Department of Microbiology, University of Guelph, Guelph,
Ontario N1G 2W1, Canada

STEPHEN J. BILLINGTON
Department of Veterinary Science and Microbiology,
University of Arizona, Tucson, AZ 85721

ALAN L. BISNO
Miami Veterans Affairs Medical Center,
1201 Northwest 16th St., #111, Miami, FL 33125-1693

GREGORY A. BOHACH
Department of Microbiology, Molecular Biology, and
Biochemistry, University of Idaho, Moscow, ID 83844-3052

DAVID BRILES
University of Alabama at Birmingham, 845 19th St. South,
Birmingham, AL 35294

REINHOLD BRÜCKNER
Mikrobiologie, Technische Universität Kaiserslautern,
D-67773 Kaiserslautern, Germany

CARMEN BUCHRIESER
Institut Pasteur, Unité de Genomique des Microorganisms
Pathogenes, 28 rue du Dr. Roux, 75724 Paris Cedex 15, France

MICHAEL CAPARON
Department of Molecular Microbiology, Box 8230,
Washington University School of Medicine,
660 S. Euclid Avenue, St. Louis, MO 63110-1093

GURSHARAN S. CHHATWAL
Department of Microbial Pathogenesis and Vaccine Research,
GBF-National Research Centre for Biotechnology,
Technical University of Braunschweig,
D-38124 Braunschweig, Germany

KYU HONG CHO
Department of Molecular Microbiology, Box 8230,
Washington University School of Medicine, 660 S. Euclid
Avenue, St. Louis, MO 63110-1093

P. PATRICK CLEARY
Department of Microbiology, University of Minnesota,
MMC 196, 420 Delaware St. SE, Minneapolis, MN 55455

L. VINCENT COLLINS
Department of Rheumatology, University of Göteborg,
Guldhedsgatan 10a, S-413 46 Göteborg, Sweden

PASCALE COSSART
Unité des Interactions Bactéries-Cellules-Unité INSERM 604,
Institut Pasteur, 75724 Paris Cedex 15, France

PATRICE COURVALIN
Unité des Agents Antibactériens, Institut Pasteur,
25 rue du Dr. Roux, 75724 Paris Cedex 15, France

MARILYN J. CRAIN
Department of Microbiology, University of Alabama at
Birmingham, Birmingham, AL 35294

DAVID CUE
Department of Microbiology, Molecular Genetics, and
Immunology, University of Kansas Medical Center,
Kansas City, KS 66160

MADELEINE W. CUNNINGHAM
Department of Microbiology and Immunology,
University of Oklahoma Health Sciences Center,
Oklahoma City, OK 73190

ANDREA C. DEDENT
Department of Microbiology, University of Chicago,
Chicago, IL 60637

DAVID W. DYER
Department of Microbiology and Immunology,
University of Oklahoma Health Sciences Center,
Oklahoma City, OK 73190

JEROME ETIENNE
INSERM E0230, Faculté de Médecine Laennec,
Rue Guillaume Paradin, 69372 Lyon Cedex 08, France

JOSEPH J. FERRETTI
University of Oklahoma Health Sciences Center,
Oklahoma City, OK 73190

NEVILLE FIRTH
School of Biological Sciences, University of Sydney,
Sydney, NSW 2006, Australia

VINCENT A. FISCHETTI
Laboratory of Bacterial Pathogenesis and Immunology,
The Rockefeller University, 1230 York Avenue,
New York, NY 10021

HENRY S. FRAIMOW
Division of Infectious Diseases,
Cooper Hospital/University Medical Center,
Camden, NJ 08103

NANCY E. FREITAG
Seattle Biomedical Research Institute, 307 Westlake Avenue
N., Suite 500, Seattle, WA 98109-5219

PHILIPPE GAUDU
Unité de Recherches Laitières et Génétique Appliquée,
Institut National de la Recherche Agronomique,
78352 Jouy en Josas, France

ALLISON F. GILLASPY
Department of Microbiology and Immunology,
University of Oklahoma Health Sciences Center,
Oklahoma City, OK 73190

MICHAEL S. GILMORE
Department of Ophthalmology, Harvard Medical School,
Boston, MA 02114

WERNER GOEBEL
Lehrstuhl für Mikrobiologie, Biozentrum, Universität
Würzburg, Am Hubland, 97074 Würzburg, Germany

RACHEL J. GORWITZ
Division of Healthcare Quality Promotion, National Center
for Infectious Diseases, Centers for Disease Control and
Prevention, Atlanta, GA 30333

ALEXANDRA GRUSS
Unité de Recherches Laitières et Génétique Appliquée,
Institut National de la Recherche Agronomique,
78352 Jouy en Josas, France

KARIN HAMMER
Technical University of Denmark, Building 301,
DK-2800 Lyngby, Denmark

LYNN E. HANCOCK
Division of Biology, Kansas State University,
Manhattan, KS 66506

JODIE S. HARING
Department of Microbiology, University of Iowa,
Iowa City, IA 52242

JOHN T. HARTY
Department of Microbiology and Interdisciplinary Program in
Immunology, University of Iowa, Iowa City, IA 52242

CHRISTINE HEILMANN
Institute of Medical Microbiology, University of Münster,
Domagkstr. 10, D-48149 Münster, Germany

DARREN E. HIGGINS
Department of Microbiology and Molecular Genetics,
Harvard Medical School, 200 Longwood Avenue,
Boston, MA 02115-5701

SUSAN K. HOLLINGSHEAD
Department of Microbiology, University of Alabama at
Birmingham, Birmingham, AL 35294

DAVID C. HOOPER
Division of Infectious Diseases, Massachusetts General
Hospital, Harvard Medical School, Boston, MA 02114-2696

JOHN J. IANDOLO
Department of Microbiology and Immunology,
University of Oklahoma Health Sciences Center,
Oklahoma City, OK 73190

HOWARD F. JENKINSON
University of Bristol, Department of Oral and Dental Science,
Lower Maudlin Street, Bristol BS1 2LY, United Kingdom

PETER RUHDAL JENSEN
Technical University of Denmark, Building 301,
DK-2800 Lyngby, Denmark

ERIC A. JOHNSON
Food Research Institute, University of Wisconsin,
1925 Willow Drive, Madison, WI 53706-1187

B. HELEN JOST
Department of Veterinary Science and Microbiology,
University of Arizona, Tucson, AZ 85721

DENNIS KASPER
Channing Lab, Department of Medicine, Brigham &
Women's Hospital, Harvard Medical School,
Boston, MA 02115

DOUGLAS S. KERNODLE
Division of Infectious Diseases, Department of Medicine,
Vanderbilt University School of Medicine, Nashville,
TN 37232-2605, and Infectious Diseases Section,
Department of Veterans Affairs Medical Center,
Nashville, TN 37212-2637

THERESA M. KOEHLER
University of Texas Health Science Center,
Department of Microbiology and Molecular Genetics,
6431 Fannin, JFB 1 765, Houston, TX 77030

JÜRGEN KREFT
Lehrstuhl für Mikrobiologie, Biozentrum,
Universität Würzburg, Am Hubland,
97074 Würzburg, Germany

HOWARD K. KURAMITSU
8518 Cahill Drive, #51, Austin, TX 78729

CHIA Y. LEE
Department of Microbiology and Immunology, University of
Arkansas for Medical Sciences, 4301 West Markham Street,
Little Rock, AR 72205

JEAN C. LEE
Channing Laboratory, Department of Medicine,
Brigham and Women's Hospital and Harvard Medical School,
Boston, MA 02115

STUART B. LEVY
Tufts University School of Medicine, 136 Harrison Avenue,
M&V 803, Boston, MA 02111

GERARD LINA
INSERM E0230, Faculté de Médecine Laennec,
Rue Guillaume Paradin, 69372 Lyon Cedex 08, France

JOHN F. LOVE
Department of Microbiology, Boston University School of
Medicine, 650 Albany Street, X830, Boston, MA 02118

FRANKLIN D. LOWY
Division of Infectious Diseases, Department of Medicine,
Columbia University, College of Physicians and Surgeons,
New York, NY 10032

DAVID M. LYERLY
TechLab, Inc., 1861 Pratt Drive, Corporate Research Center,
Blacksburg, VA 24060-6364

DENA LYRAS
Department of Microbiology, Monash University, Wellington
Road, Clayton, Victoria 3168, Australia

LAWRENCE MADOFF
Channing Lab, Department of Medicine,
Brigham & Women's Hospital, Harvard Medical School,
Boston, MA 02115

HORST MALKE
University of Oklahoma Health Sciences Center,
Oklahoma City, OK 73190

LUCIANO A. MARRAFFINI
Department of Microbiology, University of Chicago,
Chicago, IL 60637

VALÉRIO R. F. MATIAS
Department of Microbiology, University of Guelph,
Guelph, Ontario N1G 2W1, Canada

BRUCE A. McCLANE
Department of Molecular Genetics and Biology,
University of Pittsburgh School of Medicine,
E1240 Biomed Science Tower,
Pittsburgh, PA 15261

JOHN K. McCORMICK
Lawson Health Research Institute, London,
Ontario N6A 4V2, Canada

DAVID J. McMILLAN
Queensland Institute of Medical Research, Brisbane,
Queensland 4029, Australia

LAURA M. McMURRY
Tufts University School of Medicine, 136 Harrison Avenue,
M&V 803, Boston, MA 02111

W. MICHAEL McSHAN
Department of Pharmaceutical Sciences, CPB 307, 1110 N.
Stonewall Boulevard, University of Oklahoma Health
Sciences Center, Oklahoma City, OK 73190

JUDY K. MORONA
Department of Molecular Biosciences, Adelaide University,
Adelaide, South Australia 5005, Australia

STEFANIE MÜLLER-ALTROCK
Lehrstuhl für Mikrobiologie, Biozentrum, Universität
Würzburg, Am Hubland, 97074 Würzburg, Germany

JOHN R. MURPHY
Department of Microbiology, Boston University School of
Medicine, 650 Albany Street, X830, Boston, MA 02118

VICTOR NIZET
Department of Pediatrics, University of California,
San Diego, La Jolla, CA 92093

RICHARD P. NOVICK
Skirball Institute for Biomolecular Medicine,
New York University Medical Center, 550 First Avenue,
New York, NY 10016-6481

CARLOS J. ORIHUELA
Department of Infectious Diseases, St. Jude Children's
Research Hospital, Memphis, TN 38105-2794

JOSHUA ORVIS
Department of Microbiology and Immunology,
University of Oklahoma Health Sciences Center,
Oklahoma City, OK 73190

VIJAY PANCHOLI
Department of Pathology, 129 Hamilton Hall,
The Ohio State University Medical Center,
1645 Neil Avenue, Columbus, OH 43210-1218

LAWRENCE C. PAOLETTI
Channing Lab, Department of Medicine,
Brigham & Women's Hospital, Harvard Medical School,
Boston, MA 02115

JAMES C. PATON
School of Molecular and Biomedical Science, University of
Adelaide, Adelaide, SA 5005, Australia

GEORG PETERS
Institute of Medical Microbiology, University of Münster,
Domagkstr. 10, D-48149 Münster, Germany

MARNIE L. PETERSON
Department of Microbiology, University of Minnesota
Medical School, Minneapolis, MN 55455

JAVIER PIZARRO-CERDÁ
Unité des Interactions Bactéries-Cellules/Unité INSERM 604,
Institut Pasteur, 75724 Paris Cedex 15, France

DANIEL A. PORTNOY
Department of Molecular and Cell Biology and The School of
Public Health, University of California, Berkeley,
Berkeley, CA 94720

KLAUS T. PREISSNER
Institut für Biochemie, Fachbereich Humanmedizin,
Justus-Liebig-Universität, Friedrichstrasse 24,
D-35392 Giessen, Germany

RICHARD A. PROCTOR
407 SMI, Department of Medical Microbiology and
Immunology, 1300 University Avenue, Madison, WI 53706

STEVEN J. PROJAN
Biological Technologies, Wyeth Research,
87 Cambridge Park Drive, Cambridge, MA 02140

BRUCE A. ROE
Department of Microbiology and Immunology,
University of Oklahoma Health Sciences Center,
Oklahoma City, OK 73190

JULIAN I. ROOD
Australian Bacterial Pathogen Program, Department of
Microbiology, Monash University, Wellington Road,
Melbourne, Victoria 3168, Australia

RALF ROSENSTEIN
Mikrobielle Genetik, Universität Tübingen,
D-72076 Tübingen, Germany

CRAIG E. RUBENS
Department of Pediatrics, Children's Hospital &
Regional Medical Center, University of Washington,
Seattle, WA 98105

R. R. B. RUSSELL
Oral Biology/Dental School, University of Newcastle upon
Tyne, Newcastle NE2 4BW, United Kingdom

ALEXEY RUZIN
Infectious Disease, Wyeth Research, 401 North Middletown
Road, Pearl River, NY 10965

WALTER F. SCHLECH III
Department of Microbiology and Immunology, Dalhousie
University, Halifax, Nova Scotia B3H 4R2, Canada

PATRICK SCHLIEVERT
Department of Microbiology, University of Minnesota,
Minneapolis, MN 55455

OLAF SCHNEEWIND
Department of Microbiology, University of Chicago,
Chicago, IL 60637

ANNE SCHUCHAT
Centers for Disease Control and Prevention,
Atlanta, GA 30333

RONALD A. SKURRAY
School of Biological Sciences, University of Sydney,
Sydney, NSW 2006, Australia

DENNIS L. STEVENS
Infectious Disease Section, Building 45, Veterans Affairs
Medical Center, 500 West Fort Street, Boise, ID 83702-4501

BIRGIT STROMMENGER
Robert Koch Institute, Wernigerode Branch,
Burgstrasse 37, G-38855 Wernigerode, Germany

E. SWIATLO
Division of Infectious Diseases, Veterans Affairs Medical
Center, and Research and Education, 1500 E. Woodrow
Wilson Drive, Jackson, MS 39216-5199

SUSANNE R. TALAY
Department of Microbial Pathogenicity and Vaccine
Research, Technical University/GBF National Research
Centre for Biotechnology, D-38124 Braunschweig, Germany

ANDRZEJ TARKOWSKI
Department of Rheumatology and Inflammation
Research, University of Göteborg, Guldhedsgatan 10a,
S-413 46 Göteborg, Sweden

FRED C. TENOVER
Hospital Infections Program, Centers for Disease
Control and Prevention, Mailstop G-08, 1600 Clifton Road,
Atlanta, GA 30333

ALEXANDER TOMASZ
The Rockefeller University, 1230 York Avenue, #152,
New York, NY 10021

ANNE TRISTAN
INSERM E0230, Faculté de Médecine Laennec,
Rue Guillaume Paradin, 69372 Lyon Cedex 08, France

ELAINE TUOMANEN
Department of Infectious Diseases,
St. Jude's Children's Research Hospital,
322 N. Lauderdale, Memphis, TN 38105-2794

FRANÇOIS VANDENESCH
INSERM E0230, Faculté de Médecine Laennec,
Rue Guillaume Paradin, 69372 Lyon Cedex 08, France

M. MARGARET VICKERMAN
Department of Periodontics and Endodontics, School of
Dentistry, University of Buffalo, Buffalo, NY 14214

BEINAN WANG
Gonda Department of Cell and Molecular Biology,
House Ear Institute, University of Southern California,
Los Angeles, CA 90057

KEITH E. WEAVER
Department of Microbiology, School of Medicine,
University of South Dakota, 414 E. Clark Street,
Vermillion, SD 57069-2390

JEFFREY N. WEISER
Department of Medicine, University of Pennsylvania, 402A
Johnson Pavilion, Philadelphia, PA 19104-6076

JERROLD WEISS
Division of Infectious Disease, University of Iowa School of
Medicine, D158 MTF, Iowa City, IA 52242

GUIDO WERNER
Robert Koch Institute, Wernigerode Branch, Burgstraße 37,
G-38855 Wernigerode, Germany

MICHAEL R. WESSELS
Division of Infectious Diseases, Children's Hospital Boston,
300 Longwood Avenue, Boston, MA 02115

TRACY D. WILKINS
TechLab, Inc., 1861 Pratt Drive, Corporate Research Center,
Blacksburg, VA 24060-6364, and Fralin Biotech Center,
Virginia Polytechnic Institute and State University,
Blacksburg, VA 24061

WOLFGANG WITTE
Robert Koch Institute, Wernigerode Branch, Burgstraße 37,
D-38855 Wernigerode, Germany

VERONICA WORRELL
Department of Microbiology and Immunology,
University of Oklahoma Health Sciences Center,
Oklahoma City, OK 73190

YUJI YAMAMOTO
Unité de Recherches Laitières et Génétique Appliquée,
Institut National de la Recherche Agronomique,
78352 Jouy en Josas, France

MICHAEL YEAMAN
Division of Infectious Diseases, Department of Medicine,
David Geffen School of Medicine, UCLA Box 466,
Los Angeles, CA 90095

JANET YOTHER
Department of Microbiology, University of Alabama at
Birmingham, Birmingham, AL 35294

Preface

Gram-positive bacteria are structurally distinct from their gram-negative relatives. These differences, reflected in the lack of an outer membrane and related secretory systems and the presence of a thick peptidoglycan, have enabled these organisms to develop novel approaches to pathogenesis by acquiring (among others) a unique family of surface proteins, toxins, and enzymes. The initial edition of this volume was a fully referenced research compendium directed to the gram-positive bacterial pathogen at all levels. In this second edition, we have attempted, whenever possible, to include the explosion of new data generated from genomic sequencing. It includes the current theories on the mechanisms of gram-positive bacterial pathogenicity, together with current knowledge on gram-positive structure and mechanisms of antibiotic resistance. This edition emphasizes streptococci, staphylococci, listeria, and spore-forming pathogens, with chapters written by many of the leading researchers in these areas. The chapters systematically dissect these organisms biologically, genetically, and immunologically, in an attempt to understand the strategies used by these bacteria to cause human disease. It is hoped that the insights gained from understanding these strategies will lead to the rational design of novel therapeutics and vaccines to control infections caused by gram-positive bacteria.

VINCENT A. FISCHETTI
RICHARD P. NOVICK
JOSEPH J. FERRETTI
DANIEL A. PORTNOY
JULIAN I. ROOD

THE GRAM-POSITIVE CELL WALL

SECTION EDITOR: Vincent A. Fischetti

THE GRAM-POSITIVE CELL WALL structure differs in many respects from its gram-negative counterpart. While much has been reported regarding the complexity of the gram-negative envelope, until recently similar information has not been available for the gram-positive wall. The following chapters are designed to give the reader a good impression of gram-positive cell wall structure and the increasing complexity of its surface protein coat, as well as our increasing knowledge of its polymeric nature and the location of its periplasmic constituents, offering the most comprehensive understanding to date of the gram-positive envelope.

Ultrastructure of Gram-Positive Cell Walls

TERRY J. BEVERIDGE AND VALÉRIO R. F. MATIAS

1

Over the last decade there has been renewed interest in the mechanisms by which gram-positive pathogens infect humans and animals and the ways in which these pathogens interact with host tissue during pathogenesis. This renaissance has been driven by the accelerated incidence of antibiotic resistance among major pathogens such as *Mycobacterium*, *Staphylococcus*, *Streptococcus*, and *Enterococcus* spp. and by the increase in infections caused by gram-positive pathogens in immunocompromised patients with AIDS (3, 13, 15, 42, 51). Clearly, the infectivity of a disease agent is complicated and relies on both the virulence of the pathogen and the susceptibility of the host, but it is apparent that the outermost components of each pathogen and their interaction with specific host cell lines are of great importance during infection. For this reason, it is important that the polymers that make up gram-positive cell walls be correctly allocated to the natural locations in which they reside. In this way, the surface interactions between pathogen and host should be better understood.

Remarkably, after approximately 50 years of study of gram-positive bacteria by electron microscopy (14), many of the ultrastructural details still await elucidation (5). It has proven to be extremely difficult to preserve native wall ultrastructure as bacteria are prepared for electron microscopy and then subjected to the high-energy load of the electron beam (11). Yet, although many other physical techniques have been tried (e.g., selected area electron diffraction, X-ray diffraction, neutron diffraction, nuclear magnetic resonance, X-ray lithography, scanning tunneling microscopy, and atomic force microscopy [AFM]), the tried-and-true technique is still transmission electron microscopy (TEM) (11, 28). Scanning electron microscopy (SEM) has not provided nearly as much information about these cell walls as TEM, since it can reveal only the surface topography. Because gram-positive walls are poorly conducting, the wall surface is usually "gold sputtered" for SEM to provide a thin conductive layer that prevents charging (and artifactual imaging) by the electron beam (11); this external gold layer can reduce the resolution as well as compact polymeric filaments extending from the wall. Yet, if great care is taken in specimen preparation, SEM topographical preparations can reveal remarkable views of polymeric arrangements (e.g., the concentric pattern of peptidoglycan and associated polymers of the *Staphylococcus epidermidis* cell wall shown in Fig. 1). Such detailed SEM images are rare and require dedicated electron microscopy laboratories that are experienced in microbiological preparations. Most reliable high-resolution information has stemmed from TEM images garnered from a number of separate techniques, such as negative stains, shadow-castings, thin sections of conventional embeddings, and freeze-etchings of intact bacteria or isolated cell wall fragments (4). Recently, modern cryo-techniques such as freeze-substitution have given a more precise view of gram-positive walls and, even more recently, hydrated cryo-thin sections have provided our best interpretation of the polymeric structure of these difficult to decipher walls (29, 29a).

Here, frozen cells are vitrified in amorphous ice by snap-freezing within milli- to microsecond time frames, and then cryo-thin sectioned and viewed while in a frozen state in specialized cryo-TEMs. Because the cells are vitrified in amorphous ice, which resembles a "glass," the polymeric structure of the wall remains in its natural hydrated state. Just as cryo-TEM has expanded our perceptual boundaries on wall structure, so has AFM. These microscopes possess no lenses, but through an intricate system of cantilevers, tips, and piezo-ceramic manipulators (20), are capable of molecular resolution of wall polymers.

In this chapter we explain the procedures used to examine gram-positive walls by TEM and discuss the results in the context of our current view of polymeric arrangements. Often today, electron microscopy is performed by institutional nonspecialized TEM centers that have little microbiological experience, and few of these have state-of-the-art cryo-units. Because of this, we discuss the traditional techniques and their results in case this more modern equipment is not readily available so that the reader knows what to expect. We then give a more up-to-date current viewpoint provided by cryo-TEM and AFM and show how these results integrate with traditional views.

Gram-Positive Pathogens, 2nd edition, edited by Vincent A. Fischetti et al.
© 2006 ASM Press, Washington, D.C.

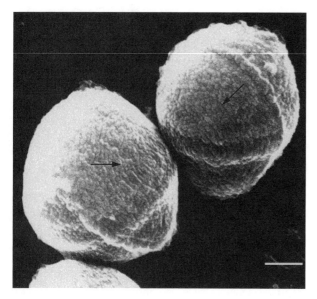

FIGURE 1 SEM micrograph of *S. epidermidis* showing the concentric circular structures of the cell wall surface (arrows) that suggest the circular arrangement of polymers in the cell wall. Bar = 250 nm. (Reproduced from *J. Electron Microsc.* **27:**147–148 [1978] with permission.)

TECHNIQUES USED TO STUDY THE ULTRASTRUCTURE OF GRAM-POSITIVE CELL WALLS

A detailed explanation of the procedures used for TEM processing of bacteria is beyond the scope of this chapter. Instead, only a general description of each technique is given, with emphasis on the reasons behind each manipulation. More detailed explanations can be found in Beveridge et al. (11), Dufrêne (20), and Koval and Beveridge (28).

Freeze-Etching

Freeze-etching is a cryo-technique in which the sample is frozen to a vitrified state, fractured, etched, and shadowed

FIGURE 3 TEM freeze-etching of a *B. licheniformis* showing a closing septum (small arrows) between two daughter cells. Again, cross fractures of the cell wall show little infrastructure within the wall's matrix. The large arrow with circle denotes shadow direction. Bar = 100 nm.

with a metal vapor so that the topography of the fracture can be seen (9). For gram-positive bacteria, the fracture usually runs through the hydrophobic domain of the plasma (cytoplasmic) membrane and reveals intramembranous proteins (4, 11). Some fractures run over and through the surface of the cell wall and usually show a relatively featureless matrix. Cross fractures, though, reveal a complex polymeric network whose texture is difficult to resolve (Fig. 2). Fractures of growing (Fig. 3) and completed (Fig. 4) septa are especially important because they reveal how the new wall polymers in gram-positive walls are laid down during division.

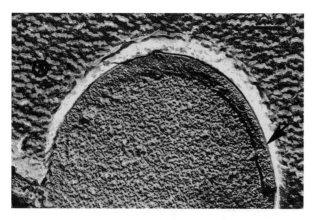

FIGURE 2 TEM micrograph of a freeze-fractured and freeze-etched *B. subtilis* showing a cross fracture of the cell wall (large arrow). Notice that little polymeric infrastructure is seen in the wall by this technique. The encircled arrow points out the shadow direction and the scale bar = 50 nm.

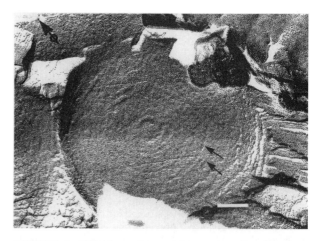

FIGURE 4 TEM freeze-fracture image of a cell wall septum from an *S. aureus* wall preparation that was previously treated with trichloroacetic acid to remove all associated teichoic acid and proteins. Because this wall now contains only peptidoglycan, the concentric circular arrangement (small arrows) must be the arrangement of peptidoglycan. The large arrow with circle denotes shadow direction. Bar = 100 nm. (Reproduced from *J. Bacteriol.* **150:**844–850 [1982] with permission.)

Negative Staining

Most information has come from negative stains of cell wall fragments derived by the physical breakage of intact bacteria. Polar caps of gram-positive rods are readily identified, and interestingly, the concentric arrangement of polymers can be identified in images of septa (9), which are similar to those seen in SEM and freeze-etching (Fig. 1 and 4). Of all TEM techniques, negative staining can provide the highest resolution of gram-positive cell walls; for example, the capsule and S-layer of *Bacillus anthracis* can be resolved above the polymeric matrix of the wall (Fig. 5).

Conventional Embeddings

Conventional embedding is a tried-and-true technique that was first used in the 1950s (14), but because this method requires chemical fixation and dehydration, it is prone to artifacts (11). Because it eventually provides thin sections of plastic-embedded bacteria, the procedure reveals how the various layers of the cell envelope are juxtaposed on top of one another (Fig. 6). This method revealed that gram-positive walls consist of an amorphous amalgamation of peptidoglycan and ancillary secondary polymers such as teichoic and teichuronic acids (4, 9, 30).

Although this technique has been used for decades, it is still among the best for discerning the distribution of cellular constituents and the juxtaposition of envelope layers. (Imagine! Here is the capability to section a 1.0 to 3.0 × 0.5 to 1.0 μm cell into ~500 Å slices so that ribosomes, DNA, membranes, and walls can be seen [8, 10, 28]! This section thickness is a little more than 500 hydrogen atoms thick [11].) Here, the cell wall in cross section is a rather featureless amorphous matrix (e.g., in *Staphylococcus aureus* [Fig. 6]), but the septum of dividing cells is clearly growing inward and the midline, where the daughter cells will eventually separate, is clearly visible (Fig. 6). This technique should be readily available to microbiologists at most universities or research centers.

Freeze-Substitution

For at least a decade, electron microscopists have realized that conventional embeddings do not provide a highly

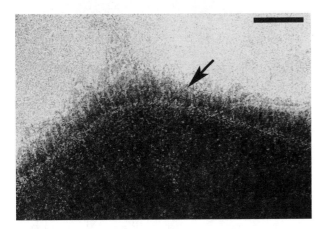

FIGURE 5 TEM negative stain of the pole of a *B. anthracis* cell showing capsular material (arrow) above the cell wall. Underneath the capsule a moiré pattern is seen, which is the S-layer (6) of this pathogen. Both of these structures are virulence factors. The arrow points to small fibrils that are the capsule of the bacterium. Bar = 100 nm.

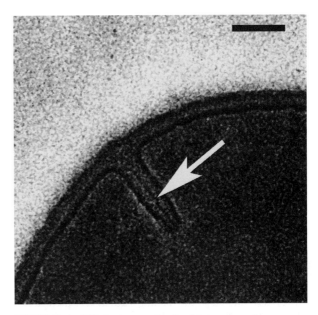

FIGURE 6 TEM micrograph of a thin section of a conventionally embedded *S. aureus* showing a growing septum (arrow) complete with a developing midline that will be used to eventually separate the two daughter cells. Note that the septum has some texture but that the older preexisting wall does not. Bar = 100 nm.

accurate picture of thin-sectioned bacteria. No matter how hard they tried to preserve cells with chemical fixatives, substantial quantities of lipid, protein, and nucleic acids were extracted during the procedure, and many structural components were compacted. It was left to the cryo-technique of freeze-substitution during the 1980s and 1990s to reformulate our perception of cell walls (7, 8, 10, 21, 25, 28, 38, 47). This technique combines rapid freezing (vitrification) and conventional embedding (22, 23, 35–37). A thin layer of cells is snap-frozen in liquid propane or ethane at close to −196°C and immediately immersed in a substitution medium consisting of 2% (wt/vol) osmium tetroxide and 2% (wt/vol) uranyl acetate dissolved in anhydrous acetone. A molecular sieve to absorb sublimed water is also added, and the entire suspension is held at −80°C for 2 days. When the cells are snap-frozen, they are encased in noncrystalline, vitreous ice in a matter of milliseconds; all molecular motion in the bacteria is immediately stopped (including the action of degrading enzymes), and delicate, hydrated structure is preserved. During substitution, the cells are fixed, stained, and dehydrated (all at the same time) at such a low temperature (−80°C) that fine structure is maintained. Once substitution is complete, the bacteria can be returned to ambient temperature and infiltrated with the same plastics used in conventional embedding. Thin sections of cell walls subjected to freeze-substitution are no longer featureless (Fig. 7). They have a dark-stained region immediately above the plasma membrane and a highly fibrous surface with an electron-translucent layer in the middle (7, 9, 10, 21, 24, 47).

Hydrated Cryo-Sections

Hydrated cryo-sectioning is one of the most advanced cryo-TEM techniques, and it is the most difficult. For this reason few studies have been done on bacteria (19, 29, 29a), and

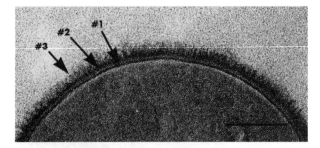

FIGURE 7 TEM micrograph of a thin section of freeze-substituted *B. subtilis* showing the tripartite fine structure in the cell wall that demonstrates cell wall turnover. Region #1 is densely stained because the newly laid down peptidoglycan is condensed and reactive to the staining reagents. Region #2 is more lightly stained because this is the stress-bearing peptidoglycan that is highly stretched and therefore not as dense or condensed as in region #1. Region #3 is undergoing hydrolysis by peptidoglycan hydrolases (autolysins) and is more fibrous. Because covalent bonds are being clipped, there are many reactive groups with which the staining reagents can interact, and this region is darkly stained. Bar = 100 nm.

only one has been published on gram-positive cell walls (29a). Like freeze-etching and freeze-substitution, cells have to be vitrified. Here, though, a relatively large quantity must be processed so that they eventually can be thin sectioned in an ultramicrotome. Because of the large biomass, vitrification can be difficult and high-pressure freezing is often a necessity. Here, the cells are subjected to a high pressure and snap-frozen with a jet of liquid nitrogen. The high pressure and cold immediately vitrifies substantial (microgram) quantities of bacteria. To aid vitrification, the bacteria are usually grown in medium containing a cryoprotectant (e.g., 15% [wt/vol] sucrose or 20% [wt/vol] dextran). (Freezing is an important step because bacteria are excellent ice nucleation particles on which ice crystals grow. If this happens, the cells are deformed and often punctured so that their native structure is lost.) Once frozen, the vitrified cell pellet is placed in a cryo-ultramicrotome and thin sectioned with a specialized diamond knife at −140 to −160°C. Once the frozen thin sections are cut, they are manipulated into a cryo-specimen holder and placed into the cryo-chamber of the TEM. Both holder and chamber are held close to liquid nitrogen temperature to ensure that the cells remain vitrified. Throughout all of these steps, it is important that all manipulations be done under low humidity because the cold thin sections act as a cryo-trap for water vapor in the air, which can immediately sublime onto the sections as obscuring ice crystals. As one can imagine, all steps in the preparation of cryo-sections are difficult and require experience, steady hands, patience, and rather exotic equipment (19, 29). But it is worth it!

The structure found in the cells is as close to the native structure as possible. In fact, if the intact cells (before cryosectioning) were thawed, they would come back to life and swim away. This is one of the very few techniques that ensures that the bacteria remain hydrated (they are embedded in vitreous ice), and it is the only technique that, at the same time, reveals internal and enveloping structure (19, 29, 29a), because of the thin section. Yet, there are still some difficulties to be overcome. Unlike conventional thin sections, the cells cannot be stained with heavy metals to provide contrast. Somehow, we must differentiate the cells

from the frozen matrix of amorphous ice. This is done by relying on the differential scattering power of the mass of the cells versus that of the ice (the C, N, S, P, and inorganic metal cations of cells scatter electrons better than the H

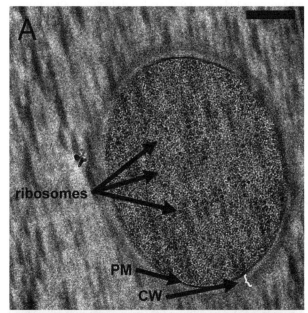

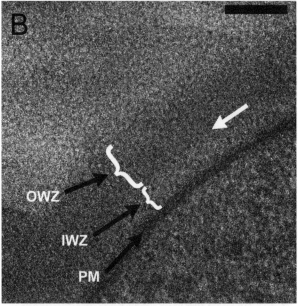

FIGURE 8 (A) Cryo-TEM micrograph of a hydrated frozen section of a *B. subtilis* cell. The ribosomes are dispersed throughout the cytoplasm (arrows), and the plasma membrane (PM) is bounded by a bipartite cell wall (CW). Unlike other TEM images, this cell is unstained, and the contrast is provided by cellular mass distribution and phase contrast. Ridges and valleys can be seen in the surrounding ice and are cutting artifacts (see reference 29 for details). Bar = 200 nm. (B) High magnification of a hydrated frozen section of the *B. subtilis* cell envelope for comparison with the freeze-substitution tripartite wall image in Fig. 7. The PM is enclosed by a low-density inner wall zone (IWZ), which is bounded by a high-density outer wall zone (OWZ). The inner surface of the OWZ (white arrow) appears to have higher contrast and therefore more innate density. Bar = 50 nm.

and O of water). This differential is enhanced by artificially under- or over-focusing the image (i.e., using the phase contrast function). Now, ribosomes, DNA, plasma membranes, and the wall can be seen (Fig. 8A), and high magnification reveals a multilayered wall (Fig. 8B) that can be correlated to that of conventional and freeze-substituted sections (Fig. 7). More details can be found in references 29 and 29a.

AFM

AFM is unlike any other microscopy we have mentioned because no lenses are used. Instead, the sample is placed on a solid surface, and an extremely sharp tip on the end of a thin cantilever is lowered until contact is made. The tip is moved in a raster in the x and y planes over the sample via a piezo-ceramic scanner. A laser is bounced off the cantilever to track the tip's lateral movement as well as its height (z) location in time via a light-detecting diode. In this way the topography of microbes can be viewed in exquisite detail (more details can be found in reference 20). There are technical difficulties to overcome when viewing bacteria because the force of the tip can dislodge the cells and deform bacterial surface layers, obscuring minute molecular detail. The great advantage of AFM is that this microscopy can be done underwater, so hydrated cell surfaces are seen. A recent study of S. aureus has shown that the newly formed wall during septation retains concentric ridges of wall material after division (Fig. 9A), which disappear as this region grows older (46). The older wall can have different surface textures, and some walls can be extremely porous (Fig. 9B), presumably because of wall turnover. This same study also demonstrates the adhesive forces of the cell's surface polymers as the tip is attracted and pulled away from it.

GENERAL CHEMISTRY OF THE GRAM-POSITIVE CELL WALL

Peptidoglycan

Most gram-positive cell walls possess an extensive meshwork of peptidoglycan that is typically 15 to 30 nm thick in conventional thin sections. Because one layer of peptidoglycan is ca. 1 nm thick (4), this implies that the cell walls of bacteria such as B. subtilis, which has a wall 25 nm thick, consist of 25 layers of peptidoglycan. A certain proportion of all peptidoglycan polymers are cross-linked to adjacent strands so that a huge macromolecular murein sacculus completely surrounds the cell. For B. subtilis, ca. 50% of the peptidoglycan fibers are cross-linked through direct bonding between the meso-diaminopimelic acid (at position C_3 along the peptide stem) and the terminal D-alanine (at position C_4) of adjacent peptide stems emanating from the N-acetyl muramyl residues; this is A1γ chemotype peptidoglycan (41). Mycobacterium tuberculosis also has A1γ peptidoglycan (41). There is a great diversity of gram-positive peptidoglycan chemotypes (i.e., more than 100 chemotypes are known (41) that rely on subtly different amino acid substituents in the peptide stems and on different linkage units between the stems. For example, S. aureus has a pentaglycine linkage unit between an L-lysine (C_3) (which replaces the diaminopimelic acid of A1α chemotype) and the terminal D-alanine (C_4); this is the A1γ chemotype. This chemotype can subtly alter when S. aureus is stressed by methicillin or high NaCl content (17, 50). (The chemotype classification system provides a wealth of chemical information about each type of peptidoglycan; the

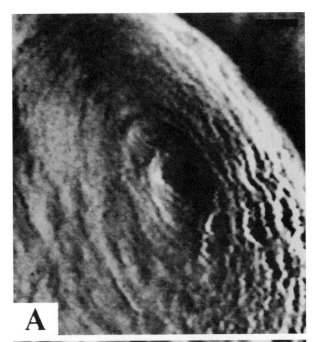

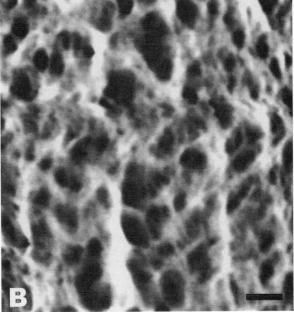

FIGURE 9 (A) AFM deflection image of a newly formed S. aureus cell wall derived from the septum showing concentric rings surrounding the central depression of the closed septum. The rings correspond to the manner in which the new wall polymers are laid down and configured during septation and cell separation. Bar = 50 nm. (Previously published in J. Bacteriol. **186:**3286–3295 [2004] and reproduced with permission.) (B) High-resolution AFM image of an old region of the S. aureus cell wall (well removed from a septal region) showing the topography of the wall surface. Notice how fibrous and porous it is; this is probably due to cell wall turnover. Bar = 50 nm.

roman capital letters [i.e., A or B] indicate the cross-linking class; the Arabic numbers [i.e., 1 to 5] indicate the type of interpeptide bridge or lack of it; and the Greek letters [i.e., α to δ] indicate the amino acid present in position C_3 of the stem peptide.)

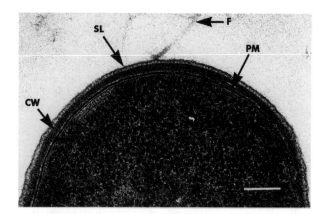

FIGURE 10 TEM thin section of a cell pole of a conventionally embedded *Bacillus thuringiensis* showing its S-layer (SL) above the cell wall (CW) and plasma membrane (PM). A flagellum (F) emanating from the surface can also be seen. Bar = 100 nm.

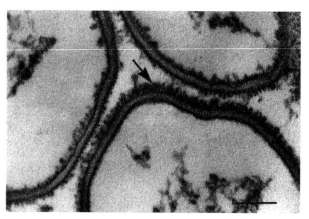

FIGURE 11 Thin section of isolated cell walls from *S. aureus* that were processed by the freeze-substitution technique after the culture was mechanically disrupted. As with the freeze-substituted *B. subtilis* cells shown in Fig. 7, these cell walls also have a fibrous surface (arrow) and tripartite structure. Bar = 100 nm. (Reproduced from *J. Bacteriol.* **169:**2482–2487 [1987] with permission.)

Secondary Polymers

Attached to the peptidoglycan, especially to the *N*-acetyl muramyl residues, are a variety of secondary polymers. In the simplest case, as exemplified by *B. subtilis* and *B. licheniformis*, these are teichoic (a highly phosphorylated polymer) and teichuronic (phosphate is substituted by uronic acid) acids (4, 34). The ratio of these two secondary polymers is modulated in *B. subtilis* by available environmental phosphate, whereas in *B. licheniformis*, a mixture of both polymers is constant. Some gram-positive bacteria, such as some *Clostridium* and *Sporosarcina* species, do not possess secondary polymers, whereas others have a greater number and more complex variety than the two polymeric acids that were previously mentioned. For example, *M. tuberculosis* possesses mycolic acids, arabinogalactans, and glycolipids that are linked to the peptidoglycan (16), as are the M protein of *Streptococcus pyogenes* (see chapter 2, this volume) and the nonteichoic acid Lancefield antigens of *Streptococcus* and *Enterococcus*. To compound the complexity, certain walls (e.g., those of *Streptococcus pneumoniae* and *Lactobacillus* spp.) contain both teichoic acids (linked to peptidoglycan muramyl moieties) and lipoteichoic acids (linked to the plasma membrane by the "lipo" substituent) (52).

ELECTRON MICROSCOPY

It is clear that the chemistry of gram-positive walls can be complicated and that many types of secondary polymers can be associated with the peptidoglycan. Yet, peptidoglycan is common to all and is the primary fabric that controls most aspects of cell shape and, as it is laid down, of growth. No matter the peptidoglycan chemotype, TEM has not found major differences in the structural formats of most gram-positive walls unless the wall has an S-layer or capsule above it (Fig. 10). Typically, the wall matrix is homogeneous with little differentiation of polymers within it (Fig. 2 and 6). We believe that the peptidoglycan strands (and associated secondary polymers) are arranged in a circumferential manner at right angles to the long cell axis (4, 48, 49). A new scaffold model has been proposed that could affect our traditional model of peptidoglycan consisting of a sequence of layers from bottom to top of the cell wall (cf. references 18 and 27). Yet, there is little hint of any arrangement

unless actively growing regions of the wall (e.g., septa) are investigated (2, 39, 46). In this case, small concentric ridges emanating out from the center of the septum are seen by freeze-etching and AFM preparations (Fig. 4 and 9A). These ridges are presumed to be the arrangement of the new wall polymers as they are fabricated and inserted into the growing septum (1, 2, 46). Once the septum is completed and the two daughter cells have separated from one another, this region becomes a new hemispherical cap (or pole) in rod-shaped bacteria or a newly formed septal wall zone as in one-half of the cell wall of a *Staphylococcus* sp. (Fig. 9). In the latter case or in *Enterococcus* spp. (9), a septal scar denotes the old from the new cell wall (46). Intuitively, the concentric arrangement of polymers seen in septa (Fig. 1, 4, and 9A) is the typical arrangement in cell walls of coccoid cells and also substantiates the circumferential arrangement believed to exist in rod-shaped bacteria (see reference 9 for concentric rings in a *Bacillus* septum).

CELL WALL TURNOVER

Gram-positive walls undergo a process of turnover for cell growth to occur. This must be carefully accomplished because there are substantial hydrostatic turgor pressures within gram-positive bacteria. In fact, these cells would lyse because of this pressure if it were not for the strong fabrication of their walls. Therefore, the "make before break" concept of wall turnover has been invoked to explain cellular growth (27). New polymers are exported, compacted, and linked both together and to preexisting wall polymers at the inner face of the cell wall by penicillin-binding proteins (e.g., see references 33 and 39). These new polymers are packed tightly together to a relatively high density and are not under stress from cellular turgor pressure. This is because the older peptidoglycan (which directly overlies these new polymers) is the stress-bearing material. Being stressed, the older peptidoglycan is stretched almost to its breaking limit and is (therefore) a low-density fabric. Directly above the stress-bearing region is the oldest peptidoglycan, which is being solubilized by its constituent autolysins.

New material is constantly being put in at the inner wall face, and old material is constantly being removed from the outer wall face, but more material is put in than is removed. This wall turnover, from inner to outer faces, combined with the expansive forces exerted by turgor pressure, ensures wall expansion and cell growth. A dynamic equilibrium between polymer renewal at the inner face and polymer hydrolysis at the outer face is highly controlled to ensure cellular growth and division. If the equilibrium is disturbed, the consequences can be severe; e.g., the antibiotics penicillin and cephalosporin bind to penicillin-binding proteins and inhibit the renewal of wall polymers at the inner face. Since the autolysins continue to hydrolyze surface peptidoglycan (e.g., see reference 44), the wall grows thinner until turgor pressure physically lyses the cell.

Freeze-substitution has captured the process of wall turnover in gram-positive bacteria (24). The wall is discriminated into three separate regions (Fig. 7). (i) The inner face is a darkly staining region because of the high density of its polymers and the abundance of reactive groups (which bind the stain). (ii) The middle region is more lightly stained and has low density because it is stress-bearing. (iii) The outer face is highly fibrous because autolysins are breaking up and solubilizing the polymeric matrix in this region. This same general format can be seen in a wide number of such cell walls (Fig. 7 and 11). It is important to recognize that because secondary polymers (e.g., teichoic acid) are attached to the peptidoglycan, they too will be removed (as muramyl fragments) by autolytic action. For example, if the isolated walls seen in Fig. 11 are treated with alkali or trichloroacetic acid to remove teichoic acids, the surface fibrils are absent (47).

Cryo-sections give us the most natural view of polymer organization in gram-positive walls, but this organization is difficult to discern. Unlike freeze-substituted walls, cryo-walls are not stained and the polymers have low contrast, which is imparted to them simply by their mass. Careful study, though, reveals subtle differences between freeze-substituted and cryo walls (cf. Fig. 7 and 8B). In cryo-sections only a hint of the outer fringe on the wall can be seen because there is so little mass to scatter electrons (29a). The darkly stained inner wall region seen in freeze-substitutions now seems more aligned to the outer face of the plasma membrane followed by a weakly scattering region that could be a periplasmic space (which was not seen by freeze-substitution). Clearly, more experimentation will be necessary before final correlation between the two techniques can be made.

MYCOBACTERIAL WALLS—AN EXAMPLE OF A MORE UNUSUAL WALL TYPE

Because mycobacterial walls possess a number of unusual and unique substances, they are difficult to preserve. These walls contain 30 to 60% (by weight) lipid (16). Substances such as glycolipids, glycopeptidolipids, glycophospholipids, lipo-oligosaccharides, and even mycolic acids resist cross-linking by chemical fixatives and are prone to extraction by the organic solvents used during the dehydration procedure of conventional embeddings. Yet, with care, conventional embeddings have shown a remarkable degree of cell wall complexity; a triple-layered structure is seen (40). The outermost wall structure consists of an irregular electron-dense outer layer, followed by an electron-translucent region (sometimes referred to as the electron-transparent zone), which sits on top of a densely stained peptidoglycan layer. The electron-translucent region, which some researchers consider to be a lipid capsule, has been especially difficult to preserve (16, 35).

There have not been many attempts to elucidate the mycobacterial wall by freeze-substitution. Because of the high lipid content in the walls, these bacteria are highly hydrophobic, which affects the freezing rate; good vitrification of the cells by freezing is difficult (35–37). Persistence and high freezing rates have finally produced good freeze-substitution images (35). Although the tripartite wall of conventional embeddings is still seen, the wall is approximately two-thirds the conventional thickness (Fig. 12). Because many lipids are extracted from mycobacterial walls during conventional embeddings, strong hydrophobic interactions (which compress the cell wall) are reduced and the

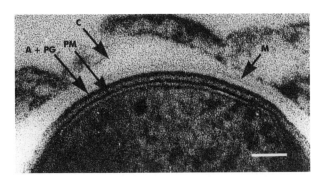

FIGURE 12 Thin section of a freeze-substituted *M. kansasii* that shows the complexity of this bacterium's cell envelope. C, capsule or electron-transparent region; M, mycolates; A + PG, arabinogalactan plus peptidoglycan; PM, plasma membrane. Bar = 50 nm. More details of this image can be found in references 14 and 28. (Reproduced with permission of the author.)

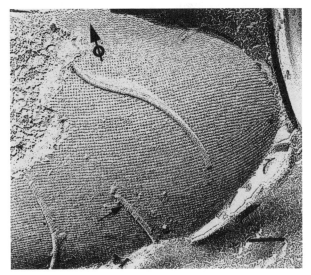

FIGURE 13 Freeze-etched surface of an *Aneurinibacillus thermoaerophilus* 10155/C$^+$ showing the S-layer with its square lattice ($p4$; $a = b = 10$ nm), which is made up of a glycoprotein. See reference 20 for more details. Flagella can be seen on top of the S-layer (small arrow). The large arrow with circle denotes shadow direction. Bar = 100 nm.

wall expands during a conventional embedding (37). Figure 12 shows a freeze-substituted wall of *Mycobacterium kansasii* and the possible arrangement of its various constituents (see references 16 and 37 for more details).

S-LAYERED WALLS

It is not unusual for gram-positive walls to have S-layers. These are paracrystalline surface arrays that self-assemble onto the surfaces of bacteria (and archaea) into oblique (*p1*, *p2*), square (*p4*), or hexagonally (*p3*, *p6*) packed lattices composed of proteins or glycoproteins with a molecular mass of 50 to 120 kDa (6, 32, 43). They are frequently found on *Bacillus*, *Clostridium*, *Lactobacillus*, and *Sporosarcina* spp. and are seen as an extra surface layer on top of the cell wall (Fig. 10). Although the S-layers of gram-positive pathogens have not been extensively studied, especially by freeze-substitution (26), it is clear that the S-layer of *B. anthracis* (Fig. 5), along with the bacterium's two toxins and poly-γ-D-glutamic acid capsule, contributes to this microorganism's virulence (31). It has also been demonstrated that *Clostridium difficile* possesses an S-layer (45). Freeze-etching has been especially useful in the detection of gram-positive S-layers because a major surface fracture plane often follows the outer face of S-layers, and etching clearly reveals the lattice type (Fig. 13).

A GRAM-POSITIVE PERIPLASM

Unlike gram-negative cell envelopes, gram-positive envelopes do not seem to have a clearly defined periplasmic space when viewed by most electron microscopic techniques (7, 10). Yet, many of the functional attributes of periplasmic components (e.g., proteases, DNases, autolysins, newly formed wall components, alkaline phosphatase) can be closely associated with the fabric of their walls. In fact, those bacilli that possess S-layers also have a pool of S-layer protein intermingled with the wall polymers that, presumably, can be called upon for the repair or new assembly of their S-layer (12). Accordingly, even without a dedicated physical space for a periplasm, gram-positive bacteria possess periplasmic components interdigitated among the polymers that make up their cell walls. By definition, then, these bacteria have a periplasm (residing within the cell wall), but unlike gram-negative bacteria, they have no clearly defined periplasmic space (7). It is possible that as we become more familiar with the interpretation of cryosections of gram-positives (Fig. 8), we will find that there is actually a defined periplasmic space within these complex cell envelopes.

We thank Kazuobu Amako and Paul Messner for supplying their excellent images [JRD6](K. A. supplied Fig. 1 and 11, and P. M. supplied Fig. 13). T.J.B.'s research was funded by National Centres of Excellence grants through the Canadian Bacterial Disease Network (CBDN) and the Advanced Food and Materials Network (AFMnet) and a grant from the Natural Sciences and Engineering Research Council of Canada (NSERC). V.R.F.M. was the recipient of a CNPq/Brazil scholarship. The NSERC Guelph Regional STEM Facility is partially funded by an NSERC-Major Facilities Access grant.

REFERENCES

1. **Amako, K., and A. Umeda.** 1977. Scanning electron microscopy of *Staphylococcus aureus*. *J. Ultrastruct. Res.* **58**:34–40.

2. **Amako, K., A. Umeda, and K. Murata.** 1982. Arrangement of peptidoglycan in the cell wall of *Staphylococcus* spp. *J. Bacteriol.* **150**:844–850.

3. **Arthur, M., and P. Courvalin.** 1993. Genetics and mechanisms of glycopeptide resistance in enterococci. *Antimicrob. Agents Chemother.* **37**:1563–1571.

4. **Beveridge, T. J.** 1981. Ultrastructure, chemistry, and function of the bacterial wall. *Int. Rev. Cytol.* **12**:229–317.

5. **Beveridge, T. J.** 1988. Wall ultrastructure: how little we know, p. 3–20. *In* P. Actor, L. Daneo-Moore, M. Higgins, M. R. J. Salton, and G. D. Shockman (ed.), *Antibiotic Inhibition of Bacterial Cell Surface Assembly and Function*. American Society for Microbiology, Washington, D.C.

6. **Beveridge, T. J.** 1994. Bacterial S-layers. *Curr. Opin. Struct. Biol.* **4**:204–212.

7. **Beveridge, T. J.** 1995. The periplasmic space and the periplasm in gram-positive and gram-negative bacteria. *ASM News* **61**:125–130.

8. **Beveridge, T. J.** 1999. Structures of gram-negative cell walls and their derived membrane vesicles. *J. Pathogens.* **181**: 4725–4733.

9. **Beveridge, T. J.** 2000. Ultrastructure of gram-positive cell walls, p. 3–10. *In* V. A. Fischetti, R. P. Novick, J. J. Ferreti, D. A. Portnoy, and J. I. Rood (ed.), *Gram-Positive Pathogens*. ASM Press, Washington, D.C.

10. **Beveridge, T. J., and L. L. Graham.** 1991. Surface layers of bacteria. *Microbiol. Rev.* **55**:684–705.

11. **Beveridge, T. J., D. Moyles, and B. Harris.** Electron microscopy. Submitted for publication in C. A. Reddy et al. (ed.), *Methods for General and Molecular Microbiology*.

12. **Breitwieser, A., K. Gruber, and U. B. Sleytr.** 1992. Evidence for an S-layer protein pool in the peptidoglycan of *Bacillus stearothermophilus*. *J. Bacteriol.* **174**:8008–8015.

13. **Brudney, K., and J. Dobkin.** 1991. Resurgent tuberculosis in New York City. Human immunodeficiency virus, homelessness and the decline of tuberculosis control programs. *Am. Rev. Respir. Dis.* **144**:745–749.

14. **Chapman, G. B., and J. Hillier.** 1953. Electron microscopy of ultrathin sections of bacteria. I. Cellular division in *Bacillus cereus*. *J. Bacteriol.* **66**:362–373.

15. **Culliton, B. J.** 1992. Drug resistant TB may bring epidemic. *Nature* **356**:473.

16. **Daffé, M., and P. Draper.** 1998. The envelope layers of mycobacteria with reference to their pathogenicity. *Adv. Microb. Physiol.* **39**:131–203.

17. **deJonge, B. L. M., Y.-S. Chang, D. Gage, and A. Tomasz.** 1992. Peptidoglycan composition of a highly methicillin-resistant *Staphylococcus aureus*. *J. Biol. Chem.* **267**:11248–11254.

18. **Dmitriev, B. A., F. V. Toukach, K. J. Schaper, O. Holst, E. T. Rietschel, and S. Ehlers.** 2003. Tertiary structure of bacterial murein: the scaffold model. *J. Bacteriol.* **185**: 3458–3468.

19. **Dubochet, J., A. W. McDowall, B. Menge, E. N. Schmid, and K. G. Lickfeld.** 1983. Electron microscopy of frozen-hydrated bacteria. *J. Bacteriol.* **155**:381–390.

20. **Dufrêne, Y. F.** 2004. Using nanotechniques to explore microbial surfaces. *Nat. Rev. Microbiol.* **2**:451–460.

21. **Graham, L. L.** 1991. Freeze-substitution studies of bacteria. *Electron Microsc. Rev.* **5**:77–103.

22. **Graham, L. L., and T. J. Beveridge.** 1990. Evaluation of freeze-substitution and conventional embedding protocols for routine electron microscopic processing of eubacteria. *J. Bacteriol.* **172**:2141–2149.

23. Graham, L. L., and T. J. Beveridge. 1990. Effect of chemical fixatives on accurate preservation of *Escherichia coli* and *Bacillus subtilis* structure in cells prepared by freeze-substitution. *J. Bacteriol.* **172:**2150–2159.

24. Graham, L. L., and T. J. Beveridge. 1994. Structural differentiation of the *Bacillus subtilis* cell wall. *J. Bacteriol.* **176:**1413–1421.

25. Hobot, J. A., E. Carlemalm, W. Villiger, and E. Kellenberger. 1984. Periplasmic gel: new concept resulting from the reinvestigation of bacterial cell envelope ultrastructure by new methods. *J. Bacteriol.* **160:**143–152.

26. Kadurugamuwa, J. L., A. Mayer, P. Messner, M. Sára, U. B. Sleytr, and T. J. Beveridge. 1998. S-layered *Aneurinibacillus* and *Bacillus* spp. are susceptible to the lytic action of *Pseudomonas aeruginosa* membrane vesicles. *J. Bacteriol.* **180:**2306–2311.

27. Koch, A. L., and R. J. Doyle. 1985. Inside-to-outside growth and turnover of the cell wall of gram-positive rods. *J. Theor. Biol.* **117:**137–157.

28. Koval, S. F., and T. J. Beveridge. 2000. Electron microscopy. *In* J. Lederberg (ed.), *Encyclopedia of Microbiology*, 2nd ed. Academic Press, Inc., New York, N.Y.

29. Matias, V. R. F., A. Al-Amoudi, J. Dubochet, and T. J. Beveridge. 2003. Cryo-transmission electron microscopy of frozen-hydrated sections of gram-negative bacteria. *J. Bacteriol.* **185:**6112–6118.

29a. Matias, V. R. F., and T. J. Beveridge. 2005. Cryo-electron microscopy reveals native polymeric cell wall structure in *Bacillus subtilis* 168 and the existence of a periplasmic space. *Mol. Microbiol.* **56:**240–251.

30. Merad, T., A. R. Archibald, I. C. Hancock, C. R. Harwood, and J. A. Hobot. 1989. Cell wall assembly in *Bacillus subtilis*: visualization of old and new wall material by electron microscopic examination of samples stained selectively for teichoic acid and teichuronic acid. *J. Gen. Microbiol.* **135:**645–655.

31. Mesnage, S., E. Tosi-Couture, M. Mock, P. Gounon, and A. Fouet. 1997. Molecular characterization of the *Bacillus anthracis* main S-layer component; evidence that it is the major cell-associated antigen. *Mol. Microbiol.* **23:**1147–1155.

32. Messner, P., and U. B. Sleytr. 1992. Crystalline bacterial cell-surface layers. *Adv. Microb. Physiol.* **33:**213–275.

33. Murray, T., D. L. Popham, and P. Setlow. 1997. Identification and characterization of *pbpA* encoding *Bacillus subtilis* penicillin-binding protein 2A. *J. Bacteriol.* **179:**3021–3029.

34. Neuhaus, F. C., and J. Baddiley. 2003. A continuum of anionic charge: structures and functions of D-alanyl-teichoic acids in gram-positive bacteria. *Microbiol. Mol. Biol. Rev.* **67:**686–723.

35. Paul, T. R., and T. J. Beveridge. 1992. Reevaluation of envelope profiles and cytoplasmic ultrastructure of mycobacteria processed by conventional embedding and freeze-substitution protocols. *J. Bacteriol.* **174:**6508–6517.

36. Paul, T. R., and T. J. Beveridge. 1993. Ultrastructure of mycobacterial surfaces by freeze-substitution. *Zentrabl. Bakteriol.* **279:**450–457.

37. Paul, T. R., and T. J. Beveridge. 1994. Preservation of surface lipids and ultrastructure of *Mycobacterium kansasii* using freeze-substitution. *Infect. Immun.* **62:**1542–1550.

38. Paul, T. R., L. L. Graham, and T. J. Beveridge. 1993. Freeze-substitution and conventional electron microscopy of medically-important bacteria. *Rev. Med. Microbiol.* **4:**65–72.

39. Paul, T. R., A. Venter, L. C. Blaszczak, T. R. Parr, H. Labishinski, and T. J. Beveridge. 1995. Localization of penicillin-binding proteins to the splitting system of *Staphylococcus aureus* septa using a mercury-penicillin V derivative. *J. Bacteriol.* **177:**3631–3640.

40. Rastogi, N., C. Fréhel, and H. L. David. 1986. Triple-layered structure of mycobacterial cell wall: evidence for the existence of a polysaccharide-rich outer layer in 18 mycobacterial species. *Curr. Microbiol.* **13:**237–242.

41. Schleifer, K. H., and O. Kandler. 1972. Peptidoglycan types of bacterial cell walls and their taxonomic implications. *Bacteriol. Rev.* **36:**407–477.

42. Sieradzki, K., and A. Tomasz. 1997. Inhibition of cell wall turnover and autolysis by vancomycin in a highly vancomycin-resistant mutant of *Staphylococcus aureus*. *J. Bacteriol.* **179:**2557–2566.

43. Sleytr, U. B., and T. J. Beveridge. 1999. Bacterial S-layers. *Trends Microbiol.* **7:**253–260.

44. Sugai, M., S. Yamada, S. Nakashima, H. Komatsuzawa, A. Matsumoto, T. Oshida, and H. Suginaka. 1997. Localized perforation of the cell wall by a major autolysin: *atl* gene products and the onset of penicillin-induced lysis of *Staphylococcus aureus*. *J. Bacteriol.* **179:**2958–2962.

45. Takeoka, A., K. Takumi, T. Koga, and T. Kawata. 1991. Purification and characterization of S-layer proteins from *Clostridium difficile* GAI 0714. *J. Gen. Microbiol.* **137:**261–267.

46. Touhami, A., M. H. Jericho, and T. J. Beveridge. 2004. Atomic force microscopy of cell growth and division in *Staphylococcus aureus*. *J. Bacteriol.* **186:**3286–3295.

47. Umeda, A., Y. Ueki, and K. Amako. 1987. Structure of the *Staphylococcus aureus* cell wall determined by the freeze-substitution method. *J. Bacteriol.* **169:**2482–2487.

48. Verwer, R. W. H., and N. Nanninga. 1976. Electron microscopy of isolated cell walls of *Bacillus subtilis* var. *niger*. *Arch. Microbiol.* **109:**195–197.

49. Verwer, R. W. H., N. Nanninga, W. Keck, and U. Schwarz. 1978. Arrangement of glycan chains in the sacculus of *Escherichia coli*. *J. Bacteriol.* **136:**723–729.

50. Vijaranakul, U., M. J. Nadakavukaren, B. deJonge, B. J. Wilkinson, and R. K. Jayaswal. 1995. Increased cell size and shortened peptidoglycan interpeptide bridge of NaCl-stressed *Staphylococcus aureus* and their reversal by glycine betaine. *J. Bacteriol.* **177:**5116–5121.

51. Woodford, N., A. P. Johnson, D. Morrison, and D. C. E. Speller. 1995. Current perspectives on glycopeptide resistance. *Clin. Microbiol. Rev.* **8:**585–615.

52. Yother, J., K. Leopold, J. White, and W. Fischer. 1998. Generation and properties of *Streptococcus pneumoniae* mutant which does not require choline or analogs for growth. *J. Bacteriol.* **180:**2093–2101.

Surface Proteins on Gram-Positive Bacteria

VINCENT A. FISCHETTI

2

As arms, legs, hair, and fur are used in higher species for their survival in the environment, surface appendages are used by bacteria for similar purposes. Surface molecules in bacteria range from complex structures such as flagella that propel the organism in aqueous environments, to less sophisticated polysaccharides and proteins. All of these molecules serve to benefit the organism for survival in a hostile environment, such as the waters of a rushing stream, the blood of an infected animal, the surface of an object, or the surface of a mucosal epithelium. Although it was previously believed that bacteria were simple single-cell organisms with little complexity, it is now becoming apparent that they are highly evolved, advanced particles that possess a wide array of surface molecules that serve to manipulate the organism in its environment. For human pathogens, surface molecules have been finely tuned to allow adherence and colonization of host surfaces, invasion of cells, evasion of the host's immune response, and persistence in infected tissues.

In an effort to emphasize the complexity of bacterial surface molecules and their use in the everyday life of the bacterium, this chapter will focus on those surface proteins found on gram-positive bacteria. For an extensive review of the subject, see reference 70.

GRAM-POSITIVE CELL WALL

From electron microscopic analysis of the gram-positive cell envelope (see chapter 1, this volume) and a large number of elegant chemical and immunological analyses, a picture of the gram-positive cell wall has emerged (Fig. 1). The structure differs significantly from the gram-negative cell wall in two ways: (i) the presence of a thicker and more cross-linked peptidoglycan and (ii) the lack of an outer membrane. Because of these differences, surface molecules in gram-positive organisms vary from those in gram-negative organisms, in which specialized systems are required to transport and anchor molecules through the outer membrane (107). In general, surface proteins in gram-positive bacteria can be separated into three categories: (i) those that anchor at their C-terminal ends (through an LPXTG motif), (ii) those that bind by way of charge or hydrophobic interactions, and (iii) those that bind via their N-terminal region (lipoproteins) (Fig. 1).

C-TERMINAL-ANCHORED PROTEINS

Decades ago, relatively harsh treatments of extraction were used to remove surface proteins from gram-positive bacteria, resulting in the isolation and characterization of molecular fragments rather than complete molecules. It was not until the early 1980s that the genes for surface proteins on gram-positive bacteria were first published (protein A from *Staphylococcus aureus* [33, 111] and M protein from *Streptococcus pyogenes* [39]). Although at first glance these molecules had little in common, it was soon realized that the C-terminal end of the protein A sequence contained a sequence error. Upon correction (111), it became apparent that this region of protein A exhibited very close sequence homologies with the M protein molecule. As other protein sequences from gram-positive bacteria were published in the ensuing years, it became obvious that there was a common theme within the structure of these C-terminal-anchored molecules. Since that time >400 proteins from gram-positive bacteria that anchor through their C-terminal region, based on the presence of the LPXTG motif, have been reported (see Table 1 for a representative list).

M Protein Structure

Of the C-terminal-anchored proteins, the M protein may be considered an archetypical molecule, a characteristic of which is the presence of regions with sequence repeats and a conserved anchor. As seen in Fig. 2, the M protein is composed of four repeat blocks, each differing in size and sequence (39). The A-repeats are each composed of 14 amino acids in which the central blocks are identical and the end blocks slightly diverge from the central consensus repeats. The B-repeats, composed of 25 amino acids each, are arranged in the same way. The C-repeats, composed of 2.5 blocks of 42 amino acids, are not as similar to each other as are the A- and B-repeats. There are also four short D-repeats that show some homology. These repeat segments make up the central helical rod region of the M molecule

Gram-Positive Pathogens, 2nd edition, edited by Vincent A. Fischetti et al.
© 2006 ASM Press, Washington, D.C.

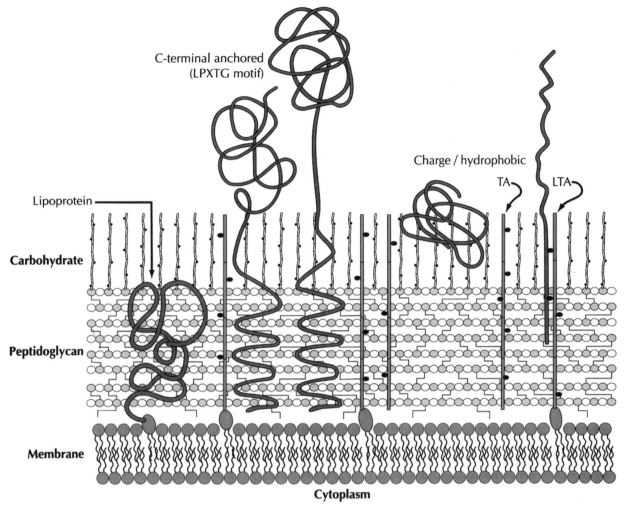

FIGURE 1 Major surface structures of the cell wall of gram-positive bacteria. Linked to the surface of the peptidoglycan, many gram-positive organisms have polysaccharide structures that in some cases are used for their immunological classification. Surface proteins are linked by three mechanisms. (i) Lipoproteins have a lipid linked through a cysteine at the N terminus. (ii) C-terminal-anchored proteins are attached and stabilized in the peptidoglycan through a C-terminal complex containing an LPXTG motif. (Most surface proteins are anchored in this way.) (iii) Certain surface proteins are attached through hydrophobic and/or charge interactions to the cell surface. (Some proteins are bound ionically to the lipoteichoic acid.) The teichoic acids (TA) are a common feature of the gram-positive cell wall. TA is usually composed of a repeating carbohydrate-phosphate polymer linked through a phosphodiester linkage to the peptidoglycan. Lipoteichoic acid (LTA) is composed of a similar polymer linked to the cytoplasmic membrane through a fatty acid (see chapter 19, this volume).

because of the high helical potential ascribed to the amino acids found within this segment, as determined by conformational analysis (25, 83).

A close examination of the sequence within the M molecule revealed a repeating seven-residue periodicity of nonpolar amino acids, a basic characteristic of α-helical coiled-coil proteins such as mammalian tropomyosin. In general, α-helical coiled-coil proteins are constructed from a reiterating seven-residue amino acid pattern $(a\text{-}b\text{-}c\text{-}d\text{-}e\text{-}f\text{-}g)_n$ in which residues in positions a and d are hydrophobic and form the core residues in the coiled-coil, and the intervening residues are primarily helix-promoting. The arrangement of

amino acids of the M6 protein within the seven-residue pattern is shown in Fig. 3. The seven-residue periodicity from amino acids 12 through 362 strongly suggests that this region is in a coiled-coil conformation and forms the helical central rod region of the molecule. Discontinuities in the heptad pattern (seen especially in the B-repeat region), which have been found in the sequence of other M proteins and other coiled-coil molecules, probably account for the flexibility of the M molecules observed in electron micrographs (83). Based on these irregularities in the heptad pattern, the central rod region is divided into three subregions that correlate with the A-, B-, and C-repeat blocks (25).

TABLE 1 C-terminal-linked sequenced surface proteins[a] from gram-positive bacteria

Name/gene	Function/name	Organism	LPXTG[b]	Reference	GenBank accession no.
1. emm6	M protein	*Streptococcus pyogenes*	**LPSTG**	39	M11338
2. emm5	M protein	*S. pyogenes*	**LPSTG**	67	M20374
3. emm12	M protein	*S. pyogenes*	**LPSTG**	86	U02342
4. emm24	M protein	*S. pyogenes*	**LPSTG**	68	M19031
5. emm49	M protein	*S. pyogenes*	**LPSTG**	34	M23689
6. emm57	M protein	*S. pyogenes*	**LPSTG**	62	X60959
7. emm2	M protein	*S. pyogenes*	**LPSTG**	5	X61276
8. emm3	M protein	*S. pyogenes*	**LPSTG**	50	Z21845
9. ARP2	IgA-binding protein	*S. pyogenes*	**LPSTG**	5	X61276
10. ARP4	IgA-binding protein	*S. pyogenes*	**LPSTG**	28	X15198
11. Mrp4	IgG/fibrinogen binding	*S. pyogenes*	**LPSTG**	76	M87831
12. FcRA	Fc-binding protein	*S. pyogenes*	**LPSTG**	35	M22532
13. Prot H	Human IgG Fc binding	*S. pyogenes*	**LPSTG**	32	M29398
14. SCP	C5a peptidase	*S. pyogenes*	**LPTTN**	11	J05229
15. T6	Protease-resistant protein	*S. pyogenes*	**LPSTG**	92	M32978
16. sof22	Serum opacity factor	*S. pyogenes*	**LPASG**	84	U02290
17. Sfb	Fibronectin binding	*S. pyogenes*	**LPATG**	105	X67947
18. ZAG	Binds $_{\gamma2}$ M, Alb, IgG	*Streptococcus zooepidemicus*	**LPTTG**	46	U02290
19. PrtF	Fibronectin binding	*S. pyogenes*	**LPATG**	95	L10919
20. PAM	Plasmin binding	*S. pyogenes*	**LPSTG**	3	Z22219
21. Prot L	Light-chain binding	*Peptostreptococcus magnus*	**LPKAG**	49	M86697
22. PAB	Human serum albumin binding	*P. magnus*	**LPEAG**	17	X77864
23. bac	IgA-binding protein	Group B streptococcus	**LPYTG**	36	X58470
24. bca	Alpha C antigen	Group B streptococcus	**LPATG**	66	M97256
25. fnbA	Fibronectin binding	*Streptococcus dysgalactiae*	**LPQTG**	61	Z22150
26. fnbB	Fibronectin binding	*S. dysgalactiae*	**LPAAG**	61	Z22151
27. Fnz	Fibronectin binding	*Streptococcus equi*	**LPQTS**	71	Y17116
28. SeM	M-like	*S. equi*	**LPSTG**	108	U73162
29. SzPSe	M-like	*S. equi*	**LPQTS**	108	U73163
30. Prot G	IgG-binding protein	Group G streptococcus	**LPTTG**	74	X06173
31. EmmG1	M protein	Group G streptococcus	**LPSTG**	15	M95774
32. DG12	Albumin-binding protein	Group G streptococcus	**LPSTG**	98	M95520
33. GfbA	Fibronectin binding	Group G streptococcus	**LPATG**	53	U31115
34. MRP	Surface protein[a]	*Streptococcus suis*	**LPNTG**	99	X64450
35. PAc	Surface protein	*Streptococcus mutans*	**LPNTG**	73	X14490
36. spaP	Surface protein	*S. mutans*	**LPNTG**	51	X17390
37. spa	Surface protein	*Streptococcus sobrinus*	**LPATG**	109	D90354
38. wapA	Wall-associated protein A	*S. mutans*	**LPSTG**	22	M19347
39. fruA	Fructosidase	*S. mutans*	**LPDTG**	9	L03358
40. Sec10	Surface protein	*Enterococcus faecalis*	**LPQTG**	48	M64978
41. Asc10	Surface protein	*Enterococcus faecalis*	**LPKTG**	48	M64978
42. asa1	Aggregation substance	*Enterococcus faecalis*	**LPQTG**	30	X17214
43. Prot A	IgG-binding protein	*Staphylococcus aureus*	**LPETG**	33	X00342
44. FnBPb	Fibronectin binding	*S. aureus*	**LPETG**	97	J04151
45. FnBPa	Fibronectin binding	*S. aureus*	**LPETG**	47	X62992
46. cna	Collagen-binding protein	*S. aureus*	**LPKTG**	82	M81736
47. clfA	Clumping factor	*S. aureus*	**LPDTG**	65	Z18852
48. EDIN	Epidermal cell inhibitor	*S. aureus*	**LPRGT**	43	NA[c]
49. prtM	Cell wall protease	*Lactobacillus paracasei*	**LPKTA**	38	M83946
50. wg2	Cell wall protease	*Streptococcus cremoris*	**LPKTG**	54	M24767
51. inlA	Internalization protein	*Listeria monocytogenes*	**LPTTG**	29	M67471
52. Fimbriae	Type 1 fimbriae	*Actinomyces viscosus*	**LPLTG**	111	M32067
53. Fimbriae	Type 2 fimbriae	*Actinomyces naeslundii*	**LPLTG**	116	M21976
54. FimA	Type 2 fimbriae	*A. naeslundii*	**LPLTG**	118	AF019629
55. Hyal1	Hyaluronidase	*Streptococcus pneumoniae*	**LPQTG**	4	L20670
56. nanA	Neuraminidase	*S. pneumoniae*	**LPETG**	10	X72967
57. glnA	Glutamine synthetase	Group B streptococcus	**LPATL**	102	U61271
58. Protein F2	Fibronectin binding	*S. pyogenes*	**LPATG**	44	U31980
59. Fbe	Fibronectin binding	*Staphylococcus epidermidis*	**LPDTG**	71	Y1711660.
60. InlC2	Internalin-like	*L. monocytogenes*	**LPTAG**	19	U77368

(Continued on next page)

TABLE 1 (*Continued*)

Name/gene	Function/name	Organism	LPXTG[b]	Reference	GenBank accession no.
61. InlD	Internalin-like	*L. monocytogenes*	**LPTAG**	19	U77368
62. InlE	Internalin-like	*L. monocytogenes*	**LPITG**	19	U77368
63. InlF	Internalin-like	*L. monocytogenes*	**LPKTG**	19	U77367
64. SfBP1	Fibronectin binding	*S. pyogenes*	**LPXTG**	87	AF071083
65. *dex*	Dextranase	*S. sobrinus*	**LPKTG**	113	M96978
66. *padl*	Pheromone	*E. faecalis*	**LPHTG**	114	X62658
67. FAI	Fibronogen/albumin/IgG	Group C streptococcus	**LPSTG**	104	NA
68. Sfbll	Fibronectin binding	*S. pyogenes*	**LPASG**	56	X83303
69. CspB	Fimbriae associated	*Streptococcus salivarius*	**LPETG**	60	AF353638
70. HarA	Haptoglobin binding	*S. aureus*	**LPKTG**	20	NA
71. RSP	Adhesin	*Erysipelothrix rhusiopathiae*	**LPN/QTG**	96	AB052682
72. SAS	Several unknown	*S. aureus*	**LPXTG**	88	NA
73. GspB	Platelet binding	*Streptococcus gordonii*	**LPRTG**	2	AAL13053

Surface proteins attached by other mechanisms

1. *ftf*	Fructosyl transferase	*S. salivarius*		85	
2. *pspA*	Surface protein	*S. pneumoniae*		42, 119	
3. *aas*	Adhesin/autolytic	*Streptococcus saprophyticus*		37	
4. FbpA	Fibrinogen binding	*S. aureus*		13	
5. FBP54	Fibronectin/fibrinogen binding	*S. pyogenes*		16	
6. Fib	Fibronectin binding	*S. aureus*		8	
7. SEN	Surface enolase	*S. pyogenes*		81	
8. SDH	Surface dehydrogenase	*S. pyogenes*		80	

[a]Surface proteins: proteins that have been identified to have a C-terminal anchor motif, but the function is unknown.
[b]L,72/72; P, 72/72; (X); T, 59/72; G, 61/72.
[c]NA, not available.

Electron and Fluorescent Microscopy

Although early electron micrographs of the surface of gram-positive bacteria revealed certain structures, it was not known whether they were proteinaceous or polysaccharides. It was not until Swanson et al. (103) published the first electron micrographs of the M protein on the surface of group A streptococci that it was realized that the molecule was an elongated structure. Although other proteins were also reported to be on the streptococcal surface, they were not apparent in these electron micrographs. Thus, direct visualization of surface molecules is limited to molecules with certain physicochemical characteristics. In the case of the M protein, it is the α-helical coiled-coil structure that allowed its visualization by electron microscopy (25).

In experiments designed to answer questions about the synthesis and placement of M protein on the cell wall, Swanson et al. (103) found that after trypsinization to remove existing M protein on living streptococci newly synthesized M protein was first seen by electron microscopy on the cell in the position of the newly forming septum. In similar experiments designed after the classical experiments of Cole and Hahn (14) using fluorescein-labeled anti-M antibodies, trypsinized streptococci placed in fresh medium for 10 min revealed M protein first at the newly forming septum (Fig. 4). On organisms examined after 40 min or more of incubation, M protein was not observed in the position of the old wall, suggesting that the M molecule is produced only where the new cell wall is synthesized. This confirmed the observation of Swanson et al. (103). Whether this is also true of the many other C-terminal-anchored surface proteins has not yet been determined; however, it is likely to be the case.

C-Terminal Anchor Region

An examination of the C-terminal end of those surface molecules that anchor at that end revealed, without exception, that all have a similar arrangement of amino acids. Up to seven charged amino acids are found at the C terminus and are composed of a mixture of both negatively and positively charged residues. Immediately N-terminal to this short charged region is a segment of 15 to 22 predominantly hydrophobic amino acids sufficient to span the cytoplasmic membrane of the bacterium, placing the charged end in the cytoplasm (Fig. 2). In all these proteins, the sequences found in the hydrophobic and charged regions are not necessarily the same, but the chemical characteristics of the amino acids used to compose them are conserved. Beginning about eight amino acids N-terminal from the hydrophobic domain is a heptapeptide with the consensus sequence LPXTG, which is extraordinarily conserved among all C-terminal-anchored proteins examined so far (26). Although several amino acid substitutions are seen in position 3 of the heptapeptide (predominantly A, Q, E, T, N, D, K, and L), positions 1, 2, 4, and 5 are nearly completely conserved (100% in the L and P positions and >85% in the T and G positions). This conservation is also maintained at the DNA level. The preservation of this heptapeptide and the high homology within the hydrophobic and charged regions suggest that the method of anchoring these molecules within the bacterial cell is also highly conserved, and the LPXTG motif may serve as an enzyme recognition sequence.

Thus, C-terminal-anchored surface proteins in gram-positive bacteria (which could number >25 different molecules in a single organism) are synthesized and exported at

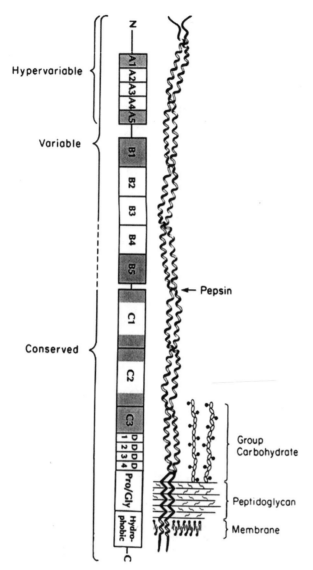

FIGURE 2 Characteristics of the complete M6 protein sequence. Blocks A, B, C, and D designate the location of the sequence repeat blocks. Block C3 is half the size of blocks C1 and C2. Shadowed blocks indicate those in which the sequence diverges from the central consensus sequence. Pro/Gly denotes the proline- and glycine-rich region likely located in the peptidoglycan. Membrane anchor is a 19-hydrophobic-amino-acid region adjacent to a 6-amino-acid charged tail. Pepsin identifies the position of the pepsin-sensitive site after amino acid 228. The C-terminal end is located within the cell wall and membrane.

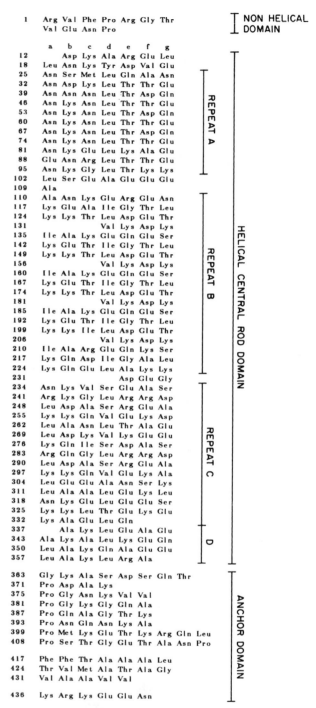

FIGURE 3 Complete M6 protein sequence arranged to highlight the seven-residue periodicity found in the helical central rod region. Region assignments are based on sequence and conformational analyses. Arrangement of the sequence is based on the position of amino acids in a seven-residue periodicity designated by letters a through g beginning at residue 12 and continuing through residue 362, with interruptions at residues 109, 131, 156, 181, 206, 231, and 337. Alignment from residue 363 to 416 is used essentially to highlight the regularity of the position of prolines in the sequence. No periodicity is found from residue 417 to the end. Three major regions are indicated (nonhelical, helical, and anchor). The pepsin-sensitive site is between Ala-228 and Lys-229.

the septum, where new cell wall is also being produced and translocated to the surface (14, 25). Therefore, to coordinate the export and anchoring of all these proteins, the C-terminal hydrophobic domain and charged tail function as a temporary stop, precisely positioning the LPXTG motif at the outer surface of the cytoplasmic membrane (Fig. 5). This motif is then cleaved, ultimately resulting in the attachment of the surface-exposed segment of the protein to a cellular substrate (94). This idea is supported by studies indicating that the C-terminal hydrophobic domain and

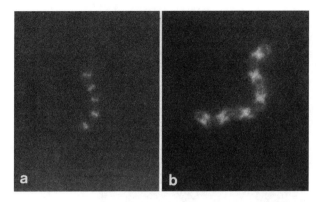

FIGURE 4 Fluorescein-labeled anti-M6 antibody analysis of the appearance of M6 protein on streptococcal cell walls. M6 streptococci were treated with trypsin to remove surface M protein. Cells, reincubated at 37°C, were removed at intervals, fixed, and stained with fluorescein-labeled anti-M6 antibody. (a) After 10 min of incubation the M protein is located within a thin band at the position of the newly forming septum. Magnification, ×5,000. (b) Location of the M protein after 40 min of incubation. Note that no fluorescein label is seen in the position of the old wall. Magnification, ×6,000.

charged tail are missing from the streptococcal M protein extracted from the streptococcal cell wall (78, 94). Studies also suggest that the substrate in the peptidoglycan to which the protein is linked is the cross bridge (e.g., glycine in *S. aureus*, alanine in *S. pyogenes*) of the pentapeptide (93).

Enzymes That Cleave the LPXTG Motif

Sortase

Sortase is a membrane-bound transpeptidase found in gram-positive bacteria (1, 6, 52, 63, 75), with the capacity to both cleave the LPXTG motif and anchor the newly formed C terminus to the nascent cross bridge in the peptidoglycan (63, 110) (for detailed review see reference 69). Mutants in the sortase gene (*SrtA*) have been shown to be defective in their ability to surface display proteins (1, 6, 52, 64). Because surface proteins are essential for the infectivity and survival of gram-positive bacteria in the host, SrtA mutants have been shown to be attenuated in colonization experiments (51) and animal infection models (64). A genome-wide analysis of sortases revealed that sortase-like proteins were present in almost all gram-positive bacterial species, and that in every gram-positive bacterium in which one sortase-like protein was identified, at least one other was also found, and in most cases several others (1, 77). Further analysis indicates that genes encoding sortase-like proteins are usually found to be clustered along with genes encoding their probable substrates. Because of their importance in surface protein display, sortase molecules would be a target for antibiotic development. Unlike current antibiotics, which affect bacterial viability, antibiotics that block the action of sortases would compromise the pathogenicity of a pathogen, yielding a new class of anti-infectives.

LPXTGase

Recently a new enzyme was identified from *S. pyogenes* that differs significantly from sortase but also cleaves within the LPXTG motif of surface proteins (59). This enzyme, termed LPXTGase, has no aromatic amino acids, is rich in alanine (particularly D-Ala [58]), is heavily glycosylated, and is 30% composed of uncommon amino acids, suggesting that it may be constructed independently of ribosomes. The enzyme, which is also found in *S. aureus*, is also important in cell surface protein display as well as cell wall assembly. Because the cleavage activity of LPXTGase is at least two orders of magnitude greater than sortase (110), LPXTGase may play an essential role in the rapid display of thousands of surface protein molecules during a short (20 to 30 min) bacterial division cycle.

Wall-Associated Region

Immediately N-terminal from the LPXTG motif is the wall-associated region, which spans about 50 to as many as 125 amino acid residues and is found in nearly all the proteins analyzed. Although this region does not exhibit a high degree of sequence identity among the known surface proteins, it is characterized by a high percentage of proline/glycine and threonine/serine residues. For some proteins (such as the M protein) the concentration of proline/glycine is significantly higher than that of threonine/serine, while in others this relationship is either reversed or nearly equal. Because of its proximity to the hydrophobic domain (which is located in the cell membrane), this region would be positioned within the peptidoglycan layer of the cell wall (78) (Fig. 1). The reason for the presence of these particular amino acids at this location is unknown. One hypothesis suggests that the prolines and glycines, with their ability to initiate bends and turns within proteins, allow the peptidoglycan to become more easily cross-linked around these folds, thus stabilizing the molecule within the cell wall (78). The function of the threonines and serines within this region is not immediately apparent.

Surface-Exposed Region

As the C-terminal region is characteristically conserved among the C-terminal-anchored surface proteins, the surface-exposed region is characteristically unique. Despite their differences, these molecules appear to fall into three groups: (i) those with several repeating sequence blocks, (ii) those with repeat blocks located close to and within the cell wall region, and (iii) those without any repeat sequences (Fig. 6). Streptococcal M protein and protein G may be considered molecules representative of those containing several sequence repeats. However, the most commonly found structures are those with repeats located close to the cell wall and an extended nonrepetitive region N-terminal to this segment.

Conformational analysis of 12 representative surface molecules by the algorithm of Garnier et al. (31) revealed that, in general, those proteins containing repeat sequences were predominantly helical within the region containing repeat segments (Fig. 7). Conversely, regions and molecules without repeat blocks were predominantly composed of amino acids exhibiting high β-sheet, β-turn, and/or random coil potential. Thus, within most of these molecules, the presence of repeat segments usually predicts the location of a helical domain. Possibly one of the pressures for the maintenance of repeat blocks is to preserve the helix potential within specific regions of these molecules, the presence of which may determine an extended protein structure, as has been shown for the M protein (83). An exception to this is found in the Sec10 protein, a predominantly helical molecule with limited repeat segments.

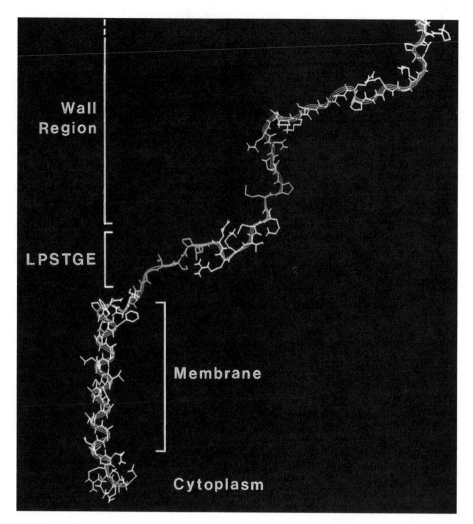

FIGURE 5 (*See the separate color insert for the color version of this illustration.*) Computer-generated model of the C-terminal end of the M protein sequence (residues 371 to 441). A comparable region is found in all C-terminal-anchored surface proteins from gram-positive bacteria (see Table 1). The predicted location of this segment of the molecule is shown in the cytoplasm, membrane, and peptidoglycan. The space between the membrane and peptidoglycan (wall region) may be considered the "periplasm" of the gram-positive bacterium. The figure was generated on a Steller computer using the Quanta 2.1A program for energy minimization.

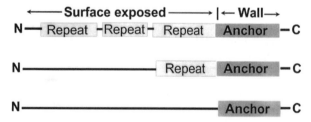

FIGURE 6 Characteristics of C-terminal-anchored surface proteins. These proteins fall into three basic categories: those with multiple repeats, single repeats, or no repeats. Proteins with a single repeat located close to the cell wall are the most common. Anchor is the region of the molecule located within the cell wall carbohydrate and peptidoglycan.

Seven-Residue Periodicity

As shown above, most of the surface molecules containing repeat sequences were found to be α-helical in those regions composed of repeats, while molecules without repeat blocks exhibited a high degree of β-sheet, β-turn, and random coil. However, little information was available to indicate that, except for M6 protein, any of these surface proteins are also in a coiled-coil conformation. To attempt to answer this question, an algorithm (Matcher) was developed (27) and used to determine if a seven-residue periodicity also exists in those molecules containing α-helical regions. When this algorithm was applied to representative sequences, extended regions of seven-residue periodicity were found in Arp4, FcRA, protein H, PAc, and spaP proteins (Fig. 7). This strongly suggested that these molecules may attain a coiled-

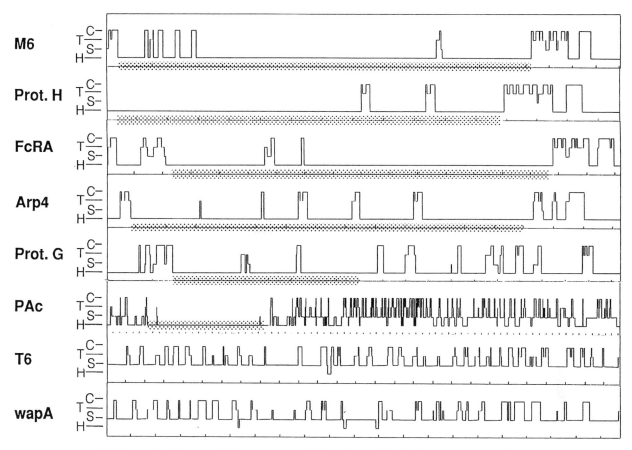

FIGURE 7 Conformational characteristics of the surface molecules from gram-positive bacteria. The sequences were analyzed by the Garnier-Robson algorithm supplied with the EuGene protein analysis package. The location within these molecules of regions exhibiting random coil (C), β-turn (T), β-sheet (S), and α-helix (H) are designated. Sequences were derived from the references listed in Table 1. Shaded areas are those containing a seven-residue periodicity based on the Matcher program. Proteins T6 and wapA exhibit no extended helical regions or seven-residue periodicity.

coil conformation within these α-helical segments. It is apparent that only those proteins that contain sequence repeats and exhibit high α-helical potential show the presence of a seven-residue pattern of hydrophobic amino acids. On the other hand, even though the protein G molecule exhibits extensive helix potential, only a limited portion was found to contain the heptad pattern. Interestingly, however, the PAc protein from *Streptococcus mutans* has only one small segment between residues 120 and 520 that displays repeat blocks, and this segment exhibits strong helix potential. This is precisely the region that was also found to contain a regular seven-residue heptad pattern. Taken together, these data strongly suggest that the coiled-coil conformation may also be a common characteristic of surface proteins found on gram-positive organisms.

Size Variation

To fully appreciate the diversity among certain surface proteins on gram-positive organisms, it is important to understand a property of the M protein molecule, namely, size variation. Using a broadly cross-reactive monoclonal antibody as a probe, the size of the M protein, extracted by solubilizing the cell wall from a number of streptococcal strains, was examined by Western blot. M protein derived from 20 different serotypes exhibited variations in size rang-

ing from 41 to 80 kDa in molecular mass, depending on the strain (25). Similar size variation was observed when streptococcal strains of the same serotype (M6) isolated over a period of 40 years were examined in a similar way. This variation in size may be explained by the observation that the M sequence contains extensive repeats at both the protein and DNA levels. Long, reiterated DNA sequences are likely to serve as substrates for recombinational events or for replicative slippage, generating deletions and duplications within the M protein gene that lead to the production of M proteins of different sizes.

Sequence analysis of the M6 gene isolates from local streptococcal outbreaks revealed that they are clonally related and not the result of separate acquisitions of unrelated M6 strains during the course of the infection (25). The observation that size variants occur among these clinical isolates suggests that the size mutant constituted the major organism in the streptococcal population at the time of isolation and that a given size mutant has a selective advantage under clinical conditions. Perhaps serological pressure as a result of a local immune response forces the appearance of the size variant.

Although sequence repeats have been reported in several surface proteins on gram-positive bacteria, size variation has not been reported or tested for all these proteins.

However, in addition to M protein, size variation has been observed in PspA from *Streptococcus pneumoniae* (112), HagA from *Porphyromonas gingivalis* (55), protein A from *S. aureus* (12), and Sfbp1 from *S. pyogenes* (87).

Multifunctional Proteins

Of the >100 proteins identified on the gram-positive surface, the great majority contain domains that bind molecules found in body secretions (Table 1), including immunoglobulins (IgA and IgG), albumin, fibronectin, and fibrinogen. Other proteins have been identified because of the presence of enzymatic domains. Based on current data, it is clear that the majority of these proteins are multifunctional, in which the identified function is limited to only one segment of the molecule. In most cases, the function of the other region(s) has not been identified. For example, as described above, it appears that many of the surface proteins contain a repeat region located close to the cell wall and a nonrepetitive segment N-terminal to it (Fig. 6). In most cases the binding domain (e.g., fibronectin, immunoglobulin) has been localized to the repeat segment while the N-terminal region exhibits no known function. In the case of streptococcal opacity factor, both domains have been defined (84); the repeats bind fibronectin while the N-terminal domain has the catalytic function that cleaves ApoA1 of high-density lipoprotein, resulting in the opacity of serum. In the case of highly repetitive proteins such as M protein and protein G, each repeat domain has a specific function. In M protein, the A-repeats bind albumin, mucin, and C4 binding protein, while the B-repeats bind fibrinogen (7, 41, 45, 90, 91), and the C-repeats bind factor H of complement (40) and keratinocytes (72). In protein G, one repeat domain binds IgG while the second binds albumin. The multifunctional characteristic of surface proteins is not limited to molecules anchored via their C-terminal region. Proteins bound through hydrophobic/charge interactions and N-terminally anchored proteins also exhibit this characteristic (see below).

Thus, in general, surface proteins on most gram-positive organisms are multifunctional molecules with at least two and in some cases three or more independent functions. Given that up to 25 or more of these surface proteins may ultimately be found on a single cell surface, the potential complexity associated with the bacterial surface becomes enormous.

PROTEINS ANCHORED BY CHARGE AND/OR HYDROPHOBIC INTERACTIONS

Although the great majority of proteins identified on the surface of gram-positive bacteria anchor through their C termini, a few have been identified to bind through less-defined charge and/or hydrophobic interactions (Table 1).

Surface Glycolytic Enzymes

Through an in-depth analysis of the individual proteins identified in a cell wall extract of group A streptococci, it was discovered that some proteins were composed of enzymes normally found in the glycolytic pathway located in the cytosol (79). In all, five enzymes have been identified (triosphosphate isomerase, glyceraldehyde-3-phosphate dehydrogenase [GAPDH], phosphoglycerate kinase, phosphoglycerate mutase, and α-enolase) that form a short contiguous segment of the glycolytic pathway. Interestingly,

they are a complex of enzymes involved in the production of ATP (Pancholi and Fischetti, unpublished data).

Support for the specific translocation of these enzymes to the cell surface instead of a nonspecific release through cell lysis comes from the inability to identify cytoplasmic markers or other glycolytic enzymes in the growth medium or on the cell surface.

These enzymes have been identified on the surface of nearly all streptococcal groups, and certain enzymes have been identified on some pathogens, the surface of fungi such as *Candida albicans*, and parasites such as trypanosomes and schistosomes (Table 2). Given the right substrate, streptococci have the capacity to produce ATP on the cell's surface, which further increases the complexity of such a bacterial surface.

Two of the best-characterized surface glycolytic enzymes on the streptococcus are GAPDH and α-enolase (79, 81). Like other surface proteins on gram-positive bacteria, GAPDH is a multifunctional protein with binding affinities for fibronectin and lysozyme as well as for cytoskeletal proteins such as actin and myosin. GAPDH has also been shown to have ADP-ribosylating activity in addition to its GAPDH activity. α-Enolase is also multifunctional in its ability to specifically also bind plasmin(ogen) (81). It is the major plasmin-binding protein for streptococci. Evidence has shown that there is one gene for each of these enzymes, which are produced in the glycolytic cytosol. A proportion (roughly 30 to 40%) is then translocated to the surface. Because these molecules are not synthesized with an N-terminal signal sequence, it is unclear how they are transported to the cell surface. Also, because they do not contain an apparent anchor motif, it is not known how they are attached to the cell surface. Because they can be removed with chaotropic agents, it is reasonable to assume that they are bound through charge and/or hydrophobic interactions (18).

Other Proteins

PspA, a surface protein and a virulence determinant for *S. pneumoniae* (115), has many of the structural characteristics found in M protein, such as sequence repeats and α-helical coiled-coil conformation, but it does not contain a C-terminal-anchor motif. It has been shown that after surface translocation, PspA anchoring requires charge interactions between the membrane-associated choline-containing lipo-

TABLE 2 Glycolytic enzymes found on the surface of microorganisms

Enzyme[a]	Bacteria	Fungus	Parasite
GAPDH	Streptococci	*Candida albicans*	Schistosome
	S. pneumoniae		Trypanosome
	S. aureus		
	Mycobacterium avium		
PGK	Streptococci	*C. albicans*	Trypanosome
TPI	Streptococci		Trypanosome
PGM	Streptococci		
α-Enolase	Streptococci	*C. albicans*	

[a]GAPDH, glyceraldehyde-3-phosphate dehydrogenase; PGK, phosphoglycerate kinase; TPI, triosphosphate isomerase; PGM, phosphoglycerate mutase.

teichoic acid and the C-terminal repeat region of the PspA molecule (Fig. 1).

N-TERMINAL-ANCHORED LIPOPROTEINS

Although lipoproteins (proteins containing a lipid covalently linked to an N-terminal cysteine) have been described in gram-negative bacteria, they have only recently been identified in gram-positive bacteria (for review, see reference 101). Although most are predicted to be lipoproteins by genetic analysis based on the consensus sequence for lipoproteins (100), they usually comprise molecules that are not surface-exposed. For example, these proteins are involved in transport systems such as the *mal* and *ami* operons of *S. pneumoniae* and the Msm system of *S. mutans*. A few of these lipoproteins have been identified by their ability to act as adhesins, allowing these organisms to adhere to a variety of substrates. SsaB, a lipoprotein from *Streptococcus sanguis* that shares sequence homology with other streptococcal adhesins, has been implicated in the interaction with a salivary receptor and the coaggregation with *Actinomyces naeslundii*. Many of these so-called adhesins have also been found to participate in various transport systems for the streptococcus. Lipoproteins have also been implicated in the spore cell cycle of *Bacillus subtilis* (21), conjugation in *Enterococcus faecalis* (106), and enzymatic functions (57, 89).

Thus, it is expected that these proteins will be located predominantly on the surface of the cell membrane, with the majority of the molecule located within the peptidoglycan (Fig. 1). In some cases it is possible that certain domains will be surface exposed, allowing them to function as adhesins or even surface enzymes.

CONCLUSION

Based on current information, the surface of the gram-positive bacterial cell wall is highly complex (Fig. 8) and could even be considered to be a distinct organelle, because it is composed of proteins with specific binding functions and enzymatic activity combined with the ability to generate energy. Considering the fact that there could be more than 25 different proteins on the cell surface, each with the potential of up to three functional domains, more than 75 independent activities could potentially be present on the cell surface, much more than had ever been previously anticipated.

Because extensive cytoplasmic domains are not present within the surface proteins thus far identified in gram-positive bacteria, it is unlikely that the binding of these molecules to specific ligands in the bacterial cell surface induces a cytoplasmic signal to activate a gene product. It is more likely that binding initiates a conformational signal on the bacterial cell surface to perform a specific function. For example, streptococci usually infect the pharynx, particularly the tonsils, through contact with contaminated saliva. Upon entering the oral cavity the organism first

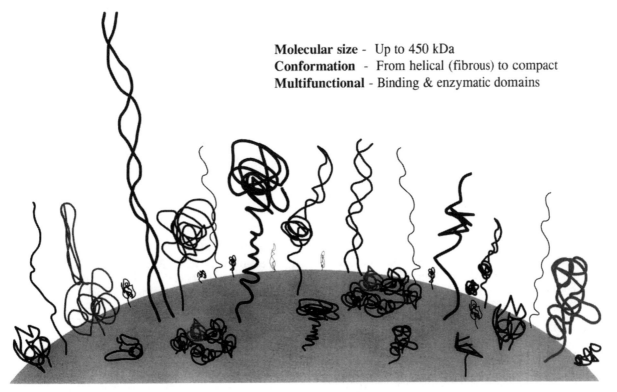

Molecular size - Up to 450 kDa
Conformation - From helical (fibrous) to compact
Multifunctional - Binding & enzymatic domains

FIGURE 8 Appearance of the surface of gram-positive bacteria exhibiting a wide array of protein molecules. Each molecule depicted may have thousands of identical copies densely packed on the surface.

encounters the mucus that coats the mucosal epithelium. Soluble components found in the mucus, such as IgA, IgG, albumin, fibronectin, etc., are able to interact with their respective binding proteins on the streptococcal surface. This binding may initiate a set of conformational events within these proteins on the surface of the bacterial particle to drive the organism through the mucus to the epithelial surface. Perhaps the energy required for this is derived from the surface glycolytic enzymes necessary to generate ATP, or an enzyme is activated to solubilize the mucin. This must all occur quickly or the organism will be swept into the gut and eliminated. Thus, the molecules necessary to initiate infection are poised and ready on the bacterial surface. Therefore, before arriving on the mucosal epithelium the organism may be considered an inert particle with a wide array of surface proteins capable of driving the organism to the mucosal surface. When stable contact is made, the organism has time to send signals into the cell to initiate infection. Attempting to sort out when and how these proteins function during the infection process will be the challenge for future studies. With this understanding, however, new strategies could be developed in controlling diseases by this class of bacteria.

Supported by U.S. Public Health Service Grant AI11822.

REFERENCES

1. **Barnett, T. C., and J. R. Scott.** 2002. Differential recognition of surface proteins in *Streptococcus pyogenes* by two sortase gene homologs. *J. Bacteriol.* **184:**2181–2191.

2. **Bensing, B. A., B. W. Gibson, and P. M. Sullam.** 2004. The *Streptococcus gordonii* platelet binding protein GspB undergoes glycosylation independently of export. *J. Bacteriol.* **186:**638–645.

3. **Berge, A., and U. Sjobring.** 1993. PAM, a novel plasminogen-binding protein from *Streptococcus pyogenes*. *J. Biol. Chem.* **268:**25417–25424.

4. **Berry, A. M., R. A. Lock, S. M. Thomas, D. P. Rajan, D. Hansman, and J. C. Paton.** 1994. Cloning and nucleotide sequence of the *Streptococcus pneumoniae* hyaluronidase gene and purification of the enzyme from recombinant *Escherichia coli*. *Infect. Immun.* **62:**1101–1108.

5. **Bessen, D. E., and V. A. Fischetti.** 1992. Nucleotide sequences of two adjacent M and M-like protein genes of group A streptococci: different RNA transcript levels and identification of a unique IgA-binding protein. *Infect. Immun.* **60:**124–135.

6. **Bierne, H., S. K. Mazmanian, M. Trost, M. G. Pucciarelli, G. Liu, P. Dehoux, the European Listeria Genome Consortium, L. Jansch, F. Garcia-Del Portillo, O. Schneewind, and P. Cossart.** 2002. Inactivation of the *srtA* gene in *Listeria monocytogenes* inhibits anchoring of surface proteins and affects virulence. *Molec. Microbiol.* **43:**869–881.

7. **Blom, A. M., K. Berggard, J. H. Webb, G. Lindahl, B. O. Villoutreix, and B. Dahlback.** 2000. Human C4b-binding protein has overlapping, but not identical, binding sites for C4b and streptococcal M proteins. *J. Immunol.* **164:**5328–5336.

8. **Boden, M. K., and J.-I. Flock.** 1994. Cloning and characterization of a gene for a 19kDa fibrinogen-binding protein from *Staphylococcus aureus*. *Mol. Microbiol.* **12:**599–606.

9. **Burne, R. A., and J. E. C. Penders.** 1993. Characterization of the *Streptococcus mutans* GS-5 *fruA* gene encoding exo-beta-D-fructosidase. *Infect. Immun.* **60:**4621–4632.

10. **Camara, M., G. J. Boulnois, P. W. Andrew, and T. J. Mitchell.** 1994. A neuraminidase from *Streptococcus pneumoniae* has the features of a surface protein. *Infect. Immun.* **62:**3688–3695.

11. **Chen, C. C., and P. P. Cleary.** 1990. Complete nucleotide sequence of the streptococcal C5a peptidase gene of *Streptococcus pyogenes*. *J. Biol. Chem.* **265:**3161–3167.

12. **Cheung, A. L., and V. A. Fischetti.** 1988. Variation in the expression of cell wall proteins of *Staphylococcus aureus* grown on solid and liquid media. *Infect. Immun.* **56:**1061–1065.

13. **Cheung, A. I., S. J. Projan, R. E. Edelstein, and V. A. Fischetti.** 1995. Cloning, expression, and nucleotide sequence of a *Staphylococcus aureus* (*fbpA*) encoding a fibrinogen-binding protein. *Infect. Immun.* **63:**1914–1920.

14. **Cole, R. M., and J. J. Hahn.** 1962. Cell wall replication in *Streptococcus pyogenes*. *Science* **135:**722–724.

15. **Collins, C. M., A. Kimura, and A. L. Bisno.** 1992. Group G streptococcal M protein exhibits structural features analogous to those of class I M protein of group A streptococci. *Infect. Immun.* **60:**3689–3696.

16. **Courtney, H. S., Y. Li, J. B. Dale, and D. L. Hasty.** 1994. Cloning, sequencing, and expression of a fibronectin/fibrinogen-binding protein from group A streptococci. *Infect. Immun.* **62:**3937–3946.

17. **de Chateau, M., and L. Bjorck.** 1994. Protein PAB, a mosaic albumin-binding bacterial protein representing the first contemporary example of module shuffling. *J. Biol. Chem.* **269:**12147–12151.

18. **Derbise, A., Y. P. Song, S. Parikh, V. A. Fischetti, and V. Pancholi.** 2004. Role of the C-terminal lysine residues of streptococcal surface enolase in Glu- and Lys-plasminogen-binding activities of group A streptococci. *Infect. Immun.* **72:**94–105.

19. **Dramsi, S., P. Dehoux, M. Lebrun, P. L. Goossens, and P. Cossart.** 1998. Identification of four new members of the internalin multigene family of *Listeria monocytogenes* EGD. *Infect. Immun.* **65:**1615–1625.

20. **Dryla, A., D. Gelbmann, and A. N. E. von Gabain.** 2003. Identification of a novel iron-regulated staphylococcal surface protein with haptoglobin-haemoglobin binding activity. *Mol. Microbiol.* **49:**37–53.

21. **Errington, J., L. Appleby, R. A. Daniel, H. Goodfellow, S. R. Partridge, and M. D. Yudkin.** 1992. Structure and function of the *spoIIIJ* gene of *Bacillus subtilis*: a vegetatively expressed gene that is essential for epsilon^{G} activity at an intermediate stage of sporulation. *J. Gen. Microbiol.* **138:** 2609–2618.

22. **Ferretti, J. J., R. R. B. Russell, and M. L. Dao.** 1989. Sequence analysis of the wall-associated protein precursor of *Streptococcus mutans* antigen A. *Mol. Microbiol.* **3:**469–478.

23. **Fischetti, V. A.** 1989. Streptococcal M protein: molecular design and biological behavior. *Clin. Microbiol. Rev.* **2:** 285–314.

24. **Fischetti, V. A., M. Jarymowycz, K. F. Jones, and J. R. Scott.** 1986. Streptococcal M protein size mutants occur at high frequency within a single strain. *J. Exp. Med.* **164:** 971–980.

25. **Fischetti, V. A., D. A. D. Parry, B. L. Trus, S. K. Hollingshead, J. R. Scott, and B. N. Manjula.** 1988. Conformational characteristics of the complete sequence

of group A streptococcal M6 protein. *Proteins: Struct. Funct. Genet.* **3:**60–69.

26. **Fischetti, V. A., V. Pancholi, and O. Schneewind.** 1990. Conservation of a hexapeptide sequence in the anchor region of surface proteins of gram-positive cocci. *Mol. Microbiol.* **4:**1603–1605.

27. **Fischetti, V. A., G. M. Landau, P. H. Sellers, and J. P. Schmidt.** 1993. Identifying periodic occurrences of a template with applications to protein structure. *Inform. Process. Lett.* **45:**11–18.

28. **Frithz, E., L.-O. Heden, and G. Lindahl.** 1989. Extensive sequence homology between IgA receptor and M protein in *Streptococcus pyogenes. Mol. Microbiol.* **3:**1111–1119.

29. **Gaillard, J. L., P. Berche, C. Frehel, E. Gouin, and P. Cossart.** 1991. Entry of *L. monocytogenes* into cells is mediated by a repeat protein analogous to surface antigens from gram-positive, extracellular pathogens. *Cell* **65:**1127–1141.

30. **Galli, D., F. Lottspeich, and R. Wirth.** 1990. Sequence analysis of *Enterococcus faecalis* aggregation substance encoded by the sex pheromone plasmid pAD1. *Mol. Microbiol.* **4:**895–904.

31. **Garnier, J., D. J. Osguthorpe, and B. Robson.** 1978. Analysis of the accuracy and implications of simple methods for predicting the secondary structure of globular proteins. *J. Mol. Biol.* **120:**97–120.

32. **Gomi, H., T. Hozumi, S. Hattori, C. Tagawa, F. Kishimoto, and L. Bjorck.** 1990. The gene sequence and some properties of protein H. *J. Immunol.* **144:**4046–4052.

33. **Guss, B., M. Uhlen, B. Nilsson, M. Lindberg, J. Sjoquist, and J. Sjodahl.** 1984. Region X, the cell-wall-attachment part of staphylococcal protein A. *Eur. J. Biochem.* **138:**413–420.

34. **Haanes, E. J., and P. P. Cleary.** 1989. Identification of a divergent M protein gene and an M protein related gene family in serotype 49 *Streptococcus pyogenes. J. Bacteriol.* **171:**6397–6408.

35. **Heath, D. G., and P. P. Cleary.** 1989. Fc-receptor and M protein genes of group A streptococci are products of gene duplication. *Proc. Natl. Acad. Sci. USA* **86:**4741–4745.

36. **Heden, L.-O., E. Frithz, and G. Lindahl.** 1991. Molecular characterization of an IgA receptor from Group B streptococci: sequence of the gene, identification of a proline-rich region with unique structure and isolation of N-terminal fragments with IgA-binding capacity. *Eur. J. Immunol.* **21:**1481–1490.

37. **Hell, W., H.-G. W. Meyer, and S. G. Gatermann.** 1998. Cloning of *aas*, a gene encoding a *Staphylococcus saprophyticus* surface protein with adhesive and autolytic properties. *Mol. Microbiol.* **29:**871–881.

38. **Holck, A., and H. Naes.** 1992. Cloning, sequencing and expression of the gene encoding the cell-envelope-associated proteinase from *Lactobacillus paracasei* subsp. *paracasei* NCDO 151. *J. Gen. Microbiol.* **138:**1353–1364.

39. **Hollingshead, S. K., V. A. Fischetti, and J. R. Scott.** 1986. Complete nucleotide sequence of type 6 M protein of the group A streptococcus: repetitive structure and membrane anchor. *J. Biol. Chem.* **261:**1677–1686.

40. **Horstmann, R. D., H. J. Sievertsen, J. Knobloch, and V. A. Fischetti.** 1988. Antiphagocytic activity of streptococcal M protein: selective binding of complement control protein Factor H. *Proc. Natl. Acad. Sci. USA* **85:**1657–1661.

41. **Horstmann, R. D., H. J. Sievertsen, M. Leippe, and V. A. Fischetti.** 1992. Role of fibrinogen in complement inhibition by streptococcal M protein. *Infect. Immun.* **60:**5036–5041.

42. **Iannelli, F., M. Oggioni, and G. Pozzi.** 2002. Allelic variation in the highly polymorphic locus *pspC* of *Streptococcus pneumoniae. Gene* **284:**63–71.

43. **Inoue, S., M. Sugai, Y. Murooka, S.-Y. Paik, Y.-M. Hong, H. Ohgai, and H. Suginaka.** 1991. Molecular cloning and sequencing of the epidermal cell differentiation inhibitor gene from *Staphylococcus aureus. Biochem. Biophys. Res. Commun.* **174:**459–464.

44. **Jaffe, J., S. Natanson-Yaron, M. G. Caparon, and E. Hanski.** 1996. Protein F2, a novel fibronectin-binding protein from *Streptococcus pyogenes*, possesses two binding domains. *Mol. Microbiol.* **21:**373–384.

45. **Johnsson, E., A. Thern, B. Dahlback, L.-O. Heden, M. Wikstrom, and G. Lindahl.** 1996. A highly variable region in members of the streptococcal M protein family binds the human complement regulator C4BP. *J. Immunol.* **157:**3021–3029.

46. **Johnsson, H., H. Lindmark, and B. Guss.** 1995. A protein G-related cell surface protein in *Streptococcus zooepidemicus. Infect. Immun.* **63:**2968–2975.

47. **Jonsson, K., C. Signas, H.-P. Muller, and M. Lindberg.** 1991. Two different genes encode fibronectin binding proteins in *Staphylococcus aureus. Eur. J. Biochem.* **202:**1041–1048.

48. **Kao, S.-M., S. B. Olmsted, A. S. Viksnins, J. C. Gallo, and G. M. Dunny.** 1991. Molecular and genetic analysis of a region of plasmid pCF10 containing positive control genes and structural genes encoding surface proteins involved in pheromone-inducible conjugation in *Enterococcus faecalis. J. Bacteriol.* **173:**7650–7664.

49. **Kastern, W., U. Sjobring, and L. Bjorck.** 1992. Structure of peptostreptococcal protein L and identification of a repeated immunoglobulin light chain-binding domain. *J. Biol. Chem.* **267:**12820–12825.

50. **Katsukawa, C.** 1994. Cloning and nucleotide sequence of type 3 M protein gene (emm3) consisting of an N-terminal variable portion and C-terminal conserved C repeat regions: relation to other genes of *Streptococcus pyogenes. J. Jpn. Assoc. Infect. Dis.* **68:**698–705.

51. **Kelly, C., P. Evans, L. Bergmeier, S. F. Lee, F. A. Progulske, A. C. Harris, A. Aitken, A. S. Bleiweis, and T. Lerner.** 1990. Sequence analysis of a cloned streptococcal surface antigen I/II. *FEBS Lett.* **258:**127–132.

52. **Kharat, A. S. and A. Tomasz.** 2003. Inactivation of the *srtA* gene affects localization of surface proteins and decreases adhesion of *Streptococcus pneumoniae* to human pharyngeal cells in vitro. *Infect. Immun.* **71:**2758–2765.

53. **Kline, J. B., S. Xu, A. L. Bisno, and C. M. Collins.** 1996. Identification of a fibronectin-binding protein (GfbA) in pathogenic group G streptococci. *Infect. Immun.* **64:**2122–2129.

54. **Kok, J., K. J. Leenhouts, A. J. Haandrikman, A. M. Ledeboer, and G. Venema.** 1988. Nucleotide sequence of the cell wall proteinase gene of *Streptococcus cremoris* Wg2. *Appl. Environ. Microbiol.* **54:**231–238.

55. **Kozarov, E., J. Whitlock, H. Dong, E. Carrasco, and A. Progulske-Fox.** 1998. The number of direct repeats in hagA is variable among *Porphyromonas gingivalis* strains. *Infect. Immun.* **66:**4721–4725.

56. **Kreikemeyer, B., S. R. Talay, and G. S. Chhatwal.** 1995. Characterization of a novel fibronectin-binding surface protein in group A streptococci. *Mol. Microbiol.* **17:**137–145.

57. Lansing, M., S. Lellig, A. Mausolf, I. Martini, F. Crescenzi, M. O'Regan, and P. Prehm. 1993. Hyaluronate synthases: cloning and sequencing of the gene from *Streptococcus* sp. *Biochem. J.* **289:**179–184.

58. Lee, S. G., and V. A. Fischetti. 2003. Presence of D-alanine in an endopeptidase from *Streptococcus pyogenes*. *J. Biol. Chem.* **278:**46649–46653.

59. Lee, S. G., V. Pancholi, and V. A. Fischetti. 2002. Characterization of a unique glycosylated anchor endopeptidase that cleaves the LPXTG sequence motif of cell surface proteins of gram-positive bacteria. *J. Biol. Chem.* **277:**46912–46922.

60. Levesque, C., C. Vadeboncoeur, and M. Frenette. 2004. The *csp* operon of *Streptococcus salivarius* encodes two predicted cell-surface proteins, one of which, CspB, is associated with the fimbriae. *Microbiology* **150:**189–198.

61. Lindgren, P.-E., M. J. McGavin, C. Signas, B. Guss, S. Gurusiddappa, M. Hook, and M. Lindberg. 1993. Two different genes coding for fibronectin-binding proteins from *Streptococcus dysgalactiae*. The complete nucleotide sequences and characterization of the binding domain. *Eur. J. Biochem.* **214:**819–827.

62. Manjula, B. N., K. M. Khandke, T. Fairwell, W. A. Relf, and K. S. Sripakash. 1991. Heptad motifs within the distal subdomain of the coiled-coil rod region of M protein from rheumatic fever and nephritis associated serotypes of group A streptococci are distinct from each other: nucleotide sequence of the M57 gene and relation of the deduced amino acid sequence of other M proteins. *J. Protein Chem.* **10:**369–383.

63. Mazmanian, S. K., G. Liu, H. Ton-That, and O. Schneewind. 1999. *Staphylococcus aureus* sortase, an enzyme that anchors surface proteins to the cell wall. *Science* **285:**760–763.

64. Mazmanian, S. K., G. Liu, E. R. Jensen, E. Lenoy, and O. Schneewind. 2000. *Staphylococcus aureus* sortase mutants defective in the display of surface proteins and in the pathogenesis of animal infections. *Proc. Natl. Acad. Sci. USA* **97:**5510–5515.

65. McDevitt, D., P. Francois, P. Vaudaux, and T. J. Foster. 1994. Molecular characterization of the clumping factor (fibrinogen receptor) of *Staphylococcus aureus*. *Mol. Microbiol.* **11:**237–248.

66. Michel, J. L., L. C. Madoff, K. Olson, D. E. Kling, D. L. Kasper, and F. M. Ausubel. 1992. Large, identical, tandem repeating units in the C protein alpha antigen gene, *bca*, of group B streptococci. *Proc. Natl. Acad. Sci. USA* **89:**10060–10064.

67. Miller, L., L. Gray, E. H. Beachey, and M. A. Kehoe. 1988. Antigenic variation among group A streptococcal M proteins: nucleotide sequence of the serotype 5 M protein gene and its relationship with genes encoding types 6 and 24 M proteins. *J. Biol. Chem.* **263:**5668–5673.

68. Mouw, A. R., E. H. Beachey, and V. Burdett. 1988. Molecular evolution of streptococcal M protein: cloning and nucleotide sequence of type 24 M protein gene and relation to other genes of *Streptococcus pyogenes*. *J. Bacteriol.* **170:**676–684.

69. Navarre, W. W., and O. Schneewind. 1994. Proteolytic cleavage and cell wall anchoring at the LPXTG motif of surface proteins in gram-positive bacteria. *Mol. Microbiol.* **14:**115–121.

70. Navarre, W. W., and O. Schneewind. 1999. Surface proteins of gram-positive bacteria and mechanisms of their targeting to the cell wall envelope. *Microbiol. Mol. Biol. Rev.* **63:**174–229.

71. Nilsson, A., L. Frykberg, J.-I. Flock, L. Pei, M. Lindberg, and B. Guss. 1998. A fibrinogen-binding protein of *Staphylococcus epidermis*. *Infect. Immun.* **66:**2666–2673.

72. Okada, N., M. K. Liszewski, J. P. Atkinson, and M. Caparon. 1995. Membrane cofactor protein (CD46) is a keratinocyte receptor for the M protein of group A streptococcus. *Proc. Natl. Acad. Sci. USA* **92:**2489–2493.

73. Okahashi, N., C. Sasakawa, S. Yoshikawa, S. Hamada, and T. Koga. 1989. Molecular characterization of a surface protein antigen gene from serotype c *Streptococcus mutans* implicated in dental caries. *Mol. Microbiol.* **3:**673–678.

74. Olsson, A., M. Eliasson, B. Guss, B. Nilsson, U. Hellman, M. Lindberg, and M. Uhlen. 1987. Structure and evolution of the repetitive gene encoding streptococcal protein G. *Eur. J. Biochem.* **168:**319–324.

75. Osaki, M., D. Takamatsu, Y. Shimoji, and T. Sekizaki. 2002. Characterization of *Streptococcus suis* genes encoding proteins homologous to sortase of gram-positive bacteria. *J. Bacteriol.* **184:**971–982.

76. O'Toole, P., L. Stenberg, M. Rissler, and G. Lindahl. 1992. Two major classes in the M protein family in group A streptococci. *Proc. Natl. Acad. Sci. USA* **89:**8661–8665.

77. Pallen, M. J., A. C. Lam, M. Antonio, and K. Dunbar. 2001. An embarrassment of sortases—a richness of substrates? *Trends Microbiol.* **9:**97–102.

78. Pancholi, V., and V. A. Fischetti. 1988. Isolation and characterization of the cell-associated region of group A streptococcal M6 protein. *J. Bacteriol.* **170:**2618–2624.

79. Pancholi, V., and V. A. Fischetti. 1992. A major surface protein on group A streptococci is a glyceraldehyde-3-phosphate dehydrogenase with multiple binding activity. *J. Exp. Med.* **176:**415–426.

80. Pancholi, V., and V. A. Fischetti. 1993. Glyceraldehyde-3-phosphate dehydrogenase on the surface of group A streptococci is also an ADP-ribosylating enzyme. *Proc. Natl. Acad. Sci. USA* **90:**8154–8158.

81. Pancholi, V., and V. A. Fischetti. 1998. α-enolase, a novel strong plasmin(ogen) binding protein of the surface of pathogenic streptococci. *J. Biol. Chem.* **273:**14503–14515.

82. Patti, J. M., H. Jonsson, B. Guss, L. M. Switalski, K. Wiberg, M. Lindberg, and M. Hook. 1992. Molecular characterizations and expression of a gene encoding a *Staphylococcus aureus* collagen adhesin. *J. Biol. Chem.* **267:**4766–4772.

83. Phillips, G. N., P. F. Flicker, C. Cohen, B. N. Manjula, and V. A. Fischetti. 1981. Streptococcal M protein: alpha-helical coiled-coil structure and arrangement on the cell surface. *Proc. Natl. Acad. Sci. USA* **78:**4689–4693.

84. Rakonjac, J. V., J. C. Robbins, and V. A. Fischetti. 1995. DNA sequence of the serum opacity factor of group A streptococci: identification of a fibronectin-binding repeat domain. *Infect. Immun.* **63:**622–631.

85. Rathsam, C., P. M. Giffard, and N. A. Jacques. 1993. The cell-bound fructosyltransferase of *Streptococcus salivarius*: the carboxyl terminus specifies attachment in a *Streptococcus gordonii* model system. *J. Bacteriol.* **175:**4520–4527.

86. Robbins, J. C., J. G. Spanier, S. J. Jones, W. J. Simpson, and P. P. Cleary. 1987. *Streptococcus pyogenes* type 12 M protein regulation by upstream sequences. *J. Bacteriol.* **169:**5633–5640.

87. Rocha, C. L., and V. A. Fischetti. 1999. Identification and characterization of a novel fibronectin-binding protein on the surface of group A streptococci. *Infect. Immun.* **67:**2720–2728.

88. Roche, F. M., R. Massey, S. J. Peacock, N. P. J. Day, L. Visai, P. Speziale, A. Lam, M. Pallen, and T. J. Foster.

2003. Characterization of novel LPXTG-containing proteins of *Staphylococcus aureus* identified from genome sequences. *Microbiology* **149**:643–654.

89. **Rothe, B., P. Roggentin, R. Frank, H. Blocker, and R. Schauer.** 1989. Cloning, sequencing and expression of sialidase gene from *Clostridium sordellii*. *J. Gen. Microbiol.* **135:** 3087–3096.

90. **Ryan, P. A., V. Pancholi, and V. A. Fischetti.** 2001. Group A streptococci bind to mucin and human pharyngeal cells through sialic acid-containing receptors. *Infect. Immun.* **69:**7402–7412.

91. **Ryc, M., E. H. Beachy, and E. Whitnack.** 1989. Ultrastructural localization of the fibrinogen-binding domain of streptococcal M protein. *Infect. Immun.* **57:**2397–2404.

92. **Schneewind, O., K. F. Jones, and V. A. Fischetti.** 1990. Sequence and structural characterization of the trypsin-resistant T6 surface protein of group A streptococci. *J. Bacteriol.* **172:**3310–3317.

93. **Schneewind, O., A. Fowler, and K. F. Faull.** 1995. Structure of the cell wall anchor of surface proteins in *Staphylococcus aureus*. *Science* **268:**103–106.

94. **Schneewind, O., P. Model, and V. A. Fischetti.** 1992. Sorting of protein A to the staphylococcal cell wall. *Cell* **70:**267–281.

95. **Sela, S., A. Aviv, A. Tovi, I. Burstein, M. G. Caparon, and E. Hanski.** 1993. Protein F: an adhesin of *Streptococcus pyogenes* binds fibronectin via two distinct domains. *Mol. Microbiol.* **10:**1049–1055.

96. **Shimoji, Y., Y. Ogawa, M. Osaki, H. Kabeya, S. Maruyama, T. Mikami, and T. Sekizaki.** 2003. Adhesive surface proteins of *Erysipelothrix rhusiopathiae* bind to polystyrene, fibronectin, and type I and IV collagens. *J. Bacteriol.* **185:** 2739–2748.

97. **Signas, C., G. Raucci, K. Jonsson, P. Lindgren, G. M. Anantharamaiah, M. Hook, and M. Lindberg.** 1989. Nucleotide sequence of the gene for fibronectin-binding protein from *Staphylococcus aureus*: use of this peptide sequence in the synthesis of biologically active peptides. *Proc. Natl. Acad. Sci. USA* **86:**699–703.

98. **Sjobring, U.** 1992. Isolation and molecular characterization of a novel albumin-binding protein from group G streptococci. *Infect. Immun.* **60:**3601–3608.

99. **Smith, H. E., U. Vecht, A. L. J. Gielkens, and M. A. Smits.** 1992. Cloning and nucleotide sequence of the gene encoding the 136-kilodalton surface protein (muramidase-released protein) of *Streptococcus suis* type 2. *Infect. Immun.* **60:**2361–2367.

100. **Sutcliffe, I. C., and D. J. Harrington.** 2002. Pattern searches for the identification of putative lipoprotein genes in gram-positive bacterial genomes. *Microbiology* **148:**2065–2077.

101. **Sutcliffe, I. C., and R. R. B. Russell.** 1995. Lipoproteins of gram-positive bacteria. *J. Bacteriol.* **177:**1123–1128.

102. **Suvorov, A. N., A. E. Flores, and P. Ferrieri.** 1997. Cloning of the glutamine synthetase gene from group B streptococci. *Infect. Immun.* **65:**191–196.

103. **Swanson, J., K. C. Hsu, and E. C. Gotschlich.** 1969. Electron microscopic studies on streptococci. I. M antigen. *J. Exp. Med.* **130:**1063–1091.

104. **Talay, S. R., M. P. Grammel, and G. S. Chhatwal.** 1996. Structure of a group C streptococcal protein that binds to fibrinogen, albumin and immunoglobulin G via overlapping modules. *Biochem. J.* **315:**577–582.

105. **Talay, S. R., P. Valentin-Weigand, P. G. Jerlstrom, K. N. Timmis, and G. S. Chhatwal.** 1992. Fibronectin-binding protein of *Streptococcus pyogenes*: sequence of the binding domain involved in adherence of streptococci to epithelial cells. *Infect. Immun.* **60:**3837–3844.

106. **Tanimoto, K., F. Y. An, and D. B. Clewell.** 1993. Characterization of the *traC* determinant of the *Enterococcus faecalis* hemolysin-bacteriocin plasmid pAD1: binding of sex pheromone. *J. Bacteriol.* **175:**5260–5264.

107. **Thanassi, D. G., E. T. Saulino, M. J. Lombardo, R. Roth, J. Heuser, and S. J. Hultgren.** 1998. The PapC usher forms an oligomeric channel: implications for pilus biogenesis across the outer membrane. *Proc. Natl. Acad. Sci. USA* **95:**3146–3151.

108. **Timoney, J. F., S. C. Artiushin, and J. S. Boschwitz.** 1997. Comparison of the sequences and functions of *Streptococcus equi* M-like proteins SeM and SzPSe. *Infect. Immun.* **65:**3600–3605.

109. **Tokuda, M., N. Okahashi, I. Takahashi, M. Nakai, S. Nagaoka, M. Kawagoe, and T. Koga.** 1991. Complete nucleotide sequence of the gene for a surface protein antigen of *Streptococcus sobrinus*. *Infect. Immun.* **59:**3309–3312.

110. **Ton-That, H., G. Liu, S. K. Mazmanian, K. F. Faull, and O. Schneewind.** 1999. Purification and characterization of sortase, the transpeptidase that cleaves surface proteins of *Staphylococcus aureus* at the LPXTG motif. *Proc. Natl. Acad. Sci. USA* **96:**12424–12429.

111. **Uhlen, M., B. Guss, B. Nilsson, S. Gatenbeck, L. Philipson, and M. Lindberg.** 1984. Complete sequence of the staphylococcal gene encoding protein A. *J. Biol. Chem.* **259:**1695–1702 & 13628.

112. **Waltman, W. D., L. S. McDaniel, B. M. Gray, and D. E. Briles.** 1990. Variation in the molecular weight of PspA (pneumococcal surface protein A) among *Streptococcus pneumoniae*. *Microb. Pathog.* **8:**61–69.

113. **Wanda, S.-Y., and R. Curtiss III.** 1994. Purification and characterization of *Streptococcus sobrinus* dextranase produced in recombinant *Escherichia coli* and sequence analysis of the dextranase gene. *J. Bacteriol.* **176:**3839–3850.

114. **Weidlich, G., R. Wirth, and D. Galli.** 1992. Sex pheromone plasmid pADI-encoded surface exclusion protein of *Enterococcus faecalis*. *Mol. Gen. Genet.* **233:** 161–168.

115. **Wu, H. Y., M. H. Nahm, Y. Guo, M. W. Russel, and D. E. Briles.** 1997. Intranasal immunization of mice with PspA (pneumococcal surface protein A) can prevent intranasal carriage, pulmonary infection, and sepsis with *Streptococcus pneumoniae*. *J. Infect. Dis.* **175:**839–846.

116. **Yeung, M. K., and J. O. Cisar.** 1988. Cloning and nucleotide sequence of a gene for *Actinomyces naeslundii* 45 type 2 fimbriae. *J. Bacteriol.* **170:**3803–3809.

117. **Yeung, M. K., and J. O. Cisar.** 1990. Sequence homology between the subunits of two immunologically and functionally distinct types of fimbriae of *Actinomyces* spp. *J. Bacteriol.* **172:**2462–2468.

118. **Yeung, M. K., J. A. Donkersloot, J. O. Cisar, and P. A. Ragsdale.** 1998. Identification of a gene involved in assembly of *Actinomyces naeslundii* T14V type 2 fimbriae. *Infect. Immun.* **66:**1482–1491.

119. **Yother, J., and D. E. Briles.** 1992. Structural properties and evolutionary relationships of PspA, a surface protein of *Streptococcus pneumoniae*, as revealed by sequence analysis. *J. Bacteriol.* **174:**601–609.

THE STREPTOCOCCUS

SECTION EDITORS: Vincent A. Fischetti
and Joseph J. Ferretti

A S A SPECIES, the streptococci are a diverse group, ranging from commensal organisms which occupy various niches of the body to
pathogens having the capacity to infect a wide range of tissue sites.
They are also diverse in their resistance to current antibiotic therapy, from
the multiresistant enterococci to the barely resistant *S. pyogenes*. It is thus
not surprising to find, when looking back over the past century of research
on pathogenic organisms, that the streptococcus appears as one of the best
studied of the gram-positive pathogens.

Although published results have steadily increased during the first three-
quarters of this period, the introduction of molecular techniques in the latter quarter of the century has seen a nearly logarithmic increase in new
information. Despite this wealth of knowledge, however, numerous questions still remain unanswered. The first part of the century relied on the
biology of the organism to answer questions of pathogenesis; in the latter
part, however, molecular techniques overshadowed the biology in a mad
scramble to understand the genetic complexity of these bacteria. Since the
last edition, the complete genome sequences of eleven different species of
streptococci have been completed. Analyses of multiple genome strains in
the same species have shown that significant differences may occur, even in
the same species. A common feature of all the streptococci is the presence
of horizontal gene transfer and recombination systems, allowing for
enhanced survival of an organism in its ecological niche, often by the emergence of new strains with increased virulence properties.

The chapters to follow are the best of a mixture of genetics, biology, and
immunology designed to both ask and answer questions pertaining to the
pathogenicity of the streptococci. The information gained from these studies and the ideas they stimulate for future experiments will allow us to
develop new tools not only to prevent infection by these organisms specifically but perhaps by gram-positive bacteria in general.

Intracellular Invasion by *Streptococcus pyogenes*: Invasins, Host Receptors, and Relevance to Human Disease

BEINAN WANG, DAVID CUE, AND P. PATRICK CLEARY

3

The human oral-nasal mucosa is the primary reservoir for *Streptococcus pyogenes* infections. Although the most common infection of consequence in temperate climates is pharyngitis, the past 15 years have witnessed a dramatic increase in invasive disease in many regions of the world. Historically, *S. pyogenes* has been associated with sepsis and fulminate systemic infections, but the mechanism by which these streptococci traverse mucosal or epidermal barriers is not understood. The discovery that *S. pyogenes* can be internalized by mammalian epithelial cells at high frequencies (17, 21, 25) and/or can open tight junctions to pass between cells (11) is a potential explanation for changes in epidemiology and the ability of this species to breach such barriers. In this chapter, the invasins and pathways used by *S. pyogenes* to reach the intracellular state are reviewed, and the relationship between intracellular invasion and human disease is discussed.

S. pyogenes has evolved a variety of both surface-bound and extracellular factors that alter the inflammatory response and impair phagocytic clearance of the bacteria. The more than 120 M genotypes use both similar and different strategies at the biochemical level to colonize their host and avoid protective defenses. These are reviewed elsewhere (16) and in other chapters of this volume. Intracellular invasion depends on at least two classes of surface proteins, the M proteins (1, 12) and fibronectin (Fn)-binding proteins (25, 31). The M proteins serve many functions in the pathogenesis of *S. pyogenes*, including resistance to phagocytosis, adherence, and intracellular invasion (16). The Fn-binding proteins PrtF and the allelic variant SfbI are also adhesins and invasins produced by 50 to 60% of the M genotypes. The function of these proteins in the context of intracellular invasion is described below.

THE IMPACT OF INTRACELLULAR INVASION ON *S. PYOGENES* INFECTION

Many geographic and temporal clusters of sepsis and toxic shock, caused by a few serotypes of *S. pyogenes*, have been reported (4, 44). An M1 subclone was associated with many such clusters (4, 27). Surprisingly, devastating soft tissue infections were often reported in patients who had not experienced previous trauma or wounds that might initiate systemic spread of the organism (44). Cleary and coworkers (5, 21) considered the possibility that intracellular invasion of mucosal surfaces provided a window for streptococci to reach deeper tissue, even the bloodstream. Indeed, a globally disseminated M1 subclone invaded A549 epithelial cells at significantly higher frequency than did other subclones of M1 streptococci. The source of M1 subclones, i.e., uncomplicated disease or more invasive disease (blood-isolated), did not correlate with efficiency of ingestion by epithelial cells (5). Only the M1 subclone that had acquired 70 kb of DNA sequence from two distinct prophages (5) was associated with high-frequency invasion of cultured cells. To date no relationship to prophages and intracellular invasion has been discovered.

Although an association between high-frequency intracellular invasion and systemic streptococcal disease is still uncertain, the intracellular state may significantly increase the capacity of this bacterium to disseminate in human populations. When strains from carriers and patients with uncomplicated pharyngitis and sepsis were compared, those from carriers were observed to be internalized by HEp2 cells at the highest frequency (24). A highly variable M1 culture could be enriched for more invasive streptococci by in vitro serial passage through human epithelial cells (6); therefore, in vivo cycling of streptococci between the interior and exterior of the mucosal epithelium may select for variants that are more efficiently internalized. From 30 to 40% of children continue to shed streptococci after treatment with penicillin (30). Brandt et al. (2) reported 80% relapse by the same initial strain following vigorous penicillin therapy when 40 consecutive isolates from 18 patients were characterized. Treatment failures and subsequent carriage of streptococci were reported to be more common when the infecting strain produced Protein F (PrtF, also termed SfbI), an invasin that promotes high-frequency ingestion of streptococci by epithelial cells (28). This correlation was not, however, confirmed by a smaller study (2). In vitro, intracellular *S. pyogenes* can resist at least 100 µg/ml of penicillin (unpublished data); therefore, penicillin treatment

---

may further select for strains that can be efficiently internalized by epithelial cells. High-frequency intracellular invasion may increase the rate of antibiotic therapy failure and therefore increase the size of the human reservoir that can disseminate the organism to others in the population. As the reservoir enlarges, the probability of serious, systemic infection will also increase. Thus, strains or serotypes that are less able to acquire an antibiotic-free niche may be less often associated with severe disease. Facinelli and colleagues (15) observed an association between genetic resistance to erythromycin and more efficient uptake of *S. pyogenes* by epithelial cells. They suggested the frightening possibility that erythromycin resistance had become genetically linked to efficient invasion of human cells and thus coselected by antibiotic therapy. A genetic relationship was not established, however; nor did the authors exclude the possibility that a single subclone had disseminated in the study population. A thorough analysis of a cluster of toxic shock in southern Minnesota showed that nearly 40% of the schoolchildren in nearby communities were carriers of a serotype M3 clone that was responsible for systemic disease in adults (7). These investigators suggested that schoolchildren served as the reservoir for *S. pyogenes* that produced toxic shock in adults with underlying physical disabilities.

ARE TONSILS A RESERVOIR FOR S. PYOGENES?

The best and most direct evidence that intracellular bacteria are an important source for dissemination of streptococci and the cause of recurrent tonsillitis is based on microscopic studies of surgically excised tonsils (29, 30). This study found that 13 of 14 tonsils from children plagued by recurrent tonsillitis harbored streptococci within epithelial cells in the tonsillar crypts. *S. pyogenes* was also observed in macrophage-like cells at high frequency in these specimens. Tonsils from control subjects who had tonsils removed for other reasons did not contain streptococci. Intracellular streptococci were also observed in cultured tonsils infected with *S. pyogenes* (29). The invasive M1 subclone was confirmed in vitro to efficiently invade primary keratinized tonsillar epithelial cells (10). In a murine intranasal infection model this M1 subclone and an M49 strain were shown to specifically invade nasal-associated lymphoid tissue known to be functionally homologous to human tonsils (34). Initially streptococci were associated with M-like cells, sporadically located along the base of the nasal epithelium, and by 24 hours postinoculation had formed microcolonies throughout the lymphoid tissue. From 1 to 10% of the streptococci, depending on the mouse, were intracellular. This study suggests that M-like cells are the primary portals for transport of *S. pyogenes* across the nasal mucosa into lymphoid tissue. Subsequent studies also demonstrated that nasal-associated lymphoid tissue is the initial, primary location for antigen-specific T-cell priming and activation following intranasal infection by *S. pyogenes* (35).

S. PYOGENES INVASINS AND HOST RECEPTORS

In many respects *S. pyogenes*' interactions with human cells are analogous to those of *Yersinia pseudotuberculosis*. This organism can invade eukaryotic cells by at least three different mechanisms, the most efficient of which is mediated by the bacterial InvA protein (14). InvA is an outer membrane protein capable of high-affinity binding to multiple β1 integrins. Integrins are a mammalian family of heterodimeric, transmembrane glycoproteins that bind to extracellular matrix proteins (e.g., Fn and laminin [Ln]) and blood-clotting proteins (e.g., fibrinogen) (20). Engagement of integrins by InvA results in activation of host signal transduction pathways and rearrangements of host cell cytoskeletal components. Invasins are a subclass of bacterial adhesin molecules required for ingestion by host cells. Typically, invasins are proteins expressed on the surfaces of bacterial cells that directly or indirectly recognize specific host cell receptors (14). In general, invasins are capable of inducing reorganization of the host cell's cytoskeleton by generating specific signals from receptors with which they interact. Integrins are often exploited for microbial entry into mammalian cells. This may be due to their ubiquitous expression and their ability to affect cytoskeletal arrangement (20).

S. pyogenes has evolved multiple mechanisms for invasion of a wide variety of mammalian cells. Several streptococcal invasins have been identified, including two related Fn-binding proteins and a variety of M proteins. The cellular receptors, ligands, and biochemical pathways vary, depending on the strain of streptococcus. Recognition that different M proteins require different soluble factors and receptors adds another layer of complexity (1).

ZIPPER AND OTHER MECHANISMS OF STREPTOCOCCAL INGESTION

Cytoskeletal events accompanying uptake of a variety of pathogenic bacteria have been studied. Ruffling of the membrane precedes the entry of *Shigella and Salmonella* spp. (14). In contrast, internalization by a zipper mechanism is mediated by interactions between surface invasins, ligands, and host cell receptors. These interactions lead to pseudopod formation, which requires the extension of actin filaments beneath the host cell membrane (12).

Scanning electron microscopy revealed that adherent M1[+] bacteria are often in close contact with host cell microvilli, suggesting that this association may be an initial step in internalization (Fig. 1A). Microvilli frequently extend across bacterial surfaces and appear to entrap streptococci. Microvilli increase in number near adherent streptococci and appear to morphologically expand into pseudopodia-like structures, although these structures may arise independently (Fig. 1B). Membrane engulfment of streptococci, suggesting a zipper mechanism of uptake, is also observed (Fig. 1C). In contrast to M1[+] streptococci, M1[−] bacteria are rarely observed to interact with microvilli, and no pseudopodia-like structures were seen (12). De novo actin polymerization is essential for uptake of these streptococci, because invasion is blocked by cytochalasin D, an inhibitor of actin polymerization (21). Confocal immunofluorescent microscopy confirmed that vacuoles containing internalized streptococci are surrounded by polymerized actin. Actin is occasionally associated with membrane-spanning chains and seems to disappear soon after entry. The M1 strain of streptococcus, like *Listeria* and *Yersinia* spp., can be internalized by a zipperlike mechanism (12, 14).

More recent microscopic studies demonstrated that physical events leading to ingestion of *S. pyogenes* vary dramatically between strains and serotypes (26, 40). Comparison of strains A40 (M12 Sfb1[+]) and A8 (M40 Sfb1[−]) revealed striking differences between them and the M1 strain 90-226 studied in the Cleary laboratory. Strain A40

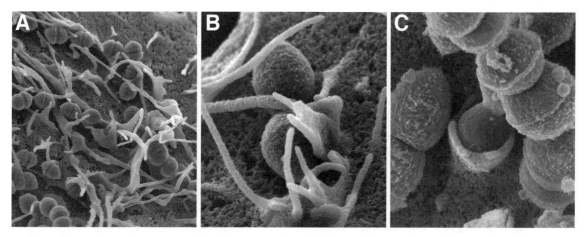

FIGURE 1 Scanning electron micrographs showing different stages of streptococcal-HeLa cell interactions. (A) A high density of microvilli surrounding and in contact with adherent streptococci. (B) A common morphological change that microvilli undergo. The cup-like structure appears to gradually engulf the bacteria and ultimately pulls them into a vacuole. (C) A streptococcal chain has been partially ingested by a mechanism that morphologically resembles a receptor-ligand interaction. It is not known whether microvilli are involved in the latter or whether streptococci are ingested by two different kinds of cytoskeletal rearrangements.

induces the accumulation of small membrane omega-shaped cavities in the host cell membrane, near adherent streptococci. Shortly thereafter, membrane invaginations are formed, into which streptococci enter. The integrity of the membrane is not disrupted and streptococci end up in a cytoplasmic vacuole. These cell membrane changes depend on Fn and α5β1 integrins, but focal adhesion complexes were not observed to contain polymerized actin (26). Further studies showed that these membrane structures are caveolae, structures previously found to mediate entry of both viruses and parasites into cells. In contrast, invasion of HEp2 cells by strain A8 (M40 Sfb1⁻) was not inhibited by an Fn peptide analogue, nor by anti-α5β1 antibody. This strain induced a dramatic aggregation of microvilli near and around attached streptococci, and ingestion of this strain was independent of caveolae and lipid rafts. In many respects, strain A8 behaves like the M1 strain 90-226; large numbers of microvilli contact streptococci and many undergo morphological changes. However, invasion of HEp2 and A549 epithelial cells by the M1 strain depends on Fn and α5β1 integrins (9, 10). It is unclear whether these differences reflect yet another mechanism of streptococcal uptake or merely depend on different experimental conditions. Irrespective of differences, this study further implicated participation of microvilli in uptake of streptococci by epithelial cells.

M PROTEINS AND OTHER Fn-BINDING PROTEINS ARE INVASINS

S. pyogenes expresses an array of cell surface proteins that facilitate bacterial colonization of a variety of human tissues (8). Some are adhesins that also function as invasins. The proteins SfbI and F1 (PrtFI) are two closely related streptococcal adhesins that bind Fn with high affinity. Approximately 50% of *S. pyogenes* isolates carry the gene encoding SfbI/PrtFI, although many do not express the gene under laboratory conditions (24, 25, 28). Prebinding of soluble Fn by PrtFI-bearing streptococci can promote bacterial adherence to cultured cells and dermal tissue. This and experimental evidence from several laboratories are consistent with SfbI/PrtFI promoting intracellular invasion via a similar mechanism.

Molinari et al. (25) demonstrated that SfbI can mediate invasion of HEp2 cells. Streptococcal invasion can be ablated by antiserum raised against SfbI or by preincubation of host cells with recombinant SfbI. Latex beads coated with recombinant SfbI readily adhere to and are efficiently ingested by epithelial cells, demonstrating that the interaction of SfbI with host cells is sufficient for internalization. Fn binding by SfbI is apparently required for invasion, because beads coated with a recombinant SfbI peptide lacking the Fn-binding domains do not efficiently adhere to HEp2 cells. Ozeri et al. (31) reported that invasion of HeLa cells by PrtFI⁺ bacteria depends on addition of serum or purified Fn. In both studies, antibodies against Fn inhibited invasion. Also, rPrtFI peptides, containing at least one Fn-binding domain, were found to inhibit PrtFI-mediated invasion. PrtFI binds to a 70-kDa N-terminal fragment of Fn. This region of Fn differs from the region bound by epithelial cells. The 70-kDa fragment can inhibit Fn-mediated invasion by competing with intact Fn for PrtFI binding. Only intact Fn molecules are capable of supporting bacterial invasion (31).

Antibody directed against the integrin β1 subunit specifically blocks Fn-PrtF1-mediated invasion of HeLa cells (31). This result suggests that one or more β1-containing integrins are involved in bacterial internalization. In contrast, invasion of GD25 (embryonic mouse stem) cells appears to be mediated by integrin αvβ3 (the major Fn-binding integrin of this cell line), because invasion is inhibited by a peptide that specifically blocks Fn binding to this integrin (31). Collectively, these results demonstrate that Fn functions as a molecular bridge between SfbI/PrtFI and host cell Fn receptors.

Expression of Sfb1/PrtF1 is not a prerequisite for high-frequency intracellular invasion. For example, serotype M1 strains typically lack the gene encoding SfbI/PrtFI, but some can invade human epithelial cells with high efficiency (5, 28). Invasion of cultured cells by the globally disseminated

M1 strain 90-226 primarily depends on expression of M1 protein. Inactivation of the M1 protein gene, *emm1*, decreased invasion of HeLa, A549, and HEp2 cells 50-fold (9, 12). Also, latex beads coated with M1 are readily ingested by HeLa cells (12). These results indicate that invasion of epithelial cells by strain 90-226 is mediated in large part by M1 protein.

Expression of M1 protein is not sufficient for invasion, however, because bacterial internalization depends on exogenous serum, the extracellular matrix protein Fn or Ln (9). M1 appears to be the major Fn-binding protein expressed by strain 90-226, as inactivation of *emm1* reduces binding by 88%. Also, purified M1 protein can bind Fn in vitro (9). However, this is not a general property of M proteins; only three types, M1, M6, and M3, are known to bind Fn (9, 42, unpublished data). Interestingly, M3 protein was also found to promote serum-dependent invasion of HEp2 cells (1).

In the presence of Fn, internalization of M1 streptococci depends on the interaction of M1, Fn, and the epithelial cell Fn receptor α5β1 (Fig. 2). A monoclonal antibody that specifically blocks Fn binding to this integrin can ablate invasion of A549, HeLa, and human tonsillar epithelial cells (9, 10). Low-molecular-weight nonpeptidyl α5β1 antagonists are also effective inhibitors of invasion of A549 and human tonsillar epithelial cells (10). The inhibitory effects of α5β1 antagonists are not due to generalized effects on either bacterial or host cells, because the inhibitors do not abrogate Ln-mediated invasion. Rather, the inhibitory effects of α5β1 antagonists are observed only when bacteria are exposed to either serum or purified Fn. Thus, as in the case of Sfb1/PrtF1[+] strains, Fn appears to function as a bridging molecule in promoting invasion by M1[+] bacteria. Ln can replace Fn as the agonist in invasion assays (Fig. 2).

Failure of Ln to promote invasion of A549 cells by M[−] mutants suggests that this extracellular matrix protein binds sufficiently to M1 protein to serve as a bridge between streptococci and α3β1 integrins.

The invasion efficiencies of M1 strains are not dictated solely by the presence of invasion agonists, however; La-Penta et al. (21) and Cleary et al. (6) reported that M1 isolates exhibit widely varying invasion efficiencies, even when experiments are performed in the presence of serum. These variations are likely due, at least in part, to varying levels of M1 protein expression. Transcription of *emm1* can vary considerably between isolates and even between different cultures of the same isolate (6). Although this possibility has yet to be thoroughly investigated, it is consistent with the finding of Ozeri et al. (31) that the number of integrin-binding sites on bacteria can greatly affect invasion efficiency. Expression of different M proteins—M6, M3, and M18—in a common streptococcal background demonstrated that M proteins can differ widely in their capacity to promote ingestion of streptococci. Serum and the cellular receptor CD46 (Fig. 2) were required for uptake of streptococci that expressed M3 protein, whereas neither was required when the recombinant expressed M6 protein (1). Efficient invasion of A549 cells by strain 90-226 also requires that the M1 protein engage both CD46 and, indirectly, α5β1 integrins (39). Thus, it appears that M proteins of different serotypes often recognize different receptors on surfaces of mammalian cells. Berkower and colleagues' (1) results suggest that the invasion potential is actually determined by the primary amino acid sequence of the respective M protein. Although it is clear that several types of M protein can facilitate invasion by *S. pyogenes*, there is no single mechanism underlying M protein-mediated ingestion of streptococci.

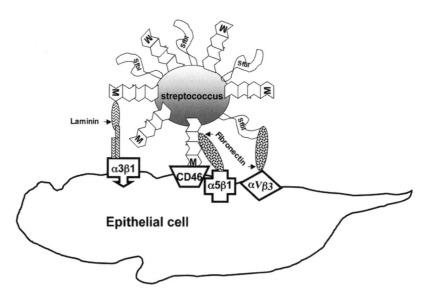

FIGURE 2 This cartoon depicts the various agonist and cellular receptors that streptococci commandeer to promote their own phagocytosis by epithelial cells. Both the M protein and high-affinity Fn proteins SfbI/PrtF1 are known invasins that depend on Fn forming a bridge between the bacterial surface and integrin receptors α5β1 or αVβ3. Although less studied, Ln can also serve as an agonist for ingestion of streptococci by bridging them to α3β1 integrins. For some strains Fn-mediated invasion of epithelial cells also requires interaction between M protein and the complement regulatory protein CD46.

STREPTOCOCCAL FOCAL ADHESION AND SIGNAL TRANSDUCTION

The complex networks of intracellular events that lead to ingestion of several microbial pathogens have begun to be unraveled. Studies have not only identified potential targets for therapeutics but in some cases revealed insight into basic cell biology questions. Uptake by host cells requires streptococci to manipulate normal outside-inside signals spawned from cellular receptors. Interactions of Fn-binding proteins and integrins trigger a cascade of signals that cause cytoskeleton rearrangement, leading to ingestion of streptococci by nonphagocytic epithelial cells. Studies of streptococcal-induced signaling pathways have focused on two major invasins and Fn-binding proteins, SfbI/PrtF1 and M1 protein. Ozeri et al. (32) reported that infection of epithelial cells with an M6 SfbI/PrtF1$^+$ strain caused integrins to be recruited to sites of bacterial adhesion and co-localization of phosphorylated focal adhesion kinase (FAK) and paxillin, common signaling and structural components of focal adhesion complexes. Invasion of epithelial cells by the M6 strain was significantly reduced when host cells were pretreated with genistein, a general inhibitor of protein tyrosine kinases, such as Src kinases and FAK. Invasion was also reduced by inhibitors of the small GTPases Rac and Cdc42 (47). Moreover, internalization of this strain by FAK$^{-/-}$ and Src$^{-/-}$ cells and cells expressing dominant negative forms of Rac1 or Cdc42 was less efficient. From these findings, Ozeri et al. proposed a model for SfbI/PrtF1-mediated *S. pyogenes* invasion that is incorporated into Fig. 3: Fn bound to surface SfbI/PrtF1 protein associates with integrins to cause integrin clustering, followed by recruitment of paxillin, FAK, and other focal adhesion proteins at the site of entry. This initiates autophosphorylation of FAK, which creates a docking site for Src kinases. Paxillin present in this complex, and additional phosphorylation sites of FAK, are then phosphorylated by Src. Ozeri et al. proposed that Rac and Cdc42 are required to orchestrate signaling events initiated by Fn bound to SfbI/PrtF1 protein (32). Then the focal adhesion complex provides an anchor for actin polymerization and cytoskeleton rearrangements required for internalization of streptococci. To date, however, the relationships among integrin-generated signals, formation of focal adhesion complex, and Rac and Cdc42 are poorly understood and unsupported experimentally. It is

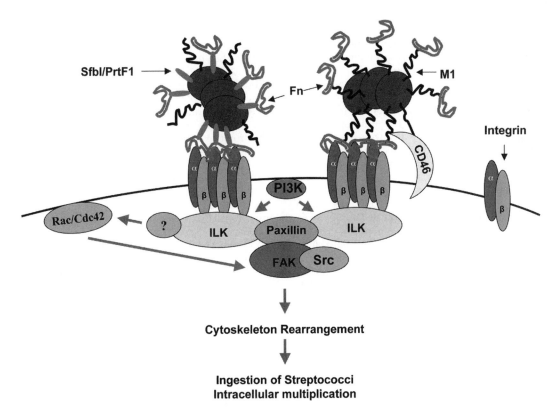

FIGURE 3 Postulated and confirmed components of streptococcal-induced focal adhesion complexes and signaling molecules required for formation of those complexes and subsequent cytoskeletal changes that lead to ingestion of streptococci by epithelial cells. Fn bound to surface M1 or SfbI/PrtF1 proteins associates with integrins to cause integrin clustering, leading to ILK activation in a PI3K-dependent manner. For some strains contact between M protein and CD46 is also required for high-level invasion. The function of CD46 in this process, however, is unknown. Following recruitment of paxillin, FAK, and other focal adhesion proteins, autophosphorylation of FAK creates a docking site for Src kinases to phosphorylate recruited paxillin and additional phosphorylation sites of FAK. Alternatively, Rac and Cdc42 are activated by ILK through an intermediate to participate in formation of focal adhesion complexes, which provide an anchor for actin polymerization and cytoskeleton rearrangement, and ultimately results in uptake of *S. pyogenes*.

also possible that streptococci interact with additional receptors, which in turn activate other mechanisms of signaling and pathways of ingestion.

M1$^+$ strains of *S. pyogenes* invade epithelial cells efficiently without expression of SfbI/PrtF1 protein. M1-mediated invasion of epithelial cells was shown to require phosphatidylinositol 3-kinase (PI3K) (37), a heterodimer consisting of the regulatory subunit p85 and the catalytic subunit p110. Activation of PI3K catalyzes phosphorylation of membrane-associated phosphatidylinositol, which binds to downstream targets, including integrin-linked kinase (ILK) (36). Unpublished studies found that ILK is also required for *S. pyogenes* invasion of HEp2 cells (48). ILK is known to be a crucial link between integrins and the cytoskeleton that is capable of phosphorylating the β_1 integrin cytoplasmic domain and can be coimmunoprecipitated with β_1 integrin in lysates of mammalian cells (19). Invasion of epithelial cells by M1$^+$ *S. pyogenes* was impaired up to 80% by a specific ILK inhibitor. Inhibition of ILK expression with siRNA, or inhibition of activity by transfection with kinase-inactive ILK gene, also significantly reduced *S. pyogenes* invasion. Controls—assays of the non-integrin-dependent invasion of epithelial cells by salmonella—confirmed that inhibitors of either PI3K or ILK did not have a general negative impact on the host cell's ability to ingest bacteria but were specific for Fn-mediated uptake of *S. pyogenes*. ILK is known to be activated upon integrin engagement in a PI3K-dependent manner and to require that its pleckstrin homology-like domain be bound to phosphatidylinositol 3,4,5-triphosphate, a product of PI3K. Therefore, it is possible that PI3K indirectly provides the cytoplasmic membrane anchor for ILK.

Invasion of epithelial cells by M6 SfbI/PrtF1$^+$ streptococcus was also reduced 70 to 80% by inhibitors of PI3K and ILK, suggesting that signaling pathways induced by this high-affinity Fn-binding protein and M1 protein overlap. ILK has been suggested to indirectly activate Rac and Cdc42 (23, 41), which can regulate actin cytoskeleton rearrangement (18), leading to various forms of cell adhesion complexes and stress fibers (48). Therefore, ILK may be an intermediate between integrins and Rac and/or Cdc42. Although M1-Fn and SfbI/PrtF1-Fn complexes may induce cellular changes by shared mechanisms, M1-mediated invasion is significantly less sensitive to genistein (48), suggesting that different mechanisms of signaling may also exist.

M1 protein can interact with multiple cellular receptors, either directly or indirectly (Fig. 2). For example, CD46, a cofactor in Factor I-mediated inactivation of complement proteins C3b and C4b (22) that is expressed on the surface of all nucleated human cells, enhances invasion of epithelial cells by M3$^+$ and M1$^+$ streptococci (1, 39). Overexpression of a mutant form of CD46 with a deletion in the cytoplasmic domain results in partial reduction of invasion. Although CD46's role in intracellular signaling and its docking partners are poorly defined, these preliminary studies suggest that M1 protein-mediated ingestion of *S. pyogenes* requires costimulation from two cellular receptors (39).

Diverse invasins, agonists, and cellular receptors predict that the signaling networks that lead to ingestion of streptococci by epithelial cells may also be varied and multiple. Although our knowledge of the signaling mechanisms responsible for ingestion of *S. pyogenes* is rudimentary, Fig. 3 attempts to interpret what is known. Association of multimeric Fn-binding protein–Fn complexes with $\alpha5\beta1$ integrins on the bacterial surface causes integrin clustering that leads to activation of ILK in a PI3K-dependent manner and

recruitment of paxillin and FAK. Autophosphorylation of FAK creates a docking site for Src kinases, resulting in phosphorylation of paxillin and additional sites of FAK. Rac and Cdc42 proteins are likely activated by ILK signaling, building focal adhesion complexes that ultimately regulate localized actin depolymerization and polymerization and the cytoskeletal changes that engulf the bacteria. Many gaps in the pathway exist, but the tools are now available to further dissect the mechanisms.

OTHER INVASINS

More than five other *S. pyogenes* proteins are known to bind Fn and/or Ln, everything from proteases to serum opacity factor. A few serotypes, including M1 strain 90-226, produce the Fn-binding protein Fba. Like M protein and the C5a peptidase (SCPA), expression of Fba is positively controlled by Mga, a multigene transcriptional activator. Mutation and complementation experiments demonstrated that Fba promotes adherence and uptake of streptococci by epithelial cells in the presence of Fn (45). More recent studies found that Fba binds the complement regulatory factor FHL-1, and that this agonist also promotes uptake of streptococci in M1$^-$ mutants of strain 90-226 (33). Elsner et al. (13) reported that mutants, which lack the lipoprotein Lbp/Lsp, showed reduced adherence and invasion of epithelial cells. However, others found that Lbp/Lsp functions as an adhesin, but mutations that eliminated expression of Lpb/Lsp had no impact on invasion of HEp2 cells (46). Our laboratory also found that mutation of Lpb/Lsp protein had no effect on invasion of A549 cells by strain 90-226. Expression of that protein on the surface of lactococcus imparted to that organism the potential to adhere to A549 cells but did not promote bacterial uptake. It is possible that this lipoprotein varies significantly between strains, or that differences in experimental conditions account for the different conclusions. Thus, binding Fn to the streptococcal surface is not in itself sufficient to promote invasion of epithelial cells. Fn density and/or conformation in the context of the streptococcal surface more than likely influences its potential to promote integrin aggregation or the signals generated by the occupied integrin.

The surface-bound protease SCPA was demonstrated to be the primary, Fn-independent invasin for *S. agalactiae* (3). Although M1 protein is the primary invasin for strain 90-226, SCPA accounts for 15 to 20% of bacteria internalized by A549 cells (38). Complete deletions or amino acid substitutions that block protease activity also reduce the capacity of SCPA to promote ingestion of *S. pyogenes*. Cheng et al. (3) postulated that a propeptide, which is autocatalytically removed during protease maturation, may also uncover the protein's cell binding site.

FATE OF INTRACELLULAR STREPTOCOCCI

Listeria spp. are able to escape from vacuoles, and subsequent actin tail formation allows spread to neighboring cells (14). Immunofluorescent microscopy did not reveal actin tails associated with streptococci at 5 or 7 h postinfection (12). Transmission electron microscopy showed streptococci enclosed in vacuoles within host cells (17, 21). The ultimate fate of internalized bacteria, however, remains controversial. Results from several studies failed to observe growth of intracellular streptococci, and infected cell lines eventually clear bacteria (17, 43). Consistent with this finding are transmission electron micrographs taken 8 h postinfection, showing degraded streptococci inside phagolysosomes (17). Österlund and Engstrand (29), however, reported that *S. pyogenes* can remain viable within HEp2 cells for up to 7 days, and we have found that M1$^+$ bacteria can be routinely recovered from infected cells after several days' exposure to antibiotics (unpublished data). Transmission electron microscopy observed that the M$^+$Sfb1$^-$ strain A8 rapidly escaped the phagosome and multiplied extensively in the cytoplasm of HEp2 cells (26). In contrast, the M$^+$ Sfb1$^+$ strain more rarely escaped the vacuole and was reported not to multiply within epithelial cells. Again the outcome of streptococcal adherence to epithelial cells depends on the strain and the spectrum of proteins expressed on the bacterial surface.

SUMMARY

S. pyogenes has evolved multiple mechanisms for invasion of epithelial cells. Strains of this species differ substantially in invasion frequency and in dependence on invasion agonists. High-frequency internalization is mediated by extracellular matrix proteins, such as Fn and Ln, that form bridges between a surface invasin and the appropriate integrin receptor on the target mammalian cell. Because adherence and invasion can be independent events, mediated by distinct surface proteins and distinct integrins, a two-receptor process is proposed for high-efficiency internalization. The spectrum of target host cells, invasion efficiency, and requirement for serum agonists is determined by the extracellular matrix-binding proteins displayed on the bacterial surface. No single invasin or surface protein accounts for high-efficiency invasion of epithelial cells by all strains of streptococci. These findings suggest that intracellular invasion may be triggered by different agonists in different tissues or at different stages of infection. Intracellular streptococci appear to account for the persistence of streptococci after penicillin therapy and to be one source of streptococci that are responsible for recurrent tonsillitis. Therefore, the capacity to invade epithelial cells may influence the prevalence of a given strain of *S. pyogenes* in human populations. A better understanding of streptococcal invasins, host cell receptors, and downstream signals may reveal why this species causes a wide spectrum of diseases, and could lead to new and improved treatment strategies.

REFERENCES

1. **Berkower, C., M. Ravins, A. E. Moses, and E. Hanski.** 1999. Expression of different group A streptococcal M proteins in an isogenic background demonstrates diversity in adherence to and invasion of eukaryotic cells. *Mol. Microbiol.* **31:**1463–1475.
2. **Brandt, C. M., F. Allerberger, B. Spellerberg, R. Holland, R. Lutticken, and G. Haase.** 2001. Characterization of consecutive *Streptococcus pyogenes* isolates from patients with pharyngitis and bacteriological treatment failure: special reference to *prt*F1 and *sic/drs. J. Infect. Dis.* **183:**670–674.
3. **Cheng, Q., D. Stafslien, S. S. Purushothaman, and P. Cleary.** 2002. The group B streptococcal C5a peptidase is both a specific protease and an invasin. *Infect. Immun.* **70:**2408–2413.
4. **Cleary, P., E. L. Kaplan, J. P. Handley, A. Wlazlo, M. H. Kim, A. R. Hauser, and P. M. Schlievert.** 1992. Clonal basis for resurgence of serious streptococcal disease in the 1980's. *Lancet* **321:**518–521.
5. **Cleary, P. P., D. LaPenta, R. Vessela, H. Lam, and D. Cue.** 1998. A globally disseminated M1 subclone of group A streptococci differs from other subclones by 70 kilobases of prophage DNA and capacity for high-frequency intracellular invasion. *Infect. Immun.* **66:**5592–5597.
6. **Cleary, P. P., L. McLandsborough, L. Ikedo, D. Cue, J. Krawczak, and H. Lam.** 1998. High-frequency intracellular infection and erythrogenic toxin A expression undergo phase variation in M1 group A streptococci. *Mol. Microbiol.* **28:**157–167.
7. **Cockerill, F. R., K. L. MacDonald, R. L. Thompson, F. Roberson, P. C. Kohner, J. Besser-Wiek, J. M. Manahan, J. M. Musser, P. M. Schlievert, J. Talbot, B. Frankfort, J. M. Steckelberg, W. R. Wilson, and M. T. Osterholm.** 1997. An outbreak of invasive group A streptococcal disease associated with high carriage rates of the invasive clone among school-aged children. *JAMA* **277:**38–43.
8. **Courtney, H. S., J. B. Dale, and D. L. Hasty.** 1997. Host cell specific adhesins of group A streptococci. *Adv. Exp. Med. Biol.* **418:**605–606.
9. **Cue, D., P. E. Dombek, H. Lam, and P. P. Cleary.** 1998. Serotype M1 *Streptococcus pyogenes* encodes multiple pathways for entry into human epithelial cells. *Infect. Immun.* **66:**4593–4601.
10. **Cue, D., S. O. Southern, P. J. Southern, J. Prabhakar, W. Lorelli, J. M. Smallheer, S. A. Mousa, and P. P. Cleary.** 2000. A nonpeptide integrin antagonist can inhibit epithelial cell ingestion of *Streptococcus pyogenes* by blocking formation of integrin alpha5beta1-fibronectin-M1 protein complexes. *Proc. Natl. Acad. Sci. USA* **97:**2858–2863.
11. **Cywes, C., and M. Wessels.** 2001. Group A Streptococcus tissue invasion by CD44-mediated cell signalling. *Nature* **414:**648–652.
12. **Dombek, P. E., J. Sedgewick, H. Lam, S. Ruschkowski, B. B. Finlay, and P. P. Cleary.** 1999. High frequency intracellular invasion of epithelial cells by serotype M1 group A streptococci: M1 protein mediated invasion and cytoskeletal rearrangements. *Mol. Microbiol.* **31:**859–870.
13. **Elsner, A., B. Kreikemeyer, A. Braun-Kiewnick, B. Spellerberg, B. A. Buttaro, and A. Podbielski.** 2002. Involvement of Lsp, a member of the LraI-lipoprotein family in *Streptococcus pyogenes*, in eukaryotic cell adhesion and internalization. *Infect. Immun.* **70:**4859–4869.
14. **Finlay, B. B., and S. Falkow.** 1997. Common themes in microbial pathogenicity revisited. *Microbiol. Mol. Biol. Rev.* **61:**136–169.
15. **Facinelli, B., C. Spinaci, G. Magi, E. Giovanetti, and P. E. Varald.** 2001. Association between erythromycin resistance and ability to enter human respiratory cells in group A streptococci. *Lancet* **358:**30–33.
16. **Fischetti, V. A.** 2000. Surface proteins on gram-positive bacteria, p. 11–24. *In* V. A. Fischetti R. P. Novick, J. J. Ferretti, D. A. Portnoy, and J. I. Rood (ed.), *Gram-Positive Pathogens.* ASM Press, Washington, D.C.
17. **Greco, R., L. De Martino, G. Donnarumma, M. P. Conte, L. Seganti, and P. Valenti.** 1995. Invasion of cultured human cells by *Streptococcus pyogenes. Res. Microbiol.* **146:**5551–5560.
18. **Hall, A.** 1998. Rho GTPases and the actin cytoskeleton. *Science* **279:**509–514.
19. **Hannigan, G. E., C. Leung-Hagesteijn, L. Fitz-Gibbon, M. G. Coppolino, G. Radeva, J. Filmus, J. C. Bell, and S. Dedhar.** 1996. Regulation of cell adhesion and anchorage-dependent growth by a new beta 1-integrin-linked protein kinase. *Nature* **379:**91–96.

20. **Hynes, R. O.** 1992. Integrins: versatility, modulation, and signaling in cell adhesion. *Cell* **69:**11–25.

21. **LaPenta, D., C. Rubens, E. Chi, and P. P. Cleary.** 1994. Group A streptococci efficiently invade human respiratory epithelial cells. *Proc. Natl. Acad. Sci. USA* **91:**12115–12119.

22. **Liszewski, M. K., and J. P. Atkinson.** 1992. Membrane cofactor protein. *Curr. Top. Microbiol. Immunol.* **178:**45–60.

23. **Mishima, W., A. Suzuki, S. Yamaji, R. Yoshimi, A. Ueda, T. Kaneko, J. Tanaka, Y. Miwa, S. Ohno, and Y. Ishigatsubo.** 2004. The first CH domain of affixin activates Cdc42 and Rac1 through alphaPIX, a Cdc42/Rac1-specific guanine nucleotide exchanging factor. *Genes Cells* **9:**193–204.

24. **Molinari, G., and G. S. Chhatwal.** 1998. Invasion and survival of *Streptococcus pyogenes* in eukaryotic cells correlates with the source of the clinical isolates. *J. Infect. Dis.* **177:**1600–1607.

25. **Molinari, G., S. R. Talay, P. Valentin-Weigand, M. Rohde, and G. S. Chhatwal.** 1997. The fibronectin-binding protein of *Streptococcus pyogenes*, SfbI, is involved in the internalization of group A streptococci by epithelial cells. *Infect. Immun.* **65:**1357–1363.

26. **Molinari, G., M. Rohde, C. A. Guzmán, and G. S. Chhatwal.** 2000. Two distinct pathways for the invasion of *Streptococcus pyogenes* in non-phagocytic cells. *Cell. Microbiol.* **2:**145–154.

27. **Musser, J. M., V. Kapur, J. Szeto, X. Pan, D. S. Swanson, and D. R. Martin.** 1995. Genetic diversity and relationships among *Streptococcus pyogenes* strains expressing serotype M1 protein: recent intercontinental spread of a subclone causing episodes of invasive disease. *Infect. Immun.* **63:**994–1003.

28. **Neeman, R., N. Keller, A. Barzilai, Z. Korenman, and S. Sela.** 1998. Prevalence of the internalization-associated gene, *prtF1*, among persisting group A streptococcus strains isolated from asymptomatic carriers. *Lancet* **352:**1974–1977.

29. **Österlund, A., and L. Engstrand.** 1995. Intracellular penetration and survival of *Streptococcus pyogenes* in respiratory epithelial cells in vitro. *Acta Otolaryngol.* **115:**685–688.

30. **Österlund, A., R. Popa, T. Nikkila, A. Scheynius, and L. Engstrand.** 1997. Intracellular reservoir of *Streptococcus pyogenes* in vivo: a possible explanation for recurrent pharyngotonsillitis. *Laryngoscope* **107:**640–647.

31. **Ozeri, V., I. Rosenshine, D. F. Mosher, R. Fassler, and E. Hanski.** 1998. Roles of integrins and fibronectin in the entry of *Streptococcus pyogenes* into cells via protein F1. *Mol. Microbiol.* **30:**625–637.

32. **Ozeri, V., I. Rosenshine, A. Ben-Ze'Ev, G. M. Bokoch, T. S. Jou, and E. Hanski.** 2001. De novo formation of focal complex-like structures in host cells by invading streptococci. *Mol. Microbiol.* **41:**561–573.

33. **Pandiripally, V., L. Wei, C. Skerka, P. F. Zipfel, and D. Cue.** 2003. Recruitment of complement factor H-like protein 1 promotes intracellular invasion by group A streptococci. *Infect. Immun.* **71:**7119–7128.

34. **Park, H. S., K. P. Francis, J. Yu, and P. P. Cleary.** 2003. Membranous cells in nasal-associated lymphoid tissue: a portal of entry for the respiratory mucosal pathogen group A streptococcus. *J. Immunol.* **171:**2532–2537.

35. **Park, H. S., M. Costalonga, R. L. Reinhardt, P. E. Dombek, M. K. Jenkins, and P. P. Cleary.** 2004. Primary induction of CD4 T cell responses in nasal-associated lymphoid tissue during group A streptococcal infection. *Eur. J. Immunol.* **34:**2843–2853.

36. **Persad, S., S. Attwell, V. Gray, N. Mawji, J. L. Deng, D. Leung, J. Yan, J. Sanghera, M. P. Walsh, and S. Dedhar.** 2001. Regulation of protein kinase B/Akt-serine 473 phosphorylation by integrin-linked kinase: critical roles for kinase activity and amino acids arginine 211 and serine 343. *J. Biol. Chem.* **276:**27462–27469.

37. **Purushothaman, S. S., B. Wang, and P. P. Cleary.** 2003. M1 protein triggers a phosphoinositide cascade for group A streptococcus invasion of epithelial cells. *Infect. Immun.* **71:**5823–5830.

38. **Purushothaman, S. S., H. S. Park, and P. P. Cleary.** 2004. Promotion of fibronectin independent invasion by C5a peptidase into epithelial cells in group A streptococcus. *Indian J. Med. Res.* **119**(Suppl):44–47.

39. **Rezcallah, M. S., K. Hodges, D. B. Gill, J. P. Atkinson, B. Wang, and P. P. Cleary.** 2005. Engagement of CD46 and α5β1 integrin by group A streptococci is required for efficient invasion of epithelial cells. *Cell. Microbiol.* **7:**645–653.

40. **Rohde, M., E. Müller, G. S. Chhatwal, and S. R. Talay.** 2003. Host cell caveolae act as an entry-port for group A streptococci. *Cell. Microbiol.* **5:**323–342.

41. **Rosenberger, G., I. Jantke, A. Gal, and K. Kutsche.** 2003. Interaction of alphaPIX (ARHGEF6) with beta-parvin (PARVB) suggests an involvement of alphaPIX in integrin-mediated signaling. *Hum. Mol. Genet.* **12:**155–167.

42. **Schmidt, K.-H., K. Mann, J. Cooney, and W. Kohler.** 1993. Multiple binding of type 3 streptococcal M protein to human fibrinogen, albumin and fibronectin. *FEMS Immunol. Med. Microbiol.* **7:**135–144.

43. **Schrager, H. M., J. G. Rheinwald, and M. R. Wessels.** 1996. Hyaluronic acid capsule and the role of streptococcal entry into keratinocytes in invasive skin infection. *J. Clin. Invest.* **98:**1954–1958.

44. **Stevens, D. L., M. H. Tanner, J. Winship, R. Swarts, K. Ries, P. Schlievert, and E. Kaplan.** 1989. Severe group A streptococcal infections associated with a toxic shock-like syndrome and scarlet fever toxin A. *N. Engl. J. Med.* **321:**1–7.

45. **Terao, Y., S. Kawabata, E. Kunitomo, J. Murakami, I. Nakagawa, and S. Hamada.** 2001. Fba, a novel fibronectin-binding protein from *Streptococcus pyogenes*, promotes bacterial entry into epithelial cells, and the *fba* gene is positively transcribed under the Mga regulator. *Mol. Microbiol.* **42:**75–86.

46. **Terao, Y., S. Kawabata, E. Kunitomo, I. Nakagawa, and S. Hamada.** 2002. Novel laminin-binding protein of *Streptococcus pyogenes*, Lbp, is involved in adhesion to epithelial cells. *Infect. Immun.* **70:**993–997.

47. **Van Aelst, L., and C. D'Souza-Schorey.** 1997. Rho GTPases and signaling networks. *Genes Dev.* **11:**2295–2322.

48. **Wang, B., Y. Yureco, S. Dedhar, and P. Cleary.** Integrin-linked kinase is an essential link between integrins and uptake of bacterial pathogens by epithelial cells. *Cell. Microbiol.*, in press.

Capsular Polysaccharide of Group A Streptococci

MICHAEL R. WESSELS

4

In her early studies of group A streptococci (GAS), Rebecca Lancefield noted an association between virulence and a distinctive appearance of the bacterial colonies on solid media. Isolates that were highly virulent for mice and that grew well in fresh human blood typically formed large colonies with a translucent, liquid appearance (mucoid) or an irregular, collapsed appearance (matte). By contrast, avirulent isolates that grew poorly in human blood formed compact, opaque colonies (glossy). Strains that grew as mucoid or matte colonies usually produced large amounts of M protein, which was given the designation "M" by Lancefield because of this association with colony morphology (44, 73). Later work by Armine Wilson (80) demonstrated that the mucoid or matte appearance of such strains was in fact due to elaboration of capsular polysaccharide, not M protein. Wilson showed that the mucoid or matte colony type was converted to a nonmucoid or glossy colony by growth on medium containing hyaluronidase, which digested the hyaluronic acid capsule, whereas growth on medium containing trypsin, which digested M protein, had no such effect. Furthermore, although many strains that produced abundant capsular polysaccharide also were rich in M protein, expression of the two surface products was not always linked; certain mucoid or highly encapsulated strains produced little or no M protein, and certain strains rich in M protein produced little or no capsule (62, 73).

Both M protein and capsule have since been shown to contribute independently to GAS virulence. The association of mucoid strains of GAS with both invasive infection and acute rheumatic fever suggested a role for the capsular polysaccharide in virulence. Anecdotal observations that linked the capsule to pathogenesis of streptococcal disease were supported by a study that characterized the colony morphology of more than 1,000 GAS clinical isolates received by a streptococcal reference lab in the 1980s. In that study, only 3% of GAS isolates from patients with pharyngitis had a mucoid colony morphology, whereas 21% of isolates from patients with invasive disease syndromes such as bacteremia, necrotizing fasciitis, or streptococcal toxic shock syndrome and 42% of isolates associated with acute rheumatic fever were mucoid (40). The overrepresentation of mucoid strains among isolates associated with invasive infection and rheumatic fever suggested that the capsule contributes to enhanced virulence. The occurrence of outbreaks of acute rheumatic fever associated with mucoid strains of GAS at several locations in the United States in the 1980s supported the same inference (49, 74, 79). These clinical and epidemiologic observations implicating the capsule in pathogenesis have been corroborated by extensive experimental studies that have demonstrated a definite role of the hyaluronic acid capsule as a virulence factor in GAS infection.

BIOCHEMISTRY AND GENETICS OF CAPSULE PRODUCTION

Hyaluronic Acid Biosynthesis

The GAS capsular polysaccharide is composed of hyaluronic acid, a high-M_r ($>10^6$ kDa) linear polymer made up of $\beta(1\rightarrow4)$-linked disaccharide repeat units of D-glucuronic acid$(1\rightarrow3)$-β-D-N-acetylglucosamine. The polysaccharide is synthesized from the nucleotide sugar precursors UDP-glucuronic acid and UDP-N-acetylglucosamine by a membrane-associated enzyme, hyaluronan synthase. High-M_r hyaluronic acid (or hyaluronan) can be produced by incubation of cell-free membrane extracts of GAS with the two substrate UDP-sugars in the presence of divalent cations (50, 69, 70).

A genetic locus required for hyaluronic acid production in GAS was located independently by three laboratories using transposon mutagenesis to produce acapsular mutants from a mucoid GAS strain. Transposon insertions that produced the acapsular phenotype were mapped to a locus that was highly conserved among GAS strains (15, 20, 78). Characterization of this region of the GAS chromosome revealed a cluster of three genes, *hasABC*, whose products are involved in hyaluronic acid biosynthesis. The first gene, *hasA*, encodes hyaluronan synthase, a protein with a predicted M_r of 47.9 kDa that includes at least four predicted membrane-spanning domains, consistent with evidence that the enzyme is membrane-associated (16, 22). The *hasA*

Gram-Positive Pathogens, 2nd edition, edited by Vincent A. Fischetti et al.

gene product shares significant similarity with the NodC protein of *Rhizobium meliloti* which is required for synthesis of an *N*-acetylglucosamine-containing polysaccharide, with chitin synthases, and with hyaluronan synthases from other microbial and higher animal species (17, 75). The GAS hyaluronan synthase has been expressed in *Escherichia coli* as a His$_6$-fusion protein. Activity of the purified enzyme is strongly stimulated by the presence of certain phospholipids, particularly cardiolipin (diphosphatidylglycerol), which appears to be associated with the enzyme in GAS (71). Cardiolipin molecules may contribute to formation of a transmembrane pore for export of the hyaluronic acid polymer, because the hyaluronan synthase itself contains only two relatively small extracellular domains (37).

The second gene, *hasB*, encodes UDP-glucose dehydrogenase, a protein with a predicted M$_r$ of 45.5 kDa that catalyzes the oxidation of UDP-glucose to UDP-glucuronic acid (21). The third gene in the cluster, *hasC*, encodes a predicted 33.7-kDa protein identified as UDP-glucose pyrophosphorylase (11). The function of this enzyme is to catalyze the condensation of UTP with glucose-1-phosphate to form UDP-glucose. Thus, the reaction catalyzed by the *hasC* product yields a substrate for UDP-glucose dehydrogenase encoded by *hasB*, whose reaction product is, in turn, a substrate for hyaluronan synthase encoded by *hasA* (Fig. 1). Although the enzyme protein encoded by *hasC* is enzymatically active, it is not required for hyaluronic acid synthesis by GAS. Selective inactivation of *hasC* resulted in no decrement in hyaluronic acid synthesis by a highly encapsulated strain of GAS; the implication is that another source of UDP-glucose is available within the cell (3). Furthermore, expression of recombinant *hasA* and *hasB* (without *hasC*) resulted in synthesis of hyaluronic acid in *E. coli* and *Enterococcus faecalis* (15, 16). Together, these observations have led to the conclusion that *hasA* and *B* are the only two genes uniquely required for bacterial synthesis of hyaluronic acid. The three genes

hasABC are transcribed as a single message of approximately 4.2 kb from a promoter immediately upstream of *hasA* (10). The identification of a potential *rho*-independent terminator at the 3′ terminus of *hasC* suggests that no additional genes are included in the operon.

The genes immediately upstream and downstream of the *has* operon appear not to be involved in capsule synthesis or surface expression (Fig. 2). Downstream of *hasC* is a small open reading frame of unknown function, then a gene with sequence similarity to the *recF* gene in other bacteria (18). Inactivation of the *recF* homolog in GAS results in increased susceptibility to UV irradiation, supporting the identification of the GAS gene product as a DNA repair enzyme. Further downstream from *recF* is an open reading frame of unknown function that resembles a membrane protein and *guaB* encoding inosine monophosphate dehydrogenase, an enzyme in the purine biosynthesis pathway (2). Upstream of *hasA* are three open reading frames of unknown function; introduction of a large deletion encompassing most of the first two open reading frames resulted in no decrement in capsule production, evidence that neither of these genes is involved in capsule expression (3). Further upstream are a homolog of phosphatidylglycerophosphate synthase (*pgsA*) and three open reading frames predicted to encode ATP-binding components and a putative permease protein of an ABC transport system (3, 58). An insertion mutation in the locus encoding one of the putative ATP-binding proteins (SPy2194 in the SF370 genome) was associated with reduced production of extracellular hyaluronic acid but no effect on hyaluronan synthase activity (58). Although this observation raises the possibility that an ABC transport system may participate in hyaluronic acid export, it requires confirmation because homologs of these genes are also found in the genome of group B streptococci in a location remote from the group B streptococcal genes involved in capsular polysaccharide synthesis.

The simplicity of the capsule gene cluster in GAS contrasts with the size and complexity of capsule synthesis loci in other encapsulated gram-positive bacteria, which generally consist of at least 12 genes (43, 54, 56, 63, 64). One exception is the type 3 pneumococcus, whose capsule production is strikingly similar to that in GAS with respect to the structure of the polysaccharide and the capsule synthesis genes; in type 3 pneumococcus, capsule synthesis involves three genes that are homologs of *hasABC* (19, 31). In both GAS and type 3 pneumococcus, it seems likely that the capsule synthesis gene cluster was acquired from another species by horizontal gene transfer. This hypothesis is also suggested by genome comparisons of GAS with group B streptococci: with the exception of the *hasABC* cluster that is absent in group B streptococci, there is a high degree of gene synteny over more than 25 kb of the corresponding chromosomal region in GAS and group B streptococci.

Regulation of Hyaluronic Acid Biosynthesis

The *has* operon is highly conserved among GAS strains. A study of nine GAS isolates that came from diverse settings and whose dates of isolation spanned 75 years found only six mutations within a 300-nucleotide central region of the *hasA* gene (17); all were single nucleotide substitutions, and none resulted in a change in the amino acid sequence of the protein. Despite the high degree of conservation of the hyaluronan synthase gene, individual GAS strains vary widely in the amount of capsule they produce. Because the *has* genes are conserved, differences in capsule expression among strains are likely to reflect differences in the regula-

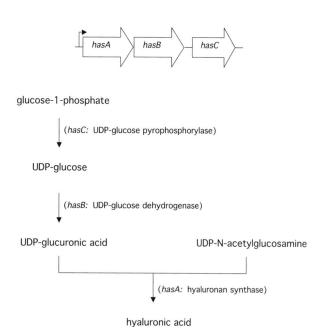

FIGURE 1 Schematic diagram of the *has* operon and the function of each gene product in the biosynthetic pathway for synthesis of hyaluronic acid in GAS.

SPy2193 2194 2195 *pgsA* SPy2197 2198   2199   **hasA** **hasB** **hasC** SPy2203 *recF* SPy2205 *guaB*

FIGURE 2 Map of the region of the GAS chromosome that includes the *has* operon encoding enzymes required for hyaluronic acid synthesis. The other genes shown appear not to be involved in capsular polysaccharide synthesis or surface expression. In some strains insertion sequence IS1239′ is present approximately 50 nucleotides upstream of the *has* operon promoter. (Adapted from the genome sequence of M1 strain SF370, GenBank accession number AE004092 [26].)

tion of *has* gene transcription. Experimental support for this hypothesis comes from the observation that *has* gene transcripts are detected during exponential growth of well-encapsulated strains of GAS, but not in strains that do not produce detectable hyaluronic acid (10). Analysis of the *has* operon promoter using a reporter system in which the promoter was fused to a promoterless chloramphenicol acetyl transferase (CAT) gene showed that full activity of the *has* promoter required no more than 12 nucleotides of chromosomal DNA upstream of the promoter's −35 site (1). This observation is consistent with the finding that many strains of GAS (10 of 23) contain the insertion element IS1239′ approximately 50 nucleotides upstream of the −35 site without any apparent effect on capsule production (3).

One factor that influences the amount of capsule produced by a particular strain is the structure of the *has* operon promoter itself. Experiments using CAT reporter gene fusions revealed threefold higher activity of the *has* promoter from a highly encapsulated M type 18 strain than that from a poorly encapsulated M type 3 strain (1). Construction of hybrid promoters by site-directed mutagenesis demonstrated that nucleotide substitutions in two regions accounted for the differences in promoter activity between the two serotypes: three nucleotides in the −35, −10 spacer region and four nucleotides in the +2 to +8 positions relative to the start site of *hasA* transcription. The characteristic fine structure of the *has* operon promoter in type 18 strains may account for the observation that M18 strains are typically mucoid (i.e., highly encapsulated) because the precise sequence of the *has* promoter appears to be serotype-specific.

Capsule production varies not only among strains but also in an individual strain under different circumstances. For example, during growth in liquid batch culture, GAS produce large capsules during exponential phase and then shed their capsules into the medium during stationary phase. Studies using a *has* promoter-CAT fusion integrated into the GAS chromosome demonstrated a rapid increase in promoter activity to a maximal level during the exponential phase of growth, with a sharp fall as the organisms entered stationary phase. The production of cell-associated capsular polysaccharide followed an identical but slightly delayed pattern, consistent with the expected lag for changes in the expression of enzyme proteins that are reflected in the rate of polysaccharide synthesis. These results indicate that the changing levels of encapsulation observed during different phases of growth are due to transcriptional regulation of *has* operon expression. Further studies on the effect of growth rate on capsule expression in a chemostat continuous culture system found that levels of CAT activity and capsule production were at least three times higher in cells maintained at a constant rapid growth rate (doubling time of 1.4 h) than in cells maintained at a slow growth rate (doubling time of 11 h) (S. Albertí, J. L. Levin, and M. R. Wessels, unpublished observations).

Up-regulation of capsule production is likely to occur during infection when the organism moves from the external environment into the pharynx or deep tissues, where both temperature and nutrient availability are optimal for bacterial growth. This hypothesis was supported by experiments using a reporter strain of GAS in which expression of green fluorescent protein was controlled by the *has* operon promoter. Introduction of the strain into the mouse peritoneum or into the baboon pharynx triggered a rapid increase in fluorescence, reflecting transcription from the *has* promoter (34). Thus, the regulation of capsule expression may be a useful adaptation to survival in host environments in which the capsule is required to defend the organism against complement-mediated phagocytic killing.

The dynamic changes in GAS capsule expression with changes in growth rate and the observation that some strains of GAS appear more highly encapsulated after animal passage have suggested that capsule gene transcription is regulated by cellular mechanisms in addition to those attributable to the intrinsic structure of the *has* operon promoter. Mga, the *trans*-acting regulatory protein that influences expression of several GAS proteins, appears not to regulate capsule expression, at least in most strains of GAS. Inactivation of Mga was associated with reduced *has* gene transcription in an M1 strain (9); however, no effect on capsule production and/or *has* transcription was observed upon Mga inactivation in strains of M6, M18, or M49 (48, 59, 60). RofA, a regulatory protein involved in control of protein F expression, also is not known to affect capsule production (29).

Transposon mutagenesis of a poorly encapsulated M3 strain of GAS resulted in the identification of a novel regulatory locus whose inactivation dramatically increased capsule production (48). The locus consists of two genes, *csrR* and *csrS* (also called *covR* and *covS*), whose predicted products resemble the regulator and sensor proteins, respectively, of bacterial two-component regulatory systems (Fig. 3). Targeted inactivation of *csrR* by allelic exchange mutagenesis resulted in a sixfold increase in capsule production. The increase in hyaluronic acid synthesis was accompanied by a parallel increase in *has* gene transcription—evidence that CsrR regulates transcription of the capsule synthesis genes. Mutation of *csrR* appeared not to affect expression of M protein or hemolytic activity, but subsequent work in several laboratories has demonstrated that the system also represses expression of several other virulence factors, including streptokinase (*ska*), streptolysin S (*sag* operon), mitogenic factor/streptodornase (*speMF*), an integrin-like protein/IgG protease (*mac* or *ideS*), and, in some strains, cysteine protease (*speB*) (25, 36, 47, 52). Microarray transcriptional profiling experiments suggested that CsrRS is involved directly or indirectly in regulation of approximately 15% of GAS genes (32). The membrane-associated CsrS protein is thought to respond to extracellular Mg^{2+} concentration

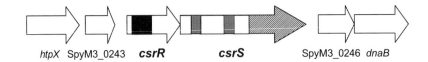

■ Predicted receiver domain
▦ Predicted membrane-spanning regions
▨ Predicted transmitter domain

FIGURE 3 Diagram of the *csrRS* chromosomal locus encoding a two-component regulatory system that regulates hyaluronic acid synthesis. Sequences corresponding to regions of the predicted proteins with properties characteristic of the response regulator (CsrR) or sensor (CsrS) components are indicated. (Adapted from the genome sequence of M3 strain MGAS315, GenBank accession number AE014074 [7].)

and perhaps other environmental signals by controlling phosphorylation of the cognate regulator CsrR (35). Phosphorylation of CsrR increases its affinity for binding to the promoter region of regulated genes to repress transcription (8, 24, 52). The likely importance of this regulatory system in virulence is supported by the markedly greater resistance to in vitro phagocytosis displayed by a (highly encapsulated) *csrR* mutant than by the wild-type parent strain and a 500-fold increase in the mutant's virulence for mice after intraperitoneal challenge (48).

THE CAPSULE AND HOST IMMUNE DEFENSES

Immunogenicity of Hyaluronic Acid

In 1937, Kendall et al. (42) reported the purification of capsular polysaccharide from mucoid strains of GAS. Their analysis revealed that an apparently identical capsular material was produced by different M types of GAS. The GAS polysaccharide contained equal amounts of *N*-acetylglucosamine and glucuronic acid as exclusive component monosaccharides; this finding suggested that it might be structurally identical to a polysaccharide isolated from bovine vitreous humor and human umbilical cord. Immunization of rabbits and horses with mucoid GAS cells failed to elicit antibodies detectable by precipitin reactions with the polysaccharide. Later studies have confirmed that the "serologically inactive" polysaccharide produced by GAS is hyaluronic acid, a high-M_r polysaccharide with a repeating unit structure identical to that of hyaluronic acid isolated from animal sources. Presumably because it is recognized as a "self" antigen, hyaluronic acid is poorly immunogenic in several animal species. Fillit and colleagues (27, 28) evoked antibodies to hyaluronic acid by immunization of rabbits with encapsulated streptococci of group A or C and by immunization of mice with hyaluronidase-treated hyaluronic acid linked to liposomes. Such antibodies, however, have not been shown to have any opsonic activity against GAS in vitro or to confer protection against GAS infection in vivo.

Capsule and Resistance to Phagocytosis

GAS, like other gram-positive pathogens, is resistant to direct bacteriolysis by complement and antibodies. There-fore, clearance of the organisms from the blood or deep tissues depends largely on uptake and killing by phagocytes. This concept was supported by the observations of Todd, Lancefield, and others that virulence of GAS isolates in both clinical settings and in experimental animal infections was closely paralleled by the capacity of the strains to resist killing by human blood leukocytes. GAS strains that produced large amounts of M protein multiplied in fresh human blood, while those that lacked M protein did not (45, 72). In addition, however, early studies demonstrated a relationship between surface expression of the hyaluronic acid capsule and resistance to phagocytosis. Mucoid or highly encapsulated strains tended to be resistant, and hyaluronidase treatment increased their susceptibility to phagocytosis in vitro, a result that suggested a protective effect of the capsule (30, 61, 68). More recent work with acapsular mutants has provided definitive evidence that the capsule is a major determinant of resistance to phagocytic killing. In several assays measuring complement-dependent phagocytic killing by human blood leukocytes, acapsular mutants of type 18 and type 24 GAS were significantly more susceptible than the corresponding encapsulated wild-type strains (14, 55, 78).

One mechanism through which capsular polysaccharides may confer resistance to phagocytosis is the inhibition of complement activation. The type III capsular polysaccharide of group B streptococci, for example, prevents deposition of C3b on the bacterial surface through inhibition of alternative complement pathway activation by sialic acid residues present in the polysaccharide (23, 51). In GAS, however, the presence of the capsular polysaccharide appears to have no effect on the amount of C3b deposited on the bacterial surface. Similar amounts of C3 were detected on acapsular mutants and on encapsulated wild-type strains after incubation of the organisms in serum (14). The increased resistance of well-encapsulated GAS strains to complement-dependent phagocytic killing, therefore, must reflect the capacity of the capsule to block access of neutrophils to opsonic complement components bound to bacterial surface rather than inhibition of C3b deposition (Fig. 4).

Resistance of GAS to phagocytosis has been assessed most often with the bactericidal test of Lancefield, in which a small inoculum of the test strain is rotated for 3 h in heparinized fresh human blood (45). Strains that exhibit a

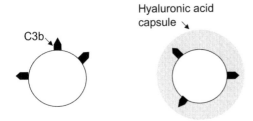

FIGURE 4 The hyaluronic acid capsule and resistance to complement-mediated phagocytosis. The capsule does not prevent deposition of C3b on the bacterial cell wall, but rather interferes with the interaction of bound C3b with phagocyte receptors.

significant increase in CFU during rotation in blood are considered resistant. An effect of the capsule is evident in this assay—acapsular mutants show less net growth than encapsulated wild-type strains—but the effect of M protein is more striking (55, 78). M protein-deficient strains not only fail to grow but typically are reduced in number after a 3-h incubation. By contrast, when resistance to phagocytosis is assessed with a different assay, the effect of capsule is more dramatic than that of M protein. In this second assay system, a larger bacterial inoculum is rotated for 1 h with human peripheral blood leukocytes in the presence of 10% normal human serum as a complement source (6, 77). In the latter assay, acapsular mutant strains and poorly encapsulated wild-type strains are killed (~90% reduction in CFU) despite the presence of M protein, while well-encapsulated strains do not change or increase in number. A mutant that is deficient in M protein but that produces wild-type amounts of capsule is killed less efficiently than a capsule-deficient mutant in the same background (55). In a study of isogenic mutants deficient in capsule, M protein, or both in the background of a mucoid type 18 GAS strain, mouse virulence correlated better with results in the second assay system than with the Lancefield whole-blood assay (55). It appears, therefore, that the role of capsule in resistance to phagocytosis as a correlate of virulence may be better assessed in the opsonophagocytic assay in 10% serum than in the whole-blood assay of Lancefield.

CAPSULE AS A VIRULENCE FACTOR

Experimental Infection Models

In the 1940s, investigators attempted to determine experimentally whether the hyaluronic acid capsule contributed to the virulence of GAS. Hirst (38) found that decapsulating GAS by treatment with leech extract containing hyaluronidase had no effect on virulence of the bacteria in mice. However, Kass and Seastone (41) found reduced virulence of decapsulated GAS in mice, perhaps because the latter authors administered hyaluronidase to the infected mice at frequent intervals in order to prevent regeneration of the capsule after challenge (41). During the 1990s, more direct proof that the capsule contributes to virulence came from several studies using acapsular mutant strains of GAS derived by transposon mutagenesis or by targeted inactivation of gene(s) required for hyaluronic acid synthesis (Table 1). In the first such study, an acapsular mutant was derived by Tn916 mutagenesis from a mucoid M type 18 strain of GAS (77). The mutant strain, TX4, did not pro-

TABLE 1 Animal models in which the hyaluronic acid capsule has been shown to enhance virulence of group A streptococci

Type of infection	Animal species	Reference(s)
Systemic infection	Mouse	77, 78
	Chicken (embryo)	65
Pharyngeal colonization	Mouse	39, 76
	Baboon	2
Pneumonia	Mouse	39
Soft tissue infection	Mouse	4, 66

duce detectable hyaluronic acid but made wild-type amounts of type 18 M protein. Loss of capsule expression in the mutant strain was associated with a >100-fold increase in 50% lethal dose in mice after intraperitoneal challenge. Further characterization of strain TX4 revealed a deletion of chromosomal DNA adjacent to the site of Tn916 insertion of approximately 9.5 kb, including the has operon required for hyaluronic acid synthesis (see above). To exclude the possibility that the reduced virulence of the mutant might be due in part to loss of an unrecognized phenotype encoded by the deleted DNA rather than entirely to the capsule synthesis genes, a second Tn916 mutant was characterized. The second mutant, TX72, harbored a single copy of Tn916 within hasA, the gene encoding hyaluronan synthase, and no associated chromosomal deletion (78). Virulence testing of TX72 in parallel with TX4 demonstrated that the two strains were similarly attenuated in the mouse model of lethal GAS infection. This result firmly established the role of the capsule in an experimental infection model. Both the simple insertion mutation in TX72 and the insertion-deletion mutation in TX4 were transduced independently to a highly encapsulated M type 24 strain (Vaughn), and both resulted in loss of capsule expression in the recipient strain. The type 24 acapsular mutants, 24-72 and 24-4, also exhibited reduced virulence in mice after intraperitoneal challenge, with respective 50% lethal dose values approximately 900 and 600 times higher, respectively, than that for the type 24 wild-type strain (76). Virulence of both acapsular mutants (TX-4 and 24-4) was similarly reduced from that of their respective parent strains (M18 and M24) in a chicken embryo model of lethal GAS infection—results indicating that the role of the capsule in virulence was not limited to a single animal species (65).

Because the capsule is required for GAS resistance to complement-mediated phagocytic killing (see above), it is not surprising that capsule expression enhances virulence in systemic infection models in which the animals succumb to overwhelming bacteremia. In addition, however, the hyaluronic acid capsule contributes in a pivotal fashion to the capacity of the organisms to produce invasive soft tissue infection. A mouse model of invasive GAS soft tissue infection has been developed in which the bacterial inoculum is delivered by superficial injection just below the skin surface, in a small volume, and with minimal trauma to the tissues (4, 66). Mice challenged with the highly encapsulated M24 strain Vaughn developed dermal necrosis with underlying purulent inflammation and secondary bacteremia, while mice challenged with the acapsular mutant strain 24-4 developed no lesion at all or minor superficial inflammation and no bacteremia (66). Similar results were observed in the same model with a moderately encapsulated M3 strain of GAS (originally isolated from a patient with necrotizing fasciitis) that was compared with two isogenic acapsular mutants derived by allelic exchange mutagenesis of the *hasA* (hyaluronan synthase) gene. In the latter experiments, the majority of animals challenged with the encapsulated M3 parent strain developed spreading, necrotic soft tissue infection and died, whereas the acapsular mutants produced only mild local infection and no deaths (4). The dramatic impact of the capsule on virulence in this model is correlated with equally striking histopathologic findings: in animals challenged with an acapsular mutant, a small focus of neutrophilic inflammation and necrosis is confined by a well-formed abscess. By contrast, after challenge with the encapsulated wild-type strain, acute inflammatory cells extend throughout the subcutaneous tissue and are associated with thrombosis of blood vessels, with infarction and necrosis of the overlying dermis—a histopathologic picture very similar to that in patients with GAS necrotizing fasciitis. Examination of the histologic sections by immunofluorescence microscopy with a GAS-specific antibody showed that the acapsular GAS cells were confined to the abscess cavity, while the encapsulated wild-type organisms were widely dispersed throughout the subcutaneous tissues.

Capsule and Mucosal Colonization

Studies by at least two groups have demonstrated a critical role for the hyaluronic acid capsule in pharyngeal colonization in mice. BALB/c mice inoculated intranasally with the stable (transposon insertion plus chromosomal deletion) acapsular mutant strain 24-4 rapidly cleared the organism from the pharynx, and no mice died (76). In contrast, throat cultures of mice challenged with the revertible (simple transposon insertion) acapsular mutant 24-72 yielded encapsulated revertants, and the rate of persistent throat colonization and lethal systemic infection was similar to that in animals challenged with the encapsulated type 24 parent strain. A study of an acapsular mutant derived from the mouse-virulent M50 strain, B514, yielded similar results: intranasal challenge with the wild-type strain resulted in chronic throat colonization in C57BL/10SnJ mice. After intranasal challenge with an acapsular mutant derived by plasmid insertion mutagenesis of the *hasA* gene, encapsulated revertants accounted for all GAS recovered from throat cultures 4 days after challenge (39). Intratracheal inoculation of C3HeB/FeJ mice with the wild-type strain produced pneumonia and secondary systemic infection in approximately 50% of the animals within 72 h, while only two animals challenged with the acapsular mutant became ill, 5 and 6 days after challenge, respectively. In both cases only encapsulated revertants were recovered from cultures of the throat, lungs, and blood (39). The findings in both studies of impaired colonization by the acapsular mutant strains and the rapid emergence of encapsulated revertants together provide strong evidence that encapsulation confers a powerful survival advantage for GAS in the upper airway. This conclusion was supported also by experiments in a primate model of GAS pharyngeal colonization: an acapsular mutant derived from an M3 strain was cleared from the pharynx of baboons more quickly than the wild-type parent strain (5).

Capsule and GAS Attachment to Epithelial Cells

The influence of the hyaluronic acid capsule on the capacity of GAS to colonize the pharyngeal epithelium may be explained, at least in part, by the protective effect of the capsule in preventing opsonophagocytic killing, although the importance of resistance to opsonophagocytosis at a mucosal site is less well defined than in blood or deep tissues. In addition, the capsule is an important factor in the process of GAS attachment to epithelial cells, both by modulating interactions of other bacterial surface molecules with the epithelium and by direct interaction of GAS capsular hyaluronic acid with eukaryotic cells. The capsule may influence adherence by interfering with the interaction of any of a variety of GAS surface molecules that have been implicated as adhesins: M and M-like proteins, fibronectin-binding proteins, lipoteichoic acid, and glyceraldehyde-3-phosphate dehydrogenase. In general, acapsular mutant strains of GAS attach in higher numbers to skin or throat keratinocytes than do the corresponding encapsulated wild-type strains. For type 24 GAS, the capsule prevents M protein-mediated adherence to keratinocytes, because inactivation of the *emm24* gene reduced adherence in the background of an acapsular mutant, but not in the background of the same strain expressing wild-type amounts of capsule (67). It is likely that the capsule exerts a similar modulating influence on bacterial adherence mediated by other GAS surface molecules.

Not only does the capsule influence the binding interactions of other GAS surface structures, there is evidence that the capsule itself can serve as a ligand for attachment of GAS to CD44 on epithelial cells. CD44 is a hyaluronic acid-binding glycoprotein found on a variety of human cells including epithelial and hematopoietic cells. The interaction of CD44 with hyaluronic acid appears to have important functions in cell migration, organization of epithelia, lymphocyte homing, and tumor metastasis. Because CD44 is expressed on the surface of tonsillar epithelial cells, it may serve as a receptor for binding of hyaluronic acid on the surface of GAS. This hypothesis was supported by the finding that a monoclonal antibody to CD44 reduced attachment of wild-type strains of GAS to human pharyngeal keratinocytes by ≃90%, but had no effect on attachment of acapsular mutant strains (67). Further evidence that CD44 acts as a receptor for GAS came from the observation that transfection of K562 cells with cDNA encoding either of two epithelial isoforms of CD44 conferred on the cells the capacity to bind GAS. Cells expressing CD44 bound a variety of GAS wild-type strains, including several in which the level of capsule expression was below the limit of detection in a dye-binding assay; however, inactivation of the hyaluronan synthase gene in such a poorly encapsulated GAS strain completely abrogated binding (67). A study of transgenic mice with defective expression of epithelial CD44 found reduced pharyngeal colonization compared to that in wild-type mice after intranasal challenge with GAS, evidence that CD44 can function as a receptor for GAS binding to the pharyngeal epithelium in vivo (13).

Capsule and CD44-Mediated Tissue Invasion

CD44 is a transmembrane protein, and its ligation by hyaluronic acid can result in signal transduction and induction of cellular events involved in cell migration. Binding of GAS to CD44 on human pharyngeal keratinocytes in vitro has been shown to trigger a similar pattern of cellular responses, including the formation of lamellipodia or membrane ruffles at the site of GAS attachment and disruption of intercellular junctions (12). These events were associated with enhanced translocation of encapsulated GAS compared to an acapsular mutant across epithelial cell layers in vitro. The results suggested that interaction of the GAS capsule with CD44 manipulates a host signaling pathway to disrupt epithelial integrity and enable bacterial penetration into underlying tissues.

Capsule and GAS Entry into Epithelial Cells

Several laboratories have documented the entry of GAS into human epithelial cells, a process that has been proposed to represent a virulence mechanism—either a step in bacterial invasion into deep tissues or a means of escape of the organism into an intracellular sanctuary shielded from antibiotics and host immune defenses (33, 46, 57, 66). However, although there is irrefutable evidence that the capsule plays a key role in invasive infection in vivo, there is equally clear evidence that the capsule prevents GAS entry into epithelial cells. In studies of GAS entry into cultured human keratinocytes from pharyngeal epithelium or external skin, poorly encapsulated strains entered cells far more efficiently than well-encapsulated strains, regardless of M type or clinical source of the isolates (66). Acapsular mutants entered more than 1,000 times more efficiently than corresponding encapsulated parent strains. The apparently opposite effects of the capsule on invasive infection in vivo compared to those on epithelial cell "invasion" in vitro suggest that bacterial entry into epithelial cells is not a step in productive infection, but rather a host response that limits the spread of poorly encapsulated strains. By contrast, virulent encapsulated strains resist ingestion by eukaryotic cells, both epithelial cells and professional phagocytes, and invade deep tissues by an extracellular route, a process that appears to be facilitated by CD44-mediated disruption of cell-cell junctions. This hypothesis is supported by the finding that clinical isolates from patients with invasive GAS infection "invade" epithelial cells at much lower rates than throat or skin isolates (53). Although it is difficult to completely discount the possibility that epithelial cells serve as a sanctuary site for intracellular GAS during some phase of colonization or latent infection, persistence of intracellular GAS for prolonged periods seems unlikely because the viability of GAS within epithelial cells falls rapidly after bacterial entry (33, 66).

CONCLUSIONS

Careful observations by microbiologists and clinicians in the 1920s and 1930s pointed to a link between the hyaluronic acid capsule and GAS disease pathogenesis. More recent studies have defined the genetic locus that directs hyaluronic acid biosynthesis and have characterized the molecular mechanisms through which the capsule enhances GAS virulence. It is now appreciated that the has capsule synthesis operon is both highly conserved and widely distributed among GAS strains, attesting to the adaptive role served by the hyaluronic acid capsule in the coevolution of GAS with the human host. As a poorly immunogenic "self" antigen, the capsular polysaccharide appears to have persisted in an invariant form in GAS, presumably because it has not been subject to the selective pressure of an effective, polysaccharide-specific, humoral immune response in the infected host that has led to emergence of multiple capsular types in other pathogenic bacteria. Studies of acapsular mutant strains have demonstrated that the capsule protects GAS from complement-mediated phagocytic killing and is essential for full virulence in a variety of experimental infection models. The GAS capsule also influences attachment of the bacteria to human epithelial cells, both by

modulating interaction of M protein and other potential adhesins and by itself serving as a ligand for attachment of GAS to CD44 on epithelial cells. The latter mechanism may be of particular importance in GAS colonization of the pharynx, as it appears to participate in adherence of diverse GAS strains including those that produce only small amounts of capsule. Finally, the successful adaptation of GAS to survival in the human host involves regulation of capsule expression that is dependent on the fine structure of the has operon promoter, on the CsrRS two-component regulatory system, and, likely, on additional mechanisms yet to be uncovered.

REFERENCES

1. **Albertí, S., C. D. Ashbaugh, and M. R. Wessels.** 1998. Structure of the has operon promoter and regulation of hyaluronic acid capsule expression in group A Streptococcus. Mol. Microbiol. **28**:343–353.
2. **Ashbaugh, C. D., and M. R. Wessels.** 1995. Identification, cloning, and expression of the group A streptococcal guaB gene encoding inosine monophosphate dehydrogenase. Gene **165**:57–60.
3. **Ashbaugh, C. D., S. Albertí, and M. R. Wessels.** 1998. Molecular analysis of the capsule gene region of group A Streptococcus: the hasAB genes are sufficient for capsule expression. J. Bacteriol. **180**:4955–4959.
4. **Ashbaugh, C. D., H. B. Warren, V. J. Carey, and M. R. Wessels.** 1998. Molecular analysis of the role of the group A streptococcal cysteine protease, hyaluronic acid capsule, and M protein in a murine model of human invasive soft-tissue infection. J. Clin. Investig. **102**:550–560.
5. **Ashbaugh, C. D., T. J. Moser, M. H. Shearer, G. L. White, R. C. Kennedy, and M. R. Wessels.** 2000. Bacterial determinants of persistent throat colonization and the associated immune response in a primate model of human group A streptococcal pharyngeal infection. Cell Microbiol. **2**:283–292.
6. **Baltimore, R. S., D. L. Kasper, C. J. Baker, and D. K. Goroff.** 1977. Antigenic specificity of opsonophagocytic antibodies in rabbit anti-sera to group B streptococci. J. Immunol. **118**:673–678.
7. **Beres, S. B., G. L. Sylva, K. D. Barbian, B. Lei, J. S. Hoff, N. D. Mammarella, M. Y. Liu, J. C. Smoot, S. F. Porcella, L. D. Parkins, D. S. Campbell, T. M. Smith, J. K. McCormick, D. Y. Leung, P. M. Schlievert, and J. M. Musser.** 2002. Genome sequence of a serotype M3 strain of group A Streptococcus: phage-encoded toxins, the high-virulence phenotype, and clone emergence. Proc. Natl. Acad. Sci. USA **99**:10078–10083.
8. **Bernish, B., and I. van de Rijn.** 1999. Characterization of a two-component system in Streptococcus pyogenes which is involved in regulation of hyaluronic acid production. J. Biol. Chem. **274**:4786–4793.
9. **Cleary, P. P., L. McLandsborough, L. Ikeda, D. Cue, J. Krawczak, and H. Lam.** 1998. High-frequency intracellular infection and erythrogenic toxin A expression undergo phase variation in M1 group A streptococci. Mol. Microbiol. **28**:157–167.
10. **Crater, D. L., and I. van de Rijn.** 1995. Hyaluronic acid synthesis operon (has) expression in group A streptococci. J. Biol. Chem. **270**:18452–18458.
11. **Crater, D. L., B. A. Dougherty, and I. van de Rijn.** 1995. Molecular characterization of hasC from an operon required for hyaluronic acid synthesis in group A streptococci. J. Biol. Chem. **270**:28676–28680.

12. **Cywes, C., and M. R. Wessels.** 2001. Group A streptococcus tissue invasion by CD44-mediated cell signalling. *Nature* **414:**648–652.

13. **Cywes, C., I. Stamenkovic, and M. R. Wessels.** 2000. CD44 as a receptor for colonization of the pharynx by group A *Streptococcus. J. Clin. Investig.* **106:**995–1002.

14. **Dale, J. B., R. G. Washburn, M. B. Marques, and M. R. Wessels.** 1996. Hyaluronate capsule and surface M protein in resistance to opsonization of group A streptococci. *Infect. Immun.* **64:**1495–1501.

15. **DeAngelis, P. L., J. Papaconstantinou, and P. H. Weigel.** 1993. Isolation of a *Streptococcus pyogenes* gene locus that directs hyaluronan biosynthesis in acapsular mutants and in heterologous bacteria. *J. Biol. Chem.* **268:**14568–14571.

16. **DeAngelis, P. L., J. Papaconstantinou, and P. H. Weigel.** 1993. Molecular cloning, identification, and sequence of the hyaluronan synthase gene from group A *Streptococcus pyogenes. J. Biol. Chem.* **268:**19181–19184.

17. **DeAngelis, P. L., N. Yang, and P. H. Weigel.** 1994. The *Streptococcus pyogenes* hyaluronan synthase: sequence comparison and conservation among various group A strains. *Biochem. Biophys. Res. Com.* **199:**1–10.

18. **DeAngelis, P. L., N. Yang, and P. H. Weigel.** 1995. Molecular cloning of the gene encoding RecF, a DNA repair enzyme, from *Streptococcus pyogenes. Gene* **156:**89–91.

19. **Dillard, J. P., M. W. Vandersea, and J. Yother.** 1995. Characterization of the cassette containing genes for type 3 capsular polysaccharide biosynthesis in *Streptococcus pneumoniae. J. Exp. Med.* **181:**973–983.

20. **Dougherty, B. A., and I. van de Rijn.** 1992. Molecular characterization of a locus required for hyaluronic acid capsule production in group A streptococci. *J. Exp. Med.* **175:**1291–1299.

21. **Dougherty, B. A., and I. van de Rijn.** 1993. Molecular characterization of *hasB* from an operon required for hyaluronic acid synthesis in group A streptococci. *J. Biol. Chem.* **10:**7118–7124.

22. **Dougherty, B. A., and I. van de Rijn.** 1994. Molecular characterization of *hasA* from an operon required for hyaluronic acid synthesis in group A streptococci. *J. Biol. Chem.* **269:**169–175.

23. **Edwards, M. S., D. L. Kasper, H. J. Jennings, C. J. Baker, and A. Nicholson-Weller.** 1982. Capsular sialic acid prevents activation of the alternative complement pathway by type III, group B streptococci. *J. Immunol.* **128:**1278–1283.

24. **Federle, M. J., and J. R. Scott.** 2002. Identification of binding sites for the group A streptococcal global regulator CovR. *Mol. Microbiol.* **43:**1161–1172.

25. **Federle, M. J., K. S. McIver, and J. R. Scott.** 1999. A response regulator that represses transcription of several virulence operons in the group A streptococcus. *J. Bacteriol.* **181:**3649–3657.

26. **Ferretti, J. J., W. M. McShan, D. Ajdic, D. J. Savic, G. Savic, K. Lyon, C. Primeaux, S. Sezate, A. N. Suvorov, S. Kenton, H. S. Lai, S. P. Lin, Y. Qian, H. G. Jia, F. Z. Najar, Q. Ren, H. Zhu, L. Song, J. White, X. Yuan, S. W. Clifton, B. A. Roe, and R. McLaughlin.** 2001. Complete genome sequence of an M1 strain of *Streptococcus pyogenes. Proc. Natl. Acad. Sci. USA* **98:**4658–4663.

27. **Fillit, H. M., M. McCarty, and M. Blake.** 1986. Induction of antibodies to hyaluronic acid by immunization of rabbits with encapsulated streptococci. *J. Exp. Med.* **164:**762–776.

28. **Fillit, H. M., M. Blake, C. MacDonald, and M. McCarty.** 1988. Immunogenicity of liposome-bound hyaluronate in mice. At least two different antigenic sites on hyaluronate are identified by mouse monoclonal antibodies. *J. Exp. Med.* **168:**971–982.

29. **Fogg, G. C., C. M. Gibson, and M. G. Caparon.** 1994. The identification of *rofA*, a positive-acting regulatory component of *prtF* expression: use of an m gamma delta-based shuttle mutagenesis strategy in *Streptococcus pyogenes. Mol. Microbiol.* **11:**671–684.

30. **Foley, S. M. J., and W. B. Wood.** 1959. Studies on the pathogenicity of group A streptococci II. The antiphagocytic effects of the M protein and the capsular gel. *J. Exp. Med.* **110:**617–628.

31. **Garcia, E., and R. Lopez.** 1997. Molecular biology of the capsular genes of *Streptococcus pneumoniae.* FEMS Microbiol. Lett. **149:**1–10.

32. **Graham, M. R., L. M. Smoot, C. A. Migliaccio, K. Virtaneva, D. E. Sturdevant, S. F. Porcella, M. J. Federle, G. J. Adams, J. R. Scott, and J. M. Musser.** 2002. Virulence control in group A Streptococcus by a two-component gene regulatory system: global expression profiling and in vivo infection modeling. *Proc. Natl. Acad. Sci. USA* **99:**13855–13860.

33. **Greco, R., L. De Martino, G. Donnarumma, M. P. Conte, L. Seganti, and P. Valenti.** 1995. Invasion of cultured human cells by *Streptococcus pyogenes. Res. Microbiol.* **146:**551–560.

34. **Gryllos, I., C. Cywes, M. H. Shearer, M. Cary, R. C. Kennedy, and M. R. Wessels.** 2001. Regulation of capsule gene expression by group A Streptococcus during pharyngeal colonization and invasive infection. *Mol. Microbiol.* **42:**61–74.

35. **Gryllos, I., J. C. Levin, and M. R. Wessels.** 2003. The CsrR/CsrS two-component system of group A Streptococcus responds to environmental Mg^{2+}. *Proc. Natl. Acad. Sci. USA* **100:**4227–4232.

36. **Heath, A., V. J. DiRita, N. L. Barg, and N. C. Engleberg.** 1999. A two-component regulatory system, CsrR-CsrS, represses expression of three Streptococcus pyogenes virulence factors, hyaluronic acid capsule, streptolysin S, and pyrogenic exotoxin B. *Infect. Immun.* **67:**5298–5305.

37. **Heldermon, C., P. L. DeAngelis, and P. H. Weigel.** 2001. Topological organization of the hyaluronan synthase from *Streptococcus pyogenes. J. Biol. Chem.* **276:**2037–2046.

38. **Hirst, G. K.** 1941. The effect of a polysaccharide splitting enzyme on streptococcal infection. *J. Exp. Med.* **73:**493–506.

39. **Husmann, L. K., D.-L. Yung, S. K. Hollingshead, and J. R. Scott.** 1997. Role of putative virulence factors of *Streptococcus pyogenes* in mouse models of long-term throat colonization and pneumonia. *Infect. Immun.* **65:**1422–1430.

40. **Johnson, D. R., D. L. Stevens, and E. L. Kaplan.** 1992. Epidemiologic analysis of group A streptococcal serotypes associated with severe systemic infections, rheumatic fever, or uncomplicated pharyngitis. *J. Infect. Dis.* **166:**374–382.

41. **Kass, E. H., and C. V. Seastone.** 1944. The role of the mucoid polysaccharide (hyaluronic acid) in the virulence of group A hemolytic streptococci. *J. Exp. Med.* **79:**319–330.

42. **Kendall, F., M. Heidelberger, and M. Dawson.** 1937. A serologically inactive polysaccharide elaborated by mucoid strains of group A hemolytic streptococcus. *J. Biol. Chem.* **118:**61–69.

43. Kolkman, M. A. B., W. Wakarchuk, P. J. M. Nuijten, and B. A. M. van der Ziejst. 1997. Capsular polysaccharide synthesis in *Streptococcus pneumoniae* serotype 14: molecular analysis of the complete *cps* locus and identification of genes encoding glycosyltransferases required for the biosynthesis of the tetrasaccharide subunit. *Mol. Microbiol.* 26:197–208.

44. Lancefield, R. C. 1940. Type specific antigens, M and T, of matt and glossy variants of group A hemolytic streptococci. *J. Exp. Med.* 71:521–537.

45. Lancefield, R. C. 1957. Differentiation of group A streptococci with a common R antigen into three serological types with special reference to the bactericidal test. *J. Exp. Med.* 107:525–544.

46. LaPenta, D., C. Rubens, E. Chi, and P. P. Cleary. 1994. Group A streptococci efficiently invade human respiratory epithelial cells. *Proc. Natl. Acad. Sci. USA* 91:12115–12119.

47. Lei, B., F. R. DeLeo, N. P. Hoe, M. R. Graham, S. M. Mackie, R. L. Cole, M. Liu, H. R. Hill, D. E. Low, M. J. Federle, J. R. Scott, and J. M. Musser. 2001. Evasion of human innate and acquired immunity by a bacterial homolog of CD11b that inhibits opsonophagocytosis. *Nat. Med.* 7:1298–1305.

48. Levin, J. C., and M. R. Wessels. 1998. Identification of *csrR/csrS*, a genetic locus that regulates hyaluronic acid capsule synthesis in group A *Streptococcus*. *Mol. Microbiol.* 30:209–219.

49. Marcon, M. J., M. M. Hribar, D. M. Hosier, D. A. Powell, M. T. Brady, A. C. Hamoudi, and E. L. Kaplan. 1988. Occurrence of mucoid M-18 *Streptococcus pyogenes* in a central Ohio pediatric population. *J. Clin. Microbiol.* 26:1539–1542.

50. Markovitz, A., J. A. Cifonelli, and A. Dorfman. 1959. The biosynthesis of hyaluronic acid by group A Streptococcus. VI. Biosynthesis from uridine nucleotides in cell-free extracts. *J. Biol. Chem.* 234:2343–2350.

51. Marques, M. B., D. L. Kasper, M. K. Pangburn, and M. R. Wessels. 1992. Prevention of C3 deposition is a virulence mechanism of type III group B *Streptococcus* capsular polysaccharide. *Infect. Immun.* 60:3986–3993.

52. Miller, A. A., N. C. Engleberg, and V. J. Dirita. 2001. Repression of virulence genes by phosphorylation-dependent oligomerization of CsrR at target promoters in S. pyogenes. *Mol. Microbiol.* 40:976–990.

53. Molinari, G., and G. S. Chhatwal. 1998. Invasion and survival of *Streptococcus pyogenes* in eukaryotic cells correlates with the source of the clinical isolates. *J. Infect. Dis.* 177:1600–1607.

54. Morona, J. K., R. Morona, and J. C. Paton. 1997. Characterization of the locus encoding the *Streptococcus pneumoniae* type 19F capsular polysaccharide biosynthetic pathway. *Mol. Microbiol.* 23:751–763.

55. Moses, A. E., M. R. Wessels, K. Zalcman, S. Albertí, S. Natanson-Yaron, T. Menes, and E. Hanski. 1997. Relative contributions of hyaluronic acid capsule and M protein to virulence in a mucoid strain of group A *Streptococcus*. *Infect. Immun.* 65:64–71.

56. Munoz, R., M. Mollerach, R. Lopez, and E. Garcia. 1997. Molecular organization of the genes required for the synthesis of type 1 capsular polysaccharide of *Streptococcus pneumoniae*: formation of binary encapsulated pneumococci and identification of cryptic dTDP-rhamnose biosynthesis genes. *Mol. Microbiol.* 25:79–92.

57. Osterlund, A., R. Popa, T. Nikkila, A. Scheynius, and L. Engstrand. 1997. Intracellular reservoir of *Streptococcus pyogenes* in vivo: a possible explanation for recurrent pharyngotonsillitis. *Laryngoscope* 107:640–647.

58. Ouskova, G., B. Spellerberg, and P. Prehm. 2004. Hyaluronan release from Streptococcus pyogenes: export by an ABC transporter. *Glycobiology* 14:931–938.

59. Perez-Casal, J., M. G. Caparon, and J. R. Scott. 1991. Mry, a *trans*-acting positive regulator of the M protein gene of *Streptococcus pyogenes* with similarity to the receptor proteins of two-component regulatory systems. *J. Bacteriol.* 173:2617–2624.

60. Podbielski, A., M. Woischnik, B. Pohl, and K. H. Schmidt. 1996. What is the size of the group A streptococcal *vir* regulon? The Mga regulator affects expression of secreted and surface virulence factors. *Med. Microbiol. Immunol.* 185:171–181.

61. Rothbard, S. 1948. Protective effect of hyaluronidase and type-specific anti-M serum on experimental group A Streptococcus infections in mice. *J. Exp. Med.* 88:325–342.

62. Rothbard, S., and R. F. Watson. 1948. Variation occurring in group A streptococci during human infection. Progressive loss of M substance correlated with increasing susceptibility to bacteriostasis. *J. Exp. Med.* 87:521–533.

63. Rubens, C. E., R. F. Haft, and M. R. Wessels. 1995. Characterization of the capsular polysaccharide genes of group B streptococci, p. 237–244. *In* J. J. Ferretti, M. S. Gilmore, T. R. Klaenhammer, and F. Brown (ed.), *Genetics of Streptococci, Enterococci, and Lactococci*, vol. 85. Karger, Basel, Switzerland.

64. Sau, S., N. Bhasin, E. R. Wann, J. C. Lee, T. J. Foster, and C. Y. Lee. 1997. The *Staphylococcus aureus* allelic genetic loci for serotype 5 and 8 capsule expression contain the type-specific genes flanked by common genes. *Microbiology* 143:2395–2405.

65. Schmidt, K.-H., E. Gunther, and H. S. Courtney. 1996. Expression of both M protein and hyaluronic acid capsule by group A streptococcal strains results in a high virulence for chicken embryos. *Med. Microbiol. Immunol.* 184:169–173.

66. Schrager, H. M., J. G. Rheinwald, and M. R. Wessels. 1996. Hyaluronic acid capsule and the role of streptococcal entry into keratinocytes in invasive skin infection. *J. Clin. Investig.* 98:1954–1958.

67. Schrager, H. M., S. Albertí, C. Cywes, G. J. Dougherty, and M. R. Wessels. 1998. Hyaluronic acid capsule modulates M protein-mediated adherence and acts as a ligand for attachment of group A *Streptococcus* to CD44 on human keratinocytes. *J. Clin. Investig.* 101:1708–1716.

68. Stollerman, G. H., M. Rytel, and O. Ortiz. 1963. Accessory plasma factors involved in the bactericidal test for type-specific antibody to group A streptococci. II. Human plasma cofactor(s) enhancing opsonization of encapsulated organisms. *J. Exp. Med.* 117:1–17.

69. Stoolmiller, A. C., and A. Dorfman. 1969. The biosynthesis of hyaluronic acid by *Streptococcus*. *J. Biol. Chem.* 244: 236–246.

70. Sugahara, K., N. B. Schwartz, and A. Dorfman. 1979. Biosynthesis of hyaluronic acid by *Streptococcus*. *J. Biol. Chem.* 254:6252–6261.

71. Tlapak-Simmons, V. L., B. A. Baggenstoss, T. Clyne, and P. H. Weigel. 1999. Purification and lipid dependence of the recombinant hyaluronan synthases from *Streptococcus pyogenes* and *Streptococcus equisimilis*. *J. Biol. Chem.* 274: 4239–4245.

72. Todd, E. W. 1927. The influence of sera obtained from cases of streptococcal septicaemia on the virulence of the homologous cocci. *Br. J. Exp. Pathol.* 8:361–368.

73. Todd, E. W., and R. C. Lancefield. 1928. Variants of hemolytic streptococci; their relation to type specific substance, virulence, and toxin. *J. Exp. Med.* **48:**751.

74. Veasy, L. G., S. E. Wiedmeier, G. S. Orsmond, H. D. Ruttenberg, M. M. Boucek, S. J. Roth, V. F. Tait, J. A. Thompson, J. A. Daly, E. L. Kaplan, and H. R. Hill. 1987. Resurgence of acute rheumatic fever in the intermountain area of the United States. *N. Engl. J. Med.* **316:** 1298–1315.

75. Weigel, P. H., V. C. Hascall, and M. Tammi. 1997. Hyaluronan synthases. *J. Biol. Chem.* **272:**13997–14000.

76 Wessels, M. R., and M. S. Bronze. 1994. Critical role of the group A streptococcal capsule in pharyngeal colonization and infection in mice. *Proc. Natl. Acad. Sci. USA* **91:**12238–12242.

77. Wessels, M. R., A. E. Moses, J. B. Goldberg, and T. J. DiCesare. 1991. Hyaluronic acid capsule is a virulence factor for mucoid group A streptococci. *Proc. Natl. Acad. Sci. USA* **88:**8317–8321.

78. Wessels, M. R., J. B. Goldberg, A. E. Moses, and T. J. DiCesare. 1994. Effects on virulence of mutations in a locus essential for hyaluronic acid capsule expression in group A streptococci. *Infect. Immun.* **62:**433–441.

79. Westlake, R., T. Graham, and K. Edwards. 1990. An outbreak of acute rheumatic fever in Tennessee. *Pediatr. Infect. Dis. J.* **9:**97–100.

80. Wilson, A. T. 1959. The relative importance of the capsule and the M antigen in determining colony form of group A streptococci. *J. Exp. Med.* **109:**257–270.

Toxins and Superantigens of Group A Streptococci

JOHN K. McCORMICK, MARNIE L. PETERSON, AND PATRICK M. SCHLIEVERT

5

Streptococcus pyogenes (group A streptococcus) is a remarkable and versatile human bacterial pathogen that is capable of producing an impressive arsenal of both surface-expressed and secreted virulence factors. This pathogen continues to generate significant morbidity by causing a variety of uncomplicated human diseases such as pharyngitis and skin infections, and more serious diseases such as acute rheumatic and scarlet fevers. In addition, group A streptococci cause some of the most devastating bacterial diseases known, such as necrotizing fasciitis/myositis and the streptococcal toxic shock syndrome (STSS). Although surface-expressed virulence factors are clearly vital and necessary for the development of disease, it is the secreted virulence factors that are likely the major causative agents of damage and toxicity seen during infection. This arsenal of secreted virulence factors is rivaled by few bacterial pathogens, and although the repertoire of the factors is highly diverse, including various enzymes, two notable membrane-active toxins, and immune system-altering toxins, classical AB toxins with intracellular targets are apparently lacking (9, 11, 45, 98, 121). In this chapter, we will focus on the true exotoxins of group A streptococci with regard to their structure, function, and genetics, as well as their roles in the pathogenesis of human disease.

BACTERIAL SUPERANTIGENS

Arguably the most impressive class of toxins produced by the group A streptococci is the superantigen family. Superantigens are among the most potent known activators of T cells, and of the available genome sequences, each strain contains between four and six distinct superantigen serotypes (9, 11, 45, 98, 121). The streptococcal superantigens belong to a larger group of structurally conserved exotoxins that are also produced by coagulase-positive staphylococci (19, 74, 84, 91). Some group C and group G beta-hemolytic streptococci also produce these toxins (6, 94, 112).

The term "superantigen" was first used to describe the massive primary T-cell response to these bacterial toxins (136). Superantigens function by causing excessive T-cell activation through simultaneous engagement of both the variable region of the T-cell antigen receptor (TCR) β-chain (referred to as "Vβ") on T cells, and different regions of major histocompatibility complex (MHC) class II molecules on antigen-presenting cells (77, 126). Superantigens use this common but very efficient mechanism of T-cell stimulation that does not represent the conventional mode for T-cell activation. Rather, in superantigen-mediated T-cell activation, the toxin binds directly to MHC class II and does not require processing of the superantigen by the antigen-presenting cell. The interaction with the TCR depends on binding to different Vβ regions, instead of peptide-specific complementarity determining regions (CDRs) from both the TCR α- and β-chains. Due to the unconventional contacts created by this interaction, superantigen-mediated T-cell activation is also not MHC-restricted, and furthermore, both CD4$^+$ and CD8$^+$ are stimulated in a Vβ-specific manner. Superantigens can thus elicit a strong primary response that is not seen with conventional peptide antigens. In the most severe case, this excessive activation of T cells results in the massive release of proinflammatory cytokines from both the T cell and the antigen-presenting cell, which can result in a very serious and highly lethal disease called toxic shock syndrome (91).

Streptococcal Superantigens and Disease

The streptococcal superantigens have historically been referred to as erythrogenic toxins or scarlet fever toxins, and the characteristic rash seen in scarlet fever is likely due to an amplified hypersensitivity reaction resulting from superantigen activity (120). The appearance of a rash was the defining feature of the erythrogenic toxins. Watson (134) first proposed the name streptococcal pyrogenic exotoxins (Spes) because their distinctive fever-producing ability initially led him to believe that the Spes belonged to a separate family of toxins. We now know that these toxins all represent the same group.

It was not until the late 1980s that descriptions of well-defined cases led to the recognition of the severely toxic streptococcal superantigen-mediated disease STSS (31, 124). The prototypical streptococcal superantigen serotypes are SpeA and SpeC, and these two toxins have been

Gram-Positive Pathogens, 2nd edition, edited by Vincent A. Fischetti et al.
© 2006 ASM Press, Washington, D.C.

recognized as the likely causative agents in most cases of STSS (26, 96, 124). SpeA and SpeC are produced by many *S. pyogenes* strains isolated from patients with STSS (19, 27, 33, 54, 59, 96). Due to the association of SpeA and SpeC with STSS, two toxoid vaccines have been generated for these toxins that, when used as vaccines in the rabbit model of STSS, provided complete protection to lethal quantities of the respective superantigen (90, 114). Six naturally occurring alleles of *speA* are known to exist (13, 99), and serotype M1 and M3 streptococcal strains expressing the *speA₂* and *speA₃* alleles caused the majority of STSS in the 1980s (99).

Multiple lines of evidence indicate that the streptococcal superantigens are significant virulence factors. Of the relatively few streptococcal superantigen serotypes that have been tested, the toxins share the properties of inducing fever (pyrogenicity) and lethal shock in the rabbit. Pure SpeA, SpeC, and SpeJ can induce symptoms of STSS without necrotizing fasciitis/myositis in this model (87, 90, 114), and pure SpeA administered to rabbits as a vaccine also conferred significant protection from STSS with necrotizing fasciitis/myositis after challenge with viable M1 and M3 streptococci (52), the two most common type M serotypes associated with STSS. Circulating superantigens have also been found in patients with STSS (122), and the lack of humoral immunity to the superantigens is a risk factor for the development of STSS (10, 40).

Although significant research on the toxicity of bacterial superantigens is conducted in the mouse model, we generally do not believe that mice are the appropriate model to evaluate the lethal effects of these toxins. Lethal effects are not seen at very high relative doses without the coadministration of a liver-damaging agent, while the rabbit is susceptible to fairly low doses when administered continuously (106). It has been clearly shown that HLA-DQ transgenic mice, but not HLA-DR, were more susceptible to SpeA-induced lethality whether administered as a pure toxin or during infection with SpeA⁺ *S. pyogenes*, whereas other work with nontransgenic mice did not show an effect (123, 133). Consistent with this, it has been convincingly shown that human MHC class II allelic variation contributes to differences in severity seen for invasive streptococcal infections and that this is due to cytokine production triggered by streptococcal superantigens (75) and that MHC class II α-chain polymorphisms affect superantigen responses in vitro (81).

Bacterial superantigens also have the peculiar ability to enhance host susceptibility to endotoxin-mediated lethal shock by up to one-million-fold (34). The amplification of susceptibility to endotoxin is mediated primarily by interferon-γ from T cells, resulting in the synergistic release of tumor necrosis factor (TNF) from macrophages and T cells as a result of exposure to superantigen (35). The dramatic enhancement of endotoxin toxicity may be the most severe symptom of STSS-like illness in which the rabbits die of a capillary leak syndrome induced by cytokines. These cytokines include interleukin-1 and TNF-α from peripheral macrophages (41, 107) and interleukin-2, TNF-β, and interferon-γ from T cells (35).

Structure/Function of Streptococcal Superantigens

Streptococcal superantigens are ribosomally synthesized, relatively low molecular mass (~22 to 28 kDa) proteins. These toxins contain classical signal peptides that are cleaved after secretion to release the mature toxin. Superantigens do not undergo major modifications after release from the cell or major structural alterations upon binding to their ligands (46, 63, 73, 78, 80, 108, 109, 127). Despite the fairly weak primary amino acid sequence homology that exists between many of the streptococcal superantigens, all of the crystal structures determined for these toxins share the "generic" superantigen fold (77). Each structure includes an N-terminal α-helix that leads into a β-barrel domain, also known as the oligosaccharide/oligonucleotide binding (OB) fold (93). A central α-helix then joins this domain to a β-sheet structure known as the β-grasp domain. In Fig. 1A we show the ribbon diagrams of all crystallized streptococcal superantigens with the OB fold domain shown in red and the β-grasp domain shown in blue. Despite the clear similarities to the overall fold, there are also significant differences in how these toxins engage their host receptors.

Co-crystallization structures were recently determined for SpeA and SpeC in contact with the mouse Vβ8.1 and human Vβ 2.1 chains, respectively (127). Earlier co-crystal structures also exist for the staphylococcal superantigens staphylococcal enterotoxin B (SEB) and staphylococcal enterotoxin C (SEC) in complex with the mouse Vβ8.1 chain (46, 78). Although highly homologous superantigens, such as SEB, SEC, and SpeA, may have similar architectures for their Vβ chain interactions, the β-chain-SpeC complex involved a significantly larger contact surface, a different orientation, and contacts with the CDR3 loop, whereas SEB, SEC, or SpeA had no intermolecular contacts with this loop (127). It also appears that there are significant differences in the binding interfaces and architectures for less similar superantigens, such as TSS toxin-1 (TSST-1) from *Staphylococcus aureus* (89), and it is likely that as other interactions are characterized, further variation will be seen.

The interaction of superantigens with MHC class II also shows considerable diversity where superantigens have two independent binding domains. One binding face occurs through the N-terminal OB fold domain and is of relatively low affinity (~10^{-5} M). Although somewhat controversial (79, 90, 130, 132), it appears likely that most, if not all, streptococcal superantigens may contain this low-affinity MHC binding site. A model of the low-affinity SpeC-MHC class II interaction predicts two major contact sites, including a polar pocket and a hydrophobic loop, and mutations within these sites have shown reductions in SpeC superantigenicity (130, 132). The architecture of this "generic," low-affinity interaction is also not likely universally conserved because the co-crystal structure for TSST-1 in complex with HLA-DR1 revealed significant contacts with the antigen peptide that did not occur in the SEB-MHC structure (63, 73). The staphylococcal enterotoxin A-MHC structure more closely resembles the SEB-MHC structure (109).

The second, zinc-dependent MHC-binding face occurs on the β-grasp domain and is of relatively high affinity (~10^{-7}). The zinc ion, shown in magenta in Fig. 1A, is coordinated through three conserved residues, whereas the fourth ligand is from a conserved residue on the MHC β-chain. Co-crystal structures for this interaction have been solved for SpeC (80) and staphylococcal enterotoxin H (108). In both structures, the superantigen contacts the α-helix of the β1 domain of MHC class II and the amino terminal of the MHC class II bound peptide. Approximately one-third of the interaction is through the peptide, indicating that different peptides likely influence binding. How-

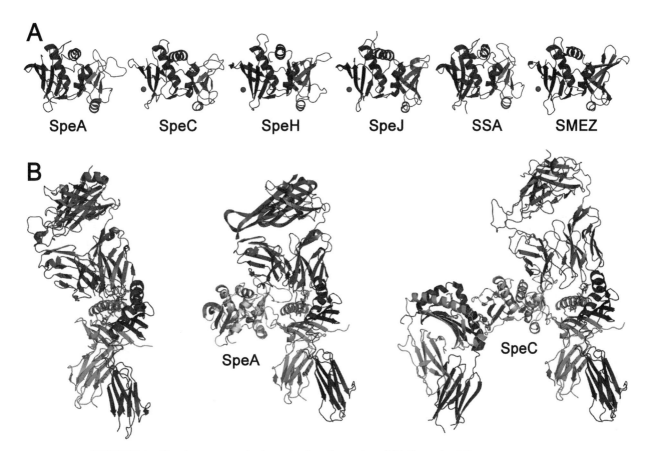

FIGURE 1 (*See the separate color insert for the color version of this illustration.*) Structure conservation and models of T-cell activation complexes for streptococcal superantigens. (A) Ribbon diagrams of the crystal structures for streptococcal pyrogenic exotoxin serotypes A, C, H, J, streptococcal superantigen (SSA), and streptococcal mitogenic exotoxin-Z_2 ($SMEZ_2$) (3, 7, 37, 115, 125). (B) Ribbon diagrams demonstrating typical antigen-mediated T-cell activation (left) (55) and modeled T-cell activation complexes for SpeA (middle) and SpeC (right). The co-crystal structures of SpeA and SpeC in complex with their respective TCR β-chains (127) and of SpeC in complex with the MHC class II through the zinc-dependent high-affinity binding domain have been determined (80). In light of recent evidence (130), it is likely that SpeC also activates T cells in a mode similar to the staphylococcal enterotoxin A model (80) where the superantigen also engages MHC class II through the generic low-affinity binding domain. The binding architecture for the generic low-affinity MHC class II binding to SpeA and SpeC is modeled using the staphylococcal enterotoxin B-MHC class II co-crystal structure (63). Note the presence of the zinc ion (magenta) coordinated in the high-affinity binding site for SpeC and that SpeA lacks this zinc site. The TCR α-chain (shown in gray) for both the SpeA and SpeC diagrams is modeled for clarity by superimposition of the α/β TCR shown on the left to the respective TCR β-chains for both superantigens. The figure was generated using Pymol (32).

ever, the amino terminal of the bound peptide is also less variable, which suggests that the high-affinity interaction binds more structurally conserved regions of the peptide-MHC complex compared with TCRs (128).

By superimposition of the high-affinity zinc-dependent MHC-SpeC structure (80) and the β-chain-SpeC structures (127), and also through orientation of the proposed low-affinity generic MHC binding face through alignment with the MHC-SEB structure (63) with both the β-chain-SpeC and β-chain-SpeA structures, models for these two activation complexes were generated (Fig. 1B). The SpeA model is analogous to the standard "wedge" model of superantigen-mediated T-cell activation (78), whereas the SpeC model is analogous to the model that has been proposed for staphylococcal enterotoxin A (80, 109). In these models,

the mechanism by which superantigens subvert the normal T-cell activation process is apparent. According to the SEB model, the antigenic peptide is displaced from the TCR, as well as all of the CDR loops except CDR2 from the TCR Vα chain (2). In the SpeC model presented here, it is predicted that this superantigen also forms a "wedge" between the TCR and MHC class II using the low-affinity interaction; thus similar unconventional contacts between MHC class II and the TCR Vα could occur. It is important to note that no TCR-superantigen-MHC complex structures have been determined, and whether the SpeC model actually occurs in vivo, where all three ligands are engaged simultaneously, is unknown. Furthermore, SEB and SpeC are not overly homologous superantigens, and because it is known that the low-affinity MHC class II interaction is variable, as for TSST-1 and SEB (63, 73), SpeC may show further variation in the orientation of MHC class II.

As indicated, not all streptococcal (or staphylococcal) superantigens contain the zinc-dependent high-affinity MHC class II binding site. As a comparative study, we have cloned, expressed from *Escherichia coli*, and purified, eight independent streptococcal superantigens (Fig. 2A) and compared their mitogenic capacity in parallel using standard human T-cell proliferation assays. As seen in Fig. 2B, each toxin produced a dose-dependent proliferative response in which for some superantigens there appeared to be some degree of activation-induced cell death at the higher concentrations. Of the eight superantigens assayed, SpeA and SSA are the only toxins that lack the predicted zinc-binding residues, and both of these toxins lost activity at ~10 pg/ml, while each of the superantigens containing the predicted zinc-binding domain lost activity at ~0.1 pg/ml (except for SpeG, which lost activity at ~1 pg/ml). From this direct comparative analysis, it appears that superantigens that contain the zinc-dependent high-affinity MHC class II binding domain are ~100-fold more potent than their counterparts that lack this domain. This correlation, however, may not be so simple, because mutation of single or multiple zinc-binding residues in SpeC drastically decreased the potency of the toxin (132). How this increased potency of wild-type toxins that contain the zinc-binding site may affect the biological and evolutionary role of these toxins in the human host is also unclear, because SpeA, which lacks this domain, has generally shown the highest association with invasive streptococcal disease and STSS. It is possible that the high-affinity binding domain represents a mechanism by which these toxins can be more efficiently sequestered and targeted at their receptor sites on antigen-presenting cells, yet their low-affinity binding domains are not as efficient as those found in superantigens lacking the high-affinity domain. Alternatively, most superantigen-induced activation by zinc-coordinating superantigens may occur independently of the low-affinity binding site.

Superantigens and Bacteriophage-Mediated Evolution of Group A Streptococci

Although there is a clear association of SpeA and SpeC with severe streptococcal disease, the true evolutionary function of the bacterial superantigens is unknown. Since early in the 20th century, the incidence of serious streptococcal disease has greatly decreased, likely because of the widespread use of antibiotics (85). However, since the mid-1980s there has been a worldwide resurgence in frequency and severity of STSS (31, 124) and other invasive diseases caused by group A streptococci (26, 97, 118), as well as rheumatic fever (64). This has raised the hypothesis that an excessively virulent strain or strains have emerged, or alternatively, that there have been major changes in host susceptibility to these infections.

Early evidence has indicated that superantigens were bacteriophage-encoded. In 1927, Frobisher and Brown (49) showed that a filterable agent from scarlet fever isolates could convert nonscarlatinal streptococci to toxigenic strains; this was again confirmed in 1964 (140). The gene encoding SpeA was later shown to be encoded on the bacteriophage T12 (65, 66), and in 1986 the structural gene was sequenced (135). McShan et al. (92) showed that the bacteriophage-encoding SpeA integrates into the serine tRNA gene in the streptococcal genome. Although we have known about the "erythrogenic" toxins for more than

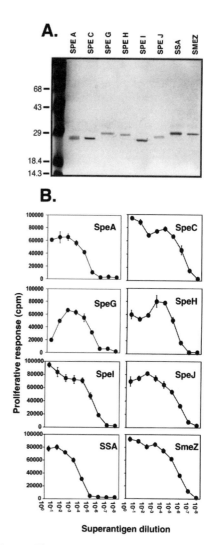

FIGURE 2 The presence of the high-affinity, zinc-dependent MHC class II binding domain is associated with approximately 100-fold higher potency of streptococcal superantigens. (A) Sodium dodecyl sulfate polyacrylamide gel demonstrating purity of eight recombinant serotypes of streptococcal superantigens expressed and purified from *E. coli*. Techniques for cloning, expression, and purification have been described (88). (B) Standard human T-cell proliferation assay using ^{3}H-thymidine of the eight recombinant streptococcal superantigens. SpeC, SpeG, SpeH, SpeI, SpeJ, and SMEZ$_2$ each contain the predicted zinc-binding motif found in the β-grasp domain. SpeA and SSA lack this motif and are no longer active at ~100-fold concentrations higher than the zinc-binding motif containing superantigens. cpm, counts per minute.

85 years, from the recent generation of complete genome sequences for different M serotypes of group A streptococci, it is now clear that there is widespread redundancy in the superantigen-producing capability of different group A streptococci strains and that most of these toxins are encoded on bacteriophage elements.

We are now aware of at least 11 different functional serotypes of streptococcal superantigens, including Spe serotypes A, C, G, H, I, J, K, L, M, SSA, and SmeZ. Note that SpeK from strain MGAS315 (11) is the same super-antigen as SpeL in strain SSI (98) and that we are using the serotype names from MGAS315. Both SpeA and SpeC are encoded on functional bacteriophage (51, 65), while the majority of other serotypes are associated with apparently complete bacteriophage genomes, or bacteriophage remnants. Collectively, the genome sequence data have revealed that the majority of genetic variation in different serotypes is mainly due to the presence or absence of lysogenic bacteriophage, or bacteriophage remnants, located at various sites within the group A streptococcal chromosome (8). The location of the superantigen gene(s) on each bacteriophage is invariably downstream of the predicted bacteriophage structural genes and adjacent to the bacteriophage attachment site. This location is likely significant in that the toxin genes may have been acquired by an aberrant excision event, and this model would predict that the superantigen gene would be retained only if it provided some type of evolutionary advantage to either the lysogenic bacteriophage or the host streptococci. Because it is difficult to imagine what advantage this may play for the bacteriophage, it is assumed that the superantigens must confer an albeit currently unrecognized benefit to group A streptococci. In addition, many of the bacteriophage elements also carry other potential virulence factors such as mitogenic factors and a phospholipase. The superantigen-encoding bacteriophage may actually represent an early event in the evolution of a pathogenicity island (52).

Because different M serotypes are associated with different disease states, it is not unreasonable to hypothesize that the genetic complement encoded on different bacteriophage elements may play an important role in different virulence properties in group A streptococci. From extensive analysis of multiple strains from similar and different M serotypes, it is clear that the multifaceted array of clinical presentations produced by group A streptococci is likely due to the acquisition of bacteriophage-encoded virulence factors, and the streptococcal superantigens are likely key virulence factors involved in this evolution (9). From comparisons of the different bacteriophage elements, nearly identical bacteriophages can harbor different superantigen gene serotypes (9, 98). This strongly implies that these bacteriophage elements have recombined to produce different bacteriophage chimeras with different toxin-producing capabilities.

SpeG and SmeZ are exceptions in that these superantigens appear to be encoded within the core chromosome (9, 11, 45, 98, 111, 121). SpeJ is also not found in association with bacteriophage genes (45), yet it is not present in the other sequenced strains. SmeZ, in particular, is known to have many different alleles, and most of the sequence changes are located on the surface of the toxin and rarely found within the predicted receptor-binding domains. This implies that SmeZ is under significant immunological pressure to alter its antigenic characteristics as a possible immune evasion strategy (111). Interestingly, SpeG has also been found in group G streptococci ($speG_{dys}$), suggesting that genetic transfer of this toxin has occurred between the two species (117). Furthermore, multiple unpublished submissions to the data bank indicated that variants of this superantigen also exist.

Superantigen redundancy is also paralleled in *S. aureus* (110), and of the characterized superantigens also found in group C or G streptococci (4, 67, 94, 112, 117), each has a close homologue among the 11 group A streptococcal serotypes. It is now generally believed that bacteriophage-

mediated transduction, which has been shown in the past to occur between group A, C, and G streptococci (29, 30), is also responsible for the transfer of superantigen genes between these different Lancefield groups. However, other than the demonstration of their activity, there is currently no clear epidemiological evidence for the role of serotypes other than SpeA or SpeC in human disease. The task of showing clear associations with the different serotypes, if they occur, may be difficult in that the repertoire of most clinical isolates is not known, and it is also likely that further serotypes will be discovered. It may be that the extreme redundancy of these toxins reflects a function of antigenic variation and that individual disease states may be tied to the lack of host immunity to the different streptococcal superantigen serotypes.

STREPTOCOCCAL PYROGENIC TOXIN TYPE B (SpeB, OR CYSTEINE PROTEASE)

speB is highly conserved and abundantly expressed in virtually all *S. pyogenes* isolates during the late exponential growth phase (25, 137). SpeB is secreted as a 40-kDa zymogen that is autocatalytically cleaved to generate an active 28-kDa proteinase (23, 36). Analysis of the amino acid sequence revealed that SpeB was nearly identical to streptococcal proteinase precursor (53). *speB* was first cloned in 1988, and protein expressed from *E. coli* was shown to retain its lymphocyte mitogenic activity, but lacked protease activity (20). Subsequent unpublished studies showed that the lack of protease activity of that isolated SpeB resulted from purifying the protein by our standard technique, including ethanol precipitation followed by isoelectric focusing (18). When ammonium sulfate was used as the precipitating agent followed by isoelectric focusing, protease activity was observed. The data suggested that alternate protein folding patterns resulted in the differential activity. The superantigenicity of the protein remains controversial, but this activity also may be dependent on the protein conformation.

SpeB is considered an important virulence factor, most likely due to the immunomodulating effects of its protease activity and possibly its lymphocyte mitogenic activity (28, 70). SpeB has broad proteolytic activity against a number of different human proteins, including interleukin-1β (70), immunoglobulins (28), fibrinogen (86), fibronectin (71), kininogens (56), and a metalloprotease (22). *S. pyogenes* proteins, such as M proteins, can also be proteolyzed by SpeB (12); however, an *S. pyogenes* surface protein G-related α2-macroglobulin-binding protein may protect some surface proteins against the proteolytic actions of SpeB by binding to plasma α2-macroglobulin, a potent protease inhibitor (113). SpeB was also recently determined to be able to directly degrade the superantigen SpeF, but not SpeA, which affected the ability of the superantigen to induce proliferation of human lymphocytes (69). A recent investigation reported that the fibronectin-dependent internalization of *S. pyogenes* into human cells was decreased by the proteolysis of protein F1, an *S. pyogenes* cell wall-attached fibronectin-binding protein, suggesting that SpeB also plays a role in the regulation of the internalization process (103). Likewise, SpeB-deficient mutants were shown to have enhanced in vitro internalization into human epithelial and endothelial cells (21).

In vivo studies supporting the role of SpeB as a virulence factor are controversial. Lukomski et al. (82) reported that a

SpeB-deficient M3 strain appeared to be less virulent in mice after intraperitoneal challenge than was the wild-type M3 strain. Additionally, Schlievert et al. reported that all nephritogenic streptococci isolates produced SpeB (119). Alternatively, a study using isogenic gene replacement strains and a mouse model of invasive soft tissue infection showed that *speB* mutants had no apparent effect on the ability of group A streptococci to cause local tissue injury and invasive infection (5). Likewise, a recent study demonstrated that an inverse relationship exists between SpeB expression and the severity of invasive group A streptococci infections (68), suggesting that *S. pyogenes* may need to differentially regulate SpeB expression, depending on the site of infection and events occurring during infection (113). Epidemiological evidence suggests a correlation between SpeB production and a preference for infections of the skin (129). SpeB was also recently reported to cause an increase in capillary permeability in the skin of guinea pigs, most likely through the induction of histamine release, whereas both SpeA and SpeC did not exhibit such activity (104). Additionally in this study, three of seven patients with STSS also possessed higher levels of plasma histamine compared to the normal subjects, suggesting a potential role of histamine induction by SpeB in streptococcal infections of skin and mucous membranes (104).

SpeB has also been shown to possess superantigen activity independent of its protease effects, although this activity is still debated (20, 39, 76, 131). Tomai et al. (131) have shown that cloned SpeB caused significant skewing (overstimulation) of T cells bearing Vβ2, consistent with the toxin's having superantigen activity.

STREPTOCOCCAL PYROGENIC TOXIN TYPE F (SpeF, OR MITOGENIC FACTOR)

SpeF was originally named mitogenic factor (138) and was later named SpeF because of its ability to induce cytokine production (102). SpeF was cloned, and no significant homology was found with any of the other Spe molecules (61). Among streptococci, *speF* appears to be present only in group A streptococci (139) and may be present in all group A streptococci (102). Purified SpeF induced cytokine production from peripheral blood mononuclear cells, including interferon-γ, and TNF-β, but not interleukin-2 or TNF-α. Also, SpeF required antigen-presenting cells for T-cell proliferation, and T cells bearing Vβ 2, 4, 8, 15, and 19 were preferentially activated (102).

Recently, SpeF was shown to have nuclease activity, and a protein containing an amino acid change from histidine to arginine at position 122 was inactive for DNase activity (62). More recently, SpeF has been shown to be DNase B, with different domains controlling superantigen and DNase activity (38). All group A streptococci produce DNases; however, aside from SpeF, these enzymes have not been well characterized.

STREPTOCOCCAL CYTOLYTIC TOXINS

Group A streptococci produce two known hemolysin exotoxins: streptolysin O (SLO) and streptolysin S (SLS). Although both are exotoxins, neither is a suspected superantigen. SLO is well characterized as a thiol-activated cytolysin that is secreted from nearly all group A streptococci (72). At least two active forms of SLO exist, with respective molecular weights between 50,000 and 70,000 and pIs between 6.0 and 6.5 and 7.0 and 7.8. SLO is active in a reduced state, but is rapidly inactivated in the presence of oxygen. SLO binds to cholesterol present in eukaryotic membranes, where it forms multimeric transmembrane pores (15, 16, 105). The ability to form these pores may result in lysis of susceptible cells (15, 16, 105). In a study characterizing the response of human keratinocytes to group A streptococci, SLO was determined to work synergistically, along with an adherence mechanism, to cause membrane damage to keratinocyte skin cells, resulting in release of proinflammatory cytokines (116). The adherence mechanism may promote an increased local concentration of SLO at the keratinocyte membrane or may allow for delivery of SLO at a specific area of the membrane. In addition to the formation of passive pores resulting in cell lysis and induction of inflammation, SLO has also been discovered recently to play an active role in the cytolysin-mediated translocation of an effector protein of streptococcal origin into the cytoplasm of host cells (116).

SLS is also produced by essentially all strains of *S. pyogenes* and is the cytolytic factor responsible for the beta-hemolytic zones used routinely in the identification of group A streptococci on sheep blood agar. This cytolytic toxin is oxygen-stable and nonimmunogenic. In addition to the lytic effect on sheep erythrocytes, SLS can damage other cell membranes, including those of lymphocytes (50) and neutrophils and platelets (60). The molecular identity of SLS was unknown until a recent analysis of an SLS-defective transposon Tn916 insertion mutant identified an operon of nine SLS-associated genes (sagA to sagI), which resembles operons encoding peptide bacteriocins in other bacterial species (14, 101). The production of active SLS involves modification and secretion of the 53-amino-acid product of the sagA gene by the products of the other sag genes, which are transcribed at much lower levels due to a partial transcription terminator between sagA and sagB (24, 101). The sag locus is conserved among group A streptococcus strains regardless of M protein type (101). A similar study using transposon mutagenesis created a mutant deficient in SLS production that remained normal with respect to other exoprotein expression, and was shown to be markedly less virulent in a mouse model of subcutaneous infection versus the isogenic wild-type strain (14). To determine the combined contributions of SLO and SLS to virulence of *S. pyogenes*, a recent study constructed single and double nonpolar deletion mutations in the slo and sagB genes of the serotype M5 group A streptococcus strain Manfredo, which lacked notable effects on the expression of other virulence factors examined (47). Analysis of these mutants in various virulence models indicated that both SLO and SLS contribute directly to the early stages of infection and the induction of necrosis in the murine dermonecrosis model, especially in those mutants containing double mutations in both the slo and sagB genes (47).

STREPTOKINASE

Streptokinase has long been recognized as a secreted factor from group A and other β-hemolytic streptococci. In addition, it is well recognized that numerous variant forms of the protein exist, and there have been suggestions that certain variant forms are associated with development of acute glomerulonephritis (68, 69, 73, 108). Streptokinase has three principal domains, α, β, and γ, of approximately equal amino acid length (146, 144, and 123 amino acids) (69). The described function of streptokinase is to bind to plasminogen and activate other plasminogen molecules by proteolysis to plasmin that then dissolves clots. The effect of streptokinase on plasminogen is nonenzymatic. This property has allowed streptokinase to be developed for therapeutic fibrinolysis associated with myocardial infarcts, strokes, and venous thrombosis. Studies indicate that all domains of streptokinase interact with the plasminogen catalytic domain and that the β domain also interacts with kringle 5 of plasminogen (87, 143). There appears to be the initial need for a rapid interaction of the β domain of streptokinase with kringle 5 (87, 143). This allows the interaction with the α and γ domains to form cooperatively an active site with plasminogen (87, 143). Substrate plasminogen appears to be recognized by the streptokinase α domain (87, 143).

More recently, studies have again been initiated to study the potential role of streptokinase in group A streptococcal virulence (32, 77, 132, 138). The studies collectively suggest that the protein is important in establishment and spread of skin infections, notably impetigo. These studies have often used mice as infection models. Since group A streptococci are not natural pathogens of mice, the investigators have begun studies to "humanize" the mice through use of both human skin transplanted to nude mice and transgenic expression of human plasminogen (132, 138). From these studies, it appears that SpeB, mentioned above, is critical for initiation of impetigo since inactivation of *speB* attenuates virulence (137). The role of streptokinase appears to be to facilitate clot dissolution that occurs through plasma influx after infection is established (77, 132, 137, 138). Dissolution of clots facilitates spread of the pathogen into adjacent tissue. The combined effects of SpeB and streptokinase in skin may also explain the results that suggest that these two factors are associated with acute glomerulonephritis, which has been linked to prior skin infections (68, 108, 126).

STREPTOCOCCAL INHIBITOR OF COMPLEMENT

Streptococcal inhibitor of complement (Sic) is a 31-kDa protein secreted by certain *S. pyogenes* strains, most notably M type 1 strains (1). Other M types may make related proteins, some of which have minimal effect on the complement system (17). Sic was originally identified as a protein that inhibits the membrane attack complex of the complement system, with its mechanism being the inhibition of C567 insertion into the membrane (44). More recently, Sic has been shown to be rapidly internalized by polymorphonuclear leukocytes and binds specifically to ezrin and moesin, proteins that function to link the actin cytoskeleton to the host cell surface (44). The effect of this is to interfere with polymorphonuclear opsonophagocytosis and intracellular killing. Sic has also been shown to inhibit directly other components of the innate immune system, including lysozyme, secretory leukocyte proteinase inhibitor, human α-defensin hNP-1, human β-defensins 1, 2, and 3, and cathelicidin LL-37, all of which can be toxic to *S. pyogenes* (42, 43, 48).

Studies have been performed that suggest Sic functions in vivo to increase streptococcal virulence. Sic is a highly variable protein, with a large number of variants arising in vivo during epidemic spread of the organism (57, 58). Such high variation in vivo suggests that the phenomenon is offering selective advantage to the microbe. Most of the variants are single amino acid changes, insertions, or deletions. Lukomski and colleagues showed that mice inoculated intranasally with M1 streptococci making Sic had higher incidence of throat colonization than mice inoculated with the Sic-negative mutants (83). It was also shown that the same Sic-negative strain was less effectively phagocytosed and killed than the wild type by epithelial cells (83). It was speculated that Sic interfered with streptococcal contact, internalization, and killing of the organism. This would promote bacterial survival and dissemination by allowing the organisms to remain longer in the extracellular environment.

CONCLUSIONS

The increased incidence of severe invasive group A streptococcal diseases and STSS remains incompletely explained; however, Spes are clearly implicated in STSS. A model for the action of Spes in STSS would involve a very complex toxic process in which death would be the final endpoint. Cytokine mediators, released in massive amounts as a result of both superantigenicity and the enhancement of susceptibility to endotoxin, and the direct toxicity of superantigens are likely to contribute to capillary leak, which is ultimately responsible for hypotension, shock, and death. When considering the design of effective therapeutic strategies to combat STSS, it is important to realize that no single streptococcal superantigen is responsible for all cases of STSS. It has been reported that uncharacterized Spes exist (100), and they have also been detected in certain group B, C, F, and G streptococcal strains that may mimic the role of the group A streptococcal Spes.

Most research on Spes has focused on their role in STSS, although these superantigens may also be involved in other important biological processes, such as colonization or circumvention of local or systemic immune responses. A recent study has suggested that SpeA may contribute to more chronic streptococcal diseases by induction of Th2-derived immunoregulatory and hematopoietic cytokines (95). Furthermore, although rabbits are the accepted animal model for STSS, the amount of Spe required to induce death appears to be 100 to 200 μg, while in humans 2 μg may be lethal. Taken together, many unanswered questions remain to be clarified to fully understand the immunobiology of Spes and their roles in human streptococcal diseases.

Finally, it is becoming clear that there are numerous additional streptococcal exotoxins and exoproteins besides the Spes, both known and novel factors, that have profound effects on the host. Although hemolysins have been known for years to possess hemolytic activity, it is now appreciated that they have numerous other effects as well. Their contributions to streptococcal disease, as well as newly identified exoproteins, are the subject of intense investigation, and the findings thus far suggest most if not all of these exotoxins and proteins alter normal immune system function, likely contributing to streptococcal persistence in the host.

We acknowledge support for this work by operating grant R3445A02 and a New Investigator Award from the Canadian Institutes of Health Research to J.K.M. and USPHS research grant HL36611 from the National Heart, Lung, and Blood Institute to P.M.S. M.L.P. was supported by USPHS training grant T32-HD07381.

REFERENCES

1. **Akesson, P., A. G. Sjoholm, and L. Bjorck.** 1996. Protein SIC, a novel extracellular protein of *Streptococcus pyo-*

genes interfering with complement function. *J. Biol. Chem.* **271**:1081–1088.

2. Andersen, P. S., P. M. Lavoie, R. P. Sekaly, H. Churchill, D. M. Kranz, P. M. Schlievert, K. Karjalainen, and R. A. Mariuzza. 1999. Role of the T cell receptor alpha chain in stabilizing TCR-superantigen-MHC class II complexes. *Immunity* **10**:473–483.

3. Arcus, V. L., T. Proft, J. A. Sigrell, H. M. Baker, J. D. Fraser, and E. N. Baker. 2000. Conservation and variation in superantigen structure and activity highlighted by the three-dimensional structures of two new superantigens from *Streptococcus pyogenes*. *J. Mol. Biol.* **299**:157–168.

4. Artiushin, S. C., J. F. Timoney, A. S. Sheoran, and S. K. Muthupalani. 2002. Characterization and immunogenicity of pyrogenic mitogens SePE-H and SePE-I of *Streptococcus equi*. *Microb. Pathog.* **32**:71–85.

5. Ashbaugh, C. D., H. B. Warren, V. J. Carey, and M. R. Wessels. 1998. Molecular analysis of the role of the group A streptococcal cysteine protease, hyaluronic acid capsule, and M protein in a murine model of human invasive soft-tissue infection. *J. Clin. Investig.* **102**:550–560.

6. Assimacopoulos, A. P., J. A. Stoehr, and P. M. Schlievert. 1997. Mitogenic factors from group G streptococci associated with scarlet fever and streptococcal toxic shock syndrome. *Adv. Exp. Med. Biol.* **418**:109–114.

7. Baker, H. M., T. Proft, P. D. Webb, V. L. Arcus, J. D. Fraser, and E. N. Baker. 2004. Crystallographic and mutational data show that the streptococcal pyrogenic exotoxin J can use a common binding surface for T-cell receptor binding and dimerization. *J. Biol. Chem.* **279**:38571–38576.

8. Banks, D. J., S. B. Beres, and J. M. Musser. 2002. The fundamental contribution of phages to GAS evolution, genome diversification and strain emergence. *Trends Microbiol.* **10**:515–521.

9. Banks, D. J., S. F. Porcella, K. D. Barbian, S. B. Beres, L. E. Philips, J. M. Voyich, F. R. DeLeo, J. M. Martin, G. A. Somerville, and J. M. Musser. 2004. Progress toward characterization of the group A *Streptococcus* metagenome: complete genome sequence of a macrolide-resistant serotype M6 strain. *J. Infect. Dis.* **190**:727–738.

10. Basma, H., A. Norrby-Teglund, Y. Guedez, A. McGeer, D. E. Low, O. El-Ahmedy, B. Schwartz, and M. Kotb. 1999. Risk factors in the pathogenesis of invasive group A streptococcal infections: role of protective humoral immunity. *Infect. Immun.* **67**:1871–1877.

11. Beres, S. B., G. L. Sylva, K. D. Barbian, B. Lei, J. S. Hoff, N. D. Mammarella, M. Y. Liu, J. C. Smoot, S. F. Porcella, L. D. Parkins, D. S. Campbell, T. M. Smith, J. K. McCormick, D. Y. Leung, P. M. Schlievert, and J. M. Musser. 2002. Genome sequence of a serotype M3 strain of group A *Streptococcus*: phage-encoded toxins, the high-virulence phenotype, and clone emergence. *Proc. Natl. Acad. Sci. USA* **99**:10078–10083.

12. Berge, A., and L. Bjorck. 1995. Streptococcal cysteine proteinase releases biologically active fragments of streptococcal surface proteins. *J. Biol. Chem.* **270**:9862–9867.

13. Bessen, D. E., M. W. Izzo, T. R. Fiorentino, R. M. Caringal, S. K. Hollingshead, and B. Beall. 1999. Genetic linkage of exotoxin alleles and emm gene markers for tissue tropism in group A streptococci. *J. Infect. Dis.* **179**:627–636.

14. Betschel, S. D., S. M. Borgia, N. L. Barg, D. E. Low, and J. C. De Azavedo. 1998. Reduced virulence of group A streptococcal Tn916 mutants that do not produce streptolysin S. *Infect. Immun.* **66**:1671–1679.

15. Bhakdi, S., J. Tranum-Jensen, and A. Sziegoleit. 1985. Mechanism of membrane damage by streptolysin-O. *Infect. Immun.* **47**:52–60.

16. Bhakdi, S., H. Bayley, A. Valeva, I. Walev, B. Walker, M. Kehoe, and M. Palmer. 1996. Staphylococcal alpha-toxin, streptolysin-O, and *Escherichia coli* hemolysin: prototypes of pore-forming bacterial cytolysins. *Arch. Microbiol.* **165**:73–79.

17. Binks, M., and K. S. Sriprakash. 2004. Characterization of a complement-binding protein, DRS, from strains of *Streptococcus pyogenes* containing the emm12 and emm55 genes. *Infect. Immun.* **72**:3981–3986.

18. Blomster-Hautamaa, D. A., and P. M. Schlievert. 1988. Preparation of toxic shock syndrome toxin-1. *Methods Enzymol.* **165**:37–43.

19. Bohach, G. A., D. J. Fast, R. D. Nelson, and P. M. Schlievert. 1990. Staphylococcal and streptococcal pyrogenic toxins involved in toxic shock syndrome and related illnesses. *Crit. Rev. Microbiol.* **17**:251–272.

20. Bohach, G. A., A. R. Hauser, and P. M. Schlievert. 1988. Cloning of the gene, speB, for streptococcal pyrogenic exotoxin type B in *Escherichia coli*. *Infect. Immun.* **56**:1665–1667.

21. Burns, E. H., Jr., S. Lukomski, J. Rurangirwa, A. Podbielski, and J. M. Musser. 1998. Genetic inactivation of the extracellular cysteine protease enhances in vitro internalization of group A streptococci by human epithelial and endothelial cells. *Microb. Pathog.* **24**:333–339.

22. Burns, E. H., Jr., A. M. Marciel, and J. M. Musser. 1996. Activation of a 66-kilodalton human endothelial cell matrix metalloprotease by *Streptococcus pyogenes* extracellular cysteine protease. *Infect. Immun.* **64**:4744–4750.

23. Bustin, M., M. C. Lin, W. H. Stein, and S. Moore. 1970. Activity of the reduced zymogen of streptococcal proteinase. *J. Biol. Chem.* **245**:846–849.

24. Carr, A., D. D. Sledjeski, A. Podbielski, M. D. Boyle, and B. Kreikemeyer. 2001. Similarities between complement-mediated and streptolysin S-mediated hemolysis. *J. Biol. Chem.* **276**:41790–41796.

25. Chaussee, M. S., E. R. Phillips, and J. J. Ferretti. 1997. Temporal production of streptococcal erythrogenic toxin B (streptococcal cysteine proteinase) in response to nutrient depletion. *Infect. Immun.* **65**:1956–1959.

26. Cleary, P. P., E. L. Kaplan, J. P. Handley, A. Wlazlo, M. H. Kim, A. R. Hauser, and P. M. Schlievert. 1992. Clonal basis for resurgence of serious *Streptococcus pyogenes* disease in the 1980s. *Lancet* **339**:518–521.

27. Cockerill, F. R., 3rd, K. L. MacDonald, R. L. Thompson, F. Roberson, P. C. Kohner, J. Besser-Wiek, J. M. Manahan, J. M. Musser, P. M. Schlievert, J. Talbot, B. Frankfort, J. M. Steckelberg, W. R. Wilson, and M. T. Osterholm. 1997. An outbreak of invasive group A streptococcal disease associated with high carriage rates of the invasive clone among school-aged children. *JAMA* **277**: 38–43.

28. Collin, M., and A. Olsen. 2003. Extracellular enzymes with immunomodulating activities: variations on a theme in *Streptococcus pyogenes*. *Infect. Immun.* **71**:2983–2992.

29. Colon, A. E., R. M. Cole, and C. G. Leonard. 1972. Intergroup lysis and transduction by streptococcal bacteriophages. *J. Virol.* **9**:551–553.

30. Colon, A. E., R. M. Cole, and C. G. Leonard. 1971. Lysis and lysogenization of groups A, C, and G streptococci by a transducing bacteriophage induced from a group G *Streptococcus*. *J. Virol.* **8**:103–110.

31. Cone, L. A., D. R. Woodard, P. M. Schlievert, and G. S. Tomory. 1987. Clinical and bacteriologic observations of a

toxic shock-like syndrome due to *Streptococcus pyogenes*. *N. Engl. J. Med.* **317:**146–149.

32. **DeLano, W. L.** 2002. The PyMOL molecular graphics system on World Wide Web. http://www.pymol.org.

33. **Demers, B., A. E. Simor, H. Vellend, P. M. Schlievert, S. Byrne, F. Jamieson, S. Walmsley, and D. E. Low.** 1993. Severe invasive group A streptococcal infections in Ontario, Canada: 1987–1991. *Clin. Infect. Dis.* **16:**792–800.

34. **Dinges, M. M., and P. M. Schlievert.** 2001. Comparative analysis of lipopolysaccharide-induced tumor necrosis factor alpha activity in serum and lethality in mice and rabbits pretreated with the staphylococcal superantigen toxic shock syndrome toxin 1. *Infect. Immun.* **69:**7169–7172.

35. **Dinges, M. M., and P. M. Schlievert.** 2001. Role of T cells and gamma interferon during induction of hypersensitivity to lipopolysaccharide by toxic shock syndrome toxin 1 in mice. *Infect. Immun.* **69:**1256–1264.

36. **Doran, J. D., M. Nomizu, S. Takebe, R. Menard, D. Griffith, and E. Ziomek.** 1999. Autocatalytic processing of the streptococcal cysteine protease zymogen: processing mechanism and characterization of the autoproteolytic cleavage sites. *Eur. J. Biochem.* **263:**145–151.

37. **Earhart, C. A., G. M. Vath, M. Roggiani, P. M. Schlievert, and D. H. Ohlendorf.** 2000. Structure of streptococcal pyrogenic exotoxin A reveals a novel metal cluster. *Protein Sci.* **9:**1847–1851.

38. **Eriksson, A., B. Eriksson, S. E. Holm, and M. Norgren.** 1999. Streptococcal DNase B is immunologically identical to superantigen SpeF but involves separate domains. *Clin. Diagn. Lab. Immunol.* **6:**133–136.

39. **Eriksson, A., and M. Norgren.** 1999. The superantigenic activity of streptococcal pyrogenic exotoxin B is independent of the protease activity. *FEMS Immunol. Med. Microbiol.* **25:**355–363.

40. **Eriksson, B. K., J. Andersson, S. E. Holm, and M. Norgren.** 1999. Invasive group A streptococcal infections: T1M1 isolates expressing pyrogenic exotoxins A and B in combination with selective lack of toxin-neutralizing antibodies are associated with increased risk of streptococcal toxic shock syndrome. *J. Infect. Dis.* **180:**410–418.

41. **Fast, D. J., P. M. Schlievert, and R. D. Nelson.** 1989. Toxic shock syndrome-associated staphylococcal and streptococcal pyrogenic toxins are potent inducers of tumor necrosis factor production. *Infect. Immun.* **57:**291–294.

42. **Fernie-King, B. A., D. J. Seilly, A. Davies, and P. J. Lachmann.** 2002. Streptococcal inhibitor of complement inhibits two additional components of the mucosal innate immune system: secretory leukocyte proteinase inhibitor and lysozyme. *Infect. Immun.* **70:**4908–4916.

43. **Fernie-King, B. A., D. J. Seilly, and P. J. Lachmann.** 2004. The interaction of streptococcal inhibitor of complement (SIC) and its proteolytic fragments with the human beta defensins. *Immunology* **111:**444–452.

44. **Fernie-King, B. A., D. J. Seilly, C. Willers, R. Wurzner, A. Davies, and P. J. Lachmann.** 2001. Streptococcal inhibitor of complement (SIC) inhibits the membrane attack complex by preventing uptake of C567 onto cell membranes. *Immunology* **103:**390–398.

45. **Ferretti, J. J., W. M. McShan, D. Ajdic, D. J. Savic, G. Savic, K. Lyon, C. Primeaux, S. Sezate, A. N. Suvorov, S. Kenton, H. S. Lai, S. P. Lin, Y. Qian, H. G. Jia, F. Z. Najar, Q. Ren, H. Zhu, L. Song, J. White, X. Yuan, S. W. Clifton, B. A. Roe, and R. McLaughlin.** 2001. Complete genome sequence of an M1 strain of *Streptococcus pyogenes*. *Proc. Natl. Acad. Sci. USA* **98:**4658–4663.

46. **Fields, B. A., E. L. Malchiodi, H. Li, X. Ysern, C. V. Stauffacher, P. M. Schlievert, K. Karjalainen, and R. A. Mariuzza.** 1996. Crystal structure of a T-cell receptor beta-chain complexed with a superantigen. *Nature* **384:**188–192.

47. **Fontaine, M. C., J. J. Lee, and M. A. Kehoe.** 2003. Combined contributions of streptolysin O and streptolysin S to virulence of serotype M5 *Streptococcus pyogenes* strain Manfredo. *Infect. Immun.* **71:**3857–3865.

48. **Frick, I. M., P. Akesson, M. Rasmussen, A. Schmidtchen, and L. Bjorck.** 2003. SIC, a secreted protein of *Streptococcus pyogenes* that inactivates antibacterial peptides. *J. Biol. Chem.* **278:**16561–16566.

49. **Frobisher, M., and J. H. Brown.** 1927. Transmissible toxigenicity of streptococci. *Bull. Johns Hopkins Hosp.* **41:**167–173.

50. **Ginsburg, I.** 1972. Mechanisms of cell and tissue injury induced by group A streptococci: relation to poststreptococcal sequelae. *J. Infect. Dis.* **126:**294–340.

51. **Goshorn, S. C., G. A. Bohach, and P. M. Schlievert.** 1988. Cloning and characterization of the gene, *speC*, for pyrogenic exotoxin type C from *Streptococcus pyogenes*. *Mol. Gen. Genet.* **212:**66–70.

52. **Hacker, J., and J. B. Kaper.** 2000. Pathogenicity islands and the evolution of microbes. *Annu. Rev. Microbiol.* **54:**641–679.

53. **Hauser, A. R., and P. M. Schlievert.** 1990. Nucleotide sequence of the streptococcal pyrogenic exotoxin type B gene and relationship between the toxin and the streptococcal proteinase precursor. *J. Bacteriol.* **172:**4536–4542.

54. **Hauser, A. R., D. L. Stevens, E. L. Kaplan, and P. M. Schlievert.** 1991. Molecular analysis of pyrogenic exotoxins from *Streptococcus pyogenes* isolates associated with toxic shock-like syndrome. *J. Clin. Microbiol.* **29:**1562–1567.

55. **Hennecke, J., A. Carfi, and D. C. Wiley.** 2000. Structure of a covalently stabilized complex of a human alphabeta T-cell receptor, influenza HA peptide and MHC class II molecule, HLA-DR1. *EMBO J.* **19:**5611–5624.

56. **Herwald, H., M. Collin, W. Muller-Esterl, and L. Bjorck.** 1996. Streptococcal cysteine proteinase releases kinins: a virulence mechanism. *J. Exp. Med.* **184:**665–673.

57. **Hoe, N. P., P. Kordari, R. Cole, M. Liu, T. Palzkill, W. Huang, D. McLellan, G. J. Adams, M. Hu, J. Vuopio-Varkila, T. R. Cate, M. E. Pichichero, K. M. Edwards, J. Eskola, D. E. Low, and J. M. Musser.** 2000. Human immune response to streptococcal inhibitor of complement, a serotype M1 group A Streptococcus extracellular protein involved in epidemics. *J. Infect. Dis.* **182:**1425–1436.

58. **Hoe, N. P., K. Nakashima, S. Lukomski, D. Grigsby, M. Liu, P. Kordari, S. J. Dou, X. Pan, J. Vuopio-Varkila, S. Salmelinna, A. McGeer, D. E. Low, B. Schwartz, A. Schuchat, S. Naidich, D. De Lorenzo, Y. X. Fu, and J. M. Musser.** 1999. Rapid selection of complement-inhibiting protein variants in group A Streptococcus epidemic waves. *Nat. Med.* **5:**924–929.

59. **Holm, S. E., A. Norrby, A. M. Bergholm, and M. Norgren.** 1992. Aspects of pathogenesis of serious group A streptococcal infections in Sweden, 1988–1989. *J. Infect. Dis.* **166:**31–37.

60. **Hryniewicz, W., and J. Pryjma.** 1977. Effect of streptolysin S on human and mouse T and B lymphocytes. *Infect. Immun.* **16:**730–733.

61. **Iwasaki, M., H. Igarashi, Y. Hinuma, and T. Yutsudo.** 1993. Cloning, characterization and overexpression of a *Streptococcus pyogenes* gene encoding a new type of mitogenic factor. *FEBS Lett.* **331:**187–192.

62. **Iwasaki, M., H. Igarashi, and T. Yutsudo.** 1997. Mitogenic factor secreted by *Streptococcus pyogenes* is a heat-stable nuclease requiring His122 for activity. *Microbiology* **143**(Pt. 7):2449–2455.

63. **Jardetzky, T. S., J. H. Brown, J. C. Gorga, L. J. Stern, R. G. Urban, Y. I. Chi, C. Stauffacher, J. L. Strominger, and D. C. Wiley.** 1994. Three-dimensional structure of a human class II histocompatibility molecule complexed with superantigen. *Nature* **368**:711–718.

64. **Johnson, D. R., D. L. Stevens, and E. L. Kaplan.** 1992. Epidemiologic analysis of group A streptococcal serotypes associated with severe systemic infections, rheumatic fever, or uncomplicated pharyngitis. *J. Infect. Dis.* **166**:374–382.

65. **Johnson, L. P., and P. M. Schlievert.** 1984. Group A streptococcal phage T12 carries the structural gene for pyrogenic exotoxin type A. *Mol. Gen. Genet.* **194**:52–56.

66. **Johnson, L. P., and P. M. Schlievert.** 1983. A physical map of the group A streptococcal pyrogenic exotoxin bacteriophage T12 genome. *Mol. Gen. Genet.* **189**:251–255.

67. **Kalia, A., and D. E. Bessen.** 2003. Presence of streptococcal pyrogenic exotoxin A and C genes in human isolates of group G streptococci. *FEMS Microbiol. Lett.* **219**:291–295.

68. **Kansal, R. G., A. McGeer, D. E. Low, A. Norrby-Teglund, and M. Kotb.** 2000. Inverse relation between disease severity and expression of the streptococcal cysteine protease, SpeB, among clonal M1T1 isolates recovered from invasive group A streptococcal infection cases. *Infect. Immun.* **68**:6362–6369.

69. **Kansal, R. G., V. Nizet, A. Jeng, W. J. Chuang, and M. Kotb.** 2003. Selective modulation of superantigen-induced responses by streptococcal cysteine protease. *J. Infect. Dis.* **187**:398–407.

70. **Kapur, V., M. W. Majesky, L. L. Li, R. A. Black, and J. M. Musser.** 1993. Cleavage of interleukin 1 beta (IL-1 beta) precursor to produce active IL-1 beta by a conserved extracellular cysteine protease from *Streptococcus pyogenes*. *Proc. Natl. Acad. Sci. USA* **90**:7676–7680.

71. **Kapur, V., S. Topouzis, M. W. Majesky, L. L. Li, M. R. Hamrick, R. J. Hamill, J. M. Patti, and J. M. Musser.** 1993. A conserved *Streptococcus pyogenes* extracellular cysteine protease cleaves human fibronectin and degrades vitronectin. *Microb. Pathog.* **15**:327–346.

72. **Kehoe, M. A., L. Miller, J. A. Walker, and G. J. Boulnois.** 1987. Nucleotide sequence of the streptolysin O (SLO) gene: structural homologies between SLO and other membrane-damaging, thiol-activated toxins. *Infect. Immun.* **55**:3228–3232.

73. **Kim, J., R. G. Urban, J. L. Strominger, and D. C. Wiley.** 1994. Toxic shock syndrome toxin-1 complexed with a class II major histocompatibility molecule HLA-DR1. *Science* **266**:1870–1874.

74. **Kotb, M.** 1995. Bacterial pyrogenic exotoxins as superantigens. *Clin. Microbiol. Rev.* **8**:411–426.

75. **Kotb, M., A. Norrby-Teglund, A. McGeer, H. El-Sherbini, M. T. Dorak, A. Khurshid, K. Green, J. Peeples, J. Wade, G. Thomson, B. Schwartz, and D. E. Low.** 2002. An immunogenetic and molecular basis for differences in outcomes of invasive group A streptococcal infections. *Nat. Med.* **8**:1398–1404.

76. **Leonard, B. A., P. K. Lee, M. K. Jenkins, and P. M. Schlievert.** 1991. Cell and receptor requirements for streptococcal pyrogenic exotoxin T-cell mitogenicity. *Infect. Immun.* **59**:1210–1214.

77. **Li, H., A. Llera, E. L. Malchiodi, and R. A. Mariuzza.** 1999. The structural basis of T cell activation by superantigens. *Annu. Rev. Immunol.* **17**:435–466.

78. **Li, H., A. Llera, D. Tsuchiya, L. Leder, X. Ysern, P. M. Schlievert, K. Karjalainen, and R. A. Mariuzza.** 1998. Three-dimensional structure of the complex between a T cell receptor beta chain and the superantigen staphylococcal enterotoxin B. *Immunity* **9**:807–816.

79. **Li, P. L., R. E. Tiedemann, S. L. Moffat, and J. D. Fraser.** 1997. The superantigen streptococcal pyrogenic exotoxin C (SPE-C) exhibits a novel mode of action. *J. Exp. Med.* **186**:375–383.

80. **Li, Y., H. Li, N. Dimasi, J. K. McCormick, R. Martin, P. Schuck, P. M. Schlievert, and R. A. Mariuzza.** 2001. Crystal structure of a superantigen bound to the high-affinity, zinc-dependent site on MHC class II. *Immunity* **14**:93–104.

81. **Llewelyn, M., S. Sriskandan, M. Peakman, D. R. Ambrozak, D. C. Douek, W. W. Kwok, J. Cohen, and D. M. Altmann.** 2004. HLA class II polymorphisms determine responses to bacterial superantigens. *J. Immunol.* **172**:1719–1726.

82. **Lukomski, S., E. H. Burns, Jr., P. R. Wyde, A. Podbielski, J. Rurangirwa, D. K. Moore-Poveda, and J. M. Musser.** 1998. Genetic inactivation of an extracellular cysteine protease (SpeB) expressed by *Streptococcus pyogenes* decreases resistance to phagocytosis and dissemination to organs. *Infect. Immun.* **66**:771–776.

83. **Lukomski, S., N. P. Hoe, I. Abdi, J. Rurangirwa, P. Kordari, M. Liu, S. J. Dou, G. G. Adams, and J. M. Musser.** 2000. Nonpolar inactivation of the hypervariable streptococcal inhibitor of complement gene (sic) in serotype M1 *Streptococcus pyogenes* significantly decreases mouse mucosal colonization. *Infect. Immun.* **68**:535–542.

84. **Marrack, P., and J. Kappler.** 1990. The staphylococcal enterotoxins and their relatives. *Science* **248**:705–711.

85. **Massell, B. F., C. G. Chute, A. M. Walker, and G. S. Kurland.** 1988. Penicillin and the marked decrease in morbidity and mortality from rheumatic fever in the United States. *N. Engl. J. Med.* **318**:280–286.

86. **Matsuka, Y. V., S. Pillai, S. Gubba, J. M. Musser, and S. B. Olmsted.** 1999. Fibrinogen cleavage by the *Streptococcus pyogenes* extracellular cysteine protease and generation of antibodies that inhibit enzyme proteolytic activity. *Infect. Immun.* **67**:4326–4333.

87. **McCormick, J. K., A. A. Pragman, J. C. Stolpa, D. Y. Leung, and P. M. Schlievert.** 2001. Functional characterization of streptococcal pyrogenic exotoxin J, a novel superantigen. *Infect. Immun.* **69**:1381–1388.

88. **McCormick, J. K., and P. M. Schlievert.** 2003. Expression, purification, and detection of novel streptococcal superantigens. *Methods Mol. Biol.* **214**:33–43.

89. **McCormick, J. K., T. J. Tripp, A. S. Llera, E. J. Sundberg, M. M. Dinges, R. A. Mariuzza, and P. M. Schlievert.** 2003. Functional analysis of the TCR binding domain of toxic shock syndrome toxin-1 predicts further diversity in MHC class II/superantigen/TCR ternary complexes. *J. Immunol.* **171**:1385–1392.

90. **McCormick, J. K., T. J. Tripp, S. B. Olmsted, Y. V. Matsuka, P. J. Gahr, D. H. Ohlendorf, and P. M. Schlievert.** 2000. Development of streptococcal pyrogenic exotoxin C vaccine toxoids that are protective in the rabbit model of toxic shock syndrome. *J. Immunol.* **165**:2306–2312.

91. **McCormick, J. K., J. M. Yarwood, and P. M. Schlievert.** 2001. Toxic shock syndrome and bacterial superantigens: an update. *Annu. Rev. Microbiol.* **55**:77–104.

92. **McShan, W. M., Y. F. Tang, and J. J. Ferretti.** 1997. Bacteriophage T12 of *Streptococcus pyogenes* integrates into the gene encoding a serine tRNA. *Mol. Microbiol.* **23:**719–728.

93. **Mitchell, D. T., D. G. Levitt, P. M. Schlievert, and D. H. Ohlendorf.** 2000. Structural evidence for the evolution of pyrogenic toxin superantigens. *J. Mol. Evol.* **51:**520–531.

94. **Miyoshi-Akiyama, T., J. Zhao, H. Kato, K. Kikuchi, K. Totsuka, Y. Kataoka, M. Katsumi, and T. Uchiyama.** 2003. *Streptococcus dysgalactiae*-derived mitogen (SDM), a novel bacterial superantigen: characterization of its biological activity and predicted tertiary structure. *Mol. Microbiol.* **47:**1589–1599.

95. **Muller-Alouf, H., D. Gerlach, P. Desreumaux, C. Leportier, J. E. Alouf, and M. Capron.** 1997. Streptococcal pyrogenic exotoxin A (SPE A) superantigen induced production of hematopoietic cytokines, IL-12 and IL-13 by human peripheral blood mononuclear cells. *Microb. Pathog.* **23:**265–272.

96. **Musser, J. M., A. R. Hauser, M. H. Kim, P. M. Schlievert, K. Nelson, and R. K. Selander.** 1991. *Streptococcus pyogenes* causing toxic-shock-like syndrome and other invasive diseases: clonal diversity and pyrogenic exotoxin expression. *Proc. Natl. Acad. Sci. USA* **88:**2668–2672.

97. **Musser, J. M., V. Kapur, J. Szeto, X. Pan, D. S. Swanson, and D. R. Martin.** 1995. Genetic diversity and relationships among *Streptococcus pyogenes* strains expressing serotype M1 protein: recent intercontinental spread of a subclone causing episodes of invasive disease. *Infect. Immun.* **63:**994–1003.

98. **Nakagawa, I., K. Kurokawa, A. Yamashita, M. Nakata, Y. Tomiyasu, N. Okahashi, S. Kawabata, K. Yamazaki, T. Shiba, T. Yasunaga, H. Hayashi, M. Hattori, and S. Hamada.** 2003. Genome sequence of an M3 strain of *Streptococcus pyogenes* reveals a large-scale genomic rearrangement in invasive strains and new insights into phage evolution. *Genome Res.* **13:**1042–1055.

99. **Nelson, K., P. M. Schlievert, R. K. Selander, and J. M. Musser.** 1991. Characterization and clonal distribution of four alleles of the *speA* gene encoding pyrogenic exotoxin A (scarlet fever toxin) in *Streptococcus pyogenes*. *J. Exp. Med.* **174:**1271–1274.

100. **Newton, D., A. Norrby-Teglund, A. McGeer, D. E. Low, P. M. Schlievert, and M. Kotb.** 1997. Novel superantigens from streptococcal toxic shock syndrome *Streptococcus pyogenes* isolates. *Adv. Exp. Med. Biol.* **418:**525–529.

101. **Nizet, V., B. Beall, D. J. Bast, V. Datta, L. Kilburn, D. E. Low, and J. C. De Azavedo.** 2000. Genetic locus for streptolysin S production by group A streptococcus. *Infect. Immun.* **68:**4245–4254.

102. **Norrby-Teglund, A., D. Newton, M. Kotb, S. E. Holm, and M. Norgren.** 1994. Superantigenic properties of the group A streptococcal exotoxin SpeF (MF). *Infect. Immun.* **62:**5227–5233.

103. **Nyberg, P., M. Rasmussen, U. Von Pawel-Rammingen, and L. Bjorck.** 2004. SpeB modulates fibronectin-dependent internalization of *Streptococcus pyogenes* by efficient proteolysis of cell-wall-anchored protein F1. *Microbiology* **150:**1559–1569.

104. **Ohkuni, H., Y. Todome, Y. Watanabe, T. Ishikaw, H. Takahashi, Y. Kannari, H. Kato, T. Uchiyama, H. Saito, V. A. Fischetti, and J. B. Zabriskie.** 2004. Studies of recombinant streptococcal pyrogenic exotoxin B/cysteine protease (rSPE B/SCP) in the skin of guinea pigs and the release of histamine from cultured mast cells and basophilic leukocytes. *Indian J. Med. Res.* **119**(Suppl.)**:**33–36.

105. **Palmer, M., A. Valeva, M. Kehoe, and S. Bhakdi.** 1995. Kinetics of streptolysin O self-assembly. *Eur. J. Biochem.* **231:**388–395.

106. **Parsonnet, J., Z. A. Gillis, A. G. Richter, and G. B. Pier.** 1987. A rabbit model of toxic shock syndrome that uses a constant, subcutaneous infusion of toxic shock syndrome toxin 1. *Infect. Immun.* **55:**1070–1076.

107. **Parsonnet, J., R. K. Hickman, D. D. Eardley, and G. B. Pier.** 1985. Induction of human interleukin-1 by toxic-shock-syndrome toxin-1. *J. Infect. Dis.* **151:**514–522.

108. **Petersson, K., M. Hakansson, H. Nilsson, G. Forsberg, L. A. Svensson, A. Liljas, and B. Walse.** 2001. Crystal structure of a superantigen bound to MHC class II displays zinc and peptide dependence. *EMBO J.* **20:**3306–3312.

109. **Petersson, K., M. Thunnissen, G. Forsberg, and B. Walse.** 2002. Crystal structure of a SEA variant in complex with MHC class II reveals the ability of SEA to crosslink MHC molecules. *Structure* (Camb.) **10:**1619–1626.

110. **Proft, T., and J. D. Fraser.** 2003. Bacterial superantigens. *Clin. Exp. Immunol.* **133:**299–306.

111. **Proft, T., S. L. Moffatt, K. D. Weller, A. Paterson, D. Martin, and J. D. Fraser.** 2000. The streptococcal superantigen SMEZ exhibits wide allelic variation, mosaic structure, and significant antigenic variation. *J. Exp. Med.* **191:**1765–1776.

112. **Proft, T., P. D. Webb, V. Handley, and J. D. Fraser.** 2003. Two novel superantigens found in both group A and group C *Streptococcus*. *Infect. Immun.* **71:**1361–1369.

113. **Rasmussen, M., and L. Bjorck.** 2002. Proteolysis and its regulation at the surface of *Streptococcus pyogenes*. *Mol. Microbiol.* **43:**537–544.

114. **Roggiani, M., J. A. Stoehr, S. B. Olmsted, Y. V. Matsuka, S. Pillai, D. H. Ohlendorf, and P. M. Schlievert.** 2000. Toxoids of streptococcal pyrogenic exotoxin A are protective in rabbit models of streptococcal toxic shock syndrome. *Infect. Immun.* **68:**5011–5017.

115. **Roussel, A., B. F. Anderson, H. M. Baker, J. D. Fraser, and E. N. Baker.** 1997. Crystal structure of the streptococcal superantigen SPE-C: dimerization and zinc binding suggest a novel mode of interaction with MHC class II molecules. *Nat. Struct. Biol.* **4:**635–643.

116. **Ruiz, N., B. Wang, A. Pentland, and M. Caparon.** 1998. Streptolysin O and adherence synergistically modulate proinflammatory responses of keratinocytes to group A streptococci. *Mol. Microbiol.* **27:**337–346.

117. **Sachse, S., P. Seidel, D. Gerlach, E. Gunther, J. Rodel, E. Straube, and K. H. Schmidt.** 2002. Superantigen-like gene(s) in human pathogenic *Streptococcus dysgalactiae*, subsp *equisimilis*: genomic localisation of the gene encoding streptococcal pyrogenic exotoxin G (speG^{dys}). *FEMS Immunol. Med. Microbiol.* **34:**159–167.

118. **Schlievert, P. M., A. P. Assimacopoulos, and P. P. Cleary.** 1996. Severe invasive group A streptococcal disease: clinical description and mechanisms of pathogenesis. *J. Lab. Clin. Med.* **127:**13–22.

119. **Schlievert, P. M., K. M. Bettin, and D. W. Watson.** 1979. Production of pyrogenic exotoxin by groups of streptococci: association with group A. *J. Infect. Dis.* **140:**676–681.

120. **Schlievert, P. M., K. M. Bettin, and D. W. Watson.** 1979. Reinterpretation of the Dick test: role of group A

streptococcal pyrogenic exotoxin. *Infect. Immun.* **26:** 467–472.

121. **Smoot, J. C., K. D. Barbian, J. J. Van Gompel, L. M. Smoot, M. S. Chaussee, G. L. Sylva, D. E. Sturdevant, S. M. Ricklefs, S. F. Porcella, L. D. Parkins, S. B. Beres, D. S. Campbell, T. M. Smith, Q. Zhang, V. Kapur, J. A. Daly, L. G. Veasy, and J. M. Musser.** 2002. Genome sequence and comparative microarray analysis of serotype M18 group A *Streptococcus* strains associated with acute rheumatic fever outbreaks. *Proc. Natl. Acad. Sci. USA* **99:**4668–4673.

122. **Sriskandan, S., D. Moyes, and J. Cohen.** 1996. Detection of circulating bacterial superantigen and lymphotoxin-alpha in patients with streptococcal toxic-shock syndrome. *Lancet* **348:**1315–136.

123. **Sriskandan, S., M. Unnikrishnan, T. Krausz, and J. Cohen.** 1999. Molecular analysis of the role of streptococcal pyrogenic exotoxin A (SPEA) in invasive soft-tissue infection resulting from *Streptococcus pyogenes*. *Mol. Microbiol.* **33:**778–790.

124. **Stevens, D. L., M. H. Tanner, J. Winship, R. Swarts, K. M. Ries, P. M. Schlievert, and E. Kaplan.** 1989. Severe group A streptococcal infections associated with a toxic shock-like syndrome and scarlet fever toxin A. *N. Engl. J. Med.* **321:**1–7.

125. **Sundberg, E., and T. S. Jardetzky.** 1999. Structural basis for HLA-DQ binding by the streptococcal superantigen SSA. *Nat. Struct. Biol.* **6:**123–129.

126. **Sundberg, E. J., P. S. Andersen, P. M. Schlievert, K. Karjalainen, and R. A. Mariuzza.** 2003. Structural, energetic, and functional analysis of a protein-protein interface at distinct stages of affinity maturation. *Structure* (Camb.) **11:**1151–1161.

127. **Sundberg, E. J., H. Li, A. S. Llera, J. K. McCormick, J. Tormo, P. M. Schlievert, K. Karjalainen, and R. A. Mariuzza.** 2002. Structures of two streptococcal superantigens bound to TCR beta chains reveal diversity in the architecture of T cell signaling complexes. *Structure* (Camb.) **10:**687–699.

128. **Sundberg, E. J., Y. Li, and R. A. Mariuzza.** 2002. So many ways of getting in the way: diversity in the molecular architecture of superantigen-dependent T-cell signaling complexes. *Curr. Opin. Immunol.* **14:**36–44.

129. **Svensson, M. D., D. A. Scaramuzzino, U. Sjobring, A. Olsen, C. Frank, and D. E. Bessen.** 2000. Role for a secreted cysteine proteinase in the establishment of host tissue tropism by group A streptococci. *Mol. Microbiol.* **38:**242–253.

130. **Swietnicki, W., A. M. Barnie, B. K. Dyas, and R. G. Ulrich.** 2003. Zinc binding and dimerization of *Streptococcus pyogenes* pyrogenic exotoxin C are not essential for T-cell stimulation. *J. Biol. Chem.* **278:**9885–9895.

131. **Tomai, M. A., P. M. Schlievert, and M. Kotb.** 1992. Distinct T-cell receptor V beta gene usage by human T lymphocytes stimulated with the streptococcal pyrogenic exotoxins and pep M5 protein. *Infect. Immun.* **60:**701–705.

132. **Tripp, T. J., J. K. McCormick, J. M. Webb, and P. M. Schlievert.** 2003. The zinc-dependent major histocompatibility complex class II binding site of streptococcal pyrogenic exotoxin C is critical for maximal superantigen function and toxic activity. *Infect. Immun.* **71:**1548–1550.

133. **Unnikrishnan, M., J. Cohen, and S. Sriskandan.** 2001. Complementation of a *speA* negative *Streptococcus pyogenes* with *speA*: effects on virulence and production of streptococcal pyrogenic exotoxin A. *Microb. Pathog.* **31:**109–114.

134. **Watson, D. W.** 1960. Host-parasite factors in group A streptococcal infections. Pyrogenic and other effects of immunologic distinct exotoxins related to scarlet fever toxins. *J. Exp. Med.* **111:**255–284.

135. **Weeks, C. R., and J. J. Ferretti.** 1986. Nucleotide sequence of the type A streptococcal exotoxin (erythrogenic toxin) gene from *Streptococcus pyogenes* bacteriophage T12. *Infect. Immun.* **52:**144–150.

136. **White, J., A. Herman, A. M. Pullen, R. Kubo, J. W. Kappler, and P. Marrack.** 1989. The V beta-specific superantigen staphylococcal enterotoxin B: stimulation of mature T cells and clonal deletion in neonatal mice. *Cell* **56:**27–35.

137. **Yu, C. E., and J. J. Ferretti.** 1991. Frequency of the erythrogenic toxin B and C genes (speB and speC) among clinical isolates of group A streptococci. *Infect. Immun.* **59:**211–215.

138. **Yutsudo, T., H. Murai, J. Gonzalez, T. Takao, Y. Shimonishi, Y. Takeda, H. Igarashi, and Y. Hinuma.** 1992. A new type of mitogenic factor produced by *Streptococcus pyogenes*. *FEBS Lett.* **308:**30–34.

139. **Yutsudo, T., K. Okumura, M. Iwasaki, A. Hara, S. Kamitani, W. Minamide, H. Igarashi, and Y. Hinuma.** 1994. The gene encoding a new mitogenic factor in a *Streptococcus pyogenes* strain is distributed only in group A streptococci. *Infect. Immun.* **62:**4000–4004.

140. **Zabriskie, J. B.** 1964. The role of temperate bacteriophage in the production of erythrogenic toxin by group A streptococci. *J. Exp. Med.* **119:**761–780.

Genetics of Group A Streptococci

KYU HONG CHO AND MICHAEL CAPARON

6

Streptococcus pyogenes (the group A streptococcus) is remarkable in terms of the large number of very different diseases it can cause in humans. These range from superficial and self-limiting diseases of the pharynx (e.g., pharyngitis, commonly known as strep throat) and skin (impetigo) to infections that involve increasingly deeper layers of tissue and are associated with increasing degrees of destruction of tissue (e.g., erysipelas, cellulitis, necrotizing fasciitis, and myositis). The organism has the ability to spread rapidly through tissue and to penetrate into the vasculature to cause lethal sepsis. Other diseases result from the production of toxins that spread through tissue or systemically from a site of local infection (scarlet fever and toxic shock syndrome). Still other diseases are the result of an immunopathological response on the part of the host that is triggered by a streptococcal infection. These diseases include rheumatic fever, acute glomerulonephritis, certain types of psoriasis, and potentially even some forms of obsessive-compulsive syndrome disorder.

S. pyogenes is even more remarkable in terms of the very large number of factors that have been identified as potential virulence determinants for these various diseases. These include surface proteins (M proteins, fibronectin-binding proteins, surface dehydrogenases, C5a peptidase), the hyaluronic acid capsule, secreted degradative enzymes (several distinct DNases, a cysteine protease, NADase, hyaluronidase), and many different secreted toxins (streptolysin S, streptolysin O, the pyrogenic exotoxins, streptococcal superantigen, streptokinase). This represents only a partial list of the different potential virulence factors this bacterium can produce, and many of these are considered in more detail in the other chapters of this volume.

Until recently, the function that any of these different potential virulence factors contributed to the pathogenesis of any specific streptococcal disease was only poorly understood. A major reason for this deficiency was the lack of sophisticated genetic systems that could be applied to the analysis of a specific virulence factor according to modern molecular criteria. A succinct statement of these criteria was outlined by Falkow in his "molecular Koch's postulates" (18). These postulates state that (i) the phenotype under investigation should be associated with pathogenic members of a species; (ii) the gene(s) associated with the virulence trait should be identified and isolated by molecular methods; (iii) specific inactivation of identified gene(s) should lead to a measurable loss in pathogenicity; and (iv) reintroduction of the unmodified wild-type gene should lead to a restoration in pathogenicity. An exciting development is that, over the past few years, the work of many different groups has contributed to the development of sophisticated group A streptococcal genetic systems that now allow the many different diseases that this organism can cause to be studied at this level of molecular resolution. The goal of this chapter will be to present an overview of these methods and their applications.

GENETIC EXCHANGE

Transduction

A key element of any genetic system involves some system of genetic exchange between different bacterial hosts which allows the construction of an altered genome in the target host which can then be subjected to an analysis of its virulence phenotypes. Unlike several other species of streptococci, the group A streptococci are not naturally competent for the uptake of exogenous DNA. Conjugative DNA transfer does occur in group A streptococci; however, this is restricted to the transfer of conjugative plasmids and conjugative transposons (see below) and there is no evidence for mobilization of chromosomal markers. There is also no evidence that any important virulence traits are encoded by these types of mobile elements, although they are undoubtedly important in the transmission of resistance to various antibiotics.

Even though there is no evidence to support genetic exchange by transformation or by conjugation, analysis of several polymorphic loci, most notably the genes which encode the M proteins, has provided considerable evidence for horizontal transfer of genetic material among natural populations of *S. pyogenes*. In this regard, considerable attention has focused on the contribution of phage to genetic

Gram-Positive Pathogens, 2nd edition, edited by Vincent A. Fischetti et al.
© 2006 ASM Press, Washington, D.C.

transfer. *S. pyogenes* strains are rich in phage, and several of these have been demonstrated to encode virulence factors, including the pyrogenic exotoxins SPE-A and SPE-C (53). Thus, it is not surprising that transduction was the first mechanism of genetic exchange that was exploited in the manipulation of the *S. pyogenes* genome. The most highly developed of these are derivatives of phage A25. This lytic phage is classified as a Bradley group B phage that recognizes peptidoglycan of groups A, C, and G streptococci as its cellular receptor. It has a 35-kb double-stranded genome with circular permutation and terminal repetition (44) and it is proficient for transducing markers in vitro. (For a detailed description of a method for transduction, see reference 8.) The most useful derivatives of A25 for use in transduction are those developed by Malke (39), who constructed a derivative with two distinct temperature-sensitive lesions (A25$_{ts1-2}$) that becomes defective for growth at 37°C. This feature is useful for transduction, because it allows the production of a transducing lysate of the donor host at the permissive temperature (30°C), but prevents the killing of transduced hosts when infection is performed at the nonpermissive temperature.

A limitation to the use of transduction for the construction of mutated chromosomes for virulence studies is that the method is restricted to the exchange of preexisting markers between different *S. pyogenes* hosts. Thus, it cannot be applied to the construction of novel mutations or in cases where a suitable selectable marker is not linked to the preexisting mutation of choice. However, transduction has been most useful in linkage analysis of transposon Tn916-generated mutations and in analysis of Tn916-generated mutations when the mutant chromosome contains multiple copies of the transposable element. In both cases, transduction is used to cross the transposon-containing locus back into a wild-type background. Because Tn916 does not transpose at high frequency during transduction, the resulting transductants arise by homologous recombination, with the result that the transposon serves as a selectable marker to cross the mutated locus back into a wild-type background (8). The phenotypes of the resulting transductants can then be analyzed and compared to that of the original mutant. In the case of a mutant with multiple insertions, it is possible to generate transductants with just one of the mutated loci so that the contribution of each individual transposon insertion to the generation of the mutant phenotype can be analyzed.

Transformation

One of the most significant advances in genetic technology for *S. pyogenes* has been the development of methods for transformation. This has allowed the introduction of heterologous DNA into an *S. pyogenes* host, and has opened the door for a large number of different techniques for mutagenesis and allelic exchange that will be described in greater detail below. The breakthrough technology that began the era of transformation for *S. pyogenes* was the introduction by several companies of reasonably priced instruments for the introduction of DNA by electroporation. As is true for most methods of transformation, success with electroporation-based transformation results from careful attention to growth conditions. Electroporation of streptococci generally requires that the cell walls be weakened. Most successful methods have adapted the techniques originally developed by Dunny and colleagues (15a) for electroporation of the enterococci. This method uses cells from the early exponential stages of growth in medium sup-

plemented with glycine. The addition of glycine is thought to contribute to a decreased level of cross-linking in the cell wall, and the exact stage of growth and concentration of glycine is determined empirically. (For a detailed description of the method, see reference 8.) Alternative conditions have been described by Simon and Ferretti (58), and a modified medium developed by Husmann et al. (32) has proven useful for transformation of strains that have difficulty growing in the more widely used media.

It is of interest to note that genomic sequences of several *S. pyogenes* strains of the M1, M3, and M5 serotypes have revealed that although they have some genes required for DNA uptake by natural transformation, they lack many genes thought to be critical for this process, notably *comAB*. However, a cluster of genes called the Sil locus, which is associated with an ability to cause disseminated infection, contains genes highly homologous to *comAB* (31). To date, the Sil genes have been found only among serotype M14 and M18 strains, and because its G+C content is markedly lower than the rest of the *S. pyogenes* chromosome, it may have been more recently acquired by these lineages from horizontal transmission (31).

PLASMID TECHNOLOGY

The development of transformation systems allowed the modification of *S. pyogenes* hosts through the introduction of plasmids proficient for episomal replication in *S. pyogenes*. The most used strategy has involved shuttle type plasmids, where the plasmids are first manipulated using standard technologies in *Escherichia coli*, purified from *E. coli*, and then used to transform an *S. pyogenes* host for analysis of the properties of the resulting strain. The main applications for this strategy have been to use the plasmids in complementation studies in which the wild-type allele of the gene of interest is introduced in *trans* into an *S. pyogenes* host that contains a defined mutation in the target gene and for expression of an *S. pyogenes* gene in a heterologous host. In this latter case, heterologous expression is utilized in functional studies to validate the contribution a given gene confers to a given phenotype. The heterologous host can be an *S. pyogenes* strain that naturally lacks the phenotype under investigation or can be another streptococcal or enterococcal species. In one of the first examples of this strategy, Scott et al. (56) introduced a copy of the gene that encodes the M6.1 protein (*emm6.1*) on a mobilizable plasmid vector, used this to transform a naturally competent streptococcal species, and then mobilized the plasmid by conjugation into an *emm*-deficient *S. pyogenes* host. Analysis of the resulting strain demonstrated that the introduction *emm6.1* converted the host strain from being sensitive to killing in a bactericidal assay to being resistant to killing in the assay. It should also be noted that using plasmid-based gene transfer techniques and the mutagenesis techniques described below, the contribution of the M6.1 protein to the phenotype of resistance to killing by phagocytic cells was the first formal application of the molecular Koch's postulates in *S. pyogenes* (49). Additional applications using transforming vectors for expression in heterologous hosts have included the demonstration that protein F is sufficient to confer a fibronectin-binding phenotype to non-fibronectin-binding *S. pyogenes* (29) and enterococcal hosts and that the *has* operon is sufficient to confer the ability to produce hyaluronate to both acapsular *S. pyogenes* and enterococcal hosts (14).

The strategies outlined above make extensive use of *E. coli* molecular biology to manipulate the plasmid prior to its introduction and analysis in *S. pyogenes*. Therefore, the most useful vectors have been those with the ability to replicate in both *E. coli* and *S. pyogenes*. The plasmid vectors that have been most commonly used have been based on one of two different replicons. The first of these are the various vectors that have been devised based on the pWV01 replicon and its relatives (35). This plasmid was isolated from a lactococcal species and is the prototype member of a family of "promiscuous replicons" that have the remarkable property that they can replicate in both gram-negative (*E. coli*) and gram-positive (*S. pyogenes*) hosts. This is a unique property that is not shared by other gram-positive or gram-negative-derived plasmids. In fact, many plasmids of gram-positive origin have not been reported to replicate reliably in *S. pyogenes*. These include vectors based on staphylococcal replicons, including pC194, pE194, and pUB110, that have been useful for genetic analysis in many other gram-positive species. The second class of commonly used vectors in *S. pyogenes* have been based on the pAMβ1 replicon originally isolated from *Enterococcus faecalis* (12). Unlike pWV01, the pAMβ1 replicon cannot replicate in *E. coli*, and this requires that the vector also contain an origin of replication that is proficient in an *E. coli* host. The plasmid pAT28 developed by Trieu-Cuot et al. (60) is typical of the pAMβ1-derived vectors and contains the high-copy pUC replicon derived from ColE1 for replication in *E. coli* and several other useful features, including the multiple cloning site and the *lacZα* reporter gene of pUC18, which allows screening for insertion of cloned DNA by α complementation in *E. coli* (60). Several other significant differences exist between the pWV01- and pAMβ1-derived vectors. Most notable among these is the fact that pWV01 replicates using a rolling-circle-type mechanism more characteristic of the single-stranded bacteriophages and likely exists in a single-stranded conformation for extended periods of time in replicating cells. In contrast, pAMβ1 replicates via a double-stranded θ-type mechanism. Also, as a general rule, pWV01-based vectors transform *S. pyogenes* at a much higher efficiency than do pAMβ1-derived vectors. Because of their different modes of replication, this difference may involve how the restriction system of *S. pyogenes* recognizes the two different plasmids. However, it should be noted that very little is currently understood about restriction in *S. pyogenes*, and that, in general, it has not posed a significant barrier to the introduction of DNA purified from any number of different *E. coli* K-12 hosts. Despite its lower transformation efficiency, the pAMβ1-derived vectors have proven very useful in obtaining expression of cloned genes that were only poorly expressed by pWV01-based vectors (21).

An alternative to using replicating plasmids to express genes has recently been developed (42). This vector system contains an *E. coli* origin of replicon to facilitate manipulation in *E. coli* and contains the integrase gene (*int*) and attachment site (*attP*) from the T12 temperate bacteriophage of *S. pyogenes*. Following its introduction into *S. pyogenes*, *int* catalyzes the site-specific recombination of the entire circular molecule into the chromosomal phage T12 attachment site (*attB*), which is located in a serine tRNA gene (42). Because the vector lacks the phage excisionase, integration is highly stable. This vector should be very useful under circumstances when it is desirable to study the gene in question at low copy number.

A fourth plasmid has been developed that has become very useful in genetic manipulation of *S. pyogenes*. This is a derivative of a pWV01-type replicon that is temperature

sensitive for replication in both *E. coli* and *S. pyogenes*. Developed by Maguin and coworkers (38), the plasmid is now known as pG+host4 and contains 4 amino-acid substitutions in the RepA protein, which is responsible for nicking one DNA strand at the plus origin to initiate replication. Other derivatives of this plasmid include pJRS233 (50) and pG+host5 (6), which respectively add low-copy (pSC101) and high-copy (ColE1) *E. coli* origins of replication so that they are not temperature sensitive for replication in *E. coli*. In addition, pJRS233 includes the multiple cloning site of pBluescriptSK+. The modification to the standard electroporation conditions described by Perez-Casal et al. (50) greatly increases the transformation efficiency of these plasmids in *S. pyogenes*. The main application of the pG+host-derived plasmids has been in the construction of in-frame deletion alleles of chromosomally encoded genes, and in the delivery of transposable elements for mutagenesis (see below).

DIRECTED MUTAGENESIS

The techniques developed for directed mutagenesis (so-called reverse genetics) in *S. pyogenes* have proven invaluable for the analysis of the contributions that specific genes make to pathogenesis. In these techniques, a defined mutation is constructed in order to inactivate a specific gene in order to construct an isogenic mutant for analysis. These techniques have become particularly useful with the availability of the first complete genome sequence of *S. pyogenes* strain ATCC 700294 (GenBank accession no. AE004092).

Allelic Replacement Using Linear DNA

The allelic replacement method of mutagenesis is straightforward and first involves cloning the gene of interest into any suitable plasmid vector in *E. coli*. With the available genome information, this has become a simple task of analyzing the sequence and designing suitable primers for PCR amplification, followed by insertion of amplification product into an *E. coli* cloning vector. As a general rule, *S. pyogenes* DNA is most stable when cloned on low-copy vectors in *E. coli*, and several different vectors have been used with great success. The only restriction is that the vector must not contain a gene that encodes a β-lactamase. Because (i) this DNA will eventually be introduced into *S. pyogenes*, (ii) no *S. pyogenes* isolate containing a β-lactamase has yet been described in nature, and (iii) β-lactam antibiotics are the drug of choice for treatment of *S. pyogenes* infections, it is unethical to introduce a β-lactamase into *S. pyogenes*. Also, as a general rule, the larger the fragment amplified, the greater the frequency with the allelic exchange will occur, although this method has been successful with fragments in the 1-kb range. Once the gene of interest has been cloned, its sequence is examined for a suitable unique restriction site. A selectable marker is then introduced into this site to construct the mutant allele. The markers that have been most successful have been those derived from gram-positive organisms that can also be used for selection in *E. coli*. They include *ermAM* (resistance to erythromycin) (GenBank accession no. M20334), *aphA3* (resistance to kanamycin) (GenBank accession no. V01547), *tetM* (resistance to tetracycline) (GenBank accession no. X92947), and *aad9* (resistance to spectinomycin) (GenBank accession no. M69221). A modified omega mutagenesis cassette constructed by introduction of the *aphA3* kanamycin-resistance determinant into the original omega interposon (ΩKm-2) has been widely used in this mutagenesis technique (48).

The advantages to the use of this element is that it has a cassettelike structure which facilitates its manipulation and strong transcription and translation termination signals so that it generates a strong polar mutation. The plasmid which contains this mutant allele is then converted to a linear molecule by digestion with a restriction endonuclease that cuts only in the vector sequences. The linear molecule is then used to transform the target *S. pyogenes* host with selection for the inserted resistance marker. Because the introduced molecule is linear, preservation of circular chromosomal structure requires that all resistant transformants arise by two homologous recombination events flanking each side of the inserted resistance marker. The end result is the exchange of the wild-type allele for the inactivated allele (Fig. 1). Although this type of recombination does not occur at high frequency, the method is usually successful because the frequency of nonhomologous recombination is typically extremely low. Insertion into the expected locus is confirmed using Southern blot and PCR-based analyses, and the resulting mutant can be subjected to functional studies. However, in interpreting the resulting functional data, it should be kept in mind that this method of mutagenesis generates strong polar mutations. Thus, expression of distal genes will also likely be affected should the gene of interest be located in a polycistronic transcriptional unit.

A variation of this technique can be used when a convenient restriction site is not located within the target gene or when it is desirable to subject a large cloned region to rapid, high-resolution mutagenesis. This is desirable in the latter case because genes for complex phenotypes and pathways are often clustered together on the chromosome. This method is a technique for shuttle mutagenesis that uses a version of the *E. coli* transposon mini-γδ (mγδ-200) that has been modified to contain a kanamycin-resistance gene that can be selected for in both *E. coli* and *S. pyogenes*. In this technique, the ability of m-γδ to easily and efficiently generate a large number of random transposon insertions into a segment of DNA cloned on a plasmid is used to obtain a series of mutations along the entire length of cloned streptococcal DNA. Mutated plasmids containing insertions at desired locations are then crossed into the streptococcal chromosome as described above. This method has been particularly useful for the identification of regulatory genes linked to a target gene of interest (22). For a detailed description of this technique, consult Hanski et al. (30).

Directed Insertional Inactivation

Perhaps the most commonly used technique for directed mutagenesis utilizes mutations that are constructed as the result of a single homologous recombination event. For this technique, instead of cloning a large segment of chromosome that includes the target gene, an internal segment of the target gene is cloned, and this segment should not include either the 5' or 3' ends of the target gene. A selectable marker for *S. pyogenes* is also included on the plasmid,

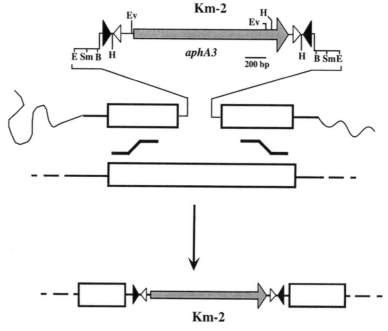

FIGURE 1 Strategy for allelic replacement mutagenesis using an omega interposon. The ΩKm-2 interposon contains a kanamycin-resistance gene that can be selected for in both gram-positive and gram-negative organisms. Additional features of ΩKm-2 include a cassettelike structure with several convenient restriction sites and strong transcription (open triangles) and translation (closed triangles) termination signals such that insertion of the element into a gene cloned in *E. coli* (top half of the figure) is strongly polar. The *E. coli* plasmid vector is converted to a linear molecule by digestion outside of the cloned streptococcal gene (vector DNA is represented by the nonstraight lines) and introduced into *S. pyogenes* with selection for resistance to kanamycin. Recombination between homologous sequences (indicated by the lines between introduced DNA and chromosome) results in the replacement of the ΩKm-2-inactivated allele for the chromosomal allele (shown below the arrow). Abbreviations: E, *Eco*RI; Sm, *Sma*I; B, *Bam*HI; H, *Hind*III; Ev, *Eco*RV.

but in this case it is located adjacent to the cloned segment rather than introduced to interrupt the cloned segment. The resulting plasmid is purified from E. coli and used to transform S. pyogenes with selection for resistance to the introduced marker. Because the commonly used E. coli replicons do not replicate in S. pyogenes, resistant transformants most frequently arise as a result of a single homologous recombination event between the internal segment of the gene introduced on the plasmid and the identical sequence on the streptococcal chromosome. The resulting chromosomal structure is generated by integration of the entire plasmid and consists of a partial duplication of the target gene, which now flanks the integrated plasmid sequences (Fig. 2A). As a consequence of using only an internal fragment of the target gene, one of the duplicated copies will lack its 3' end and the other will lack its 5' end; thus, both partial copies should be inactive. Thus, the end result is the directed insertional inactivation of the target gene. A number of modified E. coli plasmids have been developed to simplify this mutagenesis technique (46, 52, 59).

The principal advantage to this method is that single recombination events occur at higher frequency than double recombination events. In addition, this approach also requires the cloning of a much smaller segment of the chromosome, and can work with fragments as small as 0.5 kb. However, there are several disadvantages to the method. These include the fact that the mutations are not necessarily stable, because a second recombination event between the duplicated gene segments can result in excision of the plasmid vector and regeneration of a wild-type structure. Also, the mutations generated by this technique are also strongly polar. However, it is possible to modify the technique to directly address the issue of polarity. In this case, an additional control strain is constructed: instead of cloning a segment internal to the target gene, a segment is cloned that has its 5' end anchored internal to the gene but its 3' end anchored at a location distal to the 3' end of the coding region of the gene. Integration of this construction into the target gene in a wild-type strain also results in a partial duplication. However, because the cloned segment overlaps the 3' end of the gene, the first copy of the gene will be regenerated and the integrated vector will be located adjacent to this intact copy such that it will still be polar on expression of any distal genes (Fig. 2B). If the phenotype under analysis is the result of a polar effect and the loss of expression of a distal gene, then this control strain should also demonstrate the mutant phenotype, even though it has an intact and functional copy of the target gene. On the other hand, an unaltered wild-type phenotype in the control strain indicates that the mutant phenotype is solely the result of insertional inactivation of the target gene.

An additional modification of this method can be used to map the promoter and cis-acting control regions of a target operon. This is based on the technique developed by Piggot and colleagues (50a) for analysis of promoters in Bacillus subtilis and is similar in concept to the method described above for placing a polar insertion downstream of a target gene. However, in this technique, the segment of DNA cloned into the integrational vector is anchored within the coding region at its 3' end but is anchored upstream of the start of the gene at its 5' end. Integration of this construct into the target locus will again generate duplication. However, because the 3' end of the cloned segment ends within the gene, the first copy will be truncated and will be inactive. This is followed by the integrated vector and then by the second intact copy of the target gene, which is now preceded exactly by the 5' flanking region cloned into the vector. If this 5' flanking region includes the promoter and other cis-acting control regions, the target gene will be expressed. If this segment lacks these elements, the target gene will not be expressed (Fig. 2C). By using a nested set of insertions containing different lengths of the upstream control region, it is possible to map the sequences required for expression and regulation of the target gene with a high degree of precision. This method has been used to examine regulation of mga, the regulator of the genes that encode the M proteins and the C5a peptidase (46).

In-Frame Deletion

An in-frame deletion mutation has as the advantage that it is both stable and is nonpolar. It has as the disadvantage that it is somewhat more time-consuming to construct. In-frame deletions have now been constructed in the genes for many different S. pyogenes virulence factors, including the C5a peptidase (33), the M protein, the cysteine proteinase, the has operon (3), and streptolysin O (55). Most in-frame deletion mutations have been constructed by the method pioneered in the Cleary lab (33) that utilizes derivative of the temperature-sensitive plasmid pG+host4. In this method, the gene of interest is cloned into the vector in E. coli. Molecular techniques are then used to delete a large central region of the gene, but at the same time preserving its reading frame. The resulting construct is introduced into an S. pyogenes host at a temperature that is permissive for replication of the plasmid. The culture is then shifted to the nonpermissive temperature while maintaining selection for the resistance determinant of the plasmid. This selects for chromosomes into which the nonreplicating plasmid has inserted by homologous recombination between the in-frame deletion allele and the resident wild-type allele. At this stage the chromosome will now contain both the wild-type and deletion alleles. At some frequency, a second homologous recombination will occur which will result in the excision of the integrated plasmid. Depending on the recombination junctions, either the wild-type or the deletion allele will remain behind in the chromosome (Fig. 3). Shifting the culture back to the permissive temperature enriches for chromosomes from which the plasmid has excised, because replication of the integrated plasmid creates a second origin of replication for the chromosome which is usually deleterious to growth. The culture is once again shifted to the nonpermissive temperature, but this time in the absence of selection for the plasmid. This enriches for segregants that have lost the excised and now nonreplicating plasmid. These segregants are plated as single colonies, which are then examined by PCR to identify chromosomes that contain the in-frame deletion allele. For a detailed description of this technique, see Ji et al. (33).

There are several drawbacks to this technique for in-frame deletion. First, it can be time-consuming because it requires that cultures undergo multiple passages at nonpermissive and permissive temperatures. Second, multiple in vitro passage has a tendency to enrich for spontaneous mutants with reduced expression of virulence factors, including the M protein and the hyaluronic acid capsule (2). If this proves to be a significant problem for the strain under analysis, a two-step method for allelic replacement can be used by introducing the counter-selectable rpsL marker into the temperature-sensitive vector (40). This gene encodes ribosomal protein S12, and mutations in rpsL that confer resistance to streptomycin arise spontaneously in S. pyogenes at a frequency of $\sim 10^{-9}$ cell^{-1} generation^{-1} (56).

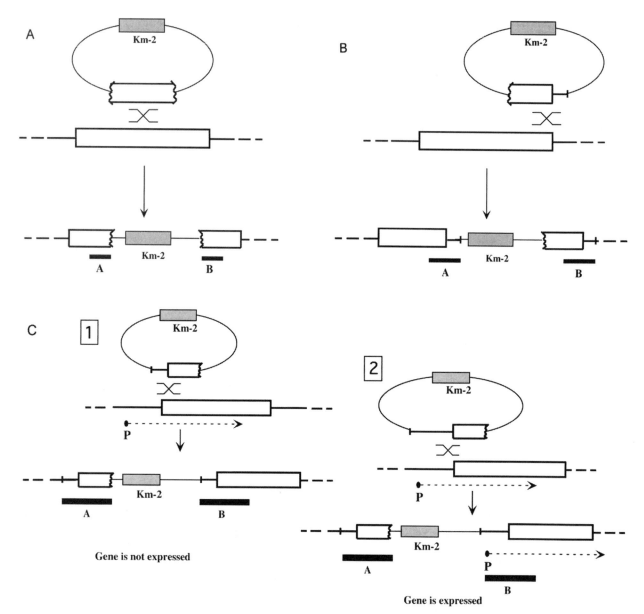

FIGURE 2 Directed insertional mutagenesis of targeted genes. (A) A DNA segment internal to the targeted gene (shown by the box enclosed by wavy lines) is cloned onto an *E. coli* plasmid that cannot replicate in *S. pyogenes*. The plasmid is introduced into an *S. pyogenes* host as a circular molecule (top of figure) with selection for a resistance marker on the plasmid (ΩKm-2). A single homologous recombination event between the chromosome and the circular molecule (shown by the "X") results in the integration of the plasmid into the chromosome and a partial duplication of the gene (the solid bars labeled A and B represent the duplicated segment) in which neither of the two copies is complete. (B) Generation of a polar insertion 39 to the target gene. If the segment of DNA cloned on the integrational plasmid contains sequences that include the 3′ terminus of the target gene, the resulting structure will also contain a partial duplication gene (the solid bars labeled A and B represent the duplicated segment), but in this case the 5′ copy will be intact and will now be flanked at its 3′ end by a polar element. This strategy can be used to test if the insertion generated in the target gene (see above) is polar on distal genes. (C) Mapping the *cis*-acting control regions of the target gene. In this strategy, the technique is modified by including different regions of DNA extending 5′ to the target gene. The plasmid is integrated into the target locus as described above, and the end product is also a partial duplication of the target gene (the solid bars labeled A and B represent the duplicated segment); however, it is the distal copy that is intact. If this region does not include the *cis*-acting control regions (represented by the broken arrow and the closed circle labeled P), the distal intact copy will not be expressed (scenario labeled 1). In contrast, if the cloned segment includes the *cis*-acting control regions, then the distal copy will be expressed (scenario labeled 2).

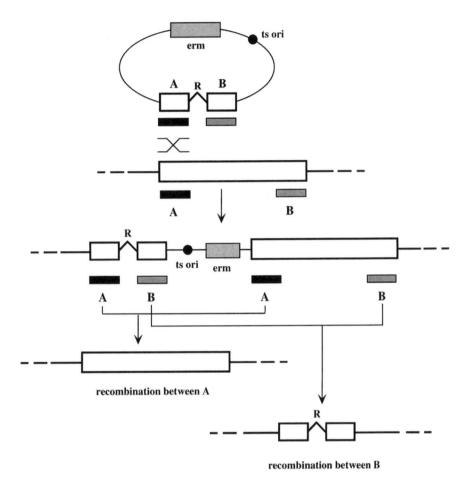

FIGURE 3 Construction of an in-frame deletion. Standard PCR-based methods are used to generate a deletion of the internal region of a copy of the target gene that has been cloned on an *E. coli*-streptococcal shuttle vector that is temperature sensitive for replication (*ts ori*). The deletion is constructed so as to maintain the reading frame of the gene (represented as the bent line connecting the 5′ region labeled A and the 3′ region labeled B). After its introduction into *S. pyogenes*, growth at a temperature nonpermissive for replication of the plasmid with selection for the antibiotic-resistant determinant of the plasmid (*erm*) selects for chromosomes in which the plasmid has integrated by homologous recombination (X). The two regions of homology flanking the deletion are represented by the solid and gray bars labeled A and B. Recombination between the A regions is shown, and the product is shown below the first arrow. A second homologous recombination event can occur between the 5′ homologous regions (labeled A) or the 3′ homologous regions (labeled B), which results in excision of the plasmid and either restoration of the wild-type structure or replacement by the deletion allele (these products are illustrated below the second set of arrows). Growth at a temperature permissive for replication of the plasmid enriches for chromosomes from which the plasmid has been excised. Presence of the wild-type or deletion allele in any one isolate is easily determined by assay for the unique restriction site engineered into the deletion allele (indicated by the R above the bent line).

However, the wild-type allele is dominant, so a partial diploid containing wild-type *rpsL* on the integrated temperature-sensitive vector and mutant *rpsL* in its chromosome will be sensitive to streptomycin. The inclusion of streptomycin during the final passage at the nonpermissive temperature directly selects for chromosomes from which the integrated plasmid has excised and segregated. This method has been used to introduce mutant alleles of the virulence regulator Mga into the chromosome for functional analyses (40).

TRANSPOSON MUTAGENESIS

The techniques described above have been extremely useful for the analysis of virulence in *S. pyogenes*. However, they required a detailed preexisting knowledge of the target gene. Although the availability of genomic information is very helpful for analysis of the chromosomal structure of known genes and for identification of homologues of known virulence genes from other organisms, the directed mutagenesis approach is most successfully applied to testing hypotheses which address the functions of previously identified genes. It is not well suited for the identification of novel genes that contribute to virulence. This is most successfully done using a traditional genetic approach, in which a virulence phenotype is established and random mutations are generated to identify genes that influence this phenotype. In *S. pyogenes*, the most successful method of random mutagenesis has used the insertion of a transposable element at multiple loci in the chromosome. This section will consider the various types of transposons that have been successfully used to identify novel genes in *S. pyogenes*.

Tn*916*

The first transposable element used to identify novel virulence genes in *S. pyogenes* was Tn916 (Fig. 4). This element is the prototype of the family of transposons discovered by the group of Clewell (12) that are now known as the conjugative transposons. Conjugative transposons have the remarkable property that they are self-transmissible and transpose from a locus in a donor chromosome to a different locus in a recipient chromosome, when these chromosomes are located in different cells. The conjugal transfer event requires cell-cell contact and can occur between very distantly related species and even between gram-positive and gram-negative hosts. This biology of these unique elements has been extensively investigated and is the subject of several excellent reviews (e.g., reference 57), and so will be only briefly mentioned here. The transposon moves by an excision-insertion type mechanism in which the first step is the excision of the element from the donor locus by a λ-like pathway that involves a recombination event between sequences adjacent to the ends of the element. However, in contrast to movement of λ in which the substrate sequences must be identical, the substrate sequences used by Tn916 are almost always nonhomologous. The circular intermediate is nicked at a discrete site and a single strand is directionally transferred into the donor cell. Transfer requires direct cell-cell contact and may involve some degree of zygote formation. This may explain how, although Tn916 does not appear to mobilize chromosomal markers, transfer of unlinked chromosomal markers can occur.

The ability of the recombination reaction to occur between nonhomologous sequences allows the element to utilize a large number of different sites for insertion. However, there are preferred target sites for insertion. These preferred sites do not necessarily share the same specific sequence, but likely share a similar conformation and are the sites of bent DNA. Interestingly, among the insertions

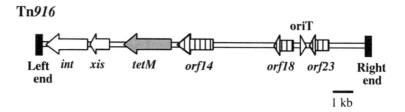

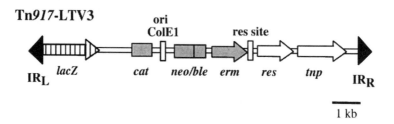

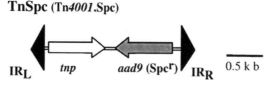

FIGURE 4 Transposon mutagenesis of *S. pyogenes*. The several transposons that have been used for mutagenesis of *S. pyogenes* are shown. Antibiotic resistance genes are represented by the gray bars; transposon ends and/or terminal inverted repeats are shown in black; genes that are essential for transposition are shown by white bars; and genes that are nonessential for transposition are represented by the striped bars. Tn916 (GenBank accession no. U09422) is the prototype conjugative transposon and contains at least 24 different open reading frames. Of these, one is required for resistance to tetracycline, two are essential for transposition, and the rest are likely involved in conjugal transfer. Tn917-LTV3 is a highly engineered derivative of the Tn3-like transposon Tn917. This element transposes via a replicative mechanism and has been modified to include a promoterless *lacZ* reporter gene to generate random transcriptional fusions and an *E. coli* ColE1 plasmid origin of replication to facilitate the cloning and analysis of inactivated loci. TnSpc is a derivative of the Tn5-like transposon Tn4001. This element transposes via a cut-and-paste mechanism and consists of the left and right inverted repeats and transposase of IS256 and a spectinomycin resistance gene.

in *S. pyogenes* that have identified virulence genes, a large percentage are not in the coding region of the gene but have been located 5′ to the coding region in the transcriptional control regions of the genes. From a practical standpoint, mutagenesis with Tn*916* requires a donor strain and an antibiotic-resistant derivative of the *S. pyogenes* host of interest. The donors can be distantly related to *S. pyogenes* and are most often either *E. faecalis* or *B. subtilis*, and it is usually a simple procedure to select a spontaneous high-level streptomycin-resistant mutant of the *S. pyogenes* strain under analysis. (For a detailed method for Tn*916* mutagenesis of *S. pyogenes*, consult Caparon and Scott [8].)

The advantage to using Tn*916* is that it can be applied to probably any isolate of *S. pyogenes*, including those that may be difficult to transform. The disadvantages are that Tn*916* is a large element for a transposon (more than 18 kb in size), does transpose into preferred sites, and frequently generates sites with multiple insertions. This latter issue can often be resolved through analysis of the resulting mutants by transduction (see above). In addition to mutagenesis, Tn*916* has been engineered as a vector for the introduction of foreign DNA into *S. pyogenes*, including promoter-reporter gene fusions for the analysis of gene expression. The transposon has served as a useful mutagenesis agent and has been used to identify novel genes involved in regulation, e.g., *mga* (7), expression of streptolysin S (4, 36, 45), and production of the capsule (15, 62), among others.

Tn*917*

Tn*917* was first described by Clewell's group in *E. faecalis* (59a), and has been highly engineered by Youngman and colleagues and applied with great success to the mutagenesis of *Listeria monocytogenes* and *B. subtilis*. This transposable element is a member of the large Tn*3* family (which also includes the γδ element of *E. coli* discussed above). The characteristic of this family is that they contain fairly large terminal repeats, although they are much smaller in size than the conjugative transposons. They produce stable insertions, generate a 5-bp target-site duplication upon insertion, and transpose at high frequency and with a high degree of randomness. Their movement involves a replicative pathway, which requires that donor and recipient molecules undergo replication, and proceeds through the cointegration of donor and host molecules, which contain two directly repeated copies of the transposon at the fusion junctions. Resolution of the cointegrate occurs via site-specific recombination catalyzed by a transposon-encoded enzyme called resolvase that recognizes a specific site in the each copy of the transposon (the *res* site). The end result is that a single copy of the transposon is located at a random site in the recipient molecule.

The requirement for replication means that the transposon must be delivered into the target *S. pyogenes* host on a replicating plasmid. A technical problem is that the plasmid must have a counter-selectable marker so that chromosomes that have obtained an insertion of the transposon can be distinguished from hosts in which the transposon remains on the plasmid. This problem was solved with the advent of temperature-sensitive plasmid vectors for *S. pyogenes*. Eichenbaum and Scott (17) developed a mutagenesis vector (pJRS290) (16) that is based on pG+host4 and contains a highly engineered derivative of Tn*917* developed by Youngman and coworkers (Tn*917*-LTV3) (6a) (Fig. 4). This derivative of Tn*917* has several useful features, including an *E. coli* plasmid origin of replication that facilitates the direct cloning of chromosomal DNA adjacent to the inserted transposon. Using this vector, Tn*917*-LTV3 was found to transpose efficiently and with a high degree of randomness in *S. pyogenes*.

Tn*4001*

The transposable element Tn*4001* was originally isolated as an agent of transmissible gentamycin-resistance in *Staphylococcus aureus*. It is a composite-type transposon and a member of the Tn*5* family. These transposable elements share a common structure in which directly repeated copies of an insertion sequence flank an antibiotic-resistance gene. Each insertion sequence itself is bounded by a short inverted repeat encoded by the transposase gene, which is the only transposon-encoded gene required for movement. The insertion sequences themselves are usually capable of transposition independent of the entire element. For Tn*4001*, two copies of IS*256* flank a central gentamycin-resistance gene. The pathway of transposition of this class of elements involves a cut-and-paste mechanism catalyzed by transposase, which recognizes as its substrates the short inverted repeats of the insertion sequences. Studies have shown that Tn*4001* has a broad host range, including staphylococci, oral streptococci, and mycoplasmas, and that it chooses its targets for insertion with a high degree of randomness. Because the cut-and-paste pathway does not require replication of the donor molecule, the element can easily be delivered into the host of choice via a nonreplicating suicide vector. However, there have been a number of problems which have limited its use. The most significant of these is that transposition of the entire element occurs at a somewhat low frequency relative to the frequency of transposition of the individual insertion sequences. As a result, a chromosome with a Tn*4001* will frequently also have multiple copies of IS*256* inserted at other loci. To address this problem, derivatives of Tn*4001* have been constructed that essentially contain only the inverted repeats of IS*256* oriented to flank the gene for transposase and a selectable marker (37). The resulting elements are relatively small in size (~2 kb) and transpose at a high frequency characteristic of an individual copy of IS*256*. The organization of the element also prevents independent transposition of the insertion sequence and ensures that the resulting population of insertions is homogeneous, although a significant percentage of the resulting mutants may have insertions at two different loci. A spectinomycin-resistant version of this transposon (called TnSpc for simplicity) (Fig. 4) was used to identify three novel genes required for expression of the cysteine protease of *S. pyogenes*, including a transcriptional regulator (*ropB*) and a chaperone (*ropA*).

The introduction of unique DNA sequences of different sizes into TnSpc has been used to produce an element for a modification of the Signature-tag mutagenesis technique christened polymorphic tag length transposon mutagenesis (31). Amplification by PCR of the tags in any pool of mutants allows the detection of specific mutant chromosomes because they will have an amplified tag of unique length. This allows for comparison between an input pool of mutants and an output pool recovered following infection to identify specific mutants that were selected against through competition with other members of the pool in vivo. This technique was used to identify the Sil locus, which can contribute to the ability of *S. pyogenes* to disseminate from a site of local tissue infection (31).

Because the terminal inverted repeats are quite short (26 bp for IS*256*), several specialized derivatives of the transposon have been developed that place a reporter gene

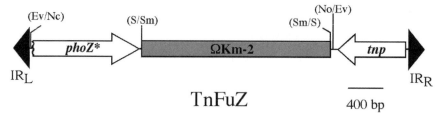

FIGURE 5 The structure of TnFuZ. The element contains the left and right inverted repeats (IR$_L$ and IR$_R$, respectively), the transposase (*tpn*) of IS256, and the *ahpA3* kanamycin resistance determinant contained on Ωkm-2. The gene encoding the *E. faecalis* alkaline phosphatase (*phoZ*) was altered by removal of the region that encoded its signal sequence, and the modified gene (*phoZ**) was introduced into the element as shown. Abbreviations: C, *Cla*I; Ev, *Eco*RV; Nc, *Nco*I; P, *Pst*I; S, *Sal*I; Sm, *Sma*I. A slash indicates a junction of two restriction fragments joined during construction of the element, and restriction sites enclosed by parentheses are inactive.

adjacent to the end. The TnFuZ (fusion to *phoZ*) element places a version of the *E. faecalis phoZ* alkaline phosphatase adjacent to the transposon inverted repeat (Fig. 5). This reporter has been modified by removal of its export signal sequence. Insertion into an open reading frame encoding an export signal that results in formation of a translation fusion with PhoZ will now direct export of the chimera if the open reading frame encodes a secretion signal. Because PhoZ is active only following its export across the cytoplasmic membrane, a simple and rapid colorimetric screen can be used to identify colonies arising from a mutant that contains an insertion in a gene that encodes a secreted protein. The TnFuZ element has been used to identify secreted proteins whose expression is enhanced during aerobic growth (25). Another example of the versatility of IS256-based elements is those that have been modified to include genes encoding a bacterial luciferase. Insertion into a transcription unit that directs transcription of the luciferase genes allows gene expression in vivo to be monitored noninvasively using specialized imaging cameras that can capture photons emitted by the streptococci in living host tissue (23).

Transposome Mutagenesis

In transposome mutagenesis, a stable complex between transposase and transposon DNA is formed in vitro and then introduced into a bacterium. The Mg^{2+} in the bacterial cytoplasm activates transposase that then catalyzes transposition of the element to some random location in the bacterial chromosome. Recent work has shown that this method can be adapted for efficient mutagenesis of *S. pyogenes* (11). In this method, a streptococcal erythromycin-resistance gene was introduced between the inverted repeats of Tn5 on the *E. coli* plasmid pMOD-2 (Epicentre Technologies) to construct pMOD-2::Ω*erm* (Fig. 6). Following digestion with a restriction endonuclease (*Pvu*II) (Fig. 6), the transposon (TmErm)-containing linear DNA fragment was purified and reacted with a modified hyperactive tranposase (EZ::TN^{TM}, Epicentre Technologies) to generate the transposome. Following transformation of *S. pyogenes* using electroporation, erythromycin-resistant colonies were obtained at frequencies between 10^3 and 10^4 CFU/μg of DNA (11). Mutations generated by this method have proven to be very stable even in the absence of erythromycin selection, possibly because the inserted transposon does not encode its own transposase.

Analysis of Transposon Mutants

Regardless of the type of transposable element utilized, a number of additional tests should be performed to ensure that the mutant phenotype is the direct result of insertion of the transposon. This involves determining the sequence of the locus into which the transposon has inserted. Because the transposons commonly used in *S. pyogenes* contain antibiotic resistance markers that can also be used for selection in *E. coli*, it is relatively simple to obtain a clone of the insertion locus. This is done by construction in *E. coli* of a plasmid library of the chromosome of the mutant with direct selection for the transposon-encoded marker. It is also possible to obtain the sequence flanking the inserted transposon using one of a number of commonly used inverse and vectorette PCR techniques or by direct sequencing of chromosomal DNA (for examples, see Lyon et al. [37] or Gibson and Caparon [25]). The availability of significant

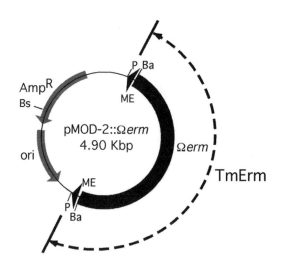

FIGURE 6 The structure of pMOD-2::Ω*erm* containing the novel transposon TmErm. The transposon TmErm contains the right and left inverted repeats (mosaic end, ME) modified from the inverted repeats of Tn5, and an erythromycin resistance marker (Ω*erm*). To apply the transposome mutagenesis to *S. pyogenes*, the DNA fragment containing transposon TmErm was purified after digestion with *Pvu*II. Thus, the ampicillin resistance determinant contained on pMOD-2::Ω*erm* was removed before the transposome mutagenesis. The purified transposon TmErm was then reacted with a transposase (EZ::TN, Epicentre Technologies) in vitro to form the transposome that is then used to transform *S. pyogenes* using electroporation. Several restriction sites in the plasmid are indicated: P, *Pvu*II; Ba, *Bam*HI; Bs, *Bsa*I.

amounts of genomic sequence means that is usually necessary to generate only a small amount of sequence data that can then be compared to the genome sequence database to obtain large amounts of sequence information. With these data in hand, the next step is to construct a mutant in the locus in the wild-type parental strain using one of the strategies for directed mutagenesis outlined above. If the original insertion is responsible for the mutant phenotype, then it is expected that all isolates obtained via directed mutagenesis will also be mutants. It should also be kept in mind when interpreting results that transposon-generated mutations are polar. It is also possible to conduct a linkage analysis of Tn916 insertions without first cloning the locus by transducing the insertion locus into a wild-type strain (see above). In the case of TnSpc, it is possible to use PCR techniques to delete transposase after the insertionally inactivated locus has been cloned into E. coli (37). This prevents the element from further transposition and allows the mutant locus to be crossed back into a wild-type background using the strategy of allelic replacement via linear DNA that was described above.

ANALYSIS OF GENE EXPRESSION

Gene regulation phenomena play a key role in pathogenesis. The interaction between the host and the microbe is dynamic and often is a progression through a number of discrete steps. Each of these steps is characterized by the expression of specific bacterial genes required for survival and multiplication and the host's response to the action of the products of these microbial genes. As a consequence, the microorganism is continually challenged to adapt to new and changing environmental conditions. Pathogens have taken advantage of the dynamic nature of this interaction and have evolved to recognize changes in specific environmental conditions as markers that define a particular host compartment or stage of infection. The pathogen makes use of this information to modulate expression of virulence genes required for survival in this host compartment. Thus, an understanding of the in vitro conditions that regulate expression of a specific virulence gene provides insight into how the gene contributes to virulence in vivo (for review, see reference 43). Regulation of virulence genes often involves control at the level of transcription. For S. pyogenes, transcriptional regulation has often been analyzed through the use of reporter genes.

The basic strategy for using reporter genes to analyze gene expression involves fusing the promoter for the virulence gene of interest to a reporter gene that encodes a gene whose product can easily be quantified. This feature is of particular utility when the assays for quantification of the product of the streptococcal gene under analysis are time consuming, expensive, or cumbersome. The reporter gene lacks its own promoter so that its product becomes an accurate relative indicator of the steady-state level of transcription initiation from the target promoter. It should be kept in mind that because the message is a chimera between the initiation signals of the target gene and the reporter gene, it will not be subject to the same posttranscriptional and posttranslational controls. Furthermore, the half-life of the reporter gene's translation product will also not reflect that of the native polypeptide's. Thus, reporter genes are most appropriately used to quantify the strength of initiation of transcription of the target promoter.

The reporter genes that have been successfully used in S. pyogenes include those encoding chloramphenicol acetyltransferase (Cat), β-galactosidase (Lac), β-glucuronidase (Gus), and alkaline phosphatase (Pho). These gene products are stable, highly active enzymes that are highly specific for their substrates. Their utility is enhanced by the availability of synthetic substrates with isotopic, colorimeteric, fluorogenic, or light-emitting properties that allow very sensitive detection. Also, the reporter genes encoding these enzymes are derived from gram-positive bacteria or have been extensively modified to optimize their expression in gram-positive hosts and differ extensively from their counterparts used for analysis in E. coli. The most successful applications of Cat have employed cat86, originally isolated from Bacillus pumilis that has been modified by the deletion of an attenuator sequence and the substitution of an ATG start codon for the native gene's TTG start codon. The advantages of Cat are that cat86 is very stably maintained in S. pyogenes and that the assays for its enzymatic acetylation of chloramphenicol are very sensitive (9, 41). Its disadvantages are that the most sensitive assays involve radioactivity, are expensive, are somewhat labor-intensive relative to other reporter genes, and require the preparation of cell-free extracts. Both Lac and Gus utilize E. coli genes that have been modified by the introduction of ribosome binding sites recognized by gram-positive bacteria (20, 34). The advantages of using these as reporter genes are that assays for their enzymatic activities are rapid and are easy to perform; many different samples can be analyzed simultaneously; a wide variety of different substrates is available; and that it is possible to analyze both permeabilized cells and cell extracts. Their disadvantages are that they are not as sensitive or as stably maintained in S. pyogenes as cat86. Nevertheless, gusA has proven to be very useful in numerous applications, including identification of regulatory sites in the hasA promoter (19) and location of a regulator (rocA) of covR expression (5).

A chimeric reporter protein based on fusion of two naturally secreted proteins has been developed. This reporter contains as its enzymatic partner the alkaline phosphatase of E. faecalis (PhoZ) (54) (Fig. 7). This enzyme is secreted from E. faecalis and is a lipoprotein that becomes tethered to the outer leaflet of the cell membrane. The enzymatic C-terminal domain, not including its lipoprotein secretion signal domain, is fused to the N-terminal domain of protein F of S. pyogenes. Protein F is a cell wall-associated fibronectin-binding adhesin whose N-terminal domain contains the sequence which targets the protein for secretion; its central domain contains the regions responsible for binding to fibronectin; its C-terminal domain is responsible for targeting the protein to a sorting pathway that covalently couples a sequence near its C terminus to the cell wall (47). Fusion of the N-terminal secretion domain of protein F to the C-terminal enzymatic domain of PhoZ results in a very stable, highly enzymatically active chimeric protein that, because it lacks the postsecretion attachment domains of either protein, is freely secreted from the streptococcal cell into the surrounding medium (27). This feature makes quantitative analysis of the secreted chimeric protein very simple and requires neither permeabilization nor preparation of cytoplasmic extracts. Because of the widespread popularity of alkaline phosphatase enzymes in a large number of histological and biochemical techniques, there is an excellent selection of sensitive substrates with a variety of useful characteristics, including those with fluorescent and light-emitting properties.

Green fluorescent protein (GFP) has rapidly become a widely used reporter of gene expression because of its ease of detection and its possible applications for analyses in real time and in vivo during infection. Numerous versions of GFP are available that have been modified to alter codon

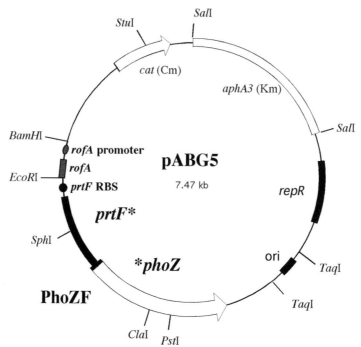

FIGURE 7 A chimeric secreted alkaline phosphatase reporter gene. The plasmid pABG5 is a pWV01-based *E. coli*-streptococcal shuttle vector whose genes encode resistance to chloramphenicol (*cat*) and kanamycin (*aphA3*). The plasmid contains a promoterless reporter gene formed by the fusion of the N-terminal region of the cell wall-associated protein F (*prtF**) to the enzymatic domain of the enterococcal alkaline phosphatase (*phoZ**). Because the chimeric protein (secreted protein F-PhoZ reporter, or PhoZF) lacks the C-terminal cell wall attachment domain of protein F and the N-terminal lipoprotein tethering domain of PhoZ, it is freely secreted from the cell. The PhoZF chimera retains the enzymatic activity of PhoZ and is easily quantified in culture supernatants. Restriction sites for *Bam*HI and *Eco*RI can be used to place the promoter of interest in an orientation to direct transcription of *phoZF*, which is then translated using the ribosome-binding site of *prtF* (RBS). The plasmid pABG5 contains the *rofA* promoter, which confers high-level expression.

usage, stability, and spectral properties (13). The GFPuv variant has been used to examine regulation of the promoter for the hyaluronic acid biosynthesis gene cluster in streptococci recovered directly from infected tissues (28). The latter study also reported an important consideration for interpreting data derived from GFP-based reporters: fluorescence is often not uniform among streptococcal cells in any population, and many cells express no detectable fluorescence (28).

Numerous strategies have been used to introduce reporter fusion constructs into *S. pyogenes* for analysis. Plasmid vectors have the advantages that they are easy to manipulate and can be used to generate and analyze a large number of permutations and variations of promoter structure to probe the *cis*- and *trans*-acting control regions. A plasmid-based reporter will be present in multiple copies, and this amplification will increase the sensitivity of detecting the activity of the promoter under analysis. This latter feature can also make plasmid-based systems sensitive to multiple-copy-number-derived artifacts, such as titration of *trans*-acting regulatory components, when these components are present in limiting quantities. Furthermore, any sequence-directed local effects of chromosomal structure on expression of a given promoter will not likely be replicated in a plasmid environment. Integration of reporter constructs into the chromosome can alleviate many of these latter types of potential artifacts but with a trade-off in ease of use and sensitivity. Integration has most often been performed by introducing the reporter construct on a nonreplicating *E. coli*-based plasmid to target integration into the target gene's native locus by homologous recombination promoted by the cloned promoter segment. The use of a reporter gene can be combined with the recombinational strategy described above for the analysis of the *cis*-acting regions of a promoter. Other reporter gene delivery vectors have included Tn*916*-based shuttle transposons and the incorporation of a promoterless LacZ near the end of Tn*917* for the generation of libraries with fusions in genes generated at random. Reporter genes have been successfully applied to the analysis of expression of many different *S. pyogenes* genes, including *mga* (24, 46), *rofA* (21), various *emm* genes (9, 41, 51), *prtF* (21, 61), and the *has* operon (1).

Real-time PCR has rapidly become the standard for analysis of gene expression in *S. pyogenes*. Although it requires expensive equipment, this disadvantage is more than offset by its advantages: it does not require the laborious construction of reporter gene fusions, it is fast and easy to perform, and it can directly analyze the abundance of any RNA for which suitable amplification primers can be developed. The utility of real-time PCR has been greatly facilitated by the development of technology for the simple and rapid purification of high-quality RNA from relatively small samples (10). This combination has made real-time PCR the method of choice for validation of data developed from

DNA microarray profiling of gene expression and for the analysis of *S. pyogenes* in tissue recovered from experimental infection (26). In real-time PCR, total RNA is purified and then converted to first-strand cDNA using reverse transcriptase. These cDNAs then serve as a template for real-time PCR using primers specific for the gene in question. Real-time thermocyclers are modified with a detection system that can continuously monitor the fluorescence of a probe that is sensitive to the quantity of DNA that is being synthesized by PCR. The amplification cycle at which a detectable amplification product is first detected is a quantitative measure of the concentration of that message in the input cDNA pool. Data are typically reported as a ratio to a reference gene whose expression is constant under the conditions of analysis.

Techniques for Heterologous and Ectopic Expression

Structure-function studies on the role of putative virulence factors require techniques for heterologous expression, both for controlling the timing and relative levels of expression of a given gene and to study the structure-function relationships of specific domains of a given virulence-associated protein. The former case requires the availability of an ectopic promoter whose activity can be tightly, easily, and quantitatively controlled by some external factor. The latter case requires a method for the expression of various domains of the polypeptide under analysis in the context of an unrelated polypeptide. This allows an independent analysis of the functionality of that domain separate from other regions of the original polypeptide. Methods to accomplish both of these expression strategies have been developed for analysis of virulence in *S. pyogenes*.

An approach for using a regulated heterologous promoter to direct expression of the gene under analysis has been developed that is based on the *nisA* promoter of *Lactococcus lactis*. In lactococci, this promoter controls the expression of a large multigene operon involved in the synthesis of the lantibiotic nisin. An interesting feature of the control of the *nisA* promoter is that its activity is tightly regulated and is stimulated in response to nisin. The signal transduction pathway is well characterized and includes the histidine protein kinase NisK and cognate response regulator. The method for using this regulated promoter involves the introduction of *nisK* and *nisR* into an *S. pyogenes* strain under the control of the constitutively active *nisR* promoter (17). The gene of interest is introduced under control of the *nisA* promoter, which now makes expression of the gene of interest sensitive to the concentration of exogenous nisin that is added to the culture. Tightly regulated expression of a *gusA* reporter gene and induction of up to 60-fold over background using a broad range of nisin concentrations have been reported for *S. pyogenes* (17).

The ability to express a domain of a target protein within the context of a heterologous protein is highly desirable for the structure-function analysis of adhesins. A technique for this strategy that uses the framework of the M6.1 protein has been developed. It is based on the work of Pozzi and coworkers (52a), who demonstrated that the central domain of the M protein could be replaced with a heterologous sequence without affecting expression and surface presentation of the resulting chimera. Essentially the N-terminal secretion and C-terminal attachment domains are used to display the introduced heterologous domain on the streptococcal cell surface. A successful application of this strategy has been

the analysis of the two distinct fibronectin-binding domains of protein F (47). For a more detailed explanation of this technique, see Hanski et al. (30).

CONCLUDING REMARKS

Rapid progress has been made in recent years in the development of sophisticated techniques for genetic analysis in *S. pyogenes*. Much of this effort has been directed at the development of methods for the mutagenesis of known genes. Much progress has also been made in the development of strategies for the identification of novel genes. It is likely that the widespread application of these techniques to the virulence properties of *S. pyogenes* will enrich our understanding of streptococcal pathogenesis with insight at the molecular level and will help to establish and clarify the contributions of specific genes. Additional use and development of methods for analysis of gene expression and heterologous expression will continue and allow analyses of virulence factors at much higher levels of resolution than previously possible.

REFERENCES

1. **Alberti, S., C. D. Ashbaugh, and M. R. Wessels.** 1998. Structure of the has operon promoter and regulation of hyaluronic acid capsule expression in group A Streptococcus. *Mol. Microbiol.* **28:**343–353.

2. **Ashbaugh, C. D., and M. R. Wessels.** 2001. Absence of a cysteine protease effect on bacterial virulence in two murine models of human invasive group A streptococcal infection. *Infect. Immun.* **69:**6683–6688.

3. **Ashbaugh, C. D., H. B. Warren, V. J. Carey, and M. R. Wessels.** 1998. Molecular analysis of the role of the group A streptococcal cysteine protease, hyaluronic acid capsule and M protein in a murine model of human invasive soft-tissue infection. *J. Clin. Invest.* **102:**550–560.

4. **Betschel, S. D., S. M. Borgia, N. L. Barg, D. E. Low, and J. C. S. De Azavedo.** 1998. Reduced virulence of group A streptococcal Tn916 mutants that do not produce streptolysin S. *Infect. Immun.* **66:**1671–1679.

5. **Biswas, I., and J. R. Scott.** 2003. Identification of rocA, a positive regulator of covR expression in the group A streptococcus. *J. Bacteriol.* **185:**3081–3090.

6. **Biswas, I., A. Gruss, S. D. Ehrlich, and E. Maguin.** 1993. High-efficiency gene inactivation and replacement system for gram-positive bacteria. *J. Bacteriol.* **175:**3628–3635.

6a. **Camilli, A., D. A. Portnoy, and P. Youngman.** 1990. Insertional mutagenesis of *Listeria monocytogenes* with a novel Tn917 derivative that allows direct cloning of DNA flanking transposon insertion. *J. Bacteriol.* **172:**3738–3744.

7. **Caparon, M. G., and J. R. Scott.** 1987. Identification of a gene that regulates expression of M protein, the major virulence determinant of group A streptococci. *Proc. Natl. Acad. Sci. USA* **84:**8677–8681.

8. **Caparon, M. G., and J. R. Scott.** 1991. Genetic manipulation of the pathogenic streptococci. *Methods Enzymol.* **204:**556–586.

9. **Caparon, M. G., R. T. Geist, J. Perez-Casal, and J. R. Scott.** 1992. Environmental regulation of virulence in group A streptococci: transcription of the gene encoding M protein is stimulated by carbon dioxide. *J. Bacteriol.* **174:**5693–5701.

10. **Cheung, A. L., K. J. Eberhardt, and V. A. Fischetti.** 1994. A method to isolate RNA from Gram-positive bacteria and mycobacteria. *Anal. Biochem.* **222:**511–524.

11. **Cho, K., and M. Caparon.** New streptococcal genes affecting SpeB expression and *Streptococcus pyogenes* virulence. Manuscript in preparation.

12. **Clewell, D. B., Y. Yagi, G. M. Dunny, and S. K. Schultz.** 1974. Characterization of three plasmid deoxyribonucleic acid molecules in a strain of *Streptococcus faecalis*: identification of a plasmid determining erythromycin resistance. *J. Bacteriol.* **117:**283–289.

13. **Crameri, A., E. A. Whitehorn, E. Tate, and W. P. Stemmer.** 1996. Improved green fluorescent protein by molecular evolution using DNA shuffling. *Nat. Biotechnol.* **14:** 315–319.

14. **DeAngelis, P. L., J. Papaconstantinou, and P. H. Weigel.** 1993. Isolation of a *Streptococcus pyogenes* gene locus that directs hyaluronan biosynthesis in acapsular mutants and in heterologous bacteria. *J. Biol. Chem.* **268:**14568–14571.

15. **Dougherty, B. A., and I. van de Rijn.** 1992. Molecular characterization of a locus required for hyaluronic acid capsule production in group A streptococci. *J. Exp. Med.* **175:**1291–1299.

15a. **Dunny, G. M., L. N. Lee, and D. J. LeBlanc.** 1991. Improved electroporation and cloning vector system for gram-positive bacteria. *Appl. Environ. Microbiol.* **57:**1194–1201.

16. **Eichenbaum, Z., and J. R. Scott.** 1997. Use of Tn917 to generate insertion mutatins in the group A streptococcus. *Gene* **186:**213–217.

17. **Eichenbaum, Z., M. J. Federle, D. Marra, W. M. de Vos, O. P. Kuipers, M. Kleerebezem, and J. R. Scott.** 1998. Use of the lactococcal nisA promoter to regulate gene expression in gram-positive bacteria: comparison of induction level and promoter strength. *Appl. Environ. Microbiol.* **64:**2763–2769.

18. **Falkow, S.** 1988. Molecular Koch's postulates applied to microbial pathogenicity. *Rev. Infect. Dis.* **10:**S274–S276.

19. **Federle, M. J., and J. R. Scott.** 2002. Identification of binding sites for the group A streptococcal global regulator CovR. *Mol. Microbiol.* **43:**1161–1172.

20. **Ferrari, E., D. J. Henner, M. Perego, and J. A. Hock.** 1988. Transcription of *Bacillus subtilis* subtilisin and expression of subtilisin in sporulation mutants. *J. Bacteriol.* **170:** 289–295.

21. **Fogg, G. C., and M. G. Caparon.** 1997. Constitutive expression of fibronectin binding in *Streptococcus pyogenes* as a result of anaerobic activation of rofA. *J. Bacteriol.* **179:**6172–6180.

22. **Fogg, G. C., C. M. Gibson, and M. G. Caparon.** 1993. Identification of rofA, a positive-acting regulatory component of prtF expression: use of a mγδ-based shuttle mutagenesis strategy in Streptococcus pyogenes. *Mol. Microbiol.* **11:**671–684.

23. **Francis, K. P., J. Yu, C. Bellinger-Kawahara, D. Joh, M. J. Hawkinson, G. Xiao, T. F. Purchio, M. Caparon, M. Lipsitch, and P. R. Contag.** 2001. Visualizing pneumococcal infections in the lungs of live mice using bioluminescent *Streptococcus pneumoniae* transformed with a novel gram-positive lux transposon. *Infect. Immun.* **69:**3350–3358.

24. **Geist, R. T., N. Okada, and M. G. Caparon.** 1993. Analysis of *Streptococcus pyogenes* promoters by using novel Tn916-based shuttle vectors for the construction of transcriptional fusions to chloramphenicol acetyltransferase. *J. Bacteriol.* **175:**7561–7570.

25. **Gibson, C. M., and M. G. Caparon.** 2002. Alkaline phosphatase reporter transposon for identification of genes encoding secreted proteins in gram-positive microorganisms. *Appl. Environ. Microbiol.* **68:**928–932.

26. **Graham, M. R., L. M. Smoot, C. A. Migliaccio, K. Virtaneva, D. E. Sturdevant, S. F. Porcella, M. J. Federle, G. J. Adams, J. R. Scott, and J. M. Musser.** 2002. Virulence control in group A Streptococcus by a two-component gene regulatory system: global expression profiling and in vivo infection modeling. *Proc. Natl. Acad. Sci. USA* **99:**13855–13860.

27. **Granok, A. B., D. Parsonage, R. P. Ross, and M. G. Caparon.** 2000. The RofA binding site in *Streptococcus pyogenes* is utilized in multiple transcriptional pathways. *J. Bacteriol.* **182:**1529–1540.

28. **Gryllos, I., C. Cywes, M. H. Shearer, M. Cary, R. C. Kennedy, and M. R. Wessels.** 2001. Regulation of capsule gene expression by group A Streptococcus during pharyngeal colonization and invasive infection. *Mol. Microbiol.* **42:**61–74.

29. **Hanski, E., P. A. Horwitz, and M. G. Caparon.** 1992. Expression of protein F, the fibronectin-binding protein of *Streptococcus pyogenes* JRS4, in heterologous streptococcal and enterococcal strains promotes their adherence to respiratory epithelial cells. *Infect. Immun.* **60:**5119–5125.

30. **Hanski, E., G. Fogg, A. Tovi, N. Okada, I. Burstein, and M. Caparon.** 1994. Molecular analysis of *Streptococcus pyogenes* adhesion. *Methods Enzymol.* **253:**269–305.

31. **Hidalgo-Grass, C., M. Ravins, M. Dan-Goor, J. Jaffe, A. E. Moses, and E. Hanski.** 2002. A locus of group A Streptococcus involved in invasive disease and DNA transfer. *Mol. Microbiol.* **46:**87–99.

32. **Husmann, L. K., D. L. Yung, S. K. Hollingshead, and J. R. Scott.** 1997. Role of putative virulence factors of *Streptococcus pyogenes* in mouse models of long-term throat colonization and pneumonia. *Infect. Immun.* **65:**1422–1430.

33. **Ji, Y., L. McLandsborough, A. Kondagunta, and P. P. Cleary.** 1996. C5a peptidase alters clearance and trafficking of group A streptococci by infected mice. *Infect. Immun.* **64:**503–510.

34. **Karow, M. L., and P. J. Piggot.** Construction of gusA transcriptional fusion vectors for *Bacillus subtilis* and their utilization for studies of spore formation. *Gene* **163:**69–74.

35. **Leenhouts, K. J., B. Tolner, S. Bron, J. Kok, G. Venema, and J. F. Seegers.** 1991. Nucleotide sequence and characterization of the broad-host-range lactococcal plasmid pWVO1. *Plasmid* **26:**55–66.

36. **Liu, S., S. Sela, G. Cohen, J. Jadoun, A. Cheung, and I. Ofek.** 1997. Insertional inactivation of Streptolysin S expression is associated with altered riboflavin metabolism in *Streptococcus pyogenes*. *Microb. Pathog.* **22:**227–234.

37. **Lyon, W., C. M. Gibson, and M. G. Caparon.** 1998. A role for Trigger Factor and an Rgg-like regulator in the transcription, secretion and processing of the cysteine proteinase of *Streptococcus pyogenes*. *EMBO J.* **17:**6263–6275.

38. **Maguin, E., P. Duwat, T. Hege, D. Ehrlich, and A. Gruss.** 1992. New thermosensitive plasmid for gram-positive bacteria. *J. Bacteriol.* **174:**5633–5638.

39. **Malke, H.** 1969. Transduction of *Streptococcus pyogenes* K 56 by temperature-sensitive mutants of the transducing phage A 25. *Z. Naturforsch. B* **24:**1556–1561.

40. **McIver, K. S., and R. L. Myles.** 2002. Two DNA-binding domains of Mga are required for virulence gene activation in the group A streptococcus. *Mol. Microbiol.* **43:**1591–1601.

41. **McIver, K. S., A. S. Heath, and J. R. Scott.** 1995. Regulation of virulence by environmental signals in group A streptococci: influence of osmolarity, temperature, gas exchange, and iron limitation on *emm* transcription. *Infect. Immun.* **63:**4540–4542.

42. **McShan, W. M., R. E. McLaughlin, A. Nordstrand, and J. J. Ferretti.** 1998. Vectors containing streptococcal bacteriophage integrases for site-specific gene insertion. *Methods Cell Sci.* **20:**51–57.

43. **Mekalanos, J. J.** 1992. Environmental signals controlling expression of virulence determinants in bacteria. *J. Bacteriol.* **174:**1–7.

44. **Moynet, D. J., A. E. Colon-Whitt, G. B. Calandra, and R. M. Cole.** 1985. Structure of eight streptococcal bacteriophages. *Virology* **142:**263–269.

45. **Nida, K., and P. P. Cleary.** 1983. Insertional inactivation of streptolysin S expression in *Streptococcus pyogenes*. *J. Bacteriol.* **155:**1156–1161.

46. **Okada, N., R. T. Geist, and M. G. Caparon.** 1993. Positive transcriptional control of *mry* regulates virulence in the group A streptococcus. *Mol. Microbiol.* **7:**893–903.

47. **Ozeri, V., A. Tovi, I. Burstein, S. Natanson-Yaron, M. G. Caparon, K. M. Yamada, S. K. Akiyama, I. Vlodavsky, and E. Hanski.** 1996. A two-domain mechanism for group A streptococcal adherence through protein F to the extracellular matrix. *EMBO J.* **15:**898–998.

48. **Perez-Casal, J., M. G. Caparon, and J. R. Scott.** 1991. Mry, a trans-acting positive regulator of the M protein gene of *Streptococcus pyogenes* with similarity to the receptor proteins of two-component regulatory systems. *J. Bacteriol.* **173:**2617–2624.

49. **Perez-Casal, J., M. G. Caparon, and J. R. Scott.** 1992. Introduction of the *emm6* gene into an *emm*-deletion strain of *Streptococcus pyogenes* restores its ability to resist phagocytosis. *Res. Microbiol.* **143:**549–558.

50. **Perez-Casal, J., E. Maguin, and J. R. Scott.** 1993. An M protein with a single C repeat prevents phagocytosis of *Streptococcus pyogenes*: use of a temperature-sensitive shuttle vector to deliver homologous sequences to the chromosome of *S. pyogenes*. *Mol. Microbiol.* **8:**809–819.

50a. **Piggot, P. J., C. A. M. Curtis, and H. DeLencastre.** 1984. Use of integrational plasmid vectors to demonstrate the poly-cistronic nature of a transcriptional unit (spoIIA) required for sporulation of *Bacillus subtilis*. *J. Gen. Microbiol.* **120:**2123–2136.

51. **Podbielski, A., J. A. Peterson, and P. P. Cleary.** 1992. Surface protein-CAT reporter fusions demonstrate differential gene expression in the *vir* regulon of *Streptococcus pyogenes*. *Mol. Microbiol.* **6:**2253–2265.

52. **Podbielski, A., B. Spellerberg, M. Woischnik, B. Pohl, and R. Lutticken.** 1996. Novel series of plasmid vectors for gene inactivation and expression analysis in group A streptococci (GAS). *Gene* **177:**137–147.

52a. **Pozzi, G., M. Contorni, M. R. Oggioni, R. Manganelli, M. Tommasino, F. Cavalieri, and V. A. Fischetti.** 1992. Delivery and expression of a heterologous antigen on the surface of streptococci. *Infect. Immun.* **60:**1902–1907.

53. **Rajo, J. V., and P. M. Schlievert.** 1998. Mechanisms of pathogenesis of staphylococcal and streptococcal superantigens. *Curr. Top. Microbiol. Immunol.* **225:**81–97.

54. **Rothschild, C. B., R. P. Ross, and A. Claiborne.** 1991. Molecular analysis of the gene encoding alkaline phosphatase in *Streptococcus faecalis* 10C1, p. 45–48. *In* G. M. Dunny, P. P. Cleary, and L. L. McKay (ed.), *Genetics and Molecular Biology of Streptococci, Lactococci, and Enterococci.* American Society for Microbiology, Washington, D.C.

55. **Ruiz, N., B. Wang, A. Pentland, and M. Caparon.** 1998. Streptolysin O and adherence synergistically modulate proinflammatory responses of keratinocytes to group A streptococci. *Mol. Microbiol.* **27:**337–346.

56. **Scott, J. R., P. C. Guenther, L. M. Malone, and V. A. Fischetti.** 1986. Conversion of an M$^-$ group A streptococcus to M$^+$ by transfer of a plasmid containing an M6 gene. *J. Exp. Med.* **164:**1641–1651.

57. **Scott, J. R., F. Bringel, D. Marra, G. Van Alstine, and C. K. Rudy.** 1994. Conjugative transposition of Tn*916*: preferred targets and evidence for conjugative transfer of a single strand and for a double-stranded circular intermediate. *Mol. Microbiol.* **11:**1099–1108.

58. **Simon, D., and J. J. Ferretti.** 1991. Electrotransformation of *Streptococcus pyogenes* with plasmid and linear DNA. *FEMS Microbiol. Lett.* **82:**219–224.

59. **Tao, L., D. J. LeBlanc, and J. J. Ferretti.** 1992. Novel streptococcal-integration shuttle vectors for gene cloning and inactivation. *Gene* **120:**105–110.

59a. **Tomich, P. K., F. Y. An, and D. B. Clewell.** 1980. Properties of erythromycin-inducible transposon Tn*917* in *Streptococcus faecalis*. *J. Bacteriol.* **141:**1366–1374.

60. **Trieu-Cuot, P., C. Carlier, C. Poyart-Salmeron, and P. Courvalin.** 1990. A pair of mobilizable shuttle vectors conferring resistance to spectinomycin for molecular cloning in *Escherichia coli* and in Gram-positive bacteria. *Nucleic Acids Res.* **18:**4296.

61. **VanHeyningen, T., G. Fogg, D. Yates, E. Hanski, and M. Caparon.** 1993. Adherence and fibronectin-binding are environmentally regulated in the group A streptococcus. *Mol. Microbiol.* **9:**1213–1222.

62. **Wessels, M. R., A. E. Moses, J. B. Goldberg, and T. J. DiCesare.** 1991. Hyaluronic acid capsule is a virulence factor for mucoid group A streptococci. *Proc. Natl. Acad. Sci. USA* **88:**8317–8321.

Cross-Reactive Antigens of Group A Streptococci

MADELEINE W. CUNNINGHAM

7

Cross-reactive antigens are molecules on the group A streptococcus that mimic host molecules and during infection induce an immune response against host tissues. Molecular mimicry is the term used to describe immunological cross-reactivity between host and bacterial antigens. Immunological cross-reactions between streptococcal and host molecules have been identified by antibodies or T cells that react with streptococcal components and tissue antigens. The advent of monoclonal antibodies and T-cell clones/hybridomas has greatly facilitated the identification of host and streptococcal antigens responsible for immunological cross-reactions associated with immunization, infection, and autoimmune sequelae. The identification of cross-reactive antigens in group A streptococci is important in our understanding of the pathogenesis of autoimmune sequelae, such as rheumatic fever and glomerulonephritis, which may occur following group A streptococcal infection.

Molecular mimicry between host and bacterial antigens was first defined as identical amino acid sequences shared between different molecules present in tissues and the bacterium (52, 53, 128). The investigation of molecular mimicry through the use of monoclonal antibodies has identified other types of molecular mimicry. The second type of mimicry involves antibody recognition of similar structures such as alpha-helical coiled-coil molecules such as streptococcal M protein and host proteins myosin, keratin, tropomyosin, vimentin, and laminin, which share regions containing 40% identity or less and whose cross-reactive sites are not completely identical (1, 4, 31, 34, 35, 38–40, 42–44, 50, 84–87). A third type of molecular mimicry is revealed in immunological cross-reactions between molecules as diverse as DNA and proteins (31, 50, 112) or carbohydrates and peptides (129–131). Studies of the cross-reactive antigens of the group A streptococcus have contributed greatly to our knowledge about molecular mimicry and the antibody molecules involved. The antibody molecules are described as polyreactive or cross-reactive to indicate recognition of multiple antigens. As described, the basis of the immunological cross-reactions may be identical or homologous amino acid sequences shared between two different proteins or may be epitopes shared between two entirely different chemical structures.

HISTORICAL PERSPECTIVE

Cross-reactive antigens and antibodies were first associated with acute rheumatic fever (ARF) and group A streptococci when it was discovered that rheumatic fever sera or anti-group A streptococcal antisera reacted with human heart or skeletal muscle tissues (143–145). Antibodies against group A streptococci in rheumatic fever serum were shown to be absorbed from the sera with human heart extracts, and, vice versa, the antiheart antibodies were shown to be absorbed with group A streptococci or streptococcal membranes (76–77). Rabbit antisera produced against group A streptococcal cell walls reacted with human heart tissue, and antibodies produced against human heart tissue reacted with group A streptococcal antigens (74–76). In rheumatic fever, heart-reactive antibodies appeared to persist in patients with rheumatic recurrences, and there appeared to be a relationship between high titers of antiheart antibodies and recurrence of rheumatic fever (144, 145). Heart-reactive antibodies were reported to decline within five years of the initial rheumatic fever attack.

To put the early work in perspective, Kaplan and colleagues (76) implicated the streptococcal cell wall containing the M protein as the cross-reactive antigen recognized by the antiheart antibodies, while studies by Zabriskie and Freimer (143, 144) implicated the streptococcal membrane as the site of the cross-reactive antigen. Beachey and Stollerman (8) and Widdowson and colleagues (139–141) reported an M-associated non-type-specific antigen and M-associated protein, respectively, which were thought to be associated with immunological cross-reactions in ARF. In 1977, van de Rijn demonstrated that highly purified peptides from group A streptococcal membranes reacted with the antiheart antibodies in rheumatic fever serum (136). Taken together, this evidence suggested that cross-reactive antigens in group A streptococci were located in both cell wall and membrane.

Gram-Positive Pathogens, 2nd edition, edited by Vincent A. Fischetti et al.
© 2006 ASM Press, Washington, D.C.

Evidence also supported the hypothesis that the group A polysaccharide was a cross-reactive antigen (57). Goldstein and colleagues showed that glycoproteins in heart valves contained the N-acetyl-glucosamine determinant, which they proposed to be responsible for immune cross-reactivity of group A streptococci with heart tissues. Dudding and Ayoub demonstrated persistence of anti-group A carbohydrate antibody in patients with rheumatic valvular disease (48). Lyampert in Russia also published studies suggesting that the group A streptococcal polysaccharide antigen induced responses against host tissues (93–95). McCarty suggested that the terminal O-linked N-acetyl-glucosamine might cross-react with antibodies against the group A carbohydrate and host tissues (103). Additional evidence suggested that the hyaluronic acid capsule of the group A streptococcus also might induce responses against joint tissues (102, 120). Administration of a peptidoglycan-polysaccharide complex prepared from group A streptococci to rats induced carditis and arthritis (22, 23, 124–127). The studies in the 1960s and 1970s left little doubt that group A streptococci induced autoantibodies against heart and other host tissues and that streptococcal components could induce inflammatory lesions resembling arthritis and carditis in animal tissues.

CROSS-REACTIVE MONOCLONAL ANTIBODIES RECOGNIZE MYOSIN AND OTHER ALPHA-HELICAL COILED-COIL MOLECULES IN TISSUES

Monoclonal antibodies (MAbs) cross-reactive with group A streptococci and human heart tissues were produced from mice immunized with streptococcal cell wall and membrane components (33, 34, 50, 87) and from rheumatic carditis patients (1, 129). Figure 1 illustrates the cross-reactivity of an antistreptococcal MAb with myocardium. In 1985, Krisher and Cunningham identified myosin as a cross-reactive host tissue antigen that provided the link between streptococci and heart (87). In 1987, it was demonstrated that cardiac myosin could induce myocarditis in genetically susceptible mice (108). Myosin is an alpha-helical coiled-coil molecule that was shown to be an important tissue target of the cross-reactive MAbs (33). Other host tissue antigens recognized by antistreptococcal mouse and human MAbs included the alpha-helical coiled-coil molecules tropomyosin (1, 50), keratin (1, 129, 132), vimentin (31), laminin (3, 4, 54), DNA (31), and N-acetyl-β-D-glucosamine (129–131), the immunodominant epitope of the group A polysaccharide. The tissue targets identified by the cross-reactive MAbs may be important in the manifestations of group A streptococcal rheumatic fever sequelae of arthritis, carditis, chorea, and erythema marginatum (41, 73). A summary of the cross-reactive human and mouse MAb specificities has been published in previous reviews (25–28).

The cross-reactive antibodies were divided into three major subsets based on their cross-reactivity with (i) myosin and other alpha-helical molecules, (ii) DNA, or (iii) N-acetyl-glucosamine. Figure 2 illustrates the subsets of cross-reactive antistreptococcal/antimyosin MAbs. All three subsets were identified among MAbs from mice immunized with group A streptococcal components, but in the human, the predominant subset reacted with the N-acetyl-glucosamine epitope and myosin and related molecules. This result is not surprising because rheumatic fever patients do not develop antinuclear antibodies during the course of their disease. Elevated levels of polyreactive antimyosin antibodies found in ARF sera (39) and in animals immunized with streptococcal membranes or walls (74, 76, 77, 144) most likely account for the reactivity of these sera with myocardium and other tissues. Similar types of cross-reactive mouse and human antibodies have been investigated by Lange (90) and by Wu et al. (142), respectively.

Polyspecific or cross-reactive autoantibodies have emerged as a theme in autoimmunity and molecular mimicry (2, 18, 31, 88). The V-D-J region genes of the human and mouse cross-reactive MAbs have been sequenced (1, 4, 104), but there is no consensus sequence to explain the molecular basis of polyspecificity and cross-reactivity. It is worth noting that the three groups of reactivities in mice do not have specific antibody V gene families or VH and VL gene combinations that correlate with a specific reactivity. However, higher avidity cross-reactive antimyosin MAb (104) was associated with reactivity with laminin and

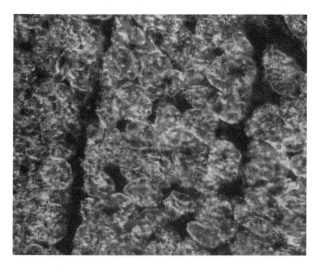

FIGURE 1 (*See the separate color insert for the color version of this illustration.*) Reaction of mouse antistreptococcal MAb with human tissue section of myocardium in indirect immunofluorescence assay. Mouse IgM (20 µg/ml) was unreactive (not shown). MAbs were tested at 20 µg/ml. (From reference 50 with permission from *Journal of Immunology*. Copyright 1989, The American Association of Immunologists, Inc.)

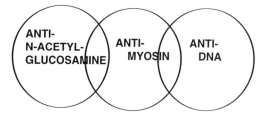

FIGURE 2 Subsets of cross-reactive antistreptococcal and antimyosin antibodies. Human and mouse antistreptococcal/antimyosin MAbs have been divided into three subsets based on their reactivity with myosin and N-acetyl-glucosamine, the dominant group A carbohydrate epitope; with DNA and the cell nucleus, a property found among mouse MAbs; and with myosin and a family of alpha-helical coiled-coil molecules. (From reference 24 with permission from Indiana University School of Dentistry Press.)

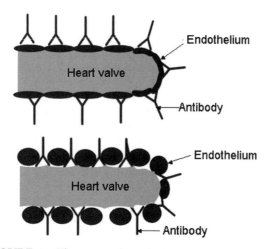

FIGURE 3 The potential mechanism of antibodies in the pathogenesis of rheumatic heart disease. Cross-reactive antibody is shown binding directly to endothelium (top) or binding to basement membrane of the valve (bottom) exposed due to shear stress or damage by antibody and complement.

complement-mediated heart cell cytotoxicity (3, 4). The V genes of cross-reactive human MAbs from rheumatic fever were encoded by a heterogeneous group of VH3 family genes (VH-3, VH-8, VH-23, VH-30) and a VH4-59 gene segment (1). Wu et al. have also reported a similar group of V gene sequences for their human antistreptococcal/antimyosin antibodies produced from Epstein-Barr virus-transformed B cell lines (142). Many sequences were found to be either in germ line configuration or highly homologous with a previously sequenced germline V gene. It has been proposed that a germ line antibody may be polyreactive due to conformational rearrangement and configurational change, permitting binding of diverse molecules (138).

Although the role of the cross-reactive antibodies in disease is not clear, cytotoxic mouse and human MAbs have identified the extracellular matrix protein laminin as a potential tissue target present in basement membrane surrounding myocardium and valve surface endothelium (3, 4, 54, 62, 63). Cross-reactive antibodies may become trapped in extracellular matrix, which may act like a sieve to capture antibody and lead to inflammation in host tissues. The cross-reactive antibodies in rheumatic heart disease target valve surface endothelium (54), as shown in Fig. 3 and 4,

3.B6 reacted with human valve.

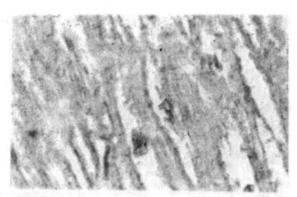

3.B6 reacted with human myocardium

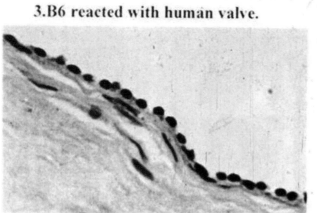

Human IgM control

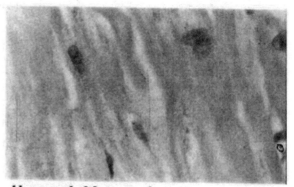

Human IgM control

FIGURE 4 (*See the separate color insert for the color version of this illustration.*) Reactivity of antistreptococcal/antimyosin MAb 3.B6 with normal human valve endothelium and myocardium. Formalin-fixed human mitral valve (top left) and myocardium (top right) were reacted with MAb 3.B6 at 10 μg/ml. MAb 3.B6 binding was detected using biotin-conjugated antihuman antibodies and alkaline phosphatase-labeled streptavidin followed by fast red substrate. Control sections (bottom) did not react with human IgM at 10 μg/ml. (With permission from the *Journal of Clinical Investigation.*)

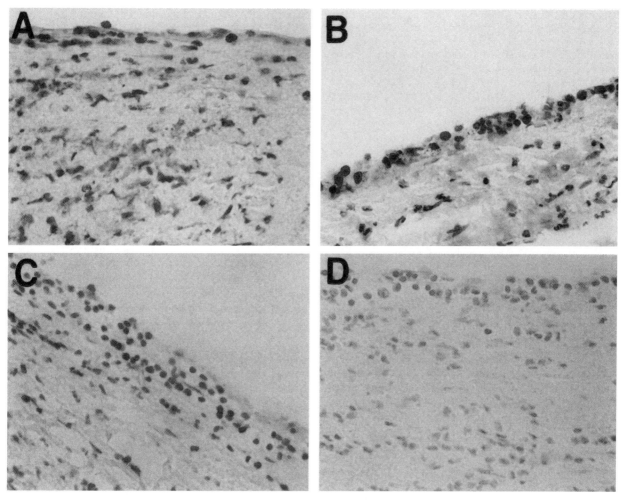

FIGURE 5   (*See the separate color insert for the color version of this illustration.*) Adhesion and extravasation of T lymphocytes into ARF valve in valvulitis. (A, B) Extravasation of CD4$^+$ lymphocytes (stained red) (original magnification, ×200 and ×400, respectively). (C) Extravasation of CD8$^+$ lymphocytes (stained red) into the valve through the valvular endothelium (magnification, ×200). An IgG$_1$ isotype control MAb (IgG$_1$) did not react with the same valve (not shown) (magnification, ×400). (With permission from reference 117, the *Journal of Infectious Diseases*, University of Chicago.)

and most likely initiate the inflammation at the endocardium where T cells infiltrate activated endothelium (117), as shown in Fig. 5, and eventually lead to scarring and neovascularization of the normally avascular valve. Vascular cell adhesion molecule 1 (VCAM-1) expression has been found on the valve surface endothelium, which attracts VLA-4 on the activated T cell (117).

MONOCLONAL ANTIBODIES IDENTIFY STREPTOCOCCAL CROSS-REACTIVE ANTIGENS

Antimyosin MAbs identified cross-reactive antigens in streptococcal membranes (31, 32) and walls (31). The streptococcal M protein was shown to react with the heart- or myosin-cross-reactive MAbs (32, 36, 50). In addition, a 60-kDa protein present in the cell membrane (5, 7) was cross-reactive with myosin in the heart, and a 67-kDa protein,

which was immunologically similar to class II major histocompatibility complex (MHC) molecules, was also identified (82). The data suggested that cross-reactive antigens were present in both the wall and membrane of the group A streptococcus. The data support previous evidence from both Kaplan and colleagues (74–78) and Zabriskie and colleagues (143–145). Although these data were once thought to be conflicting, it is clear that the cross-reactive antibodies recognize more than one antigen in the streptococcal cell. The data previously reported by Goldstein et al. (57) are also supported by the evidence that a subset of cross-reactive MAbs recognized *N*-acetyl-β-D-glucosamine, the immunodominant epitope of the group A polysaccharide. *N*-Acetylglucosamine is a major epitope of some of the cross-reactive mouse MAbs and virtually all of the human cross-reactive MAbs investigated (129–131). The studies link together the cross-reactivity of the group A carbohydrate epitope, GlcNAc, human cardiac myosin, and streptococcal M protein. Antigenic redundancy due to cross-reactivity may be an important factor in triggering disease in a susceptible host. The cross-reactive antigens are now seen as separate entities recognized by MAbs that recognize more than one antigen molecule. Thus, the previous studies, which before seemed conflicting, were all correct. The MAbs have allowed dissection of the group A streptococcal cross-reactive antigens, which would have not been possible using polyclonal sera. The following sections provide more evidence about the identification and analysis of the cross-reactive antigens of the group A streptococcus.

M Proteins and Rheumatic Fever

Investigation of the M proteins over the past 15 years has provided important information about the sequence and primary structure of the molecule. The hypothesis that M proteins and myosin have immunological similarities was supported by the structural studies of Manjula and colleagues (97, 98, 100, 109), which demonstrated the seven-amino-acid-residue periodicity common among group A streptococcal M proteins and shared with proteins such as tropomyosin, myosin, desmin, vimentin, and keratin. Figure 6 illustrates the seven-residue periodicity and the homology characteristic of alpha-helical coiled-coil proteins such as tropomyosin and myosin. Studies by Dale, Beachey, Bronze, and colleagues using polyclonal sera and affinity-purified antibodies demonstrated immunological cross-reactivity between M proteins and myosin (15–17, 42–45). Studies using cross-reactive MAbs also identified immunological cross-reactivity between streptococcal M proteins, both PepM (pepsin-extracted M protein fragment) and recombi-

nant molecules, and cardiac and skeletal myosins (31, 36, 38, 50).

Since streptococcal M proteins have been investigated for potential epitopes that were recognized by the cross-reactive MAbs, affinity-purified antimyosin antibody from ARF sera was reacted with peptides of streptococcal M5 protein. Affinity-purified antimyosin antibodies from ARF reacted with an M5 amino acid sequence (residues 184 to 188) near the pepsin cleavage site in M5 and M6 proteins (36). The epitope, located in the B-repeat region of M5 and M6 proteins, appears to be a B-cell epitope for antibody-mediated cross-reactivity with myosin (36, 40). An M5 peptide, containing the B-repeat region epitope gln-lys-ser-lys-gln (QKSKQ), was shown to inhibit antimyosin antibodies in ARF (36). Furthermore, an M5 peptide that contained the QKSKQ sequence induced antibodies in BALB/c mice against cardiac or skeletal myosins and the LMM fragment of myosin as well. This evidence further supported the previous findings indicating that the QKSKQ sequence is important in the antibody-mediated cross-reactions with myosin in ARF (36). M5 residues 164 to 197 were demonstrated to induce antibodies against sarcolemmal membrane of heart tissue (121), and M5 residues 84 to 116 induced heart-reactive antibody and reacted with antimyosin antibody purified from ARF sera (36, 121). Studies by Kraus and colleagues (85) demonstrated a vimentin cross-reactive epitope present in the M12 protein, and M protein peptides containing brain cross-reactive epitopes were localized to the M5 protein sequence 134 to 184 and the M19 sequence 1 to 24 (17, 121). A summary of the currently known myosin cross-reactive B-cell epitopes in M5 protein is shown in

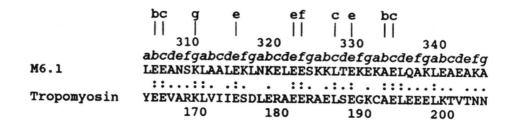

```
          bc     g        e          ef     c e     bc
          ||     |        |          ||     | |     ||
          310             320            330            340
          abcdefgabcdefgabcdefgabcdefgabcdefgabcdefg
M6.1      LEEANSKLAALEKLNKELEESKKLTEKEKAELQAKLEAEAKA
          :: ...:: .:.  .:.   :: .:.: . :::...:. ...
Tropomyosin YEEVARKLVIIESDLERAEERAELSEGKCAELEEELKTVTNN
          170            180            190            200
```

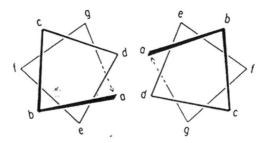

FIGURE 6 Sequence alignment of streptococcal M6 protein and human cardiac tropomyosin in a region exhibiting significant homology. Lowercase letters a to g directly above the sequence designate the position of these amino acids within the seven-residue periodicity in both segments. Lowercase letters at the top of the figure designate identities at external locations in the heptad repeat. Double dots indicate identities and single dots indicate conservative substitutions. Within this segment of the streptococcal M6 molecule, 31% homology is observed with tropomyosin. Because both molecules are alpha-helical coiled-coil proteins, they contain the seven-residue repeat pattern in which positions a and d are usually hydrophobic. Similar homologies are seen between M proteins and myosin heavy chains and any of the three laminin chains. (From reference 50 with permission from *Journal of Immunology*. Copyright 1989, The American Association of Immunologists, Inc.)

Fig. 7, which illustrates the A-, B-, and C-repeat regions of M protein.

Immunization of BALB/c mice with each of 23 peptides of M5 protein has revealed that M5 peptides NT3–NT7, B2B3B, and C1A–C3 induce anti-human cardiac myosin antibodies as shown in Fig. 8. Overlapping synthetic peptides spanning the M5 protein molecule have been reported elsewhere (40). Titers of the anti-M5 peptide sera were 10 times greater against cardiac myosin than against skeletal myosin, tropomyosin, vimentin, and laminin (40). Certain M5 peptides appeared to induce a cardiac myosin-directed response in BALB/c mice. Furthermore, mice developed mild myocarditis when immunized with M5 peptides NT4, NT5, NT6, B1A, and B3B (40). Peptides from the C-repeat region of M5 protein did not produce myocardial lesions when administered to BALB/c mice.

The C-repeat region of class I M proteins contains the class I epitope, which is identified by reactivity with anti-M protein MAb 10B6 (9, 10, 70, 71). Class I M protein serotypes were streptococcal strains associated with pharyngitis and rheumatic fever. Responses against the class I epitope were stronger in patients with rheumatic fever than in those with uncomplicated disease (11, 114). MAb 10B6, which recognizes the C-repeat epitope, also reacted with skeletal and cardiac myosins and the HMM subfragment of myosin (114). The C-repeat class I epitope in M5 contains the amino acid sequence KGLRRDLDASREAK, which shares homology with RRDL, a conserved amino acid sequence found in cardiac and skeletal myosins. The myosin sequence RRDL is located in the HMM subfragment of cardiac and skeletal myosin heavy chain (114).

It was reported that reactivity of immunoglobulin G (IgG) from ARF sera was greater to the class I peptide sequence than IgG from uncomplicated pharyngitis (114). This was confirmed in another study by Mori and colleagues (107). However, the lowered immune response to the class I epitope in uncomplicated pharyngitis was most likely because the patients with uncomplicated pharyngitis had been treated with penicillin (72). Antibodies against the class I epitope were affinity purified from ARF sera and shown to react with myosin (114). Whether the myosin-reactive antibodies against the class I epitope and C-repeat region play a role in the disease pathogenesis is not known; however, the C-repeat peptides do not produce heart lesions in animal models of carditis (40).

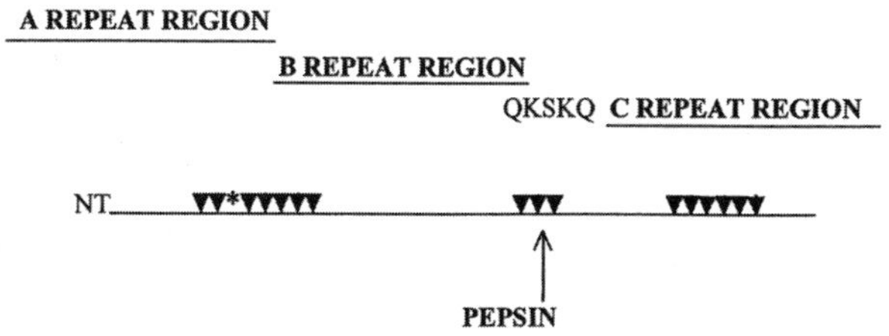

Studies by Huber and Cunningham demonstrated that streptococcal M5 peptide NT4 containing the amino acid sequence GLKTENEGLKTENEGLKTE could produce myocarditis in MRL/++ mice, an autoimmune-prone strain (66). The studies demonstrated that the myocarditis was mediated by CD4$^+$ T cells and class II MHC molecules. Antibodies against the IAk MHC molecule or antibody against CD4$^+$ T cells abrogated the myocarditis (66). Amino acid sequence homology between the NT4 peptide of M5 and human cardiac myosin demonstrates 80% identity. Figure 9 illustrates the homology between the M5 peptide NT4 sequence and cardiac myosin sequence. The mimicking sequence in NT4 is repeated four times in the M5 protein and is present in myosin only once. Repeated regions of the M protein that mimic cardiac myosin may be important in breaking tolerance in the susceptible host and producing autoimmune disease. The data support the hypothesis that epitopes in streptococcal M protein that mimic cardiac myosin may be important in breaking tolerance to this potent autoantigen. Table 1 summarizes the M protein

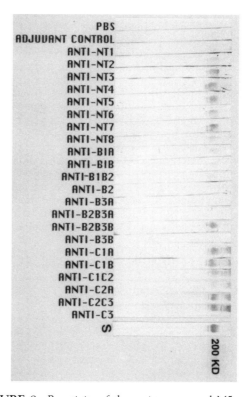

FIGURE 8 Reactivity of the antistreptococcal M5 peptide sera in the Western immunoblot of human cardiac myosin. The 200-kDa protein band of the purified human cardiac myosin, shown in the stained (S) portion of the Western blot, reacted most strongly with antipeptide sera from mice immunized with M5 peptides NT3–7, B2B3B, and C1A, C1B, C2C3, and C3. Sera were tested at a 1:1000 dilution. A control antimyosin MAb, CCM-52 (a gift from Dr. William Clark, Cardiovascular Research Institute, Michael Reese Hospital and Medical Center, Chicago, Ill.), reacted with the 200-kDa band present in our purified preparation of human cardiac myosin heavy chain. Purification of the human cardiac myosin heavy chain to homogeneity has been previously described by Dell et al. (47). The Western blot confirms the data seen in the enzyme-linked immunosorbent assay with human cardiac myosin. (From reference 40 with permission.)

A REPEAT REGION

B REPEAT REGION

QKSKQ **C REPEAT REGION**

NT_____ ▼▼*▼▼▼▼▼   ▼▼▼   ▼▼▼▼▼▼

↑

PEPSIN

FIGURE 7 Antimyosin antibody cross-reactive sites in M5 protein A-, B-, and C-repeat regions of the molecule. Arrows point to sites determined to induce human cardiac myosin cross-reactive antibody (40). The asterisk marks the location of peptide NT4 containing several repeats of an epitope in cardiac myosins that causes myocarditis in MRL/++ and BALB/c mouse strains (40, 66). Site QKSKQ is the epitope determined to react with antimyosin antibody in ARF sera (36). (From reference 4a, copyright ©2000, by Oxford University Press, Inc. Used by permission of Oxford University Press, Inc.)

<pre>
Streptococcal M Protein M5 Peptide NT4   GLKTENEG
                                              : :::
                      cardiac myosin   KLQTENGE
</pre>

FIGURE 9 Sequence homology between human cardiac myosin and peptide NT4, which causes myocarditis (40, 66). The homologous sequence repeats four times in the streptococcal M5 protein and in NT4 and once in cardiac myosin. Repeated sequences in M proteins that mimic cardiac myosin may be important in inducing inflammatory heart disease. (Adapted from reference 66 with permission from *Journal of Immunology*. Copyright 1989, The American Association of Immunologists, Inc.)

amino acid sequences observed to induce inflammatory myocardial lesions in mice.

Streptococcal M proteins not only mimic epitopes in cytoskeletal proteins, but they mimic epitopes in other strong bacterial and viral antigens, namely heat shock protein (Hsp)-65 (113) and coxsackieviral capsid proteins (38, 64, 66, 68). These immunological cross-reactions may be vital to survival of the host and suggest that antibody molecules may recognize and neutralize more than a single infectious agent. Such antibodies may be an important first line of defense and would be highly advantageous for the host. Multiple pieces of evidence from experiments using synthetic peptides have shown that sites in the M protein mimic a site(s) in the VP1 capsid protein of coxsackievirus (37, 38). Further evidence demonstrated that some of the antistreptococcal/antimyosin MAbs actually neutralized enteroviruses and were cytotoxic for heart cells (38). This was extremely interesting because coxsackieviruses cause autoimmune myocarditis in susceptible hosts (55, 65, 67). The Hsp-65 antigen has been shown to play a role in the development of arthritis and diabetes (20, 21). Cross-reactive epitopes shared between streptococcal M proteins and Hsp-65 play a role in arthritis sequelae. Antibodies against Hsp-60 have been implicated in cytotoxicity against endothelium (122). Shared epitopes among pathogens may be important in molecular mimicry, may break tolerance to cryptic host molecules, and may influence the development of autoimmune diseases.

M Proteins and Cross-Reactive T Cells

M proteins have also been shown to stimulate human T-lymphocyte responses (45, 59, 110, 111, 118, 119, 133–135), including cytotoxic T lymphocytes from ARF patients (45, 69). PepM protein has been reported to be a superantigen that stimulated Vβ2-, Vβ4,- and Vβ8-bearing T-cell subsets (133–135, 137). Other reports show that neither recombinant nor native forms of M proteins are superantigenic (46, 51, 123). A superantigenic site reported for the M5 protein has been localized to the amino acid sequence located in M5 residues 157 to 197 (KEQENKETIGTLKKILDETVKDKLAKEQKSKQNIGALKQEL) the B3-repeat region (137). The site has a significant amount of sequence homology with other superantigens. The role of M protein or other superantigens in ARF may be to activate large numbers of T cells, including some that are cross-reactive and may lead to ARF.

Studies of T-cell epitopes in rheumatic fever and in animal models focus on the M5 protein molecule because serotype M5 has been often associated with ARF outbreaks (13). T- and B-cell epitopes of the M5 protein were defined in previous studies by Robinson and colleagues (118, 119), by Pruksakorn et al. (110, 111), and in our own laboratory (40).

Guilherme et al. isolated cross-reactive T cells from mitral valves, papillary muscle, and left atrium of rheumatic fever patients previously infected with M5 group A streptococci (59). T-cell lines were responsive to several peptides of the streptococcal serotype M5 protein and proteins from heart tissue extracts. Sequences from the A- and B-repeat regions of M5 protein that stimulated the valvular T cells are shown in Table 2. In BALB/c mice using synthetic peptides, myosin-cross-reactive T-cell epitopes from the A-, B-, and C-repeat regions of the M5 protein were identified (40). Six dominant myosin cross-reactive T-cell epitopes were located in the M5 molecule (40). Table 2 summarizes (i) the dominant myosin cross-reactive T-cell epitopes of M5 protein in BALB/c mice (40), (ii) the M5 sequences recognized by T-cell clones from rheumatic heart valves (59), and (iii) M5 peptides reported by Pruksakorn et al. to stimulate human T cells from normal controls and rheumatic fever patients that were responsive to myosin peptides (110, 111). An important correlation seen in Table 2 is that M5 peptides NT4/NT5 (GLKTENEGLKTENEGLKTE and KKEHEAENDKLKQQRDTL) and B1B2/B2 (VKDKIAKEQENKETIGTL and TIGTLKKILDETVKDKIA), which were dominant cross-reactive T-cell epitopes in the BALB/c mouse, contain sequences similar to those reported to be recognized by T cells from rheumatic valves (59). Peptides NT4 and NT5 produced inflammatory infiltrates in the myocardium of animals immunized with those peptides (40). The collective evidence on cross-reactive T cells suggests that amino acid sequences in M5 protein that share homology with cardiac myosin may break tolerance and promote T-cell-mediated inflammatory heart disease in animals and man (40).

T cells isolated and cloned from human rheumatic mitral valves recognized streptococcal M5 peptides and LMM peptides (49). Mimicry between M proteins and cardiac myosin stimulates cross-reactive T cells in the peripheral blood of the host (30, 61). The T cells then traffic to the valve once the endothelium of the valve becomes activated and inflamed (117). This extravasation event at the valve endocardium allows the cross-reactive T cells to enter the valve, where they recognize and proliferate to valvular pro-

TABLE 1 M5 protein sequences that produce myocarditis in mice[a]

Peptide	Amino acid sequence	Residues
A-repeat region		
NT4	GLKTENEGLKTENEGLKTE	40–58
NT5	KKEHEAENDKLKQQRDTL	59–76
NT6	QRDTLSTQKETLEREVQN	72–89
B-repeat region		
B1A	TRQELANKQQESKENEKAL	111–129
B3B	GALKQELAKKDEANKISD	202–219

[a]NT4 produced myocarditis in BALB/c (40) and MRL/++ mice (66); the other peptides shown produced myocarditis in BALB/c mice. (From *Effects of Microbes on the Immune System* with permission from Lippincott-Williams and Wilkins.)

TABLE 2 Summary of myosin or heart cross-reactive T-cell epitopes of streptococcal M5 protein[a]

Peptide	Sequence[b]	Origin of T-cell clone or response[c]	Reference(s)
1–25	TVTRGTISDPQRAKEALDKYELENH[b]	ARF/Valve	59
81–96	DKLKQQRDTLSTQKETLEREVQN[b]	ARF/Valve	59
163–177	ETIGTLKKILDETVK[b]	ARF/Valve	59
337–356	LRRDLDASREAKKQVEKAL	Normal/PBL[c]	110
347–366	AKKQVEKALEEANSKLAALE	Mice/Normal PBL	110, 111
397–416	LKEQLAKQAEELAKLRAGKA	ARF/PBL	113
NT4 40–58	GLKTENEGLKTENEGLKTE	BALB/c/lymph node[d]	
NT5 59–76	KKEHEAENDKLKQQRDTL		
B1B2 137–154	VKDKIAKEQENKETIGTL		
B2 150–167	TIGTLKKILDETVKDKIA		
C2A 254–271	EASRKGLRRDLDASREAK		
C3 293–308	KGLRRDLDASREAKKQ		

[a]From reference 24.

[b]Amino terminal TVTRGTIS sequence (peptide 1–25) was taken from the M5 amino acid sequence published by Manjula et al. (99) and deviates from the M5 sequence published by Miller et al. (106) at positions 1 and 8. Sequences for peptides 81–96 and 163–177 were taken from the PepM5 sequence as reported by Manjula et al. (99). These two sequences are given as 67–89 and 174–188, respectively, in the sequence reported by Miller et al. (106). All other sequences shown are from the M5 gene sequence reported by Miller et al. (106).

[c]PBL, peripheral blood lymphocytes; ARF, acute rheumatic fever.

[d]BALB/c mice immunized with purified human cardiac myosin and the recovered lymph node lymphocytes were stimulated with each of the peptides in tritiated thymidine uptake assays.

teins such as vimentin (62) and which may include cardiac myosin from papillary muscle. T cells that remain in the valve survive if they continue to be stimulated by host alpha-helical proteins within the valve. The activated cross-reactive T cells, when continually stimulated within the valve, become pathogenic for the host and produce the TH1 mechanism of pathogenesis with scarring in the valve. Figure 10 is a diagram of the events in rheumatic carditis that result from immune responses against the cross-reactive antigens of the group A streptococcus (29).

M Proteins and Poststreptococcal Acute Glomerulonephritis

Sera from patients with poststreptococcal acute glomerulonephritis were shown by Kefalides and colleagues to contain antibodies against laminin, collagen, and other macromolecules found in the glomerular basement membrane (79). The epitope recognized in collagen was identified and shown to be located in the 7-S domain of type IV collagen (80). Streptococcal antigens, immunoglobulins, and complement were detected in the kidney glomeruli in acute glomerulonephritis (105). Studies by Markowitz and Lange suggested that the glomerular basement membrane shared antigens with streptococcal M12 protein (101), and others have shown immunologic cross-reactivity between the group A streptococcus and glomeruli (14, 58, 86, 89). Kraus and Beachey discovered a renal autoimmune epitope (Ile-Arg-Leu-Arg) in M protein (84). Evidence does suggest that certain M protein serotypes are associated with nephritis and that molecular mimicry or immunological cross-reactivity between glomeruli and M protein could be one mechanism by which antiglomerular antibodies are produced during infection. In animal models of nephritis induced by nephritogenic streptococci (M type 12), antiglomerular antibodies eluted from kidney glomeruli reacted with the type

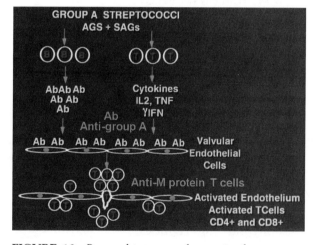

FIGURE 10 Proposed immunopathogenesis of poststreptococcal heart disease. Initially, B and T cells are activated by specific streptococcal antigens and superantigens, leading to strong responses against streptococcal and host antigens. The development of pathogenic clones of B and T lymphocytes is important in development of the disease. The antibodies against the group A carbohydrate, which is cross-reactive with the valve surface, bind to the valve surface endothelium (endocardium) and lead to damage and/or upregulation of cell adhesion molecules such as VCAM-1 on activated surface endothelium of the valve. M protein-reactive T cells enter the valve through the surface endothelium by binding to cell adhesion molecules such as VCAM-1 and extravasate into the valve (117). The formation of scar tissue in the valve, followed by neovascularization, allows the disease to continue in the valve. The specificity of the T cells entering the valve has been shown to be M protein (49, 59–61). (Reproduced from reference 29 with permission from *Frontiers in Bioscience*.)

12 streptococcal M protein (92). Furthermore, an anti-glomerular MAb reacted with the M12 protein (58). These studies support immune-mediated mechanisms in poststreptococcal acute glomerulonephritis. Streptococcal and renal antigens may be an important source of mimicking antigen present in the kidney and play a role in binding Ig and complement and in producing nephritis.

Novel Cross-Reactive Protein Antigens in Group A Streptococci

Antimyosin antibodies from ARF were used to identify group A streptococcal cross-reactive antigens other than M proteins. The cross-reactive antigens were detected in a *Streptococcus pyogenes* strain that had spontaneously lost its *emm* gene. The affinity-purified antimyosin antibodies were used to screen gene libraries as well as the antigen preparations from this M protein-deficient streptococcus. A 60-kDa streptococcal actin-like protein (6), a 60-kDa wall-membrane protein (5, 7), and a 67-kDa protein from a cloned gene product (82) were identified. All of the antigens were unique and interesting. The 60-kDa actin-like molecule behaved biochemically as an actin and produced actin-like filaments observed by electron microscopy (6). The 67-kDa cloned gene product recognized by antimyosin antibody from ARF sera was novel in that it shared enough sequence homology with MHC molecules that it reacted with sera against mouse MHC class II molecules and exhibited a hydrophilicity plot almost identical with that of MHC class II molecules (82). The gene for the 67-kDa protein was present only in pathogenic streptococci groups A, C, and G (82). The gene for the 67-kDa protein was not found in other streptococcal groups, *Escherichia coli*, or staphylococci when they were screened for hybridization with the gene probe. The role of the 67-kDa protein as a potential virulence factor in the pathogenesis of streptococcal infections or sequelae is not yet known.

IMMUNE RESPONSES TO *N*-ACETYL-β-D-GLUCOSAMINE, DOMINANT EPITOPE OF GROUP A POLYSACCHARIDE, IN THE PATHOGENESIS OF RHEUMATIC HEART DISEASE AND SYDENHAM CHOREA IN ACUTE RHEUMATIC FEVER

Immunological Cross-Reactivity between Carbohydrate and Peptides

The structure of the group A polysaccharide is a polymer of rhamnose with *N*-acetyl-D-glucosamine linked to the rhamnose backbone. The *N*-acetyl-glucosamine is the immunodominant epitope of the group A carbohydrate that distinguishes *S. pyogenes* from other streptococcal species. The *N*-acetyl-glucosamine and the group A polysaccharide have been identified in cross-reactions between streptococci and heart tissues (57, 93, 94). Recently, Shikman, Adderson, and colleagues reported that a subset of cross-reactive antimyosin/antistreptococcal MAbs from immunized mice and virtually all human MAbs from rheumatic carditis patients reacted strongly with *N*-acetyl-glucosamine epitope of the group A streptococcal carbohydrate (1, 129–131). Furthermore, synthetic peptides from keratin (129, 131), coxsackievirus (130), or human cardiac myosin (1) reacted with MAbs against *N*-acetyl-glucosamine and myosin as previously described. In addition, a synthetic cytokeratin peptide SFGSGFGGGY mimicked *N*-acetyl-glucosamine

in its reaction with antibodies to *N*-acetyl-glucosamine and lectins. Mimicry between the group A carbohydrate epitope and peptide molecules is a form of mimicry between chemically diverse molecules. Based on results with peptides altered at a single amino acid residue, it was deduced that aromatic and hydrophobic interactions were important in cross-reactive anti-carbohydrate antibody binding to the peptide. Our data, as well as that of others (91, 96), clearly link myosin and *N*-acetyl-glucosamine with antiheart cross-reactivity.

Rheumatic Heart Disease

Dudding and Ayoub described the elevation and persistence of anti-group A polysaccharide antibody in rheumatic valvulitis (48). Studies of the human antistreptococcal/antimyosin MAbs that reacted strongly with *N*-acetyl-glucosamine have revealed that one of the MAbs was cytotoxic for human endothelium and reacted with valvular endothelium in tissue sections of human valve (54). The cytotoxic MAb recognized the extracellular matrix protein laminin, which is part of the basement membrane underlying the valvular endothelium. These data suggest that human antibody cross-reactivity with valve tissues is established through antimyosin/anti-*N*-acetyl-glucosamine/anti-laminin reactivity. Antibodies that react with valve endothelium may lead to inflammation at the valve surface and promote T-cell infiltration of the valve in rheumatic heart disease. Based on our current knowledge as illustrated in Fig. 10 (28, 30), the proposed events in rheumatic valvulitis/carditis involve both antibodies and T cells. Cross-reactive antibodies target the valve and are believed to be against the group A carbohydrate and valve endothelium, based on several lines of evidence described herein (48, 54, 57, 62, 78). T cells are cross-reactive with streptococcal M proteins and homologous alpha-helical protein antigens such as myosin, laminin, tropomyosin, or vimentin and become activated and extravasate through activated endothelium into the valve, where they differentiate into CD4$^+$ TH1 cells producing gamma interferon and valvular scarring (29, 30, 49, 59–61, 117). VCAM-1 was up-regulated on valve endothelium (endocardium) in rheumatic carditis, indicating activation of the valve (117). Evidence strongly supports the cross-reactive T-cell component of the disease, in which T cells are shown to penetrate the valve endothelium into an originally avascular valve (12, 19, 30, 40, 49, 56, 59–61, 69, 81, 115–117). T cells in peripheral blood of patients and in the valve have been shown to be cross-reactive with M protein and cardiac proteins, including cardiac myosin epitopes, and they are both CD4$^+$ and CD8$^+$ phenotypes. The development of scarring in the valve is part of the pathogenesis caused by gamma interferon production. Because the scarring promotes neovascularization and development of a blood supply into normally avascular valve tissue, T cells can subsequently enter the valve through blood vessels developed in the scar. The valve becomes predisposed to cellular infiltration through both the activated valve endocardial surface and the neovascularized scar tissue.

Sydenham Chorea

In Sydenham chorea, the major neurological manifestation and movement disorder of rheumatic fever, human mAbs were produced. The chorea MAbs demonstrated specificity for mammalian lysoganglioside and the group A carbohydrate epitope *N*-acetyl-β-D-glucosamine (83). Chorea MAb and acute serum antibodies targeted the surface of human neuronal cells and induced calcium/calmodulin-dependent protein kinase activity (CAM kinase). Although the MAbs were IgM, the serum antibodies that triggered the cell signaling in neuronal cells were IgG. IgG is believed to penetrate the blood-brain barrier and reach brain tissues, where it can produce disease. The chorea antibodies were then shown to lead to production of dopamine by neuronal cells, which would ultimately trigger the chorea. The studies suggest that (i) the antibodies against streptococci and brain in Sydenham chorea may produce the central nervous system dysfunction through a neuronal signal transduction and subsequent dopamine release mechanism, and (ii) the molecular targets of the chorea antibodies include gangliosides on neuronal cells and the group A carbohydrate on the streptococci. Only the MAb with the higher affinity could induce the cell signaling and CAM kinase production in neuronal cells (83). Figure 11 is a diagram illustrating the proposed events leading to Sydenham chorea.

SUMMARY AND CONCLUSIONS

Cross-reactive antibodies and cross-reactive T cells against group A streptococci are important in the pathogenesis of autoimmune sequelae following infection. Three subsets of cross-reactive antibodies have been defined as well as host tissue alpha-helical coiled-coil antigens such as myosin, tropomyosin, keratin, vimentin, and laminin (28). Cross-reactive anti-polysaccharide antibodies recognize peptide sequences in alpha-helical proteins as well as the N-acetyl-glucosamine molecule, dominant epitope of the group A streptococcal carbohydrate (129–131).

In rheumatic heart disease, laminin, an extracellular matrix molecule present in basement membrane of the valve, may trap cross-reactive antibodies at the endocardial cell surface and lead to damage or inflammation of the endothelium. Activated endothelium will lead to subsequent extravasation of streptococcal M protein/myosin-cross-reactive T cells into the valve (54, 59–61, 117). Cross-reactive group A streptococcal antigens with apparent roles in the pathogenesis of rheumatic heart disease include the M proteins and the group A polysaccharide. Amino acid sequences of streptococcal M5 protein that have been shown to be pathogenic in animals have also been reported to be recognized by T cells from rheumatic heart valves (30, 49, 59–61). The identification of human cardiac myosin cross-reactive B- and T-cell epitopes of M5 protein has been a step forward in understanding the cross-reactive epitopes that produce disease in animals and humans (40).

In Sydenham chorea, mimicry between the N-acetyl-glucosamine molecule of the group A carbohydrate and brain gangliosides potentially leads to antibodies that bind to the surface of neuronal cells and trigger induction of dopamine in the disease (83). In the pathogenesis of chorea, the development of IgG responses and higher affinity cell surface reactive antibody that can bind strongly enough to alter host tissues may lead to disease. In carditis, T cells can enter the valve, but only those that are consistently stimulated with local antigens would retain a strong immune response and TH1 pattern against the valve triggering cytokine production and scarring.

Although the identity and pathogenic mechanisms of cross-reactive antigens and their cross-reactive antibodies and T cells are much clearer now than in the past, there are still lessons to be learned about how mimicry of host tissues leads to autoimmune sequelae following streptococcal disease.

This work was supported by grants HL35280 and HL56267 from the National Heart, Lung and Blood Institute. M.W.C is the recipient of an NHLBI Merit Award. I appreciate the important contributions of my students and fellows to this body of work.

REFERENCES

1. **Adderson, E. E., A. R. Shikhman, K. E. Ward, and M. W. Cunningham.** 1998. Molecular analysis of polyreactive monoclonal antibodies from rheumatic carditis: human anti-N-acetyl-glucosamine/anti-myosin antibody V region genes. *J. Immunol.* **161:**2020–2031.

2. **Adrezejewski, C., Jr., J. Rauch, B. D. Stollar, and R. S. Schwartz.** 1980. Antigen binding diversity and idiotypic cross-reactions among hybridoma autoantibodies to DNA. *J. Immunol.* **126:**226–231.

3. **Antone, S. M., and M. W. Cunningham.** 1992. Cytotoxicity linked to a streptococcal monoclonal antibody which recognizes laminin, p. 189–191. In G. Orefici (ed.), *New Perspectives on Streptococci and Streptococcal Infections. Proceedings of the XI Lancefield International Symposium on Streptococci and Streptococcal Diseases*, supplement 22. Gustav Fisher Verlag, New York, N.Y.

4. **Antone, S. M., E. E. Adderson, N. M. J. Mertens, and M. W. Cunningham.** 1997. Molecular analysis of V gene sequences encoding cytotoxic anti-streptococcal/anti-myosin monoclonal antibody 36.2.2 that recognizes the heart cell surface protein laminin. *J. Immunol.* **159:**5422–5430.

4a. **Ayoub, E. M., M. Kotb, and M. W. Cunningham.** 2000. Rheumatic fever pathogenesis. In D. Stevens and E. Kaplan (ed.), *Streptococcal Infections: Clinical Aspects, Microbiology, and Molecular Pathogenesis.* Oxford University Press, New York, N.Y.

5. **Barnett, L. A., and M. W. Cunningham.** 1990. A new heart-cross-reactive antigen in *Streptococcus pyogenes* is not M protein. *J. Infect. Dis.* **162:**875–882.

6. **Barnett, L. A., and M. W. Cunningham.** 1992. Evidence for actinlike proteins in an M protein-negative strain of *Streptococcus pyogenes. Infect. Immun.* **60:**3932–3936.

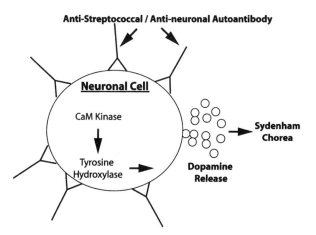

FIGURE 11 Proposed events leading to Sydenham chorea. Antibodies against the group A streptococcal carbohydrate and brain ganglioside react with the surface of neuronal cells and trigger the cell signaling event, leading to up-regulation of calcium/calmodulin-dependent protein kinase II and eventually dopamine release, which leads to the choreic movement disorder.

7. **Barnett, L. A., J. J. Ferretti, and M. W. Cunningham.** 1992. A 60-kDa acute rheumatic fever associated antigen of *Streptococcus pyogenes*. *Zentbl. Bakteriol Suppl.* 22. Gustav-Fisher-Verlag, New York, N.Y.

8. **Beachey, E. H., and G. H. Stollerman.** 1973. Mediation of cytotoxic effects of streptococcal M protein by non-type-specific antibody in human sera. *J. Clin. Investig.* **52:** 2563–2570.

9. **Bessen, D. E., and V. A. Fischetti.** 1990. Differentiation between two biologically distinct classes of group A streptococci by limited substitutions of amino acids within the shared region of M protein-like molecules. *J. Exp. Med.* **172:**1757–1764.

10. **Bessen, D., K. F. Jones, and V. A. Fischetti.** 1989. Evidence for two distinct classes of streptococcal M protein and their relationship to rheumatic fever. *J. Exp. Med.* **169:**269–283.

11. **Bessen, D. E., L. G. Veasy, H. R. Hill, N. H. Augustine, and V. A. Fischetti.** 1995. Serologic evidence for a class I group A streptococcal infection among rheumatic fever patients. *J. Infect. Dis.* **172:**1608–1611.

12. **Bhatia, R., J. Narula, K. S. Reddy, M. Koicha, A. N. Malaviya, R. B. Pothineni, R. Tandon, and M. L. Bhatia.** 1989. Lymphocyte subsets in acute rheumatic fever and rheumatic heart disease. *Clin. Cardiol.* **12:**34–38.

13. **Bisno, A. L.** 1995. Non-suppurative poststreptococcal sequelae: rheumatic fever and glomerulonephritis, p. 1799–1810. *In* G. L. Mandell, J. E. Bennett, and R. Dolin (ed.), *Principles and Practice of Infectious Diseases*, vol. 2. Churchill Livingstone, New York, N.Y.

14. **Bisno, A. L., J. W. Wood, J. Lawson, S. Roy, E. H. Beachey, and G. H. Stollermann.** 1978. Antigens in urine of patients with glomerulonephritis and in normal human serum which cross-react with group A streptococci: identification and partial characterization. *J. Lab. Clin. Med.* **91:** 500–513.

15. **Bronze, M. S., and J. B. Dale.** 1993. Epitopes of streptococcal M proteins that evoke antibodies that cross-react with human brain. *J. Immunol.* **151:**2820–2828.

16. **Bronze, M. S., E. H. Beachey, J. M. Seyer, and J. B. Dale.** 1987. Protective and heart-cross-reactive epitopes of type 19 streptococcal M protein. *Trans. Assoc. Am. Physicians* **100:**80–84.

17 **Bronze, M. S., E. H. Beachey, and J. B. Dale.** 1988. Protective and heart-cross-reactive epitopes located within the NH2 terminus of type 19 streptococcal M protein. *J. Exp. Med.* **167:**1849–1859.

18. **Carroll, P., D. Stafford, R. S. Schwartz, and B. D. Stollar.** 1985. Murine monoclonal anti-DNA antibodies bind to endogenous bacteria. *J. Immunol.* **135:**1086.

19. **Chopra, P., J. Narula, A. S. Kumar, S. Sachdeva, and M. L. Bhatia.** 1988. Immunohistochemical characterisation of Aschoff nodules and endomyocardial inflammatory infiltrates in left atrial appendages from patients with chronic rheumatic heart disease. *Int. J. Cardiol.* **20:**99–105.

20. **Cohen, I. R.** 1991. Autoimmunity to chaperonins in the pathogenesis of arthritis and diabetes. *Annu. Rev. Immunol.* **9:**567–589.

21. **Cohen, I. R., and D. B. Young.** 1991. Autoimmunity, microbial immunity, and the immunological homunculus. *Immunol. Today* **12:**105–110.

22. **Cromartie, W. J., and J. G. Craddock.** 1966. Rheumatic like cardiac lesions in mice. *Science* **154:**285–287.

23. **Cromartie, W., J. Craddock, J. H. Schwab, S. Anderle, and C. Yang.** 1977. Arthritis in rats after systemic injection of streptococcal cells or cell walls. *J. Exp. Med.* **146:** 1585–1602.

24. **Cunningham, M. W.** 1998. Presented at the Second Annual Indiana Conference, Indianapolis, Ind.

25. **Cunningham, M. W.** 1999. *Molecular Mimicry between Group A Streptococci and Myosin in the Pathogenesis of Rheumatic Fever*, vol. 2. American Registry of Pathology, Armed Forces Institute of Pathology, Washington, D.C.

26. **Cunningham, M. W.** 2000. *Group A Streptococcal Sequelae and Molecular Mimicry*. Lippincott, Williams and Wilkins, Philadelphia, Pa.

27. **Cunningham, M. W.** 2000. *Pathogenesis of Acute Rheumatic Fever*. Oxford Press, New York, N.Y.

28. **Cunningham, M. W.** 2000. Pathogenesis of group A streptococcal infections. *Clin. Microbiol. Rev.* **13:**470–511.

29. **Cunningham, M. W.** 2003. Autoimmunity and molecular mimicry in the pathogenesis of post-streptococcal heart disease. *Front. Biosci.* **8:**s533–s543.

30. **Cunningham, M. W.** 2004. T cell mimicry in inflammatory heart disease. *Mol. Immunol.* **40:**1121–1127.

31. **Cunningham, M. W., and R. A. Swerlick.** 1986. Polyspecificity of antistreptococcal murine monoclonal antibodies and their implications in autoimmunity. *J. Exp. Med.* **164:**998–1012.

32. **Cunningham, M. W., K. Krisher, and D. C. Graves.** 1984. Murine monoclonal antibodies reactive with human heart and group A streptococcal membrane antigens. *Infect. Immun.* **46:**34–41.

33. **Cunningham, M. W., N. K. Hall, K. K. Krisher, and A. M. Spanier.** 1985. A study of monoclonal antibodies against streptococci and myosin. *J. Immunol.* **136:**293–298.

34. **Cunningham, M. W., N. K. Hall, K. K. Krisher, and A. M. Spanier.** 1986. A study of anti-group A streptococcal monoclonal antibodies cross-reactive with myosin. *J. Immunol.* **136:**293–298.

35. **Cunningham, M. W., J. M. McCormack, L. R. Talaber, J. B. Harley, E. M. Ayoub, R. S. Muneer, L. T. Chun, and D. V. Reddy.** 1988. Human monoclonal antibodies reactive with antigens of the group A Streptococcus and human heart. *J. Immunol.* **141:**2760–2766.

36. **Cunningham, M. W., J. M. McCormack, P. G. Fenderson, M. K. Ho, E. H. Beachey, and J. B. Dale.** 1989. Human and murine antibodies cross-reactive with streptococcal M protein and myosin recognize the sequence GLN-LYS-SER-LYS-GLN in M protein. *J. Immunol.* **143:** 2677– 2683.

37. **Cunningham, M. W., S. M. Antone, J. M. Gulizia, and C. J. Gauntt.** 1992. Common epitopes shared between streptococcal M protein and viruses may be a link to autoimmunity, p. 534–536. *In* G. Orefici (ed.), *New Perspectives on Streptococci and Streptococcal Infections. Proceedings of the XIth Lancefield International Symposium on Streptococci and Streptococcal Diseases*, supplement 22. Gustav Fisher Verlag, New York, N.Y.

38. **Cunningham, M. W., S. M. Antone, J. M. Gulizia, B. M. McManus, V. A. Fischetti, and C. J. Gauntt.** 1992. Cytotoxic and viral neutralizing antibodies cross-react with streptococcal M protein, enteroviruses, and human cardiac myosin. *Proc. Natl. Acad. Sci. USA* **89:**1320–1324.

39. **Cunningham, M. W., S. M. Antone, J. M. Gulizia, B. A. McManus, and C. J. Gauntt.** 1993. α-helical coiled-coil molecules: a role in autoimmunity against the heart. *Clin. Immunol. Immunopathol.* **68:**129–134.

40. **Cunningham, M. W., S. M. Antone, M. Smart, R. Liu, and S. Kosanke.** 1997. Molecular analysis of human car-

diac myosin-cross-reactive B- and T-cell epitopes of the group A streptococcal M5 protein. *Infect. Immun.* **65:** 3913–3923.

41. **Dajani, A. S.** 1992. Guidelines for the diagnosis of rheumatic fever (Jones criteria, 1992 update). *JAMA* **268:** 2069–2073.

42. **Dale, J. B., and E. H. Beachey.** 1985. Epitopes of streptococcal M proteins shared with cardiac myosin. *J. Exp. Med.* **162:**583–591.

43. **Dale, J. B., and E. H. Beachey.** 1985. Multiple, heart-cross-reactive epitopes of streptococcal M proteins. *J. Exp. Med.* **161:**113–122.

44. **Dale, J. B., and E. H. Beachey.** 1986. Sequence of myosin-cross-reactive epitopes of streptococcal M protein. *J. Exp. Med.* **164:**1785–1790.

45. **Dale, J. B., and E. H. Beachey.** 1987. Human cytotoxic T lymphocytes evoked by group A streptococcal M proteins. *J. Exp. Med.* **166:**1825–1835.

46. **Degnan, B., J. Taylor, and J. A. Goodacre.** 1997. *Streptococcus pyogenes* type 5 M protein is an antigen, not a superantigen for human T cells. *Hum. Immunol.* **53:**206–215.

47. **Dell, A., S. M. Antone, C. J. Gauntt, C. A. Crossley, W. A. Clark, and M. W. Cunningham.** 1991. Autoimmune determinants of rheumatic carditis: localization of epitopes in human cardiac myosin. *Eur. Heart J.* **12:**158–162.

48. **Dudding, B. A., and E. M. Ayoub.** 1968. Persistence of streptococcal group A antibody in patients with rheumatic valvular disease. *J. Exp. Med.* **128:**1081.

49. **Fae, K., J. Kalil, A. Toubert, and L. Guilherme.** 2004. Heart infiltrating T cell clones from a rheumatic heart disease patient display a common TCR usage and a degenerate antigen recognition pattern. *Mol. Immunol.* **40:**1129–1135.

50. **Fenderson, P. G., V. A. Fischetti, and M. W. Cunningham.** 1989. Tropomyosin shares immunologic epitopes with group A streptococcal M proteins. *J. Immunol.* **142:**2475–2481.

51. **Fleischer, B., K.-H. Schmidt, D. Gerlach, and W. Kohler.** 1992. Separation of T-cell-stimulating activity from streptococcal M protein. *Infect. Immun.* **60:**1767–1770.

52. **Fujinami, R. S., and M. B. A. Oldstone.** 1985. Amino acid homology between the encephalitogenic site of myelin basic protein and virus: mechanism of autoimmunity. *Science* **230:**1043–1045.

53. **Fujinami, R. S., M. B. A. Oldstone, Z. Wroblewska, M. E. Frankel, and H. Koprowski.** 1983. Molecular mimicry in virus infections: cross reaction of measles virus phosphoprotein of herpes simplex virus protein with human intermediate filaments. *Proc. Natl. Acad. Sci. USA* **80:**2346–2350.

54. **Galvin, J. E., M. E. Hemric, K. Ward, and M. W. Cunningham.** 2000. Cytotoxic monoclonal antibody from rheumatic carditis reacts with human endothelium: implications in rheumatic heart disease. *J. Clin. Investig.* **106:** 217–224.

55. **Gauntt, C., S. Tracy, N. Chapman, H. Wood, P. Kolbeck, A. Karaganis, C. Winfrey, and M. Cunningham.** 1995. Coxsackievirus-induced chronic myocarditis in murine models. *Eur. Heart J.* (Suppl.) **16:**56–58.

56. **Gibofsky, A., S. Kerwar, and J. B. Zabriskie.** 1998. Rheumatic fever: the relationships between host, microbe and genetics. *Rheum. Dis. Clin. North Am.* **24:**237–259.

57. **Goldstein, I., B. Halpern, and L. Robert.** 1967. Immunological relationship between streptococcus A polysaccha-ride and the structural glycoproteins of heart valve. *Nature* **213:**44–47.

58. **Goroncy-Bermes, P., J. B. Dale, E. H. Beachey, and W. Opferkuch.** 1987. Monoclonal antibody to human renal glomeruli cross-reacts with streptococcal M protein. *Infect. Immun.* **55:**2416–2419.

59. **Guilherme, L., E. Cunha-Neto, V. Coelho, R. Snitcowsky, P. M. A. Pomerantzeff, R. V. Assis, F. Pedra, J. Neumann, A. Goldberg, M. E. Patarroyo, F. Pileggi, and J. Kalil.** 1995. Human heart-filtrating T cell clones from rheumatic heart disease patients recognize both streptococcal and cardiac proteins. *Circulation* **92:**415–420.

60. **Guilherme, L., N. Dulphy, C. Douay, V. Coelho, E. Cunha-Neto, S. E. Oshiro, R. V. Assis, A. C. Tanaka, P. M. Pomerantzeff, D. Charron, A. Toubert, and J. Kalil.** 2000. Molecular evidence for antigen-driven immune responses in cardiac lesions of rheumatic heart disease patients. *Int. Immunol.* **12:**1063–1074.

61. **Guilherme, L., S. E. Oshiro, K. C. Fae, E. Cunha-Neto, G. Renesto, A. C. Goldberg, A. C. Tanaka, P. M. Pomerantzeff, M. H. Kiss, C. Silva, F. Guzman, M. E. Patarroyo, S. Southwood, A. Sette, and J. Kalil.** 2001. T-cell reactivity against streptococcal antigens in the periphery mirrors reactivity of heart-infiltrating T lymphocytes in rheumatic heart disease patients. *Infect. Immun.* **69:**5345–5351.

62. **Gulizia, J. M., M. W. Cunningham, and B. M. McManus.** 1991. Immunoreactivity of anti-streptococcal monoclonal antibodies to human heart valves. Evidence for multiple cross-reactive epitopes. *Am. J. Pathol.* **138:**285–301.

63. **Gulizia, J. M., M. W. Cunningham, and B. M. McManus.** 1992. Anti-streptococcal monoclonal antibodies recognize multiple epitopes in human heart valves: cardiac myosin, vimentin, and elastin as potential valvular autoantigens, p. 267–269. *In* G. Orefici (ed.), *New Perspectives on Streptococci and Streptococcal Infections. Proceedings of the XI Lancefield International Symposium on Streptococci and Streptococcal Diseases,* supplement 22. Gustav Fisher Verlag, New York, N.Y.

64. **Huber, S., J. Polgar, A. Moraska, M. Cunningham, P. Schwimmbeck, and P. Schultheiss.** 1993. T lymphocyte responses in CVB3-induced murine myocarditis. *Scand. J. Infect. Dis.* (Suppl.) **88:**67–78.

65. **Huber, S. A., and P. A. Lodge.** 1986. Coxsackievirus B-3 myocarditis: identification of different pathogenic mechanisms in DBA/2 and BALB/c mice. *Am. J. Pathol.* **122:** 284–291.

66. **Huber, S. A., and M. W. Cunningham.** 1996. Streptococcal M protein peptide with similarity to myosin induces CD4$^+$ T cell-dependent myocarditis in MRL/++ mice and induces partial tolerance against coxsackieviral myocarditis. *J. Immunol.* **156:**3528–3534.

67. **Huber, S. A., L. P. Job, and J. F. Woodruff.** 1980. Lysis of infected myofibers by coxsackievirus B3 immune lymphocytes. *Am. J. Pathol.* **98:**681–694.

68. **Huber, S. A., A. Moraska, and M. Cunningham.** 1994. Alterations in major histocompatibility complex association of myocarditis induced by coxsackievirus B3 mutants selected with monoclonal antibodies to group A streptococci. *Proc. Natl. Acad. Sci. USA* **91:**5543–5547.

69. **Hutto, J. H., and E. M. Ayoub.** 1980. Cytotoxicity of lymphocytes from patients with rheumatic carditis to cardiac cells *in vitro,* p. 733–738. *In* S. E. Read and J. B. Zabriskie (ed.), *Streptococcal Diseases and the Immune Response.* Academic Press, New York, N.Y.

70. **Jones, K. F., and V. A. Fischetti.** 1988. The importance of the location of antibody binding on the M6 protein for opsonization and phagocytosis of group A M6 streptococci. *J. Exp. Med.* **167:**1114–1123.

71. **Jones, K. F., S. A. Khan, B. W. Erickson, S. K. Hollingshead, J. R. Scott, and V. A. Fischetti.** 1986. Immunochemical localization and amino acid sequences of cross-reactive epitopes within the group A streptococcal M6 protein. *J. Exp. Med.* **164:**1226–1238.

72. **Jones, K. F., S. S. Whitehead, M. W. Cunningham, and V. A. Fischetti.** 2000. Reactivity of rheumatic fever and scarlet fever patients' sera with group A streptococcal M protein, cardiac myosin, and cardiac tropomyosin: a retrospective study. *Infect. Immun.* **68:**7132–7136.

73. **Jones, T. D.** 1944. The diagnosis of rheumatic fever. *JAMA* **126:**481–484.

74. **Kaplan, M. H.** 1963. Immunologic relation of streptococcal and tissue antigens. I. Properties of an antigen in certain strains of group A streptococci exhibiting an immunologic cross reaction with human heart tissue. *J. Immunol.* **90:**595–606.

75. **Kaplan, M. H., and M. Meyeserian.** 1962. Immunologic cross-reaction between group-A streptococcal cells and human heart tissue. *Lancet* **1:**706–710.

76. **Kaplan, M. H., and M. L. Suchy.** 1964. Immunologic relation of streptococcal and tissue antigens. II. Cross reactions of antisera to mammalian heart tissue with a cell wall constituent of certain strains of group A streptococci. *J. Exp. Med.* **119:**643–650.

77. **Kaplan, M. H., and K. H. Svec.** 1964. Immunologic relation of streptococcal and tissue antigens. III. Presence in human sera of streptococcal antibody cross-reactive with heart tissue: association with streptococcal infection, rheumatic fever, and glomerulonephritis. *J. Exp. Med.* **119:** 651–666.

78. **Kaplan, M. H., R. Bolande, L. Rakita, and J. Blair.** 1964. Presence of bound immunoglobulins and complement in the myocardium in acute rheumatic fever. Association with cardiac failure. *N. Engl. J. Med.* **271:**637–645.

79. **Kefalides, N. A., N. T. Pegg, N. Ohno, T. Poon-King, J. B. Zabriskie, and H. Fillit.** 1986. Antibodies to basement membrane collagen and to laminin are present in sera from patients with poststreptococcal glomerulonephritis. *J. Exp. Med.* **163:**588.

80. **Kefalides, N. A., N. Ohno, C. B. Wilson, H. Fillit, J. Zabriskie, and J. Rosenbloom.** 1993. Identification of antigenic epitopes in type IV collagen by use of synthetic peptides. *Kidney Int.* **43:**94–100.

81. **Kemeny, E., T. Grieve, R. Marcus, P. Sareli, and J. B. Zabriskie.** 1989. Identification of mononuclear cells and T cell subsets in rheumatic valvulitis. *Clin. Immunol. Immunopathol.* **52:**225–237.

82. **Kil, K. S., M. W. Cunningham, and L. A. Barnett.** 1994. Cloning and sequence analysis of a gene encoding a 67-kilodalton myosin-cross-reactive antigen of *Streptococcus pyogenes* reveals its similarity with class II major histocompatibility antigens. *Infect. Immun.* **62:**2440–2449.

83. **Kirvan, C. A., S. E. Swedo, S. Heuser, and M. W. Cunningham.** 2003. Mimicry and autoantibody-mediated neuronal cell signaling in Sydenham chorea. *Nature Med.* **9:**914–920.

84. **Kraus, W., and E. H. Beachey.** 1988. Renal autoimmune epitope of group A streptococci specified by M protein tetrapeptide: Ile-Arg-Leu-Arg. *Proc. Natl. Acad. Sci. USA* **85:**4510–4520.

85. **Kraus, W., J. M. Seyer, and E. H. Beachey.** 1989. Vimentin-cross-reactive epitope of type 12 streptococcal M protein. *Infect. Immun.* **57:**2457–2461.

86. **Kraus, W., J. B. Dale, and E. H. Beachey.** 1990. Identification of an epitope of type 1 streptococcal M protein that is shared with a 43-kDa protein of human myocardium and renal glomeruli. *J. Immunol.* **145:**4089–4093.

87. **Krisher, K., and M. W. Cunningham.** 1985. Myosin: a link between streptococci and heart. *Science* **227:**413–415.

88. **Lafer, E. M., J. Rauch, C. Andrezejewski, Jr., D. Mudd, B. Furie, R. S. Schwartz, and B. D. Stollar.** 1981. Polyspecific monoclonal lupus autoantibodies reactive with both polynucleotides and phospholipids. *J. Exp. Med.* **153:** 897–909.

89. **Lange, C. F.** 1969. Chemistry of cross-reactive fragments of streptococcal cell membrane and human glomerular basement membrane. *Transplant. Proc.* **1:**959–963.

90. **Lange, C. F.** 1994. Localization of [C14] labeled anti-streptococcal cell membrane monoclonal antibodies (anti-SCM mAb) in mice. *Autoimmunity* **19:**179–191.

91. **Liao, L., R. Sindhwani, M. Rojkind, S. Factor, L. Leinwand, and B. Diamond.** 1995. Antibody-mediated autoimmune myocarditis depends on genetically determined target organ sensitivity. *J. Exp. Med.* **187:**1123–1131.

92. **Lindberg, L. H., and K. L. Vosti.** 1969. Elution of glomerular bound antibodies in experimental streptococcal glomerulonephritis. *Science* **166:**1032–1033.

93. **Lyampert, I. M., O. I. Vvedenskaya, and T. A. Danilova.** 1966. Study on streptococcus group A antigens common with heart tissue elements. *Immunology* **11:**313–320.

94. **Lyampert, I. M., L. V. Beletskrya, and G. A. Ugryumova.** 1968. The reactions of heart and other organ extracts with the sera of animals immunized with group A streptococci. *Immunology* **15:**845–854.

95. **Lyampert, I. M., L. V. Beletskaya, N. A. Borodiyuk, E. V. Gnezditskaya, B. L. Rassokhina, and T. A. Danilova.** 1976. A cross-reactive antigen of thymus and skin epithelial cells common with the polysaccharide of group A streptococci. *Immunology* **31:**47.

96. **Malkiel, S., L. Liao, M. W. Cunningham, and B. Diamond.** 2000. T-cell-dependent antibody response to the dominant epitope of streptococcal polysaccharide, N-acetyl-glucosamine, is cross-reactive with cardiac myosin. *Infect. Immun.* **68:**5803–5808.

97. **Manjula, B. N., and V. A. Fischetti.** 1980. Tropomyosin-like seven residue periodicity in three immunologically distinct streptococcal M proteins and its implications for the antiphagocytic property of the molecule. *J. Exp. Med.* **151:**695–708.

98. **Manjula, B. N., and V. A. Fischetti.** 1986. Sequence homology of group A streptococcal Pep M5 protein with other coiled-coil proteins. *Biochem. Biophys. Res. Commun.* **140:** 684–690.

99. **Manjula, B. N., A. S. Acharya, S. M. Mische, T. Fairwell, and V. A. Fischetti.** 1984. The complete amino acid sequence of a biologically active 197-residue fragment of M protein isolated from type 5 group A streptococci. *J. Biol. Chem.* **259:**3686–3693.

100. **Manjula, B. N., B. L. Trus, and V. A. Fischetti.** 1985. Presence of two distinct regions in the coiled-coil structure of the streptococcal Pep M5 protein: relationship to mammalian coiled-coil proteins and implications to its biological properties. *Proc. Natl. Acad. Sci. USA* **82:** 1064–1068.

101. Markowitz, A. S., and C. F. Lange. 1964. Streptococcal related glomerulonephritis. I. Isolation, immunocytochemistry and comparative chemistry of soluble fractions from type 12 nephritogenic streptococci and human glomeruli. *J. Immunol.* **92**:565.

102. McCarty, M. 1956. Variation in the group specific carbohydrates of group A streptococci. II. Studies on the chemical basis for serological specificity of the carbohydrates. *J. Exp. Med.* **104**:629–643.

103. McCarty, M. 1964. Missing links in the streptococcal chain leading to rheumatic fever: The Duckett Jones Memorial Lecture. *Circulation* **29**:488–493.

104. Mertens, N. M., J. E. Galvin, E. E. Adderson, and M. W. Cunningham. 2000. Molecular analysis of cross-reactive anti-myosin/anti-streptococcal mouse monoclonal antibodies. *Mol. Immunol.* **37**:901–913.

105. Michael, A. F., Jr., K. N. Drummond, R. A. Good, and R. L. Vernier. 1966. Acute poststreptococcal glomerulonephritis: immune deposit disease. *J. Clin. Investig.* **45**: 237–248.

106. Miller, L. C., E. D. Gray, E. H. Beachey, and M. A. Kehoe. 1988. Antigenic variation among group A streptococcal M proteins: nucleotide sequence of the serotype 5 M protein gene and its relationship with genes encoding types 6 and 24 M proteins. *J. Biol. Chem.* **263**:5668–5673.

107. Mori, K., N. Kamakawaji, and T. Sasazuki. 1996. Persistent elevation of immunoglobulin G titer against the C region of recombinant group A streptococcal M protein in patients with rheumatic fever. *Pediatr. Res.* **55**:502–506.

108. Neu, N., N. R. Rose, K. W. Beisel, A. Herskowitz, G. Gurri-Glass, and S. W. Craig. 1987. Cardiac myosin induces myocarditis in genetically predisposed mice. *J. Immunol.* **139**:3630–3636.

109. Phillips, G. N., P. F. Flicker, C. Cohen, B. N. Manjula, and V. A. Fischetti. 1981. Streptococcal M protein: α-helical coiled-coil structure and arrangements on the cell surface. *Proc. Natl. Acad. Sci. USA* **78**:4698.

110. Pruksakorn, S., A. Galbraith, R. A. Houghten, and M. F. Good. 1992. Conserved T and B cell epitopes on the M protein of group A streptococci. Induction of bactericidal antibodies. *J. Immunol.* **149**:2729–2735.

111. Pruksakorn, S., B. Currie, E. Brandt, C. Phornphutku, S. Hunsakunachai, A. Manmontri, J. H. Robinson, M. A. Kehoe, A. Galbraith, and M. F. Good. 1994. Identification of T cell autoepitopes that cross-react with the C-terminal segment of the M protein of group A streptococci. *Int. Immunol.* **6**:1235–1244.

112. Putterman, C., and B. Diamond. 1998. Immunization with a peptide surrogate for double stranded DNA (dsDNA) induces autoantibody production and renal immunoglobulin deposition. *J. Exp. Med.* **188**:29–38.

113. Quinn, A., T. M. Shinnick, and M. W. Cunningham. 1996. Anti-Hsp 65 antibodies recognize M proteins of group A streptococci. *Infect. Immun.* **64**:818–824.

114. Quinn, A., K. Ward, V. Fischetti, M. Hemric, and M. W. Cunningham. 1998. Immunological relationship between the class I epitope of streptococcal M protein and myosin. *Infect. Immun.* **66**:4418–4424.

115. Raizada, V., R. C. Williams, P. Chopra, N. Gopinath, K. Prakash, K. B. Sharma, K. M. Cherian, S. P. Panday, R. Arora, M. Nigam, J. B. Zabriskie, and G. Husby. 1983. Tissue distribution of lymphocytes in rheumatic heart valves as defined by monoclonal anti-T cell antibodies. *Am. J. Med.* **74**:90–96.

116. Read, S. E., H. F. Reid, V. A. Fischetti, T. Poon-King, R. Ramkissoon, M. McDowell, and J. B. Zabriskie. 1986. Serial studies on the cellular immune response to streptococcal antigens in acute and convalescent rheumatic fever patients in Trinidad. *J. Clin. Immunol.* **6**:433–441.

117. Roberts, S., S. Kosanke, S. T. Dunn, D. Jankelow, C. M. G. Duran, and M. W. Cunningham. 2001. Pathogenic mechanisms in rheumatic carditis: focus on valvular endothelium. *J. Infect. Dis.* **183**:507–511.

118. Robinson, J. H., M. C. Atherton, J. A. Goodacre, M. Pinkney, H. Weightman, and M. A. Kehoe. 1991. Mapping T-cell epitopes in group A streptococcal type 5 M protein. *Infect. Immun.* **59**:4324–4331.

119. Robinson, J. H., M. C. Case, and M. A. Kehoe. 1993. Characterization of a conserved helper T-cell epitope from group A streptococcal M proteins. *Infect. Immun.* **61**:1062–1068.

120. Sandson, J., D. Hamerman, and R. Janis. 1968. Immunologic and chemical similarities between the streptococcus and human connective tissue. *Trans. Assoc. Am. Physicians* **81**:249–257.

121. Sargent, S. J., E. H. Beachey, C. E. Corbett, and J. B. Dale. 1987. Sequence of protective epitopes of streptococcal M proteins shared with cardiac sarcolemmal membranes. *J. Immunol.* **139**:1285–1290.

122. Schett, G., Q. Xu, A. Amberger, R. Van der Zee, H. Recheis, J. Willeit, and G. Wick. 1995. Autoantibodies against heat shock protein 60 mediate endothelial cytotoxicity. *J. Clin. Investig.* **96**:2569–2577.

123. Schmidt, K.-H., D. Gerlach, L. Wollweber, W. Reichardt, K. Mann, J.-H. Ozegowski, and B. Fleischer. 1995. Mitogenicity of M5 protein extracted from *Streptococcus pyogenes* cells is due to streptococcal pyrogenic exotoxin C and mitogenic factor MF. *Infect. Immun.* **63**: 4569–4575.

124. Schwab, J. H. 1962. Analysis of the experimental lesion of connective tissue produced by a complex of C polysaccharide from group A streptococci. I. In vivo reaction between tissue and toxin. *J. Exp. Med.* **116**:17–28.

125. Schwab, J. H. 1964. Analysis of the experimental lesion of connective tissue produced by a complex of C polysaccharide from group A streptococci. II. Influence of age and hypersensitivity. *J. Exp. Med.* **119**:401–408.

126. Schwab, J. H. 1965. Biological properties of streptococcal cell wall particles. I. Determinants of the chronic nodular lesion of connective tissue. *J. Bacteriol.* **90**:1405–1411.

127. Schwab, J. H., J. Allen, S. Anderle, F. Dalldorf, R. Eisenberg, and W. J. Cromartie. 1982. Relationship of complement to experimental arthritis induced in rats with streptococcal cell walls. *Immunology* **46**:83–88.

128. Schwimmbeck, P. L., and M. B. A. Oldstone. 1989. Klebsiella pneumoniae and HLA B27-associated diseases of Reiter's syndrome and ankylosing spondylitis. *Curr. Top. Microbiol. Immunol.* **45**:45–56.

129. Shikhman, A. R., and M. W. Cunningham. 1994. Immunological mimicry between N-acetyl-beta-D-glucosamine and cytokeratin peptides. Evidence for a microbially driven anti-keratin antibody response. *J. Immunol.* **152**:4375–4387.

130. Shikhman, A. R., N. S. Greenspan, and M. W. Cunningham. 1993. A subset of mouse monoclonal antibodies

cross-reactive with cytoskeletal proteins and group A streptococcal M proteins recognizes *N*-acetyl-beta-D-glucosamine. *J. Immunol.* **151**:3902–3913.

131. **Shikhman, A. R., N. S. Greenspan, and M. W. Cunningham.** 1994. Cytokeratin peptide SFGSGFGGGY mimics *N*-acetyl-beta-D-glucosamine in reaction with antibodies and lectins, and induces in vivo anti-carbohydrate antibody response. *J. Immunol.* **153**:5593–5606.

132. **Swerlick, R. A., M. W. Cunningham, and N. K. Hall.** 1986. Monoclonal antibodies cross-reactive with group A streptococci and normal and psoriatic human skin. *J. Investig. Dermatol.* **87**:367–371.

133. **Tomai, M., M. Kotb, G. Majumdar, and E. H. Beachey.** 1990. Superantigenicity of streptococcal M protein. *J. Exp. Med.* **172**:359–362.

134. **Tomai, M., J. A. Aileon, M. E. Dockter, G. Majumbar, D. G. Spinella, and M. Kotb.** 1991. T cell receptor V gene usage by human T cells stimulated with the superantigen streptococcal M protein. *J. Exp. Med.* **174**:285–288.

135. **Tomai, M., P. M. Schlievert, and M. Kotb.** 1992. Distinct T cell receptor Vβ gene usage by human T lymphocytes stimulated with the streptococcal pyrogenic exotoxins and M protein. *Infect. Immun.* **60**:701–705.

136. **van de Rijn, I., J. B. Zabriske, and M. McCarty.** 1977. Group A streptococcal antigens cross-reactive with myocardium: purification of heart reactive antibody and isolation and characterization of the streptococcal antigen. *J. Exp. Med.* **146**:579–599.

137. **Wang, B., P. M. Schlievert, A. O. Gaber, and M. Kotb.** 1993. Localization of an immunologically functional region of the streptococcal superantigen pepsin-extracted fragment of type 5 M protein. *J. Immunol.* **151**:1419–1429.

138. **Wedemayer, G. J., P. A. Patten, L. H. Wang, P. G. Schultz, and R. C. Stevens.** 1997. Structural insights into the evolution of an antibody combining site. *Science* **276**:1665–1669.

139. **Widdowson, J. P.** 1980. The M-associated protein antigens of group A streptococci, p. 125–147. *In* S. E. Read and J. B. Zabriske (ed.), *Streptococcal Diseases and the Immune Response.* Academic Press, New York, N.Y.

140. **Widdowson, J. P., W. R. Maxted, D. L. Grant, and A. M. Pinney.** 1971. The relationship between M-antigen and opacity factor in group A streptococci. *J. Gen. Microbiol.* **65**:69–80.

141. **Widdowson, J. P., W. R. Maxted, and A. M. Pinney.** 1976. An M-associated protein antigen (MAP) of group A streptococci. *J. Hyg.* **69**:553–564.

142. **Wu, X., B. Liu, P. L. Van der Merwe, N. N. Kalis, S. M. Berney, and D. C. Young.** 1998. Myosin-reactive autoantibodies in rheumatic carditis and normal fetus. *Clin. Immunol. Immunopathol.* **87**:184–192.

143. **Zabriskie, J. B.** 1967. Mimetic relationships between group A streptococci and mammalian tissues. *Adv. Immunol.* **7**:147–188.

144. **Zabriskie, J. B., and E. H. Freimer.** 1966. An immunological relationship between the group A streptococcus and mammalian muscle. *J. Exp. Med.* **124**:661–678.

145. **Zabriskie, J. B., K. C. Hsu, and B. C. Seegal.** 1970. Heart-reactive antibody associated with rheumatic fever: characterization and diagnostic significance. *Clin. Exp. Immunol.* **7**:147–159.

Extracellular Matrix Interactions with Gram-Positive Pathogens

GURSHARAN S. CHHATWAL AND KLAUS T. PREISSNER

8

Adherence to and invasion of eukaryotic cells are the main strategies used by pathogenic bacteria for colonization, evasion of immune defenses, survival, and causing disease in mammalian hosts. Most gram-negative bacteria make use of pili to achieve adherence and invasion. Because gram-positive pathogens do not possess pili, they express specific cell surface components called adhesins that mediate their adherence to host tissues, thereby facilitating not only colonization but also invasion (61). Most of these adhesins function by recognizing and binding to various components of the host extracellular matrix (ECM) (73). ECM consists of many diverse structures and complex macromolecules that maintain the bulk of tissues and provide them with tensile strength and elasticity (64). In addition, ECM affects the cellular physiology of the organism and is critical for adhesion, migration, proliferation, and differentiation of many cell types. ECM not only serves as structural support for cells but also provides support for infiltrating pathogens to colonize and invade, particularly under conditions of injury and trauma. Cell surface adhesive components of the host are often recognized by pathogenic bacteria in a tissue- or cell-specific manner. Pathogens also contact host tissue fluids that contain a variety of adhesive components. Microbial binding may lead to structural and/or functional alterations of host proteins and to activation of cellular mechanisms that influence tissue and cell invasion of pathogens.

The interactions of bacteria with ECM, therefore, represent important pathogenicity mechanisms. The identification and characterization of host ECM molecules and complementary bacterial adhesins would contribute not only to a better understanding of the molecular aspects of pathogenesis but also to the design of novel strategies to control and manage infectious diseases. In this chapter we describe the interaction of ECM components with gram-positive pathogens. The first section describes the structure and function of ECM, and the second deals with the interactions of gram-positive pathogens with ECM components and their biological consequences. These interactions are summarized in Table 1.

EXTRACELLULAR MATRIX

The major structural components of the eukaryotic ECM are collagens that form different types of interstitial or basement membrane networks (103). The interstitial ECMs are mainly built up by fibril-forming collagens (types I, II, and III), whereas basement membranes contain a two-dimensional collagen type IV network that is often interconnected with a laminin network. Other components, such as collagenous and noncollagenous glycoproteins, elastin, proteoglycans, hyaluronan, growth factors, and proteases, become associated with ECM, giving rise to their specialized structure and function at various locations in the body (46). The main portion of interstitial ECM is produced and deposited by embedded connective tissue cells such as fibroblasts in the dermis, smooth muscle cells and fibroblasts in the vasculature, and osteoblasts and chondroblasts in bone and cartilage. In addition, melanocytes, fat, skeletal muscle, nerve and epithelial cells, and circulating blood cells, such as macrophages, granulocytes, and lymphocytes, participate in the synthesis and secretion of ECM material and determine the specific character of each tissue or organ. During embryonic development, the ECM is constantly rearranged, thereby regulating morphogenesis in an active and dynamic fashion. Likewise, the time-dependent modification and rebuilding of ECM at various locations in the body are essential for inflammation and wound healing. At these sites, the provisional ECM may provide bacterial entry and colonization, and soluble proteolytic fragments of ECM proteins are indicators of matrix turnover, particularly under disease conditions or in tumor patients (103). Specific circulating fragments of type XVIII collagen (endostatin) (88) or type IV collagen (tumstatin) (55) were shown to exert potent antiangiogenic activities on vascular cells (66). Yet, their respective interactions with microbes have not been explored.

Structure and Function of ECM Molecules

More than 25 different collagens have been defined on a molecular basis. A major subfamily of collagens is the fibril-forming types, such as I, II, III, V, and XI, produced and

Gram-Positive Pathogens, 2nd edition, edited by Vincent A. Fischetti et al.
© 2006 ASM Press, Washington, D.C.

TABLE 1 Properties of ECM components involved in interaction with gram-positive pathogens

Parameter	Collagen type I, II	Collagen type IV	Elastin	Laminin	Fibrinogen	Fibronectin	Vitronectin	Thrombospondin
Mol. mass (kDa)	Fibers	Network	Network	900	350	440	78	420
Immobilized form	Yes	Yes	Yes	Yes	Yes	Yes	Yes	Yes
Soluble form	No	No	No	No	Yes	Yes	Yes	Yes
Plasma concentration (mg/ml)					2–4	0.2–0.4	0.2–0.4	approx. 0.02
Major bacterial binding domain	A domain		30-kDa N-terminal		D fragment	29-kDa N-terminal domain/ type I module	Hemopexin-like repeats, heparin-binding site (?)	?
Interacting major gram-positive pathogens	Staphylococcus aureus	Streptococcus pyogenes	S. aureus, coagulase-negative staphylococci	S. aureus, Streptococcus sanguis, S. pyogenes, Streptococcus agalactiae, Streptococcus mutans, Streptococcus anginosus	S. aureus, A, B, C, and G streptococci, S. mutans, Staphylococcus epidermidis	S. aureus, coagulase-negative staphylococci, A, B, C, G streptococci, Streptococcus pneumoniae, S. sanguis	S. aureus, coagulase-negative staphylococci, A, C, G streptococci	S. aureus, A, C, and G streptococci

assembled during wound healing processes (78). These collagens are characterized by large stretches of triple-helical protein strands and are distinct from the fibril-associated collagens as well as a third subfamily, the network-forming collagens of which type IV and X are prominent examples. All collagen types are expressed in a tissue- and/or cell-specific manner and contribute to the appearance and mechanical properties of a given ECM, thereby determining that cellular environment.

Network-forming collagens, such as different isoforms of type IV, determine the structure of basement membranes as a specialized ECM. An intercalated laminin network that can also exist independently is linked to type IV collagen via nidogen/entactin. Heparan sulfate proteoglycans, usually perlecan, and adhesive glycoproteins are embedded into these supramolecular arrays (103). These components are in direct contact with epithelial, endothelial, striated muscle, fat, and nerve cells, which produce the individual components and deposit them in a polarized fashion toward the cellular basolateral side. Basement membranes serve quite distinct functions in the body, such as linking epithelia to the underlying interstitial ECM, covering skeletal myotubes as basal lamina, or serving as a filtration barrier in the glomerulus.

Laminins

Laminins are major constituents of basement membranes and are the first ECM proteins to be produced during embryogenesis. Because of their specific interactions with type IV collagen, proteoglycans, and other ECM components, as well as with several cell types, laminins play an important structural and functional role within basement membranes. Laminin, with a molecular mass of 900 kDa, is composed of three different chains, α, β, and γ, which are linked to each other by disulfide bridges. Owing to variations in chain composition, more than 10 different isoforms of laminin exist that are expressed in a tissue-specific manner. Laminins are in close contact with a variety of cell types, including muscle, adipocytes, neurons, and endothelial and epithelial cells. Laminins are involved in promotion of cell proliferation, attachment, and chemotaxis, as well as neurite outgrowth and enhancement of angiogenesis.

Elastin

Another important molecule of the ECM in specific tissues is elastin, which is the main component of elastic fiber. These fibers are responsible for the elasticity of lung, skin, and other tissues, particularly blood vessels. Elastin is a highly cross-linked polymer of the nonproteolytically modified, hydrophobic precursor tropoelastin. During fibrogenesis, secreted tropoelastin becomes stabilized into insoluble elastin by intermolecular cross-linking of lysines into desmosine and isodesmosine cross-links through the copper-dependent enzyme lysyl oxidase (47). The rubberlike mechanical properties of elastin result from repetitive hydrophobic domains of tropoelastin.

Adhesive Glycoproteins

Adhesive glycoproteins constitute important components of ECM. Upon vessel wall injury, particularly at sites of wound healing, initial adhesion of platelets and subsequent aggregation are dependent on adhesive glycoproteins, such as fibronectin and vitronectin, that are present in the subendothelial cell matrix and stored inside platelets and are secreted during this initial phase of hemostatic plug formation. Adhesive glycoproteins not only promote attach-

ment via their Arg-Gly-Asp (RGD)-containing epitope (77) but also perform multiple functions by interacting with other ligands, such as heparan sulfate, collagens, or mediators of humoral defense mechanisms. These proteins also exist as soluble forms, which may differ from those in the subendothelium and α-granules of platelets owing to alternative splicing, differences in the state of polymerization, different conformations, or the transition into a self-aggregating molecule. The partitioning of these proteins between humoral and cell surface or ECM phases, together with inducible receptor sites on platelets or inflammatory cells, indicates that adhesive glycoproteins are important players in tissue remodeling or defense mechanisms (77). Thus, these adhesion molecules not only provide versatile molecular links mediating adhesive processes and responsive reactions at localized sites but are also of major importance for the initial adherence phase of pathogens.

Fibronectin

Fibronectin is a ubiquitous adhesive protein that is essential for the adhesion of almost all types of cells. It is abundant in the circulation and at various ECM sites. The characteristic form of the molecule in solution is a dimer generated by disulfide bridging at the carboxy terminus of two similar subunits, each with a molecular mass of about 220 kDa (40). In addition to approximately 30 intrachain disulfide bonds, two free sulhydryl groups per subunit are involved in the formation of high-molecular-weight polymers of fibronectin, which are predominantly found in tissues. The heterogeneity observed in fibronectin molecules isolated from plasma or tissue is due to variation in both the amino acid sequence and posttranslational modifications. Although only one gene has been identified, variations in the carbohydrate content or structure, phosphorylation, sulfation, and acetylation are responsible for additional heterogeneity.

The sensitivity of fibronectin to proteolytic degradation has been used to identify the structure-function relationships of independent domains. The 30-kDa amino-terminal fragment contains the major acceptor site for factor XIIIa-mediated cross-linking and also bears the binding sites for heparin, fibrin, and bacteria, including *Staphylococcus aureus* (63). The well-known property of fibronectin to bind collagen or gelatin and to complement C1q is contained within the adjacent 40-kDa fragment, while the central portion of the molecule has no well-defined binding functions. The versatile integrin recognition sequence RGD was first recognized in the type III-11 repeat of fibronectin (74) and has been found in a large number of adhesive and nonadhesive proteins as well. Additional heparin-binding and fibrin-binding domains are located within the carboxy-terminal portion of the fibronectin molecule. Together with the RGD-containing cell attachment site, the heparin-binding domain is crucial for the establishment of stable focal adhesions, as has been demonstrated in fragment complementation assays. Synthesis and deposition of fibronectin as a self-associating fibrillar array into the growing ECM of adhesive cells occur by an active process, and the accumulation of fibronectin into an insoluble form is potentiated by disulfide bridge formation, as well as covalent cross-linking by transglutaminase/factor XIIIa.

Vitronectin

Another adhesive glycoprotein, vitronectin (75), is found in the circulation as a single-chain polypeptide with a molecular mass of 78 kDa and becomes associated with various ECM sites, particularly during tissue or vascular remodeling processes. Several immunofluorescent and histochemical studies suggest the deposition of vitronectin in a fibrillar pattern in loose connective tissue, in association with dermal elastic fibers in skin and with renal tissue (76). Moreover, the accumulation of terminal complement complexes along elastic fibers later in life, the association of vitronectin with keratin bodies during keratinocyte programmed cell death (apoptosis), and colocalization of vitronectin with deposits of the terminal complement complex in kidney tissue from patients with glomerulonephritis suggest a role for vitronectin in preventing tissue damage in proximity to local complement activation. Although the mechanism of deposition of exogenous vitronectin alone or in association with other proteins into different tissues remains unclear, it may occur in a fashion similar to that for fibronectin; thus, both proteins are prominent candidates for promoting adherence of bacteria at various accessible ECM sites. The interaction of vitronectin with glycosaminoglycans or different types of native collagens and basement membrane-associated osteonectin and the cross-linking of vitronectin by transglutaminase/factor XIIIa are reactions likely to occur in the ECM in vivo as well (48, 83).

Fibrinogen

The adhesive glycoprotein fibrinogen is a major plasma glycoprotein that serves as the predominant macromolecular substrate for thrombin in the blood-clotting cascade. The primary structure of fibrinogen (35) together with biophysical studies (32) indicate that the 350-kDa molecule is composed of two identical sets of three polypeptide chains, Aα, Bβ, and γ, which are disulfide-bridged and organized in an antiparallel fashion. Upon selective and specific proteolytic attack by α-thrombin, two pairs of fibrinopeptides, A and B, are sequentially released, and the appearance of these peptides in the circulation is an indicator of thrombin activity in vivo. Covalent stabilization of the forming fibrin clot is mediated by transglutaminase/factor XIIIa-dependent cross-linking between γ and α chains. Together with invading cells and aggregating platelets, the fibrin clot constitutes the majority of the initial provisional ECM network for the wound healing process (28).

Fibrinogen is a multifunctional protein capable of binding to collagen, fibronectin, components of the fibrinolytic system, and a variety of eukaryotic cells, as well as to bacteria. In particular, while the fibrin clot is being organized, it may already serve as a cofactor surface for tissue plasminogen activator-dependent plasminogen formation, ultimately leading to clot lysis. Cell surface receptors for fibrinogen that belong to the family of integrins have been identified on mammalian cells; the platelet integrin α_{IIb}-β_3 (GP IIb/IIIa) is principally required for platelet aggregation. Integrin α_M-β_2 (complement receptor 3) on phagocytes may also recognize fibrinogen during wound healing and defense when phagocytic clearance of fibrin(ogen)-associated clot or cell fragments is required. The recognition of the distal end of the γ-chain of fibrinogen and of two RGD-containing sites and additional epitopes by these integrins indicates that fibrinogen may serve as a bridging component between surface receptors on different cells or other extracellular sites once they become exposed.

Thrombospondin

Thrombospondin, a multifunctional glycoprotein of 420 kDa, is stored in platelet α-granules and is secreted upon stimulation (49). The released thrombospondin becomes incorporated into fibrin clots. Thrombospondin belongs to a family of structurally unrelated members of the tenascin protein family, including osteonectin and osteopontin. These matricellular proteins promote divergent cellular functions (8). Owing to highly regulated biosynthesis by various hormones and cytokines and during development, nerve regeneration, and vascular remodeling related to angiogenesis, these proteins become sequestered in the ECM and are therefore available for subsequent recruitment to different cell surface receptors, including integrins. In particular, thrombospondin plays a role in binding and modulation of cytokines and proteases. The level of thrombospondin in the circulation and other body fluids produced by vascular cells or secreted by activated platelets and its distribution in tissues appear to vary in correlation with different pathological states. Specifically, thrombospondin promotes cell attachment and spreading via αv-integrins in the vascular system but may also lead to destabilization of focal adhesions (65).

INTERACTION OF GRAM-POSITIVE PATHOGENS WITH ECM MOLECULES

Binding of Collagen to Gram-Positive Bacteria

Many staphylococcal and streptococcal species interact specifically with collagenous proteins. It has been speculated that these interactions might be important in the pathogenesis of various diseases, such as osteomyelitis and infective arthritis. Switalski et al. (93) isolated a surface protein of 135 kDa from S. aureus and tentatively identified it as a collagen receptor. Patti et al. (72) cloned and sequenced a gene, cna, encoding a collagen-binding protein from S. aureus. This collagen-binding protein exhibits all the main features of the surface protein of gram-positive bacteria. The ligand-binding domain of S. aureus is localized in the A region of collagen (72). So far, only a single gene has been identified for staphylococcal collagen-binding protein, although proteins of different molecular sizes have been reported from different strains, probably owing to strain-to-strain variation in the number of repeats (95). The collagen-binding protein from S. aureus is involved in the adherence of bacteria to cartilage. The mediation of adherence to cartilage is specific to collagen, because synthetic beads coated with collagen mediate the adherence to cartilage, whereas the beads coated with fibronectin show no adherence (95).

The binding to collagen by different streptococcal species has also been reported. A recent study showing the involvement of collagen in acute rheumatic fever, a serious autoimmune sequela of Streptococcus pyogenes, underlines the importance of this interaction (24). S. pyogenes serotype M3 and M18 isolated during outbreaks of rheumatic fever have the unique capability to bind and aggregate collagen type IV. M3 protein is identified as collagen binding factor of M3 serotypes, whereas M18 isolates bind collagen through capsule. Immunization of mice with purified M3 protein led to the generation of anticollagen type IV antibodies. These findings suggest a link between collagen binding and presence of collagen-reactive antibodies that may form a basis for rheumatic fever. Some S. pyogenes strains that do not interact with collagen directly are able to recruit collagen by prebound fibronectin (23). Fibronectin-mediated collagen recruitment represents a novel aggregation, colonization, and immune evasion mechanism of S. pyogenes. Animal pathogenic streptococci also interact with collagen type II.

Streptococcus mutans interacts with collagen type I via a 16-kDa binding protein with high affinity, as shown by adherence of bacteria to a collagen-coated surface. Furthermore, collagen mediates adherence of *S. mutans* to dentin in the oral cavity, thereby playing a role in the pathogenesis of root surface caries (94). The collagen-binding protein of *S. mutans* designated Cnm protein represents a new strain-specific member of the collagen-binding adhesion family (84). In infectious endocarditis, which is most commonly caused by streptococci and which is characterized by the formation of septic masses of platelets on the surfaces of heart valves, collagenlike platelet aggregation-associated protein of *Streptococcus sanguis* and direct interactions with host ECM collagens enhance platelet accumulation and subsequent bacterial colonization. The binding of collagen to antigen I/II family of polypeptides of oral streptococci has been reported to facilitate the invasion of bacteria into root dentinal tubes (54), which further underlines the importance of collagen-bacterial interactions.

Binding of Laminin to Gram-Positive Bacteria

Exposure of laminin to pathogenic bacteria is most frequently seen in damaged or inflamed tissues. Switalski et al. (91) described the binding of *S. pyogenes* to laminin. Human monoclonal antibodies that have been developed from rheumatic heart disease patients have been found to recognize laminin, among other proteins. These autoantibodies are generally cross-reactive with a variety of *S. pyogenes* components and are polyspecific in nature (20). It has been proposed that autoantibodies recognizing myosin in the heart muscle and laminin in the heart valve are centrally involved in the pathogenesis of rheumatic heart disease (21). Laminin therefore plays an important role in the pathogenesis of *S. pyogenes* diseases. Streptococcal-mediated damage to epithelial cell surfaces or entry through skin sores may grant the colonizing bacterium access to the laminin of the basement membrane and surrounding tissues. To date, two streptococcal laminin-binding proteins have been described. First, streptococcal pyrogenic exotoxin B has been shown to bind to laminin. This protein, which is a potent secreted and cell surface cysteine protease, degrades host ECM components, activates interleukin-1β, and is ubiquitously expressed by *S. pyogenes* strains (41). The 34-kDa laminin-binding protein Lbp was found to be present and highly conserved in all *S. pyogenes* strains examined, and mediated attachment to Hep 2 cells (101). Lmb protein from *Streptococcus agalactiae* has been identified as laminin-binding protein as well. Lmb-mediated attachment of *S. agalactiae* to laminin may be essential for the bacterial colonization of damaged epithelium and translocation of bacteria into the bloodstream (87). The binding of laminin to a *Streptococcus gordonii* strain isolated from a patient with infective endocarditis (86) was mediated by a 145-kDa cell wall protein that was regulated by the presence of laminin. The binding of laminin to *S. mutans* and *Streptococcus anginosus* has also been reported (1, 4). The sera from patients with infective carditis contained antibodies against this protein, whereas no significant recognition was seen with sera from patients with valvulopathies, suggesting an increased expression of laminin-binding protein during infective endocarditis. Although a few streptococcal strains isolated from the oral cavity, including viridan streptococci (92), expressed this protein, the majority of strains isolated from patients with endocarditis expressed binding protein(s) for laminin, indicating that this ECM component might be an important factor in the pathogenesis of viridans endo-

carditis. Binding of laminin has also been described for *S. aureus*, and a surface-localized enolase has been identified as a bacterial binding component (9).

Binding of Elastin to Gram-Positive Bacteria

The specific binding of elastin was described for a number of *S. aureus* strains (70). The binding component was identified as a 25-kDa staphylococcal surface protein, termed EbpS, capable of binding elastin with high affinity. Staphylococcal elastin binding is mediated by a discrete domain defined by a short peptide sequence in the amino terminal extracellular region of EbpS (71). EbpS is expressed at the cell surface as an integrated membrane protein. The expression of EbpS was correlated with the ability of cells to grow to a higher density in liquid cultures, indicating that EbpS may have a role in regulating cell growth (25). It has also been postulated that EbpS might contribute to infection in organs containing elastin, such as lung, skin, and blood vessels.

Fibronectin, a Matrix Protein Most Sought after by Gram-Positive Pathogens

Fibronectin not only acts as a substrate for the adhesion of eukaryotic cells but also serves as the prototype of adhesion proteins that bind specifically to microorganisms (12, 111). Because it binds to more than 16 bacterial species, the interactions with gram-positive bacteria have been characterized extensively. In particular, most *S. aureus* isolates, as well as various streptococcal strains, specifically bind and adhere to fibronectin, mediated by several different bacterial adhesins (33, 60, 96, 97) with highly homologous recognition motifs for the adhesion protein. Two genes, *fnbpA* and *fnbpB*, encoding the fibronectin-binding protein of *S. aureus* have been described. The gene products show typical features of gram-positive surface proteins, and the binding domain was localized to a 38-amino-acid repeated sequence of these proteins. FNBPA has recently been shown to induce aggregation of platelets and may play a role in *S. aureus*-associated infective endocarditis (34).

A number of fibronectin-binding proteins have also been identified in *S. pyogenes* as well as the fibronectin-binding protein SfbI of group A streptococci that mediates the adherence of bacteria with epithelial cells (96, 97). Hanski and Caparon (33) identified the fibronectin-binding protein F1 encoded by the gene *prtF*. The deduced amino acid sequences of SfbI and protein F1 indicated that both proteins were identical. SfbI protein consists of 638 amino acids and comprises five structurally distinct domains. The N-terminal signal peptide is followed by an aromatic domain and four proline-rich repeats, which are flanked by nonrepetitive spacer sequences. A second repeat region, distinct from the proline repeats, is located in the C-terminal part of the protein. Fibronectin binding was located to the C-terminal repeat region as well as to the nonrepetitive spacer sequence. Functional analysis showed that the repetitive sequence is essential for adherence, whereas both repeat and spacer sequences are required for invasion into eukaryotic cells (99). SfbI protein therefore represents a highly evolved prokaryotic molecule that exploits fibronectin not only for extracellular targeting but also for its subsequent activation, which leads to efficient cellular invasion. SfbI protein also triggers invasion through caveolae and participates in organelle formation to avoid intracellular killing (82). This might explain the mechanism of intracellular survival and persistence of *S. pyogenes*. The functional domains of SfbI were highly conserved in isolates belonging to different serotypes (98). Unlike M protein, SfbI protein did not show any cross-reactivity with host proteins (109). Besides its role in adherence, SfbI protein is also involved in the invasion of epithelial cells by group A streptococci. Synthetic beads coated with SfbI protein were readily internalized by epithelial cells, indicating that SfbI alone is sufficient for invasion (59, 61). SfbI protein, therefore, is the first defined invasin of group A streptococci. The cellular receptors responsible for protein F1 (SfbI)-mediated invasion have been identified as integrins capable of binding fibronectin (67). Pathogenic bacteria are attached to human fibronectin through tandem β-zippers, which might explain the mechanism of the integrin-dependent fibronectin-binding-protein-mediated invasion of host cells (85).

Besides its involvement in adherence and invasion, SfbI protein has also been shown to be a strong mucosal adjuvant (59). Mice immunized intranasally with ovalbumin coupled to SfbI evoked a substantially higher immunoglobulin A response in lung lavage than did mice immunized with ovalbumin alone. SfbI protein also evoked a protective immune response in mice. Animals immunized with SfbI protein were protected against the lethal challenge of group A streptococci belonging to different serotypes (31). These results underline the importance of SfbI protein in vaccine development. Kreikemeyer et al. (43) and Rakonjac et al. (79) showed that serum opacity factors of group A streptococci (SfbII and SOF22) also interacted with fibronectin specifically. Their fibronectin-binding region was located in the C-terminal repeat region of the molecule that is not involved in the serum opacity factor activity. Because SOF is mainly a secreted protein, the role of its interaction with fibronectin is not yet clear. In in vivo experiments, however, synthetic beads coated with the repeat region of SOF showed significant adherence to epithelial cells (44). Another fibronectin-binding protein, Sffbp12, was described in group A streptococci with no homology to SfbI and SfbII protein (81). Sffbp12 shares a high degree of homology with fibronectin- and fibrinogen-binding proteins from *Streptococcus dysgalactiae* and *S. aureus*, respectively. Another fibronectin-binding protein, F2, functions as a major fibronectin-binding protein adhesin in a subset of *S. pyogenes* strains and has been reported to be an important factor for both early and late pathogenetic stages of superficial infections (45). Two other fibronectin-binding proteins from *S. pyogenes*, Fba and FbaB, have been identified (100, 102). Fba has been shown to promote bacterial entry into epithelial cells, whereas FbaB is etiologically involved in the development of invasive streptococcal disease. *S. agalactiae* also binds fibronectin via a C5a peptidase, which has another function of cleaving the complement anaphylatoxin C5a (3).

Fibronectin also interacts with animal pathogenic streptococci, facilitating their adherence to host epithelial cells (106). *S. dysgalactiae*, a cattle pathogen, expresses two different binding proteins, FnbA and FnbB (53). Their fibronectin-binding domain is located in the repeat region of both proteins, whereby the repeats of FnbA and FnbB do not show any sequence homology. Despite the difference in sequence of repeats for various fibronectin-binding proteins, they were capable of cross-inhibiting the binding of fibronectin to *S. aureus*, *S. dysgalactiae*, and *S. pyogenes* (42). The repeat motif of SfbI protein, therefore, conforms to the consensus sequence previously reported for FnbA, FnbB, and FnbpA.

Fibronectin binding has also been observed with coagulase-negative staphylococci (108). Unlike other gram-positive cocci, coagulase-negative staphylococci interact only with immobilized fibronectin, which allows them to colonize artificial devices such as intraocular lenses, prosthetic cardiac valves, vascular grafts, prosthetic joints, and intravascular catheters. Fibronectin incorporated in fibrin thrombi is also the cause of infection at sites of blood clots or damaged tissue (17). Some strains of *Streptococcus pneumoniae* also interact with immobilized fibronectin via a surface-located protein, PavA (38, 110). This protein has no typical signal sequence and membrane anchor and therefore belongs to a family of anchorless proteins that get secreted by an unknown mechanism and reassociates on the bacterial surface (11). PavA plays a direct role in the pathogenesis of pneumococcal infections, because isogenic PavA mutants are more than 100-fold attenuated in virulence in a mouse sepsis model (38).

Binding of Vitronectin to Gram-Positive Bacteria

Vitronectin has equivalent effects on cellular adhesion and bacterial binding as are demonstrated for fibronectin, but owing to the deposition of vitronectin in the periphery, specific interactions with bacteria are likely to occur in damaged or altered tissues. Specific interaction of vitronectin with various strains of staphylococci as well as groups A, C, and G streptococci has been described (16). The binding of vitronectin to group G streptococci, but not to groups A and C streptococci, is inhibited by heparin, indicating the diversity of vitronectin-binding proteins among streptococci. Like for fibronectin, binding to vitronectin mediates the adherence of gram-positive bacteria to host cells (27, 106). The vitronectin-binding proteins therefore represent additional adhesins of gram-positive bacteria. The adhesin of *S. aureus* responsible for vitronectin interaction has striking similarity to a heparan sulfate-binding protein from the same staphylococcal strain, suggesting that the adhesion protein undergoes multiple interactions with different bacterial surface recognition sites (51). Although the streptococcal vitronectin-binding protein has not yet been identified, four vitronectin-binding proteins from *Staphylococcus epidermidis* have been described (50). There is an indication that interaction of vitronectin with multiple recognition sites of *S. epidermidis* surface may contribute to bacterial colonization (50). The binding of vitronectin has also been described for *S. pneumoniae*, but the pathogenic role of this interaction has not yet been elucidated (26).

Distant sequence homology exists between vitronectin and the heme-binding plasma protein hemopexin, whose ligand-binding properties strongly depend on the conformational flexibility of the protein. In addition to RGD- and heparin-dependent interactions with gram-positive bacteria, hemopexin-type repeats in vitronectin, as well as in hemopexin itself, have been identified as primary binding sites for group A streptococci (52).

Binding of Fibrinogen to Gram-Positive Bacteria

Together with invading cells and aggregating platelets, the fibrin clot represents the majority of the initial provisional ECM network for sealing a wound site. At sites of trauma, fibrinogen therefore serves as a substrate for bacterial adhesion. Various fibrinogen-binding proteins mediating bacterial colonization in wounds or catheters have been described, of which those expressed by *S. aureus* are the best characterized (57). In particular, "clumping factor" serves to recognize the carboxy-terminal portion of fibrinogen γ-chain in a manner analogous to $\alpha_{IIb}\beta_3$-integrin binding. Moreover, the homology between metal ion-dependent adhesion sites of integrin α_{IIb}- and α_M-subunits or clumping factor and an integrinlike protein from *Candida albicans* (39) indicates common mechanisms of fibrinogen binding in mammalian cells, lower eukaryotes, and prokaryotes. Staphylocoagulase serves as an additional fibrinogen-binding factor that is not involved in bacterial clumping but, owing to prothrombin binding and conversion, serves to promote fibrin formation or bacterial attachment onto fibrinogen-coated surfaces (22). In group A streptococci, M protein is a cell surface structure principally responsible for binding fibrinogen, and this binding contributes to the known antiopsonic property of M protein (112). Fibrinogen binding to streptococci of groups C and G also leads to inhibition of complement fixation and subsequent phagocytosis, indicating an important role of this interaction (14, 105). Furthermore, bacterial colonization was reduced in a mouse mastitis model by vaccination with *S. aureus* fibrinogen-binding proteins (56), providing a new concept for antimicrobial therapy. A novel fibronectin-binding protein, FbsB, has been reported in *S. agalactiae*. This protein plays an important role in the overall process of host cell entry by *S. agalactiae* (30).

The adherence of *S. aureus* to endothelial cells is mediated predominantly by fibrinogen as bridging molecule, leading to acute endovascular infections (10). Although platelet aggregation via a fibrinogen-dependent mechanism can be induced by this pathogen, this process is independent of the aforementioned principal $\alpha_{IIb}\beta_3$-integrin-binding interactions with fibrinogen (2). *S. aureus* colonizes skin lesions of more than 90% of patients with atopic dermatitis. It was shown that fibrinogen plays a major role in the enhanced binding of *S. aureus* to atopic skin (18).

In addition to staphylococcal or streptococcal surface proteins that interact with ECM components, proteins released by these bacterial species can directly influence fibrin formation or dissolution. While staphylocoagulase binds and activates prothrombin to thrombin, staphylokinase and streptokinase interact stoichiometrically with plasminogen, resulting in plasmin formation, whereby the former fibrinolytic agent acts in a fibrin-specific manner. These strategies apparently allow effective fixation and subsequent penetration of these bacteria into wound areas.

Binding of Thrombospondin to Gram-Positive Bacteria

The high-affinity binding of thrombospondin has been described for *S. aureus* strains (36). This interaction mediates the adherence of staphylococci to tissues via activated platelets during inflammation or infection. A large number of protein A-negative and -positive staphylococcus isolates adhered to thrombospondin-coated synthetic disks, indicating that adherence is significantly promoted as a function of absorbed thrombospondin. The binding of thrombospondin to coagulase-negative streptococci may contribute to bacterial adherence on biomaterial surfaces (114). Specific binding of thrombospondin has also been observed with groups A, C, and G streptococci. Like *S. aureus*, the binding component in streptococci is resistant to trypsin. The exact biological function of streptococcal-thrombospondin interaction has yet to be explored.

Degradation of ECM by Gram-Positive Pathogens

Invasion of bacteria in tissues requires the degradation of matrix proteins. This is accomplished by various bacterial enzymes as well as by host-derived plasmin (7). Tissue-type plasminogen activators produced by many eukaryotic cells, particularly under inflammatory conditions, are responsible for plasmin formation. A number of bacteria, such as groups A, C, and G streptococci, express surface receptors for plasmin(ogen) that facilitate pericellular proteolysis and invasion (68, 80). A plasmin-binding protein, PAM, has been isolated from group A streptococci. This protein belongs to the M protein family and binds to the kringle domain of plasminogen. This binding was blocked by a lysine analog, indicating that lysine residues in the M-like protein participate in the interaction (113). The interaction of streptococci with plasmin or plasminogen might contribute to streptococcal invasion. It was found that plasmin bound to streptococci could be protected from inactivation by the plasmin inhibitor α_2-antiplasmin. The structural similarity of neutrophil-derived polypeptide "defensins" to plasminogen kringle motifs (37) suggests that their antimicrobial activity can be related to interference with plasmin formation, thereby preventing the spreading of infection. Plasminogen binding to *S. pyogenes* has been shown to mediate adherence to and pericellular invasion of human pharyngeal cells (69). Plasminogen binding has also been associated with the skin infections caused by *S. pyogenes* (58, 90). The critical role of plasminogen in group A streptococcal infections has recently been demonstrated in a study using a transgene-expressing human plasminogen with markedly increased mortality in mice infected with streptococci (89). Plasminogen also influences the pneumococcal pathogenesis, because surface-located α-enolase of *S. pneumoniae* binds and activates plasminogen, leading to tissue invasion, which underlines the importance of this interaction in invasive infectious processes (5, 6).

As an acute-phase reactant in host defense, a circulating broad-spectrum proteinase inhibitor, α_2-macroglobulin, serves to eliminate complex proteinases via receptor-mediated endocytosis and may do so with microbial enzymes as well (19). Interestingly, various strains of streptococci exhibit specific binding to α_2-macroglobulin (13). A protein of 78 kDa was purified from group A streptococci that interacted specifically with native α_2-macroglobulin. Although this protein possessed no proteolytic activity, its interaction with α_2-macroglobulin led to a change in conformation similar to that obtained by α_2-macroglobulin-protease complexes (15). Through their interaction with α_2-macroglobulin, streptococci might gain access to host tissues, possibly via the α_2-macroglobulin receptor (107). A streptococcal surface protein, GRAB, interacts with α_2-macroglobulin (29). The mice infected with GRAB mutant survived longer and exhibited lower levels of bacterial dissemination, emphasizing the role of α_2-macroglobulin in the virulence of *S. pyogenes* (104).

CONCLUSIONS

Gram-positive pathogens are still a major cause of serious human diseases. The rapid emergence of resistance to antibiotics in various gram-positive cocci, and the fact that no vaccine is available against most of these organisms, makes gram-positive pathogens a major health hazard. A better understanding of the molecular mechanism of gram-positive infections is a prerequisite for development of effective vaccines and for designing novel strategies for the prevention and treatment of these infections. Adherence and invasion are the important disease-causing mechanisms of gram-positive pathogens. Because interaction of these pathogens with the components of host ECM is involved both in adherence and in invasion, the underlying mechanisms of this interaction are of utmost importance for designing novel therapeutic strategies. The ECM-binding bacterial proteins (adhesins) can be used as vaccine candidates in an antiadhesin vaccine. These strategies are, however, complicated by the fact that not all adhesin genes are present in every strain and that environmental factors can affect expression of these genes. Because each microorganism can employ multiple mechanisms of adhesion to initiate infection, effective antiadhesion drugs may contain cocktails of various inhibitors.

A classic example of the potential use of bacterial cell surface components interacting with ECM is documented with fibronectin-binding protein SfbI from S. pyogenes. SfbI protein acts both as adhesin and invasin of group A streptococci. Mice immunized with this protein not only show reduced pharyngeal colonization but are also protected against lethal streptococcal challenge. Because SfbI is highly conserved in its functional domain in streptococcal isolates from different geographical regions, it does not cross-react with the host proteins, is highly immunogenic, and represents a promising vaccine candidate. Besides its role in adhesion and invasion, SfbI protein is a strong mucosal adjuvant, and because of its nontoxic nature it has an advantage over the toxin-based adjuvants. Immunization with S. aureus fibrinogen-binding proteins led to reduced colonization in a mouse mastitis model. Synthetic peptides representing the fibronectin-binding domain of S. aureus adhesins as well as antibodies against the fibronectin-binding domain of S. aureus were successful in blocking bacterial colonization on implanted foreign material. These examples clearly demonstrate the potential use of components involved in the interaction of ECM and gram-positive pathogens and justify further research in this field.

REFERENCES

1. **Allen, B. L., B. Katz, and M. Hook.** 2002. *Streptococcus anginosus* adheres to vascular endothelium basement membrane and purified extracellular matrix proteins. *Microb. Pathog.* **32:**191-204.

2. **Bayer, A. S., P. M. Sullam, M. Ramos, C. Li, A. L. Cheung, and M. R. Yeaman.** 1995. *Staphylococcus aureus* induces platelet aggregation via a fibrinogen-dependent mechanism which is independent of principal platelet glycoprotein IIb/IIIa fibrinogen-binding domains. *Infect. Immun.* **63:**3634-3641.

3. **Beckmann, C., J. D. Waggoner, T. O. Harris, G. S. Tamura, and C. E. Rubens.** 2002. Identification of novel adhesins from Group B streptococci by use of phage display reveals that C5a peptidase mediates fibronectin binding. *Infect. Immun.* **70:**2869-2876.

4. **Beg, A. M., M. N. Jones, T. Miller-Torbert, and R. G. Holt.** 2002. Binding of *Streptococcus mutans* to extracellular matrix molecules and fibrinogen. *Biochem. Biophys. Res. Commun.* **298:**75-79.

5. **Bergmann, S., M. Rohde, G. S. Chhatwal, and S. Hammerschmidt.** 2001. alpha-Enolase of *Streptococcus pneumoniae* is a plasmin(ogen)-binding protein displayed on the bacterial cell surface. *Mol. Microbiol.* **40:**1273-1287.

6. **Bergmann, S., D. Wild, O. Diekmann, R. Frank, D. Bracht, G. S. Chhatwal, and S. Hammerschmidt.** 2003. Identification of a novel plasmin(ogen)-binding motif in surface displayed alpha-enolase of *Streptococcus pneumoniae*. *Mol. Microbiol.* **49:**411-423.

7. **Border, C. C., R. Lottenberg, G. O. von Mering, K. H. Johnston, and M. D. P. Boyle.** 1991. Isolation of a prokaryotic plasmin receptor. Relationship to a plasminogen activator produced by the same micro-organism. *J. Biol. Chem.* **266:**4922-4928.

8. **Bornstein, P.** 1995. Diversity of function is inherent in matricellular proteins: an appraisal of thrombospondin 1. *J. Cell Biol.* **130:**503-506.

9. **Carneiro, C. R., E. Postol, R. Nomizo, L. F. Reis, and R. R. Brentani.** 2004. Identification of enolase as a laminin-binding protein on the surface of *Staphylococcus aureus*. *Microbes Infect.* **6:**604-608.

10. **Cheung, A. L., M. Krishnan, E. A. Jaffe, and V. A. Fischetti.** 1991. Fibrinogen acts as a bridging molecule in the adherence of *Staphylococcus aureus* to cultured human endothelial cells. *J. Clin. Invest.* **87:**2236-2245.

11. **Chhatwal, G. S.** 2002. Anchorless adhesins and invasins of Gram-positive bacteria: a new class of virulence factors. *Trends Microbiol.* **10:**205-208.

12. **Chhatwal, G. S., and H. Blobel.** 1987. Heterogeneity of fibronectin reactivity among streptococci as revealed by binding of fibronectin fragments. *Comp. Immunol. Microbiol. Infect. Dis.* **10:**99-108.

13. **Chhatwal, G. S., H. P. Müller, and H. Blobel.** 1983. Characterization of binding of human α_2-macroglobulin to group G streptococci. *Infect. Immun.* **41:**959-964.

14. **Chhatwal, G. S., I. S. Dutra, and H. Blobel.** 1985. Fibrinogen binding inhibits the fixation of the third component of human complement on surface of groups A, B, C, and G streptococci. *Microbiol. Immunol.* **29:**973-980.

15. **Chhatwal, G. S., G. Albohn, and H. Blobel.** 1987. Novel complex formed between a nonproteolytic cell wall protein of group A streptococci and α_2-macroglobulin. *J. Bacteriol.* **169:**3691-3695.

16. **Chhatwal, G. S., K. T. Preissner, G. Müller-Berghaus, and H. Blobel.** 1987. Specific binding of the human S protein (vitronectin) to streptococci, *Staphylococcus aureus*, and *Escherichia coli*. *Infect. Immun.* **55:**1878-1883.

17. **Chhatwal, G. S., P. Valentin-Weigand, and K. N. Timmis.** 1990. Bacterial infection of wounds: fibronectin-mediated adherence of group A and C streptococci to fibrin thrombi in vitro. *Infect. Immun.* **58:**3015-3019.

18. **Cho, S. H., I. Strickland, M. Boguniewicz, and D. Y. Leung.** 2001. Fibronectin and fibrinogen contribute to the enhanced binding of *Staphylococcus aureus* to atopic skin. *J. Allergy Clin. Immunol.* **108:**269-274.

19. **Chu, C. T., and S. V. Pizzo.** 1994. Alpha-2-macroglobulin, complement, and biologic defense: antigens, growth factors, microbial proteases, and receptor ligation. *Lab. Invest.* **71:**792-812.

20. **Cunningham, M. W.** 2000. Pathogenesis of group A streptococcal infections. *Clin. Microbiol. Rev.* **13:**470-511.

21. **Cunningham, M. W.** 2003. Autoimmunity and molecular mimicry in the pathogenesis of poststreptococcal heart disease. *Front. Biosci.* **8:**533-543.

22. **Dickinson, R. B., J. A. Nagel, D. McDevitt, T. J. Foster, R. A. Proctor, and S. L. Cooper.** 1995. Quantitative comparison of clumping factor- and coagulase-mediated *Staphylococcus aureus* adhesion to surface-bound fibrinogen under flow. *Infect. Immun.* **63:**3143-3150.

23. **Dinkla, K., M. Rohde, W. T. Jansen, J. R. Carapetis, G. S. Chhatwal, and S. R. Talay.** 2003. *Streptococcus pyogenes* recruits collagen via surface-bound fibronectin: a novel colonization and immune evasion mechanism. *Mol. Microbiol.* **47:**861–869.

24. **Dinkla, K., M. Rohde, W. T. Jansen, E. L. Kaplan, G. S. Chhatwal, and S. R. Talay.** 2003. Rheumatic fever-associated *Streptococcus pyogenes* isolates aggregate collagen. *J. Clin. Invest.* **111:**1905–1912.

25. **Downer, R., F. Roche, P. W. Park, R. P. Mecham, and T. J. Foster.** 2002. The elastin-binding protein of *Staphylococcus aureus* (EbpS) is expressed at the cell surface as an integral membrane protein and not as a cell wall-associated protein. *J. Biol. Chem.* **277:**243–250.

26. **Eberhard, T., and M. Ullberg.** 2002. Interaction of vitronectin with *Haemophilus influenzae*. *FEMS Immunol. Med. Microbiol.* **34:**215–219.

27. **Filippsen, L. F., P. Valentin-Weigand, H. Blobel, K. T. Preissner, and G. S. Chhatwal.** 1990. Role of complement S protein (vitronectin) in adherence of *Streptococcus dysgalactiae* to bovine epithelial cells. *Am. J. Vet. Res.* **51:**861–865.

28. **Gailit, J., and R. A. F. Clark.** 1994. Wound repair in the context of extracellular matrix. *Curr. Opin. Cell Biol.* **6:**717–725.

29. **Godehardt, A. W., S. Hammerschmidt, R. Frank, and G. S. Chhatwal.** 2004. Binding of alpha2-macroglobulin to GRAB (protein G-related alpha2-macroglobulin-binding protein), an important virulence factor of group A streptococci, is mediated by two charged motifs in the DeltaA region. *Biochem. J.* **381:**877–885.

30. **Gutekunst, H., B. J. Eikmanns, and D. J. Reinscheid.** 2004. The novel fibrinogen-binding protein FbsB promotes *Streptococcus agalactiae* invasion into epithelial cells. *Infect. Immun.* **72:**3495–3504.

31. **Guzmán, C. A., S. R. Talay, G. Molinari, E. Medina, and G. S. Chhatwal.** 1999. Protective immune response against *Streptococcus pyogenes* in mice after intranasal vaccination with the fibronectin binding protein SfbI. *J. Infect. Dis.* **179:**901–906.

32. **Hall, C. E., and H. S. Slayter.** 1959. The fibrinogen molecule: its size, shape, and mode of polymerization. *J. Biophys. Biochem. Cytol.* **5:**11–15.

33. **Hanski, E., and M. Caparon.** 1992. Protein F, a fibronectin-binding protein, is an adhesin of the group A *Streptococcus pyogenes. Proc. Natl. Acad. Sci. USA* **89:**6172–6176.

34. **Heilmann, C., S. Niemann, B. Sinha, M. Herrmann, B. E. Kehrel, and G. Peters.** 2004. *Staphylococcus aureus* fibronectin-binding protein (FnBP)-mediated adherence to platelets, and aggregation of platelets induced by FnBPA but not by FnBPB. *J. Infect. Dis.* **190:**321–329.

35. **Henschen, A., F. Lottspeich, E. Töpfer-Petersen, and R. Warbinek.** 1979. Primary structure of fibrinogen. *Thromb. Haemost.* **41:**662–670.

36. **Herrmann, M., S. J. Suchard, L. A. Boxer, F. A. Waldvogel, and D. P. Lew.** 1991. Thrombospondin binds to *Staphylococcus aureus* and promotes staphylococcal adherence to surfaces. *Infect. Immun.* **59:**279–288.

37. **Higazi, A. A. R., I. I. Barghouti, and R. Abumuch.** 1995. Identification of an inhibitor of tissue-type plasminogen activator-mediated fibrinolysis in human neutrophils—a role for defensin. *J. Biol. Chem.* **270:**9472–9477.

38. **Holmes, A. R., R. McNab, K. W. Millsap, M. Rohde, S. Hammerschmidt, J. L. Mawdsley, and H. F. Jenkinson.** 2001. The *pavA* gene of *Streptococcus pneumoniae* encodes a fibronectin-binding protein that is essential for virulence. *Mol. Microbiol.* **41:**1395–1408.

39. **Hostetter, M. K.** 1996. An integrin-like protein in *Candida albicans:* implications for pathogenesis. *Trends Microbiol.* **4:**242–246.

40. **Hynes, R. O., and K. M. Yamada.** 1982. Fibronectins: multifunctional modular glycoproteins. *J. Cell Biol.* **95:** 369–377.

41. **Hytonen, J., S. Haataja, D. Gerlach, A. Podbielski, and J. Finne.** 2001. The SpeB virulence factor of *Streptococcus pyogenes*, a multifunctional secreted and cell surface molecule with strepadhesin, laminin-binding and cysteine protease activity. *Mol. Microbiol.* **39:**512–519.

42. **Joh, J. J., K. House-Pompeo, J. M. Patti, S. Gurusiddappa, and M. Höök.** 1994. Fibronectin receptors from Gram-positive bacteria: comparison of active sites. *Biochemistry* **33:**6086–6092.

43. **Kreikemeyer, B., S. R. Talay, and G.S. Chhatwal.** 1995. Characterization of a novel fibronectin-binding surface protein in group A streptococci. *Mol. Microbiol.* **17:**137–145.

44. **Kreikemeyer, B., D. R. Martin, and G. S. Chhatwal.** 1999. SfbII protein, a fibronectin binding surface protein of group A streptococci, is a serum opacity factor with high serotype-specific apolipoproteinase activity. *FEMS Microbiol. Lett.* **178:**305–311.

45. **Kreikemeyer, B., S. Oehmcke, M. Nakata, R. Hoffrogge, and A. Podbielski.** 2004. *Streptococcus pyogenes* fibronectin-binding protein F2: expression profile, binding characteristics, and impact on eukaryotic cell interactions. *J. Biol. Chem.* **279:**15850–15859.

46. **Kreis, T., and R. Vale.** 1993. *Guidebook to the Extracellular Matrix and Adhesion Proteins.* Oxford University Press, Oxford, United Kingdom.

47. **Labat-Robert, J., M. Bihari-Varga, and L. Robert.** 1990. Extracellular matrix. *FEBS Lett.* **268:**386–393.

48. **Lane, D. A., A. M. Flynn, G. Pejler, U. Lindahl, J. Choay, and K. T. Preissner.** 1987. Structural requirements for the neutralization of heparin-like saccharides by complement S protein/vitronectin. *J. Biol. Chem.* **262:** 16343–16349.

49. **Lawler, J. W., and R. O. Hynes.** 1987. Structure organization of the thrombospondin molecule. *Semin. Thromb. Hemost.* **13:**245–254.

50. **Li, D. Q., F. Lundberg, and A. Ljungh.** 2001. Characterization of vitronectin-binding proteins of *Staphylococcus epidermidis. Curr. Microbiol.* **42:**361–367.

51. **Liang, O. D., M. Maccarana, J. I. Flock, M. Paulsson, K. T. Preissner, and T. Wadström.** 1993. Multiple interactions between human vitronectin and *Staphylococcus aureus. Biochim. Biophys. Acta* **374:**1–7.

52. **Liang, O. D., K. T. Preissner, and G. S. Chhatwal.** 1997. The hemopexin-type repeats of human vitronectin are recognized by *Streptococcus pyogenes. Biochem. Biophys. Res. Commun.* **234:**445–449.

53. **Lindgren, P.-E., P. Speziale, M. McGavin, H.-J. Monstein, M. Höök, L. Visai, T. Kostiainen, S. Bozzini, and M. Lindberg.** 1992. Cloning and expression of two different genes from *Streptococcus dysgalactiae* encoding fibronectin receptors. *J. Biol. Chem.* **267:**1924–1931.

54. **Love, R. M., M. D. McMillan, and H. F. Jenkinson.** 1997. Invasion of dentinal tubules by oral streptococci is associated with collagen recognition mediated by the antigen I/II family of polypeptides. *Infect. Immun.* **65:**5157–5164.

55. **Maeshima, Y., A. Sudhakar, J. C. Lively, K. Ueki, S. Kharbanda, C.R. Kahn, N. Sonenberg, R. O. Hynes, and R. Kalluri.** 2002. Tumstatin, an endothelial cell-specific inhibitor of protein synthesis. *Science* **295:**140–143.

56. **Mamo, W., M. Boden, and J. I. Flock.** 1994. Vaccination with *Staphylococcus aureus* fibrinogen binding protein (FgBPs) reduces colonization of *S. aureus* in a mouse mastitis model. *FEMS Immun. Med. Microbiol.* **10:**47–53.

57. **McDevitt, D., T. Nanavaty, K. House-Pompeo, E. Bell, N. Turner, L. McIntire, T. Foster, and M. Höök.** 1997. Characterization of the interaction between the *Staphylococcus aureus* clumping factor (ClfA) and fibrinogen. *Eur. J. Biochem.* **247:**416–424.

58. **McKay, F. C., J. D. McArthur, M. L. Sanderson-Smith, S. Gardam, B. J. Currie, K. S. Sriprakash, P. K. Fagan, R. J. Towers, M. R. Batzloff, G. S. Chhatwal, M. Ranson, and M. J. Walker.** 2004. Plasminogen binding by group A streptococcal isolates from a region of hyperendemicity for streptococcal skin infection and a high incidence of invasive infection. *Infect. Immun.* **72:**364–370.

59. **Medina, E., S. R. Talay, G. S. Chhatwal, and C. A. Guzmán.** 1998. Fibronectin-binding protein I of *Streptococcus pyogenes* is a promising adjuvant for antigens delivered by mucosal route. *Eur. J. Immunol.* **28:**1069–1077.

60. **Menzies, B. E.** 2003. The role of fibronectin binding proteins in the pathogenesis of *Staphylococcus aureus* infections. *Curr. Opin. Infect. Dis.* **16:**225–229.

61. **Molinari, G., S. R. Talay, P. Valentin-Weigand, M. Rohde, and G. S. Chhatwal.** 1997. The fibronectin-binding protein of *Streptococcus pyogenes*, SfbI, is involved in the internalization of group A streptococci by epithelial cells. *Infect. Immun.* **65:**1357–1363.

62. **Molinari, G., and G. S. Chhatwal.** 1998. Invasion and survival of *Streptococcus pyogenes* in eukaryotic cells correlates with the source of the clinical isolates. *J. Infect. Dis.* **177:**1600–1607.

63. **Mosher, D. F., and R. A. Proctor.** 1980. Binding of factor XIIIa-mediated cross-linking of a 27-kilodalton fragment of fibronectin to *Staphylococcus aureus*. *Science* **209:**927–929.

64. **Mosher, D. F., J. Sottile, C. Wu, and J. A. McDonald.** 1992. Assembly of extracellular matrix. *Curr. Opin. Cell Biol.* **4:**810–818.

65. **Murphy-Ullrich, J. E., S. Gurusiddappa, W. A. Frazier, and M. Höök.** 1993. Heparin-binding peptides from thrombospondin 1 and 2 contain focal adhesion-labilizing activity. *J. Biol. Chem.* **268:**26784–26789.

66. **O'Reilly, M. S., T. Boehm, Y. Shing, N. Fukai, G. Vasios, W. S. Lane, E. Flynn, J. R. Birkhead, B. R. Olsen, and J. Folkman.** 1997. Endostatin: an endogenous inhibitor of angiogenesis and tumor growth. *Cell* **88:**277–285.

67. **Ozeri, V., I. Rosenshine, D. F. Mosher, R. Fässler, and E. Hanski.** 1998. Roles of integrins and fibronectin in the entry of *Streptococcus pyogenes* into cells via protein F1. *Mol. Microbiol.* **30:**625–637.

68. **Pancholi, V., and V. A. Fischetti.** 1998. Alpha-enolase, a novel strong plasmin(ogen) binding protein on the surface of pathogenic streptococci. *J. Biol. Chem.* **273:**14503–14515.

69. **Pancholi, V., P. Fontan, and H. Jin.** 2003. Plasminogen-mediated group A streptococcal adherence to and pericellular invasion of human pharyngeal cells. *Microb. Pathog.* **35:**293–303.

70. **Park, P. W., D. D. Roberts, L. E. Grosso, W. C. Parks, J. Rosenbloom, W. R. Abrams, and R. P. Mecham.** 1991. Binding of elastin to *Staphylococcus aureus*. *J. Biol. Chem.* **266:**23399–23406.

71. **Park, P. W., T. J. Broekelmann, B. R. Mecham, and R. P. Mecham.** 1999. Characterization of the elastin binding domain in the cell-surface 25-kDa elastin-binding protein of *Staphylococcus aureus* (EbpS). *J. Biol. Chem.* **274:**2845–2850.

72. **Patti, J. M., J. O. Boles, and M. Höök.** 1993. Identification and biochemical characterization of the ligand binding domain of the collagen adhesin from *Staphylococcus aureus*. *Biochemistry* **32:**11428–11435.

73. **Patti, J. M., B. L. Allen, M. J. McGavin, and M. Höök.** 1994. MSCRAMM-mediated adherence of microorganisms to host tissue. *Annu. Rev. Microbiol.* **48:**585–617.

74. **Pierschbacher, M. D., and E. Ruoslahti.** 1984. Cell attachment activity of fibronectin can be duplicated by small synthetic fragments of the molecule. *Nature* **309:**30–33.

75. **Preissner, K. T.** 1991. Structure and biological role of vitronectin. *Annu. Rev. Cell Biol.* **7:**275–310.

76. **Preissner, K. T., and B. Pötzsch.** 1995. Vessel wall-dependent metabolic pathways of the adhesive proteins, von-Willebrand-factor and vitronectin. *Histol. Histopathol.* **10:**239–251.

77. **Preissner, K. T., A. E. May, K. D. Wohn, M. Germer, and S. M. Kanse.** 1997. Molecular crosstalk between adhesion receptors and proteolytic cascades in vascular remodeling. *Thromb. Haemost.* **78:**88–95.

78. **Raghow, R.** 1994. The role of extracellular matrix in postinflammatory wound healing and fibrosis. *FASEB J.* **8:**823–831.

79. **Rakonjac, J. V., J. C. Robbins, and V. A. Fischetti.** 1995. DNA sequence of the serum opacity factor of group A streptococci: identification of a fibronectin-binding repeat domain. *Infect. Immun.* **63:**622–631.

80. **Ringdahl, U., M. Svensson, A. C. Wistedt, T. Renn, R. Kellner, W. Müller-Esterl, and U. Sjöbring.** 1998. Molecular co-operation between protein PAM and streptokinase for plasmin acquisition by *Streptococcus pyogenes*. *J. Biol. Chem.* **273:**6424–6430.

81. **Rocha, C. L., and V. A. Fischetti.** 1997. Identification and characterization of a new protein from *Streptococcus pyogenes* having homology with fibronectin and fibrinogen binding proteins. *Adv. Exp. Med. Biol.* **418:**737–739.

82. **Rohde, M., E. Muller, G. S. Chhatwal, and S. R. Talay.** 2003. Host cell caveolae act as an entry-port for group A streptococci. *Cell Microbiol.* **5:**323–342.

83. **Rosenblatt, S., J. A. Bassuk, C. E. Alpers, E. H. Sage, R. Timpl, and K. T. Preissner.** 1997. Differential modulation of cell adhesion by interaction between adhesive and counteradhesive proteins: characterization of the binding of vitronectin to osteonectin (BM40, SPARC). *Biochem. J.* **324:**311–319.

84. **Sato, Y., K. Okamoto, A. Kagami, Y. Yamamoto, T. Igarashi, and H. Kizaki.** 2004. *Streptococcus mutans* strains harboring collagen-binding adhesin. *J. Dent. Res.* **83:**534–539.

85. **Schwarz-Linek, U., J. M. Werner, A. R. Pickford, S. Gurusiddappa, J. H. Kim, E. S. Pilka, J. A. Briggs, T. S. Gough, M. Hook, I. D. Campbell, and J. R. Potts.** 2003. Pathogenic bacteria attach to human fibronectin through a tandem beta-zipper. *Nature* **423:**177–181.

86. **Sommer, P., C. Gleyzal, S. Guerret, J. Etienne, and J.-A. Grimaud.** 1992. Induction of a putative laminin-binding protein of *Streptococcus gordonii* in human infective endocarditis. *Infect. Immun.* **60:**360–365.

87. Spellerberg, B., E. Rozdzinski, S. Martin, J. Weber-Heynemann, N. Schnitzler, R. Lutticken, and A. Podbielski. 1999. Lmb, a protein with similarities to the LraI adhesin family, mediates attachment of *Streptococcus agalactiae* to human laminin. *Infect. Immun.* **67:**871–878.

88. Ständker, L., M. Schrader, S. M. Kanse, M. Jurgens, W. G. Forssmann, and K. T. Preissner. 1997. Isolation and characterization of the circulating form of human endostatin. *FEBS Lett.* **420:**129–133.

89. Sun, H., U. Ringdahl, J. W. Homeister, W. P. Fay, N. C. Engleberg, A. Y. Yang, L. S. Rozek, X. Wang, U. Sjobring, and D. Ginsburg. 2004. Plasminogen is a critical host pathogenicity factor for group A streptococcal infection. *Science* **305:**1283–1286.

90. Svensson, M. D., U. Sjobring, F. Luo, and D. E. Bessen. 2002. Roles of the plasminogen activator streptokinase and the plasminogen-associated M protein in an experimental model for streptococcal impetigo. *Microbiology* **148:**3933–3945.

91. Switalski, L. M., P. Speziale, M. Höök, T. Wadström, and R. Timpl. 1984. Binding of *Streptococcus pyogenes* to laminin. *J. Biol. Chem.* **259:**3734–3738.

92. Switalski, L. M., H. Murchinson, R. Timpl, R. Curtiss III, and M. Höök. 1987. Binding of laminin to oral and endocarditis strains of viridans streptococci. *J. Bacteriol.* **169:**1095–1101.

93. Switalski, L. M., P. Speziale, and M. Höök. 1989. Isolation and characterization of a putative collagen receptor from *Staphylococcus aureus* strain Cowan 1. *J. Biol. Chem.* **264:**21080–21086.

94. Switalski, L. M., W. G. Butcher, P. C. Caufield, and M. S. Lantz. 1993. Collagen mediates adhesion of *Streptococcus mutans* to human dentin. *Infect. Immun.* **61:**4119–4125.

95. Switalski, L. M., J. M. Patti, W. Butcher, A. G. Gristina, P. Speziale, and M. Höök. 1993. A collagen receptor in *Staphylococcus aureus* strains isolated from patients with septic arthritis mediates adhesion to cartilage. *Mol. Microbiol.* **7:**99–107.

96. Talay, S. R., E. Ehrenfeld, G. S. Chhatwal, and K. N. Timmis. 1991. Expression of the fibronectin-binding components of *Streptococcus pyogenes* in *Escherichia coli* demonstrates that they are proteins. *Mol. Microbiol.* **5:**1727–1734.

97. Talay, S. R., P. Valentin-Weigand, P. G. Jerlström, K. N. Timmis, and G. S. Chhatwal. 1992. Fibronectin-binding protein of *Streptococcus pyogenes*: sequence of the binding domain involved in adherence of streptococci to epithelial cells. *Infect. Immun.* **60:**3837–3844.

98. Talay, S. R., P. Valentin-Weigand, K. N. Timmis, and G. S. Chhatwal. 1994. Domain structure and conserved epitopes of Sfb protein, the fibronectin-binding adhesin of *Streptococcus pyogenes*. *Mol. Microbiol.* **13:**531–539.

99. Talay, S. R., A. Zock, M. Rohde, G. Molinari, M. Oggioni, G. Pozzi, C. A. Guzman, and G. S. Chhatwal. 2000. Co-operative binding of human fibronectin to Sfbl protein triggers streptococcal invasion into respiratory epithelial cells. *Cell Microbiol.* **2:**521–535.

100. Terao, Y., S. Kawabata, E. Kunitomo, J. Murakami, I. Nakagawa, and S. Hamada. 2001. Fba, a novel fibronectin-binding protein from *Streptococcus pyogenes*, promotes bacterial entry into epithelial cells, and the *fba*

gene is positively transcribed under the Mga regulator. *Mol. Microbiol.* **42:**75–86.

101. Terao, Y., S. Kawabata, E. Kunitomo, I. Nakagawa, and S. Hamada. 2002. Novel laminin-binding protein of *Streptococcus pyogenes*, Lbp, is involved in adhesion to epithelial cells. *Infect. Immun.* **70:**993–997.

102. Terao, Y., S. Kawabata, M. Nakata, I. Nakagawa, and S. Hamada. 2002. Molecular characterization of a novel fibronectin-binding protein of *Streptococcus pyogenes* strains isolated from toxic shock-like syndrome patients. *J. Biol. Chem.* **277:**47428–47435.

103. Timpl, R. 1989. Structure and biological activity of basement membrane proteins. *Eur. J. Biochem.* **180:**487–502.

104. Toppel, A. W., M. Rasmussen, M. Rohde, E. Medina, and G. S. Chhatwal. 2003. Contribution of protein G-related alpha2-macroglobulin-binding protein to bacterial virulence in a mouse skin model of group A streptococcal infection. *J. Infect. Dis.* **187:**1694–1703.

105. Traore, M. Y., P. Valentin-Weigand, G. S. Chhatwal, and H. Blobel. 1991. Inhibitory effects of fibrinogen on phagocytic killing of streptococcal isolates from humans, cattle and horses. *Vet. Microbiol.* **28:**295–302.

106. Valentin-Weigand, P., J. Grulich-Henn, G. S. Chhatwal, G. Müller-Berghaus, H. Blobel, and K. T. Preissner. 1988. Mediation of adherence of streptococci to human endothelial cells by complement S-protein (vitronectin). *Infect. Immun.* **56:**2851–2855.

107. Valentin-Weigand, P., M. Y. Traore, H. Blobel, and G. S. Chhatwal. 1990. Role of α2-macroglobulin in phagocytosis of group A and C streptococci. *FEMS Microbiol. Lett.* **58:**321–324.

108. Valentin-Weigand, P., K. N. Timmis, and G. S. Chhatwal. 1993. Role of fibronectin in staphylococcal colonization of fibrin thrombi and plastic surfaces. *J. Med. Microbiol.* **38:**90–95.

109. Valentin-Weigand, P., S. R. Talay, A. Kaufhold, K. N. Timmis, and G. S. Chhatwal. 1994. The fibronectin binding domain of the Sfb protein adhesin of *Streptococcus pyogenes* occurs in many group A streptococci and does not cross-react with heart myosin. *Microb. Pathog.* **17:**111–120.

110. van der Flier, M., N. Chhun, T. M. Wizemann, J. Min, J. B. McCarthy, and E. I. Tuomanen. 1995. Adherence of *Streptococcus pneumoniae* to immobilized fibronectin. *Infect. Immun.* **63:**4317–4322.

111. Vercellotti, G. M., D. Lussenhop, P. K. Peterson, L. T. Furcht, J. B. McCarthy, H. S. Jacob, and C. F. Moldow. 1984. Bacterial adherence to fibronectin and endothelial cells: a possible mechanism for bacterial tissue tropism. *J. Lab. Clin. Med.* **103:**34–43.

112. Whitnack, E., and E. H. Beachy. 1985. Inhibition of complement-mediated opsonization and phagocytosis of *Streptococcus pyogenes* by D fragments of fibrinogen and fibrin bound to cell surface M protein. *J. Exp. Med.* **162:**1983–1997.

113. Wiestedt, A. C., U. Ringdahl, W. Müller-Esterl, and U. Sjöbring. 1995. Identification of a plasminogen-binding motif in PAM, a bacterial surface protein. *Mol. Microbiol.* **18:**569–578.

114. Yanagisawa, N., D. Q. Li, and A. Ljungh. 2001. The N-terminal of thrombospondin-1 is essential for coagulase-negative staphylococcal binding. *J. Med. Microbiol.* **50:**712–719.

Streptococcus-Mediated Host Cell Signaling

VIJAY PANCHOLI

9

To be successful pathogens, bacteria must adhere to, colonize, and invade the target tissue. To successfully colonize and infect, bacteria usually need to sequentially engage (adhere) to surface-bound adhesins with complementary receptors on target cells. In this regard, group A *Streptococcus* (GAS or *Streptococcus pyogenes*) displays a remarkable tissue tropism for the human pharynx and the skin, causing a variety of diseases. The surface of GAS presents an array of proteins that perform a variety of functions (28, 71, 104). As information has become available over the last decade, it has become evident that most, if not all, surface proteins on these bacteria are multifunctional (28, 71, 104). In a given strain, one surface protein may perform several functions, or several proteins in concert may participate to perform one particular function. Since most of these proteins are multifunctional in nature and their expression is controlled by the surrounding environment, mechanisms involved in streptococcal pathogenesis have become more complex than previously believed (16, 78, 88, 89, 124). Besides this complexity, GAS is of special concern because of its ability to cause diseases such as rheumatic fever and acute poststreptococcal glomerulonephritis (157, 158). Reemergence of rheumatic fever cases, and invasive and fatal GAS infections, in the absence of any specific virulent clone suggest that host factors play a critical role in the final outcome of a variety of streptococcal diseases (19, 33, 106). Despite several years of intensive investigations into streptococcal pathogenesis, it has remained enigmatic as to why pharyngeal infection, and not other infections such as those of the skin, is a prerequisite for rheumatic fever and rheumatic heart diseases (159).

It is becoming more apparent that microorganisms, using their surface proteins, interact with host receptor molecules and regulate intracellular signaling pathways to induce their own adherence, colonization, and internalization. Although numerous studies have laid a sound foundation for the understanding of the cross-talk between gram-negative bacteria and host cells, thereby unraveling their basic virulence mechanisms (see reviews in references 11, 21, 22, 42, 61, 65, 160), the information on the cell biology of GAS-infected pharyngeal cells or skin cells has just begun to be available. Despite the fact that recent information on host cell-signaling events induced by streptococci has increased our knowledge on the mechanisms of streptococcal diseases, the field of cellular microbiology of pathogenic streptococcal infection in comparison to that by gram-negative bacterial infection has remained much less explored (17, 118, 139, 156).

As this book sees its second edition within five years, the purpose of this chapter is to incorporate all available articles that have provided direct as well as indirect evidence of the initial interactions of pathogenic streptococci with their specific host cells, resulting in the exchange of messages between them, which in turn determines the outcome of the disease.

With the availability of complete genome sequence analyses of five GAS strains, including M1, M3 (two strains), M18, and M6 (6, 8, 41, 102, 144), it is easy to predict the number of surface proteins, which may serve as potential adhesins during initial interactions with host cells. However, it is unknown how many of them remain expressed during the adherence process and serve as adhesins for specific receptors. The complex dynamic of the control of one surface protein on the others is presently unknown, because the expression of many of these proteins is environmentally regulated (16, 78, 88, 89, 124). Further, many of the surface proteins are up-regulated only when the bacteria reach the intracellular environment (52, 78, 141, 154, 155). Similarly, the interaction of bacteria with their specific host cells may also up- or down-regulate the expression of host cell receptors. The bacterial adhesin-host receptor interaction may vary with polarized cells versus nonpolarized cells, and professional phagocytes versus nonprofessional phagocytic cells. Thus, the dynamic of interaction in the early events of streptococcal infection is extremely complex. Despite many obvious missing links in our understanding of the precise signal transduction pathways evoked by these organisms in their respective host cells, the available literature suggests a common paradigm (42). Pathogenic gram-positive cocci through their surface proteins interact with specific receptors on the target cell and induce a series of biochemical signals. These signals, which are characterized by the induction of phosphokinase enzymes and phosphorylation of several intracellular proteins, ultimately target the nucleus and lead to either generalized or specific gene activation. Activation of some of these genes may result in the modulation of interleukin or cytokine expression, which may then initiate a proinflammatory response. These induced signals, and subsequent products, could have several effects on the invasion of bacteria. For example, these induced signals may modulate cytoskeletal structure and/or specific host cell receptor expression or may destroy adjoining cells and disrupt natural protective barriers in autocrine or paracrine modes. This, in turn, could facilitate bacterial entry (Fig. 1).

INITIAL INTERACTION OF GRAM-POSITIVE BACTERIAL SURFACE PROTEINS WITH MUCIN

The human respiratory epithelium is covered by a viscous layer of mucus, situated on the top of the cilia and resulting in a continuous flow towards the pharynx. By virtue of this anatomical disposition, all mucosal pathogens first interact with mucin, the main component of the respiratory mucus, before they adhere to a specific host cell receptor on the epithelial cell. Mucins are a highly complex class of glycoproteins found both in secretory and bound forms and are expressed by several genes (49, 146). The structural diversity among the mucin family also increases due to the posttranslational O-glycosylation and sulfation that contribute to more than 50% of the molecular mass of the mucin molecule (146). Although mucin may serve as an initial protective barrier from invading pathogens, overproduction of mucin has been shown to exacerbate the pathological symptoms in diseases such as bacterial pneumonia and cystic fibrosis (38, 74, 135). Although several reports have described the role of the interaction of mucin with gram-negative bacteria in the pathogenesis of cystic fibrosis and similarly with oral streptococci in oral diseases (35, 38, 74), limited information is available on the interaction of mucin with pathogenic gram-positive bacteria and its implication in the disease process (23, 33, 97, 142). The mucin-binding property of GAS is regulated by the *mga* gene (63, 133). The M protein and a 39-kDa GAS cell wall-associated surface protein are the first reported surface proteins that mediate streptococcal-specific binding to salivary/tracheobronchial mucin (134). The binding of mucin to GAS also increases streptococcal adherence to and invasion of cultured human pharyngeal cells (134). Pullulanase, a 129-kDa *mga*-regulated GAS surface protein, also binds to submaxillary mucin and serves as a surface adhesin (64). How pathogens in general and streptococci in particular traverse this mucous layer and find their way to the specific target receptor on epithelial cells is unknown. Pullulanase in this regard may perform both functions, because it also possesses carbohydrate-degrading activities, although it is not known how these two combined activities are modulated during initial interaction to bring bacteria to host cells (64). Further investigation is needed on the role of gram-positive mucolytic enzymes such as streptococcal neuraminidase (33) or other extracellular proteases (18, 83) in facilitating bacterial migration through the viscous mucin layer to target them to the epithelial surface-specific receptor before invasion. In the last decade, considerable advances have been made towards our understanding of the structure and function of mucin glycoproteins and their physiological niche (49, 148). Dohrman et al. (36) have reported that certain types of mucins such as MUC2 and MUC5AC are overexpressed in human bronchial explant and epithelial cell cultures in response to interactions with culture supernatants of gram-negative (e.g., *Pseudomonas aeruginosa*) as well as gram-positive (e.g., *S. pyogenes*, *Staphylococcus aureus*, *Staphylococcus epidermidis*) bacteria. This expression is found to be tyrosine-phosphorylation-dependent (36), indicating that a specific type of mucus production may be the result of the induction of a specific intracellular signaling event. Although respiratory cell-bound mucins such as MUC1 are thought to be involved in cell adhesion (59, 81, 149), and their cytoplasmic domains have been shown to play a role in cytoskeletal rearrangements in certain disease conditions (43), it is not clear at present whether salivary mucin or cell-bound mucin serves as adhesin or a functional receptor for bacterial surface proteins.

BINDING OF STREPTOCOCCAL ADHESINS TO HOST CELL RECEPTORS AND INTRACELLULAR SIGNALING EVENTS IN TARGET CELLS

In gram-negative bacteria, several reports have conclusively proven that the cell-cell interactions via specific receptor-adhesins elicit a variety of specific signaling events that ultimately determine disease outcome (11, 21, 22, 42, 61, 65, 160). Recent studies on streptococci have also indicated that streptococcal interaction with the host is not a static process but a highly dynamic complex process resulting in specific intracellular responses. The C-repeat region of streptococcal M protein binds to CD46, a membrane-bound complement regulatory protein related to factor H that serves as a cellular receptor for GAS on human keratinocytes (109). CD46 belongs to a family of proteins that contains four structurally related short consensus repeat domains and regulates the activation of the complement components C3b and/or C4b (132). A major M protein binding site is located within short consensus repeats 3 and 4, probably at the interface of these two domains (51). The hyaluronic acid capsule also modulates M-protein-mediated adherence and acts as a ligand for attachment of GAS to CD44 on human keratinocytes (137). CD44 is also reported to be a receptor for GAS colonization of the pharynx (30). It is also proposed that this streptococcal-host cell intimate contact, mediated by the M protein and CD46, may initiate proinflammatory responses from keratinocytes during infection (156). In fact, CD44-dependent GAS binding to polarized monolayers of human keratinocytes induces marked cytoskeletal rearrangements, resulting in membrane ruffling and disruption of intercellular junctions. Transduction of the signal induced after binding of GAS to the keratinocyte surface via CD44 involves Rac1 and the cytoskeleton linker protein ezrin as well as tyrosine phosphorylation of cellular proteins (29). Studies of bacterial translocation in two models of human skin have indicated that cell signaling triggered by the interaction of the hyaluronic capsule of *S. pyogenes* with CD44 opens intercellular junctions and promotes tissue penetration by the bacteria through a paracellular route (29).

The evidence that streptococci upon contact with their target cells such as pharyngeal cells may regulate intracellular signaling events was first suggested when a novel multifunctional protein, streptococcal surface dehydrogenase (SDH), was identified on the surface of streptococci (115). SDH, which shows binding activity to a variety of mam-

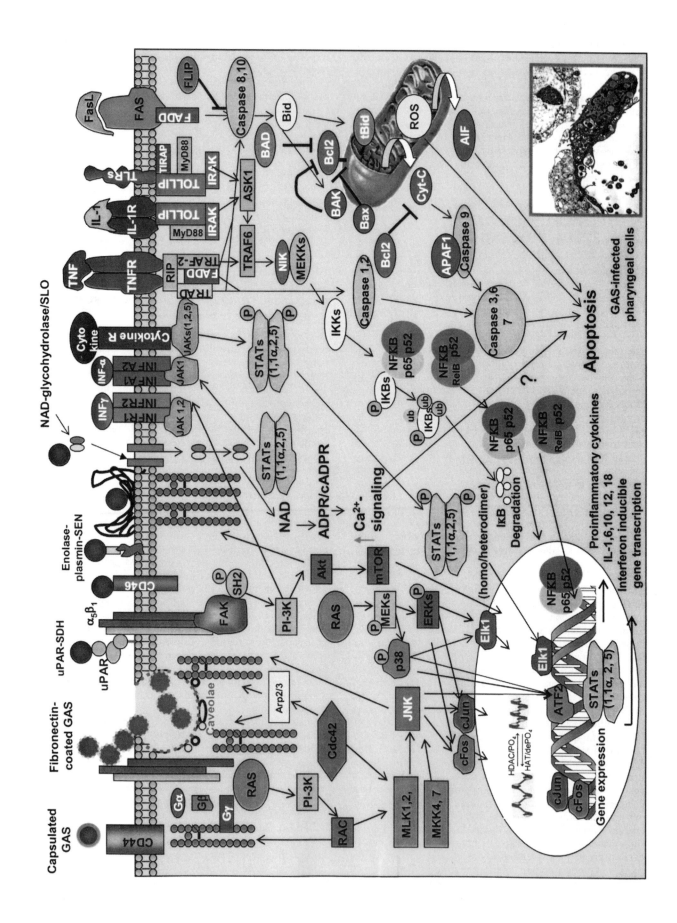

malian proteins such as fibronectin, myosin, and actin, is also an ADP-ribosylating enzyme (116). ADP-ribosylation plays an important role in intracellular signaling events and is the mechanism by which many bacterial toxins show their deleterious effect on target cells (98). Although the role of SDH as a toxin is not proven, its direct contact with human pharyngeal cells (both polarized and nonpolarized) has been shown to cause DNA condensation, a penultimate stage of apoptosis (118). Interestingly, SDH belongs to a novel class of surface-expressed, anchorless housekeeping enzymes that were recently recognized as virulence factors for pathogens (114). Like SDH, another metabolic enzyme, streptococcal surface enolase (SEN), has also been shown to be present on the surface of *S. pyogenes* and *Streptococcus pneumoniae* and functions as a strong plasminogen receptor (9, 119). Human enolase has been shown to be expressed on the surface of a variety of cells, such as hematopoetic, neuronal, endothelial, and epithelial, and serves as a receptor for plasminogen (40, 94, 120, 127). SDH binds to four

different membrane-associated proteins (118), two of which (47 kDa and 55 kDa) have been recently characterized as human enolase and urokinase plasminogen-activator receptor (uPAR), respectively (69). The recent report describing increased adherence of *S. pyogenes* to human pharyngeal cells in the presence of plasminogen indicates that this molecule may serve as a molecular bridge between human enolase expressed on the surface of pharyngeal cells and the SEN molecule expressed on the surface of *S. pyogenes* (120). SDH directly binds to human enolase (69). Since SDH is found complexed with SEN (117) and also associated with the M protein and M-like protein (34), it is conceivable that both human enolase and uPAR can serve as potential receptors for GAS-mediated signaling events. Purified SDH, like whole streptococci, is capable of activating both tyrosine and serine/threonine kinases in human pharyngeal cells (118). This activation induces serine/threonine and tyrosine phosphorylation of several cellular proteins, with the most noticeable phosphorylation of 180-,

---

FIGURE 1 (*See the separate color insert for the color version of this illustration.*) Streptococcus-mediated signaling events and their implications. The cartoon illustrates reported signaling events. Some of the pathways are hypothesized on the basis of available specific reports on streptococci-mediated signaling events and established signal transduction pathways in eukaryotes. There are at least four receptors (uPAR, enolase, CD44, CD46) that directly interact with group A *Streptococcus* (GAS) surface proteins. Indirect binding to eukaryotic cells through fibronectin is likely mediated by α5β1 integrin. GAS invades host epithelial cells by two different mechanisms, either by invagination at the point of bacterial contact with host cells or by massive induction of microvilli, which form membrane ruffling for engulfment of bacteria (filopodia). Cytoskeletal rearrangements involve induction of the IP-3 kinase pathway or RAS/CDC42/tyrosine kinase activation. Some of the GAS secretory products such as SLO and NAD glycohydrolase may cooperatively make holes in eukaryotic cells and inject bacterial product in eukaryotic cells to exploit intracellular signaling events in a manner similar to the type III secretory system of gram-negative bacteria. Intracellular GAS may then direct host cells to undergo apoptosis via caspase-dependent and -independent pathways. In GAS-mediated apoptosis mitochondria seem to play a crucial role. NAD-glycohydrolase of GAS may convert NAD to ADP-ribose and/or cyclic ADP-ribose, which may then direct the host cell to undergo apoptosis through the Ca^{2+} signaling pathway. Although not fully explored, transcriptional regulation of many inflammatory cytokines and apoptosis in the GAS-infected host cells may be mediated via NF-κB and JAK/STAT pathways. Histone-phosphorylation/dephosphorylation and histone acetylation/deacetylation also seem to play important roles in gene transcription in GAS-infected host cells. The induction of signaling pathways may vary depending on the cell lines, status of the cell (polarized versus nonpolarized), type of GAS strains, and growth phase. Abbreviations: ADPR-adenosine diphosphate-ribose; cADPR-cyclic ADPR; AIF-apoptosis inducing factor; Akt-AKT retroviral oncogene protein Ser/Thr kinase; APAF1-apoptosis protease activating factor-1 or CED-4; Arp2/3-actin-related protein 2 and 3; ASK1-apoptosis signal-regulating kinase 1 (also known as MEKK5); ATF2-activating transcription factor 2; BAD-Bcl-xL/Bcl-2 associated death promoter; Bid-BH-3 interacting domain death agonist that induces ICE-like proteases and apoptosis; Bcl2-B-cell lymphoma 2, which belongs to the Bcl-2 family of proteins and is known to inhibit apoptosis; BAK-Bcl-2 antagonist/killer. Bak is a pro-apoptotic protein; BAX-Bcl-2 associated x protein. Bax is as member of the Bcl-2 family and is pro-apoptotic; tBid-truncated Bid; CARD-caspase activation and recruitment domain; Caspase-cysteinyl aspartic acid-protease; CD44-human leukocyte differentiation receptor antigen for hyaluronate and proteoglycin serglycin; CD46-member of RCA gene family, receptor for measles virus and the M protein of GAS; Cdc42-cell division cycle 42 (GTP-binding protein); CytC-cytochrome C; c-Fos/c-Jun-transcription factor, also known as activator protein or AP-1; ELK1-Ets domain containing DNA-binding protein. Mammalian ELK-1, ELK-3 (also known as Net or SAP-2) and ELK-4 (also known as SRF accessory protein 1 [SAP-1]), which all form a ternary complex with the serum response factor (SRF); ERK-extracellular signal-related protein kinase; FADD-Fas-associated death domain; FAK-focal adhesion kinase; FAS-known as CD95 or APO-1. Fas is a member of the TNF receptor family and promotes apoptosis. FasL-Fas ligand. FasL is also known as APO-1 ligand or Apo-L; FLIP-FLICE (Fadd-like ICE [interleukin-1b converting enzyme also known as caspase-1]) inhibitory protein; $G_{\alpha,\beta,\gamma}$-G-protein α, β, and γ; HAT-histone acetyl transferase; HDAC-histone deacetylase; IKB-inhibitory IB (inhibitor of NF-κB) proteins; IKKs-IKK-1 and IKK-2 are two direct IB kinases; IL-interleukins; INF-interferon α or γ; INFR-interferon-γ receptor; IRAK-interleukin-1 receptor-associated kinases; JAK-Janus kinase; JNK-c-Jun N terminal kinase; MAPK-mitogen-activated protein kinase; MEK-MAPK activator; MEKKs-mitogen-activated protein/ERK kinase kinases; MLK1,2-CdC42-dependent kinases; MLK4,7-CdC42-dependent kinases; mTOR-mammalian target of Rapamycin (an immunosuppressant) protein; MyD88-myeloid differentiation factor 88; NAD-nicotinamide adenine dinucleotide; NFKB-nuclear factor of immunoglobulin κ locus in B cells. NF-κB activates transcription of genes in many tissues; NIK-Nck interacting kinase (interacting with MEKK) is different from NIK (NFκB-interacting kinase); PI-3K-Phosphatidyl inositol-3 kinase; RAC-small GTP-binding protein encoded by the *rac* gene; RAS-small GTP-binding protein protooncogene encoded by the *ras* gene; RIP-receptor interacting protein; ROS-reactive oxygen species; SEN, streptococcal surface enolase; SH2-Src homology region 2; SLO, streptolysin O; STAT-signal transducers and activators of transcription; TIRAP-Toll-interleukin 1 receptor domain-containing adaptor protein; TLR-Toll-like receptor; TNF-tumor necrosis factor; TNFR-tumor necrosis factor receptor; TOLL-human homologue of the *Drosophila* Toll protein; TOLLIP-Toll-interacting protein; TRAD-TNF receptor I associated death domain; TRAF-tumor necrosis factor receptor-associated factors; SDH-streptococcal surface dehydrogenase; uPAR-urokinase plasminogen activator receptor.

100-, 30-, 22-, 17-, and 8-kDa proteins. The SDH-mediated phosphorylation of certain proteins (180 and 100 kDa), and more notably the de novo phosphorylation of the 17-kDa pharyngeal protein, was inhibited only in the presence of genistein, which specifically inhibits the catalytic action of the protein tyrosine kinase (118). Intact streptococci or purified SDH are unable to induce phosphorylation of these cellular proteins or the de novo phosphorylation of the 17-kDa protein in the absence of cytoplasmic components of pharyngeal cells (118). This provides an additional proof that streptococcal- or SDH-mediated interactions occur, possibly through 37-kDa and 55-kDa (enolase and uPAR, respectively) putative surface receptors. This, in turn, may evoke a cascade of specific biochemical signals that requires the cytoplasmic milieu for intracellular signal transduction, ultimately resulting in the induced or de novo phosphorylation of cellular proteins (118). Because the primary target tissues of GAS are the human pharynx and skin, human pharyngeal cell lines such as Detroit 562 and FaDu, human keratinocyte cell line HaCat, and SCC13 may serve as ideal in vitro infection models for streptococcal studies (29, 51, 109, 118).

In the case of *S. pneumoniae*, cell wall components have been shown to activate intracellular signaling events in astrocytes; however, the specific receptor on these cells is not known (139). Phosphorylcholine of cell wall-associated teichoic acid also plays a critical role in pneumococcal binding to endothelial or lung cells, possibly through the platelet-activating factor receptor (PAF receptor) (26). Interestingly, this receptor is expressed only during the activated stage of the cell (26). Pneumococcal binding to resting cells is mediated through GalNAcβ1-4Gal and GalNAcβ1-3Gal receptors present on the cells, and their binding to de novo expressed PAF receptors and subsequent invasion of the cell is mediated through N-acetylglucosamine residues present on them (26, 47). The PAF receptor is known to participate in a variety of intracellular signaling events. However, surprisingly, pneumococcal adherence to and invasion of the target cells mediated by the PAF receptor occurs without evoking any intracellular signal transduction. Further dissection of biochemical events associated with pneumococcal adherence and invasion may provide a novel insight into the mechanism underlying the pneumococcal tissue tropism and infection (27).

Another line of evidence that pneumococci are capable of inducing signal transduction has been reported using their target microglial cells or astrocytes (139). Microglial cells and astrocytes, which are found at the site of inflammation and secrete various proinflammatory mediators, seem to play a critical role in the pneumococcal-mediated inflammation and central nervous system immune responses (152). Pneumococcal cell wall (PCW) components, such as lipopolysaccharide of gram-negative bacteria, are capable of inducing CD14-dependent signaling events in primary cultures of rat astrocytes and the human astrocytoma cell line, U373MG (139). PCW initiates tyrosine phosphorylation and the subsequent activation of mitogen-activated protein kinases (MAPK), erk-1/erk-2, c-Jun N-terminal kinase, and p38 in astrocytes in a dose-dependent manner in the presence of CD14. Since the MAPK pathway ultimately results in the induction of transcription activation factors, this study provides an important insight into the mechanism of PCW-mediated secretion of proinflammatory mediators during the disease process (140). In fact, PCW and lipoteichoic acid act synergistically to initiate signaling events in various cell types through the CD14 receptor (17). These phenomena can be demonstrated only in immature but not in permanent human monocytes, while the intact pneumococci do not use CD14 as a receptor to initiate signaling events (17). The teichoic acid-peptidoglycan network induces a variety of mammalian cells to produce tumor necrosis factor α (TNF-α), IL-1, IL-6, IL-10, IL-12, nitric oxide, NF-κB, and vascular endothelial growth factor and promotes a procoagulant state (4, 15, 44, 48, 56, 73, 80, 128, 139). Many of these effects are mediated through cell wall binding to Toll-like receptor 2 (TLR-2) (76, 86, 138).

ROLE OF INTEGRINS IN STREPTOCOCCUS-MEDIATED SIGNALING EVENTS

Integrins are large αβ heterodimeric membrane proteins found on the surface of a wide variety of mammalian cells (62, 65). These integrins seem to be involved in promoting adhesion to extracellular matrix proteins such as fibronectin, collagen, and laminin. Several fibronectin-binding proteins on the surface of GAS have been described. They include protein-F/Sfb1 (54, 95), protein F2 (68), serum opacity factor (126), protein H (45), SDH (115), Fbp54 (24), and M1 (25). Sfb1, protein F, and protein F1 are found, at least in part, to mediate streptococcal invasion of HEp2 or HeLa cells (95, 111). Similarly, the M1 protein efficiently binds to the human lung epithelial cell lines, A549, in the presence of fibronectin via α5β1 integrin (25). This integrin-mediated interaction seems to play a critical role in streptococcal invasion. While recognizing the importance of α5β1 integrin-mediated streptococcal invasion of epithelial cells, the study emphasizes that the ability of fibronectin to mediate invasion varies considerably, depending on the source of the integrin preparation (24, 25). GAS invasion is also reported to be mediated by laminin (66), which recognizes α1β1, α2β1, α3β1, α6β1, or α7β1 integrins on the cell surface (62, 65, 150). Both entry of GAS via protein F1 and uptake of fibronectin-coated beads into HeLa cells are blocked by anti-B1-antibody (96, 111). Thus, the streptococci may use one or more integrins, depending on the cell type, as receptors for invasion, and may subsequently evoke an integrin-specific signaling pathway. Bacterial invasion of a target epithelial cell is a dynamic process that involves active participation of the cytoskeletal structure (11, 20, 160). It is thus reasonable to predict that binding to integrins allows streptococci to communicate with the cytoskeletal structure of the cell and regulates the streptococcal entry into the cell. The studies showing the inhibition of streptococcal invasion of Detroit pharyngeal cells in the presence of the tyrosine and serine/threonine kinase inhibitors genistein and staurosporine indicate that the rearrangement of cytoskeletal structure, which facilitates bacterial entry, is substantially regulated by the activation of these intracellular signaling enzymes (118). In HeLa cells, a highly invasive type M1 GAS strain undergoing endocytosis has been found to be associated with polymerized actin and Lamp1 glycoprotein of lysosomal vacuole (37). The actin polymerization and hence streptococcal invasion of host cells is significantly decreased in the presence of phosphatidylinositol 3-kinase inhibitors such as LY294002 and wartmanin (125). Thus, M1 protein, either directly or indirectly via bound fibronectin, initiates signals that depend on the phosphatidylinositol 3-kinase pathway, which is crucial for the cytoskeletal rearrangement and internalization of streptococci (125). Similarly, protein F1 also mediates recruitment of de novo focal adhesion kinase, which is tyrosine phosphorylated (112). Src kinase, Yes, and Fyn proteins enhance the efficiency of GAS uptake. In addition, Rac and Cdc42 are also activated, indicating that at least two independent pathways converge to trigger actin rearrangement and endocytosis/bacterial uptake (112). The activation of different pathways for actin rearrangements is also manifested via electron microscopy. GAS strains, such as fibronectin-binding M type 12 strain A40, invade after the generation of membrane invagination at the level of the bacterial cell contact area. Non-fibronectin-binding M type 5 strain A8 recruits elongated microvilli upon bacterial contact, which then fuse to surround bacteria, leading to the formation of large pseudopod-like structures to engulf bacteria (96). The invasion of HEp-2 cells by strain A40 has recently been shown to be inhibited by treatment of cells with cholesterol-depleting agents such as methyl-β-cyclodextrin and filipin (130). Fluorescence-labeled cholera toxin B subunits have also been found to colocalize at the contact point between strain A40 and the host cell membrane (130). Interestingly, invasion by strain A8 is not decreased with either cyclodextrin or filipin. Together these results indicate that caveolae may act as an entry port for A40 strain or streptococcal strains that use fibronectin-binding-protein-mediated invasion of the host cell. Cells that possess more caveolae may show more invasion of streptococci. Whether caveolin protein serves as a receptor for streptococci is presently not known.

In all published reports, the fibronectin-binding property of GAS is ascribed a primary role in initial adherence to, colonization of, and subsequent invasion of respiratory epithelial cells (25, 67, 95, 110). However, this view, which is based solely on in vitro unpolarized tissue culture cells, needs to be reevaluated in context of the fact that in vivo streptococcal targets are polarized pharyngeal cells. Establishment of cell polarity is fundamental to differentiation and diversity of function in most cells. Extracellular matrix-binding integrin receptors, which are evenly distributed on transformed tissue culture cells, are, in fact, located at the basal side rather than on the apical side of the polarized cell (10, 39, 129). Streptococci, after entry into the host, are expected to interact with specific receptor(s) located on the apical surface before invasion. Hence, it is likely that during initial colonization and invasion of the intact epithelia, streptococci may not utilize their fibronectin-binding property because fibronectin is located at the basal side of the cell as one of the components of the extracellular matrix (Fig. 1). Once streptococci reach the extracellular matrix, their adherence to one or more of its components, such as fibronectin, may initiate signaling events in the cell via a specific class of integrins. These signaling events, in turn, may afford streptococci the ability to invade these cells further after the initial entry. It is reasonable to speculate that integrin-mediated signaling events play a critical role in streptococcal dissemination in the tissue. Recent studies have, however, changed our concept about the role of fibronectin and the fibronectin-binding ability of GAS in terms of bacterial dissemination and hence virulence (108). Using protein F1-defective/negative streptococcal strain and transgenic mice lacking plasma fibronectin, this study has conclusively shown that protein F1-expressing bacteria are less virulent to normal mice and that bacterial dissemination occurs more efficiently only in mice lacking plasma fibronectin. This demonstrates that the binding of plasma fibronectin to the bacterial surface down-regulates S. pyogenes virulence by limiting bacterial spread (108). Thus, the fibronectin-binding ability may be viewed as the bacterium's deliberate attempt to decrease its virulence and form an unassuming but more advantageous coexistence with its human host.

TARGETING THE NUCLEUS: AN ULTIMATE FUNCTION OF THE MEMBRANE RECEPTOR-ORIGINATED SIGNAL TRANSDUCTION

Unlike the M type 6 GAS strain (118), the invasive capacity of the M type 1 GAS strain (125) is unaffected by genistein, indicating that more than one signaling pathway may control streptococcal invasion. It is, therefore, likely that during streptococcal invasion of host cells, the initial bacterial interaction leads to either invagination of the host cell (caveolae) or massive accumulation of microvilli (96) (Fig. 1). The mechanisms behind these two types of invasion processes have been described in other chapters of this book. Research efforts from many laboratories over the past several years have established that within minutes after the activation of several membrane signaling molecules, there occurs a sequential stimulation of several protein kinases, including serine/threonine/tyrosine kinases and phosphatidylinositol kinases, leading to the RAS family of GTPases. These proteins then bind to and activate the mixed-lineage kinases (MLK1, 2, 4), a group of MAPKs that contain GTPase recognition motif (CRIIB) (46, 140). This leads to activation of MAPKs that regulate the c-Jun N-terminal kinase. Activation of p21-activated kinase by Rac allows activation of the inhibitory kappa B (IKB) kinase that phosphorylates IKB, the inhibitor of the nuclear factor NF-κB, promoting IKB ubiquitinylation and its degradation by the proteosome and allowing NF-κB to translocate into the nucleus and target gene expression (5, 143). In humans, activated NF-κB dimers consist mainly of the Rel family proteins p50 and p65 subunits, which then bind to responsive κB sites in the promoters and enhancers of target genes (50). Like NF-κB, STAT molecules (signal transducer in the cytoplasm and activator of transcription in the nucleus) are also latent cytoplasmic transcription factors activated by cytokine stimulation via tyrosine phosphorylation (32). The binding of cytokines to their receptors results in autophosphorylation of receptor-associated JAK kinases that phosphorylate and activate STATs (122). GAS induces NF-κB activity in human macrophage (93) and human laryngeal cell line HEp-2 (90). Similarly, group B streptococci (87, 91) and oral streptococci (105) also induce NF-κB activity in hematopoietic cells and fibroblastlike synoviocytes. S. pyogenes has also been found to induce STAT1, STAT3, and interferon (IFN) regulatory factor 1 DNA-binding activity and the formation of IFN-α-specific transcription factor complex and IFN-stimulated gene factor 3 in human primary macrophages (93). Interestingly, S. pyogenes-mediated NF-κB activation in host cells is not inhibited by cytochalasin B (90) or by a protein synthesis inhibitor, cycloheximide (93), suggesting that these bacteria could directly activate NF-κB without being internalized. STAT activation, on the other hand, is cycloheximide-sensitive (93), suggesting that prior activation and production of cytokines are required for STAT activation. These studies thus shed light on the molecular mechanisms of interaction between streptococci and host cells that lead to different inflammatory responses.

The chromatin or the nucleosome is a highly ordered structure involving DNA supercoil stabilized by octameric core histone proteins H2A, H2B, H3, and H4 (82). Acetylation/deacetylation and phosphorylation/dephosphorylation of histone proteins play an important role in relaxation or condensation of the chromatin structure (60, 161). In the last decade, it has become increasingly evident that acetylation/deacetylation and phosphorylation/dephosphorylation of core histone proteins play a crucial role in the activation and repression of gene transcription, which is important for growth and development (1, 7, 25, 60, 85, 131, 147, 161). More recently, it has been confirmed that the enzymes (acetylase and deacetylase) that are involved in the modification of core histone proteins and hence the condensation/relaxation of chromatin are, in fact, transcriptional modulators (see reviews in references 60, 123, 161). The report of streptococcal/SDH-mediated de novo phosphorylation of the 17-kDa pharyngeal cell protein and its identification as histone H3 is the first to describe the possible role of histone phosphorylation in bacterially mediated signaling in intact target cells (118) (Fig. 1). Electron microscopy of SDH-treated pharyngeal cells showing distinct condensation of chromatin in the nuclei supports the view that this effect is the direct result of the de novo phosphorylation of histone H3, although other possibilities may explain the condensation of chromatin structure (118). It is likely that the streptococcal-mediated generalized induction of protein phosphorylation of pharyngeal cells may be in part due to the upstream signaling events that target the nucleus. Implications of these findings may be viewed in terms of the ability of SDH/streptococci to induce gene transcription in the target cells. Hence to understand the role of streptococci-mediated signaling events in pathogenesis, the relationship between the streptococcal/SDH-mediated histone H3 phosphorylation and other serine/threonine and tyrosine phosphorylation events leading to transcriptional regulation must be further investigated.

PROINFLAMMATORY RESPONSE IN HOST CELLS AFTER THEIR INTERACTION WITH STREPTOCOCCI

Proinflammatory responses from the host cells in terms of the de novo or increased secretion of certain cytokines as a result of the bacterium-host cell interaction provide indirect evidence of not only signal transduction events but also the activation of specific genes and subsequent nuclear responses. Substantial in vitro and in vivo findings now provide evidence to confirm that, like lipopolysaccharide (162), pathogenic gram-positive cocci, their secretory products, and cell wall components are capable of inducing proinflammatory mediators such as interleukins and TNF-α in a CD14-dependent manner (27, 55, 91, 99, 121, 139, 156). The role of staphylococcal and streptococcal exotoxins, which function as superantigens, in the induction of mitogenic and inflammatory responses is well established and is characterized by the release of a variety of chemokines and cytokines from peripheral blood monocytes (3, 77, 107, 121). Although the release of inflammatory cytokines triggered by streptococcal superantigens plays an important role in invasive GAS disease, individuals infected with the same strain may develop very different manifestations. Kotb et al. (77) have recently proposed that leukocyte antigen class II allelic variation contributes to differences in severity of invasive streptococcal infection through the ability to regulate streptococcal superantigen-mediated cytokine responses in terms of TNF-β, IFN-γ, and IL-4 production. In particular, patients with the DRB1*1501/DQB1*0602 haplotype mount significantly reduced re-

sponses and hence are less likely to develop severe systemic disease (77). In addition to superantigens, an important role of the M protein in excessive plasma leakage from vasculature has been recently recognized, leading to life-threatening hypotension and multiorgan failure (57). Herwald et al. (57) have reported that the ability of the M protein to bind fibrinogen (2, 70) allows the M protein/fibrinogen complex to interact with β2-integrins and activate neutrophils. As a result, neutrophils release heparin-binding protein, an inflammatory mediator that induces vascular leakage and a massive inflammatory response, leading to toxic shock. More evidence is available that confirms the ability of target epithelial cells or blood monocytes to release certain interleukins and TNF as a result of their interaction with intact gram-positive bacteria in a CD14-independent manner (17, 156). The examination by Wang et al. (156) of keratinocyte responses to adherent versus nonadherent GAS has revealed a distinct pattern of expression of several proinflammatory molecules. This study has reported further that both adherent and nonadherent streptococci are capable of inducing IL-6 and prostaglandin E_2; however, only adherent streptococci are capable of inducing IL-1α, IL-1β, and IL-8. Similarly, only adherent streptococci are capable of damaging the keratinocytes. Since the production of cytokines and the damage to the keratinocytes are considerably reduced in streptolysin O-deficient GAS mutants, it is rational to conclude that to induce proinflammatory response streptococci also utilize some of their secretory products (156). In view of the fact that the primary target cell for GAS is the pharyngeal cell, it is pertinent to explore the mechanism by which streptococci are able to evoke such an inflammatory response using in vitro pharyngeal cell model.

GAS-MEDIATED APOPTOSIS OF HOST CELLS

GAS has devised several strategies for its perpetual existence as a successful extracellular pathogen. As described above, the initial events in streptococcal disease involve bacterial adherence to epithelial cells followed by invasion of cells. Subsequently, like many intracellular pathogens (53, 58, 72, 164, 165), GAS can also induce apoptotic programmed cell death (101, 151). Apoptosis is crucial for proper organismal development, maintenance of cell number homeostasis, and elimination of diseased or otherwise harmful cells (100, 145). Apoptotic cell death in response to bacterial infection is viewed to delete infected and damaged epithelial cells and to restore epithelial integrity, which is distorted during the course of infection (163). Thus, apoptosis may contribute to resolution of bacterial infection. The ability of host cells to undergo apoptosis is considered the host defense strategy to keep the pathogen at a disadvantage by not providing a target for adherence, invasion, and subsequent proliferation. However, the ability of the bacteria to direct the host cell to undergo apoptotic cell death, thus allowing it to escape from phagocytic activity, may be viewed as an important virulence factor. Apoptosis of a few epithelial cells directed by bacteria not only helps these bacteria escape from one of the host's lines of defense, but it also facilitates penetration to the next cellular barrier. Tsai et al. (151) and Nakagawa et al. (101) have shown that M type 1, 6, and 49 GAS strains induce apoptosis in 10 to 20% of the upper and lower respiratory cell lines, HEp-2 laryngeal, and A-549 lung alveolar (101, 151). In particular, the cells with internalized bacteria show increased expression of the Bax protein, which translocates to mitochondria (101), resulting in cytochrome C release. At the same time, the expression of prosurvival gene products such as Bcl-2 and Bcl-X$_L$ is significantly decreased in mitochondria. Caspase-9 activation and inhibition of apoptosis of these cells in the presence of caspase inhibitors indicate that caspase activation is likely an important mediator that triggers apoptosis (101, 151). SpeB may play a direct role in causing this limited apoptosis (151). Another study has shown that NAD$^+$-glycohydrolase of an M type 3 clinical isolate gains access to the keratinocyte cytoplasm by enhancing streptolysin O-dependent damage of keratinocyte cell membranes (14). Once inside the cell, NAD$^+$ glycohydrolase inhibits GAS internalization by keratinocytes and together with streptolysin O (SLO) induces apoptosis (14, 84). The precise signaling events are presently not known, but it is postulated that NAD$^+$ glycohydrolase may enzymatically hydrolyze NAD, resulting in the accumulation of nicotinamide, ADP-ribose (ADPR), and cyclic ADPR, which may stimulate Ca^{2+} signaling pathways, including those leading to apoptosis (14, 84, 92). With the availability of the complete human genome analysis and recent developments in high-throughput microgene array technologies, it is now possible to understand the molecular basis of apoptosis by analyzing global transcription-regulated processes in target epithelial cells and neutrophils/macrophages that occur after invasion or phagocytosis of bacterial pathogens. A recent study on global transcriptional regulation in neutrophils after phagocytosis of *Borrelia hermsii*, *Listeria monocytogenes*, *Burkholderia cepacia*, *S. aureus*, and *S. pyogenes* has shown that *S. pyogenes* induces unique changes in neutrophils' gene expression (75). More specifically, phagocytosis of GAS results in up- and down-regulation of 393 genes (168 up-regulated, 225 down-regulated), 71 of which are apoptosis-related genes that include up-regulation of AP-1, FOSL1, FOSB, JUNB, and TNFRSF5-(CD40), while at least 26 genes involved in responses to IFN are down-regulated. The down-regulated genes include IFN-responsive genes (IFIT1, IFIT4, IFITM2, IFI16, IRF1, MX1, and ADAR) and transcription factor-related genes (STAT1, STAT3, STAT5B, STAT6, JAK1, RELA, and RELB) representing the JAK-STAT survival pathway. A similar application-targeted gene array study has also shown that SDH plays an important role in GAS-mediated apoptosis of pharyngeal cells (113). This study also emphasizes that host cells are programmed to undergo cell death via more than one signaling pathway, induced as a result of GAS interactions with pharyngeal cells and also other host cells. During pneumococcal meningitis, the neuronal injury arises directly from the infecting bacteria and excessive immune responses arise in the host. In experimental meningitis, concentrations of several caspases are significantly elevated and active caspase-3 is present in dentate gyrus (153). However, intrathecal treatment with broad spectrum caspase inhibitor rescues only about 50% of the neurons of experimental mice with acute pneumococcal meningitis (12). In the microglial neuronal cell tissue culture system, however, pneumococcal infection results in rapid damage of mitochondria followed by release of cytochrome C and apoptosis-inducing factor (AIF) (13). Anti-AIF antibody also inhibits apoptosis in a significant number of cells infected with pneumococci, indicating that neuronal damage during pneumococcal meningitis occurs both as caspase-dependent and caspase-independent (mitochondrial pathway) events (12, 13). Unlike in neuronal cells, pneumococci induce apoptosis of human alveolar and broncho-epithelial cells by caspase-6 and non-caspase protease but not by caspase-3 (136).

CONCLUDING REMARKS

The direct and indirect evidence of the consequences of streptococcal interaction with target epithelial cells has suggested that the streptococcal adherence is not merely a static process but is highly regulated, dynamic, and complex. The skeleton of this dynamic process is intracellular signaling events that ultimately determine the cells' fate and subsequently that of adjacent cells (Fig. 1). While together these events allow streptococci to disseminate in tissues, the biochemical changes in affected cells determine the clinical picture of streptococcal disease. Thus, changes in intracellular signaling events because of streptococcal interactions play a critical role in the disease process.

With the availability of the complete genome sequence information of humans and of several gram-positive bacterial species, including five from four different M serotypes of GAS, our knowledge of cell-cell communication between pathogenic gram-positive cocci and their target host cells has increased significantly. Although several mechanisms of pathogenesis of streptococcal diseases have been proposed, the complexity is compounded by the fact that a variety of multifunctional proteins on the surface of streptococci have been shown to play a role in the disease process and the expression of some of these proteins is type-specific or even strain-specific. At present the details of how all these proteins, whether individually or in concert, interact with host cells and ultimately evoke intracellular signaling are unknown. In this scenario, global gene expression analyses in various relevant cell types and/or infected tissues obtained from appropriate animal models will help resolve several unanswered questions. Limited attempts have been made to discern the global gene expression profile in streptococcal-infected human neutrophils. This approach, although very powerful, is inherently expensive, labor intensive, and requires thorough computational analysis of large amounts of information. In the event of unknown functions of many human genes, application-targeted gene expression analyses may help address specific questions with regard to inflammation, involvement of cell-specific receptors, cell fate, and activation of kinases, cytokines, and/or transcription factors. This approach may help decipher distinct signal transduction pathways induced because of specific bacterial protein-host receptor interactions to accurately define the role of specific bacterial proteins and host factors in pathogenesis of various streptococcal diseases. This information can also be utilized to design appropriate and novel therapeutic agents to intervene in the disease process as has been done for many septic shock and inflammatory diseases (79, 103).

I gratefully acknowledge research grant support from NIH (AI42827).

REFERENCES

1. **Ajiro, K., and T. Nishimoto.** 1985. Specific site of histone H3 phosphorylation related to the maintenance of premature chromosome condensation. *J. Biol. Chem.* **260:**15379–15381.
2. **Akesson, P., K.-H. Schmidt, J. Cooney, and L. Bjorck.** 1994. M1 protein and protein H: IgGFc- and albumin-binding streptococcal surface proteins encoded by adjacent genes. *Biochem. J.* **300:**877–886.
3. **Alouf, J. E., and H. Muller-Alouf.** 2003. Staphylococcal and streptococcal superantigens: molecular, biological and clinical aspects. *Int. J. Med. Microbiol.* **292:**429–440.

4. **Ambrosino, D. M., N. R. Delaney, and R. C. Shamberger.** 1990. Human polysaccharide-specific B cells are responsive to pokeweed mitogen and IL-6. *J. Immunol.* **144:** 1221–1226.

5. **Baldwin, A. S., Jr.** 1996. The NF-κB and IkB proteins: new discoveries and insights. *Annu. Rev. Immunol.* **14:** 649–681.

6. **Banks, D. J., S. F. Porcella, K. D. Barbian, S. B. Beres, L. E. Philips, J. M. Voyich, F. R. DeLeo, J. M. Martin, G. A. Somerville, and J. M. Musser.** 2004. Progress toward characterization of the group A *Streptococcus* metagenome: complete genome sequence of a macrolide-resistant serotype M6 strain. *J. Infect. Dis.* **190:**727–738.

7. **Barratt, M. J., C. A. Hazzalin, E. Cano, and L. C. Mahadevan.** 1994. Mitogen-stimulated phosphorylation of histone H3 is targeted to a small hyperacetylation-sensitive fraction. *Proc. Natl. Acad. Sci. USA* **91:**4781–4785.

8. **Beres, S. B., G. L. Sylva, K. D. Barbian, B. Lei, J. S. Hoff, N. D. Mammarella, M.-Y. Liu, J. C. Smoot, S. F. Porcella, L. D. Parkins, D. S. Campbell, T. M. Smith, J. K. McCormick, D. Y. M. Leung, P. Schlievert, and J. M. Musser.** 2002. Genome sequence of a serotype M3 strain of group A *Streptococcus*: phage-encoded toxins, the high-virulence phenotype, and clone emergence. *Proc. Natl. Acad. Sci. USA* **99:**10078–10083.

9. **Bergmann, S., M. Rohde, G. S. Chhatwal, and S. Hammerschmidt.** 2001. Alpha-enolase of *Streptococcus pneumoniae* is a plasmin(ogen)-binding protein displayed on the bacterial cell surface. *Mol. Microbiol.* **40:**1273–1287.

10. **Bissell, M. J.** 1993. Introduction: form and function in the epithelia. *Sem. Cell. Bio.* **4:**157–159.

11. **Bliska, J. B., J. E. Galan, and S. Falkow.** 1993. Signal transduction in the mammalian cell during bacterial attachment and entry. *Cell* **73:**903–920.

12. **Braun, J. S., R. Novak, K. H. Herzog, S. M. Bodner, J. L. Cleveland, and E. Tuomanen.** 1999. Neuroprotection by a caspase inhibitor in acute bacterial meningitis. *Nat. Med.* **5:**298–302.

13. **Braun, J. S., R. Novak, P. J. Murray, C. M. Eischen, S. A. Susin, G. Kroemer, A. Halle, J. R. Weber, E. I. Tuomanen, and J. L. Cleveland.** 2001. Apoptosis-inducing factor mediates microglial and neuronal apoptosis caused by pneumococcus. *J. Infect. Dis.* **184:**1300–1309.

14. **Bricker, A. L., C. Cywes, C. D. Ashbaugh, and M. R. Wessels.** 2002. NAD$^+$-glycohydrolase acts as an intracellular toxin to enhance the extracellular survival of group A streptococci. *Mol. Microbiol.* **44:**257–269.

15. **Buchanan, R. M., B. P. Arulanandam, and D. W. Metzger.** 1998. IL-12 enhances antibody responses to T-independent polysaccharide vaccines in the absence of T and NK cells. *J. Immunol.* **161:**5525–5533.

16. **Caparon, M. G., R. T. Geist, J. Perez-Casal, and J. R. Scott.** 1992. Environmental regulation of virulence in group A streptococci: transcription of the gene encoding M protein is stimulated by carbon dioxide. *J. Bacteriol.* **174:**5693–5701.

17. **Cauwels, A., E. Wan, M. Leismann, and E. Tuomanen.** 1997. Coexistence of CD14-dependent and independent pathways for stimulation of human monocytes by gram-positive bacteria. *Infect. Immun.* **65:**3255–3260.

18. **Chen, C. C., and P. P. Cleary.** 1989. Cloning and expression of the streptococcal C5a peptidase gene in *Escherichia coli*: linkage to the type 12 M protein gene. *Infect. Immun.* **57:**1740–1745.

19. **Cleary, P. P., E. L. Kaplan, J. P. Handley, A. Wlazlo, M. H. Kim, A. R. Hauser, and P. M. Schlievert.** 1992. Clonal basis for resurgence of serious *Streptococcus pyogenes* disease in the 1980s. *Lancet* **339:**518–521.

20. **Cossart, P.** 1997. Subversion of the mammalian cell cytoskeleton by invasive bacteria. *J. Clin. Invest.* **99:**2307–2311.

21. **Cossart, P., P. Boquet, S. Normark, and R. Rappuoli.** 1996. Cellular microbiology emerging. *Science* **271:**315–316.

22. **Cotter, P. A., and J. F. Miller.** 1996. Triggering bacterial virulence. *Science* **273:**1183–1184.

23. **Courtney, H. S., and D. L. Hasty.** 1991. Aggregation of group A streptococci by human saliva and effect of saliva on streptococcal adherence to host cells. *Infect. Immun.* **59:**1661–1666.

24. **Courtney, H. S., Y. Li, J. B. Dale, and D. L. Hasty.** 1994. Cloning, sequencing, and expression of a fibronectin/fibrinogen-binding protein from group A streptococci. *Infect. Immun.* **62:**3937–3946.

25. **Cue, D., P. E. Dombek, H. Lam, and P. P. Cleary.** 1998. *Streptococcus pyogenes* serotype M1 encodes multiple pathways for entry into human epithelial cells. *Infect. Immun.* **66:**4593–4601.

26. **Cundell, D. R., N. P. Gerard, C. Gerard, I. Indanpaan-Heikkila, and E. I. Tuomanen.** 1995. *Streptococcus pneumoniae* anchor to activated human cells by the receptor for platelet-activating factor. *Nature* **377:**435–438.

27. **Cundell, D. R., C. Gerard, I. Idanpaan-Keikkila, E. I. Tuomanen, and N. P. Gerard.** 1996. PAf receptor anchors *Streptococcus pneumoniae* to activated human endothelial cells. *Adv. Exp. Med. Biol.* **416:**89–94.

28. **Cunningham, M. W.** 2000. Pathogenesis of a group A streptococcal infections. *Clin. Microbiol. Rev.* **13:**470–511.

29. **Cywes, C., and M. R. Wessels.** 2001. Group A *Streptococcus* tissue invasion by CD44-mediated cell signalling. *Nature* **414:**648–652.

30. **Cywes, C., I. Stamenkovic, and M. R. Wessels.** 2000. CD44 as a receptor for colonization of the pharynx by group A *Streptococcus*. *J. Clin. Invest.* **106:**995–1002.

31. **Darnell, J. E.** 1998. STATs and gene regulation. *Science* **277:**1630–1635.

32. **Davies, H. D., A. McGeer, B. Schwartz, K. Green, D. Cann, A. E. Simor, and D. E. Low.** 1996. Invasive group A streptococcal infections in Ontario, Canada. *N. Engl. J. Med.* **335:**547–554.

33. **Davis, L., M. M. Baig, and E. M. Ayoub.** 1979. Properties of extracellular neuraminidase produced by group A *Streptococcus*. *Infect. Immun.* **24:**780–786.

34. **D'Costa, S. S., T. G. Romer, and M. D. P. Boyle.** 2000. Analysis of expression of a cytosolic enzyme on the surface of *Streptococcus pyogenes*. *Biochem. Biophys. Res. Commun.* **278:**826–832.

35. **Demuth, D. R., E. E. Golub, and D. Malamud.** 1990. Streptococcal-host interactions. *J. Biol. Chem.* **265:**7120–7126.

36. **Dohrman, A., S. Miyata, M. Gallup, J.-D. Li, C. Chapelin, A. Coste, E. Escudier, J. Nadel, and A. Bashir.** 1998. Mucin gene (MUC 2 and MUC 5AC) up-regulation by gram-positive and gram-negative bacteria. *Biochim. Biophys. Acta* **1406:**251–259.

37. **Dombek, P. E., D. Cue, J. Sedgewick, H. Lam, S. Ruschkowski, B. B. Finlay, and P. P. Cleary.** 1999. High-frequency intracellular invasion of epithelial cells by serotype M1 group A streptococci: M1 protein-mediated invasion and cytoskeletal rearrangements. *Mol. Microbiol.* **31:**859–870.

38. **Doring, G., H. Obernessen, K. Botzenhart, B. Flehmig, N. Hoiby, and A. Hofmann.** 1983. Proteases of *Pseudomonas aeruginosa* in patients with cystic fibrosis. *J. Infect. Dis.* **147:**744–750.

39. **Drubin, D. G., and W. J. Nelson.** 1996. Origins of polarity. *Cell* **84:**335–344.

40. **Dudani, A. K., C. Cummings, S. Hashemi, and P. R. Ganz.** 1993. Isolation of a novel 45 kDa plasminogen receptor from human endothelial cells. *Thromb. Res.* **69:**185–196.

41. **Ferretti, J. J., W. M. McShan, D. Ajdic, D. J. Savic, G. Savic, K. Lyon, C. Primeaux, S. Sezate, A. N. Suvorov, S. Kenton, H. S. Lai, S. P. Lin, Y. Qian, H. G. Jia, F. Z. Najar, Q. Ren, H. Zhu, L. Song, J. White, X. Yuan, S. W. Clifton, B. A. Roe, and R. McLaughlin.** 2001. Complete genome sequence of an M1 strain of *Streptococcus pyogenes*. *Proc. Natl. Acad. Sci. USA* **98:**4658–4663.

42. **Finlay, B. B., and P. Cossart.** 1997. Exploitation of mammalian host cell functions by bacterial pathogens. *Science* **276:**718–725.

43. **Forstner, G.** 1995. Signal transduction, packaging and secretion of mucins. *Annu. Rev. Physiol.* **57:**585–605.

44. **Freyer, D., M. Weih, J. R. Weber, W. Burger, P. Scholz, R. Manz, A. Zeigenhorn, C. Angstwurm, and U. Dirnagl.** 1996. Pneumococcal cell wall components induce nitric oxide synthase and TNF-alpha in astroglial-enriched cultures. *Glia* **16:**1–6.

45. **Frick, I.-M., K. L. Crossin, G. M. Edelman, and L. Bjorck.** 1995. Protein H—a bacterial surface protein with affinity for both immunoglobulin and fibronectin type III domains. *EMBO J.* **14:**1674–1679.

46. **Gallo, K. A., and G. L. Johnson.** 2002. Mixed-lineage kinase control of JNK and p38 MAPK pathways. *Nat. Rev. Mol. Cell Biol.* **3:**663–671.

47. **Garcia, R. C., D. R. Cundell, E. I. Tuomanen, L. F. Kolakowski, C. Gerard, and N. P. Gerard.** 1995. The role of N-glycosylation for functional expression of the human platelet-activating factor receptor. Glycosylation is required for efficient membrane trafficking. *J. Biol. Chem.* **270:**25178–25184.

48. **Geelen, S., C. Bhattacharyya, and E. Tuomanen.** 1992. Induction of procoagulant activity on human endothelial cells by *Streptococcus pneumoniae*. *Infect. Immun.* **60:**4179–4183.

49. **Gendler, S. J., and A. P. Spicer.** 1995. Epithelial mucin genes. *Annu. Rev. Physiol.* **57:**607–634.

50. **Ghosh, S., M. J. May, and E. B. Kopp.** 1998. NF-kappaB and Rel proteins: evolutionary conserved mediators of immune responses. *Annu. Rev. Immunol.* **16:**225–260.

51. **Giannakis, E., T. S. Jokiranta, R. J. Ormsby, T. G. Duthy, D. A. Male, D. Christiansen, V. A. Fischetti, C. Bagley, B. E. Loveland, and D. L. Gordon.** 2002. Identification of the streptococcal M protein-binding site on membrane cofactor protein (CD46). *J. Immunol.* **168:**4585–4592.

52. **Graham, M. G., L. M. Smoot, C. A. Lux Migliaccio, K. Virtaneva, D. E. Sturdevant, S. F. Porcella, M. J. Federle, G. J. Adams, J. R. Scott, and J. M. Musser.** 2002. Virulence control in group A *Streptococcus* by a two-component gene regulatory system: global expression profiling and *in vivo* infection modeling. *Proc. Natl. Acad. Sci. USA* **99:**13855–13860.

53. **Guzman, C. A., E. Domann, M. Rohde, D. Bruder, A. Darji, S. Weiss, J. Wehland, T. Chakraborty, and K. N. Timmis.** 1997. Apoptosis of mouse dendritic cells is triggered by listeriolysin, the major virulence determinant of *Listeria monocytogenes*. *Mol. Microbiol.* **20:**119–126.

54. **Hanski, E., and M. Caparon.** 1992. Protein F, a fibronectin-binding protein, is an adhesin of the group A *Streptococcus, Streptococcus pyogenes*. *Proc. Natl. Acad. Sci. USA* **89:**6172–6176.

55. **Henneke, P., O. Takeuchi, J. A. van Strijp, H. K. Guttormsen, J. A. Smith, A. B. Schromm, T. A. Espevik, S. Akira, V. Nizet, D. L. Kasper, and D. T. Golenbock.** 2001. Novel engagement of CD14 and multiple toll-like receptors by group B streptococci. *J. Immunol.* **167:** 7069–7076.

56. **Hermann, C., I. Spreitzer, N. Schroeder, S. Morath, M. Lehner, W. Fischer, C. Schutt, R. Schumann, and T. Hartung.** 2002. Cytokine induction by purified lipoteichoic acids from various bacterial species-role of LBP, sCD14, CD14 and failure to induce TL-12 and subsequent IFN release. *Eur. J. Immunol.* **32:**551.

57. **Herwald, H., H. Cramer, M. Morgelin, W. Russell, U. Sollenberg, A. Norrby-Teglund, H. Flodgaard, L. Lindbom, and L. Bjorck.** 2004. M protein, a classical bacterial virulence determinant, forms complexes with fibrinogen that induce vascular leakage. *Cell* **116:**367–379.

58. **Hilbi, H., J. E. Moss, D. Hersh, Y. Chen, J. Arondel, S. Banerjee, R. A. Flavell, J. Yuan, P. J. Sansonetti, and A. Zychlinsky.** 1998. Shigella-induced apoptosis is dependent on caspase-1 which binds to IpaB. *J. Biol. Chem.* **273:** 32895–32900.

59. **Hilkens, J., M. J. L. Ligtengerg, H. L. Vos, and S. V. Litvinov.** 1992. Cell membrane-associated mucins and their adhesion-modulating property. *Trends Biol. Sci.* **17:** 359–363.

60. **Hopkin, K.** 1997. Spools, switches, or scaffolds: how might histones regulate transcription? *J. NIH Res.* **9:**34–37.

61. **Hultgren, S. J., S. Abraham, M. Caparon, P. Falk, J. W. St. Geme III, and S. Normack.** 1993. Pilus and nonpilus bacterial adhesions: assembly and function in cell regulation. *Cell* **73:**887–901.

62. **Hynes, R. O.** 1992. Integrins: versatility, modulation, and signaling in cell adhesion. *Cell* **69:**11–25.

63. **Hytonen, J., S. Haataja, P. Isomaki, and J. Finne.** 2000. Identification of a novel glycoprotein-binding activity in *Streptococcus pyogenes* regulated by the *mga* gene. *Microbiology* **146:**31–39.

64. **Hytonen, J., S. Haataja, and J. Finne.** 2003. *Streptococcus pyogenes* glycoprotein-binding strepadhesin activity is mediated by a surface-associated carbohydrate-degrading enzyme, pullulanase. *Infect. Immun.* **71:**784–793.

65. **Iseberg, R. R.** 1991. Discrimination between intracellular uptake and surface adhesion of bacterial pathogens. *Science* **252:**934–938.

66. **Iseberg, R. R., and G. T. V. Nhieu.** 1994. Binding and internalization of microorganisms by integrin receptors. *Trends Microbiol.* **2:**10–14.

67. **Jadoun, J., V. Ozeri, E. Burstein, E. Skutelsky, E. Hanski, and S. Sela.** 1998. Protein F1 is required for efficient entry of *Streptococcus pyogenes* into epithelia cells. *J. Infect. Dis.* **178:**147–158.

68. **Jaffe, J., S. Natanson-Yaron, M. G. Caparon, and E. Hanski.** 1996. Protein F2, a novel fibronectin-binding protein from *Streptococcus pyogenes*, possesses two binding domains. *Mol. Microbiol.* **21:**373–384.

69. **Jin, H., Y. P. Song, and V. Pancholi.** 2004. Urokinase plasminogen activator receptor (uPAr/CD87) is a receptor for group A streptococcal surface dehydrogenase (SDH) on the surface of human pharyngeal cells: identification of ligand-receptor binding regions, abstr. B-378, p. 99. Abstr. 104th Annu. Meet. Am. Soc. Microbiol. 2004. American Society for Microbiology, Washington, D.C.

70. **Kantor, F. S.** 1965. Fibrinogen precipitating by streptococcal M protein. I. Identity of the reactants and stoichiometry of the reaction. *J. Exp. Med.* **121:**849–859.

71. **Kehoe, M. A.** 1994. Cell-wall-associated proteins in gram-positive bacteria, p. 217–261. *In* J.-M. Ghuysen and R. Hakenbeck (ed.), *Bacterial Cell Wall.* Elsevier Science, New York, N.Y.

72. **Kemp, K., H. Bruunsgaard, P. Skinhoj, and P. B. Klarlund.** 2002. Pneumococcal infections in humans are associated with increased apoptosis and trafficking of type 1 cytokine-producing T cells. *Infect. Immun.* **70:**5019–5025.

73. **Kerr, A. R., X. Q. Wei, P. W. Andrew, and T. J. Mitchell.** 2004. Nitric oxide exerts distinct effects in local and systemic infections with *Streptococcus pneumoniae.* *Microb. Pathog.* **36:**303–310.

74. **Klinger, J., B. Tandler, C. Kiedtke, and T. Boat.** 1984. Proteinases of *Pseudomonas aeruginosa* evoke mucins release by tracheal epithelium. *J. Clin. Invest.* **74:**1669–1678.

75. **Kobayashi, S. D., K. R. Braughton, A. R. Whitney, J. M. Voyich, T. G. Schwan, J. M. Musser, and F. R. DeLeo.** 2003. Bacterial pathogens modulate an apoptosis differentiation program in human neutrophils. *Proc. Natl. Acad. Sci. USA* **100:**10948–10953.

76. **Koedel, U., B. Angele, T. Rupprecht, H. Wagner, A. Roggenkamp, H. W. Pfister, and C. J. Kirschning.** 2003. Toll-like receptor 2 participates in mediation of immune response in experimental pneumococcal meningitis. *J. Immunol.* **170:**438–444.

77. **Kotb, M., A. Norrby-Teglund, A. McGeer, H. El Sherbini, M. T. Dorak, A. Khurshid, K. Green, J. Peeples, J. Wade, G. Thomson, B. Schwartz, and D. E. Low.** 2002. An immunogenetic and molecular basis for differences in outcomes of invasive group A streptococcal infections. *Nat. Med.* **8:**1398–1404.

78. **Kreikemeyer, B., K. S. McIver, and A. Podbielski.** 2003. Virulence factors regulation and regulatory networks in *Streptococcus pyogenes* and their impact on pathogen-host interactions. *Trends Microbiol.* **11:**224–232.

79. **Levitzki, A., and A. Gazit.** 1995. Tyrosine kinase inhibition: an approach to drug development. *Science* **267:**1782–1788.

80. **Lieberman, D., S. Livnat, F. Schlaeffer, A. Porath, S. Horowitz, and R. Levy.** 1997. IL-1beta and IL-6 in community-acquired pneumonia: bacteremic pneumococcal pneumonia versus *Mycoplasma pneumoniae* pneumonia. *Infection* **25:**90–94.

81. **Livinov, S. V., and J. Hilkens.** 1993. The epithelial sialomucin, episialin, is sialylated during recycling. *J. Biol. Chem.* **268:**21364–21371.

82. **Luger, K., A. W. Mader, R. K. Richmond, D. F. Sargent, and T. J. Richmond.** 1997. Crystal structure of the nucleosome core particle at 2.8A resolution. *Nature* **389:**251–260.

83. **Lukomski, S., E. H. Burns Jr, P. R. Wyde, A. Podbielski, J. Rurangirwa, D. K. Moore-Poveda, and J. M. Musser.** 1998. Genetic inactivation of an extracellular cysteine protease (SpeB) expressed by *Streptococcus pyogenes* decreases resistance to phagocytosis and dissemination to organs. *Infect. Immun.* **66:**771–776.

84. **Madden, J. C., N. Ruiz, and M. Caparon.** 2001. Cytolysin-mediated translocation (CMT): a functional equivalent of type III secretion in Gram-positive bacteria. *Cell* **104:**143–152.

85. **Mahadevan, L. C., A. C. Willis, and M. J. Barratt.** 1991. Rapid histone H3 phosphorylation in response to growth factors, phorbol esters, okadaic acid, and protein synthesis inhibitors. *Cell* **65:**775–783.

86. **Malley, R., P. Henneke, S. C. Morse, M. J. Cieslewicz, M. Lipsitch, C. M. Thompson, E. Kurt-Jones, J. C. Paton, M. R. Wessels, and D. T. Golenbock.** 2003. Recognition of pneumolysin by Toll-like receptor 4 confers resistance to pneumococcal infection. *Proc. Natl. Acad. Sci. USA* **100:**1966–1971.

87. **Mancuso, G., A. Midiri, C. Beninati, G. Piraino, A. Valenti, G. Nicocia, D. Teti, J. Cook, and G. Teti.** 2002. Mitogen-activated protein kinases and NF-kappa B are involved in TNF-alpha responses to group B streptococci. *J. Immunol.* **169:**1401–1409.

88. **McIver, K. S., and J. R. Scott.** 1997. Role of *mga* in growth phase regulation of virulence genes of the group A streptococcus. *J. Bacteriol.* **179:**5178–5187.

89. **McIver, K. S., A. S. Heath, and J. R. Scott.** 1995. Regulation of virulence by environmental signals in group A streptococci: influence of osmolarity, temperature, gas exchange, and iron limitation of *emm* transcription. *Infect. Immun.* **63:**4540–4542.

90. **Medina, E., D. Anders, and G. S. Chhatwal.** 2002. Induction of NF-kappaB nuclear translocation in human respiratory epithelial cells by group A streptococci. *Microb. Pathog.* **33:**307–313.

91. **Medvedev, A. E., T. Flo, R. R. Ingalls, D. T. Golenbock, G. Teti, S. N. Vogel, and T. Espevik.** 1998. Involvement of CD14 and complement receptors CR3 and CR4 in nuclear factor-kappaB activation and TNF production induced by lipopolysaccharide and group B streptococcal cell walls. *J. Immunol.* **160:**4535–4542.

92. **Mehta, K., U. Sahid, and F. Malavasi.** 1996. Human CD38, a cell-surface protein with multiple function. *FASEB J.* **10:**1408–1417.

93. **Miettinen, M., A. Lehtonen, I. Julkunen, and S. Matikainen.** 2000. Lactobacilli and streptococci activate NF-kappa B and STAT signaling pathways in human macrophages. *J. Immunol.* **164:**3733–3740.

94. **Miles, L. A., C. M. Dahlberg, J. Plescia, J. Felez, K. Kato, and E. F. Plow.** 1991. Role of cell-surface lysines in plasminogen binding to cells: identification of alpha-enolase as a candidate plasminogen receptor. *Biochemistry* **30:**1682–1691.

95. **Molinari, G., S. R. Talay, P. Valentin-Weigand, M. Rohde, and G. S. Chhatwal.** 1998. The fibronectin-binding protein of *Streptococcus pyogenes,* SfbI, is involved in the internalization of group A streptococci by epithelial cells. *Infect. Immun.* **65:**1357–1363.

96. **Molinari, G., M. Rohde, C. A. Guzman, and G. S. Chhatwal.** 2000. Two distinct pathways for the invasion of *Streptococcus pyogenes* in non-phagocytic cells. *Cell. Microbiol.* **2:**145–154.

97. **Mosquera, J., A., V. N. Katiyar, J.Coello, and B. Rodriguez-Iturbe.** 1985. Neuraminidase production by streptococci from patients with glomerulonephritis. *J. Infect. Dis.* **151:**259–263.

98. **Moss, J., and M. Vaughan.** 1988. ADP-ribosylation of guanyl nucleotide-binding regulatory proteins by bacterial toxins. *Adv. Enzymol.* **61:**303–379.

99. **Muller-Alouf, H., J. E. Alouf, D. Gerlach, J.-H. Ozegowski, C. Fitting, and J.-M. Cavaillon.** 1994. Comparative study of cytokine release by human peripheral blood mononuclear cells stimulated with *Streptococcus pyogenes* superantigenic erythrogenic toxins, heat-killed streptococci and lipopolysaccharide. *Infect. Immun.* **62:**4915–4921.

100. **Nagata, S.** 1997. Apoptosis by death factor. *Cell* **88:** 355–365.

101. **Nakagawa, I., M. Nakata, S. Kawabata, and S. Hamada.** 2002. Cytochrome C-mediated caspase-9 activation triggers apoptosis in *Streptococcus pyogenes*-infected epithelial cells. *Cell. Microbiol.* **3:**395–405.

102. **Nakagawa, I., K. Kurokawa, A. Yamashita, M. Nakata, Y. Tomiyasu, N. Okahashi, S. Kawabata, K. Yamazaki, T. Shiba, T. Yasunaga, H. Hayashi, M. Hattori, and S. Hamada.** 2003. Genome sequence of an M3 strain of *Streptococcus pyogenes* reveals a large-scale genomic rearrangement in invasive strains and new insights into phage evolution. *Genome Res.* **13:**1042–1055.

103. **Natanson, C., W. D. Hoffman, A. F. Suffredini, P. Q. Eichacker, and R. L. Danner.** 1994. Selected treatment strategies for septic shock based on proposed mechanisms of pathogenesis. *Ann. Intern. Med.* **120:**771–783.

104. **Navarre, W. W., and O. Schneewind.** 1999. Surface proteins of gram-positive bacteria and mechanisms of their targeting to the cell wall envelope. *Microbiol. Molec. Biol. Rev.* **63:**174–229.

105. **Neff, L., M. Zeisel, J. Sibilia, M. Scholler-Guinard, J. P. Klein, and D. Wachsmann.** 2001. NF-kappaB and the MAP kinases/AP-1 pathways are both involved in interleukin-6 and interleukin-8 expression in fibroblast-like synoviocytes stimulated by protein I/II, a modulin from oral streptococci. *Cell. Microbiol.* **3:**703–712.

106. **Norgren, M., A. Norrby, and S. E. Holm.** 1992. Genetic diversity in T1M1 group A streptococci in relation to clinical outcome of infection. *J. Infect. Dis.* **166:**1014–1020.

107. **Norrby-Teglund, A., S. Chatellier, D. E. Low, A. McGeer, K. Green, and M. Kotb.** 2000. Host variation in cytokine responses to superantigens determine the severity of invasive group A streptococcal infection. *Eur. J. Immunol.* **30:**3247–3255.

108. **Nyberg, P., T. Sakai, K. H. Cho, M. G. Caparon, R. Fassler, and L. Bjorck.** 2004. Interactions with fibronectin attenuate the virulence of *Streptococcus pyogenes*. *EMBO J.* **23:**2166–2174.

109. **Okada, N., M. K. Liszewski, J. P. Atkinson, and M. Caparon.** 1995. Membrane cofactor protein (CD46) is a keratinocyte receptor for the M protein of group A streptococcus. *Proc. Natl. Acad. Sci. USA* **92:**2489–2493.

110. **Okada, N., M. Watarai, V. Ozeri, E. Hanski, M. Caparon, and C. Sasakawa.** 1998. A matrix form of fibronectin mediates enhanced binding of *Streptococcus pyogenes* to host tissue. *J. Biol. Chem.* **272:**26978–26984.

111. **Ozeri, V., I. Rosenshine, D. F. Mosher, R. Fassler, and E. Hanski.** 1998. Roles of integrins and fibronectin in the entry of *Streptococcus pyogenes* into cells via protein F1. *Mol. Microbiol.* **30:**625–637.

112. **Ozeri, V., I. Rosenshine, A. Ben-Ze'ev, G. M. Bokoch, T.-S. Jou, and E. Hanski.** 2001. *De novo* formation of focal complex-like structures in host cells by invading streptococci. *Mol. Microbiol.* **41:**561–573.

113. **Pancholi, V.** 2001. The regulatory role of streptococcal surface deHydrogenase (SDH) in the expression of cytokines and apoptosis related genes in group A streptococci infected human pharyngeal cells, abstr. B-177, p. 80. Abstr. 101st Annu. Meet. Am. Soc. Microbiol. 2001. American Society for Microbiology, Washington, D.C.

114. **Pancholi, V., and G. S. Chhatwal.** 2003. Housekeeping enzymes as virulence factors for pathogens. *Int. J. Med. Microbiol.* **293:**1–11.

115. **Pancholi, V., and V. A. Fischetti.** 1992. A major surface protein on group A streptococci is a glyceraldehyde-3-phosphate dehydrogenase with multiple binding activity. *J. Exp. Med.* **176:**415–426.

116. **Pancholi, V., and V. A. Fischetti.** 1993. Glyceraldehyde-3-phosphate dehydrogenase on the surface of group A streptococci is also an ADP-ribosylating enzyme. *Proc. Natl. Acad. Sci. USA* **90:**8154–8158.

117. **Pancholi, V., and V. A. Fischetti.** 1997. Identification of a glycolytic enzyme complex on the surface of group A streptococci. abstr. B-42, p. 35. Abstr. 98th Annu. Meet. Am. Soc. Microbiol. 1997. American Society for Microbiology, Washington, D.C.

118. **Pancholi, V., and V. A. Fischetti.** 1997. Regulation of the phosphorylation of human pharyngeal cell proteins by group A streptococcal surface dehydrogenase (SDH): signal transduction between streptococci and pharyngeal cells. *J. Exp. Med.* **186:**1633–1643.

119. **Pancholi, V., and V. A. Fischetti.** 1998. a-Enolase, a novel strong plasmin(ogen) binding protein on the surface of pathogenic streptococci. *J. Biol. Chem.* **273:**14503–14515.

120. **Pancholi, V., P. A. Fontan, and H. Jin.** 2003. Plasminogen-mediated group A streptococcal adherence to and pericellular invasion of human pharyngeal cells. *Microb. Pathog.* **35:**293–303.

121. **Parsonnet, J.** 1989. Mediators in the pathogenesis of toxic shock syndrome: overview. *J. Infect. Dis.* **11:**5263–5269.

122. **Pellegrini, S., and I. Dusanter-Fourt.** 1997. The structure, regulation and function of the Janus kinases (JAKs) and the signal transducers and activators of transcription (STATs). *Eur. J. Biochem.* **248:**615–633.

123. **Pennisi, E.** 1997. Opening the way to gene activity. *Science* **275:**155–157.

124. **Perez-Casal, J., M. G. Caparon, and J. R. Scott.** 1991. Mry, a trans-acting positive regulator of the M protein gene of *Streptococcus pyogenes* with similarity to the receptor proteins of two-component regulatory systems. *J. Bacteriol.* **173:**2617–2624.

125. **Purushothaman, S. S., B. Wang, and P. P. Cleary.** 2003. M1 protein triggers a phosphoinositide cascade for group A *Streptococcus* invasion of epithelial cells. *Infect. Immun.* **71:**5823–5830.

126. **Rakonjac, J. V., J. C. Robbins, and V. A. Fischetti.** 1995. DNA sequence of the serum opacity factor of group A streptococci: identification of a fibronectin-binding repeat domain. *Infect. Immun.* **63:**622–631.

127. **Redlitz, A., B. J. Fowler, E. F. Plow, and L. A. Miles.** 1995. The role of an enolase-related molecule in plasminogen binding to cells. *Eur. J. Biochem.* **227:**407–415.

128. **Rijneveld, A. W., S. Florquin, J. Branger, P. Speelman, S. J. van Deventer, and P. T. Van Der.** 2001. TNF-alpha compensates for the impaired host defense of IL-1 type I receptor-deficient mice during pneumococcal pneumonia. *J. Immunol.* **167:**5240–5246.

129. **Rodriguez-Boulan, E., and W. J. Nelson.** 1989. Morphogenesis of the polarized epithelial cell phenotype. *Science* **245:**718–725.

130. **Rohde, M., E. Muller, G. S. Chhatwal, and S. R. Talay.** 2003. Host cell caveolae act as an entry-port for group A streptococci. *Cell. Microbiol.* **5:**323–342.

131. **Roth, S. Y., and C. D. Allis.** 1996. Histone acetylation. *Cell* **87:**5–8.

132. **Russell, S.** 2004. CD46: a complement regulator and pathogen receptor that mediates links between innate and acquired immune function. *Tissue Antigens* **64:**111–118.

133. **Ryan, P. A., V. Pancholi, and V. A. Fischetti.** 1998. Binding of group A streptococci to mucin and its implications in the colonization process, abstr. B-4, p. 56. Abstr. 98th Annu. Meet. Am. Soc. Microbiol. 1998. American Society for Microbiology, Washington, D.C.

134. **Ryan, P. A., V. Pancholi, and V. A. Fischetti.** 2002. Group A streptococci bind to mucin and human pharyngeal cells through sialic acid-containing receptors. *Infect. Immun.* **69:**7402–7412.

135. **Sajjan, S. U., and J. F. Forstner.** 1998. Identification of the mucin-binding adhesin of *Pseudomonas cepacia* isolated from patients with cystic fibrosis. *Infect. Immun.* **60:** 1434–1440.

136. **Schmeck, B., R. Gross, P. D. N'Guessan, A. C. Hocke, S. Hammerschmidt, T. J. Mitchell, S. Rosseau, N. Suttorp, and S. Hippenstiel.** 2004. *Streptococcus pneumoniae*-induced caspase 6-dependent apoptosis in lung epithelium. *Infect. Immun.* **72:**4940–4947.

137. **Schrager, H. M., S. Alberti, C. Cywes, G. Dougherty, and M. R. Wessels.** 1998. Hyaluronic acid capsule modulates M protein-mediated adherence and acts as a ligand for attachment of group A *Streptococcus* to CD44 on human keratinocytes. *J. Clin. Invest.* **101:**1708–1716.

138. **Schroder, N. W., S. Morath, C. Alexander, L. Hamann, T. Hartung, U. Zahringer, U. B. Goebel, J. R. Weber, and R. R. Schumann.** 2003. Lipoteichoic acid (LTA) of *Streptococcus pneumoniae* and *Staphylococcus aureus* activates immune cells via Toll-like receptor (TLR)-2, lipopolysaccharide-binding protein (LBP), and CD14, whereas TLR-4 and MD-2 are not involved. *J. Biol. Chem.* **278:**15587–15594.

139. **Schumann, R. R., D. Pfeil, D. Freyer, W. Buerger, N. Lamping, C. J. Kirschning, U. B. Goebel, and J. R. Weber.** 1998. Lipopolysaccharide and pneumococcal cell wall components activate the mitogen activated protein kinases (MAPK) erk-1, erk-2, and p38 in astrocytes. *Glia* **22:**295–305.

140. **Seger, R., and E. G. Krebs.** 1995. The MAPK signaling cascade. *FASEB J.* **9:**726–735.

141. **Shelburne, S. A., and J. M. Musser.** 2004. Virulence gene expression in vivo. *Curr. Opin. Microbiol.* **7:**283– 289.

142. **Shuter, J., V. B. Hatcher, and F. D. Lowy.** 1998. *Staphylococcus aureus* binding to human nasal mucin. *Infect. Immun.* **64:**310–318.

143. **Siebenlist, U., G. Franzoso, and K. Brown.** 1994. Structure, regulation, and function of NF-κB. *Annu. Rev. Cell Biol.* **10:**405–455.

144. **Smoot, J. C., K. D. Barbian, J. J. Van Gompel, L. M. Smoot, M. S. Chaussee, G. L. Sylva, D. E. Sturdevant, S. M. Ricklefs, S. F. Porcella, L. D. Parkins, S. B. Beres, D. S. Campbell, T. M. Smith, Q. Zhang, V. Kapur, J. A. Daly, L. G. Veasey, and J. M. Musser.** 2002. Genome sequence and comparative microarray analysis of serotype M18 group A Streptococcus strains associated with acute rheumatic fever outbreaks. *Proc. Natl. Acad. Sci. USA* **99:**4668–4673.

145. **Steller, H.** 1995. Mechanisms and genes of cellular suicide. *Science* **267:**1445–1449.

146. **Strous, G. J., and J. Dekker.** 1992. Mucin-type glycoproteins. *Crit. Rev. Biochem. Mol. Biol.* **27:**57–92.

147. **Sweet, M. T., G. Carlson, R. G. Cook, D. Nelson, and C. D. Allis.** 1997. Phosphorylation of linker histones by a protein kinase A-like activity in mitotic nuclei. *J. Biol. Chem.* **272:**916–923.

148. **Tabak, L. A.** 1995. In defense of the oral cavity: structure, biosynthesis, and function of salivary mucins. *Annu. Rev. Physiol.* **57:**547–564.

149. **Taylor-Papadimitriou, J., and O. J. Finn.** 1997. Biology, biochemistry and immunology of carcinoma-associated mucins. *Immunol. Today* **18:**105–107.

150. **Timpl, R., and J. C. Brown.** 1998. The laminins. *Matrix Biol.* **14:**275–281.

151. **Tsai, P. J., Y. S. Lin, C. F. Kuo, H. Y. Lei, and J. J. Wu.** 1999. Group A *Streptococcus* induces apoptosis in human epithelial cells. *Infect. Immun.* **67:**4334–4339.

152. **Tuomanen, E., R. Austrian, and H. R. Masure.** 1995. Pathogenesis of pneumococcal infection. *N. Engl. J. Med.* **332:**1280–1284.

153. **von Mering, M., A. Wellmer, U. Michel, S. Bunkowski, A. Tlustochowska, W. Bruck, U. Kuhnt, and R. Nau.** 2001. Transcriptional regulation of caspases in experimental pneumococcal meningitis. *Brain Pathol.* **11:**282–295.

154. **Voyich, J. M., D. E. Sturdevant, K. R. Braughton, S. D. Kobayashi, B. Lei, K. Virtaneva, D. W. Dorward, J. M. Musser, and F. R. DeLeo.** 2003. Genome-wide protective response used by group A *Streptococcus* to evade destruction by human polymorphonuclear leukocytes. *Proc. Natl. Acad. Sci. USA* **100:**1996–2001.

155. **Voyich, J. M., K. R. Braughton, D. E. Sturdevant, C. Vuong, S. D. Kobayashi, S. F. Porcella, M. Otto, J. M. Musser, and F. R. DeLeo.** 2004. Engagement of the pathogen survival response used by group A *Streptococcus* to avert destruction by innate host defense. *J. Immunol.* **173:**1194–1201.

156. **Wang, B., N. Ruiz, A. Pentland, and M. Caparon.** 1997. Keratinocyte proinflammatory responses to adherent and nonadherent group A streptococci. *Infect. Immun.* **65:**2119–2126.

157. **Wannamaker, L. W.** 1970. Differences between streptococcal infections of the throat and of the skin. *N. Engl. J. Med.* **282:**23–31.

158. **Wannamaker, L. W.** 1970. Differences between streptococcal infections of the throat and of the skin (second of two parts). *N. Engl. J. Med.* **282:**78–85.

159. **Wannamaker, L. W.** 1973. The chains that link the throat to the heart. *Circulation* **48:**9–18.

160. **Wick, M. J., J. L. Madara, B. N. Fields, and S. J. Normark.** 1991. Molecular cross talk between epithelial cells and pathogenic microorganisms. *Cell* **67:**651–659.

161. **Wolffe, A., J. Wong, and D. Pruss.** 1997. Activators and repressors: making use of chromatin to regulate transcription. *Genes Cells* **2:**291–302.

162. **Wright, S. D., R. A. Ramos, P. S. Tobias, R. J. Ulevitch, and J. C. Mathison.** 1990. CD14, a receptor for complexes of lipopolysaccharide (LPS) and LPS binding protein. *Science* **249:**1431–1433.

163. **Zychlinsky, A., and P. J. Sansonetti.** 1997. Apoptosis as a proinflammatory event: what can we learn from bacteria-induced cell death? *Trends Microbiol.* **5:**201–204.

164. **Zychlinsky, A., K. Thirumalai, J. Arondel, J. R. Cantey, A. O. Aliprantis, and P. Sansonetti.** 1996. In vivo apoptosis in *Shigella flexneri* infection. *Infect. Immun.* **64:** 5357–5365.

165. **Zysk, G., L. Bejo, B. K. Schneider-Wald, R. Nau, and H. Heinz.** 2000. Induction of necrosis and apoptosis of neutrophil granulocytes by *Streptococcus pneumoniae*. *Clin. Exp. Immunol.* **122:**61–66.

Vaccine Approaches To Protect against Group A Streptococcal Pharyngitis

VINCENT A. FISCHETTI

10

Streptococcus pyogenes (Lancefield group A) is a human pathogen responsible for a wide range of diseases, the most common of which are nasopharyngeal infections and impetigo. More than 25 million cases of group A streptococcal infections occur each year in the United States, at a cost of more than $1 billion to the public, in addition to losses in productivity. An increase in streptococcal toxic shock and invasive infections (particularly necrotizing fasciitis) has been reported with certain strains of group A streptococci, resulting in rapid fatalities in up to 30% of the cases (53). Because there have been no reports of penicillin resistance in group A streptococci, streptococci-related diseases can be successfully treated with this antibiotic; however, with erythromycin, the second drug of choice for these bacteria, resistance is currently observed. Before treatment, group A streptococcal infections are usually associated with fever, significant discomfort, and generalized lethargy. About 3% of individuals with untreated or inadequately treated streptococcal pharyngitis develop rheumatic fever and rheumatic heart disease, a sequela of the streptococcal infection resulting in cardiac damage, particularly to the mitral valve. Although in the United States outbreaks of rheumatic fever are usually sporadic and infrequent, affecting only local areas of the country (57), in developing countries as many as 1% of school-age children are estimated to have rheumatic heart disease (20). Because of this, and the concern that penicillin-resistant strains may appear, there is a strong incentive to develop a safe and effective vaccine against group A streptococcal pharyngitis.

Although more prevalent during winter months, at any given time up to 30% of asymptomatic humans carry group A streptococci in their pharynx. Except under unusual circumstances and one serotype, there is no animal reservoir for these organisms. Thus, the eradication or significant reduction in the carriage of group A streptococci in the human pharynx would have a profound effect on the dissemination of this organism in the population and on the initiation of streptococcal disease. Even in the case of the highly invasive strains of streptococci, there is no evidence that they have their origins in other than pharyngeal sources.

Lancefield (40) showed more than 50 years ago that the surface M protein would be a prime candidate for a vaccine to protect against streptococcal infection. However, we have learned since that time that protection incurred by serum immunoglobulin G (IgG) to the M molecule is type specific, and antibodies directed to limited portions of the type-specific region may induce protection against only a limited number of strains within an M type. More recent studies have revealed that broad protection against streptococcal infection may be achieved by the induction of a local secretory response to exposed conserved sequences found within the M molecule. The approach prevents streptococcal colonization of the upper respiratory mucosa in a mouse model.

It has been known for decades that patients with rheumatic fever have serum IgG to human smooth muscle tissue at levels three to four times that found in normal serum. Although these antibodies have also been found to react with streptococcal cell wall, membrane, and M protein N-terminal determinants, their role, if any, in the pathogenesis of rheumatic fever has not yet been proven. The fact that rheumatic fever and further cardiac damage occur around the time of the streptococcal infection, and not between infections when cross-reactive antibodies are still elevated (56), argues against the direct involvement of these antibodies in the disease process. Despite this, and until proven otherwise, it will be important to minimize the induction of cross-reactive antibodies in streptococcal vaccine preparations. This chapter concentrates on the progress to date toward the development of a vaccine to protect against streptococcal nasopharyngeal infection. For a more detailed review of the subject see reference 27.

M PROTEIN STRUCTURE AND FUNCTION

Protective immunity to group A streptococcal infection is achieved through antibodies directed to the M protein (23), a major virulence factor present on the surface of all clinical isolates. M protein is a coiled-coil fibrillar protein composed of three major segments of tandem repeat sequences that

Gram-Positive Pathogens, 2nd edition, edited by Vincent A. Fischetti et al.
© 2006 ASM Press, Washington, D.C.

extends nearly 60 nm from the surface of the streptococcal cell wall (23) (Fig. 1). The A- and B-repeats located within the N-terminal half are antigenically variable among the >125 known streptococcal types, with the N-terminal nonrepetitive region and A-repeats exhibiting hypervariability. The more C-terminal C-repeats, the majority of which are surface exposed, contain epitopes that are highly conserved among the identified M proteins (35). Because of its antigenically variable N-terminal region, the M protein provides the basis for the Lancefield serological typing scheme for group A streptococci (23) and a more contemporary molecular approach to typing group A streptococci (4).

The M protein is considered the major virulence determinant because of its ability to prevent phagocytosis when present on the streptococcal surface and thus, by this definition, all clinical isolates express M protein. This function may in part be attributed to the specific binding of complement factor H to both the conserved C-repeat domain and the fibrinogen bound to the B-repeats (30), preventing the deposition of C3b on the streptococcal surface. It is proposed that when the streptococcus contacts serum, the factor H bound to the M molecule inhibits or reverses the formation of C3b,Bb complexes and helps to convert C3b to its inactive form (iC3b) on the bacterial surface, preventing C3b-dependent phagocytosis. Studies have shown that antibodies directed to the B- and C-repeat regions of the M protein are unable to promote phagocytosis. This may be the result of the ability of factor H to also control the binding of C3b to the Fc receptors on these antibodies, resulting in inefficient phagocytosis. Antibodies directed to the hypervariable N-terminal region are opsonic, perhaps because they cannot be controlled by the factor H bound to the B- and C-repeat regions. Thus, it appears that through evolu-

tion the streptococcus has devised a method to protect its conserved region from being used against itself by binding factor H to regulate the potentially opsonic antibodies that bind to these regions.

TYPE-SPECIFIC PROTECTION

The M protein has been a prime vaccine candidate to prevent group A streptococcal infections since Lancefield showed clearly that M protein-specific human and animal antibodies have the capacity to opsonize streptococci in preparation for phagocytic clearance. In general, serum IgG directed to the hypervariable NH_2-terminal portion of M protein leads to complement fixation and phagocytosis by polymorphonuclear leukocytes of streptococci of the homologous serotype (34). Even antibodies directed to whole group A streptococci will allow phagocytosis only of strains of the same M type in a phagocytic assay, suggesting that besides the M protein, no other streptococcal antigen is able to induce antibodies to override the antiphagocytic property of the M protein. Fox (26) has reviewed the early attempts in M protein vaccine development. Since the early 1970s few human trials have been realized. This is partially based on problems with hypersensitivity reactions found with the acid-extracted M protein preparations of the time and the fact that only type-specific protection was observed. In addition, repeated attempts to separate heterologous protein contaminants from the type-specific determinants proved unsuccessful. Except for one investigation, all streptococcal vaccine development since the early 1970s was based on animal studies in which the analysis of the immune response to M protein preparations was performed with and without adjuvants and in combination with other antigens.

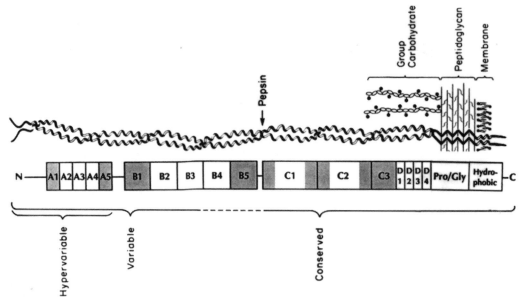

FIGURE 1 Proposed model of the M protein from M6 strain D471 (23a, 29). The coiled-coil rod region extends about 60 nm from the cell wall with a short nonhelical domain at the NH_2 terminus. The Pro/Gly-rich region of the molecule is found within the peptidoglycan (44a). The membrane-spanning segment is composed of predominantly hydrophobic amino acids, and a short charged tail extends into the cytoplasm. Data suggest that the membrane anchor may be cleaved shortly after synthesis (44a). The A-, B-, and C-repeat regions are indicated along with those segments containing conserved, variable, and hypervariable epitopes among heterologous M serotypes. Pepsin designates the position of a pepsin-susceptible site near the center of the molecule.

In 1979 Beachey et al. (1) used pepsin-extracted M24 protein (PepM24, the N-terminal half of the native M24 molecule) (Fig. 1) to immunize human volunteers. Unlike earlier acid-extracted products (26), this highly purified fragment was found to be free of non-type-specific reactivity and did not induce delayed-type hypersensitivity tests in the skin of the 37 adult volunteers. Immunization with alum-precipitated PepM24 protein led to the development of type-specific opsonic antibodies in 10 of 12 volunteers, none of whom developed heart-reactive antibodies as determined by immunofluorescence. These studies clearly indicated that M protein vaccines free of sensitizing antigens could be produced but further emphasized the type specificity of the immune response.

Using these studies as a starting point, Beachey and co-workers (2) began to develop a type-specific epitope-based vaccine strategy to protect against streptococcal disease. It was soon learned that the complete PepM24 fragment was not necessary, but peptides representing the first 20 or so N-terminal amino acids of the M24 protein also evoked type-specific opsonic antibodies to M24 streptococci. Experiments with synthetic peptides of the N terminus of M1, M5, M6, and M19 proteins resulted in the same conclusions. When a hybrid peptide was chemically synthesized representing the N-terminal sequence of both the M24 and M5 proteins and injected into animals with complete Freund's adjuvant, opsonic antibodies were produced to both M5 and M24 streptococci (3). Opsonic antibodies to three M proteins were obtained when the N-terminal sequences of three M protein sequences (M5-M6-M24) were synthesized in tandem and injected into rabbits.

MULTIVALENT TYPE-SPECIFIC VACCINE

Dale et al. (16) improved on Beachey's approach and used recombinant technology to prepare a tandem oligonucleotide array representing the N-terminal sequence of four serotypes (M24-M5-M6-M19). When the recombinant tetravalent fusion protein was purified and used to immunize rabbits, antibodies were raised against all four M proteins with variations in both the enzyme-linked immunosorbent assay titer and opsonic activity to the four respective streptococcal types. These studies were subsequently repeated using an octavalent construct (18); however, in these studies, opsonic antibodies were produced against six of the eight serotypes used in the vaccine construct (with less reactivity being observed with the serotypes represented at the ends). To help overcome this, a 26-valent vaccine was subsequently developed comprising four peptides in which the serotype at one end was duplicated at the other (31). It was designed to contain the N-terminal region of M proteins from epidemiologically important serotypes, including those responsible for pharyngitis in children, and those historically responsible for rheumatic fever. The antibodies produced in rabbits to this mixture of peptides were found to be 69% opsonic against at least one of the three rabbits immunized. Using 50% reduction in growth of streptococci in a bactericidal assay as an end point, 92% (24 of 26) of the serotypes induced an opsonic response. Whether a 50% reduction will be sufficient to block human infections remains to be tested. As with the other vaccine studies using N-terminal sequences, none of the rabbits developed tissue cross-reactive antibodies to human tissues, including cardiac.

In a more recent phase I safety study, a hexapeptide was generated containing the N-terminal region of M1, M3, M5, M6, M19, and M24 proteins and used to immunize 28 volunteers intramuscularly (38). After three immunizations, significant antibody levels were produced to all six proteins. After a 1-year follow-up, the vaccine was found to be well tolerated, with no evidence of tissue cross-reactive antibodies. These studies confirm and extend the earlier experiments of Beachey and colleagues and reveal that such an approach may be useful for the prevention of infection by specific streptococcal serotypes.

An important factor to consider in the development of a type-specific epitope-based vaccine is the potential of the streptococcus to generate new M serotypes by changing the amino-terminal portion of the M protein. For example, in the type 6 M protein of strain D471, type-specific opsonizing antibodies are directed against epitopes located both at the amino-terminal end (residues 1 through 21) and within the A-repeat block (34), which begins at amino acid 27 and continues to residue 96 in the 441-amino-acid M molecule (29). High-frequency, intragenic recombinational events within the A-repeat block can lead to a significant loss in the opsonizing ability of monospecific antibodies directed to this region (36). For instance, an opsonic antibody generated to the D471 parental strain showed some or no opsonizing activity to size-variant derivatives of this strain or other M6 streptococci isolated from patients. In addition, Harbaugh et al. (28) showed that sequence variation also occurs within the nonrepeating N-terminal end of the M protein sequence from strains of the same serotype and that these changes ultimately affect the binding of opsonic antibodies. More recently, Penney et al. (45) showed a significant degree of sequence variation within the N-terminal hypervariable region of several strains identified serologically as M1. In all, these findings strongly indicate that opsonic antibodies induced by M protein N-terminal sequences from a given vaccine strain may prove to be ineffective or weakly effective against other strains of the same serotype as a result of antigenic drift. Therefore, a type-specific vaccine necessary to protect against a streptococcal infection would require a multivalent antigen corresponding to stable immunodeterminants on serotypes that together account for the majority of the nasopharyngeal isolates prevalent within the population at a given time. Thus far, this region has not been identified.

MUCOSAL VACCINE FOR NON-TYPE-SPECIFIC PROTECTION

At its peak incidence, 50% of children between the ages of 5 and 7 years suffer from streptococcal infection each year. Furthermore, the siblings of a child with a streptococcal pharyngitis are five times more likely to acquire the organism than is one of the parents. This decreased occurrence of streptococcal pharyngitis in adults might be explained by a nonspecific age-related host factor resulting in a decreased susceptibility to streptococci. Alternatively, protective antibodies directed to antigens common to a large number of group A streptococcal serotypes might arise as a consequence of multiple infections or exposures experienced during childhood. This could result in an elevated response to conserved M protein epitopes. This latter hypothesis is partly supported by earlier studies on the immune response to the M protein, in which it was found that the B-repeat domain (Fig. 1) was clearly immunodominant. When rabbits were immunized with the whole M protein molecule, the first detectable antibodies were directed to the B-repeat region and rose steadily with time. It was only after repeated M protein immunizations that antibodies were produced against the hypervariable A- and conserved C-repeat regions.

Early human trials by D'Alessandri et al. (15) and Polly et al. (46) strongly suggested that mucosal vaccination with M protein was protective. Using highly purified acid-extracted M protein (which contains type-specific and conserved-region fragments of M protein), volunteers were immunized either intranasally or subcutaneously. When they were challenged orally with virulent streptococci, the intranasally immunized volunteers displayed lower rates of both nasopharyngeal colonization and clinical illness compared with placebo controls. Volunteers immunized subcutaneously displayed only a reduction in clinical illness and showed no reduction in nasopharyngeal colonization.

Unlike antibodies to the N-terminal region of the M molecule, it was clear from our own studies that antibodies directed to the exposed C-repeat region were not opsonic. Because of this, experiments were performed to explore whether mucosal antibodies directed only to the conserved region of M protein could be responsible for protection against streptococcal infection. Taking advantage of the pepsin site in the center of the M molecule (separating the variable and conserved regions) (Fig. 1), the recombinant M6 protein was cleaved, and the N- and C-terminal fragments were separated by sodium dodecyl sulfate-polyacrylamide gel electrophoresis and Western blotted. When the blots were reacted with different adult human sera, all adults tested had antibodies to the C-terminal conserved region, while, as expected, only sera that were opsonic for the M6 organisms reacted with the N-terminal variable region. Similar studies performed with M protein isolated from five different common serotypes (M3, M5, M6, M24, M29) revealed that sera from 10 of 17 adults tested did not have N-terminal-specific antibodies to these M types; only two sera reacted with two serotypes, and the remaining five sera reacted with only one serotype. However, all sera tested reacted to the C-terminal fragment of the M molecule. Similar results were seen when salivary IgA from adults and children was tested in enzyme-linked immunosorbent assay against the N- and C-terminal halves of the M6 molecule (unpublished data). In all, this is further evidence that the relative resistance of adults to streptococcal pharyngitis is clearly not due to the presence of type-specific antibodies to multiple M serotypes of streptococci, but may perhaps be due to the presence of antibodies to conserved determinants.

From these findings we reasoned that an immune response to the conserved region of the M molecule might afford protection by inducing a mucosal response to prevent streptococcal colonization and ultimate infection. In view of the evidence that the conserved C-repeat epitopes of the M molecule are immunologically exposed on the streptococcal surface (35), it should be possible to generate mucosal antibodies that are reactive to the majority of streptococcal types, using only the conserved-region antigen for immunization.

PASSIVE PROTECTION

Secretory IgA (sIgA) is able to protect mucosal surfaces from infection by pathogenic microorganisms despite the fact that its effector function differs from those of serum-derived immunoglobulins. When streptococci are administered intranasally to mice, they are able to cause death by first colonizing and then invading the mucosal barrier, resulting in dissemination of the organism to systemic sites. Using this model we first examined whether sIgA, delivered

directly to the mucosa, plays a role in protecting against streptococcal infection. Live streptococci were mixed with affinity-purified M protein-specific sIgA or IgG antibodies and administered intranasally to the animals (6). The results clearly showed that the anti-M-protein sIgA protected the mice against streptococcal infection and death, whereas the opsonic serum IgG administered in the same way was without effect. This indicated to us that sIgA can protect at the mucosa and may preclude the need for opsonic IgG in preventing streptococcal infection. This study was also one of the first to compare purified, antigen-specific sIgA and serum IgG for passive protection at a mucosal site.

In another laboratory, passive protection against streptococcal pharyngeal colonization was also shown by the oral administration of purified lipoteichoic acid (LTA) but not deacylated LTA before oral challenge in mice (17). The addition of anti-LTA by the same route also protected mice from oral streptococcal challenge. Although several in vitro studies showed the importance of M protein and LTA in streptococcal adherence, these in vivo studies, together with those presented above, suggest that both M protein and LTA may play key roles in the colonization of the mouse pharyngeal mucosa. However, it is uncertain whether this is also true in humans.

ACTIVE IMMUNIZATION WITH CONSERVED-REGION PEPTIDES

To determine whether a local mucosal response directed to the conserved exposed epitopes of M protein can influence the course of mucosal colonization by group A streptococci, peptides corresponding to these regions were used as immunogens in a mouse model (7). Overlapping synthetic peptides of the conserved region of the M6 protein were covalently linked to the mucosal adjuvant cholera toxin B subunit (CTB) and administered intranasally to the mice in three weekly doses. Thirty days later, animals were challenged intranasally with live streptococci (either homologous M6 or heterologous M14), and pharyngeal colonization by the challenge organism was monitored by throat swabs for the presence of streptococci for 10 to 15 days. Mice immunized with the peptide-CTB complex showed a significant reduction in colonization with either M6 or M14 streptococci compared with mice receiving CTB alone (5) (Fig. 2). Thus, despite the fact that conserved-region peptides were unable to evoke an opsonic antibody response (34), these peptides have the capacity to induce a local immune response capable of influencing the colonization of group A streptococci at the nasopharyngeal mucosa in this model system. These findings were the first to demonstrate protection against a heterologous serotype of group A streptococci with a vaccine consisting of the widely shared C-repeat region of the M6 protein.

Confirmation of these findings was later published independently using a different streptococcal serotype as the immunizing and challenge strains (10, 44). In a separate study, Pruksakorn et al. (48), using different criteria for streptococcal opsonization than those previously published (39), found that, when peptides derived from the conserved region of the M protein were used to immunize mice, they induced antibodies capable of opsonizing type 5 streptococci and streptococci isolated from Aboriginal and Thai patients with rheumatic fever. These findings are in sharp contrast to those of Jones and Fischetti (34), who showed that antibodies to the conserved region peptides of M protein are not

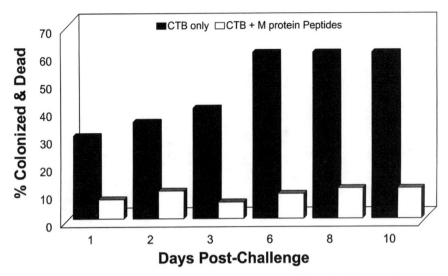

FIGURE 2 The extent of colonization and death of mice challenged with group A streptococci after oral immunization with M-protein conserved-region M6 peptides linked to CTB. Throats of orally immunized mice were swabbed each day after challenge with M14 streptococci, and the specimens were plated on blood plates to determine the extent of colonization compared with that of mice vaccinated with CTB only. Plates showing group A streptococci were scored as positive.

opsonic. Relf et al. (49), in an attempt to maintain the alpha-helical conformation of conserved-region peptides, substituted the flanking regions of these peptides with non-streptococcal sequences with high helix potential, perhaps explaining the discrepancy in opsonic activity. In an immunogenicity study these chimeric peptides were found to be immunogenic in mice immunized nasally, resulting in a mucosal IgA response and a systemic IgG response (21). When these peptides were used to immunize mice intranasally, significant protection was observed after streptococcal challenge by the same route (44), resulting in similar protection as unmodified peptides (5). Thus, in addition to a mucosal response, if these conserved-region chimeric peptides do in fact induce an opsonic response in humans (9, 48), they may offer added protection against streptococcal infection.

VACCINIA VIRUS AS A VECTOR

To further verify the validity of using the M protein conserved region as a streptococcal vaccine, experiments were repeated in a vaccinia virus vector system. In these studies, the gene coding for the complete conserved region of the M6 molecule (from the pepsin site to the C terminus) was cloned and expressed in vaccinia virus producing the recombinant VV::M6 virus (24). Tissue culture cells infected with this virus were found to produce the conserved region of the M6 molecule. Animals immunized intranasally with only a single dose of recombinant virus were significantly protected from heterologous streptococcal challenge compared with animals immunized with wild-type virus (Fig. 3). When the extent of colonization was examined in those animals immunized with wild-type or the VV::M6 recombinant, the VV::M6-immunized animals showed a marked reduction in overall colonization, indicating that mucosal immunization reduced the bacterial load on the mucosa in these animals. Animals immunized intradermally with the VV::M6 virus and challenged intranasally showed no protection.

The approaches described above proved that induction of a local immune response was critical for protection against streptococcal colonization and that the protection was not dependent upon an opsonic response. However, in the event that the streptococcus was successful in penetrating the mucosa and establishing an infection, only then would type-specific antibodies be necessary to eradicate the organism. This idea may perhaps explain why adults sporadically develop a streptococcal pharyngitis. The success of these strategies not only forms the basis of a broadly protective vaccine for the prevention of streptococcal pharyngitis but may also offer insights for the development of other vaccines. For instance, a vaccine candidate previously shown to be ineffective by the parenteral route may prove to be successful by simply changing the site of immunization. Furthermore, these results emphasize the fact that in some cases antigens need to be presented to the immune system in a specific fashion to ultimately induce a protective response.

THE USE OF GRAM-POSITIVE COMMENSALS AS VACCINE VECTORS

The CTB-linked peptide and vaccinia systems, although successful for protection against streptococcal infection, are not ideal. The CTB system requires large quantities of purified peptide and a linkage protocol that allows no more than two peptides per CTB molecule to enable the proper binding of CTB to GM_1 ganglioside (5). Although possible to achieve, these requirements make this type of vaccine relatively expensive to produce, even if recombinant technology were used to prepare the fusion molecules. Given the fact that such a vaccine ultimately must be administered in developing countries, the cost would likely be prohibitive. The vaccinia virus vector, on the other hand, is inexpensive but unlikely to gain approval from the U.S. Food and Drug Administration because oral/intranasal administration, which could result in serious complications, is required for effectiveness. Because of these limitations, we set

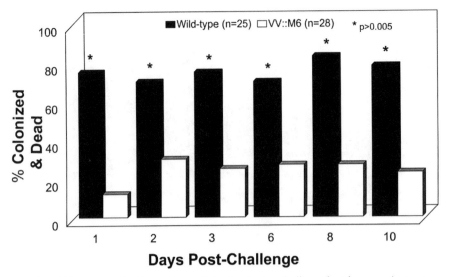

FIGURE 3 The extent of colonization and death of mice challenged with group A streptococci after oral immunization with recombinant vaccinia virus containing the gene for the whole conserved region of the M6 protein. Throats of orally immunized mice were swabbed each day after challenge with M14 streptococci, and the specimens were plated on blood plates to determine the extent of colonization compared with that of mice vaccinated with wild-type vaccinia only. Plates showing group A streptococci were scored as positive.

out to develop a mucosal vaccine delivery system that is safe, effective, and inexpensive.

Although several nonliving systems of delivering antigen to mucosal sites have been developed (7, 12, 22, 41), live vectors may afford a better and more natural response without the need to reimmunize to gain higher antibody titers. In most instances, live antigen delivery vectors are derived from bacteria (usually gram-negative) (54) or viruses (55) that are normally considered mammalian pathogens. Perhaps this is due in part to our better understanding of these organisms, making genetic manipulations easier. Usually these organisms have been extensively engineered to reduce their pathogenicity yet maintain certain invasive qualities (e.g., to invade the M cells of the gut mucosa) to induce a mucosal immune response. To circumvent some of the safety and environmental issues inherent in the wide-scale dissemination of engineered pathogens, we developed nonpathogenic gram-positive bacteria as vaccine vectors (47). In this system, foreign antigens are displayed on the surface of gram-positive human commensal organisms that colonize the niche invaded by the pathogen (oral, intestinal, vaginal). Colonization generates both an enhanced local IgA response to the foreign antigen and systemic IgG and T-cell responses. Unlike many other live bacterial systems, in which the foreign antigen is retained in the cytoplasm, translocated to the periplasm, or in some cases secreted, the gram-positive vector anchors the foreign antigen to the cell for surface display (47). Because the cell wall peptidoglycan of the gram-positive cell is a natural adjuvant, an enhanced response is obtained when the engineered organisms are processed for antibody induction.

Our ability to accomplish this is based on the recent discovery that surface molecules from gram-positive bacteria that anchor via their C termini (of which >400 have now been sequenced) have a highly conserved C-terminal region that is responsible for cell attachment (25, 43) (see chapter 2, this volume).

A NEW GENERATION OF STREPTOCOCCAL VACCINE

Based on successful mouse studies examining the immune response to a recombinant antigen expressed on the surface of *Streptococcus gordonii*, a recombinant *S. gordonii* was engineered that comprised the C-terminal half of the M protein containing the exposed conserved region of the molecule. This segment was similar to that used successfully in the vaccinia virus experiments (see above) (42). *S. gordonii*, expressing this segment of the M protein on its surface, successfully colonized for up to 12 weeks all of the 10 rabbits immunized. During this time, the animals raised a salivary IgA (Fig. 4) and serum IgG (Fig. 5) response to the intact M protein. The amount of M protein-specific sIgA was up to 5% of the total IgA in the saliva of these animals. The IgA and IgG induced by this method did not cross-react with human heart tissue, as determined by immunofluorescence assay or with tropomyosin, myosin, or vimentin by Western blot. As an added feature of the live vector system, Byrd et al. constructed *S. gordonii* that coexpressed immunomodulatory molecules such as interleukin-2 and interferon-gamma along with the M protein conserved region (11). They found that in the coexpression constructs, the cytokines modulated the systemic immune response compared to control constructs without cytokines.

The commensal delivery system would be ideal for developing countries if proven to be effective. Because it is a live vector, it would be easy to administer and not likely to require additional doses. Also, because gram-positive bacteria are stable for long periods in the lyophilized state, a cold chain would not be required. Early studies show that, when reintroduced into the human oral cavity, *S. gordonii* is capable of persisting for more than 2 years and is transmitted to other members of the family. For a developing country this factor could be ideal, because rarely is the whole population able to be immunized. However, it remains to be determined

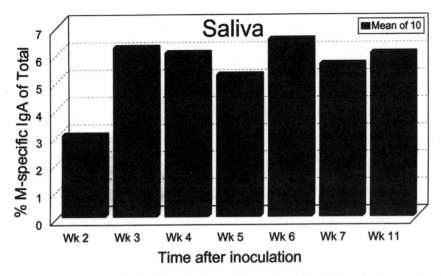

FIGURE 4 M protein-specific salivary IgA in rabbits colonized with *S. gordonii* expressing the conserved region on the cell surface. Salivary samples were taken after pylocarpine induction and tested in an enzyme-linked immunosorbent assay against the M protein.

if this approach will induce a protective immune response in humans.

NON-M-PROTEIN APPROACHES TO PROTECT AGAINST STREPTOCOCCAL INFECTION

Other approaches have been developed that may be an alternative to using M protein as a vaccine candidate (19). Cleary and coworkers (13) have identified a group A streptococcal protease that specifically cleaves the human serum chemotaxin C5a, preventing its binding to polymorphonuclear neutrophils. It was suggested that the surface-exposed streptococcal C5a peptidase prevents the influx of phagocytes to a streptococcal infection by destroying this chemo-

taxin. As proof, cleavage of C5a has been shown to reduce the influx of inflammatory cells at the site of a streptococcal infection (32). Expanding this finding, it was found that delivery of a defective form of the streptococcal C5a peptidase molecule intranasally to mice showed promise in protecting against challenge against heterologous M serotypes (33). In these studies, immunized animals cleared the challenge streptococci from the throat more rapidly than did control animals. To support this approach, human sera were tested for the presence of C5a peptidase antibodies in acute and convalescent sera from children with streptococcal pharyngitis (52). The authors found that C5a peptidase was highly immunogenic.

Using a different strategy, Musser and his group have been working on the group A streptococcal cysteine protease

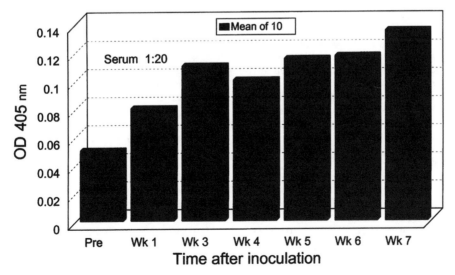

FIGURE 5 M protein-specific serum IgG in rabbits colonized with *S. gordonii* expressing the conserved region on the cell surface. Blood samples were taken at weekly intervals and tested in ELISA against the M protein.

known as streptococcal pyrogenic exotoxin B (SpeB). The gene for this protein is found in virtually all strains of group A streptococci, and in most cases its product is secreted from these organisms. Since cysteine proteases have been implicated in bacterial pathogenicity, Kapur et al. (37) found that a protease-negative SpeB mutant lost nearly all of its ability to cause death in mice when compared with wild-type organisms. They showed that mice passively immunized with rabbit IgG to cysteine protease exhibited a longer time to death than did control animals. Active immunization with cysteine protease gave the same result. Thus, although this vaccine approach prolonged the time to death, it did not prevent death.

N-Acetylglucosamine, a polysaccharide component of the streptococcal cell wall, is the group-specific antigen for group A streptococci. Because most people infected by streptococci will develop anti-N-acetylglucosamine antibodies, Zabriskie and colleagues attempted to determine if these antibodies have any effect in protecting against streptococcal infection. Using a modified in vitro phagocytic assay and a low bacterial inoculum, they found that anticarbohydrate antibodies specific for the N-acetylglucosamine were phagocytic. Comparing human and rabbit opsonic sera, they concluded that high titers of anti-N-acetylglucosamine-specific antibodies were effective in opsonization and phagocytosis of streptococci (50).

As for other candidates, serum opacity factor, a cell wall-associated and secreted protein of class II group A streptococci (8), has been shown to produce a protective immune response to heterologous streptococcal serotypes (14). The fibronectin-binding domain of SfbI, a fibronectin-binding protein of S. pyogenes, when used to immunize the upper respiratory mucosa of mice, was found to induce a protective mucosal response after challenge with streptococci by the same route (51). And finally, LTA mixed with CTB as a mucosal adjuvant and delivered nasally to mice induced an sIgA and systemic IgG response (58). When these mice were challenged with streptococci nasally, there was a significant reduction in colonization.

SUMMARY AND CONCLUSION

It would be naive to believe that a streptococcal infection is not a highly complex process, and perhaps no one single approach will control or prevent all aspects of infection by these organisms. Although we may be able to prevent infection by using a mucosal approach, these types of vaccines are not usually sterilizing, and organisms introduced at the mucosa at high doses may break through and cause sporadic infections. This is more akin to how we believe most adults naturally become resistant to streptococcal infections. One of the benefits of a successful mucosal vaccination scheme would be the reduction of streptococcal colonization in general, thus reducing the total number of these pathogens in the population. Because the main reservoir for group A streptococci for most streptococci-related illnesses is the human nasopharynx, reducing the carriage of these organisms by only 30% would have a profound impact on the dissemination of streptococci in the environment and thus a significant reduction in streptococcal disease in general. Perhaps a combination vaccine incorporating the serum opsonic power of the polyvalent type-specific approach combined with the mucosal protection offered by a mucosal vaccine will ultimately be the best way to control all pathogenic aspects of a streptococcus.

Supported in part by USPHS Grant AI11822.

REFERENCES

1. **Beachey, E. H., G. H. Stollerman, R. H. Johnson, I. Ofek, and A. L. Bisno.** 1979. Human immune response to immunization with a structurally defined polypeptide fragment of streptococcal M protein. *J. Exp. Med.* **150:**862–877.
2. **Beachey, E. H., J. M. Seyer, J. B. Dale, W. A. Simpson, and A. H. Kang.** 1981. Type-specific protective immunity evoked by synthetic peptide of *Streptococcus pyogenes* M protein. *Nature* **292:**457–459.
3. **Beachey, E. H., H. Gras-Masse, A. Tarter, M. Jolivet, F. Audibert, L. Chedid, and J. M. Seyer.** 1986. Opsonic antibodies evoked by hybrid peptide copies of types 5 and 24 streptococcal M proteins synthesized in tandem. *J. Exp. Med.* **163:**1451–1458.
4. **Beall, B., R. Facklam, and T. Thompson.** 1995. Sequencing *emm*-specific polymerase chain reaction products for routine and accurate typing of group A streptococci. *J. Clin. Microbiol.* **34:**953–958.
5. **Bessen, D., and V. A. Fischetti.** 1988. Influence of intranasal immunization with synthetic peptides corresponding to conserved epitopes of M protein on mucosal colonization by group A streptococci. *Infect. Immun.* **56:**2666–2672.
6. **Bessen, D., and V. A. Fischetti.** 1988. Passive acquired mucosal immunity to group A streptococci by secretory immunoglobulin A. *J. Exp. Med.* **167:**1945–1950.
7. **Bessen, D., and V. A. Fischetti.** 1990. Synthetic peptide vaccine against mucosal colonization by group A streptococci. I. Protection against a heterologous M serotype with shared C repeat region epitopes. *J. Immunol.* **145:**1251–1256.
8. **Bessen, D., K. F. Jones, and V. A. Fischetti.** 1989. Evidence for two distinct classes of streptococcal M protein and their relationship to rheumatic fever. *J. Exp. Med.* **169:**269–283.
9. **Brandt, E. R., W. A. Hayman, B. Currie, J. Carapetis, Y. Wood, D. C. Jackson, J. Cooper, W. D. Melrose, A. J. Saul, and M. F. Good.** 1996. Opsonic human antibodies from an endemic population specific for a conserved epitope on the M protein of group A streptococci. *Immunology* **89:**331–337.
10. **Bronze, M. S., D. S. McKinsey, E. H. Beachey, and J. B. Dale.** 1988. Protective immunity evoked by locally administered group A streptococcal vaccines in mice. *J. Immunol.* **141:**2767–2770.
11. **Byrd, C. M., T. C. Bolken, K. F. Jones, T. K. Warren, A. T. Vella, J. McDonald, D. King, Z. Blackwood, and D. E. Hruby.** 2002. Biological consequences of antigen and cytokine co-expression by recombinant *Streptoccocus gordonii* vaccine vectors. *Vaccine* **20:**2197–2205.
12. **Challacombe, S. J., D. Rahman, H. Jeffery, S. S. Davis, and D. T. O'Hagan.** 1992. Enhanced secretory IgA and systemic IgG antibody responses after oral immunization with biodegradable microparticles containing antigen. *Immunology* **76:**164–168.
13. **Cleary, P. P., U. Prahbu, J. B. Dale, D. E. Wexler, and J. Handley.** 1992. Streptococcal C5a peptidase is a highly specific endopeptidase. *Infect. Immun.* **60:**5219–5223.
14. **Courtney, H. S., D. L. Hasty, and J. B. Dale.** 2003. Serum opacity factor (SOF) of *Streptococcus pyogenes* evokes antibodies that opsonize homologous and heter-

ologous SOF-positive serotypes of group A streptococci. *Infect. Immun.* **71:**5097–5103.

15. **D'Alessandri, R., G. Plotkin, R. M. Kluge, M. K. Wittner, E. N. Fox, A. Dorfman, and R. H. Waldman.** 1978. Protective studies with group A streptococcal M protein vaccine. III. Challenge of volunteers after systemic or intranasal immunization with type 3 or type 12 group A Streptococcus. *J. Infect. Dis.* **138:**712–718.

16. **Dale, J. B., E. Y. Chiang, and J. W. Lederer.** 1993. Recombinant tetravalent group A streptococcal M protein vaccine. *J. Immunol.* **151:**2188–2194.

17. **Dale, J. B., R. W. Baird, H. S. Courtney, D. L. Hasty, and M. S. Bronze.** 1994. Passive protection of mice against group A streptococcal pharyngeal infection by lipoteichoic acid. *J. Infect. Dis.* **169:**319–323.

18. **Dale, J. B., M. Simmons, E. C. Chiang, and E. Y. Chiang.** 1996. Recombinant, octavalent group A Streptococcal M protein vaccine. *Vaccine* **14:**944–948.

19. **Dale, J. B., P. P. Cleary, V. A. Fischetti, J. M. Musser, and J. B. Zabriskie.** 1997. Group A and group B streptococcal vaccine development. A round table presentation. *Adv. Exp. Med. Biol.* **418:**863–868.

20. **Dodu, S. R. A., and S. Bothig.** 1989. Rheumatic fever and rheumatic heart disease in developing countries. *World Health Forum* **10:**203–212.

21. **Dunn, L. A., D. J. McMillan, M. Batzloff, W. Zeng, D. C. J. Jackson, J. A. Upcroft, and P. Upcroft.** 2002. Parenteral and mucosal delivery of a novel multi-epitope M protein-based group A streptococcal vaccine construct: investigation of immunogenicity in mice. *Vaccine* **20:**2635–2640.

22. **Eldridge, J. H., J. K. Staas, J. A. Meulbroek, J. R. McGhee, T. R. Tice, and R. M. Gilley.** 1991. Biodegradable microspheres as a vaccine delivery system. *Mol. Immunol.* **28:**287–294.

23. **Fischetti, V. A.** 1989. Streptococcal M protein: molecular design and biological behavior. *Clin. Microbiol. Rev.* **2:**285–314.

23a. **Fischetti, V. A., D. A. D. Parry, B. L. Trus, S. K. Hollingshead, J. R. Scott, and B. N. Manjula.** Conformational characteristics of the complete sequence of group A streptococcal M6 proteins. *Proteins* **3:**60–69.

24. **Fischetti, V. A., W. M. Hodges, and D. E. Hruby.** 1989. Protection against streptococcal pharyngeal colonization with a vaccinia: M protein recombinant. *Science* **244:**1487–1490.

25. **Fischetti, V. A., V. Pancholi, and O. Schneewind.** 1990. Conservation of a hexapeptide sequence in the anchor region of surface proteins of gram-positive cocci. *Mol. Microbiol.* **4:**1603–1605.

26. **Fox, E. N.** 1974. M proteins of group A streptococci. *Bacteriol. Rev.* **38:**57–86.

27. **Good, M. F., P. J. Cleary, J. Dale, V. A. Fischetti, K. Fuchs, H. Sabharwal, and J. Zabriskie.** 2004. Development of a vaccine to prevent infection with group A streptococci and rheumatic fever, p. 695–710. *In* M. M. Levine, J. B. Kaper, R. Rappuoli, M. A. Liu, and M. F. Good (ed.), *New Generation Vaccines.* Marcel Dekker, Inc., New York, N.Y.

28. **Harbaugh, M. P., A. Podbielski, S. Hugl, and P. P. Cleary.** 1993. Nucleotide substitutions and small-scale insertion produce size and antigenic variation in group A streptococcal M1 protein. *Mol. Microbiol.* **8:**981–991.

29. **Hollingshead, S. K., V. A. Fischetti, and J. R. Scott.** 1986. Complete nucleotide sequence of type 6 M protein

of the group A streptococcus: repetitive structure and membrane anchor. *J. Biol. Chem.* **261:**1677–1686.

30. **Horstmann, R. K., H. J. Sievertsen, M. Leippe, and V. A. Fischetti.** 1992. Role of fibrinogen in complement inhibition by streptococcal M protein. *Infect. Immun.* **60:**5036–5041.

31. **Hu, M. C., M. A. Walls, S. D. Stroop, M. A. Reddish, B. Beall, and J. B. Dale.** 2002. Immunogenicity of a 26-valent group A streptococcal vaccine. *Infect. Immun.* **70:**2171–2177.

32. **Ji, Y., L. McLandsborough, A. Kondagunta, and P. P. Cleary.** 1996. C5a peptidase alters clearance and trafficking of group A streptococci by infected mice. *Infect. Immun.* **64:**503–510.

33. **Ji, Y., B. Carlson, A. Kondagunta, and P. P. Cleary.** 1997. Intranasal immunization with C5a peptidase prevents nasopharyngeal colonization of mice by the group A *Streptococcus. Infect. Immun.* **65:**2080–2087.

34. **Jones, K. F., and V. A. Fischetti.** 1988. The importance of the location of antibody binding on the M6 protein for opsonization and phagocytosis of group A M6 streptococci. *J. Exp. Med.* **167:**1114–1123.

35. **Jones, K. F., B. N. Manjula, K. H. Johnston, S. K. Hollingshead, J. R. Scott, and V. A. Fischetti.** 1985. Location of variable and conserved epitopes among the multiple serotypes of streptococcal M protein. *J. Exp. Med.* **161:**623–628.

36. **Jones, K. F., S. K. Hollingshead, J. R. Scott, and V. A. Fischetti.** 1988. Spontaneous M6 protein size mutants of group A streptococci display variation in antigenic and opsonogenic epitopes. *Proc. Natl. Acad. Sci. USA* **85:**8271–8275.

37. **Kapur, V., J. T. Maffei, R. S. Greer, L. L. Li, G. J. Adams, and J. M. Musser.** 1994. Vaccination with streptococcal extracellular cysteine protease (interleukin-1 beta convertase) protects mice against challenge with heterologous group A streptococci. *Microb. Pathol.* **16:**443–450.

38. **Kotloff, K. L., M. Corretti, K. Palmer, J. D. Campbell, M. A. Reddish, M. C. Hu, S. S. Wasserman, and J. B. Dale.** 2004. Safety and immunogenicity of a recombinant multivalent group A streptococcal vaccine in healthy adults. *JAMA* **292:**709–715.

39. **Lancefield, R. C.** 1959. Persistence of type specific antibodies in man following infection with group A streptococci. *J. Exp. Med.* **110:**271–292.

40. **Lancefield, R. C.** 1962. Current knowledge of the type specific M antigens of group A streptococci. *J. Immunol.* **89:**307–313.

41. **McKenzie, S. J., and J. F. Halsey.** 1984. Cholera toxin B subunit as a carrier protein to stimulate a mucosal immune response. *J. Immunol.* **133:**1818–1824.

42. **Medaglini, D., G. Pozzi, T. P. King, and V. A. Fischetti.** 1995. Mucosal and systemic immune responses to a recombinant protein expressed on the surface of the oral commensal bacterium *Streptococcus gordonii* after oral colonization. *Proc. Natl. Acad. Sci. USA* **92:**6868–6872.

43. **Navarre, W. W., and O. Schneewind.** 1994. Proteolytic cleavage and cell wall anchoring at the LPXTG motif of surface proteins in gram-positive bacteria. *Mol. Microbiol.* **14:**115–121.

44. **Olive, C., T. Clair, P. Yarwood, and M. F. Good.** 2002. Protection of mice from group A streptococcal infection by intranasal immunisation with a peptide vaccine that contains a conserved M protein B cell epitope and lacks a T cell autoepitope. *Vaccine* **20:**2816–2825.

44a. **Pancholi, V., and V. A. Fischetti.** 1988. Isolation and characterization of the cell-associated region of group A streptococcal M6 protein. *J. Bacteriol.* **170:**2618–2624.

45. **Penney, T. J., D. R. Martin, L. C. Williams, S. A. de Malmanche, and P. L. Bergquist.** 1995. A single *emm* gene-specific oligonucleotide probe does not recognise all members of the *Streptococcus pyogenes* M type 1. *FEMS Microbiol. Lett.* **130:**145–150.

46. **Polly, S. M., R. H. Waldman, P. High, M. K. Wittner, A. Dorfman, and E. N. Fox.** 1975. Protective studies with a group A streptococcal M protein vaccine. II. Challenge of volunteers after local immunization in the upper respiratory tract. *J. Infect. Dis.* **131:**217–224.

47. **Pozzi, G., M. Contorni, M. R. Oggioni, R. Manganelli, M. Tommasino, F. Cavalieri, and V. A. Fischetti.** 1992. Delivery and expression of a heterologous antigen on the surface of streptococci. *Infect. Immun.* **60:**1902–1907.

48. **Pruksakorn, S., B. Currie, E. Brandt, D. Martin, A. Galbraith, C. Phornphutkul, S. Hunsakunachal, A. Manmontri, and M. F. Good.** 1994. Towards a vaccine for rheumatic fever: identification of a conserved target epitope on M protein of group A streptococci. *Lancet* **344:**639–642.

49. **Relf, W. A., J. Cooper, E. R. Brandt, W. A. Hayman, R. F. Anders, S. Pruksakorn, B. Currie, A. Saul, and M. F. Good.** 1996. Mapping a conserved conformational epitope from the M protein of group A streptococci. *Pept. Res.* **9:**12–20.

50. **Salvadori, L. G., M. S. Blake, M. McCarty, J. Y. Tai, and J. B. Zabriskie.** 1995. Group A streptococcus-liposome ELISA antibody titers to group A polysaccharide and opsonophagocytic capability to the antibodies. *J. Infect. Dis.* **171:**593–600.

51. **Schulze, K., E. Medina, G. S. Chhatwal, and C. A. Guzman.** 2003. Stimulation of long-lasting protection against *Streptococcus pyogenes* after intranasal vaccination with non adjuvanted fibronectin-binding domain of the SfbI protein. *Vaccine* **21:**1958–1964.

52. **Shet, A., E. L. Kaplan, D. R. Johnson, and P. P. Cleary.** 2003. Immune response to group A streptococcal C5a peptidase in children: implications for vaccine development. *J. Infect. Dis.* **188:**809–817.

53. **Stevens, D. L.** 1992. Invasive group A streptococcus infections. *Clin. Infect. Dis.* **14:**2–13.

54. **Tacket, C. O., D. M. Hone, R. Curtiss III, S. M. Kelly, G. Losonsky, L. Guers, A. M. Harris, R. Edelman, and M. M. Levine.** 1992. Comparison of the safety and immunogenicity of *delta-aroC delta-aroD* and *delta-cya delta-crp Salmonella typhi* strains in adult volunteers. *Infect. Immun.* **60:**536–541.

55. **Tartaglia, J., and E. Paoletti.** 1990. Live recombinant viral vaccines, p. 125–151. *In* M. H. V. van Regenmortel and A. R. Neurath (ed.), *Immunochemistry of Viruses,* II. Elsevier Science, New York, N.Y.

56. **van de Rijn, I., J. B. Zabriskie, and M. McCarty.** 1977. Group A streptococcal antigens cross-reactive with myocardium. Purification of heart-reactive antibody and isolation and characterization of the streptococcal antigen. *J. Exp. Med.* **146:**579.

57. **Veasy, L. G., S. E. Wiedmeier, G. S. Orsmond, H. D. Ruttenberg, M. M. Boucek, S. J. Roth, and V. F. Tait.** 1987. Resurgence of acute rheumatic fever in the intermountain area of the United States. *N. Engl. J. Med.* **316:**421–427.

58. **Yokoyama, Y., and Y. Harabuchi.** 2002. Intranasal immunization with lipoteichoic acid and cholera toxin evokes specific pharyngeal IgA and systemic IgG responses and inhibits streptococcal adherence to pharyngeal epithelial cells in mice. *Int. J. Pediatr. Otorhinolaryngol.* **63:**235–241.

The Bacteriophages of Group A Streptococci

W. MICHAEL McSHAN

11

Bacteriophages (phages), the ubiquitous viruses infecting almost all known species of bacteria, were discovered early in the twentieth century, apparently independently, by Twort in Britain and d'Hérelle in France (1). The ensuing years saw numerous studies in the new phenomenon of bacterial viruses, finding that phages could be isolated from water and soil as well as the exterior and interior surfaces of humans and animals. Phages could be isolated, with a few exceptions, that infected virtually all pathogenic and nonpathogenic bacteria. For a number of reasons, both medical and industrial, the scientific community has maintained a consistent level of interest in the study of bacteriophages of streptococci and the related genera. This review will focus on the influence that the phages of *Streptococcus pyogenes* (group A streptococci; GAS), both lytic and lysogenic, have on the biology and dissemination of virulence factors of this important gram-positive pathogen.

EARLY STUDIES

Evans, in a series of reports from the 1930s and 1940s, presented the first systematic studies on the bacteriophages of GAS (37–41). She initially investigated the possibility that phages could prevent or lessen the severity of streptococcal infections (37). Treatment of mice or rabbits with a virulent strain of GAS and a bacteriophage isolated from sludge did not show any protective effects for the infected animals, and in some cases, rabbits treated with antistreptococcal phages succumbed to an unusually violent course of disease.

One of the earliest facts to come to light concerning GAS bacteriophages was that the lysogenic state was very common (38), an observation that has been confirmed by numerous subsequent studies (24, 54, 63, 64, 102, 108). Another key observation made during the early days of GAS streptococcal investigations was that sterile culture filtrates from "scarlatina" (erythrogenic toxin A-producing) strains could cause toxigenic conversion of non-toxin-producing GAS (22, 45). The association of bacteriophages with this phenomenon was not made at that time because the discovery of lysogeny would not occur until several decades later. The link between toxigenic conversion of GAS and temperate phages was finally made by Zabriskie in 1964 (111) and is now recognized as an important, perhaps key, factor for virulence in these bacteria.

LYTIC BACTERIOPHAGES

General Characteristics

Bacteriophages may typically be grouped by their life cycle into two categories: the lytic phages and the temperate (lysogenic) phages. The lytic bacteriophages infect their specific host bacterium, replicate their genome and assemble new virions, and then rupture the host to release the newly formed phages. Lytic phages do not enter into any extended relationship with their host cell following infection, and they do not directly alter the genotype of the host by the introduction of a novel gene, for example. However, lytic phages can play important roles in the shaping of the biology of their GAS hosts, through elimination of the phage-susceptible members of a population consisting of more than one strain, selection for the rare phage-resistant variants in a mostly homogenous population, or by being the vectors of genetic exchange through generalized transduction.

Bacteriophage A25

Bacteriophage A25, the best studied lytic phage of GAS, was originally isolated from Paris sewage by N. A. Boulgakov and described by Maxted (78, 79). Phage A25 was one of a series of phages isolated together; its name is derived from the letter of the Lancefield group of its propagating strain combined with the type number (M25). Because this bacteriophage is able to mediate generalized transduction in GAS (67), a number of studies have addressed various aspects of its biology, and although much is still unknown about A25, it remains the best-characterized virulent GAS phage. Phage A25 is also referred to in the literature as phage 12204, its designation by the American Type Collection. The role of A25 in mediating generalized transduction is discussed in the section dedicated to that topic.

One-step growth experiments showed that phage A25 has an average burst size of approximately 30 PFU per cell

Gram-Positive Pathogens, 2nd edition, edited by Vincent A. Fischetti et al.
© 2006 ASM Press, Washington, D.C.

(71). Although the cell culture temperature does not alter the burst size, the latent period during which phage replication occurs is strongly influenced, increasing from 39 min at 37°C to more than 100 min at 26°C.

Only a few electron micrograph studies have been done on the bacteriophages of GAS, and those that have been studied, both lytic and temperate, have a similar structure with isometric heads and long noncontractile tails (Bradley's class B) (72, 89, 112). By the more recent classification scheme, these virions would belong to the large family of Siphoviridae (1). One virulent phage similar to phage A25 was examined in detail and shown to have an isometric, octahedral head measuring 58 to 60 nm across with a long flexible tail that measures 180 to 190 nm in length and 10 nm in diameter (72, 112). The tail of this phage is composed of 8-nm circular subunits and terminates in a transverse plate with a single projecting spike that is about 20 nm long (89, 112). Within the capsid is contained the linear, double-stranded DNA genome of 34.6 kb length (87).

Fischetti and Zabriskie found that phage A25 will not adsorb to mechanically disrupted GAS or isolated GAS cell walls but will adsorb to intact, living GAS (43). In contrast, Cleary and coworkers showed that heat-killed streptococci adsorbed A25 almost as well as living cells. They attributed the discrepancy with Fischetti and Zabriskie's finding to differences in growth and suspending media as well as the number of PFU used in the assay (26). The receptor for A25 is the peptidoglycan of GAS, and treatment of the cells with muralytic enzymes such as lysozyme or the group C streptococcus (GCS) phage C1 lysin destroyed the receptor activity. Additionally, phage A25 will also adsorb to streptococci from groups C, G, O, and K. Group G, in addition to absorbing phage A25, can actually become infected, and with passage of the phage through group G organisms, the phage can be adapted to efficient propagation (26). Wannamaker et al. (103) demonstrated that A25 was able to infect and produce plaques on 34 of 71 GCS strains (48%); the frequency of plaque formation by the lytic group C phage, phage C1, on the same strains was 70%. The other GAS virulent phages, A6, A12, and A27, were also able to infect these group C strains at frequencies ranging from 34 to 47%. Because the group C-specific carbohydrate and peptidoglycan are present in all members of that group, there must be surface molecules that either block or enhance access to those molecules, modulating the binding of phages A25 and C1.

The hyaluronic acid capsule of GAS provides an effective barrier against infection by many lytic bacteriophages, preventing attachment of the phage to the bacterial surface (78). In natural infections, the state of encapsulation by GAS varies according to the growth phase of the cells, with transcription of the hyaluronic acid synthesis (has) operon dropping to very low levels once stationary phase is reached (33). Maxted observed that exogenous hyaluronidase must be added to achieve efficient infection of GAS by lytic phages such as A25 in vitro (78). Temperate streptococcal phages, by contrast, often encode a hyaluronidase and thus do not require the exogenous addition of this enzyme (62). One report observed trace amounts of hyaluronidase associated with phage A25 virions (2.7 × 10^{-16} units/PFU [9]), but whether this small amount of activity was due to a phage-specified protein or a streptococcal hyaluronidase was not determined.

Although exogenous hyaluronidase is required for infection of GAS by lytic phages such as A25, release of the newly synthesized virions is mediated by phage-encoded lysins that disrupt the thick, gram-positive bacterial cell wall. Hill and Wannamaker (50) isolated a phage-associated lysin not

associated with the viral particle from cultures of *S. pyogenes* K56 that had lysed following infection with phage A25. In addition to its lytic activity against GAS, it also showed activity against streptococci from groups C, G, and H.

Lytic Phages and Streptococcal Virulence

Although the toxin-carrying temperate phages of GAS have been long associated with increasing the virulence of the lysogen by toxigenic conversion, lytic phages may also play a role in selecting populations of streptococci better adapted to resisting host defenses. Maxted found that exposure of many strains of GAS from a variety of M-types selected for the appearance of mucoid variants that were resistant to subsequent lytic phage infection (79). These mucoid variants had greatly increased hyaluronic acid production (often undetectable in the parental strain) and in some cases, also increased M protein expression. Although some of the GAS strains were able to grow in normal human blood, those strains that ordinarily did not grow all gave phage-surviving progeny that were able to survive in the presence of phagocytes. Parallel studies of mouse virulence produced similar results with mucoid phage-survivors of nonmucoid, avirulent strains acquiring the ability to survive and grow in mice (79). Evans (37) had previously reported that when GAS-infected mice were treated with bacteriophage in an effort to rescue them, the phage-treated mice died slightly more rapidly.

An important question that remained unanswered by the studies of Maxted concerned the appearance of the mucoid, phage-resistant progeny. Although the standard fluctuation test of Luria and Delbrück was applied by Maxted to determine if random mutations in the population were sufficient to explain the appearance of the mucoid colonies, the results were inconclusive (78). In light of the later findings that temperate phages (93) and multigene activators such as the mga gene product (28) can influence the expression of M protein and the hyaluronic acid capsule, it is conceivable that other factors, perhaps in addition to spontaneous mutations, influence the final population phenotype. Recent studies of the two-component regulator CovR/S have shown that inactivation of this system results in the induction of the mucoid phenotype through the overproduction of the hyaluronic acid capsule; a number of prophage genes are also up-regulated (47). Indeed, it is possible that A25 is a virulent mutant of a temperate phage that no longer is able to enter the lysogenic cycle but still can recombine with endogenous prophages in host GAS to generate novel bacteriophages that can mediate phenotypic changes in their host. This notion is strengthened by the observation that a temperate phage from group G streptococci, GT-234, is serologically related to phage A25 (30). The findings (7, 8) that some streptococcal prophages prevent A25 from replicating in their host strain is similar to the prevention of phage superinfection by homoimmune temperate phages. Possibly more than one mechanism of inhibition of A25 by the different lysogens is in operation; some prophages apparently inhibit A25 by initiating premature lysis of the superinfected cell while others appear to prevent some step in phage replication.

TEMPERATE BACTERIOPHAGES AND GENOMIC PROPHAGES

General Characteristics

The temperate or lysogenic bacteriophages are characterized by their ability to integrate their DNA into the host bacterium's chromosome via site-specific recombination, becoming a stable genetic element. Far from being a molecular curiosity, these phages often have great impact upon the pathogenic potential of their hosts by introducing phage-associated virulence genes. Studies from the pregenomics era had predicted that the frequency of lysogeny in GAS was very high (>90%) (54, 108, 109), and genome sequencing has confirmed these results. At present, six complete genomes of S. pyogenes are available: five published (M1, two isolates of M3, M6, and M18 [6, 10, 42, 85, 91]) and one unpublished (M5; http://www.sanger.ac.uk). One striking characteristic of every one of these genomes is the presence of multiple prophages and prophage remnants, with prophage DNA responsible for ~10% of the total DNA per genome. Further, most of these prophages, either through the carriage of virulence genes or positional effects, would be predicted to affect the phenotype of the host streptococcus. Phage-associated virulence factors have been observed to occur in a number of species of bacteria, including Corynebacterium diphtheriae, Clostridium botulinum, Staphylococcus aureus, Escherichia coli, Vibrio cholerae, and S. pyogenes (12, 18). Interestingly, the conversion of nontoxigenic GAS by cultural filtrates from toxigenic strains to scarlet fever toxin (erythrogenic toxin) production may have been the earliest reported example of phage-mediated virulence factors (22, 45), although the actual link to a bacteriophage was not made until many years later (111).

The linkage of GAS phages to toxin genes focused attention on these viruses for several decades, and the evidence suggested that the carriage of these phage-associated virulence genes was very high. In one pregenomics survey, the frequency of two exotoxin genes in strains isolated from cases of rheumatic heart disease, scarlet fever, or general infections was found to be 30% for speA and 50% for speC (108–110). However, when each group of strains was examined, speA was present in only 15% of the general strains but present in 50% of both the scarlet fever and rheumatic disease strains. The frequency of speC in all three groups was about 50%. Of the total number of strains examined (512), 30% carried neither speA nor speC. Strains carrying both toxins were rare, with only 5% of the general strains and 9% of the total strains having both speA and speC. The completion of multiple GAS genomes has confirmed these early studies, showing multiple phages per genome. When the six genome strains are considered, speC is present in two strains (SF370 and Manfredo), speA in the two M3 strains (SSI-1 and MGAS315), and both genes in the remaining strains (MGAS8232 and MGAS10394). However, all of the genomes carry multiple other phage-associated toxin or virulence genes in addition to speA or speC.

The first temperate GAS bacteriophage to be studied in detail was phage T12. T12 was first identified by Zabriskie (111) as being linked to the gene for the erythrogenic toxin SpeA, and subsequent studies confirmed the phage linkage and characterized this toxin gene (55, 56, 104, 105). The complete DNA sequence of T12 was determined as well as a preliminary description of its gene structure and expression (81; unpublished results). T12 has a 35,066-bp genome (%G+C = 38.5) with divergent transcriptional directions for the regulatory genes and the structural genes similar to other gram-positive temperate phages, with the toxin gene positioned between the lysis genes and the right attachment site of the integrated prophage. In the host GAS for T12, strain T25$_3$, a second prophage is co-induced with T12 upon mitomycin C treatment. This phage, having a genome of about 36,000 bases, appears to require T12 as a helper phage for replication or packaging (107, 110).

Besides the erythrogenic toxin SpeA, other toxins were found to be associated with GAS phages. The association between pyrogenic exotoxin C (SpeC) and a bacteriophage was made by Colon-Whitt et al. after inducing a strain of GAS to release phage that subsequently could convert a nontoxigenic strain to SpeC production (32). Phage CS112, the prototypic speC$^+$ temperate bacteriophage, was later identified and found to have a circular genome of 40,800 bp that appears to be terminally redundant and circularly permuted. The gene for speC is located near the phage attachment site, as was found to be true in the speA$^+$ phages (46, 57).

Morphologically, the temperate phages of GAS resemble the lytic phages, having a distinct head and a long noncontractile tail (62, 86, 98). The individual phages that have been characterized have linear or circular genomes that range from 32 to 42 kb in length (46, 94, 110). Burst sizes are usually low, being in the range of 10 to 30 PFU, although occasional strains produce burst sizes around 100 PFU (61, 107). Although the role played by lytic streptococcal phages in pathogenesis may be indirect, acting as vehicles of genetic exchange through generalized transduction, lysogeny by GAS bacteriophages can directly enhance the pathogenic potential of the host streptococcus through toxigenic conversion.

Genome Distribution and Bacteriophage Attachment Sites

The distribution of prophages in the completed streptococcal genomes is shown in Fig. 1 (the genome backbone is based upon the M1 SF370 sequence). Although prophages are found in each genome quadrant, the majority are clustered toward the terminus region on the lagging strand side (relative to oriC). Perhaps significantly, no phages target the cluster of virulence genes surrounding emm (shaded region). Some sites are clearly favored as targets for prophage integration; the 3' end of the histonelike protein is occupied in all six streptococcal genomes.

Prophage integration occurs via a Campbell-type homologous exchange between a common sequence shared between the phage and host chromosomes (attP and attB, respectively). These duplications are of varying lengths, ranging from a few bases to nearly 100 bp, and often represent coding regions of the host genome (19, 48). Once integrated, the junction regions between the prophage and host genomes are identified as the attL (the left attachment site on the integrase end of the prophage) and attR (the right attachment site on the distal end near the lysis genes). In the S. pyogenes prophages, the identifiable duplications range from 12 bp in phage MGAS10394.1 to about 96 bp in T12 and the phages that share its attachment site. In most bacterial species, the majority of duplications occur at the 3' end of some host target gene and integration either leaves the target gene intact (via the duplication) or, in at least one case, provides an alternative carboxy terminus for the specified protein (19, 20). In the S. pyogenes prophages, integration and duplication occur frequently at either the 5' or 3' end of genes (Table 1). Bacteriophage T12 integrates by site-specific recombination into what was initially identified as a gene for a serine tRNA (attB) (82), but after the completion of genome sequencing is now correctly identified as a tmRNA gene (Kelly Williams, personal communication). The 3' ends of tmRNA genes serve as attB sites for phages or cryptic prophages in a range of bacterial species, including E. coli, V. cholerae, and Dichelobacter nodosus (106). Besides the tmRNA gene, other 3'-gene targets used by

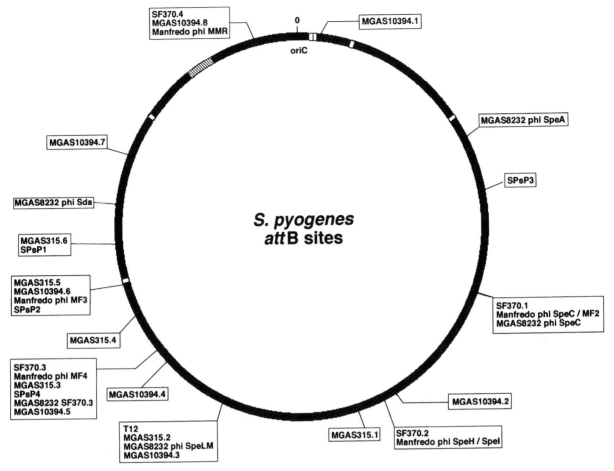

FIGURE 1 Bacteriophage attachment sites in the *S. pyogenes* genome. The locations of the genome prophages on the *S. pyogenes* genome are shown as a generalized GAS backbone based upon the M1 genome; prophages that share the same attachment site are boxed together. The rRNA operons are indicated as white blocks; the cluster of virulence genes flanking *emm* is hatched. The origin of replication is indicated (*oriC*).

TABLE 1 Prophages and major prophage remnants identified in the sequenced *S. pyogenes* genomes

Genome	M type	Prophage	*attB* sequence duplication	Gene duplication	Toxins
sf370 (42)	M1	sf370.1	AGTATAGTTG TACATG	5′ end of dipeptidase	*speC*, MF2
		sf370.2	AAACTCAAGA AGTGATTAAA TAAAACATT AAAGAACCTT GTCATATCA	3′ end of dTDP-glucose-4,6-dehydratase	*speH*, *speI*
		sf370.3	AAAGACGCTG TTAAATAATT	3′ end of histone-like protein	Mf3
		sf370.4	CAATAATGTT TGTCATAATT	5′ end of *mutL*	-
		R1930	TTAACGTTTT GAGAATTGTG T AAGCTTTACG AGCTTT	3′ end of 30S ribosomal subunit protein S9	-
MGAS315 (10)	M3	MGAS315.1	GTTTTAGAGC TATG	Intergenic region	
		MGAS315.2	TAACGGTAAC TCATGAACAA GACAAAAAGA AAAAACCTTG ATGTAACAAG GTTTTAGTAA GTTATGATTA CTTACGGTAA GCATTGATGG AGCCGGTGGG AGT	3′ end of tmRNA	SSA

(Continued on next page)

TABLE 1 (*Continued*)

Genome	M type	Prophage	*attB* sequence duplication	Gene duplication	Toxins
		MGAS315.3	AAGACGCTGT TAAATAATT	3′ end of histone-like protein	DNase
		MGAS315.4	TCTGATATAA TATAAGA	Promoter of *yesN*	*speK*
		MGAS315.5	ATACTTTGAT TATA	5′ end of conserved hypothetical protein	*speA*
		MGAS315.6	TTTATGCTAT AAT	5′ end of gamma-glutamyl kinase (*proB*)	
SSI-1 (85)	M3	SPsP1	TTACTACGTT TTTTACTACG TT	Promoter region of putative gamma-glutamyl kinase	*sdn*
		SPsP2	ATACTTTGAT TATA	5′ of *recX*	*speA*
		SPsP3	TCTGATATAA TAA	Promoter of iron-dependent repressor SPy0450	*speL*
		SPsP4	AAAGACGCTG TTAAATAATT	3′ end of histone-like protein	MF4
Manfredo	M5	phi SpeC/MF2	AGTATAGTTG TACATG	5′ end of dipeptidase	*speC*, MF2
		phi SpeH/SpeI	AAACTCAAGA AGRGATTAAA TAAAACATTA AAGAACCTTG TCATATCA	3′ end of dTDP-glucose-4,6-dehydratase	*speH*, *speI*
		phi MF3	TTTCTTTTCA ATTTTAGTGA TTTTCATA	5′ of *recX*	MF3
		phi MF4	AAAGACGCTG TTAAATAATT	3′ end of histone-like protein	MF4
		phi MutL	CAATAATGTT TGTCATAATT T	5′ end of *mutL*	-
MGAS10394 (6)	M6	MGAS10394.1	TATGCTATAA TA	3′ end of recombination protein *recO*	*sdn*
		MGAA10394.2	Not identified	Possibly tRNA-Arg	*speA* variant, *speI* fragment
		MGAS10394.3	AACTCATGAA CAAGACAAAA AGAAAAAACC TTGATGTAAC AAGGTTTTAG TAAGTTATGA TTACTTACGG TAAGCATTGA TGGAGCCGGT GGGAGT	3′ end of tmRNA	*speC* variant, *speK*, *sla*
		MGAS10394.4	Not identified; transposon	Between *comE* operon proteins 2 and 3	*mefA*
		MGAS10394.5	AATTATTTAA CAGCGTCTTT	3′ end of histone-like protein	*speC*, MF2
		MGAS10394.6	Not identified	Displaces Spy1607	*sda*
		MGAS10394.7	Not identified	Promoter of excinuclease ABC subunit A	MF3
		MGAS10394.8	CAATAATGTT TGTCATAATTT	5′ end of *mutL*	-
MGAS8232 (91)	M18	phi SpeA	CTTGATTTTA GTGACATG	3′ immediate downstream of putative SNF helicase	*speA*
		phi SpeC	TAATGGCATGTACAACTATACT	5′ end of dipeptidase	*speC*
		phi SpeLM	GGTTCGACTC CCACCGGCTC CATCAATGCT TACCGTAAGT AATCATAACT TACTAAAACC TTGTTACATC AAGGTTTTTT CTTTTTGTCT TGTTCATGAG TT	3′ end of tmRNA	*speL*, *speM*,
		phi SF370.3	TAGACGAATT ATTTAACAGC GTC	3′ end of histone-like protein	DNase
		phi Sda	TGGAGCAAAA TTTGCTAT	5′ end of HAD-like hydrolase	MF3

[a] The complete prophage and major prophage remnants from each of the six completed GAS genomes were analyzed for the site of attachment, *attB* duplication sequence, and identifiable toxin genes or other virulence genes. In general, nomenclature used in the annotation of a genome was adopted. Since an annotation for the M6 Manfredo strain has not been yet released, a descriptive nomenclature was adopted based upon the virulence gene associated with the phage or, when lacking such a gene, the site of phage attachment.

genome prophages include the histonelike protein, dTDP-glucose-4,6-dehydratase, a putative SNF helicase, and recombination protein recO. In all of these phages, integration does not alter the targeted coding region.

A number of the *S. pyogenes* genome prophages have adopted a more unusual strategy, integrating into the 5′ end of a gene or its predicted promoter. For example, a predicted polycistronic message contains the genes essential for methyl-directed mismatch repair (MMR), *mutS* and *mutL*, as well as several additional DNA repair genes (Fig. 2). Phage SF370.4 integrates into the 5′ end of *mutL*, thus separating it from the shared promoter upstream of *mutS*. The presence of the phage could then inactivate MMR and lead to a mutator phenotype. Integration at the 5′ end of a host gene is also seen in other *S. pyogenes* genome prophages, including a dipeptidase gene, recombination protein *recX*, *proB*, a HAD-like hydrolase, and a conserved hypothetical protein (Table 1). Other prophages appear to target the promoter region immediately upstream of the actual open reading frame, including the promoter regions of *yesN*, gamma-glutamyl kinase, and an iron-dependent repressor. Similarly, the transposon MGAS10394.4 with some prophage features separates the *comE* operon proteins 2 and 3, probably creating a polar mutation that silences protein 3. In all of these cases, the insertion of a prophage at the 5′ end of a gene has the potential for altering host gene expression by introducing a polar mutation or an alternative promoter, and such transcription-altering prophages may represent an important class of genetic regulatory elements in *S. pyogenes* in modifying the biology of their hosts.

Phage Diversity and Modular Exchange

At the DNA sequence level, the genome prophages of *S. pyogenes* form groups of more or less closely related members (Fig. 3 and Table 2), with phylogenetic analysis suggesting that nine groups have been identified (Fig. 4 and Table 3). However, the challenge of constructing meaningful phylogenetic trees for phages is compounded by the apparent frequency of recombination between phage genomes (18, 49),

and thus, the phylogeny of individual functional modules (such as toxin genes) may be unrelated to other regions of the genome (such as the lysogeny module). Pregenomic studies suggested that considerable diversity existed in GAS phages, with phages sharing the *speA* gene integrating into the bacterial chromosome at multiple sites; further, these sites were also used by non-*speA* phages (80, 82, 108, 110). Botstein proposed that the genome diversity seen in phages resulted from the recombination between members of a phage family so that functional "modules" would be reassorted to generate novel phages, largely through shared linker sequences (13). However, more recent studies have found that rather than homologous recombination between shared sequences driving module exchange, nonhomologous (illegitimate) recombination appears to be the principal means of exchange (36, 44, 58, 69, 83). While the results of these recombinational events may be seen in the genomes of the temperate phages of *S. pyogenes*, the actual driving force behind nonhomologous recombination of these genomes is poorly understood.

At whatever frequency, nonhomologous recombination would be likely to generate defective phage genomes as well as successful combinations, and further, such a recombinational event might occur at any point between two phage genomes. Natural selection has favored the preservation of overall gene order, with modules for integration, regulation, replication, DNA packaging, structural proteins, and host lysis arranged sequentially. Further, there appears to be a strong selection for the maintenance groups of genes together within a module, at least at the level of phages that share a common host species such as *S. pyogenes*. As an example, a group of prophages related to phage SF370.1 shares a conserved region that has been recombined with a variety of integrative and virulence modules in the generation of phage diversity (Fig. 5). Positionally, each of the six phages is integrated into different sites in the bacterial chromosome even though they may share the same virulence genes (e.g., SF370.1 and Manfredo phi MF2 both have *speC* and MF2 but integrate into different *attB* sites). The conserved

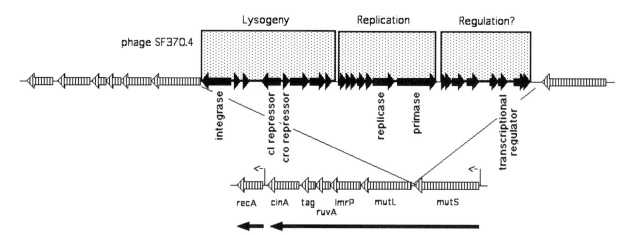

FIGURE 2 5′-integration of phage SF370.4 interrupts MMR. The potential for gene interruption by 5′-integration is illustrated by comparing the MMR region from strain SF370 (upper) with that from MGAS315 (lower). In strains lacking a prophage at this site, the genes for *mutS*, *mutL*, *lmrP*, *ruvA*, and *tag* are predicted to be expressed on a polycistronic mRNA (indicated below by a heavy arrow) with the promoter position upstream of *mutS* (←). The presence of phage SF370.4 separates the *mutL* and the downstream genes from *mutS* and the promoter, potentially creating a polar mutation.

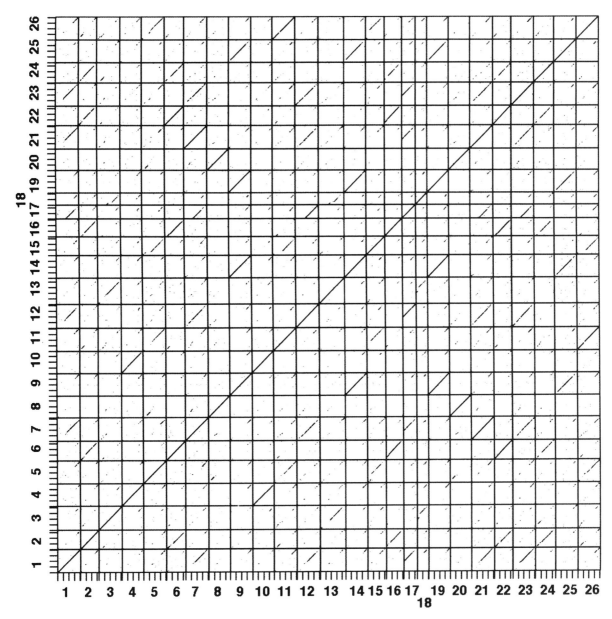

FIGURE 3 Multiple alignment of the genome prophages of *S. pyogenes* (see Table 2). The genomes of the major prophages found in the six completed genomes were aligned by dot plot analysis using a window size of 11. The three phages that integrate at *mutL* and the transposon/phage element MGAS10394.4 were excluded from this analysis.

central region found in each phage genome encodes the genes for DNA packaging, structural proteins for phage heads and tails, and lysis genes. The preservation of this module may reflect selection to maintain the regulatory elements that control gene expression or biophysical constraints in the assembly of the structural proteins. However, each individual phage has some level of variation within this region, showing insertions, deletions, or replacements relative to the other members.

Other notable clusters of prophages are observed in the genomes. One group of phages is identified by the use of the histonelike protein as the site of integration; remarkably, a member of this group is present in every sequenced genome. The dot plot analysis in Fig. 6 shows a high degree of DNA

sequence colinearity over much of these genomes; however, the region that sometimes diverges is the virulence gene: although it is most commonly occupied by one of the DNase genes (MF3 or MF4) in MGAS10394.5, it is occupied by *speC* and MF2. Why this phage is so common is uncertain. Integration occurs in the 3′ end of the histonelike protein gene and thus should not appreciably alter expression of that protein; thus, some modulation of host gene expression by integration is not anticipated. The presence of some variant of a DNase placed at this region of the chromosome may be linked to some higher level of genome organization relating to chromosomal DNA replication and global gene expression. A third family of closely related genes is found in three of the six genomes (SF370, MGAS10394, and Manfredo)

TABLE 2 Key to Fig. 3

No.	Prophage
1.	SF370.1
2.	SF370.3
3.	SF370.2
4.	MGAS315.1
5.	MGAS315.2
6.	MGAS315.3
7.	MGAS315.4
8.	MGAS315.5
9.	MGAS315.6
10.	MGAS8283 phi SpeA
11.	MGAS8283 phi SpeC
12.	MGAS8283 phi Sda
13.	MGAS8283 phi SpeLM
14.	MGAS10394.1
15.	MGAS10394.3
16.	MGAS10394.5
17.	MGAS10394.6
18.	MGAS10394.7
19.	SPsP1
20.	SSPsP2
21.	SPsP3
22.	SPsP4
23.	Manfredo phi MF
24.	Manfredo phi MF4
25.	Manfredo phi SpeH/SpeI
26.	Manfredo phi SpeC/MF2

and is remarkable for several aspects (Fig. 7). As mentioned above, these phages integrate into the 5′ end of *mutL*, and their presence is therefore predicted to interrupt MMR. Numerous studies have demonstrated that the frequency of MMR defects in pathogenic bacteria is much higher than anticipated (65, 66), and the resulting mutator phenotype may be advantageous under certain circumstances (76, 77, 90, 92). The three MMR phages are similar in structure, having the genes for lysogeny, DNA replication, and regulation but lacking all of the late genes for DNA packaging, structural proteins, and host lysis. Preliminary studies have found that although phage SF370.4 lacks the genes to complete its lytic cycle, its genome is excised during logarithmic growth and apparently reintegrated in the stationary phase (96). Thus, these MMR-inactivating prophages may represent a novel means of achieving a mutator phenotype in GAS that is altered in response to growth.

The exchange of genome modules by nonhomologous recombination may be an efficient mechanism for the generation of new phages and promote the dissemination of toxin and other virulence genes, but it complicates any attempt to create a meaningful phylogenetic tree because different genes in a particular phage, such as the integrase and a superantigen, may have different evolutionary histories. Such conflicting phylogenetic signals make it difficult to construct a unique evolutionary tree. The method of split decomposition attempts to address this problem, generating phylogenetic trees with networks rather than simple branches when the data suggest the presence of different

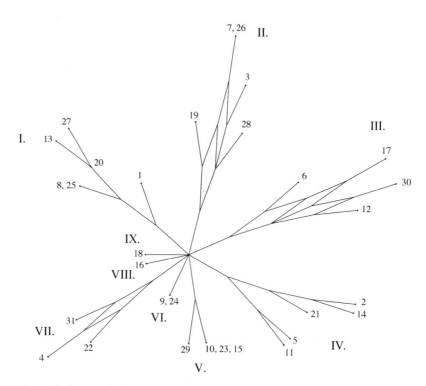

FIGURE 4 Phylogram of the *S. pyogenes* identified genome prophages. The phylogenetic tree with split decomposition analysis of the GAS genome phages shows probable groups with related evolutionary histories. Multiple sequence alignment of the prophage genomes was done using CLUSTALX (97). Phylogenetic analysis was done using the split decomposition method (3), and the software program SplitsTree (51) was used to generate the final tree.

TABLE 3 Key to Fig. 4

No.	Prophage
1	SF370.1
2	SF370.2
3	SF370.3
4	SF370.4
5	MGAS315.1
6	MGAS315.2
7	MGAS315.3
8	MGAS315.4
9	MGAS315.5
10	MGAS315.6
11	MGAS8283 phi SpeA
12	MGAS8283 phi SpeC
13	MGAS8283 phi Sda
14	MGAS8283 phi SpeLM
15	MGAS10394.1
16	MGAS10394.2
17	MGAS10394.3
18	MGAS10394.4
19	MGAS10394.5
20	MGAS10394.6
21	MGAS10394.7
22	MGAS10394.8
23	SSI-1 SPsP1
24	SSI-1 SPsP2
25	SSI-1 SPsP3
26	SSI-1 SPsP4
27	Manfredo phi Mf3
28	Manfredo phi Mf4
29	Manfredo phi SpeH/SpeI
30	Manfredo phi SpeC/Mf2
31	Manfredo phi MutL

and possibly conflicting phylogenies (3). Split decomposition analysis of the genome prophages illustrates that a number of evolutionary tree branches show complex networks, suggesting possible evidence for different and conflicting phylogenies within the genomes (Fig. 4, groups II, III, and VII). Further, because split decomposition does not attempt to force data onto a tree, it can provide a good indication of how treelike given data are, showing that some of the other groups have less complicated, and possibly more direct, evolutionary histories (groups I, IV, and V).

Late Phage Genes

Studies of the often closely related phages from *Streptococcus thermophilus* have identified gene cassettes in the phage late regions for DNA packaging that distinguish the *pac*-type phages from the *cos*-phages (34, 35). In the DNA packaging module, the *pac*-type phages have the typical gene order of small terminase, large terminase subunit, minor head protein, scaffold protein, and major head protein. The *cos*-type phages, by contrast, have the gene order of small terminase subunit, large terminase subunit, *clpP* protease, and the major head protein; the scaffold protein is generated by ClpP protease cleavage of the head protein (34, 35). Many prophages identified in Fig. 4 have genome organizations similar to the *pac*-style phages, but the members of two groups (III and IV) are predicted to be similar to the *S. thermophilus cos*-type phages. The first *cos* group (Fig. 4, group III) shows evi-

dence of considerable recombination within the group, perhaps reflecting a relatively recent emergence of these phages. The second group of *cos* phages (group IV) is marked by deletions and point mutations in the late gene cluster (Fig. 8): phage SF370.3 has a mutation in the portal protein (36), MGAS8283 phi SpeLM has one in the tape measure protein (minor tail protein), and phage MGAS10394.3 has one in the probable head protein. Phage MGAS10394.7 is even further decayed, being a fragment containing only some late genes and the MF3 streptodornase gene. As discussed below, such mutations and deletions result in the fixation of the phage-associated virulence genes in the bacterial genome. The number of phages in this group with defects suggests that this group may be relatively ancient, having decayed through long-term association with their host bacteria.

The possession of a hyaluronidase gene differentiates many temperate GAS phages from the lytic GAS phages. Kjems first observed that hyaluronidase was released by a lysogenic strain of GAS during phage multiplication (61), with at least some of the activity being associated with the isolated phage particles (62). Analysis of the prophage genomes has shown that this enzyme is contained in the late gene cluster and probably forms part of the tail fibers. Initially, two phage hyaluronidase genes were identified: *hylP*, isolated from a type 49 nephritic GAS strain lysogenic for phage H4489A (53), and *hylP2*, isolated from a type 22 GAS (54). These two hyaluronidase genes are closely related, having 66.5% identity and 77.5% similarity at the nucleotide sequence level. The *hylP* gene specifies a region in the hyaluronidase protein that contains a series of 10 Gly-X-Y amino acid triplets, a structure reminiscent of collagen. Interestingly, this region is absent in *hylP2* and is obviously dispensable for hyaluronidase activity. However, the presence of a collagenlike motif could potentially lead to the induction of antibodies that cross-react with host tissues as is seen in the polyarthritis that sometimes accompanies rheumatic fever. Further allelic variants of these phage hyaluronidase genes, possibly generated by recombination, have been identified (75). Most of the complete or nearly complete genome prophages contain an allele of hyaluronidase; however, the presence of this protein is not universal: the closely related prophages MGAS315.5 and SPsP2 lack this enzyme. The lack of hyaluronidase suggests that these phages must rely upon exogenous enzyme from the host or a "helper" phage, attach at some point in the cell cycle when little capsule is expressed (68), or have devised some alternative means for circumventing the capsule barrier.

The bacterial surface receptors for toxigenic phage have not been yet identified, but some evidence suggests that the sites for specific phages may be present on some strains and not others, the receptors possibly being related to M-type. For example, in a survey by Yu and Ferretti (110), GAS M1 strains had sequences associated with T12 carriage but never those associated with a different *speA*+ phage (phage 49), and *speA* in general was found with highest frequency in M1, M3, M49, T1, and T3/13 strains. However, 89% of the M3 strains that were *speA*+ showed evidence for either double lysogeny or the presence of some recombinant phage, hybridizing to both probes specific for phages T12 and 49; the remaining 11% hybridized only to the phage 49 probe (107, 110). In a separate survey, strains of GAS with the highest frequency of *speC* carriage were found to be M2, M4, M6, M12, T2, T4, T6, and T28 strains; the only serotype that frequently contains both *speA* and *speC* genes is M3 (46).

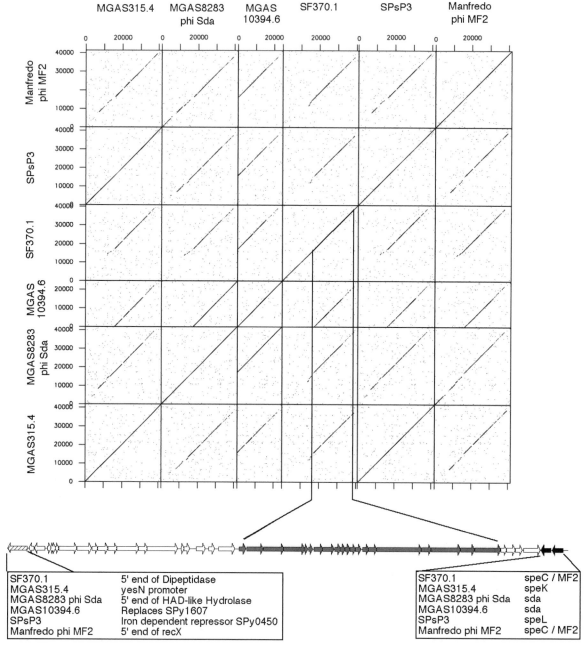

FIGURE 5 The cluster of prophages sharing a conserved region with phage SF370.1. The prophages with a conserved region encompassing the late genes for the phage structural proteins are compared by dot plot analysis. The genome of phage SF370.1 is shown below with the predicted open reading frames of the conserved region shaded in gray. The integrase gene (hatched) and virulence genes (black) are positioned at the flanking ends of the prophage genome; the box below the integrase gene indicates the site of attachment for each phage in the group, and the one below the virulence genes indicates the toxin or virulence genes associated with each prophage.

Temperate Bacteriophages and Virulence

The association of temperate phages and virulence genes is widespread in pathogenic bacteria (12, 101). Certainly, one of the most striking characteristics of any of the GAS genomes is the number of phage-associated virulence genes, and indeed, a majority of these genes were unknown prior to genome sequencing. As shown in Table 1, a considerable range of genes is carried by GAS prophages, including superantigens (*speA*, *speC*, *speG*, *speH*, *speI*, *speJ*, *speK*, *speL*, SSA, and variants), DNases (MF2, MF3, and MF4), phospholipase A2 (*sla*), and macrolide resistance (*mefA*). The known and suspected GAS phage-associated virulence genes are all positioned in the phage genomes in the region be-

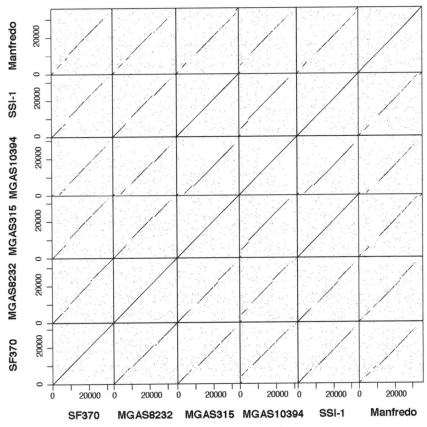

FIGURE 6 The prophages integrated at the histonelike protein gene. The prophages from each of the six genomes that use the histonelike gene as *attB* are aligned by dot plot analysis. The prophages are identified by the name of the GAS genome that they occupy.

tween the lysis module and the right attachment site with the bacterial chromosome. Recently, the term "morons" has been proposed to describe genetic additions to phage genomes that may not directly change "normal" phage functions such as integration or the assembly of capsids but possibly increase the fitness of the phage (often indirectly by increasing the fitness of the host) (58). Transcriptionally, morons are predicted to be independent of the phage genetic programs associated with lysogeny or lysis. However, recent evidence shows that the expression of these toxins may be influenced by bacterial signals, including ones resulting from the interaction of the streptococcus with the human host.

The timing and control of the expression of the phage-encoded virulence genes are a topic of great interest, given that the clinical outcome of infection can vary greatly from individual to individual. For example, a strain of GAS that was a toxin nonexpresser in artificial media was stimulated to secrete pyrogenic exotoxin SpeC and a phage-encoded DNase (Spd1 or MF2) after coculture with human pharyngeal cells, and this event was mediated by a pharyngeal cell soluble factor (16, 17). Similarly, the coculture of M3 strain MGAS315 with human epithelial pharyngeal cells induced the prophage-encoded proteins *speA* and phospholipase A(2), although simultaneous induction of the associated prophages was seen in one but not both (5). In a study of clinical M1 strains, it was similarly shown that SpeA production was very low in about half the strains. However, culture of these strains in Teflon chambers implanted into

mice for 5 days induced the expression of SpeA. Interestingly, isolates recovered from the chambers continued to produce SpeA, suggesting that an altered genetic program controlling toxin expression had been induced. The induction of SpeA expression occurred with a simultaneous inhibition of the chromosomally encoded cysteine protease SpeB, possibly preventing the degradation of the phage-encoded protein (2, 23, 59). All these studies point out that the regulation of these toxin genes is an area that still needs intense investigation to determine the relative roles played by the human host, the bacterium, and the phage. It is particularly unclear whether the responses that lead to prophage induction represent a stress that causes the phage to "abandon ship" after damage to its host or a response that might benefit the host (and therefore the phage) in terms of population survival. The in vivo release of toxigenic phage and infection of a new host that has been recently demonstrated suggests that the infection, lysogeny, and production of new phages is a highly dynamic process (15).

Temperate phages may play other roles in the modulation of host biology in addition to the carriage of toxin and other virulence genes. The expression of the major cell-surface antigen of *S. pyogenes*, the M protein (*emm*), is phenotypically variable, with some strains giving rise to frequent M⁻ colonial variants. This spontaneous variation in M protein expression was related to the presence of an extrachromosomal element by Cleary and his coworkers, who demonstrated that growth of GAS under conditions that typically cure bacteria of plasmids or phage resulted in high numbers of

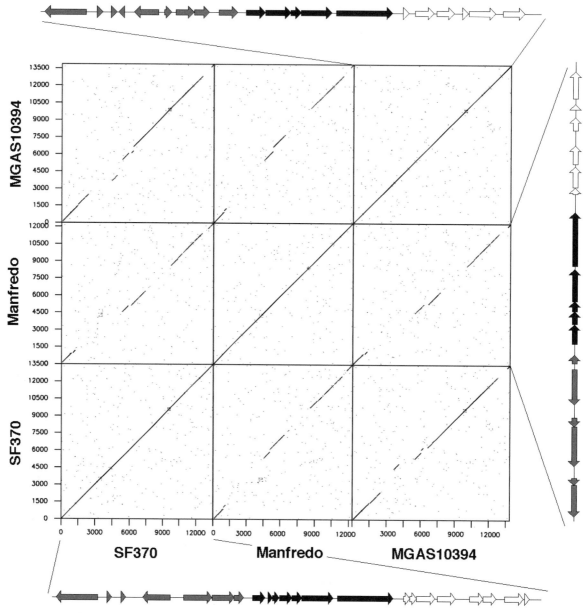

FIGURE 7 The MMR-converting phages. The prophages from SF370, MGAS10394, and Manfredo that integrate into the 5′ end of *mutL* are compared by dot plot analysis. The open reading frames for the phages are shown flanking the plot: SF370.4 (below), Manfredo (right), and MGAS10394.8 (above) with the lysogeny (gray), replication (black), and probable regulation (white) modules indicated. Although no structure genes are present, all three prophages have identifiable *attL* and *attR* sites.

M⁻ variants (25, 94). Subsequently, a temperate bacteriophage, SP24, was isolated, then, when introduced into an M⁻ *emm76* strain, resulted in *emm* expression to return to levels similar to the wild type (93), suggesting that the phage introduces a transcription regulator.

The Origins of Virulence Factors

Toxin genes and other phage-associated virulence genes are often positioned on the phage genome next to the genetic elements involved with site-specific integration of the phage into the bacterial chromosome, suggesting that such toxin genes may have originally become associated with the phage chromosome as a result of aberrant excision of the phage following induction that resulted in the acquisition of a bacterial gene (12). The fact that these toxin genes appear to play no role in the replication of the phage, often being secreted proteins with signal peptides, strengthens the notion that such elements were acquired at some point late in the phage's evolutionary history. However, the actual origins of phage-associated toxin genes are difficult to trace,

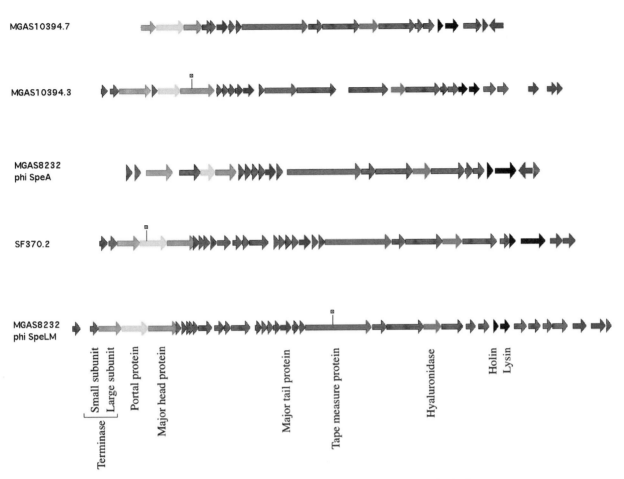

FIGURE 8 (*See the separate color insert for the color version of this illustration.*) The late gene regions of the group IV prophages. Some of the identifiable genes are listed below the MGAS8232 phi SpeLM prophage genome; the corresponding color scheme is used to identify related genes in the other prophages. The locations of probable mutations are indicated by the box symbol.

because the original genetic material may have originated from a different bacterial species and become established in the currently known system only following the appearance of a host-range variant of a phage. Some molecular evidence suggests that such events have indeed contributed to the dissemination of phage-associated toxin genes; for example, the *speA* gene of *S. pyogenes* and the enterotoxins B and C1 of *S. aureus* show a significant degree of homology and thus may share a common origin (105).

A frequent observation has been that these virulence genes (morons) differ in nucleotide content from the remainder of the phage genome. For example, the %G+C of the region containing *speA* (30%) varies from that of the rest of the phage genome, suggesting that the toxin gene may have originated from a different source than the rest of the phage genome. The phage-associated toxins found in the strain SF370 genome also showed similar divergence (42). Did these toxin genes all originate from some unknown bacteria with very low %G+C? The apparent divergence in %G+C content in these toxin genes may be misleading. If the coding region for *speA* is rewritten with the most frequently used codons in the *S. pyogenes* SF370 genome, then the resulting "optimized" open reading frame actually has a slightly lower %G+C than the native *speA* gene (Fig. 9A). Further, the basic pattern of %G+C over

the gene is very similar, suggesting that the sequence composition is at least somewhat influenced by the required codon usage to build the SpeA toxin protein. Indeed, if the process is repeated, rebuilding the *speA* gene with optimized codon usage from a number of bacterial species, the great majority of residues are conserved (Fig. 9B). The uncertainty of the origin of these virulence-related genes is compounded by the possibility that their evolutionary origin may be quite distant. Although related genes for lysogeny, replication, packaging, morphogenesis, and lysis may be identified in the phages of other species (105), no nonphage source for these toxin genes has been yet discovered. Cryptic and sometimes expressed coding regions often are found in the genomes of bacteriophages, even from those infecting nonpathogenic bacteria (14, 99, 100), and it may be that the origin of some virulence factors may have occurred by a process of selection for function from preexisting phage genes. The relationships seen between some prophages of *S. pyogenes* and *S. thermophilus* raise the possibility that prototoxin genes may have arisen in a separate genera and reached GAS by horizontal transfer (21, 36). Further, the positioning of such genes at the end of the integrated prophage may also reflect the results of selection for a genome arrangement in which toxin transcription may function independently of the phage promoters (which may

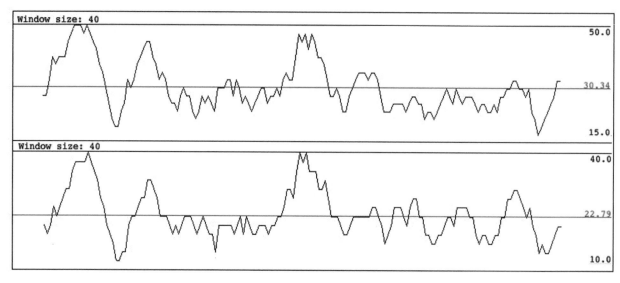

FIGURE 9 %G+C and codon usage of the native *speA* gene and "species-optimized" variants. (above) The %G+C plot of the native *speA* (upper) is shown compared to the plot of a reverse-translated version (lower) created using a codon table with the most frequently used codons from *S. pyogenes* SF370 (window size = 40). (Next page) A multiple alignment is shown comparing the native *speA* with reverse-translated *speA* variants created with optimized codon usage from a number of bacterial species: *S. pyogenes, Streptococcus pneumoniae, Lactococcus lactis, Enterococcus faecalis, S. aureus, Bacillus subtilis,* and *E. coli.* Regions of identity (>25%) are shaded in gray. Reverse translation was done using a program written in PERL with species-specific codon tables.

be silent during lysogeny). Certainly, the origin of phage virulence genes remains poorly understood, and whether the source of these genes was some unknown bacterial pool or they arose through adaptation of phage genetic material, it seems probable that toxin genes have had a long evolutionary history with GAS that helps obscure their past.

The Decay of Prophage Genomes and the Fixation of Virulence Genes

An intact, functional temperate phage is expected to exist in a dynamic state between the integrated prophage and the mature phage particle. At some frequency, this process may be interrupted by random mutations, rearrangements, or deletions that damage an integrated prophage so that the replication can no longer be completed, thus fixing the prophage and any associated virulence genes in the bacterial chromosome. The long-term maintenance of such genetic elements in the bacterial chromosome would depend upon any selective advantage conferred upon the host. The presence of a prophage in the bacterial chromosome places a burden on the host in terms of energy required to replicate the additional DNA; the presence of multiple prophages, as in the case of the GAS genome strains, further exacerbates this problem. It has been proposed (18) that the presence of any phage-associated gene that enhances the bacterium's ability to survive and gather resources helps relieve this energetic cost and favors maintenance of the prophage. There appears to be considerable strain-to-strain variation in GAS, however, as to the maintenance of these prophages as intact, function virions as opposed to noninducible remnants. For example, in strain SF370, phage SF370.1 is inducible while phages SF370.2 and SF370.3 are not, having defects that interfere with the completion of the lytic cycle (21, 36, 42). In contrast, all of the toxin-associated pro-

phages from MGAS315 are inducible by a variety of means, either by spontaneous release or by chemical agent (hydrogen peroxide or mitomycin C) (5). In a study under more physiological conditions, *speC*-containing phages were induced following coculture with human pharyngeal cells, resulting in the release of approximately 10^5 phage particles during a 3-h period. A pharyngeal cell soluble factor released during coculture probably mediates the phage induction (16).

The inactivation of a prophage while maintaining the associated virulence gene may be a frequent and evolutionarily favored event, preserving the benefit to the streptococcus conferred by the toxin while eliminating the danger of prophage induction and subsequent lysis. In the GAS prophages, frequent examples can be found in which either a genetic defect (inactivation of the portal protein in phage SF370.2) or large genome deletion (MGAS10394.7) has fixed the associated toxin gene as a permanent element of the host. Homologous recombination with another phage genome could rescue the disabled prophage or recombine to generate a novel, functional phage. If the forces that lead to prophage decay are constant over time, then inactivated prophages may be long-term residents of their host, and conversely, the fully inducible phages may have been recently acquired. The number of inducible prophage genomes in the M3 MGAS315 strain may reflect this strain's status as a recently emerged clone (10). The emergence of this highly virulent strain has been proposed to have resulted from the sequential addition of prophages (MGAS315.5, acquired in the 1920s; MGAS315.2, acquired around 1940; and MGAS315.4, acquired around 1980) to generate the current M3 clone (4). The inducible status of these phages suggests that the timeframe for maintenance of intact prophages in GAS may be measured in decades. The M6 MGAS10394 strain has, by contrast, a large number of phage

SF370	1	...	72
Native speA	1	...	72
L. lactis	1	...	72
S. pneumoniae	1	...	72
E. faecalis	1	...	72
S. aureus	1	...	72
B. subtilis	1	...	72
E. coli	1	...	72
SF370	73	...	144
Native speA	73	...	144
L. lactis	73	...	144
S. pneumoniae	73	...	144
E. faecalis	73	...	144
S. aureus	73	...	144
B. subtilis	73	...	144
E. coli	73	...	144
SF370	145	...	216
Native speA	145	...	216
L. lactis	145	...	216
S. pneumoniae	145	...	216
E. faecalis	145	...	216
S. aureus	145	...	216
B. subtilis	145	...	216
E. coli	145	...	216
SF370	217	...	288
Native speA	217	...	288
L. lactis	217	...	288
S. pneumoniae	217	...	288
E. faecalis	217	...	288
S. aureus	217	...	288
B. subtilis	217	...	288
E. coli	217	...	288
SF370	289	...	360
Native speA	289	...	360
L. lactis	289	...	360
S. pneumoniae	289	...	360
E. faecalis	289	...	360
S. aureus	289	...	360
B. subtilis	289	...	360
E. coli	289	...	360
SF370	361	...	432
Native speA	361	...	432
L. lactis	361	...	432
S. pneumoniae	361	...	432
E. faecalis	361	...	432
S. aureus	361	...	432
B. subtilis	361	...	432
E. coli	361	...	432
SF370	433	...	504
Native speA	433	...	504
L. lactis	433	...	504
S. pneumoniae	433	...	504
E. faecalis	433	...	504
S. aureus	433	...	504
B. subtilis	433	...	504
E. coli	433	...	504
SF370	505	...	576
Native speA	505	...	576
L. lactis	505	...	576
S. pneumoniae	505	...	576
E. faecalis	505	...	576
S. aureus	505	...	576
B. subtilis	505	...	576
E. coli	505	...	576
SF370	577	...	648
Native speA	577	...	648
L. lactis	577	...	648
S. pneumoniae	577	...	648
E. faecalis	577	...	648
S. aureus	577	...	648
B. subtilis	577	...	648
E. coli	577	...	648
SF370	649	...	702
Native speA	649	...	702
L. lactis	649	...	702
S. pneumoniae	649	...	702
E. faecalis	649	...	702
S. aureus	649	...	702
B. subtilis	649	...	702
E. coli	649	...	702

remnants, suggesting that the toxins associated with these prophages have been long-established host features. The group of *cos* phages (group IV) discussed above may be an older group of prophages, their long-term utility to the host preserving the distal, toxin-containing region of their genomes. However, implications of a whole group of phages that is marked by defects are that (i) few related intact phages may be present in the current circulating phage pools to provide genetic material to rescue these damaged prophages, and (ii) the group as a whole may be in the process of being supplanted by other phage families.

In addition to the pressures that lead to prophage decay in response to host stabilizing selection, other factors may lead to the loss of prophage DNA. In the case of a bacterium-phage relation that relies upon the selective effect of a single phage-encoded gene, such as *C. diphtheriae* and the diphtheria toxin, a stable maintenance of the phage may be favored. However, in *S. pyogenes* multiple phage-encoded morons may be present in a single strain and often belong to the same category (superantigen, mitogenic factor, or DNase). In such a case, a competition between morons may occur, leading to the preferential establishment of one moron and the decay and elimination of another. The prevalence of some phage-associated toxin genes (such as *speA* and *speC*) may reflect the superior advantage conferred to the host streptococcus over other toxin genes that are then eliminated to reduce the genetic burden on the bacterium. Examples of fragments of superantigen genes are observed in SF370 (R1930 with integrase containing several point mutations but intact *attL* and *attR* sites [21] or the superantigen domain next to transposase/integrase fragment SPy0858) and in MGAS10394 (*speI* fragment in phage MGAS10394.2).

TRANSDUCTION

The three common methods of genetic exchange within a species of bacteria are transformation, conjugation, and transduction. In the GAS, neither transformation nor conjugation has been observed to occur naturally. However, transduction occurs in GAS and is mediated by both lytic and temperate bacteriophages. An important observation of infections caused by GAS is the periodic shift in severity and in the predominant clinical syndromes (88). The genetic basis for these phenotypic shifts in GAS natural populations has been investigated in several studies, and the results have shown that considerable allelic variation exists at the DNA level both in bacterial- and phage-associated genes as well as in linkage between specific genes and clinical syndromes (11, 24, 60, 84, 109). The allelic variation is more than would be expected to result from accumulated random mutations between genetically isolated individuals. Because transduction is the only known natural means of genetic exchange, it is likely that this mechanism and its associated bacteriophages play an important role in the genetic shifts seen in GAS.

Transduction among GAS was first reported by Leonard, Colón, and Cole in 1968 (67). They identified five phages, two temperate and three lytic, that were able to transduce streptomycin resistance from a type 6 donor strain at varying frequencies to strain K56 (M-type 12). Of these phages that mediated transduction, phage A25 transduced at the highest frequency. Transfer of streptomycin resistance from an A25 lysate of a resistant host to the sensitive K56 donor strain was insensitive to DNase treatment of the lysate but could be prevented by phage-specific antisera. Both obser-

vations supported transduction as the mechanism of transfer, and subsequent work confirmed the ability of A25 (and several related phages) to mediate generalized transduction.

The presence of the hyaluronic capsule of GAS prevents infection by phage A25; it must be removed for transduction to occur. Increased levels of transductants can be obtained by the use of specific A25 antiphage serum to block any unabsorbed or progeny phage from infecting the transductants resulting from the initial absorption (73) or by use of temperature-sensitive mutants of phage A25 to infect decapsulated K56 (71). No enhancement in transduction frequency was obtained by combining treatment with antiphage serum and the use of the temperature-sensitive mutant phages. The efficiency of A25-mediated transduction also may be increased by UV irradiation of the transducing lysates prior to adsorption to the recipient streptococci (29, 70). As UV dose to the phage A25 lysate increases, the number of transductants increases, reaching the maximum value when a 90 to 99% reduction in PFU is achieved (73). Interestingly, no enhancement in transduction frequency was observed when two temperate transducing phages were UV treated. Twenty-six GAS recipient strains were tested for transduction by A25, A5, and A6 before and after UV irradiation of the phages. Thirteen of these strains, including M-types 6, 8, 12, 13, 19, and nontypeable strains, were efficiently transduced only after UV treatment (29).

GAS strain T253, the laboratory host strain for the erythrogenic toxin A-carrying phage T12, can serve as donor for A25 transduction; however, T253-derived lysates produce six to eight times fewer transducing particles than those derived from K56. This observed difference is not due to a diminished phage burst size because A25 grown on T253 produces approximately the same number of PFU per cell as propagation on strain K56. Interestingly, the transducing lysate from a streptomycin derivative of T253 was able to transfer this antibiotic resistance to a homologous strain at a lower frequency than to a heterologous strain (74).

Temperate GAS bacteriophages are also capable of mediating the transfer of antibiotic resistance by transduction. Ubukata and coworkers (98) observed the transfer of resistance to tetracycline, chloramphenicol, macrolides, lincomycin, and clindamycin by induced temperate phages from T-12 strains. Antibiotic resistance was transferred either alone or in combination, suggesting that the donor strain may have been harboring a genetic element conferring multiple drug resistance, such as the transposon Tn916. The transfer of erythromycin and streptomycin resistance by temperate phages induced by mitomycin C treatment has been also observed (52).

The transfer of genetic material from one streptococcal group to another by transduction is another potential means of the dissemination of genes that could affect virulence. Bacteriophages isolated from streptococcal groups A, E, and G could lyse streptococci from serogroups A, C, G, H, and L, and in some cases, these lytic phages were also capable of propagation in one or more of the other serogroups (31). The same study also showed that phage A25 as well as two other lytic GAS phages could also transduce a number of group G strains to streptomycin resistance. The same workers, Colon et al., subsequently isolated a temperate phage from a group G streptococcus (phage GT-234) that was capable of propagating on streptococci from serogroups A and C (30). Phage GT-234, which is serologically related to and morphologically indistinguishable from the lytic GAS phage A25, was able to transduce strains of serogroups A, C, and G to streptomycin resistance. Wannamaker et al. (103) demonstrated that genetic information (streptomycin resistance) could be transferred into group A strains by a transducing phage from GCS. This was a temperate phage induced from a clinical isolate of a GCS, and although most of the temperate phages induced from other strains of GCS did not plaque on GAS, this phage was unusual in that it was able to infect 40% of the GAS strains tested.

Horizontal transfer of genetic material in GAS has been shown by numerous studies to be important in the dissemination of genes in natural populations. The impact of this genetic flow is readily appreciated when shifts in the prevalence and severity of streptococcal diseases occur (27, 95). The role in transduction and bacteriophages in horizontal transfer is assumed, yet no modern studies have directly addressed the question. The majority of the studies on streptococcal transduction were done before the advent of many of the current techniques of molecular biology, so the time may have come to reexamine this phenomenon. A better understanding of streptococcal transduction may prove key to understanding the flow of genetic information in natural populations of GAS and the horizontal transfer of information from other genera.

REFERENCES

1. **Ackermann, H.-W., and M. S. DuBow.** 1987. *Viruses of Prokaryotes: General Properties of Bacteriophages*, vol. 1. CRC Press, Boca Raton, Fla.
2. **Aziz, R. K., M. J. Pabst, A. Jeng, R. Kansal, D. E. Low, V. Nizet, and M. Kotb.** 2004. Invasive M1T1 group A Streptococcus undergoes a phase-shift in vivo to prevent proteolytic degradation of multiple virulence factors by SpeB. *Mol. Microbiol.* **51:**123–134.
3. **Bandelt, H., and A. Dress.** 1992. Split decomposition: a new and useful approach to phylogenetic analysis of distance data. *Mol. Phylogenet. Evol.* **1:**242–252.
4. **Banks, D. J., S. B. Beres, and J. M. Musser.** 2002. The fundamental contribution of phages to GAS evolution, genome diversification and strain emergence. *Trends Microbiol.* **10:**515–521.
5. **Banks, D. J., B. Lei, and J. M. Musser.** 2003. Prophage induction and expression of prophage-encoded virulence factors in group A Streptococcus serotype M3 strain MGAS315. *Infect. Immun.* **71:**7079–7086.
6. **Banks, D. J., S. F. Porcella, K. D. Barbian, S. B. Beres, L. E. Philips, J. M. Voyich, F. R. DeLeo, J. M. Martin, G. A. Somerville, and J. M. Musser.** 2004. Progress toward characterization of the group A Streptococcus metagenome: complete genome sequence of a macrolide-resistant serotype M6 strain. *J. Infect. Dis.* **190:**727–738.
7. **Behnke, D., and H. Malke.** 1978. Bacteriophage interference in *Streptococcus pyogenes*. I. Characterization of prophage—host systems interfering with the virulent phage A25. *Virology* **85:**118–128.
8. **Behnke, D., and H. Malke.** 1978. Bacteriophage interference in *Streptococcus pyogenes*. II. A25 mutants resistant to prophage-mediated interference. *Virology* **85:**129–136.
9. **Benchetrit, L. C., E. D. Gray, and L. W. Wannamaker.** 1977. Hyaluronidase activity of bacteriophages of group A streptococci. *Infect. Immun.* **15:**527–532.
10. **Beres, S. B., G. L. Sylva, K. D. Barbian, B. Lei, J. S. Hoff, N. D. Mammarella, M. Y. Liu, J. C. Smoot, S. F. Porcella, L. D. Parkins, D. S. Campbell, T. M. Smith, J. K. McCormick, D. Y. Leung, P. M. Schlievert, and J. M. Musser.** 2002. Genome sequence of a serotype M3 strain of group A Streptococcus: phage-encoded toxins, the high-virulence phenotype, and clone emergence. *Proc. Natl. Acad. Sci. USA* **99:**10078–10083.
11. **Bessen, D. E., and S. K. Hollingshead.** 1994. Allelic polymorphism of *emm* loci provides evidence for horizontal gene spread in group A streptococci. *Proc. Natl. Acad. Sci. USA* **91:**3280–3284.
12. **Bishai, W. R., and J. R. Murphy.** 1988. Bacteriophage gene products that cause human disease, p. 683–724. *In* R. Calendar (ed.), *The Bacteriophages*, vol. 2. Plenum Press, New York, N.Y.
13. **Botstein, D.** 1980. A theory of modular evolution for bacteriophages. *Ann. N.Y. Acad. Sci.* **354:**484–491.
14. **Boyce, J. D., B. E. Davidson, and A. J. Hillier.** 1995. Identification of prophage genes expressed in lysogens of the *Lactococcus lactis* bacteriophage BK5-T. *Appl. Environ. Microbiol.* **61:**4099–4104.
15. **Broudy, T. B., and V. A. Fischetti.** 2003. In vivo lysogenic conversion of Tox(−) *Streptococcus pyogenes* to Tox(+) with lysogenic streptococci or free phage. *Infect. Immun.* **71:**3782–3786.
16. **Broudy, T. B., V. Pancholi, and V. A. Fischetti.** 2001. Induction of lysogenic bacteriophage and phage-associated toxin from group a streptococci during coculture with human pharyngeal cells. *Infect. Immun.* **69:**1440–1443.
17. **Broudy, T. B., V. Pancholi, and V. A. Fischetti.** 2002. The in vitro interaction of *Streptococcus pyogenes* with human pharyngeal cells induces a phage-encoded extracellular DNase. *Infect. Immun.* **70:**2805–2811.
18. **Brussow, H., C. Canchaya, and W. D. Hardt.** 2004. Phages and the evolution of bacterial pathogens: from genomic rearrangements to lysogenic conversion. *Microbiol. Mol. Biol. Rev.* **68:**560–602.
19. **Campbell, A. M.** 1992. Chromosomal insertion sites for phages and plasmids. *J. Bacteriol.* **174:**7495–7499.
20. **Campbell, A., S. J. Schneider, and B. Song.** 1992. Lambdoid phages as elements of bacterial genomes. *Genetica* **86:**259–267.
21. **Canchaya, C., F. Desiere, W. McShan, J. Ferretti, J. Parkhill, and H. Brussow.** 2002. Genome analysis of an inducible prophage and prophage remnants integrated in the *Streptococcus pyogenes* strain SF370. *Virology* **302:**245–2 58.
22. **Cantacuzene, J., and O. Boncieu.** 1926. Modifications subies pare des streptocoques d'origine non-scarlatineuse que contact des produits scarlatineux filtres. *C. R. Acad. Sci.* **182:**1185.
23. **Chatellier, S., N. Ihendyane, R. G. Kansal, F. Khambaty, H. Basma, A. Norrby-Teglund, D. E. Low, A. McGeer, and M. Kotb.** 2000. Genetic relatedness and superantigen expression in group A streptococcus serotype M1 isolates from patients with severe and nonsevere invasive diseases. *Infect. Immun.* **68:**3523–3534.
24. **Chaussee, M. S., J. Liu, D. L. Stevens, and J. J. Ferretti.** 1996. Genetic and phenotypic diversity among isolates of *Streptococcus pyogenes* from invasive infections. *J. Infect. Dis.* **173:**901–908.
25. **Cleary, P. P., Z. Johnson, and L. Wannamaker.** 1975. Genetic instability of M protein and serum opacity factor of group A streptococci: evidence suggesting extrachromosomal control. *Infect. Immun.* **12:**109–118.
26. **Cleary, P. P., L. W. Wannamaker, M. Fisher, and N. Laible.** 1977. Studies of the receptor for phage A25 in group A streptococci: the role of peptidoglycan in reversible adsorption. *J. Exp. Med.* **145:**578–593.

27. Cleary, P. P., E. L. Kaplan, J. P. Handley, A. Wlazlo, M. H. Kim, A. R. Hauser, and P. M. Schlievert. 1992. Clonal basis for resurgence of serious *Streptococcus pyogenes* disease in the 1980s. *Lancet* **339**:518–521.

28. Cleary, P. P., L. McLandsborough, L. Ikeda, D. Cue, J. Krawczak, and H. Lam. 1998. High-frequency intracellular infection and erythrogenic toxin A expression undergo phase variation in M1 group A streptococci. *Mol. Microbiol.* **28**:157–167.

29. Colon, A. E., R. M. Cole, and C. G. Leonard. 1970. Transduction in group A streptococci by ultraviolet-irradiated bacteriophages. *Can. J. Microbiol.* **16**:201–202.

30. Colon, A. E., R. M. Cole, and C. G. Leonard. 1971. Lysis and lysogenization of groups A, C, and G streptococci by a transducing bacteriophage induced from a group G Streptococcus. *J. Virol.* **8**:103–110.

31. Colon, A. E., R. M. Cole, and C. G. Leonard. 1972. Intergroup lysis and transduction by streptococcal bacteriophages. *J. Virol.* **9**:551–553.

32. Colon-Whitt, A., R. S. Whitt, and R. M. Cole. 1979. Production of an erythrogenic toxin (streptococcal pyrogenic exotoxin) by a non-lysogenised group-A streptococcus, p. 64–65. *In* M. T. Parker (ed.), *Pathogenic Streptococci.* Reedbooks Ltd., Chertsey, England.

33. Crater, D. L., and I. van de Rijn. 1995. Hyaluronic acid synthesis operon (has) expression in group A streptococci. *J. Biol. Chem.* **270**:18452–18458.

34. Desiere, F., S. Lucchini, and H. Brussow. 1999. Comparative sequence analysis of the DNA packaging, head, and tail morphogenesis modules in the temperate cos-site *Streptococcus thermophilus* bacteriophage Sfi21. *Virology* **260**:244–253.

35. Desiere, F., R. D. Pridmore, and H. Brussow. 2000. Comparative genomics of the late gene cluster from Lactobacillus phages. *Virol.* **275**:294–305.

36. Desiere, F., W. M. McShan, D. van Sinderen, J. J. Ferretti, and H. Brussow. 2001. Comparative genomics reveals close genetic relationships between phages from dairy bacteria and pathogenic streptococci: evolutionary implications for prophage-host interactions. *Virology* **288**:325–341.

37. Evans, A. C. 1933. Inactivation of antistreptococcus bacteriophage by animal fluids. *Public Health Rep.* **48**:411–426.

38. Evans, A. C. 1934. The prevalence of streptococcus bacteriophage. *Science* **80**:40–41.

39. Evans, A. C. 1934. Streptococcus bacteriophage: a study of four serological types. *Public Health Rep.* **49**:1386–1401.

40. Evans, A. C. 1940. The potency of nascent streptococcus bacteriophage B. *J. Bacteriol.* **39**:597–604.

41. Evans, A. C., and E. M. Stockrider. 1942. Another serologic type of streptococcic bacteriophage. *J. Bacteriol.* **42**:211–214.

42. Ferretti, J. J., W. M. McShan, D. Ajdic, D. J. Savic, G. Savic, K. Lyon, C. Primeaux, S. Sezate, A. N. Suvorov, S. Kenton, H. Lai, S. Lin, Y. Qian, H. G. Jia, F. Z. Najar, Q. Ren, H. Zhu, L. Song, J. White, X. Yuan, S. W. Clifton, B. A. Roe, and R. McLaughlin. 2001. Complete genome sequence of an M1 strain of *Streptococcus pyogenes. Proc. Natl. Acad. Sci. USA* **98**:4658–4663.

43. Fischetti, V. A., and J. B. Zabriskie. 1968. Studies on streptococcal bacteriophages. II. Adsorption studies on group A and group C streptococcal bacteriophages. *J. Exp. Med.* **127**:489–505.

44. Ford, M. E., G. J. Sarkis, A. E. Belanger, R. W. Hendrix, and G. F. Hatfull. 1998. Genome structure of mycobacteriophage D29: implications for phage evolution. *J. Mol. Biol.* **279**:143–164.

45. Frobisher, M., and J. H. Brown. 1927. Transmissible toxicogenicity of streptococci. *Bull. Johns Hopkins Hosp.* **41**:167–173.

46. Goshorn, S. C., and P. M. Schlievert. 1989. Bacteriophage association of streptococcal pyrogenic exotoxin type C. *J. Bacteriol.* **171**:3068–3073.

47. Graham, M. R., L. M. Smoot, C. A. Migliaccio, K. Virtaneva, D. E. Sturdevant, S. F. Porcella, M. J. Federle, G. J. Adams, J. R. Scott, and J. M. Musser. 2002. Virulence control in group A Streptococcus by a two-component gene regulatory system: global expression profiling and in vivo infection modeling. *Proc. Natl. Acad. Sci. USA* **99**:13855–13860.

48. Groth, A. C., and M. P. Calos. 2004. Phage integrases: biology and applications. *J. Mol. Biol.* **335**:667–678.

49. Hendrix, R. W., J. G. Lawrence, G. F. Hatfull, and S. Casjens. 2000. The origins and ongoing evolution of viruses. *Trends Microbiol.* **8**:504–508.

50. Hill, J. E., and L. W. Wannamaker. 1981. Identification of a lysin associated with a bacteriophage (A25) virulent for group A streptococci. *J. Bacteriol.* **145**:696–703.

51. Huson, D. H. 1998. SplitsTree: analyzing and visualizing evolutionary data. *Bioinformatics* **14**:68–73.

52. Hyder, S. L., and M. M. Streitfeld. 1978. Transfer of erythromycin resistance from clinically isolated lysogenic strains of *Streptococcus pyogenes* via their endogenous phage. *J. Infect. Dis.* **138**:281–286.

53. Hynes, W. L., and J. J. Ferretti. 1989. Sequence analysis and expression in *Escherichia coli* of the hyaluronidase gene of *Streptococcus pyogenes* bacteriophage H4489A. *Infect. Immun.* **57**:533–539.

54. Hynes, W. L., L. Hancock, and J. J. Ferretti. 1995. Analysis of a second bacteriophage hyaluronidase gene from *Streptococcus pyogenes*: evidence for a third hyaluronidase involved in extracellular enzymatic activity. *Infect. Immun.* **63**:3015–3020.

55. Johnson, L. P., and P. M. Schlievert. 1983. A physical map of the group A streptococcal pyrogenic exotoxin bacteriophage T12 genome. *Mol. Gen. Genet.* **189**:251–255.

56. Johnson, L. P., and P. M. Schlievert. 1984. Group A streptococcal phage T12 carries the structural gene for pyrogenic exotoxin type A. *Mol. Gen. Genet.* **194**:52–56.

57. Johnson, L. P., P. M. Schlievert, and D. W. Watson. 1980. Transfer of group A streptococcal pyrogenic exotoxin production to nontoxigenic strains of lysogenic conversion. *Infect. Immun.* **28**:254–257.

58. Juhala, R. J., M. E. Ford, R. L. Duda, A. Youlton, G. F. Hatfull, and R. W. Hendrix. 2000. Genomic sequences of bacteriophages HK97 and HK022: pervasive genetic mosaicism in the lambdoid bacteriophages. *J. Mol. Biol.* **299**:27–51.

59. Kazmi, S. U., R. Kansal, R. K. Aziz, M. Hooshdaran, A. Norrby-Teglund, D. E. Low, A. B. Halim, and M. Kotb. 2001. Reciprocal, temporal expression of SpeA and SpeB by invasive M1T1 group a streptococcal isolates in vivo. *Infect. Immun.* **69**:4988–4995.

60. Kehoe, M. A., V. Kapur, A. M. Whatmore, and J. M. Musser. 1996. Horizontal gene transfer among group A streptococci: implications for pathogenesis and epidemiology. *Trends Microbiol.* **4**:436–443.

61. Kjems, E. 1958. Studies on streptococcal bacteriophages. 2. Adsorption, lysogenization, and one-step growth experiments. *Acta Pathol. Microbiol. Scand.* **42**:56–66.

62. **Kjems, E.** 1958. Studies on streptococcal bacteriophages. 3. Hyaluronidase produced by the streptococcal phage-host cell system. *Acta Pathol. Microbiol. Scand.* **44:**429–439.

63. **Kjems, E.** 1960. Studies on streptococcal bacteriophages. 5. Serological investigation of phages isolated from 91 strains of group A haemolytic streptococci. *Acta Pathol. Microbiol. Scand.* **49:**205–212.

64. **Krause, R. M.** 1957. Studies on bacteriophages of hemolytic streptococci. I. Factors influencing the interaction of phage and susceptible host cell. *J. Exp. Med.* **106:**365–383.

65. **LeClerc, J. E., and T. A. Cebula.** 2000. Pseudomonas survival strategies in cystic fibrosis. *Science* **289:**391–392.

66. **LeClerc, J. E., B. Li, W. L. Payne, and T. A. Cebula.** 1996. High mutation frequencies among *Escherichia coli* and *Salmonella* pathogens. *Science* **274:**1208–1211.

67. **Leonard, C. G., A. E. Colón, and R. M. Cole.** 1968. Transduction in group A streptococcus. *Biochem. Biophys. Res. Commun.* **30:**130–135.

68. **Levin, J., and M. Wessels.** 1998. Identification of csrR/csrS, a genetic locus that regulates hyaluronic acid capsule synthesis in group A Streptococcus. *Mol. Microbiol.* **30:**209–219.

69. **Lucchini, S., F. Desiere, and H. Brussow.** 1999. Similarly organized lysogeny modules in temperate Siphoviridae from low GC content gram-positive bacteria. *Virology* **263:**427–435.

70. **Malke, H.** 1969. Transduction of *Streptococcus pyogenes* K 56. *Microbiol. Genet. Bull.* **31:**23.

71. **Malke, H.** 1969. Transduction of *Streptococcus pyogenes* K 56 by temperature-sensitive mutants of the transducing phage A25. *Z. Naturforsch. Teil B* **24:**1556–1561.

72. **Malke, H.** 1970. Characteristics of transducing group A streptococcal bacteriophages A 5 and A 25. *Arch. Gesamte Virusforsch.* **29:**44–49.

73. **Malke, H.** 1972. Transduction in group A streptococci. *In* L. W. Wannamaker and J. M. Matsen (ed.), *Streptococci and Streptococcal Diseases: Recognition, Understanding, and Management.* Academic Press, New York, N.Y.

74. **Malke, H.** 1973. Phage A25-mediated transfer induction of a prophage in *Streptococcus pyogenes*. *Mol. Gen. Genet.* **125:**251–264.

75. **Marciel, A. M., V. Kapur, and J. M. Musser.** 1997. Molecular population genetic analysis of a *Streptococcus pyogenes* bacteriophage-encoded hyaluronidase gene: recombination contributes to allelic variation. *Microb. Pathog.* (England) **22:**209–217.

76. **Matic, I., C. Rayssiguier, and M. Radman.** 1995. Interspecies gene exchange in bacteria: the role of SOS and mismatch repair systems in evolution of species. *Cell* **80:**507–515.

77. **Matic, I., F. Taddei, and M. Radman.** 2000. No genetic barriers between *Salmonella enterica* serovar typhimurium and *Escherichia coli* in SOS-induced mismatch repair-deficient cells. *J. Bacteriol.* **182:**5922–5924.

78. **Maxted, W. R.** 1952. Enhancement of streptococcal bacteriophage lysis by hyaluronidase. *Nature* (London) **170:**1020–1021.

79. **Maxted, W. R.** 1955. The influence of bacteriophage on *Streptococcus pyogenes*. *J. Gen. Microbiol.* **12:**484–495.

80. **McShan, W. M., and J. J. Ferretti.** 1997. Genetic diversity in temperate bacteriophages of *Streptococcus pyogenes*: identification of a second attachment site for phages carrying the erythrogenic toxin A gene. *J. Bacteriol.* **179:**6509–6511.

81. **McShan, W. M., and J. J. Ferretti.** 1997. Genetic studies of erythrogenic toxin carrying temperate bacteriophages of *Streptococcus pyogenes*, p. 971–973. *In* T. Horaud, A. Bouvet, R. Leclercq, H. D. Montclos, and M. Sicard (ed.), *Streptococci and the Host.* Plenum Publishing, New York, N.Y.

82. **McShan, W. M., Y.-F. Tang, and J. J. Ferretti.** 1997. Bacteriophage T12 of *Streptococcus pyogenes* integrates into the gene for a serine tRNA. *Mol. Microbiol.* **23:**719–728.

83. **Monod, C., F. Repoila, M. Kutateladze, F. Tetart, and H. M. Krisch.** 1997. The genome of the pseudo T-even bacteriophages, a diverse group that resembles T4. *J. Mol. Biol.* **267:**237–249.

84. **Musser, J. M., V. Kapur, J. Szeto, X. Pan, D. S. Swanson, and D. R. Martin.** 1995. Genetic diversity and relationships among serotype M1 strains of *Streptococcus pyogenes*. *Dev. Biol. Stand.* **85:**209–213.

85. **Nakagawa, I., K. Kurokawa, A. Yamashita, M. Nakata, Y. Tomiyasu, N. Okahashi, S. Kawabata, K. Yamazaki, T. Shiba, T. Yasunaga, H. Hayashi, M. Hattori, and S. Hamada.** 2003. Genome sequence of an M3 strain of *Streptococcus pyogenes* reveals a large-scale genomic rearrangement in invasive strains and new insights into phage evolution. *Genome Res.* **13:**1042–1055.

86. **Niemann, H., A. Birch-Andersen, E. Kjems, B. Mansa, and S. Stirm.** 1976. Streptococcal bacteriophage 12/12-borne hyaluronidase and its characterization as a lyase (EC 4.2.99.1) by means of streptococcal hyaluronic acid and purified bacteriophage suspensions. *Acta Pathol. Microbiol. Scand. Sect. B* **84:**145–153.

87. **Pomrenke, M. E., and J. J. Ferretti.** 1989. Pysical maps of the streptococcal bacteriophage A25 and C1 genomes. *J. Basic Microbiol.* **29:**395–398.

88. **Quinn, R. W.** 1989. Comprehensive review of morbidity and mortality trends for rheumatic fever, streptococcal disease, and scarlet fever: the decline of rheumatic fever. *Rev. Infect. Dis.* **11:**928–953.

89. **Read, S. E., and R. W. Reed.** 1972. Electron microscopy of the replicative events of A25 bacteriophages in group A streptococci. *Can. J. Microbiol.* **18:**93–96.

90. **Richardson, A. R., and I. Stojiljkovic.** 2001. Mismatch repair and the regulation of phase variation in *Neisseria meningitidis*. *Mol. Microbiol.* **40:**645–655.

91. **Smoot, J. C., K. D. Barbian, J. J. Van Gompel, L. M. Smoot, M. S. Chaussee, G. L. Sylva, D. E. Sturdevant, S. M. Ricklefs, S. F. Porcella, L. D. Parkins, S. B. Beres, D. S. Campbell, T. M. Smith, Q. Zhang, V. Kapur, J. A. Daly, L. G. Veasy, and J. M. Musser.** 2002. Genome sequence and comparative microarray analysis of serotype M18 group A Streptococcus strains associated with acute rheumatic fever outbreaks. *Proc. Natl. Acad. Sci. USA* **99:**4668–4673.

92. **Sniegowski, P.** 1998. Mismatch repair: origin of species? *Curr. Biol.* **8:**R59–R61.

93. **Spanier, J. G., and P. P. Cleary.** 1980. Bacteriophage control of antiphagocytic determinants in group A streptococci. *J. Exp. Med.* **152:**1393–1406.

94. **Spanier, J. G., and P. P. Cleary.** 1983. A restriction map and analysis of the terminal reduncacy in the Group A streptococcal bacteriophage SP 24. *Virology.* **130:**502–513.

95. **Stevens, D. L., M. H. Tanner, J. Winship, R. Swarts, K. M. Ries, P. M. Schlievert, and E. Kaplan.** 1989. Severe group A streptococcal infections associated with a toxic shock-like syndrome and scarlet fever toxin A. *N. Engl. J. Med.* **321:**1–7.

96. **Thompson, P., and W. M. McShan.** 2003. Bacteriophage regulation of mismatch repair in *Streptococcus pyogenes* SF370. Abstr. 103rd Annu. Meet. Am. Soc. Microbiol. 2003. American Society for Microbiology, Washington, D.C.

97. **Thompson, J. D., D. G. Higgins, and T. J. Gibson.** 1994. CLUSTAL W: improving the sensitivity of progressive multiple sequence alignment through sequence weighting, positions-specific gap penalties and weight matrix choice. *Nucleic Acids Res.* **22:**4673–4680.

98. **Ubukata, K., M. Konno, and R. Fujii.** 1975. Transduction of drug resistance to tetracycline, chloramphenicol, macrolides, lincomycin and clindamycin with phages induced from *Streptococcus pyogenes. J. Antibiot.* **28:**681–688.

99. **Ventura, M., S. Foley, A. Bruttin, S. C. Chennoufi, C. Canchaya, and H. Brussow.** 2002. Transcription mapping as a tool in phage genomics: the case of the temperate *Streptococcus thermophilus* phage Sfi21. *Virology* **296:**62–76.

100. **Ventura, M., C. Canchaya, M. Kleerebezem, W. M. de Vos, R. J. Siezen, and H. Brussow.** 2003. The prophage sequences of *Lactobacillus plantarum* strain WCFS1. *Virology* **316:**245–255.

101. **Wagner, P. L., and M. K. Waldor.** 2002. Bacteriophage control of bacterial virulence. *Infect. Immun.* **70:**3985–3993.

102. **Wannamaker, L. W., S. Skjold, and W. R. Maxted.** 1970. Characterization of bacteriophages from nephritogenic group A streptococci. *J. Infect. Dis.* **121:**407–418.

103. **Wannamaker, L. W., S. Almquist, and S. Skjold.** 1973. Intergroup phage reactions and transduction between group C and group A streptococci. *J. Exp. Med.* **137:** 1338–1353.

104. **Weeks, C. R., and J. J. Ferretti.** 1984. The gene for type A streptococcal exotoxin (erythrogenic toxin) is located in bacteriophage T12. *Infect. Immun.* **46:**531–536.

105. **Weeks, C. R., and J. J. Ferretti.** 1986. Nucleotide sequence of the type A streptococcal exotoxin gene (erythrogenic toxin) from *Streptococcus pyogenes* bacteriophage T12. *Infect. Immun.* **52:**144–150.

106. **Williams, K. P.** 2002. Integration sites for genetic elements in prokaryotic tRNA and tmRNA genes: sublocation preference of integrase subfamilies. *Nucleic Acids Res.* **30:**866–875.

107. **Yu, C.-E.** 1990. Molecular characterization of new *speA* gene-containing bacteriophages and epidemiologic analysis of erythrogenic toxin genes among clinical group A streptococcal strains. Ph.D. dissertation. University of Oklahoma, Oklahoma City, Okla.

108. **Yu, C.-E., and J. J. Ferretti.** 1989. Molecular epidemiologic analysis of the type A streptococcal exotoxin (erythrogenic toxin) gene (*speA*) in clinical *Streptococcus pyogenes* strains. *Infect. Immun.* **57:**3715–3719.

109. **Yu, C. E., and J. J. Ferretti.** 1991. Frequency of the erythrogenic toxin B and C genes (speB and speC) among clinical isolates of group A streptococci. *Infect. Immun.* **59:**211–215.

110. **Yu, C.-E., and J. J. Ferretti.** 1991. Molecular characterization of new group A streptococcal bacteriophages containing the gene for streptococcal erythrogenic toxin A (*speA*). *Mol. Gen. Genet.* **231:**161–168.

111. **Zabriskie, J. B.** 1964. The role of temperate bacteriophage in the production of erythrogenic toxin by group A streptococci. *J. Exp. Med.* **119:**761–779.

112. **Zabriskie, J. B., S. E. Read, and V. A. Fischetti.** 1972. Lysogeny in streptococci, p. 99–118. *In* L. W. Wannamaker and J. M. Matsen (ed.), *Streptococci and Streptococcal Diseases: Recognition, Understanding, and Management.* Academic Press, New York, N.Y.

Molecular Epidemiology, Ecology, and Evolution of Group A Streptococci

DEBRA E. BESSEN AND SUSAN K. HOLLINGSHEAD

12

Within a bacterial species, there are often strains that differ from another in important biological properties. This is certainly the case for *Streptococcus pyogenes* (i.e., group A streptococci; GAS), whose members can cause a wide variety of human diseases, yet there does not appear to exist a single omnipotent strain. To better understand strain differences and their relevance to human disease, stable markers have been identified within organisms of this species. Epidemiological markers are useful for investigating outbreaks of disease, and they can also provide a reference point for deciphering the genetic organization of a bacterial population. The epidemiology of a microbial disease is often, in large part, a reflection of the ecology and evolution of the causative agent.

SEROLOGICAL MARKERS

During the 1920s, Dr. Rebecca Lancefield began work aimed at understanding the basis for protective immunity to GAS infection. Antibodies raised to extractable surface antigens, called M proteins, led to opsonophagocytosis of the strain from which the M protein was derived (44). However, the antibodies directed to the M protein of one organism often failed to protect against many other GAS isolates tested. A serological typing scheme arose through the development of antibodies directed to M proteins of different isolates. More than 80 distinct M-types have now been identified, and protective immunity to GAS infection, as measured in vitro, is M-type-specific.

Subsequent analysis of M proteins demonstrated that they form hairlike fibrils that extend about 60 nm from the surface of the bacterial cell. The M and M-like proteins are extremely complex in both structure and function. The determinants of serological type lie at the amino-termini (i.e., distal fibril tips) of many, but not all M or M-like proteins. A review of M protein structure is presented in chapter 2 of this book.

M protein serotyping is often concordant with a second typing scheme that can be used for a major subgroup of GAS: those that produce opacity factor (OF). OF is a protein, either bound to the cell surface or secreted, that displays apolipoproteinase and fibronectin-binding activities (62). OF-typing is based on neutralization of apolipoproteinase activity by specific antibody, and it correlates with M-type to a large extent, although there are strains of a single M-type that are associated with different OF-types, and vice versa (2). The genes encoding M protein (*emm*) and OF (*sof*) map ~12 kb apart on the GAS chromosome, and both genes are transcriptionally regulated by the product of the *mga* gene (35, 54).

A third serological typing scheme is based on the T-antigen, defined by its resistance to trypsin. There appear to be far more distinct M-types than T-types. The *tee* gene of a T6-type strain has been mapped to a highly recombinatorial region of the GAS genome, positioned ~300 kb from the *emm*⁻ and *sof*⁻ regions (5, 68). However, the genetic basis for most other T-antigen types remains unknown.

With the M-typing scheme in place, decades of epidemiological studies revealed that some serotypes of GAS have a strong tendency to cause throat infection, but not skin infection, and similarly, other serotypes are often seen with impetigo, but much less often with pharyngitis (10, 78). This observation gave rise to the concept of distinct throat and skin types. Similarly, only certain M-types were frequently observed in association with outbreaks of rheumatic fever (rheumatogenic types), and acute glomerulonephritis could be attributed to a small subset of M-types (nephritogenic types) as well (60). Thus, the M serotypic markers identify organisms as having a strong tendency to cause some disease, but not all disease, suggesting that there exists a degree of specialization among strains belonging to this species.

ECOLOGY AND PRIMARY INFECTION

GAS are free-living organisms; however, their ecological niche appears to be quite narrow—their only known biological host is the human. The primary sites for colonization of GAS involve two tissues: the nasal and oropharyngeal mucosal epithelium of the upper respiratory tract (URT) and the superficial layers of the epidermis. Whether the surface of unbroken skin allows GAS to colonize and reproduce or if tiny breaks in the stratum corneum are required

Gram-Positive Pathogens, 2nd edition, edited by Vincent A. Fischetti et al.
© 2006 ASM Press, Washington, D.C.

for gaining a foothold is unclear. GAS have no known environmental reservoir, and transmission is almost always from person to person. Thus, the throat and skin of the human host represent the primary habitats for GAS. It is at these two tissue sites that GAS can undergo successful reproductive growth and transmission to the next host.

Once colonization is established in the URT, the organism can cause a symptomatic pharyngitis, which can be cleared in two weeks by protective antibody that is mounted by the host in response to infection. Alternatively, a clinically inapparent infection can arise whereby the host lacks obvious clinical symptoms of illness, yet there is a significant immune response. A third possibility is asymptomatic carriage, in which a vigorous host immune response to many GAS antigens is lacking (39). The carrier state can persist for months, and the organism remains capable of transmission to a new host. However, relative to acute pharyngitis, the number of GAS present during carriage may be reduced and GAS are often in altered state, signified by a relative lack of M protein expressed on the surface.

When GAS enter a prolonged stationary phase in vitro, there is decreased synthesis of several virulence factors, including M protein and capsule (45). Similar conditions leading to the down-regulation of virulence factors might also apply to the long-term survival of GAS during asymptomatic carriage and when residing intracellularly. Intracellular invasion of epithelial cells by GAS may contribute to persistent infection, and may also be a prerequisite for gaining access to the bloodstream, whereupon it can cause severe invasive disease (13). The relative contribution of cell wall-deficient L forms of GAS to persistent infection, if any, is not understood.

At the skin, GAS survive and replicate just below the dry keratinized layer, provoking a strong inflammatory response that gives rise to the purulent lesions of impetigo (i.e., pyoderma). It is not clear whether there exists a true carrier state at the skin, and the uncertainty appears to stem from the duration of bacterial colonization at the skin in the absence of inflammation (i.e., disease), which may be only transient (47). Like GAS, the scabies mite infests the superficial layers of the skin, just below the stratum corneum; scabies can cause intense itching and has been implicated as a risk factor in the development of streptococcal pyoderma in many tropical communities (49, 60).

The relative incidence of GAS disease varies throughout the world, in accordance with both season and locale. In the temperate regions of North America and Europe, pharyngitis is highly prevalent during the winter months and impetigo (although less common) is most often encountered during warmer weather. In many tropical regions, such as the Northern Territory of Australia, GAS impetigo is far more common than pharyngeal infection among the aboriginal population, and there are no discrete seasonal peaks in incidence of disease (49). Climate and the level of hygiene strongly influence the prevalence of impetigo, whereas crowding is a major risk factor for nasopharyngeal infection.

Streptococcal pharyngitis and impetigo are superficial, self-limiting infections that usually cause only a mild illness. The reason why GAS are a major public health concern is because of their ability to cause severe invasive disease and trigger nonsuppurative sequelae of an autoimmune-like nature. In severe invasive disease, such as necrotizing fasciitis, GAS gain access to deep tissue that is normally sterile. Although invasive disease is associated with high morbidity and mortality, its frequency of occurrence is low. Population-based surveys conducted during the 1990s estimated

the annual incidence of invasive GAS disease at <10 cases per 100,000 (16). This is in sharp contrast to non-life-threatening GAS infections, where between the ages of five and seven years, an average of half of all children experience one GAS infection annually (11). The superficial nature of pharyngitis and impetigo provides the bacterium with an easy exit for transmission to new hosts.

STRUCTURE AND EVOLUTION OF *emm* GENES

Bacteria encounter new environments whenever they infect a new host, infect the same host at a new tissue site, or the local ecology changes as a consequence of the ongoing host-pathogen interaction. During the course of infection, there are often shifts in the availability of nutrients and in the innate and specific immune response. Environmental signals can be sensed by the bacterium and transduced into alterations of gene expression. The ability of a single strain to survive and reproduce in a host environment, such as the throat or skin, suggests that it has both the requisite genes and the mechanisms for regulating their expression.

The ecology of the oropharyngeal mucosa and epidermal tissue is different in many ways. Adaptation of a bacterium to a new environment evolves through genetic change. A reproductive advantage (i.e., increased fitness) by individual progeny of a large bacterial population leads to their natural selection. Over the long time scale of evolution, alleles conferring a strong biological advantage under particular ecological conditions can achieve fixation, giving rise to a subpopulation of specialists. The finding of strong tissue site preferences for GAS of many M protein serotypes is indicative of genotypic differences between the so-called throat and skin strains, and suggests that there exist genes that are responsible for conferring these adaptations.

Tracing the evolutionary history of the genes encoding serological markers provides a starting point for trying to better understand the genetic basis for the biological properties that correlate with serological type. This approach is perhaps more meaningful for GAS than it is for some other bacterial species because the M-serotypic markers are the translational products of genes, rather than the biosynthetic products of an enzymatic process, as is the case for the polysaccharide capsular serotypes of pneumococci and the lipopolysaccharide O-antigen serotypes of *Enterobacteriaceae*.

Nearly all GAS have one, two, or three *emm* or *emm*-like genes arranged in tandem on the chromosome, and they are usually 200 to 300 bp apart from one another (Fig. 1) (28, 32, 59). The multiple *emm* and *emm*-like genes of a single isolate differ from one another in sequence. The *emm* gene family is characterized by high levels of sequence identity for portions encoding the leader peptide (extreme 5′ end) and cell-associated region (3′ end). To understand the evolutionary history of the *emm* gene family, it is useful to separately analyze the appropriate *emm* gene segments, because recombinational events involving *emm* genes can obscure their phylogenetic relationships.

About two-thirds of the mature protein products of *emm* and *emm*-like genes are exposed on the bacterial cell surface, with the amino-terminus lying at the distal fibril tip. The determinants of M-serotype are encoded by the 5′ end of the central *emm* gene, just downstream from the leader peptide-encoding region. A phylogenetic tree based on the 5′-end nucleotide sequence encoding serotypic determinants has been constructed (80). However, phylogenetic re-

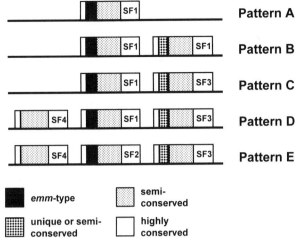

FIGURE 1 Structural features of *emm* and *emm*-like genes of GAS. Each large box represents an *emm* or *emm*-like gene, which are arranged in tandem on the chromosome, usually 200 to 300 bp apart. The extent of nucleotide sequence homologies among subregions within *emm* and *emm*-like genes ranges from highly conserved (open; for regions encoding 5′-end leader peptide and 3′-end cell-associated domains) to highly variable (filled; for regions encoding the determinants of M-serotype). SF forms of the peptidoglycan-spanning coding regions are indicated; five major patterns (A through E) for the content and arrangement of *emm* and *emm*-like SF gene forms are shown. Other nomenclature that is often used in reference to *emm*-like genes includes *mrp* or *fcrA* (for the SF4 *emm* gene), and *enn* or *emmL* (for the downstream *emm*-like genes of patterns B through E).

lationships are difficult to infer from 5′ end-based *emm* gene trees because the extensive sequence heterogeneity casts uncertainty as to the exact position of many branches. The recent implementation of a widely used *emm*-sequence typing scheme, which closely corresponds to serological typing, has led to the identification of >160 *emm*-types exhibiting <95% sequence identity at their 5′ ends (1, 21).

Protective antibodies produced by the host and directed to the M fibril tips exert strong negative selection pressure against immunoreactive M-types. Thus, genetic change leading to the loss of critical antigenic epitopes can be favorable, and provide the mutant with a selective advantage and the opportunity for immune escape (4, 19). The *emm* genes appear to be highly mutable loci, displaying an increased frequency of random mutations that promote phenotypic variation. Many *emm*-subtypes, defined by small differences in amino acid sequence near the amino-terminus, have been described (48). The molecular mechanisms underlying genetic change in *emm* may be similar to strategies used by many bacterial virulence genes for adaptive evolution in the face of shifting environmental conditions (55). Frameshift and compensatory mutations near the 5′ end are one mechanism for generating antigenic diversity in M proteins, as are short in-frame deletions and duplications (27, 29, 31, 65). The frequency of genetic change may be facilitated by short stretches of sequence repeats or polypyrimidine tracts. An accumulation of spontaneous point mutations, by itself, might also provide enough antigenic change for positive selection. In contrast to immune escape, the localization of a common binding site for complement C4b-

binding protein at the amino-terminal ends of M protein molecules derived from several serotypes (37) suggests that functional constraints also shape the natural selection of *emm*-sequence types.

Immediately downstream from the highly heterogeneous 5′ ends of the central *emm* gene are semiconserved regions of discrete blocks of nucleotide sequence that are shared among a few to many distinct M-serotypes (Fig. 1). The flanking *emm*-like genes also have semiconserved regions encompassing the bulk of their surface-exposed encoded portion. High-affinity binding activities for a variety of human-derived tissue and plasma proteins, such as components of the complement and coagulation-fibrinolytic pathways, have been mapped to products of several of the semiconserved domains. Some of the discrete domains are actually a series of DNA repeats. The domains give rise to a mosaic-like arrangement that is highly suggestive of extensive recombinational events (72). Recombination can mask phylogenetic relationships, and the evolutionary history of *emm* genes is not readily apparent from the domain structure. However, there does appear to be some degree of nonrandomness within the central *emm* gene region, because certain domains have a high tendency to occur in conjunction with other domains (7, 73). The need to bind specific tissue and plasma components of the human host will presumably shape the content of the binding domains found within a given strain. Recombinational events leading to new mosaic combinations appear to be less frequent than changes at the 5′ ends, and may occur only on a longer term evolutionary scale.

Within-host genetic change of the infecting GAS population is most apparent in the form of deletion and duplication of blocks of DNA sequence repeats contained within *emm* genes, occurring by homologous recombination and/or DNA slipped-strand mispairing (24, 31). These genetic events are evidenced in sequential isolates collected from the oropharynx of long-term carriers in the preantibiotic era. Isolates collected from various individuals involved in outbreaks of short duration in institutional settings (e.g., military, child daycare) display deletion and duplication of sequence repeats within their *emm* genes. A loss of antigenic epitopes that are targets of protective antibody is one consequence of these genetic changes, perhaps by inducing conformational alterations that affect the amino-terminal end (38).

In contrast to the portion of *emm* genes encoding for the surface-exposed part of the M protein molecule, the 3′ end is relatively conserved among different M-serotypes. A comparison of the nucleotide sequences of the 3′ ends of *emm* and *emm*-like genes, encoding for the cell-associated regions of the M and M-like proteins, indicates that they constitute four major phylogenetic lineages; presumably, each lineage arose through duplication and subsequent divergence of an ancestral *emm* gene (32, 33). The sequence divergence observed at the 3′ ends is largely contained within the peptidoglycan-spanning domain-coding region, which is flanked by highly conserved sequences both upstream and downstream, corresponding to the coding regions for the putative group A carbohydrate- and cytoplasmic membrane-spanning domains, respectively. The four principal peptidoglycan-spanning domain forms differ not only in amino acid sequence, but also in their lengths, ranging from 36 to 55 amino acid residues. The four phylogenetic lineages each represent a subfamily (SF) of *emm* genes, designated SF1 through SF4. Conceivably, distinct peptidoglycan structures are required in order to accommodate the assembly of each of the *emm* SF gene products on the cell surface.

ECOLOGICAL MARKERS

The content and relative chromosomal arrangements of *emm* SF genes are found to exist in five basic patterns (A through E), with very few exceptions (Fig. 1). Pattern A, D, and E strains are the most common. In nearly all instances, isolates of the same *emm*-type share the same *emm* pattern grouping (53). Pattern A–C, D, and E strains account for 17, 39, and 41%, respectively, of the >150 *emm*-types examined. Despite their importance in GAS disease in the United States and western Europe, the pattern A–C subpopulation of strains accounts for only a small percentage of the total *emm*-types.

A strong correlation between *emm* pattern and tissue site of isolation was first shown for a genetically diverse set of isolates (6). *emm* pattern A–C strains have a high tendency to be recovered from pharyngitis, whereas pattern D strains are usually isolated from impetigo lesions. As a group, pattern E strains are readily found at both tissues. These correlations have been confirmed in several population-based studies. Reasonable inferences can be made on *emm* pattern based on the *emm*-type (53). Nearly all pharyngitis isolates (>99%) collected from hospitals in Rome, Italy, are of *emm*-types typically found in patterns A–C or E, whereas <1% are of *emm*-types associated with pattern D (17). Similarly, in a recent report by Shulman et al. on >1,900 GAS isolates collected from cases of pharyngitis in the United States over 2 years (70), 53% were of *emm*-types typical of pattern A–C strains, 47% were typical of pattern E, and <1% were typical of pattern D. In a rural aboriginal community in tropical Australia where impetigo is hyperendemic, no cases of pharyngitis were detected during a 25-month surveillance period; of the impetigo isolates recovered, 46 and 41%, respectively, were *emm* pattern D or E (9). Although differences in *emm* pattern distribution among throat and skin isolates are highly significant by statistical tests, the associations between *emm* patterns A–C and the throat, and *emm* pattern D and the skin, are not absolute. Nevertheless, the data suggest that pattern A–C and D strains are niche specialists, having strong preference for just one tissue site. Pattern E strains are readily recovered from both tissues and are considered generalists. Interestingly, nearly all pattern E strains are OF producers, and conversely, nearly all OF producers are pattern E, suggesting that their distinction as a unique subpopulation extends beyond their *emm* chromosomal arrangement.

The existence of stable *emm* genotypes associated with strong tissue site preferences allows one to address several interesting issues related to horizontal gene transfer (HGT) and speciation.

GENETIC VARIATION AND EVOLUTION

Evolution is a two-step process: genetic change followed by natural selection. There are multiple sources of genetic variation in GAS. These include point mutation, slipped-stranded DNA mispairing, and numerous forms of recombination, both homologous and site-specific. Notably, GAS are loaded with bacteriophage, and some strains possess conjugative transposons and plasmids (3, 66). There is suggestive evidence that GAS can undergo natural transformation (30). It is less clear whether GAS undergo homology-directed illegitimate recombination, whereby very short stretches of homology can lead to insertion of foreign DNA, as observed for *Streptococcus pneumoniae* (61). High-frequency genetic changes are observed in the gene encoding the Sic protein, as evidenced by intrahost variation (51). Within-host genetic changes are also observed for the *emm* gene, involving slipped-stranded mispairing of DNA encoding blocks of sequence repeats (31). Genes encoding the fibronectin-binding protein F1 and the collagenlike protein SclB are also rich in repetitive sequences (63, 71).

Multilocus sequence typing (MLST) based on housekeeping (HK) loci is a powerful tool used to understand genetic relationships among strains of a bacterial species. HK genes encode proteins vital to the cell, such as metabolic enzymes; they are used because they are (assumed to be) selectively neutral. Multiple loci are used because no single locus can accurately reflect the ancestral relationships between strains if recombination rates are high. MLST schemes have been developed, or are currently under development, for >25 species of medically important microbial pathogens (http://www.mlst.net). For MLST of GAS, the nucleotide sequence is determined for an internal portion of seven HK loci (18).

Bacterial species vary widely in their relative rates of genetic recombination, arising from HGT between strains. Recombination can be measured by statistical tests for congruency between HK gene tree topologies (23). Congruent trees signify low recombination rates; incongruent trees indicate that recombination has been sufficiently frequent to eliminate the phylogenetic signal. For GAS, there is a complete lack of HK gene tree congruency, and evolution occurs by a netlike process. Along with *S. pneumoniae*, GAS are the most highly recombinogenic bacteria among six species examined (23).

Niche differentiation is an early key step in speciation (15). Sequence divergence among neutral HK genes can be useful for identifying discrete ecological populations. Pattern A–C and D strains, representing the throat and skin specialist subpopulations, can be regarded as distinct ecological populations because they exploit different niches. However, despite some niche separation created by distinct epidemiological trends and innate tissue tropisms, there is no evidence for neutral HK gene divergence between strains of differing *emm* patterns, or between isolates known to be recovered from the throat versus impetigo lesions. HGT between throat and skin specialists may occur during throat carriage or occasional coinfections (12). Alternatively, the pattern E generalist subpopulation may serve as a genetic shuttle between pattern A–C and D strains (8).

The strong linkage observed between *emm* chromosomal pattern and tissue site preference, against a background of random associations between HK alleles, suggests that the *emm* gene products themselves (or physically linked genes) have a direct role in conferring host tissue tropisms.

Despite high levels of recombination among GAS, the relationship between *emm*-type and clone is stable over some spatial scales. A one-to-one relationship was found between nearly all *emm*-types and genotypes defined by MLST for 137 GAS isolates collected from a remote aboriginal island community in tropical Australia (52). When these isolates are compared to >200 isolates obtained from elsewhere in the world, there are many examples of isolates from aboriginal and non-Australian subjects sharing the same *emm*-type, but in about 50% of *emm*-types, the multilocus genotypes of isolates of the same *emm*-type from the two regions are very different. A single *emm*-type may typically define a single clone within the United States (18), and on the remote island (52), but in many cases their genetic backgrounds are drastically different from one another. This finding has implications for attempts to make global associations between *emm*-types and certain disease manifestations.

A striking feature of GAS is their lack of pathogenicity islands (PAI), a subset of genomic islands (large blocks of DNA acquired from another species) that can confer a "quantum leap in phenotype." Several other important bacterial pathogens also have a dearth of PAIs, and there appears to be a strong inverse relationship between the number of PAIs (based on the literature) (67) and the relative degree of recombination (based on congruency of HK gene tree topologies and other measures) (22, 23, 77). Acquisition of a PAI is probably a very rare event. It is possible that GAS acquired PAIs deep in their evolutionary history, but they were quickly broken up by recombination and are no longer recognized as such.

Perhaps the closest genetic relative of GAS is the human commensal-like group C and G streptococcus (GCS/GGS; *Streptococcus dysgalactiae* subsp. *equisimilus*). Although a common inhabitant of the URT, on occasion GCS/GGS are recovered in association with disease. There is strong evidence that the gene encoding streptokinase, a plasminogen activator, was acquired from GCS/GGS by some strains of GAS following interspecies HGT (41). The degree to which orthologous gene replacements shape the population biology of GAS remains largely unexplored.

An important evolutionary advantage of recombination is that it allows for quick exploration of a wide array of genotypes and phenotypes. The genome plasticity evidenced by high levels of recombination may provide the organism with a selective advantage under a wide variety of environmental conditions.

SELECTION PRESSURES AND TRANSMISSION DYNAMICS

For a bacterial species whose world is largely confined to the human population, the selective pressures that most profoundly shape its genetic structure are intrinsic to the human condition. Local microenvironmental conditions can shift during the course of infection within a single host. Also, ecological conditions can vary widely from host to host, and not all exposed hosts are susceptible to infection by a given strain. In general terms, the basic reproductive rate (i.e., fitness) of an organism is directly proportional to its successful transmission to new hosts (46). From the evolutionary standpoint, invasive disease is a dead end for the infecting GAS organism because the severely ill patient becomes immobilized; thus, opportunities for its transmission to new hosts are diminished. If virulence is too high and the primary host is killed off quickly, leaving insufficient time for transmission to a new host, then the bacterium will die along with the infected host.

In the absence of a GAS vaccine, all protective immunity is acquired through prior infection. Following infection with at least some GAS strains, protective immunity appears to be lifelong (43). Presumably, circumstances exist in which an organism undergoes a genetic change that leads to immune escape during the course of an infection. However, there are few, if any, well-documented examples of this occurring, perhaps because it is not a particularly high-frequency event and infected patients are promptly treated. Signatures of host immune selection may lie in the high

levels of amino acid sequence polymorphism that are observed for several surface proteins, such as M protein, OF, and protein F1 (2, 74).

A distinguishing feature of GAS that is also shared with *Staphylococcus aureus* lies in the diversity of the T-lymphocyte superantigens (SAgs) that are produced by different strains. For GAS, more than 10 distinct molecular forms exhibit SAg activity, leading to the release of inflammatory cytokines by the host. Many SAg genes are encoded by bacteriophage; thus, there is extensive strain-to-strain variation in SAg content. SAgs are highly specific in their binding to different forms of both the HLA class II molecule and the β portion of the T-cell receptor. Of course, the human host population displays tremendous diversity in HLA class II haplotype and also in the relative amounts of each β form. In severe systemic GAS disease, certain HLA class II haplotypes confer strong protection from disease, whereas other haplotypes increase the risk of severe disease (42). It remains unknown whether GAS SAgs can provide an advantage to the bacterium during primary infection at the URT or skin, and whether this capacity varies in accordance with the genetic susceptibility of the host. The diversity in SAg structural forms and their specificities appears to be a footprint of a strong selective pressure.

Bacteriocin-like inhibitory substances produced by *Streptococcus salivarius*, a commensal of the human oral cavity, are toxic to at least some strains of GAS (75). *S. salivarius* produces a linear peptide bacteriocin (salivaricin A) that leads to bacteriostatic growth inhibition of GAS in vitro. *S. salivarius* can also produce a second lantibiotic-type bacteriocin (salivaricin B) that has bactericidal activity towards GAS. Thus, the composition of the normal flora may selectively influence GAS survival. Bacteriocin-like genes have also been identified within the GAS genome, such as streptolysin S, a cytolysin known to target mammalian cells (57). Direct killing between GAS and other bacterial species, and between different GAS strains, is a potentially important factor in shaping the population genetic structure of GAS via intense direct competition.

Antibiotics can act as a potent selective force. Penicillin is the drug of choice for treating most GAS infections; however, resistance to penicillin is not known to have occurred in a natural setting (34). Macrolides and lincosamides are the primary treatment for GAS infections in patients with β-lactam hypersensitivity or chronic, recurrent pharyngitis due to prior treatment failure (40). Resistance to these drugs is on the rise. In recent years, resistance to macrolides in GAS has exceeded 30% in some locales (50). Perhaps not surprising, there is a strong correlation between macrolide consumption and resistance in GAS (25, 26, 69). The three primary resistance genes found in GAS—*erm(B)*, *erm(A)*, *mef(A)*—are carried by mobile genetic elements. Extensive genetic diversity is observed among macrolide-resistant clones of GAS (64).

Although there is no true resistance to penicillin, treatment failures do occur. GAS recovered from the URT from cases of antibiotic treatment failure disproportionately harbor *prtF1*, encoding the fibronectin-binding Protein F1 that facilitates intracellular invasion and thereby promotes a carrier state (56). Presumably, intracellular localization protects GAS from β-lactam action (58). In vitro studies show that the ability of GAS to persist in the throat following antibiotic therapy corresponds with their capacity to adhere to and be internalized by epithelial cells.

The proportion of *prtF1*-positive GAS is significantly higher for macrolide-resistant than for macrolide-sensitive isolates (20). In San Francisco, macrolide resistance is even more common in GAS isolates recovered from patients with invasive disease than cases of pharyngitis (81). Because invasive disease most likely originates from a clinically inapparent URT infection, together the findings support the idea that macrolide-resistant GAS are highly adept at causing a prolonged, inapparent infection that on occasion can evolve into serious invasive disease. Clindamycin is the drug of choice for the treatment of severe systemic GAS disease because it blocks exotoxin synthesis and has good tissue penetration. A failure to block exotoxin synthesis following clindamycin treatment could lead to increased morbidity and mortality due to soft tissue infection by macrolide-resistant GAS.

Human behavior is a factor in the evolution of microorganisms. For the past 50 years, the GAS infection that triggers symptoms of sore throat and high fever is quickly eradicated in most of the developed world, because patients actively seek medical treatment in the form of antibiotic therapy. But do variants of GAS exist that are of lowered virulence because of a diminished capacity to cause overt clinical symptoms, and if so, does the low-virulence phenotype provide a selective advantage that allows for their emergence as highly prevalent clones? Only if the symptoms of acute pharyngitis specifically enhance the transmission of GAS to a new host (i.e., virulence and transmission are positively coupled) will the advantage of evading near certain death by antibiotics be counterbalanced by a decrease in transmission. In several recent large outbreaks of rheumatic fever in communities across the United States, a substantial proportion of cases was associated with clinically inapparent GAS infection that went unnoticed and untreated in the weeks prior to acute autoimmune attack (76a).

During a 10-week period in early 1995, there was a confined outbreak of seven cases of severe invasive disease attributed to a single clone of M-type 3 GAS. Six of the seven patients resided in one of two communities in southeastern Minnesota, and five had underlying medical conditions or predisposing factors (14). The M3 clone accounted for 78% of the GAS detected through screening elementary school-children for pharyngeal carriage (92% participated); in all, 25% of the children harbored the M3 clone in the absence of any illness that would have otherwise prevented them from attending school (14). In contrast, among GAS-associated sore throat cases identified in community clinics, only 28% were identical to the M3 clone. The findings on asymptomatic carriage suggest that the M3 clone is highly prevalent and highly transmissible. Although the M3 clone can give rise to acute pharyngitis, it was not nearly as well represented among pharyngitis cases as compared to cases of invasive disease and carriage. Is the M3 clone best described as a "super killer" or as a "super transmitter"? Are the high levels of virulence (i.e., invasive disease) strictly dependent on host factors? Is its transmission largely independent of acute pharyngitis?

The answers to these questions are not fully understood. In a more recent report, an invasive index value was calculated, defined as the ratio of isolates of a given *emm*-type from cases of invasive disease to cases of pharyngitis (48, 70). Each set of disease-associated isolates was collected by population-based surveillance at multiple sites throughout the United States. As in the Minnesota outbreak, *emm*-type 3 isolates had an invasive index value >1; however, several *emm*-types associated with a high proportion of invasive disease cases had indices <1. The *emm*-type associated with the highest proportion of invasive disease cases (*emm1*, 18.2%) had an invasive index value close to 1 (0.96), indicating that it accounts for a nearly identical proportion of invasive and pharyngitis isolates. Missing from the assessment is a complete picture of the prevalence of each clone in the community. Because GAS can be transmitted to new hosts in the absence of overt disease, the number of hosts with asymptomatic carriage and clinically inapparent infection should be considered. Knowledge of the level of exposure of the total host population to GAS is essential for a complete understanding of the epidemiological patterns, and for the optimal design of therapeutic and intervention strategies directed to problematic clones.

There may also be intrinsic properties of the bacterium that can enhance its rate of transmission. For example, it has been proposed that expression of the fibronectin-binding protein F1 facilitates colonization during the earliest stage of GAS throat infection (76). In numerous epidemics of GAS pharyngitis during which there was a high rate of person-to-person transmission, the causative agent was often found to be highly enriched for hyaluronic acid capsule (36, 79). Overall, we have incomplete knowledge of the biological requirements that must be met by the GAS for their successful journey into their next human incubator.

REFERENCES

1. **Beall, B.** 2004. http://www.cdc.gov/ncidod/biotech/strep/emmtypes.htm.

2. **Beall, B., G. Gherardi, M. Lovgren, B. Forwick, R. Facklam, and G. Tyrrell.** 2000. Emm and sof gene sequence variation in relation to serological typing of opacity factor positive group A streptococci. *Microbiology* **146:** 1195–1209.

3. **Beres, S. B., G. L. Sylva, K. D. Barbian, B. Lei, J. S. Hoff, N. D. Mammarella, M. Y. Liu, J. C. Smoot, S. F. Porcella, L. D. Parkins, D. S. Campbell, T. M. Smith, J. K. McCormick, D. Y. Leung, P. M. Schlievert, and J. M. Musser.** 2002. Genome sequence of a serotype M3 strain of group A *Streptococcus*: phage-encoded toxins, the high-virulence phenotype, and clone emergence. *Proc. Natl. Acad. Sci. USA* **99:**10078–10083.

4. **Beres, S. B., G. L. Sylva, D. E. Sturdevant, C. N. Granville, M. Y. Liu, S. M. Ricklefs, A. R. Whitney, L. D. Parkins, N. P. Hoe, G. J. Adams, D. E. Low, F. R. DeLeo, A. McGeer, and J. M. Musser.** 2004. Genome-wide molecular dissection of serotype M3 group A Streptococcus strains causing two epidemics of invasive infections. *Proc. Natl. Acad. Sci. USA* **101:**11833–11838.

5. **Bessen, D. E., and A. Kalia.** 2002. Genomic localization of a T-serotype locus to a recombinatorial zone encoding for extracellular matrix-binding proteins in *Streptococcus pyogenes. Infect. Immun.* **70:**1159–1167.

6. **Bessen, D. E., C. M. Sotir, T. L. Readdy, and S. K. Hollingshead.** 1996. Genetic correlates of throat and skin isolates of group A streptococci. *J. Infect. Dis.* **173:**896–900.

7. **Bessen, D. E., M. W. Izzo, E. J. McCabe, and C. M. Sotir.** 1997. Two-domain motif for IgG-binding activity by group A streptococcal emm gene products. *Gene* **196:** 75–82.

8. **Bessen, D. E., M. W. Izzo, T. R. Fiorentino, R. M. Caringal, S. K. Hollingshead, and B. Beall.** 1999. Genetic linkage of exotoxin alleles and emm gene markers for tissue tropism in group A streptococci. *J. Infect. Dis.* **179:** 627–636.

9. **Bessen, D. E., J. R. Carapetis, B. Beall, R. Katz, M. Hibble, B. J. Currie, T. Collingridge, M. W. Izzo, D. A. Scaramuzzino, and K. S. Sriprakash.** 2000. Contrasting molecular epidemiology of group A streptococci causing tropical and non-tropical infections of the skin and throat. *J. Infect. Dis.* **182:**1109–1116.

10. **Bisno, A. L., and D. Stevens.** 2000. *Streptococcus pyogenes* (including streptococcal toxic shock syndrome and necrotizing fasciitis), p. 2101–2117. In G. L. Mandell, R. G. Douglas, and R. Dolin (ed.), *Principles and Practice of Infectious Diseases*, 5th ed., vol. 2. Churchill Livingstone, Philadelphia, Pa.

11. **Breese, B. B., and C. B. Hall.** 1978. *Beta Hemolytic Streptococcal Diseases.* Houghton Mifflin, Boston, Mass.

12. **Carapetis, J., D. Gardiner, B. Currie, and J. D. Mathews.** 1995. Multiple strains of *Streptococcus pyogenes* in skin sores of Aboriginal Australians. *J. Clin. Microbiol.* **33:**1471–1472.

13. **Cleary, P. P., L. McLandsborough, L. Ikeda, D. Cue, J. Krawczak, and H. Lam.** 1998. High-frequency intracellular infection and erythrogenic toxin A expression undergo phase variation in M1 group A streptococci. *Mol. Microbiol.* **28:**157–167.

14. **Cockerill, F. R., K. L. MacDonald, R. L. Thompson, F. Roberson, P. C. Kohner, J. Besser-Wiek, J. M. Manahan, J. M. Musser, P. M. Schlievert, J. Talbot, B. Frankfort, J. M. Steckelberg, W. R. Wilson, and M. T. Osterholm.** 1997. An outbreak of invasive group A streptococcal disease associated with high carriage rates of the invasive clone among school-aged children. *JAMA* **277:**38–43.

15. **Cohan, F. M.** 2002. What are bacterial species? *Annu. Rev. Microbiol.* **56:**457–487.

16. **Davies, H. D., A. McGeer, B. Schwartz, K. Green, D. Cann, A. E. Simor, and D. E. Low.** 1996. Invasive group A streptococcal infections in Ontario, Canada. *N. Engl. J. Med.* **335:**547–554.

17. **Dicuonzo, G., G. Gherardi, G. Lorino, S. Angeletti, M. DeCesaris, E. Fiscarelli, D. E. Bessen, and B. Beall.** 2001. Group A streptococcal genotypes from pediatric throat isolates in Rome, Italy. *J. Clin. Microbiol.* **39:**1687–1690.

18. **Enright, M. C., B. G. Spratt, A. Kalia, J. H. Cross, and D. E. Bessen.** 2001. Multilocus sequence typing of *Streptococcus pyogenes* and the relationship between emm-type and clone. *Infect. Immun.* **69:**2416–2427.

19. **Eriksson, B. K. G., A. Villasenor-Sierra, M. Norgren, and D. L. Stevens.** 2001. Opsonization of T1M1 group A Streptococcus: dynamics of antibody production and strain specificity. *Clin. Infect. Dis.* **32:**E24–E30.

20. **Facinelli, B., C. Spinaci, G. Magi, E. Giovanetti, and P. E. Varaldo.** 2001. Association between erythromycin resistance and ability to enter human respiratory cells in group A streptococci. *Lancet* **358:**30–33.

21. **Facklam, R., B. Beall, A. Efstratiou, V. Fischetti, E. Kaplan, P. Kriz, M. Lovgren, D. Martin, B. Schwartz, A. Totolian, D. Bessen, S. Hollingshead, F. Rubin, J. Scott, and G. Tyrrell.** 1999. Report on an international workshop: demonstration of emm typing and validation of provisional M-types of group A streptococci. *Emerg. Infect. Dis.* **5:**247–253.

22. **Falush, D., C. Kraft, N. S. Taylor, P. Correa, J. G. Fox, M. Achtman, and S. Suerbaum.** 2001. Recombination and mutation during long-term gastric colonization by *Helicobacter pylori*: estimates of clock rates, recombination size, and minimal age. *Proc. Natl. Acad. Sci. USA* **98:**15056–15061.

23. **Feil, E. J., E. C. Holmes, D. E. Bessen, M.-S. Chan, N. P. J. Day, M. C. Enright, R. Goldstein, D. Hood, A. Kalia, C. E. Moore, J. Zhou, and B. G. Spratt.** 2001. Recombination within natural populations of pathogenic bacteria: short-term empirical estimates and long-term phylogenetic consequences. *Proc. Natl. Acad. Sci. USA* **98:**182–187.

24. **Fischetti, V. A., M. Jarymowycz, K. F. Jones, and J. R. Scott.** 1986. Streptococcal M protein size mutants occur at high frequency within a single strain. *J. Exp. Med.* **164:**971–980.

25. **Freeman, A. F., and S. T. Shulman.** 2002. Macrolide resistance in group A Streptococcus. *Pediatr. Infect. Dis. J.* **21:**1158–1160.

26. **Fujita, K., K. Murono, M. Yoshikawa, and T. Murai.** 1994. Decline of erythromycin resistance of group A streptococci in Japan. *Pediatr. Infect. Dis. J.* **13:**1075–1078.

27. **Gardiner, D. L., A. M. Goodfellow, D. R. Martin, and K. S. Sriprakash.** 1998. Group A streptococcal Vir types are M protein gene (emm) sequence type specific. *J. Clin. Microbiol.* **36:**902–907.

28. **Haanes, E. J., and P. P. Cleary.** 1989. Identification of a divergent M protein gene and an M protein related gene family in serotype 49 *Streptococcus pyogenes*. *J. Bacteriol.* **171:**6397–6408.

29. **Harbaugh, M. P., A. Podbielski, S. Hugl, and P. P. Cleary.** 1993. Nucleotide substitutions and small-scale insertion produce size and antigenic variation in group A streptococcal M1 protein. *Mol. Microbiol.* **8:**981–991.

30. **Hidalgo-Grass, C., M. Ravins, M. Dan-Goor, J. Jaffe, A. E. Moses, and E. Hanski.** 2002. A locus of group A Streptococcus involved in invasive disease and DNA transfer. *Mol. Microbiol.* **46:**87–99.

31. **Hollingshead, S. K., V. A. Fischetti, and J. R. Scott.** 1987. Size variation in group A streptococcal M protein is generated by homologous recombination between intragenic repeats. *Mol. Gen. Genet.* **207:**196–203.

32. **Hollingshead, S. K., T. L. Readdy, D. L. Yung, and D. E. Bessen.** 1993. Structural heterogeneity of the emm gene cluster in group A streptococci. *Mol. Microbiol.* **8:**707–717.

33. **Hollingshead, S. K., T. Readdy, J. Arnold, and D. E. Bessen.** 1994. Molecular evolution of a multi-gene family in group A streptococci. *Mol. Biol. Evol.* **11:**208–219.

34. **Horn, D., J. Zabriskie, R. Austrian, P. Cleary, J. Ferretti, V. Fischetti, E. Gotschlich, E. Kaplan, M. McCarty, S. Opal, R. Roberts, A. Tomasz, and Y. Wachtfogel.** 1998. Why have group A streptococci remained susceptible to penicillin? Report on a symposium. *Clin. Infect. Dis.* **26:**1341–1345.

35. **Jeng, A., V. Sakota, Z. Y. Li, V. Datta, B. Beall, and V. Nizet.** 2003. Molecular genetic analysis of a group A Streptococcus operon encoding serum opacity factor and a novel fibronectin-binding protein, SfbX. *J. Bacteriol.* **185:**1208–1217.

36. **Johnson, D. R., D. L. Stevens, and E. L. Kaplan.** 1992. Epidemiological analysis of group A streptococcal serotypes associated with severe systemic infections, rheumatic fever, or uncomplicated pharyngitis. *J. Infect. Dis.* **166:**374–382.

37. **Johnsson, E., A. Thern, B. Dahlback, L. O. Heden, M. Wikstrom, and G. Lindahl.** 1996. A highly variable region in members of the streptococcal M protein family binds the human complement regulator C4BP. *J. Immunol.* **157:**3021–3029.

38. Jones, K. F., S. K. Hollingshead, J. R. Scott, and V. A. Fischetti. 1988. Spontaneous M6 protein size mutants of group A streptococci display variation in antigenic and opsonogenic epitopes. *Proc. Natl. Acad. Sci. USA* **85:**8271–8275.

39. Kaplan, E. L. 1980. The group A streptococcal upper respiratory tract carrier state: an enigma. *J. Pediatr.* **97:**337–345.

40. Kaplan, E. L., and D. R. Johnson. 2001. Unexplained reduced microbiological efficacy of intramuscular benzathine penicillin G and of oral penicillin V in eradication of group A streptococci from children with acute pharyngitis. *Pediatrics* **108:**1180–1186.

41. Kapur, V., S. Kanjilal, M. R. Hamrick, L.-L. Li, T. S. Whittam, S. A. Sawyer, and J. M. Musser. 1995. Molecular population genetic analysis of the streptokinase gene of *Streptococcus pyogenes* mosaic alleles generated by recombination. *Mol. Microbiol.* **16:**509–519.

42. Kotb, M., A. Norrby-Teglund, A. McGeer, H. El-Sherbini, M. T. Dorak, A. Khurshid, K. Green, J. Peeples, J. Wade, G. Thomson, B. Schwartz, and D. E. Low. 2002. An immunogenetic and molecular basis for differences in outcomes of invasive group A streptococcal infections. *Nat. Med.* **8:**1398–1404.

43. Lancefield, R. C. 1959. Persistence of type specific antibodies in man following infection with group A streptococci. *J. Exp. Med.* **110:**271–292.

44. Lancefield, R. C. 1962. Current knowledge of the type specific M antigens of group A streptococci. *J. Immunol.* **89:**307–313.

45. Leonard, B. A. B., M. Woischnik, and A. Podbielski. 1998. Production of stabilized virulence factor-negative variants by group A streptococci during stationary phase. *Infect. Immun.* **66:**3841–3847.

46. Levin, B. R. 1996. The evolution and maintenance of virulence in microparasites. *Emerg. Infect. Dis.* **2:**93–102.

47. Leyden, J. J., R. Stewart, and A. M. Kligman. 1980. Experimental infections with group A streptococci in humans. *J. Invest. Dermatol.* **75:**196–201.

48. Li, Z. Y., V. Sakota, D. Jackson, A. R. Franklin, and B. Beall. 2003. Array of M protein gene subtypes in 1064 recent invasive group A streptococcus isolates recovered from the active bacterial core surveillance. *J. Infect. Dis.* **188:**1587–1592.

49. Martin, D. R., and K. S. Sriprakash. 1996. Epidemiology of group A streptococcal disease in Australia and New Zealand. *Rec. Adv. Microbiol.* **4:**1–40.

50. Martin, J. M., M. Green, K. A. Barbadora, and E. R. Wald. 2002. Erythromycin-resistant group A streptococci in schoolchildren in Pittsburgh. *N. Engl. J. Med.* **346:**1200–1206.

51. Matsumoto, M., N. P. Hoe, M. Y. Liu, S. B. Beres, G. L. Sylva, C. M. Brandt, G. Haase, and J. M. Musser. 2003. Intrahost sequence variation in the streptococcal inhibitor of complement gene in patients with human pharyngitis. *J. Infect. Dis.* **187:**604–612.

52. McGregor, K., N. Bilek, A. Bennett, A. Kalia, B. Beall, J. Carapetis, B. Currie, K. Sriprakash, B. Spratt, and D. Bessen. 2004. Group A streptococci from a remote community have novel multilocus genotypes but share emmtypes and housekeeping alleles. *J. Infect. Dis.* **189:**717–723.

53. McGregor, K. F., B. G. Spratt, A. Kalia, A. Bennett, N. Bilek, B. Beall, and D. E. Bessen. 2004. Multi-locus sequence typing of *Streptococcus pyogenes* representing most

known emm-types and distinctions among sub-population genetic structures. *J. Bacteriol.* **186:**4285–4294.

54. McLandsborough, L. A., and P. P. Cleary. 1995. Insertional inactivation of virR in *Streptococcus pyogenes* M49 demonstrates that VirR functions as a positive regulator of ScpA, FcRA, OF, and M protein. *FEMS Microbiol. Lett.* **128:**45–52.

55. Moxon, E. R., P. B. Rainey, M. A. Nowak, and R. E. Lenski. 1994. Adaptive evolution of highly mutable loci in pathogenic bacteria. *Curr. Biol.* **4:**24–33.

56. Neeman, R., N. Keller, A. Barzilai, Z. Korenman, and S. Sela. 1998. Prevalence of internalisation-associated gene, prtF1, among persisting group-A streptococcus strains isolated from asymptomatic carriers. *Lancet* **352:**1974–1977.

57. Nizet, V., B. Beall, D. J. Bast, V. Datta, L. Kilburn, D. E. Low, and J. C. De Azavedo. 2000. Genetic locus for streptolysin S production by group A streptococcus. *Infect. Immun.* **68:**4245–4254.

58. Osterlund, A., R. Popa, T. Nikkila, A. Scheynius, and L. Engstrand. 1997. Intracellular reservoir of *Streptococcus pyogenes* in vivo: a possible explanation for recurrent pharyngotonsillitis. *Laryngoscope* **107:**640–647.

59. Podbielski, A. 1993. Three different types of organization of the vir regulon in group A streptococci. *Mol. Gen. Genet.* **237:**287–300.

60. Potter, E. V., M. Svartman, I. Mohammed, R. Cox, T. Poon-King, and D. P. Earle. 1978. Tropical acute rheumatic fever and associated streptococcal infections compared with concurrent acute glomerulonephritis. *J. Pediatr.* **92:**325–333.

61. Prudhomme, M., V. Libante, and J. P. Claverys. 2002. Homologous recombination at the border: insertion-deletions and the trapping of foreign DNA in *Streptococcus pneumoniae*. *Proc. Natl. Acad. Sci. USA* **99:**2100–2105.

62. Rakonjac, J. V., J. C. Robbins, and V. A. Fischetti. 1995. DNA sequence of the serum opacity factor of group A streptococci: identification of a fibronectin-binding repeat domain. *Infect. Immun.* **63:**622–631.

63. Rasmussen, M., and L. Bjorck. 2001. Unique regulation of SclB—a novel collagen-like surface protein of *Streptococcus pyogenes*. *Mol. Microbiol.* **40:**1427–1438.

64. Reinert, R. R., R. Lutticken, J. A. Sutcliffe, A. Tait-Kamradt, M. Y. Cil, H. M. Schorn, A. Bryskier, and A. Al-Lahham. 2004. Clonal relatedness of erythromycin-resistant *Streptococcus pyogenes* isolates in Germany. *Antimicrob. Agents Chemother.* **48:**1369–1373.

65. Relf, W. A., D. R. Martin, and K. S. Sriprakash. 1994. Antigenic diversity within a family of M proteins from group A streptococci: evidence for the role of frameshift and compensatory mutations. *Gene* **144:**25–30.

66. Santagati, M., F. Iannelli, C. Cascone, F. Campanile, M. R. Oggioni, S. Stefani, and G. Pozziz. 2003. The novel conjugative transposon Tn1207.3 carries the macrolide efflux gene mef(A) in *Streptococcus pyogenes*. *Microb. Drug Resist.* **9:**243–247.

67. Schmidt, H., and M. Hensel. 2004. Pathogenicity islands in bacterial pathogenesis. *Clin. Microbiol. Rev.* **17:**14–56.

68. Schneewind, O., K. F. Jones, and V. A. Fischetti. 1990. Sequence and structural characterization of the trypsin-resistant T6 surface protein of group A streptococci. *J. Bacteriol.* **172:**3310–3317.

69. Seppala, H., T. Klaukka, J. Vuopio-Varkila, A. Muotiala, H. Helenius, K. Lager, and P. Huovinen. 1997. The effect of changes in the consumption of macrolide antibiotics on erythromycin resistance in group A streptococci

in Finland. Finnish Study Group for Antimicrobial Resistance. *N. Engl. J. Med.* **337:**441–446.

70. **Shulman, S.** 2004. Group A streptococcal pharyngitis serotype surveillance in North America, 2000–2002. *Clin. Infect. Dis.* **39:**325–332.

71. **Spinaci, C., G. Magi, C. Zampaloni, L. A. Vitali, C. Paoletti, M. R. Catania, M. Prenna, L. Ferrante, S. Ripa, P. E. Varaldo, and B. Facinelli.** 2004. Genetic diversity of cell-invasive erythromycin-resistant and -susceptible group A streptococci determined by analysis of the RD2 region of the prtF1 gene. *J. Clin. Microbiol.* **42:**639–644.

72. **Stenberg, L., P. W. O'Toole, J. Mestecky, and G. Lindahl.** 1994. Molecular characterization of protein Sir, a streptococcal cell surface protein that binds both immunoglobulin A and immunoglobulin G. *J. Biol. Chem.* **269:** 13458–13464.

73. **Svensson, M. D., U. Sjöbring, and D. E. Bessen.** 1999. Selective distribution of a high-affinity plasminogen binding site among group A streptococci associated with impetigo. *Infect. Immun.* **67:**3915–3920.

74. **Towers, R. J., P. K. Fagan, S. R. Talay, B. J. Currie, K. S. Sriprakash, M. J. Walker, and G. S. Chhatwal.** 2003. Evolution of sfbI encoding streptococcal fibronectin-binding protein I: horizontal genetic transfer and gene mosaic structure. *J. Clin. Microbiol.* **41:**5398–5406.

75. **Upton, M., J. R. Tagg, P. Wescombe, and H. F. Jenkinson.** 2001. Intra- and interspecies signaling between *Streptococcus salivarius* and *Streptococcus pyogenes* mediated by

SalA and SalA1 lantibiotic peptides. *J. Bacteriol.* **183:** 3931–3938.

76. **VanHeyningen, T., T. Fogg, D. Yates, E. Hanski, and M. Caparon.** 1993. Adherence and fibronectin-binding are environmentally regulated in the group A streptococcus. *Mol. Microbiol.* **9:**1213–1222.

76a. **Veasy, L. G., L. Y. Tani, and H. R. Hill.** 1994. Persistence of acute rheumatic fever in the intermountain area of the United States. *J. Pediatr.* **124:**9–16.

77. **Viscidi, R. P., and J. C. Demma.** 2003. Genetic diversity of *Neisseria gonorrhoeae* housekeeping genes. *J. Clin. Microbiol.* **41:**197–204.

78. **Wannamaker, L. W.** 1970. Differences between streptococcal infections of the throat and of the skin. *N. Engl. J. Med.* **282:**23–31.

79. **Wessels, M. R., and M. S. Bronze.** 1994. Critical role of the group A streptococcal capsule in pharyngeal colonization and infection in mice. *Proc. Natl. Acad. Sci. USA* **91:**12238–12242.

80. **Whatmore, A. M., V. Kapur, D. J. Sullivan, J. M. Musser, and M. A. Kehoe.** 1994. Non-congruent relationships between variation in emm gene sequences and the population genetic structure of group A streptococci. *Molec. Microbiol.* **14:**619–631.

81. **York, M. K., L. Gibbs, F. Perdreau-Remington, and G. F. Brooks.** 1999. Characterization of antimicrobial resistance in *Streptococcus pyogenes* isolates from the San Francisco Bay area of northern California. *J. Clin. Microbiol.* **37:**1727–1731.

B. Group B Streptococci

Pathogenic Mechanisms and Virulence Factors of Group B Streptococci

VICTOR NIZET AND CRAIG E. RUBENS

13

Although group B streptococci (GBS) commonly colonize the lower gastrointestinal tract and vaginal epithelium of healthy adults, they remain a potentially devastating pathogen to susceptible infants. Because the newborn is quantitatively and qualitatively deficient in host defenses, including phagocytes, complement, and specific antibody, an environment exists in which a variety of potential GBS virulence factors are unveiled. The complex interactions between the bacterium and the newborn host that lead to disease manifestation can be divided into several important categories (Table 1). This chapter will review GBS pathogenic mechanisms involved in adherence to epithelial surfaces, cellular invasion of epithelial and endothelial barriers, direct injury to host tissues, avoidance of immunologic clearance, and induction of the sepsis syndrome. Special attention will be focused on recent molecular genetic discoveries, including the sequencing of three complete GBS genomes, which have led to the identification of specific virulence determinants implicated in the pathogenesis of newborn infection.

ADHERENCE TO EPITHELIAL SURFACES

Essential to acquisition of the organism by the newborn, the first step in the pathogenesis of GBS disease is asymptomatic colonization of the female genital tract. In comparison to other microorganisms, GBS bind very efficiently to human vaginal cells (151), with maximal adherence at the acidic pH characteristic of vaginal mucosa (158, 187). GBS also adhere effectively to human cells from a variety of fetal, infant, and adult tissues, including placental membranes (42), buccal, and pharyngeal mucosa (17, 68), alveolar epithelium (160) and endothelium (75), and brain endothelium, each potentially relevant to vertical transmission and development of invasive disease in the infant (109). These findings indicate that either (i) the specific component of the host cell surface to which GBS attach is widely distributed among human tissues or (ii) GBS can adhere to multiple cell surface components.

Both ionic and hydrophobic interactions contribute to GBS host cell adherence (18, 160). Epithelial cell binding is inhibited by pretreatment of the GBS with a variety of proteases, preincubating the cells with hydrophobic GBS surface proteins, or preincubating the GBS with antibodies to these proteins (18, 103, 160, 179). The GBS polysaccharide capsule itself attenuates the adherence potential of the organism (75, 160). Potential mechanisms include steric hindrance of a high-affinity adhesin-receptor interaction or a decrease in repulsive forces generated between negatively charged sialic acid residues on the GBS capsule and the surface of the host cell.

A nonproteinaceous ligand involved in GBS adherence is lipoteichoic acid (LTA), an amphiphilic glycolipid polymer that extends through the cell wall and is known to mediate host cell attachment by other gram-positive pathogens (7, 180). Low-affinity binding of GBS to human buccal epithelial cells is competitively inhibited by preincubation with purified LTA (105, 163). GBS with high levels of LTA bind more efficiently to buccal cells of fetal rather than adult origin, suggesting a potential link to the age-restricted susceptibility to GBS disease (105). Enzymatic treatment of the fetal buccal epithelial cells with trypsin or periodate abolishes LTA binding, indicating the presence of a glycoprotein receptor(s) on the fetal cells absent from the adult cells (106). Interestingly, topical administration of LTA or glycerol phosphate, a subunit component of LTA putatively involved in binding, may decrease GBS vaginal colonization of pregnant mice (27).

Recent investigations indicate that GBS effectively bind the extracellular matrix components fibronectin, fibrinogen, and laminin. These proteins are known to interact with host cell-anchored proteins such as integrins that have been demonstrated to mediate adherence of related gram-positive pathogens to epithelial cells (142). GBS bind to immobilized fibronectin to facilitate mucosal colonization (159), but not to soluble fibronectin that may serve as an opsonin for phagocyte recognition (20). Recently, a genomewide phage display technique was employed to identify GBS genes that mediate the selective fibronectin adherence (8). Discovered in this screen was a fibronectin-binding property associated with the surface-anchored GBS C5a peptidase, ScpB. Decreased fibronectin binding was observed in isogenic scpB mutants, and a direct interaction

Gram-Positive Pathogens, 2nd edition, edited by Vincent A. Fischetti et al.
© 2006 ASM Press, Washington, D.C.

TABLE 1 Virulence attributes of group B streptococci implicated in neonatal infection

Pathogenic categories	Specific mechanisms	GBS determinants
Adherence to epithelial surfaces	Colonization of vaginal mucosa Attachment to respiratory epithelium Fibronectin binding Fibrinogen binding Laminin binding	Lipoteichoic acid, αC protein, C5a peptidase (ScpB),[a] FbsA protein, LraI protein
Cellular invasion of epithelial and endothelial barriers	Chorionic invasion/transcytosis Lung epithelial invasion Lung endothelial invasion Brain endothelial invasion/transcytosis	αC protein,[a] C5a peptidase,[a] β-hemolysin/cytolysin,[a] Spb1, IagA anchoring of LTA[a]
Direct injury to host tissues	Injury to placental membranes Injury to lung epithelial cells Injury to lung endothelial cells Injury and apoptosis of hepatocytes Injury to blood-brain barrier endothelium	β-Hemolysin/cytolysin,[a] proteases, collagenase, CAMP factor, hyaluronate lyase
Avoidance of immunologic clearance	Resistance to opsonophagocytosis Blocking neutrophil recruitment Nonimmune antibody binding Resistance to antimicrobial peptides Resistance to oxidative burst killing	Polysaccharide capsule,[a] C5a peptidase,[a] αC protein[a] D-alanylation of LTA,[a] Pbp1a,[a] protease CspA,[a] carotenoid pigment,[a] SOD[a]
Induction of the sepsis syndrome	Cytokine release (IL-1, TNF-α) Nitric oxide synthase induction CNS inflammation	Peptidoglycan, β-hemolysin/ cytolysin,[a] cell wall lipids

[a] Role in virulence demonstrated by animal studies with isogenic, factor-deficient mutants.

was demonstrated between recombinant ScpB and solid-phase fibronectin (8, 24).

Similar targeted mutagenesis studies have demonstrated that adherence of GBS to laminin involves Lmb, a homologue of the Lra1 adhesin family (153); attachment of GBS to fibrinogen is mediated by repetitive motifs within surface-anchored protein FbsA (141). The transcriptional regulator RogB positively regulates the ability of GBS to bind fibrinogen and fibronectin by increasing expression of downstream genes with extracellular matrix binding motifs as well as *fbs*A (54). GBS fibronectin binding is dependent upon cellular glutamine transport encoded by the *gln*PQ operon (161). Finally, the surface protein Rib confers protective immunity and is expressed by most invasive GBS isolates (155); Rib is closely related to the R28 protein of group A streptococcus (GAS) that promotes epithelial cell binding (156).

INVASION OF EPITHELIAL AND ENDOTHELIAL CELLS

GBS have been shown to penetrate and survive within several human cell types—a phenomenon that has been termed cellular invasion. This phenotypic property may account for the ability of GBS to traverse a number of host cellular obstacles, including the placental membranes, the alveoli of the infant lung, and the neonatal blood-brain barrier. The pathogenic consequence of bacterial passage through the respective barriers may be significant: amnionitis and overwhelming fetal exposure; bacteremia and systemic spread; and entry into the central nervous system (CNS) and meningitis. The electron micrographs in Fig. 1 show ultrastructural evidence of GBS cellular invasion from a series of in vivo and in vitro studies. In all cell types examined, intracellular GBS were found within membrane-bound vacuoles, suggesting that the organism somehow elicits its own endocytotic uptake.

It has long been suspected that GBS can penetrate into the amniotic cavity through intact placental membranes, because fulminant early-onset disease develops in some infants delivered by cesarean section with no identifiable obstetric risk factors (35, 40). Rapid death and advanced lung inflammatory changes on autopsy of such patients strongly imply that the onset of infection occurred in utero (77). Experimentally, GBS are seen to migrate through freshly isolated chorioamniotic membranes, binding to the maternal surface by 2 h and appearing on the fetal surface within 8 h of inoculation (42). The nature of GBS placental penetration has been dissected further in studies using primary chorion and amnion cell cultures (184). GBS are highly invasive for chorion cells and are capable of transcytosing through intact chorion cell monolayers without disruption of intracellular tight junctions. However, although GBS adhere to amnion cells, they fail to invade these cells under a variety of assay conditions. The amnion may thus constitute a formidable host barrier against infection of the fetus, and GBS penetration into the amniotic cavity may require direct or indirect injury to amnion cells (see below) rather than cellular invasion and transcytosis (184).

Following aspiration of infected amniotic fluid, the initial focus of GBS infection is the newborn lung. Indeed, pneumonia with marked respiratory distress is a hallmark of early-onset infection. To disseminate from the alveolar space and gain access to the systemic circulation, GBS must traverse three host barriers: the alveolar epithelium, the pulmonary interstitium, and the pulmonary endothelium. In vivo evidence for GBS invasion of these host cells comes from a primate model in which early-onset pneumonia and septicemia were established following intraamniotic inoculation of the

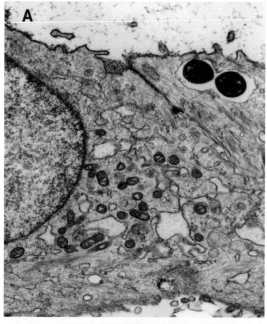

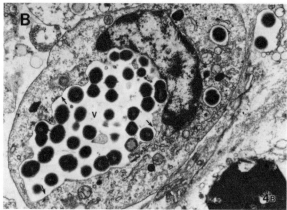

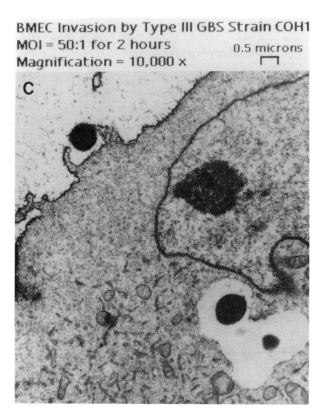

FIGURE 1 Intracellular invasion of host cells by GBS. (A) Primary cultures of human chorionic epithelial cells. (B) Respiratory epithelial cells of an infant macaque following infection by intra-amniotic inoculation. (C) Immortalized cultures of human brain microvascular endothelial cells. In each case, intracellular GBS are found within membrane-bound vacuoles (v). Host cytoskeletal changes in panel C suggest that the organism elicits its own endocytotic uptake.

organism. Electron microscopy of lung tissue from the infant macaques demonstrated GBS within membrane-bound vacuoles of type I and II alveolar epithelial cells, interstitial fibroblasts, and capillary endothelial cells (46, 134). Subsequent tissue culture studies demonstrated morphologically comparable GBS invasion of human lung epithelial cells (135) and human umbilical vein or piglet lung microvascular endothelial cells (46, 47).

GBS isolates of capsular serotypes Ia, Ib, Ia/c, II, and III all invade alveolar epithelial cells, although a small degree of strain variation exists in the magnitude of invasion (66, 135). GBS invade more efficiently into pulmonary microvascular endothelial cells than endothelial cells of pulmonary artery origin (47), suggesting a tropism for particular endothelial surfaces. In order to produce meningitis, GBS circulating in the bloodstream must subsequently penetrate human brain microvascular endothelial cells (BMEC), the single-cell layer that constitutes the blood-brain barrier.

GBS invasion and transcytosis of intact, polar tissue culture monolayers of human BMEC have been demonstrated (109). Serotype III strains, which account for the majority of CNS isolates, invaded BMEC more efficiently than strains from other common GBS serotypes.

Time course studies in epithelial and endothelial cells indicate that GBS survive intracellularly for up to 24 h after invasion, without appreciable bacterial replication (109, 135). Addition of specific inhibitors demonstrates that active bacterial DNA, RNA, and protein synthesis are necessary for invasion and that the host endocytotic mechanism involves actin microfilaments (109, 135, 184). Invasion of BMEC appears to require microtubular cytoskeletal elements and active host protein synthesis as well (109). Uptake of GBS is associated with induction of protein-kinase signal transduction pathways in the epithelial cell (170). These pathways may be dependent upon calmodulin, as entry of GBS into epithelial cells is inhibited in a dose-dependent manner when the extracellular calcium concentration is reduced (170).

The polysaccharide capsule itself does not appear to be essential for GBS cellular invasion. Rather, an isogenic nonencapsulated mutant invades epithelial and endothelial cells more efficiently than the type III clinical isolate from which it was derived (46, 66, 109). As discussed above, the GBS capsule decreases adherence of the organism to alveolar epithelium (160), presumably through steric interference of certain receptor-ligand interactions. Should initial adherence to the epithelial cell be a requirement for subsequent invasion, inhibition of invasion by the presence of capsule would be anticipated. Likewise, capsule may mask other "invasin" molecules on the GBS surface that promote epithelial cell uptake independent of adherence.

Some experimental evidence exists to indicate that cellular invasion is a crucial component in the pathogenesis of neonatal GBS disease. When tested in the tissue culture model, GBS isolates from the blood of infected neonates are significantly more invasive for respiratory epithelia than isolates from vaginal carriers or colonized neonates without clinical symptoms (169). These results indicate that in vitro invasion of human epithelial cell monolayers is a marker for the ability of a GBS to produce invasive disease in vivo. Genetic phenotyping of type III GBS strains identified a particular restriction digest pattern, III-3, characteristic of the vast majority of isolates from invasive neonatal infection (157). Subsequent subtractive hybridization studies identified a gene unique to restriction digest pattern III-3 strains encoding the surface-anchored protein Spb1, required for maximal epithelial cell invasion (2). Similarly, experiments employing targeted mutagenesis demonstrated that removal of the genes encoding the fibronectin-binding C5a peptidase ScpB or the alpha C surface protein each significantly reduced GBS epithelial cell invasion (15, 24). Further, the alpha protein of the c-protein complex that is present on the surface of some but not all serotypes has been implicated in binding to glycoprotein receptors on the surface of a vaginal epithelial cell line and mediates transcytosis across confluent monolayers (6).

In the case of GBS brain endothelial cell invasion and the pathogenesis of meningitis, a novel role of surface-expressed LTA has been identified (30). Two independent random mutants with decreased brain endothelial cell invasion were found to possess a disruption in the same gene (iagA, invasion-associated gene), encoding a glycosyltransferase homologue. In-frame allelic replacement of iagA in the GBS chromosome confirmed the observed phenotype, as the ΔiagA mutant exhibited a fourfold decrease in BMEC invasion. The glycolipid diglucosyldiacylglycerol, an anchor for LTA and predicted product of the IagA glycosyltransferase, was absent in the ΔiagA mutant, which consequently released increased amounts of LTA into the media. When injected intravenously in mice, the ΔiagA mutant established an initial bacteremia comparable to wild-type GBS. However, the ΔiagA mutant produced less mortality, a finding associated with less blood-brain barrier penetration and fewer animals with culture or histopathologic evidence of meningitis (30).

DIRECT INJURY TO HOST CELLULAR BARRIERS

Localized bacterial proliferation with associated injury to host tissues is apparent in pathologic specimens from patients with GBS disease, in particular the placenta (chorioamnionitis), lung (pneumonia), and brain (meningitis). GBS products, in particular its β-hemolysin/cytolysin (β-H/C), have been shown to be directly cytotoxic to human cells and thus represent potential virulence determinants of the organism. Disruption of epithelial and endothelial cell barriers through direct cell injury could theoretically facilitate placental penetration and systemic spread of the organism in the infected infant. Note that intracellular invasion and transcytosis by GBS, described in the previous section, can occur without significant injury to host cells in vitro. The relative contribution of intracellular invasion versus direct tissue injury to GBS pathogenesis remains to be determined—it is likely that they act in concert to produce the clinical spectrum of invasive disease. In fulminant early-onset disease and pneumonia, autopsy findings of alveolar hemorrhage and proteinaceous exudate (1, 122) attest to significant pulmonary epithelial and endothelial cell destruction, through which GBS would have direct access to the circulation. By contrast, in late-onset GBS disease, a subacute clinical presentation without pneumonia is the rule. Intracellular invasion and transcytosis of the alveolar lining by colonizing bacteria may represent the principal mechanism of bacterial spread to the bloodstream. Subsequent invasion and transcytosis of BMEC by circulating GBS could result in meningitis, a frequent clinical complication.

The great majority (>98%) of GBS clinical isolates demonstrate β-hemolysis when plated on sheep blood agar (36). The GBS β-H/C itself has yet to be purified to homogeneity, owing largely to the fact that its activity is unstable. High-molecular-weight carrier molecules such as albumin are required to preserve β-H/C activity in GBS culture supernatants (38, 93, 167). A role for β-H/C in lung epithelial and endothelial injury was first recognized through the development of GBS transposon mutants expressing either a nonhemolytic (NH) or hyperhemolytic (HH) phenotype (48, 107). When monolayers of human alveolar epithelial cells are exposed to log-phase GBS or stabilized β-H/C extracts of GBS cultures, cellular injury can be measured by lactate dehydrogenase release or by trypan blue nuclear staining. Whereas NH strains produce no detectable injury beyond baseline (media alone), β-H/C-producing strains induce injury to lung epithelial and endothelial cells in direct correlation to their ability to lyse sheep erythrocytes (48, 107). Electron microscopic studies of the injured lung epithelial cells reveal global loss of microvillus architecture, disruption of cytoplasmic and nuclear membranes, and marked swelling of the cytoplasm and organelles (107)—findings which suggest that the β-H/C acts as a pore-forming cytolysin. Radiolabeled rubidium (^{86}Rb+) and hemoglobin exhibit identical efflux kinetics from sheep red blood cells exposed to GBS β-H/C, an indication that the toxin produces membrane lesions of large size (94).

In vitro, GBS injury to cultured lung epithelial and endothelial cells can be documented at bacterial concentrations of 10^6 (most hemolytic clinical isolate) to 10^8 (least hemolytic clinical isolate) GBS per ml (48, 107). When GBS pneumonia is induced in newborn primates, bacterial density reaches 10^9 to 10^{11} per g of lung tissue (134), indicating that an ample local reservoir exists for β-H/C production and host cell injury. GBS β-H/C expression is associated with dose-dependent increases in albumin transit across polar endothelial cell monolayers, suggesting that it may contribute to pulmonary vascular leakage (48). Protein efflux from pulmonary vessels into the terminal air spaces is consistent with the clinical pattern of alveolar congestion and hyaline membrane formation seen in early-onset pneumonia.

The extent of alveolar epithelial cell injury produced by HH strains is reduced in a stepwise fashion when dipalmitoyl phosphatidylcholine, the major component of human surfactant, is added in concentrations corresponding to the physiologic increase in alveolar fluid dipalmitoyl phosphatidylcholine during the third trimester of pregnancy (107). This finding provides a theoretical rationale for the increased incidence of severe GBS pneumonia in premature, surfactant-deficient neonates (107, 162).

Several virulence studies suggest a role for β-H/C expression in GBS disease acquired via the pulmonary route. Using intranasal inoculation, a comparison of NH and HH GBS mutants shows that increased β-H/C activity is associated with a decreased LD_{50} and an earlier time to death for a given inoculum (175). When the bacterial inoculum is delivered directly into the rat lung by transthoracic puncture, a ΔβH/C mutant exhibits a 1,000-fold increased LD_{50} compared to wild-type (108). In a newborn rabbit model of GBS pulmonary infection, those pups infected intratracheally with wild-type GBS developed focal pneumonia and by 18 h had 100-fold more bacteria in lung tissues than pups infected with a ΔβH/C mutant (64). Mortality and the development of bacteremia were significantly higher in the animals challenged with wild-type GBS compared to the ΔβH/C mutant. Lung compliance during mechanical ventilation was impaired upon infection with wild-type GBS but not with the ΔβH/C strain (64).

The locus of the GBS chromosome that encodes the β-H/C activity was first discovered by Spellerberg and colleagues using a negative selection strategy (152). Several transposon insertion sites of NH mutants all clustered within a 7-kb region of the GBS chromosome that was named *cyl*, in reference to the cytolytic activity exhibited by the β-H/C. This study and subsequent reports of the GBS *cyl* gene cluster identified twelve separate open reading frames (ORFs) organized in an operon structure (120, 152, 154). The association of the *cyl* locus with GBS β-H/C production was corroborated through independent positive screening of a plasmid library of GBS chromosomal DNA in *Escherichia coli* with the aim of identifying hemolytic transformants (120). One such clone yielded a recombinant plasmid containing the GBS *cylE* and *cylF* genes. Precise in-frame allelic replacement of *cylE* with an antibiotic cassette yields an NH mutant in which β-H/C activity can be restored by reintroduction of *cylE* on a plasmid vector. When expressed in *E. coli* as a recombinant fusion protein, *cylE* alone is sufficient to confer a robust beta-hemolytic phenotype after 48 h incubation (120). The combination of the mutagenesis, complementation, and heterologous expression studies strongly suggests that *cylE* encodes the structural determinant of the GBS β-H/C, but does not exclude the possibility that additional GBS genes are involved in its processing, activation, or export to the cell surface. The predicted product of the *cylE* ORF is a 78-kDa protein without significant homology to any other proteins in the GenBank databases. It remains to be determined whether the full-length protein or a derivative represents the mature toxin, and the lack of sequence homologies precludes further inference on the precise mechanism of β-H/C action at this time.

Generation of GBS mutants that exhibit an HH phenotype suggests the existence of regulatory pathways controlling the level of toxin production. Recently, one clear negative regulator of GBS *cyl* operon expression was identified, a homologue of the *covS-covR* two-component global transcriptional regulator that is so well studied in GAS (82). A GBS *covS/R* knockout mutant showed markedly increased β-H/C activity, and *covR* was shown to bind directly to the promoter motif at the head of the *cyl* operon (82).

Mechanisms by which GBS may promote placental membrane rupture and premature delivery are being examined. Isolated chorioamniotic membranes exposed to GBS have decreased tensile strength, elasticity, and work to rupture (140). Peptide fragments released from these membranes suggest that the organism is producing one or more proteases that degrade the placental tissue. GBS proteolytic activity has been identified in culture supernatants, but no correlation exists between the proteolytic activity of a given strain and its virulence in a mouse model (32). A cell-associated collagenase activity of GBS has been postulated (67). Antibodies raised against collagenase from *Clostridium histolyticum* cross-react with cell-associated proteins produced by GBS and inhibit GBS hydrolysis of a synthetic peptide collagen analog (67). Disruption of collagen fibrils could theoretically play a role in GBS penetration of the chorioamnion and premature rupture of membranes. However, when the gene for this enzyme (*pepB*) was cloned, sequenced, and expressed, it was found incapable of solubilizing a film of reconstituted rat tail collagen (88). Rather it appears to be a zinc metallopeptidase capable of degrading a variety of small bioactive peptides (e.g., bradykinin, neurotensin).

GBS may induce placental membrane rupture indirectly by alteration of host cell processes. For example, the presence of GBS within the lower uterine cavity or cervix activates the maternal decidua cell peroxidase-H_2O_2-halide system, which could promote oxygen-radical-induced damage to adjacent fetal membranes (138). Filtered extracts of GBS modify the arachidonic acid metabolism of cultured human amnion cells, favoring production of prostaglandin E_2 (9, 81). High local concentrations of this compound are known to stimulate the onset of normal labor, and may also be a mechanism for initiation of premature labor (50). GBS stimulate macrophage inflammatory protein-1a and interleukin-8 (IL-8) production from human chorion cells; these chemokines are important mediators signaling migration of inflammatory cells and may also contribute to the pathogenesis of infection-associated preterm labor (31).

GBS secrete a protein that degrades hyaluronic acid, an important component of the extracellular matrix in higher organisms. The GBS hyaluronate lyase gene has been cloned and expressed in *E. coli*, and shares 50.7% amino acid identity with the pneumococcal hyaluronidase (87). Hyaluronate lyase is expressed in increased levels by type III GBS isolates (101), and strains obtained from neonates with bloodstream infections produced higher levels of the enzyme than strains from asymptomatically colonized infants (100) or adults with noninvasive disease (78). The biological role of the hyaluronate lyase in GBS pathogenesis remains uncertain. Theoretically, breakdown of hyaluronic acid in the extracellular matrix could facilitate tissue spread by the organism (119). The gradual increase in hyaluronic acid content of the placental membrane during gestation may enhance the nutritional or pathogenic value of expressing this enzyme during placental membrane infection (149). Further analysis of the hyaluronate lyase will now be possible with solution of its crystal structure and the elucidation of the mechanism of its catalytic process (98).

Finally, the CAMP phenomenon refers to synergistic hemolytic zones produced by colonies of GBS streaked adjacent to colonies of *Staphylococcus aureus* on sheep blood agar (25). GBS CAMP factor is an extracellular protein of

23.5 kDa that further destabilizes and lyses erythrocyte membranes pretreated with β-toxin, a staphylococcal sphingomyelinase (10). Human erythrocytes, like sheep erythrocytes, have not been observed to undergo hemolysis with CAMP factor alone. The CAMP factor gene (*cfb*) has been cloned and expressed in *E. coli* (139), and the recombinant protein elicits antibodies that inhibit the CAMP phenomenon (115). Recent electron microscopy and chemical cross-linking studies have shown that CAMP factor oligomerizes in the target membrane to form discrete pores and trigger cell lysis (83). The limited evidence for direct toxicity comes from experiments in which partially purified CAMP factor preparation produces mortality in rabbits when injected intravenously (146–148).

AVOIDANCE OF IMMUNE CLEARANCE

Upon penetration of GBS into the lung tissue or bloodstream of the newborn infant, an immunologic response is recruited to clear the organism. The central elements of this response are host phagocytic cells of the neutrophil, and to a lesser extent, monocyte-macrophage cell lines. However, as is the case with most other pathogenic bacteria, effective phagocytosis of GBS by neutrophils and macrophages requires opsonization. Without the participation of specific antibodies and serum complement, phagocytosis of GBS is dramatically reduced (3, 33, 144). The predilection of certain neonates to suffer invasive GBS disease may thus reflect quantitative or qualitative deficits in (i) phagocytic cell function, (ii) serotype-specific anti-GBS immunoglobulin, and/or (iii) the classic and alternate complement pathways. In addition to these host factors, GBS possess a number of unique virulence attributes that interfere with effective opsonophagocytosis; chief among them is the type-specific polysaccharide capsule.

GBS associated with human disease are almost invariably encapsulated, belonging to one of the nine recognized capsule serotypes: Ia, Ib, and II through VIII. With minor exceptions, the various GBS capsular polysaccharide antigens are composed of the same four component monosaccharides: glucose, galactose, N-acetylglucosamine, and sialic acid (type VI lacks N-acetylglucosamine and type VIII contains rhamnose). However, serotype-specific epitopes of each polysaccharide are created by differences in the arrangement of component sugars into a unique repeating unit (176). Hyperimmune rabbit antisera directed against a given type-specific polysaccharide antigen provide passive protection to mice from lethal challenge with virulent strains from the homologous but not heterologous serotypes. A low level of human maternal anticapsular immunoglobulin G (IgG) is a major risk factor for invasive GBS infections in the neonate (5).

The biochemistry and immunology of GBS capsular polysaccharide have been studied most thoroughly in serotype III organisms. The native type III capsular polysaccharide is a high-molecular-weight polymer composed of more than 100 repeating pentasaccharide units. Each pentasaccharide unit contains a trisaccharide backbone of galactose, glucose, and N-acetylglucosamine with a side chain of galactose and a terminal sialic acid moiety (69, 176). Added complexity is produced by variable levels of O-acetylation modifications of the capsular sialic acid observed in several GBS serotypes (85). Sialic acid is known to be a critical element in the epitope of type III GBS polysaccharide capsule that confers protective immunity. After treatment with sialidase, the altered capsular polysaccharide fails to elicit protective antibodies

against GBS infection. Moreover, protective antibodies derived from native type III capsule do not bind to the altered (asialo) capsule backbone structure (76). Human infants who possess antibodies that react only to the desialylated capsule remain at high risk for invasive disease (76).

A correlation between the sialic acid component of type III GBS capsule and animal virulence was first noted in studies employing chemical modification or spontaneous but genetically uncharacterized mutants. Organisms treated with sialidase are opsonized more effectively by complement through the alternative pathway, and are consequently more readily phagocytosed by human neutrophils in vitro (34). Sialidase treatment of type III GBS resulted in diminished lethality of the organism upon intravenous administration to neonatal rats (145). Serial subculture of a wild-type GBS strain in the presence of type III-specific antiserum allowed identification of mutants lacking the terminal sialic acid of the polysaccharide capsule. These mutants possessed a 1,000-fold greater LD_{50} following tail-vein injection in mice (185).

Direct proof for the role of type III GBS capsule in virulence was provided by the construction of isogenic capsule-deficient mutants by means of Tn916 (or Tn916ΔE) mutagenesis (131, 132). Libraries of GBS::Tn916 transconjugates were screened by immunoblot analysis for alterations in capsule expression. Two major types of type III GBS capsule mutants have been identified by this method. The first mutant phenotype completely lacked evidence of capsular material by immune electron microscopy, and failed to react with antisera to type III GBS or to type 14 pneumococcus (which recognizes the asialo-core structure of type III capsule) (131, 177). The second mutant phenotype reacted only with pneumococcal type 14 antisera, and has been shown by structural carbohydrate chemistry to specifically lack the terminal sialic acid residues of the native type III capsule (178).

Interference of effective C3 deposition by sialylated polysaccharide capsule appears to be an important virulence mechanism of GBS. Situated at the convergence of the classical and alternative complement pathways, deposition of C3 on the bacterial surface, with subsequent cleavage and degradation to opsonically active fragments C3b and iC3b, is a pivotal element in host defense against invasive infections. C3 deposition and degradation occur on the surface of GBS representing a variety of serotypes (21). However, the extent of C3 deposition by the alternative pathway is inversely related to the size and density of the polysaccharide capsule present on the surface of type Ib and type III GBS strains (95, 150). Isogenic type III mutant strains expressing a sialic acid-deficient capsule, or lacking capsular polysaccharide altogether, bind 8 to 16 times more C3 than the parent strain (95). Moreover, C3 fragments bound to the acapsular mutant are predominantly in the active form, C3b, whereas the inactive form, C3bi, is predominantly bound to the surface of the parent strain. In comparison to parent strains, the isogenic acapsular or asialo type III capsule mutants were susceptible to opsonophagocytosis in the presence of complement and peripheral blood neutrophils (95, 131, 177).

The type III GBS capsule mutants were also significantly less virulent in animal models of GBS infection. In a model of neonatal GBS pneumonia and bacteremia, neonatal rats were inoculated with either the parent strain or an acapsular mutant by intratracheal injection. In animals that received the acapsular mutant, fewer GBS were recovered per gram of lung, more bacteria were associated with resident alveolar macrophages, and the animals became significantly less bacteremic than animals that received the parent strain (96). Subcutaneous injection of the acapsular or asialo mutants in neonatal rats resulted in similar LD$_{50}$ values that were at least 100-fold greater than those obtained with the parent strain (131, 177). Together these data provided compelling evidence that the capsule protects the organism from phagocytic clearance during the initial pulmonary phase and later bacteremic phase of early-onset GBS infection. A new discovery finds complement binding also to be critical for the humoral immune response to GBS capsular polysaccharide. C3$^{-/-}$ mice failed to uptake GBS capsular polysaccharide into marginal zone B cells or dendritic cells and consequently produced low levels of specific anticapsular IgM and IgG antibodies (118).

The transposon insertion sites for both the asialo- and noncapsular GBS mutants mapped to the same 20-kb region of type III chromosome (80, 133). Southern blot analyses with genomic DNA samples from other clinical isolates indicated that the entire capsule gene region is highly conserved among several GBS strains representing a variety of serotypes (80). It is now recognized that GBS capsule biosynthesis is encoded in the single long transcript of a 16- (or more) gene operon now fully sequenced from all of the serotypes (Fig. 2). The capsule genes are organized according to their role in synthesis: regulation, chain length, export, repeating unit synthesis, and sialic acid synthesis/activation. Recently, elegant experiments have shown that the heterologous expression of a single polymerase gene (cpsH) from this operon can cause a type GBS Ia strain to express type III capsule epitopes, and vice versa (22). The genes involved in sialic acid metabolism (neuA-neuD) are unique to GBS among gram-positive organisms, sharing

similarity and function with sialic acid genes from E. coli K1 (55). The sialyltransferase gene encoded by cpsK, which adds sialic acid to the terminal side chain, shares homology to genes in Haemophilus and is even functional in Haemophilus ducreyi (23). Indeed, the low G+C content of the sialic acid genes from E. coli K1 may suggest that the origin of these genes is from gram-positive bacteria. The other genes share significant homology to polysaccharide genes from type 14 capsule genes from Streptococcus pneumoniae (SPN) and other organisms (Fig. 2). The evolution of the different serotype capsules appears to represent the acquisition of new glycosyltransferase genes by horizontal gene transfer or through gene conversion mechanisms (26). The common arrangement of the polysaccharide synthesis genes and their significant homology suggest important conservation of the gene structure among gram-positive bacteria that produce complex polysaccharides on their surface. For GBS, this conservation is also driven by the need to maintain this important host evasion factor.

Two divergently transcribed ORFs, cpsX and cpsY, separated by a common regulatory region were identified upstream of the cpsA–D genes involved in polysaccharide capsule biosynthesis in GBS (79, 186). These genes are implicated in the regulation of capsule expression in GBS, because the CpsX protein shares sequence similarities with LytR of Bacillus subtilis, an attenuator of transcription; cpsY has similarity to genes encoding members of the lysR family of transcriptional regulators. However, when mutagenized, cpsY did not have an impact on cps expression; rather this gene, now called mtaR, was found to be involved in regulating methionine transport (143).

GBS are known to regulate the degree of capsular polysaccharide expression with cell growth rate (111). Organ-

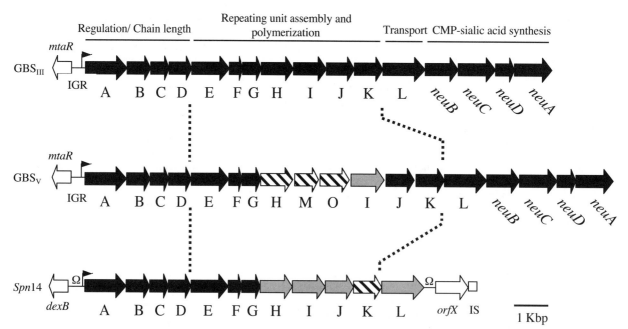

FIGURE 2 CPS synthesis operons of type III and V GBS and S. pneumoniae type 14 (Spn14). Functional organization of the loci is indicated at top. Block arrows indicate ORFs with their gene designations below. Promoter locations are indicated (→), as are transcriptional terminators (Ω), intergenic regions (IGR), and insertion elements (IS). For the type V and Spn14 sequences, solid black indicates >40% identity and solid gray indicates 20 to 40% identity with the respective type III GBS homologue. Diagonal hatching, no homology; white, unrelated to CPS production.

isms passed on solid growth media become less encapsulated (phase shifting) than when they are passed in animals (52). Mouse passage of various serotypes of GBS is followed by increases in sialylated capsule content that correlate with increased virulence (110). Subpopulations of heavily or poorly encapsulated GBS can be subcultured from strains isolated from infected infants (56). It is interesting to speculate that GBS may regulate capsule expression in response to the host environment at different stages in the pathogenic process. Nonencapsulated mutants are more adherent and significantly more invasive for respiratory epithelium than the parent strain (66, 160). These findings suggest that the capsule hinders interaction with the epithelial cell surface and/or partially masks important bacterial surface ligands that recognize receptors on the epithelial surface (158). In mucosal colonization and the early stages of newborn infection, a dynamic balance may exist between producing enough capsule to avoid immune clearance but not so much as to prevent bacterial-epithelial interaction (130). Once having invaded tissues or circulating in the bloodstream, an up-regulation of GBS capsule expression may then be favored as a means of preventing rapid opsonophagocytic clearance. In vitro, it appears that only brief periods of fast growth are required for up-regulation of *cps* gene-specific mRNAs and expression of high levels of cell-associated capsule (129).

The c antigen of GBS is a protein complex consisting of two distinct components: α, which is sensitive to the protease trypsin, and β, which is trypsin-resistant (11, 181). Overall, one or both components of the c protein are found in approximately 60% of GBS clinical isolates, but in notably less than 1% of type III strains (11, 71). The α component is expressed as a series of proteins that demonstrate a laddering pattern on immunoblots and vary greatly in molecular weight (14 to 145 kDa) among GBS strains (39, 41, 91). Large identical, tandem repeat units make up 74% of the DNA sequence in the α antigen gene, and are likely to play a role in generating the size diversity observed in α antigen expression (99). The β component of c protein is typically expressed as a single 130-kDa protein capable of binding human IgA (16, 137). Determination of the nucleotide sequence of the complete β antigen gene shows that it encodes a polypeptide typical of other gram-positive cell wall proteins (70).

Increasing evidence suggests that c protein may contribute to GBS virulence. Type II strains possessing both the α and β components of c protein produce significantly higher mortality in the infant rat model than type II strains lacking c protein (39). In vitro studies showed that type II strains without c protein were more easily killed than c protein-positive strains in an opsonophagocytic bactericidal assay (112). This limitation in phagocytosis of α- and β-containing type II strains is not due to differential uptake by the neutrophils, because the organisms are efficiently internalized, but rather to an apparent defect in intracellular killing (113). Protection against phagocytosis may reflect nonimmune binding of IgA by the β component (137), which appears to interfere sterically with deposition of opsonically active complement protein C3 on the GBS surface (113). Inactivation of the α antigen gene (*bca*) by allelic replacement resulted in a five- to sevenfold attenuation in lethality in the mouse model when compared with the isogenic wild-type strain (86). Finally, antibodies to the α component of c protein did not passively protect neonatal mice from lethal challenge with the *bca* gene-knockout mutant, suggesting that the antigen is a target for protective

immunity (86). Deletion in the repeat region of the α component of c protein enhances the pathogenicity of GBS in immune mice by (i) loss of a protective (conformational) epitope(s) and (ii) loss of antibody binding due to a decrease in antigen size relative to other cell surface components (51).

In addition to opsonization, the serum complement system contributes to host defense through generation of soluble chemotactic factors that promote neutrophil mobilization, in particular C5a. A hallmark feature of severe early-onset GBS disease is the poor influx of host neutrophils into sites of tissue infection (60, 134). GBS contribute to poor neutrophil mobilization by production of an enzyme that specifically cleaves and inactivates human C5a (65). GBS inactivate C5a by proteolytic cleavage between histidine-67 and lysine-68 near the C terminus (12). This finding is consistent with the known critical role of the C terminus of C5a in activation of human neutrophils as demonstrated by site-directed mutagenesis (104).

The normally cell-associated GBS C5a peptidase activity has been purified to apparent homogeneity by chromatographic methods, revealing an enzyme with a molecular weight of approximately 120 kDa (13). C5a peptidase appears to be a serine esterase based on sensitivity to the inhibitor di-isopropyl fluorophosphate (13). The enzymatic activity of soluble C5a peptidase is completely neutralized by serum from normal human adults, in large part due to naturally occurring IgG antibodies (65). GBS expression of C5a peptidase reduces neutrophil recruitment to the lungs of C5-deficient mice reconstituted with human C5a (14). A related GBS cell-surface protease produced by all serotype strains, named CspA, targets host fibrinogen, producing adherent fibrin-like cleavage products that coat the bacterial surface and interfere with opsonophagocytic clearance (58).

The GBS β-H/C is directly cytolytic to macrophages and neutrophils (90). As a consequence, Δβ-H/C mutants are more rapidly killed by host phagocytes during in vitro phagocytic assays and more quickly cleared from the bloodstream in animal infection models (90, 121, 125). GBS can trigger macrophage apoptosis through β-H/C-mediated pore formation and calcium influx or by β-H/C independent processes that may include modification of phosphatidylserine metabolism (19, 90, 168).

Curiously, although streptococci are commonly thought of as extracellular pathogens, GBS have been shown to survive for prolonged periods within the phagolysosome of macrophages and to be >10-fold more resistant to hydrogen peroxide killing than catalase-positive *S. aureus* (183). One GBS defense against oxidative stress is the enzyme superoxide dismutase, as evidenced by the fact that a *sodA* mutant is highly susceptible to macrophage killing and survives poorly in vivo (116). Another factor allowing GBS to survive inside phagocytes is the production of an orange carotenoid pigment, a property unique among hemolytic streptococci and genetically linked to the *cyl* operon encoding the GBS β-H/C (154). The free-radical scavenging properties of the carotenoid neutralize hydrogen peroxide and singlet oxygen, thereby providing a shield against the key elements of phagocyte oxidative burst killing (90).

CAMP factor is released from log-phase GBS cultures into the media, and is capable of binding weakly to the Fc region of human IgG and IgM (74). The complete amino acid sequence of CAMP factor has been determined, and partial sequence homology exists with the Fc-binding regions of *S. aureus* protein A (136). In the mouse model, data suggest a role for CAMP factor in resistance to immune clearance. Mice treated with a sublethal dose of GBS but then injected with purified CAMP factor developed septicemia and died (37). However, neither CAMP factor alone nor various peptide fragments of the CAMP factor coadministered with the bacteria resulted in similar mortality. Interference of normal immune system function by hyaluronate lyase is also possible, because human macrophages, neutrophils, and lymphocytes express the CD44 receptor that binds specifically to hyaluronan (59). Lymphocyte adherence to endothelial cells is inhibited by bovine testicular and fungal hyaluronidases (102), and presumably would also be sensitive to the GBS hyaluronate lyase (119).

The application of signature-tagged mutagenesis for in vivo screening in a GBS neonatal rat sepsis model has helped identify unexpected virulence genes encoding factors that help the bacterium survive immune clearance (72). For example, *ponA*, which codes for an extracytoplasmic penicillin-binding protein (PBP1a), promotes resistance to phagocytic killing independent of capsule (73). GBS mutants with deletion of PBP1a are less virulent following both lung and systemic challenge, which is correlated to an increased susceptibility to cationic antimicrobial peptides (defensins, cathelicidins) produced by host epithelial cells and phagocytes (57). Another way GBS avoids antimicrobial peptide clearance is through the D-alanylation of LTA in the bacterial cell wall, a process encoded by genes in the *dlt* operon. A GBS *dlt*A mutant exhibited decreased negative surface charge that impeded cationic host defense peptides from reaching their cell membrane target of action (117).

INDUCTION OF THE SEPSIS SYNDROME

If failures in epithelial barrier function and immunologic clearance allow GBS to establish bacteremia in the neonate, development of the sepsis syndrome, and in many cases profound septic shock, may be the consequence. Severe early-onset GBS disease is clinically indistinguishable from septic shock associated with gram-negative endotoxemia. Findings include systemic hypotension, persistent pulmonary hypertension, tissue hypoxemia and acidosis, temperature instability, disseminated intravascular coagulation, neutropenia, and ultimately, multiple-organ system failure. Because infusion of GBS produces similar pathophysiological changes in neonatal animal models of sepsis, several investigations have begun to elucidate the patterns in which GBS activate host inflammatory mediators to induce sepsis syndrome and circulatory shock.

Animal models in which GBS are infused intravenously demonstrate a biphasic host inflammatory response (45, 61, 127). The acute phase (<1 h) is manifest by increased pulmonary artery pressure and decreased arterial oxygenation, and is associated with a rise in serum levels of thromboxane B_2, a stable metabolite marker of the pulmonary vasoconstrictor thromboxane A_2. Pulmonary hypertension and hypoxemia persist through the late phase (2 to 4 h), in which a progressive pattern of systemic hypotension, decreased cardiac output, and metabolic acidosis develops together with hematologic abnormalities and organ system dysfunction. Inflammatory markers of the late phase include increases in serum thromboxane B_2, tumor necrosis factor alpha (TNFα), and 6-keto-prostaglandin F1α (a stable metabolite of prostacyclin). Experiments employing specific antagonists confirm the importance of these compounds in producing the hemodynamic alterations of GBS sepsis, and

demonstrate the involvement of still other mediators in the host inflammatory cascade.

As a known stimulator of both the cyclooxygenase and lipooxygenase pathways, IL-1 may occupy a proximal position in the cytokine cascade of septic shock (28). Treatment with an IL-1 receptor antagonist improves cardiac output and mean arterial pressure and increases the length of survival in piglets receiving a continuous infusion of GBS (172). In mice, GBS induce a "Th1-like" cytokine response (IL-2, interferon-γ, IL-12) in the absence of cytokines important in B-cell help (IL-4, IL-5, IL-10) (128). This pattern of response may allow GBS to evade antibody production important for clearance. The cytokine IL-12 may also play a particularly important role in systemic GBS infection. IL-12 elevation is seen 12 to 72 h after GBS challenge in the neonatal rat. Pretreatment with a monoclonal antibody against IL-12 results in greater mortality and levels of bacteremia, whereas therapeutic administration of IL-12 results in lower mortality and bloodstream bacterial counts (92).

Some ambiguity exists as to the precise role played by TNFα in GBS septicemia of the newborn. One can frequently detect TNFα in the blood, urine, or cerebrospinal fluid of infants with invasive GBS disease (182). Human mixed mononuclear cell cultures exposed to GBS release TNFα in a dose- and time-dependent manner; moreover, neonatal monocytes exhibit a larger TNFα response than adult cells (182). GBS infusion in piglets is associated with TNFα release during the late phase of hemodynamic response, but the TNFα inhibitor pentoxifylline has only modest effects on the ongoing pulmonary hypertension, hypoxemia, and systemic hypotension (44). Marked improvement in these hemodynamic parameters is seen when pentoxifylline treatment is combined with indomethicin inhibition of thromboxane B_2 and prostacyclin synthesis (164). Serum TNFα levels in the mouse and rat also rise after challenge with GBS; however, administration of polyclonal or monoclonal anti-TNFα antibody does not affect overall mortality in these models (165, 164).

In contrast to gram-negative pathogens and endotoxin, the specific nature of the GBS component(s) that trigger the host cytokine cascade is less well understood. Cell wall preparations of GBS cause TNFα release from human monocytes in a manner requiring CD14 and complement receptors type 3 and 4 (97). The group B polysaccharide and peptidoglycan appear to be significantly greater stimulators of TNFα release from monocytes than LTA or type-specific capsular polysaccharide (171). IL-1 and IL-6 release are also stimulated by soluble GBS cell wall antigens (174).

Studies using isogenic type III GBS mutants lacking polysaccharide capsule have shown that the presence of capsule has no effect on production of TNFα by human mononuclear cells in vitro (182), and does not change the degree of pulmonary hypertension observed in vivo (43). In contrast, capsule-deficient mutants of a type Ib GBS strain actually produce a greater degree of pulmonary hypertension than the parent strain in the piglet model (114). The latter finding implies that the type-specific Ib capsular polysaccharide may actually cloak the GBS cell wall component responsible for triggering the early phase of the host inflammatory response. This property of GBS capsule may be an important virulence attribute, allowing the organism to multiply and spread beyond a pulmonary focus before adequate host clearance mechanisms are recruited.

Incubation of piglet mesenteric arteries with heat-killed GBS produces a marked hyporesponsiveness to noradrenaline, the endothelial cell-derived vasoconstrictor ET-1, and the synthetic thromboxane A_2 analog U46619. This hyporesponsiveness appears to result from enhanced release of nitric oxide (NO), suggesting a role for nitric oxide synthetase (iNOS) induction in the systemic hypotension in GBS sepsis (173). GBS-treated pulmonary arteries also exhibited NO-mediated hyporesponsiveness to noradrenaline and ET-1, but responded normally to U46619 (173). Absence of the thromboxane A_2-induced component of NO-mediated pulmonary hyporesponsiveness might help explain the coexistence of pulmonary hypertension with systemic hypotension during GBS sepsis syndrome. Studies employing isogenic HH and NH mutants demonstrate that the GBS β-H/C is involved in transcriptional induction of iNOS in murine macrophages in a dose- and time-dependent manner (126). Intravenous challenge of rabbits with HH mutants resulted in significantly higher mortality, higher median NO serum levels, and greater organ injury and disseminated intravascular coagulation when compared to wild-type or NH mutants (126). These findings corroborate earlier studies in which administration of filter-purified GBS β-H/C extracts to rabbits or rats produced dose-dependent hypotensive changes and a limited number of deaths due to shock, findings not seen with streptolysin S from *Streptococcus pyogenes* (53).

Recently, the importance of complement components in amplifying GBS TNFα induction was corroborated when reduced levels of the cytokine were observed in the blood of C3 or C3 receptor-deficient mice stimulated with GBS (84). Knockout mouse studies indicate that GBS cell wall peptidoglycan-induced activation of p38 and NF-κB depends upon the cytoplasmic toll-like receptor (TLR) adaptor protein MyD88, but does not proceed via the well-studied TLR2 and/or TLR4 (63). However, an additional as yet undiscovered secreted GBS factor appears to activate phagocytes via TLR2 and TLR6 (62). GBS β-H/C activity is correlated directly to systemic hypotension and liver necrosis in a rabbit model of septicemia (125). Recently, it was discovered that GBS β-H/C and cell wall components act synergistically to induce macrophage production of iNOS and generation of NO (124), a potent factor in the sepsis cascade.

One location in which an overexuberant host inflammatory response to GBS may be particularly unwelcome is the confines of the brain and CNS. In the infant rat model, early GBS meningitis is characterized by acute neutrophilic inflammation in the subarachnoid space and ventricles, vasculopathy, and neuronal injury. The initiation of the CNS inflammatory response is triggered by the blood-brain barrier endothelium, which activates a specific pattern of gene transcription for neutrophil recruitment, including production of chemokines (e.g., IL-8, Groα), endothelial receptors, and neutrophil activators (29). The principal provocative factor for the blood-brain barrier inflammatory gene response is the GBS β-H/C, and this toxin has also been shown to contribute to the development of meningitis (29) and neuronal apoptosis (J. Weber and V. Nizet, unpublished).

The GBS terminal α2→3 Neu5Ac capsular component is identical to a sugar epitope widely displayed on the surface of all mammalian cells (4). Furthermore, the terminal α2→3-linked Neu5Ac is overexpressed in humans who, in evolution, have lost the genes to produce the alternative sialic acid Neu5Gc. One can hypothesize that GBS is a particularly vexsome human pathogen because its sialylated surface capsule has undergone selection to resemble host "self" and avoid immune recognition. In fact, compared to wild-type strains, isogenic capsule-deficient mutants of GBS

elicit greater degrees of proinflammatory cytokine release from human brain endothelial cells (29) and produce greater brain inflammation when injected intracisternally in newborn piglets (89). These observations likely reflect a combination of (i) decreased immune recognition as a result of GBS capsule molecular mimicry of host epitopes and (ii) increased access of host pattern recognition molecules (e.g., TLRs) to the cell wall components LTA and peptidoglycan hidden beneath the physical cloak of the polysaccharide capsule.

INSIGHT FROM COMPLETE GENOME SEQUENCES

Two complete GBS genomes, one a serotype III strain (49) and a second a serotype V strain (166), were sequenced in 2002. An additional serotype Ia genome sequence is available through online databases. Each ~2.2-Mb chromosome contains >2,100 predicted protein-encoding ORFs. Revealed in these analyses are several new candidate GBS pathogenicity factors identified (i) as orthologs to known virulence determinants in other streptococcal pathogens, e.g., GAS and SPN, and/or (ii) by their possession of the classic C-terminal LP(X)TG sorting signal of gram-positive cell wall-anchored proteins. Examples would include predicted GBS surface proteins with significant homology to the fibronectin-binding protein PFPB of GAS or the adhesin PspC of SPN. Other interesting targets for future analyses of GBS regulation of virulence include more than 100 candidate transcriptional regulators, including at least 17 two-component systems (sensor histidine kinase + response regulator) and paralogs of the Rgg, RofA, and Nra global transcriptional regulators of GAS (49). Overall, slightly more than half of the genes in the GBS genomes could be considered to share orthologs with either GAS or SPN.

Comparative hybridization experiments using whole genome microarrays were performed between the sequence serotype V strain and 19 other GBS strains of varying serotypes, revealing a surprising degree of genetic heterogeneity even among strains of the same serotype (166). Many predicted virulence genes present in GBS but not in GAS or SPN are associated with mobile genetic elements such as bacteriophages or transposon insertion sequences, suggesting horizontal acquisition of virulence factors by GBS from other bacterial species (166). Most of the known and putative GBS virulence genes are found clustered within some of 14 chromosomal "islands" of unique GBS sequence. Acquisition of these genetic sequences could theoretically have occurred through phage transduction and conjugation, or alternatively through natural transformation. Although GBS have not classically been considered naturally transformable, orthologs of several genes necessary for competence in SPN have been identified in the GBS genome (49).

CONCLUSIONS

GBS infection represents a complex interaction between the bacterial pathogen and the susceptible newborn. Human infants exhibit several well-documented deficiencies in host defense mechanisms, creating an environment in which a variety of potential GBS virulence factors are revealed. Chief among these is the sialylated polysaccharide surface capsule, whose clear-cut role in pathogenesis derives from its ability to confer resistance to opsonophagocytosis upon the organism. The multifunctional GBS β-H/C toxin contributes to disease pathogenesis through direct tissue injury, promotion of intracellular invasion, triggering of apoptosis, resistance to phagocytosis, and activation of host inflammatory pathways. The α-component of the c protein antigen has been cloned and sequenced, and targeted mutagenesis demonstrates that the gene product contributes to virulence and is also a target of protective immunity. Another pathogenicity factor, the C5a peptidase, has been characterized genetically and biochemically. It has been shown to impair neutrophil recruitment, and it is recognized to function as a cellular adhesin and invasin.

The pathogen-host relationship between GBS and human newborns becomes increasingly complex when one considers that bacterial attributes that promote virulence at a given stage in pathogenesis may attenuate virulence at other stages. For example, presence of the antiphagocytic polysaccharide capsule has an inhibitory effect on GBS invasion of host epithelial and endothelial cells. However, some capsule must be synthesized on the mucosal surface to prevent clearance by resident macrophages. Furthermore, by cloaking proinflammatory cell wall components, encapsulated GBS strains are associated with less pulmonary hypertension and meningeal irritation than unencapsulated mutants. These observations, plus those observed between maternal and newborn isolates, suggest that GBS may need to up- or down-regulate expression of capsular polysaccharide and other potential virulence factors at specific stages of the infectious process.

The recent availability of well-characterized isogenic mutants lacking individual virulence determinants has allowed precise hypothesis testing in vitro and in vivo. Increased knowledge of the molecular basis of GBS pathogenesis will help clarify the ontogeny of effective innate immunity in the early stages of human life, and provide novel targets for chemotherapy or immunoprophylaxis of GBS infections. For example, the recognition of a novel serine-threonine kinase and phosphatase pathway that is involved in cell morphology, growth, and virulence could present several biochemical targets for antimicrobial development (123).

Efficacious vaccines based on protein-conjugated GBS capsular polysaccharides are poised for phase III clinical trials, and recent studies have highlighted the potential for conserved GBS surface proteins (e.g., ScpB, Sip, BSP) to serve as serotype-independent vaccine. The next generation of GBS research, accelerated by the power of genomics, promises to enhance the status of this complex pathogen as a model organism for molecular microbiology and immunology investigations.

REFERENCES

1. **Ablow, R. C., S. G. Driscoll, E. L. Effmann, I. Gross, C. J. Jolles, R. Uauy, and J. B. Warshaw.** 1976. A comparison of early-onset group B streptococcal neonatal infection and the respiratory-distress syndrome of the newborn. *N. Engl. J. Med.* **294**:65–70.
2. **Adderson, E. E., S. Takahashi, Y. Wang, J. Armstrong, D. V. Miller, and J. F. Bohnsack.** 2003. Subtractive hybridization identifies a novel predicted protein mediating epithelial cell invasion by virulent serotype III group B *Streptococcus agalactiae*. *Infect. Immun.* **71**:6857–6863.
3. **Anderson, D. C., B. J. Hughes, M. S. Edwards, G. J. Buffone, and C. J. Baker.** 1983. Impaired chemotaxigenesis by type III group B streptococci in neonatal sera: relationship to diminished concentration of specific anticapsu-lar antibody and abnormalities of serum complement. *Pediatr. Res.* **17**:496–502.
4. **Angata, T., and A. Varki.** 2002. Chemical diversity in the sialic acids and related alpha-keto acids: an evolutionary perspective. *Chem. Rev.* **102**:439–469.
5. **Baker, C. J., and D. L. Kasper.** 1976. Correlation of maternal antibody deficiency with susceptibility to neonatal group B streptococcal infection. *N. Engl. J. Med.* **294**:753–756.
6. **Baron, M. J., G. R. Bolduc, M. B. Goldberg, T. C. Auperin, and L. C. Madoff.** 2004. Alpha C protein of group B *Streptococcus* binds host cell surface glycosaminoglycan and enters cells by an actin-dependent mechanism. *J. Biol. Chem.* **279**:24714–24723.
7. **Beachey, E. H.** 1975. Binding of group A streptococci to human oral mucosal cells by lipoteichoic acid. *Trans. Assoc. Am. Physicians* **88**:285–292.
8. **Beckmann, C., J. D. Waggoner, T. O. Harris, G. S. Tamura, and C. E. Rubens.** 2002. Identification of novel adhesins from Group B streptococci by use of phage display reveals that C5a peptidase mediates fibronectin binding. *Infect. Immun.* **70**:2869–2876.
9. **Bennett, P. R., M. P. Rose, L. Myatt, and M. G. Elder.** 1987. Preterm labor: stimulation of arachidonic acid metabolism in human amnion cells by bacterial products. *Am. J. Obstet. Gynecol.* **156**:649–655.
10. **Bernheimer, A. W., R. Linder, and L. S. Avigad.** 1979. Nature and mechanism of action of the CAMP protein of group B streptococci. *Infect. Immun.* **23**:838–844.
11. **Bevanger, L., and O. J. Iversen.** 1983. The Ibc proteins of group B streptococci: trypsin extracted alpha antigen and detection of the alpha and beta antigens. *Acta Pathol. Microbiol. Immunol. Scand.* **91**:75–81.
12. **Bohnsack, J. F., K. W. Mollison, A. M. Buko, J. C. Ashworth, and H. R. Hill.** 1991. Group B streptococci inactivate complement component C5a by enzymic cleavage at the C-terminus. *Biochem. J.* **273**:635–640.
13. **Bohnsack, J. F., X. N. Zhou, P. A. Williams, P. P. Cleary, C. J. Parker, and H. R. Hill.** 1991. Purification of the proteinase from group B streptococci that inactivates human C5a. *Biochim. Biophys. Acta* **1079**:222–228.
14. **Bohnsack, J. F., K. Widjaja, S. Ghazizadeh, C. E. Rubens, D. R. Hillyard, C. J. Parker, K. H. Albertine, and H. R. Hill.** 1997. A role for C5 and C5a-ase in the acute neutrophil response to group B streptococcal infections. *J. Infect. Dis.* **175**:847–855.
15. **Bolduc, G. R., M. J. Baron, C. Gravekamp, C. S. Lachenauer, and L. C. Madoff.** 2002. The alpha C protein mediates internalization of group B *Streptococcus* within human cervical epithelial cells. *Cell Microbiol.* **4**:751–758.
16. **Brady, L. J., and M. D. Boyle.** 1989. Identification of non-immunoglobulin A-Fc-binding forms and low-molecular-weight secreted forms of the group B streptococcal beta antigen. *Infect. Immun.* **57**:1573–1581.
17. **Broughton, R. A., and C. J. Baker.** 1983. Role of adherence in the pathogenesis of neonatal group B streptococcal infection. *Infect. Immun.* **39**:837–843.
18. **Bulgakova, T. N., K. B. Grabovskaya, M. Ryc, and J. Jelinkova.** 1986. The adhesin structures involved in the adherence of group B streptococci to human vaginal cells. *Folia Microbiol.* (Prague) **31**:394–401.
19. **Buratta, S., K. Fettucciari, M. Mambrini, I. Fetriconi, P. Marconi, and R. Mozzi.** 2002. Group B *Streptococcus* (GBS) modifies macrophage phosphatidylserine metabolism during induction of apoptosis. *FEBS Lett.* **520**:68–72.

20. Butler, K. M., C. J. Baker, and M. S. Edwards. 1987. Interaction of soluble fibronectin with group B streptococci. *Infect. Immun.* **55:**2404–2408.

21. Campbell, J. R., C. J. Baker, and M. S. Edwards. 1992. Influence of serotype of group B streptococci on C3 degradation. *Infect. Immun.* **60:**4558–4562.

22. Chaffin, D. O., S. B. Beres, H. H. Yim, and C. E. Rubens. 2000. The serotype of type Ia and III group B streptococci is determined by the polymerase gene within the polycistronic capsule operon. *J. Bacteriol.* **182:**4466–4477.

23. Chaffin, D. O., K. McKinnon, and C. E. Rubens. 2002. CpsK of *Streptococcus agalactiae* exhibits alpha2,3-sialyltransferase activity in *Haemophilus ducreyi*. *Mol. Microbiol.* **45:**109–122.

24. Cheng, Q., D. Stafslien, S. S. Purushothaman, and P. Cleary. 2002. The group B streptococcal C5a peptidase is both a specific protease and an invasin. *Infect. Immun.* **70:**2408–2413.

25. Christie, R., N. E. Atkins, and E. Munch-Peterson. 1944. A note on lytic phenomenon shown by group B streptococci. *Aust. J. Exp. Biol. Med. Sci.* **22:**197–200.

26. Cieslewicz, M. J., D. O. Chaffin, G. Glusman, D. L. Kasper, A. Madan, S. Rodrigues, J. Fahey, M. Wessels, and C. E. Rubens. 2005. Structural and genetic diversity of group B streptococcus capsular polysaccharides. *Infect. Immun.* **73:**3096–3103.

27. Cox, F., L. Taylor, E. K. Eskew, and S. J. Mattingly. 1993. Prevention of group B streptococcal colonization and bacteremia in neonatal mice with topical vaginal inhibitors. *J. Infect. Dis.* **167:**1118–1122.

28. Dinarello, C. A. 1992. The role of interleukin-1 in host responses to infectious diseases. *Infect. Agents Dis.* **1:**227–236.

29. Doran, K. S., G. Y. Liu, and V. Nizet. 2003. Group B streptococcal beta-hemolysin/cytolysin activates neutrophil signaling pathways in brain endothelium and contributes to development of meningitis. *J. Clin. Invest.* **112:**736–744.

30. Doran, K. S., E. J. Engelson, A. Khosravi, H. C. Maisey, I. Fedtke, O. Equils, K. S. Michelsen, M. Arditi, A. Peschel, and V. Nizet. 2005. Blood-brain barrier invasion by group B *Streptococcus* depends upon proper cell-surface anchoring of lipoteichoic acid. *J. Clin. Invest.* **115:**2499–2507.

31. Dudley, D. J., S. S. Edwin, J. Van Wagoner, N. H. Augustine, H. R. Hill, and M. D. Mitchell. 1997. Regulation of decidual cell chemokine production by group B streptococci and purified bacterial cell wall components. *Am. J. Obstet. Gynecol.* **177:**666–672.

32. Durham, D. L., S. J. Mattingly, T. I. Doran, T. W. Milligan, and D. C. Straus. 1981. Correlation between the production of extracellular substances by type III group B streptococcal strains and virulence in a mouse model. *Infect. Immun.* **34:**448–454.

33. Edwards, M. S., A. Nicholson-Weller, C. J. Baker, and D. L. Kasper. 1980. The role of specific antibody in alternative complement pathway-mediated opsonophagocytosis of type III, group B *Streptococcus*. *J. Exp. Med.* **151:**1275–1287.

34. Edwards, M. S., D. L. Kasper, H. J. Jennings, C. J. Baker, and A. Nicholson-Weller. 1982. Capsular sialic acid prevents activation of the alternative complement pathway by type III, group B streptococci. *J. Immunol.* **128:**1278–1283.

35. Eickhoff, T. C., J. O. Klein, A. K. Daly, P. Ingal, and M. Finland. 1964. Neonatal sepsis and other infections due to group B beta-hemolytic streptococci. *N. Engl. J. Med.* **271:**1221–1228.

36. Facklam, R. R., J. F. Padula, E. C. Wortham, R. C. Cooksey, and H. A. Rountree. 1979. Presumptive identification of group A, B, and D streptococci on agar plate media. *J. Clin. Microbiol.* **9:**665–672.

37. Fehrenbach, F. J., D. Jurgens, J. Ruhlmann, B. Sterzik, and M. Ozel. 1988. Role of CAMP-factor (protein B) in virulence, p. 351–357. *In* F. J. Feherenbach (ed.), *Bacterial Protein Toxins*. Gustav Fischer, Stuttgart, Germany.

38. Ferrieri, P. 1982. Characterization of a hemolysin isolated from group-B streptococci, p. 142–143. *In* S. E. Holm and P. Christensen (ed.), *Basic Concepts of Streptococci and Streptococcal Diseases*. Reedbooks, Surrey, United Kingdom.

39. Ferrieri, P. 1988. Surface-localized protein antigens of group B streptococci. *Rev. Infect. Dis.* **10**(Suppl. 2):S363–S366.

40. Ferrieri, P., P. P. Cleary, and A. E. Seeds. 1977. Epidemiology of group-B streptococcal carriage in pregnant women and newborn infants. *J. Med. Microbiol.* **10:**103–114.

41. Flores, A. E., and P. Ferrieri. 1996. Molecular diversity among the trypsin resistant surface proteins of group B streptococci. *Zentbl. Bakteriol.* **285:**44–51.

42. Galask, R. P., M. W. Varner, C. R. Petzold, and S. L. Wilbur. 1984. Bacterial attachment to the chorioamniotic membranes. *Am. J. Obstet. Gynecol.* **148:**915–928.

43. Gibson, R. L., G. J. Redding, W. E. Truog, W. R. Henderson, and C. E. Rubens. 1989. Isogenic group B streptococci devoid of capsular polysaccharide or beta-hemolysin: pulmonary hemodynamic and gas exchange effects during bacteremia in piglets. *Pediatr. Res.* **26:**241–245.

44. Gibson, R. L., G. J. Redding, W. R. Henderson, and W. E. Truog. 1991. Group B streptococcus induces tumor necrosis factor in neonatal piglets. Effect of the tumor necrosis factor inhibitor pentoxifylline on hemodynamics and gas exchange. *Am. Rev. Respir. Dis.* **143:**598–604.

45. Gibson, R. L., W. E. Truog, W. R. Henderson, Jr., and G. J. Redding. 1992. Group B streptococcal sepsis in piglets: effect of combined pentoxifylline and indomethacin pretreatment. *Pediatr. Res.* **31:**222–227.

46. Gibson, R. L., M. K. Lee, C. Soderland, E. Y. Chi, and C. E. Rubens. 1993. Group B streptococci invade endothelial cells: type III capsular polysaccharide attenuates invasion. *Infect. Immun.* **61:**478–485.

47. Gibson, R. L., C. Soderland, W. R. Henderson, Jr., E. Y. Chi, and C. E. Rubens. 1995. Group B streptococci (GBS) injure lung endothelium in vitro: GBS invasion and GBS-induced eicosanoid production is greater with microvascular than with pulmonary artery cells. *Infect. Immun.* **63:**271–279.

48. Gibson, R. L., V. Nizet, and C. E. Rubens. 1999. Group B streptococcal beta-hemolysin promotes injury of lung microvascular endothelial cells. *Pediatr. Res.* **45:**626–634.

49. Glaser, P., C. Rusniok, C. Buchrieser, F. Chevalier, L. Frangeul, T. Msadek, M. Zouine, E. Couve, L. Lalioui, C. Poyart, P. Trieu-Cuot, and F. Kunst. 2002. Genome sequence of *Streptococcus agalactiae*, a pathogen causing invasive neonatal disease. *Mol. Microbiol.* **45:**1499–1513.

50. Gomez, R., F. Ghezzi, R. Romero, H. Munoz, J. E. Tolosa, and I. Rojas. 1995. Premature labor and intra-amniotic infection. Clinical aspects and role of the cytokines in diagnosis and pathophysiology. *Clin. Perinatol.* **22:**281–342.

51. Gravekamp, C., B. Rosner, and L. C. Madoff. 1998. Deletion of repeats in the alpha C protein enhances the pathogenicity of group B streptococci in immune mice. *Infect. Immun.* **66:**4347–4354.

52. Gray, B. M., and D. G. Pritchard. 1992. Phase variation in the pathogenesis of group B streptococcal infections,

p. 452–454. In G. Orefici (ed.), *New Perspectives on Streptococci and Streptococcal Infections*. Gustav Fisher Verlag, New York, N.Y.

53. **Griffiths, B. B., and H. Rhee.** 1992. Effects of haemolysins of groups A and B streptococci on cardiovascular system. *Microbios* **69:**17–27.

54. **Gutekunst, H., B. J. Eikmanns, and D. J. Reinscheid.** 2003. Analysis of RogB-controlled virulence mechanisms and gene repression in *Streptococcus agalactiae*. *Infect. Immun.* **71:**5056–5064.

55. **Haft, R. F., M. R. Wessels, M. F. Mebane, N. Conaty, and C. E. Rubens.** 1996. Characterization of *cps*F and its product CMP-*N*-acetylneuraminic acid synthetase, a group B streptococcal enzyme that can function in K1 capsular polysaccharide biosynthesis in *Escherichia coli*. *Mol. Microbiol.* **19:**555–563.

56. **Hakansson, S., S. E. Holm, and M. Wagner.** 1987. Density profile of group B streptococci, type III, and its possible relation to enhanced virulence. *J. Clin. Microbiol.* **25:**714–718.

57. **Hamilton, A., A. L. Jones, and R. H. Needham.** 2004. Role of GBS penicillin-binding protein 1a in evasion of the host immune response. Abstr. 104th Annu. Meet. Am. Soc. Microbiol. 2004. American Society for Microbiology, Washington, D.C.

58. **Harris, T. O., D. W. Shelver, J. F. Bohnsack, and C. E. Rubens.** 2003. A novel streptococcal surface protease promotes virulence, resistance to opsonophagocytosis, and cleavage of human fibrinogen. *J. Clin. Invest.* **111:**61–70.

59. **Haynes, B. F., M. J. Telen, L. P. Hale, and S. M. Denning.** 1989. CD44—a molecule involved in leukocyte adherence and T-cell activation. *Immunol. Today* **10:**423–428.

60. **Hemming, V. G., D. W. McCloskey, and H. R. Hill.** 1976. Pneumonia in the neonate associated with group B streptococcal septicemia. *Am. J. Dis. Child.* **130:**1231–1233.

61. **Hemming, V. G., W. F. O'Brien, G. W. Fischer, S. M. Golden, and S. F. Noble.** 1984. Studies of short-term pulmonary and peripheral vascular responses induced in oophorectomized sheep by the infusion of a group B streptococcal extract. *Pediatr. Res.* **18:**266–269.

62. **Henneke, P., O. Takeuchi, J. A. van Strijp, H. K. Guttormsen, J. A. Smith, A. B. Schromm, T. A. Espevik, S. Akira, V. Nizet, D. L. Kasper, and D. T. Golenbock.** 2001. Novel engagement of CD14 and multiple toll-like receptors by group B streptococci. *J. Immunol.* **167:**7069–7076.

63. **Henneke, P., O. Takeuchi, R. Malley, E. Lien, R. R. Ingalls, M. W. Freeman, T. Mayadas, V. Nizet, S. Akira, D. L. Kasper, and D. T. Golenbock.** 2002. Cellular activation, phagocytosis, and bactericidal activity against group B streptococcus involve parallel myeloid differentiation factor 88-dependent and independent signaling pathways. *J. Immunol.* **169:**3970–3977.

64. **Hensler, M. E., G. Y. Liu, S. Sobczak, K. Benirshcke, V. Nizet, and G. P. Heldt.** 2005. Virulence role of group B Streptococcus beta-hemolysin/cytolysin and carotenoid pigment function in a neonatal rabbit model of early-onset pulmonary infection. *J. Infect. Dis.* **191:**1287–1291.

65. **Hill, H. R., J. F. Bohnsack, E. Z. Morris, N. H. Augustine, C. J. Parker, P. P. Cleary, and J. T. Wu.** 1988. Group B streptococci inhibit the chemotactic activity of the fifth component of complement. *J. Immunol.* **141:**3551–3556.

66. **Hulse, M. L., S. Smith, E. Y. Chi, A. Pham, and C. E. Rubens.** 1993. Effect of type III group B streptococcal capsular polysaccharide on invasion of respiratory epithelial cells. *Infect. Immun.* **61:**4835–4841.

67. **Jackson, R. J., M. L. Dao, and D. V. Lim.** 1994. Cell-associated collagenolytic activity by group B streptococci. *Infect. Immun.* **62:**5647–5651.

68. **Jelinkova, J., K. B. Grabovskaya, M. Ryc, T. N. Bulgakova, and A. A. Totolian.** 1986. Adherence of vaginal and pharyngeal strains of group B streptococci to human vaginal and pharyngeal epithelial cells. *Zentbl. Bakteriol. Mikrobiol. Hyg. A* **262:**492–499.

69. **Jennings, H. J., C. Lugowski, and D. L. Kasper.** 1981. Conformational aspects critical to the immunospecificity of the type III group B streptococcal polysaccharide. *Biochemistry* **20:**4511–4518.

70. **Jerlstrom, P. G., S. R. Talay, P. Valentin-Weigand, K. N. Timmis, and G. S. Chhatwal.** 1996. Identification of an immunoglobulin A binding motif located in the beta-antigen of the c protein complex of group B streptococci. *Infect. Immun.* **64:**2787–2793.

71. **Johnson, D. R., and P. Ferrieri.** 1984. Group B streptococcal Ibc protein antigen: distribution of two determinants in wild-type strains of common serotypes. *J. Clin. Microbiol.* **19:**506–510.

72. **Jones, A. L., K. M. Knoll, and C. E. Rubens.** 2000. Identification of *Streptococcus agalactiae* virulence genes in the neonatal rat sepsis model using signature-tagged mutagenesis. *Mol. Microbiol.* **37:**1444–1455.

73. **Jones, A. L., R. H. Needham, A. Clancy, K. M. Knoll, and C. E. Rubens.** 2003. Penicillin-binding proteins in *Streptococcus agalactiae*: a novel mechanism for evasion of immune clearance. *Mol. Microbiol.* **47:**247–256.

74. **Jurgens, D., B. Sterzik, and F. J. Fehrenbach.** 1987. Unspecific binding of group B streptococcal cocytolysin (CAMP factor) to immunoglobulins and its possible role in pathogenicity. *J. Exp. Med.* **165:**720–732.

75. **Kallman, J., J. Schollin, S. Hakansson, A. Andersson, and E. Kihlstrom.** 1993. Adherence of group B streptococci to human endothelial cells in vitro. *APMIS* **101:**403–408.

76. **Kasper, D. L., C. J. Baker, R. S. Baltimore, J. H. Crabb, G. Schiffman, and H. J. Jennings.** 1979. Immunodeterminant specificity of human immunity to type III group B streptococcus. *J. Exp. Med.* **149:**327–339.

77. **Katzenstein, A. L., C. Davis, and A. Braude.** 1976. Pulmonary changes in neonatal sepsis to group B beta-hemolytic *Streptococcus*: relation of hyaline membrane disease. *J. Infect. Dis.* **133:**430–435.

78. **Kjems, E., B. Perch, and J. Henrichsen.** 1980. Serotypes of group B streptococci and their relation to hyaluronidase production and hydrolysis of salicin. *J. Clin. Microbiol.* **11:**111–113.

79. **Koskiniemi, S., M. Sellin, and M. Norgren.** 1998. Identification of two genes, *cps*X and *cps*Y, with putative regulatory function on capsule expression in group B streptococci. *FEMS Immunol. Med. Microbiol.* **21:**159–168.

80. **Kuypers, J. M., L. M. Heggen, and C. E. Rubens.** 1989. Molecular analysis of a region of the group B streptococcus chromosome involved in type III capsule expression. *Infect. Immun.* **57:**3058–3065.

81. **Lamont, R. F., M. Rose, and M. G. Elder.** 1985. Effect of bacterial products on prostaglandin E production by amnion cells. *Lancet* **2:**1331–1333.

82. **Lamy, M.-C., M. Zouine, J. Fert, M. Vergassola, E. Couve, E. Pellegrini, P. Glaser, F. Kunst, T. Msadek, P. Trieu-Cuot, and C. Poyart.** 2004. CovS/CovR of group B streptococcus: a two-component global regulatory system involved in virulence. *Mol. Microbiol.* **54:**1250–1268.

83. Lang, S., and M. Palmer. 2003. Characterization of *Streptococcus agalactiae* CAMP factor as a pore-forming toxin. *J. Biol. Chem.* **278**:38167–38173.

84. Levy, O., R. M. Jean-Jacques, C. Cywes, R. B. Sisson, K. A. Zarember, P. J. Godowski, J. L. Christianson, H. K. Guttormsen, M. C. Carroll, A. Nicholson-Weller, and M. R. Wessels. 2003. Critical role of the complement system in group B streptococcus-induced tumor necrosis factor alpha release. *Infect. Immun.* **71**:6344–6353.

85. Lewis, A. L., V. Nizet, and A. Varki. 2004. Discovery and characterization of sialic acid O-acetylation in group B *Streptococcus. Proc. Natl. Acad. Sci. USA* **101**:11123–11128.

86. Li, J., D. L. Kasper, F. M. Ausubel, B. Rosner, and J. L. Michel. 1997. Inactivation of the alpha C protein antigen gene, *bca*, by a novel shuttle/suicide vector results in attenuation of virulence and immunity in group B *Streptococcus. Proc. Natl. Acad. Sci. USA* **94**:13251–13256.

87. Lin, B., S. K. Hollingshead, J. E. Coligan, M. L. Egan, J. R. Baker, and D. G. Pritchard. 1994. Cloning and expression of the gene for group B streptococcal hyaluronate lyase. *J. Biol. Chem.* **269**:30113–30116.

88. Lin, B., W. F. Averett, J. Novak, W. W. Chatham, S. K. Hollingshead, J. E. Coligan, M. L. Egan, and D. G. Pritchard. 1996. Characterization of PepB, a group B streptococcal oligopeptidase. *Infect. Immun.* **64**:3401–3406.

89. Ling, E. W., F. J. Noya, G. Ricard, K. Beharry, E. L. Mills, and J. V. Aranda. 1995. Biochemical mediators of meningeal inflammatory response to group B streptococcus in the newborn piglet model. *Pediatr. Res.* **38**:981–987.

90. Liu, G. Y., K. S. Doran, T. Lawrence, N. Turkson, M. Puliti, L. Tissi, and V. Nizet. 2004. Sword and shield: linked group B streptococcal β-hemolysin/cytolysin and carotenoid pigment function to subvert host phagocyte defense. *Proc. Natl. Acad. Sci. USA* **101**:14491–14496.

91. Madoff, L. C., S. Hori, J. L. Michel, C. J. Baker, and D. L. Kasper. 1991. Phenotypic diversity in the alpha C protein of group B streptococci. *Infect. Immun.* **59**:2638–2644.

92. Mancuso, G., V. Cusumano, F. Genovese, M. Gambuzza, C. Beninati, and G. Teti. 1997. Role of interleukin 12 in experimental neonatal sepsis caused by group B streptococci. *Infect. Immun.* **65**:3731–3735.

93. Marchlewicz, B. A., and J. L. Duncan. 1980. Properties of a hemolysin produced by group B streptococci. *Infect. Immun.* **30**:805–813.

94. Marchlewicz, B. A., and J. L. Duncan. 1981. Lysis of erythrocytes by a hemolysin produced by a group B *Streptococcus* sp. *Infect. Immun.* **34**:787–794.

95. Marques, M. B., D. L. Kasper, M. K. Pangburn, and M. R. Wessels. 1992. Prevention of C3 deposition by capsular polysaccharide is a virulence mechanism of type III group B streptococci. *Infect. Immun.* **60**:3986–3993.

96. Martin, T. R., J. T. Ruzinski, C. E. Rubens, E. Y. Chi, and C. B. Wilson. 1992. The effect of type-specific polysaccharide capsule on the clearance of group B streptococci from the lungs of infant and adult rats. *J. Infect. Dis.* **165**:306–314.

97. Medvedev, A. E., T. Flo, R. R. Ingalls, D. T. Golenbock, G. Teti, S. N. Vogel, and T. A. Espevik. 1998. Involvement of CD14 and complement receptors CR3 and CR4 in nuclear factor-kappa B activation and TNF production induced by lipopolysaccharide and group B streptococcal cell walls. *J. Immunol.* **160**:4535–4542.

98. Mello, L. V., B. L. De Groot, S. Li, and M. J. Jedrzejas. 2002. Structure and flexibility of *Streptococcus agalactiae*

hyaluronate lyase complex with its substrate. Insights into the mechanism of processive degradation of hyaluronan. *J. Biol. Chem.* **277**:36678–36688.

99. Michel, J. L., L. C. Madoff, K. Olson, D. E. Kling, D. L. Kasper, and F. M. Ausubel. 1992. Large, identical, tandem repeating units in the C protein alpha antigen gene, *bca*, of group B streptococci. *Proc. Natl. Acad. Sci. USA* **89**:10060–10064.

100. Milligan, T. W., C. J. Baker, D. C. Straus, and S. J. Mattingly. 1978. Association of elevated levels of extracellular neuraminidase with clinical isolates of type III group B streptococci. *Infect. Immun.* **21**:738–746.

101. Milligan, T. W., S. J. Mattingly, and D. C. Straus. 1980. Purification and partial characterization of neuraminidase from type III group B streptococci. *J. Bacteriol.* **144**:164–171.

102. Miyake, K., C. B. Underhill, J. Lesley, and P. W. Kincade. 1990. Hyaluronate can function as a cell adhesion molecule and CD44 participates in hyaluronate recognition. *J. Exp. Med.* **172**:69–75.

103. Miyazaki, S., O. Leon, and C. Panos. 1988. Adherence of *Streptococcus agalactiae* to synchronously growing human cell monolayers without lipoteichoic acid involvement. *Infect. Immun.* **56**:505–512.

104. Mollison, K. W., W. Mandecki, E. R. Zuiderweg, L. Fayer, T. A. Fey, R. A. Krause, R. G. Conway, L. Miller, R. P. Edalji, M. A. Shallcross, et al. 1989. Identification of receptor-binding residues in the inflammatory complement protein C5a by site-directed mutagenesis. *Proc. Natl. Acad. Sci. USA* **86**:292–296.

105. Nealon, T. J., and S. J. Mattingly. 1984. Role of cellular lipoteichoic acids in mediating adherence of serotype III strains of group B streptococci to human embryonic, fetal, and adult epithelial cells. *Infect. Immun.* **43**:523–530.

106. Nealon, T. J., and S. J. Mattingly. 1985. Kinetic and chemical analyses of the biologic significance of lipoteichoic acids in mediating adherence of serotype III group B streptococci. *Infect. Immun.* **50**:107–115.

107. Nizet, V., R. L. Gibson, E. Y. Chi, P. E. Framson, M. Hulse, and C. E. Rubens. 1996. Group B streptococcal beta-hemolysin expression is associated with injury of lung epithelial cells. *Infect. Immun.* **64**:3818–3826.

108. Nizet, V., R. L. Gibson, and C. E. Rubens. 1997. The role of group B streptococci beta-hemolysin expression in newborn lung injury. *Adv. Exp. Med. Biol.* **418**:627–630.

109. Nizet, V., K. S. Kim, M. Stins, M. Jonas, E. Y. Chi, D. Nguyen, and C. E. Rubens. 1997. Invasion of brain microvascular endothelial cells by group B streptococci. *Infect. Immun.* **65**:5074–5081.

110. Orefici, G., S. Recchia, and L. Galante. 1988. Possible virulence marker for *Streptococcus agalactiae* (Lancefield Group B). *Eur. J. Clin. Microbiol. Infect. Dis.* **7**:302–305.

111. Paoletti, L. C., R. A. Ross, and K. D. Johnson. 1996. Cell growth rate regulates expression of group B *Streptococcus* type III capsular polysaccharide. *Infect. Immun.* **64**: 1220–1226.

112. Payne, N. R., and P. Ferrieri. 1985. The relation of the Ibc protein antigen to the opsonization differences between strains of type II group B streptococci. *J. Infect. Dis.* **151**:672–681.

113. Payne, N. R., Y. K. Kim, and P. Ferrieri. 1987. Effect of differences in antibody and complement requirements on phagocytic uptake and intracellular killing of "c" protein-positive and -negative strains of type II group B streptococci. *Infect. Immun.* **55**:1243–1251.

114. Philips, J. B. D., J. X. Li, B. M. Gray, D. G. Pritchard, and J. R. Oliver. 1992. Role of capsule in pulmonary

hypertension induced by group B streptococcus. *Pediatr. Res.* **31**:386–390.

115. **Podbielski, A., O. Blankenstein, and R. Lutticken.** 1994. Molecular characterization of the cfb gene encoding group B streptococcal CAMP-factor. *Med. Microbiol. Immunol.* **183**:239–256.

116. **Poyart, C., E. Pellegrini, O. Gaillot, C. Boumaila, M. Baptista, and P. Trieu-Cuot.** 2001. Contribution of Mn-cofactored superoxide dismutase (SodA) to the virulence of *Streptococcus agalactiae*. *Infect. Immun.* **69**:5098–5106.

117. **Poyart, C., E. Pellegrini, M. Marceau, M. Baptista, F. Jaubert, M. C. Lamy, and P. Trieu-Cuot.** 2003. Attenuated virulence of *Streptococcus agalactiae* deficient in D-alanyl-lipoteichoic acid is due to an increased susceptibility to defensins and phagocytic cells. *Mol. Microbiol.* **49**:1615–1625.

118. **Pozdnyakova, O., H. K. Guttormsen, F. N. Lalani, M. C. Carroll, and D. L. Kasper.** 2003. Impaired antibody response to group B streptococcal type III capsular polysaccharide in C3- and complement receptor 2-deficient mice. *J. Immunol.* **170**:84–90.

119. **Pritchard, D. G., B. Lin, T. R. Willingham, and J. R. Baker.** 1994. Characterization of the group B streptococcal hyaluronate lyase. *Arch. Biochem. Biophys.* **315**:431–437.

120. **Pritzlaff, C. A., J. C. Chang, S. P. Kuo, G. S. Tamura, C. E. Rubens, and V. Nizet.** 2001. Genetic basis for the beta-haemolytic/cytolytic activity of group B *Streptococcus*. *Mol. Microbiol.* **39**:236–247.

121. **Puliti, M., V. Nizet, C. von Hunolstein, F. Bistoni, P. Mosci, G. Orefici, and L. Tissi.** 2000. Severity of group B streptococcal arthritis is correlated with beta-hemolysin expression. *J. Infect. Dis.* **182**:824–832.

122. **Quirante, J., R. Ceballos, and G. Cassady.** 1974. Group B beta-hemolytic streptococcal infection in the newborn. I. Early onset infection. *Am. J. Dis. Child.* **128**:659–665.

123. **Rajagopal, L., A. Clancy, and C. E. Rubens.** 2003. A eukaryotic type serine/threonine kinase and phosphatase in *Streptococcus agalactiae* reversibly phosphorylate an inorganic pyrophosphatase and affect growth, cell segregation, and virulence. *J. Biol. Chem.* **278**:14429–14441.

124. **Ring, A., J. S. Braun, V. Nizet, W. Stremmel, and J. L. Shenep.** 2000. Group B streptococcal beta-hemolysin induces nitric oxide production in murine macrophages. *J. Infect. Dis.* **182**:150–157.

125. **Ring, A., J. S. Braun, J. Pohl, V. Nizet, W. Stremmel, and J. L. Shenep.** 2002. Group B streptococcal beta-hemolysin induces mortality and liver injury in experimental sepsis. *J. Infect. Dis.* **185**:1745–1753.

126. **Ring, A., C. Depnering, J. Pohl, V. Nizet, J. L. Shenep, and W. Stremmel.** 2002. Synergistic action of nitric oxide release from murine macrophages caused by group B streptococcal cell wall and beta-hemolysin/cytolysin. *J. Infect. Dis.* 2002 **186**:1518–1521.

127. **Rojas, J., L. E. Larsson, C. G. Hellerqvist, K. L. Brigham, M. E. Gray, and M. T. Stahlman.** 1983. Pulmonary hemodynamic and ultrastructural changes associated with Group B streptococcal toxemia in adult sheep and newborn lambs. *Pediatr. Res.* **17**:1002–1008.

128. **Rosati, E., K. Fettucciari, L. Scaringi, P. Cornacchione, R. Sabatini, L. Mezzasoma, R. Rossi, and P. Marconi.** 1998. Cytokine response to group B streptococcus infection in mice. *Scand. J. Immunol.* **47**:314–323.

129. **Ross, R. A., L. C. Madoff, and L. C. Paoletti.** 1999. Regulation of cell component production polysaccharide

by growth rate in the group B *Streptococcus*. *J. Bacteriol.* **181**:5389–5394.

130. **Rubens, C. E.** 1994. Type III capsular polysaccharide of group B streptococci: role in virulence and the molecular basis of capsule expression, p. 327–329. *In* V. L. Miller, J. B. Kaper, D. A. Portnoy, and R. R. Isberg (ed.), *Molecular Genetics of Bacterial Pathogenesis.* American Society for Microbiology, Washington, D.C.

131. **Rubens, C. E., M. R. Wessels, L. M. Heggen, and D. L. Kasper.** 1987. Transposon mutagenesis of type III group B *Streptococcus*: correlation of capsule expression with virulence. *Proc. Natl. Acad. Sci. USA* **84**:7208–7212.

132. **Rubens, C. E., M. R. Wessels, J. M. Kuypers, D. L. Kasper, and J. N. Weiser.** 1990. Molecular analysis of two group B streptococcal virulence factors. *Semin. Perinatol.* **14**:22–29.

133. **Rubens, C. E., J. M. Kuypers, L. M. Heggen, D. L. Kasper, and M. R. Wessels.** 1991. Molecular analysis of the group B streptococcal capsule genes, p. 179–183. *In* G. M. Dunny, P. P. Cleary, and L. L. McKay (ed.), *Genetics and Molecular Biology of Streptococci, Lactococci and Enterococci.* American Society for Microbiology, Washington, D.C.

134. **Rubens, C. E., H. V. Raff, J. C. Jackson, E. Y. Chi, J. T. Bielitzki, and S. L. Hillier.** 1991. Pathophysiology and histopathology of group B streptococcal sepsis in *Macaca nemestrina* primates induced after intraamniotic inoculation: evidence for bacterial cellular invasion. *J. Infect. Dis.* **164**:320–330.

135. **Rubens, C. E., S. Smith, M. Hulse, E. Y. Chi, and G. van Belle.** 1992. Respiratory epithelial cell invasion by group B streptococci. *Infect. Immun.* **60**:5157–5163.

136. **Ruhlmann, J., B. Wittmann-Liebold, D. Jurgens, and F. J. Fehrenbach.** 1988. Complete amino acid sequence of protein B. *FEBS Lett.* **235**:262–266.

137. **Russell-Jones, G. J., E. C. Gotschlich, and M. S. Blake.** 1984. A surface receptor specific for human IgA on group B streptococci possessing the Ibc protein antigen. *J. Exp. Med.* **160**:1467–1475.

138. **Sbarra, A. J., G. B. Thomas, C. L. Cetrulo, C. Shakr, A. Chaudhury, and B. Paul.** 1987. Effect of bacterial growth on the bursting pressure of fetal membranes in vitro. *Obstet. Gynecol.* **70**:107–110.

139. **Schneewind, O., K. Friedrich, and R. Lutticken.** 1988. Cloning and expression of the CAMP factor of group B streptococci in *Escherichia coli*. *Infect. Immun.* **56**:2174–2179.

140. **Schoonmaker, J. N., D. W. Lawellin, B. Lunt, and J. A. McGregor.** 1989. Bacteria and inflammatory cells reduce chorioamniotic membrane integrity and tensile strength. *Obstet. Gynecol.* **74**:590–596.

141. **Schubert, A., K. Zakikhany, M. Schreiner, R. Frank, B. Spellerberg, B. J. Eikmanns, and D. J. Reinscheid.** 2002. A fibrinogen receptor from group B *Streptococcus* interacts with fibrinogen by repetitive units with novel ligand binding sites. *Mol. Microbiol.* **46**:557–569.

142. **Schwartz-Linek, U., M. Hook, and J. R. Potts.** 2004. The molecular basis of fibronectin-mediated bacterial adherence to host cells. *Mol. Microbiol.* **52**:631–641.

143. **Shelver, D., L. Rajagopal, T. O. Harris, and C. E. Rubens.** 2003. MtaR, a regulator of methionine transport, is critical for survival of group B streptococcus in vivo. *J. Bacteriol.* **185**:6592–6599.

144. **Shigeoka, A. O., R. T. Hall, V. G. Hemming, C. D. Allred, and H. R. Hill.** 1978. Role of antibody and com-

plement in opsonization of group B streptococci. *Infect. Immun.* **21:**34–40.

145. **Shigeoka, A. O., N. S. Rote, J. I. Santos, and H. R. Hill.** 1983. Assessment of the virulence factors of group B streptococci: correlation with sialic acid content. *J. Infect. Dis.* **147:**857–863.

146. **Shyur, S. D., H. V. Raff, J. F. Bohnsack, D. K. Kelsey, and H. R. Hill.** 1992. Comparison of the opsonic and complement triggering activity of human monoclonal IgG1 and IgM antibody against group B streptococci. *J. Immunol.* **148:**1879–1884.

147. **Siegel, J. D., G. H. McCracken, Jr., N. Threlkeld, B. Milvenan, and C. R. Rosenfeld.** 1980. Single-dose penicillin prophylaxis against neonatal group B streptococcal infections. A controlled trial in 18,738 newborn infants. *N. Engl. J. Med.* **303:**769–775.

148. **Skalka, B., and J. Smola.** 1981. Lethal effect of CAMP-factor and UBERIS-factor—a new finding about diffusible exosubstances of *Streptococcus agalactiae* and *Streptococcus uberis. Zentralbl. Bakteriol.* **249:**190–194.

149. **Skinner, S. J., and G. C. Liggins.** 1981. Glycosaminoglycans and collagen in human amnion from pregnancies with and without premature rupture of the membranes. *J. Dev. Physiol.* **3:**111–121.

150. **Smith, C. L., D. G. Pritchard, and B. M. Gray.** 1991. Program Abstr. 31st Intersci. Conf. Antimicrob. Agents Chemother., abstr.

151. **Sobel, J. D., P. Myers, M. E. Levison, and D. Kaye.** 1982. Comparison of bacterial and fungal adherence to vaginal exfoliated epithelial cells and human vaginal epithelial tissue culture cells. *Infect. Immun.* **35:**697–701.

152. **Spellerberg, B., B. Pohl, G. Haase, S. Martin, J. Weber-Heynemann, and R. Lutticken.** 1999. Identification of genetic determinants for the hemolytic activity of *Streptococcus agalactiae* by ISS1 transposition. *J. Bacteriol.* **181:**3212–3219.

153. **Spellerberg, B., E. Rozdzinski, S. Martin, J. Weber-Heynemann, N. Schnitzler, R. Lutticken, and A. Podbielski.** 1999. Lmb, a protein with similarities to the LraI adhesin family, mediates attachment of *Streptococcus agalactiae* to human laminin. *Infect. Immun.* **67:**871–878.

154. **Spellerberg, B., S. Martin, C. Brandt, and R. Lutticken.** 2000. The *cyl* genes of *Streptococcus agalactiae* are involved in the production of pigment. *FEMS Microbiol. Lett.* **188:**125–128.

155. **Stalhammar-Carlemalm, M., L. Stenberg, and G. Lindahl.** 1993. Protein rib: a novel group B streptococcal cell surface protein that confers protective immunity and is expressed by most strains causing invasive infections. *J. Exp. Med.* **177:**1593–1603.

156. **Stalhammar-Carlemalm, M., T. Areschoug, C. Larsson, and G. Lindahl.** 1999. The R28 protein of *Streptococcus pyogenes* is related to several group B streptococcal surface proteins, confers protective immunity and promotes binding to human epithelial cells. *Mol. Microbiol.* **33:**208–219.

157. **Takahashi, S., E. E. Adderson, Y. Nagano, N. Nagano, M. R. Briesacher, and J. F. Bohnsack.** 1998. Identification of a highly encapsulated, genetically related group of invasive type III group B streptococci. *J. Infect. Dis.* **177:**1116–1119.

158. **Tamura, G. S., and C. E. Rubens.** 1994. Host-bacterial interactions in the pathogenesis of group B streptococcal infections. *Curr. Opin. Infect. Dis.* **7:**317–322.

159. **Tamura, G. S., and C. E. Rubens.** 1995. Group B streptococci adhere to a variant of fibronectin attached to a solid phase. *Mol. Microbiol.* **15:**581–589.

160. **Tamura, G. S., J. M. Kuypers, S. Smith, H. Raff, and C. E. Rubens.** 1994. Adherence of group B streptococci to cultured epithelial cells: roles of environmental factors and bacterial surface components. *Infect. Immun.* **62:**2450–2458.

161. **Tamura, G. S., A. Nittayajarn, and D. L. Schoentag.** 2002. A glutamine transport gene, *glnQ*, is required for fibronectin adherence and virulence of group B streptococci. *Infect. Immun.* **70:**2877–2885.

162. **Tapsall, J. W., and E. A. Phillips.** 1991. The hemolytic and cytolytic activity of group B streptococcal hemolysin and its possible role in early onset group B streptococcal disease. *Pathology* **23:**139–144.

163. **Teti, G., F. Tomasello, M. S. Chiofalo, G. Orefici, and P. Mastroeni.** 1987. Adherence of group B streptococci to adult and neonatal epithelial cells mediated by lipoteichoic acid. *Infect. Immun.* **55:**3057–3064.

164. **Teti, G., G. Mancuso, F. Tomasello, and M. S. Chiofalo.** 1992. Production of tumor necrosis factor-alpha and interleukin-6 in mice infected with group B streptococci. *Circ. Shock* **38:**138–144.

165. **Teti, G., G. Mancuso, and F. Tomasello.** 1993. Cytokine appearance and effects of anti-tumor necrosis factor alpha antibodies in a neonatal rat model of group B streptococcal infection. *Infect. Immun.* **61:**227–235.

166. **Tettelin, H., V. Masignani, M. J. Cieslewicz, J. A. Eisen, S. Peterson, M. R. Wessels, I. T. Paulsen, K. E. Nelson, I. Margarit, T. D. Read, L. C. Madoff, A. M. Wolf, M. J. Beanan, L. M. Brinkac, S. C. Daugherty, R. T. DeBoy, A. S. Durkin, J. F. Kolonay, R. Madupu, M. R. Lewis, D. Radune, N. B. Fedorova, D. Scanlan, H. Khouri, S. Mulligan, H. A. Carty, R. T. Cline, S. E. Van Aken, J. Gill, M. Scarselli, M. Mora, E. T. Iacobini, C. Brettoni, G. Galli, M. Mariani, F. Vegni, D. Maione, D. Rinaudo, R. Rappuoli, J. L. Telford, D. L. Kasper, G. Grandi, and C. M. Fraser.** 2002. Complete genome sequence and comparative genomic analysis of an emerging human pathogen, serotype V *Streptococcus agalactiae. Proc. Natl. Acad. Sci. USA* **99:**12391–12396.

167. **Tsaihong, J. C., and D. E. Wennerstrom.** 1983. Effect of carrier molecules on production and properties of extracellular hemolysin produced by *Streptococcus agalactiae. Curr. Microbiol.* **31:**5–9.

168. **Ulett, G. C., J. F. Bohnsack, J. Armstrong, and E. E. Adderson.** 2003. Beta-hemolysin-independent induction of apoptosis of macrophages infected with serotype III group B streptococcus. *J. Infect. Dis.* **188:**1049–1053.

169. **Valentin-Weigand, P., and G. S. Chhatwal.** 1995. Correlation of epithelial cell invasiveness of group B streptococci with clinical source of isolation. *Microb. Pathog.* **19:**83–91.

170. **Valentin-Weigand, P., H. Jungnitz, A. Zock, M. Rohde, and G. S. Chhatwal.** 1997. Characterization of group B streptococcal invasion in HEp-2 epithelial cells. *FEMS Microbiol. Lett.* **147:**69–74.

171. **Vallejo, J. G., C. J. Baker, and M. S. Edwards.** 1996. Roles of the bacterial cell wall and capsule in induction of tumor necrosis factor alpha by type III group B streptococci. *Infect. Immun.* **64:**5042–5046.

172. **Vallette, J. D., Jr., R. N. Goldberg, C. Suguihara, T. Del Moral, O. Martinez, J. Lin, R. C. Thompson, and E. Bancalari.** 1995. Effect of an interleukin-1 receptor antagonist on the hemodynamic manifestations of group B streptococcal sepsis. *Pediatr. Res.* **38:**704–708.

173. **Villamor, E., F. Perez Vizcaino, J. Tamargo, and M. Moro.** 1996. Effects of group B *Streptococcus* on the responses to U46619, endothelin-1, and noradrenaline in

isolated pulmonary and mesenteric arteries of piglets. *Pediatr. Res.* **40:**827–833.

174. **von Hunolstein, C., A. Totolian, G. Alfarone, G. Mancuso, V. Cusumano, G. Teti, and G. Orefici.** 1997. Soluble antigens from group B streptococci induce cytokine production in human blood cultures. *Infect. Immun.* **65:**4017–4021.

175. **Wennerstrom, D. E., J. C. Tsaihong, and J. T. Crawford.** 1985. Evaluation of the role of hemolysin and pigment in the pathogenesis of early onset group B streptococcal infection, p. 155–156. *In* Y. Kimura, S. Kotami, and Y. Shiokowa (ed.), *Recent Advances in Streptococci and Streptococcal Diseases.* Reedbooks, Bracknell, United Kingdom.

176. **Wessels, M. R., V. Pozsgay, D. L. Kasper, and H. J. Jennings.** 1987. Structure and immunochemistry of an oligosaccharide repeating unit of the capsular polysaccharide of type III group B Streptococcus. A revised structure for the type III group B streptococcal polysaccharide antigen. *J. Biol. Chem.* **262:**8262–8267.

177. **Wessels, M. R., C. E. Rubens, V. J. Benedi, and D. L. Kasper.** 1989. Definition of a bacterial virulence factor: sialylation of the group B streptococcal capsule. *Proc. Natl. Acad. Sci. USA* **86:**8983–8987.

178. **Wessels, M. R., V. J. Benedi, D. L. Kasper, L. M. Heggen, and C. E. Rubens.** 1991. The type III capsule and virulence of group B Streptococcus, p. 219–223. *In* G. M. Dunny, P. P. Cleary, and L. L. McKay (ed.), *Genetics and Molecular Biology of Streptococci, Lactococci, and Enterococci.* American Society for Microbiology, Washington, D.C.

179. **Wibawan, I. T., C. Lammler, and F. H. Pasaribu.** 1992. Role of hydrophobic surface proteins in mediating adher-

ence of group B streptococci to epithelial cells. *J. Gen. Microbiol.* **138:**1237–1242.

180. **Wicken, A. J., and K. W. Knox.** 1975. Lipoteichoic acids: a new class of bacterial antigen. *Science* **187:**1161–1167.

181. **Wilkinson, H. W., and R. G. Eagon.** 1971. Type-specific antigens of group B type Ic streptococci. *Infect. Immun.* **4:**596–604.

182. **Williams, P. A., J. F. Bohnsack, N. H. Augustine, W. K. Drummond, C. E. Rubens, and H. R. Hill.** 1993. Production of tumor necrosis factor by human cells in vitro and in vivo, induced by group B streptococci. *J. Pediatr.* **123:**292–300.

183. **Wilson, C. B., and W. M. Weaver.** 1985. Comparative susceptibility of group B streptococci and *Staphylococcus aureus* to killing by oxygen metabolites. *J. Infect. Dis.* **152:**323–329.

184. **Winram, S. B., M. Jonas, E. Chi, and C. E. Rubens.** 1998. Characterization of group B streptococcal invasion of human chorion and amnion epithelial cells *in vitro.* *Infect. Immun.* **66:**4932–4941.

185. **Yeung, M. K., and S. J. Mattingly.** 1983. Isolation and characterization of type III group B streptococcal mutants defective in biosynthesis of the type-specific antigen. *Infect. Immun.* **42:**141–151.

186. **Yim, H. H., A. Nittayarin, and C. E. Rubens.** 1997. Analysis of the capsule synthesis locus, a virulence factor in group B streptococci. *Adv. Exp. Med. Biol.* **418:**995–997.

187. **Zawaneh, S. M., E. M. Ayoub, H. Baer, A. C. Cruz, and W. N. Spellacy.** 1979. Factors influencing adherence of group B streptococci to human vaginal epithelial cells. *Infect. Immun.* **26:**441–447.

Surface Structures of Group B Streptococci Important in Human Immunity

Lawrence C. Madoff, Lawrence C. Paoletti, and Dennis L. Kasper

14

With uncanny ability to interpret results obtained with a handful of immunological techniques that would be considered insensitive by today's standards, Rebecca Lancefield and coworkers in the 1930s set the stage for decades of research on *Streptococcus agalactiae*, also known as group B streptococcus (GBS). In contrast to group A streptococcus (GAS), which was isolated mainly from humans, the first strains of GBS studied by Lancefield were isolated from cows with mastitis or from "normal milk." At that time, GBS was a well-known cause of bovine mastitis and a concern mainly of the dairy industry (72).

The first report of GBS in humans appeared in 1938 (36). However, not until the 1960s, when several reports appeared in the clinical literature on the presence of hemolytic GBS among human newborns with sepsis and meningitis, did this pathogen receive serious attention from researchers (23, 32, 48). By the 1970s, GBS had eclipsed *Escherichia coli* as the most common bacterial cause of sepsis among newborns, with an incidence of early-onset disease of 2.9 per 1,000 births and a case-fatality rate of 46% (2). Despite variability among surveillance sites (reviewed in reference 116), rates of early-onset GBS disease remained high throughout the 1970s and 1980s, prompting the Institute of Medicine to declare in 1985 that the prevention of perinatal GBS disease was a national health priority (49).

Today, colonization with GBS—a risk factor for neonatal illness—has been reported in pregnant and nonpregnant adults and the elderly, not only in the United States (3, 116) but also worldwide (121). This situation accentuates the need to complete vaccine development in this country while considering a global vaccine initiative against this life-threatening pathogen. This chapter highlights critical advances in our understanding of the role(s) of GBS surface antigens (namely, the group B carbohydrate, the type-specific capsular polysaccharides [CPSs], and proteins) in immunity and their application as components of experimental vaccines.

CARBOHYDRATE ANTIGENS OF GBS

Group B Carbohydrate

The group B carbohydrate is an antigen common to all strains and serotypes of GBS. Positioned proximal to the cell wall (126), the group B carbohydrate is composed of rhamnose, galactose, *N*-acetylglucosamine, and glucitol. These sugars form four different oligosaccharide units (97) linked by phosphodiester bonds to create a complex and highly branched tetra-antennary structure (Fig. 1). The terminal position and abundance of rhamnose suggested that this sugar would constitute or be part of an immunodominant epitope; this hypothesis was proven by inhibition studies using a number of derivative oligosaccharides with polyclonal and monoclonal antibodies to the group B carbohydrate (98).

Lancefield et al. (77) were the first to demonstrate the inability of antibody to the group B carbohydrate to protect mice from lethal challenge with viable GBS. Indeed, the presence of maternal antibody to the group B carbohydrate was associated with poor neonatal outcome despite good agreement between levels of specific immunoglobulin G (IgG) in maternal and cord sera (1). This result confirmed in humans what was already known in animals: that the group B carbohydrate was not important to natural immunity.

Still, investigators wondered whether a group B carbohydrate vaccine would be effective in eliciting functional antibody if the carbohydrate's immunogenicity were enhanced. Rainard (109) demonstrated an improved immune response in cows when the group B carbohydrate was coupled to ovalbumin. Antibody induced to this conjugate vaccine was of the IgG class, was specific for the major sugars of the group B carbohydrate, and was opsonically active against an unencapsulated GBS strain of bovine origin (109). That high-titered IgG specific for the group B carbohydrate was less active against highly encapsulated GBS strains of human origin was discerned in studies of a group B

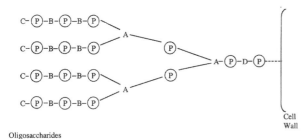

Oligosaccharides

A: α-L-Rhap-(1→3)- α-D-Galp-(1→3)-ß-D-Glcp NAc-(1→4)- α-L-Rhap-(1→2)-
[α-L-Rhap (1→3) α-D- Galp-(1→3)-ß-D-Glcp NAc-(1→4)-]- α-L-Rhap-(1→2)-
α-L-Rhap-(1→1')-D Glucitol-(3'-1)- α-L-Rhap.

B: α-L-Rhap-(1→2)-[α-L-Rhap-(1→3) α-D- Galp-(1→3)-ß-D-Glcp NAc-(1→4)-]-
α-L-Rhap (1→2)- α-L-Rhap-(1→1')-D-Glucitol-(3'-1)- α-L-Rhap.

C: α-L-Rhap-(1→2)- α-L-Rhap-(1→2)- α-L-Rhap-(1→1')-D-Glucitol-(3'-1)- α-L-
Rhap.

D: α-L-Rhap-(1→3)- α-D-Galp-(1→3)-ß-D-Glcp NAc-(1→3)- α-L-Rhap-(1→3)- α-
L-Rhap (1→3)- α-L-Rhap-(1→3)-ß-L-Rhap-(1→4)-D-GlcNAc.

FIGURE 1 Proposed structure of the group B carbohydrate antigen, as modified from reference 98. The dashed line indicates that the linkage between the group B carbohydrate and the cell wall has not been discerned.

carbohydrate-tetanus toxoid conjugate vaccine (92). Adsorption studies using rabbit antisera to this vaccine with GBS strains of human origin that were highly encapsulated or with an isogenic mutant strain that lacked CPS showed CPS interference with specific antibody binding (92); this finding confirmed the inability of group B-specific antibody to bind to the group B antigen on highly encapsulated GBS cells. Not surprisingly, antibody to bovine isolates that possess low levels of CPS may be effective in combating GBS disease in the cow, whereas the same antibody would not be effective against more highly encapsulated GBS of human origin. The ability of group B antibody to bind to poorly encapsulated GBS deserves further attention, especially because the expression of CPS by a GBS strain (of human origin) has been shown to be regulated by the cell's rate of growth (106), whereas the expression of the group B carbohydrate is not influenced by growth rate (112).

Chemistry of Group B Streptococcal CPSs

With few exceptions, all strains of GBS isolated from humans are encapsulated (Fig. 2A) and can be classified on the basis of serology and CPS structure. Nine distinct GBS serotypes have thus far been identified: Ia, Ib, II, III, IV, V, VI, VII, and VIII. In the past, serotypes Ia, Ib, II, and III were equally prevalent in normal vaginal carriage and early-onset sepsis (i.e., that developing at less than 7 days of age) (3). However, type V strains have emerged as an important cause of GBS infection (111), and strains of types VI and VIII have become prevalent among Japanese women (68). The chair models in Fig. 3 highlight the heterogeneity among the structures of GBS CPSs—this despite the fact that each repeating unit contains four of only five sugars (glucose, galactose, N-acetylglucosamine, rhamnose, and sialic acid). Recent analyses revealed that some of the sialic acid residues were O-acetylated (at position C7, C8, or C9)

in addition to being N-acetylated at C5 (79). Although not exhaustively studied, the percentage of O-acetylation on 10 GBS strains tested ranged from ~5 to ~55% (79), a result that should be considered when viewing the structures in Fig. 3.

Types Ia and Ib

Lancefield (73) was the first to notice serological cross-reactions between strains bearing Ia and Ib antigens, although she did not know what cellular structure contained the differing epitopes. Nuclear magnetic resonance analysis, performed 45 years after Lancefield's observation, showed that the native CPSs of GBS types Ia and Ib indeed are structurally similar, differing only in the linkage of the side chain galactose to N-acetylglucosamine (52); the type Ia CPS has a β-(1→4) linkage, and the type Ib CPS has a β-(1→3) linkage in this position. These linkages are critical to the immunospecificity of these CPSs. The repeating unit of Ia and Ib CPSs is a pentasaccharide with a disaccharide backbone and a trisaccharide side chain. Like all GBS CPSs, each possesses an α-(2→3)-linked sialic acid as a terminal side-chain saccharide.

Type II

The native type II CPS repeating unit structure is composed of galactose, glucose, N-acetylglucosamine, and sialic acid in a 3:2:1:1 molar ratio (53). This repeating unit structure has two side chains—a sialic acid residue and a galactose residue—linked separately to the pentasaccharide backbone.

Type III

The type III CPS has a trisaccharide backbone of glucose, N-acetylglucosamine, and galactose, and a disaccharide side chain of sialic acid and galactose linked β-(1→4) to the backbone N-acetylglucosamine (128). Although sialic acid per se is not an immunodominant saccharide, the negatively charged carboxylate group on this sugar influences the conformation and thus the immunodeterminant epitopes of

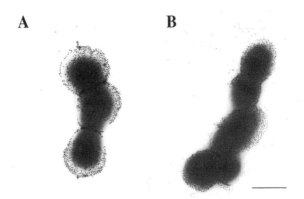

FIGURE 2 Immunogold-labeled electron micrographs showing surface localization of type Ia CPS (A) and the alpha C protein (B) of GBS strain A909. For CPS type Ia staining, rabbit antiserum to the CPS, followed by 20-nm-diameter-gold-labeled protein A, was used. For alpha C protein, rabbit antiserum to the purified one-repeat alpha C protein, followed by 15-nm-diameter-gold-labeled protein A, was used. Bar, 500 nm. (Reproduced from *Infection and Immunity* [41] with permission.)

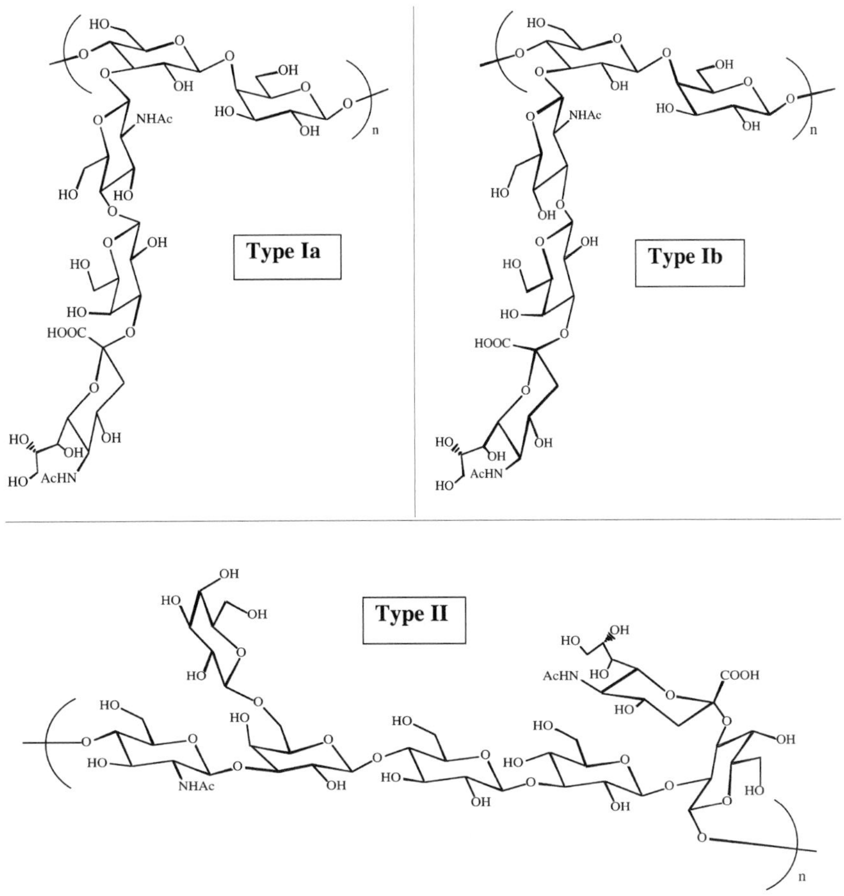

FIGURE 3   Chair models of the repeating units of GBS capsular polysaccharides.

the entire polymer (51, 127). Removal of the terminal sialic acid from the repeating unit of the type III CPS with mild acid or neuraminidase results in a CPS with chemical and immunological identity to the pneumococcal type 14 CPS (59)—a finding that has enhanced our understanding of the roles of sialic acid in GBS immunity, as discussed further below.

Type IV

The GBS type IV CPS contains galactose, glucose, N-acetylglucosamine, and sialic acid in a 2:2:1:1 molar ratio (129). The presence of sialic acid does not appear to be a critical immunodeterminant for type IV CPS, as it is for other GBS CPSs (129).

Type V

The repeating unit structure of serotype V CPS is composed of seven sugars. The CPS has a trisaccharide backbone consisting of glucose, galactose, and glucose, as well as two side chains. One side chain consists of glucose linked β-(1→3) to the backbone galactose. The other side chain is a trisaccharide of N-acetylglucosamine, galactose, and sialic acid. This side chain is linked β-(1→6) to a backbone glucose and terminates with sialic acid (132).

Type VI

The type VI CPS repeating unit comprises galactose, glucose, and sialic acid in a 2:2:1 molar ratio (63, 124). Type

VI was the first GBS CPS described that lacked N-acetylglucosamine.

Type VII

The GBS type VII repeating unit contains a trisaccharide backbone and a trisaccharide side chain with a molar ratio of 2:2:1:1 for galactose, glucose, N-acetylglucosamine, and sialic acid (63a).

Type VIII

The structure of GBS CPS type VIII is the most unusual and simplest structure of all GBS serotypes. Like type VI CPS, the type VIII CPS lacks N-acetylglucosamine; yet unlike any other GBS CPS this tetrasaccharide contains rhamnose, a sugar previously associated exclusively with the group B carbohydrate. The type VIII repeating unit contains glucose, galactose, rhamnose, and sialic acid in 1:1:1:1 molar ratio, with the sialic acid as the monosaccharide side chain (64).

Role of Sialic Acid as a Structural Feature and a Virulence Factor

Hot HCl or cold trichloroacetic acid (TCA) extraction methods used by Lancefield and Freimer (75) resulted in GBS antigens that were incomplete immunochemically. These investigators surmised that the missing component contributed to the serological specificity and negative charge of the CPS antigen. That GBS type-specific CPSs

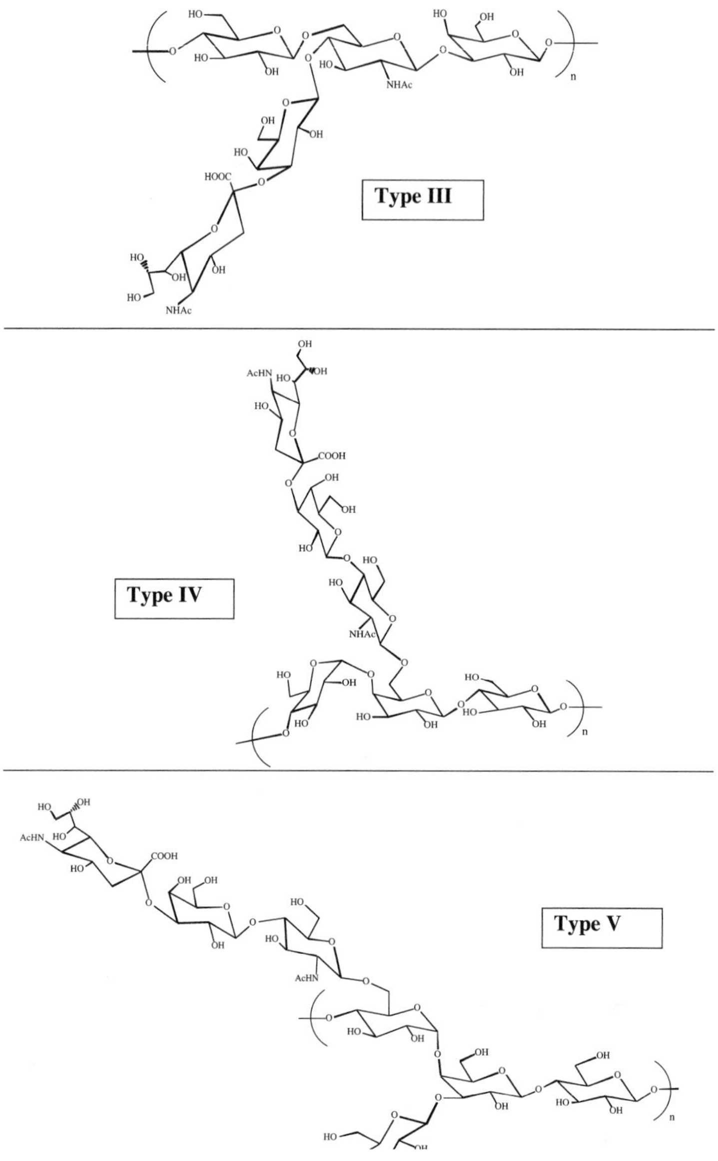

FIGURE 3 (*Continued*)

contain sialic acid was cited as a personal communication to Lancefield by Liu in 1972—a discovery that verified sialic acid indeed "represents the missing antigenic component of the carbohydrate in the TCA antigen" (74). The acidic nature of the TCA extraction procedure resulted in the purification of an incomplete antigen because the α-(2→3) ketosidic linkage between sialic acid and galactose—a bond that is universal among all GBS type-specific CPSs—is acid labile.

Structural analysis of GBS type III CPS showed that the desialylated or core structure was identical to that of *Streptococcus pneumoniae* type 14 CPS. This serendipitous observation helped to define the importance of sialic acid on type III CPS in the formation of a conformationally dependent epitope (127, 128). Specifically, carboxylate groups on sialic acid residues form intramolecular hydrogen bonds with the galactose residues in the backbone of the type III CPS (51). Reduction of the carboxylate groups to hydroxymethyl groups destroyed the CPS antigenicity with antisera to the native type III CPS, while oxidation of the carboxylate groups with sodium *meta*-periodate to remove carbons 8 and 9 did not affect antigenicity. The finding that GBS CPS retains its antigenicity upon oxidation with sodium periodate became important in the design of a coupling strategy to produce GBS conjugate vaccines. The requirement for sialic acid to impart antigenicity on GBS CPSs is not universal, however. Whereas the sialic acid residue on type Ia CPS is required for complete antigenicity, the binding of specific antisera to the structurally related type Ib does not depend on sialic acid (115).

The presence of sialic acid on GBS creates a surface that does not activate the alternative pathway of complement (31). In contrast, GBS organisms grown in the presence of neuraminidase—a condition that results in a lower degree of surface sialylation—support complement activation in the presence or absence of specific antibody. Marques et al. (91) reported lower amounts of C3 deposition on GBS organisms that possessed a sialylated as opposed to a desialy-

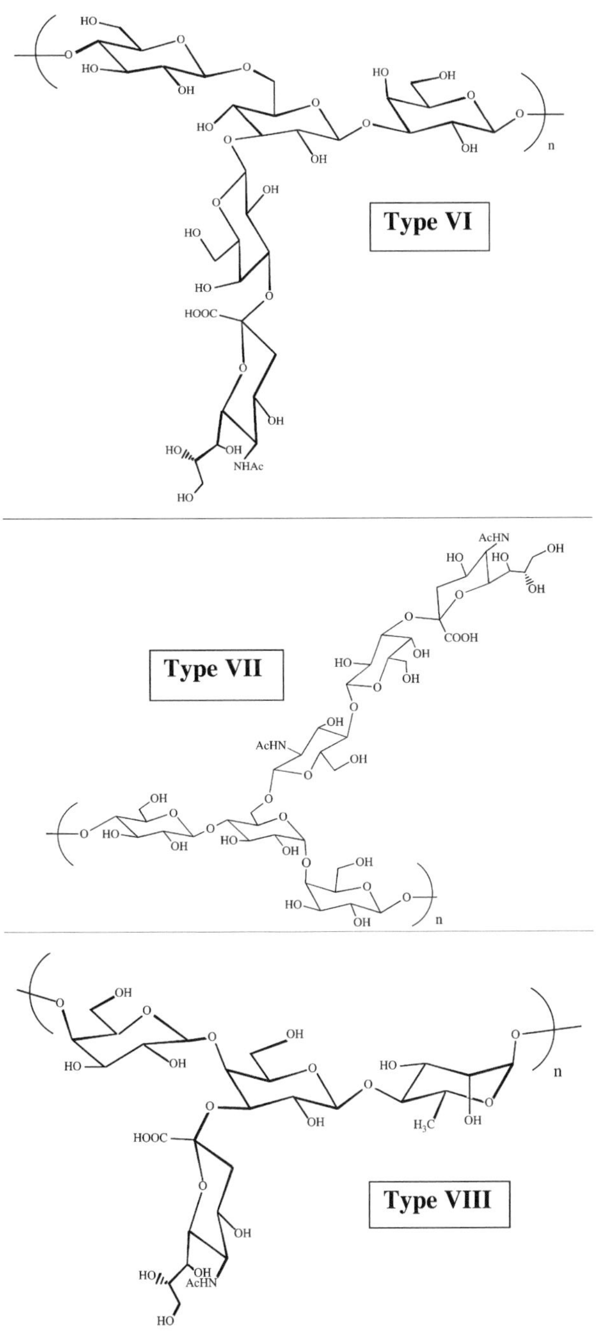

Type VI

Type VII

Type VIII

FIGURE 3 *(Continued)*

The role of sialic acid as a virulence factor of GBS was confirmed in studies with a transposon mutant of type III strain COH 31r/s that lacked the ability to synthesize sialylated CPS (130). GBS strain COH 31-21 produced an asialo type III CPS that bound to pneumococcal type 14 CPS but not native type III CPS antisera. The virulence of the asialo mutant was two orders of magnitude lower than that of the wild-type parent strain and equal to that of another transposon mutant, COH 31-15, that completely lacked CPS (130).

Genetics of CPS Antigens

The transposon mutants described above were the first used to map the genes important in GBS CPS expression (69). Since then, complete genomic sequences for the CPS biosynthesis regions for all of the known serotypes of GBS have been reported (Fig. 4) (28). As in many encapsulated bacteria, genes conserved across diverse capsular serotypes flank genes encoding enzymes unique to a specific capsule serotype. In GBS, a central group of serotype-specific glycosyltransferases and polymerases is flanked on one side (right in Fig. 4) by genes encoding enzymes that synthesize and activate sialic acid, the terminal sugar on the side chain of all nine serotypes (27, 44). Flanking the other side of the serotype-specific glycosyltransferases (left side of Fig. 4) is a group of genes hypothesized to function in export of the polysaccharide capsule (27). In analyzing the GBS capsular biosynthesis clusters of the different GBS capsular types, Cieslewicz et al. found strong evidence for horizontal gene transfer resulting in both intra- and interspecies recombination events (28). Although diversity in GBS capsules was thought to be due to immune pressure, the relatively limited number of distinct capsule types might be related to structural elements required for pathogenicity.

PROTEIN ANTIGENS OF GBS

Although the importance of surface proteins in immunity to *Streptococcus pyogenes* has long been appreciated, their role in immunity to GBS was recognized only in the 1970s,

lated CPS. In all, these findings suggested a critical role for sialic acid in evading the host's natural immune mechanisms and thus in modulating GBS virulence.

The role of the complement pathway in the immune response to sialylated antigen (GBS type III CPS) and desialylated antigen (*S. pneumoniae* type 14 CPS) was examined in untreated mice and in mice treated with cobra venom factor to cause complement depletion (90). Complement depletion did not affect the immune response to the sialylated antigen, but this condition markedly reduced the specific antibody response to desialylated CPS. These results demonstrated the importance of the CPS structure in complement activation and in the generation of a specific antibody response (90).

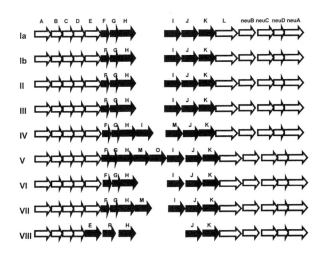

FIGURE 4 Schematic of *cps* gene clusters in all nine GBS serotypes. Conserved genes are depicted by white arrows. Variable genes are depicted by black arrows. The variable gap in the center of the *cps* clusters was introduced to permit alignment of homologous genes (28).

when Lancefield et al. (77) described the cross-serotype protection afforded by rabbit antiserum to protease-sensitive epitopes. The intervening years have witnessed considerable research in this area. This work has led to a new appreciation of the diversity of GBS surface proteins and an increased awareness of their roles in immunity and pathogenicity. It is now apparent that most GBS strains express antigenic surface proteins (Table 1). Much of the interest in these proteins arises from a desire to identify protective antigens for inclusion in GBS vaccines. Discussed here are proteins that may be of interest for GBS vaccines because they elicit protective immunity in models of infection. None have yet been tested in humans.

Alphalike Proteins: Tandem Repeat-Containing Proteins of GBS

Alpha C Protein

Wilkinson and Eagon (136) described a protein extract obtained from a serotype Ic strain of GBS. On the basis of this description, Lancefield recognized that the previously designated serotype Ic in fact comprised the Ia polysaccharide and the Ibc protein. In seminal experiments, Lancefield and coworkers (77) demonstrated that antibodies to the protein component protected mice from lethal challenge with either Ia or Ib capsular serotype GBS strains that bore the Ibc proteins. Even in their initial work, Wilkinson and Eagon (136) recognized that the protein extract from the Ic GBS strain contained two moieties, one sensitive to digestion with trypsin, the other resistant. Bevanger and Maeland (16) described the trypsin-resistant component as the alpha antigen of the Ibc protein and the trypsin-sensitive component as the beta antigen. The term C proteins was coined to clarify the distinction between the proteins and the carbohydrate antigens; the presence of the C protein(s) was denoted by the letter following the capsular serotype (e.g., Ia/c) (47). Later work by a number of investigators identified the alpha and beta components as distinct proteins, independently expressed in different strains, and encoded by separate genes (84, 85, 94, 113, 114). Generally present in strains of Ia and Ib capsular serotype and frequently present in type II strains, these antigens are occasionally found in GBS of all serotypes. The trypsin-resistant antigen is now referred to as the alpha C protein (Fig. 2B) and the trypsin-sensitive antigen as the beta C protein. In

one survey, these proteins were found in 59% of strains in a collection of 785 clinical isolates from the University of Minnesota (58). All Ib strains expressed C proteins, and 84% expressed both alpha and beta; 91% of Ia strains and 81% of type II strains expressed the alpha C protein. In a similar survey, Bevanger (15) found that 52% of human isolates of GBS expressed the alpha C protein and 25% expressed the beta C protein.

Monoclonal antibodies have been used to identify the alpha C protein and to define it as a mouse-protective epitope in GBS (85). In Western blot analysis, monoclonal antibody reacted with the vast majority of strains of GBS type Ia (all of 14 strains), type Ib (11 of 15 strains), and type II (all of 12 strains). Western blots of surface extracts from GBS strains also demonstrated an unusual phenotypic characteristic of this protein: a pattern of regularly spaced bands with a periodicity of approximately 8 kDa. Also evident on Western blots was size heterogeneity of the expressed protein from strain to strain, with the size of the largest band varying from <70 kDa to >160 kDa in a panel of 37 isolates. An interesting observation was that the protein sizes did not appear to be randomly distributed but clustered around 100 kDa, with relatively few isolates expressing either a very large or a very small alpha C protein.

The gene encoding the alpha C protein was cloned from the prototype Ia/c strain A909 from the Lancefield collection (95). This gene, *bca*, consists of 3,063 nucleotides and encodes several identifiable regions. A signal sequence with homology to other gram-positive surface proteins was 56 amino acids in length (119). This sequence was followed by an amino-terminal domain of 170 amino acids with a predicted molecular mass of 18.8 kDa. The most striking feature of the gene was a series of nine long tandem repeats that had 246 nucleotides each (encoding 82 amino acids, or 7.9 kDa) and that were identical at the nucleotide level. The repeat region composed >75% of the mature protein, and the correspondence between the size of the repeat region and the distance between the bands on the Western blot suggested that the repeats gave rise to the ladder pattern seen on Western blots. Antibodies to both the repeat region and the amino-terminal domain have been shown to be protective in an animal model and to mediate opsonophagocytic killing in vitro (62). The carboxyl terminus showed homology to those of other gram-positive surface proteins, including an LPXTG domain (identified as a peptidoglycan-binding motif), a hydrophobic membrane-spanning region, and a charged tail thought to be involved in orienting and anchoring the protein to the cell wall (95).

A size discrepancy between the alpha C proteins of GBS isolates from mothers and those from their neonates suggested that a loss of repeats within the protein gene may occur during colonization or infection with GBS (89). This phenomenon was confirmed experimentally in an immune mouse model. Although spontaneous deletions within the alpha C protein occur at a low rate (approximately 10^{-4}), passage through mice given antiserum to the alpha C protein gave rise to deletion rates approaching 50%. Moreover, the mutants bound poorly to the antiserum and were more poorly opsonized for phagocytic killing than the wild-type strain. In vitro experiments showed that purified alpha C proteins containing 9 or 16 repeats bound with high affinity to antibodies to wild-type (9-repeat) alpha C protein, whereas 1- and 2-repeat alpha C proteins bound with lower overall affinity (39). These differences were not due simply to a reduced number of antibody-binding sites but appeared to reflect a conformational dependence of the protein on

TABLE 1 Characteristics of selected group B streptococcal protein antigens

Designation	Antisera protective?	Gene	Associated GBS serotype(s)
Alpha-like proteins			
Alpha	Yes	*bca*	Ia, Ib, II
Alpha-like protein 1 (epsilon)		*alp1*	Ia
Rib protein (R4)	Yes	*rib*	III
Alp2	Yes	*alp2*	V, VIII
Alp3 (R28)		*alp3*	V, VIII
Beta C protein	Yes	*bac*	Ib
C5a peptidase	Yes	*scpb*	All
Sip	Yes	*sip*	All
BPS	Yes	*sar5*	All

repeat number. In contrast, antibodies to the 1-repeat protein appeared to bind to all of the different constructs with similar overall affinity. Antibodies to the single- or double-repeat proteins showed greater protective efficacy in vivo and tended to allow fewer deletion mutants to escape. Further studies demonstrated that immunogenicity in mice was inversely related to repeat number (40): 9- and 16-repeat alpha C proteins evoked lower titers of antibody, particularly to the amino-terminal domain, than did the 1- and 2-repeat variants. These findings suggested that modulation of repeat number may be a mechanism for interaction with the host's immune system.

A genetically constructed null mutant of the alpha C protein in GBS strain A909 exhibited virulence that was modestly attenuated (by <1 $\log_{10}$) relative to that of the wild-type in neonatal mice (80). The decreased mortality evident in the first 24 h after infection suggested a role in the early stages of pathogenesis. A single-repeat alpha C protein mutant in the same strain, derived by passage through immune mice, showed equivalent pathogenicity in naive mice but displayed enhanced lethality (>2 $\log_{10}$) in mice first immunized against the protein (41). Recent evidence indicates that the alpha C protein acts to promote internalization into cervical epithelial cells via association with a host glycosaminoglycan (13, 19). Experimentally, this interaction was inhibited by the isolated amino-terminal domain of the alpha C protein, suggesting that the activity lies within this domain.

Rib Protein

Although the alpha C protein is present mainly in GBS strains of types Ia, Ib, and II, a phenotypically similar but immunologically distinct protein, designated Rib, has been discovered in a type III GBS strain (119). This protein, present in 31 of 33 clinical type III isolates studied, showed a laddering phenotype on sodium dodecyl sulfate-polyacrylamide gel electrophoresis (SDS-PAGE), resistance to trypsin digestion, and variability in molecular size among clinical isolates. Antiserum to the protein protected between 62 and 90% of mice from challenge with GBS but did not cross-protect the animals against infection by alpha-positive GBS. Amino-terminal sequencing of the protein demonstrated homology to the alpha C protein with identity in 6 of 12 residues. When cloned and sequenced, the gene encoding this protein showed structural as well as sequence similarity to the alpha C protein (126). Like *bca*, the *rib* gene contains a signal sequence, an amino-terminal region, a series of tandem repeats, and a carboxy-terminal membrane anchor region (Fig. 5). Thus, a family of genetically similar, tandem repeat-containing proteins was identified in GBS. Rib has been shown to be identical to the previously recognized protein R4 (see below).

Other Alphalike Proteins

During the 1980s, the previously obscure GBS serotype V emerged as an important pathogen in the United States. A novel protein identified in many type V strains was phenotypically similar to the alpha C and Rib proteins (66). Antiserum to the purified protein recognized a homologous protein (of variable size) in 25 of 41 clinical isolates. This antiserum protected 78% of neonatal mice from GBS challenge. Thus, the protein exhibited the characteristic features of this family of proteins: (i) protective efficacy against infection with a homologous GBS strain in a mouse model; (ii) laddering phenotype on SDS-PAGE with variable molecular size among clinical isolates; and (iii) resistance to trypsin digestion. In addition, this protein displayed amino acid sequence similarity both to the alpha C protein (17 of 18 evaluable residues) and to the Rib protein (5 of 10 residues). Further analysis revealed two further alphalike genes, Alp2 and Alp3, found most often in GBS of serotypes V and VIII. These displayed mosaicism—each contains elements of other GBS proteins while preserving the overall structure found in alpha C protein and Rib of an amino-terminal domain, a series of long, tandem repeats, and a highly conserved carboxy-terminal domain (69). Alp3, in fact, was identical to R28, a protein described to act as an adhesin in GAS (120). Both proteins contained amino termini with high degrees of similarity to alpha and rib and repeating units that were identical to those found in Rib. These findings strongly suggested inter- and intraspecies recombination of genetic material, a phenomenon subsequently noted in the capsular biosynthesis gene cluster.

Brady et al. (21) recognized a gamma C protein when polyclonal antiserum to the prototype C-protein strain A909 was absorbed with another alpha C protein-positive strain. Residual activity against surface proteins was recognized in several strains. Gamma protein reactivity was found only in alpha-positive strains. Michel et al. (96) identified a protein in some alpha-positive GBS strains that reacted with monoclonal antibody to alpha C protein but varied in restriction digest pattern at the genetic level and in morphology on SDS-PAGE. Partial nucleotide sequencing of the gene encoding this alpha C protein variant showed sequence divergence from *bca* in the amino-terminal domain, and the variant was designated Alp1 or epsilon. It appears likely that the gamma protein reactivity identified by Brady et al. was, in fact, due to differences between alpha C protein and Alp1 and was seen when antiserum raised to the prototype alpha C protein was absorbed with an Alp1-positive GBS strain (21). The reactivity was probably a marker for antigenic epitopes present in the original alpha C protein but not in Alp1. Further sequencing of alp1 reveals repeats identical to those found in alpha C protein.

R Proteins

R proteins, also identified by their resistance to trypsin digestion, were initially recognized in *S. pyogenes* by Lancefield and Perlmann (76). Wilkinson (135) identified four immunologically distinct R proteins (R1, R2, R3, and R4) occurring in various combinations in streptococci of groups A, B, and C. Flores and Ferrieri (35) found R proteins of

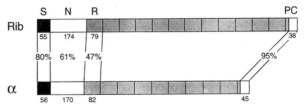

FIGURE 5 Comparison of the Rib and alpha C proteins. Overall structure of Rib from strain BM110 and alpha C protein from strain A909 and degree of amino acid residue identity between different regions of the proteins. S, signal peptide; N, NH₂-terminal region; R, one repeat; P, partial repeat; C, COOH-terminal region. The number of amino acids in each region is indicated. The Rib protein has 12 repeats of 79 amino acids, and the alpha C protein has 9 repeats of 82 amino acids. (Reproduced from the *Journal of Biological Chemistry* [127] with permission.)

subtypes R1 and R4 in 37% (49 of 131) of GBS clinical isolates. The proteins were present predominantly in type II and III strains. Linden (82) showed that rabbit antiserum to an affinity-purified R protein from type III GBS protected mice against invasive infection by type II R-positive strains but not against that by type III R-positive strains. The presence of R protein antibodies in human serum correlated with the absence of invasive infection by R protein-expressing GBS strains in human neonates (83).

The R proteins appear to share two important phenotypic characteristics with the members of the tandem repeat-containing family of GBS surface proteins: trypsin resistance and protective capacity. Bevanger et al. (17) isolated a protein from a serotype III strain, designated it an R-associated protein, and found that it was biochemically and immunologically highly similar to both R4 and Rib. Similarly, Lachenauer and Madoff (66) showed that Alp2 from a GBS type V strain was reactive with antiserum to R1. Further characterization of the R proteins will probably reveal that they are members of the same family as the other tandem repeat-containing GBS proteins and that the genes encoding them possess a similar structure. Indeed, Smith et al. concluded that Rib was identical to R4 based upon partial amino acid and genetic sequences (118).

Beta C Protein

The beta C protein was initially recognized as the trypsin-susceptible component of the C protein (136). Subsequent work has demonstrated its expression on the surface of approximately 10 to 25% of GBS strains, with predominance in Ib strains (15, 58, 84). Russell-Jones et al. (116) recognized that this antigen was a 130-kDa protein and that it bound specifically to human IgA. This binding occurs via the Fc portion of the IgA heavy chain and is of high affinity, with an affinity constant of 3.5×10^8 M^{-1} (81). The affinity is considerably higher for serum than for secretory IgA. The gene for the beta C protein was cloned by several groups (30, 46, 55, 94), and antibody to the cloned gene product was demonstrated to have protective capacity against infection in mice (94). Two complete nucleotide sequences for the gene (*bac*) have been published and differ minimally; one is 1,134 amino acids in length (46) and the other 1,164 amino acids (55). Other than a single amino acid difference, the two sequences differ only in the number of short XPZ repeats found in the carboxy-terminal half of the protein. Both contain consensus gram-positive signal sequences and carboxy-terminal membrane anchor regions that have extensive homology to other gram-positive surface proteins (including the GBS tandem-repeating proteins discussed above). Both groups of researchers studying the beta C protein detected IgA-binding capacity in the amino-terminal half of the molecule; however, Jerlstrom et al. (55) detected IgA binding within two separate nonoverlapping subclones, whereas Heden et al. (46) detected only a single IgA-binding region. Jerlstrom et al. (56) later localized IgA binding to a 73-amino-acid region containing a MLKKIE sequence and found that this sequence was required for IgA binding. These observations suggest that only one IgA-binding region exists (as is the case with other immunoglobulin-binding proteins) and that Jerlstrom's earlier finding may have resulted from the use of polyclonal IgA, which reacted in an immune fashion with a portion of the protein not involved in IgA binding.

Antibodies elicited to the beta C protein appear to be highly protective (37, 86, 94). Immunization of mice with this protein protected their neonatal pups from invasive GBS infection at >90% efficacy, and antibodies facilitated bacterial opsonization for phagocytic killing by polymorphonuclear leukocytes (86). One group has found that antibodies to the beta C protein—unlike those to any other GBS antigen—are actually bactericidal in vitro, i.e., are able to kill GBS in the absence of polymorphonuclear leukocytes (37). The beta C protein has also served as an effective carrier protein in conjugate vaccines. A glycoconjugate containing equal quantities of beta C protein and type III CPS linked by reductive amination was broadly protective against invasive GBS infection in neonatal mice challenged with either type III GBS or a type Ia/C beta-positive strain (88). Rabbit antiserum to the conjugate vaccine was opsonic against multiple strains containing either type III CPS or beta C protein. Similar conjugates with beta C protein have also been made with type Ia, II, and III CPSs and have elicited antibody titers comparable to those elicited by CPS-tetanus toxoid (TT) conjugates (99). Lachenauer et al. (70) compared naturally acquired beta C protein-specific IgG levels in mothers colonized with beta-positive GBS with those found in mothers who gave birth to infants that developed invasive illness with beta-positive strains. In contrast to findings with antibody to several type-specific polysaccharides, no correlation was found between beta-specific IgG level and protection from invasive disease; however, only a small number of cases was available (70).

Other Surface Proteins

Sip Protein

Surface immunogenic protein (Sip), a surface-localized protein of approximately 45 kDa, has been shown to be present in GBS of all serotypes. Immunization of mice with native Sip elicited protection against challenge with several GBS serotypes (22). In a neonatal mouse protection assay, Martin and colleagues used recombinantly derived Sip to immunize mouse dams (93). Survival of neonatal pups whose dams had been immunized ranged from 75 to 98% compared with 0 to 12% of controls across a broad range of serotypes. No function has yet been assigned to the Sip protein.

BPS Protein

Erdogan et al. described a 105-kDa protein in a GBS strain and initially designated it as R5 because it was derived from an R-positive GBS isolate (34). Although initially designated an R-like protein, it bears no phenotypic relationship to other R proteins and has been renamed BPS for group B protective surface protein. In a mouse protection assay it was able to elicit antibodies that protected >80% of mice from lethal challenge with a single GBS strain, compared with 20% survival in controls.

C5a Peptidase

C5a peptidase is a surface-localized protease of approximately 125 kDa, present in both GAS and GBS, that has been shown to inactivate human C5a, an important component of complement-mediated neutrophil chemoattraction (29). In addition, C5a peptidase may play a role in bacterial invasion of epithelial cells and may bind to fibronectin (14, 26). There is evidence that human serum contains antibodies to this protein (18); in GAS, immunization with C5a peptidase elicited antibodies in mice and reduced colonization of the nasopharynx by several GAS strains (57). Although C5a peptidase antibodies do not appear to be opsonic against GAS, there is evidence of opsonic activity against GBS

(24). The potential for a C5a peptidase vaccine is discussed further with conjugate GBS vaccines below.

X Proteins

Many bovine GBS isolates express X protein. This protein has not been well characterized; however, in one study, an immunoblot of the protein showed multiple laddering bands that suggested homology with the tandem repeat-containing proteins (78). Immunization with an X protein elicited opsonic antibodies in cows (110).

Glutamine Synthetase and Surface Enolase

The glutamine synthetase of GBS, a 52-kDa protein that complemented this enzymatic function in *E. coli*, was found in various serotypes of GBS (122). Evidence suggesting surface expression of this protein included the finding of amino-terminal signal sequence and carboxy-terminal membrane anchor homology with other surface proteins and the presence of antibodies to the protein in serum raised to whole bacteria. Pancholi and Fischetti (101) described a 45-kDa plasminogen-binding protein—streptococcal surface enolase—on the surface of many streptococcal species, including GBS, and hypothesized a role for this protein in virulence. It has not been determined whether either of these proteins elicits protective antibody.

Genomic Identification of GBS Surface Proteins

The availability of complete genomic sequence for GBS isolates has allowed the systematic exploration of surface-expressed proteins to identify possible protective immunogens. To date, sequences for type III and type V isolates have been published (38, 123). Each revealed an AT-rich genome with approximately 2 million nucleotides and approximately 2,100 open reading frames. As expected, the GBS genome possesses great similarity to those of *S. pyogenes* and *S. pneumoniae*; the type V GBS genome shares 1,060 of its 2,144 predicted genes with both organisms and 401 genes with one of the species (but not both). Approximately 650 proteins were predicted to be membrane-associated or secreted in the type V genome, including 24 proteins containing an LPXTG (that predict surface exposure via a cell wall-anchoring motif), 51 lipoproteins, and 177 proteins carrying a signal sequence. Two hundred ninety-one of these genes were successfully expressed and used to raise antibodies in mice. Of these, 139 recognized a predominant band in Western blot and 55 appeared to be surface-expressed in GBS on the basis of fluorescence-activated cell sorting. In the type III genome, 21 genes encoded LPXTG proteins. In addition, 36 lipoproteins, many of which are likely to be surface-exposed, were predicted on the basis of characteristic sequences. Surface exposure of these proteins was not tested experimentally. These methods hold promise to transform the process of discovery of novel vaccine antigens. Such a reverse vaccinology approach has been utilized for other pathogens (43).

GBS VACCINES

CPS Vaccines

Clinical data and experimental observations support an important role for CPS-specific antibodies in the prevention or control of GBS disease. These data include the following: (i) GBS produces type-specific CPSs that are targets of protective immunity (59, 73). (ii) Low-level type-specific maternal antibody is a risk factor for disease in infants; conversely, high specific maternal IgG is associated with protection in the neonate (5). (iii) Type-specific antibodies are opsonically active in vitro (12) and are protective in animal models of infection (88). (iv) IgG to the type-specific CPSs can cross the placenta (7), a result that is critical to the concept of a maternal vaccine against neonatal GBS disease (6).

Native GBS type Ia, Ib, II, and III CPSs have been tested as vaccines in adults. The first clinical trial with GBS vaccine enrolled 33 healthy adults to evaluate the safety and immunogenicity of type III CPS extracted by two methods (6). Although both preparations were well tolerated, the EDTA extraction method produced a type III CPS that was more immunogenic than that obtained by TCA extraction. By the 1980s, more than 300 healthy adults, including 40 third-trimester pregnant women, were safely vaccinated with type III CPS or other GBS CPS vaccines at CPS doses ranging from 10 to 150 μg (4, 7). Although well tolerated, these antigens failed to induce a strong specific antibody response in the target population, i.e., subjects with low levels of preexisting CPS-specific antibody. However, type-specific antibody levels among responders (i.e., mainly those with preexisting specific antibody levels >2 μg/ml) increased sharply 4 weeks after CPS vaccination. High levels of CPS-specific antibody persisted for at least 2 years, and declining levels were restored to their peak upon revaccination with CPS. The class of human antibody among responders to GBS type III CPS (4) and to GBS type II CPS (33) was predominantly IgG. The response to CPS antigens differed with the serotype tested; native type II CPS was the most immunogenic (88% rate of response) and type Ia CPS the least (40%) (4). GBS CPS vaccines induced antibodies that were active against homologous GBS serotypes in opsonophagocytic assays and were passively protective in animal studies (4). By the late 1980s, preclinical and clinical results showed that, although the CPS structures were critical vaccine components, they were not sufficiently immunogenic by themselves to be effective GBS vaccines. Researchers then began to focus on the development of GBS CPS vaccines whose immunogenicity was enhanced by methods that preserved their native antigenic structures.

Animal models are useful and often vital in the development of new vaccines. A mouse model of maternal vaccination-neonatal GBS disease (Fig. 6) has been used extensively to evaluate the efficacy of actively administered GBS vaccines or of passively administered type-specific immune sera (88). This model, which focuses primarily on neonatal survival, has also been adapted to test the therapeutic potential of human antisera to GBS vaccines (107).

Polysaccharide-Protein Conjugate Vaccines

Conjugation technology has yielded an abundance of materials for use as reagents in experimental immunology or as potential vaccines. Without doubt, the clinical success of conjugate vaccines against *Haemophilus influenzae* type b infections spurred interest in adapting this technology to improve the immunogenicity of other poorly immunogenic carbohydrates, including those of GBS.

The first GBS CPS conjugate vaccines were prepared in 1990 by three different coupling strategies (Table 2). Native type III CPS was coupled to TT (III-TT) using adipic acid dihydrazide as a spacer molecule (71) or directly via aldehydes formed on a selected number of sialic acid residues (131). In a different approach, coupling of a type III oligosaccharide of 14 pentasaccharide repeating units to TT using a synthetic 6-C spacer molecule resulted in a single-site attachment of the reducing end of the oligosaccharide (102). The chemical methods used to create these and other bacterial conjugate vaccines have been reviewed in detail by Jennings and Sood (50). Despite differences in coupling chemistries, CPS size, purification methods, and the adjuvant used in the development and administration of these vaccines, all three conjugates were better immunogens in laboratory animals than was uncoupled type III CPS. Moreover, antibodies elicited by these GBS conjugate vaccines were of the IgG class and were functionally active when tested in vitro and in vivo.

With the coupling chemistry well in hand, progress was then made in developing conjugate vaccines against GBS serotypes other than type III (104, 133). Immunity in mice to multiple GBS serotypes was demonstrated in studies of a mixture of Ia-TT, Ib-TT, II-TT, and III-TT conjugates administered as a single tetravalent vaccine (105). The lack of interference of one serotype with another in the tetravalent vaccine opened the way to the development of a multivalent GBS vaccine for use in humans. However, clinical observations of an increased prevalence of serotype V dictated its inclusion in a multivalent vaccine (45). Preclinical studies of a serotype V-TT conjugate vaccine have paved the way for this innovation (134).

To broaden coverage with a single vaccine construct, the beta C protein of GBS (discussed in detail above) has been used both as a carrier for type III CPS and as an immunogen in a conjugate vaccine eliciting protective antibodies to GBS strains of serotype III as well as to non-type III GBS

strains that contain this protein (88). Although the results with this vaccine proved that broader coverage could be attained with a relevant carrier protein, the alpha C protein of GBS (see above) would be a better carrier for GBS CPSs because it is present on a larger number of GBS serotypes (87). A GBS type III-alpha C protein conjugate vaccine could potentially provide coverage against up to 90% of disease-causing strains of GBS. Conjugate vaccines utilizing both the full-length 9-repeat alpha C protein and 2-repeat alpha C protein (see below) as carriers for type III polysaccharide elicited antibodies in mice that protected neonatal mouse pups against lethal intraperitoneal challenge with either alpha-positive or type III GBS (42) C5a peptidase as a potential carrier protein in conjugate vaccines. A conjugate vaccine using the C5a peptidase coupled to type III polysaccharide was able to reduce the severity of pneumonia in an intranasally inoculated mouse model of GBS infection using a single challenge strain of GBS with a different capsular serotype (25).

Because protective antibodies to type III GBS CPS appear to recognize a conformational epitope, all conjugate vaccines for clinical use were prepared using a coupling method known to preserve both the native repeating unit structure and the conformation of the CPS (51). This conjugation method (54) is essentially a two-step process: (i) GBS CPS is subjected to mild oxidation with sodium periodate, which creates free aldehyde groups on a limited number of sialic acid residues on the CPS. (ii) The newly created aldehyde groups on the CPS serve as sites for irreversible coupling to free amino groups present on carrier proteins by a process called reductive amination. The end

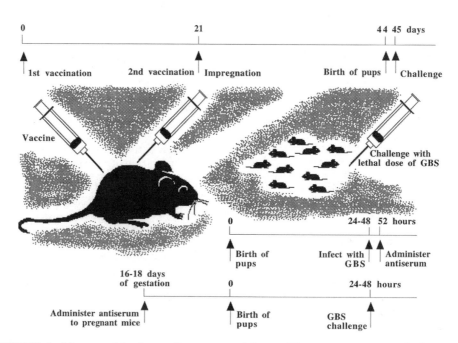

FIGURE 6 Mouse models of group B streptococcal disease. The immunogenicity and efficacy of GBS vaccines have been evaluated with the maternal vaccination-neonatal mouse model by actively vaccinating female mice, mating these mice, and subsequently challenging their offspring with GBS (top timeline). In a passive vaccination-protection model (bottom timeline), newborn pups born to dams that received immune sera during pregnancy are challenged with GBS as a means of measuring the functional capacity of IgG. The therapeutic potential of GBS vaccine-induced serum (middle timeline) has been ascertained by inoculating naive pups with GBS 4 h before administration of immune sera. In all three models, pups are infected or challenged with GBS within 24 to 48 h of birth, and survival is assessed 48 h after challenge.

TABLE 2 Characteristics of group B streptococcal conjugate vaccines

Capsular serotype(s)	Carrier protein(s)	Spacer	Coupling method	Linkage	Study	Reference(s)
III	TT	ADH[a]	CR[b]	Multisite, random	Preclinical	71
III oligosaccharide[c]	TT[d]	6-C[e]	RA[f]	Single, end-linked	Preclinical	102
III	TT	None	RA	Multisite, direct[g]	Preclinical	131
III oligosaccharide[h]	TT	6-C	RA	Single, end-linked	Preclinical	103
II	TT	None	RA	Multisite, direct	Preclinical	104
Ia	TT	None	RA	Multisite, direct	Preclinical	133
III	β-C[i]	None	RA	Multisite, direct	Preclinical	88
III	α-C[j]	None	RA	Multisite, direct	Preclinical	87
Ib	TT	None	RA	Multisite, direct	Preclinical	105
V	TT	None	RA	Multisite, direct	Preclinical	134
III	TT	None	RA	Multisite, direct	Clinical	60
Ia, Ib	TT	None	RA	Multisite, direct	Clinical	8
Ia, II, III	TT and β-C	None	RA	Multisite, direct	Preclinical	99
II	TT	None	RA	Multisite, direct	Clinical	9
V	TT	None	RA	Multisite, direct	Clinical	11
V	CRM$_{197}$[k]	None	RA	Multisite, direct	Clinical	11
III	rCTB[l]	None	RA	Multisite, direct	Preclinical	108a
III	rdHBcAg[m]	None	RA	Multisite, direct	Preclinical	108a
IV	TT	None	RA	Multisite, direct	Preclinical	101a
VII	TT	None	RA	Multisite, direct	Preclinical	101a

[a]Adipic acid dihydrazide.
[b]Carbodiimide reduction.
[c]14 pentasaccharide repeating units.
[d]Monomeric tetanus toxoid.
[e]6-Aminohexyl-1-β-D-galactopyranoside.
[f]Reductive amination using sodium cyanoborohydride.
[g]Via aldehydes formed on a selected number of sialic acid residues.
[h]6 and 25 pentasaccharide repeating units.
[i]GBS beta C protein.
[j]GBS alpha C protein.
[k]Cross-reactive material, a diphtheria mutant toxoid isolated from *Clostridium diphtheriae*.
[l]Recombinant cholera B toxin.
[m]Recombinant duck hepatitis B core antigen.

result is a conjugate vaccine that contains CPS covalently linked directly to the protein. The requirement for a covalent linkage between the CPS and carrier protein was necessary to the formation of efficacious GBS vaccines (Table 3).

Conjugate vaccines for use in phase 1 and phase 2 clinical trials have been individually prepared by the reductive amination method with GBS CPSs Ia, Ib, II, III, and V. Each CPS was individually coupled to TT, and a second type V conjugate vaccine was prepared with the mutant diphtheria toxoid CRM197 (108). Four uncoupled GBS CPS preparations (types Ia, Ib, II, and III) were used as control vaccines in these clinical trials.

Clinical Trials with III-TT Vaccine

The initial clinical trials with the second-generation (i.e., conjugate) GBS vaccines were designed to compare the safety and immunogenicity of conjugated and uncoupled type III CPS and to determine an optimal immunogenic dose (60). Both III-TT (at any CPS dose: 58, 14.5, 3.6 μg) and III CPS were well tolerated, with minimal reactogenicity reported among the 90 female recipients. The geometric mean concentration (GMC) of III CPS-specific IgG among the 30 recipients of the 58-μg dose (as CPS) of III-TT rose from 0.09 μg/ml before immunization to 4.89 μg/ml 2 weeks thereafter; during this same interval, the GMC of specific IgG among the 30 recipients of III CPS increased from 0.21 μg/ml to only 1.30 μg/ml ($P < 0.05$). The relative immuno-

genicity of III-TT appeared to be dose dependent, with a peak GMC of type-specific IgG of 2.72 μg/ml in the 14.5-μg group and 1.10 μg/ml in the 3.6-μg group (60). Moreover, antibodies elicited by immunization with III-TT recognized a conformationally dependent epitope of the CPS, promoted in vitro opsonophagocytosis and killing of GBS, and were functional in vivo in a neonatal mouse protection model (60).

Preliminary results indicate that a III-TT vaccine is also safe and immunogenic when administered to healthy women at 30 to 32 weeks of gestation (10). Maternally derived type III CPS-specific IgG measured in serum from 1- and 2-month-old infants was functionally active in vitro.

Clinical Trials with Ia-TT and Ib-TT Vaccines

The Ia-TT and Ib-TT conjugate vaccines were also well tolerated and were more immunogenic than uncoupled homologous CPS in nonpregnant, healthy women (8). The GMC of type Ia-specific IgG among recipients of a 60-μg dose (as CPS) of Ia-TT rose from 0.5 μg/ml to a peak of 26.2 μg/ml 8 weeks after vaccination, whereas during the same interval antibody levels among recipients of Ia CPS rose from 0.2 to only 2.4 μg/ml. Levels of type Ia CPS-specific IgG remained high in recipients of the Ia-TT vaccine at 1 and 2 years after vaccination, a result demonstrating the durability of the response. The response to Ia-TT vaccines was dose dependent: 60-μg and 15-μg doses elicited higher antibody levels than did a 3.75-μg dose (8).

TABLE 3 Effect of covalent attachment of polysaccharide to protein on the efficacy in mice of group B streptococcal polysaccharide conjugate vaccines

Vaccine	No. of pups surviving/no. challenged	% Survival
III-TT[a]	27/29	93
III+TT	0/33	0
III CPS	0/22	0
TT	0/10	0

[a]Female outbred mice received a priming dose of vaccine emulsified with complete Freund's adjuvant and were mated 2 weeks later. A booster dose of vaccine was administered with imcomplete Freund's adjuvant 3 weeks after the priming dose. Newborn pups (<36 h old) were challenged with an ordinarily lethal dose of GBS type III strain M781, and survival was assessed 48 h later. All groups received 1 μg per dose of type III capsular polysaccharide, either uncoupled (III CPS) or covalently coupled (III-TT) to or admixed with (III+TT) tetanus toxoid.

Like the other GBS conjugate vaccines, Ib-TT vaccine was safer and more immunogenic than uncoupled homologous CPS (8). The prevaccination GMC of type-specific antibody among recipients of Ib-TT was 0.4 μg/ml, a level that rose to a peak of 14.2 μg/ml 4 weeks after vaccination with a single 63-μg dose (as CPS). In contrast, among recipients of uncoupled type Ib CPS, the prevaccination type-specific antibody GMC of 0.4 μg/ml rose to only 4.4 μg/ml 4 weeks after vaccination. The response to Ib-TT vaccine was dose dependent, with the highest (63 μg) and intermediate (15.75 μg) doses resulting in the greatest magnitude of antibody response (8).

Clinical Trials with II-TT Vaccine

Coupling purified GBS type II CPS to TT also resulted in a safe and highly immunogenic vaccine in healthy, nonpregnant adults (9). Responses to the conjugate vaccine were dose dependent, with the lowest CPS dose of 3.6 μg evoking a peak type-specific IgG of ~15 μg/ml 2 weeks after vaccination. As expected, immune sera were functionally active in an in vitro opsonophagocytosis assay against viable type II GBS. Interestingly, high concentrations of type II CPS-specific IgM and IgA were also elicited in a vaccine dose-dependent manner. These antibody isotypes were not elicited in humans who received the other (Ia-TT, Ib-TT, and III-TT) GBS CPS conjugate vaccines, and although one can speculate that the difference may lie with the structural features (described above) of the type II CPS, the answer awaits further experimentation.

Clinical Trials with V-TT and V-CRM197 Vaccines

As reviewed in chapter 15 by Schuchat and Balter, type V GBS emerged as an important human pathogen during the 1980s. In response to the emergence of this serotype, conjugates were prepared with purified type V CPS. Two conjugate vaccines, one prepared with GBS type V CPS and coupled to TT, and the other coupled to CRM197 (Table 2), were randomly administered to healthy nonpregnant women aged 18 to 45 years (11). Both conjugate vaccines were well tolerated with no serious adverse reactions reported. Immune responses to V-TT and V-CRM197 were similar to those observed with the other GBS conjugates; type-specific IgG reached peak GMC levels between 6.5 and 8.9 μg/ml at 2 to 4 weeks after vaccination with no statistical differences observed between these two preparations. Surprisingly, both conjugate vaccines elicited high GMCs of type V

CPS-specific IgM and IgA, a response pattern similar to that measured from recipients of the II-TT vaccine (11). Both type II and type V CPSs have two side chains, one of which is composed of a single monosaccharide (Fig. 3). Could this or other CPS structural feature(s) influence the immune response to conjugates prepared with these antigens or are there host factors, such as prior immune status, that influence the antibody isotype? While answers to these questions await experimentation, a type V CPS coupled to either carrier protein is shown to be safe and immunogenic, and elicits functionally active antibody in healthy adults.

Because of the increased rate of GBS among the elderly, Palazzi and coworkers examined the immune response of a GBS type V-TT in 22 healthy adults aged 65 to 85 years (100). The antibody isotype distribution of V-TT in this age group was similar to that measured in 18- to 45-year-old adults and GMCs of type V CPS-specific IgG, IgA, and IgM ranged between 2 and 3 μg/ml 4 weeks after vaccination; concentrations were reduced only by approximately 50% over 1 year. This was the first study that demonstrated the potential of GBS conjugate vaccines to stem disease in the elderly.

SUMMARY

The seminal findings by Rebecca Lancefield and coworkers on the role of carbohydrate and protein antigens in GBS immunity have led a generation of researchers not only to a better understanding of these antigens in immunity but also toward the development of effective vaccines. Today, GBS remains a major bacterial cause of disease among neonates and a common cause of bovine mastitis (61). This organism is increasingly common in the nonpregnant population in the United States (45), especially among the elderly (reviewed in reference 116). However, as the twenty-first century begins, we have the knowledge necessary to produce and formulate effective vaccines against GBS disease.

We thank Julia Wang for assistance with the chair models of the capsular polysaccharides, Claudia Gravekamp for providing electron micrographs of GBS, Michael Wessels for Figure 4, Thomas DiCesare for illustration of the mouse model, and Julie McCoy for editorial assistance.

Most of the GBS research conducted at the Channing Laboratory and cited in this review has been funded through the National Institute of Allergy and Infectious Diseases of the NIH.

ADDENDUM IN PROOF

The shift of GBS disease from newborns to elderly adults, current colonization rates, and serotype-specific immunity to GBS among the healthy elderly population are reported in recently published articles by Edwards and coworkers (30a, 31a). These findings broaden the target population that would receive a GBS vaccine and also strengthen the call for a multivalent vaccine. Toward that end, a "reverse vaccinology" approach was used to reveal four new GBS proteins that, in combination, provided specific immunity to 12 GBS strains representing six of the nine known serotypes (89a).

REFERENCES

1. **Anthony, B. F., N. F. Concepcion, and K. F. Concepcion.** 1985. Human antibody to the group-specific polysaccharide of group B *Streptococcus. J. Infect. Dis.* **151:**221–226.

2. **Baker, C. J., and F. F. Barrett.** 1973. Transmission of group B streptococci among parturient women and their neonates. *J. Pediatr.* **83:**919–925.

3. Baker, C. J., and M. S. Edwards. 1995. Group B streptococcal infections, p. 980–1054. *In* J. S. Remington and J. O. Klein (ed.), *Infectious Diseases of the Fetus and Newborn Infant*, 4th ed. W. B. Saunders Co., Philadelphia, Pa.

4. Baker, C. J., and D. L. Kasper. 1985. Group B streptococcal vaccines. *Rev. Infect. Dis.* **7:**458–467.

5. Baker, C. J., D. L. Kasper, I. B. Tager, A. Paredes, S. Alpert, W. M. McCormack, and D. Goroff. 1977. Quantitative determination of antibody to capsular polysaccharide in infection with type III strains of group B *Streptococcus. J. Clin. Invest.* **59:**810–818.

6. Baker, C. J., M. S. Edwards, and D. L. Kasper. 1978. Immunogenicity of polysaccharides from type III group B *Streptococcus. J. Clin. Invest.* **61:**1107–1110.

7. Baker, C. J., M. A. Rench, M. S. Edwards, R. J. Carpenter, B. M. Hays, and D. L. Kasper. 1988. Immunization of pregnant women with a polysaccharide vaccine of group B *Streptococcus. N. Engl. J. Med.* **319:**1180–1220.

8. Baker, C. J., L. C. Paoletti, M. R. Wessels, H.-K. Guttormsen, M. A. Rench, M. E. Hickman, and D. L. Kasper. 1999. Safety and immunogenicity of capsular polysaccharide-tetanus toxoid conjugate vaccines for group B streptococcal types Ia and Ib. *J. Infect. Dis.* **179:**142–150.

9. Baker, C. J., L. C. Paoletti, M. A. Rench, H. K. Guttormsen, V. J. Carey, M. E. Hickman, and D. L. Kasper. 2000. Use of capsular polysaccharide-tetanus toxoid conjugate vaccine for type II group B *Streptococcus* in healthy women. *J. Infect. Dis.* **182:**1129–1138.

10. Baker, C. J., M. A. Rench, and P. McInnes. 2003. Immunization of pregnant women with group B streptococcal type III capsular polysaccharide-tetanus toxoid conjugate vaccine. *Vaccine* **21:**3468–3472.

11. Baker, C. J., L. C. Paoletti, M. A. Rench, H. K. Guttormsen, M. S. Edwards, and D. L. Kasper. 2004. Immune response of healthy women to 2 different group B streptococcal type V capsular polysaccharide-protein conjugate vaccines. *J. Infect. Dis.* **189:**1103–1112.

12. Baltimore, R. S., D. L. Kasper, C. J. Baker, and D. K. Goroff. 1977. Antigenic specificity of opsonophagocytic antibodies in rabbit anti-sera to group B streptococci. *J. Immunol.* **118:**673–678.

13. Baron, M. J., G. R. Bolduc, M. B. Goldberg, T. C. Auperin, and L. C. Madoff. 2004. Alpha C protein of group B Streptococcus binds host cell surface glycosaminoglycan and enters cells by an actin-dependent mechanism. *J. Biol. Chem.* **279:**24714–24723.

14. Beckmann, C., J. D. Waggoner, T. O. Harris, G. S. Tamura, and C. E. Rubens. 2002. Identification of novel adhesins from Group B streptococci by use of phage display reveals that C5a peptidase mediates fibronectin binding. *Infect. Immun.* **70:**2869–2876.

15. Bevanger, L. 1983. Ibc proteins as serotype markers of group B streptococci. *Acta Pathol. Microbiol. Immunol. Scand. Sect. B* **91:**231–234.

16. Bevanger, L., and J. A. Maeland. 1979. Complete and incomplete Ibc protein fraction in group B streptococci. *Acta Pathol. Microbiol. Immunol. Scand. Sect. B* **87:**51–54.

17. Bevanger, L., A. I. Kvam, and J. A. Maeland. 1995. A *Streptococcus agalactiae* R protein analysed by polyclonal and monoclonal antibodies. *APMIS* **103:**731–736.

18. Bohnsack, J. F., X. N. Zhou, J. N. Gustin, C. E. Rubens, C. J. Parker, and H. R. Hill. 1992. Bacterial evasion of the antibody response: human IgG antibodies neutralize soluble but not bacteria-associated group B streptococcal C5a-ase. *J. Infect. Dis.* **165:**315–321.

19. Bolduc, G. R., M. J. Baron, C. Gravekamp, C. S. Lachenauer, and L. C. Madoff. 2002. The alpha C protein mediates internalization of group B Streptococcus within human cervical epithelial cells. *Cell Microbiol.* **4:**751–758.

20. Brady, L. J. Personal communication, unpublished data.

21. Brady, L. J., U. D. Daphtary, E. M. Ayoub, and M. D. P. Boyle. 1988. Two novel antigens associated with group B streptococci identified by a rapid two-stage radioimmunoassay. *Infect. Immun.* **158:**965–973.

22. Brodeur, B. R., M. Boyer, I. Charlebois, J. Hamel, F. Couture, C. R. Rioux, and D. Martin. 2000. Identification of group B streptococcal Sip protein, which elicits cross-protective immunity. *Infect. Immun.* **68:**5610–5618.

23. Butter, M. N. W., and C. E. de Moor. 1967. *Streptococcus agalactiae* as a cause of meningitis in the newborn, and of bacteraemia in adults. Differentiation of human and animal varieties. *J. Microbiol. Serol.* **33:**439–450.

24. Cheng, Q., B. Carlson, S. Pillai, R. Eby, L. Edwards, S. B. Olmsted, and P. Cleary. 2001. Antibody against surface-bound C5a peptidase is opsonic and initiates macrophage killing of group B streptococci. *Infect. Immun.* **69:**2302–2308.

25. Cheng, Q., S. Debol, H. Lam, R. Eby, L. Edwards, Y. Matsuka, S. B. Olmsted, and P. P. Cleary. 2002. Immunization with C5a peptidase or peptidase-type III polysaccharide conjugate vaccines enhances clearance of group B streptococci from lungs of infected mice. *Infect. Immun.* **70:**6409–6415.

26. Cheng, Q., D. Stafslien, S. S. Purushothaman, and P. Cleary. 2002. The group B streptococcal C5a peptidase is both a specific protease and an invasin. *Infect. Immun.* **70:**2408–2413.

27. Cieslewicz, M. J., D. L. Kasper, Y. Wang, and M. R. Wessels. 2001. Functional analysis in type Ia group B streptococcus of a cluster of genes involved in extracellular polysaccharide production by diverse species of streptococci. *J. Biol. Chem.* **276:**139–146.

28. Cieslewicz, M. J., D. Chaffin, G. Glusman, D. L. Kasper, S. R. Madan, J. Fahey, M. R. Wessels, and C. E. Rubens. 2005. Structural and genetic diversity of group B Streptococcus capsular polysaccharides. *Infect. Immun.* **73:**3096–3103.

29. Cleary, P. P., J. Handley, A. N. Suvorov, A. Podbielski, and P. Ferrieri. 1992. Similarity between the group B and A streptococcal C5a peptidase genes. *Infect. Immun.* **60:**4239–4244.

30. Cleat, P. H., and K. N. Timmis. 1987. Cloning and expression in *Escherichia coli* of the Ibc protein genes of group B streptococci: binding of human immunoglobulin A to the beta antigen. *Infect. Immun.* **55:**1151–1155.

30a. Edwards, M. S., and C. J. Baker. 2005. Group B streptococcal infections in elderly adults. *Clin. Infect. Dis.* **41:**839–847.

31. Edwards, M. S., W. A. Nicholson, C. J. Baker, and D. L. Kasper. 1980. The role of specific antibody in alternative complement pathway-mediated opsonophagocytosis of type III, group B *Streptococcus. J. Exp. Med.* **151:**1275–1287.

31a. Edwards, M. S., M. A. Rench, D. L. Palazzi, and C. J. Baker. 2005. Group B streptococcal colonization and serotype-specific immunity in healthy elderly persons. *Clin. Infect. Dis.* **40:**352–357.

32. Eickhoff, T. C., J. O. Klein, A. K. Daly, D. Ingall, and M. Finland. 1964. Neonatal sepsis and other infections due to group B beta-hemolytic streptococci. *N. Engl. J. Med.* **271:**1221–1228.

33. **Eisenstein, T. K., B. J. De Cueninck, D. Resavy, G. D. Shockman, R. B. Carey, and R. M. Swenson.** 1983. Quantitative determination in human sera of vaccine-induced antibody to type-specific polysaccharides of group B streptococci using an enzyme-linked immunosorbent assay. *J. Infect. Dis.* **147:**847–856.

34. **Erdogan, S., P. K. Fagan, S. R. Talay, M. Rohde, P. Ferrieri, A. E. Flores, C. A. Guzman, M. J. Walker, and G. S. Chhatwal.** 2002. Molecular analysis of group B protective surface protein, a new cell surface protective antigen of group B streptococci. *Infect. Immun.* **70:**803–811.

35. **Flores, A. E., and P. Ferrieri.** 1989. Molecular species of R-protein antigens produced by clinical isolates of group B streptococci. *J. Clin. Microbiol.* **27:**1050–1054.

36. **Fry, R. M.** 1938. Fatal infections by haemolytic streptococcus group B. *Lancet* **i:**199–201.

37. **Fusco, P. C., J. W. Perry, S. M. Liang, M. S. Blake, F. Michon, and J. Y. Tai.** 1997. Bactericidal activity elicited by the beta C protein of group B streptococci contrasted with capsular polysaccharides. *Adv. Exp. Med. Biol.* **418:**841–845.

38. **Glaser, P., C. Rusniok, C. Buchrieser, F. Chevalier, L. Frangeul, T. Msadek, M. Zouine, E. Couve, L. Lalioui, C. Poyart, P. Trieu-Cuot, and F. Kunst.** 2002. Genome sequence of *Streptococcus agalactiae*, a pathogen causing invasive neonatal disease. *Mol. Microbiol.* **45:**1499–1513.

39. **Gravekamp, C., D. S. Horensky, J. L. Michel, and L. C. Madoff.** 1996. Variation in repeat number within the alpha C protein of group B streptococci alters antigenicity and protective epitopes. *Infect. Immun.* **64:**3576–3583.

40. **Gravekamp, C., D. L. Kasper, J. L. Michel, D. E. Kling, V. Carey, and L. C. Madoff.** 1997. Immunogenicity and protective efficacy of the alpha C protein of group B streptococci are inversely related to the number of repeats. *Infect. Immun.* **65:**5216–5221.

41. **Gravekamp, C., B. Rosner, and L. C. Madoff.** 1998. Deletion of repeats in the alpha C protein enhances the pathogenicity of group B streptococci in immune mice. *Infect. Immun.* **66:**4347–4354.

42. **Gravekamp, C., D. L. Kasper, L. C. Paoletti, and L. C. Madoff.** 1999. Alpha C protein as a carrier for type III capsular polysaccharide and as a protective protein in group B streptococcal vaccines. *Infect. Immun.* **67:**2491–2496.

43. **Grifantini, R., E. Bartolini, A. Muzzi, M. Draghi, E. Frigimelica, J. Berger, G. Ratti, R. Petracca, G. Galli, M. Agnusdei, M. M. Giuliani, L. Santini, B. Brunelli, H. Tettelin, R. Rappuoli, F. Randazzo, and G. Grandi.** 2002. Previously unrecognized vaccine candidates against group B meningococcus identified by DNA microarrays. *Nat. Biotechnol.* **20:**914–921.

44. **Haft, R. F., M. R. Wessels, M. F. Mebane, N. Conaty, and C. E. Rubens.** 1996. Characterization of *cpsF* and its product CMP-*N*-acetylneuraminic acid synthetase, a group B streptococcal enzyme that can function in K1 capsular polysaccharide biosynthesis in *Escherichia coli*. *Mol. Microbiol.* **19:**555–563.

45. **Harrison, L. H., J. A. Elliott, D. M. Dwyer, J. P. Libonati, P. Ferrieri, L. Billmann, and A. Schuchat.** 1998. Serotype distribution of invasive group B streptococcal isolates in Maryland: implications for vaccine formulation. *J. Infect. Dis.* **177:**998–1002.

46. **Heden, L. O., E. Frithz, and G. Lindahl.** 1991. Molecular characterization of an IgA receptor from group B streptococci: sequence of the gene, identification of a proline-rich region with unique structure and isolation of N-terminal fragments with IgA-binding capacity. *Eur. J. Immunol.* **21:**1481–1490.

47. **Henricksen, J. P., J. Ferrieri, J. Jelinkova, W. Koehler, and W. R. Maxted.** 1984. Nomenclature of antigens of group B streptococci. *Int. J. Syst. Bacteriol.* **34:**500.

48. **Hood, M., A. Janney, and G. Dameron.** 1961. Beta hemolytic *Streptococcus* group B associated with problems of the perinatal period. *Am. J. Obstet. Gynecol.* **82:**809–818.

49. **Institutes of Medicine, Division of Health Promotion and Disease Prevention.** 1985. Appendix P: prospects for immunizing against *Streptococcus* group B, p. 424–438. *In Diseases of Importance in the United States*, vol. I. *New Vaccine Development: Establishing Priorities*. National Academy Press, Washington, D.C.

50. **Jennings, H. J., and R. K. Sood.** 1994. Synthetic glyco-conjugates as human vaccines, p. 325–361. *In* Y. C. Lee and R. T. Lee (ed.), *Neoglycoconjugates: Preparation and Applications*. Academic Press, Inc., New York, N.Y.

51. **Jennings, H. J., C. Lugowski, and D. L. Kasper.** 1981. Conformational aspects critical to the immunospecificity of the type III group B streptococcal polysaccharide. *Biochemistry* **20:**4511–4518.

52. **Jennings, H. J., E. Katzenellenbogen, C. Lugowski, and D. L. Kasper.** 1983. Structure of native polysaccharide antigens of type Ia and type Ib group B *Streptococcus*. *Biochemistry* **22:**1258–1264.

53. **Jennings, H. J., K. G. Rosell, E. Katzenellenbogen, and D. L. Kasper.** 1983. Structural determination of the capsular polysaccharide antigen of type II group B *Streptococcus*. *J. Biol. Chem.* **258:**1793–1798.

54. **Jennings, H. J., C. Lugowski, and F. E. Ashton.** 1984. Conjugation of meningococcal lipopolysaccharide R-type oligosaccharides to tetanus toxoid as a route to a potential vaccine against group B *Neisseria meningitidis*. *Infect. Immun.* **43:**407–412.

55. **Jerlstrom, P. G., G. S. Chhatwal, and K. N. Timmis.** 1991. The IgA-binding beta antigen of the C protein complex of group B streptococci: sequence determination of its gene and detection of two binding regions. *Mol. Microbiol.* **5:**843–849.

56. **Jerlstrom, P. G., S. R. Talay, P. Valentin-Weigand, K. N. Timmis, and G. S. Chhatwal.** 1996. Identification of an immunoglobulin A binding motif located in the beta-antigen of the c protein complex of group B streptococci. *Infect. Immun.* **64:**2787–2793.

57. **Ji, Y., B. Carlson, A. Kondagunta, and P. P. Cleary.** 1997. Intranasal immunization with C5a peptidase prevents nasopharyngeal colonization of mice by the group A Streptococcus. *Infect. Immun.* **65:**2080–2087.

58. **Johnson, D. R., and P. Ferrieri.** 1984. Group B streptococcal Ibc protein antigen: distribution of two determinants in wild-type strains of common serotypes. *J. Clin. Microbiol.* **19:**506–510.

59. **Kasper, D. L., C. J. Baker, R. S. Baltimore, J. H. Crabb, G. Schiffman, and H. J. Jennings.** 1979. Immunodeterminant specificity of human immunity to type III group B *Streptococcus*. *J. Exp. Med.* **149:**327–339.

60. **Kasper, D. L., L. C. Paoletti, M. R. Wessels, H. K. Guttormsen, V. J. Carey, H. J. Jennings, and C. J. Baker.** 1996. Immune response to type III group B streptococcal polysaccharide-tetanus toxoid conjugate vaccine. *J. Clin. Invest.* **98:**2308–2314.

61. **Keefe, G. P.** 1997. *Streptococcus agalactiae* mastitis: a review. *Can. Vet. J.* **38:**429–437.

62. **Kling, D. E., C. Gravekamp, L. C. Madoff, and J. L. Michel.** 1997. Characterization of two distinct opsonic and protective epitopes within the alpha C protein of the group B *Streptococcus*. *Infect. Immun.* **65:**1462–1467.

63. Kogan, G., D. Uhrin, J.-R. Brisson, L. C. Paoletti, D. L. Kasper, C. von Hunolstein, G. Orefici, and H. J. Jennings. 1994. Structure of the type VI group B *Streptococcus* capsular polysaccharide determined by high resolution NMR spectroscopy. *J. Carbohydr. Chem.* **13**:1071–1078.

63a. Kogan, G., J. R. Brisson, D. L. Kasper, C. von Hunolstein, G. Orefici, and H. J. Jennings. 1995. Structural elucidation of the novel type VII group B *Streptococcus* capsular polysaccharide by high resolution NMR spectroscopy. *Carbohydr. Res.* **277**:1–9.

64. Kogan, G., D. Uhrin, J. R. Brisson, L. C. Paoletti, A. E. Blodgett, D. L. Kasper, and H. J. Jennings. 1996. Structural and immunochemical characterization of the type VIII group B *Streptococcus* capsular polysaccharide. *J. Biol. Chem.* **271**:8786–8790.

65. Kuypers, J. M., L. M. Heggen, and C. E. Rubens. 1989. Molecular analysis of a region of the group B *Streptococcus* chromosome involved in type III capsule expression. *Infect. Immun.* **57**:3058–3065.

66. Lachenauer, C. S., and L. C. Madoff. 1996. A protective surface protein from type V group B streptococci shares N-terminal sequence homology with the alpha C protein. *Infect. Immun.* **64**:4255–4260.

67. Lachenauer, C. S., and L. C. Madoff. 1997. Cloning and expression in *Escherichia coli* of a protective surface protein from type V group B streptococci. *Adv. Exp. Med. Biol.* **418**:615–618.

68. Lachenauer, C. S., D. L. Kasper, J. Shimada, Y. Ichiman, H. Ohtsuka, M. Kaku, L. C. Paoletti, and L. C. Madoff. 1999. Serotypes VI and VIII predominate among group B streptococci isolated from pregnant Japanese women. *J. Infect. Dis.* **179**:1030–1033.

69. Lachenauer, C., R. Creti, J. Michel, and L. Madoff. 2000. Mosaicism in the alpha-like protein genes of group B streptococci. *Proc. Natl. Acad. Sci. USA* **97**:9630–9635.

70. Lachenauer, C. S., C. J. Baker, M. J. Baron, D. L. Kasper, C. Gravekamp, and L. C. Madoff. 2002. Quantitative determination of immunoglobulin G specific for group B streptococcal beta C protein in human maternal serum. *J. Infect. Dis.* **185**:368–374.

71. Lagergard, T., J. Shiloach, J. B. Robbins, and R. Schneerson. 1990. Synthesis and immunological properties of conjugates composed of group B *Streptococcus* type III capsular polysaccharide covalently bound to tetanus toxoid. *Infect. Immun.* **58**:687–694.

72. Lancefield, R. C. 1934. A serological differentiation of specific types of bovine hemolytic streptococci (group B). *J. Exp. Med.* **59**:441–458.

73. Lancefield, R. C. 1938. Two serological types of group B hemolytic streptococci with related, but not identical, type-specific substances. *J. Exp. Med.* **67**:25–40.

74. Lancefield, R. C. 1972. Cellular antigens of group B streptococci, p. 57–65. *In* L. W. Wannamaker and J. M. Matsen (ed.), *Streptococci and Streptococcal Diseases, Recognition, Understanding and Management.* Academic Press, Inc., New York, N.Y.

75. Lancefield, R. C., and E. H. Freimer. 1966. Type-specific polysaccharide antigens of group B streptococci. *J. Hyg. Camb.* **64**:191–203.

76. Lancefield, R. C., and G. E. Perlmann. 1952. Preparation and properties of a protein (R antigen) occurring in streptococci of group A, type 28 and in certain streptococci of other serological groups. *J. Exp. Med.* **96**:83–97.

77. Lancefield, R. C., M. McCarty, and W. N. Everly. 1975. Multiple mouse-protective antibodies directed against group B streptococci. Special reference to antibodies effective against protein antigens. *J. Exp. Med.* **142**:165–179.

78. Lautrou, Y., P. Rainard, B. Poutrel, M. S. Zygmunt, A. Venien, and J. Dufrenoy. 1991. Purification of the protein X of *Streptococcus agalactiae* with a monoclonal antibody. *FEMS Microbiol. Lett.* **64**:141–145.

79. Lewis, A. L., V. Nizet, and A. Varki. 2004. Discovery and characterization of sialic acid O-acetylation in group B Streptococcus. *Proc. Natl. Acad. Sci. USA* **101**:11123–11128.

80. Li, J., D. L. Kasper, F. M. Ausubel, B. Rosner, and J. L. Michel. 1997. Inactivation of the alpha C protein antigen gene, *bca*, by a novel shuttle/suicide vector results in attenuation of virulence and immunity in group B *Streptococcus.* *Proc. Natl. Acad. Sci. USA* **94**:13251–13256.

81. Lindahl, G., B. Akerstrom, J. P. Vaerman, and L. Stenberg. 1990. Characterization of an IgA receptor from group B streptococci: specificity for serum IgA. *Eur. J. Immunol.* **20**:2241–2247.

82. Linden, V. 1983. Mouse-protective effect of rabbit anti-R-protein antibodies against group B streptococci type II carrying R-protein. Lack of effect on type III carrying R-protein. *Acta Pathol. Microbiol. Immunol. Scand. Sect. B* **91**:145–151.

83. Linden, V., K. K. Christensen, and P. Christensen. 1983. Correlation between low levels of maternal IgG antibodies to R protein and neonatal septicemia with group B streptococci carrying R protein. *Int. Arch. Allergy Appl. Immunol.* **71**:168–172.

84. Madoff, L. C., S. Hori, J. L. Michel, C. J. Baker, and D. L. Kasper. 1991. Phenotypic diversity in the alpha C protein of group B streptococcus. *Infect. Immun.* **59**:2638–2644.

85. Madoff, L. C., J. L. Michel, and D. L. Kasper. 1991. A monoclonal antibody identifies a protective C-protein alpha-antigen epitope in group B streptococci. *Infect. Immun.* **59**:204–210.

86. Madoff, L. C., J. L. Michel, E. W. Gong, A. K. Rodewald, and D. L. Kasper. 1992. Protection of neonatal mice from group B streptococcal infection by maternal immunization with beta C protein. *Infect. Immun.* **60**:4989–4994.

87. Madoff, L. C., L. C. Paoletti, J. L. Michel, E. W. Gong, and D. L. Kasper. 1994. Synthesis of a type III polysaccharide-recombinant alpha C protein conjugate vaccine for prevention of group B streptococcal infection. *Clin. Infect. Dis.* **19**:602.

88. Madoff, L. C., L. C. Paoletti, J. Y. Tai, and D. L. Kasper. 1994. Maternal immunization of mice with group B streptococcal type III polysaccharide-beta C protein conjugate elicits protective antibody to multiple serotypes. *J. Clin. Invest.* **94**:286–292.

89. Madoff, L. C., J. L. Michel, E. W. Gong, D. E. Kling, and D. L. Kasper. 1996. Group B streptococci escape host immunity by deletion of tandem repeat elements of the alpha C protein. *Proc. Natl. Acad. Sci. USA* **93**:4131–4136.

89a. Maione, D., I. Margarit, C. D. Rinaudo, V. Masignani, M. Mora, M. Scarselli, H. Tettelin, C. Brettoni, E. T. Iacobini, R. Rosini, N. D' Agostino, L. Miorin, S. Buccato, M. Mariani, G. Galli, R. Nogarotto, V. N. Dei, F. Vegni, C. Fraser, G. Mancuso, G. Teti, L. C. Madoff, L. C. Paoletti, R. Rappuoli, D. L. Kasper, J. L. Telford, and G. Grandi. 2005. Identification of a universal Group B streptococcus vaccine by multiple genome screen. *Science* **309**:148-150.

90. Markham, R. B., W. A. Nicholson, G. Schiffman, and D. L. Kasper. 1982. The presence of sialic acid on two related bacterial polysaccharides determines the site of the

primary immune response and the effect of complement depletion on the response in mice. *J. Immunol.* **128:** 2731–2733.

91. **Marques, M. B., D. L. Kasper, M. K. Pangburn, and M. R. Wessels.** 1992. Prevention of C3 deposition by capsular polysaccharide is a virulence mechanism of type III group B streptococci. *Infect. Immun.* **60:**3986–3993.

92. **Marques, M. B., D. L. Kasper, A. Shroff, F. Michon, H. J. Jennings, and M. R. Wessels.** 1994. Functional activity of antibodies to the group B polysaccharide of group B streptococci elicited by a polysaccharide-protein conjugate vaccine. *Infect. Immun.* **62:**1593–1599.

93. **Martin, D., S. Rioux, E. Gagnon, M. Boyer, J. Hamel, N. Charland, and B. R. Brodeur.** 2002. Protection from group B streptococcal infection in neonatal mice by maternal immunization with recombinant Sip protein. *Infect. Immun.* **70:**4897–4901.

94. **Michel, J. L., L. C. Madoff, D. E. Kling, D. L. Kasper, and F. M. Ausubel.** 1991. Cloned alpha and beta C-protein antigens of group B streptococci elicit protective immunity. *Infect. Immun.* **59:**2023–2028.

95. **Michel, J. L., L. C. Madoff, K. Olson, D. E. Kling, D. L. Kasper, and F. M. Ausubel.** 1992. Large identical tandem repeating units in the C protein alpha antigen gene, bca, of group B streptococci. *Proc. Natl. Acad. Sci. USA* **89:** 10060–10065.

96. **Michel, J. L., B. D. Beseth, L. C. Madoff, S. K. Olken, D. L. Kasper, and F. M. Ausubel.** 1994. Genotypic diversity and evidence for two distinct classes of the C protein alpha antigen of group B *Streptococcus*, p. 331–332. *In* A. Totolian (ed.), *Pathogenic Streptococci: Present and Future*. Lancer Publications, St. Petersburg, Russia.

97. **Michon, F., E. Katzenellenbogen, D. L. Kasper, and H. J. Jennings.** 1987. Structure of the complex group-specific polysaccharide of group B *Streptococcus*. *Biochemistry* **26:** 476–486.

98. **Michon, F., R. Chalifour, R. Feldman, M. Wessels, D. L. Kasper, A. Gamian, V. Pozsgay, and H. J. Jennings.** 1991. The alpha-L-(1→2)-trirhamnopyranoside epitope on the group-specific polysaccharide of group B streptococci. *Infect. Immun.* **59:**1690–1696.

99. **Michon, F., P. C. Fusco, A. J. D'Ambra, M. Laude-Sharp, K. Long-Rowe, M. S. Blake, and J. Y. Tai.** 1997. Combination conjugate vaccines against multiple serotypes of group B streptococci. *Adv. Exp. Med. Biol.* **418:** 847–850.

100. **Palazzi, D. L., M. A. Rench, M. S. Edwards, and C. J. Baker.** 2004. Use of type V group B streptococcal conjugate vaccine in adults 65–85 years old. *J. Infect. Dis.* **190:**558–564.

101. **Pancholi, V., and V. A. Fischetti.** 1998. Alpha-enolase, a novel strong plasmin(ogen) binding protein on the surface of pathogenic streptococci. *J. Biol. Chem.* **273:**14503–14515.

101a.**Paoletti, L. C., and D. L. Kasper.** 2002. Conjugate vaccines against group B Streptococcus type IV and VII. *J. Infect. Dis.* **186:**123–126.

102. **Paoletti, L. C., D. L. Kasper, F. Michon, J. DiFabio, K. Holme, H. J. Jennings, and M. R. Wessels.** 1990. An oligosaccharide-tetanus toxoid conjugate vaccine against type III group B *Streptococcus*. *J. Biol. Chem.* **265:**18278–18283.

103. **Paoletti, L. C., D. L. Kasper, F. Michon, J. DiFabio, H. J. Jennings, T. D. Tosteson, and M. R. Wessels.** 1992. Effects of chain length on the immunogenicity in rabbits of group B Streptococcus type III oligosaccharide-tetanus toxoid conjugates. *J. Clin. Invest.* **89:**203–209.

104. **Paoletti, L. C., M. R. Wessels, F. Michon, J. DiFabio, H. J. Jennings, and D. L. Kasper.** 1992. Group B *Streptococcus* type II polysaccharide-tetanus toxoid conjugate vaccine. *Infect. Immun.* **60:**4009–4014.

105. **Paoletti, L. C., M. R. Wessels, A. K. Rodewald, A. A. Shroff, H. J. Jennings, and D. L. Kasper.** 1994. Neonatal mouse protection against infection with multiple group B streptococcal (GBS) serotypes by maternal immunization with a tetravalent GBS polysaccharide-tetanus toxoid conjugate vaccine. *Infect. Immun.* **62:** 3236–3243.

106. **Paoletti, L. C., R. A. Ross, and K. D. Johnson.** 1996. Cell growth rate regulates expression of group B *Streptococcus* type III capsular polysaccharide. *Infect. Immun.* **64:**1220–1226.

107. **Paoletti, L. C., J. Pinel, A. K. Rodewald, and D. L. Kasper.** 1997. Therapeutic potential of human antisera to group B streptococcal glycoconjugate vaccines in neonatal mice. *J. Infect. Dis.* **175:**1237–1239.

108. **Paoletti, L. C., C. J. Baker, and D. L. Kasper.** 1998. Neonatal group B streptococcal disease: progress towards a multivalent maternal vaccine, abstr. P-16, p. 43. First Annual Conference on Vaccine Research. National Foundation for Infectious Diseases, Washington, D.C.

108a.**Paoletti, L. C., D. L. Peterson, R. Legmann, and R. J. Collier.** 2001. Preclinical evaluation of group B streptococcal polysaccharide conjugate vaccines prepared with a modified diphtheria toxin and a recombinant duck hepatitis B core antigen. *Vaccine* **20:**370–376.

109. **Rainard, P.** 1992. Isotype antibody response in cows to *Streptococcus agalactiae* group B polysaccharide-ovalbumin conjugate. *J. Clin. Microbiol.* **30:**1856–1862.

110. **Rainard, P., Y. Lautrou, P. Sarradin, and B. Poutrel.** 1991. Protein X of *Streptococcus agalactiae* induces opsonic antibodies in cows. *J. Clin. Microbiol.* **29:**1842–1846.

111. **Rench, M. A., and C. J. Baker.** 1993. Neonatal sepsis caused by a new group B streptococcal serotype. *J. Pediatr.* **122:**638–640.

112. **Ross, R. A., L. C. Madoff, and L. C. Paoletti.** 1999. Regulation of cell component production by growth rate in group B *Streptococcus*. *J. Bacteriol.* **181:**5389–5394.

113. **Russell-Jones, G. J., and E. C. Gotschlich.** 1984. Identification of protein antigens of group B streptococci, with special reference to the Ibc antigens. *J. Exp. Med.* **160:** 1476–1484.

114. **Russell-Jones, G. J., E. C. Gotschlich, and M. S. Blake.** 1984. A surface receptor specific for human IgA on group B streptococci possessing the Ibc protein antigen. *J. Exp. Med.* **160:**1467–1475.

115. **Schifferle, R. E., H. J. Jennings, M. R. Wessels, E. Katzenellenbogen, R. Roy, and D. L. Kasper.** 1985. Immunochemical analysis of the types Ia and Ib group B streptococcal polysaccharides. *J. Immunol.* **135:**4164–4170.

116. **Schuchat, A.** 1998. Epidemiology of group B streptococcal disease in the United States: shifting paradigms. *Clin. Microbiol. Rev.* **11:**497–513.

117. **Shen, X., T. Lagergard, Y. Yang, M. Lindblad, M. Fredriksson, G. Wallerstrom, and J. Holmgren.** 2001. Effect of pre-existing immunity for systemic and mucosal immune responses to intranasal immunization with group B Streptococcus type III capsular polysaccharide-cholera toxin B subunit conjugate. *Vaccine* **19:**3360–3368.

118. **Smith, B. L., A. Flores, J. Dechaine, J. Krepela, A. Bergdall, and P. Ferrieri.** 2004. Gene encoding the group B streptococcal protein R4, its presence in clinical refer-

ence laboratory isolates and R4 protein pepsin sensitivity. *Indian J. Med. Res.* **119**(Suppl)**:**213–220.

119. **Stalhammar-Carlemalm, M., L. Stenberg, and G. Lindahl.** 1993. Protein rib: a novel group B streptococcal cell surface protein that confers protective immunity and is expressed by most strains causing invasive infections. *J. Exp. Med.* **177:**1593–1603.

120. **Stalhammar-Carlemalm, M., T. Areschoug, C. Larsson, and G. Lindahl.** 1999. The R28 protein of Streptococcus pyogenes is related to several group B streptococcal surface proteins, confers protective immunity and promotes binding to human epithelial cells. *Mol. Microbiol.* **33:** 208–219.

121. **Stoll, B. J., and A. Schuchat.** 1998. Maternal carriage of group B streptococci in developing countries. *Pediatr. Infect. Dis. J.* **17:**499–503.

122. **Suvorov, A. N., A. E. Flores, and P. Ferrieri.** 1997. Cloning of the glutamine synthetase gene from group B streptococci. *Infect. Immun.* **65:**191–196.

123. **Tettelin, H., V. Masignani, M. J. Cieslewicz, J. A. Eisen, S. Peterson, M. R. Wessels, I. T. Paulsen, K. E. Nelson, I. Margarit, T. D. Read, L. C. Madoff, A. M. Wolf, M. J. Beanan, L. M. Brinkac, S. C. Daugherty, R. T. DeBoy, A. S. Durkin, J. F. Kolonay, R. Madupu, M. R. Lewis, D. Radune, N. B. Fedorova, D. Scanlan, H. Khouri, S. Mulligan, H. A. Carty, R. T. Cline, S. E. Van Aken, J. Gill, M. Scarselli, M. Mora, E. T. Iacobini, C. Brettoni, G. Galli, M. Mariani, F. Vegni, D. Maione, D. Rinaudo, R. Rappuoli, J. L. Telford, D. L. Kasper, G. Grandi, and C. M. Fraser.** 2002. Complete genome sequence and comparative genomic analysis of an emerging human pathogen, serotype V Streptococcus agalactiae. *Proc. Natl. Acad. Sci. USA* **99:**12391–12396.

124. **von Hunolstein, C., S. D'Ascenzi, B. Wagner, J. Jelinková, G. Alfarone, S. Recchia, M. Wagner, and G. Orefici.** 1993. Immunochemistry of capsular type polysaccharide and virulence properties of type VI *Streptococcus agalactiae* (group B streptococci). *Infect. Immun.* **61:** 1272–1280.

125. **Wagner, M., B. Wagner, and V. R. Kubin.** 1980. Immunoelectron microscopic study of the location of group-specific and type-specific polysaccharide antigens on isolated walls of group B streptococci. *J. Gen. Microbiol.* **120:**369–376.

126. **Wastfelt, M., M. Stalhammar-Carlemalm, A. M. Delisse, T. Cabezon, and G. Lindahl.** 1996. Identification of a family of streptococcal surface proteins with extremely repetitive structure. *J. Biol. Chem.* **271:**18892–18897.

127. **Wessels, M. R., A. Munoz, and D. L. Kasper.** 1987. A model of high-affinity antibody binding to type III group B *Streptococcus* capsular polysaccharide. *Proc. Natl. Acad. Sci. USA* **84:**9170–9174.

128. **Wessels, M. R., V. Pozsgay, D. L. Kasper, and H. J. Jennings.** 1987. Structure and immunochemistry of an oligosaccharide repeating unit of the capsular polysaccharide of type III group B Streptococcus. A revised structure for the type III group B streptococcal polysaccharide antigen. *J. Biol. Chem.* **262:**8262–8267.

129. **Wessels, M. R., W. J. Benedí, H. J. Jennings, F. Michon, J. L. DiFabio, and D. L. Kasper.** 1989. Isolation and characterization of type IV group B *Streptococcus* capsular polysaccharide. *Infect. Immun.* **57:**1089–1094.

130. **Wessels, M. R., C. E. Rubens, V. J. Benedí, and D. L. Kasper.** 1989. Definition of a bacterial virulence factor: sialylation of the group B streptococcal capsule. *Proc. Natl. Acad. Sci. USA* **86:**8983–8987.

131. **Wessels, M. R., L. C. Paoletti, D. L. Kasper, J. L. DiFabio, F. Michon, K. Holme, and H. J. Jennings.** 1990. Immunogenicity in animals of a polysaccharide-protein conjugate vaccine against type III group B *Streptococcus*. *J. Clin. Invest.* **86:**1428–1433.

132. **Wessels, M. R., J. L. DiFabio, V. J. Benedí, D. L. Kasper, F. Michon, J. R. Brisson, J. Jelinková, and H. J. Jennings.** 1991. Structural determination and immunochemical characterization of the type V group B streptococcus capsular polysaccharide. *J. Biol. Chem.* **266:** 6714–6719.

133. **Wessels, M. R., L. C. Paoletti, A. K. Rodewald, F. Michon, J. DiFabio, H. J. Jennings, and D. L. Kasper.** 1993. Stimulation of protective antibodies against type Ia and Ib group B streptococci by a type Ia polysaccharide-tetanus toxoid conjugate vaccine. *Infect. Immun.* **61:** 4760–4766.

134. **Wessels, M. R., L. C. Paoletti, J. Pinel, and D. L. Kasper.** 1995. Immunogenicity and protective activity in animals of a type V group B streptococcal polysaccharide-tetanus toxoid conjugate vaccine. *J. Infect. Dis.* **171:** 879–884.

135. **Wilkinson, H. W.** 1972. Comparison of streptococcal R antigens. *Appl. Microbiol.* **24:**669–670.

136. **Wilkinson, H. W., and R. G. Eagon.** 1971. Type-specific antigens of group B type Ic streptococci. *Infect. Immun.* **4:**596–604.

137. **Yim, H. H.** 1998. Regulation of capsular polysaccharide production in group B *Streptococcus*, abstr. 27, p. 36. Am. Soc. Microbiol. Conference on Streptococcal Genetics. American Society for Microbiology, Washington, D.C.

Epidemiology of Group B Streptococcal Infections

ANNE SCHUCHAT AND SHARON BALTER

15

Group B streptococcus (GBS), or *Streptococcus agalactiae*, was first classified in the 1930s by Lancefield and Hare (51) in their studies on the serologic differentiation of streptococci. Lancefield's studies established that most puerperal infections of the day were due to group A streptococci, although they identified GBS from the vaginal cultures of asymptomatic women. GBS was better known at the time as the cause of bovine mastitis. GBS was first described as a human pathogen in a 1938 report of three women with fatal puerperal sepsis due to GBS (42). However, it was only rarely reported as a human pathogen until the 1960s, when it was increasingly seen as an adult pathogen and emerged as the leading cause of neonatal sepsis (36). Before this time, sepsis in newborns had been principally caused by *Escherichia coli* (36). The reasons for the emergence of GBS disease are unclear.

GBS DISEASE IN NONPREGNANT ADULTS

Incidence

Although GBS is commonly thought of as a cause of disease in neonates and pregnant women, it causes substantial morbidity and mortality among nonpregnant adults (Fig. 1) and appears to be increasing in incidence in that population. Incidence rates vary by area; a study in metropolitan Atlanta in 1992 to 1993 found the incidence of invasive disease among nonpregnant adults to be 5.9 cases per 100,000 population, a 37% increase from 1989 to 1990 (20). A similar study in Maryland in 1992 to 1993 found the incidence among nonpregnant adults to be 6.5/100,000 (47). Incidence varies by age, with rates of 3.6/100,000 persons 15 to 64 years old compared with 20.1/100,000 persons 65 years and older (71). Reported case-fatality rates range from 9 to 70% (19, 30, 43); multistate population-based surveillance recently estimated a case-fatality rate of 15% in persons 65 years and older (71). The overall rate of invasive GBS disease is slightly lower for women, about 3.3/100,000, compared with 4.9/100,000 for men, but women account for more than half the cases among the elderly (38).

Of nine known GBS serotypes, the most common reported among adults are types III and V. Serotype V was first reported in 1985 and initially appeared to be a rare cause of infection (20). However, a population-based study found it to be the most common serotype causing disease in nonpregnant adults and the second most common serotype in pregnant women (47). The emergence of serotype V appears to be a relatively recent phenomenon. Although serotype V isolates would previously have been identified as nontypeable, studies done before serotype V was identified reported few nontypeable isolates; in the mid-1970s, only 4% of adults had nontypeable strains, whereas serotype V now accounts for 29% of adult cases (47). Reports of serotype VIII were originally restricted to studies in Japanese populations, but bloodstream isolates of serotype VIII from patients in Denmark have now been reported (37).

Syndromes

The most common syndromes caused by GBS in adults are skin, soft tissue, and bone infections (38, 76). These infections are often complications of chronic diabetes or decubitus ulcers (38). Cellulitis, foot ulcers, and abscesses are the most common manifestations, but other infections, including necrotizing fasciitis, have occasionally been reported (35).

Another common presentation of GBS infection in adults is bacteremia without any identified focus. In one study, about a quarter of such patients had underlying hepatic or renal failure (38), although malignancy and diabetes also were common underlying conditions. Patients with indwelling catheters are at higher risk for GBS bacteremia (49). Polymicrobial bacteremia, often with *Staphylococcus aureus*, is identified in 26 to 30% of patients with GBS bacteremia (38, 49).

GBS can also cause pneumonia (38, 76). Pneumonia is a particularly severe form of GBS disease and has a high mortality rate (49, 81). In one small case series, all seven patients with GBS pneumonia died (81). Patients with GBS pneumonia appear to have more neurologic disease, dementia, or tracheoesophageal fistulas, suggesting that GBS pneumonia may be related to aspiration of the organism (38, 81).

Gram-Positive Pathogens, 2nd edition, edited by Vincent A. Fischetti et al.
© 2006 ASM Press, Washington, D.C.

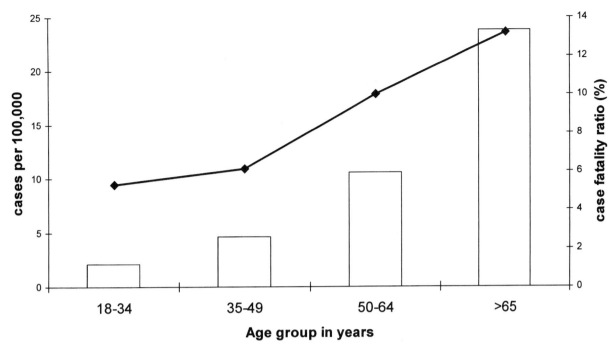

FIGURE 1 Incidence of invasive group B streptococcal disease by age (cases per 100,000) (open bars) and case fatality ratio (%) (black line) in persons ≥18 years of age. (Active Bacterial Core surveillance 2002; data available at http://www.cdc.gov/abcs.)

In the series of seven patients, six had severe underlying neurologic conditions, and one had cancer of the esophagus with tracheoesophageal fistula. Chest radiographs showed bilateral or lobar infiltrates, and although infections were polymicrobial, GBS was the predominant organism (81).

The epidemiology of GBS endocarditis has undergone a shift since the preantibiotic era. Previously, endocarditis was one of the most common syndromes caused by GBS infection, but GBS endocarditis is much rarer today (38, 71, 76). In the 1930s and 1940s, GBS endocarditis was predominantly an acute disease of pregnant women and most often affected the mitral valve (52, 76). A comparison of a series of patients with endocarditis from 1938 to 1945 with a series after 1962 found that in the latter series, about half the patients were men, and although the mitral valve remained the most common site of infection, aortic and tricuspid valve disease was also seen (52). Cases of GBS endocarditis in the latter series were both acute and subacute. Large, friable vegetations are a common feature of the disease, often leading to embolization and requiring surgical intervention (52). Recent series confirm a high mortality rate for GBS endocarditis in the modern era (69). Endocarditis in pregnancy still occurs, with the tricuspid valve most often involved (29).

Less common GBS infections such as arthritis, urinary tract infections, meningitis, and peritonitis also occur. GBS arthritis is most commonly monoarticular, although it can be polyarticular (62, 76). Diabetes, osteoarthritis, and underlying joint disease, including prosthetic joints, are common predisposing factors. In a case series, patients with GBS arthritis required both antibiotic therapy and drainage for recovery and, despite therapy, only about 50% completely recovered (35). GBS urinary tract infections occur in nonpregnant adults as well as pregnant women (34, 52). In men, GBS urinary tract infections appear to be associated with prostatic infections (34), although they may also occur with stones or other abnormalities of the urinary tract and in patients with diabetes (38, 52).

Risk Factors

Adults with chronic diseases are at higher risk for GBS infection. Conditions that increase the risk of GBS disease include diabetes, malignancy, liver disease, neurologic impairment, renal failure, other forms of immune impairment such as human immunodeficiency virus infection, and steroid use (34, 38, 49, 52, 76, 81). In one study, patients with any type of cancer had a 16-fold-higher risk of developing invasive GBS disease than did healthy adults (76), although another study found that this increased risk declined with age and that the elderly with cancer were at no greater risk than the elderly without cancer (38). A third study, which used multivariate analysis, found that only cancer of the breast independently increased the risk of invasive GBS disease. (49)

Diabetes is a commonly reported condition among patients with GBS disease (34, 38, 49, 52, 76). In a population-based study of risk factors, patients with diabetes had 10.5 times the risk of the general population (76). The precise mechanism underlying this increased risk is unclear. Diabetes increases the risk for GBS disease for both insulin-dependent and non-insulin-dependent patients across all age groups (38). Persons with diabetes do not have an increased rate of nose, throat, skin, or rectal GBS colonization compared with those without diabetes (25), but some evidence indicates that abnormalities in immune function, such as neutrophil phagocytosis or intracellular killing, may contribute to the increased risk for diabetic patients (80).

In addition to underlying disease, other risk factors for invasive GBS disease in adults include advanced age and black race (38, 76). Although the disease occurs in all age groups, in-hospital mortality is higher among older patients with GBS infections (38, 49). The mean age of GBS patients in several population-based studies was about 65 years (38, 49, 76). The explanation for the higher risk of GBS disease among black persons is unclear, although black race may be a surrogate for other factors. In one study, race was not an independent risk factor after controlling for chronic medical conditions (49). In a population-based study in metropolitan Atlanta, the incidence of invasive infection among blacks was twice as high as that among whites, but most of the disparity was associated with residence in the urban center; whites residing in the same area also had an increased risk compared with whites in other communities (38). The higher prevalence of diabetes and renal disease may also contribute to higher rates of GBS disease in this population (38).

GBS infection may be nosocomially acquired. One study found that among patients acquiring the infection in the hospital, GBS was independently associated with congestive heart failure, dementia, seizure disorder, and placement of a central venous line (49). Clustering of invasive GBS disease among hospitalized patients is rarely reported, suggesting that the underlying illness of the hospitalized population or invasive procedures introducing endogenous flora into patients predisposes them to disease, rather than the disease being transmitted from patient to patient. Because the gastrointestinal tract is thought to be a reservoir for GBS (7), there has been concern that sigmoidoscopy and colonoscopy can lead to disease postprocedure. Arthritis and pacemaker wire infections (35) due to GBS following sigmoidoscopy or colonoscopy have been reported.

GBS IN PREGNANT WOMEN AND NEWBORNS

Infections in Pregnant Women

GBS causes substantial morbidity in pregnant women. Of 307 cases of invasive GBS disease in adults identified in a population-based active surveillance study in 1990, 11% were in pregnant women (86). For the period 1993 through 1998, 345 (6.3%) of the 5,463 cases that occurred in adults were in pregnant women (71). In one study from the 1980s, the attack rate of GBS infections in pregnant women was approximately 2/1,000 deliveries (64). Multistate surveillance suggests that the incidence of invasive GBS disease in pregnant women declined from 0.29 per 1,000 live births in 1993 to 0.23 per 1,000 births in 1998 ($P < 0.03$) (71), consistent with increases in preventive antibiotics during this time period. GBS causes a variety of perinatal infections in pregnant women, including both symptomatic and asymptomatic bacteriuria, endometritis, amnionitis, meningitis, pyelonephritis, and postpartum wound infections (64).

It has also been suggested that GBS urinary tract infections or urinary tract, rectal, or genital colonization in pregnant women may lead to late-term abortions and preterm and low-birth-weight infants. A study comparing 150 women who presented with signs of threatened abortion with 100 women of similar gestational age demonstrated that women who aborted were more likely to be colonized with GBS than those who did not (31). The association was seen for both cervical colonization and urinary colonization. Smaller studies have both confirmed (60) and disputed (59) the association between GBS in the urinary tract and preterm delivery. In a large study done by the Vaginal Infections and Prematurity Group, 13,914 women were screened for GBS colonization during pregnancy. This study showed an increase in the risk of low birth weight and premature infants among women who were heavily colonized with GBS. Women with light colonization had pregnancy outcomes similar to those of the uncolonized women (68).

Sites of Maternal Colonization

Many pregnant women are colonized with GBS but are asymptomatic. However, GBS colonization in pregnant women is important because of the risk for transmission to their newborns. Newborns born to colonized mothers are more likely to develop GBS infection (6). Women may be colonized at multiple sites, including the rectum, vagina, cervix, and throat (10, 67). Rectal carriage is more common than vaginal carriage (10, 33). Cervical colonization is relatively less common, and throat colonization is even rarer (10, 33, 67). The Vaginal Infections and Prematurity Study, the largest of its kind, found the overall carriage rate among 13,914 pregnant women to be 21%, although only vaginal cultures were collected in this study (68).

The ability to detect GBS colonization depends both on the sites cultured and on the medium used. Using selective broth medium will result in higher recovery of GBS; one study demonstrated 50% improvement in the recovery of GBS with the use of selective broth medium compared with standard blood agar (65). In this study, which evaluated the impact of culture site, 94 women were cultured both vaginally and rectally; 29 women were positive for GBS. Of those, 3 (10%) were positive on vaginal culture only, 12 (41.4%) were positive on rectal culture only, and 14 (48.3%) were positive on both rectal and vaginal culture (65).

Duration of Maternal Colonization

GBS carriage may be chronic, intermittent, or transient (23, 33). In a large study of pregnant women who were screened for GBS once during a prenatal visit from the late first trimester through the third trimester, 67% of women who were GBS carriers and 8.5% of women who were not colonized with GBS at their prenatal visit were culture positive for GBS at delivery (23). The value of prenatal culture for predicting colonization at delivery increased with shorter intervals between prenatal sampling and delivery. All (16 of 16) carriers who delivered within 5 weeks of their prenatal culture remained positive at the time of delivery. In a similar study of 754 women who were sequentially cultured for rectal and vaginal colonization with GBS in the second and third trimesters, 31% were carriers of GBS in the second trimester and 28% were carriers of GBS in the third trimester; only 17% were positive in both trimesters. Serotyping of the isolates indicated that persistence of the same type was most common in anorectal-vaginal carriers and least likely in women who were vaginal carriers only; anorectal-vaginal carriers rarely acquired a new type. Of women who initially did not carry GBS, 16% acquired GBS between the second and third trimesters (33). Another study of 826 women at 35 to 36 weeks' gestation showed that the sensitivity of late antenatal cultures for identifying colonization status at delivery was 87% (84).

Risk Factors for Maternal Carriage

Efforts have been made to characterize groups with higher rates of GBS carriage. As part of the Vaginal Infections and Prematurity Study, 8,049 pregnant women underwent vaginal and cervical cultures for GBS colonization at 23 to 26 weeks' gestation (67). In a multivariate analysis, Caribbean Hispanics and blacks were more likely to carry GBS than were whites and Mexican Hispanics. The same study also found that GBS colonization was associated with increasing age, lower parity, and fewer years of education; current smoking was associated with a decreased risk of colonization. Colonization was only associated with "extreme" increases in sexual activity, defined as those women who had both frequent intercourse and multiple partners. Although carriage was associated with concurrent colonization with *Candida* spp., GBS carriage was not associated with concurrent infection with sexually transmitted pathogens, including *Neisseria gonorrhoeae*, *Trichomonas vaginalis*, or *Chlamydia trachomatis* (67). A lack of association between GBS colonization and sexually transmitted diseases has been observed in other studies (14). A study of nonpregnant college women showed that women with any sexual experience were more likely to be colonized than were those without sexual experience. That study also suggested that intrauterine devices may be associated with carriage (14). It did not demonstrate an association between GBS colonization and oral contraceptives, although this has been reported (70).

Infant Colonization

Infants born to GBS-colonized women are more likely to be colonized with GBS. In a study of 802 women in which investigators obtained endocervical and vaginal cultures at the 36th week of pregnancy and again at delivery, cultures were also obtained from infants' anterior nares, external ear canals, and the base of their umbilicus within 3 h of birth. Of infants born to mothers who had positive cultures at delivery, 37% were colonized with GBS at delivery and 49% were colonized by the time they were discharged. In contrast, of the 689 infants born to culture-negative mothers, only 1% acquired GBS colonization. All but one of the 31 GBS-colonized infants in this study were colonized with the same serotype as their mother (41).

Although infant colonization is presumed to occur in the uterus or birth canal, even if a mother is colonized in the rectum, and not the vagina, there is still a 17% rate of vertical transmission of GBS to the newborn (6). GBS can cross intact membranes, and neonatal acquisition from the mother can occur in newborns delivered by cesarean section. In a longitudinal study of vaginally delivered infants whose mothers were colonized with GBS, 61% acquired GBS colonization; of infants who were delivered by cesarean section to mothers who were colonized with GBS, 40% acquired GBS colonization with the same serotype as the mother. Infants whose mothers were heavily colonized were more likely to become colonized than those whose mothers were only lightly colonized (95% versus 31%) (6).

Infections in Newborns

GBS is a leading cause of neonatal bacterial disease in the United States. Before the use of prevention measures became widespread in the early and mid-1990s, GBS caused an estimated 7,600 cases of serious illness and 310 deaths among U.S. infants aged ≤90 days (86). Between 1993 and 2003, the overall annual incidence of early-onset GBS disease declined from 1.7 cases per 1,000 live-born infants to 0.3 cases per 1,000 live-born infants, as a result of increased use of preventive strategies (Fig. 2) (27, 28, 71). Early mortality from neonatal sepsis also declined disproportionately during the 1990s, in comparison with neonatal mortality from other causes or late-sepsis mortality (57).

Infections in newborns commonly present as bacteremia, meningitis, or pneumonia. There are two distinct syndromes: early-onset disease (EOD), which appears in the first week of life, usually within the first 24 h, and late-onset disease (LOD), which occurs on or after 7 days of age (12). Infants with EOD most commonly have sepsis or pneumonia; meningitis and bone and soft tissue infections can also oc-

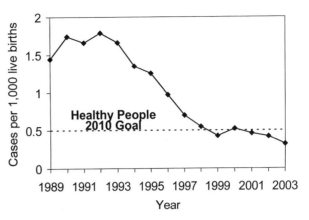

FIGURE 2 Incidence of early-onset group B streptococcal disease (cases per 1,000 live births), 1989–2003 (Active Bacterial Core surveillance) (from reference 28) compared with Healthy People 2010 goal of 0.5 per 1,000 live births.

cur. EOD is due to ascending infection from the genital tract or colonization of the infant during delivery. Amniotic infection can occur even when the membranes are intact (31), although both premature and prolonged rupture of membranes increase the risk of transmission. Case-fatality rates for EOD are estimated at 6.5% (28).

LOD is more likely to present with meningitis than is EOD, although LOD also can present as bone and soft tissue infections, urinary tract infections, or pneumonia (82). Although the disease is often thought of as occurring between the first 7 days and 12 weeks of life, in one study at a tertiary care center, 20% of cases occurred in infants older than 3 months of age (82). Infants with LOD are less likely to be severely ill upon presentation, and their deliveries are less likely to be characterized by predisposing obstetric complications (13). Case-fatality rates for LOD are estimated at 2.8% (71).

Among patients with EOD, serotypes III and Ia are most commonly identified, although serotypes V and II are also seen. The majority of patients with LOD have serotype III, although serotype Ia is also fairly common (47).

Risk Factors for Early-Onset Neonatal Disease

Maternal carriage is clearly a risk factor for neonatal GBS disease (23). Infants born to heavily colonized mothers are more likely to develop invasive disease than are infants born to mothers who are lightly colonized or not colonized at all (45, 53). Without preventive measures, about 1 to 2% of neonates born to mothers with rectovaginal colonization will develop early-onset GBS disease (23).

Although the majority of cases of GBS disease occur in full-term infants, prematurity and low birth weight are risk factors for the development of invasive neonatal GBS disease (74). Although premature infants may be more susceptible to invasive disease, it is also possible that prematurity and low birth weight are the result of GBS colonization in the mother (58, 68).

Maternal antibodies to the capsular polysaccharides of GBS protect against GBS disease. Antibodies are passively transferred from a mother to her fetus during pregnancy, offering protection against invasive disease. One study found lower antibody concentrations in women whose infants developed GBS disease than in colonized women whose infants did not (15); another found that women whose infants developed GBS disease had lower antibody titers than the general population (46). The concentration of serum antibody against GBS increases with increasing age (8); this could explain in part why young maternal age is a risk factor for invasive neonatal GBS disease (74, 75). Previous delivery of an infant with GBS disease also increases risk that EOD will occur in a subsequent pregnancy (72).

Obstetric factors that have been associated with early-onset GBS disease include intrapartum fever (23, 75), rupture of membranes before labor (75), group B streptococcal bacteriuria during pregnancy (56, 72, 75), prolonged labor (75), and prolonged rupture of membranes (23, 39). It has also been suggested that multiple gestation may be a risk factor for EOD, although this observation may be due to the fact that such infants are more likely to be low birth weight and premature (63).

Other risk factors for EOD include black race (74, 75), Hispanic ethnicity (85), and history of previous miscarriage (74). One study identified internal monitoring for more than 12 h as a risk factor for GBS disease (83), although this was not an independent risk factor in other studies (75).

Risk Factors for LOD

Risk factors for late-onset GBS disease have recently been characterized (55), though they are less well understood than those for EOD. Obstetric factors do not appear to be as important. A cohort study in metropolitan Atlanta found that black infants had 35 times the rate of LOD than infants of other races. The only other risk factor found in that study was young maternal age (74). Other reports suggested that breast-feeding by mothers with GBS mastitis may play a role in transmission of GBS, but this route of infection probably accounts for a minority of cases (9, 17). A recent hospital matched case-control study of risk factors for LOD conducted in Houston identified prematurity as the major risk factor for late-onset GBS disease, with maternal GBS colonization and black race as other independent factors associated with higher risk of LOD (55).

Although vertical transmission accounts for the colonization of most infants, nosocomial or horizontal transmission of GBS may account for some LOD. Anthony et al. (6) found evidence of nosocomial spread of colonization in 2 of 10 cohorts of infants studied. In one cohort, an infant acquired GBS colonization of the same serotype as the infant's mother; five other infants born on the same or next day and cared for in the same half of the nursery acquired GBS colonization of the same serotype, although their mothers had consistently negative cultures. Outbreaks of GBS disease in nurseries have rarely been reported (17).

TREATMENT

Group B streptococci are uniformly susceptible to penicillin in vitro, and penicillin G is the treatment of choice for established cases of GBS. The organism is generally susceptible in vitro to ampicillin, vancomycin, teicoplanin, imipenem, and first-, second-, and third-generation cephalosporins, although the activity varies with each agent (35). Several studies have been reported on antimicrobial resistance among GBS. Reports from 1998 through 2004 found resistance to erythromycin in 7 to 25% of isolates and resistance to clindamycin in 3 to 15% of isolates (5, 40, 48, 54). For newborns with presumptive GBS sepsis, the initial treatment should be penicillin G or ampicillin plus an aminoglycoside. Once GBS has been identified, penicillin G alone can be given.

For infants with GBS meningitis, the recommended dosage of penicillin G is 250,000 to 450,000 U/kg/day given intravenously in three divided doses for infants 7 days or younger; infants older than 7 days should receive 450,000 U/kg/day given intravenously in four divided doses. Ampicillin can also be used; the dose is 200 mg/kg/day given intravenously in three divided doses for infants 7 days or younger, and 300 mg/kg/day in four to six divided doses for infants over 7 days of age (2). For uncomplicated meningitis, 14 to 21 days of therapy is recommended, although longer treatment may be necessary in complicated cases (2). Infants with bacteremia of undefined focus require treatment for at least 10 days. For osteomyelitis, treatment is required for 4 weeks or more (2).

In adults, penicillin G is the treatment of choice. Ten days of treatment is recommended for bacteremia, soft tissue infections, pneumonia, and pyelonephritis, and 14 to 21 days is recommended for meningitis. For osteomyelitis, 3 to 4 weeks of therapy is required. Endocarditis patients should receive 4 to 6 weeks of therapy, and surgical debridement or valve replacement may be necessary (35).

PREVENTION

Strategies for EOD Prevention

Substantial progress in prevention of EOD has occurred in the past decade, as a result of implementation of intrapartum chemoprophylaxis. Over the 1990s, increased use of intrapartum prophylaxis led to declines in early-onset cases and deaths. In the United States, the American College of Obstetricians and Gynecologists, American Academy of Pediatrics, American Academy of Family Physicians, American College of Nurse Midwives, and Centers for Disease Control and Prevention all recommend universal screening for group B streptococcus and intrapartum prophylaxis for women identified as GBS carriers (4, 27).

Intrapartum antimicrobial prophylaxis involves the use of antibiotics during labor or after the onset of membrane rupture for selected women. Several studies indicated that intrapartum chemoprophylaxis of colonized women reduced both neonatal colonization and EOD (26). A study that randomized only those carriers who had risk factors for neonatal disease (premature labor or prolonged rupture of membranes) demonstrated a significant decrease in both neonatal colonization and disease among those who received intrapartum ampicillin (21). A study of 30,197 women who were screened for GBS at 32 weeks' gestation and treated with intrapartum penicillin showed 16 EOD infections in the study group compared with 27 in the control group of 26,195 women (0.5 per 1,000 births versus 1 per 1,000 births; $P = 0.04$) (44). In 1996, the first uniform recommendations for GBS EOD prevention were issued by the Centers for Disease Control and Prevention, the American College of Obstetricians and Gynecologists, and the American Academy of Pediatrics (1, 3, 26). The guidelines recommended use of either a screening-based strategy or a risk-based approach to selection of women for intrapartum antimicrobial prophylaxis. A retrospective cohort study conducted among infants born in 1998 and 1999 demonstrated that prenatal screening led to 50% fewer early-onset cases compared with the risk-based strategy (72). These data provided the technical basis for revised U.S prevention guidelines issued in 2002, which recommended universal prenatal screening (4, 27). The 2002 guidelines recommend collection of screening cultures at 35 to 37 weeks' gestation and use of intrapartum prophylaxis for all carriers as well as those women with threatened preterm delivery in whom a negative culture result is not available. To improve sensitivity of prenatal screening in predicting colonization status at the time of delivery, the guidelines specify that swabs should be collected from both vagina and rectum, and inoculated into selective broth medium.

The accuracy of prenatal screening cultures is improved if they are collected late in pregnancy. In a study of 826 women, the sensitivity of antenatal cultures collected between 35 and 36 weeks was 87% and the specificity was 96%; among patients with cultures collected 6 or more weeks before delivery, the sensitivity was only 43% and the specificity was 85% (84). Because carriage of GBS can be transient and recurrent, the best time to collect prenatal cultures has been a concern. Although culturing late in pregnancy ensures the reliability of the findings at delivery, women who deliver preterm and are therefore at a higher risk of transmission of neonatal disease are missed. To address the gaps involved with prenatal collection of screening specimens, investigators have explored development of rapid tests for detection of GBS at the time of labor. Although efforts initially had substantial limitations compared with the gold standard selective broth medium (11), the recent commercialization of a sensitive and specific real-time PCR test has greatly improved test performance (18, 32). Although this approach is quite promising, especially for facilities where many women present to labor and delivery without prenatal care, implementation of use of the PCR outside the research setting will benefit from ongoing evaluation (73).

Some hospitals also propose postnatal penicillin prophylaxis administered to newborns by intramuscular injection immediately after birth. Only one randomized trial was done, in which blood cultures were collected from all newborns before chemoprophylaxis; no differences were observed among the treated and untreated groups of low-birth-weight infants in this study (66). The mortality rates were also similar in both groups. Another study in a single hospital found a reduction in early-onset GBS infections, but there was no difference in mortality between the penicillin-treated group and the control group, and an increase in mortality associated with penicillin-resistant infections was initially identified in the treatment group, although it was not sustained

(78). A long-term observational study in the same hospital did find lower rates of EOD among groups in which postnatal prophylaxis was used (0.25 and 0.63 per 1,000 births) versus those in which it was not (1.19, 1.59, and 1.95 per 1,000) (77). Most of the infants who got sick despite postnatal penicillin did so in the first 4 h after birth. Postnatal prophylaxis may not be effective against early-onset GBS infections that are acquired in utero well before birth.

Investigators have also studied the use of the topical microbicide chlorhexidine to cleanse the birth canal during labor and reduce transmission of GBS or other bacteria from mother to newborn. This strategy reduced admissions to neonatal intensive care units in a trial in Sweden (24), but has had conflicting results elsewhere. A study in Malawi designed to test the effectiveness of chlorhexidine against perinatal transmission of HIV found reduced neonatal clinical sepsis and deaths as an unexpected but significant outcome (79). Conflicting data and concerns about the design of individual studies of chlorhexidine completed to date have limited the widespread uptake of this approach, although clinicians in some parts of Europe do use vaginal douching or cleansing with chlorhexidine in routine practice. Additional study of this approach is needed to clarify its value, particularly for developing country settings.

Current Recommendations for Prevention of Early-Onset Neonatal Disease

Building on earlier prevention research and extensive evaluation of the 1996 recommendations for use of either of two strategies for intrapartum antimicrobial prophylaxis (1, 3, 26), the Centers for Disease Control and Prevention, the American College of Obstetricians and Gynecologists, and the American Academy of Pediatrics issued updated prevention guidelines in 2002 (4, 27). The statements recommend a universal screening-based approach. All women are screened at 35 to 37 weeks' gestation for rectovaginal colonization with GBS (Fig. 3). Intrapartum chemoprophylaxis is offered to all women who have at least one of the following: GBS bacteriuria during pregnancy, delivery of a previous infant with GBS disease, or positive prenatal culture. If

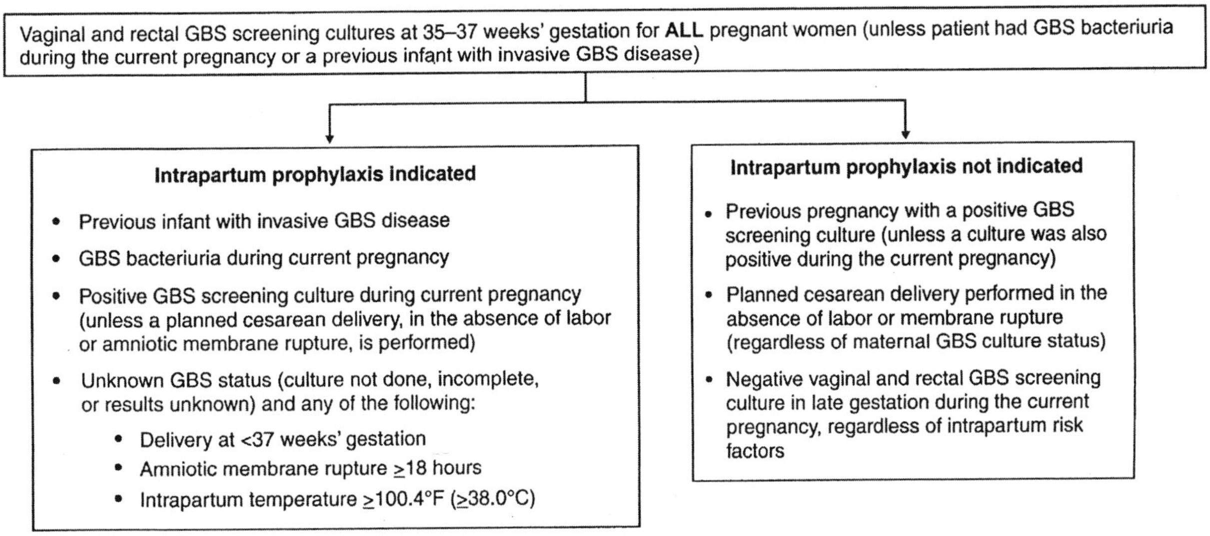

Vaginal and rectal GBS screening cultures at 35–37 weeks' gestation for **ALL** pregnant women (unless patient had GBS bacteriuria during the current pregnancy or a previous infant with invasive GBS disease)

Intrapartum prophylaxis indicated

- Previous infant with invasive GBS disease
- GBS bacteriuria during current pregnancy
- Positive GBS screening culture during current pregnancy (unless a planned cesarean delivery, in the absence of labor or amniotic membrane rupture, is performed)
- Unknown GBS status (culture not done, incomplete, or results unknown) and any of the following:
  - Delivery at <37 weeks' gestation
  - Amniotic membrane rupture ≥18 hours
  - Intrapartum temperature ≥100.4°F (≥38.0°C)

Intrapartum prophylaxis not indicated

- Previous pregnancy with a positive GBS screening culture (unless a culture was also positive during the current pregnancy)
- Planned cesarean delivery performed in the absence of labor or membrane rupture (regardless of maternal GBS culture status)
- Negative vaginal and rectal GBS screening culture in late gestation during the current pregnancy, regardless of intrapartum risk factors

FIGURE 3 Algorithm for prevention of perinatal group B streptococcal disease, 2002 guidelines (from reference 27).

results of prenatal culture are not known at delivery, intrapartum chemoprophylaxis should be given if the patient is <37 weeks' gestation, has duration of membrane rupture ≥18 h, or has a temperature of ≥100.4°F (≥38°C). To ensure accurate results, both vaginal and rectal swabs should be collected and incubated in a single selective culture broth.

The recommended intrapartum regimen is penicillin, 5-mU intravenous load, and then 2.5 mU every 4 h until delivery, or, alternatively, ampicillin, 2-g load, and then 1 g intravenously every 4 h until delivery. The increase in resistance to clindamycin and erythromycin among GBS isolates is sufficient that empiric use of those drugs for penicillin-allergic women is no longer appropriate in the United States, and new recommendations have been developed for the penicillin-allergic woman with GBS colonization. For penicillin-allergic women at low risk of anaphylaxis, use of cephazolin (2 g intravenous initial dose, then 1 g intravenously every 8 h until delivery) is recommended. For penicillin-allergic women at high risk of anaphylaxis, antimicrobial susceptibility testing is recommended. If GBS isolates are susceptible to both clindamycin and erythromycin, then clindamycin, 900 mg intravenously every 8 h until delivery, or erythromycin, 500 mg intravenously every 6 h until delivery, may be used. If the organism is resistant to either clindamycin or erythromycin, then vancomycin (1 g intravenously every 12 h until delivery) is recommended. Intrapartum chemoprophylaxis should begin as soon as possible during labor for GBS-positive women.

THE FUTURE

GBS disease is a major concern for pregnant women and their newborns and a problem among the growing population of adults with chronic diseases. Although most neonatal GBS disease can be prevented through intrapartum prophylaxis, currently available strategies are not ideal. The currently recommended strategy involves giving intrapartum antimicrobial agents to approximately 25% of pregnant women, which may result in adverse reactions and possible emergence of antimicrobial-resistant organisms (61).

Development of a vaccine against GBS may provide a better long-term solution than chemoprophylaxis. A vaccine could be given to women during pregnancy or to adolescent girls; transfer of antibodies across the placenta late in pregnancy would confer protection to the newborn. In addition, a vaccine could be effective against LOD and disease in pregnant women and nonpregnant adults. A vaccine for GBS is in development (50, 16). Although substantial progress has been achieved with the use of intrapartum prophylaxis, as evidenced by the 80% decline in incidence of EOD, concern regarding emerging resistance and lack of impact on LOD suggests that ongoing work to deliver on the promise of GBS vaccines is extremely worthwhile.

REFERENCES

1. **American Academy of Pediatrics.** 1997. Revised guidelines for prevention of early-onset group B streptococcal (GBS) infection. *Pediatrics* **99:**489–495.
2. **American Academy of Pediatrics, Committee on Infectious Diseases, G. Peter, C. B. Hall, N. A. Halsey, S. M. Marcy, and L. K. Pickering.** 1997. p. 494–501. *1997 Red Book: Report of the Committee on Infectious Diseases*, 24th ed. American Academy of Pediatrics, Elk Grove Village, Ill.
3. **American College of Obstetricians and Gynecologists, Committee on Obstetric Practice.** 1996. p. 1–7. *Prevention of Early-Onset Group B Streptococcal Disease in Newborns*, no. 173. American College of Obstetricians and Gynecologists, Washington, D.C.
4. **American College of Obstetricians and Gynecologists, Committee on Obstetric Practice.** 2002. *Prevention of Early-Onset Group B Streptococcal Disease in Newborns*. American College of Obstetricians and Gynecologists, Washington, D.C.
5. **Andrews J. I., D. J. Diekema, S. K. Hunter, P. R. Rhomberg, M. A. Pfaller, R. N. Jones, and G. V. Doern.** 2000. Group B streptococci causing neonatal bloodstream infection: antimicrobial susceptibility and serotyping results from SENTRY centers in the Western Hemisphere. *Amer. J. Obstetr. Gynecol.* **183:**859–862.
6. **Anthony, B. F., D. M. Okada, and C. J. Hobel.** 1979. Epidemiology of the group B *Streptococcus*: maternal and nosocomial sources for infant acquisitions. *J. Pediatr.* **95:**431–436.
7. **Anthony, B. F., R. Eisenstadt, J. Carter, S. K. Kwang, and C. J. Hobel.** 1981. Genital and intestinal carriage of group B streptococci during pregnancy. *J. Infect. Dis.* **143:**761–766.
8. **Anthony, B. F., I. E. Concepcion, N. F. Concepcion, C. M. Vadheim, and J. Tiwari.** 1994. Relation between maternal age and serum concentration of IgG antibody to type III group B streptococci. *J. Infect. Dis.* **170:**717–719.
9. **Arias-Camison, J. M.** 2003. Late onset group B streptococcal infection from maternal expressed breast milk in a very low birth weight infant. *J. Perinatol.* **23:**691–692.
10. **Badri, M. S., S. Zawaneh, A. C. Cruz, G. Mantilla, H. Baer, W. N. Spellacy, and E. M. Ayoub.** 1977. Rectal colonization with group B *Streptococcus*: relation to vaginal colonization of pregnant women. *J. Infect. Dis.* **135:**308–312.
11. **Baker, C. J.** 1996. Inadequacy of rapid immunoassays for intrapartum detection of group B streptococcal carriers. *Obstet. Gynecol.* **88:**51–55.
12. **Baker, C. J., and M. S. Edwards.** 1995. Group B streptococcal infections, p. 980–1054. *In* J. S. Remington and O. J. Klein (ed.), *Infectious Disease of the Fetus and Newborn Infant*, 4th ed. W. B. Saunders Co., Philadelphia, Pa.
13. **Baker, C. J., F. F. Barrett, R. C. Gordon, and M. D. Yow.** 1973. Suppurative meningitis due to streptococci of Lancefield group B: a study of 33 infants. *J. Pediatr.* **82:**724–729.
14. **Baker, C. J., D. K. Goroff, S. Alpert, V. A. Crockett, S. H. Zinner, J. R. Evrard, B. Rosner, and W. M. McCormack.** 1977. Vaginal colonization with group B streptococcus: a study of college women. *J. Infect. Dis.* **135:**392–397.
15. **Baker, C. J., D. L. Kasper, I. B. Tager, A. Paredes, S. Alpert, W. M. McCormack, and D. Goroff.** 1977. Quantitative determination of antibody to capsular polysaccharide in infection with type III strains of group B *Streptococcus*. *J. Clin. Invest.* **59:**810–818.
16. **Baker C. J., L. C. Paoletti, M. A. Rench, H. K. Guttormsen, M. S. Edwards, and D. L. Kasper.** 2004. Immune response of healthy women to 2 different group B streptococcal type V capsular polysaccharide-protein conjugate vaccines. *J. Infect. Dis.* **189:**1103–1112.
17. **Band, J. D., H. W. Clegg, P. S. Hayes, R. R. Facklam, J. Stringer, and R. E. Dixon.** 1981. Transmission of group B streptococci traced by use of multiple epidemiologic markers. *Am. J. Dis. Child.* **135:**355–358.

18. **Bergeron, M. G., D. Ke, C. Menard, F. J. Picard, M. Gagnon, M. Bernier, M. Ouellette, P. H. Roy, S. Marcoux, and W. D. Fraser.** 2000. Rapid detection of group B streptococci in pregnant women at delivery. *N. Engl. J. Med.* **343:**175–179.

19. **Blancas, D., M. Santin, M. Olmo, F. Alcaide, J. Carratala, and F. Gudiol.** 2004. Group B streptococcal disease in nonpregnant adults: incidence, clinical characteristics, and outcome. *Eur. J. Clin. Microbiol. Infect. Dis.* **23:**168–173.

20. **Blumberg, H. M., D. S. Stephens, M. Modansky, M. Erwin, J. Elliot, R. R. Facklam, A. Schuchat, W. Baughman, and M. M. Farley.** 1996. Invasive group B streptococcal disease: the emergence of serotype V. *J. Infect. Dis.* **173:**365–373.

21. **Boyer, K. M., and S. P. Gotoff.** 1986. Prevention of early-onset neonatal group B streptococcal disease with selective intrapartum chemoprophylaxis. *N. Engl. J. Med.* **314:** 1665–1669.

22. **Boyer, K. M., C. A. Gadzala, L. I. Burd, D. E. Fisher, J. B. Paton, and S. P. Gotoff.** 1983. Selective intrapartum chemoprophylaxis of neonatal group B streptococcal early-onset disease. I. Epidemiologic rationale. *J. Infect. Dis.* **148:** 795–801.

23. **Boyer, K. M., C. A. Gadzala, P. D. Kelly, L. I. Burd, and S. P. Gotoff.** 1983. Selective intrapartum chemoprophylaxis of neonatal group B streptococcal early-onset disease. II. Predictive value of prenatal cultures. *J. Infect. Dis.* **148:** 802–809.

24. **Burman, L. G., P. Christensen, K. Christensen, B. Fryklund, A. Helgesson, N. W. Svenningsen, K. Tullus, and the Swedish Chlorhexidine Study Group.** 1992. Prevention of excess neonatal morbidity associated with group B streptococci by vaginal chlorhexidine disinfection during labour. *Lancet* **340:**65–69.

25. **Casey, J. I., S. Maturlo, J. Albin, and S. C. Edberg.** 1982. Comparison of carriage rates of group B *Streptococcus* in diabetic and nondiabetic persons. *Am. J. Epidemiol.* **116:** 704–708.

26. **Centers for Disease Control and Prevention.** 1996. Prevention of perinatal group B streptococcal disease: a public health perspective. *Morbid. Mortal. Wkly. Rep.* **45:**1–24.

27. **Centers for Disease Control and Prevention.** 2002. Prevention of perinatal group B streptococcal disease: revised recommendations from CDC. *Morbid. Mortal. Wkly. Rep.* **51**(No. RR-11)**:**1–22.

28. **Centers for Disease Control and Prevention.** 2004. Diminishing racial disparities in early-onset neonatal group B streptococcal disease–United States, 2000–2003. *Morbid. Mortal. Wkly. Rep.* **53:**502–505.

29. **Crespo, A., A. S. Retter, and B. Lorber.** 2003. Group B streptococcal endocarditis in obstetric and gynecologic practice. *Infect. Dis. Obstet. Gynecol.* **11:**109–115.

30. **Dahl, M. S., I. Tessin, and B. Trollfors.** 2003. Invasive group B streptococcal infections in Sweden: incidence, predisposing factors and prognosis. *Int. J. Infect. Dis.* **7:**113–119.

31. **Daugaard, H. O., A. C. Thomsen, U. Henriques, and A. Ostergaard.** 1988. Group B streptococci in the lower urogenital tract and late abortions. *Am. J. Obstet. Gynecol.* **158:**28–31.

32. **Davies, H. D., M. A. Miller, S. Faro, D. Gregson, S. C. Kehl, and J. A. Jordan.** 2004. Multi-center study of a rapid molecular-based assay for the diagnosis of group B streptococcal colonization in pregnant women. *Clin. Infect. Dis.* **39:**1129–1135.

33. **Dillon, H. C., E. Gray, M. A. Pass, and B. M. Gray.** 1982. Anorectal and vaginal carriage of group B streptococci during pregnancy. *J. Infect. Dis.* **145:**794–799.

34. **Duma, R. J., A. N. Weinberg, T. F. Medrek, and L. J. Kunz.** 1969. Streptococcal infections: a bacteriologic and clinical study of streptococcal bacteremia. *Medicine* **48:** 87–107.

35. **Edwards, M. S., and C. J. Baker.** 1995. *Streptococcus agalactiae* (group B *Streptococcus*), p. 1835–1845. *In* G. L. Mandell, J. E. Bennett, and R. Dolin (ed.), *Principles and Practice of Infectious Diseases,* 4th ed. Churchill Livingstone, Ltd., New York, N.Y.

36. **Eickhoff, T. C., J. O. Klein, A. K. Daly, D. Ingall, and M. Finland.** 1964. Neonatal sepsis and other infections due to group B beta-hemolytic streptococci. *N. Engl. J. Med.* **271:**1221–1228.

37. **Ekelund, K., H. C. Slotved, H. U. Nielsen, M. S. Kaltoft, and H. B. Konradsen.** 2003. Emergence of invasive serotype VIII group B streptococcal infections in Denmark. *J. Clin. Microbiol.* **41:**4442–4444.

38. **Farley, M. M., R. C. Harvey, T. Stull, J. D. Smith, A. Schuchat, J. D. Wenger, and D. S. Stephens.** 1993. A population-based assessment of invasive disease due to group B *Streptococcus* in nonpregnant adults. *N. Engl. J. Med.* **328:**1807–1811.

39. **Faxelius, G., K. Bremme, K. Kvist-Christensen, P. Christensen, and S. Ringertz.** 1988. Neonatal septicemia due to group B streptococci—perinatal risk factors and outcome of subsequent pregnancies. *J. Perinat. Med.* **16:** 423–430.

40. **Fernandez, M., M. Hickman, and C. J. Baker.** 1998. Antimicrobial susceptibilities of group B streptococci isolated between 1992 and 1996 from patients with bacteremia or meningitis. *Antimicrob. Agents Chemother.* **42:**1517–1519.

41. **Ferrieri, P., P. P. Cleary, and A. E. Seeds.** 1976. Epidemiology of group B streptococcal carriage in pregnant women and newborn infants. *J. Med. Microbiol.* **10:**114.

42. **Fry, R. M.** 1938. Fatal infections by haemolytic streptococcus group B. *Lancet* **i:**199–201.

43. **Gallagher, P. G., and C. Watanakunakorn.** 1985. Group B streptococcal bacteremia in a community teaching hospital. *Am. J. Med.* **78:**795–800.

44. **Garland, S. M., and J. R. Fliegner.** 1991. Group B streptococcus and neonatal infections: the case for intrapartum chemoprophylaxis. *Aust. N.Z. Obstet. Gynaecol.* **31:**119–122.

45. **Gerards, L. J., B. P. Cats, and J. A. A. Hoogkamp-Korstanje.** 1985. Early neonatal group B streptococcal disease: degree of colonisation as an important determinant. *J. Infect.* **11:**119–124.

46. **Grubb, R., K. K. Christensen, P. Christensen, and V. Linden.** 1982. Association between maternal Gm allotype and neonatal septicemia with group B streptococci. *Immunogenetics* **9:**143–147.

47. **Harrison, L. H., J. Elliott, D. M. Dwyer, J. P. Libonati, P. Ferrieri, L. Billmann, A. Schuchat, and the Maryland Emerging Infections Program.** 1998. Serotype distribution of invasive group B streptococcal isolates in Maryland: implications for vaccine formulation. *J. Infect. Dis.* **177:**998–1002.

48. **Heelan, J. S., M. E. Hasenbein, and A. J. McAdam.** 2004. Resistance of group B streptococcus to selected antibiotics, including erythromycin and clindamycin. *J. Clin. Microbiol.* **42:**1263–1264.

49. **Jackson, L. A., R. Hilsdon, M. M. Farley, L. H. Harrison, A. L. Reingold, B. D. Plikaytis, J. D. Wenger, and**

A. Schuchat. 1995. Risk factors for group B streptococcal disease in adults. *Ann. Intern. Med.* **123:**415–420.

50. Kasper, D. L., L. C. Paoletti, M. R. Wessels, H. K. Guttormsen, V. J. Carey, H. J. Jennings, and C. J. Baker. 1996. Immune response to type III group B streptococcal polysaccharide-tetanus toxoid conjugate vaccine. *J. Clin. Invest.* **98:**2308–2314.

51. Lancefield, R. C., and R. Hare. 1935. The serological differentiation of pathogenic and non-pathogenic strains of hemolytic streptococci from parturient women. *J. Exp. Med.* **61:**335–349.

52. Lerner, P. I., K. V. Gopalakrishna, E. Wolinsky, M. C. McHenry, J. S. Tan, and M. Rosenthal. 1977. Group B *Streptococcus (S. agalactiae)* bacteremia in adults: analysis of 32 cases and review of the literature. *Medicine* **56:**457–473.

53. Lim, D. V., K. S. Kanarek, and M. E. Peterson. 1982. Magnitude of colonization and sepsis by group B streptococci in newborn infants. *Curr. Microbiol.* **7:**99–101.

54. Lin, F. Y., P. H. Azimi, L. E. Weisman, J. B. Philips III, J. Regan, P. Clark, G. G. Rhoads, J. Clemens, J. Troendle, E. Pratt, R. A. Brenner, and V. Gill. 2000. Antibiotic susceptibility profiles for group B streptococci isolated from neonates, 1995–1998. *Clin. Infect. Dis.* **31:**76–79.

55. Lin, F. Y., L. E. Weisman, J. Troendle, and K. Adams. 2003. Prematurity is the major risk factor for late-onset group B streptococcus disease. *J. Infect. Dis.* **188:**267–271.

56. Liston, T. E., R. E. Harris, S. Foshee, and D. M. Null. 1979. Relationship of neonatal pneumonia to maternal urinary and neonatal isolates of group B streptococci. *South. Med. J.* **72:**1410–1412.

57. Lukacs, S. L., K. C. Schoendorf, and A. Schuchat. 2004. Trends in sepsis-related neonatal mortality in the United States, 1985–1998. *Pediatr. Infect. Dis. J.* **23:**599–603.

58. McDonald, H., R. Vigneswaran, and J. A. O'Loughlin. 1989. Group B streptococcal colonization and preterm labor. *Aust. N.Z. J. Obstet. Gynaecol.* **29:**291–293.

59. McKenzie, H., M. L. Donnet, P. N. Howie, N. B. Patel, and D. T. Benvie. 1994. Risk of preterm delivery in pregnant women with group B streptococcal urinary infections or urinary antibodies to group B streptococcal and *E. coli* antigens. *Br. J. Obstet. Gynaecol.* **101:**107–113.

60. Moller, M., A. C. Thomsen, K. Borch, K. Dinesen, and M. Zdravkovic. 1984. Rupture of fetal membranes and premature delivery associated with group B streptococci in urine of pregnant women. *Lancet* **ii:**69–70.

61. Moore, M. R., S. J. Schrag, and A. Schuchat. 2003. Effects of intrapartum antimicrobial prophylaxis for prevention of group-B-streptococcal disease on the incidence and ecology of early-onset neonatal sepsis. *Lancet Infect. Dis.* **3:**201–213.

62. Nolla, J. M., C. Gomez-Vaquero, X. Corbella, S. Ordonez, C. Garcia-Gomez, A. Perez, J. Cabo, J. Valverde, and J. Ariza. 2003. Group B streptococcus (*Streptococcus agalactiae*) pyogenic arthritis in nonpregnant adults. *Medicine* **82:**119–128.

63. Pass, M. A., S. Khare, and H. C. Dillon. 1980. Twin pregnancies: incidence of group B streptococcal colonization and disease. *J. Pediatr.* **97:**635–637.

64. Pass, M. A., B. M. Gray, and H. C. Dillon. 1982. Puerperal and perinatal infections with group B streptococci. *Am. J. Obstet. Gynecol.* **143:**147–152.

65. Philipson, E. H., D. A. Palermino, and A. Robinson. 1995. Enhanced antenatal detection of group B streptococcus colonization. *Obstet. Gynecol.* **85:**437–439.

66. Pyati, S. P., R. S. Pildes, N. M. Jacobs, R. S. Ramamurthy, T. F. Yeh, D. S. Raval, L. D. Lilien, P. Amma, and W. I. Metzger. 1983. Penicillin in infants weighing two kilograms or less with early-onset group B streptococcal disease. *N. Engl. J. Med.* **308:**1383–1388.

67. Regan, J. A., M. A. Klebanoff, R. P. Nugent, and Vaginal Infections and Prematurity Study Group. 1991. The epidemiology of group B streptococcal colonization in pregnancy. *Obstet. Gynecol.* **77:**604–610.

68. Regan, J. A., M. A. Klebanoff, R. P. Nugent, D. A. Eshenbach, W. C. Blackwelder, Y. Lou, R. S. Gibbs, P. J. Rettig, D. H. Martin, and R. Edelman. 1996. Colonization with group B streptococci in pregnancy and adverse outcome. *Am. J. Obstet. Gynecol.* **174:**1354–1360.

69. Rollan, M. J., J. A. San Roman, I. Vilacosta, C. Sarria, J. Lopez, M. Acuna, and J. L. Bratos. 2003. Clinical profile of *Streptococcus agalactiae* native valve endocarditis. *Am. Heart J.* **146:**1095–1098.

70. Schauf, V., and V. Hlaing. 1976. Group B streptococcal colonization in pregnancy. *Obstet. Gynecol.* **47:**719–721.

71. Schrag, S. J., S. Zywicki, M. M. Farley, A. L. Reingold, L. H. Harrison, L. B. Lefkowitz, J. L. Hadler, R. Danila, P. R. Cieslak, and A. Schuchat. 2000. Group B streptococcal disease in the era of intrapartum antibiotic prophylaxis. *N. Engl. J. Med.* **342:**15–20.

72. Schrag, S. J., E. R. Zell, R. Lynfield, A. Roome, K. E. Arnold, A. Craig, L. H. Harrison, A. L. Reingold, K. Stefonek, G. Smith, M. Gamble, and A. Schuchat. 2002. A population-based comparison of strategies to prevent early-onset group B streptococcal disease in neonates. *N. Engl. J. Med.* **347:**233–239.

73. Schrag, S. J. 2004. The past and future of perinatal group B streptococcal disease prevention. *Clin. Infect. Dis.* **39:**1136–1138.

74. Schuchat, A., M. J. Oxtoby, S. L. Cochi, A. K. Sikes, A. Hightower, B. Plikaytis, and C. V. Broome. 1990. Population-based risk factors for neonatal group B streptococcal disease: results of a cohort study in metropolitan Atlanta. *J. Infect. Dis.* **162:**672–677.

75. Schuchat, A., K. Deaver-Robinson, B. D. Plikaytis, K. Zangwill, J. Mohle-Boetani, and J. D. Wenger. 1994. Multistate case-control study of maternal risk factors for neonatal group B streptococcal disease. *Pediatr. Infect. Dis. J.* **13:**623–629.

76. Schwartz, B., A. Schuchat, M. J. Oxtoby, S. L. Cochi, A. Hightower, and C. V. Broome. 1991. Invasive group B streptococcal disease in adults. *JAMA* **266:**1112–1114.

77. Siegel, J. D., and N. B. Cushion. 1996. Prevention of early-onset group B streptococcal disease: another look at single-dose penicillin at birth. *Obstet. Gynecol.* **87:**692–698.

78. Siegel, J. D., G. H. J. McCracken, N. Threldkeld, B. M. DePasse, and C. R. Rosenfeld. 1982. Single-dose penicillin prophylaxis of neonatal group B streptococcal disease. *Lancet* **i:**1426–1430.

79. Taha, T. E., R. J. Biggar, R. L. Broadhead, L. A. R. Mtimavalye, A. B. Justesen, G. N. Liomba, J. D. Chiphangwi, and P. G. Miotti. 1997. Effect of cleansing the birth canal with antiseptic solution on maternal and newborn morbidity and mortality in Malawi: clinical trial. *Br. Med. J.* **315:**216–219.

80. Tan, J. S., J. L. Anderson, C. Watanakunakorn, and J. P. Phair. 1975. Neutrophil dysfunction in diabetes mellitus. *J. Lab. Clin. Med.* **85:**26–33.

81. Verghese, A., S. L. Berk, L. J. Boelen, and J. K. Smith. 1982. Group B streptococcal pneumonia in the elderly. *Arch. Intern. Med.* **142:**1642–1645.

82. **Yagupsky, P., M. A. Menegus, and K. R. Powel.** 1991. The changing spectrum of group B streptococcal disease in infants: an eleven-year experience in a tertiary care hospital. *Pediatr. Infect. Dis. J.* **10:**801–808.

83. **Yancey, M. K., P. Duff, P. Kubilis, P. Clark, and B. H. Frentzen.** 1996. Risk factors for neonatal sepsis. *Obstet. Gynecol.* **87:**188–194.

84. **Yancey, M. K., A. Schuchat, L. K. Brown, V. L. Ventura, and G. R. Markenson.** 1996. The accuracy of late antenatal screening cultures in predicting genital group B streptococcal colonization at delivery. *Obstet. Gynecol.* **88:**811–815.

85. **Zaleznik, D. F., M. A. Rench, S. Hillier, M. A. Krohn, R. Platt, M. L. Lee, A. E. Flores, P. Ferrieri, and C. J. Baker.** 2000. Invasive disease due to group B streptococcus in pregnant women and neonates from diverse population groups. *Clin. Infect. Dis.* **30:**276–281.

86. **Zangwill, K. M., A. Schuchat, and J. D. Wenger.** 1992. Group B streptococcal disease in the United States, 1990: report from a multistate active surveillance system. *Morbid. Mortal. Wkly. Rep. CDC Surveill. Summ.* **41:**25–32.

Genetics and Pathogenicity Factors of Group C and G Streptococci

HORST MALKE

16

The Lancefield group C (GCS) and group G streptococci (GGS) carry the immunodeterminant residues N-acetyl-galactosamine and rhamnose, respectively, on the oligosaccharide side chains of their cell wall carbohydrate antigens. These organisms are distributed in both humans and animals. They are isolated as opportunistic commensals from the skin, nose, throat, vagina, and gastrointestinal tract, but may also be associated with clinically important infections of these sites and with hospital outbreaks. Serious diseases caused by human GCS and GGS often resemble those due to their closest genetic relatives, the group A streptococci (GAS), and include septicemia, pharyngitis, cellulitis, otitis media, septic arthritis, meningitis, infective endocarditis, multiple organ abscesses, necrotizing fasciitis, and toxic shock syndrome (17, 23, 48, 54, 129, 147, 165). The genetic relationships of GCS and GGS strains are complicated and incompletely resolved due to the diversity of species within the serogroups or the present uncertainties in assigning species names in particular to the human GGS strains. The current classification (Table 1) relies on habitats, pathogenicity properties, physiological characteristics, and relationships of informational macromolecules, with serological grouping being useful for differentiating infraspecific biotypes. Before the advent of large-scale DNA sequencing and gene cloning technologies, GCS or GGS had rarely been subjected to genetic studies. However, studies of the host range of bacteriophages isolated from GAS, GCS, and GGS, and of their transducing potential, provided evidence for intergroup phage reactions and intergroup transduction between strains belonging to different Lancefield groups, thus amending the original notion of the strict group specificity of streptococcal phage-host interactions (136, 161). An attempt to elucidate the genetic relationships between GAS, GCS, and GGS based on the nucleotide sequence of selectively neutral housekeeping genes supported events of interspecies gene transfer that predominantly occurred from GAS donors to GCS and GGS recipients (67). The application of recombinant DNA techniques has advanced our understanding of GCS and GGS in diverse areas, and this chapter concentrates on the structure and function of pathogenetically relevant genes and proteins studied at the molecular level in recent years.

MOLECULAR TAXONOMIC APPROACHES TO THE CLASSIFICATION OF GCS AND GGS

On the basis of 16S rRNA comparative sequence analysis, GCS and GGS fall into two species groups, the pyogenic and the anginosus group; the latter is also known as "Streptococcus milleri" group (7, 25, 68, 164). Chromosomal DNA-DNA hybridization studies combined with biochemical tests indicate that the pyogenic streptococci previously named Streptococcus dysgalactiae and "Streptococcus equisimilis," together with those belonging to the large-colony-forming GGS and serogroup L strains, exhibit high levels of DNA sequence identity and, on the basis of the commonly accepted species-level hybridization (≥70%), constitute a single species, S. dysgalactiae (28, 157). Whole-cell protein electrophoresis revealed two subpopulations within this species, leading to differentiation between S. dysgalactiae subsp. equisimilis, comprising GCS and GGS strains of human origin, and S. dysgalactiae subsp. dysgalactiae, which harbors animal GCS and serogroup L strains (155). GGS isolated from animals always appear to qualify as Streptococcus canis. On grounds of DNA sequence similarity, "Streptococcus zooepidemicus" is closely related to Streptococcus equi, warranting a similar subdivision of this species into the S. equi subsp. zooepidemicus and S. equi subsp. equi, with the latter having evolved as a clonal descendent of the former (15, 63). Recent epidemiological studies have increasingly relied on genotypic classification methods such as restriction endonuclease digestion of total cell DNA followed by conventional (95) or pulsed-field gel electrophoretic analysis of macrorestriction fragment patterns (10, 48, 54), analysis of restriction fragment length polymorphism of rDNA (ribotyping) (128), randomly amplified polymorphic DNA (RAPD) fingerprinting (9, 40), multilocus enzyme electrophoresis (MLEE) (8, 157), and typing based on comparative sequence analyses of the fast-evolving nonfunctional 16S-23S rRNA intergenic spacer region (15, 32, 49). Fluo-

TABLE 1 Classification of GCS and GGS[a]

Taxon	Lancefield group	Hosts	Specific distinguishing reactions
S. dysgalactiae subsp. *equisimilis*	C, G[b]	Humans[b]	Streptokinase activity on human plasminogen; C5a peptidase production; large-colony-forming; GGS bind IgG Fc fragment; no α- or β-D-galactosidase production; pyrrolidonylarylamidase negative
S. dysgalactiae subsp. *dysgalactiae*	C	Various animals	Alpha-hemolytic; no streptokinase activity on human plasminogen; no C5a peptidase production
S. equi subsp. *equi*	C	Horses	Trehalose, sorbitol, and ribose fermentation negative
S. equi subsp. *zooepidemicus*	C	Various animals, humans	Trehalose, sorbitol, and ribose fermentation positive
S. canis	G	Dogs, cows, cats	Beta-hemolytic; α- and β-D-galactosidase production
"*S. milleri*"	C, G, F, A	Humans	Hemolysis reaction varies between species; small-colony-forming; GGS do not bind IgG Fc fragment; Voges-Proskauer reaction positive

[a]Adapted from references cited in the text.
[b]Uncommonly, this subspecies may also comprise serogroup A and L strains and may also be found in animals.

rescent in situ hybridization has also been used for the fast identification of GCS associated with necrotizing fasciitis (135). Finally, sequencing an internal fragment of the superoxide dismutase gene has yielded a phylogenetic tree, the topology of which agreed with that obtained from 16S rRNA sequences, rendering this method suitable for species-level identification of various streptococci (122). Taken together, the results obtained, by and large, corroborate the species classification given in Table 1 and, moreover, testify to the utility of these methods for subspecific typing. For example, RAPD fingerprinting has proven useful in epidemiological investigations designed to delineate new and persistent *S. dysgalactiae* strains that caused intramammary infections in dairy cows (113). Comparison of the discriminatory power of different genomic typing methods revealed that RAPD exceeds MLEE for GGS (and GAS) typing (9), and that pulsed-field gel electrophoretic analysis is even more efficacious than MLEE and RAPD, as indicated by its potential to identify 93 distinct types among a total of 41 GAS and 58 GCS or GGS strains, with no types common to strains of different species (10). Although a comprehensive battery of species-specific gene probes for identification of GCS and GGS has yet to be developed, additional attempts to approach this goal have focused on oligonucleotides designed from 16S and 23S rRNA sequences (145). Alternatively, future differential genome analysis applied to GCS and GGS might identify genes that are responsible for species-specific features.

CELL SURFACE-ASSOCIATED PROTEINS

One hallmark of the gram-positive pathogens is the synthesis of specific cell wall-associated proteins that enable them to interact in various ways with proteins present in the body fluids or extracellular tissue matrix of their mammalian hosts. Such interactions may facilitate colonization, lead to molecular host mimicry, or interfere with various host defenses against invasion.

M and M-Like Proteins

The major bona fide virulence determinant of GAS, M protein, is generally accepted to comprise polymorphic, fibrillar surface-exposed polypeptides that share common structural and functional features, among them a high propensity to form an alpha helix with a periodicity indicative of a coiled-coil structure, the capability of inhibiting phagocytosis by human neutrophils, and the potential to elicit opsonic antibodies that provide type-specific protection. Several studies conducted mainly in the 1980s provided highly suggestive evidence based on serological cross-reactions with typing sera or monoclonal antibodies raised to M protein from GAS, phagocytosis inhibition, and DNA hybridization analysis that M proteins or M-like proteins are also encoded and expressed by GCS and GGS (30). Subsequent cloning and sequence determination of the corresponding genes as well as analysis of immunoreactivity afforded conclusive evidence in support of this notion (38, 39, 107, 148, 149, 156). The first M protein gene, *emmG1*, to be isolated from a human GGS strain encodes a product with structural features characteristic of GAS class I M proteins that are epidemiologically associated with rheumatogenic, opacity factor-negative serotypes (18). As expected, structural similarity between MG1 and the GAS class I M proteins was significantly greater in the C-terminal regions (containing the anchor domains, and the C and D repeats) than in the N-terminal portions (containing the A and B repeats) of the mature proteins (86 to 94% vs. 31 to 35% amino acid identities), an observation consistent with a difference in antigenicity between MG1 and the GAS M types. An *emmG1* probe encoding the variable region of MG1 detected homologous DNA in seven additional M-positive GGS strains but also revealed at least four distinct *emm* alleles harbored by these strains. The essential aspects of these findings were confirmed and expanded by others (127, 133, 134, 139) using a greater and epidemiologically unrelated number of GGS isolates that also included strains recovered from animal sources. It is noteworthy that none of the DNA specimens isolated from GGS strains originating from animals hybridized with appropriate *emm* gene probes from GAS, whereas all DNAs from human GGS strains (large-colony-forming isolates) did (127, 133). Probes designed to differentiate between GAS M class I and M class II genes detected only GGS *emm* genes related to GAS M class I genes (127). The complete sequence of an additional GGS *emm* gene (*emmLG593*), together with the sequences of the 5′ ends of 30 additional GGS *emm* genes, revealed six distinct "genetic types," which in paired comparisons shared <95% sequence identity in the *emm* gene region coding for the mature N termini (127). This observation adds further support to the genetic heterogeneity of the M proteins of GGS and, if consistent with future supplementary opsonophagocytotic studies, may be used for the development of GGS M typing schemes for epidemiological purposes. Most interestingly, in the latter studies (127) as well as those conducted independently by others (134, 139), certain GGS *emm* genes were identified, defined portions of which exhibit exceedingly high levels of sequence identity to the variable regions of well-established GAS class I M protein genes, most notably *emm12* and *emm57*. Inasmuch as the results of total genomic DNA-DNA hybridization, chromosomal restriction endonuclease profiling, and MLEE all indicate that GGS are appreciably divergent from GAS in overall genomic character and, hence, have no recent GAS ancestors, the most likely explanation for the high local homology of certain *emm* genes in the two serogroups is intergroup horizontal genetic exchange followed by homologous recombination events (134). Although several scenarios can be envisaged to suggest possible modes of DNA transfer, the actual transfer mechanisms are entirely left in the dark.

In GCS isolated from horses, several M or M-like proteins have been characterized at the molecular level: SeM (nearly identical to FgBP) (101) produced by *S. equi* subsp. *equi*, and SzP and SzPSe produced by *S. equi* subsp. *zooepidemicus* (148, 149). SzP and SzPSe are structurally only marginally related to SeM. They are antigenically cross-reactive and are coded by allelic variants of the same hypervariable gene. Functionally, however, all three proteins resemble M proteins of GAS in being capable of inhibiting phagocytosis, eliciting serum opsonic and protective responses, and binding (equine) fibrinogen (101, 132, 149). The complete sequences now available for these proteins are consistent with earlier serological and other observations, indicating that *S. equi* subsp. *equi* is antigenically conserved whereas equine isolates of *S. equi* subsp. *zooepidemicus* exhibit great antigenic variation (63). These observations, together with data derived from comparative MLEE, suggest that *S. equi* subsp. *zooepidemicus* is the ancestral species from which a clone now named *S. equi* subsp. *equi* has emanated (15, 63). Unlike the SzP protein family, SeM is highly conserved, and appears to be the stronger opsonizing antigen in small laboratory animals (149). Therefore, and because of its proven involvement in pathogenicity, SeM has been made the main component of commercial strangles vaccines (147). Analysis of truncated FgBP/SeM fragments has shown that binding to fibrinogen requires a fairly large stretch of the protein reaching from the vicinity of the N terminus to about residue 350 of the 534 total amino acids (102). FgBP also uses distinct structural features to bind immunoglobulin G (IgG)-Fc (103), and acquires thermal stability owing to an extensive α-helical coiled-coil structure situated in the C-terminal portion of FgPB (104). Of note, in the regions external to the cell wall, the SeM and SzP

family proteins share no obvious sequence homology with GAS M proteins (149). However, M antigens closely related to the GAS class I M proteins and to those of the human GGS strains have also been detected on *S. dysgalactiae* subsp. *dysgalactiae* associated with bovine mastitis (156) and GCS isolated from humans (11, 38, 120). It is noteworthy that sequencing the DNA upstream of these M protein genes identified co-oriented open reading frames designated *dmgA* (156) and *mgc* (38), respectively, with a high degree of sequence similarity to *mga*, the regulator of *emm* gene expression in GAS. In this connection, the FAI protein found in a GCS isolate from a horse (146) deserves special attention. FAI consists of overlapping functional modules responsible for fibrinogen, IgG, and albumin binding. Its three C repeat copies share 40 to 50% sequence identity with the B repeats of GAS M type 49 and the C repeats of M12, but exhibit no sequence similarity to any portions of SeM or the SzP proteins. Because GAS are usually not distributed in hosts other than primates, it seems doubtful whether the strain producing FAI is a genuine horse strain and whether the *fai* gene results from intergroup DNA exchange (149).

IgG-Binding Proteins

The most thoroughly studied cell surface protein produced by many human GGS and GCS strains is protein G, which binds the heavy chains of IgG Fc and Fab with an affinity comparable to that of antigen. Conventional thinking implies contribution to adherence, inhibition of phagocytosis, and host mimicry in the role that protein G might have in pathogenesis; however, strong evidence in favor of this notion is missing. Although the structure of protein G is reminiscent of staphylococcal protein A, the two polypeptides share, except for their C termini, little sequence identity (26, 46). The protein G genes (*spg*) of distinct *S. dysgalactiae* strains may differ in their actual nucleotide sequence and size. However, allowing for processes such as duplication of internal repeats, and recombination and replication slippages involving repeats (26, 114), different *spg* alleles exhibit a very high level of sequence identity, making protein G a unique type among the diverse group of gram-positive Ig-binding proteins. Compared to protein A, protein G binds not only to human, rabbit, mouse, and guinea pig IgG but also to IgG from cow, horse, sheep, and goat, which protein A does not bind. In addition to showing wider IgG species reactivity, protein G, unlike protein A, also binds human subclass IgG_3. Finally, protein G's reactivity is strictly limited to IgG (protein A also binds IgM) (46). Because of these properties, protein G has become one of the most versatile tools in immunochemistry. Dissection of the molecule by enzymatic and genetic means followed by functional analysis of defined fragments confirmed early observations indicating that protein G is multifunctional in binding IgG, serum albumin, and α_2-macroglobulin (α_2-M) (2, 14, 46, 106, 111). Thus, as schematically shown in Fig. 1, IgG Fc-binding resides in the C-terminal portion of the molecule and is mediated by two or more repeated domains, whereas the albumin binding activity is a separate function of repeated domains in the N-terminal half. α_2-M binding resides in the N-terminal tip of the mature protein that lacks repeated sequence elements. Functional remnants of protein G have been retained in the protein G-related α_2-M binding protein of GAS (124), which, however, has not yet been detected in GGS or GCS.

The three-dimensional structures of the 56-residue B1 and 57-residue B2 IgG Fc-binding domains of protein G from GGS strain GX7809 (26) have been determined by

C1

FIGURE 1 Schematic representation of the domain organization of protein G. S, signal sequence; E, α_2-macroglobulin-binding; A and B, human serum albumin-binding; C, IgG-binding; W, cell wall-associated region; M, cell membrane-associated region. (Adapted from reference 124.)

nuclear magnetic resonance spectroscopy (44) and X-ray diffraction analysis (1), respectively. The two approaches yielded essentially identical overall structures characterized by a four-stranded β sheet and a helix lying diagonally across the sheet and connecting the outer two β strands, β2 and β3. The IgG-binding interface of this structure is formed by negative charges distributed around the helix and β3 (43). The entire folding motif is very compact and shows high thermal stability owing to extensive hydrogen bonding and tight packing of its buried hydrophobic core. Taking advantage of these properties, the folding thermodynamics and mechanism of the B1 segment have been studied (131) and an M13 phage-B1 domain display system has been developed to pinpoint primary structure requirements for thermodynamic stability of the B1 domain (115). Interestingly, the IgG-binding domains also bind, in a nonimmune manner, heavy-chain constant-region sequences of the Fab portion of IgG but use disparate parts of their structure to bind the heavy chains of Fc and Fab. The crystal structure of the complex between Fab and the third IgG-binding domain of protein G shows that the outer β2 strand of the molecule aligns in an antiparallel fashion with the last β strand of the constant heavy chain domain of Fab, CH1 (22). In this manner, the β sheet structure is extended into protein G. Because the CH1 domain is highly conserved in the IgG Fab fragments from diverse species, these findings neatly explain the broad specificity of the protein G-IgG interaction. Moreover, provided that backbone-backbone hydrogen bonding between the interacting β strands is preserved, this arrangement allows Fab binding in the presence of considerable primary structure variability.

Fragmentation of human serum albumin located its single protein G binding site to disulfide loops 6 to 8, with a 5.5-kDa pepsin fragment showing complete inhibition of the interaction between protein G and the full-length albumin (27). Recently, the crystal structure for a complex between human serum albumin and the albumin-binding domain (the GA module) of the PAB protein from *Finegoldia magna* (formerly *Peptostreptococcus magnus*) has been solved (76). As a result of an intergeneric sequence transfer event, the GA module exhibits high sequence similarity to the albumin-binding domain of protein G. The human serum abumin/GA complex sheds some light on the amino acid residues involved in conferring broader albumin-binding specificity on protein G than on PAB, which preferentially binds only primate albumins, in accord with the human pathogenic lifestyle of *F. magna* (76).

In recent years, IgG-binding proteins have been found that show interesting functional differences from the prototype protein G molecules. The *mag* gene cloned from an *S. dysgalactiae* strain codes for a polypeptide with three discrete domains corresponding to those of the trifunctional protein G (60). However, the albumin-binding spectrum of Mag is

broader than that of protein G, including albumins from at least nine mammalian species (61). The *mig* gene isolated from a case of bovine mastitis specifies a protein with binding activity to IgG, which resides in five C-terminal repeats, and to the α_2-M-proteinase complex. The latter activity is located in the N-terminal portion of the Mig protein, which exhibits only weak sequence similarity to the α_2-M-binding domain of the Mag protein. Furthermore, protein Mig has no albumin binding activity (59). Finally, the Zag protein, isolated from an *S. dysgalactiae* subsp. *zooepidemicus* strain, shows a molecular architecture that is similar to that of the Mag protein, but its N-terminal α_2-M-binding domain represents a third unique type of this module with little sequence similarity to the corresponding regions of either the Mag or Mig proteins. However, inhibition assays showed that all three proteins compete for the same or overlapping sites in the α_2-M molecule (62). The pathogenetic role of these proteins remains to be established; however, evidence has been published showing that proteinase-complexed α_2-M bound to animal GCS inhibits phagocytosis in vitro (154).

Fibronectin-Binding Proteins

It was not until the beginning of the 1990s that surface proteins of GAS, GCS, or GGS that bind to fibronectin were characterized at the molecular level. The primary function of this adhesive matrix protein is to mediate substrate adhesion of eukaryotic cells via integrin receptors that bind to central parts of the molecule containing the RGD sequence motif. Among the best-studied GCS and GGS proteins responsible for fibronectin binding are those encoded by the *S. dysgalactiae* genes *fnbA*, *fnbB* (cloned from the same strain) (77, 78), and *gfbA* (70); the *S. dysgalactiae* subsp. *equisimilis fnb* gene (79); and the *S. equi* subsp. *zooepidemicus fnz* gene of the animal ZV strain, which also produces the Zag protein (see above) (81). The mature proteins appear to occur in two size classes of about 60 and 120 kDa, but they all consist of C-terminal repetitive domains that are close to the cell wall-spanning regions and mediate binding to the 29-kDa N-terminal segment of fibronectin. Comparison between the C-terminal repeats reveals a consensus sequence consisting of two adjacent glycines surrounded by highly acidic tripeptides that are required for binding activity, as indicated by the inhibitory effect of synthetic peptides that mimic individual repeats and show a high level of cross-reactivity (57, 96). In addition to these interactions, the sequence motif LAGESGET separating the first and the second repeat of the Fnz protein has been identified as an important part of a fibronectin-binding domain which does not bind the 29-kDa fibronectin fragment (81). The fibronectin-binding proteins F1 and F2 of GAS also have at least two domains, each responsible for binding. The F1 domain N-terminal to the repetitive domain contains the

LAGESGET motif (56). The F2 protein, which is homologous to FnbB and Fnb, contains a nonrepeated domain, designated UFBD, which lies upstream of the repeat region and exhibits fibronectin binding independently of the repeat domain (56). UFBD displays a high degree of sequence similarity to FnbB and Fnb regions N-terminal to their repetitive domains, strongly suggesting that fibronectin-binding proteins generally use multiple domains for ligand binding (58, 158). This also includes Fnz (82) and FnbA, in which a previously unidentified region, Au, upstream of the repeats is capable of independent fibrinogen binding (138). Interestingly, a monoclonal antibody raised to FnbA enhances, rather than inhibits, fibronectin binding to protein fragments or synthetic peptides containing Au, and recognizes only the Au-fibrinogen complex (138). Circular dichroism analysis suggested that the ligand binding site containing Au has little secondary structure, but on binding fibronectin, is induced to form a predominantly β sheet structure that is seen by the antibody (52). The infected host might respond to such ligand-induced conformational changes with antibodies directed against the complex, thus stabilizing, rather than blocking, tissue adherence of the pathogen (138). That fibronectin-binding proteins do contribute to adherence is shown, e.g., by the correlation between the presence of the gfbA gene (present in 36% of GGS strains tested) and the ability of such cells to adhere to human skin fibroblasts, as well as by the transformability of a nonadherent strain to the adherence phenotype upon introduction of gfbA (70). There also exists evidence that GCS and GGS, like their GAS relatives, are capable of invading nonphagocytic host cells (14, 105, 137), a potential that is dependent on fibronectin-binding proteins. A recent model for the internalization process based on structural studies implies that the originally disordered primary C-terminal fibronectin-binding repeats form an extended tandem β-zipper upon binding to the string of F1 modules contained in the N-terminal 29-kDa fibronectin domain. The binding of multiple fibronectin molecules results in integrin clustering on the host cell membrane and subsequent endocytosis of the streptococcal cells (130).

Apart from the microbial surface components recognizing adhesive matrix molecules characterized above, a novel fibronectin-binding protein, Sfs, displaying sequence similarity to collagen, was identified in many strains of both subspecies of *S. equi* (80). Sfs, which also has a counterpart in GAS, binds to the 30- to 40-kDa collagen-binding domain of fibronectin and, accordingly, inhibits the binding of collagen to fibronectin.

Plasmin(ogen)-Binding Proteins

Besides GAS, certain strains of human GCS and GGS as well as bovine GGS have been found to bind native human plasminogen (PG) and, with lesser affinity, an N-terminal PG fragment containing kringles 1 to 3 (153). However, in these experiments, the nature of the proteins responsible for binding remained unknown. Sequence information is available, however, for two related M-like proteins from human GCS and GGS strains capable of binding PG, fibrinogen, and serum albumin to distinct sites of these proteins (6). Moreover, the GGS protein showed partial sequence identity to the PG-binding M protein from GAS termed PAM (6). PG binding and activation are also common properties of bovine mastitis-associated group C strains, with features indicating species specificity regarding the origin of the plasma proteins (75).

The surprising discovery in GAS that the glycolytic enzyme glyceraldehyde-3-phosphate dehydrogenase (GAPDH), which possesses none of the hallmarks of gram-positive cell surface proteins, can be localized to the cell wall to function as a major binding protein for a variety of mammalian proteins including plasmin (PM) (83, 116) prompted the isolation and functional analysis of the corresponding gene (gapC) in the human group C strain H46A (36). In this strain, gapC is an essential gene that occurs in single copy and is abundantly transcribed. Its product shares 95% sequence identity with *Streptococcus pyogenes* GAPDH. Moreover, the number and position of lysine residues shown to be involved in plasmin(ogen) binding (36, 83) are identical in the two proteins. The binding parameters for human PG and PM of purified GapC protein were determined by real-time biospecific interaction analysis, with results indicating that the protein binds PM and PG with about 10-fold different affinities (K_d values, 25 nM vs. 220 nM, respectively). Thus, the zymogen and active enzyme appear to possess low-affinity binding sites for the gapC gene product, a result at variance with that ascribed to GAS GAPDH, whose plasmin-binding activity was reported to be three orders of magnitude higher (13). On reexamination, the GAS strain involved turned out to display other PM-binding substances on the surface in addition to GAPDH (21). A candidate substance could well be the glycolytic enzyme α-enolase, which had subsequently been identified as a novel ubiquitous streptococcal surface protein with strong plasmin(ogen) binding capacity (117). Given the 2-μM concentration of PG in plasma, even the high K_d value for the association of this protein with GapC would suggest that, in the vasculature, the great majority of the *S. equisimilis* H46A PG binding sites are occupied, even if GapC is the only PG-binding protein. A streptokinase (SK)-negative mutant of H46A, isolated to eliminate endogenous PG-activating capacity, has indeed been shown to bind PG, as indicated by the generation of cell surface-associated PM activity following exposure to exogenous PG activators (92). A novel aspect characterizing the interaction of plasmin(ogen) with surface GAPDHs relates to the finding that limited prior treatment with PM greatly increases the subsequent PG-binding capacity of these proteins (36). Presumably, the underlying mechanism involves the exposure of new PG binding sites on GAPDH by PM, which preferentially catalyzes the hydrolysis of α-amino-substituted lysine (and arginine) esters. In the infected host, streptococci capable of transporting GAPDH (and α-enolase) to the cell wall and secreting SK would thus appear to have evolved a remarkable system to generate and amplify cell-surface protease activity. Host PG bound with low affinity can be converted to the active enzyme directly on the cell surface. The cell-bound PM may serve to generate additional PG-binding sites on GAPDH to start new cycles of PG binding and activation, resulting in amplification of the tissue-invasive potential of the pathogen. Although there exist strains producing proteins unrelated to GAPDH that display high-affinity plasmin(ogen) binding sites (6), the special role of GAPDH (and possibly α-enolase) is due to its general occurrence on the surface of the pyogenic streptococci (116, 117). A major challenge to what is presently understood of the PG-binding and -activating capacity of pathogenic streptococci relates to the identification of possible factors involved in the modulation of cell surface protease activity that may be necessary to avoid destructive proteolysis of surface structures important for colonization and invasion. It is interesting to note that a screen designed to identify

Escherichia coli genes expressed preferentially by natural isolates when incubated in nutrient-poor aquatic medium identified a streptococcal *gapC* homolog (85% sequence homology), suggesting acquisition of the gene by horizontal transfer from a gram-positive organism and possible involvement in the preparation of cells for the colonization of a new host (24).

Complement Fragment C5a Peptidase

Complement fragment C5a functions as a chemotaxin that mediates phagocyte recruitment to the site of infection. GAS and human GGS have been found to produce a highly specific cell surface endopeptidase that cleaves C5a, thereby limiting the activity of this chemotactic signal. In Southern hybridization experiments, probes specific for the GAS C5a peptidase gene, *scpA*, detected homologous *scpG* sequences in all human GGS strains tested (16, 54) but failed to detect the gene in GGS isolated from animals (16). The *scpA* and *scpG* genes appear to be very similar and their products cross-react serologically (16). However, the *scpG* and *scpC* genes in GGS and GCS, respectively, show different linkage relationships from *scpA* because, in either serogroup, they are not located 3' adjacent to *emm* or *emm*-like genes, as is the case for *scpA* in the *mga* regulon of GAS (16, 38, 139) (see also Fig. 2). Whether *scpG/C* are members of the *mgc* regulon (38) is an open question; loss of synteny of the component genes of the *mga* regulon in GGS/GCS may be taken to indicate recent acquisition of individual genes by horizontal gene transfer. In fact, in *Streptococcus agalactiae*, *scpB* is localized on a composite transposon flanked by insertion sequences, thus highlighting the possibility of lateral transfer involved in the shuffling of *scp* genes (34).

CYTOPLASMIC MEMBRANE-ASSOCIATED ENZYMES

Hyaluronan Synthase

The hyaluronan (hyaluronic acid) capsule of the pyogenic streptococci has an important role in resistance to phagocytosis and in virulence (3, 147). In GAS, the three genes required for hyaluronan synthesis, *hasA*, *hasB* and *hasC*, constitute an operon (20). For efficient hyaluronan export, an ATP-binding cassette transporter system is required, the genes of which are located in opposite orientation adjacent to the *has* operon (115a). A *hasA* homolog encoding hyaluronan synthase, which catalyzes the alternate addition of UDP-acetylglucosamine and UDP-glucuronic acid to form the capsule polymer, has been cloned from *S. equi* subsp. *equisimilis* (73). The GCS enzyme (seHAS) sequence is 72% identical to that of its homolog from GAS (spHAS)

and, accordingly, the two proteins cross-react serologically. Both proteins are processive enzymes, and the size distributions of the hyaluronan chains they synthesize are similar. However, seHAS shows a twofold faster chain elongation rate than spHAS, making it the most active hyaluronan synthase described thus far (73). Furthermore, the membrane topology of seHAS is likely to be identical to that experimentally deduced for spHAS. This enzyme exhibits four membrane-spanning domains and two membrane-associated domains, with the termini of the protein and several central regions situated intracellularly (50). The hyaluronan synthases seHAS and spHAS were overexpressed as His$_6$-tagged membrane proteins in *E. coli*, purified to homogeneity, and extensively characterized kinetically. They represent lipid-dependent enzymes, the activity of which is strongly stimulated by cardiolipin and phosphatidylserine (150–152). In contrast, sulfhydryl-modifying reagents and the lipid-soluble 4-methylumbelliferone proved to be inhibitors of seHAS activity (64, 74). Despite suggestions to the contrary (108), no phosphoryl modification of streptococcal HAS has been detected (64). Although the 6.2-kb restriction fragment from the encapsulated GCS strain that served to clone *hasA* (73) is potentially large enough also to accommodate *hasB* and *hasC* (20), definitive proof for the existence and functioning of this operon in GCS or GGS has not been reported. Operon organization of the three genes required for capsule synthesis cannot be taken for granted, as evidenced by *Streptococcus uberis*, in which *hasC* is not situated 3'-adjacent to *hasAB* (162).

Cytoplasmic Membrane Lipoprotein Acid Phosphatase

Attempts to elucidate the linkage relationships of the *S. dysgalactiae* subsp. *equisimilis* H46A *gapC* gene resulted in the discovery of a novel streptococcal gene 3' adjacent to, and co-oriented with, *gapC* (37). This gene, designated *lppC*, codes for a membrane lipoprotein as indicated by the presence of a lipoprotein-specific signal sequence cleavage and lipidation site (VTG↓C), by globomycin-sensitivity of signal sequence processing, and by tight association of the protein with the streptococcal cytoplasmic membrane or the outer membrane of *E. coli* when heterologously expressed in this organism. Southern, Northern, and Western analyses revealed that *lppC* has homologs in GAS and is transcribed independently of *gapC* as monocistronic mRNA from a σ^{70}-like consensus promoter (37). Lipoproteins cross-reacting serologically with LppC have also been detected in disparate strains of both subspecies of *S. equi* (47). Moreover, database searching shows that *lppC* homologs are widely distributed in other pathogenic species as well. Among those, of particular interest is the *hel* gene of *Haemophilus influenzae*, which encodes the major outer membrane antigen *e* (P4),

1 kb

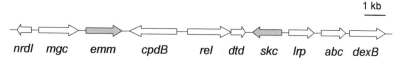

FIGURE 2 Organization of the streptokinase-M protein gene region in the *S. dysgalactiae* subsp. *equisimilis* H46A chromosome based on references 38 and 98. Arrows represent the genes and their orientation: *nrdI*, ribonucleotide reductase; *mgc*, multigene regulator of GCS; *emm*, M protein; *cpdB*, 2',3'-cyclic nucleotide 2'-phosphodiesterase; *rel*, bifunctional (p)ppGppase and (p)ppGpp synthetase; *dtd*, D-tyrosyl-tRNA^{Tyr} deacylase; *skc*, streptokinase; *lrp*, leucine-rich protein of unknown function; *abc*, ATP-binding cassette transporter; *dexB*, α-glucosidase.

exhibiting 58% sequence similarity to LppC. Biochemical, serological, and genetic evidence has been provided to show that LppC represents a novel class of gram-positive lipoproteins in functioning as an acid phosphatase (85). This enzymatic activity was subsequently found to be true of the *e* (P4) antigen as well and led to a reevaluation of its primary role as an outer membrane protein (125). In search of a physiological role for LppC, various physiological substrates were tested and efficient phosphohydrolase activity was found to be restricted to nucleoside and 2′-deoxynucleoside 3′- and 5′-monophosphates (88). The enzyme also showed transphosphorylating activity to compounds with free hydroxyl groups, explaining the failure of hydrolyzable phosphomonoesters to support the growth of strain H46A in media lacking free inorganic orthophosphate. Growth studies also showed that this strain is capable of de novo purine synthesis but depends on salvage pathways for pyrimidine acquisition. However, although UMP could compensate for U deficiency in both the wild type and an *lppC* mutant, CMP utilization specifically required the functional *lppC* gene (88).

EXTRACELLULAR PROTEINS

Streptokinase

The first SK gene (*skc*) to be cloned (86) and sequenced (91) originates from the most potent and commercially exploited producer of this PG activator, the human *S. dysgalactiae* subsp. *equisimilis* strain H46A. Later, an SK gene (*skg*) from a human GGS strain was also characterized (159), and the two encoded proteins were found to share 98% sequence identity. To date, SK alleles belong to those streptococcal genes that have most frequently been sequenced from diverse strains, and research in this area has appreciably expanded our knowledge of this important streptococcal gene and its product in various areas. Part of this work has been reviewed previously (93, 94); here I focus on recent studies done to elucidate the expression control, structure-function relationships, and aspects of adaptive evolution of SK.

Expression Control of the SK Gene

Because gene regulation is influenced by the spatial association of genes on the chromosome, it is appropriate to point out that the SK genes show homology of synteny in the genomes of GAS, and of human GCS and GGS (33, 38, 98) (Fig. 2). Conservation of linkage appears also to be true of an equine group C SK gene that hybridizes with *skc* (33) but encodes an SK that preferentially activates horse PG and antigenically differs from the proteins produced by human strains (110). *skc*, *skg*, and *ska*, collectively designated *skn*, are transcribed as 1.3-kb monocistronic mRNA from a highly conserved chromosomal region in which they are interspersed among unrelated genes transcribed in the opposite direction (Fig. 2). Transcription is terminated by a hypersymmetrical intrinsic terminator with bidirectional activity shared by the oppositely oriented *skn* and *rel-dtd* transcription units (140, 141). S1 nuclease mapping has identified the core promoter with an extended −10 region that directs transcription initiation predominantly at a G located 32 bases upstream of the *skc* translational start site (35). Circular permutation analysis combined with determination of the activity of nested deletions in the DNA upstream of the −35 region identified an intrinsic DNA bending locus that has a pivotal role in *skn* expression (45, 94). Deletions covering the bending locus greatly decrease the core promoter strength in strain H46A but not in *E. coli* (42, 94), consistent with the observation in the latter species that promoters with a TG motif 1 bp upstream of the −10 region function activator-independently, even in the absence of a recognizable −35 region. Region 2.5 of σ^{70} implicated in contacting the TG motif (5) is highly conserved in the streptococcal homolog of σ^{70} and, in particular, does carry the residues H455 and E458 responsible for recognition of the TG extension (Fig. 3). Nevertheless, as shown by the above findings, the streptococcal system is much more stringent with respect to supplementary structural requirements for full *skn* promoter activity and displays strain-specific *skn* expression levels. Thus, using reporter gene constructs in allele swap experiments between GCS and GAS strains revealed that the host genetic background dictates *skn* expression, suggesting the existence of trans-acting factor(s) with strain-specific activity that modulate *skn* expression. Evidence for the action of such factors was provided after the discovery in GAS of two independent two-component signal transduction systems, *covRS* (29) and *fasCAX* (71), which regulate *skn* expression negatively and positively, respectively (89, 144). Thus, assessing the apparent SK activity of streptococcal strains requires the dissection of the opposing activities of the *cov* and *fas* systems. Specifically, sequence analysis showed that wild-type H46A is actually a derepressed mutant for Skc (and streptolysin S [SLS]) synthesis, carrying a K102amber mutation in *covR* (144). In general, knowledge of the FasA response regulator is less advanced than that of CovR. The bootstrapped phylogenetic tree presented in Fig. 4 places FasA in a widely distributed family of streptococcal response regulators (the FasA-BlpR-ComE family) that is involved in behavioral processes, such as quorum sensing, and that uses the C-terminal LytTR domain for DNA binding (89, 109). The FasA proteins, which are distributed predominantly among the pyogenic streptococci, form a well-supported cluster within this protein family. Underscoring their distinct position, the *fasX* gene, thought to encode a nontranslated RNA functioning as the terminal effector of

FIGURE 3 Sequence comparison of *E. coli* and *S. pyogenes* σ factors (5) between region 2.4 and the start of region 3. Identical and similar amino acids are marked by asterisks and dots, respectively. The residues identified in *E. coli* as contacting the TG extension of the −10 promoter hexamer are indicated by open boxes.

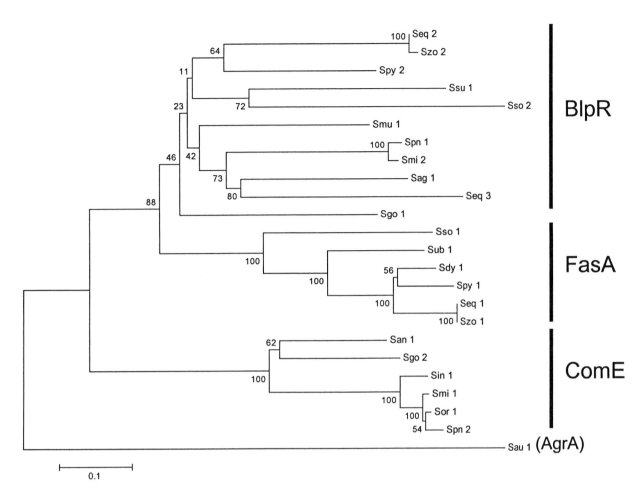

FIGURE 4 Bootstrapped (1,000 trials) neighbor-joining phylogenetic tree of the FasA-BlpR-ComE family of streptococcal response regulators graphed by the MEGA2 software (72) based on ClustalX multiple alignment (http://www.ncbi.nlm.nih.gov). Species abbreviations and sources of the proteins are as follows: Sag1, *S. agalactiae*, AF390107; San1, *S. anginosus*, AJ000864; Sau1, identical to AgrA from *Staphylococcus aureus*, AY082629; Sdy1, *S. dysgalactiae* subsp. *equisimilis*, AY075107; Seq1, Seq2, Seq3, *S. equi* subsp. *equi*, http://www.sanger.ac.uk/Projects/S_equi; Sgo1, *S. gordonii*, http://www.tigr.org/tdb/mdb/mdbinprogress.html; Sgo2, X98109; Sin1, *S. infantis*, AF498313; Smi1, *S. mitis*, AJ000871; Smi2, http://www.tigr.org/tdb/mdb/mdbinprogress.html; Smu1, *S. mutans*, AE015016; Sor1, *S. oralis*, AJ240794; Spn1, *S. pneumoniae*, AJ278302; Spn2, AJ240793; Spy1, *S. pyogenes*, AE009971; Spy2, AE009991; Sso1, Sso2, *S. sobrinus*, http://www.tigr.org/tdb/mdb/mdbinprogress.html; Ssu1, *S. suis*, AY125957; Sub1, *S. uberis*, http://www.sanger.ac.uk/Projects/S_uberis; Szo1, Szo2, *S. equi* subsp. *zooepidemicus*, http://www.sanger.ac.uk/Projects/S_zooepidemicus. When sets of sequences >90% identical were identified in the same species, only one sequence was included to avoid analysis of duplicate sequences. For taxonomic reasons, the exception was *S. equi*, for which both subspecies are included in spite of Seq1 and Szo1 on one hand, and Seq2 and Szo2 on the other, being >90% identical. Bootstrap confidence estimates (%) are given next to the tree nodes: groups found in >95% of trials are considered well supported, those found in >50% are considered suggestive.

the *fasCAX* operon (71, 144), has no homologs in the *comE* and *blpR* clusters.

Structure-Function Relationships of SK

Streptokinase (SK) produced by *S. dysgalactiae* subsp. *equisimilis* strain H46A was predominantly used in the great number of structural studies reported in the literature. This protein has at least three functions in the process of PG activation: (i) high-affinity PG binding to form a 1:1 stoichiometric PG:SK complex; (ii) interacting with PG to

structure the active site, i.e., forming the virgin enzyme with amidolytic activity (conformational activation, i.e, PG*:SK formation); (iii) binding and processing substrate PG molecules, resulting in cleavage of the activating R561-V562 peptide bond to generate the active enzyme, PM. Solution structural studies have presented evidence that the 414-amino-acid mature SK is an elongated globular molecule that consists of three independent folding domains of roughly equal sizes (19, 118, and references therein). The crystal structure of SK complexed with microplasmin (μPM, i.e., PG residues 542 to 791 containing the catalytic site) illustrates these domains denoted α (residues 1 to 150), β (151 to 287), and γ (288 to 414), interconnected by flexible strands (160). Each of these domains contacts μPM, with the β domain having the fewest interactions and the SK termini appearing to be mobile. The unstructured termini could explain early findings showing that SK muteins lacking up to 15 N-terminal or 41 C-terminal residues or that truncated SKs fused to foreign proteins continue to activate PG, albeit with diminished activity (55, 69, 87, 90, 166). Studies of binding interactions between SK and PG domains using surface plasmon resonance (84) showed that the β domain, in addition to contacting the PG catalytic domain, interacts with PG kringle 5. Based on these studies, the mechanistic steps involved in PG activation include (i) binding of all Sk domains to the PG catalytic domain; (ii) cooperation of the α and γ domains to induce PG*:SK formation; and (iii) recognition of substrate PG by PG*:SK through interactions mediated by the α domain. A unified hypothesis for the mechanism of coupling between conformational and proteolytic PG activation based on parabolic time-course assays implies that PG*:SK initially binds PG in the substrate mode to form a ternary PG*:SK:PG complex, from which free PM is produced. PM then binds strongly to free SK to form a catalytic PM:SK complex that leads to sequestering of SK at concentrations of PG ≫SK. The PM:SK complex on its part binds free PG in the substrate mode, which ultimately results in full conversion of free PG to PM (12). Taken together, these studies suggest that except for the very ends of the polypeptide chain, most of the SK sequence is essential for efficient functionality, suggesting that any attempts at decreasing the size of the protein to reduce antigenicity may detrimentally affect PG activation.

Adaptive Evolution of SK Alleles

It has been known for a long time that SKs exploited by streptococci to infect mammalians show species-specific activity (Table 1). Recent studies based on knowledge gained about the mechanism of PG activation have provided some molecular insights into this phenomenon (41). SKs produced by different human streptococcal strains show a strong conservation of amino acid residues that contact sites in the loop residues of human μPM. On the other hand, such residues lack conservation among SKs from animal streptococci that activate the corresponding animal PGs. Conversely, different animal PGs show greater sequence diversity from human μPM in loops that contact SKs produced by human-pathogenic streptococcal strains. Streptococci may thus evolve to produce SKs that match their evolving target PGs, whereas PGs evolve to escape from such interactions. Such coevolutionary patterns in PG activation would thus appear to be of advantage for both sides and make SK a determinant of host species specificity for streptococcal infections (41). In fact, a transgenic mouse line expressing human PG exhibited significantly increased mortality upon infection with GAS (145b).

A study designed to uncover molecular evolutionary mechanisms underlying tissue-specific streptococcal infections analyzed the sequences of skn genes in skin-tropic strains known to bind host PG via the PAM protein and to depend on both types of genes for virulence at the skin (66). Exploiting the well-known fact that the β domain of SK shows the highest level of sequence heterogeneity, a tight association was found between the presence of PAM and a well-supported subcluster 2b ska allele. This observation indicated coinheritance of the corresponding genotypes maintained by coselective pressure due to epistasis rather than physical genetic linkage of the ska and PAM-encoding emm loci. Interestingly, compared to the β-domain-encoding region of the ska alleles, the skn alleles from GCS and GGS show a high degree of sequence homogeneity and are most closely related to subcluster 2a alleles of ska. This subcluster is predominantly associated with throat-tropic GAS strains, consistent with the lifestyle of GCS and GGS, which primarily inhabit the upper respiratory tract. Although it remains an open question whether subcluster 2a skn alleles facilitate colonization of the throat, this finding suggests that an skn allele from GCS or GGS was transferred from either group to GAS, where it both stayed as a subcluster 2a allele and evolved into a subcluster 2b allele. Lateral transfer in the opposite direction seems to be less likely because the housekeeping genes of GCS and GGS strains carrying a subcluster 2a allele show great heterogeneity (66).

Streptolysins and Streptodornase

Based on the well-known observation that a positive antistreptolysin O test is not specific for recent GAS disease, successful attempts were made to clone and sequence the streptolysin O (slo) genes of GCS (S. dysgalactiae subsp. equisimilis) and GGS (S. canis) (112). The slo gene proved to be highly conserved in the three serogroups, with sequence identity of the protein exceeding 98%. Expectedly, the slo gene was not detected in either subspecies of S. equi (31, 112). However, an SLS-like activity has been characterized in S. equi subsp. equi and shown to be subject to transposon inactivation at a single locus (31). The production of this important cytotoxin is presumably encoded in the same nine-gene sag operon that has been characterized in GAS and S. dysgalactiae subsp. equisimilis (human GGS) associated with severe necrotizing soft tissue infections (53).

In contrast to the highly conserved slo and sag genes, the streptodornase gene, sdc, first cloned from S. dysgalactiae subsp. equisimilis strain H46A (163), encodes a DNase that exhibits only 48% sequence identity to a type D DNase encoded by the sdaD gene later isolated from a GAS strain (121). However, establishing the intergroup sequence relationships of the different streptococcal DNases must await the determination of the enzyme types that occur in GCS and GGS as well as the molecular characterization of the three remaining types from GAS. Although for a long time implicated as potential virulence factors, the streptococcal DNases have only recently been directly shown to be involved in pathogenesis (145a).

Pyrogenic Exotoxins

Pyrogenic exotoxins (Spe) are believed to play an important role in severe invasive diseases caused by streptococci. Although the first reports of such infections involving non-GAS isolates appeared in the 1990s, it was not until the beginning of the 2000s that speA, speC, and speG as characterized before in GAS were also detected in human GCS and GGS (65, 126). Similarly, S. equi subsp. equi strains were found that harbored speH, speI, speL, and speM genes (4, 123). Although speH and speI were consistently found in all isolates, they were absent from S. equi subsp. zooepidemicus (4). Moreover, these two genes were linked, like their orthologs in GAS, to a bacteriophage genome, strongly suggesting that intergeneric phage-mediated transfer is responsible for their spread.

THE STRINGENT AND RELAXED RESPONSES OF *S. DYSGALACTIAE* SUBSP. *EQUISIMILIS*

Sequence analysis of *S. dysgalactiae* subsp. *equisimilis* chromosomal DNA 3′-adjacent to the *skc* gene (98) led to the unexpected discovery of the 739-codon *rel* gene (Fig. 2), the first homolog from a gram-positive organism of the paralogous *E. coli relA* and *spoT* genes to be cloned, sequenced, and functionally characterized in some detail (97, 99). In *E. coli*, these genes are involved in the metabolism of guanosine 5′,3′-polyphosphates [(p)ppGpp], which accumulate in response to starvation for amino acids or glucose exhaustion. The accumulated (p)ppGpp mediates a broad pleiotropic response hallmarked by a rapid shutdown of stable RNA (rRNA and tRNA) synthesis (the stringent response) and, in addition, by an increase or decrease of the synthesis of about half of the cellular proteins. As revealed by in vivo complementation experiments with appropriate *E. coli* mutants and verified by in vitro enzyme activity assays, the *rel* gene of *S. dysgalactiae* subsp. *equisimilis*, designated rel_{Seq} here, encodes a strong (p)ppGpp 3′-pyrophosphohydrolase activity [(p)ppGppase] and a weaker ribosome-independent (p)ppGpp synthetic activity resulting from its ATP:GTP 3′-pyrophosphoryltransferase capacity (99). In vivo, the net effect of the opposing Rel_{Seq} activities favors (p)ppGpp degradation, resulting in the failure of rel_{Seq} to complement the absence of *relA* and, hence, in its functional likening to *spoT*. Structural support for this inference provided the detection of immunological cross-reactivity of Rel_{Seq} with polyclonal antibodies to SpoT but not RelA and, vice versa, of SpoT but not RelA with antibodies to Rel_{Seq}. However, despite these clear observations, disruption of the chromosomal rel_{Seq} gene at codon 216 by insertion mutagenesis abolished the (p)ppGpp accumulation response following amino acid starvation of wild-type *S. dysgalactiae* subsp. *equisimilis*, a phenotype characteristic of *E. coli relA* but not *spoT* mutants. Further functional characterization of rel_{Seq} showed that amino acid deprivation of *S. dysgalactiae* subsp. *equisimilis* triggers a rel_{Seq}-dependent stringent response characterized by rapid (p)ppGpp accumulation at the expense of GTP and abrupt cessation of net RNA accumulation in the wild type but not in the insertion mutant, which displays a relaxed response (97). Thus, one of the most interesting results of this work is the finding that the Rel_{Seq} protein is bifunctional and shows activities that reside separately in the paralogous SpoT and RelA proteins from *E. coli*. This establishes an important variation of the *E. coli* paradigm for regulating (p)ppGpp levels and presumably the expression of genes controlled by (p)ppGpp. In fact, evidence accumulated subsequently for other gram-positive genera supports the observations made in streptococcus and implicates a single bifunctional Rel/Spo enzyme in the metabolism of (p)ppGpp, unaccompanied by a sister *rel/spo* paralog. Cloning and expression of defined rel_{Seq} fragments followed by catalytic activity assays of the purified protein fragments revealed that the two enzymatic activities reside in the N-terminal half-protein (residues 1 to 385) and that the removal of the C-terminal half-protein has reciprocal regulatory effects on the opposing activities of the N-terminal domain (100). Further dissection of the bifunctional N-terminal domain led to the characterization of defined monofunctional subdomains, with hydrolase and synthetase activities residing in residues 1 to 224 and 79 to 385, respectively (100). The 2.1-Å crystal structure of Rel_{Seq}1–385, the first X-ray structure to be reported for an enzyme involved in the metabolism of (p)ppGpp (51), fully explains the solution biochemical studies in trapping the enzyme in two distinct conformations that typify the opposing (p)ppGpp hydrolase-OFF/synthetase-ON and hydrolase-ON/synthetase-OFF states (Fig. 5 A, B). The individual active sites are located >30 Å apart and, being mutually exclusive, demonstrate how reciprocal regulation of the two opposing activities can prevent futile cycling in the metabolism of (p)ppGpp. The conclusions derived from the crystal structure are fully supported by extensive point mutagenesis (Fig. 5C) or enzyme inhibition experiments (51). However, further work is necessary to fully understand the involvement of the stringent response in the biology of streptococcus in general and in how nutrient limitation in particular may drive the expression of virulence traits. Concomitant with prevention of wasteful macromolecular synthesis by a rapid shutdown of futile RNA synthesis following the failure of aminoacyl-tRNA pools to keep up with the demands of protein synthesis, the functional significance of the streptococcal stringent factor consists of strongly supporting cell survival under nutritional stress conditions (142). The short half-lives of 20 and 60 s for pppGpp and ppGpp, respectively, ensure quick recovery from the stasislike survival state when the nutritional conditions improve (97). Importantly, transcriptional analyses have shown that in addition to the stringent response, streptococci are capable of mounting a *rel*-independent response to amino acid deprivation that involves transcriptional modulation of a wide array of housekeeping genes as well as accessory and dedicated virulence genes. For example, among the genes responding with elevated transcript levels to starvation conditions are those encoding aminoacyl-tRNA synthetases, oligopeptide permease, two-component regulators (*cov* and *fas*), and members of the SLS operon. Based on these observations, a regulatory network with feedback mechanisms has been proposed that counteracts the stringent response, links the levels of key rate-limiting enzymes to virulence gene expression, and enables streptococci in a dynamic way to take advantage of protein-rich host environments (142, 143).

CONCLUSIONS AND PROSPECTS

It is apparent from the preceding account that molecular biological investigations into GCS and GGS have preferentially sought to identify and characterize genes and proteins putatively involved in pathogenicity. Although in many cases, this work has been aided by knowledge obtained previously from studies of the counterparts of such determinants in GAS, investigation of GCS and GGS has also produced independent knowledge of significance for the streptococcal field or bacteriology in general. This work led to the realization that GCS and GGS share many of the cell surface-associated as well as extracellular pathogenicity factors with GAS but also exhibit unique traits that merit their further investigation. With few exceptions, the analysis of genetic linkage relationships has rarely been pursued in human GCS and GGS. Although there are ongoing genome projects for both subspecies of *S. equi* (http://www.sanger.ac.uk/Projects/Microbes), no such projects have apparently been initiated yet for human GCS or GGS. The little knowledge available (33, 37, 38, 98) suggests that both homology of synteny and drastic rearrangement of gene order will be observed in future work aimed at the development of a comparative genomics of GAS, GCS, and GGS. The elucidation of the phylogenetic relationships of these organisms may be complicated given the clear indications of horizontal gene transfer following speciation of the species to

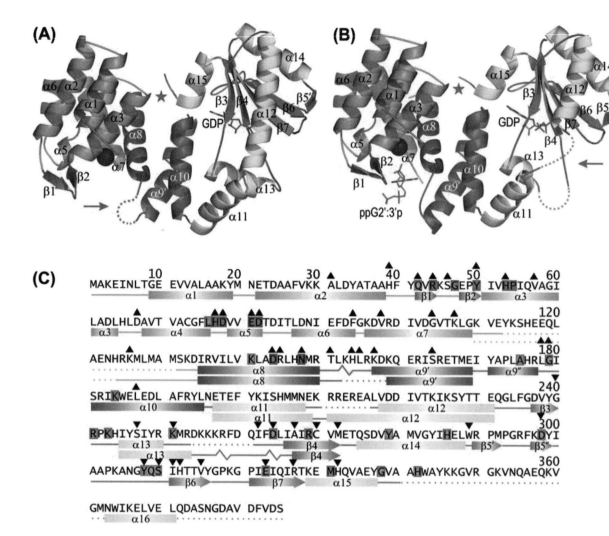

FIGURE 5 (*See the separate color insert for the color version of this illustration.*) Structure of $\text{Rel}_{Seq}1$-385. (A) ppGpp Hydrolase-OFF/synthetase-ON conformation, in complex with Mn^{2+} (blue sphere) and GDP (stick rendering). The hydrolase domain is highlighted in green α-helices and blue β-strands, the synthetase domain is in yellow α-helices and orange β-strands, and the central 3-helix bundle is in red. Part of the substrate-binding cleft comprising the hydrolase site (down-regulated) is disordered (red arrow). The small synthetase/hydrolase interdomain contact interface involved in signal transmission is labeled with a red star. (B) Hydrolase-ON/synthetase-OFF conformation, in complex with Mn^{2+}, ppG':3'p (which locks the enzyme in the hydrolase-ON/synthetase-OFF conformation) and GDP. The coloring and rendering schemes are the same as for (A). The disordered synthetase site (down-regulated) is illustrated as dashed lines with a red arrow. (C) Primary and secondary structure of $\text{Rel}_{Seq}1$-385. Secondary structure is color-coded according to (A) and (B). Unique secondary structure assignments for (A) are placed immediately below the corresponding assignments for (B). Residues absolutely conserved throughout the mono- and bifunctional RelA and SpoT homologs are underlaid with blue boxes. Residues conserved in SpoT and the bifunctional enzymes but mutated in the hydrolase-incompetent RelA homologs are underlaid with red boxes. Upright and inverted black triangles above the sequence indicate residues which, when substituted experimentally by missense mutations, lead to defective hydrolase and synthetase activities, respectively. (Reproduced from reference 51, with permission.)

be compared. Furthermore, the available evidence of linkage disequilibrium between a limited number of genes will require future work to delve into problems of genetic linkage, recombination rates, and epistasis to provide explanations for the development of constrained functions and the evolution of coadapted genes in closely related species (66). Detailed knowledge about the responses of streptococci to

the environmental stresses encountered at sites of infection is scant. For example, tissue colonization involves, first of all, a quest for food, and monomeric building blocks for macromolecular syntheses appear to be limiting and need to be generated. Conversely, the stasislike carrier state requiring survival functions may characterize cells that not only adhere to host cells but also succeed in invading them. The ability to persist under these conditions may well depend on the stringent response, and the now available structural knowledge of the streptococcal stringent factor (51) makes this enzyme a viable target for the design of antibacterial agents. Studying how ppGpp exerts its action on the streptococcal transcription apparatus will likely reveal unexpected variations on emerging notions concerning the importance of the secondary channel of RNA polymerase as a regulatory entrance for transcription factors (e.g., DksA-like proteins), substrate nucleotides (NTPs), and inhibitory molecules (ppGpp). Recent structural studies have shown that all three of these factors bind to RNA polymerase and that ppGpp and the DksA protein act synergistically to regulate transcription (119). However, although widely distributed in bacteria, database searches have failed so far to identify DksA homologs at the level of amino acid sequence among the pathogenic streptococci.

Space constraints sometimes obliged me to cite reviews or other secondary literature referring to primary research papers relevant to this chapter. Work by me and my coworkers mentioned here was supported by grants from the Deutsche Forschungsgemeinschaft, the Fonds der Chemischen Industrie, and the Thüringer Ministerium für Wissenschaft, Forschung und Kunst.

REFERENCES

1. **Achari, A., S. P. Hale, A. J. Howard, G. M. Clore, A. M. Gronenborn, K. D. Hardman, and M. Whitlow.** 1992. 1.67-A X-ray structure of the B2 immunoglobulin-binding domain of streptococcal protein G and comparison to the NMR structure of the B1 domain. *Biochemistry* **31:**10449–10457.

2. **Akerström, B., E. Nielsen, and L. Björck.** 1987. Definition of IgG- and albumin-binding regions of streptococcal protein G. *J. Biol. Chem.* **262:**13388–13391.

3. **Anzai, T., J. F. Timoney, Y. Kuwamoto, Y. Fujita, R. Wada, and T. Inoue.** 1999. In vivo pathogenicity and resistance to phagocytosis of *Streptococcus equi* strains with different levels of capsule expression. *Vet. Microbiol.* **67:**277–286.

4. **Artiushin, S. C., J. F. Timoney, A. S. Sheoran, and S. K. Muthupalani.** 2002. Characterization and immunogenicity of pyrogenic mitogens SePE-H and SePE-I of *Streptococcus equi.* *Microb. Pathog.* **32:**71–85.

5. **Barne, K. A., J. A. Bown, S. J. W. Busby, and S. D. Minchin.** 1997. Region 2.5 of the *Escherichia coli* RNA polymerase σ^{70} subunit is responsible for the recognition of the 'extended-10' motif at promoters. *EMBO J.* **16:**4034–4040.

6. **Ben Nasr, A., A. Wistedt, U. Ringdahl, and U. Sjöbring.** 1994. Streptokinase activates plasminogen bound to human group C and G streptococci through M-like proteins. *Eur. J. Biochem.* **222:**267–276.

7. **Bentley, R. W., J. A. Leigh, and M. D. Collins.** 1991. Intrageneric structure of *Streptococcus* based on comparative analysis of small-subunit rRNA sequences. *Int. J. Syst. Bacteriol.* **41:**487–494.

8. **Bert, F., B. Picard, N. Lambert-Zechovsky, and P. Goullet.** 1995. Identification and typing of pyogenic streptococci by enzyme electrophoretic polymorphism. *J. Med. Microbiol.* **42:**442–451.

9. **Bert, F., B. Picard, C. Branger, and N. Lambert-Zechovsky.** 1996. Analysis of genetic relationships among strains of groups A, C and G streptococci by random amplified polymorphic DNA analysis. *J. Med. Microbiol.* **45:**278–284.

10. **Bert, F., C. Branger, and N. Lambert-Zechovsky.** 1997. Pulsed-field gel electrophoresis is more discriminating than multilocus enzyme electrophoresis and random amplified polymorphic DNA analysis for typing pyogenic streptococci. *Curr. Microbiol.* **34:**226–229.

11. **Bisno, A. L., C. M. Collins, and J. C. Turner.** 1997. M proteins of group C streptococci isolated from patients with acute pharyngitis, p. 745–748. *In* T. Horaud, M. Sicard, A. Bouvet, R. Leclerq, and H. DeMonclos (ed.), *Streptococci and the Host.* Plenum Press, New York, N.Y.

12. **Boxrud, P. D., and P. E. Bock.** 2004. Coupling of conformational and proteolytic activation in the kinetic mechanism of plasminogen activation by streptokinase. *J. Biol. Chem.* **279:**36642–36649.

13. **Broeseker, T. A., M. D. Boyle, and R. Lottenberg.** 1988. Characterization of the interaction of human plasmin with its specific receptor on a group A streptococcus. *Microb. Pathog.* **5:**19–27.

14. **Calvinho, L. F., R. A. Almeida, and S. P. Oliver.** 1998. Potential virulence factors of *Streptococcus dysgalactiae* associated with bovine mastitis. *Vet. Microbiol.* **61:**93–110.

15. **Chanter, N., N. Collin, N. Holmes, M. Binns, and J. Mumford.** 1997. Characterization of the Lancefield group C streptococcus 16S-23S RNA gene intergenic spacer and its potential for identification and sub-specific typing. *Epidemiol. Infect.* **118:**125–135.

16. **Cleary, P. P., J. Peterson, C. Chen, and C. Nelson.** 1991. Virulent human strains of group G streptococci express a C5a peptidase enzyme similar to that produced by group A streptococci. *Infect. Immun.* **59:**2305–2310.

17. **Cohen-Poradosu, R., J. Jaffe, D. Lavi, S. Grisariu, R. Nir-Paz, L. Valinsky, M. Dan-Goor, C. Block, B. Beall, and A. E. Moses.** 2004. Group G streptococcal bacteremia in Jerusalem. *Emerg. Infect. Dis.* **10:**1455–1460.

18. **Collins, C. M., A. Kimura, and A. L. Bisno.** 1992. Group G streptococcal M protein exhibits structural features analogous to those of class I M protein of group A streptococci. *Infect. Immun.* **60:**3689–3696.

19. **Conejero-Lara, F., J. Parrado, A.I. Azuaga, C. M. Dobson, and C. P. Ponting.** 1998. Analysis of the interactions between streptokinase domains and human plasminogen. *Protein Sci.* **7:**2190–2199.

20. **Crater, D. L., and I. van de Rijn.** 1995. Hyaluronic acid synthesis operon (*has*) expression in group A streptococci. *J. Biol. Chem.* **270:**18452–18458.

21. **D'Costa, S. S., H. Wang, D. W. Metzger, and M. D. P. Boyle.** 1997. Group A streptococcal isolate 64/14 expresses surface plasmin-binding structures in addition to Plr. *Res. Microbiol.* **148:**559–572.

22. **Derrick, J. P., and D. B. Wigley.** 1992. Crystal structure of a streptococcal protein G domain bound to an Fab fragment. *Nature* **359:**752–754.

23. **Efstratiou, A.** 1997. Pyogenic streptococci of Lancefield groups C and G as pathogens in man. *J. Appl. Microbiol. Symp. Suppl.* **83:**72S–79S.

24. **Espinosa-Urgel, M., and R. Kolter.** 1998. *Escherichia coli* genes expressed preferentially in an aquatic environment. *Mol. Microbiol.* **28:**325–332.

25. **Facklam, R.** 2002. What happened to the streptococci: overview of taxonomic and nomenclature changes. *Clin. Microbiol. Rev.* **15:**613–630.

26. **Fahnestock, S. R., P. Alexander, J. Nagle, and D. Filpula.** 1986. Gene for an immunoglobulin-binding protein from a group G streptococcus. *J. Bacteriol.* **167:**870–880.

27. **Falkenberg, C., L. Björck, and B. Akerström.** 1992. Localization of the binding site for streptococcal protein

G on human serum albumin. Identification of a 5.5-kilo-dalton protein G binding albumin fragment. *Biochem.* **31:**1451–1457.

28. **Farrow, J. A. E., and M. D. Collins.** 1984. Taxonomic studies on streptococci of serological groups C, G and L and possibly related taxa. *Syst. Appl. Microbiol.* **5:**483–493.

29. **Federle, M. J., K. S. McIver, and J. R. Scott.** 1999. A response regulator that represses transcription of several virulence operons in the group A streptococcus. *J. Bacteriol.* **181:**3649–3657.

30. **Fischetti, V. A.** 1989. Streptococcal M protein: molecular design and biological behavior. *Clin. Microbiol. Rev.* **2:** 285–314.

31. **Flanagan, J., N. Collin, J. Timoney, T. Mitchell, J. A. Mumford, and N. Chanter.** 1998. Characterization of the haemolytic activity of *Streptococcus equi. Microb. Pathog.* **24:**211–221.

32. **Forsman, P., A. Tilsala-Timisjärvi, and T. Alatossava.** 1997. Identification of staphylococcal and streptococcal causes of bovine mastitis using 16S-23S rRNA spacer regions. *Microbiology* **143:**3491–3500.

33. **Frank, C., K. Steiner, and H. Malke.** 1995. Conservation of the organization of the streptokinase gene region among pathogenic streptococci. *Med. Microbiol. Immunol.* **184:** 139–146.

34. **Franken, C., G. Haase, C. Brandt, J. Weber-Heynemann, S. Martin, C. Lämmler, A. Podbielski, R. Lütticken, and B. Spellerberg.** 2001. Horizontal gene transfer and host specificity of beta-haemolytic streptococci: the role of a putative composite transposon containing *scpB* and *lmb. Mol. Microbiol.* **41:**925–935.

35. **Gase, K., T. Ellinger, and H. Malke.** 1995. Complex transcriptional control of the streptokinase gene of *Streptococcus equisimilis* H46A. *Mol. Gen. Genet.* **247:**749–758.

36. **Gase, K., A. Gase, H. Schirmer, and H. Malke.** 1996. Cloning, sequencing and functional overexpression of the *Streptococcus equisimilis* H46A *gapC* gene encoding a glyceraldehyde-3-phosphate dehydrogenase that also functions as a plasmin(ogen)-binding protein: purification and biochemical characterization of the protein. *Eur. J. Biochem.* **239:**42–51.

37. **Gase, K., G. Liu, A. Bruckmann, K. Steiner, J. Ozegowski, and H. Malke.** 1997. The *lppC* gene of *Streptococcus equisimilis* encodes a lipoprotein that is homologous to the *e* (P4) outer membrane protein from *Haemophilus influenzae. Med. Microbiol. Immunol.* **186:**63–73.

38. **Geyer, A., and K.-H. Schmidt.** 2000. Genetic organization of the M protein region in human isolates of group C and G streptococci: two types of multigene regulator-like (*mgrC*) regions. *Mol. Gen. Genet.* **262:**965–976.

39. **Geyer, A., A. Roth, S. Vettermann, E. Günther, A. Groh, E. Straube, and K.-H. Schmidt.** 1999. M protein of a *Streptococcus dysgalactiae* human wound isolate shows multiple binding to different plasma proteins and shares epitopes with keratin and human cartilage. *FEMS Immunol. Med. Microbiol.* **26:**11–24.

40. **Gillespie, B. E., B. M. Jayarao, and S. P. Oliver.** 1997. Identification of *Streptococcus* species by randomly amplified polymorphic deoxyribonucleic acid fingerprinting. *J. Dairy Sci.* **80:**471–476.

41. **Gladysheva, I. P., R. B. Turner, I. Y. Sazonova, L. Liu, and G. L. Reed.** 2003. Coevolutionary patterns in plasminogen activation. *Proc. Natl. Acad. Sci. USA* **100:** 9168–9172.

42. **Gräfe, S., T. Ellinger, and H. Malke.** 1996. Structural dissection and functional analysis of the complex promoter of the streptokinase gene from *Streptococcus equisimilis* H46A. *Med. Microbiol. Immunol.* **185:**11–17.

43. **Gronenborn, A. M., and G. M. Clore.** 1993. Identification of the contact surface of a streptococcal protein G domain complexed with a human Fc fragment. *J. Molec. Biol.* **233:**331–335.

44. **Gronenborn, A. M., D. R. Filpula, N. Z. Essig, A. Achari, M. Whitlow, P. T. Wingfield, and G. M. Clore.** 1991. A novel, highly stable fold of the immunoglobulin binding domain of streptococcal protein G. *Science* **253:**657–660.

45. **Gross, S., K. Gase, and H. Malke.** 1996. Localization of the sequence-determined DNA bending center upstream of the streptokinase gene *skc. Arch. Microbiol.* **166:**116–121.

46. **Guss, B., M. Eliasson, A. Olsson, M. Uhlén, A. K. Frej, H. Jörnvall, J. I. Flock, and M. Lindberg.** 1986. Structure of the IgG-binding regions of streptococcal protein G. *EMBO J.* **5:**1567–1575.

47. **Hamilton, A., D. Harrington, and I. C. Sutcliffe.** 2000. Characterization of acid phosphatase activities in the equine pathogen *Streptococcus equi. Syst. Appl. Microbiol.* **23:**325–329.

48. **Hashikawa, S., Y. Iinuma, M. Furushita, T. Ohkura, T. Nada, K. Torii, T. Hasegawa, and M. Ohta.** 2004. Characterization of group C and G streptococcal strains that cause streptococcal toxic shock syndrome. *J. Clin. Microbiol.* **42:**186–192.

49. **Hassan, A. A., I. U. Khan, A. Abdulmawjood, and C. Lämmler.** 2003. Inter- and intraspecies variations of the 16S-23S rDNA intergenic spacer region of various streptococcal species. *Syst. Appl. Microbiol.* **26:**97–103.

50. **Heldermon, C., P. L. DeAngelis, and P. H. Weigel.** 2001. Topological organization of the hyaluronan synthase from *Streptococcus pyogenes. J. Biol. Chem.* **276:**2037–2046.

51. **Hogg, T., U. Mechold, H. Malke, M. Cashel, and R. Hilgenfeld.** 2004. Conformational antagonism between opposing active sites in a bifunctional RelA/SpoT homolog modulates (p)ppGpp metabolism during the stringent response. *Cell* **117:**57–68.

52. **House-Pompeo, K., Y. Xu, D. Joh, P. Speziale, and M. Höök.** 1996. Conformational changes in the fibronectin binding MSCRAMMs are induced by ligand binding. *J. Biol. Chem.* **271:**1379–1384.

53. **Humar, D., V. Datta, D. J. Bast, B. Beall, J. C. Azavedo, and V. Nizet.** 2002. Streptolysin S and necrotising infections produced by group G streptococcus. *Lancet* **359:**124–129.

54. **Ikebe, T., S. Murayama, K. Saitoh, S. Yamai, R. Suzuki, J. Isobe, D. Tanaka, C. Katsukawa, A. Tamaru, A. Katayama, Y. Fujinaga, K. Hoashi, and H. Watanabe.** 2004. Surveillance of severe invasive group-G streptococcal infections and molecular typing of the isolates in Japan. *Epidemiol. Infect.* **132:**145–149.

55. **Jackson, K. W., H. Malke, D. Gerlach, J. J. Ferretti, and J. Tang.** 1986. Active streptokinase from the cloned gene in *Streptococcus sanguis* is without the carboxyl-terminal 32 residues. *Biochemistry* **25:**108–114.

56. **Jaffe, J., S. Natanson-Yaron, M. G. Caparon, and E. Hanski.** 1996. Protein F2, a novel fibronectin-binding protein from *Streptococcus pyogenes*, possesses two binding domains. *Mol. Microbiol.* **21:**373–384.

57. **Joh, H. J., K. House-Pompeo, J. M. Patti, S. Gurusiddappa, and M. Höök.** 1994. Fibronectin receptors from

gram-positive bacteria: comparison of active sites. *Biochemistry* **33:**6086–6092.

58. Joh, D., P. Speziale, S. Gurusiddappa, J. Manor, and M. Höök. 1998. Multiple specificities of the staphylococcal and streptococcal fibronectin-binding microbial surface components recognizing adhesive matrix molecules. *Eur. J. Biochem.* **258:**897–905.

59. Jonsson, H., and H. P. Müller. 1994. The type-III Fc receptor from *Streptococcus dysgalactiae* is also an α_2-macroglobulin receptor. *Eur. J. Biochem.* **220:**819–826.

60. Jonsson, H., L. Frykberg, L. Rantamäki, and B. Guss. 1994. MAG, a novel plasma protein receptor from *Streptococcus dysgalactiae*. *Gene* **143:**85–89.

61. Jonsson, H., C. Burtsoff-Asp, and B. Guss. 1995. Streptococcal protein MAG—a protein with broad albumin binding specificity. *Biochim. Biophys. Acta* **1249:**65–71.

62. Jonsson, H., H. Lindmark, and B. Guss. 1995. A protein G-related cell surface protein in *Streptococcus zooepidemicus*. *Infect. Immun.* **63:**2968–2975.

63. Jorm, L. R., D. N. Love, G. D. Bailey, G. M. McKay, and D. A. Briscoe. 1994. Genetic structure of populations of β-haemolytic Lancefield group C streptococci from horses and their association with disease. *Res. Vet. Sci.* **57:**292–299.

64. Kakizaki, I., K. Takagaki, Y. Endo, D. Kudo, H. Ikeya, T. Miyoshi, B.A. Baggenstoss, V.L. Tlapak-Simmons, K. Kumari, A. Nakane, P.H. Weigel, and M. Endo. 2002. Inhibition of hyaluronan synthesis in *Streptococcus equi* FM100 by 4-methylumbelliferone. *Eur. J. Biochem.* **269:**5066–5075.

65. Kalia, A., and D. E. Bessen. 2003. Presence of streptococcal pyrogenic exotoxin A and C genes in human isolates of group G streptococci. *FEMS Microbiol. Lett.* **219:**291–295.

66. Kalia, A., and D. E. Bessen. 2004. Natural selection and evolution of streptococcal virulence genes involved in tissue-specific adaptations. *J. Bacteriol.* **186:**110–121.

67. Kalia, A., M. C. Enright, B. G. Spratt, and D. E. Bessen. 2001. Directional gene movement from human-pathogenic to commensal-like streptococci. *Infect. Immun.* **69:**4858–4869.

68. Kawamura, Y., X. G. Hou, F. Sultana, H. Miura, and T. Ezaki. 1995. Determination of 16S rRNA sequences of *Streptococcus mitis* and *Streptococcus gordonii* and phylogenetic relationships among members of the genus *Streptococcus*. *Int. J. Syst. Bacteriol.* **45:**406–408.

69. Klessen, C., K. H. Schmidt, J. J. Ferretti, and H. Malke. 1988. Tripartite streptokinase gene fusion vectors for gram-positive and gram-negative procaryotes. *Mol. Gen. Genet.* **212:**295–300.

70. Kline, J. B., S. Xu, A. L. Bisno, and C. M. Collins. 1996. Identification of a fibronectin-binding protein (GfbA) in pathogenic group G streptococci. *Infect. Immun.* **64:**2122–2129.

71. Kreikemeyer, B., M. D. P. Boyle, B. A. Leonard Buttaro, M. Heinemann, and A. Podbielski. 2001. Group A streptococcal growth phase-associated virulence factor regulation by a novel operon (Fas) with homologies to two-component-type regulators requires a small RNA molecule. *Mol. Microbiol.* **39:**392–406.

72. Kumar, S., K. Tamura, I. B. Jakobsen, and M. Nei. 2001. MEGA2: molecular evolutionary genetics analysis software. *Bioinformatics* **17:**1244–1245.

73. Kumari, K., and P. H. Weigel. 1997. Molecular cloning, expression, and characterization of the authentic hyaluro-

nan synthase from group C *Streptococcus equisimilis*. *J. Biol. Chem.* **272:**32539–32546.

74. Kumari, K., V. L. Tlapak-Simmons, B. A. Baggenstoss, and P. H. Weigel. 2002. The streptococcal hyaluronan synthases are inhibited by sulfhydryl-modifying reagents, but conserved cysteine residues are not essential for enzyme function. *J. Biol. Chem.* **277:**13943–13951.

75. Leigh, J. A., S. M. Hodgkinson, and R. A. Lincoln. 1998. The interaction of *Streptococcus dysgalactiae* with plasmin and plasminogen. *Vet. Microbiol.* **61:**121–135.

76. Lejon, S., I.-M. Frick, L. Björck, M. Wikström, and S. Svensson. 2004. Crystal structure and biological implications of a bacterial albumin-binding module in complex with human serum albumin. *J. Biol. Chem.* **279:**42924–42928.

77. Lindgren, P. E., P. Speziale, M. McGavin, H. J. Monstein, M. Höök, L. Visai, T. Kostiainen, S. Bozzini, and M. Lindberg. 1992. Cloning and expression of two different genes from *Streptococcus dysgalactiae* encoding fibronectin receptors. *J. Biol. Chem.* **267:**1924–1931.

78. Lindgren, P. E., M. J. McGavin, C. Signäs, B. Guss, S. Gurusiddappa, M. Höök, and M. Lindberg. 1993. Two different genes coding for fibronectin-binding proteins from *Streptococcus dysgalactiae*: the complete nucleotide sequences and characterization of the binding domains. *Eur. J. Biochem.* **214:**819–827.

79. Lindgren, P. E., C. Signäs, L. Rantamäki, and M. Lindberg. 1994. A fibronectin-binding protein from *Streptococcus equisimilis*: characterization of the gene and identification of the binding domain. *Vet. Microbiol.* **41:**235–247.

80. Lindmark, H., and B. Guss. 1999. SFS, a novel fibronectin-binding protein from *Streptococcus equi*, inhibits the binding between fibronectin and collagen. *Infect. Immun.* **67:**2383–2388.

81. Lindmark, H., K. Jacobsson, L. Frykberg, and B. Guss. 1996. Fibronectin-binding protein of *Streptococcus equi* subsp. *zooepidemicus*. *Infect. Immun.* **64:**3993–3999.

82. Lindmark, H., M. Nilsson, and B. Guss. 2001. Comparison of the fibronectin-binding protein FNE from *Streptococcus equi* subspecies *equi* with FNZ from *Streptococcus equi* subspecies *zooepidemicus* reveals a major and conserved difference. *Infect. Immun.* **69:**3159–3163.

83. Lottenberg, R., C. C. Broder, M. D. Boyle, S. J. Kain, B. L. Schroeder, and R. Curtiss III. 1992. Cloning, sequence analysis, and expression in *Escherichia coli* of a streptococcal plasmin receptor. *J. Bacteriol.* **174:**5204–5210.

84. Loy, J. A., X. Lin, M. Schenone, F. J. Castellino, X. C. Zhang, and J. Tang. 2001. Domain interactions between streptokinase and human plasminogen. *Biochemistry* **40:**14686–14695.

85. Malke, H. 1998. Cytoplasmic membrane lipoprotein LppC of *Streptococcus equisimilis* functions as an acid phosphatase. *Appl. Environ. Microbiol.* **64:**2439–2442.

86. Malke, H., and J. J. Ferretti. 1984. Streptokinase: cloning, expression, and excretion by *Escherichia coli*. *Proc. Natl. Acad. Sci. USA* **81:**3557–3561.

87. Malke, H., and J. J. Ferretti. 1991. Expression and properties of hybrid streptokinases extended by N-terminal plasminogen kringle domains, p. 184–189. *In* G. M. Dunny, P. P. Cleary, and L. L. McKay (ed.), *Genetics and Molecular Biology of Streptococci, Lactococci, and Enterococci*. American Society for Microbiology, Washington, D.C.

88. Malke, H., and K. Steiner. 2000. Functional properties of the cytoplasmic membrane lipoprotein acid phosphatase of

Streptococcus dysgalactiae subsp. *equisimilis*, p. 879–883. *In* D. R. Martin and J. R. Tagg (ed.), *Streptococci and Strepto-coccal Diseases—Entering the New Millennium*. Securacopy, Porirua, New Zealand.

89. **Malke, H., and K. Steiner.** 2004. Control of streptokinase gene expression in group A & C streptococci by two-component regulators. *Indian J. Med. Res.* **119**(Suppl.):4–56.

90. **Malke, H., D. Lorenz, and J. J. Ferretti.** 1987. Streptokinase: expression of altered forms, p. 143–149. *In* J. J. Ferretti and R. Curtiss III (ed.), *Streptococcal Genetics*. American Society for Microbiology, Washington, D.C.

91. **Malke, H., B. Roe, and J. J. Ferretti.** 1985. Nucleotide sequence of the streptokinase gene from *Streptococcus equisimilis* H46A. *Gene* **34**:357–362.

92. **Malke, H., U. Mechold, K. Gase, and G. Gerlach.** 1994. Inactivation of the streptokinase gene prevents *Streptococcus equisimilis* H46A from acquiring cell-associated plasmin activity in the presence of plasminogen. *FEMS Microbiol. Lett.* **116**:107–112.

93. **Malke, H., K. Steiner, K. Gase, U. Mechold, and T. Ellinger.** 1995. The streptokinase gene: allelic variation, genomic environment, and expression control. *Dev. Biol. Stand.* **85**:183–193.

94. **Malke, H., K. Steiner, K. Gase, and C. Frank.** 2000. Expression and regulation of the streptokinase gene. *Methods* **21**:111–124.

95. **Martin, N. J., E. L. Kaplan, M. A. Gerber, M. A. Menegus, M. Randolph, K. Bell, and P. P. Cleary.** 1990. Comparison of epidemic and endemic group G streptococci by restriction enzyme analysis. *J. Clin. Microbiol.* **28**:1881–1886.

96. **McGavin, M., S. Gurusiddappa, P. E. Lindgren, M. Lindberg, G. Raucci, and M. Höök.** 1993. Fibronectin receptors from *Streptococcus dysgalactiae* and *Staphylococcus aureus*. *J. Biol. Chem.* **268**:23946–23953.

97. **Mechold, U., and H. Malke.** 1997. Characterization of the stringent and relaxed responses of *Streptococcus equisimilis*. *J. Bacteriol.* **179**:2658–2667.

98. **Mechold, U., K. Steiner, S. Vettermann, and H. Malke.** 1993. Genetic organization of the streptokinase region of the *Streptococcus equisimilis* H46A chromosome. *Mol. Gen. Genet.* **241**:129–140.

99. **Mechold, U., M. Cashel, K. Steiner, D. Gentry, and H. Malke.** 1996. Functional analysis of a *relA/spoT* gene homolog from *Streptococcus equisimilis*. *J. Bacteriol.* **178**:1401–1411.

100. **Mechold, U., H. Murphy, L. Brown, and M. Cashel.** 2002. Intramolecular regulation of the opposing (p)ppGpp catalytic activities of Rel$_{Seq}$, the Rel/Spo enzyme from *Streptococcus equisimilis*. *J. Bacteriol.* **184**:2878–2888.

101. **Meehan, M., P. Nowlan, and P. Owen.** 1998. Affinity purification and characterization of a fibrinogen-binding protein complex which protects mice against lethal challenge with *Streptococcus equi* subsp. *equi*. *Microbiology* **144**:993–1003.

102. **Meehan, M., D. A. Muldowney, N. J. Watkins, and P. Owen.** 2000. Localization and characterization of the ligand-binding domain of the fibrinogen-binding protein (FgBP) of *Streptococcus equi* subsp. *equi*. *Microbiology* **146**:1187–1194.

103. **Meehan, M., Y. Lynagh, C. Woods, and P. Owen.** 2001. The fibrinogen-binding protein (FgBP) of *Streptococcus equi* subsp. *equi* additionally binds IgG and contributes to virulence in a mouse model. *Microbiology* **147**:3311–3322.

104. **Meehan, M., S. M. Kelly, N. C. Price, and P. Owen.** 2002. The C-terminal portion of the fibrinogen-binding protein of *Streptococcus equi* subsp. *equi* contains extensive α-helical coiled-coil structure and contributes to thermal stability. *FEMS Microbiol. Lett.* **206**:81–86.

105. **Molinari, G., and G. S. Chhatwal.** 1999. Streptococcal invasion. *Curr. Opin. Microbiol.* **2**:56–61.

106. **Müller, H.-P. and L. K. Rantamäki.** 1995. Binding of native alpha 2-macroglobulin to human group G streptococci. *Infect. Immun.* **63**:2833–2839.

107. **Nicholson, M. L., L. Ferdinand, J. S. Sampson, A. Benin, S. Balter, S. W. L. Pinto, S. F. Dowell, R. R. Facklam, G. M. Carlone, and B. Beall.** 2000. Analysis of immunoreactivity to a *Streptococcus equi* subsp. *zooepidemicus* M-like protein to confirm an outbreak of poststreptococcal glomerulonephritis, and sequences of M-like proteins from isolates obtained from different host species. *J. Clin. Microbiol.* **38**:4126–4130.

108. **Nickel, V., S. Prehm, M. Lansing, A. Mausolf, A. Podbielski, J. Deutscher, and P. Prehm.** 1998. An ectoprotein kinase of group C streptococci binds hyaluronan and regulates capsule formation. *J. Biol. Chem.* **273**:23668–23673.

109. **Nikolskaya, A. N., and M. Y. Galperin.** 2002. A novel type of conserved DNA-binding domain in the transcriptional regulators of the AlgR/AgrA/LytR family. *Nucleic Acids Res.* **30**:2453–2459.

110. **Nowicki, S. T., D. Minning-Wenz, K. H. Johnston, and R. Lottenberg.** 1994. Characterization of a novel streptokinase produced by *Streptococcus equisimilis* of non-human origin. *Thromb. Haemost.* **72**:595–603.

111. **Nygren, P. A., C. Ljungquist, H. Thromborg, K. Nustad, and M. Uhlén.** 1990. Species-dependent binding of serum albumins to the streptococcal receptor protein G. *Eur. J. Biochem.* **193**:143–148.

112. **Okumura, K., A. Hara, T. Tanaka, I. Nishiguchi, W. Minamide, H. Igarashi, and T. Yutsudo.** 1994. Cloning and sequencing the streptolysin O genes of group C and group G streptococci. *DNA Sequence* **4**:325–328.

113. **Oliver, S. P., B. E. Gillespie, and B. M. Jayarao.** 1998. Detection of new and persistent *Streptococcus uberis* and *Streptococcus dysgalactiae* intramammary infections by polymerase chain reaction-based DNA fingerprinting. *FEMS Microbiol. Lett.* **160**:69–73.

114. **Olsson, A., M. Eliasson, B. Guss, B. Nilsson, U. Hellman, M. Lindberg, and M. Uhlén.** 1987. Structure and evolution of the repetitive gene encoding streptococcal protein G. *Eur. J. Biochem.* **168**:319–324.

115. **O'Neil, K. T., R. H. Hoess, D. P. Raleigh, and W. F. DeGrado.** 1995. Thermodynamic genetics of the folding of the B1 immunoglobulin-binding domain from streptococcal protein G. *Proteins* **21**:11–21.

115a. **Ouskova, G., B. Spellerberg, and P. Prehm.** 2004. Hyaluronan release from *Streptococcus pyogenes*: export by an ABC transporter. *Glycobiology* **14**:931–938.

116. **Pancholi, V., and V. A. Fischetti.** 1992. A major surface protein on group A streptococci is a glyceraldehyde-3-phosphate-dehydrogenase with multiple binding activity. *J. Exp. Med.* **176**:415–426.

117. **Pancholi, V., and V. A. Fischetti.** 1998. α-enolase, a novel strong plasmin(ogen) binding protein on the surface of pathogenic streptococci. *J. Biol. Chem.* **273**:14503–14515.

118. **Parrado, J., F. Conejero-Lara, R. A. G. Smith, J. M. Marshall, C. P. Ponting, and C. M. Dobson.** 1996. The

domain organization of streptokinase: nuclear magnetic resonance, circular dichroism, and functional characterization of proteolytic fragments. *Protein Sci.* **5**:693–704.

119. **Perederina, A., V. Svetlov, M. N. Vassylyeva, T. H. Tahirov, S. Yokoyama, I. Artsimovitch, and D. G. Vassylyev.** 2004. Regulation through the secondary channel—structural framework for ppGpp-DksA synergism during transcription. *Cell* **118**:297–309.

120. **Podbielski, A., B. Melzer, and R. Lütticken.** 1991. Application of the polymerase chain reaction to study the M protein(-like) gene family in beta-hemolytic streptococci. *Med. Microbiol. Immunol.* **180**:213–227.

121. **Podbielski, A., I. Zagres, A. Flosdorff, and J. Weber-Heynemann.** 1996. Molecular characterization of a major serotype M49 group A streptococcal DNase gene (*sdaD*). *Infect. Immun.* **64**:5349–5356.

122. **Poyart, C., G. Quesne, S. Coulon, P. Berche, and P. Trieu-Cuot.** 1998. Identification of streptococci to species level by sequencing the gene encoding the manganese-dependent superoxide dismutase. *J. Clin. Microbiol.* **36**:41–47.

123. **Proft, T., P. D. Webb, V. Handley, and J. D. Fraser.** 2003. Two novel superantigens found in both group A and group C *Streptococcus. Infect. Immun.* **71**:1361–1369.

124. **Rasmussen, M., H.-P. Müller, and L. Björck.** 1999. Protein GRAB of *Streptococcus pyogenes* regulates proteolysis at the bacterial surface by binding α$_2$-macroglobulin. *J. Biol. Chem.* **274**:15336–15344.

125. **Reilly, T. J., B. A. Green, G. W. Zlotnick, and A. L. Smith.** 2001. Contribution of the DDDD motif of *Haemophilus influenzae* e(P4) to phosphomonoesterase activity and heme transport. *FEBS Lett.* **494**:19–23.

126. **Sachse, S., P. Seidel, D. Gerlach, E. Günther, J. Rödel, E. Straube, and K.-H. Schmidt.** 2002. Superantigen-like gene(s) in human pathogenic *Streptococcus dysgalactiae* subsp. *equisimilis*: genomic localisation of the gene encoding streptococcal pyrogenic exotoxin G (*speG*^{dys}). *FEMS Immunol. Med. Microbiol.* **34**:159–167.

127. **Schnitzler, N., A. Podbielski, G. Baumgarten, M. Mignon, and A. Kaufhold.** 1995. M or M-like protein gene polymorphisms in human group G streptococci. *J. Clin. Microbiol.* **33**:356–363.

128. **Schnitzler, N., G. Haase, A. Podbielski, A. Kaufhold, C. Lämmler, and R. Lütticken.** 1997. Human isolates of large colony-forming β-hemolytic group G streptococci form a distinct clade upon 16S rRNA gene analysis, p. 363–365. *In* T. Horaud, A. Bouvet, R. Leclercq, H. de Montclos, and M. Sicard (ed.), *Streptococci and the Host.* Plenum Press, New York, N.Y.

129. **Sharma, M., R. Khatib, and M. Fakih.** 2002. Clinical characteristics of necrotizing fasciitis by group G *Streptococcus*: case report and review of the literature. *Scand. J. Infect. Dis.* **34**:468–471.

130. **Schwarz-Linek, U., M. Höök, and J. R. Potts.** 2004. The molecular basis of fibronectin-mediated bacterial adherence to host cells. *Mol. Microbiol.* **52**:631–641.

131. **Sheinerman, F. B., and C. L. Brooks.** 1998. Calculations on folding of segment B1 of streptococcal protein G. *J. Molec. Biol.* **278**:439–456.

132. **Sheoran, A. S., B. T. Sponseller, M. A. Holmes, and J. F. Timoney.** 1997. Serum and mucosal antibody isotype responses to M-like protein (SeM) of *Streptococcus equi* in convalescent and vaccinated horses. *Vet. Immunol. Immunopathol.* **59**:239–251.

133. **Simpson, W. J., J. C. Robbins, and P. P. Cleary.** 1987. Evidence for group A-related M protein genes in human but not animal-associated group G streptococcal pathogens. *Microb. Pathog.* **3**:339–350.

134. **Simpson, W. J., J. M. Musser, and P. P. Cleary.** 1992. Evidence consistent with horizontal transfer of the gene (*emm12*) encoding serotype M12 protein between group A and group G pathogenic streptococci. *Infect. Immun.* **60**:1890–1893.

135. **Sing, A., K. Trebesius, and J. Heesemann.** 2001. Diagnosis of *Streptococcus dysgalactiae* subsp. *equisimilis* (group C streptococci) associated with deep soft tissue infections using fluorescent in situ hybridization. *Eur. J. Clin. Microbiol. Infect Dis.* **20**:146–149.

136. **Skjold, S. A., H. Malke, and L. W. Wannamaker.** 1979. Transduction of plasmid-mediated erythromycin resistance between group-A and -G streptococci, p. 274–275. *In* M. T. Parker (ed.), *Pathogenic Streptococci*, Reedbooks Ltd., Chertsey, Surrey, England.

137. **Song, X. M., J. Perez-Casal, and A. A. Potter.** 2004. The Mig protein of *Streptococcus dysgalactiae* inhibits bacterial internalization into bovine mammary gland epithelial cells. *FEMS Microbiol. Lett.* **231**:33–38.

138. **Speziale, P., D. Joh, L. Visai, S. Bozzini, K. House-Pompeo, M. Lindberg, and M. Höök.** 1996. A monoclonal antibody enhances ligand binding of fibronectin MSCRAMM (adhesin) from *Streptococcus dysgalactiae. J. Biol. Chem.* **271**:1371–1378.

139. **Sriprakash, K. S., and J. Hartas.** 1996. Lateral genetic transfers between group A and G streptococci for M-like genes are ongoing. *Microb. Pathog.* **20**:275–285.

140. **Steiner, K., and H. Malke.** 1995. Transcription termination of the streptokinase gene of *Streptococcus equisimilis* H46A: bidirectionality and efficiency in homologous and heterologous hosts. *Molec. Gen. Genet.* **246**:374–380.

141. **Steiner, K., and H. Malke.** 1997. Primary structure requirements for in vivo activity and bidirectional function of the transcription terminator shared by the oppositely oriented *skc/rel-orf1* genes of *Streptococcus equisimilis. Molec. Gen. Genet.* **255**:611–618.

142. **Steiner, K., and H. Malke.** 2000. Life in protein-rich environments: the *relA*-independent response of *Streptococcus pyogenes* to amino acid starvation. *Mol. Microbiol.* **38**:1004–1016.

143. **Steiner, K., and H. Malke.** 2001. *relA*-independent amino acid starvation response network of *Streptococcus pyogenes. J. Bacteriol.* **183**:7354–7364.

144. **Steiner, K., and H. Malke.** 2002. Dual control of streptokinase and streptolysin S production by the *covRS* and *fasCAX* two-component regulators in *Streptococcus dysgalactiae* subsp. *equisimilis. Infect. Immun.* **70**:3627–3636.

145. **Sultana, F., Y. Kawamura, X. G. Hou, S. E. Shu, and T. Ezaki.** 1998. Determination of 23S rRNA sequences from members of the genus *Streptococcus* and characterization of genetically distinct organisms previously identified as members of the *Streptococcus anginosus* group. *FEMS Microbiol. Lett.* **158**:223–230.

145a. **Sumby, P., K. D. Barbian, D. J. Gardner, A. R. Whitney, D. M. Welty, R. D. Long, J. R. Bailey, M. J. Parnell, N. P. Hoe, and G. G. Adams, F. R. DeLeo, and J. M. Musser.** 2005. Extracellular deoxyribonuclease made by group A *Streptococcus* assists pathogenesis by enhancing evasion of the innate immune response. *Proc. Natl. Acad. Sci. USA* **102**:1679–1684.

145b. **Sun, H., U. Ringdahl, J. W. Homeister, W. P. Fay, N. C. Engleberg, A. Y. Yang, L. S. Rozek, X. Wang, U. Sjöbring, and D. Ginsburg.** 2004. Plasminogen is a critical host pathogenicity factor for group A streptococcal infection. *Science* **305:**1283–1286.

146. **Talay, S. R., M. P. Grammel, and G. S. Chhatwal.** 1996. Structure of a group C streptococcal protein that binds to fibrinogen, albumin and immunoglobulin G via overlapping modules. *Biochem. J.* **315:**577–582.

147. **Timoney, J. F.** 2004. The pathogenic equine streptococci. *Vet. Res.* **35:**397–409.

148. **Timoney, J. F., J. Walker, M. Zhou, and J. Ding.** 1995. Cloning and sequence analysis of a protective M-like protein gene from *Streptococcus equi* subsp. *zooepidemicus*. *Infect. Immun.* **63:**1440–1445.

149. **Timoney, J. F., S. C. Artiushin, and J. S. Boschwitz.** 1997. Comparison of the sequences and functions of *Streptococcus equi* M-like proteins SeM and SzPSe. *Infect. Immun.* **65:**3600–3605.

150. **Tlapak-Simmons, V. L., B. A. Baggenstoss, T. Clyne, and P. H. Weigel.** 1999. Purification and lipid dependence of the recombinant hyaluronan synthases from *Streptococcus pyogenes* and *Streptococcus equisimilis*. *J. Biol. Chem.* **274:**4239–4245.

151. **Tlapak-Simmons, V. L., B. A. Baggenstoss, K. Kumari, C. Heldermon, and P. H. Weigel.** 1999. Kinetic characterization of the recombinant hyaluronan synthases from *Streptococcus pyogenes* and *Streptococcus equisimilis*. *J. Biol. Chem.* **274:**4246–4253.

152. **Tlapak-Simmons, V. L., C. A. Baron, and P. H. Weigel.** 2004. Characterization of the purified hyaluronan synthase from *Streptococcus equisimilis*. *Biochemistry* **43:**9234–9242.

153. **Ullberg, M., I. Karlsson, B. Wiman, and G. Kronvall.** 1992. Two types of receptors for human plasminogen on group G streptococci. *Acta Pathol. Microbiol. Immunol. Scand.* **100:**21–28.

154. **Valentin-Weigand, P., M. Y. Traore, H. Blobel, and G. S. Chhatwal.** 1990. Role of α2-macroglobulin in phagocytosis of group A and C streptococci. *FEMS Microbiol. Lett.* **70:**321–324.

155. **Vandamme, P., B. Pot, E. Falsen, K. Kersters, and L. A. Devriese.** 1996. Taxonomic study of Lancefield streptococcal groups C, G, and L (*Streptococcus dysgalactiae*) and proposal of *S. dysgalactiae* subsp. *equisimilis* subsp. nov. *Int. J. System. Bacteriol.* **46:**774–781.

156. **Vasi, J., L. Frykberg, L. E. Carlsson, M. Lindberg, and B. Guss.** 2000. M-like proteins of *Streptococcus dysgalactiae*. *Infect. Immun.* **68:**294–302.

157. **Vieira, V. V., L. M. Teixeira, V. Zahner, H. Momen, R. R. Facklam, A. G. Steigerwalt, D. J. Brenner, and A. C. Castro.** 1998. Genetic relationships among the different phenotypes of *Streptococcus dysgalactiae* strains. *Int. J. Syst. Bacteriol.* **48:**1231–1243.

158. **Visai, L., E. De Rossi, V. Valtulina, F. Casolini, S. Rindi, P. Guglierame, G. Pietrocola, V. Bellotti, G. Riccardi, and P. Speziale.** 2003. Identification and characterization of a new ligand-binding site in FnbB, a fibronectin-binding adhesin from *Streptococcus dysgalactiae*. *Biochim. Biophys. Acta* **1646:**173–183.

159. **Walter, F., M. Siegel, and H. Malke.** 1989. Nucleotide sequence of the streptokinase gene from a group G-*Streptococcus*. *Nucleic Acids Res.* **17:**1262.

160. **Wang, X., X. Lin, J. A. Loy, J. Tang, and X. C. Zhang.** 1998. Crystal structure of the catalytic domain of human plasmin complexed with streptokinase. *Science* **281:**1662–1665.

161. **Wannamaker, L. W., S. Almquist, and S. Skjold.** 1973. Intergroup phage reactions and transduction between group C and group A streptococci. *J. Exp. Med.* **137:**1338–1353.

162. **Ward, P. N., T. R. Field, W. G. F. Ditcham, E. Maguin, and J. A. Leigh.** 2001. Identification and disruption of two discrete loci encoding hyaluronic acid capsule bio-synthesis genes *hasA*, *hasB*, and *hasC* in *Streptococcus uberis*. *Infect. Immun.* **69:**392–399.

163. **Wolinowska, R., P. Ceglowski, J. Kok, and G. Venema.** 1991. Isolation, sequence and expression in *Escherichia coli*, *Bacillus subtilis* and *Lactococcus lactis* of the Dnase (streptodornase)-encoding gene from *Streptococcus equisimilis* H46A. *Gene* **106:**115–119.

164. **Woo, P. C. Y., A. M. Y. Fung, S. K. P. Lau, S. S. Y. Wong, and K.-Y. Yuen.** 2001. Group G beta-hemolytic streptococcal bacteremia characterized by 16S ribosomal RNA sequencing. *J. Clin. Microbiol.* **39:**3147–3155.

165. **Zaoutis, T., M. Attia, R. Gross, and J. Klein.** 2004. The role of group C and group G streptococci in acute pharyngitis in children. *Clin. Microbiol. Infect.* **10:**37–40.

166. **Zhai, P., N. Wakeham, J. A. Loy, and X. C. Zhang.** 2003. Functional roles of streptokinase C-terminal flexible peptide in active site formation and substrate recognition in plasminogen activation. *Biochemistry* **42:**114–120.

Pathogenicity Factors in Group C and G Streptococci

GURSHARAN S. CHHATWAL, DAVID J. McMILLAN, AND SUSANNE R. TALAY

17

Group C and G streptococci constitute a heterogeneous complex of streptococcal species that reside as apathogenic commensals in humans and animals or act as causative agents of severe infection and organ damage associated with high mortality rates. In this chapter we give a short overview of the various group C and group G streptococcal species, the diseases they cause, and the major pathogenicity factors that contribute to the virulence of these organisms.

TAXONOMY AND DISEASE SPECTRUM OF GROUP C AND GROUP G STREPTOCOCCI

Taxonomic classification of group C and G streptococci has proven to be a complex issue, because the traditional Lancefield system is not consistent with molecular and phenotypical classification systems. Group G streptococci and group C streptococci are currently divided into five species, *Streptococcus dysgalactiae*, *Streptococcus equi*, *Streptococcus canis*, the *Streptococcus anginosus* group, and *Streptococcus phocae* (30). *S. dysgalactiae* is further divided into the subspecies *S. dysgalactiae* subsp. *equisimilis* and *S. dysgalactiae* subsp. *dysgalactiae* (hereafter referred to as *S. equisimilis* and *S. dysgalactiae*). *Streptococcus equi* is divided into the subspecies *S. equi* subsp. *equi* and *S. equi* subsp. *zooepidemicus* (hereafter referred to as *S. equi* and *S. zooepidemicus*). On the basis of genetic evidence, *S. dysgalactiae*, *S. equi*, and *S. canis* are more closely related to each other than to the *S. anginosus* group, and constitute species with the "large-colony" phenotype. Data now suggest that *S. equi* may simply be a subclone of *S. zooepidemicus* (21). *S. phocae* is a new species expressing the group C antigen thus far isolated only from seals (98, 115). These species may also contain strains that express the Lancefield group A or group F antigens.

Infections caused by large-colony-forming group C streptococci in humans are mostly due to *S. equisimilis* and, to a lesser extent, *S. zooepidemicus*. Infections with *S. dysgalactiae*, which is uniquely non-beta-hemolytic, have not been recorded in humans (30). Infection with *S. equisimilis* can cause cellulitis, peritonitis, septic arthritis (29, 36), pneumonia, endocarditis, acute pharyngitis (5, 110, 121), bac-

teremia (15, 20), and toxic shock syndrome (54, 57). Generally, large-colony-forming group C streptococcal bacteremia in humans is either community-acquired or found to be associated with exposure to animals or their products. Although occurring less frequently than bacteremia caused by streptococcal serotypes A and B, it is characterized by high mortality rates (up to 25%) as well as other major sequelae that reflect the potentially high level of virulence of these organisms. Infection with *S. zooepidemicus* results in a similar spectrum of disease, but is also associated with poststreptococcal nephritis in humans (4, 80).

In domestic animals the subspecies of *S. equi* and *S. dysgalactiae* are of clinical relevance. *S. equisimilis* can cause pneumonia, arthritis, septicemia, and abscesses. *S. dysgalactiae* is mainly responsible for bovine mastitis. *S. equi*, a pathogen primarily restricted to horses and donkeys, causes strangles, a highly contagious disease characterized by purulent discharges from the respiratory tract and the development of abscesses (40). This organism is not considered to be part of the normal flora because of its close association with disease. In contrast, *S. zooepidemicus* is an opportunistic commensal that colonizes mucosal surfaces and causes rhinopharyngitis, pneumonia (120), endometritis, neonatal septicemia, and wound infections in horses (117) as well as disease in other domestic animals such as cattle, sheep, pigs, and chicken. As mentioned above, *S. zooepidemicus* is also responsible for a range of zoonotic infections in humans.

The beta-hemolytic group G streptococcus was first identified by Lancefield and Hare (58) in 1935. Group G streptococcal strains with the large-colony phenotype are the most frequently isolated from human infections and belong to *S. equisimilis*. Another large-colony-forming species, *S. canis*, normally an animal pathogen, has also been reported to cause disease in humans (8). Some *S. anginosus* strains have also been reported to express group G carbohydrate, although these isolates are not as frequent as *S. anginosus* strains expressing group F or group C antigen (30). *S. equisimilis* expressing group G carbohydrate was initially considered nonpathogenic in humans because the organism was found as part of the normal flora of the skin, pharynx, gastrointestinal tract, and vagina. The pathogen is commonly isolated from the throat, skin lesions, and the bloodstream, depending on the kind of infection. Group G streptococci can cause epidemic pharyngitis, bacteremia, meningitis, puerperal sepsis, peritonitis, cellulitis, arthritis, wound infection, septicemia, infective endocarditis, necrotizing fasciitis, toxic shock syndrome, and glomerulonephritis (42, 82, 119). Group G streptococci and group C streptococci have also been linked to the poststreptococcal sequelae rheumatic heart disease (39). Group G streptococci also reside in domestic animals such as cattle, sheep, cats, and dogs that either constitute healthy carriers of the bacterium or develop diseases such as bovine mastitis.

The small-colony phenotype of group C and G streptococci is expressed by the *S. anginosus* group, formerly known as *Streptococcus milleri*. The group contains three species: *S. anginosus*, *Streptococcus constellatus*, and *Streptococcus intermedius* (30). Strains belonging to the *S. anginosus* group express group antigens F, C, A, and G, or no antigen. These strains are also likely to be non-beta-hemolytic. Members of these species are recognized as common commensal organisms of the human oral cavity, gastrointestinal tract, and genitourinary tract. They are also associated with abscess formation in the mouth and other bodily sites (23, 89), as well as pharyngitis (118) and endocarditis (55, 56). It has recently been reported that *S. anginosus*, along with *Streptococcus mitis* and *Treponema denticola*, can be isolated from esophageal cancer tissue and by initiating inflammation in the cancerous tissue has a role in the development or progression of these cancers (76, 78).

In the following section, we discuss the major factors that enable group C and G streptococci to infect their hosts and cause disease. Such factors include adhesive structures that initiate the infection process, antiphagocytic factors that enable the bacterium to evade the host's immune system, factors that are potentially involved in spreading in tissues, and factors that specifically bind, degrade, or damage host components (Table 1).

MECHANISMS FOR ADHERENCE

Adhesion of microorganisms to host tissues represents a critical phase in the development of infection. It is therefore unsurprising that microorganisms have evolved dedicated mechanisms for attachment and adherence to extracellular matrix (ECM) components of the host (46, 84). Of these components the high-molecular-weight glycoprotein fibronectin appears to be the major attachment target of gram-positive cocci, including group G streptococci and group C streptococci. Fibronectin itself is responsible for substrate adhesion of eukaryotic cells via specific cell surface factors of the integrin family. It also specifically interacts with other matrix components, such as collagen, fibrin, and sulfated glycosaminoglycans, demonstrating that this molecule fulfills multifunctional roles within the extracellular network (43). Because it is present in the ECM of most tissues, as well as in plasma and other body fluids in its soluble form, fibronectin represents an exquisite target for bacteria to exploit the cell attachment properties of this molecule by linking the pathogen to specific target cells. Epithelial cells of the upper respiratory tract of humans are bathed in secretions containing fibronectin in its soluble form. Once bound to the bacterial surface, it will enable the pathogen to attach and subsequently colonize the primary site of infection.

Fibronectin-binding proteins (FBPs) were first identified in *Staphylococcus aureus* and *Streptococcus pyogenes* (93).

TABLE 1 Pathogenicity factors of group C and group G streptococci

Pathogenicity factors	Organism	Reference(s)
Adherence factors		
Fibronectin-binding proteins		
FnbA	*S. dysgalactiae*	61
FnbB	*S. dysgalactiae*	61
FnB	*S. equi*	62
GfbA	*S. dysgalactiae*	14
FNZ	*S. zooepidemicus*	64
SFS	*S. equi*	63
FNE	*S. equi*	65
Collagen-binding proteins		
57-kDa collagen-binding protein	*S. zooepidemicus*	113
CNE	*S. equi* spp.	59
Plasminogen-binding proteins		
M-like proteins	*S. dysgalactiae* spp.	6
Glyceraldehydes-3-phosphate dehydrogenase	*S. equisimilis*	35
Laminin-binding proteins		
PLBP	*S. anginosus*	1
Immune evasion factors		
M-like proteins	*S. dysgalactiae*	25, 91, 112
	S. equisimilis	
DemA	*S. dysgalactiae*	112
FgBP/SeM	*S. equi* spp.	70, 108
SzPSe/SzP	*S. equi* spp.	109
Others		
C5a peptidase	*S. dysgalactiae*	24, 103
Immunoglobulin-binding proteins		
Protein G	*S. dysgalactiae*	38
MIG	*S. dysgalactiae*	48
MAG	*S. dysgalactiae*	49
ZAG	*S. zooepidemicus*	51
Toxins		
Streptokinase	*S. dysgalactiae* spp.	6
Streptolysin O	*S. dysgalactiae* spp.	83
Streptolysin S	*S. dysgalactiae* spp.	33, 42
	S. equi spp.	
Superantigens		
SpeG^{dys}	*S. equisimilis*	90
SePE-H	*S. equisimilis*	3
SePE-I	*S. equisimilis*	3
Sepe-l	*S. equi*	86
Sepe-m	*S. equi*	86
speM	*S. equisimilis*	44
ssa	*S. equisimilis*	44
smeZ	*S. equisimilis*	44
	S. canis	
SDM	*S. dysgalactiae*	74
SpeA	*S. dysgalactiae*	52
SpeC	*S. dysgalactiae*	52

Binding by these bacteria to eukaryotic cells via fibronectin is also an important preliminary event prior to the invasion of these cells (26, 47, 75). The FBPs from streptococci and staphylococcus share a common architecture, with a putative signal sequence at the N terminus and a wall- and membrane-spanning region. The major fibronectin-binding domains are located within the C-terminal part of the proteins and are composed of three to five repetitive units that consist of 35 to 37 amino acid residues and bind to the 29-kDa N-terminal fragment of fibronectin. Further binding studies of SfbI from *S. pyogenes* and FnBPA in *S. aureus* also indicate the presence of secondary fibronectin-binding sites upstream of the repeat regions (93).

A number of FBPs have now been identified in group C and G streptococci (14, 60–64). The first to be identified were FnbA and FnbB from *S. dysgalactiae* (61). These two genes exhibit low sequence homology, suggesting that they did not evolve via gene duplication but evolved independently. Even so, both FnbA and FnbB have primary sequence architecture resembling SfbI and FnBPA and bind fibronectin through C-terminal repeat regions (46). FnbA has not been shown to be expressed under standard conditions but is structurally homologous to streptococcal fibronectin binding II from *S. pyogenes* (53). Like SfbII, FnbA also has a lipoproteinase activity. FnbB is a 122-kDa protein that shows significant homology to Protein F2 from *S. pyogenes*. In addition to the fibronectin-binding repeats, the protein also has an upstream fibronectin-binding domain similar to the Du unit of FnBPA of *S. aureus* (114).

Although the modular organization of FNZ, an *S. zooepidemicus* FBP, is similar to those from other species, homology at the amino acid level is weak (64). Fibronectin binding is mediated through a repeat region and a second upstream region containing the amino acid motif LAGESGET. This motif is also present in the secondary binding domain of SfbI (106), where it acts independently from the repeat region in binding to fibronectin (94). This domain is also present in GfbA, the homologue of SfbI in *S. dysgalactiae*. Although direct interaction between GfbA and fibronectin has not been demonstrated, the observation that transference of the *gfbA* gene to another strain resulted in transference of the fibronectin-binding phenotype strongly demonstrated the function of this protein (14).

fne is the homologous gene to *fnz* in *S. equi* (65). However, FNE is unique in that a single nucleotide deletion in *fne* gene has resulted in the truncation of the protein. PCR and sequence analysis of several strains confirmed the deletion was a conserved change when compared to *fnz*. As a consequence of the deletion, FNE is expressed without a cell wall-binding motif and is secreted into the surrounding environment. The truncated FNE also lacks the classic fibronectin-binding repeats found in the C-terminal region of FNZ, but is able to bind fibronectin through an unidentified domain in the amino half of the molecule. The amino half of FNZ was subsequently shown to have a similar binding capacity. SFS is an FBP unique to *S. equi*, without homology to other FBPs, but with homology to collagen (63). A study of the distribution of *sfs* found the gene to be present in all 50 strains of *S. equi* subsp. *equi* examined as well as 41 of 48 *S. equi* subsp. *zooepidemicus* isolates. To date expression of SFS by *S. equi* subsp. *equi* has not been demonstrated in in vitro experiments. SFS contains a signal peptide, but does not contain cell wall-binding motifs, or the traditional fibronectin-binding motifs. SFS, but not FNZ, inhibits binding between fibronectin and collagen. As *S. equi* isolates express FNE, binding of fibronectin by FNE on the bacterial surface may allow the subsequent binding of SFS through fibronectin. This may in turn inhibit binding of fibronectin to collagen, which could have several physiological consequences (63).

An alternative by which group C and G streptococci adhere to host cells is via binding to other ECM molecules, including vitronectin, collagen, plasminogen, and laminin (2, 87). Vitronectin is a multifunctional serum protein that affects the humoral immune system by binding to and inhibiting the complement membrane attack complex (85) and is also a major matrix-associated adhesive glycoprotein that regulates blood coagulation. The ability of group C and G streptococci to specifically interact with vitronectin (22) and mediate the adherence to both epithelial and endothelial cells (31, 111) was demonstrated some time ago. However, a specific vitronectin-binding protein has never been identified in streptococci. Collagen is a matrix protein that influences the structure of tissues and is also involved in cell attachment, proliferation, and cell differentiation. In *S. pyogenes*, binding to collagen has recently been demonstrated through the preliminary binding of fibronectin (28), and this mechanism may also operate in group C and G streptococci. A 57-kDa collagen type II-binding protein isolated from *S. zooepidemicus* that has the potential to mediate attachment to collagen-rich tissues has also been described (113). More recently, a new collagen-binding protein (CNE) from *S. equi* and *S. zooepidemicus* has been described (59). CNE is homologous to the collagen-binding CNA protein in *S. aureus*, with the greatest level of similarity occurring in the A repeats, the domain that mediates collagen binding. Amino acid comparisons between CNE and the *S. equi* collagen-binding protein previously described by Visai et al. (112) suggested that the two proteins were unique. Plasminogen binding to group G or C streptococci through M-like proteins (6) and surface-exposed glyceraldehydes-3-phosphate dehydrogenase has been documented (35). A laminin-binding protein has been identified in *S. anginosus* but requires further characterization (1). To date proteins other than FBPs involved in binding ECM components are not well characterized on the molecular level and await further investigation.

ANTIPHAGOCYTIC FACTORS

A feature of pathogenic streptococci that strongly contributes to their virulence is the ability to resist phagocytosis. The streptococcal M protein, first identified and characterized in *S. pyogenes*, is the major antiphagocytic factor of *S. pyogenes* (32). It is a fibrillar surface-exposed molecule that forms a coiled-coil secondary structure extending from the surface of the bacterium. Binding of complement factors or fibrinogen and inhibition of C3b deposition on the bacterial surface are mechanisms by which the M protein can inhibit opsonization of the organism by the alternative complement pathway, thus evading the host's nonspecific immune defense mechanism. As a consequence, bacteria carrying the M protein are capable of surviving and multiplying in blood.

As in *S. pyogenes*, human-pathogenic strains of group G and C streptococci carry M proteins. Protein MG1, the first group G streptococcal M-like protein characterized on the molecular level (25), exhibits typical structural and biological features of M proteins, such as coiled-coil structure and the ability to generate type-specific opsonizing antibodies. Protein MG1 shares highly homologous sequences with the C-terminal repeat region of class I M proteins, which are frequently associated with rheumatic fever. M proteins of group G streptococci are also responsible for conferring resistance to phagocytosis (19). Studies on recurrently infected mice, however, showed that the animals neither acquired M-specific protective immunity against the cellulitis-causing group G streptococcus (11) nor had opsonic antibodies, an effect of the disease that is in contrast with the results observed with *S. pyogenes*. Extended epidemiological characterization of *emm* genes in human-pathogenic group G streptococci revealed that they all exhibited features of class I *emm* genes, whereas animal-pathogenic strains failed to contain homologous sequences (91, 112). The high similarity of the *emm* genes from both species gave support to the hypothesis that in group G streptococci the region containing a class I *emm* gene has been acquired through horizontal transfer (96, 102) and that mosaiclike polymorphism in the flanking regions is the result of subsequent rearrangements. It is thus not surprising that homologous M protein sequences can also be demonstrated in group C streptococci of human origin.

Characterization of the *emm* genes of acute pharyngitis-associated isolates of *S. equi* has also revealed the presence of genes highly homologous to those found in group G streptococci (12). The 58-kDa FgBP, also known as SeM, appears to be the most predominant M-like protein in *S. equi* (70, 108, 109). Analysis of 200 *S. zooepidemicus* isolates failed to identify any containing a protein reactive with FgBP antisera, demonstrating absence of FgBP in this species. FgBP contains a signal sequence domain, a cell wall-anchoring domain, and two degenerate repeat domains. Although it does not display primary sequence homology with *S. pyogenes*, computer algorithms of M proteins predict FgBP to contain an α-helical coiled-coil secondary structure (70) and to share functional features such as the ability to bind fibrinogen, prevent phagocytosis (13), and act as a protective antigen against *S. equi* challenge. The fibrinogen-binding region is not found in the repeats, as is the case in *S. pyogenes* M proteins, but has been mapped to a region extending over the amino-terminal half of the protein (71). An FgBP knockout strain that is unable to bind equine fibrinogen or immunoglobulin G (IgG) has a reduced capacity to inhibit opsonization in horse blood, and is attenuated in virulence in mouse models, indicating the importance of this molecule to the pathogenesis of *S. equi* (72).

DemA is a *S. dysgalactiae* protein with homology to FgBP identified by screening of a phagemid expression library (112). The mature DemA protein is 54 kDa in size, contains a signal sequence and cell wall-binding domain, and is predicted to have a coiled-coil secondary structure. DemA shows greatest homology to the FgBP at the C-terminal end in a region that does not participate in fibrinogen binding. The amino acid motif VSKDLADKL is present with the repeat units of both DemA and FgBP, suggesting that the sequence may have an important biological function. DemA is able to bind IgG from various animal sources, reminiscent of type IIa Ig-binding proteins of *S. pyogenes* in a domain distinct from the fibrinogen-binding domain. Nucleotide sequencing of the *demA* locus identified an open reading frame upstream of *demA* homologous to *mga*, a positive regulator of M-protein expression in *S. pyogenes*.

A second M protein designated SzPSe in *S. zooepidemicus* and SzP in *S. equi* has also been described. The two proteins are highly homologous at the amino acid level, but are clearly distinct from FgBP. SzP/SzPSe bind fibrinogen and have an α-helical structure. Although SzPSe is highly conserved within isolates separated both temporally and geographically, SzP shows variation in both amino-terminal sequences and number of repeats (109). SzPSe can therefore be considered a variant of SzP existing only in *S. equi*. These data are consistent with other studies suggesting that *S. equi* is a clonal variant of *S. zooepidemicus* (21).

C5a peptidase represents an additional factor that contributes to the antiphagocytic properties of pathogenic streptococci by specifically destroying the host chemotaxin C5a, leading to a limited recruitment of polymorphonuclear leukocytes to the site of infection (24). The gene encoding C5a peptidase is found in group G streptococci (*scpG*) isolated from human but not animal infections, *S. pyogenes* (*scpA*) and *Streptococcus agalactiae* (*scpB*). Pathogenic studies have not been carried out on contribution to virulence of ScpB. However, an ScpA-negative *S. pyogenes* isolate was shown to be cleared more rapidly than wild-type *S. pyogenes* from an air-sac model of murine infection (45). The *scpA* gene in *S. pyogenes*, but not *scpG* in group G streptococci, is found in the same locus, and is regulated conjointly, with the *emm* gene (102). In *S. agalactiae scpB* is located on a putative composite transposon (34). The level of homology of *scp* between species and their different chromosomal locations has led to the hypothesis that these genes were acquired by lateral gene transfer (34).

IMMUNOGLOBULIN-BINDING PROTEINS

Streptococcal protein G, a surface molecule associated with the majority of group C and G streptococcal isolates of human origin, represents one of the best-characterized pathogenic factors from these species. Protein G interactions with immunoglobulins and other host proteins have been the subject of detailed reviews (68, 79, 97). Protein G is defined as a type III IgG Fc receptor; it interacts with a wide species range of immunoglobulins, as well as human serum albumin, kininogen, and α2-macroglobulin. Protein G exhibits a modular structure in which the binding sites for IgG are located within the C-terminal repeat region (38, 79, 97). The central A/B-repeat region constitutes the binding domain for serum albumin, and the N-terminal E region is responsible for interacting with the native form of α2-macroglobulin (77). In contrast to human pathogenic strains of group G streptococci that exclusively bind to the native (slow) form of α2-macroglobulin via protein G, animal-derived isolates of bovine and equine origin bind the proteinase-complexed (fast) form of the molecule. The B1 domain of protein G involved in Ig binding consists of an α-helix and four β-strand sheets. This domain has been used as a model structure in numerous biochemical studies examining protein folding, protein interaction, and synthetic protein design (17, 27, 37).

Two protein G-related proteins from a mastitis-causing *S. dysgalactiae* strain, MIG (48) and MAG (49, 50), as well as ZAG (51) from *S. zooepidemicus*, have been characterized on the molecular level. Like protein G, MAG and ZAG exhibit serum albumin and type III Fc receptor activity, whereas protein MIG lacks albumin-binding activity. However, MAG is able to bind to immunoglobulins from a greater number of animal sources than protein G. MAG, ZAG, and MIG also bind to the fast form of α2-macroglobulin. The α2-macroglobulin-binding region in MIG is not homologous to those of protein G, MAG, ZAG, and GRAB. In phagocytosis assays a MIG isogenic mutant strain of *S. dysgalactiae* was not as resistant to opsonization by bovine neutrophils as the parental strain (99). MIG has also recently been shown to bind to bovine IgA (100) and can inhibit bacterial internalization into host cells (101).

ENZYMES AND TOXINS

After colonization, adherence, and evasion of host immune responses, the dissemination of pathogenic streptococci in tissues is regarded to be an important step for the onset and development of an invasive disease. One of the factors involved in this process is streptokinase, a protein found in groups C, G, and A streptococci (6). Streptokinase catalyzes the conversion of plasminogen to its active form, plasmin. The current model invokes the formation of a streptokinase/plasminogen complex, resulting in the exposure of the plasminogen-active site, which then catalyzes the conversion of other plasminogen molecules into plasmin (66). Plasmin, a serine protease, is a key enzyme in the fibrinolytic system that is able to break down tissue barriers, thereby enabling the dissemination of streptococci. M proteins coordinately interact with the secreted streptokinase by binding either fibrinogen (116) or plasminogen (107). Examination of the streptokinase genes of human group C and G streptococci reveals allelic variation of the latter (107). Some of the identified alleles were previously found in *S. pyogenes*, whereas others were unique for the tested strains. Although believed to have implications for the virulence potential of the organism, the biological significance of the variation is still unclear. Streptokinases have been isolated from both human and animal group C and group G isolates, and have specificity for the plasminogens of their respective hosts (69). Thus, streptokinases from human group A, C, and G streptococci isolates have greater homology to each other than to animal isolates of group C and G streptococci. A study by Caballero et al. (18) found the amino acid homology between streptokinases from human and animal *S. equisimilis* isolates to be only 35%. Homology between the streptokinase from *S. equisimilis* from equine and porcine origins was only 21%. The ability of streptokinases to cleave plasminogen from specific species may therefore be a critical factor in determining the host range of individual streptococcal strains (92).

Streptolysin O (SLO) is the prototype of a family of thiol-activated cytolysins produced by the genus *Streptococcus* as well as by other gram-positive bacteria, including *Bacillus*, *Clostridium*, and *Listeria* species (10). The genes coding for SLO of group C and G streptococci (*S. dysgalactiae*) are almost identical to that of *S. pyogenes* (83). SLO homologues have not been described in *S. equi*. SLO has been shown to act synergistically with streptococcal cysteine protease in producing lung injuries in a rat model (95) as well as with an adhesin in modulating signaling responses of keratinocytes during in vitro infection (88). The cytotoxic activity of SLO is based on its ability to form large oligomeric hydrophilic transmembrane pores in eukaryotic cells. The passive flux of ions and macromolecules through these pores is believed to be one mechanism by which SLO catalyzes the lysis of eukaryotic cells (9). Recent data from studies in *S. pyogenes* indicate that SLO may have an active role in translocation of effector molecules into the cytoplasm of eukaryotic cells in a manner functionally equivalent to the type III secretion pathway found in gram-negative bacteria. Translocation of an effector molecule NAD-glycohydrolase (SPN) into keratinocytes was shown to require SLO pore formation, and resulted in increased cellular cytotoxicity (16, 67). The functional domains required for pore formation and SPN translocation are located on different parts of the SLO molecule. The observation that a similar translocation domain is absent on a cytolysin from another bacterial species provides further evidence that SLO translocation of molecules into host cells is an active process (73). Because SLO can also modulate cytokine and inflammatory responses in some cell types, it has been suggested that this toxin may have an important role in subverting host immune responses during streptococcal infection (10, 88, 104).

Streptolysin S (SLS) is another cytolysin secreted by groups A, C, and G streptococci, including the animal-pathogenic *S. equi* species (33). SLS belongs to a distinct group of hemolytic toxins that are characterized by their resistance to oxidation and sensitivity to trypan blue. SLS activity results in damage to membranes of various cell types as well as subcellular organelles (81). In contrast to the 57-kDa SLO, SLS is a small 57-amino-acid protein. The gene encoding SLS, *sagA*, is part of a locus containing nine open reading frames that have significant homology with genes from bacteriocin loci (42, 82). In *S. equisimilis*, SLS expression is under the control of both the *covRS* and *fasCAX* two-component regulatory systems, which in *S. pyogenes* have been shown to be involved in the regulation of multiple virulence factors (105). In a mouse infection model, *S. equisimilis* expressing SLS proliferates and induces necrotic lesions at the site of infection, whereas SLS-negative strains do not, suggesting that SLS is an important factor in the development of necrotizing fasciitis (42). Like many other streptococcal pathogenicity factors, the precise mechanism by which SLS contributes to disease remains unknown.

Group C and G streptococci are now recognized as causative agents of invasive disease, including streptococcal toxic shock syndrome and necrotizing fasciitis. Streptococcal pyrogenic exotoxins (Spes) expressed by *S. pyogenes* are believed to be important in the initiation and progression of these diseases (7). These proteins belong to the superantigen class of proteins, which bind to major histocompatibility complex class II on antigen-presenting cells and T-cell receptor molecules, leading to proliferation of T cells and the massive release of inflammatory cytokines. The first identification of a superantigen gene (speG^{dys}) in a human *S. equisimilis* isolate was provided by Sachse et al. (90). SePE-H and SePE-I were identified from *S. equisimilis* animal isolates in the same year (3). To date, superantigen alleles have not been identified in *S. zooepidemicus* (3, 41). Two novel superantigens, *sepe-l* and *sepe-m*, were discovered by genomic sequencing of *S. equi* prior to identification in *S. pyogenes*, but appear to be infrequent in both species (86). In a panel of 20 invasive human *S. equisimilis* isolates Igwe et al. (44) demonstrated the presence of genes encoding *speM*, *ssa*, and *smeZ*, but could not detect *speL*, *speH*, and *speJ* alleles. The same study reported the presence of a SmeZ allele in *S. canis*. Another superantigen, SDM, with amino acid similarity to SpeL and SpeM, is found in *S. dysgalactiae*, but not *S. pyogenes* (74). SpeA and SpeC, the major *S. pyogenes* superantigens, are also found in *S. dysgalactiae* (52). Consistent with the genetic profile of superantigens in *S. pyogenes*, most superantigens found in group C and group G streptococci are associated with prophages. Acquisition of these pathogenicity factors from streptococci of the same or different species can therefore occur through horizontal gene transfer.

CONCLUSIONS

Group G and C streptococci represent a heterogeneous but exquisitely host-adapted range of streptococcal isolates that span the variety of professional commensals, opportunistic pathogens, and exclusive pathogens for many mammalian species, including humans. Virulence factors that enable these bacteria to colonize the host, avoid immune responses, and cause disease have been characterized in some detail in *S. dysgalactiae* and *S. equi*. Molecular characterization of virulence factors from the *S. anginosus* group is not as advanced. Analysis of the bacterial factors that specifically interact with components of the host has allowed insight into the biochemical principles as well as some functional strategies of these streptococci; analysis has also revealed interesting evolutionary aspects, including convergent development, horizontal spread, and module shuffling. As group C and group G streptococci-associated diseases are coming under greater scrutiny it has become apparent that in many instances the virulence factors of these species have close homologues in *S. pyogenes*. It is therefore unsurprising that these pathogens cause a similar spectrum of diseases in humans. Future genomic comparisons between these species will identify new factors involved in disease and help to ascertain the real degree of difference between these species. Similar studies between human and animal pathogens will also shed light on factors defining the host range of individual streptococcal species.

We thank K. Mummenbrauer and H. Brink for their help during manuscript preparation. David McMillan is the recipient of an Alexander Von Humboldt Fellowship.

REFERENCES

1. **Allen, B. L., and M. Hook.** 2002. Isolation of a putative laminin binding protein from *Streptococcus anginosus*. *Microb. Pathog.* **33:**23–31.

2. **Allen, B. L., B. Katz, and M. Hook.** 2002. *Streptococcus anginosus* adheres to vascular endothelium basement membrane and purified extracellular matrix proteins. *Microb. Pathog.* **32:**191–204.

3. **Artiushin, S. C., J. F. Timoney, A. S. Sheoran, and S. K. Muthupalani.** 2002. Characterization and immunogenicity of pyrogenic mitogens SePE-H and SePE-I of *Streptococcus equi*. *Microb. Pathog.* **32:**71–85.

4. **Balter, S., A. Benin, S. W. Pinto, L. M. Teixeira, G. G. Alvim, E. Luna, D. Jackson, L. LaClaire, J. Elliott, R. Facklam, and A. Schuchat.** 2000. Epidemic nephritis in Nova Serrana, Brazil. *Lancet* **355:**1776–1780.

5. **Barnham, M., J. Kerby, R. S. Chandler, and M. R. Millar.** 1989. Group C streptococci in human infection: a study of 308 isolates with clinical correlations. *Epidemiol. Infect.* **102:**379–390.

6. **Ben Nasr, A., A. Wistedt, U. Ringdahl, and U. Sjobring.** 1994. Streptokinase activates plasminogen bound to human group C and G streptococci through M-like proteins. *Eur. J. Biochem.* **222:**267–276.

7. **Bernal, A., T. Proft, J. D. Fraser, and D. N. Posnett.** 1999. Superantigens in human disease. *J. Clin. Immunol.* **19:**149–157.

8. **Bert, F., and N. Lambert-Zechovsky.** 1997. Septicemia caused by *Streptococcus canis* in a human. *J. Clin. Microbiol.* **35:**777–779.

9. **Bhakdi, S., J. Tranum-Jensen, and A. Sziegoleit.** 1985. Mechanism of membrane damage by streptolysin-O. *Infect. Immun.* **47:**52–60.

10. **Billington, S. J., B. H. Jost, and J. G. Songer.** 2000. Thiol-activated cytolysins: structure, function and role in pathogenesis. *FEMS Microbiol. Lett.* **182:**197–205.

11. **Bisno, A. L., and J. M. Gaviria.** 1997. Murine model of recurrent group G streptococcal cellulitis: no evidence of protective immunity. *Infect. Immun.* **65:**4926–4930.

12. **Bisno, A. L., C. M. Collins, and J. C. Turner.** 1996. M proteins of group C streptococci isolated from patients with acute pharyngitis. *J. Clin. Microbiol.* **34:**2511–2515.

13. **Boschwitz, J. S., and J. F. Timoney.** 1994. Inhibition of C3 deposition on *Streptococcus equi* subsp. *equi* by M protein: a mechanism for survival in equine blood. *Infect. Immun.* **62:**3515–3520.

14. **Bradford Kline, J., S. Xu, A. L. Bisno, and C. M. Collins.** 1996. Identification of a fibronectin-binding protein (GfbA) in pathogenic group G streptococci. *Infect. Immun.* **64:**2122–2129.

15. **Bradley, S. F., J. J. Gordon, D. D. Baumgartner, W. A. Marasco, and C. A. Kauffman.** 1991. Group C streptococcal bacteremia: analysis of 88 cases. *Rev. Infect. Dis.* **13:**270–280.

16. **Bricker, A. L., C. Cywes, C. D. Ashbaugh, and M. R. Wessels.** 2002. NAD+-glycohydrolase acts as an intracellular toxin to enhance the extracellular survival of group A streptococci. *Mol. Microbiol.* **44:**257–269.

17. **Byeon, I. J., J. M. Louis, and A. M. Gronenborn.** 2003. A protein contortionist: core mutations of GB1 that induce dimerization and domain swapping. *J. Mol. Biol.* **333:**141–152.

18. **Caballero, A. R., R. Lottenberg, and K. H. Johnston.** 1999. Cloning, expression, sequence analysis, and characterization of streptokinases secreted by porcine and equine isolates of *Streptococcus equisimilis*. *Infect. Immun.* **67:**6478–6486.

19. **Campo, R. E., D. R. Schultz, and A. L. Bisno.** 1995. M proteins of group G streptococci: mechanisms of resistance to phagocytosis. *J. Infect. Dis.* **171:**601–606.

20. **Carmeli, Y., and K. L. Ruoff.** 1995. Report of cases of and taxonomic considerations for large-colony-forming Lancefield group C streptococcal bacteremia. *J. Clin. Microbiol.* **33:**2114–2117.

21. **Chanter, N., N. Collin, N. Holmes, M. Binns, and J. Mumford.** 1997. Characterization of the Lancefield group C streptococcus 16S-23S RNA gene intergenic spacer and its potential for identification and sub-specific typing. *Epidemiol. Infect.* **118:**125–135.

22. **Chhatwal, G. S., K. T. Preissner, G. Muller-Berghaus, and H. Blobel.** 1987. Specific binding of the human S protein (vitronectin) to streptococci, *Staphylococcus aureus*, and *Escherichia coli*. *Infect. Immun.* **55:**1878–1883.

23. **Claridge, J. E., 3rd, S. Attorri, D. M. Musher, J. Hebert, and S. Dunbar.** 2001. *Streptococcus intermedius*, *Streptococcus constellatus*, and *Streptococcus anginosus* ("*Streptococcus milleri* group") are of different clinical importance and are not equally associated with abscess. *Clin. Infect. Dis.* **32:**1511–1515.

24. **Cleary, P. P., J. Peterson, C. Chen, and C. Nelson.** 1991. Virulent human strains of group G streptococci express a C5a peptidase enzyme similar to that produced by group A streptococci. *Infect. Immun.* **59:**2305–2310.

25. **Collins, C. M., A. Kimura, and A. L. Bisno.** 1992. Group G streptococcal M protein exhibits structural features analogous to those of class I M protein of group A streptococci. *Infect. Immun.* **60:**3689–3696.

26. **Cue, D., P. E. Dombek, H. Lam, and P. P. Cleary.** 1998. *Streptococcus pyogenes* serotype M1 encodes multiple pathways for entry into human epithelial cells. *Infect. Immun.* **66:**4593–4601.

27. **Ding, K., J. M. Louis, and A. M. Gronenborn.** 2004. Insights into conformation and dynamics of protein GB1 during folding and unfolding by NMR. *J. Mol. Biol.* **335:**1299–1307.

28. **Dinkla, K., M. Rohde, W. T. Jansen, J. R. Carapetis, G. S. Chhatwal, and S. R. Talay.** 2003. *Streptococcus pyogenes* recruits collagen via surface-bound fibronectin: a novel colonization and immune evasion mechanism. *Mol. Microbiol.* **47:**861–869.

29. **Dubost, J. J., M. Soubrier, C. De Champs, J. M. Ristori, and B. Sauvezie.** 2004. Streptococcal septic arthritis in adults. A study of 55 cases with a literature review. *Joint Bone Spine* **71:**303–311.

30. **Facklam, R.** 2002. What happened to the streptococci: overview of taxonomic and nomenclature changes. *Clin. Microbiol. Rev.* **15:**613–630.

31. **Filippsen, L. F., P. Valentin-Weigand, H. Blobel, K. T. Preissner, and G. S. Chhatwal.** 1990. Role of complement S protein (vitronectin) in adherence of *Streptococcus dysgalactiae* to bovine epithelial cells. *Am. J. Vet. Res.* **51:**861–865.

32. **Fischetti, V. A.** 1991. Streptococcal M protein. *Sci. Am.* **264:**58–65.

33. **Flanagan, J., N. Collin, J. Timoney, T. Mitchell, J. A. Mumford, and N. Chanter.** 1998. Characterization of the haemolytic activity of *Streptococcus equi. Microb. Pathog.* **24:**211–221.

34. **Franken, C., G. Haase, C. Brandt, J. Weber-Heynemann, S. Martin, C. Lammler, A. Podbielski, R. Lutticken, and B. Spellberg.** 2001. Horizontal gene transfer and host specificity of beta-haemolytic streptococci: the role of a putative composite transposon containing *scpB* and *lmb. Mol. Microbiol.* **41:**925–935.

35. **Gase, K., A. Gase, H. Schirmer, and H. Malke.** 1996. Cloning, sequencing and functional overexpression of the *Streptococcus equisimilis* H46A *gapC* gene encoding a glyceraldehyde-3-phosphate dehydrogenase that also functions as a plasmin(ogen)-binding protein. Purification and biochemical characterization of the protein. *Eur. J. Biochem.* **239:**42–51.

36. **Gonzalez Teran, B., M. P. Roiz, T. Ruiz Jimeno, J. Rosas, and J. Calvo-Alen.** 2001. Acute bacterial arthritis caused by group C streptococci. *Semin. Arthritis Rheum.* **31:**43–51.

37. **Gronenborn, A. M., D. R. Filpula, N. Z. Essig, A. Achari, M. Whitlow, P. T. Wingfield, and G. M. Clore.** 1991. A novel, highly stable fold of the immunoglobulin binding domain of streptococcal protein G. *Science* **253:**657–661.

38. **Guss, B., M. Eliasson, A. Olsson, M. Uhlen, A. K. Frej, H. Jornvall, J. I. Flock, and M. Lindberg.** 1986. Structure of the IgG-binding regions of streptococcal protein G. *EMBO J.* **5:**1567–1575.

39. **Haidan, A., S. R. Talay, M. Rohde, K. S. Sriprakash, B. J. Currie, and G. S. Chhatwal.** 2000. Pharyngeal carriage of group C and group G streptococci and acute rheumatic fever in an aboriginal population. *Lancet* **356:**1167–1169.

40. **Harrington, D. J., I. C. Sutcliffe, and N. Chanter.** 2002. The molecular basis of *Streptococcus equi* infection and disease. *Microbes Infect.* **4:**501–510.

41. **Hashikawa, S., Y. Iinuma, M. Furushita, T. Ohkura, T. Nada, K. Torii, T. Hasegawa, and M. Ohta.** 2004. Characterization of group C and G streptococcal strains that cause streptococcal toxic shock syndrome. *J. Clin. Microbiol.* **42:**186–192.

42. **Humar, D., V. Datta, D. J. Bast, B. Beall, J. C. De Azavedo, and V. Nizet.** 2002. Streptolysin S and necrotising infections produced by group G streptococcus. *Lancet* **359:**124–129.

43. **Hynes, R. O., and K. M. Yamada.** 1982. Fibronectins: multifunctional modular glycoproteins. *J. Cell Biol.* **95:**369–377.

44. **Igwe, E. I., P. L. Shewmaker, R. R. Facklam, M. M. Farley, C. van Beneden, and B. Beall.** 2003. Identification of superantigen genes *speM*, *ssa*, and *smeZ* in invasive strains of beta-hemolytic group C and G streptococci recovered from humans. *FEMS Microbiol. Lett.* **229:**259–264.

45. **Ji, Y., L. McLandsborough, A. Kondagunta, and P. P. Cleary.** 1996. C5a peptidase alters clearance and trafficking of group A streptococci by infected mice. *Infect. Immun.* **64:**503–510.

46. **Joh, D., P. Speziale, S. Gurusiddappa, J. Manor, and M. Hook.** 1998. Multiple specificities of the staphylococcal and streptococcal fibronectin-binding microbial surface components recognizing adhesive matrix molecules. *Eur. J. Biochem.* **258:**897–905.

47. **Joh, D., E. R. Wann, B. Kreikemeyer, P. Speziale, and M. Hook.** 1999. Role of fibronectin-binding MSCRAMMs in bacterial adherence and entry into mammalian cells. *Matrix Biol.* **18:**211–223.

48. **Jonsson, H., and H. P. Muller.** 1994. The type-III Fc receptor from *Streptococcus dysgalactiae* is also an alpha 2-macroglobulin receptor. *Eur. J. Biochem.* **220:**819–826.

49. **Jonsson, H., L. Frykberg, L. Rantamaki, and B. Guss.** 1994. MAG, a novel plasma protein receptor from *Streptococcus dysgalactiae. Gene* **143:**85–89.

50. **Jonsson, H., C. Burtsoff-Asp, and B. Guss.** 1995. Streptococcal protein MAG—a protein with broad albumin binding specificity. *Biochim. Biophys. Acta* **1249:**65–71.

51. **Jonsson, H., H. Lindmark, and B. Guss.** 1995. A protein G-related cell surface protein in *Streptococcus zooepidemicus. Infect. Immun.* **63:**2968–2975.

52. **Kalia, A., and D. E. Bessen.** 2003. Presence of streptococcal pyrogenic exotoxin A and C genes in human isolates of group G streptococci. *FEMS Microbiol. Lett.* **219:**291–295.

53. **Katerov, V., P. E. Lindgren, A. A. Totolian, and C. Schalen.** 2000. Streptococcal opacity factor: a family of bifunctional proteins with lipoproteinase and fibronectin-binding activities. *Curr. Microbiol.* **40:**149–156.

54. **Keiser, P., and W. Campbell.** 1992. 'Toxic strep syndrome' associated with group C Streptococcus. *Arch. Intern. Med.* **152:**882, 884.

55. **Kitada, K., M. Inoue, and M. Kitano.** 1997. Experimental endocarditis induction and platelet aggregation by *Streptococcus anginosus*, *Streptococcus constellatus* and *Streptococcus intermedius. FEMS Immunol. Med. Microbiol.* **19:**25–32.

56. **Kitada, K., M. Inoue, and M. Kitano.** 1997. Infective endocarditis-inducing abilities of "*Streptococcus milleri*" group. *Adv. Exp. Med. Biol.* **418:**161–163.

57. **Korman, T. M., A. Boers, T. M. Gooding, N. Curtis, and K. Visvanathan.** 2004. Fatal case of toxic shock-like syndrome due to group C streptococcus associated with superantigen exotoxin. *J. Clin. Microbiol.* **42:**2866–2869.

58. **Lancefield, R. C., and R. Hare.** 1935. The serological differentiation of pathogenic and nonpathogenic streptococci from parturient women. *J. Exp. Med.* **61:**335–349.

59. **Lannergard, J., L. Frykberg, and B. Guss.** 2003. CNE, a collagen-binding protein of *Streptococcus equi. FEMS Microbiol. Lett.* **222:**69–74.

60. **Lindgren, P. E., P. Speziale, M. McGavin, H. J. Monstein, M. Hook, L. Visai, T. Kostiainen, S. Bozzini, and M. Lindberg.** 1992. Cloning and expression of two different genes from *Streptococcus dysgalactiae* encoding fibronectin receptors. *J. Biol. Chem.* **267:**1924–1931.

61. **Lindgren, P. E., M. J. McGavin, C. Signas, B. Guss, S. Gurusiddappa, M. Hook, and M. Lindberg.** 1993. Two different genes coding for fibronectin-binding proteins from *Streptococcus dysgalactiae*. The complete nucleotide sequences and characterization of the binding domains. *Eur. J. Biochem.* **214:**819–827.

62. **Lindgren, P. E., C. Signas, L. Rantamaki, and M. Lindberg.** 1994. A fibronectin-binding protein from *Streptococcus equisimilis*: characterization of the gene and identification of the binding domain. *Vet. Microbiol.* **41:**235–247.

63. **Lindmark, H., and B. Guss.** 1999. SFS, a novel fibronectin-binding protein from *Streptococcus equi* inhibits the

binding between fibronectin and collagen. *Infect. Immun.* **67:**2383–2388.

64. **Lindmark, H., K. Jacobsson, L. Frykberg, and B. Guss.** 1996. Fibronectin-binding protein of *Streptococcus equi* subsp. *zooepidemicus. Infect. Immun.* **64:**3993–3999.

65. **Lindmark, H., M. Nilsson, and B. Guss.** 2001. Comparison of the fibronectin-binding protein FNE from *Streptococcus equi* subspecies *equi* with FNZ from *S. equi* subspecies *zooepidemicus* reveals a major and conserved difference. *Infect. Immun.* **69:**3159–3163.

66. **Lottenberg, R., D. Minning-Wenz, and M. D. Boyle.** 1994. Capturing host plasmin(ogen): a common mechanism for invasive pathogens? *Trends Microbiol.* **2:**20–24.

67. **Madden, J. C., N. Ruiz, and M. Caparon.** 2001. Cytolysin-mediated translocation (CMT): a functional equivalent of type III secretion in gram-positive bacteria. *Cell* **104:**143–152.

68. **Malke, H.** 2000. Genetics and pathogenicity factors of group C and group G streptococci, p. 163–176. *In* V. A. Fischetti, R. P. Novick, J. J. Ferretti, D. A. Portnoy, and J. I. Rood (ed.), *Gram-Positive Pathogens.* ASM Press, Washington, D.C.

69. **McCoy, H. E., C. C. Broder, and R. Lottenberg.** 1991. Streptokinases produced by pathogenic group C streptococci demonstrate species-specific plasminogen activation. *J. Infect. Dis.* **164:**515–521.

70. **Meehan, M., P. Nowlan, and P. Owen.** 1998. Affinity purification and characterization of a fibrinogen-binding protein complex which protects mice against lethal challenge with *Streptococcus equi* subsp. *equi. Microbiology* **144:**993–1003.

71. **Meehan, M., D. A. Muldowney, N. J. Watkins, and P. Owen.** 2000. Localization and characterization of the ligand-binding domain of the fibrinogen-binding protein (FgBP) of *Streptococcus equi* subsp. *equi. Microbiology* **146:**1187–1194.

72. **Meehan, M., Y. Lynagh, C. Woods, and P. Owen.** 2001. The fibrinogen-binding protein (FgBP) of *Streptococcus equi* subsp. *equi* additionally binds IgG and contributes to virulence in a mouse model. *Microbiology* **147:**3311–3322.

73. **Meehl, M. A., and M. G. Caparon.** 2004. Specificity of streptolysin O in cytolysin-mediated translocation. *Mol. Microbiol.* **52:**1665–1676.

74. **Miyoshi-Akiyama, T., J. Zhao, H. Kato, K. Kikuchi, K. Totsuka, Y. Kataoka, M. Katsumi, and T. Uchiyama.** 2003. *Streptococcus dysgalactiae*-derived mitogen (SDM), a novel bacterial superantigen: characterization of its biological activity and predicted tertiary structure. *Mol. Microbiol.* **47:**1589–1599.

75. **Molinari, G., S. R. Talay, P. Valentin-Weigand, M. Rohde, and G. S. Chhatwal.** 1997. The fibronectin-binding protein of *Streptococcus pyogenes*, SfbI, is involved in the internalization of group A streptococci by epithelial cells. *Infect. Immun.* **65:**1357–1363.

76. **Morita, E., M. Narikiyo, A. Yano, E. Nishimura, H. Igaki, H. Sasaki, M. Terada, N. Hanada, and R. Kawabe.** 2003. Different frequencies of *Streptococcus anginosus* infection in oral cancer and esophageal cancer. *Cancer Sci.* **94:**492–496.

77. **Muller, H. P., and L. K. Rantamaki.** 1995. Binding of native alpha 2-macroglobulin to human group G streptococci. *Infect. Immun.* **63:**2833–2839.

78. **Narikiyo, M., C. Tanabe, Y. Yamada, H. Igaki, Y. Tachimori, H. Kato, M. Muto, R. Montesano, H. Sakamoto, Y. Nakajima, and H. Sasaki.** 2004. Frequent and preferential infection of *Treponema denticola*, *Streptococcus mitis*, and *Streptococcus anginosus* in esophageal cancers. *Cancer Sci.* **95:**569–574.

79. **Navarre, W. W., and O. Schneewind.** 1999. Surface proteins of gram-positive bacteria and mechanisms of their targeting to the cell wall envelope. *Microbiol. Mol. Biol. Rev.* **63:**174–229.

80. **Nicholson, M. L., L. Ferdinand, J. S. Sampson, A. Benin, S. Balter, S. W. Pinto, S. F. Dowell, R. R. Facklam, G. M. Carlone, and B. Beall.** 2000. Analysis of immunoreactivity to a *Streptococcus equi* subsp. *zooepidemicus* M-like protein to confirm an outbreak of poststreptococcal glomerulonephritis, and sequences of M-like proteins from isolates obtained from different host species. *J. Clin. Microbiol.* **38:**4126–4130.

81. **Nizet, V.** 2002. Streptococcal beta-hemolysins: genetics and role in disease pathogenesis. *Trends Microbiol.* **10:**575–580.

82. **Nizet, V., B. Beall, D. J. Bast, V. Datta, L. Kilburn, D. E. Low, and J. C. De Azavedo.** 2000. Genetic locus for streptolysin S production by group A streptococcus. *Infect. Immun.* **68:**4245–4254.

83. **Okumura, K., A. Hara, T. Tanaka, I. Nishiguchi, W. Minamide, H. Igarashi, and T. Yutsudo.** 1994. Cloning and sequencing the streptolysin O genes of group C and group G streptococci. *DNA Seq.* **4:**325–328.

84. **Patti, J. M., B. L. Allen, M. J. McGavin, and M. Hook.** 1994. MSCRAMM-mediated adherence of microorganisms to host tissues. *Annu. Rev. Microbiol.* **48:**585–617.

85. **Preissner, K. T.** 1991. Structure and biological role of vitronectin. *Annu. Rev. Cell Biol.* **7:**275–310.

86. **Proft, T., P. D. Webb, V. Handley, and J. D. Fraser.** 2003. Two novel superantigens found in both group A and group C Streptococcus. *Infect. Immun.* **71:**1361–1369.

87. **Rantamaki, L. K., and H. P. Muller.** 1995. Phenotypic characterization of *Streptococcus dysgalactiae* isolates from bovine mastitis by their binding to host derived proteins. *Vet. Microbiol.* **46:**415–426.

88. **Ruiz, N., B. Wang, A. Pentland, and M. Caparon.** 1998. Streptolysin O and adherence synergistically modulate proinflammatory responses of keratinocytes to group A streptococci. *Mol. Microbiol.* **27:**337–346.

89. **Ruoff, K. L.** 1988. *Streptococcus anginosus* ("*Streptococcus milleri*"): the unrecognized pathogen. *Clin. Microbiol. Rev.* **1:**102–108.

90. **Sachse, S., P. Seidel, D. Gerlach, E. Gunther, J. Rodel, E. Straube, and K. H. Schmidt.** 2002. Superantigen-like gene(s) in human pathogenic *Streptococcus dysgalactiae*, subsp *equisimilis*: genomic localisation of the gene encoding streptococcal pyrogenic exotoxin G (speG$^{(dys)}$). *FEMS Immunol. Med. Microbiol.* **34:**159–167.

91. **Schnitzler, N., A. Podbielski, G. Baumgarten, M. Mignon, and A. Kaufhold.** 1995. M or M-like protein gene polymorphisms in human group G streptococci. *J. Clin. Microbiol.* **33:**356–363.

92. **Schroeder, B., M. D. Boyle, B. R. Sheerin, A. C. Asbury, and R. Lottenberg.** 1999. Species specificity of plasminogen activation and acquisition of surface-associated proteolytic activity by group C streptococci grown in plasma. *Infect. Immun.* **67:**6487–6495.

93. **Schwarz-Linek, U., M. Hook, and J. R. Potts.** 2004. The molecular basis of fibronectin-mediated bacterial adherence to host cells. *Mol. Microbiol.* **52:**631–641.

94. **Sela, S., A. Aviv, A. Tovi, I. Burstein, M. G. Caparon, and E. Hanski.** 1993. Protein F: an adhesin of *Streptococcus pyogenes* binds fibronectin via two distinct domains. *Mol. Microbiol.* **10:**1049–1055.

95. **Shanley, T. P., D. Schrier, V. Kapur, M. Kehoe, J. M. Musser, and P. A. Ward.** 1996. Streptococcal cysteine protease augments lung injury induced by products of group A streptococci. *Infect. Immun.* **64:**870–877.

96. **Simpson, W. J., J. M. Musser, and P. P. Cleary.** 1992. Evidence consistent with horizontal transfer of the gene (*emm12*) encoding serotype M12 protein between group A and group G pathogenic streptococci. *Infect. Immun.* **60:**1890–1893.

97. **Sjobring, U., L. Bjorck, and W. Kastern.** 1991. Streptococcal protein G. Gene structure and protein binding properties. *J. Biol. Chem.* **266:**399–405.

98. **Skaar, I., P. Gaustad, T. Tonjum, B. Holm, and H. Stenwig.** 1994. *Streptococcus phocae* sp. nov., a new species isolated from clinical specimens from seals. *Int. J. Syst. Bacteriol.* **44:**646–650.

99. **Song, X. M., J. Perez-Casal, A. Bolton, and A. A. Potter.** 2001. Surface-expressed Mig protein protects *Streptococcus dysgalactiae* against phagocytosis by bovine neutrophils. *Infect. Immun.* **69:**6030–6037.

100. **Song, X. M., J. Perez-Casal, M. C. Fontaine, and A. A. Potter.** 2002. Bovine immunoglobulin A (IgA)-binding activities of the surface-expressed Mig protein of *Streptococcus dysgalactiae.* *Microbiology* **148:**2055–2064.

101. **Song, X. M., J. Perez-Casal, and A. A. Potter.** 2004. The Mig protein of *Streptococcus dysgalactiae* inhibits bacterial internalization into bovine mammary gland epithelial cells. *FEMS Microbiol. Lett.* **231:**33–38.

102. **Sriprakash, K. S., and J. Hartas.** 1996. Lateral genetic transfers between group A and G streptococci for M-like genes are ongoing. *Microb. Pathog.* **20:**275–285.

103. **Sriprakash, K. S., and J. Hartas.** 1997. Genetic mosaic upstream of *scpG* in human group G streptococci contains sequences from group A streptococcal virulence regulon. *Adv. Exp. Med. Biol* **418:**749–751.

104. **Stassen, M., C. Muller, C. Richter, C. Neudorfl, L. Hultner, S. Bhakdi, I. Walev, and E. Schmitt.** 2003. The streptococcal exotoxin streptolysin O activates mast cells to produce tumor necrosis factor alpha by p38 mitogen-activated protein kinase- and protein kinase C-dependent pathways. *Infect. Immun.* **71:**6171–6177.

105. **Steiner, K., and H. Malke.** 2002. Dual control of streptokinase and streptolysin S production by the *covRS* and *fasCAX* two-component regulators in *Streptococcus dysgalactiae* subsp. *equisimilis.* *Infect. Immun.* **70:**3627–3636.

106. **Talay, S. R., P. Valentin-Weigand, P. G. Jerlstrom, K. N. Timmis, and G. S. Chhatwal.** 1992. Fibronectin-binding protein of *Streptococcus pyogenes*: sequence of the binding domain involved in adherence of streptococci to epithelial cells. *Infect. Immun.* **60:**3837–3844.

107. **Tewodros, W., I. Karlsson, and G. Kronvall.** 1996. Allelic variation of the *streptokinase* gene in beta-hemolytic streptococci group C and G isolates of human origin. *FEMS Immunol. Med. Microbiol.* **13:**29–34.

108. **Timoney, J. F., J. Walker, M. Zhou, and J. Ding.** 1995. Cloning and sequence analysis of a protective M-like protein gene from *Streptococcus equi* subsp. *zooepidemicus.* *Infect. Immun.* **63:**1440–1445.

109. **Timoney, J. F., S. C. Artiushin, and J. S. Boschwitz.** 1997. Comparison of the sequences and functions of *Streptococcus equi* M-like proteins SeM and SzPSe. *Infect. Immun.* **65:**3600–3605.

110. **Turner, J. C., F. G. Hayden, M. C. Lobo, C. E. Ramirez, and D. Murren.** 1997. Epidemiologic evidence for Lancefield group C beta-hemolytic streptococci as a cause of exudative pharyngitis in college students. *J. Clin. Microbiol.* **35:**1–4.

111. **Valentin-Weigand, P., J. Grulich-Henn, G. S. Chhatwal, G. Muller-Berghaus, H. Blobel, and K. T. Preissner.** 1988. Mediation of adherence of streptococci to human endothelial cells by complement S protein (vitronectin). *Infect. Immun.* **56:**2851–2855.

112. **Vasi, J., L. Frykberg, L. E. Carlsson, M. Lindberg, and B. Guss.** 2000. M-like proteins of *Streptococcus dysgalactiae.* *Infect. Immun.* **68:**294–302.

113. **Visai, L., S. Bozzini, G. Raucci, A. Toniolo, and P. Speziale.** 1995. Isolation and characterization of a novel collagen-binding protein from *Streptococcus pyogenes* strain 6414. *J. Biol. Chem.* **270:**347–353.

114. **Visai, L., E. De Rossi, V. Valtulina, F. Casolini, S. Rindi, P. Guglierame, G. Pietrocola, V. Bellotti, G. Riccardi, and P. Speziale.** 2003. Identification and characterization of a new ligand-binding site in FnbB, a fibronectin-binding adhesin from *Streptococcus dysgalactiae.* *Biochim. Biophys. Acta* **1646:**173–183.

115. **Vossen, A., A. Abdulmawjood, C. Lammler, R. Weiss, and U. Siebert.** 2004. Identification and molecular characterization of beta-hemolytic streptococci isolated from harbor seals (*Phoca vitulina*) and grey seals (*Halichoerus grypus*) of the German North and Baltic Seas. *J. Clin. Microbiol.* **42:**469–473.

116. **Wang, H., R. Lottenberg, and M. D. Boyle.** 1995. A role for fibrinogen in the streptokinase-dependent acquisition of plasmin(ogen) by group A streptococci. *J. Infect. Dis.* **171:**85–92.

117. **Welsh, R. D.** 1984. The significance of *Streptococcus zooepidemicus* in the horse. *Equine Practice* **6:**6–16.

118. **Whiley, R. A., L. M. Hall, J. M. Hardie, and D. Beighton.** 1999. A study of small-colony, beta-haemolytic, Lancefield group C streptococci within the *anginosus* group: description of *Streptococcus constellatus* subsp. *pharyngis* subsp. nov., associated with the human throat and pharyngitis. *Int. J. Syst. Bacteriol.* **49:**1443–1449.

119. **Woo, P. C., A. M. Fung, S. K. Lau, S. S. Wong, and K. Y. Yuen.** 2001. Group G beta-hemolytic streptococcal bacteremia characterized by 16S ribosomal RNA gene sequencing. *J. Clin. Microbiol.* **39:**3147–3155.

120. **Yoshikawa, H., T. Yasu, H. Ueki, T. Oyamada, H. Oishi, T. Anzai, M. Oikawa, and T. Yoshikawa.** 2003. Pneumonia in horses induced by intrapulmonary inoculation of *Streptococcus equi* subsp. *zooepidemicus.* *J. Vet. Med. Sci.* **65:**787–792.

121. **Zaoutis, T., M. Attia, R. Gross, and J. Klein.** 2004. The role of group C and group G streptococci in acute pharyngitis in children. *Clin. Microbiol. Infect.* **10:**37–40.

Group C and Group G Streptococcal Infections: Epidemiologic and Clinical Aspects

GIO J. BARACCO AND ALAN L. BISNO

18

Streptococci possessing Lancefield group C and G cell wall carbohydrates are heterogeneous in regard to biochemical reactions, hemolytic characteristics, predilection for host species, and clinical illnesses produced in humans and animals. These organisms are found as commensals in the throat, skin, and occasionally the female genitourinary tract, and their epidemiologic patterns and clinical manifestations reflect this distribution.

TAXONOMY

The designations group C and group G streptococci refer to the classification established by Lancefield in 1933. Using DNA-DNA reassociation, 16S rDNA gene sequencing, and other modern molecular techniques, the taxonomy of streptococci has moved to a classification based on genera and species (25, 58). Numerous species of streptococci may express Lancefield group C and G carbohydrates. Conversely, a single species may display group C, group G, or another Lancefield group determinant (Table 1).

In this chapter we will, to the extent possible, focus on the species-based nomenclature of streptococci. However, many of the studies reviewed, particularly those in the older literature, employed the Lancefield classification, making it impossible to know the causative species.

The great majority of human group C or G infections are due to strains of *Streptococcus dysgalactiae* or to members of the anginosus group. The former form large colonies (≥5 mm in diameter) that resemble those of *Streptococcus pyogenes* when cultivated on sheep blood agar plates, while the latter form small or "minute" colony types (<5 mm). In addition to colony size, there are differences in biochemical reactions between large colony strains and the minute colony forms carrying the C and G polysaccharide antigens.

The *Streptococcus anginosus* group (S. *anginosus*, *Streptococcus constellatus*, *Streptococcus intermedius*, and various subspecies), formerly known as either S. *anginosus* or *Streptococcus milleri*, is a source of much taxonomic confusion. This group (particularly the non-beta-hemolytic strains) is commonly classified with the "viridans" streptococcal group. They may possess group A, C, G, or F antigen, or no group antigen at all. They are more commonly alpha-hemolytic, but all three species have beta-hemolytic strains (Table 1).

Bert et al. (7) analyzed 54 human isolates and 33 animal isolates of S. *dysgalactiae* by pulse-field gel electrophoresis, and identified two major clusters that correlated very strongly with the source host. DNA-DNA reassociation studies and multilocus enzyme electrophoresis, as well as phenotypic studies, confirmed the necessity to divide S. *dysgalactiae* into two subspecies: S. *dysgalactiae* subsp. *dysgalactiae* and S. *dysgalactiae* subsp. *equisimilis* (74). S. *dysgalactiae* subsp. *equisimilis* is a pyogenic, beta-hemolytic streptococcal species associated with human disease. It may express either group C or G antigen. It has also been found to display group A or group L antigen. S. *dysgalactiae* subsp. *dysgalactiae*, *Streptococcus equi* subsp. *equi*, S. *equi* subsp. *zooepidemicus*, and *Streptococcus phocae* also express the group C polysaccharide, but they are primarily veterinary pathogens. Likewise, *Streptococcus canis*, primarily a dog pathogen, contains the group G antigen.

EPIDEMIOLOGY

The S. *anginosus* group members are normal commensals in the oral cavity. All three species, but especially S. *intermedius*, have been found in gingival crevices, dental plaque, dental root canals, and the naso- and oropharynx. S. *anginosus* isolates are also commonly isolated from urogenital and gastrointestinal sources (25, 58). S. *dysgalactiae* subsp. *equisimilis* is a common commensal of humans (61). For this reason, transmission, when it occurs, is more likely to be from person to person, and most cases are sporadic in nature, rather than associated with common source outbreaks. When outbreaks do occur, they are generally associated with close personal contact or perhaps with environmental contamination. By contrast, S. *equi* subsp. *zooepidemicus* is generally associated with exposure to animals or to common source outbreaks, especially consumption of contaminated dairy products.

Gram-Positive Pathogens, 2nd edition, edited by Vincent A. Fischetti et al.
© 2006 ASM Press, Washington, D.C.

TABLE 1 Human and animal streptococci expressing Lancefield C and G cell-wall antigens[a]

Species	Lancefield antigen(s)	Hemolytic reaction(s)	Comments
S. dysgalactiae subsp. equisimilis	C, G[b]	β	Pyogenic; formerly named S. equisimilis. Can be differentiated from β-hemolytic anginosus group strains with the C or G antigen by the formation of relatively large colonies and other phenotype traits. Agent of respiratory and deep tissue infections, cellulitis, and septicemia.
Anginosus species group	A, C, F, G, or no detectable antigen	α, β, γ	"Viridans" streptococcal group composed of three species, S. anginosus, S. constellatus, and S. intermedius. Formerly known as "S. milleri." Beta-hemolytic strains form small colonies compared to those of pyogenic β-hemolytic group A, C, and G streptococci and also differ in other phenotypic traits. Agents of purulent infections.
S. dysgalactiae subsp. dysgalactiae	C, L	α, β, γ	Pathogen of domesticated animals. Participation in human infections not well documented.
S. equi subsp. equi	C	β	Agent of equine strangles. Participation in human infections not well documented.
S. equi subsp. zooepidemicus	C	β	Agent of bovine mastitis and infection in other domesticated animals. Implicated in outbreaks of nephritis in humans.
S. canis	G	β	Dogs and other animals are the usual hosts. Documented as an infrequent human pathogen.

[a]Reprinted from reference 58 with permission of Elsevier.
[b]Isolates with the group A antigen have also been described.

ANIMAL INFECTIONS

S. dysgalactiae subsp. dysgalactiae is a major cause of mastitis in cows, dromedary camels, and various infections in lambs (24, 28, 31). S. equi subsp. equi is occasionally found in the upper respiratory tract of normal horses and is the causative agent of equine strangles. This acute, contagious, and deadly respiratory disease has led to explosive epidemics in horse stables and has serious potential economic consequences for horse fanciers (32). S equi subsp. zooepidemicus is a cause of infection in a variety of animal species, including horses, cows, dogs, rabbits, llamas, alpacas, and swine (1, 17, 24, 28, 34, 64).

HUMAN INFECTIONS

The S. anginosus group is usually classified and described with the viridans group of streptococci. These species are characterized for their disposition to form abscesses in various organs, including lung, pleural space, brain, oral and abdominal cavities, skin and soft tissues, and genitourinary tract with or without associated bacteremia (47, 63, 77). Although less commonly than other viridans streptococci, S. anginosus may also cause endocarditis (43, 58, 80). Many invasive anginosus strains, however, belong to serogroups other than C or G or are nongroupable (59, 77).

This review will focus on the more common infections caused by S. dysgalactiae subsp. equisimilis, as well as on the few reported human cases caused by S. equi subsp. zooepidemicus, S. equi subsp. equi, S. dysgalactiae subsp. dysgalactiae, and S. canis. Strains of these streptococci have been associated with infections of many body sites (2, 14, 18, 23, 28, 61, 72, 76).

Pharyngitis and Acute Glomerulonephritis

The role of S. dysgalactiae subsp. equisimilis in sporadic cases of pharyngitis remains somewhat controversial. Although occasionally isolated from patients with pharyngitis, strep-

tococci of this species are also often cultured from the throats of healthy individuals. There are a few large studies examining the relationship of S. dysgalactiae subsp. equisimilis to acute, sporadic pharyngitis. Turner et al. (71) studied students reporting to a college health service with acute pharyngitis and compared them with controls without infectious problems. Group C streptococci were cultured at a higher rate from those with pharyngitis than from the control group. Patients with positive cultures for group C streptococci were more likely to have features suggestive of a bacterial infection, such as exudative tonsillitis and anterior cervical lymphadenopathy, than were those with negative cultures. Furthermore, these group C strains resisted phagocytosis in human blood and contained genomic DNA encoding an M protein similar in structure to that of group A streptococci, providing further evidence of possible human virulence (8). In a later study, these authors described 265 students with exudative pharyngitis and compared them with 75 patients with rhinovirus infection and 162 students with noninfectious problems. S. dysgalactiae subsp. equisimilis was isolated significantly more frequently from patients with exudative pharyngitis than from either control group (70). Twenty-two cases of pharyngitis from which group C streptococci were isolated occurred during the fall of 1974 in a school for boys with learning disabilities (6). Although it is likely in this epidemiologic setting that the infecting strains were S. dysgalactiae subsp. equisimilis, they were unfortunately not speciated.

Streptococci expressing group G antigen have clearly been linked to outbreaks of pharyngitis. Many of these outbreaks have been related to a common source, usually a food product. In one such outbreak during a single week in 1968, 176 students at a college were evaluated for pharyngitis (35). The attack rate in the student body was 31%. Signs and symptoms were similar to those characteristic of group A streptococcal pharyngitis, suggesting an etiologic role for the organism. Epidemiologic investigation linked the outbreak to contaminated egg salad. In another common source outbreak, 72 persons who attended a convention developed pharyngitis, with group G streptococci isolated from most who had cultures performed (67). All of the patients had consumed chicken salad prepared by a single cook whose throat culture was positive for the organism. An epidemic of group G streptococcal pharyngitis involving 68 students occurred over a 1-week period at a North Carolina college (45). Because no common food source could be identified, the author concluded that the mode of spread was most likely person to person. The very sharp epidemic curve and brief duration of the outbreak suggest, however, that contamination of a common food vehicle is more likely. In support of this conclusion are the facts that all students interviewed had eaten in the campus cafeteria in the week preceding illness and that one student with a positive culture was a food handler. A communitywide outbreak of group G streptococcal pharyngitis, unrelated to any common source, was documented among private pediatric patients in the winter and spring of 1986 to 1987 in Connecticut (29).

Epidemics and clusters of pharyngitis cases due to *S. equi* subsp. *zooepidemicus* are related to a common source, usually consumption of unpasteurized dairy products. A remarkable feature of such outbreaks is their association with poststreptococcal acute glomerulonephritis. From December 1997 to July 1998, 253 cases of acute glomerulonephritis due to *S. dysgalactiae* subsp. *zooepidemicus* occurred in Nova Serrana, Brazil, among persons who had consumed locally produced unpasteurized cheese. Ten patients required dialysis and three died (4). A follow-up two years later of 134 of these patients showed that five of them required chronic dialysis. Of the 69 patients from that cohort who could be found and reevaluated, 42% had hypertension and there was a high proportion of patients (up to 30%) with persistent renal function abnormalities (52). Duca et al. (21) described 85 patients with pharyngitis due to *S. equi* subsp. *zooepidemicus* following the ingestion of improperly pasteurized milk. Eighty-seven percent of the patients were adults. Approximately one-third of the patients developed acute glomerulonephritis, generally in the second or third week of illness. In a smaller outbreak, five members of a family developed an upper respiratory infection related to *S. equi* subsp. *zooepidemicus* after consuming unpasteurized milk (5). Three of the five family members subsequently developed poststreptococcal glomerulonephritis, which was confirmed in one case by renal biopsy.

Infections of Skin and Soft Tissue

Strains of *S. dysgalactiae* subsp. *equisimilis* bearing either group C or G antigens not infrequently cause skin and soft tissue infections, and the skin is often the portal of entry for serious invasive disease and bacteremia. These infections can manifest as pyoderma, cellulitis, erysipelas, surgical wound infections, abscesses, and pyomyositis (11, 20, 50, 78). Infection due to these organisms may complicate ulcers associated with diabetes mellitus, immobility (61), or venous and lymphatic compromise of any cause. Recurrent cellulitis may occur, for example, in the saphenous venectomy limb of patients who have undergone coronary artery bypass grafting (3) or in the extremities of individuals who have had axillary, pelvic, or femoral node dissection for cancer. Soft tissue abscesses and even necrotizing fasciitis can occur as well, usually following puncture wounds or other trauma (14, 61). The majority of patients with serious group G streptococcal skin and soft tissue infections have underlying diseases. These are most commonly malignancy,

cardiovascular disease, alcoholism, and diabetes mellitus (51, 72). Injectable-drug users seem to be at increased risk for cellulitis and skin abscesses due to group G streptococci, and the skin is the usual source of bacteremia in such patients (20). Burn patients are also at risk for skin infections with this organism; such individuals accounted for 8% of cutaneous group G streptococci infections in one series (11). Group G streptococci have been linked to skin graft infection with subsequent loss of the graft (55).

In contrast to *S. dysgalactiae* subsp. *equisimilis*, the rare cases of skin and soft tissue infection due to *S. equi* subsp. *zooepidemicus* and *S. equi* subsp. *equi* usually involve exposure to animals. One case of cellulitis with bacteremia due to *S. equi* subsp. *zooepidemicus* was reported in a renal transplant patient who was exposed to horses at a show (46). A case of severe facial cellulitis due to *S. equi* subsp. *equi* was reported in another man who also had equine exposure (12).

Joint and Bone Infections

Gonzalez Teran et al. (30) reviewed 24 patients with group C streptococcal arthritis, two from their experience and the other 22 from the literature. Twelve (50%) cases were caused by *S. dysgalactiae* subsp. *equisimilis*, three by *S. equi* subsp. *zooepidemicus*, and the other nine were not speciated. Nine of their patients had polyarticular involvement. One-third (eight patients) had a preexisting arthropathy (two had rheumatoid arthritis, one had gout and seronegative arthropathy, three had osteoarthritis, one had a neuropathic arthropathy, and one had osteochondromatosis). Four of the 24 patients were immunosuppressed, two had human immunodeficiency virus infection, and two were on chemotherapy. Only two patients, one infected with *S. equi* subsp. *zooepidemicus*, and the other with *S. dysgalactiae* subsp. *equisimilis*, had a history of animal exposure (horses in both cases).

Numerous cases of infectious arthritis due to group G streptococci have been reported. Serious medical illnesses and previous joint disease were common features in these patients. In five cases of group G streptococcal infectious arthritis from the University of California-Los Angeles hospital system, all patients had prior joint disease and two had infected prostheses (48). In a series of seven patients, only one patient had no underlying systemic or rheumatologic illness (62). The remaining six patients had prior trauma, surgery, or inflammation at the affected joint. Four patients also had underlying medical conditions, including diabetes mellitus, alcoholism, and cardiovascular disease. In a review of 50 previously reported cases of group G streptococcal arthritis, more than one-third of patients had chronic joint disease, while just under half of the patients had one of four underlying conditions: malignancy, alcoholism, diabetes mellitus, or injectable-drug use (13). Osteomyelitis also occurs but is reported less frequently than infectious arthritis. In these cases, there is also often a significant underlying disease (11, 72).

Maternal and Neonatal Infections

Although *S. dysgalactiae* subsp. *equisimilis* can be found as part of the normal female genitourinary flora, its presence in clinical specimens often indicates infection. There have been at least two outbreaks of puerperal fever caused by *S. dysgalactiae* subsp. *equisimilis*. Thirty-three confirmed cases in England were caused by a single strain of *S. dysgalactiae* subsp. *equisimilis*. Clinical features included fever and signs of perineal infection. Sources of infection were postulated by the authors to be environmental because the organism was cultured from toilet seats and bath plug holes. The organism was, however, also cultured from the throats of many of the nursing staff (68). In another outbreak, also in England, though 4 years later, seven women developed puerperal fever due to *S. dysgalactiae* subsp. *equisimilis*. Interestingly, the isolates were identical to the strain responsible for the first outbreak. Though the microorganism was not isolated from the environment, the authors speculated that transmission may have occurred through use of a common toilet seat (27). These epidemiologic and microbiologic data suggesting transmission by fomites must be interpreted with caution. The role of environmental contamination versus nosocomial person-to-person transmission in such outbreaks remains to be determined.

Neonatal group C streptococcal infection is rare. In one case, meningitis due to *S. dysgalactiae* subsp. *equisimilis* developed in an infant whose mother was being treated for chorioamnionitis at the time of delivery (26). In another case, a preterm infant developed meningitis due to *S. dysgalactiae* subsp. *dysgalactiae*; the source of infection was not determined, as the mother was not ill and the organism could not be cultured from her (54).

Although clinical infection is rare, colonization of neonates with group G streptococci seems to be a common finding. In one study, cultures were taken from the nose and umbilicus of more than 3,000 neonates over a 1-year period at the New York Hospital (22). The monthly incidence of positive cultures for group G streptococci ranged from 41 to 76%. Seven cases of neonatal sepsis due to this organism were diagnosed over the same time period. Five of the seven cases occurred in the setting of complications of pregnancy or childbirth. In a larger review encompassing this series, premature or prolonged rupture of the amniotic membranes was the most common risk factor associated with group G streptococcal infection (16).

Bacteremia, Endocarditis, and Other Serious Invasive Diseases

S. dysgalactiae subsp. *equisimilis*, *S. equi* subsp. *zooepidemicus*, and other streptococci with either group C or G antigen have been reported to cause bacteremia, both primary and secondary to a variety of sources, most commonly cellulitis (15, 18). Many (up to 70% in some series) patients affected by these organisms have serious underlying disease, especially malignancy, cardiovascular disease, diabetes mellitus, immunosuppression, and alcohol or injectable drug use (2, 15, 44, 72, 76).

Bacteremia due to *S. dysgalactiae* subsp. *equisimilis* may be primary in approximately 20% of cases, or secondary to a focal site of infection, most often from the skin or soft tissues. Auckenthaler et al. (2) reported 38 patients who were bacteremic with group G streptococci at the Mayo Clinic-affiliated hospitals, representing 0.25% of all patients with positive blood cultures over a 10-year period. Seventy percent of the patients acquired the infection in the community, and the skin was the portal of entry in approximately three-quarters of the patients. Most of the hospital-acquired bacteremias involved a postoperative wound or a transcutaneous procedure. The patients tended to be older, with most being in the sixth to eighth decades. Many patients had venous insufficiency, lymphedema, or another cause of chronic lower extremity edema. Carmeli et al. (15) reported 10 cases of group C streptococcal bacteremia in Israel and reviewed several other case series. In this review, some pa-

tients had primary bacteremia, or bacteremia secondary to pharyngitis, epiglottitis, pericarditis, pneumonia, skin and soft tissue infection, endocarditis, or infected aneurysm. In a review from Boston University, 29 patients with group G streptococcal bacteremia were identified over a 3-year period (76). The median age of the affected patients was 68 years, and one-half had a skin infection as the primary source of the bacteremia. In another series, six cases of bacteremia occurred in injectable-drug users (20). The portal of entry for these patients was the skin. All of the infected patients had injected drugs for at least 10 years.

Some series have noted a high rate of relapsing or recurring bacteremias caused specifically by organisms carrying the group G antigen. A series of 84 cases of group G *S. dysgalactiae* subsp. *equisimilis* in Israel (18) included six patients (7%) with recurrent bacteremia, ranging from two to four episodes per patient. Two of the six patients had the same clone isolated from their subsequent bacteremias, and the other four had different isolates in each incident. Another series in Singapore reported a rate of recurrent bacteremia of 5.8% (69). This fact suggests that although group G streptococci contain M proteins (19), infections with these organisms may not induce solid protective immunity. This assumption is supported by studies in a murine model of group G streptococcal cellulitis (9).

On the other hand, bacteremia caused by *S. equi* subsp. *zooepidemicus* is generally associated with animal contact, and tends to occur in outbreaks associated with exposure to animals or animal products. In 1999, Bradley et al. (10) reviewed 88 cases of bacteremia caused by group C streptococci reported in the literature. Twenty-one of these patients reported exposure to animals or animal products, and as expected, most of these had bacteremia due to *S. equi* subsp. *zooepidemicus*. Ten patients had consumed unpasteurized milk, four patients were farmers, one was a butcher, and several had other contact with animals. In the same series, 24 patients with definite or probable endocarditis were described. Of these, five cases were due to *S. equi* subsp. *zooepidemicus*, four were due to *S. dysgalactiae* subsp. *equisimilis*, and the remainder was unspecified. Animal exposure was noted only in patients with infection due to *S. equi* subsp. *zooepidemicus* or unspeciated organisms. Underlying cardiac disease was seen in 60% of the patients for whom adequate information was available. Edwards et al. (23) described an outbreak of 11 cases of bacteremia due to *S. equi* subsp. *zooepidemicus* in West Yorkshire. Presentations included primary septicemia, endocarditis, infected aneurysm, and meningitis. All 11 patients had consumed unpasteurized milk from the same source. Yuen et al. (81) reported 11 cases of *S. equi* subsp. *zooepidemicus* bacteremia with sepsis over a 4-year period in Hong Kong. The patients had a variety of presenting syndromes, and 55% had a serious underlying illness. None of the patients reported exposure to animals or animal products. After further investigation, it was felt that the infections were acquired from ingestion of undercooked pork. Furthermore, condemned septicemic pigs were found to be infected with *S. equi* subsp. *zooepidemicus* strains whose DNA fingerprints were identical to the human isolates.

Endocarditis due to group G streptococci is uncommon. Like bacteremia, the disease tends to occur in older patients with serious underlying conditions. Preexisting valvular disease is noted in about one-half of all patients. In a review of 40 cases (65), the average age was 56 years and the overall mortality was 36%. Underlying disease was present in about one-half of the patients; six patients had a malignancy, six were diabetic, four were alcoholics, and three were injectable-drug users. Also, one-half of the patients had known preexisting valvular disease, with mitral regurgitation being the most common abnormality. Three cases occurred in patients with prosthetic valves. In a series of seven cases not included in the above review, the average age of the patients was 72 years, and only one patient was younger than 60 (73). Underlying medical conditions and/or preexisting valvular disease were noted in most cases.

S. dysgalactiae subsp. equisimilis, with both group C and G antigens, has been identified as the causative agent of the streptococcal toxic shock syndrome (STSS) (33). Two cases of STSS caused by S. equi subsp. zooepidemicus have also been described (33, 40). Several other case reports document the ability of group C and G streptococci to produce STSS (39, 41, 49, 75). Although some had evidence of superantigen production, "classical" S. pyogenes superantigens were absent in these organisms, and the factors responsible for the pathogenesis of these infections are still unidentified.

Sachse et al. (60) studied 24 pathogenic isolates of S. dysgalactiae subsp. equisimilis and found a gene encoding the streptococcal pyrogenic exotoxin G (speG^{dys}), which demonstrates that this species has the potential to produce superantigenlike proteins. Sachse et al. failed, however, to show the presence of genes related to the superantigens SPEA, SPEC, SPEZ (SMEZ), SPEH, and SPEI. Two other studies, however, have demonstrated that some strains of S. dysgalactiae subsp. equisimilis carry bacteriophage-associated genes speA, speC, speM, ssa, or smeZ identical to their counterparts in S. pyogenes. This suggests that these genes may be transferred from one species to the other, conferring enhanced pathogenicity (36, 37).

TREATMENT

S. dysgalactiae subsp. equisimilis and the other large-colony group C and G streptococci are susceptible to penicillin, which is considered the drug of choice. Their range of minimal inhibitory concentrations to penicillin G is between 0.03 and 0.06 µg/ml (56). Other cell wall-active antibiotics, such as broad-spectrum penicillins, cephalosporins, and glycopeptides, are active in vitro, as are erythromycin, clindamycin, and chloramphenicol (42). As with group A streptococci, some strains of group C and G streptococci are resistant in vitro to macrolides, and rarely to clindamycin as well (38, 79). Given the increasing use of the newer macrolides in clinical practice, the prevalence of such strains is likely to increase over time.

Lam and Bayer found that bacterial killing by penicillin is impaired when high concentrations ($>10^8$ CFU/ml) of organisms are found (42), a phenomenon well described for S. pyogenes, which down-regulates the production of penicillin-binding proteins during its stationary phase of growth, resulting in a paucity of targets for penicillin (the "Eagle effect") (66).

Another observation has been that in certain patients, particularly those with endocarditis or septic arthritis, there is a poor or delayed response to therapy. The reason for this suboptimal response is unclear but is likely, at least in part, due to the underlying conditions present. The addition of gentamicin to a cell wall-active antibiotic has been shown to be synergistic in vitro (53, 57), and some have suggested its use in certain patients with severe and invasive group G streptococcal infections, provided there is no contraindication to use of an aminoglycoside antimicrobial agent. There are no definitive clinical data to support that recommendation at this time.

SUMMARY

Although most research on beta-hemolytic streptococci has focused on group A organisms, streptococci carrying serogroup C and G antigens are being increasingly recognized as important human pathogens. Members of the S. anginosus group, usually classified with the viridans group of streptococci, are associated with a variety of pyogenic infections. Most human infections due to S. dysgalactiae subsp. equisimilis are caused by person-to-person transmission, but infections due to S. equi subsp. zooepidemicus (and, rarely, to S. equi subsp. equi) are zoonoses. Transmission of these latter species occurs by animal contact or by contamination of food products and has been associated with the development of poststreptococcal glomerulonephritis. S. dysgalactiae subsp. equisimilis is similar in morphology and often in clinical expression to S. pyogenes. These streptococci cause infections of throat and skin and soft tissues. Moreover, they may invade the bloodstream and disseminate widely to many deep tissue sites, including endocardium. Life-threatening invasive infections, including STSS, due to streptococci of groups C and G occur most frequently in patients with severe underlying medical diseases. Treatment with penicillin is adequate under most circumstances.

REFERENCES

1. **Aubry, P., T. M. Swor, C. V. Lohr, A. Tibary, and G. M. Barrington.** 2000. Septic orchitis in an alpaca. Can. Vet. J. **41:**704–706.
2. **Auckenthaler, R., P. E. Hermans, and J. A. Washington II.** 1983. Group G streptococcal bacteremia: clinical study and review of the literature. Rev. Infect. Dis. **5:**196–204.
3. **Baddour, L. M., and A. L. Bisno.** 1985. Non-group A beta-hemolytic streptococcal cellulitis: association with venous and lymphatic compromise. Am. J. Med. **79:**155–159.
4. **Balter, S., A. Benin, S. W. Pinto, L. M. Teixeira, G. G. Alvim, E. Luna, D. Jackson, L. Laclaire, J. Elliott, R. Facklam, and A. Schuchat.** 2000. Epidemic nephritis in Nova Serrana, Brazil. Lancet **355:**1776–1780.
5. **Barnham, M., T. J. Thornton, and K. Lange.** 1983. Nephritis caused by Streptococcus zooepidemicus (Lancefield group C). Lancet **1(8331):**945–948.
6. **Benjamin, J. T., and V. A. J. Perriello.** 1976. Pharyngitis due to group C hemolytic streptococci in children. J. Pediatr. **89:**254–256.
7. **Bert, F., C. Branger, B. Poutrel, and N. Lambert-Zechovsky.** 1997. Differentiation of human and animal strains of Streptococcus dysgalactiae by pulsed-field gel electrophoresis. FEMS Microbiol. Lett. **150:**107–112.
8. **Bisno, A. L., and J. M. Gaviria.** 1997. Murine model of recurrent group G streptococcal cellulitis: no evidence of protective immunity. Infect. Immun. **65:**4926–4930.
9. **Bisno, A. L., C. M. Collins, and J. C. Turner.** 1996. M proteins of group C streptococci isolated from patients with acute pharyngitis. J. Clin. Microbiol. **34:**2511–2515.
10. **Bradley, S. F., J. J. Gordon, D. D. Baumgartner, W. A. Marasco, and C. A. Kauffman.** 1991. Group C streptococcal bacteremia: analysis of 88 cases. Rev. Infect. Dis. **13:**270–280.

11. **Brahmadathan, K. N., and G. Koshi.** 1989. Importance of group G streptococci in human pyogenic infections. *J. Trop. Med. Hyg.* **92:**35–38.

12. **Breiman, R. F., and F. J. Silverblatt.** 1986. Systemic *Streptococcus equi* infection in a horse handler—a case of human strangles. *West. J. Med.* **145:**385–386.

13. **Bronze, M. S., S. Whitby, and D. R. Schaberg.** 1997. Group G streptococcal arthritis: case report and review of the literature. *Am. J. Med. Sci.* **313:**239–243.

14. **Carmeli, Y., and K. L. Ruoff.** 1995. Report of cases of and taxonomic considerations for large-colony-forming Lancefield group C streptococcal bacteremia. *J. Clin. Microbiol.* **33:**2114–2117.

15. **Carmeli, Y., J. M. Schapiro, D. Neeman, A. M. Yinnon, and M. Alkan.** 1995. Streptococcal group C bacteremia: survey in Israel and analytical review. *Arch. Intern. Med.* **155:**1170–1176.

16. **Carstensen, H., C. Pers, and O. Pryds.** 1988. Group G streptococcal neonatal septicemia: two case reports and review of the literature. *Scand. J. Infect. Dis.* **20:**407–410.

17. **Chalker, V. J., H. W. Brooks, and J. Brownlie.** 2003. The association of *Streptococcus equi* subsp. *zooepidemicus* with canine infectious respiratory disease. *Vet. Microbiol.* **95:**149–156.

18. **Cohen-Paradosu, R., J. Jaffe, D. Lavi, S. Grisariu-Greenzaid, R. Nir-Paz, L. Valinsky, M. Dan-Goor, C. Block, B. Beall, and A. Moses.** 2004. Group G streptococcal bacteremia in Jerusalem. *Emerg. Infect. Dis.* **10:**1455–1460.

19. **Collins, C. M., A. Kimura, and A. L. Bisno.** 1992. Group G streptococcal M protein exhibits structural features analogous to class I M protein of group A streptococci. *Infect. Immun.* **60:**3689–3696.

20. **Craven, D. E., A. I. Rixinger, A. L. Bisno, T. A. Goularte, and W. R. McCabe.** 1986. Bacteremia caused by group G streptococci in parenteral drug abusers: epidemiological and clinical aspects. *J. Infect. Dis.* **153:**988–992.

21. **Duca, E., G. R. Teodorovici, C. Radu, A. Vita, P. Talasman-Niculescu, E. Bernescu, C. Feldi, and V. Rosca.** 1969. A new nephritogenic streptococcus. *J. Hyg.* **67:**691–698.

22. **Dyson, A. E., and S. E. Read.** 1981. Group G streptococcal colonization and sepsis in neonates. *J. Pediatr.* **99:**944–947.

23. **Edwards, A. T., M. Roulson, and M. J. Ironside.** 1988. A milkborne outbreak of serious infection due to *Streptococcus zooepidemicus* (Lancefield group C). *Epidemiol. Infect.* **101:**43–51.

24. **Efstratiou, A., G. Colman, G. Hahn, J. F. Timoney, J. M. Boeufgras, and D. Monget.** 1994. Biochemical differences among human and animal streptococci of Lancefield group C or group G. *J. Med. Microbiol.* **41:**145–148.

25. **Facklam, R.** 2002. What happened to the streptococci: overview of taxonomic and nomenclature changes. *Clin. Microbiol. Rev.* **15:**613–630.

26. **Faix, R. G., E. I. Soskolne, and R. E. Schumacher.** 1997. Group C streptococcal infection in a term newborn infant. *J. Perinatol.* **17:**79–82.

27. **Galloway, A., I. Noel, A. Efstratiou, E. Saint, and D. R. White.** 1994. An outbreak of group C streptococcal infection in a maternity unit. *J. Hosp. Infect.* **28:**31–37.

28. **Gaviria, J. M., and A. L. Bisno.** 2000. Group C and G streptococci, p. 238–254. *In* E. L. Kaplan and D. L. Stevens (ed.), *Streptococcal Infections.* Oxford University Press, New York, N.Y.

29. **Gerber, M. A., M. F. Randolph, N. J. Martin, M. F. Rizkallah, P. P. Cleary, E. L. Kaplan, and E. M. Ayoub.** 1991. Community-wide outbreak of group G streptococcal pharyngitis. *Pediatrics* **87:**598–603.

30. **Gonzalez Teran, B., M. P. Roiz, J. T. Ruiz, J. Rosas, and J. Calvo-Alen.** 2001. Acute bacterial arthritis caused by group C streptococci. *Semin. Arthritis Rheum.* **31:**43–51.

31. **Guliye, A. Y., C. Van Creveld, and R. Yagil.** 2002. Detection of subclinical mastitis in dromedary camels (*Camelus dromedarius*) using somatic cell counts and the N-acetyl-beta-D-glucosaminidase test. *Trop. Anim. Health Prod.* **34:**95–104.

32. **Harrington, D. J., I. C. Sutcliffe, and N. Chanter.** 2002. The molecular basis of *Streptococcus equi* infection and disease. *Microbes Infect.* **4:**501–510.

33. **Hashikawa, S., Y. Iinuma, M. Furushita, T. Ohkura, T. Nada, K. Torii, T. Hasegawa, and M. Ohta.** 2004. Characterization of group C and G streptococcal strains that cause streptococcal toxic shock syndrome. *J. Clin. Microbiol.* **42:**186–192.

34. **Hewson, J., and C. K. Cebra.** 2001. Peritonitis in a llama caused by *Streptococcus equi* subsp. *zooepidemicus. Can. Vet. J.* **42:**465–467.

35. **Hill, H. R., G. G. Caldwell, E. Wilson, D. Hager, and R. A. Zimmerman.** 1969. Epidemic of pharyngitis due to streptococci of Lancefield group G. *Lancet* **2**(616):371–374.

36. **Igwe, E. I., P. L. Shewmaker, R. R. Facklam, M. M. Farley, C. Van Beneden, and B. Beall.** 2003. Identification of superantigen genes speM, ssa, and smeZ in invasive strains of beta-hemolytic group C and G streptococci recovered from humans. *FEMS Microbiol. Lett.* **229:**259–264.

37. **Kalia, A., and D. E. Bessen.** 2003. Presence of streptococcal pyrogenic exotoxin A and C genes in human isolates of group G streptococci. *FEMS Microbiol. Lett.* **219:**291–295.

38. **Kataja, J., H. Seppala, M. Skurnik, H. Sarkkinen, and P. Huovinen.** 1998. Different erythromycin resistance mechanisms in group C and group G streptococci. *Antimicrob. Agents Chemother.* **42:**1493–1494.

39. **Keiser, P., and W. Campbell.** 1992. "Toxic strep syndrome" associated with group C streptococcus. *Arch. Intern. Med.* **152:**882–883.

40. **Korman, T. M., A. Boers, T. M. Gooding, N. Curtis, and K. Visvanathan.** 2004. Fatal case of toxic shock-like syndrome due to group C streptococcus associated with superantigen exotoxin. *J. Clin. Microbiol.* **42:**2866–2869.

41. **Kugi, M., H. Tojo, I. Haraga, T. Takata, K. Handa, and K. Tanaka.** 1998. Toxic shock-like syndrome caused by group G Streptococcus. *J. Infect.* **37:**308–309.

42. **Lam, K., and A. S. Bayer.** 1983. Serious infections due to group G streptococci: report of 15 cases with in vitro-in vivo correlations. *Am. J. Med.* **75:**561–570.

43. **Lefort, A., O. Lortholary, P. Casassus, C. Selton-Suty, L. Guillevin, and J. L. Mainardi.** 2002. Comparison between adult endocarditis due to beta-hemolytic streptococci (serogroups A, B, C, and G) and *Streptococcus milleri*: a multicenter study in France. *Arch. Intern. Med.* **162:**2450–2456.

44. **Liu, C. E., T. N. Jang, F. D. Wang, L. S. Wang, and C. Y. Liu.** 1995. Invasive group G streptococcal infections: a review of 37 cases. *Zhonghua Yi Xue Za Zhi* (Taipei) **56:**173–178.

45. **McCue, J. D.** 1982. Group G streptococcal pharyngitis: analysis of an outbreak at a college. *JAMA* **248:**1333–1336.

46. **McKeage, M. J., M. W. Humble, and R. B. Morrison.** 1990. *Streptococcus zooepidemicus* cellulitis and bacteremia in a renal transplant recipient. *Aust. N. Z. J. Med.* **20:**177–178.

47. **Molina, J. M., C. Leport, A. Bure, M. Wolff, C. Michon, and J. L. Vilde.** 1991. Clinical and bacterial features of infections caused by *Streptococcus milleri. Scand. J. Infect. Dis.* **23:**659–666.

48. **Nakata, M. M., J. H. Silvers, and L. George.** 1983. Group G streptococcal arthritis. *Arch. Intern. Med.* **143:** 1328–1330.

49. **Natoli, S., C. Fimiani, N. Faglieri, L. Laurenzi, A. Calamaro, A. M. Frasca, and E. Arcuri.** 1996. Toxic shock syndrome due to group C streptococci. A case report. *Intensive Care Med.* **22:**985–989.

50. **Nohlgard, C., A. Bjorklind, and H. Hammar.** 1992. Group G streptococcal infections on a dermatological ward. *Acta Derm. Venereol.* **72:**128–130.

51. **Packe, G. E., D. F. Smith, T. M. S. Reid, and C. C. Smith.** 1991. Group G streptococcal bacteremia—a review of thirteen cases of grampian. *Scott. Med. J.* **36:**42–44.

52. **Pinto, S. W., R. Sesso, E. Vasconcelos, Y. J. Watanabe, and A. M. Pansute.** 2001. Follow-up of patients with epidemic poststreptococcal glomerulonephritis. *Am. J. Kidney Dis.* **38:**249–255.

53. **Portnoy, D., J. Prentis, and G. K. Richards.** 1981. Penicillin tolerance of human isolates of group C streptococci. *Antimicrob. Agents Chemother.* **20:**235–238.

54. **Quinn, R. J. M., A. F. Hallett, P. C. Appelbaum, and R. C. Cooper.** 1978. Meningitis caused by *Streptococcus dysgalactiae* in a preterm infant. *Am. J. Clin. Pathol.* **70:**948–950.

55. **Rider, M. A., and J. C. McGregor.** 1994. Group G streptococcus—an emerging cause of graft loss? *Br. J. Plast. Surg.* **47:**346–348.

56. **Rolston, K. V., J. L. LeFrock, and R. F. Schell.** 1982. Activity of nine antimicrobial agents against Lancefield group C and group G streptococci. *Antimicrob. Agents Chemother.* **22:**930–932.

57. **Rolston, K. V. I., P. H. Chandrasekar, and J. L. LeFrock.** 1984. Antimicrobial tolerance in group C and group G streptococci. *J. Antimicrob. Chemother.* **13:**389–392.

58. **Ruoff, K. L., and A. L. Bisno.** 2005. Classification of streptococci, p. 2360–2362. *In* G. L. Mandell, R. Dolin, and J. E. Bennett (ed.), *Principles and Practice of Infectious Diseases.* Churchill Livingstone, Philadelphia, Pa.

59. **Ruoff, K. L., L. J. Kunz, and M. J. Ferraro.** 1985. Occurrence of *Streptococcus milleri* among beta-hemolytic streptococci isolated from clinical specimens. *J. Clin. Microbiol.* **22:**149–151.

60. **Sachse, S., P. Seidel, D. Gerlach, E. Gunther, J. Rodel, E. Straube, and K. H. Schmidt.** 2002. Superantigen-like gene(s) in human pathogenic *Streptococcus dysgalactiae*, subsp. *equisimilis*: genomic localization of the gene encoding streptococcal pyrogenic exotoxin G (speG(dys)). *FEMS Immunol. Med. Microbiol.* **34:**159–167.

61. **Salata, R. A., P. I. Lerner, D. M. Shlaes, K. V. Gopalakrishna, and E. Wolinsky.** 1989. Infections due to Lancefield group C streptococci. *Medicine* **68:**225–239.

62. **Schattner, A., and K. L. Vosti.** 1998. Bacterial arthritis due to beta-hemolytic streptococci of serogroups A, B, C, F, and G. Analysis of 23 cases and a review of the literature. *Medicine* **77:**122–139.

63. **Singh, K. P., A. Morris, S. D. Lang, D. M. MacCulloch, and D. A. Bremner.** 1988. Clinically significant *Streptococcus anginosus* (*Streptococcus milleri*) infections: a review of 186 cases. *N. Z. Med. J.* **101:**813–816.

64. **Smith, K. C., A. S. Blunden, K. E. Whitwell, K. A. Dunn, and A. D. Wales.** 2003. A survey of equine abortion, stillbirth and neonatal death in the UK from 1988 to 1997. *Equine Vet. J.* **35:**496–501.

65. **Smyth, E. G., A. P. Pallett, and R. N. Davidson.** 1988. Group G streptococcal endocarditis: two cases reports, a review of the literature and recommendations for treatment. *J. Infect.* **16:**169–176.

66. **Stevens, D. L., S. Yan, and A. E. Bryant.** 1993. Penicillin-binding protein expression at different growth stages determines penicillin efficacy in vitro and in vivo: an explanation for the inoculum effect. *J. Infect. Dis.* **167:**1401–1405.

67. **Stryker, W. S., D. W. Fraser, and R. R. Facklam.** 1982. Foodborne outbreak of group G streptococcal pharyngitis. *Am. J. Epidemiol.* **116:**533–540.

68. **Teare, E. L., R. D. Smithson, A. Efstratiou, W. R. Devenish, and N. D. Noah.** 1989. An outbreak of puerperal fever caused by group C streptococci. *J. Hosp. Infect.* **13:** 337–347.

69. **Tee, W. S., P. K. Lieu, and C. C. Ngan.** 2002. Epidemiology of beta-haemolytic group G streptococcal bacteraemia in Singapore (1996 to 1998). *Ann. Acad. Med. Singapore* **31:**86–91.

70. **Turner, J. C., G. F. Hayden, D. Kiselica, J. Lohr, C. F. Fishburne, and D. Murren.** 1990. Association of group C beta-hemolytic streptococci with endemic pharyngitis among college students. *JAMA* **264:**2644–2647.

71. **Turner, J. C., F. G. Hayden, M. C. Lobo, C. E. Ramirez, and D. Murren.** 1997. Epidemiologic evidence for Lancefield group C beta-hemolytic streptococci as a cause of exudative pharyngitis in college students. *J. Clin. Microbiol.* **35:**1–4.

72. **Vartian, C., P. I. Lerner, D. M. Shlaes, and K. V. Gopalakrishna.** 1985. Infections due to Lancefield group G streptococci. *Medicine* **64:**75–88.

73. **Venezio, F. R., R. M. Gullberg, G. O. Westenfelder, J. P. Phair, and F. V. Cook.** 1986. Group G streptococcal endocarditis and bacteremia. *Am. J. Med.* **81:**29–34.

74. **Vieira, V. V., L. M. Teixeira, V. Zahner, H. Momen, R. R. Facklam, A. G. Steigerwalt, D. J. Brenner, and A. C. Castro.** 1998. Genetic relationships among the different phenotypes of *Streptococcus dysgalactiae* strains. *Int. J. Syst. Bacteriol.* **48**(Pt. 4):1231–1243.

75. **Wagner, J. G., P. M. Schlievert, A. P. Assimacopoulos, J. A. Stoehr, P. J. Carson, and K. Komadina.** 1996. Acute group G streptococcal myositis associated with streptococcal toxic shock syndrome: case report and review. *Clin. Infect. Dis.* **23:**1159–1163.

76. **Watsky, K. L., N. Kollisch, and P. Densen.** 1985. Group G streptococcal bacteremia: the clinical experience at Boston University Medical Center and a critical review of the literature. *Arch. Intern. Med.* **145:**58–61.

77. **Weightman, N. C., M. R. Barnham, and M. Dove.** 2004. *Streptococcus milleri* group bacteraemia in North Yorkshire, England (1989–2000). *Indian J. Med. Res.* **119**(Suppl): 164–167.

78. **Woo, P. C., J. L. Teng, S. K. Lau, P. N. Lum, K. W. Leung, K. L. Wong, K. W. Li, K. C. Lam, and K. Y. Yuen.** 2003. Analysis of a viridans group strain reveals a case of bacteremia due to Lancefield group G alpha-hemolytic *Streptococcus dysgalactiae* subsp. *equisimilis* in a patient with pyomyositis and reactive arthritis. *J. Clin. Microbiol.* **41:** 613–618.

79. **Woo, P. C., A. P. To, H. Tse, S. K. Lau, and K. Y. Yuen.** 2003. Clinical and molecular epidemiology of erythromycin-resistant beta-hemolytic Lancefield group G streptococci causing bacteremia. *J. Clin. Microbiol.* **41:**5188–5191.

80. **Woo, P. C., H. Tse, K. M. Chan, S. K. Lau, A. M. Fung, K. T. Yip, D. M. Tam, K. H. Ng, T. L. Que, and K. Y.** Yuen. 2004. "Streptococcus milleri" endocarditis caused by *Streptococcus anginosus. Diagn. Microbiol. Infect. Dis.* **48:**81–88.

81. **Yuen, K. Y., W. H. Seto, C. H. Choi, W. Ng, S. W. Ho, and P. Y. Chau.** 1990. *Streptococcus zooepidemicus* (Lancefield group C) septicaemia in Hong Kong. *J. Infect.* **21:**241–250.

The Cell Wall of *Streptococcus pneumoniae*

ALEXANDER TOMASZ AND WERNER FISCHER

19

Since the last edition of this book, in 2000, a number of important publications have appeared in the literature, the inclusion of which made it necessary to limit the topics of this chapter to information that has bearing on the biochemical and genetic aspects of covalently linked components of the pneumococcal cell wall. Information on proteins noncovalently attached to the cell wall (chapter 24), cell walls and phase variation (chapter 22), and inflammatory activity of cell walls (chapter 21) are reviewed separately.

Historically, studies of the pneumococcal cell wall were motivated by such unique features as the presence of choline in the teichoic acids (TAs), the pleiomorphic changes that accompany removal or alteration of choline residues, and structural changes in peptidoglycan that are associated with penicillin resistance. Most recent contributions to the field include immunofluorescence microscopic localization of cell wall synthetic enzymes at sites of wall synthesis (36), identification of genetic determinants and enzymes that are involved with the synthesis of muropeptide branches (8), sortase-dependent attachment of proteins (31), removal of N-acetyl groups from N-acetyl hexosamine residues in the cell wall glycan chains (54), and removal of phosphoryl choline residues from TAs (6, 55).

FUNCTIONAL ANATOMY OF THE PNEUMOCOCCAL CELL WALL

The overwhelming majority of "natural" isolates of pneumococci are enwrapped on their outermost surface by one of the 90 chemically different capsular polysaccharides that this bacterial species is capable of synthesizing (30). Under these diverse structures lies the cell wall, which, as far as the resolution of currently used analytical techniques can tell, is much more uniform in its chemistry: it is composed of a peptidoglycan (21) covalently linked to chains of an unusually complex TA (29, 49), which contain as structural components phosphoryl choline residues (4, 51). The phosphoryl choline residues play multiple roles in the physiology and virulence of the pneumococcus (see below). The two polymers, peptidoglycan and TA, make up the bulk of the cell wall in roughly equal (mg to mg) proportions (38). In most but not all cases, chains of the capsular polysaccharide are attached by covalent bonds to the underlying peptidoglycan (49) (Fig. 1).

In electron microscopic thin sections prepared by the method of Kellenberger (53) (osmium tetraoxide and glutaraldehyde fixation followed by uranylacetate and lead citrate staining), the cell wall of *Streptococcus pneumoniae* strain R36A appears as a band of uniform width composed of two electron-dense lines (3 to 4 nm each) enclosing a wider low-density layer (6 to 8 nm). The distribution of TA chains appears to be uniform within this wall layer (49), and this is presumably also true for the peptidoglycan. Some anatomically differentiated areas may be identified through electron microscopy. These are (i) the equatorial areas where cell wall growth becomes centripetal (formation of septum or crosswall) and which represent the "growth zones" of the entire wall; (ii) in dividing cells a circumferential thickening ("hump") of the cell wall appears at the place of the incipient septa. (iii) Parallel, or perhaps just prior to the beginning of the formation of septum, the hump appears to be split at the center, and the two halves begin to "move" on the cell surface symmetrically to the left and to the right of the in-growing septum coupled to the growth and eventual division of the cell into two daughter cells. Due to the conservative mode of replication of the pneumococcal cell wall (see below), these two half-humps are morphological age markers: they divide the cell wall of each pneumococcal cell into two hemispheres, which differ in age by one cell generation. The functional correlates of these morphological changes began to be identified by the use of immunofluorescence microscopy, which allowed the localization of the various pneumococcal high-molecular-weight penicillin-binding proteins (PBPs) and the FtsZ-ring to equatorial and septal areas of the bacteria in various stages of cell division (36). A special role of PBP3 activity in these processes was also proposed (37). (iv) The final stage of cell division, the separation of daughter cells, may be inhibited in pneumococci by several means, resulting in the formation of long chains of bacteria. Under these conditions one can observe by electron microscopy a thin bridge of cell wall material connecting neighboring cells to one

Gram-Positive Pathogens, 2nd edition, edited by Vincent A. Fischetti et al.
© 2006 ASM Press, Washington, D.C.

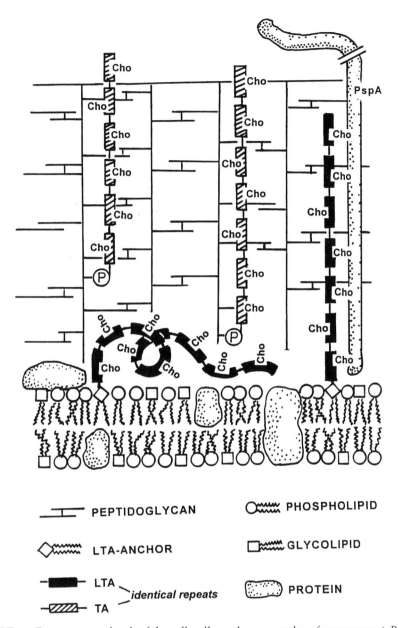

FIGURE 1 Diagrammatic sketch of the cell wall membrane complex of pneumococci. Pneumococcal surface protein (PspA) is shown as an example of a surface protein with a choline-binding signature.

another. A murein hydrolase, LytB (an endo-β-1,4-*N*-acetylglucosaminidase), has been identified as an enzyme essential for the terminal separation of daughter cells at the end of cell division (18).

Growth Zone and Cell Wall Segregation

Similarly to other streptococci, pneumococci incorporate new cell wall units into the preexisting wall material at a single growth zone located at the cellular equator. Peptidoglycan, TA units, and capsular polysaccharide (57) enter the pneumococcal surface at this growth zone, which could be visualized by exploiting the unique selectivity of a pneumococcal enzyme, the murein hydrolyase LytA, for choline-containing segments of the cell wall (33). Pneumococci require choline for growth, and the design of this experi-

ment was based on the observation that the choline component of the wall TA can be replaced by structural analogs such as ethanolamine. Pneumococci grown in ethanolamine-containing medium show several striking abnormalities: unlike the choline-grown bacteria, pneumococci utilizing ethanolamine grow in long chains and are completely resistant to the cell wall-degrading activity of LytA (52). Upon addition of trace amounts of radiolabeled choline to a culture grown on ethanolamine, the bacteria immediately shifted to the utilization of choline so that the nascent wall units that began to incorporate into the cell surface contained choline residues in the TA component of the nascent cell wall and produced regions that were susceptible to hydrolysis by exogenous LytA enzyme added to the medium. It was possible to show by electron microscopy that under these conditions the LytA enzyme performed an enzymatic "microsurgery" on the bacteria: it selectively removed a thin, equatorially located band of cell wall, thus identifying the anatomical site of wall incorporation and growth zone (33).

Another abnormality of the ethanolamine-grown pneumococci, the complete inhibition of cell separation, has allowed the design of experiments to test the mode of inheritance of pneumococcal cell walls. Pneumococci labeled in their wall by titrated choline were shifted to an ethanolamine-containing medium in which the bacteria continued to grow in the form of chains of cells, i.e., "linear clones" in which the distribution of radioactively labeled bacteria could be followed (by autoradiography) as a function of cell generations in the ethanolamine-containing medium. Because the TA choline does not exhibit turnover during growth, the localization of radioactively labeled cells within the chains of bacteria could provide clues as to the mode of wall segregation. The finding was that the radioactive label remained in large clusters in association with cells that were located either at the tips or at the center of chains. The results demonstrate the conservation of large hemispherical segments of the cell wall that are passed on intact to daughter cells during cell division (2).

STRUCTURE OF PEPTIDOGLYCAN

Purified cell walls of the nonencapsulated strain R6St (streptomycin resistant) of S. pneumoniae were hydrolyzed by the pneumococcal amidase (N-acetyl muramic acid L-alanine amidase; the product of lytA gene) under conditions in which this enzyme can quantitatively release the peptide units of cell wall muropeptides.

A method that uses muramidase digestion of the pneumococcal cell wall followed by high-performance liquid chromatography (HPLC) separation of muropeptides, through an adaptation of the method of Glauner and Schwarz, has been described (25). In the amidase method, the family of peptides were separated by HPLC; size fractionation, determination of amino acid composition and NH$_2$ termini, partial sequencing of the peptides generated by HPLC, and analysis by time-of-flight mass spectrometry have allowed the identification of a surprisingly large number of monomeric, dimeric, and trimeric peptides (21) showing a diversity comparable to that seen among the muropeptide species identified in Escherichia coli. In the structural assignments it was assumed that the amino acids within the stem peptides had the usual, alternating sequence of L and D amino acids beginning with L-alanine in position 1, followed by a D-isoglutamine, and then by L-lysine. The carboxy terminus in the stem peptide is occupied by two consecutive D-alanine residues; however, such muropeptides with intact pentapeptide residue are rare in pneumococci cultivated under normal growth conditions. Extension of this analytical technique to cell walls of clinical isolates from a large variety of isolation sites and dates and expressing a variety of different capsules has led to the proposition that the cell wall muropeptide composition of S. pneumoniae grown in the commonly used semisynthetic media and harvested in the late exponential phase is constant and characteristic of the species. The most abundant monomer of the pneumococcal peptidoglycan was a tripeptide, the most frequent dimer a directly cross-linked tri-tetrapeptide. Interestingly, the representation of carboxy-terminal alanine was extremely rare, suggesting the presence of powerful D,D and D,L carboxypeptidases. This was confirmed by testing the wall peptide composition in a PBP3-defective mutant (42) and in the laboratory strain R36A growing in the presence of subinhibitory concentrations of clavulanic acid, a selective inhibitor of PBP3 in this bacterium (46). The peptidoglycan produced under these conditions showed accumulation of peptide species that retained carboxy-terminal D-alanine residues. Pneumococci grown in the presence of clavulanate also showed abnormal physiological properties: premature induction of stationary-phase autolysis, hypersensitivity to lysozyme, and reduced MICs for deoxycholate and penicillin.

An interesting feature of the peptide network was the presence of both directly and indirectly cross-linked components. In the latter, alanyl-serine or alanyl-alanine dipeptides formed the cross-link. In terms of cross-linking mode, the pneumococcal cell wall may be classified as either A1α or A3α, depending on which dimer one chooses. A massive distortion of peptidoglycan composition in the direction of preponderance of indirectly cross-linked components was demonstrated in several penicillin-resistant clinical isolates (41).

PEPTIDOGLYCAN COMPOSITION AND PENICILLIN RESISTANCE

The first series of highly penicillin-resistant clinical isolates examined by the HPLC method was from South Africa. It was in these isolates that the mechanism of resistance, namely reduction in antibiotic "affinity" of PBPs, was identified for the first time (60). It was also in genetic crosses with these isolates used as DNA donors that the stepwise nature of penicillin resistance (i.e., the sequential reduction in the penicillin affinity of several high-molecular-weight PBPs in parallel with the gradually increasing penicillin MIC) was recognized (60).

Analysis of the penicillin-resistant South African clinical isolates revealed that they produced cell walls of a radically different composition from the one seen in the penicillin-susceptible and nonencapsulated laboratory isolate (20). When the HPLC analysis was extended to the walls of several penicillin-susceptible clinical isolates and several resistant strains (all but one from South Africa), the striking shift towards indirectly cross-linked wall peptide composition in the resistant isolates was fully confirmed. A link between resistance to penicillin and abnormal wall composition was also suggested by the analysis of genetic crosses: a shift towards the distorted wall composition of the resistant DNA donor was observed in genetic transformants above certain MIC. It was suggested that the anomalous wall peptide composition reflected the altered substrate preference of the penicillin-resistant PBPs, a shift from the linear to the branched wall peptide precursors (20).

The murMN Operon

A considerable clarification concerning determinants of cell wall structure and its relationship to penicillin resistance was obtained by the recent identification of the murMN operon, which encodes enzymes involved in the synthesis of branched structured muropeptides in the pneumococcal peptidoglycan (8). The same determinants were also described independently and named fib by another group (58). The murMN operon shows homology to the femXAB genes of Staphylococcus aureus that are involved with the addition of the pentaglycine branches to the epsilon amino group of lysine residues in the staphylococcal peptidoglycan (40). The in vivo substrate of MurM is the

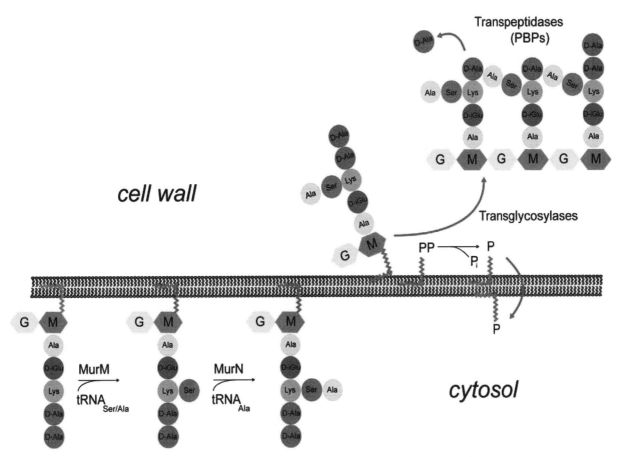

FIGURE 2 Role of MurM and MurN in cell wall branching. The substrate of the MurM- and MurN-catalyzed branching reaction is lipid II, which is composed of *N*-acetylated disaccharide units of glucosamine (hexagon with G) and muramic acid (hexagon with M) with the pentapeptide chain attached to the M residues. Lipid II is anchored on the plasma membrane through the carrier lipid bactoprenyl pyrophosphate (the zig-zag line). Attachment of the completed precursor to the pre-existing cell wall occurs on the outer surface of the plasma membrane by the activity of transglyco-sylases and transpeptidases. (Reproduced with permission from reference 16.)

lipid-carrier-linked disaccharide pentapeptide, and the MurM protein catalyzes the addition of a serine or alanine residue (depending on the particular *murM* allele) to the free amino group of lysine in the stem peptide, in a reaction in which the amino acid donor is presumed to be an amino aceyl tRNA (11). This reaction is followed by the MurM-catalyzed addition of an alanine to complete the dipeptide branch (see Fig. 2).

Inactivation of the *murM* operon or inactivation of the *murM* gene alone did not interfere with growth of the bacteria but caused the production of a peptidoglycan composed exclusively of linear muropeptides. Another consequence of inactivation was the complete loss of the penicillin-resistant phenotype. Examination of a large number of *S. pneumoniae* clinical isolates representing different genetic lineages has identified several distinct *murM* alleles carried by penicillin-resistant strains that also showed different and unique muropeptide composition in their peptidoglycan (10). Analysis of the *murM* alleles from penicillin-resistant isolates showed that they differed from one another and from the *murM* carried by penicillin-susceptible strains in regions of considerable sequence diversity that

were distributed as heterologous "patches" along the *murM* gene sequence. Different *murM* alleles from several penicillin-resistant *S. pneumoniae* strains, each with a characteristic branched peptide composition, were introduced on a plasmid into a common penicillin-susceptible laboratory strain. All transformants remained penicillin susceptible, but their cell wall composition changed in directions that corresponded to the muropeptide pattern of the strain from which the *murM* allele was derived. This observation suggests that the muropeptide composition of *S. pneumoniae* is determined by the particular *murM* carried by the strain (11, 12) (Fig. 3A and B and Table 1).

The relationship between *murM* alleles and the penicillin-resistant phenotype is less clear. In genetic transformation of high-level penicillin resistance it was shown that successful expression of the resistant phenotype required that the transformants received not only the particular mosaic *pbp* genes of the DNA donor but also the *murM* allele carried by the donor strain (47). However, in other experiments the linkage between penicillin resistance and abnormal wall composition was lost in transformation experiments in which the first round of transformation was

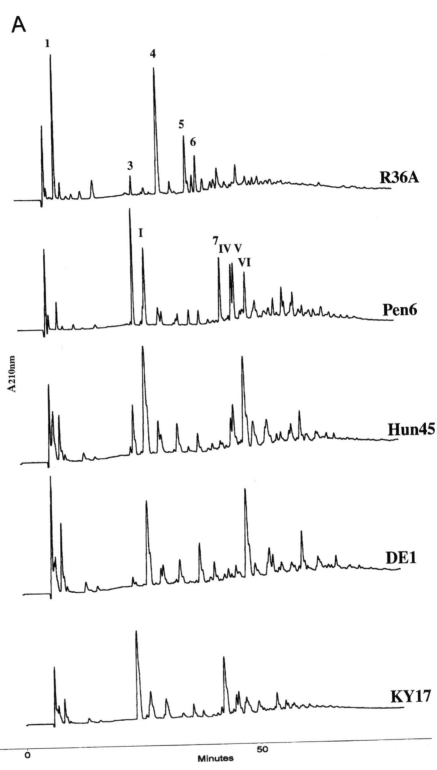

FIGURE 3 HPLC elution profiles (A) and structures (B) of stem peptides of the peptidoglycan from the penicillin-susceptible strain R36A and several penicillin-resistant strains that carry different abnormal *mur*M alleles. Structures of cell wall stem peptides identified in the pneumococcal peptidoglycan of penicillin-susceptible and -resistant strains of pneumococci. (Reproduced with permission from reference 10.)

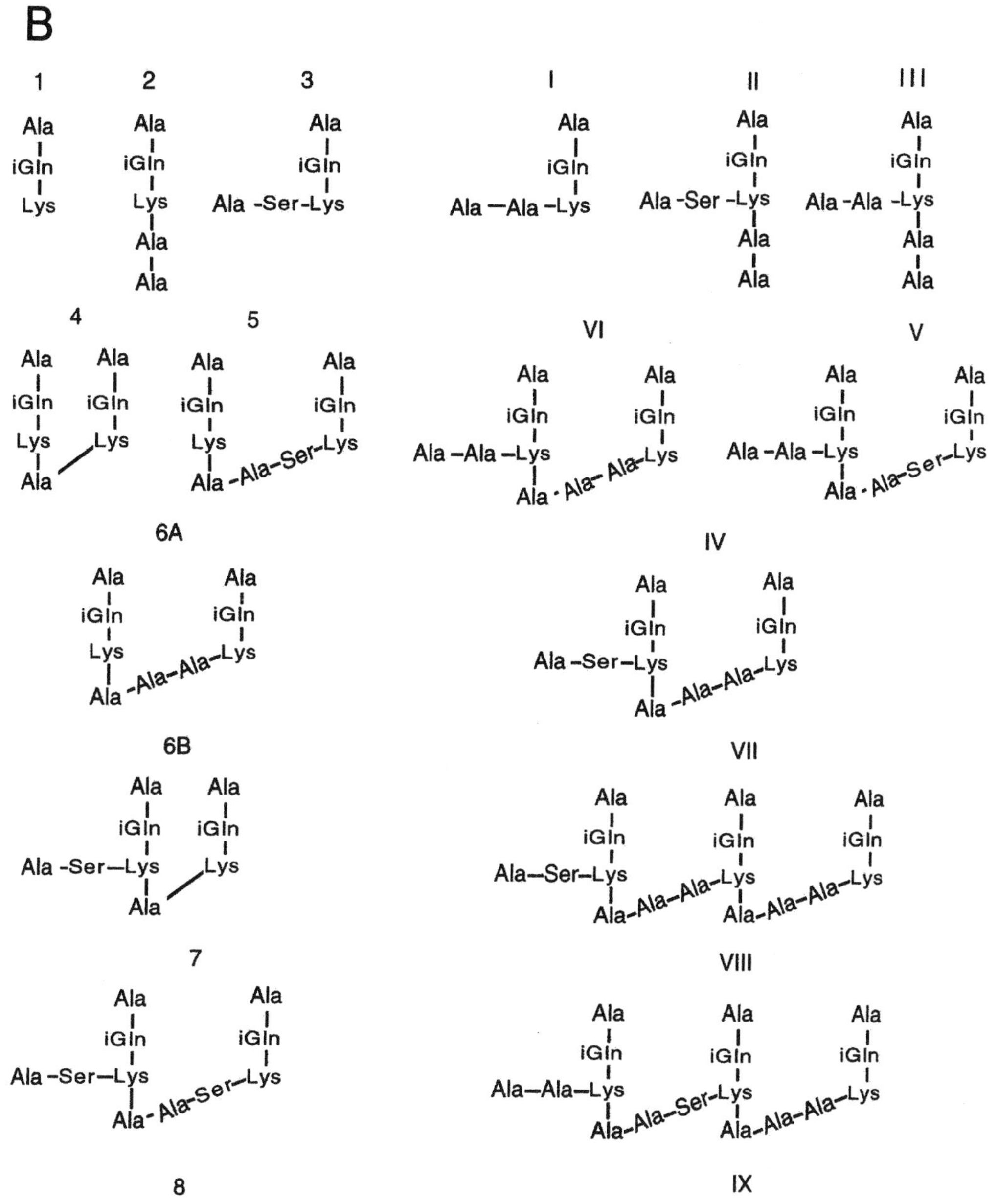

B

1
Ala
|
iGln
|
Lys

2
Ala
|
iGln
|
Lys
|
Ala
|
Ala

3
Ala
|
iGln
|
Ala -Ser-Lys

I
Ala
|
iGln
|
Ala —Ala —Lys

II
Ala
|
iGln
|
Ala -Ser -Lys
|
Ala
|
Ala

III
Ala
|
iGln
|
Ala -Ala -Lys
|
Ala
|
Ala

4
Ala   Ala
| |
iGln iGln
| |
Lys   Lys
| /
Ala

5
Ala   Ala
| |
iGln   iGln
| |
Lys   Lys
|
Ala -Ala-Ser-Lys

VI
Ala   Ala
| |
iGln   iGln
| |
Ala –Ala –Lys   Ala-Ala–Lys
              |
    Ala · Ala–Ala

V
Ala   Ala
| |
iGln   iGln
| |
Ala –Ala –Lys   Ala –iGln –Ser-Lys
              |
    Ala · Ala-Ser

6A
Ala   Ala
| |
iGln   iGln
| |
Lys
Ala -Ala-Ala-Lys

IV
Ala   Ala
| |
iGln   iGln
| |
Ala -Ser–Lys   Ala
              |
    Ala -Ala-Ala–Lys

6B
Ala   Ala
| |
iGln   iGln
| |
Ala -Ser—Lys   Lys
              /
    Ala

VII
Ala   Ala   Ala
| | |
iGln   iGln   iGln
| | |
Ala-Ser-Lys   Ala   Ala-Ala-Ala-Lys
           Ala-Ala-Ala-Lys

7
Ala   Ala
| |
iGln   iGln
| |
Ala -Ser—Lys   Ala-Ala-Ser-Lys
           |
    Ala -Ala-Ser

VIII
Ala   Ala   Ala
| | |
iGln   iGln   iGln
| | |
Ala—Ala—Lys   Ala-Ser-Lys   Ala-Ala-Ala-Lys
           Ala-Ala-Ser-Lys

8

IX

FIGURE 3   *(Continued)*

followed by a second backcross (44). Interestingly, in this particular experiment the penicillin-resistant secondary transformants began to show "fitness" defects: defective growth and premature autolysis in antibiotic-free medium, similar to the defective physiology observed in laboratory isolates of penicillin-resistant pneumococci (43).

Genetic analysis and molecular modeling of the *murM* protein have identified amino acid residues that appear to be critical for the specificity of this protein and also pinpointed domains that interact with the aminoacyl tRNA and the bactoprenyl-linked substrate of the *murM*-catalyzed reaction (16).

The Pneumococcal C-Polysaccharide and F-Antigen

In 1930, long before lipoteichoic acids (LTAs) and TAs had been discovered and defined, pneumococcal TA was de-scribed as pneumococcal C-polysaccharide by Tillet and coworkers (50). Thirteen years later, pneumococcal LTA was isolated by Goebel and his colleagues in 1943 and named lipocarbohydrate or pneumococcal F-antigen owing to its fatty acid content and immunological properties (23). In these early studies, a structural relationship between C-polysaccharide and lipocarbohydrate was suggested, and in contrast to the various strain-specific capsular polysaccha-rides, lipocarbohydrate and C-polysaccharide were consid-ered pneumococcal common antigens. This was confirmed by serological methods, which showed that all 90 known capsular types of *S. pneumoniae* possess C-polysaccharide and F-antigen (49). The two polymers differ immunologi-cally, as Forssman antigenicity is associated with the LTA (1, 3). As shown by immunoelectron microscopy, C-poly-saccharide is uniformly distributed on both the inside and outside of the cell walls, and LTA is located on the surface of the cytoplasmic membrane (48).

TABLE 1 Cell wall peptide composition of several strains of *S. pneumoniae*[a]

Peptide	% Peptide in strain:				
	R36A (*murMA*)	Pen6 (*murMB2*)	Hun45 (*murMB1*)	DE1 (*murMB3*)	KY1 (*murMB5*)
1	13.3	2.6	4.5	10.8	3.4
2	3.4	0.7	0.5	1.4	0.6
3	2.8	14.3	7.2	1.0	28.5
I	1.8	14.4	24.1	23.4	11.2
II	3.4	2.7	5.1	3.1	0.8
4	21.0	3.2	4.4	6.0	7.2
III	1.3	3.5	6.7	6.9	2.1
5	12.7	2.2	0.9	0.5	2.9
6	7.1	2.5	2.8	9.3	1.9
7	7.4	11.1	1.4	0.4	19.7
IV	2.6	6.5	4.7	0.6	4.1
V	2.8	12.3	8.1	3.7	5.2
8	7.8	2.5	1.4	0.6	5.8
VI	6.6	10.1	19.1	27.5	2.0
9	6.0	11.6	9.0	4.9	4.5
Total	100	100	100	100	100
Monomers	26	38	48	47	47
Multimers	74	62	52	53	53
Linear	50	12	14	23	17
Branched	50	88	86	77	83
Branched/linear	1.0	7.4	6.4	3.3	4.9

[a]Reproduced from reference 10 with permission.

Choline, the surface signature of pneumococci, was identified as a component of TA and LTA (4, 51). The complex structures of TA and LTA could not be unraveled before modern analytical techniques became available (see review by Fischer [13]). In 1980, Jennings and coworkers published the first complete structure of pneumococcal TA (29). The structure of pneumococcal LTA was clarified in 1992 (1), and subsequent reinvestigation of the TA, isolated from the same strain from which the LTA had been isolated, revealed that both polymers possess identical chain structures (15).

STRUCTURE OF TA AND LTA

The chains of LTA and TA are made up by identical repeats consisting of D-glucose, the rare positively charged amino sugar 2-acetamido-4-amino-2,4,6-trideoxy-D-galactose (AAT-Gal), two *N*-acetyl-D-galactosaminyl residues, phosphocholine, and ribitol 5-phosphate. The repeats are joined together by phosphodiester bonds between O5 of the ribitol and O6 of the glucosyl residue of adjacent repeats. The functionally important phosphocholine residues are phosphodiester-linked to O6 of the *N*-acetyl-D-galactosaminyl residues. The number of phosphocholine residues per repeat is strain-specific: in LTA and TA of strain R6 the majority of repeats carry two phosphocholine residues, whereas most of the repeats in strain Rx1 are substituted with one.

In addition to the unusual complexity of the chemical structure of TA, pneumococci are also unique because their TA and LTA possess identical repeat and chain structures, whereas in other gram-positive bacteria TAs and LTAs are structurally and biosynthetically distinct entities.

The chain of LTA is phosphodiester-linked to O6 of the terminal sugar residue of D-Glc*p*-(β1-4)-D-AATGal*p*-(β1-3)- D-Glc*p*(α1-3)acyl$_2$Gro, a unique glycolipid that contains the positively charged AATGal residue intercalated between two D-glucopyranosyl residues (Fig. 2). In LTA and TA the AATGal residues are of conformational importance because the positively charged amino groups interact electrostatically or by hydrogen bonding with the negatively charged phosphate groups on the adjacent glucosyl residues (13, 32).

Whereas in other gram-positive bacteria the chain of LTA is linked to a membrane glycolipid, the lipid anchor of pneumococcal LTA does not occur in the free state in the membrane, where Gal(α1-2)Glc(α1-3)acyl$_2$Gro is the major glycolipid (13). From the molar ratio of intrachain-phosphate to glycerol an average of six to seven repeats per chain is calculated for both strains.

Microheterogeneity of LTA became apparent by hydrophobic interaction chromatography and mass spectral analysis (1). The chain of LTA may vary in length between two and eight repeats, with the range differing from sample to sample. On sodium dodecyl sulfate-polyacrylamide gel electrophoresis LTA yields a ladderlike pattern of up to six bands, each differing from the next by one repeat. Species with one phosphocholine per repeat are distinguished from species with two by higher mobility of the individual bands.

Pneumococcal TA is linked to the peptidoglycan by a phosphodiester to O6 of some of the MurNAc-residues, which is demonstrated by the release of MurN-6-*P* on HCl hydrolysis of cell walls (34) and of muropeptides. It is estimated that, depending on the particular strain, between 15 and 30% of the muramic acid residues carry TA chains (17, 38). Although the release of MurNAc-6-*P* indicates the presence of a linkage unit, which in other gram-positive bacteria connects the TA chain and muramic acid by a phosphodiester bond, the typical components of already known linkage units, namely, glycerophosphate and *N*-acetyl-mannosamine, could not be detected either in the hydrolysate of pneumococcal cell walls or of TA-containing muropeptides. Because no other sugar was found either, an acid-degradable sugar such as AATGal is suggested as a component of the linkage unit (13).

COVALENT MODIFICATIONS OF THE *S. PNEUMONIAE* CELL WALL

Peptidoglycan *N*-Acetylglucosamine Deacetylase

An unusually high proportion of hexosamine units in the glycan strands of the pneumococcal cell wall is not *N*-acetylated, explaining the resistance of the peptidoglycan to the hydrolytic action of lysozyme, a muramidase that cleaves in the glycan backbone. A gene, *pgdA*, was identified as encoding for the peptidoglycan *N*-acetylglucosamine deacetylase A with amino acid sequence similarity to fungal chitin deacetylases and rhizobial NodB chitooligosaccharide deacetylases. Pneumococci in which *pgdA* was inactivated by insertion duplication mutagenesis produced fully *N*-acetylated glycan and became hypersensitive to exogenous lysozyme in the stationary phase of growth. The *pgdA* gene may contribute to pneumococcal virulence by providing protection against host lysozyme. A pneumococcal strain expressing capsular type II with inactivated *pgdA* showed reduced virulence in the mouse model of intraperitoneal infection (56).

Sortase A-Dependent Covalent Attachment of Proteins to the Pneumococcal Cell Wall

Most virulence-related pneumococcal proteins studied so far have been shown to be attached to the cell surface through the choline residues of TAs, and pneumococci have been described as a paradigm for the display of virulence proteins through specific but noncovalent associations with the cell surface (5). Such choline-binding proteins include PspA, PsaA (see chapters 21 and 22), the enzyme phosphorylcholine esterase (PCE) (6, 55), and LytA, in which a 20-amino-acid repeat was shown to recognize choline residues in wall or membrane TAs (19). The observations described in this report demonstrate that the mechanism of surface display of proteins does not depend on the choline-binding paradigm alone; pneumococci also use covalent anchoring for at least some surface proteins. Data available from the genome of *S. pneumoniae* strain R6 (28) indicate that there are at least 23 proteins carrying the LPXTG motif, and 15 of these have this recognition sequence at the C terminus, as expected for proteins that are processed by a typical sortase.

Inactivation of sortase gene *srtA* in *S. pneumoniae* strain R6 caused the release of β-galactosidase and neuraminidase A from the cell wall into the surrounding medium. Both of these surface proteins contain the LPXTG motif in the C-terminal domain. Complementation with plasmid-borne *srtA* reversed protein release. Deletion of *murM*, a gene involved in the branching of pneumococcal peptidoglycan, also caused partial release of β-galactosidase, suggesting preferential attachment of the protein to branched muropeptides in the cell wall. Inactivation of *srtA* caused decreased adherence to human pharyngeal cells in vitro but had no effect on the virulence of a capsular type III strain of *S. pneumoniae* in the mouse intraperitoneal model. The ob-

servations suggest that—as in other gram-positive bacteria—sortase-dependent display of proteins occurs in *S. pneumoniae* and that some of these proteins may be involved in colonization of the human host (31).

PCE Enzymatic Removal of Phosphoryl Choline Residues from the Pneumococcal Cell Wall

A gene, *pce*, encoding for a TA PCE, with enzymatic activity capable of removing phosphorylcholine residues from the cell wall TA and LTA, was identified independently by two research groups (6, 55). PCE carries an N-terminal signal sequence and contains a C-terminal choline-binding domain with 10 homologous repeating units similar to those found in other pneumococcal surface proteins. The catalytic activity of PCE is localized on the N-terminal part of the protein. The mature protein was overexpressed in *E. coli* and purified in a one-step procedure by choline-affinity chromatography. The product of the enzymatic digestion of ³H-choline-labeled cell walls was shown to be phosphorylcholine. Inactivation of the *pce* gene in *S. pneumoniae* strains by insertion-duplication mutagenesis caused a unique change in colony morphology and a striking increase in virulence of a capsular type III strain in the intraperitoneal mouse model. PCE may be a regulatory element involved with the interaction of *S. pneumoniae* with its human host. Sequence comparison indicates that PCE, the protein originally identified through its unique enzymatic activity (26), is identical to choline-binding protein E—described by Masure and colleagues as a protein implicated in the attachment of pneumococci to nasopharyngeal cells (24).

MULTIPLE FUNCTIONS OF CELL WALL CHOLINE RESIDUES

Nutritional Requirement for Choline

Choline is an essential growth factor for all natural isolates of pneumococci, which have to import this nutrient from the growth medium (39). Choline may be replaced by other amino alcohols such as ethanolamine, which can incorporate into LTA and TA at the same positions as phosphocholine (W. Fischer, R. Hartmann, and G. Pohlentz, unpublished data), but it cannot replace phosphocholine functionally. Ethanolamine-grown cells show a number of striking abnormalities, which include inhibition of cell separation (growth in long chains); inhibition of DNA uptake in genetic transformation; lack of autolysis during treatment with penicillin and other wall inhibitors and detergents (52); production of an "immature, enzymatically inactive" form of the autolytic enzyme LytA (35); and production of a cell wall that cannot absorb LytA (22) and is completely resistant to the hydrolytic action of LytA (27). However, the requirement of LytA for built-in phosphocholine residues is no longer seen when solubilized cell wall polymers are degraded to muropeptides (48) or when the TA is removed from peptidoglycan (45). These observations may be interpreted to indicate that TA prevents the access of autolysin to its cell wall substrate and that this effect is overcome by binding of the enzyme to the phosphocholine residues.

There are several proteins in pneumococci, including LytA and LytB, that specifically recognize and bind to phosphocholine residues. These proteins contain distinct N-terminal domains that are responsible for their specific biological activities, whereas the C-terminal domains are highly homologous and contain 6 to 10 choline-recognizing repeats of 20 amino acids each. The structure of crystallized LytA and its interaction with choline residues in the cell wall have been described recently (7).

Choline-Independent Strains

A choline-independent strain, R6Cho⁻, was recovered from a heterologous cross with DNA from *Streptococcus oralis* (45). *S. oralis* also incorporates phosphocholine into its cell wall TA, but, unlike pneumococci, has no nutritional requirement for choline. Another choline-independent strain, JY2190, is a mutant that was generated by serial passage of strain Rx1 in chemically defined medium containing decreasing concentrations of ethanolamine with each passage (59). Neither of the two strains had acquired the capability to synthesize choline or ethanolamine, because phosphorylated amino alcohols could not be detected on TA and LTA. In spite of the absence of phosphocholine, there was no alteration either in the structure of LTA and TA or in cell wall composition, including the stem peptide profile. Only the phosphate content of cell walls was reduced, consistent with the absence of phosphocholine. In vivo, the lack of active autolysins became apparent by impaired cell separation at the end of cell division and by resistance against stationary-phase and penicillin-induced lysis. Due to the absence of choline from LTA, PspA was lost into the surrounding medium, whereas, in spite of choline-free TA, the amidase was retained on the cells. Both choline-independent strains retained the capacity to incorporate nutritional choline into TAs: when grown in the presence of choline, PspA was retained, cells separated normally and became penicillin-sensitive, and phosphocholine was discovered on LTA and TA.

The metabolic background to dispense with the nutritional requirement for choline is not yet understood. A hint may come from in vitro experiments showing that the synthesis of peptidoglycan was also inhibited by choline deprivation (14). It is likely that the pneumococcal TA, like the TAs of other gram-positive bacteria, is synthesized in linkage to polyprenol phosphate and that only completed phosphocholine- or phosphoethanolamine-substituted TA is transferred through the plasma membrane to the peptidoglycan. Polyprenol-linked TA lacking these substituents would trap polyprenol phosphate, rendering it unavailable for peptidoglycan synthesis. In this hypothesis, the nutritional requirement for choline resides in a recognition site of the transferase for phospho-amino-alcohols on TA, which may have been altered in the choline-independent strains.

REFERENCES

1. **Behr, T., W. Fischer, J. Peter Katalinic, and H. Egge.** 1992. The structure of pneumococcal lipoteichoic acid. Improved preparation, chemical and mass spectrometric studies. *Eur. J. Biochem.* **207:**1063–1075.
2. **Briles, E. B., and A. Tomasz.** 1970. Radioautographic evidence for equatorial wall growth in a gram positive bacterium: segregation of choline ³H-labeled teichoic acid. *J. Cell Biol.* **47:**786–790.
3. **Briles, E. B., and A. Tomasz.** 1975. Membrane lipoteichoic acid is not a precursor to wall teichoic acid in pneumococci. *J. Bacteriol.* **122:**335–337.
4. **Brundish, D. E., and J. Baddiley.** 1968. Pneumococcal C-substance, a ribitol teichoic acid containing choline phosphate. *Biochem. J.* **110:**573–582.
5. **Cossart, P., and R. Jonquieres.** 2000. Sortase, a universal target for therapeutic agents against gram-positive bacteria? *Proc. Natl. Acad. Sci. USA* **97:**5013–5015.
6. **de las Rivas, B., J. L. Garcia, R. Lopez and P. Garcia.** 2001. Molecular characterization of the pneumococcal teichoic acid phosphorylcholine esterase. *Microb. Drug Resist.* **7:**213–222.
7. **Fernandez-Tornero, C., E. Garcia, R. Lopez, G. Gimenez-Gallego, and A. Romero.** 2002. Two new crystal forms of the choline-binding domain of the major pneumococcal autolysin: insights into the dynamics of the active homodimer. *J. Mol. Biol.* **321:**163–173.
8. **Filipe, S. R., and A. Tomasz.** 2000. Inhibition of the expression of penicillin resistance in *Streptococcus pneumoniae* by inactivation of cell wall muropeptide branching genes. *Proc. Natl. Acad. Sci. USA* **97:**4891–4896.
9. **Filipe, S. R., M. G. Pinho, and A. Tomasz.** 2000. Characterization of the *murMN* operon involved in the synthesis of branched peptidoglycan peptides in *Streptococcus pneumoniae.* *J. Biol. Chem.* **275:**27768–27774.
10. **Filipe, S. R., E. Severina, and A. Tomasz.** 2000. Distribution of the mosaic structured *murM* genes among natural populations of *Streptococcus pneumoniae.* *J. Bacteriol.* **182:**6798–6805.
11. **Filipe, S. R., E. Severina, and A. Tomasz.** 2001. Functional analysis of *Streptococcus pneumoniae* MurM reveals the region responsible for its specificity in the synthesis of peptidoglycan branched peptides. *J. Biol. Chem.* **276:**39618–39628.
12. **Filipe, S. R., E. Severina, and A. Tomasz.** 2002. The *murMN* operon: a functional link between antibiotic resistance and antibiotic tolerance in *Streptococcus pneumoniae.* *Proc. Natl. Acad. Sci. USA* **99:**1550–1555.
13. **Fischer, W.** 1990. Bacterial phosphoglycolipids and lipoteichoic acids, p. 123–234. *In* M. Kates (ed.), *Handbook of Lipid Research*, Vol. 6, *Glycolipids, Phosphoglycolipids, and Sulfoglycolipids.* Plenum Press, New York, N.Y.
14. **Fischer, H., and A. Tomasz.** 1985. Peptidoglycan cross-linking and teichoic acid attachment in *Streptococcus pneumoniae.* *J. Bacteriol.* **163:**46–54.
15. **Fischer, W., T. Behr, R. Hartmann, J. P. Katalinic, and H. Egge.** 1993. Teichoic acid and lipoteichoic acid of *Streptococcus pneumoniae* possess identical chain structures. A reinvestigation of teichoic acid (C polysaccharide). *Eur. J. Biochem.* **215:**851–857.
16. **Fiser, A., S. R. Filipe, and A. Tomasz.** 2003. Cell wall branches, penicillin resistance and the secrets of the MurM protein. *Trends Microbiol.* **11:**547–553.
17. **Garcia, P., J. L. Garcia, E. Garcia, and R. Lopez.** 1989. Purification and characterization of the autolysin glycosidase of *Streptococcus pneumoniae.* *Biochem. Biophys. Res. Commun.* **158:**251–256.
18. **Garcia, P., M. P. Gonzalez, E. Garcia, R. Lopez, and J. L. Garcia.** 1999. LytB, a novel pneumococcal murein hydrolase essential for cell separation. *Mol. Microbiol.* **31:**1275–1277.
19. **Garcia, J. L., A. R. Sanchez-Beato, F. J. Medrano, and R. Lopez.** 2000. Versatility of choline-binding domain, p. 231–245. *In* A. Tomasz (ed.), *Streptococcus pneumoniae, Molecular Biology and Mechanisms of Disease.* Mary Ann Liebert Inc., Larchmont, N.Y.
20. **Garcia-Bustos, J., and A. Tomasz.** 1990. A biological price of antibiotic resistance: major changes in the peptidoglycan structure of penicillin-resistant pneumococci. *Proc. Natl. Acad. Sci. USA* **87:**5414–5419.

21. **Garcia-Bustos, J. F., B. T. Chait, and A. Tomasz.** 1987. Structure of the peptide network of pneumococcal peptidoglycan. *J. Biol. Chem.* **262:**15400–15405.

22. **Giudicelli, S., and A. Tomasz.** 1984. Attachment of pneumococcal autolysin to wall teichoic acids, an essential step in enzymatic wall degradation. *J. Bacteriol.* **158:**1188–1190.

23. **Goebel, W. F., T. Shedlovsky, G. I. Lavin, and M. H. Adams.** 1943. The heterophil antigen of *Pneumococcus*. *J. Biol. Chem.* **148:**1–15.

24. **Gosink, K. K., E. R. Mann, C. Guglielmo, E. I. Tuomanen, and H. R. Masure.** 2000. Role of novel choline binding proteins in virulence of *Streptococcus pneumoniae*. *Infect. Immun.* **68:**5690–5695.

25. **Hakenbeck, R., A. Konig, I. Kern, M. van der Linden, W. Keck, D. Billot-Klein, R. Legrand, B. Schoot, and L. Gutmann.** 1998. Acquisition of five high-M_r penicillin-binding protein variants during transfer of high-level β-lactam resistance from *Streptococcus mitis* to *Streptococcus pneumoniae*. *J. Bacteriol.* **180:**1831–1840.

26. **Hoeltje, J.-V., and A. Tomasz.** 1974. Teichoic acid phosphorylcholine esterase. A novel enzyme activity in pneumococcus. *J. Biol. Chem.* **249:**7032–7034.

27. **Holtje, J.-V., and A. Tomasz.** 1975. Specific recognition of choline residues in the cell wall teichoic acid by the *N*-acetylmuramyl-L-alanine amidase of pneumococcus. *J. Biol. Chem.* **250:**6072–6076.

28. **Hoskins, J., W. E. Alborn, Jr., J. Arnold, L. C. Blaszczak, S. Burgett, B. S. DeHoff, S. T. Estrem, L. Fritz, D. J. Fu, W. Fuller, C. Geringer, R. Gilmour, J. S. Glass, H. Khoja, A. R. Kraft, R. E. Lagace, D. J. LeBlanc, L. N. Lee, E. J. Lefkowitz, J. Lu, P. Matsushima, S. M. McAhren, M. McHenney, K. McLeaster, C. W. Mundy, T. I. Nicas, F. H. Norris, M. O'Gara, R. B. Peery, G. T. Roberson, P. Rockey, P. M. Sun, M. E. Winkler, Y. Yang, M. Young-Bellido, G. Zhao, C. A. Zook, R. H. Baltz, S. R. Jaskunas, P. R. Rosteck, Jr., P. L. Skatrud, and J. I. Glass.** 2001. Genome of the bacterium *Streptococcus pneumoniae* strain R6. *J. Bacteriol.* **183:**5709–5717.

29. **Jennings, H. J., C. Lugowski, and N. M. Young.** 1980. Structure of the complex polysaccharide C-substance from *Streptococcus pneumoniae* type 1. *Biochemistry* **19:**4712–4719.

30. **Kamerling, J. P.** 2000. Pneumococcal polysaccharides: a chemical view, p. 81–115. *In* A. Tomasz (ed.), Streptococcus pneumoniae, *Molecular Biology and Mechanisms of Disease.* Mary Ann Liebert Inc., Larchmont, N.Y.

31. **Kharat, A. S., and A. Tomasz.** 2003. Inactivation of the srtA gene affects localization of surface proteins and decreases adhesion of *Streptococcus pneumoniae* to human pharyngeal cells in vitro. *Infect. Immun.* **71:**2758–2765.

32. **Klein, R. A., R. Hartmann, H. Egge, T. Behr, and W. Fischer.** 1996. The aqueous solution structure of a lipoteichoic acid from *Streptococcus pneumoniae* strain R6 containing 2,4-diamino-2,4,6-trideoxy-galactose: evidence for conformational mobility of the galactopyranose ring. *Carbohyd. Res.* **281:**79–98.

33. **Laitinen, H., and A. Tomasz.** 1990. Changes in composition of peptidoglycan during maturation of the cell wall in pneumococci. *J. Bacteriol.* **172:**5961–5967.

34. **Liu, T.-Y., and E. C. Gotschlich.** 1967. Muramic acid phosphate as a component of the muropeptide of Gram-positive bacteria. *J. Biol. Chem.* **242:**471–476.

35. **Lopez, R., E. Garcia, P. Garcia, and J. L. Garcia.** 2000. The pneumococcal cell wall degrading enzymes: a modular design to creat new lysins?, p. 197–211. *In* A. Tomasz (ed.), Streptococcus pneumoniae, *Molecular Biology and Mechanisms of Disease.* Mary Ann Liebert Inc., Larchmont, N.Y.

36. **Morlot, C., A. Zapun, O. Dideberg, and T. Vernet.** 2003. Growth and division of *Streptococcus pneumoniae*: localization of the high molecular weight penicillin-binding proteins during the cell cycle. *Mol. Microbiol.* **50:**845–855.

37. **Morlot, C., M. Noirclerc-Savoye, A. Zapun, O. Dideberg, and T. Vernet.** 2004. The D,D-carboxypeptidase PBP3 organizes the division process of *Streptococcus pneumoniae*. *Mol. Microbiol.* **51:**1641–1648.

38. **Mosser, J. L., and A. Tomasz.** 1970. Choline-containing teichoic acid as a structural component of pneumococcal cell wall and its role in sensitivity to lysis by an autolytic enzyme. *J. Biol. Chem.* **245:**287–298.

39. **Rane, L., and Y. Subbarow.** 1940. Nutritional requirements of the pneumococcus. 1. Growth factors for types I, II, V, VII, VIII. *J. Bacteriol.* **40:**695–704.

40. **Rohrer, S., and B. Berger-Bachi.** 2003. FemABX peptidyl transferases: a link between branched-chain cell wall peptide formation and beta-lactam resistance in gram-positive cocci. *Antimicrob. Agents Chemother.* **47:**837–846.

41. **Severin, A., and A. Tomasz.** 1996. Naturally occurring peptidoglycan variants of *Streptococcus pneumoniae*. *J. Bacteriol.* **178:**168–174.

42. **Severin, A., C. Schuster, R. Hakenbeck, and A. Tomasz.** 1992. Altered murein composition in a DD-carboxypeptidase mutant of *Streptococcus pneumoniae*. *J. Bacteriol.* **174:**5152–5155.

43. **Severin, A., M. V. Vaz Pato, A. M. Sa Figueiredo, and A. Tomasz.** 1995. Drastic changes in the peptidoglycan composition of penicillin resistant laboratory mutants of *Streptococcus pneumoniae*. *FEMS Microbiol. Lett.* **130:**31–35.

44. **Severin, A., A. M. S. Figueiredo, and A. Tomasz.** 1996. Separation of abnormal cell wall composition from penicillin resistance through genetic transformation of *Streptococcus pneumoniae*. *J. Bacteriol.* **178:**1788–1792.

45. **Severin, A., D. Horne, and A. Tomasz.** 1997. Autolysis and cell wall degradation in a choline-independent strain of *Streptococcus pneumoniae*. *Microb. Drug Resist.* **3:**391–400.

46. **Severin, A., E. Severina, and A. Tomasz.** 1997. Abnormal physiological properties and altered cell wall composition in *Streptococcus pneumoniae* grown in the presence of clavulanic acid. *Antimicrob. Agents Chemother.* **41:**504–510.

47. **Smith, A. M., and K. P. Klugman.** 2000. Non-penicillin-binding protein mediated high-level penicillin and cephalosporin resistance in a Hungarian clone of *Streptococcus pneumoniae*. *Microb. Drug Resist.* **6:**105–110.

48. **Sørensen, U. B. S., and J. Henrichsen.** 1987. Cross reaction between pneumococci and other streptococci due to C polysaccharide and F antigen. *J. Clin. Microbiol.* **25:**1854–1859.

49. **Sørensen, U. B. S., J. Blom, A. Birch-Andersen, and J. Henrichsen.** 1988. Ultrastructural localization of capsules, cell wall polysaccharide, cell wall proteins, and F antigen in pneumococci. *Infect. Immun.* **56:**1890–1896.

50. **Tillet, W. S., W. F. Goebel, and O. T. Avery.** 1930. Chemical and immunological properties of a species-specific carbohydrate of pneumococci. *J. Exp. Med.* **52:**895–900.

51. **Tomasz, A.** 1967. Choline in the cell wall of a bacterium: novel type of polymer-linked choline in pneumococcus. *Science* **157:**694–697.

52. **Tomasz, A.** 1968. Biological consequences of the replacement of choline by ethanolamine in the cell wall of pneu-

mococcus: chain formation, loss of transformability, and loss of autolysis. *Proc. Natl. Acad. Sci. USA* **59:**86–93.

53. **Tomasz, A.** 2000. *Streptococcus pneumoniae*: functional anatomy, p. 9–23. *In* A. Tomasz (ed.), Streptococcus pneumoniae: *Molecular Biology and Mechanisms of Disease*. Mary Ann Liebert, Inc., New York, N.Y.

54. **Vollmer, W., and A. Tomasz.** 2000. The *pgdA* gene encodes for a peptidoglycan *N*-acetylglucosamine deacetylase in *Streptococcus pneumoniae. J. Biol. Chem.* **275:**20496–20501.

55. **Vollmer, W., and A. Tomasz.** 2001. Identification of the teichoic acid phosphorylcholine esterase in *Streptococcus pneumoniae. Mol. Microbiol.* **39:**1610–1622.

56. **Vollmer, W., and A. Tomasz.** 2002. Peptidoglycan *N*-acetylglucosamine deacetylase (PgdA)—a putative virulence factor in *Streptococcus pneumoniae. Infect. Immun.* **70:**7176–7178.

57. **Wagner, M.** 1964. Studies with fluorescent antibodies of growing bacteria. I. Regeneration of the cell wall in *Diplococcus pneumoniae. Zentbl. Bakteriol. Parasitenkd.* **195:**87–93.

58. **Weber, B., K. Ehlert, A. Diehl, P. Reichmann, H. Labischinski, and R. Hakenbeck.** 2000. The fib locus in *Streptococcus pneumoniae* is required for peptidoglycan cross-linking and PBP-mediated beta-lactam resistance. *FEMS Microbiol. Lett.* **188:**81–85.

59. **Yother, J., K. Leopold, J. White, and W. Fischer.** 1998. Generation and properties of a *Streptococcus pneumoniae* mutant which does not require choline or analogs for growth. *J. Bacteriol.* **180:**2093–2101.

60. **Zighelboim, S., and A. Tomasz.** 1980. Penicillin-binding proteins of the multiply antibiotic-resistant South African strains of *Streptococcus pneumoniae. Antimicrob. Agents Chemother.* **17:**434–442.

Streptococcus pneumoniae Capsular Polysaccharide

JAMES C. PATON AND JUDY K. MORONA

20

The presence of what is now recognized as the polysaccharide capsule on the surface of *Streptococcus pneumoniae* was noted by Pasteur in the first published description of the organism in 1880, and since that time it has been the direct or indirect focus of intensive investigation (reviewed by Austrian [6, 7]). Studies during the first three decades of the 20th century demonstrated the existence of multiple capsular types of *S. pneumoniae* and the fact that antibodies to the capsule conferred type-specific protection against challenge in laboratory animals. The capsular material itself was isolated by Dochez and Avery in 1917 (28), but the fact that it was immunogenic led them to believe that this "soluble substance of the pneumococcus" was proteinaceous in nature. It was not until 1925 that Avery and colleagues (10, 11) demonstrated that the pneumococcal capsule consisted of polysaccharide, the first nonprotein antigen to be recognized.

The capsule forms the outermost layer of all fresh clinical isolates of *S. pneumoniae*. It is approximately 200 to 400 nm thick (81) and, with the exception of type 3, appears to be covalently attached to the outer surface of the cell wall peptidoglycan (82). A total of 90 structurally and serologically distinct capsular polysaccharide (CPS) types have been recognized to date (38), and the chemical structures of the repeat units for approximately 60% of these have been determined (reviewed by van Dam et al. [86]). The simplest CPS types are linear polymers with repeat units comprising two or more monosaccharides. The more complicated structural types are branched polysaccharides with repeat unit backbones composed of one to six monosaccharides plus additional side chains. Two nomenclature systems for CPS serotypes have been developed, but the Danish system, which combines antigenically cross-reacting types into groups, is now preferred to the American system, which lists serotypes in chronological order of discovery.

The polysaccharide capsule is considered to be a sine qua non of pneumococcal virulence (7). As mentioned above, all fresh clinical isolates of *S. pneumoniae* are encapsulated, and spontaneous nonencapsulated derivatives of such strains are almost completely avirulent. Indeed, as early as 1931, Avery and Dubos (9) demonstrated that enzymic depolymerization of the CPS of a type 3 pneumococcus in-creased its 50% lethal dose more than 10^5-fold. More recently, similar virulence defects have been reported for type 2 and 3 pneumococci with defined mutations in genes essential for CPS synthesis (14, 37, 58, 71). The clear morphological distinction between encapsulated ("smooth") and nonencapsulated ("rough") pneumococci, as well as the massive difference in virulence, facilitated early studies on the phenomenon of "capsular transformation." This was first demonstrated by Griffith (34), who found that a proportion of mice injected with a mixture of live rough and killed smooth pneumococci died. Smooth pneumococci expressing the same capsular serotype as the killed smooth strain were isolated from the blood of these mice. The transformation reaction was subsequently performed in vitro and led to the seminal discovery that the "transforming principle," the carrier of genetic information, was in fact DNA (12).

ROLE IN VIRULENCE

The precise manner in which the pneumococcal capsule contributes to virulence is not fully understood, although it is known to have strong antiphagocytic properties in nonimmune hosts. The majority of CPS serotypes are highly charged at physiological pH, and this may directly interfere with interactions with phagocytes (53). Pneumococcal cell wall teichoic acid (often referred to as C polysaccharide) is capable of activating the alternative complement pathway. In addition, antibodies to this and other cell surface constituents (e.g., surface proteins), which are found in most adult sera, may result in activation of the classical complement pathway, as does interaction of the teichoic acid with C-reactive protein. However, the capsule forms an inert shield, which appears to prevent either the Fc region of immunoglobulin G or iC3b fixed to deeper cell surface structures from interacting with receptors on phagocytic cells (74, 95). Recent data suggest that the capsule may also reduce the total amount of complement deposited on the bacterial surface (2). Pneumococci belonging to different CPS serotypes vary in their capacity to resist phagocytosis in vitro and also in their ability to elicit a humoral immune

Gram-Positive Pathogens, 2nd edition, edited by Vincent A. Fischetti et al.
© 2006 ASM Press, Washington, D.C.

response (86), which no doubt accounts in large part for the fact that certain types are far more commonly associated with human disease (7). Otherwise isogenic pneumococci expressing different CPS serotypes, generated by in vitro or in vivo transformation, also exhibit marked differences in virulence for mice (44, 75). However, other factors contribute to some extent; Kelly et al. (44) demonstrated that virulence is also influenced by the recipient strain. The difference in virulence between pneumococcal serotypes is clearly a function of the biological properties of the CPS itself and is not simply related to the thickness of capsule. For example, type 3 and type 37 pneumococci both produce very thick capsules, but only the former is of high virulence for humans or laboratory animals (7). Notwithstanding this, analysis of a series of mutants producing different amounts of CPS has demonstrated that within a given strain and serotype, virulence of S. pneumoniae is directly related to capsular thickness (57).

CPS-BASED VACCINES

In the immune host, binding of specific antibody to the CPS results in opsonization and rapid clearance of the invading pneumococci. For this reason a polyvalent pneumococcal CPS vaccine was licensed in the late 1970s. The vaccine provides serotype-specific protection, and the current formulation contains CPS purified from 23 of the 90 recognized types. However, the distribution of S. pneumoniae serotypes varies both temporally and geographically. The current vaccine covers approximately 85 to 90% of disease-causing serotypes in the United States or Europe, but in parts of Asia, coverage is <60% (53). Furthermore, serotype prevalence data are scanty for many developing countries, and so vaccine coverage is uncertain. The CPS vaccine is undoubtedly protective in healthy adults (against invasive infections caused by those serotypes included in the vaccine), but the efficacy is much lower in other groups at high risk of pneumococcal infection, such as the elderly; patients with underlying pulmonary, cardiac, or renal disease; and immunocompromised patients. Efficacy is poorest in young children, for whom the existing formulation has little or no demonstrable clinical benefit. Polysaccharides are T-cell-independent antigens and so are poorly immunogenic in children under 5 years. This is particularly so for the five pneumococcal CPS types that most commonly cause invasive disease in children (29).

Poor immunogenicity of polysaccharide antigens can be overcome by conjugation to protein carriers, converting them into T-cell-dependent antigens, which are far more immunogenic, and a highly efficacious pneumococcal PS-protein conjugate vaccine has been licensed (46). However, the protection is still serotype-specific, and because of the high cost, the number of types covered has been reduced. The new vaccine is seven-valent, and second-generation conjugate formulations currently undergoing clinical trials cover 11 types at most. Thus, although these vaccines may provide improved protection, it is against a much more limited serotype range (78). Furthermore, studies of nasopharyngeal colonization with S. pneumoniae during trials of these conjugate vaccines have shown that although carriage of vaccine types was reduced, the vacated niche was promptly occupied by nonvaccine serotypes potentially capable of causing disease in humans (59, 76). Nasopharyngeal carriage of S. pneumoniae is generally accepted as a prerequisite for pneumococcal disease, and serotypes being carried usually correlate with those causing disease in a community. Consistent with this principle, a trial of the pneumococcal PS-protein conjugate vaccine in Finland demonstrated that although the vaccine reduced the incidence of otitis media caused by vaccine types, this was largely offset by a 33% increase in cases of otitis media caused by nonvaccine serotypes (30). Thus, widespread introduction of such vaccines may alter the serotype distribution of pneumococcal disease rather than reducing its overall impact.

CPS BIOSYNTHESIS GENES

The frequency of transformation of pneumococci from one CPS type to another observed during the classical studies of Griffith, Avery, and others (12, 34) suggested that at least those genes encoding the serotype-specific components of the CPS biosynthetic machinery were closely linked on the chromosome. Additional genetic and biochemical studies by Austrian, Bernheimer, and colleagues (8, 16) demonstrated that during transformation, CPS biosynthesis genes are transferred as a cassette. In the last decade, advances in gene cloning and DNA technology have resulted in publication of annotated sequences of the complete genetic loci encoding CPS biosynthesis for about 20% of the 90 known S. pneumoniae serotypes (4, 27, 35, 39, 40, 42, 48–50, 54, 55, 63–67, 72, 73, 79, 87, 88). Characterization of the loci from the remaining types is in progress at the Sanger Institute, and sequences for 82 serotypes are currently available (80). Once these sequences have been annotated, analysis of the genes present should increase our understanding of how the pneumococcus acquired the capacity to synthesize such a vast array of CPS serotypes. The type 3 locus (designated cps3 [27] or cap3 [31]) and the type 19F locus (designated cps19f [35]) were the first to be completely sequenced and were shown to be located at the same position in the chromosome, between dexB and aliA (4, 27, 35, 63). The cps loci from all other serotypes, except type 37, have also been localized to the same position on the chromosome. Type 37 has a cryptic copy of the type 33F cps locus located between dexB and aliA; it contains many deletions and point mutations and is not involved in type 37 CPS production. Instead, type 37 CPS biosynthesis is directed by one gene, tts, located elsewhere on the chromosome (55). The mechanism of CPS biosynthesis that occurs in type 37 is similar to that in type 3, and this mechanism is distinct from that in all other serotypes analyzed to date, as described below.

CPS Biosynthesis in Types 3 and 37

Type 3 CPS has a simple disaccharide repeat unit comprising Glc and GlcA. The cps3/cap3 locus contains only three intact genes, which are transcribed as a single unit (4, 27). The first gene (cps3D or cap3A) encodes the UDP-Glc dehydrogenase required for the synthesis of UDP-GlcA (3). The second gene (cps3S or cap3B) encodes the type 3 synthase, a processive β-glycosyltransferase that links the alternating Glc and GlcA moieties via distinct glycosidic bonds (4, 5, 27). There is a significant degree of amino acid sequence similarity between Cps3S/Cap3B and other bacterial polysaccharide synthases, including HasA, which synthesizes the hyaluronic acid capsule of group A streptococci (24). These synthases have a common predicted architecture, with four transmembrane domains and a large central cytoplasmic domain. This latter region is believed to contain two distinct catalytic sites capable of forming the two different glycosidic linkages (43). The final complete gene in the *cps3/cap3* locus (*cps3U* or *cap3C*) encodes a Glc-1-phosphate uridylyltransferase. However, Cps3U/Cap3C is not essential for CPS biosynthesis in type 3 pneumococci, as inactivation of the gene has no effect on CPS production (27). This enzyme is also encoded by the *galU* gene, which is located elsewhere in the chromosome, and its product (UDP-Glc) is required for CPS biosynthesis in all pneumococci, regardless of serotype (60).

Type 37 CPS is a branched polysaccharide with a linear backbone of Glc with a monosaccharide Glc side chain; a single gene, *tts*, is responsible for its biosynthesis. Tts is a processive β-glucosyltransferase, which has sequence similarities to a group of plant and bacterial cellulose synthases, especially to the highly conserved motifs thought to be critical for catalysis and/or binding of UDP-Glc (55). The presence of the Cps3S/Cap3B and Tts synthases is sufficient for type 3 and type 37 CPS production, respectively. However, the expression of type 3 CPS also relies on the capacity of the cell to synthesize UDP-GlcA, which is a substrate for Cps3S/Cap3B (18, 56). CPS synthesized by processive transferases does not need a dedicated export system, as the C-terminal hydrophobic domains of these proteins have been predicted to form a pore in the membrane, through which the growing polysaccharide chain is extruded as it is synthesized (43).

CPS Biosynthesis in Other Pneumococcal Serotypes

The CPS structures for most other serotypes are more complex than types 3 and 37, with an oligosaccharide repeat unit consisting of at least three sugars (86). These repeat units may also contain phosphodiester linkages, pyruvate, glycerol, phosphoryl choline, and/or ribitol. In some serotypes one or more sugars may also be acetylated (86). The other *cps* loci analyzed to date are also much more complex than those of types 3 and 37, which is suggestive of a more elaborate biosynthetic mechanism. The loci vary from approximately 13 to 30 kb in length and consist of 10 to more than 20 genes that appear to be arranged as a single transcriptional unit (Fig. 1). Functions for their gene products have been proposed on the basis of amino acid sequence similarities with known proteins and with reference to the structure of their CPS repeat unit. The genes in each *cps* locus are arranged in a common blockwise fashion. The first four genes (*cpsA* to *D*) comprise the 5′ portion of each *cps* locus; their products have all been implicated in regulation of CPS production and are discussed in detail below. The central portion of each *cps* locus encodes the glycosyltransferases responsible for assembly of oligosaccharide repeat units, the repeat unit transporter (Wzx), and the polysaccharide polymerase (Wzy). The 3′ regions of the various *cps* loci encode enzymes for synthesis of activated monosaccharide precursors (Fig. 1).

Biosynthesis of CPS in these serotypes occurs by a Wzx/Wzy-dependent polymerization pathway, analogous to group 1 CPS biosynthesis in *Escherichia coli* and O-antigen assembly in gram-negative bacteria (94). The initial step involves transfer of a sugar moiety to a lipid carrier on the cytoplasmic face of the cell membrane. In serotypes containing glucose, this step is carried out by the glucose-1-phosphate transferase CpsE (47). These glycosyltransferases are membrane-associated, which facilitates interaction with the lipid carrier (35). Other glycosyltransferases then catalyze the sequential transfer of the other component monosaccharide precursors (synthesized in the cytoplasm by the activated monosaccharide synthesis genes) to form the polysaccharide

repeat unit. These lipid-linked repeat units are then translocated from the cytoplasmic to the extracellular side of the cell membrane by the repeat unit transporter and polymerized in a blockwise fashion by the polysaccharide polymerase, extending the polysaccharide at the reducing terminus. CpsC and D have been proposed to regulate CPS chain length and possibly export via tyrosine phosphorylation as described below. Pneumococcal CPS is believed to be linked covalently to the cell wall peptidoglycan (82). Although the precise nature of this linkage and the enzyme(s) responsible are unknown, it may be analogous to that in type III *Streptococcus agalactiae*, in which the CPS is linked via an additional oligosaccharide and a phosphodiester bond to GlcNAc residues on the peptidoglycan (25).

COMPARISON OF *cps* LOCI FROM DIFFERENT SEROTYPES

The common blockwise arrangement of genes in the *cps* loci analyzed to date (Fig. 1) suggests that comparison of their genes may provide insights into the mechanism of generation of capsular diversity. The first five genes *cpsA* to *E* are highly conserved in most *cps* loci, except in serotypes that do not contain glucose in their CPS, such as types 1, 4, 5, and 10. These serotypes contain the *cpsA* to *D* genes, but not the *cpsE* gene, which encodes a glucose-1-phosphate transferase (47, 66, 72). The *cpsA* to *E* genes exist as two distinct classes, suggesting that all pneumococcal serotypes have evolved from two distinct ancestral *cps* types (66). Members of class I include types 1, 5, 7F, 9N, 10, 14, 15B, 15C, 18C, 19F, 19B, and 19C; class II includes types 2, 6A, 6B, 8, 9V, 19A, 23F, and 33F. Within one class, the *cpsA* to *E* genes share greater than 95% sequence identity, but share only 70% sequence identity with those from the other class. Some serotypes, such as type 4, contain a hybrid *cps* locus, where recombination between a class I and a class II locus has occurred. Additionally, the type 3 and type 37 *cps* loci, which do not require the *cpsA* to *E* gene products for CPS biosynthesis, still retain defective copies of these class I and class II sequences, respectively, suggesting that they also have a common ancestry with other pneumococci of the same class.

The central part of the *cps* locus is serotype-specific, as it encodes glycosyltransferases that exhibit a high degree of substrate specificity and form distinct glycosidic linkages. Accordingly, serotypes have transferases in common only if they have identical glycosidic linkages between identical sugar moieties in their CPS. For example, closely related glycosyltransferases present in the *cps* loci of types 19F, 9N, and 9V are predicted to join ManNAc via a (β1→4) linkage to Glc (63). This central region also encodes the repeat unit transporter and the polysaccharide polymerase. Closely related polymerase and transporter genes only occur between serotypes with near-identical repeat unit backbones, such as types 19F and 19A, types 19B and 19C, and types 14, 15B, and 15C (66, 67, 88). Indeed, the specificity of these genes is enabling definitive determination of serotype by multiplex PCR rather than by classical serological methods (52).

Activated monosaccharide synthesis genes, located at the 3′ end of the *cps* locus, are more widely distributed among different *S. pneumoniae* serotypes than genes encoding specific glycosyltransferases. Even so, there is substantial deviation among functional homologs. For example, the genes encoding UDP-Glc dehydrogenases from serotypes 1, 2, and 3 exhibit 60 to 90% amino acid sequence identity to

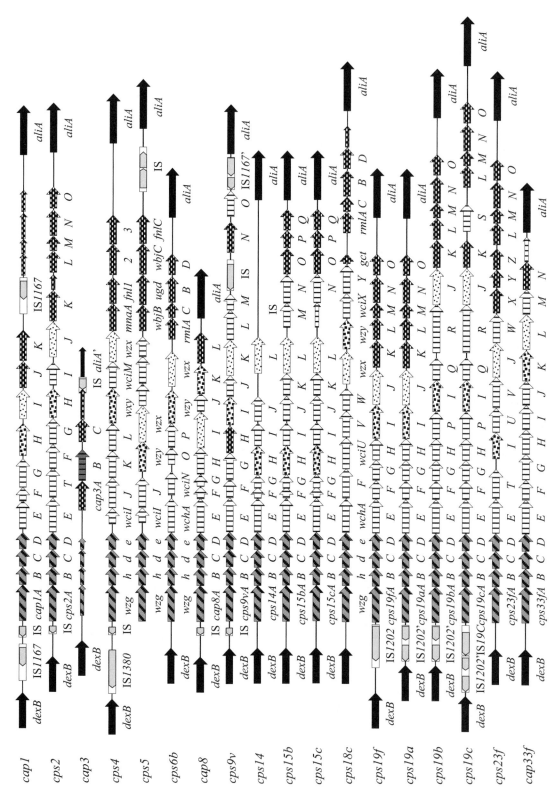

each other. Also, the type 19F and type 4 genes encoding UDP-*N*-acetylglucosamine-2-epimerases are only 66% identical. However, the four genes encoding synthesis of dTDP-rhamnose from Glc-1-phosphate have been shown to be closely related, with greater than 95% amino acid sequence identity in all serotypes tested that contain rhamnose in their CPS (63). Interestingly, defective copies of these genes have also been identified at the 3′ end of *cps* loci from serotypes not containing rhamnose in their CPS, such as type 1. Their presence suggests that the ancestor of type 1 pneumococci may have been a serotype containing rhamnose (72).

COMPARISON OF *cps* LOCI FROM CLOSELY RELATED SEROTYPES

Group 19 comprises serotypes 19F, 19A, 19B, and 19C, and the structures of their CPS repeat units are shown in Fig. 2. Types 19F and 19A have an identical trisaccharide repeat unit and differ only in the nature of the glycosidic linkage formed during polymerization. Both of these types are important causes of human disease, particularly in young children. Types 19B and 19C are only occasionally associated with human disease; they have more complicated CPS repeat units than types 19F and 19A, but differ from each other only by one additional glucose side chain in type 19C. The *cps* loci of all four members of group 19 are closely related (35, 63, 64, 67). Thirteen genes (*cps19A* to *H* and *K* to *O*) are conserved in all four serogroup members. Nearly all of the common genes from types 19F, 19B, and 19C are >95% identical to each other, whereas those from the type 19A *cps* locus are more divergent. Although the 19A and 19F loci are identical in terms of the number and arrangement of the genes present, sequence homology between individual *cps19a* and *cps19f* genes varies from 70 to 99% (at both the DNA and the deduced amino acid sequence level). This sequence divergence is surprising given that only an alteration in *cps19fI*, such that an α(1→3) rather than an α(1→2) linkage is formed during polymerization of the repeat units, is required to change a type 19F pneumococcus into type 19A (67). This suggests either that the type 19F and 19A loci diverged in the distant evolutionary past or that their component genes originated from different sources.

The type 19B and 19C *cps* loci are nearly identical and contain five genes between *cps19H* and *cps19K* unrelated to the genes in the type 19F and 19A *cps* loci (Fig. 1). These genes encode a repeat unit transporter and a polysaccharide polymerase that are specific for type 19B and 19C CPS, as well as two additional putative glycosyltransferases and an unknown protein that might possibly be involved in synthesis of an activated ribose precursor. These five genes encode all of the functions required to convert a type 19F pneumococcus to type 19B (64). The type 19C *cps* locus differs from the type 19B locus only in the insertion of a putative glucosyl transferase gene (*cps19cS*) between *cps19cK*

and *cps19cL* (Fig. 1). The presence of this gene accounts for the additional Glc side chain in the otherwise identical repeat unit structures (Fig. 2).

The structures of the type 6A and 6B CPS repeat units differ only in the linkage between the rhamnose and ribitol moieties (Fig. 2), and as expected, the type 6A and 6B *cps* loci are very closely related (42, GenBank accession number AY078347). The sequences for the genes encoding the rhamnosyltransferase differ only by a few bases and result in very few amino acid changes. However, these changes are clearly sufficient to alter the serotype-determining linkage.

Serotypes 14, 15B, and 15C have closely related CPS structures, as shown in Fig. 2. They differ only in the structure of their side chains, although type 15B is also O-acetylated at an as yet undetermined site on 70% of its repeating units (41). Similarly, their *cps* loci are closely related, containing 11 common genes. These encode the four transferases responsible for synthesis of the repeat unit backbone and the Gal side chain, as well as the repeat unit transporters and polysaccharide polymerases. An additional gene encoding the galactosyltransferase that adds a second Gal residue to the side chain is also present in the *cps* loci of all three serotypes, but the copy present in type 14 has mutated and encodes a nonfunctional truncated protein (88). The type 15B and 15C *cps* loci contain an additional five genes encoding an acetyltransferase, a glycerol-2-phosphotransferase, and three enzymes required for biosynthesis of CDP-2-glycerol. A defective copy of the glycerol-2-phosphotransferase gene is also found at the 3′ end of the type 14 *cps* locus (49). The presence of two defective genes in the type 14 *cps* locus, which are functional in the type 15B and 15C *cps* loci, suggests that these serotypes share a common ancestor. The presence of the genes involved in the addition of glycerol-2-phosphate to the polysaccharide repeat unit was unexpected, as the CPS structure determination for types 15B and 15C had identified the presence of choline, not glycerol-2-phosphate (Fig. 2) (41). However, one type 15B strain containing glycerol-2-phosphate has been reported in the literature (1). These results suggest that further investigation of the type 15B and 15C CPS structures is needed to confirm the presence of glycerol-2-phosphate. The difference between types 15B and 15C is due to the acetylation of the type 15B CPS, and is dependent upon the acetyltransferase gene. This gene is phase-variable and its activity is dependent upon the number of TA repeats present in the central region of the gene (88). The type 15B isolates examined, in which the coding sequence was in-frame, contained eight TA repeats, whereas the type 15C isolates contained either seven or nine TA repeats, resulting in a frame shift and premature termination of translation, inactivating the encoded enzyme (88). Spontaneous switching between types 15B and 15C has been demonstrated in the laboratory at a frequency of up to 1 in 250, and is likely to occur during natural infection, as these serotypes are often isolated together from children with otitis media (89).

FIGURE 1 Organization of the *cps* loci from various *S. pneumoniae* serotypes. The organizations are based on published data for types 1 (72), 2 (39), 3 (4, 27), 4 (40, 42), 5 (GenBank accession no. AY336008), 6B (42), 8 (42, 73), 9V (87), 14 (48, 49), 15B (88), 15C (88), 18C (42), 19F (35, 63), 19A (66, 67), 19B (64), 19C (67), 23F (65, 79), and 33F (54). Gene and locus designations are as published. Open reading frames (ORFs) within the DNA sequence are indicated by large boxed arrows. Highly conserved ORFs, or those encoding proteins belonging to a particular functional group, are identified as shown in the legend below the figure. Assignment of an ORF to a given function-related group has been based on the published information for each locus, as well as on additional database comparisons for some of the ORFs. The narrow boxed arrows represent cryptic ORFs not required for CPS biosynthesis in the respective serotype.

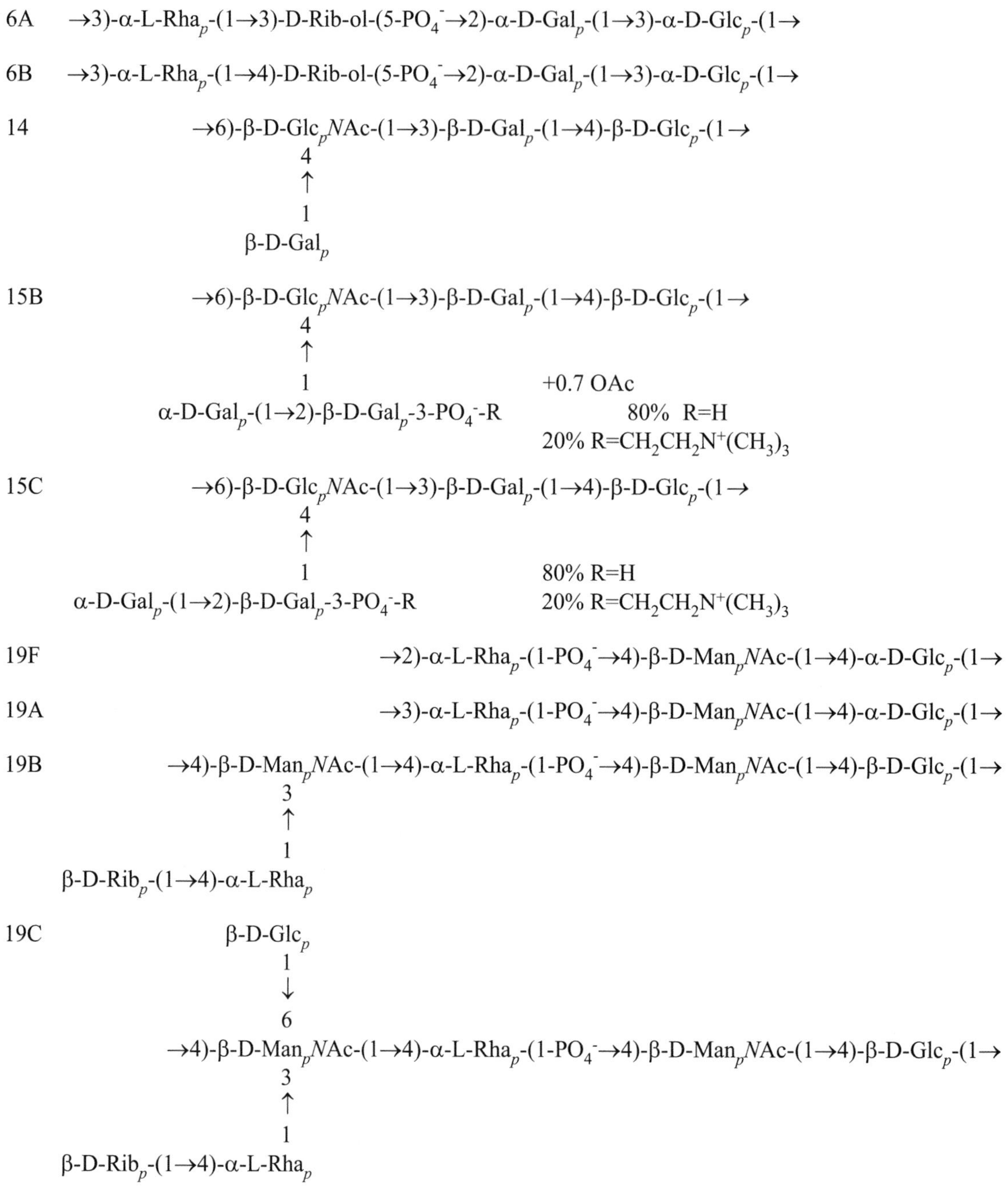

6A   →3)-α-L-Rha$_p$-(1→3)-D-Rib-ol-(5-PO$_4^-$→2)-α-D-Gal$_p$-(1→3)-α-D-Glc$_p$-(1→

6B   →3)-α-L-Rha$_p$-(1→4)-D-Rib-ol-(5-PO$_4^-$→2)-α-D-Gal$_p$-(1→3)-α-D-Glc$_p$-(1→

14   →6)-β-D-Glc$_p$NAc-(1→3)-β-D-Gal$_p$-(1→4)-β-D-Glc$_p$-(1→
                        4
                        ↑
                        1
                  β-D-Gal$_p$

15B   →6)-β-D-Glc$_p$NAc-(1→3)-β-D-Gal$_p$-(1→4)-β-D-Glc$_p$-(1→
                        4
                        ↑
                        1   +0.7 OAc
      α-D-Gal$_p$-(1→2)-β-D-Gal$_p$-3-PO$_4^-$-R   80% R=H
                                         20% R=CH$_2$CH$_2$N$^+$(CH$_3$)$_3$

15C   →6)-β-D-Glc$_p$NAc-(1→3)-β-D-Gal$_p$-(1→4)-β-D-Glc$_p$-(1→
                        4
                        ↑
                        1   80% R=H
      α-D-Gal$_p$-(1→2)-β-D-Gal$_p$-3-PO$_4^-$-R   20% R=CH$_2$CH$_2$N$^+$(CH$_3$)$_3$

19F   →2)-α-L-Rha$_p$-(1-PO$_4^-$→4)-β-D-Man$_p$NAc-(1→4)-α-D-Glc$_p$-(1→

19A   →3)-α-L-Rha$_p$-(1-PO$_4^-$→4)-β-D-Man$_p$NAc-(1→4)-α-D-Glc$_p$-(1→

19B   →4)-β-D-Man$_p$NAc-(1→4)-α-L-Rha$_p$-(1-PO$_4^-$→4)-β-D-Man$_p$NAc-(1→4)-β-D-Glc$_p$-(1→
             3
             ↑
             1
      β-D-Rib$_p$-(1→4)-α-L-Rha$_p$

19C   β-D-Glc$_p$
                     1
                     ↓
                     6
      →4)-β-D-Man$_p$NAc-(1→4)-α-L-Rha$_p$-(1-PO$_4^-$→4)-β-D-Man$_p$NAc-(1→4)-β-D-Glc$_p$-(1→
             3
             ↑
             1
      β-D-Rib$_p$-(1→4)-α-L-Rha$_p$

FIGURE 2   Comparison of the CPS biological repeat unit structures of *S. pneumoniae* serotypes 6A, 6B, 14, 15B, 15C, 19F, 19A, 19B, and 19C. These are based on published chemical repeat unit structures (17, 41, 86), adjusting for the fact that glucose is the first sugar of the biological repeat unit.

CAPSULAR TRANSFORMATION IN VIVO

There is growing evidence that the phenomenon of capsular transformation first observed by Griffith (34) is a common event in vivo. Application of modern molecular typing techniques has resulted in the detection of otherwise genetically indistinguishable pneumococci expressing different capsular types. This has been particularly evident in clonal groups that are resistant to multiple antibiotics. In-

deed, derivatives of a highly successful, multiresistant type 23F clone (which originated in Spain) expressing type 3, 9N, 14, 19A, and 19F capsules have been isolated (13, 20, 22, 75). The sequences of the regions from *dexB* to *cpsB* and from *cpsL* to *aliA* have been determined for the parental multiresistant type 23F strain and eight otherwise identical type 19F clinical isolates. Examination of polymorphisms in the conserved regions of the two *cps* loci indicated that in each case, the 5′ recombination occurred upstream of *dexB*.

In six of the eight type 19F strains, the 3' crossover point was downstream of *aliA*. However, in the other two, a recombination crossover point between the introduced type 19F sequences and the type 23F chromosome was identified; this was in *cps*M in one strain and *cps*N in the other (21). Thus, capsule switching involves exchange of very large DNA fragments, ranging from at least 15 kb to more than 22.5 kb. The existence of multiple crossover points as well as additional minor polymorphisms within the type 19F-derived *cps* genes also indicated that the eight multiresistant type 19F strains that were studied arose as a consequence of a minimum of four independent transformation events involving different type 19F donors. It therefore appears that these capsule switching events may be relatively common among pneumococci in nature (21). Multiple serotypes of *S. pneumoniae* are frequently carried concurrently in the human nasopharynx (7), providing ample opportunity for exchange of DNA between types. In addition to enhancing the spread of drug resistance among diverse capsular types, these exchanges may also provide a mechanism for evasion of serotype-specific host immune defenses, such as those resulting from immunization with pneumococcal CPS-protein conjugate vaccines, which provide cover against a limited range of serotypes.

Complete or partial insertion sequences have been located adjacent to many of the *cps* loci sequenced to date, and in some serotypes, the locus is flanked at both the 5' and 3' ends by such elements. The *cps*-flanking regions appear to be common targets for insertion elements, and this has led to the suggestion that they may play a role in horizontal transfer of *cps* genes (49, 72). There are precedents for this in other bacteria; the *Haemophilus influenzae* type b capsule genes, for example, are located on a 17-kb compound transposon (51). However, the identification of crossover points within the *cps* loci of the type 19F derivatives of the multiresistant type 23F *S. pneumoniae* clone referred to above confirms that at least in these cases, capsular exchange occurred as a consequence of homologous recombination rather than by transposition. Nevertheless, Muñoz et al. (72) have demonstrated that IS*1167* sequences flanking the *cap1* locus could direct ectopic integration of these genes into copies of IS*1167* located elsewhere in the pneumococcal chromosome, resulting in genetically binary strains.

REGULATION OF CPS PRODUCTION

Colonization of the nasopharyngeal mucosa is an essential first step in the pathogenesis of pneumococcal disease and is presumed to involve interaction between pneumococcal adhesins and specific receptors on host epithelial cells. In a proportion of cases, asymptomatic carriage progresses to invasive disease, and although the events involved are poorly understood, it is clearly a watershed in the bacteria-host relationship. Evidence has emerged that this transition involves a major switch in expression of important virulence determinants, as the pneumococcus adapts to the altered microenvironment (85). Maximal expression of capsule is clearly essential for systemic virulence, but the degree of exposure of other important pneumococcal surface structures, such as the adhesins, may also be influenced by capsular thickness. Nonencapsulated pneumococci exhibit higher adherence to human respiratory epithelial (A549) cells in vitro than otherwise isogenic derivatives expressing either type 3 or type 19F capsules (84). Thus, the very feature (encapsulation) that is absolutely essential for systemic virulence of *S. pneumoniae* could be disadvantageous during the

colonization phase. Pneumococci have been shown to undergo a bidirectional phase variation between two distinct colonial morphologies, described as opaque and transparent. The transparent phenotype exhibits increased in vitro adherence to buccal epithelial cells and cytokine-activated A549 cells relative to opaque variants of the same strain, as well as an enhanced capacity to colonize the nasopharynx of infant rats (23, 92). On the other hand, the opaque form is associated with massively increased virulence in animal models of systemic disease, and this correlates with increased production of CPS relative to cell wall teichoic acid, compared with the transparent phenotype (45). Phase variation also correlated with alteration in levels of several surface proteins, but the molecular mechanism involved has yet to be elucidated. A distinct "phase-variation" phenomenon has also been reported in type 3, 8, and 37 pneumococci grown on sorbarod biofilms. These strains exhibited switching from encapsulated to nonencapsulated phenotypes as a result of spontaneous tandem duplications of up to 239 bp in the *cap3A*, *cap8E*, and *tts* genes, respectively, which exhibited low-frequency reversion (90, 91). However, the in vivo relevance of this phenomenon is questionable, because at least some level of encapsulation appears to be required for pneumococci to stably colonize the nasopharynx (58).

Clearly, the capacity to regulate CPS production, at either the transcriptional or translational level, is important for the survival of the pneumococcus in different host environments. To date, no transcriptional control elements have been identified in association with the pneumococcal σ^{70} *cps* promoter, located about 20 nucleotides upstream of the initiation codon of the first *cps* gene (72). However, environmental conditions affecting the level of transcription of the *cps* locus have been identified. There is some evidence to suggest that the level of transcription of the *cps* locus differs in the transparent and opaque phase variants, as more CpsD could be detected by Western immunoblotting in the latter (93). This is consistent with observations that the opaque form expresses higher levels of CPS than the transparent form, thereby favoring survival in the lungs and bloodstream of the host (45). Direct evidence for transcriptional regulation of the *cps* locus in the host environment has been obtained using quantitative reverse transcriptase PCR. The level of *cps* mRNA relative to 16S rRNA in pneumococci isolated from the blood of infected mice was approximately fourfold higher compared to that in pneumococci grown in vitro (77). RegM, a homolog of the staphylococcal catabolite control protein CcpA involved in regulation of sugar metabolism pathways, was also shown to affect transcription of the *cps* locus (measured using a *cps::lacZ* transcriptional fusion in wild-type and *regM*-deleted pneumococci) (32). This study suggested that carbon source may also influence capsular expression (32).

The first four genes of the *cps* locus, *cps*A to D, are common to all pneumococcal serotypes, except types 3 and 37, and the products encoded by these genes, CpsA to D, have been shown to affect the level of CPS expression (68). Homologs of *cps*B to D are found in most gram-positive capsule loci, but *cps*A homologs have been identified only in the *cps* loci of other members of the genus *Streptococcus*. Interestingly, although CpsA resembles a *Bacillus subtilis* transcriptional attenuator, it appears to function as a transcriptional enhancer in group B streptococci. Reverse transcriptase PCR analysis indicated that a mutant in which the *cps1aA* gene was deleted expressed only 41% of wild-type *cps* mRNA levels, and also produced only 24% of wild-type CPS levels

(19). To date, *cpsA* has not been shown to affect *cps* transcription in *S. pneumoniae*. Type 19F and type 2 mutants in which the *cpsA* gene had been deleted produced reduced levels of CPS (approximately two-thirds that of the respective wild-type strains) (68, 70, 71). However, there were no obvious differences in the amounts of CpsB and CpsD proteins detected by Western blotting in any of the strains (15, 68).

CpsB, CpsC, and CpsD have all been implicated in regulation of CPS production at a posttranscriptional level. CpsC and CpsD belong to the PCP2b family of polysaccharide copolymerases and are predicted to function together in polymerization and export of CPS in a fashion similar to PCP2a proteins, such as Wzc and ExoP, in production of CPS and exopolysaccharide (EPS) in gram-negative bacteria (33, 35, 62, 94). CpsD has similarities to the C-terminal domain of PCP2a proteins and is an autophosphorylating protein-tyrosine kinase (62, 68). These kinases contain Walker A and B ATP-binding motifs and a tyrosine-rich region at the C terminus that becomes phosphorylated at multiple tyrosine residues, initially via the ATP-binding domain, and also via transphosphorylation, which occurs independently of the ATP-binding domain (14, 68, 70). In *S. pneumoniae*, the tyrosine-rich region of CpsD is arranged as an ordered [YGX] motif varying from two to four repeats in different serotypes, with four being the most prevalent. CpsC is a membrane protein with two membrane-spanning hydrophobic domains; the N and C termini are located in the cytoplasm, while the central portion is exposed on the external side of the cell membrane. It has similarities to PCP1/Wzz proteins associated with polymerization of O-antigen in gram-negative bacteria and to the N-terminal domain of PCP2a proteins (62). CpsC is required for initial CpsD tyrosine-autophosphorylation, but is not required for trans-phosphorylation between CpsD proteins (14, 68). CpsB has been identified as a manganese-dependent phosphotyrosine-protein phosphatase belonging to the polymerase and histidinol phosphatase family of phosphoesterases and is required to dephosphorylate CpsD (69). CpsB is also thought to be able to bind CpsD and prevent transphosphorylation between CpsD proteins (14). In the avirulent *S. pneumoniae* strain Rx1-19F, *cpsD* mutants containing either point mutations affecting the ATP-binding domain (Walker A motif) or a deletion of the C-terminal tyrosine-rich domain eliminated CPS production, indicating that both of these regions are important for function of CpsD (68, 70). Western immunoblotting studies of strains containing either CpsD proteins in which various combinations of the four tyrosine residues in the C-terminal [YGX]₄ repeat domain were changed to phenylalanine, or truncated CpsD proteins in which individual [YGX] repeats were deleted, showed that although the tyrosine in the first repeat is less efficiently phosphorylated than the other three, all four tyrosines are capable of being phosphorylated (70). Analysis of the CPS produced by these mutants indicated that CpsD required at least two intact [YGX] repeats or the presence of at least three of the four tyrosine residues in the [YGX]₄ repeat domain to produce wild-type levels of CPS. When CpsD contained either one intact [YGX] repeat or less than three tyrosine residues within the [YGX]₄ repeat domain, the mutant strains produced reduced amounts of a "mucoid" CPS (70). Phosphorylation of the tyrosines in the tyrosine-rich C-terminal region of PCP2a and PCP2b proteins may induce a conformational change that modulates their activity in CPS production. Replacing tyrosine with phenylalanine may have affected the conformation of CpsD, thereby interfering with its normal function and resulting in the expression of a mucoid phenotype. Strains in which *cpsB* had either been deleted or mutated to eliminate the phosphotyrosine-protein phosphatase activity of CpsB had high levels of phosphorylated CpsD protein detected by Western immunoblotting using anti-phosphotyrosine monoclonal antibody. Additionally, these strains exhibited a rough phenotype and produced very little CPS, suggesting that the phosphorylated form of CpsD was inactive (68, 70). Additionally, the loss of CpsB activity did not affect CPS production in strains that contained a CpsD protein that could no longer be phosphorylated due to mutation of all four tyrosines to phenylalanine in the [YGX]₄ repeat domain; these strains still expressed "mucoid" CPS. Thus, autophosphorylation of CpsD at tyrosine attenuates its activity and negatively regulates CPS production in Rx1-19F (68, 70).

Recently, the function of CpsB and CpsD in CPS production has been somewhat clouded by conflicting results obtained for the mouse virulent *S. pneumoniae* type 2 strain D39 (15, 71). D39 mutants in which the *cpsD* gene was deleted and mutants in which one, two, or all three of the tyrosines in the [YGX]₃ repeat domain of Cps2D were changed to phenylalanine exhibited the same phenotype as their Rx1-19F counterparts described above (15, 71). However, when the *cpsB* gene was deleted, the phenotype observed differed in the two studies. In one study the D39 *cpsB* deletion mutant, D39-BΔ, was rough, as was the Rx1-19F *cpsB* deletion mutant, while in the other study the *cpsB* deletion mutant, MB526, was smooth and produced more CPS than the parent D39 strain. In both strains a high proportion of the CpsD present in the cells was phosphorylated, indicating the loss of phosphotyrosine-protein phosphatase activity in these mutants (15, 71). Recent repeated attempts to construct a D39 *cpsB* deletion mutant by transformation with an overlap-extension PCR product in which an erythromycin-resistance gene replaces *cpsB* were unsuccessful, whereas similar *cpsD* deletion mutants were obtained readily (61). However, when DNA from the existing rough D39-BΔ strain was used to amplify the overlap-extension PCR product, mutants in which an erythromycin-resistance gene replaces *cpsB* were obtained. Interestingly, whereas most mutants exhibited a rough phenotype such as D39-BΔ, approximately 10% exhibited a smooth phenotype such as MB526 (61). The fact that these D39 *cpsB* deletion transformants exhibited both phenotypes suggests that the difference between them is most probably due to minor changes (e.g., point mutations) in the *cpsA*, *cpsC*, or *cpsD* genes, because only this region of the *cps* locus was transformed into D39 in these experiments. The exact function of any interaction between CpsA and CpsB/CpsC/CpsD has not been defined. Although deletion of *cpsA* results in reduced expression of CPS in D39, the mechanism by which this occurs and the effects of specific amino acid changes in CpsA are unknown. Amino acid changes in either CpsC or CpsD could affect the conformation of these proteins and/or affect protein-protein interactions, such that CpsC/CpsD activity becomes independent of tyrosine phosphorylation and CPS production is permanently switched on or off. Such point mutations have been described for other regulatory proteins such as *Salmonella enterica* Typhimurium PhoQ, in which a single amino acid change resulted in a constitutive phenotype (36).

These results make it difficult to define the role of CpsD tyrosine phosphorylation in regulation of CPS production in D39. Clearly, there is a difference in the role of CpsB regulation of CpsD tyrosine phosphorylation in the mouse-virulent D39 strain compared to the avirulent Rx1 background. This could be due to other regulatory factors acting on CpsB and/or CpsD in D39 that are no longer active in Rx1, or the effects of another protein that is dephosphorylated by CpsB in D39 but is absent in Rx1. However, regulation of CPS expression via phosphorylation of tyrosine in CpsD is important for virulence of *S. pneumoniae* in mice. Three different D39 CpsD mutants expressing mucoid CPS were able to cause systemic disease in mice after intraperitoneal challenge, but not via the intranasal route. These three mucoid strains exhibited wild-type capacity to colonize the lungs, but were never detected in the blood (71). This suggests that loss of regulation via phosphorylation of tyrosine in CpsD affects the ability of *S. pneumoniae* to translocate from the lungs into the bloodstream.

The availability of oxygen may be one environmental trigger in regulation of tyrosine phosphorylation of CpsD and hence, CPS production. When pneumococci in the highly encapsulated opaque phase were grown anaerobically, the amount of phosphorylated CpsD detected in the cell by Western immunoblotting increased, as did the amount of CPS expressed, compared to that seen under aerobic conditions (93). However, oxygen availability had no effect on CPS expression in transparent phase pneumococci. The changes observed in the opaque phase were not due to transcriptional regulation of the *cps* locus, because the amount of CpsD detected by Western immunoblotting was similar in both aerobic and anaerobic environments (93). These findings, together with the previously described high levels of phosphorylated CpsD in the highly encapsulated D39 *cpsB* delete strain MB526, have led to speculation that there is a positive correlation between tyrosine phosphorylation and CPS production and that the phosphorylated form of CpsD may be the active form in D39 (15). However, this hypothesis overlooks the fact that a D39 strain expressing a CpsD protein that cannot be phosphorylated due to replacement of the tyrosines in the [YGX] repeat domain with phenylalanine still produces significant amounts of CPS (68, 70). Thus, further studies are needed to clarify the precise role of tyrosine phosphorylation in the regulation of CPS production in D39.

Two proteins involved in sugar metabolism have also been shown to affect CPS production. PGM is the phosphoglucomutase that catalyzes the conversion of Glc-6-P to Glc-1-P, and GalU is a Glc-1-phosphate uridylyltransferase that catalyzes the formation of UDP-Glc from Glc-1-P. Pneumococcal mutants in which either the *galU* or the *pgm* genes were disrupted produced almost no CPS and also exhibited growth defects (37, 60). Additionally, strains in which the *pgm* gene had defined point mutations, which significantly reduced but did not eliminate the enzymatic activity of PGM, still produced reduced amounts of CPS even though they no longer exhibited growth defects. These strains also exhibited reduced virulence in mice compared to the wild-type parent strain (37). Both PGM and GalU are required for synthesis of UDP-Glc, which is a precursor for the biosynthesis of all 90 pneumococcal CPS types, as well as for other cellular structures such as teichoic acid. Thus, limiting the supply of this precursor would be expected to impact heavily upon CPS production in the pneumococcus. Indirect modulation of CPS production by controlled availability of precursors or cofactors may be one of the regulatory mechanisms used by the pneumococcus. For example, the mechanism of biosynthesis of the pneumococcal cell wall teichoic acid (C-polysaccharide) is likely to be similar to that of CPS. Accordingly, if teichoic acid synthesis is subject to direct regulation, this may influence the availability of common precursors (e.g., UDP-Glc or the lipid carrier) for CPS biosynthesis. Such a phenomenon would be consistent with the observation that phase variation in *S. pneumoniae* has opposite effects on the total amounts of CPS and teichoic acid (45). Saturation of a common lipid carrier pool with the disaccharide precursor for peptidoglycan synthesis has previously been proposed to explain blockade of exopolysaccharide production in *Streptococcus thermophilus* with a mutation in *pbp2b*, one of its penicillin-binding protein genes (83).

CONCLUDING REMARKS

In this chapter we have attempted to summarize the current state of knowledge on the CPS of *S. pneumoniae*, with particular reference to the genes encoding biosynthesis of this most important of all pneumococcal surface antigens. Notwithstanding the insights gained from the elegant classical genetic studies on CPS production carried out before the early 1970s, research in recent years has been revolutionized by the advent of modern methods for gene cloning, DNA amplification, and sequence analysis. The complete nucleotide sequences for the *cps* loci for all of the 90 known serotypes will soon be available. The functions of many of the individual genes in these loci await confirmation by conventional biochemical and genetic analysis. Nevertheless, access to the enormous body of information now available on sequence databases, combined with knowledge of the chemical structures for many of the CPS repeat units, has enabled accurate predictions of function for a significant proportion of these genes. It has also been possible to predict the mechanisms of CPS biosynthesis in pneumococci by analogy with those operating in gram-negative bacteria. The existence of two distinct mechanisms for CPS biosynthesis in *S. pneumoniae* has already been recognized. However, much remains to be learned about the precise molecular events involved in both of these processes, and about how CPS production in pneumococci is regulated. Further biochemical and mutational analyses are also required to elucidate the precise functions of the four genes at the 5′ end of the *cps* loci, which clearly encode important common steps in polysaccharide biosynthesis in pneumococci, as well as in other gram-positive genera. Given the importance of capsules to the virulence of *S. pneumoniae* and several other gram-positive pathogens, such conserved components of the CPS biosynthesis machinery may prove to be useful targets for novel antimicrobial strategies.

We are grateful to Renato Morona for many helpful discussions. Work in our laboratory is supported by a Program Grant from the National Health and Medical Research Council of Australia.

REFERENCES

1. **Abeygunawardana, C., T. C. Williams, J. S. Sumner, and J. P. Hennessey, Jr.** 2000. Development and validation of an NMR-based identity assay for bacterial polysaccharides. *Anal. Biochem.* **279:**226–240.

2. **Abeyta, M., G. G. Hardy, and J. Yother.** 2003. Genetic alteration of capsule type but not PspA type affects accessibility of surface-bound complement and surface antigens of *Streptococcus pneumoniae*. *Infect. Immun.* **71:**218–225.

3. **Arrecubieta, C., R. López, and E. Gárcía.** 1994. Molecular characterization of *cap3A*, a gene from the operon required for the synthesis of the capsule of *Streptococcus pneumoniae* type 3: sequencing of mutations responsible for

the unencapsulated phenotype and localization of the capsular cluster on the pneumococcal chromosome. *J. Bacteriol.* **176:**6375–6383.

4. **Arrecubieta, C., E. García, and R. López.** 1995. Sequence and transcriptional analysis of a DNA region involved in the production of capsular polysaccharide in *Streptococcus pneumoniae* type 3. *Gene* **167:**1–7.

5. **Arrecubieta, C., R. López, and E. García.** 1996. Type 3-specific synthase of *Streptococcus pneumoniae* (Cap3B) directs type 3 polysaccharide biosynthesis in *Escherichia coli* and in pneumococcal strains of different serotypes. *J. Exp. Med.* **184:**449–455.

6. **Austrian, R.** 1981. Pneumococcus: the first one hundred years. *Rev. Infect. Dis.* **3:**183–189.

7. **Austrian, R.** 1981. Some observations on the pneumococcus and on the current status of pneumococcal disease and its prevention. *Rev. Infect. Dis.* **3**(Suppl.)**:**S1–S17.

8. **Austrian, R., H. P. Bernheimer, E. E. B. Smith, and G. T. Mills.** 1959. Simultaneous production of two capsular polysaccharides by pneumococcus. II. The genetic and biochemical bases of binary capsulation. *J. Exp. Med.* **110:**585–602.

9. **Avery, O. T., and R. Dubos.** 1931. The protective action of a specific enzyme against type III pneumococcus infections in mice. *J. Exp. Med.* **54:**73–89.

10. **Avery, O. T., and M. Heidelberger.** 1925. Immunological relationships of cell constituents of pneumococcus. *J. Exp. Med.* **42:**367–376.

11. **Avery, O. T., and, H. J. Morgan.** 1925. Immunological reactions of the isolated carbohydrate and protein of pneumococcus. *J. Exp. Med.* **42:**347–353.

12. **Avery, O. T., C. M. MacLeod, and M. McCarty.** 1944. Studies on the chemical nature of the substance inducing transformation of pneumococcal types. Induction of transformation by a desoxyribonucleic acid fraction isolated from pneumococcus type III. *J. Exp. Med.* **79:**137–158.

13. **Barnes, D. M., S. Whittier, P. H. Gilligan, S. Soares, A. Tomasz, and F. W. Henderson.** 1995. Transmission of multidrug-resistant serotype 23F *Streptococcus pneumoniae* in group day care: evidence suggesting capsular transformation of the resistant strain *in vivo. J. Infect. Dis.* **171:**890–896.

14. **Bender, M. H., and J. Yother.** 2001. CpsB is a modulator of capsule-associated tyrosine kinase activity in *Streptococcus pneumoniae. J. Biol. Chem.* **276:**47966–47974.

15. **Bender, M. H., R. T. Cartee, and J. Yother.** 2003. Positive correlation between tyrosine phosphorylation of CpsD and capsular polysaccharide production in *Streptococcus pneumoniae. J. Bacteriol.* **185:**6057–6066.

16. **Bernheimer, H. P., I. E. Wermundsen, and R. Austrian.** 1967. Qualitative differences in the behavior of pneumococcal deoxyribonucleic acids transforming to the same capsular type. *J. Bacteriol.* **93:**320–333.

17. **Beynon, L. M., J. C. Richards, M. B. Perry, and P. J. Kniskern.** 1991. Antigenic and structural relationships within group 19 *Streptococcus pneumoniae:* chemical characterization of the specific capsular polysaccharides of type 19B and 19C. *Can. J. Chem.* **70:**218–232.

18. **Cartee, R.T., W. T. Forsee, J. W. Jensen, and J. Yother.** 2001. Expression of the *Streptococcus pneumoniae* type 3 synthase in *Escherichia coli.* Assembly of type 3 polysaccharide on a lipid primer. *J. Biol. Chem.* **276:**48831–48839.

19. **Cieslewicz, M. J., D. L. Kasper, Y. Wang, and M. R. Wessels.** 2001. Functional analysis in type Ia group B Streptococcus of a cluster of genes involved in extracellular polysaccharide production by diverse species of streptococci. *J. Biol. Chem.* **276:**139–146.

20. **Coffey, T. J., C. G. Dowson, M. Daniels, J. Zhou, C. Martin, B. G. Spratt, and J. M. Musser.** 1991. Horizontal gene transfer of multiple penicillin-binding protein genes and capsular biosynthetic genes in natural populations of *Streptococcus pneumoniae. Mol. Microbiol.* **5:**2255–2260.

21. **Coffey, T. J., M. C. Enright, M. Daniels, J. K. Morona, R. Morona, W. Hryniewicz, J. C. Paton, and B. G. Spratt.** 1998. Recombinational exchanges at the capsular polysaccharide biosynthetic locus lead to frequent serotype changes among natural isolates of *Streptococcus pneumoniae. Mol. Microbiol.* **27:**73–83.

22. **Coffey, T. J., M. C. Enright, M. Daniels, P. Wilkinson, S. Berron, A. Fenoll, and B. G. Spratt.** 1998. Serotype 19A variants of the Spanish serotype 23F multiresistant clone of *Streptococcus pneumoniae. Microb. Drug Resist.* **4:**51–55.

23. **Cundell, D. R., J. N. Weiser, J. Shen, A. Young, and E. I. Tuomanen.** 1995. Relationship between colonial morphology and adherence of *Streptococcus pneumoniae. Infect. Immun.* **63:**757–761.

24. **DeAngelis, P. L., J. Papaconstantinou, and P. H. Weigel.** 1993. Molecular cloning, identification, and sequence of the hyaluronan synthase gene from group A *Streptococcus pyogenes. J. Biol. Chem.* **268:**19181–19184.

25. **Deng, L., D. L. Kasper, T. P. Krick, and M. R. Wessels.** 2000. Characterization of the linkage between the type III capsular polysaccharide and the bacterial cell wall of group B Streptococcus. *J. Biol. Chem.* **275:**7497–7504.

26. **Dillard, J. P., and J. Yother.** 1994. Genetic and molecular characterization of capsular polysaccharide biosynthesis in *Streptococcus pneumoniae* type 3. *Mol. Microbiol.* **12:**959–972.

27. **Dillard, J. P., M. W. Vandersea, and J. Yother.** 1995. Characterization of the cassette containing genes for type 3 capsular polysaccharide biosynthesis in *Streptococcus pneumoniae. J. Exp. Med.* **181:**973–983.

28. **Dochez, A. R., and O. T. Avery.** 1917. The elaboration of specific soluble substance by pneumococcus during growth. *J. Exp. Med.* **26:**477–493.

29. **Douglas, R. M., J. C. Paton, S. J. Duncan, and D. Hansman.** 1983. Antibody response to pneumococcal vaccination in children younger than five years of age. *J. Infect. Dis.* **148:**131–137.

30. **Eskola, J., T. Kilpi, A. Palmu, J. Jokinen, J. Haapakoski, E. Herva, A. Takala, H. Kayhty, P. Karma, R. Kohberger, G. Siber, P. H. Makela, and the Finnish Otitis Media Study Group.** 2001. Efficacy of a pneumococcal conjugate vaccine against acute otitis media. *N. Engl. J. Med.* **344:**403–409.

31. **García, E., P. García, and R. López.** 1993. Cloning and sequencing of a gene involved in the synthesis of the capsular polysaccharide of *Streptococcus pneumoniae* type 3. *Mol. Gen. Genet.* **239:**188–195.

32. **Giammarinaro, P., and J. C. Paton.** 2002. Role of RegM, a homologue of the catabolite repressor protein CcpA, in the virulence of *Streptococcus pneumoniae. Infect. Immun.* **70:**5454–5461.

33. **Glucksmann, M. A., T. L. Reuber, and G. C. Walker.** 1993. Genes needed for the modification, polymerisation, export, and processing of succinoglycan by *Rhizobium meliloti:* a model for succinoglycan biosynthesis. *J. Bacteriol.* **175:**7045–7055.

34. **Griffith, F.** 1928. The significance of pneumococcal types. *J. Hyg.* **27**:113–159.

35. **Guidolin, A., J. K. Morona, R. Morona, D. Hansman, and J. C. Paton.** 1994. Nucleotide sequence of an operon essential for capsular polysaccharide biosynthesis in *Streptococcus pneumoniae* type 19F. *Infect. Immun.* **62**: 5384–5396.

36. **Gunn, J. S., E. L. Hohmann, and S. I. Miller.** 1996. Transcriptional regulation of *Salmonella* virulence: a PhoQ periplasmic domain mutation results in increased net phosphotransfer to PhoP. *J. Bacteriol.* **178**:6369–6373.

37. **Hardy, G. G., A. D. Magee, C. L. Ventura, M. J. Caimano, and J. Yother.** 2001. Essential role for cellular phosphoglucomutase in virulence of type 3 *Streptococcus pneumoniae. Infect. Immun.* **69**:2309–2317.

38. **Henrichsen, J.** 1995. Six newly recognized types of *Streptococcus pneumoniae. J. Clin. Microbiol.* **33**:2759–2762.

39. **Iannelli, F., B. J. Pearce, and G. Pozzi.** 1999. The type 2 capsule locus of *Streptococcus pneumoniae. J. Bacteriol.* **181**:2652–2654.

40. **Institute for Genomic Research.** http://www.tigr.org/pub/data/s[ru5,.4]pneumoniae/

41. **Jansson, P. E., B. Lindberg, U. Lindquist, and J. Ljungberg.** 1987. Structural studies of the capsular polysaccharide from *Streptococcus pneumoniae* types 15B and 15C. *Carbohydr. Res.* **162**:111–116.

42. **Jiang, S. M., L. Wang, and P. R. Reeves.** 2001. Molecular characterization of *Streptococcus pneumoniae* type 4, 6B, 8, and 18C capsular polysaccharide gene clusters. *Infect. Immun.* **69**:1244–1255.

43. **Keenleyside, W. J., and C. Whitfield.** 1996. A novel pathway for O-polysaccharide biosynthesis in *Salmonella enterica* serovar Borreze. *J. Biol. Chem.* **271**:28581–28592.

44. **Kelly, T., J. P. Dillard, and J. Yother.** 1994. Effect of genetic switching of capsular type on virulence of *Streptococcus pneumoniae. Infect. Immun.* **62**:1813–1819.

45. **Kim, J. O., and J. N. Weiser.** 1998. Association of intrastrain phase variation in quantity of capsular polysaccharide and teichoic acid with the virulence of *Streptococcus pneumoniae. J. Infect. Dis.* **177**:368–377.

46. **Klein, D. L.** 2000. Pneumococcal disease and the role of conjugate vaccines, p. 467–477. *In* A. Tomasz (ed.), Streptococcus pneumoniae: *Molecular Biology and Mechanisms of Disease.* Mary Ann Liebert Inc., Larchmont, N.Y.

47. **Kolkman, M. A. B., D. A. Morrison, B. A. M. van der Zeijst, and P. J. M. Nuijten.** 1996. The capsule polysaccharide synthesis locus of *Streptococcus pneumoniae* serotype 14: identification of the glycosyltransferase gene *cps14E. J. Bacteriol.* **178**:3736–3741.

48. **Kolkman, M. A. B., B. A. M. van der Zeijst, and P. J. M. Nuijten.** 1997. Functional analysis of glycosyltransferases encoded by the capsular polysaccharide locus of *Streptococcus pneumoniae* serotype 14. *J. Biol. Chem.* **272**: 19502–19508.

49. **Kolkman, M. A. B., W. Wakarchuk, P. J. M. Nuijten, and B. A. M. van der Zeijst.** 1997. Capsular polysaccharide synthesis in *Streptococcus pneumoniae* serotype 14: molecular analysis of the complete *cps* locus and identification of genes encoding glycosyltransferases required for the biosynthesis of the tetrasaccharide subunit. *Mol. Microbiol.* **26**:197–208.

50. **Kolkman, M. A. B., B. A. M. van der Zeijst, and P. J. M. Nuijten.** 1998. Diversity of capsular polysaccharide synthesis gene clusters in *Streptococcus pneumoniae. J. Biochem.* **123**:937–945.

51. **Kroll, J. S., B. M. Loynds, and E. R. Moxon.** 1991. The *Haemophilus influenzae* capsulation gene cluster: a compound transposon. *Mol. Microbiol.* **5**:1549–1560.

52. **Lawrence, E. R., D. B. Griffiths, S. A. Martin, R. C. George, and L. M. Hall.** 2003. Evaluation of semiautomated multiplex PCR assay for determination of *Streptococcus pneumoniae* serotypes and serogroups. *J. Clin. Microbiol.* **41**:601–607.

53. **Lee, C.-J., S. D. Banks, and J. P. Li.** 1991. Virulence, immunity and vaccine related to *Streptococcus pneumoniae. Crit. Rev. Microbiol.* **18**:89–114.

54. **Llull, D., R. López, E. García, and R. Muñoz.** 1998. Molecular structure of the gene cluster responsible for the synthesis of the polysaccharide capsule of *Streptococcus pneumoniae* type 33F. *Biochim. Biophys. Acta* **1449**: 217–224.

55. **Llull, D., R. Muñoz, R. López, and E. García.** 1999. A single gene (*tts*) located outside the *cap* locus directs the formation of *Streptococcus pneumoniae* type 37 capsular polysaccharide. Type 37 pneumococci are natural, genetically binary strains. *J. Exp. Med.* **190**:241–251.

56. **Llull, D., E. García, and R. López.** 2001. Tts, a processive beta-glucosyltransferase of *Streptococcus pneumoniae*, directs the synthesis of the branched type 37 capsular polysaccharide in Pneumococcus and other gram-positive species. *J. Biol. Chem.* **276**:21053–21061.

57. **MacLeod, C. M., and M. R. Krauss.** 1950. Relation of virulence of pneumococcal strains for mice to the quantity of capsular polysaccharide formed in vitro. *J. Exp. Med.* **92**:1–9.

58. **Magee, A. D., and J. Yother.** 2001. Requirement for capsule in colonization by *Streptococcus pneumoniae. Infect. Immun.* **69**:3755–3761.

59. **Mbelle, N., R. E. Huebner, A. D. Wasas, A. Kimura, I. Chang, and K. P. Klugman.** 1999. Immunogenicity and impact on nasopharyngeal carriage of a nonavalent pneumococcal conjugate vaccine. *J. Infect. Dis.* **180**: 1171–1176.

60. **Mollerach, M., R. López, and E. García.** 1998. Characterization of the *galU* gene of *Streptococcus pneumoniae* encoding a uridine diphosphoglucose pyrophosphorylase: a gene essential for capsular polysaccharide biosynthesis. *J. Exp. Med.* **188**:2047–2056.

61. **Morona, J. K., R. Morona, and J. C. Paton.** Unpublished observations.

62. **Morona, R., L. Van Den Bosch, and C. Daniels.** 2000. Evaluation of Wzz/MPA1/MPA2 proteins based on the presence of coiled-coil regions. *Microbiology* **146**:1–4.

63. **Morona, J. K., R. Morona, and J. C. Paton.** 1997. Characterization of the locus encoding the *Streptococcus pneumoniae* type 19F capsular polysaccharide biosynthetic pathway. *Mol. Microbiol.* **23**:751–763.

64. **Morona, J. K., R. Morona, and J. C. Paton.** 1997. Molecular and genetic characterization of the capsule biosynthesis locus of *Streptococcus pneumoniae* type 19B. *J. Bacteriol.* **179**:4953–4958.

65. **Morona, J. K., D. C. Miller, T. J. Coffey, C. J. Vindurampulle, B. G. Spratt, R. Morona, and J. C. Paton.** 1999. Molecular and genetic characterization of the capsule biosynthesis locus of *Streptococcus pneumoniae* type 23F. *Microbiology* **145**:781–789.

66. **Morona, J. K., R. Morona, and J. C. Paton.** 1999. Analysis of the 5′ portion of the type 19A capsule locus identifies two classes of *cpsC*, *cpsD*, and *cpsE* genes in *Streptococcus pneumoniae. J. Bacteriol.* **181**:3599–3605.

67. **Morona, J. K., R. Morona, and J. C. Paton.** 1999. Comparative genetics of capsular polysaccharide biosynthesis in *Streptococcus pneumoniae* types belonging to serogroup 19. *J. Bacteriol.* **181:**5355–5364.

68. **Morona, J. K., J. C. Paton, D. C. Miller, and R. Morona.** 2000. Tyrosine phosphorylation of CpsD negatively regulates capsular polysaccharide biosynthesis in *Streptococcus pneumoniae. Mol. Microbiol.* **35:**1431–1442.

69. **Morona, J. K., R. Morona, D. C. Miller, and J. C. Paton.** 2002. *Streptococcus pneumoniae* capsule biosynthesis protein CpsB is a novel manganese-dependent phosphotyrosine-protein phosphatase. *J. Bacteriol.* **184:**577–583.

70. **Morona, J. K., R. Morona, D. C. Miller, and J. C. Paton.** 2003. Mutational analysis of the carboxy-terminal (YGX)$_4$ repeat domain of CpsD, an autophosphorylating tyrosine kinase required for capsule biosynthesis in *Streptococcus pneumoniae. J. Bacteriol.* **185:**3009–3019.

71. **Morona, J. K., D. C. Miller, R. Morona, and J. C. Paton.** 2004. The effect that mutations in the conserved capsular polysaccharide biosynthesis genes *cpsA, cpsB* and *cpsD* have on virulence of *Streptococcus pneumoniae. J. Infect. Dis.* **189:**1905–1913.

72. **Muñoz, R., M. Mollerach, R. López, and E. García.** 1997. Molecular organization of the genes required for the synthesis of type 1 capsular polysaccharide of *Streptococcus pneumoniae:* formation of binary encapsulated pneumococci and identification of cryptic dTDP-rhamnose biosynthesis genes. *Mol. Microbiol.* **25:**79–92.

73. **Muñoz, R., M. Mollerach, R. López, and E. García.** 1999. Characterization of the type 8 capsular gene cluster of *Streptococcus pneumoniae. J. Bacteriol.* **181:**6214–6219.

74. **Musher, D. M.** 1992. Infections caused by *Streptococcus pneumoniae:* clinical spectrum, pathogenesis, immunity and treatment. *Clin. Infect. Dis.* **14:**801–807.

75. **Nesin, M., M. Ramirez, and A. Tomasz.** 1998. Capsular transformation of a multidrug-resistant *Streptococcus pneumoniae in vivo. J. Infect. Dis.* **177:**707–713.

76. **Obaro, S. K., R. A. Adegbola, W. A. Banya, and B. M. Greenwood.** 1996. Carriage of pneumococci after pneumococcal vaccination. *Lancet* **348:**271–272.

77. **Ogunniyi, A. D., P. Giammarinaro, and J. C. Paton.** 2002. The genes encoding virulence-associated proteins and the capsule of *Streptococcus pneumoniae* are upregulated and differentially expressed *in vivo. Microbiology* **148:**2045–2053.

78. **Paton, J. C.** 2004. New pneumococcal vaccines: basic science developments, p. 382–402. *In* E. I. Tuomanen, T. J. Mitchell, D. A. Morrison, and B. G. Spratt (ed.), *The Pneumococcus.* ASM Press, Washington, D.C.

79. **Ramirez, M., and A. Tomasz.** 1998. Molecular characterization of the complete 23F capsular polysaccharide locus of *Streptococcus pneumoniae. J. Bacteriol.* **180:**5273–5278.

80. **Sanger Institute.** http://www.sanger.ac.uk/Projects/S_pneumoniae/CPS/

81. **Sorensen, U. B. S., J. Blom, A. Birch-Andersen, and J. Henrichsen.** 1988. Ultrastructural localization of capsules, cell wall polysaccharide, cell wall proteins, and F antigen in pneumococci. *Infect. Immun.* **56:**1890–1896.

82. **Sorensen, U. B. S., J. Henrichsen, H.-C. Chen, and S. C. Szu.** 1990. Covalent linkage between the capsular polysaccharide and the cell wall peptidoglycan of *Streptococcus pneumoniae* revealed by immunochemical methods. *Microb. Pathog.* **8:**325–334.

83. **Stingele, F., and B. Mollet.** 1996. Disruption of the gene encoding penicillin-binding protein 2b (*pbp2b*) causes altered cell morphology and cease in exopolysaccharide production in *Streptococcus thermophilus* Sfi6. *Mol. Microbiol.* **22:**357–366.

84. **Talbot, U., A. W. Paton, and J. C. Paton.** 1996. Uptake of *Streptococcus pneumoniae* by respiratory epithelial cells. *Infect. Immun.* **64:**3772–3777.

85. **Tuomanen, E. I., and H. R. Masure.** 1997. Molecular and cellular biology of pneumococcal infection. *Microb. Drug Resist.* **3:**297–308.

86. **van Dam, J. E. G., A. Fleer, and H. Snippe.** 1990. Immunogenicity and immunochemistry of *Streptococcus pneumoniae* capsular polysaccharides. *Antonie Leeuwenhoek* **58:**1–47.

87. **van Selm, S., M. A. Kolkman, B. A. van der Zeijst, K. A. Zwaagstra, W. Gaastra, and J. P. van Putten.** 2002. Organization and characterization of the capsule biosynthesis locus of *Streptococcus pneumoniae* serotype 9V. *Microbiology* **148:**1747–1755.

88. **van Selm, S., L. M. van Cann, M. A. Kolkman, B. A. van der Zeijst, and J. P. van Putten.** 2003. Genetic basis for the structural difference between *Streptococcus pneumoniae* serotype 15B and 15C capsular polysaccharides. *Infect. Immun.* **71:**6192–6198.

89. **Venkateswaren, P. S., N. Stanton, and R. Austrian.** 1983. Type variation of strains of *Streptococcus pneumoniae* in capsular serogroup 15. *J. Infect. Dis.* **147:**1041–1054.

90. **Waite, R. D., J. K. Struthers, and C. G. Dowson.** 2001. Spontaneous sequence duplication within an open reading frame of the pneumococcal type 3 capsule locus causes high-frequency phase variation. *Mol. Microbiol.* **42:**1223–1232.

91. **Waite, R. D., D. W. Penfold, J. K. Struthers, and C. G. Dowson.** 2003. Spontaneous sequence duplications within capsule genes *cap8E* and *tts* control phase variation in *Streptococcus pneumoniae* serotypes 8 and 37. *Microbiology* **149:**497–504.

92. **Weiser, J. N., R. Austrian, P. K. Sreenivasan, and H. R. Masure.** 1994. Phase variation in pneumococcal opacity: relationship between colonial morphology and nasopharyngeal colonization. *Infect. Immun.* **62:**2582–2589.

93. **Weiser, J. N., D. Bae, H. Epino, S. B. Gordon, M. Kapoor, L. A. Zenewicz and M. Shchepetov.** 2001. Changes in availability of oxygen accentuate differences in capsular polysaccharide expression by phenotypic variants and clinical isolates of *Streptococcus pneumoniae. Infect. Immun.* **69:**5430–5439.

94. **Whitfield, C., and A. Paiment.** 2003. Biosynthesis and assembly of Group 1 capsular polysaccharides in *Escherichia coli* and related extracellular polysaccharides in other bacteria. *Carbohydr. Res.* **338:**2491–2502.

95. **Winkelstein, J. A.** 1981. The role of complement in the host's defense against *Streptococcus pneumoniae. Rev. Infect. Dis.* **3:**289–298.

Streptococcus pneumoniae: Invasion and Inflammation

CARLOS J. ORIHUELA AND ELAINE TUOMANEN

21

Streptococcus pneumoniae (the pneumococcus) is the leading cause of otitis media (OM), community-acquired pneumonia, and bacterial meningitis. Solely a human pathogen, the pneumococcus colonizes the nasopharynx and is spread between humans by aerosol. Colonization of the nasopharynx with the pneumococcus is typically asymptomatic (4). Rates of carriage vary from 5 to 10% of healthy adults to 20 to 40% of healthy children (75, 89). Risk factors associated with higher rates of carriage include race (Native Americans and Australian aboriginals among the highest rates of carriage [23, 113]), infancy (43, 44), time of the year (carriage is higher during winter months [44]), and daycare attendance (40 to 60% of children who attend daycare are colonized [26]). Duration of colonization decreases with age and varies from 2 weeks to 4 months (43, 101).

Invasive pneumococcal disease (IPD) results from the spread of the pneumococcus from the nasopharynx to the inner ear, lungs, bloodstream, and brain. Infants, the elderly, and individuals who are immunocompromised are at risk for developing IPD (15). Pneumococcal models of invasive disease must account for the commensal nature of the bacteria, yet also take into account the wide spectrum of disease the pneumococcus is capable of causing. Although attack rates are low, such a large number of individuals are colonized that the resulting morbidity and mortality associated with the pneumococcus are tremendous. Worldwide, it is estimated that *S. pneumoniae* is responsible for 15 cases of IPD per 100,000 persons per year (29), and over a million deaths annually. As of 1997, in the United States, the pneumococcus was responsible for greater than 6 million cases of OM, 100,000 cases of pneumonia, 60,000 cases of bacteremia, and 3,300 cases of meningitis annually (17). Mortality associated with bacteremia and meningitis is high, particularly at the extremes of age (132). In the elderly, mortality associated with pneumococcal bacteremia may be as high as 60%, and as high as 80% for meningitis (65).

In this chapter we will first review the molecular mechanisms that allow the pneumococcus to colonize and spread from one anatomical site to the next. We will then discuss the mechanisms of inflammation and cytotoxicity during pneumococcal infection.

TRAFFICKING OF PNEUMOCOCCI THROUGH THE BODY

Interactions with Epithelial Cells of the Nasopharynx

For over a century, *S. pneumoniae* has been categorized by serology with distinct serotypes identified on the basis of the >90 immunologically and chemically distinct polysaccharide capsules that surround and protect the bacterium from phagocytosis. Studies examining the contribution of the capsule to virulence have demonstrated that only a small subset of capsular types cause most of the IPD (33). Among these serotypes 4, 6A, 6B, 14, 23F, 19F, 9V, and 18C account for approximately 80% of the isolates from children 2 to 5 years of age in the United States. The vast majority of pneumococci colonize the nasopharynx for up to 6 weeks and are then cleared with no systemic symptoms in the host (4, 20). In instances in which IPD does occur, it has been determined that IPD occurs most frequently early after the acquisition of a new capsular serotype. Attack rates are higher for serotypes that are carried for shorter periods of time versus those that colonize for extended periods (20). Thus an inverse relationship exists between the propensity of a serotype to cause invasive disease and its ability to colonize long term. Examples of this include isolates of serotype 1, which are commonly recovered from individuals with IPD but are rarely isolated from healthy carriers (86), and isolates of serotype 6, which have lower attack rates yet are carried for longer periods of time (101).

S. pneumoniae undergoes spontaneous phase variation, alternating between a transparent and opaque phenotype (126). The distinct phenotypes are discernible by subtle changes in colony morphology and opacity when the colonies are visualized on transparent media with oblique transmitted light. It has been determined that the transparent phenotype is the predominant phase isolated from the nasopharynx of infected animals (126), whereas the opaque is the predominant phase isolated from the blood (63). The transparent phenotype expresses increased amounts of phosphorylcholine (ChoP) (64) and choline-binding protein A

Gram-Positive Pathogens, 2nd edition, edited by Vincent A. Fischetti et al.
© 2006 ASM Press, Washington, D.C.

(CbpA) (93), both of which function as adhesins and contribute to the ability of the bacteria to colonize the nasopharynx. The opaque phenotype expresses increased levels of capsule and pneumococcal surface protein A (PspA). Currently, it is thought that phase variation is a mechanism by which the pneumococcus alternates between an adhesive phenotype that is able to colonize the nasopharynx and an opsonophagocytosis-resistant phenotype that can survive in areas where phagocytes are present.

ChoP on the cell wall (Fig. 1) acts as a docking station for a unique group of 15 secreted proteins, termed choline-binding proteins (CBPs) (Fig. 2). CBPs noncovalently bind to the choline moiety of the cell wall, thereby reversibly snapping proteins with diverse functions onto the bacterial surface (31). Among CBPs, CbpA is the major pneumococcal adhesin (40) expressed predominantly in the transparent phenotype (93). Pneumococci lacking CbpA are not only defective in colonization of the nasopharynx but also

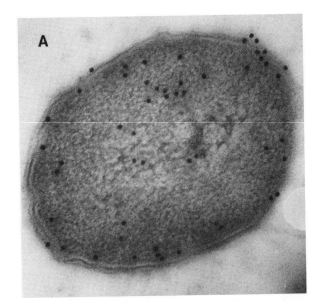
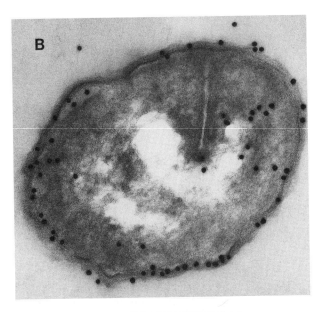
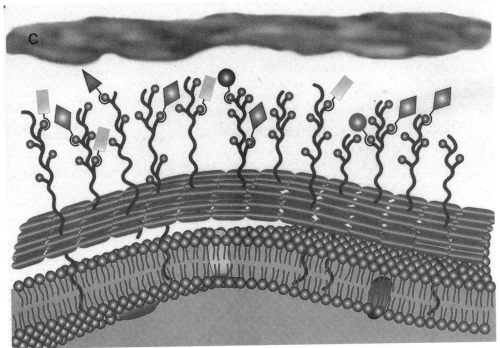

FIGURE 1 (*See the separate color insert for the color version of this illustration.*) Immunohistochemical and schematic depiction of the choline biology of the pneumococcal surface. Immunogold-labeling of pneumococci with (A) TEPC-15 antibody recognizing free choline and (B) antiautolysin antibody. These two images contrast free (A) versus CBP-bound (B) choline. (C) Schematic view of the capsule (blue), cell wall (green), and membrane (red). The teichoic and lipoteichoic acids are indicated as dark blue lines bearing choline (circles). A proportion of these are capped by CBPs. (Courtesy of Dr. K. G. Murti, St. Jude Electron Microscopy Core Facility.)

A.

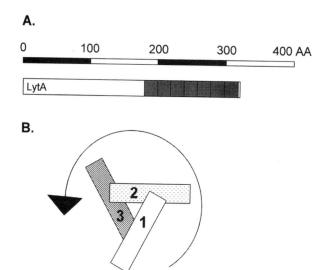

FIGURE 2 (A) Graphical representation of the CBP LytA (36). Boxes indicate individual choline-binding repeats (CBR), each consisting of the consensus sequence GWVKD-NGTWYYLNSSGAMAT. CBPs attach noncovalently to ChoP on the bacterial surface through the choline-binding domain (CBD), which typically consists of multiple CBRs. Crystal structure analysis of the CBD of LytA, the major autolysin, indicates that each 20-amino-acid CBR forms a small hairpin consisting of two antiparallel β-strands connected by a short internal loop. (B) These hairpins are connected by 8 to 10 residues that rotate each hairpin in a 120° counterclockwise direction such that a left-handed superhelix is formed, similar to a spiral staircase. ChoP is subsequently bound by hydrophobic cavities present in the grooves of the surface of the CBD steps (31).

have a diminished capacity to colonize the lower respiratory tract and cause pneumonia (80, 93). In vitro, CbpA mutants fail to bind either a nasopharyngeal epithelial cell line or activated type II human lung cells (40). Similarly, the CBPs LytB, LytC, CbpD, CbpE, and CbpG also contribute to nasopharyngeal colonization when tested in rat pups. LytB functions as a glucosaminidase (37), whereas LytC is a lysozyme (38) and CbpE is a choline esterase (121). CbpG is a serine protease that in addition to contributing to nasopharyngeal colonization is required for development of high-grade bacteremia (41).

Other virulence factors affecting the capacity of the pneumococcus to colonize the nasopharynx include sIgA1 proteases that cleave the Fc portion of human IgA1 (129). Secretory IgA1 (sIgA1) proteases neutralize the capacity of sIgA to aggregate and opsonize the pneumococcus (62). Surprisingly, hydrolyzed sIgA1 also contributes to adhesion. Fab fragments remaining on the surface of the bacterium after sIgA proteolysis neutralize the inhibitory effect of the negatively charged capsule and enhance adherence (129).

The two-component system CiaR/H is required for efficient colonization and regulates gene expression in response to oxidative stress (99). Among the genes regulated by CiaR/H is htrA. HtrA, or high temperature requirement A, is a heat-inducible serine protease that has both proteolytic and chaperone activities. HtrA assists in the survival against environmental stresses such as oxidative stress, osmotic stress, and elevated temperatures (57). Loss of the CiaR/H operon results in a 1,000-fold decrease in the num-

ber of bacteria colonizing the nasopharynx and a 25-fold decrease in levels of htrA expression. Deletion of htrA alone results in a 100-fold decrease in nasopharyngeal colonization, indicating that CiaR/H regulates other factors that contribute to colonization (99). Finally, microarray analysis has now also been used to examine gene expression during bacterial adhesion to epithelial cells in vitro. Pneumococcal genes with enhanced expression in response to pneumococcal adhesion included CbpA and HtrA (81).

Ascension into the Middle Ear

OM is a highly prevalent pediatric disease and the primary cause of physician visits by small children. Along with *Moraxella catarrhalis* and *Haemophilus influenzae,* the pneumococcus is a primary cause of OM and is isolated in approximately 30% of culture-positive middle ear infusions (9). OM results when pneumococci in the nasopharynx ascend the eustachian tube and gain access to the middle ear. Typically, OM is marked by earache, fever, nasal congestion, a feeling of fullness in the ear, and muffled hearing.

Among the virulence determinants responsible for development of OM, it is thought that the neuraminidase plays an important role in permitting the pneumococci to ascend the eustachian tubes and invade the middle ear (9, 68). *S. pneumoniae* encodes two neuraminidases, NanA and NanB, with NanA being the major neuraminidase (8, 104). Neuraminidases cleave N-acetylneuraminic acid from glycoproteins and glycolipids on the eukaryotic cell surface. It has been proposed that the neuraminidase cleaves mucin, reducing the viscosity of this barrier and permitting the bacteria to access the epithelium. Once on the surface of the host cell, it is thought that neuraminidase cleaves oligosaccharides on the surface of the cell, thus exposing cryptic receptors that enhance bacterial adherence (8, 104). Evidence supporting this hypothesis is clearance of a neuraminidase-deficient mutant twice as quickly as wild-type *S. pneumoniae* from the middle ear of chinchillas, and structural changes in the cell surface carbohydrates of eustachian tube epithelial cells following infection with wild-type pneumococci versus a mutant lacking NanA (111, 112).

Once in the middle ear, inflammation is the result of pneumolysin and cell wall components (115). Pneumolysin is a potent cytotoxic hemolysin that activates complement via the classical pathway. In contrast, cell wall components activate complement via the alternative pathway. The contribution of these determinants to inflammation will be discussed in more detail later. It is pneumolysin that is responsible for hearing loss and cochlear damage during OM (19, 131). Guinea pigs infected with wild-type *S. pneumoniae* and a mutant deficient in neuraminidase demonstrated significant damage to the reticular lamina, sensory hair cells, and supporting cells of the organ of Corti. Guinea pigs infected with a pneumolysin-deficient mutant had no visible damage.

CbpA/Polymeric Immunoglobulin Receptor (pIgR)-Mediated Invasion in the Upper Respiratory Tract

Once attached to epithelial cells, the pneumococcus is able to translocate across the mucosal barrier by co-opting the pIgR (133). Mucosal epithelial cells transport immunoglobulin A (IgA) and IgM to the apical surface by binding the immunoglobulins on the basolateral surface to pIgR. IgA attached to pIgR crosses the epithelial cell in a vesicle. On the apical surface pIgR is cleaved and immunoglobulins attached to the cleaved portion of the pIgR receptor (secretory component) are released into the lumen. Unassociated and uncleaved pIgR is subsequently shuttled back to the baso-lateral surface (59).

S. pneumoniae translocates through epithelial cells by binding to pIgR on the apical side of the host cell. As the receptor is endocytosed and recycled to the basolateral sur-face, the pneumococcus is carried with it. The peptide mo-tif YRNYPT found on CbpA mediates binding of the pneu-mococcus to the pIgR, and pretreatment of cells with the hexapeptide is sufficient to block invasion (69). In vitro CbpA alone is sufficient to mediate translocation across a cell, as latex beads coated with CbpA are endocytosed (133). Other experiments supporting a physiological role for this process include the decreased nasopharyngeal colo-nization observed in mice lacking pIgR transport and the reduced capacity for S. pneumoniae mutants lacking CbpA to enter the bloodstream (70). It should be noted that pIgR-mediated invasion occurs only in the upper respiratory tract, because pIgR is not expressed in the lower respiratory tract.

Accessing the Lower Respiratory Tract

Development of pneumococcal pneumonia is contingent on the ability of S. pneumoniae to establish a lower respira-tory tract infection despite the host defenses that either kill or clear inhaled microorganisms. Mutants deficient in neu-raminidase do not cleave mucin efficiently and have a di-minished capacity to adhere and establish a lower respira-tory tract infection. Mutants deficient in NanA have a reduced capacity to bind to chinchilla tracheas ex vivo (110) and are attenuated in their ability to cause a lower respiratory tract infection following intranasal challenge (80). Pneumococcal adhesion to eukaryotic cells is a two-step process that initially entails a loose interaction with glycoconjugates, followed by a tighter, more secure interac-tion with protein receptors. During initial stages of infection, S. pneumoniae and other respiratory tract pathogens such as Pseudomonas aeruginosa and H. influenzae bind to the carbo-hydrate determinant N-acetylgalactosamine β1-3 galactose (66). Neuraminidases expose N-acetylgalactosamine β1-3 galactose and other ligands on the host cell surface that are obscured by sialic acid (55, 112). In vitro, pneumococci ad-here more efficiently to tissue culture cells treated with neu-raminidase (110). Enhanced adherence mediated by neu-raminidase may also explain the synergism observed between the pneumococcus and influenza virus (84). It has long been known that influenza primes the lungs for the develop-ment of a secondary bacterial infection. This synergism has been successfully modeled in mice, where mice pretreated with influenza develop fulminant pneumococcal pneumo-nia and death despite challenge with a dose that is insuffi-cient to infect a mouse not treated with influenza (74). Presumably, neuraminidase expressed on the surface of the influenza virion strips sialic acid from the mucosa and exposes bacterial receptors. Oseltamivir, a neuraminidase inhibitor, has been shown to prevent pneumococcal super-infection in this postinfluenza pneumococcal challenge model.

Interactions in the Alveoli

Pneumococcal pneumonia is classically characterized by an intense inflammatory response that results in consolidation of the infected alveoli and the affected lobe. Infected lung tissue progresses through stages of engorgement and red he-patization during which capillaries and epithelial cells be-come inflamed; fluid, erythrocytes, and neutrophils accumu-late in the alveoli; and a fibrin mesh develops (red hepatiza-tion). Subsequently, the lungs darken (gray hepatization) as leukocytes enter the lesion and the bacteria are engulfed by macrophages. Resolution continues for several days as capsule-specific antibodies provide efficient opsonization and in-flammatory mediators dissipate (116).

Pneumococcal cell wall, pneumolysin, and hydrogen peroxide are the virulence determinants that mediate the inflammation and cytotoxicity that is observed in the lungs (6, 17, 30). Challenge of mice with purified pneumolysin or cell wall products is sufficient to cause edema and influx of neutrophils that result in pneumonia (96, 118). Multiple studies clearly demonstrate that deletion of the genes that encode autolysin, pneumolysin, or the enzyme that pro-duces hydrogen peroxide greatly attenuates the ability of the bacteria to survive and replicate in the lungs (6, 7, 16, 80, 96, 102). These findings suggest that the ability of the pneumococcus to cause inflammation is not incidental but is instead a requirement for its survival in the lungs. The contribution of these virulence products is discussed in greater detail below.

Access to the Bloodstream

During red hepatization, infected alveoli are overflowing with bacteria, erythrocytes, and leukocytes enwrapped in a fibrin mesh. Fibrin strands pass through the pores of Kohn from one alveolus to the next and the lymphatics are dilat-ed and filled with cells and fibrin. It is at this stage that mice become bacteremic. Access to the bloodstream by S. pneu-moniae may occur through several pathways, including via the lymphatics, via cell damage to the epithelial and en-dothelial cells, and via direct invasion of endothelial cells (Fig. 3). Most likely all three pathways contribute to blood-stream invasion in an infected animal.

In comparison to classically invasive bacteria such as Salmonella and Shigella spp., in which greater than 2 to 3% of the inoculum invades host cells, the pneumococcus is a low-efficiency invader: ~0.2% of the inoculum invades cells (21). Invasion is dependent on activation of the host cell by pneumococcal cell wall components and pneumoly-sin and results in de novo expression of platelet activating factor receptor (PAFr), C3, and other factors. PAFr, a recep-tor abundant on lung cells, recognizes ChoP either on its natural ligand—platelet activating factor (PAF)—or on the surface of the bacteria. COS cells expressing PAFr bind more bacteria than COS cells not expressing PAFr (21). Likewise, studies have colocalized PAFr with adherent bac-teria on the surface of human cells (87). PAFr binding is not limited to the pneumococcus; other respiratory pathogens such as Haemophilus spp., Neisseria spp., and Pseudomonas spp. also express ChoP on their surfaces in a phase-variable manner (128, 127). One reason for the conserved phase-dependent expression of ChoP in respiratory pathogens may be because ChoP binds to C-reactive protein (CRP) (42). CRP is an acute-phase reactant that activates complement and opsonizes the bacteria. Thus, phase variation of ChoP serves as a mechanism by which the pneumococcus switch-es from a more adherent (high ChoP) form to one that is more resistant to phagocytosis (low ChoP).

Unlike PAF, the binding of pneumococcus to the PAFr does not result in the activation of a G-protein-mediated signal transduction pathway (87). Rather, pneumococcal uptake requires activation of ERK kinases consistent with activation by β-arrestin rather than PAFr. Uptake of the pneumococcus into a vacuole involves clathrin followed by recruitment of (β-arrestin scaffold, Rab5, then Rab7, and

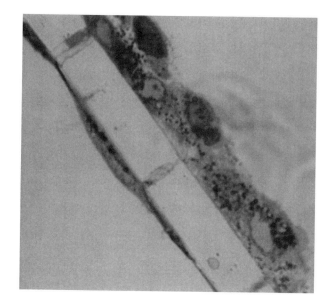

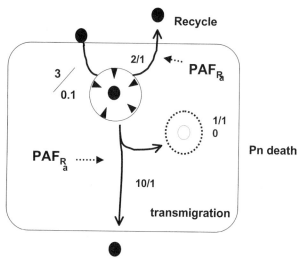

FIGURE 3 Microscopic and schematic depictions of pneumococcal invasion. (Top) Pneumococci (dark blue gram positive) bind to and are internalized into type II A549 lung cells on top of a microporous filter. Upon exiting the base of the lung cell the bacteria pass through the filter pores and invade and transcytose across primary human endothelial cells under the filter. Courtesy of Dr. C. Rosenow, Rockefeller University. (Bottom) The schematic illustrates the differences in the transcytosis process between opaque and transparent phase variants (transparent/opaque). Three fates are: (i) entry and recycling to the apical surface favored for the transparent bacteria and inhibitable by PAF receptor antagonist (PAF Ra); (ii) entry and death within the vacuole favored for opaque bacteria; and (iii) entry and transmigration across the cell, overwhelmingly favored by transparent bacteria. (Modified from reference 92.)

Rab11. Rab 5 is involved in early endocytosis; Rab 7 is found in the late endosome. Rab 11 is responsible for vacuole recycling (98). Overexpression of arrestin in endothelial cells enhances colocalization of the bacteria with Rab 7 and Rab 11 and increases survival of the pneumococci normally killed by the lysozome. Thus, it is currently thought that association of β-arrestin with the PAFr vacuole com-

plex contributes to the successful translocation of the bacteria away from the lysozome (87).

Survival in the Bloodstream

Once in the bloodstream, capsular polysaccharide (CPS) is by far the most important virulence determinant of the pneumococcus and is responsible for inhibiting phagocytosis. Data from multiple experiments suggest that the chemical structure (serotype) and amount of CPS present on the surface of the bacteria contribute to the differential ability of different serotypes and phase variants to survive in the blood (1, 12, 61, 71). Unencapsulated mutants are essentially avirulent (71), requiring 10,000- to 100,000-fold more bacteria to kill a mouse than the encapsulated parent strain following intraperitoneal injection. It is believed that CPS inhibits phagocytosis by preventing serum components such as complement, CRP, mannose-binding proteins, and antibodies attached to cell wall components from interacting with their respective receptors on the phagocytes (32, 53, 126). This inhibition is thought to be the result of steric hindrance and the negative charge of the CPS, the former physically covering the opsonin and preventing its interaction with the phagocytes, and the latter repelling the close association of bacteria with leukocytes. Formation of antibody to the serotype-specific CPS marks clearance of the infection, as antibodies to CPS are highly opsonic and are protective against subsequent pneumococcal challenge with the same serotype (17).

Pneumococcal proteins also contribute to resistance to defenses in the serum. PspA has been demonstrated to inhibit complement activation mediated via the classical pathway on the bacterial surface (90, 91). Moreover, mutants deficient in PspA are cleared more rapidly from the bloodstream of XID mice than wild type, with clearance correlating with the amount of complement on the bacterial surface (114). PspA also protects bacteria from the bactericidal effects of apolactoferrin, while antibodies to PspA enhance apolactoferrin killing (100). CbpA also inhibits complement deposition (22, 28). CbpA binds to C3 without cleavage and binds to Factor H. Binding of C3 allows the bacteria to degrade C3, whereas Factor H is a negative regulator of the alternative pathway interfering with formation of the C3 convertase.

Interactions at the Blood-Brain Barrier

Pneumococcal meningitis is by far the most devastating complication of IPD. Approximately 30% of affected individuals die (27), and half of the survivors suffer from permanent neurological sequelae (24). The majority of damage occurs in the hippocampus, particularly in the dentate gyrus, with survivors suffering from hippocampal atrophy (34). Half of neuronal damage is mediated by the host defense (10), as cell wall components are detected by host cells and in particular leukocytes (35, 46). Inflammation results in the recruitment of leukocytes into the cerebrospinal fluid (CSF), which exacerbate neuronal damage by releasing matrix metalloproteases such as MMP-9 (76), neurotoxins such as peroxynitrate (60), and proinflammatory cytokines that recruit more leukocytes (35). Ultimately the host response triggers caspase-dependent apoptosis of neurons (10). Blockage of leukocyte entry into the CSF during meningitis decreases damage by approximately 50% (119, 123) and as such is the basis of treating individuals with pneumococcal meningitis with dexamethasone prior to antibiotic therapy. The other half of neuronal damage is due directly to cytotoxic compounds such as pneumolysin and

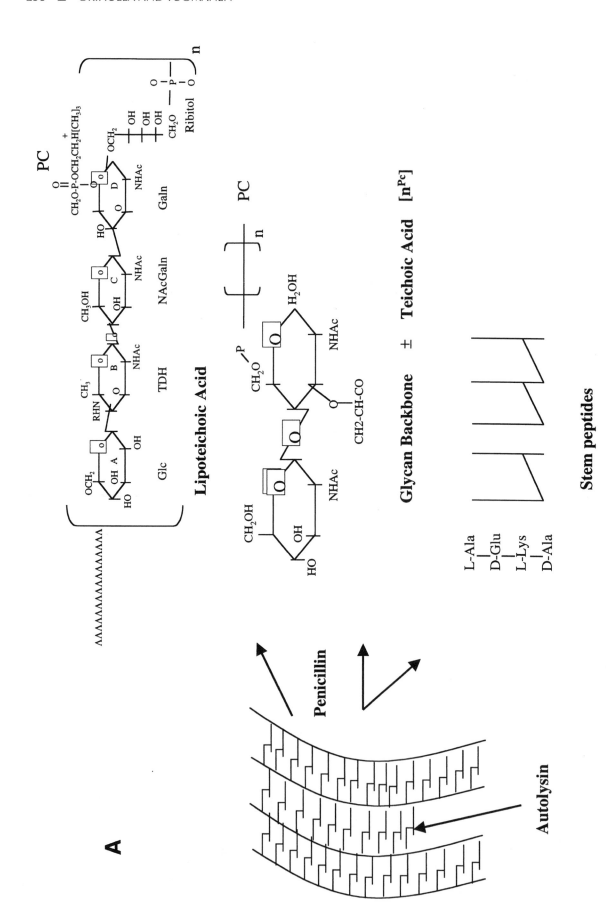

hydrogen peroxide that damage the mitochondria and cause the release of apoptosis-inducing factor (11). These factors are reviewed later in the text.

Pneumococcal invasion into the CSF is thought to occur either in the choroid plexus or in the cerebral capillaries that traverse the subarachnoid space. Cell invasion and transcytosis through the endothelial cells of the blood-brain barrier are thought to be the sole mechanisms responsible for CSF invasion. This is in contrast to the lungs/blood, where capillary leakage and lymphatics also contribute. Studies in mice have determined that invasion is dependent on activation of the endothelial cells, expression of PAFr, and CbpA expression on the bacteria (92). PAFr-knockout mice were resistant to development of meningitis (87). Likewise, CbpA mutants were unable to cross the blood-brain barrier despite bacterial titers in the blood of 10^8 CFU/ml (80).

INFLAMMATION AND CYTOTOXICITY

Intense inflammation is a hallmark of pneumococcal disease, and the pneumococcus serves as a prototype for understanding the molecular mechanisms of inflammation in response to gram-positive bacteria (108, 120). Inflammatory components released by the pneumococcus include peptidoglycan, teichoic acid, pneumolysin, secreted proteins, and hydrogen peroxide. Alone, several of these factors

have been shown to trigger inflammation and are cytotoxic (see above). In concert, these factors trigger inflammation through multiple inflammatory cascades, including the chemokine/cytokine cascade, the complement cascade, and the coagulation cascade. In a naïve host lacking serotype-specific antibodies, these cascades hasten the accumulation of leukocytes, which are ineffective for phagocytosis in the face of CPS. These white blood cells in turn release more cytokines and inflammatory mediators that serve to intensify the inflammation at the site without efficiently clearing the infection.

Inflammation

The key surface component recognized by the innate immune system is the cell wall (Fig. 4). Pneumococcal cell wall is composed of peptidoglycan with teichoic acid attached to roughly every third *N*-acetylmuramic acid residue (78). Challenge with peptidoglycan or teichoic acid elicits inflammation and re-creates many of the symptoms of pneumonia, OM, and meningitis when tested in animal models (108, 117). CPS, in contrast, is inert (130). Cell wall components mediate inflammation through multiple pathways. At the cellular level, peptidoglycan and teichoic acid bind to pattern recognition molecules such as lipopolysaccharide-binding protein, peptidoglycan recognition proteins, and CD14. These complexes in turn bind to Toll-like receptor 2 on the surface of epithelial cells, monocytes, and

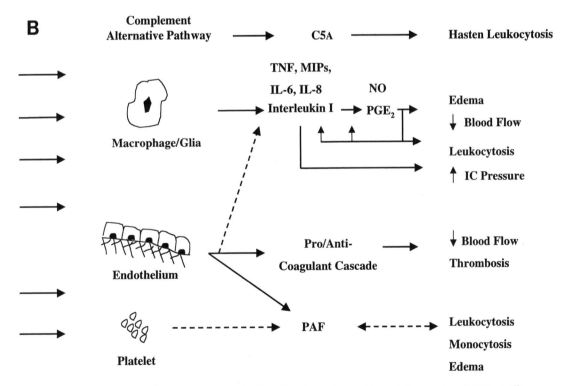

FIGURE 4 Structure of the pneumococcal cell wall and its relationship to inflammation. (A) Penicillin induces cell wall degradation by the autolysin, releasing cell wall fragments such as lipoteichoic acid, glycan polymers with and without teichoic acid, and small stem peptides. All teichoicated species contain ChoP, a key component increasing inflammatory activity. (B) All of these components interact with a variety of human cells, which in turn produce inflammatory mediators. Particularly important in this response are PAF and IL-1. These mediators combine to produce the symptomatology of pneumococcal infection, including changes in blood flow, fluid balance in the tissue, and leukocytosis. Glc: glucose; TDH: trideoxyhexose; NAcGaln: *N*-acetylgalactosamine; Galn: galactosamine; L-Ala: L-alanine; D-Glu: D-glucose; L-Lys: L-lysine; TNF: tumor necrosis factor; NO: nitric oxide; PGE₂: prostaglandin E2; IC: intracranial.

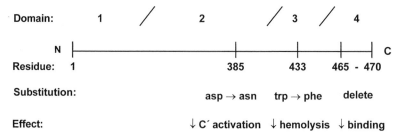

FIGURE 5 Domain structure of pneumolysin. Pneumolysin has three functionally separate domains: one activating complement, one causing hemolysis, and the other binding to cholesterol. Site-specific mutations alter these properties individually. (Compiled from references 94 and 97.)

macrophages (82, 124). Cross-linking of Toll-like receptor 2 triggers intracellular signaling that activates transcriptional regulators such as NF-κB. NF-κB expression then results in activation of the cell, including production of proinflammatory cytokines such as interleukin-1β (IL-1β), IL-6, IL-8, and tumor necrosis factor (103). Nod receptors in the cytoplasm also bind to intracellular peptidoglycan and may modulate inflammation (40, 122). Activation of epithelial and endothelial cells results in recruitment of effector cells such as neutrophils and macrophages, altered vascular permeability, and creation of a serous exudate.

In addition to activating eukaryotic cells, cell wall components activate complement that directly contributes to inflammation. Antibodies specific to bacterial proteins on the cell surface activate the classical pathway (91). Similarly, CPS and cell walls bind to hydrolyzed C3 and activate the alternative complement pathway (54, 56, 130). Either pathway results in release of C3a and C5a, both of which are potent anaphylactic molecules. Cell wall also activates complement via the lectin-binding pathway. Mannose-binding lectin, a member of the collectin family, binds to carbohydrates such as mannose and N-acetylglucosamine, a constituent of peptidoglycan (125). Binding by mannose-binding lectin results in formation of a C3 convertase on the surface of the bacteria and deposition of C3b. Mannose-binding lectin deficiency is associated with an increased risk of invasive pneumococcal infection (95). Finally, CRP bound to cell wall ChoP also activates the classical complement cascade by binding C1q (51). This results in additional opsonization and release of the potent anaphylactic molecules C3a and C5a.

Complement activation is not limited to the surface of the bacteria; cell wall fragments released by the bacteria following lysis and during cell wall turnover activate complement apart from viable bacteria (79). Likewise, pneumolysin released into the milieu also activates complement (77). Pneumolysin binds to the Fc portion of immunoglobulins and activates the classical pathway. It has been suggested that cell wall components and pneumolysin released by the bacteria serve to deplete complement (2, 3). This would have the most direct impact during bloodstream infections and in the lungs. Release of cell wall components is mediated by the murein hydrolase, LytA. LytA is responsible for pneumococcal lysis in stationary phase as well as in the presence of antibiotics (109). Autolysin-mediated lysis is responsible for the spike in inflammation observed immediately following antibiotic treatment of meningitis. Autolysin-negative mutants have reduced virulence as com-

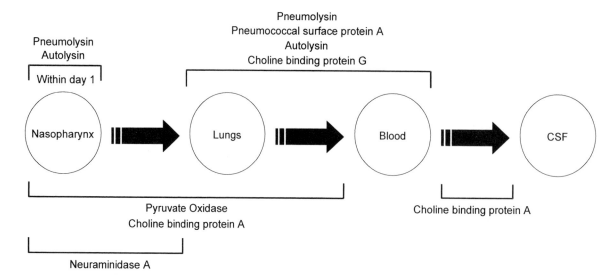

FIGURE 6 Site-specific contribution of pneumococcal virulence factors to survival and transitions from one anatomical site to the next. (Reproduced with permission from reference 80.)

TABLE 1 Anatomical site specific-expression of *Streptococcus pneumoniae* genes as determined by microarray analysis (78)[a]

Gene name	TIGR annotation	Blood[b] Fold change	Blood[b] P value	CSF[b] Fold change	CSF[b] P value	ECC[b] Fold change	ECC[b] P value	Cellular role
Virulence determinants								
blpU; bacteriocin	SP0041			−3.1	3E-07			Toxin production and resistance
cps4C	SP0348			3.9	2E-12			Capsular polysaccharide biosynthesis
pspA; pneumococcal surface protein A	SP0117[c]	3.4	2E-04	50.3	1E-08			Inhibits complement deposition
cbpJ; choline-binding protein J	SP0378					4.0	1E-04	Adhesin
cbpG; choline-binding protein G	SP0390	1.7	5E-03			3.6	9E-07	Adhesin, putative protease
cbpG; choline-binding protein F	SP0391	2.3	2E-03					Unknown
blpT	SP0524			−16.1	5E-09			Toxin production and resistance
blpA; ATP-binding protein, authentic frameshift	SP0530[c]			−11.5	1E-08			BlpC ABC transporter
blpK; bacteriocin	SP0533			−2.5	3E-06			Toxin production and resistance
blpY; immunity protein	SP0545			−16.8	8E-13			Toxin production and resistance
prtA; protective antigen A	SP0641[c,d]	15.5	7E-06	9.6	8E-13	3.3	1E-02	Serine protease
spxB; pyruvate oxidase A	SP0730			−2.6	6E-06			Hydrogen peroxide synthesis
lmb; adhesion lipoprotein	SP1002	3.4	2E-05	−5.2	1E-09			Laminin-binding protein
Conserved hypothetical protein	SP1003[c]	8.8	3E-05	−4.6	7E-09			Unknown
Conserved hypothetical protein	SP1004	2.7	1E-04					Unknown
xseA; exodeoxyribonuclease VII, large subunit	SP1207			−2.2	3E-07			DNA degradation
nanA; neuraminidase A, authentic frameshift	SP1693			3.6	2E-07			Sialidase
ply; pneumolysin	SP1923[c]			−5.6	5E-06	−9.8	3E-03	Cytolytic toxin, complement activator
lytA; autolysin	SP1937			−7.6	9E-13			Cell wall amidase, autolysis
cbpA; choline-binding protein A	SP2190[c]					2.3	2E-02	Adhesin
pcpA; choline-binding protein PcpA	SP2136[e]	7.8	7E-05					Unknown
pcpD; choline-binding protein D	SP2201[c]			−12.0	2E-12			Adhesin
htrA	SP2239[c]					8.2	4E-05	Serine protease
spoJ	SP2240[d]					3.9	9E-05	Homologous to SpoJ of B. subtilis
cps4A	SP0346					2.2	2E-03	CPS biosynthesis

[a] Bacterial RNA for microarrays was obtained from infected blood of mice (blood), CSF from meningitic rabbits (CSF), and pneumoocci attached to epithelial cells in vitro (ECC)

[b] Values within shaded boxes indicate genes with decreased transcription.

[c] Identified by STM (47).

[d] Identified by DFI (72).

[e] Confirmed by animal infection (47).

pared to wild type, most likely as a result of the inability of the mutant to release cell wall components and intracellular pneumolysin (58).

Finally, it must be noted that while complement serves to opsonize the bacteria, it is not by itself sufficient to clear the pneumococcus from the lungs. The presence of CPS inhibits the ability of phagocytes to engulf bacteria. The thick peptidoglycan cell wall is sufficiently robust to form a barrier between complement attack complexes and the bacterial cell membrane.

Cytotoxicity

In addition to the pathology derived from inflammation, the pneumococcus can directly damage eukaryotic cells. The principal mediators of cytotoxicity are pneumolysin and hydrogen peroxide (6, 11, 19, 30, 106). Pneumolysin has long been recognized as a principal virulence factor of the pneumococcus (Fig. 5). Studies using a variety of challenge routes and animal models have convincingly demonstrated that pneumolysin-deficient mutants are drastically attenuated (6, 80, 96, 106). It was initially thought that pneumolysin was released only as a result of bacterial lysis. It is now known that pneumolysin is also secreted during early exponential phase in a manner that is independent of autolysin (5). The mechanism responsible for pneumolysin transport is unknown. Pneumolysin is a pore-forming toxin that kills cells directly via necrosis (105). The toxin binds to cholesterol on the surface of the host cell and oligomerizes to form pores as large as 30 nm in diameter (39). At lower concentrations the toxin has a variety of effects on different cell types. Pneumolysin has been demonstrated to slow ciliary beating of epithelial cells (105), disrupt tight junctions (88), and inhibit the capacity of neutrophils and macrophages to kill by inhibiting oxidative burst (18, 83). Disruption of the alveoli-capillary barrier contributes to the leakage that allows serous exudates to enter the lungs and the bacteria to cross into the bloodstream (73). In the middle ear, pneumolysin is responsible for damage to the cochlea and hair cells, contributing to hearing loss (19). During meningitis pneumolysin causes neuronal damage mediated by apoptosis (11). Apoptosis is the result of an influx of extracellular calcium, presumably the result of pores in the membrane, which triggers the release of apoptosis-inducing factor (AIF) from mitochondria. Chelating extracellular calcium inhibits the release of AIF and protects cells from pneumolysin-induced apoptosis in vitro.

Hydrogen peroxide (H_2O_2) is a major product of pneumococcal metabolism and also damages host tissues. H_2O_2 is the result of the enzyme pyruvate oxidase, which decarboxylates pyruvate to produce acetyl phosphate plus H_2O_2 and CO_2 (80, 102). Mutation of pyruvate oxidase results in reduced adherence to all cell types and dramatically attenuates virulence in the respiratory tract but not in the bloodstream (80). Studies looking at the cytotoxic effects of H_2O_2 are not as comprehensive as those that have looked at pneumolysin. Nonetheless, H_2O_2 also contributes to the release of AIF from the mitochondria of neurons, resulting in apoptosis (11), and inhibits beating of ciliated ependymal cells lining the ventricular system of the brain and cerebral aqueducts (49, 50).

GENOMIC VIEW OF PATHOGENESIS

In 2001, the complete annotated genome of *S. pneumoniae* strains TIGR4 and R6 (52, 107) was published alongside a draft sequence of the strain G54 (25). R6 has the smallest genome at 2.04 Mb encoding 2,043 predicted protein coding regions; G54 is next at 2.1 Mb encoding 2,046 predicted proteins; and TIGR4 is the largest sequenced *S. pneumoniae* genome at 2.16 Mb encoding 2,236 predicted proteins. Sequence analysis of the three genomes followed by BLAST analysis of the predicted proteins was consistent with metabolism of a nutritionally fastidious lactic acid facultative anaerobe with a G + C content of 40%. An inordinate number of genes are devoted to carbohydrate metabolism. Most striking, however, was the apparent plasticity of the genome, with up to 10% of genes altered between isolates and a reference strain (45). These differences are credited to the natural ability of the pneumococcus to become competent and undergo DNA transformation. Gene shuffling has resulted in 3 to 5% of the genome being composed of insertion sequences and small dispersed DNA repeats known as BOX and RUP elements and truncation of 2% of the genes in comparison to orthologs in other bacteria.

Since release of the annotated genomes, large-scale identification of *S. pneumoniae* virulence determinants at the genomic level has been attempted (47, 67, 72, 85). Although major virulence determinants such as CPS, pneumolysin, and CBPs have been determined to have clearly established roles in pathogenesis (58) (Fig. 6), the search for new virulence determinants is ongoing. Currently, 36 to 45% of the predicted coding regions on the genome have no known function. As such, studies such as signature-tagged mutagenesis have used transposons and suicide vectors to identify mutants unable to survive passage through an animal and have identified genes required for virulence (47, 67, 85). Examples of genes identified by signature-tagged mutagenesis include the *pit1* and *pit2* loci that contribute to iron transport and are believed to be part of a pathogenicity island (13); likewise signature-tagged mutagenesis identified *mgrA*, a gene encoding a protein homologous to Mga, a transcriptional regulator of the *rlr* pathogenicity islet that contributes to adhesion (48). More recently, investigators have attempted to examine gene expression in vivo at the transcriptome level using microarrays (81) (Table 1). Among important features seen, the genes encoding CbpA, HtrA, and PspA were identified as having enhanced expression during pneumococcal attachment to epithelial cells. Likewise *pspA* and *cbpG* transcription were enhanced in the blood. Interestingly, expression of autolysin, pneumolysin, and pyruvate oxidase was reduced in the CSF. Although this study is by far the most comprehensive to date, others are required, as ideally, characterization of pneumococcal virulence determinants should include analysis of gene expression during invasive disease. Finally, the genomic content of invasive versus noninvasive strains remains to be ascertained. Evidence is accumulating indicating that particular clonotypes are more invasive than others (14). This analysis will potentially identify new targets for intervention.

REFERENCES

1. **Abeyta, M., G. G. Hardy, and J. Yother.** 2003. Genetic alteration of capsule type but not PspA type affects accessibility of surface-bound complement and surface antigens of *Streptococcus pneumoniae*. *Infect. Immun.* **71:**218–225.
2. **Alcantara, R. B., L. C. Preheim, and M. J. Gentry.** 1999. Role of pneumolysin's complement-activating activity during pneumococcal bacteremia in cirrhotic rats. *Infect. Immun.* **67:**2862–2866.

3. **Alcantara, R. B., L. C. Preheim, and M. J. Gentry-Nielsen.** 2001. Pneumolysin-induced complement depletion during experimental pneumococcal bacteremia. *Infect. Immun.* **69:**3569–3575.

4. **Austrian, R.** 1986. Some aspects of the pneumococcal carrier state. *J. Antimicrob. Chemother.* **18**(Suppl. A):35–45.

5. **Balachandran, P., S. K. Hollingshead, J. C. Paton, and D. E. Briles.** 2001. The autolytic enzyme LytA of *Streptococcus pneumoniae* is not responsible for releasing pneumolysin. *J. Bacteriol.* **183:**3108–3116.

6. **Berry, A. M., and J. C. Paton.** 2000. Additive attenuation of virulence of *Streptococcus pneumoniae* by mutation of the genes encoding pneumolysin and other putative pneumococcal virulence proteins. *Infect. Immun.* **68:**133–140.

7. **Berry, A. M., J. C. Paton, and D. Hansman.** 1992. Effect of insertional inactivation of the genes encoding pneumolysin and autolysin on the virulence of *Streptococcus pneumoniae* type 3. *Microb. Pathog.* **12:**87–93.

8. **Berry, A. M., R. A. Lock, and J. C. Paton.** 1996. Cloning and characterization of nanB, a second *Streptococcus pneumoniae* neuraminidase gene, and purification of the NanB enzyme from recombinant *Escherichia coli. J. Bacteriol.* **178:**4854–4860.

9. **Block, S. L.** 1997. Causative pathogens, antibiotic resistance and therapeutic considerations in acute otitis media. *Pediatr. Infect. Dis. J.* **16:**449–456.

10. **Braun, J. S., R. Novak, K. H. Herzog, S. M. Bodner, J. L. Cleveland, and E. I. Tuomanen.** 1999. Neuroprotection by a caspase inhibitor in acute bacterial meningitis. *Nat. Med.* **5:**298–302.

11. **Braun, J. S., J. E. Sublett, D. Freyer, T. J. Mitchell, J. L. Cleveland, E. I. Tuomanen, and J. R. Weber.** 2002. Pneumococcal pneumolysin and H(2)O(2) mediate brain cell apoptosis during meningitis. *J. Clin. Invest.* **109:**19–27.

12. **Briles, D. E., M. J. Crain, B. M. Gray, C. Forman, and J. Yother.** 1992. Strong association between capsular type and virulence for mice among human isolates of *Streptococcus pneumoniae. Infect. Immun.* **60:**111–116.

13. **Brown, J. S., S. M. Gilliland, and D. W. Holden.** 2001. A *Streptococcus pneumoniae* pathogenicity island encoding an ABC transporter involved in iron uptake and virulence. *Mol. Microbiol.* **40:**572–585.

14. **Brueggemann, A. B., T. E. Peto, D. W. Crook, J. C. Butler, K. G. Kristinsson, and B. G. Spratt.** 2004. Temporal and geographic stability of the serogroup-specific invasive disease potential of *Streptococcus pneumoniae* in children. *J. Infect. Dis.* **190:**1203–1211.

15. **Butler, J. C.** 2004. Epidemiology of pneumococcal disease, p. 148–168. *In* E. I. Tuomanen, T. J. Mitchell, D. A. Morrison, and B. G. Spratt (ed.), *The Pneumococcus.* ASM Press, Washington, D.C.

16. **Canvin, J. R., A. P. Marvin, M. Sivakumaran, J. C. Paton, G. J. Boulnois, P. W. Andrew, and T. J. Mitchell.** 1995. The role of pneumolysin and autolysin in the pathology of pneumonia and septicemia in mice infected with a type 2 pneumococcus. *J. Infect. Dis.* **172:**119–123.

17. **Centers for Disease Control and Prevention.** 1997. Prevention of pneumococcal disease: recommendations of the Advisory Committee on Immunization Practices (ACIP). *Morb. Mortal. Wkly. Rep.* **46:**1–25.

18. **Cockeran, R., A. J. Theron, C. Feldman, T. J. Mitchel, and R. Anderson.** 2004. Pneumolysin potentiates oxidative inactivation of alpha-1-proteinase inhibitor by activated human neutrophils. *Respir. Med.* **98:**865–871.

19. **Comis, S. D., M. P. Osborne, J. Stephen, M. J. Tarlow, T. L. Hayward, T. J. Mitchell, P. W. Andrew, and G. J. Boulnois.** 1993. Cytotoxic effects on hair cells of guinea pig cochlea produced by pneumolysin, the thiol activated toxin of *Streptococcus pneumoniae. Acta Otolaryngol.* **113:**152–159.

20. **Crook, D. W., A. B. Brueggemann, K. L. Sleeman, and T. E. A. Peto.** 2004. Pneumococcal carriage, p. 136–147. *In* E. I. Tuomanen, T. J. Mitchell, D. A. Morrison, and B. G. Spratt (ed.), *The Pneumococcus.* ASM Press, Washington, D.C.

21. **Cundell, D. R., N. P. Gerard, C. Gerard, I. Idanpaan-Heikkila, and E. I. Tuomanen.** 1995. *Streptococcus pneumoniae* anchor to activated human cells by the receptor for platelet-activating factor. *Nature* **377:**435–438.

22. **Dave, S., M. K. Pangburn, C. Pruitt, and L. S. McDaniel.** 2004. Interaction of human factor H with PspC of *Streptococcus pneumoniae. Indian J. Med. Res.* **119** (Suppl.):66–73.

23. **Davidson, M., A. J. Parkinson, L. R. Bulkow, M. A. Fitzgerald, H. V. Peters, and D. J. Parks.** 1994. The epidemiology of invasive pneumococcal disease in Alaska, 1986–1990—ethnic differences and opportunities for prevention. *J. Infect. Dis.* **170:**368–376.

24. **de Gans, J., and D. van de Beek.** 2002. Dexamethasone in adults with bacterial meningitis. *N. Engl. J. Med.* **347:**1549–1556.

25. **Dopazo, J., A. Mendoza, J. Herrero, F. Caldara, Y. Humbert, L. Friedli, M. Guerrier, E. Grand-Schenk, C. Gandin, M. de Francesco, A. Polissi, G. Buell, G. Feger, E. Garcia, M. Peitsch, and J. F. Garcia-Bustos.** 2001. Annotated draft genomic sequence from a *Streptococcus pneumoniae* type 19F clinical isolate. *Microb. Drug Resist.* **7:**99–125.

26. **Dunais, B., C. Pradier, H. Carsenti, M. Sabah, G. Mancini, E. Fontas, and P. Dellamonica.** 2003. Influence of child care on nasopharyngeal carriage of *Streptococcus pneumoniae* and *Haemophilus influenzae. Pediatr. Infect. Dis. J.* **22:**589–592.

27. **Durand, M. L., S. B. Calderwood, D. J. Weber, S. I. Miller, F. S. Southwick, V. S. Caviness, Jr., and M. N. Swartz.** 1993. Acute bacterial meningitis in adults. A review of 493 episodes. *N. Engl. J. Med.* **328:**21–28.

28. **Duthy, T. G., R. J. Ormsby, E. Giannakis, A. D. Ogunniyi, U. H. Stroeher, J. C. Paton, and D. L. Gordon.** 2002. The human complement regulator factor H binds pneumococcal surface protein PspC via short consensus repeats 13 to 15. *Infect. Immun.* **70:**5604–5611.

29. **Fedson, D. S., D. M. Musher, and J. Eskola.** 1998. Pneumococcal vaccine, p. 32–48. *In* S. A. Plotkin and W. A. Ordenstein (ed.), *Vaccines*, 3rd ed. WB Saunders, Philadelphia, Pa.

30. **Feldman, C., R. Anderson, R. Cockeran, T. Mitchell, P. Cole, and R. Wilson.** 2002. The effects of pneumolysin and hydrogen peroxide, alone and in combination, on human ciliated epithelium in vitro. *Respir. Med.* **96:**580–585.

31. **Fernandez-Tornero, C., R. Lopez, E. Garcia, G. Gimenez-Gallego, and A. Romero.** 2001. A novel solenoid fold in the cell wall anchoring domain of the pneumococcal virulence factor LytA. *Nat. Struct. Biol.* **8:**1020–1024.

32. **Fine, D. P.** 1975. Pneumococcal type-associated variability in alternate complement pathway activation. *Infect. Immun.* **12:**772–778.

33. **Fiore, A. E., O. S. Levine, J. A. Elliott, R. R. Facklam, and J. C. Butler.** 1999. Effectiveness of pneumococcal

polysaccharide vaccine for preschool-age children with chronic disease. *Emerg. Infect. Dis.* **5:**828–831.

34. **Free, S. L., L. M. Li, D. R. Fish, S. D. Shorvon, and J. M. Stevens.** 1996. Bilateral hippocampal volume loss in patients with a history of encephalitis or meningitis. *Epilepsia* **37:**400–405.

35. **Freyer, D., R. Manz, A. Ziegenhorn, M. Weih, K. Angstwurm, W. D. Docke, A. Meisel, R. R. Schumann, G. Schonfelder, U. Dirnagl, and J. R. Weber.** 1999. Cerebral endothelial cells release TNF-alpha after stimulation with cell walls of *Streptococcus pneumoniae* and regulate inducible nitric oxide synthase and ICAM-1 expression via autocrine loops. *J. Immunol.* **163:**4308–4314.

36. **Garcia, E., J. L. Garcia, C. Ronda, P. Garcia, and R. Lopez.** 1985. Cloning and expression of the pneumococcal autolysin gene in *Escherichia coli. Mol. Gen. Genet.* **201:**225–230.

37. **Garcia, P., M. P. Gonzalez, E. Garcia, R. Lopez, and J. L. Garcia.** 1999. LytB, a novel pneumococcal murein hydrolase essential for cell separation. *Mol. Microbiol.* **31:**1275–1277.

38. **Garcia, P., M. Paz Gonzalez, E. Garcia, J. L. Garcia, and R. Lopez.** 1999. The molecular characterization of the first autolytic lysozyme of *Streptococcus pneumoniae* reveals evolutionary mobile domains. *Mol. Microbiol.* **33:**128–138.

39. **Gilbert, R. J., J. L. Jimenez, S. Chen, I. J. Tickle, J. Rossjohn, M. Parker, P. W. Andrew, and H. R. Saibil.** 1999. Two structural transitions in membrane pore formation by pneumolysin, the pore-forming toxin of *Streptococcus pneumoniae. Cell* **97:**647–655.

40. **Girardin, S. E., I. G. Boneca, L. A. Carneiro, A. Antignac, M. Jehanno, J. Viala, K. Tedin, M. K. Taha, A. Labigne, U. Zahringer, A. J. Coyle, P. S. DiStefano, J. Bertin, P. J. Sansonetti, and D. J. Philpott.** 2003. Nod1 detects a unique muropeptide from gram-negative bacterial peptidoglycan. *Science* **300:**1584–1587.

41. **Gosink, K. K., E. R. Mann, C. Guglielmo, E. I. Tuomanen, and H. R. Masure.** 2000. Role of novel choline binding proteins in virulence of *Streptococcus pneumoniae. Infect. Immun.* **68:**5690–5695.

42. **Gould, J. M., and J. N. Weiser.** 2001. Expression of C-reactive protein in the human respiratory tract. *Infect. Immun.* **69:**1747–1754.

43. **Gray, B. M., G. M. Converse III, and H. C. Dillon, Jr.** 1980. Epidemiologic studies of *Streptococcus pneumoniae* in infants: acquisition, carriage, and infection during the first 24 months of life. *J. Infect. Dis.* **142:**923–933.

44. **Gray, B. M., M. E. Turner, and H. C. Dillon, Jr.** 1982. Epidemiologic studies of *Streptococcus pneumoniae* in infants. The effects of season and age on pneumococcal acquisition and carriage in the first 24 months of life. *Am. J. Epidemiol.* **116:**692–703.

45. **Hakenbeck, R., N. Balmelle, B. Weber, C. Gardes, W. Keck, and A. de Saizieu.** 2001. Mosaic genes and mosaic chromosomes: intra- and interspecies genomic variation of *Streptococcus pneumoniae. Infect. Immun.* **69:**2477-2486.

46. **Hanisch, U. K., M. Prinz, K. Angstwurm, K. G. Hausler, O. Kann, H. Kettenmann, and J. R. Weber.** 2001. The protein tyrosine kinase inhibitor AG126 prevents the massive microglial cytokine induction by pneumococcal cell walls. *Eur. J. Immunol.* **31:**2104–2115.

47. **Hava, D. L., and A. Camilli.** 2002. Large-scale identification of serotype 4 *Streptococcus pneumoniae* virulence factors. *Mol. Microbiol.* **45:**1389–1406.

48. **Hemsley, C., E. Joyce, D. L. Hava, A. Kawale, and A. Camilli.** 2003. MgrA, an orthologue of Mga, acts as a transcriptional repressor of the genes within the rlrA pathogenicity islet in *Streptococcus pneumoniae. J. Bacteriol.* **185:**6640–6647.

49. **Hirst, R. A., A. Rutman, K. Sikand, P. W. Andrew, T. J. Mitchell, and C. O'Callaghan.** 2000. Effect of pneumolysin on rat brain ciliary function: comparison of brain slices with cultured ependymal cells. *Pediatr. Res.* **47:**381–384.

50. **Hirst, R. A., K. S. Sikand, A. Rutman, T. J. Mitchell, P. W. Andrew, and C. O'Callaghan.** 2000. Relative roles of pneumolysin and hydrogen peroxide from *Streptococcus pneumoniae* in inhibition of ependymal ciliary beat frequency. *Infect. Immun.* **68:**1557–1562.

51. **Holzer, T. J., K. M. Edwards, H. Gewurz, and C. Mold.** 1984. Binding of C-reactive protein to the pneumococcal capsule or cell wall results in differential localization of C3 and stimulation of phagocytosis. *J. Immunol.* **133:**1424–1430.

52. **Hoskins, J., W. E. Alborn, Jr., J. Arnold, L. C. Blaszczak, S. Burgett, B. S. DeHoff, S. T. Estrem, L. Fritz, D. J. Fu, W. Fuller, C. Geringer, R. Gilmour, J. S. Glass, H. Khoja, A. R. Kraft, R. E. Lagace, D. J. LeBlanc, L. N. Lee, E. J. Lefkowitz, J. Lu, P. Matsushima, S. M. McAhren, M. McHenney, K. McLeaster, C. W. Mundy, T. I. Nicas, F. H. Norris, M. O'Gara, R. B. Peery, G. T. Robertson, P. Rockey, P. M. Sun, M. E. Winkler, Y. Yang, M. Young-Bellido, G. Zhao, C. A. Zook, R. H. Baltz, S. R. Jaskunas, P. R. Rosteck, Jr., P. L. Skatrud, and J. I. Glass.** 2001. Genome of the bacterium *Streptococcus pneumoniae* strain R6. *J. Bacteriol.* **183:**5709–5717.

53. **Hostetter, M. K.** 1986. Serotypic variations among virulent pneumococci in deposition and degradation of covalently bound C3b: implications for phagocytosis and antibody production. *J. Infect. Dis.* **153:**682–693.

54. **Hostetter, M. K., M. L. Thomas, F. S. Rosen, and B. F. Tack.** 1982. Binding of C3b proceeds by a transesterification reaction at the thiolester site. *Nature* **298:**72–75.

55. **Howie, A. J., and G. Brown.** 1985. Effect of neuraminidase on the expression of the 3-fucosyl-N-acetyllactosamine antigen in human tissues. *J. Clin. Pathol.* **38:**409–416.

56. **Hummell, D. S., R. W. Berninger, A. Tomasz, and J. A. Winkelstein.** 1981. The fixation of C3b to pneumococcal cell wall polymers as a result of activation of the alternative complement pathway. *J. Immunol.* **127:**1287–1289.

57. **Ibrahim, Y. M., A. R. Kerr, J. McCluskey, and T. J. Mitchell.** 2004. Role of HtrA in the virulence and competence of *Streptococcus pneumoniae. Infect. Immun.* **72:**3584–3591.

58. **Jedrzejas, M. J.** 2001. Pneumococcal virulence factors: structure and function. *Microbiol. Mol. Biol. Rev.* **65:**187–207.

59. **Kaetzel, C. S.** 2001. Polymeric Ig receptor: defender of the fort or Trojan horse? *Curr. Biol.* **11:**R35–R38.

60. **Kastenbauer, S., U. Koedel, and H. W. Pfister.** 1999. Role of peroxynitrite as a mediator of pathophysiological alterations in experimental pneumococcal meningitis. *J. Infect. Dis.* **180:**1164–1170.

61. **Kelly, T., J. P. Dillard, and J. Yother.** 1994. Effect of genetic switching of capsular type on virulence of *Streptococcus pneumoniae. Infect. Immun.* **62:**1813–1819.

62. **Kilian, M., J. Mestecky, and M. W. Russell.** 1988. Defense mechanisms involving Fc-dependent functions of

immunoglobulin A and their subversion by bacterial immunoglobulin A proteases. *Microbiol. Rev.* **52:**296–303.

63. **Kim, J. O., and J. N. Weiser.** 1998. Association of intrastrain phase variation in quantity of capsular polysaccharide and teichoic acid with the virulence of *Streptococcus pneumoniae. J. Infect. Dis.* **177:**368–377.

64. **Kim, J. O., S. Romero-Steiner, U. B. Sorensen, J. Blom, M. Carvalho, S. Barnard, G. Carlone, and J. N. Weiser.** 1999. Relationship between cell surface carbohydrates and intrastrain variation on opsonophagocytosis of *Streptococcus pneumoniae. Infect. Immun.* **67:**2327–2333.

65. **Kramer, M. R., B. Rudensky, I. Hadas-Halperin, M. Isacsohn, and E. Melzer.** 1987. Pneumococcal bacteremia—no change in mortality in 30 years: analysis of 104 cases and review of the literature. *Isr. J. Med. Sci.* **23:**174–180.

66. **Krivan, H. C., D. D. Roberts, and V. Ginsburg.** 1988. Many pulmonary pathogenic bacteria bind specifically to the carbohydrate sequence GalNAc beta 1-4Gal found in some glycolipids. *Proc. Natl. Acad. Sci. USA* **85:**6157–6161.

67. **Lau, G. W., S. Haataja, M. Lonetto, S. E. Kensit, A. Marra, A. P. Bryant, D. McDevitt, D. A. Morrison, and D. W. Holden.** 2001. A functional genomic analysis of type 3 *Streptococcus pneumoniae* virulence. *Mol. Microbiol.* **40:**555–571.

68. **Leibovitz, E.** 2003. Acute otitis media in pediatric medicine: current issues in epidemiology, diagnosis, and management. *Paediatr. Drugs* **5**(Suppl.):1–12.

69. **Lu, L., M. E. Lamm, H. Li, B. Corthesy, and J. R. Zhang.** 2003. The human polymeric immunoglobulin receptor binds to *Streptococcus pneumoniae* via domains 3 and 4. *J. Biol. Chem.* **278:**48178–48187.

70. **Luton, F., M. Verges, J. P. Vaerman, M. Sudol, and K. E. Mostov.** 1999. The SRC family protein tyrosine kinase p62yes controls polymeric IgA transcytosis in vivo. *Mol. Cell* **4:**627–632.

71. **Magee, A. D., and J. Yother.** 2001. Requirement for capsule in colonization by *Streptococcus pneumoniae. Infect. Immun.* **69:**3755–3761.

72. **Marra, A., J. Asundi, M. Bartilson, S. Lawson, F. Fang, J. Christine, C. Wiesner, D. Brigham, W. P. Schneider, and A. E. Hromockyj.** 2002. Differential fluorescence induction analysis of *Streptococcus pneumoniae* identifies genes involved in pathogenesis. *Infect. Immun.* **70:**1422–1433.

73. **Maus, U. A., M. Srivastava, J. C. Paton, M. Mack, M. B. Everhart, T. S. Blackwell, J. W. Christman, D. Schlondorff, W. Seeger, and J. Lohmeyer.** 2004. Pneumolysin-induced lung injury is independent of leukocyte trafficking into the alveolar space. *J. Immunol.* **173:**1307–1312.

74. **McCullers, J. A., and K. C. Bartmess.** 2003. Role of neuraminidase in lethal synergism between influenza virus and *Streptococcus pneumoniae. J. Infect. Dis.* **187:**1000–1009.

75. **Melegaro, A., N. J. Gay, and G. F. Medley.** 2004. Estimating the transmission parameters of pneumococcal carriage in households. *Epidemiol. Infect.* **132:**433–441.

76. **Meli, D. N., S. Christen, and S. L. Leib.** 2003. Matrix metalloproteinase-9 in pneumococcal meningitis: activation via an oxidative pathway. *J. Infect. Dis.* **187:**1411–1415.

77. **Mitchell, T. J., P. W. Andrew, F. K. Saunders, A. N. Smith, and G. J. Boulnois.** 1991. Complement activation and antibody binding by pneumolysin via a region of the

toxin homologous to a human acute-phase protein. *Mol. Microbiol.* **5:**1883–1888.

78. **Moreillon, P., and P. A. Majcherczyk.** 2003. Proinflammatory activity of cell-wall constituents from gram-positive bacteria. *Scand. J. Infect. Dis.* **35:**632–641.

79. **Nau, R., and H. Eiffert.** 2002. Modulation of release of proinflammatory bacterial compounds by antibacterials: potential impact on course of inflammation and outcome in sepsis and meningitis. *Clin. Microbiol. Rev.* **15:**95–110.

80. **Orihuela, C. J., G. Gao, K. P. Francis, J. Yu, and E. I. Tuomanen.** 2004. Tissue-specific contributions of pneumococcal virulence factors to pathogenesis. *J. Infect. Dis.* **190:**1661–1669.

81. **Orihuela, C. J., J. N. Radin, J. E. Sublett, G. Gao, D. Kaushal, and E. Tuomanen.** 2004. Microarray analysis of pneumococcal gene expression during invasive disease. *Infect. Immun.* **72:**5582–5596.

82. **Ozinsky, A., D. M. Underhill, J. D. Fontenot, A. M. Hajjar, K. D. Smith, C. B. Wilson, L. Schroeder, and A. Aderem.** 2000. The repertoire for pattern recognition of pathogens by the innate immune system is defined by cooperation between toll-like receptors. *Proc. Natl. Acad. Sci. USA* **97:**13766–13771.

83. **Paton, J. C., and A. Ferrante.** 1983. Inhibition of human polymorphonuclear leukocyte respiratory burst, bactericidal activity, and migration by pneumolysin. *Infect. Immun.* **41:**1212–1216.

84. **Peltola, V. T., and J. A. McCullers.** 2004. Respiratory viruses predisposing to bacterial infections: role of neuraminidase. *Pediatr. Infect. Dis. J.* **23:**S87–S97.

85. **Polissi, A., A. Pontiggia, G. Feger, M. Altieri, H. Mottl, L. Ferrari, and D. Simon.** 1998. Large-scale identification of virulence genes from *Streptococcus pneumoniae. Infect. Immun.* **66:**5620–5629.

86. **Porat, N., R. Trefler, and R. Dagan.** 2001. Persistence of two invasive *Streptococcus pneumoniae* clones of serotypes 1 and 5 in comparison to that of multiple clones of serotypes 6B and 23F among children in southern Israel. *J. Clin. Microbiol.* **39:**1827–1832.

87. **Radin, J. N., C. J. Orihuela, G. Murti, C. Guglielmo, P. J. Murray, and E. Tuomanen.** 2005. B-arrestin 1 determines the traffic pattern of PAFr-mediated endocytosis of *Streptococcus pneumoniae. Infect. Immun.* **73:**7827–7835.

88. **Rayner, C. F., A. D. Jackson, A. Rutman, A. Dewar, T. J. Mitchell, P. W. Andrew, P. J. Cole, and R. Wilson.** 1995. Interaction of pneumolysin-sufficient and -deficient isogenic variants of *Streptococcus pneumoniae* with human respiratory mucosa. *Infect. Immun.* **63:**442–447.

89. **Regev-Yochay, G., M. Raz, R. Dagan, N. Porat, B. Shainberg, E. Pinco, N. Keller, and E. Rubinstein.** 2004. Nasopharyngeal carriage of *Streptococcus pneumoniae* by adults and children in community and family settings. *Clin. Infect. Dis.* **38:**632–639.

90. **Ren, B., A. J. Szalai, O. Thomas, S. K. Hollingshead, and D. E. Briles.** 2003. Both family 1 and family 2 PspA proteins can inhibit complement deposition and confer virulence to a capsular serotype 3 strain of *Streptococcus pneumoniae. Infect. Immun.* **71:**75–85.

91. **Ren, B., A. J. Szalai, S. K. Hollingshead, and D. E. Briles.** 2004. Effects of PspA and antibodies to PspA on activation and deposition of complement on the pneumococcal surface. *Infect. Immun.* **72:**114–122.

92. **Ring, A., J. N. Weiser, and E. I. Tuomanen.** 1998. Pneumococcal trafficking across the blood-brain barrier. Mole-

cular analysis of a novel bidirectional pathway. *J. Clin. Invest.* **102:**347–360.

93. Rosenow, C., P. Ryan, J. N. Weiser, S. Johnson, P. Fontan, A. Ortqvist, and H. R. Masure. 1997. Contribution of novel choline-binding proteins to adherence, colonization and immunogenicity of *Streptococcus pneumoniae. Mol. Microbiol.* **25:**819–829.

94. Rossjohn, J., S. C. Feil, W. J. McKinstry, R. K. Tweten, and M. W. Parker. 1997. Structure of a cholesterol-binding, thiol-activated cytolysin and a model of its membrane form. *Cell* **89:**685–692.

95. Roy, S., K. Knox, S. Segal, D. Griffiths, C. E. Moore, K. I. Welsh, A. Smarason, N. P. Day, W. L. McPheat, D. W. Crook, and A. V. Hill. 2002. MBL genotype and risk of invasive pneumococcal disease: a case-control study. *Lancet* **359:**1569–1573.

96. Rubins, J. B., D. Charboneau, J. C. Paton, T. J. Mitchell, P. W. Andrew, and E. N. Janoff. 1995. Dual function of pneumolysin in the early pathogenesis of murine pneumococcal pneumonia. *J. Clin. Invest.* **95:** 142–150.

97. Saunders, F. K., T. J. Mitchell, J. A. Walker, P. W. Andrew, and G. J. Boulnois. 1989. Pneumolysin, the thiol-activated toxin of *Streptococcus pneumoniae,* does not require a thiol group for in vitro activity. *Infect. Immun.* **57:**2547–2552.

98. Seachrist, J. L., and S. S. Ferguson. 2003. Regulation of G protein-coupled receptor endocytosis and trafficking by Rab GTPases. *Life Sci.* **74:**225–235.

99. Sebert, M. E., L. M. Palmer, M. Rosenberg, and J. N. Weiser. 2002. Microarray-based identification of htrA, a *Streptococcus pneumoniae* gene that is regulated by the CiaRH two-component system and contributes to nasopharyngeal colonization. *Infect. Immun.* **70:**4059–4067.

100. Shaper, M., S. K. Hollingshead, W. H. Benjamin, Jr., and D. E. Briles. 2004. PspA protects *Streptococcus pneumoniae* from killing by apolactoferrin, and antibody to PspA enhances killing of Pneumococci by apolactoferrin. *Infect. Immun.* **72:**5031–5040.

101. Smith, T., D. Lehmann, J. Montgomery, M. Gratten, I. D. Riley, and M. P. Alpers. 1993. Acquisition and invasiveness of different serotypes of *Streptococcus pneumoniae* in young children. *Epidemiol. Infect.* **111:**27–39.

102. Spellerberg, B., D. R. Cundell, J. Sandros, B. J. Pearce, I. Idanpaan-Heikkila, C. Rosenow, and H. R. Masure. 1996. Pyruvate oxidase, as a determinant of virulence in *Streptococcus pneumoniae. Mol. Microbiol.* **19:**803–813.

103. Spellerberg, B., C. Rosenow, W. Sha, and E. I. Tuomanen. 1996. Pneumococcal cell wall activates NF-kappa B in human monocytes: aspects distinct from endotoxin. *Microb. Pathog.* **20:**309–317.

104. Stahl, W. L., and R. D. O'Toole. 1972. Pneumococcal neuraminidase: purification and properties. *Biochim. Biophys. Acta* **268:**480–487.

105. Steinfort, C., R. Wilson, T. Mitchell, C. Feldman, A. Rutman, H. Todd, D. Sykes, J. Walker, K. Saunders, P. W. Andrew, et al. 1989. Effect of *Streptococcus pneumoniae* on human respiratory epithelium in vitro. *Infect. Immun.* **57:**2006–2013.

106. Stringaris, A. K., J. Geisenhainer, F. Bergmann, C. Balshusemann, U. Lee, G. Zysk, T. J. Mitchell, B. U. Keller, U. Kuhnt, J. Gerber, A. Spreer, M. Bahr, U. Michel, and R. Nau. 2002. Neurotoxicity of pneumolysin, a major pneumococcal virulence factor, involves calcium influx and depends on activation of p38 mitogen-activated protein kinase. *Neurobiol. Dis.* **11:**355–368.

107. Tettelin, H., K. E. Nelson, I. T. Paulsen, J. A. Eisen, T. D. Read, S. Peterson, J. Heidelberg, R. T. DeBoy, D. H. Haft, R. J. Dodson, A. S. Durkin, M. Gwinn, J. F. Kolonay, W. C. Nelson, J. D. Peterson, L. A. Umayam, O. White, S. L. Salzberg, M. R. Lewis, D. Radune, E. Holtzapple, H. Khouri, A. M. Wolf, T. R. Utterback, C. L. Hansen, L. A. McDonald, T. V. Feldblyum, S. Angiuoli, T. Dickinson, E. K. Hickey, I. E. Holt, B. J. Loftus, F. Yang, H. O. Smith, J. C. Venter, B. A. Dougherty, D. A. Morrison, S. K. Hollingshead, and C. M. Fraser. 2001. Complete genome sequence of a virulent isolate of *Streptococcus pneumoniae. Science* **293:** 498–506.

108. Tomasz, A., and K. Saukkonen. 1989. The nature of cell wall-derived inflammatory components of pneumococci. *Pediatr. Infect. Dis. J.* **8:**902–903.

109. Tomasz, A., A. Albino, and E. Zanati. 1970. Multiple antibiotic resistance in a bacterium with suppressed autolytic system. *Nature* **227:**138–140.

110. Tong, H. H., M. A. McIver, L. M. Fisher, and T. F. DeMaria. 1999. Effect of lacto-N-neotetraose, asialoganglioside-GM1 and neuraminidase on adherence of otitis media-associated serotypes of *Streptococcus pneumoniae* to chinchilla tracheal epithelium. *Microb. Pathog.* **26:** 111–119.

111. Tong, H. H., L. E. Blue, M. A. James, and T. F. DeMaria. 2000. Evaluation of the virulence of a *Streptococcus pneumoniae* neuraminidase-deficient mutant in nasopharyngeal colonization and development of otitis media in the chinchilla model. *Infect. Immun.* **68:**921–924.

112. Tong, H. H., I. Grants, X. Liu, and T. F. DeMaria. 2002. Comparison of alteration of cell surface carbohydrates of the chinchilla tubotympanum and colonial opacity phenotype of *Streptococcus pneumoniae* during experimental pneumococcal otitis media with or without an antecedent influenza A virus infection. *Infect. Immun.* **70:**4292–4301.

113. Torzillo, P. J., J. N. Hanna, F. Morey, M. Gratten, J. Dixon, and J. Erlich. 1995. Invasive pneumococcal disease in central Australia. *Med. J. Aust.* **162:**182–186.

114. Tu, A. H., R. L. Fulgham, M. A. McCrory, D. E. Briles, and A. J. Szalai. 1999. Pneumococcal surface protein A inhibits complement activation by *Streptococcus pneumoniae. Infect. Immun.* **67:**4720–4724.

115. Tuomanen, E. I. 2000. Pathogenesis of pneumococcal inflammation: otitis media. *Vaccine* **19**(Suppl. 1):S38–S40.

116. Tuomanen, E. 2004. Attachment and invasion of the respiratory tract, p. 221–237. *In* E. Tuomanen, T. Mitchell, D. A. Morrison, and B. G. Spratt (ed.), *The Pneumococcus.* ASM Press, Washington, D.C.

117. Tuomanen, E., A. Tomasz, B. Hengstler, and O. Zak. 1985. The relative role of bacterial cell wall and capsule in the induction of inflammation in pneumococcal meningitis. *J. Infect. Dis.* **151:**535–540.

118. Tuomanen, E., R. Rich, and O. Zak. 1987. Induction of pulmonary inflammation by components of the pneumococcal cell surface. *Am. Rev. Respir. Dis.* **135:**869–874.

119. Tuomanen, E. I., K. Saukkonen, S. Sande, C. Cioffe, and S. D. Wright. 1989. Reduction of inflammation, tissue damage, and mortality in bacterial meningitis in rabbits treated with monoclonal antibodies against adhesion-

promoting receptors of leukocytes. *J. Exp. Med.* **170:**959–969.

120. **Tuomanen, E. I., R. Austrian, and H. R. Masure.** 1995. Pathogenesis of pneumococcal infection. *N. Engl. J. Med.* **332:**1280–1284.

121. **Vollmer, W., and A. Tomasz.** 2001. Identification of the teichoic acid phosphorylcholine esterase in *Streptococcus pneumoniae. Mol. Microbiol.* **39:**1610–1622.

122. **Watanabe, T., A. Kitani, P. J. Murray, and W. Strober.** 2004. NOD2 is a negative regulator of Toll-like receptor 2-mediated T helper type 1 responses. *Nat. Immunol.* **5:** 800–808.

123. **Weber, J. R., K. Angstwurm, W. Burger, K. M. Einhaupl, and U. Dirnagl.** 1995. Anti ICAM-1 (CD 54) monoclonal antibody reduces inflammatory changes in experimental bacterial meningitis. *J. Neuroimmunol.* **63:**63–68.

124. **Weber, J. R., D. Freyer, C. Alexander, N. W. Schroder, A. Reiss, C. Kuster, D. Pfeil, E. I. Tuomanen, and R. R. Schumann.** 2003. Recognition of pneumococcal peptidoglycan: an expanded, pivotal role for LPS binding protein. *Immunity* **19:**269–279.

125. **Weis, W. I., K. Drickamer, and W. A. Hendrickson.** 1992. Structure of a C-type mannose-binding protein complexed with an oligosaccharide. *Nature* **360:**127–134.

126. **Weiser, J. N., R. Austrian, P. K. Sreenivasan, and H. R. Masure.** 1994. Phase variation in pneumococcal opacity: relationship between colonial morphology and nasopharyngeal colonization. *Infect. Immun.* **62:**2582–2589.

127. **Weiser, J. N., M. Shchepetov, and S. T. Chong.** 1997. Decoration of lipopolysaccharide with phosphoryl-choline: a phase-variable characteristic of *Haemophilus influenzae. Infect. Immun.* **65:**943–950.

128. **Weiser, J. N., J. B. Goldberg, N. Pan, L. Wilson, and M. Virji.** 1998. The phosphorylcholine epitope undergoes phase variation on a 43-kilodalton protein in *Pseudomonas aeruginosa* and on pili of *Neisseria meningitidis* and *Neisseria gonorrhoeae. Infect. Immun.* **66:**4263–4267.

129. **Weiser, J. N., D. Bae, C. Fasching, R. W. Scamurra, A. J. Ratner, and E. N. Janoff.** 2003. Antibody-enhanced pneumococcal adherence requires IgA1 protease. *Proc. Natl. Acad. Sci. USA* **100:**4215–4220.

130. **Winkelstein, J. A., and A. Tomasz.** 1978. Activation of the alternative complement pathway by pneumococcal cell wall teichoic acid. *J. Immunol.* **120:**174–178.

131. **Winter, A. J., S. D. Comis, M. P. Osborne, M. J. Tarlow, J. Stephen, P. W. Andrew, J. Hill, and T. J. Mitchell.** 1997. A role for pneumolysin but not neuraminidase in the hearing loss and cochlear damage induced by experimental pneumococcal meningitis in guinea pigs. *Infect. Immun.* **65:**4411–4418.

132. **Zervos, M. J., S. Dembinski, T. Mikesell, and D. R. Schaberg.** 1986. High-level resistance to gentamicin in *Streptococcus faecalis:* risk factors and evidence for exogenous acquisition of infection. *J. Infect. Dis.* **153:** 1075–1083.

133. **Zhang, J. R., K. E. Mostov, M. E. Lamm, M. Nanno, S. Shimida, M. Ohwaki, and E. Tuomanen.** 2000. The polymeric immunoglobulin receptor translocates pneumococci across human nasopharyngeal epithelial cells. *Cell* **102:**827–837.

Phase Variation of *Streptococcus pneumoniae*

JEFFREY N. WEISER

22

Streptococcus pneumoniae undergoes spontaneous, reversible phenotypic variation, or phase variation, which is readily visualized as differences in colony morphology. Each isolate, therefore, is a heterogeneous population of organisms. It is now clear that these different phases are fundamentally distinct phenotypes that differ in their metabolism and multiple characteristics. Moreover, different strains may vary in different characteristics although a few common themes, including a predilection for variation of cell surface features, have emerged. For example, cell surface features that vary in amount in association with colony opacity include capsular polysaccharide and the choline-containing teichoic acid. Opaque variants, which express more capsular polysaccharide and less teichoic acid, are more virulent in animal models of sepsis but colonize the nasopharynx poorly. In contrast, transparent variants, which have less capsular polysaccharide and more teichoic acid, colonize the nasopharynx in animal models more efficiently but are relatively avirulent. This suggests that phase variation generates a mixed population that may allow for selection of organisms in vivo with characteristics permissive for either carriage or systemic infection. The genetic switch controlling these multiple phenotypic differences remains unknown.

PHENOTYPIC VARIATION IN GRAM-POSITIVE BACTERIA

The spontaneous, reversible, on-and-off switching or phase variation of virulence determinants is a well-recognized property of many gram-negative pathogens. This ability allows for the selection of variants with optimal characteristics for an individual host or distinct host environment and appears to be particularly common among the nonenteric pathogens such as *Haemophilus influenzae*, *Neisseria gonorrhoeae*, and *Neisseria meningitidis* (11, 15, 43, 56). In contrast, the enteric pathogens seem to depend more on sensing of their environment followed by programmed alterations in gene expression. There have been occasional reports of phase variation in gram-positive bacteria, but the

molecular mechanisms involved and the precise role in host-pathogen interaction are in general not well understood (3, 20, 26, 28, 34, 35).

We have described phase variation in *S. pneumoniae*, the pneumococcus, and characterized its relationship to colonization and the pathogenesis of infection (52). In particular, the focus of our laboratory has been the identification of variably expressed cell surface components as a means of gaining insight into the pathogenesis of pneumococcal disease at a molecular level. *S. pneumoniae* is highly proficient at colonization of its human host. Despite its narrow host range, it is capable of considerable flexibility, as demonstrated by the ability of different strains to synthesize a vast repertoire of at least 90 unique capsular polysaccharides. The pneumococcus has, in addition, the capacity to thrive in a number of diverse host environments, including the bloodstream and the mucosal surface of the nasopharynx. As is the case for other respiratory tract pathogens that frequently cause invasive infection, the ability of the pneumococcus to adapt to these varied environments requires changes in the expression of specific cell surface molecules.

CHARACTERISTICS OF PHENOTYPIC VARIATION IN THE PNEUMOCOCCUS

Phenotypic variation in the pneumococcus can be appreciated by detailed examination of colony morphology (52). Because a colony is an array of closely approximated organisms, differences in their physical characteristics may affect the packing of organisms within the colony. In some cases these differences may alter the passage of light through the colony, resulting in altered colony appearance. When viewed with oblique, transmitted light and magnification on transparent medium, it is possible to observe opaque and transparent colony forms in colonies derived from the same isolate (Fig. 1). Variation in colony morphology appears to be common to all strains, although it is more readily appreciated in isolates of certain serotypes, possibly because some capsule types may act to obscure phenotypic differences because of their thickness or other properties. Opacity variation in the pneumococcus is obscured on nontranslucent

Gram-Positive Pathogens, 2nd edition, edited by Vincent A. Fischetti et al.
© 2006 ASM Press, Washington, D.C.

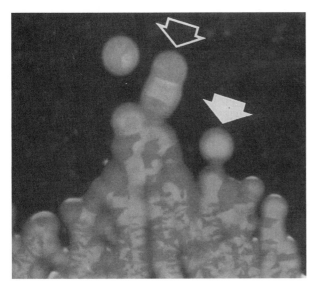

FIGURE 1 Colonies of a type 18C clinical isolate of *S. pneumoniae* showing phenotypic variation between opaque (solid arrow) and transparent (open arrow) colony forms when viewed with oblique, transmitted illumination on a transparent surface. Magnification, ×240.

medium, such as blood agar, which probably accounts for why these differences had not previously been described.

There is spontaneous, reversible variation between colony phenotypes. It is possible to detect sectored colonies resulting from phase variation during the clonal expansion of a single organism as it forms a colony. The frequency of switching is highly variable from isolate to isolate (range 10^{-3} to 10^{-6}/generation) and appears to be independent of in vitro growth conditions, including pH, temperature, and osmolarity. Under standard culture conditions, pneumococcal isolates are highly heterogeneous populations. It is possible, however, to separate many strains into nearly uniform populations of opaque and transparent and, in some cases, intermediate forms for direct comparison.

CORRELATION BETWEEN OPACITY VARIATION AND PNEUMOCOCCAL INFECTION

Animal models were used to determine whether differences in colony morphology correlate with a difference in the ability of the pneumococcus to colonize and infect a host. The relative ability of opaque and transparent variants to colonize the nasopharynx, the initial step in the pathogenesis of pneumococcal disease, was assessed in an infant rat model (52). This is a convenient model for obtaining washes of the nasopharynx to determine the quantity as well as the phenotype of colonizing organisms. Following a single intranasal inoculum, the number of organisms in the nasopharynx expands rapidly and the pups remain heavily colonized for at least several weeks. Pneumococcal carriage by infant rats, furthermore, does not appear to be limited to certain serotypes. When equal inocula (10^3 CFU) of relatively uniform populations of opaque or transparent variants of the same strain were compared in this model, only the transparent organisms were able to establish dense and stable colonization of the mucosal surface of the nasophar-

ynx. After challenge with a large inoculum (10^7 CFU) of an opaque variant, the pups gradually became colonized, but by the end of 7 days the opaque variant had been cleared and there was heavy colonization with transparent forms. This suggested that because such a large inoculum contained a small number of transparent variants, these were selected for from among the heterogeneous inoculum.

Evidence that the transparent phenotype is selected for during nasopharyngeal colonization left in question the biological role of the opaque phenotype. This was addressed by using intraperitoneal rather than intranasal inoculation in order to bypass the requirements of colonization (21). Because infant rats are not highly susceptible to invasive pneumococcal infection, it was necessary to perform these experiments in an adult mouse model of sepsis. This required the use of mouse-virulent serotypes in which the opaque and transparent forms are easily distinguished. Equal intraperitoneal inocula of relatively uniform populations of opaque or transparent variants of the same strain (10^7 CFU) were compared in this model. All mice receiving opaque organisms died of sepsis. In contrast, of the few mice that expired following inoculation with transparent organisms, splenic cultures revealed only organisms that had reverted to a more opaque phenotype. This suggests that during invasive infection, there is a strong selection for organisms with the opaque phenotype. Similarly, opaque variants show enhanced survival compared to transparent variants of the same strain during experimental middle ear infection (otitis media) in the setting of inflammation induced by concomitant influenza A virus infection (47).

Based on the animal experiments, it appears that the pneumococcus phase varies between at least two forms: one adapted for nasopharyngeal colonization and the other for events following colonization. The animal experiments, therefore, demonstrated the relevance of opacity variation to the pathogenesis of pneumococcal infection. Observations in animals, furthermore, correlate with phenotypic differences in type-matched isolates simultaneously cultured from the nasopharynx (transparent phenotype predominant) and bloodstream (opaque phenotype predominant) of the same individual with pneumococcal infection (49).

OPACITY VARIATION AND CARBOHYDRATE-CONTAINING STRUCTURES

The biochemical comparison of pneumococcal variants has focused on two major cell surface structures, the capsular polysaccharide and C polysaccharide, or cell wall teichoic acid. The capsule is a well-recognized virulence determinant, and relatively small differences in the amount of capsular polysaccharide have been shown to have a major effect on the ability of the pathogen to cause infection (27). Opacity variation is present in unencapsulated mutants, which suggests that differences in the amount or composition of the capsule are not the sole factor responsible for phenotypic variation. Variants of the same strain are indistinguishable using type-specific antiserum in a quellung reaction, suggesting that their capsules are antigenically related. Immunoelectron microscopy demonstrated a larger and denser zone of immunoreactive capsular material in opaque pneumococci (22). A capture enzyme-linked immunosorbent assay with type-specific monoclonal antibodies (MAbs) was then used to quantify amounts of capsular polysaccharide in organisms of each phenotype (21). This procedure showed that opaque pneumococci have 1.2- to

22-fold more capsular polysaccharide than the related transparent variants. Expression of capsular polysaccharide in opaque variants was further accentuated under conditions of reduced oxygen (49).

Differences in the other carbohydrate-containing structure on the cell surface, the teichoic acids, were analyzed by taking advantage of the unique structure of these molecules, which contain choline in the form of covalently bound choline phosphate or phosphorylcholine (ChoP) (2, 8). Choline, which is obtained exclusively from the nutrient medium, is essential for growth of both transparent and opaque variants. It is estimated that approximately 90% of the choline incorporated into cells is localized to the cell wall teichoic acid, with the remainder in the lipoteichoic acid (pneumococcal Forssman antigen), which is anchored in the plasma membrane (46). The teichoic acids consist of polymers that contain two to eight identical repeating units linked by ribitol phosphate, with each unit containing two ChoP residues.

The quantity of teichoic acid per cell was compared in opaque and transparent variants by measuring the incorporation of [³H]choline from the culture medium (21). After growth to the same density, transparent variants incorporated three- to eightfold more of the radiolabel per cell than did opaque variants of the same strain. Cells grown in the presence of [³H]choline were fractionated to determine the cellular location of the increased choline incorporated into organisms with the transparent phenotype. The majority of the label was found in the sodium dodecyl sulfate-insoluble cell fraction, which includes the crude cell wall and the cell wall-associated teichoic acid. These differences in the amount of cellular ChoP were also demonstrated by a quantitative enzyme-linked immunosorbent assay using a natural murine immunoglobulin A (IgA) MAb, TEPC-15, that reacts specifically to ChoP and has been shown to bind to ChoP on the pneumococcus (25). Finally, opaque and transparent phase variants were compared by flow cytometry using the MAb TEPC-15 (51). Although both phenotypes reacted with the MAb, the fivefold greater intensity of reactivity for organisms of the transparent form suggests that there is more surface-exposed ChoP associated with this phenotype. Additional evidence that teichoic acid structure mediates changes in colony opacity comes from mutational analysis of the genes involved in choline incorporation. S. pneumoniae contains two copies of licD, shown in H. influenzae to function as an NDP-choline transferase, which correlates with the expression of two ChoP residues/teichoic acid repeat unit (59). Loss of licD2 was associated with decreased [³H]choline uptake and an altered colony phenotype.

These findings would suggest that either the structure of the teichoic acid or the amount of teichoic acid per cell is subject to variation, because the only significant reservoir of cellular choline is the teichoic acids. Differences in teichoic acid structure could account for these observations if the amount of choline per teichoic acid chain varied owing to either differences in the average number of repeating units per chain or the number of ChoP residues per repeating unit. Either of these differences in structure would be expected to affect the size distribution of teichoic acid chains. Western analysis using the MAb TEPC-15 was used to visualize lipoteichoic acid (sodium dodecyl sulfate-soluble fraction). The lipoteichoic acid showed a ladderlike array of doublet bands, which represents chains with differing numbers of repeating units (51). When samples from equivalent numbers of opaque and transparent organisms were compared, there was no appreciable difference in the migration of each chain or in the average chain length. The intensity of reactivity for each band, however, was greater for the transparent variant. This observation suggests that the higher amount of choline in transparent pneumococci may be a result of increased numbers of teichoic acid residues per cell rather than variation in structure or average number of repeating units per chain. Because the cell wall-associated teichoic acid is covalently attached to muramic acid residues in the peptidoglycan and only a fraction of the muramic acid residues appear to be linked to a teichoic acid chain, the differences in the amount of teichoic acid observed in this study could be a consequence of differing proportions of muramic acid residues having an attached teichoic acid chain (7). It has been proposed that capsular polysaccharide is covalently linked to cell wall-associated teichoic acid in the pneumococcus (41). Therefore, differences in amounts of capsular polysaccharide could be a consequence of variation in cell wall-associated teichoic acid.

VARIATION IN THE EXPRESSION OF CELL SURFACE PROTEINS

The relationship of several previously identified cell surface proteins to opacity variation has been examined. The role of cell lysis or autolysis in this phenomenon was considered because of several observations: (i) Opaque colonies are dome-shaped, whereas transparent colonies are umbilicated after equivalent incubation conditions. (ii) Electron micrographs of organisms grown under identical conditions revealed that there was breakdown of the cell wall structures only in the transparent cells (52). (iii) Opaque organisms undergo spontaneous lysis more slowly and are more resistant to the detergent deoxycholate (38). Autolysis occurs once the bacteria reach stationary phase through the enzymatic degradation of the cell wall primarily by the major murein amidase, autolysin (LytA). Differences in rates of autolysis could result from variations in the peptidoglycan substrate or the expression of the amidase. The hydrolysis of purified cell walls of opaque and transparent organisms as determined by high-pressure liquid chromatography analysis of stem peptides released by treatment with amidase was indistinguishable (50). This result made it unlikely that differences in autolysis could be attributed wholly to variations in the cell wall. The expression of LytA in opaque and transparent variants was compared using antiserum to LytA (provided by R. Lopez). Low levels of autolysis in opaque variants correlated with decreased levels of immunoreactive LytA on colony immunoblots and Western analysis. Mutants in which the lytA gene has been interrupted have altered colony morphology but are still capable of displaying phenotypic variation. This indicated that LytA is only one factor contributing to opacity variation. The question of whether LytA, which is present in higher amounts on the cell surface of transparent variants, has a role in the more efficient colonization by this phenotype was examined in the infant rat model of pneumococcal carriage. LytA⁻ mutants in encapsulated strains (provided by J. Paton) were indistinguishable when compared with LytA⁺ parent strains in their ability to colonize the infant rat nasopharynx.

The hypothesis that other cell surface proteins are expressed in higher amounts in transparent variants, such as LytA, and contribute to colonization was examined as follows. LytA is known to anchor to the cell by binding to ChoP on the teichoic acid. Antiserum to pneumococcal proteins that adhered to a choline column (provided by R. Masure) was used to compare the differential expression of choline-binding proteins. In addition to LytA, this anti-serum recognized at least two other proteins that could be eluted from cells by incubation in high concentrations of choline. One of these proteins was the pneumococcal surface protein PspA, which was present in greater quantities in opaque variants (21). PspA binds to lactoferrin and has been shown to contribute to systemic pneumococcal infection in mice (14, 17, 29). The other choline-binding protein, designated CbpA, was present in higher amounts in transparent variants. Mutagenesis of the gene encoding this protein eliminated its expression and resulted in diminished adherence to type II pneumocytes and a more than 2-log decrease in the number of organisms colonizing the infant rat nasopharynx (37). CbpA, therefore, may contribute to the enhanced ability of transparent pneumococci to colonize the mucosal surface of the nasopharynx. Two other reports have suggested additional functions: the binding to secretory IgA, factor H, and complement component C3 to what appears to be the identical or closely related pneumococcal protein also referred to as SpsA, Hic, PbcA, or PspC (6, 16, 19, 37). More recently, analysis of the pneumococcal genome sequence has revealed a number of additional genes whose translation products have sequences related to the choline-binding domain (10, 58). Thus, the family of choline-binding proteins may be more extensive, and differences in their expression on the cell surface seem likely considering the variation in the ChoP anchor.

Additional evidence for variation in surface proteins was obtained from microarray analysis, which revealed 24 open reading frames that demonstrated differences in expression greater than twofold between opacity variants of independent strains (24). Twenty-one of these showed increased expression in the transparent variants, including 11 predicted to be involved in sugar utilization. Of these 11, at least two are sortase-anchored surface exoglycosidases. A single genomic region contains seven of these loci, including the gene that encodes the neuraminidase, NanA. In contrast to previous studies, there was no contribution of NanA to adherence of *S. pneumoniae* to epithelial cells or colonization in an animal model. However, NanA-dependent desialylation of human airway components that bind to the organism and may mediate bacterial clearance was observed. Targets of desialylation included human lactoferrin, secretory component, and IgA, which were shown to be present on the surface of the pneumococcus in vivo during pneumococcal pneumonia. The efficiency of desialylation was increased several-fold in the transparent variants and enhanced for host proteins binding to the surface of *S. pneumoniae*. Because deglycosylation affects the function of many host proteins, NanA may contribute to a protease-independent mechanism to modify bound targets and facilitate enhanced survival of the bacterium during colonization. Other gene products known to vary with colony opacity did not show differences in mRNA levels in the microarray analysis, suggesting that transcriptional regulation is not the primary mechanism controlling opacity phenotypes.

OTHER CHARACTERISTICS THAT VARY IN *S. PNEUMONIAE*

Using high-resolution two-dimensional protein electrophoresis, Overweg et al. showed that in comparison with transparent variants, the opaque variants reduced the expression of two proteins and overexpressed one protein (31). The proteins were identified by mass spectrometric analysis. The protein overexpressed in the opaque phenotype was similar to elongation factor Ts of *Helicobacter pylori*. One of the two proteins that were underexpressed in the opaque variants was homologous to the proteinase maturation protein PrtM of *Lactobacillus paracasei*, a member of the family of peptidyl-prolyl *cis/trans* isomerases. A consensus lipoprotein signal sequence suggests that the putative proteinase maturation protein A, designated PpmA, is located at the surface of the pneumococcus and may play a role in the maturation of surface or secreted proteins. The second underexpressed protein was identified as pyruvate oxidase, SpxB. The lower SpxB expression in opaque variants most probably explains the reduced production of hydrogen peroxide, a reaction product of SpxB, in this variant (33). Although it was subsequently shown that production of hydrogen peroxide varies consistently with colony morphology, one of the strains analyzed using high-resolution two-dimensional protein electrophoresis was found to have a frameshift mutation in *spxB*, resulting in loss of its function (32, 33). Because *spxB* is the main source of hydrogen peroxide generation and this byproduct inhibits the organism's growth as colonies, some variations in colony morphology may be consequences of differences in SpxB or its activity (42).

Recently, membrane characteristics of phenotypic variants were compared (1). These studies indicated higher microviscosity (increased 27 to 38%) of the membrane of bacterial cells in opaque variants. Membrane fatty acid composition, determined by gas chromatography, revealed that the two variants carry the same types of fatty acids but in different proportions with a lower degree of unsaturated fatty acids in the opaque variants.

One of the most thoroughly studied aspects of pneumococcal biology is the ability of the organism to take up and incorporate exogenous DNA. Comparison of the level of competence showed that transparent variants were transformed at 9- to 670-fold higher rates than the opaque variants (54). These differences were independent of the DNA incorporated and inversely correlated with amounts of capsular polysaccharide. Genetic transformation, therefore, tends to select for a less-encapsulated subpopulation with a transparent phenotype.

ROLE OF ChoP IN THE BIOLOGY OF *S. PNEUMONIAE*

Because teichoic acid is a major component of the cell surface, differences in amounts of ChoP that distinguish opaque and transparent phase variants might have numerous effects on the cell. Evidence that the cell surface expression of multiple choline-binding proteins and amounts of capsular polysaccharide vary in association with variation of the ChoP anchor has been discussed. In addition, ChoP on the pneumococcus may interact directly with host molecules. Choline, which is generally not present in prokaryotes, is a major component of eukaryotic membrane lipids in the form of phosphatidylcholine and is found on many host structures.

Pneumococci have been shown to adhere to human buccal epithelial cells, type II pneumocytes, and human lung epithelial cells, as well as vascular endothelial cells. For each of these cell types, transparent organisms bind in greater numbers than the related opaque organisms (4). This result correlates with the enhanced ability of the transparent variant to colonize the nasopharynx (52). The adherence of transparent, but not opaque, pneumococci is augmented by stimulation of resting human cells with the cytokines interleukin-1 or tumor necrosis factor. Adherence to stimulated cells correlates with the ability of transparent variants to bind to cells transfected with the receptor for platelet-activating factor (rPAF) (5). Inflammatory cytokines activate the expression of the rPAF, which has been identified on many host tissues. It has been proposed that the pneumococcus interacts with the rPAF by structural mimicry of platelet-activating factor. Because both the cell surface of the pneumococcus and PAF contain ChoP, this structure may be crucial to the binding of the bacteria to this host cell target, an interaction that is inhibited by rPAF antagonists and high concentrations of exogenous choline. This interaction has also been shown to require the presence of choline on the organism, because pneumococci grown in the presence of ethanolamine in lieu of choline are poorly adherent. It is possible that the enhanced adherence to activated cells by transparent organisms is due to the increased expression of ChoP associated with this phenotype. An alternative explanation is that differences in adherence between phenotypes are due to changes in the cell surface expression of capsular polysaccharide or choline-binding adhesins rather than a direct result of differences in content of ChoP. Phase variation in the display of the amount of ChoP and subsequent differences in choline-binding proteins may allow the organism to switch from an adherent to a nonadherent state. This may be critical to the organism's ability to exist both in the nasopharynx, where attachment to cells may be beneficial for prolonged carriage, and in the bloodstream, where adherence to cells may confer a disadvantage for survival.

Interaction with host cells may also be a factor in the ability of the pneumococcus to breach the blood-brain barrier, as occurs in the pathogenesis of meningitis. This process has been modeled using human brain microvascular endothelial cells in culture (36). Pneumococci are able to invade these cells and transcytose to the basal surface through a mechanism that involves the rPAF. The process of invasion is inhibited by the presence of capsular polysaccharide and is approximately sixfold more efficient for transparent variants that have smaller capsules.

CONTRIBUTION OF PHENOTYPIC VARIATION TO HOST CLEARANCE MECHANISMS

The expression of elevated amounts of ChoP associated with the transparent phenotype may, in addition, promote clearance of the organism by the host. Human serum contains abundant natural antibody to ChoP, and a serum protein, C-reactive protein (CRP), recognizes this structure and has been shown to promote the in vivo clearance of the pneumococcus (30, 44, 45). CRP, an acute-phase reactant whose levels may rise 1,000-fold during inflammation, is capable of activating the classical pathway of complement by binding to C1q (48). CRP expression has also been described in leukocytes and in the human respiratory epithelium (12). Only transparent pneumococci appear to bind significant amounts of human CRP (23). Binding of CRP effectively inhibits rPAF-mediated adherence to epithelial cells (13). These factors may lead to a selection of pneumococci with less cell surface ChoP, i.e., of the opaque phenotype, once an organism has bypassed the mucosal barrier.

The contribution of the capsular polysaccharide to the pathogenesis of pneumococcal infection has been ascribed to its antiphagocytic properties. The possibility that the increased virulence of opaque pneumococci could be due to the antiphagocytic effect of the higher amount of capsular polysaccharide associated with this phenotype was addressed in an opsonophagocytosis assay (23). For variants from an individual isolate, the amount of immune serum required for opsonophagocytosis by HL-60 cells in culture was proportional to the quantity of capsular polysaccharide. The average titer of serum required to achieve 50% killing was significantly greater for opaque compared with transparent variants. In addition, natural IgG2 antibody against ChoP is opsonophagocytic and more effective against transparent variants, which express increased amounts of the ChoP antigen and less antiphagocytic capsular polysaccharide (9). This suggests that the ability of opaque organisms to cause sepsis may be a result of diminished clearance by the combined activity of antibody, CRP, complement, and phagocytes.

RELATIONSHIP TO OTHER PATHOGENS OF THE HUMAN RESPIRATORY TRACT

Phase variation in the expression of ChoP is not unique to the pneumococcus, although this structure is distinctly uncommon in prokaryotes. Our laboratory has reported that another pathogen of the human respiratory tract, *H. influenzae*, a gram-negative bacterium that has a life cycle with many similarities to that of *S. pneumoniae*, decorates its cell surface with ChoP (57). In the case of *H. influenzae*, ChoP is found on the lipopolysaccharide (56). In fact, MAb TEPC-15 cross-reacts with ChoP on the surface glycolipids of both *H. influenzae* (lipopolysaccharide) and *S. pneumoniae* (teichoic acid). *H. influenzae* also acquires choline from the growth medium, although, unlike for the pneumococcus, it is not a nutritional requirement. The expression of ChoP on the lipopolysaccharide requires a previously identified chromosomal locus containing four genes, *licA* to *D*, that are also present in *S. pneumoniae*, where they contribute to teichoic acid biosynthesis (59). Phase variation is mediated by a molecular switch based on multiple tandem repeats of the sequence 5'-CAAT'-3' within the open reading frame of *licA* that are not present in the homolog in *S. pneumoniae* (55). The gene product of *licA* has homology to eukaryotic choline kinases, suggesting that the bacterial pathway for choline incorporation has common features as that of eukaryotes.

Other species that contain a *licA* homolog include some members of the genera *Neisseria* and *Mycoplasma*. Because *Neisseria*, *Mycoplasma*, *Haemophilus*, and *Streptococcus* species all share the ability to infect the respiratory tract, there may be a common mechanism involving ChoP in their pathogenesis. Interestingly, TEPC-15 binding has been detected on pili of *N. meningitidis* and *N. gonorrhoeae*, whereas in other members of the same genus ChoP is expressed on lipopolysaccharide by a mechanism similar to that of *H. influenzae* (39, 40, 53). Recently, ChoP has been confirmed as one of the posttranslational modifications of the pili of pathogenic neisseriae (18). Because pili in *N. meningitidis* also undergo phase variation, it appears that many of the major pathogens that may infect the human respiratory tract have the ChoP epitope on their cell surface and display phase variation in either the presence or amount of this unusual prokaryotic structure. *H. influenzae* decorated with ChoP was efficiently killed by CRP and anti-ChoP human IgG2 in the presence of complement (9, 56). This finding confirmed the role of this innate host defense mechanism in targeting organisms containing cell surface ChoP and provides a possible explanation for why successful pathogens such as the pneumococcus must downregulate their expression.

REFERENCES

1. **Aricha, B., I. Fishov, Z. Cohen, N. Sikron, S. Pesakhov, I. Khozin-Goldberg, R. Dagan, and N. Porat.** 2004. Differences in membrane fluidity and fatty acid composition between phenotypic variants of *Streptococcus pneumoniae. J. Bacteriol.* **186:**4638–4644.

2. **Behr, T., W. Fischer, J. Peter-Katalinic, and H. Egge.** 1992. The structure of pneumococcal lipoteichoic acid. *Eur. J. Biochem.* **207:**1063–1075.

3. **Christensen, G., L. Baddour, B. Madison, J. Parisi, S. Abraham, D. Hasty, J. Lowrance, J. Josephs, and W. Simpson.** 1990. Colonial morphology of staphylococci on Memphis agar: phase variation of slime production, resistance to beta-lactam antibiotics, and virulence. *J. Infect. Dis.* **161:**1153–1169.

4. **Cundell, D. R., J. N. Weiser, J. Shen, A. Young, and E. I. Tuomanen.** 1995. Relationship between colonial morphology and adherence of *Streptococcus pneumoniae. Infect. Immun.* **63:**757–761.

5. **Cundell, D., C. Gerard, I. Idanpaan-Heikkila, E. Tuomanen, and N. Gerard.** 1996. PAf receptor anchors *Streptococcus pneumoniae* to activated human endothelial cells. *Adv. Exp. Med. Biol.* **416:**89–94.

6. **Duthy, T., R. Ormsby, E. Giannakis, A. Ogunniyi, U. Stroeher, J. Paton, and D. Gordon.** 2002. The human complement regulator factor H binds pneumococcal surface protein PspC via short consensus repeats 13 to 15. *Infect. Immun.* **70:**5604–5611.

7. **Fischer, H., and A. Tomasz.** 1985. Peptidoglycan crosslinking and teichoic acid attachment in *Streptococcus pneumoniae. J. Bacteriol.* **163:**46–54.

8. **Fischer, W., T. Behr, R. Hartmann, K. C. J. Peter, and H. Egge.** 1993. Teichoic acid and lipoteichoic acid of *Streptococcus pneumoniae* possess identical chain structures. A reinvestigation of teichoid acid (C polysaccharide). *Eur. J. Biochem.* **215:**851–857.

9. **Goldenberg, H., T. McCool, and J. Weiser.** 2004. Human IgG2 recognizing phosphorylcholine shows cross-reactivity and evidence for protection against major bacterial pathogens of the human respiratory tract. *J. Infect. Dis.* **190:**1254–1263.

10. **Gosink, K., E. Mann, C. Guglielmo, E. Tuomanen, and H. Masure.** 2000. Role of novel choline binding proteins in virulence of *Streptococcus pneumoniae. Infect. Immun.* **68:**5690–5695.

11. **Gotschlich, E. C.** 1994. Genetic locus for the biosynthesis of the variable portion of *Neisseria gonorrhoeae* lipooligosaccharide. *J. Exp. Med.* **180:**2181–2190.

12. **Gould, J., and J. Weiser.** 2001. Expression of C-reactive protein in the human respiratory tract. *Infect. Immun.* **69:**1747–1754.

13. **Gould, J., and J. Weiser.** 2002. The inhibitory effect of C-reactive protein on bacterial phosphorylcholine-platelet activating factor receptor mediated adherence is blocked by surfactant. *J. Infect. Dis.* **186:**361–371.

14. **Hakansson, A., H. Roche, S. Mirza, L. S. McDaniel, A. Brooks-Walters, and D. Briles.** 2001. Characterization of binding of human lactoferrin to pneumococcal surface protein A. *Infect. Immun.* **69:**3372–3381.

15. **Hammerschmidt, S., A. Muller, H. Sillmann, M. Muhlenhoff, R. Borrow, A. Fox, J. van Putten, W. D. Zollinger, R. Gerardy-Schahn, and M. Frosch.** 1996. Capsule phase variation in *Neisseria meningitidis* serogroup B by slipped-strand mispairing in the polysialyltransferase gene (*siaD*): correlation with bacterial invasion and the outbreak of meningococcal disease. *Mol. Microbiol.* **20:** 1211–1220.

16. **Hammerschmidt, S., S. R. Talay, P. Brandtzaeg, and G. S. Chhatwal.** 1997. SpsA, a novel pneumococcal surface protein with specific binding to secretory immunoglobulin A and secretory component. *Mol. Microbiol.* **25:**1113–1124.

17. **Hammerschmidt, S., G. Bethe, P. Remane, and G. Chhatwal.** 1999. Identification of pneumococcal surface protein A as a lactoferrin-binding protein of *Streptococcus pneumoniae. Infect. Immun.* **67:**1683–1687.

18. **Hegge, F., P. Hitchen, F. Aas, H. Kristiansen, C. Lovold, W. Egge-Jacobsen, M. Panico, W. Leong, V. Bull, M. Virji, H. Morris, A. Dell, and M. Koomey.** 2004. Unique modifications with phosphocholine and phosphoethanolamine define alternate antigenic forms of *Neisseria gonorrhoeae* type IV pili. *Proc. Natl. Acad. Sci. USA* **101:** 10798–10803.

19. **Janulczyk, R., F. Iannelli, A. Sjöholm, G. Pozzi, and L. Björck.** 2000. Hic, a novel surface protein of *Streptococcus pneumoniae* that interferes with complement function. *J. Biol. Chem.* **275:**37257–37263.

20. **Jones, G., D. Clewell, L. Charles, and M. Vickerman.** 1996. Multiple phase variation in haemolytic, adhesive and antigenic properties of *Streptococcus gordonii. Microbiology* **142:**181–189.

21. **Kim, J., and J. Weiser.** 1998. Association of intrastrain phase variation in quantity of capsular polysaccharide and teichoic acid with the virulence of *Streptococcus pneumoniae. J. Infect. Dis.* **177:**368–377.

22. **Kim, J., W. J. H. Wani, H. Sørensen, J. Blom, and J. Weiser.** 1998. Characterization of phenotypic variants of *Streptococcus pneumoniae*, abstr. B-2, p. 56. Abstr. 98th Gen. Meet. Am. Soc. Microbiol. 1998. American Society for Microbiology, Washington, D.C.

23. **Kim, J. O., S. Romero-Steiner, U. Sørensen, J. Blom, M. Carvalho, S. Barnardi, G. Carlone, and J. N. Weiser.** 1999. Relationship between cell-surface carbohydrates and intrastrain variation on opsonophagocytosis of *Streptococcus pneumoniae. Infect. Immun.* **67:**2327–2333.

24. **King, S., K. Hippe, J. Gould, D. Bae, S. Peterson, R. Cline, C. Fasching, E. Janoff, and J. Weiser.** 2004. Phase variable desialylation of host proteins that bind to *Streptococcus pneumoniae* in vivo and protect the airway. *Mol. Microbiol.* **54:**159–171.

25. **Leon, M. A., and N. M. Young.** 1971. Specificity for phosphorylcholine of six murine myeloma proteins reactive with pneumococcus C polysaccharide and beta-lipoprotein. *Biochemistry* **10:**1424–1429.

26. **Lukomski, S., K. Nakashima, I. Abdi, V. Cipriano, B. Shelvin, E. Graviss, and J. Musser.** 2001. Identification and characterization of a second extracellular collagen-like protein made by group A Streptococcus: control of production at the level of translation. *Infect. Immun.* **69:** 1729–1738.

27. **MacLeod, C. M., and M. R. Krauss.** 1950. Relation of virulence of pneumococcal strains for mice to the quantity of capsular polysaccharide formed in vitro. *J. Exp. Med.* **92:**1–9.

28. **McCarty, M.** 1966. The nature of the opaque colony variation in group A streptococci. *J. Hyg. Camb.* **64:**185–190.

29. McDaniel, L. S., J. Yother, M. Vijayakumar, L. McGarry, W. R. Guild, and D. E. Briles. 1987. Use of insertional inactivation to facilitate studies of biological properties of pneumococcal surface protein A (PspA). *J. Exp. Med.* **165**:381–394.

30. Mold, C., S. Nakayama, T. Holzer, H. Gewurz, and T. Du Clos. 1981. C-reactive protein is protective against *Streptococcus pneumoniae* infection in mice. *J. Exp. Med.* **154**:1703–1708.

31. Overweg, K., C. Pericone, G. Verhoef, J. Weiser, H. Meiring, A. De Jong, R. De Groot, and P. Hermans. 2000. Differential protein expression in phenotypic variants of *Streptococcus pneumoniae*. *Infect. Immun.* **68**:4604–4610.

32. Pericone, C., D. Bae, M. Shchepetov, T. McCool, and J. Weiser. 2002. Short-sequence tandem and nontandem DNA repeats and endogenous hydrogen peroxide production contribute to genetic instability of *Streptococcus pneumoniae*. *J. Bacteriol.* **184**:4392–4399.

33. Pericone, C., S. Park, J. Imlay, and J. Weiser. 2003. Factors contributing to hydrogen peroxide resistance in *Streptococcus pneumoniae* include pyruvate oxidase (SpxB) and avoidance of the toxic effects of the Fenton reaction. *J. Bacteriol.* **185**:6815–6825.

34. Pincus, S. H., R. L. Cole, M. R. Wessels, M. D. Corwin, E. Kamanga-Sollo, S. F. Hayes, W. Cieplak, and J. Swanson. 1992. Group B streptococcal variants. *J. Bacteriol.* **174**:3739–3749.

35. Pincus, S. H., R. L. Cole, E. Kamanga-Sollo, and S. H. Fischer. 1993. Interaction of group B streptococcal opacity variants with the host defense system. *Infect. Immun.* **91**:3761–3768.

36. Ring, A., J. N. Weiser, and E. I. Tuomanen. 1998. Pneumococcal penetration of the blood-brain barrier: molecular analysis of a novel re-entry path. *J. Clin. Invest.* **102**:347–360.

37. Rosenow, C., P. Ryan, J. N. Weiser, S. Johnson, P. Fontan, A. Ortqvist, and H. R. Masure. 1997. Contribution of novel choline-binding proteins to adherence, colonization and immunogenicity of *Streptococcus pneumoniae*. *Mol. Microbiol.* **25**:819–829.

38. Saluja, S. K., and J. N. Weiser. 1995. The genetic basis of colony opacity in *Streptococcus pneumoniae*: evidence for the effect of box elements on the frequency of phenotypic variation. *Mol. Microbiol.* **16**:215–227.

39. Serino, L., and M. Virji. 2000. Phosphorylcholine decoration of lipopolysaccharide differentiates commensal Neisseriae from pathogenic strains: identification of licA-type genes in commensal Neisseriae. *Mol. Microbiol.* **35**:1550–1559.

40. Serino, L., and M. Virji. 2002. Genetic and functional analysis of the phosphorylcholine moiety of commensal Neisseria lipopolysaccharide. *Mol. Microbiol.* **43**:437–448.

41. Sorensen, U. B. S., J. Henrichsen, H.-C. Chen, and S. C. Szu. 1990. Covalent linkage between the capsular polysaccharide and cell wall peptidoglycan of *Streptococcus pneumoniae* revealed by immunochemical methods. *Microb. Pathog.* **8**:325–334.

42. Spellberg, B., D. R. Cundell, J. Sandros, B. J. Pearce, I. Idanpaan-Heikkila, C. Rosenow, and H. R. Masure. 1996. Pyruvate oxidase, as a determinant of virulence in *Streptococcus pneumoniae*. *Mol. Microbiol.* **19**:803–813.

43. Stern, A., and T. F. Meyer. 1987. Common mechanism controlling phase and antigenic variation in pathogenic neisseriae. *Mol. Microbiol.* **1**:5–12.

44. Szalai, A. J., D. E. Briles, and J. E. Volanakis. 1996. Role of complement in C-reactive-protein-mediated protection of mice from *Streptococcus pneumoniae*. *Infect. Immun.* **64**:4850–4853.

45. Szalai, A. J., A. Agrawal, T. J. Greenhough, and J. E. Volanakis. 1997. C-reactive protein. *Immunol. Res.* **16**:127–136.

46. Tomasz, A. 1981. Surface components of *Streptococcus pneumoniae*. *Rev. Infect. Dis.* **3**:190–210.

47. Tong, H., J. Weiser, M. James, and T. DeMaria. 2001. Effect of influenza A virus infection on nasopharyngeal colonization and otitis media induced by transparent or opaque phenotypic variants of *Streptococcus pneumoniae* in the chinchilla model. *Infect. Immun.* **69**:602–606.

48. Volanakis, J. E., and M. H. Kaplan. 1974. Interaction of C-reactive protein complexes with the complement system. II. Consumption of guinea pig complement by CRP complexes. Requirement for human C1q. *J. Immunol.* **113**:9–17.

49. Weiser, J., D. Bae, H. Epino, S. Gordon, M. Kapoor, L. Zenewicz, and M. Shchepetov. 2001. Changes in availability of oxygen accentuate differences in capsular polysaccharide expression by phenotypic variants and clinical isolates of *Streptococcus pneumoniae*. *Infect. Immun.* **69**:5430–5439.

50. Weiser, J., Z. Markiewicz, E. Tuomanen, and J. Wani. 1996. Relationship between phase variation in colony morphology, intrastrain variation in cell wall physiology and nasopharyngeal colonization by *Streptococcus pneumoniae*. *Infect. Immun.* **64**:2240–2245.

51. Weiser, J. N. 1998. Phase variation in colony opacity. *Microb. Drug Resist.* **4**:129–145.

52. Weiser, J. N., R. Austrian, P. K. Sreenivasan, and H. R. Masure. 1994. Phase variation in pneumococcal opacity: relationship between colonial morphology and nasopharyngeal colonization. *Infect. Immun.* **62**:2582–2589.

53. Weiser, J. N., J. B. Goldberg, N. Pan, L. Wilson, and M. Virji. 1998. The phosphorylcholine epitope undergoes phase variation on a 43 kD protein in *Pseudomonas aeruginosa* and on pili of pathogenic Neisseria. *Infect. Immun.* **66**:4263–4267.

54. Weiser, J. N., and M. Kapoor. 1999. Effect of intrastrain variation in amount of capsular polysaccharide on the genetic transformation of *Streptococcus pneumoniae*: implications for virulence studies of encapsulated strains. *Infect. Immun.* **67**:3690–3692.

55. Weiser, J. N., J. M. Love, and E. R. Moxon. 1989. The molecular mechanism of phase variation of *H. influenzae* lipopolysaccharide. *Cell* **59**:657–665.

56. Weiser, J. N., N. Pan, K. L. McGowan, D. Musher, A. Martin, and J. C. Richards. 1998. Phosphorylcholine on the lipopolysaccharide of *Haemophilus influenzae* contributes to persistence in the respiratory tract and sensitivity to serum killing mediated by C-reactive protein. *J. Exp. Med.* **187**:631–640.

57. Weiser, J. N., M. Shchepetov, and S. T. H. Chong. 1997. Decoration of lipopolysaccharide with phosphorylcholine; a phase-variable characteristic of *Haemophilus influenzae*. *Infect. Immun.* **65**:943–950.

58. Yother, J., and J. M. White. 1994. Novel surface attachment mechanism of the *Streptococcus pneumoniae* protein PspA. *J. Bacteriol.* **176**:2976–2985.

59. Zhang, J.-R., I. Idanpaan-Heikkila, W. Fischer, and E. Tuomanen. 1999. Pneumococcal licD2 gene is involved in phosphorylcholine metabolism. *Mol. Microbiol.* **31**:1477–1488.

Genetics of *Streptococcus pneumoniae*

JANET YOTHER AND SUSAN K. HOLLINGSHEAD

23

The origins of genetics in *Streptococcus pneumoniae* can be traced to studies that began in the late 1800s with the isolation of nonencapsulated variants and ultimately led to the discovery of bacterial gene transfer by Griffith in 1928 and the identification of DNA as the genetic material by Avery, MacLeod, and McCarty in 1944 (6, 34, 126). Transformation was the hallmark of these and many other studies analyzing the transfer of capsular polysaccharide biosynthetic genes, as well as nutritional and antibiotic markers. Today, it remains the single most important technique in the genetic analysis of *S. pneumoniae*. Many of the *S. pneumoniae* strains commonly used in genetic studies are descended from isolates obtained by Avery in 1916 and first described in 1928 by Dawson (109, 119). In fact, all of the nonencapsulated isolates used to study the transformation process are derived from a single type 2 clinical isolate and its nonencapsulated variants used in the 1944 studies of Avery et al. Table 1 provides the lineage for a number of these strains, along with properties pertinent to this review. Also listed are the strains for which the complete genome sequences have been or will soon be available, including the strain R6 from within the D39 lineage. This chapter highlights much of the current information regarding *S. pneumoniae* genetics, and the reader is referred to references within those cited for further details of the elegant earlier studies that laid the foundations for our present knowledge.

GENOMES AND DIVERSITY

The DNA sequences for three *S. pneumoniae* chromosomes were published in 2001 (27, 42, 118). The first of these was an annotated draft genome for strain G54, a serotype 19F strain (27), which was published as a near-complete genome in 31 annotated DNA assemblages. Shortly thereafter, two completed genomes were published, one for strain TIGR4 (118), a virulent serotype 4 clinical isolate, and one for R6 (42), a nonencapsulated strain in the historical D39 lineage. The sequences of two further genomes, for strains Spain[6B]-2 and Spain[23F]-1, are currently being edited and annotated at The Institute for Genomic Research (46) and The Sanger Center (101), respectively. These latter two genomes represent lineages that have been important in the global spread of antibiotic resistance.

The genome size is 2,161 kb in TIGR4 and 2,039 kb in R6, and the average percent G+C content is 39.7. There is a strong bias in the orientation of genes, with a large majority being transcribed in a direction that is away from the origin of replication. This feature is common to all of the low-G+C gram-positive bacteria. The genome sequences provide an inventory of genes that can give some insight into the metabolic and pathogenic capacities of the pneumococcus. Notable in metabolism is the lack of the tricarboxylic acid cycle and the indication that fermentation of carbohydrates is the preferred energy source, based in part on the higher percentage of transporters in this category.

Similarly, the positions of genes in different strains and the presence or absence of orthologs can provide insight into the nature of forces that operate in the natural evolution of *S. pneumoniae*. A comparison of the three genome sequences has revealed mosaic chromosomes, which are marked by regions for which the gene content and/or the gene order differs from one genome to another (12, 117). In one-on-one comparisons, the three strains each share about 90% of their genes, while about 10% of the genes are not shared. Nonshared DNA is mostly associated with gene clusters unique to one or the other of the strains. Microarray hybridizations also show that the clusters detected from whole genome comparisons are of variable presence in other strains (12, 117). These gene clusters are spaced around the entire genome, and, in several cases, the same genome location contains one cluster of genes in R6 and a different cluster in TIGR4. This suggests the possibility of sites permissive for gene transfer leading to genome plasticity. The GC content of the gene clusters is often aberrant from that of the total genome, an indication that they may have originated from outside the species. In these genome comparisons, the cluster boundaries rarely have the associated genetic hallmarks of pathogenicity islands in other bacteria. Most clusters are near one or more repetitive DNA sequences, but this may be circumstantial, as ~5% of the total DNA is assigned to the repetitive DNA class (5, 117).

Gram-Positive Pathogens, 2nd edition, edited by Vincent A. Fischetti et al.
© 2006 ASM Press, Washington, D.C.

TABLE 1 Derivation and properties of commonly used *S. pneumoniae* strains

Strain	Capsule[a]	Derivation and relevant properties[b]
D39	Type 2[+]	Clinical isolate (1928); virulent; pDP1[+]; *comC1*
R36	Type 2[−]	D39 passaged 36 × in anti-type 2 serum (1944); does not revert to Cps[+] (see R36A, below); Hex[+]; pDP1[+]
R36N	Type 2[−]	Colony variant of R36 (1944); nontransformable; pDP1[+]
R36NC	Type 2[−]	Transformable derivative of R36N (1947); *Dpn*I; Hex[+]; pDP1[+]
R36A	Type 2[−]	Colony variant of R36 (1944); deletions demonstrated in type 2 capsule locus (reference 45); GenBank accession number AF029368); Hex[+]; pDP1[+]
R6	Type 2[−]	Single-colony isolate of R36A (1962); *Dpn*I; Hex[+]; pDP1[−]; genome sequenced; GenBank accession number AE007317
R6x	Type 2[−]	R6 transformed to the Rx Hex[−] phenotype (1973)
Rx	Type 3*	R36A transformed to type 3 encapsulation (1949) and subsequently isolated as a spontaneous "nonencapsulated" derivative (1959) (but see Rx1, below); Hex[−] (*hexB*[−]); pDP1[−]
Rx1	Type 3*	Highly transformable derivative of Rx; contains the tpe 3 *cps* locus and, consequently, a 5′ deletion of *aliA* (*plpA*); produces small amounts of type 3 capsule (~20% of that observed with most type 3 strains); Cps phenotype due to point mutation in *cps3D*; virulence is not restored by transformation to normal type 3 encapsulation (unlike transformation of D39 to type 3); *Dpn*0; Hex[−] (*hexB*[−]); pDP1[−]; *comC1*
A66	Type 3[+]	Clinical isolate (1928); virulent; donor in 1944 transformation studies of Avery et al.; *Dpn*I; *comC2*
TIGR4	Type 4[+]	Clinical isolate, virulent; genome sequenced; GenBank accession number AE005672; *Dpn*I; *comC2*; MLST type 205
G54	Type 19F[+]	Clinical isolate; macrolide and tetracycline resistant, not mouse virulent; *Dpn*II; *comC1*; genome draft annotation; GenBank accession number AL449923-64, 31 contigs, MLST type 63, lineage most often carries type 15 capsule
670, Spain[6B]−2	Type 6B[+]	Clinical isolate; penicillin and macrolide resistant, not mouse virulent; *Dpn*II; *comC1*; genome sequence under way at TIGR, MLST type 90
Spain[23F]−1	Type 23F[+]	Clinical isolate; penicillin, chloramphenicol, and tetracycline resistant, not mouse virulent; *comC2*; genome sequence under way at Sanger, MLST type 81

[a]The *cps* genotype is given, and the symbols indicate whether the strain is encapsulated (+), nonencapsulated (−), or intermediate (*). Of the type 3* strains, only Rx1 has been experimentally demonstrated to produce capsular polysaccharide (23, 25).

[b]Properties are listed for strains in which they have been demonstrated, and may be the same for progenitors. Lineages and original references are described for most D39 derivatives in references 109 and 119. Numbers in parentheses indicate date of original description. Except as noted, all strains are avirulent. *Dpn* phenotypes are from reference 75 and the genome sequences. The *Dpn*0 phenotype in Rx1 is apparently due to a mutation in the *Dpn*I-encoding locus. Hex phenotypes are from references 92 and 119 and the genome sequences. *comC* alleles are from reference 90 or the genome sequences. All D39 derivatives are expected to carry the *comC1* allele.

Repetitive DNA includes mobile elements such as transposons and insertion sequences that can often indicate regions of genome evolution. Even nonmobile repetitive DNA can support gene transfer and genome remodeling, however, by serving as reiterated sites of homology between genomes. The properties of a number of insertion sequences identified in the chromosomes of *S. pneumoniae* isolates are summarized in Table 2. In addition to these are two smaller, nonmobile elements known as RUP (repeat unit of pneumococcus) (78) and BOX (63). As evidenced from the genome sequences, these elements are present in more than 100 copies and are found almost exclusively outside of coding regions. Their functions are unknown, although RUP elements are thought to be mobilizable by the IS630-Spn1. The BOX element is composed of three subunits of 59, 45, and 50 bp (box A, box B, and box C, respectively), although not all are present in every copy.

GENE TRANSFER

Both transformation and conjugation have been described in *S. pneumoniae*. Transformation serves as the primary, and perhaps sole, means of transferring chromosomal genes. Conjugation occurs with plasmids that are capable of self-transfer or mobilization and with conjugative transposons that are integrated into the chromosome. The latter mediate their own transfer but do not cotransfer linked chromosomal markers. Conjugative transposons and chromosomal

genes transferred by transformation are the major mechanisms known to be involved in the spread of antibiotic resistance in *S. pneumoniae*. Generalized transduction has not been observed to occur, although a process termed pseudotransduction, which involves properties of both transduction and transformation, has been described for one pneumococcal bacteriophage.

Transformation

Natural transformation in *S. pneumoniae* can be divided into several stages: (i) induction of competence, (ii) binding of double-stranded DNA to the cell surface, (iii) digestion of bound DNA into discrete double-stranded fragments, (iv) uptake of single-stranded linear DNA, (v) binding of the internalized DNA by single-stranded binding proteins, and (vi) recombination of homologous DNA into the chromosome by a single-strand replacement mechanism. Successful transformation of plasmids requires the uptake of two complementary, overlapping DNA strands. Essentially any DNA can be taken up, as no specific recognition sequences are required. Most DNA that is not homologous to the pneumococcal chromosome or that cannot replicate independently is degraded and becomes part of the nucleotide precursor pool. Occasionally, DNA that is not homologous to the chromosome but that is physically linked to DNA that is homologous can be inserted into the chromosome by a process known as homology-assisted illegitimate recombination (91).

TABLE 2 Insertion sequences found in *S. pneumoniae*

Element[a]	Size (bp)	Terminal inverted repeats	Target (bp)	Copies	Distribution[c]	Reference[d]
IS*1202*	1,747	23 bp, imperfect	27	1-5 (0)[b]	Mainly type 19	70
IS*1167*	1,435	24 bp, imperfect	8	3-12 (22; 14t)	Multiple, 11/22	129
IS*1381*	846	20 bp, imperfect	7	5-7 (12t)	Multiple, 8/8	100
IS*1515*	871	12 bp, imperfect	3	2-13 (1)	Type 1, 17/17; others, 3/20	76
IS*3-Spn*	1,359		4	(14t)	Unknown	
IS*1380-Spn*	1,703	13/15[e] bp, imperfect	2	(12; 1t)	Unknown	
IS*630-Spn1*	896	22 bp, imperfect		(12t)	Unknown	
IS*1239*	1,046	7/15 bp, imperfect		(14t)	Unknown	
IS*66*	2,484	18/20 bp, imperfect	8	(2t)	Unknown	
IS*200*	747	0 bp		(3; 2t)	Unknown	

[a]Each element contains an open reading frame or frames with homology to transposase sequences.
[b]Numbers in parentheses represent the number of copies in TIGR4 and the number of truncated or frameshifted copies (t).
[c]The capsular serotypes of strains in which the element occurs; numbers indicate the fraction of tested strains containing the element.
[d]Except as noted, information is from references 15 and 118.
[e]Alternative number of base pairs, depending on the copy.

Induction of Competence

S. pneumoniae becomes competent for transformation for only a brief period during the early exponential phase of growth (121). The specific point at which competence is induced is determined by both the cell density and the culturing conditions. Maximal transformation efficiencies are usually observed in a rich medium containing calcium and serum or albumin, at a starting pH of about 7.0 to 7.2. Under optimum culture conditions, transformation of a given marker from a donor chromosomal DNA preparation can be observed in 1% or more of the recipients. Alterations in starting pH affect both the timing and efficiency of transformation (18). A low pH tends to prevent the development of competence, whereas higher pH values result in multiple waves of competence that occur over a broader range of cell densities and rarely reach the peak efficiency. Under optimum conditions, cotransformation of two unlinked chromosomal markers occurs near the frequency expected for two independent events, indicating that essentially all cells in the culture become competent (87). Cotransformation of unlinked markers during subsequent, postpeak waves of competence occurs at a frequency higher than expected, indicating that only a fraction of these cells are actually capable of being transformed (18). These differing outcomes make it possible to manipulate the culture conditions, depending on the desired outcome of the transformation. In addition, the high transformation frequencies observed under optimum conditions allow for the introduction of DNA for which there is no selectable marker but for which a suitable screen exists. Indeed, the use of isolated, homogeneous DNA fragments, such as that obtained from cloned DNA or PCR amplification, has been observed to result in transformation frequencies as high as 50% (25, 39, 111).

The development of competence is mediated by the competence factor (121), a small peptide that induces expression of genes encoding transformation-related proteins. Different *S. pneumoniae* isolates contain one of two alleles of *comC*, the gene that encodes the competence-stimulating peptide (CSP) (90). The 41-amino-acid translation products from *comC1* and *comC2* each contain an identical 24-amino-acid N-terminal leader sequence that possesses a Gly-Gly processing site common to peptide bacteriocins. The remaining 17 amino acids, which represent the mature, active peptides of CSP-1 and CSP-2, differ at eight posi-

tions. Sequence analysis of encapsulated strains representing multiple capsular serotypes has shown that each has either a *comC1* or *comC2* allele and that the respective alleles are highly conserved. In general, however, transformation of encapsulated isolates is not readily observed under the growth conditions described above. In many cases, the failure to transform is related to the presence of the capsule, which impedes the induction of competence (25, 127). Noncompetent cultures, as well as many encapsulated strains, can be induced to competence by the addition of exogenous competence factor isolated from competent cells of the nonencapsulated R6 strain, by supernatants of competent cultures of the Rx1 strain, and by the addition of synthetic CSP-1 or CSP-2, depending on the allele present in the recipient strain (90, 121, 127). The conditions for inducing competence are less strict under these circumstances than for natural induction, and cultures at essentially any cell density can be induced. The optimum CSP concentration for inducing competence is strain-dependent. In addition, because competence is a transient state, the time of exposure to the competence factor prior to DNA addition is critical, usually ranging from 10 to 20 min, and is also strain-dependent.

The identification of the *comC1* and *comC2* alleles and their respective peptides has made possible the transformation of a large number of encapsulated isolates. The efficiencies of transformation can vary widely, however, and there remain many strains (approximately 50% of those tested) that have not been transformed. In addition, some strains transform more efficiently when competence is induced using supernatants from competent Rx1 cultures than when using the appropriate synthetic CSP. These observations suggest that other factors are also important in the induction of competence and the ability of recipients to be transformed.

Competence develops through a quorum-sensing mechanism in which CSP serves as the pheromone signal. The induction of competence leads to a transient restriction of most cellular protein synthesis, with high-level expression of about 14 proteins expected to be involved in transformation (71). In microarray analyses, a larger number of genes whose expression is altered by CSP was detected, but among 124 such genes, only 23 are clearly necessary for transformation (84, 96). The genes can be divided into those with early, late, and delayed induction in response to CSP, as well as a group that exhibits repression (4, 85). Early genes include those involved in quorum sensing, whereas late genes include those necessary for DNA processing. Heat shock proteins, proteases, and chaperones occur among the delayed genes, possibly reflective of a stress response. Repressed genes include ones encoding ribosomal proteins and carbohydrate and amino acid metabolism enzymes.

Recent genetic studies have characterized some of the events and genetic loci involved in detecting and responding to the CSP signal (summarized in Fig. 1). Export of CSP is mediated by an ABC transporter encoded by the *comAB* locus, which is not closely linked to *comC* (43). *comC* is the first gene in an operon with *comD* and *comE*, which encode the histidine kinase sensor and response regulator components, respectively, of a two-component signal transduction system (82). In *Streptococcus gordonii*, which contains homologs of ComCDE and undergoes competence in a manner similar to *S. pneumoniae*, ComD has been shown to be the CSP receptor (41). It likely serves the same function in *S. pneumoniae*. Expression of *comCDE* is enhanced approximately 40-fold upon addition of exogenous CSP (82). Maximal expression of *comAB* and *comCDE* occurs within 5 to

10 min after addition of CSP and returns to near basal levels by 15 to 20 min (4, 84). Loss of *comE* abolishes competence induction and *comCDE* expression (82). Binding of CSP to ComD is predicted to activate ComE via phosphorylation, and mutations in ComD that apparently alter its conformation and lead to constitutive phosphorylation also result in constitutive competence (50). The activated ComE binds to conserved target sequences, consisting of two 9-bp direct repeats separated by 12 bp, located upstream of the −10 regions of the *comAB*, *comCDE*, and *comX* promoters (53, 124). *comX* encodes the alternative sigma factor ComX, which allows core RNA polymerase to recognize competence-specific promoters (53, 60). These promoters contain a com- or cin-box (TACGAATA) located −10 from the transcription start site and a T-rich region at −25 (14). They are located upstream of a number of loci, including *cin-rec*, *cel*, *cfl*, *cgl*, *coi*, and *ssb*, that are involved in the later stages of competence and whose expression is induced in parallel (14, 83, 84). ComX thus serves as the link between the quorum-sensing part of the competence pathway (*comAB* and *comCDE*) and the DNA-processing components (53). It is an unstable protein that is not detected in noncompetent cultures but is apparent by 7.5 min after

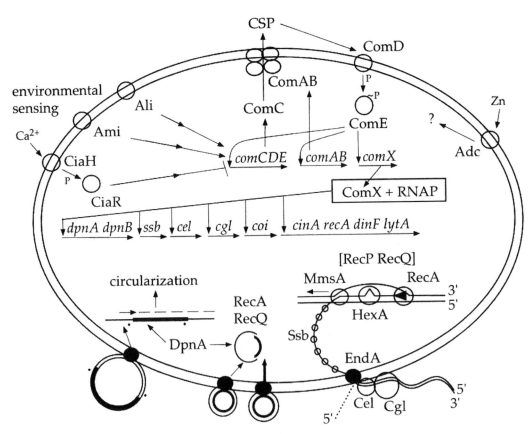

FIGURE 1 Transformation. The model is based on information discussed in the text. In some cases, the putative regulation of gene expression may be either direct or indirect. Phosphorylation of response regulator molecules by their cognate histidine kinase sensors is indicated by P. Uptake and integration of linear DNA are shown in the lower right portion, with monomer and dimer plasmid uptake shown in the central and left lower portions, respectively. Plasmid uptake will occur by the same mechanism as linear DNA uptake. For monomers, complementary overlapping strands will pair, and DNA synthesis will complete the double-stranded circular molecule. For the dimer, a small fragment of the complementary strand can serve as a primer for DNA synthesis, and circularization can occur as described in the text. Dots on the dimer molecule indicate homologous sites.

induction with CSP and about 3 min before the expression of ComX-dependent genes is detected (60). During competence, levels of the major sigma factor, Sigma A (RpoD), are decreased relative to ComX, and the molar amount of ComX exceeds that of Sigma A by approximately twofold. ComX may replace Sigma A during competence, thus explaining the preferential expression of competence-related genes during this time (60, 71). An additional factor, ComW, is necessary to maintain adequate levels of ComX, through an effect on either translation or stability, and may also increase the activity of ComX in transcription of the late competence-specific genes (58, 59).

CiaH and CiaR are the sensor and response regulator components, respectively, of a two-component signal transduction system that may be involved in calcium sensing (33, 36). CiaR is necessary to repress competence expression, although it is not clear whether the activated or non-activated form is involved. Mutations that eliminate CiaR result in constitutive *comCDE* expression and competence (21, 64, 66), whereas point mutations in *ciaH* result in no CSP being exported and a failure to respond to exogenously added CSP (36). Enhanced expression of stress response genes occurs in CSP-induced CiaR mutants, along with growth restriction and stationary-phase autolysis (21). Mutations that affect stress response proteins can also lead to competence expression (17, 97), and it has been suggested that CSP might be an alarmone that signals environmental stress (64). The Cia system, as well as the oligopeptide permeases encoded by the *ami* and *ali* (also referred to as *plp*) loci, may be involved in sensing the environment. Mutations in the *ami* and *ali* loci, and especially those in more than one locus, affect the spontaneous development of competence but do not alter the response to exogenous CSP induction (4, 81).

There has long been evidence for a macromolecular inhibitor that causes the cessation of the competent phase (121). The mechanism by which competence is abruptly shut off remains unknown, but mRNAs of the ComX-dependent loci decay in parallel, and the process requires ComX (53, 84).

Other genes or loci necessary for specific steps in transformation have been described and are discussed in the appropriate sections below. Mutant analyses have also identified several additional loci whose precise points of involvement are not entirely clear. For example, Adc mutants transform poorly unless Zn is added, and the addition of CSP in the absence of Zn does not induce competence. Therefore, the effect is likely downstream of induction and may be related to transmission of the signal from ComD to ComE (26).

DNA Release, Binding, and Uptake

CSP induction results in the release of transformable DNA from competent cells (74, 79, 113, 114). The release is dependent on the major autolysin, LytA, and the autolytic lysozyme, LytC. Double-stranded DNA can bind to competent cells at multiple locations (50 binding sites have been estimated [30]), where it is subjected to single-strand nicks. A constitutively expressed membrane-associated endonuclease (EndA) converts the nicks into double-strand cuts, resulting in fragments with a median length of about 7 kb (48, 93). Nucleolytic digestion of one strand occurs in the 5′ to 3′ direction, while the other strand is taken into the cell in the 3′ to 5′ direction (69). Entry and degradation occur at equivalent rates of ~100 nucleotides per sec at 30°C (69). Binding and subsequent uptake of a single large mole-cule may occur at multiple independent sites. Hence, even though digestion results in fragments of relatively small size, linkage of markers hundreds of kilobases apart can be detected using carefully prepared DNA. The average small size also does not preclude the uptake and integration of large DNA fragments, or of large heterologous regions flanked by sequences homologous to the recipient chromosome, as evidenced by the exchange of genetic cassettes encoding the capsular polysaccharide biosynthetic enzymes (reviewed in reference 126).

It has been proposed that EndA is part of a DNA uptake complex (93). The predicted products encoded by several genes induced by CSP and required for efficient transformation are homologous to proteins involved in DNA binding and uptake, and the assembly of DNA transport machinery in *Bacillus subtilis*. These genes are contained in the *S. pneumoniae* loci originally referred to as *cel* or *cilE*, *cgl* or *cilD*, *cilC* or *cclA*, *cilB*, and *cfl* (14, 53, 83). They have since been referred to in a manner consistent with their *B. subtilis* homologs (9). The *B. subtilis* ComG proteins are primarily membrane-associated, but some may be processed and translocated to the cell wall. They appear to have a role in allowing the DNA to traverse the peptidoglycan and reach the DNA-binding protein, ComEA (28). In *S. pneumoniae*, inactivation of ComGA (*cilD*, *cglA*) eliminated DNA binding and degradation, whereas inactivation of ComEA (*cilE*, *celA/B*) eliminated degradation and internalization of DNA, but low-efficiency binding still occurred. EndA thus cannot access DNA in the absence of these proteins, and they are postulated to be necessary for the surface attachment and deep binding, respectively, of the DNA (9). In these same studies, ComEC (*cilE*, *celA/B*) and ComFA (*cfl*) mutants were found to bind and degrade DNA, but not internalize it, suggesting a role in membrane transport.

Chromosomal Integration

Once inside the cell, the DNA is contained in an eclipse complex, where it is protected from cellular nucleases by interaction with a 19.5-kDa single-strand binding protein and no longer has transforming activity (123). Homologous recombination into the chromosome proceeds by a single-strand replacement mechanism that involves RecA activity and is affected by the Hex mismatch repair system (described below). Purified RecA mediates ssDNA-dependent NTP hydrolysis and NTP-dependent DNA strand exchange (112). CSP induces expression of the *recA* operon, which consists of *cinA-recA-dinF-lytA* (73, 80). In competent cells, all of these genes are induced from a promoter located upstream of *cinA*, whereas in noncompetent cells, lower levels of transcription proceed from other promoters. The induction of *lytA*, which encodes the major pneumococcal autolysin, is consistent with the observation of DNA release and autolytic activity during the competence phase. LytA is not, however, necessary for transformation (73). The role of the other genes in the *rec* locus has not been fully determined, but the *Escherichia coli dinF* is expressed in response to DNA-damaging agents (73), and CinA (referred to by the authors as CogA) may have a role in facilitating the localization of RecA to the membrane (67).

Mutations in at least four other loci are known to affect chromosomal integration. *mmsA* encodes a helicase (RecG homolog) that is proposed to be involved in branch migration of the three-stranded heteroduplex integration intermediate (65). *recP* is required for chromosomal but not plasmid transformation, although it is not clear how its predicted product, a transketolase, is involved. *recQ* is required for both chromosomal and plasmid transformation, but its role has not been further characterized (72, 95). Finally, genetic screens have identified a competence-induced gene necessary for efficient transformation that encodes a predicted 14.8-kDa protein with homology to single-stranded DNA-binding proteins (14). It has not been determined whether this is the eclipse complex protein.

Mismatch Repair

Incoming DNA that contains either transition mutations or small (<6 bp) deletions or insertions is usually corrected to the recipient DNA sequence by the action of the Hex mismatch repair system (reviewed in reference 19). Incoming markers can thus be observed to transform at high efficiency (transversions and large deletions/insertions), low efficiency (transitions and small deletions/insertions), or intermediate efficiency (transversions and single base deletions/insertions). The substrate for repair is the heteroduplex integration intermediate formed between the donor and recipient DNA. Recognition of a single-strand break in the donor DNA, followed by recognition of the mismatch, triggers activity of a 3' to 5' exonuclease that creates a gap of up to several kilobases in the donor DNA. The gap is then repaired using the recipient DNA as a template. Thus, a high-efficiency marker located adjacent to a low-efficiency marker may also be repaired. The same system functions to repair errors in DNA replication and is analogous to the Mut system of E. coli. Unlike the Mut system, however, the Hex system does not require recognition of unmethylated sequences in hemimethylated DNA. In S. pneumoniae, the products of at least two unlinked genes, hexA and hexB, are essential for the repair system. HexA (homologous to MutS) recognizes the mismatched base pair, but the role of HexB has not been demonstrated. By analogy with its E. coli homolog, MutL, it is likely to interact with HexA and other proteins in the mismatch repair complex and stimulate their activities (37, 92). In Hex⁻ cells, mismatches are not repaired, and high-efficiency transformation of all markers, as well as a high spontaneous mutation rate, are observed. The high transformation efficiency of many of the commonly used S. pneumoniae strains is due, in part, to the fact that they are Hex⁻, as the result of a mutation in hexB (92) (Table 1).

Expression of the hex genes is not enhanced during competence, and saturation of the system can occur with an excess number of mismatches, leading to the incorporation of low-efficiency markers. In addition, saturation is observed using donor DNA from the related Streptococcus mitis and Streptococcus oralis. Thus, the Hex system does not serve as a barrier to interspecies gene exchange, and its role in transformation, unlike that in DNA replication repair, has been proposed to be fortuitous (44).

S. pneumoniae also encodes a MutY-like adenine glycosylase, which repairs A/G mismatches to CG. MutY mutants exhibit a mutator phenotype, in which the frequency of AT to CG transversions is enhanced (99).

DNA Repair

S. pneumoniae generally lacks a UmuDC-dependent SOS response involved in error-prone repair of UV-damaged DNA (32), although transposon-encoded UmuC and UmuD homologs can mediate this response in cells harboring Tn5252 (77). Despite the lack of a LexA homolog in the S. pneumoniae chromosome, the purified S. pneumoniae RecA is able to cleave the E. coli LexA (112). Uptake of either homologous or heterologous DNA during transfor-

mation results in lysogenic induction of the temperate bacteriophage HB-746 in RecA⁺ but not RecA⁻ cells. Similarly, prophage induction by mitomycin C is dependent on a functional RecA. Both observations are consistent with an SOS-like repair system that is controlled by RecA (62). The spontaneous mutation frequency is significantly increased in cells undergoing transformation with homologous or closely related DNA but is unaffected when nonhomologous DNA is transformed (35). These observations further suggest the existence of an error-prone SOS-like repair system. The repair of double-stranded DNA breaks requires RexAB, the functional homologs of the E. coli RecBC (38).

Plasmid Transformation

Transformation of plasmids proceeds by the same pathway as that for chromosomal DNA. The efficiency (based on donor DNA concentration) is lower, however, because two separate uptake events that involve complementary and overlapping single strands must occur in order to obtain a circular, partially double-stranded molecule that can then be completed through DNA repair synthesis (Fig. 1). Consequently, plasmid DNA linearized at a unique site cannot be recircularized and established in the recipient cell. The frequency of plasmid transformation is enhanced by the use of dimers as opposed to monomers (104). Indeed, the highest frequencies of plasmid transformations are obtained using RecA⁺ E. coli strains as a source of multimeric plasmids. In these situations, only a small fragment of the complementary strand is necessary to provide a primer for synthesis of a molecule that can be more than one monomeric unit in length when using the dimer/multimer as a template. A recombination event between homologous regions can then restore the circular plasmid. Alternatively, as suggested by Saunders and Guild (104), intramolecular circularization may occur by pairing of a single-strand region of the incoming DNA with a newly synthesized complementary strand and displacement of the template strand. A ligation reaction would restore the circular molecule. The first possibility would be expected to be RecA-dependent, whereas the latter likely would not. Whether the transformation of dimers is dependent on RecA has not been determined. Defects in RecA do, however, decrease the efficiency of monomer transformation (30). In addition, RecQ is necessary for efficient plasmid transformation, but RecP is not (72, 95).

The establishment of plasmids can also be facilitated by the presence in the recipient cell of regions homologous to the incoming plasmid (111). Thus, plasmids containing cloned fragments of S. pneumoniae DNA are readily transformed, as the homologous chromosomal regions can serve as primers for DNA repair synthesis, if the original double-strand cut occurred within the cloned region of the plasmid. Likewise, plasmids with homology to endogenous plasmids are transformed at high frequency.

Artificial Transformation

Replicating plasmids can be introduced into S. pneumoniae by electroporation. Unlike plasmids taken up via the natural transformation pathway, they are subject to restriction by the Dpn system (see below), indicating that double-stranded DNA is transformed (54). Recombination of markers into the chromosome is not observed, even when competent cells are used as recipients or when DNA is made single-stranded prior to transformation. Electrotransformation of plasmids containing regions homologous to the S. pneumoniae chromosome is, however, enhanced in competent cells in a RecA-dependent manner, indicating that pairing between homologous DNAs does occur during this process (55).

Conjugation

S. pneumoniae can serve as both a donor and recipient in the conjugation of plasmid DNA and conjugative transposons (107, 110). Transfers are DNase-resistant and require stable cell-to-cell contact, such as that provided during filter matings. Encapsulated strains can serve as donors in conjugation but, in limited studies, their effectiveness as recipients appears variable (13, 29). Although no indigenous conjugative plasmids have been identified, several plasmids from other streptococci, including pIP501 of *Streptococcus agalactiae*, have successfully been conjugated into *S. pneumoniae* and then retransferred to other pneumococcal and streptococcal recipients. In addition, nonconjugative plasmids such as pMV158 can be mobilized both to and from *S. pneumoniae* (110).

Conjugative transposons that can insert at either multiple sites or only at specific sites have been identified in *S. pneumoniae*. Both are transferred in a process that likely involves excision to form a circular intermediate that is nicked and then transferred as single-stranded DNA to recipient cells. Tn*1545*, a 25.3-kb conjugative transposon first detected in *S. pneumoniae*, encodes resistance to tetracycline (*tet*M), kanamycin, and related aminoglycosides (*aphA-3*), and macrolide-lincosamide-streptogramin B type antibiotics (*erm*AM). It is closely related to Tn*916* from *Enterococcus faecalis* (20, 106). Precise excision of Tn*1545* from the chromosome is mediated by the transposase-encoded proteins Xis-Tn and Int-Tn, which are structurally and functionally similar to the Xis and Int proteins of bacteriophage λ. Excision results in a nonreplicating, circular intermediate that can integrate at another site in the resident chromosome through the action of Int-Tn or can mediate its own transfer to other recipients (89). Numerous other gram-positive bacteria can serve as both recipients and donors for Tn*1545* transfer (20). Transposition occurs in these organisms, as well as in *E. coli* in which the transposon has been introduced on a plasmid. Integration involves site-specific recombination within a short region of nonhomologous DNA and can occur at multiple locations. Specificity is for AT-rich regions and may be related to conformational aspects of the target DNA sequence (106).

Tn*5253* [originally referred to as Ω(*cat-tet*)], the first conjugative transposon identified in *S. pneumoniae*, is a 65.5-kb element that is composed of two independent conjugative transposons, Tn*5251* and Tn*5252*. Tn*5251* is 18 kb in size, encodes tetracycline resistance (*tet*M), and is a member of the Tn*1545*/Tn*916* family. It is inserted within the central region of the 47.5-kb Tn*5252*, which encodes resistance to chloramphenicol (7, 107). Conjugative transposition of either the entire Tn*5253* or the independent transposons can occur. Unlike the Tn*1545*-like transposons, integration of Tn*5252* into the pneumococcal chromosome occurs within a single 72-bp target site (*att*B) (122). Conjugative transfer of Tn*5252* has been demonstrated from *S. pneumoniae* to other streptococci, where transposition also appears to involve a unique target site. Tn*1207.1*, a transposon possibly related to Tn*5252*, carries the macrolide efflux gene *mef*(A) and specifically integrates into *celB*, thus impairing competence (22, 103). Tn*1207.1* is 7.2 kb and lacks transposition ability, but contains several open reading frames with homology to Tn*5252*. It appears to be part of the larger, 52-kb conjugative transposon Tn*1207.3* of *Streptococcus pyogenes*, which can be transferred from *S. pyogenes* to *S. pneumoniae* and inserts at a single specific site in each organism (*comEC* and *celB*, respectively) (102).

Pseudotransduction

Although true generalized transduction has not been observed with pneumococcal phages, Porter et al. (88) described a process that involved phage packaging of pneumococcal DNA and subsequent transfer to recipient cells via the transformation pathway. Termed pseudotransduction, the process included several features consistent with generalized transduction. The phage, PG24, could package different chromosomal regions into DNase-resistant phage particles that adsorbed to recipient bacteria. Gene transfer was inhibited by blocking adsorption with phage-specific antiserum or by trypsin treatment of the phage. Successful gene transfer, however, was dependent on the ability of the recipient cells to become competent and on the presence of EndA, which is required for DNA processing and entry via the transformation pathway. Frequency of transfer was affected by the Hex phenotype of the recipient, indicating entry of single-stranded DNA and heteroduplex formation between the donor and recipient DNAs. DNase sensitivity of the process occurred at the time of competence development, further indicating that the bacterial DNA was released from the phage at the outside surface of the cell and not injected into it, as occurs for the phage DNA. Transfer of DNA by pseudotransduction rather than standard transformation permitted the entry of longer fragments, presumably owing to protection of the DNA from excessive cuts by EndA while maintained at the cell surface in the phage head.

PLASMIDS AND BACTERIOPHAGE

Plasmids have been detected in a small number of *S. pneumoniae* isolates. Most are either identical or very similar to pDP1, a cryptic 3-kb plasmid first identified in the capsule type 2 strain D39 and a number of its descendants (109). Approximately 3% of more than 600 strains examined carry a member of this plasmid family (see reference 108 and references therein). The plasmids are of low copy number (less than 10), are not amplifiable, and belong to a family of rolling-circle plasmids that includes pC194 (105, 108, 109). Small plasmids unrelated to pDP1 have occasionally been noted (108), but, as with pDP1, no specific functions have been detected. The presence of pDP1 or other plasmids is unrelated to capsular serotype, restriction system, antibiotic resistance profile, clinical infection, or other phenotypes. Plasmids from other gram-positive bacteria are capable of replicating in *S. pneumoniae* and have been used for both transformation studies and cloning (25, 104, 111, 127). The addition of reporter genes, such as *gfp*, has allowed characterization of gene regulation as well as in vivo adhesion and invasion (1, 8, 47).

Bacteriophage that infect *S. pneumoniae* can be readily isolated from clinical samplings. The ω, Cp, and Dp groups contain virulent phage that were isolated directly from human throat swabs (31, 68, 86, 120). The bacterial origins of these phage are not known, but the ω phages do not infect other species of streptococci present in the oral cavity (120). Temperate phage appear to be present in the majority of pneumococcal isolates and occur in a wide variety of capsular serotypes (10, 94). However, as shown using isogenic derivatives of encapsulated and nonencapsulated strains, the presence of the capsule prevents infection by virulent ω phages under laboratory conditions (11), suggesting that natural infections may occur when capsule production is reduced, as is expected to occur during colonization of the nasopharyngeal cavity (61, 125). Pneumococcal virulence properties encoded by phage have not yet been identified.

Properties of some of the better-characterized pneumococcal phage are shown in Table 3. Some important similarities between many of the phages have been noted, and the complete nucleotide sequences have been determined for the genomes of phages Cp-1, Dp-1, EJ-1, and MM1 (reviewed in references 31 and 57). Cp-1 and HB-3, as well as their related phages, contain a terminal protein covalently linked to the 5′ end of each DNA strand. In the case of Cp-1, and possibly the other phage, the terminal protein serves as a primer in initiation of DNA replication for the linear molecule. The mechanism is like that found in adenovirus and B. subtilis phage φ29, with which Cp-1 has structural similarity. As in these other viruses, the DNA of Cp-1 and its related phage contains inverted terminal repeat structures. A number of the pneumococcal phages are also related by the structural organization of their lytic enzymes. Cp-1, Cp-9, Dp-1, and HB-3 require choline in the pneumococcal cell wall for adsorption and cell lysis. The lytic enzyme in each phage contains a C-terminal choline-binding domain that is also present in the pneumococcal autolysin and numerous other choline-binding proteins of S. pneumoniae. The N-terminal region of HB-3 encodes an amidase that is virtually identical to that of the host autolysin, whereas the same region in Cp-1 and Cp-9 encodes a lysozyme. In Dp-1, a hydrolase similar to that found in the Lactococcus lactis phage BK5-T is encoded. The lysozyme found in Cp-1 and Cp-9 is also encoded by the N-terminal region of the Cp-7 lytic enzyme, but the C-terminal region of this protein does not encode a choline-binding domain, and hence this phage does not require choline-containing cell walls for infection. These enzymes represent examples of modular evolution.

THE Dpn RESTRICTION SYSTEM

Most S. pneumoniae isolates contain the genes necessary for expression of either the DpnI or DpnII restriction system. The genes for each of these systems are located at the same site in the respective chromosomes, thus forming genetic cassettes that can be exchanged as units between strains (51). The DpnI endonuclease (DpnC) cleaves double-stranded DNA containing the methylated sequence 5′-GmeATC-3′, whereas the DpnII endonuclease (DpnB) cleaves unmethylated DNA of the same sequence. In addition to their respective endonucleases, the DpnI gene cassette encodes a second protein (DpnD) of unknown function, while the DpnII cassette encodes the adenine methyltransferases, DpnA and DpnM, that are necessary for methylating the host DNA to prevent restriction (98). DpnM acts only on double-stranded DNA, whereas DpnA can methylate both double- and single-stranded DNA (16).

The major function of the Dpn system is to protect the cell against entry of phage DNA. Neither of the Dpn endonucleases attacks single-stranded DNA or hemimethylated double-stranded DNA. Thus, natural transformation of chromosomal DNA is not affected. In contrast, plasmid transformation is affected to some extent, with an approximate 50% reduction observed for transfers between strains of differing Dpn types as compared with transfers between strains of the same type (16). The lack of complete restriction of methylated plasmids by DpnI cells apparently results from the mechanism of plasmid uptake, which can give rise to large segments of hemimethylated DNA following DNA synthesis to restore both strands of the incoming plasmid. Complete restriction of nonmethylated plasmids by the DpnII system is prevented by the action of DpnA, which methylates incoming single-stranded DNA. Expression of dpnA is induced by CSP and initiates at a com-box sequence (49). In the absence of such methylation, restriction of the newly restored plasmid that is not methylated on either strand apparently occurs faster than methylation of the double-stranded DNA by DpnM. Single-stranded, unmethylated plasmid DNA introduced by conjugation into

TABLE 3 Properties of some S. pneumoniae bacteriophage[a]

Phage	Structure	DNA	% GC	Burst size	N-terminal[b]	C-terminal[c]
ω8 (virulent)	Octahedral hexagonal head (60 nm) Flexible tail (180 × 10 nm) Tail fiber (90 nm)	50 kb, ds, linear	49	100		
Cp-1 (virulent)	Irregular hexagonal head (60 × 45 nm) Tail (20 × 15 nm) Neck appendages Head fibers	19.3 kb, ds, linear terminal protein	42	10	Lysozyme	Choline
Dp-1 (virulent)	Polyhedral head (67 nm) Tail (155 nm) Lipid envelope	56.5 kb, ds, linear	27	100	Amidase	Choline
EJ-1 (temperate)	Isometric head (57 nm) Tail (130 nm)	42.9 kb, ds, linear	39.6		Amidase	Choline
HB-3 (temperate)	Head (65 nm) Tail (156 × 1 nm)	40 kb, ds, linear terminal protein[d]			Amidase	Choline
MM-1 (temperate)	Icosahedral head (60 nm) Tail (160 nm)	40.2 kb, ds, linear terminal protein[d]	38.4		Amidase	Choline

[a]Information is from citations in the text and references therein.
[b]Activity of the N-terminal domain of the lytic enzyme.
[c]Substrate bound by the C-terminal domain of the lytic enzyme.
[d]The roles of the terminal proteins in these phage have not been determined.

*Dpn*II cells is also severely restricted, presumably because synthesis of the complementary strand by enzymes transferred with the DNA occurs more rapidly than protective methylation of the single strand by DpnA (16). The introduction of plasmids by electroporation, or other methods that involve the uptake of double-stranded DNA, is also strongly affected by the *Dpn* system (56).

PROMOTERS

The existence of extended −10 sites in promoter sequences was first noted in the *Dpn*II locus and was subsequently found to be relatively common in *S. pneumoniae*, in contrast to the relatively rare occurrence of such sequences in *E. coli*. The promoter of the *Dpn*II operon contains an extended −10 site (TATGGTATAAT) but does not have a −35 site. Analysis of 35 additional known or putative *S. pneumoniae* promoters revealed that 60% had the complete extended −10 consensus sequence (TNTGNTATAAT) and 71% had at least the TG motif of the extension. None lacked a −35 site, and those lacking a −10 extension tended to have −35 sequences close to the *E. coli* consensus (98). Numerous *S. pneumoniae* sequences exhibit promoter activity in *E. coli* but not in *S. pneumoniae*, apparently as a result of the AT richness of the *S. pneumoniae* DNA and of more stringent sequence recognition requirements of the *S. pneumoniae* RNA polymerase. Such requirements could be fulfilled by the extended −10 sequences. Excessive promoter activity of *S. pneumoniae* DNA in *E. coli* has been used to explain difficulties encountered in cloning pneumococcal sequences. Although it is clear that difficulties exist, sequences with promoter activity strong enough to destabilize cloning vectors appear to be rare in the *S. pneumoniae* chromosome (24). The expression of toxic products, the use of large fragments or high-copy-number plasmids, and sequence-specific effects unrelated to promoter activity appear to be more likely causes.

THE GENERATION AND ANALYSIS OF MUTANTS

The ability of *S. pneumoniae* to be naturally transformed makes possible a number of techniques that are not always easily performed in other bacteria. That same property, however, demands that special care be taken in the conclusions drawn from mutagenesis experiments. As noted above, the frequency of spontaneous mutations increases during transformation with homologous DNA, and mutations other than those desired are often observed to occur. In addition, a large amount of DNA that is both linked and unlinked to the gene of interest can be incorporated during chromosomal transformations. Consequently, before drawing conclusions regarding the effect of the intended mutation, it is essential either to observe a consistent phenotype among multiple, independently isolated mutants or to specifically repair the known mutation and observe reversion to the parental phenotype. The latter is preferable, because similar compensatory mutations conferring a specific growth advantage may consistently be selected for even in multiple, independent mutants. In the case of chromosomal transformations between nonisogenic strains, mutants are backcrossed several times to the parent strain to reduce the amount of extraneous donor DNA.

The most frequently used mutagenesis technique in *S. pneumoniae* involves insertion-duplication. In this procedure, a fragment of *S. pneumoniae* DNA is cloned into a plasmid that is able to replicate in *E. coli* but not *S. pneumoniae*. The clone can thus be propagated in *E. coli* and used to transform *S. pneumoniae*, where homologous recombination into the chromosome at the target site is necessary to select for an antibiotic resistance encoded by the plasmid vector. Target sequences of less than 200 bp can be used successfully, but larger fragments increase the frequency of transformation, owing in part to the higher likelihood of digestion of the surface-bound DNA within the cloned fragment (see Plasmid Transformation, above). As noted above, RecA⁺ *E. coli* strains can also be used to enhance transformation frequencies, if recombination within the cloned fragment is not expected to be an issue. Transformation and subsequent integration into the *S. pneumoniae* chromosome will result in gene disruption if the cloned fragment is an internal portion of a gene, or restoration of the entire gene if the fragment overlaps one end of the gene (Fig. 2). In the latter case, a mutant phenotype should not be observed, unless the target sequence is part of an operon. Thus, a convenient and essential control when using an internal fragment for gene disruption is to use a 3′-overlapping fragment that restores the complete gene and should, in the absence of polar effects, have the parental phenotype. Integration of the clone results in a duplication of the target sequence. Consequently, precise excision of insertions occurs at low frequency, and resolved plasmids can be recovered by transformation into *E. coli*. Unlike many clones containing large *S. pneumoniae* DNA fragments, those containing fragments of the size necessary for generating insertion-duplication mutations are generally stable in *E. coli*. Hence, libraries can be used to mutagenize the *S. pneumoniae* chromosome (52), and unknown target fragments can easily be cloned, as can surrounding DNA following digestion, ligation, and transformation to *E. coli*, using the integration vector for replication. Despite the ability of the insertions to resolve, they are relatively stable, and resolution usually occurs at a frequency of <1/10⁵. Thus, reversion is not generally apparent in the absence of a strong selection for the revertant phenotype.

Insertion-duplication using vectors containing reporter genes, including *cat* and *lacZ*, has been used successfully to analyze gene expression in *S. pneumoniae* (3, 14, 25, 83). In the case of the latter, the endogenous β-galactosidase activity must be eliminated through specific mutation of *bgaA* (128) for expression studies. Integration with vectors permitting PhoA translational fusions is used to study exported proteins (81). Most of the vectors used for insertion-duplication, as well as most replicating vectors, encode erythromycin resistance for selection in both *S. pneumoniae* and *E. coli* (3, 14, 25, 83, 127). Other commonly used antibiotic markers are chloramphenicol, kanamycin, spectinomycin, and tetracycline (83, 111, 116). Plasmids carrying β-lactamase-encoding genes, whether expressed or not, should not be used because naturally occurring resistance due to β-lactamase has not yet appeared in *S. pneumoniae*.

Transposon mutagenesis has been used with limited success in *S. pneumoniae*, but in vitro mariner mutagenesis has proven to be a useful technique (2, 64). Here, linear DNA is mutagenized in vitro using the *Himar*1 transposase and a minitransposon containing an antibiotic resistance. Gaps generated at the insertion site are repaired, and the linear fragments are transformed to *S. pneumoniae* with selection for the antibiotic marker. Site specificity is minimal, and saturation mutagenesis appears possible.

For the introduction of mutations that do not confer a selectable phenotype, direct transformation of a PCR- or

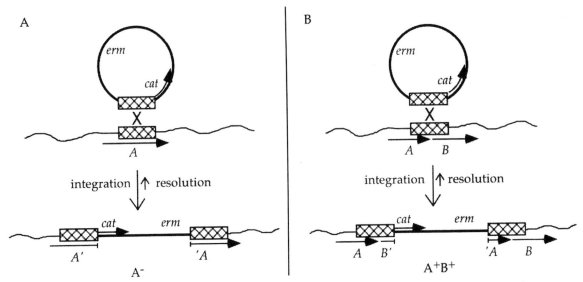

FIGURE 2 Insertion-duplication mutagenesis and restoration. (A) The effect of using an internal gene fragment to direct insertion of a nonreplicating plasmid into the chromosome. Duplication of the target fragment occurs, and the gene is disrupted by the plasmid insertion, resulting in an insertion-duplication mutation. (B) The target fragment overlaps the ends of two genes. Insertion results in duplication of the target fragment, but both genes are completely reconstructed and the result is an insertion-duplication restoration. Both genes should be functional, unless they form part of an operon, in which case the plasmid insertion would be polar on the downstream gene. The figure shows a selectable erythromycin-resistance gene (*erm*) and a promoterless chloramphenicol-resistance reporter gene (*cat*), as discussed in reference 25.

cloned fragment containing the mutation, followed by screening using PCR or other suitable method, can be used to identify mutants resulting from double crossover events (39, 40). Alternatively, a gene replacement technique involving a negative selection has been described (115). Here, the Janus cassette, which contains a selectable kanamycin-resistant and counterselectable *rpsL*, which confers dominant streptomycin sensitivity in a streptomycin-resistant background, is first introduced by homologous recombination at the site of interest. DNA containing the desired mutation is then introduced, and mutants in which recombination has occurred to replace the Janus cassette are identified by their streptomycin-resistant and kanamycin-sensitive phenotypes.

Mapping of point or other small chromosomal mutations utilizes cotransformation with known selectable markers for linkage analyses. With carefully prepared DNA, it is possible to observe linkage with markers that are hundreds of kilobases apart owing to binding and uptake of a single DNA molecule at multiple sites on the cell surface. More precise mapping, even in the absence of a linked marker, can be achieved using restriction enzyme-digested chromosomal DNA and demonstrating that transfer of the mutant or linked phenotype occurs with smaller fragments. Electrophoretic separation of the restricted DNA allows individual fractions, either purified or contained in melted slices from a low-gelling temperature agarose gel, to be transformed directly to *S. pneumoniae*. The small number of fragments contained in the fraction can then usually be cloned and tested individually for the ability to transfer the phenotype. As noted earlier, cloned fragments normally yield high transformation frequencies owing to the large concentration of homogeneous DNA. They can be transformed either as linear restriction fragments or as the intact clone without selection for the antibiotic marker. In the latter case, the double crossover event to introduce or repair the mutation in the recipient chromosome occurs about 500 times more frequently than the single crossover event necessary to integrate the plasmid and cause gene disruption. Fragments of less than 200 bp can be used for mapping, but efficiency increases with increasing fragment size. PCR fragments can similarly be used as a source of homogeneous DNA for transformation mapping.

Work in our laboratories is supported by Public Health Service grants AI28457 and GM53017 (J.Y.) and AI53749 (S.K.H.).

REFERENCES

1. **Acebo, P., C. Nieto, M. A. Corrales, M. Espinosa, and P. Lopez.** 2000. Quantitative detection of *Streptococcus pneumoniae* cells harbouring single or multiple copies of the gene encoding the green fluorescent protein. *Microbiology* **146:**1267–1273.

2. **Akerley, B. J., E. J. Rubin, A. Camilli, D. J. Lampe, H. M. Robertson, and J. J. Mekalanos.** 1998. Systematic identification of essential genes by *in vitro mariner* mutagenesis. *Proc. Natl. Acad. Sci. USA* **95:**8927–8932.

3. **Alloing, G., C. Granadel, D. A. Morrison, and J. P. Claverys.** 1996. Competence pheromone, oligopeptide permease, and induction of competence in *Streptococcus pneumoniae*. *Mol. Microbiol.* **21:**471–478.

4. **Alloing, G., B. Martin, C. Granadel, and J. P. Claverys.** 1998. Development of competence in *Streptococcus pneumoniae*: pheromone autoinduction and control of quorum sensing by the oligopeptide permease. *Mol. Microbiol.* **29:** 75–83.

5. **Aras, R. A., J. Kang, A. I. Tschumi, Y. Harasaki, and M. J. Blaser.** 2003. Extensive repetitive DNA facilitates prokaryotic genome plasticity. *Proc. Natl. Acad. Sci. USA* **100:**13579–13584.

6. **Avery, O. T., C. M. MacLeod, and M. McCarty.** 1944. Studies on the chemical nature of the substance inducing transformation of pneumococcal types. Induction of transformation by a desoxyribonucleic acid fraction isolated from pneumococcus Type III. *J. Exp. Med.* **79:**137–158.

7. **Ayoubi, P., A. O. Kilic, and M. N. Vijayakumar.** 1991. Tn*5253*, the pneumococcal W(*cat tet*) BM6001 element, is a composite structure of two conjugative transposons, Tn*5251* and Tn*5252*. *J. Bacteriol.* **173:**1617–1622.

8. **Bartilson, M., A. Marra, J. Christine, J. S. Asundi, W. P. Schneider, and A. E. Hromockyj.** 2001. Differential fluorescence induction reveals *Streptococcus pneumoniae* loci regulated by competence stimulatory peptide. *Mol. Microbiol.* **39:**126–135.

9. **Berge, M., M. Moscoso, M. Prudhomme, B. Martin, and J. P. Claverys.** 2002. Uptake of transforming DNA in Gram-positive bacteria: a view from *Streptococcus pneumoniae*. *Mol. Microbiol.* **45:**411–421.

10. **Bernheimer, H. P.** 1979. Lysogenic pneumococci and their bacteriophages. *J. Bacteriol.* **138:**618–624.

11. **Bernheimer, H. P., and J. G. Tiraby.** 1976. Inhibition of phage infection by pneumococcus capsule. *Virology* **73:**308–309.

12. **Bruckner, R., M. Nuhn, P. Reichmann, B. Weber, and R. Hakenbeck.** 2004. Mosaic genes and mosaic chromosomes—genomic variation in *Streptococcus pneumoniae*. *Int. J. Med. Microbiol.* **294:**157–168.

13. **Buu-Hoi, A., and T. Horodniceanu.** 1980. Conjugative transfer of multiple antibiotic resistance markers in *Streptococcus pneumoniae*. *J. Bacteriol.* **143:**313–320.

14. **Campbell, E. A., S. Y. Choi, and H. R. Masure.** 1998. A competence regulon in *Streptococcus pneumoniae* revealed by genomic analysis. *Mol. Microbiol.* **27:**929–939.

15. **Centre National de la Recherche Scientifique.** http://www.is.biotoul.fr/is.html

16. **Cerritelli, S., S. S. Springhorn, and S. A. Lacks.** 1989. DpnA, a methylase for single-strand DNA in the *Dpn* II restriction system, and its biological function. *Proc. Natl. Acad. Sci. USA* **86:**9223–9227.

17. **Chastanet, A., M. Prudhomme, J. P. Claverys, and T. Msadek.** 2001. Regulation of *Streptococcus pneumoniae clp* genes and their role in competence development and stress survival. *J. Bacteriol.* **183:**7295–7307.

18. **Chen, J. D., and D. A. Morrison.** 1987. Modulation of competence for genetic transformation in *Streptococcus pneumoniae*. *J. Gen. Microbiol.* **133:**1959–1967.

19. **Claverys, J. P., and S. A. Lacks.** 1986. Heteroduplex deoxyribonucleic acid base mismatch repair in bacteria. *Microbiol. Rev.* **50:**133–165.

20. **Courvalin, P., and C. Carlier.** 1986. Transposable multiple antibiotic resistance in *Streptococcus pneumoniae*. *Mol. Gen. Genet.* **205:**291–297.

21. **Dagkessamanskaia, A., M. Moscoso, V. Henard, S. Guiral, K. Overweg, M. Reuter, B. Martin, J. Wells, and J. P. Claverys.** 2004. Interconnection of competence, stress and CiaR regulons in *Streptococcus pneumoniae*: competence triggers stationary phase autolysis of *ciaR* mutant cells. *Mol. Microbiol.* **51:**1071–1086.

22. **Del Grosso, M., F. Iannelli, C. Messina, M. Santagati, N. Petrosillo, S. Stefani, G. Pozzi, and A. Pantosti.** 2002. Macrolide efflux genes *mef*(A) and *mef*(E) are carried by different genetic elements in *Streptococcus pneumoniae*. *J. Clin. Microbiol.* **40:**774–778.

23. **Dillard, J. P., M. W. Vandersea, and J. Yother.** 1995. Characterization of the cassette containing genes for type 3 capsular polysaccharide biosynthesis in *Streptococcus pneumoniae*. *J. Exp. Med.* **181:**973–983.

24. **Dillard, J. P., and J. Yother.** 1991. Analysis of *Streptococcus pneumoniae* sequences cloned into *Escherichia coli*: effect of promoter strength and transcription terminators. *J. Bacteriol.* **173:**5105–5109.

25. **Dillard, J. P., and J. Yother.** 1994. Genetic and molecular characterization of capsular polysaccharide biosynthesis in *Streptococcus pneumoniae* type 3. *Mol. Microbiol.* **12:**959–972.

26. **Dintilhac, A., G. Alloing, C. Granadel, and J. P. Claverys.** 1997. Competence and virulence of *Streptococcus pneumoniae*: Adc and PsaA mutants exhibit a requirement for Zn and Mn resulting from inactivation of putative ABC metal permeases. *Mol. Microbiol.* **25:**727–739.

27. **Dopazo, J., A. Mendoza, J. Herrero, F. Caldara, Y. Humbert, L. Friedli, M. Guerrier, E. Grand-Schenk, C. Gandin, M. de Francesco, A. Polissi, G. Buell, G. Feger, E. Garcia, M. Peitsch, and J. F. Garcia-Bustos.** 2001. Annotated draft genomic sequence from a *Streptococcus pneumoniae* type 19F clinical isolate. *Microb. Drug Resist.* **7:**99–125.

28. **Dubnau, D.** 1999. DNA uptake in bacteria. *Ann. Rev. Microbiol.* **53:**217–244.

29. **Engel, H. W., N. Soedirman, J. A. Rost, W. J. van Leeuwen, and J. D. van Embden.** 1980. Transferability of macrolide, lincomycin, and streptogramin resistances between group A, B, and D streptococci, *Streptococcus pneumoniae*, and *Staphylococcus aureus*. *J. Bacteriol.* **142:**407–413.

30. **Fox, M., and R. D. Hotchkiss.** 1957. Initiation of bacterial transformation. *Nature* **179:**1322–1325.

31. **Garcia, P., A. C. Martin, and R. Lopez.** 1997. Bacteriophages of *Streptococcus pneumoniae*: a molecular approach. *Microb. Drug Resist.* **3:**165–176.

32. **Gasc, A. M., N. Sicard, J. P. Claverys, and A. M. Sicard.** 1980. Lack of SOS repair in *Streptococcus pneumoniae*. *Mutat. Res.* **70:**157–165.

33. **Giammarinaro, P., M. Sicard, and A. M. Gasc.** 1999. Genetic and physiological studies of the CiaH-CiaR two-component signal-transducing system involved in cefotaxime resistance and competence of *Streptococcus pneumoniae*. *Microbiology* **145:**1859–1869.

34. **Griffith, F.** 1928. The significance of pneumococcal types. *J. Hyg.* **27:**113–159.

35. **Grist, R. W., and L. O. Butler.** 1983. Effect of transforming DNA on growth and frequency of mutation of *Streptococcus pneumoniae*. *J. Bacteriol.* **153:**153–162.

36. **Guenzi, E., A. M. Gasc, M. A. Sicard, and R. Hakenbeck.** 1994. A two-component signal-transducing system is involved in competence and penicillin susceptibility in laboratory mutants of *Streptococcus pneumoniae*. *Mol. Microbiol.* **12:**505–515.

37. **Hall, M., and S. Matson.** 1999. The *Escherichia coli* MutL protein physically interacts with MutH and stimulates the MutH-associated endonuclease activity. *J. Biol. Chem.* **274:**1306–1312.

38. **Halpern, D., A. Gruss, J. P. Claverys, and M. El-Karoui.** 2004. *rexAB* mutants in *Streptococcus pneumoniae*. *Microbiology* **150:**2409–2414.

39. **Hardy, G. G., M. J. Caimano, and J. Yother.** 2000. Capsule biosynthesis and basic metabolism in *Streptococcus*

pneumoniae are linked through the cellular phosphogluco-mutase. *J. Bacteriol.* **182:**1854–1863.

40. **Hardy, G. G., A. D. Magee, C. L. Ventura, M. J. Caimano, and J. Yother.** 2001. Essential role for cellular phosphoglucomutase in virulence of type 3 *Streptococcus pneumoniae. Infect. Immun.* **69:**2309–2317.

41. **Havarstein, L. S., P. Gaustad, I. F. Nes, and D. A. Morrison.** 1996. Identification of the streptococcal competence-pheromone receptor. *Mol. Microbiol.* **21:**863–869.

42. **Hoskins, J., W. E. Alborn, Jr., J. Arnold, L. C. Blaszczak, S. Burgett, B. S. DeHoff, S. T. Estrem, L. Fritz, D. J. Fu, W. Fuller, C. Geringer, R. Gilmour, J. S. Glass, H. Khoja, A. R. Kraft, R. E. Lagace, D. J. LeBlanc, L. N. Lee, E. J. Lefkowitz, J. Lu, P. Matsushima, S. M. McAhren, M. McHenney, K. McLeaster, C. W. Mundy, T. I. Nicas, F. H. Norris, M. O'Gara, R. B. Peery, G. T. Robertson, P. Rockey, P. M. Sun, M. E. Winkler, Y. Yang, M. Young-Bellido, G. Zhao, C. A. Zook, R. H. Baltz, S. R. Jaskunas, P. R. Rosteck, Jr., P. L. Skatrud, and J. I. Glass.** 2001. Genome of the bacterium *Streptococcus pneumoniae* strain R6. *J. Bacteriol.* **183:**5709–5717.

43. **Hui, F. M., L. Zhou, and D. A. Morrison.** 1995. Competence for genetic transformation in *Streptococcus pneumoniae*: organization of a regulatory locus with homology to two lactococcin A secretion genes. *Gene* **153:**25–31.

44. **Humbert, O., M. Prudhomme, R. Hakenbeck, C. G. Dowson, and J. P. Claverys.** 1995. Homologous recombination and mismatch repair during transformation in *Streptococcus pneumoniae*: saturation of the Hex mismatch repair system. *Proc. Natl. Acad. Sci. USA* **92:**9052–9056.

45. **Iannelli, F., B. J. Pearce, and G. Pozzi.** 1999. The type 2 capsule locus of *Streptococcus pneumoniae. J. Bacteriol.* **181:**2652–2654.

46. **Institute for Genomic Research.** http://www.tigr.org

47. **Kadioglu, A., J. A. Sharpe, I. Lazou, C. Svanborg, C. Ockleford, T. J. Mitchell, and P. W. Andrew.** 2001. Use of green fluorescent protein in visualisation of pneumococcal invasion of broncho-epithelial cells in vivo. *FEMS Microbiol. Lett.* **194:**105–110.

48. **Lacks, S., and M. Neuberger.** 1975. Membrane location of a deoxyribonuclease implicated in the genetic transformation of *Diplococcus pneumoniae. J. Bacteriol.* **124:**1321–1329.

49. **Lacks, S. A., S. Ayalew, A. G. de la Campa, and B. Greenberg.** 2000. Regulation of competence for genetic transformation in *Streptococcus pneumoniae*: expression of *dpnA*, a late competence gene encoding a DNA methyltransferase of the DpnII restriction system. *Mol. Microbiol.* **35:**1089–1098.

50. **Lacks, S. A., and B. Greenberg.** 2001. Constitutive competence for genetic transformation in *Streptococcus pneumoniae* caused by mutation of a transmembrane histidine kinase. *Mol. Microbiol.* **42:**1035–1045.

51. **Lacks, S. A., B. M. Mannarelli, S. S. Springhorn, and B. Greenberg.** 1986. Genetic basis of the complementary DpnI and DpnII restriction systems of *S. pneumoniae*: an intercellular cassette mechanism. *Cell* **46:**993–1000.

52. **Lee, M. S., B. A. Dougherty, A. C. Madeo, and D. A. Morrison.** 1999. Construction and analysis of a library for random insertional mutagenesis in *Streptococcus pneumoniae*: use for recovery of mutants defective in genetic transformation and for identification of essential genes. *Appl. Environ. Microbiol.* **65:**1883–1890.

53. **Lee, M. S., and D. A. Morrison.** 1999. Identification of a new regulator in *Streptococcus pneumoniae* linking quorum sensing to competence for genetic transformation. *J. Bacteriol.* **181:**5004–5016.

54. **LeFrancois, J., A. M. Gasc, and M. Sicard.** 1997. Electrotransformation of *Streptococcus pneumoniae*: evidence for restriction of the DNA on entry. *Microb. Drug Resist.* **3:**101–104.

55. **Lefrancois, J., M. M. Samrakandi, and A. M. Sicard.** 1998. Electrotransformation and natural transformation of *Streptococcus pneumoniae*: requirement of DNA processing for recombination. *Microbiology* **144:**3061–3068.

56. **Lefrancois, J., and A. M. Sicard.** 1997. Electrotransformation of *Streptococcus pneumoniae*: evidence for restriction of DNA on entry. *Microbiology* **143:**523–526.

57. **Lopez, R., and E. Garcia.** 2004. Recent trends on the molecular biology of pneumococcal capsules, lytic enzymes, and bacteriophage. *FEMS Microbiol. Rev.* 28:553–580.

58. **Luo, P., H. Li, and D. A. Morrison.** 2003. ComX is a unique link between multiple quorum sensing outputs and competence in *Streptococcus pneumoniae. Mol. Microbiol.* **50:**623–633.

59. **Luo, P., H. Li, and D. A. Morrison.** 2004. Identification of ComW as a new component in the regulation of genetic transformation in *Streptococcus pneumoniae. Mol. Microbiol.* **54:**172–183.

60. **Luo, P., and D. A. Morrison.** 2003. Transient association of an alternative sigma factor, ComX, with RNA polymerase during the period of competence for genetic transformation in *Streptococcus pneumoniae. J. Bacteriol.* **185:**349–358.

61. **Magee, A. D., and J. Yother.** 2001. Requirement for capsule in colonization by *Streptococcus pneumoniae. Infect. Immun.* **69:**3755–3761.

62. **Martin, B., P. Garcia, M.-P. Castanié, and J.-P. Claverys.** 1995. The *recA* gene of *Streptococcus pneumoniae* is part of a competence-induced operon and controls lysogenic induction. *Mol. Microbiol.* **15:**367–379.

63. **Martin, B., O. Humbert, M. Camara, E. Guenzi, J. Walker, T. Mitchell, P. Andrew, M. Prudhomme, G. Alloing, R. Hakenbeck, D. A. Morrison, G. J. Boulnois, and J.-P. Claverys.** 1992. A highly conserved repeated DNA element located in the chromosome of *Streptococcus pneumoniae. Nucleic Acids Res.* **20:**3479–3483.

64. **Martin, B., M. Prudhomme, G. Alloing, C. Granadel, and J. P. Claverys.** 2000. Cross-regulation of competence pheromone production and export in the early control of transformation in *Streptococcus pneumoniae. Mol. Microbiol.* **38:**867–78.

65. **Martin, B., G. J. Sharples, O. Humbert, R. G. Lloyd, and J. P. Claverys.** 1996. The *mmsA* locus of *Streptococcus pneumoniae* encodes a RecG-like protein involved in DNA repair and in three-strand recombination. *Mol. Microbiol.* **19:**1035–1045.

66. **Mascher, T., D. Zahner, M. Merai, N. Balmelle, A. B. de Saizieu, and R. Hakenbeck.** 2003. The *Streptococcus pneumoniae cia* regulon: CiaR target sites and transcription profile analysis. *J. Bacteriol.* **185:**60–70.

67. **Masure, H., B. Pearce, H. Shio, and B. Spellerberg.** 1998. Membrane targeting of RecA during genetic transformation. *Mol. Microbiol.* **27:**845–852.

68. **McDonnell, M., C. Ronda-Lain, and A. Tomasz.** 1975. "Diplophage": a bacteriophage of *Diplococcus pneumoniae. Virology* **63:**577–582.

69. **Mejean, V., and J. P. Claverys.** 1993. DNA processing during entry in transformation of *Streptococcus pneumoniae. J. Biol. Chem.* **268:**5594–5599.

70. **Morona, J. K., A. Guidolin, R. Morona, D. Hansman, and J. C. Paton.** 1994. Isolation, characterization, and

nucleotide sequence of IS*1202*, an insertion sequence of *Streptococcus pneumoniae*. *J. Bacteriol.* **176:**4437–4443.

71. **Morrison, D. A., and M. F. Baker.** 1979. Competence for genetic transformation in pneumococcus depends on synthesis of a small set of proteins. *Nature* **282:**215–217.

72. **Morrison, D. A., S. A. Lacks, W. R. Guild, and J. M. Hageman.** 1983. Isolation and characterization of three new classes of transformation-deficient mutants of *Streptococcus pneumoniae* that are defective in DNA transport and genetic recombination. *J. Bacteriol.* **156:**281–290.

73. **Mortier-Barriere, I., A. de Saizieu, J. P. Claverys, and B. Martin.** 1998. Competence-specific induction of *recA* is required for full recombination proficiency during transformation in *Streptococcus pneumoniae*. *Mol. Microbiol.* **27:**159–170.

74. **Moscoso, M., and J. P. Claverys.** 2004. Release of DNA into the medium by competent *Streptococcus pneumoniae*: kinetics, mechanism and stability of the liberated DNA. *Mol. Microbiol.* **54:**783–794.

75. **Muckerman, C. C., S. S. Springhorn, B. Greenberg, and S. A. Lacks.** 1982. Transformation of restriction endonuclease phenotype in *Streptococcus pneumoniae*. *J. Bacteriol.* **152:**183–190.

76. **Muñoz, R., R. Lopez, and E. Garcia.** 1998. Characterization of IS*1515*, a functional insertion sequence in *Streptococcus pneumoniae*. *J. Bacteriol.* **180:**1381–1388.

77. **Munoz-Najar, U., and M. N. Vijayakumar.** 1999. An operon that confers UV resistance by evoking the SOS mutagenic response in streptococcal conjugative transposon Tn*5252*. *J. Bacteriol.* **181:**2782–2788.

78. **Oggioni, M. R., and J. P. Claverys.** 1999. Repeated extragenic sequences in prokaryotic genomes: a proposal for the origin and dynamics of the RUP element in *Streptococcus pneumoniae*. *Microbiology* **145:**2647–2653.

79. **Ottolenghi, E., and R. Hotchkiss.** 1960. Appearance of genetic transforming activity in pneumococcal cultures. *Science* **132:**1257–1258.

80. **Pearce, B., A. Naughton, E. Campbell, and H. Masure.** 1995. The *rec* locus, a competence-induced operon in *Streptococcus pneumoniae*. *J. Bacteriol.* **177:**86–93.

81. **Pearce, B. J., A. M. Naughton, and H. R. Masure.** 1994. Peptide permeases modulate transformation in *Streptococcus pneumoniae*. *Mol. Microbiol.* **12:**881–892.

82. **Pestova, E. V., L. S. Havarstein, and D. A. Morrison.** 1996. Regulation of competence for genetic transformation in *Streptococcus pneumoniae* by an auto-induced peptide pheromone and a two-component regulatory system. *Mol. Microbiol.* **21:**853–862.

83. **Pestova, E. V., and D. A. Morrison.** 1998. Isolation and characterization of three *Streptococcus pneumoniae* transformation-specific loci by use of a *lacZ* reporter insertion vector. *J. Bacteriol.* **180:**2701–2710.

84. **Peterson, S., R. T. Cline, H. Tettelin, V. Sharov, and D. A. Morrison.** 2000. Gene expression analysis of the *Streptococcus pneumoniae* competence regulons by use of DNA microarrays. *J. Bacteriol.* **182:**6192–6202.

85. **Peterson, S. N., C. K. Sung, R. Cline, B. V. Desai, E. C. Snesrud, P. Luo, J. Walling, H. Li, M. Mintz, G. Tsegaye, P. C. Burr, Y. Do, S. Ahn, J. Gilbert, R. D. Fleischmann, and D. A. Morrison.** 2004. Identification of competence pheromone responsive genes in *Streptococcus pneumoniae* by use of DNA microarrays. *Mol. Microbiol.* **51:**1051–1070.

86. **Porter, R. D., and W. R. Guild.** 1976. Characterization of some pneumococcal bacteriophages. *J. Virol.* **19:**659–667.

87. **Porter, R. D., and W. R. Guild.** 1969. Number of transformable units per cell in *Diplococcus pneumoniae*. *J. Bacteriol.* **97:**1033–1035.

88. **Porter, R. D., N. B. Shoemaker, G. Rampe, and W. R. Guild.** 1979. Bacteriophage-associated gene transfer in pneumococcus: transduction or pseudotransduction? *J. Bacteriol.* **137:**556–567.

89. **Poyart-Salmeron, C., P. Trieu-Cuot, C. Carlier, and P. Courvalin.** 1990. The integration-excision system of the conjugative transposon Tn*1545* is structurally and functionally related to those of lambdoid phages. *Mol. Microbiol.* **4:**1513–1521.

90. **Pozzi, G., L. Masala, F. Iannelli, R. Manganelli, L. S. Havarstein, L. Piccoli, D. Simon, and D. A. Morrison.** 1996. Competence for genetic transformation in encapsulated strains of *Streptococcus pneumoniae*: two allelic variants of the peptide pheromone. *J. Bacteriol.* **178:** 6087–6090.

91. **Prudhomme, M., V. Libante, and J. P. Claverys.** 2002. Homologous recombination at the border: insertion-deletions and the trapping of foreign DNA in *Streptococcus pneumoniae*. *Proc. Natl. Acad. Sci. USA* **99:**2100–2105.

92. **Prudhomme, M., B. Martin, V. Mejean, and J. P. Claverys.** 1989. Nucleotide sequence of the *Streptococcus pneumoniae hexB* mismatch repair gene: homology of HexB to MutL of *Salmonella typhimurium* and to PMS1 of *Saccharomyces cerevisiae*. *J. Bacteriol.* **171:**5332–5338.

93. **Puyet, A., B. Greenberg, and S. A. Lacks.** 1990. Genetic and structural characterization of *endA*. A membrane-bound nuclease required for transformation of *Streptococcus pneumoniae*. *J. Mol. Biol.* **213:**727–738.

94. **Ramirez, M., E. Severina, and A. Tomasz.** 1999. A high incidence of prophage carriage among natural isolates of *Streptococcus pneumoniae*. *J. Bacteriol.* **181:**3618–3625.

95. **Reizer, J., A. Reizer, A. Bairoch, and M. H. Saier, Jr.** 1993. A diverse transketolase family that includes the RecP protein of *Streptococcus pneumoniae*, a protein implicated in genetic recombination. *Res. Microbiol.* **144:** 341–347.

96. **Rimini, R., B. Jansson, G. Feger, T. C. Roberts, M. de Francesco, A. Gozzi, F. Faggioni, E. Domenici, D. M. Wallace, N. Frandsen, and A. Polissi.** 2000. Global analysis of transcription kinetics during competence development in *Streptococcus pneumoniae* using high density DNA arrays. *Mol. Microbiol.* **36:**1279–1292.

97. **Robertson, G. T., W. L. Ng, J. Foley, R. Gilmour, and M. E. Winkler.** 2002. Global transcriptional analysis of *clpP* mutations of type 2 *Streptococcus pneumoniae* and their effects on physiology and virulence. *J. Bacteriol.* **184:** 3508–3520.

98. **Sabelnikov, A. G., B. Greenberg, and S. A. Lacks.** 1995. An extended −10 promoter alone directs transcription of the DpnII operon of *Streptococcus pneumoniae*. *J. Mol. Biol.* **250:**144–155.

99. **Samrakandi, M. M., and F. Pasta.** 2000. Hyperrecombination in *Streptococcus pneumoniae* depends on an atypical *mutY* homologue. *J. Bacteriol.* **182:**3353–3360.

100. **Sanchez-Beato, A. R., E. Garcia, R. Lopez, and J. L. Garcia.** 1997. Identification and characterization of IS*1381*, a new insertion sequence in *Streptococcus pneumoniae*. *J. Bacteriol.* **179:**2459–2463.

101. **Sanger Center.** http://www.sanger.ac.uk

102. **Santagati, M., F. Iannelli, C. Cascone, F. Campanile, M. R. Oggioni, S. Stefani, and G. Pozzi.** 2003. The novel conjugative transposon Tn*1207.3* carries the macrolide efflux gene *mef*(A) in *Streptococcus pyogenes*. *Microb. Drug Resist.* **9:**243–247.

103. **Santagati, M., F. Iannelli, M. R. Oggioni, S. Stefani, and G. Pozzi.** 2000. Characterization of a genetic element carrying the macrolide efflux gene *mef*(A) in

Streptococcus pneumoniae. Antimicrob. Agents Chemother. **44:** 2585–2587.

104. **Saunders, C. W., and W. R. Guild.** 1981. Monomer plasmid DNA transforms *Streptococcus pneumoniae. Mol. Gen. Genet.* **181:**57–62.

105. **Schuster, C., M. van der Linden, and R. Hakenbeck.** 1998. Small cryptic plasmids of *Streptococcus pneumoniae* belong to the pC194/pUB110 family of rolling circle plasmids. *FEMS Microbiol. Lett.* **164:**427–431.

106. **Scott, J., and G. Churchward.** 1995. Conjugative transposition. *Ann. Rev. Microbiol.* **49:**367–397.

107. **Shoemaker, N. B., M. D. Smith, and W. R. Guild.** 1979. Organization and transfer of heterologous chloramphenicol and tetracycline resistance genes in pneumococcus. *J. Bacteriol.* **139:**432–441.

108. **Sibold, C., Z. Markiewicz, C. Latorre, and R. Hakenbeck.** 1991. Novel plasmids in clinical strains of *Streptococcus pneumoniae. FEMS Microbiol. Lett.* **61:**91–95.

109. **Smith, M. D., and W. R. Guild.** 1979. A plasmid in *Streptococcus pneumoniae. J. Bacteriol.* **137:**735–739.

110. **Smith, M. D., N. B. Shoemaker, V. Burdett, and W. R. Guild.** 1980. Transfer of plasmids by conjugation in *Streptococcus pneumoniae. Plasmid* **3:**70–79.

111. **Stassi, D. L., P. Lopez, M. Espinosa, and S. A. Lacks.** 1981. Cloning of chromosomal genes in *Streptococcus pneumoniae. Proc. Natl. Acad. Sci. USA* **78:**7028–7032.

112. **Steffen, S. E., and F. R. Bryant.** 2000. Purification and characterization of the RecA protein from *Streptococcus pneumoniae. Arch. Biochem. Biophys.* **382:**303–309.

113. **Steinmoen, H., E. Knutsen, and L. S. Havarstein.** 2002. Induction of natural competence in *Streptococcus pneumoniae* triggers lysis and DNA release from a subfraction of the cell population. *Proc. Natl. Acad. Sci. USA* **99:**7681–7686.

114. **Steinmoen, H., A. Teigen, and L. S. Havarstein.** 2003. Competence-induced cells of *Streptococcus pneumoniae* lyse competence-deficient cells of the same strain during cocultivation. *J. Bacteriol.* **185:**7176–7183.

115. **Sung, C. K., H. Li, J. P. Claverys, and D. A. Morrison.** 2001. An *rpsL* cassette, Janus, for gene replacement through negative selection in *Streptococcus pneumoniae. Appl. Environ. Microbiol.* **67:**5190–5196.

116. **Tao, L., D. J. LeBlanc, and J. J. Ferretti.** 1992. Novel streptococcal-integration vectors for gene cloning and inactivation. *Gene* **120:**105–110.

117. **Tettelin, H., and S. K. Hollingshead.** 2004. Comparative genomics of *S. pneumoniae*, p. 15–30. *In* E. I. Tuomanen, T. J. Mitchell, D. A. Morrison, and B. G. Spratt (ed.), *The Pneumococcus.* ASM Press, Washington, D.C.

118. **Tettelin, H., K. E. Nelson, I. T. Paulsen, J. A. Eisen, T. D. Read, S. Peterson, J. Heidelberg, R. T. DeBoy, D. H. Haft, R. J. Dodson, A. S. Durkin, M. Gwinn,** J. F. Kolonay, W. C. Nelson, J. D. Peterson, L. A. Umayam, O. White, S. L. Salzberg, M. R. Lewis, D. Radune, E. Holtzapple, H. Khouri, A. M. Wolf, T. R. Utterback, C. L. Hansen, L. A. McDonald, T. V. Feldblyum, S. Angiuoli, T. Dickinson, E. K. Hickey, I. E. Holt, B. J. Loftus, F. Yang, H. O. Smith, J. C. Venter, B. A. Dougherty, D. A. Morrison, S. K. Hollingshead, and C. M. Fraser. 2001. Complete genome sequence of a virulent isolate of *Streptococcus pneumoniae. Science* **293:**498–506.

119. **Tiraby, G., M. S. Fox, and H. Bernheimer.** 1975. Marker discrimination in deoxyribonucleic acid-mediated transformation of various *Pneumococcus* strains. *J. Bacteriol.* **121:**608–618.

120. **Tiraby, J. G., E. Tiraby, and M. S. Fox.** 1975. Pneumococcal bacteriophages. *Virology* **68:**566–569.

121. **Tomasz, A., and R. D. Hotchkiss.** 1964. Regulation of the transformability of pneumococcal cultures by macromolecular cell products. *Proc. Natl. Acad. Sci. USA* **51:** 480–487.

122. **Vijayakumar, M. N., and S. Ayalew.** 1993. Nucleotide sequence analysis of the termini and chromosomal locus involved in site-specific integration of the streptococcal conjugative transposon Tn*5252. J. Bacteriol.* **175:** 2713–2719.

123. **Vijayakumar, M. N., and D. A. Morrison.** 1983. Fate of DNA in eclipse complex during genetic transformation in *Streptococcus pneumoniae. J. Bacteriol.* **156:** 644–648.

124. **Ween, O., P. Gaustad, and L. S. Havarstein.** 1999. Identification of DNA binding sites for ComE, a key regulator of natural competence in *Streptococcus pneumoniae. Mol. Microbiol.* **33:**817–827.

125. **Weiser, J. N., R. Austrian, P. K. Sreenivasan, and H. R. Masure.** 1994. Phase variation in pneumococcal opacity: relationship between colonial morphology and nasopharyngeal colonization. *Infect. Immun.* **62:**2582–2589.

126. **Yother, J.** 1999. Common themes in the genetics of streptococcal capsular polysaccharides, p. 161–184. *In* J. B. Goldberg (ed.), *Genetics of Bacterial Polysaccharides.* CRC Press, Boca Raton, Fla.

127. **Yother, J., L. S. McDaniel, and D. E. Briles.** 1986. Transformation of encapsulated *Streptococcus pneumoniae. J. Bacteriol.* **168:**1463–1465.

128. **Zahner, D., and R. Hakenbeck.** 2000. The *Streptococcus pneumoniae* beta-galactosidase is a surface protein. *J. Bacteriol.* **182:**5919–5921.

129. **Zhou, L., F. M. Hui, and D. A. Morrison.** 1995. Characterization of IS*1167*, a new insertion sequence in *Streptococcus pneumoniae. Plasmid* **33:**127–138.

Pneumococcal Vaccines

D. E. BRILES, J. C. PATON, E. SWIATLO, AND M. J. CRAIN

24

With the discovery of the pneumococcus in 1881 it became apparent that this gram-positive pathogen was a major cause of serious and often fatal pneumonia (124). It is also a major cause of meningitis and otitis media in children. Pneumococci are the largest cause of community-acquired pneumonia in the developed world. In the United States the combined rates of invasive pneumococcal bacteremia, pneumonia, and meningitis in the elderly are $\geq 50/100,000$ (49), while in children 6 to 23 months of age rates range from about $150/100,000$ to $500/100,000$, depending on race and ethnic background (36). The rate of meningitis in children in the United States is about 1 to 2 cases per 100,000 children, with a fatality rate of about 18% (51). In the developing world pneumococci are an important cause of childhood deaths due to bacterial respiratory infection following viral disease. Globally, such infections kill an estimated one million children annually (4). Within recent years about one-third to one-half of pneumococci recovered from humans have become at least partially resistant to penicillin, and penicillin-resistant strains are frequently also resistant to other common antibiotics (79). The rise of antibiotic resistance among pneumococci has already complicated treatment, especially of meningitis (69), and threatens to greatly increase the morbidity and mortality caused by pneumococci unless new control measures are developed.

It has long been recognized that the best management of most infectious disease is prevention. Vaccines offer the prospect of a highly cost-effective means of preventing morbidity and mortality caused by pneumococci. This chapter provides a concise summary of issues critical to the development and application of pneumococcal vaccines. Several relatively recent reviews address this topic in more detail (32, 45, 50, 121).

In the preantibiotic era, vaccination attempts utilized whole killed pneumococci injected parenterally. Although such vaccines were sometimes protective in humans, they were also highly reactogenic. These killed vaccines were mainly used to elicit antibody in animals for passive treatment of infected humans (60). In 1933 it was clearly demonstrated that antibody to type-specific capsular polysaccharides (PS) could be highly protective (6). However, it

soon became apparent that different strains of *Streptococcus pneumoniae* each expressed one of many different PS. Most subsequent vaccine attempts focused on the use of mixtures of the isolated PSs to elicit protection (50, 110). However, the inability of young children to make adequate responses to most soluble PS led to the development and licensing of an immunogenic PS-protein conjugate vaccine for children (18, 121). However, the wide diversity of pneumococcal PS types, the failure of PS vaccines to protect against strains with PSs not in the vaccines, the complexity of production, and the high expense of conjugate vaccines have led to investigations of other pneumococcal antigens (primarily proteins) as potential vaccine candidates.

PNEUMOCOCCAL CAPSULAR PS VACCINE

During the 1930s capsular PS was shown to be able to elicit protective antibodies. At that time it was generally assumed that capsular PS was the target of all, or most, protective antibodies (6, 124). Currently, at least 90 distinct serological types of capsular PS have been identified for pneumococci (61). Each strain can make only one PS and changes require a genetic transformation by DNA of another strains. Only about one-third of the 90 serotypes occur with any frequency in adult infections, and the distribution of serotypes is even more restricted in children (56, 121). Following the identification of capsular PSs as protection-eliciting molecules, vaccines composed of relatively pure PSs were used successfully in nursing homes and among military personnel in the 1940s. However, during this time antibiotics were becoming widely available, and these drugs diminished the perceived impact of pneumococcal infections. Two commercially available pneumococcal vaccines were withdrawn from production for lack of demand (50).

Despite a growing selection of antibiotics and increasingly sophisticated critical care technology, morbidity and mortality from invasive pneumococcal infections have remained high. Antibiotics and supportive care alone were not sufficient to completely eliminate the impact of pneumococcal disease (5). During the 1970s renewed interest in pneumococcal vaccines led to clinical trials of a PS vaccine in adults at high risk for invasive pneumococcal disease in South Africa and New Guinea. Based largely on encouraging results from these trials, a 14-valent PS vaccine was licensed in the United States in 1977. The vaccine was expanded to 23 PS serotypes in 1983, and is currently the only approved formulation available for adults in the United States (50, 121).

Clinical trials examining the efficacy of PS vaccine continued after licensure, with somewhat mixed, and controversial, results (50). A significant impediment to conducting the most rigorous controlled clinical trials has been ethical concern of denying at-risk groups an approved and recommended vaccine. A large case-controlled study of invasive pneumococcal disease in adults showed that immunization with PS vaccine had an overall protective efficacy of 56% for preventing invasive disease by serotypes in the vaccine (118). Subsequently, many smaller studies have shown variable efficacy of the PS vaccine in preventing invasive pneumococcal infections (37, 41). Although the PS vaccine is cost-effective even when considering only bacteremic disease, this vaccine is not generally accepted to reduce the incidence of pneumococcal pneumonia or localized upper respiratory infections such as acute bronchitis, sinusitis, or otitis media (48).

A major drawback to the PS vaccine is the poor immunogenicity of PS antigens in children and elderly adults. Response to PS varies by age and with the presence of certain underlying chronic medical conditions, most notably infection with human immunodeficiency virus. Certain capsular PS serotypes in the 23-valent vaccine, such as types 6B, 9V, 19F, and 23F, induce relatively low amounts of specific antibody, which fall to prevaccination levels within 3 years (117). In addition to a reduced quantity of antibody, a significant number of elderly adults have impaired functional antibody responses to PS as defined by in vitro avidity and opsonophagocytosis assays (116).

PS-PROTEIN CONJUGATE VACCINES

Children less than 2 years of age are at increased risk for infections caused by encapsulated bacteria; consequently, this group has the highest incidence of invasive pneumococcal infections (116). Children in this age group do not consistently respond to PS antigens, which directly cross-link surface immunoglobulin (Ig) on B cells and do not have an absolute requirement for T-cell help. The molecular basis for this reduced capacity to respond to T-independent antigens is not completely understood, although it may involve reduced expression of B-cell coreceptors such as complement receptor 2 (108, 109). Whatever the underlying mechanisms, children less than 2 years respond poorly to many of the common pneumococcal capsular types (94), and alternatives to purified PS antigens are needed to protect this susceptible population.

The ability to make antibodies to protein antigens appears quite rapidly after birth, and T cell-dependent antibody production is highly efficient and induces memory B cells. The development of PS-protein conjugate vaccines for children takes advantage of the ability of young children to make T cell-dependent responses to PS covalently attached to immunogenic proteins. This strategy has proven highly successful in preventing both invasive disease and colonization with *Haemophilus influenzae* type b (47). The process of synthesizing conjugate vaccines requires chemical reactions that covalently attach PS to protein; nonconjugated mixtures of peptides and PS do not induce T cell-dependent PS antibodies (8).

Immunization with PS conjugate vaccines induces effective memory antibody responses in infants and children and antibody levels can be boosted with repeated immunization (1). A large clinical trial in infants using a four-dose schedule of immunization with a heptavalent conjugate vaccine proved this strategy to be highly effective in preventing invasive (bacteremic) disease (16, 17, 125). However, this conjugate vaccine has shown somewhat more modest efficacy in preventing pneumonia (18) and otitis media (46). The heptavalent pneumococcal conjugate vaccine was licensed in 2000 in the United States for use in all children under 2 years of age and for high-risk children less than 5 years of age (125). Because the vaccine serotypes that colonize and infect children include most of the serotypes associated with antibiotic resistance, and because children carry pneumococci at a much higher frequency than adults, there are expectations that childhood immunization may reduce the transmission of antibiotic-resistant pneumococci in the entire population (20, 43, 53, 78). It has been observed, however, that significant replacement carriage and otitis media occur with capsular types not present in the vaccine (20, 46, 78). It remains to be determined if nonvaccine serotypes will eventually emerge as a cause of significant invasive disease in children, or if many of the nonvaccine type PSs are intrinsically able to attenuate pneumococci. Also of concern are observations that immunization with the conjugate vaccine alters the bacterial flora of the upper respiratory tract in children and has affected the pathogens recovered from children with carriage and acute otitis media (19, 21). A potentially worrisome trend has been noted in increased rates of carriage of *Staphylococcus aureus* in children immunized with the pneumococcal conjugate vaccine (21, 105).

The distribution of pneumococcal serotypes commonly causing invasive disease in adults is much larger than for children. Consequently, a strategy of conjugate vaccine development for adults that covered as many serotypes as the 23-valent PS vaccine would involve a very complex and costly vaccine preparation. Although conjugate vaccines have been found to be safe and immunogenic in elderly adults, an improvement in protective efficacy has not yet been observed (91, 103, 120).

Strategies to enhance the immunogenicity of PS have been tested in small animal and human studies, but none have yet reached clinical application. A conjugate of type 6B PS and tetanus toxoid reduces carriage in mice when administered intranasally with the cholera toxin B subunit (126). T-cell help during response to protein antigens involves binding of CD40 on B cells by CD154 expressed on T cells. Administration of anti-CD40 antibodies along with capsular PS results in predominantly high-affinity IgG antibodies, which is a hallmark of responses to protein antigens (44). Small amounts of interleukin-12 have been shown to be a potent adjuvant for pneumococcal PS in mice (35). Coadministration of immune modulators may prove to be an effective and practical strategy to enhance immune responses to PS.

NONCAPSULAR PS VACCINES

Overview

Polyvalent pneumococcal vaccines based on purified PS have been available for two decades, but their clinical efficacy has been limited by poor immunogenicity in high-risk groups (particularly young children). As discussed above, the problem of poor vaccine immunogenicity in children is being addressed by conjugation of the PS to protein carriers, thereby converting the PS from T-cell-independent to T-cell-dependent antigens. However, serotype coverage will be limited, as it is unlikely that more than 11 serotypes will be included in such conjugate formulations. Nasopharyngeal colonization with S. pneumoniae is thought to be a prerequisite for invasive disease. Some data from the conjugate vaccine trials indicate that, although carriage of vaccine types was reduced, the vacated niche was promptly occupied by nonvaccine serotypes known to cause invasive disease in humans (45). In addition, the cost of the conjugate vaccines is likely to be high; thus, their use may be low in third-world countries, where the need for effective pediatric vaccines is the greatest. In view of this, much recent attention has focused on the possibility of developing vaccines based on pneumococcal protein antigens common to all serotypes (32, 33). Such proteins, being T-cell-dependent antigens, are likely to be highly immunogenic in human infants and able to elicit immunological memory. The pneumococcal proteins also have potential as carriers for the PS in the conjugate vaccines.

The discovery of cross-reactive proteins was preceded by studies demonstrating that antibodies directed against the phosphocholine epitopes of teichoic and lipoteichoic acids could be protective against pneumococcal infections in mice (26, 31, 33, 70). This observation made it clear that the capsule was not the only pneumococcal antigen that could elicit a protective response. Because antibodies to the phosphocholine epitopes of teichoic acids are not made in young children, this discovery by itself did not provide a means of developing an efficacious noncapsular pneumococcal vaccine. Subsequent studies have identified several protein antigens that are immunogenic in young children and that have also been shown to elicit protective immune responses in mice. These antigens include the pneumococcal surface protein PspA; autolysin (lytA), an enzyme on the pneumococcal cell wall; and pneumolysin, a cytoplasmic protein that is released when pneumococci are autolysed. These proteins and several others are described below.

PspA and PspC

PspA is produced by all pneumococci (42). The protein is important for virulence (84, 107) and has been shown to interfere with activation of C3 and its deposition on pneumococci (106, 107). Antibodies to PspA are able to enhance complement deposition (106). PspA may also play a role on mucosal surfaces because it blocks killing by apolactoferrin (86). Antibody to PspA reverses this blockage and enhances killing by apolactoferrin (86). Although serologically variable when examined with monoclonal antibody, PspA is highly cross-reactive when examined with polyclonal sera (40, 42). Monoclonal and polyclonal antibodies to PspA can passively protect mice from otherwise fatal bacteremia and sepsis caused by pneumococci (30, 80). Parenteral immunization with PspA has been able to protect mice against fatal infections, and this protection has been observed to be highly cross-protective regardless of PspA type (27, 30, 112). Intranasal immunization with PspA has also been shown to protect against nasopharyngeal carriage in an adult mouse carriage model, raising the possibility that vaccines could be developed to prevent carriage and transmission of pneumococci (25, 126).

A number of pneumococcal proteins, including PspA, PspC, and LytA, are able to bind choline, and share similar, and sometimes indistinguishable, choline-binding domains near the C-terminal ends of the proteins (24, 76, 127, 128).

The N-terminal end of PspA is composed of a largely coiled-coil α-helical sequence (81, 127) that is responsible for most of the cross-protective immunity elicited by PspA (82, 111). Paradoxically, this region also contains significant structural and serological variability within the PspA molecule (62, 82). In spite of their variability at the amino acid sequence level PspAs are very cross-reactive (30, 42) and can be divided into families based on their cross-reactivity and amino acid sequence (40, 62). Ninety-five percent or more of clinical isolates are composed of families 1 and 2 (23, 40, 62). Immunization of humans with a single recombinant family 1 PspA leads to antibodies that are able to protect mice from otherwise fatal infection with strains expressing either family 1 or family 2 PspAs (27). However, other data indicate that the best vaccine would probably contain PspAs from both family 1 and family 2 (90, 111, 112).

The serum of virtually all adults, and most children over 7 months of age, contains detectable antibody to PspA (104, 123). The levels of antibody are higher in adults than in children. It is thus possible that natural antibodies to PspA will contribute more to the immunity to pneumococcal infection in adults than in young children. It is assumed that immunization to elicit high levels of antibodies to PspA will be able to enhance protection of young children as well as adults from infections with pneumococci.

PspC is a protein that has some similarity to PspA in its proline-rich and choline-binding domains (29). The gene for this protein was originally identified through its close similarity to PspA (83). However, the α-helical domain of PspC is more complex than that of PspA and is present as several very distinct alleles with distinct combinations of functions (34, 65). This protein has been independently discovered by others based on its ability to bind secretory IgA (SpsA), (58) choline (CbpA) (113), and factor H (Hic) (66). It has potential roles in colonization, adherence, and invasion (7, 34, 58, 64, 113). Immunity to PspC is able to protect against pneumococcal infection and carriage (7, 34).

Pneumolysin

All pneumococci produce pneumolysin, a potent 53-kDa thiol-activated pore-forming cytolysin, which can attack any cell that has cholesterol in its plasma membrane (98). In addition to its cytotoxic properties, pneumolysin is capable of directly activating the classical complement pathway in the absence of specific antibody, with a concomitant reduction in serum opsonic activity (102). This latter property is mediated by its capacity to bind the Fc region of human IgG (87).

Structure-function analysis of pneumolysin has indicated that a domain toward the C terminus of the toxin (amino acids 427 to 437), which includes a unique cysteine residue, is critical for cytotoxicity (22). This cysteine motif is highly conserved among other members of the thiol-activated cytolysin family. Several single-amino-acid substitutions within this region (and other regions involved in cell binding and pore formation) reduce the cytotoxicity of pneumolysin by up to 99.9%. A separate region, which has a degree of amino acid homology with human C-reactive protein, is responsible for IgG binding and complement activation, and a mutation in this domain (Asp385→Asn) interferes with both properties (87).

Pneumolysin has a variety of detrimental effects on cells and tissues in vitro that provide clues to its role in the pathogenesis of pneumococcal disease (97, 98). Complete inactivation of the pneumolysin gene in either a type 2 or type 3 pneumococcus has been shown to reduce virulence for mice challenged by both the intranasal and intraperitoneal routes (13, 14). Compared with the wild-type strain, intranasal challenge with the pneumolysin-negative pneumococci resulted in a less severe inflammatory response, a reduced rate of multiplication within the lung, a reduced capacity to injure the alveolar-capillary barrier, and a delayed onset of bacteremia (38, 115). Pneumolysin also plays a critical role after bacteremia has developed, by modulating the elicited inflammation to reduce its ability to keep bacteremia in check (9). Pneumolysin's ability to modulate inflammation may be related to its abilities to stimulate Toll-like receptor 4 (77) and may contribute to chemotaxis of CD4 T cells (68). By using S. pneumoniae derivatives in which the wild-type pneumolysin gene was replaced by mutated genes encoding toxins with point mutations affecting either or both of the cytotoxic and complement activation properties, it has been possible to confirm distinct roles for the two toxin activities in the pathogenesis of pneumococcal pneumonia (2, 10, 114).

These studies establish the importance of pneumolysin in the pathogenesis of pneumococcal disease and identify it as a target for vaccination. It has been known for many years that immunization with purified pneumolysin protects mice against challenge with highly virulent pneumococci (100). Potential problems of toxicity have been overcome by site-directed mutagenesis of regions of the toxin essential for cytotoxicity and complement activation, as described above. Genes encoding these recombinant pneumolysin toxoids (pneumolysoids) have also been inserted into *Escherichia coli* expression vectors, permitting large-scale production of antigens at low cost (101). Sequence analysis of pneumolysin genes from a wide range of S. pneumoniae serotypes has confirmed that there is very little variation in primary amino acid sequence (>99% identity) (74, 88), and so a single vaccine antigen should provide coverage against all pneumococci regardless of serotype. Indeed, immunization of mice with a pneumolysoid carrying a Trp_{433}-Phe mutation resulting in >99% reduction in cytotoxicity (designated PdB) provided a significant degree of protection against all nine serotypes of S. pneumoniae that were tested (3). Moreover, mouse monoclonal antibodies to pneumolysin could protect against otherwise fatal infection following intranasal inoculation of S. pneumoniae (52). Humans are known to mount an antibody response to pneumolysin as a result of natural exposure to S. pneumoniae (63, 89). The expectation that human antibodies to pneumolysin may be protective is supported by evidence that purified human antibody to pneumolysin passively protects mice from challenge with virulent pneumococci (89) and by evidence that a lack of high levels of serum antibody to pneumolysin appears to predispose patients to pneumococcal pneumonia (63). Thus, it is anticipated that immunization of humans with pneumolysoid will increase resistance to pneumococcal infection. It is expected that this protection would be the result of toxin-neutralizing antibodies and not direct opsonization because pneumolysin is not surface attached. The protection elicited by a pneumolysin-containing vaccine may be enhanced by incorporation of other protection-eliciting antigens such as PspA, immunity to which enhances complement deposition and blood clearance (28, 93, 106).

Pneumolysin has also shown promise as a carrier for the otherwise poorly immunogenic PS in experimental conjugate vaccine formulations. Immunization of mice with pneumolysoid conjugated to type 19F PS elicited a strong and boostable antibody response to both protein and PS moieties and provided infant mice with a high degree of protection against challenge with S. pneumoniae (72, 101). Similar results have been reported for conjugates of native pneumolysin with type 18C PS (71). A comparison of tetravalent pneumolysoid-PS or tetanus toxoid-PS conjugate vaccines (incorporating PS types 6B, 14, 19F, and 23F) demonstrated that pneumolysoid was at least as good a carrier protein as tetanus toxoid and, in the case of type 23F, superior (85). Clearly, such antigens have the potential to evoke a significant anti-PS response, as well as an antivirulence-protein response, thereby conferring comprehensive protection against pneumococcal disease in humans.

Other Noncapsular Protein Antigens

Several other putative pneumococcal virulence proteins, including LytA, two neuraminidases (NanA and NanB), and hyaluronidase, have been examined for vaccine potential (reviewed by Paton et al. [99]). Of these, the most effective was LytA, which elicited a level of protection similar to that achieved by immunization with pneumolysin. This protective effect appeared to be largely attributable to blockade of the release of pneumolysin; the toxin is located in the cytoplasm of growing pneumococci and is released into the external medium only when the cells undergo autolysis. Immunization with a combination of pneumolysin and LytA did not increase the degree of protection over that obtained using either antigen alone. Also, immunization with LytA provided no protection against challenge with high doses of pneumolysin-negative pneumococci (73). Immunization of mice with either of the two neuraminidases has been shown to confer low-level but statistically significant protection. In both cases protection obtained was poorer than that achieved by immunization with pneumolysin, and there was no additive protective effect. The efficacy of parenteral immunization with both NanA and NanB has yet to be determined. In studies with chinchillas (122) it has been shown that NanA is important for nasal carriage. Studies with mice have resulted in a similar observation for NanA but have not revealed a role for NanB in nasal colonization (J. Watt, D. E. Briles, and F. van Ginkel, unpublished data). That NanA may be an effective mucosal immunogen to prevent carriage has gained some support from a recent demonstration that immunization with NanA affords protection against experimental otitis media in chinchillas (75).

In other studies the relatively conserved pneumococcal surface proteins BVH-3 and BVH-11 were shown to be able to elicit protection against infection with certain capsular type 3 strains (57). PcpA is a surface choline-binding protein of unknown function that appears to be important for invasive pneumonia and bacteremia (59). Proteinase maturation protein A (PpmA) is a surface lipoprotein that shows more expression in invasive opaque phase pneumococci than in less invasive transparent phase pneumococci (96). The putative PpmA is reported to be required for full virulence in a lung infection model and to be able to elicit opsonophagocytic antibody (95). However, convincing evidence that immunity to PpmA can elicit protection is still lacking (55). PsaA is a protein that is part of a manganese transport system and is important for virulence (12, 67). Although PsaA has been considered as a vaccine candidate to elicit protection against carriage, more recent studies indicate that it is not sufficiently surface-exposed to play a role in adherence or to be an effective target of protective antibodies (28, 55, 67). Additional studies with PpmA and PsaA are needed. Other proteins such as the zinc metalloproteases ZmpB and ZmpC are virulence factors for invasive infection (39, 92) and may also be potential vaccine candidates.

Phosphocholine of Teichoic Acids

Mouse and human antibodies to pneumococcal cell teichoic and lipoteichoic acids, in particular those to the phosphocholine epitope, have long been known to protect mice from otherwise fatal pneumococcal infection (26, 31, 33). More recent studies have shown that human antibodies to these antigens can be opsonic (54) and that the levels of antibodies are inversely associated with susceptibility to invasive pneumococcal disease (129). Interestingly, antibodies to phosphocholine also react with oxidized low-density lipoprotein in human blood and atherosclerotic plaques. The recent observation that these antibodies reduce atherosclerosis in a mouse model (15, 119) suggests that a vaccine that elicited these antibodies in humans might protect against heart disease and stroke as well as pneumococcal pneumonia.

The Case for a Multivalent Pneumococcal Protein Vaccine

Virtually all of the pneumococcal proteins under consideration as vaccine antigens are directly or indirectly involved in the pathogenesis of pneumococcal disease. Mutagenesis of some combinations of virulence factor genes, for example those encoding pneumolysin and either PspA or PspC, or PspA and PspC, has been shown to synergistically attenuate pneumococcal virulence in animal models, implying that the respective proteins function independently in the pathogenic process (7, 11). This strongly suggests that immunization with combinations of these antigens might provide additive protection. Moreover, there may be differences in the relative protective capacities of the individual antigens against particular S. pneumoniae strains, particularly for surface-exposed antigens that exhibit some degree of sequence variation. Thus, a combined pneumococcal protein vaccine may elicit a higher degree of protection against a wider variety of strains than any single antigen. To date only a limited number of combination experiments have been performed. Immunization of mice with a combination of the pneumolysoid PdB and PspA provided significantly increased protection against intraperitoneal challenge than immunization with either protein alone. However, combining either protein with PsaA did not result in enhanced protection (93). The potential benefits of combination protein vaccines are also well illustrated using a recently developed mouse model of nonbacteremic pneumonia that closely reflects the commonest form of pneumococcal respiratory disease in humans (28). In this system, subcutaneous immunization (using alum adjuvant) with either PdB or PspA, but not PsaA, significantly reduced numbers of S. pneumoniae in the lungs 7 days after challenge. A significant additional reduction in bacterial load was achieved by immunization with a combination of PdB and PspA, but not when either protein was combined with PsaA (28). Thus, optimum vaccine formulation may be dependent upon the willingness of vaccine developers to combine more than one antigen to develop optimal protection. Such combinations could be different free antigens or conjugates of C-carbohydrate, phosphocholine, or capsular antigens to immunogenic pneumococcal proteins.

CONCLUSIONS

The ongoing high global morbidity and mortality associated with pneumococcal disease and the complications caused by increasing rates of resistance to antimicrobials have underpinned extensive efforts in recent years to develop more effective vaccination strategies against S. pneumoniae. These efforts have benefited from a better understanding of the mechanisms of pathogenesis of pneumococcal disease, and the advances made possible by the advent of recombinant DNA technology and access to genome sequence data. The polyvalent PS vaccines have doubtless prevented a large number of deaths from invasive disease in recipients belonging to those patient groups for whom this vaccine is currently recommended. The newer PS-protein conjugate formulations confer a very high degree of protection in young children against included serotypes, and may also have an impact on prevalence of drug-resistant strains. However, there is now general acceptance that this vaccination approach is not without its drawbacks, and as explained above, the initially substantial clinical benefits that are expected to be derived from widespread use of conjugate vaccines may diminish with time. It will take many years for the overall impact of conjugate vaccines on disease burden and the population biology of S. pneumoniae to become apparent. At the very least, use of the conjugate vaccines will buy time for development of cheaper, non-serotype-specific vaccines based on combinations of protein antigens. It must be emphasized, however, that the success of these protein vaccines is not dependent upon real or perceived failure of the conjugates. Rather, the two approaches should be viewed as complementary, each having an important role to play in global prevention of pneumococcal disease. Neither should development of parenteral protein vaccines impede future research on mucosal- or DNA-based delivery systems, which may further improve presentation of protective antigens to the immune system, thereby optimizing host responses and herd immunity.

REFERENCES

1. **Ahman, H., H. Kayhty, H. Lehtonen, O. Leroy, J. Froeschle, and J. Eskola.** 1998. Streptococcus pneumoniae capsular polysaccharide-diphtheria toxoid conjugate vaccine is immunogenic in early infancy and able to induce immunologic memory. Pediatr. Infect. Dis. J. **17:**211–216.
2. **Alexander, J. E., A. M. Berry, J. C. Paton, J. B. Rubins, P. W. Andrew, and T. J. Mitchell.** 1998. Amino acid changes affecting the activity of pneumolysin alter the behavior of pneumococci in pneumonia. Microb. Pathog. **24:**167–174.
3. **Alexander, J. E., R. A. Lock, C. C. A. M. Peeters, J. T. Poolman, P. W. Andrew, T. J. Mitchell, D. Hansman, and J. C. Paton.** 1994. Immunization of mice with pneumolysin toxoid confers a significant degree of protection against at least nine serotypes of Streptococcus pneumoniae. Infect. Immun. **62:**5683–5688.
4. **Anonymous.** 1999. Pneumococcal vaccines. WHO position paper. Wkly. Epidemiol. Rec. **74:**177–183.
5. **Austrian, R., and J. Gold.** 1964. Pneumococcal bacteremia with special reference to bacteremic pneumococcal pneumonia. Ann. Intern. Med. **60:**759–776.
6. **Avery, O. T., and W. F. Goebel.** 1933. Chemoimmunological studies of the soluble specific substance of pneumococcus. I. The isolation and properties of the acetyl polysaccharide of pneumococcus type I. J. Exp. Med. **58:**731–755.
7. **Balachandran, P., A. Brooks-Walter, A. Virolainen-Julkunen, S. K. Hollingshead, and D. E. Briles.** 2002. The role of pneumococcal surface protein C (PspC) in nasopharyngeal carriage and pneumonia and its ability to elicit protection against carriage of Streptococcus pneumoniae. Infect. Immun. **70:**2526–2534.

8. Barbour, M. L., R. T. Mayon-White, C. Coles, D. W. Crook, and E. R. Moxon. 1995. The impact of conjugate vaccine on carriage of *Haemophilus influenzae* type b. *J. Infect. Dis.* **171:**93–98.

9. Benton, K., and D. E. Briles. 1994. Pneumolysin facilitates pneumococcal sepsis by interfering with an antipneumococcal inflammatory response, p. 104. Abstr. 94th Annu. Meet. Am. Soc. Microbiol. 1994. American Society for Microbiology, Washington, D.C.

10. Benton, K. A., J. C. Paton, and D. E. Briles. 1997. The hemolytic and complement-activating properties of pneumolysin do not contribute individually to virulence in a pneumococcal bacteremia model. *Microb. Pathog.* **23:**201–209.

11. Berry, A. M., and J. Paton, C. 2000. Additive attenuation of virulence of *Streptococcus pneumoniae* by mutation of the genes encoding pneumolysin and other putative pneumococcal virulence proteins. *Infect. Immun.* **68:**133–140.

12. Berry, A. M., and J. C. Paton. 1996. Sequence heterogeneity of PsaA, a 37-kilodalton putative adhesin essential for virulence of *Streptococcus pneumoniae. Infect. Immun.* **64:**5255–5262.

13. Berry, A. M., J. C. Paton, and D. Hansman. 1992. Effect of insertional inactivation of the genes encoding pneumolysin and autolysin on the virulence of *Streptococcus pneumoniae* type 3. *Microb. Pathog.* **12:**87–93.

14. Berry, A. M., J. Yother, D. E. Briles, D. Hansman, and J. C. Paton. 1989. Reduced virulence of a defined pneumolysin-negative mutant of *Streptococcus pneumoniae. Infect. Immun.* **57:**2037–2042.

15. Binder, C. J., S. Horkko, A. Dewan, M. K. Chang, E. P. Kieu, C. S. Goodyear, P. X. Shaw, W. Palinski, J. L. Witztum, and G. J. Silverman. 2003. Pneumococcal vaccination decreases atherosclerotic lesion formation: molecular mimicry between *Streptococcus pneumoniae* and oxidized LDL. *Nat. Med.* **9:**736–743.

16. Black, S., H. Shinefield, R. Cohen, D. Floret, J. Gaudelus, C. Olivier, and P. Reinert. 2004. Clinical effectiveness of seven-valent pneumococcal conjugate vaccine (Prevenar) against invasive pneumococcal diseases: prospects for children in France. *Arch. Pediatr.* **11:**843–853.

17. Black, S., H. Shinefield, B. Fireman, E. Lewis, P. Ray, J. R. Hansen, L. Elvin, K. M. Ensor, J. Hackell, G. Siber, F. Malinoski, D. Madore, I. Chang, R. Kohberger, W. Watson, R. Austrian, and K. Edwards. 2000. Efficacy, safety and immunogenicity of heptavalent pneumococcal conjugate vaccine in children. Northern California Kaiser Permanente Vaccine Study Center Group. *Pediatr. Infect. Dis. J.* **19:**187–195.

18. Black, S. B., H. R. Shinefield, S. Ling, J. Hansen, B. Fireman, D. Spring, J. Noyes, E. Lewis, P. Ray, J. Lee, and J. Hackell. 2002. Effectiveness of heptavalent pneumococcal conjugate vaccine in children younger than five years of age for prevention of pneumonia. *Pediatr. Infect. Dis. J.* **21:**810–815.

19. Block, S. L., J. Hedrick, C. J. Harrison, R. Tyler, A. Smith, R. Findlay, and E. Keegan. 2004. Community-wide vaccination with the heptavalent pneumococcal conjugate significantly alters the microbiology of acute otitis media. *Pediatr. Infect. Dis. J.* **23:**829–833.

20. Bogaert, D., R. De Groot, and P. W. Hermans. 2004. *Streptococcus pneumoniae* colonisation: the key to pneumococcal disease. *Lancet Infect. Dis.* **4:**144–154.

21. Bogaert, D., A. van Belkum, M. Sluijter, A. Luijendijk, R. de Groot, H. C. Rumke, H. A. Verbrugh, and P. W. Hermans. 2004. Colonisation by *Streptococcus pneumoniae* and *Staphylococcus aureus* in healthy children. *Lancet* **363:**1871–1872.

22. Boulnois, G. J., J. C. Paton, T. J. Mitchell, and P. W. Andrew. 1991. Structure and function of pneumolysin, the multifunctional, thiol-activated toxin of *Streptococcus pneumoniae. Mol. Microbiol.* **5:**2611–2616.

23. Brandileone, M. C., A. L. Andrade, E. M. Teles, R. C. Zanella, T. I. Yara, J. L. Di Fabio, and S. K. Hollingshead. 2004. Typing of pneumococcal surface protein A (PspA) in *Streptococcus pneumoniae* isolated during epidemiological surveillance in Brazil: towards novel pneumococcal protein vaccines. *Vaccine* **22:**3890–3896.

24. Briese, T., and R. Hakenbeck. 1985. Interaction of the pneumococcal amidase with lipoteichoic acid and choline. *Eur. J. Biochem.* **146:**417–427.

25. Briles, D. E., E. Ades, J. C. Paton, J. S. Sampson, G. M. Carlone, R. C. Huebner, A. Virolainen, E. Swiatlo, and S. K. Hollingshead. 2000. Intranasal immunization of mice with a mixture of the pneumococcal proteins PsaA and PspA is highly protective against nasopharyngeal carriage of *Streptococcus pneumoniae. Infect. Immun.* **68:**796–800.

26. Briles, D. E., C. Forman, J. C. Horowitz, J. E. Volanakis, W. H. Benjamin, Jr., L. S. McDaniel, J. Eldridge, and J. Brooks. 1989. Antipneumococcal effects of C-reactive protein and monoclonal antibodies to pneumococcal cell wall and capsular antigens. *Infect. Immun.* **57:**1457–1464.

27. Briles, D. E., S. K. Hollingshead, J. King, A. Swift, P. A. Braun, M. K. Park, L. M. Ferguson, M. H. Nahm, and G. S. Nabors. 2000. Immunization of humans with rPspA elicits antibodies, which passively protect mice from fatal infection with *Streptococcus pneumoniae* bearing heterologous PspA. *J. Infect. Dis.* **182:**1694–1701.

28. Briles, D. E., S. K. Hollingshead, J. C. Paton, E. W. Ades, L. Novak, F. W. van Ginkel, and W. H. J. Benjamin. 2003. Immunizations with pneumococcal surface protein A and pneumolysin are protective against pneumonia in a murine model of pulmonary infection with *Streptococcus pneumoniae. J. Infect. Dis.* **188:**339–348.

29. Briles, D. E., S. K. Hollingshead, E. Swiatlo, A. Brooks-Walter, A. Szalai, A. Virolainen, L. S. McDaniel, K. A. Benton, P. White, K. Prellner, A. Hermansson, P. C. Aerts, H. Van Dijk, and M. J. Crain. 1997. PspA and PspC: their potential for use as pneumococcal vaccines. *Microb. Drug Resist.* **3:**401–408.

30. Briles, D. E., G. S. Nabors, A. Brooks-Walter, J. Paton, C., and S. Hollingshead. 2001. The potential for using protein vaccines to protect against otitis media caused by *Streptococcus pneumoniae. Vaccine* **19:**S87–S95.

31. Briles, D. E., M. Nahm, K. Schroer, J. Davie, P. Baker, J. Kearney, and R. Barletta. 1981. Antiphosphocholine antibodies found in normal mouse serum are protective against intravenous infection with type 3 *Streptococcus pneumoniae. J. Exp. Med.* **153:**694–705.

32. Briles, D. E., J. C. Paton, and S. K. Hollingshead. 2004. Pneumococcal common proteins and other vaccine strategies, p. 459–469. *In* M. M. Levine, J. B. Kaper, R. Rappuoli, M. Liu, and M. F. Good (ed.), *New Generation Vaccines*, 3rd ed. Marcel Dekker, Inc., New York, N.Y.

33. Briles, D. E., J. C. Paton, M. H. Nahm, and E. Swiatlo. 2000. Immunity to *Streptococcus pneumoniae*, p. 263–280. *In* M. Cunningham and R. S. Fujinami (ed.), *Effect of Microbes on the Immune System.* Lippincott-Raven, Philadelphia, Pa.

34. Brooks-Walter, A., D. E. Briles, and S. K. Hollingshead. 1999. The *pspC* gene of *Streptococcus pneumoniae* encodes

a polymorphic protein PspC, which elicits cross-reactive antibodies to PspA and provides immunity to pneumococcal bacteremia. *Infect. Immun.* **67:**6533–6542.

35. **Buchanan, R. M., D. E. Briles, B. P. Arulanandam, M. A. Westerink, R. H. Raeder, and D. W. Metzger.** 2001. IL-12-mediated increases in protection elicited by pneumococcal and meningococcal conjugate vaccines. *Vaccine* **19:**2020–2028.

36. **Butler, J. C.** 2004. Epidemiology of pneumococcal disease, p. 148–168. *In* E. I. Tuomanen (ed.), *The Pneumococcus.* ASM Press, Washington, D.C.

37. **Butler, J. C., R. F. Breiman, J. F. Campbell, H. B. Lipman, C. V. Broome, and R. R. Facklam.** 1993. Pneumococcal polysaccharide vaccine efficacy. An evaluation of current recommendations. *JAMA* **270:**1826–1831.

38. **Canvin, J. R., A. P. Marvin, M. Sivakumaran, J. C. Paton, G. J. Boulonois, P. W. Andrew, and T. J. Mitchell.** 1995. The role of pneumolysin and autolysin in the pathology of pneumoniae and septicemia in mice infected with a type 2 pneumococcus. *J. Infect. Dis.* **172:**119–123.

39. **Chiavolini, D., G. Memmi, T. Maggi, F. Iannelli, G. Pozzi, and M. R. Oggioni.** 2003. The three extra-cellular zinc metalloproteinases of *Streptococcus pneumoniae* have a different impact on virulence in mice. *BMC Microbiol.* **3:**14.

40. **Coral, M. C. V., N. Fonseca, E. Castaneda, J. L. Di Fabio, S. K. Hollingshead, and D. E. Briles.** 2001. Families of pneumococcal surface protein A (PspA) of *Streptococcus pneumoniae* invasive isolates recovered from Colombian children. *Emerg. Infect. Dis.* **7:**832–836.

41. **Cornu, C., D. Yzebe, P. Leophonte, J. Gaillat, J. P. Boissel, and M. Cucherat.** 2001. Efficacy of pneumococcal polysaccharide vaccine in immunocompetent adults: a meta-analysis of randomized trials. *Vaccine* **19:**4780–4790.

42. **Crain, M. J., W. D. Waltman, J. S. Turner, J. Yother, D. E. Talkington, L. M. McDaniel, B. M. Gray, and D. E. Briles.** 1990. Pneumococcal surface protein A (PspA) is serologically highly variable and is expressed by all clinically important capsular serotypes of *Streptococcus pneumoniae.* *Infect. Immun.* **58:**3293–3299.

43. **Dagan, R., R. Melamed, M. Muallem, L. Piglansky, D. Greenburg, O. Abramson, P. M. Mendalman, N. Bohidar, and P. Yagupsky.** 1996. Reduction of nasopharyngeal carriage of pneumococci during the second year of life by a heptavalent conjugate pneumococcal vaccine. *J. Infect. Dis.* **174:**1271–1278.

44. **Dullforce, P., D. C. Sutton, and A. W. Heath.** 1998. Enhancement of T cell-independent immune responses in vivo by CD40 antibodies. *Nat. Med.* **4:**88–91.

45. **Eskola, J., S. Black, and H. Shinefield.** 2004. Pneumococcal conjugate vaccines, p. 589. *In* S. A. Plotkin and W. A. Orenstein (ed.), *Vaccines.* W. B. Saunders, Philadelphia, Pa.

46. **Eskola, J., T. Kilpi, A. Palmu, J. Jokinen, J. Haapakoski, E. Herva, A. Takala, H. Kayhty, P. Karma, R. Kohberger, G. Siber, and P. H. Makela.** 2001. Efficacy of a pneumococcal conjugate vaccine against acute otitis media. *N. Engl. J. Med.* **344:**403–409.

47. **Eskola, J., A. K. Takala, and H. Kayhty.** 1993. *Haemophilus influenzae* type b polysaccharide-protein conjugate vaccines in children. *Curr. Opin. Pediatr.* **5:**55–59.

48. **Esposito, S., R. Droghetti, N. Faelli, A. Lastrico, C. Tagliabue, L. Cesati, C. Bianchi, and N. Principi.** 2003. Serum concentrations of pneumococcal anticapsular antibodies in children with pneumonia associated with *Streptococcus pneumonia* infection. *Clin. Infect. Dis.* **37:**1261–1264.

49. **Fedson, D. S., J. Anthony, and G. Scott.** 1999. The burden of pneumococcal disease among adults in developed and developing countries: what is and is not known. *Vaccine* **17**(Suppl.)**1:**S11–S18.

50. **Fedson, D. S., and D. M. Musher.** 2004. Pneumococcal polysaccharide vaccine, p. 529–588. *In* S. A. Plotkin and W. A. Orenstein (ed.), *Vaccines.* W. B. Saunders, Philadelphia, Pa.

51. **Fedson, D. S., E. D. Shapiro, F. M. LaForce, M. A. Mufson, D. M. Musher, J. S. Spika, R. F. Breiman, and C. V. Broome.** 1994. Pneumococcal vaccine after 15 years of use. Another view. *Arch. Intern. Med.* **154:**2531–2535.

52. **Garcia-Suarez Mdel, M., M. D. Cima-Cabal, N. Florez, P. Garcia, R. Cernuda-Cernuda, A. Astudillo, F. Vazquez, J. R. De los Toyos, and F. J. Mendez.** 2004. Protection against pneumococcal pneumonia in mice by monoclonal antibodies to pneumolysin. *Infect. Immun.* **72:**4534–4540.

53. **Ghaffar, F., T. Barton, J. Lozano, L. S. Muniz, P. Hicks, V. Gan, N. Ahmad, and G. H. McCracken, Jr.** 2004. Effect of the 7-valent pneumococcal conjugate vaccine on nasopharyngeal colonization by *Streptococcus pneumoniae* in the first 2 years of life. *Clin. Infect. Dis.* **39:**930–938.

54. **Goldenberg, H. B., T. L. McCool, and J. N. Weiser.** 2004. Cross-reactivity of human immunoglobulin G2 recognizing phosphorylcholine and evidence for protection against major bacterial pathogens of the human respiratory tract. *J. Infect. Dis.* **190:**1254–1263.

55. **Gor, D. O., D. Xuedong, D. E. Briles, M. R. Jacobs, and N. S. Greenspan.** Relationship between surface accessibility for PpmA, PsaA, PspA and antibody mediated immunity to systemic infection by *Streptococcus pneumoniae. Infect. Immun.*, in press.

56. **Gray, B. M., G. M. Converse III, and H. C. Dillon.** 1979. Serotypes of *Streptococcus pneumoniae* causing disease. *J. Infect. Dis.* **140:**979–983.

57. **Hamel, J., N. Charland, I. Pineau, C. Ouellet, S. Rioux, D. Martin, and B. R. Brodeur.** 2004. Prevention of pneumococcal disease in mice immunized with conserved surface-accessible proteins. *Infect. Immun.* **72:**2659–2670.

58. **Hammerschmidt, S., S. R. Talay, P. Brandtzaeg, and G. S. Chhatwal.** 1997. SpsA, a novel pneumococcal surface protein with specific binding to secretory immunoglobulin A and secretory component. *Mol. Microbiol.* **25:**1113–1124.

59. **Hava, D. L., and A. Camilli.** 2002. Large-scale identification of serotype 4 *Streptococcus pneumoniae* virulence factors. *Mol. Microbiol.* **45:**1389–1406.

60. **Heffron, R.** 1939. *Pneumonia.* The Commonwealth Fund, New York, N.Y.

61. **Henrichsen, J.** 1995. Six newly recognized types of *Streptococcus pneumoniae. J. Clin. Microbiol.* **33:**2759–2762.

62. **Hollingshead, S. K., R. S. Becker, and D. E. Briles.** 2000. Diversity of PspA: mosaic genes and evidence for past recombination in *Streptococcus pneumoniae. Infect. Immun.* **68:**5889–5900.

63. **Huo, Z., O. Spencer, J. Miles, J. Johnson, R. Holliman, J. Sheldon, and P. Riches.** 2004. Antibody response to pneumolysin and to pneumococcal capsular polysaccharide in healthy individuals and *Streptococcus pneumoniae* infected patients. *Vaccine* **22:**1157–1161.

64. **Iannelli, F., D. Chiavolini, S. Ricci, M. R. Oggioni, and G. Pozzi.** 2004. Pneumococcal surface protein C contributes to sepsis caused by *Streptococcus pneumoniae* in mice. *Infect. Immun.* **72:**3077–3080.

65. **Iannelli, F., M. R. Oggioni, and G. Pozzi.** 2002. Allelic variation in the highly polymorphic locus pspC of *Streptococcus pneumoniae. Gene* **284:**63–71.

66. Janulczyk, R., F. Iannelli, A. G. Sjoholm, G. Pozzi, and L. Bjorck. 2000. Hic, a novel surface protein of *Streptococcus pneumoniae* that interferes with complement function. *J. Biol. Chem.* **275:**37257–37263.

67. Johnston, J. W., L. E. Myers, M. M. Ochs, W. H. Benjamin, Jr., D. E. Briles, and S. K. Hollingshead. 2004. Lipoprotein PsaA in virulence of *Streptococcus pneumoniae*: surface accessibility and role in protection from superoxide. *Infect. Immun.* **72:**5858–5867.

68. Kadioglu, A., W. Coward, M. J. Colston, C. R. Hewitt, and P. W. Andrew. 2004. CD4-T-lymphocyte interactions with pneumolysin and pneumococci suggest a crucial protective role in the host response to pneumococcal infection. *Infect. Immun.* **72:**2689–2697.

69. Kaplan, S. L., and E. O. Mason, Jr. 2002. Mechanisms of pneumococcal antibiotic resistance and treatment of pneumococcal infections in 2002. *Pediatr. Ann.* **31:**250–260.

70. Kenny, J. J., G. Guelde, R. T. Fisher, and D. L. Longo. 1994. Induction of phosphocholine-specific antibodies in X-linked immune deficient mice: in vivo protection against a *Streptococcus pneumoniae* challenge. *Int. Immunol.* **6:**561–568.

71. Kuo, J., M. Douglas, H. K. Ree, and A. A. Lindberg. 1995. Characterization of a recombinant pneumolysin and its use as a protein carrier for pneumococcal type 18C conjugate vaccine. *Infect. Immun.* **63:**2706–2713.

72. Lee, C.-J., R. A. Lock, P. W. Andrew, T. J. Mitchell, D. Hansman, and J. C. Paton. 1994. Protection in infant mice from challenge with *Streptococcus pneumoniae* type 19F by immunization with a type 19F polysaccharide-pneumolysoid conjugate. *Vaccine* **12:**875–877.

73. Lock, R. A., D. Hansman, and J. C. Paton. 1992. Comparative efficacy of autolysin and pneumolysin as immunogens protecting mice against infection by *Streptococcus pneumoniae*. *Microb. Pathog.* **12:**137–143.

74. Lock, R. A., Q. Q. Zhang, A. M. Berry, and J. C. Paton. 1996. Sequence variation in the *Streptococcus pneumoniae* pneumolysin gene affecting haemolytic activity and electrophoretic mobility of the toxin. *Microb. Pathog.* **21:**71–83.

75. Long, J. P., H. H. Tong, and T. F. DeMaria. 2004. Immunization with native or recombinant *Streptococcus pneumoniae* neuraminidase affords protection in the chinchilla otitis media model. *Infect. Immun.* **72:**4309–4313.

76. Lopez, R., and E. Garcia. 2004. Recent trends on the molecular biology of pneumococcal capsules, lytic enzymes, and bacteriophage. *FEMS Microbiol. Rev.* **28:**553–580.

77. Malley, R., P. Henneke, S. C. Morse, M. J. Cieslewicz, M. Lipsitch, C. M. Thompson, E. Kurt-Jones, J. C. Paton, M. R. Wessels, and D. T. Golenbock. 2003. Recognition of pneumolysin by Toll-like receptor 4 confers resistance to pneumococcal infection. *Proc. Natl. Acad. Sci. USA* **100:**1966–1971.

78. Mbelle, N., R. E. Huebner, A. D. Wasas, A. Kimura, I. Chang, and K. P. Klugman. 1999. Immunogenicity and impact on nasopharyngeal carriage of a nonavalent pneumococcal conjugate vaccine. *J. Infect. Dis.* **180:**1171–1176.

79. McCormick, A. W., C. G. Whitney, M. M. Farley, R. Lynfield, L. H. Harrison, N. M. Bennett, W. Schaffner, A. Reingold, J. Hadler, P. Cieslak, M. H. Samore, and M. Lipsitch. 2003. Geographic diversity and temporal trends of antimicrobial resistance in *Streptococcus pneumoniae* in the United States. *Nat. Med.* **9:**424–430.

80. McDaniel, L. S., W. H. Benjamin, Jr., C. Forman, and D. E. Briles. 1984. Blood clearance by anti-phosphocholine antibodies as a mechanism of protection in experimental pneumococcal bacteremia. *J. Immunol.* **133:**3308–3312.

81. McDaniel, L. S., D. O. McDaniel, S. K. Hollingshead, and D. E. Briles. 1998. Comparison of the PspA sequence from *Streptococcus pneumoniae* EF5668 to the previously identified PspA sequence from strain Rx1 and ability of PspA from EF5668 to elicit protection against pneumococci of different capsular types. *Infect. Immun.* **66:**4748–4754.

82. McDaniel, L. S., B. A. Ralph, D. O. McDaniel, and D. E. Briles. 1994. Localization of protection-eliciting epitopes on PspA of *Streptococcus pneumoniae* between amino acid residues 192 and 260. *Microb. Pathog.* **17:**323–337.

83. McDaniel, L. S., J. S. Sheffield, E. Swiatlo, J. Yother, M. J. Crain, and D. E. Briles. 1992. Molecular localization of variable and conserved regions of *pspA*, and identification of additional *pspA* homologous sequences in *Streptococcus pneumoniae*. *Microb. Pathog.* **13:**261–269.

84. McDaniel, L. S., J. Yother, M. Vijayakumar, L. McGarry, W. R. Guild, and D. E. Briles. 1987. Use of insertional inactivation to facilitate studies of biological properties of pneumococcal surface protein A (PspA). *J. Exp. Med.* **165:**381–394.

85. Michon, F., P. C. Fusco, C. A. S. A. Minetti, M. Laude-Sharp, S. Moore, D. P. Remeta, I. Heron, and M. S. Blake. 1998. Multivalent pneumococcal capsular polysaccharide conjugate vaccines employing genetically detoxified pneumolysin as a carrier protein. *Vaccine* **16:**1732–1741.

86. Mirza, S., S. K. Hollingshead, J. H. Williams, Jr., and D. E. Briles. 2004. PspA protects *Streptococcus pneumoniae* from killing by apolactoferrin and antibody to PspA enhances killing of pneumococci by apolactoferrin. *Infect. Immun.* **72:**5031–5040.

87. Mitchell, T. J., P. W. Andrew, F. K. Saunders, A. N. Smith, and G. J. Boulnois. 1991. Complement activation and antibody binding by pneumolysin via a region of the toxin homologous to a human acute-phase protein. *Mol. Microbiol.* **5:**1883–1888.

88. Mitchell, T. J., F. Mendez, J. C. Paton, P. W. Andrew, and G. J. Boulnois. 1990. Comparison of pneumolysin genes and proteins from *Streptococcus pneumoniae* types 1 and 2. *Nucleic Acids Res.* **18:**4010.

89. Musher, D. M., H. M. Phan, and R. E. Baughn. 2001. Protection against bacteremic pneumococcal infection by antibody to pneumolysin. *J. Infect. Dis.* **183:**827–830.

90. Nabors, G. S., P. A. Braun, D. J. Herrmann, M. L. Heise, D. J. Pyle, S. Gravenstein, M. Schilling, L. M. Ferguson, S. K. Hollingshead, D. E. Briles, and R. S. Becker. 2000. Immunization of healthy adults with a single recombinant pneumococcal surface protein A (PspA) variant stimulates broadly cross-reactive antibodies. *Vaccine* **18:**1743–1754.

91. Nieminen, T., J. Eskola, and H. Kayhty. 1998. Pneumococcal conjugate vaccination in adults: circulating antibody secreting cell response and humoral antibody responses in saliva and in serum. *Vaccine* **16:**630–636.

92. Oggioni, M. R., G. Memmi, T. Maggi, D. Chiavolini, F. Iannelli, and G. Pozzi. 2003. Pneumococcal zinc metalloproteinase ZmpC cleaves human matrix metalloproteinase 9 and is a virulence factor in experimental pneumonia. *Mol. Microbiol.* **49:**795–805.

93. Ogunniyi, A. D., R. L. Folland, D. B. Briles, S. K. Hollingshead, and J. C. Paton. 2000. Immunization of mice with combinations of pneumococcal virulence proteins elicits enhanced protection against challenge with *Streptococcus pneumoniae*. *Infect. Immun.* **68:**3028–3033.

94. Overturf, G. D. 2002. Pneumococcal vaccination of children. *Semin. Pediatr. Infect. Dis.* **13:**155–164.

95. Overweg, K., A. Kerr, M. Sluijter, M. H. Jackson, T. J. Mitchell, A. P. de Jong, R. de Groot, and P. W. Hermans. 2000. The putative proteinase maturation protein A of *Streptococcus pneumoniae* is a conserved surface protein with potential to elicit protective immune responses. *Infect. Immun.* **68:**4180–4188.

96. Overweg, K., C. D. Pericone, G. G. C. Verhoef, G. N. Weiser, H. D. Meiring, A. P. J. M. De Jong, R. De Groot, and P. W. M. Hermans. 2000. Differential protein expression in phenotypic variants of *Streptococcus penumoniae*. *Infect. Immun.* **68:**4604–4610.

97. Paton, J. C. 1996. The contribution of pneumolysin to the pathogenicity of *Streptococcus pneumoniae*. *Trends Microbiol.* **4:**103–106.

98. Paton, J. C., P. W. Andrew, G. J. Boulnois, and T. J. Mitchell. 1993. Molecular analysis of the pathogenicity of *Streptococcus pneumoniae*: the role of pneumococcal proteins. *Annu. Rev. Microbiol.* **47:**89–115.

99. Paton, J. C., A. M. Berry, and R. A. Lock. 1997. Molecular analysis of putative pneumococcal virulence proteins. *Microb. Drug. Resist.* **3:**3–10.

100. Paton, J. C., R. A. Lock, and D. C. Hansman. 1983. Effect of immunization with pneumolysin on survival time of mice challenged with *Streptococcus pneumoniae*. *Infect. Immun.* **40:**548–552.

101. Paton, J. C., R. A. Lock, C.-J. Lee, J. P. Li, A. M. Berry, T. J. Mitchell, P. W. Andrew, D. Hansman, and G. J. Bulnois. 1991. Purification and immunogenicity of genetically obtained pneumolysin toxoids and their conjugation to *Streptococcus pneumoniae* type 19F polysaccharide. *Infect. Immun.* **59:**2297–2304.

102. Paton, J. C., B. Rowan-Kelly, and A. Ferrante. 1984. Activation of human complement by the pneumococcal toxin pneumolysin. *Infect. Immun.* **43:**1085–1087.

103. Powers, D. C., E. L. Anderson, K. Lottenbach, and C. M. Mink. 1996. Reactogenicity and immunogenicity of a protein-conjugated pneumococcal oligosaccharide vaccine in older adults. *J. Infect. Dis.* **173:**1014–1018.

104. Rapola, S., V. Jantti, R. Haikala, R. Syrjanen, G. M. Carlone, J. S. Sampson, D. E. Briles, J. C. Paton, A. K. Takala, T. M. Kilpi, and H. Kayhty. 2000. Natural development of antibodies to pneumococcal surface protein A, pneumococcal surface adhesin A, and pneumolysin in relation to pneumococcal carriage and acute otitis media. *J. Infect. Dis.* **182:**1146–1152.

105. Regev-Yochay, G., R. Dagan, M. Raz, Y. Carmeli, B. Shainberg, E. Derazne, G. Rahav, and E. Rubinstein. 2004. Association between carriage of *Streptococcus pneumoniae* and *Staphylococcus aureus* in children. *JAMA* **292:**716–720.

106. Ren, B., A. J. Szalai, S. K. Hollingshead, and D. E. Briles. 2004. Effects of PspA and antibodies to PspA on activation and deposition of complement on the pneumococcal surface. *Infect. Immun.* **72:**114–122.

107. Ren, B., A. J. Szalai, O. Thomas, S. K. Hollingshead, and D. E. Briles. 2003. Both family 1 and family 2 PspAs can inhibit complement deposition and confer virulence to a capsular 3 serotype *Streptococcus pneumoniae*. *Infect. Immun.* **71:**75–85.

108. Rijkers, G. T., E. A. Sanders, M. A. Breukels, and B. J. Zegers. 1998. Infant B cell responses to polysaccharide determinants. *Vaccine* **16:**1396–1400.

109. Rijkers, G. T., E. A. M. Sanders, M. A. Breukels, and B. J. M. Zegers. 1996. Responsiveness of infants to cap-

110. Robbins, J. B., R. Austrian, C.-J. Lee, S. C. Rastogi, G. Schiffman, J. Henrichsen, P. H. Makela, C. V. Broome, R. R. Facklam, R. H. Tiesjema, and J. C. Parke, Jr. 1983. Considerations for formulating the second-generation pneumococcal capsular polysaccharide vaccine with emphasis on the cross-reactive types within groups. *J. Infect. Dis.* **148:**1136–1159.

111. Roche, H., A. Hakansson, S. K. Hollingshead, and D. E. Briles. 2003. Regions of PspA/EF3296 best able to elicit protection against *Streptococcus pneumoniae* in a murine infection model. *Infect. Immun.* **71:**1033–1041.

112. Roche, H., B. Ren, L. S. McDaniel, A. Hakansson, and D. E. Briles. 2003. Relative roles of genetic background and variation in PspA in the ability of antibodies to PspA to protect against capsular type 3 and 4 strains of *Streptococcus pneumoniae*. *Infect. Immun.* **71:**4498–4505.

113. Rosenow, C., P. Ryan, J. N. Weiser, S. Johnson, P. Fontan, A. Ortqvist, and H. R. Masure. 1997. Contribution of novel choline-binding proteins to adherence, colonization and immunogenicity of *Streptococcus pneumoniae*. *Mol. Microbiol.* **25:**819–829.

114. Rubins, J. B., D. Charboneau, C. Fashing, A. M. Berry, J. C. Paton, J. E. Alexander, P. W. Andrew, T. J. Mitchell, and E. N. Janoff. 1996. Distinct roles for pneumolysin's cytotoxic and complement activities in pathogenesis of pneumococcal pneumonia. *Am. J. Respir. Crit. Care Med.* **153:**1339–1346.

115. Rubins, J. B., D. Charboneau, J. C. Paton, T. J. Mitchell, and P. W. Andrew. 1995. Dual function of pneumolysin in the early pathogenesis of murine pneumococcal pneumonia. *J. Clin. Investig.* **95:**142–150.

116. Rubins, J. B., A. K. G. Puri, D. Carboneau, R. MacDonald, N. Opstad, and E. N. Janoff. 1998. Magnitude, duration, quality, and function of pneumococcal vaccine responses in elderly adults. *J. Infect. Dis.* **178:**431–440.

117. Sankilampi, U., P. O. Honkanen, A. Bloigu, and M. Leinonen. 1997. Persistence of antibodies to pneumococcal capsular polysaccharide vaccine in the elderly. *J. Infect. Dis.* **176:**1100–1104.

118. Shapiro, E. D., A. T. Berg, R. Austrian, D. Schroeder, V. Parcells, A. Margolis, R. K. Adair, and J. D. Clemmens. 1991. Protective efficacy of polyvalent pneumococcal polysaccharide vaccine. *N. Engl. J. Med.* **325:**1453–1460.

119. Shaw, P. X., S. Horkko, M. K. Chang, L. K. Curtiss, W. Palinski, G. J. Silverman, and J. L. Witztum. 2000. Natural antibodies with the T15 idiotype may act in atherosclerosis, apoptotic clearance, and protective immunity. *J. Clin. Investig.* **105:**1731–1740.

120. Shelly, M. A., H. Jacoby, G. J. Riley, B. T. Graves, M. Pichichero, and J. J. Treanor. 1997. Comparison of pneumococcal polysaccharide and CRM_{197} conjugated pneumococcal oligosaccharide vaccines in young and elderly adults. *Infect. Immun.* **65:**242–247.

121. Siber, G. R. 1994. Pneumococcal disease: prospects for a new generation of vaccines. *Science* **265:**1385–1387.

122. Tong, H. H., L. E. Blue, M. A. James, and T. F. DeMaria. 2000. Evaluation of the virulence of a *Streptococcus pneumoniae* neuraminidase-deficient mutant in nasopharyngeal colonization and development of otitis media in the chinchilla model. *Infect. Immun.* **68:**921–924.

123. Virolainen, A., W. Russell, M. J. Crain, S. Rapola, H. Kaythy, and D. E. Briles. 2000. Human antibodies to

pneumococcal surface protein A, PspA. *Pediatr. Infect. Dis. J.* **19:**134–138.

124. **White, B.** 1938. *The Biology of Pneumococcus.* The Commonwealth Fund, New York, N.Y.

125. **Whitney, C. G., M. M. Farley, J. Hadler, L. H. Harrison, N. M. Bennett, R. Lynfield, A. Reingold, P. R. Cieslak, T. Pilishvili, D. Jackson, R. R. Facklam, J. H. Jorgensen, and A. Schuchat.** 2003. Decline in invasive pneumococcal disease after the introduction of protein-polysaccharide conjugate vaccine. *N. Engl. J. Med.* **348:** 1737–1746.

126. **Wu, H.-Y., M. Nahm, Y. Guo, M. Russell, and D. E. Briles.** 1997. Intranasal immunization of mice with PspA (pneumococcal surface protein A) can prevent intranasal carriage, pulmonary infection, and sepsis with *Streptococcus pneumoniae. J. Infect. Dis.* **175:**839–846.

127. **Yother, J., and D. E. Briles.** 1992. Structural properties and evolutionary relationships of PspA, a surface protein of *Streptococcus pneumoniae*, as revealed by sequence analysis. *J. Bacteriol.* **174:**601–609.

128. **Yother, J., and J. M. White.** 1994. Novel surface attachment mechanism for the *Streptococcus pneumoniae* protein PspA. *J. Bacteriol.* **176:**2976–2985.

129. **Zysk, G., G. Bethe, R. Nau, D. Koch, V. C. Grafin Von Bassewitz, H. P. Heinz, and R. R. Reinert.** 2003. Immune response to capsular polysaccharide and surface proteins of *Streptococcus pneumoniae* in patients with invasive pneumococcal disease. *J. Infect. Dis.* **187:**330–333.

E. ENTEROCOCCI

Pathogenicity of Enterococci

LYNN E. HANCOCK AND MICHAEL S. GILMORE

25

INTRODUCTION

The origin of the term "enterococcus" dates to the end of the 19th century, when Thiercelin described a saprophytic gram-positive coccus of intestinal origin capable of pathogenesis (128, 129). In the same year, a more detailed picture of enterococcal pathogenesis was reported by MacCallum and Hastings (80), who isolated and characterized an organism (now known to be a cytolytic *Enterococcus faecalis*) from a lethal case of acute endocarditis. From these early descriptions, the paradigm of enterococcal pathogenesis as that of a commensal opportunist has been established. Now more than a century later, enterococci are prominent nosocomial pathogens (109).

Infections caused by the genus *Enterococcus* (most notably *E. faecalis* and *Enterococcus faecium*, which account for ~95% of all infections) include urinary tract infections (UTI), bacteremia, intraabdominal infections, and endocarditis (62, 84). Enterococci are now the leading cause of surgical site infection, rank second only to staphylococci as a cause of hospital-acquired bacteremia, and rank as the third leading cause of nosocomial UTI (109). In addition to nosocomial infections, enterococci are responsible for 5 to 20% of community-acquired endocarditis (82). Disease caused by enterococci is compounded by the reduced ability to treat enterococcal infections due to their intrinsic and acquired antibiotic resistance.

A number of studies have examined resistance trends for infections caused by *E. faecalis* and *E. faecium* (62, 143). Data compiled for 15,000 isolates over a 3-year period (1995 to 1997) showed that although resistance to ampicillin and vancomycin is relatively uncommon (<2%) among *E. faecalis* isolates, *E. faecium* showed a general trend towards increasing resistance to both ampicillin (83%) and vancomycin (52%). A more recent study of nosocomial bloodstream infections over a 7-year period from 1995 to 2002 showed that enterococci account for 9% of bloodstream infections (143). Somewhat surprisingly, the percentage of *E. faecalis* isolates resistant to vancomycin continued to remain low (~2%), whereas more than 60% of *E. faecium* isolates are now vancomycin resistant (143). His-

torically, an isolate with more antibiotic resistances was thought to equate to greater virulence and an increase in length of hospital stay, cost, and mortality (23). More recent studies that adjusted for severity of illness, however, found no correlation between the status of antibiotic resistance and an increase in mortality risk (34, 74). Resistance alone cannot serve as a predictor of disease frequency and outcome. *E. faecalis*, which remains sensitive to vancomycin and ampicillin, continues to be the most frequently encountered enterococcal isolate, accounting for 79% of enterococcal infections (62). Reasons for the disparity in the number of infections caused by *E. faecalis* and *E. faecium* are not well known, although the ratio of *E. faecalis* to *E. faecium* infections appears to be decreasing in recent years (69), likely attributable to the prevalence of vancomycin resistance in *E. faecium* as compared to *E. faecalis*. The emergence of more pathogenic *E. faecium* lineages has also been described (13, 76, 142), and it remains unclear as to whether resistance drives the evolution of pathogenic traits, or whether these events happen independently. One explanation for the overrepresentation of *E. faecalis* among clinical isolates may simply relate to natural abundance at commensal sites. Several studies indicate that *E. faecalis* is both more prevalent and abundant in the human gastrointestinal tract (26, 96) compared to *E. faecium*. In one such study (96), the numbers of *E. faecalis* were on average 100-fold higher than those of *E. faecium*. In another study (26), although the mean numbers of bacteria per gram of feces were not significantly different, it was found that *E. faecalis* was more prevalent in the population, being detected in 82% of subjects, whereas *E. faecium* was only found in 36% of subjects. An alternative explanation is that *E. faecalis* is intrinsically more pathogenic (62, 88) (Table 1). The emergence of the complete genome sequence for *E. faecalis* V583 (99) and the near complete sequence for *E. faecium* TX0016 should provide additional clues as to the common ground shared by these organisms and enable a clearer understanding of how these organisms collectively cause disease.

This review focuses on the mechanisms by which enterococci cause human disease, with particular attention to advances in the field since the last edition of this text (48).

Gram-Positive Pathogens, 2nd edition, edited by Vincent A. Fischetti et al.
© 2006 ASM Press, Washington, D.C.

TABLE 1 Enterococcal virulence factors

Virulence factor	Description	Reference(s)
Cytolysin[a]	Hemolysin/bacteriocin; lyses a broad range of eukaryotic and gram-positive cells	10, 11, 64, 67
Gelatinase[a]	Secreted zinc metalloprotease	24, 105, 119
Serine protease[a]	Secreted serine protease	24, 105, 119
Hyaluronidase[b]	Degrades hyaluronic acid	107
Esp[a, b]	Cell wall-anchored protein; enhances biofilm formation and colonization of bladder epithelium	76, 116
Aggregation substance[a,b]	Cell wall-anchored protein; involved in conjugation and adhesion to eukaryotic cells	10, 54, 70, 114
Ace[a,b]	Cell wall-anchored protein; collagen-binding protein	92, 94, 108, 131
Epa[a,b]	Enterococcal polysaccharide antigen; antiphagocytic cell wall polysaccharide	126
Cps[a,b]	Capsular polysaccharide; antiphagocytic cell wall polysaccharide	47, 56
LTA[a,b]	Lipoteichoic acid; enterococcal group antigen; binding substance for conjugation	114, 135
Toxic metabolites[a,b]	Reactive oxygen species; extracellular superoxide, hydrogen peroxide	60, 87

[a]Virulence factors identified in *Enterococcus faecalis*.
[b]Virulence factors identified in *Enterococcus faecium*.

Because of chapter format, only representative works will be cited. More comprehensive citations may be found in several recent reviews on the topic of enterococcal pathogenesis (41, 102, 125), as well as a complete volume on enterococcal biology (40).

Biology and Epidemiology

Enterococci were formerly classified as group D streptococci (72), but were given genus status in 1984 (113), based on nucleic acid hybridization studies, which showed a more distant relationship to the streptococci.

Enterococci are considered commensals of the gastrointestinal tract of a variety of organisms, including humans. They are found in a number of environments, due to dissemination in animal excrement and environmental persistence. Members of this genus are endowed with intrinsic properties that contribute to their ability to adopt a commensal lifestyle in the harsh confines of the intestinal tract, as well as aid in their ability to survive, persist, and spread in the hospital. Enterococci are able to grow over wide temperature and pH ranges, survive dessication, and grow in the presence of 6.5% NaCl and 40% bile salts (40).

Environmental Persistence

Epidemiology involving enterococci has proven person-to-person transmission in the hospital (9, 63), suggesting an ability to persist for extended periods in the hospital environment. This transmission typically occurs via the hands of health care personnel, but has also been shown to involve inanimate objects such as bedrails, nursing station keyboards, hospital drapery, and clinical instruments, such as ear-probe thermometers (103). The ability to survive nutrient-poor environments, as well as dessication, has led some to suggest that enterococci enter a viable but nonculturable state as an adaptation to these poor growth conditions (25, 53, 77), but the mechanisms of how enterococci adapt to exist in such a state are still poorly understood. Several studies have also examined the nature of the environmental ruggedness of *E. faecalis* (27–30, 37–39, 73, 110) . *E. faecalis* was adapted to the presence of lethal levels of bile salts and detergents, such as sodium dodecyl sulfate, when first cultured under sublethal levels (27). The ability of enterococci to adapt and persist in the presence of detergents may allow them to survive typical cleaning regimens employed in most hospitals as part of an infection control program. More re-

cently, regulatory systems have been identified that appear to allow an adaptive response to environmental insults, such as heat and detergent stresses (sodium dodecyl sulfate) (50, 75). In a study by Hancock and Perego (50), inactivation of a response regulator, designated RR06, led to an increased sensitivity to heat stress (growth at 46°C), as well as detergent stress (0.003% sodium dodecyl sulfate). Le Breton et al. (75) showed that another response regulator, RR10, negatively regulates the heat shock proteins DnaK and GroEL. Inactivation of the *rr*10 gene resulted in a more acid-sensitive phenotype, but also resulted in enhanced survival at 50°C. Clearly, enterococci possess the ability to alter gene expression through mechanisms involving environmental sensing and the subsequent relay of information to the appropriate adaptive response. Genome scanning of the prepublished *E. faecium* genome has allowed the identification of close orthologs of these regulatory systems in *E. faecium*, suggesting that the response to environmental stresses is likely shared by all enterococci.

PATHOGENIC MECHANISMS

For enterococci to cause disease several barriers must first be overcome. An initial barrier that these organisms confront is the ability to colonize the intestinal tract, where they must compete for nutrient resources in an intestinal milieu including several hundred unique bacterial species. From the site of colonization, the organism must translocate to infectious sites, evade host clearance, and ultimately produce pathologic changes in the host through direct toxic activity, or indirectly by inducing an inflammatory response (68).

Colonization and Translocation

Enterococci normally colonize the gastrointestinal tract of humans. They are found in relative abundance in human feces (10^5 to 10^7 organisms per gram) (96), but represent less than 1% of the total microbial population, which is primarily composed of anaerobic microflora (123). This barrier to colonization, sometimes referred to as "colonization resistance," keeps the number of enterococci in the gut at relatively innocuous levels. It is thought that the production of metabolic byproducts, such as short-chain fatty acids (butyric acid), by obligate anaerobes is inhibitory to facultative anaerobes, such as enterococci (123). Another barrier to colonization is the gastric pH, which is inhospitable for most microorganisms, including enterococci. In essence, a pH gradient is established from the stomach to the lower bowel, and it is only near the large intestine (colon) that a more neutral pH range is observed to allow growth of microorganisms. In general, these obstacles to colonization keep enterococci confined to life as a commensal in the colon; it is only when these barriers are removed, typically by medical intervention, that enterococci move from commensal to rogue opportunist. In the intensive care unit setting in particular, patients are placed on H2-receptor antagonists as prophylaxis for treating stress ulcers; this action results in a gastric pH increase from pH 2 to pH 3.5 to 5.3, depending on the antagonist employed (16). In a study by Basaran et al. (3), H2-receptor antagonists were shown to promote colonization of the small bowel, as well as translocation of enteric bacteria from the intestinal tract to extraintestinal tissues. This study would also suggest that access to the small bowel by enteric bacteria promotes their translocation.

The barrier of colonization resistance can also be removed by medical intervention through the use of broad-spectrum antibiotics that possess little or no antienterococcal activity (17, 18, 20). In a study designed to examine the persistence and density of colonization by vancomycin-resistant E. faecium in a murine model, Donskey et al. (19) showed that suppression of the anaerobic microflora by the use of antimicrobials that target this microbial population resulted in increased colonization density and prolonged persistence of vancomycin-resistant E. faecium. This observation again highlights the importance of the anaerobic flora in suppressing enterococcal growth within the intestinal microenvironment. As prior antibiotic therapy appears to be a predisposing factor for enterococcal infection, multidrug-resistant isolates would appear to have an advantage in their ability to colonize a patient predisposed by antibiotic therapy. For therapies involving antibiotics that possess potent antienterococcal activity (i.e., vancomycin) for commensal strains, this may well be the case. What is somewhat puzzling is that therapy with little or no antienterococcal activity also predisposes the patient to colonization with exogenously acquired multidrug-resistant strains, suggesting that these acquired strains outcompete the native enterococcal population. The alternative explanation is that hospital strains are endowed with properties that allow them to colonize niches in the intestine that are not well suited for colonization by commensal isolates. The recently described pathogenicity island of E. faecalis encodes a bile salt hydrolase that may allow colonization of intestinal sites inhospitable for commensal strains (115). Several studies have documented patient colonization following hospital admission, and have shown that colonization with multiresistant strains is a predisposing factor for subsequent infection (62, 89). To colonize the lower bowel, enterococci must survive transit through the low pH of the stomach. Studies have examined the acid tolerance of E. faecalis (29, 122). Flahaut et al. (29) demonstrated that exposure of E. faecalis to a sublethal pH level (pH 4.8) for 15 to 30 min protected the organism from a normally lethal challenge at pH 3.2, suggesting an adaptive response to changing acid stress. Suzuki and colleagues (122) have shown that an E. faecalis mutant defective in F_1-F_0 H^+-ATPase activity was unable to grow at pH<6. The H^+-ATPase is used to regulate the cytoplasmic pH of E. faecalis by proton extrusion. This enzyme has been shown to be activated at low pH (122). More recently, Teng et al. (127) identified a two-component regulatory system, termed EtaRS, that appears

to be involved in acid stress response and virulence. Inactivation of the response regulator, EtaR, results in increased acid sensitivity and decreased virulence in a murine peritonitis model. The genes controlled by this system have yet to be identified, and it will be of interest to understand how this system senses and adapts to acid stress. Enterococci possess the ability to withstand the low gastric pH, which would facilitate colonization. This attribute may be critical in the ability of multidrug-resistant enterococcal strains to colonize the intestinal tract and cause hospital ward outbreaks. Whether infection-derived enterococcal isolates from infections show enhanced acid tolerance is yet to be determined.

In an effort to identify conditions that would promote enterococcal translocation, Wells et al. induced intestinal E. faecalis overgrowth by antibiotic treatment and demonstrated that organisms can adhere to epithelial surfaces of the ileum, cecum, and colon (139, 140). These same studies also showed that enterococci possess the ability to translocate from the intestinal lumen into the mesenteric lymph nodes, liver, and spleen (139, 140).

The mechanisms responsible for enterococcal translocation are not clearly defined. One hypothesis is that enterococci are phagocytosed by tissue macrophages or intestinal epithelial cells and are transported across the intestinal wall to the underlying lymph system. Failure to kill the phagocytosed organism could then lead to systemic spread (141). Gentry-Weeks et al. (36) have shown that E. faecalis possesses the ability to survive for up to 72 h in peritoneal macrophages. Olmsted et al. (98) examined the role of the plasmid-encoded surface protein, aggregation substance, in the ability of E. faecalis to be internalized by cultured intestinal epithelial cells (HT-29). The presence of aggregation substance was found to significantly augment E. faecalis internalization by HT-29 cells. However, it was also observed that in contrast to the one order of magnitude in enhanced uptake efficiency conferred by aggregation substance, three orders of magnitude of difference were observed between various enterococcal strains tested, indicative that additional unknown features of this species play major roles as determinants of internalization efficiency. To further dissect the functional domains of aggregation substance involved in internalization by HT-29 cells, Waters et al. (137, 138) examined the various domains of aggregation substance to determine their roles in this process. Surprisingly, the aggregation domain and not the N-terminal RGD motif of aggregation substance was shown to be important for internalization by HT-29 cells (138). An additional finding from this study was the apparent requirement for a cofactor (likely lipoteichoic acid [LTA]), because expression of aggregation substance in the binding substance mutant INY3000 failed to efficiently internalize. In a subsequent study it was shown that the amino-terminal aggregation domain of the protein could bind efficiently to purified LTA, as well as LTA purified from INY3000, suggesting that another cell component other than LTA was responsible for the cooperative internalization of E. faecalis by HT-29 cells (137). Because INY3000 possesses multiple Tn916 transposon insertions, further elucidation of the component required for aggregation substance-mediated internalization of E. faecalis by HT-29 cells will be required.

To identify additional enterococcal components involved in the translocation process, a recent study by Zeng et al. (146) examined the ability of E. faecalis strains to transcytose across monolayers of human colon carcinoma-derived T84 cells. In this assay, the plasmid-free strain OG1RF and eight other isolates (two endocarditis isolates, one urine isolate, and five fecal isolates) showed translocation, while six clinical isolates (three endocarditis and three urine isolates), along with strain JH2-2, and the control, *Escherichia coli* DH5α, had no detectable translocation. A correlation regarding virulence properties between the strains that could translocate and those that could not was not mentioned in this study, and it still remains uncertain as to what enterococcal factors are important in this process.

Biofilm Formation

In addition to translocation, additional mechanisms of disease pathogenesis get their start in the intestinal tract. The intestinal tract is an important reservoir for nosocomial pathogens such as enterococci and allows them access to infectious sites through other means. Disruption of the natural barriers to intestinal colonization, namely gastric acidity and colonization resistance, promotes overgrowth of the pathogens. Fecal incontinence and diarrhea among hospitalized patients contribute to the subsequent dissemination of pathogens into the hospital environment (17). From these external sources, enterococci can transiently colonize the skin of patients. Subsequent disruption of the skin barrier through surgery or placement of indwelling lines, such as urinary catheters or intravenous lines, then predisposes the patient to infection. It is estimated that greater than 60% of nosocomial infections arise from indwelling devices (112). The ability of organisms to form biofilms on indwelling devices then serves as a reservoir to continually seed organisms into the blood. The concept of biofilm formation by microorganisms is a relatively new concept that has emerged in the last decade. Biofilm formation represents a developmental switch from a free-living state, the so-called planktonic state, to a surface-attached, ordered community of organisms encased in an exopolymer matrix. This microbial lifestyle represents a therapeutic challenge because biofilms are more resistant to antimicrobials than their planktonic counterparts, and are more resistant to clearance by effectors of the host immune system (14).

Although it has been known that enterococci are capable of forming biofilms (2, 31), a more detailed picture of the factors involved in this process has only recently begun to emerge (49, 58, 71, 85, 100, 124, 130). Several studies have identified a role for the novel enterococcal surface protein, Esp, in *E. faecalis* biofilm formation (85, 124, 130). The gene encoding this protein has been localized to a recently described pathogenicity island (115), and the protein shares apparent structural similarity to a surface protein from *Staphylococcus aureus*, termed Bap (biofilm-associated protein), which has also been shown to be important in biofilm formation by *S. aureus* (15). Analysis of the relationship between the presence of the Esp-encoding gene (*esp*) and the biofilm formation capacity in *E. faecalis* demonstrated that the presence of the *esp* gene is highly associated (P < 0.0001) with the capacity of *E. faecalis* to form a biofilm on a polystyrene surface, as 93.5% of the *E. faecalis esp*-positive isolates were capable of forming a biofilm. Moreover, none of the *E. faecalis esp*-deficient isolates were biofilm producers (130). Seemingly contradictory data were also reported in the same study as insertional mutagenesis of the *esp* gene caused a complete loss of the biofilm formation phenotype in most strain backgrounds; however, in other genetic lineages no apparent phenotypic defect was observed. This suggests that although highly associated with the ability of *E. faecalis* to form biofilms, the absence of a functional Esp does not preclude isolates from forming a biofilm. The role of Esp in biofilm formation appears to affect primary attachment to

the abiotic surface, likely through hydrophobic interactions, because Esp was found to increase the cell's hydrophobicity as reflected by partitioning with n-hexadecane (130). A subsequent study by Tendolkar et al. (124) confirmed that the presence of Esp enhances biofilm formation by *E. faecalis* isolates by increasing the overall biomass of the biofilm. Furthermore, Esp-dependent biofilm formation appears to be medium-dependent, as glucose concentrations greater than 0.5% were required for maximal biofilm formation (124). In confirmation of a role for a carbohydrate source in biofilm formation, Hufnagel et al. (58) identified a putative sugar-binding transcriptional regulator involved in biofilm production by an *E. faecalis* isolate. Numerous studies have examined the contribution of the *fsr* quorum sensing locus in biofilm formation (49, 85, 100). These studies involved Esp-deficient isolates, so multiple routes to biofilm formation in *E. faecalis* must exist. Pillai et al. (100) also confirmed a role for glucose in the medium for enhancing biofilm formation by Fsr and gelatinase-positive isolates. Studies by Kristich et al. (71) and Hancock and Perego (49) have demonstrated that the role for *fsr* in biofilm formation is mediated primarily through the production of the secreted zinc metalloprotease, gelatinase. Hancock and Perego (49) demonstrated the rescue of a biofilm-negative strain *E. faecalis* FA2-2 to biofilm formation by the addition of purified gelatinase to the culture. The amount added covered the known physiologic range for gelatinase as has previously been determined (81), and appears to function in a dose-dependent manner. The substrate that gelatinase acts upon, which leads to biofilm formation, remains to be discovered. Waters et al. (136) recently identified a role for the protease in cell chain length formation and the degradation of misfolded surface proteins. Whether any of these processes relate to biofilm formation remains to be discovered. Clearly, biofilm formation by *E. faecalis*, as well as *E. faecium*, should remain a hot topic of the enterococcal research community for the coming years. Some questions for future consideration are: (i) Do Esp-positive *E. faecium* isolates form enhanced biofilms relative to their Esp-negative counterparts? (ii) What does the biofilm exopolymer matrix consist of, and does it relate to any of the known cell wall polysaccharides? (iii) Do factors such as Esp or gelatinase also contribute to in vivo biofilms, where free glucose concentrations would presumably be limiting?

Bacteremia

Nosocomial surveillance data for 2002 rank enterococci second only to staphylococci as the most common cause of nosocomial bacteremia, accounting for 11.8% of infections in U.S. hospitals (69). *E. faecalis* accounted for 8.3% of the total nosocomial infections or 71% of the enterococcal bloodstream infections, with the remaining 29% coming from *E. faecium*. This compares favorably with epidemiologic data from previous years in which enterococci accounted for 12.8% of all isolates (95). The translocation of enterococci across an intact intestinal epithelial barrier is thought to lead to many bacteremias with no identifiable source (66, 141). Other identifiable sources for enterococcal bacteremia include intravenous lines, abscesses, and UTIs (66). The risk factors for mortality associated with enterococcal bacteremia include severity of illness (based on APACHE II scores), patient age, and use of third-generation cephalosporins or metronidazole (121). Huycke et al. (63) showed that patients infected with hemolytic, gentamicin-resistant *E. faecalis* strains had a fivefold-increased risk for death within three weeks compared to patients with nonhemolytic, gentamicin-susceptible strains. Moreover, mode of treatment was not associated with outcome, discounting the contribution of aminoglycoside resistance to this enhanced lethality of infection. In a more recent study, Caballero-Granado et al. (8) analyzed the clinical outcome, including mortality, for bacteremia caused by *Enterococcus* spp. with and without high-level gentamicin resistance. Mortality associated with high-level gentamicin resistance (29%) was not significantly different from gentamicin-susceptible strains (28%). In addition, Caballero-Granado et al. (8) found no significant difference in the length of hospitalization after acquisition of enterococcal bacteremia. Taken together, these studies suggest that high-level aminoglycoside resistance does not affect clinical outcome and that the presence of the *E. faecalis* cytolysin (hemolysin) may confer enhanced toxicity to infection. The *E. faecalis* cytolysin has been shown to result in acute toxicity in a number of animal models. Cytolysin significantly lowers the 50% lethal dose of the infecting strain for mice (22, 64, 83). As will be discussed later, cytolysin also contributes to the acute toxicity of lupine endocarditis and endophthalmitis models (10, 67), and was shown to contribute to nematode killing in *Caenorhabditis elegans* (35).

UTI

Enterococci have been estimated to account for 110,000 UTIs annually in the United States (62). A few studies have been aimed at understanding the interaction of enterococci with uroepithelial tissue (43, 70). Kreft et al. (70) showed a potential role for the plasmid-encoded aggregation substance in the adhesion of enterococci to renal epithelial cells. *E. faecalis* harboring the pheromone-responsive plasmid pAD1 or various isogenic derivatives was better able to bind to the cultured pig renal tubular cell line, LLC-PK, than plasmid-free cells. Their findings also showed that a synthetic peptide containing the fibronectin motif, Arg-Gly-Asp-Ser, could inhibit binding. This structural motif mediates the interaction between fibronectin and eukaryotic surface receptors of the integrin family (33).

Guzman and coworkers (43) analyzed the ability of strains of *E. faecalis* isolated from either UTIs or endocarditis to adhere to urinary tract (UT) epithelial cells and to the Girardi heart cell line. UT infection isolates adhered to the UT epithelial cells in vitro, whereas strains from endocarditis adhered efficiently to the Girardi heart cell line, suggesting a role for environmental adaptation to facilitate interactions with host tissues.

Shankar et al. (116) have shown a role for the aforementioned *E. faecalis* Esp surface protein in colonization of the bladder, but not the kidneys, in an ascending UTI model, suggesting that this protein has tissue specificity for bladder epithelium. Shiono and Ike (118) have shown that some *E. faecalis* clinical isolates adhere efficiently to human bladder carcinoma cells, as well as to human bladder epithelial cells, and that such interaction is inhibited by pretreating the bacteria with fibronectin or trypsin. The conclusion is that these strains possess a protease-sensitive substance that recognizes fibronectin on the surface of the bladder epithelium. In a follow-up to this study, Tomita and Ike (131) recently showed that these highly adherent strains also recognize the extracellular matrix proteins fibronectin, laminin, and collagen type I, II, IV, and V. One of these strains, AS14, was subjected to transposon mutagenesis with Tn916, and mutants with altered collagen IV and laminin binding were identified. Out of 14 single transposon insertion mutants, 13 mapped to the recently identified Ace protein (92–94, 108, 120). The level of adherence to collagen IV and laminin was one to two orders of magnitude lower than the number observed for the wild-type strain, but the level of adherence to fibronectin remained the same as that of the wild-type strain. The collagen-binding proteins Ace (*E. faecalis*) and Acm (*E. faecium*) were originally identified using emerging genome sequence data and querying with the amino-terminal end of Cna, the collagen-binding protein from *S. aureus* (94, 108). The ability of most *E. faecalis* Ace-positive strains to adhere to collagen and laminin was highly dependent on growth temperature, with binding occurring at 46°C, but not at 37°C (92). No plausible explanation existed for this unusual binding phenotype until the recent observations by Tomita and Ike (131), which showed that gelatinase-positive *E. faecalis* strains adhered poorly to collagen and laminin due to the ability of this protease to cleave these proteins. Interestingly, the strains originally examined by Nallapareddy et al. (92) were all gelatinase-positive, and the plausible explanation for binding at 46°C and not at 37°C is that gelatinase is thermal labile at the higher temperature. Thus, in addition to degrading gelatin and casein as substrates, it would appear that the *E. faecalis* gelatinase can use collagen and laminin as substrates. What implications this has for enterococcal spread and toxicity remain to be determined.

In addition to the Esp and Ace surface proteins, Sillanpaa et al. (120) recently identified 17 additional *E. faecalis* proteins that globally resemble microbial surface component-recognizing adhesive matrix molecules. Further characterization of nine of these proteins revealed that two of the nine proteins selected for further study appeared to represent cell wall-anchored enzymes. The remaining seven proteins constitute a family of structurally related proteins potentially interacting with proteins of the host. Interestingly, all of these seven proteins are expressed in the host, as antisera from infected individuals recognize recombinant versions of the enterococcal proteins. Perhaps the most interesting finding from this study was the fact that three of the most seroreactive proteins (EF1091, EF1092, and EF1093) are encoded by an operon adjacent to a gene encoding a putative sortase (EF1094). This gene arrangement is similar to that recently described for the pilin assembly locus of *Corynebacterium diphtheriae* (132–134). The observation that these *E. faecalis* surface proteins (EF1091 to EF1093) are the most seroreactive proteins, and that they are arranged similarly to the recently described *C. diphtheria* pilin assembly locus, leads to the tentative speculation that *E. faecalis* possesses a pilus. Whether such a structure exists and the role that it may play in enterococcal biology are worthy of future investigation.

Endocarditis

Of the infections caused by enterococci, infective endocarditis is one of the most therapeutically challenging (82). Enterococci are the third leading cause of infective endocarditis, accounting for 5 to 20% of cases of native valve infective endocarditis, and 6 to 7% of prosthetic valve endocarditis (82). As noted above, enterococci cultured in serum exhibit enhanced binding to Girardi heart cells. This interaction is inhibited by periodate treatment of the bacterial cell, as well as competitive inhibition of binding, by prior incubation of the target cells with specific sugar residues, including D-galactose and L-fucose (44). This suggests that a carbohydrate antigen mediates the adherence of enterococci to cultured heart cells derived from the right auricular appendage (Girardi heart).

The presence of the pheromone-responsive plasmid pAD1 enhances vegetation formation in enterococcal endocarditis (10). By comparing endocarditis caused by isogenic mutants in either cytolysin (hemolysin) production or aggregation substance, it was observed that the presence of the cytolysin contributed to overall lethality (6/11 animals killed compared to 2/13 in the noncytolytic mutant, $P < 0.01$), whereas the presence of aggregation substance led to a twofold increase in mean vegetation weight. It was noted, however, that all strains tested were able to produce endocarditis, even plasmid-free controls. These data suggest that the presence of auxiliary genetic elements, such as plasmids and/or pathogenicity islands, can enhance the pathogenicity of the organism, but may not be essential in establishing the infection.

Serum from a patient with *E. faecalis* endocarditis was used to identify an *E. faecalis* antigen selectively expressed in serum but not in broth culture (79). This protein antigen, designated EfaA, had a predicted molecular mass of 34,768 Da. Database homology searches revealed extensive sequence similarity with several streptococcal adhesins. The authors hypothesized that this surface antigen might function as an important adhesin in endocarditis, but there is little published data to support this. More recently, Shepard and Gilmore (117) showed that of all the suspected *E. faecalis* virulence factors, EfaA resulted in the most dramatic increase in expression when comparing serum-grown versus broth-grown *E. faecalis*. In light of more recent findings (78), this is probably not all that surprising. It is now known that EfaA forms part of putative ABC-type transporter specific for manganese, along with EfaB and EfaC. Because manganese is tightly sequestered within the mammalian host, it is not unexpected that *E. faecalis* grown in serum would require the induction of a high-affinity manganese transporter, EfaABC. The induction of the *efaCBA* operon occurs at relatively low manganese levels, because under high manganese levels the product of the *efaR* gene, a manganese-responsive transcriptional regulator, represses transcription from the *efaCBA* operon (78). Apart from its induction in serum to import manganese, no other functional role has been ascribed to the EfaABC proteins.

Endophthalmitis

Colonization of host tissue may play a role in the pathogenesis of endophthalmitis. Enterococci are among the most destructive agents that cause this postoperative complication of cataract surgery (46, 67). Although designed to determine whether aggregation substance targeted *E. faecalis* to alternate anatomical structures within the eye, it was shown that enterococci attach to membranous structures in the vitreous, but that such adherence is not dependent on the presence of aggregation substance (65). In summary, the preponderance of data indicates that *E. faecalis* adhesion to host tissues is complex and appears to involve multiple adhesins, including cell surface proteins and surface carbohydrates.

Immune Evasion

For enterococci to maintain an infection, they must successfully evade both specific and nonspecific host defense mechanisms. Other gram-positive pathogens possess attributes that allow them to survive in the host in spite of powerful nonspecific host defenses mediated primarily by professional phagocytes, i.e., neutrophils, monocytes, and macrophages. These factors include antiphagocytic polysaccharide capsules, antiphagocytic surface proteins, such as the group A streptococcal M protein, and various secreted toxins with direct phagocytic cell toxicity.

Studies designed to characterize the host response to enterococcal infection have been conducted (1, 32, 52, 97). Previous studies (1, 52) concluded that neutrophil-mediated killing of enterococci was largely a function of complement, with antibody playing a less essential but potentially important role. However, a more recent study by Gaglani et al. (32) demonstrated that effective neutrophil-mediated killing was highly dependent on both antibody and complement. The discrepancy in these disparate studies potentially lies in the fact that so-called "normal serum" contains sufficient antienterococcal antibodies to mediate effective opsonization; only after absorption of these antibodies does their importance prove evident.

In a case report, Bottone and colleagues (6) isolated three highly mucoid encapsulated strains of *E. faecalis* from patients with UTIs. These mucoid isolates appeared to persist longer in the mouse following intraperitoneal injection compared to nonmucoid controls, but crude mortality was not changed, suggesting a role in preventing or delaying phagocytic clearance (5).

Several recent studies have examined the diversity of cell wall polysaccharides in *E. faecalis* (47, 55–57). Based on genetic and serologic data, *E. faecalis* capsular polysaccharides fall into four serogroups (57). Antibodies to *E. faecalis* capsular polysaccharides do, however, cross-protect across the species boundary (56), as antibodies to *E. faecalis* capsular antigens are reactive with some *E. faecium* capsular antigens and mediate opsonophagocytosis. The genetic basis for *E. faecalis* capsular biosynthesis has been discovered for two of the four serogroups, and is encoded by the *cpsC-K* operon (47, 51). In addition to capsular antigens, Murray and coworkers have identified another cell wall polysaccharide whose synthesis is encoded by the *epa* locus, enterococcal polysaccharide antigen (126, 144, 145). This pathway appears to be widespread and conserved within *E. faecalis*, and it has been designated the *E. faecalis* group antigen (47), analogous to the streptococcal group antigens. Antibodies to Epa have also been shown to promote opsonophagocytic clearance by neutrophils (126), but this prospect remains controversial (56). The extent to which capsular polysaccharides and/or Epa could be used as a potential enterococcal vaccine awaits further study.

Pathologic Tissue Damage

Following adhesion to host cell surfaces and evasion of the host immune response, the last step in the pathogenesis of infection is the production of pathologic changes in the host. Such changes can be induced by the host inflammatory cascade or by direct damage as a result of secreted toxins or proteases, as well as the production of toxic metabolic byproducts. Each of these mechanisms has been observed in studies of *E. faecalis* pathogenesis.

Indirect Tissue Damage

Enterococcal LTA, also known as the group D streptococcal antigen, has been implicated in a variety of biological processes (86). Some properties ascribed to LTA include modulation of the host immune response, as well as mediating the adherence of enterococci to host cells. Bhakdi et al. (4) found the LTA from enterococci to be as inflammatory as lipopolysaccharide of gram-negative bacteria, although more recent findings dispute this claim (7). LTA may also contribute to the ability of enterococci to exchange and rapidly disseminate genetic information, a subject that has been investigated extensively (21) (chapter 26, this volume).

The role of LTA and aggregation substance in cardiac infections has been examined (114). Strains of *E. faecalis* defective in enterococcal binding substance, which is at least partially derived from LTA, and also defective in the protein adhesin aggregation substance, did not induce clinical signs of illness when injected intraventricularly at levels of 10^8 CFU/ml. However, EBS$^+$AS$^-$ or EBS$^-$AS$^+$ strains showed signs of illness and pericardiac inflammation. All rabbits injected with the EBS$^+$AS$^+$ strain developed illness and died. Surprisingly little inflammation was observed in rabbits injected with the EBS$^+$AS$^+$ strain despite the lethality observed. The authors state that such observations are consistent with the presence of a superantigen. The presence of LTA (EBS) and AS together may mediate effects on the host immune response that differ from those seen when either component acts alone.

Direct Tissue Damage

The enterococcal cytolysin and two proteases, a zinc metalloprotease (gelatinase) and a serine protease, are secreted factors well suited to contribute to disease severity (10, 42, 64, 67, 105). The enterococcal cytolysin is well established in several independent labs and animal models (10, 64, 67). The presence of the cytolysin has also been shown to promote the appearance of *E. faecalis* in the bloodstream (62). The genetics and biology of the enterococcal cytolysin were recently reviewed (11). The toxin is genetically and structurally related to the lantibiotic family of bacteriocins, but diverges from this family in that it possesses both bacteriocin activity as well as toxicity for a variety of mammalian cell types. The cytolysin operon is typically encoded on large pheromone-responsive plasmids, but has also been shown to reside within an *E. faecalis* pathogenicity island (115). The operon consists of six genes, designated *cylL*$_L$, *cylL*$_S$, *cylM*, *cylB*, *cylA*, and *cylI*, with each gene product playing an essential role in toxin synthesis, modification, secretion, activation, and immunity. Transcription of the operon is repressed by the products of a divergent operon encoding CylR1 and CylR2, and repression is alleviated by the accumulation of fully processed CylL$_S$, which serves as a quorum-sensing molecule to prime the synthesis of more toxin components as cell density increases (45).

The most direct and quantitative evidence for pathologic damage attributable to the cytolysin was obtained using a rabbit model of endophthalmitis (67). This model was selected because a robust infection can be established with as few as 10 organisms due to the natural aberrations in the intraocular immune response. This limited response provides the offending bacterium an opportunity to adapt to in vivo growth conditions and environmental cues. Moreover, highly sensitive and quantitative measurements of the evolution of disease can be made. A role for the cytolysin in tissue pathology was unambiguously demonstrated both by a reduction in β-wave response and complete destruction of retinal architecture 24 hours postinfection.

Because of its broad protease specificity, the enterococcal gelatinase may also play a measurable role in systemic disease (22), as well as in a caries model using germ-free rats (42). Using germ-free rats, Gold et al. (42) showed that a proteolytic (Gel+) strain exhibited cariogenic activity, whereas three nonproteolytic strains exhibited little cariogenecity. Dupont et al. (22) showed a reduced 50% lethal dose for mice injected with gelatinase-producing (Gel+) strains, and these findings were subsequently confirmed using isogenic strains defective in protease production (105). On the same transcript as the *gelE* gene resides the *sprE*

gene, encoding a serine protease. Both proteases have been shown to play important roles in *C. elegans* nematode killing (119), lupine endophthalmitis (24), and murine peritonitis (105). As discussed previously, gelatinase has also been shown to be important in *E. faecalis* biofilm formation (49, 71) and appears to play a role in the degradation of the host extracellular matrix proteins collagen and laminin (131).

Both proteases are regulated at the transcriptional level by a quorum-sensing signal transduction system termed *fsr* (104). The *fsr* locus encodes a response regulator, FsrA, which is thought to be activated by phosphorylation on a conserved aspartyl residue by the phosphotransfer from a histidine kinase, FsrC. The kinase is thought to undergo autophosphorylation upon sensing the accumulation of a peptide lactone quorum molecule (90), which is coded for at the carboxy terminus of FsrB. The only genes known to be transcriptionally controlled by activated FsrA are *fsrB* and C and *gelE* and *sprE*. Most if not all the phenotypic effect of inactivating the *fsr* locus resides in the lack of protease production. There are, however, studies that suggest that additional genes might be controlled by the *fsr* locus (24, 85, 119). Further exploration of this topic will likely depend on microarray or two-dimensional gel analysis for the answer.

There is some indication that clinical isolates may be enriched for the proteolytic trait, because greater than 50% of isolates from both endocarditis and other clinical sources exhibited gelatinase activity, whereas only 27% of community fecal isolates possessed this trait (12). More recent studies that have explored the presence of both *fsr* genes and *gelE* in epidemiology have found widely differing results (101, 111). Pillai et al. (101) found that 100% (12/12) of endocarditis isolates were *fsr* positive, whereas only 53% (10/19) of fecal isolates were *fsr* positive. However, Roberts et al. (111) found that the number of *fsr*- and *gelE*-positive isolates remained equally distributed between 60 and 65% for both infection-derived isolates and isolates colonizing healthy individuals. The discrepancy could simply result from a smaller sample size in the former study and the fact that the probe for detection of the *fsr* locus in the former study included all three *fsr* genes. It has recently been demonstrated that in a large percentage of gelatinase-negative clinical isolates, a 23.9-kb chromosomal deletion has occurred, resulting in the deletion of both *fsrA* and *fsrB*, and a null protease phenotype (91). Because this represents a defined deletion, it is tempting to speculate why *E. faecalis* would delete important virulence factors. Analogous to the description of the *E. faecalis* pathogenicity island, which is known to be modulated through the insertion and deletion of genetic information (115), the organism may simply modulate its virulence so that it does not overtly offend its host.

Although *E. faecium* appears to lack both proteolytic and cytolytic activities, it has recently been shown to encode a hyaluronidase (107) that appears to be enriched in nonstool isolates. Hyaluronidase activity has yet to be demonstrated in these isolates, but it is conceivable that this trait is emerging within the *E. faecium* population. An *E. faecium* pathogenicity island was recently described that contains an ortholog of the *E. faecalis esp* gene (76). Apart from the *esp* gene and an *araC*-like transcriptional regulator, none of the remaining open reading frames resemble those present in the *E. faecalis* pathogenicity island. Clearly both *E. faecalis* and *E. faecium* are evolving into more complex pathogens.

Toxic Metabolites

In addition to secreted proteins, *E. faecalis* and *E. faecium* have recently been shown to produce toxic oxygen metabolites that have resulted in cell or organ damage (60, 87). Using a worm model, Moy et al. (87) demonstrated that *E. faecium* produces hydrogen peroxide at levels that cause cellular damage. *E. faecium* transposon insertion mutants were identified that altered *C. elegans* killing activity and displayed altered levels of hydrogen peroxide production. Mutation of an NADH oxidase-encoding gene eliminated nearly all NADH oxidase activity and reduced hydrogen peroxide production and killing, whereas mutation of a gene encoding an NADH peroxidase resulted in enhanced levels of hydrogen peroxide and more rapid nematode killing. Depending on the culture conditions, *E. faecium* is able to produce hydrogen peroxide, and the level of hydrogen peroxide production appears to correlate with nematode survival.

The vast majority (87/91) of *E. faecalis* isolates produce superoxide (O_2^-), whereas *E. faecium* isolates (5/13) do so less frequently (61). The authors of this study speculated that membrane-damaging effects of oxygen radicals may potentiate cellular damage to nearby intestinal epithelial cells. In proof, it was recently shown that the production of extracellular superoxide and hydrogen peroxide damages colonic epithelial cell DNA (60). The answer to why *E. faecalis* produces extracellular oxygen radicals stems from the fact that this oxygen radical is a byproduct of an incomplete respiratory chain. *E. faecalis* in the presence of exogenous heme is capable of reconstituting a cytochrome complex (59), enabling it to effectively respire. The reconstituted cytochrome complex suppresses extracellular superoxide production. The benefits of respiration in the presence of heme have recently been examined for *Lactococcus lactis* (106), and similar findings probably apply to why *E. faecalis* likely benefits from heme-dependent respiration. These benefits include enhanced growth yield and improved long-term survival, as well as less protein and DNA damage attributable to a reduction in oxidative and acidic stresses. The fact that enterococci generate a byproduct that damages their host appears to be simply an unintended consequence of incomplete respiration.

CONCLUSIONS AND FUTURE PERSPECTIVES

Enterococci are well adapted for survival and persistence in a variety of adverse environments, including sites of infection and inanimate hospital surfaces. The rapid emergence of antimicrobial resistance among enterococci undoubtedly also contributes to their emergence as prominent nosocomial pathogens, making them among the most difficult to treat. The recent reports of *E. faecalis* and *E. faecium* isolates containing pathogenicity islands suggest that traits in addition to antibiotic resistance move through the enterococcal gene pool and that such traits contribute to their spread in the hospital setting. Our understanding of how these organisms continually evolve will need to keep pace with their evolution or we may potentially be on the wrong end of a public health nightmare, in which isolates evolve to higher levels of antibiotic resistance and ever increasing virulence.

Portions of the work described were supported by NIH grants EY08289 and AI041108. The contributions of many who have been associated with the laboratory over the past 20 years are gratefully acknowledged.

REFERENCES

1. **Arduino, R. C., B. E. Murray, and R. M. Rakita.** 1994. Roles of antibodies and complement in phagocytic killing of enterococci. *Infect. Immun.* **62:**987–993.
2. **Baldassarri, L., R. Cecchini, L. Bertuccini, M. G. Ammendolia, F. Iosi, C. R. Arciola, L. Montanaro, R. Di Rosa, G. Gherardi, G. Dicuonzo, G. Orefici, and R. Creti.** 2001. *Enterococcus* spp. produces slime and survives in rat peritoneal macrophages. *Med. Microbiol. Immunol.* **190:**113–120.
3. **Basaran, U. N., S. Celayir, N. Eray, R. Ozturk, and O. F. Senyuz.** 1998. The effect of an H2-receptor antagonist on small-bowel colonization and bacterial translocation in newborn rats. *Pediatr. Surg. Int.* **13:**118–120.
4. **Bhakdi, S., T. Klonisch, P. Nuber, and W. Fischer.** 1991. Stimulation of monokine production by lipoteichoic acids. *Infect. Immun.* **59:**4693–4697.
5. **Bottone, E. J.** 1999. Encapsulated *Enterococcus faecalis:* role of encapsulation in persistence in mouse peritoneum in absence of mouse lethality. *Diagn. Microbiol. Infect. Dis.* **33:**65–68.
6. **Bottone, E. J., L. Patel, P. Patel, and T. Robin.** 1998. Mucoid encapsulated *Enterococcus faecalis:* an emerging morphotype isolated from patients with urinary tract infections. *Diagn. Microbiol. Infect. Dis.* **31:**429–430.
7. **Bruserud, O., O. Wendelbo, and K. Paulsen.** 2004. Lipoteichoic acid derived from *Enterococcus faecalis* modulates the functional characteristics of both normal peripheral blood leukocytes and native human acute myelogenous leukemia blasts. *Eur. J. Haematol.* **73:**340–350.
8. **Caballero-Granado, F. J., J. M. Cisneros, R. Luque, M. Torres-Tortosa, F. Gamboa, F. Diez, J. L. Villanueva, R. Perez-Cano, J. Pasquau, D. Merino, A. Menchero, D. Mora, M. A. Lopez-Ruz, and A. Vergara.** 1998. Comparative study of bacteremias caused by *Enterococcus* spp. with and without high-level resistance to gentamicin. *J. Clin. Microbiol.* **36:**520–525.
9. **Chenoweth, C., and D. Schaberg.** 1990. The epidemiology of enterococci. *Eur. J. Clin. Microbiol. Infect. Dis.* **9:**80–89.
10. **Chow, J. W., L. A. Thal, M. B. Perri, J. A. Vazquez, S. M. Donabedian, D. B. Clewell, and M. J. Zervos.** 1993. Plasmid-associated hemolysin and aggregation substance production contribute to virulence in experimental enterococcal endocarditis. *Antimicrob. Agents Chemother.* **37:**2474–2477.
11. **Coburn, P. S., and M. S. Gilmore.** 2003. The *Enterococcus faecalis* cytolysin: a novel toxin active against eukaryotic and prokaryotic cells. *Cell Microbiol.* **5:**661–669.
12. **Coque, T. M., J. E. Patterson, J. M. Steckelberg, and B. E. Murray.** 1995. Incidence of hemolysin, gelatinase, and aggregation substance among enterococci isolated from patients with endocarditis and other infections and from feces of hospitalized and community-based persons. *J. Infect. Dis.* **171:**1223–1229.
13. **Coque, T. M., R. Willems, R. Canton, R. Del Campo, and F. Baquero.** 2002. High occurrence of *esp* among ampicillin-resistant and vancomycin-susceptible *Enterococcus faecium* clones from hospitalized patients. *J. Antimicrob. Chemother.* **50:**1035–1038.
14. **Costerton, J. W., P. S. Stewart, and E. P. Greenberg.** 1999. Bacterial biofilms: a common cause of persistent infections. *Science* **284:**1318–1322.
15. **Cucarella, C., C. Solano, J. Valle, B. Amorena, I. Lasa, and J. R. Penades.** 2001. Bap, a *Staphylococcus aureus* surface protein involved in biofilm formation. *J. Bacteriol.* **183:**2888–2896.

16. **Darlong, V., T. S. Jayalakhsmi, H. L. Kaul, and R. Tandon.** 2003. Stress ulcer prophylaxis in patients on ventilator. *Trop. Gastroenterol.* **24:**124–128.

17. **Donskey, C. J.** 2004. The role of the intestinal tract as a reservoir and source for transmission of nosocomial pathogens. *Clin. Infect. Dis.* **39:**219–226.

18. **Donskey, C. J., T. K. Chowdhry, M. T. Hecker, C. K. Hoyen, J. A. Hanrahan, A. M. Hujer, R. A. Hutton-Thomas, C. C. Whalen, R. A. Bonomo, and L. B. Rice.** 2000. Effect of antibiotic therapy on the density of vancomycin-resistant enterococci in the stool of colonized patients. *N. Engl. J. Med.* **343:**1925–1932.

19. **Donskey, C. J., J. A. Hanrahan, R. A. Hutton, and L. B. Rice.** 1999. Effect of parenteral antibiotic administration on persistence of vancomycin-resistant *Enterococcus faecium* in the mouse gastrointestinal tract. *J. Infect. Dis.* **180:**384–390.

20. **Donskey, C. J., M. S. Helfand, N. J. Pultz, and L. B. Rice.** 2004. Effect of parenteral fluoroquinolone administration on persistence of vancomycin-resistant *Enterococcus faecium* in the mouse gastrointestinal tract. *Antimicrob. Agents Chemother.* **48:**326–328.

21. **Dunny, G. M., B. A. Leonard, and P. J. Hedberg.** 1995. Pheromone-inducible conjugation in *Enterococcus faecalis*: interbacterial and host-parasite chemical communication. *J. Bacteriol.* **177:**871–876.

22. **Dupont, H., P. Montravers, J. Mohler, and C. Carbon.** 1998. Disparate findings on the role of virulence factors of *Enterococcus faecalis* in mouse and rat models of peritonitis. *Infect. Immun.* **66:**2570–2575.

23. **Edmond, M. B., J. F. Ober, J. D. Dawson, D. L. Weinbaum, and R. P. Wenzel.** 1996. Vancomycin-resistant enterococcal bacteremia: natural history and attributable mortality. *Clin. Infect. Dis.* **23:**1234–1239.

24. **Engelbert, M., E. Mylonakis, F. M. Ausubel, S. B. Calderwood, and M. S. Gilmore.** 2004. Contribution of gelatinase, serine protease, and fsr to the pathogenesis of *Enterococcus faecalis* endophthalmitis. *Infect. Immun.* **72:**3628–3633.

25. **Figdor, D., J. K. Davies, and G. Sundqvist.** 2003. Starvation survival, growth and recovery of *Enterococcus faecalis* in human serum. *Oral Microbiol. Immunol.* **18:**234–239.

26. **Finegold, S. M., V. L. Sutter, and G. E. Mathisen.** 1983. Normal indigenous intestinal flora, p. 3–31. *In* D. J. Hentges (ed.), *Human Intestinal Microflora in Health and Disease.* Academic Press, New York, N.Y.

27. **Flahaut, S., J. Frere, P. Boutibonnes, and Y. Auffray.** 1996. Comparison of the bile salts and sodium dodecyl sulfate stress responses in *Enterococcus faecalis*. *Appl. Environ. Microbiol.* **62:**2416–2420.

28. **Flahaut, S., A. Hartke, J. Giard, and Y. Auffray.** 1997. Alkaline stress response in *Enterococcus faecalis*: adaptation, cross-protection, and changes in protein synthesis. *Appl. Environ. Microbiol.* **63:**812–814.

29. **Flahaut, S., A. Hartke, J. C. Giard, A. Benachour, P. Boutibonnes, and Y. Auffray.** 1996. Relationship between stress response toward bile salts, acid and heat treatment in *Enterococcus faecalis*. *FEMS Microbiol. Lett.* **138:**49–54.

30. **Flahaut, S., J. M. Laplace, J. Frere, and Y. Auffray.** 1998. The oxidative stress response in *Enterococcus faecalis*: relationship between H_2O_2 tolerance and H_2O_2 stress proteins. *Lett. Appl. Microbiol.* **26:**259–264.

31. **Foley, I., and P. Gilbert.** 1997. In-vitro studies of the activity of glycopeptide combinations against *Enterococcus faecalis* biofilms. *J. Antimicrob. Chemother.* **40:**667–672.

32. **Gaglani, M. J., C. J. Baker, and M. S. Edwards.** 1997. Contribution of antibody to neutrophil-mediated killing of *Enterococcus faecalis*. *J. Clin. Immunol.* **17:**478–484.

33. **Galli, D., F. Lottspeich, and R. Wirth.** 1990. Sequence analysis of *Enterococcus faecalis* aggregation substance encoded by the sex pheromone plasmid pAD1. *Mol. Microbiol.* **4:**895–904.

34. **Garbutt, J. M., M. Ventrapragada, B. Littenberg, and L. M. Mundy.** 2000. Association between resistance to vancomycin and death in cases of *Enterococcus faecium* bacteremia. *Clin. Infect. Dis.* **30:**466–472.

35. **Garsin, D. A., C. D. Sifri, E. Mylonakis, X. Qin, K. V. Singh, B. E. Murray, S. B. Calderwood, and F. M. Ausubel.** 2001. A simple model host for identifying Gram-positive virulence factors. *Proc. Natl. Acad. Sci. USA* **98:**10892–10897.

36. **Gentry-Weeks, C. R., R. Karkhoff-Schweizer, A. Pikis, M. Estay, and J. M. Keith.** 1999. Survival of *Enterococcus faecalis* in mouse peritoneal macrophages. *Infect. Immun.* **67:**2160–2165.

37. **Giard, J. C., A. Hartke, S. Flahaut, A. Benachour, P. Boutibonnes, and Y. Auffray.** 1996. Starvation-induced multiresistance in *Enterococcus faecalis* JH2-2. *Curr. Microbiol.* **32:**264–271.

38. **Giard, J. C., A. Hartke, S. Flahaut, P. Boutibonnes, and Y. Auffray.** 1997. Glucose starvation response in *Enterococcus faecalis* JH2-2: survival and protein analysis. *Res. Microbiol.* **148:**27–35.

39. **Giard, J. C., J. M. Laplace, A. Rince, V. Pichereau, A. Benachour, C. Leboeuf, S. Flahaut, Y. Auffray, and A. Hartke.** 2001. The stress proteome of *Enterococcus faecalis*. *Electrophoresis* **22:**2947–2954.

40. **Gilmore, M. S., D. B. Clewell, P. Courvalin, G. M. Dunny, B. E. Murray, and L. B. Rice.** 2002. *The Enterococci: Pathogenesis, Molecular Biology, and Antibiotic Resistance.* ASM Press, Washington, D.C.

41. **Gilmore, M. S., P. S. Coburn, S. R. Nallapareddy, and B. E. Murray.** 2002. Enterococcal virulence, p. 301–354. *In* M. S. Gilmore, D. B. Clewell, P. Courvalin, G. M. Dunny, B. E. Murray, and L. B. Rice (ed.), *The Enterococci: Pathogenesis, Molecular Biology, and Antibiotic Resistance.* ASM Press, Washington, D.C.

42. **Gold, O. G., H. V. Jordan, and J. van Houte.** 1975. The prevalence of enterococci in the human mouth and their pathogenicity in animal models. *Arch. Oral Biol.* **20:**473–477.

43. **Guzman, C. A., C. Pruzzo, G. LiPira, and L. Calegari.** 1989. Role of adherence in pathogenesis of *Enterococcus faecalis* urinary tract infection and endocarditis. *Infect. Immun.* **57:**1834–1838.

44. **Guzman, C. A., C. Pruzzo, M. Plate, M. C. Guardati, and L. Calegari.** 1991. Serum dependent expression of *Enterococcus faecalis* adhesins involved in the colonization of heart cells. *Microb. Pathog.* **11:**399–409.

45. **Haas, W., B. D. Shepard, and M. S. Gilmore.** 2002. Two-component regulator of *Enterococcus faecalis* cytolysin responds to quorum-sensing autoinduction. *Nature* **415:**84–87.

46. **Han, D. P., S. R. Wisniewski, L. A. Wilson, M. Barza, A. K. Vine, B. H. Doft, and S. F. Kelsey.** 1996. Spectrum and susceptibilities of microbiologic isolates in the Endophthalmitis Vitrectomy Study. *Am. J. Ophthalmol.* **122:**1–17.

47. **Hancock, L. E., and M. S. Gilmore.** 2002. The capsular polysaccharide of *Enterococcus faecalis* and its relationship to other polysaccharides in the cell wall. *Proc. Natl. Acad. Sci. USA* **99:**1574–1579.

48. **Hancock, L. E., and M. S. Gilmore.** 2000. Pathogenicity of enterococci, p. 251–258. *In* V. Fischetti, R. Novick, J. Ferretti, D. Portnoy, and J. Rood (ed.), *Gram-Positive Pathogens.* American Society for Microbiology, Washington, D.C.

49. **Hancock, L. E., and M. Perego.** 2004. The *Enterococcus faecalis fsr* two-component system controls biofilm development through production of gelatinase. *J. Bacteriol.* **186:**5629–5639.

50. **Hancock, L. E., and M. Perego.** 2004. Systematic inactivation and phenotypic characterization of two-component signal transduction systems of *Enterococcus faecalis* V583. *J. Bacteriol.* **186:**7951–7958.

51. **Hancock, L. E., B. D. Shepard, and M. S. Gilmore.** 2003. Molecular analysis of the *Enterococcus faecalis* serotype 2 polysaccharide determinant. *J. Bacteriol.* **185:** 4393–4401.

52. **Harvey, B. S., C. J. Baker, and M. S. Edwards.** 1992. Contributions of complement and immunoglobulin to neutrophil-mediated killing of enterococci. *Infect. Immun.* **60:**3635–3640.

53. **Heim, S., M. M. Lleo, B. Bonato, C. A. Guzman, and P. Canepari.** 2002. The viable but nonculturable state and starvation are different stress responses of *Enterococcus faecalis,* as determined by proteome analysis. *J. Bacteriol.* **184:** 6739–6745.

54. **Hirt, H., P. M. Schlievert, and G. M. Dunny.** 2002. In *vivo* induction of virulence and antibiotic resistance transfer in *Enterococcus faecalis* mediated by the sex pheromone-sensing system of pCF10. *Infect. Immun.* **70:**716–723.

55. **Huebner, J., A. Quaas, W. A. Krueger, D. A. Goldmann, and G. B. Pier.** 2000. Prophylactic and therapeutic efficacy of antibodies to a capsular polysaccharide shared among vancomycin-sensitive and -resistant enterococci. *Infect. Immun.* **68:**4631–4636.

56. **Huebner, J., Y. Wang, W. A. Krueger, L. C. Madoff, G. Martirosian, S. Boisot, D. A. Goldmann, D. L. Kasper, A. O. Tzianabos, and G. B. Pier.** 1999. Isolation and chemical characterization of a capsular polysaccharide antigen shared by clinical isolates of *Enterococcus faecalis* and vancomycin-resistant *Enterococcus faecium. Infect. Immun.* **67:**1213–1219.

57. **Hufnagel, M., L. E. Hancock, S. Koch, C. Theilacker, M. S. Gilmore, and J. Huebner.** 2004. Serological and genetic diversity of capsular polysaccharides in *Enterococcus faecalis. J. Clin. Microbiol.* **42:**2548–2557.

58. **Hufnagel, M., S. Koch, R. Creti, L. Baldassarri, and J. Huebner.** 2004. A putative sugar-binding transcriptional regulator in a novel gene locus in *Enterococcus faecalis* contributes to production of biofilm and prolonged bacteremia in mice. *J. Infect. Dis.* **189:**420–430.

59. **Huycke, M. M.** 2002. Physiology of enterococci, p. 133–176. *In* M. S. Gilmore, D. B. Clewell, P. Courvalin, G. M. Dunny, B. E. Murray, and L. B. Rice (ed.), *The Enterococci: Pathogenesis, Molecular Biology, and Antibiotic Resistance.* ASM Press, Washington, D.C.

60. **Huycke, M. M., V. Abrams, and D. R. Moore.** 2002. *Enterococcus faecalis* produces extracellular superoxide and hydrogen peroxide that damages colonic epithelial cell DNA. *Carcinogenesis* **23:**529–536.

61. **Huycke, M. M., W. Joyce, and M. F. Wack.** 1996. Augmented production of extracellular superoxide by blood isolates of *Enterococcus faecalis. J. Infect. Dis.* **173:**743–746.

62. **Huycke, M. M., D. F. Sahm, and M. S. Gilmore.** 1998. Multiple-drug resistant enterococci: the nature of the prob-

lem and an agenda for the future. *Emerg. Infect. Dis.* **4:**239–249.

63. **Huycke, M. M., C. A. Spiegel, and M. S. Gilmore.** 1991. Bacteremia caused by hemolytic, high-level gentamicin-resistant *Enterococcus faecalis. Antimicrob. Agents Chemother.* **35:**1626–1634.

64. **Ike, Y., H. Hashimoto, and D. B. Clewell.** 1984. Hemolysin of *Streptococcus faecalis* subspecies *zymogenes* contributes to virulence in mice. *Infect. Immun.* **45:**528–530.

65. **Jett, B. D., R. V. Atkuri, and M. S. Gilmore.** 1998. *Enterococcus faecalis* localization in experimental endophthalmitis: role of plasmid-encoded aggregation substance. *Infect. Immun.* **66:**843–848.

66. **Jett, B. D., M. M. Huycke, and M. S. Gilmore.** 1994. Virulence of enterococci. *Clin. Microbiol. Rev.* **7:**462–478.

67. **Jett, B. D., H. G. Jensen, R. E. Nordquist, and M. S. Gilmore.** 1992. Contribution of the pAD1-encoded cytolysin to the severity of experimental *Enterococcus faecalis* endophthalmitis. *Infect. Immun.* **60:**2445–2452.

68. **Johnson, A. P.** 1994. The pathogenicity of enterococci. *J. Antimicrob. Chemother.* **33:**1083–1089.

69. **Karlowsky, J. A., M. E. Jones, D. C. Draghi, C. Thornsberry, D. F. Sahm, and G. A. Volturo.** 2004. Prevalence and antimicrobial susceptibilities of bacteria isolated from blood cultures of hospitalized patients in the United States in 2002. *Ann. Clin. Microbiol. Antimicrob.* **3:**7–14.

70. **Kreft, B., R. Marre, U. Schramm, and R. Wirth.** 1992. Aggregation substance of *Enterococcus faecalis* mediates adhesion to cultured renal tubular cells. *Infect. Immun.* **60:** 25–30.

71. **Kristich, C. J., Y. H. Li, D. G. Cvitkovitch, and G. M. Dunny.** 2004. Esp-independent biofilm formation by *Enterococcus faecalis. J. Bacteriol.* **186:**154–163.

72. **Lancefield, R. C.** 1933. A serological differentiation of human and other groups of hemolytic streptococci. *J. Exp. Med.* **57:**571–595.

73. **Laplace, J. M., M. Thuault, A. Hartke, P. Boutibonnes, and Y. Auffray.** 1997. Sodium hypochlorite stress in *Enterococcus faecalis:* influence of antecedent growth conditions and induced proteins. *Curr. Microbiol.* **34:**284–289.

74. **Lautenbach, E., W. B. Bilker, and P. J. Brennan.** 1999. Enterococcal bacteremia: risk factors for vancomycin resistance and predictors of mortality. *Infect. Control Hosp. Epidemiol.* **20:**318–323.

75. **Le Breton, Y., G. Boel, A. Benachour, H. Prevost, Y. Auffray, and A. Rince.** 2003. Molecular characterization of *Enterococcus faecalis* two-component signal transduction pathways related to environmental stresses. *Environ. Microbiol.* **5:**329–337.

76. **Leavis, H., J. Top, N. Shankar, K. Borgen, M. Bonten, J. van Embden, and R. J. Willems.** 2004. A novel putative enterococcal pathogenicity island linked to the *esp* virulence gene of *Enterococcus faecium* and associated with epidemicity. *J. Bacteriol.* **186:**672–682.

77. **Lleo, M. M., M. C. Tafi, and P. Canepari.** 1998. Nonculturable *Enterococcus faecalis* cells are metabolically active and capable of resuming active growth. *Syst. Appl. Microbiol.* **21:**333–339.

78. **Low, Y. L., N. S. Jakubovics, J. C. Flatman, H. F. Jenkinson, and A. W. Smith.** 2003. Manganese-dependent regulation of the endocarditis-associated virulence factor EfaA of *Enterococcus faecalis. J. Med. Microbiol.* **52**(Pt. 2)**:** 113–119.

79. **Lowe, A. M., P. A. Lambert, and A. W. Smith.** 1995. Cloning of an *Enterococcus faecalis* endocarditis antigen:

homology with adhesins from some oral streptococci. *Infect. Immun.* **63:**703–706.

80. **MacCallum, W. G., and T. W. Hastings.** 1899. A case of acute endocarditis caused by *Micrococcus zymogenes* (nov. spec.), with a description of the microorganism. *J. Exp. Med.* **4:**521–534.

81. **Makinen, P. L., D. B. Clewell, F. An, and K. K. Makinen.** 1989. Purification and substrate specificity of a strongly hydrophobic extracellular metalloendopeptidase ("gelatinase") from *Streptococcus faecalis* (strain OG1-10). *J. Biol. Chem.* **264:**3325–3334.

82. **Megran, D. W.** 1992. Enterococcal endocarditis. *Clin. Infect. Dis.* **15:**63–71.

83. **Miyazaki, S., A. Ohno, I. Kobayashi, T. Uji, K. Yamaguchi, and S. Goto.** 1993. Cytotoxic effect of hemolytic culture supernatant from *Enterococcus faecalis* on mouse polymorphonuclear neutrophils and macrophages. *Microbiol. Immunol.* **37:**265–270.

84. **Moellering, R. C. J.** 1995. *Enterococcus* species, *Streptococcus bovis*, and *Leuconostac* species, p. 1826–1835. *In* G. L. Mandell, J. E. Bennett, and R. Dolin (ed.), *Principles and Practices of Infectious Diseases*, 4th ed. Churchill Livingston, New York, N.Y.

85. **Mohamed, J. A., W. Huang, S. R. Nallapareddy, F. Teng, and B. E. Murray.** 2004. Influence of origin of isolates, especially endocarditis isolates, and various genes on biofilm formation by *Enterococcus faecalis. Infect. Immun.* **72:**3658–3663.

86. **Montravers, P., J. Mohler, L. Saint Julien, and C. Carbon.** 1997. Evidence of the proinflammatory role of *Enterococcus faecalis* in polymicrobial peritonitis in rats. *Infect. Immun.* **65:**144–149.

87. **Moy, T. I., E. Mylonakis, S. B. Calderwood, and F. M. Ausubel.** 2004. Cytotoxicity of hydrogen peroxide produced by *Enterococcus faecium. Infect. Immun.* **72:**4512–4520.

88. **Mundy, L. M., D. F. Sahm, and M. Gilmore.** 2000. Relationships between enterococcal virulence and antimicrobial resistance. *Clin. Microbiol. Rev.* **13:**513–522.

89. **Murray, B. E.** 1998. Diversity among multidrug-resistant enterococci. *Emerg. Infect. Dis.* **4:**37–47.

90. **Nakayama, J., Y. Cao, T. Horii, S. Sakuda, A. D. L. Akkermans, W. M. de Vos, and H. Nagasawa.** 2001. Gelatinase biosynthesis-activating pheromone: a peptide lactone that mediates a quorum sensing in *Enterococcus faecalis. Mol. Microbiol.* **41:**145–154.

91. **Nakayama, J., R. Kariyama, and H. Kumon.** 2002. Description of a 23.9-kilobase chromosomal deletion containing a region encoding *fsr* genes which mainly determines the gelatinase-negative phenotype of clinical isolates of *Enterococcus faecalis* in urine. *Appl. Environ. Microbiol.* **68:**3152–3155.

92. **Nallapareddy, S. R., X. Qin, G. M. Weinstock, M. Hook, and B. E. Murray.** 2000. *Enterococcus faecalis* adhesin, *ace*, mediates attachment to extracellular matrix proteins collagen type IV and laminin as well as collagen type I. *Infect. Immun.* **68:**5218–5224.

93. **Nallapareddy, S. R., K. V. Singh, R. W. Duh, G. M. Weinstock, and B. E. Murray.** 2000. Diversity of *ace*, a gene encoding a microbial surface component recognizing adhesive matrix molecules, from different strains of *Enterococcus faecalis* and evidence for production of Ace during human infections. *Infect. Immun.* **68:**5210–5217.

94. **Nallapareddy, S. R., G. M. Weinstock, and B. E. Murray.** 2003. Clinical isolates of *Enterococcus faecium* exhibit strain-specific collagen binding mediated by Acm, a new

member of the MSCRAMM family. *Mol. Microbiol.* **47:**1733–1747.

95. **NNIS.** 1997. National Nosocomial Infections Surveillance (NNIS) report, data summary from October 1986–April 1997, issued May 1997. *Am. J. Infect. Control* **25:**477–487.

96. **Noble, C. J.** 1978. Carriage of group D streptococci in the human bowel. *J. Clin. Pathol.* **31:**1182–1186.

97. **Novak, R. M., T. J. Holzer, and C. R. Libertin.** 1993. Human neutrophil oxidative response and phagocytic killing of clinical and laboratory strains of *Enterococcus faecalis. Diagn. Microbiol. Infect. Dis.* **17:**1–6.

98. **Olmsted, S. B., G. M. Dunny, S. L. Erlandsen, and C. L. Wells.** 1994. A plasmid-encoded surface protein on *Enterococcus faecalis* augments its internalization by cultured intestinal epithelial cells. *J. Infect. Dis.* **170:**1549–1556.

99. **Paulsen, I. T., L. Banerjei, G. S. Myers, K. E. Nelson, R. Seshadri, T. D. Read, D. E. Fouts, J. A. Eisen, S. R. Gill, J. F. Heidelberg, H. Tettelin, R. J. Dodson, L. Umayam, L. Brinkac, M. Beanan, S. Daugherty, R. T. DeBoy, S. Durkin, J. Kolonay, R. Madupu, W. Nelson, J. Vamathevan, B. Tran, J. Upton, T. Hansen, J. Shetty, H. Khouri, T. Utterback, D. Radune, K. A. Ketchum, B. A. Dougherty, and C. M. Fraser.** 2003. Role of mobile DNA in the evolution of vancomycin-resistant *Enterococcus faecalis. Science* **299:**2071–2074.

100. **Pillai, S. K., G. Sakoulas, G. M. Eliopoulos, R. C. Moellering, Jr., B. E. Murray, and R. T. Inouye.** 2004. Effects of glucose on *fsr*-mediated biofilm formation in *Enterococcus faecalis. J. Infect. Dis.* **190:**967–970.

101. **Pillai, S. K., G. Sakoulas, H. S. Gold, C. Wennersten, G. M. Eliopoulos, R. C. Moellering, Jr., and R. T. Inouye.** 2002. Prevalence of the *fsr* locus in *Enterococcus faecalis* infections. *J. Clin. Microbiol.* **40:**2651–2652.

102. **Pillar, C. M., and M. S. Gilmore.** 2004. Enterococcal virulence—pathogenicity island of *E. faecalis. Front. Biosci.* **9:**2335–2346.

103. **Porwancher, R., A. Sheth, S. Remphrey, E. Taylor, C. Hinkle, and M. Zervos.** 1997. Epidemiological study of hospital-acquired infection with vancomycin-resistant *Enterococcus faecium*: possible transmission by an electronic ear-probe thermometer. *Infect. Control Hosp. Epidemiol.* **18:**771–773.

104. **Qin, X., K. V. Singh, G. M. Weinstock, and B. E. Murray.** 2001. Characterization of *fsr*, a regulator controlling expression of gelatinase and serine protease in *Enterococcus faecalis* OG1RF. *J. Bacteriol.* **183:**3372–3382.

105. **Qin, X., K. V. Singh, G. M. Weinstock, and B. E. Murray.** 2000. Effects of *Enterococcus faecalis fsr* genes on production of gelatinase and a serine protease and virulence. *Infect. Immun.* **68:**2579–2586.

106. **Rezaiki, L., B. Cesselin, Y. Yamamoto, K. Vido, E. van West, P. Gaudu, and A. Gruss.** 2004. Respiration metabolism reduces oxidative and acid stress to improve long-term survival of *Lactococcus lactis. Mol. Micribiol.* **53:**1331–1342.

107. **Rice, L. B., L. Carias, S. Rudin, C. Vael, H. Goossens, C. Konstabel, I. Klare, S. R. Nallapareddy, W. Huang, and B. E. Murray.** 2003. A potential virulence gene, *hylEfm*, predominates in *Enterococcus faecium* of clinical origin. *J. Infect. Dis.* **187:**508–512.

108. **Rich, R. L., B. Kreikemeyer, R. T. Owens, S. LaBrenz, S. V. Narayana, G. M. Weinstock, B. E. Murray, and M. Hook.** 1999. Ace is a collagen-binding MSCRAMM

from *Enterococcus faecalis*. *J. Biol. Chem.* **274**:26939–26945.

109. **Richards, M. J., J. R. Edwards, D. H. Culver, and R. P. Gaynes.** 2000. Nosocomial infections in combined medical-surgical intensive care units in the United States. *Infect. Control Hosp. Epidemiol.* **21**:510–515.

110. **Rince, A., S. Flahaut, and Y. Auffray.** 2000. Identification of general stress genes in *Enterococcus faecalis*. *Int. J. Food Microbiol.* **55**:87–91.

111. **Roberts, J. C., K. V. Singh, P. C. Okhuysen, and B. E. Murray.** 2004. Molecular epidemiology of the *fsr* locus and of gelatinase production among different subsets of *Enterococcus faecalis* isolates. *J. Clin. Microbiol.* **42**:2317–2320.

112. **Safdar, N., and D. G. Maki.** 2002. The commonality of risk factors for nosocomial colonization and infection with antimicrobial-resistant *Staphylococcus aureus*, *Enterococcus*, gram-negative bacilli, *Clostridium difficile*, and *Candida*. *Ann. Intern. Med.* **136**:834–844.

113. **Schleifer, K. H., and R. Kilpper-Balz.** 1984. Transfer of *Streptococcus faecalis* and *Streptococcus faecium* to the genus *Enterococcus* nom. rev. as *Enterococcus faecalis* comb. nov. and *Enterococcus faecium* comb. nov. *Int. J. Sys. Bacteriol.* **34**:31–34.

114. **Schlievert, P. M., P. J. Gahr, A. P. Assimacopoulos, M. M. Dinges, J. A. Stoehr, J. W. Harmala, H. Hirt, and G. M. Dunny.** 1998. Aggregation and binding substances enhance pathogenicity in rabbit models of *Enterococcus faecalis* endocarditis. *Infect. Immun.* **66**:218–223.

115. **Shankar, N., A. S. Baghdayan, and M. S. Gilmore.** 2002. Modulation of virulence within a pathogenicity island in vancomycin-resistant *Enterococcus faecalis*. *Nature* **417**:746–750.

116. **Shankar, N., C. V. Lockatell, A. S. Baghdayan, C. Drachenberg, M. S. Gilmore, and D. E. Johnson.** 2001. Role of *Enterococcus faecalis* surface protein Esp in the pathogenesis of ascending urinary tract infection. *Infect. Immun.* **69**:4366–4372.

117. **Shepard, B. D., and M. S. Gilmore.** 2002. Differential expression of virulence-related genes in *Enterococcus faecalis* in response to biological cues in serum and urine. *Infect. Immun.* **70**:4344–4352.

118. **Shiono, A., and Y. Ike.** 1999. Isolation of *Enterococcus faecalis* clinical isolates that efficiently adhere to human bladder carcinoma T24 cells and inhibition of adhesion by fibronectin and trypsin treatment. *Infect. Immun.* **67**:1585–1592.

119. **Sifri, C. D., E. Mylonakis, K. V. Singh, X. Qin, D. A. Garsin, B. E. Murray, F. M. Ausubel, and S. B. Calderwood.** 2002. Virulence effect of *Enterococcus faecalis* protease genes and the quorum-sensing locus *fsr* in *Caenorhabditis elegans* and mice. *Infect. Immun.* **70**:5647–5650.

120. **Sillanpaa, J., Y. Xu, S. R. Nallapareddy, B. E. Murray, and M. Hook.** 2004. A family of putative MSCRAMMs from *Enterococcus faecalis*. *Microbiology* **150**:2069–2078.

121. **Stroud, L., J. Edwards, L. Danzing, D. Culver, and R. Gaynes.** 1996. Risk factors for mortality associated with enterococcal bloodstream infections. *Infect. Control Hosp. Epidemiol.* **17**:576–580.

122. **Suzuki, T., C. Shibata, A. Yamaguchi, K. Igarashi, and H. Kobayashi.** 1993. Complementation of an *Enterococcus hirae* (*Streptococcus faecalis*) mutant in the alpha subunit of the H+ -ATPase by cloned genes from the same and different species. *Mol. Microbiol.* **9**:111–118.

123. **Tannock, G. W., and G. Cook.** 2002. Enterococci as members of the intestinal microflora of humans, p. 101–132. *In* M. S. Gilmore, D. B. Clewell, P. Courvalin, G. M. Dunny, B. E. Murray, and L. B. Rice (ed.), *The Enterococci: Pathogenesis, Molecular Biology, and Antibiotic Resistance*. ASM Press, Washington, D.C.

124. **Tendolkar, P. M., A. S. Baghdayan, M. S. Gilmore, and N. Shankar.** 2004. Enterococcal surface protein, Esp, enhances biofilm formation by *Enterococcus faecalis*. *Infect. Immun.* **72**:6032–6039.

125. **Tendolkar, P. M., A. S. Baghdayan, and N. Shankar.** 2003. Pathogenic enterococci: new developments in the 21st century. *Cell. Mol. Life Sci.* **60**:2622–2636.

126. **Teng, F., K. D. Jacques-Palaz, G. M. Weinstock, and B. E. Murray.** 2002. Evidence that the enterococcal polysaccharide antigen gene (*epa*) cluster is widespread in *Enterococcus faecalis* and influences resistance to phagocytic killing of *E. faecalis*. *Infect. Immun.* **70**:2010–2015.

127. **Teng, F., L. Wang, K. V. Singh, B. E. Murray, and G. M. Weinstock.** 2002. Involvement of PhoP-PhoS homologs in *Enterococcus faecalis* virulence. *Infect. Immun.* **70**:1991–1996.

128. **Thiercelin, M. E.** 1899. Sur un diplocoque saprophyte de l'intestin susceptible de devenir pathogene. *C. R. Soc. Biol.* **5**:269–271.

129. **Thiercelin, M. E.** 1899. Morphologie et modes de reproduction de l'enterocoque. *C. R. Seances Soc. Biol. Fil.* **11**:551–553.

130. **Toledo-Arana, A., J. Valle, C. Solano, M. J. Arrizubieta, C. Cucarella, M. Lamata, B. Amorena, J. Leiva, J. R. Penades, and I. Lasa.** 2001. The enterococcal surface protein, Esp, is involved in *Enterococcus faecalis* biofilm formation. *Appl. Environ. Microbiol.* **67**:4538–4545.

131. **Tomita, H., and Y. Ike.** 2004. Tissue-specific adherent *Enterococcus faecalis* strains that show highly efficient adhesion to human bladder carcinoma T24 cells also adhere to extracellular matrix proteins. *Infect. Immun.* **72**:5877–5885.

132. **Ton-That, H., L. A. Marraffini, and O. Schneewind.** 2004. Sortases and pilin elements involved in pilus assembly of *Corynebacterium diphtheriae*. *Mol. Microbiol.* **53**:251–261.

133. **Ton-That, H., and O. Schneewind.** 2004. Assembly of pili in Gram-positive bacteria. *Trends Microbiol.* **12**:228–234.

134. **Ton-That, H., and O. Schneewind.** 2003. Assembly of pili on the surface of *Corynebacterium diphtheriae*. *Mol. Microbiol.* **50**:1429–1438.

135. **Trotter, K. M., and G. M. Dunny.** 1990. Mutants of *Enterococcus faecalis* deficient as recipients in mating with donors carrying pheromone-inducible plasmids. *Plasmid* **24**:57–67.

136. **Waters, C. M., M. H. Antiporta, B. E. Murray, and G. M. Dunny.** 2003. Role of the *Enterococcus faecalis* GelE protease in determination of cellular chain length, supernatant pheromone levels, and degradation of fibrin and misfolded surface proteins. *J. Bacteriol.* **185**:3613–3623.

137. **Waters, C. M., H. Hirt, J. K. McCormick, P. M. Schlievert, C. L. Wells, and G. M. Dunny.** 2004. An amino-terminal domain of *Enterococcus faecalis* aggregation substance is required for aggregation, bacterial internalization by epithelial cells and binding to lipoteichoic acid. *Mol. Microbiol.* **52**:1159–1171.

138. **Waters, C. M., C. L. Wells, and G. M. Dunny.** 2003. The aggregation domain of aggregation substance, not the RGD motifs, is critical for efficient internalization by HT-29 enterocytes. *Infect. Immun.* **71:**5682–5689.

139. **Wells, C. L., and S. L. Erlandsen.** 1991. Localization of translocating *Escherichia coli, Proteus mirabilis,* and *Enterococcus faecalis* within cecal and colonic tissues of mono-associated mice. *Infect. Immun.* **59:**4693–4697.

140. **Wells, C. L., R. P. Jechorek, and S. L. Erlandsen.** 1990. Evidence for the translocation of *Enterococcus faecalis* across the mouse intestinal tract. *J. Infect. Dis.* **162:**82–90.

141. **Wells, C. L., M. A. Maddaus, and R. L. Simmons.** 1988. Proposed mechanisms for the translocation of intestinal bacteria. *Rev. Infect. Dis.* **10:**958–979.

142. **Willems, R. J., W. Homan, J. Top, M. van Santen-Verheuvel, D. Tribe, X. Manzioros, C. Gaillard, C. M. Vandenbroucke-Grauls, E. M. Mascini, E. van Kregten, J. D. van Embden, and M. J. Bonten.** 2001. Variant *esp* gene as a marker of a distinct genetic lineage of van-comycin-resistant *Enterococcus faecium* spreading in hospitals. *Lancet* **357:**853–855.

143. **Wisplinghoff, H., T. Bischoff, S. M. Tallent, H. Seifert, R. P. Wenzel, and M. B. Edmond.** 2004. Nosocomial bloodstream infections in US hospitals: analysis of 24,179 cases from a prospective nationwide surveillance study. *Clin. Infect. Dis.* **39:**309–317.

144. **Xu, Y., B. E. Murray, and G. M. Weinstock.** 1998. A cluster of genes involved in polysaccharide biosynthesis from *Enterococcus faecalis* OG1RF. *Infect. Immun.* **66:** 4313–4323.

145. **Xu, Y., K. V. Singh, X. Qin, B. E. Murray, and G. M. Weinstock.** 2000. Analysis of a gene cluster of *Enterococcus faecalis* involved in polysaccharide biosynthesis. *Infect. Immun.* **68:**815–823.

146. **Zeng, J., F. Teng, G. M. Weinstock, and B. E. Murray.** 2004. Translocation of *Enterococcus faecalis* strains across a monolayer of polarized human enterocyte-like T84 cells. *J. Clin. Microbiol.* **42:**1149–1154.

Enterococcal Genetics

KEITH E. WEAVER

26

The majority of interest in enterococcal genetics has been generated in response to three landmark discoveries: (i) identification of the first conjugative plasmids whose transfer systems are induced by an identifiable signal (57), (ii) identification of the first "transposons" capable of intercellular (conjugative) transposition (77, 82), and (iii) the acquisition of vancomycin resistance (123, 200). Because the most prevalent vancomycin resistance genes are located on plasmids and transposons, most work on enterococcal genetics has focused on mobile genetic elements. Examination of the complete sequence of a vancomycin-resistant clinical isolate of *Enterococcus faecalis* reaffirmed the importance of such elements in the evolution of this species, revealing that over a quarter of the genome consists of mobile and/or exogenously acquired DNA (156). In addition to the previously identified plasmids and transposons, bacteriophage, pathogenicity islands, and integrons must now be added to the list of mobile elements contributing to the evolution of *E. faecalis* and probably the other enterococci. However, understanding of the basic mechanisms of DNA replication and repair, chromosomal segregation, cell division, and transcription in this genus remains limited. The chapter will begin with a review of what is known or can be discerned from the genome sequence about these basic genetic mechanisms. Much of the information presented in this section was derived from information and tools provided for examining the *E. faecalis* and *E. faecium* genomes at the Web sites of The Institute for Genomic Research and the Joint Genome Institute, respectively (196a). The rest of the chapter will focus on the known mobile genetic elements, both old friends and new acquaintances, which seem to play such a significant role in the evolution of the enterococci.

CHROMOSOMAL REPLICATION, REPAIR, AND INHERITANCE

In a comparison of genes surrounding the replication origins of *Bacillus subtilis* and *Pseudomonas putida*, Ogasawara and Yoshikawa (148) concluded that gene order in this region is highly conserved among eubacteria. Comparison of the region surrounding the *dnaA* genes of *B. subtilis* (115) and *E. faecalis* revealed that the order of at least 10 genes is nearly identical between the two species. The conserved gene order is *jag-spoIIIJ-rnpA-rpmH-dnaA-dnaN-yaaA-recF-gyrB-gyrA*. The *spoIIIJ* and *yaaA* designations are from *B. subtilis* and are not designated as such in the *E. faecalis* genome, but sequence analysis shows that the syntenic *E. faecalis* genes belong to the same conserved orthologous group. Beyond the *gyrA* gene both species encode genes involved in ribosome structure (ribosomal RNA genes in *B. subtilis*, ribosomal proteins in *E. faecalis*), but beyond that gene order is not conserved. The gene order downstream of *dnaA* in *E. faecium* is identical to that of *E. faecalis*. Interestingly, this gene order is well conserved in two *Lactobacillus* genome sequences but not in those of *Lactococcus lactis* or the streptococci. Examination of the *E. faecalis* genome sequence immediately adjacent to the *dnaA* gene identified eight DnaA boxes, four on each side of the gene, indicating that the origin of replication is probably located nearby.

The *E. faecalis* genome includes homologs of all known components of the typical gram-positive DNA polymerase III (Pol III) holoenzyme, including a replicative DNA helicase (which is named DnaB according to the *Escherichia coli* convention rather than DnaC as in *B. subtilis*), helicase loaders, primase, clamp and clamp loaders, and DnaE- and PolC-type catalytic alpha subunits. Most have also been identified in the *E. faecium* genome. It has been proposed that PolC may be the primary leading strand DNA polymerase while DnaE plays dual roles as the lagging strand polymerase and as an error-prone polymerase during DNA repair in *B. subtilis* and *Streptococcus pyogenes* (25, 121). The *E. faecalis* PolC and DnaE proteins have been purified and found to have characteristics similar to the *B. subtilis* homologs, even cross-reacting with antibodies generated against *B. subtilis* PolC and DnaE (71), suggesting that *E. faecalis* and *B. subtilis* are functionally similar.

What is less clear is whether the *E. faecalis* DNA Pol III is localized as it is in *B. subtilis*. It has been shown that DNA Pol III holoenzyme forms a stationary focus at midcell during DNA replication, dividing near the end of the cell cycle and reforming at the cell quarter positions that become the

Gram-Positive Pathogens, 2nd edition, edited by Vincent A. Fischetti et al.
© 2006 ASM Press, Washington, D.C.

midpoint of daughter cells (124). This has led to a factory model of DNA replication in which the chromosomal DNA is pulled through a membrane-associated DNA polymerase factory rather than the polymerase traveling along a stationary chromosome. What anchors the polymerase at midcell is not clear. It would certainly be of interest to localize the DNA Pol III factory in cocci to determine what kind of movement occurs in cells that have no distinct quarter positions. In *E. faecalis* cells, cell growth involves the splitting of centrally located cell wall bands, deposition of new cell wall material between the two new bands, and migration of the bands to the center of the new daughter cells (99). It would be interesting to determine if movement of the DNA Pol III factory is coordinated with band movement.

In addition to DNA Pol III, the *E. faecalis* chromosome also encodes a DNA Pol I homolog, two putative phage-encoded DNA polymerases, and one member of the mutagenic Y-family polymerases (149) that have been shown to be involved in lesion bypass in other organisms. Y-family polymerases are commonly found on bacterial plasmids (160), and, in fact, all three of the plasmids in *E. faecalis* V583 encode at least one member of the family. The previously described *uvrA* gene of *E. faecalis* plasmid pAD1 (152) also belongs to this group and was shown to impart increased resistance to UV exposure.

Other genes associated with DNA repair present in the *E. faecalis* genome include homologs required for transcription-coupled DNA repair (*mfd*), nucleotide excision repair (*uvrA*, *uvrB*, and *uvrC*), mismatch repair (*mutS* and *mutL* homologs annotated as *hexA* and *hexB*, respectively), and numerous other nucleotide repair proteins, including seven potential *mutT* homologs. It should be noted that the previously designated *uvrABC* genes located on plasmid pAD1 (152) are not homologs of the excision repair enzymes that usually bear this name, but are instead associated with a lesion bypass polymerase. Multiple homologs of genes involved in recombination repair were also identified, including genes for a LexA-like regulator, RecA, a RexAB DNA processing enzyme, a RuvAB Holliday junction helicase, a PriA replication restart primase, a RecFOR complex, and a variety of other recombination proteins. Although a recombination-deficient mutant of *E. faecalis* has been isolated with a phenotype similar to other *recA* mutants (217), it has not been characterized genetically and the mutation site or sites are unknown. The other putative recombinational repair proteins have not been characterized.

The *E. faecalis* genome includes topoisomerases I to IV, an SMC homolog, and an HU homolog, suggesting that the chromosomal DNA is organized and compacted as in other bacteria. No replication terminator protein is annotated in the genome and no identifiable homolog of the *B. subtilis* Rtp terminator could be identified by BLAST search. In fact, Rtp appears to be limited to the *Bacillus* genus, suggesting that proteins involved in termination of replication are poorly conserved. A single XerD homolog, involved in chromosome dimer resolution in other bacteria, is annotated in the *E. faecalis* genome, but several other integrase/recombinases have been identified that could play the role of the XerCD resolution recombinase. The C terminus of the cell division protein FtsK has been implicated in organizing both the XerCD dimer resolution complex and the topoisomerase IV decatenation complex near the chromosomal replication terminus for efficient separation of daughter chromosomes in *E. coli* (64, 108). The C-terminal domain of *E. faecalis* FtsK is 42% identical to the C terminus of *E. coli* FtsK, but the N-termi-

nal domains of the two proteins are not similar, suggesting that they may play similar roles at the end of chromosomal replication but different roles in cell division.

ParAB proteins have been implicated in the partition of plasmid and chromosomal DNA in both gram-positive and gram-negative organisms (83). Several ParA and ParB homologs are encoded on the *E. faecalis* chromosome, but many of these likely reside on the remnants of integrated plasmids. However, one ParAB pair (EF3299 and EF3298, respectively) is located between *gidB* and *obg* homologs, genes commonly located on bacterial chromosomes. This ParAB pair is also highly homologous to the *B. subtilis* Soj (66% identical) and SpoOJ (51% identical) proteins, which are required for chromosome segregation and sporulation (109). Whether the *E. faecalis* ParAB homologs are involved in chromosome segregation and what other proteins might also be involved are currently unknown.

The division/cell wall cluster of *E. faecalis* was previously identified in strain A24836 (167) and appears to be identical in the fully sequenced V583 genome (note that *divIB* and *ftsQ* are orthologs). This cluster includes *ftsZ*, the gene for the bacterial tubulin homolog essential for contraction of the cell membrane during cell division, two other cell division genes *ftsA* and *ftsQ*, and several genes involved in peptidoglycan synthesis. There are also multiple *ftsW*- and *ftsK*-like genes in the V583 genome that may be involved in cell division. In *B. subtilis*, the DivIVA protein functions with cell division inhibitory MinCD proteins to ensure proper septum placement (61). Although the *E. faecalis* genome contains a strong *divIVA* homolog, there are no apparent *minCD* homologs. Because *E. faecalis* is not rod-shaped, the absence of a septum placement system might not be surprising, although it should be noted that at least some gram-negative cocci retain the *minCDE* system and use it for division site placement (170a). How the division site is chosen in gram-positive cocci is unknown. Although some recent work has focused on the coordination of cell wall synthesis with division in *Streptococcus pneumoniae* (138, 139), much more needs to be done to determine division pathways in gram-positive cocci.

REGULATION OF GENE EXPRESSION

Most of what is known about the regulation of gene expression in enterococci has been determined by the examination of mobile genetic elements and genes important in pathogenesis. This information is discussed either below in sections relating to individual mobile elements or in chapter 25. In this section we will focus on general features that can be discerned from the genomic sequence data and/or from the limited experimental data.

Regulation of antigenic surface proteins often results in near random on-off transitions, referred to as antigenic or phase variation. This type of regulation is frequently the result of "slippage" of the DNA Pol III enzyme during replication of homopolymeric stretches or iterative nucleotide repeats located within an open reading frame or between the -10 and -35 boxes of a promoter (128). Of 134 putative surface proteins identified within the *E. faecalis* genome, 65 were found to have repeats that could potentially lead to replication slippage, suggesting that this mechanism may play a major role in regulating the antigenic surface features of enterococcal cells (156).

Genes for the RNA polymerase α, β, and β′ subunits, as well as the ω assembly factor and the δ specificity factor (127), are present in the *E. faecalis* genome. Four recognizable sigma factor genes are also present, encoding a putative housekeeping sigma factor (referred to as *rpoD* or *sigA*), two homologs of the extracytoplasmic family sigma factors, and a σ^{54}-like sigma factor (*rpoN*). Although this is certainly a smaller complement of alternative sigma factors than *B. subtilis* or *E. coli*, it is more than any other annotated genome from a lactic acid bacterium. σ^{54} appears to be particularly rare in this group of bacteria, found only in *Lactobacillus plantarum* and *E. faecalis*. σ^{54} sigma factors are unique among bacterial sigma factors because they are unable to form open complexes and initiate transcription without the help of an ATP-dependent activator protein (26). Such activators are highly conserved, and several have been identified in the *E. faecalis* genome (97). Four full-length genes for σ^{54}-RNA polymerase activator proteins were identified in the *E. faecalis* genome, and all were homologous to a subfamily of activators that include a DEAH helicase motif. Immediately downstream of each of these genes was a sequence similar to known σ^{54} promoters and an operon encoding components of a PTS permease system. Mutation of one of the regulators as well as either of two genes proposed to be regulated by it resulted in the same phenotype (increased sensitivity to a bacteriocin), suggesting that the regulator actually controlled expression of the downstream genes. Mutation of σ^{54} itself resulted in increased sensitivity to the same bacteriocin (49). This work implicates *E. faecalis* σ^{54} in regulation of PTS permeases, although more work is required to define the promoter sequences recognized and the interactions with its regulators. Interestingly, the partially completed *E. faecium* genome encodes five activator proteins in the same subfamily as the *E. faecalis* σ^{54} activators.

Two-component signal transduction systems (TCSTS) have emerged as a ubiquitous mechanism used by bacteria to regulate gene expression in response to environmental conditions (103). The canonical TCSTSs are composed of a sensor histidine kinase (HK) and a response regulator (RR), which is commonly a transcriptional regulator. Analysis of the *E. faecalis* genome revealed the presence of 17 HK/RR pairs and 1 orphan RR. Classification and similarity to previously described TCSTSs are described in reference 92, and we will use those authors' nomenclature to discuss the systems characterized in enterococci. The best-characterized enterococcal TCSTSs are the Van systems (HK-RR 11) required for regulation of vancomycin resistance genes; these will be discussed in more detail below. The Fsr system (HK-RR 15) regulates production of two potential virulence factors, gelatinase and a serine protease (63, 142, 168, 169). Fsr is a quorum-sensing system that responds to the accumulation of a peptide lactone, designated GBAP, that is processed from the C-terminal end of FsrB, a protein also required for processing and export of GBAP (143). The *fsr* system appears to be analogous to the well-studied *agr* system of *Staphylococcus aureus* (162). The CroRS system (HK-RR 05) is essential for PBP5-mediated intrinsic cephalosporin resistance of *E. faecalis* (45). This system is induced by any inhibitor of peptidoglycan synthesis and is autoregulated, but deletion of *croRS* does not affect PBP5 production, and supply of ectopic PBP5 under the control of a heterologous promoter does not rescue cephalosporin resistance. Mutations in *croR* also fail to induce transcription of the general stress gene *sagA* (120). This suggests that CroRS regulates some gene that acts upstream of PBP5 to enhance resistance. A possible link between *sagA* expression and PBP5-dependent cephalosporin resistance has not been explored. The Eta system (HK-RR 10) is involved in virulence as well as resistance to low pH

and high temperature (196). Disruption of RR 04, 08, and 18 each resulted in sensitivity to heat, and transcription of HK-RR 07, 13, and 14 was induced by heat, suggesting that these TCSTSs may also play a role in stress resistance (120).

In *E. faecium* one TCSTS, EntKR, involved in expression of the bacteriocin enterocin A, has been described (150). EntKR is a quorum-sensing system, responding to a small peptide produced from the same locus, designated EntF. There appear to be no close homologs to the EntKR system in *E. faecalis*. Clear homologs of the *E. faecalis* HK-RR 01, 03, 04, 05 (CroRS), 07, 09, and 10 are present in the *E. faecium* genome sequence. In addition, a system with linked homologs of the *S. aureus agrA*, *agrB*, and *agrC* genes is present in *E. faecium*, but this system is not closely related to the *E. faecalis fsr* system discussed above. The *E. faecium* genome contains 10 more annotated RRs and nine more HKs that either have no strong homology to counterparts in the *E. faecalis* genome or are no more closely related to *E. faecalis* genes than genes from any other low G+C grampositive bacterium. Therefore, *E. faecalis* and *E. faecium* have about the same number of TCSTSs (17 and 18, respectively, with one orphan RR in each), but not necessarily the same complement of homologs.

The *E. faecalis cyl* cytolysin-producing operon is induced by a quorum-sensing mechanism operating through a "two-component" system, but in this case the sensor and regulator show no homology to HKs or RRs of typical TCSTS and the inducer is one of the components of the toxin itself (91).

The *E. faecalis* genome sequence includes numerous other putative regulatory proteins in a variety of other classes. Few of these regulators have been characterized. The *E. faecalis* version of the catabolite control protein gene *ccpA* has been mutationally inactivated and shown to regulate a variety of glucose starvation proteins and was also shown to complement a *B. subtilis ccpA* mutant (119). Putative regulators of arginine catabolism (15) and a manganese-inducible virulence operon *efaCBA* (129) have been identified, but the effects of mutations have not been tested. Several other operons appear to be regulated by various environmental conditions, but the mechanism of regulation and the regulators have not been identified (50, 93, 140, 186). A potential transcriptional activator responsive to oxidative stress, designated *hypR*, was recently described (201), and the presence of an OxyR homolog that regulates the gene for NADH peroxidase has been postulated on the basis of indirect evidence (155).

Perhaps the best-studied non-TCSTS regulatory system in enterococci is the *cop* copper homeostasis system of *Enterococcus hirae*, recently reviewed in reference 189. The four-gene operon consists of the *copA* and *copB* genes, which encode proteins for copper uptake and efflux, respectively; *copY*, which encodes a copper-responsive repressor; and *copZ*, which encodes a copper chaperone that delivers copper from CopA to CopY. At low copper levels, CopY in complex with Zn(II) binds to DNA sequences overlapping the promoter as a dimer and turns down transcription presumably by interfering with RNA polymerase binding. At high copper levels, CopZ delivers Cu(I) to CopY, displacing Zn(II) and causing dissociation from the promoter. CopZ complexed with Cu(I) is subject to proteolysis, preventing excessive accumulation of intracellular copper, which can damage the cell.

Postinitiation regulation by alternate RNA secondary structures is a common theme in the regulation of bacterial genes involved in amino acid and nucleotide biosynthesis. Such mechanisms have not been studied in detail in the enterococci, although the *pyr* operon of *E. faecalis* appears to be regulated by PyrR via an attenuation mechanism similar to that described for *B. subtilis* (85). Bioinformatic analysis suggests that T-box regulation is also present in *E. faecalis* and *E. faecium* (153). A transcription attenuation mechanism for regulation of the *tet*M gene of Tn916 was proposed based on the observation of (i) a short leader peptide encoded between the promoter and the TetM-encoding open reading frame, (ii) the existence of multiple alternative RNA secondary structures surrounding the leader peptide coding sequence, and (iii) the apparent extension of a short transcript upon induction with tetracycline (190). The details of this proposed mechanism remain to be verified.

Very little is known about posttranscriptional regulation in *E. faecalis* aside from the identification of genes involved in translation and RNA decay. Two antisense RNA systems and a translational stimulatory RNA have been described on the pAD1 and pCF10 plasmids and will be discussed in the relevant sections below.

VANCOMYCIN RESISTANCE

Enterococci are intrinsically resistant to a variety of commonly used antibiotics and have acquired resistance to many others either by chromosomal mutation or acquisition of mobile genetic elements. As such, the enterococci may provide a collection and distribution center for the dissemination of antibiotic resistance genes to other, potentially more pathogenic, gram-positive bacteria. Dramatic evidence of this was recently provided by the characterization of the first high-level vancomycin-resistant *S. aureus* (VRSA) isolate (67, 213). Examination of the plasmid content of the VRSA isolate and co-isolated vancomycin-resistant *E. faecalis* (VRE) and vancomycin-sensitive *S. aureus* (VSSA) strains suggested that a *van*-carrying transposon, Tn1546, was delivered from the VRE to the VSSA on a conjugative plasmid. The Tn1546 then transposed to a plasmid resident in the VSSA, and the donating plasmid was subsequently lost. Because the new VRSA Tn1546-carrying plasmid encodes a pSK41-like conjugation system, it is likely that these genes will continue to be disseminated.

Vancomycin is a cell wall-active antibiotic that represents the last line of defense against many multiply-resistant gram-positive pathogens. For this reason, the emergence of VRE has caused great concern in the medical community. Vancomycin and related glycopeptide antibiotics inhibit cell wall synthesis by binding to the terminal D-Ala-D-Ala dipeptide of peptidoglycan precursors, preventing their transfer from the lipid carrier to the cell wall by transglycosidases. Transpeptidases and D,D-carboxypeptidases involved in cell wall synthesis are also inhibited. VRE circumvents vancomycin action by substituting D-Ala-D-Lac or D-Ala-D-Ser for the D-Ala-D-Ala dipeptide, decreasing the affinity of peptidoglycan precursors for glycopeptides.

Unlike resistance to most other antibiotics, resistance to vancomycin involves the acquisition of a suite of enzymes required to reprogram peptidoglycan synthesis. Six genotypically and phenotypically distinct systems designated VanA-E and G have been identified in VRE. All six systems include three components: (i) a TCSTS, VanRS, required for sensing the presence of the antibiotic and inducing the resistance operon; (ii) an essential three-gene core encoding enzymes responsible for producing the D-Lac (VanH) or D-Ser (VanT) precursors, ligating those precursors to D-Ala (VanA-G), and eliminating D-Ala-D-Ala dipeptides (VanX, VanY, and/or a VanXY fusion); and (iii) accessory genes

performing various and sometimes unknown roles. Although the core and accessory genes are generally transcribed as an operon, the TCSTS is transcribed from a separate promoter (Fig. 1). The VanC, VanE, and VanG systems all encode low-level vancomycin resistance by producing peptidoglycan peptides terminating in D-Ala-D-Ser. All encode a VanXY bifunctional dipeptidase/carboxypeptidase for degrading D-Ala-D-Ala, a VanT serine racemase for producing D-Ser, and their respective ligases. The VanC systems are responsible for the intrinsic resistance of motile enterococci, are chromosomally encoded, and do not appear to be on mobile elements. Although VanC systems appear to encode a TCSTS, resistance is expressed either constitutively or only sluggishly induced (6, 60). The VanE system is very similar to VanC in sequence and gene organization but is present in an *E. faecalis* strain (154). Although likely acquired from one of the intrinsically resistant enterococci, it is chromosomally located and could not be transferred to a vancomycin-susceptible strain. VanE is inducible in spite of a nonsense mutation in the gene for the VanS HK, suggesting some level of cross-talk with other cellular TCSTSs. The VanG system is also chromosomally located in an *E. faecalis* strain, but resides on a large (~240 kbp) mobile element that also encodes an *erm*B erythromycin resistance gene (55). The VanG system is inducible, probably due to the presence of a functional VanRS TCSTS.

The VanA, VanB, and VanD systems all produce peptidoglycan precursors terminating in D-Ala-D-Lac. The VanA and VanB systems are widespread in both *E. faecalis* and *E. faecium*. In general, vancomycin resistance is more prevalent in *E. faecium* than in *E. faecalis* (179). Phenotypically, VanA-type resistance is characterized by high-level resistance to both vancomycin and teicoplanin, while VanB-type resistance provides variable levels of resistance to vancomycin only. Susceptibility to teicoplanin in the VanB system is due to a failure to induce the resistance genes in response to this antibiotic. VanB cells induced with vancomycin are resistant to both antibiotics. Genotypically, the VanA and VanB systems are quite similar (Fig. 1). Each encodes a VanRS TCSTS, a VanH pyruvate dehydrogenase for production of D-Lac, their respective ligase VanA or VanB, and separate VanX dipeptidase and VanY carboxypeptidase enzymes. Only the *van*H, *van*A/B, and *van*X genes are essential for resistance. In addition, the VanA system encodes an additional gene, *van*Z, which confers low-level resistance to teicoplanin by an unknown mechanism. The VanB system encodes a gene, *van*W (also present in VanG), of unknown function.

The VanRS TCSTS of the VanA and VanB systems have been extensively investigated and have been recently reviewed (10). In vitro the VanRS system functions much like other TCSTSs. The HK VanS autophosphorylates its conserved histidine residue and then transfers the phosphate to the VanR RR (216). Phosphorylated VanR (VanR~P) produces a larger footprint and binds with higher affinity to its target promoters than unphosphorylated VanR, consistent with its in vivo role as a transcriptional activator (105). However, in the absence of VanS, dephosphorylation of VanR~P is unusually slow compared to related RRs, and VanS stimulates dephosphorylation (216). In addition, VanR is capable of stimulating its own phosphorylation in the absence of VanS using acetyl phosphate. These in vitro results are consistent with the observation in vivo that inactivation of VanS results in constitutive activation of vancomycin resistance rather than failure to induce expression. Apparently, VanR~P is the default condition of the RR in vivo, with

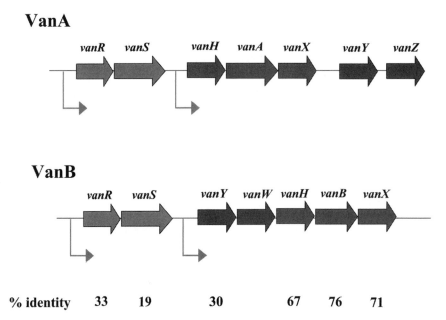

FIGURE 1   (*See the separate color insert for the color version of this illustration.*) Organization of the VanA and VanB vancomycin resistance systems. Gene functions are color-coded: green, sensory and regulatory; red, essential resistance genes; blue, auxiliary resistance genes. Detailed functions of each gene are described in the text. Green arrows show the position of the VanR-inducible promoters. The percent identity of the VanB genes to their VanA homologs is given below the individual VanB genes. Homologs have identical names in both systems except for the *vanA* and *vanB* ligases.

phosphorylation occurring either by acetyl phosphate or by cross-talk with other cellular HKs. The primary role of VanS, then, is to dephosphorylate VanR in the absence of the inducing signal (7, 8). The nature of the inducing signal remains unknown. Given the differences in induction patterns and the poor conservation of VanS, the signal may be different for the VanA and VanB systems. Interestingly, single amino acid substitutions in the signaling domain of VanS$_B$ allow induction by teicoplanin (14).

Both the VanA and VanB gene clusters are frequently, if not always, associated with mobile genetic elements. VanA is located on a Tn3-type transposon Tn*1546* and is frequently found on conjugative plasmids. In one *E. faecium* outbreak, a unique mechanism of facilitating horizontal transfer of VanA was observed (96). In this case, a nonconjugative Tn*1546*-bearing plasmid was observed to integrate into a pheromone-responsive plasmid via a mechanism involving a series of insertion sequences located outside of Tn*1546*. Conjugative transfer was enhanced by the fact that plasmid fusion occurred within a homolog of *prgX*, a key negative regulator of the pheromone response in pCF10-like plasmids (see below). The resulting insertional inactivation of *prgX* derepressed conjugation functions, and transfer occurred in the absence of pheromone. Tn*1546*-like elements have also been identified on plasmids related to the conjugative non-pheromone-responsive plasmid pMG1 (198). VanB has been localized on a variety of mobile elements. In some strains, conjugative transfer of VanB has been associated with the movement of large (90 to 250 kb), chromosomally located elements that may be complex conjugative transposons (65). VanB was further localized on one of these elements to a 64-kb composite transposon, Tn*1547*, capable of transposition to plasmids under laboratory conditions. In other cases, VanB resistance was associ-

ated with conjugative plasmids (215), but the identification of similar genes on a variety of plasmids also suggested a transposon location. A VanB gene cluster has been localized on a Tn*916*-like element designated Tn*5382* (31). Although independent mobility of this element was not demonstrated, it appears to be associated with a larger, 130- to 160-kb chromosomally located conjugative element. A similar element, designated Tn*1549*, was identified on a pAD1-like pheromone-responsive plasmid (81).

Unlike VanA and VanB, VanD systems provide constitutive vancomycin resistance due to various lesions in the *vanS* gene (20, 32, 56). In spite of the presence of an apparently intact *vanX* gene, VanD systems produce very little VanX D-Ala-D-Ala dipeptidase activity. Nevertheless, peptidoglycan precursors in VanD strains contain very little or no D-Ala-D-Ala terminal peptides due to mutations in the chromosomally encoded *ddl* gene responsible for producing the D-Ala-D-Ala precursors (20, 32, 157). Thus, a combination of constitutive expression of VanD and mutation of the chromosomal *ddl* ensures vancomycin resistance. VanD systems have been found only in *E. faecium* and do not appear to be mobile.

The complex nature of the vancomycin resistance systems argues against their acquisition by accumulation of mutations or piecemeal acquisition of individual genes. More than likely they were acquired en masse from a single source. In this context it is interesting to note that the gene organization and sequence of the VanA and VanB systems are highly similar to self-protection systems present in glycopeptide-producing species (136, 137).

Resistance to Other Antibiotics

In addition to vancomycin, enterococci have acquired resistance to a wide variety of other antibiotics. Ampicillin resistance in enterococcal species is due to the overproduction of PBP5, which has a low affinity for the drug. Overproduction was believed to be due to disruption of a putative negative regulator encoded by the *psr* gene (126), but recent data have cast doubt on this interpretation (173, 180). Specific combinations of amino acid substitutions in PBP5 also affect affinity and antibiotic resistance (171). Interestingly, in an ampicillin-resistant isolate of *E. faecium* the *pbp5* gene was located on the same large chromosomal mobile element as the *vanB*-containing Tn*5382* (see above), explaining the close association of these two resistance markers among clinical *E. faecium* isolates (31). High-level aminoglycoside resistance is provided by a variety of aminoglycoside-modifying enzymes, and at least some of these are located on plasmids and transposons (34). The most common mechanism for macrolide/licosamide/streptogramin B resistance is methylation of 23S rRNA by the *ermB* gene (164), which can be encoded on Tn3-like transposon Tn*917* (185). A plasmid was isolated in *E. faecium* that encoded *ermB*, *vanA*, and *vatD*, encoding resistance to the streptogramin synergistic combination quinupristin/dalfopristin (21). Chloramphenicol resistance is frequently encoded on *cat* genes common on broad host range, conjugative Inc18 plasmids such as pRE25, which also encodes *ermB* and several aminoglycoside-modifying enzymes (181). Tetracycline resistance is most commonly encoded by *tetM* on Tn*916*-like conjugative transposons (see below), but Tn*916*-like elements can also carry other resistance determinants (53, 117). Resistance to linezolid, evernimicin, and quinolones has so far been restricted to mutations in their chromosomal targets.

PHEROMONE-RESPONSIVE PLASMIDS

The enterococci play host to a wide variety of plasmids. Three classes of plasmids commonly found in enterococci have been studied in detail. Two of these, the rolling-circle replicating plasmids and the theta replicating Inc18 plasmids, are not restricted to the enterococci and have been reviewed extensively elsewhere (24, 54, 112–114, 147, 208) and will not be discussed further here. The third class, the pheromone-responsive plasmids, replicate primarily in *E. faecalis* and remain one of only two plasmid systems in which a specific signal for induction of conjugation has been identified (the quorum-sensing system of *Agrobacterium* Ti plasmids is the other [80]). These plasmids will be discussed below, but the reader is referred to recent reviews that discuss the replication (208) and conjugation (38, 42) of these elements in more detail than is possible here. Three other enterococcal plasmids of interest, pAMα1, pMG1, and pRUM, will be discussed briefly at the end of this section.

The pheromone-responsive plasmids range in size from 37 to 92 kb and are maintained at a low copy number in their native host. In general, the pheromone-responsive conjugation system works as follows. Conjugation in cells containing pheromone-responsive plasmids is normally repressed in the absence of an appropriate recipient. When cocultivated with or exposed to culture filtrates of plasmid-free cells, plasmid-containing cells respond by producing a plasmid-encoded surface adhesin called aggregation substance (AS). AS binds to a ligand called enterococcal-binding substance that is present on the surface of both donor and recipient cells, leading to the formation of macroscopic cellular aggregates that provide the contact required for plasmid transfer. Once transfer is complete, production of the pheromone specific for the transferred plasmid is shut down in recipient cells. However, production of pheromone(s) specific for other plasmids continues, allowing the recipient to induce a mating response in other potential donors and to continue to collect plasmids. A cartoon showing the relevant players involved in the pheromone response, how they interact, and the organization of the plasmid-encoded regulatory genes is shown in Fig. 2.

All sex pheromones characterized to date are linear, hydrophobic peptides of seven or eight amino acids. The genetic determinants responsible for production of five pheromones were identified in the *E. faecalis* V583 genome, and all were found to be encoded within signal peptide coding sequences of different lipoprotein genes (37). Involvement of these genes in production of pheromones specific for pAD1 (cAD1), pCF10 (cCF10), and pAM373 (cAM373) was confirmed by cloning and mutagenesis experiments (2, 5, 69). Mutations of the mature lipoprotein-encoding portion of the genes for cAD1 and cAM373, *cad* and *camE*, respectively, had no effect on pheromone production or mating potential, suggesting that the lipoproteins themselves are not involved in conjugation. The cCF10 gene, *ccfA* (a *spoIIIJ* homolog), could not be disrupted and is apparently essential. All of the pheromone sequences except for cCF10 are at the C terminus of the lipoprotein signal peptide so presumably signal peptidase II is required for processing, although this has not been proven. Processing at the pheromone N terminus requires the Eep protein for all pheromones except cAM373 (4). Eep belongs to a family of intramembrane processing proteins involved in regulated intramembrane proteolysis in organisms as diverse as bacteria and humans (23). Interestingly, both cAD1 and cAM373 activities can be detected in some *S. aureus* strains. In both cases the responsible peptides are produced from the signal sequences of lipoproteins unrelated to those that produce the same peptides in *E. faecalis* (19, 69). Linear synthetic peptides of the appropriate sequence function as effective inducing signals for the cognate plasmids, indicating that no further processing is necessary.

Pheromones are imported into plasmid-containing cells by the chromosomally encoded oligopeptide permease system in combination with a plasmid-encoded oligopeptide permease A homolog that provides increased sensitivity and specificity for the cognate pheromone (125). The plasmid-encoded facilitators are referred to as TraC in pAD1, pAM373, and pPD1 and as PrgZ in pCF10 (144, 178, 193). Inside the cell, pheromone is bound by a key negative regulatory protein TraA/PrgX (78, 144). Binding of cAD1 by the pAD1-encoded TraA repressor leads to a loss of affinity for the P_0 promoter and consequent derepression of the TraE1 positive regulatory protein and downstream conjugation genes (78). The mechanism of action of TraE1 is unknown, but it appears to function at a posttranscriptional level (141, 197). The effect of pheromone binding on pCF10 PrgX appears to be more subtle, apparently changing the conformation of the protein on its promoter rather than significantly affecting binding affinity, and only affecting downstream expression about twofold (12). The downstream positive regulators in the pCF10 system include three small proteins and the Q_L RNA, which appear to interact with ribosomes to force antitermination of transcription into the downstream conjugation genes (18, 35). Superimposed on TraA/PrgX regulation is an antisense RNA-controlled negative regulatory circuit. mD and Qa antisense RNAs have been identified and described in pAD1 and pCF10, respectively, and their sequences and location ("aR" in Fig. 2) are highly conserved in other pheromone plasmids. In both cases the antisense RNA has been shown to stimulate termination at t1, perhaps by stabilizing a termination-proficient conformation of the P_0 transcript (13, 197). Interestingly, PrgX and Qa are transcribed

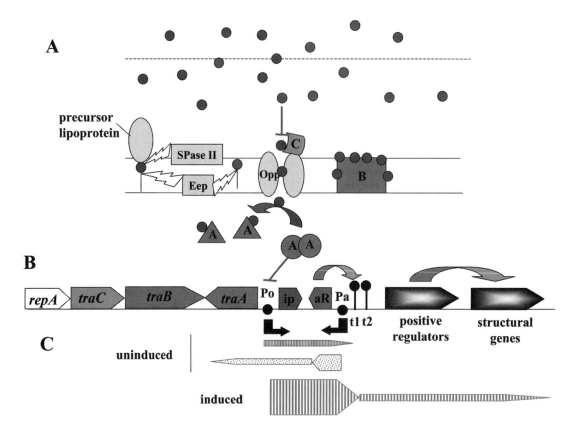

FIGURE 2 (*See the separate color insert for the color version of this illustration.*) Regulatory circuitry of the pheromone response. (A) Events occurring at the cell surface. Important chromosomally encoded determinants are colored yellow, with the exception of the pheromone itself, which is shown as blue circles. Plasmid-encoded determinants are color-coordinated with their genes shown in B. The inhibitor peptide (ip) is shown as red circles and competitively inhibits pheromone binding to TraC. TraB inhibition of pheromone secretion is depicted here as sequestration, but this is only one possible mechanism and has not been proven. (B) Events occurring at the DNA level. Binding of pheromone by TraA links events at the cell surface with events at the DNA level. A conformational change in TraA due to pheromone binding is indicated by the change in shape of the molecule and change in dimerization state, although the precise changes are still unknown. Pheromone-free TraA binds P_0 and inhibits transcription. Direction of transcription from P_0 and P_a is indicated by the arrows. The antisense RNA, generalized to aR in this figure, stimulates termination at t1, indicated by the green arrow. Positive regulatory elements vary between different pheromone-responsive plasmids, and their mechanisms of inducing downstream transcription of the conjugation structural genes may also vary. For simplicity, the more common gene names and order are used. It should be noted that the gene order of the *traC* and *traB* genes is reversed in pAD1. The RepA gene is shown to orient the reader relative to Fig. 3. (C) Events at the RNA level. Relative levels of RNA produced from the P_0 and P_a promoters under uninduced (red) and induced (green) conditions are depicted by the size of the arrows.

from the same promoter, P_a, and PrgX is required both for positive autoregulation of P_a and proper processing of Qa (12). In both the pAD1 and pCF10 systems, pheromone induction results in a significant decrease in antisense RNA production (11, 16, 17). The current model of pheromone-mediated regulation proposes that in the absence of pheromone TraA/PrgX reduces but does not eliminate transcription from the P_0 promoter. Under these conditions the level of antisense RNA is sufficient to terminate all transcription from P_0 at t1. Pheromone induction alters TraA/PrgX interaction with P_0, leading to an increase in transcription that both reduces the level of antisense RNA, perhaps by transcriptional interference, and titrates the remaining antisense RNA, allowing read-through of t1 and t2 into the

positive regulatory genes and further up-regulation of the conjugation structural genes. Many details of this system remain to be worked out.

Two plasmid-encoded factors are required for pheromone shutdown, an integral membrane protein that appears to interfere with secretion of cognate pheromone and an inhibitor peptide that acts as a competitive inhibitor of pheromone activity. The inhibitor peptides ("ip" in Fig. 2) are biochemically similar to the pheromones, and each specifically inhibits only the pheromone signal for the plasmid on which it is encoded. Inhibitor peptides are located at the C-terminal end of 21- to 23-amino-acid precursors that have features of isolated signal sequences (41), and processing requires the Eep protease (2). The relative importance of the secretion inhibitor and the inhibitor peptide varies among the pheromone plasmids. For example, in pAD1-containing cells only the iAD1 inhibitor peptide can be detected in the cell supernatant, indicating that the TraB secretion inhibitor effectively blocks cAD1 secretion (1, 205). In contrast, in pCF10-containing cells, the PrgY secretion inhibitor reduces the amount of cCF10 associated with the cell wall but does not reduce the level of secreted cCF10, suggesting that the amount of exogenous pheromone is precisely balanced by the production of inhibitor (29). Interestingly, pAM373 appears to control extracellular pheromone levels without a clear TraB/PrgY homolog (52).

Downstream of the positive regulatory elements of pAD1, transcription of a contiguous region of ~20 kb is inducible by pheromone (163). In most pheromone plasmids that have been examined, the first two genes of this region encode a surface exclusion protein and AS (102). The surface exclusion proteins limit conjugation between strains harboring the same pheromone plasmid (59, 212). AS is essential for the cellular aggregation response to pheromone and efficient mating in broth, but is not required for plasmid transfer on solid surfaces, suggesting that it is not an integral part of the mating pore (43). AS is a cell wall-anchored protein and is highly conserved in all pheromone plasmids except pAM373. AS facilitates intercellular aggregation by binding to enterococcal-binding substance that is present on the surface of both donor and recipient cells. Lipoteichoic acid (LTA) appears to be at least one component of enterococcal-binding substance because free LTA inhibits aggregation (62) and cells containing mutations affecting LTA are poor recipients (199). Recent mutagenesis and in vitro binding analysis of the AS from pCF10 have identified two distinct aggregation domains, one of which binds tightly to LTA in vitro (202, 203). However, pCF10 AS also binds LTA from E. hirae, which does not aggregate with AS-expressing E. faecalis cells, suggesting that although LTA binding may be essential for aggregation it is not sufficient. pAM373 does not encode an entry exclusion function (51), and the putative gene encoding the pAM373 AS is half the size of other AS-encoding genes and shows only local homology to them (52).

Very little is known about the pheromone-inducible genes downstream of AS. The complete sequences of pAD1 and pAM373 have been reported (52, 75), and the V583 genomic sequence includes an apparent pheromone-responsive plasmid, pTEF2 (156). All three of these plasmids contain a stretch of genes downstream of their respective AS genes that show >95% homology, indicating that even in the divergent pAM373 many of the structural genes required for conjugation are conserved. Within this conserved region is a short segment of localized reduced homology. In pAD1, this region has been shown to encode a functional origin of transfer (oriT) located between a relaxase gene, traX, and a gene with homology to coupling proteins from other conjugative plasmids, traW. A similarly located oriT was identified in pAM373 and the two oriTs were found to be mobilized only by their cognate plasmids. Plasmid-specific single-strand nicking by the putative relaxase was demonstrated within a large, conserved inverted repeat in oriT. Specificity was imparted by nonconserved direct repeats, presumed to be the relaxase-binding site, adjacent to the conserved inverted repeat (73).

In all of the pheromone plasmids that have been studied, the genes responsible for positive and negative regulation of the pheromone response are present in a cluster immediately upstream of the structural genes, as shown in Fig. 2. In pAD1 and pCF10, the genes required for replication and stable inheritance of the plasmid are located immediately adjacent to this regulatory cluster (98, 204). Homologs of these genes have been identified on pPD1 (79) and pAM373 (52). The genes of the basic replicon include a replication initiator protein RepA (PrgW in pCF10), a partition system RepBC (PrgPO in pCF10), and an antisense RNA-regulated toxin-antitoxin (TA) module designated par (Fig. 3). The repA gene belongs to a family of replication initiator genes that encode a centrally located series of repeats within the initiator protein coding sequence. Such genes have been identified in a number of narrow host-range plasmids native to low G+C gram-positive bacteria (66). In several of these plasmids, including pAD1, the internal repeats have been implicated as an origin of replication (oriV) (74, 84, 116, 192). Recent work on pAD1 has shown that (i) oriV is sufficient to support plasmid replication if RepA protein is supplied in trans, (ii) the repA gene itself is sufficient to support plasmid replication if RepA is overexpressed from an artificial promoter, and (iii) RepA binds specifically to oriV in double-stranded DNA and nonspecifically to single-stranded DNA (74). In addition, it was found that a previously identified, relatively weak origin of transfer within repA (3) was functionally separable from the origin of replication. The pCF10 initiator protein, PrgW, binds cCF10 pheromone, and cCF10 is required for plasmid maintenance. Furthermore, provision of cCF10 in a host not normally permissive for plasmid replication, L. lactis, allows plasmid establishment (58). This feature may be unique to pCF10 because mutation of the cAD1 coding sequence does not affect pAD1 replication (2).

The mechanisms of strand opening, initiation of replication, and regulation of initiation frequency have yet to be determined for the pheromone-responsive plasmids. Expression of the pSK41 RepA homolog is regulated by an antisense RNA (116), but this possibility has not been examined in pAD1. Interestingly, the DNA-binding features of pAD1 RepA—specific binding to the origin and strong nonspecific binding to single-stranded DNA—are shared with the unrelated RepE protein of pAMβ1 (122). Whether this means these two unrelated plasmids use a similar mechanism to initiate replication is not known.

The pheromone-responsive plasmid partition systems appear to be variants of the previously described Type Ib class (83). All Type I systems encode a well-conserved Walker-type ATPase designated ParA, a poorly conserved DNA-binding protein designated ParB, and a centromere-like site to which ParB binds. In Type Ib systems, the ParA component lacks a DNA-binding domain and the centromere-like site is located upstream of the parA-parB operon where binding of the ParB component can regulate transcription as well as direct partition. The RepB/PrgP and RepC/PrgO proteins correspond to Type Ib ParA and ParB proteins, respectively. In all four pheromone-responsive plasmids examined, including pAD1, pCF10, pPD1, and pAM373, a series of repeats is present upstream of the repB/prgP gene with at least one repeat overlapping a putative promoter sequence, suggesting that protein binding to the repeats could suppress transcription. Unlike other Type Ib systems, however, a series of similar repeats is also located downstream of the repC/prgO gene (208). Each plasmid has repeats with distinct DNA sequence and organization. For example, pAD1 contains 26 TAGTARRR (R = purine) repeats upstream of repB in two clusters of 13 and 12 tandem repeats with a single isolated repeat overlapping the promoter. Three more tandem repeats are located downstream

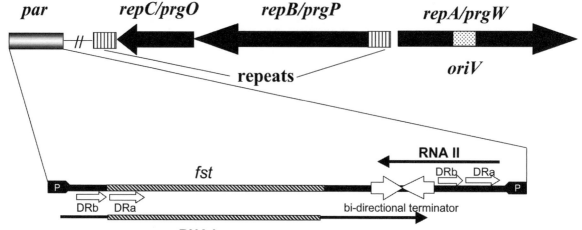

FIGURE 3 Genetic organization of a general pheromone-responsive plasmid replicon. Gene names are given as pAD1/pCF10. The stippled box within *repA/prgW* represents the origin of replication (*oriV*) to which the *repA/prgW* product binds. The lined boxes at each end of the *repB/prgP-repC/prgO* operon represent the likely centromere-like sites to which the *repC/prgO* products bind and, with the *repB/prgP* product, direct plasmid partition. The organization of the *par* region is enlarged below the replicon. Arrowheads at each end of the locus marked P represent the promoters for the RNA I and RNA II transcripts. The extent and direction of transcription of the transcripts are shown above and below the genes for RNA II and RNA I, respectively. The *fst* gene is designated by the diagonally lined box and the sequence of the peptide is shown at the bottom. DRa and DRb are direct repeats in the DNA sequence that provide complementarity between RNA I and RNA II when transcribed in opposite directions. Overlapping transcription at the bidirectional transcriptional terminator also provides complementarity.

of *repC*. pCF10 contains six tandem ATATATNNN (N=any nucleotide) repeats upstream of *prgP* with a single isolated repeat overlapping the promoter and two sets of four tandem repeats downstream of *prgO*. Recent results indicate that the pAD1 RepBC system has many features consistent with other partition systems. Thus, (i) *repBC* with the flanking repeats is capable of stabilizing heterologous plasmids, (ii) the upstream repeats act as a strong incompatibility determinant, (iii) RepC binds specifically and with high affinity to the repeats, (iv) RepB binds the repeats only in the presence of RepC and ATP, and (v) RepC down-regulates its own expression (76). Spontaneous expansion of the pAD1 *repB* upstream repeats has also been associated with loss of regulation of the conjugative response and constitutive aggregation (95). The molecular basis of this phenomenon is unknown but may not be directly related to the normal function of the repeats.

TA modules are ubiquitous on bacterial plasmids and function to stabilize plasmids within a host cell population by programming for death any cell that loses the plasmid (94). To accomplish this, addiction modules encode a stable toxin and an unstable antitoxin; as long as the plasmid is retained antitoxin is continuously replenished and the cell is protected, but if the plasmid is lost the antitoxin is degraded and the toxin kills the cell. In most cases both toxin and antitoxin are proteins, but in some cases the antitoxin is an antisense RNA that binds to the toxin-encoding mRNA and inhibits translation. The pAD1 *par* system remains the only antisense RNA-regulated TA system characterized in gram-positive bacteria (210) and is regulated by a unique antisense RNA mechanism. *par* encodes two convergently transcribed RNAs, RNAI and RNAII, initiating from each

end of *par* and terminating at a common bidirectional rho-independent terminator (207, 209) (Fig. 3). RNAI is the toxic component of the system, encoding a 33-amino-acid peptide designated Fst. RNA II is a 65-nucleotide untranslated RNA that is the *par* antitoxin. Although the RNAI and RNAII genes overlap only at the terminator, each reads in opposite directions across a pair of direct repeats. Therefore, RNAII contains regions of complementarity with both the 3′ and 5′ ends of RNAI. Furthermore, the RNAI repeats are positioned such that interaction with RNAII sequesters the translation initiation region of *fst*, preventing ribosome binding and translation (88, 90). Formation of the RNA I-RNA II complex proceeds in at least two steps with the interaction initiating at a U-turn motif in the complementary terminator loops at the 3′ ends of each RNA and then extending to the repeats at the 5′ end that suppress translation (89) (Fig. 4). An intramolecular stem-loop within RNA I that partially sequesters the *fst* ribosome-binding site may inhibit translation until RNA II binding can be completed (88). Unlike most antisense RNA systems, binding of RNA I by RNA II results in formation of a highly stable complex rather than specifically targeting the complex for degradation (206). Formation of such a stable complex would allow the toxin-encoding RNA I to persist in plasmid-containing cells. Over time, however, RNA II is less stable than RNA I, suggesting that it is eventually removed from the complex to allow translation of *fst*. The Fst toxin is a surface-active peptide that makes cells permeable and has effects on chromosome partitioning and cell division (211). The toxin sequence is shown in Fig. 3. The precise mechanism of action of the toxin is still unknown.

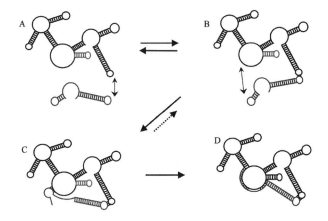

FIGURE 4 (*See the separate color insert for the color version of this illustration.*) Model of RNA I-RNA II interaction. RNA I is the larger black structure and RNA II is the smaller blue structure. Stems, loops, and bulges are depicted in their approximate locations and sizes as determined by experimentation. The red stem-loop structure within RNA I sequesters the ribosome-binding site and prevents translation until a complex is formed. The initial interaction occurs at a U-turn motif in the terminator loop of RNA I (A), followed by interaction between the complementary repeats in the 5′ end of each RNA (B and C). Because RNA II-mediated protection from the RNA I-encoded toxin occurs in vivo even when one of the repeats is mutated, the structure shown in panel C is apparently sufficient to prevent translation in vivo. Once complex formation is complete (D), the structure is extremely stable in vivo and in vitro, perhaps due to the gap between the interacting repeats.

Non-Pheromone-Responsive Enterococcal Plasmids

pAMα1 is an ~10-kb multicopy tetracycline resistance plasmid that is a composite of two replicons (44, 158), an RCR replicon that functions in B. subtilis but not in E. faecalis and carries the tetracycline resistance gene, and a putative theta-replicon that supports replication in E. faecalis. Each component plasmid also encodes a putative relaxase gene, and the junction sites contain potential oriT sequences recognized by them, suggesting a possible mechanism for their fusion. Under tetracycline selection, cells accumulate pAMα1 derivatives containing tandem repeats of the replicon carrying the tetracycline resistance gene. Initial amplification is RecA-independent and requires at least one of the relaxase genes and both oriT sequences, suggesting that amplification occurs by site-specific recombination between multiple copies of the RCR replicon catalyzed by the relaxase protein (72).

pMG1 is an ~65-kb conjugative plasmid originally isolated from E. faecium and encoding high-level gentamicin resistance. Like the pheromone-responsive plasmids, pMG1 is capable of high-frequency transfer in broth, but unlike pheromone plasmids pMG1 transfer is not induced by exposure of donor cells to culture filtrates from potential recipient cells, conjugating broth cultures do not form macroscopic cellular aggregates, and transfer occurs freely between E. faecium and E. faecalis cells. Furthermore, no homology between pMG1 and the conjugative systems of pheromone-responsive plasmids or Inc18 plasmids was observed by Southern hybridization (107). These results suggest that pMG1 encodes a unique high-frequency conjugation sys-

tem. So far, one pMG1-encoded gene, designated traA, has been shown to be induced immediately after mixing of donor and recipient cells, and mutations in this gene reduce the frequency of mating in broth but not on solid surfaces (194). Vancomycin-resistant clinical isolates have been shown to carry pMG1-like plasmids with Tn1546-like vancomycin-resistant transposons (198).

pRUM is an ~25-kb nonconjugative multiresistance plasmid isolated from an E. faecium clinical isolate (86). Sequence analysis identified one full-length putative replication initiator protein that was most closely related to the RepA initiators of the pheromone-responsive plasmids. The plasmid also encodes a proteic TA system, designated Axe-Txe, that functions to stabilize pRUM and heterologous plasmids. Unlike the Fst toxin of the pAD1 par system, whose toxicity appears to be limited to low G+C gram-positive bacteria, the Txe toxin has a broad spectrum of activity and is active on E. coli cells, where it causes filamentation, suggesting that it inhibits cell division. Homologs of the Axe-Txe system are widespread on bacterial chromosomes and plasmids.

CONJUGATIVE TRANSPOSONS

The enterococci carry the full range of transposon varieties found in other bacteria. Two Tn3-type transposons, Tn917 encoding erythromycin resistance and Tn1546 encoding vancomycin resistance, have been extensively studied (9, 159), and Tn917 and its derivatives have been used for transposon mutagenesis in a number of gram-positive organisms. The genome sequence of E. faecalis strain V583 revealed the presence of 38 insertion sequence elements with ISEf1, IS256, and IS1216 being the most frequent (156). Both IS256 and IS1216 have been associated with simple composite transposons (96, 104) and large, complex, mosaic elements (170, 172). Many of these elements contain antibiotic resistance genes and some are conjugative by means that have yet to be fully explained. For a more detailed description of these elements see reference 208.

The E. faecalis transposon Tn916 was the first well-characterized self-conjugative transposon (77, 82), although large, conjugative chromosomal elements had been previously identified in other streptococci (188). Conjugative transposons were later found to be widespread in the lactic acid bacteria and mechanistically similar but unrelated elements were identified in Bacteroides species (187). In spite of the apparent broad host range of the Tn916-like elements (39) conjugative transposons did not appear to be widespread in eubacteria. However, characterization of new elements from both gram-positive and gram-negative organisms and the reinterpretation of data on older elements formerly classified as integrative plasmids have revealed that elements with characteristics similar to conjugative transposons (i.e., that excise from their chromosomal locus, circularize, transfer to a new host cell by conjugation, and insert in the recipient cell's chromosome without extensive replication) are probably as ubiquitous in bacteria as plasmids. Because the mobility of most of these elements is dependent on a tyrosine or serine recombinase rather than a DDE-type transposase, and because many of them have relatively site-specific integration preferences, the appropriateness of the term "conjugative transposons" has been questioned and alternative names have been suggested, the most inclusive of which is integrative and conjugative elements (ICE) (27). Because the focus here is enterococcal genetics the discussion below will concentrate primarily on the best-studied elements, Tn916 and the closely related Tn*1545* (48). A map of Tn*916* is shown in Fig. 5.

Unlike transposons using a DDE-type transposase, integrated conjugative transposons are not flanked by direct repeats of target sequence. Preferred insertion sites tend to be AT-rich and possess intrinsic curvature, but the DNA sequence is not conserved and flanking sequences on each end are heterologous (106, 131). Two transposon-encoded proteins, Int and Xis, are required for excision (134, 166). Int is a tyrosine recombinase (87) with N-terminal and C-terminal DNA-binding domains (130). The N-terminal domain (Int-N) binds via a three-stranded β-sheet DNA-binding domain to direct repeat sequences near each end of Tn*916* (46, 214). The C-terminal domain (Int-C) binds to the transposon termini as well as to target sequences. Int-C contains the conserved active site residues involved in nicking-closing. Like other tyrosine recombinases, Int produces staggered cuts at each end of the integrated transposon. One cut is made flush with the end of the transposon, producing a recessed 3′ end covalently joined to Int via a phosphotyrosine linkage. The second cut is made six nucleotides into the host sequence, leaving a free 5′ OH that attacks the phosphotyrosine linkage at the other end of the transposon, resulting in circularization (195). Because the joint is derived from chromosomal DNA at each end that is not complementary, a 6-bp heteroduplex is formed in the circular intermediate called the coupling sequence (30). The fact that Int must bind and cleave at the coupling sequences, which vary from insert to insert, may explain the wide range of conjugation frequencies (from $<10^{-9}$ to $>10^{-4}$ per donor cell) observed using different donor strains (110). How the coupling sequences affect conjugation frequency is not known, but it does not appear to be due to effects on integrase affinity (161) or on excision frequency (133).

Xis binds adjacent to the Int-N binding sites on both ends of the transposon (176) and stimulates Int-mediated excision of Tn*916* in *E. coli* (166, 191), *L. lactis* (135), *B. subtilis* (134), and in vitro (177). In *E. coli*, Xis binding to the left end is thought to play an architectural role similar to λ-Xis in excision (47). HU protein plays a significant role in excision in *E. coli*, but no host factors affecting excision have been identified in gram-positive hosts. Interestingly, Xis binding to the right end of Tn*916* inhibits excision, perhaps by competing with Int-N for binding, suggesting that Xis has dual roles in excision, both stimulating and negatively regulating the process (101).

Once the circular intermediate is formed the transposon is transferred to the recipient cell by conjugation. The molecular details of this process are poorly understood. The coupling sequence from the left end of the transposon is preferentially transferred to the recipient, and this has been taken as evidence that only a single strand is transferred as in plasmid conjugation (182). However, similar results would be obtained if one strand of the coupling sequence was preferentially repaired in the donor (132). A functional *oriT* has been identified on Tn*916* that bears similarities to IncP-type origins of transfer, again implicating a plasmid-like transfer mechanism, but neither the nicking site nor the relaxase have been unambiguously identified (111). An adjacent open reading frame, ORF21, has been shown to exhibit nonspecific nicking activity (38), but it also shows homology to FtsK/SpoIIIE proteins that are known to function as motor proteins for double-stranded DNA and have been implicated in the conjugative transfer of double-stranded DNA in another ICE element, pSAM2 of *Streptomyces ambofaciens* (165). A combination of genetic analyses and DNA sequence determination has identified a number of genes involved in Tn*916* conjugation (70, 183), but the

A

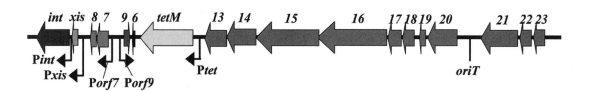

B

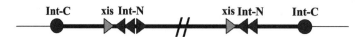

FIGURE 5 (*See the separate color insert for the color version of this illustration.*) Genetic organization of Tn*916* and the relative positions of Int and Xis binding sites. (A) Functions of Tn*916* genes are color-coded: blue, recombination; green and red, positive and negative regulation, respectively; yellow, tetracycline resistance; magenta, conjugation. Gene names and open reading frame numbers are shown above the individual genes. The origin of transfer is designated by the line between *orfs* 20 and 21 labeled *oriT*. Promoters and the direction of transcription are designated by labeled arrows below the gene line. (B) Binding sites for the Int-C and Int-N DNA-binding domains and Xis are designated by marked circles and triangles. Int and Xis binding sites are color-coordinated with the genes shown in panel A.

precise roles of none of these genes have been experimentally determined.

Tn916 does not mobilize plasmids into which it has integrated, nor does it mobilize chromosomal genes adjacent to its integration site during intercellular transposition. This suggests that some mechanism suppresses utilization of the *oriT* site until after the transposon has excised from its host genome and circularized. Two mechanisms provide this control. First, induced expression of the transfer genes requires a functional *int* gene and transcription is initiated from the *orf7* promoter which, because it is located at the opposite end of the transposon, is linked to the transfer genes only in the circular intermediate (33). Activation of the *orf7* promoter requires *orf7* and *orf8* and is repressed by *orf9*, but the mechanisms of regulation are unknown. Tetracycline induction of the *tetM* gene may also affect the *orf7* promoter. Second, Int has been shown to bind to Tn916 *oriT* by a mechanism that requires both the N- and C-terminal binding domains (100). This binding may serve to suppress conjugation until a complex functional for excision has been formed.

Once a copy of the transposon is transferred, integration into the recipient chromosome requires Int but not Xis (134). Because the known tyrosine recombinases function only on double-stranded DNA, if conjugation involves the transfer of single-stranded DNA, complementary strand synthesis must precede integration. How this is accomplished is not known. Experiments with *int-* versions of Tn916 in which Int was provided in *trans* from a plasmid in the donor revealed that *int* gene expression was not required in the recipient for integration, suggesting that the Int protein might be transported with the transposon DNA (22). Recipients frequently acquire multiple copies of the transposon, and there is no evidence that Tn916 expresses either an entry exclusion function or transposon immunity (40, 146). Interestingly, in cells harboring more than one copy of the transposon, excision of one element transactivates excision and mobility of other elements located elsewhere in the genome (68).

Sequence analysis of known ICE elements and putative ICE elements in sequenced genomes suggests that these mobile elements are modular and that the modules may have different evolutionary histories (27, 28, 174). For example, Tn5397 from *Clostridium difficile* is greater than 84% identical at the DNA sequence level to Tn916 across all of the conjugation-related genes and in the *tetM* gene, but Tn916 Int is a tyrosine recombinase while the recombinase protein of Tn5397 is a serine recombinase. Conversely, the recombination module of the VanB transposon Tn1549 is closely related to the Tn916 *int/xis* genes but most of the conjugation genes are unrelated (81). Putative ICE elements identified in the *E. faecalis* V583 genome sequence EfaC2 and EfaD2 include conjugation-related genes apparently acquired from pAD1 and pAM373 (27). A more global analysis including plasmids, bacteriophage, and genomic islands, as well as ICE elements, implies that the degree of exchange between these elements is very high, suggesting that rather than representing distinct groups, all of these elements may represent a continuum of structures that maximize mobility under various circumstances (151).

Other Mobile Genetic Elements

DNA sequence analysis of the *E. faecalis* V583 genome identified seven regions probably derived from integrated bacteriophage. Whether these represent functional prophage or remnants of decayed phage is not known. Little work has been done to genetically characterize enterococcal phage, and no transduction system has been described.

Pathogenicity islands have been identified in both *E. faecalis* (184) and *E. faecium* (118), whose virulence characteristics are described in more detail in chapter 25. The *E. faecalis* pathogenicity island includes multiple insertion sequences, a region containing genes similar to pheromone-responsive plasmid genes, and an integrase gene at one end most likely responsible for mobility of the element. The *E. faecium* pathogenicity island contains at least two of the same genes as the *E. faecalis* pathogenicity island but has many distinguishing characteristics. The island has yet to be completely sequenced, so the nature of its ends is not known and it is not clear whether it includes an integrase gene.

Integrons are gene collecting systems that include a tyrosine recombinase integrase and an outward-reading promoter that transcribes a variable suite of antibiotic resistance gene cassettes flanked by recognition sites for the integrase (see reference 175 for a review). Integrons are generally considered to be relatively restricted to gram-negative bacteria, but a recent screen for integron-associated genes in poultry litter suggests that integrons may be much more common in gram-positive bacteria than previously appreciated (145). An integron containing an *aadA* gene encoding resistance to streptomycin and spectinomycin located on an ~80-kb plasmid was previously reported in an *E. faecalis* clinical isolate (36). The role of integrons in the evolution of antibiotic resistance in the enterococci may be underappreciated.

CONCLUDING REMARKS

The completion of the genomic sequence of *E. faecalis* has provided a treasure trove of information to investigators and has once again focused the spotlight on mobile genetic elements. In addition to providing critical information on the better-known mobile elements (e.g., the identification of the genes encoding the conjugative pheromones and evidence for mosaicism of ICE elements), the genomic sequence also highlighted the importance of several elements that, until recently, have received little attention (e.g., pathogenicity islands and bacteriophage). Undoubtedly, completion of the *E. faecium* genome sequence and comparison to *E. faecalis* will provide even greater insight into the role of mobile elements in the evolution of the enterococci.

It is also important to note, however, that there are several questions of basic enterococcal biology that remain unaddressed and for which the genomic sequence data provide only a rudimentary framework for understanding. For example, although *E. faecalis* and *E. faecium* carry many of the same genes involved in cell division, chromosomal replication, and cell wall biosynthesis as the better-studied rod-shaped organisms, the pattern of cell growth and division is much different in the enterococci. Is the spherical form of growth simply a default position due to the absence of certain genes (e.g., MinCD, MreB, Mbl), or is it due to specific variant functions and interactions of common genes present in both rods and cocci? Are the paradigms of spatial association of DNA replication and resolution functions with the cell division apparatus established in *E. coli* and *B. subtilis* retained in the spherical, chain-forming bacteria? The study of gene regulation in the enterococci also remains in its infancy, and the genomic data have served primarily to emphasize how much we do not know, rather than answer many significant questions. For example, a variety of TC-STS have been identified in both *E. faecalis* and *E. faecium*, but the signals sensed by most of these systems are unknown. Even the signal for the intensively studied VanRS systems remains a mystery.

Work continues on the best-studied enterococcal mobile elements, the pheromone-responsive conjugative plasmids and the conjugative transposons. Although much has been learned concerning the sensing, transduction, synthesis, and inhibition of the pheromone signal, many of the regulatory details, especially with respect to the ultimate induction of the conjugative structural genes, remain to be worked out. Although the replication initiator and origin of replication have been identified, the mechanisms for attracting the host DNA Pol III, strand opening, and replication direction are unknown. The composition of the centromere-like site utilized in plasmid partition, particularly with respect to the downstream repeats, is unclear. Likewise, the mechanism of activation of the *par* RNA I-RNA II complex for translation and the mechanism of the toxin remain largely undefined. With the exception of processing at *oriT* in pAD1, virtually nothing is known about the genes that are essential for DNA transfer during conjugation of either the pheromone plasmids or conjugative transposons. In the case of Tn*916* questions persist about the mechanism by which the DNA is processed at *oriT*, and even whether the DNA is transferred as a single or double strand.

The discovery of integrons, pathogenicity islands, bacteriophage, and new plasmids and ICE elements makes it clear that our understanding of the complement of the enterococcal mobile genome is still far from complete. The role of bacteriophage and alternative modes of conjugation (e.g., for pMG1) in moving DNA about in this genus has been particularly neglected. Hopefully, work in these areas will continue to broaden our understanding of the complex process of evolution at work in these organisms.

REFERENCES

1. **An, F. Y., and D. B. Clewell.** 1994. Characterization of the determinant (*traB*) encoding sex pheromone shutdown by the hemolysin/bacteriocin plasmid pAD1 in *Enterococcus faecalis. Plasmid* **31**:215–221.
2. **An, F. Y., and D. B. Clewell.** 2002. Identification of the cAD1 sex pheromone precursor in *Enterococcus faecalis. J. Bacteriol.* **184**:1880–1887.
3. **An, F. Y., and D. B. Clewell.** 1997. The origin of transfer (oriT) of the enterococcal, pheromone-responding, cytolysin plasmid pAD1 is located within the repA determinant. *Plasmid* **37**:87–94.
4. **An, F. Y., M. C. Sulavik, and D. B. Clewell.** 1999. Identification and characterization of a determinant (*eep*) on the *Enterococcus faecalis* chromosome that is involved in production of the peptide sex pheromone cAD1. *J. Bacteriol.* **181**:5915–5921.
5. **Antiporta, M. H., and G. M. Dunny.** 2002. ccfA, the genetic determinant for the cCF10 peptide pheromone in *Enterococcus faecalis* OG1RF. *J. Bacteriol.* **184**:1155–1162.
6. **Arias, C. A., P. Courvalin, and P. E. Reynolds.** 2000. *vanC* cluster of vancomycin-resistant *Enterococcus gallinarum* BM4174. *Antimicrob. Agents Chemother.* **44**:1660–1666.
7. **Arthur, M., F. Depardieu, and P. Courvalin.** 1999. Regulated interactions between partner and non-partner sensors and response regulators that control glycopeptide resistance gene expression in enterococci. *Microbiology* **145**: 1849–1858.
8. **Arthur, M., F. Depardieu, G. Gerbaud, M. Galimand, R. Leclercq, and P. Courvalin.** 1997. The VanS sensor negatively controls VanR-mediated transcriptional activation of glycopeptide resistance genes of Tn*1546* and related elements in the absence of induction. *J. Bacteriol.* **179**: 97–106.
9. **Arthur, M., C. Molinas, F. Depardieu, and P. Courvalin.** 1993. Characterization of Tn1546, a Tn3-related transposon conferring glycopeptide resistance by synthesis of depsipeptide peptidoglycan precursors in *Enterococcus faecium* BM4147. *J. Bacteriol.* **175**:117–127.
10. **Arthur, M., and R. Quintiliani, Jr.** 2001. Regulation of VanA- and VanB-type glycopeptide resistance in enterococci. *Antimicrob. Agents Chemother.* **45**:375–381.
11. **Bae, T., and G. M. Dunny.** 2001. Dominant-negative mutants of *prgX*: evidence for a role for PrgX dimerization in negative regulation of pheromone-inducible conjugation. *Mol. Microbiol.* **39**:1307–1320.
12. **Bae, T., B. Kozlowicz, and G. M. Dunny.** 2002. Two targets in pCF10 DNA for PrgX binding: their role in production of Qa and prgX mRNA and in regulation of pheromone-inducible conjugation. *J. Mol. Biol.* **315**:995–1007.
13. **Bae, T., B. K. Kozlowicz, and G. M. Dunny.** 2004. Characterization of *cis*-acting *prgQ* mutants: evidence for two distinct repression mechanisms by Qa RNA and PrgX protein in pheromone-inducible enterococcal plasmid pCF10. *Mol. Microbiol.* **51**:271–281.
14. **Baptista, M., F. Depardieu, P. Reynolds, P. Courvalin, and M. Arthur.** 1997. Mutations leading to increased levels of resistance to glycopeptide antibiotics in VanB-type enterococci. *Mol. Microbiol.* **25**:93–105.
15. **Barcelona-Andres, B., A. Marina, and V. Rubio.** 2002. Gene structure, organization, expression, and potential regulatory mechanisms of arginine catabolism in *Enterococcus faecalis. J. Bacteriol.* **184**:6289–6300.
16. **Bastos, M. C. de Freire, K. Tanimoto, and D. B. Clewell.** 1997. Regulation of transfer of the *Enterococcus faecalis* pheromone-responding plasmid pAD1: temperature-sensitive transfer mutants and identification of a new regulatory determinant, *traD. J. Bacteriol.* **179**:3250–3259.
17. **Bastos, M. C. de Freire, H. Tomita, K. Tanimoto, and D. B. Clewell.** 1998. Regulation of the *Enterococcus faecalis* pAD1-related sex pheromone response: analyses of *traD* expression and its role in controlling conjugation functions. *Mol. Microbiol.* **30**:381–392.
18. **Bensing, B. A., D. A. Manias, and G. M. Dunny.** 1997. Pheromone cCF10 and plasmid pCF10-encoded regulatory molecules act post-transcriptionally to activate expression of downstream conjugation functions. *Mol. Microbiol.* **24**: 285–294.
19. **Berg, T., N. Firth, S. Apisiridej, A. Hettiaratchi, A. Leelaporn, and R. A. Skurray.** 1998. Complete nucleotide sequence of pSK41: evolution of staphylococcal conjugative multiresistance plasmids. *J. Bacteriol.* **180**:4350–4359.
20. **Boyd, D. A., J. Conly, H. Dedier, G. Peters, L. Robertson, E. Slater, and M. R. Mulvey.** 2000. Molecular characterization of the vanD gene cluster and a novel insertion element in a vancomycin-resistant enterococcus isolated in Canada. *J. Clin. Microbiol.* **38**:2392–2394.
21. **Bozdogan, B., R. Leclercq, A. Lozniewski, and M. Weber.** 1999. Plasmid-mediated coresistance to streptogramins and vancomycin in *Enterococcus faecium* HM1032. *Antimicrob. Agents Chemother.* **43**:2097–2098.
22. **Bringel, F., G. Van Alstine, and J. Scott.** 1992. Conjugative transposition of Tn916: the transposon int gene is required only in the donor. *J. Bacteriol.* **174**:4036–4041.

23. **Brown, M. S., J. Ye, R. B. Rawson, and J. L. Goldstein.** 2000. Regulated intramembrane proteolysis: a control mechanism conserved from bacteria to humans. *Cell* **100**: 391–398.

24. **Bruand, C., E. Le Chatelier, S. Ehrlich, and L. Janniere.** 1993. A fourth class of theta-replicating plasmids: the pAM beta 1 family from gram-positive bacteria. *Proc. Natl. Acad. Sci. USA* **90**:11668–11672.

25. **Bruck, I., M. F. Goodman, and M. O'Donnell.** 2003. The essential C family DnaE polymerase is error-prone and efficient at lesion bypass. *J. Biol. Chem.* **278**:44361–44368.

26. **Buck, M., M.-T. Gallegos, D. J. Studholme, Y. Guo, and J. D. Gralla.** 2000. The bacterial enhancer-dependent sigma 54 (sigma N) transcription factor. *J. Bacteriol.* **182**: 4129–4136.

27. **Burrus, V., G. Pavlovic, B. Decaris, and G. Guedon.** 2002. Conjugative transposons: the tip of the iceberg. *Mol. Microbiol.* **46**:601–610.

28. **Burrus, V., G. Pavlovic, B. Decaris, and G. Guedon.** 2002. The ICESt1 element of *Streptococcus thermophilus* belongs to a large family of integrative and conjugative elements that exchange modules and change their specificity of integration. *Plasmid* **48**:77–97.

29. **Buttaro, B. A., M. H. Antiporta, and G. M. Dunny.** 2000. Cell-associated pheromone peptide (cCF10) production and pheromone inhibition in *Enterococcus faecalis*. *J. Bacteriol.* **182**:4926–4933.

30. **Caparon, M., and J. R. Scott.** 1989. Excision and insertion of the conjugative transposon Tn916 involves a novel recombination mechanism. *Cell* **59**:1027–1034.

31. **Carias, L. L., S. D. Rudin, C. J. Donskey, and L. B. Rice.** 1998. Genetic linkage and cotransfer of a novel, vanB-containing transposon (Tn5382) and a low-affinity penicillin-binding protein 5 gene in a clinical vancomycin-resistant *Enterococcus faecium* isolate. *J. Bacteriol.* **180**:4426–4434.

32. **Casadewall, B., P. E. Reynolds, and P. Courvalin.** 2001. Regulation of expression of the vanD glycopeptide resistance gene cluster from *Enterococcus faecium* BM4339. *J. Bacteriol.* **183**:3436–3446.

33. **Celli, J., and P. Trieu-Cuot.** 1998. Circularization of Tn916 is required for expression of the transposon-encoded transfer functions: characterization of long tetracycline-inducible transcripts reading through the attachment site. *Mol. Microbiol.* **28**:103–117.

34. **Chow, J. W., V. Kak, I. You, S. J. Kao, J. Petrin, D. B. Clewell, S. A. Lerner, G. H. Miller, and K. J. Shaw.** 2001. Aminoglycoside resistance genes aph(2'')-Ib and aac(6')-Im detected together in strains of both *Escherichia coli* and *Enterococcus faecium*. *Antimicrob. Agents Chemother.* **45**:2691–2694.

35. **Chung, J., B. Bensing, and G. Dunny.** 1995. Genetic analysis of a region of the *Enterococcus faecalis* plasmid pCF10 involved in positive regulation of conjugative transfer functions. *J. Bacteriol.* **177**:2107–2117.

36. **Clark, N. C., O. Olsvik, J. M. Swenson, C. A. Spiegel, and F. C. Tenover.** 1999. Detection of a streptomycin/spectinomycin adenylyltransferase gene (aadA) in *Enterococcus faecalis*. *Antimicrob. Agents Chemother.* **43**:157–160.

37. **Clewell, D. B., F. Y. An, S. E. Flannagan, M. Antiporta, and G. M. Dunny.** 2000. Enterococcal sex pheromone precursors are part of signal sequences for surface lipoproteins. *Mol. Microbiol.* **35**:246–247.

38. **Clewell, D. B., and G. Dunny.** 2002. Conjugation and genetic exchange in enterococci, p. 265–300. *In* M. S. Gilmore, D. B. Clewell, P. Courvalin, G. Dunny, B. E. Murray, and L. B. Rice (ed.), *The Enterococci: Pathogenesis, Molecular Biology, and Antibiotic Resistance.* ASM Press, Washington, D.C.

39. **Clewell, D. B., S. E. Flannagan, D. D. Jaworski, and D. B. Clewell.** 1995. Unconstrained bacterial promiscuity: the Tn916-Tn1545 family of conjugative transposons. *Trends Microbiol.* **3**:229–236.

40. **Clewell, D. B., S. E. Flannagan, L. A. Zitzow, Y. A. Su, P. He, E. Senghas, and K. E. Weaver.** 1991. Properties of conjugative transposon Tn916, p. 39–44. *In* G. M. Dunny, P. Cleary, and L. McKay (ed.), *Genetics and Molecular Biology of Streptococci, Lactococci, and Enterococci.* American Society for Microbiology, Washington, D.C.

41. **Clewell, D. B., L. T. Pontius, F. Y. An, Y. Ike, A. Suzuki, and J. Nakayama.** 1990. Nucleotide sequence of the sex pheromone inhibitor (iAD1) determinant of *Enterococcus faecalis* conjugative plasmid pAD1. *Plasmid* **24**:156–161.

42. **Clewell, D. B., M. Victoria Francia, S. E. Flannagan, and F. Y. An.** 2002. Enterococcal plasmid transfer: sex pheromones, transfer origins, relaxases, and the *Staphylococcus aureus* issue. *Plasmid* **48**:193–201.

43. **Clewell, D. B., and K. E. Weaver.** 1989. Sex pheromones and plasmid transfer in *Enterococcus faecalis*. *Plasmid* **21**: 175–184.

44. **Clewell, D. B., Y. Yagi, G. M. Dunny, and S. K. Schultz.** 1974. Characterization of three plasmid deoxyribonucleic acid molecules in a strain of *Streptococcus faecalis*: identification of a plasmid determining erythromycin resistance. *J. Bacteriol.* **117**:283–289.

45. **Comenge, Y., R. Quintiliani, Jr., L. Li, L. Dubost, J.-P. Brouard, J.-E. Hugonnet, and M. Arthur.** 2003. The CroRS two-component regulatory system is required for intrinsic β-lactam resistance in *Enterococcus faecalis*. *J. Bacteriol.* **185**:7184–7192.

46. **Connolly, K. M., U. Ilangovan, J. M. Wojciak, M. Iwahara, and R. T. Clubb.** 2000. Major groove recognition by three-stranded [beta]-sheets: affinity determinants and conserved structural features. *J. Mol. Biol.* **300**:841–856.

47. **Connolly, K. M., M. Iwahara, and R. T. Clubb.** 2002. Xis protein binding to the left arm stimulates excision of conjugative transposon Tn916. *J. Bacteriol.* **184**:2088–2099.

48. **Courvalin, P., and C. Carlier.** 1986. Transposable multiple antibiotic resistance in *Streptococcus pneumoniae*. *Mol. Gen. Genet.* **205**:291–297.

49. **Dalet, K., C. Briand, Y. Cenatiempo, and Y. Hechard.** 2001. The rpoN gene of *Enterococcus faecalis* directs sensitivity to subclass IIa bacteriocins. *Cur. Microbiol.* **41**:441–443.

50. **Day, A. M., J. H. Cove, and M. K. Phillips-Jones.** 2003. Cytolysin gene expression in *Enterococcus faecalis* is regulated in response to aerobiosis conditions. *Mol. Genet. Genomics* **269**:31–39.

51. **DeBoever, E. H., and D. B. Clewell.** 2001. The *Enterococcus faecalis* pheromone-responsive plasmid pAM373 does not encode an entry exclusion function. *Plasmid* **45**: 57–60.

52. **DeBoever, E. H., D. B. Clewell, and C. M. Frasier.** 2000. *Enterococcus faecalis* conjugative plasmid pAM373: complete nucleotide sequence and genetic analyses of sex pheromone response. *Mol. Microbiol.* **37**:1327–1341.

53. **Del Grosso, M., A. Scotto d'Abusco, F. Iannelli, G. Pozzi, and A. Pantosti.** 2004. Tn2009, a Tn916-like element containing mef(E) in *Streptococcus pneumoniae*. *Antimicrob. Agents Chemother.* **48**:2037–2042.

54. **del Solar, G., and M. Espinosa.** 2000. Plasmid copy number control: an ever-growing story. *Mol. Microbiol.* **37**:492–500.

55. **Depardieu, F., M. G. Bonora, P. E. Reynolds, and P. Courvalin.** 2003. The *vanG* glycopeptide resistance operon from *Enterococcus faecalis* revisited. *Mol. Microbiol.* **50:**931–948.

56. **Depardieu, F., P. E. Reynolds, and P. Courvalin.** 2003. VanD-type vancomycin-resistant *Enterococcus faecium* 10/96A. *Antimicrob. Agents Chemother.* **47:**7–18.

57. **Dunny, G. M., B. L. Brown, and D. B. Clewell.** 1978. Induced cell aggregation and mating in *Streptococcus faecalis:* evidence for a bacterial sex pheromone. *Proc. Natl. Acad. Sci. USA* **75:**3479–3483.

58. **Dunny, G. M., and B. A. B. Leonard.** 1997. Cell-cell communication in gram-positive bacteria. *Annu. Rev. Microbiol.* **51:**527–564.

59. **Dunny, G. M., D. L. Zimmerman, and M. L. Tortorello.** 1985. Induction of surface exclusion (entry exclusion) by *Streptococcus faecalis* sex pheromones: use of monoclonal antibodies to identify an inducible surface antigen involved in the exclusion process. *Proc. Natl. Acad. Sci. USA* **82:**8582–8586.

60. **Dutta, I., and P. E. Reynolds.** 2002. Biochemical and genetic characterization of the vanC-2 vancomycin resistance gene cluster of *Enterococcus casseliflavus* ATCC 25788. *Antimicrob. Agents Chemother.* **46:**3125–3132.

61. **Edwards, D. H., and J. Errington.** 1997. The *Bacillus subtilis* DivIVA protein targets to the division septum and controls the site specificity of cell division. *Mol. Microbiol.* **24:**905–915.

62. **Ehrenfeld, E. E., R. E. Kessler, and D. B. Clewell.** 1986. Identification of pheromone-induced surface proteins in *Streptococcus faecalis* and evidence of a role for lipoteichoic acid in formation of mating aggregates. *J. Bacteriol.* **168:**6–12.

63. **Engelbert, M., E. Mylonakis, F. M. Ausubel, S. B. Calderwood, and M. S. Gilmore.** 2004. Contribution of gelatinase, serine protease, and fsr to the pathogenesis of *Enterococcus faecalis* endophthalmitis. *Infect. Immun.* **72:**3628–3633.

64. **Espeli, O., C. Lee, and K. J. Marians.** 2003. A physical and functional interaction between *Escherichia coli* FtsK and topoisomerase IV. *J. Biol. Chem.* **278:**44639–44644.

65. **Evers, S., J. Quintiliani, R., and P. Courvalin.** 1996. Genetics of glycopeptide resistance in enterococci. *Microb. Drug Resist.* **2:**219–223.

66. **Firth, N., S. Apisiridej, T. Berg, B. A. O'Rourke, S. Curnock, K. G. H. Dyke, and R. A. Skurray.** 2000. Replication of staphylococcal multiresistance plasmids. *J. Bacteriol.* **182:**2170–2178.

67. **Flannagan, S. E., J. W. Chow, S. M. Donabedian, W. J. Brown, M. B. Perri, M. J. Zervos, Y. Ozawa, and D. B. Clewell.** 2003. Plasmid content of a vancomycin-resistant *Enterococcus faecalis* isolate from a patient also colonized by *Staphylococcus aureus* with a VanA phenotype. *Antimicrob. Agents Chemother.* **47:**3954–3959.

68. **Flannagan, S. E., and D. B. Clewell.** 1991. Conjugative transfer of Tn916 in *Enterococcus faecalis:* trans activation of homologous transposons. *J. Bacteriol.* **173:**7136–7141.

69. **Flannagan, S. E., and D. B. Clewell.** 2002. Identification and characterization of genes encoding sex pheromone cAM373 activity in *Enterococcus faecalis* and *Staphylococcus aureus. Mol. Microbiol.* **44:**803–817.

70. **Flannagan, S. E., L. A. Zitzow, Y. A. Su, and D. B. Clewell.** 1994. Nucleotide sequence of the 18-kb conjugative transposon Tn916 from *Enterococcus faecalis. Plasmid* **32:**350–354.

71. **Foster, K. A., M. H. Barnes, R. O. Stephenson, M. M. Butler, D. J. Skow, W. A. LaMarr, and N. C. Brown.** 2003. DNA polymerase III of *Enterococcus faecalis:* expression and characterization of recombinant enzymes encoded by the polC and dnaE genes. *Protein Expr. Purif.* **27:**90–97.

72. **Francia, M. V., and D. B. Clewell.** 2002. Amplification of the tetracycline resistance determinant of pAM alpha 1 in *Enterococcus faecalis* requires a site-specific recombination event involving relaxase. *J. Bacteriol.* **184:**5187–5193.

73. **Francia, M. V., and D. B. Clewell.** 2002. Transfer origins in the conjugative *Enterococcus faecalis* plasmids pAD1 and pAM373: identification of the pAD1 *nic* site, a specific relaxase and a possible TraG-like protein. *Mol. Microbiol.* **45:**375–395.

74. **Francia, M. V., S. Fujimoto, P. Tille, K. E. Weaver, and D. B. Clewell.** 2004. Replication of *Enterococcus faecalis* pheromone-responding plasmid pAD1: location of the minimal replicon and oriV Site and RepA involvement in initiation of replication. *J. Bacteriol.* **186:**5003–5016.

75. **Francia, M. V., W. Haas, R. Wirth, E. Samberger, A. Mscholl-Silberhorn, M. S. Gilmore, Y. Ike, K. E. Weaver, F. Y. An, and D. B. Clewell.** 2001. Completion of the nucleotide sequence of the *Enterococcus faecalis* conjugative virulence plasmid pAD1 and identification of a second transfer origin. *Plasmid* **46:**117–127.

76. **Francia, M. V., K. E. Weaver, and D. B. Clewell.** Unpublished observations.

77. **Franke, A. E., and D. B. Clewell.** 1981. Evidence for a chromosome-borne resistance transposon (Tn916) in *Streptococcus faecalis* that is capable of "conjugal" transfer in the absence of a conjugative plasmid. *J. Bacteriol.* **145:** 494–502.

78. **Fujimoto, S., and D. B. Clewell.** 1998. Regulation of the pAD1 sex pheromone response of *Enterococcus faecalis* by direct interaction between the cAD1 peptide mating signal and the negatively regulating, DNA-binding TraA protein. *Proc. Natl. Acad. Sci. USA* **95:**6430–6435.

79. **Fujimoto, S., H. Tomita, E. Wakamatsu, K. Tanimoto, and Y. Ike.** 1995. Physical mapping of the conjugative bacteriocin plasmid pPD1 of *Enterococcus faecalis* and identification of the determinant related to the pheromone response. *J. Bacteriol.* **177:**5574–5581.

80. **Fuqua, W., and S. Winans.** 1994. A LuxR-LuxI type regulatory system activates *Agrobacterium* Ti plasmid conjugal transfer in the presence of a plant tumor metabolite. *J. Bacteriol.* **176:**2796–2806.

81. **Garnier, F., S. Taourit, P. Glaser, P. Courvalin, and M. Galimand.** 2000. Characterization of transposon Tn1549, conferring VanB-type resistance in *Enterococcus* spp. *Microbiology* **146:**1481–1489.

82. **Gawron-Burke, C., and D. B. Clewell.** 1982. A transposon in *Streptococcus faecalis* with fertility properties. *Nature* **300:**281–284.

83. **Gerdes, K., J. Møller-Jensen, and R. B. Jensen.** 2000. Plasmid and chromosome partitioning: surprises from phylogeny. *Mol. Microbiol.* **37:**455–466.

84. **Gering, M., F. Gotz, and R. Bruckner.** 1996. Sequence and analysis of the replication region of the *Staphylococcus xylosus* plasmid pSX267. *Gene* **182:**117–122.

85. **Ghim, S.-Y., C. C. Kim, E. R. Bonner, J. N. D'Elia, G. K. Grabner, and R. L. Switzer.** 1999. The *Enterococcus faecalis pyr* operon is regulated by autogenous transcriptional attenuation at a single site in the 5′ leader. *J. Bacteriol.* **181:**1324–1329.

86. **Grady, R., and F. Hayes.** 2003. Axe-Txe, a broad spectrum proteic toxin-antitoxin system specified by a multidrug-resistant, clinical isolate of *Enterococcus faecium*. *Mol. Microbiol.* **47**:1419–1432.

87. **Grainge, I., and M. Jayaram.** 1999. The integrase family of recombinases: organization and function of the active site. *Mol. Microbiol.* **33**:449–456.

88. **Greenfield, T. J., E. Ehli, T. Kirshenmann, T. Franch, K. Gerdes, and K. E. Weaver.** 2000. The antisense RNA of the *par* locus of pAD1 regulates the expression of a 33-amino-acid toxic peptide by an unusual mechanism. *Mol. Microbiol.* **37**:652–660.

89. **Greenfield, T. J., T. Franch, K. Gerdes, and K. E. Weaver.** 2001. Antisense RNA regulation of the *par* post-segregational killing system: structural analysis and mechanism of binding of the antisense RNA, RNAII and its target, RNAI. *Mol. Microbiol.* **42**:527–537.

90. **Greenfield, T. J., and K. E. Weaver.** 2000. Antisense RNA regulation of the pAD1 *par* post-segregational killing system requires interaction at the 5' and 3' ends of the RNAs. *Mol. Microbiol.* **37**:661–670.

91. **Haas, W., B. D. Shepard, and M. S. Gilmore.** 2002. Two-component regulator of *Enterococcus faecalis* cytolysin responds to quorum-sensing autoinduction. *Nature* **415**:84–87.

92. **Hancock, L., and M. Perego.** 2002. Two-component signal transduction in *Enterococcus faecalis*. *J. Bacteriol.* **184**:5819–5825.

93. **Hancock, L. E., B. D. Shepard, and M. S. Gilmore.** 2003. Molecular analysis of the *Enterococcus faecalis* serotype 2 polysaccharide determinant. *J. Bacteriol.* **185**:4393–4401.

94. **Hayes, F.** 2003. Toxins-antitoxins: plasmid maintenance, programmed cell death, and cell cycle arrest. *Science* **301**:1496–1499.

95. **Heath, D., F. An, K. Weaver, and D. Clewell.** 1995. Phase variation of *Enterococcus faecalis* pAD1 conjugation functions relates to changes in iteron sequence region. *J. Bacteriol.* **177**:5453–5459.

96. **Heaton, M. P., L. F. Discotto, M. J. Pucci, and S. Handwerger.** 1996. Mobilization of vancomycin resistance by transposon-mediated fusion of a VanA plasmid with an *Enterococcus faecium* sex pheromone-response plasmid. *Gene* **171**:9–17.

97. **Héchard, Y., C. Pelletier, Y. Cenatiempo, and J. Frère.** 2001. Analysis of σ54-dependent genes in *Enterococcus faecalis*: a mannose PTS permease (EIIMan) is involved in sensitivity to a bacteriocin, mesentericin Y105. *Microbiology* **147**:1575–1580.

98. **Hedberg, P. J., B. A. B. Leonard, R. E. Ruhfel, and G. M. Dunny.** 1996. Identification and characterization of the genes of *Enterococcus faecalis* plasmid pCF10 involved in replication and in negative control of pheromone-inducible conjugation. *Plasmid* **35**:46–57.

99. **Higgins, M. L., and G. D. Shockman.** 1976. Study of a cycle of cell wall assembly in *Streptococcus faecalis* by three-dimensional reconstruction of thin sections of cells. *J. Bacteriol.* **127**:1346–1358.

100. **Hinerfeld, D., and G. Churchward.** 2001. Specific binding of integrase to the origin of transfer (oriT) of the conjugative transposon Tn916. *J. Bacteriol.* **183**:2947–2951.

101. **Hinerfeld, D., and G. Churchward.** 2001. Xis protein of the conjugative transposon Tn916 plays dual opposing roles in transposon excision. *Mol. Microbiol.* **41**:1459–1467.

102. **Hirt, H., R. Wirth, and A. Muscholl.** 1996. Comparative analysis of 18 sex pheromone plasmids from *Enterococcus faecalis*: detection of a new insertion element on pPD1 and implications for the evolution of this plasmid family. *Mol. Gen. Genet.* **252**:640–647.

103. **Hoch, J. A., and T. J. Silhavy.** 1995. *Two-Component Signal Transduction*. American Society for Microbiology, Washington, D.C.

104. **Hodel-Christian, S. L., and B. E. Murray.** 1991. Characterization of the gentamicin resistance transposon Tn5281 from *Enterococcus faecalis* and comparison to staphylococcal transposons Tn4001 and Tn4031. *Antimicrob. Agents Chemother.* **35**:1147–1152.

105. **Holman, T. R., Z. Wu, B. L. Wanner, and C. T. Walsh.** 1994. Identification of the DNA-binding site for the phosphorylated VanR protein required for vancomycin resistance in *Enterococcus faecium*. *Biochemistry* **33**:4625–4631.

106. **Hosking, S., M. Deadman, E. Moxon, J. Peden, N. Saunders, and N. High.** 1998. An in silico evaluation of Tn916 as a tool for generalized mutagenesis in *Haemophilus influenzae* Rd. *Microbiology* **144**:2525–2530.

107. **Ike, Y., K. Tanimoto, H. Tomita, K. Takeuchi, and S. Fujimoto.** 1998. Efficient transfer of the pheromone-independent *Enterococcus faecium* plasmid pMG1 (Gmr) (65.1 kilobases) to *Enterococcus* strains during broth mating. *J. Bacteriol.* **180**:4886–4892.

108. **Ip, S. C. Y., M. Bregu, F.-X. Barre, and D. J. Sherratt.** 2003. Decatenation of DNA circles by FtsK-dependent Xer site-specific recombination. *EMBO J.* **22**:6399–6407.

109. **Ireton, K., N. Gunther, 4th, and A. Grossman.** 1994. spo0J is required for normal chromosome segregation as well as the initiation of sporulation in *Bacillus subtilis*. *J. Bacteriol.* **176**:5320–5329.

110. **Jaworski, D., and D. Clewell.** 1994. Evidence that coupling sequences play a frequency-determining role in conjugative transposition of Tn916 in *Enterococcus faecalis*. *J. Bacteriol.* **176**:3328–3335.

111. **Jaworski, D., and D. Clewell.** 1995. A functional origin of transfer (oriT) on the conjugative transposon Tn916. *J. Bacteriol.* **177**:6644–6651.

111a. **Joint Genome Institute.** *Enterococcus faecium* genome. http://genome.jgi-psf.org/draft_microbes/entfa/entfa.home.html

112. **Khan, S.** 1997. Rolling-circle replication of bacterial plasmids. *Microbiol. Mol. Biol. Rev.* **61**:442–455.

113. **Khan, S. A.** 2003. DNA-protein interactions during the initiation and termination of plasmid pT181 rolling-circle replication. *Prog. Nucleic Acid Res. Mol. Biol.* **75**:113–137.

114. **Khan, S. A.** 2000. Plasmid rolling-circle replication: recent developments. *Mol. Microbiol.* **37**:477–484.

115. **Kunst, F., N. Ogasawara, I. Moszer, A. M. Albertini, G. Alloni, V. Azevedo, M. G. Bertero, P. Bessieres, A. Bolotin, S. Borchert, R. Borriss, L. Boursier, A. Brans, M. Braun, S. C. Brignell, S. Bron, S. Brouillet, C. V. Bruschi, B. Caldwell, V. Capuano, N. M. Carter, S.-K. Choi, J.-J. Codani, I. F. Connerton, N. J. Cummings, R. A. Daniel, F. Denizot, K. M. Devine, A. Dusterhoft, S. D. Ehrlich, P. T. Emmerson, K. D. Entian, J. Errington, C. Fabret, E. Ferrari, D. Foulger, C. Fritz, M. Fujita, Y. Fujita, S. Fuma, A. Galizzi, N. Galleron, S.-Y. Ghim, P. Glaser, A. Goffeau, E. J. Golightly, G. Grandi, G. Guiseppi, B. J. Guy, K. Haga,**

J. Haiech, C. R. Harwood, A. Henaut, H. Hilbert, S. Holsappel, S. Hosono, M.-F. Hullo, M. Itaya, L. Jones, B. Joris, D. Karamata, Y. Kasahara, M. Klaerr-Blanchard, C. Klein, Y. Kobayashi, P. Koetter, G. Koningstein, S. Krogh, M. Kumano, K. Kurita, A. Lapidus, S. Lardinois, J. Lauber, V. Lazarevic, S.-M. Lee, A. Levine, H. Liu, S. Masuda, C. Mauel, C. Medigue, N. Medina, R. P. Mellado, M. Mizuno, D. Moestl, S. Nakai, M. Noback, D. Noone, M. O'Reilly, K. Ogawa, A. Ogiwara, B. Oudega, S.-H. Park, V. Parro, T. M. Pohl, D. Portetelle, S. Porwollik, A. M. Prescott, E. Presecan, P. Pujic, B. Purnelle, et al. 1997. The complete genome sequence of the gram-positive bacterium *Bacillus subtilis*. *Nature* **390**:249–256.

116. **Kwong, S. M., R. A. Skurray, and N. Firth.** 2004. *Staphylococcus aureus* multiresistance plasmid pSK41: analysis of the replication region, initiator protein binding and antisense RNA regulation. *Mol. Microbiol.* **51:** 497–509.

117. **Lancaster, H., A. P. Roberts, R. Bedi, M. Wilson, and P. Mullany.** 2004. Characterization of Tn916S, a Tn916-like element containing the tetracycline resistance determinant tet(S). *J. Bacteriol.* **186:**4395–4398.

118. **Leavis, H., J. Top, N. Shankar, K. Borgen, M. Bonten, J. van Embden, and R. J. L. Willems.** 2004. A novel putative enterococcal pathogenicity island linked to the esp virulence gene of *Enterococcus faecium* and associated with epidemicity. *J. Bacteriol.* **186:**672–682.

119. **Leboeuf, C., L. Leblanc, Y. Auffray, and A. Hartke.** 2000. Characterization of the ccpA gene of *Enterococcus faecalis*: identification of starvation-inducible proteins regulated by CcpA. *J. Bacteriol.* **182:**5799–5806.

120. **Le Breton, Y., G. Boel, A. Benachour, H. Prevost, Y. Auffray, and A. Rince.** 2003. Molecular characterization of *Enterococcus faecalis* two-component signal transduction pathways related to environmental stresses. *Environ. Microbiol.* **5:**329–337.

121. **Le Chatelier, E., O. J. Becherel, E. d'Alencon, D. Canceill, S. D. Ehrlich, R. P. P. Fuchs, and L. Janniere.** 2004. Involvement of DnaE, the second replicative dna polymerase from *Bacillus subtilis*, in DNA mutagenesis. *J. Biol. Chem.* **279:**1757–1767.

122. **Le Chatelier, E., L. Janniere, S. D. Ehrlich, and D. Canceill.** 2001. The RepE initiator is a double-stranded and single-stranded DNA-binding protein that forms an atypical open complex at the onset of replication of plasmid pAMbeta 1 from gram-positive bacteria. *J. Biol. Chem.* **276:**10234–10246.

123. **Leclercq, R., E. Derlot, J. Duval, and P. Courvalin.** 1988. Plasmid-mediated resistance to vancomycin and teicoplanin in *Enterococcus faecium*. *N. Engl. J. Med.* **319:** 157–161.

124. **Lemon, K. P., and A. D. Grossman.** 1998. Localization of bacterial DNA polymerase: evidence for a factory model of replication. *Science* **282:**1516–1519.

125. **Leonard, B. A. B., A. Podbielski, P. J. Hedberg, and G. M. Dunny.** 1996. *Enterococcus faecalis* pheromone binding protein, PrgZ, recruits a chromosomal oligopeptide permease system to import sex pheromone cCF10 for induction of conjugation. *Proc. Natl. Acad. Sci. USA* **93:**260–264.

126. **Ligozzi, M., F. Pittaluga, and R. Fontana.** 1993. Identification of a genetic element (psr) which negatively controls expression of *Enterococcus hirae* penicillin-binding protein 5. *J. Bacteriol.* **175:**2046–2051.

127. **Lopez de Saro, F. J., A.-Y. Moon Woody, and J. D. Helmann.** 1995. Structural analysis of the *Bacillus subtilis*[delta] factor: a protein polyanion which displaces RNA from RNA polymerase. *J. Mol. Biol.* **252:**189–202.

128. **Lovett, S. T.** 2004. Encoded errors: mutations and rearrangements mediated by misalignment at repetitive DNA sequences. *Mol. Microbiol.* **52:**1243–1253.

129. **Low, Y. L., N. S. Jakubovics, J. C. Flatman, H. F. Jenkinson, and A. W. Smith.** 2003. Manganese-dependent regulation of the endocarditis-associated virulence factor EfaA of *Enterococcus faecalis*. *J. Med. Microbiol.* **52:**113–119.

130. **Lu, F., and G. Churchward.** 1994. Conjugative transposition: Tn916 integrase contains two independent DNA binding domains that recognize different DNA sequences. *EMBO J.* **13:**1541–1548.

131. **Lu, F., and G. Churchward.** 1995. Tn916 target DNA sequences bind the C-terminal domain of integrase protein with different affinities that correlate with transposon insertion frequency. *J. Bacteriol.* **177:**1938–1946.

132. **Manganelli, R., S. Ricci, and G. Pozzi.** 1997. The joint of Tn916 circular intermediates is a homoduplex in *Enterococcus faecalis*. *Plasmid* **38:**71–78.

133. **Marra, D., B. Pethel, G. G. Churchward, and J. R. Scott.** 1999. The frequency of conjugative transposition of Tn916 is not determined by the frequency of excision. *J. Bacteriol.* **181:**5414–5418.

134. **Marra, D., and J. R. Scott.** 1999. Regulation of excision of the conjugative transposon Tn916. *Mol. Microbiol.* **31:**609–621.

135. **Marra, D., J. G. Smith, and J. R. Scott.** 1999. Excision of the conjugative transposon Tn916 in *Lactococcus lactis*. *Appl. Environ. Microbiol.* **65:**2230–2231.

136. **Marshall, C. G., G. Broadhead, B. K. Leskiw, and G. C. Wright.** 1997. D-ala-D-ala ligases from glycopeptide antibiotic-producing organisms are highly homologous to the enterococcal vancomycin-resistance ligases VanA and VanB. *Proc. Natl. Acad. Sci. USA* **94:**6480–6483.

137. **Marshall, C. G., I. A. D. Lessard, I. S. Park, and G. D. Wright.** 1998. Glycopeptide antibiotic resistance in glycopeptide-producing organisms. *Antimicrob. Agents Chemother.* **42:**2215–2220.

138. **Morlot, C., M. Noirclerc-Savoye, A. Zapun, O. Dideberg, and T. Vernet.** 2004. The D,D-carboxypeptidase PBP3 organizes the division process of *Streptococcus pneumoniae*. *Mol. Microbiol.* **51:**1641–1648.

139. **Morlot, C., A. Zapun, O. Dideberg, and T. Vernet.** 2003. Growth and division of *Streptococcus pneumoniae*: localization of the high molecular weight penicillin-binding proteins during the cell cycle. *Mol. Microbiol.* **50:** 845–855.

140. **Murata, T., I. Yamato, K. Igarashi, and Y. Kakinuma.** 1996. Intracellular Na+ regulates transcription of the ntp operon encoding a vacuolar-type Na+-translocating ATPase in *Enterococcus hirae*. *J. Biol. Chem.* **271:**23661–23666.

141. **Muscholl-Silberhorn, A. B.** 2000. Pheromone-regulated expression of sex pheromone plasmid pAD1-encoded aggregation substance depends on at least six upstream genes and a cis-acting, orientation-dependent factor. *J. Bacteriol.* **182:**3816–3825.

142. **Mylonakis, E., M. Engelbert, X. Qin, C. D. Sifri, B. E. Murray, F. M. Ausubel, M. S. Gilmore, and S. B. Calderwood.** 2002. The *Enterococcus faecalis* fsrB gene, a key component of the fsr quorum-sensing system, is asso-

ciated with virulence in the rabbit endophthalmitis model. *Infect. Immun.* **70**:4678–4681.

143. **Nakayama, J., Y. Cao, T. Horii, S. Sakuda, A. D. L. Akkermans, W. M. deVos, and H. Nagasawa.** 2001. Gelatinase biosynthesis-activating pheromone: a peptide lactone that mediates a quorum sensing in *Enterococcus faecalis. Mol. Microbiol.* **41**:145–154.

144. **Nakayama, J., Y. Takanami, T. Horii, S. Sakuda, and A. Suzuki.** 1997. Molecular mechanism of peptide-specific pheromone signaling in *Enterococcus faecalis:* function of pheromone receptor TraA and pheromone binding protein TraC encoded by pPD1. *J. Bacteriol.* **180:** 449–456.

145. **Nandi, S., J. J. Maurer, C. Hofacre, and A. O. Summers.** 2004. Gram-positive bacteria are a major reservoir of Class 1 antibiotic resistance integrons in poultry litter. *Proc. Natl. Acad. Sci. USA* **101**:7118–7122.

146. **Norgren, M., and J. R. Scott.** 1991. The presence of conjugative transposon Tn916 in the recipient strain does not impede transfer of a second copy of the element. *J. Bacteriol.* **173**:319–324.

147. **Novick, R. P.** 1998. Contrasting lifestyles of rolling-circle phages and plasmids. *Trends Biochem. Sci.* **23:**434–438.

148. **Ogasawara, N., and H. Yoshikawa.** 1992. Genes and their organization in the replication origin region of the bacterial chromosome. *Mol. Microbiol.* **6**:629–634.

149. **Ohmori, H., E. C. Friedberg, R. P. Fuchs, M. F. Goodman, F. Hanaoka, D. Hinkle, T. A. Kunkel, C. W. Lawrence, Z. Livneh, T. Nohmi, L. Prakash, S. Prakash, T. Todo, G. C. Walker, Z. Wang, and R. Woodgate.** 2001. The Y-family of DNA polymerases. *Molec. Cell* **8**:7–9.

150. **O'Keeffe, T., C. Hill, and R. P. Ross.** 1999. Characterization and heterologous expression of the genes encoding enterocin a production, immunity, and regulation in *Enterococcus faecium* DPC1146. *Appl. Environ. Microbiol.* **65:** 1506–1515.

151. **Osborn, A. M., and D. Boltner.** 2002. When phage, plasmids, and transposons collide: genomic islands, and conjugative- and mobilizable-transposons as a mosaic continuum. *Plasmid* **48**:202–212.

152. **Ozawa, Y., K. Tanimoto, S. Fujimoto, H. Tomita, and Y. Ike.** 1997. Cloning and genetic analysis of the UV resistance determinant (uvr) encoded on the *Enterococcus faecalis* pheromone-responsive conjugative plasmid pAD1. *J. Bacteriol.* **179**:7468–7475.

153. **Panina, E. M., A. G. Vitreschak, A. A. Mironov, and M. S. Gelfand.** 2003. Regulation of biosynthesis and transport of aromatic amino acids in low-GC gram-positive bacteria. *FEMS Microbiol. Lett.* **222**:211–220.

154. **Patino, L. A., P. Courvalin, and B. Perichon.** 2002. vanE gene cluster of vancomycin-resistant *Enterococcus faecalis* BM4405. *J. Bacteriol.* **184**:6457–6464.

155. **Paul Ross, R., and A. Claiborne.** 1997. Evidence for regulation of the NADH peroxidase gene (npr) from *Enterococcus faecalis* by OxyR. *FEMS Microbiol. Lett.* **151**:177–183.

156. **Paulsen, I. T., L. Banerjei, G. S. A. Myers, K. E. Nelson, R. Seshadri, T. D. Read, D. E. Fouts, J. A. Eisen, S. R. Gill, J. F. Heidelberg, H. Tettelin, R. J. Dodson, L. Umayam, L. Brinkac, M. Beanan, S. Daugherty, R. T. DeBoy, S. Durkin, J. Kolonay, R. Madupu, W. Nelson, J. Vamathevan, B. Tran, J. Upton, T. Hansen, J. Shetty, H. Khouri, T. Utterback, D. Radune, K. A. Ketchum, B. A. Dougherty, and C. M. Fraser.** 2003. Role of mobile DNA in the evolution of vancomycin-resistant *Enterococcus faecalis. Science* **299**:2071–2074.

157. **Perichon, B., B. Casadewall, P. Reynolds, and P. Courvalin.** 2000. Glycopeptide-resistant *Enterococcus faecium* BM4416 is a VanD-type strain with an impaired D-alanine:D-alanine ligase. *Antimicrob. Agents Chemother.* **44:** 1346–1348.

158. **Perkins, J. B., and P. Youngman.** 1983. *Streptococcus* plasmid pAMα1 is a composite of two separable replicons, one of which is closely related to *Bacillus* plasmid pBC16. *J. Bacteriol.* **155**:607–615.

159. **Perkins, J. B., and P. J. Youngman.** 1984. A physical and functional analysis of Tn917, a *Streptococcus* transposon in the Tn3 family that functions in *Bacillus. Plasmid* **12:** 119–138.

160. **Permina, E. A., A. A. Mironov, and M. S. Gelfand.** 2002. Damage-repair error-prone polymerases of eubacteria: association with mobile genome elements. *Gene* **293:** 133–140.

161. **Pethel, B., and G. Churchward.** 2000. Coupling sequences flanking Tn916 do not determine the affinity of binding of integrase to the transposon ends and adjacent bacterial DNA. *Plasmid* **43**:123–129.

162. **Podbielski, A., and B. Kreikemeyer.** 2004. Cell density-dependent regulation: basic principles and effects on the virulence of gram-positive cocci. *Int. J. Infect. Dis.* **8**:81–95.

163. **Pontius, L. T., and D. B. Clewell.** 1991. A phase variation event that activates conjugation functions encoded by the *Enterococcus faecalis* plasmid pAD1. *Plasmid* **26:** 172–185.

164. **Portillo, A., F. Ruiz-Larrea, M. Zarazaga, A. Alonso, J. L. Martinez, and C. Torres.** 2000. Macrolide resistance genes in *Enterococcus* spp. *Antimicrob. Agents Chemother.* **44**:967–971.

165. **Possoz, C., C. Ribard, J. Gagnat, J.-L. Pernodet, and M. Guerineau.** 2001. The integrative element pSAM2 from *Streptomyces:* kinetics and mode of conjugal transfer. *Mol. Microbiol.* **42**:159–166.

166. **Poyart-Salmeron, C., P. Trieu-Cuot, C. Carlier, and P. Courvalin.** 1989. Molecular characterization of two proteins involved in the excision of the conjugative transposon Tn1545: homologies with other site-specific recombinases. *EMBO J.* **8**:2425–2433.

167. **Pucci, M., J. Thanassi, L. Discotto, R. Kessler, and T. Dougherty.** 1997. Identification and characterization of cell wall-cell division gene clusters in pathogenic gram-positive cocci. *J. Bacteriol.* **179**:5632–5635.

168. **Qin, X., K. V. Singh, G. M. Weinstock, and B. E. Murray.** 2001. Characterization of fsr, a regulator controlling expression of gelatinase and serine protease in *Enterococcus faecalis* OG1RF. *J. Bacteriol.* **183**:3372–3382.

169. **Qin, X., K. V. Singh, G. M. Weinstock, and B. E. Murray.** 2000. Effects of *Enterococcus faecalis* fsr genes on production of gelatinase and a serine protease and virulence. *Infect. Immun.* **68**:2579–2586.

170. **Quintiliani, R., Jr., and P. Courvalin.** 1996. Characterization of Tn1547, a composite transposon flanked by the IS16 and IS256-like elements, that confers vancomycin resistance in *Enterococcus faecalis* BM4281. *Gene* **172**:1–8.

170a. **Ramirez-Arcos, S., J. Szeto, J.-A. R. Dillon, and W. Margolin.** 2002. Conservation of dynamic localization among MinD and MinE orthologs: oscillation of *Neisseria gonorrhoeae* proteins in *Escherichia coli. Mol. Microbiol.* **46**:493–504.

171. **Rice, L. B., S. Bellais, L. L. Carias, R. Hutton-Thomas, R. A. Bonomo, P. Caspers, M. G. Page, and L. Gutmann.** 2004. Impact of specific pbp5 mutations on expression of β-lactam resistance in *Enterococcus faecium. Antimicrob. Agents Chemother.* **48**:3028–3032.

172. **Rice, L. B., and L. L. Carias.** 1998. Transfer of Tn5385, a composite, multiresistance chromosomal element from *Enterococcus faecalis. J. Bacteriol.* **180:**714–721.

173. **Rice, L. B., L. L. Carias, R. Hutton-Thomas, F. Sifaoui, L. Gutmann, and S. D. Rudin.** 2001. Penicillin-binding protein 5 and expression of ampicillin resistance in *Enterococcus faecium. Antimicrob. Agents Chemother.* **45:**1480–1486.

174. **Roberts, A. P., P. A. Johanesen, D. Lyras, P. Mullany, and J. I. Rood.** 2001. Comparison of Tn5397 from *Clostridium difficile*, Tn916 from *Enterococcus faecalis* and the CW459tet(M) element from *Clostridium perfringens* shows that they have similar conjugation regions but different insertion and excision modules. *Microbiology* **147:** 1243–1251.

175. **Rowe-Magnus, D. A., and D. Mazel.** 2001. Integrons: natural tools for bacterial genome evolution. *Curr. Opin. Microbiol.* **4:**565–569.

176. **Rudy, C., J. Scott, and G. Churchward.** 1997. DNA binding by the Xis protein of the conjugative transposon Tn916. *J. Bacteriol.* **179:**2567–2572.

177. **Rudy, C., K. Taylor, D. Hinerfeld, J. Scott, and G. Churchward.** 1997. Excision of a conjugative transposon in vitro by the Int and Xis proteins of Tn916. *Nucleic Acids Res.* **25:**4061–4066.

178. **Ruhfel, R., D. Manias, and G. Dunny.** 1993. Cloning and characterization of a region of the *Enterococcus faecalis* conjugative plasmid, pCF10, encoding a sex-pheromone-binding function. *J. Bacteriol.* **175:**5253–5259.

179. **Sahm, D. F., M. K. Marsilio, and P. G. Piazza.** 1999. Antimicrobial resistance in key bloodstream bacterial isolates: electronic surveillance with the Surveillance Network Database—USA. *Clin. Infect. Dis.* **29:**259–263.

180. **Sapunaric, F., C. Franssen, P. Stefanic, A. Amoroso, O. Dardenne, and J. Coyette.** 2003. Redefining the role of psr in β-lactam resistance and cell autolysis of *Enterococcus hirae. J. Bacteriol.* **185:**5925–5935.

181. **Schwarz, F. V., V. Perreten, and M. Teuber.** 2002. Sequence of the 50-kb conjugative multiresistance plasmid pRE25 from *Enterococcus faecalis* RE25. *Plasmid* **46:**170–187.

182. **Scott, J. R., F. Bringel, D. Marra, G. Van Alstine, and C. K. Rudy.** 1994. Conjugative transposition of Tn916: preferred targets and evidence for conjugative transfer of a single strand and for a double-stranded circular intermediate. *Mol. Microbiol.* **11:**1099–1108.

183. **Senghas, E., J. M. Jones, M. Yamamoto, C. Gawron-Burke, and D. B. Clewell.** 1988. Genetic organization of the bacterial conjugative transposon Tn916. *J. Bacteriol.* **170:**245–249.

184. **Shankar, N., A. S. Baghdayan, and M. S. Gilmore.** 2002. Modulation of virulence within a pathogenicity island in vancomycin-resistant *Enterococcus faecalis. Nature* **417:**746–750.

185. **Shaw, J. H., and D. B. Clewell.** 1985. Complete nucleotide sequence of macrolide-lincosamide-streptogramin B-resistance transposon Tn917 in *Streptococcus faecalis. J. Bacteriol.* **164:**782–796.

186. **Shepard, B. D., and M. S. Gilmore.** 1999. Identification of aerobically and anaerobically induced genes in *Enterococcus faecalis* by random arbitrarily primed PCR. *Appl. Environ. Microbiol.* **65:**1470–1476.

187. **Shoemaker, N. B., R. D. Barber, and A. A. Salyers.** 1989. Cloning and characterization of a *Bacteroides* conjugal tetracycline-erythromycin resistance element by using a shuttle cosmid vector. *J. Bacteriol.* **171:**1294–1302.

188. **Shoemaker, N. B., M. D. Smith, and W. R. Guild.** 1980. DNase-resistant transfer of chromosomal cat and tet insertions by filter mating in *Pneumococcus. Plasmid* **3:**80–87.

189. **Solioz, M., and J. V. Stoyanov.** 2003. Copper homeostasis in *Enterococcus hirae. FEMS Microbiol. Rev.* **27:**183–195.

190. **Su, Y., P. He, and D. Clewell.** 1992. Characterization of the tet(M) determinant of Tn916: evidence for regulation by transcription attenuation. *Antimicrob. Agents Chemother.* **36:**769–778.

191. **Su, Y. A., and D. B. Clewell.** 1993. Characterization of the left 4 kb of conjugative transposon Tn916: determinants involved in excision. *Plasmid* **30:**234–250.

192. **Tanaka, T., and M. Ogura.** 1998. A novel *Bacillus natto* plasmid pLS32 capable of replication in *Bacillus subtilis. FEBS Lett.* **422:**243–246.

193. **Tanimoto, K., F. An, and D. Clewell.** 1993. Characterization of the traC determinant of the *Enterococcus faecalis* hemolysin-bacteriocin plasmid pAD1: binding of sex pheromone. *J. Bacteriol.* **175:**5260–5264.

194. **Tanimoto, K., and Y. Ike.** 2002. Analysis of the conjugal transfer system of the pheromone- independent highly transferable enterococcus plasmid pMG1: identification of a tra gene (traA) up-regulated during conjugation. *J. Bacteriol.* **184:**5800–5804.

195. **Taylor, K., and G. Churchward.** 1997. Specific DNA cleavage mediated by the integrase of conjugative transposon Tn916. *J. Bacteriol.* **179:**1117–1125.

196. **Teng, F., L. Wang, K. V. Singh, B. E. Murray, and G. M. Weinstock.** 2002. Involvement of phoP-phoS homologs in *Enterococcus faecalis* virulence. *Infect. Immun.* **70:**1991–1996.

196a. **The Institute for Genome Research.** *Enterococcus faecalis* genome. http://www.tigr.org/tigr-scripts/CMR2/GenomePage3.spl?database=gef.

197. **Tomita, H., and D. B. Clewell.** 2000. A pAD1-encoded small RNA molecule, mD, negatively regulates *Enterococcus faecalis* pheromone response by enhancing transcription termination. *J. Bacteriol.* **182:**1062–1073.

198. **Tomita, H., K. Tanimoto, S. Hayakawa, K. Morinaga, K. Ezaki, H. Oshima, and Y. Ike.** 2003. Highly conjugative pMG1-like plasmids carrying Tn1546-like transposons that encode vancomycin resistance in *Enterococcus faecium. J. Bacteriol.* **185:**7024–7028.

199. **Trotter, K. M., and G. M. Dunny.** 1990. Mutants of *Enterococcus faecalis* deficient as recipients in mating with donors carrying pheromone-inducible plasmids. *Plasmid* **24:**57–67.

200. **Uttley, A. H., C. H. Collins, J. Naidoo, and R. C. George.** 1988. Vancomycin-resistant enterococci. *Lancet* **1:**57–58.

201. **Verneuil, N., M. Sanguinetti, Y. Le Breton, B. Posteraro, G. Fadda, Y. Auffray, A. Hartke, and J.-C. Giard.** 2004. Effects of the *Enterococcus faecalis* hypR gene encoding a new transcriptional regulator on oxidative stress response and intracellular survival within macrophages. *Infect. Immun.* **72:**4424–4431.

202. **Waters, C. M., and G. M. Dunny.** 2001. Analysis of functional domains of the *Enterococcus faecalis* pheromone-induced surface protein aggregation substance. *J. Bacteriol.* **183:**5659–5667.

203. **Waters, C. M., H. Hirt, J. K. McCormick, P. M. Schlievert, C. L. Wells, and G. M. Dunny.** 2004. An

amino-terminal domain of *Enterococcus faecalis* aggregation substance is required for aggregation, bacterial internalization by epithelial cells and binding to lipoteichoic acid. *Mol. Microbiol.* **52:**1159–1171.

204. **Weaver, K., D. Clewell, and F. An.** 1993. Identification, characterization, and nucleotide sequence of a region of *Enterococcus faecalis* pheromone-responsive plasmid pAD1 capable of autonomous replication. *J. Bacteriol.* **175:**1900–1909.

205. **Weaver, K. E., and D. B. Clewell.** 1991. Control of *Enterococcus faecalis* sex pheromone cAD1 elaboration: effects of culture aeration and pAD1 plasmid-encoded determinants. *Plasmid* **25:**177–189.

206. **Weaver, K. E., E. A. Ehli, J. S. Nelson, and S. Patel.** 2004. Antisense RNA regulation by stable complex formation in the *Enterococcus faecalis* plasmid pAD1 par addiction system. *J. Bacteriol.* **186:**6400–6408.

207. **Weaver, K. E., K. D. Jensen, A. Colwell, and S. I. Sriram.** 1996. Functional analysis of the *Enterococcus faecalis* plasmid pAD1-encoded stability determinant par. *Mol. Microbiol.* **20:**53–63.

208. **Weaver, K. E., L. R. Rice, and G. Churchward.** 2002. Plasmids and transposons, p. 219–263. *In* M. S. Gilmore, D. B. Clewell, P. Courvalin, G. Dunny, B. E. Murray, and L. B. Rice (ed.), *The Enterococci: Pathogenesis, Molecular Biology, and Antibiotic Resistance.* ASM Press, Washington, D.C.

209. **Weaver, K. E., and D. J. Tritle.** 1994. Identification and characterization of an *Enterococcus faecalis* plasmid pAD1-encoded stability determinant which produces two small RNA molecules necessary for its function. *Plasmid* **32:**168–181.

210. **Weaver, K. E., K. D. Walz, and M. S. Heine.** 1998. Isolation of a derivative of *Escherichia coli-Enterococcus faecalis* shuttle vector pAM401 temperature sensitive for maintenance in *E. faecalis* and its use in evaluating the mechanism of pAD1 par-dependent plasmid stabilization. *Plasmid* **40:**225–232.

211. **Weaver, K. E., D. M. Weaver, C. L. Wells, C. M. Waters, M. E. Gardner, and E. A. Ehli.** 2003. *Enterococcus faecalis* plasmid pAD1-encoded Fst toxin affects membrane permeability and alters cellular responses to lantibiotics. *J. Bacteriol.* **185:**2169–2177.

212. **Weidlich, G., R. Wirth, and D. Galli.** 1992. Sex pheromone plasmid pAD1-encoded surface exclusion protein of *Enterococcus faecalis. Mol. Gen. Genet.* **233:**161–168.

213. **Weigel, L. M., D. B. Clewell, S. R. Gill, N. C. Clark, L. K. McDougal, S. E. Flannagan, J. F. Kolonay, J. Shetty, G. E. Killgore, and F. C. Tenover.** 2003. Genetic analysis of a high-level vancomycin-resistant isolate of *Staphylococcus aureus. Science* **302:**1569–1571.

214. **Wojciak, J. M., K. M. Connolly, and R. T. Clubb.** 1999. NMR structure of the Tn916 integrase-DNA complex. *Nat. Struct. Biol.* **6:**366–373.

215. **Woodford, N., D. Morrison, A. P. Johnson, A. C. Bateman, J. G. Hastings, T. S. Elliott, and B. D. Cookson.** 1955. Plasmid-mediated vanB glycopeptide resistance in enterococci. *Microb. Drug Resist.* **1:**235–240.

216. **Wright, G. D., T. R. Holman, and C. T. Walsh.** 1993. Purification and characterization of VanR and the cytosolic domain of VanS: a two-component regulatory system required for vancomycin resistance in *Enterococcus faecium* BM4147. *Biochemistry* **32:**5057–5063.

217. **Yagi, Y., and D. B. Clewell.** 1980. Recombination-deficient mutant of *Streptococcus faecalis. J. Bacteriol.* **143:**966–970.

Pathogenesis of Oral Streptococci

R. R. B. RUSSELL

27

LIFE IN THE MOUTH

The oral cavity represents an environment that is warm, moist, and well provided with nutrients in the form of a steady supply of host-derived macromolecules, with the added bonus of several meals a day supplemented by whatever intermittent snacks a particular individual may fancy. Small wonder then, that the oral cavity supports a rich and abundant microflora. Within the mouth, however, there is a range of habitats differing in such properties as the supply of oxygen, nutrient flow, pH, and the nature of substratum—hard (teeth) or soft (mucosal tissues). Any colonizing bacterium must therefore have evolved the ability to overcome the hazards encountered, such as fluctuations in the composition of nutrient supply, local oxygen availability, shear forces due to saliva flow and mastication, and a range of host defense mechanisms.

Species of streptococci are well represented among the bacteria found in the oral cavity, which has been estimated to harbor around 500 different species of bacteria, though there remain many taxa of uncertain status and many microscopically observable microbes that have not yet been isolated in laboratory culture. These oral streptococci seem to be ubiquitous among all the human populations studied. When they have been sought, identical or closely related streptococci have also been found in a wide variety of animal species, so streptococci are clearly part of the normal commensal flora of mammals; the purpose of this chapter is to consider the problems that arise when this commensal relationship breaks down and the oral streptococci become opportunistic pathogens.

PROBLEMS WITH TAXONOMY

The question of how to classify the oral streptococci has been tackled since the start of the 20th century, but even today one can still find textbooks in which all are lumped together under the descriptive term "viridans streptococci," which refers to the greening reaction, or alpha-hemolysis, produced on blood agar. Such a classification based on hemolysis is not, however, truly interchangeable with "oral streptococci" because some of these may show beta-hemoly-

sis, and the alpha-hemolytic phenotype is not restricted to species of streptococci found in the oral cavity. In a comprehensive review of the history of the developments in classification of the oral streptococci, Whiley and Beighton (70) listed 18 recognized species and described the changes in taxonomic position and nomenclature that have taken place up to 1998. Facklam (19) has also recently reviewed the changes in taxonomy and nomenclature. A clearer taxonomic picture has gradually emerged as new technical approaches have been introduced—biotyping, serotyping, DNA-DNA hybridization, and, most recently, sequencing of genes for ribosomal RNA or conserved enzymes, all leading to shifts in our understanding. The relationship between a wide range of species determined by rRNA gene sequence analysis by Kawamura et al. (31) is shown in Fig. 1. Similar phylogenetic trees that show overall agreement abut the major clusters, but uncertainties about the positioning of individual species have been produced based on genes for superoxide dismutase (52), D-alanine:D-alanine ligase (23), RNase P (64), or transfer DNA intergenic spacer length (16). Further species have been proposed in the last few years, and all fit into one of the major groupings. Table 1 lists the species regularly isolated from the human oral cavity, with the names changed to comply with Latin grammar (65).

The mitis group contains the largest number of named species of oral streptococci. rRNA data show that *Streptococcus pneumoniae* also lies in this group, though its habitat is considered to be the nasopharynx rather than the oral cavity. *S. pneumoniae* is closely related to *Streptococcus mitis* and *Streptococcus oralis*, and there is now a substantial body of evidence showing that there is extensive exchange of genetic information between the species. This was first reported for a penicillin-binding protein when genes in penicillin-resistant isolates of *S. pneumoniae* were shown to have a mosaic structure, with segments clearly identical to genes of *S. oralis*, but has since been extended to a number of different genes, including those for immunoglobulin A (IgA) protease and the *comCDE* loci required for competence (25, 34, 69). The commensal oral streptococci thus offer a pool of genetic material that can undergo gene shuffling with an

Gram-Positive Pathogens, 2nd edition, edited by Vincent A. Fischetti et al.

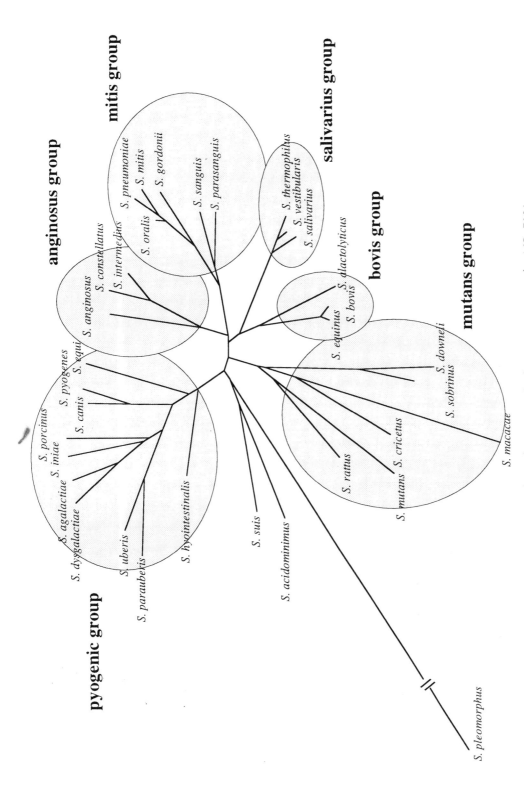

FIGURE 1 Phylogenetic relationships among 34 *Streptococcus* species by 16S rRNA gene sequence analysis. (Reproduced from Kawamura et al. [31] with permission.)

TABLE 1 Streptococci commonly found in the human mouth

Group	Species	Occurrence	Virulence properties
Mitis group	S. mitis S. australis S. cristatus S. oralis S. gordonii S. oligofermentans S. parasanguinis S. peroris S. sanguinis S. sinensis	Most common group, pioneer species in plaque colonization, most frequent cause of endocarditis	Survival in bloodstream; attachment to heart valves; induction of tissue damage
Mutans group	S. mutans S. sobrinus	Mainly found in plaque, associated with dental caries	Colonization of plaque; acidogenesis; aciduricity; extracellular glucan synthesis;
Salivarius group	S. salivarius S. vestibularis	Colonize principally mucosal surfaces	Rarely pathogenic
Anginosus group	S. anginosus S. constellatus S. intermedius	Found in gingival crevice, associated with abscesses	Survival in anaerobic conditions; resistance to host defenses; tissue damage by degradative enzymes

important pathogen and lead to the emergence of resistant strains (53). There is thus reason to think that there may be extensive mixing of other genetic traits within the mitis group, because all are naturally competent and share a common response mechanism to a competence-stimulating pheromone (25). This raises the question as to whether the currently defined species can be separated by clear boundaries, or whether they represent a continuum, with many mosaic isolates displaying a mixture of properties now regarded as characteristic of individual species. This is particularly so for the S. mitis/S. oralis/S. pneumoniae group, into which the newly described Streptococcus australis, Streptococcus infantis, Streptococcus oligofermentans, Streptococcus peroris, and Streptococcus sinensis all seem to fit. It is possible that the existence of a high level of natural competence and consequent ready exchange of genes between strains sharing a common pherotype may also explain the observed heterogeneity within the anginosus group (25, 70, 72). In contrast, species boundaries appear to be clearly delineated in the mutans and salivarius groups.

ACQUISITION OF ORAL STREPTOCOCCI

The oral cavity is sterile in utero, but although during birth the neonate is exposed to all the complex microflora of the birth canal, these organisms fail to colonize, illustrating the highly selective environment of the mouth. The oral flora is rapidly established soon after birth, with streptococci being numerically dominant, particularly S. salivarius, S. mitis, and S. oralis, which colonize the mucosal surfaces and dorsum of the tongue. From where do these infecting streptococci come? There is little evidence for the source of the pioneer species, but it is generally considered to be the mother. The transmission of S. mutans has been studied in some detail. S. mutans preferentially colonizes hard surfaces in the mouth and hence its emergence in significant numbers is delayed until eruption of the first teeth, with the majority of infants acquiring it during a "window of infectivity" around a median age of 26 months (13). There is a correlation between the levels of mutans streptococci in the

mouths of infants and in their mothers, and fingerprinting methods have been used to demonstrate that the most common source of infection is the mother (38). Although an infant may initially carry only one clonal line of S. mutans (identified by ribotyping or restriction fragment length polymorphism), by adulthood he or she may have up to seven different clonotypes, though these seem to be remarkably stable once established (1). In marked contrast, there is a great degree of diversity in the population structure of S. mitis in any one individual, and there is evidence of rapid turnover of dominant clonal types, with any one type being resident for only a few weeks (22, 27, 28). It seems probable that the diversity of types of S. mitis is due to its natural competence and the ready exchange of genetic information between strains. In other words, new clonotypes emerge by recombination between residents rather than superinfection with new strains. This free mixing and shuffling of DNA is also seen in the Neisseria spp., which similarly colonize mucosal surfaces, and it has been proposed that it may represent an evolved mechanism for generating antigenic diversity that helps the bacteria to evade the host immune defenses. Although this remains to be tested, it is interesting to note that the mutans streptococci, which are equally successful long-term colonizers of the oral cavity, seem to be genetically stable.

MECHANISM OF COLONIZATION

The immediate challenge for any bacterium entering the mouth is to avoid being spat out again or swallowed. Early work by Gibbons established both that oral streptococci can adhere to surfaces in the mouth and that they exhibit a specificity of adhesion that may at least partly explain why they colonize particular sites. Most work has been done on the formation of dental plaque, which complies with the classical model for formation of a biofilm: a cleaned tooth surface rapidly becomes coated with a conditioning layer of host-derived and bacterial macromolecules to which particular pioneer species attach, with streptococci (S. mitis, S. oralis, and S. sanguinis) being predominant. Plaque then develops by a combination of growth of attached pioneers to form microcolonies and the accretion of further bacteria by interspecies linking. This coaggregation again involves specific binding between bacterial surface adhesins, usually proteins, and receptors, which are generally carbohydrates. There is now an extensive literature on the specificity of these interactions, and elegant proposals have been put forward to explain how a network of binding interactions between different species could lead to the stabilization of a multispecies biofilm. The molecular basis of these interbacterial binding reactions and bacterial binding to host salivary or tissue macromolecules has been reviewed (30, 73). However, despite the detailed laboratory knowledge, there remains limited direct evidence of the significance of the contribution of coaggregation to colonization or pathogenesis. One study, in which enamel chips coated with coaggregating and noncoaggregating streptococci were placed in the mouth, found no difference in the subsequent accumulation of species predicted to bind to the target organisms (60), while an in vitro study with a defined set of bacteria concluded that coaggregation was not important in the early stages of biofilm formation (24). Recently, however, confocal microscopy has been used to demonstrate the occurrence of coaggregation between *S. gordonii* and *Actinomyces naeslundii* (50) in dental plaque.

IMMUNOLOGICAL PROCESSES IN THE MOUTH

Antibodies to the oral streptococci can be detected in blood or saliva from an early age, though the pattern of response against different antigens shows great variability and studies using Western blot analysis to characterize the response are often difficult to interpret because of cross-reaction between antigens of different species. Despite numerous studies, however, there has emerged no convincing evidence that the immune response controls the developing microflora. This remains one of the great puzzles, not just for the oral cavity but for other permanently colonized mucosal sites— how is the balance between commensal bacteria and host response maintained in equilibrium? Attention has been focused on the contribution of the major secretory antibody, IgA, and of the IgA1 proteases that cleave it (34, 44). Such a protease is produced by *S. sanguinis*, *S. mitis*, and *S. oralis*, three species that are among the most successful pioneer species in plaque formation. In the context of the mixed flora of plaque, the protease produced by these species may also benefit neighboring cells of other species (66); again, this effect has yet to be demonstrated in vivo. If IgA does influence the oral microflora, it is clearly unable to eliminate streptococci, and the interplay between the immune system and the oral streptococci must be very subtle. On the other hand, there is now substantial evidence that a strongly enhanced response induced by immunization or by passive application of antibody can influence the ability of streptococci to colonize and/or cause disease (36, 41, 55).

METABOLISM OF DENTAL PLAQUE

All streptococci in the oral cavity are faced with a fluctuating environment, but most is known about those species found in dental plaque. There, depending on the site in the mouth and the stage of development of plaque, wide variation is experienced in availability of nutrients, oxygen, and pH. The streptococci, as facultative anaerobes, are of central importance in the metabolism of plaque, not only because they are the organisms that are well equipped to survive in the fluctuating conditions but because they facilitate the survival of other, strictly anaerobic species (7, 51). The various species of streptococci differ both in their ability to generate acid from dietary carbohydrate and in their acid tolerance. A consequence of this is that frequent exposure to carbohydrate and the consequent fall in pH, to values that may go below pH 4.0 in carious lesions, serve to enrich the population of aciduric species and particularly the mutans streptococci (5, 8, 46, 47, 63). This observation of a population shift serves to underscore the importance of considering dental plaque as a dynamic ecosystem in which a consortium of bacterial species coexists in a variety of adhesive and metabolic interactions, which means that the community flourishes in a way that individual species cannot (6, 7, 10, 24, 46). Furthermore, the properties of bacteria growing in a mixed culture biofilm can differ substantially from those shown in pure culture either as a biofilm or planktonic culture, with different proteins being expressed (4, 68) and changes in properties such as acid tolerance and resistance to inhibitors (48).

ORAL STREPTOCOCCI AS PATHOGENS

The oral streptococci are normal commensals of the human mouth and as such play a beneficial role in colonization resistance, excluding potentially pathogenic species. Their association with disease is as opportunistic pathogens; their opportunities occur under two broad groups of conditions: (i) when environmental changes result in an ecological shift that favors an overgrowth of a particular species of streptococci, and (ii) when they escape their normal habitat and establish at another site, usually as a consequence of some breakdown in normal host defenses. Diseases caused by oral streptococci can affect the hard or soft tissues of the oral cavity, as well as remote body sites.

DENTAL CARIES

The disease of dental caries manifests as localized dissolution of the enamel surface of the tooth, which may advance to reach the underlying dentine and even penetrate as far as the tooth pulp. Caries lesions most commonly occur in pits and fissures, or in the area of contact between adjacent teeth; i.e., at stagnation sites where plaque has the chance to accumulate undisturbed. Decay is most frequently found in the occlusal (biting) surfaces of molar teeth but under certain circumstances may show a different distribution, for example, in the rapidly progressing "nursing bottle caries" of infants put to sleep with their incisor teeth effectively bathed in a sugar-rich solution from the feeding bottle, dummy, or even the mother's breast. The central pathogenic process of caries, in which plaque bacteria utilize dietary carbohydrate to generate the acid that demineralizes enamel, has been apparent for over a century, but the complexity of the plaque microflora rendered identification of causative organisms a difficult task. Debate raged between proponents of the specific plaque hypothesis, which proposed that particular species were responsible, and the nonspecific plaque hypothesis, which advanced the view that all plaque, regardless of its composition, was bad. The most recent variant, advanced by Marsh (46), is the ecological plaque hypothesis, which applies our understanding of the properties of the various plaque species to explain the shifts in the bacterial population under the environmental stress of frequent doses of sucrose or other fermentable carbohydrate. Thus, the drops in pH following sugar exposure favor the bacteria that survive and multiply well under these conditions. If these species are themselves good acid producers, there is a ratchet effect whereby the increased number of bacteria produce even more acid at the next sugar exposure, so there is a shift in the composition of the plaque population. Conversely, a plaque population containing a significant proportion of arginolytic or ureolytic bacteria that generate alkaline conditions will not be conducive to dental caries (9, 35).

Of all the species of bacteria found in dental plaque, only the mutans streptococci have been shown to have a clear association with the initiation of dental caries. Cross-sectional and longitudinal studies, carried out principally in the United States and Scandinavia, show that increased levels of mutans streptococci are strong predictors of current or future decay (40). Because S. mutans is the most common of the mutans group, being found in the vast majority of individuals, it seems clear that it is the principal etiological agent. However, most studies have relied upon selective media that have not distinguished between S. mutans and Streptococcus sobrinus, so, although the latter is much less frequent, its contribution to caries in individual cases is unclear. Although S. mutans is now generally accepted as being of prime importance, it has long been recognized that caries can occur even when this species cannot be detected. This indicates that there are other bacteria, or combinations of bacterial species, that may fill the same ecological niche as S. mutans and hence generate the same conditions in plaque that lead to sufficient acid production to cause decay. Not surprisingly, bacteria that are isolated from carious lesions and that share physiological properties with S. mutans are also mostly streptococci, belonging to a variety of species (39, 43, 57, 63), though the features of these isolates of non-mutans streptococci that make them more acidogenic and aciduric than is normally characteristic of their species remain to be discovered.

The known association between diet and dental caries should lead to caution in extrapolating results from one human population to another. Surveys in a number of African countries where sugar consumption is low have revealed that high levels of mutans streptococci can be found but with no correspondingly high incidence of dental caries (11, 12, 26). This may be due to consumption of a starch-rich diet and underscores the importance of diet in creating the local conditions that give the oral streptococci the opportunity to become pathogens, in particular, the availability of a carbon source that is readily fermentable to acid. S. mutans can produce acid from a wide range of carbohydrates, but sucrose shows a particular association with dental caries. This is in part explained by the widespread use of sucrose in foodstuffs and particularly snack foods, but besides entering glycolysis, sucrose is also the sole substrate for extracellular fructosyltransferases and glucosyltransferases of streptococci. The polymers (fructans and glucans) produced by these enzymes are important to overall plaque economy by serving as energy store, altering the permeability of plaque, and stabilizing the plaque biofilm by bacteria-bacteria and bacteria-surface adhesion (14).

PERIODONTAL DISEASE

Periodontal diseases affect the soft tissues and bone supporting the tooth and are acknowledged to have a microbial etiology, though the nature of the host response plays a powerful role in the occurrence and severity of disease. Streptococci are prominent members of the plaque flora above the gingival margin but subgingivally are displaced by gram-negative anaerobic rods. Investigations into the microbiology of periodontal disease have indicated that high numbers of streptococci at a site are generally an indicator of health, though a report found that Streptococcus constellatus was found at higher prevalence in patients with refractory periodontitis (15). Conditions in the periodontal pocket are alkaline as a result of proteolysis of blood and tissue components, so it is not surprising that species such as S. mutans are not favored. Why S. constellatus may flourish in disease, and not the other members of the anginosus group that are also normally present in the gingival crevice, is not clear.

DENTAL ABSCESSES

The pulp of the tooth and the periapical region around the root are normally sterile, but when bacteria gain entry, e.g., because of extensive caries or trauma to the tooth, they may multiply within the confined space to produce an abscess. A wide range of species can be isolated from dental abscesses, but individual abscesses generally harbor only three or four species (37). Among the species isolated, streptococci are the most common, particularly S. anginosus, which in one study was found in 33% of acute dentoalveolar abscesses (21). All three species of the anginosus group can be isolated from the oral cavity, particularly in the gingival crevice around the margins of the teeth. However, in contrast to the other oral streptococci, which are rarely found in other body sites, they also appear to be normal inhabitants of the gastrointestinal tract and vagina. The three species are also frequently isolated from purulent infections of internal organs, including the brain, liver, lungs, and spleen, with some indication that each shows a predilection for particular sites (71). Most abscesses are mixed infections, and it has been proposed that the facultatively anaerobic streptococci are important in the early stages of abscess formation, establishing local conditions suitable for subsequent invasion by strict anaerobes (59). Further development of the abscesses may depend upon synergistic interactions in which certain species benefit others by inactivation of host defenses or a number of species form a nutritional "consortium" for the efficient sequential degradation of host macromolecules (particularly glycoproteins). In this regard, the production by the anginosus group of a spectrum of degradative enzymes allows them to multiply in a location where they are remote from any diet and must obtain nutrients by breakdown of host tissue. Enzymes produced include glycosidases, nucleases, hyaluronidase, sialidase, and chondroitin sulfate depolymerase (29). There has been a report of a novel cytotoxin, intermedilysin, produced by Streptococcus intermedius (42, 49). This cytotoxin shows some homology to pneumolysin but has a different spectrum of action, being specific for human cells. This is the only example of a cytotoxin produced by one of the oral streptococci.

SYSTEMIC INFECTIONS

The systemic disease with which oral streptococci are most clearly associated is infective endocarditis (sometimes referred to as subacute bacterial endocarditis). Approximately half of all cases of infective endocarditis are attributable to streptococci, the most common species identified being S. sanguinis, S. oralis, and S. gordonii (17). It is for this reason that prophylactic antibiotics are employed before dental procedures in at-risk patients (i.e., those with a predisposing valvular defect) though the validity of this approach is the subject of considerable debate (58). Electrophoretic fingerprinting techniques have demonstrated a correspondence between oral isolates and isolates from heart valve lesions (20), blood (33, 54), and infected prosthetic joint replacements (3). Even very mild manipulations in the mouth can result in transfer of bacteria into the bloodstream, while procedures such as dental extraction (especially if there is an abscess underlying the extracted tooth) will produce a massive shower of organisms, including streptococci, into the circulation. In healthy individuals this poses little problem, but the consequences can be severe if the defense mechanisms are defective, for example in cancer patients with neutropenia where S. mitis has been reported to be the species most frequently isolated, with a disturbingly high incidence of antibiotic-resistant strains (2, 45).

Members of the anginosus group of streptococci are frequently isolated from abscesses in the brain, lung, and liver (71), but because the anginosus streptococci are naturally found at several body sites, there is no a priori case to consider the oral cavity the source of infection for abscesses elsewhere in the body. There is the strong possibility that streptococci from a periapical abscess would gain entry to the bloodstream, but there have been no definitive reports, rather than circumstantial evidence, linking an oral infection to a distant abscess.

VIRULENCE FACTORS OF ORAL STREPTOCOCCI

The specificity of preventive or therapeutic strategies such as vaccines relies upon the identification of unique targets, which, in the case of pathogens, are commonly ones associated with virulence. Do the oral streptococci have specific virulence factors as opposed to normal properties that contribute to virulence? The only textbook virulence factor is the intermedilysin toxin produced by S. intermedius, but its importance in pathogenesis is unknown. Dental caries is a consequence of acid production by the universal process of glycolysis; endocarditis is a consequence of streptococci forming colonies on heart valves; tissue damage in abscesses is due simply to streptococci being present in increased numbers and carrying out their normal metabolic activities, degrading macromolecules to derive a source of carbon and nitrogen. These normal metabolic processes thus all can contribute to the virulence of the streptococci. Little is yet known about how the oral streptococci interact with the cells of an infected host, and we are only beginning to gather information on how they can induce or suppress cellular functions. It is now clear, however, that strong cellular responses can be elicited by surface components of oral streptococci (18, 61, 62, 67), and further study of these responses should yield new insights into pathogenic mechanisms in abscess formation and valve damage in endocarditis.

A driving force for the identification of streptococcal virulence factors has been the search for suitable targets for interventions that would moderate plaque formation or the progression of dental caries (55). This is considered in more detail by Kuramitsu (chapter 28) and Jenkinson and Vickerman (chapter 29) in this volume, but most interest has focused on surface proteins of S. mutans that may be involved in adhesion, the glucosyltransferases that form sticky glucans from sucrose, and the glucan-binding proteins. These have formed the basis for experimental vaccines (36, 56), and it has been proposed that bacterial control could be achieved by provision of an excess of soluble adhesin or receptor molecules, perhaps as a mouthwash, which could disrupt or prevent the establishment of strepto-

cocci at their normal attachment site (32, 55). Our knowledge of the properties of individual species is beginning to benefit from genome sequencing projects that are identifying novel genes that may contribute to virulence, as well as insights into the population structure and genomic plasticity. However, the challenges for the future are not just to explore the molecular pathogenesis of the various streptococcal species but also to understand the complex interactions in the plaque ecosystem, the balance between harmless and harmful species, the means by which both these types of bacteria maintain a balance with the protective factors of the host, and the factors that can tip the balance away from health towards the disease state.

REFERENCES

1. **Alaluusua, S., S. J. Alaluusua, J. Karjalainen, M. Saarela, T. Holttinen, M. Kallio, P. Holtta, H. Torkko, P. Relander, and S. Asikainen.** 1994. The demonstration by ribotyping of the stability of oral *Streptococcus mutans* infection over 5 to 7 years in children. *Arch. Oral. Biol.* **39:** 467–471.

2. **Alcaide, F., J. Carratala, J. Linares, F. Gudiol, and R. Martin.** 1996. In vitro activities of eight macrolide antibiotics and RP-59500 (quinupristin-dalfopristin) against viridans group streptococci isolated from blood of neutropenic cancer patients. *Antimicrob. Agents Chemother.* **40:**2117–2120.

3. **Bartzokas, C. A., R. Johnson, M. Jane, M. V. Martin, P. K. Pearce, and Y. Saw.** 1994. Relation between mouth and hematogenous infection in total joint replacements. *Brit. Med. J.* **309:**506–508.

4. **Black, C., I. Allan, S. K. Ford, M. Wilson, and R. McNab.** 2004. Biofilm-specific surface properties and protein expression in oral *Streptococcus sanguis*. *Arch. Oral Biol.* **49:**295–304.

5. **Bowden, G. H. W., and I. R. Hamilton.** 1998. Survival of oral bacteria. *Crit. Rev. Oral Biol. Med.* **9:**54–85.

6. **Bradshaw, D. J., K. A. Homer, P. D. Marsh, and D. Beighton.** 1994. Metabolic cooperation in oral microbial communities during growth on mucin. *Microbiology* **140:**3407–3412.

7. **Bradshaw, D. J., P. D. Marsh, G. K. Watson, and C. Allison.** 1997. Oral anaerobes cannot survive oxygen stress without interacting with facultative/aerobic species as a microbial community. *Lett. Appl. Microbiol.* **25:**385–387.

8. **Brailsford, S. R., B. Shah, D. Simons, S. Gilbert, D. Clark, I. Ines, S. E. Adams, C. Allison, and D. Beighton.** 2001. The predominant aciduric microflora of root-caries lesions. *J. Dent. Res.* **80:**1828–1833.

9. **Burne, R. A., and R. E. Marquis.** 2000. Alkali production by oral bacteria and protection against dental caries. *FEMS Microbiol. Lett.* **193:**1–6.

10. **Byers, H. L., E. Tarelli, K. A. Homer, and D. Beighton.** 1999. Sequential deglycosylation and utilization of the N-linked, complex-type glycans of human alpha1-acid glycoprotein mediates growth of *Streptococcus oralis*. *Glycobiol.* **9:**469–479.

11. **Carlsson, P.** 1989. Distribution of mutans streptococci in populations with different levels of sugar consumption. *Scand. J. Dent. Res.* **97:**120–125.

12. **Carlsson, P., I. A. Gandour, B. Olsson, B. Rickardsson, and K. Abbas.** 1987. High prevalence of mutans streptococci in a population with extremely low prevalence of dental caries. *Oral Microbiol. Immunol.* **2:**121–124.

13. **Caufield, P. W., G. R. Cutter, and A. P. Dasanayake.** 1993. Initial acquisition of mutans streptococci by infants:

evidence for a discrete window of infectivity. *J. Dent. Res.* **72:**37–45.

14. **Colby, S. M., and R. R. B. Russell.** 1997. Sugar metabolism by mutans streptococci. *J. Appl. Microbiol.* **83:**S80–S88.

15. **Colombo, A. P., A. D. Haffajee, F. E. Dewhirst, B. J. Paster, C. M. Smith, M. A. Cugini, and S. S. Socransky.** 1998. Clinical and microbiological features of refractory periodontitis subjects. *J. Clin. Periodontol.* **25:**169–180.

16. **De Gheldre, Y., P. Vandamme, H. Goossens, and M. J. Struelens.** 1999. Identification of clinically relevant viridans streptococci by analysis of transfer DNA intergenic spacer length polymorphism. *Int. J. Syst. Bacteriol.* **49:**1591–1598.

17. **Douglas, C. W. I., J. Heath, K. K. Hampton, and F. E. Preston.** 1993. Identity of viridans streptococci isolated from cases of infective endocarditis. *J. Med. Microbiol.* **39:**179–182.

18. **Elting, L. S., G. P. Bodey, and B. H. Keefe.** 1992. Septicemia and shock syndrome due to viridans streptococci—a case-control study of predisposing factors. *Clin. Infect. Dis.* **14:**1201–1207.

19. **Facklam, R.** 2002. What happened to the streptococci: overview of taxonomic and nomenclature changes. *Clin. Microbiol. Rev.* **15:**613–630.

20. **Fiehn, N. E., E. Gutschik, T. Larsen, and J. M. Bangsborg.** 1995. Identity of streptococcal blood isolates and oral isolates from 2 patients with infective endocarditis. *J. Clin. Microbiol.* **33:**1399–1401.

21. **Fisher, L. E., and R. R. B. Russell.** 1993. The isolation and characterization of milleri group streptococci from dental periapical abscesses. *J. Dent. Res.* **72:**1191–1193.

22. **Fitzsimmons, S., M. Evans, C. Pearce, M. J. Sheridan, R. Wientzen, G. Bowden, and M. F. Cole.** 1996. Clonal diversity of *Streptococcus mitis* biovar 1 isolates from the oral cavity of human neonates. *Clin. Diagn. Lab. Immunol.* **3:**517–522.

23. **Garnier, F., G. Gerbaud, P. Courvalin, and M. Galimand.** 1997. Identification of clinically relevant viridans group streptococci to the species level by PCR. *J. Clin. Microbiol.* **35:**2337–2341.

24. **Guggenheim, M., S. Shapiro, R. Gmur, and B. Guggenheim.** 2001. Spatial arrangements and associative behavior of species in an in vitro oral biofilm model. *Appl. Environ. Microbiol.* **67:**1343–1350.

25. **Havarstein, L. S., R. Hakenbeck, and P. Gaustad.** 1997. Natural competence in the genus *Streptococcus*: evidence that streptococci can change pherotype by interspecies recombinational exchanges. *J. Bacteriol.* **179:**6589–6594.

26. **Helderman, W. H. V., M. I. N. Matee, J. S. van der Hoeven, and F. H. M. Mikx.** 1996. Cariogenicity depends more on diet than the prevailing mutans streptococcal species. *J. Dent. Res.* **75:**535–545.

27. **Hohwy, J., and M. Kilian.** 1995. Clonal diversity of the *Streptococcus mitis* biovar 1 population in the human oral cavity and pharynx. *Oral Microbiol. Immunol.* **10:**19–25.

28. **Hohwy, J., J. Reinholdt, and M. Kilian.** 2001. Population dynamics of *Streptococcus mitis* in its natural habitat. *Infect. Immun.* **69:**6055–6063.

29. **Homer, K. A., L. Denbow, R. A. Whiley, and D. Beighton.** 1993. Chondroitin sulfate depolymerase and hyaluronidase activities of viridans streptococci determined by a sensitive spectrophotometric assay. *J. Clin. Microbiol.* **31:**1648–1651.

30. **Jenkinson, H. F., and R. J. Lamont.** 1997. Streptococcal adhesion and colonization. *Crit. Rev. Oral Biol. Med.* **8:**175–200.

31. **Kawamura, Y., X. G. Hou, F. Sultana, H. Miura, and T. Ezaki.** 1995. Determination of 16S r-RNA sequences of *Streptococcus mitis* and *Streptococcus gordonii* and phylogenetic relationships among members of the genus *Streptococcus. Int. J. Syst. Bacteriol.* **45:**406–408.

32. **Kelly, C. G., J. S. Younson, B. Y. Hikmat, S. M. Todryk, M. Czisch, P. I. Haris, I. R. Flindall, C. Newby, A. I. Mallet, J. K. C. Ma, and T. Lehner.** 1999. A synthetic peptide adhesion epitope as a novel antimicrobial agent. *Nature Biotechnol.* **17:**42–47.

33. **Kennedy, H. F., D. Morrison, D. Tomlinson, B. E. S. Gibson, J. Bagg, and C. G. Gemmell.** 2003. Gingivitis and toothbrushes: potential roles in viridans streptococcal bacteraemia. *J. Infect.* **46:**67–70.

34. **Kilian, M., J. Reinholdt, H. Lomholt, K. Poulsen, and E. V. G. Frandsen.** 1996. Biological significance of IgA1 proteases in bacterial colonization and pathogenesis: critical evaluation of experimental evidence. *APMIS* **104:**321–338.

35. **Kleinberg, I.** 2002. A mixed-bacteriae ecological approach to understanding the role of the oral bacteria in dental caries causation: an alternative to *Streptococcus mutans* and the specific-plaque hypothesis. *Crit. Rev. Oral Biol. Med.* **13:**108–125.

36. **Koga, T., T. Oho, Y. Shimazaki, and Y. Nakano.** 2002. Immunization against dental caries. *Vaccine* **20:**2027–2044.

37. **Lewis, M. A. O., T. W. Macfarlane, and D. A. McGowan.** 1990. A microbiological and clinical review of the acute dentoalveolar abscess. *Brit. J. Oral Maxillofac. Surg.* **28:**359–366.

38. **Li, Y., and P. W. Caufield.** 1995. The fidelity of initial acquisition of mutans streptococci by infants from their mothers. *J. Dent. Res.* **74:**681–685.

39. **Lingstrom, P., F. O. van Ruyven, J. van Houte, and R. Kent.** 2000. The pH of dental plaque in its relation to early enamel caries and dental plaque flora in humans. *J. Dent. Res.* **79:**770–777.

40. **Loesche, W. J.** 1986. Role of *Streptococcus mutans* in human dental decay. *Microbiol. Rev.* **50:**353–380.

41. **Ma, J. K. C., B. Y. Hikmat, K. Wycoff, N. D. Vine, D. Chargelegue, L. Yu, M. B. Hein, and T. Lehner.** 1998. Characterization of a recombinant plant monoclonal secretory antibody and preventive immunotherapy in humans. *Nature Med.* **4:**601–606.

42. **Macey, M. G., R. A. Whiley, L. Miller, and H. Nagamune.** 2001. Effect on polymorphonuclear cell function of a human-specific cytotoxin, intermedilysin, expressed by *Streptococcus intermedius. Infect. Immun.* **69:**6102–6109.

43. **Marchant, S., S. R. Brailsford, A. C. Twomey, G. J. Roberts, and D. Beighton.** 2001. The predominant microflora of nursing caries lesions. *Caries Res.* **35:**397–406.

44. **Marcotte, H., and M. C. Lavoie.** 1998. Oral microbial ecology and the role of salivary immunoglobulin A. *Microbiol. Mol. Biol. Rev.* **62:**71–109.

45. **Marron, A., J. Carratala, F. Alcaide, A. Fernandez-Sevilla, and F. Gudiol.** 2001. High rates of resistance to cephalosporins among viridans group streptococci causing bacteraemia in neutropenic cancer patients. *J. Antimicrob. Chemother.* **47:**87–91.

46. **Marsh, P. D.** 2003. Are dental diseases examples of ecological catastrophes? *Microbiology* **149:**279–294.

47. **Marsh, P. D.** 2004. Dental plaque as a microbial biofilm. *Caries Res.* **38:**204–211.

48. **McNeill, K., and I. R. Hamilton.** 2003. Acid tolerance response of biofilm cells of *Streptococcus mutans. FEMS Microbiol. Lett.* **221:**25–30.

49. Nagamune, H., C. Ohnishi, A. Katsuura, K. Fushitani, R. A. Whiley, A. Tsuji, and Y. Matsuda. 1996. Intermedilysin, a novel cytotoxin specific for human cells, secreted by *Streptococcus intermedius* UNS46 isolated from a human liver abscess. *Infect. Immun.* **64:**3093–3100.

50. Palmer, R. J., S. M. Gordon, J. O. Cisar, and P. E. Kolenbrander. 2003. Coaggregation-mediated interactions of streptococci and actinomyces detected in initial human dental plaque. *J. Bacteriol.* **185:**3400–3409.

51. Palmer, R. J., K. Kazmerzak, M. C. Hansen, and P. E. Kolenbrander. 2001. Mutualism versus independence: Strategies of mixed-species oral biofilms in vitro using saliva as the sole nutrient source. *Infect. Immun.* **69:**5794–5804.

52. Poyart, C., G. Quesne, S. Coulon, P. Berche, and P. Trieu-Cuot. 1998. Identification of streptococci to species level by sequencing the gene encoding the manganese-dependent superoxide dismutase. *J. Clin. Microbiol.* **36:**41–47.

53. Reichmann, P., A. Konig, J. Linares, F. Alcaide, F. C. Tenover, L. McDougal, S. Swidsinski, and R. Hakenbeck. 1997. A global gene pool for high-level cephalosporin resistance in commensal *Streptococcus* species and *Streptococcus pneumoniae. J. Infect. Dis.* **176:** 1001–1012.

54. Richard, P., G. A. Delvalle, P. Moreau, N. Milpied, M. P. Felice, T. Daeschler, J. L. Harousseau, and H. Richet. 1995. Viridans streptococcal bacteremia in patients with neutropenia. *Lancet* **345:**1607–1609.

55. Russell, R. R. B. 1994. Control of specific plaque bacteria. *Adv. Dent. Res.* **8:**285–290.

56. Russell, R. R. B., and N. W. Johnson. 1987. The prospects for vaccination against dental-caries. *Brit. Dent. J.* **162:**29–34.

57. Sansone, C., J. Van Houte, K. Joshipura, R. Kent, and H. C. Margolis. 1993. The association of mutans streptococci and non-mutans streptococci capable of acidogenesis at a low pH with dental caries on enamel and root surfaces. *J. Dent. Res.* **72:**508–516.

58. Seymour, R. A., R. Lowry, J. M. Whitworth, and M. V. Martin. 2000. Infective endocarditis, dentistry and antibiotic prophylaxis; time for a rethink? *Brit. Dent. J.* **189:** 610–616.

59. Shinzato, T., and A. Saito. 1994. A mechanism of pathogenicity of streptococcus-milleri group in pulmonary infection—synergy with an anaerobe. *J. Med. Microbiol.* **40:**118–123.

60. Skopek, R. J., W. F. Liljemark, C. G. Bloomquist, and J. D. Rudney. 1993. Dental plaque development on defined streptococcal surfaces. *Oral Microbiol. Immunol.* **8:**16–23.

61. Soares, R., P. Ferreira, M. M. G. Santarem, M. Teixeira da Silva, and M. Arala-Chaves. 1990. Low T-cell and B-cell reactivity is an apparently paradoxical request for murine immunoprotection against *Streptococcus mutans*—murine protection can be achieved by immunization against a B-cell mitogen produced by these bacteria. *Scand. J. Immunol.* **31:**361–366.

62. Stinson, M. W., R. McLaughlin, S. H. Choi, Z. E. Juarez, and J. Barnard. 1998. Streptococcal histone-like protein: Primary structure of *hlpA* and protein binding to lipoteichoic acid and epithelial cells. *Infect. Immun.* **66:** 259–265.

63. Svensater, G., M. Borgstrom, G. H. W. Bowden, and S. Edwardsson. 2003. The acid-tolerant microbiota associated with plaque from initial caries and healthy tooth surfaces. *Caries Res.* **37:**395–403.

64. Tapp, J., M. Thollesson, and B. Herrmann. 2003. Phylogenetic relationships and genotyping of the genus *Streptococcus* by sequence determination of the RNase P RNA gene, *rnpB. Int. J. Syst. Evol. Microbiol.* **53:**1861–1871.

65. Truper, H. G., and L. DeClari. 1997. Taxonomic note: necessary correction of specific epithets formed as substantives (nouns) "in apposition." *Int. J. Syst. Bacteriol.* **47:** 908–909.

66. Tyler, B. M., and M. F. Cole. 1998. Effect of IgA1 protease on the ability of secretory IgA1 antibodies to inhibit the adherence of *Streptococcus mutans. Microbiol. Immunol.* **42:**503–508.

67. Vernier, A., M. Diab, M. Soell, G. Haan-Archipoff, A. Beretz, D. Wachsmann, and J. P. Klein. 1996. Cytokine production by human epithelial and endothelial cells following exposure to oral viridans streptococci involves lectin interactions between bacteria and cell surface receptors. *Infect. Immun.* **64:**3016–3022.

68. Welin, J., J. C. Wilkins, D. Beighton, and G. Svensater. 2004. Protein expression by *Streptococcus mutans* during initial stage of biofilm formation. *Appl. Environ. Microbiol.* **70:**3736–3741.

69. Whatmore, A. M., A. Efstratiou, A. P. Pickerill, K. Broughton, G. Woodard, D. Sturgeon, R. George, and C. G. Dowson. 2000. Genetic relationships between clinical isolates of *Streptococcus pneumoniae*, *Streptococcus oralis*, and *Streptococcus mitis*: characterization of "atypical" pneumococci and organisms allied to *S. mitis* harboring *S. pneumoniae* virulence factor-encoding genes. *Infect. Immun.* **68:**1374–1382.

70. Whiley, R. A., and D. Beighton. 1998. Current classification of the oral streptococci. *Oral Microbiol. Immunol.* **13:** 195–216.

71. Whiley, R. A., D. Beighton, T. G. Winstanley, H. Y. Fraser, and J. M. Hardie. 1992. *Streptococcus intermedius*, *Streptococcus constellatus*, and *Streptococcus anginosus* (the *Streptococcus milleri* group)—association with different body sites and clinical infections. *J. Clin. Microbiol.* **30:** 243–244.

72. Whiley, R. A., L. M. C. Hall, J. M. Hardie, and D. Beighton. 1997. Genotypic and phenotypic diversity within *Streptococcus anginosus. Int. J. Syst. Bacteriol.* **47:**645–650.

73. Whittaker, C. J., C. M. Klier, and P. E. Kolenbrander. 1996. Mechanisms of adhesion by oral bacteria. *Ann. Rev. Microbiol.* **50:**513–552.

The Virulence Properties of *Streptococcus mutans*

HOWARD K. KURAMITSU

28

It is now well established that *Streptococcus mutans* is an important etiological agent in the development of dental caries, one of the most prevalent human infectious diseases. Following the initial isolation of this organism in 1924, an extensive literature on the virulence properties of *S. mutans* has developed (4, 22, 32, 41). However, it is now recognized that a number of distinct species, commonly referred to collectively as the mutans streptococci, comprise the organisms originally classified as *S. mutans*. As with other pathogenic bacteria, developments in the area of molecular genetics have played a major role in defining the pathogenicity of these oral bacteria. In this regard, the recent completion of the sequencing of the genome of *S. mutans* UA159 (1) has already accelerated efforts to further define the genetic basis for the cariogenicity of these organisms. This review will focus primarily on developments in this area of research over the past five years since the first edition of this volume was published. The reader is urged to consult other reviews for major developments prior to 2000 (10, 32, 41, 54).

IDENTIFICATION OF THE VIRULENCE FACTORS OF *S. MUTANS* PRIOR TO THE SEQUENCING OF ITS GENOME

Using primarily biochemical approaches, it was established almost thirty years ago that three principal properties distinguished *S. mutans* strains from the other oral streptococci isolated from the human oral cavity: (i) their ability to synthesize insoluble adhesive glucans from sucrose; (ii) their relative acid tolerance (aciduricity); and (iii) their rapid production of lactic acid from dietary sugars (41). Furthermore, the importance of these properties relative to cariogenicity was subsequently confirmed utilizing genetic approaches with defined mutants and rat model systems (33). The development of recombinant DNA techniques as well as gene inactivation strategies was crucial in this regard. These approaches identified a number of genes of the mutans streptococci that influenced the virulence of these organisms (Table 1), including the *gtf* genes coding for glucosyltransferases (Gtfs) (33), the *gbpA* and *gbpC* genes (24, 44) encoding glucan-binding proteins, *spaP* expressing a cell surface adhesin (13), and the *glgR* gene involved in intracellular polysaccharide storage (23). In addition, a number of other genes that have been shown to affect potential virulence properties in vitro were also characterized, including some involved in the stress responses of *S. mutans* (*ffh, dgk, gbpB,* and an apurinic-apyrimidinic endonuclease gene) (3, 21, 30, 43). However, mutants defective in these genes have not yet been examined in animal model systems.

RECENT DEVELOPMENTS IN THE GENETIC ANALYSIS OF *S. MUTANS*

The isolation of gene fragments of potential virulence genes of *S. mutans* by recombinant DNA techniques has allowed for the construction of numerous monospecific mutants for characterization in in vitro and in vivo systems (33). More recently, a rapid PCR-based ligation technique has been developed for the same purpose in *S. mutans* and related transformable streptococci (35). This procedure should allow for convenient inactivation of multiple genes in this organism in a relatively short time span. Therefore, such an approach would be ideal for inactivating multiple genes detected by examining the genome sequence of *S. mutans* for preliminary identification of virulence phenotypes.

Random mutagenesis strategies have also been beneficial in identifying genes that could play a role in the virulence of *S. mutans* (33). Several studies involving the utilization of the transposons Tn916 and Tn917 have identified genes involved in aciduricity (20, 61) as well as in bacteriocin production (11). In addition, the utilization of suicide plasmid libraries in these organisms has identified a number of genes involved in biofilm formation (62). Very recently, a modification of the mariner mutagenesis system has also been adapted for use in *S. mutans* to identify the genetic basis for a cold agglutination phenotype (51). Therefore, several approaches are now available to overcome some of the limitations of using transposons (preferential insertion of the transposons into "hot spots") for generating mutant libraries.

Gram-Positive Pathogens, 2nd edition, edited by Vincent A. Fischetti et al.
© 2006 ASM Press, Washington, D.C.

TABLE 1 Virulence factors of *S. mutans* identified in vivo

Gene	Proposed role
gbpA[a]	Glucan-mediated aggregation
gbpC	Glucan-mediated aggregation and stress responses
gtfB	Glucan-mediated attachment
gtfC	Glucan-mediated attachment
gtfD	Glucan-mediated attachment
spaP	Attachment
glgR	Intracellular polysaccharide accumulation

[a]See text for references.

The random mutagenesis libraries do not allow, however, for the isolation of mutants that are altered in essential genes of an organism. The recent development of an antisense RNA strategy for use in *S. mutans* now makes it possible to examine the function of some of these genes. For example, this strategy was utilized to examine the essential *sgp* gene in *S. mutans* (3) and demonstrated the role of this gene in sucrose-independent biofilm formation (63). In addition, this strategy could also be used to identify novel essential genes in *S. mutans*, as utilized recently in *Staphylococcus aureus* (26).

As with other bacteria, the utilization of reporter genes (principally *lacZ* and *cat* fusions) to monitor gene expression in *S. mutans* has yielded important information on the effects of environmental conditions on the expression of potential virulence traits (10, 33). In addition, a procedure to randomly insert the *lacZ* gene into the chromosome of *S. mutans* using transposon Tn917 has been developed (17). Furthermore, the green fluorescent protein has been expressed in *S. mutans* (64) and not only allows for quantitation of gene expression but also for localization of these strains in biofilms. Although fluorescence in these constructs is not as strong as with some other bacteria, it was still possible to both quantitate gene expression and visualize the cells under confocal microscopy.

SIGNALING PATHWAYS IN *S. MUTANS*

One of the major conceptual advances in prokaryotic biology in the past decade has been the recognition that individual bacteria undergo metabolic changes in response to the presence of other homologous or heterogeneous organisms. This phenomenon of "quorum sensing" may be employed to monitor the presence of either "self" or "nonself" organisms (6). As with other streptococci (2), it appears that *S. mutans* can monitor its own relative density via a peptide signaling system mediated by the competence-stimulating peptide (CSP) (14). Extracellular CSP binds to a histidine kinase membrane receptor, ComD, which in turns phosphorylates a response regulator, ComE, to subsequently modulate the expression of a number of genes. Some of these genes are involved with DNA uptake and transformation, aciduricity, and biofilm formation (16, 40). More recently, it has been demonstrated that this regulatory network also modulates bacteriocin production in strains GS5 and BM71 of *S. mutans* (62a). It has also been proposed that CSP regulates at least one other regulatory pathway independent of the ComDE proteins (40). Therefore, it is likely that CSP may modulate other properties of *S. mutans*, with some of these potentially involved in virulence.

It has been suggested that another signaling system involving a furanone-like inducer, AI-2, which is the product of the *luxS* gene, is utilized by a number of different bacteria, including *S. mutans* (45, 61), to monitor cellular functions (6). In addition, it has been proposed that this system may be important in sensing the metabolic state of bacteria (7). In this regard, mutations in the *luxS* gene have been demonstrated to have marked effects on sucrose-dependent biofilm formation by *S. mutans* GS5 (15, 61) but not on sucrose-independent biofilm development (63). These differential effects are dependent, in part, upon the regulation of the genes involved in insoluble glucan synthesis, *gtfB* and *gtfC*, by AI-2 (Yoshida and Kuramitsu, unpublished results). Furthermore, *luxS* mutations affected several other genes involved in the pathogenicity of these organisms, including fructanase, the *brp* gene implicated in biofilm formation, and one or more genes involved in acid sensitivity (60). However, the target receptor for AI-2 in *S. mutans* has not yet been identified. In addition, it is not known if the synthesis of AI-2 by other oral plaque bacteria (19) can affect the virulence of *S. mutans* in vivo.

BIOFILM FORMATION BY *S. MUTANS*

Another important conceptual advance of the past decade is the recognition that bacteria growing in attached structures called biofilms are physiologically distinct from the same organisms growing in culture. Although investigators working with *S. mutans* and other dental plaque bacteria have been examining biofilms for many decades, it was not generally recognized prior to the formulation of the "biofilm concept" (12) that sessile bacteria were phenotypically different from the same organisms growing freely in culture (planktonic state). This later recognition has led to increasing emphasis on the need to examine pathogenic bacteria in the biofilm state.

Because the virulence of *S. mutans* is primarily dependent upon its ability to initially colonize tooth surfaces as a component of a biofilm (dental plaque), it is of interest to determine the molecular basis for this property. As mentioned above, recombinant DNA techniques have been utilized to confirm the important roles of the *gtf* and *spaP* genes of *S. mutans* in dental plaque formation and associated dental caries. When implanted into rats fed sucrose diets, it was clearly shown that these genes strongly influenced cariogenesis, especially on the smooth surfaces of teeth. These effects were likely modulated by decreased biofilm formation by these mutants, but this was not directly established.

More recently, utilizing a variety of different approaches, a number of specific genes have been identified that appear to modulate biofilm formation in vitro. In the case of two of these genes, *gbpA* and *gbpC* (24, 44), it was further established that these genes also affect biofilm formation in rat model systems. Because these genes affect a variety of properties in *S. mutans*, these results suggest that, as with other bacteria, biofilm formation is a complex process that can be modulated by a variety of physiological changes in each organism. Whether a unique biofilm-specific pathway exists still remains to be determined.

Another interesting feature of several of the genes that influence biofilm formation is that several of these, including *sgp* and *gbpC*, are implicated in the stress response of *S. mutans* (3, 43, 50). This is compatible with the hypothesis that biofilm formation may represent another stress response of microorganisms, triggering the development of this unique phenotype.

It was of interest that one genetic locus that affected biofilm formation in S. mutans is the *com* quorum-sensing system (40, 63). The effects on sucrose-independent biofilm formation appear to vary depending upon the strains examined as well as the biofilm assays utilized. For example, *comC* mutants of strain NG8 form the same mass of biofilms as the wild-type strain on polystyrene surfaces (40), while the same mutant in strain GS5 forms reduced levels of biofilms (63). However, the three-dimensional structure of the former mutants relative to the parental strain NG8 appears to be distinct. Interestingly, the *com* genes also appear to be required for biofilm formation in *Streptococcus gordonii* (42). Nevertheless, the role of the competence regulon in modulating biofilm formation has not yet been defined. These results suggest that it may be possible to develop specific inhibitors of the quorum-sensing system in S. mutans to attenuate biofilm formation. Such inhibitors may have an advantage over presently utilized antimicrobial agents in that there should be a decreased probability of developing resistance against biofilm antagonists.

INTERACTIONS BETWEEN S. MUTANS AND OTHER PLAQUE BACTERIA

The association of S. mutans in dense biofilms on the teeth suggests that these organisms may affect, and can be influenced by, other plaque constituents. Earlier studies have revealed that there is an inverse relationship between the quantity of S. mutans in dental plaque and the presence of S. sanguis strains (some of these strains have now been reclassified as S. gordonii or *Streptococcus sanguinis*) (29). S. mutans may antagonize the growth of the latter organisms via either acid production or the elaboration of bacteriocins (58). However, antagonistic effects of *Streptococcus sanguis* strains on S. mutans have not been elucidated previously. In this regard, very recently it has been demonstrated that S. gordonii Challis can attenuate a number of the CSP-mediated properties of S. mutans GS5 and BM71, including transformation, bacteriocin production, and acid tolerance (57a). These effects were mediated, in part, by the production of a serine-protease of strain Challis that inactivated the CSP of S. mutans. Other oral streptococci also attenuated the quorum sensing-dependent properties of S. mutans. These results might help explain some of the antagonistic effects of the S. sanguis family on S. mutans.

It is likely that genetic exchange between bacteria may play a pivotal role in the evolution of these organisms as well as provide a means for disseminating antibiotic resistance genes. In the latter regard, it is not unreasonable to speculate that some resistance genes in plaque bacteria could be transferred to organisms that normally reside outside the oral cavity. However, such gene transfer has not yet been documented. Previously, intrageneric as well as intergeneric exchange of DNA between S. mutans and other oral streptococci has been demonstrated in broth cultures (34). Recently, such exchange has been demonstrated in biofilms (Wang and Kuramitsu, unpublished results). However, transformation of S. mutans strains was shown to be attenuated in the presence of other oral streptococci such as S. gordonii, as indicated above. Thus, the transfer of chromosomal or plasmid markers from S. mutans to S. gordonii Challis could be readily demonstrated in biofilms, but the reverse transfer was severely restricted relative to transformation of monospecies S. mutans biofilms. This suggests that in vivo genetic exchange into naturally transformable strains of S. mutans in dental plaque may occur but could

be limited in those biofilms containing relatively high proportions of nonmutans streptococci.

SEQUENCING OF THE S. MUTANS GENOME

Recently, the sequencing of the genome of S. mutans UA159 was reported (1). Even prior to the publication of the annotated sequence of the UA159 genome, the availability of the unannotated sequence on the Internet proved to be very valuable in isolating a number of genes from S. mutans (8). Of 1,963 open reading frames identified, almost 300 appear to be unique to S. mutans. Sixty open reading frames correspond to ABC transporters, which suggests a variety of efflux and transport systems that have not yet been characterized in these organisms. An examination of the putative open reading frames of S. mutans has already revealed several interesting possibilities. For example, although three genes for glucan-binding proteins have been previously isolated from these organisms, genome sequencing has uncovered a potential fourth gene, *gbpD*, coding for this activity (52). In addition, a number of insertion sequence elements are present on the UA159 genome. Some of these elements may be responsible for gene rearrangements in this species. In addition, very recently an operon for a lantibiotic-like bacteriocin from strain GS5 has been identified and appears to be part of a transposon (62a). These genes are absent in strain UA159 but are present on the strain BM71 chromosome and may have been transferred into different strains following transposition.

Furthermore, the availability of the complete genome sequence of S. mutans has now made it possible to conduct global regulatory studies on this organism using both DNA microarray and proteomic approaches. One laboratory (R. Quivey, personal communication) has constructed such microarrays and is currently investigating acid tolerance responses in these organisms. In addition, The Institute of Genomic Research in conjunction with the National Institute of Dental and Craniofacial Research has now made available S. mutans microarrays to the international research community. Despite some of the limitations in using such arrays (56), it is likely that their utilization will shortly yield valuable information regarding the global regulation of the S. mutans genome as a function of pH, biofilm formation, quorum-sensing regulation, etc. This information may eventually be utilized to identify S. mutans-specific genes coding for virulence functions that could be targeted to modulate human dental caries.

Likewise, proteomic approaches have recently been undertaken in an effort to examine global regulation of gene expression in S. mutans. One earlier investigation (53) examined protein expression as a function of mature (3-day) biofilm formation with two-dimensional gels and matrix-assisted laser desorption ionization–time of flight mass spectrometry. It was demonstrated that almost 20% of the 694 proteins examined were either up- or down-regulated during biofilm formation. Many of the up-regulated genes appear to code for biosynthetic (protein, fatty acid) capacities. In contrast, several of the down-regulated proteins are involved in catabolic processes. However, very recently it was demonstrated that carbohydrate catabolic pathways appear to be up-regulated during very early biofilm formation (two hours) (59). These contrasting results with early and mature biofilms indicate that it is important to investigate differential protein expression at different time periods during biofilm maturation. In addition, utilizing a similar approach, Lemos et al.

(37) demonstrated that the *hrcA* (a gene in a stress-related operon containing *dnaK*) mutation in strain UA159 induced the expression of at least a dozen proteins, including GroEL and GroES. Another important property of *S. mutans* that has been examined utilizing a proteomic approach is the acid tolerance of these organisms. A very recent report has demonstrated the complexity of this virulence-related property and indicated that several stress response proteins (38) as well as metabolic changes resulting in more alkaline end products (39) were induced at acidic pH. Thus, such global approaches are proving to be valuable in discerning overall cellular adaptations that would be difficult to uncover using a "gene by gene" approach.

CURRENT ISSUES IN DEFINING THE VIRULENCE OF *S. MUTANS*

As summarized above, there is a consensus now emerging that we currently have a basic understanding of the major pathogenic properties of *S. mutans* and some other members of the mutans streptococci. Nevertheless, a number of significant questions in this regard still remain unanswered. For example, although the importance of glucan synthesis by the mutans streptococci has been recognized for almost four decades now, the specific roles of each of the multiple Gtfs expressed in an individual strain are still unclear. Recent data suggest that the relative proportions of each of these enzymes play an important role in cariogenesis (46). However, it is not known how these proportions translate into distinct chemical properties of the glucans, i.e., what is the chemical nature of "adhesive" glucans? Furthermore, several laboratories (46, 55) have reported on the important role of one of the enzymes, GtfC, on sucrose-dependent colonization of smooth surfaces. Nevertheless, the function of this enzyme in this process still needs to be clarified.

There is also evidence that the glucan-binding proteins of *S. mutans* play significant roles in cariogenesis (5). The interaction of these proteins with the glucans synthesized by these organisms could modify the structures of the polysaccharides and mediate aggregation of the organisms in the presence of sucrose. It is also intriguing that *gbpA* mutants modulate the expression of insoluble glucan-synthetic activity in strain UA130 (24), underlining the relationship between the Gtfs and the glucan-binding proteins. Whether the other glucan-binding proteins of these organisms also have similar influences on the Gtfs still remains to be determined. In addition to glucan-binding proteins, it has been suggested that the dextranase activity of *S. mutans* may play a role in modifying the structure of the glucans and hence the cariogenicity of these organisms (18).

Because *S. mutans* can colonize teeth in the absence of sucrose (41), there has also been much interest in the mechanisms of sucrose-independent biofilm formation by these organisms. It is likely that such interactions are mediated by multiple mechanisms, including nonspecific trapping into retentive sites, interaction of *S. mutans* surface adhesins such as SpaP with the tooth pellicle, interaction of the organisms with plaque glucans, and ionic and hydrophobic interactions between the organisms and salivary components (32). The utilization of *spaP* mutants in a rat caries model has confirmed a role for this adhesin in cariogenesis (13). However, despite in vitro evidence suggesting that such mutants are attenuated in attachment to tooth surfaces (9), it was somewhat surprising that in vivo colonization of rat teeth did not appear to differ between the *spaP* mutant and the parental strain. Therefore, it is not clear to what extent the SpaP adhesin or other potential surface adhesins mediate *S. mutans* attachment to teeth in the human oral cavity.

One issue that has received some attention in pathogenic microbiology is that concerning the variation in virulence of different strains of a pathogen. An examination of the literature has suggested that a variety of potential virulence factors of *S. mutans* can vary in different strains. Some of these differences may be due to inherent genetic differences that are beginning to be revealed by comparison of different genetic loci in *S. mutans* strains with the UA159 genome. Other differences may be due to passage of a particular strain in the laboratory, as previously suggested (15). For example, both the *spaP* and *gbpC* genes coding for potential virulence factors of strain GS5 have been demonstrated to contain chain-terminating nonsense mutations (49). It is possible that such mutations occurred during passage in the laboratory and may be selected for during in vitro culturing. In addition, as exemplified by recent studies with *S. mutans* mutants examined for biofilm formation (36, 40, 63), strain differences were noted, and these may also be dependent upon the environmental conditions used in the assay systems (media, relative anaerobiosis, composition of the solid surfaces) as well as the mutagenesis strategy utilized. Therefore, some caution should be exercised in drawing conclusions regarding virulence factors in *S. mutans* based upon studies with a single strain or one assay system.

Another factor influencing the cariogenicity of *S. mutans* is the environmental influences of the host and other plaque organisms. Technical limitations have made this important question difficult to address. However, current (47) and potential future advances in the utilization of reporter gene constructs and confocal microscopy have now suggested possible approaches for examining this important question. Recent in vitro studies (see above) have suggested that interactions with other oral streptococci can attenuate some of the virulence properties of *S. mutans* and have suggested that this is a potentially promising area of research. However, the complexity of the microbiota in dental plaque makes this a very challenging issue for the future.

Another basic question that has not yet been resolved is that of the relative importance of *S. mutans* in human cariogenesis. Is this the major etiological agent of dental caries or have we overlooked other important cariogenic organisms? Is it really the presence of a single organism such as *S. mutans* that is the determining factor in caries formation, or is it the relative abundance of different caries-promoting and caries-inhibiting organisms (27)? Obviously, we do not have data from sufficient numbers of plaque samples to answer such complex questions. However, the development of metagenomic techniques (31, 57) may eventually prove valuable in answering such questions. For example, using 16S rRNA sequencing of plaque samples it should be possible to generate sufficient data to statistically catalogue the differences between cariogenic and healthy plaque in greater detail. Ultimately, the utilization of anticaries approaches specifically targeting *S. mutans* in the human oral cavity will allow for the quantitative determination of its contribution to dental caries.

PROSPECTS FOR REDUCING THE INCIDENCE OF DENTAL CARIES

Despite several decades of intensive research on the etiology of dental caries, it is clear that the fruits of such labor have not yet been adequately translated into the clinical setting. Dental caries has declined significantly in most industrialized societies over the past few decades, due primarily to fluoridation and improved dental hygiene (54). Nevertheless, this disease is still highly prevalent, especially in medically underserved communities. Therefore, novel approaches for further reducing the incidence of dental caries are of interest. Research to develop an effective anticaries vaccine has been in progress for almost four decades but has not yet resulted in any human trials of candidate vaccines. Despite a number of animal studies that have demonstrated the effectiveness of anti-*S. mutans* vaccines in reducing the incidence of caries (48), several ethical and economic issues have prevented their testing in humans. Whether these can be resolved still remains to be determined. As an alternative to active immunization protocols, it might be possible to use passive immunization to protect against human dental caries (28). However, this approach still needs to be further evaluated before testing in human subjects.

Another potential approach for controlling human dental caries is replacement therapy (25). The use of a genetically modified strain of *S. mutans* that is incapable of producing lactic acid has been demonstrated to protect rats against subsequent challenge with wild-type strains of the organism. It is possible that this strategy will be tested in humans in the near future. Another possible future approach for attenuating the role of *S. mutans* in cariogenesis virulence may be to exploit newly emerging information on biofilm formation by these organisms. For example, it may be possible to design antagonists targeted specifically against the *com* signal transduction system, which regulates several virulence properties of these organisms (40). Alternatively, oral commensal organisms could be genetically engineered to produce factors that can antagonize the cariogenicity of *S. mutans* (alkali-generating systems, glucanases, mutacins, etc.).

These strategies, like that of the utilization of vaccines, are predicated on the hypothesis that reducing the in vivo levels of the oral cariogenic *S. mutans* species will have a significant impact on the development of dental caries. However, as suggested by Kleinberg (27), this might not be an optimal approach for controlling dental caries given the proposed complexity of the etiology of this disease. Nevertheless, given the present emphasis on the role of *S. mutans* in cariogenesis, it appears that the interactions of these organisms with other plaque organisms will likely be a major focus of future research.

This review is dedicated to the memory of Toshihiko Koga, who made a number of significant contributions to our understanding of the physiology of S. mutans. *Work in my laboratory cited in this review was primarily supported by NIH grant DE03258.*

REFERENCES

1. **Ajdic, D., W. M. McShan, R. E. McLaughlin, G. Savic, J. Chang, M. B. Carson, C. Primeaux, R. Tian, S. Kenton, H. Jia, S. Lin, Y. Qian, S. Li, H. Zhu, F. Najar, H. Lai, J. White, B. A. Roe, and J. J. Ferretti.** 2002. Genome sequence of *Streptococcus mutans* UA159, a cariogenic dental pathogen. *Proc. Natl. Acad. Sci. USA* **99:**14434–14439.
2. **Alloing, G., B. Martin, C. Granadel, and J. P. Claverys.** 1998. Development of competence in *Streptococcus pneumoniae*: pheromone autoinduction and control of quorum sensing by the oligopeptide permease. *Mol. Microbiol.* **29:** 75–83.
3. **Baev, D., R. England, and H. K. Kuramitsu.** 1999. Stress induced membrane association of *Streptococcus mutans* GTP-binding protein, an essential G-protein, and investigation of its physiological role by utilizing an antisense RNA strategy. *Infect. Immun.* **67:**4510–4516.
4. **Banas, J. A.** 2004. Virulence properties of *Streptococcus mutans*. *Front. Biosci.* **9:**1267–1277.
5. **Banas, J. A., and M. M. Vickerman.** 2003. Glucan-binding proteins of oral streptococci. *Crit. Rev. Oral Biol. Med.* **14:** 89–99.
6. **Bassler, B.** 2002. Small talk. Cell-to-cell communication in bacteria. *Cell* **109:**421–424.
7. **Beeston, A. L., and M. G. Surette.** 2002. *pfs*-Dependent regulation of autoinducer-2 production in *Salmonella enteritica* serovar Typhimurium. *J. Bacteriol.* **184:**3450–3456.
8. **Bhagwat, S. P., J. Nary, and R. A. Burne.** 2001. Effects of mutating putative two-component systems on biofilm formation by *Streptococcus mutans* UA159. *FEMS Microbiol. Lett.* **205:**225–230.
9. **Bowen, W. H., K. M. Schilling, E. Giertsen, S. Person, S. F. Lee, A. S. Bleiweis, and D. Beeman.** 1991. Role of a cell surface-associated protein in adherence and dental caries. *Infect. Immun.* **59:**4606–4609.
10. **Burne, R. A.** 1998. Oral streptococci... products of their environment. *J. Dent. Res.* **77:**445–452.
11. **Caufield, P. W., G. R. Shah, and S. K. Hollingshead.** 1990. Use of transposon Tn916 to inactivate and isolate a mutacin-associated gene from *Streptococcus mutans*. *Infect. Immun.* **58:**4126–4135.
12. **Costerton, J. W., Z. Lewandowski, D. E. Caldwell, K. R. Kerber, and H. M. Lappin-Scott.** 1995. Microbial biofilms. *Annu. Rev. Microbiol.* **49:**711–745.
13. **Crowley, P. J., L. J. Brady, S. M. Michalek, and A. S. Bleiweis.** 1999. Virulence of a *spaP* mutant of *Streptococcus mutans* in a gnotobiotic rat model. *Infect. Immun.* **67:** 1201–1206.
14. **Cvitkovitch, D. G.** 2001. Genetic competence and transformation in oral streptococci. *Crit. Rev. Oral Biol. Med.* **12:**217–243.
15. **Cvitkovitch, D. G., and I. R. Hamilton.** 1994. Biochemical change exhibited by oral streptococci resulting from laboratory subculturing. *Oral Microbiol. Immunol.* **9:**209–217.
16. **Cvitkovitch, D. G., Y. H. Li, and R. P. Ellen.** 2003. Quorum sensing and biofilm formation in streptococcal infections. *J. Clin. Investig.* **112:**1626–1632.
17. **Cvitkovitch, D. G., J. A. Gutierrez, J. Behari, P. J. Youngman, J. E. Wetz, P. J. Crowley, J. D. Hillman, L. J. Brady, and A. S. Bleiweis.** 2000. Tn917-lac mutagenesis of *Streptococcus mutans* to identify environmentally regulated genes. *FEMS Microbiol. Lett.* **182:**149–154.
18. **Freedman, M. L., J. M. Tanzer, and A. L. Coykendall.** 1981. The use of genetic variants in the study of dental caries, p. 247–261. *In* J. M. Tanzer, (ed.), *Animal Models in Cariology*. Information Retrieval Inc., Washington, D.C.
19. **Frias, J., E. Olle, and M. Alsina.** 2001. Periodontal pathogens produce quorum sensing signals. *Infect. Immun.* **69:**3431–3434.
20. **Gutierrez, J. A., P. A. Crowley, D. P. Brown, J. D. Hillman, P. Youngman, and A. S. Bleiweis.** 1996. Insertional mutagenesis and recovery of interrupted genes of *Streptococcus mutans* by using transposon Tn917: preliminary characterization of mutants displaying acid sensitivity and nutritional requirements. *J. Bacteriol.* **178:**4166–4175.
21. **Hahn, K., R. C. Faustoferri, and R. G. Quivey, Jr.** 1999. Induction of an AP endonuclease activity in *Streptococcus mutans* during growth at low pH. *Mol. Microbiol.* **31:** 1489–1498.

22. **Hamada, S., and H. D. Slade.** 1980. Biology, immunology, and cariogenicity of *Streptococcus mutans*. *Microbiol. Rev.* **44:**331–384.

23. **Harris, G. S., S. M. Michalek, and R. Curtiss III.** 1992. Cloning of a locus involved in *Streptococcus mutans* intracellular polysaccharide accumulation and virulence testing of an intracellular polysaccharide-deficient mutant. *Infect. Immun.* **60:**3175–3185.

24. **Hazlett, K. R. O., S. M. Michalek, and J. A. Banas.** 1998. Inactivation of the *gbpA* gene of *Streptococcus mutans* increases virulence and promotes in vivo accumulation of recombinations between glucosyltransferases B and C genes. *Infect. Immun.* **66:**2180–2185.

25. **Hillman, J. D.** 2002. Genetically modified *Streptococcus mutans* for the prevention of dental caries. *Antonie Leeuwenhoek* **82:**361–366.

26. **Ji, Y., B. Zhang, S. F. Van Horn, P. Warren, G. Woodnutt, M. K. R. Burnham, and M. Rosenberg.** 2001. Identification of critical staphylococcal genes using conditional phenotypes generated by antisense RNA. *Science* **293:**2266–2269.

27. **Kleinberg, I.** 2002. A mixed-bacteria ecological approach to understanding the role of the oral bacteria in dental caries causation: an alternative to *Streptococcus mutans* and the specific plaque hypothesis. *Crit. Rev. Oral Biol. Med.* **13:**108–125.

28. **Koga, T., T. Oho, Y. Shimazaki, and Y. Nakano.** 2002. Immunization against dental caries. *Vaccine* **20:**2027–2044.

29. **Kolenbrander, P. E.** 1988. Intergeneric coaggregation among human oral bacteria and ecology of dental plaque. *Annu. Rev. Microbiol.* **42:**627–656.

30. **Kremer, B. H., M. van der Kraan, P. J. Crowley, I. R. Hamilton, L. J. Brady, and A. S. Bleiweis.** 2001. Characterization of the *sat* operon in *Streptococcus mutans*: evidence for a role of Ffh in acid tolerance. *J. Bacteriol.* **183:**2543–2552.

31. **Kroes, I., P. W. Lepp, and D. A. Relman.** 1999. Bacterial diversity within the human subgingival crevice. *Proc. Natl. Acad. Sci. USA* **96:**14547–14552.

32. **Kuramitsu, H. K.** 1993. Virulence factors of mutans streptococci: role of molecular genetics. *Crit. Rev. Oral Biol. Med.* **4:**159–176.

33. **Kuramitsu, H. K.** 2000. *Streptococcus mutans*: molecular genetic analysis, p. 280–286. *In* V. Fischetti, R. P. Novick, J. J. Ferretti, D. A. Portnoy, and J. I. Rood (ed.), *Gram-Positive Pathogens*. ASM Press, Washington, D.C.

34. **Kuramitsu, H. K., and V. Trappa.** 1984. Genetic exchange between oral streptococci during mixed growth. *J. Gen. Microbiol.* **130:**2497–2500.

35. **Lau, P. C., C. K. Sung, J. H. Lee, D. A. Morrison, and D. G. Cvitkovitch.** 2002. PCR ligation mutagenesis in transformable streptococci: application and efficiency. *J. Microbiol. Methods.* **49:**193–205.

36. **Lemos, J. A., T. A. Brown, Jr., and R. A. Burne.** 2004. Effects of RelA in key virulence properties of planktonic and biofilm populations of *Streptococcus mutans*. *Infect. Immun.* **72:**1431–1440.

37. **Lemos, J. A., Y. Y. Chen, and R. A. Burne.** 2001. Genetic and physiological analysis of the *groE* operon and role of the HrcA repressor in stress gene regulation and acid tolerance in *Streptococcus mutans*. *J. Bacteriol.* **183:**6074–6084.

38. **Len, A. C., D. W. Harty, and N. A. Jacques.** 2004a. Stress-responsive proteins are upregulated in *Streptococcus mutans* during acid tolerance. *Microbiology* **150:**1339–1351.

39. **Len, A. C., D. W. Harty, and N. A. Jacques.** 2004b. Proteomic analysis of *Streptococcus mutans* metabolic phenotype during acid tolerance. *Microbiology* **150:**1353–1366.

40. **Li, Y. H., P. C. Lau, N. Tang, G. Svensater, R. P. Ellen, and D. G. Cvitokovtich.** 2002. Novel two-component regulatory system involved in biofilm formation and acid resistance in *Streptococcus mutans*. *J. Bacteriol.* **184:**6333–6342.

41. **Loesche, W. J.** 1986. Role of *Streptococcus mutans* in human dental decay. *Microbiol. Rev.* **50:**353–380.

42. **Loo, C. Y., D. A. Corliss, and N. Ganeshkumar.** 2000. *Streptococcus gordonii* biofilm formation: identification of genes which code for biofilm phenotypes. *J. Bacteriol.* **182:**1374–1382.

43. **Mattos-Graner, R. O., S. Jin, W. F. King, T. Chen, D. J. Smith, and M. J. Duncan.** 2001. Cloning of the *Streptococcus mutans* gene encoding glucan binding protein B and analysis of genetic stability and protein production in clinical isolates. *Infect. Immun.* **69:**6931–6941.

44. **Matsumura, M., T. Izumi, M. Matsumoto, M. Tsuji, T. Fujiwara, and T. Ooshima.** 2003. The role of glucan-binding proteins in the cariogenicity of *Streptococcus mutans*. *Microbiol. Immunol.* **47:**213–215.

45. **Merritt, J. E., F. Qi, S. D. Goodman, M. H. Anderson, and W. Shi.** 2003. Mutation of *luxS* affects biofilm formation in *Streptococcus mutans*. *Infect. Immun.* **71:**1972–1979.

46. **Ooshima, T., M. Matsumura, T. Hoshino, S. Kawabata, S. Sobue, and T. Fujiwara.** 2001. Contributions of three glucosyltransferases to sucrose-dependent adherence of *Streptococcus mutans*. *J. Dent. Res.* **80:**1672–1677.

47. **Palmer, R., Jr., S. M. Gordon, J. O. Cisar, and P. E. Kolenbrander.** 2003. Coaggregation-mediated interactions of streptococcal actinomyces detected in human dental plaque. *J. Bacteriol.* **185:**3400–3409.

48. **Russell, M. W., N. K. Childers, S. M. Michalek, D. J. Smith, and M. A. Taubman.** 2004. A caries vaccine? The state of the science of immunization against dental caries. *Caries Res.* **38:**230–235.

49. **Sato, Y., K. Okamoto, and H. Kizaki.** 2002. *gbpC* and *pac* gene mutations detected in *Streptococcus mutans* GS-5. *Oral Microbiol. Immunol.* **17:**263–266.

50. **Sato, Y., Y. Yamamoto, and H. Kizaki.** 2000. Xylitol-induced elevated expression of the *gbpC* gene in a population of *Streptococcus mutans* cells. *Eur. J. Oral Sci.* **108:**538–545.

51. **Sato, Y., K. Okamoto, A. Kagami, Y. Yamamoto, K. Ohta, T. Igarashi, and H. Kizaki.** 2004. Application of *in vitro* mutagenesis to identify the gene responsible for cold agglutination phenotype of *Streptococcus mutans*. *Microbiol. Immunol.* **48:**444–456.

52. **Shah, D. S., and R. R. Russell.** 2004. A novel glucan-binding protein with lipase activity from the oral pathogen *Streptococcus mutans*. *Microbiology* **150:**1947–1956.

53. **Svensater, G., J. Welin, J. C. Wilkins, D. Beighton, and I. R. Hamilton.** 2001. Protein expression by planktonic and biofilm cells of *Streptococcus mutans*. *FEMS Microbiol. Lett.* **205:**139–146.

54. **Tanzer, J. M.** 1992. Microbiology of dental caries, p. 377–424. *In* J. Slots and M. A. Taubman (ed.), *Contemporary Oral Microbiology and Immunology*. Mosby Year Book, St. Louis, Mo.

55. **Tsumori, H., and H. K. Kuramitsu.** 1997. The role of *Streptococcus mutans* glucosyltransferases in the sucrose-dependent attachment to smooth surfaces: essential role of the GtfC enzyme. *Oral Microbiol. Immunol.* **12:**274–280.

56. **Vasil, M. L.** 2003. DNA microarrays in analysis of quorum sensing: strengths and limitations. *J. Bacteriol.* **185:**2061–2065.

57. **Venter, J. C., K. Remington, J. F. Heidelberg, A. L. Halpern, D. Rusch, J. A. Eisen, D. Wu, I. Paulsen, K. E. Nelson, W. Nelson, D. E. Fouts, S. Levy, A. H. Knap, M. W. Lomas, K. Nealson, O. White, J. Peterson, J. Hoffman, R. Parsons, H. Baden-Tillson, C. Pfannkoch, Y. H. Rogers, and H. O. Smith**. 2004. Environmental genome shotgun sequencing of the Sargasso Sea. *Science* **304:**66–74.

57a. **Wang, B. Y., and H. K. Kuramitsu.** 2005. Interactions between oral bacteria: inhibition of *Streptococcus mutans* bacteriocin production by *Streptococcus gordonii*. *Appl. Environ. Microbiol.* **71:**354–362.

58. **Weerkamp, A., L. Larik-Bongaerts, and D. G. Vogel.** 1977. Bacteriocins as factors in the in vitro interaction between oral streptococci in plaque. *Infect. Immun.* **16:**773–780.

59. **Weilin, J., J. C. Wilkins, D. Beighton, and G. Svensater.** 2004. Protein expression by *Streptococcus mutans* during initial stage of biofilm formation. *Appl. Environ. Microbiol.* **70:**3736–3741.

60. **Wen, Z. T., and R. A. Burne.** 2003. Functional genomics approach to identify genes required for biofilm development by *Streptococcus mutans*. *Appl. Environ. Microbiol.* **68:**1196–1203.

61. **Wen, Z. T., and R. A. Burne.** 2004. LuxS-mediated signaling in *Streptococcus mutans* is involved in regulation of acid and oxidative stress tolerance and biofilm formation. *J. Bacteriol.* **186:**2682–2691.

62. **Yamashita, Y., T. Takehara, and H. K. Kuramitsu.** 1993. Molecular characerization of a *Streptococcus mutans* mutant altered in environmental stress responses. *J. Bacteriol.* **175:**6220–6228.

62a. **Yonezawa, H., and H. K. Kuramitsu.** 2005. Genetic analysis of a unique bacteriocin, Smb, produced by *Streptococcus mutans* GS5. *Antimicrob. Agents Chemother.* **49:**541–548.

63. **Yoshida, A., and H. K. Kuramitsu.** 2002a. Multiple *Streptococcus mutans* genes are involved in biofilm formation. *Appl. Environ. Microbiol.* **68:**6283–6291.

64. **Yoshida, A., and H. K. Kuramitsu.** 2002b. *Streptococcus mutans* biofilm formation: utilization of a *gtfB* promoter-green fluorescent protein (P*gtfB*::*gfp*) construct to monitor development. *Microbiology* **148:**3388–3394.

Genetics of *sanguinis* Group Streptococci

HOWARD F. JENKINSON AND M. MARGARET VICKERMAN

29

Viridans streptococci make up the highest proportion of streptococci found within the human oral cavity, colonizing the hard (dental) and soft (epithelial) tissues. These organisms are non-beta-hemolytic, catalase-negative, leucine aminopeptidase-positive, and show no growth in 6.5% NaCl broth. However, the term viridans streptococci is not entirely satisfactory because many of the included species do not express the characteristic alpha-hemolysis (greening) on blood agar plates, attributable to hydrogen peroxide production (2). The viridans streptococci may be referred to as oral streptococci, but it should be noted that some of the species originate from the gastrointestinal and genitourinary tracts (17). Within the oral streptococci grouping it is possible to distinguish the *mutans* group organisms, e.g., *Streptococcus mutans*, *Streptococcus sobrinus*, etc., which ferment mannitol, from the *anginosus*, *mitis*, and *salivarius* group organisms, which do not. Bacteria from within these latter three groups, unlike the *mutans* group organisms, generally cause little or no dental decay in laboratory animals. Species classified as *mitis* group organisms demonstrate 95% 16S rDNA sequence homology (32) and include *mitis*, *pneumoniae*, *oralis*, *infantis*, *australis*, *parasanguinis* (previously *parasanguis*), *cristatus* (*crista*), *sanguinis* (*sanguis*), *peroris*, *gordonii*, and *sinensis* (see chapter 27). The most closely related organisms, *Streptococcus mitis*, *Streptococcus oralis*, and *Streptococcus pneumoniae*, show 99% 16S rDNA sequence homology. Three species, *Streptococcus gordonii*, *Streptococcus parasanguinis*, and *Streptococcus sanguinis*, can be separated from the other *mitis* group organisms by their positive reactions in the arginine and esculin tests. On this basis they may be constituted as the *sanguinis* group streptococci. There are many similarities between the species *gordonii* and *sanguinis*, but they can be differentiated on the basis that only *S. gordonii* binds α-amylase (33). *S. parasanguinis* can be distinguished in several ways from the other two species, one of which is by its inability to produce extracellular polysaccharide from sucrose (17).

The *mitis* group bacteria are commensals of the human oral cavity and nasopharynx. *S. mitis* and *S. oralis* are major pioneer species colonizing the oral mucosa in neonates. Production of immunoglobulin A1 (IgA1) protease (50), hexosaminidases, and neuraminidase by these organisms may provide a colonization advantage under exposure of secretory IgA antibodies in mother's milk and permit growth of bacteria on sugars released from host glycoconjugates. Following tooth eruption, the oral microflora becomes more highly complex, and there are increases in the isolation frequencies of all streptococcal species, especially *S. gordonii*, *S. parasanguinis*, and *S. sanguinis*. These organisms are found at most adult oral sites and have a high affinity for binding to salivary glycoprotein pellicles as well as for binding to other oral bacteria (reviewed in reference 37). Hence, streptococci are believed to form the foundation layers on teeth and other oral surfaces to which other organisms bind, and therefore are highly significant in the development of oral bacterial communities. Because, for the most part, we live in commensal harmony with these oral streptococci, there is an underlying notion that they are generally nonpathogenic. However, outside of the oral environment, *sanguinis* group streptococci are among the most common bacterial pathogens associated with an increasingly prevalent and serious endovascular condition, infective endocarditis. Historically, the characteristics of *sanguinis* group organisms that have been most studied are competence for DNA-mediated transformation and adhesion to saliva-coated surfaces. With the recent completion of the genome sequence of *S. gordonii* Challis (CH1), and the recognition of streptococci as critical components in the development of oral biofilms, the main focus of this chapter will be to consider the genomics, genetic control, and molecular mechanisms of processes associated with intercellular communication (mediated by peptide pheromones and quorum-sensing molecules), adhesion to surfaces, and host colonization.

MITIS GROUP STREPTOCOCCAL GENOMES

The genomes of *S. pneumoniae* TIGR4 (ATCC BAA-334), a pathogenic encapsulated strain (60), and nonencapsulated strain R6 (25) have been sequenced, allowing comparison of genes present in virulent versus avirulent pneumococcal isolates. The complete genome of *S. gordonii* Challis (CH1) has recently been sequenced (S. R. Gill, unpublished); the genome of *S. mitis* NCTC 12261 has been randomly sequenced to six times coverage and assembled into a pseudomolecule for preliminary annotation (H. Tettelin, personal communication). Comparative summaries of partial and complete genomes of oral streptococci are shown in Table 1. Functional assignments of open reading frames (ORFs) have been based on published experimental data, and on sequence similarities of ORFs detected by algorithms designed to identify functional domains. As is common among prokaryotic genomes, the *mitis* group streptococcal genomes have only 62 to 70% of identified ORFs with assigned functions. The remaining ORFs encode hypothetical proteins, with or without conserved domains, or are genes of unknown function. Owing to the natural competence of these strains for uptake of DNA and transformation, and the multiplicity of species cohabiting the oral cavity, it is not surprising that there is evidence for gene acquisition through horizontal transfer of DNA. The pneumococcal genomes show the variety of genes expected in a naturally competent species occupying the nutritionally diverse oral environment. These include a large number of genes (~140) encoding membrane transporters, consistent with the known metabolic traits of the organism. The *S. gordonii* and *S. mitis* genomes show similar large numbers of substrate transporters, in particular for a wide range of carbohydrates. Various genes encoding partial or complete insertion sequence (IS) elements, transposases, segments of prophages, and other elements associated with mobile extrachromosomal functions have integrated onto the chromosomes. However, none of the *mitis* group streptococci genomes sequenced to date provides evidence of intact prophage or replicative plasmids; rather each genome consists of a single DNA molecule. The *S. gordonii* and *S. mitis* genomes contain many regions of direct and inverted DNA repeats, which are potential hot spots for recombination. In addition, the genomes contain relatively large numbers of genes (about 20% in *S. gordonii*) with atypical G+C content, significantly higher or lower than the overall average percentage, suggesting their foreign origins (S. R. Gill, unpublished data). In summary, it is probable that commensal *sanguinis* group streptococci have a more diverse set of genes, allowing them to thrive in a wider range of environmental conditions compared with the more selective conditions that are exploited by cariogenic *mutans* group streptococci. Comparisons of genes found among oral and nonoral streptococcal species will have epidemiological significance and provide insights into the biological functions necessary or specific for determining survival in the oral environment.

COMPETENCE DEVELOPMENT AND DNA-MEDIATED TRANSFORMATION

Competence Pheromones

Many *mitis* group streptococci are naturally transformable and attain a special developmental state (competence) at some point in their growth cycle during which DNA may be taken up. Development of competence in *S. gordonii* has long been known to depend upon the production of an extracellular competence factor that is heat-resistant and trypsin-sensitive and is produced in a defined window of growth in batch culture, usually early exponential phase (38). Major advances in understanding of the competence and transformation processes in *mitis* group streptococci came about following purification of the *S. pneumoniae* competence factor (designated competence-stimulating peptide or pheromone [CSP]) (22). Based upon the sequence of natural competence factor,

a 17-amino-acid (aa) residue peptide, CSP, was synthesized and shown to stimulate competence development by pneumococcal cells, more or less irrespective of their growth phase, thus confirming competence factor as an unmodified oligopeptide pheromone (22). The gene encoding CSP (*comC*) in *S. pneumoniae* and in *S. gordonii* is located immediately upstream of *comD* and *comE* genes encoding, respectively, the sensor histidine kinase and response regulator protein of a two-component signal transduction system (TCS) (23). The ComD protein is the receptor for the *comC* gene product and senses extracellular levels of CSP. The *comCDE* locus, which is flanked by two conserved tRNA genes, is also present in *S. mitis*, *S. oralis*, *S. sanguinis*, and a number of other oral streptococcal species (64).

The streptococcal CSPs belong to a class of peptides, including bacteriocins, that are synthesized as precursors containing, within the N-terminal leader peptide, a conserved double-glycine processing site. A dedicated ABC-type transporter is responsible for proteolytic cleavage of the leader and translocation of the C-terminal peptide out across the cytoplasmic membrane. At least four *comC* alleles are present within the species *S. pneumoniae* (64). CSP-1 and CSP-2, each encoding 17-aa residue peptides that are processed from 41-aa residue prepeptides, are the two major competence pherotypes in pneumococci. They will induce competence only in strains of the corresponding pherotype, e.g., the ComD receptor of *S. pneumoniae* R6 (pherotype CSP-1) does not efficiently recognize CSP-2. Three *comC* alleles have been found among different strains of *S. gordonii*; two of these, designated *comC1* and *comC2*, encode 19-aa residue CSPs that contain only five identical residues, but which are processed from 50-aa residue prepeptides that carry identical leader sequences. The *S. gordonii* CSP is completely inactive on *S. pneumoniae*. The specificity of CSP recognition may increase the likelihood that DNA taken up during competence comes from closely related bacteria (65).

Genetic and Environmental Regulation of Competence Development

The CSP product of the *comC* gene is proposed to be a central signaling molecule that mediates quorum sensing in cultures of *S. gordonii* and *S. pneumoniae*. Experimental evidence suggests that the mechanisms involved in competence development and transformation are probably fundamentally similar in these two species of streptococci (40). The ComC precursor is processed and secreted by an ABC transporter and accessory protein encoded by the *comA* and *comB* genes. ComA proteins from *S. pneumoniae* and *S. gordonii* are 85% identical and have significant homology to bacteriocin transporters. Downstream of *comB* in *S. gordonii* is a region that has the potential to encode three different peptides, one of which is a basic 52-aa residue peptide that might be required for competence independently of CSP (40). It is apparent now that CSP is not the only signaling molecule involved in streptococcal competence development and that other environmentally responsive pathways impact on the processes (see below).

Activation of the competent state occurs when the external concentration of CSP reaches a threshold level (about 10 ng/ml in *S. pneumoniae* cultures) (22). The sensor domain of ComD responds to this by autophosphorylating and subsequently activating (by phosphorylation) the response regulatory protein ComE. Activated ComE (ComE ~P) binds upstream of *comC*, and to the promoter region of *comAB*, thus activating transcription of these operons, amplifying production of and response to the pheromone, and

TABLE 1 Summary of oral streptococcal genomes

Species	bp	% G+C	No. of ORFs	% Coding
S. gordonii Challis CH1	2,197,628	40.5	2,090	88.4
S. mitis NCTC 12261[a]	2,045,857	40.4	2,149	85.8
S. pneumoniae TIGR4	2,160,837	39.6	2,236	87.3
S. pneumoniae R6	2,038,615	39.6	2,219	86.3
S.mutansUA159	2,030,921	36.7	2,096	86.4
S. sobrinus 6715[a]	2,409,145	42.5	2,489	81.3

[a]Data for *S. mitis* and *S. sobrinus* are estimates based on unfinished genome sequences from The Institute for Genomic Research (www.tigr.org). Each has been assembled into a pseudomolecule from six to eight times sequence coverage

synchronizing development of competence within the culture. ComE~P also induces transcription of the *comX* gene, which encodes an alternate sigma factor (41). This replaces the primary sigma factor from the RNA polymerase holoenzyme and activates transcription of the so-called late competence genes, the products of which are involved in DNA uptake, processing, and recombination. Thus, ComE could be considered the regulatory master of the transformation process. This is supported by the observation that in a constitutively competent strain of *Streptococcus infantis*, which lacks *comC*, competence is lost by disruption of *comE* (65).

Although the ComABCDE pathway accounts for how the competent state may be subject to autoinduction, a more complex regulatory network is envisaged because a number of additional intrinsic and environmental factors affect transformation. A recently discovered pathway BlpA-BCSRH in *S. pneumoniae* regulates production of several class II bacteriocins and their immunity proteins. This system bears strong resemblance to ComABCDE in that BlpC encodes a bacteriocin-inducing peptide that is processed and exported by a secretion complex (BlpAB), and is recognized by the TCS BlpSRH (34). Although bacteriocin-inducing peptide-1 does not activate competence, transcription of an operon encoding the QsrAB transporter (a putative sodium pump) is positively regulated by both BlpR and ComE (34). Thus, two independent quorum-sensing systems, responding to different pheromones, can activate transcription of the same target gene. Another TCS in *S. pneumoniae*, designated CiaRH, is involved in regulating competence as well as virulence (12). Although it is not known how CiaRH and competence regulation are interconnected, recent evidence suggests that levels of HtrA, a heat shock protein, are under positive control of CiaRH. In the competence-induced transcriptome (12, 49) a number of genes with delayed-onset expression are associated with stress responses, for example *groES*, *groEL*, and *dnaK*. Several phenotypes associated with deficiency in CiaRH can be related to decreases in HtrA levels, including resistance to autolysis, increased sensitivity to oxidative stress, and decreased transformation frequency (26). Orthologs of the *ciaRH* genes are present in *S. gordonii*, *S. mutans*, and *Streptococcus pyogenes*. Although transcriptome analysis of *S. gordonii* has not yet been undertaken, orthologs to many of the competence-associated *S. pneumoniae* genes are present in the *S. gordonii* genome (S. R. Gill, unpublished results).

In addition to TCS control of competence development there is evidence for peptide regulation of competence mediated via peptide transport. Mutations in the peptide-binding protein genes of the *ami* locus, encoding an oligopeptide transport system in pneumococcus, alter the timing of competence development (1). The orthologous ABC-type binding protein-dependent oligopeptide permease designated Hpp in *S. gordonii*, which preferentially transports hexa- and heptapeptides (29), is also involved in competence development. To account for these observations, additional regulatory circuits must operate to provide temporal controls on the competence pathways. The activity of the Hpp oligopeptide transporter also influences expression of the *S. gordonii cshA* gene, which encodes a cell wall-linked surface protein adhesin (46). Transcript levels from the *cshA* gene normally increase in late exponential growth phase and appear to be sensitive to an extracellular signal sensed via Hpp. It may be of ecological significance, then, that the activities of this oligopeptide transporter can modulate both adhesion and competence, considering that oral streptococci grow within the human host principally in biofilms. Under these conditions the regulated expression of bacterial cell surface adhesins would be important for the development and maintenance of communities responsive to intercellular peptide signals (see below).

Transformation

Natural transformation is a mechanism whereby related species acquire genetic diversity. Growth and survival of recombinants in vivo are determined by host environmental selective pressures, such as the availability of nutrients and sites for adhesion, and the level and specificity of immune defenses. Genes for many streptococcal cell surface proteins contain amino acid repeat blocks (see chapter 2), and it is probable that the DNA sequences encoding these are preferred sites for recombination, thus generating diversity in substrate-binding functions and antigenicity. In the laboratory, transformation has been utilized successfully in many genetic studies of oral streptococci. To generate mutants, gene replacement techniques in particular have been employed. In the preferred method, linear DNA comprising a selectable marker (e.g., antibiotic resistance) inserted within a cloned fragment of chromosomal DNA is transformed into the host strain, and the marker is integrated into the chromosome by homologous recombination (double crossover), resulting in allelic replacement of the target site. This is an efficient mutagenesis technique in streptococci that are naturally competent. However, in using this protocol, it should be duly recognized that insertion of a replacement allele with its own transcriptional terminator may have polar effects on expression of cotranscribed genes downstream. Insertional mutagenesis of streptococcal genes may also be achieved with intact, nonreplicating suicide vectors carrying a selectable marker adjacent to a cloned region of chromosomal homology. Plasmid integration and duplication of the chromosomal target are accomplished by a Campbell-like recombination event between homologous

DNA (59). This procedure may also be used to replace a parental allele with a mutant allele with engineered base pair changes. Growing a transformant with a chromosomally integrated plasmid containing the mutated allele in the absence of antibiotic selective pressure can result in looping out of the vector through homologous recombination across the flanking DNA. Some of the resulting transformants will contain the mutant allele with base pair changes. This method is useful for studying transcriptional and translational effects on gene expression (55). Integration of a selectable marker into *S. pneumoniae* and *S. gordonii* by double-crossover homologous recombination with linear DNA is a more efficient process than insertion-duplication with transforming circular double-stranded DNA (40). Replicative plasmid transformation of *S. gordonii* is highly efficient and has been utilized for complementation analysis, expression of heterologous genes, and the isolation of selectable markers such as TetM, *ermAM*, *aad9*, and *aphA3* encoding tetracycline, erythromycin, spectinomycin, and kanamycin resistances, respectively. Plasmids containing beta-lactamase determinants conferring resistance to penicillins are not used in oral streptococcal genetics because of concerns about the possibility of introducing penicillin resistance into organisms in which naturally occurring beta-lactamases are absent. Although allelic replacement with linear transforming DNA can be readily achieved in naturally transformable *S. gordonii*, it appears to be accomplished in electrocompetent *S. parasanguinis* only by transforming with circular double-stranded plasmid DNA (19).

ADHESION

Multiple Adhesive Interactions

Insertional mutagenesis methods (as described above) in combination with functional analyses of purified native, or recombinant, streptococcal surface proteins have revealed much information about the molecular basis of adhesion. The *sanguinis* group organisms, being predominant colonizers of salivary pellicle, carry a diverse range of adhesins for salivary molecules. They also express adhesins that bind host cells and other microbial cells (37). Multiple adhesins can be theorized to confer numerous advantages to the streptococci. Adhesins of differing specificities will result in more avid binding, increase the probability of an individual cell engaging an oral cavity receptor, and allow binding to the wide range of ligands present at different body sites. Interbacterial adhesion is especially relevant in the complex microbiota of the oral cavity. Many of the later bacterial arrivals in the development of plaque communities include pathogenic gram-negative species such as *Fusobacterium nucleatum*, *Porphyromonas gingivalis*, and *Treponema denticola* that bind directly to streptococci (35). Interbacterial binding is mainly mediated by protein-carbohydrate recognitions (37), and the best characterized of these interactions are those occurring between different streptococcal species, and between streptococci and *Actinomyces naeslundii*, another major component of dental plaque (48). Strains of most *mitis* group streptococci that exhibit lactose-sensitive binding to *A. naeslundii* produce antigenically diverse linear cell wall phosphopolysaccharides containing the hostlike recognition motifs GalNAcβ1→3Gal or Galβ1→3GalNAc (67). These motifs are recognized by the fimbrial lectins of *A. naeslundii* (10) and by GalNAc-sensitive lectins present on other streptococcal cells, *Haemophilus parainfluenzae* and *Prevotella loescheii* (37).

Surface Protein Adhesins

Many of the streptococcal surface proteins that have been implicated in adhesion and colonization are covalently linked to cell wall peptidoglycan through sortase enzyme recognition of the C-terminal anchor motif LPXTG (see chapter 2). The genome sequence of *S. gordonii* contains 25 ORFs encoding proteins with LPXTG motifs, as well as genes encoding sortase A (5) and sortase B (S. R. Gill, unpublished data). Several of these potential cell surface proteins contain conserved collagen or cell wall-binding domains, suggesting their importance as colonization determinants. Cell surface polypeptides contribute to the range of surface structures that are displayed by cells of *S. gordonii*, *S. parasanguinis*, and *S. sanguinis*. A majority of *S. sanguinis* strains carry peritrichous fibrils of approximate lengths of 50 to 70 nm, while some isolates produce tufts of fibrils and some strains carry fimbriae (21). It has proved difficult to show definitive associations between production of surface appendages and adhesion. However, mutations in *S. gordonii cshA* or *S. parasanguinis fap1* genes affect adhesion and production of fibrils or fimbriae, respectively, in these species. CshA is a 259-kDa cell wall-associated protein that forms surface fibrils on *S. gordonii* (43) and on *S. sanguinis* (15). The nonrepetitive N-terminal region of the CshA polypeptide, spanning 836 aa residues, contains sequences that mediate adhesion of *S. gordonii* cells to immobilized fibronectin and *A. naeslundii* cells (45). Isogenic *cshA* mutants of *S. gordonii* are deficient in binding these receptors, exhibit reduced hydrophobicity, and are impaired in ability to invade endothelial cells (53). Fap1 is a >200-kDa glycosylated surface protein in *S. parasanguinis* with sequence regions rich in threonine and serine, and is related to the *S. gordonii* sialic acid-binding proteins Hsa and GspB (see below). It is proposed that cross-linking of Fap1 precursor molecules, possibly involving sortase enzyme activity (62), leads to fimbrial assembly on *S. parasanguinis* (66).

Possibly the best characterized of the oral streptococcal adhesins are those protein members of the antigen I/II family. The genes encoding these polypeptides were first described in *S. mutans*, but similar genes have now been detected in most indigenous oral streptococcal species (30), with 14 complete sequences now available. *S. gordonii* is unusual in expressing two antigen I/II polypeptides, designated SspA and SspB (Table 2), that are the products of tandemly arranged and independently expressed chromosomal genes (13). These polypeptides bind salivary agglutinin glycoprotein (also designated gp340), found in parotid saliva (51) and type I collagen (24), and mediate adhesion of streptococci to *A. naeslundii* (14) and to *P. gingivalis* (36). The multiple binding properties are attributed to discrete functional domains formed by least three regions of the polypeptide (30, 36, 61). The antigen I/II polypeptides may be capable of distinguishing between immobilized or soluble forms of salivary glycoproteins, thus allowing streptococci to colonize oral surfaces coated with salivary components, despite the presence of excess soluble forms of the components in saliva.

Sialic acid (NeuNAc)-binding adhesins are present on most strains of *S. gordonii* and *S. sanguinis*. The major Ca^{2+}-dependent lectin activity for α2-3-linked sialic acid-containing receptors in *S. gordonii* Challis is associated with the protein Hsa (Table 2). This forms surface fibrillar structures of >200 kDa composed of protein and wheat germ agglutinin-reactive carbohydrate that mediate hemagglutination of human erythrocytes (56). *S. gordonii* M99 does not contain the *hsa* gene but rather an alternative allele *gspB*, which encodes a highly similar serine-rich repeat protein

TABLE 2   Polypeptide adhesion and colonization factors of *S. gordonii*

Polypeptide	No. of amino acid residues[a]	Surface retention[b]	Properties or functions	Distribution among streptococcal species[c]	Reference(s)
CshA	2,508	CWA	Forms surface fibrils, determinant of cell surface hydrophobicity; binds immobilized fibronection, *S. oralis*, *A. naeslundii*, *Candida albicans*	*S. sanguinis*, *S. oralis*	15, 43, 45
CshB	2,292	CWA	CshA homolog (~60% identity over first 863 aa residues); cooperative in expression of CshA	*S. sanguinis*, *S. oralis*	15, 43, 45
SspA	1,575	CWA	Binds salivary agglutinin (gp340), collagen type I, *A. naeslundii*, *C. albicans*	*S. agalactiae*, *S. cristatus*, *S. intermedius*, *S. mutans*, *S. oralis*, *S. sanguinis*, *S. sobrinus*	13, 14, 24, 30, 36
SspB	1,499	CWA	SspA homologue (~65% aa residue identity overall); binds gp340, collagen type I, *A. naeslundii*, *C. albicans*, *P. gingivalis*	*S. agalactiae*, *S. cristatus*, *S. intermedius*, *S. mutans*, *S. oralis*, *S. sanguinis*, *S. sobrinus*	13, 14, 24, 30, 36
Hsa[d]	2,178	CWA	Glycoprotein; binds sialic acid receptors on blood cells and mucins	*S. agalactiae*, *S. cristatus*, *S. parasanguinis*, *S. pneumoniae*, *S. sanguinis*	56, 66
GspB	3,072	CWA	Hsa ortholog, glycoprotein; binds human platelets	*S. agalactiae*, *S. cristatus*, *S. parasanguinis*, *S. pneumoniae*, *S. sanguinis*	3, 57, 66
AbpA	195	SA	Binds salivary α-amylase	*S. pyogenes*	58
AbpB	652	LPP	Unrelated to AbpA: binds salivary α-amylase, probable dipeptidase		58
FbpA	550	SA	Required for CshA expression; binds immobilized fibronectin	*S. mutans*, *S. pneumoniae*, *S. pyogenes*	9
GtfG	1,577	SA/EP	Glucan production from sucrose; biofilm formation	*S. mutans*, *S. salivarius*, *S. sanguinis*, *S. sobrinus*	31, 55, 63
ScaA	310	LPP	Metal ion (preferentially Mn^{2+}) transport; oxidative stress tolerance; biofilm formation	All species	6, 27, 28
HppA	667	LPP	Oligopeptide-binding protein; binds serum proteins, *A. naeslundii*; biofilm formation	*S. pneumoniae*, *S. mitis*, *S. oralis*	29, 46

[a] Precursor polypeptide, data from GenBank entries and from genomic sequence available at www.tigr.org.
[b] CWA, cell wall-anchored; SA, cell surface-associated; EP, extracellular (released) protein; LPP, lipoprotein.
[c] Species in which orthologs identified, but not necessarily expressed by all strains.
[d] Strains of *S. gordonii* carry either the *hsa* or *gspB* allele (N. S. Jakubovics, unpublished data).

(57). Hsa and GspB polypeptides represent a new family of serine-rich repeat proteins, including Fap1 from *S. parasanguinis*, that are glycosylated prior to export via a specialized secretory system (3). Potential receptors for these polypeptides within the host include salivary mucins, secretory IgA1, platelets, and leukosialin on polymorphonuclear leukocytes (56, 57).

S. gordonii and several other *mitis* group species are able to avidly bind the most abundant salivary enzyme, α-amylase. In *S. gordonii* this is mediated by a cell surface 20-kDa polypeptide designated AbpA (Table 2). Amylase binding might be a means by which a surface enzymic activity is acquired by bacteria for generation of oligosaccharides for uptake and metabolism. In addition, α-amylase present in salivary pellicle could serve as a receptor for adhesion (58). Adhesive and nutritional functions of surface proteins are probably closely linked. Adhesins may act as capture proteins to provide substrates for degradation by cell surface enzyme complexes, while the solute-binding components of nutrient uptake systems may possibly also function as adhesins (37).

Many oral streptococci produce glucosyltransferases that hydrolyze sucrose and polymerize the glucose into glucans (see chapter 27). *S. gordonii* carries a single *gtfG* gene encoding an enzyme that synthesizes a mixed α-1,3- and α-1,6-linked glucan that may facilitate interbacterial interactions and host colonization. The level of glucosyltransferase activity is controlled by environmental signals and by genetic elements. Variants expressing only 30% or less of wild-type glucosyltransferase activity do not accumulate in significant numbers within glucans on surfaces. Other phenotypes, such as adhesion to salivary pellicle, binding to *A. naeslundii*, and β-hemolysin production, are also subject to phase variation (31). Therefore, generation of phase variants in vivo might aid dispersal of cells to colonize different sites. A positive regulator, Rgg, controls *gtfG* transcription in *S. gordonii* (55). Rgg-like proteins, which occur in both oral and nonoral streptococci, regulate exoprotein production (63) and may be involved in multigene regulatory cascades (8).

BIOFILMS

Oral Microbial Communities

Biofilm development is a multistage process. Initial attachment to a surface involves bacterial adhesins. Adhered bacteria then undergo cell division, colonize the surface, produce extracellular materials, and provide for further adhesion, accumulation, and incorporation of other bacteria to form a community. Quorum-sensing circuits are intimately involved in biofilm formation, and therefore, it is envisaged that bacteria coordinate group behavior to accomplish activities that individual cells cannot achieve. Given therefore the importance of adhesion and cell signaling in biofilm development, these are currently areas of extensive study in oral microbiology.

In at least 40 species of bacteria, LuxS encodes autoinducer 2 (AI-2) synthase necessary for production of AI-2, which is a signal for interspecies communication (18). In *S. gordonii*, AI-2 is produced maximally at late exponential phase of growth. *S. gordonii luxS* mutants are altered in ability to form mono-species (4) and mixed-species (44) biofilms. This may be related to AI-2 regulation of carbohydrate metabolism, because alterations in pathways of sugar uptake and utilization are associated with *S. gordonii* biofilm formation (39). A number of TCS functions are also crucial

for community development. The ComCDE pathway plays a role in early biofilm formation in *S. mutans*, during which cells are induced to competence and maintain competence in the biofilm (11). Although a function for ComABCDE in *S. gordonii* biofilm development has not been reported, another TCS designated BrfAB has been postulated to be critical for *S. gordonii* biofilm formation (68). Multiple extracellular signals are thus sensed by biofilm-forming cells and are integrated into establishing a successful community. Expression of adhesins is undoubtedly subject to modulation by interbacterial signals, and by host-derived signals (24), but information on this aspect of community development is currently lacking. Determining the molecular mechanisms involved in physical and chemical communication between biofilm community organisms is clearly a major challenge for the future.

Infective Endocarditis

Upon gaining access to the bloodstream, *sanguinis* group streptococci can infect the heart valves and endocardium. At these sites, thrombotic vegetations are formed consisting of bacteria cross-linked with platelets and plasma proteins. These unique biofilms are major pathogenic determinants in endocarditis and occlusive vascular disease. Strains of *sanguinis* group streptococci vary in their abilities to adhere to platelets and to induce platelet aggregation. In *S. sanguinis* it is proposed that a class I adhesin mediates initial attachment of bacteria to platelets, and then aggregation is induced by an antigenically distinct class II adhesin designated platelet aggregation-associated protein (16). The platelet-interactive domain of platelet aggregation-associated protein has been identified as a heptapeptide PGEQGPK shared by the platelet-binding domains of collagen types I and III. In *S. gordonii*, platelet binding is mediated, at least in part, by the glycosylated serine-rich repeat proteins Hsa or GspB (57). Little is known currently about how expression of these host interactive polypeptides might be influenced by bacterial signaling molecules or by host molecules in vivo.

A virulence factor implicated in *S. parasanguinis*-associated endocarditis is the FimA lipoprotein (6). This polypeptide is orthologous to ScaA, a putative manganese (Mn^{2+})-binding protein in *S. gordonii* that is induced under Mn^{2+}-limited conditions and linked to an ABC-type uptake system (27). *fimA* mutants are deficient in binding to fibrin clots in vitro and are reduced in virulence in an animal model of endocarditis (6). ScaA has been shown to be essential for Mn^{2+}-dependent oxidative stress tolerance in *S. gordonii* (28). It is suggested that FimA (or ScaA) production might be up-regulated under metal ion-depleted and aerobic conditions at endocardial sites, thus promoting bacterial adhesion, survival, and development of vegetations. Other molecules released by the bacteria, such as histonelike proteins, may contribute fortuitously to virulence by enhancing immune complex formation at local or distal sites (54).

HETEROLOGOUS GENE EXPRESSION

In the quest to develop safer vaccines, there has been much interest in generating live recombinant bacteria as vaccine antigen delivery vehicles. In particular, the use of lactic acid bacteria, which normally colonize mucosal surfaces, has been advanced as a means for engendering local IgA antibody responses to an antigen, as well as conventional systemic IgG antibody and T-cell responses. In part because *S. gordonii* Challis is so efficiently transformable, this organism has been promoted as a pilot organism in feasibility studies of commensal-mediated human or animal immunization. By utilizing a chromosomally integrated recombinant gene cassette, consisting of an in vivo active *S. gordonii* promoter situated upstream from a segment of group A streptococcal M protein carrying the cell wall anchorage region, a number of sequences encoding various heterologous antigens have been engineered and expressed on the *S. gordonii* cell surface (47). Mice immunized with killed recombinant organisms develop significant antigen-specific IgA responses at mucosal surfaces as well as serum IgG responses. Simultaneous display of an immunomodulating protein along with a heterologous protein antigen on the surface of *S. gordonii* can be used to potentiate immune responses to the coexpressed antigen (7, 42). In a development of this type of presentation system, therapeutic proteins have been expressed on the surface of *S. gordonii* and delivered to mucosal sites. These have included interleukin-1 receptor antagonist, for control of interleukin-1-dependent inflammation (52), and cyanovirin-N, which is a potent human immunodeficiency virus-inactivating protein, for capturing human immunodeficiency virus virions (20).

For these novel delivery strategies to be put into practice, many issues still need to be addressed, particularly with regard to the safety of using recombinant commensal bacteria in therapeutics. Crucial information will be acquired through more appreciation of the molecular mechanisms by which pathogenically unpredictable organisms such as *S. gordonii* colonize human body sites. Genetic studies of *sanguinis* group streptococci have significantly advanced our understanding of the processes involved in transformation, adhesion, and virulence. This knowledge will ultimately lead to new ways of controlling infections caused by these bacteria and by related streptococci. Paradoxically, it is conceivable that organisms such as *S. gordonii* may, in the future, be considered therapeutic products in novel strategies to combat other infectious diseases.

We thank many colleagues for their special insights that have helped in the production of this chapter, including D. B. Clewell, G. M. Dunny, P. M. Fives-Taylor, S. R. Gill, P. S. Handley, M. C. Herzberg, M. Kilian, P. E. Kolenbrander, R. J. Lamont, and H. Tettelin. Research in the authors' laboratories was supported by the British Heart Foundation, Health Research Council of New Zealand, National Institutes of Health (USPHS grant DE11090), and The Wellcome Trust.

REFERENCES

1. **Alloing, G., B. Martin, C. Granadel, and J.-P. Claverys.** 1998. Development of competence in *Streptococcus pneumoniae*: pheromone autoinduction and control of quorum sensing by the oligopeptide permease. *Mol. Microbiol.* **29:** 75–83.
2. **Barnard, J. P., and M. W. Stinson.** 1996. The alpha-hemolysin of *Streptococcus gordonii* is hydrogen peroxide. *Infect. Immun.* **64:**3853–3857.
3. **Bensing, B. A., and P. M. Sullam.** 2002. An accessory *sec* locus of *Streptococcus gordonii* is required for export of the surface protein GspB and for normal levels of binding to human platelets. *Mol. Microbiol.* **44:**1081–1094.
4. **Blehert, D. S., R. J. Palmer, Jr., J. B. Xavier, J. S. Almeida, and P. E. Kolenbrander.** 2003. Autoinducer 2 production by *Streptococcus gordonii* DL1 and the biofilm phenotype of a *luxS* mutant are influenced by nutritional conditions. *J. Bacteriol.* **185:**4851–4860.
5. **Bolken, T. C., C. A. Franke, K. F. Jones, G. O. Zeller, C. H. Jones, E. K. Dutton, and D. E. Hruby.** 2001. Inactivation of the *srtA* gene in *Streptococcus gordonii* inhibits cell wall anchoring of surface proteins and decreases in vitro and in vivo adhesion. *Infect. Immun.* **69:**75–80.
6. **Burnette-Curley, D., V. Wells, H. Viscount, C. L. Munro, J. C. Fenno, P. Fives-Taylor, and F. L. Macrina.** 1995. FimA, a major virulence factor associated with *Streptococcus parasanguis* endocarditis. *Infect. Immun.* **63:** 4669–4674.
7. **Byrd, C. M., T. C. Bolken, K. F. Jones, T. K. Warren, A. T. Vella, J. McDonald, D. King, Z Blackwood, and D. E. Hruby.** 2002. Biological consequences of antigen and cytokine co-expression by recombinant *Streptococcus gordonii* vaccine vectors. *Vaccine* **20:**2197–2205.
8. **Chausssee, M.S., G.A. Somerville, L Reitzer, and J.M. Musser.** 2003. Rgg coordinates virulence factor synthesis and metabolism in *Streptococcus pyogenes*. *J. Bacteriol.* **185:** 6016–6024.
9. **Christie, J., R. McNab, and H. F. Jenkinson.** 2002. Expression of fibronectin-binding protein FbpA modulates adhesion in *Streptococcus gordonii*. *Microbiology* **148:**1615–1625.
10. **Cisar, J. O., A. L. Sandberg, G. P. Reddy, C. Abeygunawardana, and C. A. Bush.** 1997. Structural and antigenic types of cell wall polysaccharides from viridans group streptococci with receptors for oral actinomyces and streptococcal lectins. *Infect. Immun.* **65:**5035–5041.
11. **Cvitkovitch, D. G., L. Ynug-Hua, and R. P. Ellen.** 2003. Quorum sensing and biofilm formation in streptococcal infections. *J. Clin. Investig.* **112:**1626–1632.
12. **Dagkessamanskaia, A., M. Moscoso, V. Henard, S. Guiral, K. Overweg, M. Reuter, B. Martin, J. Wells, and J. P. Claverys.** 2004. Interconnection of competence, stress and CiaR regulons in *Streptococcus pneumoniae*: competence triggers stationary phase autolysis of *ciaR* mutant cells. *Mol. Microbiol.* **51:**1071–1086.
13. **Demuth, D. R., Y. Duan, W. Brooks, A. R. Holmes, R. McNab, and H. F. Jenkinson.** 1996. Tandem genes encode cell-surface polypeptides SspA and SspB which mediate adhesion of the oral bacterium *Streptococcus gordonii* to human and bacterial receptors. *Mol. Microbiol.* **20:**403–413.
14. **Egland, P. G., L. D. Du, and P. E. Kolenbrander.** 2001. Identification of independent *Streptococcus gordonii* SspA and SspB functions in coaggregation with *Actinomyces naeslundii*. *Infect. Immun.* **69:**7512–7516.
15. **Elliott, D., E. Harrison, P. S. Handley, S. K. Ford, E. Jaffray, N. Mordan, and R. McNab.** 2003. Prevalence of Csh-like fibrillar surface proteins among mitis group oral streptococci. *Oral. Microbiol. Immunol.* **18:**114–120.
16. **Erickson, P. R., and M. C. Herzberg.** 1993. The *Streptococcus sanguis* platelet aggregation-associated protein. Identification and characterization of the minimal platelet-interactive domain. *J. Biol. Chem.* **268:**1646–1649.
17. **Facklam, R.** 2002. What happened to the streptococci: overview of taxonomic and nomenclature changes. *Clin. Microbiol. Rev.* **15:**613–630.
18. **Federle, M. J., and B. L. Bassler.** 2003. Interspecies communication in bacteria. *J. Clin. Investig.* **112:**1291–1299.
19. **Fenno, J. C., A. Shaikh, and P. Fives-Taylor.** 1993. Characterization of allelic replacement in *Streptococcus parasanguis*: transformation and homologous recombination in a "nontransformable" streptococcus. *Gene* **130:**81–90.
20. **Giomarelli, B., R. Provvedi, F. Meacci, D. Medaglini, G. Pozzi, T. Mori, J. B. McMahon, R. Gardella, and M. R. Boyd.** 2002. The microbicide cyanovirin-N expressed on

the surface of commensal bacterium *Streptococcus gordonii* captures HIV-1. *AIDS* **16:**1351–1356.

21. **Handley, P. S., P. L. Carter, J. E. Wyatt, and L. M. Hesketh.** 1985. Surface structures (peritrichous fibrils and tufts of fibrils) found on *Streptococcus sanguis* strains may be related to their ability to coaggregate with other oral genera. *Infect. Immun.* **47:**217–227.

22. **Håvarstein, L. S., G. Coomaraswamy, and D. A. Morrison.** 1995. An unmodified hexadecapeptide pheromone induces competence for genetic transformation in *Streptococcus pneumoniae. Proc. Natl. Acad. Sci. USA* **92:**11140–11144.

23. **Håvarstein, L. S., P. Gaustaud, I. F. Nes, and D. A. Morrison.** 1996. Identification of the streptococcal competence-pheromone receptor. *Mol. Microbiol.* **21:**863–869.

24. **Heddle, C., A. H. Nobbs, N. S. Jakubovics, M. Gal, J. P. Mansell, D. Dymock, and H. F. Jenkinson.** 2003. Host collagen signal induces antigen I/II adhesin and invasion gene expression in oral *Streptococcus gordonii. Mol. Microbiol.* **50:**597–607.

25. **Hoskins, J., W. E. Alborn, Jr., J. Arnold, L. C. Blaszczak, S. Burgett, B. S. DeHoff, S. T. Estrem, L. Fritz, D. J. Fu, W. Fuller, C. Geringer, R. Gilmour, J. S. Glass, H. Khoja, A. R. Kraft, R. E. Lagace, D. J. LeBlanc, L. N. Lee, E. J. Lefkowitz, J. Lu, P. Matsushima, S. M. McAhren, M. McHenney, K. McLeaster, C. W. Mundy, T. I. Nicas, F. H. Norris, M. O'Gara, R. B. Peery, G T. Robertson, P. Rockey, P. M. Sun, M. E. Winkler, Y. Yang, M. Young-Bellido, G. Zhao, C. A. Zook, R. H. Baltz, S. R. Jaskunas, P. R. Rosteck, Jr., P. L. Skatrud, and J. I. Glass.** 2001. Genome of the bacterium *Streptococcus pneumoniae* strain R6. *J. Bacteriol.* **183:**5709–5717.

26. **Ibrahim, Y. M., A. R. Kerr, J. McCluskey, and T. J. Mitchell.** 2004. Control of virulence by the two-component system Ciar/H is mediated via HtrA, a major virulence factor of *Streptococcus pneumoniae. J. Bacteriol.* **186:**5258–5266.

27. **Jakubovics, N. S., A. W. Smith, and H. F. Jenkinson.** 2000. Expression of the virulence-related Sca (Mn^{2+}) permease in *Streptococcus gordonii* is regulated by a diphtheria toxin metallorepressor-like protein ScaR. *Mol. Microbiol.* **38:**140–153.

28. **Jakubovics, N. S., A. W. Smith, and H. F. Jenkinson.** 2002. Oxidative stress tolerance is manganese (Mn^{2+}) regulated in *Streptococcus gordonii. Microbiology* **148:**3255–3263.

29. **Jenkinson, H. F., R. A. Baker, and G. W. Tannock.** 1996. A binding-lipoprotein-dependent oligopeptide transport system in *Streptococcus gordonii* essential for uptake of hexa- and heptapeptides. *J. Bacteriol.* **178:**68–77.

30. **Jenkinson, H. F., and D. R. Demuth.** 1997. Structure, function and immunogenicity of streptococcal antigen I/II polypeptides. *Mol. Microbiol.* **23:**183–190.

31. **Jones, G. W., D. B. Clewell, L. G. Charles, and M. M. Vickerman.** 1996. Multiple phase variation in haemolytic, adhesive and antigenic properties of *Streptococcus gordonii. Microbiology* **142:**181–189.

32. **Kawamura, Y., X. G. Hou, F. Sultana, H. Miura, and T. Ezaki.** 1995. Determination of 16S rRNA sequences of *Streptococcus mitis* and *Streptococcus gordonii* and phylogenetic relationships among members of the genus *Streptococcus. Int. J. Syst. Bacteriol.* **45:**406–408.

33. **Kilian, M., L. Mikkelsen, and J. Henrichsen.** 1989. Taxonomic study of viridans streptococci: description of *Streptococcus gordonii* sp. nov. and emended descriptions of *Streptococcus sanguis* (White and Niven 1946), *Streptococ-*

cus oralis (Bridge and Sneath 1982), and *Streptococcus mitis* (Andrewes and Horder 1906). *Int. J. Syst. Bacteriol.* **39:**471–484.

34. **Knutsen, E., O. Ween, and L. S. Håvarstein.** 2004. Two separate quorum-sensing systems upregulate transcription of the same ABC transporter in *Streptococcus pneumoniae. J. Bacteriol.* **186:**3078–3085.

35. **Kolenbrander, P. E., R. N. Andersen, D. S. Blehert, P. G. Egland, J. S. Foster, and R. J. Palmer, Jr.** 2002. Communication among oral bacteria. *Microbiol. Mol. Biol. Rev.* **66:**485–505.

36. **Lamont, R. J., A. El-Sabaeny, Y. Park, G. S. Cook, J. W. Costerton, and D. R. Demuth.** 2002. Role of the *Streptococcus gordonii* SspB protein in the development of *Porphyromonas gingivalis* biofilms on streptococcal substrates. *Microbiology* **148:**1627–1636.

37. **Lamont, R. J., and H. F. Jenkinson.** 2000. Adhesion as an ecological determinant in the oral cavity, p. 131–168. *In* H. K. Kuramitsu and R. P. Ellen (ed.), *Oral Bacterial Ecology: The Molecular Basis.* Horizon Scientific Press, Wymondham, United Kingdom.

38. **Leonard, G. C.** 1973. Early events in development of streptococcal competence. *J. Bacteriol.* **114:**1198–1205.

39. **Loo, C. Y., K. Mitrakul, I. B. Voss, C. V. Hughes, and N. Ganeshkumar.** 2003. Involvement of an inducible fructose phosphotransferase operon in *Streptococcus gordonii* biofilm formation. *J. Bacteriol.* **185:**6241–6254.

40. **Lunsford, R. D.** 1998. Streptococcal transformation: essential features and applications of a natural gene exchange system. *Plasmid* **39:**10–20.

41. **Luo, P., H. Li, and D. A. Morrison.** 2003. ComX is a unique link between multiple quorum sensing outputs and competence in *Streptococcus pneumoniae. Mol. Microbiol.* **50:**623–633.

42. **Maggi, T., M. Spinosa, S. Ricci, D. Medaglini, G. Pozzi, and M. R. Oggioni.** 2002. Genetic engineering of *Streptococcus gordonii* for the simultaneous display of two heterologous proteins at the bacterial surface. *FEMS Microbiol. Lett.* **210:**135–141.

43. **McNab, R., H. Forbes, P. S. Handley, D. M. Loach, G. W. Tannock, and H. F. Jenkinson.** 1999. Cell wall-anchored CshA polypeptide (259 kilodaltons) in *Streptococcus gordonii* forms surface fibrils that confer hydrophobic and adhesive properties. *J. Bacteriol.* **181:**3087–3095.

44. **McNab, R., S. K. Ford, A. El-Sabaeny, B. Barbieri, G. S. Cook, and R. J. Lamont.** 2003. Lux-S based signaling in *Streptococcus gordonii*: autoinducer 2 controls carbohydrate metabolism and biofilm formation with *Porphyromonas gingivalis. J. Bacteriol.* **185:**274–284.

45. **McNab, R., A. R. Holmes, J. M. Clarke, G. W. Tannock, and H. F. Jenkinson.** 1996. Cell surface polypeptide CshA mediates binding of *Streptococcus gordonii* to other oral bacteria and to immobilized fibronectin. *Infect. Immun.* **64:**4204–4210.

46. **McNab, R., and H. F. Jenkinson.** 1998. Altered adherence properties of a *Streptococcus gordonii* hppA (oligopeptide permease) mutant result from transcriptional effects on *cshA* adhesin gene expression. *Microbiology* **144:**127–136.

47. **Medaglini, D., G. Pozzi, T. P. King, and V. A. Fischetti.** 1995. Mucosal and systemic immune responses to a recombinant protein expressed on the surface of the oral commensal bacterium *Streptococcus gordonii* after oral colonization. *Proc. Natl. Acad. Sci. USA* **92:**6868–6872.

48. **Palmer, R. J., Jr., S. M. Gordon, J. O. Cisar, and P. E. Kolenbrander.** 2003. Coaggregation-mediated interac-

tions of streptococci and actinomyces detected in initial human dental plaque. *J. Bacteriol.* **185:**3400–3409.

49. Peterson, S. N., C. K. Sung, R. Cline, B. V. Desai, E. C. Snesrud, P. Luo, J. Walling, H. Li, M. Mintz, G. Tsegaye, P. C. Burr, Y. Do, S. Ahn, J. Gilbert, R. D. Fleischmann, and D. A. Morrison. 2004. Identification of competence pheromone responsive genes in *Streptococcus pneumoniae* by use of DNA microarrays. *Mol. Microbiol.* **51:**1051–1070.

50. Poulsen, K., J. Reinholdt, C. Jespersgaard, K. Boye, T. A. Brown, M. Hauge, and M. Kilian. 1998. A comprehensive genetic study of streptococcal immunoglobulin A1 proteases: evidence for recombination within and between species. *Infect. Immun.* **66:**181–190.

51. Prakobphol, A., F. Xu, V. M. Hoang, T. Larsson, J. Bergstrom, I. Johansson, L. Frangsmyr, U. Holmskov, H. Leffler, C. Nilsson, T. Boren, J. R. Wright, N. Strömberg, and S. J. Fisher. 2000. Salivary agglutinin, which binds *Streptococcus mutans* and *Helicobacter pylori*, is the lung scavenger receptor cysteine-rich protein gp-340. *J. Biol. Chem.* **275:**3980–3986.

52. Ricci, S., G. Macchia, P. Ruggiero, P. Bossu, L. Xu, D. Medaglini, A. Tagliabue, L. Hammarstrom, G. Pozzi, and D. Boraschi. 2003. In vivo mucosal delivery of bioactive human interleukin 1 receptor antagonist produced by *Streptococcus gordonii*. *BMC Biotechnol.* **3:**15.

53. Stinson, M. W., S. Alder, and S. Kumar. 1998. Invasion and killing of human endothelial cells by viridans group streptococci. *Infect. Immun.* **71:**2365–2372.

54. Stinson, M. W., R. McLaughlin, S. H. Choi, Z. E. Juarez, and J. Barnard. 1998. Streptococcal histone-like protein: primary structure of *hlpA* and protein binding to lipoteichoic acid and epithelial cells. *Infect. Immun.* **66:**259–265.

55. Sulavik, M. C., G. Tardif, and D. B. Clewell. 1992. Identification of a gene, *rgg*, which regulates expression of glucosyltransferase and influences the Spp phenotype of *Streptococcus gordonii* Challis. *J. Bacteriol.* **174:**3577–3586.

56. Takahashi, Y., K. Konishi, J. O. Cisar, and M. Yoshikawa. 2002. Identification and characterization of *hsa*, the gene encoding the sialic acid-binding adhesin of *Streptococcus gordonii* DL1. *Infect. Immun.* **70:**1209–1218.

57. Takamatsu, D., B. A. Bensing, and P. M. Sullam. 2004. Genes in the accessory *sec* locus of *Streptococcus gordonii* have three functionally distinct effects on the expression of the platelet-binding protein GspB. *Mol. Microbiol.* **52:**189–203.

58. Tanzer, J. M., L. Grant, A. Thompson, L. Li, J. D. Rogers, E. M. Haase, and F. A. Scannapieco. 2003. Amy-

lase-binding proteins A (AbpA) and B (AbpB) differentially affect colonization of rats' teeth by *Streptococcus gordonii*. *Microbiology* **149:**2653–2660.

59. Tao, L., D. J. LeBlanc, and J. J. Ferretti. 1992. Novel streptococcal-integration shuttle vectors for gene cloning and inactivation. *Gene* **120:**105–110.

60. Tettelin, H., K. E. Nelson, I. T. Paulsen, J. A. Eisen, T. D. Read, S. Peterson, J. Heidelberg, R. T. DeBoy, D. H. Haft, R. J. Dodson, A. S. Durkin, M. Gwinn, J. F. Kolonay, W. C. Nelson, J. D. Peterson, L. A. Umayam, O. White, S. L. Salzberg, M. R. Lewis, D. Radune, E. Holtzapple, H. Khouri, A. M. Wolf, T. R. Utterback, C. L. Hansen, L. A. McDonald, T. V. Feldblyum, S. Angiuoli, T. Dickinson, E. K. Hickey, I. E. Holt, B. J. Loftus, F. Yang, H. O. Smith, J. C. Venter, B. A. Dougherty, D. A. Morrison, S. K. Hollingshead, and C. M. Fraser. 2001. Complete genome sequence of a virulent isolate of *Streptococcus pneumoniae*. *Science* **293:**498–506.

61. Troffer-Charlier, N., J. Ogier, D. Mopras, and J. Cavarelli. 2002. Crystal structure of the V-region of *Streptococcus mutans* antigen I/II at 2.4 Å resolution suggests a sugar performed binding site. *J. Mol. Biol.* **318:**179–188.

62. Ton-That, H., L. A. Marraffini, and O. Schneewind. 2004. Sortases and pilin elements involved in pilus assembly of *Corynebacterium diphtheriae*. *Mol. Microbiol.* **53:**251–261.

63. Vickerman, M. M., M. Wang, and L. J. Baker. 2003. An amino acid change near the carboxyl terminus of the *Streptococcus gordonii* regulatory protein Rgg affects its abilities to bind DNA and influence expression of the glucosyltransferase gene *gtfG*. *Microbiology* **149:**399–406.

64. Whatmore, A. M., V. A. Barcus, and C. G. Dowson. 1999. Genetic diversity of the streptococcal competence (*com*) gene locus. *J. Bacteriol.* **181:**3144–3154.

65. Ween, O., S. Teigen, P. Gaustad, M. Kilian, and L. S. Håvarstein. 2002. Competence without a competence pheromone in a natural isolate of *Streptococcus infantis*. *J. Bacteriol.* **184:**3426–3432.

66. Wu, H., and P. M. Fives-Taylor. 1999. Identification of dipeptide repeats and a cell wall sorting signal in the fimbriae-associated adhesin, Fap1, of *Streptococcus parasanguis*. *Mol. Microbiol.* **34:**1070–1081.

67. Xu, D. Q., J. Thompson, and J. O. Cisar. 2003. Genetic loci for coaggregation receptor polysaccharide biosynthesis in *Streptococcus gordonii* 38. *J. Bacteriol.* **185:**5419–5430.

68. Zhang, Y., Y. Lei, A. Khammanivong, and M. C. Herzberg. 2004. Identification of a novel two-component system in *Streptococcus gordonii* V288 involved in biofilm formation. *Infect. Immun.* **72:**3489–3494.

Genetics of Lactococci

PHILIPPE GAUDU, YUJI YAMAMOTO, PETER RUHDAL JENSEN, KARIN HAMMER,
AND ALEXANDRA GRUSS

30

Lactococci have been used for centuries in dairy fermentations. These gram-positive, generally nonpathogenic, nonmotile, and nonsporulating bacteria are members of the *Streptococcaceae* family, which includes food, commensal, and virulent species (Fig. 1). *L. lactis* is a relatively simple bacterium, with a 2.4-Mb genome. Many of its functions of interest are nonredundant, which facilitates functional genetic studies. Lactococci are also presumed to be devoid of virulence factors (although isolated cases did report lactococcus as the infectious agent) (see reference 3 and references therein). The objective of this chapter is to present current information in different areas of lactococcal genetics, keeping in mind (where possible) pertinence of findings to related pathogens. Indeed, although streptococcal pathogens are mainly considered in light of factors needed for virulence, work on lactococci has gone deeper in characterizing basic metabolic properties, nutrient uptake, and survival, all of which may underlie virulence mechanisms. Numerous streptococcal genes potentially involved in metabolism were identified from global screenings (e.g., signature-tagged mutagenesis or transcriptional screening); however, in-depth studies questioning their roles in bacterial maintenance are rare. The overall nonvirulence of lactococci could be useful in determining how virulence factors may also participate in bacterial "everyday life," outside the animal host.

This chapter, organized in four sections, highlights major recent work in lactococci, including surprising metabolic capacities, physiology, stress response, and studies leading to novel successful uses of lactococci for protein delivery:

- Basic features of lactococci. A primary description of *Lactococcus* and its place among other *Streptococcaceae* are presented.
- Metabolic options for lactococci. As "industrial" bacteria, one focus in lactococci is on their metabolic processes during fermentation. *Lactococcus* was recently discovered to shift to a respiration metabolism when provided with an exogenous heme source. Other streptococci can also respire. These new metabolic studies show the flexibility of lactococcal metabolism, and provide a valuable prototype for the lifestyle of certain streptococcal pathogens.

- How lactococci can respond to stress. In lactococci, numerous and diverse factors have been identified as being involved in helping a population establish itself in a crowded or hostile environment. Genetic escape strategies have been identified by using simple selective systems.
- Applications and genetic tools of lactococci. Lactococci have been the object of intense study as delivery vectors for biomolecules. The use of protein expression systems in lactococci is gradually gaining an important position in protein delivery strategies, for the production of vaccines, and for drug delivery. Tools developed in lactococci are often applicable to other low %G+C gram-positive bacteria. In addition, as a simple nonpathogen, *L. lactis* is a useful host in which specific factors from pathogenic bacteria are introduced in order to analyze their roles in virulence. Several of these studies are reaching fruition and could lead to viable medical treatments.

BASIC FEATURES OF LACTOCOCCI

What Is *L. lactis*?

Lactic acid bacteria (LAB) are named for their ability to produce lactic acid via a fermentation metabolism. *L. lactis* is a mesophilic LAB with an optimal growth temperature of ~30°C (Table 1). It is the most extensively characterized LAB, which comprise a highly diverse group, including various cocci and bacilli. However, the term "LAB" is misleading; although LAB generally refers to bacteria used in food fermentation, lactic acid producers also include opportunists and pathogens, including streptococci and enterococci; in the case of *Enterococcus faecalis*, its status is unclear, as it is reportedly beneficial for fermentation, but clinical isolates are also a cause of infection (87). Among sequenced relatives, *L. lactis* appears to be most closely related to *Streptococcus mutans* (Fig. 1).

The 2.4-Mb genome of *L. lactis* IL1403 strain (12) is intermediate in size between streptococcal pathogens such as *Streptococcus pneumoniae* or *Streptococcus pyogenes* (reported

Gram-Positive Pathogens, 2nd edition, edited by Vincent A. Fischetti et al.
© 2006 ASM Press, Washington, D.C.

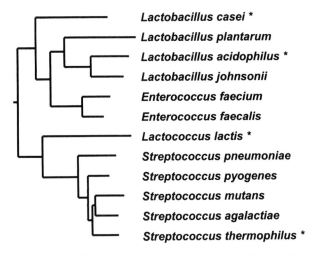

Lactobacillus casei *
Lactobacillus plantarum
Lactobacillus acidophilus *
Lactobacillus johnsonii
Enterococcus faecium
Enterococcus faecalis
Lactococcus lactis *
Streptococcus pneumoniae
Streptococcus pyogenes
Streptococcus mutans
Streptococcus agalactiae
Streptococcus thermophilus *

FIGURE 1 Phylogenetic tree reveals similarities between lactococci and streptococcal pathogens. Asterisks indicate species commonly found in dairy foods. The tree is based on alignments using the conserved *recA* gene (http://prodes.toulouse.inra.fr/multalin).

as 2.1 and 1.9 Mb, respectively) and the commensal bacterium *E. faecalis* (3.4 Mb).

Phylogenetic trees based on conserved genes, ribosomal 16S rRNA homologies, or genome alignments classify *L. lactis* as being within the family *Streptococcaceae*. Such comparisons indicate that *L. lactis* and the pathogenic streptococci may have a common origin. It is also clear that their relatedness surpasses by far that of lactococci with many other LAB (which also include many *Lactobacillus* species; Fig. 1). Streptococci also produce lactic acid, and so may be considered as a branch of LAB. The *Streptococcaceae* family diverged by unknown selective pressures to generate groups of pathogens, commensals, and food bacteria, which seem to be adapted to very different environments.

Varied Lifestyle of LAB

L. lactis and LAB in general seem to have a varied lifestyle. Lactococci are isolated from plants, and are likely to be ingested by grazing animals, together with milk, in the case of calves. Coingestion could explain how lactococci ended up in milk. The need for several plasmid- or transposon-encoded characteristics for growth in milk (e.g., enzymes for sugar and protein metabolism, DNA restriction and abortive phage infection functions, and bacteriocins) supports the hypothesis that milk is not the original habitat of LAB.

Genome Plasticity

Studies of lactococci have focused on two organisms, IL1403, a subsp. *lactis* strain, and MG1363, an *L. lactis* subsp. *cremoris* strain. Despite ~80% sequence identity between these genomes, they are not collinear, as they differ by a large chromosomal inversion (56). Furthermore, even closely related isolates of MG1363 show considerable polymorphism, corresponding to large rearrangements (55) that might be mediated by mobile elements. Studies of artificial chromosomal rearrangements suggest that lactococci can tolerate certain large genomic inversions if the origin and terminus regions are not disturbed (17).

Genome transfer and rearrangements may occur in lactococci via conjugation, transposons, insertion sequences, and phage, as suggested from identification of common elements in different species, or by experimental systems in which natural DNA transfer occurs (e.g., see references 5, 14, 71). Some *L. lactis* strains contain a conjugative factor that mediates genome transfer and can be manipulated for strain construction (38, 115). DNA transfer occurs in some bacteria by natural competence (19). *L. lactis* encodes homologs of all late competence genes, as characterized in *Bacillus subtilis*, suggesting that these organisms might have a natural competence state. Evidence for a competent state was recently provided for another LAB, *Leuconostoc mesenteroides* (41).

L. lactis IL1403 contains 40 copies of insertion elements, 14 of which correspond to an IS element similar to IS*1070* from *Leuconostoc lactis* (117), as well as at least four prophage elements (11). *L. lactis* and *Streptococcus thermophilus* share common integrative conjugative elements, as well as highly conserved regions coding, for example, exopolysaccharide synthesis (14). Existence of functional genes that have been transferred to *L. lactis* is suggested by the presence of atypical DNA regions that differ structurally from their context.

Close interactions between microorganisms, e.g., in the gastrointestinal or vaginal mucosa of animals, or in industrial milk fermentation processes, could lead to horizontal genetic exchange. Close physical contact was visualized microscopically between lactococci and little-related bacteria (S. Kulakauskas, personal communication). Contact between these bacteria is also suggested by the existence of nearly identical genes, e.g., in lactococci and in *S. thermophilus* (14).

METABOLIC OPTIONS FOR LACTOCOCCI

Lactococcal metabolism has been intensively studied for its industrial importance in fermentation processes, with a

TABLE 1 Characteristics of *L. lactis*

Classification	Gram positive, 38% G + C; genome, 2.4 Mb; nonpathogenic food microorganism (referred to as a lactic acid bacterium)
Closest neighbors	Streptococci (food, commensal, pathogen)
Optimal growth temp	30°C
Growth medium	Plants, milk, food, silage
Environmental contacts	Plants (environmental niche); grazing animals; milk and other foods[a]; gastrointestinal tract (transit)
Metabolism	Fermentation; can undergo aerobic respiration if heme is supplied in medium
Survival	Poor if grown by fermentation; good long-term survival after growth by respiration

[a]Requires plasmid-encoded factors.

focus on metabolic pathways and their engineering (see reference 22 for review). However, basic metabolic functions may have far-reaching effects, and as will be described below, metabolic shifts can result in dramatic changes in *L. lactis* growth characteristics and survival.

Five years ago, an essentially overlooked metabolic process in lactococci came to the limelight. Researchers confirmed and developed a 1970 study showing that lactococci not only ferment sugars, but are also capable of forming an active electron transport chain to generate respiration metabolism (10, 28, 36, 110). But there is a catch: for respiration to occur, an external heme source must be supplied. This provocative finding went even further. Respiration growth leads to a remarkably better bacterial survival, such that cells that undergo respiration can survive conservation for weeks, and even months longer than fermenting bacteria (28, 36, 97). Respiration metabolism and its consequences are described below. We speculate on the possible relevance of respiration metabolism for certain pathogenic streptococci.

The two energy metabolism options, fermentation and respiration, are presented. Pathways for nitrogen and nucleotide metabolism, which may be common to both means of energy metabolism, are also described.

Energy Option 1: Fermentation in *L. lactis*

From a simplistic viewpoint, lactococci seem to use sugars to provide energy and amino acids to synthesize proteins (Fig. 2). Dairy lactococci have multiple nutritional requirements for amino acids and vitamins, probably resulting from their adaptation to a life in milk. Lactose is the major sugar source in milk, and through its uptake and degradation, lactococci generate energy in glycolysis. Casein, the major protein component in milk, is degraded to provide the major carbon source for anabolism. The flow of carbon for energy production is therefore almost separable from the flow of carbon for anabolism in these bacteria, making them ideally suited for metabolic studies.

Sugars

All species belonging to the genus *Lactococcus* produce acid from glucose, fructose, mannose, and *N*-acetylglucosamine. The species *L. lactis* used for dairy fermentation is known for its mainly homolactic fermentation of lactose and other sugars. *L. lactis* subsp. *lactis* strains are more versatile than subsp. *cremoris* in their use of diverse sugar sources, including maltose, ribose, and trehalose (116). The sugar may be transported by a plasmid-encoded phosphotransferase system in dairy strains or, albeit at a slower rate, by a permease in nondairy strains (20).

Fermentation of carbohydrate may be shifted from homolactic (lactate production) to mixed acid fermentation (acetic acid, formic acid, CO_2, and ethanol produced in addition to lactate). Two very different sets of conditions may accompany this change, namely, (i) altered redox state created by increased aeration during growth or (ii) reduced flow of the sugar used for energy production. Further explanation of both sets of conditions follows.

Oxygen appears to be involved in maintaining the NADH/NAD ratio, which itself seems to regulate the switch between homolactic and mixed acid fermentation (20); aerobic conditions result in oxidation of NADH to NAD^+ (catalyzed by NADH oxidases), thereby reducing the NADH/NAD^+ ratio in the cell. Lactate dehydrogenase (LDH) is active at high NADH/NAD^+ ratios (i.e., low oxy-

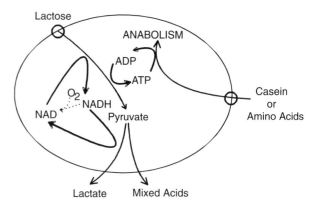

FIGURE 2 Carbon metabolism in *L. lactis*. Fermentation of sugar results in ATP production, which in turn is used for anabolism. During anaerobic conditions and rapid sugar flux, all sugar is converted to lactate (homolactic fermentation). When the sugar flux is slower, or the oxygen concentration high, mixed acid fermentation is observed. The latter two conditions are characterized by lower NADH/NAD^+ ratios than those found during homolactic fermentations. Note that the major part of the carbon for anabolism is derived from amino acids (or casein) supplied in the medium.

gen), while glyceraldehyde-3P dehydrogenase (GAPDH) is inhibited. Increasing the amounts of NADH oxidase from a nisin-inducible promoter was shown to decrease the NADH/NAD ratio and to decrease the in vivo activity of LDH. The increased pool size of pyruvate was in this case directed to acetoin and the flavor compound diacetyl (reviewed in reference 22). Other types of engineering (e.g., through mutations that block specific pathways [22]) can also result in altered flavor properties in lactococcal fermentation. Note that LDH is essential in *S. mutans*, but not in *L. lactis*, possibly because *S. mutans* lacks alternative pathways for reoxidation of NADH that are present in *L. lactis*. However, three LDH-like open reading frames are potentially present on the *L. lactis* IL1403 genome, and studies suggest that at least two genes can be activated (13).

Sugar flow may be decreased if galactose is used as an energy source or when lactose is transported by a permease (34). All sugar-carbon utilization produces pyruvate regardless of the growth conditions, and the pathway used for further conversion determines if fermentation is homolactic or mixed acid. Decreased sugar flow thus favors activity of enzymes giving rise to mixed fermentation products. One hypothesis is that when carbon fluxes are high, GAPDH is a bottleneck in glycolysis, resulting in high pools of intermediates upstream of GAPDH. These pools would then inhibit pyruvate formate-lyase in one of the fermentation pathways from pyruvate, resulting in homolactic fermentation under anaerobic conditions.

The extent to which GAPDH is a bottleneck in lactococci has been recently examined; strains were constructed with GAPDH activities ranging from 13 to 210% of the normal activity, and metabolic fluxes in these mutants were measured (45, 111). Surprisingly, GAPDH was found to be in large excess, even when carbon fluxes were high; in terms of flux control, GAPDH seems not to be the controlling factor (Fig. 3). Moreover, the fermentation pattern remains homolactic even after fourfold reduction in GAPDH activity, which shows that GAPDH has no control on mixed acid flux either. Similar studies were carried out to determine the importance of other glycolytic enzymes on the fermentation pattern. Phosphofructokinase had no control on glycolytic flux, nor on mixed acid flux (50), despite that this enzyme is present in very limited excess (2). LDH had no control on glycolytic flux either, but did exert a strong negative control on mixed acid flux (1). An intriguing possibility is that the excess glycolytic enzymes are somehow shuttled to the surface, where they could play other roles in bacterial dissemination; glycolytic enzymes have been reported to be present on the surface of numerous *Streptococcaceae* (83).

The genetic organization of enzymes involved in sugar utilization may reveal regulation at the transcriptional level. *L. lactis* appears to coordinate the expression of three genes, each involved in key but distinct steps in fermentation: *pfk*, *pyk*, and *ldh* (encoding phosphofructokinase, pyruvate kinase, and LDH, respectively), by having them present in one operon (called *las*). The existence of this operon may prevent unwanted accumulation of glycolytic intermediates. In *S. pyogenes* and *S. pneumoniae*, genomic analyses show that the *ldh* gene is not within this operon, suggesting that a common regulation of the three genes may be unique to lactococci. The *las* operon is induced in the presence of glucose via the catabolite control protein CcpA (64). Expression of the entire *las* operon has also been modulated around the normal level to determine its importance for metabolic flux. The flux was highest when expression was at its normal level, and decreased rapidly when the expression was reduced or increased (50).

In *Escherichia coli*, glycolytic metabolic flux was recently shown to be controlled almost exclusively by the demand for ATP (52, 82). In *L. lactis*, the ATP demand also controls flux in slowly or nongrowing cells but not in quickly growing cells (51, 85).

The importance of individual enzymes for system properties such as fluxes, metabolite concentrations, frequencies of infection, survival rates, etc., can be accessed by modulating the activities of the respective components. A new method using synthetic promoter libraries allows for accurate tuning of gene expression in most biological systems (45, 111).

Energy Option 2: Respiration in *L. lactis*

Components of the Electron Transport Chain

An active respiration chain comprises three elements: (i) an electron donor, supplied by NADH dehydrogenases; (ii) quinones (a nonprotein component of the respiration chain), which deliver electrons to the terminal oxidoreductases; and (iii) terminal oxidoreductases, which contain heme as an essential cofactor and use oxygen as the final electron acceptor (89, 98). Although *L. lactis* reportedly undergoes fermentation in a rich medium, addition of a heme source to aerated medium was found to activate respiration metabolism in *L. lactis* (28, 110). Existence of an electron transport chain (Fig. 4) was supported by genetic studies (28, 36) and demonstration of heme-dependent membrane NADH oxidase activity (10). Under these conditions, the cell biomass doubled (in keeping with greater energy production by respiration activity), the pH was increased, and cells produced very large amounts of acetoin rather than lactic acid (28).

Like other gram-positive bacteria that grow via a respiratory metabolism, *L. lactis* utilizes only menaquinones as electron carrier to the terminal oxidoreductase. Inactivation of *menB*, which encodes a menaquinone biosynthesis enzyme (dihydroxynaphthonic acid synthase), totally abolished both quinone production and respiration in *L. lactis* (L. Rezaiki, G. Lamberet, A. Gruss, and P. Gaudu, in preparation). However, although other respiring bacteria have several oxidoreductases that ensure respiration under different conditions, *L. lactis* uses a single enzyme, the cytochrome bd quinol oxidase (encoded by *cydAB*) (28, 36). Interestingly, this kind of oxidase has a high affinity for oxygen (25). Moreover, cytochrome bd quinol oxidase may contribute to bacterial virulence; in gram-negative pathogens, *cydAB* inactivation attenuates pathogenicity (30, 118).

Respiration in *L. lactis* requires uptake of heme (iron is insufficient), suggesting the presence of a heme transporter. Genetic studies indicate that the *fhu* operon mediates heme uptake in *L. lactis* (35).

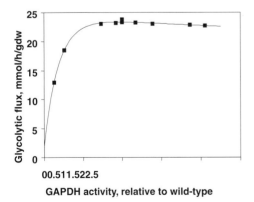

FIGURE 3 Increases in GAPDH activity do not increase glycolytic flux in steadily growing *L. lactis* MG1363. Levels of GAPDH activity were modulated by driving expression using different strength promoters. At the wild-type enzyme level (set to 1), GAPDH has zero control on the flux. gdw, grams dry weight. (Reproduced with permission from reference 111a.)

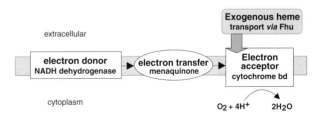

FIGURE 4 Activation of respiration metabolism in *L. lactis*. Lactococci were found capable of respiration metabolism. An active respiration chain requires three components: an electron donor (NADH dehydrogenase, potentially encoded by *noxA* and *noxB* genes; it expulses H⁺ and transfers e⁻); quinone electron transfer molecules (synthesized by enzymes encoded by the *men* genes; it further transfers e⁻); and an oxidoreductase (*cydAB*-encoded cytochrome bd quinol oxidase; it transfers e⁻ to its final acceptor, oxygen, which then reacts with H⁺ to produce water). An ATP synthase presumably recovers expulsed H⁺ to produce ATP during its entry.

Respiration Is Regulated

Respiration appears to be dominant late in growth, when examined in rich, glucose-containing medium. During initial growth, fermentation metabolism using glucose is favored. After glucose is consumed, respiration appears to predominate, and *cydAB* expression is increased (28). Lactate, produced by glucose catabolism, is then consumed. Recent experiments show that the catabolite control protein CcpA acts on (i) heme uptake regulation and (ii) the carbon source used during full respiration (35). As long as glucose is present at high concentrations, CcpA represses heme uptake by the *fhu* operon, probably by maintaining high levels of the FhuR repressor. Furthermore, inactivation of *fhuR*, the putative repressor, results in precocious heme uptake (35). When glucose levels decline, heme uptake is increased. As heme is required for cytochrome oxidase function, its uptake is suggested to be a limiting factor for activating respiration (35).

What Is the Natural Heme Source for Lactococci?

Heme, an iron porphyrin molecule, is needed for lactococci to respire. Although heme is found in blood hemoglobin, this seems an unlikely heme source in view of the natural environment of lactococci, which is thought to be in plants or in milk. Numerous explanations are possible. For example, as with all microorganisms, *Lactococcus* is likely to be in contact with heme-producing bacteria, which could supply their heme to surrounding bacteria. Plants produce heme-like molecules, including chlorophyll and, in the case of leguminous plants, leghemoglobin (from root nodules), both of which may be sources of the heme porphyrin needed for respiration. Dead animals and insects in the ground may be a source of heme. Intriguingly, respiration metabolism results in the generation of large amounts of acetoin (28), which reportedly stimulates plant growth (100), possibly suggesting the existence of synergistic interactions between these organisms, based on bacterial respiration.

Respiration Metabolism Increases Survival of Lactococci

L. lactis survival is greatly extended after respiration, as compared to fermentation growth (Fig. 5). Under storage conditions at 4°C, near 100% survival can be maintained for several months, whereas aerated cells without heme die off quickly (28, 97). Two effects of respiration contribute to improved survival; one is that respiring cells produce less acid than fermenting cells. As such, the self-destructive effects of acidification are alleviated. Second, respiration-grown cells efficiently eliminate oxygen from the medium. This is attributed to the high affinity for oxygen of the cytochrome quinol oxidase, as reported for this enzyme in *E. coli* (25). Both DNA damage and spontaneous mutation frequencies are decreased in stationary phase, compared to aerobically grown fermenting cells (97). Interestingly, respiration-grown cells exert an effect in *trans*. Cells incapable of respiration show better long-term survival when grown in the presence of respiration-competent cells. It thus appears that respiration metabolism is an important feature in lactococcal survival in oxidative conditions.

L. lactis Is a Respiration Prototype

The respiration phenotype of lactococci is not like that of known aerobic bacteria, e.g., *E. coli* or *B. subtilis*. It differs in that it respires using a restricted electron transport chain, and only if heme is provided. Comparative studies reveal that certain but not all *Streptococcaceae* have the capacity to respire when provided with the right cofactors (36). They include *Streptococcus agalactiae* (120), *E. faecalis* (43, 119), and potentially *Streptococcus uberis* (as predicted from genome).

Nitrogen Metabolism

In a milk medium, lactococci derive amino acids from casein via hydrolysis by the extracellular protease PrtP, transport of some of the generated peptides, and further degradation by a multitude of intracellular peptidases (see reference 53 for review). The amino acids that are readily available in a milk medium are used both directly as amino acid building blocks, and also as a general carbon supply for other forms of anabolism in lactococci. Extracellular proteases, like other functions needed for optimal growth in milk, are plasmid-encoded.

Peptidases are of key importance for amino acid utilization. There are at least 14 peptidases having different specificities. These intracellular enzymes play important roles in bacterial survival. Although no particular phenotype is observed when these mutants are examined in laboratory media, lactococci grow poorly or die in milk fermentation conditions.

Dairy lactococci differ from plant lactococci in that they require several amino acids for growth. Surprisingly, however, strains of both origins appear to have the necessary genes for biosynthesis. However, dairy lactococci require Ile, Leu, Val, and His, and sometimes Arg, Met, Pro, and/or Glu (116). These amino acid requirements in dairy strains resulted from multiple mutations in the structural genes, but not in deletions of these genes (11). This may suggest that mutations accumulated as an economic measure in strains maintained in a dairy environment. Similar results are reported for *S. thermophilus*, a dairy bacterium related to *S. pyogenes*.

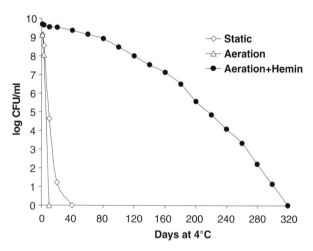

FIGURE 5 Respiration metabolism increases survival capacity of lactococci. When supplemented with hemin, aerobically grown lactococci can undergo respiration metabolism. As a result, cells stored at 4°C show a markedly better survival, as compared to cells grown aerobically in the absence of hemin or in static conditions. Improved survival was also observed when cells are maintained at 30°C. Experiment shown was performed by Karin Vido (URLGA, INRA, France).

Nucleotide Metabolism

Mutants in nucleotide metabolism may display many different phenotypes, because nucleotides are not only substrates for DNA and RNA polymerases, but are also substrates or allosteric effectors for many enzymes, and furthermore constitute parts of different coenzymes. Modulation of nucleotide pools in the cell is also influenced by the presence of exogenous nucleobases or nucleosides in the medium. Therefore, knowledge of the pathways of uptake and utilization of these compounds (the so-called salvage pathways, which vary in different organisms) is important to interpret or predict the bacterial response to changes in the medium and to increased intracellular degradation of nucleic acids.

As seen above for operons involved in carbon metabolism, gene organization has unique characteristics in lactococci. For example, the genes encoding the pyrimidine biosynthesis pathway leading to the formation of UMP are organized in five different operons in L. lactis, and in a single operon in all other investigated gram-positive bacteria. Similarly, purine biosynthesis genes involved in the first ten steps leading to IMP production are located in five separate operons in L. lactis, and in a single operon in B. subtilis.

Pyrimidine biosynthesis is regulated by attenuation and antitermination by PyrR. The PyrR protein binds to PyrR "boxes" present on the 5' ends of the untranslated mRNAs on four of the identified pyrimidine biosynthetic operons (12, 69). In the presence of high intracellular concentrations of UMP or UTP, PyrR binding blocks the formation of an antiterminator structure, resulting in transcriptional termination, similar to what was observed in B. subtilis (63). Mutation of the pyrR gene results in increased levels of the pyrimidine biosynthetic enzymes (69).

Purine biosynthesis in L. lactis is positively controlled by PurR, and purR mutants are purine auxotrophs. PurR binds to a Pur box consensus sequence (48) to activate transcription of the purC and purD operons (interestingly, the homologous PurR in B. subtilis works as a repressor). Data suggest that the PurR activating effector is 5-phosphoribosyl-1-pyrophosphate. Levels of this compound are reduced when purine is added, as they are used to synthesize purine mononucleotides; thus, purine addition may indirectly result in decreased gene expression (49). Proteomic studies show that all identified proteins induced by purine starvation belong to the purR regulon (8). Furthermore, in operons activated by PurR, the distance between the Pur box and the -10 region was 57 to 58 bp. Using this information 11 putative PurR-regulated promoters were found on the L. lactis IL1403 genome. Except for a membrane protein with unknown function, all PurR-regulated genes were found to be involved in purine biosynthesis or uptake.

The ability to utilize exogenous nucleobases or nucleosides present in the medium or formed from intracellular degradation of nucleic acids via the salvage pathways can reflect how well cells can survive under stress conditions. This capacity seems to differ between microorganisms. The nucleobases uracil, guanine, and adenine are taken up and converted to nucleotide monophosphates, while cytosine is not utilized by lactococci (68). Furthermore, all nucleosides except for cytidine can be degraded to the corresponding nucleobase. The pyrimidine nucleosides may also be directly converted to nucleotides, as the corresponding nucleoside kinases (udk and tdk) were shown to be functional in lactococci. Two high-affinity nucleoside transport systems have been identified by uptake assays and mutual inhibition of growth of pyrimidine- or purine-requiring mutants. One

is uridine-specific, while the other takes up cytidine and purine nucleosides, as well as the corresponding deoxyribonucleosides (70).

Low-nucleotide pool sizes may serve as internal stress signals that provoke expression of stress response genes in L. lactis (26, 94). One intriguing possibility comes from the recent identification of metabolites (including nucleosides) as cofactors in riboswitch regulation (79).

HOW LACTOCOCCI CAN RESPOND TO STRESS

As industrial bacteria, lactococci are submitted to harsh conditions during growth and conditioning for different processes. For example, bacteria may be exposed to osmotic and oxidative stress (in starter preparation), low pH (by lactic acid production during growth), and temperature variations (between 8 and 40°C in industrial processes). Of particular relevance are oxidation, acidification, and temperature shifts. Stress conditions contribute to DNA, protein, and/or lipid damage and can also produce other cellular changes, all of which can eventually lead to cell death.

Conserved Stress-Response Functions

Stress-response pathways are overlapping in bacteria, and thus a single stress condition may induce an active response against multiple stress conditions. Exposure of L. lactis to pH 4 for 3 h can result in $\sim 10^4$-fold drop in viability; preadaptation at an intermediate pH (pH 5) can result in complete survival, as well as improved resistance to other stress conditions (40, 95). Similarly, nonlethal heat shock results in a 2- to 100-fold induction of specific mRNA transcripts corresponding to heat shock genes (4). Among the conserved stress-response genes identified in lactococci are those encoding chaperones (DnaK, DnaJ, GrpE, and GroESL) (4) or proteases (HtrA, HflB, and ClpP) (33, 77, 91). Several two-component (sensor- regulator) systems, likely needed to react to specific environmental signals, were partially characterized in L. lactis (81). Expression of superoxide dismutase, needed for removal of toxic radicals, is acid inducible (102).

Strategies for Survival in Stress Conditions

Novel means of surviving in the presence of salt, toxins, and oxygen have been characterized in lactococci, as exemplified here.

Salt and acid induce expression of GadB and C, which are putatively involved in glutamate transport by an antiporter; glutamate transport presumably involves efflux of H^+, thereby maintaining intracellular pH (101).

Toxic products such as bile, quaternary compounds, and antibiotics may be actively pumped out of the cell by specialized transport functions, which may allow bacteria to tolerate their environment. Among the numerous transport systems that shuttle metabolites in and out of the cell, some mediate drug expulsion, and consequently, can confer drug tolerance (see reference 88 for review). In L. lactis, one multidrug pump having specificity for a wide range of amphiphilic, cationic drugs (including antibiotics, quaternary ammonium compounds, aromatic dyes, and phosphonium ions) is LmrA. The lmrA gene encodes an efflux pump that is responsible for the export of toxic molecules such as ethidium in exchange for H^+ influx. Surprisingly, LmrA (590 amino acids) is very similar to the human multidrug resistance p-glycoprotein (LmrA is 32% identical to half of the p-glycoprotein, particularly within known functional domains). Remarkably, LmrA is functional in eukaryotic cells, and is able to replace p-glycoprotein defects, thus making *L. lactis* an excellent model to study drug extrusion (88). Note that sequence comparisons predict an LmrA homolog in *S. pneumoniae* (an open reading frame with ~30% identity over 539 amino acids is present).

Production of lactic acid may be a means of dominating bacterial competitors or of inactivating phage. In addition, lactococci, like other *Streptococcaceae*, produce toxic hydrogen peroxide under aerobic fermentation conditions (97). Another type of bacterial "warfare" may be realized by bacteriocins, which have widely different host spectra (see reference 24 for review); lysis of heterologous bacteria by bacteriocins may provide needed nutrients for the producer cells.

Finally, basic cell metabolism can determine how well the bacterium copes with oxidative conditions. Respiration metabolism presents a clear advantage to lactococci in an aerobic environment (97).

Selections for Improved Adaptation to Environmental Stress Situations

Another ubiquitous stress is oxygen. If not eliminated from the cell, reactive oxygen derivatives provoke cell damage that can be lethal. One means of reducing oxygen-related damage is by removing oxygen. In fermenting cells, H_2O-forming NADH oxidases do eliminate oxygen during growth (62), although cells are sensitive to oxygen-related damage in stationary phase (27). Overproduction of H_2O-forming NADH oxidase could not only change metabolic end products (62), but also might improve survival in an oxidizing environment, due to oxygen removal. An alternative means of creating a more reducing environment is by adding glutathione, a redox peptide (59), or dithiothreitol, a reducing agent (35), to the medium. Lactococci lack catalase, which eliminates hydrogen peroxide in many aerobic bacteria. Hydrogen peroxide has been effectively removed by cloning catalase in *L. lactis* (99b). Results from our laboratory demonstrate that respiration metabolism in lactococci is an efficient means of eliminating oxygen, compared to fermentation, leading to good survival in stationary phase (28, 97).

Although acidification is generated by the fermentation growth process itself, and as such is self-inflicted, cell survival is handicapped by acid accumulation (95). Significantly, acids may accumulate if cells are grown under immobilized conditions (e.g., biofilms). In this case, cell growth and/or survival may be limited. This situation may provide a natural selection for strains to escape from a constrained environment. Immobilization studies using a semiliquid medium have been used to impose a natural selection for mutants that can more readily escape a constraining environment (73). In one case, bacteria that make chains sediment more slowly than single cells (Fig. 6). "Dechaining" mutants, affected in penicillin-binding protein and cell wall-synthesis enzyme PBP1A, were isolated; the mutants no longer formed chains, and were able to sediment more readily in the semiliquid medium (73). In *S. agalactiae*, interruption of *ponA* results in reduced virulence (46). In view of the identified role of PBP1A in lactococci, it is tempting to speculate that the *ponA S. agalactiae* mutant may be defective in its chain-forming ability, and consequently, its in vivo localization. Use of semiliquid medium for selections might prove effective in examining factors that are at work when bacterial pathogens are immobilized in their host (61a).

Transposon insertional mutagenesis was also used to select for stress-resistant strains of lactococci (26, 94, 95). Three examples are given. (i) A combination of stress conditions is lethal for lactococci (and possibly for other organisms), although each condition alone may be nonlethal (94, 95). Simultaneous high temperature (37°C), oxygen, and either low pH (i.e., similar to conditions encountered when lactococci are ingested) or a *recA* background give rise to mutants of which many seem to affect intracellular metabolic pools of guanosine-phosphate and phosphate in stress response (26, 94). Low intracellular levels of these metabolite pools in the mutants may constitute a starvation signal to induce a stress response. These mutant strains show better long-term survival than their nonmutated parents. It thus appears that a general stress response is induced in *L. lactis* when intracellular guanosine-phosphate and phosphate pools are low. (ii) Hydrogen peroxide is toxic to lactococci. An H_2O_2-resistant mutant was isolated at high temperature (37°C), and although its resistance was 1,000-fold greater than that of the parental strain, it displayed no other stress resistance phenotypes (99a). One point of interest of this type of strain is its capacity to live in coculture with strains producing millimolar amounts of H_2O_2-producing strains, e.g., some lactobacilli (80) and streptococci (106). Use of H_2O_2-resistant lactococci could lead to the development of new fermented products. Furthermore, more efficient growth of lactococci could improve the hygiene of food products. (iii) Mutants were selected for increased resistance to dithiothreitol, a reducing agent, at elevated temperature (Cesselin, Duwat, Djae, Gruss, and El Karoui, in preparation). By preventing the formation of disulfide bonds, dithiothreitol disables a part of the oxidative stress response pathway (numerous lactococcal stress-response proteins contain one or more CXXC motifs). Eighteen dithiothreitol-resistant mutants all mapped within a single operon, *pst*, involved in phosphate transport. Results indicate that this operon may control a pathway of oxidative stress response genes, including superoxide dismutase.

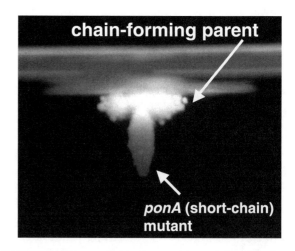

FIGURE 6 Bacterial root formation in semiliquid medium. Bacterial chains (here, an *acmA* mutant of *L. lactis*; "parental strain") sediment slowly in a semiliquid (0.035% agar) medium. A bacterial "dechained" mutant sediments more quickly to form a "root." In such experiments, all the roots corresponded to independent mutants in the same gene, *ponA*, encoding PBP1A. (Photograph kindly provided by S. Kulakauskas, URLGA, INRA, France.)

Stress-resistant lactococci are potentially valuable in dairy fermentation. Their greater resistance to stress may overcome survival variability as seen in conventional strains. The spe-cifically acid-resistant strains may provide resistance to extreme acid pH conditions, or may be better at maintaining a neutral internal pH. Stress-resistant strains may survive longer in fermentation, and may also be more resistant to harsh storage conditions (like freezing and lyophilization). Second, as such strains may survive better in the harsh environments encountered in the gut, they may be attractive for probiotic uses. Third, LAB are potentially valuable candidates for production of molecules with medical or biotechnological uses; lactococci are nontoxic and can be used to express and export proteins or other molecules of interest, either for industrial production or in the gut. The use of LAB with improved survival properties can have specific advantages over conventional strains.

APPLICATIONS AND TOOLS OF LACTOCOCCI

Some of the most spectacular applications of lactococci concern their use in "bioprotein" delivery, i.e., for antigen or enzyme delivery in vivo. These applications are based on the idea that L. lactis can act as a nonpathogenic carrier, which can be administered orally or nasally without provoking a reaction other than that induced by the bioprotein of interest. Several reports give encouraging results in the areas of (i) prophylaxis to prevent bacterial infection (37, 44, 67, 99), (ii) treatment of inflammatory bowel disease (113), (iii) prophylaxis and/or treatment of virally induced tumors (7), and (iv) allergy prevention (18, 96).

Another use of lactococci is in the effort to identify the roles of virulence factors. The complexity of the virulence process has led to an approach in which genes encoding individual factors are expressed separately in a nonpathogenic host (here, L. lactis) to examine their properties and roles in virulence (92, 93).

The recent numerous applications using lactococci result from an intense development of genetic tools, which in some cases perform as well as what is available in E. coli. Some tools developed in L. lactis are adaptable to other gram-positive bacteria.

Transposition and Gene Replacement

The development of transposition (using insertion element IS*S1*) and gene replacement systems based on a thermosensitive replicon (reference 66 and references therein) has opened up the way for selections and screenings of new phenotypes. The plasmid vector, pG$^+$host, a derivative of broad host-range plasmid pWV01 (see reference 47 for review), is thermosensitive in L. lactis (off at 37.5°C), in streptococci and staphylococci, and in some lactobacilli. Modifications in this system allow for random mutagenesis systems based on homologous integration via cloning of chromosomal DNA fragments (58). Furthermore, the existence of an E. coli strain bearing a chromosomal copy of rep (58), the plasmid-encoded replication protein, facilitates gene cloning, as well as characterization of pG$^+$host insertions.

Conjugative Chromosomal Transfer

A system analogous to Hfr transfer in E. coli has been discovered and developed in L. lactis (38). It is based upon the natural occurrence of an integrated sex factor in the model strain MG1363; this factor can transfer chromosomal DNA

at frequencies inversely correlated to the distance from the transfer site, in an orientation-directed manner. Conjugal crosses greatly facilitate strain construction in lactococci, and allow for the construction of strains combining multiple mutations. An inducible conjugation system was recently developed (115).

Site-Specific Single-Copy Integration

Lactococcal bacteriophages were initially studied with the goal of controlling starter culture lysis during fermentations. These phage studies led to the characterization of numerous bacterial strategies to abort phage activity. Phage studies have also been exploited to develop a site-specific integration system. Integration of temperate phages makes use of a phage-specified integrase that catalyzes insertion of the phage at a specific bacterial target, which is often localized at or near a tRNA gene. Using elements of lactococcal bacteriophage TP901-1, a site-specific integrative vector was designed to obtain chromosomal single-copy integration (15). This system should allow stable insertion and expression of foreign genes, and can also be used to study expression of genes in single copy under different growth conditions. A lactococcal intron was also targeted as a site for efficient and stable insertion of genes without the need for selection (32).

Regulated and Optimized Promoters

Studies in lactococci have greatly benefited from in-depth dissection and application of the highly regulated nisin biosynthesis pathway (nisin is a bacteriocin encoded by a conjugative transposon [42]) (108, 109). The promoter for nisin biosynthesis gene, nisA, is regulated by nisR and nisK gene products. In the absence of nisin or of nisR and nisK genes, promoter activity is very low or off. Addition of sublethal amounts of nisin results in strong induction of promoter activity, as demonstrated in innumerable applications of this system. This system has been shown to be functional in other gram-positive bacteria (29).

Recent studies of metabolic and stress-response pathways have led to the development of novel promoter systems, which may be advantageous in some cases over the highly useful nisin system. Novel controlled-expression systems have arisen from studies of pH, salt, metal, or sugar-regulated promoters (61, 65, 74, 103, 104). These other systems may overcome two potentially important limitations of the nisin system: (i) the requirement for either a specific host strain or a second plasmid (to provide nisR and nisK genes), and (ii) effects of nisin on the membrane, which may in particular interfere with studies of membrane proteins.

Constitutive expression of promoters at fixed levels can be valuable for quantitative physiological studies or for fine-tuning of gene expression in biotechnology. A set of synthetic promoters that differ by the sequence and length of spacers between the consensus sequences was designed and allows a broad range of constitutive activities (45, 111).

Protein Export Reporter

Protein export reporters can be precious tools in studies of membrane protein structure (54) for identifying export signals (90). One such reporter, the nuclease of Staphylococcus aureus, is a stable, well-characterized protein that is active when present as an amino- or carboxy-terminal fusion to other peptides, and is faithful in reporting export events and in determining membrane protein topology (90). It has already been used in L. lactis to follow expression of exported proteins under different environmental conditions. The major advantages of using the nuclease over previously described export reporters are that it rapidly assumes its conformation and so avoids degradation, and as few as ~300 nuclease molecules per cell can be detected in colony assays (90).

Protein Secretion and Anchoring Signals

Export signals and vectors have been developed to deliver proteins of interest in the extracellular medium or on the bacterial surface. The export signal of Usp45, a secreted *L. lactis* protein (116a), or signals from identified secreted native or heterologous proteins are routinely used. Secretion efficiency can be improved by introducing (if necessary) an overall negative charge at the N-terminal end of the mature, translocated secreted protein (57).

Anchoring of exported proteins via C-terminal LPXTG motifs is widely documented, particularly with respect to virulence factors, such as *S. pyogenes* M-protein (31, 75, 105). *L. lactis* encodes several anchored proteins and at least two sortaselike proteins (24a). Expression of the anchoring motif of the *S. pyogenes* M6 protein has been adopted to express recombinant proteins on the lactococcal surface (86). The AcmA autolysin is also cell wall-associated, due to the presence of a three-times repeated LysM motif (112). Anchoring via the AcmA binding motif was found to effectively present antigens at the lactococcal surface; a spacer between the anchoring domain and the protein of interest (a fimbrial protein) facilitates its access to target epithelial cells (60).

Expression Strain

Lactococci are remarkable for their simple genomes, compared to complex bacterial models, such as *B. subtilis* or *E. coli*. It is common for *L. lactis* to encode a single gene to ensure a function, while more complex bacteria encode several genes. An interesting example is HtrA, which is the only surface protease in *L. lactis*. In contrast, *B. subtilis* and *E. coli* both encode numerous exported proteases (91). An *htrA* mutation results in a temperature-sensitive strain that allows stable production of exported proteins at 30°C (91).

Cell Lysis Systems

Controlled cell lysis is a potentially powerful means of arresting cellular and metabolic activity; in fermentation, it may additionally result in a coordinated release of enzymes, which could accelerate product maturation. The host autolysin, AcmA (16), or bacteriophage-encoded lysins and holins (which allow lysin release) are good candidates for this purpose. This application is potentially useful in controlling cell growth in fermented dairy products, as well as for enzyme release (103). Expression of lysin and holin by a nisin-induced promoter does indeed appear to accelerate cheese ripening (21). Note that this same approach has obvious applications to growth control of bacterial pathogens; an *S. pyogenes*-specific lysin was shown to effectively prevent or treat infection in mice (76).

The cell envelope is an important barrier protecting the cell from stress situations. Cell wall damage via autolytic enzymes can render cells more sensitive to environmental conditions. Bacteria that have undergone even partial cell wall damage were found to be permeable to small labeled probes used in standard in situ hybridization methods, while undamaged cells are not (9). The method, developed for studies in lactococci to estimate the cell wall state, can be used for different gram-positive bacteria.

Containment and Food-Grade Strains

Inactivation of the *L. lactis* thymidylate synthase *thyA* gene results in a requirement for thymine or thymidine (84, 114). In *L. lactis*, a *thyA* mutant was exploited as a means of strain containment, i.e., strains can grow in a thymine-containing environment, such as the gut, but not in more limiting environments (114).

Nonsense suppressor strains are classically used in *E. coli* genetics to analyze the phenotypes of point mutations. A nonsense suppressor strain of *L. lactis* was also isolated. Plasmids carrying the suppressor genes could suppress an otherwise lethal nonsense mutation in the cell. This property of suppressors was exploited to construct and establish a food-grade plasmid (i.e., no foreign DNA) containing the suppressor genes in a suppressible purine auxotroph; this plasmid is stable in a milk medium, which cannot sustain growth of a purine auxotroph (23).

Higher Antigenicity

Studies have compared effectiveness of presentation of different antigens as bacterial cytoplasmic, surface-anchored, or secreted proteins. Cell wall-associated antigens seem to induce greater immune response than secreted or cytoplasmically expressed proteins (6, 78). Furthermore, immune response was reportedly enhanced in mutant strains with cell wall defects due to mutation of the alanine racemase (39).

CONCLUSIONS

L. lactis is perhaps the microorganism most eaten by humans. It is a member of a family composed of pathogens (e.g., *S. pneumoniae*, *S. pyogenes*), commensal microorganisms (e.g., *Streptococcus gordonii*, *S. mutans*), and food microorganisms (e.g., *S. thermophilus* and *L. lactis*). Studies of *L. lactis* may allow us to understand by how little pathogens and nonpathogens differ. As a bacterium that acidifies its own medium, *L. lactis* may have a high capacity for stress resistance when preadapted; stress-resistant mutants with constitutive stress resistance can be selected. As a food microorganism, *L. lactis* may come into close contact with other bacteria in both the food environment and in the gut. As a nontoxic bacterium that secretes relatively few proteins in quantity, *L. lactis* may also be an organism of choice for oral vaccine or protein delivery design (see references 72, 107 for review) and for biotechnological uses. The development of surface display systems in LAB will be potentially useful in the development of oral vaccines based on the nontoxic LAB. As an organism present on plants, in milk, in dairy products, and in the gut, *L. lactis* may be the organism of choice for studies on the influence of environmental stress on evolution.

We are grateful to our laboratory colleagues K. Vido, S. Kulakauskas, L. Rezaiki, L. Bermudez, I. Poquet, P. Langella, J.C. Piard, and M. El Karoui for their insights on lactococci, which were incorporated into this chapter.

REFERENCES

1. **Andersen, H. W., M. B. Pedersen, K. Hammer, and P. R. Jensen.** 2001. Lactate dehydrogenase has no control on lactate production but has a strong negative control on formate production in *Lactococcus lactis*. *Eur. J. Biochem.* **268:**6379–6389.

2. **Andersen, H. W., C. Solem, K. Hammer, and P. R. Jensen.** 2001. Twofold reduction of phosphofructokinase activity in *Lactococcus lactis* results in strong decreases in

growth rate and in glycolytic flux. *J. Bacteriol.* **183:** 3458–3467.

3. **Antolin, J., R. Ciguenza, I. Saluena, E. Vazquez, J. Hernandez, and D. Espinos.** 2004. Liver abscess caused by *Lactococcus lactis cremoris*: a new pathogen. *Scand. J. Infect. Dis.* **36:**490–491.

4. **Arnau, J., K. I. Sorensen, K. F. Appel, F. K. Vogensen, and K. Hammer.** 1996. Analysis of heat shock gene expression in *Lactococcus lactis* MG1363. *Microbiology* **142:** 1685–1691.

5. **Belhocine, K., I. Plante, and B. Cousineau.** 2004. Conjugation mediates transfer of the Ll.LtrB group II intron between different bacterial species. *Mol. Microbiol.* **51:** 1459–1469.

6. **Bermudez-Humaran, L. G., N. G. Cortes-Perez, Y. Le Loir, J. M. Alcocer-Gonzalez, R. S. Tamez-Guerra, R. M. de Oca-Luna, and P. Langella.** 2004. An inducible surface presentation system improves cellular immunity against human papillomavirus type 16 E7 antigen in mice after nasal administration with recombinant lactococci. *J. Med. Microbiol.* **53:**427–433.

7. **Bermudez-Humaran, L. G., P. Langella, N. G. Cortes-Perez, A. Gruss, R. S. Tamez-Guerra, S. C. Oliveira, O. Saucedo-Cardenas, R. Montes de Oca-Luna, and Y. Le Loir.** 2003. Intranasal immunization with recombinant *Lactococcus lactis* secreting murine interleukin-12 enhances antigen-specific Th1 cytokine production. *Infect. Immun.* **71:**1887–1896.

8. **Beyer, N. H., P. Roepstorff, K. Hammer, and M. Kilstrup.** 2003. Proteome analysis of the purine stimulon from *Lactococcus lactis. Proteomics* **3:**786–797.

9. **Bidnenko, E., C. Mercier, J. Tremblay, P. Tailliez, and S. Kulakauskas.** 1998. Estimation of the state of the bacterial cell wall by fluorescent in situ hybridization. *Appl. Environ. Microbiol.* **64:**3059–3062.

10. **Blank, L. M., B. J. Koebmann, O. Michelsen, L. K. Nielsen, and P. R. Jensen.** 2001. Hemin reconstitutes proton extrusion in an H(+)-ATPase-negative mutant of *Lactococcus lactis. J. Bacteriol.* **183:**6707–6709.

11. **Bolotin, A., S. Mauger, K. Malarme, S. D. Ehrlich, and A. Sorokin.** 1999. Low-redundancy sequencing of the entire *Lactococcus lactis* IL1403 genome. *Antonie Leeuwenhoek* **76:**27–76.

12. **Bolotin, A., P. Wincker, S. Mauger, O. Jaillon, K. Malarme, J. Weissenbach, S. D. Ehrlich, and A. Sorokin.** 2001. The complete genome sequence of the lactic acid bacterium *Lactococcus lactis* ssp. *lactis* IL1403. *Genome Res.* **11:**731–753.

13. **Bongers, R. S., M. H. Hoefnagel, M. J. Starrenburg, M. A. Siemerink, J. G. Arends, J. Hugenholtz, and M. Kleerebezem.** 2003. IS981-mediated adaptive evolution recovers lactate production by ldhB transcription activation in a lactate dehydrogenase-deficient strain of *Lactococcus lactis. J. Bacteriol.* **185:**4499–4507.

14. **Bourgoin, F., A. Pluvinet, B. Gintz, B. Decaris, and G. Guedon.** 1999. Are horizontal transfers involved in the evolution of the *Streptococcus thermophilus* exopolysaccharide synthesis loci? *Gene* **233:**151–161.

15. **Brondsted, L., and K. Hammer.** 1999. Use of the integration elements encoded by the temperate lactococcal bacteriophage TP901-1 to obtain chromosomal single-copy transcriptional fusions in *Lactococcus lactis. Appl. Environ. Microbiol.* **65:**752–758.

16. **Buist, G., H. Karsens, A. Nauta, D. van Sinderen, G. Venema, and J. Kok.** 1997. Autolysis of *Lactococcus lactis* caused by induced overproduction of its major autolysin, AcmA. *Appl. Environ. Microbiol.* **63:**2722–2728.

17. **Campo, N., M. J. Dias, M. L. Daveran-Mingot, P. Ritzenthaler, and P. Le Bourgeois.** 2004. Chromosomal constraints in Gram-positive bacteria revealed by artificial inversions. *Mol. Microbiol.* **51:**511–522.

18. **Chatel, J. M., S. Nouaille, K. Adel-Patient, Y. Le Loir, H. Boe, A. Gruss, J. M. Wal, and P. Langella.** 2003. Characterization of a *Lactococcus lactis* strain that secretes a major epitope of bovine beta-lactoglobulin and evaluation of its immunogenicity in mice. *Appl. Environ. Microbiol.* **69:**6620–6627.

19. **Claverys, J. P., M. Prudhomme, I. Mortier-Barriere, and B. Martin.** 2000. Adaptation to the environment: *Streptococcus pneumoniae*, a paradigm for recombination-mediated genetic plasticity? *Mol. Microbiol.* **35:**251–259.

20. **Cocaign-Bousquet, M., C. Garrigues, P. Loubiere, and N. D. Lindley.** 1996. Physiology of pyruvate metabolism in *Lactococcus lactis. Antonie Leeuwenhoek* **70:**253–267.

21. **de Ruyter, P. G., O. P. Kuipers, W. C. Meijer, and W. M. de Vos.** 1997. Food-grade controlled lysis of *Lactococcus lactis* for accelerated cheese ripening. *Nat. Biotechnol.* **15:**976–979.

22. **de Vos, W. M., and J. Hugenholtz.** 2004. Engineering metabolic highways in lactococci and other lactic acid bacteria. *Trends Biotechnol.* **22:**72–79.

23. **Dickely, F., D. Nilsson, E. B. Hansen, and E. Johansen.** 1995. Isolation of *Lactococcus lactis* nonsense suppressors and construction of a food-grade cloning vector. *Mol. Microbiol.* **15:**839–847.

24. **Diep, D. B., and I. F. Nes.** 2002. Ribosomally synthesized antibacterial peptides in Gram positive bacteria. *Curr. Drug Targets* **3:**107–122.

24a. **Dieye, Y.** 2002. Ph.D. thesis. Université Paris XI, Paris, France.

25. **D'Mello, R., S. Hill, and R. K. Poole.** 1996. The cytochrome bd quinol oxidase in *Escherichia coli* has an extremely high oxygen affinity and two oxygen-binding haems: implications for regulation of activity in vivo by oxygen inhibition. *Microbiology* **142(Pt 4):**755–763.

26. **Duwat, P., S. D. Ehrlich, and A. Gruss.** 1999. Effects of metabolic flux on stress response pathways in *Lactococcus lactis. Mol. Microbiol.* **31:**845–858.

27. **Duwat, P., S. D. Ehrlich, and A. Gruss.** 1995. The recA gene of *Lactococcus lactis*: characterization and involvement in oxidative and thermal stress. *Mol. Microbiol.* **17:**1121–1131.

28. **Duwat, P., S. Sourice, B. Cesselin, G. Lamberet, K. Vido, P. Gaudu, Y. Le Loir, F. Violet, P. Loubiere, and A. Gruss.** 2001. Respiration capacity of the fermenting bacterium *Lactococcus lactis* and its positive effects on growth and survival. *J. Bacteriol.* **183:**4509–4516.

29. **Eichenbaum, Z., M. J. Federle, D. Marra, W. M. de Vos, O. P. Kuipers, M. Kleerebezem, and J. R. Scott.** 1998. Use of the lactococcal nisA promoter to regulate gene expression in gram-positive bacteria: comparison of induction level and promoter strength. *Appl. Environ. Microbiol.* **64:**2763–2769.

30. **Endley, S., D. McMurray, and T. A. Ficht.** 2001. Interruption of the cydB locus in *Brucella abortus* attenuates intracellular survival and virulence in the mouse model of infection. *J. Bacteriol.* **183:**2454–2462.

31. **Fischetti, V. A., V. Pancholi, and O. Schneewind.** 1990. Conservation of a hexapeptide sequence in the anchor region of surface proteins from gram-positive cocci. *Mol. Microbiol.* **4:**1603–1605.

32. **Frazier, C. L., J. San Filippo, A. M. Lambowitz, and D. A. Mills.** 2003. Genetic manipulation of *Lactococcus lactis* by

using targeted group II introns: generation of stable insertions without selection. *Appl. Environ. Microbiol.* **69:** 1121–1128.

33. **Frees, D., and H. Ingmer.** 1999. ClpP participates in the degradation of misfolded protein in *Lactococcus lactis. Mol. Microbiol.* **31:**79–87.

34. **Garrigues, C., P. Loubiere, N. D. Lindley, and M. Cocaign-Bousquet.** 1997. Control of the shift from homolactic acid to mixed-acid fermentation in *Lactococcus lactis*: predominant role of the NADH/NAD+ ratio. *J. Bacteriol.* **179:**5282–5287.

35. **Gaudu, P., G. Lamberet, S. Poncet, and A. Gruss.** 2003. CcpA regulation of aerobic and respiration growth in *Lactococcus lactis. Mol. Microbiol.* **50:**183–192.

36. **Gaudu, P., K. Vido, B. Cesselin, S. Kulakauskas, J. Tremblay, L. Rezaiki, G. Lamberret, S. Sourice, P. Duwat, and A. Gruss.** 2002. Respiration capacity and consequences in *Lactococcus lactis. Antonie Leeuwenhoek* **82:**263–269.

37. **Gilbert, C., K. Robinson, R. W. Le Page, and J. M. Wells.** 2000. Heterologous expression of an immunogenic pneumococcal type 3 capsular polysaccharide in *Lactococcus lactis. Infect. Immun.* **68:**3251–3260.

38. **Godon, J. J., C. J. Pillidge, K. Jury, C. A. Shearman, and M. J. Gasson.** 1995. Molecular analysis of the *Lactococcus lactis* sex factor. *Dev. Biol. Stand.* **85:**423–430.

39. **Grangette, C., H. Muller-Alouf, P. Hols, D. Goudercourt, J. Delcour, M. Turneer, and A. Mercenier.** 2004. Enhanced mucosal delivery of antigen with cell wall mutants of lactic acid bacteria. *Infect. Immun.* **72:**2731–2737.

40. **Hartke, A., S. Bouche, J. C. Giard, A. Benachour, P. Boutibonnes, and Y. Auffray.** 1996. The lactic acid stress response of *Lactococcus lactis* subsp. *lactis. Curr. Microbiol.* **33:**194–199.

41. **Helmark, S., M. E. Hansen, B. Jelle, K. I. Sorensen, and P. R. Jensen.** 2004. Transformation of *Leuconostoc carnosum* 4010 and evidence for natural competence of the organism. *Appl. Environ. Microbiol.* **70:**3695–3699.

42. **Horn, N., S. Swindell, H. Dodd, and M. Gasson.** 1991. Nisin biosynthesis genes are encoded by a novel conjugative transposon. *Mol. Gen. Genet.* **228:**129–135.

43. **Huycke, M. M., D. Moore, W. Joyce, P. Wise, L. Shepard, Y. Kotake, and M. S. Gilmore.** 2001. Extracellular superoxide production by *Enterococcus faecalis* requires demethylmenaquinone and is attenuated by functional terminal quinol oxidases. *Mol. Microbiol.* **42:**729–740.

44. **Iwaki, M., N. Okahashi, I. Takahashi, T. Kanamoto, Y. Sugita-Konishi, K. Aibara, and T. Koga.** 1990. Oral immunization with recombinant *Streptococcus lactis* carrying the *Streptococcus mutans* surface protein antigen gene. *Infect. Immun.* **58:**2929–2934.

45. **Jensen, P. R., and K. Hammer.** 1998. The sequence of spacers between the consensus sequences modulates the strength of prokaryotic promoters. *Appl. Environ. Microbiol.* **64:**82–87.

46. **Jones, A. L., R. H. Needham, A. Clancy, K. M. Knoll, and C. E. Rubens.** 2003. Penicillin-binding proteins in *Streptococcus agalactiae*: a novel mechanism for evasion of immune clearance. *Mol. Microbiol.* **47:**247–256.

47. **Khan, S. A.** 1997. Rolling-circle replication of bacterial plasmids. *Microbiol. Mol. Biol. Rev.* **61:**442–455.

48. **Kilstrup, M., S. G. Jessing, S. B. Wichmand-Jorgensen, M. Madsen, and D. Nilsson.** 1998. Activation control of pur gene expression in *Lactococcus lactis*: proposal for a consensus activator binding sequence based on deletion analysis and site-directed mutagenesis of purC and purD promoter regions. *J. Bacteriol.* **180:**3900–3906.

49. **Kilstrup, M., and J. Martinussen.** 1998. A transcriptional activator, homologous to the *Bacillus subtilis* PurR repressor, is required for expression of purine biosynthetic genes in *Lactococcus lactis. J. Bacteriol.* **180:**3907–3916.

50. **Koebmann, B. J., H. W. Andersen, C. Solem, and P. R. Jensen.** 2002. Experimental determination of control of glycolysis in *Lactococcus lactis. Antonie Leeuwenhoek* **82:**237–248.

51. **Koebmann, B. J., C. Solem, M. B. Pedersen, D. Nilsson, and P. R. Jensen.** 2002. Expression of genes encoding F(1)-ATPase results in uncoupling of glycolysis from biomass production in *Lactococcus lactis. Appl. Environ. Microbiol.* **68:**4274–4282.

52. **Koebmann, B. J., H. V. Westerhoff, J. L. Snoep, D. Nilsson, and P. R. Jensen.** 2002. The glycolytic flux in *Escherichia coli* is controlled by the demand for ATP. *J. Bacteriol.* **184:**3909–3916.

53. **Kunji, E. R., I. Mierau, A. Hagting, B. Poolman, and W. N. Konings.** 1996. The proteolytic systems of lactic acid bacteria. *Antonie Leeuwenhoek* **70:**187–221.

54. **Kunji, E. R., D. J. Slotboom, and B. Poolman.** 2003. *Lactococcus lactis* as host for overproduction of functional membrane proteins. *Biochim. Biophys. Acta* **1610:**97–108.

55. **Le Bourgeois, P., M. L. Daveran-Mingot, and P. Ritzenthaler.** 2000. Genome plasticity among related ++Lactococcus strains: identification of genetic events associated with macrorestriction polymorphisms. *J. Bacteriol.* **182:** 2481–2491.

56. **Le Bourgeois, P., M. Lautier, L. van den Berghe, M. J. Gasson, and P. Ritzenthaler.** 1995. Physical and genetic map of the *Lactococcus lactis* subsp. *cremoris* MG1363 chromosome: comparison with that of *Lactococcus lactis* subsp. *lactis* IL 1403 reveals a large genome inversion. *J. Bacteriol.* **177:**2840–2850.

57. **Le Loir, Y., S. Nouaille, J. Commissaire, L. Bretigny, A. Gruss, and P. Langella.** 2001. Signal peptide and propeptide optimization for heterologous protein secretion in *Lactococcus lactis. Appl. Environ. Microbiol.* **67:**4119–4127.

58. **Leenhouts, K., G. Buist, A. Bolhuis, A. ten Berge, J. Kiel, I. Mierau, M. Dabrowska, G. Venema, and J. Kok.** 1996. A general system for generating unlabelled gene replacements in bacterial chromosomes. *Mol. Gen. Genet.* **253:**217–224.

59. **Li, Y., J. Hugenholtz, T. Abee, and D. Molenaar.** 2003. Glutathione protects *Lactococcus lactis* against oxidative stress. *Appl. Environ. Microbiol.* **69:**5739–5745.

60. **Lindholm, A., A. Smeds, and A. Palva.** 2004. Receptor binding domain of *Escherichia coli* F18 fimbrial adhesin FedF can be both efficiently secreted and surface displayed in a functional form in *Lactococcus lactis. Appl. Environ. Microbiol.* **70:**2061–2071.

61. **Llull, D., and I. Poquet.** 2004. New expression system tightly controlled by zinc availability in *Lactococcus lactis. Appl. Environ. Microbiol.* **70:**5398–5406.

61a. **Llull, D., P. Veiga, J. Tremblay, and S. Kulakauskas.** 2005. Immobilization-based isolation of capsule-negative mutants of *Streptococcus pneumoniae. Microbiology* **151:**1911–1917.

62. **Lopez de Felipe, F., and J. Hugenholtz.** 1999. Pyruvate flux distribution in NADH-oxidase-overproducing *Lactococcus lactis* strain as a function of culture conditions. *FEMS Microbiol. Lett.* **179:**461–466.

63. **Lu, Y., and R. L. Switzer.** 1996. Evidence that the *Bacillus subtilis* pyrimidine regulatory protein PyrR acts by binding to pyr mRNA at three sites in vivo. *J. Bacteriol.* **178:** 5806–5809.

64. Luesink, E. J., R. E. van Herpen, B. P. Grossiord, O. P. Kuipers, and W. M. de Vos. 1998. Transcriptional activation of the glycolytic las operon and catabolite repression of the gal operon in *Lactococcus lactis* are mediated by the catabolite control protein CcpA. *Mol. Microbiol.* **30:** 789–798.

65. Madsen, S. M., J. Arnau, A. Vrang, M. Givskov, and H. Israelsen. 1999. Molecular characterization of the pH-inducible and growth phase-dependent promoter P170 of *Lactococcus lactis. Mol. Microbiol.* **32:**75–87.

66. Maguin, E., H. Prevost, S. D. Ehrlich, and A. Gruss. 1996. Efficient insertional mutagenesis in lactococci and other gram-positive bacteria. *J. Bacteriol.* **178:**931–935.

67. Mannam, P., K. F. Jones, and B. L. Geller. 2004. Mucosal vaccine made from live, recombinant *Lactococcus lactis* protects mice against pharyngeal infection with *Streptococcus pyogenes. Infect. Immun.* **72:**3444–3450.

68. Martinussen, J., and K. Hammer. 1995. Powerful methods to establish chromosomal markers in *Lactococcus lactis*: an analysis of pyrimidine salvage pathway mutants obtained by positive selections. *Microbiology* **141:**1883–1890.

69. Martinussen, J., J. Schallert, B. Andersen, and K. Hammer. 2001. The pyrimidine operon pyrRPB-carA from *Lactococcus lactis. J. Bacteriol.* **183:**2785–2794.

70. Martinussen, J., S. L. Wadskov-Hansen, and K. Hammer. 2003. Two nucleoside uptake systems in *Lactococcus lactis*: competition between purine nucleosides and cytidine allows for modulation of intracellular nucleotide pools. *J. Bacteriol.* **185:**1503–1508.

71. McKay, L. L., K. A. Baldwin, and J. D. Efstathiou. 1976. Transductional evidence for plasmid linkage of lactose metabolism in streptococcus lactis C2. *Appl. Environ. Microbiol.* **32:**45–52.

72. Mercenier, A., H. Muller-Alouf, and C. Grangette. 2000. Lactic acid bacteria as live vaccines. *Curr. Issues Mol. Biol.* **2:**17–25.

73. Mercier, C., C. Durrieu, R. Briandet, E. Domakova, J. Tremblay, G. Buist, and S. Kulakauskas. 2002. Positive role of peptidoglycan breaks in lactococcal biofilm formation. *Mol. Microbiol.* **46:**235–243.

74. Miyoshi, A., E. Jamet, J. Commissaire, P. Renault, P. Langella, and V. Azevedo. 2004. A xylose-inducible expression system for *Lactococcus lactis. FEMS Microbiol. Lett.* **239:**205–212.

75. Navarre, W. W., and O. Schneewind. 1999. Surface proteins of gram-positive bacteria and mechanisms of their targeting to the cell wall envelope. *Microbiol. Mol. Biol. Rev.* **63:**174–229.

76. Nelson, D., L. Loomis, and V. A. Fischetti. 2001. Prevention and elimination of upper respiratory colonization of mice by group A streptococci by using a bacteriophage lytic enzyme. *Proc. Natl. Acad. Sci. USA* **98:**4107–4112.

77. Nilsson, D., A. A. Lauridsen, T. Tomoyasu, and T. Ogura. 1994. A *Lactococcus lactis* gene encodes a membrane protein with putative ATPase activity that is homologous to the essential *Escherichia coli* ftsH gene product. *Microbiology* **140**(Pt 10):2601–2610.

78. Norton, P. M., H. W. Brown, J. M. Wells, A. M. Macpherson, P. W. Wilson, and R. W. Le Page. 1996. Factors affecting the immunogenicity of tetanus toxin fragment C expressed in *Lactococcus lactis. FEMS Immunol. Med. Microbiol.* **14:**167–177.

79. Nudler, E., and A. S. Mironov. 2004. The riboswitch control of bacterial metabolism. *Trends Biochem. Sci.* **29:**11–17.

80. Ocana, V. S., A. A. Pesce de Ruiz Holgado, and M. E. Nader-Macias. 1999. Selection of vaginal H_2O_2-generat-

ing *Lactobacillus* species for probiotic use. *Curr. Microbiol.* **38:**279–284.

81. O'Connell-Motherway, M., D. van Sinderen, F. Morel-Deville, G. F. Fitzgerald, S. D. Ehrlich, and P. Morel. 2000. Six putative two-component regulatory systems isolated from *Lactococcus lactis* subsp. *cremoris* MG1363. *Microbiology* **146:**935–947.

82. Oliver, S. 2002. Metabolism: demand management in cells. *Nature* **418:**33–34.

83. Pancholi, V., and V. A. Fischetti. 1992. A major surface protein on group A streptococci is a glyceraldehyde-3-phosphate-dehydrogenase with multiple binding activity. *J. Exp. Med.* **176:**415–426.

84. Pedersen, M. B., P. R. Jensen, T. Janzen, and D. Nilsson. 2002. Bacteriophage resistance of a deltathyA mutant of *Lactococcus lactis* blocked in DNA replication. *Appl. Environ. Microbiol.* **68:**3010–3023.

85. Pedersen, M. B., B. J. Koebmann, P. R. Jensen, and D. Nilsson. 2002. Increasing acidification of nonreplicating *Lactococcus lactis* deltathyA mutants by incorporating ATPase activity. *Appl. Environ. Microbiol.* **68:**5249–5257.

86. Piard, J. C., R. Jimenez-Diaz, V. A. Fischetti, S. D. Ehrlich, and A. Gruss. 1997. The M6 protein of *Streptococcus pyogenes* and its potential as a tool to anchor biologically active molecules at the surface of lactic acid bacteria. *Adv. Exp. Med. Biol.* **418:**545–550.

87. Pillar, C. M., and M. S. Gilmore. 2004. Enterococcal virulence—pathogenicity island of E. faecalis. *Front. Biosci.* **9:**2335–2346.

88. Poelarends, G. J., P. Mazurkiewicz, and W. N. Konings. 2002. Multidrug transporters and antibiotic resistance in *Lactococcus lactis. Biochim. Biophys. Acta* **1555:**1–7.

89. Poole, R. K., and G. M. Cook. 2000. Redundancy of aerobic respiratory chains in bacteria? Routes, reasons and regulation. *Adv. Microb. Physiol.* **43:**165–224.

90. Poquet, I., S. D. Ehrlich, and A. Gruss. 1998. An export-specific reporter designed for gram-positive bacteria: application to *Lactococcus lactis. J. Bacteriol.* **180:**1904–1912.

91. Poquet, I., V. Saint, E. Seznec, N. Simoes, A. Bolotin, and A. Gruss. 2000. HtrA is the unique surface housekeeping protease in *Lactococcus lactis* and is required for natural protein processing. *Mol. Microbiol.* **35:**1042–1051.

92. Purushothaman, S. S., B. Wang, and P. P. Cleary. 2003. M1 protein triggers a phosphoinositide cascade for group A Streptococcus invasion of epithelial cells. *Infect. Immun.* **71:**5823–5830.

93. Que, Y. A., P. Francois, J. A. Haefliger, J. M. Entenza, P. Vaudaux, and P. Moreillon. 2001. Reassessing the role of *Staphylococcus aureus* clumping factor and fibronectin-binding protein by expression in *Lactococcus lactis. Infect. Immun.* **69:**6296–6302.

94. Rallu, F., A. Gruss, S. D. Ehrlich, and E. Maguin. 2000. Acid- and multistress-resistant mutants of *Lactococcus lactis*: identification of intracellular stress signals. *Mol. Microbiol.* **35:**517–528.

95. Rallu, F., A. Gruss, and E. Maguin. 1996. *Lactococcus lactis* and stress. *Antonie Leeuwenhoek* **70:**243–251.

96. Repa, A., C. Grangette, C. Daniel, R. Hochreiter, K. Hoffmann-Sommergruber, J. Thalhamer, D. Kraft, H. Breiteneder, A. Mercenier, and U. Wiedermann. 2003. Mucosal co-application of lactic acid bacteria and allergen induces counter-regulatory immune responses in a murine model of birch pollen allergy. *Vaccine* **22:**87–95.

97. **Rezaiki, L., B. Cesselin, Y. Yamamoto, K. Vido, E. van West, P. Gaudu, and A. Gruss.** 2004. Respiration metabolism reduces oxidative and acid stress to improve long-term survival of *Lactococcus lactis*. *Mol. Microbiol.* **53:** 1331–1342.

98. **Richardson, D. J.** 2000. Bacterial respiration: a flexible process for a changing environment. *Microbiology* **146** (Pt 3):551–571.

99. **Robinson, K., L. M. Chamberlain, K. M. Schofield, J. M. Wells, and R. W. Le Page.** 1997. Oral vaccination of mice against tetanus with recombinant *Lactococcus lactis*. *Nat. Biotechnol.* **15:**653–657.

99a. **Rochat, T., J. J. Gratadoux, G. Corthier, B. Coqueran, M. E. Nader-Macias, A. Gruss, and P. Langella.** 2005. *Lactococcus lactis* Spox spontaneous mutants: a family of oxidative stress-resistant dairy strains. *Appl. Environ. Microbiol.* **71:**2782–2788.

99b. **Rochat, T., A. Miyoshi, J. J. Gratadoux, P. Duwat, S. Sourice, V. Azevedo, and P. Langella.** 2005. High-level resistance to oxidative stress in *Lactococcus lactis* conferred by *Bacillus subtilis* catalase KatE. *Microbiology* **151:** 3011–3018.

100. **Ryu, C. M., M. A. Farag, C. H. Hu, M. S. Reddy, H. X. Wei, P. W. Pare, and J. W. Kloepper.** 2003. Bacterial volatiles promote growth in Arabidopsis. *Proc. Natl. Acad. Sci. USA* **100:**4927–4932.

101. **Sanders, J. W., K. Leenhouts, J. Burghoorn, J. R. Brands, G. Venema, and J. Kok.** 1998. A chloride-inducible acid resistance mechanism in *Lactococcus lactis* and its regulation. *Mol. Microbiol.* **27:**299–310.

102. **Sanders, J. W., K. J. Leenhouts, A. J. Haandrikman, G. Venema, and J. Kok.** 1995. Stress response in *Lactococcus lactis*: cloning, expression analysis, and mutation of the lactococcal superoxide dismutase gene. *J. Bacteriol.* **177:**5254–5260.

103. **Sanders, J. W., G. Venema, and J. Kok.** 1997. A chloride-inducible gene expression cassette and its use in induced lysis of *Lactococcus lactis*. *Appl. Environ. Microbiol.* **63:**4877–4882.

104. **Sanders, J. W., G. Venema, J. Kok, and K. Leenhouts.** 1998. Identification of a sodium chloride-regulated promoter in *Lactococcus lactis* by single-copy chromosomal fusion with a reporter gene. *Mol. Gen. Genet.* **257:**681–685.

105. **Schneewind, O., D. Mihaylova-Petkov, and P. Model.** 1993. Cell wall sorting signals in surface proteins of gram-positive bacteria. *EMBO J.* **12:**4803–4811.

106. **Seki, M., K. Iida, M. Saito, H. Nakayama, and S. Yoshida.** 2004. Hydrogen peroxide production in *Streptococcus pyogenes*: involvement of lactate oxidase and coupling with aerobic utilization of lactate. *J. Bacteriol.* **186:**2046–2051.

107. **Shanahan, F.** 2004. Making microbes work for mankind—clever trick or a glimpse of the future for IBD treatment? *Gastroenterology* **127:**667–668.

108. **Siegers, K., and K. D. Entian.** 1995. Genes involved in immunity to the lantibiotic nisin produced by *Lactococcus lactis* 6F3. *Appl. Environ. Microbiol.* **61:**1082–1089.

109. **Siegers, K., S. Heinzmann, and K. D. Entian.** 1996. Biosynthesis of lantibiotic nisin. Posttranslational modification of its prepeptide occurs at a multimeric membrane-associated lanthionine synthetase complex. *J. Biol. Chem.* **271:**12294–12301.

110. **Sijpesteijn, A. K.** 1970. Induction of cytochrome formation and stimulation of oxidative dissimilation by hemin in *Streptococcus lactis* and *Leuconostoc mesenteroides*. *Antonie Leeuwenhoek* **36:**335–348.

111. **Solem, C., and P. R. Jensen.** 2002. Modulation of gene expression made easy. *Appl. Environ. Microbiol.* **68:**2397–2403.

111a. **Solem, C., B. J. Koebmann, and P. R. Jensen.** 2003. Glyceraldehyde-3-phosphate dehydrogenase has no control on the glycolytic flux in *Lactococcus lactis* MG1363. *J. Bacteriol.* **185:**1564–1571.

112. **Steen, A., G. Buist, K. J. Leenhouts, M. El Khattabi, F. Grijpstra, A. L. Zomer, G. Venema, O. P. Kuipers, and J. Kok.** 2003. Cell wall attachment of a widely distributed peptidoglycan binding domain is hindered by cell wall constituents. *J. Biol. Chem.* **278:**23874–23881.

113. **Steidler, L., W. Hans, L. Schotte, S. Neirynck, F. Obermeier, W. Falk, W. Fiers, and E. Remaut.** 2000. Treatment of murine colitis by *Lactococcus lactis* secreting interleukin-10. *Science* **289:**1352–1355.

114. **Steidler, L., S. Neirynck, N. Huyghebaert, V. Snoeck, A. Vermeire, B. Goddeeris, E. Cox, J. P. Remon, and E. Remaut.** 2003. Biological containment of genetically modified *Lactococcus lactis* for intestinal delivery of human interleukin 10. *Nat. Biotechnol.* **21:**785–789.

115. **Stentz, R., K. Jury, T. Eaton, M. Parker, A. Narbad, M. Gasson, and C. Shearman.** 2004. Controlled expression of CluA in *Lactococcus lactis* and its role in conjugation. *Microbiology* **150:**2503–2512.

116. **Teuber, M.** 1995. The genus *Lactococcus*, p. 173–234. *In* B. J. B. Wood and W. H. Holzapfel (ed.), *The Genera of Lactic Acid Bacteria*. Blackie Academic and Professional, Glasgow, United Kingdom.

116a. **van Asseldonk, M., W. M. de Vos, and G. Simons.** 1993. Functional analysis of the *Lactococcus lactis* usp45 secretion signal in the secretion of a homologous proteinase and a heterologous alpha-amylase. *Mol. Gen. Genet.* **240:** 428–434.

117. **Vaughan, E. E., and W. M. de Vos.** 1995. Identification and characterization of the insertion element IS1070 from *Leuconostoc lactis* NZ6009. *Gene* **155:**95–100.

118. **Way, S. S., S. Sallustio, R. S. Magliozzo, and M. B. Goldberg.** 1999. Impact of either elevated or decreased levels of cytochrome bd expression on *Shigella flexneri* virulence. *J. Bacteriol.* **181:**1229–1237.

119. **Winstedt, L., L. Frankenberg, L. Hederstedt, and C. von Wachenfeldt.** 2000. *Enterococcus faecalis* V583 contains a cytochrome bd-type respiratory oxidase. *J. Bacteriol.* **182:** 3863–3866.

120. **Yamamoto, Y., C. Poyart, P. Trieu-Cuot, G. Lamberet, A. Gruss, and P. Gaudu.** 2005. Respiration metabolism of Group B *Streptococcus* is activated by environmental haem and quinine and contributes to virulence. *Mol. Microbiol.* **56:**525–534.

THE STAPHYLOCOCCUS

SECTION EDITOR: Richard P. Novick

W ITH THE COMPLETION AND ANNOTATION OF NINE staphylococcal genomes, including seven *S. aureus* strains and two *S. epidermidis*, our understanding of these important bacteria has taken an exponential leap forward. The staphylococci have taken their place among the well-characterized microorganisms, and one may now appreciate their overall metabolic organization as well as those features of the genome that are related to disease causation and that continue to be the focus of this section. Epidemiology and typing now have a very strong footing, and we have a much better appreciation of the contribution of mobile genetic elements to pathogenesis. As in the first edition, this section devotes considerable attention to the non-*aureus* staphylococci. Overall, it covers the cell surface and its associated proteins and capsules, extracellular proteins, including toxins and degradative enzymes, regulatory aspects of pathogenesis and of carbohydrate catabolism, mobile elements, their transfer and gene content, experimental infection models, interactions with host cells, and the role of host defenses and antibiotic resistance. The literature on staphylococci seems to have expanded exponentially, and there has been a corresponding expansion of citations in these chapters, which will provide excellent access to the overall literature.

Diagnostics, Typing, and Taxonomy

WOLFGANG WITTE, BIRGIT STROMMENGER, AND GUIDO WERNER

31

DIAGNOSTICS

Speciation in bacterial systematics is based on a comparison of organismal characteristics in order to arrange microorganisms in groups sharing common properties. The basic taxonomic unit, the species, is defined by the most characteristic properties necessary for reliable identification of a particular organism as belonging to this group. The purpose of speciation is identification of a microorganism as belonging to a basic taxon, which, for example, has a particular ecological or clinical significance. The classification of different species into groups leads to genera with a number of key characteristics.

Besides the species characteristics, organisms exhibit a number of other properties that can be used for a further differentiation into "types" below the species level. Ideally, typing would identify the clonal progeny of an organism within a species population.

In general, the purpose of typing is the study of population dynamics and the spread of microorganisms that undergo clonal (nonsexual) reproduction. Thus typing is an important tool of epidemiology for tracing the spread of particular strains (clones) and discovering routes of transmission and reservoirs.

Bacterial taxonomy still joins gram-positive, catalase-positive cocci such as staphylococci and micrococci in the family of *Micrococcaceae*, although the genera of this family exhibit a number of very different molecular features (e.g., GC content, cell wall composition) (Table 1). Further analysis of 16S and 23S rRNA sequences will answer the question whether staphylococci form a quite separate taxonomic group.

As with other microorganisms, speciation in staphylococci was first based on a data set of morphological characteristics, physiological properties, and chemical composition of the cell wall (17). A problem with this scheme was that some of the taxonomic characteristics, such as acid formation from carbohydrates and the formation of cell wall-associated proteins, extracellular enzymes, and toxins, were unstable owing to genomic rearrangements or the acquisition or loss of accessory genetic elements.

Later DNA-DNA hybridization studies have shown that staphylococci form a well-defined genus, which can be sub-divided into several species groups (39) (Fig. 1). The results of the DNA homology studies have been confirmed at the epigenetic level (comparative characterization of catalases, aldolases, and L- and D-lactate dehydrogenases [39]) and for a number of staphylococcal species, as well as by 16S and 23S rRNA sequence analysis (22, 33).

Phylogenetic relationships based on 16S rRNA sequence analysis for most species confirm the data from DNA-DNA hybridization (Fig. 2).

A more recent investigation revealed that the sequence polymorphism in a "hot spot" stretch of the heat shock protein 60 gene (*hsp60*) is species-specific in staphylococci (21). In contrast to the variable parts of the 16S rRNA gene sequence, there are at least two, in some cases even more, mismatches. Therefore, *hsp60* is of special interest for DNA-based species diagnostics. Another study revealed a species-specific polymorphism of the glycerol aldehyde phosphatase (*gap*) gene, which can be detected by PCR and AluI-digestion of the amplimers (50). A further approach has used internal transcribed spacer PCR in the rRNA operon to identify staphylococcal species (4).

For routine purposes *Staphylococcus aureus*, the most significant pathogen among staphylococci, can be easily identified by demonstration of the free coagulase enzyme or the clumping factor, a cell surface-associated fibrinogen-binding protein. Clumping factor causes agglutination in the presence of human plasma and is therefore very easy to detect. However, *Staphylococcus schleiferi* also exhibits clumping activity when suspended in human blood plasma, and certain epidemic clones of methicillin-resistant *S. aureus* (MRSA) do not. The accuracy of commercial slide agglutination tests has recently been improved by the addition of antibodies against capsular antigens, which provides sufficient sensitivity for speciation of MRSA.

The identification of coagulase-negative staphylococcal (CNS) species in clinical bacteriology is primarily based on a set of different biochemical characteristics (17). At present, panels used in clinical routine include about 30 metabolic characteristics as well as resistance to novobiocin as characteristics of the *Staphylococcus saprophyticus* group and resistance to furazolidon for separation from micrococci.

Gram-Positive Pathogens, 2nd edition, edited by Vincent A. Fischetti et al.
© 2006 ASM Press, Washington, D.C.

TABLE 1 Characteristics of the genus *Staphylococcus* in comparison to other genera classified as members of the family *Micrococcaceae*

Parameter	Characteristic for the genus:			
	Staphylococcus	*Micrococcus*	*Planococcus*	*Stomatococcus*
GC content of DNA (mol% G+C)	30–35	70–75	40–51	56–60
Cell wall composition (more than 2 mol of glycin per mol of glutaminic acid in peptidoglycan)	+	−	−	−
Type of fructose-1,6-diphosphate aldolase	I	II	ND[a]	II
Cytochrome *c*	−	+	ND	+
Character used in bacteriological diagnostics				
Sensitivity to lysostaphin	+	−	−	−
Sensitivity to furazolidon	−	+	ND	ND

[a]ND, not determined.

The distribution frequency of metabolic characteristics used for speciation of staphylococci is based on an empirically established data bank. Attribution of particular reaction profiles to a species can be performed by a code book or more quickly and precisely by computerized probability analysis.

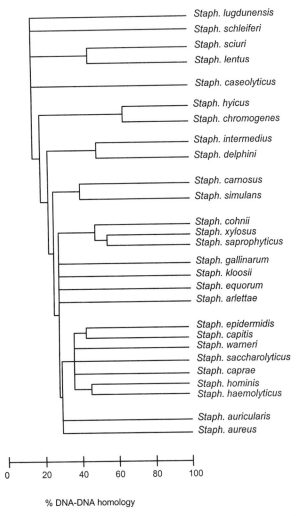

FIGURE 1 Phylogenetic relations among staphylococcal species as deduced from DNA-DNA hybridization. (Reprinted with permission from reference 39.)

The results of automated or semiautomated systems based on the above-mentioned biochemical reaction profiles do not necessarily correlate with conventional tests on other characteristics of *S. aureus*, such as coagulase. Particular MRSA strains may possess metabolic deficiencies (e.g., lack of urease or of pyrrolidonylarylaminidase) that can lead to their incorrect identification as a CNS species.

For rapid and reliable discrimination of *S. aureus* from CNS, PCR reactions have been established for detection of the following *S. aureus*-specific sequences: *nuc*, the gene for heat-stable DNase (1, 2), specific sequences in rRNA (38), and a DNA sequence specific for *S. aureus* as identified by subtractive hybridization (24). Also, PCR demonstration of *fem* genes, such as *femA*, has been recommended (18); there are, however, doubts that a sufficient number of isolates from different CNS species have been checked for this characteristic.

EPIDEMIOLOGICAL TYPING

General Aspects

Outbreaks of *S. aureus* infections are in most cases due to the clonal expansion of a particular strain. CNS from colonization and infections in a particular hospital unit are mostly polyclonal; epidemic clonal spread was rarely described until now. Typing in epidemiology has the goal of discriminating the epidemic clone by characteristics that differ from those of epidemiologically unrelated strains. The ideal typing characteristic should be stable within the epidemic strain and sufficiently diverse within the species population. Diversity results from random, nonlethal genetic events that are assayed by the typing system (e.g., neutral mutations influencing the charge of an enzyme of primary metabolism and thus its motility in multilocus enzyme electrophoresis; deletions or insertions influencing the distance between restriction sites and thus also restriction endonuclease cleavage patterns). Genetic diversity among strains possessing clinical significance with regard to pathogenicity or antibiotic resistance or selection factors will be limited when a particular clonal group that is optimally adapted to the hospital situation has been disseminated. The development of more sensitive molecular techniques during the past 10 years has led to an increased ability to detect subsets of evolutionary lineages among related groups of bacterial strains.

Requirements for Typing Systems

Characteristics used for typing should be demonstrable in nearly all of the isolates within a species to guarantee a high

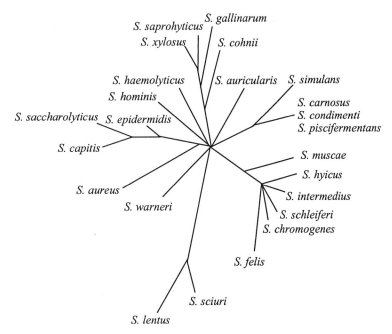

FIGURE 2 Phylogenetic relationship of staphylococci reflected by the 16S rRNA-based tree. (Reprinted from reference 33, with permission.)

degree of typeability. Results of typing have to be reproducible in repeated laboratory testing of the same isolate as well as during the course of an epidemic and between different laboratories. A particular typing method should generate a sufficient number of types (discriminatory power), and the frequency of the most frequent type should be less than 5% of the number of isolates of the species population investigated.

Phenotypic Typing Methods for *S. aureus*

The oldest phenotypic typing method is serotyping. Although its typeability and reproducibility seemed to have been fairly good, serotyping was later replaced by phage typing, which has a much higher discriminatory power. In phage typing isolates are discriminated by their susceptibility or resistance to lysis by each member of a set of typing phages. In general terms the specificity of phage patterns is based on receptor specificity in adsorption, DNA restriction, superinfection immunity of lysogenic cells, and further mechanisms of postadsorption interference. The phage typing system of *S. aureus* was established empirically. Receptor specificity is obviously not a factor in phage typing of *S. aureus*. Later studies revealed that DNA restriction and modification are probably the most common mechanism of variation (42), but superinfection immunity is also a determinant of typing patterns.

Ideally, each unrelated strain in a population should exhibit a specific phage pattern. For a number of strains this is really found, and here phage patterns correspond to specific molecular typing patterns (e.g., most *S. aureus* strains that produce toxic shock syndrome toxin, exfoliative toxin, or exhibit borderline methicillin resistance have specific phage patterns; strains of types 94, 95, or 96 also represent specific genotypes [35, 46]).

Most strains of *S. aureus*, however, are lysogenic. Prophages influence phage patterns by superinfection immunity and DNA restriction, and variations in lysogeny occur frequently. Due to this variability phage typing is of very limited

use, especially for multiply-resistant strains. Another problem is nontypeability, which requires expansion of the bacteriophage panel.

Bacteriocinogenicity typing, which is based on capacity to form bacteriocins active against different indicator strains, was also tried in *S. aureus* and was abandoned because of unsatisfactory reproducibility and discriminatory power. This is not surprising because bacteriocins are very often plasmid-determined and have broad host ranges.

As with other bacterial pathogens, exoprotein profiles have been used as an epidemiological marker for typing CNS and *S. aureus*. In *S. aureus* this approach is impaired by unsatisfactory resolution of the various proteins over a broad molecular mass range in polyacrylamide gel electrophoresis. This was overcome using immunoblots (25). One has, however, to remember that genes for exoproteins can be located on accessory genetic elements (e.g., on prophages for proteins SEA and SAK) and that lysogenization can influence the expression of other exoproteins, as is the case for *geh* and *hlb*.

Besides resistance to antibiotics, clinical isolates of *S. aureus* often exhibit resistance to heavy metal ions and different cationic substances. These characters have been also used for typing (resistotyping). Most of them are encoded by plasmids and therefore may be lost or newly acquired, which compromises stability of typing results, a problem also encountered with typing by plasmid profiles (see below).

Genotypic Typing Methods

In general terms genotypic typing is a means of sampling the genome but must focus on regions that have a desirable degree of variability. How one decides on the appropriate degree of variability is clearly the key question in attempting to develop typing schemes. As a general rule for epidemiology, several levels of discrimination seem to be required. The most sensitive level required for epidemiology would generate a single pattern for a set of obviously clonal strains from a specific outbreak, different from patterns for epidemiologically unrelated strains. A second level would be able to differentiate from unrelated strains the widely disseminated progeny of such a clone, and a third level would be used to characterize strains common to particular clinical conditions, which might or might not have a definable common genotype.

In the following the different genotyping methods applied to S. aureus are described in historical order.

Plasmid Profile Analysis

This method became applicable after the introduction of microliter techniques for plasmid isolation (13). Staphylococcal strains, especially CNS, often differ in the number and size of their plasmids, easily detectable in agarose gels. The discriminatory power is low, but can be improved by determining restriction endonuclease cleavage patterns of the plasmids. The value of plasmid profile analysis is limited because plasmids can be spontaneously lost or newly acquired. The plasmids of hospital strains can be related, as is the case, for example, for conjugative plasmids carrying aminoglycoside resistance determinants from the United States or Australia.

Plasmid profile analysis is still in use for a quick comparison of isolates of CNS from several blood cultures derived from the same patients to exclude contaminants.

Multilocus Enzyme Electrophoresis

Multilocus enzyme electrophoresis reflects indirectly genotypic polymorphisms by differences in the electrophoretic mobilities of individual, soluble enzymes of primary metabolism. Differences in electrophoretic mobility are due to mutations affecting the charge of an enzyme. This method is rather laborious and therefore not broadly used in epidemiology. It also has the drawback that mutations not affecting the charge of enzymes are missed; therefore, it is not a true genotyping method. For studies of population genetics, however, it has been extremely useful, e.g., for showing clonal relatedness of S. aureus exhibiting phage pattern 95 (35) and clonal diversity among MRSA (27, 28).

Restriction Fragment Length Polymorphism

Restriction fragment length polymorphism typing is based on the random distribution of restriction endonuclease cleavage sites on the bacterial genome, which is reflected by fragment length. Large fragments generated with infrequently cleaving enzymes are resolved by pulsed-field gel electrophoresis. Differences between strains seen in size and number of fragments derived from digestion of genomic DNA can be due to different genetic events such as point mutations, which change the endonuclease recognition site, insertions, deletions, inversions, or transpositions. Thus, a single point mutation leading to loss or gain of one particular restriction site results in a three-fragment difference between the mutant and the parental strain. A three-fragment difference is also seen after an insertion or deletion of a transposable element or a prophage containing one recognition site, whereas translocation or inversion of DNA containing one recognition site would cause a four-fragment difference (43).

Fragment patterns derived from digestion with frequently cutting enzymes cannot be sufficiently separated by ordinary electrophoresis. A particular local DNA sequence with variable cleavage sites can readily be resolved by hybridizing specific probes for this sequence to whole cellular DNA digests. The most common application of this is ribotyping because S. aureus possesses up to eight copies of the rRNA operon, having length polymorphisms for the rRNA gene spacers. There are also polymorphisms with regard to the location of restriction endonuclease cleavage sites on rRNA gene spacers and on DNA sequences neighboring the rRNA gene operons. In comparative trials, though, ribotyping was found less discriminative than SmaI-macrorestriction patterns (44); it was successfully used for the discrimination of MRSA from different continents (15), and can also be used for speciation.

Southern blot hybridization with probes based on chromosomal accessory genetic elements has also been used effectively for strain discrimination. One example is methicillin resistance, based on acquisition of the mecA gene, which is located on "mec-associated DNA," a DNA element of about 50 kb, often including Tn554. The mecA gene contains a cleavage site for the ClaI enzyme. Hybridization of probes for mecA and for Tn554 to ClaI digests of genomic DNA of MRSA detects polymorphisms within mec-associated DNA and neighboring regions; it revealed a probably ancestral relatedness of "classical" MRSA from worldwide sources (20).

Field inversion gel electrophoresis and pulsed-field gel electrophoresis allow a sufficient separation of fragments with comparably high molecular masses (macrorestriction patterns). Because the conditions for lysis of the staphylococcal cells in agarose blocks and deproteinization of DNA, cleavage, and electrophoresis are well standardized, the results are highly reproducible. For S. aureus the enzyme SmaI has been most useful, yielding 8 to 20 fragments ranging from 8 to 800 kb in size. A harmonized protocol proved extremely useful for tracing spread of particular epidemic MRSA in Europe (26).

For application to epidemiological tracing, staphylococcal isolates are regarded as different strains if their SmaI-macrorestriction patterns differ in four or more bands (43). In other words, the occurrence of two or more genetic events influencing the distribution of restriction sites is taken to indicate an epidemiological difference. This criterion is rather arbitrary and could lead to the wrong conclusion: for example, translocation or inversion involving DNA containing one or more sites will affect four or more fragments. Although SmaI-macrorestriction patterns are rather stable during outbreaks of nosocomial infections with MRSA, genomic rearrangements have been observed along with the dissemination of particular MRSA clones beyond a particular site or region (47). When SmaI-macrorestriction patterns are interpreted in terms of epidemiological relatedness, one to three fragment differences (probably due to one genetic event) are considered as variation within the outbreak clonal population, four to six differences indicate a subclone of the outbreak population, and more than seven, an unrelated isolate (43).

Typing by PCR

PCR is able to produce millions of copies from a particular DNA stretch within a short time. Polymorphisms detected by PCR can be based on genetic events taking place between the location of primer binding sequences, thus leading to different length of amplimers. Length polymorphisms in repetitive DNA sequences can also easily be detected by PCR.

Arbitrarily Primed PCR

A prerequisite to the use of PCR for typing is the availability of DNA sequences for the whole bacterial chromosome or at least for particular genes. Until the DNA sequence of the *S. aureus* chromosome has been analyzed for more repetitive sequences useful for PCR typing, arbitrary primers are used as genetic markers, as has been done with other organisms (45). The use of single short primers of arbitrary nucleotide sequence leads to random amplified polymorphic DNA (RAPD) PCR.

RAPD primers consist of 8 to 10 nucleotides; a RAPD PCR is performed at comparably low annealing temperatures. Thus these primers also recognize sequences only partly complementary homologous to them. This can lead to problems of reproducibility, even in repetition of the same typing experiment, especially when more than one method of template DNA preparation or more than one batch of the polymerase is used (see also reference 44).

PCR of Repetitive Sequences

All bacterial genomes harbor repetitive sequences that can be used for epidemiological typing. In most cases these motifs occur throughout the entire chromosome but rarely within genes. For *S. aureus*, repeats like the enterobacterial repetitive intergenic consensus in *Enterobacteriaceae* have not yet been described. Consequently, the enterobacterial repetitive intergenic consensus-2 primers have been used for *S. aureus*, but at rather low annealing temperature, as with RAPD. This situation has been improved by the use of intragenic repetitive regions that have been found in the coagulase (*coa*) and protein A (*spa*) genes (11, 40). Although the polymorphism in the *coa* gene can lead to amplimers of different length and with different *AluI*-cleavage sites, the *coa* polymorphism is not as discriminative as *SmaI*-macrorestriction patterns (40). PCR for length polymorphisms due to variable number of tandem repeats in general proved less discriminative than *SmaI*-macrorestriction patterns. This can be overcome by assessing different variable number of tandem repeat loci with multiplex PCR. For *S. aureus* this was established for *sdr*, *clfA*, *clfB*, *ssp*, *coa*, and *spa* (36). The discriminatory power and reproducibility correspond to *SmaI*-macrorestriction patterns. However, when analyzing gels, the appearing fragments cannot always be traced back to single loci; therefore, hierarchical clustering is not possible.

PCR of rRNA Gene Spacer Sequences

With the exception of a few bacterial species, rRNA operons are present in 2 to 11 copies per prokaryotic chromosome. Because of homologous sequences of the rRNA operon, intrachromosomal recombinational events can lead to polymorphisms. Thus, rRNA gene spacers can differ with regard to sequences and length in the same chromosome and between strains. Length variations can be assessed by PCR with primers binding to the nonvariable sequences of the 16S and 23S rRNA genes. In *S. aureus* this PCR is much less discriminatory than *SmaI*-macrorestriction patterns (5, 48).

PCR for DNA Stretches Flanked by Insertion Elements (IS)

IS257 seems to be widely disseminated, and has a significant role in the evolution of multiply-resistant plasmids; IS256 is associated with Tn4001, carrying genes for gentamicin resistance and efflux of cationic compounds. Despite the mobility of IS257, chromosomal insertions are stable within the time frame of most epidemiological analyses. PCR for length polymorphism of inter-IS256 sequences has been found to be useful for typing multiply-resistant MRSA (7).

PCR for DNA Stretches Flanked by Other Known Sequences

The *S. aureus* chromosome contains the attachment region of Tn916 at several locations. Nucleotide sequences neighboring ribosomal binding sites are obviously rather conserved (29). If in the right orientation, PCR with primer sequences for both known sequences (tar916-shida PCR) amplifies fragments of different length and can be applied for typing *S. aureus* (6).

Amplified Fragment Length Polymorphism

The amplified fragment length polymorphism method is based on selective PCR amplification of restriction fragments from a total digest of whole cellular DNA: restriction of DNA is followed by ligation of oligonucleotide adaptors, selective amplification of sets of restriction fragments and analysis, most on a polyacrylamide capillary. Selective amplification is achieved by use of primers, which extend beyond the restriction site. Although amplified fragment length polymorphism had been applied to a number of different bacterial pathogens, its use for staphylococci remained limited (e.g., typing of *Staphylococcus epidermidis* from blood cultures [41]), probably because of problems with reproducibility of complete digestion of whole cellular DNA.

Sequence-Based Typing

Introduction of and advances in automated sequencing technology gave broad access to affordable, precise sequencing. Analysis of genomic polymorphisms by sequencing has a big advantage compared to "fragment-based" typing techniques because of unambiguous interpretation of results and their portability.

Multilocus Sequence Typing (MLST)

MLST was first introduced into molecular population studies on *Neisseria meningitidis* (23). MLST analyzes housekeeping genes, which are thought to be selectively neutral. Ideally, genes used for MLST should exhibit on average 40 alleles per locus. As in other current MLST schemes also in *S. aureus* seven gene fragments of about 450 bp each are generated by PCR and subjected to sequencing in both directions (8). The corresponding genes are carbamate kinase (*arcC*), shikimate dehydrogenase (*aro*), glycerol kinase (*glp*), guanylate kinase (*gmk*), phosphatase acetyltransferase (*pta*), triosesphosphate isomerase (*tpi*), and acetyl coenzyme A acetyltransferase (*yqi*). The resulting sequence is compared to previously characterized alleles for each via a database provided at the MLST website. Results of MLST can be presented as a UPGMA (unweighted pair group method with arithmetic mean) dendrogram derived from pair-wise differences in allelic profiles or by BURST (which stands for based upon related sequence types) (10). Traditional dendrograms get difficult to interpret when large numbers of isolates have to be examined. Furthermore, relationships between isolates having a high degree of alleles in common (e.g., four or five alleles of seven) are not correctly represented in UPGMA-based dendrograms (3). BURST has been introduced for typing bacteria with an epidemic population structure. BURST requires a list of strain or clone numbers (defined by a particular allelic profile) and allelic profiles and groups them into clonal complexes that share at least five alleles in common. The ancestor of each clonal complex is assumed to be the genotype that had the largest number of single-locus variants. Single-locus variants are presented in a concentric

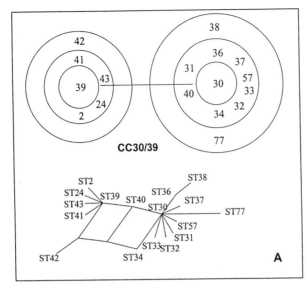

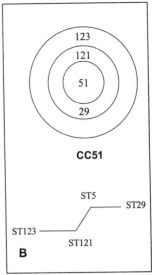

FIGURE 3 Examples of BURST analysis according to Feil et al. (10). (A) Clonal lineages derived from ancestor lineage ST30; ST36 is EMRSA (epidemic MRSA) 16 in United Kingdom. (B) Clonal lineages derived from ancestor ST51.

circle surrounding the ancestor; double-locus variants possess another ring outside of that. BURST has been very successfully used in studies on the evolutionary history of MRSA by M. Enright and coworkers (9). An example of BURST analysis is shown in Fig. 3.

spa Sequence Typing

MLST distinguishes major clonal lineages of the species S. aureus. Its discriminatory power is not sufficient for distinguishing strains within a clonal group for local epidemiological or outbreak investigation purposes. With respect to genomic polymorphisms it has been shown for S. aureus that point mutations by far exceed recombination events. Therefore, single-locus DNA sequencing of repeat regions of the coa (coagulase) and the spa gene (protein A) became of particular interest (12, 40) (for structure of the spa gene, see Fig. 4). The X region of spa is subject to spontaneous mutations as well as loss and gain of repeats. When the repeat structure was assigned an alphanumerical code and spa types were deduced according to the order of repeats, a good

correlation between clonal groupings by MLST, cluster analysis of SmaI-macrorestriction patterns, and spa types was found. Using a microarray based on 2,817 open reading frames within each of 36 different MRSA, it has been shown that spa sequence typing performed better than MLST, pulsed-field gel electrophoresis, and coa sequence typing with regard to discriminatory power and degree of agreement with sequences on the microarray (19). The development of a novel software tool with rapid repeat determination, data management, and retrieval allows the broader application of spa sequence typing to nosocomial epidemiology (14). An example of numerical coding of spa types is shown in Table 2.

Characterization of Staphylococcal Cassette Chromosome (SCC*mec*)

Long-term studies of the epidemiology of MRSA and of the evolutionary relationships among these bacteria have shown that a complete characterization of MRSA lineages not only requires identification of the core genome but also an identification of the large elements carrying the mecA gene (mec-associated DNA [16]).

Until now the complete structure of at least five types of mec elements, also referred to as SCCmec, has been elucidated (Fig. 5). Main components of these elements are the ccr-complex (chromosome cassette recombinase), of which at least four types are now known, and the mec-complex, containing mecA and associated regulatory genes. Class A and B mec differ by integration of IS1272 and deletion of mecI and a part of mecR. SCCmec types II and III contain integrated plasmids as puB110 and pT181 or transposable elements as Tn554 and IS431 (30).

For identification of SCCmec-types in epidemiological studies different PCR systems have been proposed, which detect sequences specific for each type. The approach by Ito et al. (16) identifies the ccr-complex, the mec class, as well as sequences specific for type IVa and IVb. The multiplex PCR established by Oliveira and de Lencastre (31) focuses on a few sequences specific for each type. Type IV SCCmec

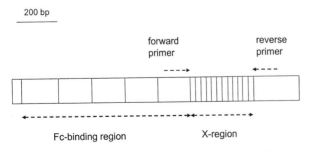

FIGURE 4 Representation of the spa gene coding for protein A. Five and 12 repeats have been indicated in the Fc-binding region and the X region, respectively. The locations of the forward and reverse primers, used to amplify and sequence the X region, are shown.

TABLE 2 Example of numerical coding of *spa* types[a]

spa type or repeat	Repeats or nucleotide sequence
01	r26 r30 r17 r34 r17 r20 r17 r12 r17 r16
02	r26 r23 r17 r34 r17 r20 r17 r12 r17 r16
r26	GAG GAA GAC AAC AAA AAA CCT GGT
r23	AAA GAA GAC GGC AAC AAA CCT GGC
r30	AAA GAA GAC GGC AAC AAA AAA CCT GGT
r17	AAA GAA GAC GGC AAC AAG CCT GGT
r34	AAA GAA GAC AAC AAA AAA CCT GGT
r20	AAA GAA GAC AAC AAC AAA CCT GCC
r12	AAA GAA GAC AAC AAC AAG CCT GGT
r16	AAA GAA GAC GGC AAC AAA CCT GGT

[a]Reprinted with permission from reference 14.

elements are identified in this system by a sequence also present in types I and II and by lack of sequences specific for types I, II, and III.

Different clonal lineages have obviously acquired different types of SCC*mec* at different times and occasions. Thus, it is likely that a clonal lineage of MLST type ST5 has independently acquired SCC*mec* I (Belgium), SCC*mec* II (Japan, Central Europe), and SCC*mec* IV (United States).

The more recently emerging epidemic MRSA, which contain only few resistance determinants as ST22 (widely disseminated in the United Kingdom and in Germany) and ST45 (in Germany, Finland, and the Netherlands), contain type IV SCC*mec*, also described for community-acquired MRSA.

Combined Use of Different Molecular Typing Methods

In addition to outbreak analysis, epidemiological surveillance of *S. aureus* must be able to track interhospital dissemination and the evolution of multiply-resistant strains. Strains exhibiting one particular macrorestriction or PCR pattern may not necessarily be identical or related to a common ancestor. However, a strong match of the results of several typing techniques based on different genomic polymorphisms indicates clonal relatedness. This approach has been above all applied to the epidemiology of MRSA. For example, extraregional dissemination of a particular MRSA clone in Spain and in Portugal was clearly demonstrated by *Sma*I-macrorestriction analysis and patterns of *mecA* and Tn554 hybridization to *Cla*I digests and Tn554 to *Sma*I-digests (37). There is large congruence of clusters of MRSA exhibiting related *Sma*I-macrorestriction patterns and lineages defined by MLST and by *spa*-sequence typing (19, 32).

The attribution of particular epidemic MRSA to genotypically related clusters by *Sma*I-macrorestriction patterns in Central Europe had been confirmed by independent genomic polymorphisms detected by PCR (48). With the application of MLST to *S. aureus* and in particular to MRSA and with further insights into the structure of SCC*mec* elements, resolution strain phylogeny became possible (34). These studies have shown that methicillin resistance has emerged on multiple occasions in a few phylogenetic lineages that had been widely disseminated (Table 3). Strain identification by *Sma*I-macrorestriction patterns followed by MLST and characterization of SCC*mec* proved extremely useful in following emergence and dissemination of community-acquired MRSA in different parts of the world

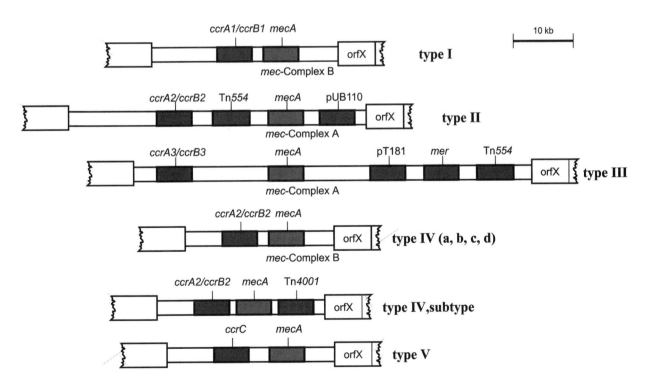

FIGURE 5 (*See the separate color insert for the color version of this illustration.*) Different types of SCC*mec* elements demonstrated in MRSA. orfX, open reading frame X of the chromosome. (Adapted with permission from references 16 and 30.)

TABLE 3 Epidemic MRSA with wide geographical dissemination

MLST type	Allelic profile	SCC*mec*	International dissemination
ST247	3-3-1-12-4-4-16	I	Belgium, Germany, Finland, France, Portugal, Spain, Sweden, United Kingdom, United States
ST239	3-3-1-1-4-4-3	III	Australia, Brazil, Germany, Finland, Greece, Italy, Netherlands, Portugal, Slovenia, Sweden, United Kingdom, United States
ST254	3-32-1-1-4-4-3	I	Denmark, Germany, United Kingdom
ST22	7-6-1-5-8-8-6	IV	Germany, Ireland, Sweden, United Kingdom, Germany
ST45	10-14-8-6-10-3-2	IV	Belgium, Germany, Finland, Netherlands, Sweden
ST228	1-4-1-4-12-24-29	I	Austria, Belgium, Germany, Slovenia, Switzerland
ST5	1-4-4-12-1-10	I	United Kingdom
		II	Japan
		IV	Germany, United States
ST1[a]	1-1-1-1-1-1-1	IV	United States, Germany
ST8[a]	3-3-1-1-4-4-3	IV	United States, New Zealand
ST30[a]	2-2-2-2-6-3-2	IV	Australia
ST80[a]	1-3-1-14-11-51-10	IV	France, Germany, Switzerland

[a]Community-acquired MRSA.

(30). Although there are different clonal lineages of community-acquired MRSA in different continents, there is obviously also a wide dissemination of particular community-acquired MRSA clones, such as ST80 in Central Europe, ST8 in the United States and in Australia, and ST1 in the United States and in Germany (49).

Comparison of Different Genotypic Typing Systems

Which of the available typing systems (Table 4) to choose for practical epidemiology depends upon the aim at different levels of investigation. Typing by means of PCR is comparatively cheap and provides results that are sufficient for first quick information about relatedness of nosocomial isolates. If sequencing capacities are accessible, the same can be achieved by *spa* sequence typing, with the advantage that results from sequence-based typing are highly reproducible, portable, and can easily be stored in databases. Most probably *spa* sequence typing will replace SmaI-

macrorestriction patterns, which are highly discriminatory but rather laborious. MLST requires the highest workload and is not discriminatory enough for epidemiological typing, but is highly predicative in analysis of evolutionary relationships.

REFERENCES

1. **Brakstad, O. G., J. A. Maeland, and Y. Tveten.** 1993. Multiplex polymerase chain reaction for detection of genes for *Staphylococcus aureus* thermonuclease and methicillin resistance and correlation of oxacillin resistance. *Acta Path. Microbiol. Scand.* **101:**681–688.
2. **Brakstad, O. O., K. Aasbakk, and J. A. Maeland.** 1992. Detection of *Staphylococcus aureus* by polymerase chain reaction amplification of the *nuc* gene. *J. Clin. Microbiol.* **30:**1654–1660.
3. **Claus, H., C. Cuny, B. Pasemann, and W. Witte.** 1998. A database system for fragment patterns of genomic DNA of *Staphylococcus aureus*. *Zentbl. Bakteriol.* **287:**105–116.

TABLE 4 Comparison of genotyping systems applied to *S. aureus*

Typing systems	Reproducibility	Discriminatory power
Multilocus enzyme electrophoresis	High to moderate/good	Good
Multilocus sequence typing	High	Good
spa sequencing	High	Good
Restriction fragment length polymorphisms		
SmaI macrorestriction patterns	High	Excellent
Ribotyping	Good	Moderate
Hybridization of *mecA* and Tn554 probes to genomic digests	High	Good
PCR-detected polymorphisms		
rRNA gene spacer	High	Moderate
RAPD (ERIC-2 and BG2 primers)	Moderate (some PCR products inconsistently seen)	Good
tar916-shida	Moderate (some PCR products inconsistently seen)	Good
Inter-IS256	Moderate (some PCR products inconsistently seen)	High for gentamicin-resistant isolates
coa repetitive regions	High	Moderate
spa repetitive regions	High	Moderate
Plasmid profiles	High	Weak

4. **Couto, I., S. Pereira, M. Miragaia, I. Santos Sanchez, and H. de Lencastre.** 2001. Identification of clinical isolates from humans by internal transcribed spacer PCR. *J. Clin. Microbiol.* **39:**3099–3103.

5. **Cuny, C., H. Claus, and W. Witte.** 1996. Discrimination of *S. aureus* by PCR for r-RNA gene spacer size polymorphisms and comparison to *SmaI* macrorestriction patterns. *Zentbl. Bakteriol.* **283:**466–476.

6. **Cuny, C., and W. Witte.** 1996. Typing of *Staphylococcus aureus* by PCR for DNA sequences flanked by transposon Tn916 target region and ribosomal binding site. *J. Clin. Microbiol.* **34:**1502–1505.

7. **Deplano, A. M., O. Vaneechoutte, G. Verschraegen, and M. J. Struelens.** 1997. Typing of *Staphylococcus aureus* and *Staphylococcus epidermidis* strains by PCR analysis of inter-IS256 spacer length polymorphism. *J. Clin. Microbiol.* **35:**2580–2587.

8. **Enright, M. C., N. P. J. Day, C. E. Davies, S. J. Peacock, and B. G. Spratt.** 2000. Multilocus sequence typing for characterization of methicillin-resistant and methicillin-susceptible clones of *Staphylococcus aureus. J. Clin. Microbiol.* **38:**1008–1015.

9. **Enright, M. C., D. A. Robinson, G. Randle, E. J. Feil, H. Grundmann, and B. G. Spratt.** 2002. The evolutionary history of methicillin-resistant *Staphylococcus aureus* (MRSA). *Proc. Natl. Acad. Sci. USA* **99:**7687–7692.

10. **Feil, E. J., J. E. Cooper, H. Grundmann, D. Ashley-Robinson, M. C. Enright, T. Berendt, S. J. Peacock, J. Maynard Smith, M. Murphy, B. G. Spratt, C. E. Moore, and N. P. J. Day.** 2003. How clonal is *Staphylococcus aureus? J. Bacteriol.* **185:**3307–3316.

11. **Frenay, H. M., A. E. Bunschoten, L. M. Schouls, W. J. van Leeuwen, C. M. Vandenbroucke-Grauls, J. Verhoef, and F. R. Mooi.** 1996. Molecular typing of methicillin-resistant *Staphylococcus aureus* on the basis of protein A gene polymorphism. *Eur. J. Clin. Microbiol. Infect. Dis.* **15:**60–64.

12. **Frenay, H. M. E., J. P. G. Theelen, L. M. Schouls, M. J. E. Vandenbroucke-Grauls, J. Verhoef, W. J. Van Leeuwen, and F. P. Mooi.** 1994. Discrimination of epidemic and nonepidemic methicillin-resistant *Staphylococcus aureus* strains in the basis of protein A gene polymorphism. *J. Clin. Microbiol.* **32:**846–847.

13. **Goering, R.V., and E. A. Ruff.** 1993. Comparative analysis of conjugative plasmids mediating gentamicin resistance in *Staphylococcus aureus. Antimicrob. Agents Chemother.* **24:**450–452.

14. **Harmsen, D., H. Claus, W. Witte, J. Rotganger, D. Turnwald, and U. Vogel.** 2003. Typing of methicillin-resistant *Staphylococcus aureus* in a university hospital setting by using novel software for *spa* repeat determination and database management. *J. Clin. Microbiol.* **41:**5442–5448.

15. **Hiramatsu, K. I.** 1995. Molecular evolution of MRSA. *Microbiol. Immunol.* **39:**531–543.

16. **Ito, T., Y. Katayama, K. Asada, N. Mori, K. Tsutsumimoto, C. Tiensasitorn, and K. Hiramatsu.** 2001. Structural comparison of three types of staphylococcal cassette chromosome *mec* integrated in the chromosome in methicillin-resistant *Staphylococcus aureus. Antimicrob. Agents Chemother.* **45:**1323–1336.

17. **Kloos, W. E., and D. W. Lambe.** 1991. *Staphylococcus*, chapter 28. *In* A. Balows, W. J. Hausler, Jr., K. L. Herrmann, H. D. Isenberg, and H. J. Shadomy (ed.), *Manual of Clinical Microbiology*, 5th ed. American Society for Microbiology, Washington, D.C.

18. **Kobayashi, N., H. Wu, K. Kojma, K. Taniguchi, S. Urasawa, N. Kehara, Y. Omizi, Y. Kishi, A. Yagihashi, and I. Kurokawa.** 1994. Detection of *mecA, femA* and *femB* genes in clinical strains of staphylococci using polymerase chain reaction. *Epidemiol. Infect.* **113:**259–266.

19. **Koreen, L., S. V. Ramaswamy, E. A. Graviss, S. Naidick, J. M. Musser, and B. N. Kreiswirth.** 2004. *spa*-typing method for discriminating among *Staphylococcus aureus* isolates: implications for use of a single marker to detect genetic micro- and macrovariation. *J. Clin. Microbiol.* **42:**792–799.

20. **Kreiswirth, B., J. Kornblum, R. D. Arbeit, W. Eisner, J. N. Maslow, A. McGeer, D. E. Low, and R. P. Novick.** 1993. Evidence for a clonal origin of methicillin resistance in *Staphylococcus aureus. Science* **259:**227–230.

21. **Kwok, A., Y. Shei-Siang, R. P. Reynolds, S. J. Bay, Y. Av-Gay, N. J. Dovichi, and A. W. Dhow.** 1999. Species identification and phylogenetic relationship based on partial HSP60 gene sequences within the genus *Staphylococcus. Int. J. System. Bacteriol.* **49:**1181–1192.

22. **Ludwig, W., and H. H. Schleifer.** 1994. Bacterial phylogeny based on 16S and 23S rRNA sequence analysis. *FEMS Microbiol. Rev.* **15:**155–173.

23. **Maiden, M. C., J. A. Bygraves, E. Feil, G. Morelli, J. E. Russell, R. Urwin, Q. Zhang, J. Zhou, K. Zurth, D. A. Caugant, I. M. Feavers, M. Achtman, and B. G. Spratt.** 1998. Multilocus sequence typing: a portable approach to the identification of clones within populations of pathogenic microorganisms. *Proc. Natl. Acad. Sci. USA* **95:**3140–3145.

24. **Martineau, F., F. Picard, P. H. Roy, M. Oulette, and M. Bergeon.** 1998. Species specific and ubiquitons. DNA-based assays for rapid identification of *Staphylococcus aureus. J. Clin. Microbiol.* **36:**618–623.

25. **Mulligan, M., R. Y. Kwok, D. M. Citron, J. F. John, and P. B. Smith.** 1988. Immunoblots, antimicrobial resistance, and bacteriophage typing of oxacillin resistant *Staphylococcus aureus. J. Clin. Microbiol.* **26:**2395–2401.

26. **Murchan, S., M. E. Kaufmann, A. Deplano, R. de Ryck, M. Struelens, C. E. Zinn, V. Fussing, S. Salmenlinna, J. Vuopio-Varkila, N. El Solh, C. Cuny, W. Witte, P. T. Tassios, N. Legakis, W. van Leeuwen, A. van Belkulm, A. Vindel, I. Laconcha, J. Garaizar, S. Haeggman, B. Olsson-Liljequist, U. Ransjo, G. Coombes, and B. Cookson.** 2003. Harmonization of pulsed-field gel electrophoresis protocols for epidemiological typing of strains of methicillin-resistant *Staphylococcus aureus*: a single approach developed by consensus in 10 European laboratories and its application for tracing the spread of related strains. *J. Clin. Microbiol.* **41:**1574–1585.

27. **Musser, J., and R. K. Selander.** 1990. Genetic analysis of natural populations of *Staphylococcus aureus. In* R. P. Novick (ed.), *Molecular Biology of the Staphylococci*. VCH Publisher, New York, N.Y.

28. **Musser, J. M.** 1996. Molecular population genetic analysis of emerged bacterial pathogens: selected insights. *Emerg. Infect. Dis.* **2:**1–17.

29. **Novick, R. P.** 1990. The *Staphylococcus aureus* as a molecular genetic system. *In* R. P. Novick (ed.), *Molecular Biology of the Staphylococci*. VCH Publisher, New York, N.Y.

30. **Okuma, K., K. Iwakawa, J. D. Turnridge, W. B. Grubb, J. M. Bell, F. G. O'Brien, G. W. Coombs, J. W. Pearman, F. C. Tenover, M. Kapi, C. Tiensasitorn, T. Ito, and K. Hiramatsu.** 2002. Dissemination of new methicillin-resistant *Staphylococcus aureus* clones in the community. *J. Clin. Microbiol.* **40:**4289–4294.

31. **Oliveira, D., and H. de Lencastre.** 2002. Multiplex PCR strategy for rapid identification of structural types and variants of the *mec* element in methicillin-resistant *Staphylococcus aureus*. *Antimicrob. Agents Chemother.* **46:** 2155–2161.

32. **Peacock, S. J., G. D. de Silva, A. Justice, A. Cowland, C. E. Moore, C. G. Winearls, and N. P. Day.** 2002. Comparison of multilocus sequence typing and pulsed-field gel electrophoresis as tools for typing *Staphylococcus aureus* isolates in a microepidemiological setting. *J. Clin. Microbiol.* **40:**3764–3770.

33. **Probst, A., C. Hertel, L. Richter, L. Wassill, W. Ludwig, and W. Hammes.** 1998. *Staphylococcus condimenti* sp. nov., from soy sauce mash, and *Staphylococcus carnosus* (Schleifer and Fischer, 1982) subsp. *utilis* subsp. nov. *Int. J. Syst. Bacteriol.* **48:**651–658.

34. **Robinson, A., and M. C. Enright.** 2003. Evolutionary models of the emergence of methicillin-resistant *Staphylococcus aureus*. *Antimicrob. Agents Chemother.* **47:**3926–3934.

35. **Rosdahl, W. T., W. Witte, J. M. Musser, and J. O. Jarlow.** 1994. *Staphylococcus aureus* of type 95, spread of a single clone. *Epidemiol. Infect.* **113:**463–470.

36. **Sabat, A., J. Kryszton-Russjan, W. Strzalka, R. Filipek, K. Kosowska, W. Hryniewicz, J. Travis, and J. Potempa.** 2003. New method for typing *Staphylococcus aureus* strains: multiple–locus variable–number tandem repeat analysis of polymorphism and genetic relationships of clinical isolates. *J. Clin. Microbiol.* **41:**1801–1804.

37. **Santos-Sanches, I., M. A. De Sousa, S. I. Calheiros, L. Felicito, I. Pedra, and H. De Lencastre.** 1995. Multidrug-resistant Iberian epidemic clone of methicillin-resistant *Staphylococcus aureus* endemic in a hospital in Northern Portugal. *Microb. Drug Res.* **1:**299–306.

38. **Saruta, K., T. Matsungana, M. Kono, S. Hoshina, S. Ikawa, O. Sakai, and K. Machida.** 1997. Rapid identification and typing of *Staphylococcus aureus* by nested PCR amplified ribosomal DNA spacer region. *FEMS Microbiol. Lett.* **146:**271–278.

39. **Schleifer, K. H., and R. M. Kroppenstedt.** 1990. Chemical and molecular classification of staphylococci. *J. Appl. Bacteriol. Symp. Ser.* **19:**9S-45S.

40. **Schwarzkopf, A., and H. Karch.** 1994. Genetic variation in *Staphylococcus aureus* coagulase genes: potential and limit for use as epidemiological marker. *J. Clin. Microbiol.* **32:**2407–2412.

41. **Sloos, J. H., P. Janssen, C. P. A. van Boven, and L. Dijkshoorn.** 1998. AFLP™ typing of *Staphylococcus epidermidis* in multiple sequential blood cultures. *Res. Microbiol.* **149:** 221–228.

42. **Stobberingh, E. E., and K. C. Winkler.** 1977. Restriction deficient mutants of *Staphylococcus aureus*. *J. Gen. Microbiol.* **90:**359–367.

43. **Tenover, F. C., R. D. Arbeit, and R. V. Goering.** 1997. How to select and interpret molecular strain typing methods for epidemiological studies of bacterial infections: a review for healthcare epidemiologists. *Infect. Control Hosp. Epidemiol.* **18:**426–439.

44. **Van Belkum, A., R. Bax, and G. Prevost.** 1994. Comparison of four genotyping assays for epidemiological study of methicillin-resistant *Staphylococcus aureus*. *Eur. J. Clin. Microbiol. Infect. Dis.* **13:**420–424.

45. **Versalovic, J., T. Koeuth, and J. R. Lupski.** 1991. Distribution of repetitive DNA sequences in eubacteria and application to fingerprinting of bacterial genomes. *Nucleic Acids Res.* **19:**6823–6831.

46. **Witte, W., C. Cuny, and H. Claus.** 1993. Clonal relatedness of *Staphylococcus aureus* strains from infections in humans as deduced from genomic DNA fragment patterns. *Med. Microbiol. Lett.* **2:**72–79.

47. **Witte, W., C. Cuny, O. Zimmermann, R. Ruchel, M. Hopken, R. Fischer, and S. Wagner.** 1994. Stability of genomic DNA fragment patterns in methicillin-resistant *Staphylococcus aureus* during the course of intra- and interhospital spread. *Eur. J. Epidemiol.* **10:**743–748.

48. **Witte, W., M. Kresken, C. Braulke, and C. Cuny.** 1997. Increasing incidence and widespread dissemination of methicillin-resistant *Staphylococcus aureus* (MRSA) in hospitals in central Europe, with special reference to German hospitals. *Clin. Microbiol. Infect.* **3:**414–422.

49. **Witte, W., C. Braulke, C. Cuny, B. Strommenger, G. Werner, D. Heuck, U. Jappe, C. Wendt, H. J. Linde, and D. Harmsen.** 2005. Emergence of MRSA with Panton-Valentine Leukocidin genes in Central Europe. *Eur. J. Clin. Microbiol. Infect. Dis.* **24:**1–5.

50. **Yugens, S., A. Temprano, M. Sanchez, J. M. Luengo, and G. Nahamo.** 2001. Identification of *Staphylococcus* spp. by PCR-restriction fragment length polymorphism of *gap* gene. *J. Clin. Microbiol.* **39:**3693–3695.

The *Staphylococcus aureus* NCTC 8325 Genome

ALLISON F. GILLASPY, VERONICA WORRELL, JOSHUA ORVIS, BRUCE A. ROE,
DAVID W. DYER, AND JOHN J. IANDOLO

32

Staphyloccocus aureus strain NCTC 8325 was used to develop the original genome map of *S. aureus*, which owes much of its development to the efforts of Peter A. Pattee and colleagues (17, 28). By focusing on the *S. aureus* phage group III strain NCTC 8325, Pattee's laboratory utilized transduction and transformation to identify a series of linkage groups that were organized into a rough genetic map. Once physical techniques for genome mapping (restriction endonuclease digestion and pulsed field gel electrophoresis) became feasible, Pattee and coworkers attempted to fit the genetic linkage data to the physical maps of the genome. The map continued to evolve, and in 2000 an effort was undertaken by our group to sequence and annotate the entire genome of strain 8325 (http://microgen.ouhsc.edu). The genome sequence and annotation are now complete and will be the main focus of this chapter. In addition to strain NCTC 8325, the genome sequence and annotation for at least six other *S. aureus* strains (COL, N315, Mu50, MW2, MRSA252, MSSA476) have been completed in recent years (3, 16, 20). Information for these genomes is available on websites at The Institute for Genomic Research (http://www.tigr.org) and the National Center for Biotechnology Information (http://www.ncbi.nlm.nih.gov).

To date, there are more complete genome sequences available for *S. aureus* than for any other microbial species. The availability of several genomes has led to detailed comparative analyses between clinical isolates and prototype strains as well as strains multiply-resistant and sensitive to several antibiotics (16, 23). These comparisons have confirmed earlier suspicions that the majority of the differences among strains are traceable to mobile genetic elements, including temperate bacteriophages, transposons, insertion sequences, and pathogenicity islands. This relationship can be seen in colinearity analysis of strain NCTC 8325 and COL using a dot-plot program (Fig. 1). Indeed, this plot shows that the major differences between these two strains, evident by the gaps in the graph, are at the sites of prophage and other element insertions for each genome. In fact, the most recent in silico comparative analysis concluded that the "core genome" makes up ~75% of any *S. aureus* genome (23), with the similarity and colinearity of the different sequenced isolates being close to 100%. In this regard, we have done a BLASTp analysis comparing the entire predicted coding sequence from each of the sequenced genomes to the coding sequence of NCTC 8325. Based on bitsum scores, we were able to determine that NCTC 8325 is most closely related to strain COL followed by MW2, MSSA476, Mu50, N315, and MRSA252, respectively (Fig. 2). This was also confirmed with data from Holden and colleagues (16) using concatenated sequences from seven loci used in multilocus sequence typing (MLST). Overall, the seven *S. aureus* strains sequenced to date all cluster together on a phylogenetic tree and show little diversity.

THE GENOME MAP OF NCTC 8325

The complete circular genome map of NCTC 8325 shows the position of each predicted open reading frame (ORF) within the genome and the predicted functional role for each coding sequence (Fig. 3, Table 1). The annotation of the ~2.82-Mb genome contains 2,892 predicted ORFs, 61 tRNA genes, three structural RNAs, and five complete ribosomal RNA operons. The genome has a G+C content of 32.9% and an average gene length of 824 nucleotides with 84.5% of the genome consisting of coding sequence, similar to the other sequenced *S. aureus* strains. Among the 2,892 ORFs, 85.4% have a predicted methionine (ATG) start codon, while the remaining 14.6% have a predicted valine (GTC) or leucine (TTG) start.

More than half of the genome (~61%) encodes proteins of unknown function. The majority of these are similar to predicted ORFs in the sequences of other strains of *S. aureus* or other microorganisms and have therefore been designated as "conserved hypothetical proteins." Approximately 5.5% of the NCTC 8325 genome is strain-specific and predicted to encode hypothetical proteins. This is similar to other sequenced *S. aureus*, which range from 5 to 6% unique content.

The current map for the NCTC 8325 genome (Fig. 3) shows the position of each of the 2,892 ORFs designated by color based on their predicted functional roles. It illustrates a strong coding bias for each strand. Approximately half of the coding sequences are located predominantly on one

Gram-Positive Pathogens, 2nd edition, edited by Vincent A. Fischetti et al.
© 2006 ASM Press, Washington, D.C.

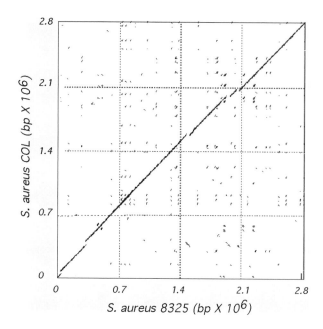

FIGURE 1 Genome comparison of strain NCTC 8325 and COL using the DNAMAN (Lynnon Corp.) dot-plot comparison program. Genomes were compared at the nucleotide level every 50 bp across the entire sequence. Similarity at each position is illustrated by the presence of a black dot. These two genomes are considered collinear due to the high degree of identity along the entire sequence. Numbers on the *x* axis represent nucleotide positions in the NCTC 8325 genome while numbers on the *y* axis represent nucleotide positions in the COL genome.

replichore (8), and the second half of the coding sequences are located predominantly on the other replichore. Interestingly, this is seen in all sequenced *S. aureus* genomes (3, 16, 20) and other low-G+C gram-positive organisms such as *Streptococcus pyogenes* (6, 13), *Streptococcus mutans* (1), *Streptococcus agalactiae* (30), and *Staphylococcus epidermidis* (36). This phenomenon was first observed in the *Escherichia*

coli K-12 genome and appears to be related to the GC skew and the relationship of each region to the origin of replication (8). Nevertheless, the extent of the coding bias in the lower-GC, gram-positive organisms listed above appears to be much more dramatic and warrants additional study.

The functional role categories used in the NCTC 8325 annotation are those proposed by The Institute for Genomic Research (http//:www.tigr.org). The distribution for each of the 18 role categories of NCTC 8325 ORFs is shown in Fig. 4. The total percentage of ORFs on this graph is greater than 100% because some of the predicted genes were assigned multiple roles during annotation. For example, there are 221 ORFs encoded within the prophage regions in NCTC 8325. Most have no known function, but are found within other bacteriophage genomes. These ORFs are designated as "conserved hypothetical prophage genes." In total, there are 80 ORFs encoded within the three temperate bacteriophage genomes that have this designation. They

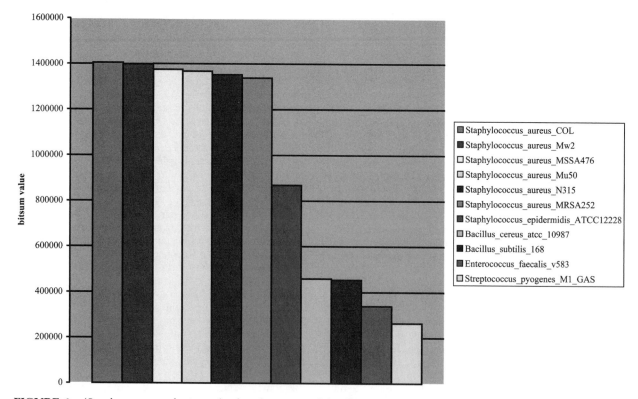

FIGURE 2 (*See the separate color insert for the color version of this illustration.*) Whole genome comparisons using the bitsum method. Coding regions from each of the six *S. aureus* sequenced strains were used in a BLASTp analysis with ORFs from NCTC 8325. The bit scores for these BLASTp comparisons were then summed (the bitsum) and used to plot the similarity of each genome to strain 8325. Coding regions from the completed genomes for other gram-positive organisms are shown for comparison.

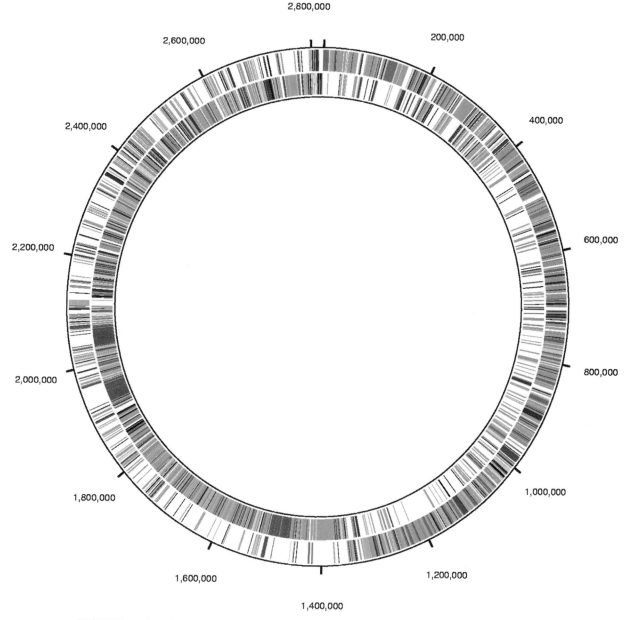

FIGURE 3 *(See the separate color insert for the color version of this illustration.)* Circular map of the NCTC 8325 genome. The ORFs present on the sense strand are represented on the outermost circle, the ORFs on the antisense strand are shown on the second circle, and the innermost circle is the GC skew on each coding strand. Nucleotide positions are labeled every 200,000 bp, and each of the predicted ORFs is colored based on the assigned functional role category in the final annotation. Colors and designations for each role category can be found in the legend for Fig. 4. As is customary, the first ORF represented is *dnaA*.

have therefore been counted as both "conserved hypothetical proteins" and the "mobile and extrachromosomal elements" (Fig. 4). Further discussion of the bacteriophages and their annotation is found later in this chapter.

Many ORFs encoding housekeeping functions in *S. aureus* are highly conserved sequences with an organization similar to that in other bacteria. In particular, ribosomal protein operons, ribosomal RNA operons, and genes flanking the origin of replication are as observed in the *Bacillus subtilis* (2) and *Escherichia coli* (4, 8) genomes. Collectively,

the staphylococcal genomes also carry a variable number of mobile genetic elements, including insertion sequences, transposons, prophages, and uncharacterized foreign DNA inserts. The additional DNA sequences are often responsible for virulence properties, such as enterotoxin production, production of toxic shock syndrome toxin, or resistance to methicillin. In the NCTC 8325 genome, these types of elements contain approximately 8% of the total ORFs, with the highest number being prophage-encoded. The variable modular elements and other interesting features of NCTC

Role Category	#	%
Amino acid biosynthesis	69	2%
Biosynthesis of cofactors, prosthetic groups, and carriers	70	2%
Cell envelope	98	3%
Cellular processes	82	3%
Central intermediary metabolism	34	1%
Conserved hypothetical proteins	1636	51%
DNA metabolism	70	2%
Energy metabolism	177	6%
Fatty acid and phospholipid metabolism	32	1%
Hypothetical proteins / Unknown / Unclassified	201	6%
Mobile and extrachromosomal element functions	236	7%
Protein fate	58	2%
Protein synthesis	114	4%
Purines, pyrimidines, nucleosides, and nucleotides	57	2%
Regulatory functions	65	2%
Signal transduction	31	1%
Transcription	21	1%
Transport and binding proteins	160	5%
Total	3211	

FIGURE 4 (*See the separate color insert for the color version of this illustration.*) Graphical representation of the functional role category assignments for each of the 2,892 predicted ORFs for NCTC 8325. Numbers of genes and the percentage of the genome they represent are shown for each category. Percentage and the number of ORFs total more than 100% due to multiple role assignments for some ORFs.

8325 and other staphylococcal genomes are summarized below. A tabular listing of all ORFs assigned a putative function can be found in Table 1.

Insertion Sequences and Transposons

Insertion sequences and transposons have been identified in a variety of strains of *S. aureus*. Strain 8325 appears to contain components of at least nine potential insertion elements (IS). There are at least two copies are of an IS*1181*-like element, and two copies are highly similar to IS*1272*. The remainder are truncated and degenerate IS-like fragments. However, these degenerate IS elements appear to be present in the genome of several strains, including COL, Mu50, and N315. None of the 8325 complement of IS elements are associated with antibiotic resistance, toxins, or pathogenicity islets.

A variety of insertion sequences and transposons has been found in the other sequenced strains. IS*256* is a component of Tn*4001*, and multiple copies of this element have been detected on the chromosomes of certain clinical *S. aureus* strains (9). IS*431* (IS*257*) is a component of Tn*4003* and is found on a number of antibiotic resistance-encoding and conjugative plasmids of *S. aureus* (5, 29). This insertion sequence is also found in the methicillin-resistance region of the chromosome. IS*1181* was initially identified on an *S. aureus* plasmid, and copies of this element are present in the 8325 genome (12).

Transposons identified in other *S. aureus* strains include Tn*551*, encoding macrolide-lincosamide-streptogramin B resistances, and Tn*4001*, encoding erythromycin, gentamicin, neomycin, and kanamycin resistance (25). Both of these transposons were utilized by Pattee in his initial mapping of the chromosome. Additional transposons include Tn*4003*, encoding trimethoprim resistance, and Tn*4002* and Tn*4201*, encoding β-lactamase (15, 21). Site-specific transposons include Tn*552* (encoding β-lactamase), Tn*554* (encoding resistance to erythromycin and spectinomycin), and Tn*3582* (which is closely related to Tn*554*) (26, 30, 33). These transposable elements have preferred primary insertion sites in the chromosome. For Tn*552* this consists of an integrated copy of *bin-res*, and for Tn*554*, a unique site, *att554*. The latter transposon can insert at low frequency at secondary sites if the primary site is deleted (26). Interestingly, Tn*554*, Tn*552*, and a Tn*916*-like element are found in the highly antibiotic-resistant strains N315, Mu50, and MRSA252, but not in NCTC 8325.

Prophages

S. aureus NCTC 8325 is known to carry three temperate bacteriophages, φ11, φ12, and φ13. The φ11 and φ13 prophages are located in relative close proximity to one another in the *Sma*I-F fragment of the genome with only ~100 kb of DNA separating the two elements, while the φ12 prophage is inserted in the *Sma*I-A fragment and separated from the others by approximately 415 kb. Bacteriophages φ11 and φ12 insert intergenically and their lysogens have no known phenotype other than repressing infection by similar phages. The *attB* site of both phages lies between conserved hypothetical proteins; the former is between ORF 2016 and ORF 2017 and the latter between ORF 1513 and ORF 1514. φ13 inserts intragenically into ORF 2240 (*hlb*), inactivating the gene function. Coding sequences for the genes of all three temperate phages are found predominantly on the strand opposite the direction of replication. Sequencing of the NCTC 8325 genome resulted in the first complete sequence available for each of these bacteriophages, and a complete annotation and discussion of these were published separately in 2002 (18).

Temperate staphylococcal bacteriophages also carry a number of virulence factors such as enterotoxin A production, staphylokinase, and leukocidin (35), as well as certain transposons (7, 11, 19). They insert both inter- and intragenically into the genome and therefore have the capacity to alter the genetic organization of the host genome and gene expression. Interruption of lipase and beta-toxin production are the best known examples. φL54a is a negative converting phage whose *att* site is in the lipase gene (*geh*), and lysogeny by it inactivates glycerol ester hydrolase (*geh*) production (10, 11, 22). Among converting phages, φ13 is unique because it carries the staphylokinase gene (*sak*) and

TABLE 1 Tabular listing of all predicted ORFs in NCTC 8325 listed by functional role category based on TIGR roles

Amino acid biosynthesis category	Functional role	Gene
Aspartate family		
ORF00338[a]	5-Methyltetrahydropteroyltriglutamate–homocysteine S-methyltransferase	metE[b]
ORF00340	trans-Sulfuration enzyme family protein	
ORF00422	trans-Sulfuration enzyme family protein, putative	
ORF00849	Aminotransferase, class V superfamily, putative	
ORF01319	Aspartate kinase, putative	
ORF01320	Homoserine dehydrogenase, putative	hom
ORF01321	Threonine synthase	thrC
ORF01322	Homoserine kinase	thrB
ORF01394	Aspartokinase II, putative	
ORF01395	Aspartate-semialdehyde dehydrogenase	asd
ORF01396	Dihydrodipicolinate synthase	dapA
ORF01397	Dihydrodipicolinate reductase	dapB
ORF01398	2,3,4,5-Tetrahydropyridine-2-carboxylate N-succinyltransferase, putative	dapD
ORF01401	Diaminopimelate decarboxylase	lysA
ORF01845	Formate-tetrahydrofolate ligase, putative	fhs
Glutamate family		
ORF00147	Acetylglutamate kinase, putative	
ORF00148	Glutamate N-acetyltransferase/amino-acid acetyltransferase	argJ
ORF00149	N-Acetylglutamate gamma-semialdehyde dehydrogenase, putative	
ORF00150	Ornithine aminotransferase, putative	
ORF00434	Transcriptional regulator, lysR family, putative	
ORF00435	Glutamate synthase, large subunit, putative	gltB
ORF00436	NADH-glutamate synthase small subunit, putative	gltD
ORF00895	Glutamate dehydrogenase, NAD-specific, putative	gluD
ORF00898	Argininosuccinate lyase	argH
ORF00899	Argininosuccinate synthase	argG
ORF01128	Ornithine carbamoyltransferase	argF
ORF01287	Glutamine synthetase, type I	glnA
ORF01597	Pyrroline-5-carboxylate reductase	proC
ORF01617	Arginine repressor	argR
ORF02760	Glutamate synthase alpha subunit, putative	gltB
ORF02968	Ornithine carbamoyltransferase	argF
Pyruvate family		
ORF00536	Branched-chain amino acid aminotransferase	ilvE
ORF00914	2-Isopropylmalate synthase, putative	
ORF01451	Threonine dehydratase	ilvA
ORF02281	Dihydroxy-acid dehydratase	ilvD
ORF02282	Acetolactate synthase, large subunit, biosynthetic type	ilvB
ORF02284	Ketol-acid reductoisomerase	ilvC
ORF02285	2-Isopropylmalate synthase	leuA
ORF02286	3-Isopropylmalate dehydrogenase	leuB
ORF02287	3-Isopropylmalate dehydratase, large subunit	leuC
ORF02288	3-Isopropylmalate dehydratase, small subunit	leuD
ORF02289	Threonine dehydratase, putative	
Serine family		
ORF00422	trans-Sulfuration enzyme family protein, putative	
ORF00510	Serine acetyltransferase, putative	cysE
ORF01833	D-3-Phosphoglycerate dehydrogenase	serA
ORF02354	Serine hydroxymethyltransferase, putative	glyA
Aromatic amino acid family		
ORF00832	3-Dehydroquinate dehydratase, type I	aroD
ORF01364	Prephenate dehydrogenase, putative	
ORF01366	Anthranilate synthase component I	trpE
ORF01367	Anthranilate synthase component II, putative	
ORF01368	Anthranilate phosphoribosyltransferase	trpD
ORF01369	Indole-3-glycerol phosphate synthase, putative	
ORF01370	N-(5′-Phosphoribosyl)anthranilate isomerase, putative	trpF

(Continued on next page)

Amino acid biosynthesis category	Functional role	Gene
ORF01371	Tryptophan synthase, beta subunit	*trpB*
ORF01372	Tryptophan synthase, alpha subunit	*trpA*
ORF01481	3-Phosphoshikimate 1-carboxyvinyltransferase	*aroA*
ORF01482	3-Dehydroquinate synthase	*aroB*
ORF01483	Chorismate synthase	*aroC*
ORF01635	Shikimate kinase, putative	*aroK*
ORF01699	Shikimate 5-dehydrogenase	*aroE*
Histidine family		
ORF00733	Histidinol-phosphate aminotransferase	*hisC*
ORF01845	Formate-tetrahydrofolate ligase, putative	*fhs*
ORF03008	Imidazole glycerol phosphate synthase subunit hisF, putative	*hisF*
ORF03009	Phosphoribosylformimino-5-aminoimidazole carboxamide ribotide isomerase, putative	*hisA*
ORF03010	Imidazole glycerol phosphate synthase, subunit H (hisH), putative	*hisH*
ORF03011	Imidazoleglycerol-phosphate dehydratase, putative	*hisB*
ORF03012	Histidinol-phosphate aminotransferase, putative	
ORF03013	Histidinol dehydrogenase	*hisD*
ORF03014	ATP phosphoribosyltransferase, putative	

Biosynthesis of cofactors, prosthetic groups, and carriers

Molybdopterin		
ORF02536	Molybdopterin cofactor biosynthesis protein A, putative	
ORF02537	Molybdopterin-guanine dinucleotide biosynthesis mobA, putative	
ORF02538	Molybdopterin-converting factor, subunit 1	*moaD*
ORF02540	Molybdopterin-converting factor moa, putative	*gphA*
ORF02541	Molybdopterin-guanine dinucleotide biosynthesis protein MobB	*mobB*
ORF02542	Molybdopterin biosynthesis protein moeA, putative	
ORF02543	Molybdenum cofactor biosynthesis protein C	*moaC*
ORF02544	Molybdopterin precursor biosynthesis moaB, putative	
Menaquinone and ubiquinone		
ORF00983	2-Succinyl-6-hydroxy-2,4-cyclohexadiene-1-carboxylic acid synthase/2-oxoglutarate decarboxylase	*menD*
ORF00985	Enoyl-CoA hydratase/isomerase family protein, putative	
ORF01487	Menaquinone biosynthesis methyltransferase, putative	*ubiE*
ORF01618	Geranyltranstransferase, putative	*ispA*
ORF01916	*o*-Succinylbenzoic acid–CoA ligase, putative	
Lipoate		
ORF00861	Lipoic acid synthetase	*lipA*
Heme, porphyrin, and cobalamin		
ORF01065	Heme A synthase, putative	
ORF01066	Protoheme IX farnesyltransferase	
ORF01771	Glutamate-1-semialdehyde-2,1-aminomutase	*hemL*
ORF01772	Porphobilinogen synthase	*hemB*
ORF01773	Uroporphyrinogen-III synthase, putative	*hemD*
ORF01774	Porphobilinogen deaminase	*hemC*
ORF01776	Glutamyl-tRNA reductase	*hemA*
ORF01960	Protoporphyrinogen oxidase	*hemG*
ORF01961	Ferrochelatase	*hemH*
ORF01962	Uroporphyrinogen decarboxylase	*hemE*
ORF02000	Glutamate-1-semialdehyde-2,1-aminomutase, putative	*hemL*
ORF02945	Siroheme synthase, putative	
Other		
ORF00466	4-Diphosphocytidyl-2C-methyl-D-erythritol kinase	*ispE*
ORF00577	Mevalonate kinase, putative	*mvaK1*
ORF01486	Heptaprenyl diphosphate syntase component II, putative	*hepT*
ORF02139	Pyrazinamidase/nicotinamidase, putative	

(*Continued on next page*)

TABLE 1 (*Continued*)

Amino acid biosynthesis category	Functional role	Gene
ORF02860	HMG-CoA synthase, putative	
Pantothenate and coenzyme A		
ORF01075	Pantetheine-phosphate adenylyltransferase	*coaD*
ORF01178	Phosphopantothenoylcysteine decarboxylase/phosphopantothenate–cysteine ligase	*coaBC*
ORF01795	Dephospho-CoA kinase	
ORF01845	Formate-tetrahydrofolate ligase, putative	*fhs*
ORF02739	2-Dehydropantoate 2-reductase	*panE*
ORF02916	Aspartate 1-decarboxylase	*panD*
ORF02918	Pantoate–beta-alanine ligase	*panC*
ORF02919	3-Methyl-2-oxobutanoate hydroxymethyltransferase	*panB*
ORF02920	2-Dehydropantoate 2-reductase	*pane*
Pyridoxine		
ORF00499	Pyridoxine biosynthesis protein	
Glutathione		
ORF00171	Gamma-glutamyltranspeptidase, putative	*ggt*
Biotin		
ORF02712	6-Carboxyhexanoate–CoA ligase, putative	
ORF02714	Biotin synthase	*bioB*
ORF02715	Adenosylmethionine-8-amino-7-oxononanoate aminotransferase	*bioA*
ORF02716	Dethiobiotin synthase, putative	*bioD*
Folic acid		
ORF00489	Dihydropteroate synthase	*folP*
ORF00490	Dihydroneopterin aldolase	*folB*
ORF00491	2-Amino-4-hydroxy-6-hydroxymethyldihydropteridine pyrophosphokinase	*folK*
ORF00722	*para*-Aminobenzoate synthase, glutamine amidotransferase, component II, putative	*pabB*
ORF00723	*para*-Aminobenzoate synthase, component I, putative, authentic frameshift	
ORF01007	FolD bifunctional protein, putative	*folD*
ORF01434	Dihydrofolate reductase	*folA*
ORF01766	Folylpolyglutamate synthase/dihydrofolate synthase, putative	*folC*
ORF02354	Serine hydroxymethyltransferase, putative	*glyA*
Pyridine nucleotides		
ORF00284	5′-Nucleotidase, lipoprotein e(P4) family	
ORF00943	ATP-NAD kinase, putative	
ORF01697	Nicotinate (nicotinamide) nucleotide adenylyltransferase	*nadD*
ORF02132	NAD$^+$ synthetase	*nadE*
ORF02133	Nicotinate phosphoribosyltransferase, putative	
Thiamine		
ORF00562	Phosphomethylpyrimidine kinase	*thiD*
ORF01824	Thiamine biosynthesis protein ThiI	*thiI*
ORF02328	Thiamine-phosphate pyrophosphorylase	*thiE*
ORF02329	Hydroxyethylthiazole kinase, putative	*thiM*
ORF02330	Phosphomethylpyrimidine kinase	*thiD*
Riboflavin, flavin mononucleotide, and flavin adenine dinucleotide		
ORF01249	Riboflavin biosynthesis protein RibF	*ribF*
ORF01886	Riboflavin synthase, beta subunit	*ribH*
ORF01887	Riboflavin biosynthesis protein, putative	*ribAB*
ORF01888	Riboflavin synthase, alpha subunit	*ribE*
ORF01889	Riboflavin biosynthesis protein RibD	*ribD*

Cell envelope

Biosynthesis of murein sacculus and peptidoglycan

ORF00222	*tagB* protein, putative	*tagB*

(*Continued on next page*)

TABLE 1 Tabular listing of all predicted ORFs in NCTC 8325 listed by functional role category based on TIGR roles (*Continued*)

Amino acid biosynthesis category	Functional role	Gene
ORF00223	Teichoic acid biosynthesis protein F, putative	*tagF*
ORF00248	Peptidoglycan hydrolase, putative	*lytM*
ORF00641	Teichoic acid translocation ATP-binding protein, putative	*rfbI*
ORF00643	*tagB* protein, putative	*tagB*
ORF00644	*tagX* protein, putative	*tagX*
ORF00645	Glycerol-3-phosphate cytidylyltransferase	*tagD*
ORF00646	Penicillin-binding protein 4, putative	*pbp4*
ORF00752	UDP-*N*-acetylenolpyruvoylglucosamine reductase, putative	*murB*
ORF00869	D-Alanine-activating enzyme	*dltA*
ORF00871	D-Alanyl carrier protein	*dltC*
ORF00954	UDP-*N*-acetylmuramoylalanyl-D-glutamate–2,6-diaminopimelate ligase	*murE*
ORF00998	*fmt* protein, putative	*fmt*
ORF01106	Glutamate racemase	*murI*
ORF01145	Penicillin-binding protein 1	*pbpA*
ORF01146	Phospho-*N*-acetylmuramoyl-pentapeptide-transferase	*mraY*
ORF01147	UDP-*N*-acetylmuramoylalanine–D-glutamate ligase	*murD*
ORF01400	Alanine racemase, putative	
ORF01424	UDP-*N*-acetylglucosamine–*N*-acetylmuramyl-(pentapeptide) pyrophosphoryl-undecaprenol *N*-acetylglucosamine transferase	*murG*
ORF01467	Penicillin-binding protein 2	*pbp2*
ORF01652	Penicillin-binding protein 3	*pbp3*
ORF01759	Rod shape-determining protein MreC	*mreC*
ORF01856	UDP-*N*-acetylmuramate–alanine ligase	*murC*
ORF02107	UDP-*N*-acetylmuramyl tripeptide synthetase, putative	
ORF02305	Alanine racemase	*alr*
ORF02317	UDP-*N*-acetylmuramoylalanyl-D-glutamyl-2,6-diaminopimelate–D-alanyl-D-alanyl ligase	*murF*
ORF02318	D-Alanine–D-alanine ligase	
ORF02337	UDP-*N*-acetylglucosamine 1-carboxyvinyltransferase	*murA*
ORF02365	UDP-*N*-acetylglucosamine 1-carboxyvinyltransferase	*murA*
ORF02405	Phosphoglucosamine mutase	*glmM*

Surface structures

ORF00812	Clumping factor	*clfA*
ORF00816	Extracellular matrix and plasma binding protein, putative	
ORF01501	Elastin-binding protein	*ebpS*
ORF02161	MHC class II analog protein	*map*
ORF02802	Fibronectin-binding protein B, putative	*fnbB*
ORF02803	Fibronectin-binding protein precursor, putative	*fnbA*
ORF02963	Clumping factor B, putative	*clfB*

Biosynthesis and degradation of surface polysaccharides and lipopolysaccharides

ORF00114	Capsular polysaccharide biosynthesis protein, putative	
ORF00115	Capsular polysaccharide synthesis enzyme Cap5B	*capB*
ORF00116	Capsular polysaccharide synthesis enzyme Cap8C	*capC*
ORF00117	Capsular polysaccharide biosynthesis protein Cap5D, putative	*capD*
ORF00118	Capsular polysaccharide biosynthesis protein Cap5E, putative	*capE*
ORF00119	Capsular polysaccharide synthesis enzyme Cap8F	*capF*
ORF00120	UDP-*N*-acetylglucosamine 2-epimerase	
ORF00121	Capsular polysaccharide synthesis enzyme O-acetyl transferase Cap5H, putative	*capH*
ORF00122	Capsular polysaccharide biosynthesis protein Cap5I, putative	*capI*
ORF00123	Capsular polysaccharide biosynthesis protein Cap5J, putative	*capJ*
ORF00124	Capsular polysaccharide biosynthesis protein Cap5K, putative	*capK*
ORF00125	Cap5L protein/glycosyltransferase, putative	*capL*
ORF00126	Capsular polysaccharide biosynthesis protein Cap8M	*capM*
ORF00127	Cap5N protein/UDP-glucose 4-epimerase, putative	*capN*
ORF00128	Cap5O protein/UDP-*N*-acetyl-D-mannosaminuronic acid dehydrogenase	*capO*
ORF00129	UDP-*N*-acetylglucosamine 2-epimerase	
ORF00427	Autolysin precursor, putative	
ORF00471	UDP-*N*-acetylglucosamine pyrophosphorylase	*glmU*
ORF00642	Teichoic acid biosynthesis protein, putative	*tagG*

(*Continued on next page*)

TABLE 1 (*Continued*)

Amino acid biosynthesis category	Functional role	Gene
ORF00870	*dltB* protein, putative	*dltB*
ORF00872	Extramembranal protein	*dltD*
ORF00974	Glycosyl transferase, group 1	
ORF00994	Bifunctional autolysin precursor, putative	*atl*
ORF01219	Cell wall hydrolase, putative	*lytN*
ORF01237	Undecaprenyl diphosphate synthase	*uppS*
ORF01871	Polysaccharide biosynthesis protein, putative	
ORF02352	UDP-*N*-acetylglucosamine 2-epimerase	
ORF02423	UDP-*N*-acetylglucosamine pyrophosphorylase, putative	
ORF02580	*N*-Acetylmuramoyl-L-alanine amidase, putative	
ORF02801	UTP-glucose-1-phosphate uridylyltransferase	*galU*
ORF02998	Capsular polysaccharide biosynthesis protein, Cap5C	*capC*
ORF02999	Capsular polysaccharide biosynthesis protein Cap5B	*capB*
ORF03000	Capsular polysaccharide biosynthesis, *capA*, putative	*capA*
ORF03004	*icaB* protein, putative	*icaB*

Other

ORF00074	Periplasmic binding protein, putative	
ORF00105	Phosphonate ABC transporter, substrate-binding protein, putative	
ORF00183	Membrane protein, putative	
ORF00200	Membrane protein, putative	
ORF00544	*sdrC* protein, putative	*sdrC*
ORF00545	*sdrD* protein, putative	*sdrD*
ORF00613	Iron compound ABC transporter, substrate-binding protein, putative	
ORF00918	Truncated MHC class II analog protein	
ORF00927	Oligopeptide ABC transporter, substrate-binding protein, putative	
ORF00928	Oligopeptide ABC transporter, substrate-binding protein, putative	
ORF01079	Neurofilament protein	
ORF01114	Fibrinogen-binding protein	*fib*
ORF01389	Phosphate ABC transporter, periplasmic phosphate-binding protein, putative	*phoX*
ORF02466	Truncated MHC class II analog protein	
ORF02696	*fmhA* protein, putative	
ORF02742	Amino acid transporter, putative	
ORF02767	Peptide ABC transporter, peptide-binding protein, putative	
ORF02834	Sortase, putative	*srtA*
ORF03006	Lipase	*geh*

Cellular processes

Adaptations to atypical conditions

ORF00942	GTP pyrophosphokinase, putative	
ORF01403	Cold shock protein, putative	
ORF01742	GTP pyrophosphokinase	*relA*
ORF02298	Sigma factor B, putative	*sigB*
ORF02932	Choline dehydrogenase	*betA*
ORF02933	Betaine aldehyde dehydrogenase	*betB*
ORF03045	Cold shock protein, putative	

DNA transformation

ORF01691	DNA internalization-related competence protein ComEC/Rec2	

Detoxification

ORF00093	Superoxide dismutase, putative	*sod*
ORF00364	Alkyl hydroperoxide reductase, subunit F, putative	*ahpF*
ORF00365	Alkyl hydroperoxide reductase	
ORF00802	Carboxylesterase precursor, putative	*estA*
ORF01327	Catalase	
ORF01653	Superoxide dismutase, Mn, putative	*sodA*
ORF01891	Arsenate operon regulator	*arsR*
ORF01893	Arsenical pump membrane protein subfamily	*arsB*
ORF01894	Arsenate reductase	*arsC*

(*Continued on next page*)

Amino acid biosynthesis category	Functional role	Gene
Cell division		
ORF00229	ScdA protein, putative	*scdA*
ORF00342	Chromosome partioning protein, ParB family, putative	
ORF01148	Cell division protein, putative	
ORF01149	Cell division protein, putative	
ORF01150	Cell division protein FtsZ	*ftsZ*
ORF01204	SMC family, C-terminal domain family	
ORF01668	GTP-binding protein Era	*era*
Other		
ORF00233	Holin-like protein LrgB, putative	*lrgB*
ORF02375	Autoinducer-2 production protein LuxS, putative	*luxS*
Toxin production and resistance		
ORF00214	Phosphoenolpyruvate-dependent sugar phosphotransferase system, EIIA 2, putative	
ORF00246	Drug transporter, putative	
ORF00464	Dimethyladenosine transferase	*ksgA*
ORF00691	Undecaprenol kinase, putative	
ORF00703	Quinolone resistance norA protein, putative	
ORF00884	Multiple resistance and pH regulation protein F (MrpF/PhaF) family, putative	
ORF01121	Alpha-hemolysin precursor	*hla*
ORF01373	Methicillin resistance factor, FemA, putative	*femA*
ORF01374	Methicillin resistance factor, putative	*femB*
ORF01705	Enterotoxin family protein	*sea*
ORF01879	Virulence factor regulator protein	*rot*
ORF01951	Epidermin biosynthesis protein EpiC, authentic point mutation	*epiC*
ORF01954	Leukotoxin, LukD [pathogenicity island SaPIn3]	*lukD*
ORF02003	ABC transporter, ATP-binding/permease protein, putative	
ORF02098	DNA-binding response regulator VraR, putative	*vraR*
ORF02099	Histidine kinase, putative	
ORF02260	Delta-hemolysin precursor	*hld*
ORF02609	Fosfomycin resistance protein, putative	*fosB*
ORF02629	Drug resistance transporter, EmrB/QacA subfamily, putative	
ORF02696	*fmhA* protein, putative	*fmhA*
ORF02708	Gamma-hemolysin h-gamma-ii subunit, putative	*hlgA*
ORF02709	Leukocidin s subunit precursor, putative	*lukS*
ORF02710	Leukocidin f subunit precursor	*lukF*
ORF02740	Drug transporter, putative	
Pathogenesis		
ORF00069	Protein A	*spa*
ORF00192	Coagulase	*coa*
ORF00620	Staphylococcal accessory regulator T, putative	*sarT*
ORF00813	Truncated secreted von Willebrand factor-binding protein	
ORF00814	Truncated secreted von Willebrand factor-binding protein (coagulase)VWbp, putative	
ORF00816	Extracellular matrix and plasma binding protein, putative	
ORF00986	Cysteine protease, putative	
ORF00987	Cysteine protease precursor, putative	
ORF00988	Glutamyl endopeptidase precursor, putative	
ORF01585	Staphylococcal respiratory response protein SrrB, putative	*srrB*
ORF01879	Virulence factor regulator protein	*rot*
ORF01935	Serine protease SplF, putative	*splF*
ORF01936	Serine protease SplE	*splE*
ORF01938	Serine protease SplD	*splD*
ORF01939	Serine protease SplC	*splC*
ORF01941	Serine protease SplB	*splB*
ORF01942	Serine protease SplA	*splA*
ORF02127	Staphopain thiol proteinase	
ORF02261	Accessory gene regulator protein B	*agrB*
ORF02264	Accessory gene regulator protein C	*agrC*
ORF02265	Accessory gene regulator protein A	*agrA*

(*Continued on next page*)

TABLE 1 (*Continued*)

Amino acid biosynthesis category	Functional role	Gene
ORF02463	Hyaluronate lyase	
ORF02571	Secretory antigen precursor, putative	*ssaA*
ORF02576	Secretory antigen precursor SsaA, putative	*ssaA*
ORF02706	Immunoglobulin G-binding protein Sbi, putative	*sbi*
ORF02803	Fibronectin-binding protein precursor, putative	*fnbA*
ORF02963	Clumping factor B, putative	*clfB*
ORF02971	Aureolysin, putative	*aur*
ORF03005	Intercellular adhesion protein C, putative	*icaC*
Cell adhesion		
ORF02161	MHC class II analog protein	*map*
Central intermediary metabolism		
Nitrogen metabolism		
ORF02558	Urease, gamma subunit	*ureA*
ORF02559	Urease, beta subunit, putative	*ureB*
ORF02561	Urease, alpha subunit	*ureC*
ORF02562	Urease accessory protein UreE, putative	*ureE*
ORF02563	Urease accessory protein UreF, putative	*ureF*
ORF02564	Urease accessory protein UreG	*ureG*
ORF02565	Urease accessory protein UreD, putative	*ureD*
ORF02679	Respiratory nitrate reductase, delta subunit, putative	*narJ*
ORF02682	Uroporphyrin-III C-methyltransferase, putative	*nasF*
ORF02683	Assimilatory nitrite reductase [NAD(P)H], small subunit, putative	*nirD*
ORF02684	Assimilatory nitrite reductase [NAD(P)H], large subunit, putative	*nirB*
Sulfur metabolism		
ORF02947	Sulfite reductase (NADPH) flavoprotein alpha-component, putative	*cysJ*
Amino sugars		
ORF00295	*N*-Acetylneuraminate lyase subunit, putative	
ORF00710	*N*-Acetylglucosamine-6-phosphate deacetylase	*nagA*
ORF02399	Glucosamine–fructose-6-phosphate aminotransferase, isomerizing	*glmS*
Other		
ORF00153	Indolepyruvate decarboxylase, putative	
ORF00243	NADH dehydrogenase subunit, putative	
ORF00574	Phosphate acetyltransferase	*pta*
ORF00577	Mevalonate kinase, putative	*mvaK1*
ORF00578	Diphosphomevalonate decarboxylase	*mvaD*
ORF00579	Phosphomevalonate kinase	
ORF00849	Aminotransferase, class V superfamily, putative	
ORF01103	Succinate dehydrogenase cytochrome b-558 subunit, putative	*sdhC*
ORF01632	Glycine cleavage system P-protein subunit ll, putative	
ORF01633	Glycine cleavage system P-protein subunit I, putative	
ORF01702	MTA/SAH nucleosidase	
ORF01909	S-Adenosylmethionine synthetase	*metK*
ORF01916	*o*-Succinylbenzoic acid–CoA ligase, putative	
ORF02860	HMG-CoA synthase, putative	
ORF02877	Squalene synthase, putative	*crtN*
ORF02879	Squalene desaturase, putative	
DNA metabolism		
Degradation of DNA		
ORF00818	Thermonuclease precursor	*nuc*
ORF01619	Exodeoxyribonuclease VII, small subunit	*xseB*
ORF01620	Exodeoxyribonuclease VII, large subunit	*xseA*
DNA replication, recombination, and repair		
ORF00001	Chromosomal replication initiator protein DnaA	*dnaA*

(*Continued on next page*)

TABLE 1 Tabular listing of all predicted ORFs in NCTC 8325 listed by functional role category based on TIGR roles (*Continued*)

Amino acid biosynthesis category	Functional role	Gene
ORF00002	DNA polymerase III, beta subunit	*dnaN*
ORF00004	DNA replication and repair protein, putative	*recF*
ORF00005	DNA gyrase, B subunit	*gyrB*
ORF00006	DNA gyrase, A subunit	*gyrA*
ORF00018	Replicative DNA helicase	*dnaB*
ORF00109	Replication initiation protein, putative	
ORF00110	Truncated replication initiation protein, putative	
ORF00429	MutT/nudix family protein, putative	
ORF00442	DNA polymerase III, gamma and tau subunits, putative	*dnaX*
ORF00445	Recombination protein RecR	*recR*
ORF00454	DNA polymerase III, delta prime subunit, putative	*holB*
ORF00477	Transcription-repair coupling factor	*mfd*
ORF00507	DNA repair protein RadA	*radA*
ORF00564	Uracil-DNA glycosylase	*ung*
ORF00612	Endonuclease III, putative	
ORF00699	Deoxyribodipyrimidine photolyase, putative	
ORF00730	ATP-dependent DNA helicase RecQ	*recQ*
ORF00776	Excinuclease ABC, B subunit	*uvrB*
ORF00779	Excinuclease ABC subunit B, putative	*uvrB*
ORF00780	Excinuclease ABC, A subunit	*uvrA*
ORF00904	Exonuclease RexB, putative	*rexB*
ORF00905	ATP-dependent nuclease subunit A, putative	*addA*
ORF01095	Ribonuclease HIII	*rnhC*
ORF01098	DNA-dependent DNA polymerase beta chain, putative	
ORF01099	DNA mismatch repair MutS2 family protein	*mutS2*
ORF01102	Excinuclease ABC, C subunit	*uvrC*
ORF01179	Primosomal protein N	*priA*
ORF01194	ATP-dependent DNA helicase RecG	*recG*
ORF01222	DNA topoisomerase I	*topA*
ORF01224	Site-specific recombinase, putative	*xerC*
ORF01241	DNA polymerase III, alpha subunit, gram-positive type	*polC*
ORF01261	Competence-damage inducible protein cinA	*cinA*
ORF01262	Recombination protein RecA, putative	*recA*
ORF01272	DNA mismatch repair protein MutS	*mutS*
ORF01273	DNA mismatch repair protein HexB, putative	*hexB*
ORF01333	LexA repressor	*lexA*
ORF01351	DNA topoisomerase IV, B subunit	*parE*
ORF01352	DNA topoisomerase IV, A subunit	*parC*
ORF01469	Endonuclease III, putative	*nth*
ORF01472	DnaQ family exonuclease/DinG family helicase, putative	
ORF01502	ATP-dependent DNA helicase RecQ, putative	*recQ*
ORF01591	Integrase/recombinase XerD	*xerD*
ORF01593	NUDIX domain protein	
ORF01615	DNA repair protein RecN	*recN*
ORF01658	Endonuclease IV, putative	*nfo*
ORF01663	DNA primase, putative	*dnaG*
ORF01750	Holliday junction DNA helicase RuvB	*ruvB*
ORF01751	Holliday junction DNA helicase RuvA	*ruvA*
ORF01796	Formamidopyrimidine-DNA glycosylase	*mutM*
ORF01797	DNA polymerase I	*polA*
ORF01811	DNA polymerase III, alpha subunit superfamily	*dnaE*
ORF02005	A/G-specific adenine glycosylase, putative	*mutY*
ORF02122	DNA ligase, NAD-dependent	*ligA*
ORF02123	ATP-dependent DNA helicase PcrA	*pcrA*
ORF02276	MutS domain V protein	
ORF02517	DNA topoisomerase III, putative	*topB*
ORF02621	DNA-3-methyladenine glycosylase, putative	
ORF02791	Pyrophosphohydrolase, putative	
ORF02861	Methylated-DNA–protein-cysteine methyltransferase, putative	*ogt*

Restriction/modification

ORF00162	Type I site-specific deoxyribonuclease, HsdR family, putative	

(*Continued on next page*)

TABLE 1 *(Continued)*

Amino acid biosynthesis category	Functional role	Gene
ORF00397	Type I restriction-modification system, M subunit	*hsdM*
ORF00398	Restriction modification system specificity subunit, putative	*hsdS*
ORF01932	Type I restriction-modification enzyme, S subunit, EcoA family, putative	*hsdS*
ORF01933	Type I restriction-modification system, M subunit	*hsdM*

Chromosome-associated proteins

ORF01490	DNA-binding protein HU, putative	*hup*

Energy metabolism

Amino acids and amines

ORF00008	Histidine ammonia-lyase	*hutH*
ORF00076	Ornithine cyclodeaminase, putative	
ORF00836	Glycine cleavage system H protein	*gcvH*
ORF00894	Ornithine aminotransferase, putative	
ORF01129	Carbamate kinase	*arcC*
ORF01452	Alanine dehydrogenase	*ald*
ORF01497	L-Asparaginase, putative	*ansA*
ORF01611	2-Oxoisovalerate dehydrogenase, E2 component, dihydrolipoamide acetyltransferase, putative	*bfmBB*
ORF01612	2-Oxoisovalerate dehydrogenase, E1 component, beta subunit, putative	*bfmB1b*
ORF01613	2-Oxoisovalerate dehydrogenase, E1 component, alpha subunit, putative	*bfmB1a*
ORF01632	Glycine cleavage system P-protein subunit ll, putative	
ORF01633	Glycine cleavage system P-protein subunit I, putative	
ORF01634	Glycine cleavage system T protein	*gcvT*
ORF01818	Alanine dehydrogenase	*ald*
ORF01867	D-Amino acid aminotransferase	*dat*
ORF02374	Aminobenzoyl-glutamate utilization protein B, putative	*abgB*
ORF02409	Arginase	*roc*
ORF02606	Imidazolonepropionase	*hutI*
ORF02607	Urocanate hydratase	*hutU*
ORF02610	Formiminoglutamase	*hutG*
ORF02839	L-Serine dehydratase, iron-sulfur-dependent, alpha subunit	*sdhA*
ORF02840	L-Serine dehydratase, iron-sulfur-dependent, beta subunit	*sdhB*
ORF02869	Delta-1-pyrroline-5-carboxylate dehydrogenase, putative	
ORF02965	Carbamate kinase	*arcC*
ORF02969	Arginine deiminase	*arcA*

Biosynthesis and degradation of polysaccharides

ORF00236	6-Phospho-beta-glucosidase, putative	*bglA*
ORF00438	Alpha amylase family protein, putative	
ORF01601	Alpha-glucosidase, putative	*malA*
ORF02268	Sucrose-6-phosphate dehydrogenase, putative	*cscA*
ORF02449	6-Phospho-beta-galactosidase	*lacG*
ORF02452	Tagatose 1,6-diphosphate aldolase	*lacD*
ORF02453	Tagatose-6-phosphate kinase	*lacC*
ORF02454	Galactose-6-phosphate isomerase, LacB subunit	*lacB*
ORF02455	Galactose-6-phosphate isomerase, LacA subunit	*lacA*

Other

ORF00100	Deoxyribose-phosphate aldolase	*deoC*
ORF01276	Glycerol kinase	*glpK*
ORF01362	4-Oxalocrotonate tautomerase, putative	
ORF01491	Glycerol-3-phosphate dehydrogenase, NAD-dependent, putative	*gpsA*
ORF01846	Acetyl-CoA synthetase, putative	*acs*
ORF02379	Deoxyribose-phosphate aldolase	*deoC*
ORF02723	Glycerate kinase, putative	
ORF02849	Pyruvate oxidase, putative	
ORF02859	Hydroxymethylglutaryl-CoA reductase, degradative	
ORF02929	Acetyl-CoA synthetase, putative	

Aerobic

ORF00320	NADPH-dependent FMN reductase, putative	

(Continued on next page)

TABLE 1 Tabular listing of all predicted ORFs in NCTC 8325 listed by functional role category based on TIGR roles (*Continued*)

Amino acid biosynthesis category	Functional role	Gene
Anaerobic		
ORF00142	Formate dehydrogenase, NAD-dependent, putative	
ORF00188	Pyruvate formate-lyase 1 activating enzyme, putative	*pflA*
ORF00206	L-Lactate dehydrogenase	
ORF02550	Formate dehydrogenase family accessory protein FdhD	*fdhD*
ORF02582	Formate dehydrogenase, alpha subunit, putative	
ORF02678	Respiratory nitrate reductase, gamma subunit	*narI*
ORF02679	Respiratory nitrate reductase, delta subunit, putative	*narJ*
ORF02680	Nitrate reductase, beta subunit	*narH*
ORF02681	Nitrate reductase, alpha subunit	
ORF02922	L-Lactate dehydrogenase	
ATP-proton motive force interconversion		
ORF02134	Nitric oxide synthase, oxygenase domain, putative	
ORF02340	ATP synthase F1, epsilon subunit	*atpC*
ORF02341	ATP synthase F1, beta subunit	*atpD*
ORF02343	ATP synthase F1, gamma subunit	*atpG*
ORF02345	ATP synthase F1, alpha subunit	*atpA*
ORF02346	ATP synthase F1, delta subunit	*atpH*
ORF02347	ATP synthase F0, B subunit	*atpF*
ORF02349	ATP synthase F0, C subunit	*atpE*
ORF02350	ATP synthase F0, A subunit	*atpB*
Tricarboxylic acid cycle		
ORF00243	NADH dehydrogenase subunit, putative	
ORF00866	D-Isomer specific 2-hydroxyacid dehydrogenase, NAD binding domain protein	
ORF01103	Succinate dehydrogenase cytochrome b-558 subunit, putative	*sdhC*
ORF01104	Succinate dehydrogenase, flavoprotein chain TC0881, putative	*frdA*
ORF01105	Iron-sulfur subunit of succinate dehydrogenase, putative	*sdhB*
ORF01216	Succinyl-CoA synthetase, beta subunit, putative	*sucC*
ORF01218	Succinyl-CoA synthetase, alpha subunit, putative	*sucD*
ORF01347	Aconitate hydratase 1	*acnA*
ORF01416	2-Oxoglutarate dehydrogenase, E2 component, dihydrolipoamide succinyltransferase	*sucB*
ORF01418	2-Oxoglutarate dehydrogenase, E1 component	*sucB*
ORF01801	Isocitrate dehydrogenase, NADP-dependent	*icd*
ORF01810	NADP-dependent malic enzyme, putative	
ORF01983	Fumarate hydratase, class II	*fumC*
ORF02134	Nitric oxide synthase, oxygenase domain, putative	
ORF02647	Malate:quinone-oxidoreductase	*mqo*
ORF02927	Malate:quinone-oxidoreductase	*mqo*
Sugars		
ORF00217	Sorbitol dehydrogenase, putative	
ORF00239	Ribokinase, putative	*rbsK*
ORF00291	*pfkB* family carbohydrate kinase family, putative	
ORF00534	L-Ribulokinase	*araB*
ORF02267	Fructokinase, putative	
ORF02385	Mannose-6-phosphate isomerase, class I	*manA*
ORF02403	Mannitol-1-phosphate 5-dehydrogenase, putative	*mtlD*
ORF02614	Aldose 1-epimerase precursor, putative	*mro*
ORF02808	Gluconate kinase	*gntK*
ORF02976	Mannose-6-phosphate isomerase, class I	*manA*
Pentose phosphate pathway		
ORF00866	D-Isomer specific 2-hydroxyacid dehydrogenase, NAD binding domain protein	
ORF01189	Ribulose-phosphate 3-epimerase	*rpe*
ORF01337	Transketolase	*tkt*
ORF01599	Glucose-6-phosphate 1-dehydrogenase	*zwf*
ORF01605	6-Phosphogluconate dehydrogenase, decarboxylating	*gnd*
ORF02612	Ribose 5-phosphate isomerase, putative	*rpiA*

(*Continued on next page*)

TABLE 1 (*Continued*)

Amino acid biosynthesis category	Functional role	Gene
Pyruvate dehydrogenase		
ORF01040	Pyruvate dehydrogenase complex, E1 component, alpha subunit, putative	*pdhA*
ORF01041	Pyruvate dehydrogenase complex, E1 component, pyruvate dehydrogenase beta subunit, putative	*pdhB*
ORF01042	Dihydrolipoamide S-acetyltransferase component of pyruvate dehydrogenase complex E2, putative	*pdhC*
ORF01043	Dihydrolipoamide dehydrogenase	*lpdA*
ORF01614	Dihydrolipoamide dehydrogenase	*lpdA*
Glycolysis/gluconeogenesis		
ORF00206	L-Lactate dehydrogenase	
ORF00243	NADH dehydrogenase subunit, putative	
ORF00504	ATP:guanido phosphotransferase, C-terminal catalytic domain protein	
ORF00707	Fructose 1-phosphate kinase, putative	*lacC*
ORF00795	Glyceraldehyde-3-phosphate dehydrogenase, type I	*gap*
ORF00796	Phosphoglycerate kinase, putative	*pgk*
ORF00797	Triosephosphate isomerase	*tpiA*
ORF00798	Phosphoglycerate mutase, 2,3-bisphosphoglycerate-independent	*gpmA*
ORF00799	Enolase	*eno*
ORF00866	D-Isomer specific 2-hydroxyacid dehydrogenase, NAD binding domain protein	
ORF00900	Glucose-6-phosphate isomerase	*pgi*
ORF01064	Pyruvate carboxylase	*pyc*
ORF01103	Succinate dehydrogenase cytochrome b-558 subunit, putative	*sdhC*
ORF01646	Glucokinase, putative	
ORF01794	Glyceraldehyde-3-phosphate dehydrogenase, type I	*gap*
ORF01806	Pyruvate kinase	*pyk*
ORF01807	6-Phosphofructokinase, putative	*pfkA*
ORF01910	Phosphoenolpyruvate carboxykinase (ATP)	*pckA*
ORF02703	Phosphoglycerate mutase, putative	*gpm*
ORF02839	L-Serine dehydratase, iron-sulfur-dependent, alpha subunit	*sdhA*
ORF02922	L-Lactate dehydrogenase	
ORF02926	Fructose-bisphosphate aldolase class-I, putative	
Methanogenesis		
ORF00866	D-Isomer specific 2-hydroxyacid dehydrogenase, NAD binding domain protein	
Fermentation		
ORF00113	Alcohol dehydrogenase, iron-containing, putative	
ORF00132	Aldehyde dehydrogenase, putative	
ORF00187	Formate acetyltransferase	*pfl*
ORF00188	Pyruvate formate-lyase 1 activating enzyme, putative	*pflA*
ORF00221	Alcohol dehydrogenase, zinc-containing	
ORF00243	NADH dehydrogenase subunit, putative	
ORF00574	Phosphate acetyltransferase	*pta*
ORF00608	Alcohol dehydrogenase I, putative	*adh1*
ORF00866	D-Isomer specific 2-hydroxyacid dehydrogenase, NAD binding domain protein	
ORF01103	Succinate dehydrogenase cytochrome b-558 subunit, putative	*sdhC*
ORF01820	Acetate kinase	*ackA*
ORF02142	Aldehyde dehydrogenase, putative	
ORF02363	Aldehyde dehydrogenase, putative	*gabD*
ORF02467	Alpha-acetolactate decarboxylase	*budA*
ORF02468	Acetolactate synthase, putative	*ilvK*
ORF02830	D-Lactate dehydrogenase, putative	
ORF02875	D-Lactate dehydrogenase, putative	
ORF02921	Alpha-acetolactate decarboxylase	*budA*
Electron transport		
ORF00243	NADH dehydrogenase subunit, putative	
ORF00320	NADPH-dependent FMN reductase, putative	
ORF00366	NAD(P)H-flavin oxidoreductase, putative	
ORF00412	NADH dehydrogenase subunit 5, putative	*ndhF*
ORF00771	Peptide chain release factor 2, putative	
ORF00785	Thioredoxin reductase	*trxB*

(*Continued on next page*)

TABLE 1 Tabular listing of all predicted ORFs in NCTC 8325 listed by functional role category based on TIGR roles (*Continued*)

Amino acid biosynthesis category	Functional role	Gene
ORF00834	Thioredoxin, putative	
ORF00908	Coenzyme A disulfide reductase, putative	
ORF00999	Quinol oxidase, subunit IV, putative	
ORF01000	Cytochrome c oxidase subunit III, putative	
ORF01001	Quinol oxidase, subunit I	
ORF01002	Quinol oxidase AA3, subunit II, putative	
ORF01031	Cytochrome d ubiquinol oxidase, subunit I, putative	
ORF01032	Cytochrome d ubiquinol oxidase, subunit II, putative	
ORF01100	Thioredoxin	*trx*
ORF01103	Succinate dehydrogenase cytochrome b-558 subunit, putative	*sdhC*
ORF01504	Ferredoxin, putative	*fer*
ORF02527	FmhB protein, putative	
ORF02550	Formate dehydrogenase family accessory protein FdhD	*fdhD*
ORF02679	Respiratory nitrate reductase, delta subunit, putative	*narJ*
ORF02829	NAD(P)H-flavin oxidoreductase, putative	

Fatty acid and phospholipid metabolism

Degradation

ORF00051	1-Phosphatidylinositol phosphodiesterase precursor, putative	
ORF00086	3-Ketoacyl-acyl carrier protein reductase, putative	*fabG*
ORF00300	Lipase precursor	*geh*
ORF00802	Carboxylesterase precursor, putative	*estA*
ORF01071	Glycerophosphoryl diester phosphodiesterase, putative	
ORF02962	Tributyrin esterase, putative	
ORF03006	Lipase	*geh*

Biosynthesis

ORF00195	Acetyl-CoA acetyltransferase, putative	*atoB*
ORF00336	Acetyl-CoA acetyltransferase, putative	*atoB*
ORF00504	ATP:guanido phosphotransferase, C-terminal catalytic domain protein	
ORF00558	Acetyl-CoA acetyltransferase, putative	*atoB*
ORF00921	3-Oxoacyl-synthase, putative	*fabF-1*
ORF00947	Enoyl-(acyl-carrier-protein) reductase	*fabI*
ORF01197	Fatty acid/phospholipid synthesis protein PlsX	*plsX*
ORF01198	Malonyl CoA-acyl carrier protein transacylase	*fabD*
ORF01199	3-Oxoacyl-(acyl-carrier-protein) reductase, putative	*fabG*
ORF01201	Acyl carrier protein	*acpP*
ORF01260	CDP-diacylglycerol–glycerol-3-phosphate 3-phosphatidyltransferase	*pgsA*
ORF01310	Cardiolipin synthetase, putative	*cls*
ORF01623	Acetyl-CoA carboxylase, biotin carboxylase	*accC*
ORF01624	Acetyl-CoA carboxylase, biotin carboxyl carrier protein	
ORF01671	Diacylglycerol kinase, putative	*dgkA*
ORF01709	Acetyl-CoA carboxylase, biotin carboxylase, putative	*accC*
ORF01710	Acetyl-CoA carboxylase, biotin carboxyl carrier protein, putative	
ORF01808	Acetyl-CoA carboxylase, carboxyl transferase, alpha subunit	*accA*
ORF01809	Acetyl-CoA carboxylase, carboxyl transferase, beta subunit	*accD*
ORF02306	Holo-(acyl-carrier-protein) synthase	*acpS*
ORF02323	Cardiolipin synthetase, putative	*cls*
ORF02336	Beta-hydroxyacyl-ACP dehydratase, putative	*fabZ*
ORF02791	Pyrophosphohydrolase, putative	

Other

ORF00173	Acyl carrier phosphodiesterase	
ORF01406	Acylphosphatase	

Hypothetical proteins

Domain

ORF00186	Lipoprotein, putative	
ORF00204	Globin domain protein	

(Continued on next page)

TABLE 1 (*Continued*)

Amino acid biosynthesis category	Functional role	Gene
ORF00329	*mttA*/Hcf106 family protein-related protein	
ORF00352	Integrase-like protein	
ORF00483	S1 RNA binding domain protein	
ORF00516	Preprotein translocase SecE subunit	
ORF01046	ABC transporter domain protein	
ORF01161	Truncated transposase	
ORF01837	1-Acyl-*sn*-glycerol-3-phosphate acyltransferases domain protein	
ORF01945	Membrane protein, putative	
ORF01947	Membrane protein, putative	
ORF01950	Flavoprotein, *epiD*, putative	
ORF01952	Lantibiotic epidermin biosynthesis protein EpiB, putative	
ORF01953	Gallidermin superfamily *epiA*, putative	
ORF01955	Leukotoxin, LukE [pathogenicity island SaPIn3], putative	*lukE*
ORF01992	Phosphotransferase system, EIIC domain protein	
ORF02012	Transglycosylase domain protein	
ORF02426	Membrane protein, putative	
ORF02430	ABC transporter periplasmic binding protein, putative	
ORF02577	D-Isomer specific 2-hydroxyacid dehydrogenase, NAD binding domain protein	
ORF02639	Truncated transposase	
ORF02650	Lipoprotein, putative	
ORF02658	Membrane protein, putative	
ORF02726	Positive transcriptional activator, putative	
ORF02733	Membrane protein, putative	
ORF02753	Membrane protein, putative	
ORF02821	Membrane spanning protein, putative	
ORF02855	LysM domain protein	
ORF02883	LysM domain protein	
ORF03002	Intercellular adhesion protein A, putative	
ORF03023	Drp35	
ORF03042	Integrase/recombinase, core domain family	
ORF03046	Helix-turn-helix domain protein	

Other categories

Transposon functions

ORF00112	Transposase, IS*200* family, putative	
ORF00245	Truncated transposase	
ORF01390	Truncated transposase	
ORF01409	Truncated transposase	
ORF01410	Truncated transposase	
ORF01804	Transposase, putative	
ORF01805	Truncated transposase, putative	
ORF01905	Truncated transposase, putative	
ORF01906	Truncated transposase, putative	
ORF01911	Transposase, IS*200* family	
ORF01927	Transposase, IS*3* family, truncation-related protein, putative	
ORF01928	Transposase family protein, putative	
ORF01993	Transposase, putative	*tra*
ORF02392	Truncated resolvase	
ORF02410	Transposase, putative	*tra*

Prophage functions

ORF00349	Bacteriophage L54a, single-stranded DNA binding protein	*ssb*
ORF00624	Integrase/recombinase, putative	
ORF01281	Host factor 1 protein, putative	
ORF01515	Peptidoglycan hydrolase, putative	
ORF01516	Holin protein	
ORF01519	SLT orf 129-like protein	
ORF01520	SLT orf 488-like protein	
ORF01521	SLT orf 636-like protein	
ORF01523	SLT orf 527-like protein	
ORF01524	Holin-like protein	

(*Continued on next page*)

TABLE 1 Tabular listing of all predicted ORFs in NCTC 8325 listed by functional role category based on TIGR roles (*Continued*)

Amino acid biosynthesis category	Functional role	Gene
ORF01525	Phage tail tape measure protein, TP901 family, core region domain protein	
ORF01528	Bacteriophage L54a, bacterial Ig-like domain group 2 family protein	
ORF01529	Major tail protein	
ORF01531	SLT orf 123-like protein	
ORF01532	SLT orf 110-like protein	
ORF01536	Scaffolding protease	
ORF01537	Phage portal protein, HK97 family	
ORF01538	Phage terminase, large subunit, putative	
ORF01539	Terminase-small subunit	
ORF01540	Bacteriophage L54a, HNH endonuclease family protein	
ORF01542	SNF2 family N-terminal domain protein	
ORF01543	phi related protein	
ORF01549	Transcriptional activator *rinB*-related protein	
ORF01552	Bacteriophage L54a, deoxyuridine 5-triphosphate nucleotidohydrolase	*cut*
ORF01553	PVL orf 52-like protein	
ORF01554	PV83 orf 27-like protein	
ORF01556	PVL orf 52-like protein	
ORF01558	PVL orf 51-like protein	
ORF01561	PVL orf 50-like protein	
ORF01562	ETA orf 26-like protein-related protein	
ORF01563	Phage-encoded DNA polymerase I	
ORF01565	Phage-related protein	
ORF01566	phi APSE P51-like protein	
ORF01570	PVL orf 37-like protein	
ORF01571	SLT orf 71-like protein-related protein	
ORF01574	Helix-turn-helix domain protein	
ORF01575	Helix-turn-helix domain protein	
ORF01576	Exonuclease family	
ORF01580	phi PVL orf 30 homologue	
ORF01582	Bacteriophage integrase	
ORF02019	Autolysin	
ORF02020	Holin	
ORF02021	phi ETA orf 63-like protein	
ORF02022	Phage tail fiber protein, putative	
ORF02023	Bifunctional autolysin precursor	*atl*
ORF02025	phi SLT orf 99-like protein	
ORF02026	phi ETA orf 58-like protein-related protein	
ORF02027	SLT orf 129-like protein	
ORF02028	phiETA orf 57-like protein	
ORF02029	phi ETA orf 56-like protein	
ORF02030	phi ETA orf 55-like protein	
ORF02033	Phage tape measure protein	
ORF02036	Phage structural protein, putative	
ORF02038	Phage protein, HK97 gp10 family	
ORF02041	phi Mu50B-like protein	
ORF02042	phi Mu50B-like protein	
ORF02043	Phage head protein, putative	
ORF02047	Phage putative head morphogenesis protein, SPP1 gp7 family domain protein	
ORF02048	Phage portal protein, SPP1 family	
ORF02049	Phage terminase, large subunit, PBSX family	
ORF02050	Terminase small subunit, putative	
ORF02051	*int* gene activator RinA	
ORF02053	Transcriptional activator rinb-related protein	
ORF02057	dUTP pyrophosphatase (EC 3.6.1.23)	
ORF02059	phi PVL orf 52-like protein	
ORF02060	phi PVL orf 51-like protein	
ORF02061	phi PVL orf 50-like protein	
ORF02062	Helix-turn-helix DNA binding protein	
ORF02063	PV83 orf 23-like protein-related protein	
ORF02064	phi ETA orf 25-like protein	
ORF02067	Bacteriophage L54a, DnaB-like helicase family protein	

(*Continued on next page*)

TABLE 1 (*Continued*)

Amino acid biosynthesis category	Functional role	Gene
ORF02069	phi PV83 orf 20-like protein	
ORF02070	phi PV83 orf 19-like protein	
ORF02071	Single-strand DNA-binding protein, putative	*ssb*
ORF02074	phi PVL orf 39-like protein	
ORF02076	phi PVL orf 38-like protein-related protein	
ORF02077	phi PV83 orf 12-like protein-related protein	
ORF02078	phi PV83 orf 10-like protein	
ORF02080	Bacteriophage L54a, antirepressor, putative	
ORF02083	Bacteriophage L54a, Cro-related protein	
ORF02084	Phage repressor protein, putative	
ORF02086	PV83 orf 4-like protein-related protein	
ORF02088	Excisionase-related protein	
ORF02089	Integrase, phage family, putative	
ORF02170	Peptidoglycan hydrolase, putative	
ORF02171	Staphylokinase precursor, putative	*sak*
ORF02173	Amidase	
ORF02174	Holin, phage phi LC3 family	
ORF02178	phi PVL orf 22-like protein	
ORF02180	Phage minor structural protein, N-terminal region domain protein	
ORF02181	phi PVL orfs 18-19-like protein	
ORF02182	Tail length tape measure protein	
ORF02184	phi PVL orf 14-like protein	
ORF02185	phi PVL orf 13-like protein	
ORF02186	phi PVL orf 12-like protein	
ORF02187	Phage protein, HK97 gp10 family	
ORF02188	Phage head-tail adaptor, putative	
ORF02191	Phage major capsid protein, HK97 family	
ORF02193	Prohead protease	
ORF02194	Phage portal protein, HK97 family	
ORF02195	phi PVL orf 3-like protein-related protein	
ORF02196	Phage terminase, large subunit, putative	
ORF02197	Phage terminase, small subunit, putative	
ORF02199	phi PVL orf 62-like protein	
ORF02207	phi PVL orf 52-like protein	
ORF02208	PV83 orf 27-like protein	
ORF02210	phi PVL orf 51-like protein	
ORF02211	phi PVL orf 50-like protein	
ORF02213	phi ETA orf 25-like protein	
ORF02216	Phage DnaC-like protein	*dnaC*
ORF02217	phi ETA orf 22-like protein	
ORF02219	phi ETA orf 20-like protein	
ORF02220	phi ETA orf 18-like protein	
ORF02223	phi PVL orf 39-like protein	
ORF02224	phi PVL orf 38-like protein-related protein	
ORF02229	Anti repressor	
ORF02231	Anti repressor	
ORF02232	phi PVL orf 33-like protein	
ORF02233	phi PVL orf 32-like protein	
ORF02234	Repressor-like protein-related protein	
ORF02235	Repressor	
ORF02238	phi PVL ORF 30 homologue	
ORF02239	Integrase	
ORF02240	Truncated beta-hemolysin	*hlb*
ORF02250	Phage terminase, small subunit, putative	
ORF02334	Bacteriophage L54a, single-stranded DNA binding protein	*ssbB*
ORF02978	Phage infection protein, putative	
ORF03040	Integrase/recombinase	

Protein fate

Protein and peptide secretion and trafficking

ORF00769	Preprotein translocase, SecA subunit	*secA*

(*Continued on next page*)

TABLE 1   Tabular listing of all predicted ORFs in NCTC 8325 listed by functional role category based on TIGR roles (*Continued*)

Amino acid biosynthesis category	Functional role	Gene
ORF00801	Preprotein translocase, SecG subunit	*secG*
ORF00902	Signal peptidase IA, putative	
ORF00903	Signal peptidase IB, putative	
ORF01162	Lipoprotein signal peptidase	*lspA*
ORF01205	Signal recognition particle-docking protein FtsY	*ftsY*
ORF01207	Signal recognition particle protein	*ffh*
ORF01746	Protein-export membrane protein SecDF	*secF*
ORF01747	Preprotein translocase, YajC subunit	*yajC*
ORF01779	Trigger factor	*tig*
ORF01972	Protein export protein PrsA, putative	*prsA*
ORF02491	Preprotein translocase, SecY subunit, putative	*secY*
ORF02985	Preprotein translocase, SecA subunit, putative	*secA*
Protein folding and stabilization		
ORF00891	Peptidyl-prolyl *cis-trans* isomerase, cyclophilin-type, putative	
ORF01225	Heat shock protein HslV, putative	*hslV*
ORF01226	Heat shock protein HslVU, ATPase subunit HslU	*hslU*
ORF01682	DnaJ protein	*dnaJ*
ORF01683	DNAk protein, putative	*dnaK*
ORF01684	Co-chaperone GrpE	*grpE*
ORF01779	Trigger factor	*tig*
ORF02254	Chaperonin, 60 kDa, GrpEL, putative	*groEL*
ORF02255	Chaperonin, 10 kDa, GroES, putative	*groES*
Protein modification and repair		
ORF00782	Prolipoprotein diacylglyceryl transferase	*lgt*
ORF00963	Lipoyltransferase and lipoate-protein ligase, putative	
ORF01038	Polypeptide deformylase	*def*
ORF01182	Polypeptide deformylase, putative	*def*
ORF01360	Peptide methionine sulfoxide reductase	*msrA*
ORF01432	Peptide methionine sulfoxide reductase	*msrA*
ORF01473	BirA bifunctional protein, putative	*birA*
ORF02102	Methionine aminopeptidase, type I	*map*
ORF02996	Peptide methionine sulfoxide reductase, putative	*msrA*
Degradation of proteins, peptides, and glycopeptides		
ORF00505	Endopeptidase, putative	
ORF00757	Peptidase T, putative	*pepT*
ORF00790	ATP-dependent Clp protease, proteolytic subunit ClpP	*clpP*
ORF00879	Probable cytosol aminopeptidase	
ORF00912	ATP-dependent Clp protease, ATP-binding subunit ClpB	*clpB*
ORF00937	Oligoendopeptidase F	*pepF*
ORF00958	Serine protease HtrA, putative	*htrA*
ORF00987	Cysteine protease precursor, putative	
ORF00988	Glutamyl endopeptidase precursor, putative	
ORF01606	Peptidase T, putative	*pepT*
ORF01626	Proline dipeptidase, putative	
ORF01778	ATP-dependent Clp protease, ATP-binding subunit ClpX	*clpX*
ORF01935	Serine protease SplF, putative	*splF*
ORF01936	Serine protease SplE	*splE*
ORF01938	Serine protease SplD	*splD*
ORF01939	Serine protease SplC	*splC*
ORF01941	Serine protease SplB	*splB*
ORF01942	Serine protease SplA	*splA*
ORF01949	Intracellular serine protease, putative	
ORF02092	Aminopeptidase PepS, putative	*pepS*
ORF02127	Staphopain thiol proteinase	
ORF02277	O-sialoglycoprotein endopeptidase, putative	*gcp*
ORF02300	STAS domain, putative	
ORF02862	ATP-dependent Clp protease, ATP-binding subunit ClpC, putative	*clpC*
ORF02971	Aureolysin, putative	*aur*
ORF03025	Pyrrolidone-carboxylate peptidase	*pcp*

(*Continued on next page*)

TABLE 1 *(Continued)*

Amino acid biosynthesis category	Functional role	Gene
Protein synthesis		
Other		
ORF00475	Peptidyl-tRNA hydrolase	*pth*
ORF00804	SsrA-binding protein	*smpB*
Translation factors		
ORF00529	Translation elongation factor G	*fusA*
ORF00530	Translation elongation factor Tu	*tuf*
ORF00956	Peptide chain release factor 3	*prfC*
ORF01234	Translation elongation factor Ts	*tsf*
ORF01236	Ribosome recycling factor	*frr*
ORF01246	Translation initiation factor IF-2	*infB*
ORF01625	Translation elongation factor P	*efp*
ORF01786	Translation initiation factor IF-3	*infC*
ORF02359	Peptide chain release factor 1	*prfA*
ORF02489	Translation initiation factor IF-1	*infA*
Ribosomal proteins: synthesis and modification		
ORF00017	Ribosome protein L9	*rplI*
ORF00324	Ribosome-protein-serine acetyltransferase, putative	
ORF00348	Ribosome protein S6	*rpsF*
ORF00350	Ribosome protein S18	*rpsR*
ORF00474	Ribosome 5S rRNA E-loop binding protein Ctc/L25/TL5	
ORF00518	Ribosome protein L11	*rplK*
ORF00519	Ribosome protein L1	*rplA*
ORF00520	Ribosome protein L10	*rplJ*
ORF00521	Ribosome protein L7/L12	*rplL*
ORF00527	Ribosome protein S12	*rpsL*
ORF00528	Ribosome protein S7	*rpsG*
ORF00771	Peptide chain release factor 2, putative	
ORF01078	Ribosome protein L32	*rpmF*
ORF01191	Ribosome protein L28	*rpmB*
ORF01208	Ribosome protein S16	*rpsP*
ORF01211	Ribosome protein L19	*rplS*
ORF01232	Ribosome protein S2	*rpsB*
ORF01250	Ribosome protein S15	*rpsO*
ORF01328	Ribosome protein L33	*rpmG*
ORF01329	Ribosome protein S14p/S29e	*rpsN*
ORF01493	30S ribosomal protein S1, putative	
ORF01587	Ribosomal large subunit pseudouridine synthase B, putative	*rluB*
ORF01651	Ribosomal protein L33	*rpmG*
ORF01678	Ribosomal protein S21	*rpsU*
ORF01681	Ribosomal protein L11 methyltransferase	*prmA*
ORF01689	Ribosomal protein S20	*rpsT*
ORF01755	Ribosomal protein L27	*rpmA*
ORF01757	Ribosomal protein L21	*rplU*
ORF01784	Ribosomal protein L20	*rplT*
ORF01785	Ribosomal protein L35	*rpmI*
ORF01829	Ribosomal protein S4	*rpsD*
ORF02278	Ribosomal-protein-alanine acetyltransferase	*rimI*
ORF02361	Ribosomal protein L31	*rpmE*
ORF02477	Ribosomal protein S9, putative	*rpsI*
ORF02478	Ribosomal protein L13	*rplM*
ORF02484	Ribosomal protein L17	*rplQ*
ORF02486	Ribosomal protein S11, putative	*rpsK*
ORF02488	Ribosomal protein L36	*rpmJ*
ORF02492	Ribosomal protein L15	*rplO*
ORF02493	Ribosomal protein L30	*rpmD*
ORF02494	Ribosomal protein S5	*rpsE*
ORF02495	Ribosomal protein L18	*rplR*
ORF02496	Ribosomal protein L6, putative	*rplF*

(Continued on next page)

TABLE 1 Tabular listing of all predicted ORFs in NCTC 8325 listed by functional role category based on TIGR roles (*Continued*)

Amino acid biosynthesis category	Functional role	Gene
ORF02498	Ribosomal protein S8, putative	*rpsH*
ORF02499	Ribosomal protein S14p/S29e, putative	*rpsN*
ORF02500	50S ribosomal protein L5, putative	*rplE*
ORF02501	Ribosomal protein L24	*rplX*
ORF02502	Ribosomal protein L14	*rplN*
ORF02503	30S ribosomal protein S17, putative	*rpsQ*
ORF02504	Ribosomal protein L29	*rpmC*
ORF02505	Ribosomal protein L16	*rplP*
ORF02506	Ribosomal protein S3	*rpsC*
ORF02507	Ribosomal protein L22	*rplV*
ORF02508	Ribosomal protein S19	*rpsS*
ORF02509	Ribosomal protein L2	*rplB*
ORF02510	Ribosomal protein L23, putative	*rplW*
ORF02512	Ribosomal protein L3, putative	*rplC*
ORF02527	FmhB protein, putative	
ORF03055	Ribosomal protein L34	*rpmH*
tRNA aminoacylation		
ORF00009	Seryl-tRNA synthetase	*serS*
ORF00461	Methionyl-tRNA synthetase, putative	*metS*
ORF00493	Lysyl-tRNA synthetase	*lysS*
ORF00509	Glutamyl-tRNA synthetase	*gltX*
ORF00511	Cysteinyl-tRNA synthetase	*cysS*
ORF00611	Arginyl-tRNA synthetase	*argS*
ORF00933	Tryptophanyl-tRNA synthetase	*trpS*
ORF01092	Phenylalanyl-tRNA synthetase, alpha subunit	*pheS*
ORF01093	Phenylalanyl-tRNA synthetase, beta subunit	*pheT*
ORF01159	Isoleucyl-tRNA synthetase	*ileS*
ORF01183	Methionyl-tRNA formyltransferase	*fmt*
ORF01240	Prolyl-tRNA synthetase	*proS*
ORF01471	Asparaginyl-tRNA synthetase, putative	*asnS*
ORF01666	Glycyl-tRNA synthetase	*glyS*
ORF01722	Alanyl-tRNA synthetase	*alaS*
ORF01737	Aspartyl-tRNA synthetase	*aspS*
ORF01738	Histidyl-tRNA synthetase	*hisS*
ORF01741	D-Tyrosyl-tRNA(Tyr) deacylase	*dtd*
ORF01767	Valyl-tRNA synthetase	*valS*
ORF01788	Threonyl-tRNA synthetase	*thrS*
ORF01839	Tyrosyl-tRNA synthetase	*tyrS*
ORF01845	Formate-tetrahydrofolate ligase, putative	*fhs*
ORF01875	Leucyl-tRNA synthetase	*leuS*
ORF02116	Glutamyl-tRNA(Gln) amidotransferase, B subunit	*gatB*
ORF02117	Glutamyl-tRNA(Gln) amidotransferase, A subunit	*gatA*
ORF02118	Glutamyl-tRNA(Gln) amidotransferase, C subunit	*gatC*
tRNA and rRNA base modification		
ORF00464	Dimethyladenosine transferase	*ksgA*
ORF01184	Sun protein	*sun*
ORF01210	tRNA (guanine-N1)-methyltransferase	*trmD*
ORF01269	tRNA-i(6)A37 modification enzyme MiaB	*miaB*
ORF01280	tRNA delta(2)-isopentenylpyrophosphate transferase	*misA*
ORF01725	tRNA methyl transferase, putative	
ORF01726	(5-Methylaminomethyl-2-thiouridylate)-methyltransferase	
ORF01748	Queuine tRNA-ribosyltransferase	*tgt*
ORF01749	S-Adenosylmethionine:tRNA ribosyltransferase-isomerase	*queA*
ORF01870	Ribosomal small subunit pseudouridine synthase A, putative	*rsuA*
ORF01982	Ribosomal large subunit pseudouridine synthase, RluD subfamily, putative	*rluD*
ORF02113	RNA methyltransferase, TrmA family, putative	
ORF02480	tRNA pseudouridine synthase A	*truA*
ORF03053	tRNA modification GTPase TrmE	*trmE*

(*Continued on next page*)

TABLE 1 *(Continued)*

Amino acid biosynthesis category	Functional role	Gene
Purines, pyrimidines, nucleosides, and nucleotides		
2′-Deoxyribonucleotide metabolism		
ORF00741	Ribonucleoside-diphosphate reductase 2, putative	
ORF00742	Ribonucleotide-diphosphate reductase alpha chain, putative	*nrdE*
ORF00743	Ribonucleotide-disphosphate reductase beta chain, putative	*nrdF*
ORF01435	Thymidylate synthase, putative	*thyA*
ORF02942	Anaerobic ribonucleoside-triphosphate reductase, putative	*nrdD*
Pyrimidine ribonucleotide biosynthesis		
ORF01166	Aspartate carbamoyltransferase	*pyrB*
ORF01168	Dihydroorotase	*pyrC*
ORF01169	Carbamoyl-phosphate synthase, small subunit	*carA*
ORF01170	Carbamoyl-phosphate synthase, large subunit	*carB*
ORF01171	Orotidine 5′-phosphate decarboxylase	*pyrF*
ORF01172	Orotate phosphoribosyltransferase	*pyrE*
ORF02368	CTP synthase	*pyrG*
ORF02791	Pyrophosphohydrolase, putative	
ORF02909	Dihydroorotate dehydrogenase	*pyrD*
Salvage of nucleosides and nucleotides		
ORF00097	Purine nucleoside phosphorylase	*deoD*
ORF00372	Xanthine phosphoribosyltransferase	*xpt*
ORF00485	Hypoxanthine phosphoribosyltransferase	*hpt*
ORF01670	Cytidine deaminase	*cdd*
ORF01702	MTA/SAH nucleosidase	
ORF01715	Uridine kinase	*udk*
ORF01743	Adenine phosphoribosyltransferase	*apt*
ORF02353	Uracil phosphoribosyltransferase	*upp*
ORF02377	Pyrimidine nucleoside phosphorylase, putative	*pyn*
ORF02380	Purine nucleoside phosphorylase	*deoD*
Other		
ORF00100	Deoxyribose-phosphate aldolase	*deoC*
ORF00101	Phosphopentomutase	*deoB*
ORF00860	5′-Nucleotidase family protein	
ORF02379	Deoxyribose-phosphate aldolase	*deoC*
Purine ribonucleotide biosynthesis		
ORF00019	Adenylosuccinate synthetase	*purA*
ORF00374	Inosine-5′-monophosphate dehydrogenase	*guaB*
ORF00375	GMP synthase, putative	*guaA*
ORF00467	*pur* operon repressor	*purR*
ORF00472	Ribose-phosphate pyrophosphokinase, putative	*prsA*
ORF01008	Phosphoribosylaminoimidazole carboxylase, catalytic subunit	*purE*
ORF01009	Phosphoribosylaminoimidazole carboxylase, ATPase subunit	*purK*
ORF01010	Phosphoribosylaminoimidazole-succinocarboxamide synthase	*purC*
ORF01011	Phosphoribosylformylglycinamidine synthase, PurS protein	*purS*
ORF01012	Phosphoribosylformylglycinamidine synthase I	*purQ*
ORF01013	Phosphoribosylformylglycinamidine synthase II	*purL*
ORF01014	Amidophosphoribosyltransferase	*purF*
ORF01015	Phosphoribosylformylglycinamidine cyclo-ligase	*purM*
ORF01016	Phosphoribosylglycinamide formyltransferase, putative	*purN*
ORF01017	Phosphoribosylaminoimidazolecarboxamide formyltransferase/IMP cyclohydrolase	*purH*
ORF01018	Phosphoribosylamine-glycine ligase	*purD*
ORF01485	Nucleoside diphosphate kinase, putative	*ndk*
ORF01845	Formate-tetrahydrofolate ligase, putative	*fhs*
ORF02126	Adenylosuccinate lyase	*purB*
ORF02354	Serine hydroxymethyltransferase, putative	*glyA*
Nucleotide and nucleoside interconversions		
ORF00451	Thymidylate kinase	*tmk*

(Continued on next page)

TABLE 1 Tabular listing of all predicted ORFs in NCTC 8325 listed by functional role category based on TIGR roles (*Continued*)

Amino acid biosynthesis category	Functional role	Gene
ORF01176	Guanylate kinase	*gmk*
ORF01235	Uridylate kinase, putative	*pyrH*
ORF01330	Guanosine monophosphate reductase	*guaC*
ORF01485	Nucleoside diphosphate kinase, putative	*ndk*
ORF01496	Cytidylate kinase	*cmk*
ORF02360	Thymidine kinase, putative	
ORF02490	Adenylate kinase, putative	*adk*

Regulatory functions

RNA interactions

ORF01164	PyrR bifunctional protein	*pyrR*
ORF01274	Glycerol uptake operon antiterminator regulatory protein, putative	*glpP*
ORF01356	Transcriptional antiterminator	*glcT*
ORF01668	GTP-binding protein Era	*era*

DNA interactions

ORF00070	Staphylococcal accessory regulator homolog	*sar*
ORF00096	Transcriptional regulator, GntR family, putative	
ORF00434	Transcriptional regulator, lysR family, putative	
ORF00467	*pur* operon repressor	*purR*
ORF00620	Staphylococcal accessory regulator T, putative	*sarT*
ORF00715	Response regulator, putative	*drrA*
ORF00992	Transcriptional regulator, MarR family, putative	
ORF01333	LexA repressor	*lexA*
ORF01592	Transcriptional regulator, Fur, putative	*fur*
ORF01602	Transcriptional regulator, putative	*lacI*
ORF01617	Arginine repressor	*argR*
ORF01685	Heat-inducible transcription repressor HrcA	*hrcA*
ORF01800	Two-component response regulator, putative	
ORF01850	Catabolite control protein A	*ccpA*
ORF01891	Arsenate operon regulator	*arsR*
ORF01980	DNA-binding response regulator, putative	*vraR*
ORF01997	Ferric uptake regulator homolog, putative	
ORF02098	DNA-binding response regulator VraR, putative	*vraR*
ORF02269	Sucrose operon repressor, putative	*scrR*
ORF02315	DNA-binding response regulator, putative	
ORF02456	Lactose phosphotransferase system repressor, putative	*lacR*
ORF02457	Transcriptional regulator, Sir2 family, putative	
ORF02461	Transcriptional regulator, merR family, putative	
ORF02599	Hex regulon repressor, putative	
ORF02643	DNA-binding response regulator, putative	
ORF02956	DNA-binding response regulator, putative	

Protein interactions

ORF00021	Sensory box histidine kinase VicK, putative	*vicK*
ORF00184	Response regulator receiver domain protein	
ORF00231	Two-component response regulator, putative	
ORF00714	Sensor histidine kinase SaeS, putative	*saeS*
ORF00715	Response regulator, putative	*drrA*
ORF00781	HPr(Ser) kinase/phosphatase	*hprK*
ORF01585	Staphylococcal respiratory response protein SrrB, putative	*srrB*
ORF01586	DNA-binding response regulator, putative	
ORF01980	DNA-binding response regulator, putative	*vraR*
ORF02098	DNA-binding response regulator VraR, putative	*vraR*
ORF02099	Histidine kinase, putative	
ORF02301	SigmaB regulation protein RsbU, putative	*rsbU*
ORF02302	SigmaB regulation protein RsbU, putative	*rsbU*
ORF02314	Sensor protein KdpD, putative	*kdpD*
ORF02315	DNA-binding response regulator, putative	
ORF02643	DNA-binding response regulator, putative	

(*Continued on next page*)

TABLE 1 *(Continued)*

Amino acid biosynthesis category	Functional role	Gene
ORF02644	Sensor histidine kinase, putative	
ORF02955	Sensor histidine kinase, putative	
ORF02956	DNA-binding response regulator, putative	
Other		
ORF00020	Two-component response regulator, putative	
ORF00230	Two-component sensor histidine kinase, putative	*lytS*
ORF00504	ATP:guanido phosphotransferase, C-terminal catalytic domain protein	
ORF00517	Transcription antitermination protein, putative	*nusG*
ORF00771	Peptide chain release factor 2, putative	
ORF00794	Glycolytic operon regulator	*gapR*
ORF01361	Transcriptional regulator, putative	
ORF01384	PhoU family, putative	
ORF01879	Virulence factor regulator protein	*rot*
ORF02261	Accessory gene regulator protein B	*agrB*
ORF02264	Accessory gene regulator protein C	*agrC*
ORF02265	Accessory gene regulator protein A	*agrA*
ORF02390	Lytic regulatory protein, putative	
ORF02527	FmhB protein, putative	
ORF02583	Transcriptional regulator, putative	
ORF02799	Staphylococcal accessory regulator T, putative	*sarT*
ORF03001	*ica* operon transcriptional regulator IcaR, putative	*icaR*

Signal transduction

Two-component systems

ORF00021	Sensory box histidine kinase VicK, putative	*vicK*
ORF00184	Response regulator receiver domain protein	
ORF00231	Two-component response regulator, putative	
ORF00714	Sensor histidine kinase SaeS, putative	*saeS*
ORF00715	Response regulator, putative	*drrA*
ORF01313	Histidine kinase-, DNA gyrase B-, and HSP90-like ATPase domain protein	
ORF01420	DNA-binding response regulator, putative	
ORF01585	Staphylococcal respiratory response protein SrrB, putative	*srrB*
ORF01586	DNA-binding response regulator, putative	
ORF01799	Histidine kinase-, DNA gyrase B-, and HSP90-like ATPase domain protein	
ORF01800	Two-component response regulator, putative	
ORF01980	DNA-binding response regulator, putative	*vraR*
ORF02098	DNA-binding response regulator VraR, putative	*vraR*
ORF02099	Histidine kinase, putative	
ORF02314	Sensor protein KdpD, putative	*kdpD*
ORF02315	DNA-binding response regulator, putative	
ORF02643	DNA-binding response regulator, putative	
ORF02644	Sensor histidine kinase, putative	
ORF02955	Sensor histidine kinase, putative	
ORF02956	DNA-binding response regulator, putative	

PTS

ORF00209	PTS system, glucose-specific IIBC component, putative	*ptsG*
ORF00216	PTS system component	
ORF00708	Fructose-specific permease, putative	*fruA*
ORF02400	PTS system, mannitol-specific component, putative	
ORF02402	PTS system, mannitol-specific IIa component, putative	
ORF02456	Lactose phosphotransferase system repressor, putative	*lacR*
ORF02597	PTS system component, putative	
ORF02661	PTS system sucrose-specific IIBC component, putative	
ORF02662	PTS system sucrose-specific IIBC component	
ORF02848	PTS system glucose-specific IIABC component	
ORF02975	PTS system, fructose-specific IIABC component, putative	

Transcription

Other

ORF02316	ATP-dependent RNA helicase, DEAD box family, putative	

(Continued on next page)

TABLE 1 Tabular listing of all predicted ORFs in NCTC 8325 listed by functional role category based on TIGR roles (*Continued*)

Amino acid biosynthesis category	Functional role	Gene
DNA-dependent RNA polymerase		
ORF00524	RNA polymerase beta chain, putative	*rpoB*
ORF00525	DNA-directed RNA polymerase beta-prime chain, putative	*rpoC*
ORF02369	DNA-directed RNA polymerase, delta subunit, putative	*rpoE*
ORF02485	DNA-directed RNA polymerase alpha chain, putative	*rpoA*
RNA processing		
ORF01203	Ribonuclease III, putative	*rnc*
ORF01247	Ribosome-binding factor A	*rbfA*
ORF03054	Ribonuclease P protein component	*rnpA*
Transcription factors		
ORF00517	Transcription antitermination protein, putative	*nusG*
ORF01243	Transcription termination-antitermination factor, putative	*nusA*
ORF01621	N utilization substance protein B, putative	*nusB*
ORF01662	RNA polymerase sigma factor, putative	*rpoD*
ORF01714	Transcription elongation factor GreA	*greA*
ORF02298	Sigma factor B, putative	
ORF02299	Anti-sigma B factor, putative	*rsbW*
ORF02362	Transcription termination factor Rho	*rho*
ORF02664	Transcriptional regulator, putative	
ORF02809	Gluconate operon transcriptional repressor, putative	*gntR*
Degradation of RNA		
ORF00803	Ribonuclease R, putative	*vacB*
ORF01251	Polyribonucleotide nucleotidyltransferase, putative	*pnp*
Transport and binding proteins		
Nucleosides, purines, and pyrimidines		
ORF00373	Xanthine permease, putative	*pbuX*
ORF01165	Uracil permease, putative	*uraA*
Carbohydrates, organic alcohols, and acids		
ORF00067	L-Lactate permease	
ORF00114	Capsular polysaccharide biosynthesis protein, putative	
ORF00155	PTS system, glucose-specific component	*glcA*
ORF00175	Multiple sugar-binding transport ATP-binding protein, putative	*ugpC*
ORF00177	Maltose ABC transporter, permease protein, putative	*malF*
ORF00178	Maltose ABC transporter, permease protein	*malG*
ORF00209	PTS system, glucose-specific IIBC component, putative	*ptsG*
ORF00214	Phosphoenolpyruvate-dependent sugar phosphotransferase system, EIIA 2, putative	
ORF00215	PTS system component, putative	
ORF00216	PTS system component	
ORF00240	Ribose ABC transporter protein, putative	*rbsD*
ORF00708	Fructose-specific permease, putative	*fruA*
ORF00781	HPr(Ser) kinase/phosphatase	*hprK*
ORF01028	Phosphocarrier protein hpr, putative	*ptsI*
ORF01029	Phosphoenolpyruvate-protein phosphotransferase	
ORF02400	PTS system, mannitol-specific component, putative	
ORF02402	PTS system, mannitol-specific IIa component, putative	
ORF02450	PTS system, lactose-specific IIBC component, putative	
ORF02451	PTS system lactose-specific IIA component, putative	*lacF*
ORF02520	Sugar transporter, putative	
ORF02597	PTS system component, putative	
ORF02648	L-Lactate permease	
ORF02661	PTS system sucrose-specific IIBC component, putative	
ORF02662	PTS system sucrose-specific IIBC component	
ORF02687	Formate/nitrite transporter, putative	
ORF02806	Gluconate permease, putative	*gntP*
ORF02848	PTS system glucose-specific IIABC component	*malX*

(*Continued on next page*)

TABLE 1 *(Continued)*

Amino acid biosynthesis category	Functional role	Gene
ORF02943	Citrate transporter, putative	*citN*
ORF02975	PTS system, fructose-specific IIABC component, putative	
Anions		
ORF01385	Phosphate ABC transporter ATP-binding protein, putative	*pstB*
ORF01386	Phosphate ABC transporter, permease protein, putative	*pstA*
ORF01389	Phosphate ABC transporter, periplasmic phosphate-binding protein, putative	*phoX*
ORF02546	Molybdenum transport ATP-binding protein ModC, putative	*modC*
ORF02547	Molybdenum ABC transporter, permease protein, putative	*modB*
ORF03030	Sodium, sulfate symporter, putative	
Other		
ORF00102	Phosphonates ABC transporter, permease protein CC0363, putative	
ORF00103	Phosphonates ABC transporter, permease protein CC0363, putative	
ORF00104	Amino acid ABC transporter, ATP-binding protein, putative	
ORF00105	Phosphonate ABC transporter, substrate-binding protein, putative	
ORF00246	Drug transporter, putative	
ORF00284	5'-Nucleotidase, lipoprotein e(P4) family	
ORF00317	Glycerol-3-phosphate transporter	*glpT*
ORF00556	Proline/betaine transporter, putative	*proP*
ORF00641	Teichoic acid translocation ATP-binding protein, putative	*rfbI*
ORF00703	Quinolone resistance norA protein, putative	
ORF01893	Arsenical pump membrane protein subfamily	*arsB*
ORF02003	ABC transporter, ATP-binding/permease protein, putative	
ORF02629	Drug resistance transporter, EmrB/QacA subfamily, putative	
ORF02740	Drug transporter, putative	
Amino acids, peptides, and amines		
ORF00151	Branched-chain amino acid transport system II carrier protein	*brnQ*
ORF00167	Peptide ABC transporter, ATP-binding protein, putative	
ORF00169	Peptide ABC transporter, permease protein, putative	
ORF00282	Branched-chain amino acid transport system II carrier protein	*brnQ*
ORF00732	Amino acid ABC transporter, permease protein, putative	*proX*
ORF00926	Oligopeptide ABC transporter, ATP-binding protein, putative	*oppF*
ORF00927	Oligopeptide ABC transporter, substrate-binding protein, putative	
ORF00928	Oligopeptide ABC transporter, substrate-binding protein, putative	
ORF00929	Oligopeptide ABC transporter, ATP-binding protein, putative	
ORF00930	Oligopeptide ABC transporter, ATP-binding protein, putative	
ORF00931	Oligopeptide ABC transporter, permease protein, putative	
ORF00932	Oligopeptide ABC transporter, permease protein, putative	*oppC*
ORF01047	Spermidine/putrescine ABC transporter, permease protein, putative	
ORF01048	Spermidine/putrescine ABC transporter, permease protein, putative	*potC*
ORF01049	Spermidine/putrescine ABC transporter, spermidine/putrescine-binding protein, putative	*potD*
ORF01346	Glycine betaine transporter, putative	*opuD*
ORF01354	Sodium:alanine symporter family protein, putative	*alsT*
ORF01377	Oligopeptide ABC transporter, ATP-binding protein, putative	
ORF01378	Peptide ABC transporter, ATP-binding protein, putative	
ORF01379	Oligopeptide transporter putative membrane permease domain	
ORF01380	Oligopeptide transporter putative membrane permease domain	
ORF01411	Branched-chain amino acid transport system II carrier protein	*brnQ*
ORF01990	Amino acid ABC transporter, ATP-binding protein, putative	*glnQ*
ORF02119	Proline uptake protein [validated]	
ORF02444	Osmoprotectant transporter, BCCT family, opuD homolog, putative	*opuD*
ORF02557	Urea transporter, putative	
ORF02622	Sodium/glutamate symporter	*gltS*
ORF02697	Amino acid ABC transporter, ATP-binding protein, putative	
ORF02698	Amino acid ABC transproter, permease protein, putative	
ORF02729	Amino acid ABC transporter homolog, putative	
ORF02741	Amino acid ABC transporter, permease protein, putative	*proW-1*
ORF02742	Amino acid transporter, putative	
ORF02743	Amino acid ABC transporter, permease protein, putative	

(Continued on next page)

TABLE 1 Tabular listing of all predicted ORFs in NCTC 8325 listed by functional role category based on TIGR roles (*Continued*)

Amino acid biosynthesis category	Functional role	Gene
ORF02744	Amino acid ABC transporter, ATP-binding protein, putative	
ORF02763	Peptide ABC transporter, ATP-binding protein, putative	
ORF02764	Peptide ABC transporter, ATP-binding protein, putative	
ORF02765	Nickel ABC transporter, permease protein, putative	
ORF02766	Peptide ABC transporter, permease protein, putative	
ORF02767	Peptide ABC transporter, peptide-binding protein, putative	
ORF02937	Choline transporter, putative	*cudT*
ORF02967	Arginine/ornithine antiporter, putative	*arcD*
Unknown substrate		
ORF00176	Bacterial extracellular solute-binding protein, putative	
ORF00287	ABC transporter, ATP-binding protein, putative	
ORF00333	ABC transporter, ATP-binding protein, putative	
ORF00424	ABC transporter, permease protein, putative	
ORF00426	ABC transporter, substrate-binding protein, putative	
ORF00634	ABC transporter, substrate-binding protein, putative	
ORF00636	Iron (chelated) ABC transporter, permease protein, putative	*psaC*
ORF00667	ABC transporter ATP-binding protein, putative	*vraF*
ORF00668	ABC transporter permease, putative	*vraG*
ORF00681	Major facilitator superfamily protein superfamily	
ORF00729	ABC transporter, ATP-binding protein	
ORF00731	ABC transporter domain protein	
ORF00842	ABC transporter, ATP-binding protein, putative	
ORF00847	ABC transporter, ATP-binding protein, putative	
ORF00970	ABC transporter, ATP-binding protein, putative	
ORF01311	ABC transporter, ATP-binding protein, putative	
ORF01392	ABC transporter, ATP-binding protein, putative	
ORF01430	Phosphotransferase system enzyme IIA, putative	
ORF01657	ABC transporter, putative	*mreA*
ORF01948	ABC transporter domain protein	
ORF01967	ABC transporter, ATP-binding protein, putative	
ORF01991	ABC transporter, permease protein, putative	
ORF02137	Sodium-dependent transporter (huNaDC-1), putative	
ORF02152	ABC transporter, ATP-binding protein, putative	
ORF02154	ABC transporter, ATP-binding protein, putative	
ORF02274	ABC transporter, ATP-binding protein, putative	
ORF02397	ABC transporter, ATP-binding protein, putative	
ORF02482	ABC transporter, ATP-binding protein, putative	
ORF02483	ABC transporter, ATP-binding protein, putative	
ORF02754	ABC transporter, ATP-binding protein, putative	
ORF02773	Transporter, putative	
ORF02954	ABC transporter, ATP-binding protein, putative	
ORF03019	ABC transporter, ATP-binding protein, putative	
ORF03036	ABC transporter, ATP-binding protein, putative	
ORF03037	Permease, putative	
Cations		
ORF00071	Lipoprotein, SirC, putative	*sirC*
ORF00072	Lipoprotein, SirB, putative	*sirB*
ORF00613	Iron compound ABC transporter, substrate-binding protein, putative	
ORF00652	Iron compound ABC transporter, ATP-binding protein	
ORF00653	Ferrichrome transport permease fhuB, putative	*fhuB*
ORF00654	Ferrichrome ABC transporter (permease), putative	*fhuG*
ORF00883	Monovalent cation/proton antiporter, MnhG/PhaG subunit subfamily, putative	
ORF00886	Na$^+$/H$^+$ antiporter, MnhD component, putative	*mnhD*
ORF00887	Na$^+$/H$^+$ antiporter subunit, putative	*mnhC*
ORF00888	Monovalent cation:proton antiporter subfamily, putative	
ORF00889	Na$^+$/H$^+$ antiporter subunit, putative	*mnhA*
ORF00945	Magnesium transporter	*mgtE*
ORF01086	Iron compound ABC transporter, permease protein, putative	
ORF01087	Iron compound ABC transporter, permease protein	
ORF01997	Ferric uptake regulator homolog, putative	

(*Continued on next page*)

TABLE 1 (*Continued*)

Amino acid biosynthesis category	Functional role	Gene
ORF02108	Ferritin, putative	
ORF02137	Sodium-dependent transporter (huNaDC-1), putative	
ORF02247	Cation transport protein, putative	
ORF02270	Ammonium transporter	*amt*
ORF02300	STAS domain, putative	
ORF02310	Potassium-transporting ATPase, C subunit	*kdpC*
ORF02311	Potassium-translocating P-type ATPase, B subunit, putative	*kdpB*
ORF02312	Potassium-transporting ATPase, A subunit	*kdpA*
ORF02389	Cation efflux family protein, putative	
ORF02481	Cobalt transport protein, putative	
ORF02549	Molybdenum ABC transporter, periplasmic molybdate-binding protein	*modA*
ORF02573	Na^+/H^+ antiporter NhaC	*nhaC*
ORF02601	Na^+/H^+ antiporter, putative	
ORF02864	Ferrous iron transport protein B	*feoB*
ORF02873	Cation-transporting ATPase, E1-E2 family, putative	*copA*
ORF02874	Cation-transporting ATPase, E1-E2 family, putative	
ORF03033	High-affinity nickel transporter, putative	

Unknown function

Enzymes of unknown specificity

ORF00158	PTS system component	
ORF00450	Orn/Lys/Arg decarboxylase, putative	
ORF00555	Hydrolase, haloacid dehalogenase-like	
ORF00720	6-Pyruvoyl tetrahydropterin synthase superfamily, putative	
ORF00724	Chorismate binding enzyme, putative	
ORF00893	FMN oxidoreductase, putative	
ORF01055	Inositol monophosphatase family protein, putative	
ORF01143	S-Adenosyl-methyltransferase MraW	
ORF01185	Radical SAM enzyme, Cfr family	
ORF01279	Hydrolase, alpha/beta fold family domain protein	
ORF02111	ImpB/MucB/SamB family superfamily	
ORF02737	Epimerase/dehydratase, putative	
ORF02958	Alkaline phosphatase III precursor, putative	

General

ORF00174	M23/M37 peptidase domain protein	
ORF00296	ROK family protein	
ORF00359	Phosphoglycerate mutase family protein	
ORF00503	UvrB/uvrC motif domain protein	
ORF00538	Haloacid dehalogenase-like hydrolase, putative	
ORF00554	SIS domain protein	
ORF00615	Haloacid dehalogenase-like hydrolase, putative	
ORF00671	Secretory antigen SsaA homologue	
ORF00773	LysM domain protein	
ORF01058	GTP-binding protein TypA, putative	
ORF01110	Fibrinogen-binding protein-related	
ORF01175	Fibronectin-binding protein A-related	
ORF01223	*gid* protein	
ORF01431	Methionine sulfoxide reductase, putative	
ORF01598	AtsA/ElaC family protein	
ORF01649	Rhomboid family protein	
ORF01665	CBS domain protein	
ORF01752	ACT domain protein	
ORF01753	GTP-binding protein	
ORF01777	GTP-binding protein	
ORF01828	GAF domain protein	
ORF01840	Transglycosylase domain protein	
ORF01981	Sensor histidine kinase, putative	
ORF02256	Abortive infection protein	
ORF02297	S1 RNA-binding domain protein	

(*Continued on next page*)

TABLE 1 (*Continued*)

Amino acid biosynthesis category	Functional role	Gene
ORF02333	*sceD* protein, putative	
ORF02358	Modification methylase, HemK family, putative	
ORF02641	Permease, putative domain protein	
ORF02887	Immunodominant antigen A, putative	
ORF02953	Permease, putative domain protein	
ORF03051	Glucose-inhibited division protein B	
ORF03052	Glucose inhibited division protein A, putative	

[a]Note that a single ORF may be listed in more than one role category.

[b]Gene designations are based on those previously defined for *S. aureus* or *B. subtilis*, if available. Duplicate gene names may occur as a result of multiple alleles encoding homologous genes or unresolved frameshifts. For lack of space, genes designated as hypothetical or conserved hypothetical are not listed. A complete listing is available on our website (http://microgen.ouhsc.edu).

lysogeny results in both negative and positive conversion. It inserts intragenically into *hlb*, inactivating the gene encoding beta-toxin while encoding *sak*, the gene for staphylokinase expression.

Variable Genetic Elements

All of the sequenced staphylococcal isolates contain DNA sequences for which there is no allelic equivalent in other strains. These sequences represent important components of the *S. aureus* genome and can be studied more easily as the number of sequenced genomes increases. *S. aureus* NCTC 8325 appears to contain fewer of the previously identified genomic islands, pathogenicity islands, or transposons than other sequenced *S. aureus* strains. Specifically, NCTC 8325 does not contain an SCC*mec* region, which would encode the *ccrA* and *ccrB* recombinase genes present in methicillin-resistant strains COL, MRSA252, Mu50, and N315 (16, 20). This is not unexpected because NCTC 8325 is not resistant to this antibiotic. In addition, the *S. aureus* strains that have been sequenced contain a variety of pathogenicity islands (SaPI) (see chapters 33 and 41, this volume) that seem to occur more often in more virulent strains. Nine of these form a family of closely related mobile elements, each encoding an integrase and, usually, two or more superantigen toxins, including, in various combinations, toxic shock syndrome toxin-1, enterotoxin B, C, G, I(?), K, L, and M(?) (14, 24, 27). Their overall properties are summarized in Table 2. The SaPI variants have somewhat dissimilar integrases and reside at different sites within the chromosome.

In the interest of uniformity, we have adopted the nomenclature originally used by the Novick group (24), and we propose its use be adopted officially. NCTC 8325 contains none of these nine SaPIs. However, it does contain two other (putative) pathogenicity islands present in all the other genomes sequenced thus far. One of these, *v*Saα (3), encoding enterotoxin-like (*set*) and lipoprotein-like (*lpl*) ORFs, is present in NCTC 8325 (at coordinates 383358-408063 and contains ORF 00376 to 00405). Variations of this island differ primarily in the number and type of their enterotoxin-like ORFs (20). The second, *v*Saβ, usually carries determinants for enterotoxin-like, serine protease-like leukotoxin, components of a type I restriction-modification system, and the operon for the staphylococcin, epidermin. The epidermin genes are apparently inactive because

TABLE 2 The SaPI family

Element (strain)	Description	*att* site core sequence	Virulence genes	Reference
SaPI1 (RN4282)	15.2 kb; excised, replicated and transduced at high frequency by phage 80α	TTATTTAGCAGGAATAA	*tst, ent*	24
SaPI2 (RN3984)	Size? Excised, replicated, and transduced at high frequency by phage 80		*tst*	31
SaPI3 (COL)	15.6 kb; probably excised, replicated, and transduced at high frequency by phage 29	TTATTTAGCAGGAATAA	*seb sel sek*	27
SaPI4	Size? att-int		*tst, sec-3*	27
SaPIbov 1 (RF122)	15.8 kb; excised spontaneously; also excised, replicated, and transduced at high frequency by 80α and φ11	AAAAAAGGCTGGAAACCGCGTA ATTACGGTAACTCCAGCCTATC ATTTGCTATATATAATTA TTCCCACTCAAT	*tst, sel, sek*	14
SaPIbov2	21 kb, excised spontaneously	TAATTATTCCCACTCGAT	*bap*	34
SaPIn1 (n315) SaPIm 1 (mu50)	(Identical) 15 kb	GTTTTACCATCATTCCCGGCAT	*tst, sel, sec-3*	20
*v*Sa3 (SaGIm)	Excised spontaneously	TCCCGCCGTCTCCAT	*fhuD*	20

NCTC8325 does not produce epidermin. The NCTC 8325 version lacks the enterotoxin cluster (*ent*), but contains the leukotoxin cluster (*lukDE*) and several putative secretory serine proteases (*splA-F*). Finally, a novel island, identified in MRSA252 and not present in NCTC 8325, displays synteny with previously described SPI1 islands (16) and contains homologues of pathogenicity island proteins plus several hypothetical proteins with no similarity to other characterized genes. This island is found inserted downstream of the ribosomal protein *rpsR*. In addition to the information presented here, a more detailed discussion of these elements is found in chapters 33 and 41, this volume. Additional understanding of such elements will continue to increase as the analysis of genomic data from NCTC 8325 and the other staphylococcal genomes continues.

NOTE: The completed sequence and annotation (Table 1) as well as circular and linear maps of strain NCTC 8325 are available on our website (http://microgen.ouhsc.edu), under "projects." The GenBank sequence file and annotation also can be accessed through the National Center for Biotechnology Information microbial genomes database (http://ncbi.nlm.nih.gov).

This project was supported by NIH grant AI43568 from the National Institute of Allergy and Infectious Diseases and by a grant from the Merck Genome Research Institute.

REFERENCES

1. **Ajdic, D., W. M. McShan, R. E. McLaughlin, G. Savic, J. Chang, M. B. Carson, C. Primeaux, R. Tian, S. Kenton, H. Jia, S. Lin, Y. Qian, S. Li, H. Zhu, F. Najar, H. Lai, J. White, B. A. Roe, and J. J. Ferretti.** 2002. Genome sequence of *Streptococcus mutans* UA159, a cariogenic dental pathogen. *Proc. Natl. Acad. Sci. USA* **99:** 14434–14439.

2. **Anagnostopoulos, C., P. J. Piggot, and J. A. Hoch.** 1993. The genetic map of *Bacillus subtilis*, p. 425–462. *In* A. L. Sonenshein, J. A. Hoch, and R. Losick (ed.), Bacillus subtilis *and Other Gram-Positive Bacteria: Biochemistry, Physiology, and Molecular Genetics.* American Society for Microbiology, Washington, D.C.

3. **Baba, T., F. Takeuchi, M. Kuroda, H. Yuzawa, K. Aoki, A. Oguchi, Y. Nagai, N. Iwama, K. Asano, T. Naimi, H. Kuroda, L. Cui, K. Yamamoto, and K. Hiramatsu.** 2002. Genome and virulence determinants of high virulence community-acquired MRSA. *Lancet* **359:**1819–1827.

4. **Bachmann, B. J.** 1983. Linkage map of *Escherichia coli* K-12, edition 7. *Microbiol. Rev.* **47:**180–230.

5. **Barberis-Maino, L., B. Berger-Bächi, H. Weber, W. D. Beck, and F. H. Kayser.** 1987. IS*431*, a staphylococcal insertion sequence-like element related to IS26 from *Proteus vulgaris. Gene* **59:**107–113.

6. **Beres, S. B., G. L. Sylva, K. D. Barbian, B. Lei, J. S. Hoff, N. D. Mammarella, M. Y. Liu, J. C. Smoot, S. F. Porcella, L. D. Parkins, D. S. Campbell, T. M. Smith, J. K. McCormick, D. Y. Leung, P. M. Schlievert, and J. M. Musser.** 2002. Genome sequence of a serotype M3 strain of group A *Streptococcus*: phage-encoded toxins, the high-virulence phenotype, and clone emergence. *Proc. Natl. Acad. Sci. USA* **99:**10078–10083.

7. **Betley, M. J., and J. J. Mekalanos.** 1988. Nucleotide sequence of the type A staphylococcal enterotoxin gene. *J. Bacteriol.* **170:**34–41.

8. **Blattner, F. R., G. Plunkett III, C. A. Bloch, N. T. Perna, V. Burland, M. Riley, J. Collado-Vides, J. D. Glasner, C. K. Rode, G. F. Mayhew, J. Gregor, N. W. Davis, H. A. Kirkpatrick, M. A. Goeden, D. J. Rose, B. Mau, and Y. Shao.** 1997. The complete genome sequence of *Escherichia coli* K-12. *Science* **277:**1453–1474.

9. **Byrne, M. E., D. A. Rouch, and R. A. Skurray.** 1989. Nucleotide sequence analysis of IS256 from the *Staphylococcus aureus* gentamicin-tobramycin-kanamycin-resistance transposon Tn*4001. Gene* **81:**361–367.

10. **Coleman, D. C., J. P. Arbuthnott, H. M. Pomeroy, and T. H. Birkbeck.** 1986. Cloning and expression in *Escherichia coli* and *Staphylococcus aureus* of the beta-lysin determinant from *Staphylococcus aureus*: evidence that bacteriophage conversion of beta-lysin activity is caused by insertional inactivation of the beta-lysin determinant. *Microb. Pathog.* **1:**549–564.

11. **Coleman, D. C., D. J. Sullivan, R. J. Russell, J. P. Arbuthnott, B. F. Carey, and H. M. Pomeroy.** 1989. *Staphylococcus aureus* bacteriophages mediating the simultaneous lysogenic conversion of beta-lysin, staphylokinase and enterotoxin A: molecular mechanism of triple conversion. *J. Gen. Microbiol.* **135:**1679–1697.

12. **Derbise, A., K. G. Dyke, and N. el Solh.** 1994. Isolation and characterization of IS*1181*, an insertion sequence from *Staphylococcus aureus. Plasmid* **31:**251–264.

13. **Ferretti, J. J., W. M. McShan, D. Ajdic, D. J. Savic, G. Savic, K. Lyon, C. Primeaux, S. Sezate, A. N. Suvorov, S. Kenton, H. S. Lai, S. P. Lin, Y. Qian, H. G. Jia, F. Z. Najar, Q. Ren, H. Zhu, L. Song, J. White, X. Yuan, S. W. Clifton, B. A. Roe, and R. McLaughlin.** 2001. Complete genome sequence of an M1 strain of *Streptococcus pyogenes. Proc. Natl. Acad. Sci. USA* **98:**4658–4663.

14. **Fitzgerald, J. R., S. R. Monday, T. J. Foster, G. A. Bohach, P. J. Hartigan, W. J. Meaney, and C. J. Smyth.** 2001. Characterization of a putative pathogenicity island from bovine *Staphylococcus aureus* encoding multiple superantigens. *J. Bacteriol.* **183:**63–70.

15. **Gillespie, M. T., B. R. Lyon, and R. A. Skurray.** 1988. Structural and evolutionary relationships of beta-lactamase transposons from *Staphylococcus aureus. J. Gen. Microbiol.* **134:**2857–2866.

16. **Holden, M. T., E. J. Feil, J. A. Lindsay, S. J. Peacock, N. P. Day, M. C. Enright, T. J. Foster, C. E. Moore, L. Hurst, R. Atkin, A. Barron, N. Bason, S. D. Bentley, C. Chillingworth, T. Chillingworth, C. Churcher, L. Clark, C. Corton, A. Cronin, J. Doggett, L. Dowd, T. Feltwell, Z. Hance, B. Harris, H. Hauser, S. Holroyd, K. Jagels, K. D. James, N. Lennard, A. Line, R. Mayes, S. Moule, K. Mungall, D. Ormond, M. A. Quail, E. Rabbinowitsch, K. Rutherford, M. Sanders, S. Sharp, M. Simmonds, K. Stevens, S. Whitehead, B. G. Barrell, B. G. Spratt, and J. Parkhill.** 2004. Complete genomes of two clinical *Staphylococcus aureus* strains: evidence for the rapid evolution of virulence and drug resistance. *Proc. Natl. Acad. Sci. USA* **101:**9786–9791.

17. **Iandolo, J. J., J. P. Bannantine, and G. C. Stewart.** 1997. Genetic and physical map of the chromosome of *Staphylococcus aureus*, p. 39–54. *In* G. L. Archer and K. Crosley (ed.), *Staphylococci and Staphylococcal Diseases.* Churchill-Livingstone, New York, N.Y.

18. **Iandolo, J. J., V. Worrell, K. H. Groicher, Y. Qian, R. Tian, S. Kenton, A. Dorman, H. Ji, S. Lin, P. Loh, S. Qi, H. Zhu, and B. A. Roe.** 2002. Comparative analysis of the genomes of the temperate bacteriophages phi 11, phi 12 and phi 13 of *Staphylococcus aureus* 8325. *Gene* **289:**109–118.

19. **Kondo, I., and K. Fujise.** 1977. Serotype B staphylococcal bacteriophage singly converting staphylokinase. *Infect. Immun.* **18:**266–272.

20. **Kuroda, M., T. Ohta, I. Uchiyama, T. Baba, H. Yuzawa, I. Kobayashi, L. Cui, A. Oguchi, K. Aoki, Y. Nagai, J. Lian, T. Ito, M. Kanamori, H. Matsumaru, A. Maruyama, H. Murakami, A. Hosoyama, Y. Mizutani-Ui, N. K. Takahashi, T. Sawano, R. Inoue, C. Kaito, K. Sekimizu, H. Hirakawa, S. Kuhara, S. Goto, J. Yabuzaki, M. Kanehisa, A. Yamashita, K. Oshima, K. Furuya, C. Yoshino, T. Shiba, M. Hattori, N. Ogasawara, H. Hayashi, and K. Hiramatsu.** 2001. Whole genome sequencing of meticillin-resistant *Staphylococcus aureus. Lancet* **357:**1225–1240.

21. **Lee, C. Y.** 1995. Association of staphylococcal type-1 capsule-encoding genes with a discrete genetic element. *Gene* **167:**115–119.

22. **Lee, C. Y., and J. J. Iandolo.** 1986. Lysogenic conversion of staphylococcal lipase is caused by insertion of the bacteriophage L54a genome into the lipase structural gene. *J. Bacteriol.* **166:**385–391.

23. **Lindsay, J. A., and M. T. Holden.** 2004. *Staphylococcus aureus*: superbug, super genome? *Trends Microbiol.* **12:**378–385.

24. **Lindsay, J. A., A. Ruzin, H. F. Ross, N. Kurepina, and R. P. Novick.** 1998. The gene for toxic shock toxin is carried by a family of mobile pathogenicity islands in *Staphylococcus aureus. Mol. Microbiol.* **29:**527–543.

25. **Lyon, B. R., J. W. May, and R. A. Skurray.** 1984. Tn*4001*: a gentamicin and kanamycin resistance transposon in *Staphylococcus aureus. Mol. Gen. Genet.* **193:**554–556.

26. **Murphy, E., S. Phillips, I. Edelman, and R. P. Novick.** 1981. Tn*554*: isolation and characterization of plasmid insertions. *Plasmid* **5:**292–305.

27. **Novick, R. P., P. Schlievert, and A. Ruzin.** 2001. Pathogenicity and resistance islands of staphylococci. *Microbes Infect.* **3:**585–594.

28. **Pattee, P. A.** 1993. The genetic map of *Staphylococcus aureus*, p. 489–496. *In* A. L. Sonenshein, J. A. Hoch, and R. Losick (ed.), Bacillus subtilis *and Other Gram-Positive Bacteria: Biochemistry, Physiology, and Molecular Genetics.* American Society for Microbiology, Washington, D.C.

29. **Rouch, D. A., L. J. Messerotti, L. S. Loo, C. A. Jackson, and R. A. Skurray.** 1989. Trimethoprim resistance transposon Tn*4003* from *Staphylococcus aureus* encodes genes for a dihydrofolate reductase and thymidylate synthetase flanked by three copies of IS*257. Mol. Microbiol.* **3:**161–175.

30. **Rowland, S. J., and K. G. Dyke.** 1989. Characterization of the staphylococcal beta-lactamase transposon Tn*552. EMBO J.* **8:**2761–2773.

31. **Ruzin, A., J. Lindsay, and R. P. Novick.** 2001. Molecular genetics of SaPI1—a mobile pathogenicity island in *Staphylococcus aureus. Mol. Microbiol.* **41:**365–377.

33. **Townsend, D. E., S. Bolton, N. Ashdown, D. I. Annear, and W. B. Grubb.** 1986. Conjugative, staphylococcal plasmids carrying hitch-hiking transposons similar to Tn*554*: intra- and interspecies dissemination of erythromycin resistance. *Aust. J. Exp. Biol. Med. Sci.* **64:**367–379.

34. **Ubeda, C., M. A. Tormo, C. Cucarella, P. Trotonda, T. J. Foster, I. Lasa, and J. R. Penades.** 2003. Sip, an integrase protein with excision, circularization and integration activities, defines a new family of mobile *Staphylococcus aureus* pathogenicity islands. *Mol. Microbiol.* **49:**193–210.

35. **van der Vijver, J. C., M. van Es-Boon, and M. F. Michel.** 1972. Lysogenic conversion in *Staphylococcus aureus* to leucocidin production. *J. Virol.* **10:**318–319.

36. **Zhang, Y. Q., S. X. Ren, H. L. Li, Y. X. Wang, G. Fu, J. Yang, Z. Q. Qin, Y. G. Miao, W. Y. Wang, R. S. Chen, Y. Shen, Z. Chen, Z. H. Yuan, G. P. Zhao, D. Qu, A. Danchin, and Y. M. Wen.** 2003. Genome-based analysis of virulence genes in a non-biofilm-forming *Staphylococcus epidermidis* strain (ATCC 12228). *Mol. Microbiol.* **49:**1577–1593.

Genetics: Accessory Elements and Genetic Exchange

NEVILLE FIRTH AND RONALD A. SKURRAY

33

As has been the case for other bacterial genera, studies into the genetic basis of staphylococcal pathogenicity and antimicrobial resistance revealed the presence of determinants not uniformly represented in the genomes of all strains. Although not constituting fundamental requirements for survival per se, such accessory elements usually encode functions required to meet the demands of a particular environmental niche. The extent of the accessory DNA component of the genome is now being clarified by comparative analysis of whole *Staphylococcus aureus* genome sequences, which indicate that it can constitute in excess of 20% of a strain's genetic makeup (9, 29, 42, 51, 56). It is clear that the acquisition, maintenance, and dissemination of accessory elements have been central to the ongoing success of staphylococci as pathogens. Staphylococci represent a salient illustration of the adaptability afforded to microorganisms by access to additional functions through gene transfer mechanisms.

MECHANISMS OF GENETIC EXCHANGE

Although DNA can be introduced into staphylococci in the laboratory via each of the three traditional bacterial gene transfer mechanisms—transformation, transduction, and conjugation—the latter two are believed to be the most significant mediators of natural genetic exchange. Transformation is very inefficient, has a curious cofactor requirement that can be satisfied by a component of phage 55C, and is thought to be limited by extracellular nucleases and/or restriction systems encoded by staphylococci (58, 72). In addition to their own transfer, staphylococcal conjugative plasmids (see below) also facilitate the transmission of other (nonconjugative) plasmids, either by mobilization, if the other plasmid encodes a specialized relaxation system (79), or more generally via cointegrate formation, and potentially subsequent resolution, in a process termed conduction (7, 59). Conjugation is also thought to represent the most likely basis for the apparent transmission of large DNA segments that correspond to approximately 10 to 20% of the chromosome (83). A novel but poorly understood mechanism of genetic exchange, named mixed-culture transfer or phage-mediated conjugation, has also been identified in staphylococci (52, 58). Although phage, or perhaps components of phage, play a role in mixed-culture transfer, the process has been shown to be mechanistically distinct from transduction. There have been several reports of subinhibitory levels of some antibiotics enhancing the efficiency of DNA transfer between staphylococci; however, the basis of this stimulation remains unclear (1, 10, 23).

Identical or nearly identical accessory genes, elements, and plasmids have been detected in different staphylococcal species and other bacterial genera, such as enterococci and streptococci (28, 71, 75). Such observations suggest that, directly and/or indirectly, the gene transfer mechanisms operating in staphylococci facilitate not only intraspecific transfer, but also interspecific and intergeneric exchange, and hence access to an extended and shared reservoir of determinants (28, 75).

STAPHYLOCOCCAL PLASMIDS

One or more plasmids are usually found in clinical isolates of *S. aureus* and coagulase-negative staphylococci (72). Most staphylococcal plasmids can be categorized as one of three main classes based on physical/genetic organization and functional characteristics (71, 75), although another group, the pSK639 family plasmids (55), should be considered a fourth class. A number of other plasmids that do not appear to fall within these classes await further investigation. Fifteen plasmid incompatibility groups have so far been identified in staphylococci (71, 102); however, plasmid classification is increasingly based on DNA sequence data, particularly of replication regions. Whereas some staphylococcal plasmids are phenotypically cryptic, most that have been described encode antimicrobial resistance determinants; several have also been attributed other clinically significant properties, such as toxin production (108, 109). Although our knowledge is skewed by the medical bias of staphylococcal research, it is likely that plasmids are prevalent in all staphylococcal species.

Small Rolling-Circle Plasmids

The most thoroughly characterized staphylococcal plasmids are those that utilize asymmetric rolling-circle (RC) replica-

Gram-Positive Pathogens, 2nd edition, edited by Vincent A. Fischetti et al.
© 2006 ASM Press, Washington, D.C.

tion via a single-stranded DNA intermediate (for a detailed review of RC plasmid replication, see Khan [50]). Probably reflecting constraints imposed by this replication strategy, RC (sometimes alternatively called single-stranded DNA) plasmids are normally less than 5 kb in size and are rarely found to contain transposable elements (71). This restriction in RC plasmid size probably results from increased production of high-molecular-weight DNA by larger plasmids. Thought to represent a defect in the termination step of RC replication, accumulation of high-molecular-weight DNA impairs cell growth, leading to counterselection against host cells and hence the plasmid itself (39). RC plasmids are usually maintained at 10 to 60 copies/cell and are phenotypically cryptic or carry only a single resistance gene, although there are examples of the carriage of two determinants (72).

Staphylococcal RC plasmids can be subdivided into four families, exemplified by pT181, pC194, pE194, and pSN2, based on replication region sequence similarity (71). Representative RC plasmids are shown in Fig. 1. These families contain plasmids from different staphylococcal species and other bacterial genera, such as *Bacillus*, *Lactobacillus*, and *Streptococcus*, attesting to the horizontal transmission of RC plasmids (39). Some of these plasmids encode a mobilization system (e.g., *mobABC* on pC221) (Fig. 1). A locus consisting of a gene, *pre*, and site, RS$_A$, originally identified as a site-specific recombination function on plasmids such as pT181 and pE194 (Fig. 1) (71), may function as a second type of mobilization system, possibly involving conduction

(35, 37, 78). RC plasmids, including those in different families, often share highly similar DNA segments, such that they appear to be mosaic structures consisting of discrete functional cassettes encoding replication, resistance, and sometimes recombination and/or mobilization functions (71, 80). Exchange of these segments is thought to be promoted by the replication strategy of RC plasmids and their extended regions of sequence similarity (36). Site-specific recombination functions such as *pre*, which favor cointegrate formation rather than multimer resolution (71), may also be involved.

pSK639 Family Plasmids

A group of plasmids related to a prototype, pSK639, were initially identified in strains of *Staphylococcus epidermidis* (55), and subsequently in *S. aureus* (68, 69). In size and characteristics, these plasmids can be considered to fall between the RC and multiresistance groups of staphylococcal plasmids. Like RC plasmids, pSK639 is relatively small (8 kb) and confers only trimethoprim resistance, mediated by a trimethoprim-insensitive dihydrofolate reductase encoded by the *dfrA* gene (Fig. 2) (2, 87). Moreover, most pSK639 family plasmids possess mobilization regions closely related to those of the pT181 family RC plasmids pC221 (Fig. 1), pC223, and pS194 (2). However, pSK639-type plasmids are proposed to utilize a replication region related to that of the theta-replicating *Lactococcus lactis* plasmid pWV02 (2, 39). This family is also distinguished from the RC plasmids by the capacity to

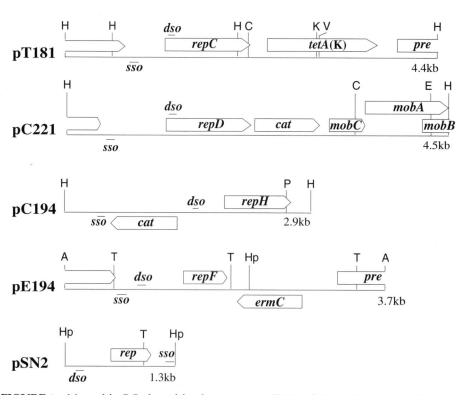

FIGURE 1 Maps of the RC plasmid family prototypes, pT181, pC194, pE194, and pSN2, and the mobilizable pT181 family plasmid, pC221 (75); see text for additional references. Plasmid sizes are shown on the right. Genes (arrowed boxes) and loci encoding the following functions are indicated: *cat*, chloramphenicol resistance; *dso*, double-stranded origin of DNA replication; *ermC*, erythromycin resistance; *mobA/mobB/mobC*, plasmid mobilization; *pre*, plasmid recombination/mobilization; *rep/repC/repD/repF/repH*, initiation of plasmid replication; *sso*, single-stranded origin of DNA replication; *tetA*(K), tetracycline resistance. Restriction sites shown: A, *AluI*; C, *ClaI*; E, *EcoRI*; H, *HindIII*; Hp, *HpaII*; K, *KpnI*; P, *PvuII*; T, *TaqI*; V, *EcoRV*.

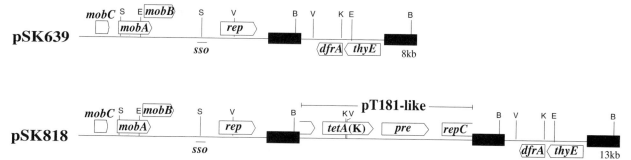

FIGURE 2 Maps of representative pSK639 family plasmids, pSK639 and pSK818 (2, 55). The locations of IS257 copies are marked by solid boxes, and the position of the cointegrated pT181-like plasmid within pSK818 is indicated. Plasmid sizes are shown on the right. Genes (arrowed boxes) and loci encoding the following functions are indicated: *dfrA*, trimethoprim resistance; *mobA/mobB/mobC*, plasmid mobilization; *pre*, plasmid recombination/mobilization; *rep/repC*, initiation of plasmid replication; *sso*, single-stranded origin of DNA replication; *tetA*(K), tetracycline resistance; *thyE*, thymidylate synthetase. Restriction sites shown: B, *Bgl*II; E, *Eco*RI; K, *Kpn*I; S, *Sal*I; V, *Eco*RV.

carry insertion sequence (IS) elements, such as copies of IS257 (55). Indeed, there are examples of these elements mediating the cointegrative fusion of a pSK639-type plasmid and an RC plasmid, such as pT181 in the case of pSK818 (Fig. 2), to form hybrid multiresistance plasmids (55). Similarly, the composite transposon-like structure designated Tn4003 (87) and related elements evident on a number of S. aureus multiresistance plasmids (see below) represent cointegrated remnants of a pSK639-like plasmid (28, 55). Other pSK639 family plasmids include the S. epidermidis plasmids pIP1629 and pIP1630 that confer resistance to type A streptogramins, and in the case of pIP1630, trimethoprim resistance (8), and the bacteriocin-encoding S. aureus plasmids pRJ6 and pRJ9 (68, 69).

Multiresistance Plasmids

Staphylococcal multiresistance plasmids are usually 15 to 40 kb in size and exist at approximately five copies per cell (72). These plasmids typically encode several antimicrobial resistance determinants, often in association with transposable elements (28, 75). Because of their size, these plasmids are presumed to replicate via the theta mode (39), a notion supported by limited experimental (93) and sequence data (25, 31). Two groups of multiresistance plasmids have been recognized, the β-lactamase/heavy metal resistance plasmids and the pSK1 family.

S. aureus strains from the 1960s and 1970s commonly contained plasmids that conferred resistance to β-lactam antibiotics and heavy metals or other inorganic ions. These plasmids characteristically carried a Tn552-like β-lactamase-encoding transposon or a derivative thereof (22, 88), and frequently possessed a composite structure designated Tn4004 conferring resistance to mercuric ions, and operons encoding resistance to arsenical and/or cadmium ions (Fig. 3) (58, 91). Additionally, some β-lactamase/heavy metal resistance plasmids carry the transposons Tn4001 or Tn551, conferring resistance to aminoglycosides (33, 58) and macrolide/lincosamide/streptogramin type B (MLS) antibiotics (73), respectively, and/or a *qacA* or *qacB* gene mediating multidrug resistance to antiseptics and disinfectants (Fig. 3) (58).

Based on structural similarities, five families (α, β, γ, δ, and orphan) of β-lactamase/heavy metal resistance plasmids have previously been described (91). The α and γ families are closely related, and plasmids that appear to represent hybrids of members from different families have also been isolated (Fig. 3) (33, 91). The replication initiation regions from two γ family β-lactamase/heavy metal resistance plasmids, pI9789::Tn552 (Fig. 3) from S. aureus (24) and pSX267 from Staphylococcus xylosus (31), have been characterized and found to possess a high degree of identity at the nucleotide level. The deduced replication initiation proteins of these plasmids exhibit amino acid sequence similarity to those from enterococcal pheromone-response plasmids such as pAD1, Lactobacillus plasmids such as pLJ1, and the Bacillus natto plasmid pLS32 (24).

The plasmids pMU50 (also referred to as pVRSA), pN315, pSAS, and pMW2, characterized recently in the course of genome sequencing projects (9, 42, 46, 51), appear to belong to the β-lactamase/heavy metal resistance plasmid group. However, pN315, pSAS, and pMW2 all contain a second replication gene related to that of pSK639; in the latter two at least this may be the active *rep* gene because only remnants remain of their pI9789-like *rep* genes.

Plasmids related to the prototype pSK1 (Fig. 4) were first identified in epidemic S. aureus and coagulase-negative staphylococcal strains isolated in Australia during the 1980s and later in isolates from Europe (75). In addition to a *qacA* gene mediating multidrug resistance to antiseptics/disinfectants carried ubiquitously by pSK1 family plasmids (86, 97), members of this family variously contain Tn4001, which confers resistance to gentamicin and other aminoglycosides (85), a Tn552-like β-lactamase transposon, Tn4002 (32), and/or a composite structure designated Tn4003 that mediates trimethoprim resistance (Fig. 4) (87). Tn4003 is now thought to represent a vestige of a cointegrated pSK639-like plasmid (see above) (28, 55). The complete nucleotide sequence of pSK1 has been determined (25), revealing that its replication initiation region is related to those of the β-lactamase/heavy metal resistance plasmids pI9789::Tn552 and pSX267 (24). Antimicrobial resistance genes and associated elements account for slightly less than half of pSK1; the remainder corresponds to a 15-kb DNA segment conserved on all known members of the pSK1 family (96). In addition to the replication region, this conserved segment contains a number of open reading frames, the functions of which remain to be established.

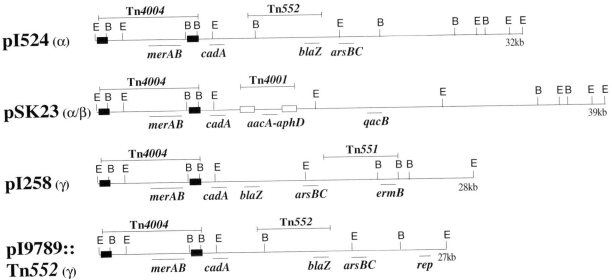

FIGURE 3 Maps of representative β-lactamase/heavy metal multiresistance plasmids pI524, pSK23, pI258, and pI9789:: Tn552 (75); see text for additional references. The family to which each plasmid belongs is shown in parentheses on the left, and plasmid sizes are indicated on the right. The positions of transposons are shown above each map. Open and solid boxes denote copies of IS256 and IS257, respectively. Selected loci encoding the following functions are indicated beneath the maps: *aacA-aphD*, aminoglycoside resistance; *arsBC*, arsenic resistance; *blaZ*, penicillin resistance; *cadA*, cadmium resistance; *ermB*, erythromycin resistance; *merAB*, mercury resistance; *qacB*, multidrug resistance to antiseptics and disinfectants; *rep*, initiation of plasmid replication. Restriction sites shown: B, *Bgl*II; E, *Eco*RI.

Conjugative Plasmids

The largest staphylococcal multiresistance plasmids (>30 kb) are those that encode their own conjugative transfer (72, 75). Only one family of conjugative plasmids has been examined in detail and includes *S. aureus* plasmids such as pSK41 (11, 27), pGO1 (65), and pJE1 (Fig. 5) (23). Structurally related conjugative plasmids have been found in coagulase-negative staphylococci (5, 48), and interspecific transfer has been demonstrated (3, 47). Conjugative transfer of pSK41 family plasmids occurs only on solid surfaces, and does so at comparatively low frequencies, in the range of 10^{-5} to 10^{-7} transconjugants per donor cell (59). Members of this plasmid family were first identified in association with gentamicin resistance in some strains in the mid-1970s (59, 75).

The complete nucleotide sequence of pSK41 is available (11). The pSK41 replication initiation region is related to those of the staphylococcal multiresistance plasmids described above (24). Conjugation by pSK41 family plasmids is mediated by a transfer system consisting of approximately 15 genes (11, 14, 27, 65, 92); two of these are adjacent to the origin of transfer, *oriT*, whereas the remainder are located within a 14-kb transfer-associated region, *tra*, which is flanked by copies of IS257 (Fig. 5). Despite the evolutionary relationship between the replication region of pSK41 and that of transmissible enterococcal pheromone-response plasmids such as pAD1, the pSK41-type conjugation system is instead related to those of the *L. lactis* plasmid pMRC01 (21), and the broad-host-range plasmid pIP501 (11), originally identified in *Streptococcus agalactiae*. This relatedness is manifest by similarity in *tra* product amino acid sequences and genetic organization (26, 34). Furthermore, several of the products from these gram-positive conjugation systems

possess similarity to proteins from gram-negative transfer systems, such as the *Escherichia coli* F plasmid conjugation system and the *Agrobacterium tumefaciens* Ti and Tra DNA transfer systems (27, 65). These relationships indicate that the pSK41-like conjugation systems belong to the type IV secretion system superfamily (34). Although the details of staphylococcal conjugation are yet to be elucidated, the basic process is likely to resemble that of the more extensively characterized gram-negative counterparts, viz., transfer of a single strand of plasmid DNA via direct cell-cell contact. Mobilization of relaxed coresident plasmids (79) and the demonstrated nicking of pGO1 *oriT* (14) are consistent with this notion. However, unlike the related gram-negative DNA transfer systems, no pilus-like structure seems to be associated with staphylococcal conjugation, perhaps reflecting the distinction in the cell envelope organizations and accounting for the solid surface requirement for conjugative transfer.

Multiple copies of IS257 are a feature of pSK41-like plasmids. For example, pGO1 contains nine directly repeated copies and a single copy in the opposite orientation (Fig. 5). Several of these elements delimit DNA segments that carry resistance genes, and it is now known that in most instances these correspond to cointegrated copies of smaller plasmids (11, 95). Such determinants include the aminoglycoside adenyltransferase gene, *aadD*, and the bleomycin resistance gene, *ble*, on a copy of pUB110, and the small multidrug resistance determinant, *smr* (formerly *qacC/D*), on another RC plasmid (Fig. 5). A third cointegrated RC plasmid, to which no phenotype has yet been ascribed, is also present adjacent to the cointegrated pUB110 on some of these conjugative plasmids. Plasmids such as pJE1, and probably pGO1, also contain a Tn4003-like structure now known to represent a fourth integrated plasmid, evidently a pSK639-

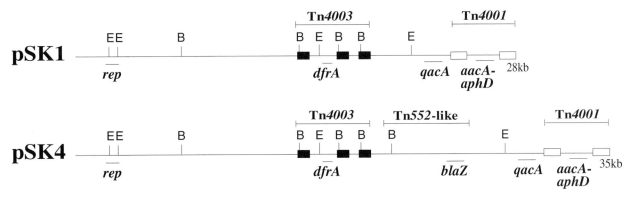

FIGURE 4 Maps of representative pSK1 family multiresistance plasmids, pSK1 and pSK4 (75); see text for additional references. Plasmid sizes are indicated on the right. The positions of transposons are indicated above each map. Open and solid boxes denote copies of IS256 and IS257, respectively. Selected loci encoding the following functions are indicated beneath the maps: *aacA-aphD*, aminoglycoside resistance; *blaZ*, penicillin resistance; *dfrA*, trimethoprim resistance; *qacA*, multidrug resistance to antiseptics and disinfectants; *rep*, initiation of plasmid replication. Restriction sites shown: B, *Bgl*II; E, *Eco*RI.

like trimethoprim-resistance plasmid (see above). In each case so far examined, the replication functions of the smaller cointegrated plasmids appear to have been inactivated either by insertion of IS257 during fusion of the two replicons or by subsequent deletions adjacent to this element (11). Similar IS257-associated events are presumably responsible for the Tn4001–IS257 hybrid structures found on most pSK41 family plasmids, thereby rendering them transpositionally defective (12). Analysis of the pSK41 sequence suggests that this family of plasmids has arisen by accretion of resistance determinants by preexisting conjugative plasmids, rather than via the en bloc incorporation of a conjugation system into a resistance plasmid (11).

Significantly, vancomycin resistance is the most recent addition to the determinants carried by pSK41-like plasmids, in the form of pLW1043 (Fig. 5). This plasmid was identified in the first reported high-level vancomycin-resistant *S. aureus* (VRSA) strain, isolated in Michigan in 2002 (104). Evidence indicates that pLW1043 resulted from the transposition of the VanA glycopeptide resistance transposon Tn1546 (see chapter 63, this volume) into a copy of IS257 within the pSK41-like plasmid pAM829 (77, 104). The broad-host-range conjugative VanA plasmid pAM830, carried by an *Enterococcus faecalis* strain coisolated with the VRSA, probably mediated the intergeneric transfer event (30, 104). The second reported VRSA was also found to contain Tn1546-like VanA transposon, but in this case it was harbored by a plasmid of approximately 120 kb. Glycopeptide resistance associated with this plasmid differs from that mediated by pLW1043 in that it confers only moderate levels of vancomycin resistance, which is phenotypically unstable (77, 98). The size of the plasmid hints that it may represent a cointegrate consisting of enterococcal and staphylococcal plasmids (98).

TRANSPOSONS AND INSERTION SEQUENCES

Transposable elements detected in staphylococci are listed in Table 1 together with relevant features and associations. For many, designation as a transposable element is based on possession of diagnostic characteristics, such as terminal inverted repeats (TIRs), flanking insertion sequences and/or target duplications, presence of an open reading frame encoding a transposase homolog, and/or detection in varied genetic contexts, rather than formal demonstration of mobility in a *recA* recombination-defective host. As outlined for specific elements below, in addition to facilitating the translocation of genes between replicons, IS elements are increasingly being shown to play more subtle roles in phenotypic expression and genome evolution.

Inverted copies of IS256 flank the composite transposon Tn4001 (Fig. 3 and 4) (85). Tn4001 and elements related to it are thought to be largely responsible for the emergence of linked resistance to the aminoglycosides, gentamicin, tobramycin, and kanamycin in staphylococci, having been found on the chromosome and multiresistance plasmids (Fig. 3, 4, and 5) in clinical *S. aureus* and coagulase-negative staphylococcal isolates from around the world (58, 75). Indeed, some of the earliest reported gentamicin-resistant strains, dating from 1975, have been shown to carry a Tn4001-like structure on the chromosome (105). Resistance is mediated by the gene *aacA-aphD*, which encodes a bifunctional enzyme possessing both aminoglycoside acetyltransferase and phosphotransferase activities (75). Transposons closely related to Tn4001 are also evident in enterococci (41) and streptococci (43). In enterococci, IS256 and related IS are also associated with elements encoding resistance to vancomycin (81) and erythromycin (82).

Roles in gene expression and regulation have been attributed to IS256, via both transcriptional promotion and insertional inactivation. Transcription of the *aacA–aphD* resistance gene of Tn4001 is directed by a complete promoter located within the end of the upstream copy of IS256, and it has also been found that a small flanking deletion can result in a −35-like sequence at the terminus of that element forming part of a hybrid promoter that mediates a higher level of aminoglycoside resistance (62). Additionally, roles for IS256 hybrid promoters have been demonstrated in the expression of methicillin resistance (17, 61). Insertional inactivation of genes by IS256 has been shown to modulate expression of several phenotypes, including virulence and glycopeptide resistance. Alternating IS256 insertions and excisions in the *ica* operon responsible for the synthesis of the polysaccharide

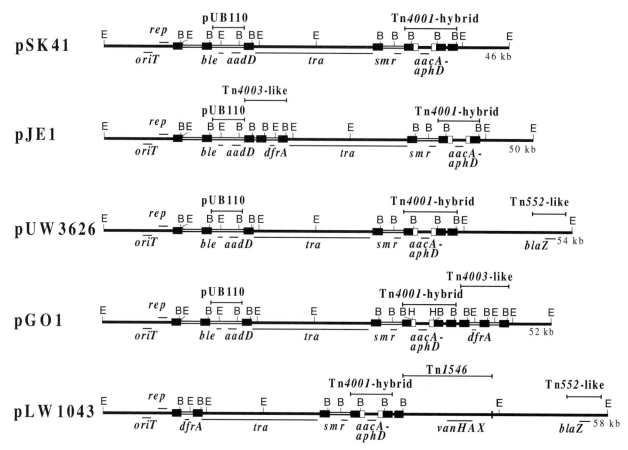

FIGURE 5 Maps of representative conjugative multiresistance plasmids pSK41, pJE1, pUW3626, pGO1, and pLW1043 (75, 104); see text for additional references. Plasmid sizes are indicated on the right. The positions of transposons and an integrated copy of the RC plasmid, pUB110, are indicated above each map; the latter and other small cointegrated plasmids are denoted by double lines. Solid boxes denote copies of IS257, including the element interrupted by Tn1546 in pLW1043, and open boxes represent truncated copies of IS256. Selected loci encoding the following functions are indicated beneath the maps: aadD/aacA-aphD, aminoglycoside resistances; blaZ, penicillin resistance; ble, bleomycin resistance; dfrA, trimethoprim resistance; oriT, origin of conjugative DNA transfer; rep, initiation of plasmid replication; smr, multidrug resistance to antiseptics and disinfectants; tra, conjugative transfer; vanHAX, glycopeptide resistance. Restriction sites shown: B, BglII; E, EcoRI. The position of the BglII site within IS257 indicates the orientation of the element.

intercellular adhesin mediate phase-variable biofilm formation in *S. epidermidis* strains responsible for indwelling device infections (110). Notably, biofilm phase variation has also been attributed to IS256 insertions that affect global regulatory circuits, including the *sarA* global regulator, the quorum-sensing *agr* system, and *rsbU*, which positively regulates the σ^B stress response factor (16, 103). IS256 inactivation of the *tca* locus was found to increase the teicoplanin resistance of a glycopeptide intermediate-resistant *S. aureus* strain (60).

IS257 (also known as IS431) has been found in diverse genetic contexts in staphylococci, variously associated with a number of determinants (Table 1). The prevalence of resistance genes flanked by copies of this element resulted in the designation of several such composite structures as transposons, viz., Tn4003 and Tn4004. Although the possibility that these structures can behave as legitimate transposons cannot be excluded, such an organization is probably a consequence of the transposition mechanism of IS257. This IS element is now thought to undergo nonresolved

replicative transposition (28, 95). The expected outcome of such an event involving two replicons is cointegration, so that the replicons become fused with a directly repeated copy of IS257 at each junction, as has been experimentally demonstrated by Needham et al. (67). Many IS257-flanked segments represent plasmids cointegrated into larger plasmids or the chromosome (28, 95). Frequently observed deletion events adjacent to IS257 may result from intramolecular transposition of IS257. However, replicon fusions and sequence deletions have also been shown to have resulted from homologous recombination between preexisting copies of IS257 (11).

The implications of the activities of IS257 are best illustrated by the pSK41 family plasmids (Fig. 5). It would seem that IS257-mediated cointegration has enabled these conjugative plasmids to collect functions, in the form of smaller plasmids, as they move horizontally through bacterial populations. Insertions and flanking deletions mediated by this element provide a mechanism for the inactivation or re-

TABLE 1 Staphylococcal insertion sequences and transposons[a]

Element	Size (kb)	Associated resistance(s)/ other phenotype(s)[b]	Relevant gene(s)	TIR (bp)/ flanking IS[c]	Target duplication (bp)	Location(s)[d]
IS256	1.3	Gentamicin/kanamycin/ tobromycin	aacA-aphA	26	8	C, P
IS257[e]	0.8	Antiseptics/disinfectants	smr	27	8	P
		Bleomycin	ble			C, P
		Cadmium	cadD			P
		Gentamicin/kanamycin/ tobromycin	aacA-aphA			C, P
		Kanamycin/neomycin/ paromomycin/tobromycin	aadD			C, P
		Mercury	merA, merB			C, P
		Methicillin	mecA			C
		Mupirocin	mupA			P
		Tetracycline	tetA(K)			C, P
		Trimethoprim	dfrA			P
		Virginiamicin	vgb			P
		Preprolysostaphin	lss			P
		Lysostaphin immunity factor	lif			P
IS1181	2.0	Kanamycin/neomycin	aphA-3	23	8	C
		Streptomycin	aadE			C
IS1182	1.9	Kanamycin/neomycin	aphA-3	33	8	C
		Streptomycin	aadE			C
IS1272	1.9	Methicillin	mecA	16	Unknown	C
IS1293	1.3	Preprolysostaphin	lss	26	Unknown	P
		Lysostaphin immunity factor	lif			P
Tn551	5.3	Macrolides/lincosamides/ streptogramin B (MLS)	ermB	40	5	P
Tn552[f]	6.1	Penicillins	blaZ	116	6/7	C, P
Tn554[g]	6.7	Macrolides/lincosamides/ streptogramin B (MLS)	ermA	Absent	Absent	C, P
		Spectinomycin	spc			
Tn1546	10.8	Vancomycin/teicoplanin	vanHAX	38	5	P
Tn3854	4.5	Kanamycin/neomycin/ streptomycin	Unknown	Unknown	Unknown	P
Tn4001[h]	4.7	Gentamicin/kanamycin/ tobromycin	aacA-aphD	IS256 (I)[i]	8	C, P
Tn4003	4.7	Trimethoprim	dfrA	IS257 (D)[j]	8	P
Tn4004	7.8	Mercury	merA, merB	IS257 (D)	8	P
Tn5404	16	Kanamycin/neomycin	aphA-3	116	6	C
		Streptomycin	aadE			
Tn5405	12	Kanamycin/neomycin	aphA-3	IS1182 (I)	8	C
		Streptomycin	aadE			
Tn5406	5.5	Streptogramin A	vgaAv	Absent	Absent	C, P
Tn5801	25.8	Tetracycline/minocycline	tetM	Absent	11	C

[a]See Paulsen et al. (75), Firth and Skurray (28), and the text for references.
[b]In the case of IS, association is based on probable involvement in the acquisition, dissemination, and/or expression of the resistance/phenotype.
[c]TIR, terminal inverted repeat. (I), inverted orientation; (D), direct orientation.
[d]C, chromosome; P, plasmid.
[e]IS257 is also known as IS431.
[f]Tn3852, Tn4002, and Tn4201 are likely to be similar or identical to Tn552.
[g]Tn3853 is likely to be similar or identical to Tn554.
[h]Tn3851 and Tn4031 are likely to be similar or identical to Tn4001.
[i](I), inverted orientation.
[j](D), direct orientation.

moval of deleterious sequences, such as redundant replication functions. Similarly, deletion of transposition functions has immobilized the Tn4001-IS257-hybrid elements on these plasmids (Fig. 5), thereby enhancing the maintenance of the plasmids under aminoglycoside selection.

A central role for IS257 has been suggested in the evolution of staphylococcal trimethoprim resistance, including the probable capture of the resistance gene, dfrA, from the chromosome of an S. epidermidis strain (28, 95). Furthermore, an IS257-hybrid promoter is responsible for transcription of this determinant (54, 95). In some plasmids, the high-level trimethoprim resistance typically conferred by this determinant has been moderated as a consequence of IS257-associated flanking deletions. An analogous IS257-hybrid promoter transcribes the tetA(K) tetracycline resistance gene of a cointegrated copy of pT181 within type III staphylococcal chromosome cassette *mec* (SCC*mec*) elements (Fig. 6), and several other genes are likely to be similarly promoted (94).

IS*1272* is thought to have contributed to the evolution of methicillin resistance in *S. aureus* and coagulase-negative staphylococci (4, 6, 40). A remnant of this element, ψIS*1272*, is evident adjacent to a deletion within the methicillin resistance (*mec*) region of some SCC*mec* elements (Fig. 6 and see below). This deletion removed *mecI* and part of *mecR1*, regulatory genes that control transcription of the divergent *mecA* gene, thereby resulting in higher levels of methicillin resistance. Multiple intact copies of IS*1272* have been found on the chromosomes of *S. aureus* and coagulase-negative staphylococcal strains, and it is particularly prevalent in *Staphylococcus haemolyticus* (6, 42).

In addition to the high-level glycopeptide resistance VanA element, Tn*1546*, of enterococcal origin (104) described above, three other Tn*3*-type transposons have been detected in staphylococci, viz., Tn*551*, Tn*552*, and Tn*5404* (28, 75). The MLS resistance transposon, Tn*551*, is closely related to Tn*917* from *E. faecalis* but confers constitutive rather than inducible MLS resistance (72). Tn*551* was originally identified on the β-lactamase/heavy metal resistance plasmid, pI258 (Fig. 3), but has subsequently been shown to transpose to numerous sites in other large plasmids and the chromosome. Despite a tendency to "hot-spot," it has

proven an invaluable tool for mapping and mutagenesis studies (44, 72).

In contrast to Tn*551*, the β-lactamase transposon Tn*552* seems to be restricted to a very limited set of insertion sites on the chromosome and plasmids (75). In multiresistance (Fig. 3 and 4) and conjugative plasmids (Fig. 5), the insertion sites of Tn*552* and Tn*552*-like elements are located within resolution sites upstream of resolvase genes encoded by each of the plasmids (11, 46, 76, 89). Accordingly, Tn*552* belongs to a group of transposons termed "*res* site hunters" (63). Despite such insertional specificity, Tn*552*-type transposons are thought to represent the source of all staphylococcal β-lactamase genes (22). However, in many plasmids, only remnants of such transposons remain, presumably as a consequence of multiple insertion events and/or rearrangements promoted by the recombinase systems present on these elements and plasmids.

The aminoglycoside resistance transposon Tn*5404* appears to have resulted from the transposition of a chromosomal element to a plasmid in a clinical *S. aureus* strain (20). This transposon shares an invertible segment with an adjacent copy of Tn*552*, such that an entire copy of one or the other of these elements is generated depending on the orientation of the segment (20). Another invertible segment within Tn*5404* represents a composite structure designated Tn*5405* (20), which is bounded by copies of IS*1182*, an ele-

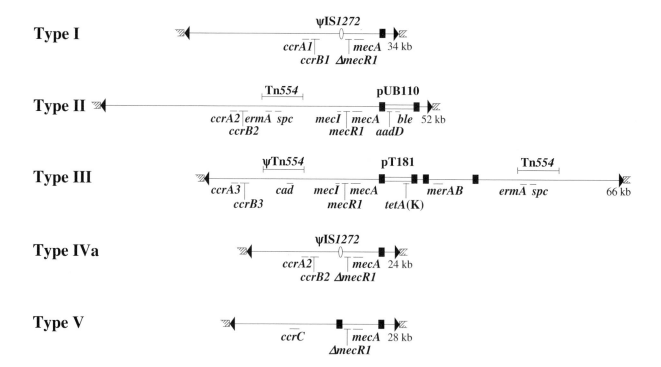

FIGURE 6 Maps of representative SCC*mec* elements (45, 74). The sizes of the elements are indicated on the right. The positions of transposons Tn*554* and ψTn*554* and cointegrated copies of the RC plasmids pUB110 and pT181 are shown; the plasmids are denoted by double lines. Solid boxes denote copies of IS*257*, and open ovals indicate the position of ψIS*1272*, an IS*1272* remnant located adjacent to a deletion that removed *mecI* and part of *mecR1* (Δ*mecR1*). Arrowheads denote TIRs. Non-SCC*mec* chromosomal DNA is shown hatched. Selected loci encoding the following functions are indicated: *aadD*, aminoglycoside resistance; *ble*, bleomycin resistance; *cad*, cadmium resistance; *ccrA1/ccrA2/ccrA3/ccrB1/ccrB2/ccrB3/ccrC*, recombinases; *ermA*, MLS resistance; *mecA*, methicillin resistance; *mecI/mecR1*, regulation of methicillin resistance; *spc*, spectinomycin resistance; *tetA*(K), tetracycline resistance.

ment related to IS*1272* (see above) (19). Comparison of this pair of IS*1182* elements with an independent chromosomal copy revealed that the presumptive transposase gene of the former contains a point mutation (19). Moreover, one of the IS*1182* elements flanking Tn*5405* is interrupted by a copy of IS*1181*, which is both prevalent and active in *S. aureus* (13, 19, 51).

The central region of Tn*5405* has been detected in the absence of flanking IS*1182* elements. In addition to *aadE* and *aphA–3* aminoglycoside resistance genes, such regions sometimes encode a functional *sat4* streptothricin resistance determinant and appear to have been disseminated among other genera, including enterococci and *Campylobacter* (18). Elucidation of any relationship between Tn*5405* and Tn*3854*, which confers the same resistance phenotype, awaits characterization of the latter.

An incomplete vestige of an element related to IS*1181* has been detected on the plasmid pACK1, from *Staphylococcus simulans* biovar *staphylolyticus* ATCC 1362. This element, designated IS*1293*, is located at one end of a segment that contains the genes encoding preprolysostaphin (*lss*) and lysostaphin immunity factor (*lif*) (99). It is likely that the *lss–lif* segment of this plasmid, which possesses a truncated remnant of IS*257* at its other end, has been acquired through horizontal transfer (99).

The MLS and spectinomycin resistance transposon, Tn*554*, is an unusual transposable element that lacks TIRs and generates no target sequence duplications upon insertion (66). Tn*554* transposes at an efficiency approaching 100% into a unique site in the *S. aureus* and *S. epidermidis* chromosomes, termed *att554*, with a preference for one orientation. Secondary insertions can be generated, albeit at much lower frequency, if *att554* is absent or occupied by a preexisting copy of Tn*554*. However, natural isolates have been identified that carry this or related elements, such as ψTn*554*, encoding cadmium resistance, at different secondary sites, including some located in SCC*mec* regions (see below) (46, 51). A Tn*554*-like element, Tn*3853*, has also been detected on a conjugative plasmid from *S. epidermidis*, pWG25 (100). Another element related to Tn*554*, Tn*5406*, encodes a variant of the *vga* gene, *vgaAv*, that confers resistance to streptogramin A but not mixtures of streptogramin A and B (38). Tn*5406* has been found at chromosomal sites equivalent to those of Tn*554*, and also on large transmissible plasmids.

Several new putative transposable elements have been identified in *S. aureus* genome sequences. These include IS-like elements ISX and ISYZ and Tn*916*-like conjugative transposons, such as Tn*5801*, which encodes *tet*(M)-mediated tetracycline and minocycline resistance (42, 46).

In staphylococci, the reassortment of resistance genes is largely limited to the activities of IS elements and transposons because no mechanisms related to the integrons of gram-negative bacteria have been detected. Rather, resistance gene clusters in staphylococci have commonly been assembled via sequential cointegrative capture of small resistance plasmids mediated by IS*257*, a process that resembles, particularly in its outcome, the accretion of resistance genes within integrons.

STAPHYLOCOCCAL CHROMOSOME CASSETTES

Most methicillin-resistant *S. aureus* isolates contain a 20- to 70-kb DNA segment, generically termed the *mec* region, not present in methicillin-sensitive strains. These segments are a type of mobile genetic element called SCC*mec* (46).

Methicillin resistance is mediated, at least in part, by the *mecA* gene, which encodes the 76-kDa low-affinity penicillin-binding protein, PBP2′ (also known as PBP2a) (see chapter 62). Two regulatory loci, *mecI* and *mecR1*, are commonly found upstream of, and transcribed divergently to, *mecA* (however, see IS*1272*, above). The *mec* region appears to act as a chromosomal hot spot for the insertion of additional antimicrobial resistance determinants, often in association with transposable elements (40). These include Tn*554* or related elements encoding resistance to erythromycin and spectinomycin, or cadmium, and IS*257*-flanked segments conferring resistance to mercury, tetracycline, and/or aminoglycosides and bleomycin; the latter two segments are known to be cointegrated copies of the plasmids pT181 and pUB110, respectively (Fig. 6) (28, 95). SCC*mec* are bounded by TIRs and usually encode two recombinase genes, *ccrA* and *ccrB*. The *ccr* genes mediate the excision and circularization of the cassette and its site- and orientation-specific insertion into the *attB*$_{SCC}$ site of the *S. aureus* chromosome and homologous sites in other staphylococcal species (46).

Various structural variants of SCC*mec* have been characterized and to date these have been classified into five types (Fig. 6), based on the organization of the *mecA* gene complex and sequence class of the *ccr* gene(s) present (45, 46). SCC*mec* are found in coagulase-negative staphylococcal strains, and the ubiquitous carriage of an *mecA* homolog by *Staphylococcus sciuri* has led to the suggestion that this or a closely related coagulase-negative staphylococcus represents the origin of the *mec* determinants found in other species (106). Current evidence indicates that the five major epidemic methicillin-resistant *S. aureus* lineages have arisen through at least 20 independent SCC*mec* acquisition events (84).

The SSC*mec* resistance islands are a subset of a SCC element family because related structures have been found that do not encode a methicillin resistance determinant (46, 57, 64). These include SCC*cap1* from *S. aureus*, which contains a gene cluster for production of type I capsular polysaccharide (57), and SSC$_{12263}$ from *S. hominis*, which has been suggested to represent a primordial SCC vehicle (49). The mechanism responsible for horizontal transmission of SCC elements remains unclear.

BACTERIOPHAGES

Most clinical *S. aureus* isolates are lysogenic, commonly containing multiple prophages (72); the *S. aureus* genomes sequenced to date all contain at least one prophage (9, 42, 51). Superinfection immunity and other factors, including restriction-modification systems, modulate the susceptibility of a strain to phage infection. This phenomenon has formed the basis of a phage-typing system, employing an international reference set of phages, used for epidemiological and evolutionary analysis of *S. aureus* strains (see chapter 31, this volume).

In addition to facilitating generalized transduction, phage infection has been found to alter the staphylococcal genome in other ways. Temperate phages can mediate lysogenic conversion of infected hosts through the carriage of several virulence determinants. Genes identified on lysogenic phage include those for enterotoxin A and a variant P (*sea* and *sep*), homologs of enterotoxins G and K (*seg2* and *sek2*), exfoliative toxin A (*eta*), staphylokinase (*sak*), and Panton-Valentine leukocidin (*lukSF*-PV) (9, 42, 51, 107). In contrast, prophage integration has been shown to inactivate *S.*

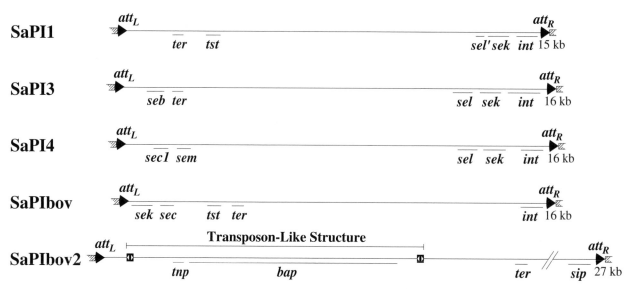

FIGURE 7 Maps of representative SaPI elements (70, 101). The sizes of the elements are indicated on the right. Arrowheads denote left and right directly repeated attachment sequences (*att*_L and *att*_R). Non-SaPI DNA is shown hatched. The position of the transposon-like structure in SaPIbov2 is shown; the 149-bp direct repeats that bound the structure are denoted as boxes that contain white arrowheads representing IS257-like TIRs. Selected loci encoding the following functions are indicated: *bap*, biofilm-associated protein; *int*, integrase; *seb*, enterotoxin B; *sec*, enterotoxin C; *sec1*, enterotoxin C1; *sek*, enterotoxin K; *sel*, enterotoxin L; *sel'*, enterotoxin L remnant; *sem*, enterotoxin M; *sip*, staphylococcal integrase protein; *ter*, phage-like encapsidation terminase; *tnp*, IS257-like transposase; *tst*, toxic shock toxin 1.

aureus chromosomal virulence genes responsible for production of beta-toxin (β-hemolysin) and lipase (glycerol ester hydrolase) due to the presence of phage attachment sites within their respective structural genes, *hlb* and *geh* (15, 53).

PATHOGENICITY ISLANDS

Bacteriophages also appear to play an important role in the transmission a family of 15- to 27-kb chromosomally integrating DNA segments found to commonly encode various superantigen genes, termed *S. aureus* pathogenicity islands (SaPIs) (Fig. 7) (70). Demonstrated and putative superantigen genes carried by SaPIs include those for toxic shock syndrome toxin 1 (*tst*) and enterotoxins B, C, K, L, and M (*seb*, *sec*, *sek*, *sel*, *sem*, respectively). SaPIs also encode a phagelike integrase, *int*, which mediates site-specific insertion into the bacterial chromosome. Some SaPIs, such as SaPI1 and SaPI3, insert at the same *att* site, although seemingly only one at a time, whereas others, such as SaPIbov, utilize a distinct *att* site located elsewhere in the genome (70); to date, SaPIs have been found in at least four locations in the *S. aureus* chromosome (9).

Laboratory studies suggest that transmission of a particular SaPI is associated with only certain staphylococcal phage (70). The prototype element, SaPI1, is transduced at high frequency as small phagelike particles only by phage 80α, which is required for excision and replication of the element but not *recA*-independent site-specific integration. Integration and excision of SaPI1 are proposed to occur via classical Campbell recombination because the circular intermediate contains a copy of the duplicated 17-bp sequence that flanks the integrated form (90). The transmission of another element, SaPIbov2, might not necessarily

be mediated by phage because its *sip*-encoded integrase is able to both integrate and excise the element (101). This element encodes no recognized superantigens, instead carrying within a transposonlike structure a gene called *bap*, for a large biofilm-associated protein involved in colonization of the bovine mammary gland.

In addition to the integrase-containing SaPI family of pathogenicity islands described above, distinct genomic pathogenicity islands are evident at two locations in all *S. aureus* chromosomes sequenced thus far. Although carriage of these seems ubiquitous at this stage and the islands found at each site are clearly related, they have been distinguished from highly conserved surrounding "core" sequences by virtue of the notable extent to which they vary between strains. The νSaα family of islands found at one site represents clusters of probable *set*-like exotoxin and lipoprotein genes, whereas the νSaβ family found at the other site encodes clusters of enterotoxins and serine proteases (9, 42, 51). Mobility of these island types is yet to be established.

PERSPECTIVES

Gene transfer mechanisms, together with accessory elements such as plasmids, transposable elements, prophages, and pathogenicity and resistance islands, serve as catalysts for microbial evolution by providing access to a shared reservoir of niche-adaptive functions. In the first edition of this chapter we predicted that genome sequencing projects would yield previously undetected accessory elements, and this has indeed been the case. However, just as importantly, the combination of comparative genomics and sequence-based strain typing is clarifying the relationships between clinical *S. aureus* strains. These studies are providing a new perspective on the scope and importance of the accessory genome, demonstrating that variation in pathogenic potential largely does not reside in the core genome, but rather that it is primarily attributable to the complement of mobile accessory elements that are present. We are confident that the application of these approaches to staphylococci from other niches will further expand the known repertoire of accessory elements and our appreciation of their significance.

Work in the laboratory of N.F. and R.A.S. on staphylococcal genetics is supported by National Health and Medical Research Council (Australia) Project Grant 153816 and Australian Research Council Grant DP0346013.

REFERENCES

1. **Al-Masaudi, S. B., M. J. Day, and A. D. Russell.** 1991. Effect of some antibiotics and biocides on plasmid transfer in *Staphylococcus aureus. J. Appl. Bacteriol.* **71:**239–243.

2. **Apisiridej, S., A. Leelaporn, C. D. Scaramuzzi, R. A. Skurray, and N. Firth.** 1997. Molecular analysis of a mobilizable theta-mode trimethoprim resistance plasmid from coagulase-negative staphylococci. *Plasmid* **38:**13–24.

3. **Archer, G. L., D. R. Dietrick, and J. L. Johnston.** 1985. Molecular epidemiology of transmissible gentamicin resistance among coagulase-negative staphylococci in a cardiac surgery unit. *J. Infect. Dis.* **151:**243–251.

4. **Archer, G. L., D. M. Niemeyer, J. A. Thanassi, and M. J. Pucci.** 1994. Dissemination among staphylococci of DNA sequences associated with methicillin resistance. *Antimicrob. Agents Chemother.* **38:**447–454.

5. **Archer, G. L., and J. Scott.** 1991. Conjugative transfer genes in staphylococcal isolates from the United States. *Antimicrob. Agents Chemother.* **35:**2500–2504.

6. **Archer, G. L., J. A. Thanassi, D. M. Niemeyer, and M. J. Pucci.** 1996. Characterization of IS*1272*, an insertion sequence-like element from *Staphylococcus haemolyticus. Antimicrob. Agents Chemother.* **40:**924–929.

7. **Archer, G. L., and W. D. Thomas, Jr.** 1990. Conjugative transfer of antimicrobial resistance genes between staphylococci, p. 115–122. *In* R. P. Novick (ed.), *Molecular Biology of the Staphylococci.* VCH Publishers, New York, N.Y.

8. **Aubert, S., K. G. H. Dyke, and N. El Solh.** 1998. Analysis of two *Staphylococcus epidermidis* plasmids coding for resistance to streptogramin A. *Plasmid* **40:**238–242.

9. **Baba, T., F. Takeuchi, M. Kuroda, H. Yuzawa, K. Aoki, A. Oguchi, Y. Nagai, N. Iwama, K. Asano, T. Maimi, H. Kuroda, L. Cui, K. Yamamoto, and K. Hiramatsu.** 2002. Genome and virulence determinants of high virulence community-acquired MRSA. *Lancet* **359:**1819–1827.

10. **Barr, V., K. Barr, M. R. Millar, and R. W. Lacey.** 1986. β-lactam antibiotics increase the frequency of plasmid transfer in *Staphylococcus aureus. J. Antimicrob. Chemother.* **17:**409–413.

11. **Berg, T., N. Firth, S. Apisiridej, A. Hettiaratchi, A. Leelaporn, and R. A. Skurray.** 1998. Complete nucleotide sequence of pSK41: evolution of staphylococcal conjugative plasmids. *J. Bacteriol.* **180:**4350–4359.

12. **Byrne, M. E., M. T. Gillespie, and R. A. Skurray.** 1990. Molecular analysis of a gentamicin resistance transposon-like element on plasmids isolated from North American *Staphylococcus aureus* strains. *Antimicrob. Agents Chemother.* **34:**2106–2113.

13. **Chesneau, O., R. Lailler, A. Derbise, and N. El Solh.** 1999. Transposition of IS*1181* in the genomes of *Staphylococcus* and *Listeria. FEMS Microbiol. Lett.* **177:**93–100.

14. **Climo, M. W., V. K. Sharma, and G. L. Archer.** 1996. Identification and characterization of the origin of conjugative transfer (*oriT*) and a gene (*nes*) encoding a single-stranded endonuclease on the staphylococcal plasmid pGO1. *J. Bacteriol.* **178:**4975–4983.

15. **Coleman, D., J. Knights, R. Russell, D. Shanley, T. H. Birkbeck, G. Dougan, and I. Charles.** 1991. Insertional inactivation of the *Staphylococcus aureus* β-toxin by bacteriophage φ13 occurs by site- and orientation-specific integration of the φ 13 genome. *Mol. Microbiol.* **5:**933–939.

16. **Conlon, K. M., H. Humphreys, and J. P. O'Gara.** 2004. Inactivations of *rsbU* and *sarA* by IS*256* represent novel mechanisms of biofilm phenotypic variation in *Staphylococcus epidermidis. J. Bacteriol.* **186:**6208–6219.

17. **Couto, I., S. W. Wu, A. Tomasz, and H. de Lencastre.** 2003. Development of methicillin resistance in clinical isolates of *Staphylococcus sciuri* by transcriptional activation of the *mecA* homologue native to the species. *J. Bacteriol.* **185:**645–653.

18. **Derbise, A., G. de Cespedes, and N. El Solh.** 1997. Nucleotide sequence of the *Staphylococcus aureus* transposon, Tn*5405*, carrying aminoglycosides resistance genes. *J. Basic Microbiol.* **37:**1–6.

19. **Derbise, A., K. G. H. Dyke, and N. El Solh.** 1996. Characterization of a *Staphylococcus aureus* transposon, Tn*5405*, located within Tn*5404* and carrying the aminoglycoside resistance genes, *aphA-3* and *aadE. Plasmid* **35:**174–188.

20. **Derbise, A., K. G. H. Dyke, and N. El Solh.** 1995. Rearrangements in the staphylococcal β-lactamase-encoding plasmid, pIP1066, including a DNA inversion that generates two alternative transposons. *Mol. Microbiol.* **17:**769–779.

21. **Dougherty, B. A., C. Hill, J. F. Weidman, D. R. Richardson, J. C. Venter, and R. P. Ross.** 1998. Sequence and analysis of the 60 kb conjugative, bacteriocin-producing plasmid pMRC01 from *Lactococcus lactis* DPC3147. *Mol. Microbiol.* **29:**1029–1038.

22. **Dyke, K., and P. Gregory.** 1997. Resistance mediated by β-lactamase, p. 139–157. *In* K. B. Crossley and G. L. Archer (ed.), *The Staphylococci in Human Disease.* Churchill Livingstone, New York, N.Y.

23. **Evans, J., and K. G. H. Dyke.** 1988. Characterization of the conjugation system associated with the *Staphylococcus aureus* plasmid pJE1. *J. Gen. Microbiol.* **134:**1–8.

24. **Firth, N., S. Apisiridej, T. Berg, B. A. O'Rourke, S. Curnock, K. G. H. Dyke, and R. A. Skurray.** 2000. Replication of staphylococcal multiresistance plasmids. *J. Bacteriol.* **182:**2170–2178.

25. **Firth, N., S. Apisiridej, and R. A. Skurray.** Unpublished data.

26. **Firth, N., T. Berg, and R. A. Skurray.** 1999. Evolution of conjugative plasmids from Gram-positive bacteria. *Mol. Microbiol.* **31:**1598–1599.

27. **Firth, N., K. P. Ridgway, M. E. Byrne, P. D. Fink, L. Johnson, I. T. Paulsen, and R. A. Skurray.** 1993. Analysis of a transfer region from the staphylococcal conjugative plasmid pSK41. *Gene* **136:**13–25.

28. **Firth, N., and R. A. Skurray.** 1998. Mobile elements in the evolution and spread of multiple-drug resistance in staphylococci. *Drug Resist. Updates* **1:**49–58.

29. **Fitzgerald, J. R., D. E. Sturdevant, S. M. Mackie, S. R. Gill, and J. M. Musser.** 2001. Evolutionary genomics of *Staphylococcus aureus*: insights into the origin of methicillin-resistant strains and the toxic shock syndrome epidemic. *Proc. Natl. Acad. Sci. USA* **98:**8821–8826.

30. **Flannagan, S. E., J. W. Chow, S. M. Donabedian, W. J. Brown, M. B. Perri, M. J. Zervos, Y. Ozawa, and D. B.**

Clewell. 2003. Plasmid content of a vancomycin-resistant *Enterococcus faecalis* isolate from a patient also colonized by *Staphylococcus aureus* with a VanA phenotype. *Antimicrob. Agents Chemother.* **47:**3954–3959.

31. **Gering, M., F. Gütz, and R. Bruckner.** 1996. Sequence and analysis of the replication region of the *Staphylococcus xylosus* plasmid pSX267. *Gene* **182:**117–122.

32. **Gillespie, M. T., B. R. Lyon, and R. A. Skurray.** 1988. Structural and evolutionary relationships of β-lactamase transposons from *Staphylococcus aureus. J. Gen. Microbiol.* **134:**2857–2866.

33. **Gillespie, M. T., and R. A. Skurray.** 1986. Plasmids in multiresistant *Staphylococcus aureus. Microbiol. Sci.* **3:**53–58.

34. **Grohmann, E., G. Muth, and M. Espinosa.** 2003. Conjugative plasmid transfer in gram-positive bacteria. *Microbiol. Mol. Biol. Rev.* **67:**277–301.

35. **Grohmann, E., E. L. Zechner, and M. Espinosa.** 1997. Determination of specific DNA strand discontinuities with nucleotide resolution in exponentially growing bacteria harboring rolling circle-replicating plasmids. *FEMS Microbiol. Lett.* **152:**363–369.

36. **Gruss, A., and S. D. Ehrlich.** 1989. The family of highly interrelated single-stranded deoxyribonucleic acid plasmids. *Microbiol. Rev.* **53:**231–241.

37. **Guzmán, L. M., and M. Espinosa.** 1997. The mobilization protein, MobM, of the streptococcal plasmid pMV158 specifically cleaves supercoiled DNA at the plasmid *oriT. J. Mol. Biol.* **266:**688–702.

38. **Haroche, J., J. Allignet, and N. El Solh.** 2002. Tn*5406*, a new staphylococcal transposon conferring resistance to streptogramin A and related compounds including dalfopristin. *Antimicrob. Agents Chemother.* **46:**2337–2343.

39. **Helinski, D. R., A. E. Toukdarian, and R. P. Novick.** 1996. Replication control and other stable maintenance mechanisms of plasmids, p. 2295–2324. *In* F. C. Neidhardt, R. Curtiss III, J. L. Ingraham, E. C. C. Lin, K. B. Low, Jr., B. Magasanik, W. Reznikoff, M. Riley, M. Schaechter, and H. E. Umbarger (ed.), Escherichia coli *and* Salmonella: *Cellular and Molecular Biology.* ASM Press, Washington, D.C.

40. **Hiramatsu, K., L. Cui, M. Kuroda, and T. Ito.** 2001. The emergence and evolution of methicillin-resistant *Staphylococcus aureus. Trends Microbiol.* **9:**486–493.

41. **Hodel-Christian, S. L., and B. E. Murray.** 1992. Comparison of the gentamicin resistance transposon Tn*5281* with regions encoding gentamicin resistance in *Enterococcus faecalis* isolates from diverse geographic locations. *Antimicrob. Agents Chemother.* **36:**2259–2264.

42. **Holden, M. T., E. J. Feil, J. A. Lindsay, S. J. Peacock, N. P. Day, M. C. Enright, T. J. Foster, C. E. Moore, L. Hurst, R. Atkin, A. Barron, N. Bason, S. D. Bentley, C. Chillingworth, T. Chillingworth, C. Churcher, L. Clark, C. Corton, A. Cronin, J. Doggett, L. Dowd, T. Feltwell, Z. Hance, B. Harris, H. Hauser, S. Holroyd, K. Jagels, K. D. James, N. Lennard, A. Line, R. Mayes, S. Moule, K. Mungall, D. Ormond, M. A. Quail, E. Rabbinowitsch, K. Rutherford, M. Sanders, S. Sharp, M. Simmonds, K. Stevens, S. Whitehead, B. G. Barrell, B. G. Spratt, and J. Parkhill.** 2004. Complete genomes of two clinical *Staphylococcus aureus* strains: evidence for the rapid evolution of virulence and drug resistance. *Proc. Natl. Acad. Sci. USA* **101:**9786–9791.

43. **Horaud, T., G. de Cespédès, and P. Trieu-Cuot.** 1996. Chromosomal gentamicin resistance transposon Tn*3706* in *Streptococcus agalactiae* B128. *Antimicrob. Agents Chemother.* **40:**1085–1090.

44. **Iandolo, J. J., J. P. Bannantine, and G. C. Stewart.** 1997. Genetic and physical map of the chromosome of *Staph-*

ylococcus aureus, p. 39–53. *In* K. B. Crossley and G. L. Archer (ed.), *The Staphylococci in Human Disease.* Churchill Livingstone, New York, N.Y.

45. **Ito, T., X. X. Ma, F. Takeuchi, K. Okuma, H. Yuzawa, and K. Hiramatsu.** 2004. Novel type V staphylococcal cassette chromosome *mec* driven by a novel cassette chromosome recombinase, *ccrC. Antimicrob. Agents Chemother.* **48:**2637–2651.

46. **Ito, T., K. Okuma, X. X. Ma, H. Yuzawa, and K. Hiramatsu.** 2003. Insights on antibiotic resistance of *Staphylococcus aureus* from its whole genome: genomic island SCC. *Drug Resist. Updates* **6:**41–52.

47. **Jaffe, H. W., H. M. Sweeney, C. Nathan, R. A. Weinstein, S. A. Kabins, and S. Cohen.** 1980. Identity and interspecific transfer of gentamicin-resistance plasmids in *Staphylococcus aureus* and *Staphylococcus epidermidis. J. Infect. Dis.* **141:**738–747.

48. **Jaffe, H. W., H. M. Sweeney, R. A. Weinstein, S. A. Kabins, C. Nathan, and S. Cohen.** 1982. Structural and phenotypic varieties of gentamicin resistance plasmids in hospital strains of *Staphylococcus aureus* and coagulase-negative staphylococci. *Antimicrob. Agents Chemother.* **21:** 773–779.

49. **Katayama, Y., F. Takeuchi, T. Ito, X. X. Ma, Y. Ui-Mizutani, I. Kobayashi, and K. Hiramatsu.** 2003. Identification in methicillin-susceptible *Staphylococcus hominis* of an active primordial mobile genetic element for the staphylococcal cassette chromosome *mec* of methicillin-resistant *Staphylococcus aureus. J. Bacteriol.* **185:**2711–2722.

50. **Khan, S. A.** 1997. Rolling-circle replication of bacterial plasmids. *Microbiol. Mol. Biol. Rev.* **61:**442–455.

51. **Kuroda, M., T. Ohta, I. Uchiyama, T. Baba, H. Yuzawa, I. Kobayashi, L. Cui, A. Oguchi, K. Aoki, Y. Nagai, J. Lian, T. Ito, M. Kanamori, H. Matsumaru, A. Maruyama, H. Murakami, A. Hosoyama, Y. Mizutani-Ui, N. K. Takahashi, T. Sawano, R. Inoue, C. Kaito, K. Sekimizu, H. Hirakawa, S. Kuhara, S. Goto, J. Yabuzaki, M. Kanehisa, A. Yamashita, K. Oshima, K. Furuya, C. Yoshino, T. Shiba, M. Hattori, N. Ogasawara, H. Hayashi, and K. Hiramatsu.** 2001. Whole genome sequencing of methicillin-resistant *Staphylococcus aureus. Lancet* **357:**1225–1240.

52. **Lacey, R. W.** 1980. Evidence for two mechanisms of plasmid transfer in mixed cultures of *Staphylococcus aureus. J. Gen. Microbiol.* **119:**423–435.

53. **Lee, C. Y., and J. J. Iandolo.** 1986. Lysogenic conversion of staphylococcal lipase is caused by insertion of the bacteriophage L54a genome into the lipase structural gene. *J. Bacteriol.* **166:**385–391.

54. **Leelaporn, A., N. Firth, M. E. Byrne, E. Roper, and R. A. Skurray.** 1994. Possible role of insertion sequence IS*257* in dissemination and expression of high- and low-level trimethoprim resistance in staphylococci. *Antimicrob. Agents Chemother.* **38:**2238–2244.

55. **Leelaporn, A., N. Firth, I. T. Paulsen, and R. A. Skurray.** 1996. IS*257*-mediated cointegration in the evolution of a family of staphylococcal trimethoprim resistance plasmids. *J. Bacteriol.* **178:**6070–6073.

56. **Lindsay, J. A., and M. T. Holden.** 2004. *Staphylococcus aureus:* superbug, super genome? *Trends Microbiol.* **12:**378–385.

57. **Luong, T. T., S. Ouyang, K. Bush, and C. Y. Lee.** 2002. Type 1 capsule genes of *Staphylococcus aureus* are carried in a staphylococcal cassette chromosome genetic element. *J. Bacteriol.* **184:**3623–3629.

58. **Lyon, B. R., and R. Skurray.** 1987. Antimicrobial resistance of *Staphylococcus aureus:* genetic basis. *Microbiol. Rev.* **51:**88–134.

59. **Macrina, F. L., and G. L. Archer.** 1993. Conjugation and broad host range plasmids in streptococci and staphylococci, p. 313–329. *In* D. B. Clewell (ed.), *Bacterial Conjugation.* Plenum Press, New York, N.Y.

60. **Maki, H., N. McCallum, M. Bischoff, A. Wada, and B. Berger-Bächi.** 2004. tcaA inactivation increases glycopeptide resistance in *Staphylococcus aureus. Antimicrob. Agents Chemother.* **48:**1953–1959.

61. **Maki, H., and K. Murakami.** 1997. Formation of potent hybrid promoters of the mutant *llm* gene by IS256 transposition in methicillin-resistant *Staphylococcus aureus. J. Bacteriol.* **179:**6944–6948.

62. **Matsuo, H., M. Kobayashi, T. Kumagai, M. Kuwabara, and M. Sugiyama.** 2003. Molecular mechanism for the enhancement of arbekacin resistance in a methicillin-resistant *Staphylococcus aureus. FEBS Lett.* **546:**401–406.

63. **Minakhina, S., G. Kholodii, S. Mindlin, O. Yurieva, and V. Nikiforov.** 1999. Tn5053 family transposons are *res* hunters sensing plasmidal *res* sites occupied by cognate resolvases. *Mol. Microbiol.* **33:**1059–1068.

64. **Mongkolrattanothai, K., S. Boyle, T. V. Murphy, and R. S. Daum.** 2004. Novel non-*mecA*-containing staphylococcal chromosomal cassette composite island containing *pbp4* and *tagF* genes in a commensal staphylococcal species: a possible reservoir for antibiotic resistance islands in *Staphylococcus aureus. Antimicrob. Agents Chemother.* **48:**1823–1836.

65. **Morton, T. M., D. M. Eaton, J. L. Johnston, and G. L. Archer.** 1993. DNA sequence and units of transcription of the conjugative transfer gene complex (*trs*) of *Staphylococcus aureus* plasmid pGO1. *J. Bacteriol.* **175:**4436–4447.

66. **Murphy, E.** 1990. Properties of the site-specific transposable element Tn554, p. 123–135. *In* R. P. Novick (ed.), *Molecular Biology of the Staphylococci.* VCH Publishers, New York, N.Y.

67. **Needham, C., W. C. Noble, and K. G. H. Dyke.** 1995. The staphylococcal insertion sequence IS257 is active. *Plasmid* **34:**198–205.

68. **Netz, D. J. A., R. Pohl, A. G. Beck-Sickinger, T. Selmer, A. J. Pierik, M. C. F. Bastos, and H.-G. Sahl.** 2002. Biochemical characterisation and genetic analysis of aureocin A53, a new, atypical bacteriocin from *Staphylococcus aureus. J. Mol. Biol.* **319:**745–756.

69. **Netz, D. J. A., H.-G. Sahl, R. Marcolino, J. S. Nascimento, S. S. Oliveira, M. B. Soares, and M. C. F. Bastos.** 2001. Molecular characterisation of aureocin A70, a multipeptide bacteriocin isolated from *Staphylococcus aureus. J. Mol. Biol.* **311:**939–949.

70. **Novick, R. P.** 2003. Mobile genetic elements and bacterial toxinoses: the superantigen-encoding pathogenicity islands of *Staphylococcus aureus. Plasmid* **49:**93–105.

71. **Novick, R. P.** 1989. Staphylococcal plasmids and their replication. *Ann. Rev. Microbiol.* **43:**537–565.

72. **Novick, R. P.** 1990. The *Staphylococcus* as a molecular genetic system, p. 1–37. *In* R. P. Novick (ed.), *Molecular Biology of the Staphylococci.* VCH Publishers, New York, N.Y.

73. **Novick, R. P., I. Edelman, M. D. Schwesinger, A. D. Gruss, E. C. Swanson, and P. A. Pattee.** 1979. Genetic translocation in *Staphylococcus aureus. Proc. Natl. Acad. Sci. USA* **76:**400–404.

74. **Okuma, K., K. Iwakawa, J. D. Turnidge, W. B. Grubb, J. M. Bell, F. G. O'Brien, G. W. Coombs, J. W. Pearman, F. C. Tenover, M. Kapi, C. Tiensasitorn, T. Ito, and K. Hiramatsu.** 2002. Dissemination of new methicillin-resistant *Staphylococcus aureus* clones in the community. *J. Clin. Microbiol.* **40:**4289–4294.

75. **Paulsen, I. T., N. Firth, and R. A. Skurray.** 1997. Resistance to antimicrobial agents other than β-lactams, p. 175–212. *In* K. B. Crossley and G. L. Archer (ed.), *The Staphylococci in Human Disease.* Churchill Livingstone, New York, N.Y.

76. **Paulsen, I. T., M. T. Gillespie, T. G. Littlejohn, O. Hanvivatvong, S. J. Rowland, K. G. H. Dyke, and R. A. Skurray.** 1994. Characterisation of *sin*, a potential recombinase-encoding gene from *Staphylococcus aureus. Gene* **141:**109–114.

77. **Perichon, B., and P. Courvalin.** 2004. Heterologous expression of the enterococcal vanA operon in methicillin-resistant *Staphylococcus aureus. Antimicrob. Agents Chemother.* **48:**4281–4285.

78. **Projan, S. J.** Unpublished data.

79. **Projan, S. J., and G. L. Archer.** 1989. Mobilization of the relaxable *Staphylococcus aureus* plasmid pC221 by the conjugative plasmid pGO1 involves three pC221 loci. *J. Bacteriol.* **171:**1841–1845.

80. **Projan, S. J., and R. Novick.** 1988. Comparative analysis of five related staphylococcal plasmids. *Plasmid* **19:**203–221.

81. **Quintiliani, R., Jr., and P. Courvalin.** 1996. Characterization of Tn1547, a composite transposon flanked by the IS16 and IS256-like elements, that confers vancomycin resistance in *Enterococcus faecalis* BM4281. *Gene* **172:**1–8.

82. **Rice, L. B., L. L. Carias, and S. H. Marshall.** 1995. Tn5384, a composite enterococcal mobile element conferring resistance to erythromycin and gentamicin whose ends are directly repeated copies of IS256. *Antimicrob. Agents Chemother.* **39:**1147–1153.

83. **Robinson, D. A., and M. C. Enright.** 2004. Evolution of *Staphylococcus aureus* by large chromosomal replacements. *J. Bacteriol.* **186:**1060–1064.

84. **Robinson, D. A., and M. C. Enright.** 2003. Evolutionary models of the emergence of methicillin-resistant *Staphylococcus aureus. Antimicrob. Agents Chemother.* **47:**3926–3934.

85. **Rouch, D. A., M. E. Byrne, Y. C. Kong, and R. A. Skurray.** 1987. The aacA-aphD gentamicin and kanamycin resistance determinant of Tn4001 from *Staphylococcus aureus*: expression and nucleotide sequence analysis. *J. Gen. Microbiol.* **133:**3039–3052.

86. **Rouch, D. A., D. S. Cram, D. DiBerardino, T. G. Littlejohn, and R. A. Skurray.** 1990. Efflux-mediated antiseptic resistance gene qacA from *Staphylococcus aureus*: common ancestry with tetracycline- and sugar-transport proteins. *Mol. Microbiol.* **4:**2051–2062.

87. **Rouch, D. A., L. J. Messerotti, L. S. Loo, C. A. Jackson, and R. A. Skurray.** 1989. Trimethoprim resistance transposon Tn4003 from *Staphylococcus aureus* encodes genes for a dihydrofolate reductase and thymidylate synthetase flanked by three copies of IS257. *Mol. Microbiol.* **3:**161–175.

88. **Rowland, S., and K. G. H. Dyke.** 1990. Tn552, a novel transposable element from *Staphylococcus aureus. Mol. Microbiol.* **4:**961–975.

89. **Rowland, S.-J., W. M. Stark, and M. R. Boocock.** 2002. Sin recombinase from *Staphylococcus aureus*: synaptic complex architecture and transposon targeting. *Mol. Microbiol.* **44:**607–619.

90. **Ruzin, A., J. Lindsay, and R. P. Novick.** 2001. Molecular genetics of SaPI1—a mobile pathogenicity island in *Staphylococcus aureus. Mol. Microbiol.* **41:**365–377.

91. **Shalita, Z., E. Murphy, and R. P. Novick.** 1980. Penicillinase plasmids of *Staphylococcus aureus*: structural and evolutionary relationships. *Plasmid* **3:**291–311.

92. **Sharma, V. K., J. L. Johnston, T. M. Morton, and G. L. Archer.** 1994. Transcriptional regulation by TrsN of conjugative transfer genes on staphylococcal plasmid pGO1. *J. Bacteriol.* **176:**3445–3454.

93. **Sheehy, R. J., and R. P. Novick.** 1975. Studies on plasmid replication. V. Replicative intermediates. *J. Mol. Biol.* **93:** 237–253.

94. **Simpson, A. E., R. A. Skurray, and N. Firth.** 2000. An IS257-derived hybrid promoter directs transcription of a *tetA*(K) tetracycline resistance gene in the *Staphylococcus aureus* chromosomal *mec* region. *J. Bacteriol.* **182:**3345–3352.

95. **Skurray, R. A., and N. Firth.** 1997. Molecular evolution of multiply-antibiotic-resistant staphylococci. *Ciba Found. Symp.* **207:**167–183.

96. **Skurray, R. A., D. A. Rouch, B. R. Lyon, M. T. Gillespie, J. M. Tennent, M. E. Byrne, L. J. Messerotti, and J. W. May.** 1988. Multiresistant *Staphylococcus aureus:* genetics and evolution of epidemic Australian strains. *J. Antimicrob. Chemother.* **21:**19–38.

97. **Tennent, J. M., B. R. Lyon, M. Midgley, I. G. Jones, A. S. Purewal, and R. A. Skurray.** 1989. Physical and biochemical characterization of the *qacA* gene encoding antiseptic and disinfectant resistance in *Staphylococcus aureus. J. Gen. Microbiol.* **135:**1–10.

98. **Tenover, F. C., L. M. Weigel, P. C. Appelbaum, L. K. McDougal, J. Chaitram, S. McAllister, N. Clark, G. Killgore, C. M. O'Hara, L. Jevitt, J. B. Patel, and B. Bozdogan.** 2004. Vancomycin-resistant *Staphylococcus aureus* isolate from a patient in Pennsylvania. *Antimicrob. Agents Chemother.* **48:**275–280.

99. **Thumm, G., and F. Götz.** 1997. Studies on prolysostaphin processing and characterization of the lysostaphin immunity factor (Lif) of *Staphylococcus simulans* biovar *staphylolyticus. Mol. Microbiol.* **23:**1251–1265.

100. **Townsend, D. E., S. Bolton, N. Ashdown, D. I. Annear, and W. B. Grubb.** 1986. Conjugative, staphylococcal plasmids carrying hitch-hiking transposons similar to Tn554: intra- and interspecies dissemination of erythromycin resistance. *Aust. J. Exp. Biol. Med. Sci.* **64:**367–379.

101. **Úbeda, C., M. A. Tormo, C. Cucarella, P. Trotonda, T. J. Foster, I. Lasa, and J. R. Penadés.** 2003. Sip, an integrase protein with excision, circularization and integration activities, defines a new family of mobile Staph-ylococcus aureus pathogenicity islands. *Mol. Microbiol.* **49:**193–210.

102. **Udo, E. E., and W. B. Grubb.** 1991. A new incompatibility group plasmid in *Staphylococcus aureus. FEMS Microbiol. Lett.* **62:**33–36.

103. **Vuong, C., S. Kocianova, Y. Yao, A. B. Carmody, and M. Otto.** 2004. Increased colonization of indwelling medical devices by quorum-sensing mutants of *Staphylococcus epidermidis* in vivo. *J. Infect. Dis.* **190:**1498–1505.

104. **Weigel, L. M., D. B. Clewell, S. R. Gill, N. C. Clark, L. K. McDougal, S. E. Flannagan, J. F. Kolonay, J. Shetty, G. E. Killgore, and F. C. Tenover.** 2003. Genetic analysis of a high-level vancomycin-resistant isolate of *Staphylococcus aureus. Science* **302:**1569–1571.

105. **Wright, C. L., M. E. Byrne, N. Firth, and R. A. Skurray.** 1998. A retrospective molecular analysis of gentamicin resistance in *Staphylococcus aureus* strains from UK hospitals. *J. Med. Microbiol.* **47:**173–178.

106. **Wu, S., C. Piscitelli, H. de Lencastre, and A. Tomasz.** 1996. Tracking the evolutionary origin of the methicillin resistance gene: cloning and sequencing of a homologue of *mecA* from a methicillin susceptible strain of *Staphylococcus sciuri. Microb. Drug Resist.* **2:**435–441.

107. **Yamaguchi, T., T. Hayashi, H. Takami, K. Nakasone, M. Ohnishi, K. Nakayama, S. Yamada, H. Komatsuzawa, and M. Sugai.** 2000. Phage conversion of exfoliative toxin A production in *Staphylococcus aureus. Mol. Microbiol.* **38:**694–705.

108. **Yamaguchi, T., T. Hayashi, H. Takami, M. Ohnishi, T. Murata, K. Nakayama, K. Asakawa, M. Ohara, H. Komatsuzawa, and M. Sugai.** 2001. Complete nucleotide sequence of a *Staphylococcus aureus* exfoliative toxin B plasmid and identification of a novel ADP-ribosyltransferase, EDIN-C. *Infect. Immun.* **69:**7760–7771.

109. **Zhang, S., J. J. Iandolo, and G. C. Stewart.** 1998. The enterotoxin D plasmid of *Staphylococcus aureus* encodes a second enterotoxin determinant (*sej*). *FEMS Microbiol. Lett.* **168:**227–233.

110. **Ziebuhr, W., V. Krimmer, S. Rachid, I. Lossner, F. Götz, and J. Hacker.** 1999. A novel mechanism of phase variation of virulence in *Staphylococcus epidermidis:* evidence for control of the polysaccharide intercellular adhesin synthesis by alternating insertion and excision of the insertion sequence element IS256. *Mol. Microbiol.* **32:**345–356.

Carbohydrate Catabolism: Pathways and Regulation

REINHOLD BRÜCKNER AND RALF ROSENSTEIN

34

The central pathways of carbon metabolism are conserved in virtually all organisms, but details of specific biosynthetic and degradative pathways vary considerably between bacteria, plants, and animals. In addition, metabolic capacities differ widely among bacteria. As a consequence, generalization and even comparative approaches do not appear to be sufficient for a complete understanding of the physiology of a certain bacterium. For that end, biosynthetic potential and nutritional requirements have to be defined for each bacterial strain individually. On the other hand, the comparison of related organisms, e.g., AT-rich gram-positive bacteria, may reveal common principles in physiology and regulation. A prominent example would perhaps be carbon catabolite repression, which is executed in AT-rich gram-positive organisms by a different mechanism than in enteric bacteria (8, 48).

In *Staphylococcus aureus* and other staphylococcal species, relatively few molecular details are known about carbohydrate utilization, biosynthetic pathways, and nutritional requirements. The limited knowledge on sugar utilization systems is especially surprising, because *S. aureus* was the first gram-positive bacterium in which the phosphoenolpyruvate (PEP)-dependent carbohydrate phosphotransferase system (PTS) was described. Numerous biochemical and physiological studies, carried out before gene isolation techniques became common, have not been pursued to the genetic level. Therefore, transcriptional regulation of genes involved in catabolic or biosynthetic pathways has been analyzed for a rather limited number of systems. In other bacteria, especially in *Escherichia coli* and *Bacillus subtilis*, these studies have led to fundamental concepts of molecular biology and gene regulation. Considering the need of pathogenic bacteria to multiply in a host organism to cause disease, there are still good reasons to investigate cellular metabolism in the staphylococci. Because earlier work on sugar utilization and other aspects of intermediary metabolism has been reviewed elsewhere (4, 31), we will concentrate on the latest experimental data from various staphylococcal species, and we will evaluate their significance for *S. aureus* considering the recently published genome sequences of the strains N315, Mu50, MW2, MRSA252, and MSSA476 (1, 25, 36). If ap-

propriate, we will also take into account the genome sequences of two apathogenic staphylococcal species—one of a nonbiofilm, noninfection associated *Staphylococcus epidermidis* strain (ATCC 1228) (58) and the other, recently completed by one of us (45), of *Staphylococcus carnosus* TM300, a species applied to meat fermentation (49).

GLYCOLYTIC PATHWAYS

The central routes of glucose metabolism in *S. aureus* are the Embden-Meyerhof-Parnas (EMP) pathway (Fig. 1) and the pentose phosphate cycle. The Entner-Doudoroff pathway does not appear to exist. Depending on the growth conditions, the majority of glucose, about 85%, is catabolized via the EMP pathway. The predominant end product of anaerobic glucose metabolism is lactate, while under aerobic growth conditions acetate and CO_2 are produced (4). Glucose enters the EMP pathway as glucose-6-phosphate, which may be produced by PTS-mediated transport and concomitant phosphorylation or by the activity of a glucose kinase, when glucose enters the cells in unmodified form. Various other carbohydrates, hexoses, hexitols, or disaccharides, such as fructose, mannitol, sucrose, or maltose, are fed into the EMP pathway by the activity of peripheral sugar-specific enzymes, such as PTS permeases, hydrolases, kinases, and dehydrogenases. In the *S. aureus* genomes the genes encoding all components of the EMP pathway are clearly detectable.

THE PHOSPHOTRANSFERASE SYSTEM

The PEP:carbohydrate PTS catalyzes the concomitant uptake and phosphorylation of a great variety of carbohydrates (42). The system consists of carbohydrate-specific PTS permeases, referred to as enzymes II (EII), and two proteins, enzyme I (EI) and histidine-containing protein (HPr), that are needed for all PTS-dependent transport processes. The latter proteins have, therefore, been designated the general PTS proteins. The sugar-specific permeases are composed of up to four protein domains (EIIA, B, C, D), at least one of which is membrane-bound. These protein domains may exist

Gram-Positive Pathogens, 2nd edition, edited by Vincent A. Fischetti et al.
© 2006 ASM Press, Washington, D.C.

as separate proteins or may be fused as a single polypeptide chain. The general PTS proteins are soluble, cytoplasmic proteins. The phosphoryl-transfer chain begins with EI and PEP and proceeds via the phosphocarrier protein HPr to the EIIA and EIIB domains of the PTS permeases. Subsequently the incoming sugar is phosphorylated by the EIIB domain (Fig. 1).

As already mentioned, *S. aureus* was the first grampositive species in which PTS activity has been demonstrated. The uptake of glucose, mannose, mannitol, glucosamine, *N*-acetylglucosamine, sucrose, lactose, galactose, and β-glucosides is reported to be PTS-dependent (31). The PTS dependency of certain sugars may vary within the staphylococci. The PTS-independent utilization of lactose by

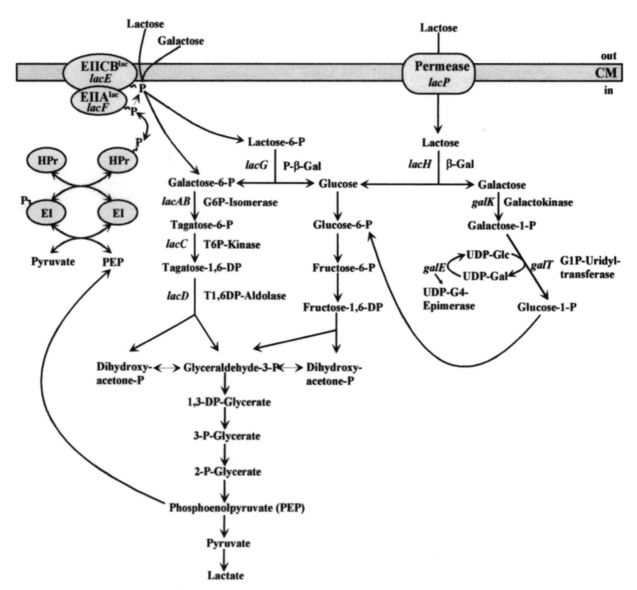

FIGURE 1 Alternative lactose catabolic pathways in staphylococci. Transport of lactose and galactose and their catabolism are shown. In *S. aureus*, lactose and galactose are transported by the PEP:carbohydrate PTS. Internalized lactose-6-phosphate is hydrolyzed by a phospho-β-galactosidase to galactose-6-phosphate and glucose. Galactose-6-phosphate is catabolized through the tagatose-6-phosphate pathway. This pathway most likely exists in staphylococci exhibiting lactose PTS activity. In *S. xylosus*, *S. carnosus*, and probably other staphylococcal species that do not possess a lactose PTS, a permease is responsible for the transport of lactose. Galactose uptake has not been studied in these species. Nonphosphorylated lactose is hydrolyzed by a β-galactosidase to yield glucose and galactose. Galactose is likely catabolized through the Leloir pathway. Glucose-6-phosphate, produced by a glucose kinase, enters the EMP pathway, the main glycolytic pathway in staphylococci. Only the galactoside-specific genes and their encoded products are mentioned in the pathways. Abbreviations: CM, cytoplasmic membrane; EI, enzyme I; EIIA^{lac}, lactose-specific enzyme IIA; EIICB^{lac}, lactose-specific enzyme IICB; HPr, histidine-containing protein; β-Gal, β-galactosidase; P-β-Gal, phospho-β-galactosidase; G6P-Isomerase, galactose-6-phosphate isomerase; G1P-Uridyltransferase, galactose-1-phosphate uridyltransferase; T6P-Kinase, tagatose-6-phosphate kinase; T1,6DP-Aldolase, tagatose-1,6-diphosphate aldolase; UDP-Gal, UDP-galactose; UDP-Glc, UDP-glucose; UDP-G4-Epimerase, UDP-galactose-4 epimerase; PEP, phosphoenolpyruvate; P, phosphate; DP, diphosphate.

Staphylococcus xylosus (2) in contrast to the PTS dependency in *S. aureus* would be an example of this variability. Glucose utilization does not completely rely on PTS activity, because HPr or EI mutants of *S. aureus* still utilize glucose (44). The same is also true for *S. xylosus* (30) and *S. carnosus* (7); dual entry of glucose may thus be a general phenomenon. The disaccharide maltose is mainly taken up independently of PTS, but additional minor PTS-mediated transport cannot be ruled out (44). On the other hand, utilization of glycerol is clearly PTS-independent. Several other carbohydrates such as cellobiose, raffinose, xylose, and arabinose do not serve as carbon sources for *S. aureus*. Metabolism of others, such as trehalose, has not been investigated. The patterns of sugar utilization determined experimentally are generally supported by the *S. aureus* genome sequences, although unambiguous identification of the substrates of PTS transporters is not always possible. The utilization of the pentose xylose, a trait found in only a few staphylococcal species (31), does not involve the PTS system in *S. xylosus* (50, 51).

At the molecular level, information on genes encoding the general PTS proteins is available from *S. aureus* (1, 25, 36), *S. carnosus* (16, 34), *S. epidermidis* (58), and *S. xylosus* (28). The primary structure of *S. aureus* HPr was initially determined by sequencing the purified protein (3), before the corresponding DNA was sequenced. In all staphylococcal species analyzed so far, the HPr gene, *ptsH*, is followed by *ptsI*, the gene encoding EI. A promoter in front of *ptsH* has been mapped in *S. xylosus* (28). Subcloning experiments showed that *ptsI* of *S. xylosus* is cotranscribed with *ptsH*, forming the *ptsHI* operon. Moreover, translation of *ptsI* is at least partially coupled to *ptsH* translation due to the lack of an appropriately spaced Shine-Dalgarno sequence. Because the *ptsHI* genes show a high degree of similarity, the findings in *S. xylosus* most likely also apply to the other staphylococcal species.

GLUCOSE UTILIZATION

Two genes, *glcU* and *glkA*, that are responsible for PTS-independent glucose utilization in *S. xylosus* have been isolated and characterized (17, 56). They encode the glucose uptake protein GlcU and the glucose kinase GlkA. The genes, located at two distinct loci in the genome, were detected by transposon Tn917 insertions, resulting in a defect in glucose-specific catabolite repression of the lactose operon. The corresponding proteins constitute a PTS-independent glucose utilization system that substantially contributes to glucose catabolism in *S. xylosus*. The presence of *glcU/glkA* sufficiently explains the ability of a *ptsI* mutant to utilize glucose as carbon source. In the completed staphylococcal genome sequences homologs of *glcU* and *glkA* are present, strongly suggesting a role in PTS-independent glucose uptake and phosphorylation. Although *glcU* is followed by a glucose dehydrogenase gene *gdh* in *S. xylosus*, *S. carnosus*, and *S. epidermidis*, *gdh* is missing at this position in *S. aureus*.

Data concerning staphylococcal PTS-dependent glucose uptake are so far restricted to *S. carnosus*. In this staphylococcal species, two genes located right next to each other, *glcA* and *glcB*, have been cloned, both of which complemented an *E. coli* mutant strain deficient in glucose uptake (10). The GlcA/B proteins are highly similar to each other (69% identity) and to glucose-specific EII proteins from *B. subtilis* or *E. coli*. The two *S. carnosus* PTS permeases had fused EII domains in the order EIICBA. Subsequent analysis of their substrate specificity indicated that both used glu-

cose as the primary substrate, but various glucosides were recognized differently (11). Immediately upstream of *glcA*, a gene, *glcT*, was detected, whose deduced amino acid sequence shows a high degree of similarity to bacterial regulators involved in antitermination (52). A putative transcriptional terminator partially overlapping an inverted repeat, which could be the target site for the antiterminator protein GlcT, is found in the *glcT-glcA* intergenic region. This organization resembles the *glcT-ptsG* region of *B. subtilis*, encoding the GlcT antiterminator protein and the glucose-specific EII, respectively (53). Studies of *S. carnosus* GlcT activity in the heterologous host *B. subtilis* indicated that the protein is indeed able to cause antitermination (32). *S. aureus* also possesses two glucose-specific PTS permeases, whose genes are not arranged in tandem. The gene most likely encoding the antiterminator GlcT is at a third locus in the *S. aureus* genome. In contrast to *S. aureus* and *S. carnosus*, *S. epidermidis* appears to have only one glucose-specific PTS permease and a *glcT* gene, which is also not linked. Interestingly, these genes have been implicated in the regulation of biofilm formation in a biofilm-producing *S. epidermidis* strain (33).

LACTOSE AND GALACTOSE UTILIZATION

Utilization of lactose and galactose in *S. aureus* relies on the PTS-dependent uptake and phosphorylation of the sugars, resulting in lactose-6-phosphate and galactose-6-phosphate, respectively. Glucose and galactose-6-phosphate are produced from intracellular lactose-6-phosphate by a phospho-β-galactosidase (23). Glucose is metabolized via the EMP pathway, whereas galactose-6-phosphate is degraded via the tagatose-6-phosphate pathway (Fig. 1). The corresponding genes are clustered and are arranged in a heptacistronic operon (*lacABCDFEG*). The locus was originally defined by mutations abolishing lactose utilization. The genes *lacFE* code for the galactoside-specific PTS permease, EIIA and EIICB (5). The last gene, *lacG*, encodes the phospho-β-galactosidase (6). The *lacABCD* cluster specifies the enzymes of the tagatose-6-phosphate pathway, with *lacAB* coding for galactose-6-phosphate isomerase, *lacC* for tagatose-6-phosphate kinase, and *lacD* for tagatose-1, 6-diphosphate aldolase (46). Upstream of the *lacABCDFEG* operon and arranged in the same orientation is *lacR*, the gene encoding the lactose operon repressor (40). LacR resembles several lactose repressors from other gram-positive bacteria (12) that belong to the DeoR regulator family, named after the repressor of the deoxyribonucleotide operon in *E. coli* (54).

Supplementation of the growth medium with lactose or galactose results in the induction of the *lac* genes, with galactose-6-phosphate being the intracellular inducer. Apart from this sugar-specific regulation, the lactose operon of *S. aureus* is also subject to global carbon catabolite repression, but the nature of the regulatory mechanism remains unclear (40). It is conceivable that part of the described catabolite repression of the *lac* operon of *S. aureus* is actually caused by glucose-mediated inducer exclusion (44, 47).

The lactose metabolism pathway described for *S. aureus* is not universal for all staphylococci. In *S. xylosus*, for example, lactose is taken up in nonphosphorylated form by a lactose permease, a member of the GPH protein family that transports galactosides, pentoses, and hexuronides (41). The lactose permease is encoded by *lacP*, which is the first gene of the lactose operon (2). The following *lacH* gene encodes the β-galactosidase of *S. xylosus* (2). After lactose hydrolysis, glucose is phosphorylated by a glucose kinase and catabolized through the EMP pathway and galactose, likely by the Leloir pathway (Fig. 1) (37). The galactokinase gene needed to produce galactose-1-phosphate (Fig. 1) has been isolated from an *S. xylosus* library but was not further characterized (9). The *lacR* gene, upstream and in the opposite orientation of the *lacPH* operon of *S. xylosus*, encodes an activator belonging to the AraC/XylS family (21). The *lacPH* operon is induced by addition of lactose to the culture medium, but not by galactose (2). A similar lactose utilization gene cluster is also present in *S. carnosus*. In this cluster, *lacR* is located in between *lacP* and *lacH* (45).

MANNITOL UTILIZATION

The mannitol PTS has been described for *S. carnosus*. The system consists of an EIICB enzyme, encoded by *mtlA*, and EIIA, encoded by *mtlF*, which together form the mannitol-specific PTS permease (18, 19). The *mtlF* and *mtlA* genes are separated by about 2 kb, but the DNA sequence of the intervening region had not been determined. Downstream of *mtlA* is *mtlD*, the gene for mannitol-1-phosphate dehydrogenase, which produces fructose-6-phosphate. In *S. aureus*, the gene order is the same, with a gene encoding a putative antiterminator protein between *mtlF* and *mtlA*, strongly suggesting control of the mannitol utilization operon by antitermination. Inspection of the region between *mtlF* and *mtlA* in *S. carnosus* also revealed a gene encoding an antiterminator.

SUCROSE UTILIZATION

The sucrose PTS permease, analyzed in *S. xylosus* and encoded by *scrA*, is composed of fused EIIBC domains (55). The EIIA domain, which is essential for PTS-mediated sucrose uptake, has not been identified in *S. xylosus*. Based on the analysis of sucrose utilization in *E. coli* and *B. subtilis*, it appears questionable whether a separate sucrose-specific EIIA protein exists. In both organisms, the EIIA domain specific for glucose, either as an independent enzyme or fused to the IICB domains of the glucose permease, serves as phosphoryl donor for the sucrose-specific EII enzymes. The GlcA/B proteins mentioned above would thus be good candidates for being involved in sucrose uptake.

Internalized sucrose-6-phosphate is cleaved by sucrose-phosphate hydrolase, the gene product of *scrB*, to yield glucose-6-phosphate and fructose (9). Fructose is subsequently phosphorylated by a fructokinase encoded by *scrK* (30). The genes *scrB* and *scrK* are located next to each other and form an operon. However, the sucrose permease gene *scrA* is not located in close vicinity to the *scrBK* genes. Expression of *scrA* as well as *scrB* is induced by sucrose in the medium. Regulation is dependent on the LacI/GalR-type repressor ScrR, whose gene *scrR* is found upstream of *scrB* (22). Imperfect palindromic sequences in the promoter region of *scrA* and *scrB* serve as binding sites for *scrR* (22). In addition to sucrose-specific ScrR-dependent regulation, the sucrose-permease gene *scrA* is subject to carbon catabolite repression by the catabolite control protein CcpA. An *scr* gene cluster, *scrR*, *scrB*, *scrK*, is also present in *S. aureus*. It is located next to *agrA*, the gene encoding the response regulator of the accessory gene regulator system (*agr*) (39). The *scrA* gene is also located elsewhere in the *S. aureus* genome.

MALTOSE UTILIZATION

Maltose utilization in *S. xylosus* is dependent on an α-glucosidase or maltase, whose gene, *malA*, is the second gene of the *malRA* operon. Besides maltose, the enzyme can also cleave short maltodextrins. The second gene of the operon, *malR*, encodes a regulator belonging to the LacI/GalR family (15), which was found to control maltose uptake but not its own expression. Maltose uptake genes have not been identified in *S. xylosus*. The same enzymatic activity as mediated by MalA in *S. xylosus* has been characterized in *S. aureus*, and the bicistronic *malRA* operon is present in this organism. In addition, genes encoding a maltose/maltodextrin transport system are found in *S. aureus*. Therefore, all components of non-PTS maltose utilization are present in *S. aureus*.

In addition to non-PTS maltose utilization, minor PTS-mediated maltose transport is detectable in *S. aureus*. Transposon mutagenesis in *S. xylosus* indicated that inactivation of a trehalose PTS also resulted in a slight defect in maltose utilization. Considering the presence of a trehalose PTS in *S. aureus*, it appears likely that minor maltose utilization is mediated by this system.

XYLOSE UTILIZATION

Utilization of xylose is scarce among staphylococci. It has been analyzed in *S. xylosus*, representing the only molecular information on staphylococcal pentose utilization. Three *S. xylosus* genes, *xylA*, *xylB*, and *xylR*, involved in xylose catabolism have been identified (50, 51). They encode the xylose isomerase XylA, the xylulose kinase XylB, and the regulator XylR. XylA and XylB generate xylulose-5-phosphate, which enters the pentose phosphate cycle. Genes encoding proteins for xylose transport have not yet been identified, but it is established that xylose enters the cells in nonmodified form by a PTS-independent route. Transcriptional analysis showed that *xylAB* is transcribed from one promoter in front of *xylA*, while *xylR* is transcribed independently. Transcription initiated at the *xylA* promoter is inducible by the addition of xylose to the growth medium, whereas *xylR* expression is constitutive with respect to xylose. Regulation of *xylAB* expression is dependent on a functional XylR protein, which acts as a repressor. Like the sucrose system, xylose utilization is also subject to carbon catabolite repression.

CARBON CATABOLITE REPRESSION

The availability of carbohydrates, especially of glucose, leads to regulatory processes often referred to as glucose effect or carbon catabolite repression (47, 48). In *S. aureus*, numerous publications describe the influence of glucose on a variety of cellular processes, such as utilization of alternative carbon sources, production of extracellular enzymes, activity of glycolytic enzymes, or cytochrome content. The production of potential virulence factors attracted considerable attention. However, the mechanism by which glucose exerts its regulatory effect has not been elucidated so far. The analysis of catabolite repression of inducible systems is quite often complicated by the inability of the inducer to enter the cells in appreciable amounts when glucose is also present in the medium, a process referred to as inducer exclusion (47). The molecular mechanisms leading to inducer exclusion in gram-positive bacteria are not yet completely understood (13). Inducer expulsion, the rapid removal of internalized inducer, has been described for some gram-positive organisms (47), but does not appear to be operative in *S. aureus* (47). Its general physiological significance remains elusive (8).

In gram-positive bacteria one form of carbon catabolite repression relies on a transcriptional regulator, termed CcpA (24), a member of the GalR-LacI family of transcription

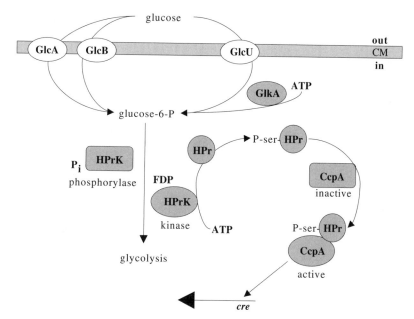

FIGURE 2 Glucose-mediated carbon catabolite repression by the catabolite control protein CcpA in staphylococci. The components involved in regulation are: CcpA, carbon catabolite control protein; GlkA, glucose kinase; GlcU, PTS-independent glucose uptake protein; HPr, general PTS phosphocarrier protein; HPrK, HPr kinase/P-HPr phosphorylase; GlcA/GlcB, glucose-specific PTS permeases; FDP, fructose-1,6-diphosphate. The general PTS protein EI, essential for the transport of all PTS sugars, is omitted for clarity. The thick arrow represents a gene that is subject to carbon catabolite repression by CcpA via *cre* (catabolite-responsive element) interaction. The position of the CcpA binding site *cre* in the promoter region is indicated. The double line marked with CM represents the cytoplasmic membrane. The fact that some proteins may only function as dimers or multimers is not depicted. *S. epidermidis* ATCC 1228 has apparently only one glucose-specific PTS permease. Glucose may be internalized by PTS-dependent or -independent transport. The glycolytic intermediate FDP activates kinase activity of HPrK, which produces P-ser-HPr by ATP-dependent phosphorylation. P-ser-HPr acts as a corepressor for CcpA, enabling the regulator to bind specifically to *cre* sites. When P_i prevails over ATP and FDP, HPrK dephosphorylates P-ser-HPr.

factors (57). CcpA shows a relatively weak affinity for its cognate operator sites, catabolite responsive elements (*cre*) (26), and must, therefore, be activated in order to bind efficiently to *cre*. The corepressor enabling CcpA to bind to DNA is serine-46 phosphorylated HPr (P-ser-HPr), the phosphocarrier protein of the PTS. The bifunctional HPr kinase/phosphorylase HPrK phosphorylates HPr in the presence of ATP and the glycolytic intermediate fructose-1,6-diphosphate (20, 35, 43). Therefore, HPrK is the sensor enzyme for catabolite repression in AT-rich gram-positive bacteria linking glycolytic activity to transcriptional regulation. When phosphate prevails over ATP and fructose-1,6-diphosphate, HPrK dephosphorylates P-ser-HPr, leading to the formation of HPr and pyrophosphate (38). Pyrophosphate is then most likely removed by the pyrophosphatase YvoE, whose gene is located in the *hprK* operon (38).

Among the staphylococci, CcpA- and HPrK-dependent carbon catabolite repression has been analyzed in some detail in *S. xylosus* (14, 27–29). Inactivation of the *ccpA* gene by a resistance cassette led to a pleiotropic loss of transcriptional regulation at several promoters, including one of the two *ccpA* promoters. Therefore, *ccpA* expression is under autogenous control, apparently reflecting the need of *S. xylosus* to carefully balance CcpA regulation (14). Loss of HPr kinase/phosphorylase activity by inactivation of the *hprK* gene abolished CcpA-dependent catabolite repression, as

well as repression that is most likely caused by inducer exclusion (27).

In addition to the loss of carbon catabolite repression, the *S. xylosus* HPr kinase/phosphorylase-deficient strain showed an unexpected phenotype. Its growth was inhibited by glucose. The mutant strain transported glucose at much higher rates than the wild type, and it also produced methylglyoxal, indicating unbalanced glucose metabolism. Therefore, carbon catabolite repression is apparently needed to balance carbohydrate uptake and catabolic capacities to protect the cell from adverse effects of unlimited carbohydrate utilization (8).

Considering the staphylococcal genome sequences determined so far, it is very likely that carbon catabolite repression proceeds in *Staphylococcus* as determined for *S. xylosus*. An outline of glucose-mediated catabolite repression is presented in Fig. 2.

Work on sugar utilization and carbon catabolite repression in S. xylosus was sponsored by the European Union Biotech Program and the Deutsche Forschungsgemeinschaft.

REFERENCES

1. Baba, T., F. Takeuchi, M. Kuroda, H. Yuzawa, K. Aoki, A. Oguchi, Y. Nagai, N. Iwama, K. Asano, T. Naimi, H. Kuroda, L. Cui, K. Yamamoto, and K. Hiramatsu.

2002. Genome and virulence determinants of high viru- lence community-acquired MRSA. *Lancet* **359:**1819– 1827.

2. **Bassias, J., and R. Brückner.** 1998. Regulation of lactose utilization genes in *Staphylococcus xylosus*. *J. Bacteriol.* **180:**2273–2279.

3. **Beyreuther, K., H. Raufuss, O. Schrecker, and W. Hengstenberg.** 1977. The phosphoenolpyruvate- dependent phosphotransferase system of *Staphylococcus au- reus*. 1. Amino-acid sequence of the phosphocarrier pro- tein HPr. *Eur. J. Biochem.* **75:**275–286.

4. **Blumenthal, H. J.** 1972. Glucose catabolism in staphylo- cocci, p. 111–135. *In* J. O. Cohen (ed.), *The Staphylococci.* Wiley Interscience, New York, N.Y.

5. **Breidt, F., Jr., W. Hengstenberg, U. Finkeldei, and G. C. Stewart.** 1987. Identification of the genes for the lactose- specific components of the phosphotransferase system in the *lac* operon of *Staphylococcus aureus. J. Biol. Chem.* **262:** 16444–16449.

6. **Breidt, F., Jr., and G. C. Stewart.** 1986. Cloning and ex- pression of the phospho-beta-galactosidase gene of *Staph- ylococcus aureus* in *Escherichia coli. J. Bacteriol.* **166:**1061– 1066.

7. **Brückner, R.** 1997. Gene replacement in *Staphylococcus carnosus* and *Staphylococcus xylosus. FEMS Microbiol. Lett.* **151:**1–8.

8. **Brückner, R., and F. Titgemeyer.** 2002. Carbon catabolite repression in bacteria: choice of the carbon source and au- toregulatory limitation of sugar utilization. *FEMS Micro- biol. Lett.* **209:**141–148.

9. **Brückner, R., E. Wagner, and F. Götz.** 1993. Characteri- zation of a sucrase gene from *Staphylococcus xylosus. J. Bac- teriol.* **175:**851–857.

10. **Christiansen, I., and W. Hengstenberg.** 1996. Cloning and sequencing of two genes from *Staphylococcus carnosus* coding for glucose-specific PTS and their expression in *Escherichia coli* K-12. *Mol. Gen. Genet.* **250:**375–379.

11. **Christiansen, I., and W. Hengstenberg.** 1999. Staphylo- coccal phosphoenolpyruvate-dependent phosphotransfer- ase system—two highly similar glucose permeases in *Staphylococcus carnosus* with different glucoside specificity: protein engineering *in vivo? Microbiology* **145:**2881– 2889.

12. **de Vos, W. M., and E. E. Vaughan.** 1994. Genetics of lac- tose utilization in lactic acid bacteria. *FEMS Microbiol. Rev.* **15:**217–237.

13. **Dossonnet, V., V. Monedero, M. Zagorec, A. Galinier, G. Pérez-Martínez, and J. Deutscher.** 2000. Phosphoryla- tion of HPr by the bifunctional HPr Kinase/P-Ser-HPr phosphatase from *Lactobacillus casei* controls catabolite re- pression and inducer exclusion but not inducer expulsion. *J. Bacteriol.* **182:**2582–2590.

14. **Egeter, O., and R. Brückner.** 1996. Catabolite repression mediated by the catabolite control protein CcpA in *Staph- ylococcus xylosus. Mol. Microbiol.* **21:**739–749.

15. **Egeter, O., and R. Brückner.** 1995. Characterization of a genetic locus essential for maltose-maltotriose utilization in *Staphylococcus xylosus. J. Bacteriol.* **177:**2408–2415.

16. **Eisermann, R., R. Fischer, U. Kessler, A. Neubauer, and W. Hengstenberg.** 1991. Staphylococcal phosphoenol- pyruvate-dependent phosphotransferase system. Purifica- tion and protein sequencing of the *Staphylococcus carnosus* histidine-containing protein, and cloning and DNA se- quencing of the *ptsH* gene. *Eur. J. Biochem.* **197:**9–14.

17. **Fiegler, H., J. Bassias, I. Jankovic, and R. Brückner.** 1999. Identification of a gene in *Staphylococcus xylosus* encoding a novel glucose uptake protein. *J. Bacteriol.* **181:** 4929–4936.

18. **Fischer, R., R. Eisermann, B. Reiche, and W. Hengsten- berg.** 1989. Cloning, sequencing and overexpression of the mannitol-specific enzyme-III-encoding gene of *Staphylo- coccus carnosus. Gene* **82:**249–257.

19. **Fischer, R., and W. Hengstenberg.** 1992. Mannitol- specific enzyme II of the phosphoenolpyruvate-dependent phosphotransferase system of *Staphylococcus carnosus. Eur. J. Biochem.* **204:**963–969.

20. **Galinier, A., M. Kravanja, R. Engelmann, W. Hengsten- berg, M. C. Kilhoffer, J. Deutscher, and J. Haiech.** 1998. New protein kinase and protein phosphatase families me- diate signal transduction in bacterial catabolite repression. *Proc. Natl. Acad. Sci. USA* **95:**1823–1828.

21. **Gallegos, M.-T., R. Schleif, A. Bairoch, K. Hofmann, and J. L. Ramos.** 1997. AraC/XylS family of transcription- al regulators. *Microbiol. Mol. Biol. Rev.* **61:**393–410.

22. **Gering, M., and R. Brückner.** 1996. Transcriptional regu- lation of the sucrase gene of *Staphylococcus xylosus* by the repressor ScrR. *J. Bacteriol.* **178:**462–469.

23. **Hengstenberg, W., D. Kohlbrecher, E. Witt, R. Kruse, I. Christiansen, D. Peters, R. Pogge von Strandmann, P. Stadtler, B. Koch, and H. R. Kalbitzer.** 1993. Structure and function of proteins of the phosphotransferase system and of 6-phospho-beta-glycosidases in gram-positive bac- teria. *FEMS Microbiol. Rev.* **12:**149–163.

24. **Henkin, T. M.** 1996. The role of CcpA transcriptional regulator in carbon metabolism in *Bacillus subtilis. FEMS Microbiol. Lett.* **135:**9–15.

25. **Holden, M. T., E. J. Feil, J. A. Lindsay, S. J. Peacock, N. P. Day, M. C. Enright, T. J. Foster, C. E. Moore, L. Hurst, R. Atkin, A. Barron, N. Bason, S. D. Bentley, C. Chillingworth, T. Chillingworth, C. Churcher, L. Clark, C. Corton, A. Cronin, J. Doggett, L. Dowd, T. Feltwell, Z. Hance, B. Harris, H. Hauser, S. Holroyd, K. Jagels, K. D. James, N. Lennard, A. Line, R. Mayes, S. Moule, K. Mungall, D. Ormond, M. A. Quail, E. Rabbi- nowitsch, K. Rutherford, M. Sanders, S. Sharp, M. Sim- monds, K. Stevens, S. Whitehead, B. G. Barrell, B. G. Spratt, and J. Parkhill.** 2004. Complete genomes of two clinical *Staphylococcus aureus* strains: evidence for the rapid evolution of virulence and drug resistance. *Proc. Natl. Acad. Sci. USA* **101:**9786–9791.

26. **Hueck, C. J., W. Hillen, and M. H. Saier, Jr.** 1994. Analysis of a *cis*-active sequence mediating catabolite re- pression in gram-positive bacteria. *Res. Microbiol.* **145:** 503–518.

27. **Huynh, P. L., I. Jankovic, N. F. Schnell, and R. Brückner.** 2000. Characterization of an HPr kinase mutant of *Staph- ylococcus xylosus. J. Bacteriol.* **182:**1895–1902.

28. **Jankovic, I., and R. Brückner.** 2002. Carbon catabolite repression by the catabolite control protein CcpA in *Staphylococcus xylosus. J. Mol. Microbiol. Biotechnol.* **4:** 309–314.

29. **Jankovic, I., O. Egeter, and R. Brückner.** 2001. Analysis of catabolite control protein A-dependent repression in *Staphylococcus xylosus* by a genomic reporter gene system. *J. Bacteriol.* **183:**580–586.

30. **Jankovic, I., J. Meyer, and R. Brückner.** 2003. Catabolite control protein CcpA-dependent glucose repression in *Staphylococcus xylosus*: efficient activation of CcpA by glu- cose transported independently from the phosphotrans- ferase system, p. 141–146. *In* P. Dürre and B. Friedrich (ed.), *Regulatory Networks in Prokaryotes.* Horizon Scien- tific Press, Norfolk, Va.

31. Kloos, W. E., K.-H. Schleifer, and F. Götz. 1991. The genus *Staphylococcus*, p. 1369–1420. *In* A. Balows, H. G. Trüper, M. Dwarkin, W. Harder, and K. H. Schleifer (ed.), *The Procaryotes*. Springer Verlag, Heidelberg, Germany.

32. Knezevic, I., S. Bachem, A. Sickmann, H. E. Meyer, J. Stülke, and W. Hengstenberg. 2000. Regulation of the glucose-specific phosphotransferase system (PTS) of *Staphylococcus carnosus* by the antiterminator protein glcT. *Microbiology* **146:**2333–2342.

33. Knobloch, J. K., M. Nedelmann, K. Kiel, K. Bartscht, M. A. Horstkotte, S. Dobinsky, H. Rohde, and D. Mack. 2003. Establishment of an arbitrary PCR for rapid identification of Tn917 insertion sites in *Staphylococcus epidermidis*: characterization of biofilm-negative and nonmucoid mutants. *Appl. Environ. Microbiol.* **69:**5812–5818.

34. Kohlbrecher, D., R. Eisermann, and W. Hengstenberg. 1992. Staphylococcal phophoenolpyruvate-dependent phosphotransferase system: molecular cloning and nucleotide sequence of the *Staphylococcus carnosus* ptsI gene and expression and complementation studies of the gene product. *J. Bacteriol.* **174:**2208–2214.

35. Kravanja, M., R. Engelmann, V. Dossonnet, M. Bluggel, H. E. Meyer, R. Frank, A. Galinier, J. Deutscher, N. Schnell, and W. Hengstenberg. 1999. The hprK gene of *Enterococcus faecalis* encodes a novel bifunctional enzyme: the HPr kinase/phosphatase. *Mol. Microbiol.* **31:**59–66.

36. Kuroda, M., T. Ohta, I. Uchiyama, T. Baba, H. Yuzawa, I. Kobayashi, L. Cui, A. Oguchi, K. Aoki, Y. Nagai, J. Lian, T. Ito, M. Kanamori, H. Matsumaru, A. Maruyama, H. Murakami, A. Hosoyama, Y. Mizutani-Ui, N. K. Takahashi, T. Sawano, R. Inoue, C. Kaito, K. Sekimizu, H. Hirakawa, S. Kuhara, S. Goto, J. Yabuzaki, M. Kanehisa, A. Yamashita, K. Oshima, K. Furuya, C. Yoshino, T. Shiba, M. Hattori, N. Ogasawara, H. Hayashi, and K. Hiramatsu. 2001. Whole genome sequencing of methicillin-resistant *Staphylococcus aureus*. *Lancet* **357:**1225–1140.

37. Maxwell, E. S., K. Kurahashi, and H. M. Kalckar. 1962. Enzymes of the Leloir pathway. *Methods Enzymol.* **5:**174–189.

38. Mijakovic, I., S. Poncet, A. Galinier, V. Monedero, S. Fieulaine, J. Janin, S. Nessler, J. A. Marquez, K. Scheffzek, S. Hasenbein, W. Hengstenberg, and J. Deutscher. 2002. Pyrophosphate-producing protein dephosphorylation by HPr kinase/phosphorylase: a relic of early life? *Proc. Natl. Acad. Sci. USA* **99:**13442–13447.

39. Novick, R. P., S. J. Projan, J. Kornblum, H. F. Ross, G. Ji, B. Kreiswirth, F. Vandenesch, and S. Moghazeh. 1995. The agr P2 operon: an autocatalytic sensory transduction system in *Staphylococcus aureus*. *Mol. Gen. Genet.* **248:**446–458.

40. Oskouian, B., and G. C. Stewart. 1990. Repression and catabolite repression of the lactose operon of *Staphylococcus aureus*. *J. Bacteriol.* **172:**3804–3812.

41. Poolman, B., J. Knol, C. van der Does, P. J. F. Henderson, W.-J. Liang, G. Leblanc, T. Pourcher, and I. Mus-Veteau. 1996. Cation and sugar selectivity determinants in a novel family of transport proteins. *Mol. Microbiol.* **19:**911–922.

42. Postma, P. W., J. W. Lengeler, and G. R. Jacobson. 1993. Phosphoenolpyruvate: carbohydrate phosphotransferase systems of bacteria. *Microbiol. Rev.* **57:**543–594.

43. Reizer, J., C. Hoischen, F. Titgemeyer, C. Rivolta, R. Rabus, J. Stülke, D. Karamata, M. H. Saier, Jr., and W. Hillen. 1998. A novel protein kinase that controls carbon catabolite repression in bacteria. *Mol. Microbiol.* **27:**1157–1169.

44. Reizer, J., S. L. Sutrina, M. H. Saier, Jr., G. C. Stewart, A. Peterkofsky, and P. Reddy. 1989. Mechanistic and physiological consequences of HPr(ser) phosphorylation on the activities of the phosphoenolpyruvate:sugar phosphotransferase system in gram-positive bacteria: studies with site-specific mutants of HPr. *EMBO J.* **8:**2111–2120.

45. Rosenstein, R., and F. Götz. Unpublished results.

46. Rosey, E. L., B. Oskouian, and G. C. Stewart. 1991. Lactose metabolism by *Staphylococcus aureus*: characterization of lacABCD, the structural genes of the tagatose 6-phosphate pathway. *J. Bacteriol.* **173:**5992–5998.

47. Saier, M. H., Jr., S. Chauvaux, G. M. Cook, J. Deutscher, I. T. Paulsen, J. Reizer, and J.-J. Ye. 1996. Catabolite repression and inducer control in Gram-positive bacteria. *Microbiology* **142:**217–230.

48. Saier, M. H., Jr., S. Chauvaux, J. Deutscher, J. Reizer, and J.-J. Ye. 1995. Protein phosphorylation and regulation of carbon metabolism in gram-negative versus gram-positive bacteria. *Trends Biochem. Sci.* **20:**267–271.

49. Schleifer, K. H., and U. Fischer. 1982. Description of a new species of the genus *Staphylococcus*: *Staphylococcus carnosus*. *Int. J. Syst. Bacteriol.* **32:**153–156.

50. Sizemore, C., E. Buchner, T. Rygus, C. Witke, F. Götz, and W. Hillen. 1991. Organization, promoter analysis and transcriptional regulation of the *Staphylococcus xylosus* xylose utilization operon. *Mol. Gen. Genet.* **227:**377–384.

51. Sizemore, C., B. Wieland, F. Götz, and W. Hillen. 1992. Regulation of *Staphylococcus xylosus* xylose utilization genes at the molecular level. *J. Bacteriol.* **174:**3042–3048.

52. Stülke, J., M. Arnaud, G. Rapoport, and I. Martin-Verstraete. 1998. PRD—a protein domain involved in PTS-dependent induction and carbon catabolite repression of catabolic operons in bacteria. *Mol. Microbiol.* **28:**865–874.

53. Stülke, J., I. Martin-Verstraete, M. Zagorec, M. Rose, A. Klier, and G. Rapoport. 1997. Induction of the *Bacillus subtilis* ptsGHI operon by glucose is controlled by a novel antiterminator, GlcT. *Mol. Microbiol.* **25:**65–78.

54. Valentin-Hansen, P., P. Hojrup, and S. Short. 1985. The primary structure of the DeoR repressor from *Escherichia coli* K-12. *Nucleic Acids Res.* **13:**5927–5936.

55. Wagner, E., F. Götz, and R. Brückner. 1993. Cloning and characterization of the scrA gene encoding the sucrose-specific enzyme II of the phosphotransferase system from *Staphylococcus xylosus*. *Mol. Gen. Genet.* **241:**33–41.

56. Wagner, E., S. Marcandier, O. Egeter, J. Deutscher, F. Götz, and R. Brückner. 1995. Glucose kinase-dependent catabolite repression in *Staphylococcus xylosus*. *J. Bacteriol.* **177:**6144–6152.

57. Weickert, M. J., and S. Adhya. 1992. A family of bacterial regulators homologous to Gal and Lac repressors. *J. Biol. Chem.* **267:**15869–15874.

58. Zhang, Y. Q., S. X. Ren, H. L. Li, Y. X. Wang, G. Fu, J. Yang, Z. Q. Qin, Y. G. Miao, W. Y. Wang, R. S. Chen, Y. Shen, Z. Chen, Z. H. Yuan, G. P. Zhao, D. Qu, A. Danchin, and Y. M. Wen. 2003. Genome-based analysis of virulence genes in a non-biofilm-forming *Staphylococcus epidermidis* strain (ATCC 12228). *Mol. Microbiol.* **49:**1577–1593.

Respiration and Small-Colony Variants of *Staphylococcus aureus*

RICHARD A. PROCTOR

35

STAPHYLOCOCCAL RESPIRATION

Staphylococcus aureus respiration and carbohydrate metabolism follow patterns very similar to those in *Escherichia coli* (3), with glycolysis in staphylococci following the nearly universal patterns found in all biological systems. Glycolysis results in the formation of some ATP. However, under aerobic conditions, much more ATP is formed when the pyruvate formed from the breakdown of glucose or fructose enters the citric acid cycle, where it is oxidized to CO_2 and H_2O. Similarly, lipids undergo β-oxidation to produce acetyl-CoA for the Krebs cycle; this also produces large quantities of NADH and ATP. The oxidation of these substrates requires that electrons flow through the electron transport chain. The reducing equivalents (electrons plus hydrogen ions) result from the oxidation of carbohydrates, amino acids, and lipids with the simultaneous reduction of NAD and FAD to NADH and $FADH_2$, respectively. Electrons flow from NADH and $FADH_2$ to menaquinone and cytochromes (Fig. 1) as these substrates are oxidized. The synthesis of ATP associated with electron transport occurs in the large transmembrane F_0F_1 ATPase (22, 23, 29, 44, 62). In *S. aureus*, there is only one quinone, menaquinone; in *E. coli*, menaquinone is used for aerobic electron transport, while fumarate and ubiquinone are involved in anaerobic metabolism (8, 9, 15). Glucose and fructose follow the glycolytic pathway to produce ATP when electron transport is interrupted. Uptake of glucose, mannitol, mannose, lactose, sucrose, maltose, and glycerol requires ATP and complex regulation via the phosphoenolpyruvate transferase system (59), thus the net quantity of ATP production is small when compared to electron transport-generated ATP production.

In addition to energy balance, redox and pH balance must also be maintained. The Krebs cycle uses the largest quantity of NAD in the cell and releases large quantities of NADH. When NADH levels are high, several Krebs enzymes are inactivated by NADH through negative allosteric regulation, e.g., aconitase and succinyl dehydrogenase (68, 69). In addition, entry of acetyl groups into the Krebs cycle is also decreased because one of the pyruvate dehydrogenase complex

enzymes is susceptible to allosteric regulation by NADH, and it is involved in acetyl-CoA production. Thus, when electron transport is interrupted, levels of NADH rise, which shuts down the Krebs cycle. The importance of this allosteric regulation is emphasized by our recent observation that mannose, but not mannitol, will stimulate growth of electron transport-deficient *S. aureus* (75). Mannose and mannitol are metabolized to fructose-6-phosphate, and this leads to ATP production via glycolysis. Consequently, the difference between these compounds is not due to differences in net ATP production. However, in the first metabolic step in mannitol metabolism, NAD is reduced to NADH as mannose is produced from mannitol, thereby increasing NADH levels. High NADH levels slow growth of respiratory-deficient bacteria. Reactions that are favored in respiratory-deficient small-colony variants (SCVs) are shown in Table 1, and it is apparent that these variants are optimizing reactions that produce ATP and oxidize NADH by non-oxygen-dependent mechanisms. In addition, because of decreased Krebs cycle activity in SCVs, large amounts of lactic acid cannot be metabolized (38); therefore, reactions that alkalinize the medium are favored, e.g., the arginine deiminase pathway. This pathway has the added benefit of producing ATP.

Respiratory-defective mutants of *S. aureus* have been associated with unusually persistent, recurrent, and antibiotic-resistant infections (1, 33, 34, 37, 47, 49, 50, 53–56, 60, 61, 64–66, 76, 77, 79, 81). These respiratory variants have been found to have defects in hemin and menaquinone biosynthesis, resulting in defective electron transport and reduced quantities of ATP (Fig. 1). Because the majority of the ATP produced by *S. aureus* goes toward cell wall biosynthesis, interruption of electron transport leads to slow growth and small colonies; hence, they are classified as SCVs. Although mutations in any essential gene can cause slow growth and small colonies, this chapter considers a subset of these mutants that was first observed in patients who had unusually persistent infections and were later found to have defects in oxidative metabolism and electron transport. The SCV acronym as used here refers only to slow-growing mutants defective in electron transport. These electron-transport variants produce reduced amounts

Gram-Positive Pathogens, 2nd edition, edited by Vincent A. Fischetti et al.
© 2006 ASM Press, Washington, D.C.

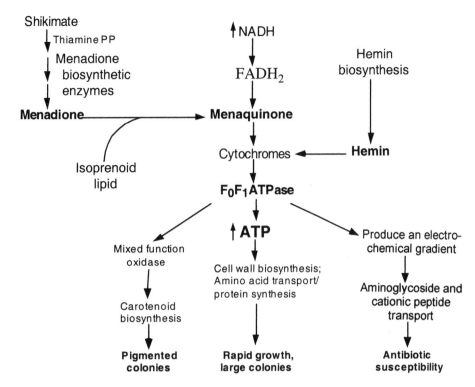

FIGURE 1 Relationship between electron transport and the small-colony variant phenotype in *S. aureus*.

of lytic toxins and are able to persist within host cells (4, 6, 33, 34, 38, 49, 50, 53–56, 60, 64–66, 74, 76–79, 81).

HISTORY OF SMALL-COLONY VARIANTS

The first report of SCVs was in 1910 and concerned *Salmonella typhi* (19). Initially, these were called gonidial mutants, a term that was abbreviated to G forms or colonies (5, 27). The idea that G forms were part of the natural growth cycle of bacteria arose because they were frequently found among many bacterial genera (49). Further work proved that they were not part of the growth cycle when it was demonstrated that many of these variants readily reverted to the large-colony form and that their small colony size related to more exacting metabolic requirements (reviewed in references 49, 50, 54–56). Subsequent studies revealed that a variety of host or environmental factors could select for mutants with this phenotype. Studies dating from the 1930s and 1940s reported decreased respiratory rates, restricted carbohydrate utilization, low ATP levels, reduced fermentation rates, and diminished dye reduction in SCVs from multiple bacterial species (49, 50, 54–56). It was also observed by a number of early investigators that SCVs produced no pigment and were auxotrophic for hemin or menadione. All of the phenotypic changes that were reported for these SCVs can be related to an interruption of electron transport (4).

MULTIPLE PHENOTYPIC CHANGES ASSOCIATED WITH INTERRUPTED ELECTRON TRANSPORT

A comparison of staphylococcal SCVs with parental wild type reveals a highly characteristic phenotype. A large number of phenotypic changes are apparent—small, non-pigmented, nonhemolytic colonies with organisms that fail to ferment mannitol and are resistant to aminoglycoside antibiotics. All aspects of this unique phenotype can be attributed to the loss of one or another of the enzymes critical to the biosynthesis of electron transport chain components (Fig. 1).

The initial interest in electron transport arose because many of the clinical and laboratory isolates of *S. aureus* SCVs were found to be hemin or menadione auxotrophs (reviewed in references 49, 50, 54–56). Failure to produce hemin blocks the synthesis of cytochromes. Because this is often the result of mutations in genes controlling porphyrin biosynthesis, the SCV phenotype can be reversed by giving hemin or its precursors. *S. aureus* is easier to study than *E. coli* because it is permeable to hemin (32). Hemin auxotrophy may also result from the inability to insert ferric iron into the pyrrole structure used for cytochrome biosynthesis, but, interestingly, such auxotrophs can insert ferrous iron into catalase, suggesting that the defect is in the ferrochelatase (36). Similarly, mutations in the menadione biosynthetic genes result in a loss of menaquinone. Adding hemin or menadione completely reverses the SCV phenotype in clinical isolates, laboratory-generated strains, and site-directed *hemB* and *menD* mutants (4, 6, 33, 34, 37, 38, 49, 50, 54–56, 65, 66, 74, 76–79). Anaerobic growth also reduplicates the SCV phenotype in *S. aureus* (4, 55), which is not surprising because menaquinone is not synthesized under anaerobic conditions (24). In contrast, *E. coli* has two quinones: menaquione is produced under anaerobic conditions, whereas ubiquinone is used when oxygen is present (8). Consequently, anaerobic growth in *E. coli* does not reduplicate the SCV phenotype in *E. coli*, but *E. coli* SCVs do grow slowly and are more resistant to aminoglycosides (21).

TABLE 1 Redox and ATP-producing reactions in SCVs

Reaction	Enzyme[a]	Gene
Favorable		
(1) Pyruvate + NADH = lactate + NAD$^+$	LDH	lctE
(2) Pyruvate + NO$_2$ + NADH = NH$_4$ + NAD$^+$	HMP	hmp
(3) Pyruvate = acetate aldehyde + NADH = ethanol + NAD$^+$	PdhB	pdhB + SA0182[b]
	ADH	adhE
(4) Phospho-enolpyruvate + CO$_2$ + ADP = ATP + oxaloacetate	PycA	pycA
(5) Oxaloacetate + NADH = malate + NAD$^+$	Mqo2	mqo2
(6) Arginine = citruline + NH$_4$	ArcA	arcA
citruline = ornithine + carbamoy-P	ArcB	arcB
carbamoy-P + ADP = ATP + NH$_4$	ArcC	arcC
(7) Pyruvate = formate = CO$_2$	PflB/Fdh	pflB/fdh
(8) Acetoacetyl-CoA + NADH =	FadB	SA0224[b]
3-OH-butanoyl-CoA + NAD$^+$		
(9) Glucose or fructose + ADP = 3 carbon + ATP	Glycolytic enzymes	
Unfavorable		
(1) Mannitol-1-P + NAD$^+$ = mannose-6-P + NADH	MltD	mltDA
(2) Acetyl-CoA to citrate via Krebs cycle yields	CitB	citB
4 NADH + FAHD$_2$ + GTP	Other Krebs cycle enzymes	
(3) β-oxidation of fatty acid = 1 NADH/acetyl group	Fatty acid oxidation enzymes	

[a]Abbreviations: ADH, alcohol dehydrogenase; ArcA = arginine deiminase; ArcB = ornithine transcarbamolyase; ArcC = carbamoylate kinase; CitB = aconitase; FadB = 3-hydroxyacyl-CoA dehydrogenase: HMP, hydroxymonocarboxylic acid phosphatase (flavohemoprotein or pyruvate/nitrite reductase); LDH = lactate dehydrogenase; MltD = mannitol-1-phosphate 5-dehydrogenase; Mqo2 = malate:quinone oxidoreductase; PycA = pyruvate carboxylase; PdhB = pyruvate dehydrogenase El component beta subunit; PflB = pyruvate formate lyase.

[b]SA0182 and SA0224 refer to gene numbers in the S. aureus N315 genome.

Slow growth may occur due to several factors when electron transport is inoperative. One direct effect of interrupting electron transport is reduced synthesis of teichoic acid, because the biosynthesis of phosphatidyl glycerol, a teichoic acid precursor, requires electron transport (26, 36, 40). Also, ATP is required for many cellular functions, such as cell wall biosynthesis and uptake of amino acids and carbohydrates; hence, organisms interrupted in electron transport have a lower capacity to produce ATP and show slower growth, resulting in small colonies. We found that the interdivision time for S. aureus 6850, a clinical isolate (parent strain), was 20 min versus 180 min for S. aureus JB-1, a menadione auxotrophic SCV, when grown in tryptic soy broth (4). Hale et al. (27) found similar growth rates with an S. aureus parent versus SCV (20 min versus 120 min), respectively, in nutrient broth. Massey and Peacock found division times of five independent strains to be 20 min versus 210 min in SCVs selected by gentamicin exposure (41, 42). In contrast, a hemin auxotrophic SCV had a 131- to 173-min generation time versus its parent's time of 53 min (40). Although there are clearly differences in growth rates of several respiratory-defective strains of S. aureus, all of these variants produce small colonies and fulfill the definition of an SCV, i.e., an organism that forms colonies that are 10-fold smaller than its parent.

The failure of SCVs to ferment carbohydrates, such as mannitol, can also be related to defects in electron transport (49, 50, 54–56, 75). As noted above, interruption of electron transport results in the accumulation of NADH, which down-regulates the citric acid cycle enzymes, thereby making the fermentation of mannitol and other such carbohydrates unfavorable.

Electron transport is also required for carotenoid biosynthesis. Electrons are shuttled through the electron transport system to the mixed function oxidase system, which is linked to pigment formation (31). Hence, SCVs that are defective in electron transport form nonpigmented colonies.

Resistance of SCVs to aminoglycosides and other antibacterial substances is a further consequence of defects in electron transport (5, 7, 39, 40–43, 74, 76–78, and reviewed in reference 50). A transmembrane potential (negative on the inside) is formed as electrons flow through the F$_0$F$_1$ ATPase (7, 22, 23, 29, 39, 44, 62). This charge gradient enhances the accumulation of aminoglycosides that are positively charged and associate with the negatively charged S. aureus membrane when a sufficiently large electrochemical gradient is present (18, 25, 40, 50). Consequently, menadione and hemin auxotrophs are readily recovered when S. aureus is exposed to gentamicin at relatively high concentrations (four to eight times the MIC) (4, 41–43, 50) because they are resistant to the drug due to decreased uptake. In addition, S. aureus SCVs create a highly acidic pH in the culture medium (7, 38), which greatly decreases the activity of gentamicin against S. aureus (41, 42). Some lantibiotics and host cationic proteins are most active when the bacterial membrane has a strongly negative transmembrane potential (7, 12, 39, 49, 50, 54, 60). Thus, the development of electron transport-defective S. aureus SCVs in the absence of antibiotic pressure might be explained by selection for resistance to positively charged antibacterial host peptides or proteins (7, 12, 39, 49, 50, 54, 60, 73). Only those compounds requiring a large membrane electrochemical gradient for uptake will be affected, because 60 to 70% of the normal membrane potential in S. aureus can be established in the absence of electron transport through ATP production via the glycolytic pathway when glucose or fructose is present (11, 22, 23, 25, 62). Glucose allows the F$_0$F$_1$ ATPase to break down ATP as protons are transported out of the cell, thereby producing a membrane potential (7, 22, 23, 29, 39, 44, 62). In contrast, daptomycin is a negatively charged antibiotic, and *S. aureus* increases its membrane potential to resist this antibiotic (67). One would predict that daptomycin would not induce electron transport SCVs.

Other early observations about the SCV phenotype, such as a decreased ability to reduce dyes and to utilize oxygen by *S. aureus* SCVs, are also consistent with decreased electron transport (49, 50, 54–56, 75). Thus, most of the phenotypic features reported for clinical and laboratory-generated *S. aureus* SCVs can be explained by the model shown in Fig. 1.

OTHER CONSIDERATIONS ABOUT ELECTRON TRANSPORT AND SCV

Some *S. aureus* SCVs have primary defects in unsaturated fatty acid biosynthesis, and this also may cause defects in electron transport (26, 36, 72). Menaquinone is formed from menadione plus a repeating isoprenoid (usually eight repeating units in *S. aureus*) (26, 72). This lipid tail is required for association with the bacterial membrane, and the cytochromes subsequently nucleate around the quinone to form the electron transport chain (24). A similar situation was seen with some serovar Typhimurium SCVs, which were able to form normal colonies when the medium was supplemented with oleic acid (82). Hence, some *S. aureus* SCVs may be electron transport-deficient because they cannot synthesize unsaturated fatty acids that are used in the biosynthesis of menaquinone.

Inhibition or mutation of the membrane complex that synthesizes ATP and creates the membrane potential, the F_0F_1 ATPase, would also be expected to produce SCVs. Mutations in the genes involved in the biosynthesis of the F_0F_1 ATPase in several bacterial species result in the formation of SCVs (22, 23, 29, 44, 62). By using Tn551 mutagenesis in the presence of hemin, menadione, and gentamicin, we isolated *S. aureus* mutants that show a typical SCV phenotype with reduced ATP production on minimal medium and increased resistance to gentamicin, but no auxotrophy for menadione or hemin (unpublished data). Menadione and hemin auxotrophs were eliminated because they are susceptible to gentamicin under these conditions. These *S. aureus* mutants resemble F_0F_1 ATPase mutants in other species showing increased oxygen utilization. High oxygen use is due to increased activity of dehydrogenases and type b cytochromes, which remove reducing equivalents formed during glycolysis (29, 31, 62). Other non-oxygen-dependent enzymes are increased as well to oxidize NADH, e.g., menaquinol NADH oxidoreductase, lactate dehydrogenase, and alcohol dehydrogenase (38, 75). In addition, a mutation in *ctaA*, heme A synthase, also shows many features typical of the SCV phenotype (13). We have also found that bacterial electron transport inhibitors Z69 and Z90 are able to reproduce the SCV phenotype (52). These observations give further support for the model in Fig. 1. Stable *S. aureus* SCVs with metabolic characteristics suggestive of electron transport variants have been recovered by growth in medium containing lithium chloride (83). Although other salts (e.g., barium nitrite) selected for SCVs, these strains were very unstable. This is of some interest because lithium chloride is known to interfere with the F_0F_1 ATP-like ATPase in *Ilyobacter tartaricus* (44). Perhaps, lithium chloride damages organisms with an intact electron transport chain, hence selecting for SCVs.

S. aureus subspecies *anaerobius*, an animal pathogen, is commonly isolated from abscesses in sheep, forms small colonies, and is deficient in porphyrin synthesis, but its characteristics are different from those of *S. aureus* SCVs (17, 57). This species is unable to grow in an aerobic environment, fails to produce catalase when supplemented with hemin, and produces a violet pigment. This organism has been isolated only once from humans (39), but its distinctive characteristics should readily separate it from *S. aureus* SCVs.

WELL-CHARACTERIZED HEMIN- OR MENAQUINONE-DEFICIENT *S. AUREUS* STRAINS

S. aureus JB-1 is a menadione auxotrophic SCV that was selected with gentamicin from its parent, *S. aureus* 6850, an organism isolated from a patient with bacteremia and multiple metastatic sites of infection (4). Strain 6850 is a particularly stable strain with regard to its production of toxins, pigment, and exoproteins. *S. aureus* JB-1 produces less ATP, transmembrane potential, pigment, alpha-toxin, and coagulase; fails to ferment mannitol; uses less oxygen than its parent strain; is more resistant than its parent to aminoglycosides and to lantibiotics that require a large membrane potential for uptake; is able to survive on biomaterial surfaces despite massive concentrations of antibiotics; and is able to persist within cultured endothelial cells (4, 7, 12, 39, 53). This complete SCV phenotype is reversed by the addition of menadione or o-succinylbenzoate but not by an earlier menadione biosynthetic precursor, chorismate (4), suggesting that it has a mutation in the menadione pathway prior to the o-succinylbenzoate intermediate.

Some hemin auxotrophs selected by resistance to gentamicin have also been characterized in detail (4, 40). These fail to ferment lactose and mannitol; have a uniform ability to grow on glucose, but not on oxidizable but nonfermentable substrates such as succinate, α-ketoglutarate, lactate, or malate; are negative for pigment formation, DNase activity, beta-hemolysis, catalase, benzidine reduction (suggesting the absence of cytochromes), and tetrazolium dye reduction (an assay for electron transport activity); show little O_2 utilization during exponential phase growth; accumulate lactate; have increased pyruvate kinase and phosphohexose isomerase (suggesting the use of glycolysis for growth); have decreased uptake of tritiated tobramycin but are susceptible to novobiocin and chloramphenicol; and persist within cultured endothelial cells (4, 40). The SCV phenotype is reversed by hemin and in some cases hemin precursors such as Δ-aminolevulinic acid. Some of these SCVs are gram-negative and coagulase-negative (40), but this is not uniformly found (4, 35, 49, 50, 53–56).

Standard transformational and linkage analysis of known mutations (auxotrophies) of the hemin biosynthetic pathway allowed mapping of the hemin biosynthetic pathway in *S. aureus* in 1968 (72). These mutants were selected by resistance to kanamycin, and they grew as small colonies. More recently, it was found that mutations in this pathway produce the unique SCV phenotype (4, 40), as do transposon mutations (32). Nevertheless, the fact that single point mutations could have such a profoundly pleiotropic effect was not generally accepted until one of the *hem* genes, *hemB*, encoding Δ-aminolevulinic acid dehydratase, had been cloned, sequenced, and insertionally inactivated (78). The *S. aureus hemB* knockout mutant had the entirely predictable and complete SCV phenotype (4, 32) as described above, which was complemented by the cloned *hemB* and reversed by the addition of hemin to the culture medium.

The *S. aureus hemB* is quite closely related to the corresponding gene from other species (*Bacillus subtilis*, 62% identity; *E. coli*, 55%; mouse, 54%; pea, 55%; and spinach, 54%) and can complement the *E. coli hemB* mutant (32, 78). The *hemB* (porphobilinogen synthase) mutant behaves like clinical SCVs in that it persisted within cultured endothelial cells because it enters these cells, but it does not lyse them due to decreased alpha-toxin production (78). Both the protein and mRNA for alpha-toxin are reduced. Complementation of the mutant in *trans* or addition of hemin reversed the phenotype and allowed the organism to lyse the endothelial cells. Further characterization of the *hemB* mutant has provided further information about this mutant. An extensive proteomic and biochemical analysis shows that a *hemB* mutant increases pathways that are associated with NADH oxidation by non-electron transport-dependent mechanisms and decreased production of protein A, coagulase, and hemolysins, but increased production of proteases (38). Examination of the *hemB* mutant in an arthritis model demonstrated that it was able to produce similar joint damage to the parent strain (30), probably due to increased expression of matrix adhesins (73).

We have also created a *menD* mutant and found that it also reproduces the clinical phenotype seen with SCVs (6). The *hemB* and *menD* mutants were examined in the rabbit endocarditis model (6). Both mutants, despite their slow growth, were able to colonize the damaged heart valves, which is consistent with their increased expression of fibronectin-binding protein (73). Also, the mutants showed a trend toward increased resistance to antibiotic killing on the heart valve, but this did not reach a statistical difference (6). However, the *menD* mutant was found to persist in tissues despite antibiotic therapy. The inability of the *hemB* mutant to persist was attributed to its ability to obtain heme from the lysed erythrocytes within the emboli from the heart valve. These data are consistent with our clinical studies, wherein hemin auxotrophic strains are more frequently isolated from infections on relatively avascular tissues, e.g., bone or pulmonary mucosal surfaces, whereas menadione auxotrophs are found equally distributed in all tissues.

CLINICAL STUDIES OF SCVs

In recent studies, *S. aureus* SCVs have been shown to produce persistent, recurrent, and antibiotic-resistant infections (1, 2, 33, 34, 37, 43, 51, 55, 60, 61, 65, 66, 70, 76, 77, 79). Patients infected with SCVs often have a history of long-term treatment with antibiotics given either parenterally (especially aminoglycosides) or locally in the form of antibiotic-impregnated beads. They may have prolonged disease-free intervals (as long as 53 years), but during recurrences, they show unusual resistance to apparently active antibiotic therapy (1, 2, 33, 34, 37, 43, 51, 55, 60, 61, 65, 66, 70, 76, 77, 79). The organism may be misidentified as viridans streptococci or coagulase-negative staphylococci (35, 50, 51, 55). These studies are consistent with laboratory models and clinical anecdotes. For example, persistence of SCVs occurred in the kidneys of mice despite increased susceptibility to complement (81), and one of the first reported persistent and recurrent human SCV cases occurred in a patient treated with penicillin and streptomycin whose *S. aureus* aortic valve endocarditis had a late recurrence with the same strain (47). The ability of SCVs to persist within host cells helps to shield them from some antibiotics (4, 34, 76). Taken together, SCVs present a challenge for treatment because the intracellular location,

slow growth, and decreased antibiotic uptake reduce the efficacy of antibiotic treatment (1, 2, 14, 16, 33, 34, 37, 40, 43, 51, 55, 60, 61, 65, 66, 70, 76, 77, 79). Placement of gentamicin beads for the treatment of *S. aureus* osteomyelitis may select for SCVs that are likely to produce infections that are recurrent and difficult to cure with antibiotics (77).

The clinical microbiologist must be alerted to look for SCVs, because their slow growth means that they may either be missed entirely or may be overgrown by more rapidly dividing strains or revertants. Furthermore, they are easily misidentified because of their unusual colony morphology and biochemical profile (34, 35, 55, 66, 76, 79). Missing an SCV in a mixed culture may have serious consequences because it is more antibiotic resistant than the parent strain, leading to a major reporting error if it is not identified. Also, failure to identify the SCVs as a variant subpopulation may cause the clinician to believe that a new, rather than recurrent, infection is present, thereby hiding an antibiotic failure. Care should be used when selecting media, as some contain high concentrations of menadione and/or hemin (e.g., brain heart infusion and Schaedler's broths), whereas other media will allow the expression of the menadione and hemin auxotrophic SCV phenotype (tryptic soy and Mueller Hinton broths are low in hemin and menadione).

More recent clinical reports are consistent with the animal model data for the *hemB* knockout wherein the SCV shows pathogenic potential. This concept has evolved from the idea that SCVs produced only persistent, but innocuous, infections. A report by Salgado et al. showed that *S. aureus* SCVs in an intensive care unit setting can be highly virulent and difficult to clear by intensive antibiotic therapy (61). Similarly, occurrences of very persistent and aggressive infections have been reported in brain and hip abscesses, wounds, endocarditis, prostheses, joints, and skin in normal hosts and patients with AIDS (2, 37, 64, 65, 70, 79). *S. aureus* SCVs are particularly prevalent in the sputums of patients with cystic fibrosis (33, 34, 76). SCVs appear to play a role in the persistent colonization of cystic fibrosis patients and may produce increased severity and rate of progression of lung dysfunction (33, 34).

MENADIONE AND HEMIN AUXOTROPHS PREDOMINATE AMONG CLINICAL ISOLATES

Although many different mutations might produce slow growth, menadione and hemin auxotrophs predominate among clinical *S. aureus* SCV isolates (49, 50, 52–56). There may be several reasons for this. First, the mutation must result in a variant that has some survival advantage. Mutations causing defects in electron transport produce variants that are more resistant to positively charged antimicrobials, that are resistant to host cationic peptides, and that are able to persist within host cells because of reduced toxin production (discussed below). Second, the mutation must not produce a defect that can be readily reversed by compounds naturally present in the host. Both menadione and hemin are at low concentrations in mammalian cells, whereas other substances, e.g., tryptophan, would be supplied by the intracellular milieu. Third, the mutation must not be lethal to the microorganism. Because *S. aureus* can survive anaerobically, electron transport is not absolutely essential for survival. Taken together, these conditions favor the selection of mutants in menadione and hemin biosynthesis. SCVs can readily be selected at exceptionally high rates, i.e., $1/10^3$, under laboratory conditions by culturing within endothelial cells (76) or by exposure to gentamicin (41, 42).

INSTABILITY OF THE SCV PHENOTYPE: A REQUIREMENT FOR REVERSION TO A TOXIN-PRODUCING PHENOTYPE

One of the earlier observations concerning SCVs from clinical specimens was that the phenotype was often unstable (49, 54). For descriptive purposes, we consider strains "highly unstable" when they lose their SCV phenotype on the first passage (common), "stable" when the phenotype is maintained through 10 passes on solid medium (unusual), and "highly stable" when it can be repeatedly passed (rare). The instability of SCVs has made studying these variants difficult.

Slow-growing variants are often seen following penicillin treatment; however, these strains have been proved to be quite unstable (64). In contrast, streptomycin-selected SCVs are stable and have characteristics very similar to those of the now well-characterized SCVs that are defective in electron transport (5). The precise defects in penicillin-selected SCVs have not been defined, but the aminoglycoside-resistant SCVs are due to mutations in genes relating to the biosynthesis of components used in the electron transport chain. The mechanism behind the instability is unknown.

SCVs AND ANTIBIOTIC RESISTANCE

S. aureus SCVs are more resistant to aminoglycosides and cationic proteins such as thrombin-induced platelet microbicidal protein or nisin (7, 25, 39, 40–43, 50, 51, 60). Also, the relatively slow growth of SCVs makes them less susceptible to β-lactam antibiotics (10, 20, 21, 50, 64, 80). The level of resistance may be as small as two to three doubling dilutions in MIC testing, but it can be as high as 20-fold increased (42). However, due to intracellular location of the SCVs and the relatively lower concentrations of antibiotics within the intracellular compartment, this reaches clinically relevant resistance. One particularly intriguing and worrisome observation was that even when the SCV state was not more stable, a transient SCV state was needed for the development of fluoroquinolone resistance in *S. aureus* (48). A major concern is that the SCV state might represent a stage when the organism was able to undergo more frequent mutations. One of the major systems for correcting mutations is the RecA system, which requires ATP to bind single-stranded DNA. Because SCVs have low levels of ATP, it is plausible that the RecA system may not be functioning optimally. Thus, SCVs are not only more resistant to antimicrobial agents, but they may be involved in the development of resistance.

When bacteria adhere to biomaterial surfaces, they undergo dramatic metabolic changes such as (i) slow growth, (ii) decreased oxidative metabolism, and (iii) enhanced resistance to antibiotics (10, 12, 21, 28, 80). Although SCVs and surface-adherent bacteria show similar metabolic profiles (49), the SCVs become even more resistant to antibiotics when surface adherent. Indeed, surface-adherent *S. aureus* SCVs that enter into stationary phase when in a biofilm become resistant to antibiotics at MIC >1,000 times the parent strain under the same conditions (12). Hence, biofilm SCVs are virtually untreatable, and this may represent part of the reason for antibiotic failure to clear device-related *S. aureus* infections.

ELECTRON TRANSPORT, RESPIRATION, AND TOXIN PRODUCTION IN *S. AUREUS*

Connections between toxin production and environmental conditions have been noted in *S. aureus* for many years. For example, anaerobic growth results in markedly decreased hemolysis and results in the typical SCV phenotype (4). In addition, exogenous glucose, nonacidic pH, or Mg^{2+} at >10 to 12 μg/ml results in decreased carotenoid pigment, alpha-toxin, and toxic shock syndrome toxin production (45, 58). A consistent reduction in alpha-toxin production is seen in clinical *S. aureus* SCVs (4, 33, 34, 37, 43, 55, 56, 60, 65, 66, 76) as well as in *hemB* mutants (6, 32, 38, 78). This occurs even though a *hemB* mutant is able to grow over seven orders of magnitude and enter into stationary phase, a situation in which *agr* is normally strongly activated and toxins are produced (38, 78). These data suggest a link between respiratory metabolism and toxin production. In support of this concept, we have found that two electron transport inhibitors (Z69 and Z90) at sub-MICs produce typical SCVs in *S. aureus* that are unable to hemolyze rabbit erythrocytes or lyse cultured endothelial cells (52).

The signaling pathway(s) that decrease toxin production (alpha-toxin, enterotoxins), but not other exoproteins such as proteases or fibronectin-binding protein (30, 75), are unknown. Clearly, the pattern of exoprotein production is very different from that seen with mutations in *sarA* or *agr* (see chapter 41, this volume). Changes in the redox state may influence some regulators that are sensitive to NADH, ATP, or GTP levels. Characterizing the signals in the pathways that link metabolism and toxin production may provide opportunities for modifying signaling with drugs to prevent disease, as has been suggested by work with glycerol monolaurate, wherein *agr* and the toxins it regulates are decreased (63) by the laurate portion of the molecule (46).

REFERENCES

1. **Abele-Horn, M., B. Schupfner, P. Emmerling, H. Waldner, and H. Goring.** 2000. Persistent wound infection after herniotomy associated with small-colony variants of *Staphylococcus aureus. Infection* **28:**53–54.
2. **Baddour, L. M., and G. D. Christensen.** 1987. Prosthetic valve endocarditis due to small-colony staphylococcal variants. *Rev. Infect. Dis.* **9:**1168–1174.
3. **Baldwin, J. E., and H. Krebs.** 1981. The evolution of metabolic cycles. *Nature* **291:**381–382.
4. **Balwit, J. M., P. van Langevelde, J. M. Vann, and R. A. Proctor.** 1994. Gentamicin-resistant, menadione and hemin auxotrophic *Staphylococcus aureus* persist within cultured endothelial cells. *J. Infect. Dis.* **170:**1033–1037.
5. **Barbour, R. G. H.** 1950. Small colony variants ("G" forms) produced by *Staphylococcus pyogenes* during the development of resistance to streptomycin. *Aust. J. Exp. Biol. Med. Sci.* **28:**411–421.
6. **Bates, D., C. von Eiff, G. Peters, A. Bayer, and R. A. Proctor.** 2003. A *menD*, but not *hemB*, mutant persists within the kidney in the rabbit endocarditis model. *J. Infect. Dis.* **187:**1654–1661.
7. **Baumert, N., C. von Eiff, F. Schaaff, G. Peters, R. A. Proctor, and H. G. Sahl.** 2002. Physiology and antibiotic susceptibility of *Staphylococcus aureus* small colony variants. *Microb. Drug Resist.* **8:**253–260.

8. **Bentley, R., and R. Meganathan.** 1982. Biosynthesis of vitamin K (menaquinone) in bacteria. *Microbiol. Rev.* **46:** 241–280.

9. **Bishop, D. H. L., K. D. Pandya, and H. K. King.** 1962. Ubiquinone and vitamin K in bacteria. *Biochem. J.* **83:** 606–614.

10. **Brown, M. R., P. J. Collier, and P. Gilbert.** 1990. Influence of growth rate on susceptibility to antimicrobial agents: modification of the cell envelope and batch and continuous culture studies. *Antimicrob. Agents Chemother.* **34:**1623–1628.

11. **Chinn, B. D.** 1936. Characteristics of small colony variants with special reference to *Shigella paradysenteriae sonne.* *J. Infect. Dis.* **59:**137–157.

12. **Chuard, C., P. E. Vaudaux, R. A. Proctor, and D. P. Lew.** 1997. Decreased susceptibility to antibiotic killing of small colony variants of *Staphylococcus aureus* in fluid phase and on fibronectin-coated surfaces. *J. Antimicrob. Chemother.* **39:**603–608.

13. **Clements, M. O., S. P. Watson, R. K. Poole, and S. J. Foster.** 1999. CtaA of *Staphylococcus aureus* is required for starvation survival, recovery, and cytochrome biosynthesis. *J. Bacteriol.* **181:**501–507.

14. **Clerch, B., E. Rivera, and M. Llagostera.** 1996. Identification of a pKM101 region which confers a slow growth rate and interferes with susceptibility to quinonolone in *Escherichia coli* AB1157. *J. Bacteriol.* **178:**5568–5572.

15. **Collins, M. D., and D. Jones.** 1981. Distribution of isoprenoid quinone structure types in bacteria and their taxonomic implications. *Microbiol. Rev.* **45:**316–354.

16. **Darouiche, R. O., and R. J. Hamill.** 1994. Antibiotic penetration of and bactericidal activity within endothelial cells. *Antimicrob. Agents Chemother.* **38:**1059–1064.

17. **de la Fuente, R., K. H. Schleifer, F. Götz, and H.-P. Köst.** 1986. Accumulation of porphyrins and pyrrole pigments by *Staphylococcus aureus* ssp. *anaerobius* and its aerobic mutant. *FEMS Microbiol. Lett.* **45:**183–188.

18. **Eisenberg, E. S., L. J. Mandel, H. R. Kaback, and M. H. Miller.** 1984. Quantitative association between electrical potential across the cytoplasmic membrane and early gentamicin uptake and killing in *Staphylococcus aureus.* *J. Bacteriol.* **157:**863–867.

19. **Eisenberg, P.** 1910. Untersuchungen über die variablen Typhusstamm (*Bacterium typhi* mutabile), sowie über eine eigentumliche hemmende Wirkung des gewöhnlichen Agar, verursacht durch Autoklavierung, abstr. 56. *Zentbl. Bakteriol. Abt.,* **1:**208.

20. **Eng, R. H. K., F. T. Padberg, S. M. Smith, E. N. Tan, and C. E. Cherubin.** 1991. Bactericidal effects of antibiotics on slowly and nongrowing bacteria. *Antimicrob. Agents Chemother.* **35:**1824–1828.

21. **Evans, D. J., D. G. Allison, M. R. W. Brown, and P. Gilbert.** 1991. Susceptibility of *Pseudomonas aeruginosa* and *Escherichia coli* biofilms toward ciprofloxacin: effect of specific growth rate. *J. Antimicrob. Chemother.* **27:**177–184.

22. **Fillingame, R. H.** 1997. Coupling H^+ transport and ATP synthesis in F_0F_1 ATP synthases: glimpses of interacting parts in a dynamic molecular machine. *J. Exp. Biol.* **200:** 217–224.

23. **Fillingame, R. H., P. C. Jones, W. Jiang, F. I. Valiyaveetil, and O. Y. Dmitriev.** 1998. Subunit organization and structure in the F_0 sector of *Escherichia coli* F_0F_1 ATP synthase. *Biochim. Biophys. Acta* **1365:**135–142.

24. **Frerman, F. E., and D. C. White.** 1967. Membrane lipid changes during formation of a functional electron transport system in *Staphylococcus aureus.* *J. Bacteriol.* **94:**1868–1874.

25. **Gilman, S., and V. Saunders.** 1986. Accumulation of gentamicin by *Staphylococcus aureus:* the role of transmembrane electrical potential. *J. Antimicrob. Chemother.* **17:** 37–44.

26. **Goldenbaum, P. E., and D. C. White.** 1974. Role of lipid in the formation and function of the respiratory system of *Staphylococcus aureus.* *Ann. N.Y. Acad. Sci.* **236:**115–123.

27. **Hale, J. H.** 1947. Studies on staphylococcal mutation: characteristics of the "G" (gonidial) variant and factors concerned in its production. *Br. J. Exp. Pathol.* **28:**202–210.

28. **Huang, C. T., F. P. Yu, G. A. McFeters, and P. S. Stewart.** 1995. Nonuniform spatial patterns of respiratory activity within biofilms during disinfection. *Appl. Environ. Microbiol.* **61:**2252–2256.

29. **Jensen, P. R., and O. Michelsen.** 1992. Carbon and energy metabolism of *atp* mutants of *Escherichia coli.* *J. Bacteriol.* **174:**7635–7641.

30. **Jonsson, I.-M., C. von Eiff, R. A. Proctor, G. Peters, C. Ryden, and A. Tarkowski.** 2003. Virulence of a *hemB* mutant *Staphylococcus aureus* small colony variant in a murine model of septic arthritis. *Microbial Pathog.* **34:**73–79.

31. **Joyce, G. H., and D. C. White.** 1971. Effect of benso(a) pyrene and piperonyl butoxide on formation of respiratory system, phospholipids, and carotenoids of *Staphylococcus aureus.* *J. Bacteriol.* **106:**403–411.

32. **Kafala, B., and A. Sasarman.** 1994. Cloning and sequence analysis of the *hemB* gene of *Staphylococcus aureus.* *Can. J. Microbiol.* **40:**651–657.

33. **Kahl, B, A. Duebbers, G. Lubritz, J. Haeberle, H. G. Koch, B. Ritzerfeld, M. Reilly, E. Harms, R. A. Proctor, M. Herrmann, and G. Peters.** 2003. Population dynamics and persistence of *Staphylococcus aureus* infection in cystic fibrosis patients. *J. Clin. Microbiol.* **41:**4424–4427.

34. **Kahl, B., R. A. Proctor, A. Schulze-Everding, M. Herrmann, H. G. Koch, I. Harms, and G. Peters.** 1998. Persistent infection with small colony variant strains of *Staphylococcus aureus* in patients with cystic fibrosis. *J. Infect. Dis.* **177:**1023–1029.

35. **Kahl, B., C. von Eiff, M. Herrmann, G. Peters, and R. A. Proctor.** 1996. Staphylococcal small colony variants present a challenge to clinicians and clinical microbiologists. *Antimicrob. Infect. Dis. Newsl.* **15:**59–63.

36. **Kaplan, M. W., and W. E. Dye.** 1976. Growth requirements of some small-colony-forming variants of *Staphylococcus aureus.* *J. Clin. Microbiol.* **4:**343–348.

37. **Kipp, F., W. Ziebuhr, K. Becker, V. Krimmer, N. Hobeta, G. Peters, and C. von Eiff.** 2003. Detection of *Staphylococcus aureus* by 16S rRNA directed in situ hybridisation in a patient with a brain abscess caused by small colony variants. *J. Neurol. Neurosurg. Psychiatry* **74:**1000–1002.

38. **Kohler, C., C. von Eiff, G. Peters, R. A. Proctor, M. Hecker, and S. Engelmann.** 2003. Physiological characterization of a heme-deficient mutant of *Staphylococcus aureus* by a proteomic approach. *J. Bacteriol.* **185:**6928–6937.

39. **Koo, S. P., A. S. Bayer, H.-G. Sahl, R. A. Proctor, and M. R. Yeaman.** 1996. Staphylocidal action of thrombin-induced platelet microbicidal protein (tPMP) is not solely dependent on transmembrane potential ($\Delta\psi$). *Infect. Immun.* **64:**1070–1074.

40. Lewis, L. A., K. Li, M. Bharosay, M. Cannella, V. Jorgenson, R. Thomas, D. Pena, M. Velez, B. Pereira, and A. Sassine. 1990. Characterization of gentamicin-resistant respiratory-deficient (Res⁻) variant strains of *Staphylococcus aureus*. *Microbiol. Immunol.* **34:**587–605.

41. Massey, R. C., A. Buckling, and S. J. Peacock. 2001. Phenotypic switching of antibiotic resistance circumvents permanent costs in *Staphylococcus aureus*. *Curr. Biol.* **11:**1810–1814.

42. Massey, R. C., and S. J. Peacock. 2002. Antibiotic-resistant sub-populations of the pathogenic bacterium *Staphylococcus aureus* confer population-wide resistance. *Curr. Biol.* **12:**R686–R687.

43. Musher, D. M., R. E. Baughn, G. B. Templeton, and J. N. Minuth. 1977. Emergence of variant forms of *Staphylococcus aureus* after exposure to gentamicin and infectivity of the variants in experimental animals. *J. Infect. Dis.* **136:**360–369.

44. Neumann, S., U. Matthey, G. Kaim, and P. Dimroth. 1998. Purification and properties of the F_0F_1 ATPase of *Ilyobacter tartaricus*, a sodium ion pump. *J. Bacteriol.* **180:**3312–3316.

45. Novick, R. 1993. *Staphylococcus*, p. 17–33. *In* A. L. Sonenshein, J. A. Hoch, and R. Losick (ed.), Bacillus subtilis and Other Gram-Positive Bacteria. American Society for Microbiology, Washington, D.C.

46. Novick, R. P. 2003. Autoinduction and signal transduction in the regulation of staphylococcal virulence. *Mol. Microbiol.* **48:**1429–1449.

47. Nydahl, B. C., and W. L. Hall. 1965. The treatment of staphylococcal infection with nafcillin with a discussion of staphylococcal nephritis. *Ann. Intern. Med.* **63:**27–43.

48. Pan, X. S., P. J. Hamlyn, R. Talens-Visconti, F. L. Alovero, R. H. Manzo, and L. M. Fisher. 2002. Small-colony mutants of *Staphylococcus aureus* allow selection of gyrase-mediated resistance to dual-target fluoroquinolones. *Antimicrob. Agents Chemother.* **46:**2498–2506.

49. Proctor, R. A. 1994. Microbial pathogenic factors: small colony variants, p. 77–90. *In* A. L. Bisno and F. A. Waldvogel (ed.), *Infections Associated with Indwelling Medical Devices*, 2nd ed. American Society for Microbiology, Washington, D.C.

50. Proctor, R. A. 1998. Bacterial energetics and antimicrobial resistance. *Drug Resist. Updates* **1:**227–235.

51. Proctor, R. A., D. M. Bates, and P. J. McNamara. 2001. Electron transport-deficient *Staphylococcus aureus* small-colony variants as emerging pathogens, p. 95–110. *In* W. Craig (ed.), *Emerging Infections*, vol. 5. ASM Press, Washington, D.C.

52. Proctor, R. A., S. Dalal, B. Kahl, D. Brar, G. Peters, and W. W. Nichols. 2002. Two diarylurea electron transport inhibitors reduce *Staphylococcus aureus* hemolytic activity and protect cultured endothelial cells from lysis. *Antimicrob. Agents Chemother.* **46:**2333–2336.

53. Proctor, R. A., J. M. Balwit, and O. Vesga. 1994. Variant subpopulations of *Staphylococcus aureus* can cause persistent and recurrent infections. *Infect. Agents Dis.* **3:**302–312.

54. Proctor, R. A., and G. Peters. 1998. Small colony variants in staphylococcal infections: diagnostic and therapeutic implications. *Clin. Infect. Dis.* **27:**419–423.

55. Proctor, R. A., P. van Langevelde, M. Kristjansson, J. N. Maslow, and R. D. Arbeit. 1995. Persistent and relapsing infections associated with small colony variants of *Staphylococcus aureus*. *Clin. Infect. Dis.* **20:**95–102.

56. Proctor, R. A., O. Vesga, M. F. Otten, S.-P. Koo, M. R. Yeaman, H.-G. Sahl, and A. S. Bayer. 1996. *Staphylococcus aureus* small colony variants cause persistent and resistant infections. *Chemotherapy* (Basel) **42**(Suppl. 2):47–52.

57. Ray, S. C., R. Schulick, D. Flayhart, and J. Dick. 1997. First report of infection with *Staphylococcus aureus* subspecies *anaerobius*, abstr. 567. *In* Proc. 35th Infect. Dis. Soc. Am. Meet. Infectious Disease Society of America, Alexandria, Va.

58. Regassa, L. B., R. P. Novick, and M. J. Betley. 1992. Glucose and nonmaintained pH decrease expression of the accessory gene regulator (*agr*) in *Staphylococcus aureus*. *Infect. Immun.* **60:**3381–3388.

59. Reizere, J., M. H. Saier, Jr., J. Deutscher, F. Grenier, J. Thompson, and W. Hengstenberg. 1988. The phosphoenolpyruvate:sugar phosphotransferase system in gram-positive bacteria: properties, mechanism, and regulation. *Crit. Rev. Microbiol.* **15:**297–338.

60. Sadowska, B., A. Bonar, C. von Eiff, R. A. Proctor, M. Chmiela, W. Rudnicka, and B. Rozalska. 2002. Characteristics of *Staphylococcus aureus*, isolated from airways of cystic fibrosis patients, and their small colony variants. *FEMS Immunol. Med. Microbiol.* **32:**191–197.

61. Salgado, D. R., F. A. Boza, M. Pinto, and J. Sampaio. 2001. Outbreak with small colony variants of methicillin-resistant *Staphylococcus aureus* in an ICU, abstr. K-1226. *In* Proc. 41st Intersci. Conf. Antimicrob. Agents Chemother. American Society for Microbiology, Washington, D.C.

62. Santana, M., M. S. Ionescu, A. Vertes, R. Longin, F. Kunst, A. Danchin, and P. Glaser. 1994. *Bacillus subtilis* F_0F_1 ATPase: DNA sequence of the *atp* operon and characterization of *atp* mutants. *J. Bacteriol.* **176:**6802–6811.

63. Schlievert, P. M., J. R. Dringer, M. H. Kim, S. J. Projan, and R. P. Novick. 1992. Effect of glycerol monolaurate on bacterial growth and toxin production. *Antimicrob. Agents Chemother.* **36:**626–631.

64. Schnitzer, R. J., L. J. Camagni, and M. Buck. 1943. Resistance of small colony variants (G forms) of a staphylococcus toward the bacteriostatic activity of penicillin. *Proc. Soc. Exp. Biol. Med.* **53:**75–89.

65. Seifert, H., C. von Eiff, and G. Fatkenheuer. 1999. Fatal case due to methicillin-resistant *Staphylococcus aureus* small colony variants in an AIDS patient. *Emerg. Infect. Dis.* **5:**450–453.

66. Seifert, H., H. Wisplinghoff, P. Schnabel, and C. von Eiff. 2003. Small colony variants of *Staphylococcus aureus* and pacemaker-related infection. *Emerg. Infect. Dis.* **9:**1316–1318.

67. Silverman, J. A., N. G. Perlmutter, and H. M. Shapiro. 2003. Correlation of daptomycin bactericidal activity and membrane depolarization in *Staphylococcus aureus*. *Antimicrob. Agents Chemother.* **47:**2538–2544.

68. Somerville, G. A., M. S. Chausse, C. I. Morgan, J. R. Fitzgerald, D. W. Dorward, L. J. Reitzer, and J. M. Musser. 2002. *Staphylococcus aureus* aconitase inactivation unexpectedly inhibits post-exponential-phase growth and enhances stationary-phase survival. *Infect. Immun.* **70:**6373–6382.

69. Somerville, G. A., B. Said-Salim, J. M. Wickman, S. J. Raffel, B. N. Kreiswirth, and J. M. Musser. 2003. Correlation of acetate catabolism and growth yield in *Staphylococcus aureus*: implications for host-pathogen interactions. *Infect. Immun.* **71:**4724–4732.

70. **Spearman, P., D. Lakey, S. Jotte, A. Chernowitz, S. Claycomb, and C. Stratton.** 1996. Sternoclavicular joint septic arthritis with small-colony variant *Staphylococcus aureus. Diagn. Microbiol. Infect. Dis.* **26:**13–15.

71. **Taber, H. W.** 1993. Respiratory chains, p. 199–212. *In* A. L. Sonenshein, J. A. Hoch, and R. Losick (ed.), Bacillus subtilis *and Other Gram-Positive Bacteria.* American Society for Microbiology, Washington, D.C.

72. **Tien, W., and D. C. White.** 1968. Linear sequential arrangement of genes for the biosynthetic pathway of heme in *Staphylococcus aureus. Proc. Natl. Acad. Sci. USA* **61:**1392–1398.

73. **Vaudaux, P., P. Francois, C. Bisognano, W. L. Kelley, D. P. Lew, J. Schrenzel, R. A. Proctor, P. J. McNamara, G. Peters, and C. von Eiff.** 2002. Increased expression of clumping factor and fibronectin-binding proteins by small colony variants of *Staphylococcus aureus. Infect. Immun.* **70:**5428–5437.

74. **Vesga, O., J. M. Vann, D. Brar, and R. A. Proctor.** 1996. *Staphylococcus aureus* small colony variants are induced by the endothelial cell intracellular milieu. *J. Infect. Dis.* **173:**739–742.

75. **von Eiff, C., P. McNamara, D. Bates, K. Becker, X. Lei, M. Ziman, B. Bochner, G. Peters, and R. A. Proctor.** Phenotype MicroArray of *S. aureus menD* and *hemB* mutants with small colony variant (SCV) phenotype. *In* Proc. 11th Int. Symp. Staphylococci Staphylococcal Infect., Charleston, S.C.

76. **von Eiff, C., K. Becker, D. Metze, G. Lubritz, J. Hockmann, T. Schwarz, and G. Peters.** 2001. Intracellular persistence of *Staphylococcus aureus* small-colony variants within keratinocytes: a cause for antibiotic treatment failure in a patient with Darier's disease. *Clin. Infect. Dis.* **32:**1643–1647.

77. **von Eiff, C., D. Bettin, R. A. Proctor, B. Rolauffs, N. Lindner, W. Winkelmann, and G. Peters.** 1997. Recovery of small colony variants of *Staphylococcus aureus* following gentamicin bead placement for osteomyelitis. *Clin. Infect. Dis.* **25:**1250–1251.

78. **von Eiff, C., C. Heilmann, R. A. Proctor, C. Woltz, G. Peters, and F. Götz.** 1997. A site-directed *Staphylococcus aureus hemB* mutant is a small-colony variant which persists intracellularly. *J. Bacteriol.* **179:**4706–4712.

79. **von Eiff, C., P. Vaudaux, B. C. Kahl, D. Lew, S. Emler, A. Schmidt, G. Peters, and R. A. Proctor.** 1999. Bloodstream infections caused by small-colony variants of coagulase-negative staphylococci following pacemaker implantation. *Clin. Infect. Dis.* **29:**932–934.

80. **Widmer, A. F., A. Wiestner, R. Frei, and W. Zimmerli.** 1991. Killing of nongrowing and adherent *Escherichia coli* determines drug efficacy in device-related infections. *Antimicrob. Agents Chemother.* **35:**741–746.

81. **Wise, R. I., and W. W. Spink.** 1954. The influence of antibiotics on the origin of small colonies (G variants) of *Micrococcus pyogenes* var. *aureus. J. Clin. Invest.* **33:**1611–1622.

82. **Xu, K., J. Delling, and T. Elliott.** 1992. The genes required for heme synthesis in *Salmonella typhimurium* include those encoding alternative functions for aerobic and anaerobic coproporphyrinogen oxidation. *J. Bacteriol.* **174:**3953–3963.

83. **Youmans, G. P., and E. Delves.** 1942. The effect of inorganic salts on the production of small colony variants by *Staphylococcus aureus. J. Bacteriol.* **44:**127–136.

The Staphylococcal Cell Wall

ALEXANDER TOMASZ

36

INTRODUCTION

Historical Overview

Ever since the recognition of the cell wall in the late 1940s and early 1950s as a unique anatomical component of all eubacterial cells, *Staphylococcus aureus* has often served as the gram-positive model in wall-related studies. One of the first demonstrations that bacterial cell walls can be isolated as physical entities with the size and shape of the whole bacterium was with *S. aureus*. It was in penicillin-treated *S. aureus* that the UDP-linked amino-sugar-containing wall precursor peptides were discovered, providing the first insights into the unique building blocks of cell wall biosynthesis. The history of interest in the staphylococcal cell wall also reflects the history of success and failure of the antibiotic era. The clue that eventually led to the discovery of penicillin (and later to the autolytic enzymes) was provided by the lysis of staphylococcal colonies in the vicinity of a mold contaminant on an agar plate in Fleming's laboratory.

Elucidation of the mode of action of several important antibiotics in the 1960s and 1970s has been intimately linked to studies on the biosynthesis of staphylococcal cell walls. This included studies on the mode of action of penicillin and other β-lactam antibiotics as specific inhibitors (acylating agents) directed against the active site of penicillin-binding proteins (PBPs)—transpeptidases (TPases) with or without an additional transglycosylase (TGase) function, which catalyze terminal stages in the assembly of the bacterial cell wall. It was mainly from studies in *S. aureus* and from parallel studies in *Escherichia coli* that by the early 1980s a coherent picture emerged about the biosynthetic pathway that leads to the formation of the lipid-linked disaccharide pentapeptide, which, with some structural variations, is the universal building block of cell wall peptidoglycan both in gram-positive and gram-negative bacteria. Reviews and references summarizing various aspects of studies on the cell walls of staphylococci up to the late 1980s are available (29, 33).

The Changing Image of Cell Walls

During the early era biochemists and microbiologists working with cell walls viewed these as more or less inert exoskeletons essential to withstand the turgor pressure of cytoplasm. However, these views are changing rapidly with the recognition of the complexity of chemical structure and biosynthetic pathways and the large number of genetic determinants involved with the synthesis of cell wall. Cell walls are also intimately involved with host-related functions of staphylococci (see below and chapters 40 and 42). The new image of cell walls emerging is that of a dynamic and very live structure.

The assembly and replication of cell walls in dividing cells pose some of the most challenging questions of microbial cell biology, surpassing in complexity the questions of chromosome replication. How is the flow of precursor molecules and their polymerization on the outer surface of the plasma membrane controlled? What controls the unique species-specific chemical composition of peptidoglycan? What principles govern the organization of wall polymers into supermolecular "sacculi" that have the same size and shape as the particular bacterium? How and by what mechanism does the nascent innermost layer of this envelope mature while moving outward towards the cell surface? How and why and through what signals and by what catalysts are the outermost layers of cell wall shed into the medium during wall turnover? And how is the spatial and temporal accuracy of wall synthesis coordinated with cell division?

A recent elegant series of studies began to address such questions through the combination of specific bacterial cell division mutants and the use of fluorescence staining procedures and confocal microscopy of dividing staphylococci. Pinho and Errington (66) used fluorescence-labeled vancomycin in combination with a transient modification of the chemistry of the carboxy-terminal residue of the lipid-linked wall precursor to localize sites of wall synthesis. Other reagents, such as fluorescence-labeled wheat germ agglutinin and an antibody against PBP2, allowed the parallel localization of newly made cell wall material and the subcellular sites where PBP2 resides in dividing bacteria (66) (Fig. 1).

About This Review

In addition to the reemergence of interest in cell walls in the context of modern microbial cell biology, two approach-

Gram-Positive Pathogens, 2nd edition, edited by Vincent A. Fischetti et al.
© 2006 ASM Press, Washington, D.C.

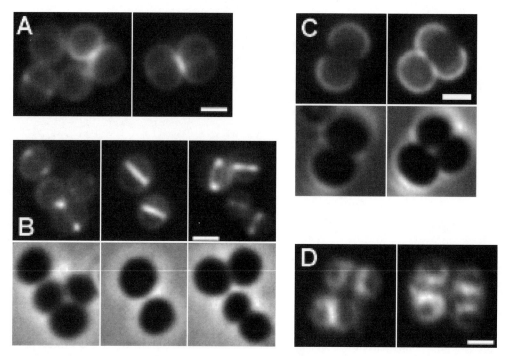

FIGURE 1 Localization of cell wall synthesis in *S. aureus*. (A) Van-fluorescence labeling of RN4220 cells. The entire cell wall, including the septum, has been labeled. (B) Van-fluorescence labeling of RN4220 cells after transient incubation with an excess of D-serine. This appears to result in specific labeling of new peptidoglycan. Different stages of septum formation can be observed. Left panel: Cells have two fluorescent spots that presumably correspond to a ring of new peptidoglycan at the division site. Middle panel: Cells have a fluorescent line across the middle, which should correspond to a complete disk of new peptidoglycan—the closed septum. Right panel: A tilted cell allows visualization of an entire ring of new peptidoglycan in an incomplete septum. (C) Wheat germ agglutinin labeling of RN4220 cells, followed by incubation in the absence of the dye. Recently synthesized or uncovered wall material appears as a nonfluorescent region that constitutes the new hemisphere of each daughter cell. (D) Immunofluorescence imaging of PBP2 in RN4220 cells. Scale bars, 1 μm. Phase-contrast microscopy images are shown below each fluorescence image in panels B and C. (Reproduced with permission from reference 65.)

es have been making great impact on discoveries in this field: the introduction of high-resolution analytical techniques (high-pressure liquid chromatography [HPLC] and mass spectrometry) and the increasing application of molecular genetic approaches. A comprehensive coverage of all data in a period of such rapid expansion of a field would be difficult. The purpose of an updated review may be better served by putting the interested reader "on track" of some of the most recent findings and emerging trends by quoting relevant literature. After a brief reminder of the anatomy of staphylococcal cell walls, new information will be reviewed in the chapter under four headings: High-Resolution Analysis of the *S. aureus* Peptidoglycan; Variations in Peptidoglycan Composition; Genetic Determinants and Enzymes in Cell Wall Synthesis; and Complex Functions of Cell Walls.

Anatomy of the *S. aureus* Cell Wall

Most clinical isolates of *S. aureus* express on their outermost surface one of the 11 chemically different capsular polysaccharides that have been identified so far (58). The chemical subunit structures of these important and often antiphagocytic carbohydrate polymers have been elucidated, and rapid advances have been made in the identification of genetic determinants and organization of capsular loci (75) (for more

detailed coverage, see chapter 37, this volume). Underneath these somewhat variant surface layers is the staphylococcal cell wall. In electron microscopic thin sections followed by heavy-metal staining, the cell wall appears as a triple layer: a diffusely staining middle layer sandwiched between two electron-dense lines (Fig. 2). The electron micrograph in Fig. 2 also shows the effect of inhibited cell wall turnover (Fig. 2B) resulting in the accumulation of large amounts of unstructured cell wall material on the cell surface of a vancomycin-resistant clinical isolate of *S. aureus* (82). The photograph of the isogenic vancomycin-susceptible parental cell recovered from the same clinical source is also shown (Fig. 2A). Cell wall thickening is often the phenotype of *S. aureus* with vancomycin-insensitive *S. aureus* (VISA)-type vancomycin resistance (14).

Unlike in streptococci, in *S. aureus* consecutive cell divisions occur in three division planes, each at right angles to one another, and proper orientation of cell wall septa must involve a complex and superbly controlled mechanism. Cyclic morphological alterations of cell walls during growth, division, and separation of daughter cells, under conditions of normal growth and also under exposure to a variety of antimicrobial agents, have been documented by the elegant electron microscopic studies of Giesbrecht and his colleagues (30).

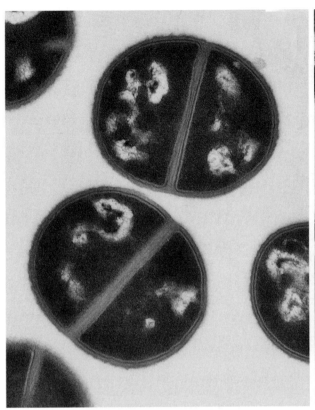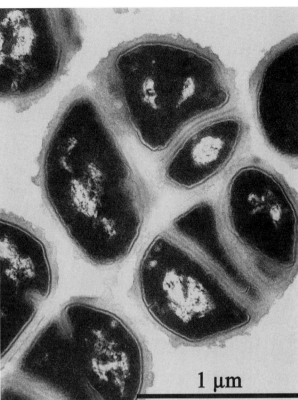

1 μm

FIGURE 2 The anatomy of cell walls in normal (left) and vancomycin-resistant (right) *S. aureus*. (Reproduced with permission from reference 81.)

Earlier Studies Established Basic Compositional/Structural Features of the *S. aureus* Cell Wall

The *S. aureus* cell wall is composed of a highly cross-linked A3α-type peptidoglycan with pentaglycine oligopeptide units connecting the ε-amino group of the lysine component of one muropeptide to the penultimate D-alanine of another. This peptidoglycan, together with ribitol-type teichoic acid chains (which are attached to the 6-hydroxyl groups of some of the N-acetylmuramic acid residues of the glycan chain) (1, 40), surrounds the *S. aureus* cell in the form of a multilayered envelope (Fig. 3). This review chapter will present information only on the peptidoglycan component and will not address teichoic acids.

HIGH-RESOLUTION ANALYSIS OF THE *S. AUREUS* PEPTIDOGLYCAN

Progress in the high-resolution chemistry of the *S. aureus* cell wall came from the introduction of HPLC and mass spectrometric methods (97) for the analysis of the peptidoglycan. First, studies with gel permeation HPLC established the presence of muropeptide oligomers in lengths extending to nanometers and beyond (85). This was followed by the adaptation of the reverse-phase HPLC system (originally developed for the analysis of *E. coli* cell walls) in combination with mass spectrometry for the analysis of strains of *S. aureus* (17, 18). The primary motivation in these studies was to understand better the mechanism of methicillin resistance in the highly and homogeneously methicillin-resistant *S. aureus* (MRSA) strain COL and its large number

of transposon mutants, but the results obtained are valid for both resistant and susceptible strains of *S. aureus*. Together with studies on the cell walls of *Streptococcus pneumoniae* (see chapter 19, this volume), analysis of the peptidoglycan of the MRSA provided the first high-resolution view of the complexity of cell wall structures in gram-positive bacteria.

Enzymatic hydrolysis (with the M1 muramidase) of the staphylococcal peptidoglycan (isolated from purified cell walls after removal of the teichoic acid chains with hydrofluoric acid) was followed by borohydride reduction and separation of the muropeptide components on a reverse-phase HPLC column. This method resolved the hydrolysate to more than 21 distinct UV-absorbing peaks plus a "hump" of unresolved material of longer retention times, which made up more than 25% of the muropeptides (Fig. 4).

Analysis of muropeptide components obtained after digestion with muramidase, or with the combination of muramidase plus lysostaphin (a bacteriolytic endopeptidase that attacks the pentaglycine bridge), for amino acid and amino sugar composition and for molecular mass (fast atom bombardment mass spectrometry), revealed the main structural principle staphylococci use in building the peptidoglycan. The major monomeric building block is a disaccharide pentapeptide carrying D-isoglutamine in position 2, an intact D-alanyl-D-alanine carboxy terminal, most frequently with a pentaglycine substituent attached to the ε-amino group of the lysine residue (muropeptide 5) or, occasionally, without it (muropeptide 1). These monomeric units make up about 6% of all muropeptides. Another 20% consists of dimers in which the pentaglycine substituent of one muropeptide unit is cross-linked to the penultimate D-alanine

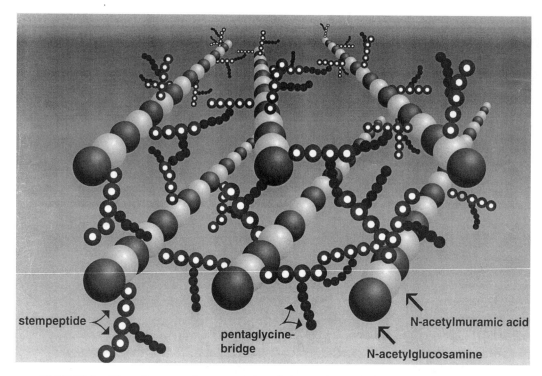

FIGURE 3 Three-dimensional structure of staphylococcal peptidoglycan. The straight lines of large bowls represent the sugar moieties of the peptidoglycan. Each globe in these lines symbolizes an amino sugar, N-acetylglucosamine (black globe), or N-acetylmuramic acid (white globe). Stem-peptides, branching from N-acetylmuramic acid, are characterized by small dark globes with a white center. The connecting interpeptide bridges (pentaglycines) between the stem-peptides are shown as small black globes. Schematic drawing by Peter Giesbrecht, Thomas Kersten, Heiner Maidhof, and Jorg Wecke, Robert-Koch Institute, Berlin, Germany.

of a neighboring one (muropeptide 11). About 40% consists of higher oligomers containing three (muropeptide 15) to nine muropeptide units generated by the same cross-linking principle. Still higher oligomers account for an additional 15 to 25% of the muropeptide units (the "hump" of unresolved components eluting with retention times higher than 110 min).

The high resolving power of the HPLC technique is also illustrated in Fig. 4. Muropeptide components differing by only one mass unit can be clearly separated. The particular case in Fig. 4 involves an MRSA mutant in which some of the isoglutamine residues of the stem-peptides in the parental strain (Fig. 4B) are replaced by free glutamic acid residues in the mutant (Fig. 4A). The change in molecular size involves the replacement of an amino group with a hydroxyl group (+1 mass unit) (Table 1) (61). Using this technique it was possible to identify the structure of an unusual—doubly cross-linked—muropeptide dimer that was accumulating in a cefotaxime-resistant mutant (6).

A variation in the preparative technique also allows determination of the structure of the glycan chains. In this case, peptidoglycan is first hydrolyzed with lysostaphin in order to disrupt the pentaglycine bridges connecting neighboring muropeptides followed by removal of stem-peptides by the pneumococcal amidase (N-acetylmuramyl-L-alanine amidase), which hydrolyzes the covalent bond connecting the L-alanine residue to the acetyl muramic acid residue of the glycan chain. After removal of the stem-peptides, the size distribution and composition of the glycan chains are determined by reverse-phase HPLC. Application of this method has resolved the glycan strands of S. aureus to a family of major peaks representing oligosaccharides composed of repeating dissaccharide units (N-acetylglucosamine beta-1,4-N-acetylmuramic acid) with different degrees of polymerization and terminating with N-acetylmuramic acid residues at the reducing ends. The method allowed separation of strands up to 23 to 26 disaccharide units. Minor satellite peaks were also present throughout the HPLC elution profile, most likely representing products of an N-acetylglucosaminidase activity (Fig. 5) (7).

Stability of Muropeptide Composition

Analysis of peptidoglycan prepared from a large number of S. aureus isolates of different clonal types (as defined by multilocus enzyme analysis), both susceptible or resistant to methicillin (either homogeneous or heteroresistant), contemporary as well as preantibiotic-era S. aureus isolates, all showed virtually identical HPLC muropeptide patterns, provided that the cells were grown in the same medium and in a "balanced" state of growth and harvested in the late exponential phase of growth (Ornelas-Soares, de Lencastre, and Tomasz, unpublished observations). Selective Tn551 inactivation of the mecA gene did not cause any detectable change in muropeptide composition (17). These data suggest that similarly to other bacteria, the muropeptide composition of S. aureus cell walls is specific for the species. The mechanism responsible for maintaining this specificity is unknown.

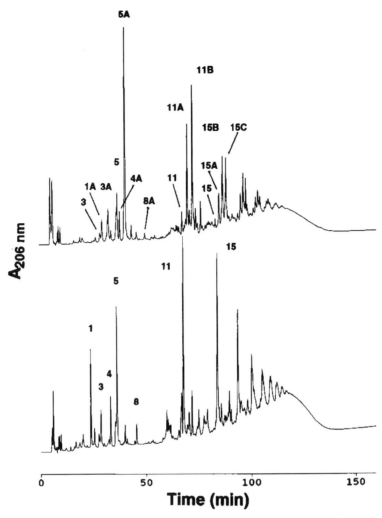

FIGURE 4 Analysis of the peptidoglycan of a methicillin-resistant strain of *S. aureus* (bottom) and its *femC* mutant (top). (Reproduced with permission from reference 60.)

VARIATIONS IN PEPTIDOGLYCAN COMPOSITION

Effect of Antibiotics

Growth of MRSA strain COL with high methicillin resistance in a wide range of subinhibitory concentrations of methicillin caused striking changes in muropeptide composition: the representation of trimeric plus higher oligomeric components was drastically reduced (from a combined representation of close to 50% to less than 10%) while the proportion of monomeric and dimeric components increased (from about 15% up to about 50%). This major compositional change was virtually complete at methicillin concentrations of about 5 μg/ml (i.e., at antibiotic concentrations that would fully acylate the four native PBPs in membrane preparations) and did not change further even in the presence of 750 μg of antibiotic per ml in the medium (the MIC value for methicillin in this strain was 1,600 μg/ml). It was suggested that the abruptness of this change in peptidoglycan composition may represent the switching over from the normal wall biosynthetic system (the four native PBPs) to another one (PBP2A, encoded by the methicillin-resistance gene *mecA*) capable of functioning in the presence of high concentrations of methicillin (17). The anomalous composition of the peptidoglycan produced under these conditions would then reflect the limited capacity of PBP2A for cross-linking more than single monomeric muropeptides. Nevertheless, similar albeit much less abrupt compositional changes were also observed when methicillin-susceptible staphylococci or an isogenic derivative of strain COL (with inactivated *mecA*) was grown in sub-MIC concentrations of methicillin.

Effect of Growth Phase and Medium Composition

Secondary modification of the staphylococcal peptidoglycan (O-acetylation of muramic acid residues and changes in the N-acetylation of the amino sugars) has been described earlier, as a function of the culture's growth phase or during treatment with chloramphenicol. Recent studies followed up earlier observations on the effect of exogenous glycine on the structure of staphylococcal peptidoglycan (16). Growth of MRSA strain COL with high methicillin resistance in the presence of high concentrations of glycine (between 0.06 and 0.25 M), D-serine, or other D-amino acids caused dramatic distortions of the HPLC pro-

TABLE 1 Structure of muropeptide components in the methicillin-resistant parental strain and its *femC* mutant 208[a]

	Muropeptide	Amino acid analysis				Mass spec (M + H)	HPLC retention time (min)
		Glx	Lys	Ala	Gly		
Parental	5	1	1.0	3.1	4.6	1254	36.2
Mutant 208	5	ND[b]	ND	ND	ND	1253.7	36.2
	5A	1	1.0	3.3	4.7	1255.4(+1)	40.1
Parental	11	1	1.0	2.9	4.3	2418	67.5
	11	1	1.0	2.6	5.0	2418.4	67.6
Mutant 208	11A	1	1.0	2.6	4.7	2418.8(+1)	70.1
	11B	1	1.0	2.8	4.8	2419.9(+2)	72.5
Parental	15	1	1.0	2.4	4.5	(3582)	83.3
Mutant 208	15B	1	1.0	2.5	5.0	3584.2(+2)	86.9
	15C	1	1.0	2.5	4.8	3585.3(+3)	88.5

```
G-M                          G-M                          G-M
 |                            |                            |
Ala          G-M            Ala            G-M            Ala
 |            |              |              |              |
iGlu         Ala           iGlu    G-M     Ala           iGlu
 |            |              |      |       |              |
Lys–(Gly)5   iGlu         Lys–(Gly)5 Ala   iGlu         Lys–(Gly)5
 |            |                    |        |              |
Ala        Lys–(Gly)5–Ala        iGlu   Lys–(Gly)5–Ala  Lys–(Gly)5–Ala
 |            |                    |        |
Ala          Ala                 Lys–(Gly)5–Ala
             |                    |
             Ala                  Ala
                                  |
                                  Ala
```

Muropeptide 5   Muropeptide 11   Muropeptide 15
(MW = 1,253)   (MW = 2,417)   (MW = 3,581)

[a]Reproduced with permission from reference 60.
[b]ND, not determined.

file of muropeptides. There was extensive reduction in cross-linkage, and major monomeric, dimeric, trimeric, and some higher oligomeric muropeptide components were gradually replaced (parallel with the increasing glycine or D-amino acid concentration) by novel components that were similar to the normal muropeptides except that their carboxy-terminal D-alanine residues were replaced by glycine or the other amino acids added to the medium. Such abnormally terminating muropeptides appear to be poor substrates for TPases because oligomers higher than pentameric compounds were rare in the peptidoglycan of cells grown in high concentrations of glycine (16, 19).

Effect of Vancomycin and Teicoplanin Resistance

Laboratory mutants with increased MIC values for glycopeptide antibiotics have been isolated (5, 65, 79, 80). The cell wall composition was determined in a series of isogenic laboratory mutants of the MRSA strain COL with high methicillin resistance (81). The peptidoglycan of the mutants showed distorted muropeptide composition that paralleled the increasing vancomycin MIC values. There was a gradual increase in the molar proportion of the major monomeric muropeptides carrying the intact D-alanyl-D-alanine carboxy termini and reduction in the representation of dimeric and oligomeric components. It was suggested that some of the unusual properties of these mutants and the mechanism of resistance are related to the enrichment

of the cell walls in muropeptide monomers. Muropeptides terminating in the D-alanyl-D-alanine residues are known to form the binding sites for glycopeptide antibiotics. The mechanism of resistance may then be related to trapping the glycopeptides in the mature layer of the peptidoglycan enriched for muropeptide monomers, thus preventing the antibiotic molecules from reaching sites of wall biosynthesis at the plasma membrane (79, 84). Another structural change is that increased glycan chain length was also detected in a vancomycin-resistant mutant (45).

VISA Isolates

Clinical isolates of *S. aureus* with reduced susceptibility to vancomycin have been repeatedly described in the literature and in at least some of these so-called VISA isolates alterations in peptidoglycan composition have been observed, including increased representation of muropeptides with non-amidated glutamic acid residues and increased monomeric and decreased oligomeric components (8, 9, 13, 15, 34, 82, 83).

Vancomycin-Resistant *S. aureus*

In 2002 highly vancomycin-resistant strains of *S. aureus* were recovered from clinical specimen. These bacteria acquired the vancomycin-resistant gene complex through the enterococcal transposon Tn*1546*. Expression of Tn*1546* produced drastic changes in the cell wall composition: all pentapeptides were replaced by tetrapeptides, and the pepti-

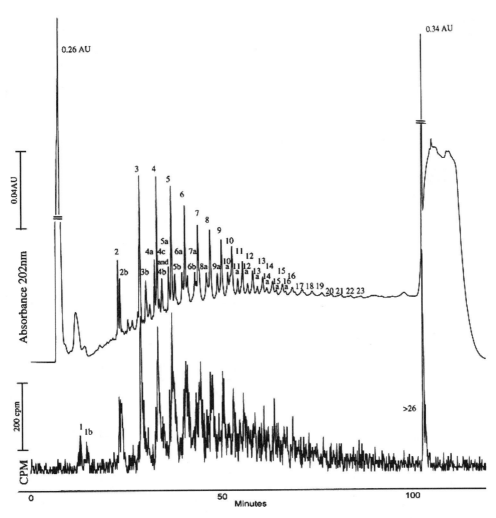

FIGURE 5 Distribution of glycan chain length in *S. aureus*. (Top) Tracing of the HPLC profile by UV absorbance. (Bottom) Tracing of the same HPLC chromatogram through radioactivity due to the labeling of the glycan with H³ N-acetylglucosamine. (Reproduced with permission from reference 8.)

doglycan contained at least 22 novel muropeptide species that frequently showed a deficit or complete absence of pentaglycine branches. The UDP-Mur-NAc pentapeptide, the major component of the cell wall precursor pool in vancomycin-sensitive cells, was replaced by UPD-MurNAc depsipeptide in the resistant bacteria, and PBP2A, the product of the methicillin-resistance gene *mecA*, was unable to utilize such precursors for wall synthesis. The MRSA strain COL carrying both *mecA* and the *vanA* genes had extremely high MICs both for beta-lactam antibiotics and for vancomycin (77). The major PBP responsible for the assembly of the abnormally structured peptidoglycan produced in the vancomycin-resistant cells was PBP2 (76).

GENETIC DETERMINANTS AND ENZYMES IN CELL WALL SYNTHESIS

"Methicillin-Conditional" Mutants

The central genetic determinant of MRSA strains is the heterologous gene *mecA*, acquired from an unknown donor. Inactivation of *mecA* results in complete loss of resistance. Nevertheless, reduction in resistance level from modest de-

crease to a virtually complete loss of resistance can also be the consequence of inactivation of a surprisingly large number of domestic genes that have been referred to as *fem* or auxiliary genes (22, 23). The transposon mutants of MRSA that led to the identification of auxiliary genes were originally isolated in order to clarify some complex genetic features of staphylococcal methicillin resistance. The first Tn*551* mutant (later named "*femA*") was isolated in 1983 (4), even before the basic genetic mechanism of methicillin resistance was recognized. Subsequently, additional transposon mutants with reduced methicillin resistance were isolated (48, 59), followed by a more systematic effort that identified more than twenty auxiliary genes (21). Biochemical analysis indicated that in several of these mutants the inactivated genes were involved with staphylococcal cell wall synthesis. This finding suggested that reduction in methicillin-resistance level may provide a relatively easily selectable phenotype for the identification of genes in cell wall synthesis and metabolism (21). As a working model, it was proposed that the methicillin MIC value in such "methicillin-conditional" mutants may primarily reflect the "success" of the *mecA* gene product (PBP2A), positioned close to the end of a long and complex metabolic pathway, to continue catalyzing peptidoglycan incorporation even when this metabolic pathway is disrupted by an inhibited auxiliary gene. In this model quantitative fluctuation of the methicillin MIC value was supposed to serve as a sensitive gauge to register errors, perturbations, or interference with any step in wall synthesis that precedes (or in any way assists) the terminal wall synthetic reaction catalyzed by PBP2A. Such errors may include the production of structurally incorrect cell wall precursors that may not compete successfully with the antibiotic for the active site of PBP2A (20), or such incorrectly structured muropeptides may not be able to perform some as yet unidentified effector function in wall synthesis (analogous perhaps to regulatory functions of cell wall metabolites in the control of β-lactamase expression) (see reference 37). Abnormally structured stempeptides that are incorporated into the cell wall may create localized structural defects or may serve as signals for murein hydrolases. It was assumed that any of these defects may cause a decrease in the MIC for methicillin.

Reports on new auxiliary mutants continue to appear in the literature. A catalogue and partial characterization of a large group of such mutants may be found in reference 22. Tn551 MRSA mutants selected for reduced MIC (such as femXA/B, gluM, murE, llm, fmtA through C, murF, murE, etc.) have led to insights into hitherto unknown or not fully characterized steps in staphylococcal wall precursor biosynthesis (27, 31, 39, 41–44, 46, 47, 52, 54, 62, 72, 73, 86, 87, 95, 96).

femAX/B

The femX (Fmhb), femA, and femB genes are part of an operon involved with the synthesis and attachment of the pentaglycine branches to the epsilon amino group of the lysine residue of the S. aureus muropeptides. The synthesis of pentaglycine bridges occurs in three steps. The protein product of Fmhb catalyzes the addition of the first glycine residue to the muropeptides. This gene is essential for bacterial survival and growth (74). The protein product of femA is involved with the addition of the first and second glycine residues to the cross bridge, and the protein product of femB with the addition of the third and fourth glycine residues (71–73). The crystal structure of the FemA protein has been determined (3) and—in an elegant series of studies—the entire sequence of pentaglycine branch synthesis and attachment has been determined and reproduced in vitro using bactoprenol-linked disaccharide pentapeptide (92) as acceptor and tRNA-linked glycine as the source of the amino acid residues (76, 90).

Insights into Regulation

Up-regulation of the transcription of pbpB, the structural gene of PBP2, by exposure of S. aureus cells to oxacillin, vancomycin, and other cell wall inhibitors, has been reported (10), and a two-component regulatory system VraSR involved with modulation of the cell wall biosynthetic pathway was identified (50). The use of DNA microarrays for transcriptional profiling of S. aureus exposed to wall inhibitors was also described (91).

Penicillin-Binding Proteins

Tentative roles for the four staphylococcal PBPs in wall synthesis were originally assigned on the basis of morphological/biochemical effects of β-lactams that showed more or less selective binding to individual PBPs (28). More recent studies used genetic techniques to clarify the functions of PBP.

Genetic work using allelic replacement has clearly established the essential nature of PBP1 (93). However, the specific contribution of this protein to wall synthesis remains to be determined.

The structural gene of PBP2, pbpB, was cloned and sequenced (32, 60). The protein was shown to be composed of a TPase and a TGase domain, the former being essential for growth of the bacteria. Studies with conditional mutants of pbpB demonstrated that this essential function of PBP2 may be replaced by PBPB2A, the product of the imported drug resistance gene mecA (60, 69). The TGase domain of PBP2 was shown to cooperate functionally with PBP2A when the bacteria were growing in the presence of antibiotics (68). Critically located point mutations in pbpB were associated with methicillin resistance in both laboratory mutants and among some clinical isolates that did not carry the mecA gene but showed low-level β-lactam resistance (89). Involvement of PBP2 with vancomycin resistance was suggested by the apparent overproduction of the protein in a staphylococcal mutant (15).

The structural gene of PBP3, pbpC, was cloned and sequenced (67). The gene encodes for a class B high-molecular-weight protein with a C-terminal penicillin-binding domain showing the conserved motifs characteristic of TPases and an N-terminal nonbinding domain of unknown function. The pbpC was inactivated and the intact gene was introduced into a lab mutant lacking PBP3. Analysis of such constructs with the fluorographic technique for PBPs clearly showed that the cloned and sequenced pbpC was indeed the structural gene of PBP3 of S. aureus. Unlike in the cases of pbpA and pbpB, inactivation of pbpC in either methicillin-susceptible or -resis-tant backgrounds allowed growth of the bacteria, and there was no change in the muropeptide composition. The only observable alteration of such mutants was a reduction in autolysis rates. However, when mutants were grown in the presence of sub-MIC concentrations of methicillin the bacteria showed frequent aberrant morphologies (67).

The structural gene of PBP4, pbpD, was cloned and sequenced (24, 25, 35). Bacteria with inactivated PBP4 grew normally but produced a peptidoglycan in which the highly oligomeric components were substantially reduced. Inactivation of PBP4 had no effect on the level of methicillin resistance in drug-resistant strains. An extensive deletion in the promoter region was identified in β-lactam-resistant laboratory isolates that showed increase in peptidoglycan cross-linking (36). Inactivation of pbpD was demonstrated in a highly vancomycin-resistant lab mutant that produced peptidoglycan with greatly reduced cross-linking (84). These findings confirm earlier proposals that PBP4 can act as a secondary TPase. The possibility that expression of the PBP4 determinant is controlled via the upstream ABC transporter (which may be part of an operon) is being considered (25).

PBP2A—the protein product of the resistance gene mecA—was crystallized, allowing the identification of structural features responsible for the extremely low affinity of this protein to most beta-lactam antibiotics (51, 90).

The first monofunctional glycosyl transferase of S. aureus was identified and characterized (57). How and under what conditions this enzyme shares the TGase activity with that of the bifunctional PBP2 (68) are not known.

The first division and cell wall (DCW) cluster of cell division and cell wall-related genes (70) was described for S. aureus. As in the E. coli DCW cluster, it contains a PBP gene (pbpA for PBP1) at one end, followed by some determinants of muropeptide biosynthesis (mreY and murD) and then by cell division-related genes (*div1B*, *ftsA*, *ftsZ*). However, in contrast to *E. coli*, the genes were not overlapping and most of the muropeptide synthesis genes were absent. One of these (*murE*) was identified at a distance from the DCW (52).

COMPLEX FUNCTIONS OF CELL WALLS

The capacity of staphylococcal cell walls, peptidoglycan, and teichoic acid to induce the production of proinflammatory cytokines has been repeatedly documented (53, 56, 88, 94). Two global regulatory genes—*agr* and *sar*—that control the production and release of several staphylococcal virulence-related factors were shown to influence staphylococcal autolysis (induced by penicillin or Triton X-100) in an opposing fashion (26). The carboxy-terminal end of staphylococcal surface proteins (such as protein A) was shown to be covalently bound to the oligoglycine cross bridge of peptidoglycan muropeptide units, through a threonine residue in the protein (generated after cleaving of the wall sorting signal LPXTG), which is then linked to the amino-terminal glycine residue of the muropeptide (49,

99). The structure of these cross bridges is under the control of the *fem* gene complex. Therefore, the already pleiomorphic *femA/B* mutant (showing defective wall structure, reduced methicillin and lysostaphin resistance, and slow wall turnover) may also be affected in virulence.

The active involvement of the cell wall in complex functions is also being recognized in studies on the transport (release and uptake) of large molecules (23). Increased susceptibility of *femA/B* null mutants to some antibacterial agents other than β-lactams may be related to increased porosity of the peptidoglycan (87). Increased extractability of proteins was noted in a highly teicoplanin-resistant staphylococcal mutant with greatly decreased peptidoglycan cross-linking. Both resistance and extractability (wall porosity?) were abolished in a Tn*551* mutant selected for reduced resistance to teicoplanin (80).

Murein Hydrolases in Cell Division and Antibiotic Resistance

The cell biological functions of staphylococcal murein hydrolases and their regulation are being intensively investigated, and these studies benefited greatly from the introduction of

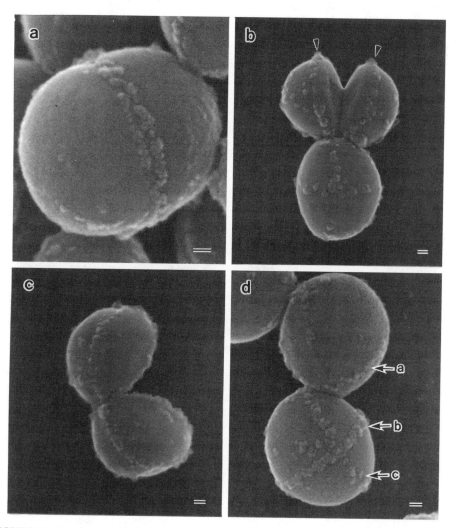

FIGURE 6 Localization of *atl* gene products on the cell surface of *S. aureus* during the division cycle as determined by scanning electron microscopy. Panels a to d show the immunogold labeling patterns on cells at different stages of the cell cycle. Scale bar, 100 nm. (Reproduced with permission from Yamada et al. [98].)

the sodium dodecyl sulfate-polyacrylamide gel electrophoresis method for profiling bacteriolytic enzymes and from genetic work on the cloning and sequencing of murein hydrolases (38, 55). Regulatory loci affecting autolysis as well as morphology were identified (11, 12). Oshida and his colleagues identified the autolysin gene *atl* encoding a bifunctional protein of deduced molecular size of 137,381 Da (63, 64). The protein was shown to contain an amidase as well as an endo-β-*N*-acetylglucosaminidase domain that undergo proteolytic processing at the cell surface. The availability of detailed sequence information on *atl* has led to some extremely interesting observations. The still "double-headed" *atl* gene product containing the fused amidase and glucosaminidase proteins was found to be localized by immunogold labeling on the plasma membrane as a circumferential ring at the future cell divisional site, and this localization precedes the appearance of centripetally growing cell wall (98) (Fig. 6). The findings suggest that this enzyme, known to be involved with cell separation, may function in conjunction with some staphylococcal *fts*-type protein in targeting the site of future cell division. The molecular "signature" that directs the autolysin gene to its site of action at the cellular equator was shown (by site-directed mutagenesis) to reside within the three repeat elements of the *atl* gene (2).

Alteration of autolytic properties (reduction or increase in rates) has been consistently seen in staphylococcal mutants isolated for changed β-lactam resistance. Stepwise increase in autolytic rates (and cell wall turnover rates) in parallel with the increase in resistance level was demonstrated in cefotaxime-resistant laboratory mutants (6). Reduced rates of wall turnover (and autolysis) were observed in *femA*, *femB*, and *femC* mutants. The reverse situation— major reduction (and heterogeneity) in methicillin resistance—was observed in several, but not all, transductants generated by crossing the inactivated *atl* into the highly methicillin-resistant strain COL (63).

SUMMARY

Through these findings the staphylococcal cell wall is beginning to emerge not only as a critically important exoskeleton responsible for the structural integrity of the bacterium, but also as an organelle of multiple functions: it is intimately involved with interactions with the system of innate immunity of the host; provides covalent attachment sites for a number of host-related surface proteins; and occupies center stage in resistance to β-lactam and glycopeptide antibiotics. Some observations also begin to throw light on how the control of degradative and synthetic activities and morphogenetic principles come together in the assembly and replication of cell walls.

I thank P. Giesbrecht, M. Sugai, M. Pinho, and J. Errington for allowing the reproduction of their illustrations in this review.

REFERENCES

1. **Archibald, A. R.** 1972. The chemistry of staphylococcal cell walls, p. 75. *In* J. O. Cohen (ed.), *The Staphylococci.* Wiley-Interscience, New York, N.Y.
2. **Baba, T., and O. Schneewind.** 1998. Targeting of muralytic enzymes to the cell division site of gram-positive bacteria: repeat domains direct autolysin to the equatorial surface ring of *Staphylococcus aureus. EMBO J.* **17:**4639–4646.
3. **Benson, T. E., D. B. Prince, V. T. Mutchler, K. A. Curry, A. M. Ho, R. W. Sarver, J. C. Hagadorn, G. H. Choi,** and **R. L. Garlick.** 2002. X-ray crystal structure of *Staphylococcus aureus* FemA. *Structure* **10:**1107–1115.
4. **Berger-Bächi, B.** 1983. Insertional inactivation of staphylococcal methicillin resistance by Tn*551. J. Bacteriol.* **154:** 479–487.
5. **Bischoff, M., M. Roos, J. Putnik, A. Wada, P. Glanzmann, P. Giachino, P. Vaudaux,** and **B. Berger-Bächi.** 2001. Involvement of multiple genetic loci in *Staphylococcus aureus* teicoplanin resistance. *FEMS Microbiol. Lett.* **194:**77–82.
6. **Boneca, I. G., N. Xu, D. A. Gage, B. L. M. de Jonge,** and **A. Tomasz.** 1997. Structural characterization of an abnormally cross-linked muropeptide dimer that is accumulated in the peptidoglycan of methicillin- and cefotaxime-resistant mutants of *Staphylococcus aureus. J. Biol. Chem.* **272:**29053–29059.
7. **Boneca, I. G., Z.-H. Huang, D. A. Gage,** and **A. Tomasz.** 2000. Characterization of *Staphylococcus aureus* cell wall glycan strands, evidence for a new endo-β-*N*-acetylglucosaminidase activity. *J. Biol. Chem.* **275:**9910– 9918.
8. **Boyle-Vavra, S., H. Labischinski, C. C. Ebert, K. Ehlert,** and **R. S. Daum.** 2001. A spectrum of changes occurs in peptidoglycan composition of glycopeptide-intermediate clinical *Staphylococcus aureus* isolates. *Antimicrob. Agents Chemother.* **45:**280–287.
9. **Boyle-Vavra, S., R. B. Carey,** and **R. S. Daum.** 2001. Development of vancomycin and lysostaphin resistance in a methicillin-resistant *Staphylococcus aureus* isolate. *J. Antimicrob. Chemother.* **48:**617–625.
10. **Boyle-Vavra, S., S. Yin, M. Challapalli,** and **R. S. Daum.** 2003. Transcriptional induction of the penicillin-binding protein 2 gene in *Staphylococcus aureus* by cell wall-active antibiotics oxacillin and vancomycin. *Antimicrob. Agents Chemother.* **47:**1028–1036.
11. **Brunskill, E. W.,** and **K. W. Bayles.** 1996. Identification of LytSR-regulated genes from *Staphylococcus aureus. J. Bacteriol.* **178:**5810–5812.
12. **Brunskill, E. W., B. L. de Jonge,** and **K. W. Bayles.** 1997. The *Staphylococcus aureus scdA* gene: a novel locus that affects cell division and morphogenesis. *Microbiology* **143:** 2877–2882.
13. **Cui, L., H. Murakami, K. Kuwahara-Arai, H. Hanaki,** and **K. Hiramatsu.** 2000. Contribution of a thickened cell wall and its glutamine nonamidated component to the vancomycin resistance expressed by *Staphylococcus aureus* Mu50. *Antimicrob. Agents Chemother.* **44:**2276–2285.
14. **Cui, L., X. Ma, K. Sato, K. Okuma, F. C. Tenover, E. M. Mamizuka, C. G. Gemmell, M. N. Kim, M. C. Ploy, N. El-Solh, V. Ferraz,** and **K. Hiramatsu.** 2003. Cell wall thickening is a common feature of vancomycin resistance in *Staphylococcus aureus. J. Clin. Microbiol.* **41:**5–14.
15. **Daum, R. S., S. Gupta, R. Sabbagh,** and **W. M. Milewski.** 1992. Characterization of *Staphylococcus aureus* isolates with decreased susceptibility to vancomycin and teicoplanin: isolation and purification of a constitutively produced protein associated with decreased susceptibility. *J. Infect. Dis.* **166:**1066–1072.
16. **De Jonge, B. L. M., D. Gage,** and **N. Xu.** 2002. The carboxyl terminus of peptidoglycan stem peptides is a determinant for methicillin resistance in *Staphylococcus aureus. Antimicrob. Agents Chemother.* **46:**3151–3155.
17. **De Jonge, B. L. M., Y.-S. Chang, D. Gage,** and **A. Tomasz.** 1992. Peptidoglycan composition of a highly methicillin-resistant *Staphylococcus aureus* strain: the role of penicillin binding protein 2A. *J. Biol. Chem.* **267:**11248–11254.

18. **De Jonge, B. L. M., Y.-S. Chang, D. Gage, and A. Tomasz.** 1992. Peptidoglycan composition in heterogeneous Tn551 mutants of a methicillin-resistant *Staphylococcus aureus* strain. *J. Biol. Chem.* **267:**11255–11259.

19. **De Jonge, B. L. M., Y.-S. Chang, N. Xu, and D. Gage.** 1996. Effect of exogenous glycine on peptidoglycan composition and resistance in a methicillin-resistant *Staphylococcus aureus* strain. *Antimicrob. Agents Chemother.* **40:** 1498–1503.

20. **De Lencastre, H., B. L. M., de Jonge, P. R. Matthews, and A. Tomasz.** 1994. Molecular aspects of methicillin resistance in *Staphylococcus aureus*. *J. Antimicrob. Chemother.* **33:**7–24.

21. **De Lencastre, H., and A. Tomasz.** 1994. Reassessment of the number of auxiliary genes essential for expression of high-level methicillin resistance in *Staphylococcus aureus*. *Antimicrob. Agents Chemother.* **38:**2590–2598.

22. **De Lencastre, H., S. W. Wu, M. G. Pinho, A. M. Ludovice, S. R. Filipe, S. Gardete, R. Sobral, S. Gill, M. Chung, and A. Tomasz.** 1999. Antibiotic resistance as a stress response: complete sequence of a large number of chromosomal loci in *Staphylococcus aureus* strain COL that impact on the expression of resistance to methicillin. *Microb. Drug Resist.* **5:**163–175.

23. **Dijkstra, A. J., and W. Keck.** 1996. Peptidoglycan as a barrier to transenvelope transport. *J. Bacteriol.* **178:**5555–5562.

24. **Domanski, T. L., and K. W. Bayles.** 1995. Analysis of *Staphylococcus aureus* genes encoding penicillin-binding protein 4 and an ABC-type transporter. *Gene* **167:**111–113.

25. **Domanski, T. L., B. L. M., de Jonge, and K. W. Bayles.** 1997. Transcription analysis of the *Staphylococcus aureus* gene encoding penicillin-binding protein 4. *J. Bacteriol.* **179:**2651–2657.

26. **Fujimoto, D. F., and K. W. Bayles.** 1998. Opposing roles of the *Staphylococcus aureus* virulence regulators, Agr and Sar, in Triton X-100- and penicillin-induced autolysis. *J. Bacteriol.* **180:**3724–3726.

27. **Fujimura, T., and K. Murakami.** 1997. Increase of methicillin resistance in *Staphylococcus aureus* caused by deletion of a gene whose product is homologous to lytic enzymes. *J. Bacteriol.* **179:**6294–6301.

28. **Georgopapadakou, N. H., B. A. Dix, and Y. R. Mauriz.** 1986. Possible physiological functions of penicillin-binding proteins in *Staphylococcus aureus*. *Antimicrob. Agents Chemother.* **29:**333–336.

29. **Ghuysen, J.-M., and R. Hakenbeck (ed.).** 1994. *Bacterial Cell Wall.* Elsevier Science B.V., Amsterdam, The Netherlands.

30. **Giesbrecht, P., T. Kersten, H. Maidhof, and J. Weeke.** 1998. The staphylococcal cell wall: morphogenesis and fatal variations in the presence of penicillin. *Microbiol. Mol. Biol. Rev.* **62:**1371–1414.

31. **Gustafson, J., A. Strässle, H. Hächler, F. H. Kayser, and B. Berger-Bächi.** 1994. The femC locus of *Staphylococcus aureus* required for methicillin resistance includes the glutamine synthetase operon. *J. Bacteriol.* **176:**1460–1467.

32. **Hackbarth, C. J., T. Kocagoz, S. Kocagoz, and H. F. Chambers.** 1995. Point mutations in *Staphylococcus aureus* PBP2 gene affect penicillin-binding kinetics and are associated with resistance. *Antimicrob. Agents Chemother.* **39:** 103–106.

33. **Hakenbeck, R., J.-V. Höltje, and H. Labischinski (ed.).** 1983. *The Target of Penicillin.* Walter de Gruyter, Berlin, Germany.

34. **Hanaki, H., H. Labischinski, Y. Inaba, N. Kondo, H. Murakami, and K. Hiramatsu.** 1998. Increase in glutamine-non-amidated muropeptides in the peptidoglycan of vancomycin-resistant *Staphylococcus aureus* strain Mu50. *J. Antimicrob. Chemother.* **42:**315–320.

35. **Henze, U. U., and B. Berger-Bächi.** 1995. *Staphylococcus aureus* penicillin-binding protein 4 and intrinsic β-lactam resistance. *Antimicrob. Agents Chemother.* **39:**2415–2422.

36. **Henze, U. U., and B. Berger-Bächi.** 1996. Penicillin-binding protein 4 overproduction increases β-lactam resistance in *Staphylococcus aureus*. *Antimicrob. Agents Chemother.* **40:**2121–2125.

37. **Jacobs, C., J.-M. Frère, and S. Normark.** 1997. Cytosolic intermediates for cell wall biosynthesis and degradation control inducible β-lactam resistance in gram-negative bacteria. *Cell* **88:**823–832.

38. **Jayaswal, R. K., Y. I. Lee, and B. J. Wilkinson.** 1990. Cloning and expression of a *Staphylococcus aureus* gene encoding a peptidoglycan hydrolase activity. *J. Bacteriol.* **172:**5783–5788.

39. **Jolly, L., S. Wu, J. van Heijenoort, H. de Lencastre, D. Mengin-Lecreulx, and A. Tomasz.** 1997. The femR315 gene from *Staphylococcus aureus*, the interruption of which results in reduced methicillin resistance, encodes a phosphoglucosamine mutase. *J. Bacteriol.* **179:**5321–5325.

40. **Kojima, N., Y. Araki, and E. Ito.** 1985. Structure of the linkage units between ribitol teichoic acids and peptidoglycan. *J. Bacteriol.* **161:**299–306.

41. **Komatsuzawa, H., G. H. Choi, T. Fujiwara, Y. Huang, K. Ohta, M. Sugai, and H. Suginaka.** 2000. Identification of a fmtA-like gene that has similarity to other PBPs and beta-lactamases in *Staphylococcus aureus*. *FEMS Microbiol. Lett.* **188:**35–39.

42. **Komatsuzawa, H., K. Ohta, M. Sugai, T. Fujiwara, P. Glanzmann, B. Berger-Bachi, and H. Suginaka.** 2000. Tn551-mediated insertional inactivation of the fmtB gene encoding a cell wall-associated protein abolishes methicillin resistance in *Staphylococcus aureus*. *J. Antimicrob. Chemother.* **45:**421–431.

43. **Komatsuzawa, H., T. Fujiwara, H. Nishi, S. Yamada, M. Ohara, N. McCallum, B. Berger-Bachi, and M. Sugai.** 2004. The gate controlling cell wall synthesis in *Staphylococcus aureus*. *Mol. Microbiol.* **53:**1221–1231.

44. **Komatsuzawa, H., K. Ohta, H. Labischinski, M. Sugai, and H. Suginaka.** 1999. Characterization of fmtA, a gene that modulates the expression of methicillin resistance in *Staphylococcus aureus*. *Antimicrob. Agents Chemother.* **43:** 2121–2125.

45. **Komatsuzawa, H., K. Ohta, S. Yamada, K. Ehlert, H. Labischinski, J. Kajimura, T. Fujiwara, and M. Sugai.** 2002. Increased glycan chain length distribution and decreased susceptibility to moenomycin in a vancomycin-resistant *Staphylococcus aureus* mutant. *Antimicrob. Agents Chemother.* **46:**75–81.

46. **Komatsuzawa, H., K. Ohta, T. Fujiwara, G. H. Choi, H. Labischinski, and M. Sugai.** 2001. Cloning and sequencing of the gene, fmtC, which affects oxacillin resistance in methicillin-resistant *Staphylococcus aureus*. *FEMS Microbiol. Lett.* **203:**49–54.

47. **Komatsuzawa, H., M. Sugai, K. Ohta, T. Fujiwara, S. Nakashima, J. Suzuki, C. Y. Lee, and H. Suginaka.** 1997. Cloning and characterization of the fmt gene which affects the methicillin resistance level and autolysis in the presence of Triton X-100 in methicillin-resistant *Staphylococcus aureus*. *Antimicrob. Agents Chemother.* **41:**2355–2361.

48. **Kornblum, J., B. J. Hartman, R. P. Novick, and A. Tomasz.** 1986. Conversion of a homogeneously methicillin resistant strain of *Staphylococcus aureus* to heterogeneous resistance by Tn551-mediated insertional inactivation. *Eur. J. Clin. Microbiol.* **5:**714–718.

49. **Kruger, R. G., B. Otvos, B. A. Frankel, M. Bentley, P. Dostal and D. G. McCafferty.** 2004. Analysis of the substrate specificity of the *Staphylococcus aureus* sortase transpeptidase SrtA. *Biochemistry* **43:**1541–1551.

50. **Kuroda, M., H. Kuroda, T. Oshima, F. Takeuchi, H. Mori, and K. Hiramatsu.** 2003. Two-component system VraSR positively modulates the regulation of cell-wall biosynthesis pathway in *Staphylococcus aureus*. *Mol. Microbiol.* **49:**807–821.

51. **Lim, D., and N. C. J. Strynadka.** 2002. Structural basis for the β-lactam resistance of PBP2a from methicillin-resistant *Staphylococcus aureus*. *Nat. Struct. Biol.* **9:**789–885.

52. **Ludovice, A. M., S. Wu, and H. de Lencastre.** 1998. Molecular cloning and DNA sequencing of the *Staphylococcus aureus* UDP-N-acetylmuramyl tripeptide synthetase (*murE*) gene, essential for the optimal expression of methicillin resistance. *Microb. Drug Resist.* **4:**85–90.

53. **Luker, F. I., D. Mitchell, and H. P. Laburn.** 2000. Fever and motor activity in rats following day and night injections of *Staphylococcus aureus* cell walls. *Am. J. Physiol. Regul. Integr. Comp. Physiol.* **279:**R610–R616.

54. **Maki, H., T. Yamaguchi, and K. Murakami.** 1994. Cloning and characterization of a gene affecting the methicillin resistance level and the autolysis rate in *Staphylococcus aureus*. *J. Bacteriol.* **176:**4993–5000.

55. **Mani, N., P. Tobin, and R. K. Jayaswal.** 1993. Isolation and characterization of autolysis-defective mutants of *Staphylococcus aureus* created by Tn917-lacZ mutagenesis. *J. Bacteriol.* **175:**1493–1499.

56. **Matsui, K., R. Motohashi, and A. Nishikawa.** 2000. Cell wall components of *Staphylococcus aureus* induce interleukin-5 production in patients with atopic dermatitis. *J. Interferon Cytokine Res.* **20:**321–324.

57. **May Wang, Q., R. B Perry, R. B. Johnson, W. E. Alborn, W.-K. Yeh, and P. L. Skatrud.** 2001. Identification and characterization of a monofunctional glycosyltransferase from *Staphylococcus aureus*. *J. Bacteriol.* **183:**4779–4785.

58. **Moreau, M., J. C. Richards, J. M. Fournier, R. A. Byrd, W. W. Karakawa, and W. F. Vann.** 1990. Structure of the type-5 capsular polysaccharide of *Staphylococcus aureus*. *Carbohydr. Res.* **201:**285–297.

59. **Murakami, K., and A. Tomasz.** 1989. Involvement of multiple genetic determinants in high-level methicillin-resistance in *Staphylococcus aureus*. *J. Bacteriol.* **171:**874–879.

60. **Murakami, K., T. Fujimura, and M. Doi.** 1994. Nucleotide sequence of the structural gene for the penicillin-binding protein 2 of *Staphylococcus aureus* and the presence of a homologous gene in other staphylococci. *FEMS Microbiol. Lett.* **117:**131–136.

61. **Ornelas-Soares, A., H. de Lencastre, B. de Jonge, D. Gage, Y.-S. Chang, and A. Tomasz.** 1993. The peptidoglycan composition of a *Staphylococcus aureus* mutant selected for reduced methicillin resistance. *J. Biol. Chem.* **268:**26268–26272.

62. **Ornelas-Soares, A., H. de Lencastre, B. L. M. de Jonge, and A. Tomasz.** 1994. Reduced methicillin resistance in a new *Staphylococcus aureus* transposon mutant that incorpo-rates muramyl dipeptides into the cell wall peptidoglycan. *J. Biol. Chem.* **269:**27246–27250.

63. **Oshida, T., and A. Tomasz.** 1992. Isolation and characterization of a Tn551-autolysis mutant of *Staphylococcus aureus*. *J. Bacteriol.* **174:**4952–4959.

64. **Oshida, T., M. Sugai, H. Komatsuzawa, Y.-M. Hong, H. Suginaka, and A. Tomasz.** 1995. A *Staphylococcus aureus* autolysin that has an N-acetylmuramoyl-L-alanine amidase domain and an endo-β-N-acetylglucosaminidase domain: cloning, sequence analysis and characterization. *Proc. Natl. Acad. Sci. USA* **92:**285–289.

65. **Pfeltz, R. F., V. K. Singh, J. L. Schmidt, M. A. Batten, C. S. Baranyk, M. J. Nadakavukaren, R. K. Jayaswal, and B. J. Wilkinson.** 2000. Characterization of passage-selected vancomycin-resistant *Staphylococcus aureus* strains of diverse parental backgrounds. *Antimicrob. Agents Chemother.* **44:**294–303.

66. **Pinho, M. G., and J. Errington.** 2003. Dispersed mode of *Staphylococcus aureus* cell wall synthesis in the absence of the division machinery. *Mol. Microbiol.* **50:**871–881.

67. **Pinho, M. G., H. de Lencastre, and A. Tomasz.** 2000. Cloning, characterization, and inactivation of the gene *pbpC*, encoding penicillin-binding protein 3 of *Staphylococcus aureus*. *J. Bacteriol.* **182:**1074–1079.

68. **Pinho, M. G., H. de Lencastre, and A. Tomasz.** 2001. An acquired and a native penicillin-binding protein cooperate in building the cell wall of drug-resistant staphylococci. *Proc. Natl. Acad. Sci. USA* **98:**10886–10891.

69. **Pinho, M. G., S. Filipe, H. de Lencastre, and A. Tomasz.** 2001. Complementation of the essential peptidoglycan transpeptidase function of penicillin-binding protein (PBP) 2 by the drug resistance protein PBP2A in *Staphylococcus aureus*. *J. Bacteriol.* **183:**6525–6531.

70. **Pucci, M. J., J. A. Thanassi, L. F. Discotto, R. E. Kessler, and T. J. Dougherty.** 1998. Identification and characterization of cell wall-cell division gene clusters in pathogenic gram-positive cocci. *J. Bacteriol.* **179:**5632–5635.

71. **Rohrer, S., and B. Berger-Bachi.** 2003. Application of a bacterial two-hybrid system for the analysis of protein-protein interactions between FemABX family proteins. *Microbiology* **149:**2733–2738.

72. **Rohrer, S., H. Maki, and B. Berger-Bachi.** 2003. What makes resistance to methicillin heterogeneous? *J. Med. Microbiol.* **52:**605.

73. **Rohrer, S., and B. Berger-Bachi.** 2003. FemABX peptidyl transferases: a link between branched-chain cell wall peptide formation and beta-lactam resistance in gram-positive cocci. *Antimicrob. Agents Chemother.* **47:**837–846.

74. **Rohrer, S., K. Ehlert, M. Tschierske, H. Labischinski, and B. Berger-Bachi.** 1999. The essential *Staphylococcus aureus* gene fmhB is involved in the first step of peptidoglycan pentaglycine interpeptide formation. *Proc. Natl. Acad. Sci. USA* **96:**9351–9356.

75. **Sau, S., N. Bhasin, E. R. Wann, J. C. Lee, T. J. Foster, and C. Y. Lee.** 1997. The *Staphylococcus aureus* allelic genetic loci for serotype 5 and 8 capsule expression contain the type-specific genes flanked by common genes. *Microbiology* **143:**2395–2405.

76. **Schneider, T., M. M. Senn, B. Berger-Bachi, A. Tossi, H. G. Sahl, and I. Wiedemann.** 2004. In vitro assembly of a complete, pentaglycine interpeptide bridge containing cell wall precursor (lipid II-Gly5) of *Staphylococcus aureus*. *Mol. Microbiol.* **53:**675–685.

77. **Severin, A., K. Tabei, F. Tenover, M. Chung, N. Clarke, and A. Tomasz.** 2004. High level oxacillin and van-

comycin resistance and altered cell wall composition in *Staphylococcus aureus* carrying both the staphylococcal *mecA* gene and the enterococcal *vanA* gene complex. *J. Biol. Chem.* **279:**3398–3407.

78. **Severin, A., S. W. Wu, K. Tabei, and A. Tomasz.** 2004. Penicillin-binding protein 2 is essential for the expression of high level vancomycin resistance and cell wall synthesis in vancomycin-resistant *S. aureus* carrying the enterococcal *vanA* gene complex. *Antimicrob. Agents Chemother.* **48:**4566–4573.

79. **Sieradzki, K., and A. Tomasz.** 1997. Inhibition of cell wall turnover and autolysis by vancomycin in a highly vancomycin-resistant mutant of *Staphylococcus aureus. J. Bacteriol.* **179:**2557–2566.

80. **Sieradzki, K., and A. Tomasz.** 1998. Suppression of glycopeptide resistance in a highly teicoplanin resistant mutant of *Staphylococcus aureus* by transposon inactivation of genes involved in cell wall synthesis. *Microb. Drug Resist.* **4:**159–168.

81. **Sieradzki, K., and A. Tomasz.** 1999. Gradual alterations in cell wall structure and metabolism in vancomycin-resistant mutants of *Staphylococcus aureus. J. Bacteriol.* **181:**7566–7570.

82. **Sieradzki, K., and A. Tomasz.** 2003. Alterations of cell wall structure and metabolism accompany reduced susceptibility to vancomycin in an isogenic series of clinical isolates of *Staphylococcus aureus. J. Bacteriol.* **185:**7103–7110.

83. **Sieradzki, K., L. Borio, J. Dick, and A. Tomasz.** 2003. Evolution of VISA strain *in vivo*: multiple genetic changes in a single lineage of MRSA under the impact of antibiotics administered for chemotherapy. *J. Clin. Microbiol.* **41:**1687–1693.

84. **Sieradzki, K., M. G. Pinho, and A. Tomasz.** 1999. Inactivated *pbp*4 in highly glycopeptide-resistant laboratory mutants of *Staphylococcus aureus. J. Biol. Chem.* **274:**18942–18946.

85. **Snowden, M. A., and H. R. Perkins.** 1990. Peptidoglycan cross-linking in *Staphylococcus aureus*: an apparent random polymerisation process. *Eur. J. Biochem.* **191:**373–377.

86. **Sobral, R. G., A. M. Ludovice, S. Gardete, K. Tabei, H. de Lencastre, and A. Tomasz.** 2003. Normally functioning *murF* is essential for the optimal expression of methicillin resistance in *Staphylococcus aureus. Microb. Drug Resist.* **9:**231–241.

87. **Strandén, A. M., K. Ehlert, H. Labischinski, and B. Berger-Bächi.** 1997. Cell wall monoglycine cross-bridges and methicillin hypersusceptibility in a *femAB* null mutant of methicillin-resistant *Staphylococcus aureus. J. Bacteriol.* **179:**9–16.

88. **Timmerman, C. P., E. Mattson, L. Martinez-Martinez, L. de Graaf, J. A. G. van Strijp, V. H. Verbrugh, J. Verhoef, and A. Fleer.** 1993. Induction of release of tumor necrosis factor from human monocytes by staphylococci and staphylococcal peptidoglycans. *Infect. Immun.* **61:**4167–4172.

89. **Tomasz, A., H. B. Drugeon, H. M. de Lencastre, D. Jabes, L. McDougall, and J. Bille.** 1989. New mechanism for methicillin resistance in *Staphylococcus aureus*: clinical isolates that lack the PBP 2a gene and contain normal penicillin-binding proteins with modified penicillin-binding capacity. *Antimicrob. Agents Chemother.* **33:**1869–1874.

90. **Toney, J. H., G. G. Hammond, B. Leiting, K.-A. D. Pryor, J. K. Wu, G. C. Cuca, and D. L. Pompliano.** 1998. Soluble penicillin-binding protein 2a: β-lactam binding and inhibition of non-β-lactams using a 96-well format. *Anal. Biochem.* **255:**113–119.

91. **Utaida, S., P. M. Dunman, D. Macapagal, E. Murphy, S. J. Projan, V. K. Singh, R. K. Jayaswal, and B. J. Wilkinson.** 2003. Genome-wide transcriptional profiling of the response of *Staphylococcus aureus* to cell-wall-active antibiotics reveals a cell-wall-stress stimulon. *Microbiology* **149:** 2719–2732.

92. **VanNieuwenhze, M. S., S. C. Mauldin, M. Zia-Ebrahimi, B. E. Winger, W. J. Hornback, S. L. Saha, J. A. Aikins, and L. C. Blaszczak.** 2002. The first total synthesis of lipid II: the final monomeric intermediate in bacterial cell wall biosynthesis. *J. Am. Chem. Soc.* **124:**3656–3660.

93. **Wada, A., and H. Watanabe.** 1998. Penicillin-binding protein 1 of *Staphylococcus aureus* is essential for growth. *J. Bacteriol.* **180:**2759–2765.

94. **Wray, G. M., S. J. Foster, C. J. Hinds, and C. Thiemermann.** 2001. A cell wall component from pathogenic and non-pathogenic gram-positive bacteria (peptidoglycan) synergises with endotoxin to cause the release of tumour necrosis factor-alpha, nitric oxide production, shock, and multiple organ injury/dysfunction in the rat. *Shock* **5:** 135–142.

95. **Wu, S., H. de Lencastre, A. Sali, and A. Tomasz.** 1996. A phosphoglucomutase-like gene essential for the optimal expression of methicillin resistance in *Staphylococcus aureus*: molecular cloning and DNA sequencing. *Microb. Drug Resist.* **2:**277–286.

96. **Wu, S., H. de Lencastre, and A. Tomasz.** 1996. Sigma-B, a putative operon encoding alternate sigma factor of *Staphylococcus aureus* RNA polymerase: molecular cloning and DNA sequencing. *J. Bacteriol.* **178:**6036–6042.

97. **Xu, N., Z.-H. Huang, B. L. M. de Jonge, and D. A. Gage.** 1997. Structural characterization of peptidoglycan muro-peptides by matrix-assisted laser desorption ionization mass spectrometry and postsource decay analysis. *Anal. Biochem.* **248:**7–14.

98. **Yamada, S., M. Sugai, H. Komatsuzawa, S. Nakashima, T. Oshida, A. Matsumoto, and H. Suginaka.** 1996. An autolysin ring associated with cell separation of *Staphylococcus aureus. J. Bacteriol.* **178:**1565–1571.

99. **Zong, Y., S. K. Mazmanian, O. Schneewind, and S. V. Narayana.** 2004. The structure of sortase B, a cysteine transpeptidase that tethers surface protein to the *S. aureus* cell wall. *Structure* **12:**105–112.

Staphylococcal Capsule

CHIA Y. LEE AND JEAN C. LEE

37

Capsular polysaccharides (CPs) are bacterial cell surface components that contribute to microbial virulence by promoting evasion of or interference with the host immune system. Capsules produced by *Staphylococcus aureus* have traditionally been divided into 11 serotypes, although evidence to support all 11 serotypes is lacking. In fact, only four of the CPs (serotypes 1, 2, 5, and 8) have been chemically defined (Fig. 1).The capsule loci from serotypes 1, 5, and 8 have been identified and genetically characterized. Most *S. aureus* strains isolated from humans produce either serotype 5 (CP5) or serotype 8 (CP8) capsules. Serotype 1 and 2 strains, though very rare, produce copious amounts of capsule that result in mucoid colonies on agar plates. On the other hand, serotype 5 and 8 strains produce less capsule under routine laboratory conditions. Therefore, type 5 and 8 strains have been referred to as microencapsulated, a designation that is a misnomer because these strains may produce abundant capsule in vivo or under appropriate in vitro conditions (Fig. 2). This chapter will focus on major advances in staphylococcal capsule research since the previous review (28). For a comprehensive study, readers are referred to a recent review by O'Riordan and Lee (46).

GENETIC ANALYSIS OF CAPSULE EXPRESSION

The *cap5* and *cap8* genes required for the synthesis of CP5 and CP8, respectively, have been cloned and characterized (54, 55). The two loci are allelic, and each contains 16 genes in an operon (Fig. 3). Twelve of the 16 genes are almost identical, and they are responsible for synthesis of the sugar residues that are shared between CP5 and CP8. The remaining four genes are specific to each serotype; they dictate the unique linkages between the sugars, the position of N-acetyl-mannosaminuronic acid (ManNAcA) O-acetylation, and polysaccharide transport and polymerization. It is also noteworthy that the specific genes are located in the center of the operon and are flanked by the common genes. This organization suggests that serotype conversion occurs by homologous recombination between the conserved regions. Indeed, a serotype 8 strain was converted to produce

CP5 by transduction from a type 5 strain containing a Tn917 insertion near the specific *cap5* genes (65). Thus, it is conceivable that serotype conversion could occur in a host with appropriate selective pressure.

A cluster of 13 *cap1* genes forming an operon was initially identified for synthesis of the type 1 capsule (26, 33). However, two more genes were identified later upon further sequencing of the downstream region (41). Interestingly, the entire *cap1* operon is located in a staphylococcal cassette chromosome (SCC) genetic element similar to the methicillin resistance SCC*mec* elements (20). These cassette chromosome elements are located at the same specific site in the *S. aureus* chromosome, and thus the SCC*cap1* and SCC*mec* elements are allelic. The SCC*cap1* element is defective in mobilization due to mutations within the recombinase genes. Nonetheless, the attachment sites are functional because excision of the element can be demonstrated by complementation with functional recombinase genes from an SCC*mec* element. Although the SCC*cap1* element itself and the left flanking regions contain numerous mutations and DNA rearrangements, the *cap1* genes are intact and functional. This observation suggests that SCC*cap1* inserted into the SCC insertion site a long time ago, and that the *cap1* genes confer a survival advantage on the host strain (41).

The fact that *cap1* is located at the SCC locus implies that a strain could possess two capsule loci in its chromosome and simultaneously produce two different capsule types. In fact, the serotype 1 strain M has been shown to carry the *cap1* genes at the SCC locus and a gene cluster homologous to the *cap8* genes at the *cap5(8)* locus. Deletion of the *cap1* genes from the bacterial chromosome resulted in a mutant that produced a different capsule type (55). Whether the wild-type strain M produces two CPs at the same time has not yet been determined. The mucoid serotype 2 strain Smith also contains two capsule loci in its chromosome, as demonstrated by Southern hybridization experiments (27). It will be of interest to assess whether the *cap2* gene cluster is located in an SCC genetic element similar to the *cap1* genes.

Approximately 20% of *S. aureus* isolates from humans are nontypeable (NT), i.e., they are nonreactive with anti-

Gram-Positive Pathogens, 2nd edition, edited by Vincent A. Fischetti et al.
© 2006 ASM Press, Washington, D.C.

CP1:   → 4)-α-D-GalNAcA*p*-(1→ 4)-α-D-GalNAcA*p*-(1→ 3)-α-D-FucNAc*p*-(1→
         (A taurine residue is linked by an amide bond to every 4th D-GalNAcA*p* residue)

CP2:   → 4)-β-D-GlcNAcA*p*-(1→ 4)-β-D-GlcNAcA*p*-(L-alanyl)-(1→

CP5:   → 4)-β-D-ManNAcA*p*-(1→ 4)-α-L-FucNAc*p*-(3OAc)-(1→ 3)-β-D-FucNAc*p*-(1→

CP8:   → 3)-β-D-ManNAcA*p*-(4OAc)-(1→ 3)-α-L-FucNAc*p*-(1→ 3)-α-D-FucNAc*p*-(1→

FIGURE 1 Structures of staphylococcal capsules. Abbreviations: GalNAcA, *N*-acetyl-galac-tosaminuronic acid; GlcNAcA, *N*-acetyl-glucosaminuronic acid; O Ac, O-acetyl.

bodies to CP types 1, 2, 5, and 8. The majority of these strains carry an intact copy of the *cap5(8)* locus and produce a *cap5(8)* transcript, indicating that these NT isolates fail to express CP due to point mutations within genes essential for CP production (6a). Some NT isolates produce reduced *cap5(8)* transcript, and this phenotype is correlated with mutations within the *cap5(8)A* promoter or defects within regulatory genes. The *cap5(8)* locus was replaced by IS257 in a small subset of bovine isolates of *S. aureus* from Argentina (6a). Our findings reveal that lack of capsule expression in NT *S. aureus* can be explained by multiple mechanisms, and these data argue against the existence of capsule serotypes other than 1, 2, 5, and 8.

Recently, the genomes of four coagulase-negative staphylococcal strains (two strains of *S. epidermidis* [14, 67] and one strain each of *S. haemolyticus* [60a] and *S. saprophyticus* [24a]) have been sequenced. Both *S. haemolyticus* and *S. saprophyticus* carry a capsule gene locus that is partially homologous to the *S. aureus cap5(8)* locus. In contrast, *cap*-homologous genes have not been identified in the two *S. epidermidis* genomes. The CP from *S. haemolyticus* JCHC1435 was recently purified and characterized (S. Flahaut and J. C. Lee, unpublished data), and an analysis of the expression of the *S. saprophyticus* capsule genes is under way.

REGULATION

Both *cap1* and *cap8* operons are transcribed by a principal promoter upstream of the operon, although weak internal promoters in both operons exist (47, 56). The *cap1* promoter is much stronger than the *cap8* promoter in the same *S. aureus* genetic background. The *cap1* genes are apparently constitutively expressed because no *cis*-acting elements exist upstream of the −35 promoter region of *cap1*. Nonetheless, the mucoid phenotype of both type 1 and type 2 strains is unstable (26, 58). For the type 1 strain M this instability was shown to be due to random mutations in the *cap1* genes (34). However, it is not known whether the high mutation rate occurs specifically at the *cap1* locus or whether it occurs nonspecifically throughout the genome.

In contrast to the serotype 1 and 2 strains, serotype 5 and 8 capsule production is highly regulated by environmental cues. Iron limitation, carbon dioxide, pH, nutrient availability, oxygen tension, and an in vivo environment have all been shown to influence the level of capsule expression (9, 15, 16, 32, 51, 59, 60). The difference in CP8 production in vivo and in vitro has been estimated to be as high as 300-fold (32). Analyses of the *cap8* promoter showed that a 10-bp inverted repeat just upstream of the −35 region is required for optimal expression of the *cap8* genes. A mutant strain containing a chromosomal mutation within this repeat region showed a drastic reduction in CP8 production (48). A cellular factor has been implicated in binding to the 10-bp repeat, but repeated efforts to isolate the factor from whole cell extracts failed when the DNA fragment containing the repeat region was used as bait. More recent experiments using enriched cellular fractions have identified a 28-kDa putative binding protein. The gene encoding the

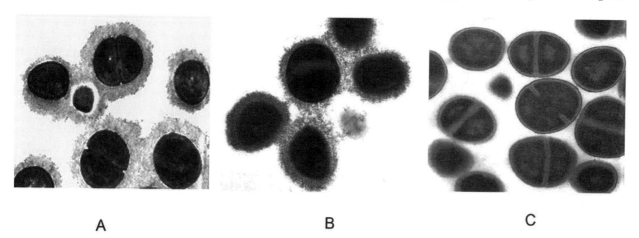

A   B   C

FIGURE 2 Transmission electron micrographs of *S. aureus* producing CP5 (A), CP8 (B), or no capsule (C). Prior to dehydration and embedding, the bacteria were incubated with capsular antibodies to stabilize and visualize the capsule.

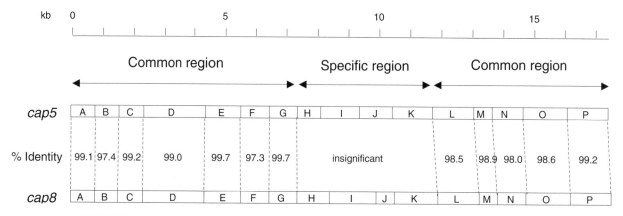

FIGURE 3 Comparison of *cap5* and *cap8* gene clusters. Gene designations are shown in boxes. Percent identity indicates the amino acid identity of the deduced proteins between the two clusters. The genes are transcribed from left to right.

protein was identified as *codY*, a pleiotropic transcriptional regulator conserved in low-G+C gram-positive bacteria. The His$_6$-CodY protein efficiently binds to the *cap8* promoter (37). Ongoing experiments will determine whether the 28-kDa protein binds to the 10-bp inverted repeat.

The virulence genes in *S. aureus* are regulated by several global regulators that form a very complex regulatory network. A central player in the network is the two-component regulatory system *agr*, which senses cell density through an autoinducing peptide pheromone (18). The *agr* system was the first global regulator to be implicated in controlling capsule expression (8). Another well-studied regulator is SarA, which has been shown to be a DNA-binding transcriptional regulator. SarA regulates its target genes either by activating the *agr* system or by an *agr*-independent pathway (6). Regulation of capsule expression by *agr* and *sar* was studied in detail by constructing single and double mutants that were evaluated for capsule production, *cap*-specific mRNA expression, and promoter activity analyzed with reporter gene fusions (35). Results from this study showed that *agr* was the major positive regulator for capsule production and that regulation was exerted at the transcriptional level. Positive regulation of capsule production by *agr* was also observed in an in vivo rabbit endocarditis model (64). SarA played a minor role in regulating capsule expression, and its effect most likely occurred at the posttranslational level (35).

A second two-component regulator that influences CP5 and CP8 production is *arlRS* (13). The *arl* system has been shown to be a key regulator of autolysis and the synthesis of several *S. aureus* extracellular virulence factors. ArlRS regulation of capsule production was first identified by genetic screening whereby a reporter gene (*xylE*) fused to the *cap8* promoter was used to identify transposon mutations affecting *cap* gene expression (38). Further studies indicated that *arlR* was a strong transcriptional activator of *S. aureus* CP production (6a, 38). Recently, the *sae* two-component system was reported to negatively regulate CP5 and CP8 expression at the transcriptional level (58a). The *sae* locus comprises four genes, including the two-component regulatory genes *saeR* and *saeS* that were originally reported (45a, 58a). How *sae* regulates capsule expression has not yet been studied in detail.

Screening the transposon mutant library using the *xylE* reporter gene resulted in the identification of two related genes with similarity to the *sbcC* and *sbcD* recombinase genes of *Bacillus*. These two genes are adjacent, and inter-ruption of either gene increased capsule production two- to fourfold, which suggests that these genes are repressors of *S. aureus* capsule production. Promoter fusion analyses indicate that the genes affect transcription of the *cap5(8)* promoter (5). Whether these *sbcCD* homologues are involved in recombination in *S. aureus* is unknown.

Recently, a global regulator named *mgrA*, distantly related to *sarA*, was found to have a major effect on capsule production (40). Three different laboratories independently identified *mgrA* using different strategies (17, 40, 62). MgrA has been shown to be a DNA-binding protein (17, 42) that binds to the *cap5(8)* promoter outside of the 10-bp inverted repeat region. The *mgrA* gene is transcribed from two promoters, and each transcript contains a considerable length of untranslated region that is essential for maximal *mgrA* expression (36). Thus, modulating *mgrA* expression can affect capsule production.

SigB is an *S. aureus* alternative stress sigma factor that regulates many staphylococcal genes, including some involved in bacterial virulence. Recent transcriptional profiling analysis of the alternative sigma factor regulon showed that the *cap5* genes were upregulated by *sigB* (3). However, no consensus *sigB* promoter sequence was found in the *cap5(8)* promoter region, indicating that *sigB* indirectly activates capsule expression.

Taken together, these data showed that *cap5(8)* genes are under the control of a complex regulatory network that includes: *agr*, *arlRS*, *sae*, *mgrA*, *sigB*, a *codY*-like gene, *sbcCD*-like genes, and, to a lesser extent, *sarA*. A detailed examination of the interaction between these regulators in controlling capsule production has not been carried out. Nonetheless, experiments using other target genes have indicated that interactions among these regulators are very complicated. For example, *agr* and *arl* have been shown to mutually repress each other (13); *mgr* and *agr* have been shown to activate each other (17, 37); and *mgr* has been shown to activate *arl* but also repress *arl* through *sarV* (42). Unraveling the intricacies of the diverse regulatory pathways that influence *S. aureus* capsule production will be a challenging task.

BIOSYNTHESIS

Sequence comparison of *S. aureus* capsule genes with proteins in the databases led to the assignment of putative functions for most of the 16 *cap5(8)* genes (29, 46, 54). Eleven of

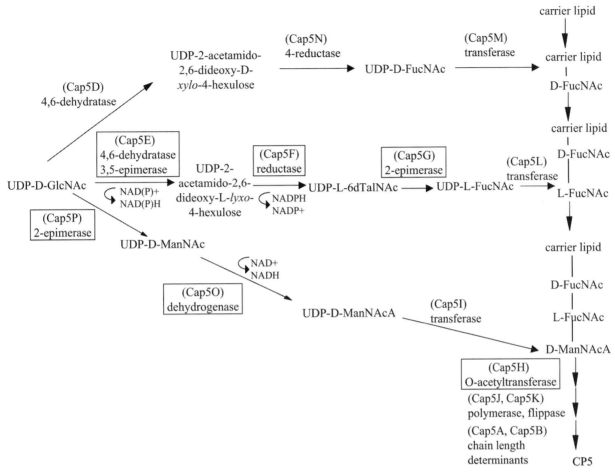

FIGURE 4 Proposed pathway for the biosynthesis of *S. aureus* CP5. The gene products for which functions have been experimentally confirmed are shown in boxes.

the 16 genes have been shown to be essential for capsule production (56). The predicted functions of six of the capsule genes have been confirmed through a combination of genetic and biochemical approaches. These six genes are involved in the synthesis of the nucleotide-activated sugars UDP-*N*-acetyl L-fucosamine (UDP-L-FucNAc) and UDP-*N*-acetyl D-mannosaminuronic acid (UDP-D-ManNAcA) from the common precursor UDP-*N*-acetyl D-glucosamine (UDP-D-GlcNAc). An updated version of the proposed CP5 biosynthetic pathway is depicted in Fig. 4.

The putative gene products of *cap5P* and *cap5O* bear homology to enzymes that are involved in the biosynthesis of UDP-ManNAcA from UDP-GlcNAc (54). In fact, the *S. aureus cap5P* and *cap5O* genes complemented mutations in the *Escherichia coli* homologs *rffE* and *rffD*, respectively, which are involved in biosynthesis of the enterobacterial common antigen (23, 52). The *rffE* gene encodes a UDP-GlcNAc-2-epimerase, and the *rffD* gene encodes a UDP-ManNAc dehydrogenase. Both genes are required for the biosynthesis of UDP-ManNAcA, a component in the synthesis of the enterobacterial common antigen.

The gene product of the *S. aureus cap5P* gene was overexpressed in *E. coli*, purified, and shown to have 2-epimerase activity, converting UDP-GlcNAc to UDP-ManNAc (21). CP5 expression was not affected by insertional inactivation of *cap5P* due to the presence of a second UDP-GlcNAc-2-

epimerase, *mnaA*, on the staphylococcal chromosome outside of the *cap5(8)* locus (21). Portoles et al. overexpressed Cap5O in *E. coli*, purified the enzyme, and demonstrated that it oxidized UDP-ManNAc to UDP-ManNAcA in an NAD⁺-dependent reaction (50). A *cap5O* mutation created in *S. aureus* Reynolds rendered the bacterium negative for CP5; capsule expression was restored when *cap5O* was provided to the mutant in *trans*.

The *S. aureus cap5H* gene is responsible for O-acetylation of the CP5 ManNAcA residue. A mutant with a Tn*918* insertion in the *cap5H* gene produced wild-type levels of O-deacetylated CP5 as determined by nuclear magnetic resonance and immunological methods (2). When *cap5H* was provided to the mutant in *trans*, it fully restored CP5 O-acetylation. Similarly, *cap8J* is most likely the O-acetylation gene in the *cap8* gene cluster, based on sequence homology and on preliminary experiments showing that a monoclonal antibody specific to O-acetylated CP8 failed to react with a *cap8J* mutant. The *cap5H* and *cap8J* genes are located in the type-specific region in their respective operons, consistent with the fact that CP5 and CP8 are O-acetylated on different sugar residues (18a).

The gene products of *S. aureus cap5E*, *cap5F*, and *cap5G* are involved in the synthesis of UDP-L-FucNAc. Cap5E, Cap5F, and Cap5G were overexpressed in *E. coli*, purified, and shown to synthesize UDP-L-FucNAc in vitro from the precursor UDP-D-GlcNAc (24). Cap5E is a multifunctional enzyme that catalyzes the 4,6-dehydration and 3,5-epimerization of UDP-GlcNAc to yield a mixture of three keto-deoxy sugars. One of the intermediates (UDP-2-acetamido-2,6-dideoxy-β-L-*lyxo*-4-hexulose) is reduced at C-4 by Cap5F to yield UDP-2-acetamido-2,6-dideoxy-L-talose (UDP-L-6dTalNAc). Incubation of UDP-L-6dTalNAc with Cap5G resulted in a product that was consistent with the 2-epimerization of UDP-L-6dTalNAc to UDP-L-FucNAc (24).

Mutations in *cap5E*, *cap5F*, or *cap5G* abolished CP5 production (24, 65). Sequence analysis of the genome from *S. aureus* NCTC 8325 revealed that this strain carries a point mutation in *cap5E* that results in a single amino acid change (M134 to R134) in the SMK domain, a catalytic triad that is critical for Cap5E enzymatic activity (24). Wann et al. showed that providing *cap5E* to strain 8325 in *trans* restored capsule production (65). Similarly, plasmid-borne *cap5F* and *cap5G*, provided in *trans*, restored CP5 expression to *cap5F* or *cap5G* mutants, respectively (24).

Putative functions have been ascribed to the remaining capsule genes based on sequence homology (29, 46, 54), although experimental evidence is lacking to support their actual activities. Cap5(8)B is likely a regulator of polysaccharide chain length, because CP8 produced by a *cap8B* mutant exhibited a lower molecular mass than that synthesized by the wild-type strain (29). Cap5D and Cap5N are involved in the synthesis of UDP-D-FucNAc. Based on the observation that a cloned *cap5D* gene complemented a mutation in *Pseudomonas aeruginosa* WbpM (4), we predict that Cap5D has 4,6 dehydratase activity, converting the precursor UDP-GlcNAc to UDP-2-acetamido-2,6 dideoxy-D-*xylo*-4-hexulose. This intermediate could then be stereospecifically reduced at C4 to UDP-D-FucNAc by Cap5N. Cap5N shares considerable identity (40%) at the amino acid level with various UDP-Glc-4-epimerases, but it also has two motifs that are characteristic of proteins with reductase activity—a GxxGxxG nucleotide binding motif and an SYK domain. Work is under way in our laboratory to experimentally verify this pathway. The functions of these genes and the *cap5(8)* genes involved with putative transferase, polymerase, and flippase activities are still unconfirmed.

ROLE IN VIRULENCE

Evidence that CP5 and CP8 promote staphylococcal virulence has come from in vitro experiments, as well as from studies performed in a variety of animal models of *S. aureus* infection and colonization. Several studies have correlated *S. aureus* capsule production with resistance to in vitro phagocytic uptake and killing (2, 19, 61, 66). Human neutrophils phagocytose capsule-negative mutants in the presence of nonimmune serum with complement activity, whereas serotype 5 isolates require both capsule-specific antibodies and complement for optimal opsonophagocytic killing. Luong and Lee (39) showed that a CP8-overproducing mutant was more resistant to in vitro opsonophagocytic killing by human neutrophils than the parental strain Becker. Nilsson et al. (45) reported that mouse macrophages phagocytosed significantly greater numbers of a CP5-negative mutant than of the parental strain Reynolds. Once phagocytosed, the CP5-positive strain survived intracellularly to a greater extent than the mutant strain.

Cunnion et al. (7) compared opsonization of isogenic *S. aureus* strains and demonstrated that the CP5-positive strain bound 42% less serum complement (C3) than the acapsular mutant. Additional in vitro studies have indicated that purified CP8 activated rat and human CD4+ T cells in vitro. Transfer of CP8-activated T cells (plus adjuvant) into the peritoneal cavity of naïve rats resulted in peritoneal abscess formation (63).

Encapsulated *S. aureus* strains have also been shown to be more virulent than acapsular mutants in animal models of staphylococcal infection. The CP5-positive strain Reynolds produced higher bacteremia levels in mice and resisted host clearance to a greater extent than two capsule-deficient mutants (61). Likewise, a CP8-overproducing strain persisted longer in the bloodstream, liver, and spleen of infected mice than the parental strain Becker (39). Strain Reynolds was more virulent than an acapsular mutant in rodent models of renal infection or abscess formation (50, 63). In fact, purified CP5 and CP8 provoked intraabdominal abscess formation in animals when given with a sterile cecal contents adjuvant (63). Mice inoculated with the serotype 5 *S. aureus* strain developed more frequent and severe arthritis, demonstrated greater weight loss, and showed a higher mortality rate than mice infected with capsule-negative mutants (45). An encapsulated strain showed more persistent colonization than a capsule-deficient mutant in a mouse model of nasal carriage (22).

To investigate whether CP5 and CP8 differ in their biological activities, Watts et al. (66) created isogenic mutants of *S. aureus* Reynolds that expressed CP5, CP8, or no capsule. The CP5$^+$ strain was more virulent in a mouse model of infection than was the CP8$^+$ strain, and this finding was correlated with greater survival of the serotype 5 strain in phagocytic killing assays.

In contrast, CP1, CP5, and CP8 were shown to attenuate bacterial virulence in a rat model of catheter-induced endocarditis (1, 44). Staphylococcal adherence to the damaged heart valve is critical to initiate infection in the endocarditis model of infection. The inverse correlation between capsule production and infectivity in this model suggests that the capsule may be masking adhesins that have been shown to be important determinants of virulence in endocarditis (25, 43). Experiments to evaluate in vitro adherence of *S. aureus* to endothelial cells (49) indicate that binding is maximal under conditions in which CP5 production is minimal. Moreover, an acapsular mutant adhered to endothelial cells in greater numbers than the wild-type strain. Restoration of CP5 production by genetic complementation resulted in an *S. aureus* strain with a poorly adherent phenotype. Similarly, we have observed that both CP5 and CP8 expression diminishes *S. aureus* clumping factor A-mediated binding to fibrinogen and platelets (53). Taken together, the results of these studies indicate that the capsule is important in staphylococcal virulence, particularly in blood-borne infections. However, CP may impede the interaction not only between the staphylococci and phagocytes, but also between staphylococci and other mammalian cells.

CAPSULAR VACCINES

Several published reports have indicated that antibodies elicited to the CPs of *S. aureus* are protective in animal models of experimental staphylococcal infection (12, 30, 31). Because polysaccharides are poorly immunogenic and generally elicit a T-cell-independent antibody response, Fattom et al. (11) conjugated CP5 and CP8 to recombinant *P. aeruginosa* exotoxoid A. The conjugate vaccines were highly immunogenic in mice and humans and induced antibodies that opsonized encapsulated *S. aureus* for phagocytosis (10, 11). Antibodies elicited by immunization with the CP5 and CP8 conjugate vaccines were protective in a mouse model of *S. aureus* lethality and disseminated infection (12). Similarly, administration of antibodies to the *S. aureus* conjugate vaccine protected rats against infection in a catheter-induced model of staphylococcal endocarditis if the animals were challenged by the intraperitoneal route (30).

Nabi Biopharmaceuticals (Boca Raton, Fla.) combined the CP5- and CP8-conjugate vaccines into a bivalent vaccine called StaphVAX™ for immunization of humans at high risk for *S. aureus* infection. A phase III clinical trial completed in 2000 assessed the safety, immunogenicity, and efficacy of StaphVAX in preventing *S. aureus* bacteremia. Approximately 1,800 patients with end-stage renal disease who were undergoing hemodialysis were given the CP5/CP8 conjugate or a placebo (57). Hemodialysis patients are at high risk for staphylococcal infection, with 3 to 4 of every 100 patients infected with *S. aureus* per year. At the end of the study period (54 weeks), the vaccine reduced the incidence of bacteremia in the study population by only 26% (not significant). However, the vaccine efficacy was 57% ($P = 0.02$) at 40 weeks, at which point the antibody levels in the vaccinated patients declined, mirroring the decline in efficacy of the vaccine. A confirmatory phase III clinical trial involving 3,600 hemodialysis patients was recently completed. Although the details of the study results have not yet been released, no protection against bacteremia in vaccine recipients was observed. These findings indicate that other staphylococcal components such as adhesins or toxoids that have shown protective efficacy in animal models of staphylococcal infection should be included in a multicomponent vaccine to protect against *S. aureus* infections.

CONCLUSION

Since our last review of this subject, substantial progress has been achieved in three major areas of study relevant to *S. aureus* capsule production. The first area is deciphering the regulation of *cap5* and *cap8* genes. Several regulators that control capsule expression have been identified and characterized, whereas others need further investigation. Regulation of CP expression is complex, consistent with the fact that the CP plays an intricate role in different *S. aureus* infectious processes. Second, our understanding of the biosynthesis of CP5 and CP8 has been advanced. At least six genes have been functionally characterized, and others are under investigation. The third area of advancement is a better understanding of the role that CP5 and CP8 play in the pathogenesis of staphylococcal infections. This has been accomplished by independent studies from different laboratories, putting an end to the controversies over the relevance of *S. aureus* CP production that arose in the 1990s. Nonetheless, further studies are warranted not only to elucidate the basic biology involved in capsule synthesis and regulation, but also to improve strategies to utilize these polysaccharides as vaccine candidates to prevent staphylococcal infections.

This work was supported by NIH grants AI37027 and AI54607 (to C.Y.L.) and AI29040 and AI23244 (to J.C.L.). We thank members of our laboratories for their significant contributions to this work.

REFERENCES

1. **Baddour, L. M., C. Lowrance, A. Albus, J. H. Lowrance, S. K. Anderson, and J. C. Lee.** 1992. *Staph-*
ylococcus aureus microcapsule expression attenuates bacterial virulence in a rat model of experimental endocarditis. *J. Infect. Dis.* **165:**749–753.

2. **Bhasin, N., A. Albus, F. Michon, P. J. Livolsi, J.-S. Park, and J. C. Lee.** 1998. Identification of a gene essential for O-acetylation of the *Staphylococcus aureus* type 5 capsular polysaccharide. *Mol. Microbiol.* **27:**9–21.

3. **Bischoff, M., P. Dunman, J. Kormanec, D. Macapagal, E. Murphy, W. Mounts, B. Berger-Bachi, and S. Projan.** 2004. Microarray-based analysis of the *Staphylococcus aureus* sigma B regulon. *J. Bacteriol.* **186:**4085–4099.

4. **Burrows, L. L., R. V. Urbanic, and J. S. Lam.** 2000. Functional conservation of the polysaccharide biosynthetic protein WbpM and its homologues in *Pseudomonas aeruginosa* and other medically significant bacteria. *Infect. Immun.* **68:**931–936.

5. **Chen, Z., T. T. Luong, and C. Y. Lee.** Unpublished results.

6. **Chien, Y., A. C. Manna, S. J. Projan, and A. L. Cheung.** 1999. SarA, a global regulator of virulence determinants in *Staphylococcus aureus*, binds to a conserved motif essential for *sar*-dependent gene regulation. *J. Biol. Chem.* **274:**37169–37176.

6a. **Cocchiaro, J. L., M. I. Gomez, A. Risley, R. M. Solinga, D. O. Sordelli, and J. C. Lee.** Molecular characterization of the capsule locus from nontypeable *Staphylococcus aureus*. *Mol. Microbiol.*, in press.

7. **Cunnion, K. M., J. C. Lee, and M. M. Frank.** 2001. Capsule production and growth phase influence binding of complement to *Staphylococcus aureus*. *Infect. Immun.* **69:**6796–6803.

8. **Dassy, B., T. Hogan, T. J. Foster, and J. M. Fournier.** 1993. Involvement of the accessory gene regulator (*agr*) in expression of type-5 capsular polysaccharide by *Staphylococcus aureus*. *J. Gen. Microbiol.* **139:**1301–1306.

9. **Dassy, B., W. T. Stringfellow, M. Lieb, and J. M. Fournier.** 1991. Production of type 5 capsular polysaccharide by *Staphylococcus aureus* grown in a semi-synthetic medium. *J. Gen. Microbiol.* **137:**1155–1162.

10. **Fattom, A., R. Schneerson, D. C. Watson, W. W. Karakawa, D. Fitzgerald, I. Pastan, X. Li, J. Shiloach, D. A. Bryla, and J. B. Robbins.** 1993. Laboratory and clinical evaluation of conjugate vaccines composed of *Staphylococcus aureus* type 5 and type 8 capsular polysaccharides bound to *Pseudomonas aeruginosa* recombinant exoprotein A. *Infect. Immun.* **61:**1023–1032.

11. **Fattom, A., R. Schneerson, S. C. Szu, W. F. Vann, J. Shiloach, W. W. Karakawa, and J. B. Robbins.** 1990. Synthesis and immunologic properties in mice of vaccines composed of *Staphylococcus aureus* type 5 and type 8 capsular polysaccharides conjugated to *Pseudomonas aeruginosa* exotoxin A. *Infect. Immun.* **58:**2367–2374.

12. **Fattom, A. I., J. Sarwar, A. Ortiz, and R. Naso.** 1996. A *Staphylococcus aureus* capsular polysaccharide (CP) vaccine and CP-specific antibodies protect mice against bacterial challenge. *Infect. Immun.* **64:**1659–1665.

13. **Fournier, B., A. Klier, and G. Rapoport.** 2001. The two-component system ArlS-ArlR is a regulator of virulence gene expression in *Staphylococcus aureus*. *Mol. Microbiol.* **41:**247–261.

14. **Gill, S. R., D. E. Fouts, G. L. Archer, E. F. Mongodin, R. T. DeBoy, J. Ravel, I. T. Paulsen, J. F. Kolonay, L. Brinkac, M. Beanan, R. J. Dodson, S. C. Daugherty, R. Madupu, S. V. Angiuoli, A. S. Durkin, D. H. Haft, J. Vamathevan, H. Khouri, T. Utterback, C. Lee, G. Dimitrov, L. Jiang, H. Qin, J. Weidman, K. Tran, K. Kang,**

I. R. Hance, K. E. Nelson, and C. M. Fraser. 2005. Insights on evolution of virulence and resistance from the complete genome analysis of an early methicillin-resistant *Staphylococcus aureus* strain and a biofilm-producing methicillin-resistant *Staphylococcus epidermidis* strain. *J. Bacteriol.* **187:**2426–2438.

15. Herbert, S., S. W. Newell, C. Lee, K. P. Wieland, B. Dassy, J. M. Fournier, C. Wolz, and G. Doring. 2001. Regulation of *Staphylococcus aureus* type 5 and type 8 capsular polysaccharides by CO_2. *J. Bacteriol.* **183:**4609–4613.

16. Herbert, S., D. Worlitzsch, B. Dassy, A. Boutonnier, J.-M. Fournier, G. Bellon, A. Dalhoff, and G. Doring. 1997. Regulation of *Staphylococcus aureus* capsular polysaccharide type 5: CO_2 inhibition in vitro and in vivo. *J. Infect. Dis.* **176:**431–438.

17. Ingavale, S. S., W. Van Wamel, and A. L. Cheung. 2003. Characterization of RAT, an autolysis regulator in *Staphylococcus aureus*. *Mol. Microbiol.* **48:**1451–1466.

18. Ji, G. Y., R. C. Beavis, and R. P. Novick. 1995. Cell density control of staphylococcal virulence mediated by an octapeptide pheromone. *Proc. Natl. Acad. Sci. USA* **92:**12055–12059.

18a. Jones, C. 2005. Revised structures for the capsular polysaccharides from *Staphylococcus aureus* types 5 and 8, components of novel glycoconjugate vaccines. *Carbohydr. Res.* **340:**1097–1106.

19. Karakawa, W. W., A. Sutton, R. Schneerson, A. Karpas, and W. F. Vann. 1988. Capsular antibodies induce type-specific phagocytosis of capsulated *Staphylococcus aureus* by human polymorphonuclear leukocytes. *Infect. Immun.* **56:**1090–1095.

20. Katayama, Y., T. Ito, and K. Hiramatsu. 2000. A new class of genetic element, staphylococcus cassette chromosome *mec*, encodes methicillin resistance in *Staphylococcus aureus*. *Antimicrob. Agents Chemother.* **44:**1549–1555.

21. Kiser, K. B., N. Bhasin, L. Deng, and J. C. Lee. 1999. *Staphylococcus aureus cap5P* encodes a UDP-N-acetylglucosamine 2-epimerase with functional redundancy. *J. Bacteriol.* **181:**4818–4824.

22. Kiser, K. B., J. M. Cantey-Kiser, and J. C. Lee. 1999. Development and characterization of a *Staphylococcus aureus* nasal colonization model in mice. *Infect. Immun.* **67:**5001–5006.

23. Kiser, K. B., and J. C. Lee. 1998. *Staphylococcus aureus cap5O* and *cap5P* genes functionally complement mutations affecting enterobacterial common antigen biosynthesis in *Escherichia coli*. *J. Bacteriol.* **180:**403–406.

24. Kneidinger, B., K. O'Riordan, J. Li, J. R. Brisson, J. C. Lee, and J. S. Lam. 2003. Three highly conserved proteins catalyze the conversion of UDP-N-acetyl-D-glucosamine to precursors for the biosynthesis of O antigen in *Pseudomonas aeruginosa* O11 and capsule in *Staphylococcus aureus* type 5. Implications for the UDP-N-acetyl-L-fucosamine biosynthetic pathway. *J. Biol. Chem.* **278:**3615–3627.

24a. Kuroda, M., A Yamashita, H. Hirakawa, M. Kumano, K. Morikawa, M. Higashide, A. Maruyama, Y. Inose, K. Matoba, H. Toh, S. Kuhara, M. Hattori, and T. Ohta. 2005. Whole genome sequence of *Staphylococcus saprophyticus* reveals the pathogenesis of uncomplicated urinary tract infection. *Proc. Natl. Acad. Sci. USA* **102:**13272–13277.

25. Kuypers, J. M., and R. A. Proctor. 1989. Reduced adherence to traumatized rat heart valves by a low-fibronectin-binding mutant of *Staphylococcus aureus*. *Infect. Immun.* **57:**2306–2312.

26. Lee, C. Y. 1992. Cloning of genes affecting capsule expression in *Staphylococcus aureus* strain M. *Mol. Microbiol.* **6:**1515–1522.

27. Lee, C. Y. Unpublished data.

28. Lee, C. Y., and J. C. Lee. 2000. Staphylococcal capsule, p. 361–366. *In* V. A. Fischetti, R. P. Novick, J. J. Ferretti, D. A. Portnoy, and J. I. Rood (ed.), *Gram-Positive Pathogens*. ASM Press, Washington, D.C.

29. Lee, J. C., and C. Y. Lee. 1999. Capsular polysaccharides of *Staphylococcus aureus*, p. 185–205. *In* J. B. Goldberg (ed.), *Genetics of Bacterial Polysaccharides*. CRC Press, Boca Raton, Fla.

30. Lee, J. C., J.-S. Park, S. E. Shepherd, V. Carey, and A. Fattom. 1997. Protective efficacy of antibodies to the *Staphylococcus aureus* type 5 capsular polysaccharide in a modified model of endocarditis in rats. *Infect. Immun.* **65:**4146–4151.

31. Lee, J. C., N. E. Perez, C. A. Hopkins, and G. B. Pier. 1988. Purified capsular polysaccharide-induced immunity to *Staphylococcus aureus* infection. *J. Infect. Dis.* **157:**723–730.

32. Lee, J. C., S. Takeda, P. J. Livolsi, and L. C. Paoletti. 1993. Effects of in vitro and in vivo growth conditions on expression of type-8 capsular polysaccharide by *Staphylococcus aureus*. *Infect. Immun.* **61:**1853–1858.

33. Lin, W. S., T. Cunneen, and C. Y. Lee. 1994. Sequence analysis and molecular characterization of genes required for the biosynthesis of type 1 capsular polysaccharide in *Staphylococcus aureus*. *J. Bacteriol.* **176:**7005–7016.

34. Lin, W. S., and C. Y. Lee. 1996. Instability of type 1 capsule production in *Staphylococcus aureus*, abstr. B-236. Abstr. 96th Gen. Meet. Am. Soc. Microbiol. 1996. American Society for Microbiology, Washington, D.C.

35. Luong, T., S. Sau, M. Gomez, J. C. Lee, and C. Y. Lee. 2002. Regulation of *Staphylococcus aureus* capsular polysaccharide expression by *agr* and *sarA*. *Infect. Immun.* **70:**444–450.

36. Luong, T. T., C. Chen, and C. Y. Lee. 2004. Promoter analysis of the global regulator *mgrA* in *Staphylococcus aureus*, abstr. B-043. Abstr. 104th Gen. Meet. Am. Soc. Microbiol. 2004. American Society for Microbiology, Washington, D.C.

37. Luong, T. T., and C. Y. Lee. Unpublished results.

38. Luong, T. T., and C. Y. Lee. 2002. Identification of genes that regulate type 5 and 8 capsular polysaccharides in *Staphylococcus aureus*, abstr. B-323. Abstr. 102nd Gen. Meet. Am. Soc. Microbiol. 2002. American Society for Microbiology, Washington, D.C.

39. Luong, T. T., and C. Y. Lee. 2002. Overproduction of type 8 capsular polysaccharide augments *Staphylococcus aureus* virulence. *Infect. Immun.* **70:**3389–3395.

40. Luong, T. T., S. W. Newell, and C. Y. Lee. 2003. Mgr, a novel global regulator in *Staphylococcus aureus*. *J. Bacteriol.* **185:**3703–3710.

41. Luong, T. T., S. Ouyang, K. Bush, and C. Y. Lee. 2002. Type 1 capsule genes of *Staphylococcus aureus* are carried in a staphylococcal cassette chromosome genetic element. *J. Bacteriol.* **184:**3623–3629.

42. Manna, A. C., S. S. Ingavale, M. Maloney, W. van Wamel, and A. L. Cheung. 2004. Identification of *sarV* (SA2062), a new transcriptional regulator, is repressed by SarA and MgrA (SA0641) and involved in the regulation of autolysis in *Staphylococcus aureus*. *J. Bacteriol.* **186:**5267–5280.

43. Moreillon, P., J. M. Entenza, P. Francioli, D. McDevitt, T. J. Foster, P. Francois, and P. Vaudaux. 1995. Role of

Staphylococcus aureus coagulase and clumping factor in pathogenesis of experimental endocarditis. *Infect. Immun.* **63:**4738–4743.

44. **Nemeth, J., and J. C. Lee.** 1995. Antibodies to capsular polysaccharides are not protective against experimental *Staphylococcus aureus* endocarditis. *Infect. Immun.* **63:** 375–380.

45. **Nilsson, I.-M., J. C. Lee, T. Bremell, C. Ryden, and A. Tarkowski.** 1997. The role of staphylococcal polysaccharide microcapsule expression in septicemia and septic arthritis. *Infect. Immun.* **65:**4216–4221.

45a. **Novick, R. P., and D. Jiang.** 2003. The staphylococcal *saeRS* system coordinates environmental signals with *agr* quorum sensing. *Microbiology* **149:**2709–2717.

46. **O'Riordan, K., and J. C. Lee.** 2004. *Staphylococcus aureus* capsular polysaccharides. *Clin. Microbiol. Rev.* **17:**218–234.

47. **Ouyang, S., and C. Y. Lee.** 1997. Transcriptional analysis of type 1 capsule genes in *Staphylococcus aureus. Mol. Microbiol.* **23:**473–482.

48. **Ouyang, S., S. Sau, and C. Y. Lee.** 1999. Promoter analysis of the *cap8* operon, involved in type 8 capsular polysaccharide production in *Staphylococcus aureus. J. Bacteriol.* **181:**2492–2500.

49. **Pohlmann-Dietze, P., M. Ulrich, K. B. Kiser, G. Doring, J. C. Lee, J. M. Fournier, K. Botzenhart, and C. Wolz.** 2000. Adherence of *Staphylococcus aureus* to endothelial cells: influence of the capsular polysaccharide, the global regulator *agr*, and the bacterial growth phase. *Infect. Immun.* **68:**4865–4871.

50. **Portoles, M., K. B. Kiser, N. Bhasin, K. H. N. Chan, and J. C. Lee.** 2001. *Staphylococcus aureus* Cap5O has UDP-ManNAc dehydrogenase activity and is essential for capsule expression. *Infect. Immun.* **69:**917–923.

51. **Poutrel, B., F. B. Gilbert, and M. Lebrun.** 1995. Effects of culture conditions on production of type 5 capsular polysaccharide by human and bovine *Staphylococcus aureus* strains. *Clin. Diagn. Lab. Immunol.* **2:**166–171.

52. **Rick, P. D., and R. P. Silver.** 1996. Enterobacterial common antigen and capsular polysaccharides, p. 104–122. *In* F. C. Neidhardt, R. Curtiss III, J. L. Ingraham, E. C. C. Lin, K. B. Low, B. Magasanik, W. S. Reznikoff, M. Riley, M. Schaechter, and H. E. Umbarger (ed.), Escherichia coli *and* Salmonella: *Cellular and Molecular Biology*, 2nd ed. ASM Press, Washington, D.C.

53. **Risley, A., C. Cywes, T. Foster, and J. C. Lee.** 2004. Production of capsular polysaccharide masks clumping factor A-mediated adherence of *Staphylococcus aureus* to fibrinogen and platelets, abstr. D-254. Abstr. 104th Gen. Meet. Am. Soc. Microbiol. 2004. American Society for Microbiology, Washington, D.C.

54. **Sau, S., N. Bhasin, E. R. Wann, J. C. Lee, T. J. Foster, and C. Y. Lee.** 1997. The *Staphylococcus aureus* allelic genetic loci for serotype 5 and 8 capsule expression contain the type-specific genes flanked by common genes. *Microbiology* **143:**2395–2405.

55. **Sau, S., and C. Y. Lee.** 1996. Cloning of type 8 capsule genes and analysis of gene clusters for the production of different capsular polysaccharides in *Staphylococcus aureus. J. Bacteriol.* **178:**2118–2126.

56. **Sau, S., J. Sun, and C. Y. Lee.** 1997. Molecular characterization and transcriptional analysis of type 8 capsule genes in *Staphylococcus aureus. J. Bacteriol.* **179:**1614–1621.

57. **Shinefield, H., S. Black, A. Fattom, G. Horwith, S. Rasgon, J. Ordonez, H. Yeoh, D. Law, J. B. Robbins, R. Schneerson, L. Muenz, S. Fuller, J. Johnson, B. Fireman, H. Alcorn, and R. Naso.** 2002. Use of a *Staphylococcus aureus* conjugate vaccine in patients receiving hemodialysis. *N. Engl. J. Med.* **346:**491–496.

58. **Smith, R. M., J. T. Parisi, L. Vidal, and J. N. Baldwin.** 1977. Nature of the genetic determinant controlling encapsulation in *Staphylococcus aureus* Smith. *Infect. Immun.* **17:**231–234.

58a. **Steinhuber, A., C. Goerke, M. G. Bayer, G. Doring, and C. Wolz.** 2003. Molecular architecture of the regulatory locus *sae* of *Staphylococcus aureus* and its impact on expression of virulence factors. *J. Bacteriol.* **185:**6278–6286.

59. **Stringfellow, W. T., B. Dassy, M. Lieb, and J. M. Fournier.** 1991. *Staphylococcus aureus* growth and type 5 capsular polysaccharide production in synthetic media. *Appl. Environ. Microbiol.* **57:**618–621.

60. **Sutra, L., P. Rainard, and B. Poutrel.** 1990. Phagocytosis of mastitis isolates of *Staphylococcus aureus* and expression of type-5 capsular polysaccharide are influenced by growth in the presence of milk. *J. Clin. Microbiol.* **28:**2253–2258.

60a. **Takeuchi, F., S. Watanabe, T. Baba, H. Yuzawa, T. Ito, Y. Morimoto, M. Kuroda, L. Cui, M. Takahashi, A. Ankai, S. Baba, S. Fukui, J. C. Lee, and K. Hiramatsu.** 2005. Whole-genome sequencing of *Staphylococcus haemolyticus* uncovers extreme plasticity of its genome and evolution of human-colonizing staphylococcal species. *J. Bacteriol.* **187:**7292–7308.

61. **Thakker, M., J.-S. Park, V. Carey, and J. C. Lee.** 1998. *Staphylococcus aureus* serotype 5 capsular polysaccharide is antiphagocytic and enhances bacterial virulence in a murine bacteremia model. *Infect. Immun.* **66:**5183–5189.

62. **Truong-Bolduc, Q. C., X. Zhang, and D. C. Hooper.** 2003. Characterization of NorR protein, a multifunctional regulator of *norA* expression in *Staphylococcus aureus. J. Bacteriol.* **185:**3127–3138.

62a. **Tuchscherr, L. P. N., F. R. Buzzola, L. P. Alvarez, R. L. Caccuri, J. C. Lee, and D. O. Sordelli.** 2005. Capsule-negative *Staphylococcus aureus* induces chronic experimental mastitis in mice. *Infect. Immun.* **73:**7932–7937.

63. **Tzianabos, A. O., J. Y. Wang, and J. C. Lee.** 2001. Structural rationale for the modulation of abscess formation by *Staphylococcus aureus* capsular polysaccharides. *Proc. Natl. Acad. Sci. USA* **98:**9365–9370.

64. **van Wamel, W., Y. Q. Xiong, A. S. Bayer, M. R. Yeaman, C. C. Nast, and A. L. Cheung.** 2002. Regulation of *Staphylococcus aureus* type 5 capsular polysaccharides by *agr* and *sarA in vitro* and in an experimental endocarditis model. *Microb. Pathog.* **33:**73–79.

65. **Wann, E. R., B. Dassy, J. M. Fournier, and T. J. Foster.** 1999. Genetic analysis of the *cap5* locus of *Staphylococcus aureus. FEMS Microbiol. Lett.* **170:**97–103.

66. **Watts, A., D. Ke, Q. Wang, A. Pillay, A. Nicholson-Weller, and J. C. Lee.** 2005. *Staphylococcus aureus* strains that express serotype 5 or serotype 8 capsular polysaccharides differ in virulence. *Infect. Immun.* **73:**3502–3511.

67. **Zhang, Y. Q., S. X. Ren, H. L. Li, Y. X. Wang, G. Fu, J. Yang, Z. Q. Qin, Y. G. Miao, W. Y. Wang, R. S. Chen, Y. Shen, Z. Chen, Z. H. Yuan, G. P. Zhao, D. Qu, A. Danchin, and Y. M. Wen.** 2003. Genome-based analysis of virulence genes in a non-biofilm-forming *Staphylococcus epidermidis* strain (ATCC 12228). *Mol. Microbiol.* **49:**1577–1593.

Staphylococcus aureus Exotoxins

GREGORY A. BOHACH

38

For the purpose of this review, *Staphylococcus aureus* exotoxins fall into three general groups: (i) membrane-active agents, (ii) pyrogenic toxin superantigens (PTSAgs), and (iii) exfoliative toxins (ETs). The membrane-active agents are hemolysins, cytolysins, or leukocidins that contribute to general pathogenesis of several staphylococcal illnesses but no specific illness, except possibly for certain cases of pneumonia. *S. aureus* toxins with undisputable superantigen (SAg) activity include the pyrogenic toxin family. These toxins share a set of immunomodulatory activities as a result of their SAg function that promote ability of many or most pyrogenic toxins to induce toxic shock syndrome (TSS). Some SAgs, such as the staphylococcal enterotoxins, have additional activities that endow them with the ability to induce other diseases such as staphylococcal food poisoning (SFP). The ETs have protease activity that enables them to cause specific pathologies on the skin.

MEMBRANE-ACTIVE AGENTS

Alpha-Toxin

Alpha-toxin is a cytolytic pore-forming toxin and is one of the most potent bacterial toxins known (reviewed in reference 13). Alpha-toxin is especially toxic for rabbits (50% lethal dose = 1.3 μg), and rabbit erythrocytes are also very sensitive to lysis by this toxin. Human erythrocytes are approximately 1,000-fold less sensitive. It is suspected that two cell-binding mechanisms exist. Cells such as rabbit erythrocytes are suspected to have high-affinity binding sites; other cells such as human erythrocytes have only low-affinity sites (53). The low-affinity binding mechanism is responsible for alpha-toxin damage to protein-free liposomes at high toxin concentrations. The identity of the high-affinity binding site is unknown, but it is probably a membrane protein.

Most *S. aureus* isolates possess *hla*, the structural gene for alpha-toxin, but are variable in its expression or in the amount of toxin expressed. TSS isolates are often nonhemolytic and less inflammatory and often have a nonsense mutation in *hla*. Work with mutants in animal models indicated that strains expressing alpha-toxin are more virulent

than isogenic derivatives (40). Although adults usually have antibody, high-affinity binding sites may allow the toxin to affect susceptible cells at very low concentrations. In vitro and in vivo, alpha-toxin is hemolytic, cytotoxic, dermonecrotic, and lethal. Several cell types, including erythrocytes, mononuclear immune cells, epithelial and endothelial cells, and platelets, can be killed by alpha-toxin. Cell death directly from membrane damage has been attributed to impaired osmoregulation, cation and small molecule influx/efflux, necrosis, and apoptosis (34, 61). Sublethal quantities of alpha-toxin can allow Ca^{2+} influx, activating phospholipases and arachidonic acid metabolism in addition to neutrophil-induced vasocontraction and endothelial cell dysfunction (17). Prostaglandins and/or leukotrienes can be generated, resulting in vasoactive effects that augment the direct lethality of alpha-toxin for endothelial cells. Additionally, proinflammatory cytokines and procoagulatory compounds released from activated monocytes and platelets, respectively, contribute to effects of alpha-toxin systemically, particularly for the cardiovascular system and the lungs. The induction of inflammation by alpha-toxin is mediated largely by interleukin-1a (IL-1a) and IL-6, and is enhanced by a synergism with other bacterial cellular and cell-free components (87).

Alpha-toxin promotes cell lysis and release of internalized *S. aureus*. Staphylococcal small-colony variants, linked to persistent and recurrent infections, can reside inside host cells partly as a result of dramatically reduced alpha-toxin expression (128). In a variety of cells, including professional and nonprofessional phagocytes, alpha-toxin has been shown to be the major inducer of caspase activation and apoptosis (52). Following exposure to alpha-toxin expressed either by intracellular or extracellular staphylococci, the toxin induces cell death with components of both apoptosis and necrosis (34, 75). Alpha-toxin and beta-toxin (discussed below) are suspected of promoting staphylococcal virulence by activating host cell shedding of syndecan-1, the major heparin sulfate proteoglycan of host cell surfaces (93). Interestingly, a role for alpha-toxin has been proposed in *S. aureus* biofilm production; an *hla* mutant was shown to have reduced capacity to colonize plastic surfaces (18).

Gram-Positive Pathogens, 2nd edition, edited by Vincent A. Fischetti et al.
© 2006 ASM Press, Washington, D.C.

Alpha-toxin is secreted as a monomer of 293 residues. Upon binding to the membrane, the monomer oligomerizes to form a ring-shaped pore (Fig. 1). Although originally interpreted to be a hexamer, structural studies have provided evidence for both hexameric and heptameric rings. Only part of the toxin penetrates the bilayer; the bulk of it remains on the surface. The pore is a water-filled channel, and a model for toxin assembly (Fig. 2) was deduced from mutagenesis and chemical modification experiments (124). Toxin at various stages of assembly differs in protease susceptibility. The amino latch region (residues 1 to 20) is susceptible to protease in forms α1, α1*, and α7*, but is resistant in the heptameric pore of α7, suggesting that folding of the latch is part of the last stage of toxin-pore assembly. A membrane-associated glycine-rich region of largely β structure (residues 110 to 148) is protease-sensitive only in the monomer α1, suggesting that it is hidden from protease attack, even though membrane penetration is the last stage. Evidence in formulating the model also came from cysteine-scanning mutagenesis. Single cysteine substitutions were derivatized with a polarity-sensitive fluorescence probe and reacted with liposomes. Residues in the glycine-rich region, particularly residues 126 to 140, showed a marked shift in fluorescence, indicating that they are embedded in a hydrophobic lipid environment.

The crystal structure of an alpha-toxin heptamer is mushroom-shaped and forms a channel ranging in diameter from 14 to 46 Å (115). The lower half of a 14-stranded

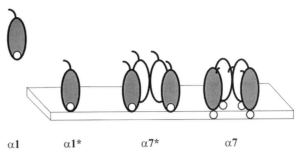

FIGURE 2 Model for alpha-toxin assembly based on crystallographic data and structure-function experiments. The model depicts the formation of a heptameric ring (α7 and α7*). A cross-section of the ring revealing only four monomers is shown so that the proposed structural alterations are visible. In this model, alpha-toxin is expressed and secreted as a monomer (α1). α1, bound to the target membrane (designated α1*), promotes assembly of the heptamer (α7). In the final stage of assembly, β-sheets in each monomer (depicted as small circles) insert into the membrane, forming a channel, and the N-terminal latches contact adjacent monomers, rendering them resistant to proteolysis.

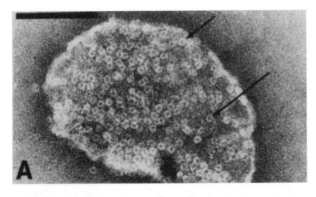

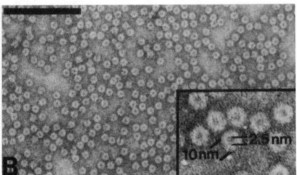

FIGURE 1 Membrane pore formation by alpha-toxin. (A) Rabbit erythrocyte membrane fragment negatively stained following lysis with alpha-toxin. Arrows designate representative ring-shaped structures (10 nm) on the membrane. (B) Ring-shaped alpha-toxin multimers isolated in detergent solution. (Inset) The rings are magnified so that the internal channel (2.5 nm) and ring perimeter (10 nm) are clearly visible.

antiparallel β-barrel penetrates and spans the lipid bilayer to form the pore. Each subunit contributes two amphipathic β-strands encompassing residues 118 to 140, with residue 128 forming the turning point at the membrane face. At the top of the cap the channel diameter is 28 Å and is lined by the N-terminal amino latches. The channel widens to 46 Å before narrowing again to 15 Å at the junction of the cap and stem. The stem pore varies from 14 to 24 Å, depending on side chains in the lumen. Within the stem are two bands of hydrophobic residues. Both ends are defined by rings of acidic and basic residues. At the top of the cap domain, the amino latch makes extensive contacts with its clockwise-related neighbor. In the body of the cap, adjacent protomers make contact through β-sandwiches.

Beta-Toxin

In 1935 Glenny and Stevens differentiated beta-toxin from alpha-toxin by antibody neutralization and showed that it lysed sheep, but not rabbit, erythrocytes (reviewed in reference 9). Beta-toxin is also unique in its ability to induce hot-cold lysis and is the staphylococcal hemolysin that participates in the CAMP reaction for identification of group B streptococci (43). Incubation of sensitive erythrocytes with beta-toxin at 37°C results in little or no lysis. However, lysis ensues if treated erythrocytes are chilled below 10°C. On the basis of this activity, it has been concluded that beta-toxin production is more frequent with animal (approximately 88%) than with human isolates (11 to 45%). We have noted, however, that use of antisera is a more sensitive means of detecting expression of the toxin. Thus, many human isolates that do not exhibit a hot-cold reaction express beta-toxin that is detectable by immunoblotting. Similar results were reported by Aarestrup et al. (1).

The *S. aureus* beta-toxin gene *hlb* (103) carries the attachment site for serological group F converting phages (22). In lysogens, the gene is disrupted by prophage integration, and phenotypic alterations through this mechanism occur frequently during infections (46). Beta-toxin is very basic and is often the most abundant protein in culture supernatants, facilitating its purification from *S. aureus* and

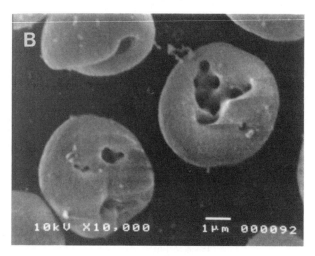

A

PHOSPHORYLCHOLINE

CERAMIDE

SPHINGOMYELIN

B

10kV X10,000   1μm 000092

FIGURE 3 Properties of staphylococcal beta-toxin. (A) Sphingomyelin chemical formula showing beta-toxin cleavage site resulting in generation of phosphorylcholine and ceramide. (B) Scanning electron micrograph showing lesions in human erythrocyte membranes caused by beta-toxin after shifting the temperature to 4°C.

Staphylococcus intermedius (30). Based on sequencing data, *S. aureus* beta-toxin is a 33,742-Da protein with two cysteines that are likely to form a disulfide linkage required for activity. The *S. intermedius* toxin is similar but not identical to *S. aureus* beta-toxin. Both proteins are related to sphingomyelinase of *Bacillus cereus* and to a similar enzyme produced by *Leptospira interrogans*. Structure-function studies on staphylococcal beta-toxin are limited, although it is likely that its mode of action mimics the *B. cereus* enzyme, which has been better characterized (123).

Beta-toxin displays species-dependent activity. Sheep, cow, and goat erythrocytes are most sensitive. Human red blood cells are intermediate in sensitivity, whereas murine and canine erythrocytes are resistant. Beta-toxin is a neutral sphingomyelinase, and the degree of erythrocyte sensitivity depends on membrane sphingomyelin content. Beta-toxin acts as a type C phosphatase, hydrolyzing sphingomyelin to phosphorylcholine and ceramide (Fig. 3A). Beta-toxin-induced hemolysis and sphingomyelin hydrolysis require divalent cations. Mg^{2+} is most effective, although either Co^{2+} or Mn^{2+} can enhance activity. Ca^{2+} and Zn^{2+} are

inhibitory. Beta-toxin enzyme activity occurs maximally at or around 37°C but is negligible at 4°C. This suggests that hot-cold lysis has two stages, including sphingomyelin hydrolysis followed by physical disruption of the membrane at lower temperatures. Unlike lesions induced by pore-forming toxins, beta-toxin causes invaginations of selected regions of the membrane (Fig. 3B). Sphingomyelin is located in membrane outer leaflet patches. Cohesive forces are apparently sufficient to hold the ceramide hydrolysis product in position in the membrane. On cooling, a phase separation occurs with condensation of ceramide into pools and collapse of the bilayer (70).

Results of early toxicity studies with beta-toxin should be interpreted with caution because toxin preparations were often contaminated with alpha-toxin. Subcutaneous administration of purified beta-toxin causes erythema in rabbits and, in general, beta-toxin is at least 10- to 160-fold less toxic than alpha-toxin in animal models. The use of isogenic *S. aureus* strains has shown that beta-toxin contributes to pathogenesis in a murine mastitis model (40), and that it has a small role to play in ocular keratitis (84). Beta-toxin is leukotoxic to a variety of cells, although morphological alterations observed microscopically are less pronounced than with erythrocytes (72). Leukocytes are also susceptible to hot-cold lysis, although sublytic concentrations or incubation at physiological temperatures results in abnormal function. Beta-toxin inhibits monocyte migration and stimulates release of IL-1β, IL-6r, and soluble CD14 from human monocytes (129). The effects on other leukocytes are less pronounced. Although a number of investigations have failed to demonstrate hot-cold lysis of neutrophils, impaired chemotaxis and Fc-binding were reported. Free ceramide is a potent second messenger and is involved in a number of cascade reactions leading to kinase and phosphatase induction, thereby promoting apoptosis. Although beta-toxin induces apoptosis in human leukemic cell lines and murine fibrosarcoma cell lines (60), Walev et al. (129) could not demonstrate a role for ceramide in the effects of beta-toxin on monocytes. Beta-toxin reduces ciliary activity of cultured rabbit nasal epithelial cells and induces sinusitis when administered to maxillary sinus in vivo (64).

Delta-Toxin

Williams and Harper proposed the existence of delta-toxin as the fourth cytolytic *S. aureus* toxin in 1947 (131). Delta-toxin is unique because it is small, heat stable, has possible surfactive properties, and is lytic towards many types of membranes from most animal species, including those on erythrocytes, other cells, organelles, and even bacterial protoplasts (reviewed in reference 9). Also, cytoplasmic leakage and lysis of cells exposed to delta-toxin occur without a demonstrable lag, similar to treatment with Triton X-100 and other detergents (109).

Delta-toxin, a 26-residue peptide, is encoded by the *hld* gene located near the 5′ end of the Agr RNAIII transcript. It is produced by nearly all *S. aureus* isolates and by a high percentage of other staphylococcal species. At least two variants of delta-toxin exist among coagulase-positive staphylococci; toxins expressed by human and canine strains of *S. aureus* are only 62% identical and are immunologically distinct. Other staphylococcal species express related peptides. Molecular modeling suggested that delta-toxin is a helical amphipathic peptide with its hydrophobic and hydrophilic residues on opposing sides of the helix. High-resolution nuclear magnetic resonance studies have yielded similar structures, but differences exist between solution and membrane-bound toxin (29, 120). In solution, residues 2 to 20 form a stable helix, whereas the C-terminal residues are more flexible. Bound to lipid micelles, residues 5 to 23 form an extended helix. The exact location and orientation of the peptide in relation to the membrane are still unclear. It is likely that the toxin inserts at least partly into the lipid bilayer, disordering lipid chain dynamics. Initially, theoretical modeling predictions suggested that the mode of lysis involves formation of channels in membranes composed of aggregates of six molecules of delta-toxin (74). In contrast, kinetics studies recently conducted by Pokorny et al. (99) did not find evidence for a stable membrane-inserted pore. Instead, they proposed a "sinking raft" model in which a critical number of toxin monomers accumulate on the surface in parallel orientation to the membrane. The monomers rapidly cross the membrane, causing release of the contents in a rapid and reversible manner.

Because delta-toxin exhibits activity toward a broad spectrum of cells, it is potentially cytotoxic for tissues and could have an adverse effect on leukocytes. Although delta-toxin induces dermonecrosis when administered intradermally into the skin of rabbits, extremely high concentrations are required to cause lethality in laboratory animals (133). It is only minimally immunogenic, but is inhibited by binding to proteins, cholesterol, and phospholipids in serum.

Sommerville et al. (114) reported that delta-toxin deformylation is linked to tricarboxylic acid cycle activity, when levels of iron are limited. They proposed that staphylococcal regulation of peptide deformylase, which requires iron, is a mechanism by which the bacterium can interfere with neutrophil chemotaxis promoted by peptides containing formyl-methionine. Delta-toxin is related to several other peptides, including the bee venom mellitin. However, delta-toxin diverges from this peptide family by its lack of antibacterial activity. Structure-function studies have identified likely residues that endow members of this family with distinct and separable hemolytic and antibacterial activities (28).

Gamma-Toxin, Leukocidin, and Other Bicomponent Toxins

Gamma-toxin, leukocidin, and other bicomponent toxins are a family of toxins encoded by several genetic loci; *hlg*, *luk-PV*, and *lukDE* are the best characterized (Fig. 4A) (47, 76). Its family members contain two synergistically acting proteins: one S component (LukS-PV, HlgA, HlgC, LukE, or LukM) and one F component (LukF-PV, LukD, or HlgB), designated on the basis of their mobility (slow or fast) in ion-exchange chromatography. Panton-Valentine leukocidin (PVL) and gamma-toxin are two prototypic bicomponent toxins. The classical PVL S and F components are LukS-PV and LukF-PV (101), but in some strains, LukS-PV is replaced by LukM-PV. Gamma-toxin contains HlgA and HlgB, its S and F components, respectively.

In most *S. aureus* strains, the *hlg* locus encodes three polypeptides. Two (HlgA and HlgC) are S components and are related to LukS-PV, encoded by the *luk-PV* locus (Fig. 4). HlgB is similar to LukF-PV (25). More than 99% of clinical isolates carry the *hlg* locus; only 2% of clinical isolates carry the *luk* genes and express PVL. The latter isolates, which also contain *hlg*, produce all three S components and both F components. Any S component can combine with each of the F proteins, leading to formation of numerous possible toxin combinations in PVL-producing strains. Only two possible combinations exist (HlgB+HlgA or

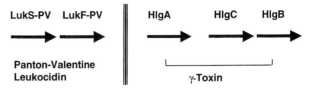

FIGURE 4 Molecular aspects of staphylococcal bicomponent toxins. Organization of bicomponent toxin genes in a strain harboring both the *hlg* and *luk-PV* loci. Any S and F component may combine to generate a unique bicomponent toxin. The two prototype bicomponent toxins, PVL and delta-toxin, are composed of LukS-PV+LukF-PV and HlgA+HlgB, respectively.

HlgB+HlgC) in typical strains harboring only the *hlg* locus. Only PVL and gamma-toxin have been specifically named. Other bicomponent toxins are designated by listing their two components. A bicomponent toxin variant (LukE+LukD) expressed by *S. aureus* Newman, which harbors the *LukDE* locus, has been reported to possess dermonecrotic and weakly leukotoxic activities but is not hemolytic (47). The PVL structural genes were shown to be harbored on temperate phages in three strains—V8, P83, and the clinical isolate A980470 (82, 140). The phage genome in each of the strains tested contained the gene for LukF-PV plus that for either LukS-PV or LukM-PV. These phages, designated ΦPVL, ΦSLT, and ΦPV83, respectively, constitute a heterogeneous group of phages that appear to have the same attachment site. ΦSLT was capable of lysogenizing several clinical isolates as well as RN4220 (82).

PVL can stimulate and lyse neutrophils and macrophages. Other normal cells are not affected by PVL, which is nonhemolytic (100). Gamma-toxin is strongly hemolytic and 90-fold less leukotoxic. Similarly to PVL, the LukF-PV component paired with either HlgC or HlgB promotes leukotoxic activity, but has poor hemolytic activity. HlgB+HlgC is weakly hemolytic but is leukotoxic, albeit 30-fold less so than PVL. PVL is active against human and rabbit leukocytes but not murine, ovine, or guinea pig cells. Of 11 species of animals tested, erythrocytes from rabbits were the only cells lysed efficiently by gamma-toxin. Humans cells were lysed but were 1,000-fold less sensitive than rabbit cells (6). Despite this, additional toxicity may be displayed in vivo. Animal studies have shown that mice, rabbits, and guinea pigs are all sensitive to gamma-toxin.

Synergistic function involves sequential binding of the F and S components. The action of gamma-toxin on erythrocytes requires initial binding of HlgB (F) followed by HlgA (S) (91) and subsequent generation of a pore. Ring-shaped structures (2.1 to 2.4 nm diameter) purified from erythrocytes consist of a complex (150 to 250 kDa) of HlgA and HlgB in an average 1:1 ratio (62, 117). Other evidence suggests formation of heteroheptameric structures in the membrane with F:S ratios of 3:4 and 4:3 (118). Similarly, the two components of the PVL also bind sequentially to human neutrophils, although most reports have indicated that the 32-kDa S component binds first (23). Although the stoichiometry of S- and F-related subunits on interactions with human neutrophils is not clear presently, it is believed that PVL forms pores. Miles et al. (77) provided evidence for an octomeric transmembrane pore, consistent with its comparatively high conductance. The number of toxin molecules bound per cell and the pore diameter are affected by divalent cations, especially Ca^{2+}. At physiological Ca^{2+} levels, the pores are permeable to small divalent ions but not to ethidium bromide. Pores formed in the absence of Ca^{2+} are larger (at least 0.78 nm in diameter) and allow the passage of ethidium bromide (36). In vitro studies have shown an affinity of S components of both PVL and gamma-toxin for GM1 gangliosides (83, 92), although the molecular nature of the receptor has not been conclusively determined.

Although the three-dimensional crystal structure has not been solved for a complete bicomponent toxin hetero-complex, the structures of LukF-PV and LukS-PV have been reported (48, 76). These two PVL subunits share considerable structural features with each other and with alpha-toxin, consistent with a conserved mechanism of action by the staphylococcal β-barrel pore-forming toxins. The most significant sequence divergence occurs in the amino latch and stem β-strands, which undergo conformational changes during oligomerization.

Exposure of neutrophils to PVL in vitro leads to swelling and rounding of the cells and their nuclei. Cell lysis is preceded by degranulation and nuclear rupture (83). One of the major effects of PVL is a rapid accumulation of intracellular free Ca^{2+}, possibly by activation of an endogenous Ca^{2+} channel (24). PVL has been associated with certain types of cutaneous infections, in particular furunculosis, cellulitis, and cutaneous abscesses (69). Antibodies provide some protection and are associated with less severe disease (134). Recently, PVL production has been associated with community-acquired methicillin-resistant strains (125) and with severe necrotizing pneumonia in children and young adults (44). The latter illness is characterized by a high lethality and extensive necrosis of the tracheal and bronchial mucosa and interalveolar septa.

PYROGENIC TOXIN SUPERANTIGENS

Classical pyrogenic toxins include staphylococcal enterotoxins (SEs) types A to E, TSS toxin-1 (TSST-1), and streptococcal pyrogenic exotoxins (14). Recently, additional SEs and SE-like (SEl) toxins with unconfirmed enterotoxic activity (68), types G to R and U (67, 81, 85, 139), have been identified. Original classification of the SEs (SEA to SEE) was based largely on antigenic differences (Table 1). Minor molecular variants are common and were first observed with SEC and TSST-1. Genomic analysis has facilitated the identification and molecular characterization of previously unrecognized SEs and SEls at a rapid rate. As a result, classification is now usually based on molecular relatedness (68). Unfortunately, the rapid rate of new toxin discovery has resulted in more than one SE being given the same designation in the literature (59, 88, 89). Therefore, it is now recommended that nomenclature for new PTSAgs be assigned by the International Nomenclature Committee for Staphylococcal Superantigens prior to publication (68). The PTSAgs may be expressed by veterinary isolates as well as human strains (32, 113).

PTSAg molecular genetics are complex. Most of the PTSAgs are regulated by the *agr* and *sar* loci. Depending on the toxin, the structural gene may be harbored on potentially mobile elements, including plasmids (SED and SEls J and R), bacteriophages (SEs A and D, and SElP), or pathogenicity islands (SEs B and C, SEls K, L, and Q, and TSST-1) (86). The *seg* and *sei* genes are located on a large operon with additional SE genes and pseudogenes (59, 79). This gene cluster has been termed the enterotoxin gene cluster (*egc*) (59). In addition to SEG and SEI, this operon also en-

TABLE 1 Reported, confirmed, or potential staphylococcal PTSAgs

Toxin[a]	Reference(s)	Human Vβ specificity[b]
SEA	79	1.1, 5.3, 6.3, 6.4, 6.9, 7.3, 7.4, 9.1, 18
SEB	27	3, 12, 13.2, 14, 15, 17, 20
SEC1	27	3, 12, 13.2, 14, 15, 17, 20
SEC2	27	3, 12, 13.1, 13.2, 14, 17, 20
SEC3_FRI909	27	3, 12, 13.1, 13.2, 14, 17, 20
SEC3_FRI909	27	3, 12, 13.1, 13.2, 14, 17, 20
SEC_bovine	26	3, 12, 13.2, 14, 15, 17, 20
SEC_ovine	26	3, 12, 13.2, 14, 15, 17, 20
SEC_canine	32	3, 12, 13.2, 14, 15, 17, 20
SED	12, 63	5.1, 12
SEE	78	5.1, 6.3, 6.4, 6.9, 8.1, 18
SEG	59, 81	3, 12, 13a, 14
SElH	95, 107	None[h]
SEl[c]	59, 81	1, 5.1, 5.2, 5.3, 23
SElJ	139	ND
SElK	89	5.1, 5.2, and 6.7
SElL[d]	59, 88	5.1, 5.2, 6.7, 16, 22
SElM[e]	59	9
SElN[f]	59	5.1, 7, 9, 22
SElO[g]	59	6a, 6b, 8, 9, 18, 21.3
SElP	42	ND
SElQ[d]	90	2, 5.1, 21.3
SElR	85	3, 11, 12, 13.2, 14
SElU	67	ND
TSST-1	21	2

[a]SE nomenclature used in this table is that recommended by the International Nomenclature Committee for Staphylococcal Superantigens (68). Toxins either lacking activity or not yet tested for emetic activity in the primate oral feeding assay are designated staphylococcal enterotoxin-like toxins (SEls), according to standard nomenclature (68).

[b]ND, not determined or reported.

[c]Weakly emetic (81).

[d]Nonemetic.

[e]Originally designated SEK (59).

[f]Originally designated SEL (59).

[g]Originally designated SEM (59).

[h]T-cell stimulation reportedly results from Vα stimulation (95).

codes SElM, SElN, and SElO. In some strains, recombination of a pseudogene results in generation of the SElU structural gene (67).

Williams et al. (132) identified a family of proteins that is expressed by a set of clustered genes harbored on a pathogenicity island (37). Although these proteins are distantly related to PTSAgs and were initially designated staphylococcal enterotoxin-like proteins, their toxicity in vitro has not been demonstrated. Because of a lack of confirmed toxicity and minimal relatedness to the SEs, and to prevent confusion when staphylococcal enterotoxin type T is eventually identified and reported, the International Nomenclature Committee for Staphylococcal Superantigens has renamed them staphylococcal superantigen-like proteins (68).

The SEs are best known for causing SFP, which results from ingestion of food contaminated with toxin. SFP is the leading cause of food-borne microbial intoxication worldwide and is usually linked to improper storage of food (54). Most cases are self-limiting and have a mean duration of 23.6 h. Patients typically present with emesis after a short incubation period (mean of 4.4 h). Other symptoms, including nausea, diarrhea, and abdominal pain or cramping, are also common. The SEs are stable in the gastrointestinal tract and indirectly stimulate the emetic reflex center. Although the vagus nerve is involved, molecular events involved are unclear (33). There is mounting evidence that mast cell activation occurs and that synthesis of inflammatory mediators and neuropeptide substance P occurs upon SE activity in the gastrointestinal tract and elsewhere (4, 16, 111). TSST-1 is not implicated in SFP.

TSS is an acute systemic illness. The criteria for defining TSS cases were established in 1981 (106). Patients present with hypotension, fever, rash, and desquamation during convalescence and have involvement of at least three additional organ systems. Some have proposed implementing a revised case definition to include patients with less severe disease, such as those whose syndrome was attenuated as a result of early treatment (94). Staphylococcal TSS may manifest in either of two general forms, menstrual or nonmenstrual. Menstrual TSS usually occurs in women whose vaginal/cervical mucosae are colonized by TSST-1-producing *S. aureus*. Tampon use is a risk factor in menstrual TSS, and a correlation between tampon absorbency and risk of developing TSS has been established. Nonmenstrual TSS may result from *S. aureus* infection elsewhere in the body. Either SEs or TSST-1 may mediate the nonmenstrual form.

Shared biological properties attributed to the PTSAgs include induction of fever, hypotension, T-cell proliferation, immunosuppression, enhanced susceptibility to endotoxin shock, and induction of cytokines (14). Key reports in 1989 (41, 108, 130) helped elucidate the mechanisms by which PTSAgs function. They interact in a unique way with the immune system: (i) with rare exceptions, they cause polyclonal proliferation of T cells bearing certain T-cell receptor (TCR) Vβ elements, (ii) they bind specifically to major histocompatibility complex class II (MHC-II) on antigen-presenting cells, and this binding is required for maximum T-cell stimulation, and (iii) this interaction typically results in deletion of stimulated Vβ-expressing T cells. Molecules with these properties were termed T-cell SAgs and the PTSAgs are prototypes of this family.

SAgs bind unprocessed to MHC-II at a region distinct from the peptide-binding groove (58, 65). Because the MHC-II-bound toxin stimulates T cells based on their Vβ sequences rather than antigen specificity, many T cells (up to 30%) are activated. This type of cellular stimulation, plus elevated cytokine levels, adversely affects the host. Combined, they cause fever, lethal shock, immunosuppression, and several clinical syndromes such as TSS and possibly certain autoimmune diseases. The major cytokines induced initially include IL-1, tumor necrosis factors alpha and beta, interferon-γ, and IL-2. The roles of these cytokines and secondary proinflammatory mediators in pathogenesis of TSS have been reviewed in detail (79).

It is proposed that *S. aureus* benefits from immunosuppression induced by the PTSAgs. Immunosuppression may be demonstrated in vitro and in patients. For example, recurrent TSS has been attributed to failure to generate neutralizing antitoxin antibodies. Systemic exposure to TSST-1 or SEs causes a decline in function and levels of certain lymphocyte populations (130). Apoptosis mediated by Fas/Fas ligand is one mechanism by which SAgs delete certain populations. $CD8^+$ cells suppress $CD4^+$ cell responses by inducing apoptosis of $CD4^+$ cells through ligation with Fas. In addition, SAg-activated T cells can undergo anergy, in which they fail to proliferate or secrete IL-2 (39, 71). High doses of SAg also promote B-cell apoptosis and the down-

regulation of immunoglobulin-secreting B cells (116). Much of the immunosuppression induced by PTSAgs in vivo is likely due to induction of immunoregulatory T cells. Interestingly, mice treated with SAgs developed both $CD4^+;CD25^+$ and $CD4^+;CD25^-$ T cells. Both subpopulations were more suppressive than naturally occurring $CD4^+;CD25^+$ regulatory T cells (45).

Crystal structures have been reported for numerous PTSAgs (79). They are all, including TSST-1, compact ellipsoidal proteins folded into two domains composed of mixed α/β structures (Fig. 5A). The smaller domain of the SEs, domain 1, contains most of the N-terminal half of the protein, excluding the N terminus. This domain is similar in topology to the well-known oligonucleotide/oligosaccharide-binding motif, although there is presently no evidence that binding to carbohydrates or nucleic acids occurs or is required for function. The top of domain 1 of SEs also contains the disulfide loop, a conserved feature of most SEs. The larger domain 2 has a β-grasp motif and is composed of C-terminal residues, plus the N terminus extending over the top. The interface between the two domains is delineated by a short shallow cavity at the top of the molecule, and by a large groove extending all the way along the back of the molecule. Despite lacking significant sequence homology with the SEs, TSST-1 shares a similar domain structural organization. Four key features of most SEs are absent in TSST-1: (i) the SE N terminus, which folds back over domain 2, (ii) an α-helix in the loop at the base of domain 1, (iii) a second α-helix positioned in the groove between both domains, and (iv) the disulfide loop. SEs A, D, and E and SElH require Zn^{+2} to stimulate T cells most efficiently (119). Zn^{2+} in SEA is bound to the external portion of domain 2, near the N terminus, through residues H187, H225, and D227. Although not apparently required for activity, the SECs bind a Zn^{2+} atom at the base of the groove on the back side of the molecule (15). This binding site, including a HEXXH motif, resembles thermolysin, the prototype of one class of metalloenzymes. SED has a second Zn^{2+}-binding site in a position similar to that of SECs. The role of this second site in these toxins may be to stabilize a local structural configuration (119).

Each PTSAg has a different affinity for and recognizes a particular repertoire of MHC-II molecules. The toxins have evolved a variety of modes for binding to this receptor (20). The first general method is displayed by SEB. An SEB:HLA-DR1 crystal structure (58) showed that SEB interacts with the HLA-DR1 α1 chain at a concave surface on the receptor adjacent to but outside of the peptide-binding groove. This interaction requires 19 SEB residues on the edge and top of domain 1, in and near the cysteine loop. Binding to the receptor in this manner orients domain 2 away from the receptor α-chain. The crystal structure of HLA-DR1 complexed with TSST-1 (65) was similar to the SEB:HLA-DR1 structure. However, TSST-1 extends further over the top of the receptor, contacting the bound peptide and the β-chain.

Binding of other SEs to MHC-II is more complex. Some, such as SEA, have a low-affinity MHC-II-binding site that overlaps that of SEB and TSST-1, plus a high-affinity site on the outside of domain 2 near the N terminus (56). The SEA high-affinity site involves coordination of Zn^{2+} through three toxin residues and residue 81 of the MHC-II β chain. Two MHC-II-binding sites per toxin allow two MHC-II molecules to be cross-linked (122). Similar modes of action, including formation of homodimers by some

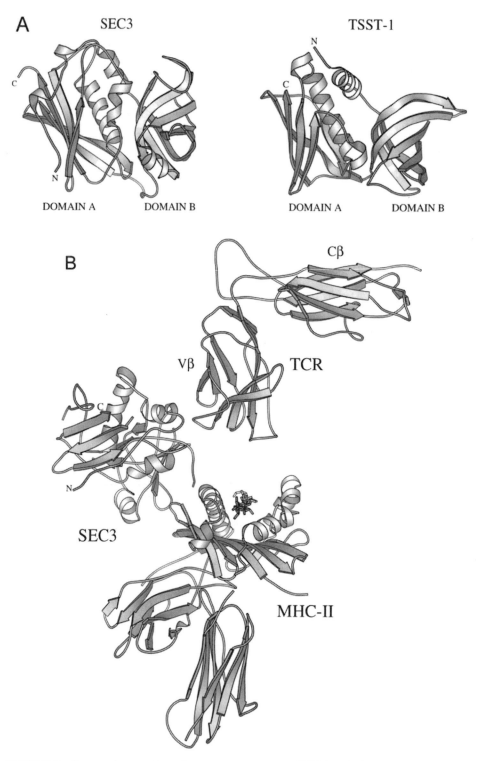

FIGURE 5 Structural properties and receptor interactions of PTSAgs. (A) A structural comparison of SEC3 and TSST-1. Ribbon diagrams shown are based on crystal structures published for the two toxins (15, 31). The two structures are oriented so that the TCR-binding cavity in each is located at the top and the cysteine loop, unique to the SEs, is on the upper right-hand corner of SEC3. Both toxins possess a similar domain organization and an overall topology despite having several important differences as discussed in the text. (B) A model of the trimolecular complex with SEB or SEC bound to TCR and MHC-II (adapted from results of references 25 and 37). In this model, SAgs orient the two receptors away from each other, inducing an aberrant mechanism of T-cell activation. Note that antigenic peptide associated with MHC-II is positioned away from the TCR-binding site.

toxins, have been proposed for other SEs with high-affinity Zn^{2+}-binding sites on the outer face of domain 2 (SED, SEE) (5, 119).

SElH contains only the high-affinity site; the crystal structure has been solved for this toxin complexed with HLA-DR1 (96). The MHC-II interaction of SElH, and presumably other PTSAgs, with high-affinity zinc-binding sites is more extensive than that for SEB. SElH binds predominantly to the β-chain and the peptide and covers a contact region approximately twice that of SEB. This feature, plus additional stability provided by the zinc bridge, are likely responsible for the high affinity between SElH and HLA-DR1.

Most PTSAgs interact with a defined TCR repertoire (Table 1), determined by sequences of TCR Vβ and by toxin residues in the shallow cavity at the top of the molecule (27). A crystal structure of SEC3 complexed with part of the murine TCR β chain (35) revealed that the toxin binds residues in the CDR1, CDR2, and HV4 loops of Vβ. Superimposing crystal structures from (i) the Vβ:SEC3 complex, (ii) the SEB:HLA-DR1 complex, and (iii) a TCR Vα homodimer led to a trimolecular complex model representing SAg binding to the T cell and antigen-presenting cell (Fig. 5B). In this model the PTSAg wedges between the TCR and the MHC-II. This orients the peptide-binding cleft away from the TCR in a configuration greatly different from typical antigen presentation. The overall affinity of the entire complex determines the effectiveness of the stimulation (66). Binding by toxins with low-affinity for the TCR can be compensated for by stronger binding to MHC-II, and vice versa.

SElH appears to have the most divergent mechanism of MHC-II-binding and subsequent T-cell stimulation. SElH binds overtop MHC-II in a mechanism that does not promote interactions between the toxin and the TCR Vβ, or between MHC-II and Vβ. Petersson et al. (95) could not demonstrate Vβ-specific expansion by SElH. Instead, SElH stimulated T cells by interacting with the TCR Vα (Vα10).

The roles of SAg activity and T-cell proliferation in SFP have received significant amounts of investigation, and some evidence exists that SAg function alone is not significant to induce this illness. For example, studies with various SEs have generated mutants that dissociate T-cell proliferation from emesis. Substitution of residue 25 in the SEA TCR-binding cavity dramatically reduces T-cell proliferation with no demonstrable effect on emesis (51). Similar conclusions were drawn from studies with SEB and SEC (3, 55). Although the disulfide bond and cysteine loop have been proposed to be involved in emesis, the contents of the loops are quite variable. Therefore, it is unlikely that the cysteines themselves or the loop residues play a key direct role in emesis. It is more likely that the disulfide bond maintains a conformation and orientation of other residues crucial for emesis (15, 55). It is suspected that a highly conserved stretch of residues in the SE small domain, immediately downstream from the disulfide bond, may mediate emesis and overlap or affect a site required for T-cell proliferation (51, 55). This is consistent with the recent demonstration that toxins lacking at least one cysteine residue (SEI, SElL, and SElQ) are minimally emetic or nonemetic when administered orally to monkeys (90).

It is generally agreed that the massive cytokines coinciding with T-cell proliferation contribute significantly to PTSAg-induced lethal shock, although work with TSST-1 site-specific mutants has suggested that induction of lethality is more complex than originally suspected. The cause of a reduced or absent in vivo toxicity for most TSST-1 mutants can be traced to their defect in T-cell proliferative ability. Generally, the defect is attributed to an amino acid alteration that interferes with their ability to bind to either the TCR or MHC-II. However, there is substantial evidence to suggest that an additional property of PTSAgs, separate from their T-cell proliferative ability, is sufficient to induce lethality. One interesting mutant, Q136A, retains the ability to induce T-cell proliferation but is devoid of lethal function in vivo (31). Residue 136 in TSST-1 is located in a largely buried site in the toxin molecule. The crystal structure of the Q136A mutant revealed that substitution of glutamine with alanine at position 136 caused a dramatic change in the conformation of the β7 to β9 loop covering the back of the central α-helix (Fig. 5A).

EXFOLIATIVE (EPIDERMOLYTIC) TOXINS

The ETs have been conclusively implicated in staphylococcal scalded-skin syndrome (SSSS). SSSS as defined by Melish and Glasgow (73) includes a spectrum of staphylococcal illnesses in patients, predominantly children, characterized by formation of bullae or skin blisters and a potential for widespread peeling. According to their designation, SSSS includes Ritter's disease, toxic epidermal necrosis, bullous impetigo, and certain erythema cases. Widespread skin peeling occurs in patients lacking ET antibodies, but localized lesions as in bullous impetigo develop if antibody is present in the serum. Melish and Glasgow (73) also found that inoculating SSSS isolates in newborn mice generated a positive Nikolsky sign and sterile skin lesions similar to those in humans with SSSS. They proposed that a soluble toxin, identified shortly thereafter, was responsible for effects in the mouse model. This model is still standard methodology for assessing the epidermolytic effects of ETs.

SSSS occurs predominantly in infants and children. Until recently, the prevailing explanation was that the adaptive immune response in immunologically mature individuals neutralizes toxin disseminating systemically. This is also consistent with results in the murine model showing that susceptibility declines between days 7 and 8, during which maturation of the adaptive response occurs. However, Plano et al. (97) demonstrated that the innate immune response is not required for protection against SSSS symptoms. Instead, susceptibility seemed to correlate with less efficient clearance from the circulation in neonatal animals.

Two antigenically distinct forms, designated ETA and ETB (reviewed in reference 10), are the best characterized ETs and are produced most frequently by phage group II by *S. aureus* isolates; strains expressing ETs constitute *agr* group IV staphylococcal isolates (57). The two proteins share greater than 40% identity over 242 and 246 residues in ETA and ETB, respectively. In addition, both proteins are approximately 25% identical to the *S. aureus* serine (V8) protease and possess the serine protease catalytic triad of H72, D120, and S195 (102). Two additional *S. aureus* molecular variants (ETC and ETD) have been reported (137), and several unique ETs have been identified in other staphylococcal species (*Staphylococcus chromogenes*, *Staphylococcus hyicus*, and *S. intermedius*) associated with swine exudative dermatitis and canine pyoderma (2, 110, 121).

The ETA structural gene (*eta*) is chromosomally located and in some strains has been shown to be harbored on a phage that can be mobilized and convert *eta*⁻ strains to toxin producers (135, 138). *etb* and *etd* are located on plasmids and pathogenicity islands, respectively; both types of elements also contain EDIN proteins, which are ADP-ribosyltransferases capable of modifying Rho GTPases (136). The significance of EDINs or their association with ETs is not clear presently.

Lesions in SSSS and mice are characterized by separation of stratum granulosa cells causing intraepidermal skin peeling. With sufficient doses, ETs can induce effects in as little as 10 min. The physical separation of the skin coincides with desmosome degeneration along a plane that eventually marks the cleavage plane. Histological observations preceding desmosome degradation include gap formations between cells and distended intercellular spaces. The mechanism of exfoliative activity has been the focus of considerable investigation at the cellular and subcellular levels (112). Although esterase activity had been confirmed (11), and experimental evidence long suggested that ET protease activity is responsible for skin lesions in SSSS (105), direct observation of protease activity remained elusive until recently. Rago et al. (104) first demonstrated ET proteolytic activity toward melanocyte-stimulating factors. The susceptible bond is located on the C-terminal side of a conserved glutamic acid residue. Although the physiological relevance of cleaving melanocyte-stimulating factor is unclear, blister formation in SSSS is now known to result from the cleavage of a homologous glutamic acid target in desmoglein-1 (Dsg1), a cadherin-like cell-to-cell desmosomal adhesion molecule (49). Related molecules such as Dsg3 and E-cadherin are not affected by ETs (8). The target residue is Glu 381, located between domains 3 and 4 (49), and its susceptibility is dependent upon Ca^{2+} (50). Proteolysis of Dsg1 in this manner results in a loss of function and cell separation, leading to blister formation. It has been proposed that this effect allows the organism to spread under the stratum corneum (7).

Crystal structures have been solved for ETA and ETB (19, 126, 127). Structurally, both toxins resemble other serine proteases, of which chymotrypsin is the prototype, containing two perpendicular β-barrel domains and C-terminal α-helix (Fig. 6). ETA possesses a unique N-terminal 15-residue α-helix that interacts with loop D adjacent to the catalytic site. In this conformation, the peptide bond between loop residues P92 and G93 is flipped 180° compared to other serine proteases so that substrates would not have access to the active site. The putative latent protease might be activated by binding to a cellular receptor or substrate via the N-terminal helix, releasing loop D and opening the active site.

The ETs have also been reported to function as SAgs, although this activity is controversial. Compared to PTSAgs, the ETs are 100-fold less potent in inducing T-cell proliferation and less toxic in rabbits (80). Initial reports (21) suggested that ETs cause stimulation of human T cells bearing Vβ2. In contrast, Fleischer and Bailey (38) suggested that the ETs were not SAgs, but that the putative activity resulted from SAg contamination of the ET preparations. They were able to purify commercial ETA such that fractions with the ability to stimulate human Vβ2 T cells were separated from the toxin. Monday et al. (80) appeared to clarify this issue and confirmed that both ETs are, in fact, SAgs. However, highly purified and recombinant ETs expressed in either *S. aureus* or *Escherichia coli* selectively expanded human T cells expressing Vβs 3, 12, 13.2, 14, 15, and 17, but not Vβ2. Skin peeling ability and T-cell proliferation are separable by mutagenesis. One ETA mutant with an altered active site serine (S195C) is unable to induce skin peeling despite inducing T-cell proliferation similar to the native

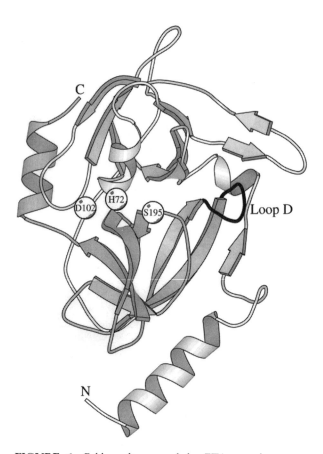

FIGURE 6 Ribbon diagram of the ETA crystal structure showing important functional features. Similar to other chymotrypsinlike proteases, ETA has two β-barrel domains and a C-terminal α-helix. The N-terminal domain, which includes a highly charged α-helix, is unique and is suspected to be involved in receptor binding. The positions of residues H72, D102, and S195, comprising the putative catalytic triad, are superimposable with the analogous residues of α-thrombin. D164 in loop D controls access of substrate to the protease active site by hydrogen bonding to G193. This causes the P192-G193 peptide bond to flip 180 degrees compared to that seen in other serine proteases and may explain the lack of demonstrable proteolytic activity in vitro. Binding of the N-terminal α-helix to its receptor has been proposed to cause a shift in the position of loop D and thereby the P192-G193 peptide bond, allowing access to the active site in vivo (127).

toxin (127). Plano et al. (98) were unable to demonstrate that recombinant ETA has superantigen activity.

We thank T. Foster, S. Bhakdi, D. Boehm, K. Dziewanowska, Y. Piemont, D. Ohlendorf, P. Schlievert, David Terman, and G. Vath for helpful discussions, review of portions of the manuscript, and, in some cases, providing figures. C. Deobald and M. Marshall assisted in preparation of the manuscript and figures. This work was supported by U.S. Public Health Service grants U54AI57141 and P20-RR15587.

REFERENCES

1. **Aarestrup, F. M., H. D. Larsen, N. H. Eriksen, C. S. Elsberg, and N. E. Jensen.** 1999. Frequency of alpha- and beta-haemolysin in *Staphylococcus aureus* of bovine and human origin. A comparison between pheno- and geno-

type and variation in phenotypic expression. *APMIS* **107:** 425–430.

2. **Ahrens, P., and L. O. Andresen.** 2004. Cloning and sequence analysis of genes encoding *Staphylococcus hyicus* exfoliative toxin types A, B, C, and D. *J. Bacteriol.* **186:** 1833–1837.

3. **Alber, G., D. K. Hammer, and B. Fleischer.** 1990. Relationship between enterotoxic- and T lymphocyte-stimulating activity of staphylococcal enterotoxin B. *J. Immunol.* **144:**4501–4506.

4. **Alber, G., P. H. Scheuber, B. Reck, B. Sailer-Kramer, A. Hartmann, and D. K. Hammer.** 1989. Role of substance P in immediate-type skin reactions induced by staphylococcal enterotoxin B in unsensitized monkeys. *J. Allergy Clin. Immunol.* **84:**880–885.

5. **Al-Daccak, R., K. Mehindate, F. Damdoumi, P. Etongue-Mayer, H. Nilsson, P. Antonsson, M. Sundstrom, M. Dohlsten, R. P. Sekaly, and W. Mourad.** 1998. Staphylococcal enterotoxin D is a promiscuous superantigen offering multiple modes of interactions with the MHC class II receptors. *J. Immunol.* **160:**225–232.

6. **Alouf, J. E.** 1977. Cell membranes and cytolytic bacterial toxins, p. 220–270. *In* P. Cuatrecasas (ed.), *Receptors and Recognition*, series B, vol. 1. *The Specificity and Action of Animal, Bacterial and Plant Toxins.* Chapman and Hall Ltd., London, United Kingdom

7. **Amagai, M., N. Matsuyoshi, Z. H. Wang, C. Andl, and J. R. Stanley.** 2000. Toxin in bullous impetigo and staphylococcal scalded-skin syndrome targets desmoglein 1. *Nat. Med.* **6:**1275–1277.

8. **Amagai, M., T. Yamaguchi, Y. Hanakawa, K. Nishifuji, M. Sugai, and J. R. Stanley.** 2002. Staphylococcal exfoliative toxin B specifically cleaves desmoglein 1. *J. Invest. Dermatol.* **118:**845–850.

9. **Arbuthnott, J. P.** 1982. Bacterial cytolysins (membrane-damaging toxins), p. 107–129. *In* P. Cohen and S. van Heyningen (ed.), *Molecular Action of Toxins and Viruses.* Elsevier Biomedical Press, Amsterdam, The Netherlands.

10. **Bailey, C. J., B. P. Lockhart, M. B. Redpath, and T. P. Smith.** 1995. The epidermolytic (exfoliative) toxins of *Staphylococcus aureus. Med. Microbiol. Immunol.* **184:**53–61.

11. **Bailey, C. J., and M. B. Redpath.** 1992. The esterolytic activity of epidermolytic toxins. *Biochem. J.* **284:**177–180.

12. **Bayles, K. W., and J. J. Iandolo.** 1989. Genetic and molecular analyses of the gene encoding staphylococcal enterotoxin D. *J. Bacteriol.* **171:**4799–4806.

13. **Bhakdi, S., and J. Tranum-Jensen.** 1991. Alpha-toxin of *Staphylococcus aureus. Microbiol. Rev.* **55:**733–751.

14. **Bohach, G. A., D. J. Fast, R. D. Nelson, and P. M. Schlievert.** 1990. Staphylococcal and streptococcal pyrogenic toxins involved in toxic shock syndrome and related illnesses. *Crit. Rev. Microbiol.* **17:**251–272.

15. **Bohach, G. A., L. M. Jablonski, C. F. Deobald, Y. I. Chi, and C. V. Stauffacher.** 1995. Functional domains of staphylococcal enterotoxins, p. 339–356. *In* M. Ecklund, J. L. Richard, and K. Mise (ed.), *Molecular Approaches to Food Safety: Issues Involving Toxic Microorganisms.* Alaken, Inc., Fort Collins, Colo.

16. **Boyle, T., V. Lancaster, R. Hunt, P. Gemski, and M. Jett.** 1994. Method for simultaneous isolation and quantitation of platelet activating factor and multiple arachidonate metabolites from small samples: analysis of effects of *Staphylococcus aureus* enterotoxin B in mice. *Anal. Biochem.* **216:** 373–382.

17. **Buerke, M., U. Sibelius, U. Grandel, U. Buerke, F. Grimminger, W. Seeger, J. Meyer, and H. Darius.** 2002. *Staph-ylococcus aureus* alpha toxin mediates polymorphonuclear leukocyte-induced vasocontraction and endothelial dysfunction. *Shock* **17:**30–35.

18. **Caiazza, N. C., and G. A. O'Toole.** 2003. Alpha-toxin is required for biofilm formation by *Staphylococcus aureus. J. Bacteriol.* **185:**3214–3217.

19. **Cavarelli, J., G. Prevost, W. Bourguet, L. Moulinier, B. Chevrier, B. Delagoutte, A. Bilwes, L. Mourey, S. Rifai, Y. Piemont, and D. Moras.** 1997. The structure of *Staphylococcus aureus* epidermolytic toxin A, an atypic serine protease, at 1.7 Å resolution. *Structure* **5:**813–824.

20. **Chintagumpala, M. M., J. A. Mollick, and R. R. Rich.** 1991. Staphylococcal toxins bind to different sites on HLA-DR. *J. Immunol.* **147:**3876–3881.

21. **Choi, Y., B. Kotzin, L. Herron, J. Callahan, P. Marrack, and J. Kappler.** 1989. Interaction of *Staphylococcus aureus* toxin superantigens with human T cells. *Proc. Natl. Acad. Sci. USA* **86:**8941–8945.

22. **Coleman, D. C., J. P. Arbuthnott, H. M. Pomeroy, and T. H. Birkbeck.** 1986. Cloning and expression in *Escherichia coli* and *Staphylococcus aureus* of the beta-lysin determinant from *Staphylococcus aureus*: evidence that bacteriophage conversion of beta-lysin activity is caused by insertional inactivation of the beta-lysin determinant. *Microb. Pathog.* **1:**549–564.

23. **Colin, D. A., I. Mazurier, S. Sire, and V. Finck-Barbancon.** 1994. Interaction of the two components of leukocidin from *Staphylococcus aureus* with human polymorphonuclear leukocyte membranes: sequential binding and subsequent activation. *Infect. Immun.* **62:**3184–3188.

24. **Colin, D. A., O. Meunier, L. Staali, H. Monteil, and G. Prevost.** 1996. Action mode of two components poreforming leucotoxins from *Staphylococcus aureus. Med. Microbiol. Immunol.* **185:**107–114.

25. **Cooney, J., Z. Kienle, T. J. Foster, and P. W. O'Toole.** 1993. The gamma-hemolysin locus of *Staphylococcus aureus* comprises three linked genes, two of which are identical to the genes for the F and S components of leukocidin. *Infect. Immun.* **61:**678–771.

26. **Deringer, J. R., R. J. Ely, S. R. Monday, C. V. Stauffacher, and G. A. Bohach.** 1997 Vβ-dependent stimulation of bovine and human T cells by host-specific staphylococcal enterotoxins. *Infect. Immun.* **65:**4048–4054.

27. **Deringer, J. R., R. J. Ely, C. V. Stauffacher, and G. A. Bohach.** 1996. Subtype-specific interactions of type C staphylococcal enterotoxins with the T-cell receptor. *Mol. Microbiol.* **22:**523–534.

28. **Dhople, V. M., and R. Nagaraj.** 2005. Conformation and activity of delta-lysin and its analogs. *Peptides* **26:**217–225.

29. **Dufourc, E. J., J. Dufourcq, T. H. Birkbeck, and J. H. Freer.** 1990. δ-Haemolysin from *Staphylococcus aureus* and model membranes. A solid-state ^{2}H-NMR and ^{31}P-NMR study. *Eur. J. Biochem.* **187:**581–587.

30. **Dziewanowska, K., V. E. Edwards, J. R. Deringer, G. A. Bohach, and D. J. Guerra.** 1996. Comparison of the β-toxins from *Staphylococcus aureus* and *Staphylococcus intermedius. Arch. Biochem. Biophys.* **335:**102–108.

31. **Earhart, C. A., D. T. Mitchell, D. L. Murray, D. M. Pinheiro, M. Matsumura, P. M. Schlievert, and D. H. Ohlendorf.** Structures of five mutants of toxic shock syndrome toxin-1 with reduced biological activity. *Biochemistry* **37:**7194–7202.

32. **Edwards, V. M., J. R. Deringer, S. D. Callantine, C. F. Deobald, P. H. Berger, V. Kapur, C. V. Stauffacher, and G. A. Bohach.** 1997. Characterization of the canine type C enterotoxin produced by *Staphylococcus intermedius* pyoderma isolates. *Infect. Immun.* **65:**2346–2352.

33. **Elwell, M. R., C. T. Liu, R. O. Spertzel, and W. R. Beisel.** 1975. Mechanisms of oral staphylococcal enterotoxin B-induced emesis in the monkey. *Proc. Soc. Exp. Biol. Med.* **148:**424–427.

34. **Essmann, F., H. Bantel, G. Totzke, I. H. Engels, B. Sinha, K. Schulze-Osthoff, and R. U. Janicke.** 2003. *Staphylococcus aureus* alpha-toxin-induced cell death: predominant necrosis despite apoptotic caspase activation. *Cell Death Differ.* **10:**1260–1272.

35. **Fields, B. A., E. L. Malchiodi, H. Li, X. Ysern, C. V. Stauffacher, P. M. Schlievert, K. Karjalainen, and R. A. Mariuzza.** 1996. Crystal structure of a T-cell receptor beta-chain complexed with a superantigen. *Nature* **384:**188–192.

36. **Finck-Barbancon, V., G. Duportail, O. Meunier, and D. A. Colin.** 1993. Pore formation by two-component leukocidin from *Staphylococcus aureus* within the membrane of human polymorphonuclear leukocytes. *Biochim. Biophys. Acta* **1182:**275–282.

37. **Fitzgerald, J. R., S. D. Reid, E. Ruotsalainen, T. J. Tripp, M. Liu, R. Cole, P. Kuusela, P. M. Schlievert, A. Jarvinen, and J. M. Musser.** 2003. Genome diversification in *Staphylococcus aureus*: molecular evolution of a highly variable chromosomal region encoding the Staphylococcal exotoxin-like family of proteins. *Infect. Immun.* **71:**2827–2838.

38. **Fleischer, B., and C. J. Bailey.** 1992. Recombinant epidermolytic (exfoliative) toxin A of *Staphylococcus aureus* is not a superantigen. *Med. Microbiol. Immunol.* **180:**273–279.

39. **Florquin, S., and L. Aaldering.** 1997. Superantigens: a tool to gain new insight into cellular immunity. *Res. Immunol.* **148:**373–386.

40. **Foster, T. J., M. O'Reilly, P. Phonimdaeng, J. Cooney, A. H. Patel, and A. J. Bramley.** 1990. Genetic studies of virulence factors of *Staphylococcus aureus*. Properties of coagulase and gamma-toxin and the role of alpha-toxin, beta-toxin and protein A in the pathogenesis of *S. aureus* infections, p. 403–417. In R. P. Novick (ed.), *Molecular Biology of the Staphylococci*. VCH, Cambridge, New York, N.Y.

41. **Fraser, J. D.** 1989. High-affinity binding of staphylococcal enterotoxins A and B to HLA-DR. *Nature* **339:**221–223.

42. **Furoda, M., T. Ohta, I. Uchiyama, T. Baba, H. Yuzawa, I. Kobayashi, L. Cui, A. Oguchi, K. Aoki, Y. Nagai, J. Lian, T. Ito, M. Kanamori, H. Matsumaru, A. Maruyama, H. Murakami, A. Hosoyama, Y. Mizutani-Ui, N. K. Takahashi, T. Sawano, R. Inoue, C. Kaito, K. Sekimizu, H. Hirakawa, S. Kuhara, S. Goto, J. Yabuzaki, M. Kanehisa, A. Yamashita, K. Oshima, K. Furuya, C. Yoshino, T. Shiba, M. Hattori, N. Ogasawara, H. Hayashi, and K. Hiramatsu.** 2001. Whole genome sequencing of methicillin-resistant *Staphylococcus aureus*. *Lancet* **357:**1225–1240.

43. **Gase, K., J. J. Ferretti, C. Primeaux, and W. M. McShan.** 1999. Identification, cloning, and expression of the CAMP factor gene (cfa) of group A streptococci. *Infect. Immun.* **67:**4725–4731.

44. **Gillet, Y., B. Issartel, P. Vanhems, J. C. Fournet, G. Lina, M. Bes, F. Vandenesch, Y. Piemont, N. Brousse, D. Floret, and J. Etienne.** 2002. Association between *Staphylococcus aureus* strains carrying gene for Panton-Valentine leukocidin and highly lethal necrotising pneumonia in young immunocompetent patients. *Lancet* **359:**753–759.

45. **Grundstrom, S., L. Cederbom, A. Sundstedt, P. Scheipers, and F. Ivars.** 2003. Superantigen-induced regulatory T cells display different suppressive functions in the presence or absence of natural CD4+CD25+ regulatory T cells in vivo. *J. Immunol.* **170:**5008–5017.

46. **Goerke, C., S. Matias y Papenberg, S. Dasbach, K. Dietz, R. Ziebach, B. C. Kahl, and C. Wolz.** 2004. Increased frequency of genomic alterations in *Staphylococcus aureus* during chronic infection is in part due to phage mobilization. *J. Infect. Dis.* **189:**724–734.

47. **Gravet, A., D. Colin, R. Keller, H. Giradot, H. Monteil, and G. Prevost.** 1998. Characterization of a novel structural member, LukE-LukD, of the bi-component staphylococcal leucotoxins family. *FEBS Lett.* **436:**202–208.

48. **Guillet, V., P. Roblin, S. Werner, M. Coraiola, G. Menestrina, H. Monteil, G. Prevost, and L. Mourey.** 2004. Crystal structure of leucotoxin S component: new insight into the staphylococcal beta-barrel pore-forming toxins. *J. Biol. Chem.* **279:**41028–41037.

49. **Hanakawa, Y., N. M. Schechter, C. Lin, L. Garza, H. Li, T. Yamaguchi, Y. Fudaba, K. Nishifuji, M. Sugai, M. Amagai, and J. R. Stanley.** 2002. Molecular mechanisms of blister formation in bullous impetigo and staphylococcal scalded skin syndrome. *J. Clin. Invest.* **110:**53–60.

50. **Hanakawa, Y., T. Selwood, D. Woo, C. Lin, N. M. Schechtern, and J. R. Stanley.** 2003. Calcium-dependent conformation of desmoglein 1 is required for its cleavage by exfoliative toxin. *J. Invest. Dermatol.* **121:**383–389.

51. **Harris, T. O., and M. J. Betley.** 1995. Biological activities of staphylococcal enterotoxin type A mutants with N-terminal substitutions. *Infect. Immun.* **63:**2133–2140.

52. **Haslinger, B., K. Strangfeld, G. Peters, K. Schulze-Osthoff, and B. Sinha.** 2003. *Staphylococcus aureus* alpha-toxin induces apoptosis in peripheral blood mononuclear cells: role of endogenous tumour necrosis factor-alpha and the mitochondrial death pathway. *Cell Microbiol.* **5:**729–741.

53. **Hildebrand, A., M. Roth, and S. Bhakdi.** 1991. *Staphylococcus aureus* alpha-toxin: dual mechanisms of binding to target cells. *J. Biol. Chem.* **266:**17195–17200.

54. **Holmberg, S. D., and P. A. Blake.** 1984. Staphylococcal food poisoning in the United States. New facts and old misconceptions. *JAMA* **251:**487–489.

55. **Hovde, C. J., J. C. Marr, M. L. Hoffmann, S. P. Hackett, Y. I. Chi, K. K. Crum, D. L. Stevens, C. V. Stauffacher, and G. A. Bohach.** 1994. Investigation of the role of the disulphide bond in the activity and structure of staphylococcal enterotoxin C1. *Mol. Microbiol.* **13:**897–909.

56. **Hudson, K. R., R. E. Tiedemann, R. G. Urban, S. C. Lowe, J. L. Strominger, and J. D. Fraser.** 1995. Staphylococcal enterotoxin A has two cooperative binding sites on major histocompatibility complex class II. *J. Exp. Med.* **182:**711–720.

57. **Jarraud, S., G. J. Lyon, A. M. Figueiredo, L. Gerard, F. Vandenesch, J. Etienne, T. W. Muir, and R. P. Novick.** 2000. Exfoliatin-producing strains define a fourth agr specificity group in *Staphylococcus aureus*. *J. Bacteriol.* **182:**6517–6522.

58. **Jardetsky, T. S., J. H. Brown, J. C. Gorga, L. J. Stern, R. G. Urban, Y. I. Chi, C. V. Stauffacher, J. L. Strominger, and D. C. Wiley.** 1994. Three-dimensional structure of a human class II histocompatibility molecule complexed with superantigen. *Nature* **368:**711–718.

59. **Jarraud, S., M. A. Peyrat, A. Lim, A. Tristan, M. Bes, C. Mougel, J. Etienne, F. Vandenesch, M. Bonneville, and G. Lina.** 2001. egc, a highly prevalent operon of enterotoxin gene, forms a putative nursery of superantigens in *Staphylococcus aureus*. *J. Immunol.* **166:**669–677.

60. **Jarvis, W. D., R. N. Kolesnick, F. A. Fornari, R. S. Traylor, D. A. Gewirtz, and S. Grant.** 1994. Induction of apoptotic DNA damage and cell death by activation of the sphingomyelin pathway. *Proc. Natl. Acad. Sci. USA* **91:** 73–77.

61. **Jonas, D., I. Walev, T. Berger, M. Liebetrau, M. Palmer, and S. Bhakdi.** 1994. Novel path to apoptosis: small transmembrane pores created by staphylococcal alpha-toxin in T lymphocytes evoke internucleosomal DNA degradation. *Infect. Immun.* **62:**1304–1312.

62. **Kaneko, J., O. Toshiko, T. Tomita, and Y. Kamio.** 1997. Sequential binding of staphylococcal γ-hemolysin to human erythrocytes and complex formation of the hemolysin on the cell surface. *Biosci. Biotechnol. Biochem.* **61:**846–851.

63. **Kappler, J., B. Kotzin, L. Herron, E. W. Gelfand, R. D. Bigler, A. Boylston, S. Carrel, D. N. Posnett, Y. Choi, and P. Marrack.** 1989. V beta-specific stimulation of human T cells by staphylococcal toxins. *Science* **244:**811–813.

64. **Kim, C. S., S. Y. Jeon, Y. G. Min, C. Rhyoo, J. W. Kim, J. B. Yun, S. W. Park, and T. Y. Kwon.** 2000. Effects of beta-toxin of *Staphylococcus aureus* on ciliary activity of nasal epithelial cells. *Laryngoscope* **110:**2085–2088.

65. **Kim, J., R. G. Urban, J. L. Strominger, and D. C. Wiley.** 1994. Toxic shock syndrome toxin-1 complexed with a class II major histocompatibility molecule HLA-DR1. *Science* **266:**1870–1874.

66. **Leder, L., A. Llera, P. M. Lavoie, M. I. Lebedeva, H. Li, R. P. Sekaly, G. A. Bohach, P. J. Gahr, P. M. Schlievert, K. Karjalainen, and R. A. Mariuzza.** 1998. A mutational analysis of the binding of staphylococcal enterotoxins B and C3 to the T cell receptor beta chain and major histocompatibility complex class II. *J. Exp. Med.* **187:**823–833.

67. **Letertre, C., S. Perelle, F. Dilasser, and P. Fach.** 2003. Identification of a new putative enterotoxin SEU encoded by the *egc* cluster of *Staphylococcus aureus*. *J. Appl. Microbiol.* **95:**38–43.

68. **Lina, G., G. A. Bohach, S. P. Nair, K. Hiramatsu, E. Jouvin-Marche, and R. Mariuzza, International Nomenclature Committee for Staphylococcal Superantigens.** 2004. Standard nomenclature for the superantigens expressed by *Staphylococcus*. *J. Infect. Dis.* **189:**2334–2336.

69. **Lina, G., Y. Piemont, F. Godail-Gamot, M. Bes, M. O. Peter, V. Gauduchon, F. Vandenesch, and J. Etienne.** 1999. Involvement of Panton-Valentine leukocidin-producing *Staphylococcus aureus* in primary skin infections and pneumonia. *Clin. Infect. Dis.* **29:**1128–1132.

70. **Low, D. K. R., and J. H. Freer.** 1977. Biological effects of highly purified β-lysin (sphingomyelinase C) from *Staphylococcus aureus*. *FEMS Microbiol. Lett.* **2:**133–138.

71. **Mahlknecht, U., M. Herter, M. K. Hoffmann, D. Niethammer, and G. E. Dannecker.** 1996. The toxic shock syndrome toxin-1 induces anergy in human T cells in vivo. *Hum. Immunol.* **45:**42–45.

72. **Marshall, M. J., G. A. Bohach, and D. F. Boehm.** 2000. Characterization of *Staphylococcus aureus* beta-toxin induced leukotoxicity. *J. Nat. Toxins* **9:**125–138.

73. **Melish, M. E., and L. A. Glasgow.** 1970. The staphylococcal scalded skin syndrome: development of an experimental model. *N. Engl. J. Med.* **282:**1114–1119.

74. **Mellor, I. R., D. H. Thomas, and M. S. P. Sansom.** 1988. Properties of ion channels formed by *Staphylococcus aureus* δ-toxin. *Biochim. Biophys. Acta* **942:**280–294.

75. **Mempel, M., C. Schnopp, M. Hojka, H. Fesq, S. Weidinger, M. Schaller, H. C. Korting, J. Ring, and D. Abeck.** 2002. Invasion of human keratinocytes by *Staphylococcus*

76. **Menestrina, G., M. D. Serra, and G. Prevost.** 2001. Mode of action of beta-barrel pore-forming toxins of the staphylococcal alpha-hemolysin family. *Toxicon* **39:**1661–1672.

77. **Miles, G., L. Movileanu, and H. Bayley.** 2002. Subunit composition of a bicomponent toxin: staphylococcal leukocidin forms an octameric transmembrane pore. *Protein Sci.* **11:**894–902.

78. **Monday, S. R., and G. A. Bohach.** 1999. Genetic, structural, biological, pathophysiological and clinical aspects of *Staphylococcus aureus* enterotoxins and toxic shock syndrome toxin-1, p. 589–610. *In* J. E. Alouf and J. H. Freer (ed.), *Sourcebook of Bacterial Protein Toxins.* Academic Press, London, United Kingdom.

79. **Monday, S. R., and G. A. Bohach.** 2000. Genes encoding staphylococcal enterotoxins are linked and separated by DNA related to other staphylococcal enterotoxins. *J. Nat. Toxins* **10:**1–8.

80. **Monday, S. R., G. M. Vath, W. A. Ferens, C. Deobald, J. V. Rago, P. J. Gahr, D. Monie, J. J. Iandolo, S. K. Chapes, W. C. Davis, D. H. Ohlendorf, P. M. Schlievert, and G. A. Bohach.** 1999. Unique superantigen activity of staphylococcal exfoliative toxins. *J. Immunol.* **181:**4550–4559.

81. **Munson, S. H., M. T. Tremaine, M. J. Betley, and R. A. Welch.** 1998. Identification and characterization of staphylococcal enterotoxin types G and I from *Staphylococcus aureus*. *Infect. Immun.* **66:**3337–3348.

82. **Narita, S., J. Kaneko, J. Chiba, Y. Piemont, S. Jarraud, J. Etienne, and Y. Kamio.** 2001. Phage conversion of Panton-Valentine leukocidin in *Staphylococcus aureus*: molecular analysis of a PVL-converting phage, ΦSLT. *Gene* **268:**195–206.

83. **Noda, M., and I. Kato.** 1991. Leukocidal toxins, p. 243–251. *In* J. E. Alouf and J. H. Freer (ed.), *Sourcebook of Bacterial Protein Toxins.* Academic Press, London, United Kingdom.

84. **O'Callaghan, R. J., M. C. Callegan, J. M. Moreau, L. C. Green, T. J. Foster, O. M. Hartford, L. S. Engel, and J. M. Hill.** 1997. Specific roles of alpha-toxin and beta-toxins during *Staphylococcus* corneal infection. *Infect. Immun.* **65:**1571–1578.

85. **Omoe, K., K. Imanishi, D. L. Hu, H. Kato, H. Takahashi-Omoe, A. Nakane, T. Uchiyama, and K. Shinagawa.** 2004. Biological properties of staphylococcal enterotoxin-like toxin type R. *Infect. Immun.* **72:**3664–3667.

86. **Omoe, K., D. L. Hu, H. Takahashi-Omoe, A. Nakane, and K. Shinagawa.** 2003. Identification and characterization of a new staphylococcal enterotoxin-related putative toxin encoded by two kinds of plasmids. *Infect. Immun.* **71:**6088–6094.

87. **Onogawa, T.** 2002. Staphylococcal alpha-toxin synergistically enhances inflammation caused by bacterial components. *FEMS Immunol. Med. Microbiol.* **33:**15–21.

88. **Orwin, P. M., J. Fitzgerald, D. Y. Leung, J. A. Gutierrez, G. A. Bohach, and P. M. Schlievert.** 2003. Characterization of *Staphylococcus aureus* enterotoxin L. *Infect. Immun.* **71:**2916–2919.

89. **Orwin, P. M., D. Y. Leung, D. H. Donahue, R. P. Novick, and P. M. Schlievert.** 2001. Biochemical and biological properties of Staphylococcal enterotoxin K. *Infect. Immun.* **69:**360–366.

90. **Orwin, P. M., D. Y. Leung, T. J. Tripp, G. A. Bohach, C. A. Earhart, D. H. Ohlendorf, and P. M. Schlievert.**

aureus and intracellular bacterial persistence represent haemolysin-independent virulence mechanisms that are followed by features of necrotic and apoptotic keratinocyte cell death. *Br. J. Dermatol.* **146:**943–951.

2002. Characterization of a novel staphylococcal enterotoxin-like superantigen, a member of the group V subfamily of pyrogenic toxins. *Biochemistry* **41:**14033–14040.

91. **Ozawa, T., J. Kaneko, and Y. Kamio.** 1995. Essential binding of LukF of staphylococcal γ-hemolysin followed by the binding of HγII for the hemolysis of human erythrocytes. *Biosci. Biotech. Biochem.* **559:**1181–1183.

92. **Ozawa, T., J. Kaneko, H. Narija, K. Izaki, and Y. Kamio.** 1994. Inactivation of the γ-hemolysin HγII component by addition of monoganglioside G_{M1} to human erythrocyte. *Biosci. Biotech. Biochem.* **58:**602–605.

93. **Park, P. W., T. J. Foster, E. Nishi, S. J. Duncan, M. Klagsbrun, and Y. Chen.** 2004. Activation of syndecan-1 ectodomain shedding by *Staphylococcus aureus* alpha-toxin and beta-toxin. *J. Biol. Chem.* **279:**251–258.

94. **Parsonnet, J.** 1998. Case definition of staphylococcal TSS: a proposed revision incorporating laboratory findings, p. 15. *In* F. Arbuthnott and B. Furman (ed.), *Proceedings of the European Conference on Toxic Shock Syndrome.* The Royal Society of Medicine Limited, London, United Kingdom.

95. **Petersson, K., H. Pettersson, N. J. Skartved, B. Walse, and G. Forsberg.** 2003. Staphylococcal enterotoxin H induces V alpha-specific expansion of T cells. *J. Immunol.* **170:**4148–4154.

96. **Petersson, K., M. Hakansson, H. Nilsson, G. Forsberg, L. A. Svensson, A. Liljas, and B. Walse.** 2001. Crystal structure of a superantigen bound to MHC class II displays zinc and peptide dependence. *EMBO J.* **20:**3306–3312.

97. **Plano, L. R., B. Adkins, M. Woischnik, R. Ewing, and C. M. Collins.** 2001. Toxin levels in serum correlate with the development of staphylococcal scalded skin syndrome in a murine model. *Infect. Immun.* **69:**5193–5197.

98. **Plano, L. R., D. M. Gutman, M. Woischnik, and C. M. Collins.** 2000. Recombinant *Staphylococcus aureus* exfoliative toxins are not bacterial superantigens. *Infect. Immun.* **68:**3048–3052.

99. **Pokorny, A., T. H. Birkbeck, and P. F. Almeida.** 2002. Mechanism and kinetics of delta-lysin interaction with phospholipid vesicles. *Biochemistry* **41:**11044–11056.

100. **Prevost, G., P. Coupie, P. Prevost, S. Gayet, P. Petiau, B. Cribier, H. Monteil, and Y. Piemont.** 1995. Epidemiological data on *Staphylococcus aureus* strains producing synergohymenotropic toxins. *J. Med. Microbiol.* **42:**237–245.

101. **Prevost, G., B. Cribier, P. Couppie, P. Petiau, G. Supersac, V. Finck-Barbancon, H. Monteil, and Y. Piemont.** 1995. Panton-Valentine leucocidin and gamma-hemolysin from *Staphylococcus aureus* ATCC 49775 are encoded by distinct genetic loci and have different biological activities. *Infect. Immun.* **63:**4121–4129.

102. **Prevost, G., S. Rifai, M. L. Chaix, S. Meyer, and Y. Piemont.** 1992. Is the His72, Asp120, Ser195 constitutive of the catalytic site of staphylococcal exfoliative toxin A? p. 488–489. *In* B. Witholt (ed.), *Bacterial Protein Toxins.* Fischer, Stuttgart, Germany.

103. **Projan, S. J., J. Kornblum, B. Kreiswirth, S. L. Moghazeh, W. Eisner, and R. P. Novick.** 1989. Nucleotide sequence: the β-hemolysin gene of *Staphylococcus aureus.* *Nucleic Acids Res.* **17:**3305.

104. **Rago, J. V., G. M. Vath, T. J. Tripp, G. A. Bohach, D. H. Ohlendorf, and P. M. Schlievert.** 2000. Staphylococcal exfoliative toxins cleave alpha- and beta-melanocyte-stimulating hormones. *Infect. Immun.* **68:**2366–2368.

105. **Redpath, M. B., T. J. Foster, and C. J. Bailey.** 1991. The role of the serine protease active site in the mode of action of epidermolytic toxin of *Staphylococcus aureus.* *FEMS Microbiol. Lett.* **81:**151–156.

106. **Reingold, A. L., N. T. Hargrett, K. N. Shands, B. B. Dan, G. P. Schmid, B. Y. Strickland, and C. V. Broome.** 1982. Toxic shock syndrome surveillance in the United States, 1980 to 1981. *Ann. Intern. Med.* **96:**875–880.

107. **Ren, K., J. D. Bannan, V. Pancholi, A. L. Cheung, J. C. Robbins, V. A. Fischetti, and J. B. Zabriskie.** 1994. Characterization and biological properties of a new staphylococcal exotoxin. *J. Exp. Med.* **180:**1675–1683.

108. **Rich, R. R., J. A. Mollick, and R. G. Cook.** 1989. Superantigens: interaction of staphylococcal enterotoxins with MHC class II molecules. *Trans. Am. Clin. Climatol. Assoc.* **101:**195–204.

109. **Rogalsky, M.** 1979. Nonenteric toxins of *Staphyloccus aureus.* *Microbiol. Rev.* **43:**320–360.

110. **Sato, H., K. Hirose, R. Terauchi, S. Abe, I. Moromizato, S. Kurokawa, and N. Maehara.** 2004. Purification and characterization of a novel *Staphylococcus chromogenes* exfoliative toxin. *J. Vet. Med. B Infect. Dis. Vet. Public Health* **51:**116–122.

111. **Scheuber, P. H., C. Denzlinger, D. Wilker, G. Beck, D. Keppler, and D. K. Hammer.** 1987. Staphylococcal enterotoxin B as a nonimmunological mast cell stimulus in primates: the role of endogenous cysteinyl leukotrienes. *Int. Arch. Allergy. Appl. Immunol.* **82:**289–291.

112. **Smith, T. P., D. A. John, and C. J. Bailey.** 1987. The binding of epidermolytic toxin from *Staphylococcus aureus* to mouse epidermal tissue. *Histochem. J.* **19:**137–149.

113. **Smyth, D. S., P. J. Hartigan, W. J. Meaney, J. R. Fitzgerald, C. F. Deobald, G. A. Bohach, and C. J. Smyth.** 2005. Superantigen genes encoded by the *egc* cluster and SaPIbov are predominant among *Staphylococcus aureus* isolates from cows, goats, sheep, rabbits and poultry. *J. Med. Microbiol.* **54:**401–411.

114. **Somerville, G. A., A. Cockayne, M. Durr, A. Peschel, M. Otto, and J. M. Musser.** 2003. Synthesis and deformylation of *Staphylococcus aureus* delta-toxin are linked to tricarboxylic acid cycle activity. *J. Bacteriol.* **185:**6686–6694.

115. **Song, L., M. R. Hobaugh, C. Shustak, S. Cheley, H. Bayley, and J. E. Gouaux.** 1996. Structure of the staphylococcal α-hemolysin, a heptameric transmembrane pore. *Science* **274:**1859–1865.

116. **Stohl, W., J. E. Elliott, D. H. Lynch, and P. A. Kiener.** 1998. CD95 (Fas)-based, superantigen-dependent, CD4+ T cell-mediated down-regulation of human in vitro immunoglobulin responses. *J. Immunol.* **160:**5231–5238.

117. **Sugawara, N., T. Tomita, and Y. Kamio.** 1997. Assembly of γ-hemolysin into a pore-forming ring-shaped complex on the surface of human erythrocytes. *FEBS Lett.* **410:**333–337.

118. **Sugawara-Tomita, N., T. Tomita, and Y. Kamio.** 2002. Stochastic assembly of two-component staphylococcal gamma-hemolysin into heteroheptameric transmembrane pores with alternate subunit arrangements in ratios of 3:4 and 4:3. *J. Bacteriol.* **184:**4747–4756.

119. **Sundstrom, M., L. Abrahmsen, P. Antonsson, K. Mehindate, W. Mourad, and M. Dohlsten.** 1996. The crystal structure of staphylococcal enterotoxin type D reveals Zn2+-mediated homodimerization. *EMBO J.* **15:**6832–6840.

120. **Tappin, M. J., A. Pastore, R. S. Norton, J. H. Freer, and I. D. Campbell.** 1988. High-resolution ^{1}H NMR study of the solution structure of δ-hemolysin. *Biochemistry* **27:**1643–1647.

121. **Terauchi, R., H. Sato, Y. Endo, C. Aizawa, and N. Maehara.** 2003. Cloning of the gene coding for *Staphylococcus intermedius* exfoliative toxin and its expression in *Escherichia coli. Vet. Microbiol.* **94**:31–38.

122. **Tiedemann, R. E., and J. D. Fraser.** 1996. Cross-linking of MHC class II molecules by staphylococcal enterotoxin A is essential for antigen-presenting cell and T cell activation. *J. Immunol.* **157**:3958–3966.

123. **Tomita, T., Y. Ueda, H. Tamura, R. Taguchi, and H. Ikezawa.** 1993. The role of acidic amino-acid residues in catalytic and adsorptive sites of *Bacillus cereus* sphingomyelinase. *Biochim. Biophys. Acta* **1203**:85–92.

124. **Valeva, A., A. Weisser, B. Walker, M. Kehoe, H. Bayley, S. Bhakdi, and M. Palmer.** 1996. Molecular architecture of a toxin pore: a 15 residue sequence lines the transmembrane channel of staphylococcal alpha-toxin. *EMBO J.* **15**:1857–1864.

125. **Vandenesch, F., T. Naimi, M. C. Enright, G. Lina, G. R. Nimmo, H. Heffernan, N. Liassine, M. Bes, T. Greenland, M. E. Reverdy, and J. Etienne.** 2003. Community-acquired methicillin-resistant *Staphylococcus aureus* carrying Panton-Valentine leukocidin genes: worldwide emergence. *Emerg. Infect. Dis.* **9**:978–984.

126. **Vath, G. M., C. A. Earhart, D. D. Monie, J. J. Iandolo, P. M. Schlievert, and D. H. Ohlendorf.** 1999. The crystal structure of exfoliative toxin B: a superantigen with enzymatic activity. *Biochemistry* **38**:10239–10246.

127. **Vath, G. M., C. A. Earhart, J. V. Rago, M. H. Kim, G. A. Bohach, P. M. Schlievert, and D. H. Ohlendorf.** 1997. The structure of the superantigen exfoliative toxin A suggests a novel regulation as a serine protease. *Biochemistry* **36**:1559–1566.

128. **von Eiff, C., R. A. Proctor, and G. Peters.** 2000. Small colony variants of staphylococci: a link to persistent infections. *Berl. Munch. Tierarztl. Wochenschr.* **113**:321–325.

129. **Walev, I., U. Weller, S. Strauch, T. Foster, and S. Bhakdi.** 1996. Selective killing of human monocytes and cytokine release provoked by sphingomyelinase (beta toxin) of *Staphylococcus aureus. Infect. Immun.* **64**:2974–2979.

130. **White, J., A. Herman, A. M. Pullen, R. Kubo, J. W. Kappler, and P. Marrack.** 1989. The V beta-specific superantigen staphylococcal enterotoxin B: stimulation of mature T cells and clonal deletion in neonatal mice. *Cell* **56**:27–35.

131. **Williams, R. E. O., and G. H. Harper.** 1947. Staphylococcal haemolysins on sheep blood agar with evidence for a fourth haemolysin. *J. Pathol. Bacteriol.* **59**:69–78.

132. **Williams, R. J., J. M. Ward, B. Henderson, S. Poole, B. P. O'Hara, M. Wilson, S. P. Nair.** 2000. Identification of a novel gene cluster encoding staphylococcal exotoxin-like proteins: characterization of the prototypic gene and its protein product, SET1. *Infect. Immun.* **68**:4407–4415.

133. **Wiseman, G. M.** 1975. The hemolysins of *Staphylococcus aureus. Bacteriol. Rev.* **39**:317–344.

134. **Woodin, A. M.** 1970. Staphylococcal leukocidin, p. 327–355. *In* T. C. Montie, S. Kadis, and S. J. Ajl (ed.), *Microbial Toxins*. Academic Press, New York, N.Y.

135. **Yamaguchi, T., T. Hayashi, H. Takami, K. Nakasone, M. Ohnishi, K. Nakayama, S. Yamada, H. Komatsuzawa, and M. Sugai.** 2000. Phage conversion of exfoliative toxin A production in *Staphylococcus aureus. Mol. Microbiol.* **38**:694–705.

136. **Yamaguchi, T., T. Hayashi, H. Takami, M. Ohnishi, T. Murata, K. Nakayama, K. Asakawa, M. Ohara, H. Komatsuzawa, and M. Sugai.** 2001. Complete nucleotide sequence of a *Staphylococcus aureus* exfoliative toxin B plasmid and identification of a novel ADP-ribosyltransferase, EDIN-C. *Infect. Immun.* **69**:7760–7771.

137. **Yamaguchi, T., K. Nishifuji, M. Sasaki, Y. Fudaba, M. Aepfelbacher, T. Takata, M. Ohara, H. Komatsuzawa, M. Amagai, and M. Sugai.** 2002. Identification of the *Staphylococcus aureus etd* pathogenicity island which encodes a novel exfoliative toxin, ETD, and EDIN-B. *Infect. Immun.* **70**:5835–5845.

138. **Yoshizawa, Y., J. Sakurada, S. Sakurai, K. Machida, I. Kondo, and S. Masuda.** 2000. An exfoliative toxin A-converting phage isolated from *Staphylococcus aureus* strain ZM. *Microbiol. Immunol.* **44**:189–191.

139. **Zhang, S., J. J. Iandolo, and G. C. Stewart.** 1998. The enterotoxin D plasmid of *Staphylococcus aureus* encodes a second enterotoxin determinant (*sej*). *FEMS Microbiol. Lett.* **168**:227–233.

140. **Zou, D., J. Kaneko, S. Narita, and Y. Kamio.** 2000. Prophage, ΦPV83-pro, carrying panton-valentine leucocidin genes, on the *Staphylococcus aureus* P83 chromosome: comparative analysis of the genome structures of ΦPV83-pro, ΦPVL, Φ11, and other phages. *Biosci. Biotechnol. Biochem.* **64**:2631–2643.

Extracellular Enzymes

STAFFAN ARVIDSON

39

Staphylococcus aureus produces a large number of extracellular enzymes, many of which are regarded as important virulence factors. These enzymes can degrade organic macromolecules in the environment to provide low-molecular-weight nutrients for the bacterium. Degradation of tissue constituents is also an essential part of staphylococcal pathogenicity and can promote spread of the bacterium in host tissues. Some extracellular enzymes contribute to the virulence of staphylococci by attacking molecules involved in host defenses against infection. Searches in the genome sequences of several *S. aureus* strains identified 22 genes coding for extracellular enzymes (32) (Table 1). This list of exoenzymes does not contain secreted enzymes with toxic activities such as the epidermolytic toxins and beta-hemolysin, which are described elsewhere in this volume. On the other hand, coagulase and staphylokinase (Sak), which have no enzyme activity by themselves but serve as cofactors for certain enzymes in the animal host, have been included.

COAGULASE

Coagulase production is the principal criterion used in the clinical microbiology laboratory for the identification of *S. aureus*. Although a few strains of *S. aureus* do not produce detectable amounts of coagulase, all strains seem to possess a coagulase gene (*coa*) (66).

Coagulase binds with human prothrombin in a 1:1 molar ratio to form a complex named staphylothrombin, which can convert fibrinogen to fibrin (29). Unlike the physiological activation of prothrombin, formation of staphylothrombin does not involve proteolytic cleavage of prothrombin. Instead, zymogen activation is triggered by insertion of the N terminus of coagulase into the "activation" pocket of prothrombin (19). The N terminus of coagulase thus serves the same purpose as the new N terminus formed through proteolytic activation of prothrombin.

Coagulase also binds to fibrinogen (4). A fraction of coagulase, which is associated with the bacterial cell surface, was therefore thought to be responsible for the clumping of bacterial cells when mixed with plasma. However, clumping factor is a distinct fibrinogen-binding protein that can promote binding of bacteria to solid-phase fibrinogen, in contrast to coagulase, which binds only soluble fibrinogen (43). Coagulase does not possess a cell wall-anchoring sequence (LPXTG) but seems to be associated with the bacterial cells by other means. A fibrinogen-binding protein (FbpA) belonging to the coagulase family, with a unique sequence, LPXSITG, in the middle of the molecule, has been characterized (9). Although this potential anchoring signal is not followed by a typical membrane-spanning region and a charged tail, it is speculated that it could participate in anchoring this new type of coagulase to the cell wall.

Eight serotypes of coagulase have been identified by neutralization tests. The primary structures of coagulases belonging to serotypes I, II, and III have been compared (47). Four distinct segments of coagulase can be recognized: (i) a typical signal peptide of 26 amino acid residues; (ii) a highly variable N-terminal region of 150 to 270 amino acid residues, which are only 50% identical between the serotypes; (iii) a central region with more than 90% identical residues; and (iv) a C-terminal region composed of 5, 6, or 8 tandem repeats of 27 amino acid residues. The prothrombin-binding region of coagulase is in the variable N terminus, while fibrinogen binds to the C-terminal repeat region (42).

Although the 81-bp tandem repeats encoding the C-terminal region of coagulase are well conserved, individual repeats differ in the presence or absence of *Alu*I and *Cfo*I restriction endonuclease sites. Based on the restriction fragment length polymorphism of the 3' end of the coagulase gene, it has been possible to discriminate between isolates of *S. aureus* in epidemiological studies (21, 23).

Production of coagulase is negatively regulated by *agr*. Consistent with the regulatory model, production of coagulase in wild-type bacteria is maximized during early exponential growth, while in *agr* mutants production of coagulase continues throughout growth. However, the levels of *coa*-specific transcript and extracellular coagulase activity were much lower in the *agr* mutant than in the wild type, indicating that *agr* also has a positive effect on *coa* expression (33). Transcription of *coa* is also affected by the *sae*

Gram-Positive Pathogens, 2nd edition, edited by Vincent A. Fischetti et al.
© 2006 ASM Press, Washington, D.C.

TABLE 1 Extracellular enzymes and enzyme activators from *S. aureus*

Enzyme/name	Activity/substrate	Gene name	Accession numbers[a]
Coagulase	Prothrombin activator	*coa*	X17679, SA0222
Staphylokinase (Sak)	Plasminogen activator	*sak*	U77328, SA1758
V8 protease, serine protease	Glutamic acid-specific protease	*sppA*	P04188, SA0901
Serine protease-like proteases	Not known	*splA-F*	SA1627-SA1631
Staphopain A (cysteine protease)	Cleaves most peptide bonds	*scpA*	SA1725
Staphopain B (cysteine protease)	Protease, unknown specificity	*sspB*	P81297, SA0900
Metalloprotease	Cleaves before bulky hydrophobic amino acids	*aur*	SA2430
Lipase (true)	Cleaves long-chain glycerol esters	*geh*	M12715, SA0309
Lipase, esterase	Cleaves short-chain glycerol esters	*lip*	M90693, SA2463
PI-phospholipase C (PI-PLC)	Specific for phosphatidylinositol	*plc*	L19298, SA0091
Fatty acid modifying enzyme (FAME)	Esterification of fatty acids to cholesterol		
Hyaluronate lyase	Depolymerization of hyaluronan	*hysA*	U21221, SA2003
Nuclease, thermonuclease	Degrades double- and single-stranded DNA and RNA	*nuc*	J01785, SA1160
β-Lactamase	Inactivates certain β-lactam antibiotics	*blaZ*	M15526
Endo-β-*N*-acetylglucosaminidase	Autolytic murein hydrolase	*atl*	D17366, SA0905
N-Acetylmuramyl-L-alanine amidase	Autolytic murein hydrolase	*atl*	D17366, SA0905

[a]SA numbers are open reading frame numbers in *S. aureus* strain N315.

locus. In an *agr:sae* double mutant, *coa*-specific mRNA levels were markedly reduced compared with the *agr* single mutant, suggesting that *sae* is a positive regulator (20).

The role of coagulase in staphylococcal pathogenesis is unclear. One possible function could be that fibrin clotting around a focal infection protects the bacteria from various host defense mechanisms. Several reports have indicated that site-specific inactivation of the coagulase gene does not impair virulence in experimental endocarditis, subcutaneous, or mammary infections of mice. However, coagulase-negative mutants were less virulent than the parental strain in a mouse model of blood-borne staphylococcal pneumonia (59). This suggests that coagulase may be more important in some types of infections than in others.

STAPHYLOKINASE (Sak)

Sak is a potent activator of plasminogen, the precursor of the fibrinolytic protease plasmin. Plasmin is formed from plasminogen through activator-catalyzed cleavage of a single peptide bond (R561-V562) in the zymogen, resulting in two polypeptide chains that are held together by two disulfide bonds. The present knowledge about the mechanism of action of Sak and its use as a thrombolytic agent has been reviewed (12).

Sak is not an enzyme but forms a 1:1 stoichiometric complex with plasmin that has high plasminogen-activating activity. Sak also binds to plasminogen, but this complex is inactive and must be converted to Sak-plasmin to become active. During the process of active complex formation, the 10 N-terminal amino acid residues of Sak are removed to expose Lys11 as the new N terminus. Deletion of Lys 11, or substitution with Cys, eliminates the plasminogen activator potential. The three-dimensional structure of Sak has been resolved at 1.8 Å, and the putative site of interaction with plasminogen has been identified (52).

In contrast to streptokinase, a similar nonenzymatic plasminogen activator produced by *Streptococcus pyogenes*, Sak is able to induce thrombolysis without causing systemic plasminogen activation, thereby decreasing the risk of severe bleeding. The explanation for this seems to be that the circulating Sak-plasmin complex is rapidly inhibited by α_2-antiplasmin, while the complex that is bound to fibrin is

>100-fold less sensitive to the inhibitor. The streptokinase-plasmin complex, on the other hand, is insensitive to inhibition by α_2-antiplasmin. Sak also has a higher affinity for plasmin(ogen) bound to fibrin than for free plasmin(ogen). The fibrin-selective activity of Sak makes it an attractive alternative to streptokinase as a thrombolytic agent. Clinical trials have shown that Sak can cause recanalization of the occluded arteries in more than 80% of patients with myocardial infarction. However, most patients develop high titers of neutralizing immunoglobulin G within 2 weeks after administration of Sak. Variants of Sak with a reduced immunogenicity have been developed by site-directed mutagenesis (12).

Sak is produced by lysogenic strains of *S. aureus* carrying certain prophages. Three different groups of phages that carry the Sak gene and mediate lysogenic conversion of Sak have been described. Some serotype B phages cause positive lysogenic conversion of Sak production alone (30). Simultaneous positive conversion of Sak, or Sak plus enterotoxin A, and negative conversion of beta-hemolysin can be mediated by different serotype F phages (11). The negative conversion of beta-hemolysin production is due to insertion of the phage DNA into the beta-hemolysin gene (*hlb*).

The genes of four natural variants of Sak have been characterized. They differ at six nucleotide positions within the coding region (163 codons), leading to amino acid exchanges at positions 33, 34, 36, and 43 of the mature proteins.

Sak belongs to the group of extracellular proteins that are positively regulated by *agr* (1). As a consequence, Sak is generally not produced at the same time as coagulase, which is negatively regulated by *agr*. A possible role of Sak in an infection would therefore be to facilitate the release and spread of bacteria trapped in fibrin clots, or in abscesses surrounded by fibrin. Sak also induces secretion of defensins from human neutrophilic granulocytes, and more importantly, Sak neutralizes the bactericidal effect of α-defensins by complex binding (25).

PROTEASES

Most strains of *S. aureus* produce extracellular proteolytic activity. Enzymes belonging to the families of serine-, metallo-, and cysteine-(thiol) proteases have been characterized (1).

Although the proteolytic activity varies considerably among *S. aureus* strains most human isolates possess genes coding for one serine protease (V8 protease), one metalloprotease (aureolysin), and two different cysteine proteases (staphopain A and staphopain B) (27). The proteases are secreted as proenzymes with little or no enzyme activity, which need to be proteolytically cleaved to gain full activity. Maturation proceeds as a cascade, starting with the activation of aureolysin (Fig. 1).

Serine Proteases

S. aureus serine protease (SspA), or V8 protease (named after strain V8), belongs to a small group of enzymes that preferentially cleave glutamoyl peptide bonds. A weak activity is also seen against peptide bonds following aspartic acid residues. Other enzymes of this group are produced by *Bacillus licheniformis*, *Bacillus subtilis*, *Streptomyces griseus*, and *Streptomyces fradiae* (3). Because of their narrow substrate specificity, V8 protease and the other Glu-specific proteases have been widely used for site-specific fragmentation of proteins before amino acid sequencing. Although the sequence identity between the V8 protease and the other Glu-specific proteases is only on the order of 20 to 25%, the catalytic triad residues, His-51, Asp-93, and Ser-169 (mature V8 protease numbering), and the amino acids around these residues are well conserved. The conformation of the active site is identical to that of the epidermolytic toxins A and B from *S. aureus* and trypsin (51).

The gene coding for V8 protease (*sspA*) is part of an operon together with two other genes, *sspB* and *sspC*, coding for a thiol protease (staphopain B) and a staphopain B inhibitor, respectively (40) (see below). The *ssp* operon has been found in all *S. aureus* complete genomes sequenced so far. Expression of the *ssp* operon is positively controlled by *agr* and negatively controlled by *sarA*.

V8 protease is synthesized as a preproenzyme with a typical N-terminal signal peptide of 29 amino acids, followed by a 39-amino-acid propeptide. The proenzyme, which is enzymatically inactive, is activated through proteolytic cleavage by the extracellular metalloprotease, aureolysin, and some as yet unidentified protease (Fig. 1) (60). This is different from other Glu-specific proteases produced by *Bacillus* and *Streptomyces* species, which are self-processed. A special feature of the V8 protease is the presence of

tandemly repeated tripeptides, Pro-Asn/Asp-Asn, at the COOH-terminal end of the mature protein. In the prototype enzyme from strain V8 the tripeptide is repeated 12 times, whereas the published sequences of five other strains revealed between 9 and 19 repeats. In all strains the repeat region is followed by identical sequences of 12 amino acids. The role of the tripeptide region is unknown.

Bacterial extracellular proteases are generally regarded as digestive enzymes providing the bacterium with low-molecular-weight nutrients. Because of its restricted substrate specificity, V8 protease would have only a limited utility in this respect unless it works together with other proteolytic enzymes. However, during an infection, V8 protease may promote bacterial survival and spread through inactivation of important host proteins. V8 protease can cleave the heavy chains of all human immunoglobulin classes in a way that could impair host defense against the bacterium. It also inactivates human α_1-proteinase inhibitor, which is the major inhibitor of elastase released from neutrophilic granulocytes during phagocytosis of invading microorganisms (50). The uncontrolled activity of host elastase might contribute to tissue damage and degradation of proteins involved in host defense. However, a nonpolar inactivation of *sspA* in the prototype *S. aureus* strain 8325-4 did not significantly alter its virulence in a murine abscess model of infection (54). On the other hand, a transposon mutation of *sspA* in the same parental strain was attenuated in three separate animal infection models (13).

V8 protease also seems to play a role in the bacterial transition from an adhesive to a nonadhesive phenotype by rapidly degrading fibronectin-binding proteins and other staphylococcal cell surface proteins (28, 44).

A cluster of four to six genes coding for serine protease-like enzymes (*splA* through *splF*) has been found in all *S. aureus* genomes completed so far. The Spl genes are cotranscribed and subject to *agr*-dependent positive regulation (53), and negative regulation by *sarA* (7). The Spl operon is located on the νSaα pathogenicity island that also carries several enterotoxin genes and leukocidin genes, *lukD* and *lukE* (32). The Spl proteins show 33 to 36% amino acid sequence identity with V8 protease, and contain the catalytic amino acid triad typical of trypsin-like serine proteases. Unlike V8 protease the Spl proteins seem to lack a propeptide that needs to be cleaved off for the enzyme to be active. So far only SplB and SplC have been shown to have proteolytic activity.

Cysteine Proteases

S. aureus secretes two extracellular cysteine proteases, staphopain A and staphopain B. The first characterized enzyme was staphopain A (originally named protease II), which is typically inactivated by thiol-reacting agents and has a broad substrate specificity, similar to papain (1). It also has strong activity against elastin, suggesting a role in staphylococcal pathogenesis (48). Staphopain A is efficiently inhibited by plasma α_2-macroglobulin, but is insensitive to inhibition by human cystatin C or kininogens, suggesting that some of the tissue damages related to *S. aureus* infection might be caused by this enzyme. The three-dimensional crystal structure of staphopain A, in particular its active site, is very similar to that of papain, although the enzymes show little similarity at the amino acid level (22). Staphopain A is synthesized as a preproenzyme of 388 amino acids with a typical N-terminal signal peptide for secretion, followed by a propeptide of 198 amino acid residues that is thought to keep the enzyme in an inactive form until

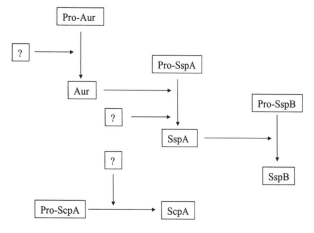

FIGURE 1 Cascade of maturation of the extracellular proteases of *S. aureus*. Question marks represent unidentified proteases.

it is removed through proteolytic cleavage. As mature enzyme was produced in mutant strains lacking V8 protease, aureolysin, or staphopain B, the enzyme responsible for the maturation process remains to be identified (60). Possibly prostaphopain A can undergo autocatalytic maturation like many papain-like proteases.

The gene coding for staphopain A, *scpA*, is cotranscribed with a gene (*scpB*) coding for staphostatin A, a specific inhibitor of staphopain A (57). Staphostatin A lacks a typical signal peptide for secretion and is most likely localized to the bacterial cytosol, where it may protect cytosolic proteins from prematurely activated staphopain A.

The second cysteine protease, staphopain B, is encoded by *sspB*, which is part of an operon together with *sspA*, coding for V8 protease, and *sspC*, coding for an inhibitor of staphopain B, named staphostatin B. The amino acid sequence of mature staphopain B is 47% identical to that of staphopain A, and contains a typical cysteine protease histidine active site consensus pattern (54). Like other enzymes belonging to the papain family of proteases, staphopain B requires cysteine for its enzyme activity, is susceptible to cysteine protease inhibitor E64, and cleaves preferentially peptides with a hydrophobic amino acid residue at P2 (40). However, the strict requirement for an arginine at position P1 makes staphopain B differ from other members of the papain family, which generally accept several other amino acids at this position. Interestingly, staphopain B could cleave human fibronectin, fibrinogen, and kininogen at sites generally recognized by plasma serine proteases such as plasmin and kallicrein. These effects of staphopain B suggest that it could play a role in *S. aureus* infection by promoting spread of the organism. This is consistent with the observation that *sspB* mutants are significantly attenuated in virulence (13, 60).

Staphopain B is secreted as a 357-amino-acid proenzyme that appears to have a low but significant proteolytic activity against gelatin and that must be cleaved by the V8 protease at glutamic acid residue number 183 to attain full activity (Fig. 1) (40, 54).

As in the case of staphopain A, a specific inhibitor, staphostatin B, controls the intracellular activity of staphopain B. The staphostatins A and B, which are of similar size (12.6 and 12.9 kDa), show the same three-dimensional eight-stranded β-barrel structure, in spite of a low amino acid sequence similarity (16, 17). Together with homologous proteins from *Staphylococcus epidermidis* and *Staphylococcus warnerii* they seem to form a completely new class of cysteine protease inhibitors, which bind to the active site of their corresponding enzymes in a substratelike fashion.

Metalloprotease

Aureolysin is a 33-kDa, calcium-binding, zinc-requiring endopeptidase with optimum activity at neutral pH. Like other neutral metalloproteases, it cleaves peptide bonds on the N-terminal side of bulky hydrophobic amino acid residues. Although zinc appears to be required for the catalytic activity, calcium ions seem to stabilize the protein against proteolytic degradation. Nucleotide sequence data together with the amino acid sequence of the mature enzyme (301 amino acids) indicate that aureolysin is synthesized as a preproenzyme with an N-terminal signal peptide for secretion followed by a profragment of 181 amino acid residues (58). The three-dimensional structure of aureolysin is very similar to that of thermolysin from *Bacillus thermoproteolyticus*, neutral protease from *Bacillus cereus*, and elastase from *Pseudomonas aeruginosa* (2). All the amino acid residues involved in substrate binding and the catalytic activity are conserved among these proteases.

Little is known about the role of aureolysin in the pathogenicity of *S. aureus*. In contrast to most members of the thermolysin family of metalloproteases, aureolysin lacks activity against elastin, a feature that could be explained by the presence of a loop peptide on top of its active site cleft (2). On the other hand, aureolysin has been shown to cleave plasma proteinase inhibitors that control host proteases, including elastase and cathepsin G from neutrophilic granulocytes, indicating a role in pathogenesis of staphylococcal disease (49, 50). Indirectly, aureolysin might have a role in pathogenicity by modulating the activity of other virulence factors produced by the pathogen. Aureolysin is involved in the processing and activation of V8 protease (15, 60), which in turn activates staphopain B, and participates in shedding of proteins from the bacterial surface, e.g., fibronectin-binding protein and protein A. Aureolysin is also directly involved in the shedding of clumping factor (41) and in processing of extracellular lipase zymogen (55).

LIPASE

Glycerol Ester Hydrolase

True lipases are glycerol ester hydrolases that degrade water-insoluble long-chain triacylglycerols at the lipid-water interface. Interaction with the substrate interface leads to an increased enzymatic activity, known as interfacial activation. Most lipases are also active against acyl *p*-nitrophenylesters, Tweens (polyoxyethylenesorbitan), and sometimes phospholipids. Related enzymes hydrolyzing preferentially water-soluble glycerol esters are also called lipases but should properly be referred to as short-chain glycerol ester hydrolases, or esterases. Both types of enzyme belong to the family of serine esterases, possessing a catalytic triad consisting of a serine, a histidine, and either glutamic acid or aspartic acid.

Both types of enzyme are produced by *S. aureus*. Esterase activity toward water-soluble glycerol esters is produced by nearly all strains, while production of true lipases has been reported to be less common (55). A possible explanation would be that production of lipase is subject to negative lysogenic conversion by bacteriophage L54a, which can insert in the coding sequence of the staphylococcal true lipase, *geh* (34). The *geh* gene is present in all *S. aureus* complete genome sequences published so far. The lipase encoded by *geh* hydrolyzes long-chain triacylglycerols as well as water-soluble triacylglycerols and Tweens. It has a pH optimum for activity at 8.0 to 8.5 and is stimulated by calcium ions (31). Lipase is secreted with a propeptide of 258 amino acid residues, which is cleaved by an extracellular metalloprotease, presumably aureolysin (55). However, prolipase and mature lipase show the same enzymatic activity. The propeptide, which is highly hydrophilic, has therefore been suggested to stimulate secretion of the more hydrophobic lipase and to protect it from proteolytic degradation (56).

The "lipase" produced by strain NCTC8530 cleaves preferentially short-chain triacylglycerols, while most long-chain lipids are not hydrolyzed at all (62). Maximum activity is toward butyryl esters at pH 6.5. It does not show interfacial activation and should therefore be classified as a short-chain glycerol ester hydrolase. Because of its high activity towards butyryl esters I will use the name butyryl esterase henceforth. The gene coding for this enzyme was denoted *lip* in the GenBank database, but has unfortunately also been named *geh* (46). All *S. aureus* genomes analyzed up to now contain both *lip* and *geh*.

Like the true lipase, encoded by *geh*, the butyryl esterase is synthesized as a large proenzyme (77 kDa) that possesses the same enzymatic activity as the mature enzyme (46 kDa) (46). The proregion of the butyryl esterase is only 14% identical to that of the true lipase. However, the hydropathy and flexibility plots of the proregions are nearly superimposable, suggesting a structural rather than a catalytic role of the propeptide. Gene fusion experiments indicate that the proregion is important for the secretion of butyryl esterase or other heterologous proteins (35). The mature esterase is 54% identical to the true lipase, with the highest homology around the active-site residues, Ser-412, Asp-603, and His-645; it is 83% identical to lipase (GehSE1) from *S. epidermidis*, which is a typical butyryl esterase (63).

The function of lipases is not fully understood, but they may be important for bacterial nutrition. A role in virulence has also been suggested, based on the observation that staphylococcal lipase impairs granulocyte function (55). Free fatty acids, resulting from lipase activity, are also known to impair the immune system. On the other hand, long-chain fatty acids are bacteriocidal and would therefore seem to interfere with pathogenicity. However, fatty acids are detoxified by fatty acid-modifying enzyme (FAME), which is secreted by most lipase-producing strains of *S. aureus* (see below).

It has been reported that expression of lipase is positively regulated by *agr* and negatively regulated by *sar* (10). However, because lipase activity was assayed using substrates that do not discriminate between the true lipase and the butyryl esterase, it is not known which enzyme is regulated.

PI-Phospholipase C

Two different phospholipases C are produced by *S. aureus*. One is the hemolytic sphingomyelinase referred to as beta-hemolysin, which is described in chapter 38, this volume. The other is a phosphatidylinositol-specific (PI) phospholipase, PI-PLC, that was identified in *S. aureus* culture supernatants more than 35 years ago. Because it did not show cytotoxicity, it was long overlooked as a potential virulence factor. However, PI-PLC can degrade membrane-associated inositol phospholipids and release glycan-PI-anchored cell surface proteins, thereby interfering with important eukaryotic cell functions (39). More than 150 proteins are known to be bound to the cell surface via glycan-PI. Among these are several hydrolytic ectoenzymes, adhesion molecules, and various receptors and surface antigens. Inositol phospholipids are also involved in many signal transduction processes. Thus, PI-PLC has the potential to compromise host cell functions in a number of ways that could contribute to staphylococcal pathogenicity.

The gene encoding PI-PLC has been cloned and sequenced (Table 1), and the gene product has been characterized (14). The *plc* gene encodes a mature protein with a calculated molecular mass of 34,107 Da, which is in good agreement with the apparent molecular mass of purified PI-PLC (32 kDa). Northern blotting revealed a *plc*-specific mRNA of approximately 1 kb, indicating that the transcript is monocistronic. A typical signal peptide of 26 amino acid residues was removed from the N terminus during secretion. The mature enzyme showed high specificity for PI with optimum activity at pH 5.5 to 6.0. Although PI-PLC from *S. aureus* does not contain any cysteine residues, enzyme activity is inhibited by low concentrations of thiol-reactive

agents such as HgCl$_2$ and *p*-chloromercuribenzoate. The reason for this is not understood.

All fresh clinical isolates of *S. aureus* were found to produce PI-PLC, although the amount of enzyme produced varied 30-fold between strains (14). Interestingly, when frozen stock cultures of these strains were tested several months later, most of the strains produced substantially less PI-PLC. Specific mRNA could only be detected in those strains producing high amounts of PI-PLC, suggesting that the wide range of PI-PLC production between stains might reflect differences at the transcriptional level. Analysis of PI-PLC production in isogenic pairs of wild-type and *agr* mutant strains revealed 80% reduced levels of PI-PLC in the mutants as compared with the parental strains, indicating that expression of *plc* is positively regulated by an *agr*-dependent mechanism. Different levels of *agr* activity could thus explain the differences in PI-PLC production between clinical isolates, as has been shown for other *agr*-regulated genes.

FATTY ACID-MODIFYING ENZYME (FAME)

Staphylococcal abscesses contain long-chain free fatty acids and other neutral lipids that are bacteriocidal to *S. aureus*. FAME, which is found in culture supernatants of about 80% of *S. aureus* strains, can inactivate these bacteriocidal lipids by catalyzing the esterification of these lipids to alcohols, preferably cholesterol (26, 37). Bacteriocidal lipids are released from glycerides in the abscess, presumably by the action of staphylococcal lipase. This is consistent with the observation that most strains that produce lipase also produce FAME. It has been shown that FAME-producing strains of *S. aureus* are more virulent in a murine model (45).

Saturated fatty acids with 15 to 19 carbons are most efficiently esterified by FAME. The enzyme has a pH optimum between 5.5 and 6.0 and a temperature optimum of about 40°C. Enzyme activity is inhibited by tri- and diglycerides with unsaturated fatty acid chains. FAME has also been identified in culture supernatants from *S. epidermidis* (5). The gene(s) coding for FAME activity has not been identified. Among the genes with a typical signal peptide for secretion identified in the *S. aureus* genome sequences are two probable lipases (32), which might be involved in this detoxification process.

Production of FAME was markedly reduced in *agr* mutant strains as compared with the corresponding parental strains (6). Consistent with positive regulation of FAME by *agr*, maximum production of FAME appeared during the postexponential phase of growth in wild-type strains. Production of FAME was also reduced in a *sarA* mutant strain, which is consistent with the reduced *agr* activity observed in *sarA* mutants.

HYALURONATE LYASE

Hyaluronic acid is a ubiquitous component of the extracellular matrix of vertebrates. Extracellular enzymes that could hydrolyze hyaluronic acid were therefore among the first enzymes to be implicated in bacterial pathogenesis. Hyaluronic acid is a linear polysaccharide composed of repeating units of D-glucuronic acid(1-β-3)*N*-acetyl-D-glucosamine(1-β-4). Three classes of hyaluronidases are recognized based on their different mechanisms of action. Hyaluronidases of bacterial origin, including that from *S. aureus*, that degrade hyaluronic acid by a β-elimination mechanism yielding disaccharides that contain glucuronosyl residues with a double bond are named hyaluronate lyases.

The reading frame of staphylococcal hyaluronate lyase, *hysA*, encodes a protein of 807 amino acid residues with a calculated molecular mass of 92 kDa (18). An N-terminal signal sequence of 40 amino acids ending with a typical signal peptidase cleavage site suggests that the molecular mass of the secreted enzyme would be 87.4 kDa. This is close to the reported value of 84 kDa. *hysA* is present in all published *S. aureus* genome sequences. The amino acid sequence of staphylococcal hyaluronate lyase shares homology with the hyaluronate lyases from *Streptococcus pneumoniae*, *Streptococcus agalactiae*, and *Propionibacterium acne* (64). The greatest similarity is seen with the pneumococcal enzyme (36% identical). A histidine residue (His-479) in hyaluronate lyase from *S. agalactiae*, which is essential for catalytic activity, is present in a region that is highly conserved in the enzymes from the four different species (36). Several other regions, rich in basic residues that are thought to be involved in hyaluronate binding, are also conserved. Earlier work has shown that staphylococcal hyaluronate lyase activity is inhibited by thiol-reactive agents, indicating that free SH groups are essential. The presence of two cysteine residues in the staphylococcal enzyme is consistent with these observations. However, the homologous hyaluronate lyases from *S. pneumoniae* and *S. agalactiae* lack cysteine residues, suggesting that cysteines are not part of the active site.

A role of hyaluronate lyase in staphylococcal pathogenicity was recently demonstrated in a subcutaneous abscess model of infection in mice (38). In particular, the wild-type strain of *S. aureus* producing hyaluronate lyase caused much more tissue damage than the isogenic *hysA* mutant. Like several other exoenzyme genes (e.g., *sspA*, *aur*, *scpA*, *geh*), expression of *hysA* is positively regulated by *agr* and negatively regulated by *sarA*.

NUCLEASE

Thermostable nuclease (TNase, also called staphylococcal nuclease) is produced by nearly all strains of *S. aureus* and has been used as a diagnostic criterion for this species. However, highly related thermonucleases are also produced by *Staphylococcus hyicus* (68% identity to *S. aureus* TNase) and *Staphylococcus intermedius* (45% identity) (8). TNase hydrolyzes single- or double-stranded DNA and RNA at the 5' position of phosphodiester bonds by a calcium-dependent mechanism. It is one of the most extensively studied enzymes in terms of protein structure and catalytic properties (24, 67). Because the enzyme can refold spontaneously after thermal unfolding, it has been widely used as a model to study protein-folding mechanisms (61). The mature extracellular form of the enzyme consists of 149 amino acids (molecular mass 16.8 kDa) and has a globular form. Amino acid residues Asp-21 and Asp-40 are involved in binding a single Ca^{2+}, while Arg-35, Glu-43, and Arg-87, together with Ca^{2+}, are involved in substrate binding and catalysis.

The nuclease gene (*nuc*) encodes a protein of 228 amino acid residues. An unusually long signal peptide (60 amino acids), containing two highly hydrophobic stretches of approximately the same length, separated by a region containing three basic residues, is cleaved off during secretion both in *S. aureus* and in heterologous systems (i.e., *B. subtilis* and *Corynebacterium glutamicum*). The secreted form of the enzyme (nuclease B) has a 19-residue N-terminal propeptide, which is cleaved by an extracellular protease. The propeptide appears to function as a secretion enhancer (65). The propeptide did not stimulate secretion when it was attached to another protein, indicating that the propeptide acts as a specific secretion enhancer for the nuclease. The propeptides of other staphylococcal exoenzymes may have a similar effect (35).

These investigations have been supported by grant 4513 from the Swedish Research Council.

REFERENCES

1. **Arvidson, S. O.** 1983. Extracellular enzymes from *Staphylococcus aureus*, p. 745–808. *In* C. S. F. Easmon and C. Adlam (ed.), *Staphylococci and Staphylococcal Infections*, vol. 2. Academic Press Inc., London, United Kingdom.

2. **Banbula, A., J. Potempa, J. Travis, C. Fernandez-Catalan, K. Mann, R. Huber, W. Bode, and F. Medrano.** 1998. Amino-acid sequence and three-dimensional structure of the *Staphylococcus aureus* metalloproteinase at 1.72 Å resolution. *Structure* **6:**1185–1193.

3. **Birktoft, J. J., and K. Breddam.** 1994. Glutamyl endopeptidases. *Methods Enzymol.* **244:**114–126.

4. **Boden, M. K., and J. I. Flock.** 1989. Fibrinogen-binding protein/clumping factor from *Staphylococcus aureus*. *Infect. Immun.* **57:**2358–2363.

5. **Chamberlain, N. R., and S. A. Brueggemann.** 1997. Characterisation and expression of fatty acid modifying enzyme produced by *Staphylococcus epidermidis*. *J. Med. Microbiol.* **46:**693–697.

6. **Chamberlain, N. R., and B. Imanoel.** 1996. Genetic regulation of fatty acid modifying enzyme from *Staphylococcus aureus*. *J. Med. Microbiol.* **44:**125–129.

7. **Chan, P. F., and S. J. Foster.** 1998. Role of SarA in virulence determinant production and environmental signal transduction in *Staphylococcus aureus*. *J. Bacteriol.* **180:**6232–6241.

8. **Chesneau, O., and N. el Solh.** 1994. Primary structure and biological features of a thermostable nuclease isolated from *Staphylococcus hyicus*. *Gene* **145:**41–47.

9. **Cheung, A. I., S. J. Projan, R. E. Edelstein, and V. A. Fischetti.** 1995. Cloning, expression, and nucleotide sequence of a *Staphylococcus aureus* gene (fbpA) encoding a fibrinogen-binding protein. *Infect. Immun.* **63:**1914–1920.

10. **Cheung, A. L., J. M. Koomey, C. A. Butler, S. J. Projan, and V. A. Fischetti.** 1992. Regulation of exoprotein expression in *Staphylococcus aureus* by a locus (*sar*) distinct from *agr*. *Proc. Natl. Acad. Sci. USA* **89:**6462–6466.

11. **Coleman, D. C., D. J. Sullivan, R. J. Russell, J. P. Arbuthnott, B. F. Carey, and H. M. Pomeroy.** 1989. *Staphylococcus aureus* bacteriophages mediating the simultaneous lysogenic conversion of beta-lysin, staphylokinase and enterotoxin A: molecular mechanism of triple conversion. *J. Gen. Microbiol.* **135**(Pt 6)**:**1679–1697.

12. **Collen, D.** 1998. Staphylokinase: a potent, uniquely fibrin-selective thrombolytic agent. *Nat. Med.* **4:**279–284.

13. **Coulter, S. N., W. R. Schwan, E. Y. Ng, M. H. Langhorne, H. D. Ritchie, S. Westbrock-Wadman, W. O. Hufnagle, K. R. Folger, A. S. Bayer, and C. K. Stover.** 1998. *Staphylococcus aureus* genetic loci impacting growth and survival in multiple infection environments. *Mol. Microbiol.* **30:**393–404.

14. **Daugherty, S., and M. G. Low.** 1993. Cloning, expression, and mutagenesis of phosphatidylinositol-specific phospholipase C from *Staphylococcus aureus*: a potential staphylococcal virulence factor. *Infect. Immun.* **61:**5078–5089.

15. **Drapeau, G. R.** 1978. Role of metalloprotease in activation of the precursor of staphylococcal protease. *J. Bacteriol.* **136:**607–613.

16. **Dubin, G.** 2003. Defense against own arms: staphylococcal cysteine proteases and their inhibitors. *Acta Biochim. Pol.* **50:**715–724.

17. **Dubin, G., M. Krajewski, G. Popowicz, J. Stec-Niemczyk, M. Bochtler, J. Potempa, A. Dubin, and T. A. Holak.** 2003. A novel class of cysteine protease inhibitors: solution structure of staphostatin A from *Staphylococcus aureus.* *Biochemistry* **42:**13449–13456.

18. **Farrell, A. M., D. Taylor, and K. T. Holland.** 1995. Cloning, nucleotide sequence determination and expression of the *Staphylococcus aureus* hyaluronate lyase gene. *FEMS Microbiol. Lett.* **130:**81–85.

19. **Friedrich, R., P. Panizzi, P. Fuentes-Prior, K. Richter, I. Verhamme, P. J. Anderson, S. Kawabata, R. Huber, W. Bode, and P. E. Bock.** 2003. Staphylocoagulase is a prototype for the mechanism of cofactor-induced zymogen activation. *Nature* **425:**535–539.

20. **Giraudo, A. T., A. L. Cheung, and R. Nagel.** 1997. The sae locus of *Staphylococcus aureus* controls exoprotein synthesis at the transcriptional level. *Arch. Microbiol.* **168:**53–58.

21. **Goh, S. H., S. K. Byrne, J. L. Zhang, and A. W. Chow.** 1992. Molecular typing of *Staphylococcus aureus* on the basis of coagulase gene polymorphisms. *J. Clin. Microbiol.* **30:**1642–1645.

22. **Hofmann, B., D. Schomburg, and H. J. Hecht.** 1993. Crystal structure of a thiol proteinase from *Staphylococcus aureus* V-8 in the E-64 inhibitor complex. *Acta Crystallogr.* **49**(Suppl.)**:**102.

23. **Hookey, J. V., J. F. Richardson, and B. D. Cookson.** 1998. Molecular typing of *Staphylococcus aureus* based PCR restriction fragment polymorphism and DNA sequence analysis of the coagulase gene. *J. Clin. Microbiol.* **36:**1083–1089.

24. **Hynes, T. R., and R. O. Fox.** 1991. The crystal structure of staphylococcal nuclease refined at 1.7 Å resolution. *Proteins* **10:**92–105.

25. **Jin, T., M. Bokarewa, T. Foster, J. Mitchell, J. Higgins, and A. Tarkowski.** 2004. *Staphylococcus aureus* resists human defensins by production of staphylokinase, a novel bacterial evasion mechanism. *J. Immunol.* **172:**1169–1176.

26. **Kapral, F. A., S. Smith, and D. Lal.** 1992. The esterification of fatty acids by *Staphylococcus aureus* fatty acid modifying enzyme (FAME) and its inhibition by glycerides. *J. Med. Microbiol.* **37:**235–237.

27. **Karlsson, A., and S. Arvidson.** 2002. Variation in extracellular protease production among clinical isolates of *Staphylococcus aureus* due to different levels of expression of the protese repressor *sarA. Infect. Immun.* **70:**4239–4246.

28. **Karlsson, A., P. Saravia-Otten, K. Tegmark, E. Morfeldt, and S. Arvidson.** 2001. Decreased amounts of cell wall-associated protein A and fibronectin-binding proteins in *Staphylococcus aureus sarA* mutants due to up-regulation of extracellular proteases. *Infect. Immun.* **69:**4742–4748.

29. **Kawabata, S., T. Morita, S. Iwanaga, and H. Igarashi.** 1985. Enzymatic properties of staphylothrombin, an active molecular complex formed between staphylocoagulase and human prothrombin. *J. Biochem.* (Tokyo) **98:**1603–1614.

30. **Kondo, I., and K. Fujise.** 1977. Serotype B staphylococcal bacteriophage singly converting staphylokinase. *Infect. Immun.* **18:**266–272.

31. **Kotting, J., H. Eibl, and F. J. Fehrenbach.** 1988. Substrate specificity of *Staphylococcus aureus* (TEN5) lipases with isomeric oleoyl-sn-glycerol ethers as substrates. *Chem. Phys. Lipids* **47:**117–122.

32. **Kuroda, M., T. Ohta, I. Uchiyama, T. Baba, H. Yuzawa, I. Kobayashi, L. Cui, A. Oguchi, K. Aoki, Y. Nagai, J. Lian, T. Ito, M. Kanamori, H. Matsumaru, A. Maruyama, H. Murakami, A. Hosoyama, Y. Mizutani-Ui, N. K. Takahashi, T. Sawano, R. Inoue, C. Kaito, K. Sekimizu, H. Hirakawa, S. Kuhara, S. Goto, J. Yabuzaki, M. Kanehisa, A. Yamashita, K. Oshima, K. Furuya, C. Yoshino, T. Shiba, M. Hattori, N. Ogasawara, H. Hayashi, and K. Hiramatsu.** 2001. Whole genome sequencing of methicillin-resistant *Staphylococcus aureus. Lancet* **357:**1225–1240.

33. **Lebeau, C., F. Vandenesch, T. Greenland, R. P. Novick, and J. Etienne.** 1994. Coagulase expression in *Staphylococcus aureus* is positively and negatively modulated by an agr-dependent mechanism. *J. Bacteriol.* **176:**5534–5536.

34. **Lee, C. Y., and J. J. Iandolo.** 1986. Lysogenic conversion of staphylococcal lipase is caused by insertion of the bacteriophage L54a genome into the lipase structural gene. *J. Bacteriol.* **166:**385–391.

35. **Liebl, W., and F. Gotz.** 1986. Studies on lipase directed export of *Escherichia coli* beta-lactamase in *Staphylococcus carnosus. Mol. Gen. Genet.* **204:**166–173.

36. **Lin, B., W. F. Averett, and D. G. Pritchard.** 1997. Identification of a histidine residue essential for enzymatic activity of group B streptococcal hyaluronate lyase. *Biochem. Biophys. Res. Commun.* **231:**379–382.

37. **Long, J. P., J. Hart, W. Albers, and F. A. Kapral.** 1992. The production of fatty acid modifying enzyme (FAME) and lipase by various staphylococcal species. *J. Med. Microbiol.* **37:**232–234.

38. **Makris, G., J. D. Wright, E. Ingham, and K. T. Holland.** 2004. The hyaluronate lyase of *Staphylococcus aureus*—a virulence factor? *Microbiology* **150:**2005–2013.

39. **Marques, M. B., P. F. Weller, J. Parsonnet, B. J. Ransil, and A. Nicholson-Weller.** 1989. Phosphatidylinositol-specific phospholipase C, a possible virulence factor of *Staphylococcus aureus. J. Clin. Microbiol.* **27:**2451–2454.

40. **Massimi, I., E. Park, K. Rice, W. Muller-Esterl, D. Sauder, and M. J. McGavin.** 2002. Identification of a novel maturation mechanism and restricted substrate specificity for the SspB cysteine protease of *Staphylococcus aureus. J. Biol. Chem.* **277:**41770–41777.

41. **McAleese, F. M., E. J. Walsh, M. Sieprawska, J. Potempa, and T. J. Foster.** 2001. Loss of clumping factor B fibrinogen binding activity by *Staphylococcus aureus* involves cessation of transcription, shedding and cleavage by metalloprotease. *J. Biol. Chem.* **276:**29969–29978.

42. **McDevitt, D., P. Francois, P. Vaudaux, and T. J. Foster.** 1994. Molecular characterization of the clumping factor (fibrinogen receptor) of *Staphylococcus aureus. Mol. Microbiol.* **11:**237–248.

43. **McDevitt, D., P. Vaudaux, and T. J. Foster.** 1992. Genetic evidence that bound coagulase of *Staphylococcus aureus* is not clumping factor. *Infect. Immun.* **60:**1514–1523.

44. **McGavin, M. J., C. Zahradka, R. Kelly, and J. E. Scott.** 1997. Modification of the *Staphylococcus aureus* fibronectin binding phenotype by V8 protease. *Infect. Immun.* **65:**2621–2628.

45. **Mortensen, J. E., T. R. Shryock, and F. A. Kapral.** 1992. Modification of bactericidal fatty acids by an enzyme of *Staphylococcus aureus. J. Med. Microbiol.* **36:**293–298.

46. **Nikoleit, K., R. Rosenstein, H. M. Verheij, and F. Gotz.** 1995. Comparative biochemical and molecular analysis of the *Staphylococcus hyicus*, *Staphylococcus aureus* and a hybrid lipase. Indication for a C-terminal phospholipase domain. *Eur. J. Biochem.* **228:**732–738.

47. **Phonimdaeng, P., M. O'Reilly, P. W. O'Toole, and T. J. Foster.** 1988. Molecular cloning and expression of the coagulase gene of *Staphylococcus aureus* 8325-4. *J. Gen. Microbiol.* **134:**75–83.

48. **Potempa, J., A. Dubin, G. Korzus, and J. Travis.** 1988. Degradation of elastin by a cysteine proteinase from *Staphylococcus aureus*. *J. Biol. Chem.* **263:**2664–2667.

49. **Potempa, J., D. Fedak, A. Dubin, A. Mast, and J. Travis.** 1991. Proteolytic inactivation of α_1-anti-chymotrysin. Sites of cleavage and generation of chemotactic activity. *J. Biol. Chem.* **266:**21482–21487.

50. **Potempa, J., W. Watorek, and J. Travis.** 1986. The inactivation of human plasma alpha 1-proteinase inhibitor by proteinases from *Staphylococcus aureus*. *J. Biol. Chem.* **261:** 14330–14334.

51. **Prasad, L., Y. Leduc, K. Hayakawa, and L. T. Delbaere.** 2004. The structure of a universally employed enzyme: V8 protease from *Staphylococcus aureus*. *Acta Crystallogr. D Biol. Crystallogr.* **60:**256–259.

52. **Rabijns, A., H. L. De Bondt, and C. De Ranter.** 1997. Three-dimensional structure of staphylokinase, a plasminogen activator with therapeutic potential. *Nat. Struct. Biol.* **4:**357–360.

53. **Reed, S. B., C. A. Wesson, L. E. Liou, W. R. Trumble, P. M. Schlievert, G. A. Bohach, and K. W. Bayles.** 2001. Molecular characterization of a novel *Staphylococcus aureus* serine protease operon. *Infect. Immun.* **69:**1521–1527.

54. **Rice, K., R. Peralta, D. Bast, J. de Azavedo, and M. J. McGavin.** 2001. Description of staphylococcus serine protease (*ssp*) operon in *Staphylococcus aureus* and nonpolar inactivation of *sspA*-encoded serine protease. *Infect. Immun.* **69:**159–169.

55. **Rollof, J., and S. Normark.** 1992. In vivo processing of *Staphylococcus aureus* lipase. *J. Bacteriol.* **174:**1844–1847.

56. **Rosenstein, R., and F. Gotz.** 2000. Staphylococcal lipases: biochemical and molecular characterization. *Biochimie* **82:** 1005–1014.

57. **Rzychon, M., A. Sabat, K. Kosowska, J. Potempa, and A. Dubin.** 2003. Staphostatins: an expanding new group of proteinase inhibitors with a unique specificity for the regulation of staphopains, *Staphylococcus* spp. cysteine proteinases. *Mol. Microbiol.* **49:**1051–1066.

58. **Sabat, A., K. Kosowska, K. Poulsen, A. Kasprowicz, A. Sekowska, B. Van den Burg, J. Travis, and J. Potempa.** 2000. Two allelic forms of the aureolysin gene (*aur*) within *Staphylococcus aureus*. *Infect. Immun.* **68:**973–976.

59. **Sawai, T., K. Tomono, K. Yanagihara, Y. Yamamoto, M. Kaku, Y. Hirakata, H. Koga, T. Tashiro, and S. Kohno.** 1997. Role of coagulase in a murine model of hematogenous pulmonary infection induced by intravenous injection of *Staphylococcus aureus* enmeshed in agar beads. *Infect. Immun.* **65:**466–471.

60. **Shaw, L., E. Golonka, J. Potempa, and S. J. Foster.** 2004. The role and regulation of the extracellular proteases of *Staphylococcus aureus*. *Microbiology* **150:**217–228.

61. **Shortle, D., Y. Wang, J. R. Gillespie, and J. O. Wrabl.** 1996. Protein folding for realists: a timeless phenomenon. *Protein Sci.* **5:**991–1000.

62. **Simons, J. W., H. Adams, R. C. Cox, N. Dekker, F. Gotz, A. J. Slotboom, and H. M. Verheij.** 1996. The lipase from *Staphylococcus aureus*. Expression in *Escherichia coli*, large-scale purification and comparison of substrate specificity to *Staphylococcus hyicus* lipase. *Eur. J. Biochem.* **242:**760–769.

63. **Simons, J. W., M. D. van Kampen, S. Riel, F. Gotz, M. R. Egmond, and H. M. Verheij.** 1998. Cloning, purification and characterisation of the lipase from *Staphylococcus epidermidis*—comparison of the substrate selectivity with those of other microbial lipases. *Eur. J. Biochem.* **253:**675–683.

64. **Steiner, B., S. Romero-Steiner, D. Cruce, and R. George.** 1997. Cloning and sequencing of the hyaluronate lyase gene from *Propionibacterium acnes*. *Can. J. Microbiol.* **43:**315–321.

65. **Suciu, D., and M. Inouye.** 1996. The 19-residue pro-peptide of staphylococcal nuclease has a profound secretion-enhancing ability in *Escherichia coli*. *Mol. Microbiol.* **21:** 181–195.

66. **Vandenesch, F., C. Lebeau, M. Bes, D. McDevitt, T. Greenland, R. P. Novick, and J. Etienne.** 1994. Coagulase deficiency in clinical isolates of *Staphylococcus aureus* involves both transcriptional and post-transcriptional defects. *J. Med. Microbiol.* **40:**344–349.

67. **Weber, D. J., A. G. Gittis, G. P. Mullen, C. Abeygunawardana, E. E. Lattman, and A. S. Mildvan.** 1992. NMR docking of a substrate into the X-ray structure of staphylococcal nuclease. *Proteins* **13:**275–287.

Staphylococcal Sortases and Surface Proteins

ANDREA C. DEDENT, LUCIANO A. MARRAFFINI, AND OLAF SCHNEEWIND

40

The cell wall envelope of *Staphylococcus aureus* can be viewed as a surface organelle with anchored proteins that interact with the bacterial environment (65). Most notably during staphylococcal colonization or invasion of host tissues, cell wall-anchored surface proteins fulfill a wide spectrum of functions (24). These include surface protein binding to host tissues, interaction with adhesive matrix molecules, immune evasive strategies, and binding of heme proteins for bacterial iron scavenging during infection (19, 52, 58, 95). Surface proteins of *S. aureus* have been studied for more than 50 years. The prototype surface molecule, protein A, was characterized in atomic detail because of its ability to bind the Fc portion of human and animal immunoglobulin (Ig) (13, 14). Protein A was also used to analyze the anchoring of surface protein to the cell wall envelope, which led to the identification and characterization of sortase enzyme (57, 89). In this chapter, we review briefly what is known about surface proteins of *S. aureus*, their mechanisms of anchoring to the cell wall envelope, and their contributions to the pathogenesis of staphylococcal infections.

PROTEIN A

In 1958 Jensen reported that *S. aureus* causes precipitation of Ig (41). Fourteen years later Sjöquist and colleagues solubilized protein A from the bacterial envelope by treatment of peptidoglycan with lysostaphin, a glycyl-glycine endopeptidase that cleaves staphylococcal cell wall cross bridges (93). Protein A amino acid sequence, gene sequence, and three-dimensional nuclear magnetic resonance and X-ray diffraction structures revealed a molecule comprised of five nearly identical Ig-binding domains as well as the molecular elements involved in binding Ig (14, 29, 92, 106). The five repeat domains are tethered to region X of protein A, a segment with disordered structure and a variable number of eight amino acid repeats (30). The C-terminal end of protein A is linked covalently to staphylococcal cell wall cross bridges via an amide bond between the T of the LPXTG motif (see below) and the amino group of pentaglycine cross bridges attached to wall peptides (87, 99). Mutations in the protein A gene (*spa*) cause significant defects in the pathogenesis of *S. aureus* infections. For example, reduced bacterial survival in blood or in the presence of macrophages is likely due to the inability of these variants to sequester Ig via Fc binding (74). However, the observed phenotypes may also be attributed to defects in the binding of protein A to von Willebrand factor, a serum polypeptide that promotes physiological homeostasis of human or animal blood, or to protein A binding to tumor necrosis factor receptor 1, a signaling molecule involved in proinflammatory cytokine responses and innate immunity (28, 33).

ANCHORING PROTEIN A TO THE CELL WALL

Protein A is synthesized as a precursor protein bearing an N-terminal signal peptide and a C-terminal cell wall sorting signal with an LPXTG motif, a hydrophobic domain, and a positively charged tail (21, 88). After initiation into the secretory pathway and signal peptide cleavage, the 35-amino-acid residue C-terminal cell wall sorting signal retains the polypeptide in the membrane (Fig. 1). The protein A precursor is then cleaved by sortase A at the LPXTG motif between the T and G residues and is captured as a thioester-linked acyl enzyme (64, 100). Nucleophilic attack of the amino group of pentaglycine within the lipid II cell wall biosynthetic precursor is thought to resolve the acyl enzyme, thereby regenerating the active site (79, 102). Surface protein linked to lipid II is then incorporated into the cell wall via the transpeptidation and transglycosylation reactions of cell wall synthesis (57, 59, 79, 100, 103). Because cell wall synthesis and surface protein anchoring occur at designated sites in the envelope, continuous incorporation of polypeptides into the cell wall is required to ensure the uniform distribution of protein A on the bacterial surface (9, 10). Once cell wall degradation commences at bacterial cell division sites, protein A linked to murein tetrapeptide is released into the environment (12, 26, 55, 100, 110).

SORTASE A

Sortase A, a 206-residue polypeptide, is anchored in the staphylococcal membrane via an N-terminal signal peptide/membrane anchor sequence (56). Replacement of the

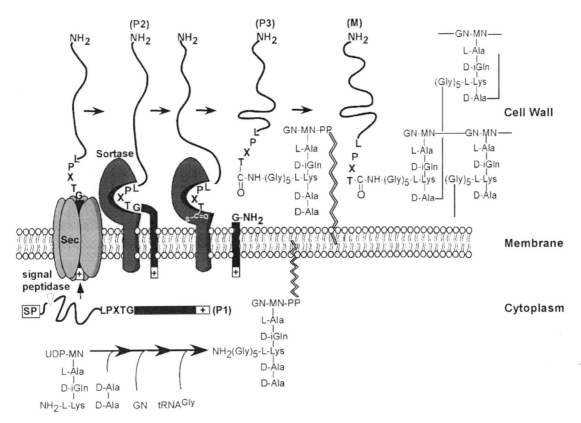

FIGURE 1 Sortase-mediated anchoring of staphylococcal surface proteins. In order to catalyze the covalent linkage of a polypeptide to the cell wall, the sortase enzyme must recognize and act on two substrates: the surface protein and the peptidoglycan. Sortase protein substrates are first synthesized in the cytoplasm as a full-length precursor (P1) containing an N-terminal signal peptide (SP) and a C-terminal sorting signal. The signal peptide directs the cellular export of the polypeptide through the Sec system and is cleaved by a signal peptidase upon secretion. This results in the generation of P2 precursor, a shorter protein containing only the sorting signal. During the export of this precursor the hydrophobic and positively charged sequences of the sorting signal act as a "stop transfer" element, leading to the retention of P2 in the bacterial membrane. Sortase contains a signal peptide/membrane anchor sequence that localizes the enzyme to the membrane. Hence the LPXTG motif of the P2 precursor is accessible to the cysteine residue of the sortase active site. The thiol group attacks the peptide bond between the T and G of the LPXTG motif, generating an acyl intermediate (i.e., sortase and the protein substrate linked through a thioester bond). Lipid II, the peptidoglycan synthesis precursor, constitutes the cell wall substrate of sortase. Lipid II contains a single peptidoglycan subunit (GlcNAc-MurNAc-L-Ala-D-iGln-L-Lys (-NH_2-Gly5)-D-Ala-D-Ala-CO_2H) associated to the bacterial membrane through an undecaprenyl phosphate group. Its synthesis occurs in the cytoplasm by the successive enzymatic addition of N-acetylmuramic acid (MN), the cell wall peptide residues (L-Ala, D-iGln, L-Lys, and D-Ala-D-Ala), N-acetylglucosamine (GN), and each of the glycines of the cross bridge via a tRNA^{Gly} donor to a UDP carrier. The peptidoglycan subunit thus generated is transferred from UDP to undecaprenyl phosphate, which in turn is translocated to the outer side of the membrane. There, lipid II is subjected to the transglycosylase and transpeptidase activities of penicillin-binding proteins during peptidoglycan polymerization. In the sorting reaction, the free amino group of the pentaglycine cross bridge of lipid II attacks the acyl intermediate, linking the C-terminal threonine of the surface protein to lipid II (P3) and regenerating the sortase active site. This constitutes the mature (M) form of the surface protein. Incorporation of this modified peptidoglycan subunit into the growing cell wall subsequently positions the surface protein in the cell envelope (M, mature form).

membrane anchor with a six histidyl tag permits purification of recombinant, soluble SrtA$_{\Delta N}$ from *Escherichia coli* lysate using nickel-nitrilotriacetic acid affinity chromatography (100). Purified SrtA$_{\Delta N}$ cleaves LPXTG motif bearing polypeptides between threonine and glycine in vitro (103). In the presence of cell wall substrates, for example, polyglycine or cell wall peptides bearing cross bridges, sortase A functions as a transpeptidase, forming an amide bond between the C-terminal threonine and the amino group of pentaglycine (103). All sortase activity can be abolished by treatment of the enzyme with methyl-methane thiosulfonate (104). These compounds form a disulfide bond with

the active site cysteine thiol of sortase A (112). Treatment with a reducing agent can reverse this reaction, restoring sortase enzymatic activity (100). Additionally, sortase transpeptidation can specifically be inhibited by treatment with hydroxylamine, a strong nucleophile that attacks the thioester bond between the enzyme and the surface protein, causing release of the latter as a polypeptide bearing a C-terminal threonine hydroxamate (104).

Nuclear magnetic resonance studies revealed a barrel-shaped solution structure of SrtA$_{\Delta N}$, which is comprised of both parallel and antiparallel beta-sheets (40). X-ray crystallography produced a high resolution structure of sortase

A, revealing an eight-stranded β-barrel structure with strands β7 and β8 forming the floor of a hydrophobic depression that encompasses the active site. The wall of this catalytic pocket is constructed by residues located in loops connecting strands β3-β4, β2-β3, β6-β7, and β7-β8. The structure of the active site suggested the presence of a cysteine-histidine ion pair located within this pocket, where the LPXTG motif of a surface protein is cleaved. Cocrystallization studies revealed the close proximity of the scissile peptide bond to the active site thiol (112). Three residues conserved among sortases, histidine 120, cysteine 184, and arginine 197, are each required for sortase activity. Although arginine 197 is in close proximity (4.5 nm) to the active site thiol, it is not yet clear whether this arginine or histidine 120, which appears to be positioned at a greater distance to the active site, functions as a general base during catalysis (39, 54, 102). The amino acid sequence of sortase A has been used as a query for BLAST searches, which by now has led to the identification of more than 200 sortase genes in genomes of gram-positive bacteria (11, 70, 101). In addition to sortase, the LPXTG motif is also cleaved by a second enzyme present in *S. aureus* termed LPXTGase (see chapter 2, this volume) (51, 102).

SORTASE A-ANCHORED SURFACE PROTEINS

Using protein A and other cell wall sorting signals as queries in bioinformatics searches, between 18 and 22 different sortase-anchored surface protein genes were identified in the genomes of *S. aureus* strains (see Table 1 for a listing of 22 surface proteins) (27, 59, 60, 84). The function of most, but not all, of these sorting signals in cell wall anchoring was demonstrated to be absolutely dependent on sortase A. Thus, by analyzing the virulence of *srtA* mutants in ani-

mal experiments, the general contribution of all cell wall-anchored surface proteins to the pathogenesis of *S. aureus* disease could be determined. Sortase A mutants displayed a 1.5-log reduction in the ability of staphylococci to cause acute lethal disease in mice (56). Further, a 3-log reduction in abscess formation in internal organs, a 3-log reduction in infectious murine arthritis, and significant defects in the development of rat infectious endocarditis were also observed (42, 109). The following describes what is known about the biochemical and physiological properties of these molecules.

Fibronectin-Binding Microbial Surface Components Recognizing Adhesive Matrix Molecules (MSCRAMMS)

Foster and Höök developed our current understanding of MSCRAMMS as bacterial elements of tissue adhesion and immune evasion (24). Several staphylococcal proteins have served as a paradigm for these studies, most notably the fibronectin-binding proteins FnbpA and FnbpB (22, 44, 85, 91). Both proteins encompass a large N-terminal domain (about 500 amino acid residues) of unknown function, followed by four to five 50-residue repeat domains responsible for binding the N-terminal domain of fibronectin. FnbpA/FnbpB interaction with fibronectin involves structural rearrangements that lead to the ordering of the Fnbp repeat domains upon ligand binding (44, 83, 108). As fibronectin is found in extracellular matrices of most tissues as well as in soluble form within body fluids, staphylococci can adhere to almost any tissue and serum protein-coated medically implanted foreign bodies (75). With such widespread binding potential, certainly one fascinating aspect of staphylococcal binding to fibronectin is the invasion of host cells and subsequent intracellular replication (3, 20).

TABLE 1 Cell-wall-anchored surface proteins of *S. aureus*

Surface protein[a]	AA[b]	Ligand[c]	Motif[d]	Reference(s)
Protein A (Spa)	508	Immunoglobulin, von Willebrand factor, TNFR	LPETG	28, 33, 106
Fibronectin-binding protein A (FnbA)	1,018	Fibronectin, fibrinogen, elastin	LPETG	91
Fibronectin-binding protein B (FnbB)	914	Fibronectin, fibrinogen, elastin	LPETG	44
Clumping factor A (ClfA)	933	Fibrinogen	LPDTG	61
Clumping factor B (ClfB)	913	Fibrinogen, keratin	LPETG	68
Collagen adhesion (Cna)	1,183	Collagen	LPKTG	77
SdrC	947	Unknown	LPETG	45
SdrD	1,315	Unknown	LPETG	45, 46
SdrE	1,166	Unknown	LPETG	45
Pls	1,637	Unknown	LPDTG	59, 60
SasA	2,261	Unknown	LPDTG	59, 60
SasB	937	Unknown	LPDTG	59, 60
SasC	2,186	Unknown	LPNTG	59, 60
SasD	241	Unknown	LPAAG	59, 60
SasE/IsdA	354	Heme	LPKTG	59, 60, 96
SasF	637	Unknown	LPKAG	59, 60
SasG/Aap	1,117	Unknown	LPKTG	38, 59, 60
SasH	308	Unknown	LPKTG	59, 60
SasI/HarA/IsdH	895	Haptoglobin	LPKTG	19, 59, 60, 96
SasJ/IsdB	645	Hemoglobin, heme	LPQTG	59, 60, 96
SasK	211	Unknown	LPKTG	59, 60
IsdC	227	Heme	NPQTN	52, 58

[a]Surface protein, identified as containing a C-terminal cell-wall sorting signal.
[b]AA, protein length in amino acids.
[c]Ligand, molecular component(s) recognized and bound by protein.
[d]Motif, consensus motif recognized by sortase and present in C-terminal cell-wall sorting signal.

Collagen-Binding MSCRAMMS

Staphylococcal strains causing connective tissue infections or osteomyelitis regularly express the collagen adhesion protein, Cna (77, 98). A large N-terminal domain encompasses the binding site for collagen, the A domain, which assumes a jelly roll fold with two β-sheets connected by a short α-helix (82). The resulting molecular trench can accommodate the collagen triple helices. Positioning of the N-terminal A domain for collagen binding is structurally similar but inverse to that of Igs (15, 82). The contribution of collagen adhesion to the pathogenesis of staphylococcal infectious arthritis has been demonstrated (76, 98).

Fibrinogen-Binding MSCRAMMS

S. aureus strains clump in the presence of plasma; this phenomenon, which has been exploited for diagnostic purposes, is the product of a molecular interaction between two MSCRAMMS, clumping factor A and B, with fibrinogen (25, 61, 68). ClfA and ClfB are structurally related and comprise a large N-terminal A domain and a repeat or R domain, which is composed exclusively of serine-aspartate repeats (32, 45). The ligand binding sites of ClfA and ClfB have been mapped to residues 220 to 559; however, it is surprising that the ligand binding sites of the two homologs are only 27% identical, suggesting overlapping but perhaps not completely redundant function (62). The ligand binding sites of ClfA assume an IgG-like fold (16, 62, 78, 107). An elegant molecular mechanism of fibrinogen substrate binding, coined "dock, lock and latch," has recently been demonstrated for SdrG, a fibrinogen-binding *Staphylococcus epidermidis* MSCRAMM that also encompasses repeat domains (80). A cleft 30 Å in length between two IgG-like folds of SdrG constitutes the fibrinogen-binding site with at least 62 contacts between the two molecules that occlude the cleavage sites for thrombin. In the model, the C-terminal segment of the protein "locks" on to the "docked" fibrinogen by covering it and latching onto the next domain (80). The IgG domain structural fold is shared between several cell wall-anchored surface molecules and appears to be a general feature. Further, molecular binding and occlusion mechanisms in fibrinogen binding are thought to aid staphylococcal spread in infected tissues, a notion that has been corroborated by experiments demonstrating the requirement for ClfA in a rat endocarditis model of infection (107).

Sdr Proteins

Both *S. aureus* and *S. epidermidis* strains encode for multiple cell wall-anchored surface proteins with large serine-aspartate repeat (Sdr) domains (32, 45, 80). The B domains of Sdr proteins contain high affinity calcium binding sites that adopt an EF hand fold, a common structure observed in other calcium-binding proteins (46, 105). Although it seems highly likely that these proteins are involved in binding host factors, such interactions have thus far not been demonstrated for the majority of the Sdr proteins.

SORTASE B

S. aureus expresses a second sortase, SrtB, under iron-starvation conditions, including during host infection (60). The *srtB* gene is located in the *isd* locus (iron-regulated surface determinants) of *S. aureus*, which is comprised of three transcriptional units, *isdA*, *isdB*, and *isdCDEF-srtB-isdG* (58). IsdC, a cell wall-anchored polypeptide with an NPQTN sorting signal, is the only known substrate of sortase B. Purified sortase B (SrtB$_{\Delta N}$, lacking an N-terminal signal peptide/membrane anchor) cleaves NPQTN peptide in vitro (58, 60, 96). After cleavage between threonine and asparagine of the NPQTN motif in vivo, the C-terminal threonine of mature IsdC polypeptide is tethered to the pentaglycine cross bridge of the staphylococcal cell wall (53). IsdC is not displayed on the staphylococcal surface but remains buried within the cell wall envelope, as shown by proteinase K protection assays. Fusion of the IsdC sorting signal to the C terminus of enterotoxin B also leads to cell wall anchoring and burial within the envelope, whereas sortase A-mediated anchoring results in surface display of both native substrates and an equivalent fusion between enterotoxin B and protein A cell wall sorting signal (53). Biochemical analysis of cell wall anchor structures revealed significant differences between sortase A and sortase B anchor products. Sortase A anchor structures include polymerized glycan strands with variable length (4 to 16 disaccharides) as well as cross-linked cell wall peptides (2 to 12 linked wall peptides) (66, 67, 99). In contrast, sortase B anchor peptides are tethered to glycan strands with 1 to 7 disaccharides and 1 to 7 wall peptides (53). One plausible mechanistic explanation for these differences is that sortase B may anchor IsdC to assembled cell wall with a low degree of cross-linking, while sortase A may generate lipid II linked proteins that are ubiquitously incorporated into the envelope (53). However, this hypothesis remains speculative and additional work is needed to unravel the mechanistic details of positioning polypeptides at discrete locations in the envelope. The X-ray diffraction structure of purified SrtB$_{\Delta N}$ with and without pentaglycine substrate has been determined (111, 113). In general, the enzyme assumes a similar fold and positioning of its active site residues (Cys184, His120, Arg197—sortase A nomenclature) to that of sortase A with the amino group of the cell wall cross bridge located in close proximity to the active site thiol (111, 113).

SORTASE-ANCHORED PROTEINS AND HEME IRON TRANSPORT

Four Isd proteins (iron-regulated surface determinants) are involved in binding heme or hemoproteins and appear to play a role in iron scavenging during staphylococcal host infection (Fig. 2). HarA/IsdH is encoded by a gene outside of the *isd* locus and has been shown to bind haptoglobin/hemoglobin complexes (19). IsdB, on the other hand, binds to hemoglobin, and all four proteins, IsdA, IsdB, IsdC, and IsdH/HarA, bind heme (58, 96). Although it seems likely that these proteins may be involved in capturing hemoproteins on the bacterial surface, liberating heme and then promoting heme transport across the bacterial cell wall envelope, this has not yet been demonstrated biochemically (96). Heme transport across the plasma membrane involves multiple ABC type transporters, and once within the bacterial cytoplasm, iron is released from the tetrapyrrol ring via the staphylococcal heme-cleaving enzymes IsdG and IsdI (94). Several staphylococcal hemolysins may be responsible for the release of hemoproteins from erythrocytes or other host cells via their membrane lytic properties (17).

CELL WALL-ASSOCIATED SURFACE PROTEINS

In addition to the subset of *S. aureus* sortase-anchored cell wall surface proteins that are covalently attached to the cell wall, there are also a number of surface proteins that lack a

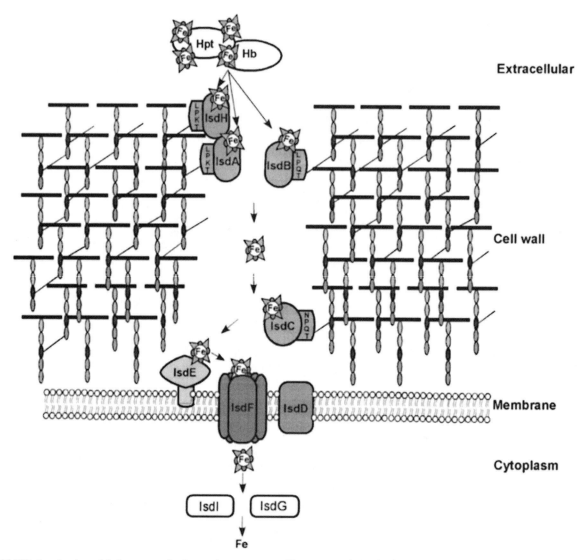

FIGURE 2 Anchored Isd proteins facilitate the transport of heme-iron through the staphylococcal cell wall. The *isd* (iron-regulated surface determinants) locus in *S. aureus* is composed of three transcriptional units, *isdA*, *isdB*, and *isdCDEF-srtB-isdG* (*srtB*, sortase B). IsdA and IsdB contain an LPXTG motif, are anchored to the cell wall by sortase A, and are exposed in the outside of the cell. IsdC contains an NPQTN motif and is anchored by sortase B. In contrast to sortase A substrates, IsdC is buried in the cell wall, and it is anchored to poorly cross-linked peptidoglycan. IsdD is an integral membrane protein, IsdE is a lipoprotein, and IsdF represents a subunit of an oligomeric ABC transporter. IsdG is a heme-oxygenase. The current model for the uptake of heme-iron by the Isd proteins states that IsdA, IsdB, and IsdH would interact with host hemoproteins such as hemoglobin (Hb), haptoglobin (Hpt), and/or hemopexin, which are released after erythrocyte lysis. The interaction would result in the sequestration of heme-iron by IsdA, IsdB, and IsdH. IsdC would serve as an intermediate heme-iron receptor between the outer cell surface and membrane components of the Isd system, thus occupying a central role in the passage of heme-iron across the staphylococcal cell wall. IsdEDF would facilitate the passage of heme through the bacterial membrane. In the cytoplasm, IsdG and IsdI would degrade the porphyrine ring and release the iron for use by the bacterium.

C-terminal cell wall sorting signal, yet remain in one way or another cell wall associated (Table 2) (27, 47). These proteins are involved in a variety of processes, including immunomodulatory properties, colonization, and dissemination through interactions with the host extracellular matrix (ECM) and plasma proteins. One of these proteins, the extracellular adherence protein Eap, seems to play a variety of roles as evidenced by the diverse interactions in which it is involved (31). Eap binds to at least seven different ECM and plasma proteins, to itself, to components of the cell wall, and to intercellular adhesion molecule 1 present on

endothelial cells and antigen-presenting cells (23, 49, 63, 71). This last interaction blocks leukocyte extravasation and interactions with T cells (7). The protein contains six repeat domains displaying homology to the N-terminal β-chain of major histocompatibility complex class II molecules (44). In addition to Eap, the *S. aureus* extracellular fibrinogen-binding protein (Fib/Efb), ECM- and plasma-binding protein (Emp), and cell wall-associated fibronectin-binding protein (Ebh) all act as adhesins, binding to one or more components of the ECM and/or plasma proteins (4, 5, 8, 37, 42). A fifth ECM-binding protein, elastin-binding protein (EbpS),

TABLE 2 Cell-wall-associated surface proteins of *S. aureus*

Surface-associated protein[a]	AA[b]	Ligand[c]	Cell wall association[d]	Signal peptide[e]	Reference(s)
Extracellular matrix and plasma-binding protein (Emp)	341	Fibronectin, fibrinogen, collagen, and vitronectin	SDS extractable	Yes	37
Cell wall-associated fibronectin-binding protein (Ebh)	10,498	Fibronectin	SDS extractable	Yes	8
Extracellular fibrinogen-binding protein (Fib/Efb)	165	Fibrinogen and complement factor C3b	Weakly associated, most found in extracellular milieu	Yes	4, 37, 50, 72
Elastin-binding protein (EbpS)	486	Elastin	Integral membrane protein	No	18, 73
Extracellular adherence protein (Eap/Map)	689	Extracellular matrix and plasma proteins	LiCl extractable	Yes	43, 63
Bifunctional autolysin (Atl)	1,256	Unknown cell wall component	LiCl extractable, R1, R2, R3 targeting domains	Yes	2, 35, 48, 69, 110
Lysostaphin	389	Unknown cell wall component	Cell wall targeting domain	Yes	1, 6, 81, 86
Φ11 hydrolase	481	Unknown cell wall component	Cell wall targeting domain	No	67, 90

[a]Surface-associated protein, localization to cell wall, but lacking a C-terminal cell-wall sorting signal.
[b]AA, protein length in amino acids.
[c]Ligand, molecular component(s) recognized and bound by protein.
[d]Cell wall association, interaction with and targeting to cell wall.
[e]Signal peptide, presence or absence of an N-terminal leader peptide directing protein to SEC pathway.

that binds to elastin was originally reported to be cell wall associated, specifically due to its protease susceptibility (73). This protein differs somewhat from the other four in that it has since been demonstrated to be an integral membrane protein (18). Eap, Efb, Emp, and Ebh are all closely associated with the cell surface, requiring chaotrophic agents for release (4, 8, 37, 73). Despite what is known about these proteins, the exact nature of the cell wall attachment for each remains unclear.

In addition to the cell wall-associated proteins with adhesion-type binding properties, there are also proteins that specifically bind and modify the cell wall peptidoglycan. Among these are the bifunctional autolysin (Atl) (69), the glycyl-glycine endopeptidase lysostaphin (86), and Φ11 hydrolase (67), all of which target and remain closely associated with the cell wall. Atl contains both an amidase domain and an endo-β-*N*-acetylglucosaminidase domain important for degradation of the *S. aureus* cell wall (69). Atl is localized to the cell poles, where it is thought to be involved in cell wall separation and turnover during cell division (48, 110). Three basic repeats in the central region of the protein are involved in the binding and targeting of the enzyme to the septal regions of the cell surface (2). Lysostaphin, another cell wall-associated protein, is produced by *Staphylococcus simulans* (36, 81). *S. simulans* is resistant to the lysostaphin glycyl-glycine hydrolyzing activity (34). Hence, it is thought that the enzyme may provide a selective advantage to *S. simulans* in mixed bacterial populations, killing competing staphylococci such as *S. aureus* (97). Lysostaphin contains a targeting domain at its C terminus that allows the protein to bind to the cell wall of *S. aureus* cells (1). A third protein, the bacteriophage muralytic enzyme Φ11 hydrolase, binds to and degrades the cell wall envelope through its D-alanyl-glycyl endopeptidase and *N*-acetylmuramyl-L-alanyl amidase activities, thus allowing the release of Φ11 phage progeny from the cytoplasm. The C-terminal region of the Φ11 hydrolase is similar in sequence to that of the lysostaphin cell wall targeting domain, presumably fulfilling the same role in targeting the hydrolase to the *S. aureus* cell wall (67).

The study and characterization of these and other cell wall-associated proteins has provided much insight into our understanding of the interactions of *S. aureus* with its environment. Not only does the cell wall of *S. aureus* provide the attachment point with which a wide array of host protein interactions and immunomodulation can occur, but it also serves as a battleground, with microbes in the surrounding environment targeting and binding their own arsenal of proteins for assault. Notwithstanding all that is already known about the function and envelope targeting of cell wall-associated surface proteins, a complete listing of these molecules or the identification of their envelope ligands has not yet been revealed, and further work is needed to unravel these fascinating mechanisms.

Work in our laboratory is supported by funding from the National Institute of Allergy and Infectious Diseases, Infectious Disease Branch (AI38897 and AI52474).

REFERENCES

1. **Baba, T., and O. Schneewind.** 1996. Target cell specificity of a bacteriocin molecule: a C-terminal signal directs lysostaphin to the cell wall of *Staphylococcus aureus*. EMBO J. **15**:4789–4797.

2. **Baba, T., and O. Schneewind.** 1998. Targeting of muralytic enzymes to the cell division site of Gram-positive bacteria: repeat domains direct autolysin to the equatorial surface ring of *Staphylococcus aureus*. EMBO J. **17**:4639–4646.

3. **Bayles, K. W., C. A. Wesson, L. E. Liou, L. K. Fox, G. A. Bohach, and W. R. Trumble.** 1998. Intracellular *Staphylococcus aureus* escapes the endosome and induces apoptosis in epithelial cells. Infect. Immun. **66**:336–342.

4. **Boden, M. K., and J. I. Flock.** 1994. Cloning and characterization of a gene for a 19 kDa fibrinogen-binding protein from *Staphylococcus aureus*. Mol. Microbiol. **12**:599–606.

5. **Boden Wastfelt, M. K., and J. I. Flock.** 1995. Incidence of the highly conserved fib gene and expression of the

fibrinogen-binding (Fib) protein among clinical isolates of *Staphylococcus aureus. J. Clin. Microbiol.* **33:**2347–2352.

6. **Browder, H. P., W. A. Zygmunt, J. R. Young, and P. A. Tavormina.** 1965. Lysostaphin: enzymatic mode of action. *Biochem. Biophys. Res. Commun.* **19:**383–389.

7. **Chavakis, T., M. Hussain, S. M. Kanse, G. Peters, R. G. Bretzel, J. I. Flock, M. Herrmann, and K. T. Preissner.** 2002. *Staphylococcus aureus* extracellular adherence protein serves as anti-inflammatory factor by inhibiting the recruitment of host leukocytes. *Nat. Med.* **8:**687–693.

8. **Clarke, S. R., L. G. Harris, R. G. Richards, and S. J. Foster.** 2002. Analysis of Ebh, a 1.1-megadalton cell wall-associated fibronectin-binding protein of *Staphylococcus aureus. Infect. Immun.* **70:**6680–6687.

9. **Cole, R. M.** 1965. Bacterial cell wall replication followed by immunofluorescence. *Bacteriol. Rev.* **29:**326–344.

10. **Cole, R. M., and J. J. Hahn.** 1962. Cell wall replication in *Streptococcus pyogenes. Science* **135:**722–724.

11. **Comfort, D., and R. T. Clubb.** 2004. A comparative genome analysis identifies distinct sorting pathways in gram-positive bacteria. *Infect. Immun.* **72:**2710–2722.

12. **de Jonge, B. L. M., H. de Lencastre, and A. Tomasz.** 1991. Suppression of autolysis and cell wall turnover in heterogeneous Tn551 mutants of a methicillin-resistant *Staphylcoccus aureus* strain. *J. Bacteriol.* **173:**1105–1110.

13. **Deisenhofer, J.** 1981. Crystallographic refinement and atomic models of a human Fc fragment and its complex with fragment B of protein A from *Staphylococcus aureus* at 2.9- and 2.8-A resolution. *Biochemistry* **20:**2361–2370.

14. **Deisenhofer, J., T. A. Jones, R. Huber, J. Sjödahl, and J. Sjöquist.** 1978. Crystallization, crystal structure analysis and atomic model of the complex formed by a human Fc fragment and fragment B of protein A from *Staphylococcus aureus. Hoppe-Seyl. Zeitsch. Physiol. Chem.* **359:**975–985.

15. **Deivanayagam, C. C., R. L. Rich, M. Carson, R. T. Owens, S. Danthuluri, T. Bice, M. Höök, and S. V. Narayana.** 2000. Novel fold and assembly of the repetitive B region of the *Staphylococcus aureus* collagen-binding surface protein. *Structure Fold Des.* **8:**67–78.

16. **Deivanayagam, C. C., E. R. Wann, W. Chen, M. Carson, K. R. Rajashankar, M. Höök, and S. V. Narayana.** 2002. A novel variant of the immunoglobulin fold in surface adhesins of *Staphylococcus aureus:* crystal structure of the fibrinogen-binding MSCRAMM, clumping factor A. *EMBO J.* **21:**6660–6672.

17. **Dinges, M. M., P. M. Orwin, and P. M. Schlievert.** 2000. Exotoxins of *Staphylococcus aureus. Clin. Microbiol. Rev.* **13:**16–34.

18. **Downer, R., F. Roche, P. W. Park, R. P. Mecham, and T. J. Foster.** 2002. The elastin-binding protein of *Staphylococcus aureus* (EbpS) is expressed at the cell surface as an integral membrane protein and not as a cell wall-associated protein. *J. Biol. Chem.* **277:**243–250.

19. **Dryla, A., D. Gelbmann, A. von Gabain, and E. Nagy.** 2003. Identification of a novel iron regulated staphylococcal surface protein with haptoglobin-haemoglobin binding activity. *Mol. Microbiol.* **49:**37–53.

20. **Dziewanowska, K., P. J.M., C. F. Deobald, B. K.W., W. R. Trumble, and G. A. Bohach.** 1999. Fibronectin binding protein and host cell tyrosine kinase are required for internalization of *Staphylococcus aureus* by epithelial cells. *Infect. Immun.* **67:**4673–4678.

21. **Fischetti, V. A., V. Pancholi, and O. Schneewind.** 1990. Conservation of a hexapeptide sequence in the anchor region of surface proteins from gram-positive cocci. *Mol. Microbiol.* **4:**1603–1605.

22. **Flock, J. I., G. Fröman, K. Jönsson, B. Guss, C. Signäs, B. Nilsson, G. Raucci, M. Höök, T. Wadström, and M. Lindberg.** 1987. Cloning and expression of the gene for a fibronectin-binding protein from *Staphylococcus aureus. EMBO J.* **6:**2351–2357.

23. **Flock, M., and J. I. Flock.** 2001. Rebinding of extracellular adherence protein Eap to *Staphylococcus aureus* can occur through a surface-bound neutral phosphatase. *J. Bacteriol.* **183:**3999–4003.

24. **Foster, T. J., and M. Höök.** 1998. Surface protein adhesins of *Staphylococcus aureus. Trends Microbiol.* **6:**484–488.

25. **Foster, T. J., and D. McDevitt.** 1994. Surface-associated proteins of *Staphylococcus aureus:* their possible roles on virulence. *FEMS Microbiol. Lett.* **118:**199–206.

26. **Ghuysen, J.-M., D. J. Tipper, C. H. Birge, and J. L. Strominger.** 1965. Structure of the cell wall of *Staphylococcus aureus* strain Copenhagen. VI. The soluble glycopeptide and its sequential degradation by peptidases. *Biochemistry* **4:**2245–2254.

27. **Gill, S. R., D. E. Fouts, G. L. Archer, E. F. Mongodin, R. T. Deboy, J. Ravel, I. T. Paulsen, J. F. Kolonay, L. Brinkac, M. Beanan, R. J. Dodson, S. C. Daugherty, R. Madupu, S. V. Angiuoli, A. S. Durkin, D. H. Haft, J. Vamathevan, H. Khouri, T. Utterback, C. Lee, G. Dimitrov, L. Jiang, H. Qin, J. Weidman, K. Tran, K. Kang, I. R. Hance, K. E. Nelson, and C. M. Fraser.** 2005. Insights on evolution of virulence and resistance from the complete genome analysis of an early methicillin-resistant *Staphylococcus aureus* strain and a biofilm-producing methicillin-resistant *Staphylococcus epidermidis* strain. *J. Bacteriol.* **187:**2426–2438.

28. **Gomez, M. I., A. Lee, B. Reddy, A. Muir, G. Soong, A. Pitt, A. Cheung, and A. Prince.** 2004. *Staphylococcus aureus* protein A induces airway epithelial inflammatory responses by activating TNFR1. *Nat. Med.* **10:**842–848.

29. **Gouda, H., H. Torigoe, A. Saito, M. Sato, Y. Arata, and I. Shimada.** 1992. Three-dimensional solution structure of the B domain of staphylococcal protein A: a comparison of the solution and crystal structures. *Biochemistry* **31:**9665–9672.

30. **Guss, B., M. Uhlæn, B. Nilsson, M. Lindberg, J. Sjöquist, and J. Sjödahl.** 1984. Region X, the-cell-wall-attachment part of staphylococcal protein A. *Eur. J. Biochem.* **138:**413–420.

31. **Harraghy, N., M. Hussain, A. Haggar, T. Chavakis, B. Sinha, M. Herrmann, and J. I. Flock.** 2003. The adhesive and immunomodulating properties of the multifunctional *Staphylococcus aureus* protein Eap. *Microbiology* **149:**2701–2707.

32. **Hartford, O., P. Francois, P. Vaudaux, and T. J. Foster.** 1997. The dipeptide repeat region of the fibrinogen-binding protein (clumping factor) is required for functional expression of the fibrinogen-binding domain on the *Staphylococcus aureus* cell surface. *Mol. Microbiol.* **25:**1065–1076.

33. **Hartleib, J., N. Kohler, R. Dickinson, G. Chhatwal, J. Sixma, O. Hartford, T. J. Foster, G. Peters, B. Kehrl, and M. Herrmann.** 2000. Protein A is the von Willebrand factor binding protein of *Staphylococcus aureus. Blood* **96:**2149–2156.

34. **Heath, H. E., L. S. Heath, J. D. Nitterauer, K. E. Rose, and G. L. Sloan.** 1989. Plasmid-encoded lysostaphin endopeptidase resistance of *Staphylococcus simulans* biovar *staphylolyticus. Biochem. Biophys. Res. Commun.* **160:**1106–1109.

35. Heilmann, C., M. Hussain, G. Peters, and F. Gotz. 1997. Evidence for autolysin-mediated primary attachment of *Staphylococcus epidermidis* to a polystyrene surface. *Mol. Microbiol.* **24:**1013–1024.

36. Heinrich, P., R. Rosenstein, M. Bohmer, P. Sonner, and F. Gotz. 1987. The molecular organization of the lysostaphin gene and its sequences repeated in tandem. *Mol. Gen. Genet.* **209:**563–569.

37. Hussain, M., K. Becker, C. von Eiff, J. Schrenzel, G. Peters, and M. Herrmann. 2001. Identification and characterization of a novel 38.5-kilodalton cell surface protein of *Staphylococcus aureus* with extended-spectrum binding activity for extracellular matrix and plasma proteins. *J. Bacteriol.* **183:**6778–6786.

38. Hussain, M., M. Herrmann, C. von Eiff, F. Perdreau-Remington, and G. Peters. 1997. A 140-kilodalton extracellular protein is essential for the accumulation of *Staphylococcus epidermidis* strains on surfaces. *Infect. Immun.* **65:**519–524.

39. Ilangovan, U., J. Iwahara, H. Ton-That, O. Schneewind, and R. T. Clubb. 2001. Assignment of 1H, 13C and 15N signals of sortase. *J. Biomol. NMR* **19:**379–380.

40. Ilangovan, U., H. Ton-That, J. Iwahara, O. Schneewind, and R. T. Clubb. 2001. Structure of sortase, the transpeptidase that anchors proteins to the cell wall of *Staphylococcus aureus*. *Proc. Natl. Acad. Sci. USA* **98:**6056–6061.

41. Jensen, K. 1958. A normally occurring staphylococcus antibody in human serum. *Acta Pathol. Microbiol. Scand.* **44:**421–428.

42. Jonsson, I. M., S. K. Mazmanian, O. Schneewind, T. Bremell, and A. Tarkowski. 2003. The role of *Staphylococcus aureus* sortase A and sortase B in murine arthritis. *Microb. Infect.* **5:**775–780.

43. Jönsson, K., D. McDevitt, M. H. McGavin, J. M. Patti, and M. Höök. 1995. *Staphylococcus aureus* expresses a major histocompatibility complex class II analog. *J. Biol. Chem.* **270:**21457–21460.

44. Jönsson, K., C. Signäs, H. P. Müller, and M. Lindberg. 1991. Two different genes encode fibronectin binding proteins in *Staphylococcus aureus*. The complete nucleotide sequence and characterization of the second gene. *Eur. J. Biochem.* **202:**1041–1048.

45. Josefsson, E., K. W. McCrea, D. Ní Eidhin, D. O'Connell, J. Cox, M. Höök, and T. J. Foster. 1998. Three new members of the serine-aspartate repeat protein multigene family of *Staphylococcus aureus*. *Microbiology* **144:**3387–3395.

46. Josefsson, E., D. O'Connell, T. J. Foster, I. Durussel, and J. A. Cox. 1998. The binding of calcium to the B-repeat segment of SrdD, a cell surface protein of *Staphylococcus aureus*. *J. Biol. Chem.* **273:**31145–31152.

47. Kehoe, M. A. 1994. Cell-wall-associated proteins in Gram-positive bacteria, p. 217–261. *In* J.-M. Ghuysen and R. Hakenbeck (ed.), *Bacterial Cell Wall*. Elsevier, Amsterdam, The Netherlands.

48. Komatsuzawa, H., M. Sugai, S. Yamada, A. Matsumoto, T. Oshida, and H. Suginaka. 1997. Subcellular localization of the major autolysin, ATL, and its processed proteins in *Staphylococcus aureus*. *Microbiol. Immunol.* **41:**469–479.

49. Kreikemeyer, B., D. McDevitt, and A. Podbielski. 2002. The role of the map protein in *Staphylococcus aureus* matrix protein and eukaryotic cell adherence. *Int. J. Med. Microbiol.* **292:**283–295.

50. Lee, L. Y., X. Liang, M. Höök, and E. L. Brown. 2004. Identification and characterization of the C3 binding domain of the *Staphylococcus aureus* extracellular fibrinogen-binding protein (Efb). *J. Biol. Chem.* **279:**50710–50716.

51. Lee, S. G., V. Pancholi, and V. A. Fischetti. 2002. Characterization of a unique glycosylated anchor endopeptidase that cleaves the LPXTG sequence motif of cell surface proteins of Gram-positive bacteria. *J. Biol. Chem.* **277:**46912–46922.

52. Mack, J., C. Vermeiren, D. E. Heinrichs, and M. J. Stillman. 2004. *In vivo* heme scavenging by *Staphylococcus aureus* IsdC and IsdE proteins. *Biochem. Biophys. Res. Commun.* **320:**781–788.

53. Marraffini, L. A., and O. Schneewind. 2005. Anchor structure of staphylococcal surface proteins. V. Anchor structure of the sortase B substrate IsdC. *J. Biol. Chem.* **280:**16263–16271.

54. Marraffini, L. A., H. Ton-That, Y. Zong, S. V. L. Narayana, and O. Schneewind. 2004. Anchoring of surface proteins to the cell wall of *Staphylococcus aureus*. IV. A conserved arginine residue is required for efficient catalysis of sortase A. *J. Biol. Chem.* **279:**37763–37770.

55. Matsuhashi, M. 1994. Utilization of lipid-linked precursors and the formation of peptidoglycan in the process of cell growth and division: membrane enzymes involved in the final steps of peptidoglycan synthesis and the mechanism of their regulation, p. 55–72. *In* J.-M. Ghuysen and R. Hakenbeck (ed.), *Bacterial Cell Wall*. Elsevier Biochemical Press, Amsterdam, The Netherlands.

56. Mazmanian, S. K., G. Liu, E. R. Jensen, E. Lenoy, and O. Schneewind. 2000. *Staphylococcus aureus* mutants defective in the display of surface proteins and in the pathogenesis of animal infections. *Proc. Natl. Acad. Sci. USA* **97:**5510–5515.

57. Mazmanian, S. K., G. Liu, H. Ton-That, and O. Schneewind. 1999. *Staphylococcus aureus* sortase, an enzyme that anchors surface proteins to the cell wall. *Science* **285:**760–763.

58. Mazmanian, S. K., E. P. Skaar, A. H. Gasper, M. Humayun, P. Gornicki, J. Jelenska, A. Joachimiak, D. M. Missiakas, and O. Schneewind. 2003. Passage of heme-iron across the envelope of *Staphylococcus aureus*. *Science* **299:**906–909.

59. Mazmanian, S. K., H. Ton-That, and O. Schneewind. 2001. Sortase-catalyzed anchoring of surface proteins to the cell wall of *Staphylococcus aureus*. *Mol. Microbiol.* **40:**1049–1057.

60. Mazmanian, S. K., H. Ton-That, K. Su, and O. Schneewind. 2002. An iron-regulated sortase enzyme anchors a class of surface protein during *Staphylococcus aureus* pathogenesis. *Proc. Natl. Acad. Sci. USA* **99:**2293–2298.

61. McDevitt, D., P. Francois, P. Vaudaux, and T. J. Foster. 1994. Molecular characterization of the clumping factor (fibrinogen receptor) of *Staphylococcus aureus*. *Mol. Microbiol.* **11:**237–248.

62. McDevitt, D., T. Nanavaty, K. House-Pompeo, E. Bell, N. Turner, L. McIntire, T. Foster, and M. Höök. 1997. Characterization of the interaction between the *Staphylococcus aureus* clumping factor (ClfA) and fibrinogen. *Eur. J. Biochem.* **247:**416–424.

63. McGavin, M. H., D. Krajewska-Pietrasik, C. Ryden, and M. Höök. 1993. Identification of a *Staphylococcus aureus* extracellular matrix-binding protein with broad specificity. *Infect. Immun.* **61:**2479–2485.

64. Navarre, W. W., and O. Schneewind. 1994. Proteolytic cleavage and cell wall anchoring at the LPXTG motif of surface proteins in gram-positive bacteria. *Mol. Microbiol.* **14:**115–121.

65. **Navarre, W. W., and O. Schneewind.** 1999. Surface proteins of Gram-positive bacteria and the mechanisms of their targeting to the cell wall envelope. *Microbiol. Mol. Biol. Rev.* **63:**174–229.

66. **Navarre, W. W., H. Ton-That, K. F. Faull, and O. Schneewind.** 1998. Anchor structure of staphylococcal surface proteins. II. COOH-terminal structure of muramidase and amidase-solubilized surface protein. *J. Biol. Chem.* **273:**29135–29142.

67. **Navarre, W. W., H. Ton-That, K. F. Faull, and O. Schneewind.** 1999. Multiple enzymatic activities of the murein hydrolase from staphylococcal phage Φ11. Identification of a D-alanyl-glycine endopeptidase activity. *J. Biol. Chem.* **274:**15847–15856.

68. **Ní Eidhin, D., S. Perkins, P. Francois, P. Vaudaux, M. Höök, and T. J. Foster.** 1998. Clumping factor B (ClfB), a new surface-located fibrinogen-binding adhesin of *Staphylococcus aureus. Mol. Microbiol.* **30:**245–257.

69. **Oshida, T., M. Sugai, H. Komatsuzawa, Y.-M. Hong, H. Suginaka, and A. Tomasz.** 1995. A *Staphylococcus aureus* autolysin that has an N-acetylmuramoyl-L-alanine amidase domain and an endo-b-N-acetylglucosaminidase domain: cloning, sequence analysis, and characterization. *Proc. Natl. Acad. Sci. USA* **92:**285–289.

70. **Pallen, M. J., A. C. Lam, M. Antonio, and K. Dunbar.** 2001. An embarrassment of sortases—a richness of substrates. *Trends Microbiol.* **9:**97–101.

71. **Palma, M., A. Haggar, and J. I. Flock.** 1999. Adherence of *Staphylococcus aureus* is enhanced by an endogenous secreted protein with broad binding activity. *J. Bacteriol.* **181:**2840–2845.

72. **Palma, M., D. Wade, M. Flock, and J. I. Flock.** 1998. Multiple binding sites in the interaction between an extracellular fibrinogen-binding protein from *Staphylococcus aureus* and fibrinogen. *J. Biol. Chem.* **273:**13177–13181.

73. **Park, P. W., D. D. Roberts, L. E. Grosso, W. C. Parks, J. Rosenbloom, W. R. Abrams, and R. P. Mecham.** 1991. Binding of elastin to *Staphylococcus aureus. J. Biol. Chem.* **266:**23399–23406.

74. **Patel, A. H., P. Nowlan, E. D. Weavers, and T. Foster.** 1987. Virulence of protein A-deficient and alpha-toxin-deficient mutants of *Staphylococcus aureus* isolated by allele replacement. *Infect. Immun.* **55:**3103–3110.

75. **Patti, J. M., B. L. Allen, M. J. McGavin, and M. Höök.** 1994. MSCRAMM-mediated adherence of microorganisms to host tissues. *Annu. Rev. Microbiol.* **48:**89–115.

76. **Patti, J. M., T. Bremell, D. Krajewska-Pietrasik, A. Abdelnour, A. Tarkowski, C. Ryden, and M. Höök.** 1994. The *Staphylococcus aureus* collagen adhesin is a virulence determinant in experimental septic arthritis. *Infect. Immun.* **62:**152–161.

77. **Patti, J. M., H. Jonsson, B. Guss, L. M. Switalski, K. Wiberg, M. Lindberg, and M. Höök.** 1992. Molecular characterization and expression of a gene encoding a *Staphylococcus aureus* collagen adhesin. *J. Biol. Chem.* **267:**4766–4772.

78. **Perkins, S., E. J. Walsh, C. C. Deivanayagam, S. V. Narayana, T. J. Foster, and M. Höök.** 2001. Structural organization of the fibrinogen-binding region of the clumping factor B MSCRAMM of *Staphylococcus aureus. J. Biol. Chem.* **276:**44721–44728.

79. **Perry, A. M., H. Ton-That, S. K. Mazmanian, and O. Schneewind.** 2002. Anchoring of surface proteins to the cell wall of *Staphylococcus aureus*. III. Lipid II is an *in vivo* peptidoglycan substrate for sortase-catalyzed surface protein anchoring. *J. Biol. Chem.* **277:**16241–16248.

80. **Ponnuraj, K., M. G. Bowden, S. Davis, S. Gurusiddappa, D. Moore, D. Choe, Y. Xu, M. Höök, and S. V. Narayana.** 2003. A "dock, lock, and latch" structural model for a staphylococcal adhesin binding to fibrinogen. *Cell* **115:**217–228.

81. **Recsei, P. A., A. D. Gruss, and R. P. Novick.** 1987. Cloning, sequence, and expression of the lysostaphin gene from *Staphylococcus simulans. Proc. Natl. Acad. Sci. USA* **84:**1127–1131.

82. **Rich, R. L., C. C. Deivanayagam, R. T. Owens, M. Carson, A. Hook, D. Moore, J. Symersky, V. W. Yang, S. V. Narayana, and M. Höök.** 1999. Trench-shaped binding sites promote multiple classes of interactions between collagen and the adherence receptors, alpha(1) beta(1) integrin and *Staphylococcus aureus* cna MSCRAMM. *J. Biol. Chem.* **274:**24906–24913.

83. **Roche, F. M., R. Downer, F. Keane, P. Speziale, P. W. Park, and T. J. Foster.** 2004. The N-terminal A domain of fibronectin-binding proteins A and B promotes adhesion of *Staphylococcus aureus* to elastin. *J. Biol. Chem.* **279:**38433–38440.

84. **Roche, F. M., R. Massey, S. J. Peacock, N. P. Day, L. Visai, P. Speziale, A. Lam, M. Pallen, and T. J. Foster.** 2003. Characterization of novel LPXTG-containing proteins of *Staphylococcus aureus* identified from genome sequences. *Microbiology* **149:**643–654.

85. **Ryden, C., K. Rubin, P. Speziale, M. Höök, M. Lindberg, and T. Wadstrom.** 1983. Fibronectin receptors from *Staphylococcus aureus. J. Biol. Chem.* **258:**3396–3401.

86. **Schindler, C. A., and V. T. Schuhardt.** 1964. Lysostaphin: a new bacteriolytic agent for the staphylococcus. *Proc. Natl. Acad. Sci. USA* **51:**414–421.

87. **Schneewind, O., A. Fowler, and K. F. Faull.** 1995. Structure of the cell wall anchor of surface proteins in *Staphylococcus aureus. Science* **268:**103–106.

88. **Schneewind, O., K. F. Jones, and V. A. Fischetti.** 1990. Sequence and structural characteristics of the trypsin-resistant T6 surface protein of group A streptococci. *J. Bacteriol.* **172:**3310–3317.

89. **Schneewind, O., P. Model, and V. A. Fischetti.** 1992. Sorting of protein A to the staphylococcal cell wall. *Cell* **70:**267–281.

90. **Shockman, G. D., and J.-V. Höltje.** 1994. Microbial peptidoglycan (murein) hydrolases, p. 131–166. *In* J.-M. Ghuysen and R. Hakenbeck (ed.), *Bacterial Cell Wall.* Elsevier Biochemical Press, Amsterdam, The Netherlands.

91. **Signas, C., G. Raucci, K. Jonsson, P.-E. Lindgren, G. M. Anantharamaiah, M. Höök, and M. Lindberg.** 1989. Nucleotide sequence of the gene for a fibronectin-binding protein from *Staphylococcus aureus*: use of this peptide sequence in the synthesis of biologically active peptides. *Proc. Natl. Acad. Sci. USA* **86:**699–703.

92. **Sjödahl, J.** 1977. Repetitive sequences in protein A from *Staphylococcus aureus.* Arrangement of five regions within the protein, four being highly homologous and Fc-binding. *Eur. J. Biochem.* **73:**343–351.

93. **Sjöquist, J., B. Meloun, and H. Hjelm.** 1972. Protein A isolated from *Staphylococcus aureus* after digestion with lysostaphin. *Eur. J. Biochem.* **29:**572–578.

94. **Skaar, E. P., A. H. Gaspar, and O. Schneewind.** 2004. IsdG and IsdI, heme degrading enzymes in the cytoplasm of *Staphylococcus aureus. J. Biol. Chem.* **279:**436–443.

95. **Skaar, E. P., M. Humayun, K. L. DeBord, and O. Schneewind.** 2004. Iron source preference of *Staphylococcus aureus* infections. *Science* **305:**1626–1628.

96. Skaar, E. P., and O. Schneewind. 2004. Iron-regulated surface determinants (Isd) of *Staphylococcus aureus*: stealing iron from heme. *Microbes Infect.* **6:**390–397.

97. Strauss, A., G. Thumm, and F. Gotz. 1998. Influence of Lif, the lysostaphin immunity factor, on acceptors of surface proteins and cell wall sorting efficieny in *Staphylococcus carnosus. J. Bacteriol.* **180:**4960–4962.

98. Switalski, L. M., J. M. Patti, W. Butcher, A. G. Gristina, P. Speziale, and M. Höök. 1993. A collagen receptor on *Staphylococcus aureus* strains isolated from patients with septic arthritis mediates adhesion to cartilage. *Mol. Microbiol.* **7:**99–107.

99. Ton-That, H., K. F. Faull, and O. Schneewind. 1997. Anchor structure of staphylococcal surface proteins. I. A branched peptide that links the carboxyl terminus of proteins to the cell wall. *J. Biol. Chem.* **272:**22285–22292.

100. Ton-That, H., G. Liu, S. K. Mazmanian, K. F. Faull, and O. Schneewind. 1999. Purification and characterization of sortase, the transpeptidase that cleaves surface proteins of *Staphylococcus aureus* at the LPXTG motif. *Proc. Natl. Acad. Sci. USA* **96:**12424–12429.

101. Ton-That, H., L. A. Marraffini, and O. Schneewind. 2004. Protein sorting to the cell wall envelope of Gram-positive bacteria. *Biochim. Biophys. Acta* **1694:**269–278.

102. Ton-That, H., S. K. Mazmanian, L. Alksne, and O. Schneewind. 2002. Anchoring of surface proteins to the cell wall of *Staphylococcus aureus*. II. Cysteine 184 and histidine 120 of sortase A form a thiolate imidazolium ion pair for catalysis. *J. Biol. Chem.* **277:**7447–7452.

103. Ton-That, H., S. K. Mazmanian, K. F. Faull, and O. Schneewind. 2000. Anchoring of surface proteins to the cell wall of *Staphylococcus aureus*. I. Sortase catalyzed *in vitro* transpeptidation reaction using LPXTG peptide and NH$_2$-Gly$_3$ substrates. *J. Biol. Chem.* **275:**9876–9881.

104. Ton-That, H., and O. Schneewind. 1999. Anchor structure of staphylococcal surface proteins. IV. Inhibitors of the cell wall sorting reaction. *J. Biol. Chem.* **274:**24316–24320.

105. Tung, H. S., B. Guss, U. Hellman, L. Persson, K. Rubin, and C. Ryden. 2000. A bone sialoprotein-binding protein from *Staphylococcus aureus*: a member of the staphylococcal Sdr family. *Biochem. J.* **345:**611–619.

106. Uhlén, M., B. Guss, B. Nilsson, S. Gatenbeck, L. Philipson, and M. Lindberg. 1984. Complete sequence of the staphylococcal gene encoding protein A. *J. Biol. Chem.* **259:**1695–1702 and 13628 (Corr.).

107. Vernachio, J., A. S. Bayer, T. Le, Y. L. Chai, B. Prater, A. Schneider, B. Ames, P. Syribeys, J. Robbins, and J. M. Patti. 2003. Anti-clumping factor A immunoglobulin reduces the duration of methicillin-resistant *Staphylococcus aureus* bacteremia in an experimental model of infective endocarditis. *Antimicrob. Agents Chemother.* **47:**3400–3406.

108. Wann, E. R., S. Gurusiddappa, and M. Höök. 2000. The fibronectin-binding MSCRAMM FnbPA of *Staphylococcus aureus* is a bifunctional protein that also binds to fibrinogen. *J. Biol. Chem.* **275:**13863–13871.

109. Weiss, W. J., E. Lenoy, T. Murphy, L. Tardio, P. Burgio, S. J. Projan, O. Schneewind, and L. Alksne. 2004. Effect of *srtA* and *srtB* gene expression on the virulence of *Staphylococcus aureus* in animal infection. *J. Antimicrob. Chemother.* **53:**480–486.

110. Yamada, S., M. Sugai, H. Komatsuzawa, S. Nakashima, T. Oshida, A. Matsumoto, and H. Suginaka. 1996. An autolysin ring associated with cell separation of *Staphylococcus aureus. J. Bacteriol.* **178:**1565–1571.

111. Zhang, R.-g., G. Joachimiak, R.-y. Wu, S. K. Mazmanian, D. M. Missiakas, O. Schneewind, and A. Joachimiak. 2004. Structures of sortase B from *Staphylococcus aureus* and *Bacillus anthracis* reveal catalytic amino acid triad in the active site. *Structure* **12:**1147–1156.

112. Zong, Y., T. W. Bice, H. Ton-That, O. Schneewind, and S. V. L. Narayana. 2004. Crystal structures of *Staphylococcus aureus* sortase A and its substrate complex. *J. Biol. Chem.* **279:**31383–31389.

113. Zong, Y., S. K. Mazmanian, O. Schneewind, and S. V. Narayana. 2004. The structure of sortase B, a cysteine transpeptidase that tethers surface protein to the *Staphylococcus aureus* cell wall. *Structure* **12:**105–112.

Staphylococcal Pathogenesis and Pathogenicity Factors: Genetics and Regulation

RICHARD P. NOVICK

41

INTRODUCTION: STAPHYLOCOCCAL PATHOGENESIS

As facultative pathogens, staphylococci can live freely in the inanimate environment, they can exist as external colonizers or commensals, and they can live within the tissues or cells of a host organism, causing disease. Additionally, staphylococci may form biofilms in any of their three states of existence. In the disease context, the classical staphylococcal infection is the abscess. Organisms entering the tissues of a host animal produce a series of extracellular proteins and other factors such as cell wall and capsular components that enable them to coagulate fibrinogen, adhere to the intercellular matrix, degrade tissue components, and lyse local cellular elements. This evokes a potent innate immune response that includes interleukins, opsonins, complement, and phagocytes. Additionally, humans have circulating antibodies to most staphylococcal antigens, and these will obviously participate in the initial response. These antibodies, the innate immune response, and fibrin generated by the organism wall off the lesion, creating a pocket within which a life-and-death battle between the organism and the phagocytes is waged, generating pus, which consists of dead neutrophils, necrotic tissue, living and dead bacteria, and the contents of lysed host and bacterial cells (Fig. 1).

The immunocompetent host nearly always wins this battle, but not without the considerable discomfort caused by local swelling, induration, and inflammation, which is why the typical skin abscess is called a boil, and why the contents of the abscess pocket that are released when the lesion opens and drains used to be called "laudable pus," because it indicated a successful outcome. Although everyone gets superficial skin infections, staphylococcal infections are only occasionally initiated de novo in deep tissue sites in healthy individuals. These result in deep abscesses that often require surgical intervention. *Staphylococcus aureus* can also alight on the heart valves, occasionally without visible antecedent, but much more often in intravenous drug users or in elderly people with calcification or old rheumatic valve damage. This is uniformly fatal without treatment and, even when treated aggressively, causes major damage to the valve, and can seed other tissues, generating metastatic abscesses. Heart valve lesions, known as vegetations, consist largely of platelets, fibrin, organisms, and neutrophils, and their structure is considered akin to biofilms formed on inanimate surfaces. A particularly troublesome, but not fatal, condition is osteomyelitis, which, again, can occasionally occur spontaneously, but much more often follows an open fracture. Osteomyelitis can persist more or less indefinitely, despite treatment, gradually destroying the infected bone, but not generally spreading to other sites.

A special set of pathological conditions, toxinoses, are caused by single toxins. In many cases, the purified toxin can generate all of the symptoms; in some cases, the living organism must be present, contributing, for example, the ability to adhere to the extracellular matrix or to resist eradication by the host, enabling the toxin to be produced in sufficient quantities and at the right location. *S. aureus* causes several toxinoses, including toxic shock syndrome (TSS), scalded skin syndrome (exfoliatin A and B), food poisoning (various enterotoxins), and necrotizing pneumonia (Panton-Valentine leukocidin [PVL]).

Although *S. aureus* has traditionally been regarded as an extracellular pathogen, there is increasing evidence that it may be taken up in significant numbers by cells other than professional phagocytes, such as epithelial and endothelial cells (see chapter 42, this volume). This has been implicated particularly in staphylococcal mastitis, which is seen uncommonly in humans, nearly always during lactation, but very commonly in cattle, where it has major commercial consequences. The overall clinical significance of intracellular *S. aureus* remains unclear, except in the case of the small-colony variants (see chapter 39, this volume). Several gene products have been implicated in internalization, and the function and regulation of these and other pathogenicity factors in the intracellular environment are discussed below.

Pathogenicity Factors and Their Deployment

Traditionally, bacterial pathogenicity or virulence factors are products whose role in the disease process is either clearly demonstrable, e.g., toxins, or more or less obvious on the basis of biological properties, e.g., enzymes that degrade tis-

Gram-Positive Pathogens, 2nd edition, edited by Vincent A. Fischetti et al.
© 2006 ASM Press, Washington, D.C.

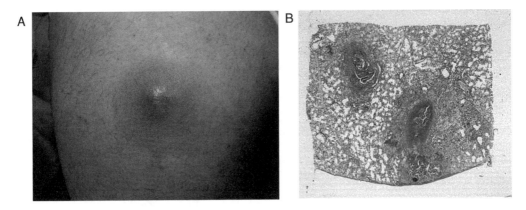

FIGURE 1 (*See the separate color insert for the color version of this illustration.*) Staphylococcal abscesses. (A) Cutaneous furuncle (Nadir Goksugur, M.D., Dermatlas; http://www.dermatlas.org). (B) Stained section of pulmonary abscess. Kindly provided by Martin Nachbar.

sue components. As these genes are actively expressed in laboratory cultures, they and their products have been relatively easy to identify and study (Table 1). The list in Table 1 is an approximation, because it is based on information from only a very few strains and because, even for these, many of the exoproteins visible as bands on stained polyacrylamide gels have not been identified. These are all accessory factors, important for adaptation to the host tissue environment but not essential for growth and cell division, and are collectively referred to as the virulon. The properties and functions of these various factors are considered in chapters 38 and 39 (this volume). This chapter outlines current understanding of their genetics and regulation.

Staphylococcal pathogenesis is multifactorial, involving three classes of factors: secreted proteins, including superantigens (SAgs), cytotoxins, and tissue-degrading enzymes; cell surface-bound proteins (the MSCRAMMs [microbial surface components recognizing adhesive matrix molecules]), including fibrinogen-binding protein, fibronectin-binding protein, collagen-binding protein, other adhesins, and antiopsonins; and cell surface components, including the polysaccharide capsule and components of the cell wall peptidoglycan. The secreted proteins are assumed to enable the organism to attack the local cellular and structural elements of diverse tissues and organs, and the MSCRAMMs enable the bacteria to adhere to tissue components, to resist phagocytosis and other host defenses (3, 108), and, in the case of fibronectin-binding protein, to facilitate internalization by epithelial and endothelial cells (40) (see chapters 39 and 40, this volume). Components of the cell wall have an important role in septic shock, inducing an excessive and detrimental host response. As the latter are encoded by constitutive housekeeping genes and are essential, they represent an exception to the rule that virulence factors are encoded by accessory genes. For present purposes, they are not regarded as components of the virulon.

In addition to the pathogenicity factors listed in Table 1 are those identified by signature-tagged mutagenesis and in vitro expression technology, many of which are of unknown function. Assignment of roles in pathogenesis to these genes, which requires ruling out polar effects, and an understanding of their biological activities, may lead to important insights into the infective process. For heuristic purposes, it is not considered useful to include in the list of pathogenicity determinants every gene that is required for the establishment and/or maintenance of an infection; although mutagenic strategies involving in vivo screening have identified many such genes, most of them are standard housekeeping genes, which may prove attractive as novel therapeutic targets but provide little if any enlightenment on the causation of disease.

Most, if not all, of the *S. aureus* virulence factors are variably expressed by one or another of the coagulase-negative species, as described in chapters 46 and 47 (this volume). For any of these species, most strains express none of these factors; a few strains express one or a few of them. It is interesting that many strains of *Staphylococcus warneri* and *Staphylococcus lugdunensis* express either β-hemolysin or staphylokinase, but not both (121), and may harbor prophages that are responsible, as in *S. aureus*.

GENETICS OF THE STAPHYLOCOCCAL VIRULON

Considering first the genetics of staphylococcal virulence factors, there would appear to be two classes—those encoded by constant chromosomal genes, present in most or all strains, and those encoded by variable genes, present in a minority of strains, and usually belonging to accessory genetic elements, including plasmids, transposons, prophages, and pathogenicity islands (SaPIs), some of which are mobile. It is probably not coincidental that all of the known staphylococcal genes responsible for toxinoses—conditions that are attributable to single proteins—are variable, whereas genes that act in concert as part of the multifactorial virulon are constant. The carriage of toxinosis-causing genes by variable genetic elements is also seen widely in other species (95).

Accessory Genetic Elements and Virulence Genes

Lysogenic Conversion

It is well known that certain staphylococcal virulence genes are encoded by prophages. These are known as converting phages, and they contribute significantly to the pathobiology of the host bacterium. The first convincing report of this in

TABLE 1 Staphylococcal extracellular accessory proteins

Gene	Location	n315[a]	Product	Activity/function	Timing	Reference(s)
			Superantigens			
sea	Phage	None[b]	Enterotoxin A	Food poisoning, TSS	xp[c]	138, 144
seb	SaPI3	None	Enterotoxin B	Food poisoning, TSS	pxp	46, 144
sec	SaPI4	SA1817	Enterotoxin C	Food poisoning, TSS	pxp	115
sed	Plasmid	None	Enterotoxin D	Food poisoning, TSS	pxp	162
eta	ETA phage	None	Exfoliatin A	Scalded skin syndrome	pxp	126
etb	Plasmid	None	Exfoliatin B	Scalded skin syndrome	pxp	
tst	SaPI1,2,bov1	SA1819	Toxic shock syndrome toxin 1	TSS	pxp	113, 144
			Cytotoxins			
hla	Chrom	SA1007	Alpha-hemolysin	Hemolysin, cytotoxin	pxp	53, 113, 117, 124, 134, 144
hlb	Chrom	SA1811	Beta-hemolysin	Hemolysin, cytotoxin	pxp	53, 113, 117
hld	Chrom	SAS065	Delta-hemolysin	Hemolysin, cytotoxin	xp	53, 113, 124, 134
hlg	Chrom	SA2207-9	Gamma-hemolysin	Hemolysin, cytotoxin	pxp	21, 117
lukS/F	PVL phage	None	P-V leukocidin	Leukolysin	pxp	117, 144
			Enzymes			
SplA-F	Chrom	SA1631-1627	Serine protease-like	Putative protease		117
sspA	Chrom	SA0901	V8 protease	Spreading factor	pxp	2, 26, 144
aur	Chrom	SA2430	Metalloprotease (aureolysin)	Processing enzyme?	pxp	2, 26
sspB	Chrom	SA0900	Cysteine protease	Processing enzyme?		2, 117
scp	Chrom	SA1725	Staphopain (protease II)	Spreading, nutrition	pxp	2
geh	Chrom	SA0309	Glycerol ester hydrolase	Spreading, nutrition	pxp	117, 129
lip	Chrom	SA2463	Lipase (butyryl esterase)	Spreading, nutrition	pxp	24
fme	Chrom	None	Fatty acid methyl ester	Fatty acid esterification	pxp	24
plc	Chrom	None	Phosphatidylinositol-specific phospholipase C		pxp	2
nuc	Chrom	SA0746	Nuclease	Nutrition	pxp	129
hal	Chrom	SA2003	Hyaluronic acid lyase	Spreading factor	pxp	80
coa	Chrom	SA0222	Coagulase	Clotting, clot digestion	exp	53, 72, 117, 150
sak	Phage	SA1758	Staphylokinase	Plasminogen activator	pxp	113
			Surface proteins			
spa	Chrom	SA0107	Protein A	Anti-immune, anti-PMN	exp	30, 53, 113, 117
cna	PT islet	None	Collagen-binding protein	Collagen binding	pxp	20
fnbA	Chrom	SA2291	Fibronectin BPA	Fibronectin binding	exp	119
fnbB	Chrom	SA2290	Fibronectin BPB	Fibronectin binding	exp	119
clfA	Chrom	SA0742	Clumping factor A	Fibrinogen binding	exp	150
clfB	Chrom	SA2423	Clumping factor B	Fibrinogen binding	exp	86, 117
hlfB	Chrom		Lactoferrin BP	Lactoferrin binding		91a
cap5	Chrom	SA0144-0159	Polysaccharide capsule type 5	Antiphagocytosis?	pxp	106
cap8	Chrom		Polysaccharide capsule type 8	Antiphagocytosis?	pxp	

[a]Designation in n315 genome.
[b]Not identified in n315 sequence or absent from n315.
[c]Xp, throughout exponential phase; exp, early exponential phase only; pxp, postexponential phase.

S. aureus involved conversion to staphylokinase production (149). Early reports of toxin conversion include PVL (140) and alpha- and delta-toxins (31, 37). PVL conversion has been amply confirmed, whereas the other two have not and are probably incorrect. Other toxin genes that have more recently been shown to be phage-encoded include enterotoxin A (15) and exfoliatin A (157).

These toxin genes are always at one end of the prophage, which has generated the hypothesis (as yet unproven) that they have been acquired by aberrant excision analogously to λdv. It is interesting that despite being phage-coded, these genes are not widespread among *S. aureus* genotypes, which one would assume is a result of narrow host range for the phages.

S. aureus is unusual in that the insertion sites for two different prophages are within the coding sequences for two important exoproteins, lipase (*geh*) (74, 116) and β-hemolysin (*hlb*) (32, 33, 149), resulting in "negative lysogenic conversion." This is remarkably widespread and is responsible for the lack of β-hemolysin production by most human clinical isolates, whereas most farm animal isolates do not carry a β-hemolysin-inactivating prophage. The pathobiological significance of negative lysogenic conversion is obscure; however, it is also true that many temperate phages insert into chromosomal genes, but these are generally not prominent in the bacterial phenotype and therefore have not generated much attention. Given that temperate phages are endosymbionts, so that their evolution is driven by forces independent of those driving evolution of their hosts, insertion into host genes is more an aspect of convenience for the phage than of biological significance for the host. Indeed, beyond providing protection against superinfection by coimmune phages, most known prophages do not contribute significantly to the host cell's phenotypes. At the same time, most β-hemolysin-inactivating phages encode staphylokinase, many also encode enterotoxin A, and some encode a recently discovered protein, CHIPS, that blocks neutrophil chemotaxis by binding to the C5a receptor (34). These genes presumably have selective value for the bacteria, so that such endosymbionts may contribute to the fitness of their hosts.

Plasmids and Transposons

Though plasmids and transposons are largely responsible for the evolution and spread of resistance to antibiotics and other toxic substances among staphylococci, they have thus far been found to have at most a minimal role in virulence. Only two plasmids (but no transposons) have been described that carry virulence factors, namely enterotoxin B (SEB) (146) and D (11). The prototypical SEB plasmid also encodes a bacteriocin and the enterotoxin D-encoding plasmid carries two additional genes for putative enterotoxins (102, 161). SEB-like toxins are also plasmid-carried in other staphylococci such as *Staphylococcus hyicus* (120).

SaPIs

TSST-1 and other SAg toxins, including SEB and enterotoxin C, are encoded by a family of novel phage-related staphylococcal pathogenicity islands (SaPIs) (99) (see chapters 32 and 33, this volume). These are ~15 kb elements that are located at specific chromosomal sites, are induced by certain temperate phages to excise and replicate during vegetative phage growth, are packaged into small-headed phage particles, and are transferred at extraordinarily high frequencies. These are the first pathogenicity islands for which mobility has been directly demonstrated. They have specific insertion sites and encode integrases and other phagelike products, and their interaction with phages seems similar to that between coliphages P2 and P4. Thus far, six of these have been described in some detail; most encode additional putative SAgs as well as those by which they were first identified. Additional SaPI-like islands have been identified in the sequenced *S. aureus* chromosomes, and other putative islands have also been identified. These differ radically from the original SaPIs, encoding predicted restriction-modification systems, transposon elements (of unknown functionality), γ-hemolysin, and predicted putative proteases and enterotoxins. They do not, however, encode integrases, they are not flanked by inverted repeats, and they have not been shown to be mobile or involved in pathogenesis. Moreover, certain of them are absolutely conserved, raising the question of whether their designation as chromosomal islands is appropriate.

Chromosomal (Constant) Virulence Genes

For present purposes, genes that have not been identified with variable genetic elements are considered to belong to the standard genetic complement of the organism. Virulence genes in this category are listed as "Chrom" in Table 1. This is typified by the genome of strain n315 (71); genes found in this strain are listed by their index numbers in Table 1, indicating, incidentally, that these are located throughout the chromosome. A considerable number of the genes in Table 1 are not found in n315, which underlines the striking genetic variability among strains of *S. aureus*. This is discussed further under "Interstrain Variations," below. Unlike the enteric pathogens, whose virulence is often determined and defined by plasmids and pathogenicity islands, *S. aureus* is fully virulent on the basis of its chromosomal complement of virulence genes, which are generally unlinked; as noted, variable genetic elements are responsible primarily for the toxinoses. Though this is not, of course, an absolute rule, it points to an interesting difference between gram-negative and -positive pathogens. One of the tasks facing any gram-negative pathogen is the export of its virulence factors—a task that is very largely accomplished by pathogenicity islands, many of which encode type III or type IV secretion systems, which are not found in gram-positive pathogens. Thus, the outer membrane of the gram-negatives is probably responsible for the location of virulence genes as well as toxin genes on variable genetic elements.

REGULATION OF THE STAPHYLOCOCCAL VIRULON

The first evidence for virulence gene regulation was the isolation of pleiotropic staphylococcal mutants defective in the production of hemolysins and other virulence factors (18). Subsequently, the isolation of a transposon insertion with this phenotype (114) led to the identification and cloning of a global regulatory locus, *agr* (accessory gene regulation) (89, 105), that is important for virulence because *agr* mutants are greatly attenuated in each of several animal models (1, 8, 27, 49) (see chapter 44, this volume). Subsequent studies have led to the demonstration of a considerable number of other regulatory genes that interact in a complex manner and are affected in turn by a variety of environmental factors, some of which may reflect conditions existing in host tissues. It has thus become clear that staphylococcal virulence is regulated by a vast and complex network of interacting genetic and environmental factors that is only now beginning to be unraveled. At this juncture, an overriding question is, what is the biological rationale underlying this remarkable regulatory strategy? One hypothesis is that facultative pathogens such as *S. aureus* express locally appropriate subsets of accessory genes as the output of this governing regulatory organization. Although a great deal of information has been obtained by in vitro analyses, definitive testing of this hypothesis will require highly sophisticated in vivo studies. In the following sections, the basic information obtained by in vitro studies is summarized, and its application to early studies of in vivo regulation is noted. Microarray technology has recently been used to define the regulatory spectra of various genetic and environmental factors (12, 17, 39, 70, 117, 147), and its continued use will be very important in deciphering the overall regulatory strategy of the organism. Note that the set of genes whose expression is controlled by a specific regulatory gene is that gene's regulon; the set controlled by a specific environmental factor is that factor's stimulon.

Time Course of In Vitro Expression

The complexity of virulon regulation in *S. aureus* can be appreciated by an examination of the time course of virulence gene expression in vitro (Fig. 2). The surface proteins appear earlier than the secreted enzymes, immunotoxins, and cytotoxins, and the above-mentioned intracellular metabolic enzymes. This sequential activation seems to be, at least in part, a function of population density. Starting with stationary phase, there would seem to be three key transition points in the in vitro growth cycle, possibly occurring in response to intracellular signals, such as GTP levels. First is the transition to exponential phase, which involves not only the revival of biosynthetic and other metabolic pathways required for growth and cell division, but also the synthesis of some surface proteins, coagulase, and possibly other accessory proteins. The synthesis of these is probably initiated during the transition from stationary phase to exponential phase and may come under the general metabolic program governing this transition. The nature of the signal(s) acting at this stage represents a key area for study. Other surface protein genes are switched on shortly after the onset of exponential growth and, as typified by *spa* (gene abbreviations are listed in Table 1), are switched off shortly thereafter, concomitantly with the appearance of *agr*-RNAIII (142). This clear reciprocity, however, is not seen with all strains and under all conditions (135), and may be related to σ^B activity (S. Herbert and R. P. Novick, unpublished data) or to growth conditions and media. The *agr* autoinducing peptide (AIP) reaches its threshold around mid-exponential phase, activating *agr* expression. In derivatives of the commonly studied NCTC8325, however, certain exoprotein genes, such as *coa*, are sharply down-regulated well

before the appearance of RNAIII, suggesting that some other inhibitory signal is responsible. The second transition, between the exponential and postexponential phases (possibly a consequence of decreasing availability of dissolved oxygen owing to increasing population density), is, in most strains, accompanied by up-regulation of the genes encoding secreted proteins. *Agr*, which, in 8325, is activated two or more hours earlier, sets the level of expression of most of these proteins, but not the timing (142); in fact, up-regulation of these genes occurs at the onset of the postexponential phase, regardless of when, or even whether, RNAIII transcription is activated (unpublished data).

Similar in vitro temporal programs have been described for other bacterial pathogens, e.g., *Bordetella pertussis* (58, 122), which activate their virulence program in response to specific signals, such as increased temperature, that are characteristic of the host tissue environment.

Regulatory Contingencies

It can be inferred from Figure 2 that regulatory elements not only up- or down-regulate their target genes but do so at particular times during the standard growth cycle. This implies that activation or inactivation of the regulators themselves is also time-dependent. Three primary mechanisms can be envisioned: specific environmental factors, autoinduction, and changes in the intracellular milieu.

Environmental Factors

During growth in typical flask cultures, pH drops from >7 to <6 owing to the fermentation of glucose. This change in pH, sensed by unknown receptors, has a profound effect on the expression of both regulatory and target genes (147). For example, it sharply down-regulates the longer *sae* transcripts (96), secondarily affecting all *sae*-regulated genes. Similarly, it is impossible to maintain O_2 saturation in flask cultures despite vigorous aeration; reduction in external

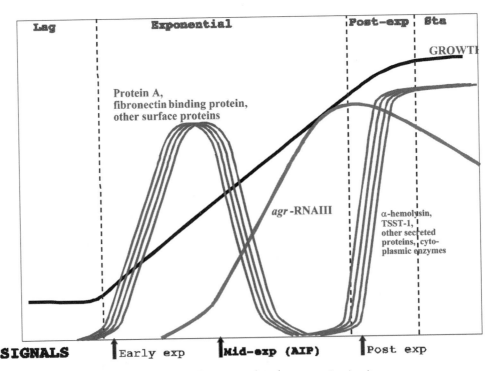

FIGURE 2 Temporal program of virulon expression in vitro.

pO_2 has major effects on several regulatory genes, including *sarA*, *sigB*, *srrAB*, and *agr* (76, 156), and consequently, profound effects on their target genes. Possible transducers of the low pO_2 response include σ^B and *srrAB*. Other environmental factors that are not automatic consequences of the in vitro growth cycle include external salt concentration, osmolarity, and metabolic inhibitors such as subinhibitory antibiotics, ethanol, and salicylate. Some of these act through (unknown) surface receptors; others penetrate the cell and act intracellularly on primary targets, including σ^B.

Autoinduction

Autoinduction is an automatic response to a low molecular weight product, which, in gram-positive bacteria, is generally a peptide that binds to the receptor of a two-component signaling (TCS) module. Early in the growth cycle, the autoinducing peptide is synthesized constitutively and accumulates during growth. As it accumulates, it increasingly activates a positive feedback loop, generating a rapidly accelerating response. *S. aureus* possesses a single known autoinduction system, the *agr* system, which regulates a large set of virulence factors and other accessory genes, acting through other regulators by a mechanism that remains to be determined.

The Intracellular Milieu

Changes in intracellular redox potential, in the levels of nucleotide tri- or polyphosphates, or in the concentrations or oxidation states of energy-dependent cofactors (NAD, flavin mononucleotide, etc.) have also been shown or postulated to have profound effects on the intracellular regulatory environment (132, 133) (see also chapter 35, this volume).

Regulatory Genes and Their Functions

Table 2 gives a summary of the known genes and environmental conditions that affect the expression of pathogenicity factors by *S. aureus* as defined above. An important proviso for most of the observations described here and elsewhere is that they are based largely on one or a very few strains and there are clearly profound strain-specific differences in regulatory modalities. Available temporal information, mostly on the *agr* system, is reviewed with reference to other regulatory factors where known. An overall concept is that the primary regulators are the TCS transduction modules, which transmit their signals either directly to target genes, or, probably more often, to intermediary transcription factors. The latter regulate target genes and often participate in feedback loops that affect one another, or one or more of the TCS modules, or both. In some cases, the transcription factors appear to act independently of input from the TCS modules. It is suggested that those environmental factors that do not penetrate the cell are sensed by signal receptors, whereas those that penetrate may be sensed directly by transcription factors.

TCS Modules

Four distinct TCS modules are presently known to be involved, *agr* (113), *saeRS* (51, 52), *srrAB* (156) (independently analyzed by Throup et al. [137] and referred to as *srhRS*), and *arlAB* (42). These four represent one-quarter of the putative TCS modules identified by examination of the *S. aureus* genome (22, 84; D. McDevitt, personal communication), and there is every reason to believe that some of the others are also involved. It is interesting that signal receptors in gram-positive bacteria typically have multiple transmembrane helices (4, 5, 36, 56, 59, 111, 130) and are activated by peptides, whereas those of gram-negative bacteria generally have one, rarely two, transmembrane helices and are activated by a variety of signals other than peptides. It is also notable that autoinducers in gram-negative bacteria are *N*-acylhomoserine lactones rather than peptides, and, with one or two exceptions, such as the *Vibrio harveyi* autoinducer (23), these act intracellularly rather than via signal transduction. A partial explanation for this difference is that peptides over a certain size do not penetrate the outer membrane of gram-negative bacteria, whereas homoserine lactones are freely diffusible.

The *agr* System

The *agr* system consists of an ~3 kb locus containing divergent transcription units, driven by promoters P2 and P3. The P2 operon encodes a TCS module and its autoinducing ligand (97). The *agr*-activating ligand is a post-translationally modified peptide (AIP) that binds to the N-terminal transmembrane domain of the *agr* signal receptor, *agrC* (63, 75, 79), activating the *agr* TCS module, of which AgrA is the response regulator. Activated AgrA binds to (66) and up-regulates promoters P2 and P3; this represents the primary function of the operon. The P3 transcript, RNAIII, rather than the response regulator, AgrA, is the intracellular effector of target gene regulation (60, 98). An RNA structure modifier, such as Hsa (Y. Fang et al., unpublished) or Hfq (F. Vandenesch, personal communication), is likely to act on RNAIII, affecting its function. *Agr* autoinduction results in a very rapid burst of activity once the autoinduction threshold has been reached. The AIP, seven to ten aminoacyl residues in length (62), is encoded by *agrD* as a propeptide and is processed, modified, and secreted by AgrB, a 26-kDa transmembrane protein (159, 160). The modification is a cyclic thiolactone bond between an internal cysteine and the C-terminal carboxyl (62, 85, 103). Structure-function studies of this peptide have shown that linear variants are inactive (85). Most changes in the primary amino acid sequence cause the peptide to become an inhibitor rather than an activator of the *agr* response. This is also the consequence of replacing the thiolactone bond with a lactone or an amide, indicating that the thiolactone bond is essential for *agr* activation.

Autoinduction of the P2 operon results in the concomitant activation of transcription from the *agr* P3 promoter, producing the 514 P3 transcript RNAIII, the effector of *agr*-specific target gene regulation (98). The P3 operon encodes one translated product, the 26-amino acid δ-hemolysin peptide, which seems not to have any role in the regulatory function of the *agr* locus (60, 98), though a nonsense mutation in codon 3 of δ-hemolysin results in the failure to activate transcription of *sspA*, the gene for serine protease (60). This may be a consequence of the effects on RNAIII secondary structure rather than any direct regulatory role for the δ-hemolysin peptide. RNAIII analogs encoded by other staphylococci vary considerably in sequence but are predicted to have similar overall secondary structure. Most, but not all, encode δ-hemolysin.

The primary regulatory function of RNAIII is at the level of transcription (98), mediated through regulatory proteins. RNAIII deletions (98), *trans*-specific RNAIIIs, and interspecies RNAIII hybrids (14, 136) have given different levels of expression for several exoproteins, suggesting that the molecule contains subdomains with separable functions, perhaps representing sites of interaction with different regulatory proteins.

TABLE 2 Regulatory genes and their roles

Regulatory unit	Role	Reference(s)	Strains with deficiencies
A. TCSs known to be involved in accessory gene regulation			
agrACDB/rnaIII	Regulates many accessory genes (encoding both extracellular and cytoplasmic proteins) in response to cell density.	94	None[a]
	Up-regulates secreted protein genes (e.g., hla, tst, sspA) and down-regulates surface protein genes (e.g., spa).		
	Up-regulates saePQ, arlRS and down-regulates srrAB		
saePQRS	Regulates many extracellular protein genes in response to different environmental signals and in response to agr RNAIII.	52	None
	Up-regulates coa and many secreted protein genes.	96	
arlRS	Regulates certain accessory genes and autolysis.	43, 44	n315 Δ16nt (frameshift)
	Down-regulates agrRNAII/RNAIII and up-regulates sarA.		
	Down-regulates spa, coa, and norA. Modulates responses to agents affecting DNA supercoiling.		
srrAB	Down-regulates agrRNAII/RNAIII, spa, and tst at low PO_2. (May provide link between respiratory metabolism, O_2 level, and regulation of virulence.)	156	None
B. Transcription factors known to be involved in accessory gene regulation			
rot	Pleiotropic regulator: affects accessory gene transcription in direction opposite that of agr.	88, 117	None
sarA	Pleiotropic repressor: represses transcription of spa and protease genes and sarT, sarS, and sarV. Affects agr RNAIII transcription under certain conditions.	28	None
sarR	Represses transcription of sarA.	81	None
sarS (sarH1)	Activates transcription of adjacent gene, spa.	135	None
tcaR	Activates transcription of spa through sarS and attenuates transcription of sasF, encoding a cell wall-associated protein.	87	NCTC8325 Δ73 AA 3′
sarT	Represses transcription of hla. Activates sarS and represses sarU transcription.	124	MRSA252 missing sarT and sarU
sarU	Activates transcription of agr RNAIII.	82	
mgrA (norR, rat)	Affects expression of many genes involved in virulence and autolysis. Activates cap5, cap8, and nuc and represses hla, spa, coa, and sspA transcription. Activates lytSR and arlRS and represses norA and sarV transcription.	77	None
sarV	Affects expression of genes involved in autolysis and virulence. Increases murein hydrolase activity.	83	None

[a]"None" indicates that the gene is conserved among the seven sequenced genomes.

Translational Regulation

RNAIII contains a series of nucleotides in its 5′ region complementary to the leader region of hla and another series of nucleotides in its 3′ region complementary to the 5′ region of spa. Deletion of the 5′ end, containing the sequences complementary to the α-hemolysin mRNA leader, abolished translation of α-hemolysin without affecting transcription of the hla gene (98), and deletion of the 3′ end, containing the sequences complementary to the spa mRNA leader, eliminated RNAIII-specific inhibition of Spa translation (unpublished data). The hla mRNA leader has a secondary structure that precludes translation; pairing with the complementary sequences in RNAIII prevents this inhibitory secondary structure from forming, thus permitting translation (90). In the secondary structure of RNAIII (13), however, the sequences that would unblock hla trans-

lation are paired and would be unavailable for complementary pairing with the hla leader. This pairing would also be expected to block α-hemolysin translation, and so we suspect that if RNAIII assumed this configuration, it would have to be unfolded in order to permit α-hemolysin translation as well as to interact with the hla leader. δ-hemolysin is not translated until about 1 h after the synthesis of RNAIII, consistent with a requirement for a posttranslational change in its conformation (7). Moreover, deletion of the 3′ half of RNAIII, which would eliminate the putative inhibitory intramolecular pairing, eliminated this delay. Presumably, such a change in conformation would also be required for the activation of δ-hemolysin translation by RNAIII. One possibility is translation of the 18 translatable codons 5′ to hld, the δ-hemolysin coding sequence. It has recently been observed that a small H-N analog, StaA, in

Escherichia coli affects RNA secondary structure and is synthesized for a short period during the exponential phase of growth (158). Such a protein could be responsible for modifying RNAIII conformation so as to permit its translation. In contrast to that of *hla*, translation of *spa* mRNA is blocked by RNAIII, owing to pairing between complementary regions of RNAIII and the translation initiation region of *spa* (94).

agr sets the level but not the timing of exoprotein gene expression. An independent timing signal is necessary for the up-regulation, at least in strains of the NCTC8325 lineage (143). *agr* activation entails a tremendous metabolic burden, resulting in frequent spontaneous *agr* mutants in the laboratory (18, 131), especially in strains that lack σ^B, which modulates the *agr* regulon (see below).

It has been proposed that ribosomal protein L2 (or RAP) is an alternative *agr* activator that is supposed to act early in growth. Unconfirmed reports claim that this protein is present in (postexponential) staphylococcal culture supernatants, although it lacks a signal peptide. The purified protein, in milligram quantities, has very weak (about twofold) *agr*-activating activity (67) and is therefore of very doubtful significance. RAP is supposed to act via TRAP, a cytoplasmic protein that is stably histidine phosphorylated in a growth-dependent manner and is partially required for *agr* activation (6). In a TRAP-defective mutant of a standard *agr*⁺ strain, RNAIII production is delayed and decreased (6). Because the AIP has no effect on *rnaIII* transcription in the absence of the *agr* signaling pathway (79), TRAP would have to be acting within that pathway.

Agr Specificity Groups

agr is conserved throughout the staphylococci with interesting variations in the *B-D-C* region (see Fig. 3A). These variations have resulted in at least four *agr* specificity groups in *S. aureus* and one or more in each of 15 other staphylococcal species examined (38, 61, 62, 104). The groups are defined by the mutual inhibition by their peptides of the *agr* response in heterologous pairings, resulting in a novel type of bacterial interference in which the *agr* regulon, rather than growth, is blocked (62). The ability of an AIP to activate its cognate receptor is highly sequence-specific; a single amino acid substitution can change group specificity (see AIPs I and IV in Fig. 4).

The divergent regions determine group specificity and must therefore have evolved in concert. Functional variants within the *agr* locus are designed for cross-group and, presumably, cross-species interference rather than cooperative communication, so that they serve to isolate populations and may represent a major determinant of strain and species divergence. *agr* groupings are broadly correlated with strain genotypes as defined by multilocus sequence typing and by amplicon fragment length polymorphisms (61), resulting in general congruence between phylogenetic trees and *agr* groups (61). These genotypes are correlated with pathotypes: most menstrual toxic shock syndrome strains belong to *agr* group III (62), as do most strains implicated in leukocidin-induced necrotizing pneumonia (49); many of the clinical isolates showing intermediate vancomycin resistance are *agr*-defectives, belonging to *agr* group II (118), and most exfoliatin-producing strains belong to *agr* group IV (61).

Although *agr* mutants are greatly attenuated, they are not avirulent; *agr*, *spa* double mutants, however, are entirely avirulent in the mouse mammary infection model (41), which is particularly noteworthy because protein A is overproduced in *agr* mutants and may compensate in some unknown way for the absence of the other factors. Microarray transcriptional profiling (39) has revealed that a number of genes encoding catabolic enzymes, such as proteases, nucleases, and lipases, belong to the global *agr* regulon. These can be regarded as nutritional factors as well as virulence factors, and may represent an overlap between the environmental and pathogenicity gene subsets. The coordinate postexponential expression of similar catabolic enzymes by soil organisms such as *Bacillus subtilis* could indicate that a catabolic enzyme regulon has been borrowed from soil bacteria by gram-positive pathogens for incorporation into their virulence regulons. *agr* also appears to down-regulate one autolysin (45) and two or more penicillin-binding proteins (106). However, the significance of this for virulence is unknown.

The *saeRS* System

saeRS (Fig. 5) represents the second major TCS module involved in global regulation of the staphylococcal virulon, controlling the production of a subset of exoproteins that partially overlaps with the *agr*-regulated set (50); in particular, it strongly up-regulates *nuc* and *coa* transcription (52). *sae* has a complex transcriptional pattern that is both autoregulated and is profoundly influenced by *agr* and by certain environmental stimuli. *sae* transcription in vitro (96) (Fig. 5) is initiated by a 2 kb transcript that disappears postexponentially. Two larger transcripts appear in midexponential phase, immediately after the onset of RNAIII synthesis. In the presence of glucose, the two longer transcripts disappear as soon as the pH falls below 6, suggesting that the traditional glucose inhibition of exoprotein synthesis is largely due to the reduction in pH that generally accompanies glucose metabolism. The longer transcripts are not seen in the *saeR* mutant or during growth at pH 5.5, and are greatly reduced in an *agr*-null strain, in a *sarA* mutant, and in the presence of 1 M NaCl or a subinhibitory concentration of clindamycin (96). As these environmental cues act independently of *agr*, *sae* probably has a major role in the integration of cell density signaling with signaling through environmental stimuli and other regulatory elements.

arlRS

arlRS is a third TCS module involved in regulation of the staphylococcal virulon (44) and, like *sae*, interacts mutually with *agr*. *agr* mutants are defective in *arlRS* expression, whereas an *arlS* mutant apparently overexpresses *agr*, especially the *agr* P2 transcript, consistent with independent regulation of P2 and P3 (44). Agr and *arlRS* thus formally represent an autorepression circuit such that *arlRS* counters *agr* autoinduction. Consistent with this is the reported down-regulation by *arlRS* of overall exoprotein synthesis, presumably consequent to down-regulation of *agr*. However, *arlRS* appears to act independently of *agr* with respect to protein A production.

srrAB (srhSR)

srrAB is the fourth TCS module evidently involved in expression of the staphylococcal virulon, especially under microaerobic condition (156) (Fig. 3B). *srrAB* is also known as *srhSR* (137); it is a homolog of the O_2-responsive *resDE* system of *B. subtilis* (16). The *srrAB/srhSR* mutants are profoundly growth defective in the absence of oxygen, although they grow normally under aerobic conditions. This TCS module appears to inhibit *agr* activation (156) and is itself down-regulated by *agr* (unpublished data; P. Schlievert,

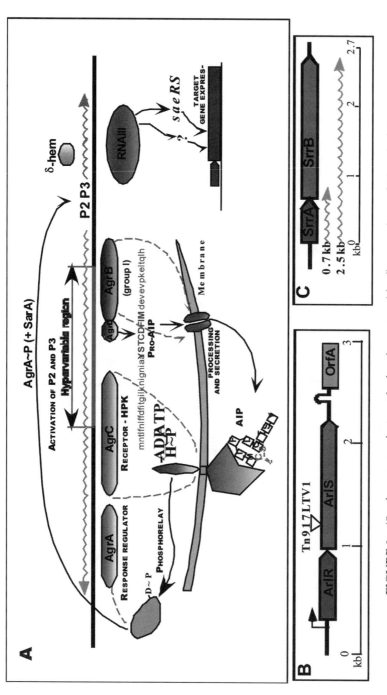

FIGURE 3 (*See the separate color insert for the color version of this illustration.*) *agr, srhRS, and arlAB. TCS modules known to affect the virulon.* (A) The *agr* system. The pro-AIP peptide is processed and secreted by AgrB, binds to an extracellular loop in the receptor-HPK, AgrC, activating autophosphorylation (or dephosphorylation), followed by phosphorylation or dephosphorylation of the response regulator, AgrA. AgrA, in conjunction with SarA, activates the two *agr* promoters, P2 and P3, leading to the production of RNAIII. RNAIII controls transcription of the target genes via one or more intracellular regulatory mediators, including a second two-component module, *saeRS*. (B) *arlRS*. The *arlRS* locus encodes a receptor-HPK (*arlS*) and a response regulator (*arlR*), driven by a single promoter and followed by a terminator stem-loop. (C) *srrAB*. The *srrAB* locus encodes a receptor-HPK (*srrB*) and a response regulator (*srrA*), driven by a single promoter that generates two transcripts whose relative significance is unknown. (Reprinted from reference 94 with the kind permission of Blackwell Publishing, Ltd.)

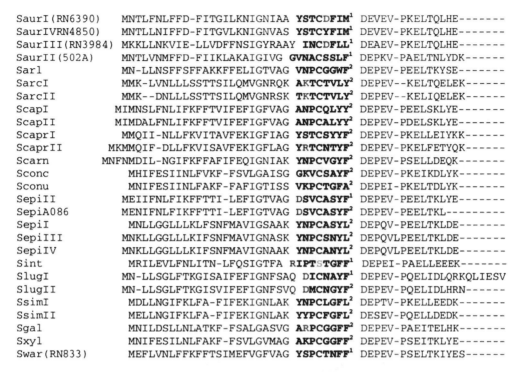

```
SaurI(RN6390)      MNTLFNLFFD-FITGILKNIGNIAA YSTCDFIM¹ DEVEV-PKELTQLHE-------
SaurIVRN4850)      MNTLLNIFFD-FITGVLKNIGNVAS YSTCYFIM¹ DEVEV-PKELTQLHE-------
SaurIII(RN3984)    MKKLLNKVIE-LLVDFFNSIGYRAAY INCDFLL¹ DEAEV-PKELTQLHE-------
SaurII(502A)       MNTLVNMFFD-FIIKLAKAIGVG GVNACSSLF¹ DEPKV-PAELTNLYDK------
Sar1               MN-LLNSFFSFFAKKFFELIGTVAG VNPCGGWF² DEPEV-PEELTKYSE-------
SarcI              MMK-LVNLLLSSTTSILQMVGNRQK AKTCTVLY² DEPEV--KELTQELEK------
SarcII             MMK--DNLLLSSTTSILQMVGNRSK TKTCTVLY² DEPEV--KELIQELEK------
ScapI              MIMNSLFNLIFKFFTVIFEFIGFVAG ANPCQLYY² DEPEV-PEELSKLYE-------
ScapII             MIMDALFNLIFKFFTVIFEFIGFVAG ANPCALYY² DEPEV-PDELSKLYE-------
ScaprI             MMQII-NLLFKVITAVFEKIGFIAG YSTCSYYF² DEPEV-PKELLEIYKK------
ScaprII            MKMMQIF-DLLFKVISAVFEKIGFLAG YRTCNTYF² DEPEV-PKELFETYQK------
Scarn              MNFNMDIL-NGIFKFFAFIFEQIGNIAK YNPCVGYF² DEPEV-PSELLDEQK-------
Sconc                 MHIFESIINLFVKF-FSVLGAISG GKVCSAYF² DEPEV-PKEIKDLYK-------
Sconu                 MNIFESIINLFAKF-FAFIGTISS VKPCTGFA² DEPEI-PKELTDLYK-------
SepiII             MEIIFNLFIKFFTTI-LEFIGTVAG DSVCASYF¹ DEPEV-PEELTKLYE-------
SepiA086           MENIFNLFIKFFTTI-LEFIGTVAG DSVCASYF² DEPEV-PEELTKL---------
SepiI              MNLLGGLLLKLFSNFMAVIGSAAK YNPCASYL² DEPQV-PEELTKLDE-------
SepiIII            MNKLLGGLLLKIFSNFMAVIGNASK YNPCSNYL² DEPQVLPEELTKLDE-------
SepiIV             MNKLLGGLLLKIFSNFMAVIGNAAK YNPCANYL² DEPQVLPEELTKLDE-------
Sint               MRILEVLFNLITN-LFQSIGTFA RIPTSTGFF¹ DEPEI-PAELLEEEK-------
SlugI              MN-LLSGLFTKGISAIFEFIGNFSAQ DICNAYF¹ DEPEV-PQELIDLQRKQLIESV
SlugII             MN-LLSGLFTKGISVIFEFIGNFSVQ DMCNGYF² DEPEV-PQELIDLHRN------
SsimI              MDLLNGIFKLFA-FIFEKIGNLAK YNPCLGFL¹ DEPTV-PKELLEEDK-------
SsimII             MELLNGIFKLFA-FIFEKIGNLAK YYPCFGFL¹ DESEV-PQELLDEDK-------
Sgal               MNILDSLLNLATKF-FSALGASVG ARPCGGFF² DEPEV-PAEITELHK-------
Sxyl               MNIFESILNLFAKF-FSVLGVMAG AKPCGGFF² DEPEV-PSEITKLYE-------
Swar(RN833)        MEFLVNLFFKFFTSIMEFVGFVAG YSPCTNFF¹ DEPEV-PSELTKIYES------
```

FIGURE 4 *agrD* peptides from various staphylococcal species. Sequences were aligned visually. Predicted AIPs are in bold and are set between spaces. Saur, *S. aureus*; Sarl, *Staphylococcus arlettae*; Sarc, *Staphylococcus auricularis*; Scap, *Staphylococcus capitis*; Scapr, *Staphylococcus caprae*; Scarn, *Staphylococcus carnosus*; Sconc, *Staphylococcus cohnii cohnii*; Sconu, *Staphylococcus cohneii urealyticus*; Sepi, *Staphylococcus epidermidis*; Sint, *Staphylococcus intermedius*; Slug, *Staphylococcus lugdunensis*; Ssim, *Staphylococcus simulans*; Sgal, *Staphylococcus gallinarum*; Sxyl, *Staphylococcus xylosus*; Swar, *Staphylococcus warneri*. [1]Sequence confirmed by in vitro synthesis or mass spectroscopy. [2]Peptide sequence predicted from nucleotide sequence. (Reprinted from reference 94 with the kind permission of Blackwell Publishing, Ltd.)

personal communication). *agr* and *srrAB/srhRS* thus represent a mutual cross-inhibition circuit that would necessarily have to respond to extrinsic regulatory inputs. *srrAB/srhRS* also regulates many genes involved in energy metabolism and evidently regulates energy transduction under anaerobic conditions (137). It may be activated by menaquinone or a derivative, one of the intermediates in the oxidative respiratory pathway. Thus, this TCS module may connect the *agr* signaling pathway with the overall energy metabolism of the cell. It appears that each of these three TCS modules exerts its effects on the virulon and other accessory genes largely, although not entirely, through its interaction with *agr*. Thus, *saeRS* and *agr* appear to be mutually up-regulatory; *srrAB* and *agr* are reported to be mutually down-regulatory; and *arlRS* and *agr* seem to constitute a mutual autorepression circuit. These different interactions, probably indirect, presumably represent a central regulatory logic, but one whose biology remains to be determined.

Transcription Factors

In general, transmission of environmental signals recognized by transmembrane and intracellular receptors to effector (or target) genes involves pleiotropic intracellular transcription factors, including the Sar (staphylococcal accessory regulation) family of homologous winged helix-turn-helix DNA binding proteins, namely SarA, R, S, T, U, V, Rot, TcaR (87), and MgrA (78). These regulatory factors, listed in Table 2 (recently reviewed by Arvidson and Tegmark [2] and by Cheung and Zhang [29]), affect a wide variety of genes, most of which encode virulence or other accessory functions. They interact with one another, with the TCS modules, and with σ^B, as well as with the target genes themselves, generating an extraordinarily complex regulatory network. Their general properties are summarized in Table 2. The DNA-binding segments of these proteins are well conserved, many containing the motif KXRXXXDER, whereas other parts of the proteins are less well conserved (see Fig. 6). The prototype, SarA, is a 14.7-kDa DNA-binding protein, distantly related to VirF of *Shigella flexneri*. SarA binds as a dimer, whereas at least three of its homologs, SarS, SarU and SarY, which bind as monomers, appear to represent duplications and are therefore intrinsically dimeric. Given the degree of structural and sequence similarity among the members of this family (29, 87), the possibility of heterodimeric combinations has been suggested (135). They appear to belong to the group of regulatory proteins that bind DNA with limited sequence specificity (e.g., H-NS and HU). The binding sites for these proteins are very AT-rich, and have been difficult to identify with any great degree of specificity, given the overall AT-richness of the staphylococcal genome. Because it is likely that more than one can bind to a given site, an in vitro functional assay would be required for any definitive assignment.

SAgs

Remarkably, at least two of the major staphylococcal SAg toxins, TSST-1 and SEB, are themselves transcription factors,

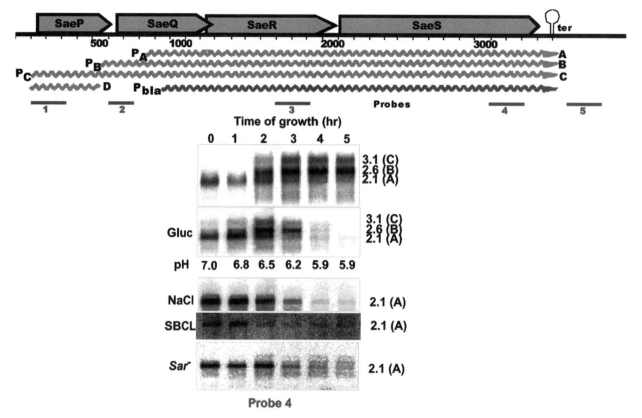

FIGURE 5 The *saeRS* system. (A) The *sae* locus, about 3.5 kb, contains four open reading frames, P, Q, R, and S. R and S form a classical TCS module. The functions of P and Q are unknown. *Sae* is transcribed from two or three promoters, one of which is active in an *agr*-null strain and the other(s) is activated by RNAIII. All three major transcripts, A, B, and C, end at ter. D may be independently transcribed or derived from C by processing. PCR probes used to map the transcripts are shown. (B) Transcription pattern (see text). (Reprinted from reference 96.)

acting as global repressors of most exoprotein genes at the level of transcription, and are also autorepressors (144). These two proteins, as well as the other staphylococcal and streptococcal SAgs, are structurally very closely related, as shown by X-ray crystallography (100). However, there is no striking sequence similarity corresponding to the inhibitory regions of the two proteins. The data establish clearly that the protein itself, rather than the mRNA or the DNA, is the inhibitor, and that a region ending at about the middle of the C-terminal half of the protein is necessary (144). As the purified toxin has no effect when added to a culture, and a deletion derivative lacking the signal peptide retains its inhibitory activity (N. Vojtov, H. Ross, and R. P. Novick, unpublished data), it is clear that an intracellular form, most likely the precursor, is the effector. It appears that neither protein binds DNA directly, presumably acting through an intermediate transcription factor. As the inhibitory effects are manifest as early as one can detect the toxins, the putative intermediary must also be present at this early time. It remains unclear precisely how this regulatory paradigm fits into the overall regulatory network; however, it is evidently of major clinical importance: in postsurgical toxic shock syndrome, resulting from a contaminated wound, the infection is often very difficult to detect, as, unlike the typical staphylococcal lesion, the wound is neither purulent nor inflamed. It is well known that TSST-1 has major effects on the production of cytokines and very prob-

ably influences the inflammatory response by this means. It is additionally hypothesized that one or more of the exoproteins, possibly lipase, the synthesis of which is inhibited by the SAg, is responsible for attracting polymorphonuclear leukocytes and stimulating the inflammatory response. It is predicted that other variable genes encoding toxins that cause toxinoses are likely to act in this manner.

Alternative Sigma Factors

A major mechanism of response to environmental stimuli is via alternative sigma factors, which are generally activated directly within the cell rather than through signal transduction. *S. aureus* possesses two of these, homologs of *B. subtilis* σ^B (35, 152) and σ^H (91), respectively. σ^B is required for the expression of one of the σ^B transcripts (152) and of genes involved in pigment synthesis, defense against oxidative stress, and other functions. The σ^B regulon, recently identified by microarray analysis (17), includes genes putatively involved in cell envelope biosynthesis and turnover, intermediary metabolism, signaling pathways, and virulence. σ^B up-regulates 198 genes and down-regulates 53 (17). Many adhesin genes are up-regulated while exoprotein and toxin genes are repressed. σ^B is activated by environmental stress and energy depletion (reduced ATP/ADP ratio), as well as by environmental stimuli such as ethanol (25) and salicylic acid (9), and its activity is regulated by a complex post-translational pathway (125). σ^B recognizes a unique promoter

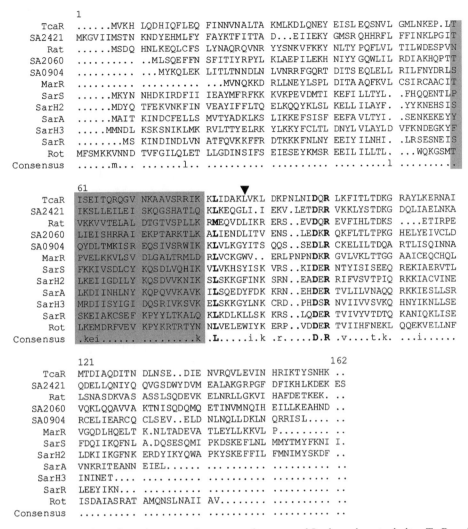

```
          1
   TcaR   ......MVKH LQDHIQFLEQ FINNVNALTA KMLKDLQNEY EISL EQSNVL GMLNKEP.LT
 SA2421   MKGVIIMSTN KNDYEHMLFY FAYKTFITTA D...EIIEKY GMSR QHHRFL FFINKLPGIT
    Rat   ......MSDQ HNLKEQLCFS LYNAQRQVNR YYSNKVFKKY NLTY PQFLVL TILWDESPVN
 SA2060   .......... ..MLSQEFFN SFITIYRPYL KLAEPILEKH NIYY GQWLIL RDIAKHQPTT
 SA0904   .......... ...MYKQLEK LITLTNNDLN LVNRRFGQRT DITS EQLELL RILFNYDRLS
   MarR   .......... .......... ...MVNQKKD RLLNEYLSPL DITA AQFKVL CSIRCAACIT
   SarS   ......MKYN NHDKIRDFII IEAYMFRFKK KVKPEVDMTI KEFI LLTYL. .FHQQENTLP
  SarH2   ......MDYQ TFEKVNKFIN VEAYIFFLTQ ELKQQYKLSL KELL ILAYF. .YYKNEHSIS
   SarA   ......MAIT KINDCFELLS MVTYADKLKS LIKKEFSISF EEFA VLTYI. .SENKEKEYY
  SarH3   ....MMNDL KSKSNIKLMK RVLTTYELRK YLKKYFCLTL DNYL VLAYLD VFKNDEGKYF
   SarR   ......MS KINDINDLVN ATFQVKKFFR DTKKKFNLNY EEIY ILNHI. .LRSESNEIS
    Rot   MFSMKKVNND TVFGILQLET LLGDINSIFS EIESEYKMSR EEIL ILLTL. ...WQKGSMT
Consensus ......m... .......l.. .......... .......... ........l .........
```

```
          61                              ▼
   TcaR   ISEITQRQGV NKAAVSRRIK KLIDAKLVKL DKPNLNIDQR LKFITLTDKG RAYLKERNAI
 SA2421   IKSLLEILEI SKQGSHATLQ KLKEQGLI.I EKV.LETDRR VKKLYSTDKG DQLIAELNKA
    Rat   VKKVVTELAL DTGTVSPLLK RMEQVDLIKR ERS..EVDQR EVFIHLTDKS ....ETIRPE
 SA2060   LIEISHRRAI EKPTARKTLK ALIENDLITV ENS..LEDKR QKFLTLTPKG HELYEIVCLD
 SA0904   QYDLTMKISR EQSIVSRWIK KLVLKGYITS QQS..SEDLR CKELILTDQA RTLISQINNA
   MarR   PVELKKVLSV DLGALTRMLD RLVCKGWV.. ERLPNPNDKR GVLVKLTTGG AAICEQCHQL
   SarS   FKKIVSDLCY KQSDLVQHIK VLVKHSYISK VRS..KIDER NTYISISEEQ REKIAERVTL
  SarH2   LKEIIGDILY KQSDVVKNIK SLSKKGFINK SRN..EADER RIFVSVTPIQ RKKIACVINE
   SarA   LKDIINHLNY KQPQVVKAVK ILSQEDYFDK KRN..EHDER TVLILVNAQQ RKKIESLLSR
  SarH3   MRDIISYIGI DQSRIVKSVK ELSKKGYLNK CRD..PHDSR NVIIVVSVKQ HNYIKNLLSE
   SarR   SKEIAKCSEF KPYYLTKALQ KLKDLKLLSK KRS..LQDER TVIVYVTDTQ KANIQKLISE
    Rot   LKEMDRFVEV KPYKRTRTYN NLVELEWIYK ERP..VDDER TVIIHFNEKL QQEKVELLNF
Consensus .kei...... ..........k .L.....i.k .r.....D.R .v....t.k. ...i......
```

```
          121                             162
   TcaR   MTDIAQDITN DLNSE..DIE NVRQVLEVIN HRIKTYSNHK ..
 SA2421   QDELLQNIYQ QVGSDWYDVM EALAKGRPGF DFIKHLKDEK ES
    Rat   LSNASDKVAS ASSLSQDEVK ELNRLLGKVI HAFDETKEK. ..
 SA2060   VQKLQQAVVA KTNISQDQMQ ETINVMNQIH EILLKEAHND ..
 SA0904   RCELIEARCQ CLSEV..ELD NLNQLLDKLN QRRISL.... ..
   MarR   VGQDLHQELT K.NLTADEVA TLEYLLKKVL P......... ..
   SarS   FDQIIKQFNL A.DQSESQMI PKDSKEFLNL MMYTMYFKNI I.
  SarH2   LDKIIKGFNK ERDYIKYQWA PKYSKEFFIL FMNIMYSKDF ..
   SarA   VNKRITEANN EIEL...... .......... .......... ..
  SarH3   ININET.... .......... .......... .......... ..
   SarR   LEEYIKN... .......... .......... .......... ..
    Rot   ISDAIASRAT AMQNSLNAII AV........ .......... ..
Consensus .......... .......... .......... .......... ..
```

FIGURE 6 Sar homologs. Amino acid sequence alignment of Sar homologs, including TcaR and MgaA, against MarR from *E. coli* and other homologous protein sequences identified from the *S. aureus* N315 genome by BLASTP analysis. The sequences of SarS (SarH1) and SarH2, which both contain two domains with homology to SarA, were truncated so that only the N-terminal domain of each was included in the alignment. The region containing the predicted helix-turn-helix motif of TcaR (Network Protein Sequence analysis [87]) and several of the other homologs is highlighted in gray. Strongly conserved residues are indicated in the consensus line and universally conserved residues are in bold type. The arrowhead represents the position (amino acid 79) at which the TcaR protein in NCTC8325-4 is truncated. (Reprinted from reference 87.)

GTTT(N14-17)GGGTAT), which has been identified for 23 different *S. aureus* genes (48), including *coa* (93) plus one of the three *sarA* (10), and one of the three *sarS* promoters (135), plus genes encoding transport functions and genes involved in generating NADH$_2$. σ^B is also required for certain genes that lack a σ^B promoter; these are presumably regulated by σ^B-dependent transcription factors. In *S. aureus*, σ^B feeds into the global regulatory network governing the expression of accessory genes, acting mostly through other regulatory genes and transcription factors, but also acting directly on those few that have σ^B-dependent promoters. Thus, it has reciprocal activities, up-regulating some exoprotein genes at a very early stage of growth, such as *coa* and *fnbB*, of which the latter does not have a σ^B-dependent promoter, and down-regulating others at the end of the exponential phase

in vitro. As many of the latter are involved in virulence, σ^B seems to be antagonistic to *agr*. A recent report suggests that σ^B is important for pathogenesis (64), though several earlier studies suggest that it is not (26, 69, 93). These latter studies, however, were done with σ^B-knockouts in strains of the NCTC8325 lineage that are already deficient in σ^B function, having a deletion in *rsbU*, a gene required for the activation of σ^B (68). Additionally, because σ^B is required for the expression of *fnbP*, an adhesin that is important for internalization, internalization is significantly reduced in a σ^B mutant (92).

σ^H does not seem to be involved in virulence but is required for the expression of several genes that are closely related to certain competence genes of *B. subtilis* (91). It is not known whether these σ^H-dependent genes are involved in the (very inefficient) *S. aureus* competence system.

Interstrain Variations

There are, naturally, major differences among strains with respect to those accessory genes that are encoded by variable genetic elements. These may occur in any imaginable combination, and may, of course, profoundly affect pathogenicity, tissue tropism, etc. In general, there is no known biological rationale for the assortment of these genes and elements. More important for present purposes is interstrain variation in the regulatory genes that control the expression of the virulon and its components (19). In addition to the above-mentioned variation in the σ^B activation pathway, characteristic of all strains of the NCTC8325 lineage, defects in several other important regulatory genes have been identified among the seven sequenced *S. aureus* genomes, as indicated in Table 2. Thus NCTC8325 strains have a stop codon in TcaR, a protein that is important for the expression of *sarS* (87), which probably accounts for the weak expression of *spa* in these strains. MRSA252 is missing *sarT* and *sarU*, which is predicted to reduce expression of the *agr* system; n315 has a 16 nucleotide gap in the C-terminal region of ArlR, causing a frameshift and almost certainly inactivating the protein. These sequence features can account for a significant portion, but are very unlikely to account for all, of the observed (and as yet unobserved) interstrain variation.

Regulatory Organization

The entire accessory gene regulatory network must be coupled to the overall energy metabolism of the cell, and it has been suggested that there must be a key coupling parameter, such as the levels of nucleotide polyphosphates or of other energy-transducing cofactors such as $NADH_2$ (R. Proctor, personal communication). Key enzymes of intermediary metabolism, such as aconitase (132), could be involved in this coupling, possibly acting through one or more TCS modules, such as *srrAB* (*srhRS*).

Environmental factors that affect expression of various components of the virulon would not be related to timing signals but would exert their effects whenever they are encountered. Certain of these (pH, O_2 tension, CO_2 concentration) would typically vary during growth in laboratory cultures and would be expected to have increased importance late in growth. Additionally, 1 M NaCl blocks RNAIII transcription (S. Herbert and R. P. Novick, unpublished data) and also blocks transcription of many *agr*-up-regulated genes such as *hla* and *sspA* (25), as well as of *agr*-down-regulated genes such as *spa* and *fnb* (25). These effects appear to be independent of RNAIII expression. Subinhibitory concentrations of antibiotics that inhibit protein synthesis (47), surfactants such as glycerol monolaurate (109), lowering of pH or O_2 tension, and limitation of essential amino acids (73) have a general inhibitory effect on exoprotein gene expression. The inhibitory effects of these environmental factors, many of which act independently of *agr*, suggest that the cell has a specific regulatory modality that selectively switches off expression of accessory genes, such as virulence factors and other exoproteins, when the organism is stressed by inhibition of protein synthesis. Consistent with this is the observation that respiration-defective mutants (small-colony variants) also show a global reduction in exoprotein synthesis (see chapter 35, this volume).

Divalent cations (25) (S. Arvidson, personal communication; S. Foster, personal communication) have also been observed to affect the expression of many of these same virulence genes (Table 1), and β-lactam antibiotics seem to

have an effect opposite to that of the protein synthesis inhibitors, namely, stimulating expression of some of the exoprotein genes, possibly owing to cross-talk between signaling systems (see, for example, references 65 and 101).

Interactions of Transcription Factors with Other Regulatory Elements

Figure 7 shows a hypothetical scheme outlining the regulatory circuitry involving SarA and its homologs. Intermediary transcription factors (see above) have been shown to affect the transcription of accessory genes, of one another, and of the TCS modules (2, 82, 123, 124), generating complex activation cascades and feedback loops and defining subsets of accessory genes. Presumably, RNAIII interacts directly or indirectly with these factors, by regulating either their synthesis or their action. For example, a recent microarray analysis has revealed that Rot has broad regulatory effects on genes belonging to the *agr* regulon and generally acts counter to *agr*, down-regulating genes encoding secreted proteins and up-regulating surface protein genes (117). In order to account for the observed effects of *agr*, one must postulate that *agr* down-regulates Rot. Because *rot* is constitutively transcribed (88), it is likely that any such down-regulation would be posttranscriptional. In the case of SarS (see Fig. 7), it is known that RNAIII blocks expression (135), perhaps through Rot. It has also been observed that SarT down-regulates *agr* (124), apparently acting via SarU (82), an *agr* up-regulator. The down-regulation of *sarT* by *agr* thus generates a negative feedback loop, as shown in Fig. 7. Additional data such as those obtained by transcript profiling (39, 117) will fill out this circuitry, perhaps defining regulatory subsets of target genes that could have biological/clinical relevance.

Expression of the Virulon in Biofilms and in the Intracellular and In Vivo Environments

Biofilm formation is increasingly recognized as an important feature of several types of staphylococcal infection, especially cystic fibrosis, osteomyelitis, endocarditis, and foreign body infections. The special environment of the biofilm has major effects on the regulation and expression of accessory genes, including virulence factors (12). *agr* expression oscillates during the development and growth of biofilm and is poorly expressed in the internal regions of the biofilm, possibly related to increased expression of adhesins (145), but is well expressed in the outer regions, which may be responsible for detachment of the organism and spread to new sites (155). Consistent with this is a deficiency in biofilm formation in SarA-defective mutants (139), possibly related to the up-regulation of adhesin determinants by SarA.

As noted above, there is increasing evidence that *S. aureus* can invade and persist in endothelial and epithelial cells (see chapter 42, this volume). A key feature of invasion is the interaction of fibronectin-binding proteins with fibronectin and with α5β1 integrin (128). Thus, *agr*-defective mutants, which overproduce FnbB, have an enhanced ability to invade mammalian cells, but do not escape the endosome and are defective for the induction of apoptosis (148). Although *agr* is expressed intracellularly (110, 127)—probably owing to accumulation of the *agr* AIP at an activating concentration (though RNAIII activation in vivo in the absence of the *agr* TCS module has been reported [138])—and the bacteria escape from the endosome and cause apoptosis, the role of specific *agr*-regulated factors remains to be determined, as does the clinical significance of intracellular growth.

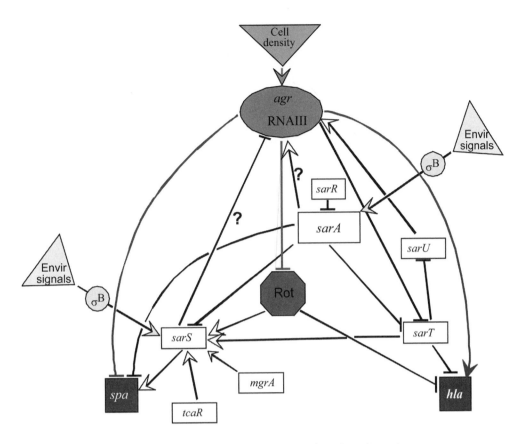

FIGURE 7 Regulatory interactions involving SarA and its homologs. Arrows represent up-regulation, bars represent down-regulation. The two outermost curved lines represent translation; the other lines represent interactions that are probably, but not always certainly, transcriptional. The interactions illustrated are based on reviews by Arvidson and Tegmark (2) and Cheung and Zhang (29) and on recent papers by Manna and Cheung (82) and Said-Salim et al. (117). Although the abbreviations are mostly in italics, on the assumption that the interactions are likely to be at the transcriptional level, there is actually very little evidence to indicate whether they are direct or indirect or at what level they occur. Question marks represent the most speculative. σ^B is shown entering the system via *sarS* and *sarA*, which have σ^B-dependent promoters and are likely to represent important intermediates in the pathways by which environmental signals are handled. (Reprinted from reference 94.)

Temporal Program

Much of the facultative gene expression system, especially including the virulon, is temporally organized so that the component genes must contain regulatory sequences that are activated combinatorially in a time-dependent manner by different incoming signals acting through intracellular response elements. The key question of how this well-defined temporal program (see above and Fig. 2) applies to the in vivo environment has recently been addressed by several laboratories. A variety of methods have been used, some involving reporter gene transcriptional fusions, and others measuring mRNAs by reverse transcriptase PCR, or gene products immunologically. Additionally, several different models have been used and a number of different strains have been examined. Gene expression is generally measured in bacteria recovered from the lesion at a single time point. Although these "snapshot" studies (see Table 3) comparing regulatory mutants with wild type do not address the temporal program, they usually show that in vivo and in vitro activities are correlated. In one time course study, using *gfp* fusions in an NCTC8325 derivative in the rabbit endocarditis model, the *agr* P2 promoter was activated before the

P3 promoter. Activation was maximal by 6 h (154) and was largely unchanged thereafter. In another study, using either an NCTC8325 or SA502A (RN6607) and an *agrp3-lux* fusion, in the mouse subcutaneous abscess model, there was a peak of activity at 3 h then a sharp decline in activity and a neutrophil-induced eclipse lasting 24 to 36 h, followed by reactivation concomitant with the presence of necrotic tissue in the mature abscess (see Fig. 8). A similar rapid rise and decline of *agrp3* activity was observed following injection of bacteria into the mouse quadricep muscle (J. Jin and R. P. Novick, unpublished data). In the subcutaneous abscess model, administration of an *agr*-inhibiting peptide delayed but did not block activation of the P3 promoter, although it blocked abscess formation. It appears that in this model, the bacteria must produce their toxic extracellular proteins within the first 3 h. In support of this, a similar abscess was produced by a sterile supernatant from an *agr*⁺ srain but not from an *agr*⁻ (151). Analysis by reverse transcriptase PCR of bacteria isolated from an indwelling foreign body (55) showed decreasing expression of RNAIII and *hla* over time. Similarly, Goerke et al. (54) also observed weak expression of the *agrp3* promoter in staphylococci

TABLE 3 Virulon expression in vivo

Strain	Mutation	Model	Method	Timing	Genes	Organs	Reference
8325-4	saeR::Tn vs. wt	Sheep peritoneum	Western	24 h snapshot	hla, hlb, coa, nuc, spa, all↓		112
Newman	agr	Rabbit endocarditis	Gfp fusion[a]	8 h 48 h	fnbA↓	Valve	153
Newman	sarA	Rabbit endocarditis	Gfp fusion	8 h 48 h	fnbA↑ ↑	Valve	
Newman	agr, sarA	Rabbit endocarditis	Gfp fusion	8 h 48 h	fnbA↑ ↑	Valve	
Newman	agr	Rabbit endocarditis	Gfp fusion	48 h	cap5↓	Valve	141
Uncharacterized CF iolates	wt	CF tissue, rat granuloma pouch	ELISA[d]	Snapshot	cap5↓[c]	Lung	57
RN6390[b]	wt	Rabbit endocarditis	Gfp fusion	0–48 h time course	RNAII RNAIII (see text)	Valve	154
Newman	wt	Rabbit endocarditis	Gfp fusion	6 h vs. 48 h	RNAII↑ RNAIII↑	Valve	154
RN6390[b] and Newman	agr sarA sae	Guinea pig, foreign body	RT PCR	144 h	hla→ hla→ hla↓	Implanted plastic tube	55

[a]Scored by flow cytometry as percent gfp-positive bacteria.
[b]Derivative of NCTC 8325.
[c]Correlated with down-regulation of cap5 by CO_2 in vitro (57).
[d]Enzyme-linked immunosorbent assay.

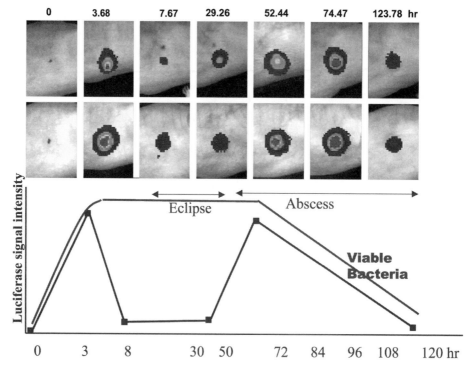

FIGURE 8 Temporal expression of agr in vivo. Agr[+] bacteria with plasmid-carried agr-P3::lux fusion. 1.5×10^7 organisms in early exponential phase injected subcutaneously with cytodex beads at time T1, three mice. Imaged with IVIS (Xenogen) system at times indicated. Images are in false color with increasing intensity from dark gray to light gray to white to medium gray (grayscale representations of blue to green to yellow to red). Signal intensity is plotted below (■) along with bacterial counts obtained by sacrificing infected mice, excising and homogenizing the lesion, and plating for viable bacteria.

isolated from cystic fibrosis sputa. Overall, results of in vivo expression studies to date suggest that the staphylococcal virulon is activated very early during infection and subsequently becomes much less active, probably representing much reduced metabolic activity of the bacteria as time goes on. This general behavior is correlated with *agr* activity; time-dependent expression of other regulatory genes remains to be measured.

REFERENCES

1. **Abdelinour, A., S. Arvidson, T. Bremell, C. Ryden, and A. Tarkowski.** 1993. The accessory gene regulator (*agr*) controls *Staphylococcus aureus* virulence in a murine arthritis model. *Infect. Immun.* **61:**3879–3885.

2. **Arvidson, S., and K. Tegmark.** 2001. Regulation of virulence determinants in *Staphylococcus aureus*. *Int. J. Med. Microbiol.* **291:**159–170.

3. **Arvidson, S. O.** 1983. Extracellular enzymes from *Staphylococcus aureus*, p. 745–808. *In Staphylococci and Staphyloccocal Infections*, vol. 2. Academic Press, London, United Kingdom.

4. **Axelsson, L., and A. Holck.** 1995. The genes involved in production of and immunity to sakacin A, a bacteriocin from *Lactobacillus sake* Lb706. *J. Bacteriol.* **177:**2125–2137.

5. **Ba-Thein, W., M. Lyristis, K. Ohtani, I. T. Nisbet, H. Hayashi, J. I. Rood, and T. Shimizu.** 1996. The virR/virS locus regulates the transcription of genes encoding extracellular toxin production in *Clostridium perfringens*. *J. Bacteriol.* **178:**2514–20.

6. **Balaban, N., T. Goldkorn, Y. Gov, M. Hirshberg, N. Koyfman, H. R. Matthews, R. T. Nhan, B. Singh, and O. Uziel.** 2001. Regulation of *Staphylococcus aureus* pathogenesis via target of RNAIII-activating protein (TRAP). *J. Biol. Chem.* **276:**2658–2667.

7. **Balaban, N., and R. P. Novick.** 1995. Translation of RNAIII, the *Staphylococcus aureus agr* regulatory RNA molecule, can be activated by a 3′-end deletion. *FEMS Microbiol. Lett.* **133:**155–161.

8. **Barg, N., C. Bunce, L. Wheeler, G. Reed, and J. Musser.** 1992. Murine model of cutaneous infection with gram-positive cocci. *Infect. Immun.* **60:**2636–2640.

9. **Bayer, A. S., A. L. Cheung, L. I. Kupferwasser, and M. R. Yeaman.** 2000. The nonsteroidal anti-inflammatory drugs (NSAIDs), salicylic acid and diclofenac, downmodulate both sar- and agr- dependent genes in Staphylococcus aureus (SA). Abstr. 100th Annu. Meet. Am. Soc. Microbiol. 2000. American Society for Microbiology, Washington, D.C.

10. **Bayer, M. G., J. H. Heinrichs, and A. L. Cheung.** 1996. The molecular architecture of the sar locus in *Staphylococcus aureus*. *J. Bacteriol.* **178:**4563–4570.

11. **Bayles, K. W., and J. J. Iandolo.** 1989. Genetic and molecular analyses of the gene encoding staphylococcal enterotoxin D. *J. Bacteriol.* **171:**4799–4806.

12. **Beenken, K. E., P. M. Dunman, F. McAleese, D. Macapagal, E. Murphy, S. J. Projan, J. S. Blevins, and M. S. Smeltzer.** 2004. Global gene expression in *Staphylococcus aureus* biofilms. *J. Bacteriol.* **186:**4665–4684.

13. **Benito, Y., F. A. Kolb, P. Romby, G. Lina, J. Etienne, and F. Vandenesch.** 2000. Probing the structure of RNAIII, the *Staphylococcus aureus* agr regulatory RNA, and identification of the RNA domain involved in repression of protein A expression. *RNA* **6:**668–679.

14. **Benito, Y., G. Lina, T. Greenland, J. Etienne, and F. Vandenesch.** 1998. trans-complementation of a *Staph-*

15. *ylococcus aureus* agr mutant by *Staphylococcus lugdunensis* agr RNAIII. *J. Bacteriol.* **180:**5780–5783.

15. **Betley, M. J., and J. J. Mekalanos.** 1985. Staphylococcal enterotoxin A is encoded by phage. *Science* **229:**185–187.

16. **Birkey, S. M., W. Liu, X. Zhang, M. F. Duggan, and F. M. Hulett.** 1998. Pho signal transduction network reveals direct transcriptional regulation of one two-component system by another two-component regulator: *Bacillus subtilis* PhoP directly regulates production of ResD. *Mol. Microbiol.* **30:**943–953.

17. **Bischoff, M., P. Dunman, J. Kormanec, D. Macapagal, E. Murphy, W. Mounts, B. Berger-Bachi, and S. Projan.** 2004. Microarray-based analysis of the *Staphylococcus aureus* sigmaB regulon. *J. Bacteriol.* **186:**4085–4099.

18. **Bjorklind, A., and S. Arvidson.** 1980. Mutants of *Staphylococcus aureus* affected in the regulation of exoprotein synthesis. *FEMS Microbiol. Lett.* **7:**203–206.

19. **Blevins, J. S., K. E. Beenken, M. O. Elasri, B. K. Hurlburt, and M. S. Smeltzer.** 2002. Strain-dependent differences in the regulatory roles of sarA and agr in *Staphylococcus aureus*. *Infect. Immun.* **70:**470–480.

20. **Blevins, J. S., A. F. Gillaspy, T. M. Rechtin, B. K. Hurlburt, and M. S. Smeltzer.** 1999. The staphylococcal accessory regulator (sar) represses transcription of the *Staphylococcus aureus* collagen adhesin gene (cna) in an agr-independent manner. *Mol. Microbiol.* **33:**317–326.

21. **Bronner, S., P. Stoessel, A. Gravet, H. Monteil, and G. Prevost.** 2000. Variable expressions of *Staphylococcus aureus* bicomponent leucotoxins semiquantified by competitive reverse transcription-PCR. *Appl. Environ. Microbiol.* **66:**3931–3938.

22. **Brunskill, E. W., and K. W. Bayles.** 1996. Identification of LytSR-regulated genes from *Staphylococcus aureus*. *J. Bacteriol.* **178:**5810–5812.

23. **Cao, J. G., Z. Y. Wei, and E. A. Meighen.** 1995. The lux autoinducer-receptor interaction in *Vibrio harveyi*: binding. *Biochem. J.* **312:**439–444.

24. **Chamberlain, N. R., and B. Imanoel.** 1996. Genetic regulation of fatty acid modifying enzyme from *Staphylococcus aureus*. *J. Med. Microbiol.* **44:**125–129.

25. **Chan, P. F., and S. J. Foster.** 1998. The role of environmental factors in the regulation of virulence-determinant expression in *Staphylococcus aureus* 8325-4. *Microbiology* **144:**2469–2479.

26. **Chan, P. F., S. J. Foster, E. Ingham, and M. O. Clements.** 1998. The *Staphylococcus aureus* alternative sigma factor sigmaB controls the environmental stress response but not starvation survival or pathogenicity in a mouse abscess model. *J. Bacteriol.* **180:**6082–6089.

27. **Cheung, A. L., K. J. Eberhardt, E. Chung, M. R. Yeaman, P. M. Sullam, M. Ramos, and A. S. Bayer.** 1994. Diminished virulence of sar⁻/agr⁻ mutant of *Staphylococcus aureus* in the rabbit model of endocarditis. *J. Clin. Investig.* **94:**1815–1822.

28. **Cheung, A. L., J. H. Heinrichs, and M. G. Bayer.** 1996. Characterization of the sar locus and its interaction with agr in *Staphylococcus aureus*. *J. Bacteriol.* **178:**418–423.

29. **Cheung, A. L., and G. Zhang.** 2002. Global regulation of virulence determinants in *Staphylococcus aureus* by the SarA protein family. *Front. Biosci.* **7:**1825–1842.

30. **Chien, Y., A. C. Manna, S. J. Projan, and A. L. Cheung.** 1999. SarA, a global regulator of virulence determinants in *Staphylococcus aureus*, binds to a conserved motif essential for sar-dependent gene regulation. *J. Biol. Chem.* **274:**37169–37176.

31. **Clecner, B., and S. Sonea.** 1966. Acquisition of the delta type of hemolytic property by lysogenic conversion in *Staphylococcus aureus*. *Rev. Can. Biol.* **25:**145–148.

32. **Coleman, D. C., J. P. Arbuthnott, H. M. Pomeroy, and T. H. Birkbeck.** 1986. Cloning and expression in *Escherichia coli* and *Staphylococcus aureus* of the beta-lysin determinant from *Staphylococcus aureus*: evidence that bacteriophage conversion of beta-lysin activity is caused by insertional inactivation of the beta-lysin determinant. *Microb. Pathog.* **1:** 549–564.

33. **Coleman, D. C., D. J. Sullivan, R. J. Russell, J. P. Arbuthnott, B. F. Carey, and H. M. Pomeroy.** 1989. *Staphylococcus aureus* bacteriophages mediating the simultaneous lysogenic conversion of b-lysin, staphylokinase and enterotoxin A: molecular mechanism of triple conversion. *J. Gen. Microbiol.* **135:**1679–1697.

34. **de Haas, C. J., K. E. Veldkamp, A. Peschel, F. Weerkamp, W. J. Van Wamel, E. C. Heezius, M. J. Poppelier, K. P. Van Kessel, and J. A. van Strijp.** 2004. Chemotaxis inhibitory protein of *Staphylococcus aureus*, a bacterial antiinflammatory agent. *J. Exp. Med.* **199:**687–695.

35. **Deora, R., T. Tseng, and T. K. Misra.** 1997. Alternative transcription factor sigmaSB of *Staphylococcus aureus*: characterization and role in transcription of the global regulatory locus sar. *J. Bacteriol.* **179:**6355–6359.

36. **Diep, D. B., L. S. Havarstein, J. Nissen-Mayer, and I. F. Nes.** 1994. The gene encoding plantarcin A, a bacteriocin from *Lactobacillus plantarum* C11, is located on the same transcription unit as an agr-like regulatory system. *Appl. Environ. Microbiol.* **60:**160.

37. **Dobardzic, R., and S. Sonea.** 1971. Hemolysin production and lysogenic conversion in *Staphylococcus aureus*. *Ann. Inst. Pasteur* (Paris) **120:**42–49.

38. **Dufour, P., S. Jarraud, F. Vandenesch, T. Greenland, R. P. Novick, M. Bes, J. Etienne, and G. Lina.** 2002. High genetic variability of the agr locus in *Staphylococcus* species. *J. Bacteriol.* **184:**1180–1186.

39. **Dunman, P. M., E. Murphy, S. Haney, D. Palacios, G. Tucker-Kellogg, S. Wu, E. L. Brown, R. J. Zagursky, D. Shlaes, and S. J. Projan.** 2001. Transcription profiling-based identification of *Staphylococcus aureus* genes regulated by the agr and/or sarA loci. *J. Bacteriol.* **183:**7341–7353.

40. **Dziewanowska, K., A. R. Carson, J. M. Patti, C. F. Deobald, K. W. Bayles, and G. A. Bohach.** 2000. Staphylococcal fibronectin binding protein interacts with heat shock protein 60 and integrins: role in internalization by epithelial cells. *Infect. Immun.* **68:**6321–6328.

41. **Foster, T. J., M. O'Reilly, P. Phonimdaeng, J. Cooney, A. H. Patel, and A. J. Bramley.** 1990. Genetic studies of virulence factors of *Staphylococcus aureus*. Properties of coagulase, alpha-toxin, beta-toxin and protein A in the pathogenesis of *S. aureus* infections, p. 403–420. *In* R. P. Novick (ed.), *Molecular Biology of the Staphylococci*. VCH Publishers, New York, N.Y.

42. **Fournier, B., and D. C. Hooper.** 2000. A new two-component regulatory system involved in adhesion, autolysis, and extracellular proteolytic activity of *Staphylococcus aureus*. *J. Bacteriol.* **182:**3955–3964.

43. **Fournier, B., and A. Klier.** 2004. Protein A gene expression is regulated by DNA supercoiling which is modified by the ArlS-ArlR two-component system of *Staphylococcus aureus*. *Microbiology* **150:**3807–3819.

44. **Fournier, B., A. Klier, and G. Rapoport.** 2001. The two-component system ArlS-ArlR is a regulator of virulence gene expression in *Staphylococcus aureus*. *Mol. Microbiol.* **41:**247–261.

45. **Fujimoto, D. F., and K. W. Bayles.** 1998. Opposing roles of the *Staphylococcus aureus* virulence regulators, Agr and Sar, in Triton X-100- and penicillin-induced autolysis. *J. Bacteriol.* **180:**3724–3726.

46. **Gaskill, M. E., and S. A. Khan.** 1988. Regulation of the enterotoxin B gene in *Staphylococcus aureus*. *J. Biol. Chem.* **263:**6276–6280.

47. **Gemmel, C. G., and A. M. A. Shibl.** 1976. The control of toxin and enzyme biosynthesis in staphylococci by antibiotics, p. 657–664. *In* J. Jeljaszewicz (ed.), *Staphylococci and Staphylococal Diseases*. Gustav Fischer Verlag, Stuttgart, Germany.

48. **Gertz, S., S. Engelmann, R. Schmid, A. K. Ziebandt, K. Tischer, C. Scharf, J. Hacker, and M. Hecker.** 2000. Characterization of the sigma(B) regulon in *Staphylococcus aureus*. *J. Bacteriol.* **182:**6983–6991.

49. **Gillet, Y., B. Issartel, P. Vanhems, J. C. Fournet, G. Lina, M. Bes, F. Vandenesch, Y. Piemont, N. Brousse, D. Floret, and J. Etienne.** 2002. Association between *Staphylococcus aureus* strains carrying gene for Panton-Valentine leukocidin and highly lethal necrotising pneumonia in young immunocompetent patients. *Lancet* **359:**753–759.

50. **Giraudo, A., G. Martinez, A. Calzolari, and R. Nagel.** 1994. Characterization of a transpositional mutant of *Staphylococcus aureus* underproducing exoproteins. *Rev. Latinoam. Microbiol.* **36:**171–176.

51. **Giraudo, A., C. Raspanti, A. Calzolari, and R. Nagel.** 1994. Characterization of a Tn551-mutant of *Staphylococcus aureus* defective in the production of several exoproteins. *Can. J. Microbiol.* **40:**677–681.

52. **Giraudo, A. T., A. Calzolari, A. A. Cataldi, C. Bogni, and R. Nagel.** 1999. The sae locus of *Staphylococcus aureus* encodes a two-component regulatory system. *FEMS Microbiol. Lett.* **177:**15–22.

53. **Giraudo, A. T., A. L. Cheung, and R. Nagel.** 1997. The sae locus of *Staphylococcus aureus* controls exoprotein synthesis at the transcriptional level. *Arch. Microbiol.* **168:** 53–58.

54. **Goerke, C., S. Campana, M. G. Bayer, G. Doring, K. Botzenhart, and C. Wolz.** 2000. Direct quantitative transcript analysis of the agr regulon of *Staphylococcus aureus* during human infection in comparison to the expression profile in vitro. *Infect. Immun.* **68:**1304–1311.

55. **Goerke, C., U. Fluckiger, A. Steinhuber, W. Zimmerli, and C. Wolz.** 2001. Impact of the regulatory loci agr, sarA and sae of *Staphylococcus aureus* on the induction of alpha-toxin during device-related infection resolved by direct quantitative transcript analysis. *Mol. Microbiol.* **40:**1439–1447.

56. **Havarstein, L. S., P. Gaustad, I. F. Nes, and D. A. Morrison.** 1996. Identification of the streptococcal competence-pheromone receptor. *Mol. Microbiol.* **21:**863–869.

57. **Herbert, S., D. Worlitzsch, B. Dassy, A. Boutonnier, J. M. Fournier, G. Bellon, A. Dalhoff, and G. Doring.** 1997. Regulation of *Staphylococcus aureus* capsular polysaccharide type 5: CO_2 inhibition in vitro and in vivo. *J. Infect. Dis.* **176:**431–438.

58. **Huh, Y. J., and A. A. Weiss.** 1991. A 23-kilodalton protein, distinct from BvgA, expressed by virulent *Bordetella pertussis* binds to the promoter region of *vir*-regulated toxin genes. *Infect. Immun.* **59:**2389–2395.

59. **Huhne, K., L. Axelsson, A. Holck, and L. Krockel.** 1996. Analysis of the sakacin P gene cluster from *Lacto-

bacillus sake Lb674 and its expression in sakacin-negative Lb. sake strains. *Microbiology* **142:**1437–1448.

60. **Janzon, L., and S. Arvidson.** 1990. The role of the delta-lysin gene (*hld*) in the regulation of virulence genes by the accessory gene regulator (*agr*) in *Staphylococcus aureus*. *EMBO J.* **9:**1391–1399.

61. **Jarraud, S., G. J. Lyon, A. M. Figueiredo, L. Gerard, F. Vandenesch, J. Etienne, T. W. Muir, and R. P. Novick.** 2000. Exfoliatin-producing strains define a fourth agr specificity group in *Staphylococcus aureus*. *J. Bacteriol.* **182:**6517–6522.

62. **Ji, G., R. Beavis, and R. P. Novick.** 1997. Bacterial interference caused by autoinducing peptide variants. *Science* **276:**2027–2030.

63. **Ji, G., R. C. Beavis, and R. P. Novick.** 1995. Cell density control of staphylococcal virulence mediated by an octapeptide pheromone. *Proc. Natl. Acad. Sci. USA* **92:**12055–12059.

64. **Jonsson, I. M., S. Arvidson, S. Foster, and A. Tarkowski.** 2004. Sigma factor B and RsbU are required for virulence in *Staphylococcus aureus*-induced arthritis and sepsis. *Infect. Immun.* **72:**6106–6111.

65. **Kernodle, D. S., P. A. McGraw, N. L. Barg, B. E. Menzies, R. K. Voladri, and S. Harshman.** 1995. Growth of *Staphylococcus aureus* with nafcillin in vitro induces alpha-toxin production and increases the lethal activity of sterile broth filtrates in a murine model. *J. Infect. Dis.* **172:**410–409.

66. **Koenig, R. L., J. L. Ray, S. J. Maleki, M. S. Smeltzer, and B. K. Hurlburt.** 2004. *Staphylococcus aureus* AgrA binding to the RNAIII-agr regulatory region. *J. Bacteriol.* **186:**7549–7555.

67. **Korem, M., A. S. Sheoran, Y. Gov, S. Tzipori, I. Borovok, and N. Balaban.** 2003. Characterization of RAP, a quorum sensing activator of *Staphylococcus aureus*(1). *FEMS Microbiol. Lett.* **223:**167–175.

68. **Kullik, I., and P. Giachino.** 1997. The alternative sigma factor s^B in *Staphylococcus aureus*: regulation of the *sigB* operon in response to growth phase and heat shock. *Arch. Microbiology* **167:**151–159.

69. **Kullik, I., P. Giachino, and T. Fuchs.** 1998. Deletion of the alternative sigma factor s^B in *Staphylococcus aureus* reveals its function as a global regulator of virulence genes. *J. Bacteriol.* **180:**4814–4820.

70. **Kuroda, M., H. Kuroda, T. Oshima, F. Takeuchi, H. Mori, and K. Hiramatsu.** 2003. Two-component system VraSR positively modulates the regulation of cell-wall biosynthesis pathway in *Staphylococcus aureus*. *Mol. Microbiol.* **49:**807–821.

71. **Kuroda, M., T. Ohta, I. Uchiyama, T. Baba, H. Yuzawa, I. Kobayashi, L. Cui, A. Oguchi, K. Aoki, Y. Nagai, J. Lian, T. Ito, M. Kanamori, H. Matsumaru, A. Maruyama, H. Murakami, A. Hosoyama, Y. Mizutani-Ui, N. K. Takahashi, T. Sawano, R. Inoue, C. Kaito, K. Sekimizu, H. Hirakawa, S. Kuhara, S. Goto, J. Yabuzaki, M. Kanehisa, A. Yamashita, K. Oshima, K. Furuya, C. Yoshino, T. Shiba, M. Hattori, N. Ogasawara, H. Hayashi, and K. Hiramatsu.** 2001. Whole genome sequencing of methicillin-resistant *Staphylococcus aureus*. *Lancet* **357:**1225–1240.

72. **Lebeau, C., F. Vandenesch, T. Greeland, R. P. Novick, and J. Etienne.** 1994. Coagulase expression in *Staphylococcus aureus* is positively and negatively modulated by an *agr*-dependent mechanism. *J. Bacteriol.* **176:**5534–5536.

73. **Leboeuf-Trudeau, T., J. de Repentigny, R. M. Frenette, and S. Sonea.** 1969. Tryptophan metabolism and toxin formation in *S. aureus* Wood 46 strain. *Can. J. Microbiol.* **15:**1–7.

74. **Lee, C. Y., and J. J. Iandolo.** 1986. Lysogenic conversion of staphylococcal lipase caused by insertion of the bacteriophage L54a genome into the lipase structural gene. *J. Bacteriol.* **166:**385–391.

75. **Lina, G., S. Jarraud, G. Ji, T. Greenland, A. Pedraza, J. Etienne, R. P. Novick, and F. Vandenesch.** 1998. Transmembrane topology and histidine protein kinase activity of AgrC, the *agr* signal receptor in *Staphylococcus aureus*. *Mol. Microbiol.* **28:**655–662.

76. **Lindsay, J. A., and S. J. Foster.** 1999. Interactive regulatory pathways control virulence determinant production and stability in response to environmental conditions in *Staphylococcus aureus*. *Mol. Gen. Genet.* **262:**323–331.

77. **Luong, T., S. Sau, M. Gomez, J. C. Lee, and C. Y. Lee.** 2002. Regulation of *Staphylococcus aureus* capsular polysaccharide expression by agr and sarA. *Infect. Immun.* **70:**444–450.

78. **Luong, T. T., S. W. Newell, and C. Y. Lee.** 2003. Mgr, a novel global regulator in *Staphylococcus aureus*. *J. Bacteriol.* **185:**3703–3710.

79. **Lyon, G. J., J. S. Wright, T. W. Muir, and R. P. Novick.** 2002. Key determinants of receptor activation in the agr autoinducing peptides of *Staphylococcus aureus*. *Biochemistry* **41:**10095–10104.

80. **Makris, G., J. D. Wright, E. Ingham, and K. T. Holland.** 2004. The hyaluronate lyase of *Staphylococcus aureus*—a virulence factor? *Microbiology* **150:**2005–2013.

81. **Manna, A., and A. L. Cheung.** 2001. Characterization of sarR, a modulator of sar expression in *Staphylococcus aureus*. *Infect. Immun.* **69:**885–896.

82. **Manna, A. C., and A. L. Cheung.** 2003. sarU, a sarA homolog, is repressed by SarT and regulates virulence genes in *Staphylococcus aureus*. *Infect. Immun.* **71:**343–353.

83. **Manna, A. C., S. S. Ingavale, M. Maloney, W. van Wamel, and A. L. Cheung.** 2004. Identification of sarV (SA2062), a new transcriptional regulator, is repressed by SarA and MgrA (SA0641) and involved in the regulation of autolysis in *Staphylococcus aureus*. *J. Bacteriol.* **186:**5267–5280.

84. **Martin, P. K., T. Li, D. Sun, D. P. Biek, and M. B. Schmid.** 1999. Role in cell permeability of an essential two-component system in *Staphylococcus aureus*. *J. Bacteriol.* **181:**3666–3673.

85. **Mayville, P., G. Ji, R. Beavis, H.-M. Yang, M. Goger, R. P. Novick, and T. W. Muir.** 1999. Structure-activity analysis of synthetic autoinducing thiolactone peptides from *Staphylococcus aureus* responsible for virulence. *Proc. Natl. Acad. Sci. USA* **96:**1218–1223.

86. **McAleese, F. M., E. J. Walsh, M. Sieprawska, J. Potempa, and T. J. Foster.** 2001. Loss of clumping factor B fibrinogen binding activity by *Staphylococcus aureus* involves cessation of transcription, shedding and cleavage by metalloprotease. *J. Biol. Chem.* **276:**29969–29978.

87. **McCallum, N., M. Bischoff, H. Maki, A. Wada, and B. Berger-Bächi.** 2004. TcaR, a putative MarR-like regulator of sarS expression. *J. Bacteriol.* **186:**2966–2972.

88. **McNamara, P. J., K. C. Milligan-Monroe, S. Khalili, and R. A. Proctor.** 2000. Identification, cloning, and initial characterization of rot, a locus encoding a regulator of virulence factor expression in *Staphylococcus aureus*. *J. Bacteriol.* **182:**3197–3203.

89. **Morfeldt, E., L. Janzon, S. Arvidson, and S. Lofdahl.** 1988. Cloning of a chromosomal locus (exp) which regulates the expression of several exoprotein genes in *Staphylococcus aureus. Mol. Gen. Genet.* **211:**435–440.

90. **Morfeldt, E., K. Tegmark, and S. Arvidson.** 1996. Transcriptional control of the *agr*-dependent virulence gene regulator, RNAIII, in *Staphylococcus aureus. Mol. Microbiol.* **21:**1227–1237.

91. **Morikawa, K., Y. Inose, H. Okamura, A. Maruyama, H. Hayashi, K. Takeyasu, and T. Ohta.** 2003. A new staphylococcal sigma factor in the conserved gene cassette: functional significance and implication for the evolutionary processes. *Genes Cells* **8:**699–712.

91a. **Naidu, A. S., M. Andersson, and A. Forsgren.** 1992. Identification of a human lactoferrin-binding protein in *Staphylococcus aureus. J. Med. Microbiol.* **36:**177–183.

92. **Nair, S. P., M. Bischoff, M. M. Senn, and B. Berger-Bachi.** 2003. The sigma B regulon influences internalization of *Staphylococcus aureus* by osteoblasts. *Infect. Immun.* **71:**4167–4170.

93. **Nicholas, R. O., T. Li, D. McDevitt, A. Marra, S. Sucoloski, P. L. Demarsh, and D. R. Gentry.** 1999. Isolation and characterization of a sigB deletion mutant of *Staphylococcus aureus. Infect. Immun.* **67:**3667–3669.

94. **Novick, R. P.** 2003. Autoinduction and signal transduction in the regulation of staphylococcal virulence. *Mol. Microbiol.* **48:**1429–1449.

95. **Novick, R. P.** 2003. Mobile genetic elements and bacterial toxinoses: the superantigen-encoding pathogenicity islands of *Staphylococcus aureus. Plasmid* **49:**93–105.

96. **Novick, R. P., and D. Jiang.** 2003. The staphylococcal saeRS system coordinates environmental signals with agr quorum sensing. *Microbiology* **149:**2709–2717.

97. **Novick, R. P., S. J. Projan, J. Kornblum, H. F. Ross, G. Ji, B. Kreiswirth, F. Vandenesch, and S. Moghazeh.** 1995. The agr P2 operon: an autocatalytic sensory transduction system in *Staphylococcus aureus. Mol. Gen. Genet.* **248:**446–458.

98. **Novick, R. P., H. F. Ross, S. J. Projan, J. Kornblum, B. Kreiswirth, and S. Moghazeh.** 1993. Synthesis of staphylococcal virulence factors is controlled by a regulatory RNA molecule. *EMBO J.* **12:**3967–3975.

99. **Novick, R. P., P. Schlievert, and A. Ruzin.** 2001. Pathogenicity and resistance islands of staphylococci. *Microbes Infect.* **3:**585–594.

100. **Ohlendorf, D. H., C. A. Earhart, D. T. Mitchell, G. M. Vath, M. Roggiani, J. V. Rago, M. H. Kim, G. A. Bohach, and P. M. Schlievert.** 1998. Structural biology of toxins associated with TSS. *Int. Congr. Symp. Ser.* **229:**89–91.

101. **Ohlsen, K., W. Ziebuhr, K. P. Koller, W. Hell, T. A. Wichelhaus, and J. Hacker.** 1998. Effects of subinhibitory concentrations of antibiotics on alpha-toxin (hla) gene expression of methicillin-sensitive and methicillin-resistant *Staphylococcus aureus* isolates. *Antimicrob. Agents Chemother.* **42:**2817–2823.

102. **Omoe, K., D. L. Hu, H. Takahashi-Omoe, A. Nakane, and K. Shinagawa.** 2003. Identification and characterization of a new staphylococcal enterotoxin-related putative toxin encoded by two kinds of plasmids. *Infect. Immun.* **71:**6088–6094.

103. **Otto, M., R. Sussmuth, G. Jung, and F. Gotz.** 1998. Structure of the pheromone peptide of the *Staphylococcus epidermidis* agr system. *FEBS Lett.* **424:**89–94.

104. **Otto, M., R. Sussmuth, C. Vuong, G. Jung, and F. Gotz.** 1999. Inhibition of virulence factor expression in *Staph-ylococcus aureus* by the *Staphylococcus epidermidis* agr pheromone and derivatives. *FEBS Lett.* **450:**257–262.

105. **Peng, H.-L., R. P. Novick, B. Kreiswirth, J. Kornblum, and P. Schlievert.** 1988. Cloning, characterization and sequencing of an accessory gene regulator (*agr*) in *Staphylococcus aureus. J. Bacteriol.* **179:**4365–4372.

106. **Piriz Duran, S., F. H. Kayser, and B. Berger-Bachi.** 1996. Impact of sar and agr on methicillin resistance in *Staphylococcus aureus. FEMS Microbiol. Lett.* **141:**255–260.

107. **Pohlmann-Dietze, P., M. Ulrich, K. B. Kiser, G. Doring, J. C. Lee, J. M. Fournier, K. Botzenhart, and C. Wolz.** 2000. Adherence of *Staphylococcus aureus* to endothelial cells: influence of capsular polysaccharide, global regulator agr, and bacterial growth phase. *Infect. Immun.* **68:**4865–4871.

108. **Projan, S., and R. Novick.** 1997. The molecular basis of virulence, p. 55–81. *In* G. Archer and K. Crossley (ed.), *Staphylococci in Human Disease.* Churchill Livingstone, New York, N.Y.

109. **Projan, S. J., S. Brown-Skrobot, P. Schlievert, F. Vandenesch, and R. P. Novick.** 1994. Glycerol monolaurate inhibits the production of b-lactamase, toxic shock syndrome toxin-1 and other staphylococcal exoproteins by interfering with signal transduction. *J. Bacteriol.* **176:**4204–4209.

110. **Qazi, S. N., E. Counil, J. Morrissey, C. E. Rees, A. Cockayne, K. Winzer, W. C. Chan, P. Williams, and P. J. Hill.** 2001. agr expression precedes escape of internalized *Staphylococcus aureus* from the host endosome. *Infect. Immun.* **69:**7074–7082.

111. **Quadri, L. E., M. Kleerebezem, O. P. Kuipers, W. M. de Vos, K. L. Roy, J. C. Vederas, and M. E. Stiles.** 1997. Characterization of a locus from *Carnobacterium piscicola* LV17B involved in bacteriocin production and immunity: evidence for global inducer-mediated transcriptional regulation. *J. Bacteriol.* **179:**6163–6171.

112. **Rampone, H., G. L. Martinez, A. T. Giraudo, A. Calzolari, and R. Nagel.** 1996. In vivo expression of exoprotein synthesis with a Sae mutant of *Staphylococcus aureus. Can. J. Vet. Res.* **60:**237–240.

113. **Recsei, P., B. Kreiswirth, M. O'Reilly, P. Schlievert, A. Gruss, and R. Novick.** 1986. Regulation of exoprotein gene expression by *agr. Mol. Gen. Genet.* **202:**58–61.

114. **Recsei, P., B. Kreiswirth, M. O'Reilly, P. Schlievert, A. Gruss, and R. P. Novick.** 1986. Regulation of exoprotein gene expression in *Staphylococcus aureus* by agar. *Mol. Gen. Genet.* **202:**58–61.

115. **Regassa, L. B., J. L. Couch, and M. J. Betley.** 1991. Steady-state staphylococcal enterotoxin type C mRNA is affected by a product of the accessory gene regulator (*agr*) and by glucose. *Infect. Immun.* **59:**955–962.

116. **Rosendal, K., P. Buelow, and O. Jessen.** 1964. Lyogenic conversion in *Staphylococcus aureus*, to a change in the production of extracellular "Tween"-splitting enzyme. *Nature* **204:**1222–1223.

117. **Said-Salim, B., P. M. Dunman, F. M. McAleese, D. Macapagal, E. Murphy, P. J. McNamara, S. Arvidson, T. J. Foster, S. J. Projan, and B. N. Kreisworth.** 2003. Global regulation of *Staphylococcus aureus* genes by rot. *J. Bacteriol.* **185:**610–619.

118. **Sakoulas, G., G. M. Eliopoulos, R. Moellering, C. Wennersten, L. Venkataraman, R. P. Novick, and H. S. Gold.** 2002. Accessory gene regulator (*agr*) locus in geographically diverse *Staphylococcus aureus* isolates with reduced susceptibility to vancomycin. *Antimicrob. Agents Chemother.* **46:**1492–1502.

119. **Saravia-Otten, P., H. P. Muller, and S. Arvidson.** 1997. Transcription of *Staphylococcus aureus* fibronectin binding protein genes is negatively regulated by agr and an agr-independent mechanism. *J. Bacteriol.* **179:**5259–5263.

120. **Sato, H., T. Watanabe, K. Higuchi, K. Teruya, A. Ohtake, Y. Murata, H. Saito, C. Aizawa, H. Danbara, and N. Maehara.** 2000. Chromosomal and extrachromosomal synthesis of exfoliative toxin from *Staphylococcus hyicus. J. Bacteriol.* **182:**4096–4100.

121. **Sawicka-Grzelak, A., A. Szymanowska, A. Mlynarczyk, and G. Mlynarczyk.** 1993. Production of staphylokinase and hemolysin by coagulase-negative staphylococcus. *Med. Dosw. Mikrobiol.* **45:**7–10.

122. **Scarlato, V., B. Arico, A. Prugnola, and R. Rappuoli.** 1991. Sequential activation and environmental regulation of virulence genes in *Bordetella pertussis. EMBO J.* **10:**3971–3975.

123. **Schmidt, K. A., A. C. Manna, and A. L. Cheung.** 2003. SarT influences sarS expression in *Staphylococcus aureus. Infect. Immun.* **71:**5139–5148.

124. **Schmidt, K. A., A. C. Manna, S. Gill, and A. L. Cheung.** 2001. SarT, a repressor of alpha-hemolysin in *Staphylococcus aureus. Infect. Immun.* **69:**4749–4758.

125. **Scott, J. M., N. Smirnova, and W. G. Haldenwang.** 1999. A Bacillus-specific factor is needed to trigger the stress-activated phosphatase/kinase cascade of sigmaB induction. *Biochem. Biophys. Res. Commun.* **257:**106–110.

126. **Sheehan, B. J., T. J. Foster, C. J. Dorman, S. Park, and G. S. Stewart.** 1992. Osmotic and growth-phase dependent regulation of the eta gene of *Staphylococcus aureus*: a role for DNA supercoiling. *Mol. Gen. Genet.* **232:**49–57.

127. **Shompole, S., K. T. Henon, L. E. Liou, K. Dziewanowska, G. A. Bohach, and K. W. Bayles.** 2003. Biphasic intracellular expression of *Staphylococcus aureus* virulence factors and evidence for Agr-mediated diffusion sensing. *Mol. Microbiol.* **49:**919–927.

128. **Sinha, B., P. P. Francois, O. Nusse, M. Foti, O. M. Hartford, P. Vaudaux, T. J. Foster, D. P. Lew, M. Herrmann, and K. H. Krause.** 1999. Fibronectin-binding protein acts as *Staphylococcus aureus* invasin via fibronectin bridging to integrin alpha5beta1. *Cell Microbiol.* **1:**101–117.

129. **Smeltzer, M. S., M. E. Hart, and J. J. Iandolo.** 1993. Phenotypic characterization of xpr, a global regulator of extracellular virulence factors in *Staphylococcus aureus. Infect. Immun.* **61:**919–925.

130. **Solomon, J., R. Magnuson, A. Sruvastavam, and A. Grossman.** 1995. Convergent sensing pathways mediate response to two extracellular competence factors in *Bacillus subtilis. Genes Dev.* **9:**547–558.

131. **Somerville, G. A., S. B. Beres, J. R. Fitzgerald, F. R. DeLeo, R. L. Cole, J. S. Hoff, and J. M. Musser.** 2002. In vitro serial passage of *Staphylococcus aureus*: changes in physiology, virulence factor production, and agr nucleotide sequence. *J. Bacteriol.* **184:**1430–1437.

132. **Somerville, G. A., M. S. Chaussee, C. I. Morgan, J. R. Fitzgerald, D. W. Dorward, L. J. Reitzer, and J. M. Musser.** 2002. *Staphylococcus aureus* aconitase inactivation unexpectedly inhibits post- exponential-phase growth and enhances stationary-phase survival. *Infect. Immun.* **70:**6373–6382.

133. **Somerville, G. A., A. Cockayne, M. Durr, A. Peschel, M. Otto, and J. M. Musser.** 2003. Synthesis and deformylation of *Staphylococcus aureus* delta-toxin are linked to tricarboxylic acid cycle activity. *J. Bacteriol.* **185:**6686–6694.

134. **Tegmark, K., A. Karlsson, and S. Arvidson.** 2000. Identification and characterization of SarH1, a new global regulator of virulence gene expression in *Staphylococcus aureus. Mol. Microbiol.* **37:**398–409.

135. **Tegmark, K., A. Karlsson, and S. Arvidson.** 2000. Identification and characterization of SarH1, a new global regulator of virulence gene expression in *Staphylococcus aureus. Mol. Microbiol.* **37:**398–409.

136. **Tegmark, K., E. Morfeldt, and S. Arvidson.** 1998. Regulation of agr-dependent virulence genes in *Staphylococcus aureus* by RNAIII from coagulase-negative staphylococci. *J. Bacteriol.* **180:**3181–3186.

137. **Throup, J. P., F. Zappacosta, R. D. Lunsford, R. S. Annan, S. A. Carr, J. T. Lonsdale, A. P. Bryant, D. McDevitt, M. Rosenberg, and M. K. Burnham.** 2001. The srhSR gene pair from *Staphylococcus aureus*: genomic and proteomic approaches to the identification and characterization of gene function. *Biochemistry* **40:**10392–10401.

138. **Tremaine, M. T., D. K. Brockman, and M. J. Betley.** 1993. Staphylococcal enterotoxin A gene (*sea*) expression is not affected by the accessory gene regulator (*agr*). *Infect. Immun.* **61:**356–359.

139. **Valle, J., A. Toledo-Arana, C. Berasain, J. M. Ghigo, B. Amorena, J. R. Penades, and I. Lasa.** 2003. SarA and not sigmaB is essential for biofilm development by *Staphylococcus aureus. Mol. Microbiol.* **48:**1075–1087.

140. **van der Vijver, J. C., M. van Es-Boon, and M. F. Michel.** 1972. Lysogenic conversion in *Staphylococcus aureus* to leucocidin production. *J. Virol.* **10:**318–319.

141. **van Wamel, W., Y. Q. Xiong, A. S. Bayer, M. R. Yeaman, C. C. Nast, and A. L. Cheung.** 2002. Regulation of *Staphylococcus aureus* type 5 capsular polysaccharides by agr and sarA in vitro and in an experimental endocarditis model. *Microb. Pathog.* **33:**73–79.

142. Reference deleted.

143. **Vandenesch, F., J. Kornblum, and R. P. Novick.** 1991. A temporal signal, independent of *agr*, is required for *hla* but not *spa* transcription in *Staphylococcus aureus. J. Bacteriol.* **173:**6313–6320.

144. **Vojtov, N., H. F. Ross, and R. P. Novick.** 2002. Global repression of exotoxin synthesis by staphylococcal superantigens. *Proc. Natl. Acad. Sci. USA* **99:**10102–10107.

145. **Vuong, C., H. L. Saenz, F. Gotz, and M. Otto.** 2000. Impact of the agr quorum-sensing system on adherence to polystyrene in *Staphylococcus aureus. J. Infect. Dis.* **182:**1688–1693.

146. **Warren, R. L.** 1980. Exfoliative toxin plasmids of bacteriophage group 2 *Staphylococcus aureus*: sequence homology. *Infect. Immun.* **30:**601–606.

147. **Weinrick, B., P. M. Dunman, F. McAleese, E. Murphy, S. J. Projan, Y. Fang, and R. P. Novick.** 2004. Effect of mild acid on gene expression in *Staphylococcus aureus. J. Bacteriol.* **186:**8407–8423.

148. **Wesson, C. A., L. E. Liou, K. M. Todd, G. A. Bohach, W. R. Trumble, and K. W. Bayles.** 1998. *Staphylococcus aureus* agr and sar global regulators influence internalization and induction of apoptosis. *Infect. Immun.* **66:**5238–5243.

149. **Winkler, K. C., J. de Waart, and C. Grootsen.** 1965. Lysogenic conversion of staphylococci to loss of b-toxin. *J. Gen. Microbiol.* **39:**321–333.

150. **Wolz, C., D. McDevitt, T. J. Foster, and A. L. Cheung.** 1996. Influence of agr on fibrinogen binding in *Staphylococcus aureus* Newman. *Infect. Immun.* **64:**3142–3147.

151. **Wright, J. S., R. Jin, and R. P. Novick.** *Proc. Natl. Acad. Sci. USA*, in press.

152. **Wu, S., H. de Lencastre, and A. Tomasz.** 1996. Sigma-B, a putative operon encoding alternate sigma factor of *Staphylococcus aureus* RNA polymerase: molecular cloning and DNA sequencing. *J. Bacteriol.* **178:**6036–6042.

153. **Xiong, Y. Q., A. S. Bayer, M. R. Yeaman, W. Van Wamel, A. C. Manna, and A. L. Cheung.** 2004. Impacts of sarA and agr in *Staphylococcus aureus* strain Newman on fibronectin-binding protein A gene expression and fibronectin adherence capacity in vitro and in experimental infective endocarditis. *Infect. Immun.* **72:**1832–1836.

154. **Xiong, Y. Q., W. Van Wamel, C. C. Nast, M. R. Yeaman, A. L. Cheung, and A. S. Bayer.** 2002. Activation and transcriptional interaction between agr RNAII and RNAIII in *Staphylococcus aureus* in vitro and in an experimental endocarditis model. *J. Infect. Dis.* **186:**668–677.

155. **Yarwood, J. M., D. J. Bartels, E. M. Volper, and E. P. Greenberg.** 2004. Quorum sensing in *Staphylococcus aureus* biofilms. *J. Bacteriol.* **186:**1838–1850.

156. **Yarwood, J. M., J. K. McCormick, and P. M. Schlievert.** 2001. Identification of a novel two-component regulatory system that acts in global regulation of virulence factors of *Staphylococcus aureus*. *J. Bacteriol.* **183:**1113–1123.

157. **Yoshizawa, Y., J. Sakurada, S. Sakurai, K. Machida, I. Kondo, and S. Masuda.** 2000. An exfoliative toxin A-converting phage isolated from *Staphylococcus aureus* strain ZM. *Microbiol. Immunol.* **44:**189–191.

158. **Zhang, A., S. Rimsky, M. E. Reaban, H. Buc, and M. Belfort.** 1996. Escherichia coli protein analogs StpA and H-NS: regulatory loops, similar and disparate effects on nucleic acid dynamics. *EMBO J.* **15:**1340–1349.

159. **Zhang, L., L. Gray, R. P. Novick, and G. Ji.** 2002. Transmembrane topology of AgrB, the protein i nvolved in the post-translational modification of AgrD in *Staphylococcus aureus*. *J. Biol. Chem.* **277:**34736–34742.

160. **Zhang, L., Lin, J., Ji, G.** 2004. Membrane anchoring of the AgrD N-terminal amphipathic region is required for its processing to produce a quorum sensing pheromone in *Staphylococcus aureus*. *J. Biol. Chem.* **279:**19448–19456.

161. **Zhang, S., J. J. Iandolo, and G. C. Stewart.** 1998. The enterotoxin D plasmid of *Staphylococcus aureus* encodes a second enterotoxin determinant (sej). *FEMS Microbiol. Lett.* **168:**227–233.

162. **Zhang, S., and G. C. Stewart.** 2000. Characterization of the promoter elements for the staphylococcal enterotoxin D gene. *J. Bacteriol.* **182:**2321–2325.

Staphylococcus aureus—Eukaryotic Cell Interactions

CARLOS ARRECUBIETA AND FRANKLIN D. LOWY

42

The pathogen *Staphylococcus aureus* causes a diversity of diseases that range from minor skin and soft tissue infections to life-threatening systemic infections. All of these infections involve bacterial interactions with host cells and tissues. *S. aureus*–endothelial cell interactions have been the most extensively studied and are among the most important events in the pathogenesis of invasive systemic disease. Systemic infections are characterized by endovascular involvement that may include cardiac valvular destruction and metastatic infections as seen in endocarditis or vasculitis, coagulopathy, and multiorgan dysfunction as seen with sepsis. Staphylococci must interact with the endothelium in order to establish these infections or must traverse the endovascular space to create tissue-based abscesses (39, 51). In vitro studies demonstrate that staphylococci colonize and invade endothelial cells and, as a consequence of this process, cause significant cellular alterations. These cellular events contribute to the pathogenesis of staphylococcal disease. While this chapter will primarily focus on *S. aureus*–endothelial cell interactions as a model of staphylococcal interaction with eukaryotic cells, reference will also be made to more recent publications describing staphylococcal interactions with other cell types.

ADHERENCE

S. aureus Adherence to Endothelial Cells

S. aureus adherence to endothelial cells is the critical first step in the invasion process. Using human and canine cardiac valvular tissue, Gould et al. (27) demonstrated that staphylococci adhere to endovascular tissue and endothelial cells grown in tissue culture more avidly than do other bacterial species. Subsequent studies confirmed these observations using human and porcine endothelial cells grown in tissue culture (40, 66, 92). In general, the bacterial species most commonly associated with acute bacterial endocarditis were also the most adherent (66, 92).

Effect of Varied Environmental Conditions on *S. aureus* Adherence to Host Tissue Surfaces

Variation in endothelial cell growth conditions altered adherence of staphylococci to endothelial cells. Cheung et al. (20) showed that endothelial cells grown in the presence of tumor necrosis factor alpha (TNF-α) were more susceptible to staphylococcal infection. The same group also reported that *S. aureus* adherence to endothelial cells was increased in the presence of fibrinogen and other blood constituents (18). Shenkman et al. (80) confirmed this observation by showing that *S. aureus* adherence to endothelial cells was enhanced by thrombin activation of the endothelial cells and that this adherence was mediated by fibrinogen and not by platelets. Finally, Blumberg et al. (13) showed that the absence of acidic fibroblast growth factor in the medium increased adherence while the addition of endotoxin and interleukin-1 (IL-1) was without effect.

Alteration of the matrix underlying endothelial cells also alters staphylococcal adherence. Alston et al. (7) demonstrated that the extracellular matrix (ECM) elaborated by *S. aureus*-infected endothelial cells signaled phenotypic changes in newly seeded endothelial cells in vitro. These cells became more susceptible to subsequent infection. This altered susceptibility appeared to be due to a reduction in the sulfation of the matrix heparan sulfate proteoglycans that resulted in a similar reduction in sulfation of the cellular heparan sulfate proteoglycans (7).

These studies suggest that subtle alterations in cellular growth conditions produce significant changes in the susceptibility of endothelial cells to infection. Similar microenvironmental changes in vivo may not produce detectable pathological changes but may nevertheless increase the possibility of infection at a particular endovascular site.

Contribution of Bacterial Factors to Adherence

Bacterial growth conditions influence staphylococcal adherence to eukaryotic cells. Staphylococci harvested during exponential growth phase are more adherent than those harvested during stationary phase, suggesting that the proteins mediating bacterial adherence are under the control of

Gram-Positive Pathogens, 2nd edition, edited by Vincent A. Fischetti et al.
© 2006 ASM Press, Washington, D.C.

global regulatory genes such as *agr* or *sar* (19, 71, 88). Shenkman et al. (79, 81) suggested that the effect of these regulatory genes seems to be exercised through adherence to fibrinogen and fibronectin. Pohlmann-Dietze et al. (72) observed that *agr* also controls expression of type 5 capsular polysaccharide, which, in turn, appears to mask the binding domain of staphylococcal adhesin(s) that mediate binding to endothelial cells.

The structurally related family of staphylococcal surface proteins designated microbial surface components recognizing adhesive matrix molecules plays a crucial role in bacterial adherence to host cells and tissues (67). In many instances the binding interaction involves ECM components such as collagen, fibronectin, and fibrinogen serving as bridging ligands connecting the bacterium and the eukaryotic cell (67, 70). In this manner, both *S. aureus* fibronectin-binding proteins A and B (FnBPA and FnBPB) mediate staphylococcal adhesion to and internalization by several mammalian cell types, including endothelial cells (2, 55, 70, 84).

Endothelial Cell Receptors for *S. aureus*

Sinha et al. (84) and Massey et al. (55) recently identified the eukaryotic membrane integrin $\alpha 5\beta 1$ as the receptor for fibronectin-mediated *S. aureus* adherence in 293 (epithelial) and endothelial cells, respectively. It has also been postulated that intercellular adhesion molecule 1 (ICAM-1) might act as the endothelial cell receptor for the staphylococcal extracellular adherence protein (Eap), which is both secreted and surface-located in *S. aureus* and has the ability to rebind to the staphylococcal surface (17, 31).

S. aureus–Platelet Interactions

The adhesion of *S. aureus* to platelets and the resulting platelet aggregation are major determinants of virulence in the pathogenesis of endocarditis (86). O'Brien et al. (62) described multiple mechanisms for stimulating platelet aggregation in vitro with the involvement of at least ClfA, ClfB, SdrE, and protein A staphylococcal surface proteins. This redundancy, as they suggest, supports the notion of its relevance in the pathogenesis of invasive diseases. *S. aureus* has long been known to adhere to platelets through a fibrinogen/fibrin-dependent mechanism (35). Recently, with the use of phage display techniques, two additional proteins, coagulase and Efb, have been proposed to mediate bacterial adherence to immobilized platelets through their ability to bind fibrinogen, which serves as a bridging molecule in that interaction (33). Similarly, FnBPA and FnBPB have more recently been described as ligands in vitro mediating staphylococcal adherence to platelets. However, adhesion processes mediated by these two proteins seem to be substantially different (34).

Nguyen et al. (62) recently identified the protein gC1qR/-p33, which can be located either intracellularly or in the eukaryotic membrane, as a cellular binding site for staphylococcal protein A. This provides an additional potential mechanism for bacterial adherence to sites of vascular injury or thrombosis.

Platelets now appear to have a dual role in the pathogenesis of endovascular infections such as infective endocarditis. Activated platelets at sites of injury provide an adhesive substrate for staphylococcal adherence. However, they also elaborate platelet microbicidal proteins that have antimicrobial properties and contribute to the host defense against infection (47, 98).

S. aureus Adherence to Other Tissue Sites

Adherence to Nasal and Respiratory Epithelial Cells

S. aureus is a part of the resident flora in the anterior nares of approximately 30% of the population (45). The basis for this colonization is not well understood (69). ClfB, a fibrinogen-binding protein, appears to mediate adhesion to desquamated nasal epithelial cells as well as to a human keratinocyte cell line, apparently by attaching to cytokeratin 10, present on the surface of these two cell types (65). SasG and Pls proteins also appear to facilitate binding to nasal epithelial cells (77). More recently, Weidenmaier et al. (94) suggested that cell wall-associated teichoic acid mediates *S. aureus* adherence to human nasal epithelial cells.

Eap has also been shown to act as an adhesion factor of *S. aureus* to various cell types such as fibroblasts and epithelial cells (29, 38, 46). Moreover, the addition of exogenous Eap was shown to increase the adherence of both the wild-type and *eap*-defective mutants to fibroblasts, suggesting that this protein might also act as a bridging ligand for other *S. aureus* surface proteins (38).

Gomez et al. (26) have shown that protein A interacts with TNF receptor 1, a receptor for TNF-α that is widely distributed on airway epithelial cells, and that this interaction, along with the subsequent signaling pathway activation, plays a role in the pathogenesis of staphylococcal pneumonia.

Effect of Flow Conditions on *S. aureus* Adherence to Endothelial Cells

Reddy and Ross (75) studied staphylococcal adherence to endothelial cells using flow assays at physiological levels of shear stress and suggested that mechanical forces present in vivo reduce the ability of *S. aureus* to bind endothelial cell surfaces. Shenkman et al. (79, 81) analyzed the adherence properties of *S. aureus* under both static and flow conditions and suggested that different adhesion mechanisms govern binding processes in these two different environments. Using similar linear flow study conditions, it has recently been shown that fluid-shear conditions that are representative of the vascular environment affect platelet–*S. aureus* interactions that are crucial to the development of bloodstream infections. In this way, peak adhesion efficiency between bacterial cells and activated platelets occurs at low shear conditions (100 s^{-1}). Furthermore, it appears to be a transition from a ClfA-dependent process at low shear rates ($\leq 400 \text{ s}^{-1}$) to a protein A/ClfA-dependent mechanism at high shears (68).

INVASION

Ogawa et al. (66) first demonstrated that following adherence, staphylococci are endocytosed into membrane-bound vacuoles in an endothelial cell-mediated process (Fig. 1). The invasion process appears dependent on initial bacterial contact followed by internalization in a eukaryotic cell-dependent process (6, 21, 37, 52). As noted, both *S. aureus* proteins FnBPA and FnBPB play a role in fibronectin-dependent adhesion and internalization by several mammalian cell types (2, 55, 70, 84). In addition, Sinha et al. (83), in a series of elegant experiments using a heterologous expression system for staphylococcal genes, demonstrated that either one of these two proteins is sufficient to elicit adhesion and internalization by epithelial cells (Fig. 2).

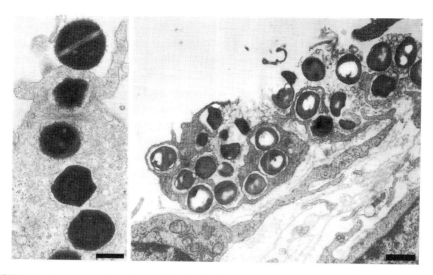

FIGURE 1 Demonstration of endothelial cell phagocytosis of *S. aureus* in vitro. (Left) Staphylococci incubated with human umbilical vein endothelial cells in tissue culture (30 min). The bacteria are phagocytized, enclosed within a membrane-bound vacuole, and transported into the cell. Bar, 0.5 μm. (Right) Section of rabbit aorta incubated with staphylococci (bacteria incubated with tissue for 30 min, then replaced with medium for a 5.5-h incubation). The endothelial cell contains a large number of bacteria enclosed within vacuoles. The cell has ruptured, releasing bacteria into the medium. Bar, 1.0 μm. (Reprinted from the *Journal of Ultrastructural and Molecular Structural Research* [52] with the permission of Academic Press.)

The staphylococcal protein Eap has been postulated to be involved in the internalization of *S. aureus* by fibroblasts (29). However, the findings of Haggar et al. (29) are in contrast with those of Kreikemeyer et al. (46), who found that Eap-defective mutant strains were better internalized than the wild-type strain. These contradictory results may reflect differences in the surface protein expression patterns between the different strains and different eukaryotic cell types used in their respective studies.

Pls, another staphylococcal surface protein found only in staphylococcal chromosomal cassette *mec* type I methicillin-resistant *S. aureus*, has been shown to have a negative effect on adhesion and internalization of *S. aureus* by 293 embryonic epithelial cells (42).

Collectively, these findings suggest that internalization of *S. aureus* occurs by receptor-mediated endocytosis, in which FnBP-fibronectin acts as a high affinity ligand for the β1-integrin host cell receptor. In this manner a model has recently been proposed for the attachment to epithelial cells taking place among FnBP-fibronectin, β1 integrin, and the heat shock protein Hsp60 (Fig. 2) (23).

Following internalization, there is fusion of the bacteria-containing phagosome with cellular lysosomes and limited, if any, intracellular bacterial replication. Through pharmacological and genetic interference assays, several studies have characterized some of the host cell factors implicated in the receptor-mediated endocytosis (4). In this way, the major role played by the cytoskeleton (microfilaments and microtubules) was demonstrated. Staphylococcal protein A, interacting with cytoskeletal actin filaments, appears to play a role in invasion of cultured human KB cells (41). In addition, the activation of the tyrosine phosphorylation cascade has been shown to be a key intracellular signal governing integrin β1-mediated internalization of *S. aureus* (4), in which

the Src family of protein-tyrosine kinases seems to play a critical role in mediating invasion-promoting signals (1).

S. aureus was considered for many years to be an extracellular organism, although it is now accepted that it has the ability to invade nonprofessional phagocytic cells (i.e., cells other than macrophages and polymorphonuclear [PMN] leukocytes) in vitro (24, 30, 66, 70, 84). Interestingly, these cell types internalize microbial pathogens using b1-integrins, whereas β2-integrins are implicated in the uptake by professional phagocytes (78).

INTRACELLULAR SURVIVAL OF *S. AUREUS*

The fact that *S. aureus* infections frequently recur, coupled with the observation that *S. aureus* cells can survive within eukaryotic cells, led to speculation that *S. aureus* may be an intracellular pathogen (30, 50). The intracellular location was hypothesized to provide a sanctuary, shielding the organism from host defenses and from antibiotics, thereby contributing to the persistence and/or recurrence of infection (73).

Studies show that *S. aureus* is internalized and survives in a variety of mammalian cells (6, 10, 11, 36, 58). In order to survive within the host cell, *S. aureus* must first escape from the encapsulating endosomal membrane and then multiply or at least survive in the host cell cytoplasm (10, 28, 37). However, the mechanism by which the endosomal membrane is breached is not completely understood.

Recent observations in cultured bovine mammary epithelial cells supported the hypothesis that the *agr* locus controls the expression of some of the genes necessary for escape from the endosome and further growth in the cytoplasm (74). Shompole et al. (82), using the same cell line, showed a correlation between activation of the Agr regulon,

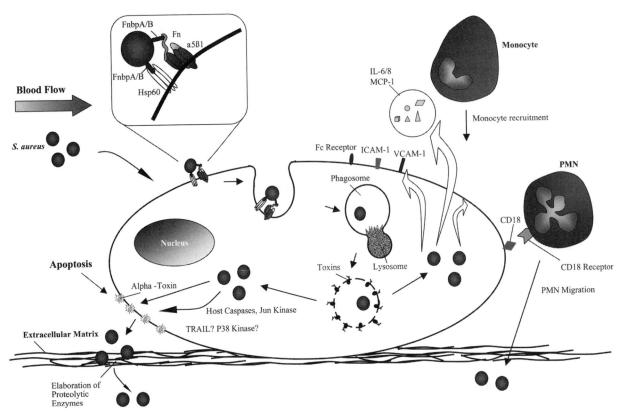

FIGURE 2 Model for *S. aureus*–endothelial cell interactions. Staphylococci adhere using fibronectin (Fn) as a bridging ligand to host cell integrin α5β1 and Hsp60. This process elicits endothelial cell-mediated phagocytosis of *S. aureus* and the subsequent phagolysosomal fusion. *S. aureus* escapes the endosome through the action of unidentified toxins. Staphylococci in the cytosol can induce host cell apoptosis via a variety of pathways involving among other components alpha-toxin, host cell caspases and Jun amino-terminal kinase. *S. aureus* cells induce expression of a wide range of both secreted and host cell surface factors. Some of these molecules are in turn involved in pathogenetic processes that include recruitment of PMN leukocytes and monocytes to sites of infection with migration across the endovascular tissue and the ECM to sites of inflammation. *S. aureus* cells also elaborate a variety of proteolytic enzymes that facilitate their translocation across the ECM to adjacent tissues.

expression of staphylococcal membrane-active toxins, and increase in permeabilization of the endosomal membrane. The high density of *S. aureus* cells within the endosome might trigger activation of the quorum-sensing Agr regulon. After escaping the endosome some strains are able to survive within the eukaryotic cytoplasm for long periods of time and therefore are potentially capable of causing persistent or recurrent infections. This feature has been associated with a particular subpopulation of mutants known phenotypically as small colony variants. Their capacity for intracellular survival is correlated with a reduced production of alpha-toxin (8, 73).

CONSEQUENCES OF STAPHYLOCOCCAL INVASION

Fc Receptor, Adhesion Molecule, and Cytokine Expression

A variety of cellular changes occurs as a result of *S. aureus* invasion (Fig. 2). Intracellular staphylococci induce eukaryotic cells to produce, and in many instances secrete, mole-

cules that play a role in the pathogenesis of staphylococcal infections. Surface expression of proteins, such as Fc receptors and adhesion molecules (11, 12, 22), as well as secretion of cytokines, all occur in response to staphylococcal invasion (87, 96, 97).

Beekhuizen et al. (11) observed that *S. aureus*-infected endothelial cells become more adhesive for monocytes and granulocytes in vitro in a staphylococcal concentration-dependent manner and that expression of ICAM-1, vascular cell adhesion molecule 1 (VCAM-1) and major histocompatibility complex class I protein was induced after infection. Moreover, granulocyte adhesion was mediated by CD11/CD18 integrins, while monocytes were bound via CD11a/CD18 and CD49d/ CD29 integrins. In addition, Moreland et al. (60) recently demonstrated that *S. aureus* cells can induce PMN leukocyte migration across a monolayer of endothelial cells and that CD18 is involved in this process. Tekstra et al. (87) showed that endothelial production of the monocyte chemotactic protein-1 (MCP-1) chemokine, which plays a role in monocyte recruitment, is induced after staphylococcal internalization. These findings strongly suggest that *S. aureus*-infected endothelial cells initiate a cascade of reactions that leads to an inflammatory response. Similar in vivo conditions resulting from infection could stimulate leukocyte migration from the vascular space into the bacterium-infected tissue.

Chavakis et al. (17) showed both in vivo and in vitro that the secreted staphylococcal protein Eap acts as an anti-inflammatory agent by blocking leukocyte recruitment due to its ability to bind to the endothelium via ICAM-1.

As noted before, endothelial cells release cytokines after staphylococcal internalization; different strains have different abilities to induce cytokine expression (87, 96, 97). Moreover, the IL-8-containing supernatant from infected endothelial cells had functional activity, attracting PMN leukocytes across an endothelial cell monolayer in an in vitro transmigration assay (97).

Some staphylococcal cellular components such as lipoteichoic acid and peptidoglycan from the cell wall appear to be capable of inducing increased chemokine secretion in endothelial cells (90). Interestingly, it has been observed that *S. aureus*, but not staphylococcal toxins, have the capacity to act as primary factors in induction of cytokine secretion by endothelial cells, whereas exotoxins might be responsible for subsequent cytokine increases in prestimulated cells (85). Stimulation of secretion of cytokines and chemokines such as IL-6, IL-12, colony-stimulating factors, and MCP-1 by *S. aureus* invasion has also been observed in osteoblasts (14, 15), a cell type not previously known to be capable of inducing protective cell-mediated immune responses.

Some of these findings have been extended to in vivo studies (44, 54). In this manner it was observed that mutations in both *sar* and *agr* regulatory loci elicited an attenuated cytokine and chemokine expression in murine brain abscesses. Furthermore, the same reduction in the levels of proinflammatory cytokines and chemokines was obtained with an alpha-toxin-defective mutant, suggesting this molecule as an important virulence factor in brain abscess development (44).

Apoptosis

S. aureus-mediated eukaryotic cell apoptosis occurs when bacteria escape from the membrane-bound vacuoles following internalization (10). This process has been observed in epithelial cells (10, 43, 95), keratinocytes (64), endothelial cells (56, 95), osteoblasts (89), fibroblasts (61), and chondrocytes (49), as well as in professional phagocytic cells such as monocytes and granulocytes (9, 32, 53).

Moreover, the mechanism responsible for the *S. aureus*-initiated eukaryotic cell apoptotic cycle has been partially elucidated. Vann and Proctor (91) used isogenic alpha-toxin-producing and -nonproducing strains to demonstrate that cellular damage was alpha-toxin-mediated. Likewise, alpha-toxin has been shown to induce apoptosis in endothelial cells (57) and in blood mononuclear cells (32).

Further studies on endothelial cells showed that *S. aureus*-induced apoptosis occurs through the activation of cellular caspases as well as the stimulation of the Jun amino-terminal kinase signaling pathway (25). Wesson et al. (95) demonstrated that host caspases 3 and 8 play a role in *S. aureus*-induced apoptosis in epithelial cells, while Weglarczyk et al. (93) reported that caspase 8 plays a role in monocyte apoptosis. *S. aureus* invasion of osteoblasts induces expression of TNF-related apoptosis-inducing ligand (3). This reaction in turn seems to mediate activation of caspase 8 and subsequent apoptosis (5). In addition, unidentified *S. aureus*-specific secreted factor(s) appears to enhance the oxidative response of neutrophils via the activation of the host

p38-mitogen-activated protein kinase, which is crucial for apoptosis (53, 63), thus explaining the potent induction of inflammation elicited by *S. aureus* compared to other staphylococcal species (63).

CLINICOPATHOLOGICAL CORRELATIONS

A prevailing concern regarding the in vitro observations of *S. aureus* invasion of eukaryotic cells has been the limited amount of supporting clinical and experimental in vivo data. In particular there has been an absence of clinicopathological data supporting the role of *S. aureus* as an intracellular pathogen (50). Several recent in vivo studies using animal models of infection have, however, provided some information on the invasion process in staphylococcal disease. These studies have demonstrated the role of different virulence factors in the development of *S. aureus* infections (44, 48, 86). In addition, staphylococcal internalization (16, 76), intracellular survival (28), and the subsequent induction of cytokine release by host cells (54) have also been analyzed in vivo and their involvement in the pathogenesis of *S. aureus* infections has been partially assessed.

A murine mastitis model was used to demonstrate the importance of FnBPs in the process of colonization and internalization (16), and observations in an embryonic chick osteoblast invasion model supported the hypothesis that invasion of osteoblasts by *S. aureus* plays a role in the etiology of osteomyelitis (76). Intracellular survival has also been suggested in vivo in a murine brain abscess model (44). Moreillon et al. (59) also reported intracellular localization of ClfA-expressing lactococci in a rat model of endocarditis.

The studies performed to date suggest a sequence of events that leads to staphylococcal colonization and infection of different tissues (Fig. 2). Adherence of staphylococci to eukaryotic tissues is a complex multifactorial process that takes place through multiple specific interactions between bacterial and host factors. These interactions seem to be closely modulated by microenvironmental conditions at the site of infection, such as fluid shear forces or modulations in the composition of the extracellular milieu. Some of these changes may be the result of tissue injury or other events that are known to facilitate the onset of staphylococcal infection. Staphylococcal internalization and its capacity for intracellular survival may help bacteria evade exposure to host defenses. Moreover, in vivo observations of *S. aureus* invasion and subsequent survival within the host cell point to a likely relevance in the development of persistent infections in different tissues. At this point, however, there is insufficient information available to speculate on which staphylococcal infections utilize eukaryotic cellular invasion or intracellular bacterial persistence as a part of the pathogenetic process. *S. aureus* has the ability to infect a wide variety of tissues, therefore causing local or systemic infections. Both local and systemic infections can metastasize by crossing a multitude of histological barriers and spread to other tissues. Finally, *S. aureus* also appears able to partially modulate the host cell-mediated immune response by eliciting or inhibiting its inflammatory response, which could account for differences in the outcomes of the infective process.

F.D.L. was supported in part by grants from the NIH—DA09656, DA11868, and DA15018, and from the CDC—CCR223380-01.

REFERENCES

1. **Agerer, F., A. Michel, K. Ohlsen, and C. R. Hauck.** 2003. Integrin-mediated invasion of *Staphylococcus aureus* into human cells requires Src family protein-tyrosine kinases. *J. Biol. Chem.* **278:**42524–42531.

2. **Ahmed, S., S. Meghji, R. J. Williams, B. Henderson, J. H. Brock, and S. P. Nair.** 2001. *Staphylococcus aureus* fibronectin-binding proteins are essential for internalization by osteoblasts but do not account for differences in intracellular levels of bacteria. *Infect. Immun.* **69:**2872–2877.

3. **Alexander, E. H., J. L. Bento, F. M. Hughes, Jr., I. Marriott, M. C. Hudson, and K. L. Bost.** 2001. *Staphylococcus aureus* and *Salmonella enterica* serovar *Dublin* induce tumor necrosis factor-related apoptosis-inducing ligand expression by normal mouse and human osteoblasts. *Infect. Immun.* **69:**1581–1586.

4. **Alexander, E. H., and M. C. Hudson.** 2001. Factors influencing the internalization of *Staphylococcus aureus* and impacts on the course of infections in humans. *Appl. Microbiol. Biotechnol.* **56:**361–366.

5. **Alexander, E. H., F. A. Rivera, I. Marriott, J. Anguita, K. L. Bost, and M. C. Hudson.** 2003. *Staphylococcus aureus*-induced tumor necrosis factor-related apoptosis-inducing ligand expression mediates apoptosis and caspase-8 activation in infected osteoblasts. *BMC Microbiol.* **3:**5.

6. **Almeida, R. A., K. R. Matthews, E. Cifrian, A. J. Guidry, and S. P. Oliver.** 1996. *Staphylococcus aureus* invasion of bovine mammary epithelial cells. *J. Dairy Sci.* **79:**1021–1026.

7. **Alston, W. K., D. A. Elliott, M. E. Epstein, V. B. Hatcher, M. Tang, and F. D. Lowy.** 1997. Extracellular matrix heparan sulfate modulates endothelial cell susceptibility to *Staphylococcus aureus*. *J. Cell Physiol.* **173:**102–109.

8. **Balwit, J. M., P. van Langevelde, J. M. Vann, and R. A. Proctor.** 1994. Gentamicin-resistant menadione and hemin auxotrophic *Staphylococcus aureus* persist within cultured endothelial cells. *J. Infect. Dis.* **170:**1033–1037.

9. **Baran, J., K. Guzik, W. Hryniewicz, M. Ernst, H. D. Flad, and J. Pryjma.** 1996. Apoptosis of monocytes and prolonged survival of granulocytes as a result of phagocytosis of bacteria. *Infect. Immun.* **64:**4242–4248.

10. **Bayles, K. W., C. A. Wesson, L. E. Liou, L. K. Fox, G. A. Bohach, and W. R. Trumble.** 1998. Intracellular *Staphylococcus aureus* escapes the endosome and induces apoptosis in epithelial cells. *Infect. Immun.* **66:**336–342.

11. **Beekhuizen, H., J. S. van de Gevel, B. Olsson, I. J. van Benten, and R. van Furth.** 1997. Infection of human vascular endothelial cells with *Staphylococcus aureus* induces hyperadhesiveness for human monocytes and granulocytes. *J. Immunol.* **158:**774–782.

12. **Bengualid, V., V. B. Hatcher, B. Diamond, E. A. Blumberg, and F. D. Lowy.** 1990. *Staphylococcus aureus* infection of human endothelial cells potentiates Fc receptor expression. *J. Immunol.* **145:**4279–4283.

13. **Blumberg, E. A., V. B. Hatcher, and F. D. Lowy.** 1988. Acidic fibroblast growth factor modulates *Staphylococcus aureus* adherence to human endothelial cells. *Infect. Immun.* **56:**1470–1474.

14. **Bost, K. L., J. L. Bento, J. K. Ellington, I. Marriott, and M. C. Hudson.** 2000. Induction of colony-stimulating factor expression following *Staphylococcus* or *Salmonella* interaction with mouse or human osteoblasts. *Infect. Immun.* **68:**5075–5083.

15. **Bost, K. L., W. K. Ramp, N. C. Nicholson, J. L. Bento, I. Marriott, and M. C. Hudson.** 1999. *Staphylococcus aureus* infection of mouse or human osteoblasts induces high levels of interleukin-6 and interleukin-12 production. *J. Infect. Dis.* **180:**1912–1920.

16. **Brouillette, E., B. G. Talbot, and F. Malouin.** 2003. The fibronectin-binding proteins of *Staphylococcus aureus* may promote mammary gland colonization in a lactating mouse model of mastitis. *Infect. Immun.* **71:**2292–2295.

17. **Chavakis, T., M. Hussain, S. M. Kanse, G. Peters, R. G. Bretzel, J. I. Flock, M. Herrmann, and K. T. Preissner.** 2002. *Staphylococcus aureus* extracellular adherence protein serves as anti-inflammatory factor by inhibiting the recruitment of host leukocytes. *Nat. Med.* **8:**687–693.

18. **Cheung, A. L., and V. A. Fischetti.** 1990. The role of fibrinogen in staphylococcal adherence to catheters in vitro. *J. Infect. Dis.* **161:**1177–1186.

19. **Cheung, A. L., J. M. Koomey, C. A. Butler, S. J. Projan, and V. A. Fischetti.** 1992. Regulation of exoprotein expression in *Staphylococcus aureus* by a locus (sar) distinct from agr. *Proc. Natl. Acad. Sci. USA* **89:**6462–6466.

20. **Cheung, A. L., J. M. Koomey, S. Lee, E. A. Jaffe, and V. A. Fischetti.** 1991. Recombinant human tumor necrosis factor-α promotes adherence of *Staphylococcus aureus* to cultured human endothelial cells. *Infect. Immun.* **59:**3827–3831.

21. **Craven, N., and J. C. Anderson.** 1979. The location of *Staphylococcus aureus* in experimental chronic mastitis in the mouse and the effect on the action of sodium cloxacillin. *Br. J. Exp. Pathol.* **60:**453–459.

22. **Drake, T. A., and M. Pang.** 1988. *Staphylococcus aureus* induces tissue factor expression in cultured human cardiac valve endothelium. *J. Infect. Dis.* **157:**749–756.

23. **Dziewanowska, K., A. R. Carson, J. M. Patti, C. F. Deobald, K. W. Bayles, and G. A. Bohach.** 2000. Staphylococcal fibronectin binding protein interacts with heat shock protein 60 and integrins: role in internalization by epithelial cells. *Infect. Immun.* **68:**6321–6328.

24. **Dziewanowska, K., J. M. Patti, C. F. Deobald, K. W. Bayles, W. R. Trumble, and G. A. Bohach.** 1999. Fibronectin binding protein and host cell tyrosine kinase are required for internalization of *Staphylococcus aureus* by epithelial cells. *Infect. Immun.* **67:**4673–4678.

25. **Esen, M., B. Schreiner, V. Jendrossek, F. Lang, K. Fassbender, H. Grassme, and E. Gulbins.** 2001. Mechanisms of *Staphylococcus aureus* induced apoptosis of human endothelial cells. *Apoptosis* **6:**431–439.

26. **Gomez, M. I., A. Lee, B. Reddy, A. Muir, G. Soong, A. Pitt, A. Cheung, and A. Prince.** 2004. *Staphylococcus aureus* protein A induces airway epithelial inflammatory responses by activating TNFR1. *Nat. Med.* **10:**842–848.

27. **Gould, K., C. H. Ramirez-Ronda, R. K. Holmes, and J. P. Sanford.** 1975. Adherence of bacteria to heart valves in vitro. *J. Clin. Invest.* **56:**1364–1370.

28. **Gresham, H. D., J. H. Lowrance, T. E. Caver, B. S. Wilson, A. L. Cheung, and F. P. Lindberg.** 2000. Survival of *Staphylococcus aureus* inside neutrophils contributes to infection. *J. Immunol.* **164:**3713–3722.

29. **Haggar, A., M. Hussain, H. Lonnies, M. Herrmann, A. Norrby-Teglund, and J. I. Flock.** 2003. Extracellular adherence protein from *Staphylococcus aureus* enhances internalization into eukaryotic cells. *Infect. Immun.* **71:**2310–2317.

30. **Hamill, R. J., J. M. Vann, and R. A. Proctor.** 1986. Phagocytosis of *Staphylococcus aureus* by cultured bovine aortic endothelial cells: model for postadherence events in endovascular infections. *Infect. Immun.* **54:**833–836.

31. **Harraghy, N., M. Hussain, A. Haggar, T. Chavakis, B. Sinha, M. Herrmann, and J. I. Flock.** 2003. The adhesive and immunomodulating properties of the multifunctional *Staphylococcus aureus* protein Eap. *Microbiology* **149:** 2701–2707.

32. **Haslinger, B., K. Strangfeld, G. Peters, K. Schulze-Osthoff, and B. Sinha.** 2003. *Staphylococcus aureus*-toxin induces apoptosis in peripheral blood mononuclear cells: role of endogenous tumour necrosis factor-α and the mitochondrial death pathway. *Cell Microbiol.* **5:**729–741.

33. **Heilmann, C., M. Herrmann, B. E. Kehrel, and G. Peters.** 2002. Platelet-binding domains in 2 fibrinogen-binding proteins of *Staphylococcus aureus* identified by phage display. *J. Infect. Dis.* **186:**32–39.

34. **Heilmann, C., S. Niemann, B. Sinha, M. Herrmann, B. E. Kehrel, and G. Peters.** 2004. *Staphylococcus aureus* fibronectin-binding protein (FnBP)-mediated adherence to platelets, and aggregation of platelets induced by FnBPA but not by FnBPB. *J. Infect. Dis.* **190:**321–329.

35. **Herrmann, M., Q. J. Lai, R. M. Albrecht, D. F. Mosher, and R. A. Proctor.** 1993. Adhesion of *Staphylococcus aureus* to surface-bound platelets: role of fibrinogen/fibrin and platelet integrins. *J. Infect. Dis.* **167:**312–322.

36. **Hess, D. J., M. J. Henry-Stanley, E. A. Erickson, and C. L. Wells.** 2003. Intracellular survival of *Staphylococcus aureus* within cultured enterocytes. *J. Surg. Res.* **114:**42–49.

37. **Hudson, M. C., W. K. Ramp, N. C. Nicholson, A. S. Williams, and M. T. Nousiainen.** 1995. Internalization of *Staphylococcus aureus* by cultured osteoblasts. *Microb. Pathog.* **19:**409–419.

38. **Hussain, M., A. Haggar, C. Heilmann, G. Peters, J. I. Flock, and M. Herrmann.** 2002. Insertional inactivation of Eap in *Staphylococcus aureus* strain Newman confers reduced staphylococcal binding to fibroblasts. *Infect. Immun.* **70:**2933–2940.

39. **Ing, M. B., L. M. Baddour, and A. S. Bayer.** 1997. Bacteremia, and infective endocarditis: pathogenesis, diagnosis, and complications, p. 331–354. *In* K. B. Crossley and G. L. Archer (ed.), *The Staphylococci in Human Disease.* Churchill Livingstone, New York, N.Y.

40. **Johnson, C. M., G. A. Hancock, and G. D. Goulin.** 1988. Specific binding of *Staphylococcus aureus* to cultured porcine cardiac valvular endothelial cells. *J. Lab. Clin. Med.* **112:**16–22.

41. **Jung, K. Y., J. D. Cha, S. H. Lee, W. H. Woo, D. S. Lim, B. K. Choi, and K. J. Kim.** 2001. Involvement of staphylococcal protein A and cytoskeletal actin in *Staphylococcus aureus* invasion of cultured human oral epithelial cells. *J. Med. Microbiol.* **50:**35–41.

42. **Juuti, K. M., B. Sinha, C. Werbick, G. Peters, and P. I. Kuusela.** 2004. Reduced adherence and host cell invasion by methicillin-resistant *Staphylococcus aureus* expressing the surface protein Pls. *J. Infect. Dis.* **189:**1574–1584.

43. **Kahl, B. C., M. Goulian, W. van Wamel, M. Herrmann, S. M. Simon, G. Kaplan, G. Peters, and A. L. Cheung.** 2000. *Staphylococcus aureus* RN6390 replicates and induces apoptosis in a pulmonary epithelial cell line. *Infect. Immun.* **68:**5385–5392.

44. **Kielian, T., A. Cheung, and W. F. Hickey.** 2001. Diminished virulence of an α-toxin mutant of *Staphylococcus aureus* in experimental brain abscesses. *Infect. Immun.* **69:** 6902–6911.

45. **Kluytmans, J., A. van Belkum, and H. Verbrugh.** 1997. Nasal carriage of *Staphylococcus aureus*: epidemiology, underlying mechanisms, and associated risks. *Clin. Microbiol. Rev.* **10:**505–520.

46. **Kreikemeyer, B., D. McDevitt, and A. Podbielski.** 2002. The role of the map protein in *Staphylococcus aureus* matrix protein and eukaryotic cell adherence. *Int. J. Med. Microbiol.* **292:**283–295.

47. **Kupferwasser, L. I., M. R. Yeaman, S. M. Shapiro, C. C. Nast, and A. S. Bayer.** 2002. In vitro susceptibility to thrombin-induced platelet microbicidal protein is associated with reduced disease progression and complication rates in experimental *Staphylococcus aureus* endocarditis: microbiological, histopathologic, and echocardiographic analyses. *Circulation* **105:**746–752.

48. **Lee, L. Y., Y. J. Miyamoto, B. W. McIntyre, M. Hook, K. W. McCrea, D. McDevitt, and E. L. Brown.** 2002. The *Staphylococcus aureus* Map protein is an immunomodulator that interferes with T cell-mediated responses. *J. Clin. Invest.* **110:**1461–1471.

49. **Lee, M. S., S. W. Ueng, C. H. Shih, and C. C. Chao.** 2001. Primary cultures of human chondrocytes are susceptible to low inocula of *Staphylococcus aureus* infection and undergo apoptosis. *Scand. J. Infect. Dis.* **33:**47–50.

50. **Lowy, F. D.** 2000. Is *Staphylococcus aureus* an intracellular pathogen? *Trends Microbiol.* **8:**341–343.

51. **Lowy, F. D.** 1998. *Staphylococcus aureus* infections. *N. Engl. J. Med.* **339:**520–532.

52. **Lowy, F. D., J. Fant, L. L. Higgins, S. K. Ogawa, and V. B. Hatcher.** 1988. *Staphylococcus aureus*—human endothelial cell interactions. *J. Ultrastruct. Mol. Struct. Res.* **98:**137–146.

53. **Lundqvist-Gustafsson, H., S. Norrman, J. Nilsson, and A. Wilsson.** 2001. Involvement of p38-mitogen-activated protein kinase in *Staphylococcus aureus*-induced neutrophil apoptosis. *J. Leukoc. Biol.* **70:**642–648.

54. **Marriott, I., D. L. Gray, S. L. Tranguch, V. G. Fowler, Jr., M. Stryjewski, L. Scott Levin, M. C. Hudson, and K. L. Bost.** 2004. Osteoblasts express the inflammatory cytokine interleukin-6 in a murine model of *Staphylococcus aureus* osteomyelitis and infected human bone tissue. *Am. J. Pathol.* **164:**1399–1406.

55. **Massey, R. C., M. N. Kantzanou, T. Fowler, N. P. Day, K. Schofield, E. R. Wann, A. R. Berendt, M. Hook, and S. J. Peacock.** 2001. Fibronectin-binding protein A of *Staphylococcus aureus* has multiple, substituting, binding regions that mediate adherence to fibronectin and invasion of endothelial cells. *Cell Microbiol.* **3:**839–851.

56. **Menzies, B. E., and I. Kourteva.** 1998. Internalization of *Staphylococcus aureus* by endothelial cells induces apoptosis. *Infect. Immun.* **66:**5994–5998.

57. **Menzies, B. E., and I. Kourteva.** 2000. *Staphylococcus aureus* α-toxin induces apoptosis in endothelial cells. *FEMS Immunol. Med. Microbiol.* **29:**39–45.

58. **Molinari, G., and G. S. Chhatwal.** 1999. Streptococcal invasion. *Curr. Opin. Microbiol.* **2:**56–61.

59. **Moreillon, P., Y. A. Que, and A. S. Bayer.** 2002. Pathogenesis of streptococcal and staphylococcal endocarditis. *Infect. Dis. Clin. North Am.* **16:**297–318.

60. **Moreland, J. G., G. Bailey, W. M. Nauseef, and J. P. Weiss.** 2004. Organism-specific neutrophil-endothelial cell interactions in response to *Escherichia coli*, *Streptococcus pneumoniae*, and *Staphylococcus aureus*. *J. Immunol.* **172:** 426–432.

61. **Murai, M., J. Sakurada, K. Seki, H. Shinji, Y. Hirota, and S. Masuda.** 1999. Apoptosis observed in BALB/3T3 cells having ingested *Staphylococcus aureus*. *Microbiol. Immunol.* **43:**653–661.

62. **Nguyen, T., B. Ghebrehiwet, and E. I. Peerschke.** 2000. *Staphylococcus aureus* protein A recognizes platelet

gC1qR/p33: a novel mechanism for staphylococcal interactions with platelets. *Infect. Immun.* **68:**2061–2068.

63. **Nilsdotter-Augustinsson, A., A. Wilsson, J. Larsson, O. Stendahl, L. Ohman, and H. Lundqvist-Gustafsson.** 2004. *Staphylococcus aureus*, but not *Staphylococcus epidermidis*, modulates the oxidative response and induces apoptosis in human neutrophils. *APMIS* **112:**109–118.

64. **Nuzzo, I., M. R. Sanges, A. Folgore, and C. R. Carratelli.** 2000. Apoptosis of human keratinocytes after bacterial invasion. *FEMS Immunol. Med. Microbiol.* **27:**235–240.

65. **O'Brien, L. M., E. J. Walsh, R. C. Massey, S. J. Peacock, and T. J. Foster.** 2002. *Staphylococcus aureus* clumping factor B (ClfB) promotes adherence to human type I cytokeratin 10: implications for nasal colonization. *Cell Microbiol.* **4:**759–770.

66. **Ogawa, S. K., E. R. Yurberg, V. B. Hatcher, M. A. Levitt, and F. D. Lowy.** 1985. Bacterial adherence to human endothelial cells in vitro. *Infect. Immun.* **50:**218–224.

67. **Patti, J. M., B. L. Allen, M. J. McGavin, and M. Hook.** 1994. MSCRAMM-mediated adherence of microorganisms to host tissues. *Annu. Rev. Microbiol.* **48:**585–617.

68. **Pawar, P., P. K. Shin, S. A. Mousa, J. M. Ross, and K. Konstantopoulos.** 2004. Fluid shear regulates the kinetics and receptor specificity of *Staphylococcus aureus* binding to activated platelets. *J. Immunol.* **173:**1258–1265.

69. **Peacock, S. J., I. de Silva, and F. D. Lowy.** 2001. What determines nasal carriage of *Staphylococcus aureus? Trends Microbiol.* **9:**605–610.

70. **Peacock, S. J., T. J. Foster, B. J. Cameron, and A. R. Berendt.** 1999. Bacterial fibronectin-binding proteins and endothelial cell surface fibronectin mediate adherence of *Staphylococcus aureus* to resting human endothelial cells. *Microbiology* **145:**3477–3486.

71. **Peng, H. L., R. P. Novick, B. Kreiswirth, J. Kornblum, and P. Schlievert.** 1988. Cloning, characterization, and sequencing of an accessory gene regulator (agr) in *Staphylococcus aureus. J. Bacteriol.* 170:4365–4372.

72. **Pohlmann-Dietze, P., M. Ulrich, K. B. Kiser, G. Doring, J. C. Lee, J. M. Fournier, K. Botzenhart, and C. Wolz.** 2000. Adherence of *Staphylococcus aureus* to endothelial cells: influence of capsular polysaccharide, global regulator agr, and bacterial growth phase. *Infect. Immun.* **68:**4865–4871.

73. **Proctor, R. A., P. van Langevelde, M. Kristjansson, J. N. Maslow, and R. D. Arbeit.** 1995. Persistent and relapsing infections associated with small-colony variants of *Staphylococcus aureus. Clin. Infect. Dis.* **20:**95–102.

74. **Qamer, S., J. A. Sandoe, and K. G. Kerr.** 2003. Use of colony morphology to distinguish different enterococcal strains and species in mixed culture from clinical specimens. *J. Clin. Microbiol.* **41:**2644–2646.

75. **Reddy, K., and J. M. Ross.** 2001. Shear stress prevents fibronectin binding protein-mediated *Staphylococcus aureus* adhesion to resting endothelial cells. *Infect. Immun.* **69:**3472–3475.

76. **Reilly, S. S., M. C. Hudson, J. F. Kellam, and W. K. Ramp.** 2000. In vivo internalization of *Staphylococcus aureus* by embryonic chick osteoblasts. *Bone* **26:**63–70.

77. **Roche, F. M., M. Meehan, and T. J. Foster.** 2003. The *Staphylococcus aureus* surface protein SasG and its homologues promote bacterial adherence to human desquamated nasal epithelial cells. *Microbiology* **149:**2759–2767.

78. **Ruoslahti, E.** 1996. RGD and other recognition sequences for integrins. *Annu. Rev. Cell Dev. Biol.* **12:**697–715.

79. **Shenkman, B., E. Rubinstein, A. L. Cheung, G. E. Brill, R. Dardik, I. Tamarin, N. Savion, and D. Varon.** 2001. Adherence properties of *Staphylococcus aureus* under static and flow conditions: roles of agr and sar loci, platelets, and plasma ligands. *Infect. Immun.* **69:**4473–4478.

80. **Shenkman, B., E. Rubinstein, I. Tamarin, R. Dardik, N. Savion, and D. Varon.** 2000. *Staphylococcus aureus* adherence to thrombin-treated endothelial cells is mediated by fibrinogen but not by platelets. *J. Lab. Clin. Med.* **135:**43–51.

81. **Shenkman, B., D. Varon, I. Tamarin, R. Dardik, M. Peisachov, N. Savion, and E. Rubinstein.** 2002. Role of agr (RNAIII) in *Staphylococcus aureus* adherence to fibrinogen, fibronectin, platelets and endothelial cells under static and flow conditions. *J. Med. Microbiol.* **51:**747–754.

82. **Shompole, S., K. T. Henon, L. E. Liou, K. Dziewanowska, G. A. Bohach, and K. W. Bayles.** 2003. Biphasic intracellular expression of *Staphylococcus aureus* virulence factors and evidence for Agr-mediated diffusion sensing. *Mol. Microbiol.* **49:**919–927.

83. **Sinha, B., P. Francois, Y. A. Que, M. Hussain, C. Heilmann, P. Moreillon, D. Lew, K. H. Krause, G. Peters, and M. Herrmann.** 2000. Heterologously expressed *Staphylococcus aureus* fibronectin-binding proteins are sufficient for invasion of host cells. *Infect. Immun.* **68:**6871–6878.

84. **Sinha, B., P. P. Francois, O. Nusse, M. Foti, O. M. Hartford, P. Vaudaux, T. J. Foster, D. P. Lew, M. Herrmann, and K. H. Krause.** 1999. Fibronectin-binding protein acts as *Staphylococcus aureus* invasin via fibronectin bridging to integrin a5b1. *Cell Microbiol.* **1:**101–117.

85. **Soderquist, B., J. Kallman, H. Holmberg, T. Vikerfors, and E. Kihlstrom.** 1998. Secretion of IL-6, IL-8 and G-CSF by human endothelial cells in vitro in response to *Staphylococcus aureus* and staphylococcal exotoxins. *APMIS* **106:**1157–1164.

86. **Sullam, P. M., A. S. Bayer, W. M. Foss, and A. L. Cheung.** 1996. Diminished platelet binding in vitro by *Staphylococcus aureus* is associated with reduced virulence in a rabbit model of infective endocarditis. *Infect. Immun.* **64:**4915–4921.

87. **Tekstra, J., H. Beekhuizen, J. S. Van De Gevel, I. J. Van Benten, C. W. Tuk, and R. H. Beelen.** 1999. Infection of human endothelial cells with *Staphylococcus aureus* induces the production of monocyte chemotactic protein-1 (MCP-1) and monocyte chemotaxis. *Clin. Exp. Immunol.* **117:**489–495.

88. **Tompkins, D. C., L. J. Blackwell, V. B. Hatcher, D. A. Elliott, C. O'Hagan-Sotsky, and F. D. Lowy.** 1992. *Staphylococcus aureus* proteins that bind to human endothelial cells. *Infect. Immun.* **60:**965–969.

89. **Tucker, K. A., S. S. Reilly, C. S. Leslie, and M. C. Hudson.** 2000. Intracellular *Staphylococcus aureus* induces apoptosis in mouse osteoblasts. *FEMS Microbiol. Lett.* **186:**151–156.

90. **van Langevelde, P., E. Ravensbergen, P. Grashoff, H. Beekhuizen, P. H. Groeneveld, and J. T. van Dissel.** 1999. Antibiotic-induced cell wall fragments of *Staphylococcus aureus* increase endothelial chemokine secretion and adhesiveness for granulocytes. *Antimicrob. Agents Chemother.* **43:**2984–2989.

91. **Vann, J. M., and R. A. Proctor.** 1988. Cytotoxic effects of ingested *Staphylococcus aureus* on bovine endothelial cells: role of *S. aureus* alpha-hemolysin. *Microb. Pathog.* **4:**443–453.

92. **Vercellotti, G. M., D. Lussenhop, P. K. Peterson, L. T. Furcht, J. B. McCarthy, H. S. Jacob, and C. F. Moldow.** 1984. Bacterial adherence to fibronectin and endothelial cells: a possible mechanism for bacterial tissue tropism. *J. Lab. Clin. Med.* **103:**34–43.

93. **Weglarczyk, K., J. Baran, M. Zembala, and J. Pryjma.** 2004. Caspase-8 activation precedes alterations of mitochondrial membrane potential during monocyte apoptosis induced by phagocytosis and killing of *Staphylococcus aureus*. *Infect. Immun.* **72:**2590–2597.

94. **Weidenmaier, C., J. F. Kokai-Kun, S. A. Kristian, T. Chanturiya, H. Kalbacher, M. Gross, G. Nicholson, B. Neumeister, J. J. Mond, and A. Peschel.** 2004. Role of teichoic acids in *Staphylococcus aureus* nasal colonization, a major risk factor in nosocomial infections. *Nat. Med.* **10:** 243–245.

95. **Wesson, C. A., J. Deringer, L. E. Liou, K. W. Bayles, G. A. Bohach, and W. R. Trumble.** 2000. Apoptosis induced by *Staphylococcus aureus* in epithelial cells utilizes a mechanism involving caspases 8 and 3. *Infect. Immun.* **68:** 2998–3001.

96. **Yao, L., V. Bengualid, F. D. Lowy, J. J. Gibbons, V. B. Hatcher, and J. W. Berman.** 1995. Internalization of *Staphylococcus aureus* by endothelial cells induces cytokine gene expression. *Infect. Immun.* **63:**1835–1839.

97. **Yao, L., F. D. Lowy, and J. W. Berman.** 1996. Interleukin-8 gene expression in *Staphylococcus aureus*-infected endothelial cells. *Infect. Immun.* **64:**3407–3409.

98. **Yeaman, M. R.** 1997. The role of platelets in antimicrobial host defense. *Clin. Infect. Dis.* **25:**951–968; quiz 969–970.

The Epidemiology of *Staphylococcus* Infections

FRED C. TENOVER AND RACHEL J. GORWITZ

43

The staphylococci are a remarkably diverse group of organisms that cause a variety of diseases, ranging from innocuous skin infections to often fatal forms of endocarditis. Staphylococci have been recognized as serious pathogens for over a century (16, 52), yet despite large-scale efforts to halt their spread, particularly in hospitals, they remain the most common cause of community- and health care-associated bacteremia. Of the approximately 2 million patients who acquire a health care-associated infection annually in the United States, approximately 230,000 will have an infection associated with *Staphylococcus aureus* (25, 59). The disease spectrum of the staphylococci includes abscesses, bacteremia, central nervous system infections, endocarditis, osteomyelitis, pneumonia, urinary tract infections, and a host of syndromes caused by exotoxins, including bullous impetigo, food poisoning, scalded skin syndrome, necrotizing pneumonia, and toxic shock syndrome. Remarkably, in addition to being the leading cause of bacteremia in the United States, staphylococci also are among the most common causes of food-borne illness and skin disease (2, 50, 85). Herein, some of the common illnesses caused by staphylococci, particularly *S. aureus*, are described.

COMMUNITY-ACQUIRED INFECTIONS

Colonization and Infection of Skin and Soft Tissues

Staphylococci colonize a sizable portion of the human population. Colonization affords organisms, such as *S. aureus*, the opportunity to gain access to skin sites, which, when infected, can serve as a source for more serious diseases, such as bacteremia, endocarditis, or toxemias (e.g., toxic shock syndrome). More commonly, however, breaches of skin sites result in either abscesses or furuncles (commonly known as boils) that can progress to carbuncles (more serious, deep-seated infections of several hair follicles). Approximately 30% of the population is stably colonized with *S. aureus*, and as many as 30 to 50% of the population may show transient colonization of the nares, axilla, perineum, or vagina (26, 45, 47, 55). Diabetics, intravenous (i.v.) drug users, pa-

tients on dialysis, and patients with AIDS have higher rates of *S. aureus* colonization (29, 62, 77, 86, 87). Hospitalized patients and health care workers are also at higher risk of becoming colonized for extended periods of time (26).

Colonization, particularly of the nares, by *S. aureus* leads to hand carriage, and from the hands, the organisms are frequently spread to other areas of the body. Thus, staphylococci often follow a nose-to-hands-to-wound route of infection (26, 87). Decreasing colonization, such as through the use of nasal ointments containing mupirocin, can aid in reducing both health care-associated and familial spread of staphylococci that can cause diseases such as recurrent folliculitis and furuncles (41, 45, 46, 86). Normally, large numbers of staphylococci are required to cause infection, but disruption of skin barriers, as seen in i.v. drug users or diabetics, increases the risk of staphylococcal infections. Dialysis patients are at particularly high risk for infection (86, 87).

Skin Infections

Approximately one-half of all skin infections are caused by *S. aureus* (24, 72). Infections include carbuncles, cellulitis, folliculitis, furuncles, hydradenitis suppurtiva, impetigo, mastitis, pyodermas, and pyomyositis. Impetigo, which involves release of epidermolytic toxins, can range from mild, recurrent infections to the more serious bullous impetigo, characterized by blisters that continually break and become infected, to the potentially life-threatening scalded skin syndrome (52, 72). Chronic *S. aureus* skin and soft tissue infections may be suggestive of a host factor disorder such as chronic granulomatous disease, an X-linked genetic disorder with frequent pyogenic infections. Although cellulitis is usually treated with antimicrobial chemotherapy, the major mode of treatment for folliculitis and abscesses is surgical drainage rather than antimicrobial agents.

Bacteremia and Endocarditis

Virtually any *S. aureus* infection can lead to bacteremia. *S. aureus* causes about 11 to 38% of community-acquired bacteremia (19, 45, 83). Mortality from *S. aureus* bacteremia ranges from 11 to 48%, a figure that has increased steadily for a number of years (56). In a review of data on adult bac-

Gram-Positive Pathogens, 2nd edition, edited by Vincent A. Fischetti et al.
© 2006 ASM Press, Washington, D.C.

teremia from three hospitals by Weinstein et al. (83), *S. aureus* was the most common cause of clinically significant bacteremia (18.9%), and non-*aureus* staphylococci (NAS) (9.2%) were the third most common cause (when organisms judged to be probable contaminants were excluded from analysis) (82). These results are similar to data from the Mayo Clinic (19), where *S. aureus* was the most common cause of bacteremia (18.4%) and NAS were the third most common cause (10.4%), with both organism groups showing a significant increase in frequency over the preceding 5-year period. In Weinstein's study, 50.6% of *S. aureus* infections were community acquired. The majority of *S. aureus* and NAS infections were related to indwelling lines or intravascular devices. However, the proportion of cases of *S. aureus* bacteremia was significantly lower in neutropenic patients, owing in part to the increase in gram-negative bacteremia arising from gastrointestinal sources. Among all causes of bacteremia, NAS showed the lowest associated mortality (5.5%); associated mortality for *S. aureus* was 11.9%. A recent meta-analysis compared the mortality associated with methicillin-resistant *S. aureus* (MRSA) and methicillin-susceptible *S. aureus* and noted a significantly higher mortality rate associated with MRSA bacteremia (20). Another study reported that mortality associated with nosocomial MRSA bacteremia is higher in those patients who were not nasal carriers (84).

Approximately 10 to 40% of community-acquired cases of *S. aureus* bacteremia progress to endocarditis (21, 44, 56). This figure is higher in i.v. drug users, often because they are heavily colonized with *S. aureus* and have frequent breaches of skin barriers, and is lower in patients with nosocomial bacteremia. Most patients have no apparent focus for the bacteremia preceding endocarditis. *S. aureus* differs from many other pathogens in that it can cause infectious endocarditis on a normal, native heart valve. *S. aureus* infectious endocarditis can present as right-sided endocarditis, primarily in i.v. drug users, left-sided native valve endocarditis, or prosthetic valve endocarditis (PVE) (37, 42).

Only 2 to 13% of cases of NAS bacteremia progress to endocarditis, although NAS are the most common cause of PVE (37, 42). PVE is divided into early PVE, i.e., those cases occurring within 60 days of valve replacement, and late PVE, which includes those infections occurring later than 60 days after valve replacement (42). Thirty to 67% of early PVE cases are caused by NAS, compared with only 20 to 28% of late PVE cases. Late PVE caused by NAS is usually community-acquired; early onset is health care-associated. In many cases of late PVE there is no obvious source. Most late PVE cases are caused by methicillin-resistant strains of *Staphylococcus epidermidis* (44), which are common in both the community and hospitalized patients.

Osteomyelitis

Staphylococcal osteomyelitis is classified as either acute or chronic (45, 80). Acute hematogenous osteomyelitis is usually a disease of children, primarily neonates, in whom it affects the long bones of the lower extremity (66, 80). The disease can be cured if detected early and treated with intensive antimicrobial therapy. Frequent therapeutic failures are seen, particularly if therapy is discontinued in less than 4 weeks (66). In adults, *S. aureus* bacteremia rarely leads to osteomyelitis of long bones; instead, vertebral bodies are more commonly affected. Chronic osteomyelitis, which can complicate open fractures or penetrating wounds, is usually an infection of at least 6 months' duration, is more commonly seen in adults, and is often refractory to cure with an-

timicrobial therapy. Chronic osteomyelitis can also develop after an initial, but failed, attempt at antimicrobial therapy. Although fluoroquinolones have shown excellent activity in vitro against gram-negative pathogens, none of the current quinolones provide optimal staphylococcal coverage even when combined with other agents such as rifampin (66, 80). In addition, quinolone usage in children is generally discouraged due to potential for damage to immature cartilage (1).

Toxin-Mediated Diseases

Several staphylococcal diseases are mediated by toxins, including impetigo (see above), food poisoning, necrotizing pneumonia, and toxic shock syndrome. Staphylococcal food poisoning is a result of ingesting one of several staphylococcal enterotoxins, the most ubiquitous of which is enterotoxin A (40, 50, 85). Because this is a true toxin-mediated disease, viable staphylococci are not always present in the contaminated vehicle. Staphylococcal food poisoning is characterized by nausea, vomiting, headache (which can be severe), and, less commonly, diarrhea (2, 40, 50, 85). Time to onset of symptoms after consuming contaminated food averages 4.4 h (40, 50). Because it is usually a self-limited disease that rarely leads to systemic infection, and because reporting by state health departments is passive and incomplete, the true incidence of staphylococcal food poisoning is unknown (2, 50).

The disease mediated by toxic shock syndrome toxin 1 (TSST-1) was first described in 1978 (76). Toxic shock syndrome has been associated with the use of highly absorbent tampons (71) but has been described in association with many types of *S. aureus* infections, some of which can be quite minor initially (22, 65). Toxic shock syndrome consists of fever above 39°C; hypotension; rash, usually followed by desquamation of the skin, especially on the palms and soles; hyperemia of mucous membranes; and involvement of multiple organ systems, as evidenced by diarrhea, thrombocytopenia, cardiopulmonary dysfunction, or a variety of other symptoms (22, 71). TSST-1 is a potent superantigen eliciting a variety of cytokines that, along with tumor necrosis factor, contribute to the severity of illness (69, 78). A single clone of *S. aureus* causes the majority of cases of menstrual toxic shock (57). Other superantigen toxins, especially enterotoxins A and B, are responsible for the majority of nonmenstrual toxic shock cases. Most adults are immune to TSST-1 and are not susceptible to TSST-1-mediated toxic shock. A small number of individuals are anergic to this toxin and suffer repeated bouts of the syndrome. Toxin-mediated diseases caused by other staphylococcal toxins, including scalded skin syndrome, are infrequent in the United States, but there are few data available on the prevalence of such diseases in other parts of the world.

Panton-Valentine leukocidin (PVL) is a bicomponent synergohymenotropic staphylococcal cytotoxin that causes leukocyte destruction and tissue necrosis (33, 51). Although its precise role in disease is unclear, it has been associated with necrotic lesions involving the skin and severe necrotizing pneumonia (33, 51). In one study of *S. aureus* isolates collected at a reference center in France between 1985 and 1998, PVL genes were detected in 93% of sampled strains associated with furunculosis and 85% of those associated with severe necrotic hemorrhagic pneumonia (51). The PVL-associated pneumonia cases were all community acquired and mainly involved immunocompetent children and young adults with previously normal lungs and a predisposing viral infection. In contrast, PVL genes were not detected in any of the S. *aureus* strains associated with hospital-acquired pneumonia. A subsequent study comparing PVL-positive and PVL-negative S. *aureus* pneumonia cases found that pneumonia cases caused by PVL-positive strains occurred in younger individuals, were more often preceded by an influenza-like illness, and were more likely to produce fever greater than 39°C, heart rate above 140 beats per min, hemoptysis, pleural effusion, and leukopenia (33). The survival rate 48 h after admission was also significantly lower for patients infected with PVL-positive S. *aureus* strains as compared to those with PVL-negative strains (63% versus 94%).

Community-Associated Pneumonia

Community-acquired pneumonia caused by S. *aureus* is not common but does occur, often as a consequence of influenza. Typically, S. *aureus* pneumonia develops in elderly patients approximately 4 to 14 days following the onset of influenza. Although such patients initially begin to improve clinically from influenza, they become increasingly ill with symptoms of high fever, productive sputum, and often respiratory failure (52). This postinfluenza pneumonia usually responds to antimicrobial agents if treated early, but can be fatal.

HEALTH CARE-ASSOCIATED INFECTIONS

Staphylococci are among the most common causes of health care-associated infections, including bacteremia, surgical site infections (SSIs), and pneumonia (25, 59, 60). Data from the National Nosocomial Infections Surveillance (NNIS) system from intensive care units (ICUs) for 2000 to 2004 show S. *aureus* to be the most common cause of nosocomial pneumonia and surgical site infections, and the third most common cause of nosocomial bloodstream infection (Centers for Disease Control and Prevention [CDC], unpublished data). Outbreaks of health care-associated infections with S. *aureus* occur in patients across the spectrum of ages, from adults in tertiary care centers to infants in hospital nurseries. The percentage of S. *aureus* infections caused by MRSA in ICUs from 1995 to 2003 is shown in Fig. 1 by age group. Overall the percentage rose from 33% in 1995 to 60% in 2003. The highest percentage was in patients >64 years of age, which in 2003 was 72%. Outbreaks in neonates usually result in skin infections and bacteremia, although more invasive diseases, such as osteomyelitis and meningitis, can occur (36). Transmission of S. *aureus* strains from health care workers to patients has resulted in mediastinitis (32), nosocomial toxic shock syndrome (48), and the spread of antimicrobial-resistant strains within an institution (23). One particularly virulent strain of S. *aureus* was shown to have spread from the index patient, who introduced the strain into Canada from a hospital in India, to multiple patients in medical centers across Canada (67).

Bloodstream Infections

Primary bloodstream infections are defined as infections in which no other primary site can be discerned (5). Diagnosis is usually based on positive blood cultures drawn from one or more venipuncture sites or from intravascular catheters. S. *aureus* was the third most frequent pathogen associated with bloodstream infections in the NNIS system from 2000 to 2003 (CDC, unpublished data) and a frequent cause of health care-associated bacteremia in other countries as well (41, 75, 79). As noted in the study by Weinstein et al. (83),

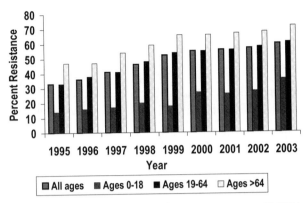

FIGURE 1 NNIS system data showing percentage of MRSA isolates among all S. *aureus* isolates causing infections in patients in U.S. intensive care units from 1995 to 2003 by patient age (CDC, unpublished data).

these infections are often related to the presence of an intravascular device. Bloodstream infections associated with these devices are serious and more often occur in patients with (i) central i.v. lines that were urgently or emergently placed, (ii) lines that have been in place for at least 72 h, and (iii) lines placed in femoral veins. Removal of the purportedly infected catheter, followed by bacteriologic culture using semiquantitative methods, can be helpful for diagnosis and a guide for antimicrobial chemotherapy.

Many medical and surgical procedures in which a device is indwelling but not implanted can lead to a device-associated infection with or without bloodstream infection. Staphylococci are often associated with this type of infection (19, 79, 83). In situations where an implantable device is left in place, NAS are the most frequent cause of infections (5, 54, 83).

Pneumonia

S. *aureus* is a common etiologic agent of pneumonia in the hospital setting (52). Risk factors for S. *aureus* pneumonia are similar to those for health care-associated pneumonia in general and include recent surgery, particularly thoracic or high abdominal surgery; chronic obstructive lung disease; ventilator therapy; older age; and immunosuppressive therapy (52). Mortality is approximately 15 to 20%. Therapy must be aggressive and usually consists of a β-lactamase-resistant penicillin, such as nafcillin, or vancomycin for several weeks.

S. *aureus* remains the most common cause of empyema, i.e., infections in the pleural space. Acute empyema usually results from direct extension from a pneumonia or lung abscess or from hematogenous seeding (52). Chest pain, fever, shortness of breath, and signs of pleural effusion are nearly always present. Thoracocentesis is required for diagnosis, but therapy nearly always requires chest tube placement and, frequently, surgery, because loculations are common complications of S. *aureus* infections.

SSIs

SSIs constituted approximately 15% of the health care-associated infections reported to the NNIS system by hospitals that collected hospitalwide surveillance data (25, 59). SSIs are a major infection control concern because they are associated with serious morbidity, mortality, and high health care cost. The risk of SSI is related to a number of factors. Among the most important are the operative procedure performed; the degree of microbiological contamination of the operative field; the duration of operation; and the intrinsic susceptibility of the patient, due to advanced age, malnutrition, trauma, loss of skin integrity (e.g., burns), or the presence of an infection at a distant site. Typically occurring 7 to 10 days following the operative procedure, SSIs have a wide range of onset times, i.e., from 2 to 30 days. The time of onset may be strongly influenced by the administration of the appropriate antimicrobial agents prophylactically. The dose, timing, and selection of the prophylactic agents of choice have been reviewed elsewhere (46).

S. *aureus* is the most common cause of SSIs among ICU patients in the United States (CDC, unpublished data). This is largely due to its common presence on the skin or mucous membranes of patients, although other factors tend to influence the development of an SSI, as noted above. The use of prosthetic materials has a strong influence on the development of an SSI with S. *aureus*. Elek and Conen (24) have shown that a subcutaneous injection of >10^5 CFU of S. *aureus* was easily controlled by host defense mechanisms, whereas 3×10^2 CFU of the same organism led to infection in the presence of prosthetic material such as a suture. The development of a glycocalyx by most strains of NAS and many strains of S. *aureus* is thought to protect these organisms from host defenses on the surface of prosthetic material (52). For effective treatment, removal of the foreign material is usually necessary, followed by antimicrobial therapy (46).

ANTIMICROBIAL RESISTANCE

Antimicrobial Resistance in the Hospital Setting

Several reports suggest that the prevalence of S. *aureus* strains resistant to methicillin, oxacillin, or nafcillin is increasing in the United States and abroad (3, 6, 41, 64, 75, 87) and that such strains can cause outbreaks (3, 23, 36, 67). NNIS data show that the pooled mean of percent MRSA in 154 ICUs was 51.6% through June 2003 (59). The pooled mean %MRSA was 42.0% among non-ICU inpatient areas (55 units) and 25.9% in outpatient areas (49 units). The MRSA rate from January to December 2002 was 57.1%, which represents a 13% increase over the MRSA rate from 1997 to 2001 (59).

S. *aureus* clinical isolates with intermediate levels of resistance to vancomycin (vancomycin MICs of 8 to 16 μg/ml), termed "vancomycin-intermediate S. *aureus*," were first identified in the late 1990s (10, 39, 70, 73). A recent case-control study compared patients infected with S. *aureus* strains for which the vancomycin MICs were 4 μg/ml (i.e., S. *aureus* with reduced susceptibility to vancomycin) to those infected with vancomycin-susceptible MRSA strains. Patients infected with S. *aureus* with reduced susceptibility to vancomycin were significantly more likely to have had antecedent vancomycin use and prior MRSA infection in the second to third month before the current infection (30).

In 2002, the first two reports of infections caused by S. *aureus* for which the vancomycin MICs were ≥32 μg/ml (i.e., vancomycin-resistant S. *aureus* or VRSA) occurred in the United States in Michigan (17) and Pennsylvania (74). A third VRSA case was identified in New York in 2004 (15). The isolates from these three infections differed in their pulsed-field gel electrophoresis (PFGE) profiles, showing them to be independent in origin, even though they were generally from the same major lineage of MRSA (17, 74). All three isolates were resistant to both vancomycin

and β-lactam agents. In contrast to the chromosomally mediated vancomycin resistance seen in the vancomycin-intermediate S. *aureus* strains (39, 73), the VRSA isolates possessed high-level vancomycin resistance conferred by the presence of the *van*A operon, acquired via conjugal transfer in vivo from a vancomycin-resistant enterococcus (81). Of particular concern, automated antimicrobial susceptibility testing methods commonly used in the United States were unable to detect vancomycin resistance in these isolates (15, 74). For this reason, the CDC recommends that laboratories performing automated susceptibility testing of S. *aureus* strains should include a vancomycin-agar screening plate containing 6 μg of vancomycin per ml and examine the plate for growth after 24 h of incubation to ensure detection of all VRSA isolates.

Antimicrobial Resistance among Community-Associated S. *aureus* Strains

The first outbreak of MRSA infections acquired in the community among individuals without traditional risk factors for health care-associated MRSA (HA-MRSA) was reported in 1980 (9). In the late 1990s, four fatal cases of MRSA sepsis in previously healthy children received national attention (11). MRSA infections in children without exposure to health care were also noted in Chicago during this same time period (38). Since then, clusters of community-associated MRSA (CA-MRSA) infections have been reported among other groups of children (8), prisoners (14), athletes (13), Native Americans (35), military recruits (88), and men who have sex with men (12).

In contrast to HA-MRSA isolates, CA-MRSA isolates tend to be susceptible to most antimicrobial agents other than β-lactams and erythromycin, although patterns of resistance vary geographically (18, 27, 58). CA-MRSA isolates also differ by PFGE and multilocus sequence type and have a different array of virulence factors when compared with HA-MRSA isolates (4, 27, 63). The gene conferring methicillin resistance in S. *aureus* (*mec*A) is carried on a mobile genetic element termed staphylococcal cassette chromosome *mec* (SCC*mec*). Of the five SCC*mec* types that have been identified thus far, CA-MRSA isolates most often contain the SCC*mec* type IV (53, 63), which is smaller, potentially more mobile, and less likely to carry other antimicrobial resistance genes than SCC*mec* types I and II, which are typically found in HA-MRSA strains (58). Furthermore, although PVL is rare among HA-MRSA strains, it is commonly present in CA-MRSA strains (4, 27, 33, 51, 58).

The clinical presentations and sites of infection of CA-MRSA strains parallel those seen with methicillin-susceptible S. *aureus* infections in the community, in contrast with the MRSA infections seen in health care settings. In a large prospective population-based study of MRSA infections, skin and soft tissue infections comprised 75% of CA-MRSA infections, as compared to only 37% of HA-MRSA infections, which were more likely than CA-MRSA strains to cause respiratory tract, urinary tract, and bloodstream infections (58).

CA-MRSA infections are more likely to occur in younger individuals, and in individuals from minority ethnic groups or of low socioeconomic status, whereas risk factors for HA-MRSA infection include older age, immunosuppression, dialysis, recent surgery, and use of invasive devices. Recent antimicrobial agent use is a risk factor for both CA-MRSA and HA-MRSA (34, 68).

There is a subset of MRSA isolates (including both HA-MRSA and CA-MRSA) that is resistant to erythromycin but susceptible to clindamycin on initial susceptibility testing, but demonstrates inducible resistance to clindamycin (28, 43). Although the clinical significance of this inducible resistance is unclear, whenever clindamycin is being considered as a therapeutic option, the isolate should be tested for inducible clindamycin resistance using the clindamycin disk induction test (D zone test) (43, 61).

CONTROL MEASURES

Infection Control in the Health Care Setting

Transmission of infection in health care settings requires three elements: a source of infecting microorganisms, a susceptible host, and a means of transmission for the microorganism (7, 31). For S. aureus infections, the human sources may be the patient, health care personnel, or, occasionally, outside visitors. These sources may be either infected or colonized with S. aureus. The role of the environment in transmitting staphylococcal infections remains controversial, although contaminated products, such as anesthetics, can clearly transmit infections (49). The host factors related to infection with S. aureus depend on the site of infection, the patient's immunologic status, and the infectious dose. Transmission may be by direct or indirect contact with the source or by contact with a common vehicle. Food-borne transmission of toxin-producing S. aureus is uncommon in health care settings.

Contact transmission is the most important and frequent mode of transmission for S. aureus (7, 31). Direct contact transmission involves contact of body surface to body surface and physical transfer of S. aureus to the host from an infected or colonized person (26). For example, a surgical resident colonized in his nares with MRSA transmitted staphylococcal infections to six patients during two distinct outbreaks (32). The organisms from the patients were shown by bacteriophage typing and antibiogram analysis to be the same strain as that harbored by the resident. Several attempts to eradicate the MRSA from the resident's nares with various antimicrobial agent combinations were initially unsuccessful. However, mupirocin was eventually successful in eradicating the organism. Kreiswirth et al. (48) reported a similar case of transmission of S. aureus from a neurosurgeon, who was a disseminator of a TSST-1-producing strain of S. aureus, to two patients who developed symptoms of toxic shock syndrome following surgical procedures. DNA probe typing confirmed the similarity of the three isolates from the neurosurgeon and the two patients. The health care-associated infections occurred 5 years apart.

Indirect contact transmission involves contact of a susceptible host with a contaminated intermediate object, usually inanimate, such as gloves that are not changed between patients. In the health care setting, indirect contact transmission, usually on the gloves or hands of health care personnel, is often the dominant mode of transmission for S. aureus. Hand washing remains the single most important measure to reduce the risk of transmitting microorganisms from one person to another (25, 31). In addition to hand washing, gloves play an important role in reducing the risks of transmission of S. aureus in health care settings. However, wearing gloves does not eliminate the need for hand washing, because gloves may have small, inapparent defects, may be torn during use, or may not be changed between patients.

Containment measures for patients with infections with S. aureus usually require standard precautions, which include hand washing after touching body fluids or contaminated items, whether or not gloves are worn; clean, nonsterile glove use before touching mucous membranes or body fluids; clean, nonsterile gown use to protect skin and prevent soiling of clothing during procedures and patient care activities that are likely to generate splashes of body fluids; and handling patient care equipment, environmental surfaces, and linens in a manner that prevents skin and mucous membrane exposures and avoids transfer of microorganisms to other patients or the environment. In general, private rooms are not needed if patients can undertake the usual personal hygiene measures and are unlikely to contaminate the environment (31). If a patient has an SSI or draining abscess with S. aureus, contact isolation may be needed. Contact isolation requires a private room, routine gown use, and dedicated patient care equipment for the duration of illness. Nurseries pose special problems for containment of MRSA infections. In one institution, continual transmission of an endemic strain of MRSA over a 3-year period was interrupted only when the institution applied an antibacterial solution of three dyes to the umbilical stumps of infants on the unit to reduce colonization with MRSA, increased the number of nurses on the unit to alleviate understaffing, and added a dedicated infection control nurse to tighten infection control efforts (36). The few cases of MRSA that occurred after these interventions were shown by PFGE typing to be caused by strains different from the endemic MRSA strain that had plagued the institution for more than 3 years. Indeed, the endemic strain had been eradicated from the medical center.

Control of S. aureus Transmission in the Community

Modes of MRSA transmission in the community are similar to those in health care settings. Clusters of MRSA infection in the community have occurred in settings where individuals are in close proximity to one another, have frequent skin-to-skin contact, engage in activities that may result in compromised skin surfaces, share clothing or equipment, or do not maintain good hygiene (12–14). Guidelines for the prevention of MRSA transmission in community settings therefore focus on maintaining good hygiene, covering skin lesions, avoiding contact with drainage of skin lesions from other individuals, cleaning shared equipment between uses, avoiding sharing personal items, and seeking medical attention promptly if skin lesions occur (12, 14). Although their effectiveness has not been established in controlled trials, other measures that have been employed and that may be helpful in an outbreak setting include the use of nasal decolonization with agents such as 2% mupirocin ointment and short-term implementation of bathing with antiseptic solutions such as 4% chlorhexidine (14).

Treatment of the S. aureus Carrier State

Staphylococcal infection and carriage occur commonly in humans. Fortunately, health care personnel serve as reservoirs (and disseminators) of S. aureus much less commonly than do patients (7). Carriage of S. aureus is most common in the anterior nares, although other skin sites may be involved. The frequency of carriage among health care personnel ranges from 20 to 90%, but fewer than 10% of health care workers are ever linked to disseminating S. aureus, probably because they are colonized with low numbers of organisms. Carriage of S. aureus in the nares has been shown to correspond to hand carriage, and persons with skin lesions caused by the organism are more likely than asymptomatic nasal carriers to disseminate the bacteria (86, 87). Culture surveys of health care personnel can detect carriers but do not indicate who among them can disseminate *S. aureus* efficiently. Thus, such surveys are not cost-effective and may subject personnel with positive cultures to unnecessary treatment and removal from duty. An epidemiologic investigation of clusters of *S. aureus* infections is needed to identify personnel who are disseminators of *S. aureus*. Such implicated personnel can then be removed from clinical duties until carriage has been eradicated.

Several antimicrobial regimens have been used to eradicate carriage of *S. aureus*. These include oral agents, i.e., ciprofloxacin, trimethoprim-sulfamethoxazole, and rifampin, or topical agents such as mupirocin (3, 41, 86, 87). Unfortunately, antimicrobial resistance to all these agents has emerged, underscoring the need for judicious antimicrobial treatment of the carrier state and avoidance of the indiscriminate culture survey and broad-scale treatment of culture-positive personnel.

Restriction from patient care activities or food handling is indicated for personnel who have draining skin lesions with *S. aureus* until they have received appropriate therapy. No work restrictions are needed for colonized personnel unless they have been epidemiologically implicated in *S. aureus* transmission in a facility.

REFERENCES

1. **American Academy of Pediatrics.** 2003. Antimicrobial agents and related therapy, p. 693–694. *In* L. K. Pickering (ed.), *Red Book: 2003 Report of the Committee on Infectious Diseases.* American Academy of Pediatrics, Elk Grove, Ill.

2. **Armstrong, G. L., J. Hollingsworth, and J. G. Morris, Jr.** 1998. Bacterial foodborne disease, p. 109–138. *In* A. S. Evan and P. S. Brachman (ed.), *Bacterial Infections of Humans. Epidemiology and Control,* 3rd ed. Plenum Medical Book Company, New York, N.Y.

3. **Aubry-Damon, H., P. Legrand, C. Brun-Buisson, A. Astier, C. J. Soussy, and R. Leclercq.** 1997. Reemergence of gentamicin-susceptible strains of methicillin-resistant *Staphylococcus aureus*: roles of an infection control program and changes in aminoglycoside use. *Clin. Infect. Dis.* **25**:647–653.

4. **Baba, T., F. Takeuchi, M. Kuroda, H. Yuzawa, K. Aoki, A. Oguchi, Y. Nagai, N. Iwama, K. Asnao, T. Naimi, H. Kuroda, L. Cui, K. Yamamoto, and K. Hiramasu.** 2002. Genome and virulence determinants of high virulence community-acquired MRSA. *Lancet* **359**:1819–1827.

5. **Banerjee, S., G. Emori, D. H. Culver, R. P. Gaynes, W. J. Martone, T. G. Emori, T. C. Horan, J. R. Edwards, W. R. Jarvis, J. S. Tolson, T. S. Henderson, J. M. Hughes, and the National Nosocomial Infections Surveillance (NNIS) System.** 1991. Trends in nosocomial bloodstream infections in the United States, 1980–1989. *Am. J. Med.* **91**:(Suppl. 3B):86S–89S.

6. **Barrett, S. P., R. V. Mummery, and B. Chattopadhyay.** 1998. Trying to control MRSA causes more problems than it solves. *J. Hosp. Infect.* **39**:85–93.

7. **Bolyard, E. A., O. C. Tablan, and W. W. Williams.** 1998. Guideline for infection control in healthcare personnel, 1998. *Am. J. Infect. Control* **26**:289–354.

8. **Buckingham, S. C., L. K. McDougal, L. Cathey, K. Comeaux, M. Brown, A. S. Craig, S. K. Fridkin, and F. C. Tenover.** 2004 Emergence of community-associated methicillin-resistant *Staphylococcus aureus* in Memphis children. *Pediatr. Infect. Dis. J.* **23**:619–624.

9. **Centers for Disease Control.** 1981. Methicillin-resistant *Staphylococcus aureus*—United States. *Morb. Mortal. Wkly. Rep.* **30**:557–559.

10. **Centers for Disease Control and Prevention.** 1997. *Staphylococcus aureus* with reduced susceptibility to vancomycin—United States, 1997. *Morb. Mortal. Wkly. Rep.* **46**:765–766.

11. **Centers for Disease Control and Prevention.** 1999. Four pediatric deaths from community–acquired methicillin-resistant *Staphylococcus aureus*—Minnesota and North Dakota, 1997–1999. *Morb. Mortal. Wkly. Rep.* **48**:707–710.

12. **Centers for Disease Control and Prevention.** 2003. Outbreaks of community-associated methicillin-resistant *Staphylococcus aureus* skin infections—Los Angeles County, California, 2002–2003. *Morb. Mortal. Wkly. Rep.* **52**:88.

13. **Centers for Disease Control and Prevention.** 2003. Methicillin-resistant *Staphylococcus aureus* infections among competitive sports participants—Colorado, Indiana, Pennsylvania, and Los Angeles County, 2000–2003. *Morb. Mortal. Wkly. Rep.* **52**:793–795.

14. **Centers for Disease Control and Prevention.** 2003. Methicillin-resistant *Staphylococcus aureus* infections in correctional facilities—Georgia, California, and Texas, 2001–2003. *Morb. Mortal. Wkly. Rep.* **52**:992–996.

15. **Centers for Disease Control and Prevention.** 2004. Vancomycin-resistant *Staphylococcus aureus*—New York, 2004. *Morb. Mortal. Wkly. Rep.* **53**:322–323.

16. **Chambers, H. F.** 2001. The changing epidemiology of *Staphylococcus aureus*? *Emerg. Infect. Dis.* **7**:178–182.

17. **Chang, S., D. M. Sievert, J. C. Hageman, M. L. Boulton, F. C. Tenover, F. P. Downes, S. Shah, J. T. Rudrik, G. R. Pupp, W. J. Brown, D. Cardo, S. K. Fridkin, and the Vancomycin-Resistant *Staphylococcus aureus* Investigative Team.** 2003. Infection with vancomycin-resistant *Staphylococcus aureus* containing the *vanA* resistance gene. *N. Engl. J. Med.* **348**:1342–1347.

18. **Charlbois, E. D., F. Perdreau-Remington, B. Kreiswirth, D. R. Bangsberg, D. Ciccarone, B. A. Diep, V. L. Ng, K. Chansky, B. Edlin, and H. F. Chambers.** 2004. Origins of community strains of methicillin-resistant *Staphylococcus aureus*. *Clin. Infect. Dis.* **39**:47–54.

19. **Cockerill, F. R., III, J. G. Hughes, E. A. Vetter, R. A. Mueller, A. L. Weaver, D. M. Ilstrup, J. E. Rosenblatt, and W. R. Wilson.** 1997. Analysis of 281,797 consecutive blood cultures performed over an eight-year period: trends in microorganisms isolated and the value of anaerobic culture of blood. *Clin. Infect. Dis.* **24**:403–418.

20. **Cosgrove, S. E., G. Sakoulas, E. N. Perencevich, M. J. Schwaber, A. W. Karchmer, and Y. Carmeli.** 2004. Comparison of mortality associated with methicillin-resistant and methicillin-susceptible *Staphylococcus aureus* bacteremia: a meta-analysis. *Clin. Infect. Dis.* **36**:53–59.

21. **Dajani, A. S., K. A. Taubert, W. Wilson, A. F. Bolger, A. Bayer, P. Ferrieri, M. H. Gewitz, S. T. Shulman, S. Nouri, J. W. Newburger, C. Hutto, T. J. Pallasch, T. W. Gage, M. E. Levison, G. Peter, and G. Zuccaro, Jr.** 1997. Prevention of bacterial endocarditis: recommendations by the American Heart Association. *Clin. Infect. Dis.* **25**:1448–1458.

22. **Davis, J. P., P. J. Chesney, P. J. Wand, M. LaVenture, and the Investigation and Laboratory Team.** 1980. Toxic-shock syndrome. *N. Engl. J. Med.* **303**:1429–1435.

23. **Dominguez, M. A., H. de Lencastre, J. Linares, and A. Tomasz.** 1994. Spread of a dominant methicillin-resistant *Staphylococcus aureus* (MRSA) clone during an outbreak of

MRSA disease in a Spanish hospital. *J. Clin. Microbiol.* **32:**2081–2087.

24. **Elek, S. D., and P. E. Conen.** 1957. The virulence of *Staphylococcus aureus* for man. A study of the problems of wound infection. *Br. J. Exp. Pathol.* **38:**573–577.

25. **Emori, T. G., and R. P. Gaynes.** 1993. An overview of nosocomial infections including the role for the microbiology laboratory. *Clin. Microbiol. Rev.* **6:**428–442.

26. **Fekety, F. R., Jr.** 1964. The epidemiology and prevention of staphylococcal infection. *Medicine* (Baltimore) **43:**593–618.

27. **Fey, P. D., B. Said-Salim, M. E. Rupp, S. H. Hinrichs, D. J. Boxrud, C. C. Davis, B. N. Kreiswirth, and P. M. Schlievert.** 2003. Comparative molecular analysis of community- or hospital-acquired methicillin-resistant *Staphylococcus aureus. Antimicrob. Agents Chemother.* **47:**196–203.

28. **Fiebelkorn, K. R., S. A. Crawford, M. L. McElmeel, and J. H. Jorgensen.** 2003. Practical disk diffusion method for detection of inducible clindamycin resistance in *Staphylococcus aureus* and coagulase-negative staphylococci. *J. Clin. Microbiol.* **41:**4740–4744.

29. **Frank, U., F. D. Daschner, G. Schulgen, and J. Mills.** 1997. Incidence and epidemiology of nosocomial infections in patients infected with human immunodeficiency virus. *Clin. Infect. Dis.* **25:**318–320.

30. **Fridkin, S. K., J. Hageman, L. K. McDougal, J. Mohammed, W. R. Jarvis, T. M. Perl, F. C. Tenover, and the Vancomycin-Intermediate *Staphylococcus aureus* Epidemiology Study Group.** 2003. Epidemiological and microbiological characterization of infections caused by *Staphylococcus aureus* with reduced susceptibility to vancomycin, United States, 1997–2001. *Clin. Infect. Dis.* **36:**429–439.

31. **Garner, J. S.** 1996. Guideline for isolation precautions in hospitals. *Infect. Control Hosp. Epidemiol.* **17:**53–80.

32. **Gaynes, R., R. Marosok, J. Mowry-Hanley, C. Laughlin, K. Foley, C. Friedman, and M. Kirsh.** 1991. Mediastinitis following coronary artery bypass surgery: a 3-year review. *J. Infect. Dis.* **163:**117–121.

33. **Gillet, Y., B. Issartel, P. Vanhems, J.-C. Fournet, G. Lina, M. Bes, F. Vandeenesch, Y. Piémont, N. Brousse, D. Floret, and J. Etienne.** 2002. Association between *Staphylococcus aureus* strain carrying gene for Panton-Valentine leukocidin and highly lethal necrotising pneumonia in young immunocompetent patients. *Lancet* **359:**753–759.

34. **Graffunder, E. M., and R. A. Venezia.** 2002. Risk factors associated with nosocomial methicillin-resistant *Staphylococcus aureus* (MRSA) infection including previous use of antimicrobials. *J. Antimicrob. Chemother.* **49:**999–1005.

35. **Groom, A. V., D. H. Wolsey, T. S. Naimi, K. Smith, S. Johnson, D. Boxrud, K. A. Moore, and J. E. Cheek.** 2001 Community-acquired methicillin-resistant *Staphylococcus aureus* in a rural American Indian community. *JAMA* **286:**2101–2105.

36. **Haley, R. W., N. B. Cushion, F. C. Tenover, T. L. Bannerman, D. Dryer, J. Ross, P. J. Sanchez, and J. D. Siegel.** 1995. Eradication of endemic methicillin-resistant *Staphylococcus aureus* infections from a neonatal intensive care unit. *J. Infect. Dis.* **171:**614–624.

37. **Harris, S. L.** 1992. Definitions and demographic characteristics, p. 1–18. *In* D. Kaye (ed.), *Infective Endocarditis*, 2nd ed. Raven Press, Ltd., New York, N.Y.

38. **Herold, B., L. C. Immergluck, M. C. Maranan, D. S. Lauderdale, R. E. Gaskin, S. Boyle-Vavra, C. D. Leitch, R. S. Daum.** 1998. Community-acquired methicillin-resistant *Staphylococcus aureus* in children with no predisposing risk. *JAMA* **279:**593–598.

39. **Hiramatsu, K., H. Hanaki, T. Ino, K. Yabuta, T. Oguri, and F. C. Tenover.** 1997. Methicillin-resistant *Staphylococcus aureus* clinical strain with reduced vancomycin susceptibility. *J. Antimicrob. Chemother.* **40:**135–136.

40. **Holmberg, S. D., and P. A. Blake.** 1984. Staphylococcal food poisoning in the United States. New facts and old misconceptions. *JAMA* **251:**487–489.

41. **Irish, D., I. Eltringham, A. Teall, H. Pickett, H. Farelly, S. Reith, and B. Cookson.** 1998. Control of an outbreak of an epidemic methicillin-resistant *Staphylococcus aureus* also resistant to mupirocin. *J. Hosp. Infect.* **39:**19–26.

42. **Ivert, T. S. A., W. E. Dismukes, C. G. Cobbs, E. H. Blackstone, J. W. Kirklin, and L. A. Bergdahl.** 1984. Prosthetic valve endocarditis. *Circulation* **69:**223–232.

43. **Jorgensen, J. H., S. A. Crawford, M. L. McElmeel, and K. R. Fiebelkorn.** 2004. Detection of inducible clindamycin resistance of staphylococci in conjunction with performance of automated broth susceptibility testing. *J. Clin. Microbiol.* **42:**1800–1802.

44. **Karchmer, A. W.** 1992. Staphylococcal endocarditis, p. 225–249. *In* D. Kaye (ed.), *Infective Endocarditis*, 2nd ed. Raven Press, Ltd., New York, N.Y.

45. **Kauffman, C. A., and S. F. Bradley.** 1997. Epidemiology of community-acquired infection, p. 287–308. *In* K. B. Crossley and G. L. Archer (ed.), *The Staphylococci in Human Disease*. Churchill Livingstone, New York, N.Y.

46. **Kernodle, D. S., and A. B. Kaiser.** 1996. Postoperative infections and antimicrobial prophylaxis, p. 2742–2756. *In* G. L. Mandell, J. E. Bennett, and R. Dolin (ed.), *Principles and Practice of Infectious Diseases*, 4th ed. Churchill Livingstone, New York, N.Y.

47. **Kluytmans, J., A. van Belkum, and H. Verbrugh.** 1997. Nasal carriage of *Staphylococcus aureus*: epidemiology, underlying mechanisms, and associated risks. *Clin. Microbiol. Rev.* **10:**505–520.

48. **Kreiswirth, B. N., G. R. Kravitz, P. M. Schlievert, and R. P. Novick.** 1986. Nosocomial transmission of a strain of *Staphylococcus aureus* causing toxic shock syndrome. *Ann. Intern. Med.* **105:**704–707.

49. **Kuehnert, M. J., R. M. Webb, E. M. Jochimsen, G. A. Hancock, M. J. Arduino, S. Hand, M. Currier, and W. R. Jarvis.** 1997. *Staphylococcus aureus* bloodstream infections among patients undergoing electroconvulsive therapy traced to breaks in infection control and possible extrinsic contamination by propofol. *Anesth. Analg.* **85:**420–425.

50. **Le Loir, Y., F. Baron, and M. Gautier.** 2003. *Staphylococcus aureus* and food poisoning. *Genet. Molec. Res.* **2:**63–76.

51. **Lina, G., Y. Piemont, F. Godail-Gamot, M. Bes, M.-O. Peter, V. Gauduchon, F. Vandensch, and J. Etienne.** 1999. Involvement of Panton-Valentine leukocidin-producing *Staphylococcus aureus* in primary skin infections and pneumonia. *Clin. Infect. Dis.* **29:**1128–1132.

52. **Lowy, F. D.** 1998. *Staphylococcus aureus* infections. *N. Engl. J. Med.* **339:**520–532.

53. **Ma, X. X., T. Ito, C. Tiensasitorn, M. Jamklang, P. Chongtrakool, S. Boyle-Vavra, R. S. Daum, and K. Hiramatsu.** 2002. Novel type of staphylococcal cassette chromosome mec identified in community-acquired methicillin-resistant *Staphylococcus aureus* strains. *Antimicrob. Agents Chemother.* **46:**1147–1152.

54. **Martin, M. A., M. A. Pfaller, and R. P. Wenzel.** 1989. Coagulase-negative staphylococcal bacteremia. Mortality and hospital stay. *Ann. Intern. Med.* **110:**9–16.

55. **Martin, R. R., V. Buttram, P. Besch, J. J. Kirkland, and G. P. Petty.** 1982. Nasal and vaginal *Staphylococcus aureus*

in young women: quantitative studies. *Ann. Intern. Med.* **96**(Pt. 2):951–953.

56. Mortara, L. A., and A. S. Bayer. 1993. *Staphylococcus aureus* bacteremia and endocarditis. *Infect. Dis. Clin. North Am.* **7:** 53–68.

57. Musser, J. M., P. M. Schlievert, A. W. Chow, P. Ewan, B. N. Kreiswirth, V. T. Rosdahl, A. S. Naidu, W. Witte, and R. K. Selander. 1990. A single clone of *Staphylococcus aureus* causes the majority of cases of toxic shock syndrome. *Proc. Natl. Acad. Sci. USA* **87:**225–229.

58. Naimi, T. S., K. H. LeDell, K. Como-Sabetti, S. M. Borchardt, D. J. Boxrud, J. Etienne, S. K. Johnson, F. Vandenesch, S. Fridkin, C. O'Boyle, R. N. Danila, and R. Lynfield. 2003. Comparison of community- and health care-associated methicillin-resistant *Staphylococcus aureus* infection. *JAMA* **290:**2976–2984.

59. National Nosocomial Infections Surveillance System. 2003. National Nosocomial Infections Surveillance (NNIS) report, data summary from January 1992 through June 2003, issued August 2003. *Am. J. Infect. Control* **31:** 481–498.

60. Náwas, T., A. Hawwari, E. Hendrix, J. Hebden, R. Edelman, M. Martin, W. Campbell, R. Naso, R. Schwalbe, and A. I. Fattom. 1998. Phenotypic and genotypic characterization of nosocomial *Staphylococcus aureus* isolates from trauma patients. *J. Clin. Microbiol.* **36:**414–420.

61. NCCLS. 2004. Performance standards for antimicrobial susceptibility testing: 14th Informational Supplement. NCCLS Document M100-S14. NCCLS, Wayne, Pa.

62. Noble, W. C., H. A. Valkenburg, and C. H. I. Wolters. 1967. Carriage of *Staphylococcus aureus* in random samples of a normal population. *J. Hyg.* (London) **65:**567–573.

63. Okuma, K., K. Iwakawa, J. D. Turnidge, W. B. Grubb, J. M. Bell, F. G. O'Brien, G. W. Coombs, J. W. Pearman, F. C. Tenover, M. Kapi, C. Tiensasitorn, T. Ito, and K. Hiramatsu. 2002. Dissemination of new methicillin-resistant *Staphylococcus aureus* clones in the community. *J. Clin. Microbiol.* **40:**4289–4294.

64. Panlilio, A. L., D. H. Culver, R. P. Gaynes, S. Banerjee, T. S. Henderson, J. S. Tolson, and W. J. Martone. 1992. Methicillin-resistant *Staphylococcus aureus* in U.S. hospitals, 1975–1991. *Infect. Control Hosp. Epidemiol.* **13:**582–586.

65. Parsonnet, J. 1996. Nonmenstrual toxic shock syndrome: new insights into diagnosis, pathogenesis, and treatment, p. 1–20. *In* J. S. Remington and M. N. Swartz (ed.), *Current Clinical Topics in Infectious Diseases.* Blackwell Science, Cambridge, United Kingdom.

66. Rissing, J. P. 1997. Antimicrobial therapy for chronic osteomyelitis in adults: role of the quinolones. *Clin. Infect. Dis.* **25:**1327–1333.

67. Roman, R. S., J. Smith, M. Walker, S. Byrne, K. Ramotar, B. Dycjk, A. Kabani, and L. E. Nicolle. 1997. Rapid geographic spread of a methicillin-resistant *Staphylococcus aureus* strain. *Clin. Infect. Dis.* **25:**698–705.

68. Sattler, C., E. O. Mason, and S. L. Kaplan. 2002. Prospective comparison of risk factors and demographic and clinical characteristics of community-acquired, methicillin-resistant versus methicillin-susceptible *Staphylococcus aureus* infection in children. *Pediatr. Infect. Dis. J.* **21:** 910–916.

69. Schlievert, P. M., K. N. Shands, B. B. Dan, G. P. Schmid, and R. D. Nishimura. 1981. Identification and characterization of an exotoxin from *Staphylococcus aureus* associated with toxic shock syndrome. *J. Infect. Dis.* **143:** 509–516.

70. Smith, T., M. L. Pearson, K. R. Wilcox, C. Cruz, M. V. Lancaster, B. Robinson-Dunn, F. C. Tenover, M. J. Arduino, M. J. Zervos, J. M. Miller, J. D. Band, and W. Jarvis. 1999. Emergence of vancomycin resistance in *Staphylococcus aureus*: epidemiology and clinical significance. *N. Engl. J. Med.* **340:**493–501.

71. Shands, K. N., G. P. Schmid, B. B. Dan, D. Blum, R. J. Guidotti, N. T. Hargrett, R. L. Anderson, D. L. Hill, C. V. Broome, J. D. Band, and D. W. Fraser. 1980. Toxic-shock syndrome in menstruating women: association with tampon use and *Staphylococcus aureus* and clinical features in 52 cases. *N. Engl. J. Med.* **303:**1436–1442.

72. Swartz, M., and A. N. Weinberg. 1987. Infections due to gram-positive bacteria, p. 2100–2121. *In* T. B. Fitzpatrick, K. A. Arndt, and W. H. Clark (ed.), *Dermatology in General Medicine*, 3rd ed. McGraw-Hill, New York, N.Y.

73. Tenover, F. C., M. V. Lancaster, B. C. Hill, C. D. Steward, S. A. Stocker, G. A. Hancock, C. M. O'Hara, S. A. McAllister, N. C. Clark, and K. Hiramatsu. 1998. Characterization of staphylococci with reduced susceptibilities to vancomycin and other glycopeptides. *J. Clin. Microbiol.* **36:**1020–1027.

74. Tenover, F. C., L. M. Weigel, P. C. Appelbaum, L. K. McDougal, J. Chaitram, S. McAllister, N. Clark, G. Killgore, C. M. O'Hara, L. Jevitt, J. B. Patel, B. Bozdogan. 2004. Characterization of a vancomycin-resistant clinical isolate of *Staphylococcus aureus* from Pennsylvania. *Antimicrob. Agents Chemother.* **48:**275–280.

75. Tiemersma, E. W., S. L. A. M. Bronzwaer, O. Lyytikäinen, J. E. Degener, P. Schrijnemakers, N. Bruinsma, J. Monen, W. Witte, H. Grundmann, and European Antimicrobial Resistance Surveillance System Participants. 2004. Methicillin-resistant *Staphylococcus aureus* in Europe 1999–2002. *Emerg. Infect. Dis.* **10:**1627–1634.

76. Todd, J., M. Fishaut, F. Kapral, and T. Welch. 1978. Toxic-shock syndrome associated with phage-group-I staphylococci. *Lancet* **2:**1116–1118.

77. Tuazon, C. U., and J. N. Sheagren. 1974. Increased rate of carriage of *Staphylococcus aureus* among narcotic addicts. *J. Infect. Dis.* **129:**725–727.

78. Uchiyama, T., X. J. Yan, K. Imanishi, and J. Yagi. 1994. Bacterial superantigens—mechanism of T cell activation by the superantigens and their role in the pathogenesis of infectious diseases. *Microbiol. Immunol.* **38:**245–256.

79. Vallúes, J., C. León, and F. Alvarez-Lerma. 1997. Nosocomial bacteremia in critically ill patients: a multicenter study evaluating epidemiology and prognosis. *Clin. Infect. Dis.* **24:**387–395.

80. Waldvogel, F. A., and H. Vasey. 1980. Osteomyelitis: the past decade. *N. Engl. J. Med.* **303:**360–363.

81. Weigel, L. M., D. B. Clewell, S. R. Gill, N. C. Clark, L. K. McDougal, S. Flannagan, G. Killgore, and F. C. Tenover. 2003. Genetic analysis of a high-level vancomycin-resistant isolate of *Staphylococcus aureus*. *Science* **302:**1569–1571.

82. Weinstein, M. P. 1996. Current blood culture methods and systems: clinical concepts, technology, and interpretation of results. *Clin. Infect. Dis.* **23:**40–46.

83. Weinstein, M. P., M. L. Towns, S. M. Quartey, S. Mirrett, L. G. Reimer, G. Parmigiani, and L. B. Reller. 1997. The clinical significance of positive blood cultures in the 1990s: a prospective comprehensive evaluation of the microbiology, epidemiology, and outcome of bacteremia and fungemia in adults. *Clin. Infect. Dis.* **24:**584–602.

84. Wertheim, H. F. L., M. C. Vos, A. Ott, A. van Belkum, A. Voss, J. A. J. W. Kluytmans, P. H. J. van Keulen,

C. M. J. E. Vandenbroucke-Grauls, M. H. M. Meester, and H. A. Verbrugh. 2004. Risk and outcome of nosocomial *Staphylococcus aureus* bacteraemia in nasal carriers versus non-carriers. *Lancet* **364:**703–705.

85. **Wieneke, A. A., D. Roberts, and R. J. Gilbert.** 1993. Staphylococcal food poisoning in the United Kingdom, 1969–1990. *Epidemiol. Infect.* **110:**519–531.

86. **Yu, V. L., A. Goetz, M. Wagnener, P. B. Smith, J. D. Rihs, J. Hanchett, and J. J. Zuravleff.** 1986. *Staphylococcus aureus* nasal carriage and infection in patients on hemodialysis: efficacy of antibiotic prophylaxis. *N. Engl. J. Med.* **315:**91–96.

87. **Zimakoff, J., F. B. Pedersen, L. Bergen, J. Baagø-Nielsen, B. Daldorph, F. Espersens, B. G. Hansen, N. Hoiby, O. B. Jepsen, P. Joffe, H. J. Kolmos, M. Klausen, K. Kristoffersen, J. Ladefoged, S. Olesen-Larsen, V. T. Rosdahl, J. Scheibel, B. Storm, and P. Tofte-Jensen.** 1996. *Staphylococcus aureus* carriage and infections among patients in four haemo- and peritoneal-dialysis centers in Denmark. *J. Hosp. Infect.* **33:**289–300.

88. **Zinderman, C. E., B. Conner, M. A. Malakooti, J. E. LaMar, A. Armstrong, and B. K. Bohnker.** 2004. Community-acquired methicillin-resistant *Staphylococcus aureus* among military recruits. *Emerg. Infect. Dis.* **10:**941–944.

Animal Models of Experimental *Staphylococcus aureus* Infection

L. VINCENT COLLINS AND ANDRZEJ TARKOWSKI

44

Staphylococcus aureus infections remain a permanent threat to humankind; they are associated with high morbidity and mortality. Staphylococci can give rise to a diverse spectrum of diseases ranging from cutaneous infections to life-threatening conditions such as sepsis, endocarditis, and arthritis. The increasing prevalence of immunocompromised subjects and the appearance of methicillin-resistant staphylococci should prompt researchers to reach a better understanding of the host-bacterium relationship, a prerequisite for better therapeutic and preventive measures. Studies of pathogenetic mechanisms in human *S. aureus* infections have met with shortcomings, most of them related to uncertainty as to the exact time of the onset of the disease as well as the ethical problems related to manipulations of the host immune system and/or the invading bacterium. Use of laboratory animals should overcome most of these problems. In this respect the mouse is the most versatile animal to use owing to (i) the development of spontaneous staphylococcal infections (13), (ii) the availability of many inbred and genetically well-characterized mouse strains, (iii) the availability of large numbers of strains either lacking a certain gene(s) of potential interest (so-called gene knockout mice) or having overexpression of a certain gene(s) (so-called transgenic mice), and (iv) last, but not least, the existing knowledge of the murine immune system. However, some diseases cannot be replicated or monitored easily in mice, and in these cases it has been necessary to develop alternative animal models.

In many instances it is clear that subjects displaying skin defects (e.g., due to eczema), damaged joints (e.g., due to rheumatoid arthritis), or defective heart valves (e.g., due to rheumatic fever) will be at risk of acquiring *S. aureus* infection. Also, patients with installed prostheses or indwelling catheters will be more prone to develop staphylococcal infections originating at these sites. Animal models to mimic these situations are particularly difficult to obtain.

The aim of this chapter is to discuss experimental models of *S. aureus* infections, including toxic shock, sepsis, endocarditis, colonization of joints and bones, mastitis, eye and skin infections, and septic arthritis. We have excluded those studies carried out primarily to test antibiotic thera-

pies and have instead concentrated on models that provide an insight into the pathogenesis of *S. aureus*. As an example of a model for studying staphylococcal disease, we present the murine model of septic arthritis and sepsis and discuss how models such as this might be used to formulate treatment and prophylaxis regimens.

ANIMAL MODELS OF STAPHYLOCOCCAL DISEASE

The various disease entities associated with *S. aureus* infections and some proposed animal models are listed in Table 1.

Toxic Shock and Sepsis

Cases of toxic shock syndrome (TSS) due to the toxic shock syndrome toxin 1 (TSST-1) of *S. aureus* are predominantly linked to tampon use in women. TSS acquired in this way is difficult to reproduce in animals, although some success has been achieved with a simulation of tampon use in rabbits (58). Attempts have also been made to induce TSS or a TSS-like syndrome by other means in laboratory animals (48, 72), but a definitive model for this disease has not been forthcoming.

Septicemia caused by *S. aureus* is usually the result of hematogenous infection from surgical intervention or from a wound site. Animal models of sepsis in which the bacteria are injected intravenously (i.v.), e.g., in chickens (22) or mice (94), closely mimic the hematogenous spread of disease seen in humans and have proved useful in defining the components of the host immune system involved in resistance to colonization and septic shock. Recently, novel nonmammalian models of infection that involve the use of the fruit fly *Drosophila melanogaster* (62), the nematode *Caenorhabditis elegans* (79), and silkworm larvae *Bombyx mori* (47) as hosts have been used to assess the in vivo virulence of *S. aureus* strains. These models are advantageous from the perspectives of low cost, ease of infection, the ability to study large numbers of infected hosts, and ethical considerations. Although these model organisms lack an acquired immune system, they reproduce many human immune responses. The fruit fly, for example, possesses elements of

Gram-Positive Pathogens, 2nd edition, edited by Vincent A. Fischetti et al.
© 2006 ASM Press, Washington, D.C.

TABLE 1 Experimental animal models used to study diseases associated with *S. aureus* infection

Disease	Animal model	Reference(s)
Toxic shock and sepsis	Rabbit	58
	Baboon	72
	Chicken	22
	Mouse	80, 94
Endocarditis	Rat	6, 49, 52, 77, 78
	Guinea pig	57
	Rabbit	18, 33, 76
Osteomyelitis	Dog	24, 25, 86
	Chicken	22, 26, 27, 83
	Rat	35, 40, 71
	Sheep	47
	Rabbit	16, 21, 93
	Guinea pig	68
Mastitis	Mouse	17, 30, 55, 56
	Rabbit	4
	Cow	84, 91
	Sheep	5, 90
Wound infection	Rat	66
	Guinea pig	10
Skin abscess/dermatitis	Mouse	15, 59, 60
Eye infection	Guinea pig	23
	Rat	7
	Rabbit	9
Septic arthritis	Rabbit	54, 73, 82
	Rat	12
	Mouse	13, 14
Kidney infection	Rabbit	29
	Mouse	31, 32

the human innate immune system, such as antimicrobial peptides and toll-like receptor expression. In addition, staphylococcal infections in these hosts are amenable to antibiotic treatment (34, 47, 62), which supports their use as model systems for screening and assessing the efficacy of antibiotic regimens in vivo.

Endocarditis

Animal models of endocarditis typically entail catheterization via the right carotid artery into the left ventricle, resulting in traumatization of the aortic valve. The formation of a mature vegetation on the valve ensues, comprising platelets, fibrin, inflammatory cells, and depositions of fibrinogen and fibronectin that serve as sites of adhesion for *S. aureus* (77). A rat model of endocarditis has been developed (33) along with a number of other models, most notably in the guinea pig (57) and the rabbit (76).

In the rat model of endocarditis, it was shown that immunization with a fibronectin fusion protein afforded protection against i.v. challenge with *S. aureus* (78). A similar rat model was used to demonstrate that passive immunization with antibodies raised against *S. aureus* type 5 capsular polysaccharide was protective against serotype 5-induced endocarditis (52).

The pivotal role played by certain staphylococcal virulence factors in endocarditis has also been elucidated through use of the rat model. These factors include the fibronectin adhesin (49), the capsular polysaccharide (6), clumping factor (61), and the collagen adhesin (36). In addition, in the rabbit model it was demonstrated that *agr*-and *sar*-regulated factors are important in bacterial colonization of valves (18).

Osteomyelitis

Colonization of the bones and joints by *S. aureus* may occur as a result of either hematogenous infection or following local trauma. The virulence factors involved in survival in blood, homing and attachment to joint tissues and bone, induction of inflammation, and subsequent joint destruction are complex, but an understanding of some of these processes is emerging from pathological observations of the disease in animal models that resemble those encountered in humans. Hematogenous, *S. aureus*-induced osteomyelitis in either the acute or chronic form can be simulated in animal models, e.g., in dogs (24, 25), chickens (26, 27, 83), and rats (35), with the associated inflammatory reactions and bone changes. Chronic osteomyelitis, on the other hand, can be induced by direct injection of *S. aureus* into the tibial marrow cavity in sheep (46), rats (71), and dogs (86).

Posttraumatic osteomyelitis following instillation of *S. aureus* into fractured leg bones has been studied in rabbits (93) and guinea pigs (68). These animals provide useful models for the development of therapeutic interventions in human bone infection following trauma.

Conventional therapies for acute hematogenous osteomyelitis involve surgical drainage and antibiotic therapy. New therapies are being investigated using biodegradable implants impregnated with antibiotics in rabbit (16, 21) and rat (40) models to treat experimental *S. aureus*-induced osteomyelitis.

Mastitis

Mastitis due to *S. aureus* in ruminants is an economically important disease against which there is as yet no reliable efficacious vaccine. The mouse model of intramammary inoculation has been used extensively in studying this disease and represents a reproducible and cheap simulation of the disease in dairy cattle (17). Various attempts have been made in the past to design vaccines directed against the bacterial surface polysaccharides and proteins (adhesins) and against the toxins of *S. aureus* and to test their efficacy in bovine, mouse, and rabbit models (28, 84). Mamo and colleagues (56) utilized the mouse model to demonstrate that immunization with a fragment of the fibronectin-binding protein reduced the incidence of severe mastitis. In a related study, mice vaccinated with the fibrinogen-binding protein showed a reduced incidence and severity of mastitis (55). Rabbits vaccinated with a toxoid derivative were protected against the lethal gangrenous form of mastitis, but there was no protection against abscess formation (4). Adjuvant whole-cell–toxoid combinations have also shown promise in the protection of heifers (91) and ewes (90) from clinical mastitis, as has a killed whole-cell, toxoid, liposome-enclosed exopolysaccharide preparation in ewes (5). More recently, a live attenuated *S. aureus* strain was used to provide protection in mice when inoculated locally during late pregnancy or early lactation (30). It is clear, therefore, that the mouse model is a valuable tool for elucidating the role of specific antigens in mastitis and for testing potential vaccine preparations.

Skin and Wound Infections

Staphylococcal infections of the skin can be minor or life threatening, depending on the integrity of the skin surface and the invasiveness of the bacterial strain. Traumatization of the skin surface is often used to investigate *S. aureus* wound infectivity. As examples, a rat model was used to show that a fibrinogen-binding protein of *S. aureus* played a key role in wound infection (66), and an invasive burn wound infection model in guinea pigs was used to study the role of host factors in combating infection of burns (10). Alternatively, bacteria are injected subcutaneously into mice along with microbeads or other foreign material, and lesions are measured after a given time period (15). This method allows comparison of the ability of different bacterial isolates to induce dermonecrosis or abscess formation associated with foreign body penetration. Recently, a murine model of *S. aureus* dermatitis has been developed. In this model, staphylococci are provided intracutaneously in the absence of foreign material (59). The ensuing infection gives rise to both local (Fig. 1) and systemic manifestations, typically disappearing within 2 weeks. However, upon depletion of host neutrophils, mice display both a considerably longer and more severe course of the infection (60).

Eye Infections

A model of staphylococcal keratitis in guinea pigs has been described (23), and a similar model in rats was used to determine the effects of antibiotic and corticosteroid treatments on the progression of *S. aureus* infection of the cornea (7). Radial incision of the rabbit conjunctiva followed by inoculation of *S. aureus* resulted in purulent conjunctivitis and a model for testing antimicrobial treatments in this disease (9).

Septic Arthritis

Well-defined human studies of infectious arthritis are almost nonexistent, as the time of induction of infection is often unknown. Furthermore, adequate tissue biopsies are rarely obtained. Experimental models of arthritis employing intra-articular injection in rabbits (54, 73, 82) have overcome some of these difficulties. However, in cases of human septic arthritis, staphylococci typically reach joints through hematogenic spreading rather than by local inoculation. Thus, the intra-articular route of inoculation bypasses the early, and potentially critical, stages of pathogenesis. During these stages the bacteria would conceivably be required to adapt to the environment within the host, to survive bactericidal components in the blood, to home to bone and synovial tissue, and to penetrate these structures to reach the joint cavity. A model of hematogenically transmitted joint and bone infection was clearly desirable. A spontaneous hematogenous arthritis in mice (13) and rats (12) was detected and described, proving that rodents may occasionally be natural hosts for staphylococci. The murine infectious model provides the undisputed advantage of using well-defined and inbred strains of mice—a prerequisite for analyzing the genetic and immunological background of the host's susceptibility (85). In this model approximately 90% of the mice develop septic arthritis, with disease initiation usually 24 h after i.v. inoculation of *S. aureus*. The characteristics of the murine model closely mirror those seen in human septic arthritis, especially showing a very high frequency and degree of periarticular bone erosivity (11). For this reason the murine model has proved extremely useful in elucidating the role of several virulence factors of *S. aureus* in septic arthritis, which include collagen adhesin (69), clumping factor (Fig. 2) (45), protein A (67), sortase A (Fig. 3)

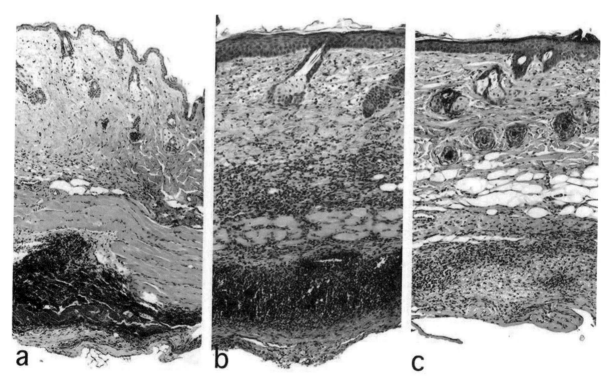

FIGURE 1 (*See the separate color insert for the color version of this illustration.*) Early and persistent infiltrations of inflammatory cells during *S. aureus* dermatitis. The micrograph shows the inflammatory infiltrate in mouse skin that was inoculated intracutaneously with 2×10^8 CFU of *S. aureus* LS-1: after 6 h (A), 48 h (B), and 1 week (C). The inflammatory infiltrate, which mainly contains macrophages, peaks at 48 h and starts to disappear 2 weeks after the inoculation (from reference 59).

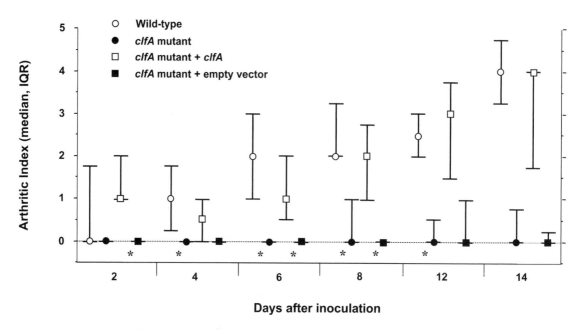

FIGURE 2 ClfA influence on septic arthritis. Development of arthritis in NMRI mice that were inoculated i.v. with 1.4 to 1.9×10^7 CFU of *S. aureus* wild-type strain Newman, *clfA* mutant DU5876 (*clfA*), *clfA* complemented mutant strain DU5898 (*clfA* mutant + *clfA*), or *clfA* mutant with vector plasmid strain DU5899 (*clfA* mutant + empty vector). Circles and squares show the median, and whiskers show the arthritic index. Two sets of statistical tests were performed: the wild-type and DU5876-infected mice were compared, and the DU5898- and DU5899-infected mice were compared. * $P < 0.001$. The severity of arthritis is markedly reduced in mice challenged i.v. with the *clfA* mutant, compared with mice infected with the wild-type strain (from reference 45).

(42), bone sialoprotein (14), small-colony variant phenotype (43), Agr-, Sar-, and sigma factor B/RsbU-regulated components (1, 44, 63), staphylococcal cell envelope teichoic acid (20) and membrane lipid modifications (70), and the polysaccharide capsule (64). The model of septic arthritis and sepsis has also been valuable in underscoring the importance of crucial components of the host immune system, e.g., neutrophils (89), T lymphocytes (2), B lymphocytes (95), and major histocompatibility complex class II expression (3) in the pathogenesis of septic arthritis.

Kidney Abscess and Nephritis

Hematogenously acquired *S. aureus* infection resulting in arthritis, as described above, is typically accompanied by chronic renal abscesses. Acute suppurative infections following i.v. injection of *S. aureus* have been reported in mice (31, 32) and rabbits (29). In the mouse model, infection leads to the formation of multiple cortical and medullar abscesses (32, 80, 81). Histopathological examination of the focal and diffuse infiltrates in the kidney cortex and medulla reveals the presence of both phagocytic cells and CD4$^+$ and CD8$^+$ lymphocytes (88). The TSST-1 (53, 88) and capsule (51) have been implicated as important virulence factors in this syndrome. A combination of corticosteroids and antibiotics was shown to be more efficacious in treating septic murine nephritis than were antibiotics alone, suggesting that T-lymphocyte infiltration plays a significant role in this disease (87).

GENE KNOCKOUT MOUSE MODELS

Cytokines play a critically important role in the pathogenesis of *S. aureus* infection, and the modulation of specific cytokines is attracting substantial interest as a means of treating disease. For instance, it has been observed that both gamma interferon (IFN-γ) mRNA and protein expression are clearly increased during the course of *S. aureus* infection in mice (96). Inactivation of the receptor for IFN-γ by gene knockout technology removes the functional IFN-γ receptor in mice. When these mice were inoculated with *S. aureus*, more frequent and severe arthritis compared with wild-type littermates was noted. Also, sepsis-triggered mortality in the IFN-γ receptor knockout mice was increased in early (but not late) stages of the infection (98). In contrast, supplementation of normal mice with extrinsic IFN-γ significantly decreased staphylococcal sepsis-triggered mortality on one hand, but enhanced the development of arthritis on the other hand (97). Interestingly, T-box transcription factor-deficient mice display significantly decreased production of IFN-γ and fail to control systemic *S. aureus* infection (39). Interleukin-4 (IL-4) is another cytokine produced by T lymphocytes. Mice not able to produce this compound are protected against septic arthritis and sepsis-triggered mortality. Our data suggest that this protection is due to the role of IL-4 as an inhibitory factor in phagocytosis, decreasing clearance of bacteria during ongoing infection (38). Yet another cytokine produced predominantly by monocytes/macrophages is tumor necrosis factor. This cytokine is known to be one of the major causative factors involved in degradation of cartilage and subchondral bone in aseptic arthritis, e.g., rheumatoid arthritis. In the case of septic arthritis, tumor necrosis factor increases joint destruction but ameliorates the severity of sepsis (37). Much more remains to be learned about the activities and concentrations of cytokines and chemokines, the synergy between different cytokines, and the responding cells in the tissues of experi-

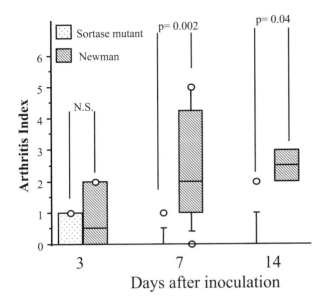

FIGURE 3 Effect of sortase mutation on the ability of *S. aureus* to cause septic arthritis. Severity of arthritis in NMRI mice inoculated with 6×10^6 CFU of *S. aureus* Newman or its isogenic sortase-deficient mutant, strain SKM3 (10 mice/group). Data shown are median and interquartile range for each group of mice. Comparisons were made using the Mann-Whitney U-test. *P* values are for the comparison between the wild-type Newman strain and the sortase mutant strain SKM3. N.S., not significant. The results show that the sortase loss significantly reduces the arthritogenicity of *S. aureus* strain Newman (from reference 42).

mentally infected animals at different stages of the disease process. In the future it may be possible, using the various animal models, to improve the outcome of *S. aureus* infection by treatments with either anti-inflammatory cytokines (e.g., IL-10, IL-4, and/or transforming growth factor β) or antagonists to proinflammatory cytokines (e.g., IL-1 receptor antagonist).

DEVELOPMENT OF NEW THERAPIES USING ANIMAL MODELS

Antistaphylococcal Therapies

The emergence of antibiotic-resistant staphylococci and in particular the methicillin-resistant *S. aureus* strains has stimulated a resurgence in the development of new antibiotics with antistaphylococcal activities and alternatives to classical antibacterial therapies. Notable examples of the latter approach are recent studies using peptides produced by bacteria to modulate staphylococcal virulence (8, 41).

Vaccination Strategies Using Animal Models

An ideal vaccine candidate should induce responses that prevent bacterial adherence, promote opsonophagocytic killing by leukocytes, and neutralize secreted toxic proteins, as suggested by Lee (50). Such a vaccine has thus far not been developed. Infection with *S. aureus* itself does not per se provide any protection against subsequent infections. Indeed, serological studies have, in general, failed to reveal a correlation between bacterial antibody titers and protective capacity. In recent times three distinct approaches to achieve immunization have been assayed. The first involves vaccination with staphylococcal polysaccharides alone or conjugated to a protein carrier. Results from experimental trials show that this approach clearly decreases the severity of infection in models of *S. aureus* peritonitis, endocarditis, bacteremia, and renal abscess formation (50). One potential obstacle in these immunizations is the multitude of different capsular polysaccharide serotypes ($n = 1$) that can be used by various staphylococcal strains. An encouraging note, however, is that two polysaccharide serotypes (numbers 5 and 8) constitute about 75% of all disease-inducing staphylococcal strains.

A second possibility that has been explored is to trigger protective immunity against staphylococcal superantigens. To induce immunity using native superantigens is not advisable because these molecules will trigger severe inflammation, potentially leading to septic shock. An alternative used by Woody and colleagues (92) is to employ enterotoxins devoid of their superantigenic properties but still expressing major antigenic determinants. Vaccination of mice with staphylococcal enterotoxin B mutated in a hydrophobic loop dominating the interface with major histocompatibility complex class II locus did not trigger an inflammatory response but induced immunity. Antibodies to this molecule were able to protect a majority of mice against lethal staphylococcal enterotoxin B challenge (92). Recently, we have shown that intranasal administration of staphylococcal enterotoxin A (SEA), but not a recombinant SEA lacking superantigenic activity (19), protected mice against lethal systemic SEA challenge (Fig. 4). The protection afforded by this treatment was superantigen-specific, because intranasal exposure to SEA did not protect against death caused by subsequent TSST-1 systemic challenge. These results show that although mucosal tolerance to a superantigen may be readily triggered by means of immunodeviation, the high number of different toxins that may be produced by various virulent staphylococcal strains hinders the development of a universal staphylococcal vaccine.

S. aureus binds to numerous components of the extracellular matrix, including collagens, laminin, bone sialoprotein, fibronectin, fibrinogen, and vitronectin. Antibodies directed to staphylococcal adhesins and thus interfering with bacterial binding to the host tissue might prevent successful colonization. Indeed, experiments involving immunization with fibronectin-binding protein provided partial protection against experimental endocarditis and mastitis in rodents (reviewed in reference 50). Our laboratory has demonstrated that immunization of mice with recombinant collagen-binding adhesin significantly alleviates the outcome of severe, life-threatening *S. aureus* sepsis (65). Here, vaccination-induced antibodies displayed protective properties, as proved by passive transfer experiments. Because collagen adhesin is expressed on the majority of staphylococci involved in invasive infections, it is another promising vaccine candidate that should merit evaluation in clinical trials.

Treatment of Ongoing Infection

Antibiotics are and will continue to be the standard treatment for systemic staphylococcal infections. However, despite adequate use of antibiotics and consequent eradication of staphylococci, continuous tissue destruction may occur. This can be clearly seen in the case of septic arthritis and is mediated by the exaggerated activation of the host immune response (see above). In addition, septic shock caused by

A. C57BL/6

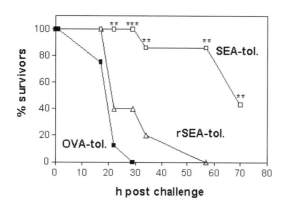

B. BALB/c

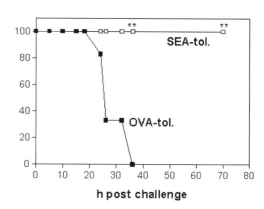

FIGURE 4 Mucosal administration of SEA protects against SEA-induced death. Mice were given SEA, recombinant nonsuperantigenic SEA (rSEA), or ovalbumin (OVA) intranasally three times 1 week apart, and challenged intraperitoneally with SEA + LPS 1 week later. Data are expressed as survival of OVA-tolerized (■), rSEA-tolerized (▲), and SEA-tolerized (□) C57BL/6 (A) and BALB/c (B) mice during the first 70 h post-SEA challenge. **, $P<0.01$; ***, $P<0.001$ compared to OVA-tolerized animals. The data show that mice that receive SEA intranasally are protected against a subsequent lethal SEA challenge (from reference 19).

S. aureus will not successfully respond to antibiotic treatment alone. Thus, there is room for improvement with respect to treatment of ongoing *S. aureus* infections. We have recently shown that concomitant corticosteroid and antibiotic treatment will, despite the inhibitory role of steroids on neutrophil function, clearly down-regulate the severity of septic arthritis and nephritis (74, 87). Alternative combinations, which include an antibiotic plus an antioxidant (free radical trap), are also effective at reducing the inflammatory and destructive effects of staphylococci in infected joints (75).

FINAL COMMENTS

During the last decade, the use of experimental models of staphylococcal infections has clarified the involvement of several bacterial virulence factors as well as many hematopoietic cell types and their products in the pathogenesis of infection. Much still remains obscure—for example, how bacterial virulence factors act in vivo and the many details of the host-bacterium interplay. More information is required about the risk factors for acquiring staphylococcal infection to properly select subjects for vaccination procedures, and new modalities should be developed to treat ongoing infections with combinations of anti-inflammatory agents, passive immunization, and antibiotics to minimize the risk of sequels. Animal models that mimic the etiology, progression, and pathology of the disease in the natural host are a crucial component in the development of therapeutic strategies.

REFERENCES

1. **Abdelnour, A., S. Arvidson, T. Bremell, C. Ryden, and A. Tarkowski.** 1993. The accessory gene regulator (*agr*) controls *Staphylococcus aureus* virulence in a murine arthritis model. *Infect. Immun.* **61:**3879–3885.
2. **Abdelnour, A., T. Bremell, R. Holmdahl, and A. Tarkowski.** 1994. Clonal expansion of T lymphocytes causes arthritis and mortality in mice infected with toxic shock syndrome toxin-1-producing staphylococci. *Eur. J. Immunol.* **24:**1161–1166.
3. **Abdelnour, A., Y. X. Zhao, R. Holmdahl, and A. Tarkowski.** 1997. Major histocompatibility complex class II region confers susceptibility to *Staphylococcus aureus* arthritis. *Scand. J. Immunol.* **45:**301–307.
4. **Adlam, C., P. D. Ward, A. C. McCartney, J. P. Arbuthnott, and C. M. Thorley.** 1977. Effect of immunization with highly purified alpha- and beta-toxins on staphylococcal mastitis in rabbits. *Infect. Immun.* **17:**259–266.
5. **Amorena, B., R. Baselga, and I. Albizu.** 1994. Use of liposome-immunopotentiated exopolysaccharide as a component of an ovine mastitis staphylococcal vaccine. *Vaccine* **12:**243–249.
6. **Baddour, L. M., C. Lowrance, A. Albus, J. H. Lowrance, S. K. Anderson, and J. C. Lee.** 1992. *Staphylococcus aureus* microcapsule expression attenuates bacterial virulence in a rat model of experimental endocarditis. *J. Infect. Dis.* **165:**749–753.
7. **Badenoch, P. R., G. J. Hay, P. J. McDonald, and D. J. Coster.** 1985. A rat model of bacterial keratitis. Effect of antibiotics and corticosteroid. *Arch. Ophthalmol.* **103:**718–722.
8. **Balaban, N., T. Goldkorn, R. T. Nhan, L. B. Dang, S. Scott, R. M. Ridgley, A. Rasooly, S. C. Wright, J. W. Larrick, R. Rasooly, and J. R. Carlson.** 1998. Autoinducer of virulence as a target for vaccine and therapy against *Staphylococcus aureus*. *Science* **280:**438–440.
9. **Behrens-Baumann, W., and T. Begall.** 1993. Reproducible model of a bacterial conjunctivitis (in German). *Ophthalmologica* **206:**69–75.
10. **Bjornson, A. B., H. S. Bjornson, N. A. Lincoln, and W. A. Altemeier.** 1984. Relative roles of burn injury, wound colonization, and wound infection in induction of alterations of complement function in a guinea pig model of burn injury. *J. Trauma* **24:**106–115.
11. **Bremell, T., A. Abdelnour, and A. Tarkowski.** 1992. Histopathological and serological progression of experimental *Staphylococcus aureus* arthritis. *Infect. Immun.* **60:**2976–2985.
12. **Bremell, T., S. Lange, R. Holmdahl, C. Ryden, G. K. Hansson, and A. Tarkowski.** 1994. Immunopathological

features of rat *Staphylococcus aureus* arthritis. *Infect. Immun.* **62:**2334–2344.

13. **Bremell, T., S. Lange, L. Svensson, E. Jennische, K. Grondahl, H. Carlsten, and A. Tarkowski.** 1990. Outbreak of spontaneous staphylococcal arthritis and osteitis in mice. *Arthritis Rheum.* **33:**1739–1744.

14. **Bremell, T., S. Lange, A. Yacoub, C. Ryden, and A. Tarkowski.** 1991. Experimental *Staphylococcus aureus* arthritis in mice. *Infect. Immun.* **59:**2615–2623.

15. **Bunce, C., L. Wheeler, G. Reed, J. Musser, and N. Barg.** 1992. Murine model of cutaneous infection with grampositive cocci. *Infect. Immun.* **60:**2636–2640.

16. **Calhoun, J. H., and J. T. Mader.** 1997. Treatment of osteomyelitis with a biodegradable antibiotic implant. *Clin. Orthop. Relat. Res.* **341:**206–214.

17. **Chandler, R. L.** 1970. Experimental bacterial mastitis in the mouse. *J. Med. Microbiol.* **3:**273–282.

18. **Cheung, A. L., K. J. Eberhardt, E. Chung, M. R. Yeaman, P. M. Sullam, M. Ramos, and A. S. Bayer.** 1994. Diminished virulence of a sar⁻/agr⁻ mutant of *Staphylococcus aureus* in the rabbit model of endocarditis. *J. Clin. Invest.* **94:**1815–1822.

19. **Collins, L. V., K. Eriksson, R. G. Ulrich, and A. Tarkowski.** 2002. Mucosal tolerance to a bacterial superantigen indicates a novel pathway to prevent toxic shock. *Infect. Immun.* **70:**2282–2287.

20. **Collins, L. V., S. A. Kristian, C. Weidenmaier, M. Faigle, K. P. M. van Kessel, J. A. G. van Strijp, F. Gotz, B. Neumeister, and A. Peschel.** 2002. *Staphylococcus aureus* strains lacking D-alanine modifications of teichoic acids are highly susceptible to human neutrophil killing and are virulence-attenuated in mice. *J. Infect. Dis.* **186:**214–219.

21. **Dahners, L. E., and C. H. Funderburk.** 1987. Gentamicin-loaded plaster of Paris as a treatment of experimental osteomyelitis in rabbits. *Clin. Orthop. Relat. Res.* **219:**278–282.

22. **Daum, R. S., W. H. Davis, K. B. Farris, R. J. Campeau, D. M. Mulvihill, and S. M. Shane.** 1990. A model of *Staphylococcus aureus* bacteremia, septic arthritis, and osteomyelitis in chickens. *J. Orthoped. Res.* **8:**804–813.

23. **Davis, S. D., L. D. Sarff, and R. A. Hyndiuk.** 1978. Staphylococcal keratitis. Experimental model in guinea pigs. *Arch. Ophthalmol.* **96:**2114–2116.

24. **Deysine, M., H. D. Isenberg, and G. Steiner.** 1983. Chronic haematogenous osteomyelitis; studies on an experimental model. *Int. Orthoped.* **7:**69–78.

25. **Deysine, M., E. Rosario, and H. D. Isenberg.** 1976. Acute hematogenous osteomyelitis: an experimental model. *Surgery* **79:**97–99.

26. **Emslie, K. R., and S. Nade.** 1983. Acute hematogenous staphylococcal osteomyelitis. A description of the natural history in an avian model. *Am. J. Pathol.* **110:**333–345.

27. **Emslie, K. R., N. R. Ozanne, and S. M. Nade.** 1983. Acute haematogenous osteomyelitis: an experimental model. *J. Pathol.* **141:**157–167.

28. **Foster, T. J.** 1991. Potential for vaccination against infections caused by *Staphylococcus aureus*. *Vaccine* **9:**221–227.

29. **Freedman, L. R.** 1960. Experimental pyelonephritis. VI. Observations on susceptibility of the rabbit kidney to infection by a virulent strain of *Staphylococcus aureus*. *Yale J. Biol. Med.* **32:**272–279.

30. **Gomez, M. I., V. E. Garcia, M. M. Gherardi, M. C. Cerquetti, and D. O. Sordelli.** 1998. Intramammary immunization with live-attenuated *Staphylococcus aureus* protects mice from experimental mastitis. *FEMS Immunol. Med. Microbiol.* **20:**21–27.

31. **Gorrill, R. H.** 1951. Experimental staphylococcal infections in mice. *Br. J. Exp. Pathol.* **32:**151–155.

32. **Gorrill, R. H.** 1958. The establishment of staphylococcal abscesses in the mouse kidney. *Br. J. Exp. Pathol.* **39:**203–212.

33. **Gutschik, E.** 1983. Experimental staphylococcal endocarditis: an overview. *Scand. J. Infect. Dis. Suppl.* **41:**87–94.

34. **Hamamoto, H., K. Kurokawa, C. Kaito, K. Kamura, I. Manitra Razanajatovo, H. Kusuhara, T. Santa, and K. Sekimizu.** 2004. Quantitative evaluation of the therapeutic effects of antibiotics using silkworms infected with human pathogenic microorganisms. *Antimicrob. Agents Chemother.* **48:**774–779.

35. **Hienz, S. A., H. Sakamoto, J. I. Flock, A. C. Morner, F. P. Reinholt, A. Heimdahl, and C. E. Nord.** 1995. Development and characterization of a new model of hematogenous osteomyelitis in the rat. *J. Infect. Dis.* **171:**1230–1236.

36. **Hienz, S. A., T. Schennings, A. Heimdahl, and J. I. Flock.** 1996. Collagen binding of *Staphylococcus aureus* is a virulence factor in experimental endocarditis. *J. Infect. Dis.* **174:**83–88.

37. **Hultgren, O., H. P. Eugster, J. D. Sedgwick, H. Korner, and A. Tarkowski.** 1998. TNF/lymphotoxin-alpha double-mutant mice resist septic arthritis but display increased mortality in response to *Staphylococcus aureus*. *J. Immunol.* **161:**5937–5942.

38. **Hultgren, O., M. Kopf, and A. Tarkowski.** 1998. *Staphylococcus aureus*-induced septic arthritis and septic death is decreased in IL-4-deficient mice: role of IL-4 as promoter for bacterial growth. *J. Immunol.* **160:**5082–5087.

39. **Hultgren, O. H., M. Verdrengh, and A. Tarkowski.** 2004. T-box transcription-factor-deficient mice display increased joint pathology and failure of infection control during staphylococcal arthritis. *Microbes Infect.* **6:**529–535.

40. **Itokazu, M., T. Ohno, T. Tanemori, E. Wada, N. Kato, and K. Watanabe.** 1997. Antibiotic-loaded hydroxyapatite blocks in the treatment of experimental osteomyelitis in rats. *J. Med. Microbiol.* **46:**779–783.

41. **Ji, G., R. C. Beavis, and R. P. Novick.** 1995. Cell density control of staphylococcal virulence mediated by an octapeptide pheromone. *Proc. Natl. Acad. Sci. USA* **92:**12055–12059.

42. **Jonsson, I.-M., S. K. Mazmanian, O. Schneewind, M. Verdrengh, T. Bremell, and A. Tarkowski.** 2002. On the role of *Staphylococcus aureus* sortase and sortase-catalyzed surface protein anchoring in murine septic arthritis. *J. Infect. Dis.* **185:**1417–1424.

43. **Jonsson, I.-M., C. von Eiff, R. A. Proctor, G. Peters, C. Ryden, and A. Tarkowski.** 2003. Virulence of a hemB mutant displaying the phenotype of a *Staphylococcus aureus* small colony variant in a murine model of septic arthritis. *Microb. Pathog.* **34:**73–79.

44. **Jonsson, I.-M., S. Arvidson, S. Foster, and A. Tarkowski.** 2004. Sigma factor B and RsbU are required for virulence in *Staphylococcus aureus*-induced arthritis and sepsis. *Infect. Immun.* **72:**6106–6111.

45. **Josefsson, E., O. Hartford, L. O'Brien, J. M. Patti, and T. Foster.** 2001. Protection against experimental *Staphylococcus aureus* arthritis by vaccination with clumping factor A, a novel virulence determinant. *J. Infect. Dis.* **184:**1572–1580.

46. **Kaarsemaker, S., G. H. Walenkamp, and A. E. van de Bogaard.** 1997. New model for chronic osteomyelitis

with *Staphylococcus aureus* in sheep. *Clin. Orthop. Relat. Res.* **339:**246– 252.

47. **Kaito, C., N. Akimitsu, H. Watanabe, and K. Sekimizu.** 2002 Silkworm larvae as an animal model of bacterial infection pathogenic to humans. *Microb. Pathog.* **32:**183–190.

48. **Kohrman, K. A., J. J. Kirkland, and P. J. Danneman.** 1989. Response of various animal species to experimental infection with different strains of *Staphylococcus aureus*. *Rev. Infect. Dis.* **11**(Suppl. 1)**:**S231–S236. (Discussion, **11** [Suppl. 1]:S236–S237.)

49. **Kuypers, J. M., and R. A. Proctor.** 1989. Reduced adherence to traumatized rat heart valves by a low-fibronectin-binding mutant of *Staphylococcus aureus*. *Infect. Immun.* **57:**2306–2312.

50. **Lee, J. C.** 1996. The prospects for developing a vaccine against *Staphylococcus aureus*. *Trends Microbiol.* **4:**162–166.

51. **Lee, J. C., M. J. Betley, C. A. Hopkins, N. E. Perez, and G. B. Pier.** 1987. Virulence studies, in mice, of transposon-induced mutants of *Staphylococcus aureus* differing in capsule size. *J. Infect. Dis.* **156:**741–750.

52. **Lee, J. C., J. S. Park, S. E. Shepherd, V. Carey, and A. Fattom.** 1997. Protective efficacy of antibodies to the *Staphylococcus aureus* type 5 capsular polysaccharide in a modified model of endocarditis in rats. *Infect. Immun.* **65:**4146–4151.

53. **Lee, J. C., N. E. Perez, and C. A. Hopkins.** 1989. Production of toxic shock syndrome toxin 1 in a mouse model of *Staphylococcus aureus* abscess formation. *Rev. Infect. Dis.* **11**(Suppl. 1)**:**S254–S259.

54. **Linhart, W. E., S. Spendel, G. Weber, and S. Zadravec.** 1990. Septic arthritis—an experimental animal model useful in free oxygen radical research. *Z. Versuchstierkd.* **33:**65–71.

55. **Mamo, W., M. Boden, and J. I. Flock.** 1994. Vaccination with *Staphylococcus aureus* fibrinogen binding proteins (FgBPs) reduces colonisation of *S. aureus* in a mouse mastitis model. *FEMS Immunol. Med. Microbiol.* **10:**47–53.

56. **Mamo, W., P. Jonsson, J. I. Flock, M. Lindberg, H. P. Muller, T. Wadstrom, and L. Nelson.** 1994. Vaccination against *Staphylococcus aureus* mastitis: immunological response of mice vaccinated with fibronectin-binding protein (FnBP-A) to challenge with *S. aureus*. *Vaccine* **12:**988–992.

57. **Maurin, M., H. Lepidi, B. La Scola, M. Feuerstein, M. Andre, J. F. Pellissier, and D. Raoult.** 1997. Guinea pig model for *Staphylococcus aureus* native valve endocarditis. *Antimicrob. Agents Chemother.* **41:**1815–1817.

58. **Melish, M. E., S. Murata, C. Fukunaga, K. Frogner, and C. McKissick.** 1989. Vaginal tampon model for toxic shock syndrome. *Rev. Infect. Dis.* **11**(Suppl. 1)**:**S238–S246. (Discussion, **11**[Suppl. 1]:S246–S247.)

59. **Molne, L., and A. Tarkowski.** 2000. An experimental model of cutaneous infection induced by superantigen-producing *Staphylococcus aureus*. *J. Invest. Dermatol.* **114:**1120–1125.

60. **Molne, L., M. Verdrengh, and A. Tarkowski.** 2000. Role of neutrophil leukocytes in cutaneous infection caused by *Staphylococcus aureus*. *Infect. Immun.* **68:**6162–6167.

61. **Moreillon, P., J. M. Entenza, P. Francioli, D. McDevitt, T. J. Foster, P. Francois, and P. Vaudaux.** 1995. Role of *Staphylococcus aureus* coagulase and clumping factor in pathogenesis of experimental endocarditis. *Infect. Immun.* **63:**4738–4743.

62. **Needham, A. J., M. Kibart, H. Crossley, P. W. Ingham, and S. J. Foster.** 2004. *Drosophila melanogaster* as a model host for *Staphylococcus aureus* infection. *Microbiology* **150:**2347–2355.

63. **Nilsson, I.-M., T. Bremell, C. Ryden, A. L. Cheung, and A. Tarkowski.** 1996. Role of the staphylococcal accessory gene regulator (*sar*) in septic arthritis. *Infect. Immun.* **64:**4438–4443.

64. **Nilsson, I.-M., J. C. Lee, T. Bremell, C. Ryden, and A. Tarkowski.** 1997. The role of staphylococcal polysaccharide microcapsule expression in septicemia and septic arthritis. *Infect. Immun.* **65:**4216–4221.

65. **Nilsson, I.-M., J. M. Patti, T. Bremell, M. Hook, and A. Tarkowski.** 1998. Vaccination with a recombinant fragment of collagen adhesin provides protection against *Staphylococcus aureus*-mediated septic death. *J. Clin. Invest.* **101:**2640–2649.

66. **Palma, M., S. Nozohoor, T. Schennings, A. Heimdahl, and J. I. Flock.** 1996. Lack of the extracellular 19-kilodalton fibrinogen-binding protein from *Staphylococcus aureus* decreases virulence in experimental wound infection. *Infect. Immun.* **64:**5284–5289.

67. **Palmqvist, N., T. Foster, A. Tarkowski, and E. Josefsson.** 2003. Protein A is a virulence factor in *Staphylococcus aureus* arthritis and septic death. *Microb. Pathog.* **33:**239–249.

68. **Passl, R., C. Muller, C. C. Zielinski, and M. M. Eibl.** 1984. A model of experimental post-traumatic osteomyelitis in guinea pigs. *J. Trauma* **24:**323–326.

69. **Patti, J. M., T. Bremell, D. Krajewska-Pietrasik, A. Abdelnour, A. Tarkowski, C. Ryden, and M. Hook.** 1994. The *Staphylococcus aureus* collagen adhesin is a virulence determinant in experimental septic arthritis. *Infect. Immun.* **62:**152–161.

70. **Peschel, A., R. W. Jack, M. Otto, L. V. Collins, P. Staubitz, G. Nicholson, H. Kalbacher, W. F. Nieuwenhuizen, G. Jung, A. Tarkowski, K. P. M. van Kessel, and J. A. G. van Strijp.** 2001. *Staphylococcus aureus* resistance to human defensins and evasion of neutrophil killing via the novel virulence factor MprF is based on modification of membrane lipids with L-lysine. *J. Exp. Med.* **193:**1067–1076.

71. **Power, M. E., M. E. Olson, P. A. Domingue, and J. W. Costerton.** 1990. A rat model of *Staphylococcus aureus* chronic osteomyelitis that provides a suitable system for studying the human infection. *J. Med. Microbiol.* **33:**189–198.

72. **Quimby, F., and H. T. Nguyen.** 1985. Animal studies of toxic shock syndrome. *Crit. Rev. Microbiol.* **12:**1–44.

73. **Riegels-Nielson, P., N. Frimodt-Moller, and J. S. Jensen.** 1987. Rabbit model of septic arthritis. *Acta Orthoped. Scand.* **58:**14–19.

74. **Sakiniene, E., T. Bremell, and A. Tarkowski.** 1996. Addition of corticosteroids to antibiotic treatment ameliorates the course of experimental *Staphylococcus aureus* arthritis. *Arthritis Rheum.* **39:**1596–1605.

75. **Sakiniene, E., and L. V. Collins.** 2002. Combined antibiotic and free radical trap treatment is effective at combating *Staphylococcus aureus* induced septic arthritis. *Arthritis Res. Ther.* **4:**196–200.

76. **Sande, M. A.** 1981. Evaluation of antimicrobial agents in the rabbit model of endocarditis. *Rev. Infect. Dis.* **3**(Suppl.)**:**S240–S249.

77. **Santoro, J., and M. E. Levison.** 1978. Rat model of experimental endocarditis. *Infect. Immun.* **19:**915–918.

78. **Schennings, T., A. Heimdahl, K. Coster, and J. I. Flock.** 1993. Immunization with fibronectin binding protein from *Staphylococcus aureus* protects against experimental endocarditis in rats. *Microb. Pathog.* **15:**227–236.

79. **Sifri, C. D., J. Begun, F. M. Ausubel, and S. B. Calder-wood.** 2003. *Caenorhabditis elegans* as a model host for *Staphylococcus aureus* pathogenesis. *Infect Immun.* **71:** 2208–2217.

80. **Smith, I. M., A. P. Wilson, E. C. Hazard, W. K. Hummer, and M. E. Dewey.** 1960. Death from staphylococci in mice. *J. Infect. Dis.* **107:**369–378.

81. **Smith, J. M., and R. J. Dubos.** 1956. The behavior of virulent and avirulent staphylococci in the tissues of normal mice. *J. Exp. Med.* **103:**87–108.

82. **Smith, R. L., G. Kajiyama, and D. J. Schurman.** 1997. Staphylococcal septic arthritis: antibiotic and nonsteroidal anti-inflammatory drug treatment in a rabbit model. *J. Orthoped. Res.* **15:**919–926.

83. **Speers, D. J., and S. M. Nade.** 1985. Ultrastructural studies of adherence of *Staphylococcus aureus* in experimental acute hematogenous osteomyelitis. *Infect. Immun.* **49:** 443–446.

84. **Sutra, L., and B. Poutrel.** 1994. Virulence factors involved in the pathogenesis of bovine intramammary infections due to *Staphylococcus aureus*. *J. Med. Microbiol.* **40:**79–89.

85. **Tarkowski, A., M. Bokarewa, L. V. Collins, I. Gjertsson, O. H. Hultgren, I.-M. Jonsson, E. Sakiniene, E. Josefsson, and M. Verdrengh.** 2002. Current status of pathogenetic mechanisms in staphylococcal arthritis. *FEMS Microbiol. Lett.* **217:**125–132.

86. **Varshney, A. C., H. Singh, R. S. Gupta, and S. P. Singh.** 1989. Experimental model of staphylococcal osteomyelitis in dogs. *Indian J. Exp. Biol.* **27:**816–819.

87. **Verba, V., E. Sakiniene, and A. Tarkowski.** 1997. Beneficial effect of glucocorticoids on the course of haematogenously acquired *Staphylococcus aureus* nephritis. *Scand. J. Immunol.* **45:**282–286.

88. **Verba, V., and A. Tarkowski.** 1996. Participation of V beta 4(+)-, V beta 7(+)-, and V beta 11(+)-T lymphocytes in haematogenously acquired *Staphylococcus aureus* nephritis. *Scand. J. Immunol.* **44:**261–266.

89. **Verdrengh, M., and A. Tarkowski.** 1997. Role of neutrophils in experimental septicemia and septic arthritis induced by *Staphylococcus aureus*. *Infect. Immun.* **65:** 2517–2521.

90. **Watson, D. L.** 1988. Vaccination against experimental staphylococcal mastitis in ewes. *Res. Vet. Sci.* **45:**16–21.

91. **Watson, D. L.** 1992. Vaccination against experimental staphylococcal mastitis in dairy heifers. *Res. Vet. Sci.* **53:** 346–353.

92. **Woody, M. A., T. Krakauer, R. G. Ulrich, and B. G. Stiles.** 1998. Differential immune responses to staphylococcal enterotoxin B mutations in a hydrophobic loop dominating the interface with major histocompatibility complex class II receptors. *J. Infect. Dis.* **177:**1013–1022.

93. **Worlock, P., R. Slack, L. Harvey, and R. Mawhinney.** 1988. An experimental model of post-traumatic osteomyelitis in rabbits. *Br. J. Exp. Pathol.* **69:**235–244.

94. **Yao, L., J. W. Berman, S. M. Factor, and F. D. Lowy.** 1997. Correlation of histopathologic and bacteriologic changes with cytokine expression in an experimental murine model of bacteremic *Staphylococcus aureus* infection. *Infect. Immun.* **65:**3889–3895.

95. **Zhao, Y. X., A. Abdelnour, R. Holmdahl, and A. Tarkowski.** 1995. Mice with the xid B cell defect are less susceptible to developing *Staphylococcus aureus*-induced arthritis. *J. Immunol.* **155:**2067–2076.

96. **Zhao, Y. X., A. Ljungdahl, T. Olsson, and A. Tarkowski.** 1996. In situ hybridization analysis of synovial and systemic cytokine messenger RNA expression in superantigen-mediated *Staphylococcus aureus* arthritis. *Arthritis Rheum.* **39:**959–967.

97. **Zhao, Y. X., I.-M. Nilsson, and A. Tarkowski.** 1998. The dual role of interferon-gamma in experimental *Staphylococcus aureus* septicaemia versus arthritis. *Immunology* **93:**80–85.

98. **Zhao, Y. X., and A. Tarkowski.** 1995. Impact of interferon-gamma receptor deficiency on experimental *Staphylococcus aureus* septicemia and arthritis. *J. Immunol.* **155:** 5736–5742.

Cellular and Extracellular Defenses against Staphylococcal Infections

JERROLD WEISS, ARNOLD S. BAYER, AND MICHAEL YEAMAN

45

Staphylococci are prominent members of the normal flora of humans that can also produce a wide variety of invasive infections (70). These include infections of wounds, prostheses and other foreign bodies, bones, endocardium, the bloodstream, and further metastatic sites. Breaches of the skin and mucosal barriers greatly increase the likelihood of invasive staphylococcal infections, affirming the importance of these peripheral barriers in maintaining a normally asymptomatic host-bacterial relationship. A cardinal sign of staphylococcal invasion, especially that of *Staphylococcus aureus*, is the acute inflammatory host response in which the influx of polymorphonuclear leukocytes (PMN) is a very prominent feature. This reflects the major role of these professional phagocytes in first-line defense against invading bacteria (26). Deficiencies in the mobilization or function of PMN are associated with increased susceptibility to infection by many extracellular bacterial pathogens, including staphylococci. However, more recent studies have revived interest in secretion-based extracellular defenses against staphylococci and other gram-positive bacteria (85, 118, 119, 133). Hallmarks of many staphylococcal infections (e.g., *S. aureus* abscesses, bacterial vegetations at injured endothelium, *Staphylococcus epidermidis* foreign body infections) include deposition of bacteria in physical states that are likely refractory to phagocytosis (55, 112). Secreted antistaphylococcal agents may act alone, providing host defense against bacteria that resist or exceed phagocyte-based defenses, and may also act in concert with resident and mobilized phagocytes to increase antibacterial cytotoxicity of host defenses (see below).

ANTISTAPHYLOCOCCAL ACTION OF PHAGOCYTES (PMN)

In general, the action of PMN at extravascular sites of infection requires a highly regulated series of PMN responses resulting in the directed migration of PMN from blood to sites of infection, sequestration of bacterial prey, and intracellular cytotoxic action (26). A gradient of bacteria- and host-derived chemotaxins and other mediators triggers

complementary changes in the surface properties of the PMN and endothelium and in PMN contractility, leading to sequential adherence and trans-endothelial migration of PMN in the immediate vicinity of underlying infectious sites. How invading staphylococci provoke acute inflammation is still not precisely known but almost certainly includes the proinflammatory action of formylated peptides derived from bacterial protein processing and derivatives or fragments of the peptidoglycan matrix of the cell wall and other cell wall-associated bacterial products such as lipoteichoic acid (LTA) (23, 48, 64). D-Alanylation of LTA, important in resistance of staphylococci to antibacterial peptides, PMN, and, especially, group IIA phospholipase A2 (PLA2; see below), also enhances the proinflammatory properties of LTA (78). Invading *S. aureus* rapidly activates complement, generating large amounts of C5a, a potent chemotactic factor for neutrophils and, to a somewhat lesser extent, monocytes (22). Other host mediators contributing to leukocyte recruitment likely include bioactive lipid metabolites (e.g., platelet-activating factor, leukotriene B4) and chemokines (e.g., interleukin-8 [IL-8]) (80). Most clinical isolates of *S. aureus* carry a phage-derived gene encoding a chemotaxis inhibitory protein (22). This secretory protein decreases PMN and monocyte recruitment and activation induced by bacterial formylated peptides and host C5a, but not that induced by IL-8, delaying mobilization of phagocytes and thus giving invading bacteria more time to multiply and adapt to the host environment before confronting PMN. An additional mechanism to blunt leukocyte chemotaxis is conferred by the extracellular adherence protein (12), which inhibits ICAM-1 dependent leukocyte mobilization.

PMN at inflammatory sites may have increased phagocytic and cytotoxic capacities reflecting pleiotropic effects of bacterial proinflammatory products as well as of host chemokines and cytokines (34). In mouse models of staphylococcal infection, introduced genetic deficiencies in either endothelial adhesion molecules (e.g., ICAM-1 [114]) or host cytokines (e.g., tumor necrosis factor alpha [51]) can increase susceptibility to infection or decrease outcome severity. These seemingly paradoxical findings are consistent

Gram-Positive Pathogens, 2nd edition, edited by Vincent A. Fischetti et al.
© 2006 ASM Press, Washington, D.C.

with the belief that normally protective acute inflammatory host responses to staphylococcal infection may also have harmful effects, depending on specific features of the infection (e.g., size of the bacterial inoculum).

Maximum efficiency of intraphagocytic destruction of staphylococci requires fluid-phase opsonins to facilitate delivery of bacteria to phagocytes and subsequent uptake. The possible contribution to intraphagocytic bacterial destruction of extracellular antistaphylococcal agents that can bind to the bacteria before phagocytosis has thus far received little attention (but see below). Plasma-derived and extracellular matrix-associated proteins (e.g., fibronectin) that can interact simultaneously with cell wall-associated bacterial proteins and surface sites on PMN may promote bacterial clearance. Antibodies to cell wall (e.g., ClfA, Fbe) proteins that engage plasma and matrix-associated host proteins are also opsonic (87, 96). However, the most potent host opsonins are cell wall- and capsule-directed antibodies and fragments of C3 derived from complement activation (56, 89, 113). Nonencapsulated bacteria can promote complement activation via both the alternative, classical and lectin-activated pathways, leading to deposition of C3b and iC3b and targeting of bacteria to complement receptors (CR3, CR4) on phagocytes. Antibodies to staphylococcal cell walls are normally present in plasma, reflecting the frequent exposure to staphylococci. These antibodies are both directly opsonic via Fcγ receptors on phagocytes and complement activating, likely contributing to the high levels of "native" resistance to invasive staphylococcal infection (66). This resistance is further enhanced by mannose-binding lectin, which uses bacterial carbohydrate recognition to activate complement and induce acute inflammation and opsonophagocytosis in response to invading S. aureus (83, 100). Bacteremic isolates of S. aureus usually have the capacity to produce capsule, but the extent to which S. aureus in the bloodstream is encapsulated is less clear (86). The presence of a capsular layer surrounding the bacterial cell wall can reduce the efficiency of antibody and complement interaction with the cell wall and of deposited C3 fragments with phagocyte receptors and, hence, the opsonic activity of normal plasma (18, 19, 109). These protective effects of capsule can be overcome by anticapsular antibodies. In certain models of infection, type-specific anticapsular antibodies are protective, presumably by promoting bacterial clearance (67). Similarly, the capsular slime layer of coagulase-negative S. epidermidis likely impedes phagocytic clearance of bacteria adherent to foreign bodies; antibodies to these surface constituents may be opsonic as well as antiadherent. In vitro studies suggest that stationary-phase bacteria may be more difficult to clear, both by favoring retention of expressed capsule and other exopolysaccharides at or near the bacterial surface and by other capsule-independent cell envelope modifications that reduce activation of C3 deposition (18).

Uptake by professional phagocytes of staphylococci and most other bacterial prey triggers activation of a cellular oxidative response (respiratory burst) and fusion of cytoplasmic granules with the phagocytic vacuole (1, 26, 30) (Fig. 1). The efficiency of these responses may be enhanced by preexposure to host and bacterial proinflammatory mediators (e.g., tumor necrosis factor alpha and lipopolysaccharide; "priming") as occurs during infection (82, 84). Both events, together with gradual vacuolar acidification, act to create an intensely noxious microenvironment in which the ingested bacterium resides. At least in PMN, however, initial cytotoxic effects precede acidification and instead occur while the phagosome is slightly alkaline (estimated

pH 7.5 to 8.0), a pH change driven by the electrogenic properties of the respiratory burst NADPH oxidase (26, 94) (Fig. 1). The extent to which PMN and mononuclear phagocytes (monocytes and macrophages) differ in their bactericidal and digestive capacities toward staphylococci has still not been rigorously examined. However, the levels of oxidase activity are higher in PMN (26, 31) and the granules of PMN appear to be enriched in substances that can convert superoxide anion and H_2O_2 to more highly reactive and cytotoxic oxidant species (e.g., myeloperoxidase [MPO]) or can exert bactericidal effects independent of oxygen (e.g., defensins, bactericidal/permeability-increasing protein, cathelicidins, cathepsin G) (30, 32, 62). Conversely, mononuclear phagocytes (especially macrophages) appear to be better equipped to digest bacteria and bacterial remnants (57, 120). These properties are consistent with specialized roles of PMN and mononuclear phagocytes in host responses to invading bacteria: PMN crucial for early arrest of bacterial multiplication and dissemination, macrophages (and dendritic cells) needed for later induction of adaptive immunity, elimination of bacterial remnants, and resolution of inflammation. Inducible expression of phagocyte nitric oxide synthase, for example by bacterial and host proinflammatory products, has a much greater role in the longer-lived macrophage where reactive nitrogen species may help limit multiplication of intracellular microbes until adaptive immune effector systems (e.g., Th1 cells→interferon-γ→macrophage activation) are mobilized (31, 81).

Oxygen-Dependent Actions of Phagocytes

In vitro killing of staphylococci by both PMN and mononuclear phagocytes is reduced, especially at higher bacterial loads, when phagocytosis is not accompanied by activation of the respiratory burst (45, 93, 98, 123). Moreover, in chronic granulomatous disease (CGD), a disorder characterized by recurrent, potentially life-threatening infections in which host phagocytes fail to mount a respiratory burst during phagocytosis, the major microbial pathogen is S. aureus; infections with S. epidermidis are also common (26). Genetically engineered mice lacking essential components of the respiratory burst oxidase (see below) also display reduced host resistance to staphylococci in vitro and in vivo (53, 90, 94). Thus, an essential role of the phagocyte oxidative response in host defense against staphylococci is well established. Experiments in mice (73) suggest an important role for nitric oxide (NO) in antistaphylococcal defenses, but the cellular origin of NO and downstream reactive nitrogen metabolites at sites of infection and their role (e.g., cytotoxicity versus signaling) and prominence in human responses to infection are not yet firmly established (45).

The prominence in CGD of infections by catalase-positive (e.g., staphylococci) but not catalase-negative (e.g., streptococci) pyogenic cocci is consistent with a prominent role of H_2O_2-derived reactive oxidant species (ROS) in intraphagocytic killing of these bacteria (26, 72), although this view has been more recently contested (76, 95). Reactive oxidants formed secondary to respiratory burst NADPH oxidase activation that are cytotoxic toward staphylococci have been identified (44, 45). However, controversy remains concerning the precise targets and roles of ROS during and after phagocytosis of S. aureus (10, 31, 94, 95). Neither O_2^- nor H_2O_2, at the levels produced, is sufficient to directly account for O_2-dependent killing of staphylococci. A variety of secondary oxidants derived from O_2^- or H_2O_2 with potent bactericidal (cytotoxic) properties has been described, but the prominence and function of specific

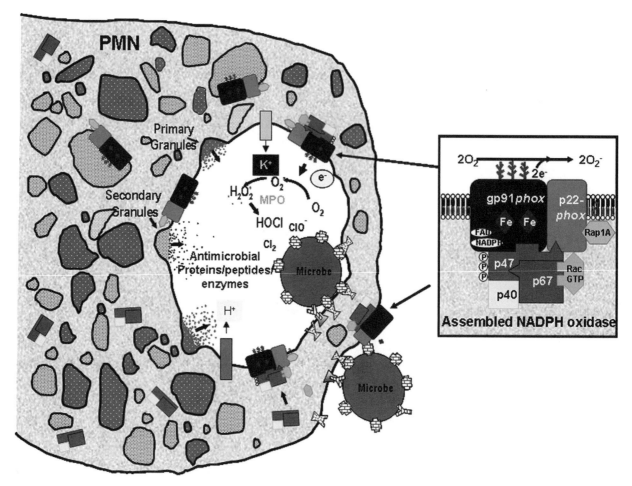

FIGURE 1 Schematic representation of PMN engaged in phagocytosis: attachment and internalization of bacterial prey into the phagocytic vacuole, fusion of cytoplasmic granules with the phagosome to deliver antibacterial peptides and proteins, and mobilization of assembled NADPH oxidase within the phagolysosome. Also indicated are K^+ and H^+ fluxes induced by electrogenic effects of oxidase activation. See text for additional details.

oxidants, either within phagolysosomes or at interfaces containing adherent bacteria, remain uncertain. Chlorination of ingested particles, including *S. aureus*, has been demonstrated and likely reflects the action of HOCl, formed by the reaction of H_2O_2 and Cl^- catalyzed by MPO (10, 44). Whether or not levels and sites of bacterial chlorination during phagocytosis are sufficient for a substantial role of HOCl-mediated chlorination in killing of ingested *S. aureus* is less clear (10, 95). MPO (and HOCl) can also react with many other anions, including nitrite. The latter may provide a mechanism whereby NO and secondary reactive nitrogen oxides (31, 45) are produced in settings (e.g., human phagocytes) where endogenous NO synthase activity may be limiting. Conversely, in settings where MPO is limiting (e.g., macrophages [26]) or at infectious sites where other host cells expressing NO synthase are prominent (e.g., endothelium), other oxidants, including peroxynitrite derived from reaction of O_2^- and NO (31, 81), may play a greater role. This may account, in part, for the much greater susceptibility to staphylococcal infection in CGD than in MPO deficiency. In patients with CGD, prophylactic gamma interferon therapy has significantly reduced the incidence and consequences of infection without remarkable changes in phagocyte oxi-

dase activity, indicating the likely importance of other host determinants of resistance to staphylococci (126).

"Oxygen-Independent" Actions of Phagocytes

The ability of PMN, especially at low bacteria/phagocyte ratios, to kill at least certain staphylococci in the absence of oxidase activation (17, 62) presumably reflects the presence of antibacterial compounds (e.g., defensins, cathelicidins, proteases) within the granules of PMN that can directly kill staphylococci (30, 68, 141). Oxygen-independent killing is greater toward mutant (e.g., *dlt*, *mprF*) *S. aureus* (17, 62) that are unable to modify prominent cell envelope polyanions (see below). This is consistent with an important role of cationic granule-associated antimicrobial proteins and peptides in oxygen-independent antistaphylococcal actions of PMN and indicates, at least in part, the molecular basis of the relative resistance of *S. aureus* to oxygen-independent killing by phagocytes. Monocytes are less active, even against the mutant strains, consistent with the apparently lower abundance of cationic antimicrobial proteins/peptides in these cells in comparison to PMN (30, 62, 71, 141).

Even cytotoxicity that is oxygen-dependent for maximum efficiency, however, may include actions of nonoxidative phagocyte weapons acting either in parallel or somehow enhanced in their actions by oxidase activity. Oxidase activation and elaboration of ROS can increase both intracellular protease activity against ingested bacteria and extracellular action of secreted proteases (50, 94, 121). These effects may be due to oxidative alterations of substrate (e.g., microbial target) and/or modifications of endogenous inhibitors that normally constrain enzyme activity. Recent studies have focused attention on the possibility that the electron flux induced by oxidase activation within the newly formed phagosome drives K^+ accumulation within the phagosome (Fig. 1), facilitating release of cationic antimicrobial proteins and peptides from the anionic sulfated proteoglycan matrix of the granules and translocation to the ingested microbe (94). Increased intraphagosomal $[K^+]$ is short-lived (94), mirroring the short half-life of active oxidase (82) and vacuolar swelling that occurs once the phagosome-associated cytoskeletal network disperses (94). This could be important if, as generally believed, interactions between released cationic proteins/peptides and the polyanionic surface of the ingested bacterium depend on electrostatic attraction (88). Although this proposed mechanism of oxidase-induced activation of granule proteins should apply to all cationic proteins and peptides, including MPO, the studies that provoked this hypothesis have suggested a specific role of cathepsin G in the antistaphylococcal cytotoxicity of (murine) PMN. How protease activation is linked, directly or indirectly, to bacterial killing is not known. Isolated cathepsin G displays only weak antistaphylococcal activity (99), suggesting either that conditions within the phagocytic vacuole are much more favorable for the antistaphylococcal activity of cathepsin G or that cathepsin G's effects on *S. aureus* during phagocytosis are less direct, perhaps involving proteolytic activation of other antibacterial systems. Comparison of the antimicrobial arsenal of different species of PMN has revealed significant species-related differences, especially among the nonoxidative cationic (poly)peptides (29, 30, 68, 141), making it even more difficult to deduce the mechanism of killing of staphylococci by human PMN based on studies with cells from knockout mice. Suggestions that all oxidase-dependent antimicrobial actions of phagocytes are "oxygen-independent," i.e., independent of ROS, seem inconsistent with emerging data that oxidative stress responses by ingested bacteria are protective (105).

To what extent bacterial killing by phagocytes or whole inflammatory exudates (see below) is accompanied by digestion of bacterial macromolecules and the determinants of that process (e.g., properties of the bacterial cell envelope, abundance and activity of host enzymes, activation of bacterial enzymes) has not, with few exceptions (33, 40, 122, 128), been extensively studied. Digestion of major bacterial structural components and enzymatic detoxification of bioactive bacterial products could be important for elimination of bacterial remnants and resolution of infection-induced inflammation. However, many images of ingested staphylococci within PMN, in vitro and in vivo, reveal relatively intact-appearing bacteria (33, 43, 74), suggesting that killing of staphylococci by PMN is not necessarily accompanied by extensive overall bacterial destruction. How and when more extensive disassembly of staphylococci is accomplished (secondary to uptake of apoptotic PMN by macrophages?) is not known and remains to be studied. The resistance of the cell wall peptidoglycan of many gram-positive bacteria to lysozyme (13) may make other bacterial macromolecules less accessible to degradation and contribute to the persistence of cell wall remnants observed under certain circumstances (40). In the case of *S. aureus*, O-acetylation at the C6-OH of muramic acid residues within cell wall peptidoglycan (7) and the apparent absence of deacylating enzymes within mammalian tissues (39) make these bacteria extremely resistant to lysozyme. It should also be noted that generally killing of staphylococci by PMN is incomplete (33, 62, 94), and, at least in certain circumstances (e.g., high MOI), includes surviving intracellular bacteria that may contribute to persistence of infection (43). Especially aggressive *S. aureus* infections (e.g., highly lethal necrotizing pneumonia) are associated with expression of the Panton-Valentine leukocidin. This toxic protein disrupts the integrity of the plasma membrane of neutrophils, leading to release of reactive oxidants and granule proteins and, if present, viable intracellular staphylococci. In contrast, interactions of *S. epidermidis* with neutrophils appear to be much less provocative, possibly promoting longer persistence of infected neutrophils (84).

ANTISTAPHYLOCOCCAL ACTION OF PLATELETS

Staphylococcal infections in CGD are overwhelmingly of extravascular nature (26). This is consistent with the retention of normal clearance function of phagocytes in this disease but also raises the possibility that other mechanisms of host defense against intravascular infections are operative. Gram-positive pathogens are, by far, the most frequent causes of endovascular infections, including infective endocarditis, vascular catheter sepsis, intravascular graft infections, and infections associated with hemodialysis shunts (52). In these clinical settings, *S. aureus*, *S. epidermidis*, and other coagulase-negative staphylococci as well as the viridans group of streptococci and the enterococci are the major etiological agents (52). A common feature in these clinical syndromes is microbial colonization of damaged endothelium. At such sites, platelets represent the initial and most abundant blood cells to arrive (133). Historically, it had been thought that platelets promote the development of endovascular infections by providing an adhesive surface upon which bacteremic organisms can dock to initiate infection. However, a body of evidence now exists emphasizing that platelets serve an antimicrobial role (Fig. 2) (133). Platelets can internalize bacteria and express on their surface a number of molecules similar to those involved in neutrophil-based host defenses (e.g., complement receptors, Fcγ receptors, and selectins). Likely to be of greater importance toward bacteria attached to surfaces is the ability of platelets to release an array of antimicrobial (poly)peptides in response to physiological stimuli present at sites of endovascular damage (133). Platelet products released during degranulation include proteins directly cytotoxic toward gram-positive and other bacteria (e.g., platelet microbicidal proteins [PMPs] (133), group IIA PLA2 (63, 118), and also molecules chemotactic for PMN, further enhancing the local mobilization of host defenses at endovascular sites of infection (133). Further emphasizing the coordination of host cellular and secretory defenses are the findings that several human and rabbit PMPs include known and novel microbicidal chemokines, or kinocidins (107, 139, 140; also see below). Locally secreted antistaphylococcal agents at sites of endovascular infection may include products of PMN, platelets, and/or the endothelium and both cytotoxic proteins and downstream metabolites of oxidative and NO metabolism (Fig. 2C). Rabbits rendered profoundly and selectively thrombocytopenic after induction of infection

A

UNACTIVATED

Dense (δ) Granule
[e.g., ADP, serotonin, Ca^{+2}, eicosanoids, TXA$_2$, PAF]

Lysosomal (λ) Granule
[e.g., lysozyme, kininogen]

Alpha (α) Granule
[e.g., fibrinogen, p-selectin, vWF, PDGF, Group IIA PLA$_2$, and likely PMPs and tPMPs]

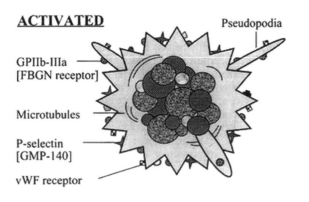

ACTIVATED

GPIIb-IIIa
[FBGN receptor]

Microtubules

P-selectin
[GMP-140]

vWF receptor

Pseudopodia

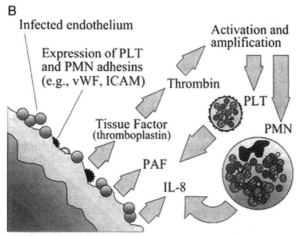

B

Infected endothelium

Expression of PLT and PMN adhesins (e.g., vWF, ICAM)

Activation and amplification

Thrombin

Tissue Factor (thromboplastin)

PAF

IL-8

PLT

PMN

C

PMN recruitment adhesion, activation

PLT adhesion, degranulation

Antimicrobial peptides (eg., tPMPs, PLA$_2$)
Toxic oxygen metabolites

Extracellular killing

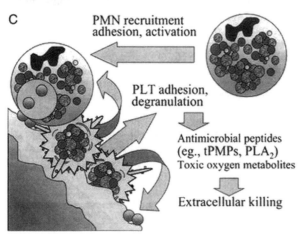

are impaired in their ability to clear viridans streptococci from damaged endocardium (106). Moreover, strains of viridans streptococci and *S. aureus* that are relatively resistant to platelet-derived antibacterial proteins (e.g., thrombin-induced PMP-1 [tPMP-1]; see below) are more virulent in animal models of infectious endocarditis, owing mainly to increased bacterial proliferation after adherence to sterile vegetations (21, 24). In vitro studies of tPMP-1 susceptibility of bacteremic isolates of *S. aureus, S. epidermidis,* and the viridans streptococci from patients with either infectious endocarditis-related or -unrelated infections revealed that isolates from patients with infective endocarditis tended to be substantially more resistant to low concentrations of tPMP-1 in vitro than were isolates not associated with infective endocarditis (129). A recent follow-up study of 60 bacteremic isolates from a single medical center (Duke University, Durham, N.C.) also revealed that strains from patients with infective endocarditis or vascular catheter infections were significantly more resistant in vitro to tPMP-1 than were isolates arising from soft tissue abscesses (2). These results suggest that reduced susceptibility to tPMP-1 provides the invading microbe with a survival advantage at sites of endovascular damage. Methicillin-resistant *S. aureus* with accessory gene regulator (*agr*) dysfunction and PMP-resistance in vitro appears to have a higher propensity to cause persistent bacteremia than methicillin-resistant *S. aureus* lacking these phenotypes (37), further suggesting a role for platelets and their products in intra- and endovascular host defense against gram-positive pathogens.

In vitro anti-*S. aureus* effects of platelets require high platelet/bacteria ratios (>1,000/1), are greater toward bacterial strains more sensitive to tPMP-1, and are accompanied by release of tPMP-1 (110). Platelet release of tPMP-1 in response to *S. aureus* was amplified by an autocrine pathway whereby released platelet ADP activates successive waves of platelet degranulation via the platelet P2X receptor. In platelets, as in PMN, there are significant species-related

FIGURE 2 (A) Schematic representation of platelets before and after activation. Note changes upon cell activation in cell architecture and surface receptors that facilitate platelet adherence and degranulation. Also note the presence of antimicrobial peptides and proteins (e.g., platelet microbicidal proteins [PMPs], thrombin-induced PMPs [tPMPs], and group IIA PLA2) believed to be stored in the alpha granule. (B) Cross-talk between vascular endothelium, platelets (PLT), and PMN in response to localized infection. Infected endothelium expresses products that either directly (e.g., platelet activating factor [PAF] and interleukin-8 [IL-8]) or indirectly (tissue factor) trigger platelet and PMN recruitment and activation and up-regulate surface receptors for (activated) platelets and PMN. (C) Recruitment, adherence, and activation of platelets and PMN (and endothelium) in juxtaposition to adherent bacteria. Degranulation and activation of respiratory burst in PMN and NO production by endothelium (not shown) lead to localized extracellular mobilization of antimicrobial peptides and proteins and toxic oxygen and nitrogen metabolites. Note that products of platelet degranulation include proteins known (e.g., platelet factor IV) or believed (e.g., tPMPs, PMPs, PLA2) to recruit and/or potentiate antimicrobial functions of PMN. Thus, secreted products may provide mechanisms for extracellular killing of (adherent) bacteria that are refractory to phagocytosis and for enhanced uptake and intracellular destruction of bacteria still susceptible to phagocytosis.

differences in the composition of granule-associated antimicrobial proteins that likely affect the antimicrobial spectrum and potency of thrombin-induced secretions and complicate the use of animal models to deduce relevant mechanisms of host defense and bacterial virulence in human endovascular infections. For example, group IIA PLA2 contributes significantly to the antistaphylococcal activity of rat and rabbit platelet acid extracts and thrombin-induced releasates, but is ~1,000-fold less prominent in human platelets.

ANTISTAPHYLOCOCCAL ACTIVITY OF BODY FLUIDS

The complement system confers upon normal plasma potent cytotoxic activity against a broad range of nonencapsulated gram-negative bacteria. In contrast, with the exception of exquisitely lysozyme-sensitive soil bacteria, gram-positive bacteria, including staphylococci, streptococci, and enterococci, are generally resistant to normal plasma and grow readily in this body fluid (28, 118). However, serum from certain animal species (e.g., rats and rabbits but not humans) is potently bactericidal toward these and many other gram-positive bacteria (28, 118). The bactericidal activity of rabbit and rat sera toward S. aureus and several other gram-positive bacteria is mainly attributable to group IIA PLA2 released from platelets during in vitro clotting (28, 119). This enzyme is naturally mobilized from a variety of intravascular and extravascular sources during inflammation (54) and thereby confers potent extracellular bactericidal activity toward many gram-positive bacteria as part of the body's response to bacterial invasion. This has been demonstrated in sterile peritoneal inflammatory exudates in rabbits induced by intraperitoneal administration of glycogen (119), plasma of baboons following intravenous bacterial challenge (118), and acute-phase human sera (54). At higher concentrations of S. aureus ($\geq 10^7$ bacteria/ml), fibrinogen-rich biological fluids rapidly induce bacterial clumping, reducing bacterial sensitivity to PLA2 (27). Depending on the extent of clumping, proteases, such as plasmin, can penetrate and dissolve protein cross-links holding bacterial clumps together and increase bacterial sensitivity to PLA2 action. Thus, the extracellular mobilization of group IIA PLA2 during inflammation, together with local activation of proteolytic activity, could provide a mechanism by which the host can control proliferation and survival of S. aureus even after bacterial clumping. In tears, where exposure to bacterial invaders is virtually constant, PLA2 levels are constitutively high and again appear to account for much of the antistaphylococcal activity of this body fluid (92).

Specialized epithelia produce a wide array of secretory (poly)peptides that, in aggregate, display broad spectrum antimicrobial activity in vitro (68, 97, 141). These secretory antimicrobial products provide, as in more primitive organisms, an early shield against microbial invasion across mucosa epithelia and skin. The ability of coagulase-negative staphylococci to colonize skin and the ability of S. aureus to colonize the nasal epithelia presumably reflect, in part, resistance of these organisms to secreted antimicrobial proteins at these sites. Human β-defensin-1 (HBD-1) is highly expressed constitutively but has low antistaphylococcal activity, especially at physiological salt concentrations (77). HBD-2, highly induced by host and bacterial proinflammatory mediators (68, 77), is more potent but also salt-sensitive (68, 77). HBD-3 and the single human cathelicidin (hCAP18→LL-37; see below), though less robustly induced by injury and infections, are more potent and less salt-sensitive and show increased potency in the presence of HBD-2 (77). In contrast to psoriatic skin lesions in which levels of HBD-2 and LL-37 are high and infection uncommon, levels of these peptides are much lower in inflamed skin from patients with atopic dermatitis (due to suppressive effects of IL-4 and IL-13) and infection with S. aureus is much more common (85). Cathelicidin-deficient mice show increased sensitivity to skin infections with group A streptococci (141).

ANTISTAPHYLOCOCCAL ACTION OF INFLAMMATORY EXUDATES

The acute inflammatory response mobilizes both PMN and extracellular antistaphylococcal activity at the site of bacterial invasion. Clumped and encapsulated S. aureus, resistant to phagocytosis, is still (partially) sensitive to extracellular group IIA PLA2 (27, 35), suggesting that the major target of this antibacterial enzyme may be invading bacteria that resist or exceed the capacity of resident and mobilized phagocytes. Primed or activated neutrophils in vitro and at inflammatory sites in vivo release granule proteins and chromatin, forming extracellular fibers that can restrict bacterial dissemination by binding bacteria (including staphylococci) and killing and partially digesting some of the trapped bacteria (8). In addition, as illustrated by the group IIA PLA2 (61, 122), the ability of extracellular antibacterial agents to bind to bacteria provides a means by which the extracellular enzyme can be delivered to phagocytosing PMN and contribute to digestion of ingested bacteria (33, 122). Degradation of phospholipids of ingested S. aureus, in fact, depends on the presence of extracellular group IIA PLA2 (Fig. 3 [33]). Effects of the extracellular enzyme on intracellular bacterial digestion occur even at enzyme concentrations below that needed for extracellular action (Fig. 3), suggesting that conditions within the phagocytic vacuole, at least initially, are more favorable for PLA2 action. Enhanced bacterial phospholipid degradation depends on activation of the respiratory burst oxidase (33). In the absence of oxidase activation, the PMN do not potentiate PLA2 action and, instead, protect the bacteria from the extracellular enzyme. The collaboration between oxygen-dependent and -independent antibacterial systems in phagocyte-dependent cytotoxicity, including both PMN-derived and soluble exogenous agents, emphasizes the complex integration of elements of innate immunity-mediating antimicrobial action that is likely to occur in PMN-rich inflammatory exudates. The synergy between extracellular group IIA PLA2 and PMN means that levels of PLA2 sufficient to cause bacterial phospholipid degradation are achieved sooner during the inflammatory response, when levels of extracellular enzyme are otherwise limiting for antistaphylococcal action. Conversely, ingestion by PMN of S. aureus before sufficient mobilization of the group IIA PLA2 may provide a protective niche that increases the probability of the persistence of bacteria and bacterial remnants (40, 43, 74). The role of oxidase activation in the action of both antibacterial protease(s) and PLA2 strongly suggests that the host defense defects in CGD include, in addition to reduced elaboration of cytotoxic ROS, reduced activity of "oxygen-independent" antibacterial enzymes whose antibacterial activity is somehow increased by oxidase activation. The dependence of several antibacterial enzymes on oxidase activation results, in CGD, in a defect in microbial digestion as well as killing that may contribute to the

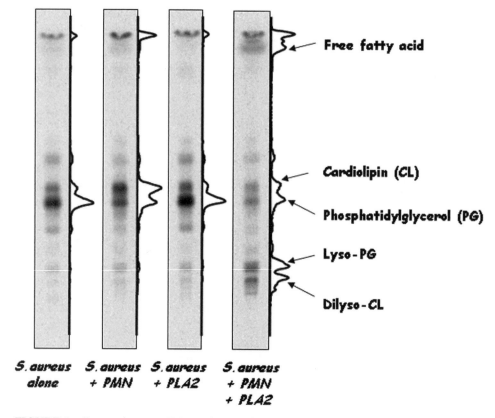

FIGURE 3 Synergy between PMN and extracellular group IIA PLA2 in digestion of phospholipids of *S. aureus* during phagocytosis by human PMN. Shown are profiles of metabolically labeled bacterial lipids (119), resolved by thin-layer chromatography, after incubation for 2 h as indicated. Note that appreciable bacterial phospholipid degradation, at concentration of PLA2 tested, occurs only in combined presence of PLA2 and PMN. Note also substantial conversion of bacterial PG→CL, indicative of a stress response induced shortly after phagocytosis. See text and reference 33 for more details.

chronic inflammation that is a hallmark of CGD (26). Even in normal individuals, oxygen limitation in deep-seated infections and necrotic tissue could, in a similar manner, create circumstances less favorable for elimination of *S. aureus* and their remnants.

MOLECULAR CHARACTERISTICS OF ANTISTAPHYLOCOCCAL HOST DEFENSE SYSTEMS

Oxygen-Dependent Systems in PMN (Phagocytes)

Respiratory Burst Oxidase

Activation of professional phagocytes, as normally occurs during phagocytosis, triggers the assembly of a multicomponent enzyme complex that catalyzes the transfer of reducing equivalents from NADPH to molecular oxygen, yielding superoxide anion (Fig. 1) (1, 82). Essential components of the respiratory burst NADPH oxidase include both transmembrane and cytosolic proteins. The former is an unusual *b*-type cytochrome (b_{558}) consisting of a 91-kDa glycoprotein (gp91-*phox*) and a 22-kDa subunit (p22-*phox*). Cytosolic components of the oxidase include p47-*phox*, p67-*phox*, p40-*phox*, and Rac1 or Rac2 and Rap1A, low-molecular-weight G proteins. gp91-*phox* has the intrinsic molecular features needed to reduce oxygen univalently via NADPH, but the other components are necessary for intracellular stability (p22-*phox*) and optimal oxidase activity. Thus, acquired gene defects in either the X-linked gp91-*phox* or in autosomal genes encoding p22-, p47-, or p67-*phox* confer a CGD phenotype, although disease severity can vary markedly (26).

Before cell activation, most cytochrome b_{558} resides within secondary granules, with additional subpools in secretory vesicles and in the plasma membrane (82). During phagocytosis, cytochrome b_{558} is mobilized from these sites to the forming phagocytic vacuole or phagolysosome, where it serves as a docking site for the cytosolic oxidase components (82). Multiple sites of contact between the cytochrome and the soluble oxidase components have been implicated, including SH3 domains in p47- and p67-*phox* and proline-rich targets of SH3 domains in p22-, p47-, and p67-*phox*. The importance of these sites in oxidase assembly and function has been illustrated in a CGD patient who had a homozygous point mutation in p22-*phox*. The mutation (P156Q) disrupted a proline-rich region in p22-*phox* normally involved in interactions with an SH3 domain of p47-*phox* during oxidase activation (26).

Under optimal conditions, full assembly of the respiratory burst oxidase occurs exclusively at membrane sites contiguous with the attached or internalized target bacterium. No assembly or oxidase activity is detected at remaining secondary granules (82). The molecular basis of this specificity is unknown. Initial membrane association of the cytosolic components is independent of flavocytochrome b_{558} and may be mediated by interactions of protein PX domains with lipid metabolites generated in situ. A role of the submembranous actin-based cytoskeleton is also possible. Based on the predicted topology of gp91-phox, O_2^- is released either within the lumen of the phagosome during phagocytosis or across the extracytoplasmic leaflet of the plasma membrane when particle attachment and cell activation are not coupled to particle uptake.

Bactericidal Reactive Oxidants

A variety of secondary oxidants derived from superoxide anion and/or H_2O_2 or NO with potent bactericidal (cytotoxic) properties have been described (31, 45, 81). These include hydroxyl radical, singlet oxygen, hypochlorous acid, N-chloramines, and peroxynitrite. Some products (e.g., hydroxyl radical, peroxynitrite) are very short-lived and have a very short range of action, whereas others (e.g., N-chloramines) are more long-lived and hence potentially able to act at a distance. Together, these products may target a wide range of cellular constituents, including DNA, lipids, protein thiols, and Fe/S centers, accounting for their cytotoxicity against virtually all microbes but potentially host cells as well. A protective role of these oxidants in host defense can be achieved by coupling oxidase activation to microbial recognition and sequestration and constraining the half-life of the oxidase and of the oxidants generated.

In several instances, pathogenic microorganisms can subvert and/or resist the actions of reactive oxidant and nitrogen species by evasion, suppression, enzymatic inactivation (e.g., catalase), scavenging, sequestration of iron, and induction of stress and repair responses (31, 81). However, the possible role of many of these specific mechanisms in interactions of important gram-positive bacterial pathogens with phagocytes is largely understudied.

Oxygen-Independent Systems

A wide variety of mammalian (poly)peptides has been described that can kill staphylococci and many other gram-positive bacteria under arbitrary in vitro conditions (30, 68, 141). The ability of these proteins, alone or in concert with other host defense systems, to exert antibacterial effects under in vitro conditions that more closely simulate conditions at extracellular or intracellular sites of infection requires much more study. We focus our description below on those agents whose mechanism of action against S. aureus has been most extensively studied.

Group IIA PLA2

The group IIA PLA2 is a member of a family of low M_r (13,000 to 18,000) "secretory" PLA2 (sPLA2) with close overall structural and functional similarity to related enzymes in the venoms of snakes and insects. Hallmarks of these proteins include a highly compact three-dimensional structure stabilized by six to eight disulfides and a Ca^{2+}-dependent catalytic machinery (103). Whereas the overall structure and catalytic properties of these enzymes are closely similar, their sites of expression and mobilization and apparent biological functions differ greatly. Among the 10 sPLA2 in humans (103), the group IIA PLA2 is unique in its high net-positive charge (ranging from +12 to +17 in five different mammalian species) and antibacterial potency. This enzyme can attack, at nanomolar concentrations, both gram-negative and gram-positive bacteria, though the former action typically requires the assistance of other host defense proteins to facilitate access of the PLA2 to phospholipids in the gram-negative bacterial envelope (30, 71, 122). Extracellular levels of the group IIA PLA2 vary widely in time and space. In plasma and presumably many other tissue fluids, group IIA PLA2 concentrations under resting conditions are subnanomolar but can increase 100- to 1,000-fold during acute inflammation as part of the acute-phase response (54, 63, 118). In contrast, at certain other sites (e.g., tears, seminal fluid), group IIA PLA2 levels are constitutively high and may exceed micromolar concentrations (92). These group IIA PLA2-rich fluids display potent antibacterial activity against S. aureus and many other gram-positive bacteria that is due largely to the presence of this enzyme (92, 118, 119). Transgenic mice overexpressing human group IIA PLA2 show increased resistance to Escherichia coli and S. aureus infections (65), demonstrating protective effects of group IIA PLA2 mobilization in vivo though under somewhat nonphysiological conditions.

The 90% lethal dose of group IIA PLA2 toward S. aureus is ca. 1 to 10 nM in artificial medium (e.g., tissue culture fluid) and in natural biological fluids. In contrast to the properties of many other host antibacterial peptides and proteins, the group IIA PLA2 is fully active in media containing physiological levels of monovalent and divalent cations. Many other species of gram-positive bacteria are highly sensitive to group IIA PLA2, including S. epidermidis, alpha- and beta-hemolytic streptococci (e.g., viridans, pneumococci, groups A and B), enterococci (Enterococcus faecalis and Enterococcus faecium), Listeria monocytogenes, and Bacillus anthracis, both encapsulated as well as nonencapsulated strains (35, 47, 92, 119). In contrast, host cells are highly resistant to group IIA PLA2, at least under normal (resting) conditions (63, 79).

Bacterial killing by PLA2 involves several stepwise interactions with the bacterial envelope, including surface binding, penetration of the cell wall, degradation of membrane phospholipids, and activation of autolysins (9, 36, 61). Mutations that either reduce surface basicity of the group IIA PLA2 or disrupt its catalytic machinery greatly reduce or eliminate antibacterial activity, reflecting the importance of both the cationic and catalytic properties in the potent antibacterial activity of group IIA PLA2 (5, 61, 119). Binding and penetration of the cell wall depend on the highly cationic properties of the group IIA PLA2 that distinguish it from all other sPLA2 (5, 61). Protoplasts fully denuded of cell wall are susceptible to all sPLA2. Initial surface binding likely involves strong electrostatic interactions with polyanionic (lipo)teichoic acids ([L]TA). PLA2 binding is antagonized by high, supraphysiological concentrations of NaCl (>0.5 M) or $MgCl_2$ (>20 mM) but not by physiological salt concentrations, consistent with the activity of this enzyme in biological fluids (118, 119). D-alanylation of (L)TA inhibits antistaphylococcal activity of a variety of cationic antimicrobial compounds (88), but effects on PLA2 activity are much greater than that on smaller cationic peptides (dlt mutant is ca. 30 to 100 times more sensitive to PLA2 [61]). In contrast to effects on the peptides, D-alanylation of (L)TA affects activity of the bound PLA2, not initial PLA2 binding to S. aureus. Initial binding is calcium-independent, but membrane phospholipid degradation and subsequent bacterial killing and lysis are calcium-dependent, reflecting the calcium-dependent catalytic properties of the PLA2. Although the K_d for calcium binding to the PLA2 is in the millimolar range (63, 103), μM concentrations of ambient Ca^{2+} satisfy PLA2 needs for antistaphylococcal action (119), suggesting that calcium bound to (L)TA in the bacterial envelope may be locally mobilized upon PLA2 binding.

The ability of bound PLA2 to penetrate the cell wall to attack phospholipids in the cell membrane is diminished in nongrowing bacteria and enhanced in bacteria treated with β-lactams, even at subinhibitory doses (36). The binding of the cationic PLA2 to anionic moieties in the cell wall may cause displacement and local activation of autolysins that are less cationic than the group IIA PLA2, and thus promote penetration of PLA2. Envelope sites of new cell wall synthesis and remodeling or of β-lactam-induced reduced cross-linking may represent sites most permissive for PLA2 penetration (36). PLA2-sensitive bacteria can tolerate degradation and loss of up to 50% of membrane phospholipids in 30 min, but a greater rate of phospholipid degradation typically leads to bacterial killing and lysis (36). Autolysin-deficient S. aureus can tolerate even greater phospholipid degradation, suggesting that killing by PLA2 is triggered when phospholipid loss exceeds the bacterium's capacity for biosynthetic repair and diffusely activates bacterial autolytic activity.

In contrast to several other sPLA2, the group IIA PLA2 displays very weak hydrolytic activity toward phospholipids in host cell membranes (79). This is attributable to substitution of a key tryptophan (Trp31) residue present on the surface of the other sPLA2 that is important for stable sPLA2 binding and limited penetration of membranes in which the outer leaflet lipids are mainly zwitterionic, as is typically the case on the surface of mammalian cells. Released membrane "blebs" from activated cells are much more sensitive targets for the group IIA PLA2, possibly reflecting increased prominence of acidic phospholipids on the outer surface and a geometry (i.e., increased radius of curvature) that more resembles bacterial cocci. Substitution of phosphatidylglycerol, the principal membrane phospholipids in S. aureus and many other gram-positive bacteria (119), with lysine has little or no effect on the activity of group IIA PLA2 toward intact bacteria or membrane protoplasts (61), indicating that cell wall, not cell membrane, polyanions are the major determinants of PLA2 activity toward S. aureus and probably other gram-positive bacteria.

PMPs and Platelet Kinocidins

Stimulation of rabbit or human platelets with thrombin, as naturally occurs at sites of endovascular damage, leads to extracellular accumulation of potent antimicrobial activity against gram-positive pathogens (133). Several antimicrobial (poly)peptides are released from platelets treated with thrombin. They were originally termed thrombodefensins because they were believed to represent the platelet version of neutrophil defensins (20). However, the major antibacterial substances toward gram-positive bacteria that are present in these secretions, the so-called PMPs and group IIA PLA2, are genetically and structurally unrelated to defensins.

The PMPs within rabbit platelets comprise an array of seven antimicrobial peptides (137). These peptides are all cationic, heat-stable, and range in molecular mass from 6.0 to 9.0 kDa, and are thus distinct in size from either defensins or group IIA PLA2. Other structural features that distinguish PMPs from defensins include the prominence of lysine residues and the presence of two to four, rather than six, cystines in the PMPs (46). Of the seven PMPs, five are iso-

lated by acid extraction of washed platelets (PMPs 1 to 5), whereas two other peptides (tPMP-1 and tPMP-2) are released following thrombin stimulation. The latter two peptides differ by only a single amino acid (glycine in tPMP-1 and arginine in tPMP-2); preliminary sequence data for PMPs 1 to 5 suggest that they also derive from distinct genes. Amino acid sequencing and mass spectroscopy have revealed that distinct N-terminal polymorphism variants of PMP-1 isolated from quiescent or thrombin-stimulated platelets arise from a common PMP-1 propeptide (139). Cloning of PMP-1 from bone marrow and characterization of its full-length cDNA confirmed that PMP-1 is translated as a 106-amino-acid precursor and is processed to yield 73-residue (8,053 Da) and 72-residue (7,951 Da) variants. Human platelets contain antimicrobial peptides analogous to PMPs 1 to 5 and the tPMPs (107). Phylogeny, sequence motifs, and three-dimensional structures show that PMP-1 shares greatest homology with human PF-4 (hPF-4). Hence, its structural and antimicrobial properties establish the identity of PMP-1 as a rabbit analogue of hPF-4. In addition to hPF-4, human PMPs (also termed thrombocidins) include platelet basic peptide and its proteolytic products such as connective tissue activating peptide 3, as well as RANTES and thymosin-b-4. Several are known CXC (β) chemokines (e.g., hPF-4, platelet basic peptide, and fragments such as connective tissue activating peptide 3), while RANTES is a known CC (α) chemokine, containing either cystine-X-cystine or cystine-cystine, respectively, at the C terminus. Therefore, in addition to direct antimicrobial properties, these peptides likely play an indirect host defense function via chemoattraction and antimicrobial potentiation of inflammatory cells (e.g., PMN). Chemokines that also exert direct microbicidal activity have been termed kinocidins (139, 140), consistent with their likely coordinating roles that bridge molecular and cellular mechanisms of antimicrobial host defense (16, 131).

Human and rabbit PMPs are active against S. aureus, streptococci, and Candida albicans in nanomolar concentrations. Interestingly, PMPs from both species exhibit conditional optima. For example, rabbit PMPs 1 to 5 are more active at acid pH, whereas tPMP-1 and -2 are more active at neutral pH. Similar themes are observed with human PMPs. Of these peptides, the antimicrobial action of tPMP-1 has been studied most intensively. Except for the enterococci, tPMP-1 is microbicidal against most of the clinically relevant gram-positive pathogens tested to date, including S. aureus, S. epidermidis, the viridans streptococci, and Candida spp. (135, 137).

Within 5 to 15 min of exposure of S. aureus to tPMP-1, perturbations of the cell membrane are manifested by ultrastructural analysis, suggesting that, like other endogenous antimicrobial peptides, tPMP-1 might target the microbial cell membrane as part of its lethal pathway (134). In support of this hypothesis, auxotrophs of S. aureus that exhibit defects in their ability to generate a normal transmembrane electric potential (Δψ) are more resistant to killing by tPMP-1 than are their parental counterparts with normal Δψ. Dietary supplementation (e.g., menadione) of the Δψ-deficient mutants to meet their auxotrophic requirements restores the Δψ to near parental levels and reconstitutes tPMP-1 susceptibility (134). The ability of tPMP-1 to increase bacterial membrane permeability to propidium iodide is also diminished in the Δψ mutants and restored by addition of menadione (134). Studies by Koo et al. (58) have suggested that the overall proton motive force (Δψ and ΔpH) is important in tPMP-1-induced microbicidal action. The dependence of tPMP-1 on Δψ for its microbicidal action further distinguishes this peptide from neutrophil defensins, which appear to kill *S. aureus* in a $\Delta\psi$-independent manner. Addition of tPMP-1 to model planar lipid bilayers causes membrane permeabilization and disruption of the bilayer in a voltage-dependent manner with maximal effects induced at a *trans*-negative voltage orientation relative to the site of addition of the peptide (60). Well-defined, voltage-gated pores formed by defensins in similar model membranes are not formed by tPMP-1. The staphylocidal activity of tPMP-1 is substantially reduced under in vitro conditions at which microbial membrane energetics are reduced (e.g., stationary phase, low temperature) (59). Finally, cytoplasmic membrane fluidity of tPMP-1-resistant strains of *S. aureus* obtained by serial passage, transposon mutagenesis, or plasmid carriage is markedly different from that of their genetically related tPMP-1-susceptible counterparts.

Although initial effects of tPMP-1 target the bacterial cell membrane, the discrepancy between early membrane effects of tPMP-1 (which occur within minutes of exposure) and the delayed microbicidal effects (which require 1 to 2 h of treatment) suggests that other targets are involved in the lethal pathway of tPMP-1. Pretreatment of tPMP-1-susceptible strains with antibiotics that block either 50S ribosome-dependent protein synthesis or DNA gyrase subunit B actions completely inhibits tPMP-1-induced microbicidal effects. Moreover, *S. aureus* strains deficient in their autolytic pathway are also less susceptible to tPMP-1, suggesting an important role for this system in the overall lethal mechanism of this peptide (130). Collectively, these data support the concept that tPMP-1-induced lethality involves intracellular targets and autolysis as well as membrane perturbations.

In addition to its direct microbicidal effects against gram-positive pathogens, tPMP-1 also causes prolonged postexposure growth-inhibitory effects against staphylococci, similar to the effects of cell wall-active antibiotics such as oxacillin and vancomycin (136). In contrast to the microbicidal effects of tPMP-1, these growth-inhibitory properties of the peptide are observed in both tPMP-1-susceptible and -resistant strains. Simultaneous exposure of staphylococci to tPMP-1 and antibiotics such as oxacillin produces synergistic bactericidal effects. In addition, preexposure of *S. aureus* or *Candida* spp. to tPMP-1 significantly reduces the capacity of these organisms to bind to platelets in vitro, an effect that can be further magnified by exposure to other antimicrobial agents (138).

The contribution of PMPs to extracellular host defenses against adherent bacteria depends on mechanisms that induce release of these antibacterial peptides. To date, three physiologically relevant stimuli have been shown to prompt the release of one or more PMPs in vitro. Each of these stimuli is associated with either endothelial damage or microbial colonization of such sites. As noted above, thrombin is a potent stimulus for the release of tPMP-1 and tPMP-2 from rabbit platelets (137). *S. aureus*-induced rabbit platelet aggregation also results in secretion from platelets of substances with biological activity and electrophoretic properties compatible with PMPs. Moreover, in the presence of staphylococcal alpha-toxin, a major secretory virulence factor of *S. aureus*, rabbit platelets are lysed coincident with the release of PMP-1 and PMP-2 (3).

Defensins

Defensins are a family of highly abundant cysteine-rich cationic peptides that contribute to host defense against bacterial, fungal, and viral infections (68). Defensins and defensinlike cysteine-rich peptides are widely expressed ancient components of antimicrobial host defenses. Three subfamilies of defensins exist: α-defensins, most prominent in neutrophils and Paneth cells of the small intestine; β-defensins, abundant at skin and mucous membranes of respiratory, gastrointestinal, and genitourinary tracts; and φ-defensins, thus far identified only in Old World monkeys, lesser apes, and orangutans. φ-Defensins are lectins with broad-spectrum antiviral properties. α- and β-defensins are 3.5 to 4.5 kDa, with distinct spacing and arrangement of three disulfides that stabilize an otherwise similar triple-stranded antiparallel β-sheet with cationic/hydrophobic (amphiphilic) character (144). φ-defensins, by contrast, are cyclic octadecapeptides, also with three disulfides, that are produced by an unusual series of posttranslational reactions.

The number and type of defensins in PMN and Paneth cells are species-dependent: human, rabbit, and rat PMN contain several α-defensins; bovine PMN contain several β-defensins; and porcine, ovine, and murine PMN contain no defensins. The four α-defensins in human PMN (human neutrophil defensins [HNP] 1 to 4) together comprise 30 to 50% of the total protein stored in the primary (azurophilic) granules. In addition, two different α-defensins (HNP-5 and -6) are expressed in Paneth cells. By contrast, murine PMN contain no defensins, whereas the Paneth cells in mice express up to 20 different defensins.

Human β-defensin-1 (HBD-1) is highly expressed in kidney, pancreas, and, to a lesser extent, other epithelial cells (143). Mucosal β-defensins are induced by inflammatory stimuli (25, 77, 97) and appear to play a role in bactericidal activity of human lung fluid (42). Of the human defensins, HBD-3 is most cationic (net charge of +9) and most active against *S. aureus* and *E. faecium*. Human α-defensin genes and many β-defensin genes are encoded within a cluster of genes on chromosome 8p22-p23.

The α-defensins are synthesized as ca. 95-residue noncytotoxic pre-proforms (39). Efficient sorting and cleavage to generate the mature cytotoxic peptides present in the granules depend on the propiece. In addition, the net-charge properties of the anionic propiece and cationic mature peptide have coevolved to maintain the charge neutrality of the proform despite wide variation in the net (+) charge of the mature defensin. This charge neutralization within the nascent polypeptide is thought to protect the producing cell (e.g., PMN) from the possible toxicity of the mature peptide during biogenesis and targeting of defensins for storage in granules (111). Abundant acidic glycosaminoglycans may serve a similar function in the granules (30, 94).

Defensins are broadly cytotoxic toward bacteria, especially gram-positive bacteria, fungi, metazoan parasites, and mammalian cells (30, 68). Hence, the cytotoxicity of defensins lacks great intrinsic specificity. However, storage of these peptides within granules and delivery to phagolysosomes (Fig. 1) should direct the peptides' cytotoxicity more selectively toward the ingested microbial prey. Independent microbicidal effects are expressed in vitro at micromolar concentrations under hypotonic conditions, but much higher concentrations of most defensins are required when physiological concentrations of salts are present. Thus, secreted (i.e., diluted) defensins are much less likely to exert cytotoxic effects unless retained within a microcompartment. Secreted defensins may also be trapped within inactive complexes by other host proteins in plasma and other body fluids, further reducing the likelihood of extracellular cytotoxic action (30, 68). However, at least certain defensins can act in synergy with other host defense agents, which may permit these peptides to contribute to antibacterial host defense at concentrations that have no independent antibacterial activity (69).

The more highly cationic (type 1) α-defensins (e.g., rabbit defensins NP-1 and NP-2) do not require target pathogens to be metabolically active to produce a microbicidal effect. In contrast, the less cationic type 2 α-defensins (e.g., HNP-1 and -2) are maximally active against metabolically active targets. These features have important implications in antimicrobial host defense, because infecting microbes may vary in their envelope charge and/or metabolic activities, particularly within distinct infection sites and different microenvironments of the host. For example, persistent *S. aureus* small-colony variants exhibit reduced oxygen consumption, ATP production, and membrane bioenergetics (91). Thus, an array of defensins exists that is likely suited to act against a broad spectrum of microbes, including those of differing metabolic status.

Defensins target and disrupt the bacterial cytoplasmic membrane (101). Staphylococcal membranes exhibit blebbing and distortion after exposure to rabbit PMN granule extracts rich in defensins, group IIA PLA2, and other cationic proteins such as cathelicidins (117). Initial interactions are likely driven by electrostatic attraction between the cationic defensins and anionic surface lipids. D-Alanylation of LTA and lysine modifications of phosphatidylglycerol produce similar (~5- to 10-fold) reduction of defensin activity against *S. aureus* (88). Insertion of bound defensins into model membranes is driven by transmembrane potential and involves the formation of multimeric pores (125). Differences in effects on model membranes between particular defensin species appear to correlate with the propensity of the defensin to form stable dimers in the context of lipid environments, simulating the bacterial cytoplasmic membrane (49). However, these differences do not obviously correlate with differences in cytotoxic activity.

Several other effects of defensins have been described in vitro that are independent of their cytotoxic activities but could contribute to a role in host responses to infection and injury. These include chemotactic activity toward monocytes (108) and T cells (11), antagonism of adrenocorticotropin action at its receptor, stimulation of epithelial cell growth, and secretion of IL-8 and inhibition of fibrinolysis; the last suggests a mechanism by which PMN could maintain a fibrin barrier at sites of infection (30, 68). Apparently subcytotoxic doses of HNP-3 have protective effects in experimental infections (including *S. aureus*) in mice, possibly by promoting leukocyte accumulation at sites of infection (124).

Cathelicidins

Cathelicidins represent a family of antimicrobial proteins with an N-terminal signal peptide, highly conserved cathelin (cathepsin L-inhibitor)-like domain (~100 amino acids), and a structurally variable C-terminal antimicrobial domain ranging from 12 to approximately 100 amino acids. Cathelicidins have been identified in all mammalian species thus far examined (141). Some species (e.g., pig, cow, sheep) produce several cathelicidin species but in humans only one cathelicidin has been identified (hCAP18 = holo-protein; LL-37 = processed antimicrobial C-terminal peptide).

In contrast to the defensins, cathelicidins are generally stored in the secondary (specific) granules of PMN in unprocessed form (32, 104). Relatively selective secretion of secondary granules, as can occur during recruitment of PMN to inflammatory sites, may in part account for the sig-

nificant levels of unprocessed cathelicidin that have been detected in extracellular fluids. The holo-protein (e.g., hCAP18) lacks independent cytotoxic activity against bacteria but can enhance the activity of other host defense proteins and inhibit the activity of bacterial proinflammatory products (e.g., lipopolysaccharide) (142). Independent cytotoxic activity requires processing of cathelicidin to release the active C-terminal peptide. Processing of PMN-derived cathelicidin is typically mediated by a protease (proteinase-3 in human PMN, elastase in porcine PMN) stored in the primary granules of PMN and thus occurs when vigorous activation of the PMN induces exocytosis of both granules, such as during phagocytosis (104, 141). Infection can also stimulate extracellular secretion and processing and contribute significantly to the extracellular antibacterial activity of an inflammatory fluid (141). The separation of the parent cathelicidin and processing enzymes in separate compartments and selective degranulation suggest distinct functions of the parent and processed forms of cathelicidin at different stages of the host inflammatory response.

hCAP18 is also widely expressed in many nonmyeloid tissues, including the epididymis, spermatids, keratinocytes, epithelial cells of airways, mouth, tongue, esophagus, intestine, cervix, and vagina (141). It is especially abundant in seminal plasma. Expression of hCAP18 is low in normal skin but induced during inflammation and recovered from wound and blister fluids, probably from activated keratinocytes and mobilized PMN (38). In vitro, antibacterial activity is apparent at micromolar concentrations against a range of gram-negative and gram-positive bacteria, including group A to C streptococci. *E. faecalis* and *S. epidermidis* are relatively resistant. The contribution of cathelicidins to host defense may be further enhanced by their ability to act in synergy with several different host defense agents (69, 102). Cathelicidins also exhibit immunomodulatory properties, including chemotaxis and induction of histamine release.

SUMMARY AND PERSPECTIVES

As this chapter reflects, much has been learned about the compositional identity of the antimicrobial arsenal of professional phagocytes (especially of PMN) and of various secretions. Continued advances in analysis of complex biological samples by proteomics may still reveal additional antimicrobial weapons of the host. How any of these cytotoxins, in isolation, kills microbial targets remains uncertain. Not surprisingly, and more importantly, how various cytotoxic products of phagocytes and their surrounding environment may function together to optimize elimination of viable microorganisms and bioactive microbial products is also largely unknown. The diversity of host antimicrobial systems may reflect the need to combat a vast array of potential microbial invaders that vary in sensitivity to individual cytotoxins as well as, perhaps, to reduce selective pressure favoring emergence and propagation of resistant microbial variants. A priori, the diversity of cytotoxins and their mechanisms of action should make possible highly cooperative actions, but the number of well-documented examples of such synergy, especially under physiologically relevant conditions, remains small. Nevertheless, perhaps the most important new insights since the first edition have been the growing recognition of the complex coordination of mechanistically and spatially (e.g., cellular and extracellular) diverse antibacterial systems. How this ultimately translates to efficient bacterial killing and clearance of bacterial remnants remains a work in progress. It is ironic that as the pendulum has swung back to a greater focus on nonoxidative cytotoxic systems, much more insight has emerged as to the genetic, biochemical, and structural basis of intrinsic resistance of staphylococci (e.g., *S. aureus*) to many of these "oxygen-independent" weapons. The possibility that some of these modifications consist of remodeling reactions responsive to distinct environmental cues and that exposure to host defenses induces these and other bacterial "stress responses" (14, 115) suggests new approaches to study of the complex dynamics of host-bacterial interactions that should help unravel what determines favorable outcome for the host and/or parasite.

REFERENCES

1. **Babior, B.** 1999. Review: NADPH oxidase: an update. *Blood* **93:**1464–1476.
2. **Bayer, A. S., D. Cheng, M. R. Yeaman, G. R. Corey, R. S. McClelland, L. J. Harrel, and V. G. Fowler, Jr.** 1998. In vitro resistance to thrombin-induced microbicidal protein among clinical bacteremic isolates of *Staphylococcus aureus* correlates with an endovascular infectious source. *Antimicrob. Agents Chemother.* **42:**3169–3172.
3. **Bayer, A. S., M. D. Ramos, B. E. Menzies, M. R. Yeaman, A. Shen, and A. L. Cheung.** 1997. Hyperproduction of alpha-toxin by *Staphylococcus aureus* results in paradoxically reduced virulence in experimental endocarditis—host defense role for platelet microbicidal proteins. *Infect. Immun.* **65:**4652–4660.
4. **Beekhuizen, H., J. S. van de Gevel, B. Olsson, I. J. van Benten, and R. van Furth.** 1997. Infection of human vascular endothelial cells with *Staphylococcus aureus* induces hyperadhesiveness for human monocytes and granulocytes. *J. Immunol.* **158:**774–782.
5. **Beers, S. A., A. G. Buckland, R. S. Koduri, W. Cho, M. H. Gelb, and D. C. Wilton.** 2002. The antibacterial properties of secreted phospholipases A2: a major physiological role for the group IIA enzyme that depends on the very high pI of the enzyme to allow penetration of the bacterial cell wall. *J. Biol. Chem.* **277:**1788–1793.
6. **Belaaouaj, A., R. McCarthy, M. Baumann, Z. Gao, T. Ley, S. Abraham, and S. Shapiro.** 1998. Mice lacking neutrophil elastase reveal impaired host defense against gram-negative bacterial sepsis. *Nat. Med.* **4:**615–618.
7. **Bera, A., S. Herbert, A. Jakob, W. Vollmer, and F. Götz.** Why are pathogenic staphylococci so lysozyme resistant? The peptidoglycan O-acetyltransferase OatA is the major determinant for lysozyme resistance of *Staphylococcus aureus.* *Mol. Microbiol.* **55:**778–787.
8. **Brinkmann, V., U. Reichard, C. Goosmann, B. Fauler, Y. Uhlemann, D. S. Weiss, Y. Weinrauch, and A. Zychlinsky.** 2004. Neutrophil extracellular traps kill bacteria. *Science* **303:**1532–1535.
9. **Buckland, A. G., and D. C. Wilton.** 2000. The antibacterial properties of secreted phospholipases A(2). *Biochim. Biophys. Acta* **14880:**71–82.
10. **Chapman, A. L., M. B. Hampton, R. Senthilmohan, C. C. Winterbourn, and A. J. Kettle.** 2002. Chlorination of bacterial and neutrophil proteins during phagocytosis and killing of *Staphylococcus aureus. J. Biol. Chem.* **277:**9757–9762.
11. **Chertov, O., D. F. Michiel, L. Xu, J. M. Wang, K. Tani, W. J. Murphy, D. L. Longo, D. D. Taub, and J. J. Oppenheim.** 1996. Identification of defensin-1, defensin-2 and CAP37/azurocidin as T-cell chemoattractant proteins released from interleukin-8-stimulated neutrophils. *J. Biol. Chem.* **271:**2935–2940.
12. **Chevakis, T., M. Hussain, S. M. Kanse, G. Peters, R. G. Bretzel, J. L. Flock, M. Herrmann, and K. T. Preissner.** 2002. *Staphylococcus aureus* extracellular adherence protein serves as anti-inflammatory factor by inhibiting the recruitment of host leukocytes. *Nat. Med.* **8:**687–693.
13. **Clarke, A. J., and C. Dupont.** 1991. O-acetylated peptidoglycan: its occurrence, pathobiologic significance, and biosynthesis. *Can. J. Microbiol.* **38:**85–91.
14. **Clements, M. O., and S. J. Foster.** 1999. Stress resistance in *Staphylococcus aureus. Trends Microbiol.* **7:**458–462.
15. **Cole, A. M., S. Tahk, A. Oren, D. Yoshioka, Y. H. Kim, A. Park, and T. Ganz.** 2001. Determinants of *Staphylococcus aureus* nasal carriage. *Clin. Diagn. Lab. Immunol.* **8:**1064–1069.
16. **Cole, A. M., T. Ganz, A. M. Liese, M. D. Burdick, L. Liu, and R. M. Strieter.** 2001. IFN-inducible ELR-CXC chemokines display defensin-like antimicrobial activity. *J. Immunol.* **167:**623–627.
17. **Collins, L. V., S. A. Kristian, C. Weidenmaier, M. Faigle, K. P. M. van Kessel, J. A. G. van Strijp, F. Götz, B. Neumeister, and A. Peschel.** 2002. *Staphylococcus aureus* strains lacking D-alanine modifications of teichoic acids are highly susceptible to human neutrophil killing and are virulence attenuated in mice. *J. Infect. Dis.* **186:**214–219.
18. **Cunnion, K. M., J. C. Lee, and M. M. Frank.** 2001. Capsule production and growth phase influence binding of complement to *Staphylococcus aureus. Infect. Immun.* **69:**6796–6803.
19. **Cunnion, K. M., H. M. Zhang, and M. M. Frank.** 2003. Availability of complement bound to *Staphylococcus aureus* to interact with membrane complement receptors influences efficiency of phagocytosis. *Infect. Immun.* **71:**656–662.
20. **Dankert, J.** 1988. Role of platelets in early pathogenesis of viridans group streptococcal endocarditis: a study of thrombodefensins. Ph.D. thesis. University of Groningen, Groningen, The Netherlands.
21. **Dankert, J., J. van der Werff, S. A. J. Zaat, W. Joldersma, D. Klein, and J. Hess.** 1995. Involvement of bactericidal factors from thrombin-stimulated platelets in clearance of adherent viridans streptococci in experimental infective endocarditis in rabbits. *Infect. Immun.* **63:**663–671.
22. **de Haas, C. J. C., K. E. Veldkamp, A. Peshel, F. Weerkamp, W. J. B. van Wamel, E. C. J. M. Heezius, M. J. J. G. Poppelier, K. P. M. van Kessel, and J. A. G. van Strijp.** 2004. Chemotaxis inhibitory protein of *Staphylococcus aureus*, a bacterial anti-inflammatory agent. *J. Exp. Med.* **199:**687–695.
23. **De Kimpe, S. J., M. Kengatharan, C. Thiemermann, and J. R. Vane.** 1995. The cell wall components peptidoglycan and lipoteichoic acid from *Staphylococcus aureus* act in synergy to cause shock and multiple organ failure. *Proc. Natl. Acad. Sci. USA* **92:**10359–10363.
24. **Dhawan, V. G., A. S. Bayer, and M. R. Yeaman.** 1998. Influence of in vitro susceptibility to thrombin-induced platelet microbicidal protein on the progression of experimental *Staphylococcus aureus* endocarditis. *Infect. Immun.* **66:**3476–3479.
25. **Diamond, G., J. P. Russell, and C. L. Bevins.** 1996. Inducible expression of an antibiotic peptide gene in lipopolysaccharide-challenged tracheal epithelial cells. *Proc. Natl. Acad. Sci. USA* **93:**5156–5160.
26. **Dinauer, M. C., W. M. Nauseef, and P. E. Newburger.** 2001. Inherited disorders of phagocyte killing, p. 4857–4887. *In* C. R. Scriver, A. L. Beaudet, W. S. Sly, D. Valle,

B. Childs, K. W. Kinzler, and B. Vogelstein (ed.), *The Metabolic and Molecular Bases of Inherited Diseases*. McGraw-Hill, New York, N.Y.

27. **Dominiecki, M. E., and J. Weiss.** 1999. Antibacterial action of extracellular mammalian group IIA phospholipase A2 against grossly clumped *Staphylococcus aureus*. *Infect. Immun.* **67:**2299–2305.

28. **Donaldson, D. M., and J. G. Tew.** 1977. Beta-lysin of platelet origin. *Bacteriol. Rev.* **41:**501–513.

29. **Eisenhauer, P. B., and R. I. Lehrer.** 1992. Mouse neutrophils lack defensins. *Infect. Immun.* **60:**3446–3447.

30. **Elsbach, P., J. Weiss, and O. Levy.** 1999. Oxygen-independent antimicrobial systems of phagocytes, p. 801–817. *In* J. I. Gallin, R. Snyderman, and C. Nathan (ed.), *Inflammation. Basic Principles and Clinical Correlates*, 3rd ed. Lippincott-Raven, New York, N.Y.

31. **Fang, F.** 2004. Antimicrobial reactive oxygen and nitrogen species: concepts and controversies. *Nat. Rev. Microbiol.* **2:**820–832.

32. **Faurschou, M., and Borregaard, N.** 2003. Neutrophil granules and secretory vesicles in inflammation. *Microbes Infect.* **5:**1317–1328.

33. **Femling, J. K., W. M. Nauseef, and J. P. Weiss.** 2005. Synergy between extracellular group IIA phospholipase A2 and phagocyte NADPH oxidase in digestion of phospholipids of *Staphylococcus aureus* ingested by human neutrophils. *J. Immunol.* **175:**4653–4661.

34. **Ferrante, A., A. J. Martin, E. J. Bates, D. H. B. Goh, D. P. Harvey, D. Parsons, D. A. Rathjen, G. Russ, and J.-M. Dayer.** 1993. Killing of *Staphylococcus aureus* by tumor necrosis factor-α-activated neutrophils. *J. Immunol.* **151:**4821–4828.

35. **Foreman-Wykert, A. K.** 1999. Determinants of the bactericidal action of mammalian 14 kDa group IIA phospholipase A2 against gram-positive bacteria, p. 113. *In* Microbiology. New York University, New York, N.Y.

36. **Foreman-Wykert, A. K., Y. Weinrauch, P. Elsbach, and J. Weiss.** 1999. Cell-wall determinants of the bactericidal action of group IIA phospholipase A2 against Gram-positive bacteria. *J. Clin. Investig.* **103:**715–721.

37. **Fowler, V. G., Jr., G. Sakoulas, L. M. McIntyre, V. G. Meka, R. D. Arbeit, C. H. Cabell, M. E. Stryjewski, G. M. Eliopoulos, L. B. Reller, G. R. Corey, T. Jones, N. Lucindo, M. R. Yeaman, and A. S. Bayer.** 2004. Persistent bacteremia due to methicillin-resistant *Staphylococcus aureus* infection is associated with *agr* dysfunction and low-level *in vitro* resistance to thrombin-induced platelet microbicidal protein. *J. Infect. Dis.* **190:**1140–1149.

38. **Frohm, M., H. Gunne, A. C. Bergman, B. Agerberth, T. Bergman, A. Boman, S. Liden, H. Jornvall, and H. G. Boman.** 1996. Biochemical and antibacterial analysis of human wound and blister fluid. *Eur. J. Biochem.* **237:**86–92.

39. **Ganz, T.** 1994. Biosynthesis of defensins and other antimicrobial peptides. *Ciba Found. Symp.* **186:**62–71.

40. **Ginsburg, I., and M. Lahav.** 1983. How are bacterial cells degraded by leukocytes in vivo? An enigma. *Clin. Immunol. Newsl.* **4:**147–153.

41. **Ginsburg, I.** 2002. The role of bacteriolysis in the pathophysiology of inflammation, infection and post-infectious sequelae. *APMIS* **110:**753–770.

42. **Goldman, M. J., G. M. Anderson, E. D. Stolzenberg, U. P. Kari, M. Zasloff, and J. M. Wilson.** 1997. Human beta-defensin-1 is a salt-sensitive antibiotic in lung that is inactivated in cystic fibrosis. *Cell* **88:**553–560.

43. **Gresham, H. D., J. H. Lowrance, T. E. Caver, B. S. Wilson, A. L. Cheung, and F. P. Lindberg.** 2000. Survival of *Staphylococcus aureus* inside neutrophils contributes to infection. *J. Immunol.* **164:**3713–3722.

44. **Hampton, M. B., A. J. Kettle, and C. C. Winterbourn.** 1996. Involvement of superoxide and myeloperoxidase in oxygen-dependent killing of *Staphylococcus aureus* by neutrophils. *Infect. Immun.* **64:**3512–3517.

45. **Hampton, M. B., A. J. Kettle, and C. C. Winterbourn.** 1998. Inside the neutrophil phagosome: oxidants, myeloperoxidase, and bacterial killing. *Blood* **92:**3007–3017.

46. **Harwig, S. S., T. Ganz, and R. I. Lehrer.** 1994. Neutrophil defensins. Purification, characterization and antimicrobial testing. *Methods Enzymol.* **236:**160–176.

47. **Harwig, S. S., L. Tan, X. D. Qu, Y. Cho, P. B. Eisenhauer, and R. I. Lehrer.** 1995. Bactericidal properties of a murine intestinal phospholipase A2. *J. Clin. Investig.* **95:**603–610.

48. **Heumann, D., C. Barras, A. Severin, M. P. Glauser, and A. Tomasz.** 1994. Gram-positive cell walls stimulate synthesis of tumor necrosis factor alpha and interleukin-6 by human monocytes. *Infect. Immun.* **62:**2715–2721.

49. **Hristova, K., M. E. Selsted, and S. H. White.** 1996. Interactions of monomeric rabbit neutrophil defensin with bilayers: comparison with dimeric human defensin HNP-2. *Biochemistry* **35:**11888–11894.

50. **Hubbard, R. C., F. Ogushi, G. A. Fells, A. M. Cantin, S. Jallat, M. Courtney, and R. G. Crystal.** 1987. Oxidants spontaneously released by alveolar macrophages of cigarette smokers can inactivate the active site of alpha 1-antitrypsin, rendering it ineffective as an inhibitor of neutrophil elastase. *J. Clin. Investig.* **80:**1289–1295.

51. **Hultgren, O., H.-P. Eugster, J. D. Sedgwick, H. Körner, and A. Tarkowski.** 1998. TNF/lymphotoxin-α double mutant mice resist septic arthritis but display increased mortality in response to *Staphylococcus aureus*. *J. Immunol.* **161:**5937–5942.

52. **Ing, M. B., L. Baddour, and A. S. Bayer.** 1997. Staphylococcal bacteremia and infective endocarditis—pathogenesis, diagnosis and complications. *In* G. Archer and K. Crossley (ed.), *Staphylococci and Staphylococcal Diseases*. Churchill-Livingstone Publishers, New York, N.Y.

53. **Jackson, S. H., J. I. Gallin, and S. M. Holland.** 1995. The p47phox mouse knock-out model of chronic granulomatous disease. *J. Exp. Med.* **182:**751–758.

54. **Kallajoki, M., and T. J. Nevalainen.** 1997. Expression of Group II phospholipase A2 in human tissues, p. 8–16. *In* W. Uhl, T. J. Nevalainen, and M. W. Buchler (ed.), *Phospholipase A2: Basic and Clinical Aspects in Inflammatory Diseases*, vol. 24. S. Karger, Basel, Switzerland.

55. **Kapral, F. A.** 1966. Clumping of *Staphylococcus aureus* in the peritoneal cavity of mice. *J. Bacteriol.* **92:**1188–1195.

56. **Karakawa, W. W., A. Sutton, R. Schneerson, A. Karpas, and W. F. Vann.** 1988. Capsular antibodies induce type-specific phagocytosis of capsulated *Staphylococcus aureus* by human polymorphonuclear leukocytes. *Infect. Immun.* **56:**1090–1095.

57. **Katz, S. S., Y. Weinrauch, R. S. Munford, P. Elsbach, and J. Weiss.** 1999. Lipopolysaccharide deacylation following extracellular or intracellular killing of *Escherichia coli* by rabbit inflammatory peritoneal exudates. *J. Biol. Chem.* **274:**36579–36584.

58. **Koo, S.-P., A. S. Bayer, R. A. Proctor, H.-G. Sahl, and M. R. Yeaman.** 1996. Staphylocidal action of platelet microbicidal protein is not solely dependent on intact transmembrane potential. *Infect. Immun.* **60:**1070–1074.

59. **Koo, S.-P., M. R. Yeaman, and A. S. Bayer.** 1996. Staphylocidal action of platelet microbicidal protein is

modified by microenvironment and target cell growth phase. *Infect. Immun.* **64:**3758–3764.

60. **Koo, S.-P., M. R. Yeaman, C. C. Nast, and A. S. Bayer.** 1997. The bacterial cell membrane is a principal target for the staphylocidal action of thrombin-induced platelet microbicidal protein. *Infect. Immun.* **65:**4795–4800.

61. **Koprivnjak, T., A. Peschel, M. H. Gelb, N. S. Liang, and J. P. Weiss.** 2002. Role of charge properties of bacterial envelope in bactericidal action of human group IIA phospholipase A2 against *Staphylococcus aureus. J. Biol. Chem.* **277:**47636–47644.

62. **Kristian, S. A., M. Dürr, J. A. G. Van Strijp, B. Neumeister, and A. Peschel.** 2003. MprF-mediated lysinylation of phospholipids in *Staphylococcus aureus* leads to protection against oxygen-independent neutrophil killing. *Infect. Immun.* **71:**546–549.

63. **Kudo, I., and M. Murakami.** 2002. Phospholipase A2 enzymes. *Prostaglandins Other Lipid Mediat.* **68–69:**3–58.

64. **Kusonoki, T., E. Hailman, T. S. C. Juan, H. S. Lichenstein, and S. D. Wright.** 1995. Molecules from *Staphylococcus aureus* that bind CD14 and stimulate innate immune responses. *J. Exp. Med.* **182:**1673–1682.

65. **Laine, V. J., D. S. Grass, and T. J. Nevalainen.** 1999. Protection by group II phospholipase A2 against *Staphylococcus aureus. J. Immunol.* **162:**7402–7408.

66. **Lee, J. C.** 1996. The prospects for developing a vaccine against *Staphylococcus aureus. Trends Microbiol.* **4:**162–166.

67. **Lee, J. C., J.-S. Park, S. E. Shepherd, V. Carey, and A. Fattom.** 1997. Protective efficacy of antibodies to the *Staphylococcus aureus* type 5 capsular polysaccharide in a modified model of endocarditis in rats. *Infect. Immun.* **65:**4146–4151.

68. **Lehrer, R. I.** 2004. Primate defensins. *Nat. Rev. Microbiol.* **2:**727–738.

69. **Levy, O., C. E. Ooi, J. Weiss, R. I. Lehrer, and P. Elsbach.** 1994. Individual and synergistic effects of rabbit granulocyte proteins on *Escherichia coli. J. Clin. Investig.* **94:**672–682.

70. **Lowy, F. D.** 1998. Medical progress: *Staphylococcus aureus* infections. *N. Engl. J. Med.* **339:**520–532.

71. **Madsen, L. M., M. Inada, and J. Weiss.** 1996. Determinants of activation by complement of group II phospholipase A2 acting against Escherichia coli. *Infect. Immun.* **64:**2425–2430.

72. **Mandell, G. L.** 1975. Catalase, superoxide dismutase, and virulence of *Staphylococcus aureus*. In vitro and in vivo studies with emphasis on staphylococcal—leukocyte interaction. *J. Clin. Investig.* **55:**561–566.

73. **McInnes, I. B., B. Leung, X. Q. Wei, C. C. Gemmell, and F. Y. Liew.** 1998. Septic arthritis following *Staphylococcus aureus* infection in mice lacking inducible nitric oxide synthase. *J. Immunol.* **160:**308–315.

74. **Melly, M. A., J. B. Thomison, and D. E. Rogers.** 1960. Fate of staphylococci within human leukocytes. *J. Exp. Med.* **112:**1121–1130.

75. **Menzies, B. E., and I. Kourteva.** 1998. Internalization of *Staphylococcus aureus* by endothelial cells induces apoptosis. *Infect. Immun.* **66:**5994–5998.

76. **Messina, C. G. M., E. P. Reeves, J. Roes, and A. W. Segal.** 2002. Catalase negative *Staphylococcus aureus* retain virulence in mouse model of chronic granulomatous disease. *FEBS Lett.* **518:**107–110.

77. **Midorikawa, K., K. Ouhara, H. Komatsuzawa, T. Kawai, S. Yamada, T. Fujiwara, K. Yamazaki, K. Sayama, M. A. Taubman, H. Kurihara, K. Hashimoto, and M. Sugai.** 2003. *Staphylococcus aureus* susceptibility to innate antimi-

crobial peptides, β-defensins and CAP18, expressed by human keratinocytes. *Infect. Immun.* **71:**3730–3739.

78. **Morath, S., A. Geyer, and T. Hartung.** 2001. Structure-activity relationship of cytokine induction by lipoteichoic acid from *Staphylococcus aureus. J. Exp. Med.* **193:**393–397.

79. **Murakami, M., Y. Nakatani, and I. Kudo.** 1996. Type II secretory phospholipase A2 associated with cell surfaces via C-terminal heparin-binding lysine residues augments stimulus-initiated delayed prostaglandin generation. *J. Biol. Chem.* **271:**30041–30051.

80. **Murdoch, C., and A. Finn.** 2000. Chemokine receptors and their role in inflammation and infectious diseases. *Blood* **95:**3032–3043.

81. **Nathan, C., and M. U. Shiloh.** 2000. Reactive oxygen and nitrogen intermediates in the relationship between mammalian hosts and microbial pathogens. *Proc. Natl. Acad. Sci. USA* **97:**8841–8848.

82. **Nauseef, W. M.** 2004. Assembly of the phagocyte NADPH oxidase. *Histochem. Cell Biol.* **122:**277–291.

83. **Neth, O., D. L. Jack, M. Johnson, N. J. Klein, and M. W. Turner.** 2002. Enhancement of complement activation and opsonophagocytosis by complexes of mannose-binding lectin with mannose-binding lectin-associated serine protease after binding to *Staphylococcus aureus. J. Immunol.* **169:**4430–4436.

84. **Nilsdotter-Augustinsson, A., A. Wilsson, J. Larsson, O. Stendahl, L. Öhman, and H. Lundqvist-Gustafsson.** 2004. *Staphylococcus aureus,* but not *Staphylococcus epidermidis,* modulates the oxidative response and induces apoptosis in human neutrophils. *APMIS* **112:**109–118.

85. **Ong, P. Y., T. Ohtake, C. Brandt, I. Strickland, M. Boguniewicz, T. Ganz, R. L. Gallo, and D. Y. M. Leung.** 2002. Endogenous antimicrobial peptides and skin infections in atopic dermatitis. *N. Engl. J. Med.* **347:**1151–1160.

86. **O'Riordan, K., and J. C. Lee.** 2004. *Staphylococcus aureus* capsular polysaccharides. *Clin. Microbiol. Rev.* **17:**218–234.

87. **Patti, J. M.** 2004. A humanized monoclonal antibody targeting *Staphylococcus aureus. Vaccine* **22S:**S39–S43.

88. **Peschel, A.** 2002. How do bacteria resist human antimicrobial peptides? *Trends Microbiol.* **10:**179–186.

89. **Peterson, P. K., B. J. Wilkinson, Y. Kim, D. Schmeling, S. D. Douglas, P. G. Quie, and J. Verhoef.** 1978. The key role of peptidoglycan in the opsonization of *Staphylococcus aureus. J. Clin. Investig.* **61:**597–609.

90. **Pollock, J. D., D. A. Williams, M. A. Gifford, L. L. Li, X. Du, J. Fisherman, S. H. Orkin, C. M. Doerschuk, and M. C. Dinauer.** 1995. Mouse model of X-linked chronic granulomatous disease, an inherited defect in phagocyte superoxide production. *Nat. Genet.* **9:**202–209.

91. **Proctor, R. A., J. M. Balwit, and O. Vesga.** 1994. Variant subpopulations of *Staphylococcus aureus* as a cause of persistent infections. *Infect. Agents Dis.* **3:**302–312.

92. **Qu, X. D., and R. I. Lehrer.** 1998. Secretory phospholipase A2 is the principal bactericide for staphylococci and other gram-positive bacteria in human tears. *Infect. Immun.* **66:**2791–2797.

93. **Quie, P. G., J. G. White, B. Holmes, and R. A. Good.** 1967. In vitro bactericidal capacity of human polymorphonuclear leukocytes: diminished activity in chronic granulomatous disease of childhood. *J. Clin. Investig.* **46:**668–679.

94. **Reeves, E. P., H. Lu, H. L. Jacobs, C. G. Messina, S. Bolsover, G. Gabella, E. O. Potma, A. Warley, J. Roes, and A. W. Segal.** 2002. Killing activity of neutrophils is mediated through activation of proteases by K+ flux. *Nature* **416:**291–297.

95. **Reeves, E. P., M. Nagl, J. Godovac-Zimmermann, and A. W. Segal.** 2003. Reassessment of the microbicidal activity of reactive oxygen species and hypochlorous acid with reference to the phagocytic vacuole of the neutrophil granulocyte. *J. Med. Microbiol.* **52:**643–651.

96. **Rennermalm, A., M. Nilsson, and J. I. Flock.** 2004. The fibrinogen binding protein of *Staphylococcus epidermidis* is target for opsonic antibodies. *Infect. Immun.* **72:**3081–3083.

97. **Schonwetter, B. S., E. D. Stolzenberg, and M. A. Zasloff.** 1995. Epithelial antibiotics induced at sites of inflammation. *Science* **267:**1645–1648.

98. **Segal, A. W., A. M. Harper, R. C. Garcia, and D. Merzbach.** 1982. The action of cells from patients with chronic granulomatous disease on *Staphylococcus aureus. J. Med. Microbiol.* **15:**441–449.

99. **Shafer, W. M., and V. C. Onunka.** 1989. Mechanism of staphylococcal resistance to non-oxidative antimicrobial action of neutrophils: importance of pH and ionic strength in determining the bactericidal action of cathepsin G. *J. Gen. Microbiol.* **135:**825–830.

100. **Shi, L., K. Takahashi, J. Dundee, S. Shahroor-Karni, S. Thie., C. Jensenius, F. Gad, M. R. Hamblin, K. N. Sastry, and R. A. B. Ezekowitz.** 2004. Mannose-binding lectin-deficient mice are susceptible to infection with *Staphylococcus aureus. J. Exp. Med.* **199:**1379–1390.

101. **Shimoda, M., K. Ohki, Y. Shimamota, and O. Kohashi.** 1995. Morphology of defensin-treated *Staphylococcus aureus. Infect. Immun.* **63:**2886–2891.

102. **Singh, P. K., B. F. Tack, P. B. McCray, Jr., and M. J. Welsh.** 2000. Synergistic and additive killing by antimicrobial factors found in human airway surface liquid. *Am. J. Physiol. Lung Cell Mol. Physiol.* **279:**L799–L805.

103. **Six, D. A., and E. A. Dennis.** 2000. The expanding superfamily of phospholipase A(2) enzymes: classification and characterization. *Biochim. Biophys. Acta* **1488:**1–19.

104. **Sorensen, O. E., P. Follin, A. H. Jonhsen, J. Calafat, G. S. Tjabringa, P. S. Hiemstra, and N. Borregaard.** 2001. Human cathelicidin, hCAP18, is processed to the antimicrobial peptide LL-37 by extracellular cleavage with proteinase 3. *Blood* **97:**3951–3959.

105. **Staudinger, B. J., M. A. Oberdoerster, P. J. Lewis, and H. Rosen.** 2002. mRNA expression profiles for *Escherichia coli* ingested by normal and phagocyte-oxidase-deficient human neutrophils. *J. Clin. Investig.* **110:**1151–1163.

106. **Sullam, P. M., U. Frank, M. G. Tauber, M. Yeaman, A. Bayer, and H. F. Chambers.** 1993. Effect of thrombocytopenia on the early course of streptococcal endocarditis. *J. Infect. Dis.* **168:**910–914.

107. **Tang, Y. Q., M. R. Yeaman, and M. E. Selsted.** 2002. Antimicrobial peptides from human platelets. *Infect. Immun.* **70:**6524–6533.

108. **Territo, M. C., T. Ganz, M. E. Selsted, and R. Lehrer.** 1989. Monocyte-chemotactic activity of defensins from human neutrophils. *J. Clin. Investig.* **84:**2017–2020.

109. **Thakker, M., J.-S. Park, V. Carey, and J. C. Lee.** 1998. *Staphylococcus aureus* serotype 5 capsular polysaccharide is antiphagocytic and enhances bacterial virulence in a murine bacteremia model. *Infect. Immun.* **66:**5183–5189.

110. **Trier, D., K. D. Gank, A. S. Bayer, and M. R. Yeaman.** 2000. *Staphylococcus aureus* elicits platelet antimicrobial responses via an ADP-dependent pathway. Program Abstr. 40th Intersci. Conf. Antimicrob. Agents Chemother. abstr. 1010. American Society for Microbiology, Washington, D.C.

111. **Valore, E. V., E. Martin, S. S. Harwig, and T. Ganz.** 1996. Intramolecular inhibition of human defensin HNP-1 by its propiece. *J. Clin. Investig.* **97:**1624–1629.

112. **Vaudaux, P. E., G. Zulian, E. Huggler, and F. A. Waldvogel.** 1985. Attachment of *Staphylococcus aureus* to polymethacrylate increases its resistance to phagocytosis in foreign body infection. *Infect. Immun.* **50:**472–477.

113. **Verbrugh, H. A., P. K. Peterson, B. Y. Nguyen, S. P. Sisson, and Y. Kim.** 1982. Opsonization of encapsulated *Staphylococcus aureus:* the role of specific antibody and complement. *J. Immunol.* **129:**1681–1687.

114. **Verdrengh, M., T. A. Springer, J.-C. Gutierrez, and A. Tarkowski.** 1996. A role of intercellular adhesion molecule 1 in pathogenesis of staphylococcal arthritis and in host defense against staphylococcal bacteremia. *Infect. Immun.* **64:**2804–2807.

115. **Voyich, J. M., K. R. Braughton, D. E. Sturdevant, C. Vuong, S. D. Kobayashi, S. F. Porcella, M. Otto, J. M. Musser, and F. R. DeLeo.** 2004. Engagement of the pathogen survival response used by group A Streptococcus to avert destruction by innate host defense. *J. Immunol.* **173:**1194–1201.

116. **Waldvogel, F. A.** 1999. New resistance in *Staphylococcus aureus. N. Engl. J. Med.* **340:**556–557.

117. **Walton, E.** 1978. The preparation, properties and action on *Staphylococcus aureus* of purified fractions from the cationic proteins of rabbit polymorphonuclear leukocytes. *Br. J. Exp. Pathol.* **59:**416–431.

118. **Weinrauch, Y., C. Abad, N. S. Liang, S. F. Lowry, and J. Weiss.** 1998. Mobilization of potent plasma bactericidal activity during systemic bacterial challenge: role of group IIA phospholipase A2. *J. Clin. Investig.* **102:**633–638.

119. **Weinrauch, Y., P. Elsbach, L. M. Madsen, A. Foreman, and J. Weiss.** 1996. The potent anti-*Staphylococcus aureus* activity of a sterile rabbit inflammatory fluid is due to a 14-kD phospholipase A2. *J. Clin. Investig.* **97:**250–257.

120. **Weinrauch, Y., S. S. Katz, R. S. Munford, P. Elsbach, and J. Weiss.** 1999. Deacylation of purified lipopolysaccharide by cellular and extracellular components of a sterile rabbit peritoneal inflammatory exudate. *Infect. Immun.* **67:**3376–3382.

121. **Weiss, J., L. Kao, M. Victor, and P. Elsbach.** 1987. Respiratory burst facilitates the digestion of Escherichia coli killed by polymorphonuclear leukocytes. *Infect. Immun.* **55:**2142–2147.

122. **Weiss, J., M. Inada, P. Elsbach, and R. M. Crowl.** 1994. Structural determinants of the action against *Escherichia coli* of a human inflammatory fluid phospholipase A2 in concert with polymorphonuclear leukocytes. *J. Biol. Chem.* **269:**26331–26337.

123. **Weiss, J., M. Victor, O. Stendahl, and P. Elsbach.** 1982. Killing of Gram-negative bacteria by polymorphonuclear leukocytes: the role of an O_2-independent bactericidal system. *J. Clin. Investig.* **69:**959–970.

124. **Welling, M. M., P. S. Hiemstra, M. T. van den Barselaar, A. Paulusma-Annema, P. H. Nibbering, E. K. J. Pauwels, and W. Calame.** 1998. Antibacterial activity of human neutrophil defensins in experimental infections in mice is accompanied by increased leukocyte accumulation. *J. Clin. Investig.* **102:**1583–1590.

125. **White, S. H., W. C. Wimley, and M. E. Selsted.** 1995. Structure, function, and membrane integration of defensins. *Curr. Opin. Struct. Biol.* **5:**521–527.

126. **Woodman, R., R. Erickson, J. Rae, H. Jaffe, and J. Curnutte.** 1992. Prolonged recombinant interferon-γ therapy in chronic granulomatous disease: evidence against enhanced neutrophil oxidase activity. *Blood* **79:** 1558–1562.

127. **Wright, G., C. E. Ooi, J. Weiss, and P. Elsbach.** 1990. Purification of a cellular (granulocyte) and an extracellular (serum) phospholipase A2 that participate in the destruction of *Escherichia coli* in a rabbit inflammatory exudate. *J. Biol. Chem.* **265:**6675–6681.

128. **Wright, G. C., J. Weiss, K.-S. Kim, H. Verheij, and P. Elsbach.** 1990. Bacterial phospholipid hydrolysis enhances the destruction of *Escherichia coli* ingested by rabbit neutrophils. Role of cellular and extracellular phospholipases. *J. Clin. Investig.* **85:**1925–1935.

129. **Wu, T., M. R. Yeaman, and A. S. Bayer.** 1994. Resistance to platelet microbicidal protein in vitro among bacteremic staphylococcal and viridans streptococcal isolates correlates with an endocarditis source. *Antimicrob. Agents Chemother.* **38:**729–732.

130. **Xiong, Y. Q., M. R. Yeaman, and A. S. Bayer.** 1999. *In vitro* antibacterial activities of platelet microbicidal protein and neutrophil defensin against *Staphylococcus aureus* are influenced by antibiotics differing in mechanism of action. *Antimicrob Agents Chemother.* **43:**1111–1117.

131. **Yang, D., Q. Chen, D. M. Hoover, P. Staley, K. D. Tucker, J. Lubkowski, and J. J. Oppenheim.** 2003. Many chemokines including CCL20 / MIP-3 alpha display antimicrobial activity. *J. Leukoc. Biol.* **74:**448–455.

132. **Yao, L., V. Bengualid, F. D. Lowy, J. J. Gibbons, V. B. Hatcher, and J. W. Berman.** 1995. Internalization of *Staphylococcus aureus* by endothelial cells induces cytokine gene expression. *Infect. Immun.* **63:**1835–1839.

133. **Yeaman, M. R.** 1997. The role of platelets in antimicrobial host defense. *Clin. Infect. Dis.* **25:**951–970.

134. **Yeaman, M. R., A. S. Bayer, S.-P. Koo, W. Foss, and P. M. Sullam.** 1998. Platelet microbicidal proteins and neutrophil defensin disrupt the *Staphylococcus aureus* cytoplasmic membrane by distinct mechanisms of action. *J. Clin. Investig.* **101:**178–187.

135. **Yeaman, M. R., A. S. Ibrahim, J. E. Edwards, A. S. Bayer, and M. A. Ghannoum.** 1993. Thrombin-induced platelet microbicidal protein is fungicidal in vitro. *Antimicrob. Agents Chemother.* **37:**546–553.

136. **Yeaman, M. R., D. C. Norman, and A. S. Bayer.** 1992. Platelet microbicidal protein enhances the bactericidal and post-antibiotic effects in *Staphylococcus aureus*. *Antimicrob. Agents Chemother.* **36:**1665–1660.

137. **Yeaman, M. R., A. Shen, Y. Tang, A. S. Bayer, and M. A. Selsted.** 1997. Isolation and antimicrobial activity of microbicidal proteins from rabbit platelets. *Infect. Immun.* **65:**1023–1031.

138. **Yeaman, M. R., P. R. Sullam, P. F. Dazin, and A. S. Bayer.** 1994. Platelet microbicidal protein alone and in combination with antibiotics reduces adherence of *Staphylococcus aureus* to platelets in vitro. *Infect. Immun.* **62:** 3416–3423.

139. **Yount, N. Y., K. D. Gank, Y. Q. Xiong, A. S. Bayer, T. Pender, W. H. Welch, and M. R. Yeaman.** 2004. Platelet microbicidal protein 1: structural themes of a multifunctional antimicrobial peptide. *Antimicrob. Agents Chemother.* **48:**4395–4404.

140. **Yount, N. Y., and M. R. Yeaman.** 2004. Multidimensional signatures in antimicrobial peptides. *Proc. Natl. Acad. Sci. USA* **101:**7363–7368.

141. **Zanetti, M.** 2004. Cathelicidin, multifunctional proteins of the innate immunity. *J. Leukoc. Biol.* **75:**39–48.

142. **Zarember, K. A., S. S. Katz, B. F. Tack, L. Doukhan, J. Weiss, and P. Elsbach.** 2002. Host defense functions of proteolytically processed and parent (unprocessed) cathelicidins of rabbit granulocytes. *Infect. Immun.* **70:**569–576.

143. **Zhao, C., I. Wang, and R. I. Lehrer.** 1996. Widespread expression of beta-defensin hBD-1 in human secretory glands and epithelial cells. *FEBS Lett.* **396:**319–322.

144. **Zimmermann, G. R., P. Legault, M. E. Selsted, and A. Pardi.** 1995. Solution structure of bovine neutrophil beta-defensin-12: the peptide fold of the beta-defensin is identical to that of the classical defensins. *Biochemistry* **34:**13663–13671.

Biology and Pathogenicity of *Staphylococcus epidermidis*

CHRISTINE HEILMANN AND GEORG PETERS

46

Non-*aureus* staphylococci (NAS) form a group of >30 defined species. They share many basic biological properties with *Staphylococcus aureus*, the predominant coagulase-positive species. However, NAS show substantial differences from *S. aureus*, especially with respect to their ecology and their pathogenic potential. The normal habitats of NAS are the skin and mucous membranes of humans and animals, and they represent a major part of the normal aerobic flora. Only 12 species are normally found in specimens of human origin, and some of these species occur only rarely. Diagnostically, human species of NAS are separated from *S. aureus* by their inability to produce free coagulase. They can further be differentiated on the basis of their novobiocin susceptibility: novobiocin-resistant NAS resemble the *Staphylococcus saprophyticus* group, and novobiocin-susceptible NAS mainly resemble the *Staphylococcus epidermidis* group (Table 1). By far the most frequently isolated species, from normal flora as well as from clinical specimens, is *S. epidermidis*.

This chapter deals with the current knowledge about *S. epidermidis*. It especially focuses on the pathogenicity of *S. epidermidis*, the underlying biological properties, and how these properties contrast with those of *S. aureus*. The basic biology is covered in other chapters, but a brief overview of the disease spectrum is given to place the studies on pathogenesis in perspective. The last part of the chapter deals with one ecological aspect, lantibiotics, which are potentially important for bacterial interference on skin and mucous membranes.

SPECTRUM OF DISEASE

NAS, particularly *S. epidermidis*, are among the most frequently isolated microorganisms in the clinical microbiology laboratory. The vast majority of infections attributed to NAS are nosocomial infections. Data from the National Nosocomial Infections Surveillance System (Centers for Disease Control and Prevention, Atlanta, Ga.) show that NAS are among the five most commonly reported pathogens in hospitals and the most frequently reported isolates in nosocomial bloodstream infections. There is a striking difference in the spectrum of clinical presentation of diseases caused by staphylococci: *S. aureus* causes a broad variety of pyogenic infections, as well as toxin-mediated diseases in the normal host, and novobiocin-resistant NAS, particularly *S. saprophyticus*, are found in urinary tract infections. In contrast, *S. epidermidis* rarely causes pyogenic infections in the normal host, except for natural valve endocarditis, and there is little to suggest any role in toxin-mediated diseases. However, when the host is compromised, *S. epidermidis* may even be the predominant cause of infection (87).

One such group is intravenous drug (heroin) abusers who develop right-sided endocarditis. It is hypothesized that endothelial microlesions caused by heroin microcrystals and repeated bacteremia with high levels of inocula of *S. epidermidis* due to nonsterile injection procedures are important causative factors. A second group is immunocompromised patients: *S. epidermidis* is the leading cause of septicemia, with the onset later than 48 h, in premature newborns. Depending on gestational age, the opsonophagocytosis system of premature newborns is not well enough developed to handle even bacteria with low pathogenic potential such as *S. epidermidis*. Patients with aplasia (neutropenia) after cytostatic and/or immunosuppressive therapy are also highly susceptible to *S. epidermidis* septicemia, owing to the low numbers of functioning polymorphonuclear neutrophils (PMNs). The sources of infection in all these patients are skin and mucous membranes, and the port of entry is very often an intravascular catheter. A third group, probably the most important, is patients with foreign bodies such as indwelling catheters or implanted polymer devices of various materials (e.g., polyethylene, polyurethane, and silicon rubber) increasingly used in diagnostic or therapeutic procedures. Infection is the major complication associated with the use of such devices; overall, NAS (mainly *S. epidermidis*) are the microorganisms most frequently isolated in these infections. Depending on the kind of device and its insertion site, different infection syndromes generate a variety of clinical presentations (Table 2) (87).

The clinical picture of *S. epidermidis* infections differs markedly from that of *S. aureus* infections. Normally, there are no fulminant signs of infection, and the clinical

Gram-Positive Pathogens, 2nd edition, edited by Vincent A. Fischetti et al.
© 2006 ASM Press, Washington, D.C.

TABLE 1 NAS found in human specimens

Novobiocin-susceptible NAS
  S. epidermidis group
    S. epidermidis
    Staphylococcus haemolyticus
    Staphylococcus hominis
    Staphylococcus capitis
    Staphylococcus warneri

  Staphylococcus simulans
  Staphylococcus auricularis

  S. lugdunensis
  Staphylococcus schleiferi

Novobiocin-resistant NAS
  Staphylococcus saprophyticus
  Staphylococcus cohnii
  Staphylococcus xylosus

course is more subacute or even chronic. Accordingly, making the diagnosis of *S. epidermidis* infection is often difficult. Therapy is especially problematic with foreign body (polymer)-associated infections: despite the use of appropriate antibiotics with proven in vitro efficacy and a generally functional host response, it is often not possible to eradicate the focus on the infected device. Thus, removal of the device and subsequent renewal become necessary (87).

Therefore, *S. epidermidis* infection contrasts with *S. aureus* infection in that *S. epidermidis* requires an especially predisposed host. Only in such a circumstance can *S. epidermidis* change from a commensal or saprophytic organism in the human cutaneous or mucocutaneous ecosystem to a pathogen.

BIOFILM FORMATION

The most important step in the pathogenesis of *S. epidermidis* foreign body-associated infectious diseases is the colonization of the polymer surface by the formation of multilayered cell clusters, which are embedded in an amorphous extracellular material (65). Infection of the polymer likely occurs by inoculation with only a few bacteria from the patient's skin or mucous membranes during implantation of the device. The colonizing bacteria, together with the extracellular material, which is mainly composed of cell wall teichoic acids (29), and host products are referred to as biofilm. The presence of large adherent biofilms on explanted intravascular catheters has been demonstrated by scanning electron microscopy (Fig. 1) (66).

Biofilm formation proceeds in two steps: rapid attachment of the bacteria to the surface is followed by a prolonged accumulation phase that involves cell proliferation and intercellular adhesion. In recent years, significant progress has been made in defining the molecular mechanisms involved in biofilm formation, which are summarized in Fig. 2.

Initial Attachment

Microbial adherence to biomaterials depends on the cell surface characteristics of the bacteria and on the nature of the polymer material. Factors involved include physico-

chemical forces such as charge, van der Waal's forces, and hydrophobic interactions. With *S. aureus*, it has been found that the colonization of abiotic surfaces depends on the charge of its teichoic acid: an *S. aureus dltA* mutant has a biofilm-negative phenotype due to a decreased initial attachment to a polystyrene or a glass surface, which is hydrophobic or negatively charged, respectively (17). DltA mediates the incorporation of the substituent D-alanine into the teichoic acid and thus renders the cell surface less negatively charged. A deduced protein sequence that is 75% identical to DltA from *S. aureus* is present in the genome of *S. epidermidis* RP62A (preliminary sequence data were obtained from The Institute for Genomic Research website at http://www.tigr.org). Thus, it may be speculated that the surface charge determined by teichoic acids may also contribute to *S. epidermidis* colonization of artificial surfaces. Cell surface hydrophobicity and initial adherence have been attributed to bacterial surface-associated proteins. With the aid of monoclonal antibodies that efficiently block adherence, the antigenically related staphylococcal surface proteins SSP-1 and SSP-2 (280 and 250 kDa, respectively) have been identified as fimbria-like polymers involved in *S. epidermidis* 354 adherence to polystyrene (86), but the genes and protein sequences are not yet available. The gene encoding the surface-associated autolysin AtlE of *S. epidermidis* O-47, which mediates primary attachment of bacterial cells to a polymer surface, has been cloned and sequenced (21). The 148-kDa AtlE shows high similarity to the *S. aureus* autolysin Atl (61% identical amino acids) and is proteolytically cleaved into two bacteriolytically active domains, a 60-kDa amidase and a 52-kDa glucosaminidase. In the central part of the protein, there are three repetitive sequences, possibly involved in the adhesive function. Another protein from *S. aureus*, the 238.7-kDa biofilm-associated protein Bap, is involved in attachment to polystyrene surface and intercellular adhesion leading to biofilm formation (9). The structural features of Bap correspond to those

TABLE 2 Foreign body (polymer)-associated *S. epidermidis* infections[a]

Septicemia, endocarditis
  Intravascular catheters
  Vascular prostheses
  Pacemaker leads
  Defibrillator systems
  Prosthetic heart valves
  Left ventricular assist devices

Peritonitis
  Ventriculoperitoneal CSF shunts
  CAPD catheter systems

Ventriculitis
  Internal and external CSF shunts

Chronic polymer-associated syndromes[b]
  Prosthetic joint (hip) loosening
  Fibrous capsular contracture syndrome after mammary
    augmentation with silicone prostheses
  Late-onset endophthalmitis after implantation of artificial
    intraocular lenses following cataract surgery

[a]CAPD, continuous ambulatory peritoneal dialysis; CSF, cerebrospinal fluid.
[b]At least in some instances, the role of *S. epidermidis* is very probable, but more data are needed.

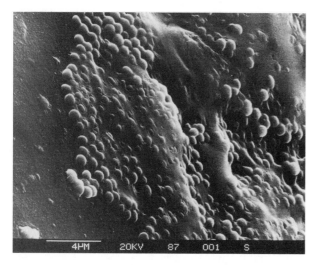

FIGURE 1 Scanning electron micrograph of an early stage of biofilm formation by *S. epidermidis* on a polyethylene surface. (Reprinted from reference 35, with permission.)

of other typical gram-positive surface proteins. The clinical significance of Bap is not clear, because it is present in only 5% of 350 bovine mastitis and absent in all 75 human clinical *S. aureus* isolates tested. However, a gene encoding a Bap-homologous protein, Bhp (M. A. Tormo et al., unpublished; GenBank database accession number AAK29746), was detected in the human clinical strain *S. epidermidis* RP62A. Bhp consists of 2,402 amino acids with a predicted molecular mass of 258 kDa and was assumed to promote biofilm formation.

Aside from proteins, a polysaccharide structure, capsular polysaccharide/adhesin (PS/A), has been associated with initial adherence and slime production (56). Tn917 mutants deficient in PS/A and initial adherence do not cause endocarditis in a rabbit model, in contrast to the isogenic

parent. Furthermore, immunization with PS/A results in protection against infection. The relationship of PS/A to other polysaccharides of *S. epidermidis* is discussed below.

Interaction of *S. epidermidis* with Extracellular Matrix Proteins

While the direct interaction between bacteria and naked polymer surfaces plays a crucial role in the early stages of the adherence process in vitro and probably also in vivo, additional factors may be important in later stages of adherence in vivo, because implanted material rapidly becomes coated with plasma and extracellular matrix proteins such as fibronectin, fibrinogen, vitronectin, thrombospondin, and von Willebrand factor. Some of these host factors could serve as specific receptors for colonizing bacteria. Indeed, adherence of clinical coagulase-positive and -negative staphylococcal isolates is significantly promoted by surface-bound fibronectin in comparison with surface-bound albumin. Adherence of all *S. aureus* strains tested is markedly promoted by immobilized fibrinogen, while adherence of *S. epidermidis* strains to immobilized fibrinogen was found to vary significantly among different strains (57).

Various genes encoding host factor-binding proteins from *S. aureus* have already been cloned and sequenced, among them fibronectin-binding proteins (FnBPs) and fibrinogen-binding proteins. Relatively few data on host factor-binding proteins of NAS are available. However, the gene encoding a fibrinogen-binding protein (Fbe) was recently cloned from *S. epidermidis* and sequenced (57). Sequence comparison revealed that the 119-kDa Fbe shows similarity to the cell wall-bound fibrinogen receptor (clumping factor; ClfA) of *S. aureus*. In contrast to ClfA, which binds to the γ chain of fibrinogen, and like ClfB, Fbe binds to the β chain of fibrinogen (63). Heterologous expression of *fbe* in *Lactococcus lactis* and analysis of an *fbe*-deficient mutant delineated the role that Fbe plays in the interaction of *S. epidermidis* with fibrinogen (18, 62). Antibodies against Fbe efficiently inhibited that adherence reaction (62, 63). Recently, Fbe was suggested as a potential

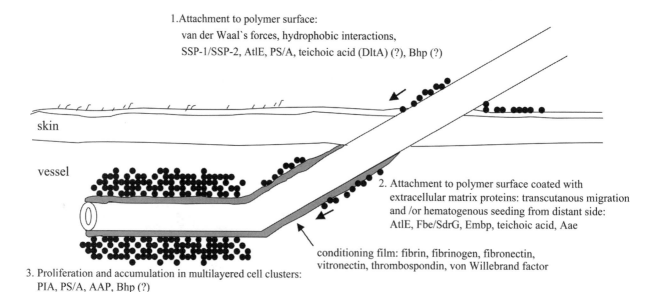

1. Attachment to polymer surface:
   van der Waal's forces, hydrophobic interactions,
   SSP-1/SSP-2, AtlE, PS/A, teichoic acid (DltA) (?), Bhp (?)

skin

vessel

2. Attachment to polymer surface coated with extracellular matrix proteins: transcutanous migration and /or hematogenous seeding from distant side: AtlE, Fbe/SdrG, Embp, teichoic acid, Aae

conditioning film: fibrin, fibrinogen, fibronectin, vitronectin, thrombospondin, von Willebrand factor

3. Proliferation and accumulation in multilayered cell clusters: PIA, PS/A, AAP, Bhp (?)

FIGURE 2 Model of different phases of *S. epidermidis* biofilm formation on a prosthetic polymer device and bacterial factors involved or potentially involved (?). Fbe/SdrG, fibrinogen-binding protein.

candidate for protein vaccination against *S. epidermidis* infections, because antibodies against Fbe significantly increased macrophage phagocytosis and reduced the severity of systemic infections (69). Moreover, antibodies against Fbe may be used to prevent or treat *S. epidermidis* infections. The overall organization of Fbe corresponds to that of other surface-associated proteins, including an LPXTG motif and the characteristic membrane-spanning region that contains serine aspartate (SD) repeats. Additional cell-surface-associated proteins containing SD repeats—i.e., SdrG (97.6 kDa; approximately 95% identical to Fbe), SdrF (179 kDa), and SdrH (50.5 kDa)—in *S. epidermidis* were identified (46). Antibodies against SdrG and SdrH proteins were found in 16 convalescent patient sera, implying that their genes are expressed during infection (46). Like Fbe, SdrG binds to the β chain in fibrinogen. More specifically, SdrG binds to the N-terminal portion (peptide β6-20) of this polypeptide, which is located proximal to the thrombin cleavage site. Furthermore, SdrG inhibits thrombin-induced fibrinogen clotting by interfering with the release of fibrinopeptide B (10). It was speculated that the reason for that binding activity of SdrG might be to prevent the release of chemotactic elements, such as fibrinopeptide B. This may reduce the influx of phagocytic neutrophils, aiding the survival of the bacteria in the host (10). For the binding mechanism of SdrG to its ligand, a "dock, lock, and latch" model has been presented after study of the crystal structures of the SdrG protein and synthetic peptides comprising the binding site in fibrinogen (67). The role of the other SD repeat proteins of *S. epidermidis* still has to be elucidated.

Recently, an FnBP designated Embp was identified in *S. epidermidis* by the phage display technique (92). The fibronectin-binding site of Embp seems to be unrelated to the fibronectin-binding site of the *S. aureus* FnBPs. Aside from proteins, cell wall teichoic acid is involved in the adherence of *S. epidermidis* to fibronectin (27). Adherence of *S. epidermidis* to immobilized fibronectin is significantly promoted in a dose-dependent fashion by teichoic acid. Preincubation of either the bacteria or fibronectin-coated surfaces with teichoic acid promoted *S. epidermidis* adherence, suggesting a possible function of teichoic acid as a bridging molecule between the bacteria and fibronectin-coated polymer material.

For the autolysin AtlE from *S. epidermidis* mediating primary attachment to a polystyrene surface (see above), vitronectin-binding activity was also found (21). There is preliminary evidence for an in vivo role of AtlE: in a central venous catheter-associated infection model, only 50% of rats challenged with an *atlE*-negative Tn917 mutant developed an infection versus 80% of rats challenged with the isogenic wild-type strain. Another autolysin/adhesin (Aas), which exhibits significant homology to Atl and AtlE, was cloned and sequenced from *S. saprophyticus* (24). Aas binds to fibronectin and to sheep erythrocytes, leading to hemagglutination. Thus, these autolysins may represent a novel class of staphylococcal adhesins. We have recently found another multifunctional autolysin/adhesin from *S. epidermidis* (designated Aae) that has bacteriolytic activity and binds to fibrinogen, fibronectin, and vitronectin in a dose-dependent and saturable fashion and with high affinity (23). Aae contains three direct repeated sequences in its N-terminal portion that are homologous to the LysM domain, which is part of several enzymes involved in cell wall metabolism and also of some adhesins.

Another example of multi- or bifunctional surface molecules of *S. epidermidis* exhibiting an enzymatic and an adhesive function is the GehD lipase: in a search for the collagen adhesin of *S. epidermidis*, the extracellular GehD lipase, which seems to partially be surface associated, was identified (2). *gehD* mutants showed reduced adherence to surface-immobilized collagen. Furthermore, recombinant GehD protein, as well as anti-GehD antibodies, was able to inhibit the adherence of *S. epidermidis* to surface-immobilized collagen, confirming its role as a collagen adhesin.

Accumulation Process

After succeeding in primary attachment to the polymer surface, bacteria proliferate and accumulate in multilayered cell clusters, which requires intercellular adhesion. Probably the same mechanisms are involved in biofilm accumulation on natural surfaces, such as the endocardium. Transposon mutants that are not able to accumulate in multilayered cell clusters lack a specific polysaccharide antigen referred to as polysaccharide intercellular adhesin (PIA). Purification and structural analysis of PIA revealed that it consists of major polysaccharide I (>80%) and minor polysaccharide II (<20%). Polysaccharide I is a linear β-1,6-linked glucosaminoglycan mainly composed of at least 130 2-deoxy-2-amino-D-glucopyranosyl residues, of which 80 to 85% are N acetylated. Polysaccharide II is structurally related to polysaccharide I but has a lower content of non-N-acetylated glucosaminyl residues and contains a small amount of phosphate and ester-linked succinyl residues. Thus, PIA represents a so far unique structure (41).

The genes (*icaABC*) that mediate cell clustering and PIA synthesis in *S. epidermidis* have been cloned and sequenced (22). Recently, an additional open reading frame (*icaD*) located between *icaA* and *icaB* and overlapping both genes was identified, and the function of the respective gene products in PIA synthesis was analyzed (15). Evidence has been provided that the proposed N-acetylglucosaminyl-transferase activity is carried out by IcaA. However, IcaA alone exhibited only low levels of transferase activity. Coexpression of *icaA* encoding the catalytic enzyme together with *icaD* led to a significant increase in activity and to synthesis of N-acetylglucosamine oligomers with a maximal length of 20 residues. Only in the presence of *icaC* does IcaAD catalyze the synthesis of long-chain oligomers reacting with PIA-specific antiserum. In a mouse model, as well as in a rat model of foreign body infections, a PIA-negative mutant was shown to be significantly less virulent than the isogenic wild-type strain; therefore, PIA represents an important pathogenicity factor (72–74). Furthermore, PIA mediates hemagglutination (12). The hemagglutinin of *S. epidermidis* has been demonstrated to be associated with biofilm production and to be a polysaccharide structure. A study investigating the pathogenic properties of strains obtained from polymer-associated septicemic disease compared with saprophytic skin and mucosal isolates demonstrated a strong correlation of biofilm formation and presence of the *ica* gene cluster essentially associated with disease isolates (93).

Formerly, another antigenic marker of slime production and accumulation designated as slime-associated antigen (SAA) was identified. Changes in the purification procedure have now shown that the composition of SAA differs from that originally described and that SAA mainly consists of N-acetylglucosamine. Hence, it has been concluded that SAA and PIA may be the same antigenic structure (1).

Recently, it has been reported that PS/A production is also determined by the *ica* gene cluster and that PS/A is chemically related to PIA (47). Both antigens are characterized by the common β-1,6-linked polyglucosamine backbone, but PS/A can be distinguished from PIA by molecular mass (>250,000 kDa for PS/A versus ~28,000 Da for PIA) and the presence of succinate groups on the majority of the amino groups of the glucosamine residues. Recently, it has been shown that the synthesis of a similar if not identical polysaccharide from *S. aureus* is mediated by a homologous *ica* gene cluster (7). However, analysis of this polysaccharide revealed that the majority of the amino groups of the glucosamine residues are acetylated, as described for PIA, rather than succinylated (42).

Proteins also seem to be essential for accumulation and biofilm formation in *S. epidermidis*. The Bap-homologous protein Bhp may be involved in biofilm accumulation (see above). A 140-kDa extracellular protein, Aap (for accumulation-associated protein), missing in the accumulation-negative mutant M7, was shown to be essential for accumulative growth in certain *S. epidermidis* strains on polymer surfaces (28, 76). An antiserum raised against the purified protein inhibited accumulation of the wild-type strain RP62A up to 98%, whereas the preimmune serum did not. Biochemical and functional properties clearly differentiate Aap from other factors known to be involved in biofilm formation. The function of Aap in the accumulation process is speculated to be the anchoring of PIA to the cell surface, because the mutant M7 produces PIA that is only loosely attached to the cell surface in contrast to the wild type (40). Recently, it was shown that Aap is able to mediate intercellular adhesion and biofilm accumulation in a completely PIA-independent manner (70). Aap, which is highly prevalent among clinical *S. epidermidis* strains (71, 85), consists of domains A and B, the latter characterized by 128 amino acid repeats. It appeared that domain B mediates cell-to-cell adhesion and biofilm accumulation. Interestingly, domain B becomes active as an intercellular adhesin only after proteolytic cleavage of the N-terminal A domain, by either endogenous staphylococcal or host proteases. This attributes a possible bimodal function to Aap during biofilm formation, including engagement of components of the innate immune response to help the organism to resist clearance by the host (70).

OTHER POTENTIAL VIRULENCE FACTORS

Extracellular Enzymes and Toxins

The establishment of an infection and the survival of the bacteria in the host depend on the ability to invade host tissues and to evade host defense systems, respectively. For this, staphylococci, in particular *S. aureus*, have developed multiple mechanisms including production of a variety of extracellular proteins and enzymes such as protein A, lipases, proteases, esterases, phospholipases, and fatty acid-modifying enzymes, as well as production of hemolysins, and toxins with superantigenic properties such as enterotoxins, exfoliative toxins, and toxic shock syndrome toxin 1. Additionally, proteases may play a role in proteolytic inactivation of host defense mechanisms such as antibodies and platelet microbicidal proteins, as well as in destruction of tissue proteins causing increased invasiveness.

In *S. epidermidis*, an extracellular metalloprotease with elastase activity has been detected, and its gene has been cloned and sequenced (82). Previously, an elastase from *S.*

epidermidis that degrades human secretory immunoglobulin A, immunoglobulin M, serum albumin, fibrinogen, and fibronectin was identified as a cysteine protease and thus assumed to be a virulence factor (80); however, the corresponding gene has not yet been cloned. An extracellular serine protease is involved in epidermin processing (14) (see below). Recently, the gene encoding a novel 27-kDa extracellular serine protease, GluSE, was cloned and characterized (58). Production of GluSE was observed in adherent cultures of *S. epidermidis* but was missing from planktonic cultures, indicating its possible role in *S. epidermidis* colonization and tissue damage. The genes of two homologous lipases (the genes encoding GehC and GehSE1 from *S. epidermidis* strains 9 and RP62A, respectively) exhibiting a high degree of similarity (97.8% identical amino acids) have been cloned and sequenced, and they have been proposed to be involved in skin colonization (11, 78). Most *S. epidermidis* strains seem to possess two lipase genes: the second lipase gene, *gehD*, encodes a mature protein of approximately 45 kDa that is 51% identical to GehC (39). The characterization and expression of fatty acid-modifying enzyme in *S. epidermidis* have been described as well (3).

In contrast to *S. aureus*, which produces all of the above-mentioned toxins in a strain-dependent manner, *S. epidermidis* is much less toxigenic. *S. epidermidis* can produce delta-toxin, which differs from the *S. aureus* delta-toxin by only three amino acids (48). The delta-toxin is encoded by *hld*, which is a component of the regulatory *agr* system (see below), and acts by formation of pores in the membrane, leading to the lysis of erythrocytes and other mammalian cells. Reports of unusual *S. epidermidis* strains producing enterotoxin C or toxic shock syndrome toxin 1 (43) are controversial.

Factors Involved in Inflammatory Reaction and Host Defense

A serious consequence of *S. epidermidis* polymer-associated infection is septicemia. In the pathophysiology of inflammatory events in septicemia, the production of cytokines such as tumor necrosis factor alpha (TNF-α), interleukin-1β (IL-1β), and IL-6 plays a major role. Peptidoglycan and teichoic acid, cell wall components purified from an *S. epidermidis* strain, stimulate human monocytes to release TNF-α, IL-1, and IL-6 in a concentration-dependent manner. Further studies revealed that human serum strongly increases peptidoglycan-induced TNF-α release by human monocytes (44).

Another feature of extracellular products of *S. epidermidis* is the interference with several neutrophil functions. Although the extracellular slime substance itself has been found to induce a significant chemotactic response in human PMNs (31), it decreases the phagocytic activity of murine peritoneal macrophages in a dose-dependent fashion (77). Moreover, when human PMNs were preincubated with increasing amounts of slime, the responsiveness to known chemotactic stimuli such as N-formyl-methionyl-leucyl-phenylalanine and zymosan-activated serum was inhibited. In addition, preincubation of PMNs with slime stimulates PMN degranulation, especially after previous treatment with cytochalasin B. This effect was dose dependent and particularly pronounced with lactoferrin. This may lead to the waste of antibacterial cellular products after contact with slime, which together with the reduced chemotactic responsiveness results in a decreased ability for intracellular killing. Indeed, in a surface opsonophagocytosis model, there was significantly less killing of the accumulation-positive strain *S. epidermidis* RP62A by human PMNs than of its accumulation-negative mutant M7 (30). Another biological active component of the extracellular products of *S. epidermidis* that protect against major components of the human innate immune system seems to be PIA (see above). Recently, it was demonstrated by analyzing an *ica* mutant strain that the production of PIA protects *S. epidermidis* against phagocytosis and killing by human polymorphonuclear leukocytes (91). Furthermore, the *ica* mutant was more susceptible to killing by human antibacterial peptides.

Extracellular slime produced by *S. epidermidis* has also been shown to reduce the blastogenic response of human peripheral mononuclear cells to T-cell mitogens (phytohemagglutinin and streptococcal blastogen A) in a dose-dependent manner (16). This was indicated by a decrease in the incorporation rate of [^{3}H]thymidine and by the number of blastic cells in the cell cultures. The underlying mechanisms are still unclear, as is the biological relevance.

Recently, the isolation and characterization of a novel proinflammatory polypeptide complex from *S. epidermidis* were reported (49). This complex had a mass of 34.5 kDa and was designated phenol-soluble modulin (PSM) based on its partition into the phenol layer on hot aqueous phenol extraction. PSM consists of three highly hydrophobic peptides (PSMα, PSMβ, and PSMγ) with 22, 44, and 25 amino acids, respectively. PSMγ was identified as a delta-toxin, while PSMα and PSMβ exhibit some similarities to previously reported staphylococcal toxins. PSM is a more potent inducer of cytokine release from monocytes and THP-1 cells than lipoteichoic acid, activates the human immunodeficiency virus type 1 long-terminal repeat in cells of macrophage lineage, and stimulates NF-κB production by these cells (49). Furthermore, PSM causes degranulation, primes for an enhanced respiratory burst, and inhibits spontaneous apoptosis in human neutrophils (37). PSM also is a chemoattractant for neutrophils and monocytes (37). Because of these pronounced proinflammatory properties, PSM is assumed to contribute markedly to sepsis caused by *S. epidermidis*. PSM synthesis is controlled by the *agr* quorum-sensing system (see below); an *agr* mutant did not produce any of the PSM components, in contrast to the wild-type strain (88). Moreover, the *agr* mutant failed to induce TNF-α production by human myeloid cells, and induction of neutrophil chemotaxis was significantly reduced with the *agr* mutant. Thus, the authors proposed that an *agr* quorum-sensing mechanism facilitates growth and survival in infected patients by adapting production of the proinflammatory PSMs to the stage of infection (88).

Iron Acquisition inside the Host

The common prerequisite for all bacterial pathogens to establish an infection is the ability to proliferate within the mammalian host. As do all bacteria, the staphylococci require iron for their growth; however, the free iron concentration (10^{-18} M) in the extracellular body fluids, owing to the presence of high-affinity iron-binding glycoproteins such as transferrin or lactotransferrin, is much too low to support staphylococcal growth.

The mechanisms by which the staphylococci acquire iron from transferrin are not well understood. In general, there are two known mechanisms for iron acquisition. The first mechanism involves the synthesis and secretion of low-molecular-mass iron chelators (siderophores), which remove iron from transferrin. The siderophore-iron complexes are then taken up by specific bacterial transport systems. The siderophores staphyloferrin A and B (481 and 448 Da, respectively), first isolated from *Staphylococcus hyicus*, were also found to be produced by *S. epidermidis* under iron-restricted conditions (38). The second mechanism by which bacteria assimilate iron depends on direct contact between the host transferrin and a bacterial surface receptor.

Both *S. epidermidis* and *S. aureus* express a number of iron-repressible cell wall- and cytoplasmic membrane-associated proteins when isolated during infection in humans and when grown in vivo in laboratory animal infections (54). These include a 42-kDa cell wall protein that functions as a receptor for human transferrin (51) and a 32-kDa cytoplasmic membrane-associated lipoprotein (5). Human transferrin is an approximately 80-kDa monomeric protein with two distinguishable homologous domains, termed the N and the C lobes, each of which contains an iron-binding site. Recent results demonstrate that the staphylococci efficiently remove iron from human transferrin sequentially from the N lobe and then from the C lobe via a receptor-mediated process that seems to be energy dependent (51). The 42-kDa transferrin receptor has been identified via N-terminal sequencing as a cell surface-associated glyceraldehyde-3-phosphate dehydrogenase (GAPDH), an enzymatic activity that is part of the glycolytic pathway (53). GAPDH is a multifunctional protein that not only is involved in the acquisition of transferrin-bound iron and catalyzes the conversion of glyceraldehyde-3-phosphate to 1,3-diphosphoglycerate, but also binds human plasmin in an enzymatically active form (53). Because plasmin is able to cleave extracellular matrix proteins and to dissolve blood clots, it was hypothesized that GAPDH-bound plasmin facilitates the crossing of tissue barriers by invasive staphylococci to cross tissue barriers and thus contributes to virulence (53). Multifunctional cell surface-bound GAPDHs have also been found in other prokaryotic and eukaryotic organisms (52). Cloning and sequencing of the DNA region encoding the 32-kDa cytoplasmic membrane-associated lipoprotein of *S. epidermidis* revealed that the corresponding gene (*sitC*) is part of a translationally coupled, iron-regulated operon (*sitABC*) that encodes an ABC-type transporter (5). It is speculated that this novel ABC transporter is involved in either siderophore- or transferrin-mediated iron uptake in *S. epidermidis*.

A newly identified mechanism by which *S. aureus* may acquire iron during infection involves the binding of hemoglobin by surface proteins, the release of heme from hemoglobin, and the transport of heme across the cell wall and plasma membrane into the cytoplasm where iron subsequently is released from heme by the activity of heme oxygenases (45). Homologs of the *S. aureus* heme oxygenases IsdG and IsdI are also encoded by the genome of *S. epidermidis* (79).

Recently, iron storage proteins in staphylococci such as ferritins, which keep intracellular levels of iron at a nontoxic concentration, have been identified (26). Ferritins from *S. aureus* and *S. epidermidis* are regulated significantly differently in response to metal ions (55).

REGULATION OF VIRULENCE FACTORS

Although studied in more detail with *S. aureus*, knowledge about the regulation of *S. epidermidis* virulence factors has been increased significantly in recent years. At least two global regulators, the *agr* (for accessory gene regulator) locus and the *sar* (for staphylococcal accessory regulator) locus, are involved in regulation of *S. epidermidis* virulence factors. The expression of both *agr* and *sar* is influenced by the alternative transcription factor σ^B, which is the global regulator of stress responses in many bacterial species.

Environmental Factors and Stress

It has been known for some time that *S. epidermidis* biofilm formation is modulated by environmental conditions and stress, such as high osmolarity, detergents, urea, ethanol, oxidative stress, and subinhibitory concentrations of antibiotics. Recent studies have demonstrated that the expression of the *icaADBC* gene cluster and thus PIA production and biofilm accumulation are influenced by such factors. External stress, such as high temperature and high osmolarity, induces *icaADBC* gene expression and biofilm formation (68). Subinhibitory concentrations of some antibiotics have a similar effect (68). The presence of ethanol or high concentrations of sodium chloride in the growth medium also creates stress conditions, which lead to increased biofilm formation and PIA production in *S. epidermidis* (34). Furthermore, the production of PIA and synthesis of the *ica*-specific mRNA are significantly enhanced under anaerobic in vitro growth conditions (8). The effect of oxygen limitation may play an important role in the in vivo situation, where anaerobiosis occurs during localized infections and leads to increased biofilm formation. The modulation of gene expression by environmental factors and stress is at least partly mediated by the alternative sigma factor, σ^B.

The *sigB* Operon

Many bacterial species express various alternative sigma factors to adapt gene expression to altered environmental conditions. Like *S. aureus*, *S. epidermidis* encodes the alternative sigma factor, σ^B (32, 34). The gene encoding σ^B is part of an operon (*sigB* [for sigma factor B] operon) that contains four genes: *rsbU*, *rsbV*, *rsbW*, and *sigB*. RsbU, RsbV, and RsbW are involved in regulation of σ^B, with RsbU being a required activator of σ^B. Tn*917* insertion into *rsbU* resulted in a biofilm-negative phenotype due to dramatically reduced PIA synthesis, strongly suggesting that these factors are σ^B dependent (34). Ethanol and salt stress are both known activators of σ^B. However, the presence of ethanol in the growth medium of the *rsbU* mutant completely restored biofilm formation, whereas salt stress did not. Thus, different regulatory mechanisms are involved in *S. epidermidis* PIA production in response to ethanol and salt stress. Recent results indicate that *icaR* may be responsible for this differential regulation (6). *icaR* is situated upstream of the *icaADBC* operon and encodes a transcriptional regulator of the *tetR* family. IcaR acts as a repressor of *icaADBC* transcription and is involved in the stimulation of *ica* operon transcription mediated by ethanol, but not by salt (6). In contrast to the above-mentioned *rsbU* mutant, a σ^B deletion mutant in strain *S. epidermidis* Tü3298 did not reveal a significant effect of σ^B on either biofilm formation or *agr* activity (32). This may be due to a nonfunctional RsbU-mediated signal transduction pathway in that particular strain.

The *agr* Locus

An *agr* homolog in *S. epidermidis* has been identified and sequenced (60, 84). DNA sequence analysis revealed a pronounced similarity between the *S. epidermidis* and *S. aureus* *agr* systems (see chapter 41). The extracellular signaling

molecule produced by a typical *S. epidermidis* strain is a cyclic octapeptide (DSVCASYF) that is encoded by *agrD* and contains a thiolester linkage between the central cysteine and the C-terminal carboxyl group (60). This octapeptide exhibits activity at nanomolar concentrations. Sequence comparison revealed no striking similarity between the signaling peptides of *S. epidermidis*, *S. aureus*, or *Staphylococcus lugdunensis* (hepta-, octa-, or nonapeptides) except for the central cysteine and its distance to the C terminus. Therefore, these conserved structural features are thought to be necessary for thiolactone formation. The AgrD proteins of *S. epidermidis* and *S. aureus* show evident similarity in the region located C terminal of the signaling peptides, suggesting that this region represents a structural element important for the modifying reaction probably mediated by AgrB. AgrB shows an overall identity of 51.3% between both species. Its location within the cytoplasmic membrane has recently been demonstrated (75). The *S. epidermidis* histidine kinase AgrC shares 50.5% identical amino acids with the *S. aureus* AgrC, with pronounced similarity in the C-terminal portion and low similarity in the N-terminal portion. These sequence data are also in agreement with the *S. lugdunensis* data, leading to the hypothesis that the N-terminal part of AgrC represents the region binding the signaling peptides that differ in sequence, while the C-terminal part interacts with the highly conserved response regulator AgrA (87.3% identity between *S. epidermidis* and *S. aureus*). In addition, an RNAIII homolog in *S. epidermidis* (560 nucleotides) was shown to regulate virulence gene expression in *S. aureus* (81). The *S. epidermidis* RNAIII had the ability to completely repress transcription of protein A and the ability to activate transcription of the alpha-toxin (*hla*) and serine protease (*ssp*) genes in an RNAIII-deficient *S. aureus* mutant. However, the stimulatory effect was reduced compared with that of *S. aureus* RNAIII. In particular, the first 50 nucleotides and last 150 nucleotides of RNAIII in *S. epidermidis* and *S. aureus* were found to be highly similar. Construction and analysis of *S. epidermidis*-*S. aureus* RNAIII hybrids showed that both the 5′ and 3′ halves of the RNA molecule are important for the regulatory function.

An *agr* mutant of *S. epidermidis* showed significantly altered protein expression: the expression of surface-associated proteins was increased, whereas the expression of extracellular proteins, as demonstrated for the exoenzymes lipase and protease, was decreased (90). Thus, the function of the *S. epidermidis* agr system in growth phase-dependent regulation of protein synthesis may correspond to that of *S. aureus* and may be summarized as follows. In an early stage of infection, cell density is low and surface-associated proteins with adhesive functions are expressed, allowing colonization of polymer material and host protein-coated material. Upon proliferation of the cells on the surface, the autoinducing signaling peptides accumulate and eventually reach the critical concentration necessary for the activation of the synthesis of RNAIII. RNAIII then down-regulates the expression of surface protein genes and up-regulates the production of tissue-degrading enzymes and other secreted proteins, which facilitates the maintenance of the infection. This potential role of *agr* in *S. epidermidis* disease suggests the use of quorum-sensing blockers as alternative antistaphylococcal drugs. However, *agr* was also found to influence biofilm formation. The *agr* mutant showed a stronger biofilm formation, primary attachment, and expression of the autolysin AtlE (reference 89 and see above). Inhibition of *agr* by quorum-sensing blockers had a similar effect. This makes it doubtful that quorum-sensing blockers will be of great benefit as anti-staphylococcal drugs (59). However, after a biofilm has formed and an *S. epidermidis* infection has been diagnosed, they might be of use in down-regulating the production of exoproteins that inhibit host defenses.

The *agr* quorum-sensing system also influences the production of lantibiotics (see below). The production of epidermin is strongly reduced in the *agr* mutant, which is not due to interference of *agr* with the transcription of the epidermin biosynthetic genes, but to interference with the extracellular processing of the leader peptide by the protease EpiP (33).

The *sar* Locus

In *S. aureus*, another global regulator, *sar*, also controls exoprotein synthesis by modulating the expression of *agr* (see chapter 41). The *sar* locus in *S. aureus* contains a major open reading frame, *sarA*, preceded by two smaller open reading frames. DNA mobility shift assays demonstrated that the *sar* gene products bind to an *agr* P2 promoter fragment, probably leading to activation of transcription of RNAII and subsequently RNAIII. A *sar* homolog of *S. epidermidis* has been cloned and sequenced (13), and it was revealed that the SarA protein of *S. epidermidis* is nearly identical (84%) to SarA of *S. aureus*. In contrast, the *sarA*-flanking DNA sequence shows only 50% identity between both strains, and the two smaller open reading frames are absent in *S. epidermidis*. Remarkably, an *S. epidermidis sar* fragment including *sarA* and the upstream flanking region interacts with an *agr* promoter fragment of *S. aureus*. Moreover, functional analysis confirmed that the *S. epidermidis sar* homolog was able to restore alpha-toxin production in an *S. aureus sar* mutant (13). Because most of the typical virulence determinants of *S. aureus* are missing in *S. epidermidis*, the genes that are under the control of *agr* and *sar* in *S. epidermidis* and other NAS must be clarified. Possible candidates include genes encoding the Fg-binding protein Fbe, SdrG, SdrF, SdrH, Bhp, Aap, autolysins, lipases, PIA and PS/A production, proteases, and delta-toxin.

Phase Variation

It has been observed for many years that biofilm-forming *S. epidermidis* strains can undergo phase variation, resulting in biofilm-negative phenotypes (4). More recently, biofilm- and autoaggregation-negative phase variants were isolated from biofilm-producing blood culture strains with Congo red agar (93). These variants occurred at a frequency of 10^{-5}, and the phenotype of the wild-type strain could be restored after repeated passages. Genetic analysis revealed that in approximately 30% of those variants, the occurrence of the biofilm-negative phenotype was due to the inactivation of either the *icaA* or the *icaC* gene by the insertion sequence element IS256 (94). This insertion is a reversible process and involves the exact excision of the IS256 element, resulting in restoration of the intact *icaA* and *icaC* genes and the biofilm-positive phenotype. It may be speculated that a switch from a biofilm-forming phenotype to a biofilm-negative variant may contribute to dissemination from an infected medical device, leading to circulation in the host organism and thereby allowing the colonization of another potential site of infection.

Fur-Like Protein

Because of the low availability of iron in the host, many bacterial pathogens use low iron concentrations as a signal to activate certain virulence factors, including toxins, adhesins, and invasins. The corresponding genes and genes involved in the biosynthesis and transport of siderophores are often regulated by Fur (for ferric uptake regulator), which has been studied extensively with gram-negative bacteria. Fur is a DNA-binding repressor protein that binds only in the presence of iron to a consensus sequence termed the Fur box located within the promoter region of the target genes. When iron levels are low, Fur does not bind, and genes are transcribed. Recently, a gene (*fur*) for a Fur-like protein in *S. epidermidis* was identified (19). Within the −35 promoter region of the *fur* gene, a sequence motif was detected with low similarity to Fur boxes. Although the *S. epidermidis* Fur protein is unable to complement an *Escherichia coli fur* mutant, the Fur protein of *E. coli* binds to the Fur box of the *S. epidermidis fur* promoter region. The role of the Fur regulatory mechanism in the expression of virulence factors in *S. epidermidis* has not yet been determined.

DtxR Homolog

An alternative iron-dependent repressor in gram-positive bacteria is DtxR (for diphtheria toxin repressor), first identified as a repressor of diphtheria toxin synthesis in *Corynebacterium diphtheriae*. Although it has a similar function, DtxR shares no homology with Fur and belongs to a newly identified family of iron-dependent repressors. DtxR homologs have been found in several bacterial species, where they regulate genes encoding iron transport systems, heme oxygenase, and virulence determinants.

Most recently, a DtxR homolog designated SirR (for staphylococcal iron regulator repressor) was identified in *S. epidermidis* by sequence analysis of the DNA region located upstream of the *sitABC* operon (reference 25 and see above). Within the *sitABC* promoter-operator region, a palindromic sequence referred to as the Sir box was found. The Sir box overlaps the transcriptional start site of *sitABC* and shows high homology to the DtxR operator consensus sequence. DNA mobility shift assays confirmed that SirR binds to the Sir box only in the presence of metal ions such as Fe^{2+} and Mn^{2+}. Southern hybridization experiments revealed that there are at least five Sir boxes in the *S. epidermidis* genome and at least three in the *S. aureus* genome, suggesting that SirR controls the expression of multiple target genes. Because an additional Sir box within the *sirR* operator is missing, SirR does not seem to be autoregulated like the *S. epidermidis* Fur-like protein.

LANTIBIOTICS

Another biological property that distinguishes *S. epidermidis* from *S. aureus* is the production of bacteriocins called lantibiotics, which are active against gram-positive bacteria. Their production may play a substantial role in bacterial interference on skin and mucous membranes and thus create an ecological niche for *S. epidermidis*. Lantibiotics are antibiotic peptides that contain the rare thioether amino acid lanthionine and/or methyllanthionine. Type A lantibiotics act by inducing the formation of pores in the cytoplasmic membrane. *S. epidermidis* and other gram-positive bacteria such as *Bacillus subtilis* and lactobacilli produce lantibiotics. Among those produced by *S. epidermidis* are the well-characterized epidermin and Pep5, as well as the newly identified epilancin K7 and epicidin 280. In general, these peptides are gene encoded and posttranslationally modified. The genes involved are organized in biosynthetic gene clusters located on plasmids. Nine genes implicated in epidermin production are encoded on the 54-kb plasmid pTü32. They include epiA as the structural gene for the epidermin precursor peptide and epiB, epiC, epiD, and epiP, which are involved in posttranslational modification. The flavoprotein EpiD catalyzes the oxidative decarboxylation of the C-terminal cysteine residue of the precursor peptide (36). EpiB and EpiC are assumed to catalyze the dehydration of serine and threonine residues and the formation of thioether bonds (35). EpiP is an extracellular serine protease that processes the 52-amino-acid epidermin precursor peptide into the mature 22-amino-acid peptide antibiotic (14). Bacteria producing bacteriocins, including lantibiotics, are always immune to their own bacteriocin. The immunity against epidermin and increased epidermin production is mediated by epiF, epiE, and epiG. These genes encode the proposed components of an ABC transporter system (64). EpiQ is a regulator that activates the transcription of epiABCD and epiFEG.

The biosynthesis of Pep5 is carried out by the pepTIAP-BC gene cluster located on the 20-kb plasmid pED503 (50). Pep5 biosynthesis involves pepA as the structural gene; the genes pepB and pepC encode putative modification enzymes; pepP encodes an intracellular leader peptidase that cleaves off the N-terminal leader sequence; pepT encodes a translocator of the ABC-type transporter, which exports the mature lantibiotic; and pepI is involved in conferring immunity (61). Most recently, the novel lantibiotic epicidin 280, whose amino acid composition exhibits 75% similarity to Pep5, has been described and its biosynthetic gene cluster has been analyzed (20).

The structural gene elkA of epilancin K7 has been cloned and sequenced (83). Preceding elkA, elkT, a gene encoding a proposed translocator protein, has been detected that possibly mediates the export of epilancin K7. The gene elkP is located downstream of elkA and encodes a putative leader peptidase essential for processing.

FUTURE ASPECTS

Our knowledge of the biology and pathogenicity of NAS, particularly S. epidermidis, has significantly increased in recent years. But we are still far away from a sufficient understanding of this versatile microorganism. We still know very little about the biology of S. epidermidis in its normal habitat, the cutaneous-mucocutaneous flora. Increased research in this area is obviously necessary to gain more insight into the complex balance mechanisms between bacterium and human host. This will help us to better understand when and how S. epidermidis can change from saprophyte to pathogen. Improvement in the armamentarium of molecular methods will enable us to analyze not only the genome but also the proteome of S. epidermidis in the future. But this must be complemented by further research on the functional level, including the development of good animal models.

REFERENCES

1. **Baldassarri, L., G. Donnelli, A. Gelosia, M. C. Voglino, A. W. Simpson, and G. D. Christensen.** 1996. Purification and characterization of the staphylococcal slime-associated antigen and its occurrence among Staphylococcus epidermis clinical isolates. Infect. Immun. **64:**3410–3415.

2. **Bowden, M. G., L. Visai, C. M. Longshaw, K. T. Holland, P. Speziale, and M. Höök.** 2002. Is the GehD lipase from Staphylococcus epidermidis a collagen binding adhesin? J. Biol. Chem. **277:**43017–43023.

3. **Chamberlain, N. R., and S. A. Brueggemann.** 1997. Characterisation and expression of fatty acid modifying enzyme produced by Staphylococcus epidermidis. J. Med. Microbiol. **46:**693–697.

4. **Christensen, G. D., L. M. Baddour, and W. A. Simpson.** 1987. Phenotypic variation of Staphylococcus epidermidis slime production in vitro and in vivo. Infect. Immun. **55:**2870–2877.

5. **Cockayne, A., P. J. Hill, N. B. Powell, K. Bishop, C. Sims, and P. Williams.** 1998. Molecular cloning of a 32-kilodalton lipoprotein component of a novel iron-regulated Staphylococcus epidermidis ABC transporter. Infect. Immun. **66:**3767–3774.

6. **Conlon, K. M., H. Humphreys, and J. P. O'Gara.** 2002. icaR encodes a transcriptional repressor involved in environmental regulation of ica operon expression and biofilm formation in Staphylococcus epidermidis. J. Bacteriol. **184:**4400–4408.

7. **Cramton, S. E., C. Gerke, N. F. Schnell, W. W. Nichols, and F. Götz.** 1999. The intercellular adhesion (ica) locus is present in Staphylococcus aureus and is required for biofilm formation. Infect. Immun. **67:**5427–5433.

8. **Cramton, S. E., M. Ulrich, F. Götz, and G. Döring.** 2001. Anaerobic conditions induce expression of polysaccharide intercellular adhesin in Staphylococcus aureus and Staphylococcus epidermidis. Infect. Immun. **69:**4079–4085.

9. **Cucarella, C., C. Solano, J. Valle, B. Amorena, I. Lasa, and J. R. Penades.** 2001. Bap, a Staphylococcus aureus surface protein involved in biofilm formation. J. Bacteriol. **183:**2888–2896.

10. **Davis, S. L., S. Gurusiddappa, K. W. McCrea, S. Perkins, and M. Höök.** 2001. SdrG, a fibrinogen-binding bacterial adhesin of the microbial surface components recognizing adhesive matrix molecules subfamily from Staphylococcus epidermidis, targets the thrombin cleavage site in the Bβ chain. J. Biol. Chem. **276:**27799–27805.

11. **Farrell, A. M., T. J. Foster, and K. T. Holland.** 1993. Molecular analysis and expression of the lipase of Staphylococcus epidermidis. J. Gen. Microbiol. **139:**267–277.

12. **Fey, P. D., J. S. Ulphani, F. Götz, C. Heilmann, D. Mack, and M. E. Rupp.** 1999. Characterization of the relationship between polysaccharide intercellular adhesin and hemagglutination in Staphylococcus epidermidis. J. Infect. Dis. **179:**1561–1564.

13. **Fluckiger, U., C. Wolz, and A. L. Cheung.** 1998. Characterization of a sar homolog of Staphylococcus epidermidis. Infect. Immun. **66:**2871–2878.

14. **Geissler, S., F. Götz, and T. Kupke.** 1996. Serine protease EpiP from Staphylococcus epidermidis catalyzes the processing of the epidermin precursor peptide. J. Bacteriol. **178:**284–288.

15. **Gerke, C., A. Kraft, R. Sussmuth, O. Schweitzer, and F. Götz.** 1998. Characterization of the N-acetylglucosaminyltransferase activity involved in the biosynthesis of the Staphylococcus epidermidis polysaccharide intercellular adhesin. J. Biol. Chem. **273:**18586–18593.

16. **Gray, E. D., G. Peters, M. Verstegen, and W. E. Regelmann.** 1984. Effect of extracellular slime substance from Staphylococcus epidermidis on the human cellular immune response. Lancet **i:**365–367.

17. **Gross, M., S. E. Cramton, F. Götz, and A. Peschel.** 2001. Key role of teichoic acid net charge in Staphylococcus aureus colonization of artificial surfaces. Infect. Immun. **69:**3423–3426.

18. **Hartford, O., L. O'Brien, K. Schofield, J. Wells, and T. J. Foster.** 2001. The Fbe (SdrG) protein of Staphylococcus

epidermidis HB promotes bacterial adherence to fibrinogen. *Microbiology* **147:**2545–2552.

19. **Heidrich, C., K. Hantke, G. Bierbaum, and H. G. Sahl.** 1996. Identification and analysis of a gene encoding a Fur-like protein of *Staphylococcus epidermidis. FEMS Microbiol. Lett.* **140:**253–259.

20. **Heidrich, C., U. Pag, M. Josten, J. Metzger, R. W. Jack, G. Bierbaum, G. Jung, and H. G. Sahl.** 1998. Isolation, characterization, and heterologous expression of the novel lantibiotic epicidin 280 and analysis of its biosynthetic gene cluster. *Appl. Environ. Microbiol.* **64:**3140–3146.

21. **Heilmann, C., M. Hussain, G. Peters, and F. Götz.** 1997. Evidence for autolysin-mediated primary attachment of *Staphylococcus epidermidis* to a polystyrene surface. *Mol. Microbiol.* **24:**1013–1024.

22. **Heilmann, C., O. Schweitzer, C. Gerke, N. Vanittanakom, D. Mack, and F. Götz.** 1996. Molecular basis of intercellular adhesion in the biofilm-forming *Staphylococcus epidermidis. Mol. Microbiol.* **20:**1083–1091.

23. **Heilmann, C., G. Thumm, G. S. Chhatwal, J. Hartleib, A. Uekötter, and G. Peters.** 2003. Identification and characterization of a novel autolysin (Aae) with adhesive properties from *Staphylococcus epidermidis. Microbiology* **149:** 2769–2778.

24. **Hell, W., H. G. Meyer, and S. G. Gatermann.** 1998. Cloning of *aas*, a gene encoding a *Staphylococcus saprophyticus* surface protein with adhesive and autolytic properties. *Mol. Microbiol.* **29:**871–881.

25. **Hill, P. J., A. Cockayne, P. Landers, J. A. Morrissey, C. M. Sims, and P. Williams.** 1998. SirR, a novel iron-dependent repressor in *Staphylococcus epidermidis. Infect. Immun.* **66:** 4123–4129.

26. **Horsburgh, M. J., M. O. Clements, H. Crossley, E. Ingham, and S. J. Foster.** 2001. PerR controls oxidative stress resistance and iron storage proteins and is required for virulence in *Staphylococcus aureus. Infect. Immun.* **69:**3744–3754.

27. **Hussain, M., C. Heilmann, G. Peters, and M. Herrmann.** 2001. Teichoic acid enhances adhesion of *Staphylococcus epidermidis* to immobilized fibronectin. *Microb. Pathog.* **31:**261–270.

28. **Hussain, M., M. Herrmann, C. von Eiff, F. Perdreau-Remington, and G. Peters.** 1997. A 140-kilodalton extracellular protein is essential for the accumulation of *Staphylococcus epidermidis* strains on surfaces. *Infect. Immun.* **65:** 519–524.

29. **Hussain, M., M. H. Wilcox, and P. J. White.** 1993. The slime of coagulase-negative staphylococci: biochemistry and relation to adherence. *FEMS Microbiol. Rev.* **10:**191–207.

30. **Johnson, G.** Personal communication.

31. **Johnson, G. M., D. A. Lee, W. E. Regelmann, E. D. Gray, G. Peters, and P. G. Quie.** 1986. Interference with granulocyte function by *Staphylococcus epidermidis* slime. *Infect. Immun.* **54:**13–20.

32. **Kies, S., M. Otto, C. Vuong, and F. Götz.** 2001. Identification of the *sigB* operon in *Staphylococcus epidermidis*: construction and characterization of a *sigB* deletion mutant. *Infect. Immun.* **69:**7933–7936.

33. **Kies, S., C. Vuong, M. Hille, A. Peschel, C. Meyer, F. Götz, and M. Otto.** 2003. Control of antimicrobial peptide synthesis by the *agr* quorum sensing system in *Staphylococcus epidermidis*: activity of the lantibiotic epidermin is regulated at the level of precursor peptide processing. *Peptides* **24:**329–338.

34. **Knobloch, J. K., K. Bartscht, A. Sabottke, H. Rohde, H. H. Feucht, and D. Mack.** 2001. Biofilm formation by

Staphylococcus epidermidis depends on functional RsbU, an activator of the *sigB* operon: differential activation mechanisms due to ethanol and salt stress. *J. Bacteriol.* **183:** 2624–2633.

35. **Kupke, T., and F. Götz.** 1996. Expression, purification, and characterization of EpiC, an enzyme involved in the biosynthesis of the lantibiotic epidermin, and sequence analysis of *Staphylococcus epidermidis epiC* mutants. *J. Bacteriol.* **178:**1335–1340.

36. **Kupke, T., C. Kempter, G. Jung, and F. Götz.** 1995. Oxidative decarboxylation of peptides catalyzed by flavoprotein EpiD. Determination of substrate specificity using peptide libraries and neutral loss mass spectrometry. *J. Biol. Chem.* **270:**11282–11289.

37. **Liles, W. C., A. R. Thomsen, D. S. O'Mahony, and S. J. Klebanoff.** 2001. Stimulation of human neutrophils and monocytes by staphylococcal phenol-soluble modulin. *J. Leukoc. Biol.* **70:**96–102.

38. **Lindsay, J. A., and T. V. Riley.** 1994. Staphylococcal iron requirements, siderophore production, and iron-regulated protein expression. *Infect. Immun.* **62:**2309–2314.

39. **Longshaw, C. M., A. M. Farrell, J. D. Wright, and K. T. Holland.** 2000. Identification of a second lipase gene, *gehD*, in *Staphylococcus epidermidis*: comparison of sequence with those of other staphylococcal lipases. *Microbiology* **146:** 1419–1427.

40. **Mack, D.** Personal communication.

41. **Mack, D., W. Fischer, A. Krokotsch, K. Leopold, R. Hartmann, H. Egge, and R. Laufs.** 1996. The intercellular adhesin involved in biofilm accumulation of *Staphylococcus epidermidis* is a linear β-1,6-linked glucosaminoglycan: purification and structural analysis. *J. Bacteriol.* **178:**175–183.

42. **Maira-Litran, T., A. Kropec, C. Abeygunawardana, J. Joyce, G. Mark, D. A. Goldmann, and G. B. Pier.** 2002. Immunochemical properties of the staphylococcal poly-*N*-acetylglucosamine surface polysaccharide. *Infect. Immun.* **70:**4433–4440.

43. **Marin, M. E., M. C. de la Rosa, and I. Cornejo.** 1992. Enterotoxigenicity of *Staphylococcus* strains isolated from Spanish dry-cured hams. *Appl. Environ. Microbiol.* **58:** 1067–1069.

44. **Mattsson, E., J. Rollof, J. Verhoef, H. Van Dijk, and A. Fleer.** 1994. Serum-induced potentiation of tumor necrosis factor alpha production by human monocytes in response to staphylococcal peptidoglycan: involvement of different serum factors. *Infect. Immun.* **62:**3837–3843.

45. **Mazmanian, S. K., E. P. Skaar, A. H. Gaspar, M. Humayun, P. Gornicki, J. Jelenska, A. Joachmiak, D. M. Missiakas, and O. Schneewind.** 2003. Passage of heme-iron across the envelope of *Staphylococcus aureus. Science* **299:**906–909.

46. **McCrea, K. W., O. Hartford, S. Davis, D. N. Eidhin, G. Lina, P. Speziale, T. J. Foster, and M. Höök.** 2000. The serine-aspartate repeat (Sdr) protein family in *Staphylococcus epidermidis. Microbiology* **146:**1535–1546.

47. **McKenney, D., J. Hübner, E. Muller, Y. Wang, D. A. Goldmann, and G. B. Pier.** 1998. The *ica* locus of *Staphylococcus epidermidis* encodes production of the capsular polysaccharide/adhesin. *Infect. Immun.* **66:**4711–4720.

48. **McKevitt, A. I., G. L. Bjornson, C. A. Mauracher, and D. W. Scheifele.** 1990. Amino acid sequence of a deltalike toxin from *Staphylococcus epidermidis. Infect. Immun.* **58:** 1473–1475.

49. **Mehlin, C., C. M. Headley, and S. J. Klebanoff.** 1999. An inflammatory polypeptide complex from *Staphylococcus*

epidermidis: isolation and characterization. *J. Exp. Med.* **189:**907–918.

50. Meyer, C., G. Bierbaum, C. Heidrich, M. Reis, J. Suling, M. I. Iglesias-Wind, C. Kempter, E. Molitor, and H. G. Sahl. 1995. Nucleotide sequence of the lantibiotic Pep5 biosynthetic gene cluster and functional analysis of PepP and PepC. Evidence for a role of PepC in thioether formation. *Eur. J. Biochem.* **232:**478–489.

51. Modun, B., R. W. Evans, C. L. Joannou, and P. Williams. 1998. Receptor-mediated recognition and uptake of iron from human transferrin by *Staphylococcus aureus* and *Staphylococcus epidermidis*. *Infect. Immun.* **66:**3591–3596.

52. Modun, B., J. Morrissey, and P. Williams. 2000. The staphylococcal transferrin receptor: a glycolytic enzyme with novel functions. *Trends Microbiol.* **8:**231–237.

53. Modun, B., and P. Williams. 1999. The staphylococcal transferrin-binding protein is a cell wall glyceraldehyde-3-phosphate dehydrogenase. *Infect. Immun.* **67:**1086–1092.

54. Modun, B. J., A. Cockayne, R. Finch, and P. Williams. 1998. The *Staphylococcus aureus* and *Staphylococcus epidermidis* transferrin-binding proteins are expressed in vivo during infection. *Microbiology* **144:**1005–1012.

55. Morrissey, J. A., A. Cockayne, K. Brummell, and P. Williams. 2004. The staphylococcal ferritins are differentially regulated in response to iron and manganese and via PerR and Fur. *Infect. Immun.* **72:**972–979.

56. Muller, E., J. Hubner, N. Gutierrez, S. Takeda, D. A. Goldmann, and G. B. Pier. 1993. Isolation and characterization of transposon mutants of *Staphylococcus epidermidis* deficient in capsular polysaccharide/adhesin and slime. *Infect. Immun.* **61:**551–558.

57. Nilsson, M., L. Frykberg, J. I. Flock, L. Pei, M. Lindberg, and B. Guss. 1998. A fibrinogen-binding protein of *Staphylococcus epidermidis*. *Infect. Immun.* **66:**2666–2673.

58. Ohara-Nemoto, Y., Y. Ikeda, M. Kobayashi, M. Sasaki, S. Tajika, and S. Kimura. 2002. Characterization and molecular cloning of a glutamyl endopeptidase from *Staphylococcus epidermidis*. *Microb. Pathog.* **33:**33–41.

59. Otto, M. 2001. *Staphylococcus aureus* and *Staphylococcus epidermidis* peptide pheromones produced by the accessory gene regulator *agr* system. *Peptides* **22:**1603–1608.

60. Otto, M., R. Süssmuth, G. Jung, and F. Götz. 1998. Structure of the pheromone peptide of the *Staphylococcus epidermidis agr* system. *FEBS Lett.* **424:**89–94.

61. Pag, U., C. Heidrich, G. Bierbaum, and H. G. Sahl. 1999. Molecular analysis of expression of the lantibiotic Pep5 immunity phenotype. *Appl. Environ. Microbiol.* **65:**591–598.

62. Pei, L., and J. I. Flock. 2001. Lack of *fbe* reduces *Staphylococcus epidermidis* adherence to Fg-coated surfaces. *Microb. Pathog.* **31:**185–193.

63. Pei, L., M. Palma, M. Nilsson, B. Guss, and J. I. Flock. 1999. Functional studies of a fibrinogen binding protein from *Staphylococcus epidermidis*. *Infect. Immun.* **67:**4525–4530.

64. Peschel, A., and F. Götz. 1996. Analysis of the *Staphylococcus epidermidis* genes epiF, -E, and -G involved in epidermin immunity. *J. Bacteriol.* **178:**531–536.

65. Peters, G., R. Locci, and G. Pulverer. 1982. Adherence and growth of coagulase-negative staphylococci on surfaces of intravenous catheters. *J. Infect. Dis.* **146:**479–482.

66. Peters, G., R. Locci, and G. Pulverer. 1981. Microbial colonization of prosthetic devices. II. Scanning electron microscopy of naturally infected intravenous catheters. *Zentralbl. Bakteriol. Mikrobiol. Hyg. B* **173:**293–299.

67. Ponnuraj, K., M. G. Bowden, S. Davis, S. Gurusiddappa, D. Moore, D. Choe, Y. Xu, and M. Höök. 2003. A "dock, lock, and latch" structural model for a staphylococcal adhesin binding to fibrinogen. *Cell* **115:**217–228.

68. Rachid, S., K. Ohlsen, W. Witte, J. Hacker, and W. Ziebuhr. 2000. Effect of subinhibitory antibiotic concentrations on polysaccharide intercellular adhesin expression in biofilm-forming *Staphylococcus epidermidis*. *Antimicrob. Agents Chemother.* **44:**3357–3363.

69. Rennermalm, A., M. Nilsson, and J.-I. Flock. 2004. The fibrinogen binding protein of *Staphylococcus epidermidis* is a target for opsonic antibodies. *Infect. Immun.* **72:**3081–3083.

70. Rohde, H., C. Burdelski, K. Bartscht, M. Hussain, F. Buck, M. A. Horstkotte, J. K.-M. Knobloch, C. Heilmann, M. Herrmann, and D. Mack. 2005. Induction of *Staphylococcus epidermidis* biofilm formation via proteolytic processing of the accumulation associated protein by staphylococcal and host proteases. *Mol. Microbiol.* **55:**1883–1895.

71. Rohde, H., M. Kalitzky, N. Kröger, S. Scherpe, M. A. Horstkotte, J. K.-M. Knobloch, A. R. Zander, and D. Mack. 2004. Detection of virulence-associated genes not useful for discriminating between invasive and commensal *Staphylococcus epidermidis* strains on a bone marrow transplant unit. *J. Clin. Microbiol.* **42:**5614–5619.

72. Rupp, M. E., P. D. Fey, C. Heilmann, and F. Götz. 2001. Characterization of the importance of *Staphylococcus epidermidis* autolysin and polysaccharide intercellular adhesin in the pathogenesis of intravascular catheter-associated infection in a rat model. *J. Infect. Dis.* **183:**1038–1042.

73. Rupp, M. E., J. S. Ulphani, P. D. Fey, K. Bartscht, and D. Mack. 1999. Characterization of the importance of polysaccharide intercellular adhesin/hemagglutinin of *Staphylococcus epidermidis* in the pathogenesis of biomaterial-based infection in a mouse foreign body infection model. *Infect. Immun.* **67:**2627–2632.

74. Rupp, M. E., J. S. Ulphani, P. D. Fey, and D. Mack. 1999. Characterization of *Staphylococcus epidermidis* polysaccharide intercellular adhesin/hemagglutinin in the pathogenesis of intravascular catheter-associated infection in a rat model. *Infect. Immun.* **67:**2656–2659.

75. Saenz, H. L., V. Augsburger, C. Vuong, R. W. Jack, F. Götz, and M. Otto. 2000. Inducible expression and cellular location of AgrB, a protein involved in the maturation of the staphylococcal quorum-sensing pheromone. *Arch. Microbiol.* **174:**452–455.

76. Schumacher-Perdreau, F., C. Heilmann, G. Peters, F. Götz, and G. Pulverer. 1994. Comparative analysis of a biofilm-forming *Staphylococcus epidermidis* strain and its adhesion-positive, accumulation-negative mutant M7. *FEMS Microbiol. Lett.* **117:**71–78.

77. Shiau, A. L., and C. L. Wu. 1998. The inhibitory effect of *Staphylococcus epidermidis* slime on the phagocytosis of murine peritoneal macrophages is interferon-independent. *Microbiol. Immunol.* **42:**33–40.

78. Simons, J. W., M. D. van Kampen, S. Riel, F. Götz, M. R. Egmond, and H. M. Verheij. 1998. Cloning, purification and characterisation of the lipase from *Staphylococcus epidermidis*—comparison of the substrate selectivity with those of other microbial lipases. *Eur. J. Biochem.* **253:**675–683.

79. Skaar, E. P., A. H. Gaspar, and O. Schneewind. 2004. IsdG and IsdI, heme-degrading enzymes in the cytoplasm of *Staphylococcus aureus*. *J. Biol. Chem.* **279:**436–443.

80. Sloot, N., M. Thomas, R. Marre, and S. Gatermann. 1992. Purification and characterisation of elastase from *Staphylococcus epidermidis*. *J. Med. Microbiol.* **37:**201–205.

81. **Tegmark, K., E. Morfeldt, and S. Arvidson.** 1998. Regulation of *agr*-dependent virulence genes in *Staphylococcus aureus* by RNAIII from coagulase-negative staphylococci. *J. Bacteriol.* **180:**3181–3186.

82. **Teufel, P., and F. Götz.** 1993. Characterization of an extracellular metalloprotease with elastase activity from *Staphylococcus epidermidis. J. Bacteriol.* **175:**4218–4224.

83. **van de Kamp, M., H. W. van den Hooven, R. N. Konings, G. Bierbaum, H. G. Sahl, O. P. Kuipers, R. J. Siezen, W. M. de Vos, C. W. Hilbers, and F. J. van de Ven.** 1995. Elucidation of the primary structure of the lantibiotic epilancin K7 from *Staphylococcus epidermidis* K7. Cloning and characterisation of the epilancin-K7-encoding gene and NMR analysis of mature epilancin K7. *Eur. J. Biochem.* **230:**587–600.

84. **Van Wamel, W. J., G. van Rossum, J. Verhoef, C. M. Vandenbroucke-Grauls, and A. C. Fluit.** 1998. Cloning and characterization of an accessory gene regulator (*agr*)-like locus from *Staphylococcus epidermidis. FEMS Microbiol. Lett.* **163:**1–9.

85. **Vandecasteele, S. J., W. E. Peetermans, R. R. Merckx, B. J. A. Rijnders, and J. Van Eldere.** 2003. Reliability of the *ica, aap,* and *atlE* genes in the discrimination between invasive, colonizing, and contaminant Staphylococcus epidermidis isolates in the diagnosis of catheter-related infections. *Clin. Microbiol. Infect.* **9:**114–119.

86. **Veenstra, G. J., F. F. Cremers, H. van Dijk, and A. Fleer.** 1996. Ultrastructural organization and regulation of a biomaterial adhesin of *Staphylococcus epidermidis. J. Bacteriol.* **178:**537–541.

87. **von Eiff, C., C. Heilmann, and G. Peters.** 1998. *Staphylococcus epidermidis*: why is it so successful? *Clin. Microbiol. Infect.* **4:**297–300.

88. **Vuong, C., M. Dürr, A. B. Carmody, A. Peschel, S. J. Klebanoff, and M. Otto.** 2004. Regulated expression of pathogen-associated molecular pattern molecules in *Staphylococcus epidermidis*: quorum-sensing determines pro-inflammatory capacity and production of phenol-soluble modulins. *Cell. Microbiol.* **6:**753–759.

89. **Vuong, C., C. Gerke, G. A. Somerville, E. R. Fischer, and M. Otto.** 2003. Quorum-sensing control of biofilm factors in *Staphylococcus epidermidis. J. Infect. Dis.* **188:**706–718.

90. **Vuong, C., F. Götz, and M. Otto.** 2000. Construction and characterization of an *agr* deletion mutant of *Staphylococcus epidermidis. Infect. Immun.* **68:**1048–1053.

91. **Vuong, C., J. M. Voyich, E. R. Fischer, K. R. Braughton, A. R. Whitney, F. R. DeLeo, and M. Otto.** 2004. Polysaccharide intercellular adhesin (PIA) protects *Staphylococcus epidermidis* against major components of the human innate immune system. *Cell. Microbiol.* **6:**269–275.

92. **Williams, R. J., B. Henderson, L. J. Sharp, and S. P. Nair.** 2002. Identification of a fibronectin-binding protein from *Staphylococcus epidermidis. Infect. Immun.* **70:**6805–6810.

93. **Ziebuhr, W., C. Heilmann, F. Götz, P. Meyer, K. Wilms, E. Straube, and J. Hacker.** 1997. Detection of the intercellular adhesion gene cluster (*ica*) and phase variation in *Staphylococcus epidermidis* blood culture strains and mucosal isolates. *Infect. Immun.* **65:**890–896.

94. **Ziebuhr, W., V. Krimmer, S. Rachid, I. Lossner, F. Götz, and J. Hacker.** 1999. A novel mechanism of phase variation of virulence in *Staphylococcus epidermidis*: evidence for control of the polysaccharide intercellular adhesin synthesis by alternating insertion and excision of the insertion sequence element IS256. *Mol. Microbiol.* **32:**345–356.

Biology and Pathogenicity of Staphylococci Other than *Staphylococcus aureus* and *Staphylococcus epidermidis*

ANNE TRISTAN, GERARD LINA, JEROME ETIENNE, AND
FRANÇOIS VANDENESCH

47

Staphylococci other than *Staphylococcus aureus* and *Staphylococcus epidermidis* correspond to a vast group of >40 species and subspecies, all of which are frequently (though erroneously, since some of them are phenotypically coagulase positive) referred to as coagulase-negative staphylococci. These bacteria are commensals on the skins of animals and humans, and some of them have become pathogens by virtue of medical progress both in human and veterinary medicine. Until the mid-1970s, identification of non-*aureus* species of staphylococci (NAS) was technically difficult and their taxonomy was ill defined, but since 1975 advances in staphylococcal systematics have enabled the description of a number of staphylococcal species. These include species whose pathogenic potential had been clearly demonstrated in humans (e.g., *S. haemolyticus*, *S. saprophyticus*, *S. lugdunensis*, *S. schleiferi*, *S. warneri*, or *S. caprae*) or in animals (e.g., *S. intermedius*, *S. hyicus*, or *S. simulans*) (61). Many new species and subspecies of NAS have now been described; their biology and pathogenic potentials are only beginning to be defined. However, it is becoming more and more apparent that various staphylococcal species produce proteins that were until recently thought to belong exclusively to *S. aureus*. This chapter will not extensively discuss the biology of all known species of staphylococci but will focus on species of NAS selected for their pathogenic potential and those for which recent discoveries have had a significant outcome on the general knowledge of staphylococcal biology. A dendrogram of staphylococcal species based on overall DNA similarity is presented in chapter 31.

STAPHYLOCOCCUS LUGDUNENSIS

S. lugdunensis was originally described in 1988. Isolates of this species can be misidentified as *S. aureus* on the basis of clumping factor production (58 to 79% positive) and from the expression (although weak) of thermo-stable DNase, two characteristics that are typical of and often considered diagnostic of *S. aureus*.

Diseases

S. lugdunensis is a human commensal more pathogenic than most other NAS, causing primary infections of the human skin and postoperative wound infections, with a predominance of the sites below the waist (109). Deep-seated infections often associated with bacteremia occur, including vascular line infection, peritonitis, prosthetic infections, septic arthritis, and infective endocarditis (51, 108, 109). Indeed, though *S. lugdunensis* can be the cause of severe and destructive infective endocarditis, the outcome with patients who have undergone valve replacement is usually more favorable (28). This suggests that any isolate of *S. lugdunensis* should be assumed to be a pathogen. With animals, *S. lugdunensis* has been associated with mastitis in goats.

Genomic Diversity and Natural Habitat

An unusual feature of *S. lugdunensis* is its high genotypic uniformity, as revealed by genomic markers such as plasmid profiling restriction endonuclease analysis of chromosomal DNA (109) or by pulsed-field gel electrophoresis (PFGE). It has been suggested that *S. lugdunensis* is a skin commensal. However, its predilection for sites of infection below the waist (108) and its high frequency of isolation from the inguinal area suggest that it is a member of the inguinal cutaneous flora (113).

Antibiotic Susceptibility

Isolates of *S. lugdunensis* are usually highly susceptible to antistaphylococcal antibiotics: only about 25% of strains produce β-lactamase (108). Resistance to methicillin has been detected, although infrequently (102). Resistance to cadmium has been detected in almost 60% of *S. lugdunensis* isolates and found to be carried by a 3.16-kb plasmid (prototype plasmid pLUG10) that is a member of the pT181 group of class I staphylococcal plasmids (23). However, the cadmium resistance genes (*cadB* and *cadX*) present on pLUG10 are usually found on plasmids of the pC194 group of class I plasmids, such as pOX6 of *S. aureus*. This suggests that cadmium resistance in staphylococci may be conferred by homologous gene sets carried by class I plasmids with different origins.

Determinants of Pathogenicity

The synthesis of an extracellular slime substance or glycocalyx, a factor interfering with phagocyte function and in-

Gram-Positive Pathogens, 2nd edition, edited by Vincent A. Fischetti et al.
© 2006 ASM Press, Washington, D.C.

volved in the pathogenesis of foreign body infections, has been demonstrated. The glycocalyx activates monocyte prostaglandin E_2 production, which in turn contributes to the inhibition of T-cell proliferation. This activation of monocytes also results in modulation of tumor necrosis factor alpha and nitric oxide production, two significant antimicrobial activities of macrophages (98). Production of other enzymes including esterase, fatty acid-modifying enzyme (FAME), and lipase (67, 73), which may act as invasion factors, has been reported (see Table 4). S. lugdunensis binds to collagen, fibronectin, vitronectin, laminin, fibrinogen, thrombospondin, plasminogen, and human immunoglobulin G (IgG) immobilized on latex beads (see Table 2) (83, 84). However, S. lugdunensis showed only moderate attachment to fibronectin-coated polymethylmethacrylate coverslips, whereas none of 10 strains of S. lugdunensis showed any significant attachment to fibrinogen (P. François and P. Vaudaux, personal communication). S. lugdunensis strains possess a von Willebrand factor-binding protein with features typical of staphylococcal cell surface proteins. It is speculated that this receptor may play a critical role in bacterial attachment to minor vascular lesions and thereby in the initial step of infective endocarditis (78).

Production of hemolytic activity synergistic with that of beta-hemolysin, protease, lipase, and esterase by S. lugdunensis has been variably observed (Table 1) (42, 68, 112). The synergistic hemolysin has been designated SLUSH (for S. lugdunensis synergistic hemolysin); it consists of three very similar 43-residue peptides highly related to the three gonococcal growth inhibitor (GGI) peptides of S. haemolyticus that also possesses synergistic haemolytic activity (29, 120). SLUSH may also exhibit some antibacterial activity in that crude supernatants from hemolytic but not nonhemolytic S. lugdunensis strains show antibacterial activity with several staphylococcal species (29). Moreover, SLUSH has no homologies with the S. aureus delta-hemolysin; unlike the delta-hemolysin gene (hld), it is not encoded within the accessory gene regulator (agr) locus that is present in S. lugdunensis (Table 1) (111). Several algorithms predict a predominantly alpha-helical structure for all three SLUSH peptides (29). SLUSH, GGI, the delta-hemolysin, and enterocin L50 from Enterococcus faecium (a broad-spectrum bacteriocin consisting of two very similar 43- and 44-residue peptides) have many features in common in that they have hemolytic and/or antimicrobial activities, are ribosomally synthesized, are secreted without a signal peptide, and do not seem to be cotranscribed with an immunity protein gene (25). A study of the distribution of hld and slush genes among 12 species of hemolytic staphylococci revealed that 5 species contained hld, apparently within the agr region, 4 contained slush, and 1 (S. caprae) contained both (Table 1) (30).

Cloning and sequencing of the entire agr locus of S. lugdunensis (agr-sl) and comparison of this sequence with that of the agr locus of S. aureus (agr-sa) revealed the presence of the four reading frames that are present in the P2 operon of agr-sa (111). The purified heptapeptide (DICNAYF) encoded by agrD-sl and released extracellularly appears to inhibit RNAIII expression of S. aureus (50) but does not increase

TABLE 1 Synergistic hemolysis and detection of hld, slush, and agr genes in Staphylococcus species

Staphylococcus species	Synergistic hemolysin test		No. of strains with positive hybridization/no. of strains tested (30) with probe			agr-RNAIII sequence available (presence of hld gene) (103, 111)
	No. (%) of strains with positive test in Donvito et al.'s study (30)	% of strains with positive test in Hébert et al.'s studies (42, 43)	hld	slush	agr	
S. aureus	9/11 (82)	33	1/1	0/1	4/4	+ (hld)
S. auricularis	0/10 (0)	NT[a]	0/3	0/3	4/4	
S. capitis subsp. capitis	11/11 (100)	93	1/4	0/4	4/4	
S. capitis subsp. ureolyticus	9/11 (82)		1/4	0/4	4/4	
S. caprae	10/11 (91)	NT	3/4	3/4	4/4	
S. epidermidis	8/11 (73)	73	4/4	0/4	4/4	+(hld)
S. haemolyticus	11/12 (92)	99	0/4	4/4	4/4	
S. hominis	8/11 (70)	53	0/4	0/4	4/4	
S. warneri	7/11 (64)	63	4/4	0/4	4/4	+ (hld × 2)
S. pasteuri	10/11 (91)	NT	4/4	0/4	4/4	
S. cohnii subsp. cohnii	9/9 (100)	100	0/4	4/4	4/4	
S. cohnii subsp. urealyticum	5/9 (56)		0/4	4/4	4/4	
S. saprophyticus	0/11 (0)	NT	0/4	0/4	4/4	
S. xylosus	4/11 (36)	54	0/4	1/4	4/4	
S. simulans	9/11 (82)	96	0/4	0/4	4/4	+(hld)
S. lugdunensis	51/53 (96)	95	0/14	14/14	4/4	+(no hld)
S. schleiferi subsp. coagulans	0/3 (0)	0	0/3	0/3	4/4	
S. schleiferi subsp. schleiferi	0/11 (0)		0/4	0/4	4/4	

[a]NT, not tested.

RNAIII expression of *S. lugdunensis* (R. P. Novick, personal communication). Besides the possible role of the *S. lugdunensis* heptapeptide in the competition with *S. aureus* for the same ecological niche, the function of this peptide in *agr-sl* functioning is not known. Similarly, the whole function of the *agr* system of *S. lugdunensis* has not yet been identified. RNAIII from *S. lugdunensis* (RNAIII-*sl*) has an estimated length of 573 nucleotides (nt); its sequence differs considerably from that of *S. aureus* RNAIII (RNAIII-*sa*) having no delta-hemolysin gene and a low overall homology despite some conservation at the 5' and 3' ends (111). These differences between RNAIII-*sa* and RNAIII from other staphylococcal species have been used as a model to study the mode of regulation of *agr*-target genes in *S. aureus* (17, 103). This is presented in detail in chapter 41 on regulation of gene expression.

STAPHYLOCOCCUS SAPROPHYTICUS

S. saprophyticus is known as a frequent cause of acute urinary tract infection among female outpatients. A new taxon designated *S. saprophyticus* subsp. *bovis* has been recently established on the basis of a study of strains isolated from the anterior nares of cows (41). Only *S. saprophyticus* subsp. *saprophyticus* has, however, been isolated from infection.

Diseases

After *Escherichia coli*, *S. saprophyticus* is the second-most-common cause of cystitis in young, healthy, sexually active women (36). *S. saprophyticus* primarily causes acute urinary tract infections including cystitis, urethritis, and pyelonephritis, occasionally complicated by infected calculus or bacteremia. It may be implicated in prostatitis. This organism has also been implicated as a cause of endophthalmitis after cataract surgery, native valve endocarditis, septicemia, and an exceptional case of nosocomial pneumonia (44).

Genomic Diversity and Natural Habitat

S. saprophyticus subsp. *saprophyticus* is isolated occasionally from the skin of humans and other mammals and as a contaminant of food of animal origin (92). Populations on human skin are usually small and transient (92). In a study including 276 women from an outpatient gynecology practice, the prevalence of colonization of the female urogenital tracts by *S. saprophyticus* was estimated to be 6.9%; the rectum, urethra, urine, and cervix were the most frequent sites of colonization, in decreasing order (36). Plasmid analysis of a series of 150 *S. saprophyticus* strains isolated from urinary tract infections in women showed that 82% of the strains harbored plasmids, some of them with complex plasmid profiles, thus allowing differentiation among strains.

Antibiotic Susceptibility

S. saprophyticus is usually susceptible to methicillin, quinolone, aminoglycoside, lincosamide, and glycopeptide. Susceptibility to benzylpenicillin, erythromycin, tetracycline, and chloramphenicol is variable (92). Some rare strains carry the *mecA* gene and possess a *femA* homolog (117). Cystitis due to vancomycin-intermediate *S. saprophyticus* has been reported (121).

Determinants of Pathogenicity

Several potential virulence factors of *S. saprophyticus* subsp. *saprophyticus* have been identified and/or characterized in the recent years. This species appears to have a higher capacity to adhere specifically to uroepithelial cells than many

other staphylococcal species. *S. saprophyticus* expresses two surface proteins. (i) One is a 95-kDa fibrillar protein designated Ssp (for *S. saprophyticus* surface-associated protein), which is involved in the interactions with uroepithelial cells (36). (ii) The other is a 160-kDa hemagglutinin which mediates binding of *S. saprophyticus* to uroepithelial cells and sheep erythrocytes; its receptor on eukaryotic cells has been identified as a protein in contrast to many other hemaglutinins. The 160-kDa hemagglutinin also acts as a fibronectin receptor in the human host. These binding properties correspond to a surface protein (designated Aas) with adhesive and autolytic properties; Aas is related but not similar to the autolysin-adhesin of *S. aureus* (Atl) and *S. epidermidis* (AtlE) (45).

Additionally, this species exhibits strong cell surface hydrophobicity that may promote adherence of the organism to uroepithelial cells (36). Treatment of bacteria with proteases reduces hydrophobicity, indicating that the structure mediating hydrophobicity is a protein. It has not been stated whether Aas was involved in this surface property.

S. saprophyticus subsp. *saprophyticus* produces a urease (see Table 4), which is an important virulence factor in that it contributes to cytopathogenicity and tissue invasiveness by inducing severe damage to the bladder tissues (35, 36). Production of other enzymes including elastase, FAME, and lipase (see Table 4) (67, 73) that may act as invasion factors has been reported. Enterotoxin production by certain strains isolated from healthy goats has been detected by enzyme-linked immunosorbent assay (ELISA) (see Table 3) (106). This has not been confirmed by emetic or mitogenic assay.

STAPHYLOCOCCUS HAEMOLYTICUS

S. haemolyticus is regarded as an emerging nosocomial pathogen with a tendency to develop antibiotic resistance.

Diseases

S. haemolyticus has been implicated in human infection including native valve endocarditis, septicemia, peritonitis, and urinary tract, wound, bone, and joint infections. Multiresistant and endemic *S. haemolyticus* strains have been frequently associated with nosocomially acquired infections including bacteremia, especially in neonatal intensive care units (55, 87).

Genomic Diversity and Natural Habitat

A part of the human normal skin flora, *S. haemolyticus* is also found in prosimians, monkeys, and domestic animals. *S. haemolyticus* colonizes drier regions of the body more successfully than most species, although its largest population is found in the axillae, perineum, and inguinal area (59).

Antibiotic Susceptibility

Susceptibility to benzylpenicillin, methicillin, tetracycline, and erythromycin is variable. Multiple resistances to antibiotics are common (55). However, some methicillin-resistant *S. haemolyticus* strains express a low-level methicillin resistance in spite of their carriage of *mecA* (53). Decreased susceptibility to glycopeptides (which involves teicoplanin more than vancomycin) was reported very early with *S. haemolyticus* (100). Cells from both clinical and laboratory-derived teicoplanin-resistant strains exhibited abnormally roughened, irregular outlines when observed by transmission electron microscopy (38). Detailed structural analysis revealed that the peptidoglycan of these strains is highly cross-linked, containing additional serine in place of glycine in their cross bridges. This alteration might interfere with the binding of glycopeptide dimers and therefore with the cooperative binding of the antibiotic to its target (19).

Determinants of Pathogenicity

S. haemolyticus binds to vitronectin, laminin, collagen, and fibronectin immobilized on latex beads (Table 2) (77, 84). S. haemolyticus produces GGI, a synergistic hemolysin which also exhibits antigonococcal activity by causing cytoplasmic leakage and eventually death of gonococcal cells (120). GGI is made of three highly homologous 44-amino-acid peptides, which are 39 to 48% homologous to the three 43-amino-acid SLUSH peptides (see the discussion above of S. lugdunensis) (29, 120). Enterotoxin production by certain strains isolated from sheep milk (15) and from healthy goats (106) has been detected by ELISA (Table 3). This has not been confirmed by mitogenic or emetic assay. S. haemolyticus produces a 45-kDa lipase, which is synthesized as a 80-kDa preproenzyme form. It has high homologies (49 to 67%) with other staphylococcal lipases (from S. aureus, S. epidermidis, and S. hyicus). Like S. aureus lipase, it shows relatively high hydrolytic activity for short-chain-length triglycerides such as tributyrin and tripropionin (79).

STAPHYLOCOCCUS SCHLEIFERI

The species (together with S. lugdunensis) was described in 1988. S. schleiferi has now been divided into two subspecies: S. schleiferi subsp. schleiferi (initially isolated from humans) and S. schleiferi subsp. coagulans (initially isolated from dog) (48). Very rare isolates of both subspecies have been reported to induce clotting of rabbit plasma, thus appearing coagulase positive (although they do not possess true coagulase), a characteristic that can lead to their mis-identification as S. aureus (110).

Diseases

S. schleiferi subsp. schleiferi has been incriminated as an opportunistic nosocomial pathogen that infrequently is the cause of surgical site infections, bacteremia, brain empyema, urinary tract infections, and infections of implanted prosthetic materials including ventricle-peritoneal shunts and pacemakers (27, 46, 65). Indeed, its isolation from the preaxillary flora at the time of pacemaker insertion is, unlike any other staphylococcal species, strongly predictive of subsequent pacemaker infection (27). The outbreak potential of S. schleiferi subsp. schleiferi infection in conjunction with cardiac surgery (65) has been also reported. S. schleiferi subsp. coagulans was obtained from strains isolated from the external meatus of dogs suffering from external otitis (48) and from one case of postsurgical wound infection in a human patient (110).

Genomic Diversity and Natural Habitat

S. schleiferi is a rather genetically homogeneous species, since a very limited polymorphism is detected by PFGE or plasmid analysis. However, ribotyping combined with PFGE appeared to be efficient in detecting intraspecific variations among S. schleiferi isolates (65). The ecological niches of this microorganism are at present not well known, except for its natural occurrence in the human preaxillary flora (27); however, it is not known whether carriage is persistent or transient.

Antibiotic Susceptibility

S. schleiferi is usually susceptible to all antistaphylococcal antibiotics including penicillin. This is unusual in light of nosocomial acquisition of infections attributed to the organism.

TABLE 2 Binding of various Staphylococcus species to extracellular matrix proteins

Species	Fn				Cn		Vn	Lm	Fg
	PAA (76)	PAA and I-PBA (83, 84)	PAA and PCR (77)	WLAB and PCR (85)	PAA (76)	PAA and I-PBA (83, 84)	PAA and I-PBA (83, 84)	PAA and I-PBA (83, 84)	PAA and I-PBA (83, 84)
S. capitis	3/6	0/2	1/2	NT	6/6	2/2	0/2	1/2	NT
S. chromogenes	4/47	NT	NT	NT	32/47	NT	NT	NT	NT
S. cohnii	5/14	2/2	NT	NT	9/14	2/2	2/2	1/2	NT
S. haemolyticus	NT	1/1	15/16	NT	NT	1/1	1/1	1/1	NT
S. hominis	10/16	2/2	6/6	NT	10/16	2/2	2/2	2/2	NT
S. hyicus	37/171	0/2	6/7	NT	127/171	1/2	0/2	0/2	NT
S. intermedius	NT	NT	1/2	NT	NT	NT	NT	NT	NT
S. lugdunensis	NT	9/11*	3/3	NT	NT	8/11	10/11*	11/11*	7/11*
S. saprophyticus	NT	0/2	1/1	3/3	NT	1/2	0/2	0/2	NT
S. schleiferi	NT	NT	NT	NT	NT	NT	NT	NT	NT
S. simulans	4/18	0/2	NT	NT	14/18	1/2	0/2	0/2	NT
S. warneri	1/4	1/2	3/5	NT	2/4	2/2	0/2	1/2	NT
S. xylosus	5/19	1/2	NT	NT	16/19	2/2	1/2	1/2	NT

^aNT, not tested; PAA, particle agglutination assays; I-PBA, iodine-labeled protein-binding assays; PCR, PCR amplification of fnb-like genes using primers based on conserved sequences of S. aureus fnbA and fnbB; WLAB, Western ligand affinity blotting; Fn, fibronectin; Vn, vitronectin; Lm, laminin; Cn, collagen; Fg, fibrinogen; *, low-affinity binding.

TABLE 3 Superantigenic toxins produced by *Staphylococcus* species other than *S. aureus* and *S. epidermidis*

| Species | Enterotoxins (SEs) or TSST-1 detected by: | | | | | | | | | | |
| | ELISA | | | | | | Gene sequence | | Emetic assay | | Mitogenic assay |
	Almazan et al. (7)	Bautista et al. (15)	Hirooka et al. (47)	Valle et al. (106)	Orden et al. (81)	Vernozy-Rozand et al. + Southern blot (118)	Edwards et al. (32)	Becker et al. (16)	Adesiyun et al. (2)	Edwards et al. (32)	Edwards et al. (32)
S. capitis	NT[a]	NT	NT	NT	NT	1/11 SEE, see	NT	NT	NT	NT	NT
S. caprae	NT	NT	NT	1/18 SEA 2/18 SEB 1/18 SEC 1/18 SED 1/18 SEE	NT	NT	NT	NT	NT	NT	NT
S. chromogenes	NT	NT	NT	3/23 SEC	NT	0/1 SEE	NT	NT	2/2; not SEA-E	NT	NT
S. cohnii	NT	1/4 SEC	NT	0/6 SEA-E	NT	0/2 SEE	NT	NT	NT	NT	NT
S. equorum	NT		NT	0/3 SEA-E	NT	2/24 SEE, see	NT	NT	NT	NT	NT
S. haemolyticus	NT	1/6 SEA 1/6 SEB 3/6 SEC 4/6 SED	NT	1/64 SEA 3/64 SEB 15/64 SEC 4/64 SEE	NT	NT	NT	NT	NT	NT	NT
S. hyicus	NT	NT	NT	2/13 SEC	NT	NT	NT	NT	3/3; not SEA-E	NT	NT
S. intermedius	2/66 SEA 1/66 SEB 13/66 SEC 4/66 SED 1/66 SEE	NT	6/73 SEC 7/73 SED 6/73 SEE 4/73 TSST-1	NT	NT	NT	sec canine	33/281 sec	NT	SEC canine	SEC canine
S. lentus	NT	NT	NT	2/3 SEE	NT	1/7 SEE, see	NT	NT	NT	NT	NT
S. saprophyticus	NT	NT	NT	1/13 SEB 3/13 SEC 2/13 SEE	NT	NT	NT	NT	NT	NT	NT
S. sciuri	NT	NT	NT	1/20 SEC 3/20 SEE	NT	0/1 SEE	NT	NT	NT	NT	NT
S. simulans	NT	NT	NT	1/45 SEA 2/45 SEB 5/45 SEC	NT	5/51 SEE, see	NT	NT	NT	NT	NT
S. warneri	NT	NT	NT		NT	0/6 SEE	NT	NT	NT	NT	NT
S. xylosus	NT	4/5 SED	NT	4/23 SEC 2/23 SEE	2/3 TSST-1	1/24 SEE	NT	NT	NT	NT	NT

[a]NT, not tested.

Determinants of Pathogenicity

S. schleiferi causes subcutaneous abscess formation in mice (68). Production of virulence factors such as glycocalyx, esterase, protease, beta-hemolysin (overall similar to the beta-toxin of *S. aureus* and *S. intermedius*), FAME, and lipase has been previously described (Table 4) (42, 68, 72, 73). In an unpublished study (P. François and P. Vaudaux, personal communication), 9 of 11 strains of *S. schleiferi* showed a moderate to strong attachment on fibronectin- and fibronogen-coated polymethylmethacrylate coverslips that was equivalent to that of *S. aureus*. No fibrinogen-binding surface components homologous to the ClfA protein of *S. aureus* were identified by either Western immunoblotting or analysis of the ClfA antigen by fluorescence-activated cell sorting (FACS), and the putative fibrinogen adhesin of *S. schleiferi* is still unknown. In contrast, *S. schleiferi* expressed one fibronectin-binding adhesin homologous to the fibronectin-binding protein(s) of *S. aureus* according to four criteria. (i) Adhesion of *S. schleiferi* to fibronectin-coated surfaces was blocked by a recombinant D-repeat peptide of *S. aureus* FnBPa. (ii) A high-molecular-weight protein band of ca. 200 kDa which comigrated with fibronectin-binding proteins of *S. aureus* was identified by fibronectin ligand affinity Western blotting. (iii) Reverse transcription-PCR of mRNAs of 10 of 11 strains of *S. schleiferi* using forward and reverse primers of a highly conserved region of *fnb* genes of *S. aureus* 8325-4 revealed a constant-size 331-bp fragment. (iv) Sequencing of the reverse transcription-PCR-amplified 331-bp fragment in two strains of *S. schleiferi* exhibiting the highest levels of fibronectin-binding activity revealed a very high similarity with a *fnb* region of *S. aureus* controls (François and Vaudaux, personal communication). Comparable results were obtained by S. J. Peacock in that Western ligand affinity blotting of cell wall-associated protein extracted from *S. schleiferi* isolates demonstrated a fibronectin-binding protein with an apparent molecular weight of 200 kDa (for the reference strain) or 180 kDa (for two clinical isolates). PCR analysis using primers designed to amplify the 345-bp fragment of *S. aureus* 8325-4 *fnbA* that encodes the D1- to D3-binding domain of FnBPA gave a product of comparable size (85).

STAPHYLOCOCCUS CAPRAE

S. caprae (together with *Staphylococcus gallinarum*) was described in 1983. Both these species were originally isolated from animals (92).

Diseases

S. caprae has usually been associated with goats, but since 1991 an increasing number of laboratories report isolating the organism from human clinical specimens of both community- and hospital-acquired infections (96). Among these, urinary tract infection, skin infection, bacteremia associated with intravenous access, endocarditis, and bone and joint postsurgical infections were reported.

Genomic Diversity and Natural Habitat

S. caprae was originally isolated from goat milk; it is the most prevalent NAS in mastitis-free goat milk and has not been isolated from cow or sheep milk. *S. caprae* demonstrates considerable conservation in chromosome structure, as indicated by PFGE analysis of *S. caprae* isolates from various geographical locations. However, human isolates of *S. caprae* were distinguished from goat isolates on the basis of SmaI PFGE profile and, to some extent, on the basis of their cellular fatty acids (61) and their ribotypes (107). The normal habitat of *S. caprae* in humans remains to be discovered, but circumstantial evidence from its association with nosocomial infections suggests that it is likely to be a commensal of skin flora.

Antibiotic Susceptibility

S. caprae is usually susceptible to most antistaphylococcal antibiotics including penicillin, but all the strains tested to date (of both human and animal origin) have been uniformly resistant to fosfomycin. Resistance to methicillin has been detected, although infrequently, with isolates from humans (96, 107).

Determinants of Pathogenicity

S. caprae possesses an autolysin-adhesin (AtlC), which binds fibronectin and possesses amidase and glusosaminidase activities (5). AtlC is a 155-kDa protein, similar to the other staphylococcal autolysin-adhesin (48 to 72% amino acid identity). An *ica* four-gene operon preceded by an *icaR* homolog, closely related to those of *S. epidermidis* and *S. aureus* (≥68% similarity DNA level, 67 to 88% amino acid identity), has been found in strains from human specimens and goat milk. However, the presence of the *ica* operon was not always associated with biofilm formation (5). Most isolates of *S. caprae* express a synergistic hemolysin resembling the delta-lysin of *S.aureus*. Southern blot hybridization using *S. aureus hld* (encoding delta-lysin) and *S. lugdunensis slush* (encoding SLUSH) probes revealed that *S. caprae* contains *hld* and *slush* homolog genes, both responsible for synergistic-hemolytic activity (Table 1) (30). Enterotoxin production by certain strains isolated from healthy goats has been detected by ELISA (Table 3) (106). This has not been confirmed by emetic or mitogenic assay. Production of enzymes including lipase, FAME, and urease (67, 73) that may act as invasion factors has been reported.

STAPHYLOCOCCUS WARNERI

S. warneri was first described in 1975 and is usually found in small populations on human skin (61). A group that is nearly a subspecies, represented by nonhuman *S. warneri*, is one of the major staphylococcal species found living on the skin and nasal membranes of various prosimians and monkeys.

Diseases

S. warneri has the usual characteristics of the NAS and can cause significant infection both in the community and in the hospital. *S. warneri* has been reported to cause bacteremia, infective endocarditis, cerebrospinal fluid shunt infection, subdural empyema, vertebral osteomyelitis, and urinary tract infection (1, 21).

Antibiotic Susceptibility

Decreased vancomycin susceptibility has been reported in a strain responsible for an outbreak in a neonatal intensive care unit (22).

Determinants of Pathogenicity

S. warneri induces abscess formation in a mouse model with a foreign body implant but less frequently than *S. schleiferi*, *S. lugdunensis*, and *S. epidermidis* (68). *S. warneri* binds to collagen and fibronectin immobilized on latex beads (Table 2) (76, 77, 84). Production of glycocalyx, lipase, protease,

TABLE 4 Other extracellular factors produced by *Staphylococcus* species other than *S. aureus* and *S. epidermidis*[a]

Species	Esterase	Lipase			FAME	SHT	Exfoliative toxin	Urease	
	Lambe et al.; activity (68)	Lambe et al.; activity (68)	Long et al.; activity (73)	Ayora et al., van Kampen et al.; gene and biochemical (9, 115)	Long et al.; activity (73)	Prevost et al.; gene and activity (86)	Ahrens and Andresen, Sato et al., Terauchi et al.; gene and activity (3, 91, 104)	Schleifer; % activity (92)	Gaterman and Marre, Jose et al.; gene (35, 52)
S. capitis	NT	NT	1/6	NT	4/6	NT	NT	<10	NT
S. caprae	NT	NT	2/6	NT	4/6	NT	NT	>90	NT
S. chromogenes	NT	NT	NT	NT	NT	NT	SCET	NT	NT
S. cohnii	NT	NT	5/6	NT	5/6	NT	NT	<10	NT
S. haemolyticus	6/6	NT	0/9	NT	0/9	NT	NT	<10	NT
S. hominis	NT	6/6	3/6	NT	3/6	NT	NT	>90	NT
S. hyicus	NT	NT	NT	*shl*	NT	NT	ExhA to D, SHETA-B	11–89	NT
S. intermedius	NT	NT	NT	NT	NT	LukS-I/LukF-I, 100%	SIET	>90	NT
S. lugdunensis	6/6	6/6	0/10	NT	0/10	NT	NT	NT	NT
S. saprophyticus	NT	NT	9/9	NT	7/9	NT	NT	>90	Cloned and sequenced
S. schleiferi	6/6	6/6	10/10	NT	8/10	NT	NT	NT	NT
S. simulans	NT	NT	1/10	NT	8/10	NT	NT	>90	NT
S. warneri	6/6	6/6	3/5	NT	3/5	NT	NT	>90	NT
S. xylosus	NT	NT	NT	NT	NT	NT	NT	>90	Cloned and sequenced

[a]SHT, synergohymenotropic toxin; NT, not tested; SCET, *S. chromogenes* exfoliative toxin; ExhA, exfoliative toxin A; SHETA, *S. hyicus* exfoliative toxin A from *S. hyicus*; SIET, *S. intermedius* exfoliative toxin.

and esterase has been previously described (68). The lipase of *S. warneri* is secreted as a protein with an apparent molecular mass of 90 kDa that is processed in the supernatant to a protein of 45 kDa. Purified lipase has a broad substrate specificity, and results of inhibition studies are consistent with the presence of a serine residue at the catalytic site (101). A synergistic hemolysin that is produced by the majority of the isolates has been attributed to *hld* homologs within RNAIII (Table 1) (30, 103). RNAIII from *S. warneri* has been cloned and sequenced. It has an estimated length of 684 nt, and its sequence shows a high degree of identity, especially in the first 50 and last 150 nt, with that of RNAI-II-*sa*. Surprisingly, it contains two nonidentical copies of delta-like hemolysin, both 25 amino acids in length, that differ by seven and five residues, respectively, from *S. aureus* delta-hemolysin (103). However, the distribution of the charged residues suggests that the variants of delta-hemolysin produced by *S. warneri* can still form amphipatic alpha-helices (103). *S. warneri* RNAIII has the ability to regulate several *agr*-dependent genes of *S. aureus* (see chapter 41 on the regulation of gene expression) (103). Enterotoxin production by certain strains isolated from healthy goats has been detected by ELISA (Table 3) (106). This has not been confirmed by emetic or mitogenic assay. Production of enzymes including esterase, lipase, FAME, and urease (67, 73, 116) that may act as invasion factors has been reported (Table 4).

STAPHYLOCOCCUS PASTEURI

S. pasteuri is a new species of coagulase-negative staphylococcus described in 1993 and isolated from human, animal, and food specimens (24). *S. pasteuri* strains are phenotypically similar to *S. warneri* strains, but a clear-cut distinction between these two species may be obtained by comparing their rRNA gene restriction patterns (24). A synergistic hemolysin that is produced by most of the isolates has been attributed to an *hld* homolog (Table 1) (30).

STAPHYLOCOCCUS SIMULANS

Described by Schleifer and Kloos in 1975, this species was named *S. simulans* for having some phenotypic similarities to *S. aureus* (92). *S. simulans* bv. staphylolyticus produces lysostaphin, a bacteriocin that is a potent lysine for *S. aureus* and certain other staphylococci.

Diseases

S. simulans may be associated with a variety of animal infections including bovine and ovine mastitis, feline and canine pyodermas, and abscesses (34). *S. simulans* has been isolated as a rare cause of human infection including urinary tract, wound, bone, and joint infections; vertebral osteomyelitis; septicemia; and native valve endocarditis (49, 89).

Genomic Diversity and Natural Habitat

This species is isolated on the skin and urethra of healthy humans, the skin of other mammals, and food including goat milk and cheese (61, 92, 118). In a study of nursing home residents, 35% of carriage sites were screened positive; the organism was most frequently carried in the perineum, but it was also found in the nose and hairline (11).

Antibiotic Susceptibility

The susceptibility of *S. simulans* isolates to benzylpenicillin, methicillin, erythromycin, tetracycline, and kanamycin is variable. Some strains carry plasmids encoding β-lactamase or a ribosomal methylase gene conferring macrolide and lincosamide resistance (10, 92). Methicillin-resistant isolates have been reported (11).

Determinants of Pathogenicity

Binding of *S. simulans* to fibronectin and collagen immobilized on latex beads has been previously described (Table 2) (76, 84). Production of enzymes including lipase, FAME, and urease (67, 73) that may act as invasion factors has been reported (Table 4). The synergistic hemolysis produced by *S. simulans* (43) has been attributed to an *hld* homolog within RNAIII that has been cloned and sequenced (Table 1) (103). RNAIII of *S. simulans* has an estimated length of 573 nt, and its sequence shows a high degree of identity, especially in the first 50 nt and last 150 nt, with that of RNAIII-*sa*. The predicted 26-residue delta-hemolysin from *S. simulans* shows 19 of 26 amino acids to be identical to those of *S. aureus* (103). *S. simulans* RNAIII has the ability to regulate several *agr*-dependent genes of *S. aureus* (see chapter 41 on the regulation of gene expression) (103). Enterotoxin production by certain strains isolated from goat milk and cheese has been detected by ELISA; the results obtained were further confirmed by Southern blotting using two oligonucleotide probes specific for the *S. aureus* enterotoxin E gene (Table 3) (118). This has not been confirmed by emetic or mitogenic assay. *S. simulans* has been observed to produce capsular polysaccharide (20).

The lysostaphin-producing strain designated *S. simulans* bv. staphylolyticus carries a large β-lactamase plasmid encoding lysostaphin (Lss) and the lysostaphin immunity factor (Lif). Lss is an extracellular glycylglycine endopeptidase that lyses other staphylococci by hydrolyzing the polyglycine interpeptide bridges in their cell wall peptidoglycan. Lss is synthesized as a 493-amino-acid preprotein, with a 36-amino-acid signal peptide, a propetide of 211 amino acids, and a mature protein of 246 amino acids. Prolysostaphin is processed in the culture supernatant by an extracellular cysteine protease. Lif, which is encoded in the opposite direction, confers lysostaphin immunity by increasing the serine/glycine ratio of the interpeptide bridges. Lif shows similarity to FemA and FemB proteins, which are involved in the biosynthesis of the glycine interpeptide bridge of staphylococcal peptidoglycan (105).

Lss has the potential to be a novel therapeutic agent. It was shown to be more effective in the treatment of experimental *S. aureus* endocarditis in rabbits than vancomycin (26). Combination of Lss with oxacillin is synergistic in vitro against oxacillin-resistant *S. epidermidis* (58). Moreover, in a rabbit model of experimental aortic valve endocarditis using an oxacillin-resistant *S. epidermidis* strain, a combination of Lss and methicillin was as effective as vancomycin alone (58). Lss has also been compared to mupirocin with a cotton rat model of nasal colonization. A single dose of Lss cream was more effective than a single dose of mupirocin ointment in eradicating *S. aureus* (66). Finally, to enhance mastitis resistance of dairy animals, Kerr et al. (56) generated transgenic mice secreting Lss into milk. These mice expressed substantial resistance to an intramammary challenge of *S. aureus* (56).

STAPHYLOCOCCUS CAPITIS

S. capitis subsp. *urealyticus* is distinguished from *S. capitis* subsp. *capitis* by its urease activity (61). *S. capitis* has been reported to cause urinary tract infections, catheter-related bacteremia, cellulitis, cerebrospinal fluid shunt infection, and infective endocarditis (69, 90). Although the scalp is the most usual habitat of S. capitis (61), it has not been unequivocally identified as the portal of entry in these cases of endocarditis. Heteroresistance to vancomycin occurs in S. capitis strains (114). All the strains tested to date have been resistant to fosfomycin.

S. capitis binds to laminin, collagen, and fibronectin immobilized on latex beads (Table 2) (76, 77, 84). Production of enzymes including lipase and FAME (67, 73) that may act as invasion factors has been reported (Table 4). A synergistic hemolysin that is produced by the majority of the isolates has been attributed to an hld homolog by Southern blot hybridization (Table 1) (30). The production of enterotoxin by strains isolated from goat milk and cheese has been reported (118). Enterotoxin was detected by ELISA and by Southern blotting with two oligonucleotide probes specific for S. aureus enterotoxin E (Table 3) (118). Enterotoxin function, however, has not been confirmed by mitogenic or emetic assay.

S. capitis strain EPK1 produces a 35-kDa glycylglycine endopeptidase (ALE-1), which hydrolyzes interpeptide pentaglycine chains of cell wall peptidoglycan of S. aureus. So far, strain EPK1 is the only producer of ALE-1 among strains of S. capitis tested (M. Sugai, personal communication). Characterization of the enzyme activity and cloning of the plasmid-encoded ale-1 gene revealed that ALE-1 is very similar to prolysostaphin produced by S. simulans bv. staphylolyticus (99). A protein homology search suggests that ALE-1 and Lss are members of a Zn^{2+} protease family with a 38-amino-acid-long motif, Tyr-X-His-X(11)-Val-X(12/20)-Gly-X(5-6)-His (99). The epr gene located upstream of and in the opposite orientation to ale-1 confers resistance to ALE-1 and Lss. As observed with Lif (responsible for lysostaphin immunity), the epr product may be involved in the addition of serine to the pentapeptide peptidoglycan precursor (99).

STAPHYLOCOCCUS HOMINIS

S. hominis is one of the major staphylococcal species inhabiting the skin of humans and is considered an opportunistic pathogen of low virulence. A new subspecies, Staphylococcus hominis subsp. novobiosepticus, isolated from human blood cultures, wound, breast abscess, cases of pneumonia, and catheter tips, has been previously described (62). S. hominis binds to vitronectin, laminin, collagen, and fibronectin immobilized on latex beads (Table 2) (76, 77, 84). A synergistic hemolysin that is produced by the majority of the isolates could not be attributed to slush or hld homologs by Southern blot hybridization (Table 1), possibly suggesting the existence of a third staphylococcal synergistic hemolysin (30). Production of enzymes including esterase, lipase, FAME, and urease (67, 73) that may act as invasion factors has been reported (Table 4).

A homolog of the staphylococcal cassette chromosome mec (SCCmec) element of methicillin-resistant S. aureus has been identified in the genome of a methicillin-susceptible S. hominis isolate. This genetic element has mosaic-like patterns of homology with extant SCCmec elements and is considered a type I SCCmec. The existence of an SCC without a mec determinant is indicative of a staphylococcal site-specific mobile genetic element that serves as a vehicle of transfer for various genetic markers between staphylococcal species (54).

STAPHYLOCOCCUS XYLOSUS

S. xylosus has hosts in all mammalian orders (59) but has rarely been associated with human infections. It is used as a starter culture in the production of dry sausage and fermented fish. S. xylosus binds to vitronectin, laminin, collagen, and fibronectin immobilized on latex beads (Table 2) (76, 84). A synergistic hemolysin that is produced by approximately one-half of the isolates has been attributed to a slush homolog (Table 1) (30). The production of enterotoxin and toxic shock syndrome toxin 1 (TSST-1) by strains isolated from goat milk and cheese (118), sheep milk (15), healthy goats (106), and from sheep, goat, and cow mastitis specimens (81) detected by ELISA has been reported. In one study, the results obtained were further confirmed by Southern blotting with two oligonucleotide probes specific for the S. aureus enterotoxin E gene (Table 3) (118). Enterotoxin function has not been confirmed by mitogenic or emetic assay. A gene encoding urease that may act as an invasion factor has been reported (Table 4) (52).

The biology of nutrition and metabolism of S. xylosus has been the subject of increasing attention. The lactose utilization genes and the mechanism of catabolite repression in S. xylosus have been previously characterized (14). A genetic locus essential for maltose-maltotriose utilization in S. xylosus has been identified, and its function has been characterized. The serine acetyltransferase gene (cysE) of S. xylosus was identified by transposon mutagenesis; it is surrounded by genes encoding glutamyl-tRNA synthetase (gltX) and cysteinyl-tRNA synthetase (cysS) in an organization identical to that found for Bacillus subtilis and Bacillus stearothermophilus (33). The sucrose-specific regulon for sucrose utilization by S. xylosus has been characterized.

S. xylosus has at least two differentially expressed catalases. S. xylosus inhibits the formation of hydrogen peroxide by its catalase activity and so could prevent color and aroma imperfections in sausage. It also contributes to aroma, mainly by modulating the level and the nature of volatile compounds coming from lipid oxidation. Antioxidant activities of S. xylosus (e.g., catalase and one cytosolic superoxide dismutase) are thought to be involved in the development of the sensorial qualities (12, 13).

STAPHYLOCOCCUS INTERMEDIUS

Diseases and Habitat

S. intermedius is the most important NAS that produces coagulase; it is, however, taxonomically distinct from S. aureus. It is the predominant coagulase-positive staphylococcus in the mouth and in skin infections of dogs. It has also been found in a wide range of other animal species including pigeons, minks, cats, foxes, raccoons, gray squirrels, goats, and horses (18). In humans, S. intermedius has rarely been found, even among individuals with frequent animal exposure, but it is a common and potentially invasive zoonotic pathogen of canine-inflicted human wounds (70). It has, however, been isolated in rare cases of non-canine-inflicted wounds, including a case of infective endocarditis, a case of catheter-related bacteremia (74), and a case of pneumonia following coronary artery bypass grafting (37). S. intermedius was also considered the etiologic agent in an outbreak of food intoxication involving butter-blend products and resulting in >265 cases in the western United States; all the isolates were reported to produce enterotoxin

A (57). In fact, the true frequency of *S. intermedius* in non-canine-inflicted wounds remains unclear, as it can be confused with *S. aureus* in medical laboratory analysis on the basis of coagulase production.

Genomic Diversity

Genomic DNA fingerprinting by PFGE of *S. intermedius* isolates has revealed a high degree of polymorphism (95). Conventional genomic DNA fingerprinting suggested that isolates from healthy dogs and those from canine pyoderma cases belong to distinct clusters (4). Ribotyping did not confirm these results; isolates from healthy dogs and cases of canine pyoderma were indistinguishable. However, pigeon and equine strains showed a variety of ribotypes, including those of the canine isolates, suggesting exchange of strains between animal species (63).

Determinants of Pathogenicity

Binding of *S. intermedius* to fibronectin immobilized on latex beads has been described (Table 2) (77). *S. intermedius* produces a 42-kDa Ig-binding protein, specific for the Fc domain, that shows close functional and antigenic similarity with protein A of *S. aureus* (40). It is expressed both in cell-wall-bound and secreted forms. *S. intermedius* produces a staphylocoagulase which resembles that of *S. aureus* in its rate and method of action on prothrombin but is antigenically distinct from the *S. aureus* coagulase (88). It has not been characterized at the molecular level. *S. intermedius* has been described as a beta- and delta-hemolysin producer, but only the beta-hemolysin has been characterized: the protein has been purified and its biochemical properties have been compared to that of *S. aureus* beta-hemolysin. Both toxins have similar enzymatic properties, belong to the class of neutral sphingomyelinases C, and have a high specificity for sphingomyelin with identical kinetic parameters. Despite these similarities, the size and amino acid compositions of the two toxins differ; there was no detectable cysteine residue in *S. intermedius* beta-hemolysin. The available N-terminal amino acid sequence of *S. intermedius* beta-hemolysin shows only 9 of 19 residues identical to *S. aureus* beta-hemolysin, confirming the lack of identity between the two toxins (31). A synergohymenotropic toxin that is produced by *S. intermedius* isolates has been characterized. It is made of two components (LukS-I and LukF-I), like the Panton-Valentine leukocidin encoded in the *lukI* operon, and is leukotoxic to polymorphonuclear leukocytes from various species. It is, however, weakly hemolytic on rabbit erythrocytes (86) (Table 4).

Production of enterotoxins A, C, D, and E and TSST-1 by strains of *S. intermedius* isolated from dog infections (7, 47) and from butter-blend products responsible for food intoxication (57) has been reported. From these toxins, only staphylococcal enterotoxin C (SEC) has been characterized at the molecular level. Called SEC canine, it is a 239-amino-acid protein with >95% sequence identity with the SEC variants produced by *S. aureus* (SEC1, SEC2, and SEC3). Purified SEC canine induces an emetic response in monkeys and induces proliferation of T cells in a Vβ-dependent manner with the same profile as that induced by SEC1 (Table 3) (32). The production of other enterotoxins detected by ELISA methods has been reported (Table 3); however, multiplex PCR targeting *sea*, *seb*, *sec*, *sed*, and *see* revealed that *sec* was the only enterotoxin gene present in 33 (11%) of 281 strains of *S. intermedius* tested (16). A 30-kDa exfoliative toxin-like toxin has been isolated from the culture filtrates of *S. intermedius*. This *S. intermedius* exfoliative toxin (SIET) is serologically different from *S. aureus* exfoliative toxin A (ETA), ETB, and ETC and from *Staphylococcus hyicus* exfoliative toxins A and B. Erythema, exfoliation, and crusting were observed in dogs injected with SIET (104). Production of urease that may act as invasion factor has been reported (Table 4).

STAPHYLOCOCCUS COHNII

S. cohnii subsp. *urealyticum* and *S. cohnii* subsp. *cohnii* are recently designated subspecies of *S. cohnii* (64). These subspecies (formerly subspecies 1 and 2) differ by phenotypic and metabolic properties and in their host range: *S. cohnii* subsp. *urealyticum* has been isolated from both humans and other primates, whereas *S. cohnii* subsp. *cohnii* has been isolated only from humans (61). *S. cohnii* has been isolated from human urinary tract and wound infections, septic arthritis, and meningitis. Susceptibility to benzylpenicillin, erythromycin, tetracycline, and chloramphenicol is variable. Methicillin-resistant strains have been isolated. A staphylococcal plasmid carrying two novel genes, *vatC* and *vgbB*, encoding resistance to streptogramin A and B antibiotics, has been recently described. (i) *vatC* encodes a 212-amino-acid acetyltransferase that inactivates streptogramin A and exhibits 58.2 to 69.8% amino acid identity with the staphylococcal Vat and VatB proteins and the *E. faecium* SatA protein. (ii) *vgbB* encodes a 295-amino-acid lactonase that inactivates streptogramin B and shows 67% amino acid identity with the staphylococcal Vgb lactonase (6).

S. cohnii binds to fibronectin, collagen, vitronectin, and laminin immobilized on latex beads (Table 2) (76, 84). A synergistic hemolysin that is produced by the majority of the isolates has been attributed to the *slush* homolog by Southern blot hybridization (Table 1) (30). The production of enterotoxin by strains isolated from sheep milk (15) detected by ELISA has been reported (Table 3). Enterotoxin function has not been confirmed by mitogenic or emetic assay. Production of enzymes including lipase and FAME (67, 73) that may act as invasion factors has been reported (Table 4).

STAPHYLOCOCCUS SCIURI

S. sciuri (*S. sciuri* subsp. *sciuri*, *S. sciuri* subsp. *carnaticus*, and *S. sciuri* subsp. *rodentium*) is commonly isolated from the skin of rodents and somewhat less frequently from the skin of ungulates, carnivores, and marsupials (60, 92). It has been shown to be an invasive pathogen for animals, causing wound infections and mastitis. However, *S. sciuri* may colonize humans; its isolation from human clinical samples such as skin, vagina, blood, urine, and central venous catheters has been reported. It has been involved in rare cases of wound infections, soft tissue infections, abscesses, boils, peritonitis, and endocarditis (75). *S. sciuri* is capable of biofilm production, displays strong proteolytic and DNase activities, produces hemolysins, and stimulates nitric oxid production by rat macrophages. The exact contribution of these potential virulence factors to pathogenesis remains to be determined (97).

A homolog of the *S. aureus* methicillin-resistance gene *mecA* was recently shown to be ubiquitous in 134 independent isolates of *S. sciuri*. Among these, isolates of *S. sciuri* subsp. *sciuri* and *S. sciuri* subsp. *carnaticus* showed only marginal, if any, resistance to methicillin (MIC, 0.75 to 6.0 µg/ml), while most isolates of S. sciuri subsp. rodentium expressed a heterogeneous methicillin resistance phenotype. Investigation of the genetic organization of the mecA region in the three subspecies revealed that S. sciuri strains can contain two different forms of mecA. The first form is virtually identical to the mecA gene of S. aureus and is present in S. sciuri subsp. rodentium; it is proposed that these strains may have acquired a mecA gene from S. aureus or S. epidermidis (122). The second form, which can coexist with the first one in S. sciuri subsp. rodentium, shows somewhat less similarity (79.5%) to mecA of S. aureus and appears to be the predominant form in S. sciuri subspecies. Strains carrying only this form are susceptible to methicillin, suggesting that it is a silent gene that may be an evolutionary relative or precursor of the mecA gene of S. aureus (122, 123). Resistance to lincosamide and streptogramin A with susceptibility to erythromycin has been observed with some strains of S. sciuri subsp. sciuri isolated from humans and could not be attributed to any of the known genes conferring resistance to macrolide-lincosamide-streptogramin B antibiotics (G. Lina, unpublished data). The S. sciuri erm gene [erm(33)] encoding inducible resistance to macrolides, lincosamides, and streptogramin B antibiotics is a product of recombination between erm(C) and erm(A). Such a recombination is a novel observation for erm genes (93). A plasmid-borne chloramphenicol resistance gene has been described for S. sciuri (94).

STAPHYLOCOCCUS HYICUS

S. hyicus (S. hyicus subsp. hyicus and S. hyicus subsp. chromogenes) is an opportunistic pathogen found in pigs and cattle (61, 92). S. hyicus binds to vitronectin, laminin, collagen, and fibronectin immobilized on latex beads (77, 84). S. hyicus strains isolated from pigs (but not from cows) produce a 42-kDa Ig-binding protein specific for the Fc domain that shows close functional and antigenic similarity with protein A of S. aureus (40).

S. hyicus strains isolated from pigs affected with exudative epidermitis produce six antigenically distinct exfoliative toxins: S. hyicus exfoliative toxin A (SHETA), SHETB, and exfoliative toxins from S. hyicus (ExhA, ExhB, ExhC, and ExhD). SHETB and ExhA to -D, together with S. aureus toxins ETA, ETB, and ETD, form a group of highly related molecules, while SHETA is quite different from the other exfoliative toxins (3). These toxins produce exfoliation by piglet and chicken skin assays. However, exfoliation was not observed with specimens from other animals (mice, rats, guinea pigs, hamsters, dogs, or cats).

S. hyicus subsp. hyicus produces a lipase (Table 4), designated SHL, that is secreted as a 86-kDa proenzyme and is processed to the mature 46-kDa enzyme by extracellular metalloprotease ShpII (9). Evidence has been presented showing that the proregion of the lipase acts as an intramolecular chaperone, which facilitates translocation of the native lipase and of a number of completely unrelated proteins fused to the propeptide. It was also observed that the proregion protects the proteins from proteolytic degradation (39). SHL is exceptional in that it displays a high level of phospholipase activity, hydrolyzes neutral lipids, and has no chain length preference. Site-directed mutagenesis and domain exchange were used to determine that in the C-terminal domain it is Ser356 that mainly determines phospholipase activity (115).

The production of enterotoxin by strains isolated from healthy goats (106), detected by ELISA, has been reported.

In another study, the production of an emetic toxin has been observed with a monkey emetic model (Table 3). However, the emetic toxin was not immunologically reactive with enterotoxins A to E (2). Production of urease that may act as invasion factors has been reported (Table 4).

STAPHYLOCOCCUS CARNOSUS

S. carnosus is used as a starter culture in the production of dry sausage and fermented fish. It is poorly pathogenic, and its natural habitat has not been determined. Physical and genetic maps of S. carnosus chromosome have been determined (119), and specific vectors have been developed for gene cloning and expression of heterologous protein in S. carnosus, including a highly efficient surface display expression system. For instance, the cholera toxin B subunit from Vibrio cholerae was properly expressed in this system; thus, the cell surface display of heterologous receptors on S. carnosus could be considered potential live bacterial vaccine delivery systems for administration by the mucosal route (71).

STAPHYLOCOCCUS GALLINARUM

S. gallinarum derives its name from its avian host and is found mainly in the skin of poultry; it may also be found in other birds (61, 92). S. gallinarum produces the bacteriocin gallidermin, which exhibits activity against propionibacteria. Like epidermin, Pep5, and epilancin K7 from S. epidermidis, gallidermin is a member of the class of lanthionine-containing peptide antibiotics also designated lantibiotics. Gallidermin is closely related to epidermin; it differs only in a Leu/Ile exchange in position 6. The biosynthesis of all of these lantibiotics proceeds from structural genes, which encode prepeptides that are enzymatically modified to give the mature peptides. Additional genes encoding transporters, immunity functions, regulatory proteins, and the modification enzymes which catalyze the biosynthesis of the rare amino acids are found in gene clusters adjacent to the structural genes (82).

STAPHYLOCOCCUS CHROMOGENES

S. chromogenes is a common coagulase-negative Staphylococcus species found on the skin of healthy pigs and is a nonpathogenic bacterium. It has been occasionally isolated together with other nonpathogenic bacteria from lymph nodes of patients suffering dermatolymphangioadenitis complicating filarial lymphedema (80). Some strains have been isolated from the skin of pigs affected with exudative epidermitis. An exfoliative toxin produced by a pig strain has been designated S. chromogenes exfoliative toxin. S. chromogenes exfoliative toxin is serologically different from S.aureus ETA, ETB, and ETC; from SHETA and SHETB; and from SIET (91).

Owing to space limitations, only a limited number of references are included in this chapter. We apologize for the inevitable omissions. Those references that were chosen generally possess extensive reference lists of their own, so that bibliographical sources are readily available, where not cited directly.

REFERENCES

1. **Abgrall, S., P. Meimoun, A. Buu-Hoi, J. P. Couetil, L. Gutmann, and J. L. Mainardi.** 2001. Early prosthetic

valve endocarditis due to *Staphylococcus warneri* with negative blood culture. *J. Infect.* **42:**166.

2. **Adesiyun, A. A., S. R. Tatini, and D. G. Hoover.** 1984. Production of enterotoxin(s) by *Staphylococcus hyicus*. *Vet. Microbiol.* **9:**487–495.

3. **Ahrens, P., and L. O. Andresen.** 2004. Cloning and sequence analysis of genes encoding *Staphylococcus hyicus* exfoliative toxin types A, B, C, and D. *J. Bacteriol.* **186:**1833–1837.

4. **Allaker, R. P., N. Garrett, L. Kent, W. C. Noble, and D. H. Lloyd.** 1993. Characterisation of *Staphylococcus intermedius* isolates from canine pyoderma and from healthy carriers by SDS-PAGE of exoproteins, immunoblotting and restriction endonuclease digest analysis. *J. Med. Microbiol.* **39:**429–433.

5. **Allignet, J., S. Aubert, K. G. Dyke, and N. El Solh.** 2001. *Staphylococcus caprae* strains carry determinants known to be involved in pathogenicity: a gene encoding an autolysin-binding fibronectin and the *ica* operon involved in biofilm formation. *Infect. Immun.* **69:**712–718.

6. **Allignet, J., N. Liassine, and N. El Solh.** 1998. Characterization of a staphylococcal plasmid related to pUB110 and carrying two novel genes, *vatC* and *vgbB*, encoding resistance to streptogramins A and B and similar antibiotics. *Antimicrob. Agents Chemother.* **42:**1794–1798.

7. **Almazan, J., R. de la Fuente, E. Gomez-Lucia, and G. Suarez.** 1987. Enterotoxin production by strains of *Staphylococcus intermedius* and *Staphylococcus aureus* isolated from dog infections. *Zentralbl. Bakteriol. Mikrobiol. Hyg. A* **264:**29–32.

8. **Andresen, L. O.** 1998. Differentiation and distribution of three types of exfoliative toxin produced by *Staphylococcus hyicus* from pigs with exudative epidermitis. *FEMS Immunol. Med. Microbiol.* **20:**301–310.

9. **Ayora, S., P. E. Lindgren, and F. Gotz.** 1994. Biochemical properties of a novel metalloprotease from *Staphylococcus hyicus* subsp. *hyicus* involved in extracellular lipase processing. *J. Bacteriol.* **176:**3218–3223.

10. **Barcs, I., and L. Janosi.** 1992. Plasmids encoding for erythromycin ribosomal methylase of *Staphylococcus epidermidis* and *Staphylococcus simulans*. *Acta Microbiol. Hung.* **39:**85–92.

11. **Barnham, M., R. Horton, J. M. P. Smith, J. Richardson, R. R. Marples, and S. Reith.** 1996. Methicillin-resistant *Staphylococcus simulans* masquerading as MRSA in a nursing home. *J. Hosp. Infect.* **34:**331–337.

12. **Barriere, C., R. Bruckner, D. Centeno, and R. Talon.** 2002. Characterisation of the *katA* gene encoding a catalase and evidence for at least a second catalase activity in *Staphylococcus xylosus*, bacteria used in food fermentation. *FEMS Microbiol. Lett.* **216:**277–283.

13. **Barriere, C., D. Centeno, A. Lebert, S. Leroy-Setrin, J. L. Berdague, and R. Talon.** 2001. Roles of superoxide dismutase and catalase of *Staphylococcus xylosus* in the inhibition of linoleic acid oxidation. *FEMS Microbiol. Lett.* **201:**181–185.

14. **Bassias, J., and R. Bruckner.** 1998. Regulation of lactose utilization genes in *Staphylococcus xylosus*. *J. Bacteriol.* **180:**2273–2279.

15. **Bautista, L., P. Gaya, M. Medina, and M. Nunez.** 1988. A quantitative study of enterotoxin production by sheep milk staphylococci. *Appl. Environ. Microbiol.* **54:**566–569.

16. **Becker, K., B. Keller, C. von Eiff, M. Bruck, G. Lubritz, J. Etienne, and G. Peters.** 2001. Enterotoxigenic potential of *Staphylococcus intermedius*. *Appl. Environ. Microbiol.* **67:**5551–5557.

17. **Benito, Y., G. Lina, T. Greenland, J. Etienne, and F. Vandenesch.** 1998. *trans*-complementation of *Staphylococcus aureus agr* mutant by *S. lugdunensis agr* RNAIII. *J. Bacteriol.* **180:**5780–5783.

18. **Biberstein, E. L., S. Jang, and D. C. Hirsh.** 1984. Species distribution of coagulase-positive staphylococci in animals. *J. Clin. Microbiol.* **19:**610–615.

19. **Billot-Klein, D., L. Gutmann, D. Bryant, D. Bell, J. Van Heijenoort, J. Grewal, and D. M. Shlaes.** 1996. Peptidoglycan synthesis and structure in *Staphylococcus haemolyticus* expressing increasing levels of resistance to glycopeptide antibiotics. *J. Bacteriol.* **178:**4696–4703.

20. **Burriel, A. R.** 1998. In vivo presence of capsular polysaccharide in coagulase-negative staphylococci of ovine origin. *New Microbiol.* **21:**49–54.

21. **Buttery, J. P., M. Easton, S. R. Pearson, and G. G. Hogg.** 1997. Pediatric bacteremia due to *Staphylococcus warneri*: microbiological, epidemiological, and clinical features. *J. Clin. Microbiol.* **35:**2174–2177.

22. **Center, K. J., A. C. Reboli, R. Hubler, G. L. Rodgers, and S. S. Long.** 2003. Decreased vancomycin susceptibility of coagulase-negative staphylococci in a neonatal intensive care unit: evidence of spread of *Staphylococcus warneri*. *J. Clin. Microbiol.* **41:**4660–4665.

23. **Chaouni, L., T. Greenland, J. Etienne, and F. Vandenesch.** 1996. Nucleic acid sequence and affiliation of pLUG10, a novel cadmium resistance plasmid from *Staphylococcus lugdunensis*. *Plasmid* **36:**1–8.

24. **Chesneau, O., A. Morvan, F. Grimont, H. Labischinski, and N. El Solh.** 1993. *Staphylococcus pasteuri* sp. nov., isolated from human, animal, and food specimens. *Int. J. Syst. Bacteriol.* **43:**237–244.

25. **Cintas, L. M., P. Casaus, H. Holo, P. E. Hernandez, I. F. Nes, and L. S. Havarstein.** 1998. Enterocins L50A and L50B, two novel bacteriocins from *Enterococcus faecium* L50, are related to staphylococcal hemolysins. *J. Bacteriol.* **180:**1988–1994.

26. **Climo, M. W., R. L. Patron, B. P. Goldstein, and G. L. Archer.** 1998. Lysostaphin treatment of experimental methicillin-resistant *Staphylococcus aureus* aortic valve endocarditis. *Antimicrob. Agents Chemother.* **42:**1355–1360.

27. **Dacosta, A., H. Lelièvre, G. Kirkorian, M. Célard, P. Chevalier, F. Vandenesch, J. Etienne, and P. Touboul.** 1998. Role of the preaxillary flora in pacemaker infections. A prospective study. *Circulation* **97:**1791–1795.

28. **De Hondt, G., M. Ieven, C. Vandermersch, and J. Colaert.** 1997. Destructive endocarditis caused by *Staphylococcus lugdunensis*. Case report and review of the literature. *Acta Clin. Belg.* **52:**27–30.

29. **Donvito, B., J. Etienne, L. Denoroy, T. Greenland, Y. Benito, and F. Vandenesch.** 1997. Synergistic hemolytic activity of *Staphylococcus lugdunensis* is mediated by three peptides encoded by a non-*agr* genetic locus. *Infect. Immun.* **65:**95–100.

30. **Donvito, B., J. Etienne, T. Greenland, C. Mouren, V. Delorme, and F. Vandenesch.** 1997. Distribution of the synergistic haemolysin genes *hld* and *slush* with respect to *agr* in human staphylococci. *FEMS Microbiol. Lett.* **151:**139–144.

31. **Dziewanowska, K., V. M. Edwards, J. R. Deringer, G. A. Bohach, and D. J. Guerra.** 1996. Comparison of the beta-toxins from *Staphylococcus aureus* and *Staphylococcus intermedius*. *Arch. Biochem. Biophys.* **335:**102–108.

32. **Edwards, V. M., J. R. Deringer, S. D. Callantine, C. F. Deobald, P. H. Berger, V. Kapur, C. V. Stauffacher, and G. A. Bohach.** 1997. Characterization of the canine type

C enterotoxin produced by *Staphylococcus intermedius* pyoderma isolates. *Infect. Immun.* **65:**2346–2352.

33. **Fiegler, H., and R. Bruckner.** 1997. Identification of the serine acetyltransferase gene of *Staphylococcus xylosus*. *FEMS Microbiol. Lett.* **148:**181–187.

34. **Fthenakis, G. C., R. R. Marples, J. F. Richardson, and J. E. Jones.** 1994. Some properties of coagulase-negative staphylococci isolated from cases of ovine mastitis. *Epidemiol. Infect.* **112:**171–176.

35. **Gatermann, S., and R. Marre.** 1989. Cloning and expression of *Staphylococcus saprophyticus* urease gene sequences in *Staphylococcus carnosus* and contribution of the enzyme to virulence. *Infect. Immun.* **57:**2998–3002.

36. **Gatermann, S. G., and K. B. Crossley.** 1997. Urinary tract infections, p. 493–508. *In* K. B. Crosssley and G. L. Archer (ed.), *The Staphylococci in Human Disease*. Churchill Livingstone, Inc., New York, N.Y.

37. **Gerstadt, K., J. S. Daly, M. Mitchell, M. Wessolossky, and S. H. Cheeseman.** 1999. Methicillin-resistant *Staphylococcus intermedius* pneumonia following coronary artery bypass grafting. *Clin. Infect. Dis.* **29:**218–219.

38. **Giovanetti, E., F. Biavasco, A. Pugnaloni, R. Lupidi, G. Biagini, and P. E. Varaldo.** 1996. An electron microscopic study of clinical and laboratory-derived strains of teicoplanin-resistant *Staphylococcus haemolyticus*. *Microb. Drug Resist.* **2:**239–243.

39. **Gotz, F., H. M. Verheij, and R. Rosenstein.** 1998. Staphylococcal lipases: molecular characterisation, secretion, and processing. *Chem. Phys. Lipids* **93:**15–25.

40. **Greene, R. T., and C. Lammler.** 1992. Isolation and characterization of immunoglobulin binding proteins from *Staphylococcus intermedius* and *Staphylococcus hyicus*. *Zentralbl. Veterinarmed. B* **39:**519–525.

41. **Hajek, V., H. Meugnier, M. Bes, Y. Brun, F. Fiedler, Z. Chmela, Y. Lasne, J. Fleurette, and J. Freney.** 1996. *Staphylococcus saprophyticus* subsp. *bovis* subsp. nov., isolated from bovine nostrils. *Int. J. Syst. Bacteriol.* **46:**792–796.

42. **Hébert, G. A.** 1990. Hemolysins and other characteristics that help differentiate and biotype *Staphylococcus lugdunensis* and *Staphylococcus schleiferi*. *J. Clin. Microbiol.* **28:**2425–2431.

43. **Hébert, G. A., C. G. Crowder, G. A. Hancock, W. R. Jarvis, and C. Thornsberry.** 1988. Characteristics of coagulase-negative staphylococci that help differentiate these species and other members of the family *Micrococcaceae*. *J. Clin. Microbiol.* **26:**1939–1949.

44. **Hell, W., T. Kern, and M. Klouche.** 1999. *Staphylococcus saprophyticus* as an unusual agent of nosocomial pneumonia. *Clin. Infect. Dis.* **29:**685-686.

45. **Hell, W., H. G. Meyer, and S. G. Gatermann.** 1998. Cloning of aas, a gene encoding a *Staphylococcus saprophyticus* surface protein with adhesive and autolytic properties. *Mol. Microbiol.* **29:**871–881.

46. **Hernandez, J. L., J. Calvo, R. Sota, J. Aguero, J. D. Garcia-Palomo, and M. C. Farinas.** 2001. Clinical and microbiological characteristics of 28 patients with *Staphylococcus schleiferi* infection. *Eur. J. Clin. Microbiol. Infect. Dis.* **20:**153–158.

47. **Hirooka, E. Y., E. E. Muller, J. C. Freitas, E. Vicente, Y. Yoshimoto, and M. S. Bergdoll.** 1988. Enterotoxigenicity of *Staphylococcus intermedius* of canine origin. *Int. J. Food Microbiol.* **7:**185–191.

48. **Igimi, S., E. Takahashi, and T. Mitsuoka.** 1990. *Staphylococcus schleiferi* subsp. *coagulans* subsp. nov., isolated from the external auditory meatus of dogs with external ear otitis. *Int. J. Syst. Bacteriol.* **40:**409–411.

49. **Jansen, B., F. Schumacher-Perdreau, G. Peters, G. Reinhold, and J. Schonemann.** 1992. Native valve endocarditis caused by *Staphylococcus simulans*. *Eur. J. Clin. Microbiol. Infect. Dis.* **11:**268–269.

50. **Ji, G., R. Beavis, and R. P. Novick.** 1997. Bacterial interference caused by autoinducing peptide variants. *Science* **276:**2027–2030.

51. **Jones, R. M., M. A. Jackson, C. Ong, and G. K. Lofland.** 2002. Endocarditis caused by *Staphylococcus lugdunensis*. *Pediatr. Infect. Dis. J.* **21:**265–268.

52. **Jose, J., S. Christians, R. Rosenstein, F. Gotz, and H. Kaltwasser.** 1991. Cloning and expression of various staphylococcal genes encoding urease in Staphylococcus carnosus. *FEMS Microbiol. Lett.* **64:**277–281.

53. **Katayama, Y., T. Ito, and K. Hiramatsu.** 2001. Genetic organization of the chromosome region surrounding *mecA* in clinical staphylococcal strains: role of IS431-mediated *mecI* deletion in expression of resistance in *mecA*-carrying, low-level methicillin-resistant *Staphylococcus haemolyticus*. *Antimicrob. Agents Chemother.* **45:**1955–1963.

54. **Katayama, Y., F. Takeuchi, T. Ito, X. X. Ma, Y. Ui-Mizutani, I. Kobayashi, and K. Hiramatsu.** 2003. Identification in methicillin-susceptible *Staphylococcus hominis* of an active primordial mobile genetic element for the staphylococcal cassette chromosome *mec* of methicillin-resistant *Staphylococcus aureus*. *J. Bacteriol.* **185:** 2711–2722.

55. **Kazembe, P., A. E. Simor, A. E. Swarney, L. G. Yap, B. Kreiswirth, J. Ng, and D. E. Low.** 1993. A study of the epidemiology of an endemic strain of *Staphylococcus haemolyticus* (TOR-35) in a neonatal intensive care unit. *Scand. J. Infect. Dis.* **25:**507–513.

56. **Kerr, D. E., K. Plaut, A. J. Bramley, C. M. Williamson, A. J. Lax, K. Moore, K. D. Wells, and R. J. Wall.** 2001. Lysostaphin expression in mammary glands confers protection against staphylococcal infection in transgenic mice. *Nat. Biotechnol.* **19:**66–70.

57. **Khambarty, F. M., R. W. Bennett, and D. B. Shah.** 1994. Application of pulsed-field gel electrophoresis to the epidemiological characterization of *Staphylococcus intermedius* implicated in a food-related outbreak. *Epidemiol. Infect.* **113:**75–81.

58. **Kiri, N., G. Archer, and M. W. Climo.** 2002. Combinations of lysostaphin with beta-lactams are synergistic against oxacillin-resistant *Staphylococcus epidermidis*. *Antimicrob. Agents Chemother.* **46:**2017–2020.

59. **Kloos, W. E.** 1986. Ecology of human skin, p. 37–50. *In* P. A. Mardh and K. H. Schleifer (ed.), *Coagulase-Negative Staphylococci*. Almqvist & Wiksell International, Stockholm, Sweden.

60. **Kloos, W. E., D. N. Ballard, J. A. Webster, R. J. Hubner, A. Tomasz, I. Couto, G. L. Sloan, H. P. Dehart, F. Fiedler, K. Schubert, H. de Lencastre, I. S. Sanches, H. E. Heath, P. A. Leblanc, and A. Ljungh.** 1997. Ribotype delineation and description of *Staphylococcus sciuri* subspecies and their potential as reservoirs of methicillin resistance and staphylolytic enzyme genes. *Int. J. Syst. Bacteriol.* **47:**313–323.

61. **Kloos, W. E., and T. L. Bannerman.** 1994. Update on clinical significance of coagulase-negative staphylococci. *Clin. Microbiol. Rev.* **7:**117–140.

62. **Kloos, W. E., C. G. George, J. S. Olgiate, P. L. Van, M. L. McKinnon, B. L. Zimmer, E. Muller, M. P. Weinstein, and S. Mirrett.** 1998. *Staphylococcus hominis* subsp. *novobiosepticus* subsp. nov., a novel trehalose- and N-acetyl-D-glucosamine-negative, novobiocin- and multiple-antibiotic-resistant sub-

species isolated from human blood cultures. *Int. J. Syst. Bacteriol.* **3:**799–812.

63. **Kloos, W. E., and J. F. Wolfshohl.** 1979. Evidence of desoxyribonucleotide sequence divergence between staphylococci living on human and other primate skin. *Curr. Microbiol.* **3:**167–172.

64. **Kloos, W. E., and J. F. Wolfshohl.** 1991. *Staphylococcus cohnii* subspecies: *Staphylococcus cohnii* subsp. *cohnii* subsp. nov. and *Staphylococcus cohnii* subsp. *urealyticum* subsp. nov. *Int. J. Syst. Bacteriol.* **41:**284–289.

65. **Kluytmans, J., H. Berg, P. Steegh, F. Vandenesch, J. Etienne, and A. van Belkum.** 1998. Outbreak of *Staphylococcus schleiferi* wound infections: strain characterization by randomly amplified polymorphic DNA analysis, PCR ribotyping, conventional ribotyping, and pulsed-field gel electrophoresis. *J. Clin. Microbiol.* **36:**2214–2219.

66. **Kokai-Kun, J. F., S. M. Walsh, T. Chanturiya, and J. J. Mond.** 2003. Lysostaphin cream eradicates *Staphylococcus aureus* nasal colonization in a cotton rat model. *Antimicrob. Agents Chemother.* **47:**1589–1597.

67. **Krzeminski, Z., and A. Raczynska.** 1990. Elastolytic activity of staphylococci isolated from human oral cavity. *Med. Dosw. Mikrobiol.* **42:**1–4.

68. **Lambe, D. W., K. P. Fergusson, J. L. Keplinger, C. G. Gemmell, and J. H. Kalbfleisch.** 1990. Pathogenicity of *Staphylococcus lugdunensis*, *Staphylococcus schleiferi*, and three other coagulase-negative staphylococci in a mouse model and possible virulence factors. *Can. J. Microbiol.* **36:**455–463.

69. **Latorre, M., P. M. Rojo, R. Franco, and R. Cisterna.** 1993. Endocarditis due to *Staphylococcus capitis* subspecies *ureolyticus*. *Clin. Infect. Dis.* **16:**343–344.

70. **Lee, J.** 1994. *Staphylococcus intermedius* isolated from dog-bite wounds. *J. Infect.* **29:**105–118.

71. **Liljeqvist, S., P. Samuelson, M. Hansson, T. N. Nguyen, H. Binz, and S. Stahl.** 1997. Surface display of the cholera toxin B subunit on *Staphylococcus xylosus* and *Staphylococcus carnosus*. *Appl. Environ. Microbiol.* **63:**2481–2488.

72. **Linehan, D., J. Etienne, and D. Sheehan.** 2003. Relationship between haemolytic and sphingomyelinase activities in a partially purified beta-like toxin from *Staphylococcus schleiferi*. *FEMS Immunol. Med. Microbiol.* **36:**95–102.

73. **Long, J. P., J. Hart, W. Albers, and F. A. Kapral.** 1992. The production of fatty acid modifying enzyme (FAME) and lipase by various staphylococcal species. *J. Med. Microbiol.* **37:**232–234.

74. **Mahoudeau, I., X. Delabranche, G. Prevost, H. Monteil, and Y. Piemont.** 1997. Frequency of isolation of *Staphylococcus intermedius* from humans. *J. Clin. Microbiol.* **35:**2153–2154.

75. **Marsou, R., M. Bes, M. Boudouma, Y. Brun, H. Meugnier, J. Freney, F. Vandenesch, and J. Etienne.** 1999. Distribution of *Staphylococcus sciuri* subspecies among human clinical specimens, and profile of antibiotic resistance. *Res. Microbiol.* **150:**531–541.

76. **Miedzobrodzki, J., A. S. Naidu, J. L. Watts, P. Ciborowski, K. Palm, and T. Wadstrom.** 1989. Effect of milk on fibronectin and collagen type I binding to *Staphylococcus aureus* and coagulase-negative staphylococci isolated from bovine mastitis. *J. Clin. Microbiol.* **27:**540–544.

77. **Minhas, T., H. A. Ludlam, M. Wilks, and S. Tabaqchali.** 1995. Detection by PCR and analysis of the distribution of a fibronectin-binding protein gene (*fbn*) among staphylococcal isolates. *J. Med. Microbiol.* **42:**96–101.

78. **Nilsson, M., J. Bjerketorp, A. Wiebensjo, A. Ljungh, L. Frykberg, and B. Guss.** 2004. A von Willebrand factor-binding protein from *Staphylococcus lugdunensis*. *FEMS Microbiol. Lett.* **234:**155–161.

79. **Oh, B., H. Kim, J. Lee, S. Kang, and T. Oh.** 1999. *Staphylococcus haemolyticus* lipase: biochemical properties, substrate specificity and gene cloning. *FEMS Microbiol. Lett.* **179:**385–392.

80. **Olszewski, W. L., S. Jamal, G. Manokaran, S. Pani, V. Kumaraswami, U. Kubicka, B. Lukomska, F. M. Tripathi, E. Swoboda, F. Meisel-Mikolajczyk, E. Stelmach, and M. Zaleska.** 1999. Bacteriological studies of blood, tissue fluid, lymph and lymph nodes in patients with acute dermatolymphangioadenitis (DLA) in course of 'filarial' lymphedema. *Acta Trop.* **73:**217–224.

81. **Orden, J. A., J. Goyache, J. Hernandez, A. Domenech, G. Suarez, and E. Gomez-Lucia.** 1992. Production of staphylococcal enterotoxins and TSST-1 by coagulase negative staphylococci isolated from ruminant mastitis. *Zentralbl. Veterinarmed. B* **39:**144–148.

82. **Ottenwalder, B., T. Kupke, S. Brecht, V. Gnau, J. Metzger, G. Jung, and F. Gotz.** 1995. Isolation and characterization of genetically engineered gallidermin and epidermin analogs. *Appl. Environ. Microbiol.* **61:**3894–3903.

83. **Paulsonn, M., C. Petersson, and A. Ljungh.** 1993. Serum and tissue protein binding and cell surface properties of *Staphylococcus lugdunensis*. *J. Med. Microbiol.* **38:**96–102.

84. **Paulsson, M., A. Ljungh, and T. Wadstrom.** 1992. Rapid identification of fibronectin, vitronectin, laminin, and collagen cell surface binding proteins on coagulase-negative staphylococci by particle agglutination assays. *J. Clin. Microbiol.* **30:**2006–2012.

85. **Peacock, S. J., G. Lina, J. Etienne, and T. J. Foster.** 1999. *Staphylococcus schleiferi* subsp. *schleiferi* expresses a fibronectin-binding protein. *Infect. Immun.* **67:**4272–4275.

86. **Prevost, G., T. Bouakham, Y. Piemont, and H. Monteil.** 1995. Characterisation of a synergohymenotropic toxin produced by *Staphylococcus intermedius*. *FEBS Lett.* **376:**135–140.

87. **Raimundo, O., H. Heussler, J. B. Bruhn, S. Suntrarachun, N. Kelly, M. A. Deighton, and S. M. Garland.** 2002. Molecular epidemiology of coagulase-negative staphylococcal bacteraemia in a newborn intensive care unit. *J. Hosp. Infect.* **51:**33–42.

88. **Raus, J., and D. N. Love.** 1990. Comparison of the staphylocoagulase activities of *Staphylococcus aureus* and *Staphylococcus intermedius* on Chromozym-TH. *J. Clin. Microbiol.* **28:**207–210.

89. **Razonable, R. R., D. G. Lewallen, R. Patel, and D. R. Osmon.** 2001. Vertebral osteomyelitis and prosthetic joint infection due to *Staphylococcus simulans*. *Mayo Clin. Proc.* **76:**1067–1070.

90. **Sandoe, J. A., K. G. Kerr, G. W. Reynolds, and S. Jain.** 1999. *Staphylococcus capitis* endocarditis: two cases and review of the literature. *Heart* **82:**e1.

91. **Sato, H., K. Hirose, R. Terauchi, S. Abe, I. Moromizato, S. Kurokawa, and N. Maehara.** 2004. Purification and characterization of a novel *Staphylococcus chromogenes* exfoliative toxin. *J. Vet. Med. B Infect. Dis. Vet. Public Health* **51:**116–122.

92. **Schleifer, K. H.** 1986. Gram-positive cocci, p. 999–1103. *In* P. H. A. Sneath, N. S. Mair, M. E. Sharpe, and J. G. Holt (ed.), *Bergey's Manual of Systematic Bacteriology*, vol. 2. Williams & Wilkins, Baltimore, Md.

93. **Schwarz, S., C. Kehrenberg, and K. K. Ojo.** 2002. *Staphylococcus sciuri* gene *erm*(33), encoding inducible resistance to macrolides, lincosamides, and streptogramin B antibiotics, is a product of recombination between

erm(C) and *erm*(A). *Antimicrob. Agents Chemother.* **46:**3621–3623.

94. **Schwarz, S., C. Werckenthin, and C. Kehrenberg.** 2000. Identification of a plasmid-borne chloramphenicol-florfenicol resistance gene in *Staphylococcus sciuri. Antimicrob. Agents Chemother.* **44:**2530–2533.

95. **Shimizu, A., H. A. Berkhoff, W. E. Kloos, C. G. George, and D. N. Ballard.** 1996. Genomic DNA fingerprinting, using pulsed-field gel electrophoresis, of *Staphylococcus intermedius* isolated from dogs. *Am. J. Vet. Res.* **57:**1458–1462.

96. **Shuttleworth, R., R. J. Behme, A. McNabb, and W. D. Colby.** 1997. Human isolates of *Staphylococcus caprae*: association with bone and joint infections. *J. Clin. Microbiol.* **35:**2537–2541.

97. **Stepanovic, S., D. Vukovicc, V. Trajkovic, T. Samardzic, M. Cupic, and M. Svabic-Vlahovic.** 2001. Possible virulence factors of *Staphylococcus sciuri.* FEMS Microbiol. *Lett.* **199:**47–53.

98. **Stout, R. D., Y. Li, A. R. Miller, and D. W. Lambe.** 1994. Staphylococcal glycocalyx activates macrophage prostaglandin E$_2$ and interleukin 1 production and modulates tumor necrosis factor alpha and nitric oxide production. *Infect. Immun.* **62:**4160–4166.

99. **Sugai, M., T. Fujiwara, K. Ohta, H. Komatsuzawa, M. Ohara, and H. Suginaka.** 1997. *epr*, which encodes glycylglycine endopeptidase resistance, is homologous to *femAB* and affects serine content of peptidoglycan cross bridges in *Staphylococcus capitis* and *Staphylococcus aureus. J. Bacteriol.* **179:**4311–4318.

100. **Tabe, Y., A. Nakamura, and J. Igari.** 2001. Glycopeptide susceptibility profiles of nosocomial multiresistant *Staphylococcus haemolyticus* isolates. *J. Infect. Chemother.* **7:** 142–147.

101. **Talon, R., N. Dublet, M. C. Montel, and M. Cantonnet.** 1995. Purification and characterization of extracellular *Staphylococcus warneri* lipase. *Curr. Microbiol.* **30:**11–16.

102. **Tee, W. S., S. Y. Soh, R. Lin, and L. H. Loo.** 2003. *Staphylococcus lugdunensis* carrying the *mecA* gene causes catheter-associated bloodstream infection in premature neonate. *J. Clin. Microbiol.* **41:**519–520.

103. **Tegmark, K., E. Morfeldt, and S. Arvidson.** 1998. Regulation of *agr*-dependent virulence genes in *Staphylococcus aureus* by RNAIII from coagulase-negative staphylococci. *J. Bacteriol.* **180:**3181–3186.

104. **Terauchi, R., H. Sato, T. Hasegawa, T. Yamaguchi, C. Aizawa, and N. Maehara.** 2003. Isolation of exfoliative toxin from *Staphylococcus intermedius* and its local toxicity in dogs. *Vet. Microbiol.* **94:**19–29.

105. **Thumm, G., and F. Gotz.** 1997. Studies on prolysostaphin processing and characterization of the lysostaphin immunity factor (Lif) of *Staphylococcus simulans* biovar *staphylolyticus. Mol. Microbiol.* **23:**1251–1265.

106. **Valle, J., E. Gomez-Lucia, S. Piriz, J. Goyache, J. A. Orden, and S. Vadillo.** 1990. Enterotoxin production by staphylococci isolated from healthy goats. *Appl. Environ. Microbiol.* **56:**1323–1326.

107. **Vandenesch, F., S. J. Eykyn, M. Bes, H. Meugnier, J. Fleurette, and J. Etienne.** 1995. Identification and ribotypes of *Staphylococcus caprae* isolates isolated as human pathogens and from goat milk. *J. Clin. Microbiol.* **33:**888–892.

108. **Vandenesch, F., S. J. Eykyn, J. Etienne, and J. Lemozy.** 1995. Skin and post-surgical wound infections due to

Staphylococcus lugdunensis. Clin. Microbiol. Infect. **1:**73–74.

109. **Vandenesch, F., S. J. Eykyn, and J. Etienne.** 1995. Infections caused by newly-described species of coagulase-negative staphylococci. *Rev. Med. Microbiol.* **6:**94–100.

110. **Vandenesch, F., C. Lebeau, M. Bes, G. Lina, B. Lina, T. Greenland, Y. Benito, Y. Brun, J. Fleurette, and J. Etienne.** 1994. Clotting activity in *Staphylococcus schleiferi* subspecies from human patients. *J. Clin. Microbiol.* **32:**388–392.

111. **Vandenesch, F., S. Projan, B. Kreiswirth, J. Etienne, and R. P. Novick.** 1993. *agr* related sequences in *Staphylococcus lugdunensis. FEMS Microbiol. Lett.* **111:**115–122.

112. **Vandenesch, F., M. J. Storrs, F. Poitevin-Later, J. Etienne, P. Courvalin, and J. Fleurette.** 1991. Delta-like haemolysin produced by *Staphylococcus lugdunensis. FEMS Microbiol. Lett.* **78:**65–68.

113. **van der Mee-Marquet, N., A. Achard, L. Mereghetti, A. Danton, M. Minier, and R. Quentin.** 2003. *Staphylococcus lugdunensis* infections: high frequency of inguinal area carriage. *J. Clin. Microbiol.* **41:**1404–1409.

114. **Van Der Zwet, W. C., Y. J. Debets-Ossenkopp, E. Reinders, M. Kapi, P. H. Savelkoul, R. M. Van Elburg, K. Hiramatsu, and C. M. Vandenbroucke-Grauls.** 2002. Nosocomial spread of a *Staphylococcus capitis* strain with heteroresistance to vancomycin in a neonatal intensive care unit. *J. Clin. Microbiol.* **40:**2520–2525.

115. **van Kampen, M., J. W. Simons, N. Dekker, M. R. Egmond, and H. M. Verheij.** 1998. The phospholipase activity of *Staphylococcus hyicus* lipase strongly depends on a single Ser to Val mutation. *Chem. Phys. Lipids* **93:** 39–45.

116. **van Kampen, M. D., R. Rosenstein, F. Gotz, and M. R. Egmond.** 2001. Cloning, purification and characterisation of *Staphylococcus warneri* lipase 2. *Biochim. Biophys. Acta* **1544:**229–241.

117. **Vannuffel, P., M. Heusterspreute, M. Bouyer, B. Vandercam, M. Philippe, and J. L. Gala.** 1999. Molecular characterization of femA from *Staphylococcus hominis* and *Staphylococcus saprophyticus*, and femA-based discrimination of staphylococcal species. *Res. Microbiol.* **150:**129–141.

118. **Vernozy-Rozand, C., C. Mazuy, G. Prevost, C. Lapeyre, M. Bes, Y. Brun, and J. Fleurette.** 1996. Enterotoxin production by coagulase-negative staphylococci isolated from goats' milk and cheese. *Int. J. Food Microbiol.* **30:**271–280.

119. **Wagner, E., J. Doskar, and F. Gotz.** 1998. Physical and genetic map of the genome of *Staphylococcus carnosus* TM300. *Microbiology* **144:**509–517.

120. **Watson, D. C., M. Yaguchi, J. G. Bisaillon, R. Beaudet, and R. Morosoli.** 1988. The amino acid sequence of a gonococcal growth inhibitor from *Staphylococcus haemolyticus. Biochem. J.* **252:**87–93.

121. **Weiss, K., D. Rouleau, and M. Laverdiere.** 1996. Cystitis due to vancomycin-intermediate *Staphylococcus saprophyticus. J. Antimicrob. Chemother.* **37:**1039–1040.

122. **Wu, S., H. de Lencastre, and A. Tomasz.** 1998. Genetic organization of the *mecA* region in methicillin-susceptible and methicillin-resistant strains of *Staphylococcus sciuri. J. Bacteriol.* **180:**236–242.

123. **Wu, S. W., H. de Lencastre, and A. Tomasz.** 2001. Recruitment of the *mecA* gene homologue of *Staphylococcus sciuri* into a resistance determinant and expression of the resistant phenotype in *Staphylococcus aureus. J. Bacteriol.* **183:**2417–2424.

Antibiotic Resistance in the Staphylococci

STEVEN J. PROJAN AND ALEXEY RUZIN

48

This chapter summarizes specific resistance mechanisms found in the staphylococci; several other chapters address resistance to various specific classes of antimicrobial agents across the gram-positive spectrum.

The importance of antimicrobial chemotherapy for the treatment of staphylococcal infections cannot be overstated. Before the widespread use of antimicrobials, *Staphylococcus aureus* bacteremia was fatal approximately 90% of the time (104). While the therapeutic use of penicillin G in the early 1940s greatly reduced mortality, the first resistant strains were described almost immediately (36). Today, we have at our disposal a large number of nominally efficacious antimicrobial agents which are active against the staphylococci. Unfortunately, we are also confronted with a dazzling array of resistance determinants and mutant strains compromising the utility of all but one of the classes of antimicrobial agents against one of the most virulent of pathogenic organisms.

THE CONSEQUENCE OF ANTIMICROBIAL RESISTANCE IN STAPHYLOCOCCI

What are the consequences of resistance with respect to the staphylococci? First and foremost, resistance effectively reduces the number of therapeutic options available to treat staphylococcal infections. While actual reports of clinical failures of antimicrobial chemotherapy for staphylococcal infections are rare, they are increasing in frequency. In addition, there are data that suggest a far greater risk to the patient infected with a resistant organism than to a patient infected with a susceptible strain of staphylococcus. In at least one report, the attributable mortality for patients infected with strains of methicillin-resistant *S. aureus* (MRSA) was 10-fold higher than for patients infected with methicillin-susceptible *S. aureus* strains (72). There have been precious few studies tracking the economic impact of resistant infections in general, much less staphylococcal infections in particular. To date, one privately funded study in 1999 remains the definitive study and clearly demonstrates the serious fiscal costs of resistant versus susceptible infections (90). However, it should be pointed out that some infections, even those

caused by susceptible staphylococci, do not respond well to antimicrobial chemotherapy; among these are osteomyelitis and bacterial endocarditis, which often require surgical intervention (122).

RESISTANCE AND THE SOURCE OF THE INFECTION

When discussing resistance among the staphylococci, it is important to draw a distinction between community-acquired versus hospital-acquired (nosocomial) infections. Almost uniformly, resistance rates among community isolates are significantly lower than resistance rates for nosocomial isolates. Ascertaining the source of an infection can, therefore, have important implications in choosing a course of antimicrobial chemotherapy. The fact that hospital isolates appear to have higher rates of resistance implies that these isolates are epidemiologically distinct from community isolates; the molecular biology of these strains certainly bears that hypothesis out. This also implies that nosocomial isolates represent microflora that are resident in the hospital and are either transferred from patient to patient via health care workers or transferred from colonized health care workers to patients. As surgical site infection is one of the nosocomial infections often attributed to staphylococci, one may wonder how such infections were prevented prior to the advent of antimicrobial chemotherapy, given our current reliance on antibiotic prophylaxis. One practice was to perform the surgery in a mist of carbolic acid (phenol), which, while not well tolerated by many participating in such surgeries, did serve to prevent a large number of infections. The modern lesson to be drawn from this is that we now use antibiotics as a substitute for good hygiene and sound surgical practice (84).

Within the last 5 years, there has been a clear increase in the number of community-acquired infections caused by strains of MRSA. These infections have been called community-onset or (more commonly) community-acquired MRSA. These strains, which have become widely disseminated, are clearly increasing in number (94). In general, these infections have mainly been less serious skin infections; however, some

Gram-Positive Pathogens, 2nd edition, edited by Vincent A. Fischetti et al.
© 2006 ASM Press, Washington, D.C.

cases of severe invasive pneumonia with an alarming mortality rate have been described (115). While they are resistant to few other antibiotics at the present time (the proposed therapy is to use the lincomycin antibiotic clindamycin) (66), it is considered only a matter of time before these strains accumulate multiple resistances and dramatically alter clinical practice.

FACTORS THAT HAVE LED TO THE EMERGENCE AND DISSEMINATION OF RESISTANCE

Discussions of the root cause(s) of bacterial drug resistance often generate more heat than light. These discussions are complicated by a lack of supporting data for many of the hypotheses brought forward to explain why we observe resistance. However, it is clear that, as stated by Levy, "Given sufficient time and drug use, antibiotic resistance will emerge" (58).

Antimicrobial resistance among the staphylococci is not a phenomenon limited to antibiotics. Strains resistant to arsenicals and mercury were identified well before what is now known as the antibiotic era, but for the purposes of this chapter the discussion will be limited to agents currently used clinically to treat staphylococcal infections.

MECHANISMS OF RESISTANCE

It is now dogma that resistance falls into one of three mechanistic classes: (i) prevention of accumulation within the bacterial cells, usually via efflux of the agent out of the bacterial cell by either dedicated or general efflux pumps (70); (ii) alteration of the molecular target of the antibiotic; and (iii) inactivation of the antibiotic. In fact, even for the above list there are some resistance genes that do not really fall neatly into any of those three categories (e.g., resistance to the semisynthetic tetracycline minocycline via ribosome protection) and in the case of some (if not most) of the classes of drugs, more than one resistance mechanism is at play (Table 1).

GENETICS OF RESISTANCE

From the genetic point of view, resistance falls into one of two classes: mutation of a bacterial gene or acquisition of a dedicated resistance gene from some other organism by some form of genetic exchange (transduction, conjugation, or transformation). In general, resistance via mutation is an alteration of the target site of the antibiotic, although increased expression of either the target (and titration of the antimicrobial agent) or a nonspecific efflux pump also can be the result of a mutation. Acquired resistance determinants are, by and large, dedicated to narrow classes of compounds, but they can run the gamut of mechanisms and are usually inducible (meaning that the gene encoding the resistance is expressed by the host bacterium when it is exposed to the drug, at concentrations insufficient to inhibit bacterial growth). Dissemination of these acquired-resistance determinants varies in frequency, depending on the type of genetic element carrying the resistance determinant.

As pointed out below (and in chapter 62), the presence of the methicillin resistance determinant by itself in a strain is not sufficient for the bacterium to be phenotypically methicillin resistant. Apparently, a series of other genetic events must take place for methicillin resistance to be manifest; although this case may be unique, this implies that acquisition of a resistance determinant by itself may not be sufficient for the expression of resistance.

TABLE 1 Agents used to treat staphylococcal infections

Agent	Molecular target
Cell wall biosynthesis inhibitors	
Beta-lactams (ampicillin)	PBPs (transpeptidation)
Glycopeptides (vancomycin, teicoplanin)	MurNac pentapeptide (transglycosylation)
Membrane-active agents	
Daptomycin	Lipoteichoic acid, cell membrane
Transcription inhibitors	
Rifampicin	RNA polymerase
DNA inhibitors	
Fluoroquinolones (ciprofloxacin)	DNA gyrase; topoisomerase IV
Protein synthesis inhibitors	
Aminoglycosides (gentamicin)	Bacterial ribosome
Tetracyclines (minocycline, doxycycline)	
Chloramphenicol	
MLS$_B$ group (erythromycin)	
Streptogramin A	
Pseudomonic acid (mupirocin)	
Fusidic acid	
Oxazolidinones (linezolid)	
Essential small molecule biosynthesis inhibitors	
Sulfonamides	Dihydropterate synthetase
Trimethoprim	Dihydrofolate reductase

BETA-LACTAM RESISTANCE

In the history of antimicrobial chemotherapy, the most useful of antistaphylococcal agents have been the beta-lactam antibiotics, the prototype of which is penicillin. These agents, which include several structural classes (e.g., penams, penems, cephamycins, cephalosporins, carbapenems, and monobactams), all contain one common structural feature: the beta-lactam ring. This four-member ring structure can be thought of as the first peptide mimetic in that the target of these drugs is the transpeptidase domain of the cell wall biosynthetic enzymes referred to as penicillin-binding proteins (PBPs). Binding of the antibiotic blocks peptidoglycan biosynthesis and often results in the subsequent induction of autolysins, leading to lysis of the bacterial cells. As a rule, the beta-lactams are far more effective in killing growing (dividing) bacteria.

Resistance

Not long after the introduction of penicillin into clinical practice, the first beta-lactamase-producing staphylococci were described. The rapid dissemination of strains carrying genes encoding the beta-lactamases, enzymes that hydrolyze the beta-lactam ring and therefore inactivate the antibiotic, had the effect of almost ending the antibiotic era before it got started. Today, >90% of clinical isolates of *S. aureus* produce beta-lactamases.

Subsequent to the discovery of the beta-lactamases in *S. aureus*, it was found that the genes encoding their production are all very similar, with only minor differences in amino acid sequence and substrate specificity. Despite these minor differences, it was possible to classify beta-lactamase by serology (117). The genes encoding both the beta-lactamase structural gene (*blaZ*) and the two regulatory genes encoding the sensor-transducer (*blaR1*) and the repressor (*blaI*) are found on a transposable element (Tn552), which itself is often found on a plasmid (89). Induction by a beta-lactam leads to the increased production of beta-lactamase, due to proteolytic inactivation of the repressor (41). More recent data provide evidence for proteolytic cleavage of both the transducer and the repressor, thus suggesting a unique signaling cascade that has not been previously described for bacteria (37, 123).

To combat resistance mediated by beta-lactamases, semisynthetic penicillins were developed (e.g., methicillin and oxacillin) which could be only very poorly hydrolyzed by the staphylococcal beta-lactamases and were, for many years, extremely effective at controlling most staphylococcal infections. In fact, it was not until the advent of the MRSA strains that antibiotic resistance among the staphylococci again became a public health concern.

Methicillin Resistance

MRSA is a something of a misnomer, because these strains can often be resistant to all clinically available beta-lactam antibiotics, rendering these once-potent agents ineffective in treating staphylococcal infections. Recently, there have been several reports of novel beta-lactams that have good levels of activity against MRSA. However, none of these are yet in clinical development. Interestingly, methicillin is no longer used therapeutically, and many researchers who long used the agent in their work are finding it difficult to obtain supplies for their studies; oxacillin is often used instead. The genetic basis of this resistance to all beta-lactam antibiotics is the *mecA* gene that encodes PBP2a, which has a low affinity for beta-lactams. Similar to organization of the *bla* gene cluster, the *mecA* operon includes *mecR1* and *mecI* genes that are homologous to *blaR1* and *blaI*, respectively. While the MecI and BlaI repressors are able to substitute for each other, the sensor-transducers MecR1 and BlaR1 can only recognize their cognate repressors, providing the specificity of induction (59, 64). The *mecA* gene is carried on staphylococcal chromosome cassette *mec* (SCC*mec*), a large mobile genetic element that apparently was introduced into *S. aureus* from an exogenous source (51). The evolutionary origin of *mecA* and the SCC*mec* element is unclear; separate studies describe either *Staphylococcus sciuri* or *Staphylococcus haemolyticus* as a possible source (7, 121).

Until recently, the model of resistance postulated that upon exposure to beta-lactam antibiotics PBP2a takes over the biosynthesis of the cell wall, whereas all native staphylococcal PBPs (PBP1, -2, -2b, -3, and -4) are completely inactivated. According to the revised model, transpeptidase function of PBP2a cooperates with transglycosylase function of the native PBP2 to catalyze cell wall biosynthesis in the presence of beta-lactams (78, 79). However, the mere presence of the *mecA* gene is not sufficient for high-level methicillin resistance. One of the most interesting phenomena among certain strains carrying the *mecA* gene is a high degree of heterotypic expression of resistance to methicillin. This heterotypy is revealed by population analyses of a genetically homogenous culture of *S. aureus*. The observation is that a large proportion of the bacteria in the culture are apparently susceptible to the antibiotic (typically, oxacillin is used), and a relatively small proportion are resistant to concentrations as high as 800 μg/ml. The apparently resistant colonies, however, maintain the same heterotypic phenotype when recultured and similarly analyzed (113).

While part of this phenomenon may be associated with expression of the *mecA* gene, which apparently requires the genetic inactivation of MecI repressor (69), mutations in other genes termed *fem* (for factors essential for expression of methicillin resistance) or *aux* (for auxiliary) genes can affect homogenous, high-level resistance (9, 10, 30, 31). At present, it is estimated there are >20 such genes. A number of them are involved in peptidoglycan biosynthesis, and others encode putative regulators, while the functions of remaining genes are still poorly understood. Among the group involved in cell wall biosynthesis are *murE*, which is required for the addition of L-lysine to the peptide moiety of cell wall precursor (61), and *femAB* operon, which together with *fmhB* is necessary for the formation of the pentaglycine cross bridge (44). Interestingly, a recent study suggested that *murE*, either directly or indirectly, controls the expression of both *pbpB* (the structural gene of PBP2) and *mecA* (38).

The genetic background of the host strain also plays a significant role in restricting the horizontal spread of *mecA*. It has been shown that expression and maintenance of the methicillin resistance determinant were facilitated in the presence of *bla* regulatory genes (14, 26, 45, 53, 93, 105). An additional factor is the selective pressure from other antibiotics. For example, one study reported that deletion of *mecA* was triggered by the acquisition of vancomycin resistance (1). Thus, the existence of host-specific and environmental barriers may account for the limited clonal nature of MRSA.

DAPTOMYCIN

This lipopeptide natural product, a drug that was originally described in 1986 and finally approved for use in 2003, is a membrane-active agent. Daptomycin was originally reported to be an inhibitor of lipoteichoic acid biosynthesis, although recent data suggest that it acts by disrupting the bacterial membrane (103). Approved narrowly for the treatment of complicated skin and skin structure infections caused by gram-positive pathogens, daptomycin has seen limited use, and there have been no reports of resistant clinical strains. However, because daptomycin is a concentration-dependent cidal agent, it may find significant utility in deep-seated, difficult-to-treat infections such as infectious endocarditis.

GLYCOPEPTIDES

Because MRSA strains are uniformly resistant to beta-lactams and are often resistant to most if not all of the agents discussed below, the last remaining line of defense lies in the glycopeptide antibiotics vancomycin and teicoplanin. These antibiotics are also cell wall biosynthesis inhibitors and block polymerization by binding to the terminal D-Ala-D-Ala of the N-acetylmuramic acid (MurNAc) pentapeptide.

Resistance

Until the late 1990s, S. aureus was uniformly susceptible to vancomycin. Later, however, a number of clinical isolates of S. aureus with reduced susceptibility to glycopeptides (vancomycin MICs of 8 to 16 µg/ml) have been described. These strains, called glycopeptide intermediate-susceptible S. aureus (GISA), have arisen in a very few patients who have been subject to long-term antimicrobial chemotherapy with vancomycin. This low-level resistance is thought to result from several, stepwise mutations yielding incrementally higher MICs. Similar mutants arise in laboratory strains, also subject to long-term selection.

The common features of GISA strains include various cell wall abnormalities and altered expression of PBPs (27, 35, 62, 75, 101, 102). While the exact mechanism of decreased glycopeptide susceptibility is still under investigation, the major role is attributed to the affinity trapping of glycopeptides due to several factors such as reduced peptidoglycan cross-linking, increased synthesis of nonamidated muropeptides, and increased cell wall thickness (27, 28). The results of another study suggested increase in transglycosylase activity as an additional contributing factor (55). In addition to the genes encoding PBPs, genetic studies identified a few other loci that may contribute to glycopeptide resistance or affect glycopeptide susceptibility in GISA. Among them are the *vra* and *tca* operons; the environmental stress-induced *sigB* operon; the quorum-sensing *agr* gene cluster; a number of other genes including *alt*, *ddh*, *mprF* (also known as *fmtC*); and genes that participate in fructose uptake, fatty acid metabolism, and purine biosynthesis (12, 15, 17, 54, 56, 63, 67, 91, 95, 96).

It should be noted that the therapeutic use of teicoplanin is somewhat controversial; it has not been approved for use in the United States, and it has been suggested that teicoplanin when used clinically may actually select for vancomycin-resistant staphylococci similar to the GISA strains described above (100).

In contrast with the low-level resistance in staphylococci that is confined to individual strains, high-level resistance to glycopeptide antibiotics (vancomycin MIC >1,000 µg/ml) is transferable and has been described in enterococci since the mid-1980s. In this case, the molecular target of glycopeptides is altered, typically due to the presence of either *vanA* or *vanB* gene clusters, both of which are carried by transposons. Detailed information on the mechanism of *van*-mediated resistance and the various types of *van* genes can be found in chapter 63.

Despite the fact that the possibility of transfer of *vanA* determinants from *Enterococcus faecalis* to S. aureus was reported by Noble and coworkers in 1992 (71), it took a decade for *vanA*-mediated resistance to emerge in S. aureus clinical strains. Since the summer of 2002, three clinical isolates of fully vancomycin-resistant S. aureus (VRSA) were reported in Michigan, Pennsylvania, and New York (19–21). Like all previously described clinical GISA strains, these new isolates were also MRSA, which raises the possibility of potentially untreatable staphylococcal infections. Although VRSA strains were susceptible to some of the old and some of the recently approved antibiotics, including trimethoprim-sulfamethoxazole, quinupristin-dalfopristin, linezolid, and daptomycin (16, 20, 22), their emergence stresses the need for the development of new antibiotics to combat multidrug resistance. Antimicrobial agents with activity against VRSA (and MRSA, for that matter) that are currently at various stages of development include glycopeptides oritavancin and dalbavancin and glycolipodepsipeptides ramoplanin and tigecycline, a novel glycylcycline. Tigecycline, which is currently approved in the United States and is in development worldwide, displayed excellent broad-spectrum antibacterial activity and complete lack of resistance among gram-positive pathogens. A novel class of glycopeptide antibiotics that deserves attention are the mannopeptimycins, which possess a unique mechanism of action and have a potent activity against a wide variety of gram-positive bacteria including MRSA, GISA, and vancomycin-resistant enterococci (43, 77, 92).

RIFAMPIN

Rifampin, a member of the rifamycin class of antibiotics, inhibits transcription by attacking the beta-subunit of RNA polymerase (118). Because RNA polymerase is required both for transcription and in the initiation of DNA replication, it is effectively a bifunctional antimicrobial; as such, it is especially potent in inhibiting the growth of nearly all bacteria, including the staphylococci.

Resistance

Despite the fact that rifampin is one of the most potent antistaphylococcal agents in vitro, with MICs of 0.03 µg/ml and less, resistance readily arises by mutation in the *rpoB* gene (which encodes the beta-subunit of RNA polymerase), resulting in a lower affinity for the antibiotic (8). Therefore, when rifampin is used therapeutically, it is often used in combination with other antibiotics (e.g., fusidic acid).

FLUOROQUINOLONES

The fluoroquinolone antimicrobials (which include ciprofloxacin and levofloxacin) are one of the few classes of antibacterial agents that are not based on a natural product. Rather, these drugs are derivatives of nalidixic acid and have found wide utility owing to their broad spectrum of activity and oral bioavailability. The fluoroquinolones specifically target the bacterial type II topoisomerases DNA gyrase and topoisomerase IV. These enzymes have double-stranded breaking and joining activity, with DNA gyrase being responsible for introducing negative superhelical turns and topoisomerase IV responsible for decatenation of the two daughter chromosomes on termination of replication. Treatment of bacteria with fluoroquinolones results in lethal accumulation of double-stranded breaks. As with the beta-lactams, the fluoroquinolones are more active against dividing bacteria. With the gram-positive bacteria, topoisomerase IV has been demonstrated to be the primary target of the fluoroquinolones while DNA gyrase has been shown to be the primary target in gram-negative bacteria.

Resistance

Resistance to the fluoroquinolones is a result of a series of stepwise mutations in the genes encoding DNA gyrase (*gyrA* and *gyrB*) and topoisomerase IV (*grlA* and *grlB*). In addition, mutations which result in the upregulation of an efflux pump (NorA) have been identified both in vitro and among clinically resistant isolates (48). It has been hypothesized that fluoroquinolones that would inhibit the activity of both DNA gyrase and topoisomerase IV equally would not be subject to the kind of bootstrapping of resistance that we currently observe (124). It has been suggested that natural products are poor starting points for the development of new antibiotics because the producing organisms (which are usually bacteria) are immune to the action of the antibiotics they produce and because the genetic information for that immunity will eventually be transferred into the target pathogenic bacteria after environmental selection with the antibiotic (112). This is undoubtedly (106) the case with tetracycline-producing strains, which harbor genes encoding tetracycline-specific efflux proteins clearly related to the efflux genes found in pathogenic bacteria (34). However, as we have seen with the fluoroquinolones, resistance can also readily arise de novo by mutation to a wholly synthetic class of agents, as well as to natural antibiotics such as rifampin.

AMINOGLYCOSIDES

The aminoglycoside class of antibiotics, as exemplified by streptomycin, are second in quantity only to the penicillins for use therapeutically. Drugs in this class include amikacin, gentamicin, netilmicin, and tobramycin. In general, they are protein synthesis inhibitors which bind irreversibly to the 30S subunit of the bacterial ribosome. The result of this binding is that aminoacyl-tRNAs are apparently unable to bind productively to the acceptor site, preventing elongation of the peptide chain. Mutations in the streptomycin-binding site can result in less-effective inhibition but can also result in a loss of fidelity in protein synthesis, with the insertion of incorrect amino acid residues into the elongating peptide. However, some aminoglycosides other than streptomycin may bind to multiple sites on the 30S subunit and are therefore less prone to mutations which alter the target of the antibiotics (97). The aminoglycosides are normally described as bactericidal which may be related to irreversible binding to their target. Aminoglycosides are also thought to be bactericidal because they are thought to have membrane disruption properties and to promote their own uptake into bacterial cells, which may contribute to their lethality. Spectinomycin, which lacks an amino sugar, has the same molecular target (the 30S ribosomal subunit) but binds reversibly and is bacteriostatic. The membrane penetrating activity of the aminoglycosides may also explain why aminoglycosides can synergize with other antimicrobials (especially the beta-lactams).

Resistance

In general, resistance to the aminoglycosides commonly used therapeutically is via enzymatic inactivation of the antibiotics (reviewed by Shaw et al. in reference 99). The plethora of aminoglycoside-modifying enzymes found in bacteria has probably derived from both producing organisms and mutation of cellular housekeeping genes (74). In the case of the staphylococci, the modifying enzymes appear to be of exogenous origin as they are found on large and small plasmids and transposons. Perhaps the most common aminoglycoside resistance genes are the *aac(6')-aph(2'')* genes found on Tn4001 (40). This gene actually encodes a single protein with two functional (and separable) domains having different enzymatic activities, *aac(6')*, with acetylating activity, and *aph(2'')*, with phosphorylating activity. Tn4001 and related transposons are often found associated with high-molecular-weight, conjugative plasmids (111) but can also be found on the staphylococcal chromosome (40). Somewhat less frequently found is *ant(4')-Ia* (encoding an adenylylase) followed by *aph(3')-IIIa*, encoding a phosphorylase. In addition, these genes (and the enzymes that they encode) are often found in combination (65). The streptomycin resistance gene on plasmid pS194, based on similarity to better-defined enzymes, is likely to be an adenylyltransferase gene (83).

Another potential resistance strategy is the formation of small-colony (or electron transport) variants. The role small-colony variants play in actual infections is still a subject of conjecture, but they can be isolated in vitro by selection for aminoglycoside resistance (see chapter 35).

TETRACYCLINES

The tetracyclines are a class of structurally related compounds, which include minocycline and doxycycline and are characterized by four interlocking six-carbon rings. These antibiotics inhibit protein synthesis by binding to the 30S subunit of the ribosome and block the entry of aminoacyl-tRNAs into the acceptor site. In general, the tetracyclines are bacteriostatic for the staphylococci (76); despite a mechanism of action similar to that of the aminoglycosides, this may be a function of the relative affinities of these drugs for their targets—i.e., aminoglycosides bind more avidly than tetracyclines.

Resistance

There are two tetracycline resistance genes that predominate among staphylococci. The *tet*(K) gene, almost exclusively found on pT181 and related plasmids, encodes a tetracycline efflux pump which is somewhat unusual in that it has 14 membrane-spanning alpha-helices as opposed to far more common 12 membrane-spanning alpha-helices seen widely in the major facilitator superfamily of transport proteins (42, 74). While *tet*(K) is relatively efficient in removing tetracycline from cells, it is not very effective in eliminating either doxycycline or minocycline. Far less frequently found in staphylococci is *tet*(L), which encodes an efflux pump similar in structure to that of *tet*(K). A third tetracycline resistance gene, *tet*(M), occurring in *S. aureus* is virtually identical to the *tet*(M) gene originally identified on the transposons Tn916 and Tn1545 in enterococci (24). However, in *S. aureus tet*(M) is apparently found at a constant chromosomal location and lacks the ability to transpose (68). While pT181 plasmids are ubiquitous, the *tet*(M) determinant is probably more clinically significant in that it encodes resistance to all tetracyclines, including minocycline and doxycycline. The expression of both *tet*(K) and *tet*(M) has been shown to be induced by subinhibitory concentrations of tetracyclines. While the mechanism of induction of *tet*(K) has not been well defined, the *tet*(M) gene is induced by a ribosome-stalling mechanism relieving transcriptional attenuation of the nascent mRNA (108).

CHLORAMPHENICOL

Chloramphenicol inhibits protein synthesis by blocking peptide bond formation. The antibiotic binds to the 50S ribosome subunit and interferes with the binding of the aminoacyl moiety of the aminoacyl-tRNA (76). Despite its potency and the fact that relatively few clinical isolates are resistant to chloramphenicol, it is rarely used in the treatment of infected patients. The reason for this is that approximately 1 in 50,000 patients treated with chloramphenicol develops aplastic anemia, which is invariably fatal. However, with the emergence of glycopeptide-resistant staphylococci, leaving only novel therapies and agents of questionable efficacy, chloramphenicol treatment may represent a viable therapeutic option for life-threatening infections.

Resistance

To date, resistance to chloramphenicol in the staphylococci is associated solely with inactivation by chloramphenicol acetyltransferases encoded by *cat* genes. The *cat* genes encoding these enzymes in the staphylococci have been found on a number of small plasmids (e.g., pC194, pC221, pC223, and pUB112). Despite the diversity of the replicons with which they are associated, these *cat* genes are all clearly related with and similar to genes encoding chloramphenicol acetyltransferases in other genera. The staphylococcal genes appear to be uniformly inducible, with the version found on pUB112 the most intensely studied. The primary mechanism of induction appears to be translational attenuation (33), which is a common theme among chloramphenicol resistance genes (60).

MLS$_B$

This group of protein synthesis inhibitors (the so-called macrolide-lincosamide-streptogramin B [MLS$_B$] group) are related both structurally and mechanistically. They function by binding to the 50S ribosomal subunit and inhibiting the peptidyl transferase reaction (13). These agents are bacteriostatic for the staphylococci unless combined with agents such as streptogramin A (see below).

Resistance

The predominant form of resistance to these agents is methylation of the 23S rRNA encoded by three related genes (*ermA*, *ermB*, and *ermC*). This alteration totally eliminates the binding of the drug to its target. While all three of these determinants have been associated with mobile genetic elements, perhaps the most widespread of these is the *ermC* gene, which has been found mostly on small plasmids such as pE194 (49) or pE5 (82). The *ermC* gene nominally confers inducible resistance to macrolides (as does *ermA*), such as erythromycin, but not to lincosamides, such as clindamycin. This means that macrolides (at low concentrations) induce the expression of resistance to any MLS$_B$ antibiotic, whereas lincosamides (like clindamycin) do not (32). Therefore, a strain harboring a plasmid carrying a

wild-type *ermC* gene would be erythromycin resistant but clindamycin susceptible. However, deletions in the mRNA leader sequence upstream from the *ermC* coding sequence render the expression of resistance constitutive. Therefore, it is not surprising that, as the use of clindamycin became more widespread, constitutively resistant strains became commonplace (116). This is actually an important observation. It has long been presumed that constitutive expression of any resistance determinant would result in an unacceptable metabolic burden to the bacterium, resulting in a competitive disadvantage and subsequent counterselection. In fact, the ability of constitutively resistant strains to be maintained even in the absence of apparent selection demonstrates that (at least in this example) the metabolic burden is not always unacceptable. This also implies that merely removing selection (by restricting the use of certain antibiotics) will not necessarily result in a significant decrease in the proportion of resistant isolates.

Recently, a gene encoding efflux pumps specific for these agents has been reported. MsrA is a macrolide-specific ABC (ATP-binding cassette) transporter encoded by *msrA* (86, 87). Inactivation by *S. aureus* has been described for a streptogramin B compound, virginiamycin B, by the product of the *vgb* gene, a hydrolase (2). There has also been a report of inactivation (120) by a single clinical isolate of *S. aureus*, but the responsible gene and/or protein has not been described as yet.

STREPTOGRAMIN COMBINATIONS

These combinations, essentially natural product antimicrobials such as virginiamycin and pristinamycin, consist of a combination of at least a streptogramin A- and a streptogramin B-type antibacterial (116). The streptogramin A compounds are bacteriostatic protein synthesis inhibitors which bind to the peptidyl transferase domain of the 50S ribosomal subunit and prevent the binding of the 3′ terminal of aminoacyl-tRNA and peptidyl-tRNAs. The streptogramin B compounds also bind to the 50S ribosomal subunit (perhaps at subunit L4), inhibit the peptidyl transferase reaction (76), and are bacteriostatic for staphylococci. However, the combination of a streptogramin A and a streptogramin B is often bactericidal (13), and a combination of semisynthetic derivatives of the pristinamycin components (referred to as quinupristin and dalfopristin individually and by the trade name Synercid when given together) has recently been introduced for the treatment of infections with gram-positive bacteria (52).

Resistance to Streptogramin A and B Combinations

Resistance to agents of the streptogramin B type is discussed above. With respect to the A-type compounds, two forms of resistance have been noted: inactivation by acetylation and efflux. Several structurally related virginiamycin acetyl transferase genes have been described by El Solh and colleagues (5). These now include at least four genes (*vatA*, *vatB*, *vatC*, and *sat4*) found in the staphylococci. To date, a two-gene complex encoding a putative ABC transporter (*vga* and *vgaB*) has also been described (4); a plasmid, pIP630, has been identified carrying *vga*, *vatA*, and *vgb* (encoding a streptogramin B hydrolase described above).

As indicated above, the streptogramin combinations are considered bactericidal; however, activity toward a strain carrying resistance to either the A or B class is bacteriostatic, and it is considered that the critical factor in resistance to the combination is the presence of a gene encoding an acetyltransferase (3).

FUSIDIC ACID

Fusidic acid, which is steroidal in structure, inhibits protein synthesis by binding to elongation factor G (EF-G), although it has been suggested that, given the excess in EF-G within bacterial cells, fusidic acid acts by other mechanisms such as blocking EF-Tu-aminoacyl-tRNA complex formation (29). However, as mutations in the gene encoding EF-G in *S. aureus* lead to resistance, it is likely that EF-G is, indeed, the target of fusidic acid. It should be noted that while fusidic acid is not approved for use in the United States, it has been used effectively (mainly as a topical agent) and in combination with other agents such as rifampin for the treatment of MRSA infections.

Resistance

Resistance to fusidic acid is mainly due to mutations in the gene encoding EF-G, the *fusA* gene. These mutations putatively decrease fusidic acid binding to its target. Plasmid-encoded fusidic acid resistance (*fusB*) has also been reported and is presumably mediated by an efflux mechanism (23).

OXAZOLIDINONES

Linezolid is the first member of a wholly synthetic class of protein synthesis inhibitors called the oxazolidinones. It has been shown to act by blocking the initiation of polypeptide synthesis and has been proven to be a valuable alternative to vancomycin treatment of MRSA (125). While resistance to linezolid is rare (107), cases have been reported where long-term exposure can eventually select for linezolid-resistant strains (119). Both in vitro and clinically derived linezolid resistance is the result of mutations in the genes encoding the target of the drug (the 50S ribosomal subunit) (109).

SULFONAMIDES AND TRIMETHOPRIM

Mechanism of Action

These agents are generally used in combination (e.g., Bactrim and cotrimoxazole). The sulfonamides are wholly synthetic analogs of *p*-aminobenzoic acid and are competitive inhibitors of dihydropteroate synthetase, an enzyme in the biosynthetic pathway for dihydrofolate. Dihydrofolate, a substrate of dihydrofolate reductase (DHFR), is reduced to tetrahydrofolate, which is a necessary cofactor for the biosynthesis of the amino acids glycine and methionine and of purines and pyrimidines. The toxic effect is probably due mainly to the depletion of thymidylate, resulting in thymineless death. Because trimethoprim, a nucleoside analog, is an inhibitor of bacterial DHFRs, the combination of sulfonamides and trimethoprim blocks two sequential steps in the pathway of several essential bacterial metabolites. Eukaryotic cells are generally not susceptible to trimethoprim because it has a much lower affinity for eukaryotic versions of DHFRs (46).

Resistance

Sulfonamide resistance in the staphylococci, which arose soon after the introduction of the sulfa drugs, is chromosomally encoded (by the *sulA* gene) and is attributed to the overproduction of *p*-aminobenzoate (57). While overpro-

duction of bacterial DHFR can provide level resistance to trimethoprim, in general high-level resistance is provided by a DHFR with a low affinity for trimethoprim. Genes encoding this enzyme are often found on large conjugative plasmids like pGO1 or pSK1 (6, 88) and together with a gene encoding a thymidylate synthase (*thyE*) are flanked by two copies of the insertion sequence IS*257*.

MUPIROCIN

Formerly known as pseudomonic acid, mupirocin blocks protein synthesis by competitively inhibiting isoleucyl-tRNA synthetase (50). Mupirocin (formulated with the trade name Bactroban) has come into wide use as a topical agent for the treatment of gram-positive infections and more recently has been employed successfully to treat nasal carriers of MRSA, especially those in chronic care settings (e.g., nursing homes) and hospital staff (85).

Resistance

Low-level resistance to mupirocin is likely due to mutations in the chromosomally encoded isoleucyl-tRNA synthetase, while high-level (and clinically relevant) resistance is clearly due to an acquired gene (*mupA*) encoding an isoleucyl-tRNA synthetase with reduced affinity for mupirocin (39, 47). To date, the *mupA* gene has been found associated with both small and large plasmids but always flanked by two copies of IS*257*. The origin of the *mupA* gene has yet to be traced.

COMPOUNDS IN DEVELOPMENT

While the increase in multiply-resistant strains of staphylococci has stimulated research directed toward new classes of antistaphylococcal drugs and improved versions of the antibiotics we currently employ, there are few promising new drugs on the horizon, a situation that has remained unchanged over the past decade (81). The good news is that daptomycin (24) and linezolid are now viable therapeutic options (see above). As has been true for over a decade, newer versions of older drugs remain in various stages of clinical development, such as new versions of glycopeptides (see below), macrolides (ketolides), cephalosporins, fluoroquinolones, and tetracyclines. Dalbavancin and oritavancin are semisynthetic glycopeptides which are structurally and mechanistically similar to vancomycin but with better pharmacokinetic properties. Both are currently under clinical investigation (114). Recently approved for clinical use is tigecycline, a derivative of the semisynthetic tetracycline minocycline, which is the first glycylcycline to demonstrate clinical efficacy (80), including activity toward resistant strains.

Some of these updated versions have excellent potency against resistant staphylococci both in vitro and in vivo. Also, some compounds previously set aside are once again being tested clinically for antistaphylococcal activity, including the cell wall biosynthesis inhibitor ramoplanin (11). The endopeptidase lysostaphin has proved extremely effective (better than any other agent in clinical use or development) as an antistaphylococcal agent in an endocarditis model (25) but still has drawn little interest among pharmaceutical companies, probably owing to its narrow spectrum of activity and an observed rapid emergence of resistance.

We thank Patricia Bradford for assistance and collegiality in the preparation of this chapter.

REFERENCES

1. **Adhikari, R. P., G. C. Scales, K. Kobayashi, J. M. B. Smith, B. Berger-Bächi, and G. M. Cook.** 2004. Vancomycin-induced deletion of the methicillin resistance gene *mecA* in *Staphylococcus aureus. J. Antimicrob. Chemother.* **54:**360–363.

2. **Allignet, J., V. Loncle, P. Mazodier, and N. el Solh.** 1988. Nucleotide sequence of a staphylococcal plasmid gene, *vgb*, encoding a hydrolase inactivating the B components of virginiamycin-like antibiotics. *Plasmid* **20:**271–275.

3. **Allignet, J., and N. el Solh.** 1995. Diversity among the gram-positive acetyltransferases inactivating streptogramin A and structurally related compounds and characterization of a new staphylococcal determinant, *vatB. Antimicrob. Agents Chemother.* **39:**2027–2036.

4. **Allignet, J., and N. El Solh.** 1997. Characterization of a new staphylococcal gene, *vgaB*, encoding a putative ABC transporter conferring resistance to streptogramin A and related compounds. *Gene* **202:**133–138.

5. **Allignet, J., N. Liassine, and N. El Solh.** 1998. Characterization of a staphylococcal plasmid related to pUB110 and carrying two novel genes, *vatC* and *vgbB*, encoding resistance to streptogramins A and B and similar antibiotics. *Antimicrob. Agents Chemother.* **42:**1794–1798.

6. **Archer, G. L., J. P. Coughter, and J. L. Johnston.** 1986. Plasmid-encoded trimethoprim resistance in staphylococci. *Antimicrob. Agents Chemother.* **29:**733–740.

7. **Archer, G. L., J. A. Thanassi, D. M. Niemeyer, and M. J. Pucci.** 1996. Characterization of IS*1272*, an insertion sequence-like element from *Staphylococcus haemolyticus. Antimicrob. Agents Chemother.* **40:**924–929.

8. **Aubry-Damon, H., C. J. Soussy, and P. Courvalin.** 1998. Characterization of mutations in the *rpoB* gene that confer rifampin resistance in *Staphylococcus aureus. Antimicrob. Agents Chemother.* **42:**2590–2594.

9. **Berger-Bächi, B., and S. Rohrer.** 2002. Factors influencing methicillin resistance in staphylococci. *Arch. Microbiol.* **178:**165–171.

10. **Berger-Bächi, B., A. Strassle, J. E. Gustafson, and F. H. Kayser.** 1992. Mapping and characterization of multiple chromosomal factors involved in methicillin resistance in *Staphylococcus aureus. Antimicrob. Agents Chemother.* **36:**1367–1373.

11. **Billot-Klein, D., D. Shlaes, D. Bryant, D. Bell, R. Legrand, L. Gutmann, and J. van Heijenoort.** 1997. Presence of UDP-*N*-acetylmuramyl-hexapeptides and -heptapeptides in enterococci and staphylococci after treatment with ramoplanin, tunicamycin, or vancomycin. *J. Bacteriol.* **179:**4684–4688.

12. **Bischoff, M., and B. Berger-Bächi.** 2001. Teicoplanin stress-selected mutations increasing σ^B activity in *Staphylococcus aureus. Antimicrob. Agents Chemother.* **45:**1714–1720.

13. **Bouanchaud, D. H.** 1997. In-vitro and in-vivo antibacterial activity of quinupristin/dalfopristin. *J. Antimicrob. Chemother.* **39**(Suppl. A)**:**15–21.

14. **Boyce, J. M., and A. A. Medeiros.** 1987. Role of beta-lactamase in expression of resistance by methicillin-resistant *Staphylococcus aureus. Antimicrob. Agents Chemother.* **31:**1426–1428.

15. **Boyle-Vavra, S., B. L. de Jonge, C. C. Ebert, and R. S. Daum.** 1997. Cloning of the *Staphylococcus aureus ddh* gene encoding NAD$^+$-dependent D-lactate dehydrogenase and insertional inactivation in a glycopeptide-resis-tant isolate. *J. Bacteriol.* **179:**6756–6763.

16. **Bozdogan, B., D. Esel, C. Whitener, F. A. Browne, and P. C. Appelbaum.** 2003. Antibacterial susceptibility of a vancomycin-resistant *Staphylococcus aureus* strain isolated at the Hershey Medical Center. *J. Antimicrob. Chemother.* **52:**864–868.

17. **Brandenberger, M., M. Tschierske, P. Giachino, A. Wada, and B. Berger-Bächi.** 2000. Inactivation of a novel three-cistronic operon *tcaR-tcaA-tcaB* increases teicoplanin resistance in *Staphylococcus aureus. Biochim. Biophys. Acta* **1523:**135–139.

18. **Canepari, P., and M. Boaretti.** 1996. Lipoteichoic acid as a target for antimicrobial action. *Microb. Drug Resist.* **2:**85–89.

19. **Centers for Disease Control and Prevention.** 2002. *Staphylococcus aureus* resistant to vancomycin—United States, 2002. *Morb. Mortal. Wkly. Rep.* **51:**565–567.

20. **Centers for Disease Control and Prevention.** 2004. Vancomycin-resistant *Staphylococcus aureus*—New York, 2004. *Morb. Mortal. Wkly. Rep.* **53:**322–323.

21. **Centers for Disease Control and Prevention.** 2002. Vancomycin-resistant *Staphylococcus aureus*—Pennsylvania, 2002. *Morb. Mortal. Wkly. Rep.* **51:**902.

22. **Chang, S., D. M. Sievert, J. C. Hageman, M. L. Boulton, F. C. Tenover, F. P. Downes, S. Shah, J. T. Rudrik, G. R. Pupp, W. J. Brown, D. Cardo, S. K. Fridkin, and the Vancomycin-Resistant Staphylococcus aureus Investigative Team.** 2003. Infection with vancomycin-resistant *Staphylococcus aureus* containing the *vanA* resistance gene. *N. Engl. J. Med.* **348:**1342–1347.

23. **Chopra, I.** 1976. Mechanisms of resistance to fusidic acid in *Staphylococcus aureus. J. Gen. Microbiol.* **96:**229–238.

24. **Clewell, D. B., S. E. Flannagan, and D. D. Jaworski.** 1995. Unconstrained bacterial promiscuity: the Tn916-Tn1545 family of conjugative transposons. *Trends Microbiol.* **3:**229–236.

25. **Climo, M. W., R. L. Patron, B. P. Goldstein, and G. L. Archer.** 1998. Lysostaphin treatment of experimental methicillin-resistant *Staphylococcus aureus* aortic valve endocarditis. *Antimicrob. Agents Chemother.* **42:**1355–1360.

26. **Cohen, S., and H. M. Sweeney.** 1973. Effect of the prophage and penicillinase plasmid of the recipient strain upon the transduction and the stability of methicillin resistance in *Staphylococcus aureus. J. Bacteriol.* **116:**803–811.

27. **Cui, L., X. Ma, K. Sato, K. Okuma, F. C. Tenover, E. M. Mamizuka, C. G. Gemmell, M. N. Kim, M. C. Ploy, N. El Solh, V. Ferraz, and K. Hiramatsu.** 2003. Cell wall thickening is a common feature of vancomycin resistance in *Staphylococcus aureus. J. Clin. Microbiol.* **41:**5–14.

28. **Cui, L., H. Murakami, K. Kuwahara-Arai, H. Hanaki, and K. Hiramatsu.** 2000. Contribution of a thickened cell wall and its glutamine nonamidated component to the vancomycin resistance expressed by *Staphylococcus aureus* Mu50. *Antimicrob. Agents Chemother.* **44:**2276–2285.

29. **Cundliffe, E.** 1972. The mode of action of fusidic acid. *Biochem. Biophys. Res. Commun.* **46:**1794–1801.

30. **De Lencastre, H., and A. Tomasz.** 1994. Reassessment of the number of auxiliary genes essential for expression of high-level methicillin resistance in *Staphylococcus aureus. Antimicrob. Agents Chemother.* **38:**2590–2598.

31. **De Lencastre, H., S. W. Wu, M. G. Pinho, A. M. Ludovice, S. Filipe, S. Gardete, R. Sobral, S. Gill, M. Chung, and A. Tomasz.** 1999. Antibiotic resistance as a stress response: complete sequencing of a large number of chromosomal loci in *Staphylococcus aureus* strain COL that impact on the expression of resistance to methicillin. *Microb. Drug Resist.* **5:**163–175.

32. **Denoya, C. D., D. H. Bechhofer, and D. Dubnau.** 1986. Translational autoregulation of *ermC* 23S rRNA methyltransferase expression in *Bacillus subtilis*. *J. Bacteriol.* **168:** 1133–1141.

33. **Dick, T., and H. Matzura.** 1990. Chloramphenicol-induced translational activation of *cat* messenger RNA in vitro. *J. Mol. Biol.* **212:**661–668.

34. **Doyle, D., K. J. McDowall, M. J. Butler, and I. S. Hunter.** 1991. Characterization of an oxytetracycline-resistance gene, *otrA*, of *Streptomyces rimosus*. *Mol. Microbiol.* **5:**2923–2933.

35. **Finan, J. E., G. L. Archer, M. J. Pucci, and M. W. Climo.** 2001. Role of penicillin-binding protein 4 in expression of vancomycin resistance among clinical isolates of oxacillin-resistant *Staphylococcus aureus*. *Antimicrob. Agents Chemother.* **45:**3070–3075.

36. **Fleming, A.** 1942. In vitro tests of penicillin potency. *Lancet* **i:**732.

37. **Garcia-Castellanos, R., G. Mallorqui-Fernandez, A. Marrero, J. Potempa, M. Coll, and F. X. Gomis-Ruth.** 2004. On the transcriptional regulation of methicillin resistance: MecI repressor in complex with its operator. *J. Biol. Chem.* **279:**17888–17896.

38. **Gardete, S., A. M. Ludovice, R. G. Sobral, S. R. Filipe, H. de Lencastre, and A. Tomasz.** 2004. Role of *murE* in the expression of beta-lactam antibiotic resistance in *Staphylococcus aureus*. *J. Bacteriol.* **186:**1705–1713.

39. **Gilbart, J., C. R. Perry, and B. Slocombe.** 1993. High-level mupirocin resistance in *Staphylococcus aureus*: evidence for two distinct isoleucyl-tRNA synthetases. *Antimicrob. Agents Chemother.* **37:**32–38.

40. **Gillespie, M. T., B. R. Lyon, L. J. Messerotti, and R. A. Skurray.** 1987. Chromosome- and plasmid-mediated gentamicin resistance in *Staphylococcus aureus* encoded by Tn4001. *J. Med. Microbiol.* **24:**139–144.

41. **Gregory, P. D., R. A. Lewis, S. P. Curnock, and K. G. Dyke.** 1997. Studies of the repressor (BlaI) of beta-lactamase synthesis in *Staphylococcus aureus*. *Mol. Microbiol.* **24:**1025–1037.

42. **Guay, G. G., S. A. Khan, and D. M. Rothstein.** 1993. The *tet*(K) gene of plasmid pT181 of *Staphylococcus aureus* encodes an efflux protein that contains 14 transmembrane helices. *Plasmid* **30:**163–166.

43. **He, H., R. T. Williamson, B. Shen, E. I. Graziani, H. Y. Yang, S. M. Sakya, P. J. Petersen, and G. T. Carter.** 2002. Mannopeptimycins, novel antibacterial glycopeptides from *Streptomyces hygroscopicus*, LL-AC98. *J. Am. Chem. Soc.* **124:**9729–9736.

44. **Henze, U., T. Sidow, J. Wecke, H. Labischinski, and B. Berger-Baechi.** 1993. Influence of *femB* on methicillin resistance and peptidoglycan metabolism in *Staphylococcus aureus*. *J. Bacteriol.* **175:**1612–1620.

45. **Hiramatsu, K., E. Suzuki, H. Takayama, Y. Katayama, and T. Yokota.** 1990. Role of penicillinase plasmids in the stability of the *mecA* gene in methicillin-resistant *Staphylococcus aureus*. *Antimicrob. Agents Chemother.* **34:**600–604.

46. **Hitchings, G. H.** 1973. Mechanism of action of trimethoprim-sulfamethoxazole I. *J. Infect. Dis.* **128:** S433.

47. **Hodgson, J. E., S. P. Curnock, K. G. Dyke, R. Morris, D. R. Sylvester, and M. S. Gross.** 1994. Molecular characterization of the gene encoding high-level mupirocin resistance in *Staphylococcus aureus* J2870. *Antimicrob. Agents Chemother.* **38:**1205–1208.

48. **Hooper, D. C.** 1995. Quinolone mode of action. *Drugs* **49**(Suppl. 2):10–15.

49. **Horinouchi, S., and B. Weisblum.** 1982. Nucleotide sequence and functional map of pE194, a plasmid that speci-

50. **Hughes, J., and G. Mellows.** 1978. Inhibition of isoleucyl-transfer ribonucleic acid synthetase in Escherichia coli by pseudomonic acid. *Biochem. J.* **176:**305–318.

51. **Ito, T., Y. Katayama, and K. Hiramatsu.** 1999. Cloning and nucleotide sequence determination of the entire *mec* DNA of pre-methicillin-resistant *Staphylococcus aureus* N315. *Antimicrob. Agents Chemother.* **43:**1449–1458.

52. **Jones, R. N., C. H. Ballow, D. J. Biedenbach, J. A. Deinhart, and J. J. Schentag.** 1998. Antimicrobial activity of quinupristin-dalfopristin (RP 59500, Synercid) tested against over 28,000 recent clinical isolates from 200 medical centers in the United States and Canada. *Diagn. Microbiol. Infect. Dis.* **31:**437–451.

53. **Katayama, Y., H. Z. Zhang, D. Hong, and H. F. Chambers.** 2003. Jumping the barrier to beta-lactam resistance in *Staphylococcus aureus*. *J. Bacteriol.* **185:**5465–5472.

54. **Koehl, J. L., A. Muthaiyan, R. K. Jayaswal, K. Ehlert, H. Labischinski, and B. J. Wilkinson.** 2004. Cell wall composition and decreased autolytic activity and lysostaphin susceptibility of glycopeptide-intermediate *Staphylococcus aureus*. *Antimicrob. Agents Chemother.* **48:** 3749–3757.

55. **Komatsuzawa, H., K. Ohta, S. Yamada, K. Ehlert, H. Labischinski, J. Kajimura, T. Fujiwara, and M. Sugai.** 2002. Increased glycan chain length distribution and decreased susceptibility to moenomycin in a vancomycin-resistant *Staphylococcus aureus* mutant. *Antimicrob. Agents Chemother.* **46:**75–81.

56. **Kuroda, M., K. Kuwahara-Arai, and K. Hiramatsu.** 2000. Identification of the up- and down-regulated genes in vancomycin-resistant *Staphylococcus aureus* strains Mu3 and Mu50 by cDNA differential hybridization method. *Biochem. Biophys. Res. Commun.* **269:**485–490.

57. **Landy, M., N. W. Larkum, E. J. Oswald, and P. Streighoff.** 1943. Increased synthesis of p-aminobenzoic acid associated with the development of sulfonamide resistance in *Staphylococcus aureus*. *Science* **97:**265.

58. **Levy, S. B.** 1998. Multidrug resistance—a sign of the times. *N. Engl. J. Med.* **338:**1376–1378.

59. **Lewis, R. A., and K. G. Dyke.** 2000. MecI represses synthesis from the beta-lactamase operon of *Staphylococcus aureus*. *J. Antimicrob. Chemother.* **45:**139–144.

60. **Lovett, P. S.** 1996. Translation attenuation regulation of chloramphenicol resistance in bacteria—a review. *Gene* **179:**157–162.

61. **Ludovice, A. M., S. W. Wu, and H. de Lencastre.** 1998. Molecular cloning and DNA sequencing of the *Staphylococcus aureus* UDP-N-acetylmuramyl tripeptide synthetase (*murE*) gene, essential for the optimal expression of methicillin resistance. *Microb. Drug Resist.* **4:**85–90.

62. **Mainardi, J. L., D. M. Shlaes, R. V. Goering, J. H. Shlaes, J. F. Acar, and F. W. Goldstein.** 1995. Decreased teicoplanin susceptibility of methicillin-resistant strains of *Staphylococcus aureus*. *J. Infect. Dis.* **171:**1646–1650.

63. **Maki, H., N. McCallum, M. Bischoff, A. Wada, and B. Berger-Bachi.** 2004. *tcaA* inactivation increases glycopeptide resistance in *Staphylococcus aureus*. *Antimicrob. Agents Chemother.* **48:**1953–1959.

64. **McKinney, T. K., V. K. Sharma, W. A. Craig, and G. L. Archer.** 2001. Transcription of the gene mediating methicillin resistance in *Staphylococcus aureus* (*mecA*) is corepressed but not coinduced by cognate *mecA* and beta-lactamase regulators. *J. Bacteriol.* **183:**6862–6868.

65. **Miller, G. H., F. J. Sabatelli, L. Naples, R. S. Hare, K. J. Shaw, et al.** 1995. The most frequently occurring aminogly-

coside resistance mechanisms—combined results of surveys in eight regions of the world. *J. Chemother.* **7**(Suppl. 2):17–30.

66. **Mongkolrattanothai, K., S. Boyle, M. D. Kahana, and R. S. Daum.** 2003. Severe *Staphylococcus aureus* infections caused by clonally related community-acquired methicillin-susceptible and methicillin-resistant isolates. *Clin. Infect. Dis.* **37**:1050–1058.

67. **Mongodin, E., J. Finan, M. W. Climo, A. Rosato, S. Gill, and G. L. Archer.** 2003. Microarray transcription analysis of clinical *Staphylococcus aureus* isolates resistant to vancomycin. *J. Bacteriol.* **185**:4638–4643.

68. **Nesin, M., P. Svec, J. R. Lupski, G. N. Godson, B. Kreiswirth, J. Kornblum, and S. J. Projan.** 1990. Cloning and nucleotide sequence of a chromosomally encoded tetracycline resistance determinant, *tetA*(M), from a pathogenic, methicillin-resistant strain of Staphylococcus aureus. *Antimicrob. Agents Chemother.* **34**:2273–2276.

69. **Niemeyer, D. M., M. J. Pucci, J. A. Thanassi, V. K. Sharma, and G. L. Archer.** 1996. Role of *mecA* transcriptional regulation in the phenotypic expression of methicillin resistance in *Staphylococcus aureus. J. Bacteriol.* **178:** 5464–5471.

70. **Nikaido, H.** 1994. Prevention of drug access to bacterial targets: permeability barriers and active efflux. *Science* **264:**382–388.

71. **Noble, W. C., Z. Virani, and R. G. Cree.** 1992. Co-transfer of vancomycin and other resistance genes from *Enterococcus faecalis* NCTC 12201 to *Staphylococcus aureus. FEMS Microbiol. Lett.* **72:**195–198.

72. **O'Kane, G. M., T. Gottlieb, and R. Bradbury.** 1998. Staphylococcal bacteraemia: the hospital or the home? A review of *Staphylococcus aureus* bacteraemia at Concord Hospital in 1993. *Aust. N. Z. J. Med.* **28:**23–27.

73. **Pao, S. S., I. T. Paulsen, and M. H. Saier, Jr.** 1998. Major facilitator superfamily. *Microbiol. Mol. Biol. Rev.* **62:**1–34.

74. **Paradise, M. R., G. Cook, R. K. Poole, and P. N. Rather.** 1998. Mutations in *aarE*, the *ubiA* homolog of *Providencia stuartii*, result in high-level aminoglycoside resistance and reduced expression of the chromosomal aminoglycoside 2'-*N*-acetyltransferase. *Antimicrob. Agents Chemother.* **42:** 959–962.

75. **Peschel, A., C. Vuong, M. Otto, and F. Gotz.** 2000. The D-alanine residues of *Staphylococcus aureus* teichoic acids alter the susceptibility to vancomycin and the activity of autolytic enzymes. *Antimicrob. Agents Chemother.* **44:**2845–2847

76. **Pestka, S.** 1971. Inhibitors of ribosome functions. *Annu. Rev. Microbiol.* **25:**487–562.

77. **Petersen, P. J., T. Z. Wang, R. G. Dushin, and P. A. Bradford.** 2004. Comparative in vitro activities of AC98-6446, a novel semisynthetic glycopeptide derivative of the natural product mannopeptimycin alpha, and other antimicrobial agents against gram-positive clinical isolates. *Antimicrob. Agents Chemother.* **48:**739–746.

78. **Pinho, M. G., H. de Lencastre, and A. Tomasz.** 2001. An acquired and a native penicillin-binding protein cooperate in building the cell wall of drug-resistant staphylococci. *Proc. Natl. Acad. Sci. USA* **98:**10886–10891.

79. **Pinho, M. G., S. R. Filipe, H. de Lencastre, and A. Tomasz.** 2001. Complementation of the essential peptidoglycan transpeptidase function of penicillin-binding protein 2 (PBP2) by the drug resistance protein PBP2A in *Staphylococcus aureus. J. Bacteriol.* **183:**6525–6531.

80. **Postier, R. G., S. L. Green, S. R. Klein, E. J. Ellis-Grosse, E. Loh, and Tigecycline 200 Study Group.** 2004. Results of a multicenter, randomized, open-label efficacy and safety study of two doses of tigecycline for complicated skin and skin-structure infections in hospitalized patients. *Clin. Ther.* **26:**704–714.

81. **Projan, S. J.** 2003. Why is big Pharma getting out of antibacterial drug discovery? *Curr. Opin. Microbiol.* **6:**427–430.

82. **Projan, S. J., M. Monod, C. S. Narayanan, and D. Dubnau.** 1987. Replication properties of pIM13, a naturally occurring plasmid found in *Bacillus subtilis*, and of its close relative pE5, a plasmid native to *Staphylococcus aureus. J. Bacteriol.* **169:**5131–5139.

83. **Projan, S. J., S. Moghazeh, and R. P. Novick.** 1988. Nucleotide sequence of pS194, a streptomycin-resistance plasmid from *Staphylococcus aureus. Nucleic Acids Res.* **16:**2179–2187.

84. **Rao, G. G.** 1998. Risk factors for the spread of antibiotic-resistant bacteria. *Drugs* **55:**323–330.

85. **Raz, R., D. Miron, R. Colodner, Z. Staler, Z. Samara, and Y. Keness.** 1996. A 1-year trial of nasal mupirocin in the prevention of recurrent staphylococcal nasal colonization and skin infection. *Arch. Intern. Med.* **156:**1109–1112.

86. **Ross, J. I., E. A. Eady, J. H. Cove, and S. Baumberg.** 1996. Minimal functional system required for expression of erythromycin resistance by *msrA* in *Staphylococcus aureus* RN4220. *Gene* **183:**143–148.

87. **Ross, J. I., E. A. Eady, J. H. Cove, W. J. Cunliffe, S. Baumberg, and J. C. Wootton.** 1990. Inducible erythromycin resistance in staphylococci is encoded by a member of the ATP-binding transport super-gene family. *Mol. Microbiol.* **4:**1207–1214.

88. **Rouch, D. A., L. J. Messerotti, L. S. Loo, C. A. Jackson, and R. A. Skurray.** 1989. Trimethoprim resistance transposon Tn4003 from *Staphylococcus aureus* encodes genes for a dihydrofolate reductase and thymidylate synthetase flanked by three copies of IS257. *Mol. Microbiol.* **3:**161–175.

89. **Rowland, S. J., and K. G. Dyke.** 1990. Tn552, a novel transposable element from *Staphylococcus aureus. Mol. Microbiol.* **4:**961–975.

90. **Rubin, R. J., C. A. Harrington, A. Poon, K. Dietrich, J. A. Greene, and A. Moiduddin.** 1999. The economic impact of *Staphylococcus aureus* infection in New York City hospitals. *Emerging Infect. Dis.* **5:**9–17.

91. **Ruzin, A., A. Severin, S. L. Moghazeh, J. Etienne, P. A. Bradford, S. J. Projan, and D. M. Shlaes.** 2003. Inactivation of *mprF* affects vancomycin susceptibility in *Staphylococcus aureus. Biochim. Biophys. Acta* **1621:**117–121.

92. **Ruzin, A., G. Singh, A. Severin, Y. Yang, R. G. Dushin, A. G. Sutherland, A. Minnick, M. Greenstein, M. K. May, D. M. Shlaes, and P. A. Bradford.** 2004. Mechanism of action of the mannopeptimycins, a novel class of glycopeptide antibiotics active against vancomycin-resistant gram-positive bacteria. *Antimicrob. Agents Chemother.* **48:** 728–738.

93. **Ryffel, C., F. H. Kayser, and B. Berger-Bachi.** 1992. Correlation between regulation of *mecA* transcription and expression of methicillin resistance in staphylococci. *Antimicrob. Agents Chemother.* **36:**25–31.

94. **Said-Salim, B., B. Mathema, and B. N. Kreiswirth.** 2003. Community-acquired methicillin-resistant Staphylococcus aureus: an emerging pathogen. *Infect. Control Hosp. Epidemiol.* **24:**451–455.

95. **Sakoulas, G., G. M. Eliopoulos, R. C. Moellering, Jr., R. P. Novick, L. Venkataraman, C. Wennersten, P. C. DeGirolami, M. J. Schwaber, and H. S. Gold.** 2003.

Staphylococcus aureus accessory gene regulator (*agr*) group II: is there a relationship to the development of intermediate-level glycopeptide resistance? *J. Infect. Dis.* **187**:929–938.

96. Sakoulas, G., G. M. Eliopoulos, R. C. Moellering, Jr., C. Wennersten, L. Venkataraman, R. P. Novick, and H. S. Gold. 2002. Accessory gene regulator (*agr*) locus in geographically diverse *Staphylococcus aureus* isolates with reduced susceptibility to vancomycin. *Antimicrob. Agents Chemother.* **46**:1492–1502.

97. Schlessinger, D., and G. Medoff. 1975. Streptomycin, dehidrostreptomycin amd the gentamicins, p. 535. *In* J. W. Corcoran and F. E. Hahn (ed.), *Antibiotics*, vol. 3. Springer-Verlag, New York, N.Y.

98. Schwarz, S., P. D. Gregory, C. Werckenthin, S. Curnock, and K. G. Dyke. 1996. A novel plasmid from Staphylococcus epidermidis specifying resistance to kanamycin, neomycin and tetracycline. *J. Med. Microbiol.* **45**:57–63.

99. Shaw, K. J., P. N. Rather, R. S. Hare, and G. H. Miller. 1993. Molecular genetics of aminoglycoside resistance genes and familial relationships of the aminoglycoside-modifying enzymes. *Microbiol. Rev.* **57**:138–163.

100. Shlaes, D. M., and J. H. Shlaes. 1995. Teicoplanin selects for *Staphylococcus aureus* that is resistant to vancomycin. *Clin. Infect. Dis.* **20**:1071–1073.

101. Sieradzki, K., M. G. Pinho, and A. Tomasz. 1999. Inactivated *pbp4* in highly glycopeptide-resistant laboratory mutants of *Staphylococcus aureus*. *J. Biol. Chem.* **274**:18942–18946.

102. Sieradzki, K., and A. Tomasz. 1997. Inhibition of cell wall turnover and autolysis by vancomycin in a highly vancomycin-resistant mutant of *Staphylococcus aureus*. *J. Bacteriol.* **179**:2557–2566.

103. Silverman, J. A., N. G. Perlmutter, and H. M. Shapiro. 2003 Correlation of daptomycin bactericidal activity and membrane depolarization in *Staphylococcus aureus*. *Antimicrob. Agents Chemother.* **47**:2538–2544.

104. Smith, I. M., and A. B. Vickers. 1960. Natural history of 338 treated and untreated patients with staphylococcal septicaemia. *Lancet* **i**:1318.

105. Stewart, G. C., and E. D. Rosenblum. 1980. Transduction of methicillin resistance in *Staphylococcus aureus*: recipient effectiveness and beta-lactamase production. *Antimicrob. Agents Chemother.* **18**:424–432.

106. Stone, M. J., and D. H. Williams. 1992. On the evolution of functional secondary metabolites (natural products). *Mol. Microbiol.* **6**:29–34.

107. Streit, J. M., R. N. Jones, H. S. Sader, and T. R. Fritsche. 2001. Assessment of pathogen occurrences and resistance profiles among infected patients in the intensive care unit: report from the SENTRY Antimicrobial Surveillance Program (North America, 2001). *Int. J. Antimicrob. Agents* **24**:111–118.

108. Su, Y. A., P. He, and D. B. Clewell. 1992. Characterization of the *tet*(M) determinant of Tn916: evidence for regulation by transcription attenuation. *Antimicrob. Agents Chemother.* **36**:769–778.

109. Swaney, S. M., H. Aoki, M. C. Ganoza, and D. L. Shinabarger. 1998. The oxazolidinone linezolid inhibits initiation of protein synthesis in bacteria. *Antimicrob. Agents Chemother.* **42**:3251–3255.

110. Thakker-Varia, S., W. D. Jenssen, L. Moon-McDermott, W. P. Weinstein, and D. T. Dubin. 1987. Molecular epidemiology of macrolides-lincosamides-streptogramin B resistance in *Staphylococcus aureus* and coagulase-negative staphylococci. *Antimicrob. Agents Chemother.* **31**:735–743.

111. Thomas, W. D., Jr., and G. L. Archer. 1989. Mobility of gentamicin resistance genes from staphylococci isolated in the United States: identification of Tn4031, a gentamicin resistance transposon from *Staphylococcus epidermidis*. *Antimicrob. Agents Chemother.* **33**:1335–1341.

112. Tomasz, A. 1994. Multiple-antibiotic-resistant pathogenic bacteria. A report on the Rockefeller University Workshop. *N. Engl. J. Med.* **330**:1247–1251.

113. Tomasz, A., S. Nachman, and H. Leaf. 1991. Stable classes of phenotypic expression in methicillin-resistant clinical isolates of staphylococci. *Antimicrob. Agents Chemother.* **35**:124–129.

114. Van Bambeke, F., Y. Van Laethem, P. Courvalin, and P. M. Tulkens. 2004 Glycopeptide antibiotics: from conventional molecules to new derivatives. *Drugs* **64**:913–936.

115. Vandenesch, F., T. Naimi, M. C. Enright, G. Lina, G. R. Nimmo, H. Heffernan, N. Liassine, M. Bes, T. Greenland, M. E. Reverdy, and J. Etienne. 2003. Community-acquired methicillin-resistant *Staphylococcus aureus* carrying Panton-Valentine leukocidin genes: worldwide emergence. *Emerging Infect. Dis.* **9**:978–984.

116. Vazquez, D. 1975. The streptogramin family of antibiotics, p. 521. *In* J. W. Corcorran and F. E. Hahn (ed.), *Antibiotics*, vol. 3. Springer-Verlag, New York, N.Y.

117. Voladri, R. K., M. K. Tummuru, and D. S. Kernodle. 1996. Structure-function relationships among wild-type variants of *Staphylococcus aureus* beta-lactamase: importance of amino acids 128 and 216. *J. Bacteriol.* **178**:7248–7253.

118. Wehrli, W., and M. Staehelin. 1971. Actions of the rifamycins. *Bacteriol. Rev.* **35**:290–309.

119. Wennersten, C., L. Venkataraman, P. C. DeGirolami, G. M. Eliopoulos, R. C. Moellering, Jr., and H. S. Gold. 2004. Linezolid resistance in sequential *Staphylococcus aureus* isolates associated with a T2500A mutation in the 23S rRNA gene and loss of a single copy of rRNA. *J. Infect. Dis.* **190**:311–317.

120. Wondrack, L., M. Massa, B. V. Yang, and J. Sutcliffe. 1996. Clinical strain of *Staphylococcus aureus* inactivates and causes efflux of macrolides. *Antimicrob. Agents Chemother.* **40**:992–998.

121. Wu, S. W., H. De Lencastre, and A. Tomasz. 2001. Recruitment of the *mecA* gene homologue of *Staphylococcus sciuri* into a resistance determinant and expression of the resistant phenotype in *Staphylococcus aureus*. *J. Bacteriol.* **183**:2417–2424.

122. Yu, V. L., G. D. Fang, T. F. Keys, A. A. Harris, L. O. Gentry, P. C. Fuchs, N. M. Wagener, and E. S. Wong. 1994. Prosthetic valve endocarditis: superiority of surgical valve replacement versus medical therapy only. *Ann. Thorac. Surg.* **58**:1073–1077.

123. Zhang, H. Z., C. J. Hackbarth, K. M. Chansky, and H. F. Chambers. 2001. A proteolytic transmembrane signaling pathway and resistance to beta-lactams in staphylococci. *Science* **291**:1962–1965.

124. Zhao, X., C. Xu, J. Domagala, and K. Drlica. 1997. DNA topoisomerase targets of the fluoroquinolones: a strategy for avoiding bacterial resistance. *Proc. Natl. Acad. Sci. USA* **94**:13991–13996.

125. Zurenko, G. E., J. K. Gibson, D. L. Shinabarger, P. A. Aristoff, C. W. Ford, and W. G. Tarpley. 2001 Oxazolidinones: a new class of antibacterials. *Curr. Opin. Pharmacol.* **1**:470–476.

THE LISTERIAE

SECTION EDITOR: Daniel A. Portnoy

L ISTERIOSIS AND *LISTERIA MONOCYTOGENES* are of interest to a broad range of investigators, ranging from food microbiologists, clinicians, immunologists, and medical microbiologists to cell biologists. Thus, while *L. monocytogenes* remains one of the leading causes of mortality from food-borne infections in the United States, it has emerged as an excellent model system to study basic aspects of bacterial pathogenesis and cell-mediated immunity. Indeed, for over 4 decades immunologists have exploited the *L. monocytogenes* model to study cell-mediated immunity. Much of our current understanding of cell-mediated immunity has derived from the murine model of listeriosis. It may be true that more is known about the murine response to *L. monocytogenes* than to any other bacterial pathogen; however, it should be noted that the vast majority of these studies have used an intravenous or intraperitoneal route of inoculation, and the relevance of these studies to the natural, oral route of infection is not always clear. Newly developed animal models, including pregnant guinea pigs and transgenic mice, are sure to have a vital impact during the next few years. *L. monocytogenes* has also been utilized as a live vector for the development of vaccines against causative agents of infectious diseases, such as human immunodeficiency virus and other malignancies. During the next few years, clinical trials will be initiated to test the efficacy of such vaccines.

In the past decade, *L. monocytogenes* has also emerged as a model system for studying basic aspects of intracellular pathogenesis. Many new genetic techniques have been developed, including transposon mutagenesis, allelic exchange, reporter systems, inducible gene expression, integration vectors, and transducing phages. The completion of the genome sequence and the subsequent development of bacterial microarrays both occurred since the publication of the first edition of this text and have resulted in the identification and characterization of many novel determinants of pathogenesis. The combination of *Listeria* genomics with mouse genomics, along with an extensive battery of genetic and cell biological tools, should make the next few years truly exciting.

Epidemiology and Clinical Manifestations of *Listeria monocytogenes* Infection

WALTER F. SCHLECH III

49

EPIDEMIOLOGY

Listeria monocytogenes is a gram-positive motile facultative anaerobe that inhabits a broad ecologic niche (81). With selective media, it can be readily isolated from soil, water, and vegetation, including raw produce designated for human consumption without further processing (29). Newer chromogenic media may offer some advantages in detection of contaminated foodstuffs (58). Contamination of meat and vegetables is on the surface and is relatively common, with up to 15% of these foods harboring the organism. In addition, the organism is a transient inhabitant of both animal and human gastrointestinal tracts (30, 35), and intermittent carriage suggests frequent exposure (31). The gut is the source for the organism in invasive listeriosis when it occurs, and large numbers of organisms may cause a noninvasive febrile gastroenteritis.

The organism is psychrophilic and enjoys a competitive advantage against other gram-positive and gram-negative microorganisms in cold environments, such as refrigerators. It may also be amplified in spoiled food products, particularly when spoilage leads to increased alkalinity. Feeding of spoiled silage with a high pH has resulted in epidemics of listeriosis in sheep or cattle (43).

Several large food-borne outbreaks of listeriosis in humans have parallels to epidemic listeriosis in animals. These outbreaks, which have been attributed to coleslaw (65), unpasteurized cheeses (10–12, 39), pasteurized milk (23), butter (44), and several meat products (13, 17), have established that human infection by *L. monocytogenes* has a food-borne origin. Many other food products have been implicated in both epidemic and sporadic disease (Table 1). Uncertainty exists as to why outbreaks of listeriosis occur in human populations, although the 50% infective dose in sporadic disease is probably high. Enhancement of organism-specific virulence factors may play a role in epidemic disease, although all isolates of *L. monocytogenes* have the constitutive ability to produce all the virulence factors characteristic of the species.

Recent evidence has suggested that most sporadic cases of listeriosis are also food borne. These reports remained anecdotal until large case-controlled studies of sporadic disease implicated food products, including cold meats, turkey franks, and delicatessen-type foods, as vehicles for development of sporadic invasive listeriosis in humans (66, 68).

Our current understanding of the epidemiology of human listeriosis suggests that the organism is a common contaminant of food products and that ingestion of small numbers of *L. monocytogenes* occurs frequently in human populations. In one prospective study, five to nine exposures per person-year were estimated (31). Amplification of the organism in biofilms or on food products undergoing processing but not pasteurization and kept at cold temperatures allows overgrowth of *L. monocytogenes*, in contrast to inhibition of other organisms. Subsequent ingestion of large numbers of the organism may overwhelm local host-defense systems in the gastrointestinal tract and reticuloendothelial systems of the liver and spleen with subsequent development of invasive disease. The annual rate of sporadic listeriosis in Europe (28) and North America (14) is usually <1/100,000 population per year. Active surveillance carried out in hospital microbiology laboratories in specific geographic regions in the United States has confirmed data obtained from passive reporting systems (14). Sporadic listeriosis appears to be more common in the spring and summer months. This could be explained by seasonal variations in the types of food products eaten by human populations, with higher-risk products eaten in the warmer months. In addition, data suggest that preexisting damage to the gastrointestinal mucosa by other microorganisms such as those associated with viral gastroenteritis may allow translocation of *L. monocytogenes* from the gastrointestinal tract, with subsequent development of invasive disease (69). These viral pathogens often have seasonal patterns that overlap with those of invasive listeriosis.

Demographic data from surveillance studies indirectly reveal several host-specific risk factors for invasive listeriosis. Infection is most commonly seen in the first 30 days of life or in patients older than 60 years. In the first instance, the fetus is infected during maternal sepsis with *L. monocytogenes* or from perivaginal and perianal colonization of the mother by transition through the birth canal. Host defense

TABLE 1 Some foods implicated in published reports of food-borne listeriosis

Coleslaw (cabbage)	Strawberries
Pasteurized whole milk	Nectarines
Chocolate milk	Patés
Mexican-style cheese	Goat cheese
Soft cheeses (different types)	Salted mushrooms
Delicatessen foods (deli meats)	Lettuce
Ice cream	Alfalfa tablets
Shrimp salad	Pork tongue in aspic
Uncooked hotdogs	Rillettes
Turkey franks	Rice salad
Fresh cream	Undercooked chicken
Baby corn	Blueberries
Smoked fish	

against listeriosis is impaired in those infants with underdeveloped macrophage and cell-mediated immune function, and invasive listeriosis is more likely to occur if colonization of the liver, respiratory tract, or gastrointestinal tract has occurred. A unique outbreak of neonatal listeriosis in Costa Rica has been described: the vehicle was *L. monocytogenes*-contaminated mineral oil used to clean infants after delivery from healthy mothers with cross-contamination of shared mineral oil (67). The index case was infected through the traditional route of maternal-fetal transmission. The increased risk of invasive listeriosis in older patients reflects the increasing incidence of immunosuppressive conditions, such as solid tumors and hematologic malignancy, in this age group. Outbreaks of food-borne listeriosis have even occurred in hospitals, emphasizing these risk factors (44). Control of early infection in humans and in animal models is highly dependent on an intact gastrointestinal mucosa and effective macrophage function in the liver, spleen, and peritoneum following bacterial translocation from the gastrointestinal tract. Both these protective events can be impaired by the primary disease or by chemotherapy or radiation-induced damage. In addition, treatment of malignancy and the use of immunosuppressive agents with a specific effect on cell-mediated immune function, such as corticosteroids or cyclosporine (64), predispose to invasive infection by diminishing *L. monocytogenes*-specific host responses that occur after the initial phase of infection.

The cell-mediated immune response to *L. monocytogenes* is normally impaired in pregnant women (75) and, accompanied by the decreased gastrointestinal motility (80) seen in pregnancy, may predispose to invasive listeriosis and subsequent transplacental infection of the infant. This results in early-onset listeriosis characterized by the delivery of an often premature and severely ill infant. Spontaneous recovery of the mother from *Listeria* sepsis normally occurs following delivery of the infant.

In late-onset listeriosis, the infant is infected through maternal gastrointestinal carriage of *L. monocytogenes* without sepsis, during transition through a colonized birth canal. In these cases, clinical disease in the infant develops 7 to 14

days later. Direct cutaneous invasion is unlikely, and it is believed that aspiration of the organism into the respiratory tract or swallowing of the organism by the infant may occur during the incubation period (67).

Several large outbreaks of a febrile gastroenteritis syndrome have further highlighted the importance of *L. monocytogenes* as a food-borne pathogen. In these outbreaks, with an average incubation period of approximately 24 h, attack rates (up to 72%) were much higher than those reported for outbreaks of invasive listeriosis. The reported vehicles for these more typical food-borne infections have included shrimp (59), rice salad (63), chocolate milk (15), corn salad (2), ready-to-eat meats (24, 71), and fresh cheese (11). The foods implicated were usually heavily contaminated ($>10^9$ CFU of *L. monocytogenes*/ml), and the amount of food ingested appeared to correlate with infection, suggesting that the high attack rates are not related to enhanced intrinsic virulence of the particular infecting strain of *L. monocytogenes*.

While a predisposition to invasive listeriosis is seen in patients with malignancy or organ transplant, recent data suggest that specific impairment of the immune system caused by human immunodeficiency virus (HIV) infection is an important factor in sporadic listeriosis (34). Several studies report that attack rates for invasive listeriosis in HIV-positive patients are 500- to 1,000-fold higher than those in the general population. In California, where active surveillance of hospital laboratories for listeriosis is carried out, a reduction in invasive listeriosis cases in HIV infection was brought about by widely promulgated dietary recommendations to prevent food-borne illness and by the use of prophylaxis for *Pneumocystis carinii* pneumonia, primarily trimethoprim-sulfamethoxazole, to which *L. monocytogenes* is susceptible (21). Reductions in the overall incidence of listeriosis in non-HIV-positive patients may also be attributed to distribution of dietary recommendations to populations at risk, including pregnant women, patients with malignancies, and organ transplant recipients (76). Perhaps more importantly, the decreased incidence of listeriosis may be due to the promotion of guidelines to promote universal awareness of the problem in the food-processing industry, which has undertaken Hazard Analysis at Critical Control Points (9, 47, 55) and Microbial Risk Assessment (49, 61) programs to reduce contamination of foods with *L. monocytogenes*, as well as with other food-borne pathogens such as *Salmonella* sp., *Campylobacter* sp., and *Escherichia coli*. These activities have provided increased protection in the face of increased public demand for fresh, unprocessed food products that may not have been cooked or pasteurized and that by definition present a greater degree of risk for food-borne illness.

In addition to Hazard Analysis at Critical Control Points programs, regulatory agencies have aggressively pursued the control of *L. monocytogenes* contamination of food. The U.S. Food and Drug Administration has a zero-tolerance policy for *L. monocytogenes* in its industry sampling programs (78). Other countries have less-stringent guidelines, allowing a small amount of contamination ($<10^2$ CFU/g) to strike a balance between protection of public health and needless condemnation of otherwise-edible food products. While invasive listeriosis may be more common in some countries in Europe than in the United States, it is not clear whether these differences can be attributed to less-stringent standards in Europe that allow more *L. monocytogenes* in the food supply. The debate continues between zero-tolerance advocates and those supporting a risk assessment approach to *Listeria* contamination of food (18).

CLINICAL DISEASE DUE TO *L. MONOCYTOGENES*

A wide variety of clinical syndromes have been associated with *L. monocytogenes* infection in both animals and humans (Table 2). The earliest descriptions of *L. monocytogenes* sepsis were of an epizootic affecting South African rodents (56) and of laboratory colonies of rabbits (52). One distinguishing characteristic of infection in rabbits was the production of monocytosis in blood, which suggested the species name *monocytogenes*. A monocytosis-producing antigen has been described as a virulence factor of *L. monocytogenes* (70), but monocytosis in the peripheral blood is not a characteristic of invasive infection in humans.

Many wild and domesticated animals are subject to invasive listeriosis. Animals acquire the organism from the environment through grazing, amplified by fecal contamination of soil and vegetation. Specific syndromes with parallels in human disease have been recognized in animals. In New Zealand in the 1930s, Gill (26) described circling disease, a rhombencephalitis of sheep that may affect flocks fed spoiled silage. *L. monocytogenes* has also been implicated as a cause of abortion and prematurity in ruminants. Intravenous and oral models of *L. monocytogenes* infection in rodents can duplicate the illness seen in the natural state in animals, including maternal sepsis and abortion (38).

The clinical syndromes associated with listeriosis in humans have been more recently elucidated. Neonatal listeriosis was initially described in postwar Europe in premature septic newborns in the former East Germany (57). This description of early-onset listeriosis was followed by reports of neonatal meningitis (late-onset listeriosis) occurring somewhat later in the postpartum period. *L. monocytogenes* as a cause of meningitis in neonates is third to group B streptococci and *E. coli* in the developed world (16). The use of antibiotic prophylaxis to prevent group B streptococcal infection may also have reduced cases of neonatal listeriosis (4). In less-developed countries, gram-negative meningitis with *E. coli* or *Salmonella* species is more common, but *Listeria* meningitis still occurs.

Early-onset neonatal listeriosis has characteristic clinical features, including prematurity, sepsis at birth, fever, a diffuse maculopapular cutaneous eruption, and evidence of significant hepatic involvement with jaundice (23, 53). The mortality rate of early-onset listeriosis, even with treatment, is very high, and stillbirth is also common in this setting. Autopsy findings in cases of early-onset listeriosis show significant chorioamnionitis in placental remnants and granulomas in multiple organs, particularly the liver and spleen, of infected infants. The original descriptions from East Germany characterized the entire syndrome as granulomatosis infantiseptica.

The mothers of these septic infants may be asymptomatic but commonly have flu-like or pyelonephritis symptoms before the early onset of labor, and their blood cultures are frequently positive for *L. monocytogenes*. Symptoms in the mother include fever, chills, and malaise, which resolve spontaneously following delivery of the infected infant and placenta (53). Anecdotal case reports suggest that early treatment of the mother who has *Listeria* sepsis can prevent transplacental infection or treat the fetus in utero with subsequent delivery of a healthy, uninfected infant. Unfortunately, this usually only happens when a community-based outbreak of *L. monocytogenes* has been identified and physicians are aware of the problem in a particular geographic region through public health alerts.

Late-onset neonatal meningitis due to *L. monocytogenes* has the typical features of the same syndrome caused by other organisms in this setting, including fever, irritability, bulging fontanelles, and meningismus (36). These symptoms usually develop 1 to 2 weeks following delivery. The mother has usually had an uncomplicated pregnancy, delivery, and postpartum course with no signs of sepsis. The clinical syndrome usually dictates a lumbar puncture, and the cerebrospinal fluid (CSF) in 50% of the cases will reveal the organism by Gram's stain. CSF cultures are usually positive, although the organism may be isolated simultaneously or only from the blood in some cases. The CSF shows other characteristics of bacterial meningitis, including a high polymorphonuclear leukocyte count, elevated protein, and low glucose levels with a decrease in the CSF-serum glucose ratio.

ADULT MENINGOENCEPHALITIS

L. monocytogenes is an uncommon cause of bacterial meningitis in adults. There are two major clinical presentations. The first is a typical subacute bacterial meningitis characterized by fever, headache, and neck stiffness (54). Because the organism is not commonly seen on Gram's stain of CSF and because the cell counts are lower than with other forms of bacterial meningitis, an initial diagnosis of viral meningitis is commonly made before culture of the organism from CSF or blood. The onset of the syndrome can occur over several days, unlike meningococcal or pneumococcal meningitis, both of which have a more abrupt onset. During epidemics of food-borne listeriosis, *Listeria* meningitis can occur in apparently healthy individuals of all ages. In sporadic disease, patients more commonly have obvious defects in cell-mediated immune function that predispose them to listeriosis.

The second form of central nervous system listeriosis in adults is a rhombencephalitis that has features characteristic of the same illness in animals described as circling disease (1). Fever, headache, nausea, and vomiting occur early, with signs of meningeal irritation less commonly present. Subsequently, patients develop multiple cranial nerve abnormalities accompanied by cerebellar dysfunction, including ataxia. Fever may not be present in up to 15% of patients, which makes the diagnosis more difficult and more suggestive of noninfectious disorders. CSF pleocytosis may be minimal, and the organism is rarely seen on Gram's stain. The diagnosis is established by culture of CSF or blood. Magnetic resonance imaging is the

TABLE 2 Major clinical syndromes associated with *L. monocytogenes*

Neonatal meningitis	Hepatitis
Meningoencephalitis in adults	Liver abscess
Rhombencephalitis	Cutaneous infections (in animal workers)
Bacteremia in infants or adults	Endophthalmitis
Native or prosthetic valve endocarditis	Febrile gastroenteritis
Spontaneous bacterial peritonitis	CAPD peritonitis
Pneumonia	Septic arthritis
Osteomyelitis	

best diagnostic study and frequently demonstrates typical multiple microabscesses of the cerebellum and diencephalon (Fig. 1). The mortality rate in this condition approaches 50%; despite treatment, residual morbidity, including permanent cranial nerve palsies and ataxia, may persist.

L. monocytogenes can also be responsible for cerebritis or typical brain abscess in the supratentorial region (19). In these cases, the typical rhombencephalitic symptoms due to microabscesses are absent. Host risk factors reflecting immune deficiency are more commonly seen in these cases, as they are with *Listeria* sepsis.

LISTERIA SEPSIS

Listeria sepsis, or bacteremia without central nervous system involvement, represents one-third of adult cases of invasive listeriosis. The symptoms are nonspecific but usually include fever and chills. As noted above, in pregnant women *Listeria* sepsis often masquerades as pyelonephritis or influenza (53). The diagnosis is often established in retrospect following delivery of an infected infant. In nonpregnant adults, *Listeria* sepsis almost always occurs in patients with malignancy, organ transplant, or other immunocompromised states (27, 60). In these settings, the presentation is also nonspecific and mimics sepsis with other gram-positive and gram-negative pathogens. The mortality rate for *Listeria* sepsis in these series is 25 to 30%.

OTHER CLINICAL SYNDROMES

Cutaneous Listeriosis

Cutaneous listeriosis is an occupational hazard of veterinary workers exposed to infected amniotic fluid or placental remnants that are removed from the birth canal of animals (49). Occasionally, cutaneous infection, including conjunctivitis, has been seen in laboratory workers. Cutaneous listeriosis is characterized by low-grade fever and multiple papulopustular lesions of the skin from which the organism can be isolated. Its appearance is similar to the rash seen in infants with early-onset disseminated listeriosis. In adults, the condition may resolve spontaneously without treatment, but the infection itself should be entirely preventable with appropriate gloving and other protective wear.

Bacterial Endocarditis Caused by *L. monocytogenes*

Bacterial endocarditis presumably follows transient bacteremia from a gastrointestinal source, with subsequent establishment of endovascular infection on an abnormal heart valve. *L. monocytogenes* is an uncommon cause of native valve endocarditis, and over 50% of cases that have been described involve prosthetic valves (22, 74). Infection with *L. monocytogenes* is usually found as part of the late prosthetic valve endocarditis syndrome. Diagnostic criteria for *Listeria* endocarditis include presence of a prosthetic valve with or without vegetation and continuous bacteremia with *L. monocytogenes*. Septic emboli and abscess formations in other organs are relatively frequent. In native valve endocarditis, *L. monocytogenes* can sometimes follow previous episodes of streptococcal bacterial endocarditis or other valvular heart disease. Patients with malignancy, diabetes, steroid therapy, and renal and liver transplantation have been diagnosed with *Listeria* endocarditis. Their presentation is nonspecific for *L. monocytogenes* and includes prolonged fever, chills, and ultimately signs of congestive heart failure. Septic embolization occurs in two-thirds of patients, and aortic and mitral valve involvement is most common. *L. monocytogenes* can also cause arterial infections that involve prosthetic abdominal and aortic grafts or native abdominal aortic aneurysms (25). The mortality of this condition approached 40% before 1985 but has been reduced to 12% with better recognition and surgical management.

Hepatitis and Liver Abscess Due to *L. monocytogenes*

L. monocytogenes has been described as a cause of acute hepatitis in several case reports (6). It occurs as acute onset of fever and jaundice accompanied by blood cultures positive for *L. monocytogenes*. The diagnosis is usually unsuspected. Severe disease with death has been described, and autopsy or liver biopsy generally reveals microabscesses and occasionally granulomas similar to those seen in severe neonatal disease. Predisposing factors include cirrhosis and liver transplantation (79), although *Listeria* hepatitis can occur in an otherwise healthy host.

Solitary and multiple liver abscesses with fever have also been described (7, 8). Bacteremia occurs in one-half of these patients. Predisposing factors include diabetes mellitus, transplantation, cirrhosis, and alcoholism. Aspiration of the abscess demonstrates the organism. The mortality rate is 50%, and postmortem examination often reveals abscesses in other organs as well. Patients with multiple abscesses appear to do worse than those with solitary abscess, despite appropriate treatment.

Listeria Peritonitis

L. monocytogenes can also cause isolated episodes of peritonitis (72). It is most commonly seen in patients undergoing continuous ambulatory peritoneal dialysis (CAPD); the organisms are isolated from the dialysate or from blood culture. The organisms presumably cause infection through translocation from the gastrointestinal tract in patients who have ingested the organism with food. This complication is extremely rare and represents <1% of all cases of CAPD peritonitis. It can also cause spontaneous bacterial peritonitis

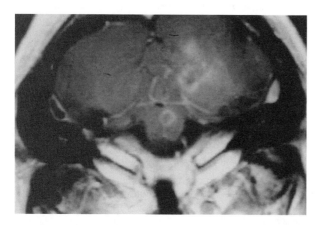

FIGURE 1 Magnetic resonance imaging of the cerebellum and midbrain of a 72-year-old, previously healthy man. Spinal fluid and blood cultures demonstrated *L. monocytogenes*. Well-formed microabscesses typical of listeriosis can be seen within the brain parenchyma.

in advanced liver disease with ascites (33). The mortality rate is low, and laboratory and clinical features are typical of spontaneous bacterial peritonitis due to other organisms. The organism can rarely cause this disease in patients who have undergone liver transplantation.

Musculoskeletal Infection

L. monocytogenes is a very uncommon cause of osteomyelitis (42). Reports of *Listeria* osteomyelitis emphasize the role of diabetes mellitus or leukemia as predisposing factors, particularly when long-term corticosteroids are administered. Relapses have been described despite effective antibiotic therapy. Septic arthritis due to *L. monocytogenes* appears to be more common than bone infection, and rheumatoid arthritis appears to be a frequent associated condition. Low-dose methotrexate therapy may predispose to this infection. Infection may follow joint injection with corticosteroids. Infection has been described in prosthetic hips and knees, as well as in native joints. The organism can also cause vertebral osteomyelitis with epidural abscess (37). With prolonged antibiotic therapy, medical treatment alone, as opposed to removal of the prosthetic joint, may be successful. Deaths are rare and are normally due to the underlying disease.

Gastroenteritis

A febrile gastroenteritis syndrome has been described for listeriosis (2, 11, 15, 24, 59, 63, 71). Gastrointestinal prodromal symptoms, such as diarrhea or abdominal pain, have been common in large outbreaks of food-borne adult listeriosis, but sepsis and meningoencephalitis have been the usual presenting syndromes. Population-based attack rates for invasive listeriosis have been low in this setting. In *Listeria* gastroenteritis, a more typical food-borne illness occurs with high attack rates (up to 72%) among the individuals exposed to the vehicle of infection. Most patients are well before development of the infection. While bacteremia has occurred in some patients, primary symptoms are diarrhea, fever, fatigue, chills, and myalgias occurring 24 h following exposure. This incubation period is considerably shorter than the 3- to 4-week period for the more usual forms of invasive listeriosis.

Pregnant women appear to be more likely to have sepsis in these outbreaks. Isolation of *L. monocytogenes* from stool with selective media has been rare, but serologic tests have been used to help define the extent of the outbreaks. In the outbreaks reported to date, rice salad, shrimp salad, chocolate milk, corn, deli meats, and fresh cheese have been the reported vehicles, and high colony counts of the organism (up to 10^9 CFU/g) appear to be present in the contaminated food. Invasive listeriosis can also be a result of loss of gastrointestinal integrity due to other gastrointestinal tract pathogens such as *Shigella* sp. (40) or to presumed viral gastroenteritis (69). This may account for some sporadic and epidemic cases of *Listeria* sepsis and meningoencephalitis.

Diagnosis of Listeriosis

Diagnosis of all forms of *L. monocytogenes* infection depends on isolation of the organism from a normally sterile site, usually blood or CSF. Gram's stain of specimens of sterile spinal fluid, peritoneal fluid, or joint aspirates occasionally, but unpredictably, reveals gram-positive coccobacilli characteristic of *L. monocytogenes*. In some forms of central nervous system infection, particularly rhombencephalitis, several samples may have to be obtained to isolate the organism.

In situations where antibiotic therapy has already been administered, isolation of the organism from a nonsterile site may support a diagnosis of listeriosis. In pregnant women, stool or vaginal cultures may be positive when selective media for *L. monocytogenes* are used for culture. In febrile gastroenteritis syndromes, where traditional pathogens have not been isolated with standard media, culture of the stool with selective media for *L. monocytogenes* may also demonstrate the organism.

Serologic tests, until recently, have not been useful in defining infection (5). In an outbreak of febrile gastroenteritis, retrospective analysis of sera from most patients has demonstrated high levels of anti-listeriolysin A antibody (11). However, serologic tests are not likely to be useful for the diagnosis of acute listeriosis. Direct detection of PCR products from *L. monocytogenes* in CSF has been studied but sensitivity appears to be low (3).

Treatment of Listeriosis

L. monocytogenes remains susceptible to most B-lactam antibiotics, with the exception of cephalosporins, to which the organism is usually resistant (32, 62). Because newer cephalosporins are commonly used for the treatment of nonspecific sepsis syndromes or for the empirical treatment of bacterial meningitis, specific therapy for listeriosis may be delayed for some patients. When listeriosis is a likely diagnosis, the use of ampicillin or, in penicillin-allergic patients, vancomycin will provide empirical coverage for *L. monocytogenes* until the diagnosis is made by culture.

A combination of ampicillin and gentamicin is the current therapy of choice for all forms of listeriosis (41, 77). Ampicillin is not bactericidal for *L. monocytogenes*, and in vitro and in vivo data suggest that an additive or synergistic effect with gentamicin may improve outcomes (45). No randomized controlled clinical trials of therapy in humans have been carried out, however.

Trimethoprim-sulfamethoxazole, with or without the addition of rifampin, is an alternative treatment regimen that has been recommended. In one retrospective study, the combination of amoxicillin and cotrimoxazole was found to be more effective than ampicillin and gentamicin (51). New quinolones may also be effective but data are limited to in vitro studies (46)

Trimethoprim-sulfamethoxazole has also been used as a prophylactic agent against a number of microorganisms, including *P. carinii*, in patients with HIV infection and in patients undergoing chemotherapy for leukemia or lymphoma. This drug would be effective in protecting against *L. monocytogenes* infections. The use of prophylaxis has been temporally associated with a decrease in the incidence of listeriosis in these compromised hosts in combination with dietary guidelines that have been issued for these patients in recent years (21).

The duration of treatment for invasive listeriosis has not been studied. Relapses appear to be uncommon, and 2 to 3 weeks of therapy with ampicillin and gentamicin is sufficient for most forms of listeriosis. Rhombencephalitis with abscess formation in the central nervous system may require more prolonged therapy, but data are not available that support treatment beyond 4 weeks (41, 77).

REFERENCES

1. **Armstrong, R. W., and P. C. Fung.** 1993. Brainstem encephalitis due to *Listeria monocytogenes*: case report and review. *Clin. Infect. Dis.* **16:**689–702.

2. **Aureli, P., G. C. Fiorucci, D. Caroli, G. Marchiaro, O. Novaro, L. Leone, and S. Salmaso.** 2000. An outbreak of febrile gastroenteritis associated with corn contaminated by *Listeria monocytogenes N. Engl. J. Med.* **342:**1236–1241.

3. **Backman, A., P.-G. Lantz, P. Radstrom, and P. Olcen.** 1999. Evaluation of an extended diagnostic PCR assay for detection and verification of the commonest causes of bacterial meningitis in CSF and other biological samples. *Mol. Cell. Probes* **13:**49–60.

4. **Baltimore, R. S., S. M. Huie, J. I. Meek, A. Schuchat, and Katherine O'Brien.** 2001. Early onset neonatal sepsis in the era of group B streptococcal prevention. *Pediatrics* **108:**1094–1098.

5. **Berche, P., K. A. Reich, M. Bonnichon, J. L. Beretti, C. Geoffroy, J. Raveneau, P. Cossart, J. L. Gaillard, P. Geslin, H. Kreis, and M. Veron.** 1990. Detection of antilisteriolysin O for serodiagnosis of human listeriosis. *Lancet* **335:**624–627.

6. **Bourgeois, N., F. Jacobs, M. L. Tavares, F. Rickaert, C. Deprez, C. Liesnard, F. Moonens, J. Van de Stadt, M. Gelin, and M. Adler.** 1993. *Listeria monocytogenes* hepatitis in a liver transplant recipient: a case report and review of the literature. *J. Hepatol.* **18:**284–289.

7. **Braun, T. I., D. Travis, R. R. Dee, and R. E. Nieman.** 1993. Liver abscess due to *Listeria monocytogenes*: case report and review. *Clin. Infect. Dis.* **17:**267–269.

8. **Bronnimann, S., H. U. Baer, R. Malinverni, and M. W. Buchler.** 1998. *Listeria monocytogenes* causing solitary liver abscess: case report and review of the literature. *Dig. Surg.* **15:**364–368.

9. **Buchanan, R. L., and R. C. Whiting.** 1998. Risk assessment: a means for linking HACCP plans and public health. *J. Food Prot.* **61:**1531–1534.

10. **Bula, C. J., J. Bille, and M. P. Glauser.** 1995. An epidemic of food-borne listeriosis in western Switzerland: description of 57 cases involving adults. *Clin. Infect. Dis.* **20:**66–72.

11. **Carrique-Mas, J. J., I. Hokeberg, Y. Andersson, M. Arneborn, W. Tham, M. L. Danielsson-Tham, B. Osterman, M. Leffler, M. Steen, E. Eriksson, G. Hedin, and J. Giesecke.** 2003. Febrile gastroenteritis after eating on-farm manufactured fresh cheese—an outbreak of listeriosis? *Epidemiol. Infect.* **130:**79–86.

12. **Centers for Disease Control and Prevention.** 2001. Outbreak of listeriosis associated with homemade Mexican-style cheese——North Carolina, October 2000–January 2001. *Morb. Mortal. Wkly. Rep.* **50:**560–562.

13. **Centers for Disease Control and Prevention.** 2002. Public health dispatch: outbreak of listeriosis—northeastern United States, 2002. *Morb. Mortal. Wkly. Rep.* **51:**950–951.

14. **Centers for Disease Control and Prevention.** 2004. Preliminary FoodNet data on the incidence of infection with pathogens transmitted commonly through food—selected sites, United States, 2003. *Morb. Mortal. Wkly. Rep.* **53:**338–343.

15. **Dalton, C. B., C. C. Austin, J. Sobel, P. S. Hayes, W. F. Bibb, L. M. Graves, B. Swaminathan, M. E. Proctor, and P. M. Griffin.** 1997. An outbreak of gastroenteritis and fever due to *Listeria monocytogenes* in milk. *N. Engl. J. Med.* **336:**100–105.

16. **Dawson, K. G., J. C. Emerson, and J. L. Burns.** 1999. Fifteen years of experience with bacterial meningitis. *Pediatr. Infect. Dis. J.* **18:**816–822.

17. **deValk, H., V. Vaillant, C. Jacquet, J. Rocourt, F. Le Querrec, F. Stainer, N. Quelquejeu, O. Pierre, V. Pierre, J.-C. Desenclos, and V. Goulet.** 2001. Two consecutive nationwide outbreaks of listeriosis in France, October 1999–February 2000. *Am. J. Epidemiol.* **154:**944–950.

18. **Donnelly, C. W.** 2001. *Listeria monocytogenes*: a continuing challenge. *Nutr. Rev.* **59:**183–194.

19. **Eckburg, P. B., J. G. Montoya, and K. L. Vosti.** 2001. Brain abscess due to *Listeria monocytogenes*: five cases and a review of the literature. *Medicine* **80:**223–235.

20. **Evans, J. R., A. C. Allen, D. A. Stinson, R. Bortolussi, and L. J. Peddle.** 1985. Perinatal listeriosis: report of an outbreak. *Pediatr. Infect. Dis.* **4:**237–241.

21. **Ewert, D. P., L. Lieb, P. S. Hayes, M. W. Reeves, and L. Mascola.** 1995. *Listeria monocytogenes* infection and serotype distribution among HIV-infected persons in Los Angeles County, 1985–1992. *J. Acquir. Immune Defic. Syndr. Hum. Retrovirol.* **8:**461–465.

22. **Fernandez Guerrero, M. L., P. Rivas, R. Rabago, A. Nunez, M. deGorgolas, and J. Martinell.** 2004. Prosthetic valve endocarditis due to *Listeria monocytogenes*. Report of two cases and reviews. *Int. J. Infect. Dis.* **8:**97–102.

23. **Fleming, D. W., S. L. Cochi, K. L. MacDonald, J. Brondum, P. S. Hayes, B. D. Plikaytis, M. B. Holmes, A. Auduria, C. V. Broome, and A. L. Reingold.** 1985. Pasteurized milk as a vehicle of infection in an outbreak of listerosis. *N. Engl. J. Med.* **312:**404–407.

24. **Frye, D. M., R. Zweig, J. Sturgeon, M. Tormey, M. LeCavalier, I. Lee, L. Lawani, and L. Mascola.** 2002. An outbreak of febrile gastroenteritis associated with delicatessen meat contaminated with *Listeria monocytogenes*. *Clin. Infect. Dis.* **35:**943–949.

25. **Gautoar, A. R., L. A. Cone, D. R. Woodard, R. J. Mahler, R. D. Lynch, and D. H. Stoltzman.** 1992. Arterial infections due to *Listeria monocytogenes*: report of four cases and review of world literature. *Clin. Infect. Dis.* **14:**23–28.

26. **Gill, D. A.** 1937. Ovine bacterial encephalitis (circling disease) and the bacterial genus Listerella. *Aust. Vet. J.* **13:**46–56.

27. **Goulet, V., and P. Marchetti.** 1996. Listeriosis in 225 non-pregnant patients in 1992. Clinical aspects and outcome in relationship to predisposing conditions. *Scand. J. Infect. Dis.* **28:**367–374.

28. **Goulet, V., H. de Valk, O. Pierre, F. Stainer, J. Rocourt, V. Vaillant, C. Jacquet, and J.-C. Desenclos.** 2001. Effect of prevention measures on incidence of human listeriosis, France, 1987–1997. *Emerg. Infect. Dis.* **7:**983–989.

29. **Graves, L. M., B. Swaminathan, G. W. Ajello, G. D. Malcolm, R. E. Weaver, R. Ransom, K. Dever, B. D. Plikaytis, A. Schuchat, J. D. Wenger, R. W. Pinner, C. V. Broome, and the Listeria Study Group.** 1992. Comparison of three selective enrichment methods for the isolation of *Listeria monocytogenes* from naturally contaminated foods. *J. Food Prot.* **55:**952–959.

30. **Grif, K., I. Hein, M. Wagner, E. Brandl, O. Mpamugo, J. McLauchlin, M. P. Dierich, and F. Allerberger.** 2001. Prevalence and characterization of *Listeria monocytogenes* in the feces of healthy Austrians. *Wien. Klin. Wochenschr.* **113:**737–742.

31. **Grif, K., G. Patscheider, M. P. Dierich, and F. Allerberger.** 2003. Incidence of fecal carriage of *Listeria monocytogenes* in three healthy volunteers: a one-year prospective stool survey. *Eur. J. Clin. Microbiol. Infec. Dis.* **22:**16–20.

32. **Hof, H., T. Nichterlein, and M. Kretschmar.** 1997. Management of listeriosis. *Clin. Microbiol. Rev.* **10:**345–357.

33. **Jayaraj, K., A. M. Di Bisceglie, and S. Gibson.** 1998. Spontaneous bacterial peritonitis caused by infection with *Listeria monocytogenes*: case report and review of the literature. *Am. J. Gastroenterol.* **93:**1556–1558.

34. Jurado, R. L., M. M. Farley, E. Pereira, R. C. Harvey, A. Schuchat, J. D. Wenger, and D. S. Stephens. 1993. Increased risk of meningitis and bacteremia due to *Listeria monocytogenes* in patients with human immunodeficiency virus infection. *Clin. Infect. Dis.* **17:**224–227.

35. Kampelmacher, E. H., and L. M. Van Noorle Jansen. 1969. Isolation of *Listeria monocytogenes* from feces of clinically healthy humans and animals. *Zentralbl. Bakteriol.* **211:**353–359.

36. Kessler, S. L., and A. S. Dajani. 1990. Listeria meningitis in infants and children. *Pediatr. Infect. Dis. J.* **9:**61–63.

37. Khan, K. M., W. Pao, and J. Kendler. 2001. Epidural abscess and vertebral osteomyelitis caused by *Listeria monocytogenes*: case report and literature review. *Scand. J. Infect. Dis.* **33:**714–716.

38. Lammerding, A. M., K. A. Glass, A. Gendron-Fitzpatrick, and M. P. Doyle. 1992. Determination of virulence of different stains of *Listeria monocytogenes* and *Listeria innocua* by oral inoculation of pregnant mice. *Appl. Bacteriol. Microbiol.* **58:**3991–4000.

39. Linnan, M. J., L. Mascola, X. D. Lou, V. Goulet, S. May, G. Salminen, D. W. Hird, M. L. Yonekura, P. Hayes, R. Weaver, A. Audurier, B. D. Plikaytis, S. L. Fannin, A. Kleks, and C. V. Broome. 1988. Epidemic listeriosis associated with Mexican-style cheese. *N. Engl. J. Med.* **319:**823–828.

40. Lorber, B. 1991. Listeriosis following shigellosis. *Rev. Infect. Dis.* **13:**865–866.

41. Lorber, B. 1997. Listeriosis. *Clin. Infect. Dis.* **24:**1–9.

42. Louthrenoo, W., and H. R. Schumacher, Jr. 1990. *Listeria monocytogenes* osteomyelitis complicating leukemia: report and literature review of *Listeria* osteoarticular infections. *J. Rheumatol.* **17:**107–110.

43. Low, J. C., and C. P. Renton. 1985. Septicemia, encephalitis and abortions in a housed flock of sheep caused by *Listeria monocytogenes* type 1/2. *Vet. Rec.* **114:**147–150.

44. Lyytikainen, O., T. Autio, R. Maijala, P. Ruutu, T. Honkanen-Buzalski, M. Miettinen, M. Hatakka, J. Mikkola, V.-J. Antilla, T. Johansson, L. Rantala, T. Aalto, H. Korkeala, and A. Siitonen. 2000. An outbreak of invasive *Listeria monocytogenes* serotype 3a infections from butter in Finland. *J. Infect. Dis.* **181:**1838–1841.

45. MacGowan, A., M. Wooton, K. Bowker, H. A. Holt, and D. Reeves. 1998. Ampicillin-aminoglycoside interaction studies using *Listeria monocytogenes*. *J. Antimicrob. Chemother.* **41:**417–418.

46. Marco, F., M. Almela, J. Nolla-Salas, P. Coll, I. Gasser, M. D. Ferrer, M. de Simon, and the Collaborative Study Group of Listeriosis of Barcelona. 2000. In vitro activity of 22 antimicrobial agents against *Listeria monocytogenes* strains isolated in Barcelona, Spain. *Diagn. Microbiol. Infect. Dis.* **38:**259–261.

47. Majewski, M. C. 1992. Food safety: the HACCP approach to hazard control. *Commun. Dis. Rep. CDR Rev.* **9:**R105–R108.

48. Mascola, L., F. Sorvillo, V. Goulet, B. Hall, R. Weaver, and M. J. Linnan. 1992. Fecal carriage of *Listeria monocytogenes*—observations during a community-wide, common-source outbreak. *Clin. Infect. Dis.* **15:**557–558.

49. McLauchlin, J., and J. C. Low. 1994. Primary cutaneous listeriosis in adults: an occupational disease in veterinarians and farmers. *Vet. Rec.* **135:**615–617.

50. McLauchlin, J., R. T. Mitchell, W. J. Smerdon, and K. Jewell. 2004. *Listeria monocytogenes* and listeriosis: a review of hazard characterization for use in microbiological risk assessment of foods. *Int. J. Food Microbiol.* **92:**15–33.

51. Merle-Melet, M., L. Dossou-Glete, P. Maurer, P. Meyer, A. Lozniewski, O. Kuntzburger, M. Weber, and A. Gerard. 1996. Is amoxicillin-cotrimoxasole the most appropriate antibiotic regimen for listeria meningoencephalitis? Review of 22 cases and the literature. *J. Infect.* **33:**79–85.

52. Murray, E. G. D., R. A. Webb, and M. B. R. Swann. 1926. A disease of rabbits characterized by large mononuclear leucocytosis, caused by a hitherto undescribed bacillus: Bacterium monocytogenes. *J. Pathol. Bacteriol.* **29:**407–439.

53. Mylonakis, E., M. Paliou, E. L. Hohmann, S. B. Calderwood, and E. J. Wing. 2002. Listeriosis during pregnancy: a case series and review of 222 cases. *Medicine* **81:**260–269.

54. Mylonakis, E., E. L. Hohmann, and S. B. Calderwood. 1998. Central nervous system infection with *Listeria monocytogenes*: 33 years' experience at a general hospital and review of 776 episodes from the literature. *Medicine* **77:**313–336.

55. Panisello, P. J., R. Rooney, P. C. Quantick, and R. Stanwell-Smith. 2000. Application of foodborne disease outbreak data in the development and maintenance of HACCP systems. *Int. J. Food Microbiol.* **59:**221–234.

56. Pirie, J. H. H. 1927. A new disease of veld rodents, "Tiger River Disease." *Publ. S. Afr. Inst. Med. Res.* **3:**163–186.

57. Potel, J. 1952. Zur granulomatosis infantiseptica. *Zentralbl. Bakteriol. Parasitenkd. Infektionskr. Hyg.* **158:**329–331.

58. Reissbrodt, R. 2004. New chromogenic plating media for detection and enumeration of pathogenic *Listeria* spp.—an overview. *Int. J. Food Microbiol.* **15:**1–9.

59. Riedo, F. X., R. W. Pinner, M. L. Tosca, M. L. Cartter, M. L. Graves, N. W. Reeves, R. E. Weaver, B. D. Plikaytis, and C. V. Broome. 1994. A point-source food borne listeriosis outbreak: documented incubation period and possible mild illness. *J. Infect Dis.* **170:**693–696.

60. Rivero, G. A., H. A. Torres, K. V. I. Rolston, and D. P. Kontoyiannis. 2003. *Listeria monocytogenes* infection in patients with cancer. *Diag. Microbiol. Infect. Dis.* **47:**393–398.

61. Rocourt, J., A. Hogue, H. Toyofuku, C. Jacquet, and J. Schlundt. 2001. *Listeria* and listeriosis: risk assessment as a new tool to unravel a multifaceted problem. *Am. J. Infect. Control* **29:**225–227.

62. Safdar, A., and D. Armstrong. 2003. Antimicrobial activities against 84 *Listeria monocytogenes* isolates from patients with systemic listeriosis at a comprehensive cancer center (1955–1997). *J. Clin. Microbiol.* **41:**483–485.

63. Salamina, G., E. Dalle Donne, A. Niccolini, G. Poda, D. Cesaroni, M. Brici, M. Maldini, R. Fini, A. Schuchat, B. Swaminathan, W. Bibb, J. Rocourt, N. Binkin, and S. Salmaso. 1996. A foodborne outbreak of gastroenteritis due to *Listeria monocytogenes*. *Epidemiol. Infect.* **117:**429–436.

64. Schlech, W. F., III. 1993. An animal model of foodborne *Listeria monocytogenes* virulence: effect of alterations in local and systemic immunity on invasive infection. *Clin. Investig. Med.* **16:**219–225.

65. Schlech, W. F., III, P. M. Lavigne, R. A. Bortolussi, A. C. Allen, E. V. Haldane, A. J. Wort, A. W. Hightower, S. E. Johnson, S. H. King, E. S. Nicholls, and C. V. Broome. 1983. Epidemic listeriosis—evidence for transmission by food. *N. Engl. J. Med.* **308:**203–206.

66. Schuchat, A., K. A. Deaver, J. D. Wenger, B. D. Plikaytis, L. Mascola, R. W. Pinner, A. L. Reingold, C. V. Broome, and the Listeriosis Study Group. 1992. Role of foods in sporadic listeriosis. I. Case-control study of dietary risk factors. *JAMA* **267:**2041–2045.

67. Schuchat, A., A. C. Lizano, C. V. Broome, B. Swaminathan, C. Kim, and K. Winn. 1991. Outbreak of neonatal listeriosis associated with mineral oil. *Pediatr. Infect. Dis. J.* **10**:183–189.

68. Schwartz, B., C. V. Broome, G. R. Brown, A. W. Hightower, C. A. Ciesielski, S. Gaventa, B. G. Gellin, L. Mascola, and the Listeriosis Study Group. 1988. Association of sporadic listeriosis with consumption of uncooked hotdogs and undercooked chicken. *Lancet* **ii**:779–782.

69. Schwartz, B., D. Hexter, C. V. Broome, A. W. Hightower, R. V. Hirschhorn, J. D. Porter, P. S. Hayes, W. F. Bibb, B. Lorber, and D. G. Faris. 1989. Investigation of an outbreak of listeriosis; new hypothesis for the etiology of epidemic *Listeria monocytogenes* infections. *J. Infect. Dis.* **159**:680–685.

70. Shuin, D. T., and S. B. Galsworthy. 1982. Stimulation of monocyte production by an endogenous mediator induced by a component from *Listeria monocytogenes*. *Immunology* **46**:343–351.

71. Sim, J., D. Hood, L. Finnie, M. Wilson, C. Graham, M. Brett, and J. A. Hudson. 2002. Series of incidents of *Listeria monocytogenes* non-invasive febrile gastroenteritis involving ready-to-eat meats. *Lett. Appl. Microbiol.* **35**:409–413.

72. Sivalingam, J. J., P. Martin, H. S. Fraimow, J. C. Yarze, and L. S. Friedman. 1992. *Listeria monocytogenes* peritonitis: case report and literature review. *Am. J. Gastroenterol.* **87**:1839–1845.

73. Skogberg, K. J., J. Syrjanen, M. Jahkola, O. V. Reinkonen, J. Paavonen, J. Ahonen, S. Kontiainen, P. Ruutu, and V. Valtonen. 1992. Clinical presentation and outcome of listeriosis in patients with and without immunosuppressive therapy. *Clin. Infect. Dis.* **14**:815–821.

74. Spyrou, N., M. Anderson, and R. Foale. 1997. Listeria endocarditis: current management and patient outcome—world literature review. *Heart* **4**:380–383.

75. Sridama, V., F. Pacni, S.-L. Wang, A. Moawad, M. Reilly, and L. J. Degroot. 1982. Decreased levels of helper T cells: a possible cause of immunodeficiency in pregnancy. *N. Engl. J. Med.* **307**:352–356.

76. Tappero, J. W., A. Schuchat, K. A. Deaver, L. Mascola, and J. D. Wenger for the Listeriosis Study Group. 1995. Reduction in the incidence of human listeriosis in the United States: effectiveness of prevention efforts? *JAMA* **273**:1118–1122.

77. Temple, M. E., and M. C. Nahata. 2000. Treatment of listeriosis. *Ann. Pharmacother.* **34**:656–661.

78. Thompson, P., P. A. Salisbury, C. Adams, and D. L. Archer. 1990. U.S. food legislation. *Lancet* **336**:1557–1559.

79. Vargas, V., C. Aleman, I. de Torres, L. Castells, J. Gavalda, C. Margarit, R. Esteban, and J. Guardia. 1998. Listeria monocytogenes-associated acute hepatitis in a liver transplant recipient. *Liver* **18**:213–215.

80. Wald, A., D. H. Vanthiel, L. Hoechstetter, J. S. Gavaler, K. M. Egler, R. Verm, L. Scott, and R. Lester. 1982. Effect of pregnancy on gastrointestinal transit. *Dig. Dis. Sci.* **27**:1015–1018.

81. Welshimer, H. J., and J. Donker-Voet. 1971. *Listeria monocytogenes* in nature. *Appl. Microbiol.* **21**:516–519.

Listeria monocytogenes Infection of Mice: an Elegant Probe To Dissect Innate and T-Cell Immune Responses

JODIE S. HARING AND JOHN T. HARTY

50

To be successful, microbial pathogens must evade the numerous immune mechanisms employed by their hosts long enough to establish a productive infection and transfer to the next host. Although infection with the intracellular bacterium *Listeria monocytogenes* can result in severe illnesses such as sepsis and meningitis in immunocompromised people, a much more common outcome is control and clearance of the organism without serious malady (135). Because *L. monocytogenes* is able to infect common laboratory animals such as the mouse (99), immunologists have been able to dissect this host-pathogen interplay and elucidate the immune functions required to contain and eliminate *L. monocytogenes* infection. Importantly, the mouse model of *L. monocytogenes* infection has yielded discoveries that are not only specific to this host-pathogen interaction but also help define fundamental concepts of innate and adaptive immunity.

MOUSE MODEL OF *L. MONOCYTOGENES* INFECTION

L. monocytogenes is a facultative intracellular bacterium (91, 93), which means it does not have to reside within a mammalian cell for its survival and replication. It is clear, however, that *L. monocytogenes* has evolved to accommodate such a lifestyle by developing mechanisms to escape the potentially lethal environment inside phagocytic vesicles (125) and accomplish cell-to-cell spread without ever leaving the protective intracellular environment (24). Because of their accessibility, short generation time, and (as shown more recently) their amenability to genetic manipulation, laboratory mice are the animals most commonly used for studying the immune system. A murine model of *L. monocytogenes* infection was first described by George Mackaness over 40 years ago (86). These initial experiments demonstrated that mice rapidly eliminated low-dose infection in an antibody-independent manner, highlighting a role for cellular immunity in clearance of *L. monocytogenes* infection. In addition, the previously infected mice were resistant to subsequent challenge with much higher numbers of bacteria, indicating that they had developed immunity as

a result of their initial infection. These groundbreaking experiments were only just the beginning of what has become an enormous literature describing perhaps the most carefully characterized model of host resistance to infectious disease. When considered together, such studies have earned *L. monocytogenes* status as an elegant probe to dissect innate and T-cell immune responses to infection.

INNATE IMMUNE RESPONSE FOLLOWING INFECTION WITH *L. MONOCYTOGENES*

Antilisterial Mechanisms in the Liver: the Primary Site of Infection

The natural route of *L. monocytogenes* infection in humans is via the gut after consumption of contaminated food products (135). However, infection of mice in this manner requires an extremely large inoculum and often results in asynchronous systemic infections, which complicate experimental design (32). In most experiments, *L. monocytogenes* is inoculated into mice intravenously (i.v.) or intraperitoneally. Within 15 min after i.v. injection, 80% of *L. monocytogenes* can be found in the liver (47, 86). The majority of the remaining inoculum is found in the spleen (86).

In the liver, the first cells that *L. monocytogenes* comes into contact with are the Kupffer cells, which are the resident tissue macrophages of this organ and line the blood vessels. These cells physically associate with and internalize *L. monocytogenes*, trapping the bacteria long enough to allow recruitment of macrophages and neutrophils from the blood that assist in killing *L. monocytogenes* (30, 33). Once associated with bacteria, Kupffer cells become activated and produce numerous cytokines, including interleukin 6 (IL-6), IL-12, IL-1β, and tumor necrosis factor (TNF), which participate in the recruitment and activation of blood-derived leukocytes and foster antimicrobial activities of other liver cells (41, 48).

During the first 6 h of infection, the majority of *L. monocytogenes* organisms in the liver are killed, most likely by neutrophils (30), which will be discussed later. After this

Gram-Positive Pathogens, 2nd edition, edited by Vincent A. Fischetti et al.
© 2006 ASM Press, Washington, D.C.

time, the remaining listeriae are found in hepatocytes, which are the main site of bacterial replication (33). Infected hepatocytes have some potential antilisterial defenses such as the production of acute-phase response proteins and reactive oxygen intermediates upon exposure to the proinflammatory cytokines (49, 123); however, these activities are only marginally protective, and the number of bacteria in hepatocytes increases exponentially over the next 3 days of infection (86).

It is critical that the replication of *L. monocytogenes* be controlled to some degree during the first 72 h of infection, until the adaptive immune response is primed and antigen (Ag)-specific T cells expand to sufficient numbers. If the replication of bacteria is not dampened, the mice will succumb to infection before the adaptive immune system even begins to participate in the antilisterial response. Neutrophils are indispensable in the early control of *L. monocytogenes* infection. A large number ($>2 \times 10^6$) of neutrophils enter the liver within the first 12 h after infection with *L. monocytogenes* and can be visualized surrounding infected hepatocytes, forming large, segregated foci of infection within this organ (104, 105). Mice treated with the monoclonal antibody RB6-8C5, which depletes neutrophils (and other immune cells) (53, 67), are unable to control *L. monocytogenes* infection and die 3 to 4 days postinfection (p.i.) from acute liver failure (33, 104). Likewise, mice deficient in IL-1 or IL-6, both of which exhibit altered neutrophil homeostasis in the blood and decreased recruitment of neutrophils into the liver following infection, are markedly more susceptible to *L. monocytogenes* than wild-type mice (see Table 1 for a list of susceptibilities to virulent *L. monocytogenes* infection of specific mouse strains discussed in this chapter) (36, 105).

Although the exact function of neutrophils in the antilisterial immune response is of some debate, these cells can internalize and kill *L. monocytogenes* (35). In addition, it has been suggested that neutrophils may lyse infected hepatocytes, thereby destroying the preferred cell for replication

TABLE 1 Relative susceptibility of specific mouse strains to virulent *L. monocytogenes* infection

Mouse strain	Relative susceptibility to infection[a]	Reference(s)
Wild type BALB/c	++++	57
IFN-$\gamma^{-/-}$	+	57
IFN-γ RI$^{-/-}$	+	63
STAT1	+	88
IFNAR1$^{-/-}$	+++++	2, 27, 94
Perforin$^{-/-}$	++++	71
IFN$\gamma^{-/-}$/perforin$^{-/-}$	+	4
TNF$^{-/-}$	++	132
TNF$^{-/-}$/perforin$^{-/-}$	++	132
IL-1$^{-/-}$	++	105
IL-6$^{-/-}$	++	36
CCR2$^{-/-}$	++	78
MyD88$^{-/-}$	+	106
SCID	++	10
RAG-2$^{-/-}$	++	72
TCR$\gamma^{-/-}$	++	72
TCR$\gamma^{-/-}$	+++	79
μMT	+++	109

[a]Intravenous infection with virulent *L. monocytogenes*.

of these bacteria and precluding efficient cell-to-cell spread (30). In other experiments it has been demonstrated that neutrophils can directly kill extracellular bacteria, perhaps released from dying hepatocytes (33) via degranulation and exposure of *L. monocytogenes* to a plethora of antibacterial enzymes and toxic oxygen metabolites. In addition to their direct antibacterial capabilities, it appears that neutrophils may be an important source of chemokines required to entice circulating lymphocytes into the liver. In a normal infection, lymphocytes begin to appear in the liver by 48 h p.i.; however, in mice depleted of neutrophils, lymphocytes are not recruited, despite or perhaps because of the significant increase in bacterial load exhibited by these mice (30).

In addition to the influx of neutrophils early after *L. monocytogenes* infection, blood-derived monocytes and natural killer (NK) cells are also recruited into the liver. The primary function of infiltrating monocytes is to phagocytose extracellular bacteria and dead or dying hepatocytes, which may contain bacteria. Several subsets of blood-derived monocytes exist in the mouse. During acute *L. monocytogenes* infection, phagocytic cells that express high levels of the surface marker Ly-6C, therefore resembling immature bone marrow-derived monocytes, predominate in the blood (117). The importance of these cells to the antilisterial immune response has not been extensively investigated; however, neutralization of macrophage colony-stimulating factor, an important chemokine in the recruitment of these cells, results in a dramatic increase in bacterial numbers at early times p.i. (33). In addition, treatment of mice with silica, which has been shown in vitro to decrease the viability and phagocytic capabilities of macrophages, results in a significant increase in bacterial replication 72 h after *L. monocytogenes* infection (120, 140).

NK cells are lytic cells whose activity is not restricted by specific Ag recognition (138). Instead, NK cells lyse cells with aberrant major histocompatibility complex class I (MHC-I) expression or after ligation of activating cell surface receptors. One of the roles NK cells may play in controlling *L. monocytogenes* infection is lysing infected hepatocytes and exposing intracellular bacteria to phagocytosis by neutrophils or macrophages. The lytic activity of NK cells is increased after in vitro exposure to *L. monocytogenes* (50).

Perhaps even more important than their lytic capacity is the ability of NK cells to make and secrete gamma interferon (IFN-γ). Upon exposure to IL-12 and TNF, most likely produced by *L. monocytogenes*-infected macrophages or dendritic cells (DCs) (40, 76), NK cells become potent producers of IFN-γ, a proinflammatory cytokine. The importance of IFN-γ in antilisterial immunity has been well documented. Mice that are deficient in IFN-γ, either of the two IFN-γ receptor chains, or signaling components downstream of the IFN-γ receptor are profoundly susceptible to *L. monocytogenes* infection (57, 63, 84, 88, 127). The 50% lethal dose for virulent *L. monocytogenes* in wild-type BALB/c mice is $10^{4.1}$, while IFN-$\gamma^{-/-}$ BALB/c mice succumb after infection with <10 bacteria. However, IFN-$\gamma^{-/-}$ mice are resistant to infection with certain attenuated strains of *L. monocytogenes* (9, 57). The roles of IFN-γ in the adaptive phase of the immune response will be discussed later in this chapter. During the early phases of the immune response, one of the primary roles of IFN-γ is to activate phagocytes to kill ingested bacteria via the induction of reactive oxygen species (reactive oxygen intermediates) and nitric oxide (NO) (40). In some studies, *L. monocytogenes* infection has been shown to be poorly controlled in mice depleted of NK cells, indicating that these cells may be an important source of IFN-γ at early

L. monocytogenes Infection in the Spleen

As mentioned previously, approximately 10 to 20% of the initial *L. monocytogenes* inoculum is taken up by the spleen (86). Within this organ, bacteria are localized in splenic marginal zone macrophages and other cells with dendritic morphology very early after infection. These cells do not appear capable of killing *L. monocytogenes*, since there is no initial decrease in bacterial numbers in the spleen. As a result, productive infection of this organ is established immediately (30). Neutrophils are recruited into the spleen in great numbers, although their accumulation is delayed compared to the liver and they do not seem to be able to control the *L. monocytogenes* infection in this organ. By day 2 p.i., large numbers of both intracellular and extracellular *L. monocytogenes* in the spleen are observed (30, 31).

A subpopulation of splenic DCs termed TipDCs have been identified recently in *L. monocytogenes*-infected mice (107). These cells begin to accumulate in the spleen 1 to 2 days p.i. with *L. monocytogenes* and produce TNF and inducible nitric oxide synthase. Recruitment of TipDCs into the spleen is aberrant in mice deficient for the chemokine receptor CCR2, which may be one reason why these mice are more susceptible to infection (78). Although these cells contribute to innate resistance to *L. monocytogenes*, TipDCs are not required for the generation of antilisterial T-cell responses (107).

The spleen is where Ag-specific CD4 and CD8 T-cell responses to *L. monocytogenes*-derived epitopes are generated; however, there is a large amount of lymphocyte death observed in the spleen 2 to 3 days after infection with *L. monocytogenes* (26, 68). In a recent study, it was demonstrated that nonspecific T cells (T cells not recognizing *L. monocytogenes*-derived epitopes) were selectively eliminated from the spleen, freeing up the necessary space for Ag-specific T cells to expand and clear the infection (68). The nonspecific T cells were shown to be dying via apoptosis (68), which may be caused by exposure to the *L. monocytogenes*-derived toxin listeriolysin O (LLO) (26). The apoptosis-inducing activity of LLO may be overridden in T cells that receive activation signals through their T-cell receptors.

The deaths of T cells and other cells in the spleen after *L. monocytogenes* infection may be important in pathogenesis and result from IFN-α/β-induced apoptosis. A collection of recent studies have demonstrated that mice deficient in IFN-α/β receptor 1 (IFNAR1) or IFN regulatory factor 3, an important signaling component downstream from IFNAR1, are surprisingly more resistant to *L. monocytogenes* infection than wild-type mice (2, 27, 94). IFNAR1-deficient mice exhibited an increase in serum IL-12p70 early after infection and only slightly decreased levels of IFN-γ compared to those in wild-type mice (2, 27). In addition, significantly less apoptosis was observed in the spleens of *L. monocytogenes*-infected IFNAR1 mice (2, 27, 94), with a particularly striking preservation of TipDCs and other macrophage-like cell subsets (2). These data present an interesting contrast between immune responses to viruses and bacteria as revealed in the *L. monocytogenes* mouse model, given that IFN-α/β is indispensable for resistance to virtually all viral infections (126).

An important innate recognition system, the Toll-like receptors (TLRs) function as a bridge between the innate and adaptive immune systems (119). TLRs are expressed on various cell types, including DCs, and recognize biological patterns that originate in a variety of microbes (i.e., double-stranded RNA, peptidoglycan, lipopolysaccharide, CpG-containing DNA, and flagellin) (119). As a consequence of TLR binding, intracellular signaling events initiated through the adaptor molecule MyD88 lead to activation of the cell (119). In DCs, stimulation through TLRs results in the upregulation of costimulatory molecules important in the activation of T cells (61). MyD88$^{-/-}$ mice are more susceptible to infection with virulent *L. monocytogenes* (106); but if they are infected with an attenuated strain (*actA* deficient) which cannot perform cell-to-cell spread, they develop robust CD8 T-cell responses (130). These data indicate that during *L. monocytogenes* infection, DCs can acquire bacterial-derived Ags and activate naïve T cells in the absence of this important innate recognition system. It remains to be determined which of the specific TLRs are critical for innate resistance to *L. monocytogenes*.

ADAPTIVE IMMUNE RESPONSES GENERATED AFTER *L. MONOCYTOGENES* INFECTION

DCs and Priming of Anti-*L. monocytogenes* T-Cell Responses

Once taken up into a cell, *L. monocytogenes* is contained within a phagosome. To survive, bacteria must escape this intensely bactericidal environment, or they will be efficiently killed in neutrophils or activated macrophages. To gain access to the more benign, nutrient-rich cytoplasm, *L. monocytogenes* employs the pore-forming protein LLO (100). This protein is a member of the cholesterol-dependent family of cytolysins and functions best at an acidic pH (45). LLO is an essential virulence factor for *L. monocytogenes*. Mutant strains deficient in production of LLO cannot establish infection and inefficiently prime antilisterial T-cell responses (12, 19).

Once in the cytoplasm, *L. monocytogenes* begins to replicate, doubling in number approximately every 40 min (99). Infected cells die somewhere around 8 h p.i., but the infection is not curtailed by death of the infected cell. *L. monocytogenes* has developed an extremely efficient and clever mechanism of cell-to-cell spread. The bacterial protein ActA contains several binding sites for host cell cytoskeletal proteins and thereby serves as a perpetual nucleation site for actin polymerization (99). In this way, *L. monocytogenes* is able to highjack the host cell's cytoskeletal machinery and use it to propel itself into neighboring cells without ever leaving the protective intracellular environment (122). Strains that are unable to make the ActA protein (*actA* deficient) are severely attenuated in vivo, highlighting how important this method of cell-cell spread is to the pathogenicity of *L. monocytogenes* (18, 75). However, *actA*-deficient strains of *L. monocytogenes* prime robust T-cell responses and are extremely useful experimental tools for studying immune responses in immunocompromised mice, which survive infection with these attenuated strains quite well (9, 57).

As presented earlier in this chapter, several components of the innate immune system are critical in controlling *L. monocytogenes* infection; however, they are not enough to accomplish clearance of virulent bacteria. T cells are required to achieve sterilizing immunity to *L. monocytogenes*. This is best demonstrated by infection of SCID or RAG-deficient mice, which possess intact innate immune systems but lack virtually all B and T lymphocytes. Infection of these mice with virulent *L. monocytogenes* results in initial control of the infection due to innate defense mechanisms, but bacteria are never cleared; the mice exhibit persistent infection and eventually succumb to *L. monocytogenes* infection (10, 72, 92).

To prime antilisterial T-cell responses, *L. monocytogenes*-derived proteins must be processed and presented on MHC-I and -II molecules by DCs to naïve Ag-specific CD8 and CD4 T cells. Both CD4 and CD8 T cells specific for *L. monocytogenes*-derived epitopes are capable of providing protective immunity independent of the other T-cell subset, although CD8 T cells are markedly better mediators of resistance to *L. monocytogenes* (14, 34, 43, 54, 55, 58, 85). Data obtained using mice depleted or made genetically deficient of DCs clearly demonstrate that these cells are essential for activating naïve CD8 T cells in response to *L. monocytogenes* and other infections (69). DCs could acquire *L. monocytogenes*-derived Ags either by being directly infected or by phagocytosing cells that contain bacteria and presenting the processed Ags via cross-presentation (60).

As an example of the potent capacity of DCs to activate naïve T cells, it has been demonstrated in recent studies that in vitro-propagated DCs matured with lipopolysaccharide and loaded with peptides derived from *L. monocytogenes* are capable of initiating protective antilisterial T-cell responses in vivo in the absence of infection (7, 51). T cells activated by peptide-loaded DCs expand, contract, and establish stable numbers of memory cells with the same kinetics as observed after infection with *L. monocytogenes*, highlighting the unique stimulatory capacity of these antigen-presenting cells (APCs) (7, 51).

Processing of *L. monocytogenes*-Derived Ags Presented on MHC-I Molecules to CD8 T Cells

Multiple *L. monocytogenes*-derived proteins contain epitopes recognized by CD8 T cells in mice. Because *L. monocytogenes* spends the majority of its life cycle in the cytoplasm of infected cells, proteins made by the bacterium are subject to surveillance by the endogenous MHC-I Ag-processing pathway (21). Newly synthesized proteins made by *L. monocytogenes* replicating in the cytoplasm are degraded by the proteosome, which generates a pool of 8- to 15-amino-acid peptides (103). These peptides are actively transported into the endoplasmic reticulum by the TAP proteins, where they are loaded onto MHC-I molecules; finally, the fully formed complex is translocated to the cell surface where it becomes accessible to CD8 T cells.

In infected BALB/c mice, the greatest number of *L. monocytogenes*-specific CD8 T cells is directed against amino acids 91 to 99 of LLO (97). Subdominant CD8 T-cell responses are generated against residues 217 to 225 and 449 to 457 of p60, a murein hydrolase, and amino acids 84 to 92 of mpl, a bacterial metalloprotease (20, 96, 111). All of these epitopes are presented by the H-2Kd MHC-I molecule.

The mechanisms that result in higher numbers of CD8 T cells against specific immunodominant epitopes are not completely defined (137). To generate an MHC-I epitope, the protein containing the appropriate peptide sequence must be degraded within the infected cell. For example, it has been demonstrated that p60, which is constitutively produced by *L. monocytogenes*, has a half-life of approximately 90 min in infected cells. Upon degradation, 30% of

p60 molecules give rise to peptide 449-457, while only 3% of p60 molecules contribute to the 217-225 epitope (111, 128). In contrast, LLO is degraded so rapidly that it is almost impossible to detect inside infected cells; however, LLO must be present at some time to facilitate escape from the phagosome, and it can be detected in cells that have been treated with proteosome inhibitors (129).

Once generated, MHC-I/peptide complexes have different half-lives, which may affect the magnitude of the T-cell response to each epitope. H-2Kd-p60$_{217-225}$ and H-2Kd-LLO$_{91-99}$ complexes each have a half-life of 6 h (22, 112, 128), whereas H-2Kd-p60$_{449-457}$ has a half-life of only 1 h (21). Interestingly, the anti-LLO$_{91-99}$ and p60$_{217-225}$ CD8 T-cell responses are larger in magnitude than the anti-p60$_{449-457}$ response. In addition to stability, specific MHC-I/peptide complexes can be present on the surface of cells in greatly differing amounts (96, 111, 113), which could greatly influence the corresponding immune response.

In addition to the number and half-life of MHC-I/peptide complexes, the magnitude of the T-cell response directed against a *L. monocytogenes*-derived epitope is governed by the number of available Ag-specific T-cell precursors present in the naïve T-cell pool. Thus, the presence of more CD8 T-cell precursors bearing surface receptors capable of recognizing the amino acid sequence of LLO$_{91-99}$ may also explain why LLO$_{91-99}$-specific CD8 T cells are present in the highest numbers after *L. monocytogenes* infection, although data to support this hypothesis are lacking.

Priming of CD8 T-cell subsets specific for an *L. monocytogenes*-derived epitope does not always generate T cells that participate in control of the infection (38, 108, 139). Proteins that are not secreted into the cytoplasm by *L. monocytogenes* contain peptides that are capable of binding MHC-I molecules and priming CD8 T-cell responses; however, these T cells do not provide protective immunity. Because professional APCs have the unique capacity to acquire and present Ags via cross-presentation (60), peptides derived from nonsecreted proteins are presented on these cells but not on non-APCs infected with *L. monocytogenes*, which can only pro-cess Ags via the endogenous MHC-I-processing pathway (56). Thus, Ag-specific CD8 T cells specific for an epitope derived from a nonsecreted protein never see their Ag on infected nonprofessional APCs, and therefore they cannot participate in the elimination of the vast majority of infected cells. This phenomenon has been demonstrated for both a model nonsecreted Ag and epitopes derived from the ActA protein (38, 108, 139).

The priming of CD8 T cells specific for the nonsecreted model epitope NP$_{118-126}$ has recently been shown to be largely dependent on neutrophils (124). The CD8 T-cell response generated after infection with *L. monocytogenes* expressing the nonsecreted form of NP$_{118-126}$ was severely impaired in mice that had been treated with two different neutrophil-depleting antibodies prior to infection. In addition, neutrophils isolated from infected mice successfully served as substrate cells for DC-mediated cross-presentation of the NP$_{118-126}$ epitope in vitro. Together, these data indicate that neutrophils, which kill *L. monocytogenes* but then die themselves, are an important source for cross-presentation of nonsecreted bacterial Ags in vivo (124).

Ags Presented on MHC-II Molecules to CD4 T Cells

In addition to Ag-specific CD8 T cells, *L. monocytogenes*-specific CD4 T cells are also capable of providing protective immunity to naïve mice (43, 58). Several epitopes generated from LLO are presented on MHC-II molecules to CD4 T cells. The largest CD4 T-cell response is observed with C57BL/6 mice and is directed against $LLO_{190-201}$ (presented by $I-A^b$) (44). Two other peptides from LLO, spanning amino acids 215 to 234 and 354 to 371, and a peptide generated from p60, residues 301 to 312, are all presented on MHC-II molecules to CD4 T cells (21).

Even though some *L. monocytogenes* bacteria are killed following internalization, the life cycle of this organism does not include a significant amount of time in the phagosome, which is where proteins would be degraded and loaded onto MHC-II complexes. Instead, it is thought that the majority of MHC-II Ag processing and presentation of *L. monocytogenes*-derived epitopes is via the cross-presentation route (114). Dying cells (neutrophils, macrophages, or infected hepatocytes) that contain bacteria or remnants of killed bacteria are taken up by DCs into phagosomes, where the bacterial proteins are efficiently degraded and made accessible for loading through fusion of the phagosome with endocytic vesicles containing MHC-II molecules.

Magnitude and Kinetics of T-Cell Responses following Infection with *L. monocytogenes*

Despite the recent development of MHC/peptide tetramers, which identify Ag-specific T cells by virtue of their T-cell response specificity, naïve Ag-specific T-cell precursors in the spleen are too few to be measured prior to infection. Studies have estimated that <1,000 precursors exist per spleen for a given epitope (15, 17, 28). Following recognition of Ag, *L. monocytogenes*-specific T cells undergo a massive expansion phase, which results in a peak number of Ag-specific T cells in the spleen at day 7 to 9 after primary infection, depending on the strain of bacteria used to infect the mice (23). Ag-specific T-cell populations specific for the more dominant *L. monocytogenes*-derived epitopes can be detected as early as 3 to 4 days p.i. by using MHC/peptide tetramers or intracellular staining for cytokines after a brief stimulation in vitro with specific peptide (5). As mentioned before, after infection of BALB/c mice, the greatest number of *L. monocytogenes*-specific CD8 T cells is directed against the LLO_{91-99} epitope, followed next by the number of $p60_{217-225}$-specific cells (9). Although the absolute numbers of T cells specific for each epitope differ quite dramatically, the kinetics with which each population of Ag-specific T cells reaches its individual peak number of cells is nearly synchronous (23). Following expansion, *L. monocytogenes*-specific T cells undergo a rapid contraction phase, during which 70 to 80% of the cells present at the peak die in 3 to 5 days. The remaining T cells represent the initial memory pool.

It was originally believed that the kinetics of T-cell expansion was governed by the duration of infection and Ag display, since the onset of the contraction phase after acute infection with virulent *L. monocytogenes* infection coincided with clearance of infection. However, several recent studies have demonstrated the ability of CD8 T cells to undergo extensive Ag-independent proliferation after only a short exposure to Ag (8, 70, 89, 136). This was best demonstrated in studies where mice were treated with ampicillin 24 h after infection with a high dose of *L. monocytogenes*. This treatment decreased the number of bacteria below the level of detection by day 3 p.i. By using a sensitive method to detect the presence of specific MHC-I/peptide complexes present in the spleen (the direct ex vivo Ag display assay), it was determined that functional Ag display was virtually absent by day 2 p.i., yet Ag-specific CD8 T cells expanded and contracted with the same kinetics as in the non-antibiotic-

treated mice (8). These experiments demonstrate that the onset of CD8 T-cell contraction is programmed by early events after *L. monocytogenes* infection and not by the duration of infection.

Although the mechanisms that regulate contraction are not well defined, one molecule that has been demonstrated to play a role in this process is IFN-γ (9, 29, 37). It was discussed earlier in this chapter that IFN-γ is essential in orchestrating innate defenses against *L. monocytogenes*; however, its role in regulating T-cell homeostasis was not predicted. CD8 T cells against *L. monocytogenes*-derived epitopes are efficiently primed in IFN-$\gamma^{-/-}$ mice infected with an *actA*-deficient strain of *L. monocytogenes* (9). These cells expand with normal kinetics but undergo an extremely protracted contraction phase, resulting in the accumulation of Ag-specific CD8 T cells <60 days p.i. IFN-γ has the capacity to cause a multitude of cellular effects by regulating transcription of hundreds of genes (16). It is currently not known if IFN-γ acts directly on Ag-specific CD8 T cells in wild-type mice to cause their contraction, perhaps by making them more sensitive to activation-induced cell death (102), or if this cytokine functions through some indirect mechanism such as altering the expression of Ag-presenting or costimulatory molecules at key times during the immune response (16). Recent work has shown that Ag-specific CD8 T cells become unresponsive to IFN-γ by downregulating the signaling component of the IFN-γ receptor very early after *L. monocytogenes* infection. In addition, these T cells upregulate IFN-γ-dependent genes, suggesting that they are receiving an IFN-γ signal during the time when they are primed (J. S. Haring and J. T. Harty, submitted for publication). It has been shown that splenic DCs produce IFN-γ after *L. monocytogenes* infection (95), which may represent the relevant cellular source of early IFN-γ. Alternatively, the source may be NK cells or memory phenotype CD8 T cells (CD44^{hi}), which make IFN-γ in an Ag-independent manner after exposure to IL-12 and IL-18 early after *L. monocytogenes* infection (13, 81).

Further evidence exists to support the hypothesis that IFN-γ signals delivered to CD8 T cells very early after infection work in concert with signals delivered through the TCR to initiate the expansion-contraction program. The results of a recent study demonstrated that CD8 T cells primed in mice treated with antibiotics prior to infection with *L. monocytogenes* expanded and developed into protective memory cells without undergoing contraction (7). If mice were treated with CpG oligonucleotides, to induce IFN-γ among other proinflammatory cytokines at the time of priming, T cells contracted normally (7). However, CpG treatment of infected IFN-$\gamma^{-/-}$ mice did not induce contraction of CD8 T cells, demonstrating a key role for the IFN-γ molecule itself.

Following infection with *L. monocytogenes*, 5 to 10% of expanded effector cells survive the contraction phase to seed the initial memory pool of Ag-specific CD8 T cells. Exactly how memory cells survive contraction is not known. IL-7 is an important cytokine for the survival of T cells, and recent experiments have suggested that cells expressing the IL-7 receptor alpha chain (CD127) at the peak of expansion following viral (131) or *L. monocytogenes* (64) infection are the cells that survive to become functional memory cells. In support of these data, the lack of CD8 T-cell contraction in the antibiotic-pretreated mice correlated with increased expression of CD127 at the peak of the response (7).

L. monocytogenes-specific memory T cells exhibit different response kinetics than primary T cells. Memory cells undergo an even more rapid expansion phase, achieve higher peak numbers, and exhibit prolonged contraction (6, 8, 46). T cells responding after secondary infection mediate clearance of bacteria much more quickly than primary cells, due to their increased numbers and their ability to perform effector functions immediately upon recognition of Ag (8).

Effector Functions of T Cells and Their Contribution to Antilisterial Immunity

As mentioned earlier, T cells are required for clearance of *L. monocytogenes*, and CD4 and CD8 T cells are both capable of providing protective immunity against *L. monocytogenes* (14, 34, 43, 54, 55, 58, 85). Much effort has been expended to demonstrate the effector functions of T cells that are required for antilisterial immunity. The primary function of CD8 T cells is to kill infected cells. This process is accomplished via the perforin/granzyme or Fas/FasL pathways, both of which initiate apoptosis of the infected cell (59). Ligating and causing the aggregation of the death receptor Fas (CD95) on the surface of an infected cell initiates a caspase cascade that ultimately results in multiple cellular effects and apoptosis (77). Perforin is released from activated CD8 T cells and facilitates the entry of granzymes A and B into the infected cell and the initiation of the caspase cascade (110).

CD8 T-cell responses are efficiently primed in perforin$^{-/-}$ mice. In fact, after primary infection, the peak numbers of Ag-specific CD8 T cells are higher in perforin$^{-/-}$ mice than in wild-type mice; thus, they establish higher numbers of memory cells (9). Perforin$^{-/-}$ mice exhibit delayed clearance of *L. monocytogenes* from the spleen but not the liver after primary infection (66); however, the 50% lethal dose of virulent *L. monocytogenes* is not lower in perforin$^{-/-}$ mice than in wild-type mice (71). Despite the enhanced number of memory CD8 T cells, immunized perforin$^{-/-}$ mice are less resistant to rechallenge with *L. monocytogenes* than immunized wild-type mice (133, 134). Subsequent studies determined that memory cells generated in perforin$^{-/-}$ mice are approximately fivefold less protective on a per-cell basis than wild-type memory CD8 T cells (90). Surprisingly, CD8 T cells that are completely incapable of lysing infected cells still provide protection from lethal *L. monocytogenes* infection. Both *L. monocytogenes*-specific wild-type and perforin$^{-/-}$ CD8 T cells can mediate bacterial clearance in Fas-deficient mice (133). Taken together, these data indicate that cytolysis of infected cells is an important, but not essential, effector function of *L. monocytogenes*-specific CD8 T cells.

In addition to killing infected cells, CD8 T cells are potent producers of IFN-γ and TNF. Production of these cytokines by T cells is not constitutive but instead is swiftly upregulated upon exposure to Ag (3, 115). Both of these cytokines are involved in activating antibacterial activities in macrophages and are essential in innate defenses against *L. monocytogenes* (57, 98). Ag-specific CD8 T cells can be efficiently generated in IFN-γ$^{-/-}$ mice (9, 57). Long-lasting immunity is generated in the absence of IFN-γ, as mice immunized with *actA*-deficient *L. monocytogenes* or DNA constructs encoding LLO are capable of resisting a subsequent challenge with virulent *L. monocytogenes* (11, 57). In addition, when Ag-specific CD8 T cells generated in IFN-γ$^{-/-}$ mice are adoptively transferred into naïve mice, they are capable of providing protective immunity against challenge with virulent *L. monocytogenes* (57). These data indicate that CD8 T-cell-derived IFN-γ is not essential for these cells to perform their antilisterial functions.

The production of TNF by CD8 T cells could not only increase the antilisterial activities of macrophages, but binding of TNF to its receptor can initiate apoptosis, which could be an additional mechanism by which CD8 T cells kill infected cells (59). Ag-specific CD8 T-cell responses can be generated from TNF$^{-/-}$ mice, and these cells provide protective immunity upon transfer into naïve mice but not TNF$^{-/-}$ mice (132). Thus, the data demonstrate that TNF is an important cytokine for innate resistance to *L. monocytogenes*, but CD8 T-cell-derived TNF is not required for protective immunity provided by these cells.

While data have been gathered that elucidate the importance of specific antibacterial effector mechanisms, to this date, no one effector function has been discovered to be absolutely required for CD8 T-cell mediated anti-*L. monocytogenes* immunity. Even Ag-specific CD8 T cells from perforin$^{-/-}$/IFN-γ$^{-/-}$ and perforin$^{-/-}$/TNF$^{-/-}$ double-deficient mice provide substantial protection against infection with virulent bacteria (4, 132). Given the high risk of a poor outcome resulting from an uncontrolled *L. monocytogenes* infection, it is beneficial to have such a multifaceted defense to prevent vulnerability if one antibacterial mechanism is incapacitated by a coinciding illness, pharmaceutical treatment, or gene mutation.

γ/δ and Nonclassical MHC-Restricted T-Cell Responses to *L. monocytogenes*

Conventional CD4 and CD8 T cells are thymically educated T-cell subsets with rearranged α/β T-cell receptors that are restricted by classical MHC-II and MHC-Ia Ag presentation molecules, respectively. In addition to these T cells, there are two other classes of T cells that participate in antilisterial immune responses. γ/δ T cells are an enigmatic population of T cells not entirely selected in the thymus. The Ags and molecules that present the putative Ags recognized by these cells are largely unknown. Some γ/δ T cells have been shown to lyse cells upon activation, but by and large their main effector function is cytokine production (25, 65). In mice genetically deficient in γ/δ T cells or if γ/δ T cells are depleted from mice prior to infection with *L. monocytogenes*, the result is prolonged infection, but eventually bacteria are cleared (62, 79, 87). It has been suggested that γ/δ T cells may serve a role in controlling inflammation in the liver after *L. monocytogenes* infection (42). One study showed that depletion of γ/δ T cells from mice prior to infection led to greater numbers of extracellular *L. monocytogenes* in the liver and an increase in the number of infiltrating neutrophils without the formation of granulomas (42). Together, these data suggest that γ/δ T cells participate in early control of *L. monocytogenes* infection by providing cytokines like IFN-γ and possibly exert a regulatory function by controlling the extent of neutrophil infiltration into the liver, but they are not required for overall survival or clearance of bacteria.

H2-M3 is a nonclassical MHC-Ib molecule that binds bacterium-derived or mitochondrial peptides containing *N*-formyl methionine on the amino terminus (83). Several *L. monocytogenes*-derived peptides are presented by this MHC-Ib molecule and stimulate Ag-specific CD8 T-cell responses (21, 80). H2-M3-restricted T cells make IFN-γ in an Ag-specific manner and, after infection, peak in numbers approximately 2 days before MHC-Ia-restricted CD8 T cells (52, 74). H2-M3-restricted T cells establish stable numbers of memory cells after either *L. monocytogenes* infection (74) or activation via peptide-coated DCs (52), and they express levels of surface activation markers such as CD44 and CD62L that are similar to levels in MHC-Ia-restricted memory cells (73). However, after a secondary infectious exposure, H2-M3-restricted T cells in *L. monocytogenes*-immune mice expand significantly less than MHC-Ia-restricted T cells (73). In contrast, H2-M3-restricted T cells in peptide-DC-primed mice expand vigorously after *L. monocytogenes* infection; however, they are restricted to CD95-mediated cytolysis and provide less protective immunity than MHC-Ia-restricted CD8 T cells (52). It was demonstrated in the latter study that the presence of memory MHC-Ia-restricted CD8 T cells curbs the expansion of memory H2-M3-restricted CD8 T cells by limiting the availability of Ag presentation on DCs (52). These data indicate that the protective capacity of H2-M3-restricted CD8 T cells is limited to primary infections, where they may function as a bridge between innate and adaptive immunity, and memory MHC-Ia-restricted CD8 T cells can suppress nonessential T-cell responses during secondary infections (52).

L. MONOCYTOGENES AS A VACCINE DELIVERY VECTOR

The extensive studies of immunity to *L. monocytogenes* in the mouse model have resulted in one of the most complete characterizations of a host-pathogen interaction in a mammalian system. Because of this body of knowledge, scientists can begin to devise ways to manipulate this interaction to create specific outcomes. One concept that is actively being pursued is using *L. monocytogenes* as a vaccine delivery vector. The ability of *L. monocytogenes* to enter phagocytic and nonphagocytic cells greatly increases the types of cells permissive to infection, which is a limitation of some viral vaccine vectors. *L. monocytogenes*, even if rendered attenuated by inhibiting its ability to accomplish cell-to-cell spread (*actA*-deficient strains), is extremely efficient at delivering proteins to the cytoplasm of cells, and this efficiently stimulates Ag-specific CD8 T-cell responses (9, 57). As an issue of safety, *L. monocytogenes* infection can be abrogated in mice and people by treatment with antibiotics, which has very little effect on the subsequent T-cell responses in mice (7, 8, 89).

Some efforts have already been made to develop strains of *L. monocytogenes* that can be used to vaccinate against proteins of interest. In one recent set of experiments, mice were given melanoma cells followed by immunization i.v. with virulent, or *actA*-deficient, strains of *L. monocytogenes* that had been engineered to express a tumor Ag. The vaccination protocol was successful at preventing tumor establishment in the immunized mice. Antibiotic treatment at 24 or 36 h postimmunization did not decrease the rate of tumor rejection. In addition, existing immunity to *L. monocytogenes* had virtually no effect on the protective capacity of the vaccine, indicating that at least one additional boost with the *L. monocytogenes* vaccine would be feasible (116). In addition to this tumor vaccination model, attenuated strains of *L. monocytogenes* that express the Gag protein from HIV have been made and efficiently prime anti-Gag CD8 T-cell responses in both adult and neonatal mice (82, 101).

A recent clinical trial has been described in which healthy human volunteers were given graded doses of attenuated *actA*-deficient *L. monocytogenes* orally and monitored for symptoms associated with the inoculation and shedding of bacteria. A small number of volunteers exhibited transient abnormal liver function tests early after inoculation; however, no serious side effects in any of the participants were detected (1). Although antibody and T-cell responses

generated against *L. monocytogenes* were not consistently detected in these subjects, this was an important preliminary study to investigate the safety of using an attenuated form of this potentially pathogenic bacterium as a safe vaccine vector in humans.

CONCLUDING REMARKS

L. monocytogenes is a robust and versatile species of bacteria. It can survive in a wide range of environments and infect animals, including humans, when provided with the opportunity. Both the innate and T-cell components of the mammalian immune system work in concert to rid the body of bacteria before the development of serious pathology. *L. monocytogenes* infection of mice has proven to be an instrumental model system in the advancement of our overall knowledge about how protective immune responses are generated. Arguably, the data amassed by dissecting the immune response to *L. monocytogenes* in mice have created the most complete understanding of a host-pathogen interaction. The next step is to translate what we have learned in total about studying *L. monocytogenes* infection into strategies for human treatments.

REFERENCES

1. **Angelakopoulos, H., K. Loock, D. M. Sisul, E. R. Jensen, J. F. Miller, and E. L. Hohmann.** 2002. Safety and shedding of an attenuated strain of *Listeria monocytogenes* with a deletion of *actA/plcB* in adult volunteers: a dose escalation study of oral inoculation. *Infect. Immun.* **70:** 3592–3601.

2. **Auerbuch, V., D. Brockstedt, N. Meyer-Morse, M. O'Riordan, and D. A. Portnoy.** 2004. Mice lacking type I interferon receptor are resistance to *Listeria monocytogenes*. *J. Exp. Med.* **200:**527–533.

3. **Badovinac, V. P., G. A. Corbin, and J. T. Harty.** 2000. Cutting edge: OFF cycling of TNF production by antigen-specific CD8+ T cells is antigen independent. *J. Immunol.* **165:**5387–5391.

4. **Badovinac, V. P., and J. T. Harty.** 2000. Adaptive immunity and enhanced CD8+ T cell response to *Listeria monocytogenes* in the absence of perforin and IFN-γ. *J. Immunol.* **164:**6444–6452.

5. **Badovinac, V. P., and J. T. Harty.** 2001. Detection and analysis of antigen-specific CD8+ T cells. *Immunol. Res.* **24:**325–332.

6. **Badovinac, V. P., K. A. Messingham, S. E. Hamilton, and J. T. Harty.** 2003. Regulation of CD8+ T cells undergoing primary and secondary responses to infection in the same host. *J. Immunol.* **170:**4933–4942.

7. **Badovinac, V. P., B. B. Porter, and J. T. Harty.** 2004. CD8$^+$ T cell contraction is controlled by early inflammation. *Nat. Immunol.* **5:**809–817.

8. **Badovinac, V. P., B. B. Porter, and J. T. Harty.** 2002. Programmed contraction of CD8$^+$ T cells after infection. *Nat. Immunol.* **3:**619–626.

9. **Badovinac, V. P., A. R. Tvinnereim, and J. T. Harty.** 2000. Regulation of antigen-specific CD8+ T cell homeostasis by perforin and interferon-γ. *Science* **290:**1354–1358.

10. **Bancroft, G. J., R. D. Schreiber, and E. R. Unanue.** 1991. Natural immunity: a T-cell-independent pathway of macrophage activation, defined in the SCID mouse. *Immunol. Rev.* **124:**5–24.

11. **Barry, R. A., H. Archie Bouwer, T. Clark, K. Cornell, and D. Hinrichs.** 2003. Protection of IFN-γ knockout

mice against *Listeria monocytogenes* challenge following intramuscular immunization with DNA vaccines encoding listeriolysin O. *Vaccine* 21:2122–2132.

12. **Barry, R. A., H. G. Bouwer, D. A. Portnoy, and D. J. Hinrichs.** 1992. Pathogenicity and immunogenicity of *Listeria monocytogenes* small-plaque mutants defective for intracellular growth and cell-to-cell spread. *Infect. Immun.* 60:1625–1632.

13. **Berg, R. E., E. Crossley, S. Murray, and J. Forman.** 2003. Memory CD8+ T cells provide innate immune protection against *Listeria monocytogenes* in the absence of cognate antigen. *J. Exp. Med.* 198:1583–1593.

14. **Bishop, D. K., and D. J. Hinrichs.** 1987. Adoptive transfer of immunity to *Listeria monocytogenes*. The influence of in vitro stimulation on lymphocyte subset requirements. *J. Immunol.* 139:2005–2009.

15. **Blattman, J. N., R. Antia, D. Sourdive, X. Wang, S. M. Kaech, K. Murali-Krishna, J. D. Altman, and R. Ahmed.** 2002. Estimating the precursor frequency of naive antigen-specific CD8 T cells. *J. Exp. Med.* 195:657–664.

16. **Boehm, U., T. Klamp, M. Groot, and J. C. Howard.** 1997. Cellular responses to interferon-γ. *Annu. Rev. Immunol.* 15:749–795.

17. **Bousso, P., A. Casrouge, J. D. Altman, M. Haury, J. Kanellopoulos, J. P. Abastado, and P. Kourilsky.** 1998. Individual variations in the murine T cell response to a specific peptide reflect variability in naive repertoires. *Immunity* 9:169–178.

18. **Brundage, R. A., G. A. Smith, A. Camilli, J. A. Theriot, and D. A. Portnoy.** 1993. Expression and phosphorylation of the *Listeria monocytogenes* ActA protein in mammalian cells. *Proc. Natl. Acad. Sci. USA* 90:11890–11894.

19. **Brunt, L. M., D. A. Portnoy, and E. R. Unanue.** 1990. Presentation of *Listeria monocytogenes* to CD8+ T cells requires secretion of hemolysin and intracellular bacterial growth. *J. Immunol.* 145:3540–3546.

20. **Busch, D. H., H. G. Bouwer, D. Hinrichs, and E. G. Pamer.** 1997. A nonamer peptide derived from *Listeria monocytogenes* metalloprotease is presented to cytolytic T lymphocytes. *Infect. Immun.* 65:5326–5329.

21. **Busch, D. H., K. Kerksiek, and E. G. Pamer.** 1999. Processing of *Listeria monocytogenes* antigens and the in vivo T-cell response to bacterial infection. *Immunol. Rev.* 172:163–169.

22. **Busch, D. H., and E. G. Pamer.** 1998. MHC class I/-peptide stability: implications for immunodominance, in vitro proliferation, and diversity of responding CTL. *J. Immunol.* 160:4441–4448.

23. **Busch, D. H., I. Pilip, and E. G. Pamer.** 1998. Evolution of a complex T cell receptor repertoire during primary and recall bacterial infection. *J. Exp. Med.* 188:61–70.

24. **Cameron, L. A., P. A. Giardini, F. S. Soo, and J. A. Theriot.** 2000. Secrets of actin-based motility revealed by a bacterial pathogen. *Nat. Rev. Mol. Cell Biol.* 1:110–119.

25. **Carding, S. R., and P. J. Egan.** 2002. γ/δ T cells: functional plasticity and heterogeneity. *Nat. Rev. Immunol.* 2:336–345.

26. **Carrero, J., B. Calderon, and E. R. Unanue.** 2004. Listeriolysin O from *Listeria monocytogenes* is a lymphocyte apoptogenic molecule. *J. Immunol.* 172:4866–4874.

27. **Carrero, J., B. Calderon, and E. R. Unanue.** 2004. Type I interferon sensitizes lymphocytes to apoptosis and reduces resistance to *Listeria* infection. *J. Exp. Med.* 200:535–540.

28. **Casrouge, A., E. Beaudoing, S. Dalle, C. Pannetier, J. Kanellopoulos, and P. Kourilsky.** 2000. Size estimate of the αβ TCR repertoire of naive mouse splenocytes. *J. Immunol.* 164:5782–5787.

29. **Chu, C., S. Wittmer, and D. Dalton.** 2000. Failure to suppress the expansion of the activated CD4 T cell population in interferon γ-deficient mice leads to exacerbation of experimental autoimmune encephalomyelitis. *J. Exp. Med.* 192:123–128.

30. **Conlan, J. W.** 1999. Early host-pathogen interactions in the liver and spleen during systemic murine listeriosis: an overview. *Immunobiology* 201:178–187.

31. **Conlan, J. W.** 1996. Early pathogenesis of *Listeria monocytogenes* infection in the mouse spleen. *J. Med. Microbiol.* 44:295–302.

32. **Conlan, J. W.** 1997. Neutrophils and tumour necrosis factor-alpha are important for controlling early gastrointestinal stages of experimental murine listeriosis. *J. Med. Microbiol.* 46:239–250.

33. **Cousens, L. P., and E. J. Wing.** 2000. Innate defenses in the liver during *Listeria* infection. *Immunol. Rev.* 174:150–159.

34. **Czuprynski, C. J., and J. F. Brown.** 1987. Dual regulation of anti-bacterial resistance and inflammatory neutrophil and macrophage accumulation by L3T4+ and Lyt 2+ *Listeria*-immune T cells. *Immunology* 60:287–293.

35. **Czuprynski, C. J., P. Henson, and P. Campbell.** 1984. Killing of *Listeria monocytogenes* by inflammatory neutrophils and mononuclear phagocytes from immune and nonimmune mice. *J. Leukoc. Biol.* 35:193–208.

36. **Dalrymple, S. A., L. A. Lucian, R. Slattery, T. McNeil, D. M. Aud, S. Fuchino, F. Lee, and R. Murray.** 1995. Interleukin-6-deficient mice are highly susceptible to *Listeria monocytogenes* infection: correlation with inefficient neutrophilia. *Infect. Immun.* 63:2262–2268.

37. **Dalton, D., L. Haynes, C. Chu, S. Swain, and S. Wittmer.** 2000. Interferon-γ eliminates responding CD4 T cells during mycobacterial infection by inducing apoptosis of activated CD4 T cells. *J. Exp. Med.* 192:117–122.

38. **Darji, A., D. Bruder, S. zur Lage, B. Gerstel, T. Chakraborty, J. Wehland, and S. Weiss.** 1998. The role of the bacterial membrane protein ActA in immunity and protection against *Listeria monocytogenes*. *J. Immunol.* 161:2414–2420.

39. **Dunn, P. L., and R. J. North.** 1991. Early gamma interferon production by natural killer cells is important in defense against murine listeriosis. *Infect. Immun.* 59:2892–2900.

40. **Edelson, B. T., and E. R. Unanue.** 2000. Immunity to *Listeria* infection. *Curr. Opin. Immunol.* 12:425–431.

41. **Ehlers, S., M. E. Mielke, T. Blankenstein, and H. Hahn.** 1992. Kinetic analysis of cytokine gene expression in the livers of naive and immune mice infected with *Listeria monocytogenes*. The immediate early phase in innate resistance and acquired immunity. *J. Immunol.* 149:3016–3022.

42. **Fu, Y., C. Roark, K. Kelly, D. Drevets, P. Campbell, R. O'Brien, and W. Born.** 1994. Immune protection and control of inflammatory tissue necrosis by γ/δ T cells. *J. Immunol.* 153:3101–3115.

43. **Geginat, G., M. Lalic, M. Kretschmar, W. Goebel, H. Hof, D. Palm, and A. Bubert.** 1998. Th1 cells specific for a secreted protein of *Listeria monocytogenes* are protective in vivo. *J. Immunol.* 160:6046–6055.

44. **Geginat, G., S. Schenk, M. Skoberne, W. Goebel, and H. Hof.** 2001. A novel approach of direct ex vivo epitope mapping identifies dominant and subdominant CD4 and CD8 T cell epitopes from *Listeria monocytogenes*. *J. Immunol.* 166:1877–1884.

45. Glomski, I. J., M. M. Gedde, A. W. Tsang, J. A. Swanson, and D. A. Portnoy. 2002. The *Listeria monocytogenes* hemolysin has an acidic pH optimum to compartmentalize activity and prevent damage to infected host cells. *J. Cell. Biol.* **156:**1029–1038.

46. Grayson, J., L. Harrington, J. Lanier, E. J. Wherry, and R. Ahmed. 2002. Differential sensitivity of naive and memory CD8+ T cells to apoptosis in vivo. *J. Immunol.* **169:**3760–3770.

47. Gregory, S. H., A. J. Sagnimeni, and E. J. Wing. 1996. Bacteria in the bloodstream are trapped in the liver and killed by immigrating neutrophils. *J. Immunol.* **157:**2514–2520.

48. Gregory, S. H., E. J. Wing, K. L. Danowski, N. van Rooijen, K. F. Dyer, and D. J. Tweardy. 1998. IL-6 produced by Kupffer cells induces STAT protein activation in hepatocytes early during the course of systemic listerial infections. *J. Immunol.* **160:**6056–6061.

49. Gregory, S. H., E. J. Wing, R. A. Hoffman, and R. L. Simmons. 1993. Reactive nitrogen intermediates suppress the primary immunologic response to *Listeria*. *J. Immunol.* **150:**2901–2909.

50. Guo, Y., D. W. Niesel, H. K. Ziegler, and G. R. Klimpel. 1992. *Listeria monocytogenes* activation of human peripheral blood lymphocytes: induction of non-major histocompatibility complex-restricted cytotoxic activity and cytokine production. *Infect. Immun.* **60:**1813–1819.

51. Hamilton, S. E., and J. T. Harty. 2002. Quantitation of CD8+ T cell expansion, memory, and protective immunity after immunization with peptide-coated dendritic cells. *J. Immunol.* **169:**4936–4944.

52. Hamilton, S. E., B. B. Porter, K. A. Messingham, V. P. Badovinac, and J. T. Harty. 2004. MHC class Ia-restricted memory T cells inhibit expansion of a nonprotective MHC class Ib (H2-M3)-restricted memory response. *Nat. Immunol.* **5:**159–168.

53. Han, Y., and J. Cutler. 1997. Assessment of a mouse model of neutropenia and the effect of an anti-candidiasis monoclonal antibody in these animals. *J. Infect. Dis.* **175:**1169–1175.

54. Harty, J. T., and M. J. Bevan. 1996. CD8 T-cell recognition of macrophages and hepatocytes results in immunity to *Listeria monocytogenes*. *Infect. Immun.* **64:**3632–3640.

55. Harty, J. T., and M. J. Bevan. 1992. CD8+ T cells specific for a single nonamer epitope of *Listeria monocytogenes* are protective in vivo. *J. Exp. Med.* **175:**1531–1538.

56. Harty, J. T., and M. J. Bevan. 1999. Responses of CD8(+) T cells to intracellular bacteria. *Curr. Opin. Immunol.* **11:**89–93.

57. Harty, J. T., and M. J. Bevan. 1995. Specific immunity to *Listeria monocytogenes* in the absence of IFNγ. *Immunity* **3:**109–117.

58. Harty, J. T., R. D. Schreiber, and M. J. Bevan. 1992. CD8 T cells can protect against an intracellular bacterium in an interferon γ-independent fashion. *Proc. Natl. Acad. Sci. USA* **89:**11612–11616.

59. Harty, J. T., A. R. Tvinnereim, and D. W. White. 2000. CD8+ T cell effector mechanisms in resistance to infection. *Annu. Rev. Immunol.* **18:**275–308.

60. Heath, W., and F. Carbone. 2001. Cross-presentation, dendritic cells, tolerance, and immunity. *Annu. Rev. Immunol.* **19:**47–64.

61. Hertz, C., S. Kiertscher, P. Godowski, D. Bouis, M. Norgand, M. Roth, and R. Modlin. 2001. Microbial lipopeptides stimulate dendritic cell maturation via Toll-like receptor 2. *J. Immunol.* **166:**2444–2450.

62. Hiromatsu, K., Y. Yoshikai, G. Matsuzaki, S. Ohga, K. Muramori, K. Matsumoto, J. Bluestone, and K. Nomoto. 1992. A protective role of γ/δ T cells in primary infection with *Listeria monocytogenes* in mice. *J. Exp. Med.* **175:**49–56.

63. Huang, S., W. Hendriks, A. Althage, S. Hemmi, H. Bluethmann, R. Kamijo, J. Vilcek, R. M. Zinkernagel, and M. Aguet. 1993. Immune response in mice that lack the IFN-γ receptor. *Science* **259:**1742–1745.

64. Huster, K., V. Busch, M. Scheimann, K. Linkermann, K. Kerksiek, H. Wagner, and D. H. Busch. 2004. Selective expression of the IL-7 receptor on memory T cells identifies early CD40L-dependent generation of distinct CD8+ memory T cell subsets. *Proc. Natl. Acad. Sci. USA* **101:**5610–5615.

65. Jameson, J., D. Witherden, and W. L. Havran. 2003. T-cell effector mechanisms: γ/δ and CD1d-restricted subsets. *Curr. Opin. Immunol.* **15:**349–353.

66. Jensen, E. R., A. A. Glass, W. R. Clark, E. J. Wing, J. F. Miller, and S. H. Gregory. 1998. Fas (CD95)-dependent cell-mediated immunity to *Listeria monocytogenes*. *Infect. Immun.* **66:**4143–4150.

67. Jensen, J., T. Warner, and E. Balish. 1993. Resistance of SCID mice to *Candida albicans* administered intravenously or colonizing the gut: role of polymorphonuclear leukocytes and macrophages. *J. Infect. Dis.* **167:**912–919.

68. Jiang, J., L. Lau, and H. Shen. 2003. Selective depletion of nonspecific T cell during the early stage of immune responses to infection. *J. Immunol.* **171:**4352–4358.

69. Jung, S., D. Unutmaz, P. Wong, G. Sano, K. De los Santos, T. Sparwasser, S. Wu, S. Vuthoori, K. Ko, F. Zavala, E. G. Pamer, D. R. Littman, and R. A. Lang. 2002. In vivo depletion of CD11c(+) dendritic cells abrogates priming of CD8(+) T cells by exogenous cell-associated antigens. *Immunity* **17:**211–220.

70. Kaech, S. M., and R. Ahmed. 2001. Memory CD8+ T cell differentiation: initial antigen encounter triggers a developmental program in naive cells. *Nat. Immunol.* **2:**415–422.

71. Kagi, D., B. Ledermann, K. Burki, H. Hengartner, and R. Zinkernagel. 1994. CD8(+) T cell-mediated protection against an intracellular bacterium by perforin-dependent cytotoxicity. *Eur. J. Immunol.* **24:**3068–3072.

72. Kaufmann, S., and C. Ladel. 1994. Role of T cell subsets in immunity against intracellular bacteria: experimental infections of knock-out mice with *Listeria monocytogenes* and *Mycobacterium bovis* BCG. *Immunobiology* **191:**509–519.

73. Kerksiek, K., A. Ploss, I. Leiner, D. H. Busch, and E. G. Pamer. 2003. H2-M3-restricted memory T cells: persistence and activation without expansion. *J. Immunol.* **170:**1862–1869.

74. Kerksiek, K. M., D. H. Busch, I. M. Pilip, S. E. Allen, and E. G. Pamer. 1999. H2-M3-restricted T cells in bacterial infection: rapid primary but diminished memory responses. *J. Exp. Med.* **190:**195–204.

75. Kocks, C., E. Gouin, M. Tabouret, P. Berche, H. Ohayon, and P. Cossart. 1992. *L. monocytogenes*-induced actin assembly requires the *actA* gene product, a surface protein. *Cell* **68:**521–531.

76. Kolb-Maurer, A., U. Kammerer, M. Maurer, I. Gentschev, E. Brocker, P. Rieckmann, and E. Kampgen. 2003. Production of IL-12 and IL-18 in human dendritic cells upon infection with *Listeria monocytogenes*. *FEMS Immunol. Med. Microbiol.* **35:**255–262.

77. Krueger, A., S. Fas, S. Baumann, and P. Krammer. 2003. The role of CD95 in the regulation of peripheral T-cell apoptosis. *Immunol. Rev.* **193**:58-69.

78. Kurihara, T., G. Warr, J. Loy, and R. Bravo. 1997. Defects in macrophage recruitment and host defense in mice lacking CCR2 chemokine receptor. *J. Exp. Med.* **186**: 1757–1762.

79. Ladel, C., C. Blum, and S. Kaufman. 1996. Control of natural killer cell-mediated innate resistance against the intracellular pathogen *Listeria monocytogenes* by γ/δ T lymphocytes. *Infect. Immun.* **64**:1744–1749.

80. Lenz, L., B. Dere, and M. J. Bevan. 1996. Identification of an H2-M3-restricted *Listeria* epitope: implications for antigen presentation by M3. *Immunity* **5**:63–72.

81. Lertmemongkolchai, G., G. Cai, C. A. Hunter, and G. J. Bancroft. 2001. Bystander activation of CD8+ T cells contributes to the rapid production of IFN-γ in response to bacterial pathogens. *J. Immunol.* **166**:1097–1105.

82. Lieberman, J., and F. R. Frankel. 2002. Engineered *Listeria monocytogenes* as an AIDS vaccine. *Vaccine* **20**:2007–2010.

83. Lindahl, K., D. Byers, V. Dabhi, R. Hovik, E. Jones, G. Smith, C. Wang, H. Xiao, and M. Yoshino. 1997. H2-M3, a full-service class Ib histocompatibility antigen. *Annu. Rev. Immunol.* **15**:851–879.

84. Lu, B., C. Ebensperger, Z. Dembic, Y. Wang, M. Kvatyuk, T. Lu, R. L. Coffman, S. Pestka, and P. B. Rothman. 1998. Targeted disruption of the interferon-γ receptor 2 gene results in severe immune defects in mice. *Proc. Natl. Acad. Sci. USA* **95**:8233–8238.

85. Lukacs, K., and R. Kurlander. 1989. Lyt-2+ T cell-mediated protection against listeriosis. Protection correlates with phagocyte depletion but not with IFN-γ production. *J. Immunol.* **142**:2879–2886.

86. Mackaness, G. B. 1962. Cellular resistance to infection. *J. Exp. Med.* **116**:381–406.

87. Matsuzaki, G., H. Yamada, K. Kishihara, Y. Yoshikai, and K. Nomoto. 2002. Mechanism of murine Vg1+ γ/δ T cell-mediated innate immune response against *Listeria monocytogenes* infection. *Eur. J. Immunol.* **32**:928–935.

88. Meraz, M., J. White, K. Sheehan, E. Bach, S. Rodig, A. Dighe, D. Kaplan, J. Riley, A. Greenlund, D. Campbell, K. Carver-Moore, R. DuBois, R. Clark, M. Aguet, and R. D. Schreiber. 1996. Targeted disruption of the Stat1 gene in mice reveals unexpected physiologic specificity in the Jak-STAT signaling pathway. *Cell* **84**:431–442.

89. Mercado, R., S. Vijh, S. E. Allen, K. Kerksiek, I. Pilip, and E. G. Pamer. 2000. Early programming of T cell populations responding to bacterial infection. *J. Immunol.* **165**:6833–6839.

90. Messingham, K. A., V. P. Badovinac, and J. T. Harty. 2003. Deficient anti-listerial immunity in the absence of perforin can be restored by increasing memory CD8+ T cell numbers. *J. Immunol.* **171**:4254–4262.

91. Milon, G. 1997. *Listeria monocytogenes* in laboratory mice: a model of short-term infectious and pathogenic processes controllable by regulated protective immune responses. *Immunol. Rev.* **158**:37–46.

92. Mocci, S., S. A. Dalrymple, R. Nishinakamura, and R. Murray. 1997. The cytokine stew and innate resistance to *L. monocytogenes*. *Immunol. Rev.* **158**:107–114.

93. Moulder, J. W. 1985. Comparative biology of intracellular parasitism. *Microbiol. Rev.* **49**:298–337.

94. O'Connell, R., S. Saha, S. Vaidya, K. W. Bruhn, G. Miranda, B. Zarnegar, A. Perry, B. Nguyen, T. Lane, T. Taniguchi, J. F. Miller, and G. Cheng. 2004.

Type I interferon enhances susceptibility to *Listeria monocytogenes* infection. *J. Exp. Med.* **200**:437–445.

95. Ohteki, T., T. Fukao, K. Suzue, C. Maki, M. Ito, M. Nakamura, and S. Koyasu. 1999. Interleukin 12-dependent interferon gamma production by CD8a+ lymphoid dendritic cells. *J. Exp. Med.* **189**:1981–1986.

96. Pamer, E. G. 1994. Direct sequence identification and kinetic analysis of an MHC class I-restricted *Listeria monocytogenes* CTL epitope. *J. Immunol.* **152**:686–694.

97. Pamer, E. G., J. T. Harty, and M. J. Bevan. 1991. Precise prediction of a dominant class I MHC-restricted epitope of *Listeria monocytogenes*. *Nature* **353**:852–855.

98. Pasparakis, M., L. Alexopoulou, V. Episkopou, and G. Kollias. 1996. Immune and inflammatory responses in TNFα-deficient mice: a critical requirement for TNFα in the formation of primary B cell follicles, follicular dendritic cell networks and germinal centers, and in the maturation of the humoral immune response. *J. Exp. Med.* **184**: 1397–1411.

99. Portnoy, D. A., V. Auerbuch, and I. J. Glomski. 2002. The cell biology of *Listeria monocytogenes* infection: the intersection of bacterial pathogenesis and cell-mediated immunity. *J. Cell Biol.* **158**:409–414.

100. Portnoy, D. A., P. S. Jacks, and D. J. Hinrichs. 1988. Role of hemolysin for the intracellular growth of *Listeria monocytogenes*. *J. Exp. Med.* **167**:1459–1471.

101. Rayevskaya, M., N. Kushnir, and F. R. Frankel. 2002. Safety and immunogenicity in neonatal mice of a hyperattenuated *Listeria* vaccine directed against human immunodeficiency virus. *J. Virol.* **76**:918–922.

102. Refaeli, Y., L. Van Parijs, S. I. Alexander, and A. K. Abbas. 2002. Interferon γ is required for activation-induced death of T lymphocytes. *J. Exp. Med.* **196**:999–1005.

103. Rock, K. L., C. Gramm, L. Rothstein, K. Clark, R. Stein, L. Dick, D. Hwang, and A. L. Goldberg. 1994. Inhibitors of the proteasome block the degradation of most cell proteins and the generation of peptides presented on MHC class I molecules. *Cell* **78**:761–771.

104. Rogers, H. W., M. P. Callery, B. Deck, and E. R. Unanue. 1996. *Listeria monocytogenes* induces apoptosis of infected hepatocytes. *J. Immunol.* **156**:679–684.

105. Rogers, H. W., C. S. Tripp, R. D. Schreiber, and E. R. Unanue. 1994. Endogenous IL-1 is required for neutrophil recruitment and macrophage activation during murine listeriosis. *J. Immunol.* **153**:2093–2101.

106. Seki, E., H. Tsutsui, N. Tsuji, N. Hayashi, K. Adachi, H. Nakano, S. Futatsugi-Yumikura, O. Takeuchi, K. Hoshino, and S. Akira. 2002. Critical roles of myeloid differentiation factor-88-dependent proinflammatory cytokine release in early phase clearance of *Listeria monocytogenes* in mice. *J. Immunol.* **169**:3863–3868.

107. Serbina, N., T. Salazar-Mather, C. Biron, W. Kuziel, and E. G. Pamer. 2003. TNF/iNOS-producing dendritic cells mediate innate immune defense against bacterial infection. *Immunity* **19**:59–70.

108. Shen, H., J. F. Miller, X. Fan, D. Kolwyck, R. Ahmed, and J. T. Harty. 1998. Compartmentalization of bacterial antigens: differential effects on priming of CD8 T cells and protective immunity. *Cell* **92**:535–545.

109. Shen, H., C. Tato, and X. Fan. 1998. *Listeria monocytogenes* as a probe to study cell-mediated immunity. *Curr. Opin. Immunol.* **10**:450–458.

110. Shresta, S., C. Pham, D. Thomas, T. Graubert, and T. Ley. 1998. How do cytotoxic lymphocytes kill their targets? *Curr. Opin. Immunol.* **10**:581–587.

111. **Sijts, A. J., A. Neisig, J. Neefjes, and E. G. Pamer.** 1996. Two *Listeria monocytogenes* CTL epitopes are processed from the same antigen with different efficiencies. *J. Immunol.* **156:**683–692.

112. **Sijts, A. J., and E. G. Pamer.** 1997. Enhanced intracellular dissociation of major histocompatibility complex class I-associated peptides: a mechanism for optimizing the spectrum of cell surface-presented cytotoxic T lymphocyte epitopes. *J. Exp. Med.* **185:**1403–1411.

113. **Skoberne, M., R. Holtappels, H. Hof, and G. Geginat.** 2001. Dynamic antigen presentation patterns of *Listeria monocytogenes*-derived CD8 T cell epitopes in vivo. *J. Immunol.* **167:**2209–2218.

114. **Skoberne, M., S. Schenk, H. Hof, and G. Geginat.** 2002. Cross-presentation of *Listeria monocytogenes*-derived CD4 T cell epitopes. *J. Immunol.* **169:**1410–1418.

115. **Slifka, M. K., F. Rodriguez, and J. L. Whitton.** 1999. Rapid on/off cycling of cytokine production by virus-specific CD8+ T cells. *Nature* **401:**76–79.

116. **Starks, H., K. W. Bruhn, H. Shen, R. A. Barry, T. W. Dubensky, D. Brockstedt, D. J. Hinrichs, D. E. Higgins, J. F. Miller, M. Giedlin, and H. G. Bouwer.** 2004. *Listeria monocytogenes* as a vaccine vector: virulence attenuation or existing antivector immunity does not diminish therapeutic efficacy. *J. Immunol.* **173:**420–427.

117. **Sunderkotter, C., T. Nikolic, M. Dillion, N. van Rooijen, M. Stehling, D. Drevets, and P. Leenen.** 2004. Subpopulations of mouse blood monocytes differ in maturation stage and inflammatory response. *J. Immunol.* **172:**4410–4417.

118. **Takada, H., G. Matsuzaki, K. Hiromatsu, and K. Nomoto.** 1994. Analysis of the role of natural killer cells in *Listeria monocytogenes* infection: relation between natural killer cells and T-cell receptor γ/δ T cells in the host defense mechanism at the early stage of infection. *Immunology* **82:**106–112.

119. **Takeda, K., T. Kaisho, and S. Akira.** 2003. Toll-like receptors. *Annu. Rev. Immunol.* **21:**335–376.

120. **Takeya, K., S. Shimotori, T. Taniguchi, and K. Nomoto.** 1977. Cellular mechanisms in the protection against infection by *Listeria monocytogenes* in mice. *J. Gen. Microbiol.* **100:**373–379.

121. **Teixeira, H., and S. Kaufmann.** 1994. Role of NK1.1+ cells in experimental listeriosis. NK1+ cells are early IFN-γ producers but impair resistance to *Listeria monocytogenes* infection. *J. Immunol.* **152:**1873–1882.

122. **Tilney, L. G., P. S. Connelly, and D. A. Portnoy.** 1990. Actin filament nucleation by the bacterial pathogen, *Listeria monocytogenes*. *J. Cell Biol.* **111:**2979–2988.

123. **Trautwein, C., K. Boker, and M. P. Manns.** 1994. Hepatocyte and immune system: acute phase reaction as a contribution to early defence mechanisms. *Gut* **35:**1163–1166.

124. **Tvinnereim, A. R., S. E. Hamilton, and J. T. Harty.** 2004. Neutrophil involvement in cross-priming CD8+ T cell responses to bacterial antigens. *J. Immunol.* **173:**1994–2002.

125. **Tweten, R. K., M. W. Parker, and A. E. Johnson.** 2001. The cholesterol-dependent cytolysins. *Curr. Top. Microbiol. Immunol.* **257:**15–33.

126. **van den Broek, M., U. Muller, S. Huang, and R. Zinkernagel.** 1995. Immune defence in mice lacking type I and/or type II interferon receptors. *Immunol. Rev.* **148:**5–18.

127. **Varinou, L., K. Ramsauer, M. Karaghiosoff, T. Kolbe, K. Pfeffer, M. Muller, and T. Decker.** 2003. Phosphorylation of the Stat1 transactivation domain is required for full-fledged IFN-γ-dependent innate immunity. *Immunity* **19:**793–802.

128. **Villanueva, M. S., P. Fischer, K. Feen, and E. G. Pamer.** 1994. Efficiency of MHC class I antigen processing: a quantitative analysis. *Immunity* **1:**479–489.

129. **Villanueva, M. S., A. J. Sijts, and E. G. Pamer.** 1995. Listeriolysin O is processed efficiently into an MHC class I-associated epitope in *Listeria monocytogenes*-infected cells. *J. Immunol.* **155:**5227–5233.

130. **Way, S., T. Kollman, A. Hajjar, and C. Wilson.** 2003. Protective cell-mediated immunity to *Listeria monocytogenes* in the absence of myeloid differentiation factor 88. *J. Immunol.* **171:**533–537.

131. **Wherry, E. J., V. Teichgraber, T. C. Becker, D. Masopust, S. M. Kaech, R. Antia, U. H. von Andrian, and R. Ahmed.** 2003. Lineage relationship and protective immunity of memory CD8 T cell subsets. *Nat. Immunol.* **4:**225–234.

132. **White, D. W., V. P. Badovinac, G. Kollias, and J. T. Harty.** 2000. Cutting edge: antilisterial activity of CD8+ T cells derived from TNF-deficient and TNF/perforin double-deficient mice. *J. Immunol.* **165:**5–9.

133. **White, D. W., and J. T. Harty.** 1998. Perforin-deficient CD8+ T cells provide immunity to *Listeria monocytogenes* by a mechanism that is independent of CD95 and IFN-γ but requires TNF-α. *J. Immunol.* **160:**898–905.

134. **White, D. W., A. MacNeil, D. H. Busch, I. M. Pilip, E. G. Pamer, and J. T. Harty.** 1999. Perforin-deficient CD8+ T cells: in vivo priming and antigen-specific immunity against *Listeria monocytogenes*. *J. Immunol.* **162:**980–988.

135. **Wing, E. J., and S. H. Gregory.** 2002. *Listeria monocytogenes*: clinical and experimental update. *J. Infect. Dis.* **185**(Suppl. 1):S18–S24.

136. **Wong, P., and E. G. Pamer.** 2001. Cutting edge: antigen-independent CD8 T cell proliferation. *J. Immunol.* **166:**5864–5868.

137. **Yewdell, J., and J. Bennink.** 1999. Immunodominance in major histocompatibility complex class I-restricted T lymphocyte responses. *Annu. Rev. Immunol.* **17:**51–88.

138. **Yokoyama, W., S. Kim, and A. French.** 2004. The dynamic life of natural killer cells. *Annu. Rev. Immunol.* **22:**405–429.

139. **Zenewicz, L., K. Foulds, J. Jiang, X. Fan, and H. Shen.** 2002. Nonsecreted bacterial proteins induce recall CD8 T cell responses but do not serve as protective antigens. *J. Immunol.* **169:**5805–5812.

140. **Zimmerman, B., B. Canono, and P. Campbell.** 1986. Silica decreases phagocytosis and bactericidal activity of both macrophages and neutrophils in vitro. *Immunology* **59:**521–525.

Genetic Tools for Use with *Listeria monocytogenes*

DARREN E. HIGGINS, CARMEN BUCHRIESER, AND NANCY E. FREITAG

51

The number and sophistication of genetic tools that have become available in recent years for the molecular characterization of *Listeria monocytogenes* have continued to increase. Plasmid vectors, reporter genes, systems designed for transposon mutagenesis, heterologous expression systems, integration vectors, and transducing phage have all greatly advanced the experimental capacity to generate, characterize, and complement mutations within *L. monocytogenes* and to define functional roles of gene products. This chapter is a brief description of some of the genetic tools currently available for use with *L. monocytogenes*. Key references are given throughout this description to provide sources for expanded details on plasmid constructions, assay conditions, and other technical aspects. The variety of genetic tools described on the following pages is meant to be representative of the resources available to those interested in *L. monocytogenes* genetics and should not be considered a complete and exhaustive list of all available plasmid vectors, transposons, etc.

THE *L. MONOCYTOGENES* CHROMOSOME AND THE IMPACT OF GENOMICS ON *LISTERIA* GENETICS

Genome sequencing projects and the advent of global approaches such as proteomics and transcriptional profiling have opened exciting new possibilities and have led to the development of a new field of biology—comparative and functional genomics. In *Listeria* research, the era of genomics started with the determination of the complete genome sequences of two members of the *Listeria* genus by a consortium of 10 European laboratories, the pathogenic species *L. monocytogenes* (strain EGD-e) and the closely related nonpathogenic species *Listeria innocua* (strain CLIP11626) (http://genolist.pasteur.fr/ ListiList/) (36). *L. monocytogenes* EGD-e is a derivative of strain EGD used by G. B. Mackaness in his pioneering studies on cell-mediated immunity and has been extensively studied in many laboratories. Complete genome comparison of *L. monocytogenes* EGD-e and *L. innocua* CLIP11626 identified a conserved genome organization and a high number of orthologous genes but also revealed the presence of 9.5% *L. monocytogenes*

EGD-e-specific genes and 5% *L. innocua* CLIP11626-specific genes, when prophages were not taken into account (12, 36). The specific genes are distributed in small regions within the *Listeria* genomes, suggesting that multiple acquisition and deletion events have led to the present genome content. To better understand virulence, host-range differences, and evolution within the genus *Listeria*, the complete genome sequences of four additional *L. monocytogenes* strains were determined: an epidemic *L. monocytogenes* isolate of serotype 4b (CLIP80459, sequenced at the Institut Pasteur) (11), two epidemic serotype 4b strains (F2365 and H7858), and one serotype 1/2a sporadic *L. monocytogenes* isolate sequenced at The Institute of Genomic Research in collaboration with the U.S. Department of Agriculture (62). The DNA sequence for strain F2365 is available at http://www.tigr.org.

The comparison of epidemic *L. monocytogenes* CLIP80459 serotype 4b and of the nonepidemic *L. monocytogenes* EGD-e serotype 1/2a isolate identified genetic divergence between the two strains, as about 6% of the sequences were found to be serotype 4b specific (23). Furthermore, preliminary comparison of the two epidemic *L. monocytogenes* serotype 4b isolates (F2365 and H7858) indicated that the two previously identified subgroups among *L. monocytogenes* 4b strains (42) showed about 2.5% differences in their genetic basis (11). The availability of these data sets allowed identification of unique marker genes for each subgroup (25, 95), as well as the development of different tools for comparative genomics like macro- and microarrays.

To gain further insight into the genetic basis for pathogenesis of the genus *Listeria*, the genomes of the second pathogenic *Listeria* species, *Listeria ivanovii* strain PAM 55 (collaboration of the Institut Pasteur, the German PathoGenoMik Network, and the group headed by J. A. Vazques Boland), and a representative of each species within the *Listeria* genus (*L. seeligeri*, *L. welshimeri*, and *L. grayi*, through the collaboration of the Institut Pasteur and the German PathoGenoMik Network), are now being sequenced. The availability of the genome sequences of all species of *Listeria*, as well as the already-available sequences of five *L. monocytogenes* isolates, will provide in-depth comparative genomics for the identification of unknown virulence determinants and further the

Gram-Positive Pathogens, 2nd edition, edited by Vincent A. Fischetti et al.
© 2006 ASM Press, Washington, D.C.

understanding of the evolution of *L. monocytogenes* as a pathogen. Furthermore, this extensive genomic knowledge, together with the availability of large, comprehensive strain collections, makes *L. monocytogenes* an ideal model for the study of virulence and bacterial evolution.

One hallmark of *L. monocytogenes* is its ability to adapt to many different environments. The bacterial transcriptome reflects an organism's immediate, ongoing, and genome-wide response to its environment. More widely used tools for studying expression profiles of bacteria are whole-genome DNA macro- and microarrays, which provide a comprehensive transcriptional analysis enabling researchers to view the organism as a system. Recently, whole genome macro- and microarrays for *L. monocytogenes*, based on the complete genome sequence of strain EGD-e (36), were developed using oligonucleotides specific for each gene within the genome (27, 40). The first study using such whole-genome macroarrays evaluated the expression profiles of wild-type *L. monocytogenes* and strains containing a deletion of a critical virulence regulatory factor known as *prfA* (60). This study revealed that PrfA can act as an activator or as a repressor and suggested that PrfA may directly or indirectly activate different sets of genes in concert with specific forms of RNA polymerase (60).

Two other reports using DNA arrays studied genes regulated by two alternative RNA polymerase sigma factors in *L. monocytogenes*, sigma 54 and sigma B (4, 50). Arous and colleagues (4) used whole-genome macroarrays to compare the global gene expression of the wild-type EGD-e strain and an *rpoN* (sigma 54) mutant. Their results suggested that sigma 54 is mainly involved in the control of carbohydrate metabolism in *L. monocytogenes*. Kazmierczak and colleagues (50) combined biocomputing and microarray-based strategies to identify sigma B-dependent genes in *L. monocytogenes*. These analyses identified a total of 55 genes with statistically significant sigma B-dependent expression including stress response genes (e.g., *gadB*, *ctc*, and the glutathione reductase gene *lmo1433*) and virulence genes (e.g., *inlA*, *inlB*, and *bsh*) (48). Finally, DNA microarrays spotted with human or mouse cDNAs have also been used to study the host transcriptional response to *L. monocytogenes* infection (6, 54).

The continued use of genomic approaches with *L. monocytogenes* will provide a basis for diverse functional studies to better understand the phenotypic and virulence differences between *L. monocytogenes* strains and will further our knowledge on the evolution of *L. monocytogenes* as a pathogen. The Pathogen Functional Genomic Resources Center and The Institute for Genomic Research have an *L. monocytogenes* microarray (http://pfgrc.tigr.org/desc.shtml) that is available for investigator-initiated projects through the National Institute of Allergy and Infectious Diseases. The guidelines for obtaining the arrays can be found at http://www.niaid.nih.gov/dmid/genomes/pfgrc/guidelines.htm.

PLASMID VECTORS

Plasmid vectors are critical tools for many aspects of bacterial genetics. The ability to complement chromosomal gene mutations in vitro and in vivo, to deliver transposable elements, to monitor gene expression via reporter-gene fusions, and to introduce specific mutations into the *L. monocytogenes* chromosome via plasmid introduction and integration has tremendously advanced molecular characterization of *L. monocytogenes* pathogenesis. In addition, the use of *L. monocytogenes* as an antigen and gene delivery ve-

hicle has been greatly facilitated by the development of specialized plasmid vectors. A number of plasmid vectors developed for use in other gram-positive bacteria have been used successfully with *L. monocytogenes*, and several of these have been improved or modified since their original constructions. Table 1 lists a sample of these plasmids and a brief summary of their utility for *L. monocytogenes*. The plasmids listed have been divided into three groups. Group I includes those plasmids that have been used for complementation studies of *L. monocytogenes* gene mutations and for general gene expression; these plasmids contain both gram-negative and gram-positive origins of replication. Group II features plasmids used for the delivery of transposons into *L. monocytogenes*, primarily derivatives of the transposon Tn917. Group III includes plasmid vectors used for allelic exchange with *L. monocytogenes* chromosomal sequences, integration of expression constructs onto the chromosome, and construction of chromosomal and plasmid reporter-gene fusions. Several of the plasmids listed in Table 1 will be described in more detail later in this section.

Introduction of Plasmid DNA into *L. monocytogenes*

Introduction of plasmid DNA into *L. monocytogenes* can be accomplished by transformation of *L. monocytogenes* protoplasts (15, 84) or by electroporation (63). Electroporation provides a simple and rapid method, and transformants are generally recovered following 2 days of growth on selective medium. The efficiency of transformation via electroporation can be increased by growing *L. monocytogenes* in the presence of small amounts of penicillin to interfere with cell wall synthesis (63); however, transformation frequencies even after penicillin treatment are generally not very high (in our hands, usually 100 to 1,000 colonies/μg of DNA, although efficiencies as high as 4×10^6 colonies/μg of DNA have been reported for some strains) (63).

Methods have also been reported that improve the transfer of plasmid DNA into *L. monocytogenes* via conjugation (52, 82). Plasmid vectors have been constructed that can be transferred directly from an *Escherichia coli* donor to *L. monocytogenes* at frequencies as high as 10^{-3} transconjugants/per donor CFU. This improved efficiency of plasmid introduction has added to the repertoire of genetic techniques available for *L. monocytogenes*.

Plasmid Vectors for Gene Expression and Complementation Studies

A variety of shuttle plasmids with broad host ranges and that are capable of replication in both *E. coli* and *L. monocytogenes* have been described (Table 1). For the expression of gene products via their own promoters or for gene complementation experiments, plasmid vectors such as pMK4 (79) and pAM401 (88) have been routinely used (Fig. 1). These plasmids contain multiple cloning sites and antibiotic resistance genes that are selectable in both gram-positive and gram-negative bacteria. Recently constructed and improved shuttle vectors include those containing origins of transfer to allow conjugation directly from *E. coli* hosts into *L. monocytogenes* and shuttle vectors that are maintained at low copy numbers in *E. coli* to facilitate the characterization of *L. monocytogenes* genes that may be toxic at high copy numbers. Such vectors include the pCON-2 and pCON2-101 plasmids (29). pCON-2 and pCON2-101 both contain antibiotic resistance genes selectable in gram-negative and

TABLE 1 Representative *L. monocytogenes* plasmid vectors

Plasmid	Size (kb)	Origin of representative gram−/gram+[a]	Origin of transfer[b]	Description	Reference(s)
Group I (general cloning and gene expression)					
pAM401	10.4	Yes/yes	No	Shuttle vector; Cm[rc]	88
pMK3	7.2	Yes/yes	No	Shuttle vector; Kn[r]	79
pMK4	5.6	Yes/yes	No	Shuttle vector; Cm[r]	79
pAT18	6.6	Yes/yes	Yes	Shuttle vector; Em[r]	81
pCON-2	7.6	Yes/yes	Yes	Shuttle vector; Cm[r]	29
pCON2-101	10.1	Yes/yes	Yes	Shuttle vector; Cm[r]; low-copy number in *E. coli*	29
pSPAC	5.7	Yes/yes	No	Expression vector, places gene under control of SPAC promoter; Cm[r]	94
Group II (delivery of transposons)					
pAM118	30.1	Yes/yes	Yes	Shuttle vector with Tn*916*; Em[r] Tet[r]	35
pLTV1	20.6	Yes/yes (ts)[d]	No	Shuttle vector with Tn*917-lac*; Em[r] Cm[r] Tet[r]	15
pLTV3	22.1	Yes/yes (ts)	No	Shuttle vector with Tn*917-lac*; Em[r] Cm[r] Tet[r] Ble[r]	15
pTV32-OK	14.1	Yes (ts)/yes (ts)	No	Shuttle vector with Tn*917-lac*; pWVO1 ts origin; Em[r]	19
Group III (allelic exchange, chromosomal gene insertion, reporter gene fusions)					
pLIV1	13.4	Yes/yes (ts)	Yes	ts integrational vector, places gene under control of SPAC/*lac*Oid promoter within *orfZ*; Cm[r] Em[r]	20
pLIV2	8.9	Yes/no	Yes	Site-specific integration vector, places gene under control of SPAC/*lac*Oid promoter within tRNA[Arg.], Cm[r]	1
pLSV1	6.3	Yes/yes (ts)	No	ts integrational vector; Em[r]	92
pKSV7	6.9	Yes/yes (ts)	No	ts integrational vector; Cm[r]	77
pCON-1	7.6	Yes/yes (ts)	Yes	ts integrational vector; Cm[r]	8, 29
pMAD	9.6	Yes/yes (ts)	No	ts vector for allelic exchange with β-galactosidase selection; Em[r]	3
pHS-LV	10.8	Yes/yes (ts)	No	ts integrational vector for insertion of genes within *orfZ*; Em[r]	74
pPL1	6.1	Yes/no	Yes	Site-specific integrational vector for insertion of genes within *comK*; Cm[r]	52
pPL2	6.1	Yes/no	Yes	Site-specific integrational vector for insertion of genes within tRNA[Arg.]; Cm[r]	52
pPL3e	8.0	Yes/no	Yes	Site-specific integrational vector for insertion of genes within tRNA[Arg.]; Em[r]	37
pHPL3	7.9	Yes/no	Yes	Site-specific integrational vector, places gene under control of Hyper SPO1 promoter within tRNA[Arg.]; Cm[r]	37
pLOV	8.0	Yes/no	Yes	Site-specific integrational vector, allows, elevated expression of genes under Hyper SPO1 within tRNA[Arg.]; Cm[r]	72
pNF579	7.6	Yes/yes (ts)	No	ts integrational vector for creating transcriptional *gfp* fusions; Cm[r]	29
pNF580	8.1	Yes/yes (ts)	No	ts integrational vector for creating transcriptional *gus* fusions; Cm[r]	29
pGF-EM	9.4	Yes/yes (ts)	Yes	ts integrational vector for gene replacement and generation of *gfp* transcriptional fusions; Em[r]	53
pAMGFP	11.1	Yes/yes	No	Shuttle vector for creating transcriptional *gfp* fusions; Cm[r]	87
pNF8	7.7	Yes/yes	Yes	Shuttle plasmid for *gfp* expression; Em[r]	26
pLCR	12	Yes/yes (ts)	No	ts integrational vector for creating transcriptional *lacZ* and *cat* fusions; Kn[r]	61
pAUL-A	9.2	Yes/yes (ts)	No	ts integrational vector for insertional mutagenesis and directional cloning; Em[r]	69
pTCV-*lac*	12	Yes/yes	Yes	Shuttle vector for construction of *lacZ* transcriptional fusions, low copy; Em[r]	67
pSB292	7.1	Yes/yes	No	Shuttle vector for construction of *luxAB* transcriptional fusions; Kn[r] Cm[r]	64

[a]Presence of plasmid origin for replication function in gram-negative (gram−) or gram-positive (gram+) bacteria.
[b]Presence or absence of origin of transfer, allowing plasmid conjugation.
[c]Antibiotic resistance markers for drug selection in *L. monocytogenes*: Cm, chloramphenicol; Kn, kanamycin; Em, erythromycin; Sp, spectinomycin; Ble, bleomycin; Tet, tetracycline.
[d]ts, temperature-sensitive plasmid origin of replication.

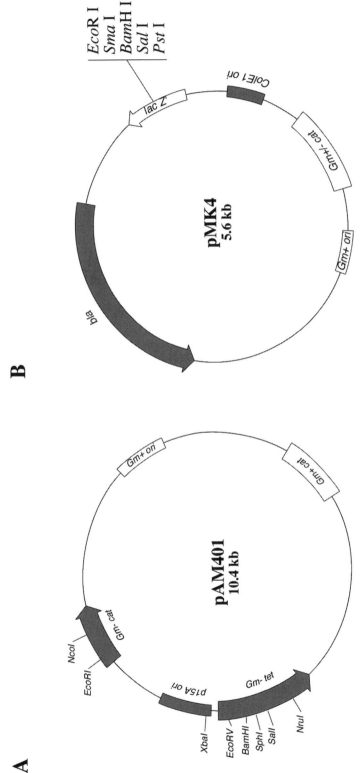

FIGURE 1 Plasmid vectors for gene expression-complementation in *L. monocytogenes*. (A) pAM401 (88) contains the p15A origin of replication, chloramphenicol (*Gm− cat*), and tetracycline (*Gm− tet*) resistance genes for selection in *E. coli*. A chloramphenicol resistance gene (*Gm+ cat*) and origin of replication (*Gm+* ori) gene allow selection in *L. monocytogenes*. Unique cloning sites present within the gram-negative replicon portion are indicated. (B) pMK4 (79) contains a ColE1 origin of replication, β-lactamase gene (*bla*), and *Gm+/− cat* resistance gene for selection in *E. coli*. The *Gm+* ori and *Gm+/− cat* resistance genes provide selection in *L. monocytogenes*. pMK4 harbors a fragment of the *E. coli lacZ* gene (*lacZ'*) that is capable of α-complementation and contains five unique restriction sites for cloning.

gram-positive bacteria, multiple cloning sites within regions capable of α-complementation (93), origins for conjugative transfer (85), and origins for plasmid replication in gram-negative and gram-positive bacteria.

pCON2-101 has two additional advantages: (i) a pSC101-derived origin that maintains the plasmid at low copy number in *E. coli* and (ii) an f1 origin derived from filamentous bacteriophage for the generation of single-stranded DNA templates.

The inducible SPAC promoter, developed for use in *Bacillus subtilis*, has also been successfully used for controlling gene expression in *L. monocytogenes* (30). The SPAC expression system, developed by Yansura and Henner (94), is based on the concept of using the *E. coli lac* repressor to control the transcription of recombinant genes in *B. subtilis*. A promoterless copy of the gene of interest is inserted downstream from a *lac* operator-regulated promoter in the presence of the *lacI* gene, which is under the control of a constitutive gram-positive promoter. Confirmation of the utility of SPAC as a method for controlling gene expression in *L. monocytogenes* was carried out by placing the green fluorescent protein (GFP) gene of *Aequorea victoria* (17) under the control of the SPAC promoter on plasmid pSPAC and by introducing the construct into *L. monocytogenes*. Expression of GFP was found to be dependent upon the addition of IPTG (isopropyl-β-D-thiogalactopyranoside) to the medium (30); thus, the SPAC system is suitable for situations where it is desirable to control the expression of a gene of interest in *L. monocytogenes*. Recent improvements to the SPAC expression system have proved useful for inducible expression of native genes in *L. monocytogenes* (20, 37, 38). The recently generated pLIV1 (20, 38) and pLIV2 vectors (Fig. 2) allow tightly regulated IPTG-inducible gene expression during extracellular growth and intracellular infection. Both plasmids are derived from vectors that facilitate chromosomal integration and provide IPTG dose-dependent complementation of *L. monocytogenes* genes in single copy (20, 38).

Plasmid Vectors for Allelic Exchange and Site-Specific Integration in *L. monocytogenes*

A powerful method allowing detailed examination of *L. monocytogenes* genetic elements is the ability to introduce specific mutations into the *L. monocytogenes* chromosome through the process of allelic exchange. First described for *L. monocytogenes* by Michel et al. (57), the replacement of wild-type chromosomal sequences with defined nucleotide substitutions, deletions, or a reporter gene construct has been since simplified by the development and use of shuttle plasmids carrying gram-positive temperature-sensitive origins of replication. The method, described by Camilli et al. (16), is depicted in Fig. 3 and proceeds as follows: plasmids are introduced into *L. monocytogenes* at temperatures permissive for plasmid replication and transformants are isolated on selective media. Plasmid-containing strains are then grown in the presence of drug selection at temperatures nonpermissive for plasmid replication; these are conditions that enrich for bacteria containing integrations of plasmid DNA into homologous regions of the *L. monocytogenes* chromosome (Fig. 3A and B). After selection of plasmid-integrant colonies, the plasmid is then excised from the chromosome and lost by growing the bacteria at temperatures permissive for plasmid replication in the absence of drug selection (Fig. 3C and D). Bacterial colonies are then screened for drug sensitivity and for the presence of the chromosomal mutation. If the plasmid construct contains homologous chromosomal sequences of approximately

equal size flanking the mutation (typically 500 bp to 1 kb) and if the mutation confers no growth defect or advantage upon bacterial colonies, then bacteria that contain the mutation of interest within their chromosomal sequences generally represent about 50% of the recovered population.

An integrational vector developed by Smith and Youngman (77) has several features that make it very useful for the process of allelic exchange in *L. monocytogenes*. Plasmid pKSV7 contains gram-positive temperature-sensitive replication functions derived from pE194ts; the copy number of this plasmid at 32°C in *B. subtilis* is about 5 (39, 83). pKSV7 also carries a ColE1 origin of replication, chloramphenicol and β-lactamase resistance genes for antibiotic selection, and a cluster of multiple cloning sites within *lacZ* sequences for insert screening by α-complementation. pKSV7 has been successfully used for the generation of single- and multiple-base substitutions, deletions, and insertions within the *L. monocytogenes* chromosome (for examples, see references 16, 31, and 48). A similar vector, pCON-1, exists for the purpose of allelic exchange: pCON-1 contains the added advantage of an origin of transfer element (8, 29). An additional vector for allelic exchange, pMAD, has been recently developed by Arnaud et al. (3). pMAD contains a copy of the *bgaB* gene encoding a thermostable β-galactosidase from *Bacillus stearothermophilus* under the control of a constitutive promoter recognized in gram-positive and gram-negative bacteria, enabling easy blue-white screening of transformants on indicator plates. A vector has also been constructed for inserting genes of interest into a nonessential region of the *L. monocytogenes* chromosome. Vector pHS-LV was designed to introduce genes or sequences by allelic exchange into the *L. monocytogenes orfZ* region, an open reading frame of unknown function located at the 3′ end of the *prfA* regulon. This vector has been used extensively for the construction of *L. monocytogenes* strains expressing foreign antigens (47, 73, 74).

A recent innovation for genetic analysis in *L. monocytogenes* has been the construction of site-specific integration vectors based upon components isolated from bacteriophage (52). Plasmids pPL1 and pPL2 (Fig. 4) are shuttle vectors that contain gram-positive and gram-negative antibiotic resistance genes, the p15A origin of replication for low-copy maintenance in *E. coli*, and an extensive multiple cloning site. However, these vectors lack a gram-positive origin of replication and do not replicate autonomously in *L. monocytogenes*. Each plasmid contains an *L. monocytogenes*-specific bacteriophage attachment site (*attPP′*) and associated integrase gene (*int*). Upon transfer of plasmid derivatives into *L. monocytogenes*, expression of the *int* gene product mediates site-specific recombination between the homologous plasmid *attP* and chromosomal *attB* sites, resulting in single-copy site-specific chromosomal integration of the plasmid vector. While the pPL1 vector has been used for integration into the nonessential *comK* locus within 10403S and SLCC5764 (also known as MacK) strains, recombination requires that the *comK* locus not already be occupied by an endogenous prophage. Therefore, use of the pPL1 vector necessitates curing of the prophage in commonly used *L. monocytogenes* strains such as 10403S, LO28, and EGD-e. The integration site for the pPL2 vector, which contains the *int* gene and *attPP′* site from the ScottA-derived PSA phage, is within the 3′ end of a tRNA^{Arg} gene. pPL2 has proven to be very useful as a site-specific integration vector for a number of applications (20, 52, 91), as the PSA prophage is less widely distributed among *L. monocytogenes* strains. Furthermore, integration within the tRNA^{Arg}

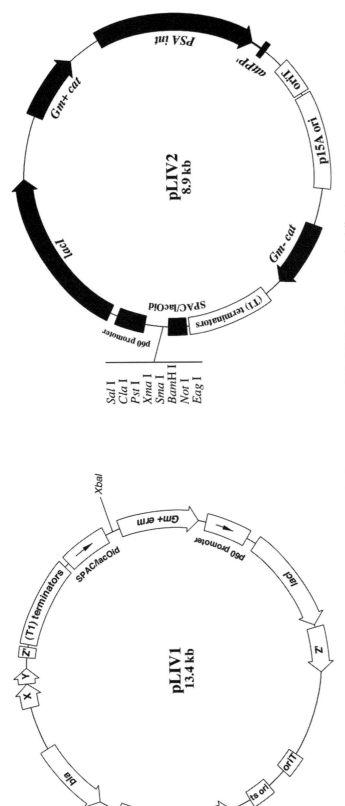

FIGURE 2 Inducible expression-integration vectors for *L. monocytogenes*. (A) pLIV1 (20) contains the following sequences: a temperature-sensitive origin of replication (ts ori) and a chloramphenicol resistance gene (Gm+ cat) for plasmid selection in *L. monocytogenes*, the ColE1 origin of replication and β-lactamase gene (bla) for cloning and selection in *E. coli*, an origin of transfer (oriT) to allow conjugal mating of plasmid derivatives from *E. coli* to *L. monocytogenes*, a unique XbaI restriction site for cloning genes under the transcriptional control of the SPAC/lacOid IPTG-inducible promoter (20), tandem copies of the *rrnB* T1 transcription terminator (T1 terminators) upstream of the SPAC/lacOid region to ensure that transcription of cloned genes initiates only from the SPAC promoter, the *L. monocytogenes* p60 gene (*iap*) promoter to allow constitutive expression of the *lac* repressor gene (*lacI*), and an erythromycin resistance determinant within the expression cassette (Gm+ *erm*) for selection of inducible constructs on the chromosome. The inducible expression cassette can be integrated into the *L. monocytogenes* chromosome within the *orfZ* gene (Z') via allelic exchange. (B) pLIV2 (Table 1) is a new inducible expression vector derived from pLIV1 and the site-specific phage integration vector pPL2 (Fig. 4) (52). pLIV2 contains the inducible expression cassette region from pLIV1 with eight unique restriction sites available for cloning. pLIV2 contains the p15A origin of replication and a chloramphenicol resistance gene (Gm− *cat*) for selection in *E. coli*. The PSA bacteriophage integrase gene (PSA *int*) and attachment site (*attPP'*) allow site-specific integration of the vector within an *L. monocytogenes* tRNA^Arg gene, following conjugal transfer from *E. coli* facilitated by the *oriT* region. The Gm+ *cat* resistance gene allows selection of integrated plasmids in *L. monocytogenes*.

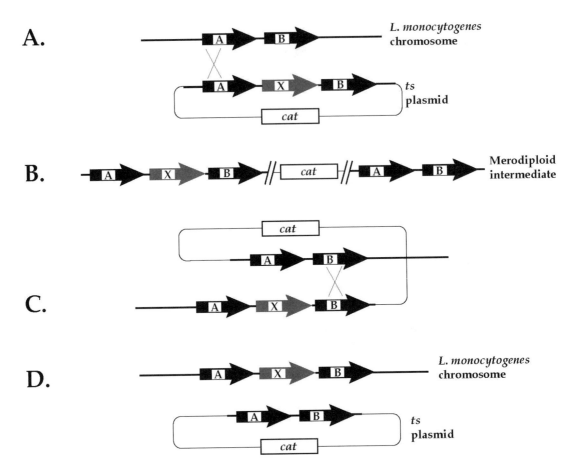

FIGURE 3 Schematic diagram for construction of chromosomal mutations in *L. monocytogenes*. The method depicted and described in the text is suitable for the introduction of insertions, deletions, or single- or multiple-nucleotide substitutions within the *L. monocytogenes* chromosome. (A) Chromosomal integration of the temperature sensitive plasmid vector by homologous recombination between plasmid encoded genes A or B and the chromosomal alleles. The designated crossover points are arbitrary and can occur on either side of gene X. (B) Growth of bacterial cultures in the presence of drug selection at temperatures nonpermissive for plasmid replication selects for the merodiploid intermediates that result from plasmid integration into the chromosome. (C) Merodiploid intermediate strains are then passed for several generations without selective pressure at temperatures permissive for plasmid replication. Spontaneous excision of the integrated plasmid from the chromosome occurs. (D) Excised plasmids are cured at temperatures nonpermissive for plasmid replication in the absence of drug selection.

locus can occur in the presence of the PSA prophage, thus alleviating the requirement of prophage curing.

Recent derivatives of the pPL2 vector have been generated that further expand the use of this site-specific integration vector system. These include pPL3e, pHPL3, and pLOV (37, 38, 72). pPL3e contains tandem copies of the *rrnB* T1 transcription terminators (20) at the beginning of the multiple cloning site of pPL2 and also includes the gram-positive *ermC* resistance gene (21) in place of the gram-positive *cat* gene of pPL2. pPL3e is useful for performing selection of a chromosomal inserted gene in conjunction with an autonomously replicating vector encoding chloramphenicol resistance such as pAM401. In addition, the transcription terminators facilitate manipulations in *E. coli* by dampening read-through expression of potentially toxic cloned genes. pHPL3 is a modification of pPL2 in which the Hyper-SPO1 promoter (68) has been inserted immediately following the

rrnB transcription terminators. The Hyper-SPO1 promoter within pHPL3 allows for constitutive expression of single copy cloned genes in *L. monocytogenes*. The pLOV vector (72) is a further modification of pHPL3 that contains sequences following the Hyper-SPO1 promoter that confer enhanced expression of cloned genes. The pLOV vector has been shown to yield up to 20-fold-higher levels of expressed protein than a construct harboring the Hyper-SPO1 promoter alone.

TRANSPOSONS

Transposable elements provide an effective means of creating random insertional mutations in bacterial populations. Several different transposons have been used in the generation of libraries of *L. monocytogenes* insertion mutants. Tn*1545* and Tn*916* are conjugative transposons that have

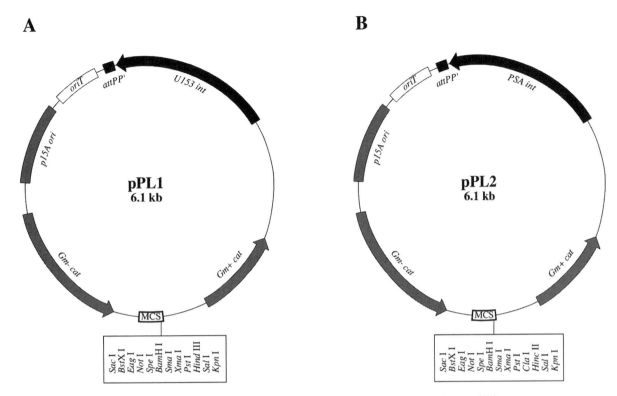

FIGURE 4 Site-specific phage integration vectors for *L. monocytogenes*. (A) pPL1 (52) contains the following features: a p15A origin of replication and chloramphenicol resistance gene (*Gm- cat*) for selection in *E. coli*, an origin of transfer (*oriT*) for conjugal mating from *E. coli* to *L. monocytogenes*, the U153 bacteriophage integrase gene (*U153 int*) and attachment site (*attPP'*) to allow site-specific integration of the vector within the *L. monocytogenes comK* gene, and a chloramphenicol resistance gene (*Gm+ cat*) for selection of plasmid maintenance in *L. monocytogenes*. A multiple cloning site region (MCS) contains 12 unique restriction sites for cloning. (B) Features of pPL2 (52) are as described for pPL1 with the exception that pPL2 contains the PSA bacteriophage integrase gene (*PSA int*) and attachment site (*attPP'*) for site-specific integration of pPL2 within an *L. monocytogenes* tRNA^{Arg} gene. pPL2 contains 13 unique restriction sites within the MCS region.

been introduced successfully into *L. monocytogenes*; however, the use of these two transposons has been restricted by both the relatively low frequencies of transposition observed and the limitations imposed on the randomness of insertion by the requirements for sequence homology between both ends of the elements and sequences surrounding the sites of integration (18, 70). Tn*1545* has, however, been successfully incorporated into experiments utilizing signature-tagged mutagenesis in *Listeria* (5). The Tn3-like transposon Tn*917* has proven more useful and has been used successfully with many gram-positive organisms including *L. monocytogenes*. Tn*917* generates extremely stable insertional mutations with a relatively high degree of randomness. Two modified forms of the Tn*917* transposon were constructed by Camilli et al. (15) and have been used in the generation of large-scale libraries of *L. monocytogenes* mutants (Fig. 5) (2, 80, 89). These modified Tn*917* elements have several advantages. (i) The transposons are carried on vectors with temperature-sensitive origins of replication that simplify the recovery of chromosomal insertions. (ii) They carry a promoterless copy of *E. coli lacZ* for the generation of transcriptional fusions. (iii) They contain several features (a ColE1 replicon, an *E. coli* selectable antibiotic resistance gene, and a polylinker cloning site) that facilitate recovery in *E. coli* of chromosomal DNA adjacent to the sites of insertion.

Other derivatives of Tn*917* have been described and utilized in *L. monocytogenes* (7, 19, 34). In addition, Tn*10*-derived transposons have been described that are active in *B. subtilis* (66) and that may be suitable for *L. monocytogenes* mutagenesis, but such use of these vectors has not yet been described. Efforts to adapt the small, random transposable Mariner element for use in *L. monocytogenes* are currently being pursued by the laboratories of Daniel Portnoy (University of California, Berkeley) and Hélène Marquis (Cornell University, Ithaca, N.Y.), with promising results.

REPORTER GENES

Several reporter genes developed for use in other systems have proven useful for monitoring *L. monocytogenes* transcriptional gene regulation. Transcriptional fusions to reporter genes such as *lacZ*, *gus*, *gfp*, *lux*, and *cat* have all been constructed in *L. monocytogenes* and have been used successfully to monitor patterns of bacterial gene expression in culture and within infected cells and animals (Table 1). The advantages and/or disadvantages of some of these reporter systems will be discussed briefly below.

Fusions to β-Galactosidase

lacZ fusions have been generated either by transposon insertion (Tn*917-lac* elements) (15, 34, 55) or via integrative

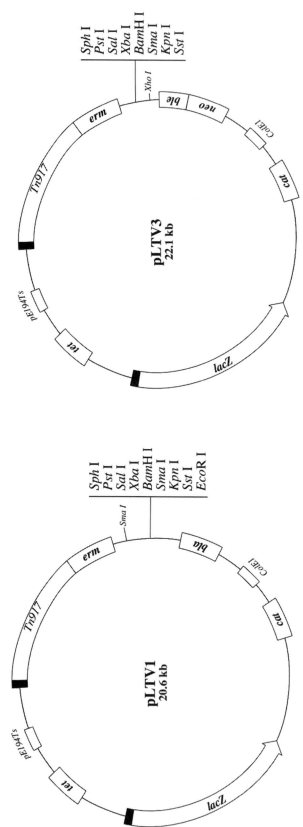

FIGURE 5 Tn917 delivery plasmids pLTV1 and pLTV3. ColE1, replication functions derived from pBR322 (10); pE194Ts, temperature-sensitive gram-positive origin of replication (83); *neo*, neomycinphosphotransferase II; *ble*, bleomycin resistance gene; *erm*, Tn917 ribosomal methyltransferase gene; *tet*, tetracycline resistance gene from pAMα1Δ1 (65); *cat*, *Staphylococcus aureus* pC194-derived chloramphenicol acetyltransferase gene (45); *bla*, pBR322 β-lactamase gene; *lacZ*, promoterless *lacZ* gene from *E. coli* with translation initiation signals derived from *B. subtilis* gene *spoVG*.

plasmids (61). β-Galactosidase activity can be measured for both extracellular and intracellularly grown bacteria (55), and the use of 4-methyl-umbelliferyl-β-D-galactopyranoside (MUG) as a substrate enables a higher degree of sensitivity than is possible with the colorimetric assay using *o*-nitrophenyl-β-D-galactopyranoside (96). MUG assays should be carried out in the presence of 0.1% Triton X-100 to enhance bacterial permeability for the MUG substrate (51, 61). Intracellular assays of bacterial β-galactosidase activity should also control for any endogenous β-galactosidase activity present in the infected cell lines. One method to reduce the background contribution of endogenous mammalian β-galactosidase activity is to adjust the pH of the assay conditions to approximately 8.0 (51, 61).

Fusions to β-Glucuronidase

gusA, the gene encoding β-glucuronidase, has been used as a reporter gene in a variety of prokaryotic systems. Transcriptional fusions to *gus* have been constructed and introduced into *L. monocytogenes* (8, 28, 58, 75, 76, 78, 90). The use of *gus* as a reporter gene has at least two advantages over *lacZ*. (i) There is very little endogenous β-glucuronidase activity in *L. monocytogenes*. (ii) *gus* fusions are far superior for assessing expression phenotypes on solid medium (using 5-bromo-4-chloro-3-indolyl glucuronide as an indicator). Colorimetric assays (46) are easily performed, and fluorometric assay substrates are also commercially available and can be used as described above for the MUG assay of β-galactosidase activity, with the substitution of 4-methylumbelliferyl-β-D-glucuronide for 4-methyl-umbelliferyl-β-D-galactopyranoside (75, 76, 90).

Fusions to *gfp*

The GFP of *A. victoria* (17) is a useful reporter gene system for the examination of gene expression in that GFP requires no cofactors for fluorescent activity and because it can be used in living systems, as well as in fixed samples. It is possible to monitor the timing of *L. monocytogenes* gene expression for individual bacteria located within infected animal cells by using single-copy chromosomally located transcriptional fusions to *gfp* (24, 30). GFP fluorescence can be detected for bacteria located either in the cytoplasm or within vacuoles of infected host cells, following fixation and microscopic examination. Work by Wilson et al. (87) used GFP and fluorescence-activated cell sorting to identify in vivo-induced promoters in *L. monocytogenes* using plasmid vector pAMGFP, a derivative of the pAM401 shuttle vector (Fig. 1) (88). Additional vectors for expression of *gfp* in *L. monocytogenes* have been developed (26, 33, 53). Constitutive expression of GFP within *L. monocytogenes* may also provide a means of distinguishing different *L. monocytogenes* strains within an infected host cell. Dietrich et al. (22) have used plasmid-borne fusions to *gfp* to monitor *L. monocytogenes* promoter activity, as well as the efficiency of plasmid delivery to infected cells, with an attenuated suicide strain of *L. monocytogenes*.

Fusions to *cat* and *lux*

Moors et al. (61) developed a plasmid vector that allows transcriptional fusions to both *cat* (encoding chloramphenicol transacetylase) and *lacZ* to be constructed in single copy in the *L. monocytogenes* chromosome. This vector, pLCR, enables both extracellular and intracellular measurement of target gene expression in *L. monocytogenes*.

Transcriptional fusions to plasmid-borne *luxAB*-encoded luciferase were originally used by Park et al. (64) to monitor

L. monocytogenes gene expression in a variety of growth media (pSB292) (Table 1). Luciferase activity was assessed by using a luminometer and by measuring the peak of light produced. A recent and exciting use of the *lux* reporter was described by Hardy et al. (41), in which they used in vivo bioluminescence imaging with an IVIS Imaging system to reveal the spatiotemporal distribution of *L. monocytogenes* during infection of mice.

HETEROLOGOUS EXPRESSION SYSTEMS FOR *L. MONOCYTOGENES* GENE PRODUCTS

In certain circumstances, it may be beneficial to use a heterologous expression system to monitor protein function or promoter activity. The naturally competent and well-studied bacterium *B. subtilis* has several advantages to recommend it as a heterologous expression system host for *L. monocytogenes* gene products. In addition to *B. subtilis* being a fellow gram-positive bacterium that is closely related to *L. monocytogenes* (36), a variety of genetic tools have been developed over the years to aid in the rapid generation of reporter gene fusions and inducible promoter constructs. *B. subtilis* has been used as a host system to demonstrate the ability of the *L. monocytogenes* listeriolysin O protein to promote lysis of phagosomal membranes in the absence of any additional *L. monocytogenes* proteins (9), to aid in the identification of the *L. monocytogenes* plcA gene product (14), to demonstrate the ability of PrfA to activate *hly* expression in the absence of any additional *L. monocytogenes* proteins (32), and to measure the relative levels of PrfA-dependent activation of target promoter sequences (71, 86). The *B. subtilis* expression systems used in each of these cases took advantage of readily available plasmid vectors and bacteriophage for rapid generation of the constructs of interest.

L. MONOCYTOGENES BACTERIOPHAGE

Perhaps the most important genetic tool recently developed for *L. monocytogenes* genetic analysis has been the isolation and use of generalized transducing bacteriophages.

Bacteriophages have been used for years as a method of typing of *L. monocytogenes* strains, and an extensive collection of bacteriophages has been assembled for this purpose. Until recently, however, little utilization was made of these bacteriophages in terms of advancing their potential as tools for *L. monocytogenes* genetics. The availability of transducing bacteriophage has several advantages for analysis of bacteria. They can be used to verify that a specific transposon insertion is associated with a mutant phenotype, they can facilitate the rapid introduction of selectable mutations into new strain backgrounds, they can be used to move point mutations closely associated with selectable markers into new strains, and they are useful for localized mutagenesis and fine structure genetic mapping.

David Hodgson of the University of Warwick has organized a collection of *L. monocytogenes* bacteriophages and tested them for transduction ability (44). A number of bacteriophages were found to be capable of transduction for *L. monocytogenes* serotype 1/2a and 4b strains; two of these are described below.

P35

A generalized transducing phage, P35 was isolated from silage and grows on serotype 1/2a strains SLCC5764, 10403S, LO28, and EGD (44). It is a temperate, citrate-

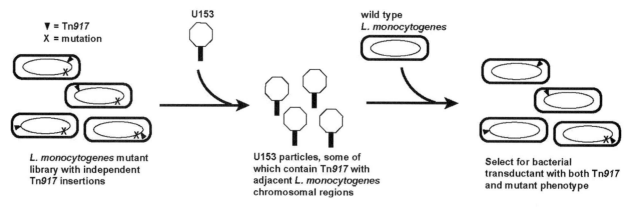

FIGURE 6 A method for isolating a transposon insertion linked to a particular mutant locus in *L. monocytogenes* (transposon tagging, as described by D. Kaiser) (49). As an example, a *L. monocytogenes* mutant strain is depicted with a single base pair change within the bacterial chromosome that confers a mutant phenotype (mutation indicated by X). A library of random transposon insertions (Tn917) is generated within the mutant strain, and a U153 bacteriophage lysate is prepared from the mutant strain transposon insertion library. Wild-type *L. monocytogenes* is incubated with the phage lysate, and transductants are selected based on the presence of the antibiotic resistance marker provided by Tn917 and the mutant phenoytpe conferred by mutation X. The Tn917 transposon is now genetically linked to mutation X, and the distance between the two can be determined based on the cotransduction frequency of Tn917 and X (see reference 59 for an experimental example and additional details). This approach can theoretically be used to map any unmarked mutation that confers a phenotype within the *L. monocytogenes* chromosome.

sensitive, chloroform-sensitive bacteriophage that only grows at room temperature. Examination of purified phage reveals a large capsid with about 38 kb of DNA and a tail of modest length. It is capable of low-frequency transduction and interstrain transduction between 10403S, EGD, and LO28, and it has been shown to infect 75% of all serotype 1/2 strains tested (43, 44).

U153

A second phage with very high transduction frequency, U153, was isolated by David Hodgson from *L. monocytogenes* strain CUSI153 (44). It has a genome of about 40 kb, and it grows on SLCC5764 and 10403S but not on LO28 or EGD. It is capable of high-frequency transduction and interstrain transduction between 10403, NCTC7973, and SLCC5764 (44).

R. Calendar and colleagues have found that *L. monocytogenes* strain 10403S carries a bacteriophage, φ10403S, that looks very similar to U153 (13). Both bacteriophages have the same attachment site specificity within *comK*, since lysogenization by U153 can cure φ10403S. φ10403S and U153 have viral interference properties: both can interfere with a bacteriophage isolated from strain EGD, whereas the bacteriophage from EGD cannot interfere with φ10403S and U153. This interference property probably explains why so few *L. monocytogenes* phages grow on 10403S (13).

P35 and U153 have been used successfully to generate new strain constructions (43, 91) and to verify the association between transposon insertions and mutant phenotypes (43). Transduction also makes it now possible to identify single-base mutations in *L. monocytogenes* via transposon tagging (Fig. 6) (43, 59, 76), following chemical mutagenesis or the use of a recently described *L. monocytogenes* mutator strain (56). The availability of transducing phage opens enormous potential for the further characterization and genetic manipulation of *L. monocytogenes*. David Hodgson,

Department of Biological Sciences, University of Warwick, is willing to send out any of the *L. monocytogenes* bacteriophages as requested. In addition, anyone who is using an unusual strain of *L. monocytogenes* may send it to him, and he will test it with his collection of transducing bacteriophages to see which are capable of growth on that strain.

SUMMARY

The expanding number of tools available for the detailed characterization and genetic manipulation of *L. monocytogenes* increases the utility of this organism as a model system for the molecular study of an intracellular pathogen. Innovative efforts by many investigators have accelerated our ability to understand multiple aspects of bacterial pathogenesis in terms of pathogen evolution, gene expression, signal transduction, and protein function outside and inside of infected host cells.

We thank P. Youngman, M. Moors, H. Marquis, P. Cossart, D. Portnoy, R. Calendar, and D. Hodgson and colleagues in their laboratories for many helpful discussions and for the communication of data before publication.

Work was supported by Public Health Service grants AI41816 and AI055651 (N.E.F.) and AI53669 (D.E.H.) from the National Institutes of Health and GPH 9 (C.B.) from the Institut Pasteur.

REFERENCES

1. **Alberti-Segui, C., and D. Higgins.** Unpublished data.
2. **Annous, B. A., L. A. Becker, D. O. Bayles, D. P. Labeda, and B. J. Wilkinson.** 1997. Critical role of anteiso-C15:0 fatty acid in the growth of *Listeria monocytogenes* at low temperatures. *Appl. Environ. Microbiol.* **63:**3887–3894.
3. **Arnaud, M., A. Chastanet, and M. Debarbouille.** 2004. New vector for efficient replacement in naturally non-transformable, low GC-content, gram-positive bacteria. *Appl. Environ. Microbiol.* **70:**6887–6891.

4. **Arous, S., C. Buchrieser, P. Folio, P. Glaser, A. Namane, M. Hebraud, and Y. Hechard.** 2004. Global analysis of gene expression in an *rpoN* mutant of *Listeria monocytogenes*. *Microbiology* **150:**1581–1590.

5. **Autret, N., I. Dubail, P. Trieu-Cuot, P. Berche, and A. Charbit.** 2001. Identification of new genes involved in the virulence of *Listeria monocytogenes* by signature-tagged transposon mutagenesis. *Infect. Immun.* **69:**2054–2065.

6. **Baldwin, D. N., V. Vanchinathan, P. O. Brown, and J. A. Theriot.** 2003. A gene-expression program reflecting the innate immune response of cultured intestinal epithelial cells to infection by *Listeria monocytogenes*. *Genome Biol.* **4:**R2.

7. **Begley, M., C. Hill, and C. G. Gahan.** 2003. Identification and disruption of *btlA*, a locus involved in bile tolerance and general stress resistance in *Listeria monocytogenes*. *FEMS Microbiol. Lett.* **218:**31–38.

8. **Behari, J., and P. Youngman.** 1998. Regulation of *hly* expression in *Listeria monocytogenes* by carbon sources and pH occurs through separate mechanisms mediated by PrfA. *Infect. Immun.* **66:**3635–3642.

9. **Bielecki, J., P. Youngman, P. Connelly, and D. A. Portnoy.** 1990. *Bacillus subtilis* expressing a haemolysin gene from *Listeria monocytogenes* can grow in mammlian cells. *Nature* **345:**175–176.

10. **Bolivar, F., P. J. Rodriquez, P. J. Greene, M. C. Betlach, H. L. Heyneker, H. W. Boyer, J. H. Crosa, and S. Falkow.** 1977. Construction and characterization of new cloning vehicles. II. A multipurpose cloning system. *Gene* **2:**95–113.

11. **Buchrieser, C.** Unpublished data.

12. **Buchrieser, C., C. Rusniok, F. Kunst, P. Cossart, and P. Glaser.** 2003. Comparison of the genome sequences of *Listeria monocytogenes* and *Listeria innocua*: clues for evolution and pathogenicity. *FEMS Immunol. Med. Microbiol.* **35:**207–213.

13. **Calendar, R.** Personal communication.

14. **Camilli, A., H. Goldfine, and D. A. Portnoy.** 1991. *Listeria monocytogenes* mutants lacking phosphatidylinositol-specific phospholipase C are avirulent. *J. Exp. Med.* **173:**751–754.

15. **Camilli, A., D. A. Portnoy, and P. Youngman.** 1990. Insertional mutagenesis of *Listeria monocytogenes* with a novel Tn917 derivative that allows direct cloning of DNA flanking transposon insertions. *J. Bacteriol.* **172:**3738–3744.

16. **Camilli, A., L. G. Tilney, and D. A. Portnoy.** 1993. Dual roles of *plcA* in *Listeria monocytogenes* pathogenesis. *Mol. Microbiol.* **8:**143–157.

17. **Chalfie, M., Y. Tu, G. Euskirchen, W. W. Ward, and D. C. Prasher.** 1994. Green fluorescent protein as a marker for gene expression. *Science* **263:**802-805.

18. **Clewell, D. B., S. E. Flannagan, Y. Ike, J. M. Jones, and C. Gawron-Burke.** 1988. Sequence analysis of termini of conjugative transposon Tn916. *J. Bacteriol.* **170:**3046-3052.

19. **Cvitkovitch, D. G., J. A. Gutierrez, J. Behari, P. J. Youngman, J. E. Wetz, P. J. Crowley, J. D. Hillman, L. J. Brady, and A. S. Bleiweis.** 2000. Tn917-lac mutagenesis of *Streptococcus mutans* to identify environmentally regulated genes. *FEMS Microbiol. Lett.* **182:**149–154.

20. **Dancz, C. E., A. Haraga, D. A. Portnoy, and D. E. Higgins.** 2002. Inducible control of virulence gene expression in *Listeria monocytogenes*: temporal requirement of listeriolysin O during intracellular infection. *J. Bacteriol.* **184:**5935–5945.

21. **Denoya, C. D., D. H. Bechhofer, and D. Dubnau.** 1986. Translational autoregulation of *ermC* 23S rRNA methyltransferase expression in *Bacillus subtilis*. *J. Bacteriol.* **168:**1133–1141.

22. **Dietrich, G., A. Bubert, I. Gentschev, Z. Sokolovic, A. Simm, A. Catic, S. H. E. Kaufmann, J. Hess, A. A. Szalay, and W. Goebel.** 1998. Delivery of antigen-encoding plasmid DNA into the cytosol of macrophages by attenuated suicide *Listeria monocytogenes*. *Nat. Biotechnol.* **16:**181–185.

23. **Doumith, M., C. Cazalet, N. Simoes, L. Frangeul, C. Jacquet, F. Kunst, P. Martin, P. Cossart, P. Glaser, and C. Buchrieser.** 2004. New aspects regarding evolution and virulence of *Listeria monocytogenes* revealed by comparative genomics and DNA arrays. *Infect. Immun.* **72:**1072–1083.

24. **Drevets, D. A., T. A. Jelinek, and N. E. Freitag.** 2001. *Listeria monocytogenes*-infected phagocytes can initiate central nervous system infection in mice. *Infect. Immun.* **69:**1344–1350.

25. **Evans, M. R., B. Swaminathan, L. M. Graves, E. Altermann, T. R. Klaenhammer, R. C. Fink, S. Kernodle, and S. Kathariou.** 2004. Genetic markers unique to *Listeria monocytogenes* serotype 4b differentiate epidemic clone II (hot dog outbreak strains) from other lineages. *Appl. Environ. Microbiol.* **70:**2383–2390.

26. **Fortinea, N., P. Trieu-Cuot, O. Gaillot, E. Pellegrini, P. Berche, and J. L. Gaillard.** 2000. Optimization of green fluorescent protein expression vectors for in vitro and in vivo detection of *Listeria monocytogenes*. *Res. Microbiol.* **151:**353–360.

27. **Frangeul, L., P. Glaser, C. Rusniok, C. Buchrieser, E. Duchaud, P. Dehoux, and F. Kunst.** 2004. CAAT-Box, Contigs-Assembly and Annotation Tool-Box for genome sequencing projects. *Bioinformatics* **20:**790–797.

28. **Fraser, K. R., D. Sue, M. Wiedmann, K. Boor, and C. P. O'Byrne.** 2003. Role of σ^B in regulating the compatible solute uptake systems of *Listeria monocytogenes*: osmotic induction of *opuC* is σ^B dependent. *Appl. Environ. Microbiol.* **69:**2015–2022.

29. **Freitag, N. E.** 2000. Genetic tools for use with *Listeria monocytogenes*, p. 488–498. *In* V. A. Fischetti, R. P. Novick, J. J. Ferretti, D. A. Portnoy, and J. I. Rood (ed.), *Gram-Positive Pathogens*, 1st ed. ASM Press, Washington, D.C.

30. **Freitag, N. E., and K. E. Jacobs.** 1999. Examination of *Listeria monocytogenes* intracellular gene expression by using the green fluorescent protein of *Aequorea victoria*. *Infect. Immun.* **67:**1844–1852.

31. **Freitag, N. E., and D. A. Portnoy.** 1994. Dual promoters of the *Listeria monocytogenes prfA* transcriptional activator appear essential in vitro but are redundant in vivo. *Mol. Microbiol.* **12:**845–853.

32. **Freitag, N. E., P. Youngman, and D. A. Portnoy.** 1992. Transcriptional activation of the *Listeria monocytogenes* hemolysin gene in *Bacillus subtilis*. *J. Bacteriol.* **174:**1293–1298.

33. **Gahan, C. G., J. O'Mahony, and C. Hill.** 2001. Characterization of the *groESL* operon in *Listeria monocytogenes*: utilization of two reporter systems (*gfp* and *hly*) for evaluating in vivo expression. *Infect. Immun.* **69:**3924–3932.

34. **Gardan, R., P. Cossart, and J. Labadie.** 2003. Identification of *Listeria monocytogenes* genes involved in salt and alkaline-pH tolerance. *Appl. Environ. Microbiol.* **69:**3137–3143.

35. **Gawron-Burke, C., and D. B. Clewell.** 1984. Regeneration of insertionally inactivated streptococcal DNA fragments after excision of transposon Tn916 in *Escherichia coli*: strategy for targeting and cloning of genes from gram-positive bacteria. *J. Bacteriol.* **159:**214–221.

36. **Glaser, P., L. Frangeul, C. Buchrieser, C. Rusniok, A. Amend, F. Baquero, P. Berche, H. Bloecker, P. Brandt, T. Chakraborty, A. Charbit, F. Chetouani, E. Couve, A. de Daruvar, P. Dehoux, E. Domann, G. Dominguez-**

Bernal, E. Duchaud, L. Durant, O. Dussurget, K. D. Entian, H. Fsihi, F. G. Portillo, P. Garrido, L. Gautier, W. Goebel, N. Gomez-Lopez, T. Hain, J. Hauf, D. Jackson, L. M. Jones, U. Kaerst, J. Kreft, M. Kuhn, F. Kunst, G. Kurapkat, E. Madueno, A. Maitournam, J. M. Vicente, E. Ng, H. Nedjari, G. Nordsiek, S. Novella, B. de Pablos, J. C. Perez-Diaz, R. Purcell, B. Remmel, M. Rose, T. Schlueter, N. Simoes, A. Tierrez, J. A. Vazquez-Boland, H. Voss, J. Wehland, and P. Cossart. 2001. Comparative genomics of *Listeria* species. *Science* **294:** 849–852.

37. **Gründling, A., L. S. Burrack, H. G. A. Bouwer, and D. E. Higgins.** 2004. *Listeria monocytogenes* regulates flagellar motility gene expression through MogR, a transcriptional repressor required for virulence. *Proc. Natl. Acad. Sci. USA* **101:** 12318–12323.

38. **Gründling, A., M. D. Gonzalez, and D. E. Higgins.** 2003. Requirement of the *Listeria monocytogenes* broad-range phospholipase PC-PLC during infection of human epithelial cells. *J. Bacteriol.* **185:**6295–6307.

39. **Gryczan, T. J., J. Hahn, S. Contente, and D. Dubnau.** 1982. Replication and incompatibility properties of plasmid pE194 in *Bacillus subtilis. J. Bacteriol.* **152:**722–735.

40. **Haas, S. A., M. Hild, A. P. Wright, T. Hain, D. Talibi, and M. Vingron.** 2003. Genome-scale design of PCR primers and long oligomers for DNA microarrays. *Nucleic Acids Res.* **31:**5576–5581.

41. **Hardy, J., K. P. Francis, M. DeBoer, P. Chu, K. Gibbs, and C. H. Contag.** 2004. Extracellular replication of *Listeria monocytogenes* in the murine gall bladder. *Science* **303:** 851–853.

42. **Herd, M., and C. Kocks.** 2001. Gene fragments distinguishing an epidemic-associated strain from a virulent prototype strain of *Listeria monocytogenes* belong to a distinct functional subset of genes and partially cross-hybridize with other *Listeria* species. *Infect. Immun.* **69:**3972–3979.

43. **Hodgson, D.** Personal communication.

44. **Hodgson, D. A.** 2000. Generalized transduction of serotype 1/2 and serotype 4b strains of *Listeria monocytogenes. Mol. Microbiol.* **35:**312–323.

45. **Horinouchi, S., and B. Weisblum.** 1982. Nucleotide sequence and functional map of pC194, a plasmid that specifies inducible chloramphenicol resistance. *J. Bacteriol.* **150:**815–825.

46. **Jefferson, R. A.** 1989. The GUS reporter gene system. *Nature* **342:**837–838.

47. **Jensen, E. R., R. Selvakumar, H. Shen, R. Ahmed, F. O. Wettstein, and J. F. Miller.** 1997. Recombinant *Listeria monocytogenes* vaccination eliminates papillomavirus-induced tumors and prevents papilloma formation from viral DNA. *J. Virol.* **71:**8467–8474.

48. **Jones, S., and D. A. Portnoy.** 1994. Characterization of *Listeria monocytogenes* pathogenesis in a strain expressing perfringolysin O in place of listeriolysin O. *Infect. Immun.* **62:**5608–5613.

49. **Kaiser, A. D.** 1984. Genetics of myxobacteria, p. 163–184. *In* E. Rosenberg (ed.), *Myxobacteria: Development and Cell Interactions.* Springer, New York, N.Y.

50. **Kazmierczak, M. J., S. C. Mithoe, K. J. Boor, and M. Wiedmann.** 2003. *Listeria monocytogenes* σ[B] regulates stress response and virulence functions. *J. Bacteriol.* **185:** 5722–5734.

51. **Klarsfeld, A., P. L. Goossens, and P. Cossart.** 1994. Five *Listeria monocytogenes* genes preferentially expressed in infected mammalian cells: *plcA, purH, purD, pyrE* and an

arginine ABC transporter gene, *arpJ. Mol. Microbiol.* **13:**585–597.

52. **Lauer, P., M. Y. Chow, M. J. Loessner, D. A. Portnoy, and R. Calendar.** 2002. Construction, characterization, and use of two *Listeria monocytogenes* site-specific phage integration vectors. *J. Bacteriol.* **184:**4177–4186.

53. **Li, G., and S. Kathariou.** 2003. An improved cloning vector for construction of gene replacements in *Listeria monocytogenes. Appl. Environ. Microbiol.* **69:**3020–3023.

54. **McCaffrey, R. L., P. Fawcett, M. O'Riordan, K. D. Lee, E. A. Havell, P. O. Brown, and D. A. Portnoy.** 2004. A specific gene expression program triggered by gram-positive bacteria in the cytosol. *Proc. Natl. Acad. Sci. USA* **101:**11386–11391.

55. **Mengaud, J., S. Dramsi, E. Gouin, J. A. Vazquez-Boland, G. Milon, and P. Cossart.** 1991. Pleiotropic control of *Listeria monocytogenes* virulence factors by a gene that is autoregulated. *Mol. Microbiol.* **5:**2273–2283.

56. **Merino, D., H. Reglier-Poupet, P. Berche, and A. Charbit.** 2002. A hypermutator phenotype attenuates the virulence of *Listeria monocytogenes* in a mouse model. *Mol. Microbiol.* **44:**877–887.

57. **Michel, E., K. A. Reich, R. Favier, P. Berche, and P. Cossart.** 1990. Attenuated mutants of the intracellular bacterium *Listeria monocytogenes* obtained by single amino acid substitutions in listeriolysin O. *Mol. Microbiol.* **4:** 2167–2178.

58. **Milenbachs, A. A., D. P. Brown, M. Moors, and P. Youngman.** 1997. Carbon-source regulation of virulence gene expression in *Listeria monocytogenes. Mol. Microbiol.* **23:**1075–1085.

59. **Milenbachs Lukowiak, A., K. J. Mueller, N. E. Freitag, and P. Youngman.** 2004. Deregulation of *Listeria monocytogenes* virulence gene expression by two distinct and semi-independent pathways. *Microbiology* **150:**321–333.

60. **Milohanic, E., P. Glaser, J. Y. Coppee, L. Frangeul, Y. Vega, J. A. Vazquez-Boland, F. Kunst, P. Cossart, and C. Buchrieser.** 2003. Transcriptome analysis of *Listeria monocytogenes* identifies three groups of genes differently regulated by PrfA. *Mol. Microbiol.* **47:**1613–1625.

61. **Moors, M. A., B. Levitt, P. Youngman, and D. A. Portnoy.** 1999. Expression of listeriolysin O and ActA by intracellular and extracellular *Listeria monocytogenes. Infect. Immun.* **67:**131–139.

62. **Nelson, K. E., D. E. Fouts, E. F. Mongodin, J. Ravel, R. T. DeBoy, J. F. Kolonay, D. A. Rasko, S. V. Angiuoli, S. R. Gill, I. T. Paulsen, J. Peterson, O. White, W. C. Nelson, W. Nierman, M. J. Beanan, L. M. Brinkac, S. C. Daugherty, R. J. Dodson, A. S. Durkin, R. Madupu, D. H. Haft, J. Selengut, S. Van Aken, H. Khouri, N. Fedorova, H. Forberger, B. Tran, S. Kathariou, L. D. Wonderling, G. A. Uhlich, D. O. Bayles, J. B. Luchansky, and C. M. Fraser.** 2004. Whole genome comparisons of serotype 4b and 1/2a strains of the food-borne pathogen *Listeria monocytogenes* reveal new insights into the core genome components of this species. *Nucleic Acids Res.* **32:** 2386–2395.

63. **Park, S. F., and G. S. A. B. Stewart.** 1990. High-efficiency transformation of *Listeria monocytogenes* by electroporation of penicillin-treated cells. *Gene* **94:**129–132.

64. **Park, S. F., G. S. A. B. Stewart, and R. G. Kroll.** 1992. The use of bacterial luciferase for monitoring the environmental regulation of expression of genes encoding virulence factors in *Listeria monocytogenes. J. Gen. Microbiol.* **138:**2619–2627.

65. **Perkins, J. B., and P. Youngman.** 1983. *Streptococcus* plasmid pAMα1 is a composite of two separate replicons, one of which is closely related to *Bacillus* plasmid pBC16. *J. Bacteriol.* **155:**607–615.

66. **Petit, M.-A., C. Bruand, L. Janniere, and S. D. Ehrlich.** 1990. Tn*10*-derived transposons active in *Bacillus subtilis*. *J. Bacteriol.* **172:**6736–6740.

67. **Poyart, C., and P. Trieu-Cuot.** 1997. A broad-host-range mobilizable shuttle vector for the construction of transcriptional fusions to beta-galactosidase in gram-positive bacteria. *FEMS Microbiol. Lett.* **156:**193–198.

68. **Quisel, J. D., W. F. Burkholder, and A. D. Grossman.** 2001. In vivo effects of sporulation kinases on mutant Spo0A proteins in *Bacillus subtilis*. *J. Bacteriol.* **183:**6573–6578.

69. **Schaferkordt, S., and T. Chakraborty.** 1995. Vector plasmid for insertional mutagenesis and directional cloning in *Listeria* spp. *Biotechniques* **19:**720–722, 724–725.

70. **Scott, J. R., P. A. Kirchman, and M. E. Caparon.** 1988. An intermediate in transposition of the conjugative transposon Tn*916*. *Proc. Natl. Acad. Sci. USA* **85:**4809–4813.

71. **Sheehan, B., A. Klarsfeld, T. Msadek, and P. Cossart.** 1995. Differential activation of virulence gene expression by PrfA, the *Listeria monocytogenes* virulence regulator. *J. Bacteriol.* **177:**6469–6476.

72. **Shen, A., and D. E. Higgins.** Unpublished data.

73. **Shen, H., J. F. Miller, X. Fan, D. Kolwyck, R. Ahmed, and J. T. Harty.** 1998. Compartmentalization of bacterial antigens: differential effects on priming of CD8 T cells and protective immunity. *Cell* **92:**535–545.

74. **Shen, H., M. K. Slifka, M. Matloubian, E. R. Jensen, R. Ahmed, and J. F. Miller.** 1995. Recombinant *Listeria monocytogenes* as a live vaccine vehicle for the induction of protective anti-viral cell-mediated immunity. *Proc. Natl. Acad. Sci. USA* **92:**3987–3991.

75. **Shetron-Rama, L. M., H. Marquis, H. G. A. Bouwer, and N. E. Freitag.** 2002. Intracellular induction of *Listeria monocytogenes* actA expression. *Infect. Immun.* **70:**1087–1096.

76. **Shetron-Rama, L. M., K. Mueller, J. M. Bravo, H. G. Bouwer, S. S. Way, and N. E. Freitag.** 2003. Isolation of *Listeria monocytogenes* mutants with high-level in vitro expression of host cytosol-induced gene products. *Mol. Microbiol.* **48:**1537–1551.

77. **Smith, K., and P. Youngman.** 1992. Use of a new integrational vector to investigate compartment-specific expression of the *Bacillus subtilis* spoIIM gene. *Biochimie* **74:**705–711.

78. **Sue, D., K. J. Boor, and M. Wiedmann.** 2003. σ^B-dependent expression patterns of compatible solute transporter genes opuCA and lmo1421 and the conjugated bile salt hydrolase gene bsh in *Listeria monocytogenes*. *Microbiology* **149:**3247–3256.

79. **Sullivan, M. A., R. E. Yasbin, and F. E. Young.** 1984. New shuttle vectors for *Bacillus subtilis* and *Escherichia coli* which allow rapid detection of inserted fragments. *Gene* **29:**21–26.

80. **Sun, A. N., A. Camilli, and D. A. Portnoy.** 1990. Isolation of *Listeria monocytogenes* small-plaque mutants defective for intracellular growth and cell-to-cell spread. *Infect. Immun.* **58:**3770–3778.

81. **Trieu-Cuot, P., C. Carlier, C. Poyart-Salmeron, and P. Courvalin.** 1991. Shuttle vectors containing a multiple cloning site and a lacZ alpha gene for conjugal transfer of DNA from *Escherichia coli* to gram-positive bacteria. *Gene* **102:**99–104.

82. **Trieu-Cuot, P., E. Derlot, and P. Courvalin.** 1993. Enhanced conjugative transfer of plasmid DNA from *Escherichia coli* to *Staphylococcus aureus* and *Listeria monocytogenes*. *FEMS Microbiol. Lett.* **109:**19–24.

83. **Villafane, R., D. H. Gechhofer, C. S. Narayanan, and D. Dubnau.** 1987. Replication control genes of plasmid pE194. *J. Bacteriol.* **169:**4822–4829.

84. **Vincente, M. F., J. C. Perez-Diaz, and F. Baquero.** 1987. A protoplast transformation system for *Listeria sp. Plasmid* **18:**89–92.

85. **Williams, D. R., D. I. Young, and M. Young.** 1990. Conjugative plasmid transfer from *Escherichia coli* to *Clostridium acetobutylicum*. *J. Gen. Microbiol.* **136:**819–826.

86. **Williams, J. R., C. Thayyullathil, and N. E. Freitag.** 2000. Sequence variations within PrfA DNA binding sites and effects on *Listeria monocytogenes* virulence gene expression. *J. Bacteriol.* **182:**837–841.

87. **Wilson, R. L., A. R. Tvinnereim, B. D. Jones, and J. T. Harty.** 2001. Identification of *Listeria monocytogenes* in vivo-induced genes by fluorescence-activated cell sorting. *Infect. Immun.* **69:**5016–5024.

88. **Wirth, R., F. Y. An, and D. B. Clewell.** 1986. Highly efficient protoplast transformation system for *Streptococcus faecalis* and a new *Escherichia coli-S. faecalis* shuttle vector. *J. Bacteriol.* **165:**831–836.

89. **Wonderling, L. D., B. J. Wilkinson, and D. O. Bayles.** 2004. The htrA (degP) gene of *Listeria monocytogenes* 10403S is essential for optimal growth under stress conditions. *Appl. Environ. Microbiol.* **70:**1935–1943.

90. **Wong, K. K., H. G. Bouwer, and N. E. Freitag.** 2004. Evidence implicating the 5′ untranslated region of *Listeria monocytogenes* actA in the regulation of bacterial actin-based motility. *Cell. Microbiol.* **6:**155–166.

91. **Wong, K. K. Y., and N. E. Freitag.** 2004. A novel mutation within the central *Listeria monocytogenes* regulator PrfA that results in constitutive expression of virulence gene products. *J. Bacteriol.* **186:**6265–6276.

92. **Wuenscher, M. D., S. Kohler, W. Goebel, and T. Chakraborty.** 1991. Gene disruption by plasmid integration in *Listeria monocytogenes*: insertional inactivation of the listeriolysin determinant lisA. *Mol. Gen. Genet.* **228:**177–182.

93. **Yanish-Perron, C., J. Vierira, and J. Messing.** 1985. Improved M13 phage cloning vectors and host strains: nucleotide sequences of the M13mp18 and pUC19 vectors. *Gene* **33:**103–109.

94. **Yansura, D. G., and D. J. Henner.** 1984. Use of the *Escherichia coli* lac repressor and operator to control gene expression in *Bacillus subtilis*. *Proc. Natl. Acad. Sci. USA* **81:**439–443.

95. **Yildirim, S., W. Lin, A. D. Hitchins, L. A. Jaykus, E. Altermann, T. R. Klaenhammer, and S. Kathariou.** 2004. Epidemic clone I-specific genetic markers in strains of *Listeria monocytogenes* serotype 4b from foods. *Appl. Environ. Microbiol.* **70:**4158–4164.

96. **Youngman, P.** 1987. Plasmid vectors for recovering and exploiting Tn*917* transpositions in *Bacillus* and other gram-positive bacteria, p. 79–103. *In* K. Hardy (ed.), *Plasmids: a Practical Approach.* IRL Press, Oxford, United Kingdom.

Regulation of Virulence Genes in Pathogenic *Listeria* spp.

WERNER GOEBEL, STEFANIE MÜLLER-ALTROCK, AND JÜRGEN KREFT

52

Regulation of virulence genes in pathogenic bacteria must occur by mechanisms allowing the coordinate and differential expression of the virulence factors during infection. Only in this way can genes, encoding a virulence factor(s) that is advantageous for a specific step in infection, be activated while those eventually disadvantageous for this step can be turned off. Physicochemical parameters that change when the pathogen enters a host, such as temperature, pH value, available supply of oxygen, essential metal ions (especially iron), or nutrients, may act directly or indirectly as signals for the differential expression of these genes.

In the genus *Listeria*, there are two pathogenic species: *Listeria monocytogenes* and *Listeria ivanovii*. While *L. ivanovii* is pathogenic only for animals, *L. monocytogenes* can infect domestic animals and humans (40, 71, 117, 125). Mainly via the food chain, *L. monocytogenes* gains access to the intestine; from there, it can spread to deeper organs and may cause an infrequent, but often fatal, disease. A third species, *Listeria seeligeri*, is considered avirulent, but interestingly it carries many of the listerial virulence genes, including that for the central virulence regulator positive regulatory factor A (PrfA). *L. monocytogenes* is a gram-positive, rod-shaped, nonsporulating, facultative intracellular bacterium with the ability to invade different mammalian cells, to survive in the cytosol of these host cells, and to spread into neighboring host cells. When equipped with the necessary virulence genes, these bacteria can move from the primary site of infection (normally the intestine) to the peripheral organs, i.e., liver and spleen, then further into the peripheral blood circuit, and eventually to the brain. During a successful infection, *L. monocytogenes* may thus encounter and invade a number of host cells and tissues. This in turn may demand the precisely timed expression of a variety of virulence factors.

REGULATORY PROTEINS OF *L. MONOCYTOGENES* AND *L. IVANOVII*

The genome sequences of several representatives of two species of the genus *Listeria*, *L. monocytogenes* and *Listeria innocua*, have been published (41, 86); the genome sequences of the other members of the genus will be available soon.

The genome sequences known so far already allowed the identification of those regulators that are specific for the pathogenic *L. monocytogenes* and hence are candidates for virulence gene control.

Of those regulators identified by this approach (summarized in Table 1), PrfA and (to some extent) SigB, although not specific for the pathogenic *Listeria* species, have been so far shown by genetic and biochemical studies to be involved in the regulation of virulence genes. The focus of this chapter will be on gene regulation by PrfA.

VIRULENCE GENES OF *L. MONOCYTOGENES* AND *L. IVANOVII*

In recent years, many virulence genes in *L. monocytogenes* and the related animal pathogen *L. ivanovii* have been identified (for recent reviews, see references 62 and 117), and the functions of the encoded virulence factors have been studied extensively. It was shown that these proteins are involved in adherence and uptake by host cells (22, 104), evasion from the primary phagosome into the host cell cytosol, intra- and intercellular motility (77), and replication efficiency in the host cell cytosol (117). Some of these genes are clustered in a kind of pathogenicity island, now termed *Listeria*-specific pathogenicity island 1 (LIPI-1) (62, 116), within the listerial chromosome, which is located always between the two housekeeping genes *ldh* and *prs* that encode lactate dehydrogenase and phosphoribosyl pyrophosphate synthetase, respectively (44, 65, 91, 107). LIPI-1 has been found in all virulent isolates of *L. monocytogenes* and *L. ivanovii* and occurs even in the avirulent *Listeria seeligeri*, albeit in a largely altered form (103). The other, similarly regulated (by PrfA) virulence genes, i.e., those for several large and small internalins (2, 14, 26, 30, 38, 92), for a hexose phosphate transporter (18), and for a bile acid hydrolase (28), are distributed on the chromosome of *L. monocytogenes*. *L. ivanovii* differs widely from *L. monocytogenes* with respect to the internalin genes and their localization on the chromosome (26, 31, 69, 103). The number of genes for small internalins is much larger in *L. ivanovii* than in *L. monocytogenes*, and many of the *i-inl* genes are clustered in a

Gram-Positive Pathogens, 2nd edition, edited by Vincent A. Fischetti et al.
© 2006 ASM Press, Washington, D.C.

TABLE 1 Regulators specific for *Listeria monocytogenes* EGD-e

Name	Description
PrfA	Transcription regulator of virulence genes
Lmo0041	Conserved hypothetical protein, putative regulator (LacI family)
Lmo0083	Similar to transcription regulator (MerR family)
Lmo0106	Similar to transcription regulator (ROK family)
Lmo0252	Similar to transcription repressor
Lmo0253	Similar to transcription antirepressor
Lmo0445	Similar to transcription regulator
Lmo0492	Similar to transcription regulator (LysR family)
Lmo0630	Similar to antiterminator (Bg1G family)
Lmo0733	Similar to transcription regulator
Lmo0734	Similar to transcription regulator (LacI family)
Lmo0753	Similar to transcription regulator (Crp/Fnr family)
Lmo1030	Similar to transcription regulator (LacI family)
Lmo1060	Putative response regulator
Lmo1116	Similar to regulatory protein (AraC/XylS family)
Lmo1974	Similar to transcription repressor (GntR family)
Lmo2144	Similar to transcription repressor (GntR family)
Lmo2773	Similar to antiterminator (BglG family)
Lmo2784	Similar to antiterminator (BglG family)
Lmo2814	Similar to transcription repressor (GntR family)
BvrA	Similar to transcription antiterminator (BglG family)

[a]Sequences are from ListiList (http://genolist.pasteur.fr/ListiList/) (41).

second *L. ivanovii*-specific pathogenicity island, now termed LIPI-2 (62).

Other differently regulated (PrfA-independent) genes with potential roles in virulence encode the extracellular protein p60, a peptidoglycan hydrolase (63, 90, 128), catalase and superoxide dismutase (7, 47, 122), the stress protease ClpC (85, 101, 102), the surface protein p104 (LAP) (51), the amidase Ami (81), and an atypical fibronectin-binding protein, FbpA (25).

REGULATION OF THE LISTERIAL VIRULENCE GENES BY ENVIRONMENTAL PARAMETERS

Genetic and biochemical studies have shown that the expression of PrfA-regulated listerial virulence genes (see below) is influenced by temperature (27, 52, 66, 96), pH (1, 23), oxygen conditions (113), iron (4), and carbon sources (1, 9, 23, 78, 79, 97).

Below 20°C, where *L. monocytogenes* is still capable of proficient growth (54), none of the virulence genes seems to be expressed in broth culture, while they are readily transcribed at 37°C and at least some of them (e.g., *hly* and *actA*) are even further induced at heat shock temperatures (110). However, in a *Drosophila* infection model (17, 74), the virulence genes were expressed even at room temperature.

Repression of *hly* by cellobiose, the phenolic beta-glucoside arbutin, but also by several readily fermentable mono- and disaccharides, including glucose, has been described (9, 57, 78, 79, 88, 89).

Starvation conditions, e.g., incubation in minimal essential medium, induce most genes of LIPI-1 and also some of the *inl* genes, including that for uptake of phosphorylated sugars (6, 30, 97, 99). Growth in a defined minimal medium containing glucose as the sole carbon source also leads to induced transcription of most of these virulence genes (75), whereas growth in rich brain heart infusion (BHI) medium normally results in rather poor expression (99). Interestingly, the addition of activated charcoal (33, 80, 99) or the XAD-4 beads (which absorb mainly hydrophobic components) to BHI medium during growth of *L. monocytogenes* yields high transcriptional activation of all PrfA-regulated virulence genes, suggesting that accumulation of a substance in the culture medium is inhibitory for the expression of these genes (see below).

There is also evidence that iron limitation may influence the expression of some virulence genes (4, 20, 21).

Not surprisingly, the intracellular environment of mammalian host cells activates PrfA and hence stimulates the expression of PrfA-regulated virulence genes and several other metabolic genes that are required for the intracellular replication cycle of *L. monocytogenes* (6, 10, 34, 57, 82, 95, 108).

PLEIOTROPIC REGULATORY FACTOR PrfA

PrfA is crucial as a transcriptional activator of most virulence genes. PrfA is encoded by the first gene of LIPI-1 (see above) and controls all five virulence genes (*plcA*, *hly*, *mpl*, *actA*, and *plcB*) (8, 91, 107) of this gene cluster and several others (for a recent review, see reference 61). Since LIPI-1 also occurs in *L. ivanovii* and *L. seeligeri* (see above), it is not surprising that these *Listeria* species carry *prfA* genes whose products share extended sequence homology (about 80% identity on the protein level) with the *L. monocytogenes* PrfA protein (Fig. 1A). PrfA of *L. monocytogenes* and *L. ivanovii* can be functionally fully exchanged, as shown in an in vitro transcription system using purified PrfA proteins (83).

PrfA is necessary for the transcription of all genes of LIPI-1. Among the internalin (*inl*) genes, PrfA-dependent transcription has been demonstrated in *L. monocytogenes* for the genes encoding the two large internalins, InlA and InlB, and for *inlC*, encoding the singular small secreted internalin of *L. monocytogenes* (30, 70, 73). In *L. ivanovii*, many more *inl* genes, especially those encoding the large number of small internalins produced by this species (31, 32, 43), are regulated by PrfA. All other *inl* genes encoding large internalins (26, 92) and identified in the genome sequences of *L. monocyto-*

genes (41, 86) are not under PrfA-mediated control. PrfA-dependent gene regulation has been also shown for the *hpt* gene, encoding a hexose phosphate transporter in both pathogenic *Listeria* species (18) and suggested for *bsh* (28), the gene for bile acid hydrolase of *L. monocytogenes*.

Recent comparative transcriptome studies set up to identify the entire PrfA regulon (80) revealed a large number of additional genes whose expression was positively or negatively affected by PrfA. A negative function of PrfA in the regulation of some genes in *L. monocytogenes*, e.g., *clpC* and the gene for a 64-kDa protein, had been claimed earlier (100, 110); a similar effect has been seen in *L. ivanovii* (65).

However, the influence of PrfA on the transcription of the latter genes is probably indirect (93); in Fig. 1B, therefore, we included only those genes as PrfA-regulated ones for which direct PrfA-dependent transcription has been convincingly shown by genetic and biochemical studies (3, 16, 64, 73, 76, 106).

PrfA IS A MEMBER OF THE Crp/Fnr FAMILY OF TRANSCRIPTION ACTIVATORS

Based on the sequence and structural (see below) similarity with cyclic AMP (cAMP)-binding protein of *Escherichia coli*

A

```
                                                           #
L. monocytogenes EGD/L028   MNAQAEEFKK YLETNGIKPK QFHKKELIFN QWDPQEYCIF LYDGITKLTS   50
L. ivanovii ATCC 19119      -D---AD--N F--------- K-K--DI--- --------V- -----A---N
L. seeligeri SLCC 3954      -STK--D--E ---S---Q-- Y-R---F--S --E-N----- -HE-VA----

   #                   #          #
ISENGTIMNL QYYKGAFVIM SGFIDTETSV GYYNLEVISE QATAYVIKIN  100
----S----- -------I-- --SL--GKPL ---------K T-----L--S
---S-D-L-- -------I-- T------K-L -------V-E --A--I---S

                                                *
ELKELLSKNL THFFYVFQTL QKQVSYSLAK FNDFSINGKL GSICGQLLIL  150
D--N-V-S-I -QL--II-A- ---------- -----V---- ----------
D----V--D- KQL--IID-- ---------- -----S---- ------F---

TYVYGKETPD GIKITLDNLT MHELGYSSGI AHSSAVSRII SKLKQEKVIV  200
-----EKN-- ----A----- -Q-------- --C------L ----K----E
A----E---N ------EK-- -Q---C---- ---------- ------N--E
                      I--  helix-turn-helix  --I

YKNSCFYVQN LDYLKRYAPK LDEWFYLACP ATWGKLN               237
--H-S----- -----KI--- ---------- S--E-F-
--D-Y--IK- IA---KI--- I--------H NS-D-F-
```

B

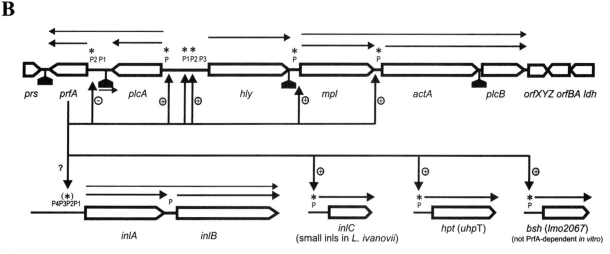

FIGURE 1 (A) Comparison of the amino acid sequences of the PrfA proteins from *L. monocytogenes*, *L. ivanovii*, and *L. seeligeri*. The pound sign (#) indicates conserved glycines at the turns of the β-roll structure; an asterisk (*) indicates the position of the G145S substitution, leading to a constitutively active PrfA (98, 118). Sequences are from references 65, 67, 76, and 103. (B) Structure and transcriptional orga-nization of the PrfA-dependent virulence gene cluster (LIPI-1) and of other PrfA-dependent virulence genes in *L. monocytogenes* and *L. ivanovii*. Genes and products: *prs*, phosphoribosyl synthase; *prfA*, positive regulatory factor A; *plcA*, phosphatidylinositol-specific phospholipase C; *hly*, listeriolysin O; *mpl*, metalloprotease; *actA*, actin-polymerization protein; *plcB*, broad-range phospholipase C (lecithinase); orfXYZ and orfBA, open reading frames of unknown function; *ldh*, lactate dehydrogenase; *inlA* and *inlB*, large cell-wall-bound internalins A and B; *inlC*, small secreted internalin C; *hpt*, hexose phosphate transporter; *bsh*, bile acid hydrolase; P, promoter. An asterisk above indicates the presence of a PrfA box within the promoter. Thin arrows above the gene symbols indicate the different transcripts. Arrows below with a plus or minus sign indicate transcriptional induction or repression, respectively, by PrfA. Filled pentagons show the additional open reading frames present in LIPI-1 of *L. seeligeri*.

(Crp or CAP), PrfA belongs to the Crp/Fnr family of transcription activators (61, 65). Crp is the major regulator in catabolite repression and regulates expression of many genes involved in carbon metabolism in *E. coli* and other gram-negative bacteria (59), while Fnr regulates the cellular response to anaerobic growth conditions in *E. coli* (111). Loss of Crp leads to a significant attenuation of virulence in *Salmonella* and *Shigella* spp.

Most members of the Crp/Fnr family have been found in gram-negative bacteria, and PrfA is one of only a few well-characterized representatives of this family in gram-positive bacteria (60), like Flp of *Lactobacillus casei* (50) and Srv of *Streptococcus pyogenes* (94).

Extensive molecular studies have been carried out with the cAMP-binding factor Crp; since PrfA shares extended sequential and structural similarity with Crp, a short overview of the most essential features of Crp might be useful for the understanding of PrfA.

MOLECULAR AND FUNCTIONAL ASPECTS OF Crp

In solution, Crp forms a homodimer with a molecular mass of 45 kDa. The monomeric Crp (209 amino acids) consists of an N-terminal domain (NTD; amino acids 1 to 129) and a C-terminal domain (CTD; amino acids 139 to 209), which are connected by a short hinge region. The NTD participates in cAMP binding and homodimerization. The region of amino acids 19 to 99 consists of short, antiparallel β-sheets (β-roll structure) that are separated by conserved glycine residues. Adjacent to this domain is a region termed helix C, which is involved in dimer formation and is also essential for cAMP binding. Detailed studies have shown that cAMP is bound to the NTD by ionic and hydrogen bonds via the amino acids G71, E72, R82, S83, R123, T127, and S128 (121). The CTD carries the helix E-turn-helix F motif between amino acids 168 and 191 that is essential for DNA binding. The two helices interact with DNA by polar side chains and positively charged residues of certain amino acids. Activation of transcription from Crp-cAMP-dependent promoters requires contact with the RNA polymerase (RNAP) involving the Crp activation regions AR1 (amino acids 156 to 164), AR2 (amino acids 19, 21, and 101), and AR3 (amino acids 52 to 58) (87, 129). Figure 2A summarizes the functionally important features of Crp.

In the absence of cAMP, Crp binds nonspecifically to DNA. Complex formation of Crp with cAMP probably influences the orientation of helix F, thereby enhancing binding of the Crp-cAMP complex to the specific target sequence. Glucose limitation leads to a cellular increase in the level of cAMP. Two molecules of cAMP bind to homodimeric Crp, thereby causing a conformational change. The Crp-cAMP complex recognizes and binds to a symmetric consensus sequence, TGTGA-N_6-TCACA, which is located at various positions (between −40 and −200 bp) upstream of the transcriptional start sites of Crp-regulated genes. Based on the location of the Crp-binding site, two well-defined classes of Crp-dependent promoters are distinguished (11–13, 29): class I promoters carry the Crp-binding site in variable positions upstream of the −10 promoter box (in *lacP1*, a typical class I promoter, it is located at position −61.5). The interaction with the CTD of the alpha subunit (alpha-CTD) of RNAP occurs in this case via a single ARI region of the Crp dimer, which forms an exposed loop structure. Class II promoters, e.g., *galP1*, carry the Crp-binding

sequence at position −41.5, thus overlapping the −35 promoter box. In this case, both subunits of Crp are in contact with RNAP. The transcriptional activation of class II promoters is more complex than that of class I promoters and requires the interaction of alpha-CTD with AR1 of the promoter-distal subunit and of the NTD of the alpha subunit (alpha-NTD) of RNAP with AR2 of the promoter-proximal Crp subunit. It is this interaction that catalyzes the transition of the closed RNAP-promoter complex into the transcription-competent open complex. There is an additional interaction of the sigma subunit of RNAP with AR3, at least under in vitro conditions. Another class of promoters, typified by the *malK* promoter, needs an additional protein for transcriptional activation (59). In this case, the Crp-binding sequence is further upstream than it is in class I or class II promoters, and its position varies.

SIMILARITIES BETWEEN PrfA AND Crp

Although a direct comparison of the amino acid sequences of Crp (209 amino acids) and PrfA (237 amino acids) shows only 20% identity, there are significant sequential and structural similarities between these two regulatory proteins, including some of the functional regions of Crp described above. The recently described three-dimensional structure of PrfA (115), depicted in Fig. 2B, indeed shows high structural similarity with Crp. The NTD of PrfA also exhibits the short β-sheets interrupted by glycine residues (β-roll structure). The amino acid residues essential for the binding of cAMP to Crp (39) and for dimerization are less well conserved in PrfA, but interestingly the exchange of G145 into serine leads to a mutant PrfA that causes a derepressed phenotype of most PrfA-regulated virulence genes (98) (see below). There are comparable mutations in helix D of Crp (G145S and A144T) that lead to a cAMP-independent active conformation of Crp (39). These data support the view that PrfA may also be converted to a transcriptional active form by binding of a cofactor.

There are also sequential similarities in PrfA to the activation region AR1 of Crp (107). The CTD of PrfA contains a helix-turn-helix (HTH) motif (amino acids 171 to 191) with substantial sequence homology to that of Crp. Moreover, Sheehan et al. (105) have provided direct evidence by site-specific mutagenesis that S183 and S184 are essential for binding of PrfA to DNA.

Unlike Crp, PrfA contains an extended C terminus with a putative leucine zipper motif (65). This structure is absent from all other members of the Crp/Fnr family, and its function is not yet known.

ALTERATIONS IN DEFINED POSITIONS OF PrfA LEAD TO FUNCTIONAL CHANGES

As previously noted (99), *L. monocytogenes* isolates differ in the expression of virulence genes when grown in standard media like BHI.

Several mutations leading to PrfA proteins (termed PrfA*) that are constitutively active under all conditions inhibiting activity of wild-type PrfA have been described (98, 109, 118, 119, 127). Some point mutations, like that leading to a G145S substitution (98), are located in a region of PrfA corresponding to helix D of Crp. Corresponding mutations in Crp affecting helix D lead to a conformation that exhibits a cAMP-independent transcriptional

A

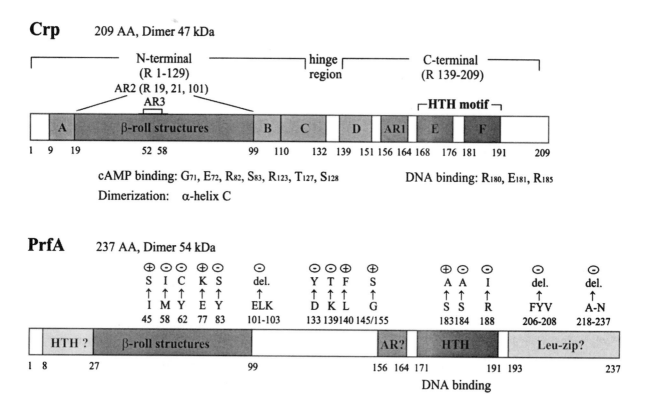

B

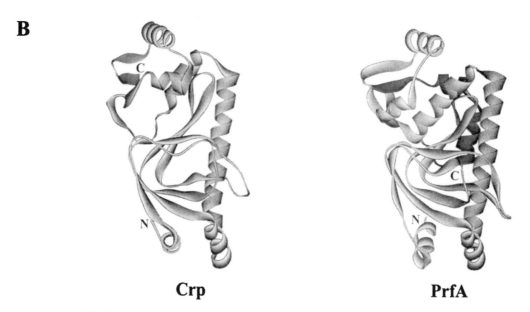

FIGURE 2 (A) Schematic comparison of Crp from *E. coli* (121) and PrfA from *L. monocytogenes*. The functionally important features of Crp (e.g., the N-terminal β-roll structures, the C-terminal HTH motif, and the activating regions AR1, AR2, and AR3) are indicated, as well as amino acids known to be involved in cAMP and DNA binding. Capital letters (A to D) denote alpha-helical regions. For PrfA, the predicted N-terminal HTH motif and the putative leucine zipper (Leu-zip) structure are marked. Amino acid replacements in PrfA that cause a decreased (−) or increased (+) level of activity are shown above. (B) Three-dimensional structures of Crp (124) and PrfA (115).

activation. Other mutations resulting in a PrfA* phenotype are located at different positions but may influence the structure of this domain anyway.

By using an assay that selects for mutations in PrfA that decrease its transcriptional activation, several point mutations were obtained (M58I, Y62C, and Y83S) in the β-roll structure (amino acids 27 to 99), in the hinge region connecting the NTD to the CTDCTD (D133Y and K139T), and in the HTH DNA-binding domain of PrfA (R188I) and several short in-frame deletions (49) (Fig. 2A). A Leu-to-Pro mutation at position 147 results in a form of the protein that is rapidly degraded (127).

An S184A exchange in the HTH DNA-binding domain also leads to decreased binding of PrfA to its target sequence and reduced expression of virulence genes, while the S183A exchange increases binding of PrfA to this site and causes enhanced expression of virulence genes (105). Recently, it has been shown that an L140F mutation in PrfA also leads to a constitutively active PrfA protein (119, 127).

cAMP is not an effector for PrfA, and amino acids essential for cAMP binding in Crp that are still conserved in PrfA may simply represent relics of a common ancestral precursor of the two regulatory proteins. As discussed below, there is, however, some evidence for the involvement of a low-molecular-weight effector(s) for PrfA that is distinct from cAMP. One may speculate that the domain of PrfA with similarity to the cAMP pocket of Crp may represent its binding site.

A deletion removing most of the PrfA-specific leucine zipperlike C terminus of PrfA (which is entirely absent in Crp and the other members of this family) also abolishes the activity of PrfA (65).

REGULATION OF PrfA SYNTHESIS

The transcription of the *prfA* gene and hence the cellular level of PrfA are subject to a complex regulation circuit. Two promoters, *prfAP1* and *prfAP2*, are positioned 113 and 30 bp, respectively, in front of the *prfA* start codon (35, 36). Transcription from these promoters leads to monocistronic *prfA* transcripts of 0.9 and 0.8 kb, respectively. While transcription starting at *prfAP1* seems to be carried out by RNAP loaded with the vegetative sigma factor (SigA), in vivo and in vitro studies show (84, 93) that the *prfAP2* promoter is transcribed by RNAP loaded with sigma factor B (SigB), suggesting a direct link between PrfA and this stress sigma factor. Indeed, a majority of genes identified in the above-mentioned transcriptome analysis (80) as being activated by PrfA are apparently transcribed from SigB-recognized promoters.

The involvement of SigB in virulence gene regulation becomes further obvious by the recent discovery that some virulence genes of *L. monocytogenes* and possibly also of *L. ivanovii*, like *inlA* and *bsh*, are transcribed by SigB-dependent promoters, in addition to the PrfA-dependent ones (56).

Recently it was shown that the monocistronic *prfA* mRNA transcribed from *prfAP1* acts as a thermosensor (52). At temperatures below 30°C, this RNA forms a secondary structure that inhibits translation of the P1 transcript, due to the inaccessibility of the ribosome-binding site. This inhibitory structure is resolved at higher temperatures.

In addition to the monocistronic transcription, *prfA* is transcribed as bicistronic mRNA together with *plcA* from a promoter located in front of *plcA*, which is activated by PrfA (76). This positive autoregulation of *prfA* transcription seems to be essential for the proper expression of the virulence genes, since the interruption of the *plcA-prfA*

operon by transposon insertions into the intergenic region leads to a significant drop in PrfA-dependent gene expression (15). A similar decrease in virulence gene expression is also observed by the deletion of the two promoters in front of *prfA*. This has led to a model for the regulation of *prfA* expression that assures that *prfA* is first transcribed from one (or both) of its own promoters. This monocistronic transcript generates enough PrfA protein to activate the PrfA-dependent promoter in front of *plcA* (35). The bicistronic *plcA-prfA* transcript leads to an increased level of cellular PrfA. It has been claimed that excess PrfA may inhibit transcription of the monocistronic *prfA* mRNA, possibly by binding of PrfA to a degenerated putative PrfA box of *prfAP2*. However, no such inhibition was seen in in vitro transcription studies (64, 93) and no effect on the expression of the PrfA-regulated virulence genes was observed in vivo when this putative PrfA binding site was removed (35). Surprisingly, it has recently been shown that the negative autoregulation of PrfA is not required for the virulence of *L. monocytogenes* (45).

Interestingly, in *L. seeligeri*, the *plcA* and *prfA* genes are interrupted by a large insertion (103); consequently, *prfA* is very weakly expressed (55).

THE PrfA-BINDING SITE (PrfA BOX) AND ITS INTERACTION WITH PrfA

All known PrfA-regulated promoters of *L. monocytogenes*, *L. ivanovii*, and *L. seeligeri* possess a conserved symmetric sequence (PrfA box) at around position −40 from the transcriptional start site. This arrangement of the PrfA box relative to the −10 promoter box is similar to that of the Crp-dependent class II promoters (see above). The consensus sequence of the 14-bp PrfA box is TTAACA-NN-TGT-TAA. This sequence is perfectly maintained in the common PrfA box of the *plcA* and *hly* promoters. The corresponding PrfA boxes of the promoters for *mpl*, *actA*, and *inlC* exhibit a 1-bp change each, while the PrfA boxes of the *inlA-inlB* promoter and of the putative PrfA-dependent *bsh* promoter are even more degenerated with two and three mismatches, respectively (5, 27, 28, 70). In *L. ivanovii*, the common PrfA box of the *plcA/ilo* and that of the *actA* and *i-inlE* promoter also show perfect symmetry, while all other identified PrfA boxes of *L. ivanovii* again show at least a 1-bp change (Fig. 3).

The symmetry of the PrfA-binding site suggests that this regulatory protein, like other members of the Crp/Fnr family, may bind as a homodimer (24).

That the PrfA box is indeed the recognition sequence for PrfA was first suggested by the effect of mutations in this sequence (37). Direct biochemical evidence came from gel retardation assays and DNaseI protection experiments (3, 4, 24). The addition of purified PrfA and RNAP to a PrfA box-containing promoter fragment leads to the formation of three complexes, which can be resolved by agarose gel electrophoresis. The fastest-migrating complex, CIII, consists of PrfA bound to its target sequence; the slowest-migrating complex, CI, contains both PrfA and RNAP. The intermediately migrating complex, CII, represents RNAP bound to the promoter region (3, 4). Formation of CII and CI requires sequences upstream and downstream of the PrfA box, including the −10 region, but even formation of CIII complex is more efficient in the presence of the −10 box (72). These bandshift data are in line with the results from the DNaseI protection assays. Purified PrfA alone protects the PrfA box and 8 to 10 nucleotides upstream of it, as has been

L. monocytogenes EGD

	PrfA-box		-10
hly	TTAACA TT TGTTAA	- 23bp -	TAGAAT
plcA	TTAACA AA TGTTAA	- 22bp -	TAAGAT
mpl	TTAACA AA TGTaAA	- 22bp -	TATAAT
actA	TTAACA AA TGTTAg	- 23bp -	TATTCT
inlC	TTAACg CT TGTTAA	- 22bp -	TAACAT
inlA	aTAACA TA aGTTAA	- 22bp -	TATTAT
hpt (uhpT)	aTAACA AG TGTTAA	- 23bp -	TATATT
bsh (lmo2067)	TTAAaA AT TtTTAA	- 23bp -	GGTAAT

L. ivanovii ATCC19119

ilo	TCTTTAACA TT TGTTAAAGA	- 22bp -	TACAAT
plcA	TCTTTAACA AA TGTTAAAGA	- 22bp -	TAAGAT
mpl	TTTAACA AA TGTcAAA	- 22bp -	TATAAT
actA	TTAACA AA TGTTAA	- 23bp -	TATTCT
i-inlC	TTAACg CT TGTTAA	- 22bp -	TAACAT
i-inlD	TTAACt TT TGTTAt	- 22bp -	TAAAAT
i-inlE	TTAACA TT TGTTAA	- 22bp -	TATGAT
i-inlF	TTAACt TT TGTTAt	- 22bp -	TAGAAT

FIGURE 3 Sequence comparison of PrfA-regulated virulence gene promoters from *L. monocytogenes* and *L. ivanovii*. Sequences are from published data or from our own unpublished results.

shown using *hly* and *inlA* promoter probes. Upon addition of partially purified RNAP from *L. monocytogenes*, the region protected from DNaseI extends down to the −10 promoter region; the protected region upstream of the PrfA box is also enlarged (3). This observation that the PrfA/RNAP complex contacts DNA regions both upstream and downstream of the PrfA-binding site resembles the situation known for class II Crp-dependent promoters of *E. coli* (12).

Purified mutant PrfA* (described above) exhibited a strongly increased affinity to the PrfA box compared to wild-type PrfA (118); however, CI complex formation by PrfA* in the presence of RNAP is not much stronger than with PrfA. Transcription initiation at different PrfA-dependent promoters measured in an in vitro transcription system shows only a slight increase with PrfA*, compared to PrfA (83). From these results, it is concluded that the strong PrfA* binding to the PrfA box does not proportionally enhance the binding of RNAP and the formation of the open complex necessary for transcription initiation.

As expected, the efficiency of PrfA binding depends on the quality of the PrfA box; e.g., binding is most efficient to the PrfA box of the *hly* promoter and weakest to that of the *inlA* promoter (24). Furthermore, it has been shown (48, 118) that the permanently activated PrfA* protein (with

the G145S exchange) binds more efficiently to the PrfA box than does the corresponding wild-type PrfA protein.

Electrophoretic mobility shift experiments with PrfA-containing cell-free extracts of *L. monocytogenes* and the short PrfA box target sequences do not yield CIII complexes, but specific slow-migrating complexes (CI and CII) are obtained with longer DNA probes containing the entire promoter sequence, including the PrfA box. Binding of purified PrfA protein to the longer target DNA results in the formation of a faster-migrating complex (CIII), which is converted into CI by the addition of *L. monocytogenes* RNAP. Binding affinity is significantly increased to DNA fragments containing the entire promoter sequence including the −10 box (48, 72).

INTRACELLULAR EXPRESSION OF PrfA-REGULATED GENES

The results described above suggest that regulation of virulence genes mediated by PrfA involves environmental parameters, as well as additional bacterial factors. Recent in vivo studies on the regulation of listerial virulence genes show that PrfA is activated when the listeriae are inside host cells; transcription of PrfA-dependent listerial virulence genes required for the various intracellular steps, like *hly*, *actA*, and *inlC*, is highly activated inside mammalian host cells (10, 34, 95, 108). Undoubtedly, the proper regulation of PrfA and the PrfA-dependent virulence genes plays a decisive role in *L. monocytogenes* infection, since a *prfA*-deficient mutant of *L. monocytogenes* is entirely avirulent in the mouse model (76); *prfA** mutants which produce permanently activated PrfA are impaired in in vitro (75) and intracellular growth (42). However, in the mouse model some of these mutants are more virulent than the wild type (109). So clearly, this question needs further investigation. Recent comparative transcriptome analyses show upregulation of most of the known PrfA-regulated virulence genes (with the exception of *inlAB*) inside host cells, in addition to several other metabolic genes whose expression may be indirectly affected by PrfA (53, 75). The products of these genes are likely to be needed for intracellular survival and/or growth of *L. monocytogenes*.

OTHER TRANSCRIPTION FACTORS INVOLVED IN THE REGULATION OF VIRULENCE

The genome sequences of *L. monocytogenes* have revealed a large number of genes encoding transcription factors (41, 86), some of which appear to be specific for this pathogenic *Listeria* species, as they are absent in the genome sequence of the apathogenic but closely related species *L. innocua* (Table 1). As already discussed above, the involvement of the alternative sigma factor, SigB, in the transcription control of several PrfA-dependent virulence genes has been clearly demonstrated in vivo (56, 84, 123) and in vitro (93). The transcriptional repressor MogR, required for *Listeria* motility, also seems to be involved in virulence (46).

REGULATION OF VIRULENCE GENE EXPRESSION ON THE POSTTRANSCRIPTIONAL LEVEL

Expression of virulence genes seems to be regulated not only on the transcriptional level but also posttranscriptionally. As previously shown, synthesis of p60 protein appears to be controlled on the translational level (58) by an as-yet-unknown mechanism. As mentioned above, translation of mono-cistronic *prfA* mRNA transcribed from the P1 promoter is only activated at temperatures above 30°C (52). The translation of the *actA* mRNA is strongly influenced by the 5′ upstream untranslated region (UTR), which can form a stable secondary structure which may inhibit accessibility of the ribosome to the Shine-Dalgarno sequence, thus blocking initiation of *actA* mRNA translation (126). A similar mechanism appears to be responsible for the translational control of internalin A and B synthesis (InlA and InlB). The 5′ UTR of the bicistronic *inlAB* transcript can also form a secondary structure(s) affecting its translation efficiency (114), and there is some evidence that anoxic conditions may act as a signal for this translation control of the *inlAB* mRNA. It is likely that the transcripts of other PrfA-regulated virulence genes, like *inlC* (30, 69) or *hpt* (18), are also subject to posttranscriptional control, as these transcripts carry also extended 5′ UTRs with the potential of forming stable secondary structures (unpublished observations). Whether small RNAs, shown to be of importance for posttranscriptional regulation of a variety of prokaryotic genes (68, 112, 120), are also involved in the translational control of these and other transcripts of virulence genes remains speculative at the moment (but see reference 19).

CONCLUSION AND OUTLOOK

Extensive studies of the molecular aspects of *L. monocytogenes* virulence have led to the identification of a variety of virulence genes involved in the various steps of the intracellular infection cycle. The differential regulation of most of these virulence genes is intimately connected with the pleiotropic regulatory protein PrfA. This transcriptional activator of the pathogenic *Listeria* species *L. monocytogenes* and *L. ivanovii* shares common properties with other members of the Crp/Fnr family to which it belongs, but it also possesses unique features. Like most regulators of this family, the PrfA protein most probably binds as a homodimer to its symmetric target sequence and may enhance binding of RNAP to PrfA-dependent promoters. There is, however, growing evidence that PrfA is embedded in other regulatory circuits of *L. monocytogenes* and *L. ivanovii* (e.g., the SigB network), which may modulate the activity as well as the amount of PrfA. Much has to be learned about the external physicochemical and metabolic parameters, as well as the additional listerial factors influencing PrfA-dependent gene regulation under extracellular and (particularly) intracellular conditions. The present data suggest that these additional parameters may affect gene regulation by PrfA in two ways: (i) by changing the conformation of PrfA, which in turn may influence binding of PrfA to the different target sites located in the corresponding promoters with positive (and possibly also negative) effects on the initiation of transcription of PrfA-regulated genes, and (ii) by inducing gene expression of *prfA*, thus increasing the amount of cellular PrfA, which again may have positive or negative effects on the transcription of PrfA-regulated genes. Finally, the participation of additional listerial (protein) factors that may interact with PrfA directly to modulate its binding to the target sites and/or to RNAP, similar to the situation of Crp-dependent promoters that require additional factors, should also be considered. The determination of the three-dimensional structure of PrfA, the complete identification of the PrfA regulon by transcriptome analyses, and the forthcoming genome sequences of additional *Listeria* species will have a strong impact on the further study of virulence gene regulation in *Listeria*. Undoubtedly, the precise knowl-

edge of the mechanisms involved in the regulation of the listerial virulence genes will be crucial for the understanding of the pathogenesis of infections by pathogenic *Listeria* spp.

REFERENCES

1. **Behari, J., and P. Youngman.** 1998. Regulation of *hly* expression in *Listeria monocytogenes* by carbon sources and pH occurs through separate mechanisms mediated by PrfA. *Infect. Immun.* **66:**3635–3642.
2. **Bergmann, B., D. Raffelsbauer, M. Kuhn, M. Götz, S. Hom, and W. Goebel.** 2002. InlA- but not InlB-mediated internalization of *Listeria monocytogenes* by nonphagocytic mammalian cells needs the support of other internalins. *Mol. Microbiol.* **43:**557–570.
3. **Böckmann, R., C. Dickneite, W. Goebel, and J. Bohne.** 2000. PrfA mediates specific binding of RNA polymerase of *Listeria monocytogenes* to PrfA-dependent virulence gene promoters resulting in a transcriptionally active complex. *Mol. Microbiol.* **36:**487–497.
4. **Böckmann, R., C. Dickneite, B. Middendorf, W. Goebel, and Z. Sokolovic.** 1996. Specific binding of the *Listeria monocytogenes* transcriptional regulator PrfA to target sequences requires additional factor(s) and is influenced by iron. *Mol. Microbiol.* **22:**643–653.
5. **Bohne, J., H. Kestler, C. Uebele, Z. Sokolovic, and W. Goebel.** 1996. Differential regulation of the virulence genes of *Listeria monocytogenes* by the transcriptional activator PrfA. *Mol. Microbiol.* **20:**1189–1198.
6. **Bohne, J., Z. Sokolovic, and W. Goebel.** 1994. Transcriptional regulation of *prfA* and PrfA-regulated virulence genes in *Listeria monocytogenes*. *Mol. Microbiol.* **11:**1141–1150.
7. **Brehm, K., A. Haas, W. Goebel, and J. Kreft.** 1992. A gene encoding a superoxide dismutase of the facultative intracellular bacterium *Listeria monocytogenes*. *Gene* **118:**121–125.
8. **Brehm, K., J. Kreft, M. T. Ripio, and J. A. Vázquez-Boland.** 1996. Regulation of virulence gene expression in pathogenic *Listeria*. *Microbiologia* **12:**219–236.
9. **Brehm, K., M. T. Ripio, J. Kreft, and J. A. Vázquez-Boland.** 1999. The *bvr* locus of *Listeria monocytogenes* mediates virulence gene repression by beta-glucosides. *J. Bacteriol.* **181:**5024–5032.
10. **Bubert, A., Z. Sokolovic, S. K. Chun, L. Papatheodorou, A. Simm, and W. Goebel.** 1999. Differential expression of *Listeria monocytogenes* virulence genes in mammalian host cells. *Mol. Gen. Genet.* **261:**323–336.
11. **Busby, S., and R. H. Ebright.** 1994. Promoter structure, promoter recognition, and transcription activation in prokaryotes. *Cell* **79:**743–746.
12. **Busby, S., and R. H. Ebright.** 1997. Transcription activation at class II CAP-dependent promoters. *Mol. Microbiol.* **23:**853–859.
13. **Busby, S., and R. H. Ebright.** 1999. Transcription activation by catabolite activator protein (CAP). *J. Mol. Biol.* **293:**199–213.
14. **Cabanes, D., O. Dussurget, P. Dehoux, and P. Cossart.** 2004. Auto, a surface associated autolysin of *Listeria monocytogenes* required for entry into eukaryotic cells and virulence. *Mol. Microbiol.* **51:**1601–1614.
15. **Camilli, A., L. G. Tilney, and D. A. Portnoy.** 1993. Dual roles of *plcA* in *Listeria monocytogenes* pathogenesis. *Mol. Microbiol.* **8:**143–157.
16. **Chakraborty, T., M. Leimeister-Wächter, E. Domann, M. Hartl, W. Goebel, T. Nichterlein, and S. Notermans.** 1992. Coordinate regulation of virulence genes in *Listeria*

monocytogenes requires the product of the *prfA* gene. *J. Bacteriol.* **174**:568–574.

17. Cheng, L. W., and D. A. Portnoy. 2003. Drosophila S2 cells: an alternative infection model for *Listeria monocytogenes. Cell. Microbiol.* **5**:875–885.

18. Chico-Calero, I., M. Suárez, B. González-Zorn, M. Scortti, J. Slaghuis, W. Goebel, and J. A. Vázquez-Boland. 2002. Hpt, a bacterial homolog of the microsomal glucose-6-phosphate translocase, mediates rapid intracellular proliferation in *Listeria. Proc. Natl. Acad. Sci. USA* **99**:431–436.

19. Christiansen, J. K., M. H. Larsen, H. Ingmer, L. Sogaard-Andersen, and B. H. Kallipolitis. 2004. The RNA-binding protein Hfq of *Listeria monocytogenes*: role in stress tolerance and virulence. *J. Bacteriol.* **186**:3355–3362.

20. Conte, M. P., C. Longhi, G. Petrone, M. Polidoro, P. Valenti, and L. Seganti. 2000. Modulation of *actA* gene expression in *Listeria monocytogenes* by iron. *J. Med. Microbiol.* **49**:681–683.

21. Conte, M. P., C. Longhi, M. Polidoro, G. Petrone, V. Buonfiglio, S. Di Santo, E. Papi, L. Seganti, P. Visca, and P. Valenti. 1996. Iron availability affects entry of *Listeria monocytogenes* into the enterocytelike cell line Caco-2. *Infect. Immun.* **64**:3925–3929.

22. Cossart, P., J. Pizarro-Cerda, and M. Lecuit. 2003. Invasion of mammalian cells by Listeria monocytogenes: functional mimicry to subvert cellular functions. *Trends Cell Biol.* **13**:23–31.

23. Datta, A. R., and M. H. Kothary. 1993. Effects of glucose, growth temperature, and pH on listeriolysin O production in *Listeria monocytogenes. Appl. Environ. Microbiol.* **59**:3495–3497.

24. Dickneite, C., R. Böckmann, A. Spory, W. Goebel, and Z. Sokolovic. 1998. Differential interaction of the transcription factor PrfA and the PrfA-activating factor (Paf) of *Listeria monocytogenes* with target sequences. *Mol. Microbiol.* **27**:915–928.

25. Dramsi, S., F. Bourdichon, D. Cabanes, M. Lecuit, H. Fsihi, and P. Cossart. 2004. FbpA, a novel multifunctional *Listeria monocytogenes* virulence factor. *Mol. Microbiol.* **53**:639–649.

26. Dramsi, S., P. Dehoux, M. Lebrun, P. L. Goossens, and P. Cossart. 1997. Identification of four new members of the internalin multigene family of *Listeria monocytogenes* EGD. *Infect. Immun.* **65**:1615–1625.

27. Dramsi, S., C. Kocks, C. Forestier, and P. Cossart. 1993. Internalin-mediated invasion of epithelial cells by *Listeria monocytogenes* is regulated by the bacterial growth state, temperature and the pleiotropic activator *prfA. Mol. Microbiol.* **9**:931–941.

28. Dussurget, O., D. Cabanes, P. Dehoux, M. Lecuit, C. Buchrieser, P. Glaser, and P. Cossart. 2002. *Listeria monocytogenes* bile salt hydrolase is a PrfA-regulated virulence factor involved in the intestinal and hepatic phases of listeriosis. *Mol. Microbiol.* **45**:1095–1106.

29. Ebright, R. H. 1993. Transcription activation at class I CAP-dependent promoters. *Mol. Microbiol.* **8**:797–802.

30. Engelbrecht, F., S. K. Chun, C. Ochs, J. Hess, F. Lottspeich, W. Goebel, and Z. Sokolovic. 1996. A new PrfA-regulated gene of *Listeria monocytogenes* encoding a small, secreted protein which belongs to the family of internalins. *Mol. Microbiol.* **21**:823–837.

31. Engelbrecht, F., C. Dickneite, R. Lampidis, M. Götz, U. DasGupta, and W. Goebel. 1998. Sequence comparison of the chromosomal regions encompassing the inter-

nalin C genes (*inlC*) of *Listeria monocytogenes* and *L. ivanovii. Mol. Gen. Genet.* **257**:186–197.

32. Engelbrecht, F., G. Dominguez-Bernal, J. Hess, C. Dickneite, L. Greiffenberg, R. Lampidis, D. Raffelsbauer, J. J. Daniels, J. Kreft, S. H. Kaufmann, J. A. Vazquez-Boland, and W. Goebel. 1998. A novel PrfA-regulated chromosomal locus, which is specific for *Listeria ivanovii*, encodes two small, secreted internalins and contributes to virulence in mice. *Mol. Microbiol.* **30**:405–417.

33. Ermolaeva, S., S. Novella, Y. Vega, M. T. Ripio, M. Scortti, and J. A. Vázquez-Boland. 2004. Negative control of *Listeria monocytogenes* virulence genes by a diffusible autorepressor. *Mol. Microbiol.* **52**:601–611.

34. Freitag, N. E., and K. E. Jacobs. 1999. Examination of *Listeria monocytogenes* intracellular gene expression by using the green fluorescent protein of *Aequorea victoria. Infect. Immun.* **67**:1844–1852.

35. Freitag, N. E., and D. A. Portnoy. 1994. Dual promoters of the *Listeria monocytogenes prfA* transcriptional activator appear essential in vitro but are redundant in vivo. *Mol. Microbiol.* **12**:845–853.

36. Freitag, N. E., L. Rong, and D. A. Portnoy. 1993. Regulation of the *prfA* transcriptional activator of *Listeria monocytogenes*: multiple promoter elements contribute to intracellular growth and cell-to-cell spread. *Infect. Immun.* **61**:2537–2544.

37. Freitag, N. E., P. Youngman, and D. A. Portnoy. 1992. Transcriptional activation of the *Listeria monocytogenes* hemolysin gene in *Bacillus subtilis. J. Bacteriol.* **174**:1293–1298.

38. Gaillard, J. L., P. Berche, C. Frehel, E. Gouin, and P. Cossart. 1991. Entry of *L. monocytogenes* into cells is mediated by internalin, a repeat protein reminiscent of surface antigens from gram-positive cocci. *Cell* **65**:1127–1141.

39. Garges, S., and S. Adhya. 1988. Cyclic AMP-induced conformational change of cyclic AMP receptor protein (CRP): intragenic suppressors of cyclic AMP-independent CRP mutations. *J. Bacteriol.* **170**:1417–1422.

40. Gellin, B. G., and C. V. Broome. 1989. Listeriosis. *JAMA* **261**:1313–1320.

41. Glaser, P., L. Frangeul, C. Buchrieser, C. Rusniok, A. Amend, F. Baquero, P. Berche, H. Bloecker, P. Brandt, T. Chakraborty, A. Charbit, F. Chetouani, E. Couvé, A. de Daruvar, P. Dehoux, E. Domann, G. Domínguez-Bernal, E. Duchaud, L. Durant, O. Dussurget, K. D. Entian, H. Fsihi, F. Garcia-del Portillo, P. Garrido, L. Gautier, W. Goebel, N. Gómez-López, T. Hain, J. Hauf, D. Jackson, L. M. Jones, U. Kaerst, J. Kreft, M. Kuhn, F. Kunst, G. Kurapkat, E. Madueno, A. Maitournam, J. M. Vicente, E. Ng, H. Nedjari, G. Nordsiek, S. Novella, B. de Pablos, J. C. Pérez-Diaz, R. Purcell, B. Remmel, M. Rose, T. Schlueter, N. Simoes, A. Tierrez, J. A. Vázquez-Boland, H. Voss, J. Wehland, and P. Cossart. 2001. Comparative genomics of *Listeria* species. *Science* **294**:849–852.

42. Götz, M. Unpublished data.

43. González-Zorn, B., G. Domínguez-Bernal, M. Suárez, M. T. Ripio, Y. Vega, S. Novella, A. Rodríguez, I. Chico, A. Tierrez, and J. A. Vázquez-Boland. 2000. SmcL, a novel membrane-damaging virulence factor in *Listeria. Int. J. Med. Microbiol.* **290**:369–374.

44. Gouin, E., J. Mengaud, and P. Cossart. 1994. The virulence gene cluster of *Listeria monocytogenes* is also present in *Listeria ivanovii*, an animal pathogen, and *Listeria seeligeri*, a nonpathogenic species. *Infect. Immun.* **62**:3550–3553.

45. **Greene, S. L., and N. E. Freitag.** 2003. Negative regulation of PrfA, the key activator of *Listeria monocytogenes* virulence gene expression, is dispensable for bacterial pathogenesis. *Microbiology* **149:**111–120.

46. **Gründling, A., L. S. Burrack, H. G. Bouwer, and D. E. Higgins.** 2004. *Listeria monocytogenes* regulates flagellar motility gene expression through MogR, a transcriptional repressor required for virulence. *Proc. Natl. Acad. Sci. USA* **101:**12318–12323.

47. **Haas, A., K. Brehm, J. Kreft, and W. Goebel.** 1991. Cloning, characterization, and expression in *Escherichia coli* of a gene encoding *Listeria seeligeri* catalase, a bacterial enzyme highly homologous to mammalian catalases. *J. Bacteriol.* **173:**5159–5167.

48. **Herler, M.** Unpublished data.

49. **Herler, M., A. Bubert, M. Goetz, Y. Vega, J. A. Vázquez-Boland, and W. Goebel.** 2001. Positive selection of mutations leading to loss or reduction of transcriptional activity of PrfA, the central regulator of *Listeria monocytogenes* virulence. *J. Bacteriol.* **183:**5562–5570.

50. **Irvine, A. S., and J. R. Guest.** 1993. *Lactobacillus casei* contains a member of the CRP-FNR family. *Nucleic Acids Res* **21:**753.

51. **Jaradat, Z. W., J. W. Wampler, and A. W. Bhunia.** 2003. A *Listeria* adhesion protein-deficient *Listeria monocytogenes* strain shows reduced adhesion primarily to intestinal cell lines. *Med. Microbiol. Immunol.* (Berlin) **192:**85–91.

52. **Johansson, J., P. Mandin, A. Renzoni, C. Chiaruttini, M. Springer, and P. Cossart.** 2002. An RNA thermosensor controls expression of virulence genes in *Listeria monocytogenes*. *Cell* **110:**551–561.

53. **Joseph, B.** Unpublished data.

54. **Junttila, J. R., S. I. Niemela, and J. Hirn.** 1988. Minimum growth temperatures of *Listeria monocytogenes* and non-haemolytic *Listeria*. *J. Appl. Bacteriol.* **65:**321–327.

55. **Karunasagar, I., R. Lampidis, W. Goebel, and J. Kreft.** 1997. Complementation of *Listeria seeligeri* with the *plcA-prfA* genes from *L. monocytogenes* activates transcription of seeligerolysin and leads to bacterial escape from the phagosome of infected mammalian cells. *FEMS Microbiol. Lett.* **146:**303–310.

56. **Kazmierczak, M. J., S. C. Mithoe, K. J. Boor, and M. Wiedmann.** 2003. *Listeria monocytogenes* σ^B regulates stress response and virulence functions. *J. Bacteriol.* **185:**5722–5734.

57. **Klarsfeld, A. D., P. L. Goossens, and P. Cossart.** 1994. Five *Listeria monocytogenes* genes preferentially expressed in infected mammalian cells: *plcA*, *purH*, *purD*, *pyrE* and an arginine ABC transporter gene, *arpJ*. *Mol. Microbiol.* **13:**585–597.

58. **Köhler, S., A. Bubert, M. Vogel, and W. Goebel.** 1991. Expression of the *iap* gene coding for protein p60 of *Listeria monocytogenes* is controlled on the posttranscriptional level. *J. Bacteriol.* **173:**4668–4674.

59. **Kolb, A., S. Busby, H. Buc, S. Garges, and S. Adhya.** 1993. Transcriptional regulation by cAMP and its receptor protein. *Annu. Rev. Biochem.* **62:**749–795.

60. **Körner, H., H. J. Sofia, and W. G. Zumft.** 2003. Phylogeny of the bacterial superfamily of Crp-Fnr transcription regulators: exploiting the metabolic spectrum by controlling alternative gene programs. *FEMS Microbiol. Rev.* **27:**559–592.

61. **Kreft, J., and J. A. Vázquez-Boland.** 2001. Regulation of virulence genes in *Listeria*. *Int. J. Med. Microbiol.* **291:**145–157.

62. **Kreft, J., J. A. Vázquez-Boland, S. Altrock, G. Domínguez-Bernal, and W. Goebel.** 2002. Pathogenicity islands and other virulence elements in *Listeria*. *Curr. Top. Microbiol. Immunol.* **264:**109–125.

63. **Kuhn, M., and W. Goebel.** 1989. Identification of an extracellular protein of *Listeria monocytogenes* possibly involved in intracellular uptake by mammalian cells. *Infect. Immun.* **57:**55–61.

64. **Lalic-Mülthaler, M., J. Bohne, and W. Goebel.** 2001. In vitro transcription of PrfA-dependent and -independent genes of *Listeria monocytogenes*. *Mol. Microbiol.* **42:**111–120.

65. **Lampidis, R., R. Gross, Z. Sokolovic, W. Goebel, and J. Kreft.** 1994. The virulence regulator protein of *Listeria ivanovii* is highly homologous to PrfA from *Listeria monocytogenes* and both belong to the Crp-Fnr family of transcription regulators. *Mol. Microbiol.* **13:**141–151.

66. **Leimeister-Wächter, M., E. Domann, and T. Chakraborty.** 1992. The expression of virulence genes in *Listeria monocytogenes* is thermoregulated. *J. Bacteriol.* **174:**947–952.

67. **Leimeister-Wächter, M., C. Haffner, E. Domann, W. Goebel, and T. Chakraborty.** 1990. Identification of a gene that positively regulates expression of listeriolysin, the major virulence factor of *Listeria monocytogenes*. *Proc. Natl. Acad. Sci. USA* **87:**8336–8340.

68. **Lenz, D. H., K. C. Mok, B. N. Lilley, R. V. Kulkarni, N. S. Wingreen, and B. L. Bassler.** 2004. The small RNA chaperone Hfq and multiple small RNAs control quorum sensing in *Vibrio harveyi* and *Vibrio cholerae*. *Cell* **118:**69–82.

69. **Lingnau, A., T. Chakraborty, K. Niebuhr, E. Domann, and J. Wehland.** 1996. Identification and purification of novel internalin-related proteins in *Listeria monocytogenes* and *Listeria ivanovii*. *Infect. Immun.* **64:**1002–1006.

70. **Lingnau, A., E. Domann, M. Hudel, M. Bock, T. Nichterlein, J. Wehland, and T. Chakraborty.** 1995. Expression of the *Listeria monocytogenes* EGD *inlA* and *inlB* genes, whose products mediate bacterial entry into tissue culture cell lines, by PrfA-dependent and -independent mechanisms. *Infect. Immun.* **63:**3896–3903.

71. **Lorber, B.** 1997. Listeriosis. *Clin. Infect. Dis.* **24:**1–11.

72. **Luo, Q., M. Herler, S. Müller-Altrock, and W. Goebel.** 2005. Supportive and inhibitory elements of a putative PrfA-dependent promoter in *Listeria monocytogenes*. *Mol. Microbiol.* **55:**986–997.

73. **Luo, Q., M. Rauch, A. K. Marr, S. Müller-Altrock, and W. Goebel.** 2004. In vitro transcription of the *Listeria monocytogenes* virulence genes *inlC* and *mpl* reveals overlapping PrfA-dependent and -independent promoters that are differentially activated by GTP. *Mol. Microbiol.* **52:**39–52.

74. **Mansfield, B. E., M. S. Dionne, D. S. Schneider, and N. E. Freitag.** 2003. Exploration of host-pathogen interactions using *Listeria monocytogenes* and *Drosophila melanogaster*. *Cell. Microbiol.* **5:**901–911.

75. **Marr, A. K., B. Joseph, and W. Goebel.** Evidence for the possible interference of PrfA, the central regulator of *L. monocytogenes* virulence, with components of the catabolite repression system. Submitted for publication.

76. **Mengaud, J., S. Dramsi, E. Gouin, J. A. Vázquez-Boland, G. Milon, and P. Cossart.** 1991. Pleiotropic control of *Listeria monocytogenes* virulence factors by a gene that is autoregulated. *Mol. Microbiol.* **5:**2273–2283.

77. **Merz, A. J., and H. N. Higgs.** 2003. *Listeria* motility: biophysics pushes things forward. *Curr. Biol.* **13:**R302–R304.

78. **Milenbachs, A. A., D. P. Brown, M. Moors, and P. Youngman.** 1997. Carbon-source regulation of virulence gene expression in *Listeria monocytogenes*. *Mol. Microbiol.* **23:**1075–1085.

79. **Milenbachs Lukowiak, A., K. J. Mueller, N. E. Freitag, and P. Youngman.** 2004. Deregulation of *Listeria monocytogenes* virulence gene expression by two distinct and semi-independent pathways. *Microbiology* **150:**321–333.

80. **Milohanic, E., P. Glaser, J. Y. Coppée, L. Frangeul, Y. Vega, J. A. Vázquez-Boland, F. Kunst, P. Cossart, and C. Buchrieser.** 2003. Transcriptome analysis of *Listeria monocytogenes* identifies three groups of genes differently regulated by PrfA. *Mol. Microbiol.* **47:**1613–1625.

81. **Milohanic, E., R. Jonquieres, P. Cossart, P. Berche, and J. L. Gaillard.** 2001. The autolysin Ami contributes to the adhesion of *Listeria monocytogenes* to eukaryotic cells via its cell wall anchor. *Mol. Microbiol.* **39:**1212–1224.

82. **Moors, M. A., B. Levitt, P. Youngman, and D. A. Portnoy.** 1999. Expression of listeriolysin O and ActA by intracellular and extracellular *Listeria monocytogenes*. *Infect. Immun.* **67:**131–139.

83. **Müller-Altrock, S., and N. Mauder.** Unpublished data.

84. **Nadon, C. A., B. M. Bowen, M. Wiedmann, and K. J. Boor.** 2002. Sigma B contributes to PrfA-mediated virulence in *Listeria monocytogenes*. *Infect. Immun.* **70:**3948–3952.

85. **Nair, S., E. Milohanic, and P. Berche.** 2000. ClpC ATPase is required for cell adhesion and invasion of *Listeria monocytogenes*. *Infect. Immun.* **68:**7061–7068.

86. **Nelson, K. E., D. E. Fouts, E. F. Mongodin, J. Ravel, R. T. DeBoy, J. F. Kolonay, D. A. Rasko, S. V. Angiuoli, S. R. Gill, I. T. Paulsen, J. Peterson, O. White, W. C. Nelson, W. Nierman, M. J. Beanan, L. M. Brinkac, S. C. Daugherty, R. J. Dodson, A. S. Durkin, R. Madupu, D. H. Haft, J. Selengut, S. Van Aken, H. Khouri, N. Fedorova, H. Forberger, B. Tran, S. Kathariou, L. D. Wonderling, G. A. Uhlich, D. O. Bayles, J. B. Luchansky, and C. M. Fraser.** 2004. Whole genome comparisons of serotype 4b and 1/2a strains of the food-borne pathogen *Listeria monocytogenes* reveal new insights into the core genome components of this species. *Nucleic Acids Res.* **32:**2386–2395.

87. **Niu, W., Y. Kim, G. Tau, T. Heyduk, and R. H. Ebright.** 1996. Transcription activation at class II CAP-dependent promoters: two interactions between CAP and RNA polymerase. *Cell* **87:**1123–1134.

88. **Park, S. F.** 1994. The repression of listeriolysin O expression in *Listeria monocytogenes* by the phenolic beta-D-glucoside, arbutin. *Lett. Appl. Microbiol.* **19:**258–260.

89. **Park, S. F., and R. G. Kroll.** 1993. Expression of listeriolysin and phosphatidylinositol-specific phospholipase C is repressed by the plant-derived molecule cellobiose in *Listeria monocytogenes*. *Mol. Microbiol.* **8:**653–661.

90. **Pilgrim, S., A. Kolb-Mäurer, I. Gentschev, W. Goebel, and M. Kuhn.** 2003. Deletion of the gene encoding p60 in *Listeria monocytogenes* leads to abnormal cell division and loss of actin-based motility. *Infect. Immun.* **71:**3473–3484.

91. **Portnoy, D. A., T. Chakraborty, W. Goebel, and P. Cossart.** 1992. Molecular determinants of *Listeria monocytogenes* pathogenesis. *Infect. Immun.* **60:**1263–1267.

92. **Raffelsbauer, D., A. Bubert, F. Engelbrecht, J. Scheinpflug, A. Simm, J. Hess, S. H. Kaufmann, and W. Goebel.** 1998. The gene cluster *inlC2DE* of *Listeria monocytogenes* contains additional new internalin genes and is im-

portant for virulence in mice. *Mol. Gen. Genet.* **260:**144–158.

93. **Rauch, M., Q. Luo, S. Müller-Altrock, and W. Goebel.** 2005. SigB-dependent in vitro transcription of *prfA* and some newly identified genes of *Listeria monocytogenes* whose expression is affected by PrfA in vivo. *J. Bacteriol.* **187:**800–804.

94. **Reid, S. D., A. G. Montgomery, and J. M. Musser.** 2004. Identification of *srv*, a PrfA-like regulator of group A *Streptococcus* that influences virulence. *Infect. Immun.* **72:** 1799–1803.

95. **Renzoni, A., P. Cossart, and S. Dramsi.** 1999. PrfA, the transcriptional activator of virulence genes, is upregulated during interaction of *Listeria monocytogenes* with mammalian cells and in eukaryotic cell extracts. *Mol. Microbiol.* **34:**552–561.

96. **Renzoni, A., A. Klarsfeld, S. Dramsi, and P. Cossart.** 1997. Evidence that PrfA, the pleiotropic activator of virulence genes in *Listeria monocytogenes*, can be present but inactive. *Infect. Immun.* **65:**1515–1518.

97. **Ripio, M. T., K. Brehm, M. Lara, M. Suárez, and J. A. Vázquez-Boland.** 1997. Glucose-1-phosphate utilization by *Listeria monocytogenes* is PrfA dependent and coordinately expressed with virulence factors. *J. Bacteriol.* **179:**7174–7180.

98. **Ripio, M. T., G. Domínguez-Bernal, M. Lara, M. Suárez, and J. A. Vázquez-Boland.** 1997. A Gly145Ser substitution in the transcriptional activator PrfA causes constitutive overexpression of virulence factors in *Listeria monocytogenes*. *J. Bacteriol.* **179:**1533–1540.

99. **Ripio, M. T., G. Domínguez-Bernal, M. Suárez, K. Brehm, P. Berche, and J. A. Vázquez-Boland.** 1996. Transcriptional activation of virulence genes in wild-type strains of *Listeria monocytogenes* in response to a change in the extracellular medium composition. *Res. Microbiol.* **147:**371–384.

100. **Ripio, M. T., J. A. Vázquez-Boland, Y. Vega, S. Nair, and P. Berche.** 1998. Evidence for expressional crosstalk between the central virulence regulator PrfA and the stress response mediator ClpC in *Listeria monocytogenes*. *FEMS Microbiol. Lett.* **158:**45–50.

101. **Rouquette, C., C. de Chastellier, S. Nair, and P. Berche.** 1998. The ClpC ATPase of *Listeria monocytogenes* is a general stress protein required for virulence and promoting early bacterial escape from the phagosome of macrophages. *Mol. Microbiol.* **27:**1235–1245.

102. **Rouquette, C., M. T. Ripio, E. Pellegrini, J. M. Bolla, R. I. Tascon, J. A. Vázquez-Boland, and P. Berche.** 1996. Identification of a ClpC ATPase required for stress tolerance and in vivo survival of *Listeria monocytogenes*. *Mol. Microbiol.* **21:**977–987.

103. **Schmid, M. W., E. Y. W. Ng, R. Lampidis, M. Emmerth, M. Walcher, J. Kreft, W. Goebel, M. Wagner, and K.-H. Schleifer.** 2005. Evolutionary history of the genus *Listeria* and its virulence genes. *Syst. Appl. Microbiol.* **28:**1–18.

104. **Schubert, W. D., and D. W. Heinz.** 2003. Structural aspects of adhesion to and invasion of host cells by the human pathogen *Listeria monocytogenes*. *ChemBioChem* **4:**1285–1291.

105. **Sheehan, B., A. Klarsfeld, R. Ebright, and P. Cossart.** 1996. A single substitution in the putative helix-turn-helix motif of the pleiotropic activator PrfA attenuates *Listeria monocytogenes* virulence. *Mol. Microbiol.* **20:**785–797.

106. **Sheehan, B., A. Klarsfeld, T. Msadek, and P. Cossart.** 1995. Differential activation of virulence gene expression by PrfA, the *Listeria monocytogenes* virulence regulator. *J. Bacteriol.* **177:**6469–6476.

107. **Sheehan, B., C. Kocks, S. Dramsi, E. Gouin, A. D. Klarsfeld, J. Mengaud, and P. Cossart.** 1994. Molecular and genetic determinants of the *Listeria monocytogenes* infectious process. *Curr. Top. Microbiol. Immunol.* **192:**187–216.

108. **Shetron-Rama, L. M., H. Marquis, H. G. Bouwer, and N. E. Freitag.** 2002. Intracellular induction of *Listeria monocytogenes actA* expression. *Infect. Immun.* **70:**1087–1096.

109. **Shetron-Rama, L. M., K. Mueller, J. M. Bravo, H. G. Bouwer, S. S. Way, and N. E. Freitag.** 2003. Isolation of *Listeria monocytogenes* mutants with high-level in vitro expression of host cytosol-induced gene products. *Mol. Microbiol.* **48:**1537–1551.

110. **Sokolovic, Z., J. Riedel, M. Wuenscher, and W. Goebel.** 1993. Surface-associated, PrfA-regulated proteins of *Listeria monocytogenes* synthesized under stress conditions. *Mol. Microbiol.* **8:**219–227.

111. **Spiro, S., and J. R. Guest.** 1990. FNR and its role in oxygen-regulated gene expression in *Escherichia coli*. *FEMS Microbiol. Rev.* **6:**399–428.

112. **Storz, G., J. A. Opdyke, and A. Zhang.** 2004. Controlling mRNA stability and translation with small, noncoding RNAs. *Curr. Opin. Microbiol.* **7:**140–144.

113. **Stritzker, J., J. Janda, C. Schoen, M. Taupp, S. Pilgrim, I. Gentschev, P. Schreier, G. Geginat, and W. Goebel.** 2004. Growth, virulence, and immunogenicity of *Listeria monocytogenes aro* mutants. *Infect. Immun.* **72:**5622–5629.

114. **Stritzker, J., C. Schoen, and W. Goebel.** 2005. Enhanced synthesis of internalin A in *aro* mutants of *Listeria monocytogenes* indicates posttranscriptional control of the *inlAB* mRNA. *J. Bacteriol.* **187:**2836–2845.

115. **Thirumuruhan, R., K. Rajashankar, A. A. Fedorov, T. Dodatko, M. R. Chance, S. C. Almo, and New York Structural Genomics Research Consortium (NYSGRC).** 2003. Crystal structure of PrfA, the transcriptional regulator in *Listeria monocytogenes*. Protein Data Bank accession code 1OMI. http://www.rcsb.org/pdb. [Online.]

116. **Vázquez-Boland, J. A., G. Domínguez-Bernal, B. González-Zorn, J. Kreft, and W. Goebel.** 2001. Pathogenicity islands and virulence evolution in *Listeria*. *Microbes Infect.* **3:**571–584.

117. **Vázquez-Boland, J. A., M. Kuhn, P. Berche, T. Chakraborty, G. Domínguez-Bernal, W. Goebel, B. González-Zorn, J. Wehland, and J. Kreft.** 2001. *Listeria* pathogenesis and molecular virulence determinants. *Clin. Microbiol. Rev.* **14:**584–640.

118. **Vega, Y., C. Dickneite, M. T. Ripio, R. Böckmann, B. González-Zorn, S. Novella, G. Domínguez-Bernal,** W. Goebel, and J. A. Vázquez-Boland. 1998. Functional similarities between the *Listeria monocytogenes* virulence regulator PrfA and cyclic AMP receptor protein: the PrfA* (Gly145Ser) mutation increases binding affinity for target DNA. *J. Bacteriol.* **180:**6655–6660.

119. **Vega, Y., M. Rauch, M. J. Banfield, S. Ermolaeva, M. Scortti, W. Goebel, and J. A. Vázquez-Boland.** 2004. New *Listeria monocytogenes prfA** mutants, transcriptional properties of PrfA* proteins and structure-function of the virulence regulator PrfA. *Mol. Microbiol.* **52:**1553–1565.

120. **Vogel, J., V. Bartels, T. H. Tang, G. Churakov, J. G. Slagter-Jäger, A. Hüttenhofer, and E. G. Wagner.** 2003. RNomics in *Escherichia coli* detects new sRNA species and indicates parallel transcriptional output in bacteria. *Nucleic Acids Res.* **31:**6435–6443.

121. **Weber, I. T., and T. A. Steitz.** 1987. Structure of a complex of catabolite gene activator protein and cyclic AMP refined at 2.5 A resolution. *J. Mol. Biol.* **198:**311–326.

122. **Welch, D. F., C. P. Sword, S. Brehm, and D. Dusanic.** 1979. Relationship between superoxide dismutase and pathogenic mechanisms of *Listeria monocytogenes*. *Infect. Immun.* **23:**863–872.

123. **Wemekamp-Kamphuis, H. H., J. A. Wouters, P. P. L. A. de Leeuw, T. Hain, T. Chakraborty, and T. Abee.** 2004. Identification of sigma factor σ^B-controlled genes and their impact on acid stress, high hydrostatic pressure, and freeze survival in *Listeria monocytogenes* EGD-e. *Appl. Environ. Microbiol.* **70:**3457–3466.

124. **White, M. A., J. C. Lee, and R. O. Fox.** 2001. Structure of Crp-Camp at 1.9 A. Protein Data Bank accession code 1I5Z. http://www.rcsb.org/pdb. [Online.]

125. **Wing, E. J., and S. H. Gregory.** 2002. *Listeria monocytogenes*: clinical and experimental update. *J. Infect. Dis.* **185**(Suppl. 1):S18–S24.

126. **Wong, K. K., H. G. Bouwer, and N. E. Freitag.** 2004. Evidence implicating the 5′ untranslated region of *Listeria monocytogenes actA* in the regulation of bacterial actin-based motility. *Cell. Microbiol.* **6:**155–166.

127. **Wong, K. K., and N. E. Freitag.** 2004. A novel mutation within the central *Listeria monocytogenes* regulator PrfA that results in constitutive expression of virulence gene products. *J. Bacteriol.* **186:**6265–6276.

128. **Wuenscher, M. D., S. Kohler, A. Bubert, U. Gerike, and W. Goebel.** 1993. The *iap* gene of *Listeria monocytogenes* is essential for cell viability, and its gene product, p60, has bacteriolytic activity. *J. Bacteriol.* **175:**3491–3501.

129. **Zhou, Y., T. J. Merkel, and R. H. Ebright.** 1994. Characterization of the activating region of *Escherichia coli* catabolite gene activator protein (CAP). II. Role at class I and class II CAP-dependent promoters. *J. Mol. Biol.* **243:**603–610.

Cell Biology of Invasion and Intracellular Growth by *Listeria monocytogenes*

JAVIER PIZARRO-CERDÁ AND PASCALE COSSART

53

Listeria monocytogenes is a facultative intracellular pathogen that grows inside mammalian cells both in vivo and in tissue culture models of infection. Its capacity to survive within macrophages was noted very early on by Mackaness (47) and believed to be one of its key properties for virulence. It is now recognized that *L. monocytogenes* grows in a broad variety of cell types in animal models and in cell culture systems. Nearly 15 years ago, the stages describing the intracellular cell cycle of *L. monocytogenes* were defined morphologically for the first time (Fig. 1) (57, 77). During the past decade, important advances directly related to the expansion of modern molecular biology—i.e., the determination of the genome sequences of *L. monocytogenes* and of the related nonpathogenic species *Listeria innocua* (26)—have permitted a broader comprehension of the determinants involved in the *L. monocytogenes* host-cell interaction. In this chapter, we will review the molecular mechanisms involved in host-cell invasion, intracellular growth, and cell-to-cell spread.

MORPHOLOGICAL DESCRIPTION OF THE ENTRY PROCESS INTO NONPHAGOCYTIC CELLS

The process of entry of *L. monocytogenes* into nonphagocytic cells has been examined by scanning and transmission electron microscopy (Fig. 2) (36, 53). Internalization of the bacteria starts by a close apposition of the plasma membrane on the bacterium, which becomes progressively enwrapped within the cell. This phenomenon is usually referred to as the zipper mechanism, in contrast to the trigger mechanism used by *Salmonella* and *Shigella* spp., a process that resembles macropinocytosis. The zipper mechanism for bacterial entry was first described in the case of invasin-mediated entry of *Yersinia* spp. (31). In that case, the surface protein invasin interacts with β1 integrins to promote uptake of yersiniae. As described below, the same type of ligand-receptor interaction occurs for *L. monocytogenes*. The situation is very different in *Salmonella* or *Shigella* spp., for which triggering of signaling events is mediated by proteins directly translocated within the mammalian cell via a type III secretion system.

BACTERIAL PROTEINS INVOLVED IN ADHESION AND ENTRY INTO HOST CELLS

Internalin (InlA): a Surface Protein Linked to the Cell Wall via a LPXTG Motif

Internalin (InlA) is an 800-amino-acid protein that displays two regions of repeats, the first consisting of 15 successive leucine-rich repeats (LRRs) of 22 amino acids (Fig. 3) (25). LRRs are motifs with a constant periodicity of leucine residues, generally found in proteins involved in strong protein-protein interactions, although the LRRs are not necessarily part of the domain of interaction. The second repeat region is made of two 70-amino-acid residues and a third partial repeat of 49 amino acids. In addition, internalin has all the features of a protein that is targeted to and exposed on the bacterial surface, i.e., a signal peptide, and a C-terminal region containing an LPXTG peptide followed by a hydrophobic sequence and a few charged residues. This type of C terminus is found in >100 gram-positive bacterial surface proteins and allows covalent linkage of the protein to the peptidoglycan after cleavage of the T-G link, as demonstrated for protein A of *Staphylococcus aureus* (67). In the *L. monocytogenes* genome, 41 genes encoding LPXTG proteins are detected (12).

Several lines of evidence indicate that internalin is sufficient for entry in cells expressing its receptor (see below). Indeed, expression of *inlA*, the gene encoding internalin, in *L. innocua* and also in the more distantly related grampositive bacterium *Enterococcus faecalis*, confers invasiveness to these noninvasive species (25, 43). Moreover, latex beads coated with internalin are readily internalized (43).

The region of internalin essential for entry into mammalian cells is the LRRs, as first shown by inhibition of entry with monoclonal antibodies raised against internalin and recognizing the LRRs and then confirmed by a deletion analysis carried out on *inlA* (52). Moreover, *L. innocua* cells expressing the internalin LRRs and the interrepeat (IR) region fused to the C-terminal part of protein A are able to invade cells expressing the internalin receptor (see below) (43).

Gram-Positive Pathogens, 2nd edition, edited by Vincent A. Fischetti et al.
© 2006 ASM Press, Washington, D.C.

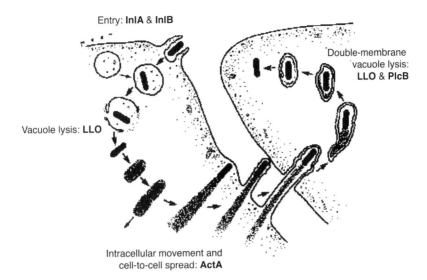

Entry: **InlA & InlB**

Double-membrane
vacuole lysis:
LLO & PlcB

Vacuole lysis: **LLO**

Intracellular movement and
cell-to-cell spread: **ActA**

FIGURE 1 Model of the *L. monocytogenes* intracellular life cycle. Bacteria induce their internalization in a phagosome that is subsequently lysed; once in the host cell cytoplasm, *L. monocytogenes* proliferates and polymerizes the host cell actin (stippled regions) to propulse itself, invading neighboring cells and starting a new infectious cycle (77). The main bacterial products implicated in each step are indicated.

Full-length internalin is expressed on the bacterial surface and can be found in various amounts in the culture supernatants, an observation that initially raised the interesting possibility that this molecule could be released to elicit entry (23). However, genetic evidence leading to truncation of the C-terminal domain of internalin or swapping of this domain with that of another surface protein has established that the surface-associated form of internalin mediates entry (39). Internalin can also be secreted in several strains as a truncated protein, owing to mutations in *inlA* (33). The recent epidemiologic survey of different *L. monocytogenes* strains of various origins (clinical strains or food isolates) indicates that clinical strains express a full-length internalin more frequently (96% of the cases) than food isolates (65% of the cases) (32), providing epidemiological evidence that internalin is a critical factor for pathogenesis in human listeriosis and that it could be used as a marker for virulence in food safety assessment programs.

InlB: an Invasion Protein with GW Modules, a Noncovalent Cell Wall Association Motif

InlB is a 630-amino-acid protein found both at the bacterial surface and in bacterial culture supernatants (19). It has a signal sequence and eight tandem LRRs very similar to those of internalin, followed by one IR region and one B repeat (Fig. 3). The 232 C-terminal amino acids of InlB are necessary and sufficient to anchor InlB to the bacterial surface (6): this region contains three tandem repeats of approximately 80 amino acids beginning with the sequence GW, hence the name given to these repeats. These residues interact with lipoteichoic acids and anchor noncovalently InlB to the bacterial cell wall (34). Similar GW repeats are also present in the C-terminal part of lysostaphin, a protein secreted by *Staphylococcus simulans* that associates with the cell wall of *S. aureus* even when exogenously added. Externally added InlB is also able to associate with *L. monocytogenes* and several other gram-positive bacteria. This external association leads to entry of a Δ*inlB* mutant and also

promotes entry of the noninvasive species *Staphylococcus carnosus* and *L. innocua*, suggesting that InlB may interact with the cell wall after secretion or release from the bacterial surface and that this interaction could contribute to invasion (6). The LLR region of InlB is sufficient for entry, as it confers invasiveness to latex beads or to noninvasive *L. innocua* (8, 58). However, the other regions of InlB (including the GW modules and the B repeat) are also required for efficient invasion of target cells (see below). The crystal structures of the full-length InlB, the LLR domains, and the LRR-IR domains have been resolved, and they suggest that InlB adopts a highly elongated structure that could accommodate several ligands (49, 50, 68).

InlC and Other Internalins: Proteins Supporting In Vivo Infection

Internalin and InlB are two members of the internalin multigene family characterized by the presence of LRRs similar to those in internalin. This family contains now 24 members (Fig. 3). The first additional member identified was InlC (or IrpA), an abundant secreted protein which does not display LPXTG or GW motifs, whose presence depends on PrfA, the transcription regulator of all known virulence genes (17, 24). Four other genes (*inlC2*, *inlD*, *inlE*, and *inlF*) encoding LPXTG proteins were then identified by screening DNA libraries of an EGD strain at low stringency, using *inlA* as a probe (22). Their expression at the transcriptional level is less efficient than that of *inlA* and *inlB*, and simple or multiple deletions have shown that these proteins do not play a direct role in invasion (22, 24). A different gene cluster (*inlG*, *inlH*, and *inlE*) was then identified in another EGD isolate (64). In the mouse model, the proteins InlE, InlF, InlG, and InlH are required for host tissue colonization (64, 68). A recent study (4) suggests that InlA needs the support of other internalins for efficient entry into nonphagocytic cells: the InlA-mediated entry is increased in the presence of InlB and InlC; in the absence of InlB, it requires the presence of InlG, InlH, and InlE.

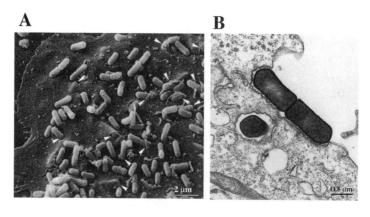

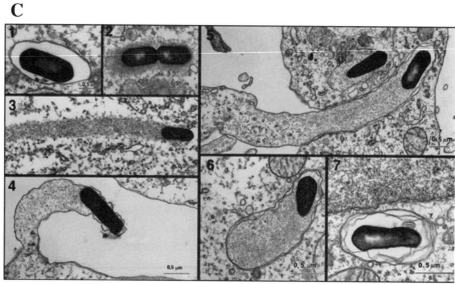

FIGURE 2 Scanning (A) and transmission (B and C) electron microscopy of different steps of the *L. monocytogenes* intracellular life cycle in the human epithelial Caco-2 cell line. (A and B) *L. monocytogenes* is internalized in host cells by a zipper phagocytosis mechanism, without triggering dramatic actin cytoskeleton rearrangements (53). (C) Bacteria are internalized in a phagosome (1) that is lysed after entry; cytoplasmic bacteria proliferate (2) and polymerize host cell actin (3) to move in the host cell cytoplasm; *L. monocytogenes* reaching the host cell plasma membrane induces the formation of protrusions (4) that can invade neighboring cells (5); bacteria are then located in a double membrane secondary vacuole (6) that is lysed (7), and the bacteria start a new infectious cycle (36).

Another internalin-like protein (Lmo2026) was recently identified in a signature-tagged transposon mutagenesis screening and it has been implicated in murine brain invasion (2). Other internalins (membrane-bound through LPXTG motifs or secreted like InlC) have been discovered through analysis of the *L. monocytogenes* sequenced genome, but their structure and their function have not yet been investigated. It should be noted that on the basis of hybridization studies at low stringency, all *L. monocytogenes* species seem to have *inl* genes (25).

p60, Ami, and Auto: Autolysins Implicated in Infection

Several autolysins have been shown to contribute to infection, and it has been hypothesized that they could have functioned as primitive colonizing factors, allowing bacteria to interact with surfaces that express molecules analogous to their natural receptors (teichoic or lipoteichoic acids).

The gene *iap* (for invasion-associated protein) encodes a 60-kDa protein (p60) that was first identified in a spontaneous mutant that was unable to invade mouse fibroblasts (37). p60 contains an amino-terminal signal sequence followed by an SH3 domain region of unknown functional relevance and several threonine-asparagine repeats; in its carboxyl-terminal domain it displays a murein hydrolase activity domain (10, 79). Deletion of *iap* leads to abnormal bacterial division and loss of actin-based motility due to an uneven distribution of ActA on the *L. monocytogenes* surface (60), suggesting that the noninvasive phenotype is an indirect consequence of the defective autolysin activity.

The gene *ami* was identified in a Southern blotting screening of chromosomal DNA at low stringency from

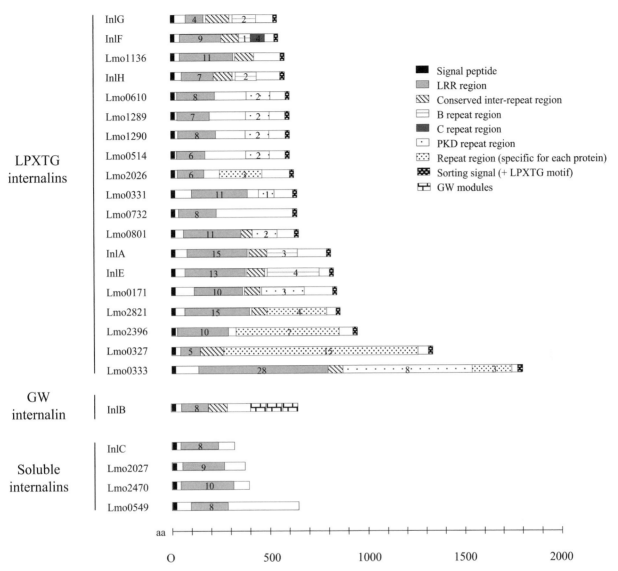

FIGURE 3 Members of the internalin family. A total of 24 proteins have been recognized in the *L. monocytogenes* genome that possess LRRs and belong to the internalin family: 19 proteins present LPXTG motifs and are covalently linked to the bacterial cell wall (including InlA), 1 protein is loosely attached to the membrane through GW motifs (InlB), and 4 proteins do not present anchor motifs and are released as soluble proteins (including InlC) (12).

wild-type EGD with the 3′ end of *inlB* containing the GW motifs as a probe (6). Ami is a 102-kDa protein that contains an amino-terminal signal sequence, followed by a domain similar to the *N*-acetylmuramoyl-L-alanine amidase domain of the Atl autolysin of *S. aureus*, and a carboxyl-terminal domain made of eight GW modules. Besides its autolytic amidase activity on *L. monocytogenes* cell walls (51), Ami has been implicated in bacterial adhesion to host cells in a Δ*inlAB* background (56): indeed, its GW modules are sufficient to restore adhesion of *ami* mutants in the Δ*inlAB* background. *ami* mutants are also attenuated in the mouse model (54). A recent study has shown that while the amino-terminal enzymatic domain of Ami from serovar 1/2a is almost identical to that of serovar 4b (which includes the strains most frequently involved in listeriosis outbreaks), the carboxy-terminal domain is less conserved be-

tween the two proteins (55). Eight GW modules are detected in Ami serovar 1/2a, while only six are present in Ami serovar 4b, and purified carboxy-terminal Ami serovar 4b is less able to bind human hepatocytic cells than carboxy-terminal Ami serovar 1/2a. This is the first evidence for a divergence between serovars 1/2a and 4b in a gene involved in adhesion of *L. monocytogenes*.

Besides Ami, seven additional proteins were subsequently discovered in the *L. monocytogenes* genome that also contain GW modules; interestingly, six of these proteins display, like Ami, autolytic activities. One of these novel autolysins, absent from the genome of the nonpathogenic species *L. innocua*, is encoded by the gene *aut* (13). Auto is necessary for entry in several epithelial and fibroblastic cell lines, but it is not sufficient, since it does not confer invasiveness when expressed by *L. innocua* (13). Inactivation of *aut* leads to a decrease in virulence, following intravenous inoculation of mice and oral infection of guinea pigs (13). Its precise role remains to be elucidated. Interestingly, *aut* is absent from the genome of *L. monocytogenes* serovar 4b.

FbpA: a Novel Multifunctional Virulence Factor

FbpA was identified through a signature-tagged mutagenesis screening designed to identify new *L. monocytogenes* virulence factors (20). *fbpA* encodes a 60-kDa surface protein with homologies to atypical fibronectin-binding proteins that is secreted through a second *secA* system (Sec A2) recently discovered in *L. monocytogenes* (45). FbpA binds to immobilized fibronectin in a dose-dependent manner and increases *L. monocytogenes* adherence to HEp-2 cells in the presence of exogenous fibronectin. FbpA also affects the expression of listeriolysin O (LLO) and InlB, but not that ActA or InlA, suggesting that this protein could behave as an escort factor. In vivo, deletion of FbpA leads to a defect in liver colonization in mice inoculated intravenously (20).

ActA and LLO: Intracellular Virulence Factors also Involved in Adhesion and Invasion

ActA is the protein responsible for the intracellular actin-based motility of *L. monocytogenes* (see below) (36). However, it has also been implicated in adhesion and invasion of target cells; indeed, an ActA-deficient mutant is impaired in attachment and entry in CHO cells and in IC-21 murine macrophages, owing to altered heparan sulfate recognition functions (1). More recently, expression of ActA in *L. innocua* was sufficient to promote bacterial entry in epithelial cells (75).

LLO is a 60-kDa pore-forming, cholesterol-dependent cytolysin involved in the escape of *L. monocytogenes* from its primary and secondary vacuoles (see below) (62). LLO oligomerizes and forms Ca^{2+}-permeable pores that lead to extracellular Ca^{2+} influx potentiating *L. monocytogenes* entry into Hep-2 cells (21).

CELLULAR RECEPTORS INVOLVED IN BACTERIAL ENTRY

E-Cadherin: the Receptor for Internalin

The internalin receptor on the human epithelial cell line Caco-2 was identified using affinity chromatography (53). It is the cell adhesion molecule E-cadherin, a 110-kDa transmembrane glycoprotein that normally mediates calcium-dependent cell-cell adhesion through homophilic interactions between extracellular domains. Cadherins are tissue specific, with E-cadherin specifically expressed in epithelial cells. These proteins play a critical role in cell sorting during development and maintenance of tissue cohesion and architecture during adult life. In polarized epithelial cells, E-cadherin is mainly expressed at the adherens junctions and on the basolateral face of the cell. Integrity of the intracytoplasmic domain of cadherins is required for optimal intercellular adhesion. This domain interacts with proteins named catenins, which in turn interact with the actin cytoskeleton, highlighting the importance of the cytoskeleton in maintaining adhesion of adjacent epithelial cells.

The ectodomain of E-cadherin, composed of five extracellular modules, is sufficient for bacterial adherence. However, deletion of the intracytoplasmic domain impairs bacterial infection (41). This intracellular domain contains a β-catenin-binding site, which in turns binds α-catenin, and

this last molecule links the whole complex to the actin cytoskeleton. A fusion molecule of the E-cadherin ectodomain and of the α-catenin actin-binding site restores invasion, suggesting that *L. monocytogenes* exploits the same molecular scaffold as the one involved in adherens junctions function to induce its entry into target cells. The unconventional myosin VIIa and its ligand vezatin, which are part of adherens junctions in polarized epithelial cells, have recently been implicated in the entry process: vezatin could recruit the myosin VIIa to the E-cadherin/actin cytoskeleton complex, and the myosin VIIa could then generate the contractile force required for bacterial internalization (Fig. 4) (74).

The interaction of InlA with E-cadherin is species specific (40). The proline at position 16 (Pro16) of the first extracellular module of the E-cadherin is crucial, as demonstrated by the fact that the human E-cadherin (which possesses a Pro16) allows InlA-dependent entry while the mouse E-cadherin (which presents a glutamic acid at the same position) does not allow invasion (40). A mouse model expressing the human E-cadherin in the intestine permits *L. monocytogenes* infection by the oral route, demonstrating the pivotal role of InlA in the crossing of the intestinal barrier (44). Recently, it was demonstrated that the InlA-E-cadherin interaction also plays a critical role in the targeting and crossing of the human maternofetal barrier by *L. monocytogenes* (42).

The crystal structure of the first extracellular domain of E-cadherin in complex with the amino-terminal region of InlA has been resolved, and it shows that the LRR region of InlA embraces and specifically recognizes the E-cadherin extracellular domain. Furthermore, the absence of one amino acid in the leucine repeat 6 of InlA (the only repeat that presents 21 instead of 22 amino acids) seems to create a hydrophobic pocket between repeats 5 and 7 that accommodates Pro16 (69).

Met, Glycosaminoglycans, and gC1q-R: Three Host-Cell Molecules That Interact with InlB

The hepatocyte growth factor-scatter factor receptor, or Met, has been identified as the main receptor for InlB (71). Met is a tyrosine kinase receptor that is required for normal growth during development, and its abnormal activation is implicated in the pathogenesis of most types of human solid tumors. Met is synthesized as a large protein that is cleaved into α and β subunits as it matures on the cell surface. Once activated, Met elicits its own phosphorylation on two critical tyrosine residues, referred to as multiple docking sites ($Y^{1349}VHV$ and $Y^{1356}VNV$), which serve to recruit downstream signaling molecules and adaptor proteins. InlB binds Met through the concave surface of the LRR region, and this interaction leads to transient Met phosphorylation (71); recruitment-phosphorylation of the adaptor proteins Cbl, Gab1, and Shc (30); and activation of the phosphatidylinositol 3-kinase (PI 3-kinase)(see below) (Fig. 4) (29).

Optimal activity of Met when activated by its natural ligand (the hepatocyte growth factor) requires glycosaminoglycans (GAGs) at the surface of target cells, probably involving oligomerization of the growth factor and/or its storage and protection from extracellular proteases. Interestingly, the GW modules of InlB also interact with GAGs, and GAGs potentiate the InlB-dependent entry of *L. monocytogenes* (35), suggesting that the released form of InlB can also play a role in invasion (Fig. 4).

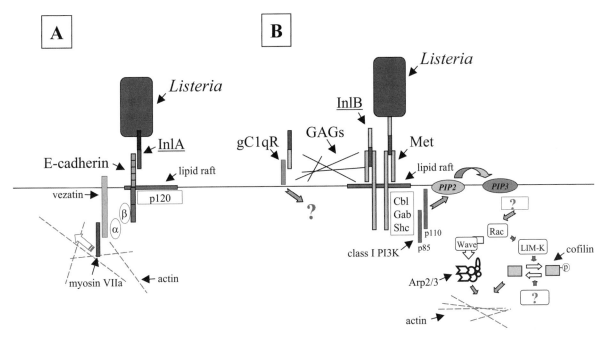

FIGURE 4 Model of *L. monocytogenes* internalization pathways in host cells. (A) InlA internalization pathway. The bacterial protein InlA binds E-cadherin located in lipid rafts, which in turns interacts with p120-, β-, and α-catenins. α-Catenin would link the whole complex to the actin cytoskeleton and to vezatin, which recruits the unconventional myosin VIIa, generating the contractile force required for internalization. (B) InlB internalization pathway. Soluble InlB can interact with gC1qr, but the signaling cascades triggered by this interaction are unknown; soluble InlB (stabilized by GAGs) or bacterial-associated InlB can interact with Met located in lipid rafts, inducing the recruitment of adaptor proteins (Cbl, Gab1, or Shc) and of class I PI 3-kinase. Conversion of PI(4,5)P$_2$ (PIP2) into PI(3,4,5)P$_3$ (PIP3) leads to translocation to the plasma membrane of an unknown factor that activates Rac; in turn, Rac indirectly activates LIM kinase, controlling a cofilin phosphocycle (that implicates an unknown phosphatase) that regulates actin polymerization-depolymerization. Rac is also implicated in the activation of the actin nucleating Arp2/3 complex, probably through a signaling cascade that includes Wave.

gC1q-R/p32, the receptor of the globular part of the complement component C1q, has also been identified as a receptor for InlB (7). The GW modules of InlB bind specifically to gC1q-R, and this interaction requires the release of InlB from the bacterial surface (49). However, since gC1q-R is devoid of a transmembrane domain or a membrane anchor, it is unclear how it can deliver intracellular signals upon InlB binding. Its subcellular localization also remains controversial.

INTEGRATING SIGNALS: THE INVASION PROCESS

Lipid Rafts: Host Membrane Microenvironment Favoring Receptor Complex Scaffolding

The formation of lipid microdomains provides a mechanism for the selection of molecular effectors into functional units at the plasma membrane for efficient signaling and sorting processes. Cholesterol is the major lipid of the plasma membrane, and it is a critical component that controls microdomain segregation by stabilizing a liquid-ordered phase in the membrane called lipid rafts. A recent work (70) showed that cholesterol depletion by methyl-β-cyclodextrin (MβCD) treatment reversibly inhibits InlA- and InlB-mediated entry of *L. monocytogenes*, suggesting that lipid rafts are required for host cell invasion. Indeed, lipid raft markers such

as glycosyl-PI-anchored proteins, myristoylated and palmitoylated peptide and the ganglioside GM1, were recruited at the bacterial entry site. More interestingly, the InlA- and InlB-dependent entry pathways were disrupted at different levels after MβCD treatment: the initial interaction between E-cadherin and InlA required the presence of E-cadherin in intact lipid microdomains. With InlB-dependent entry, only downstream signaling of Met leading to actin polymerization is microdomain dependent, while the recruitment of Met at the bacterial entry site, its phosphorylation, and the association of PI 3-kinase with phosphotyrosine proteins does not require membrane-intact lipid rafts. In particular, the closure of the phagocytic cup around InlB-coated beads is affected when cells are treated with MβCD. These results showed not only that lipid rafts are required for the invasion of target cells by *L. monocytogenes*, but also that E-cadherin and Met, two important cellular surface molecules, require intact lipid microdomains for their full activity (70).

Phosphoinositide Metabolism: Second Messengers Recruiting Critical Effectors for Entry

Phosphoinositides and products of their metabolism are important signaling molecules that play critical roles in the regulation of actin dynamics and membrane fission/fusion events. In particular, PI-4,5-bisphosphate [PI(4,5)P$_2$] and

PI-3,4,5-trisphosphate [PI(3,4,5)P$_3$] have been extensively studied and shown to modulate endocytosis or phagocytosis by interacting with cytoskeletal effectors such as dynamin-2, Wiskott-Aldrich syndrome protein (WASP), profilin, cofilin, Rho gua-nosine exchange factors, and members of the ezrin/radixin/ moesin family.

PI 3-kinase p85/p110 is required for the InlB-dependent *L. monocytogenes* entry (29). Treatment of target cells with the specific PI 3-kinase inhibitor wortmannin or expression of a dominant-negative form of the regulatory subunit p85 blocks invasion. Activation of Met by InlB causes the phosphorylation of the adaptor proteins Gab1, Cbl, and Shc, and these proteins promote the recruitment of the PI 3-kinase at the bacterial entry site (30, 71). Upon infection, the levels of PI(3,4)P$_2$ and of PI(3,4,5)P$_3$ dramatically increase in target cells (29). PI(3,4,5)P$_3$ is known to activate several exchange factors for Rho GTPases that could be involved in remodeling the actin cytoskeleton. The observation that PI(3,4,5)P$_3$ synthesis is inhibited by pretreating cells with wortmannin, but not by pretreatment with cytochalasin D, suggests that cytoskeletal rearrangements indeed occur downstream of the PI 3-kinase.

Cofilin and LIM Kinase: Organizers of Actin Reorganization during Invasion

Dynamic processes such as phagocytosis or cell migration require not only actin filament nucleation and branching (promoted by the seven protein complex Arp2/3) but also disassembly of actin filaments to reorganize membrane structures. The actin depolymerizing factor or cofilin enhances the rate of actin turnover by both severing actin filaments (thereby creating new barbed ends) and by increasing actin depolymerization at pointed ends, providing a pool of actin monomers for F-actin assembly. Filament turnover is regulated by a cofilin phosphocycle (cofilin is inactivated by phosphorylation and is reactivated by dephosphorylation) and the LIM kinase catalyzes the phosphorylation of cofilin. Bierne et al. (5) showed not only that the InlB-induced internalization of *L. monocytogenes* is dependent on the Arp2/3 complex, but also that cofilin, LIM kinase, and Rac are essential for InlB-mediated cytoskeletal rearrangements; cofilin not only stimulates actin polymerization allowing bacterial internalization, but also controls filament disassembly at the phagocytic cup, another step crucial for an efficient entry process.

BACTERIAL PROTEINS INVOLVED IN ESCAPE FROM THE VACUOLE AND INTRACELLULAR LIFE

LLO, Phosphatidylinositol Phospholipase C (PLC), and Phosphatidylcholine (PC)-PLC: Effectors of Vacuolar Lysis

Intracellular pathogens can be divided into those that reside within a host vacuole and those, like *L. monocytogenes*, that escape and grow directly in the host cytosol. This critical aspect of *L. monocytogenes* pathogenesis was not clear until electron microscopy studies were performed using tissue culture models of infection (47, 77). Before this time, most studies used primary cultures of mouse peritoneal macrophages in which the majority of bacteria were killed and hence restricted to the vacuolar compartment. In contrast, most cell lines lack bactericidal capacity and allow the majority of ingested bacteria access to the cytosol. Even

bone marrow-derived macrophages allow approximately 50% of the bacteria into the cytosol (16, 62), whereas resident peritoneal macrophages kill approximately 90% of ingested bacteria (63). Activated macrophages are extremely effective at killing *L. monocytogenes*.

The primary *L. monocytogenes* determinant responsible for escape from the vacuole and thus entrance into the cytosol is LLO, encoded by the *hly* gene. Mutants lacking LLO are avirulent and fail to escape from their vacuole. Expression of LLO by either *Bacillus subtilis* or *L. innocua* also promotes the capacity of these nonpathogenic bacteria to escape from their internalization vacuole and to grow in the cytosol. LLO is a member of a large family of pore-forming hemolysins secreted by gram-positive pathogens, including streptolysin O secreted by *Streptococcus pyogenes* and perfringolysin O secreted by *Clostridium perfringens*. All members of the family are thought to share a common mechanism of action that involves binding to membrane cholesterol followed by insertion, oligomerization of 20 to 80 monomers, and pore formation. LLO is characterized by its optimal activity between pH 5.5 and 6.0 (the pH of the early phagosome) (3), activity that can be traced to the leucine residue at position 461 (27). A PEST-like sequence in LLO was reported to target this cytolysin for degradation when present in the cytosol (15), but this observation has been challenged. Another study (46) suggests that the PEST-like sequence is required instead for optimal escape from the vacuole.

L. monocytogenes secretes two PLCs that have been also implicated in the lysis of intracellular vacuoles (28). One is specific for PI and glycosyl-PI-anchored proteins (PI-PLCs), while the PC-PLC is a broad-range PLC. Both enzymes act synergistically with LLO in lysing primary and secondary vacuoles (72). In the absence of LLO, the PC-PLC can also promote lysis of primary vacuoles in human epithelial cells (48).

Hpt: a Hexose-Uptake System Required for Intracytoplasmic Replication

L. monocytogenes relies on a PrfA-dependent hexose phosphate transporter (Hpt) to uptake glucose-1-phosphate, a sugar required for bacterial intracellular replication and proliferation in mouse organs (14). This permease is a structural and functional homolog of the eukaryote glucose-6-phosphate transporter that is responsible for glucose-6-phosphate uptake from the cytosol to the endoplasmic reticulum. However, no information is available concerning other nutritional requirements or mechanisms used by *L. monocytogenes* to obtain nutrients from the host cell cytosol.

ActA: Effector of Actin-Based Motility

Upon escape from a vacuole and arrival to the host cell cytosol, *L. monocytogenes* begins rapid growth and becomes encapsulated by host actin filaments and other actin-binding proteins (57, 77). After a couple of bacterial cell divisions, the bacteria begin to move in the cytosol at measurable rates. Movement is entirely dependent on polymerization of host actin (66, 76). New filaments form at the rear end of moving bacteria and then depolymerize. Moving bacteria have a characteristic tail composed of actin filaments and actin-binding proteins. The length of the tail reflects the rate of bacterial movement.

ActA is the only bacterial protein required for actin-based motility (9, 18, 36). Mutants lacking ActA escape normally from a vacuole and grow at wild-type growth rates in the cytosol, but they grow as microcolonies and are unable

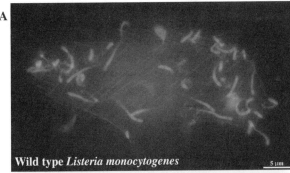

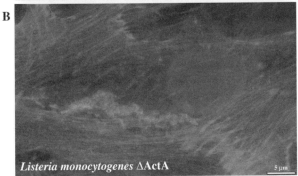

FIGURE 5 (*See the separate color insert for the color version of this illustration.*) Fluorescence microscopy of Vero cells infected with wild-type *L. monocytogenes* (A) or *L. monocytogenes* ΔActA (B). Wild-type *L. monocytogenes* induces the formation of actin comet tails that propulse the bacteria in the host cell cytoplasm (A), while *L. monocytogenes* ΔActA proliferates in microcolonies that are unable to spread from the primary infected cell (B). DNA is labeled with DAPI (4′,6′-diamidino-2-phenylindole; blue), the bacterial cell wall is labeled with an anti-*L. monocytogenes* antibody (red), and actin is labeled with fluorescent phalloidin (green).

to spread from cell to cell (Fig. 5). Importantly, ActA mutants are 1,000-fold-less virulent in the murine model of infection (61). The ActA protein can be divided into three domains (38, 59, 73). The C-terminal portion of ActA encodes a transmembrane domain that anchors ActA to the bacterial surface. The central region of ActA contains four short proline-rich repeats that bind members of the enabled/vasodilator-stimulated phosphoprotein (Ena/VASP) family of proteins, which modulate bacterial speed and directionality by recruiting profilin and providing polymerization-competent actin monomers to the ActA N-terminal domain. The N-terminal domain of ActA is essential and sufficient for actin polymerization: purified ActA or an N-terminal fragment acts as an actin nucleator when mixed 1:1 with the Arp2/3 complex (78), mimicking the Wiskott-Aldrich syndrome protein family that is the main host cell activator of the Arp2/3 complex (Fig. 6).

ActA is clearly an essential determinant of *L. monocytogenes* pathogenesis, whose role is most likely to facilitate cell-to-cell spread. Thus, actin-based motility per se is not important, but rather the fact that it promotes the movement of one infected cell to a neighboring cell without being exposed to the extracellular environment. Very few studies have addressed the mechanisms of cell-to-cell spread by *L. monocytogenes*. It has been suggested that bacteria could take advantage of normal paracytophagic behavior (engulfment of portions of neighboring cells in noninfected monolayers) to favor the intercellular passage of bacteria (65).

INTRACELLULAR LIFE: ESCAPING FROM THE HOST CELL IMMUNE SYSTEM

As discussed in chapter 50, acquired immunity to *L. monocytogenes* is largely mediated by the effector function of CD8$^+$ T cells. Antibody plays no measurable role in immunity. The cell biology of infection is consistent with the immunological perspective; i.e., once bacteria enter a cell, they are not exposed to the humoral immune response. Rather, growth in the cytosol potentially exposes bacterial proteins to the major histocompatibility complex class I pathway of antigen processing and presentation. Indeed, *L. monocytogenes* antigens presented in association with the major histocompatibility complex class I antigen H2-K^d are derived from secreted bacterial proteins. Surprisingly, the dominant antigen recognized by immune CD8$^+$ T cells is derived from LLO (11). Cells infected with *L. monocytogenes* are recognized and lysed by immune CD8$^+$ T cells. Thus, the capacity of *L. monocytogenes* to spread cell to cell allows the bacteria to escape both the humoral and cellular immune responses.

We thank people from the Cossart laboratory for helpful discussions. We apologize to those people who have contributed to the advancement of the field of listeriosis but whose work was not cited in our chapter, due to size limitations.

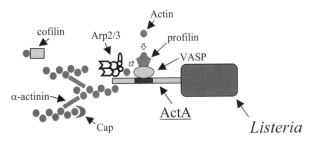

FIGURE 6 Model of actin polymerization by *L. monocytogenes* in host cells. The bacterial protein ActA recruits the host cell molecule VASP, which in turn recruits profilin to the bacterial tail; actin monomers are captured by profilin and are transferred to the Arp2/3 complex, which nucleates the polymerization of host actin, leading to the formation of actin comet tails. A capping protein inhibits the polymerization of actin on barbed ends, cofilin promotes the depolymerization of actin from pointed ends, and α-actinin stabilizes actin polymers.

REFERENCES

1. **Alvarez-Dominguez, C., J. A. Vazquez-Boland, E. Carrasco-Marin, P. Lopez-Mato, and F. Leyva-Cobian.** 1997. Host-cell heparan sulfate proteoglycans mediate attachment and entry of *Listeria monocytogenes*, and the listerial surface protein ActA is involved in heparan sulfate receptor recognition. *Infect. Immun.* **65:**78–88.

2. **Autret, N., I. Dubail, P. Trieu-Cuot, P. Berche, and A. Charbit.** 2001. Identification of new genes involved in the virulence of *Listeria monocytogenes* by signature-tagged transposon mutagenesis. *Infect. Immun.* **69:**2054–2065.

3. **Beauregard, K. E., K. Lee, R. J. Collier, and J. A. Swanson.** 1997. pH-dependent perforation of macrophage phagosomes by listeriolysin O from *L. monocytogenes*. *J. Exp. Med.* **186:**1159–1163.

4. **Bergmann, B., D. Raffelsbauer, M. Kuhn, M. Goetz, S. Hom, and W. Goebel.** 2002. InlA- but not InlB-mediated internalization of Listeria monocytogenes by non-phagocytic mammalian cells needs the support of other internalins. *Mol. Microbiol.* **43**:557–570.

5. **Bierne, H., E. Gouin, P. Roux, P. Caroni, H. L. Yin, and P. Cossart.** 2001. A role for cofilin and LIM kinase in *Listeria*-induced phagocytosis. *J. Cell Biol.* **155**:101–112.

6. **Braun, L., S. Dramsi, P. Dehoux, H. Bierne, G. Lindahl, and P. Cossart.** 1997. InlB: an invasion protein of *L. monocytogenes* with a novel type of surface association. *Mol. Microbiol.* **25**:285–294.

7. **Braun, L., B. Ghebrehiwet, and P. Cossart.** 2000. gC1q-R/p32, a C1q-binding protein, is a receptor for the InlB invasion protein of *Listeria monocytogenes. EMBO J.* **19**: 1458–1466.

8. **Braun, L., H. Ohayon, and P. Cossart.** 1998. The InlB protein of *L. monocytogenes* is sufficient to promote entry into mammalian cells. *Mol. Microbiol.* **27**:1077–1087.

9. **Brundage, R. A., G. A. Smith, A. Camilli, J. A. Theriot, and D. A. Portnoy.** 1993. Expression and phosphorylation of the *L. monocytogenes* ActA protein in mammalian cells. *Proc. Natl. Acad. Sci. USA* **90**:11890–11894.

10. **Bubert, A., M. Kuhn, W. Goebel, and S. Kohler.** 1992. Structural and functional properties of the p60 proteins from different *Listeria* species. *J. Bacteriol.* **174**:8166–8171.

11. **Busch, D. H., I. M. Pilip, S. Vijh, and E. G. Pamer.** 1998. Coordinate regulation of complex T cell populations responding to bacterial infection. *Immunity* **8**:353–362.

12. **Cabanes, D., P. Dehoux, O. Dussurget, L. Frangeul, and P. Cossart.** 2002. Surface proteins and the pathogenic potential of Listeria monocytogenes. *Trends Microbiol.* **10**: 238–245.

13. **Cabanes, D., O. Dussurget, P. Dehoux, and P. Cossart.** 2004. Auto, a surface associated autolysin of *Listeria monocytogenes* required for entry into eukaryotic cells and virulence. *Mol. Microbiol.* **51**:1601–1614.

14. **Chico-Calero, I., M. Suárez, B. González-Zorn, M. Scortti, J. Slaghuis, W. Goebel, and J. A. Vázquez-Boland.** 2002. Hpt, a bacterial homolog of the microsomal glucose-6-phosphate translocase, mediates rapid intracellular proliferation in *Listeria. Proc. Natl. Acad. Sci. USA* **99**:431–436.

15. **Decatur, A. L., and D. A. Portnoy.** 2000. A PEST-like sequence in listeriolysin O essential for Listeria monocytogenes pathogenicity. *Science* **290**:992–995.

16. **de Chastellier, C., and P. Berche.** 1994. Fate of *Listeria monocytogenes* in murine macrophages: evidence for simultaneous killing and survival of intracellular bacteria. *Infect. Immun.* **62**:543–553.

17. **Domann, E., S. Zechel, A. Lingnau, T. Hain, A. Darji, T. Nichterlein, J. Wehland, and T. Chakraborty.** 1997. Identification and characterization of a novel PrfA-regulated gene in *Listeria monocytogenes* whose product, IrpA, is highly homologous to internalin proteins, which contain leucine-rich repeats. *Infect. Immun.* **65**:101–109.

18. **Domann, E., J. Wehland, M. Rohde, S. Pistor, M. Hartl, W. Goebel, M. Leimeister-Wächter, M. Wuenscher, and T. Chakraborty.** 1992. A novel bacterial virulence gene in *L. monocytogenes* required for host cell microfilament interaction with homology to the proline-rich region of vinculin. *EMBO J.* **11**:1981–1990.

19. **Dramsi, S., I. Biswas, E. Maguin, L. Braun, P. Mastroeni, and P. Cossart.** 1995. Entry of *L. monocytogenes* into hepatocytes requires expression of InlB, a surface protein of the internalin multigene family. *Mol. Microbiol.* **16**: 251–261.

20. **Dramsi, S., F. Bourdichon, D. Cabanes, M. Lecuit, H. Fsihi, and P. Cossart.** 2004. FbpA, a novel multifunctional Listeria monocytogenes virulence factor. *Mol. Microbiol.* **53**:639–649.

21. **Dramsi, S., and P. Cossart.** 2003. Listeriolysin O-mediated calcium influx potentiates entry of *Listeria monocytogenes* into the human Hep-2 epithelial cell line. *Infect. Immun.* **71**:3614–3618.

22. **Dramsi, S., P. Dehoux, M. Lebrun, P. Goossens, and P. Cossart.** 1997. Identification of four new members of the internalin multigene family in *Listeria monocytogenes* EGD. *Infect. Immun.* **65**:1615–1625.

23. **Dramsi, S., C. Kocks, C. Forestier, and P. Cossart.** 1993. Internalin-mediated invasion of epithelial cells by *L. monocytogenes* is regulated by the bacterial growth state, temperature and the pleiotropic activator, prfA. *Mol. Microbiol.* **9**:931–941.

24. **Engelbrecht, F., S.-K. Chun, C. Ochs, J. Hess, F. Lottspeich, W. Goebel, and Z. Sokolovic.** 1996. A new PrfA-regulated gene of *L. monocytogenes* encoding a small, secreted protein which belongs to the family of internalins. *Mol. Microbiol.* **21**:823–837.

25. **Gaillard, J. L., P. Berche, C. Frehel, E. Gouin, and P. Cossart.** 1991. Entry of *L. monocytogenes* into cells is mediated by internalin, a repeat protein reminiscent of surface antigens from gram-positive cocci. *Cell* **65**:1127–1141.

26. **Glaser, P., L. Frangeul, C. Buchrieser, C. Rusniok, A. Amend, F. Baquero, P. Berche, H. Bloecker, P. Brandt, T. Chakraborty, A. Charbit, F. Chetouani, E. Couve, A. de Daruvar, P. Dehoux, E. Domann, G. Dominguez-Bernal, E. Duchaud, L. Durant, O. Dussurget, K. D. Entian, H. Fsihi, F. Garcia-del Portillo, P. Garrido, L. Gautier, W. Goebel, N. Gomez-Lopez, T. Hain, J. Hauf, D. Jackson, L. M. Jones, U. Kaerst, J. Kreft, M. Kuhn, F. Kunst, G. Kurapkat, E. Madueno, A. Maitournam, J. M. Vicente, E. Ng, H. Nedjari, G. Nordsiek, S. Novella, B. de Pablos, J. C. Perez-Diaz, R. Purcell, B. Remmel, M. Rose, T. Schlueter, N. Simoes, A. Tierrez, J. A. Vazquez-Boland, H. Voss, J. Wehland, and P. Cossart P.** 2001. Comparative genomics of *Listeria* species. *Science* **294**:849–852.

27. **Glomski, I. J., A. L. Decatur, and D. A. Portnoy.** 2003. *Listeria monocytogenes* mutants that fail to compartmentalize listeriolysin O activity are cytotoxic, avirulent, and unable to evade host extracellular defenses. *Infect. Immun.* **71**:6754–6765.

28. **Goldfine, H., and S. J. Wadsworth.** 2002. Macrophage intracellular signaling induced by *Listeria monocytogenes*. *Microbes Infect.* **4**:1335–1343.

29. **Ireton, K., B. Payrastre, H. Chap, W. Ogawa, H. Sakaue, M. Kasuga, and P. Cossart.** 1996. A role for phosphoinositide 3-kinase in bacterial invasion. *Science* **274**:780–782.

30. **Ireton, K., B. Payrastre, and P. Cossart.** 1999. The *Listeria monocytogenes* protein InlB is an agonist of mammalian phosphoinositide 3-kinase. *J. Biol. Chem.* **274**:17025–17032.

31. **Isberg, R. R., and G. Tran Van Nhieu.** 1994. Two mammalian cell internalization strategies used by pathogenic bacteria. *Annu. Rev. Genet.* **27**:395–422.

32. **Jacquet, C., M. Doumith, J. I. Gordon, P. M. Martin, P. Cossart, and M. Lecuit.** 2004. A molecular marker for evaluating the pathogenic potential of foodborne *Listeria monocytogenes. J. Infect. Dis.* **189**:2094–2100.

33. **Jonquieres, R., H. Bierne, J. Mengaud, and P. Cossart.** 1998. The *inlA* of *Listeria monocytogenes* LO28 harbors a nonsense mutation resulting in release of internalin. *Infect. Immun.* **66**:3420–3422.

34. **Jonquieres, R., H. Bierne, F. Fiedler, P. Gounon, and P. Cossart.** 1999. Interaction between the protein InlB of *Listeria monocytogenes* and lipoteichoic acid: a novel mechanism of protein association at the surface of gram-positive bacteria. *Mol. Microbiol.* **34:**902–914.

35. **Jonquieres, R., J. Pizarro-Cerda, and P. Cossart.** 2001. Synergy between the N- and C-terminal domains of InlB for efficient invasion of non-phagocytic cells by *Listeria monocytogenes. Mol. Microbiol.* **42:**955–965.

36. **Kocks, C., E. Gouin, M. Tabouret, P. Berche, H. Ohayon, and P. Cossart.** 1992. *L. monocytogenes*-induced actin assembly requires the *actA* gene product, a surface protein. *Cell* **68:**521–531.

37. **Kuhn, M., and W. Goebel.** 1989. Identification of an extracellular protein of *L. monocytogenes* possibly involved in intracellular uptake by mammalian cells. *Infect. Immun.* **57:**55–61.

38. **Lasa, I., V. David, E. Gouin, J.-B. Marchand, and P. Cossart.** 1995. The amino-terminal part of ActA is critical for the actin-based motility of *L. monocytogenes*; the central proline-rich region acts as a stimulator. *Mol. Microbiol.* **18:**425–436.

39. **Lebrun, M., J. Mengaud, H. Ohayon, F. Nato, and P. Cossart.** 1996. Internalin must be on the bacterial surface to mediate entry of *Listeria monocytogenes* into epithelial cells. *Infect. Immun.* **57:**55–61.

40. **Lecuit, M., S. Dramsi, C. Gottardi, M. Fedor-Chaiken, B. Gumbiner, and P. Cossart.** 1999. A single amino acid in E-cadherin responsible for host specificity towards the human pathogen *Listeria monocytogenes. EMBO J.* **18:**3956–3963.

41. **Lecuit, M., R. Hurme, J. Pizarro-Cerdá, H. Ohayon, B. Geiger, and P. Cossart.** 2000. A role for alpha- and beta-catenins in bacterial uptake. *Proc. Natl. Acad. Sci. USA* **97:**10008–10013.

42. **Lecuit, M., D. M. Nelson, S. D. Smith, H. Khun, M. Huerre, M. C. Vacher-Lavenu, J. I. Gordon, and P. Cossart.** 2004. Targeting and crossing of the human maternofetal barrier by *Listeria monocytogenes*: role of internalin interaction with trophoblast E-cadherin. *Proc. Natl. Acad. Sci. USA* **101:**6152–6157.

43. **Lecuit, M., H. Ohayon, L. Braun, J. Megaud, and P. Cossart.** 1997. Internalin of *Listeria monocytogenes* with an intact leucine-rich repeat region is sufficient to promote internalization. *Infect. Immun.* **65:**5309–5319.

44. **Lecuit, M., S. Vandormael-Pournin, J. Lefort, M. Huerre, P. Gounon, C. Dupuis, C. Babinet and P. Cossart.** 2001. A transgenic model for listeriosis: role of internalin in crossing the intestinal barrier. *Science* **292:**1722–1725.

45. **Lenz, L. L., S. Mohammadi, A. Geissler, and D. A. Portnoy.** 2003. SecA2-dependent secretion of autolytic enzymes promotes *Listeria monocytogenes* pathogenesis. *Proc. Natl. Acad. Sci. USA* **100:**12432–12437.

46. **Lety, M. A., C. Frehel, I. Dubail, J. L. Beretti, S. Kayal, P. Berche, and A. Charbit.** 2001. Identification of a PEST-like motif in listeriolysin O required for phagosomal escape and for virulence in *Listeria monocytogenes. Mol. Microbiol.* **39:**1124–1139.

47. **Mackaness, G. B.** 1962. Cellular resistance to infection. *J. Exp. Med.* **116:**381–406.

48. **Marquis, H., V. Doshi, and D. A. Portnoy.** 1995. The broad-range phospholipase C and a metalloprotease mediate listeriolysin O-independent escape of *L. monocytogenes* from a primary vacuole in human epithelial cells. *Infect. Immun.* **63:**4531–4534.

49. **Marino, M., M. Banerjee, R. Jonquieres, P. Cossart, and P. Ghosh.** 2002. GW domains of the *Listeria monocytogenes* invasion protein InlB are SH3-like and mediate binding to host ligands. *EMBO J.* **21:**5623–5634.

50. **Marino, M., L. Braun, P. Cossart, and P. Ghosh.** 1999. Structure of the InlB leucine-rich repeats, a domain that triggers host cell invasion by the bacterial pathogen *L. monocytogenes. Mol. Cell* **4:**1063–1072.

51. **McLaughlan, A. M., and S. J. Foster.** 1998. Molecular characterization of an autolytic amidase of *Listeria monocytogenes* EGD. *Microbiology* **144:**1359–1367.

52. **Mengaud, J., M. Lecuit, M. Lebrun, F. Nato, J.-C. Mazie, and P. Cossart.** 1996. Antibodies to the leucine-rich repeat region of internalin block entry of *Listeria monocytogenes* into cells expressing E-cadherin. *Infect. Immun.* **64:**5430–5433.

53. **Mengaud, J., H. Ohayon, P. Gounon, R. M. Mege, and P. Cossart.** 1996. E-cadherin is the receptor for internalin, a surface protein required for entry of *L. monocytogenes* into epithelial cells. *Cell* **84:**923–932.

54. **Milohanic, E., R. Jonquieres, P. Cossart, P. Berche, J. L. Gaillard.** 2001. The autolysin Ami contributes to the adhesion of *Listeria monocytogenes* to eukaryotic cells via its cell wall anchor. *Mol. Microbiol.* **39:**1212–1224.

55. **Milohanic, E., R. Jonquieres, P. Glaser, P. Dehoux, C. Jacquet, P. Berche, P. Cossart, and J. L. Gaillard.** 2004. Sequence and binding activity of the autolysin-adhesin Ami from epidemic *Listeria monocytogenes* 4b. *Infect. Immun.* **72:**4401–4409.

56. **Milohanic, E., B. Pron, P. Berche, J. L. Gaillard, et al.** 2000. Identification of new loci involved in adhesion of *Listeria monocytogenes* to eukaryotic cells. *Microbiology* **146:**731–739.

57. **Mounier, J., A. Ryter, M. Coquis-Rondon, and P. J. Sansonetti.** 1990. Intracellular and cell-to-cell spread of *Listeria monocytogenes* involves interaction with F-actin in the enterocytelike cell line Caco-2. *Infect. Immun.* **58:**1048–1058.

58. **Müller, S., T. Hain, P. Pashalidis, A. Lignau, E. Domann, T. Chakraborty, and J. Wehland.** 1998. Purification of the *inlB* gene product of *Listeria monocytogenes* and demonstration of its biological activity. *Infect. Immun.* **66:**3128–3133.

59. **Niebuhr, K., F. Ebel, R. Frank, M. Reinhard, E. Domann, U. D. Carl, U. Walter, F. B. Gertler, J. Wehland, and T. Chakraborty.** 1997. A novel proline-rich motif present in ActA of *L. monocytogenes* and cytoskeletal proteins is the ligand for the EVH1 domain, a protein module present in the Ena/VASP family. *EMBO J.* **16:**5433–5444.

60. **Pilgrim, S., A. Kolb-Maurer, I. Gentschev, W. Goebel, and M. Kuhn.** 2003. Deletion of the gene encoding p60 in *Listeria monocytogenes* leads to abnormal cell division and loss of actin-based motility. *Infect. Immun.* **71:**3473–3484.

61. **Pistor, S., T. Chakraborty, K. Niebuhr, E. Domann, and J. Wehland.** 1994. The ActA protein of *L. monocytogenes* acts as a nucleator inducing reorganization of the actin cytoskeleton. *EMBO J.* **13:**758–763.

62. **Portnoy, D. A., P. S. Jacks, and D. J. Hinrichs.** 1988. Role of hemolysin for the intracellular growth of *Listeria monocytogenes. J. Exp. Med.* **167:**1459–1471.

63. **Portnoy, D. A., R. D. Schreiber, P. Connelly, and L. G. Tilney.** 1989. Gamma interferon limits access of *L. monocytogenes* to the macrophage cytoplasm. *J. Exp. Med.* **170:**2141–2146.

64. **Raffelsbauer, D., A. Bubert, F. Engelbrecht, J. Scheinpflug, A. Simm, J. Hess, S. H. Kaufmann, and W. Goebel.** 1998. The gene cluster inlC2DE of *Listeria monocytogenes* contains

additional new internalin genes and is important for virulence in mice. *Mol. Gen. Genet.* **260:**144–158.

65. **Robbins, J. R., A. I. Barth, H. Marquis, E. L. de Hostos, W. J. Nelson, and J. A. Theriot.** 1999. *Listeria monocytogenes* exploits normal host cell processes to spread from cell to cell. *J. Cell Biol.* **146:**1333–1350.

66. **Sanger, J. M., J. W. Sanger, and F. S. Southwick.** 1992. Host cell actin assembly is necessary and likely to provide the propulsive force for intracellular movement of *Listeria monocytogenes. Infect. Immun.* **60:**3609–3619.

67. **Schneewind, O., A. Fowler, and K. F. Faull.** 1995. Structure of the cell wall anchor of surface proteins in *Staphylococcus aureus. Science* **268:**103–106.

68. **Schubert, W. D., G. Gobel, M. Diepholz, A. Darji, D. Kloer, T. Hain, T. Chakraborty, J. Wehland, E. Domann, and D. W. Heinz.** 2001. Internalins from the human pathogen *Listeria monocytogenes* combine three distinct folds into a contiguous internalin domain. *J. Mol. Biol.* **312:**783–794.

69. **Schubert, W. D., C. Urbanke, T. Ziehm, V. Beier, M. P. Machner, E. Domann, J. Wehland, T. Chakraborty, and D. W. Heinz.** 2002. Structure of internalin, a major invasion protein of *Listeria monocytogenes*, in complex with its human receptor E-cadherin. *Cell* **111:**825–836.

70. **Seveau, S., H. Bierne, S. Giroux, M. C. Prevost, and P. Cossart.** 2004. Role of lipid rafts in E-cadherin- and HGF-R/Met-mediated entry of *Listeria monocytogenes* into host cells. *J. Cell Biol.* **166:**743–753.

71. **Shen, Y., M. Naujokas, M. Park, and K. Ireton.** 2000. InlB-dependent internalization of *Listeria* is mediated by the Met receptor tyrosine kinase. *Cell* **103:**501–510.

72. **Smith, G. A., H. Marquis, S. Jones, N. C. Johnston, D. A. Portnoy, and H. Goldfine.** 1995. The two distinct phospholipases C of *Listeria monocytogenes* have overlapping roles in escape from a vacuole and cell-to-cell spread. *Infect. Immun.* **63:**4231–4237.

73. **Smith, G. A., J. A. Theriot, and D. A. Portnoy.** 1996. The tandem repeat domain in the *L. monocytogenes* ActA protein controls the rate of actin-based motility, the percentage of moving bacteria, and the localization of vasodilator-stimulated phosphoprotein and profilin. *J. Cell Biol.* **135:**647–660.

74. **Sousa, S., D. Cabanes, A. El-Amraoui, C. Petit, M. Lecuit, and P. Cossart.** 2004. Unconventional myosin VIIa and vezatin, two proteins crucial for *Listeria* entry into epithelial cells. *J. Cell Sci.* **117:**2121–2130.

75. **Suarez, M., B. Gonzalez-Zorn, Y. Vega, I. Chico-Calero, J. A. Vazquez-Boland.** 2001. A role for ActA in epithelial cell invasion by *Listeria monocytogenes. Cell. Microbiol.* **3:**853–864.

76. **Theriot, J. A., T. J. Mitchison, L. G. Tilney, and D. A. Portnoy.** 1992. The rate of actin-based motility of intracellular *L. monocytogenes* equals the rate of actin polymerization. *Nature* **357:**257–260.

77. **Tilney, L. G., and D. A. Portnoy.** 1989. Actin filaments and the growth, movement, and spread of the intracellular bacterial parasite, *L. monocytogenes. J. Cell Biol.* **109:**1597–1608.

78. **Welch, M. D., J. Rosenblatt, J. Skoble, D. A. Portnoy, and T. J. Mitchison.** 1998. Interaction of human Arp2/3 complex and the *L. monocytogenes* ActA protein in actin filament nucleation. *Science* **281:**105–108.

79. **Wuenscher, M., S. Kohler, A. Bubert, U. Gerike, and W. Goebel.** 1993. The *iap* gene of *Listeria monocytogenes* is essential for cell viability and its gene product, p60, has bacteriolytic activity. *J. Bacteriol.* **175:**3491–3501.

SPORE-FORMING PATHOGENS AND GRAM-POSITIVE MEMBERS OF THE *ACTINOMYCETALES*

SECTION EDITOR: Julian I. Rood

ISEASES SUCH AS BOTULISM, tetanus, gas gangrene, anthrax, and diphtheria are classical diseases that were recognized well before the advent of the science of microbiology and what we now regard as modern medicine. These syndromes were known before the germ theory of disease had been verified and before the concept of spontaneous generation had been disproved. Their analysis provided some of the paradigm formative studies in microbiology: consider the demonstration by Koch in 1877 that *Bacillus anthracis* was the causative agent of anthrax and the identification of diphtheria toxin in 1888 by Roux and Yersin and tetanus toxin in 1889 by Kitasato. Despite the fact that these diseases have been studied for many years, there have been some very exciting discoveries in the past decade. The crystal structures of key toxins such as botulinum toxin, diphtheria toxin, anthrax toxin components, and *Clostridium perfringens* alpha-toxin have been elucidated. The precise enzymatic role and substrate of the botulinum and tetanus metalloprotease neurotoxins have been elucidated, and definitive evidence that alpha-toxin is the essential toxin in *C. perfringens*-mediated gas gangrene has been obtained. Only now are we determining the actual mode of action of the anthrax toxins and how their various components interact. These are exciting times indeed for research on some classical bacterial diseases. Since the first edition of this volume was published, this excitement has increased with the increasing availability of complete genome sequences. Wherever possible, the chapters in this edition have been expanded to include a discussion of data obtained from recently elucidated genome sequences and of the implications of that data for our understanding of fundamental biological processes and for disease pathogenesis.

This section also deals with several diseases, the etiology or importance of which has become apparent only relatively recently. The best example is pseudomembranous colitis, which is caused by *Clostridium difficile*. This organism produces two of the largest toxins that are known, toxins whose action as monoglucosyltransferases has been elucidated only in the past few years. These chapters also deal with what can be regarded as emerging infections such as norcardiosis and actinomycosis.

The chapters are essentially self-contained and are generally focused on a particular pathogen or pathogenic genus. Hence, there are separate chapters on diphtheria, anthrax, *Nocardia*, and *Actinomyces*. The exception to this structure is a discussion of the pathogenic clostridia. A general chapter on the genetics of these organisms is followed by three chapters that are centered on pathogenesis. These chapters are organized on the basis of the mechanism of pathogenesis rather than the bacterial species. Clostridial diseases all involve the elucidation of powerful protein toxins. Irrespective of which bacterium produces these toxins, they primarily affect either the nervous system or the gastrointestinal tract or cause diseases that involve extensive tissue damage and necrosis. Therefore, it was thought that a more coherent understanding of the pathogenesis of these diseases would result from having separate chapters that discussed the neurotoxic, enterotoxic, and histotoxic clostridia.

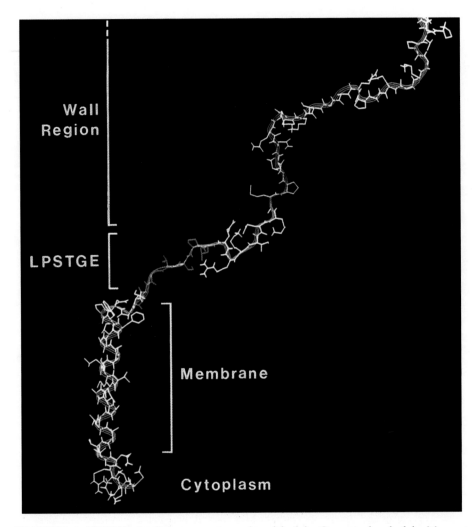

CHAPTER 2, FIGURE 5 Computer-generated model of the C-terminal end of the M protein sequence (residues 371 to 441). A comparable region is found in all C-terminal-anchored surface proteins from gram-positive bacteria (see Table 1). The predicted location of this segment of the molecule is shown in the cytoplasm, membrane, and peptidoglycan. The space between the membrane and peptidoglycan (wall region) may be considered the "periplasm'" of the gram-positive bacterium. The figure was generated on a Steller computer using the Quanta 2.1A program for energy minimization.

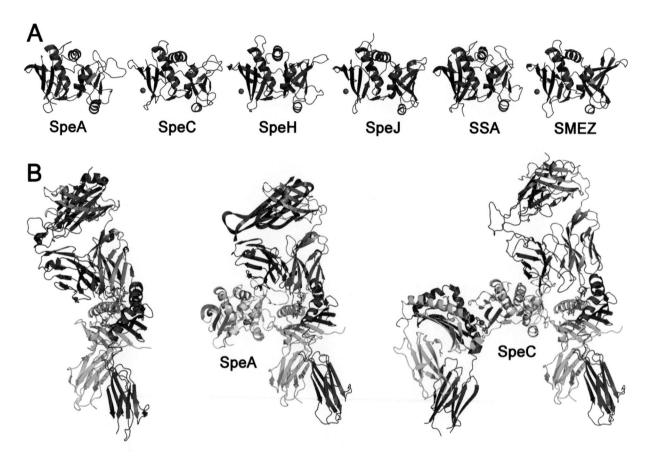

CHAPTER 5, FIGURE 1   Structure conservation and models of T-cell activation complexes for streptococcal superantigens. (A) Ribbon diagrams of the crystal structures for streptococcal pyrogenic exotoxin serotypes A, C, H, J, streptococcal superantigen (SSA), and streptococcal mitogenic exotoxin-Z_2. (SMEZ$_2$) (3, 7, 37, 115, 125). (B) Ribbon diagrams demonstrating typical antigen-mediated T-cell activation (left) (55) and modeled T-cell activation complexes for SpeA (middle) and SpeC (right). The co-crystal structures of SpeA and SpeC in complex with their respective TCR β-chains (127) and of SpeC in complex with the MHC class II through the zinc-dependent high-affinity binding domain have been determined (80). In light of recent evidence (130), it is likely that SpeC also activates T cells in a mode similar to the staphylococcal enterotoxin A model (80) where the superantigen also engages MHC class II through the generic low-affinity binding domain. The binding architecture for the generic low-affinity MHC class II binding to SpeA and SpeC is modeled using the staphylococcal enterotoxin B-MHC class II co-crystal structure (63). Note the presence of the zinc ion (magenta) coordinated in the high-affinity binding site for SpeC and that SpeA lacks this zinc site. The TCR α-chain (shown in gray) for both the SpeA and SpeC diagrams is modeled for clarity by superimposition of the α/β TCR shown on the left to the respective TCR β-chains for both superantigens. The figure was generated using Pymol (32).

CHAPTER 7, FIGURE 1 Reaction of mouse antistreptococcal MAb with human tissue section of myocardium in indirect immunofluorescence assay. Mouse IgM (20 μg/ml) was unreactive (not shown). MAbs were tested at 20 μg/ml. (From reference 50 with permission from *Journal of Immunology*. Copyright 1989, The American Association of Immunologists, Inc.)

3.B6 reacted with human valve.

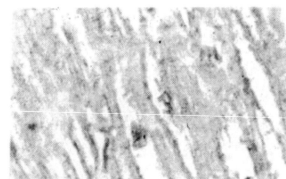

3.B6 reacted with human myocardium

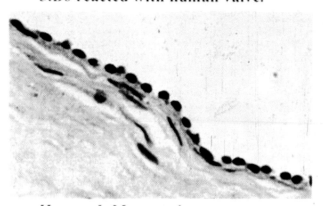

Human IgM control

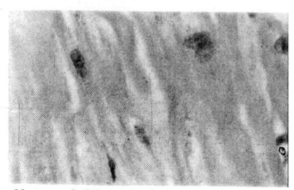

Human IgM control

CHAPTER 7, FIGURE 4   Reactivity of antistreptococcal/antimyosin MAb 3.B6 with normal human valve endothelium and myocardium. Formalin-fixed human mitral valve (top left) and myocardium (top right) were reacted with MAb 3.B6 at 10 μg/ml. MAb 3.B6 binding was detected using biotin-conjugated antihuman antibodies and alkaline phosphatase-labeled streptavidin followed by fast red substrate. Control sections (bottom) did not react with human IgM at 10 μg/ml. (With permission from the *Journal of Clinical Investigation.*)

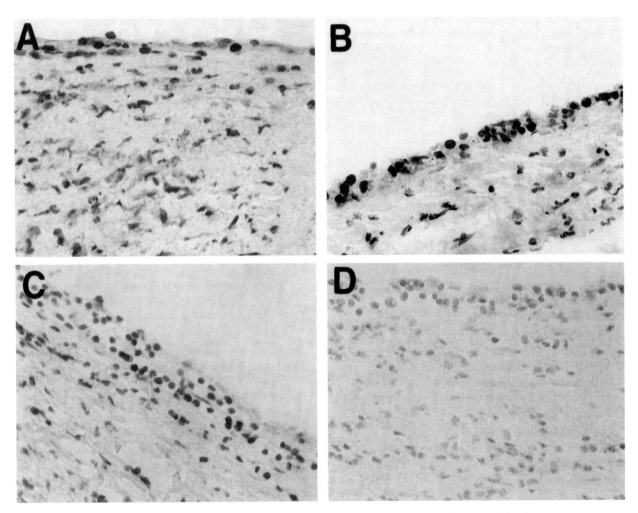

CHAPTER 7, FIGURE 5 Adhesion and extravasation of T lymphocytes into ARF valve in valvulitis. (A, B) Extravasation of CD4⁺ lymphocytes (stained red) (original magnification, ×200 and ×400, respectively). (C) Extravasation of CD8⁺ lymphocytes (stained red) into the valve through the valvular endothelium (magnification, ×200). An IgG₁ isotype control MAb (IgG₁) did not react with the same valve (not shown) (magnification, ×400). (With permission from reference 117, the *Journal of Infectious Diseases*, University of Chicago.)

CHAPTER 9, FIGURE 1 Streptococcus-mediated signaling events and their implications. The cartoon illustrates reported signaling events. Some of the pathways are hypothesized on the basis of available specific reports on streptococci-mediated signaling events and established signal transduction pathways in eukaryotes. There are at least four receptors (uPAR, enolase, CD44, CD46) that directly interact with group A *Streptococcus* (GAS) surface proteins. Indirect binding to eukaryotic cells through fibronectin is likely mediated by $\alpha5\beta1$ integrin. GAS invades host epithelial cells by two different mechanisms, either by invagination at the point of bacterial contact with host cells or by massive induction of microvilli, which form membrane ruffling for engulfment of bacteria (filopodia). Cytoskeletal rearrangements involve induction of the IP-3 kinase pathway or RAS/CDC42/tyrosine kinase activation. Some of the GAS secretory products such as SLO and NAD glycohydrolase may cooperatively make holes in eukaryotic cells and inject bacterial product in eukaryotic cells to exploit intracellular signaling events in a manner similar to the type III secretory system of gram-negative bacteria. Intracellular GAS may then direct host cells to undergo apoptosis via caspase-dependent and -independent pathways. In GAS-mediated apoptosis mitochondria seem to play a crucial role. NAD-glycohydrolase of GAS may convert NAD to ADP-ribose and/or cyclic ADP-ribose, which may then direct the host cell to undergo apoptosis through the Ca^{2+} signaling pathway. Although not fully explored, transcriptional regulation of many inflammatory cytokines and apoptosis in the GAS-infected host cells may be mediated via NF-κB and JAK/STAT pathways. Histone-phosphorylation/dephosphorylation and histone acetylation/deacetylation also seem to play important roles in gene transcription in GAS-infected host cells. The induction of signaling pathways may vary depending on the cell lines, status of the cell (polarized versus nonpolarized), type of GAS strains, and growth phase. Abbreviations: ADPR-adenosine diphosphate-ribose; cADPR-cyclic ADPR; AIF-apoptosis inducing factor; Akt-AKT retroviral oncogene protein Ser/Thr kinase; APAF1-apoptosis protease activating factor-1 or CED-4; Arp2/3-actin-related protein 2 and 3; ASK1-apoptosis signal-regulating kinase 1 (also known as MEKK5); ATF2-activating transcription factor 2; BAD-Bcl-xL/Bcl-2 associated death promoter; Bid-BH-3 interacting domain death agonist that induces ICE-like proteases and apoptosis; Bcl2-B-cell lymphoma 2, which belongs to the Bcl-2 family of proteins and is known to inhibit apoptosis; BAK-Bcl-2 antagonist/killer. Bak is a pro-apoptotic protein; BAX-Bcl-2 associated x protein. Bax is a member of the Bcl-2 family and is pro-apoptotic; tBid-truncated Bid; CARD-caspase activation and recruitment domain; Caspase-cysteinyl aspartic acid-protease; CD44-human leukocyte differentiation receptor antigen for hyaluronate and proteoglycin serglycin; CD46-member of RCA gene family, receptor for measles virus and the M protein of GAS; Cdc42-cell division cycle 42 (GTP-binding protein); CytC-cytochrome C; c-Fos/c-Jun-transcription factor, also known as activator protein or AP-1; ELK1-Ets domain containing DNA-binding protein. Mammalian ELK-1, ELK-3 (also known as Net or SAP-2) and ELK-4 (also known as SRF accessory protein 1 [SAP-1]), which all form a ternary complex with the serum response factor (SRF); ERK-extracellular signal-related protein kinase; FADD-Fas-associated death domain; FAK-focal adhesion kinase; FAS-known as CD95 or APO-1. Fas is a member of the TNF receptor family and promotes apoptosis. FasL-Fas ligand. FasL is also known as APO-1 ligand or Apo-L; FLIP-FLICE (Fadd-like ICE [interleukin-1b converting enzyme also known as caspase-1]) inhibitory protein; $G_{\alpha,\beta,\gamma}$-G-protein α, β, and γ; HAT-histone acetyl transferase; HDAC-histone deacetylase; IKB-inhibitory IB (inhibitor of NF-κB) proteins; IKKs-IKK-1 and IKK-2 are two direct IB kinases; IL-interleukins; INF-interferon α or γ; INFR-interferon-γ receptor; IRAK-interleukin-1 receptor-associated kinases; JAK-Janus kinase; JNK-c-Jun N terminal kinase; MAPK-mitogen-activated protein kinase; MEK-MAPK activator; MEKKs-mitogen-activated protein/ERK kinase kinases; MLK1,2-CdC42-dependent kinases; MLK4,7-CdC42-dependent kinases; mTOR-mammalian target of Rapamycin (an immunosuppressant) protein; MyD88-myeloid differentiation factor 88; NAD-nicotinamide adenine dinucleotide; NFKB-nuclear factor of immunoglobulin κ locus in B cells. NF-κB activates transcription of genes in many tissues; NIK-Nck interacting kinase (interacting with MEKK) is different from NIK (NFκB-interacting kinase); PI-3K-Phosphatidyl inositol-3 kinase; RAC-small GTP-binding protein encoded by the *rac* gene; RAS-small GTP-binding protein protooncogene encoded by the *ras* gene; RIP-receptor interacting protein; ROS-reactive oxygen species; SEN-streptococcal surface enolase; SH2-Src homology region 2; SLO-streptolysin O; STAT-signal transducers and activators of transcription; TIRAP-Toll-interleukin 1 receptor domain-containing adaptor protein; TLR-Toll-like receptor; TNF-tumor necrosis factor; TNFR-tumor necrosis factor receptor; TOLL-human homologue of the *Drosophila* Toll protein; TOLLIP-Toll-interacting protein; TRAD-TNF receptor I associated death domain; TRAF-tumor necrosis factor receptor-associated factors; SDH-streptococcal surface dehydrogenase; uPAR-urokinase plasminogen activator receptor.

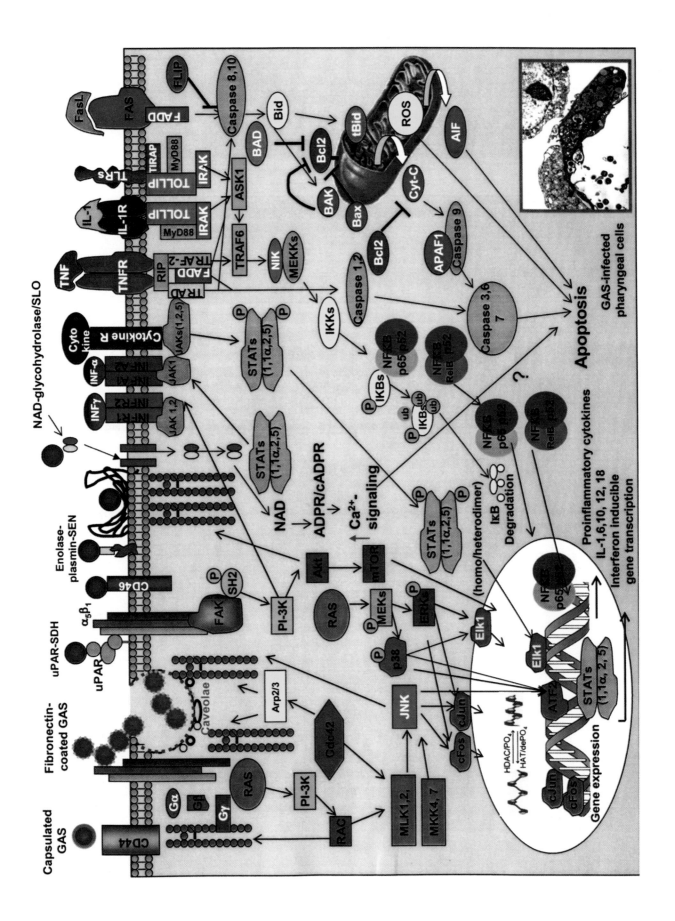

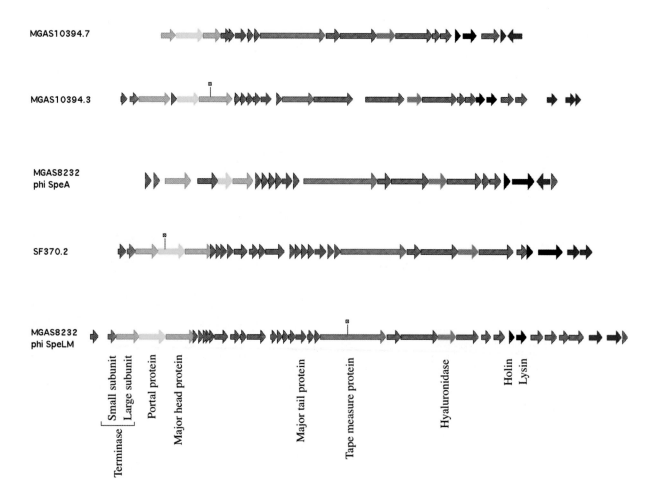

CHAPTER 11, FIGURE 8 The late gene regions of the group IV prophages. Some of the identifiable genes are listed below the MGAS8232 phi SpeLM prophage genome; the corresponding color scheme is used to identify related genes in the other prophages. The locations of probable mutations are indicated by the box symbol.

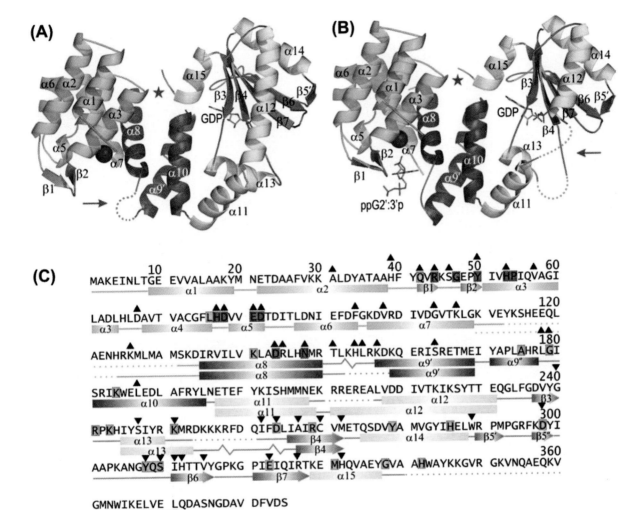

(C)

```
        10          20          30    ▲    ▲40 ▲▲▲    ▲50  ▲    ▲60
MAKEINLTGE  EVVALAAKYM  NETDAAFVKK  ALDYATAAHF  YQVRKSGEPY  IVHPIQVAGI
              α1                      α2         β1    β2      α3

                 ▲            ▲▲    ▲▲          ▲          ▲  ▲     120
LADLHLDAVT  VACGFLHDVV  EDTDITLDNI  EFDFGKDVRD  IVDGVTKLGK  VEYKSHEEQL
  α3          α4          α5          α6          α7

       ▲                        ▲▲      ▲  ▲▲▲                  ▲▲ 180
AENHRKMLMA  MSKDIRVILV  KLADRLHNMR  TLKHLRKDKQ  ERISRETMEI  YAPLAHRLGI
                          α8                      α9"         α9"
                          α8                      α9'

     ▲                                                          240
SRIKWELEDL  AFRYLNETEF  YKISHMMNEK  RREREALVDD  IVTKIKSYTT  EQGLFGDVYG
  α10         α11                     α12                      β3
              ▽           α11          ▽          α12             300
RPKHIYSIYR  KMRDKKKRFD  QIFDLIAIRC  VMETQSDVYA  MVGYIHELWR  PMPGRFKDYI
  α13         ▽  ▽        β4          α14             β5'       β5"
    α13                    β4
     ▽      ▽▽          ▽             ▽               360
AAPKANGYQS  IHTTVYGPKG  PIEIQIRTKE  MHQVAEYGVA  AHWAYKKGVR  GKVNQAEQKV
              β6          β7          α15

GMNWIKELVE  LQDASNGDAV  DFVDS
    α16
```

CHAPTER 16, FIGURE 5 Structure of Rel$_{Seq}$1-385. (A) ppGpp Hydrolase-OFF/synthetase-ON conformation, in complex with Mn^{2+} (blue sphere) and GDP (stick rendering). The hydrolase domain is highlighted in green α-helices and blue β-strands, the synthetase domain is in yellow α-helices and orange β-strands, and the central 3-helix bundle is in red. Part of the substrate-binding cleft comprising the hydrolase site (down-regulated) is disordered (red arrow). The small synthetase/hydrolase interdomain contact interface involved in signal transmission is labeled with a red star. (B) Hydrolase-ON/synthetase-OFF conformation, in complex with Mn^{2+}, ppG2':3'p (which locks the enzyme in the hydrolase-ON/synthetase-OFF conformation) and GDP. The coloring and rendering schemes are the same as for (A). The disordered synthetase site (down-regulated) is illustrated as dashed lines with a red arrow. (C) Primary and secondary structure of Rel$_{Seq}$1-385. Secondary structure is color-coded according to (A) and (B). Unique secondary structure assignments for (A) are placed immediately below the corresponding assignments for (B). Residues absolutely conserved throughout the mono- and bifunctional RelA and SpoT homologs are underlaid with blue boxes. Residues conserved in SpoT and the bifunctional enzymes but mutated in the hydrolase-incompetent RelA homologs are underlaid with red boxes. Upright and inverted black triangles above the sequence indicate residues which, when substituted experimentally by missense mutations, lead to defective hydrolase and synthetase activities, respectively. (Reproduced from reference 51, with permission.)

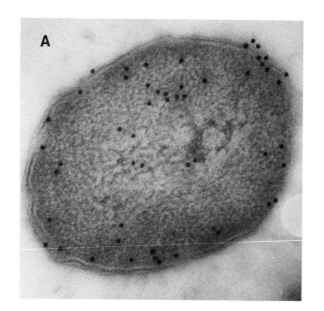

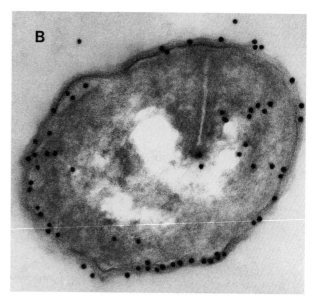

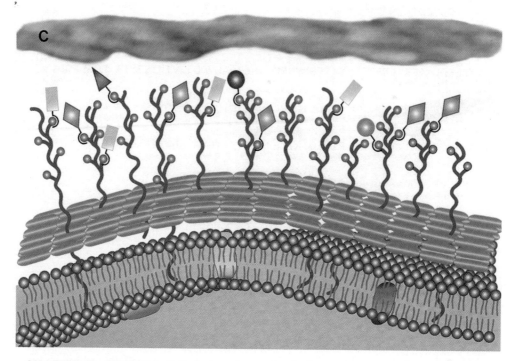

CHAPTER 21, FIGURE 1 Immunohistochemical and schematic depiction of the choline biology of the pneumococcal surface. Immunogold-labeling of pneumococci with (A) TEPC-15 antibody recognizing free choline and (B) antiautolysin antibody. These two images contrast free (A) versus CBP-bound (B) choline. (C) Schematic view of the capsule (blue), cell wall (green), and membrane (red). The teichoic and lipoteichoic acids are indicated as dark blue lines bearing choline (circles). A proportion of these are capped by CBPs. (Courtesy of Dr. K. G. Murti, St. Jude Electron Microscopy Core Facility.)

VanA

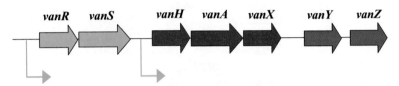

VanB

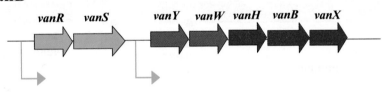

% identity   **33**   **19**   **30**   **67**   **76**   **71**

CHAPTER 26, FIGURE 1   Organization of the VanA and VanB vancomycin resistance systems. Gene functions are color-coded: green, sensory and regulatory; red, essential resistance genes; blue, auxiliary resistance genes. Detailed functions of each gene are described in the text. Green arrows show the position of the VanR-inducible promoters. The percent identity of the VanB genes to their VanA homologs is given below the individual VanB genes. Homologs have identical names in both systems except for the *vanA* and *vanB* ligases.

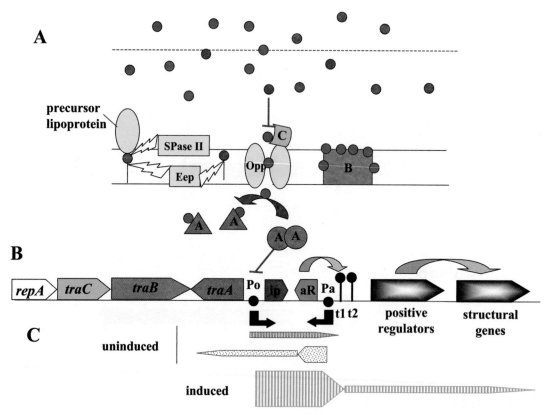

CHAPTER 26, FIGURE 2 Regulatory circuitry of the pheromone response. (A) Events occurring at the cell surface. Important chromosomally encoded determinants are colored yellow, with the exception of the pheromone itself, which is shown as blue circles. Plasmid-encoded determinants are color-coordinated with their genes shown in B. The inhibitor peptide (ip) is shown as red circles and competitively inhibits pheromone binding to TraC. TraB inhibition of pheromone secretion is depicted here as sequestration, but this is only one possible mechanism and has not been proven. (B) Events occurring at the DNA level. Binding of pheromone by TraA links events at the cell surface with events at the DNA level. A conformational change in TraA due to pheromone binding is indicated by the change in shape of the molecule and change in dimerization state, although the precise changes are still unknown. Pheromone-free TraA binds P_0 and inhibits transcription. Direction of transcription from P_0 and P_a is indicated by the arrows. The antisense RNA, generalized to aR in this figure, stimulates termination at t1, indicated by the green arrow. Positive regulatory elements vary between different pheromone-responsive plasmids, and their mechanisms of inducing downstream transcription of the conjugation structural genes may also vary. For simplicity, the more common gene names and order are used. It should be noted that the gene order of the *traC* and *traB* genes is reversed in pAD1. The RepA gene is shown to orient the reader relative to Figure 3 (see Fig. 3 in Chapter 26). (C) Events at the RNA level. Relative levels of RNA produced from the P_0 and P_a promoters under uninduced (red) and induced (green) conditions are depicted by the size of the arrows.

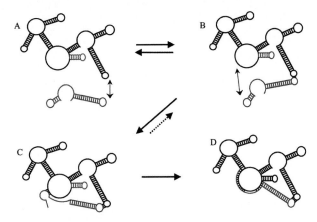

CHAPTER 26, FIGURE 4 Model of RNA I–RNA II interaction. RNA I is the larger black structure and RNA II is the smaller blue structure. Stems, loops, and bulges are depicted in their approximate locations and sizes as determined by experimentation. The red stem-loop structure within RNA I sequesters the ribosome-binding site and prevents translation until a complex is formed. The initial interaction occurs at a U-turn motif in the terminator loop of RNA I (A), followed by interaction between the complementary repeats in the 5′ end of each RNA (B and C). Because RNA II-mediated protection from the RNA I-encoded toxin occurs in vivo even when one of the repeats is mutated, the structure shown in panel C is apparently sufficient to prevent translation in vivo. Once complex formation is complete (D), the structure is extremely stable in vivo and in vitro, perhaps due to the gap between the interacting repeats.

A

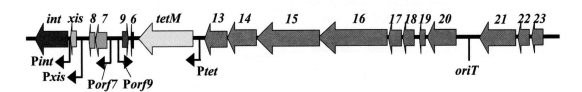

B

CHAPTER 26, FIGURE 5 Genetic organization of Tn*916* and the relative positions of Int and Xis binding sites. (A) Functions of Tn*916* genes are color-coded: blue, recombination; green and red, positive and negative regulation, respectively; yellow, tetracycline resistance; magenta, conjugation. Gene names and open reading frame numbers are shown above the individual genes. The origin of transfer is designated by the line between *orfs 20* and *21* labeled *oriT*. Promoters and the direction of transcription are designated by labeled arrows below the gene line. (B) Binding sites for the Int-C and Int-N DNA-binding domains and Xis are designated by marked circles and triangles. Int and Xis binding sites are color-coordinated with the genes shown in panel A.

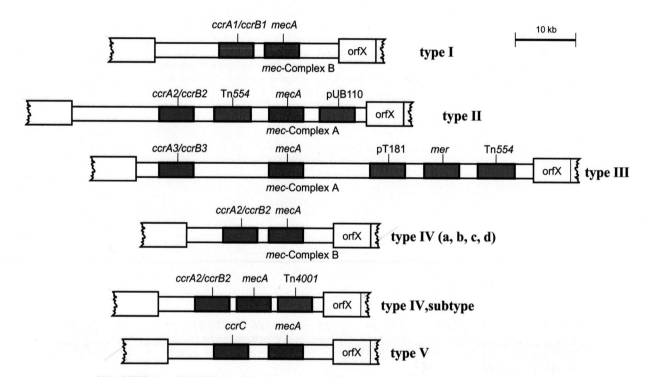

CHAPTER 31, FIGURE 5 Different types of SCC*mec* elements demonstrated in MRSA. orf X, open reading frame X of the chromosome. (Adapted with permission from references 16 and 30.)

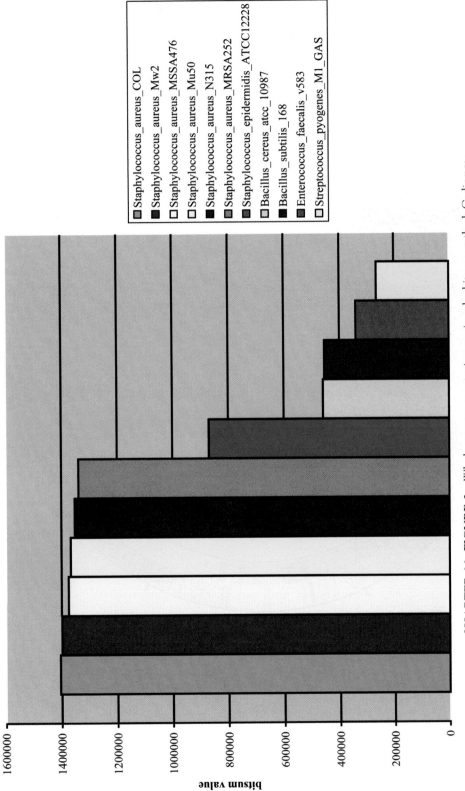

CHAPTER 32, FIGURE 2 Whole genome comparisons using the bitsum method. Coding regions from each of the six *S. aureus* sequenced strains were used in a BLASTp analysis with ORFs from NCTC 8325. The bit scores for these BLASTp comparisons were then summed (the bitsum) and used to plot the similarity of each genome to strain 8325. Coding regions from the completed genomes for other gram-positive organisms are shown for comparison.

Legend:
- Staphylococcus_aureus_COL
- Staphylococcus_aureus_Mw2
- Staphylococcus_aureus_MSSA476
- Staphylococcus_aureus_Mu50
- Staphylococcus_aureus_N315
- Staphylococcus_aureus_MRSA252
- Staphylococcus_epidermidis_ATCC12228
- Bacillus_cereus_atcc_10987
- Bacillus_subtilis_168
- Enterococcus_faecalis_v583
- Streptococcus_pyogenes_M1_GAS

bitsum value

1600000
1400000
1200000
1000000
800000
600000
400000
200000
0

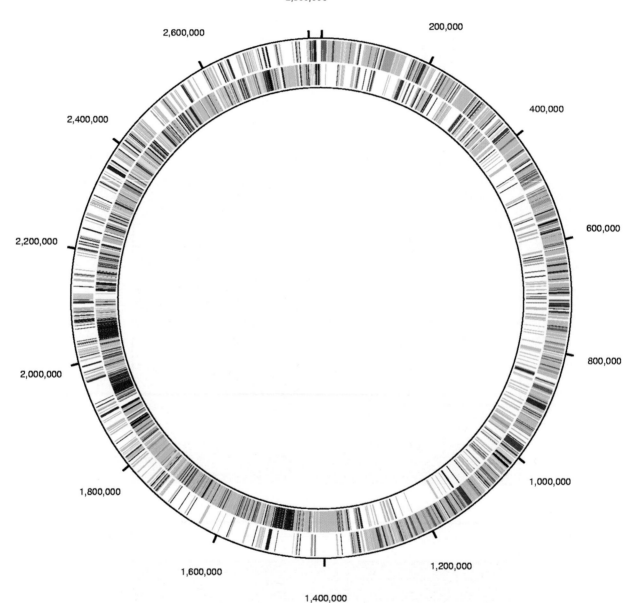

CHAPTER 32, FIGURE 3   Circular map of the NCTC 8325 genome. The ORFs present on the sense strand are represented on the outermost circle, the ORFs on the antisense strand are shown on the second circle, and the innermost circle is the GC skew on each coding strand. Nucleotide positions are labeled every 200,000 bp, and each of the predicted ORFs is colored based on the assigned functional role category in the final annotation. Colors and designations for each role category can be found in the legend for Fig. 4 (see Fig. 4, next page). As is customary, the first ORF represented is *dnaA*.

Role Category	#	%
Amino acid biosynthesis	69	2%
Biosynthesis of cofactors, prosthetic groups, and carriers	70	2%
Cell envelope	98	3%
Cellular processes	82	3%
Central intermediary metabolism	34	1%
Conserved hypothetical proteins	1636	51%
DNA metabolism	70	2%
Energy metabolism	177	6%
Fatty acid and phospholipid metabolism	32	1%
Hypothetical proteins / Unknown / Unclassified	201	6%
Mobile and extrachromosomal element functions	236	7%
Protein fate	58	2%
Protein synthesis	114	4%
Purines, pyrimidines, nucleosides, and nucleotides	57	2%
Regulatory functions	65	2%
Signal transduction	31	1%
Transcription	21	1%
Transport and binding proteins	160	5%
Total	3211	

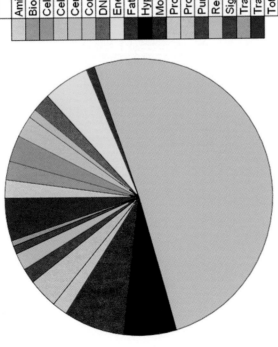

CHAPTER 32, FIGURE 4   Graphical representation of the functional role category assignments for each of the 2,892 predicted ORFs for NCTC 8325. Numbers of genes and the percentage of the genome they represent are shown for each category. Percentage totals and the total number of ORFs equal more than 100% due to multiple role assignments for some ORFs.

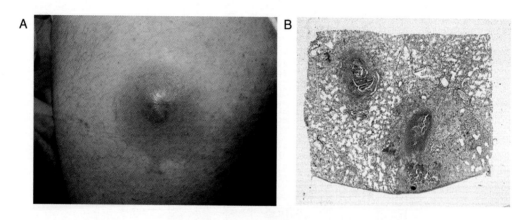

CHAPTER 41, FIGURE 1 Staphylococcal abscesses. (A) Cutaneous furuncle (Nadir Goksugur, M.D., Dermatlas; http://www.dermatlas.org). (B) Stained section of pulmonary abscess. Kindly provided by Martin Nachbar.

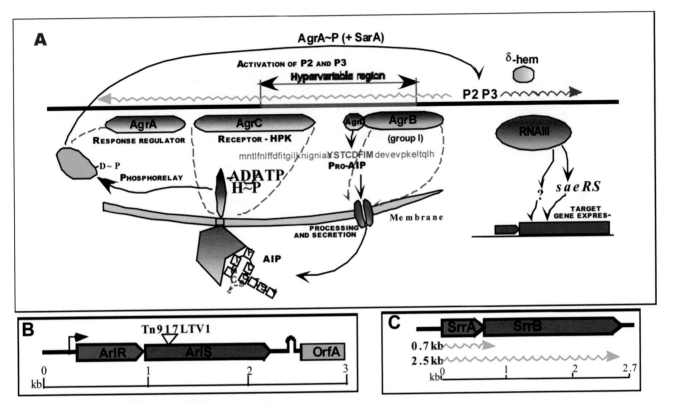

CHAPTER 41, FIGURE 3 *agr*, *srhRS*, and *arlAB*. TCS modules known to affect the virulon. (A) The *agr* system. The pro-AIP peptide is processed and secreted by AgrB, binds to an extracellular loop in the receptor-HPK, AgrC, activating autophosphorylation (or dephosphorylation), followed by phosphorylation or dephosphorylation of the response regulator, AgrA. AgrA, in conjunction with SarA, activates the two *agr* promoters, P2 and P3, leading to the production of RNAIII. RNAIII controls transcription of the target genes via one or more intracellular regulatory mediators, including a second two-component module, *saeRS*. (B) *arlRS*. The *arlRS* locus encodes a receptor-HPK (*arlS*) and a response regulator (*arlR*), driven by a single promoter and followed by a terminator stem-loop. (C) *srrAB*. The *srrAB* locus encodes a receptor-HPK (*srrB*) and a response regulator (*srrA*), driven by a single promoter that generates two transcripts whose relative significance is unknown. (Reprinted from reference 94 with the kind permission of Blackwell Publishing, Ltd.)

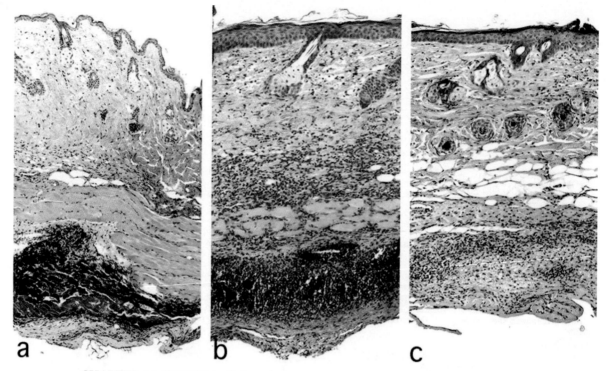

CHAPTER 44, FIGURE 1   Early and persistent infiltrations of inflammatory cells during *S. aureus* dermatitis. The micrograph shows the inflammatory infiltrate in mouse skin that was inoculated intracutaneously with 2×10^8 CFU of *S. aureus* LS-1: after 6 h (A), 48 h (B), and 1 week (C). The inflammatory infiltrate, which mainly contains macrophages, peaks at 48 h and starts to disappear 2 weeks after the inoculation (from reference 59).

CHAPTER 53, FIGURE 5   Fluorescence microscopy of Vero cells infected with wild-type *L. monocytogenes* (A) or *L. monocytogenes* ΔActA (B). Wild-type *L. monocytogenes* induces the formation of actin comet tails that propulse the bacteria in the host cell cytoplasm (A), while *L. monocytogenes* ΔActA proliferates in microcolonies that are unable to spread from the primary infected cell (B). DNA is labeled with DAPI (4′,6′-di-amidino-2-phenylindole; blue), the bacterial cell wall is labeled with an anti-*L. monocytogenes* antibody (red), and actin is labeled with fluorescent phalloidin (green).

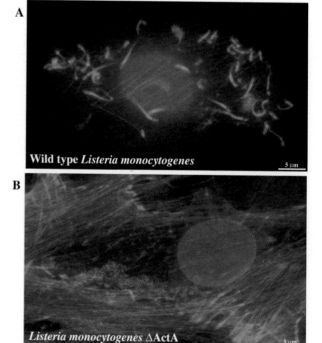

A

Wild type *Listeria monocytogenes*   5 μm

B

Listeria monocytogenes ΔActA   5 μm

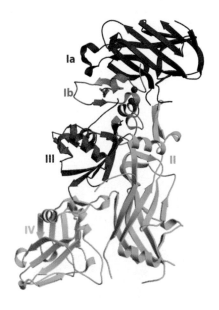

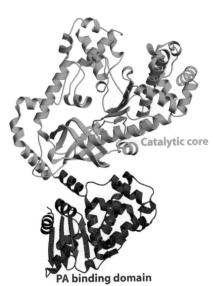

CHAPTER 54, FIGURE 1 Ribbon diagrams representing structures of the anthrax toxin proteins. (A) Monomeric PA. Ia (blue), 20-kDa fragment removed with cleavage; Ib (yellow), forms N terminus of PA_{63} and contains two structural calcium ions (red); II (green), pore formation; III (magenta), oligomerization of PA_{63}; IV (turquoise), receptor binding. (B) LF. Substrate-binding and catalytic domains (green) and PA-binding domain (magenta). (C) EF in complex with calmodulin. Catalytic core (green); PA-binding domain (magenta); helical domain (yellow) interacts with calmodulin (red). (Courtesy of W.-J. Tang.)

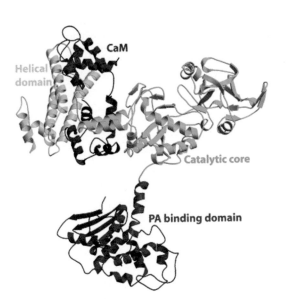

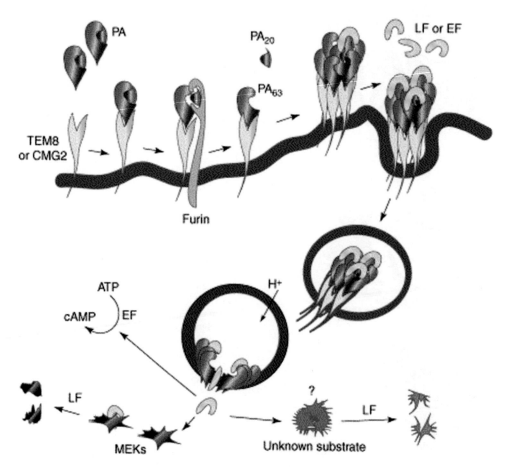

CHAPTER 54, FIGURE 2 Model of anthrax toxin action. PA binds to receptors TEM8 or CMG2. Following proteolytic cleavage of PA by furin, PA_{63} oligomerizes to form a heptameric pre-pore. EF-LF binds to the prepore, and the complex is endocytosed. Acidification of the intracellular compartment triggers translocation of EF and LF to the cytosol. EF, a calmodulin-dependent adenylate cyclase, converts ATP to cAMP. LF, a zinc-dependent protease, cleaves members of the MEK family and may also affect other targets. (Reprinted from Moayeri and Leppla [75].)

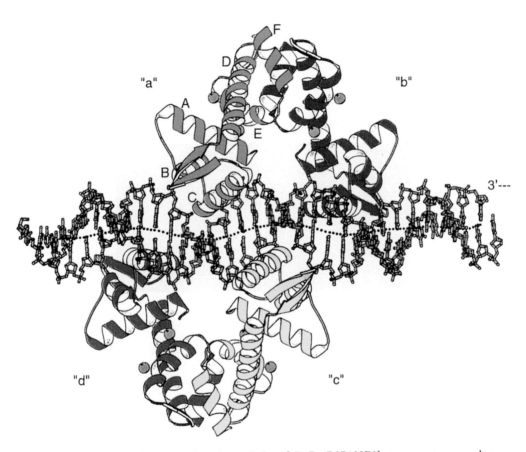

CHAPTER 59, FIGURE 2 Structure of the 2[Ni DtxR(C102D)]$_2$-*tox* operator complex. Residues 3 to 120 in each DtxR(C102D) monomer are designated "a" to "d." Ribbons and arrows are used to indicate α-helices and β-strands in each monomer. The 33-bp DNA segment carries the 27-bp interrupted palindromic *tox* operator sequence. (Adapted from White et al. [107].)

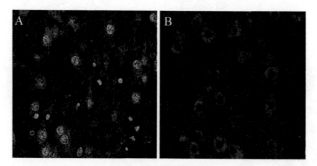

CHAPTER 61, FIGURE 6 Confocal micrographs of nocardia-induced apoptosis. (A) Apoptosis of dopaminergic neurons in the substantia nigra in a head-shake mouse 14 days after infection. The red stain localizes dopaminergic neurons. Free 3'-OH ends of DNA from apoptotic nuclei were labeled with nucleotides conjugated to fluorescein isothiocyanate (green). (B) Uninfected control. Dopaminergic neurons in the substantia nigra in a healthy uninfected mouse. The red stain localizes dopaminergic neurons. Note that there is no apoptosis. (Reproduced from reference 68.)

Bacillus anthracis

THERESA M. KOEHLER

54

ANTHRAX

Past and Present

Prior to the extraordinary interest in *Bacillus anthracis* generated by the recent bioterrorism events in the United States, much of microbiologists' awareness of the bacterium resulted from its historical significance. Accounts of a disease resembling anthrax are described in biblical and classical literature, causing many to cite anthrax as one of the oldest documented infectious diseases. In more recent history, anthrax epizootics of the 17th through the 19th centuries caused huge domestic livestock losses. A panzootic of anthrax that killed approximately one-half of the sheep in Europe in the mid-1800s provided impetus for study of the disease by early microbiologists. In 1850, Davaine described rod-shaped bodies, "bacteridia," in the blood of animals that had died of anthrax. Unequivocal proof that the anthrax bacilli cause disease is attributed to Koch, who in 1877 clearly established the etiology of anthrax by infecting animals with an organism isolated in pure culture. In 1881, Pasteur used heat-attenuated anthrax bacilli to protect sheep, cattle, and horses from a virulent strain of the anthrax bacillus, thus developing the first live attenuated bacterial vaccine (40, 110).

Immunization of livestock and improvements in environmental hygiene during the 20th century led to the global decline of anthrax (40). Nevertheless, the worldwide incidence of anthrax in animal populations and the risk of human infection associated with animal outbreaks fostered continued investigations of the bacterium and anthrax disease (47). In 1939, Sterne isolated a highly attenuated strain of *B. anthracis* that is currently used as a veterinary vaccine (103). In the 1950s, H. Smith discovered the anthrax toxin in fundamental experiments employing animal models (101). In the latter part of the 20th century, studies of the anthrax toxin proteins by a number of investigators revealed a detailed understanding of toxin structure and function. Then, in the fall of 2001, the deliberate distribution of *B. anthracis* spores via the United States postal system resulted in 22 cases of anthrax in humans. This first outbreak of bioterrorism-related anthrax in the United

States led to an unprecedented worldwide resurgence in investigations of *B. anthracis*. New information built upon the fundamental discoveries of earlier microbiologists is being gained at a rapid pace with the goal of improving the prevention, diagnosis, and treatment of anthrax disease.

Systemic and Cutaneous Diseases

B. anthracis can infect all mammals, some birds, and possibly even reptiles. Systemic anthrax, generally resulting from inhalation or ingestion of *B. anthracis* spores, has a high fatality rate. In the cases among humans in the United States in 2001, 5 of 11 patients diagnosed with inhalation anthrax died (53). The survival of the other six patients was attributed primarily to early treatment with antibiotics and aggressive medical care. Inhalation anthrax begins with influenza-like symptoms such as fever and malaise. Symptoms often include cough (minimal or productive), nausea or vomiting, and dyspnea. Onset of symptoms can range from days to weeks, depending upon the spore-containing particle size and dose. Initial symptoms are followed by sudden onset of severe respiratory distress; without early intervention, death ultimately occurs within 2 to 3 days from respiratory failure, sepsis, and shock (52). When *B. anthracis* spores are inhaled, alveolar macrophages transport the organism to regional lymph nodes, where spore germination and vegetative growth occur (40). From there, the bacteria enter the bloodstream, where concentrations can reach $>10^8$ cells per ml. Symptoms of gastrointestinal anthrax are similar to those of inhalation anthrax and are sometimes accompanied by hemorrhagic diarrhea.

A generally less-severe form of anthrax can result from entry of *B. anthracis* cells or spores via a cut or abrasion in the skin. Cutaneous anthrax infections appear initially as a small papule or pimple surrounded by a ring of vesicles. Within a few days, the central lesion ulcerates, dries, and blackens to form a characteristic eschar. The region also exhibits marked edema. Cutaneous anthrax infections generally remain localized and are readily treatable with antibiotics. Approximately 20% of untreated cases of cutaneous anthrax progress to fatal septicemia (40).

Gram-Positive Pathogens, 2nd edition, edited by Vincent A. Fischetti et al.
© 2006 ASM Press, Washington, D.C.

Epidemiology

Of all naturally acquired human cases of anthrax, >95% are cutaneous infections. Cutaneous anthrax can result from contact with dead infected animals or their products. Inhalation anthrax in humans, also known as wool-sorter's disease, was once common among textile workers who inhaled spores in dust generated while working with contaminated wool and animal hides. Humans and scavenging animals acquire gastrointestinal anthrax by ingesting spores in meat from infected animals, while herbivores ingest *B. anthracis* spores from the soil while grazing. Although virtually all mammals appear to be susceptible to anthrax infection, the infectious dose and severity of disease vary widely in different species. Domestic livestock, including sheep and cattle, and wild herbivores, such as bison, elephants, hippopotami, and kudu, are particularly susceptible to infection (113).

Natural dissemination of *B. anthracis* spores occurs in a number of ways. Blowflies transfer spores from infected carcasses to trees and shrubs that serve as food for herbivores. Vultures do not appear to be susceptible to anthrax but can carry spores to water supplies. In addition, spores can be found in the feces of carnivorous scavengers. *B. anthracis* spores can remain viable in soil, water, and contaminated animal products for decades. The most severely affected parts of the world are central Asia and western areas of Africa. Endemic anthrax in major wildlife reserves in southern Africa threatens several endangered species. A so-called anthrax belt exists from Turkey to Pakistan. In North America, anthrax is endemic in northern Alberta and the Northwest Territories of Canada. Sporadic outbreaks of anthrax occur within the United States; northwest Mississippi-southeast Arkansas and western Texas are considered areas of endemicity. The disease is also endemic throughout Mexico, Central America, and many South American countries (113).

Intentional dissemination of *B. anthracis* spores occurred in the United States in 2001. In October and November of that year, 11 inhalation and 11 cutaneous cases of anthrax in humans were traced to four envelopes containing *B. anthracis* spores in powder form. The envelopes were sent to different locations via the U.S. Postal Service, and most of the patients were either mail handlers or were exposed at worksites where contaminated letters were processed or received. Others are thought to have been exposed to cross-contaminated mail (53). The tragic event demonstrated the apparent ease with which the organism can be dispersed and solidified the placement of *B. anthracis* on the list of biothreat agents.

B. ANTHRACIS

Taxonomy, Physiology, and Morphology

B. anthracis is a member of the group 1 bacilli, which also include *Bacillus cereus*, *Bacillus thuringiensis*, *Bacillus mycoides*, and *Bacillus weihenstephanensis*. Historically, the species were grouped phylogenetically on the basis of their similar physiology. Recent genomic analyses have generally supported this grouping (49, 88, 90), but the phylogeny of these species continues to be deliberated. Although there are rare strain-specific exceptions, distinguishing phenotypic characteristics of *B. anthracis* include lack of motility, weak hemolytic activity, sensitivity to gamma bacteriophage and penicillin, and the elaboration of the specific virulence factors anthrax toxin and capsule.

B. anthracis is a facultative anaerobe and grows in most rich undefined media with a doubling time of approximately 30 min. The only absolute nutritional requirements for growth are methionine and thiamine. However, most strains grow poorly on glucose-salts medium containing only these amino acids. Commonly used minimal media contain nine or more amino acids. Uracil, adenine, guanine, and manganese stimulate growth of some strains. The optimal growth temperature for *B. anthracis* is 37°C, and cells cannot grow above 43°C (106).

Under most culture conditions, the rod-shaped vegetative cells of *B. anthracis* form long bamboo-like chains characteristic of the species. Like all *Bacillus* species, nutrient limitation can cause *B. anthracis* cells to develop into metabolically dormant spores which are resistant to environmental challenges such as UV radiation, heat, and desiccation. Sporulating cells carry elliptic, centrally located spores. In infected tissues, chains of vegetative cells are shorter than those formed during growth in vitro and spores are generally not apparent (111). *B. anthracis* cells are capsulated in infected tissues and when grown in appropriate in vitro conditions (see below).

Genome

The *B. anthracis* genome is comprised of a 5.23-Mbp chromosome and two plasmids, pXO1 (182 kb) and pXO2 (95 kb) (90), and is predicted to encode approximately 5,700 proteins. Housekeeping functions are partitioned to the chromosome, while the well-characterized virulence genes are located on the plasmids. Plasmid pXO1 carries the structural genes for the toxin proteins (78), and pXO2 harbors the biosynthetic genes for capsule formation (67). In comparison to the chromosome, the plasmids have a greater proportion of insertion sequences and genes without ascribed functions (90). Generally, plasmid content has served for facile identification of *B. anthracis* and distinction from other *B. cereus* group species. However, recent reports of closely related species harboring similar plasmids or plasmid-encoded genes indicate that speciation can be complex (46, 89).

Despite the importance of pXO1 and pXO2 in virulence, there have been relatively few reported investigations of plasmid replication and stability in *B. anthracis*. The replication initiation protein and origin of replication of pXO2 were identified recently (108), but the mechanism for replication of pXO1 is completely unknown. Neither plasmid is self transmissible, but conjugative plasmids of *B. thuringiensis* can facilitate transfer of pXO1 and pXO2 in intra- and interspecies matings (106). Spontaneous loss of pXO1 from *B. anthracis* during growth in laboratory media is rare; however, pXO1-cured isolates can be obtained following growth at 43°C (107). Unlike pXO1, pXO2 is lost spontaneously at a relatively high frequency. pXO2$^+$ strains occasionally yield rough colonies of noncapsulated cells when streaked to solid medium and incubated under appropriate conditions. In one study, approximately one-half of spontaneous capsule-negative mutants isolated had lost pXO2. The remaining mutants retained the plasmid but lost the ability to synthesize the capsule (34). Growing strains in the presence of novobiocin can induce curing of pXO2. Epidemiological investigations have revealed environmental isolates of *B. anthracis* that are pXO1$^+$ pXO2$^-$, in addition to strains harboring both virulence plasmids. The significance of plasmid pXO2-cured strains in environmental samples is subject to speculation (112).

Overall, the metabolic and transport genes of the *B. anthracis* chromosome are similar to those of the most-studied gram-positive *Bacillus*, *B. subtilis*. However, genomic analyses suggest features that may be important for the pathogenic lifestyle of *B. anthracis*. For example, compared to *B. subtilis*, *B. anthracis* has a large capacity for amino acid and peptide utilization, possibly reflecting its growth in the protein-rich environment of decaying animal matter. The *B. anthracis* genome also contains numerous genes that are important for iron acquisition, which may be important for iron scavenging during growth in a mammalian host (90). One of the most interesting revelations is the presence of several homologues of genes known to be involved in *B. cereus* and *B. thuringiensis* pathogenesis. These include genes predicted to encode various cytolysins and phospholipases. Moreover, the genome contains numerous genes for surface proteins that are potential vaccine candidates or targets for therapeutics. Many of these are homologous to *B. cereus* proteins, illustrating the relatedness of the two species.

Genetic variability within *B. anthracis* has been assessed using amplified fragment length polymorphism marker analysis and amplification of specific chromosomal and plasmid-associated loci. These studies indicate that *B. anthracis* is most closely related to *B. cereus*, yet unlike *B. cereus*, *B. anthracis* exhibits remarkably little strain diversity (43). Allelic categories have been proposed on the basis of variable DNA sequences designated variable number of tandem repeat loci. Correlations between alleles and geographic distribution have been established, indicating that the variable number of tandem repeat markers can serve as an epidemiological tool (54).

ANTHRAX TOXINS

Unique Variation of the A-B Model

B. anthracis secretes three proteins, protective antigen (PA), lethal factor (LF), and edema factor (EF), known collectively as anthrax toxin. Purified toxin proteins in binary combinations have distinct effects when injected into experimental animals. Lethal toxin (LeTx), a mixture of LF and PA, causes death while edema toxin (EdTx), a mixture of EF and PA, produces edema. PA (M_r 82,684) mediates binding and translocation of the enzymatic proteins LF (M_r 90,237) and EF (M_r 89,840) into host cells. None of the three proteins has a toxic effect when administered alone. As its name implies, PA is immunogenic when injected into susceptible animals.

The structural genes for all three toxin proteins, *pagA* (PA), *lef* (LF), and *cya* (EF), are located at distinct loci on pXO1 (78). Toxin proteins can be purified relatively easily from culture supernates of pXO1$^+$ *B. anthracis* strains and from recombinant *Escherichia coli* strains harboring the cloned genes. In recent years, a wealth of information has become available regarding the structures and functions of the toxin proteins, and there has been significant progress regarding the roles of the toxins in anthrax pathogenesis. Experimental studies of anthrax toxin are summarized below. Readers are referred to more comprehensive reviews for additional information (14, 75).

Entry into Host Cells

Two related proteins, tumor endothelial marker 8 and capillary morphogenesis protein 2, have been identified as anthrax toxin receptors (6, 97). These proteins are found on cells of many different tissues although it is not known if either receptor is relevant in vivo. The crystal structure of PA reveals four functional domains, each associated with specific steps in intoxication (Fig. 1A) (83). The current model for PA-mediated entry of EdTx and LeTx is shown in

Fig. 2. Assembly of anthrax toxin begins when PA binds to a receptor and is cleaved by furin or a furinlike protease (33), releasing a 20-kDa amino-terminal fragment (PA$_{20}$). The remaining 63-kDa receptor-bound fragment (PA$_{63}$) then oligomerizes to form a ring-shaped heptamer (73). The heptamer binds EF and LF competitively, resulting in complexes that contain one to three molecules of EF and/or LF per PA$_{63}$ heptomer (76). The receptor-toxin complex is internalized via membrane lipid rafts (1) and trafficked to the endosome where the low-pH environment induces a conformational change in the PA$_{63}$ heptamer such that it converts from a prepore to a pore (channel) in the membrane (4, 29). EF and LF associated with the heptamer translocate the membrane and contact their cytosolic targets (122).

LeTx

LeTx is the central effector of shock and death from systemic anthrax and is therefore considered to be the primary virulence factor of *B. anthracis*. Although LeTx is lethal for a number of laboratory animals, the physiological effect of the toxin is most apparent in a Fischer 344 rat model. Following intravenous injection of large amounts of LeTx, rats succumb to pulmonary edema within 40 min (61). The LF gene, *lef*, is required for virulence in a mouse model for anthrax (84).

LF is a Zn^{2+}-dependent protease that cleaves the amino terminus of mitogen-activated protein kinase kinases (MAPKKs). The crystallographic structure of LF indicates four domains (Fig. 1B). The amino-terminal domain is homologous to EF and is required for binding to PA. Domains 2, 3, and 4 create a groove that binds the 16-residue amino terminus of the substrate prior to cleavage. The catalytic zinc coordination center is in domain 4 (80). In vivo, cleavage of MAPKKs by LF inhibits the MAPK signal transduction pathway, a key signaling pathway that leads to activation of transcription factors in the nucleus and other effectors throughout the cell (22, 82, 117). Considering the fundamental role of the MAPK signaling pathway, it is likely that inhibition of MAPKK activities is important in anthrax pathogenesis. However, exactly how LeTx leads to the death of the host is not clear.

LeTx causes rapid cytolysis of certain macrophage cell lines and macrophages isolated from some inbred mouse strains (30). Thus, cytolysis of these cells serves as a convenient assay for LF activity (61). Nevertheless, some macrophages from humans and some other animals susceptible to anthrax disease are resistant to lysis by LeTx, indicating that macrophages do not play a direct role in the lethal effect of LeTx (74). Genetic studies of inbred mice with varying sensitivity to LeTx revealed a genetic polymorphism mapping to a single gene, designated *Ltx1*, on mouse chromosome 11. The *Ltx1* gene encodes a kinesin-like motor protein, Kif1C. The polymorphism in *Ltx1* influences intoxication events downstream of toxin entry into the cytosol, and it has been proposed that the polymorphism represents genetic differences in an additional substrate for LF (119, 120).

Interestingly, LeTx-resistant macrophages can be sensitized to LeTx by treatment with lipopolysaccharide (LPS) or tumor necrosis factor (TNF) (55, 81), and it has been proposed that components of the *B. anthracis* cell wall or capsule may activate and sensitize macrophages during infection (75). Results of early studies suggested that the lethality of the toxin was due to hyperproduction of cytokines by macrophages (42). However, cleavage of MAPKKs actually decreases synthesis of nitric oxide and TNF-α, and LeTx has been reported to suppress proinflammatory cytokine production in murine macrophages (23). A recent study revealed that low levels of LeTx inhibit the glucocorticoid receptor,

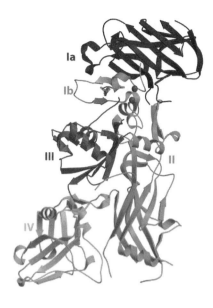

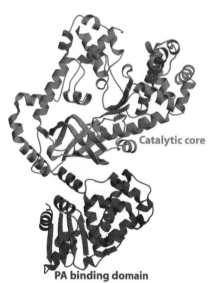

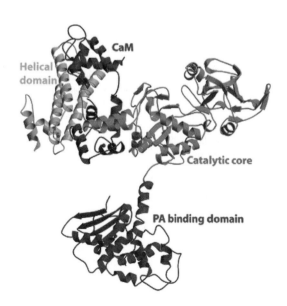

an important component of host defense against inflammation. These findings suggest that LeTx may impair the innate immune system of the host (14). In a seemingly contradictory sense, cultured macrophages develop resistance to LeTx-mediated cytolysis when they are pretreated with sublytic levels of the toxin, suggesting that macrophages are capable of adapting to and tolerating toxic doses of LeTx (95). Complete understanding of the molecular events that lead to LeTx-mediated death awaits further experimentation.

EdTx

The edema observed in cutaneous anthrax infections is attributed to the anthrax EdTx. Subcutaneous administration of EdTx gives rise to edema at the site of inoculation in laboratory animals (61). The contribution of EdTx to systemic anthrax is not clear. In a mouse model for systemic anthrax, strains deleted for the *cya* gene are attenuated 100-fold (9).

EF is a calmodulin-dependent and Ca^{+2}-dependent adenylyl cyclase. The amino-terminal part of the protein is homologous to that of LF and is required for binding to PA (Fig. 1C). Comparison of the crystallographic structures of the carboxy-terminal catalytic part of the protein in the unbound state and bound to calmodulin indicate that upon calmodulin binding, EF undergoes a conformational change in which a 15-kDa helical domain rotates away from the catalytic core of the enzyme, stabilizing a disordered loop and leading to EF activation (19). Treatment of some cell types with EdTx results in dramatic increases in cyclic AMP (cAMP), in some cases rising 1,000-fold to reach 2,000 μmol/mg of cell protein. Routine assay for EF is performed by measuring adenylyl cyclase activity (61). EF-catalyzed increases in cAMP levels disrupt normal second-messenger signaling, potentially resulting in numerous effects on cellular metabolism. EdTx appears to disrupt antibacterial responses of phagocytes. Human neutrophils treated with the toxin have reduced phagocytic and oxidative burst abilities but increased chemotactic responses to N-formyl-Met-Leu-Phe (61). Cultured monocytes treated with EdTx express reduced levels of lipopolysaccharide-inducible TNF-α but secrete elevated levels of interleukin 6 (41). As is true for other toxins that elevate cAMP, a major role of EdTx during infection may be to impair phagocytic function.

B. ANTHRACIS—HOST INTERACTIONS

Animal Models

The majority of animal models for anthrax have been used to assess pathophysiological effects of purified toxin and to test efficacy of vaccines against anthrax. Mice in particular have been used extensively for testing vaccines (51, 77, 85).

FIGURE 1 (*See the separate color insert for the color version of this illustration.*) Ribbon diagrams representing structures of the anthrax toxin proteins. (A) Monomeric PA. Ia (blue), 20-kDa fragment removed with cleavage; Ib (yellow), forms N terminus of PA_{63} and contains two structural calcium ions (red); II (green), pore formation; III (magenta), oligomerization of PA_{63}; IV (turquoise), receptor binding. (B) LF. Substrate-binding and catalytic domains (green) and PA-binding domain (magenta). (C) EF in complex with calmodulin. Catalytic core (green); PA-binding domain (magenta); helical domain (yellow) interacts with calmodulin (red). (Courtesy of W.-J. Tang.)

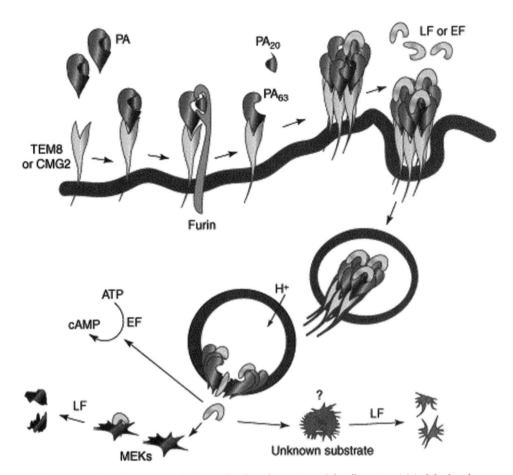

FIGURE 2 *(See the separate color insert for the color version of this illustration.)* Model of anthrax toxin action. PA binds to receptors TEM8 or CMG2. Following proteolytic cleavage of PA by furin, PA_{63} oligomerizes to form a heptameric prepore. EF-LF binds to the prepore, and the complex is endocytosed. Acidification of the intracellular compartment triggers translocation of EF and LF to the cytosol. EF, a calmodulin-dependent adenylate cyclase, converts ATP to cAMP. LF, a zinc-dependent protease, cleaves members of the MEK family and may also affect other targets. (Reprinted from Moayeri and Leppla [75].)

Most murine studies use a subcutaneous infection route. For the virulent Ames strain, the subcutaneous 50% lethal dose (LD_{50}) has been reported to be approximately 25 to 30 spores (121). The toxigenic, noncapsulated ($pXO1^+$ $pXO2^-$) Sterne strain of *B. anthracis* has been used with inbred mice to show that host genetics can influence the response to the bacterium. Most mice are resistant to large doses (i.e., $>10^7$ spores) of the Sterne strain inoculated subcutaneously, but A/J and DBA/2 mice are susceptible to doses of as little as ~1,000 spores (121). These mice are deficient in the C5 component of complement (79), which may explain their susceptibility to the noncapsulated strain. There are a few reports of intratracheal infection of mice. As might be expected, LD_{50}s are generally higher for intratracheal infection of mice than for subcutaneous infection (7, 64).

Guinea pigs have been used effectively for both subcutaneous and intranasal *B. anthracis* inoculation strategies. The LD_{50} for a subcutaneous inoculation is ~50 to 100 spores, but 100,000 spores must be delivered intranasally for a reproducible lung infection (51, 52, 62). Similarly, rabbits require 100,000 to 200,000 spores for lethal infection. In the rabbit model, the humoral response to the human anthrax vaccine AVA (for anthrax vaccine adsorbed) appears to be predictive of protection in an inhalation model (87). Finally, rhesus monkeys have been used for inhalation models as a surrogate for human infections for testing efficacy of vaccine and antibiotics against anthrax (31).

B. anthracis—Macrophage Interactions

Although much work has been done to test the effect of anthrax toxin on macrophages, interactions between *B. anthracis* spores and macrophages are not well studied. Inhaled spores reach the respiratory bronchioles and alveoli, where they are efficiently phagocytosed by alveolar macrophages via recruitment of F-actin (36, 92, 98). Infected macrophages migrate along the lymphatic channels to local lymph nodes of the mediastinum, often the first affected region in anthrax disease. Germination of *B. anthracis* spores inside alveolar macrophages is apparent from microscopic studies published many years ago (92, 98). More recent studies employing high-resolution visualization methods have confirmed these observations (36, 93). In vivo germination may be the result of specific signals generated by the macrophage. The *B. anthracis* chromosomal locus, gerS, is associated with spore germination in the presence of cultured macrophages (48).

Germinated spores colocalize with a lysosome-associated membrane protein, CD107a, indicating that spores traffic in the phagolysosomal pathway (35). The germinated spores appear to escape membrane-bound vesicles, followed by multiplication in the cytosol of the macrophage (93), but the mechanisms for release from the vesicles and whether or not macrophage death accompanies the ultimate release of vegetative cells from the host cell are not clear.

Studies of B. anthracis-macrophage interaction are limited, primarily because of challenges dictated by the physiology of this bacterium. Quantitative assessment of numbers of macrophage-associated bacteria following infection is difficult due to nonspecific but tight association of spores with macrophage surfaces, the inability to kill spores using gentamicin and other antibiotics, the developmental nature of the bacterium, and its ability to produce heterogeneous populations of chains containing vegetative cells. Currently, there are no standard methods in the literature for quantitative measurement of phagocytosis or for B. anthracis germination, viability, and proliferation in the context of the macrophage.

SURFACE STRUCTURES

Capsule

When B. anthracis is grown under appropriate conditions (see below), the outermost surface of vegetative cells is covered by a capsule. As is true for numerous pathogens, the capsule is an important virulence factor. Encapsulated cells are strongly resistant to phagocytosis by cultured macrophages (67), and a noncapsulated B. anthracis mutant is unable to disseminate from the lung to the spleen in a murine model for inhalation anthrax (21). The B. anthracis capsule is unusual in that it is a polypeptide composed of gamma-linked alpha-peptide chains of 50 to 100 D-glutamic acid residues (106). The capsule alone is relatively nonimmunogenic. However, glutamic acid polymers conjugated to various carriers can generate an immune response to glutamic acid (91, 96, 118). Moreover, a strong immune response is elicited against D-glutamic acid polymers when mice are coimmunized with agonist monoclonal antibodies to CD40 (57). These findings have renewed interest in the B. anthracis capsule as a potential target for detection, treatment, and/or prevention of anthrax disease.

The cap operon is on pXO2. The first three genes of the operon, capB, capC, and capA, encoding proteins of 44, 16, and 46 kDa, respectively, are required for capsule synthesis. Results of hydropathicity analysis and localization of the gene products in E. coli minicells indicate that all three proteins are membrane associated (65, 67). However, their functions with regard to capsule synthesis have not been elucidated. The fourth gene of the cap operon, capD (also known as dep), encodes a 51-kDa protein that catalyzes the hydrolysis of capsule material into lower-molecular-weight polymers. A capD-null mutant is avirulent in a mouse model for anthrax (66), and it has been proposed that the low-molecular-weight peptides act to inhibit host defenses (115).

S-Layer

Between the capsule and the cell wall of B. anthracis lies a paracrystalline surface layer called the S-layer. The S-layer was first observed on noncapsulated cells, and its formation is independent of capsule synthesis (69). The S-layer is composed of two 94-kDa proteins, EA1 (for extractable antigen 1) and Sap. EA1 was identified initially as a vegetative cell protein that reacted strongly with serum from animals vaccinated with a live noncapsulated B. anthracis strain (25). EA1 appears to constitute the main S-layer lattice and is the major cell-associated protein (26, 70). Sap was first identified as an abundant secreted protein in culture supernatants of noncapsulated strains. The S-layer proteins are noncovalently attached to the cell wall polysaccharide at their amino-terminal domains (27). The Sap protein is less tightly associated with the S-layer than EA1 (24). Adherence of the capsule to the cell surface is not dependent upon synthesis of the S-layer proteins. A mutant strain, carrying deletions in both S-layer genes, is fully capsulated (69).

Production and surface localization of Sap and EA1 appear to be under developmental control (15, 71). During batch culture, the proteins appear sequentially on the cell surface, with the Sap layer of the exponential phase replaced by an EA1 layer during stationary phase. Differential expression has been attributed to transcription of the sap gene by RNA polymerase containing σ^A, and transcription of the eag gene by RNA polymerase containing σ^H. Interestingly, temporal control of eag requires Sap, and EA1 is involved in strict feedback regulation of eag. Sap and EA1 proteins bind the eag promoter in vitro, suggesting that they might act as transcriptional repressors (71).

It is not known whether the B. anthracis S-layer has a specific role in virulence. S-layers are produced by many bacteria and are believed to function in molecular sieving, cell shape maintenance, and phage fixation. It has been proposed that S-layers produced by pathogenic bacteria affect binding to host cells and protect against complement-mediated killing. It is difficult to imagine such functions for the B. anthracis S-layer, because structural and immunological analyses indicate that cells are completely covered by the capsule (69). Nevertheless, it is not clear whether all B. anthracis cells are capsulated at all stages of infection and it is conceivable that the S-layer may have a protective role in noncapsulated cells.

Exosporium

The surface of the B. anthracis spore is of significant interest because it represents the first contact of the pathogen with the host. The spore coat of B. anthracis, which is similar in structure and composition to that of other Bacillus species (59), is covered by a loose-fitting diffuse layer called the exosporium (39). The exosporium appears to be unique to B. anthracis and its close relative B. cereus. It consists of a basal layer and a hair-like nap of fine filaments extending out from the basal layer (38, 58). Studies of the chemical composition of the exosporium from B. anthracis and B. cereus have revealed at least 10 proteins that are tightly associated with this outer layer. Several of these are related to morphogenetic and outer spore coat proteins of B. subtilis, but most do not have homologues in B. subtilis (12, 102, 109).

The filaments of the exosporium are largely comprised of a highly immunogenic glycoprotein called BclA (for Bacillus collagen-like protein) (104). BclA has an internal collagen-like region (CLR) of GXX repeats, which includes a large proportion of GPT triplets. In B. anthracis, the CLR is highly polymorphic, containing between 17 and 91 GXX repeats and one to eight copies of a 21-amino-acid sequence, (GPT)$_5$GDTGTT, named the BclA repeat. The length of the CLR is responsible for the variation in filament length (105). Two oligosaccharides have been reported to be attached to BclA via a GalNAc linker. Multiple copies of a 715-Da tetrasaccharide are linked to the CLR, while a 324-Da disaccharide is attached outside of this region. Interestingly, the tetrasaccharide contains a novel nonreducing terminal sugar, termed anthrose, which has not been found in spores of *B. cereus* and *B. thuringiensis* (18).

There has been little functional characterization of the exosporium structure or components. It has been speculated that the filaments and/or other components of the exosporium may function in uptake or other interactions with host cells. Two enzymes associated with the exosporium, alanine racemase and nucleoside hydrolase, have been proposed to metabolize small molecule germinants, thus limiting premature germination (109).

VIRULENCE GENE EXPRESSION

Host Cues

Toxin and capsule synthesis by *B. anthracis* represents an intriguing example of coordinate expression of virulence genes in response to host-related cues. Expression of the toxin and capsule genes by *B. anthracis* during growth in certain media is enhanced in the presence of bicarbonate or under elevated ($\geq 5\%$) atmospheric CO_2. This CO_2 effect on toxin and capsule synthesis is specific and not simply due to the buffering capacity of dissolved bicarbonate during bacterial growth or to decreased oxygen levels (106). Elevated CO_2 is postulated to be a physiologically significant signal during anthrax infection. Concentrations of bicarbonate (about 20 mM) and CO_2 (about 40 mm Hg) in humans are similar to those that activate toxin production during growth in vitro (60).

CO_2-enhanced virulence gene expression is at the level of transcription (2, 11, 56, 100, 116). In the case of the toxin genes, mutants harboring transcriptional *lacZ* and *cat* fusions have been used to monitor relative promoter activity. At late log phase, when toxin gene promoter activity and toxin protein yields are highest, expression of a *pag-lacZ* transcriptional fusion on pXO1 is induced five- to eightfold during growth in 5% CO_2 compared with growth in air (0.03% CO_2). Growth in 20% CO_2 increases the transcription up to 19-fold (56).

Growth temperature also affects expression of the toxin genes. When cells are grown in elevated CO_2, promoter activity of the three toxin genes is increased four- to sixfold during incubation at 37°C, compared with incubation at 28°C (100). Unlike CO_2, temperature does not appear to globally regulate virulence gene expression; no temperature effects on capsule gene expression have been reported.

The *atxA* Regulon

The current model for regulation of toxin and capsule gene expression is depicted in Fig. 3. All of the known *B. anthracis* virulence genes and many of their regulatory elements are located on pXO1 and pXO2. The toxin genes (*cya*, *lef*, and *pagA*) are located noncontiguously within a 30-kb region of pXO1 (78). The *cap* operon of pXO2 contains genes associated with synthesis and depolymerization of the D-glutamic acid capsule, *capB*, *capC*, *capA*, and *capD* (66). Transposon-mediated insertion mutagenesis of the pXO1⁺ pXO2⁻ Sterne strain resulted in identification of *atxA* (for anthrax toxin activator) as a *trans*-acting positive regulator of toxin gene transcription (17, 56, 114). Subsequent studies using strains carrying pXO2 revealed that *atxA* is a global regulator of gene expression in *B. anthracis*, affecting expression of numerous other plasmid- and chromosome-borne genes (5). The *atxA* gene encodes a 56-kDa protein, but the molecular mechanism for AtxA function is not known. In some cases *atxA* effects on gene expression are mediated by downstream regulators. *atxA* controls *cap* gene expression via *acpA* and *acpB*, two pXO2-borne genes with partial functional similarity (20, 37, 116). *atxA* control of the S-layer genes *eag* and *sap* is mediated by a fourth plasmid-associated regulator, *pagR*, which is cotranscribed with *pagA* (45). PagR has specific DNA-binding activity for the control regions of the *pag* operon and the S-layer genes (72).

The relationship between CO_2-enhanced virulence gene expression and regulatory gene function is not clear. *atxA* can activate toxin gene expression in air-grown cells, although the steady-state levels of *cya*, *lef*, and *pag* mRNA transcripts are significantly increased during growth in elevated CO_2 (17, 56). Expression of the *atxA* gene is unaffected by the CO_2 signal (16), indicating that CO_2 may affect the function of AtxA or the function or expression of some other regulator. Transposon mutagenesis of a pXO1⁺ strain resulted in a mutant that synthesizes all three toxin proteins in the absence of elevated CO_2. It has been proposed that the mutant lacks a repressor protein that binds to toxin gene promoter DNA in the absence of the CO_2 signal (114). However, the gene for the putative repressor has not been identified. In addition, overexpression of *atxA* appears to reduce *pag* expression, suggesting that some other factor

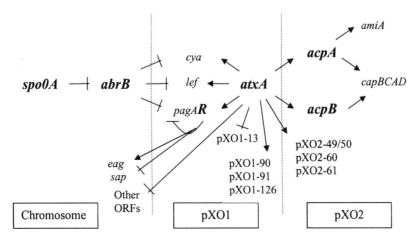

FIGURE 3 Model depicting *B. anthracis* virulence gene regulation.

may be titrated by excess AtxA. If an additional *trans*-acting factor exists, it cannot be titrated by adding more target promoter sequences. When the *pag* promoter was cloned on a multicopy plasmid and introduced into a pXO1$^+$ strain, no change in PA synthesis was observed (99). Finally, although *atxA* is required for *cap* gene expression when *B. anthracis* is cultured at physiological levels of CO_2, *atxA*-independent *cap* expression occurs when cultures are incubated in high (20%) CO_2 levels. Expression of *acpA* and *acpB* is increased under these growth conditions (20, 116), suggesting that the levels of the downstream regulators are limiting for *cap* gene expression in cells grown in air.

The *atxA*, *acpA*, and *acpB* genes have been tested for function in vivo. Subcutaneous inoculation of mice with high doses of spores of the toxigenic noncapsulated Sterne strain causes a lethal disease. An *atxA*-null mutant of the Sterne strain is avirulent in this model. Moreover, the antibody response to all three toxin proteins is decreased significantly in *atxA*-null infected mice, compared with those infected with the parent strain (17). Function of *acpA* and *acpB* has been assessed in a mouse model for inhalation anthrax employing pXO1$^+$ pXO2$^+$ strains. A mutant deleted for *acpA* and *acpB* is avirulent in this model. Furthermore, PCR analysis of tissues from infected mice indicates *acpA*-*acpB*-dependent *cap* gene expression (21). Interestingly, although *acpA* and *acpB* show functional similarity when *B. anthracis* is cultured in vitro (20), an *acpB*-null mutant is attenuated in the mouse inhalation anthrax model, whereas an *acpA*-null mutant is unaffected for virulence (21). These data suggest that *acpB* may affect additional genes important for *B. anthracis* pathogenesis.

Other Regulators

B. anthracis virulence gene expression is also linked to a complex pathway for bacterial development that is well-studied in *B. subtilis*. A *B. anthracis* homologue of the transition state regulator *abrB* controls the timing and expression levels of *pagA*, *lef*, and *cya* during growth in batch culture. An *abrB*-null mutant synthesizes elevated levels of all three toxin proteins, with transcripts more apparent during early exponential phase. In *B. subtilis*, *abrB* is negatively regulated by the response regulator *spo0A*. A *B. anthracis spo0A* mutant displays elevated *abrB* expression and decreased *pagA* expression, indicating a similar pathway in *B. anthracis*. It has been postulated that *abrB* exerts its effect on the toxin genes via control of *sigH*, as is true for many *abrB*-regulated genes in *B. subtilis*. However, the *B. subtilis* consensus sequence for recognition by SigH has not been found upstream of the toxin genes (94). Another alternative sigma factor, SigB, may be a minor regulator of virulence in *B. anthracis*. SigB has been characterized as a general stress transcription factor in *B. subtilis*. Although there is no evidence that SigB affects toxin or capsule gene expression, a *sigB* mutant exhibits reduced virulence in a mouse model for anthrax (28).

GENETIC TOOLS

Natural Genetic Exchange

DNA transfer between *B. anthracis* and the closely related species *B. cereus* and *B. thuringiensis* can occur by transduction and conjugation. Bacteriophage CP-51 can transduce various plasmids among these species and is useful for transduction of chromosomal markers between *B. anthracis* strains (106). CP-51-mediated transduction of pXO2 from

B. anthracis to *B. cereus* was instrumental in demonstrating that the plasmid was associated with capsule synthesis (34). *B. thuringiensis* plasmids pXO11 and pXO12 are self transmissible and transfer to *B. cereus*, *B. anthracis*, and other *B. thuringiensis* subspecies. *B. anthracis* strains harboring pXO12 can transfer pXO1 and pXO2 to cured *B. anthracis* strains by conduction involving the transposon Tn4430, which is found on the conjugative plasmid. Conjugative plasmid transfer between *B. subtilis* and *B. anthracis* has also been reported. *B. subtilis* (natto) plasmid pLS20 facilitates the transfer of the tetracycline-resistance plasmid pBC16 from *B. subtilis* to *B. anthracis* (106).

Generation of Recombinant Strains

B. anthracis is not naturally transformable, but plasmids can be electroporated into *B. anthracis* at frequencies up to 10^4 transformants per μg of DNA (68). Several plasmid vectors derived from *Staphylococcus aureus* plasmids pUB110, pC194, and pE194 and *B. cereus* plasmid pBC16 can replicate and express antibiotic resistance in *B. anthracis*. Most *B. anthracis* strains are sensitive to tetracycline, kanamycin (and neomycin), chloramphenicol, spectinomycin, and macrolide-lincosamide-streptogramin B antibiotics, making selection for plasmid-borne markers encoding the appropriate resistance determinants feasible. Electroporation frequencies are highest when plasmids are isolated from DNA methyltransferase-deficient *E. coli* strains or from *B. subtilis* 168, a strain that is not known to possess adenine-methylating activities. These observations suggest that *B. anthracis* has a restriction system dependent on a specific pattern of adenine methylation for the recognition of foreign DNA (68). There are no reports of electroporation of linear DNA fragments into *B. anthracis*.

It is possible to generate specific insertions and gene replacements in *B. anthracis* by homologous recombination. Integrative vectors, carrying *B. anthracis* DNA sequences and lacking a replication origin functional in *B. anthracis*, can be transferred from *E. coli* to *B. anthracis* with selection for plasmid-encoded markers by using the RK2 mating system. This strategy was used to disrupt each of the anthrax toxin genes on pXO1, creating important strains for studies of toxin gene function (84). Recombinant strains can also be constructed using temperature-sensitive plasmids created initially for use in *B. subtilis* or *E. coli*-*B. anthracis* shuttle vectors that are naturally unstable in *B. anthracis* in the absence of selection (10, 13, 17). Neither method is highly efficient. *B. anthracis* will not grow at temperatures exceeding 43°C, sometimes making plasmid curing difficult. Shuttle vectors containing a selectable marker flanked by *B. anthracis* DNA sequences corresponding to a locus of interest can be used to create gene replacement and insertion mutants, but screens for mutants resulting from a double-crossover recombination event can be tedious. Nevertheless, numerous recombinants have been isolated by these methods.

Random Mutagenesis

Point mutants of *B. anthracis* can be isolated following UV mutagenesis of spores. Rifampin- and streptomycin-resistant mutants and numerous amino acid, purine, and pyrimidine auxotrophs have been obtained by this method (3). Insertion mutations can be generated using *Enterococcus faecalis* transposons Tn916 and Tn917. Each method has disadvantages. Low electroporation and *E. coli*-*B. anthracis* mating frequencies make it difficult to screen genomic libraries for clones that complement point mutants. Tn916 insertion mutants are unstable in the absence of selection for the tetracycline-resistance gene encoded by the transposon (50). Tn917 insertion mutants are stable in the absence of selection. However, insertions can be accompanied by local deletions and rearrangements of *B. anthracis* DNA. In *B. anthracis* strains harboring either pXO1 or pXO2, the majority of Tn917 insertions map to the virulence plasmids as opposed to the chromosome (44), possibly due to the large number of insertion sequence-like elements on pXO1 and pXO2.

Measurements of Gene Expression

The availability of the *B. anthracis* genome sequence has permitted analysis of the *B. anthracis* transcriptome (5, 63). Expression of specific genes can be monitored using transcriptional fusions employing promoterless genes for β-galactosidase and chloramphenicol acetyltransferase (2, 56, 100). Measurements of specific galactosidase activity associated with these strains must be interpreted with caution. Specific enzyme activity of *B. anthracis* cells producing high levels of β-galactosidase can be highly variable when cells are grown in liquid medium. Autolysis of the cells can lead to release of inactive β-galactosidase into the culture medium (99).

ANTHRAX VACCINES

The Sterne Strain

The veterinary vaccine for anthrax is a spore preparation of an attenuated *B. anthracis* strain isolated by M. Sterne in 1937. The Sterne strain and its derivatives have been used worldwide for >50 years. This strain produces the anthrax toxin proteins but is noncapsulated, owing to the absence of pXO2. Although the vaccine is effective, there are safety concerns. The strain has a low level of virulence in some animal species, and it can occasionally result in necrosis at the injection site (40).

PA

The currently licensed human vaccine for anthrax is a noncellular vaccine composed primarily of PA. The vaccine is produced in the United States and the United Kingdom by similar methods. PA is obtained from a toxigenic noncapsulated strain by adsorption of culture supernatant to aluminum hydroxide or precipitation with alum (106). Efficacy of the vaccine has been established in studies of rhesus macaques and in a field study of an exposed and susceptible human population (86). Nevertheless, the vaccine has some shortcomings. The vaccination protocol includes a series of doses, and repeated annual boosters are required to maintain immunity. The precise composition of the vaccine is unknown and can vary in different preparations (123). There is some concern that rare adverse reactions may be attributed to residual levels of EF, LF, or other *B. anthracis*-derived enzymes. Also, in some animal models, the vaccine does not provide protection against all virulent *B. anthracis* strains (40).

NEXT-GENERATION VACCINES

Recombinant PA (rPA) is under evaluation as a second-generation vaccine for anthrax. As is true for the currently licensed vaccine, rPA provides high-level protection against aerosol challenge with a diverse panel of *B. anthracis* isolates in rhesus macaques (86). But unlike the currently used vaccine, rPA is a purified protein, free of contami-

nants. There have also been reports of genetic immunization against anthrax. DNA-based immunization against PA and LF provides protection against aerosol challenge in rabbits (32).

Although PA is considered the most important component of the current vaccine, there are considerable data indicating that *B. anthracis* antigens other than PA may supplement the immune response to PA. Results of animal studies indicate that the Sterne live spore vaccine is more protective than the human PA vaccine (40). Also, the addition of killed *B. anthracis* spores to PA results in complete protection against challenge with virulent *B. anthracis* strains in certain animals in which purified PA is infective (8). Poly-D-glutamic acid protein conjugates represent another potential addition to a PA-based vaccine. Recent reports indicate the immunogenetic potential of the glutamic acid polymers linked to carriers or coadministered with immunostimulants (57, 91, 96, 118).

FUTURE DIRECTIONS

Major advances in our understanding of structure and function of the "classic" virulence factors of *B. anthracis*, the anthrax toxin proteins and the poly-D-glutamic capsule, combined with new information regarding the anthrax toxin receptors are fueling new strategies for anthrax therapeutics and improved human vaccines. Investigations of other potential virulence factors, including adhesins, secreted enzymes, and sensors for germination, have been prompted by recent genomic proteomic analyses. As investigators turn to studies of the expression and function of known and suspected virulence factors in vivo, a more detailed picture of *B. anthracis* pathogenesis will be revealed. In particular, detailed analysis of the host response to the pathogen will fill a critical void in our knowledge of anthrax disease. Finally, molecular genetic analyses involving multiple *B. anthracis* strains will continue to facilitate epidemiological studies and development of advanced methods for detection and identification. Such investigations will also increase our understanding of the evolution of this important pathogen and its relatedness to other soil bacilli.

Work in my laboratory is supported by Public Health Service grants AI33537 and U54 AI057156 from the National Institutes of Health.

I thank members of my laboratory, C. Rick Lyons of the University of New Mexico, and Wei-Jen Tang of the University of Chicago for helpful comments on the manuscript.

REFERENCES

1. **Abrami, L., S. Liu, P. Cosson, S. H. Leppla, and F. G. van der Goot.** 2003. Anthrax toxin triggers endocytosis of its receptor via a lipid raft-mediated clathrin-dependent process. *J. Cell Biol.* **160:**321–328.
2. **Bartkus, J. M., and S. H. Leppla.** 1989. Transcriptional regulation of the protective antigen gene of *Bacillus anthracis*. *Infect. Immun.* **57:**2295–2300.
3. **Battisti, L., B. D. Green, and C. B. Thorne.** 1985. Mating system for transfer of plasmids among *Bacillus anthracis*, *Bacillus cereus*, and *Bacillus thuringiensis*. *J. Bacteriol.* **162:** 543–550.
4. **Beauregard, K. E., R. J. Collier, and J. A. Swanson.** 2000. Proteolytic activation of receptor-bound anthrax protective antigen on macrophages promotes its internalization. *Cell. Microbiol.* **2:**251–258.
5. **Bourgogne, A., M. Drysdale, S. G. Hilsenbeck, S. N. Peterson, and T. M. Koehler.** 2003. Global effects of

virulence gene regulators in a *Bacillus anthracis* strain with both virulence plasmids. *Infect. Immun.* 71:2736–2743.

6. **Bradley, K. A., J. Mogridge, M. Mourez, R. J. Collier, and J. A. Young.** 2001. Identification of the cellular receptor for anthrax toxin. *Nature* **414**:225–229.

7. **Brook, I., T. B. Elliott, R. A. Harding, S. S. Bouhaouala, S. J. Peacock, G. D. Ledney, and G. B. Knudson.** 2001. Susceptibility of irradiated mice to *Bacillus anthracis* sterne by the intratracheal route of infection. *J. Med. Microbiol.* **50**:702–711.

8. **Brossier, F., M. Levy, and M. Mock.** 2002. Anthrax spores make an essential contribution to vaccine efficacy. *Infect. Immun.* **70**:661–664.

9. **Brossier, F., M. Weber-Levy, M. Mock, and J. C. Sirard.** 2000. Role of toxin functional domains in anthrax pathogenesis. *Infect. Immun.* **68**:1781–1786.

10. **Brown, D. P., L. Ganova-Raeva, B. D. Green, S. R. Wilkinson, M. Young, and P. Youngman.** 1994. Characterization of *spo0A* homologues in diverse *Bacillus* and *Clostridium* species identifies a probable DNA-binding domain. *Mol. Microbiol.* **14**:411–426.

11. **Cataldi, A., A. Fouet, and M. Mock.** 1992. Regulation of *pag* gene expression in *Bacillus anthracis*: use of a *pag-lacZ* transcriptional fusion. *FEMS Microbiol. Lett.* **98**:89–94.

12. **Charlton, S., A. J. Moir, L. Baillie, and A. Moir.** 1999. Characterization of the exosporium of *Bacillus cereus*. *J. Appl. Microbiol.* **87**:241–245.

13. **Chen, Y., F. C. Tenover, and T. M. Koehler.** 2004. β-Lactamase gene expression in a penicillin-resistant *Bacillus anthracis* strain. *Antimicrob. Agents Chemother.* **48**:4873–4877.

14. **Collier, R. J., and J. A. Young.** 2003. Anthrax toxin. *Annu. Rev. Cell Dev. Biol.* **19**:45–70.

15. **Couture-Tosi, E., H. Delacroix, T. Mignot, S. Mesnage, M. Chami, A. Fouet, and G. Mosser.** 2002. Structural analysis and evidence for dynamic emergence of *Bacillus anthracis* S-layer networks. *J. Bacteriol.* **184**:6448–6456.

16. **Dai, Z., and T. M. Koehler.** 1997. Regulation of anthrax toxin activator gene (*atxA*) expression in *Bacillus anthracis*: Temperature, not CO_2/bicarbonate, affects AtxA synthesis. *Infect. Immun.* **65**:2576–2582.

17. **Dai, Z., J.-C. Sirard, M. Mock, and T. M. Koehler.** 1995. The *atxA* gene product activates transcription of the anthrax toxin genes and is essential for virulence. *Mol. Microbiol.* **16**:1171–1181.

18. **Daubenspeck, J. M., H. Zeng, P. Chen, S. Dong, C. T. Steichen, N. R. Krishna, D. G. Pritchard, and C. L. Turnbough, Jr.** 2004. Novel oligosaccharide side chains of the collagen-like region of BclA, the major glycoprotein of the *Bacillus anthracis* exosporium. *J. Biol. Chem.* **279**:30945–30953.

19. **Drum, C. L., S. Z. Yan, J. Bard, Y. Q. Shen, D. Lu, S. Soelaiman, Z. Grabarek, A. Bohm, and W. J. Tang.** 2002. Structural basis for the activation of anthrax adenylyl cyclase exotoxin by calmodulin. *Nature* **415**:396–402.

20. **Drysdale, M., A. Bourgogne, S. G. Hilsenbeck, and T. M. Koehler.** 2004. *atxA* controls *Bacillus anthracis* capsule synthesis via *acpA* and a newly discovered regulator, *acpB*. *J. Bacteriol.* **186**:307–315.

21. **Drysdale, M., S. Heninger, J. Hutt, Y. Chen, C. R. Lyons, and T. M. Koehler.** 2005. Capsule synthesis by *Bacillus anthracis* is required for dissemination in murine inhalation anthrax. *EMBO J.* **24**:221–227.

22. **Duesbery, N. S., C. P. Webb, S. H. Leppla, V. M. Gordon, K. R. Klimpel, T. D. Copeland, N. G. Ahn, M. K. Oskarsson, K. Fukasawa, K. D. Paull, and G. F. Vande Woude.** 1998. Proteolytic inactivation of MAP-kinase-kinase by anthrax lethal factor. *Science* **280**:734–737.

23. **Erwin, J. L., L. M. DaSilva, S. Bavari, S. F. Little, A. M. Friedlander, and T. C. Chanh.** 2001. Macrophage-derived cell lines do not express proinflammatory cytokines after exposure to *Bacillus anthracis* lethal toxin. *Infect. Immun.* **69**:1175–1177.

24. **Etienne-Toumelin, I., J. C. Sirard, E. Duflot, M. Mock, and A. Fouet.** 1995. Characterization of the *Bacillus anthracis* S-layer: cloning and sequencing of the structural gene. *J. Bacteriol.* **177**:614–620.

25. **Ezzell, J. W., Jr., and T. G. Abshire.** 1988. Immunological analysis of cell-associated antigens of *Bacillus anthracis*. *Infect. Immun.* **56**:349–356.

26. **Farchaus, J. W., W. J. Ribot, M. B. Downs, and J. W. Ezzell.** 1995. Purification and characterization of the major surface array protein from the avirulent *Bacillus anthracis* Delta Sterne-1. *J. Bacteriol.* **177**:2481–2489.

27. **Fouet, A., and S. Mesnage.** 2002. *Bacillus anthracis* cell envelope components. *Curr. Top. Microbiol. Immunol.* **271**:87–113.

28. **Fouet, A., O. Namy, and G. Lambert.** 2000. Characterization of the operon encoding the alternative σ^B factor from *Bacillus anthracis* and its role in virulence. *J. Bacteriol.* **182**:5036–5045.

29. **Friedlander, A. M.** 1986. Macrophages are sensitive to anthrax lethal toxin through an acid-dependent process. *J. Biol. Chem.* **261**:7123–7126.

30. **Friedlander, A. M., R. Bhatnagar, S. H. Leppla, L. Johnson, and Y. Singh.** 1993. Characterization of macrophage sensitivity and resistance to anthrax lethal toxin. *Infect. Immun.* **61**:245–252.

31. **Fritz, D. L., N. K. Jaax, W. B. Lawrence, K. J. Davis, M. L. Pitt, J. W. Ezzell, and A. M. Friedlander.** 1995. Pathology of experimental inhalation anthrax in the rhesus monkey. *Lab. Investig.* **73**:691–702.

32. **Galloway, D. R., and L. Baillie.** 2004. DNA vaccines against anthrax. *Expert Opin. Biol. Ther.* **4**:1661–1667.

33. **Gordon, V. M., K. R. Klimpel, N. Arora, M. A. Henderson, and S. H. Leppla.** 1995. Proteolytic activation of bacterial toxins by eukaryotic cells is performed by furin and by additional cellular proteases. *Infect. Immun.* **63**:82–87.

34. **Green, B. D., L. Battisti, T. M. Koehler, and C. B. Thorne.** 1985. Demonstration of a capsule plasmid in *Bacillus anthracis*. *Infect. Immun.* **49**:291–297.

35. **Guidi-Rontani, C., M. Levy, H. Ohayon, and M. Mock.** 2001. Fate of germinated *Bacillus anthracis* spores in primary murine macrophages. *Mol. Microbiol.* **42**:931–938.

36. **Guidi-Rontani, C., M. Weber-Levy, E. Labruyere, and M. Mock.** 1999. Germination of *Bacillus anthracis* spores within alveolar macrophages. *Mol. Microbiol.* **31**:9–17.

37. **Guignot, J., M. Mock, and A. Fouet.** 1997. AtxA activates the transcription of genes harbored by both *Bacillus anthracis* virulence plasmids. *FEMS Microbiol. Lett.* **147**:203–207.

38. **Hachisuka, Y., K. Kojima, and T. Sato.** 1966. Fine filaments on the outside of the exosporium of *Bacillus anthracis* spores. *J. Bacteriol.* **91**:2382–2384.

39. **Hachisuka, Y., S. Kozuka, and M. Tsujikawa.** 1984. Exosporia and appendages of spores of *Bacillus* species. *Microbiol. Immunol.* **28**:619–624.

40. **Hambleton, P., and P. C. Turnbull.** 1990. Anthrax vaccine development: a continuing story. *Adv. Biotechnol. Processes* **13**:105–122.

41. **Hanna, P.** 1998. Anthrax pathogenesis and host response. *Curr. Top. Microbiol. Immunol.* **225**:13–35.

42. **Hanna, P. C., D. Acosta, and R. J. Collier.** 1993. On the role of macrophages in anthrax. *Proc. Natl. Acad. Sci. USA* **90:**10198–10201.

43. **Hill, K. K., L. O. Ticknor, R. T. Okinaka, M. Asay, H. Blair, K. A. Bliss, M. Laker, P. E. Pardington, A. P. Richardson, M. Tonks, D. J. Beecher, J. D. Kemp, A. B. Kolsto, A. C. Wong, P. Keim, and P. J. Jackson.** 2004. Fluorescent amplified fragment length polymorphism analysis of *Bacillus anthracis, Bacillus cereus,* and *Bacillus thuringiensis* isolates. *Appl. Environ. Microbiol.* **70:**1068–1080.

44. **Hoffmaster, A. R., and T. M. Koehler.** 1997. The anthrax toxin activator gene *atxA* is associated with CO_2-enhanced non-toxin gene expression in *Bacillus anthracis. Infect. Immun.* **65:**3091–3099.

45. **Hoffmaster, A. R., and T. M. Koehler.** 1999. Autogenous regulation of the *Bacillus anthracis pag* operon. *J. Bacteriol.* **181:**4485–4492.

46. **Hoffmaster, A. R., J. Ravel, D. A. Rasko, G. D. Chapman, M. D. Chute, C. K. Marston, B. K. De, C. T. Sacchi, C. Fitzgerald, L. W. Mayer, M. C. Maiden, F. G. Priest, M. Barker, L. Jiang, R. Z. Cer, J. Rilstone, S. N. Peterson, R. S. Weyant, D. R. Galloway, T. D. Read, T. Popovic, and C. M. Fraser.** 2004. Identification of anthrax toxin genes in a *Bacillus cereus* associated with an illness resembling inhalation anthrax. *Proc. Natl. Acad. Sci. USA* **101:**8449–8454.

47. **Hugh-Jones, M. E., and V. de Vos.** 2002. Anthrax and wildlife. *Rev. Sci. Tech.* **21:**359–383.

48. **Ireland, J., and P. Hanna.** 2002. Amino acid- and purine ribonucleoside-induced germination of *Bacillus anthracis* ΔSterne endospores: *gerS* mediates responses to aromatic ring structures. *J. Bacteriol.* **184:**1296–1303.

49. **Ivanova, N., A. Sorokin, I. Anderson, N. Galleron, B. Candelon, V. Kapatral, A. Bhattacharyya, G. Reznik, N. Mikhailova, A. Lapidus, L. Chu, M. Mazur, E. Goltsman, N. Larsen, M. D'Souza, T. Walunas, Y. Grechkin, G. Pusch, R. Haselkorn, M. Fonstein, S. D. Ehrlich, R. Overbeek, and N. Kyrpides.** 2003. Genome sequence of *Bacillus cereus* and comparative analysis with *Bacillus anthracis. Nature* **423:**87–91.

50. **Ivins, B. E., S. L. Welkos, G. B. Knudson, and D. J. Leblanc.** 1988. Transposon Tn*916* mutagenesis in *Bacillus anthracis. Infect. Immun.* **56:**176–181.

51. **Ivins, B. E., S. L. Welkos, G. B. Knudson, and S. F. Little.** 1990. Immunization against anthrax with aromatic compound-dependent (Aro⁻) mutants of *Bacillus anthracis* and with recombinant strains of *Bacillus subtilis* that produce anthrax protective antigen. *Infect. Immun.* **58:**303–308.

52. **Ivins, B. E., S. L. Welkos, S. F. Little, M. H. Crumrine, and G. O. Nelson.** 1992. Immunization against anthrax with *Bacillus anthracis* protective antigen combined with adjuvants. *Infect. Immun.* **60:**662–668.

53. **Jernigan, D. B., P. L. Raghunathan, B. P. Bell, R. Brechner, E. A. Bresnitz, J. C. Butler, M. Cetron, M. Cohen, T. Doyle, M. Fischer, C. Greene, K. S. Griffith, J. Guarner, J. L. Hadler, J. A. Hayslett, R. Meyer, L. R. Petersen, M. Phillips, R. Pinner, T. Popovic, C. P. Quinn, J. Reefhuis, D. Reissman, N. Rosenstein, A. Schuchat, W. J. Shieh, L. Siegal, D. L. Swerdlow, F. C. Tenover, M. Traeger, J. W. Ward, I. Weisfuse, S. Wiersma, K. Yeskey, S. Zaki, D. A. Ashford, B. A. Perkins, S. Ostroff, J. Hughes, D. Fleming, J. P. Koplan, and J. L. Gerberding.** 2002. Investigation of bioterrorism-related anthrax, United States, 2001: epidemiologic findings. *Emerg. Infect. Dis.* **8:**1019–1028.

54. **Keim, P., and K. L. Smith.** 2002. *Bacillus anthracis* evolution and epidemiology. *Curr. Top. Microbiol. Immunol.* **271:**21–32.

55. **Kim, S. O., Q. Jing, K. Hoebe, B. Beutler, N. S. Duesbery, and J. Han.** 2003. Sensitizing anthrax lethal toxin-resistant macrophages to lethal toxin-induced killing by tumor necrosis factor-alpha. *J. Biol. Chem.* **278:**7413–7421.

56. **Koehler, T. M., Z. Dai, and M. Kaufman-Yarbray.** 1994. Regulation of the *Bacillus anthracis* protective antigen gene: CO_2 and a *trans*-acting element activate transcription from one of two promoters. *J. Bacteriol.* **176:**586–595.

57. **Kozel, T. R., W. J. Murphy, S. Brandt, B. R. Blazar, J. A. Lovchik, P. Thorkildson, A. Percival, and C. R. Lyons.** 2004. mAbs to *Bacillus anthracis* capsular antigen for immunoprotection in anthrax and detection of antigenemia. *Proc. Natl. Acad. Sci. USA* **101:**5042–5047.

58. **Kramer, M. J., and I. L. Roth.** 1968. Ultrastructural differences in the exosporium of the Sterne and Vollum strains of *Bacillus anthracis. Can. J. Microbiol.* **14:**1297–1299.

59. **Lai, E. M., N. D. Phadke, M. T. Kachman, R. Giorno, S. Vazquez, J. A. Vazquez, J. R. Maddock, and A. Driks.** 2003. Proteomic analysis of the spore coats of *Bacillus subtilis* and *Bacillus anthracis. J. Bacteriol.* **185:**1443–1454.

60. **Lentner, C.** 1981. *Geigy Scientific Tables: Units of Measurement, Body Fluids, Composition of the Body, Nutrition,* vol. 1 Ciba Geigy, Basel, Switzerland.

61. **Leppla, S. H.** 1995. Anthrax toxins, p. 543–572. *In* J. Moss, B. Iglewski, M. Vaughan, and A. T. Tu (ed.), *Bacterial Toxins and Virulence Factors in Disease.* Marcel Dekker, New York, N.Y.

62. **Little, S. F., and G. B. Knudson.** 1986. Comparative efficacy of *Bacillus anthracis* live spore vaccine and protective antigen vaccine against anthrax in the guinea pig. *Infect. Immun.* **52:**509–512.

63. **Liu, H., N. H. Bergman, B. Thomason, S. Shallom, A. Hazen, J. Crossno, D. A. Rasko, J. Ravel, T. D. Read, S. N. Peterson, J. Yates III, and P. C. Hanna.** 2004. Formation and composition of the *Bacillus anthracis* endospore. *J. Bacteriol.* **186:**164–178.

64. **Lyons, C. R., J. Lovchik, J. Hutt, M. F. Lipscomb, E. Wang, S. Heninger, L. Berliba, and K. Garrison.** 2004. Murine model of pulmonary anthrax: kinetics of dissemination, histopathology, and mouse strain susceptibility. *Infect. Immun.* **72:**4801–4809.

65. **Makino, S., C. Sasakawa, I. Uchida, N. Terakado, and M. Yoshikawa.** 1988. Cloning and CO_2-dependent expression of the genetic region for encapsulation from *Bacillus anthracis. Mol. Microbiol.* **2:**371–376.

66. **Makino, S., M. Watarai, H. I. Cheun, T. Shirahata, and I. Uchida.** 2002. Effect of the lower molecular capsule released from the cell surface of *Bacillus anthracis* on the pathogenesis of anthrax. *J. Infect. Dis.* **186:**227–233.

67. **Makino, S.-I., I. Uchida, N. Terakado, C. Sasakawa, and M. Yoshikawa.** 1989. Molecular characterization and protein analysis of the *cap* region, which is essential for encapsulation in *Bacillus anthracis. J. Bacteriol.* **171:**722–730.

68. **Marrero, R., and S. L. Welkos.** 1995. The transformation frequency of plasmids into *Bacillus anthracis* is affected by adenine methylation. *Gene* **152:**75–78.

69. **Mesnage, S., E. Tosi-Couture, P. Gounon, M. Mock, and A. Fouet.** 1998. The capsule and S-layer: two independent and yet compatible macromolecular structures in *Bacillus anthracis. J. Bacteriol.* **180:**52–58.

70. **Mesnage, S., E. Tosi-Couture, M. Mock, P. Gounon, and A. Fouet.** 1997. Molecular characterization of the *Bacillus*

anthracis main S-layer component: evidence that it is the major cell-associated antigen. *Mol. Microbiol.* **23:**1147–1155.

71. **Mignot, T., S. Mesnage, E. Couture-Tosi, M. Mock, and A. Fouet.** 2002. Developmental switch of S-layer protein synthesis in *Bacillus anthracis. Mol. Microbiol.* **43:**1615–1627.

72. **Mignot, T., M. Mock, and A. Fouet.** 2003. A plasmid-encoded regulator couples the synthesis of toxins and surface structures in *Bacillus anthracis. Mol. Microbiol.* **47:**917–927.

73. **Milne, J. C., D. Furlong, P. C. Hanna, J. S. Wall, and R. J. Collier.** 1994. Anthrax protective antigen forms oligomers during intoxication of mammalian cells. *J. Biol. Chem.* **269:**20607–20612.

74. **Moayeri, M., D. Haines, H. A. Young, and S. H. Leppla.** 2003. *Bacillus anthracis* lethal toxin induces TNF-alpha-independent hypoxia-mediated toxicity in mice. *J. Clin. Investig.* **112:**670–682.

75. **Moayeri, M., and S. H. Leppla.** 2004. The roles of anthrax toxin in pathogenesis. *Curr. Opin. Microbiol.* **7:**19–24.

76. **Mogridge, J., K. Cunningham, and R. J. Collier.** 2002. Stoichiometry of anthrax toxin complexes. *Biochemistry* **41:**1079–1082.

77. **Mourez, M., D. B. Lacy, K. Cunningham, R. Legmann, B. R. Sellman, J. Mogridge, and R. J. Collier.** 2002. 2001: a year of major advances in anthrax toxin research. *Trends Microbiol.* **10:**287–293.

78. **Okinaka, R. T., K. Cloud, O. Hampton, A. Hoffmaster, K. K. Hill, P. Keim, T. M. Koehler, G. Lamke, S. Kumano, J. Mahillon, D. Manter, Y. Martinez, D. Ricke, R. Svensson, and P. J. Jackson.** 1999. The sequence and organization of pXO1, the large *Bacillus anthracis* plasmid harboring the anthrax toxin genes. *J. Bacteriol.* **181:**6509–6515.

79. **Ooi, Y. M., and H. R. Colten.** 1979. Genetic defect in secretion of complement C5 in mice. *Nature* **282:**207–208.

80. **Pannifer, A. D., T. Y. Wong, R. Schwarzenbacher, M. Renatus, C. Petosa, J. Bienkowska, D. B. Lacy, R. J. Collier, S. Park, S. H. Leppla, P. Hanna, and R. C. Liddington.** 2001. Crystal structure of the anthrax lethal factor. *Nature* **414:**229–233.

81. **Park, J. M., F. R. Greten, Z. W. Li, and M. Karin.** 2002. Macrophage apoptosis by anthrax lethal factor through p38 MAP kinase inhibition. *Science* **297:**2048–2051.

82. **Pellizzari, R., C. Guidi-Rontani, G. Vitale, M. Mock, and C. Montecucco.** 2000. Lethal factor of *Bacillus anthracis* cleaves the N-terminus of MAPKKs: analysis of the intracellular consequences in macrophages. *Int. J. Med. Microbiol.* **290:**421–427.

83. **Petosa, C., R. J. Collier, K. R. Klimpel, S. H. Leppla, and R. C. Liddington.** 1997. Crystal structure of the anthrax toxin protective antigen. *Nature* **385:**833–838.

84. **Pezard, C., P. Berche, and M. Mock.** 1991. Contribution of individual toxin components to virulence of *Bacillus anthracis. Infect. Immun.* **59:**3472–3477.

85. **Pezard, C., M. Weber, J. C. Sirard, P. Berche, and M. Mock.** 1995. Protective immunity induced by *Bacillus anthracis* toxin-deficient strains. *Infect. Immun.* **63:**1369–1372.

86. **Phipps, A. J., C. Premanandan, R. E. Barnewall, and M. D. Lairmore.** 2004. Rabbit and nonhuman primate models of toxin-targeting human anthrax vaccines. *Microbiol. Mol. Biol. Rev.* **68:**617–629.

87. **Pitt, M. L., S. F. Little, B. E. Ivins, P. Fellows, J. Barth, J. Hewetson, P. Gibbs, M. Dertzbaugh, and A. M. Fried-** lander. 2001. In vitro correlate of immunity in a rabbit model of inhalational anthrax. *Vaccine* **19:**4768–4773.

88. **Priest, F. G., M. Barker, L. W. Baillie, E. C. Holmes, and M. C. Maiden.** 2004. Population structure and evolution of the *Bacillus cereus* group. *J. Bacteriol.* **186:**7959–7970.

89. **Rasko, D. A., J. Ravel, O. A. Okstad, E. Helgason, R. Z. Cer, L. Jiang, K. A. Shores, D. E. Fouts, N. J. Tourasse, S. V. Angiuoli, J. Kolonay, W. C. Nelson, A. B. Kolsto, C. M. Fraser, and T. D. Read.** 2004. The genome sequence of Bacillus cereus ATCC 10987 reveals metabolic adaptations and a large plasmid related to *Bacillus anthracis* pXO1. *Nucleic Acids Res.* **32:**977–988.

90. **Read, T. D., S. N. Peterson, N. Tourasse, L. W. Baillie, I. T. Paulsen, K. E. Nelson, H. Tettelin, D. E. Fouts, J. A. Eisen, S. R. Gill, E. K. Holtzapple, O. A. Okstad, E. Helgason, J. Rilstone, M. Wu, J. F. Kolonay, M. J. Beanan, R. J. Dodson, L. M. Brinkac, M. Gwinn, R. T. DeBoy, R. Madpu, S. C. Daugherty, A. S. Durkin, D. H. Haft, W. C. Nelson, J. D. Peterson, M. Pop, H. M. Khouri, D. Radune, J. L. Benton, Y. Mahamoud, L. Jiang, I. R. Hance, J. F. Weidman, K. J. Berry, R. D. Plaut, A. M. Wolf, K. L. Watkins, W. C. Nierman, A. Hazen, R. Cline, C. Redmond, J. E. Thwaite, O. White, S. L. Salzberg, B. Thomason, A. M. Friedlander, T. M. Koehler, P. C. Hanna, A. B. Kolsto, and C. M. Fraser.** 2003. The genome sequence of *Bacillus anthracis* Ames and comparison to closely related bacteria. *Nature* **423:**81–86.

91. **Rhie, G. E., M. H. Roehrl, M. Mourez, R. J. Collier, J. J. Mekalanos, and J. Y. Wang.** 2003. A dually active anthrax vaccine that confers protection against both bacilli and toxins. *Proc. Natl. Acad. Sci. USA* **100:**10925–10930.

92. **Ross, J. M.** 1957. Pathogenesis of anthrax following administration of spores by the respiratory route. *J. Pathol. Bacteriol.* **73:**485–494.

93. **Ruthel, G., W. J. Ribot, S. Bavari, and T. A. Hoover.** 2004. Time-lapse confocal imaging of development of *Bacillus anthracis* in macrophages. *J. Infect. Dis.* **189:**1313–1316.

94. **Saile, E., and T. M. Koehler.** 2002. Control of anthrax toxin gene expression by the transition state regulator *abrB. J. Bacteriol.* **184:**370–380.

95. **Salles, I. I., A. E. Tucker, D. E. Voth, and J. D. Ballard.** 2003. Toxin-induced resistance in *Bacillus anthracis* lethal toxin-treated macrophages. *Proc. Natl. Acad. Sci. USA* **100:**12426–12431.

96. **Schneerson, R., J. Kubler-Kielb, T. Y. Liu, Z. D. Dai, S. H. Leppla, A. Yergey, P. Backlund, J. Shiloach, F. Majadly, and J. B. Robbins.** 2003. Poly(γ-D-glutamic acid) protein conjugates induce IgG antibodies in mice to the capsule of *Bacillus anthracis*: a potential addition to the anthrax vaccine. *Proc. Natl. Acad. Sci. USA* **100:**8945–8950.

97. **Scobie, H. M., G. J. Rainey, K. A. Bradley, and J. A. Young.** 2003. Human capillary morphogenesis protein 2 functions as an anthrax toxin receptor. *Proc. Natl. Acad. Sci. USA* **100:**5170–5174.

98. **Shafa, F., B. J. Moberly, and P. Gerhardt.** 1966. Cytological features of anthrax spores phagocytized in vitro by rabbit alveolar macrophages. *J. Infect. Dis.* **116:**401–413.

99. **Sirard, J.-C., M. Mock, and A. Fouet.** 1995. Molecular tools for the study of transcriptional regulation in *Bacillus anthracis. Res. Microbiol.* **146:**729–737.

100. **Sirard, J.-C., M. Mock, and A. Fouet.** 1994. The three *Bacillus anthracis* toxin genes are coordinately regulated by bicarbonate and temperature. *J. Bacteriol.* **176:**5188–5192.

101. **Smith, H.** 2000. Discovery of the anthrax toxin: the beginning of in vivo studies on pathogenic bacteria. *Trends Microbiol.* **8:**199–200.

102. **Steichen, C., P. Chen, J. F. Kearney, and C. L. Turnbough, Jr.** 2003. Identification of the immunodominant protein and other proteins of the *Bacillus anthracis* exosporium. *J. Bacteriol.* **185:**1903–1910.

103. **Sterne, M.** 1939. The use of anthrax vaccines prepared from avirulent (uncapsulated) variants of *Bacillus anthracis. Onderstepoort J. Vet. Sci. Anim. Ind.* **13:**307–312.

104. **Sylvestre, P., E. Couture-Tosi, and M. Mock.** 2002. A collagen-like surface glycoprotein is a structural component of the *Bacillus anthracis* exosporium. *Mol. Microbiol.* **45:**169–178.

105. **Sylvestre, P., E. Couture-Tosi, and M. Mock.** 2003. Polymorphism in the collagen-like region of the *Bacillus anthracis* BclA protein leads to variation in exosporium filament length. *J. Bacteriol.* **185:**1555–1563.

106. **Thorne, C. B.** 1993. *Bacillus anthracis*, p. 113–124. *In* A. L. Sonenshein, J. A. Hoch, and R. Losick (ed.), *Bacillus subtilis and Other Gram-Positive Bacteria: Biochemistry, Physiology, and Molecular Genetics.* American Society for Microbiology, Washington, D.C.

107. **Thorne, C. B.** 1985. Genetics of *Bacillus anthracis*, p. 56–62. *In* L. Leive (ed.), *Microbiology.* American Society for Microbiology, Washington, D.C.

108. **Tinsley, E., A. Naqvi, A. Bourgogne, T. M. Koehler, and S. A. Khan.** 2004. Isolation of a minireplicon of the virulence plasmid pXO2 of *Bacillus anthracis* and characterization of the plasmid-encoded RepS replication protein. *J. Bacteriol.* **186:**2717–2723.

109. **Todd, S. J., A. J. Moir, M. J. Johnson, and A. Moir.** 2003. Genes of *Bacillus cereus* and *Bacillus anthracis* encoding proteins of the exosporium. *J. Bacteriol.* **185:**3373–3378.

110. **Turnbull, P. C.** 1991. Anthrax vaccines: past, present and future. *Vaccine* **9:**533–539.

111. **Turnbull, P. C.** 1991. Bacillus, p. 233–245. *In* S. Baron (ed.), *Medical Microbiology,* 4th ed. The University Medical Branch at Galveston, Galveston, Tex.

112. **Turnbull, P. C., R. A. Hutson, M. J. Ward, M. N. Jones, C. P. Quinn, N. J. Finnie, C. J. Duggleby, J. M.** Kramer, and J. Melling. 1992. *Bacillus anthracis* but not always anthrax. *J. Appl. Bacteriol.* **72:**21–28.

113. **Turnbull, P. C. B. (ed.).** 1996. *Proceedings of the International Workshop on Anthrax,* vol. 87, special supplement. Salisbury Medical Society, Salisbury, United Kingdom.

114. **Uchida, I., J. M. Hornung, C. B. Thorne, K. R. Klimpel, and S. H. Leppla.** 1993. Cloning and characterization of a gene whose product is a *trans*-activator of anthracis toxin synthesis. *J. Bacteriol.* **175:**5329–5338.

115. **Uchida, I., S. Makino, C. Sasakawa, M. Yoshikawa, C. Sugimoto, and N. Terakado.** 1993. Identification of a novel gene, *dep,* associated with depolymerization of the capsular polymer in *Bacillus anthracis. Mol. Microbiol.* **9:**487–496.

116. **Vietri, N. J., R. Marrero, T. A. Hoover, and S. L. Welkos.** 1995. Identification and characterization of a *trans*-activator involved in the regulation of encapsulation by *Bacillus anthracis. Gene* **152:**1–9.

117. **Vitale, G., R. Pellizzari, C. Recchi, G. Napolitani, M. Mock, and C. Montecucco.** 1998. Anthrax lethal factor cleaves the N-terminus of MAPKKs and induces tyrosine/threonine phosphorylation of MAPKs in cultured macrophages. *Biochem. Biophys. Res. Commun.* **248:**706–711.

118. **Wang, T. T., P. F. Fellows, T. J. Leighton, and A. H. Lucas.** 2004. Induction of opsonic antibodies to the gamma-D-glutamic acid capsule of *Bacillus anthracis* by immunization with a synthetic peptide-carrier protein conjugate. *FEMS Immunol. Med. Microbiol.* **40:**231–237.

119. **Watters, J. W., K. Dewar, J. Lehoczky, V. Boyartchuk, and W. F. Dietrich.** 2001. Kif1C, a kinesin-like motor protein, mediates mouse macrophage resistance to anthrax lethal factor. *Curr. Biol.* **11:**1503–1511.

120. **Watters, J. W., and W. F. Dietrich.** 2001. Genetic, physical, and transcript map of the Ltxs1 region of mouse chromosome 11. *Genomics* **73:**223–231.

121. **Welkos, S. L., T. J. Keener, and P. H. Gibbs.** 1986. Differences in susceptibility of inbred mice to *Bacillus anthracis. Infect. Immun.* **51:**795–800.

122. **Wesche, J., J. L. Elliott, P. O. Falnes, S. Olsnes, and R. J. Collier.** 1998. Characterization of membrane translocation by anthrax protective antigen. *Biochemistry* **37:**15737–15746.

123. **Whiting, G. C., S. Rijpkema, T. Adams, and M. J. Corbel.** 2004. Characterisation of adsorbed anthrax vaccine by two-dimensional gel electrophoresis. *Vaccine* **22:**4245–4251.

Clostridial Genetics

DENA LYRAS AND JULIAN I. ROOD

55

The genus *Clostridium* consists of an extremely diverse group of primarily gram-positive bacteria that have traditionally been grouped together based on their anaerobic growth requirements and their ability to produce heat-resistant endospores. Consequently, the different members of the genus are very dissimilar, and the genus lacks phylogenetic coherence. There are 120 different species within the genus, 35 of which can be considered capable of causing disease in humans or animals (78).

The most common feature of the pathogenic clostridia is that the cell and tissue damage that they cause primarily results from the production of potent extracellular toxins. Although somewhat artificial in that it crosses species boundaries, dividing the pathogenic clostridia into three major groups based upon their resultant disease pathology is useful. These groups consist of the neurotoxic clostridia, which produce toxins that affect the nervous system; the enterotoxic clostridia, which produce toxins that affect the gastrointestinal tract; and the histotoxic clostridia, whose necrotic pathology results from the production of one or more toxins that affect the structural and functional integrity of host cells located at or near the site of infection. The division into these three groups is used in the chapters of this volume on the clostridia rather than a species-specific approach because it leads to a more unified understanding of the pathogenesis process. This chapter focuses on the genetics and genomics of the pathogenic clostridia, dealing exclusively with the major clostridial pathogens *Clostridium perfringens*, *Clostridium difficile*, *Clostridium botulinum*, and *Clostridium tetani*.

PHYLOGENETICS OF THE PATHOGENIC CLOSTRIDIA

The initial application of homology studies using 23S rRNA molecules led to the division of 56 clostridial species into four major groups (37). Groups I and II consisted of well-defined species with a low-percent (24 to 32%) G+C content, group III consisted of low-percent G+C species that did not fit into the other groups, and group IV consist-

ed of high-G+C (41 to 45%) organisms. The major clostridial pathogens *C. perfringens* and *C. botulinum* belonged to group I, whereas *C. tetani* was a group II species. Subsequent studies involving sequence analysis of 16S rRNA molecules have confirmed these findings and have shown that, based on phylogenetic analysis, many members of the genus *Clostridium* are closely related to other bacteria (16, 78). The original group I organisms (37) are primarily located in 16S rRNA cluster I (Fig. 1), and it has been suggested that these organisms, which include *C. perfringens*, *Clostridium novyi*, *C. tetani*, and *C. botulinum*, are the true members of the genus *Clostridium*. Of the other major pathogens, *Clostridium septicum* and *Clostridium chauvoei* belong to cluster II, which may well remain within the genus *Clostridium* (78). However, *C. difficile* and *Clostridium sordellii* belong to cluster XI. These organisms should eventually be reclassified as belonging to another genus. However, apart from the confusion that such changes in nomenclature will cause among diagnostic microbiologists and clinicians, there is another problem in that cluster I includes the type species of the genus *Sarcina*. As the older name, *Sarcina* would have precedence, according to taxonomic convention; this is clearly an untenable situation (16, 78).

THE GENETICS OF *C. BOTULINUM*

Toxin-Encoding Bacteriophages and Plasmids of *C. botulinum*

There are seven distinct toxin types of *C. botulinum*, the causative agent of both human and animal botulism. These types are distinguished by their ability to produce antigenically distinct botulinum neurotoxins (BoNTs). Phylogenetically, these isolates represent at least three quite distinct strains (Fig. 1), which in any other genus would be classified as separate species. However, because of the disease significance of these organisms, clinical considerations have been allowed to override phylogenetics, and all of these isolates are still designated *C. botulinum*.

Gram-Positive Pathogens, 2nd edition, edited by Vincent A. Fischetti et al.
© 2006 ASM Press, Washington, D.C.

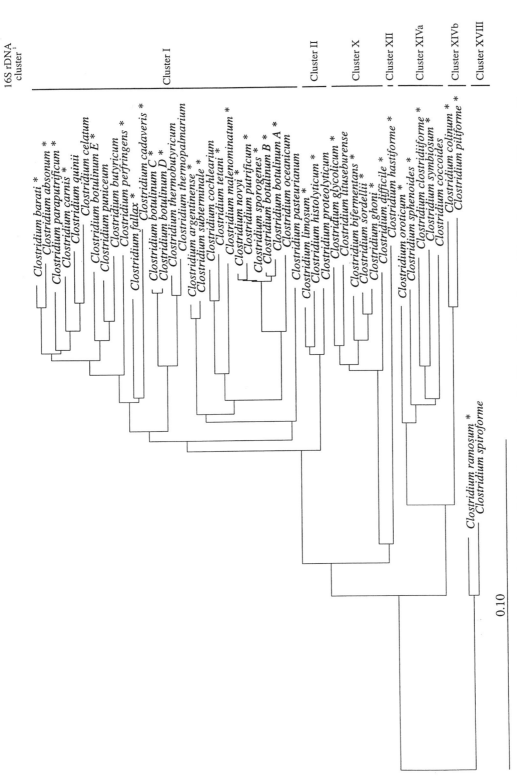

FIGURE 1 Phylogenetic relationships of the pathogenic clostridia. The scale bar indicates 10 base changes per 100 nucleotides. (Reproduced from Stackebrandt and Rainey [78] with the kind permission of the authors and Academic Press.)

The nonproteolytic *C. botulinum* types B, E, and F represent a distinct grouping of isolates that have almost 100% 16S rRNA sequence identity to toxigenic strains of *Clostridium butyricum* (16). By contrast, the proteolytic type A and F strains are virtually indistinguishable from *Clostridium putrificum* and *Clostridium sporogenes* by phylogenetic analysis. *C. botulinum* types C and D and *C. novyi* also are essentially identical species on phylogenetic grounds, whereas *C. botulinum* type G strains are quite distinct, being very closely related to *Clostridium subterminale* (16). Note that none of these *C. botulinum* toxin types are closely related to *C. tetani*.

This apparent phylogenetic chaos becomes more understandable when the location of the genes encoding the various BoNTs is analyzed. The genes encoding the various BoNT proteins are located in a gene cluster that also includes the genes encoding a hemagglutinin protein and a nontoxic nonhemagglutinin protein, as well as several other open reading frames (ORFs) and a putative regulatory gene (30). It has been known for some time that the genes encoding the BoNT type C (BoNT/C) and BoNT/D toxins and the associated gene cluster are located on pseudolysogenic bacteriophage that are not integrated into the chromosome but are maintained in an extrachromosomal state (30, 34). Curing these bacteriophages leads to the concomitant loss of the ability to produce the BoNT/C or BoNT/D toxins and their associated proteins such as hemagglutinin protein and nontoxic nonhemagglutinin proteins, a property that is restored upon reinfection. In addition, type C and type D strains of *C. botulinum* that have been cured of their toxigenic bacteriophage are indistinguishable from type A strains of *C. novyi* that have been cured of a bacteriophage that carries the *C. novyi* alpha-toxin gene. Strains can be readily interconverted between *C. botulinum* types C and D and *C. novyi* type A by infection with the respective BoNT/C-, BoNT/D-, and *C. novyi* alpha-toxin-specific bacteriophages. These data highlight the phylogenetic inconsistency of the current classification system.

C. botulinum type C and D strains also produce the C3 exoenzyme, an extracellular ADP-ribosyltransferase that acts on the mammalian Rho protein. The toxigenic bacteriophage that carries the BoNT/C and BoNT/D genes also carries the C3 structural gene. This gene is located on an unstable 21.5-kb fragment that has flanking directly repeated 6-bp Tn554-like core motifs. Deletion of the 21.5-kb region leads to the loss of one of these repeats, but it is not known if deletion results from site-specific or homologous recombination events. These observations have prompted the suggestion, for which there is no direct experimental evidence, that the C3 gene may be located on a transposable genetic element (29).

Bacteriophages have also been identified in type A, B, E, and F strains of *C. botulinum*, but in these isolates they are integrated into the chromosome and are difficult to cure (30). There is no evidence that toxin production in these toxin types is bacteriophage encoded (34).

Plasmids have been reported for almost all *C. botulinum* toxin types, but with one exception these plasmids do not carry toxin genes (34). The major exception involves type G strains, in which it has been clearly shown that the BoNT/G gene is located on a 114-kb plasmid that also encodes the hemagglutinin and nontoxic nonhemagglutinin proteins and the ability to produce the bacteriocin boticin G. In addition, although there is no evidence that the BoNT/E gene is plasmid determined in *C. botulinum*, a BoNT/E homolog that appears to be carried on a large plasmid has been identified in *C. butyricum* (28).

Genetic Manipulation of *C. botulinum*

Although methods for the genetic manipulation of *C. perfringens* are well established, it has taken some time for similar methods to be developed for the neurotoxic clostridia (36). The enterococcal tetracycline resistance transposon Tn916 has been shown to encode its own conjugative transfer from *Enterococcus faecalis* to *C. botulinum* at low but acceptable frequencies. The transposon appears to insert at several different sites in the *C. botulinum* chromosome, and multiple insertions are common. Several auxotrophic mutants have been isolated. In addition, Tn916 mutagenesis of *C. botulinum* type A strain 62A has been used to isolate mutants defective in their ability to produce BoNT/A (35).

Pulsed-field gel electrophoresis of MluI-, SmaI-, and RsrII-digested DNA has been used to determine the size of the genome of strain 62A, but the 4.04-Mb genome has not been mapped (40). Fourteen Tn916 insertion mutants of strain 62A were also mapped by pulsed-field gel electrophoresis, with 5 of 14 transconjugants containing single Tn916 insertions at different sites in the chromosome. Tn916-mediated deletions were also observed (40), and the Tn916-derived Tox⁻ mutant LNT01 has been shown to contain a deletion of ca. 70 kb in the toxin gene region (35). The complete sequence of the *C. botulinum* Hall strain A (ATCC 3502) genome has been determined but has not yet been published. The 3.89-Mb genome has a G+C content of 28.2%, and there is also a plasmid of 16.3 kb (http://www.sanger.ac.uk/Projects/C_botulinum/).

A transformation method has been developed for the plasmid-free *C. botulinum* type A strain Hall A, but it is relatively inefficient, yielding a maximum of ca. 3.4×10^3 transformants per μg of DNA (89). The process involves the electroporation of polyethylene glycol-treated mid-log-phase cells with the streptococcal vector pGK12, which encodes both erythromycin and chloramphenicol resistance. Other workers have used the *E. faecalis* shuttle vector pAT19 to transform strain 62A and to study the regulation of BoNT/A expression. These genetic experiments have provided evidence that toxin gene expression is activated by the *botR/A* gene product (52).

A versatile conjugation system has now been developed for introducing cloned genes into *C. botulinum* (9). This system involves the use of the mobilizable *C. perfringens*-*Escherichia coli* shuttle vector pJIR1457 (45) and the *E. coli* donor strain S17-1. Conjugative transfer of pJIR1457 from *E. coli* to the *C. botulinum* Hall A and 62A strains occurs at high frequency, and the resultant transconjugants stably maintain the plasmid. Therefore, this process can be routinely used to introduce cloned genes back into *C. botulinum*. For example, a pJIR1457 derivative containing a gene encoding a derivative of the BoNT/A light chain has been transferred to the toxin mutant LNT01 by conjugation, and the resultant transconjugant has been shown to produce high levels of modified light chain (9).

THE GENETICS OF *C. TETANI*

The *C. tetani* Genome and the Toxin Plasmid

The complete genome sequence of *C. tetani* strain E88, which is a variant of the vaccine strain Massachusetts, has been determined (10). The 2.8-Mb genome has a G+C content of 28.6% and carries 2,372 potential ORFs. There is also a 74.1-kb plasmid, pE88, that carries the tetanus toxin structural gene, *tetX*, and has a G+C content of only

24.5%. The genome contains very few potentially mobile genetic elements, suggesting that the *C. tetani* has a relatively stable genome that does not incorporate foreign DNA very frequently. There are several potential virulence genes located on the chromosome, including genes putatively encoding hemolysins, a fibronectin-binding protein, and surface adhesins. It is not known if any if these genes are required for virulence. The metabolic properties of *C. tetani* are reflected in the genome sequence. There are 21 genes that appear to be involved in amino acid degradation but relatively few genes involved in sugar metabolism. In addition, there are 35 genes that are postulated to be involved in Na^+-dependent transport, which in the absence of a functional F_0F_1-type ATPase are likely to be involved in providing the energy for active transport of amino acids across the cell membrane (10).

The gene encoding the tetanus neurotoxin (TeNT) has been known for some time to be plasmid determined. An initial analysis of 21 toxigenic strains of *C. tetani* revealed that they all contained a large plasmid that was absent from cured nontoxigenic strains. These observations were confirmed by subsequent studies of strain Massachusetts that led to the identification and mapping of the 75-kb plasmid pCL1, which carries the TeNT structural gene on a 16.5-kb EcoRI fragment (34). The determination of the sequence of pE88 has made a very significant contribution to our knowledge of these plasmids (10). As well as *tetX* and the *tetR* gene, which encodes the TetR-positive regulator, an alternative sigma factor (50, 52, 53, 63), pE88, encodes several other putative virulence genes. These putative products include a collagenase, ColT, several ABC-like peptide transport systems, and several regulatory proteins, including a two-component signal transduction system. Once more, it is not known if any of these factors are involved in virulence.

Genetic Manipulation of *C. tetani*

Tn916 has also been shown to encode its own transfer from *E. faecalis* to *C. tetani* (82), with the transposon inserting at multiple sites in the recipient genome. In addition, the transconjugants were able to act as donors in subsequent matings with either *C. tetani* or *E. faecalis*, the former at frequencies as high as 3.9×10^{-4} transconjugants per donor cell. However, there have been no further studies on this promising transposon mutagenesis system in *C. tetani*.

Transformation methods have also been developed for *C. tetani* (53). Electroporation was used to introduce the regulatory gene *tetR* into *C. tetani* and to show that overexpression of *tetR* leads to increased transcription of the TeNT structural gene and overexpression of TeNT. The vector was a derivative of the high-copy-number *E. faecalis* shuttle plasmid pAT19. The development of these methods is very important for future genetic studies of *C. tetani*.

THE GENETICS OF *C. PERFRINGENS*

Antibiotic Resistance in *C. perfringens*

Chloramphenicol Resistance

Chloramphenicol resistance in *C. perfringens* is not as common as erythromycin or tetracycline resistance and is mediated by the production of chloramphenicol acetyltransferase (CAT) enzymes (44). The *catP* chloramphenicol resistance gene is carried on transposons that are located on large conjugative tetracycline resistance plasmids (1). Comparative analysis of the nucleotide sequence of *catP* showed that, apart from the equivalent gene from *C. difficile*, it does not have significant similarity to other *cat* genes. However, the deduced CatP protein has significant similarity at the amino acid sequence level, including the carboxy-terminal region that contains the active site. CatP is most closely related to CAT monomers from *C. difficile*, *Vibrio anguillarum*, and *Campylobacter coli*, all of which have a four-amino-acid deletion in comparison to other CAT proteins (44). The *catP* gene was also identified in *Neisseria meningitidis* (24). Further studies have extended this finding and have shown that *catP* is present in *N. meningitidis* isolates from different geographical locations (73).

A different chloramphenicol resistance gene, *catQ*, has also been detected in a single strain of *C. perfringens* (44). The CatQ monomer is most closely related to CatB from *C. butyricum* and does not contain the four-amino-acid deletion present in the other clostridial *cat* genes. Therefore, this determinant seems to have evolved independently of *catP*, although it is possible that they share a common ancestor (44).

Erythromycin Resistance

The study of erythromycin or macrolide-lincosamide-streptogramin B (MLS) resistance in *C. perfringens* has predominantly focused on the Erm(B) determinant (44). The *erm*(B) gene is located on a 63-kb nonconjugative plasmid, pIP402, although it is not widespread in *C. perfringens*. The 738-bp gene is identical to the *erm* gene from the promiscuous *E. faecalis* plasmid pAMβ1 and has at least 98% nucleotide sequence identity to other *erm*(B) genes (8). In contrast to many of these genes, the pIP402 and pAMβ1 *erm* genes are not preceded by a leader peptide sequence and are constitutively expressed (8). An ORF of unknown function, ORF3, which is present downstream of the *erm*(B) genes from Tn917, pAMβ1, pAM77, and pIP501, is also found downstream of the pIP402 gene (Fig. 2A) (8). Mutational analysis of the *C. perfringens* Erm(B) protein has confirmed that conserved methyltransferase motifs identified by comparative sequence analysis are essential for function and has led to a greater understanding of the structure and function of this protein (23).

The *C. perfringens erm*(B) gene is located between two almost identical directly repeated sequences, DR1 (1,341 bp) and DR2 (1,340 bp) (Fig. 2A), which may encode a protein, ORF298, that has similarity to chromosomal and plasmid partitioning proteins (8). In both DR1 and DR2, there are almost identical 47-bp palindromic sequences, *palA* and *palB* (Fig. 2) (8). Comparative analysis has shown that both the pAMβ1 and pIP501 *erm* regions have DR2 but have a 975-bp internal deletion in DR1, whereas other determinants have small portions of DR1 but none have complete copies (Fig. 2A) (44). The *C. difficile* Tn5398-encoded determinant, which is unusual as it contains two *erm* genes, has a complete DR1 but an internal deletion in DR2 (21). In each case, the deletion endpoints are different, suggesting that they have arisen from separate deletion events (8).

Hybridization analyses have shown that a second *erm* gene, *erm*(Q), is also found in *C. perfringens*. This determinant represents the most common erythromycin resistance determinant in *C. perfringens*. The wider distribution of *erm*(Q) may reflect differences in the mechanisms by which the two determinants are disseminated (44).

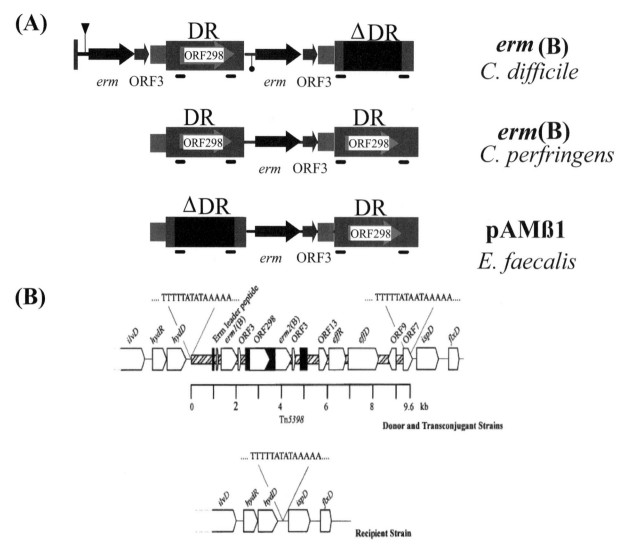

FIGURE 2 (A) Comparison of the genetic organization of the *erm*(B) gene regions. Regions of nucleotide sequence similarity are indicated by the same shading. The deletions in the pAMβ1 DR1-like sequences are indicated by the designation ΔDR. The filled arrows indicate the individual ORFs and their direction of transcription. The approximate locations of the palindromic sequences (*palA* and *palB*) but not their sizes are indicated by the boldface lines below the filled boxes. Leader peptide sequences are indicated by filled triangles. Modified from Lyras and Rood (44). (B) Genetic organization of Tn*5398*. The ORFs and their direction of transcription are represented by blocked arrows. The region encompassed by Tn*5398* is represented by the cross-hatched box and is further indicated by the scale below the diagram. Regions encompassing DR sequences are indicated by black boxes. The target site in the recipient is also shown, as is the location of each of the ends of the element and the target sequence. (Adapted from Farrow et al. [22].)

Tetracycline Resistance

Tetracycline resistance is the most common resistance phenotype found in *C. perfringens*, and resistant strains usually carry at least two distinct resistance genes (44). All tetracycline-resistant *C. perfringens* strains carry the *tetA*(P) gene. In more than half of these strains a second, overlapping resistance gene, *tetB*(P), is found downstream of *tetA*(P) (43). The Tet(P) determinant was isolated from the conjugative plasmid pCW3 and is found on all known tetracycline resistance plasmids from *C. perfringens*. It can also be encoded on nonconjugative plasmids or on the chromosome. A *tet*(M)-like gene is found in most of the isolates that do not

carry *tetB*(P), but it is not associated with the *tetA*(P) gene (43). Despite the wide distribution of the *tetB*(P) and *tet*(M) genes, no isolates have been found that carry both genes (43).

The *tetA*(P) gene encodes a 46-kDa transmembrane protein, TetA(P), which mediates the active efflux of tetracycline from the cell (74). Hydrophobicity analysis suggests that TetA(P) represents a novel type of efflux protein in comparison to other tetracycline efflux proteins. It appears to have two major hydrophilic domains that are not centrally located, whereas the prototype tetracycline efflux proteins have two large related six-transmembrane domains that are separated by a large central cytoplasmic domain (74). Site-directed mutagenesis has shown that amino acids that are predicted to be located in or near transmembrane domains, namely, Glu-52, Glu-59, and Glu-89, are essential for the active efflux of tetracycline (39). Random mutagenesis experiments have led to the identification of other residues that are required for tetracycline efflux. These residues are primarily located in the cytoplasmic loops 2 to 3 and 4 to 5 and the putative periplasmic loop 7 to 8 (6). Recent studies have involved site-directed mutagenesis of a region located within loop-joining transmembrane domains 2 and 3 of TetA(P) (7). Alteration of some residues, such as Pro-61 and Arg-71, abolished tetracycline resistance whereas alteration of other residues, such as Phe-58 and Lys-72, had no effect on the resistance phenotype. The data obtained in this study support the hypothesis that the region encoding this motif represents a distant version of the motif A region found in members of the major facilitator superfamily and other efflux proteins and that this region is essential for tetracycline resistance mediated by TetA(P) (7).

C. perfringens isolates that do not carry tetB(P) carry the tetA408(P) gene, which is 37 bp shorter than tetA(P) (43). The TetA408(P) protein has four amino acid changes within its last putative transmembrane domain, but the predicted transmembrane domain structure of the protein has not been significantly altered and the efflux phenotype is intact (43).

The tetB(P) gene encodes a putative 72.6-kDa protein that has significant amino acid sequence similarity to TetM-like cytoplasmic tetracycline resistance proteins and is therefore likely to function by a ribosomal protection mechanism (74). This gene encodes resistance to both tetracycline and minocycline, as do the tet(M)-like genes, whereas tetA(P) only confers resistance to tetracycline (43, 74).

The nucleotide sequence of the tet(M) gene from the nonconjugative C. perfringens strain CW459 has been determined. The deduced amino acid sequence of the protein encoded by this gene is almost identical (99.7%) to that from Tn916 but is less closely related (90.5%) to the Tet(M) protein from the C. difficile transposon Tn5397 (64). These results suggest that the C. perfringens tet(M) gene is more likely to be derived from a Tn916-like precursor and that the clostridial tet(M) genes seem to have evolved independently of one another (64).

Early studies showed that the tet(P) genes are found in both conjugative and nonconjugative tetracycline-resistant strains from diverse geographical locations and environmental sources (43). Conjugative transfer is usually associated with plasmids that are identical or closely related to pCW3, and in these isolates resistance has been found to be inducible (32). Interestingly, the inducible phenotype is dependent on the host strain harboring the plasmid, suggesting that induction requires a host-encoded factor that is yet to be identified (32). In nonconjugative isolates, resistance is constitutively expressed (32). Detailed transcriptional analysis of the Tet(P) determinant has shown that the tetA(P) and tetB(P) genes comprise an operon that is transcribed from a single promoter, P3, located 529 bp upstream of the tetA(P) start codon (33). A potential factor-independent terminator, T1, located approximately 390 bp downstream of the transcriptional start point but before the start of tetA(P), was also identified (33). Deletion or mutation of T1 resulted in increased read-through transcription, suggesting that this terminator may be an intrinsic control element of the tet(P) operon and may act to prevent lethal overexpression of the TetA(P) transmembrane protein (33). These studies

also showed that induction occurs at the level of initiation of transcription from the P3 promoter, suggesting the involvement of an unidentified regulatory protein that modulates P3 promoter activity (33).

The Genome Sequence of *C. perfringens*

The genome sequence of C. perfringens strain 13 has been determined (72). In addition, the sequences of two additional strains, the type strain ATCC13124 and the enterotoxin-producing strain SM101, have been completed but not yet published (http://www.tigr.org/tdb/mdb/mdbinprogress.html). The strain 13 genome is a little smaller at 3.03 Mb and has a G+C content of 28.6%. This strain also carries the 53.4 kb cryptic plasmid, pCP13. Strain 13 has 10 rRNA operons and 26 potential virulence-related genes, although of the latter genes only the plc (α-toxin) and pfoA (perfringolysin O) genes have been shown to be involved in virulence (68). The designation of many of the other virulence genes must be regarded as preliminary as there is little or no functional genomic data available for most of them.

In general, C. perfringens genes tended to be expressed in the same orientation as the direction of replication. As in C. tetani, there were very few mobile genetic elements and no pathogenicity islands (72). In keeping with its ability to ferment carbohydrates and obtain its energy by substrate level phosphorylation, there was a full complement of genes encoding glycolytic enzymes but none encoding enzymes of the TCA cycle or electron transport chain proteins. In addition, most of the genes encoding enzymes involved in amino acid biosynthesis were missing, as expected for an organism that requires many of its amino acids to be taken up from its environment. It is therefore not surprising that more than 200 transport-related genes were identified in the genome, including 52 ABC transporter systems. Finally, 20 sensor kinase/response regulator gene pairs were identified (72), including the well characterized virRS operon, the products of which regulate extracellular toxin production in C. perfringens (68).

Plasmids of *C. perfringens*

Many different types of plasmids have been found in C. perfringens, including plasmids that encode antibiotic resistance, bacteriocin production and immunity, and virulence factors or toxins (70). Many of these plasmids, especially those encoding toxins, have not been extensively studied, although several such studies are currently in progress.

C. perfringens is one of the only members of the clostridia in which conjugative antibiotic resistance plasmids have been found. Two such plasmids have been examined in detail, the 47-kb tetracycline resistance plasmid pCW3 and the 54-kb tetracycline and chloramphenicol resistance plasmid pIP401. Detailed restriction maps of both pCW3 and pIP401 have been constructed; comparison of these maps reveals that the two plasmids are very closely related. Further studies have shown that all conjugative tetracycline resistance plasmids from C. perfringens are closely related to or identical to pCW3 (44). pCW3 has now been completely sequenced, and functionally significant regions involved in conjugation and replication have been identified (T. L. Bannam, S. Teng, and J. I. Rood, unpublished data).

Loss of chloramphenicol resistance often occurs during conjugative transfer of pIP401 and is associated with the loss of approximately 6 kb from the plasmid. Subsequent studies showed that this segment comprises the transposable element Tn4451. The pIP401 replicon can therefore be considered to be a pCW3 plasmid that contains Tn4451

(1). Nonconjugative antibiotic resistance plasmids in *C. perfringens* have also been identified, the most notable being the 63-kb plasmid pIP402, which carries *erm*(B) and therefore encodes MLS resistance (69).

Phenotypically, the best-studied of all *C. perfringens* plasmids is the 10.2-kb bacteriocin plasmid pIP404, which has 10 ORFs (25). One of these genes, *bcn*, encodes the bacteriocin BCN5. Another gene, *uviA*, which is situated in an operon located downstream of *bcn*, encodes an alternative sigma factor that is related to BotR and TetR and regulates bacteriocin production (63). The origin of replication of pIP404 has been defined, copy-number control elements have been studied, and UV-inducible promoters responsible for the expression of *bcn* and the *uviAB* operon have been identified (69).

Many *C. perfringens* toxins are encoded on large plasmids (38). To analyze some of these plasmids, DNA from 16 *C. perfringens* isolates was digested with the intron-encoded endonuclease I-CeuI, which only cuts within rRNA operons found on the chromosome. Examination of these digests confirmed that the *plc* (alpha-toxin), *pfoA* (theta-toxin), *colA* (kappa-toxin), and *nagH* (sialidase) genes are chromosomally located, whereas the *etx* (epsilon-toxin), *cpb* (beta-toxin), *iap/ibp* (iota-toxin), *lam* (lambda-toxin), and urease genes are located on large extrachromosomal elements that do not contain I-CeuI sites (38). The beta2-toxin gene, *cpb2*, which does not hybridize with *cpb*, is also found on a large plasmid (26). Further work indicated that there was a significant association between *C. perfringens* isolates carrying a plasmid-carried *cpb2* gene and clostridial gastrointestinal diseases in domestic animals (86). Another study has confirmed that the *C. perfringens* urease genes, *ureABC*, are found on large plasmids that also often carry *etx*, *iap/ibp*, or the enterotoxin gene, *cpe* (20). It can be concluded that distinguishing *C. perfringens* type A isolates from type B to type E isolates clearly depends on the presence in the latter isolates of plasmids that confer the ability to produce beta-, epsilon-, or iota-toxin. Until now, very little was known about any of these large toxin plasmids; however, this situation is rapidly changing with the sequences of some of these plasmids currently being determined in several laboratories.

The nucleotide sequence of the cryptic plasmid pCP13 from the transformable *C. perfringens* strain 13 has been determined (72). This 54,310-bp plasmid has a G+C content of 25.5%, which is slightly lower than that of the chromosome at 28.6%. Determination of the nucleotide sequence showed that pCP13 carried a *cpb2* gene and a possible collagen adhesin gene, *cna* (72). Recent studies have suggested that transcription of *cpb2* and *cna* is regulated in a positive and negative manner, respectively, by the two-component VirR/VirS system (60). These studies indicate that the global regulatory cascade of the VirR/VirS system extends beyond genes that are chromosomally located to include plasmid-carried virulence genes (60).

The genetic location of the *cpe* gene is more variable than that of the other toxin genes. In *C. perfringens* strains isolated from food poisoning outbreaks, the *cpe* gene is chromosomally determined (15, 17), although a recently reported food poisoning outbreak was found to have been caused by an isolate carrying a plasmid-borne *cpe* gene (80). By contrast, in enterotoxin-producing isolates of animal origin and isolates associated with non-food-borne gastrointestinal infections, *cpe* is plasmid determined (15, 17). It has been postulated that chromosomal *cpe* genes are associated with food poisoning partly because cells and spores of these isolates are more heat resistant than isolates carrying plas-mid-carried *cpe* genes, which may enhance their survival in inadequately cooked or warmed foods (71). Other studies have shown that there is a strong correlation between human antibiotic-associated diarrhea and isolates which carry the *cpe* gene on a plasmid (75). This study further substantiated the link between isolates which carry chromosomal *cpe* genes and food poisoning (75). The *cpe*-encoding plasmid pMRS4969 has been shown to hybridize to pCW3 and to be conjugative, a finding which has significant clinical implications, since the in vivo conjugative acquisition of *cpe* by *C. perfringens* strains, which otherwise would not cause gastrointestinal disease, may make these strains pathogenic (13). Recently, a duplex PCR assay which can rapidly genotype enterotoxigenic type A isolates to determine whether these carry a chromosomal or plasmid-carried *cpe* gene has been developed, which should prove useful for epidemiologic and clinical diagnostic purposes (87).

Transposons and Insertion Sequences of *C. perfringens*

A number of transposons and insertion sequences have been identified and studied in *C. perfringens* (48). The best-characterized transposon is Tn*4451*, which carries the *catP* gene. This transposon is lost at high frequency from recombinant plasmids in *E. coli* and is also excised from its host plasmid pIP401 during conjugative transfer in *C. perfringens*. Transposition of Tn*4451* has been demonstrated in *E. coli* but not in the original host *C. perfringens* (44). The complete nucleotide sequence has been determined, and the 6,338-bp transposon has been found to encode six genes (Fig. 3) (5). Similar transposons, Tn*4452* and Tn*4453*, have been identified in *C. perfringens* and *C. difficile*, respectively.

Apart from *catP*, the two genes carried by Tn*4451* that have been studied in most detail are *tnpX* and *tnpZ*. The *tnpX* gene is required for the spontaneous excision of the element in both *E. coli* and *C. perfringens*, since a derivative of Tn*4451* with an internal deletion in *tnpX* is stable in both organisms. The provision of *tnpX* in *trans* restores the ability of the transposon to excise. Hybridization and PCR studies showed that the TnpX-mediated excision of Tn*4451* results in the formation of a circular form of the transposon. This behavior is similar to that of the conjugative transposons Tn*916* and Tn*1545*, where the first step in transposition involves the formation of a nonreplicating circular intermediate. Conjugation and transformation experiments have shown that the double stranded circular form of Tn*4451* is the transposition intermediate (47). Generation of this molecule leads to the formation of a strong promoter at the joint between the two ends of the element, which results in expression of the *tnpX* gene. The circular intermediate is not capable of replication and would be lost from host cells if it were not integrated into a replicating entity. The formation of the *tnpX* promoter increases the probability of integration of the circular form and therefore represents a transposon survival strategy. This process is particularly important when the single-stranded circular form is transferred to recipient cells via the TnpZ-mediated mobilization process discussed below. Since no endogenous TnpX is present in these cells, expression of this protein from the intermediate is crucial for transposon insertion, survival, and dissemination (47).

Sequence analysis indicated that the site-specific recombinase TnpX has a conserved resolvase/invertase domain (5) that we have shown by site-directed mutagenesis to be essential for the excision of Tn*4451* (18). Analysis of Tn*4451* target sites also showed that target site specificity is

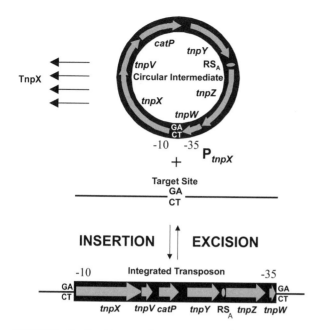

FIGURE 3 Excision and insertion model of Tn*4451*. Tn*4451* encodes six genes, as shown by the arrows, and is flanked by directly repeated GA dinucleotides. The TnpX-mediated excision of Tn*4451* leads to the formation of the circular form of the transposon, which carries one of the dinucleotides at the joint, and a deletion plasmid, in which one dinucleotide remains at the deletion site. Note the location of the components of the *tnpX* promoter. (Modified from Lyras and Rood [48].)

evident during the transposition process. The Tn*4451* target sequence was found to resemble the junction of the circular form, with insertion occurring at a GA dinucleotide and Tn*4451* insertions flanked by directly repeated GA dinucleotides that are also found at the joint of the circular form where the left and right termini of Tn*4451* are fused (18). Based on these results, a model for the excision and insertion of Tn*4451* was proposed whereby the resolvase/invertase domain of TnpX catalyzes the formation of a 2-bp staggered nick on either side of the GA dinucleotide located at the ends of Tn*4451*, at the junction of the circular form and at the insertion site (Fig. 3). Analysis of Tn*4451* derivatives with altered GA dinucleotides at the left and right ends provided the experimental evidence to support this model (18).

The development of in vivo excision and insertion assays, together with an in vitro excision assay, allowed us to determine that TnpX is the only transposon-encoded protein required for these two processes and, therefore, for transposition (49). Further, these experiments showed that TnpX catalyzes insertion by a resolvase-type mechanism, since the residues required for excision were also found to be essential for insertion (49). Significantly, the only resolvases that mediate insertion, as well as excision, belong to the large serine site-specific recombinases, which catalyze strand exchange by a nonreplicative DNA breakage and repair mechanism that involves a 2-bp staggered break across all four DNA strands and the formation of covalent phosphoserine linkages between the DNA strands and recombinase subunits (49). This group of proteins is structurally distinct from smaller resolvase proteins. Interestingly,

TnpX is most similar to the TndX protein encoded on the conjugative C. *difficile* transposon Tn*5397*, which has also been found to be the only transposon-encoded protein required for excision and insertion of this element (83), but it differs from other large resolvases, many of which are found on bacteriophage (49).

Binding studies carried out with purified TnpX have shown that the enzyme binds differentially to its transposon and target sites, which suggests that resolvase-like excision and insertion are two distinct processes (4). A combination of limited proteolysis analysis, cloning, and biochemical studies has shown that TnpX is organized into three major domains, with the N-terminal resolvase domain I (amino acids 1 to 170) not required for dimerization (4, 42). Domain II is located between amino acids 170 to 266. Domain III (amino acids 267 to 707) contains the dimerization region. A strong intramolecular interaction occurs between the N- and C-terminal regions, which are required for TnpX dimers to bind specifically to DNA. In addition, a DNA-binding site is located between residues 583 to 707. A small polypeptide consisting of amino acids 533 to 597 was found to be capable of binding specifically to the TnpX binding sites, suggesting that this is the primary DNA-binding region (42). In these studies, the DNA-binding and multimerization functions of TnpX but not those involving catalysis could be reconstituted by recombining various polypeptides encompassing the N- and C-terminal regions of the protein, suggesting that TnpX has separate catalytic, DNA-binding, and multimerization domains (42).

The *tnpZ* gene encodes a protein that has amino acid sequence similarity to a group of plasmid mobilization and recombination proteins that comprise the Mob/Pre family. These proteins interact with an upstream palindromic sequence, the *oriT* (for origin of transfer) or RS$_A$ site, resulting in plasmid nicking, which is the first step in DNA mobilization. There is an *oriT*-like sequence upstream of *tnpZ*; additional studies have shown that, in the presence of a chromosomally integrated copy of the broad-host-range IncP plasmid RP4 in E. *coli*, TnpZ promotes plasmid mobilization in *cis* and functions in *trans* to allow the mobilization of a coresident plasmid carrying an *oriT* site (19). TnpZ also facilitates the conjugative transfer of plasmids from E. *coli* to C. *perfringens*. Site-directed mutagenesis of the *oriT* site results in a significantly reduced mobilization frequency, confirming that it is required for the TnpZ-mediated mobilization process. TnpZ is unique in that it is the only known example of a Mob/Pre protein associated with a transposable genetic element (19). Furthermore, Tn*4451* is the first element to be identified from a gram-positive bacterium that is mobilizable but not self transmissible. The presence of mobilizable transposons may enable nonconjugative plasmids to be mobilized in the presence of other transfer-proficient factors. This raises the interesting possibility that TnpZ plays a role in intercellular transmission and hence the widespread dissemination of other resistance genes. Coupled with the finding that *tnpX* expression occurs from the circular intermediate via the formation of a specific promoter upon circularization, these data suggest that Tn*4451* is a highly evolved element that has acquired some properties of both conjugative and nonconjugative transposons and, in doing so, has acquired a number of mechanisms by which it optimizes its own survival and dissemination (19, 44, 47).

Finally, the function of the Tn*4451*-carried *tnpV*, *tnpY*, and *tnpW* genes is unknown. TnpV has some similarity to the Xis protein from bacteriophage λ (5) and TnpY is similar to

DNA repair proteins (49). Recent studies involving deletion of each gene from an otherwise-intact transposon showed that none of these genes are essential for excision or integration of the element (49). However, it may be that these genes are functionally required in the native clostridial hosts or that the assay systems used for these studies are not sensitive enough to detect subtle phenotypic differences. Since it is highly unlikely that all three genes are functionally redundant, further studies are required to understand what role they may play in this complex element.

A number of other insertion sequences and transposons have been identified in *C. perfringens*, namely, IS*1151*, IS*1469*, IS*1470*, and Tn*5565*. However, studies of these elements are not as advanced as those of Tn*4451*. The interesting feature of these elements is that they seem to be associated with toxin genes; although there is no direct experimental evidence, they have been implicated in the genetic transfer or mobilization of these virulence genes. Note that a Tn*916*-like conjugative transposon has been identified in *C. perfringens* strain CW459; however, studies of this element have been limited to sequence comparisons with Tn*5397* from *C. difficile*, since it appears to be nonfunctional and is therefore not capable of conjugative transfer (64).

IS*1151* is a 1,696-bp element that is located 96 bp upstream of the plasmid-carried epsilon-toxin gene, *etx*, in a *C. perfringens* type D strain. An IS*1151*-like element is located near *etx* in all type B and D isolates. IS*1151* contains two terminal 23-bp regions that constitute a perfect inverted repeat and has sequence similarity to the IS*231* family of insertion elements from *Bacillus thuringiensis*, among others (44). It was postulated that IS*1151* is derived from IS*231* and that, like IS*231* in *B. thuringiensis*, IS*1151* is involved in virulence gene transfer or mobilization in *C. perfringens*. Located upstream of IS*1151* is an ORF with significant similarity to the transposase-encoding *tnpA* gene from the Tn*3* family of transposons, which includes the cryptic Tn*4430* element from *B. thuringiensis*. IS*231* preferentially inserts at the ends of Tn*4430* in *B. thuringiensis*. The presence of IS*1151* and a region with similarity to Tn*4430* suggests that a similar situation may exist in *C. perfringens* (68).

The *cpe* gene may be chromosomal or plasmid determined and in one strain was found to be located on the same plasmid as the *etx* gene. Sequence analysis has shown that *cpe* is closely associated with a number of insertion sequences and in some chromosomal strains may be located on a transposon (12). All isolates that produce enterotoxin contain a 789-bp IS*200*-like element, IS*1469*, located 1.2 kb upstream of the *cpe* gene. In the chromosomal *cpe* strain NCTC8239, immediately upstream of IS*1469* is another insertion sequence, IS*1470A*, which belongs to the IS*30* family (Fig. 4). A second copy of IS*1470*, IS*1470B*, is found 1.1 kb downstream of *cpe* in this strain (Fig. 4). This structure appears to be that of a compound transposon, designated Tn*5565*, which contains an internal copy of IS*1469* and the *cpe* gene (11, 12). In strains isolated from food poisoning outbreaks, *cpe* is chromosomal and appears to be located on the same restriction fragment, suggesting that it is located on the same compound transposon (15). There is no evidence at this time that this putative transposon is capable of transposition. However, studies using PCR analysis have provided evidence that this element may excise to produce a circular form, which may represent a transposition intermediate (11).

In enterotoxin-producing strains of animal origin, the *cpe* gene is plasmid determined (15, 17, 81). In these strains, *cpe* is associated with IS*1469* as before, but not with IS*1470*. These plasmids also carry IS*1151*, which seems to be confined to isolates that have plasmid-borne *cpe* genes (17). In one such isolate, F3686, IS*1151* is located 260 bp downstream of the *cpe* gene, whereas in another isolate, 945P, IS*1151* is linked to the *etx* gene but not to *cpe*, although both genes are carried on the same plasmid (17). Later studies showed that *cpe* is also plasmid carried in *C. perfringens* strains associated with non-food-borne gastrointestinal infections (15).

Recent studies have determined the organization of the *cpe* locus on plasmids from 13 *C. perfringens* type A isolates (55). It was shown that, as for all chromosomal *cpe* isolates examined to date, IS*1469* sequences were present upstream of the *cpe* gene in these isolates. There is therefore a consistent association between the *cpe* gene and IS*1469* for both plasmid and chromosomally located *cpe* genes. Further analysis of these isolates suggested that sequences upstream of *cpe*, for at least 3.5 kb, are conserved in most type A isolates carrying a *cpe* plasmid but that this region differs from chromosomal isolates. The plasmid *cpe* isolates were found to lack the IS*1470* sequences found upstream of IS*1469* in the chromosomal *cpe* locus. The DNA region located downstream of *cpe* in these isolates showed somewhat more diversity, with 4 of the 13 isolates carrying IS*1151*-like sequences in this region. The remaining isolates were found to carry IS*1470* sequences in this region, but these are oppositely oriented and defective compared to the IS*1470* sequences found downstream of chromosomal *cpe*. This study therefore only found two distinctly different types of organization in the plasmid *cpe* locus in the isolates studied, suggesting that the locus has limited diversity and supporting the possibility that the chromosomal and plasmid *cpe* genes share a common origin (55). However, it is not clear how the distinctly organized

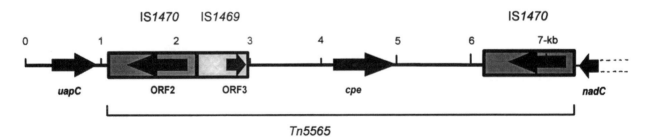

FIGURE 4 Genetic organization of the putative enterotoxin transposon Tn*5565*. The organization of the chromosomal *cpe* gene region is shown. ORFs are indicated by the arrows. The insertion elements are shown by the gray boxes. The scale is shown in kilobases. (Modified from Rood [68].)

cpe determinants may have arisen from transposition events involving Tn*5565* or the various insertion sequences.

Genetic Manipulation of *C. perfringens*

C. perfringens serves as the paradigm species for clostridial genetics. Methods for the genetic manipulation of *C. perfringens* are now well established, making this organism very amenable to genetic analysis. There are several well-characterized shuttle plasmids that can reliably and reproducibly be introduced into *C. perfringens* cells by transformation or conjugation. Methods for transposon mutagenesis and homologous recombination are also well established (67).

Although a number of *C. perfringens-E. coli* shuttle vectors have been constructed (Table 1), the most widely used shuttle vectors are derivatives of the plasmid pJIR418 (45, 67). This plasmid, which has been completely sequenced, contains the origins of replication of the *E. coli* plasmid pUC18 and the *C. perfringens* plasmid pIP404 together with the multiple cloning site and *lacZ'* gene from pUC18 (Table 1). It also contains the *erm*(B) and *catP* genes, which confer erythromycin and chloramphenicol resistance, respectively, in both *C. perfringens* and *E. coli*. Detection of recombinants can be achieved by X-Gal (5-bromo-4-chloro-3-indolyl-β-D-galactopyranoside) screening in *E. coli* or by insertional inactivation of the *erm*(B) or *catP* genes in either *E. coli* or *C. perfringens*.

Two derivatives of pJIR418, pJIR750 and pJIR751, have been constructed by deletion of the *erm*(B) and *catP* genes, respectively, thereby resulting in sequenced vectors that carry a single antibiotic resistance gene (Table 1). Subsequent studies involved the construction of derivatives of pJIR750 and pJIR751, designated pJIR1456 and pJIR1457, respectively, which can be mobilized into *C. perfringens* via conjugation from an appropriate *E. coli* donor (45).

A promoter probe plasmid based on pJIR418 has also been constructed for use in *C. perfringens*. This plasmid, pPSV (Table 1), was constructed by deletion of the *catP* gene from pJIR418, followed by the addition of a promoterless *catP* gene downstream of the multiple cloning site (54). The *catP* gene has also been used as a reporter system in the promoter probe shuttle vector pTCATT (Table 1), which contains a promoterless *catP* gene flanked by transcriptional terminators. However, PCR-mediated regeneration of the 5' terminus of the *catP* gene is required during the construction of promoter fusions in pTCATT, limiting its usefulness

(14). Other studies have attempted to develop reporter systems for use in *C. perfringens*. One system involved the construction of a plasmid that contains the *luxAB* genes from *Vibrio fischeri* under the control of the *plc* promoter. Luciferase activity and bioluminescence were obtained from this plasmid in *C. perfringens* (61). In another study, reporter plasmids using the *E. coli* gusA gene cloned into pJIR750 were constructed, and β-glucuronidase production was used successfully to monitor the sporulation-specific regulation of the *cpe* gene in *C. perfringens* (88). Recently, high-level expression of the *C. perfringens* NanI sialidase was achieved by using a ferredoxin promoter-based plasmid, pFF, for expression of this protein in *C. perfringens* (79). This system was particularly useful for purification of NanI, since the *C. perfringens* expression strain secreted the enzyme, as does the wild type, and seemed to efficiently express the protein, resulting in a substantially increased yield of 60 fold in comparison to protein isolation from an *E. coli* host. These results suggest that the pFF plasmid may be very useful for the expression of AT-rich genes from the clostridia and other bacteria in *C. perfringens* (79).

The most reliable and reproducible way of introducing shuttle plasmids into *C. perfringens* is by electroporation, although the transformation efficiency is very dependent on the host strain. A reasonable transformation frequency (3×10^5 transformants per μg of DNA) was obtained when *C. perfringens* isolate strain 13 was used as the transformation recipient. Consequently, strain 13 and derivatives of this strain are now the most widely used strains in studies of *C. perfringens* genetics (67).

Some *C. perfringens* strains cannot be transformed by electroporation (67). However, this problem can be overcome by using the shuttle plasmids pJIR1456 and pJIR1457, which carry the *oriT* region from the broad-host-range conjugative plasmid RP4 (45). These shuttle plasmids can be mobilized from the *E. coli* donor S17-1, which carries RP4 chromosomally, to a *C. perfringens* recipient at high efficiency. They can also be introduced into *C. perfringens* by the conventional electroporation process. The development of this system allows *C. perfringens* strains that cannot be transformed via electroporation to be genetically manipulated.

Methods for transposon mutagenesis in *C. perfringens* have been developed, but these methods are not very efficient. The most successful mutagenesis methods have utilized the

TABLE 1 Properties of *C. perfringens-E. coli* shuttle plasmids

Shuttle plasmid[a]	Size (kb)	*C. perfringens*		*E. coli*	
		Replicon	Selection[b]	Replicon	Selection[b]
pFF	7.0	pIP404	Em	pUC19	Ap, Em
pHR106	7.9	pJU122	Cm	pSL100	Ap, Cm
pJIR418	7.4	pIP404	Em, Cm	pUCI8	Em, Cm, XG
pJIR750	6.6	pIP404	Cm	pUC18	Cm, XG
pJIR751	6.0	pIP404	Em	pUC18	Em, XG
pJIR1456[c]	6.9	pIP404	Cm	pUC18	Cm, XG
pJIRl457[c]	6.2	pIP404	Em	pUC18	Em, XG
pPSV	6.9	pIP404	Em	pUC18	Em, XG
pTCATT	6.4	pIP404	Em	pUC18	Em

[a]For details and sources, see references 45 and 67; based on Table 5.1 from reference 67. For details on PFF, see reference 79.
[b]Ap, Cm, Em, resistance to ampicillin, chloramphenicol, and erythromycin, respectively. XG, screening for β-galactosidase production on X-Gal medium.
[c]These plasmids carry the RP4 *oriT* region and can therefore be mobilized, in the presence of plasmid RP4, to recipient cells (45).

conjugative tetracycline resistance transposon Tn916 (67). In these studies, *C. perfringens* cells were transformed with a suicide plasmid carrying Tn916, and transformants were selected on medium containing tetracycline. Hybridization analysis showed that Tn916 had inserted at different sites on the chromosome. However, multiple insertion events were common, as were deletion events. Since the *C. perfringens* transposons Tn4451 and Tn4452 have not been shown to transpose in this organism, there is still a need for more efficient transposon mutagenesis methods.

Homologous recombination and allelic exchange methods have also been used successfully for genetic analysis in *C. perfringens* (67). Homologous recombination may involve either single- or double-crossover events. For example, single-crossover events have been used to integrate suicide plasmids into the *plc* and *virR* genes, enabling an analysis of the function of these genes to be performed. Allelic exchange, achieved through a double-crossover event, has also been used to disrupt the chromosomally located *plc* and *pfoA* genes, and the plasmid-borne *tnpX* gene of Tn4451, again enabling functional analysis of these genes (67). Since two crossovers are required, allelic exchange is a less efficient process than obtaining mutants by a single-crossover event. However, the disadvantage of single-crossover events is that wild-type genes can be regenerated from the mutants by further homologous recombination events. This is not a problem with allelic exchange, which results in inherently stable mutants. It is worth noting that the efficiency by which mutants are obtained by allelic exchange depends on the size of the homologous regions that flank the gene of interest, and it is recommended that there be at least 2 kb of flanking DNA present on either side of the gene to be replaced.

THE GENETICS OF *C. DIFFICILE*

Antibiotic Resistance in *C. difficile*

Chloramphenicol Resistance

Early studies of chloramphenicol resistance in *C. difficile* W1 involved cloning of the *catP* (previously *catD*) gene. This gene is present in multiple copies on the *C. difficile* chromosome (44). Recent studies have shown that both copies are associated with transposons, designated Tn4453a and Tn4453b. These elements are very similar to Tn4451 (45). These results suggest that the chloramphenicol resistance genes and associated elements from *C. difficile* and *C. perfringens* have evolved relatively recently from a common ancestor (44).

Erythromycin Resistance

Early studies of erythromycin resistance in *C. difficile* demonstrated that the conjugative transfer of MLS resistance was common, unlike the situation in *C. perfringens*. Analysis of *C. difficile* 630 indicated that transfer appears to involve a chromosomal determinant, designated Tn5398. The erythromycin resistance gene carried on Tn5398 also belongs to the Erm(B) class and was previously designated as *ermBZ* (21). This novel determinant consists of two *erm*(B) genes, *erm1*(B) and *erm2*(B) (previously *ermBZ1* and *ermBZ2*, respectively), with ORF3 present downstream of each *erm*(B) gene. The two genes are separated by a 1.34-kb direct repeat (DR) sequence that is almost identical to direct repeats associated with the Erm(B) determinants from *C. perfringens* and *E. faecalis*. Both *erm*(B) genes are

also flanked by smaller variants of the 1.34-kb DR sequence (21). Overall, the extent of the similarity between the *erm* determinants suggests that they are derived from a common progenitor. However, it appears that the sequences that flank the *erm* genes have diverged extensively, even though the genes are very highly conserved. This conclusion is supported by the fact that many of these determinants are located on conjugative plasmids or transposons that are easily disseminated (44).

More recent studies have shown that there is a large amount of variation between *C. difficile* strains carrying *erm*(B) genes, ranging from strains carrying the simplest variant of only one *erm*(B) gene, with no association to Tn5398, to the most complex which carry two *erm*(B) genes associated with Tn5398, as found in strain 630 (22, 76). These strains were from a wide variety of geographical sources.

Several studies have identified erythromycin-resistant clinical *C. difficile* isolates that do not carry an *erm*(B) determinant (2, 22, 77). In one study, these isolates were also found to be PCR-negative for the *erm*(A), *erm*(C), *erm*(F), *erm*(Q), and *mefA* genes (77). The mechanism(s) by which these strains resist the action of erythromycin has not been elucidated.

Tetracycline Resistance

Studies of tetracycline resistance determinants from *C. difficile* are not as advanced as those from *C. perfringens*. A *tet*(M) gene has been identified in the conjugative *C. difficile* strains 630 and 662 and was found to be transferable at a low frequency. Genetic studies have indicated that this gene is located on a conjugative transposon, designated Tn5397 (56). Sequence analysis showed that the protein encoded by this gene has 90.5% amino acid sequence identity with the equivalent Tn916 gene (64). As found with Tn916, the Tn5397-carried *tet*(M) gene is inducible. However, the process is unclear how this occurs, since in Tn916 tetracycline resistance is induced by a transcriptional attenuation mechanism that involves the upstream ORF12 peptide and this ORF is partially deleted in Tn5397 (64).

Transposons of *C. difficile*

Four transposons from *C. difficile* have been described, specifically, the conjugative transposons Tn5397 and Tn5398 and the mobilizable transposons Tn4453a and Tn4453b (49). The interesting feature of these elements is that they, or the resistance determinants which they encode, have significant homology to elements from other bacteria.

The nonreplicative integrative conjugative element Tn5397 encodes a *tet*(M)-like tetracycline resistance determinant and can be transferred to *Bacillus subtilis* and then back to *C. difficile*. Interestingly, conjugative transfer of Tn5397 has been demonstrated in a model of an oral biofilm, suggesting that transfer is likely to occur in the environment of the human host (66). Tn5397 integrates into the *C. difficile* chromosome at two specific sites, whereas it integrates into *B. subtilis* at various sites. The entire transposon has been cloned, and a recombinant plasmid carrying Tn5397 has been used to transform *B. subtilis* to tetracycline resistance. The resulting transformants could transfer Tn5397 to *C. difficile* by filter mating (58).

Tn5397 is closely related to the Tn916/Tn1545 family of conjugative transposons (56, 64). Tn5397 is 20,658 bp in length and potentially encodes 21 ORFs, with four distinct gene modules that are potentially involved in conjugation, tetracycline resistance, gene regulation, and transposon excision and insertion. (64). These gene regions, with the exception of those for excision and insertion, are very closely related to equivalent Tn916 regions, as well as to a similar but defective element from C. perfringens (64). Furthermore, the Tn5397 oriT site is identical to that of Tn916. These results suggest that conjugative transfer of Tn5397 occurs in a very similar manner to that of Tn916 (64).

By contrast, the excision and insertion of Tn5397 are more similar to those of the Tn4451 family of elements than to those of Tn916, since Tn5397 relies on the large resolvase protein encoded by the tndX gene, a protein that is very similar to TnpX (59). TndX is the only transposon-encoded protein required for the excision and insertion of Tn5397 (59). Excision of the element by TndX also leads to the formation of a circular intermediate. Both TndX and TnpX act in a similar way mechanistically; however, they recognize different target sites and do not seem to be functionally interchangeable (84, 85).

Sequencing of the Tn5397 region located upstream of tet(M) indicated that the element contains a group II intron inserted into ORF14, which in Tn916 is required for intercellular transposition. Tn5397 retains its ability to conjugate, despite having this gene disrupted by the intron. Recent studies showed that this intron could undergo self splicing in vivo and that mutation of the intron-encoded protein did not impact on conjugal transfer of Tn5397, although it eliminated self splicing. This result suggests that the C-terminal portion of ORF14 is not functionally required for conjugal transfer (65).

The C. difficile isolate that carries Tn5397 also carries a second element, Tn5398, which encodes erythromycin (MLS) resistance and which is transferred independently of Tn5397. This element is transferred at low frequency to C. difficile CD37 and also to Staphylococcus aureus and B. subtilis. Transconjugants derived from the latter species can act as donors in matings using strain CD37 as the recipient. Hybridization studies revealed that, like Tn5397 and Tn916, Tn5398 integrates in a site-specific manner in C. difficile but randomly in B. subtilis (44, 58).

Tn5398 is 9.6 kb in size and contains a number of ORFs and an oriT site, as well as two identical erm(B) genes that are separated by a copy of the DR sequence found in the related C. perfringens determinant (Fig. 2B). It is therefore somewhat similar in structure to the erm(B) determinant from C. perfringens and other bacteria, although there are differences between each, as discussed previously (Fig. 2A). These results suggest that the various erm(B) determinants may have evolved by a process of gene duplication and homologous recombination events from a common progenitor (22, 47). Tn5398 is an unusual element, as it does not contain any transposase, site-specific recombinase, or conjugation genes, suggesting that it does not mediate its own excision or conjugative transfer. It may be that mobilization involves recombinase-mediated excision to form a circular intermediate followed by transfer into a recipient cell via oriT, mediated by Tn5397 or another potential conjugative element within the host. A resolvase- or recombination-mediated insertion may then serve to integrate the element into the genome of the recipient (22). Clearly, further studies of Tn5398 are required to better understand how movement of this unusual element occurs.

The chloramphenicol resistance gene from C. difficile W1 is carried on the transposons Tn4453a and Tn4453b. These elements are structurally and functionally related to Tn4451 from C. perfringens, but the three elements are distinguishable by restriction mapping (45). As with Tn4451, Tn4453a and Tn4453b are excised precisely from recombinant plasmids, generating a circular form. This process is mediated by Tn4453-carried tnpX genes. The joint of the circular form is also very similar to that of Tn4451. These results suggest that the Tn4453-encoded TnpX proteins bind to comparable target sequences and function in a manner similar to that of TnpX from Tn4451 (45). Transposition assays of E. coli showed that different transposition levels were obtained with Tn4451, Tn4453a, and Tn4453b, with Tn4453a yielding the highest levels of transposition (47). Assay systems for this family of elements have therefore been based on Tn4453a; however, the data obtained from these collective studies indicate that the elements and their genes are interchangeable and that differences obtained in functional assays are due to multiple factors, as discussed earlier in this chapter.

Genetic Manipulation of C. difficile

The genetic manipulation of C. difficile is not straightforward, but the methodology has progressed in recent years. In early studies, Tn916 or derivatives of this transposon were transferred from B. subtilis into C. difficile CD37 by filter mating. The C. difficile transconjugants each contained one copy of the transposon integrated into the same position in the genome. This site-specific insertion was also observed with Tn5397 and Tn5398. Tn916 is of limited value as a cloning vector because of its large size and the fact that it is not an autonomously replicating element. However, the site-specific insertion of Tn916 means that genes can reproducibly be introduced into a defined site in the C. difficile genome (44).

The limitations of using Tn916 as a cloning vector have been partly addressed. A plasmid that carried a segment of Tn916 was constructed and introduced into a B. subtilis strain containing a chromosomal copy of Tn916, where it integrated into this element by homologous recombination. The recombinant transposon was then transferred by conjugation to C. difficile, where it was inserted into the chromosome (57). This method was used successfully to introduce a fragment of the C. difficile toxin B gene back into the appropriate C. difficile strain. More recently, it was used to construct a sigK merodiploid in C. difficile (27). In these experiments, the Tn916::sigK derivative was incorporated into the chromosome by homologous recombination with a chromosomal sigK segment rather than by Tn916 transposition.

Electroporation is not currently an option for the introduction of DNA into C. difficile. There was an initial report of the introduction of DNA into C. difficile by electroporation (3); however, in the hands of this and many other laboratories, this approach has not been successful. A targeted gene disruption of the glycerol dehydrogenase gene, gldA, on the C. difficile chromosome has also been reported (41). A gldA gene with an internal deletion was introduced into C. difficile by conjugation on a suicide plasmid. The plasmid integrated into the gldA gene on the C. difficile chromosome, thereby inactivating the chromosomal copy of this gene. This study also has not yet been able to be repeated.

However, there are now simple conjugation methods available that allow the introduction of recombinant DNA molecules into C. difficile. A shuttle vector based on the C. perfringens pIP404 replication system, pJIR1457, can be transferred by conjugation from E. coli to C. perfringens and is stably maintained as a plasmid in both species (45). This RP4-mediated mobilization system can be used to transfer genes from E. coli to C. botulinum (12), C. septicum (D. Lyras and J. I. Rood, unpublished data), and, most importantly, to

C. difficile (51). We have successfully used this methodology to study gene regulation in *C. difficile* and were able to show that expression of the alternative sigma factor gene *txeR* was induced in stationary phase, was repressed by glucose, and was autoregulatory in the native host. Other workers have developed a similar system that allows the conjugative introduction of a range of autonomously replicating plasmids through circumvention of the restriction/methylation systems of two different toxigenic *C. difficile* strains (62). This system utilizes the native *C. difficile* plasmid pCD6. The shuttle plasmids are passaged through an *E. coli dam*[+] strain that contains an isoschizomer (M.*Sau*96I) of the *C. difficile* methylase genes; hence, the resultant plasmid is not subjected to restriction when it enters the recipient cell by conjugation from *E. coli*. In addition, potential target sites for relevant restriction endonucleases were removed. Importantly, the sequenced toxigenic *C. difficile* strain 630 has been shown to lack restriction endonucleases, despite the presence of five methylase genes, thereby making feasible the genetic analysis of virulence factors in this strain (31). Unfortunately, the analysis of this strain is complicated by the fact that it is resistant to both erythromycin and tetracycline, which are commonly used as selectable makers in clostridial genetics.

Finally, pulsed-field gel electrophoresis using the restriction enzymes SacII and NruI has been employed to generate a physical map of the 4.4-Mb chromosome of the toxigenic *C. difficile* strain ATCC 43594 (59a). This was the first step toward gaining an understanding of the genome organization of *C. difficile*. Currently, the genome sequence of strain 630 has been completed but has not been published (http://www.sanger.ac.uk/Projects/C_difficile/). The genome consists of a circular chromosome of 4.29 Mb with a G+C content of 29.06% and a circular plasmid of 7.88 kb.

We apologize to the many colleagues whose original work could not be cited. Owing to strict limitations in the number of permissible references, we have cited review articles wherever possible.

The work carried out in this laboratory was generously supported by grants from the Australian National Health and Medical Research Council.

REFERENCES

1. **Abraham, L. J., and J. I. Rood.** 1987. Identification of Tn*4451* and Tn*4452*, chloramphenicol resistance transposons from *Clostridium perfringens. J. Bacteriol.* **169:**1579–1584.
2. **Ackermann, G., A. Degner, S. H. Cohen, J. Silver, Jr., and A. C. Rodloff.** 2003. Prevalence and association of macrolide-lincosamide-streptogramin B (MLS$_B$) resistance with resistance to moxifloxacin in *Clostridium difficile. J. Antimicrob. Chemother.* **51:**599–603.
3. **Ackermann, G., Y. J. Tang, J. P. Henderson, A. C. Rodloff, J. Silver, Jr., and S. H. Cohen.** 2001. Electroporation of DNA sequences from the pathogenicity locus (PaLoc) of toxigenic *Clostridium difficile* into a nontoxigenic strain. *Mol. Cell. Probes* **15:**301–306.
4. **Adams, V., I. S. Lucet, D. Lyras, and J. I. Rood.** 2005. DNA binding properties of TnpX indicate that different synapses are formed in the excision and integration of the Tn*4451* family. *Mol. Microbiol.* **53:**1195–1207.
5. **Bannam, T. L., P. K. Crellin, and J. I. Rood.** 1995. Molecular genetics of the chloramphenicol-resistance transposon Tn*4451* from *Clostridium perfringens*:the TnpX site-specific recombinase excises a circular transposon molecule. *Mol. Microbiol.* **16:**535–551.

6. **Bannam, T. L., and J. I. Rood.** 1999. Identification of structural and functional domains of the tetracycline efflux protein TetA(P) from *Clostridium perfringens. Microbiology* **145:**2947–2955.
7. **Bannam, T. L., P. A. Johanesen, C. L. Salvado, S. J. Pidot, K. A. Farrow, and J. I. Rood.** 2004. The *Clostridium perfringens* TetA(P) efflux protein contains a functional variant of the Motif A region found in major facilitator superfamily transport proteins. *Microbiology* **150:**127–134.
8. **Berryman, D. I., and J. I. Rood.** 1995. The closely related *ermB-ermAM* genes from *Clostridium perfringens, Enterococcus faecalis* (pAMβ1), and *Streptococcus agalactiae* (pIP501) are flanked by variants of a directly repeated sequence. *Antimicrob. Agents Chemother.* **39:**1830–1834.
9. **Bradshaw, M., M. Goodnough, and E. Johnson.** 1998. Conjugative transfer of the *E. coli-C. perfringens* shuttle vector pJIR1457 to *Clostridium botulinum* type A strains. *Plasmid* **40:**233–237.
10. **Bruggeman, H., S. Baumer, W. F. Fricke, A. Wiezer, H. Liesegang, I. Decker, C. Herzberg, R. Martinez-Arias, R. Merkl, A. Henne, and G. Gottschalk.** 2003. The genome sequence of *Clostridium tetani*, the causative agent of tetanus disease. *Proc. Natl. Acad. Sci. USA* **100:**1316–1321.
11. **Brynestad, S., and P. E. Granum.** 1999. Evidence that Tn*5565*, which includes the enterotoxin gene in *Clostridium perfringens*, can have a circular form which may be a transposition intermediate. *FEMS Microbiol. Lett.* **170:**281–286.
12. **Brynestad, S., B. Synstad, and P. E. Granum.** 1997. The *Clostridium perfringens* entertotoxin gene is on a transposable genetic element in type A human food poisoning strains. *Microbiology* **143:**2109–2115.
13. **Brynestad, S., M. R. Sarker, B. A. McClane, P. E. Granum, and J. I. Rood.** 2001. Enterotoxin plasmid from *Clostridium perfringens* is conjugative. *Infect. Immun.* **69:**3483–3487.
14. **Bullifent, H. L., A. Moir, and R. W. Titball.** 1995. The construction of a reporter system and use for the investigation of *Clostridium perfringens* gene expression. *FEMS Microbiol. Lett.* **131:**99–105.
15. **Collie, R. E., and B. A. McClane.** 1998. Evidence that the enterotoxin gene can be episomal in *Clostridium perfringens* isolates associated with non-food-borne human gastrointestinal diseases. *J. Clin. Microbiol.* **36:**30–36.
16. **Collins, M. D., P. A. Lawson, A. Willems, J. J. Cordoba, J. Fernandez-Garayzabal, P. Garcia, J. Cai, H. Hippe, and J. A. Farrow.** 1994. The phylogeny of the genus *Clostridium*:proposal of five new genera and eleven new species combinations. *Int. J. Syst. Bacteriol.* **44:**812–826.
17. **Cornillot, E., B. Saint-Joanis, G. Daube, S.-I. Katayama, P. E. Granum, B. Canard, and S. T. Cole.** 1995. The enterotoxin gene (*cpe*) of *Clostridium perfringens* can be chromosomal or plasmid-borne. *Mol. Microbiol.* **15:**639–647.
18. **Crellin, P. K., and J. I. Rood.** 1997. The resolvase/invertase domain of the site-specific recombinase TnpX is functional and recognizes a target sequence that resembles the junction of the circular form of the *Clostridium perfringens* transposon Tn*4451. J. Bacteriol.* **179:**5148–5156.
19. **Crellin, P. K., and J. I. Rood.** 1998. Tn*4451* from *Clostridium perfringens* is a mobilizable transposon that encodes the functional Mob protein, TnpZ. *Mol. Microbiol.* **27:**631–642.
20. **Dupuy, B., G. Daube, M. R. Popoff, and S. T. Cole.** 1997. *Clostridium perfringens* urease genes are plasmid borne. *Infect. Immun.* **65:**2313–2320.

21. **Farrow, K. A., D. Lyras, and J. I. Rood.** 2000. The macrolide-lincosamide-streptogramin B resistance determinant from *Clostridium difficile* 630 contains two *erm*(B) genes. *Antimicrob. Agents Chemother.* **44:**411–413.

22. **Farrow, K. A., D. Lyras, and J. I. Rood.** 2001. Genomic analysis of the erythromycin resistance element Tn*5398* from *Clostridium difficile*. *Microbiology* **147:**2717–2728.

23. **Farrow, K. A., D. Lyras, G. Polekhina, K. Koutsis, M. W. Parker, and J. I. Rood.** 2002. Identification of essential residues in the Erm(B) rRNA methyltransferase of *Clostridium perfringens*. *Antimicrob. Agents Chemother.* **46:** 1253–1261.

24. **Galimand, M., G. Gerbaud, M. Guibourdenche, J.-Y. Riou, and P. Courvalin.** 1998. High-level chloramphenicol resistance in *Neisseria meningitidis*. *N. Engl. J. Med.* **339:**868–874.

25. **Garnier, T., and S. T. Cole.** 1988. Complete nucleotide sequence and genetic organization of the bacteriocinogenic plasmid, pIP404, from *Clostridium perfringens*. *Plasmid* **19:**134–150.

26. **Gibert, M., C. Jolivet-Renaud, and M. R. Popoff.** 1997. Beta2 toxin, a novel toxin produced by *Clostridium perfringens*. *Gene* **203:**65–73.

27. **Haraldsen, J. D., and A. L. Sonenshein.** 2003. Efficient sporulation in *Clostridium difficile* requires disruption of the σ^K gene. *Mol. Microbiol.* **48:**811–821.

28. **Hauser, D., M. Gibert, P. Boquet, and M. R. Popoff.** 1992. Plasmid localization of a type E botulinal neurotoxin gene homologue in toxigenic *Clostridium butyricum* strains, and absence of this gene in non-toxigenic *C. butyricum* strains. *FEMS Microbiol. Lett.* **99:**251–256.

29. **Hauser, D., M. Gibert, M. W. Eklund, P. Boquet, and M. R. Popoff.** 1993. Comparative analysis of C3 and botulinal neurotoxin genes and their environment in *Clostridium botulinum* types C and D. *J. Bacteriol.* **175:**7260–7268.

30. **Henderson, I., T. Davis, M. Elmore, and N. Minton.** 1997. The genetic basis of toxin production in *Clostridium botulinum* and *Clostridium tetani*, p. 261–294. *In* J. Rood, B. McClane, J. Songer, and R. Titball (ed.), *The Clostridia: Molecular Biology and Pathogenesis*. Academic Press, Inc., London, United Kingdom

31. **Herbert, M., T. A. O'Keeffe, D. Purdy, M. Elmore, and N. P. Minton.** 2003. Gene transfer into *Clostridium difficile* CD630 and characterisation of its methylase genes. *FEMS Microbiol. Lett.* **229:**103–110.

32. **Johanesen, P. A., D. Lyras, and J. I. Rood.** 2001. Induction of pCW3-encoded tetracycline resistance in *Clostridium perfringens* involves a host-encoded factor. *Plasmid* **46:**229–232.

33. **Johanesen, P. A., D. Lyras, T. L. Bannam, and J. I. Rood.** 2001. Transcriptional analysis of the *tet*(P) operon from *Clostridium perfringens*. *J. Bacteriol.* **183:**7110–7119.

34. **Johnson, E.** 1997. Extrachromosomal virulence determinants in the clostridia, p. 35–48. *In* J. Rood, B. McClane, J. Songer, and R. Titball (ed.), *The Clostridia:Molecular Biology and Pathogenesis*. Academic Press, Inc., London, United Kingdom

35. **Johnson, E., W. J. Lin, Y. Zhou, and M. Bradshaw.** 1997. Characterization of neurotoxin mutants in *Clostridium botulinum* type A. *Clin. Infect. Dis.* **25**(Suppl. 2):S168–S170.

36. **Johnson, E. A., and M. Bradshaw.** 2001. *Clostridium botulinum* and its neurotoxins: a metabolic and cellular perspective. *Toxicon* **39:**1703–1722.

37. **Johnson, J. L., and B. S. Francis.** 1975. Taxonomy of the clostridia: ribosomal acid homologies among the species. *J. Gen. Microbiol.* **88:** 229–244.

38. **Katayama, S., B. Dupuy, G. Daube, B. China, and S. T. Cole.** 1996. Genome mapping of *Clostridium perfringens* strains with I-*Ceu*I shows many virulence genes to be plasmid-borne. *Mol. Gen. Genet.* **251:**720–726.

39. **Kennan, R. M., L. M. McMurry, S. B. Levy, and J. I. Rood.** 1997. Glutamate residues located within putative transmembrane helices are essential for TetA(P)-mediated tetracycline efflux. *J. Bacteriol.* **179:**7011–7015.

40. **Lin, W.-J., and E. A. Johnson.** 1995. Genome analysis of *Clostridium botulinum* type A by pulsed-field gel electrophoresis. *Appl. Environ. Microbiol.* **61:**4441–4447.

41. **Liyanage, H., S. Kashket, M. Young, and E. R. Kashket.** 2001. *Clostridium beijerinckii* and *Clostridium difficile* detoxify methylglyoxal by a novel mechanism involving glycerol dehydrogenase. *Appl. Environ. Microbiol.* **67:**2004–2010.

42. **Lucet, I. S., F. E. Tynan, V. Adams, J. Rossjohn, D. Lyras, and J. I. Rood.** 2004. Identification of the structural and functional domains of the large serine recombinase TnpX from *Clostridium perfringens*. *J. Biol. Chem.* **280:** 2503–2511.

43. **Lyras, D., and J. I. Rood.** 1996. Genetic organization and distribution of tetracycline resistance determinants in *Clostridium perfringens*. *Antimicrob. Agents Chemother.* **40:** 2500–2504.

44. **Lyras, D., and J. I. Rood.** 1997. Transposable genetic elements and antibiotic resistance determinants from *Clostridium perfringens* and *Clostridium difficile*, p. 73–92. *In* J. I. Rood, B. A. McClane, J. G. Songer, and R. W. Titball (ed.), *The Clostridia: Molecular Biology and Pathogenesis*. Academic Press, Inc., London, United Kingdom.

45. **Lyras, D., and J. Rood.** 1998. Conjugative transfer of RP4-*oriT* shuttle vectors from *Escherichia coli* to *Clostridium perfringens*. *Plasmid* **39:**160–164.

46. **Lyras, D., C. Storie, A. S. Huggins, P. K. Crellin, T. L. Bannam, and J. I. Rood.** 1998. Chloramphenicol resistance in *Clostridium difficile* is encoded on Tn*4453* transposons that are closely related to Tn*4451* from *Clostridium perfringens*. *Antimicrob. Agents Chemother.* **42:**1563–1567.

47. **Lyras, D., and J. I. Rood.** 2000. Transposition of Tn*4451* and Tn*4453* involves a circular intermediate that forms a promoter for the large resolvase, TnpX. *Mol. Microbiol.* **38:**588–601.

48. **Lyras, D., and J. I. Rood.** 2005. Transposable genetic elements of the clostridia, p. 633–645. *In* P. Durre (ed.), *Handbook of the Clostridia*. CRC Press, Boca Raton, Fla.

49. **Lyras, D., V. Adams, I. Lucet, and J. I. Rood.** 2004. The large resolvase TnpX is the only transposon-encoded protein required for transposition of the Tn*4451/3* family of integrative mobilizable elements. *Mol. Microbiol.* **51:** 1787–1800.

50. **Mani, N., and B. Dupuy.** 2001. Regulation of toxin synthesis in *Clostridium difficile* by an alternative RNA polymerase sigma factor. *Proc. Natl. Acad. Sci. USA* **98:** 5844–5849.

51. **Mani, N., D. Lyras, L. Barroso, P. Howarth, T. Wilkins, J. I. Rood, A. L. Sonenshein, and B. Dupuy.** 2002. Environmental response and autoregulation of *Clostridium difficile* TxeR, a sigma factor for toxin gene expresssion. *J. Bacteriol.* **184:**5971–5978.

52. **Marvaud, J., M. Gibert, K. Inoue, Y. Fujinaga, K. Oguma, and M. Popoff.** 1998. *botR*/A is a positive regulator of botulinum neurotoxin and associated non-toxin protein genes in *Clostridium botulinum* A. *Mol. Microbiol.* **29:** 1009–10018.

53. **Marvaud, J.-C., U. Eisel, T. Binz, H. Niemann, and M. R. Popoff.** 1998. TetR is a positive regulator of the tetanus

toxin gene in *Clostridium tetani* and is homologous to BotR. *Infect. Immun.* **66:**5698–5702.

54. **Matsushita, C., O. Matsushita, M. Koyama, and A. Okabe.** 1994. A *Clostridium perfringens* vector for the selection of promoters. *Plasmid* **31:**317–319.

55. **Miyamoto, K., G. Chakrabarti, Y. Morino, and B. A. McClane.** 2002. Organization of the plasmid *cpe* locus in *Clostridium perfringens* type A isolates. *Infect. Immun.* **70:**4261–4272.

56. **Mullany, P., M. Pallen, M. Wilks, J. Stephen, and S. Tabaqchali.** 1996. A group II intron in a conjugative transposon from the gram-positive bacterium, *Clostridium difficile.* *Gene* **174:**145–150.

57. **Mullany, P., M. Wilks, L. Puckey, and S. Tabaqchali.** 1994. Gene cloning in *Clostridium difficile* using Tn916 as a shuttle conjugative transposon. *Plasmid* **31:**320–323.

58. **Mullany, P., M. Wilks, and S. Tabaqchali.** 1995. Transfer of macrolide-lincosamide-streptogramin B (MLS) resistance in *Clostridium difficile* is linked to a gene homologous with toxin A and is mediated by a conjugative transposon, Tn5398. *J. Antimicrob. Chemother.* **35:**305–315.

59. **Mullany, P., A. P. Roberts, and H. Wang.** 2002. Mechanism of integration and excision in conjugative transposons. *Cell. Mol. Life Sci.* **59:**2017–2022.

59a.**Norwood, D. A., Jr., and J. A. Sands.** 1997. Physical map of the *Clostridium difficile* chromosome. *Gene* **201:**159–168.

60. **Ohtani, K., H. I. Kawsar, K. Okumura, H. Hayashi, and T. Shimizu.** 2003. The VirR/VirS regulatory cascade affects transcription of plasmid-encoded putative virulence genes in *Clostridium perfringens* strain 13. *FEMS Microbiol. Lett.* **222:**137–141.

61. **Phillips-Jones, M. K.** 1993. Bioluminescence (*lux*) expression in the anaerobe *Clostridium perfringens.* *FEMS Microbiol. Lett.* **106:**265–270.

62. **Purdy, D., T. A. T. O'Keeffe, M. Elmore, M. Herbert, A. McLeod, M. Bokori-Brown, A. Ostrowski, and N. P. Minton.** 2002. Conjugative transfer of clostridial shuttle vectors from *Escherichia coli* to *Clostridium difficile* through circumvention of the restriction barrier. *Mol. Microbiol.* **46:**439–452.

63. **Raffestin, S., B. Dupuy, J. C. Marvaud, and M. R. Popoff.** 2005. BotR/A and TetR are alternative RNA polymerase sigma factors controlling the expression of the neurotoxin and associated protein genes in *Clostridium botulinum* type A and *Clostridium tetani.* *Mol. Microbiol.* **55:**235–249.

64. **Roberts, A. P., P. A. Johanesen, D. Lyras, P. Mullany, and J. I. Rood.** 2001. Comparison of Tn5397 from *Clostridium difficile*, Tn916 from *Enterococcus faecalis* and the CW459*tet*(M) element from *Clostridium perfringens* shows that they have similar conjugation regions but different insertion and excision modules. *Microbiology* **147:**1243–1251.

65. **Roberts, A. P., V. Braun, C. von Eichel-Streiber, and P. Mullany.** 2001. Demonstration that the group II intron from the clostridial conjugative transposon Tn5397 undergoes splicing in vivo. *J. Bacteriol.* **183:**1296–1299.

66. **Roberts, A. P., J. Pratten, M. Wilson, and P. Mullany.** 1999. Transfer of a conjugative transposon, Tn5397, in a model oral biofilm. *FEMS Microbiol. Lett.* **177:**63–66.

67. **Rood, J. I.** 1997. Genetic analysis in *C. perfringens*, p. 65–72. *In* J. I. Rood, B. A. McClane, J. G. Songer, and R. W. Titball (ed.), *The Clostridia: Molecular Biology and Pathogenesis.* Academic Press, Inc., London, United Kingdom.

68. **Rood, J. I.** 1998. Virulence genes of *Clostridium perfringens.* *Annu. Rev. Microbiol.* **52:**333–360.

69. **Rood, J. I., and S. T. Cole.** 1991. Molecular genetics and pathogenesis of *Clostridium perfringens.* *Microbiol. Rev.* **55:**621–648.

70. **Rood, J. I.** 2004. Virulence plasmids of spore-forming bacteria, p. 413–422. *In* B. E. Funnell and G. J. Phillips (ed.), *The Biology of Plasmids.* ASM Press, Washington, D.C.

71. **Sarker, M. R., R. P. Shivers, S. G. Sparks, V. K. Juneja, and B. A. McClane.** 2000. Comparative experiments to examine the effects of heating on vegetative cells and spores of *Clostridium perfringens* isolates carrying plasmid genes versus chromosomal enterotoxin genes. *Appl. Environ. Microbiol.* **66:**3234–3240.

72. **Shimizu, T., K. Ohtani, H. Hirakawa, K. Ohshima, A. Yamashita, T. Shiba, N. Ogasawara, M. Hattori, S. Kuhara, and H. Hayashi.** 2002. Complete genome sequence of *Clostridium perfringens*, an anaerobic flesh-eater. *Proc. Natl. Acad. Sci. USA* **99:**996–1001.

73. **Shultz, T. R., J. W. Tapsall, P. A. White, C. S. Ryan, D. Lyras, J. I. Rood, E. Binotto, and C. J. Richardson.** 2003. Chloramphenicol-resistant *Neisseria meningitidis* containing *catP* isolated in Australia. *J. Antimicrob. Chemother.* **52:**856–859.

74. **Sloan, J., L. M. McMurry, D. Lyras, S. B. Levy, and J. I. Rood.** 1994. The *Clostridium perfringens* TetP determinant comprises two overlapping genes: *tetA*(P) which mediates active tetracycline efflux and *tetB*(P) which is related to the ribosomal protection family of tetracycline resistance determinants. *Mol. Microbiol.* **11:**403–415.

75. **Sparks, S. G., R. J. Carmen, M. R. Sarker, and B. A. McClane.** 2001. Genotyping of enterotoxigenic *Clostridium perfringens* fecal isolates associated with antibiotic-associated diarrhea and food poisoning in North America. *J. Clin. Microbiol.* **39:**883–888.

76. **Spigaglia, P., and P. Mastrantonio.** 2002. Analysis of macrolide-lincosamide-streptogramin B (MLS$_B$) resistance determinant in strains of *Clostridium difficile.* *Microb. Drug Resist.* **8:**45–53.

77. **Spigaglia, P., and P. Mastrantonio.** 2004. Comparative analysis of *Clostridium difficile* clinical isolates belonging to different genetic lineages and time periods. *J. Med. Microbiol.* **53:**1129–1136.

78. **Stackebrandt, E., and F. A. Rainey.** 1997. Phylogenetic relationships, p. 3–19. *In* J. I. Rood, B. A. McClane, J. G. Songer, and R. W. Titball (ed.), *The Clostridia: Molecular Biology and Pathogenesis.* Academic Press, Inc., London, United Kingdom.

79. **Takamizawa, A., S. Miyata, O. Matsushita, M. Kaji, Y. Taniguchi, E. Tamai, S. Shimamoto, and A. Okabe.** 2004. High-level expression of clostridial sialidase using a ferredoxin gene promoter-based plasmid. *Protein Expr. Purif.* **36:**70–75.

80. **Tanaka, D., J. Isobe, S. Hosorogi, K. Kimata, M. Shimizu, K. Katori, Y. Gyobu, Y. Nagai, T. Yamagishi, T. Karasawa, and S. Nakamura.** 2003. An outbreak of food-borne gastroenteritis caused by *Clostridium perfringens* carrying the *cpe* gene on a plasmid. *Jpn. J. Infect. Dis.* **56:**137–139.

81. **Tanaka, D., J. Isobe, S. Hosorogi, K. Kimata, M. Shimizu, K. Katori, Y. Gyobu, Y. Nagai, T. Yamagishi, T. Karasawa, and S. Nakamura.** 2003. An outbreak of food-borne gastroenteritis caused by *Clostridium perfringens* carrying the *cpe* gene on a plasmid. *J. Infect. Dis.* **56:**137–139.

82. **Volk, W. A., B. Bizzini, K. R. Jones, and F. L. Macrina.** 1988. Inter- and intrageneric transfer of Tn916 between *Streptococcus faecalis* and *Clostridium tetani.* *Plasmid* **19:**255–259.

83. **Wang, H., and P. Mullany.** 2000. The large resolvase TndX is required and sufficient for integration and excision of derivatives of the novel conjugative transposon Tn*5397*. *J. Bacteriol.* **182**:6577–6583.

84. **Wang, H., A. P. Roberts, D. Lyras, J. I. Rood, M. Wilks, and P. Mullany.** 2000. Characterization of the ends and target sites of the novel conjugative transposon Tn*5397* from *Clostridium difficile*: excision and circularization is mediated by the large resolvase, TndX. *J. Bacteriol.* **182**:3775–3783.

85. **Wang, H., A. P. Roberts, and P. Mullany.** 2000. DNA sequence of the insertional hot spot of Tn*916* in the *Clostridium difficile* genome and discovery of a Tn*916*-like element in an environmental isolate integrated in the same hot spot. *FEMS Microbiol. Lett.* **192**:15–20.

86. **Waters, M., A. Savoie, H. S. Garmory, D. Bueschel, M. R. Popoff, J. G. Songer, R. W. Titball, B. A. Mc-** Clane, and M. R. Sarker. 2003. Genotyping and phenotyping of beta2-toxigenic *Clostridium perfringens* fecal isolates associated with gastrointestinal diseases in piglets. *J. Clin. Microbiol.* **41**:3584–3591.

87. **Wen, Q., K. Miyamoto, and B. A. McClane.** 2003. Development of a duplex PCR genotyping assay for distinguishing *Clostridium perfringens* type A isolates carrying chromosomal enterotoxin (*cpe*) genes from those carrying plasmid-borne enterotoxin (*cpe*) genes. *J. Clin. Microbiol.* **41**:1494–1498.

88. **Zhao, Y., and S. B. Melville.** 1998. Identification and characterization of sporulation-dependent promoters upstream of the enterotoxin gene (*cpe*) of *Clostridium perfringens*. *J. Bacteriol.* **180**:136–142.

89. **Zhou, Y., and E. Johnson.** 1993. Genetic transformation of *Clostridium botulinum* Hall A by electroporation. *Biotechnol. Lett.* **15**:121–126.

Neurotoxigenic Clostridia

ERIC A. JOHNSON

56

Neurotoxigenic clostridia are those species that produce characteristic neurotoxins affecting the nervous system of humans and animals (36, 45, 70). The classic neurotoxigenic diseases are botulism and tetanus, which are caused by the action of extremely toxic clostridial neurotoxins (CNTs) produced by *Clostridium botulinum* and *Clostridium tetani*, respectively. Botulinum neurotoxins (BoNTs) and tetanus neurotoxin (TeNT) are the most potent toxins known, with estimated lethal intravenous (i.v.) doses of 0.1 to 1 ng per kg of body weight (2, 41, 55, 64). Botulinum toxin is also extraordinarily poisonous by the oral route, with an estimated lethal dose of 1 µg per kg (2). The high toxicity of CNTs is primarily due to their extraordinary neurospecificity and to their catalytic cleavage of neuronal substrates at exceedingly low concentrations of CNT ($\leq 10^{-12}$ M) (55). During the past 2 decades, other species of clostridia (*Clostridium baratii* and *Clostridium butyricum*) that produce BoNTs have been recognized (32, 35, 36), indicating that the genes encoding BoNTs and associated proteins of the toxin complexes can be laterally transferred to nonpathogenic clostridia (27, 43, 59). Toxigenic clostridia have attracted considerable interest as etiologic agents of disease (2, 32, 36). The control of botulism and tetanus in humans and animals presents considerable challenges to physicians and veterinarians. The clostridia produce numerous protein toxins, and it is likely that the involvement of neurotoxins in other diseases will be revealed in the future.

The neurotoxic clostridia, *C. tetani* and *C. botulinum* (19, 44, 68, 73), were isolated >100 years ago and were demonstrated to cause true toxemias, acting solely through the action of their CNTs (for a review, see reference 44). Injection of CNTs into animals and humans will elicit disease without direct involvement of the bacteria; however, *C. tetani* and *C. botulinum* can infect wounds or colonize the intestinal tract, where they elicit their potent CNTs and possibly other virulence factors. Potential virulence factors, as well as a much greater understanding of the physiology of pathogenic clostridia, are being revealed by the analyses of their genomic sequences (10, 11, 59, 66).

The interest in BoNT has increased greatly during the past decade because of its utility as a pharmacological agent for treatment of neuronal diseases (7, 41, 64). Interest in BoNT has also been incited by awareness of its possible use in bioterrorism attacks (2). BoNTs and TeNT have become important tools in cell biology and have stimulated much interest in their basic nature and their actions on the nervous systems of humans and animals (39, 52, 55, 72). Several contemporary reviews describe the taxonomy, physiology, genetics, and diseases caused by the toxigenic clostridia, as well as the biochemistry, immunology, pharmacology, and medical uses of the CNTs (25, 39, 41, 52, 67, 72). This chapter describes the microbiological properties of *C. botulinum* and *C. tetani*, with an emphasis on pathogenesis and new findings on the organisms and their CNTs.

CLINICAL ASPECTS OF BOTULISM AND TETANUS

Botulism

Botulism is a rare but often severe paralytic disease caused by the extremely potent BoNTs produced by *C. botulinum* and certain other clostridia (3, 16, 23, 36, 46). Botulism is a true toxemia and results solely through the action of BoNT. Botulism was historically recognized as a deadly form of food poisoning occurring through the consumption of contaminated foods (19, 44, 68, 73). BoNTs can also cause disease by their entry into the circulation from wounds or from the intestinal tracts of susceptible infants and adults colonized by BoNT-producing clostridia (3, 32, 36, 46). Although there is little historical record before the 19th century, botulism was suspected to occur in ancient cultures, and certain dietary laws and food-processing methods probably evolved as a result of the disease (68). Ingestion of contaminated foods, particularly raw blood sausages, containing BoNT was noted in the 1700s to cause muscle paralysis and suffocation, with a high fatality rate exceeding 50% (14). The bacterial and toxigenic etiology was discovered from 1895 to 1897 by Emile Pierre Van Ermengem in a remarkable series of experiments (19, 73). His description of the disease, the causative organism ("*Bacillus botulinus*"), and properties of the toxin is a classic study in the field of pathogenic

Gram-Positive Pathogens, 2nd edition, edited by Vincent A. Fischetti et al.
© 2006 ASM Press, Washington, D.C.

clostridia and their toxins, and his findings have been aptly reviewed (19, 46, 73). The principles he established regarding the organism, its resistant endospores, and its potent toxin still form the foundation of knowledge for prevention of food-borne botulism. Besides classical food botulism, wound botulism was discovered in 1943 (14, 46), and infection and colonization of susceptible infant and adult gastrointestinal tracts were recognized in the 1970s and 1980s (3, 32, 35, 68). All four types of botulism are quite rare in their incidence and may consequently be misdiagnosed as more common paralytic diseases such as Guillain-Barré syndrome, myasthenia gravis, tick paralysis, stroke, or nervous system infections (3, 16). Infant botulism is the most prevalent form of botulism in the United States (3, 14), but food-borne botulism is probably the most common form of the disease in most other regions of the world (32, 46). C. botulinum is not as adept as C. tetani in colonizing wounds, and wound botulism is extremely rare compared with tetanus. However, since 1991 the numbers of cases of wound botulism have increased markedly, primarily in i.v. drug users in California and in the United Kingdom (14, 44). The genetic traits affecting colonization of wounds and the intestine have not been elucidated, but the availability of genetic tools and the genomic sequences of C. botulinum and C. tetani strains and other toxigenic clostridia (10, 11, 59) should enable elucidation of these processes. Botulism has devastating effects on wild and domestic animal populations, and it is the most prevalent type of botulism worldwide in terms of epidemics and deaths (26, 46, 68).

BoNT-producing clostridia synthesize seven serotypes of BoNTs, designated A through G (32, 36, 68). The BoNTs can be neutralized by specific antitoxins obtained by immunization of animals with toxoids (35, 46). A BoNT serotype is defined as being neutralized by polyvalent antisera to the causative toxin type but retaining toxicity on incubation with heterologous antisera (35, 46). BoNTs occur naturally as protein complexes in which the CNT is associated with nontoxic proteins and RNA (45, 49, 59, 63, 68). The complexes impart stability to the labile BoNTs and facilitate retention of toxicity during passage through the gastrointestinal tract and in other adverse environments (63). Six independent evolutionary lineages of BoNT-producing clostridia have been proposed (17), emphasizing the ability of the toxin genes to be laterally transferred among clostridia and illustrating the potential for discovery of new neurotoxic clostridial species in the future. Recently, subtypes of CNTs within a specific serotype have been detected (W. H. Tepp and E. A. Johnson, unpublished data), indicating that the evolutionary diversity is greater than previously envisioned.

Irrespective of the source of BoNT (food, wound infections, or infant or adult intestinal infections), the neuroparalytic symptoms are similar (3, 14, 16, 35). The hallmark clinical symptoms of botulism are a bilateral and descending weakening and paralysis of skeletal muscles (3, 14, 16, 70). The incubation time for onset of symptoms varies with the type of botulism, the serotype of BoNT, and the quantity of BoNT that enters the circulation. Classical food-borne botulism occurs following the consumption of food contaminated with preformed BoNT, and the vast majority of cases are caused by types A, B, and E, and (rarely) F. The most severe and long-lasting food-borne botulism occurs with type A (14, 44). The incubation time of food-borne botulism varies with the BoNT serotype and quantity of toxin ingested. In type A cases, the onset time is usually 12 to 36 h following consumption of the toxic food. The incubation peri-

od can be as short as 2 h when high quantities of toxin are ingested (14, 16) or as long as several days to weeks with type B or E BoNTs and ingestion of low quantities of toxin (14, 16, 46). Wound botulism usually has a relatively long incubation period of 4 to 14 days, reflecting the time for neurotoxigenic clostridia to colonize and produce CNTs. Infant botulism has been reported to have an incubation time from 6 to 8 h to several days (3).

The site of action of all serotypes of BoNT is the presynaptic terminals of motor neurons (16, 53, 55, 72, 67). Following binding and internalization into the presynaptic nerve terminal, the CNT causes a blockade of the release of acetylcholine and regional flaccid paralysis of muscles innervated by the intoxicated nerves (52, 55, 72). Botulinum toxin probably also triggers or inhibits physiological processes other than cholinergic neurotransmitter release, such as autonomic nerve function and release of cytokines affecting pain and inflammation (41), but these processes are only beginning to be elucidated. In most cases of botulism, cranial nerves are affected first, particularly those innervating the eyes; the first symptoms are blurred and double vision, dilated pupils, and drooping eyelids (Fig. 1). The eyes respond sluggishly to light in a darkened room. These abnormalities of cranial innervated muscles are followed by a descending and progressive paralysis affecting muscular activity and characterized by difficulty in swallowing, weakness of the neck, dry mouth, and problems in speaking. Weakness of the upper limbs and the torso can subsequently occur; in severe cases, respiratory muscle weakness generally ensues, requiring mechanical ventilation to prevent fatality by suffocation. BoNT can probably paralyze every striated muscle of the body, but there is apparently no interference with nervous activity of the central nervous system at in vivo concentrations (16, 55). Generally, the patient's hearing remains normal, consciousness is not lost, and the victim is cognizant of the progression of the disease. Certain episodes of botulism have been reported that affect autonomic and sensory functions. A patient's awareness of loss of muscle activity can lead to considerable emotional distress such as anxiety and depression (46).

The fatality rate from food-borne botulism has decreased from 70% in the 1800s and 1900s to ≤9% in recent years (14, 46, 68). Convalescence and recovery from botulism are generally prolonged, requiring weeks to months, depending on the quantity of toxin exposure and serotype, but recovery is usually complete and patients regain full normal function (3, 16). Symptoms of fatigue, muscular weakness, and constipation have been reported to persist for months in certain patients. Antibodies do not generally develop in patients who have experienced food-borne botulism but can develop in infants or adults with intestinal botulism who are exposed to toxin for continual periods of time. It has not been reported if i.v. drug users or persons with wound botulisms develop antibodies. Antibodies have been reported in individuals who have been repeatedly injected with type A botulinum toxin for therapeutic purposes, particularly in high-dose applications such as cervical dystonia (7, 64).

Electromyography can aid in detecting defective neuromuscular transmission and botulism (16), and methods to differentially diagnose botulism from other neurological diseases have been previously described (16). The definitive diagnosis of botulism depends on the detection of BoNT in the patient's serum, feces, and/or food that was consumed before onset of the disease (14, 32). Currently, the only reliable assay for BoNT is the mouse bioassay, together with neutralization of mouse toxicity with type-specific antitoxins

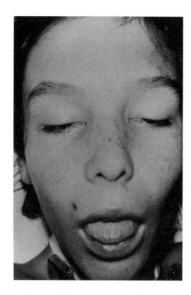

FIGURE 1 A boy suffering from food-borne botulism, showing the prominent effects of BoNT on the eyes and the descending paralysis to other regions of the face. (Photograph courtesy of Charles L. Hatheway [deceased], Centers for Disease Control and Prevention, Atlanta, Ga.)

(14, 35, 64). Enzyme-linked immunosorbent assays (ELISAs) have also been used to detect CNT antigen, and other antibody-based platforms are in development.

No specific antidote is currently available for preventing botulism or reversing paralysis once receptor binding and internalization ensue. Early administration of antibodies can decrease the severity of the disease and slow its progression, resulting in a shorter hospital stay and more rapid recovery (14, 16). Complete recovery probably requires the formation of new functional neuromuscular junctions with resumption of neurotransmission. The reason that a single exposure to exceedingly low concentrations ($\leq 10^{-12}$ M) of BoNT can cause such a long-lasting disease is currently not understood and is a very intriguing phenomenon (31, 47). The extremely long duration of action and persistence probably involves the half-life and degradation of the CNTs in neurons, as well as their effect on neuronal regeneration, which is serotype dependent (31, 47).

Botulinum toxin has been considered a potential biological warfare agent that could be administered in aerosols, food, or water (2, 16, 42). Botulinum toxin is absorbed through mucous membranes, and three cases of botulism in laboratory workers who apparently inhaled the toxin were documented (16, 46). BoNT is labile to many environmental conditions and chemicals, and preparation of an aerosol weapon would be difficult. Immunization is not feasible for protection of human populations from botulism owing to the rarity of the disease, but pentavalent (A to E) toxoid is used for immunization of researchers and for military personnel who may be exposed to BoNT in warfare. Since immunization provides only partial protection against intoxication with large quantities of toxin, immunized researchers must still follow scrupulous laboratory practices in working with botulinum toxin, including avoidance of aerosols, handling toxin in the biological safety cabinets, and use of closed containers during centrifugation and other procedures (14, 44, 53, 64). Considerable efforts in the United States and in certain other countries are being devoted to development of heptavalent vaccines and other countermeasures.

The major treatment of botulism is supportive care, with careful attention being given to respiratory status. For the treatment of adults, the only antitoxin currently available is of equine origin, and hypersensitivity reactions occur in approximately 10% of adults on exposure (14, 16). Antitoxin of human origin has been developed for treatment of infants with botulism and would be useful for the treatment of adults. Botulism immune globulin has been demonstrated to reduce both the severity of botulism in infants and the duration of hospital stays (3).

Tetanus

Tetanus is a vivid neurological disease characterized by violent and persistent spasms of the head, trunk, and limb muscles (6, 28). The harrowing symptoms of tetanus were described by Hippocrates and Aretaeus of Cappodocia thousands of years ago, but the bacterial etiology of the disease was not demonstrated until 1884 by Nicolaier and Carle and Rattone (6, 28, 44). The tetanus bacillus was isolated in pure culture by Kitasato in 1889 (28, 44). The isolation was facilitated by the heating of pus from an infected animal at 80°C for 45 to 60 min to destroy nonsporing organisms. In the late 1800s, Von Behring and Kitasato discovered the phenomena of vaccination and antitoxins, when they showed that injection of diphtheria or tetanus toxoids into animals elicited protective factors against subsequent challenges by the respective toxins (6, 28, 44).

Tetanus in adults usually occurs as a result of infection of puncture wounds or lacerations by *C. tetani* or occasionally from parenteral injection of materials contaminated with *C. tetani* (6, 28, 44). Wounds most commonly occur in the legs, as well as in the arms and hands, but in as many as 20% of the cases there is no history of injury or wound (9, 10). The incubation period for tetanus is usually 7 to 10 days but can range from 2 to 30 days (6, 28). An onset time of <1 week implies a poor prognosis, certainty of severe symptoms, and possible death (6). Neonatal tetanus is caused by infection of the fetus by passage of *C. tetani* through the separated placenta.

While tetanus has virtually been eliminated by immunization in many industrialized countries with adequate medical infrastructure, neonatal tetanus is endemic in developing countries. The worldwide incidence of tetanus has been estimated to be as many as 500,000 cases per year (6, 28). Neonatal tetanus is estimated to account for >400,000 infant deaths every year (6, 28, 44), despite the fact that it could be prevented by immunization of the mother and acquisition of passive antibodies by the fetus. Tragically, hundreds of thousands of adults and infants die every year from tetanus, and yet it is completely preventable by proper immunization. The primary recommendations for prevention of neonatal tetanus are vaccination of pregnant women and sanitary facilities for birthing.

The name tetanus is derived from the Greek *tetanos*, which means "to contract." Tetanus is characterized by exceedingly painful persistent muscular tonic spasms superimposed upon muscular rigidity (6, 28). Clinical symptoms of tetanus can be local and in close proximity to the site of injury (6, 28), or generalized, causing contractions of muscle groups, particularly those in the head and the trunk, commonly visualized by clenching of the fists, adduction of the arms, and extension of the lower extremities. Unlike certain other neurological diseases such as epilepsy, generalized tetanus is not associated with loss of consciousness and is extraordinarily painful.

During infection, TeNT is produced by *C. tetani* in the injured region, and TeNT enters the presynaptic nerve cytosol by receptor-mediated endocytosis and synaptic vesicle reuptake (6, 28, 52, 55). TeNT traverses from the nerve terminal to the nerve cell body by retrograde axonal transport (6, 28, 52), eventually reaching neurons in the spinal cord and brain stem, where it inhibits glycinergic and GABA (γ-amino-n-butyric acid)-ergic inhibitory neurotransmission (9, 6, 28, 55). It appears to not affect cholinergic synapses at physiological concentrations of 10^{-11} to 10^{-12} M (55). The presynaptic blockage of central neurons results in elevated muscle tone and hyperactive reflexes (6, 28).

Trismus may initially interfere with eating, prompting the patient to seek medical attention (6, 28). This is followed by stiffness, tetanic spasms, and violent contractions. Respiratory failure can result in death, but it may be prevented by respiratory care. Complications of generalized tetanus include acute renal failure, hyperthermia, and vertebral fractures in the elderly. Modern respiratory and supportive care has decreased the rate of mortality from >90% to about 25% (6, 28). The autonomic nervous system may also be affected in tetanus, including an episodic pulse rate, peripheral vascular constriction, sweating, and pyrexia. Diagnosis of tetanus primarily relies on presentation of the distinct clinical symptoms, but hind-limb injection of mice with wound extracts or serum can confirm the clinical diagnosis (6, 44, 74). Certain syndromes mimic tetanus, and differential diagnoses have been previously described (6, 28). Tetanus therapy includes neutralization of unbound toxin by administration of passive antitoxin, control of C. tetani proliferation by antibiotics and hygiene, excision of infected or atrophied tissue, airway management, and supportive nursing care (6, 28). Anticonvulsive therapy, application of neuromuscular blocking agents, and tracheotomy with ventilatory control are needed in severe cases. The most effective prevention measure is immunization and maintenance of immune status, with supportive care following injury.

SPECIES OF CLOSTRIDIA OTHER THAN
C. BOTULINUM AND C. TETANI IMPLICATED
IN NERVOUS SYSTEM INFECTIONS

Clostridium perfringens has a remarkable ability to infect wounds and cause gas gangrene and other wound infections, sometimes in concert with other clostridia (74). C. perfringens is a rare cause of human meningitis (67), and most cases result from infections following head injuries. *Clostridium septicum* has also been isolated from central nervous system infections (44). Neoplasm may increase the sensitivity of humans to clostridial neurological infections, possibly by enabling passage of the organisms or their toxins through membranes lining the gastrointestinal tract. C. perfringens epsilon-toxin has been shown to be toxic in the glutamatergic system of the rat hippocampus (5, 54). Pathological changes in animals exposed to epsilon-toxin mainly occur in the brain, and symptoms include neurologic features such as retraction of the head and convulsions. The relevance of this toxin to human disease is not known. Intriguing evidence has also suggested that neurotoxigenic clostridia in the gut produce toxins that may be absorbed and contribute to diseases such as autism (29).

MICROBIOLOGY OF NEUROTOXIGENIC
CLOSTRIDIA

The genus *Clostridium* is a widely diverse collection of spore-bearing anaerobic eubacteria with a gram-positive cell wall structure (13, 36, 44). Clostridia are strict anaer-

obes, obtain energy by fermentation and substrate level phosphorylation, and do not reduce sulfate to sulfide (13, 36, 44). In toxigenic species, the spores are generally wider than the vegetative organisms in which they are formed, imparting spindle shapes, the characteristic clostridial forms (36). Neurotoxigenic clostridial species include C. botulinum, C. tetani, and *Clostridium argentinense* (C. botulinum type G), as well as CNT-producing strains of C. butyricum, C. baratii, and (rarely) C. perfringens (14, 35, 44).

The isolation, identification, and maintenance of pure cultures of neurotoxigenic clostrididia present certain practical difficulties. Neurotoxigenic clostridia tend to grow as consortia, and pure cultures are often difficult to achieve and maintain (36, 44, 74). Purity of neurotoxigenic clostridia must always be rigorously ascertained by microscopy, extensive plating on nonselective media, and testing for CNT formation by the mouse bioassay. Owing to the complex nutrient requirements of neurotoxigenic clostridia, rich media are commonly used for cultivation (13, 15, 35, 64, 74). Several reference works describe preparation of these media, the use of reducing agents to lower the redox potential, and detailed methods for phenotypic and metabolic tests for identification of clostridia. The anaerobic methods required to study the clostridia have been previously described (13, 15, 36, 38, 74).

The neurotoxic clostridia are widely dispersed in nature, owing to their ability to form resistant endospores, but they have mainly been isolated from two principal habitats: soil and the intestinal tracts of humans and animals (13, 36, 44, 68, 99). Several pathogenic species appear to be common inhabitants of the intestinal tracts of humans and animals (13, 30, 36). C. tetani is frequently found in human feces, but C. botulinum is seldom isolated from this source except from intestinal botulism cases (3). The prevalence of C. tetani and types of C. botulinum vary according to the geographical region and the composition of the soil (36, 44, 74). Clostridia also frequently occur in dust, vegetable foods, and sewage and thus contaminate many environments through dispersal in dust and aerosols. C. tetani was found in 10 to 40% of fecal specimens of humans and domestic animals (74). Neurotoxigenic clostridia are saprophytic and do not have an obligatory relationship with an animal host.

Enrichment of neurotoxic clostridia is carried out by heating the samples (e.g., 80°C for 10 min) or by treating the samples with ethanol to kill vegetative organisms and enrich for spore formers (13, 36, 74). The neurotoxigenic clostridia are usually pleomorphic, and identification on a morphological basis is difficult. Identification of neurotoxigenic clostridia has traditionally been performed by utilization of select carbohydrates and assay of metabolic products, but recently the determination of genes encoding rRNA or interspatial regions has become integral for identification (30, 32, 44). Of particular importance in the identification of the pathogenic clostridia is the determination of toxins by bioassay of specimens or culture fluids using mice and specific antitoxins (14, 35). PCR and antibody methods such as ELISA have been evaluated for detection of CNTs or their structural genes without culturing the clostridia (see citations in references 32, 35, and 44). In clinical practice, the isolation of C. tetani or C. botulinum may be unnecessary, since the diseases are diagnosed by the symptoms and bioassay of CNTs. However, isolation is encouraged for further study and comparison of strains from clinical and environmental sources. C. botulinum and C. tetani may produce lower quantities of toxin after repeated laboratory culturing, and clinical or wild isolates should be preserved in liquid nitrogen or at −80°C in oxygen-impermeable containers.

Toxin titers and the identities of BoNTs and TeNT should be periodically confirmed with pure stock isolates and by mouse bioassay and specific antitoxins.

In culture, neurotoxigenic clostridia typically grow as large, rod-shaped organisms and often form filaments or chains. The vegetative cells are usually curved, their sides are parallel, and their ends are rounded (13, 36, 38). The spores of neurotoxigenic species are wider than the vegetative sporangium and swell the rod, conferring a distinctive appearance according to the position in the cell (Fig. 2). Spores of most pathogenic species can be obtained on chopped meat agar or broth or with clear complex media such as TPGY (trypticase–peptone–Bacto-peptone–glucose–yeast extract) after several days of culture (14, 36, 44). The phenotypic properties of neurotoxigenic clostridia (morphology, staining reactions, capsules, spores, and the presence of genetic elements including plasmids and bacteriophage) have been previously reviewed (32, 36, 42, 44, 59).

POTENCY AND DETECTION OF CNTs

The outstanding feature of the neurotoxigenic clostridia is their formation of a characteristic neurotoxin of extraordinary potency for humans and animals. BoNTs and TeNT are the most poisonous substances, and potencies have been estimated to be as little 0.1 to 1 ng per kg by i.v. and intramuscular (i.m.) injections (44, 64). Experiences with commercial preparations of type A BoNTs in humans, as well as nonhuman primate exposure studies, have demonstrated that the lethal i.m. dose is approximately 39 U/kg for humans (64). Assuming that a mouse 50% lethal dose is approximately 10 pg, then the lethal i.m. dose for a 70-kg human would be approximately 20 ng. Botulinum toxin is also extremely potent by the oral route, and ingestion of 70 μg of BoNT serotype A (BoNT/A) has been estimated to be

lethal for a 70-kg human (150 to 700 ng per kg) (2). The oral toxicity varies depending upon the food consumed, the association of toxin with food components, the presence of food and alcohol in the intestinal tract, and many other factors. Based on primate studies, it appears that the lethal dose ranges from 10 to 70 μg; it would obviously be much less for an infant (A. E. Larson and E. A. Johnson, unpublished data). The remarkable potency of i.m. and oral BoNTs are collectively due to their resistance in the complex form, the ability of certain serotypes to be absorbed from the intestinal tract, their extraordinary neurospecificity, and their action as zinc metalloenzymes, catalyzing the cleavage of neuronal substrates in vivo at exceedingly low concentrations of NT ($\leq 10^{-12}$ M). Due to the extreme toxicity of CNTs, careful handling and safety measures are extremely important in working with these toxins (14, 44, 53, 64).

DETECTION OF TeNT AND BoNT

Bioassays are currently the most important laboratory tests used to identify neurotoxigenic clostridial species (33, 35, 36, 69). Neutralization of toxicity by tetanus antitoxin or type-specific botulinum antitoxins ensures positive identification of toxigenic strains. Detection of CNTs is usually carried out with fecal specimens, blood (serum), suspect foods in food-borne botulism cases, and culture fluid following enrichment of the organism (3, 14, 35, 44, 69, 74). Specimens of diluted culture supernatants are injected into mice after a portion is reacted with antitoxin. TeNT is injected into the hind leg or paw, and BoNT is injected into the intraperitoneal cavity or into the tail vein. Complications in conducting and interpreting the mouse bioassay have been previously described (35).

At present, the mouse bioassay is the most important laboratory method for identification of C. botulinum and C. tetani and assay of CNTs (35, 64, 69). However, due to certain drawbacks in this assay, as well as the increased regulatory and ethical concerns in using animals for toxin determinations, there is much interest in assays not employing animals. The primary alternative assays employed have been based on immunological detection, particularly ELISA and related methods. A nonanimal assay that depends on all the steps in the intoxication mechanism including receptor binding, internalization, and catalytic cleavage of neuronal substrates, such as cell-based assays (23, 47), would be ideal for determination of toxin titers and antibodies in patients' sera. Since CNTs are zinc metallopeptidases with high specificity for their neuronal substrates, methods based on catalytic activity combined with sensitive detection methods including high-throughput fluorogenic reporters and fluorescence resonance energy transfer-sensing systems have been investigated (44). A particularly promising methodology for detection of CNTs is mass spectroscopy including high-resolution matrix-assisted laser desorption–ionization time-of-flight or electrospray to characterize the toxins and to detect reaction products from proteolytic cleavage of the neuronal substrates (4). Mass spectroscopy has considerable promise for proteomic studies based on the genomic sequences of pathogenic clostridia. Trends in assay development are also moving toward development of biosensor devices, including microfluidic and nanofluidic platforms (56). The successful development of these advanced methods will require assay of BoNTs in complex matrices such as foods and clinical samples (blood, saliva, and feces) and ideally in poisoned neurons.

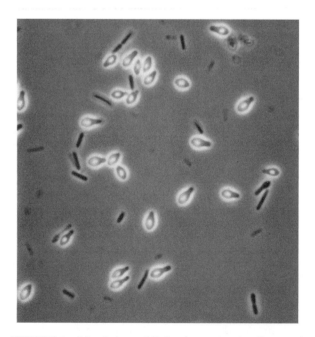

FIGURE 2 Morphology of *C. botulinum* type A cells viewed by phase-contrast microscopy. The spore-bearing cells of other serotypes of *C. botulinum* and *C. tetani* also have characterized morphologies, typically with swelling of the rod-shaped vegetative cell (36).

GENE ARRANGEMENT AND REGULATION OF TeNT AND BoNT SYNTHESIS

BoNTs and TeNT are the major virulence factors of *C. botulinum* and *C. tetani* and are responsible for the characteristic symptoms and morbidity of botulism and tetanus. However, very little is known regarding the regulation of their synthesis, particularly in vivo (58, 59). The structural gene encoding TeNT is located on a large plasmid (11, 27, 59). Depending on the serotype, the structural genes for BoNTs are located on the chromosome, bacteriophages, or plasmids (Table 1) (27, 57, 59). The genes encoding BoNT/C and BoNT/D are associated with bacteriophage, and toxin production by *C. botulinum* type C and D strains can be unstable and low titers may be produced in culture (43, 64). The gene cluster encoding the *C. botulinum* type G toxin is located on a large plasmid (43, 59), and type G toxin is also produced in very low quantities (44). With the exception of *C. botulinum* types C and D, strains of *C. botulinum* generally only produce one neurotoxin serotype (33, 36, 68). There are reports, however, of strains that produce two toxin types, one of which is produced in much higher titers than the other, such as Af, Bf, and Ba (33, 36, 68), where the major toxin produced is designated by the uppercase letter. Recent studies have demonstrated that the *C. botulinum* neurotoxin genes occur in a cluster that is highly polymorphic in its genetic structure, depending on the serotype of the subtype (Fig. 3) (21, 22, 42, 45, 49). The association of the genes for proteins in the toxin complexes with unstable genetic elements such as bacteriophage, plasmids, or transposons may contribute to their cluster heterogeneity and to lateral transfer of the toxin cluster genes (27, 43, 59).

The genes encoding the botulinum toxin progenitor within the clusters consist of two operons that are transcribed in opposite directions (45, 59, 60). One operon contains the genes for nontoxic nonhemagglutinin protein (NTNH) and BoNT; in most serotypes, a gene (*botR*) that encodes an ~21-kDa positive regulatory protein is present in this operon. Recent evidence indicates that BotR/A may function as an alternative sigma factor (60). The other divergent operon encodes hemagglutinin (HA) proteins in certain serotypes, while other serotypes and subtypes appear to express additional genes and some serotypes lack genes for HAs (21, 22, 49). The discovery of BotR/A indicates that at least one positive transcriptional regulatory protein controls expression of BoNT, but additional research is needed to further elucidate the molecular mechanisms of BoNT gene regulation, including the mechanism of repression by arginine and tryptophan in *C. botulinum* types A and E, respectively (45, 64). Sequencing of the chromosomal DNA flanking the *bont* gene clusters in type A has shown that there is considerable polymorphism in these regions and that there are differences in the presence and location of genes involved in BoNT production and maturation to the fully active CNT (8, 21, 22, 45). Defective transposons and insertion elements (IS elements) were found flanking the toxin gene clusters (21, 22, 45, 59), which could be involved in cluster heterogeneity and lateral movement. Sequence analyses have also indicated that genes encoding uncharacterized proteins are also synthesized by the promoters governing formation of CNT, NTNH, and HA in certain *C. botulinum* type A strains (22), but they have not been isolated; it is not known whether they are involved in formation or structure of the toxin complexes. In addition to the primary gene cluster occurring in all type A *C. botulinum* strains, separate clusters containing inactive neurotoxin type B gene sequences have been identified in type A *C. botulinum* strains at a surprisingly high frequency (32, 45). Nucleotide sequence analyses have shown that silent *bont/b* genes usually have high homology to the gene from authentic proteolytic type B *C. botulinum*. However, the silent genes contain stop signals and deletions that prevent functional expression (45).

The kinetics of neurotoxin complex expression during the growth cycle were recently investigated with three type A strains (8). The mRNA transcript levels carrying the genes of the type A neurotoxin complex were assayed by Northern analyses, and neurotoxin formation was determined by mouse bioassay, ELISA, and Western blotting. In all three strains, mRNA transcripts for the toxin complex genes were initially detected in early log phase, reached peak levels in early stationary phase, and rapidly decreased in mid to late stationary phase and during lysis. CNT expression and proteolytic activation of the toxin were highly

TABLE 1 Properties of botulinum (BoNT) and tetanus (TeNT) neurotoxins

Toxin	Gene location[a]	Representative species affected	Specific toxicity $(10^8)^b$	Light-chain substrate	Peptide bond cleaved[c]
BoNT/A	C	Humans, chickens	1.05–1.86	SNAP-25	Gln^{197}-Arg^{198}
BoNT/B	C	Humans, cattle, horses	0.98–1.14	VAMP	Gln^{76}-Phe^{77}
BoNT/C$_1$	B	Birds, cattle, dogs, minks	0.88	Syntaxin SNAP-25	Lys^{252}-Ala^{253} Arg^{198}-Ala^{199}
BoNT/D	B	Cattle	1.60	VAMP	Lys^{59}-Leu^{60}
BoNT/E	C	Humans, fish, aquatic birds	0.21–0.25	SNAP-25	Arg^{180}-Ile^{181}
BoNT/F	C	Humans	0.16–0.40	VAMP	Gln^{58}-Lys^{59}
BoNT/G	P	Unknown	0.1–0.3	VAMP	Ala^{81}-Ala^{82}
TeNT	P	Humans, cattle, horses, sheep, dogs, chickens, other animals		VAMP	Gln-Phe

[a] Gene location: C, chromosome; B, bacteriophage; P, plasmid. For putative chromosomal locations, this location is inferred from PCR amplification of chromosomal DNA preparations, except for type A, in which toxin gene mutations have been mapped to the chromosome (see references 45 and 59).

[b] Specific toxicity refers to toxins activated by trypsinization when necessary for maximum toxicity. Toxicities are per milligram of protein. Most of the reported data are from Sugiyama (70). The toxicities can vary considerably depending upon the strain, growth conditions, and other factors, and these specific activities are only representative of experiments conducted by Sugiyama (70). Studies (63, 70) have shown that oral doses of type A or B toxin complexes in mice are 10 to 1,000 times greater than the intraperitoneal or intravenous lethal dose depending on the size of the complex, diluent, and other factors.

[c] The specific peptide bond cleaved is shown. However, the clostridial neurotoxins require a minimum peptide length of >14 amino acid residues, depending on the serotype and a characteristic substrate tertiary structure for catalytic activity.

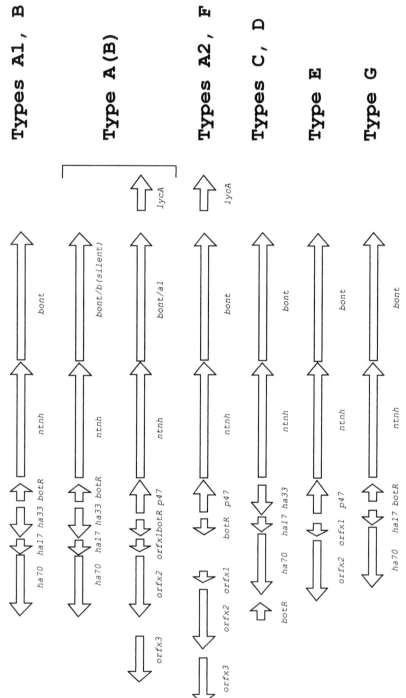

FIGURE 3 Arrangement of the genes in neurotoxin gene clusters of C. *botulinum*. (Diagram courtesy of Marite Bradshaw, University of Wisconsin.)

dependent on the strain and the growth medium (8). These results indicate that the genes of the CNT/A gene cluster are temporally expressed and regulated during the cell cycle by nutritional conditions and possibly by other mechanisms such as quorum sensing and stationary phase-expressed genes. It was shown that BoNT/A and BoNT/B from proteolytic strains of *C. botulinum* were strongly repressed by arginine and that BoNT/E from *C. botulinum* type E was repressed by tryptophan, but the molecular basis for the negative regulation has not been elucidated (45, 64). The production of BoNT is also temperature dependent in certain strains, particularly for strains that contain genes for two BoNTs (64). In *C. botulinum* type B, sporulation was enhanced by the presence of zinc in the medium, and the spore composition was distinct from that of many *Bacillus* strains (64). Recently, analysis of the genome sequences of *C. tetani* and *C. botulinum* has indicated that the sporulation process and mechanisms of control appear to be distinctly different from that of *Bacillus subtilis*, which depends on two component sensing systems, phosphorelay pathways and alternative sigma factors (37). It will be very intriguing to elucidate the mechanisms of sporulation in the clostridia.

C. tetani produces two primary virulence factors, the highly toxic TeNT and a hemolysin commonly termed tetanolysin (11, 28). As found with *C. botulinum* type A, TeNT is produced after active growth, does not have a signal sequence, and does not appear to be actively secreted during growth (60). BoNT and TeNT are possibly released during cell lysis (64), but the mechanism of CNT exit from the cell needs further study. It was shown in early work for the development of TeNT vaccine that production of TeNT is dependent on a histidine-containing peptide in the medium, suggesting that nitrogen regulation is involved in production (64), analogous to nitrogen regulation of toxin synthesis and activation observed with *C. botulinum* (45, 64). TeNT expression was recently confirmed to be regulated by peptides derived by pancreatic digestion of hormones and other polypeptides (58). In addition to nutritional regulation that probably is mediated by global regulatory mechanisms, specific regulatory proteins closely linked to the TeNT gene also regulate toxin production. A TetR-encoding gene in *C. tetani* with 50 to 65% identity to *botR* has been shown to be located immediately upstream from the TeNT structural gene (60). The gene encodes a 21.5-kDa protein that possesses features of a DNA-binding protein, including a helix-turn-helix motif and a basic pI (9.53) (60). It also showed homology with other putative regulatory genes in *Clostridium* spp., including an ~30% identity with *uviA* from *C. perfringens* and *tcdR* (previously called *txeR*) from *Clostridium difficile* (60). Overexpression of TetR in *C. tetani* increased TeNT expression and the level of the corresponding mRNA, which indicated that TetR is a positive transcriptional regulator of the TeNT gene. TetR has recently been demonstrated to have the properties of an alternative sigma factor, and the mechanism of TeNT regulation by TetR is actively being pursued (60).

Chemically defined media have been developed that support the growth of *C. tetani*, but these media give low toxin titers and may lead to strain degeneration (64). These results are analogous to those found with proteolytic strains of *C. botulinum* serotypes A and B (45, 64). TeNT is produced mainly after growth has ceased and appears to be oppositely regulated to sporulation (64). Highly toxigenic strains of both *C. tetani* and *C. botulinum* (type A) tend to sporulate poorly. These observations suggest that sporulation and toxin synthesis are governed by opposing regulatory mechanisms, and it will be intriguing to elucidate the molecular regulation of sporulation and CNT formation.

BoNT AND TeNT STRUCTURES

Crystalline type A botulinum toxin was first precipitated from culture fluid in 1908 by Somer and later purified by other researchers (64). Further studies, primarily in the United States and Japan, demonstrated that the large (900-kDa) crystalline type A molecule could be separated into toxic and nontoxic moieties (for reviews, see references 18, 64, and 70). Later, using centrifugation techniques or chromatography on ion-exchange resins, a purified single-CNT component was isolated (18, 64).

BoNTs are proteins of ~150 kDa that naturally exist as components of progenitor toxic complexes. The entire components and complete structures have not been fully elucidated for these toxin complexes. Progenitor complexes have been referred to as the M complex (ca. 300 kDa), consisting of BoNT associated with an NTNH protein of 120 to 140 kDa, or as the L and LL complexes (about 450 and 900 kDa, respectively), in which the M complex associates with HA proteins (45, 63, 64, 70). The nontoxic proteins in the complexes have been demonstrated to provide protection during manipulations and during passage through the gastrointestinal tract (63, 64). In contrast to BoNT, TeNT is produced as a single peptide of ca. 150 kDa and is not known to form complexes with nontoxic proteins. The lack of TeNT protein complexes has been suggested to be responsible for its lack of oral toxicity.

BoNTs and TeNT are produced as single-chain molecules of ca. 150 kDa that achieve their characteristic high toxicities of 10^7 to 10^8 50% lethal doses per mg (Table 1) by posttranslational proteolytic cleavage to form a di-chain molecule composed of an L chain (~50 kDa) and a heavy (H) chain (~100 kDa) linked by a disulfide bond (Fig. 4) (18, 68). Most BoNTs produced by strongly proteolytic strains of *C. botulinum* (type A) achieve full activation during culture, while proteolytic modification depends upon medium conditions, temperature, and time of incubation (18, 64, 68). Following initial cleavage into the H and L chains, more-extensive proteolytic modification may occur. In type A, an internal peptide sequence of 10 amino acids is removed, leaving residues 437 to 448, and the N-terminal Met-Ala residue is cleaved, leaving Ala as the N-terminal residue (1). In type B, a single cleavage occurs (residues 440 to 441); in type E, four amino acids are removed, leaving residues 411 to 425 (1). Less proteolytic strains of *C. botulinum* serotypes B-G generally do not produce a completely nicked neurotoxin, and full activation is attained by treatment with trypsin or another suitable proteolytic enzyme. BoNTs also undergo further proteolytic modification during storage and handling, including autocatalytic cleavage and inactivation.

BoNTs and TeNT consist of three basic functional domains (Fig. 4): (i) L chain, the catalytic domain that has endopeptidase activity on neuronal substrates; (ii) H_N, the translocation domain residing in the N-terminal region of the H chain; and (iii) H_C, the receptor-binding domain located in the C-terminal region of the H chain. The gene and amino acid sequences of BoNT and TeNT for a limited number of strains have been reviewed (50, 59). Recent studies have demonstrated that subtypes of BoNTs occur within certain serotypes, in which the 150-kDa proteins differed by 10 to 20% in amino acid sequence and associated properties such as immunogenicity and neutralization by

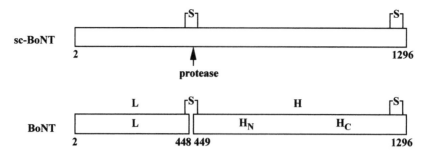

FIGURE 4 Proteolytic cleavage sites and basic domain structure of BoNT/A. After proteolytic activation, the NT, consisting of an L chain and an H chain, can be defined as three basic domains: (i) L chain, catalytic domain; (ii) HN, translocation domain; and (iii) H_C, receptor-binding domain. As described in the text, additional proteolytic modification takes place in BoNT/A following initial cleavage. (Diagram courtesy of Marite Bradshaw, University of Wisconsin.)

antitoxins (unpublished data). These findings indicate that different evolutionary lineages of CNTs exist within specific serotypes, leading to distinct BoNT structures that are differentially neutralized by antitoxins.

TeNT and BoNTs comprise a unique group of zinc proteases with unusual properties compared with other metalloproteases (9). The amino acid sequences of all seven BoNTs and TeNT have been deduced from the corresponding genes. The overall amino acid identity is about 40% for the eight CNTs whose genes have been sequenced (51, 59), and TeNT and BoNT have regions of homology, particularly in the residues defining the catalytic active site, in the translocation domain and in the two cysteine residues forming the disulfide bond connecting the H chain and the L chain (50, 59). The least degree of homology is in the carboxyl region of the H chain, which is involved in neurospecific binding. Various other regions have low similarities of amino acid sequence (50). The high homologies of amino acid sequences from physiologically and genetically distinct clostridia indicate that a single or closely related ancestral gene(s) was dispersed by lateral transfer among the clostridia, which later diverged to form the distinct lineages.

Solving the structures of BoNT/A, BoNT/B, BoNT, and TeNT domains has led to considerable insight into the structures and cellular biology of these neurotoxins (34, 51, 52, 71). The BoNT/A structure was initially solved by Stevens and colleagues, who described the 1,285-amino-acid di-chain molecule at 3.3-Å resolution (Fig. 5) (34, 51). The overall shape of BoNT/A is rectangular with dimensions of ca. 45 by 105 by 130 Å, and the molecule shows a linear arrangement of the three functional domains with no contact between the catalytic and binding domains. The three functional domains have separate and distinct structures, with the exception of an unusual loop that encircles the perimeter of the catalytic domain (34, 51). The finding of this loop was unexpected and presents a puzzle regarding catalysis, since it partly covers the catalytic active site. It is possible that the site opens on contact with the substrate, analogous to allosteric enzymes, or when internalized within neurons. Investigation of the mechanism of in vitro catalysis has demonstrated that CNT peptidase has multiple exosites that govern substrate specificity (9, 55).

Solving the BoNT/A structure revealed that the holotoxin is composed of domains of different evolutionary origins. The ganglioside-binding C-terminal subdomain has structural homology with proteins known to interact with sugars such as the H_C fragment of TeNT, serum amyloid P,

sialidase, various lectins, and the cryia and insecticidal δ-endotoxin, which bind glycoproteins and create leakage channels in membranes (34). Regions of thermolysin and leishmanolysin showed high homology to the catalytic domain of BoNT. The translocation domain was distinct in structure from bacterial pore-forming toxins and showed more resemblance to coiled-coil viral proteins such as human immunodeficiency virus type 1, gp41/GCN4, influenza HA, and the Moloney murine leukemia virus transmembrane fragment (34). BoNT appears to consist of a hybrid of varied structural motifs that evolved by combination of functional subunits to generate a highly toxic pathogenic molecule. Recent in vitro structural studies for other serotypes of BoNTs, as well as domains of the holotoxin, have shown similarity to type A and revealed interesting features that provide insight to the extreme toxicity of BoNTs (9, 52, 71). The determination of the structure of the receptor-binding fragment H_C of TeNT has provided insight into the mechanisms of neurospecific binding (52). Further studies of BoNTs, TeNT, and their domains will undoubtedly lead to increased understanding of the biology of these toxins, as well as provide a basis for improved vaccines and therapeutic agents.

The Stevens laboratory has also presented an electron density projection map of the 900-kDa botulinum type A complex at 15 Å resolution in an effort to understand the biological significance of the large complex (12). The complex was triangular in shape, had an estimated radius of about 110 Å, and possessed six distinct cylindrical lobes. The BoNT component appeared to be located in the center of the complex, where it possibly was protected from the external environment. The elucidation of the structure of the natural BoNT progenitor complexes at higher resolution would be a major accomplishment.

MECHANISMS OF INHIBITION OF NEUROTRANSMISSION BY BoNT AND TeNT

Poisoning of nerves by TeNT and BoNTs is a multistep process (55, 65, 67, 72), involving binding to gangliosides and a protein receptor(s) on presynaptic nerve cells, internalization of the CNTs by receptor-mediated endocytosis and/or synaptic vesicle reuptake, induction of channel formation, and internalization of the L catalytic domain into the presynaptic neuronal cytosol, where it specifically catalyzes neuronal protein substrate involved in vesicle traf-

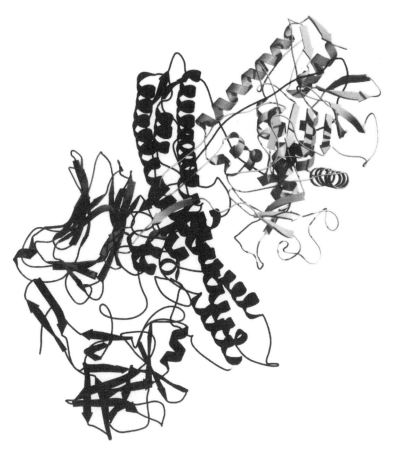

FIGURE 5   Diagram of the three-dimensional structure of BoNT/A. The catalytic domain is located at the upper right, the translocation domain in the center, the N-terminal-binding subdomain at the left, and the C-terminal-binding domain at the lower left. The catalytic zinc is represented as a ball. The overall structure is 45 by 105 by 130 Å. (Image courtesy of Ray C. Stevens, University of California, Berkeley.)

ficking. The postulated model for trafficking of the L chain into the nerve cytosol follows mechanisms elucidated for protein toxins and viruses with intracellular targets, such as diphtheria toxin and certain lipid-containing viruses. Neurospecific binding occurs by the H_C region of the H chain, probably initially to polysialogangliosides (TeNT, highest affinity to G_{T1b} and G_{D1b} and lower affinity to G_{M1} and G_{D1a}; BoNT, various affinities to G_{D1a}, G_{T1b}, and others, depending on serotype) (12, 28). Since the binding affinity to polysialogangliosides of BoNTs and TeNT is $\sim 10^{-8}$ M, a protein receptor(s) is probably also involved to bring the affinity to $\sim 10^{-12}$ M (55, 65, 67, 72). Synaptotagmin II has been identified as a protein receptor for BoNT/B (24), and other experiments are in progress to elucidate protein receptors for other serotypes.

After being incorporated within endosomes, acidification of the endosome environment has been postulated to induce pore formation and entry of the L chain into the nerve cytosol following the model for diphtheria toxin (see chapter 59). Intriguing recent work indicates that the BoNT H_C-L_C complex forms a transmembrane chaperone that prevents aggregation and enables translocation (48) by a cytosolic translocation factor complex, ATP, and the chaperone HSP 90 (61). Although initially demonstrated for diphtheria toxin, cytosolic translocation factor complex

and HSP 70 are probably also involved in translocation of CNTs through interaction with characteristic peptide sequences in the CNTs. Once internalized, the zinc-dependent catalytic domain of the seven serotypes of BoNTs interacts with and cleaves the neuronal soluble N-ethylmaleimide-sensitive fusion (NSF) protein receptor (SNARE) proteins involved in neurotransmitter vesicle transport to the membrane, vesicle-membrane fusion, and calcium-induced exocytosis (55, 65, 72). The SNAREs syntaxin and soluble NSF-attachment protein 25 (SNAP-25) are associated with the cytoplasmic face of presynaptic nerve terminals, and vesicle-associated membrane protein (VAMP) is inserted into the synaptic vesicle membrane. The three SNAREs (VAMP, SNAP-25, and syntaxin) possess a distinct three-dimensional motif (the SNARE motif) that is required for specific proteolysis by the CNTs (55, 65, 72). This nonapeptide motif occurs not only in mammals but also in *Drosophila melanogaster*, *Torpedo marmorata*, and *Arabidopsis thaliana*, but the CNTs act poorly on substrates from certain of these organisms, probably owing to mutations within or outside the SNARE motif. BoNTs and TeNT are unusual among zinc-dependent proteases in requiring a substrate at least 14 amino acids in length, probably because the CNTs recognize the shape of the substrate and not specific-peptide bonds. Recent work also has demonstrated that extensive substrate CNT contact sites exist on the surface of the L_C (9).

The use of modern molecular biology methods to investigate exocytosis, combined with electrophysiology, has led to tremendous advances in our understanding of the processes of neurotransmission and its inhibition by BoNTs and TeNT (39, 52, 67, 72). In vitro, the SNAREs SNAP-25, VAMP, and syntaxin assemble into a stable ternary complex (55, 65, 72), and disassembly requires an ATPase NSF protein together with SNAPs. These SNAREs are most susceptible to cleavage when they are not assembled in the tight ternary complex. In vivo, both fast and slow phases of membrane fusion and neurotransmitter secretion are blocked by all but one of the different BoNTs (BoNT/A) or TeNT, suggesting that the SNAREs integral for secretion must exist both in a ternary complex and free within the neuron. Although much progress has been made in understanding the mechanisms by which CNTs inhibit neurotransmission, gaps exist in our knowledge of the mechanisms of the BoNTs and TeNT. The complex mechanism of receptor-mediated endocytosis and trafficking of the toxins to the nerve cytosol is being elucidated but will require additional work. Besides being involved in exocytosis, SNAP-25 and other neuronal proteins perform several functions in the nerve, including axonal growth, maturation, and synaptogenesis (55). Exocytosis is a complex process that undoubtedly involves many molecular interactions, and the role of CNTs in altering these processes is not fully elucidated at this time. Much additional research is needed to understand their mechanisms and to validate the in vitro models using in vivo systems.

OTHER FACTORS INFLUENCING PATHOGENESIS OF *C. TETANI* AND *C. BOTULINUM*

Although BoNTs and TeNT are the primary determinants of virulence in neurotoxigenic clostridia, wound or intestinal infections also require additional virulence processes. *C. tetani* is adept at colonizing wounds, and *C. botulinum* does so relatively poorly, whereas *C. botulinum* but not *C. tetani* colonizes and causes disease in the gastrointestinal tract of infants and susceptible adults. The factors that govern the differential colonization of wounds by *C. tetani* and *C. botulinum* have not been established. It is possible that the hemolysin of *C. tetani* promotes infection by enabling the organism to acquire iron and other nutrients required for clostridial growth, but other virulence factors may be involved as well. Comparative genomic analyses should be useful in elucidating the virulence processes involved in colonization of wounds as well as intestinal colonization of humans and animals.

Sugiyama and colleagues (for a review, see reference 68) employed rodent animal models to study factors governing colonization of the intestinal tract by *C. botulinum*. The age dependence observed in humans was also duplicated in mice and rats. BoNT was produced in the colon after 2 to 3 days in 7- to 13-day-old mice challenged with *C. botulinum* type A spores. Mice 8 to 10 days old were most susceptible to this nonsymptomatic colonization. The average 50% colonizing dose of *C. botulinum* type A spores was <1,000 spores in 9-day-old mice. The age-dependent susceptibility of mice to intestinal colonization by *C. botulinum* appears to be the counterpart of the clinical infection occurring in human infants <12 months of age, but it is not due to chronological age itself, since germfree adult mice are high-

ly susceptible. BoNT was produced mostly in the large bowel, with slightly more toxin in the cecum than in the colon. This mouse model could be valuable in the study of other clostridial diseases within the intestinal environment.

The role of indigenous flora in resistance was supported when germfree mice acquired a complex gut flora after exposure to normal mice in their cages. Oral challenges with 10^5 spores were ineffective because of protection by the normal gut flora. However, the normally resistant adult mouse became susceptible when fed a kanamycin-erythromycin mixture that changed the nature of the gut flora. Pretreatment with metronidazole also sensitized mice to infection, suggesting that obligate anaerobes contributed to resistance of colonization. Interestingly, *C. perfringens* and *C. difficile* were identified in the germfree model as protecting against challenge by 10^5 spores. Two *Lactobacillus* strains, one *Bacteroides* strain, four fusiform strains, and one spirochete strain did not prevent colonization of germfree mice with 50 type A spores but did reduce the mortality rate.

Smith and Sugiyama (68) postulated that asymptomatic colonization of humans may occur more frequently than is currently believed and that overt botulism may occur only when BoNT reaches a site in the gastrointestinal tract from which it is actively absorbed. BoNT has been detected in the stools of healthy human adults, and colonization has been documented in adults whose gastrointestinal flora are altered by antibiotic therapy (46, 68). Intestinal lesions caused by *C. botulinum* in horses and other animals have been observed. The colonization of the intestinal tract by *C. tetani* has recently been postulated to cause neurological and developmental disabilities in humans, including a subset of human autism cases (29). Inhibition of neurotransmitter release by TeNT could also have a much greater impact on central nervous system function and development than previously recognized.

GENOMICS OF NEUROTOXIGENIC CLOSTRIDIA

The genomic sequences of several *Clostridium* species have recently been determined, including those of *C. tetani*, *C. perfringens*, and *Clostridium acetobutylicum* (10, 11, 57, 66); studies of others are ongoing, including those of *C. difficile* and *C. botulinum*. The genome sequence of a single strain of each species has been determined, and it is not clear if the sequences will be representative of most strains, particularly in *C. botulinum* and *C. perfringens*, in which several distinct groups of organisms occur within the species. Notwithstanding, the analyses of the genomes have revealed interesting features regarding the pathogenicity, spore formation, and metabolism of the clostridia.

The complete genome sequence of *C. tetani* E88, a nonsporulating variant of the famous Harvard or Massachusetts strain long used for toxoid production, revealed several interesting features (10, 11). This strain harbors a 78,052-bp plasmid (pE88), containing 61 open reading frames (ORFs), including genes involved in pathogenesis. The genes encoding TeNT, the positive regulatory protein TetR, and a collagenase that could destroy tissue are located on the plasmid. The analysis of pE88 also revealed various regulatory systems, including alternative sigma factor-like proteins and a two-component regulatory system located about 25 kb from the TeNT gene. pE88 had a very low G+C content of 24.5%, and it had a low number of presumptive mobile elements, including 16 transposases that were classified into four insertion sequence families. However, most of the transposases and IS elements appeared to be nonfunctional because of various classes of mutations. A number of genes encoding transporters were found on pE88, several of which are probably involved in peptide transport (10, 11).

The genomic analysis of the *C. tetani* Harvard chromosome also revealed many interesting features (10, 11). The genome was relatively small, consisting of a 2,799,251-bp chromosome encoding 2,618 ORFs and with a moles percent G+C content of 28.6%. The G+C variation within the genome was very low, and pathogenicity islands were not found. These data indicating an apparent lack of transmissible elements are intriguing, as they suggest that the *C. tetani* genome is relatively stable compared to many other bacterial groups such as the *Enterobacteriaceae*. There did exist at least one putative active mobile element, which was duplicated on pE88, that contained a complete transposase gene and a possible fragment of a transposase gene showing homology to a similar gene in *Bacillus cereus* that is present in five copies within the bacillus genome (40). *C. tetani* E88 had six rRNA operons, 54 tRNA genes, and three putative prophages (11).

Several genes were present in *C. tetani* that were postulated as potential virulence factors including those encoding collagenase, a fibronectin-binding protein, proteases, hemolysins, an internalin A homolog, a surface layer protein possibly involved in adhesion, and tetanolysin O (10, 11). Many of these genes have homologs in *C. perfringens* (66). In line with the protein-based heterotrophic metabolism of *C. tetani*, the organism contained genes with strong similarity to protein/peptide-transporting systems, several genes for amino acid decomposition, and genes for >27 peptidases, including 7 for predicted zinc metallopeptidases. The presence of many genes for degradation of amino acids but the lack of genes for sugar utilization in *C. tetani* contrasted with the genomes of *C. perfringens* (66) and *C. acetobutylicum* (57), which readily use sugars and other carbohydrates.

The complete genome sequence of a type A strain of *C. botulinum* (strain Hall ATCC 3502) has been determined by the Sanger Centre in England (http://www.sanger.ac.uk/Projects/C_botulinum/). The genome size is 3,886,916 bp and the G+C content was 28.2% (11). The strain contains a plasmid pBOT3502 that is 16.34 kb in size, has a G+C content of 26.8%, and possesses a gene for a bacteriocin analogous to a boticin gene previously isolated and characterized in my laboratory (20). The genome contains ~3,620 ORFs, ~10 rRNA operons, and ~35 putative two-component regulators. The *bont*/A and *botr*/A genes are located on the chromosome. *C. botulinum* ATCC 3502 harbors a single toxin gene cluster, encoding two polycistronic mRNAs. The toxin gene cluster was flanked by defective IS elements and transposases that are apparently nonfunctional, since they contain several mutations (11, 21, 22). Genes for two putative hemolysins are also present, and further analysis of the genome will probably reveal additional genes encoding virulence factors.

The genome appears similar to that of *C. tetani* in that there is no obvious evidence of foreign DNA acquisition (11). Interestingly, analyses of the genomes for clostridia showed that genes involved in sporulation in *Bacillus* sp., including *spo0B*, *spo0F*, *spoVf*, *spoVM*, *kinA*, *kinE*, and additional genes of the phosphorelay system and *spo/ger*, were lacking or nonfunctional in *C. perfringens*, *C. acetobutylicum*, *C. tetani*, and probably in *C. botulinum* (10, 11, 57, 66). These results suggest that the mechanism of sporulation and spore germination is significantly different from

that of *Bacillus*. Since *C. botulinum* is a very diverse species, it would be valuable to obtain the genome sequences of other strains and types and also to elucidate the proteome or the total proteins expressed by the organism. The availability of the genomic sequences of *C. botulinum* will also enable the development of microarrays for hybridization analyses of genes involved in pathogenesis, as well as genes in important processes including sporulation and stress responses and genes essential for growth in the intestinal tract, in wounds, and in foods.

SAFETY IN WORKING WITH BoNTs AND TeNT

BoNTs and TeNT are extremely toxic molecules and are considered the most potent poisons known (3, 53, 64). Toxicity of BoNT/A has been estimated as 0.2 ng/kg for humans, and the lethal dose is 1 μg or less. Because the consequences of an accidental intoxication are so severe, safety must be a primary concern of scientists studying these toxins (53, 64). The Centers for Disease Control and Prevention (CDC) recommends biosafety level 3 primary containment and personnel precautions for facilities making large quantities of CNTs. All personnel who work in the laboratory should be immunized with a pentavalent (A to E) toxoid available from the CDC. A biosafety manual should be posted in the laboratory and should contain the proper emergency phone numbers and procedures for emergency response, spill control, and decontamination. When steps in which aerosols may be created are performed, special precautions need to be taken. A class II or III biological safety cabinet or respiratory protection should be used. The use of needles and syringes for bioassays requires extreme caution. Beginning in 1997, *C. botulinum* cultures and toxins were included in a group of select agents whose transfer has been controlled by the CDC. To transfer these agents, both the person sending and the person receiving them must be registered with the CDC and exchange the appropriate approval forms.

CONCLUSIONS AND PERSPECTIVES

Remarkable advances have been achieved during the past decade in elucidating the biochemistry, structure, and pharmacological mechanisms of BoNTs and TeNT. Structural and biochemical studies of the CNTs have provided much insight into the mechanisms of substrate catalysis, neurospecific binding, and trafficking of TeNT and BoNT to their neuronal targets. These advances have certainly contributed to the remarkable success of botulinum toxin as a pharmacological agent for the treatment of various neuronal diseases and may lead to improved vaccines and countermeasures. Further information has also been attained regarding the genetic regulation and biochemical aspects of the CNTs. Paradoxically, knowledge of pathogenesis by the neurotoxic clostridia is lagging behind, such as knowledge of the virulence mechanisms required for infection of the gastrointestinal tract of susceptible infants and adults by *C. botulinum*, factors affecting the propensity of *C. tetani* compared with *C. botulinum* to infect wounds, and the role of other microorganisms and host defense mechanisms in controlling infection and neurotoxin activity. The availability of genomic sequences and comparative genomic analyses, together with the development of genetic tools such as gene replacement and vectors for controlled gene expression, will be invaluable in elucidating pathogenic mechanisms of neurotoxigenic clostridia. To prevent further human illness and deaths by neurotoxic clostridia, antidotes are urgently needed that can reverse the detrimental effects of BoNTs and TeNT once they are bound and internalized in nerves. Further knowledge of the properties of the neurotoxigenic clostridia and their CNTs will lead to improved pharmaceuticals for treatment of human disease using these most potent toxins and for needed countermeasures to botulism.

Research in my laboratory has been supported by the National Institutes of Health, the U.S. Department of Agriculture, the University of Wisconsin, and industry sponsors of the Food Research Institute, University of Wisconsin—Madison.

I am grateful to the many collaborators and mentors on various projects involving neurotoxigenic clostridia and their CNTs.

REFERENCES

1. **Anttharavally, B., W. Tepp, and B. R. DasGupta.** 1998. Status of Cys residues in the covalent structure of botulinum neurotoxin types A, B, and E. *J. Prot. Chem.* **17:**187–196.

2. **Arnon, S. S., R. Schechter, T. V. Inglesby, D. A. Henderson, J. G. Bartlett, M. Ascher, E. Eitzen, A. D. Fine, J. Hauer, M. Layton, S. Lillibridge, M. T. Osterholm, E. O'Toole, G. Parker, T. M. Perl, P. K. Russell, D. L. Swerdlow, K. Tonat, and the Working Group on Civilian Biodefense.** 2001. Botulinum toxin as a biological weapon. Medical and public health management. *JAMA* **285:**1059–1070.

3. **Arnon, S. S.** 2004. Infant botulism, p. 1758–1766. *In* R. D. Feigen, J. D. Cherry, G. J. Demmler, and S. L. Kaplan (ed.), *Textbook of Pediatric Infectious Diseases,* 5th ed. W. B. Saunders, Philadelphia, Pa.

4. **Barr, J. R., H. Moura, A. E. Boyer, A. R. Wolfitt, S. R. Kalb, A. Pavlopoulos, L. G. McWilliams, J. G. Schmidt, R. A. Martinez, and D. L. Ashley.** 2005. Botulinum neurotoxin detection and differentiation by mass spectrometry. *Emerg. Infect. Dis.* **11:**1578–1583.

5. **Barth, H., K. Aktories, M. R. Popoff, and B. G. Stiles.** 2004. Binary bacterial toxins: biochemistry, biology, and applications of common *Clostridium* and *Bacillus* proteins. *Microbiol. Mol. Biol. Rev.* **68:**373–402.

6. **Bleck, T. P.** 1991. Tetanus: pathophysiology, management, and prophylaxis, *Disease-a-Month* **37:**547–603.

7. **Borodic, G., E. Johnson, M. Goodnough, and E. Schantz.** 1996. Botulinum toxin therapy, immunlogic resistance, and problems with available materials. *Neurology* **46:**26–29.

8. **Bradshaw, M., S. S. Dineen, N. D. Maks, and E. A. Johnson.** 2004. Regulation of neurotoxin complex expression in *Clostridium botuloinum* strains 62A, Hall A-*hyper,* and NCTC 2916. *Anaerobe* **10:**321–333.

9. **Breidenbach, M. A., and A. T. Brunger.** 2004. Substrate recognition for botulinum neurotoxin serotype A. *Nature* **432:**925–929.

10. **Brüggemann, H., et al.** 2003. The genome sequence of *Clostridium tetani,* the causative agent of tetanus disease. *Proc. Natl. Acad. Sci. USA* **100:**1316–1321.

11. **Brüggeman, H., and G. Gottschalk.** 2004. Insights in metabolism and toxin production from the complete genome sequence of *Clostridium tetani. Anaerobe* **10:**53–68.

12. **Burkhard, F., F. Chen, G. M. Kuziemko, and R. C. Stevens.** 1997. Electron density projection map of the botulinum neurotoxin 900–kilodalton complex by electron crystallography. *J. Struct. Biol.* **120:**78–84.

13. **Cato, E. P., W. L. George, and S. M. Finegold.** 1986. Genus *Clostridium,* p. 1141–1200. *In* P. H. A. Sneath, N. S. Mair, M. E. Sharpe, and J. G. Holt (ed.), *Bergey's Manual of Systematic Bacteriology,* vol. 2. The Williams & Wilkins Co., Baltimore, Md.

14. **Centers for Disease Control and Prevention.** 1998. *Botulism in the United States, 1899–1996: Handbook for Epidemiologists, Clinicians, and Laboratory Workers.* Centers for Disease Control and Prevention, Atlanta, Ga.

15. **Centers for Disease Control and Prevention.** 1998. *Media for Isolation, Characterization, and Identification of Obligately Anaerobic Bacteria.* U.S. Department of Health and Human Services, Public Health Service, Centers for Disease Control and Prevention, Atlanta, Ga.

16. **Cherington, M.** 1998. Clinical spectrum of botulism. *Muscle Nerve* **21:**701–710.

17. **Collins, M. D., and A. K. East.** 1998. Phylogeny and taxonomy of the food-borne pathogen *Clostridium botulinum* and its neurotoxins. *J. Appl. Microbiol.* **84:**5–17.

18. **DasGupta, B. R.** 1989. Structure of botulinum neurotoxin, p. 53–67. *In* L. L. Simpson (ed.), *Botulinum Neurotoxin and Tetanus Toxin.* Academic Press, Inc., San Diego, Calif.

19. **Devriese, P. P.** 1999. On the discovery of *Clostridium botulinum. J. Hist. Neurosci.* **8:**43–50.

20. **Dineen, S. S., M. Bradshaw, and E. A. Johnson.** 2000. Cloning, nucleotide sequence, and expression of the gene encoding bacteriocin boticin B from *Clostridium botulinum* strain 213B. *Appl. Environ. Microbiol.* **66:**5480–5483.

21. **Dineen, S. S., M. Bradshaw, and E. A. Johnson.** 2003. Neurotoxin gene clusters in *Clostridium botulinum* type A strains: sequence comparison and evolutionary implications. *Curr. Microbiol.* **46:**345–352.

22. **Dineen, S. S., M. Bradshaw, C. Karasek, and E. A. Johnson.** 2004. Nucleotide sequence and transcriptional analysis of the type A2 neurotoxin gene cluster in *Clostridium botulinum. FEMS Microbiol. Lett.* **235:**9–16.

23. **Dong, M., W. H. Tepp, E. A. Johnson, and E. R. Chapman.** 2004. Using fluorescent sensors to detect botulinum neurotoxin activity in vitro and in living cells. *Proc. Natl. Acad. Sci. USA* **101:**14701–14706.

24. **Dong, M., D. A. Richards, M. C. Goodnough, W. H. Tepp, E. A. Johnson, and E. R. Chapman.** 2003. Synaptotagmins I and II mediate entry of botulinum neurotoxin B into cells. *J. Cell Biol.* **162:**1293–1303.

25. **Dürre, P. (ed.).** 2005. *Handbook on Clostridia.* CRC Press, Boca Raton, Fla.

26. **Eklund, M. W., and V. R. Dowell, Jr. (ed.).** 1987. *Avian Botulism: an International Perspective.* Charles C. Thomas, Springfield, Ill.

27. **Eklund, M. W., F. T. Poysky, and W. H. Habig.** 1989. Bacteriophages and plasmids in *Clostridium botulinum* and *Clostridium tetani* and their relationship to production of toxins, p. 25–51. *In* L. L. Simpson (ed.), *Botulinum Toxin and Tetanus Toxin.* Academic Press, Inc., San Diego, Calif.

28. **Finegold, S. M.** 1998. Tetanus, p. 693–722. *In* L. Collier, A. Balows, and M. Sussman (ed.), *Topley & Wilson's Microbiology and Microbial Infections,* 9th ed., vol. 3. *Bacterial Infections.* Arnold, London, United Kingdom.

29. **Finegold, S. M., D. Molitoris, Y. Song, C. Liu, M.-L. Vaisanen, E. Bolte, M. McTeague, R. Sandler, H. Wexler, E. Marlowe, M. D. Collins, P. Lawson, P. Summanen, M. Baysallar, T. Tomzynski, E. A. Johnson, R. Rolfe, H. Shah, P. Manning, and A. Kaul.** 2002. Gastrointestinal studies in late-onset autism. *Clin. Infect. Dis.* **35**(Suppl.1):S6–S16.

30. **Finegold, S. M., Y. Song, and C. Liu.** 2002. Taxonomy—general comments and update on taxonomy of clostridia and anaerobic cocci. *Anaerobe* **8:**283–285.
31. **Foran, P. G., N. Mohammed, G. O. Lisk, S. Nagwaney, G. W. Lawrence, E. Johnson, L. Smith, K. R. Aoki, and J. O. Dolly.** 2003. Evaluation of the therapeutic usefulness of botulinum neurotoxin B, C1, E, and F compared with the long lasting type A—basis for distinct durations of inhibition of exocytosis in central neurons. *J. Biol. Chem.* **278:**1363–1371.
32. **Franciosa, G., P. Aureli, and R. Schechter.** 2003. *Clostridium botulinum*, p. 61–89. *In* M. D. Miliotis and J. W. Bier (ed.), *International Handbook of Foodborne Pathogens.* Marcel Dekker, Inc., New York, N.Y.
33. **Giménez, D. F., and J. A. Giménez.** 1993. Serological subtypes of botulinal neurotoxins, p. 421–431 . *In* B. R. Dasgupta (ed.), *Botulism and Tetanus Neurotoxins: Neurotransmission and Biomedical Aspects.* Plenum Press, New York, N.Y.
34. **Hanson, M. I., and R. C. Stevens.** 2002. Structural view of botulinum neurotoxin in numerous functional states, p. 11–27. *In* M. F. Brin, M. Hallett, and J. Jankovic (ed.), *Scientific and Therapeutic Aspects of Botulinum Toxin.* Lippincott, Williams, and Wilkins, Philadelphia, Pa.
35. **Hatheway, C. L.** 1988. Botulism, p. 111–133. *In* A. Balows, J. W. J. Hausler, M. Ohashi, and A. Turano (ed.), *Laboratory Diagnosis of Infectious Diseases: Principles and Practice.* Springer-Verlag, New York, N.Y.
36. **Hatheway, C. L., and E. A. Johnson.** 1998. *Clostridium*: the spore-bearing anaerobes, p. 731–782. *In* L. Collier, A. Balows, and M. Sussman (ed.), *Topley & Wilson's Microbiology and Microbial Infections*, 9th ed., vol. 2. *Systematic Bacteriology.* Arnold, London, United Kingdom.
37. **Hoch, J. A., and M. Perego.** 1999. Two-component signal transduction in *Bacillus subtilis*: how one organism sees its world. *J. Bacteriol.* **181:**1975–1983.
38. **Holdeman, L. V., E. P. Cato, and W. E. C. Moore.** 1979. *Anaerobe Laboratory Manual*, 4th ed. Virginia Polytechic Institute and State University, Blacksburg, Va.
39. **Humeau, Y., F. Doussau, N. J. Grant, and B. Poulain.** 2000. How botulinum and tetanus neurotoxins block neurotransmitter release. *Biochimie* **82:**427–446.
40. **Ivanova, N., A. Sorokin, I. Anderson, N. Galleron, B. Candelon, V. Kapatral, A. Bhattacharya, G. Reznik, N. Mikhailova, A. Lapidus, L. Chu, M. Mazur, E. Goltzman, N. Larsen, M. D'Souza, M. Walunas, Y. Grechkin, G. Pusch, R. Haselkorn, M. Fonstein, S. D. Ehrlich, R. Overbeek, and N. Kyripides.** 2003. Genome sequence of *Bacillus cereus* and comparative analysis with *Bacillus anthracis*. *Nature* **243:**87–91.
41. **Johnson, E. A.** 1999. Clostridial toxins as therapeutic agents: benefits of nature's most toxic proteins. *Annu. Rev. Microbiol.* **53:**551–575.
42. **Johnson, E. A.** 2005. Clostridial neurotoxins, p. 491–525. *In* P. Dürre (ed.), *Handbook on Clostridia.* CRC Press, Boca Raton, Fla.
43. **Johnson, E. A.** 2005. Bacteriophages encoding botulinum and diphtheria toxin, p. 280–296. *In* M. Waldor, D. I. Friedman, and S. L. Adhya (ed.), *Phages: Their Role in Bacterial Pathogenesis and Biotechnology.* ASM Press, Washington, D.C.
44. **Johnson, E. A.** 2005. *Clostridium botulinum* and *Clostridium tetani*, p. 1035–1088. *In* S. P. Borrelio, P. R. Murray, and G. Funke (ed.), *Topley and Wilson's Microbiology and Microbial Infections*, vol. 10. Bacteriology, vol. 2. Hodder Arnold, London, United Kingdom.
45. **Johnson, E. A., and M. Bradshaw.** 2001. *Clostridium botulinum*: a metabolic and cellular perspective. *Toxicon* **39:**1703–1722.
46. **Johnson, E. A., and M. C. Goodnough.** 1998. Botulism, p. 723–741. *In* L. Collier, A. Balows, and M. Sussman (ed.), *Topley & Wilson's Microbiology and Microbial Infections*, 9th ed., vol. 2. *Systematic Bacteriology.* Arnold, London, United Kingdom.
47. **Keller, J. E., and E. A. Neale.** 2001. The role of synaptic protein Sanp-25 in the potency of botulinum neurotoxin type A. *J. Biol. Chem.* **276:**13476–13482.
48. **Koriazova, L. K., and M. Montal.** 2003. Translocation of botulinum neurotoxin light chain protease through the heavy chain channel. *Nat. Struct. Biol.* **10:**13–18.
49. **Kubota, T., N. Yonekura, Y. Hariya, E. Iosgai, H. Isogai, K. Amano, and N. Fujii.** 1998. Gene arrangement in the upstream region of Clostridium botulinum type E and Clostridium butyricum BL6340 is different from that of other types. *FEMS Microbiol. Lett.* **158:**215–221.
50. **Lacy, D. B., and R. C. Stevens.** 1999. Sequence homology and structural analysis of the clostridial neurotoxins, *J. Mol. Biol.* **291:**1091–1104.
51. **Lacy, D. B., W. Tepp, A. C. Cohen, B. R. DasGupta, and R. C. Stevens.** 1998. Crystal structure of botulinum neurotoxin type A and implications for toxicity. *Nat. Struct. Biol.* **5:**898–902.
52. **Lalli, G., S. Bohnert, K. Deinhardt, C. Verastegui, and G. Schiavo.** 2003. The journey of tetanus and botulinum neurotoxins in neurons. *Trends Microbiol.* **11:**431–437.
53. **Malizio, C. J., M. C. Goodnough, and E. A. Johnson.** 2000. Purification of botulinum type A neurotoxin. *Methods Mol. Biol.* **145:**27–39.
54. **Miyamoto, O., J. Minami, T. Toyoshima, T. Nakamura, T. Masada, S. Nagao, T. Negi, T. Itano, and A. Okabe.** 1998. Neurotoxicity of *Clostridium perfringens* epsilon-toxin for the rat hippocampus via the glutamatergic system. *Infect. Immun.* **66:**2501–2508.
55. **Montecucco, C., and G. Schiavo.** 1995. Structure and function of tetanus and botulinum neurotoxins. *Q. Rev. Biophys.* **28:**423–472.
56. **Moorthy, J., G. A. Mensing, D. Kim, S. Mohanty, D. T. Eddington, W. H. Tepp, E. A. Johnson, and D. J. Beebe.** 2004. Microfluidic tectonics platform: a colorimetric, disposable botulinum toxin enzyme-linked immunosorbent assay system. *Electrophoresis* **25:**1705–1713.
57. **Nölling, J., G. Breton, M. V. Omelchenko, K. S. Makarova, Q. Zeng, R. Gibson, H. M. Lee, J. Dubois, D. Qiu, J. Hitti, Y. I. Wolf, R. L. Tatusov, F. Sabathe, L. Doucette-Stamm, P. Soucaille, M. J. Daly, G. N. Bennett, E. V. Koonin, and D. R. Smith.** 2001. Genome sequence and comparative analysis of the solvent-producing bacterium *Clostridium acetobutylicum*. *J. Bacteriol.* **183:**4823–4828.
58. **Porfirio, Z., S. M. Prado, M. D. C. Vancetto, F. Fratelli, E. W. Alves, I. Raw, B. L. Fernandes, A. C. M. Camargo, and I. Lebrun.** 1997. Specific peptides of casein pancreatic digestion enhance the production of tetanus toxin. *J. Appl. Microbiol.* **83:**678–684.
59. **Quinn, C. P., and N. P. Minton.** 2001. Clostridial neurotoxins, p. 211–250. *In* H. Bahl and R. Dürre (ed.), *Clostridia: Biotechnology and Medical Applications.* Wiley-VCH, Weinheim, Germany.
60. **Raffestin, S., J. Christophe Marvaud, R. Cerrato, B. Dupuy, and M. R. Popoff.** 2004. Organization and regulation of the neurotoxin genes in *Clostridium botulinum* and *Clostridium tetani*. *Anaerobe* **10:**93–100.

61. **Ratts, R., H. Zeng, E. A. Berg, M. E. McComb, C. E. Costello, J. C. vanderSpek, and J. R. Murphy.** 2003. The cytosolic entry of diphtheria toxin catalytic domain requires a host cell cytosolic translocation complex. *J. Cell Biol.* **160:**1139–1150.

62. **Read, T. D., S. N. Peterson, N. Tourasse, L. W. Baille, et al.** 2003. The genome sequence of *Bacillus anthracis* and its comparison to closely related bacteria. *Nature* **423:** 81–86.

63. **Sakaguchi, G.** 1983. *Clostridium botulinum* toxins. *Pharmacol. Ther.* **19:**165–194.

64. **Schantz, E. J., and E. A. Johnson.** 1992. Properties and use of botulinum toxin and other microbial neurotoxins in medicine. *Microbiol. Rev.* **56:**80–99.

65. **Schiavo, G., M. Matteoli, and C. Montecucco.** 2000. Neurotoxins affecting neuroexocytosis. *Physiol. Rev.* **80:** 717–766.

66. **Shimizu, T., K. Ohtani, H. Hirakawa, K. Ohshima, A. Yamashita, T. Shiba, N. Ogasawara, M. Hattori, S. Kuhara, and H. Hayashi.** 2002. Complete genome sequence of *Clostridium perfringens*, an anaerobic flesh-eater. *Proc. Natl. Acad. Sci. USA* **99:**996–2001.

67. **Simpson, L. L.** 2004. Identification of the major steps in botulinum toxin action. *Annu. Rev. Pharmacol. Toxicol.* **44:**167–193.

68. **Smith, L. D. S., and H. Sugiyama.** 1988. *Botulism: the Organism, Its Toxins, the Disease.* Charles C Thomas, Springfield, Ill.

69. **Solomon, H. M., E. A. Johnson, D. T. Bernard, S. S. Arnon, and J. L. Ferreira.** 2001. *Clostridium botulinum* and its toxins, p. 317–324. *In* F. P. Downes and K. Ito (ed.), *Compendium for the Microbiological Examination of Foods,* 4th ed. American Public Health Association, Washington, D.C.

70. **Sugiyama, H.** 1980. *Clostridium botulinum* neurotoxin. *Microbiol. Rev.* **44:**419–448.

71. **Swaminathan, S., S. Eswaramoorthy, and D. Kumaran.** 2004. Structure and activity of botulinum neurotoxins. *Mov. Disord.* **19**(Suppl. 8)**:**S17–S22.

72. **Turton, K., J. A. Chaddock, and K. R. Acharya.** 2002. Botulinum and tetanus neurotoxins: structure, function, and therapeutic utility. *Trends Biochem. Sci.* **27:**552–558.

73. **Van Ermengem, E.** 1979. Classics in infectious disease. A new anaerobic bacillus and its relation to botulism. *Rev. Infect. Dis.* **1:**701–719. [Reprint, *Z. Hyg. Infektionskr.* **26:** 1–56, 1897.]

74. **Willis, A. T.** 1969. *Clostridia of Wound Infection.* Butterworths, London, United Kingdom.

Enterotoxic Clostridia: *Clostridium perfringens* Type A and *Clostridium difficile*

BRUCE A. McCLANE, DAVID M. LYERLY, AND TRACY D. WILKINS

57

The clostridia are gram-positive, anaerobic, spore-forming rods, several species of which cause gastrointestinal (GI) disease in humans or domestic animals. The involvement of some clostridia in GI disease is not surprising, given the anaerobic nature of these bacteria and their ability to produce toxins active in the GI tract. In this chapter, we discuss two enterotoxin-producing clostridia that rank among the most important enteric pathogens of humans, *Clostridium difficile* and enterotoxin-positive type A strains of *Clostridium perfringens*.

ENTEROTOXIN-PRODUCING *C. PERFRINGENS* TYPE A

The arsenal of *C. perfringens* currently contains at least 14 different toxins; however, a single isolate of this bacterium never expresses all 14 of these toxins. Based upon this observation, a commonly used classification scheme (47, 48) assigns *C. perfringens* isolates to five toxin types (A through E), depending upon their ability to produce the four major lethal toxins (LTs) (Table 1).

The major LTs are not the only biomedically important *C. perfringens* toxins; some *C. perfringens* isolates, mostly belonging to type A, express *C. perfringens* enterotoxin (CPE). CPE-positive *C. perfringens* type A strains represent only <5% of the global *C. perfringens* population (47, 48) but are nevertheless very important human GI pathogens, causing *C. perfringens* type A food poisoning, as well as non-food-borne human GI diseases such as sporadic diarrhea (diarrhea in the absence of food poisoning or antibiotic therapy) and antibiotic-associated diarrhea (AAD). Although not discussed further in this chapter, these bacteria are also an important cause of diarrhea in domestic animals (66).

C. perfringens Type A Food Poisoning

Epidemiology

C. perfringens type A food poisoning currently ranks as the third-most-common food-borne illness in the United States, where over 250,000 cases are estimated to occur each year at annual costs exceeding $120 million (47, 48). The preva-lence of this food poisoning stems, in large part, from several attributes of *C. perfringens* (47, 48). First, the ubiquitous presence of *C. perfringens* in soil, animal and human feces, river sediments, etc., provides this bacterium with ample opportunity to contaminate foods. Second, the very short (<10-min) doubling time and relative aerotolerance of *C. perfringens* permit rapid growth in foods, which is necessary for reaching the large bacterial load (>10^6 vegetative cells per gram of food) required to initiate *C. perfringens* type A food poisoning. Finally, the ability of this bacterium to survive in incompletely cooked foods is facilitated by (i) the heat tolerance of *C. perfringens* vegetative cells, which have an optimal growth temperature of 43 to 45°C but can grow at temperatures up to at least 50°C, and (ii) the even-greater heat resistance of *C. perfringens* spores (see "Virulence Factors," below).

Recognized outbreaks of *C. perfringens* type A food poisoning are usually very large, averaging about 100 cases (47, 48). The large size of these outbreaks is attributable, in part, to the association of *C. perfringens* type A food poisoning with institutional settings. Institutions are predisposed for developing this form of food poisoning because they often prepare foods in advance and then store these foods before serving, which, as described below, can facilitate the development of *C. perfringens* type A food poisoning. However, the large size of most recognized *C. perfringens* type A food poisoning outbreaks is also somewhat misleading, since outbreaks of this illness are typically overlooked by public health officials unless many people are affected.

The most important contributing factor to outbreaks of *C. perfringens* type A food poisoning is holding foods under improper storage conditions (47, 48); because of the relative heat tolerance of vegetative *C. perfringens* cells, foods should either be stored under refrigeration or at temperatures above 70°C to prevent bacterial growth. Incomplete cooking, which may allow *C. perfringens* endospores or vegetative cells to survive cooking and then (in the case of spores) to germinate in the improperly cooked food, is the second-most-common contributing factor to *C. perfringens* type A food poisoning outbreaks (47, 48). Beef roasts and poultry rank as the two most important vehicles for this

TABLE 1 Toxin typing of *C. perfringens*

C. perfringens type	Toxin(s) produced			
	Alpha	Beta	Epsilon	Iota
A	+	−	−	−
B	+	+	+	−
C	+	+	−	−
D	+	−	+	−
E	+	−	−	+

food poisoning, primarily because these large food items are difficult to cook thoroughly.

The laboratory plays an important role in identification of *C. perfringens* type A food poisoning outbreaks, and laboratory detection of this food poisoning has improved considerably in the past 10 years (48). The most reliable diagnostic approach currently available for identifying cases of *C. perfringens* type A food poisoning involves serologic detection (e.g., using commercially available enzyme-linked immunosorbent assays [ELISAs] or reverse passive latex agglutination assays) of CPE in the feces of food poisoning victims.

Pathogenesis

C. perfringens type A food poisoning is acquired by ingestion of a food item containing vegetative cells of a CPE-positive *C. perfringens* type A strain (47, 48). Most of these ingested bacteria are killed upon exposure to gastric acid; however, if the ingested food is heavily contaminated (i.e., $>10^6$ to 10^7 vegetative cells per gram of food), some vegetative cells escape into the small intestine, where they begin multiplying. After an initial multiplication period, these *C. perfringens* cells undergo sporulation, which may be triggered by their exposure to low pH in the stomach or to bile salts in the intestine. A low-M_r sporulation-promoting factor, produced by both CPE-positive and CPE-negative strains of *C. perfringens*, may also help induce in vivo sporulation (61).

CPE, which is the virulence factor considered responsible for most (or all) GI symptoms of *C. perfringens* type A food poisoning (59), is produced during in vivo sporulation (47, 48). CPE is not a secreted protein but is instead released into the intestinal lumen at the completion of sporulation, i.e., when the mother *C. perfringens* cell lyses to free its mature endospore. Once released into the intestinal lumen, CPE binds to intestinal epithelial cells and then exerts its molecular action, which causes desquamation of the intestinal epithelium. Animal model studies (47, 48) indicate that this desquamation is responsible for the onset of the GI symptoms associated with *C. perfringens* type A food poisoning. It is possible that other processes (e.g., CPE effects on intestinal paracellular permeability) also contribute to the pathogenesis of *C. perfringens* type A food poisoning (see "Virulence Factors," next column).

Animal model studies also suggest that *C. perfringens* type A food poisoning primarily involves the small intestine (47, 48). Those studies indicate that all regions of the rabbit small intestine are sensitive to CPE, with the ileum being particularly sensitive. The rabbit colon is largely unaffected by CPE, even though it binds levels of CPE similar to those in the rabbit ileum; however, it remains possible that CPE damages the human colon.

Clinically, *C. perfringens* type A food poisoning is characterized by diarrhea and abdominal cramps, which develop about 8 to 16 h after ingestion of contaminated food (47, 48). This incubation period probably results in large part from the time (~8 to 12 h) required for *C. perfringens* to complete sporulation; as previously mentioned, no CPE is released into the intestinal lumen until the completion of sporulation. In most victims, the GI symptoms of *C. perfringens* type A food poisoning typically persist for 12 to 24 h before resolving without clinical intervention. However, this disease can be fatal in elderly or debilitated individuals; estimates suggest that *C. perfringens* type A food poisoning is responsible for approximately eight deaths per year in the United States (48).

Virulence Factors

CPE-positive *C. perfringens* type A isolates associated with food poisoning typically carry their *cpe* gene on the chromosome, while CPE-positive type A isolates causing non-foodborne GI diseases usually carry their *cpe* gene on a large plasmid (18–20). For still unknown reasons, type A isolates carrying a chromosomal *cpe* gene are generally much more heat resistant (60) than type A isolates carrying a plasmid *cpe* gene. Their greater heat resistance probably favors survival of chromosomal *cpe* isolates in improperly cooked or held foods, helping to explain the strong association of these isolates with *C. perfringens* type A food poisoning.

As already mentioned, CPE is the major virulence factor responsible for *C. perfringens* type A food poisoning (47, 48). The involvement of CPE in this food-borne disease is supported by surveys demonstrating the presence of CPE in the feces of virtually all *C. perfringens* food poisoning victims, while this toxin is absent from the feces of healthy people. Furthermore, the amount of CPE present in the feces of food poisoning victims corresponds to CPE levels shown to cause significant GI effects in animal models. Also, reports indicate that CPE-positive *C. perfringens* type A isolates are more effective than CPE-negative *C. perfringens* type A isolates at eliciting ileal fluid accumulation in animal models and that this fluid accumulation can be inhibited with a CPE-specific antibody. Additionally, all GI symptoms of *C. perfringens* type A food poisoning can be reproduced by feeding human volunteers highly purified CPE. Finally, molecular Koch's postulate analyses conducted using isogenic *cpe* mutants demonstrated that CPE expression is required for a *C. perfringens* type A food poisoning derivative to cause GI effects in rabbit ileal loops (59).

CPE (M_r 35,317; pI 4.3) is a single polypeptide of 319 amino acids (47, 48). Except for some limited homology with the Antp70/C1 protein of *Clostridium botulinum*, the significance of which is unclear, the CPE amino acid sequence appears to be unique. The secondary structure of CPE apparently consists of ~80% β-pleated sheets and ~20% random coils, while the tertiary structure of this enterotoxin remains unknown. Compared with the staphylococcal enterotoxins, CPE is a relatively labile protein that loses biological activity when heated to 56°C for 5 min or when exposed to a pH of <5 or >10.

CPE action (Fig. 1) initiates when the toxin binds to one or more protein receptors that are present on many, but not all, mammalian cells. The total repertoire of CPE receptors has not yet been resolved, but certain members of the claudin family of tight junction proteins can serve as functional CPE receptors. Specifically, rat fibroblasts (which are naturally CPE insensitive because they cannot bind the enterotoxin) become highly CPE sensitive when transfected to express claudin-3, -4, -6, -7, -8, and -14 but not when transfected to express claudin-1, -2, -5, or -10 (24, 34, 35).

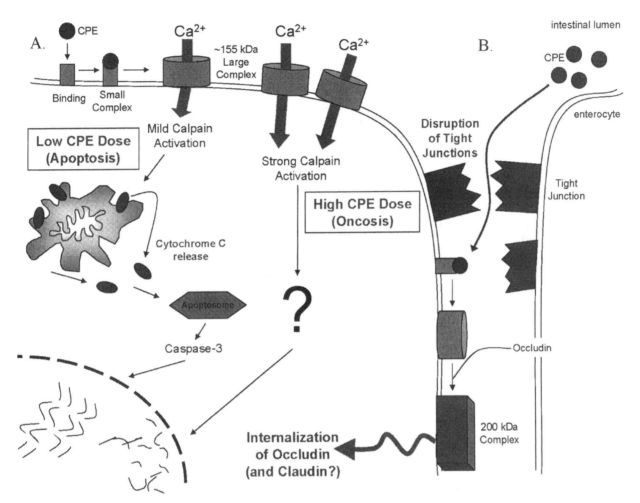

FIGURE 1 Model for the mechanism of action of CPE. (A) CPE binds to receptors, forming a small complex. At 37°C, the small complex interacts with other proteins to form an ~155-kDa large complex. The ~155-kDa complex is a pore that allows Ca^{2+} influx. With high CPE doses, massive Ca^{2+} influx occurs that triggers oncosis; with low CPE doses, there is a more moderate Ca^{2+} influx that triggers apoptosis. Activation of either cell death pathway causes morphologic damage that exposes receptors on the basolateral surface of the intoxicated cell and adjacent cells to still-unbound CPE. This allows additional formation of the ~155-kDa large complex and also permits bound CPE to interact with occludin to form an ~200-kDa complex. Formation of those two large CPE complexes triggers internalization of tight junction proteins, which damages the tight junction and leads to paracellular permeability alterations that contribute to CPE-induced diarrhea.

It has not yet been demonstrated that CPE interacts with claudins in naturally sensitive cells such as enterocytes.

Once bound to cells or isolated mammalian membranes, CPE first localizes in a small, sodium dodecyl sulfate (SDS)-sensitive complex of ~90 kDa (47, 48). Antibody probe and pronase challenge studies (38, 70) indicate that when sequestered in small complex, much (if not all) of the CPE protein remains exposed on the membrane surface. At 4°C, CPE action does not progress beyond small complex formation, and these cells never exhibit cytotoxic effects (47, 48).

Above ~20°C, the small CPE complex quickly interacts with additional (still-unknown) proteins to form higher-molecular-weight, SDS-resistant complexes (63, 64). When CPE is applied to the apical surface of a polarized epithelium (as occurs during GI disease), the initial SDS-resistant CPE complex formed is ~155 kDa in mass. Formation of that complex permeabilizes a mammalian cell to small

(<200-Da) molecules (63, 64). Two lines of evidence suggest this CPE complex may represent a pore in eukaryotic membranes, including the following. (i) Plasma membranes offer CPE substantial protection from pronase digestion when the toxin is sequestered in the ~155-kDa complex, a finding consistent with the possibility that a portion of the CPE molecule becomes inserted into the lipid bilayer of the plasma membrane when localized in this complex (70). (ii) At very high concentrations, even pure CPE can form channels and/or pores in artificial membranes lacking receptor proteins (29). Once initiated, CPE-induced changes in small molecule permeability rapidly kill a mammalian cell (17). Calcium influx appears to be the key trigger for cell death pathway activation in a CPE-treated cell, with higher CPE doses causing a massive calcium influx that triggers oncosis but with lower CPE doses causing a more modest calcium influx that induces a classical apoptosis (16). Both of those CPE-induced cell death pathways involve the cytosolic calcium-binding proteins calmodulin and calpain (16).

CPE-induced cell death exposes receptors on the basolateral surface of both dying cells and adjacent healthy cells to CPE. CPE binds to those newly exposed basolateral receptors and then interacts with the tight junction protein occludin to form an SDS-resistant, 200-kDa complex (63, 64). Formation of that complex triggers the internalization of occludin and perhaps claudins, thereby damaging the tight junction and (probably) increasing paracellular permeability (46), which likely contributes to the intestinal fluid-electrolyte secretion that manifests as diarrhea during GI disease.

How does the CPE protein mediate the effects described above? The presence of receptor-binding activity in the C-terminal half of the enterotoxin was initially suggested by studies using chemically cleaved native CPE; this tentative assignment was then unambiguously confirmed when a recombinant CPE$_{171-319}$ fragment was shown to exhibit similar binding characteristics as native CPE (47, 48). Subcloning experiments then further localized receptor-binding activity to the 30 C-terminal amino acids, e.g., studies (47, 48) using recombinant C-terminal CPE deletion fragments demonstrated that removing the five C-terminal amino acids from CPE eliminates both receptor-binding and cytotoxic activity. Besides confirming the importance of the extreme C-terminal region of CPE for receptor-binding activity, this result also provides strong evidence that a second, independent receptor-binding domain does not exist elsewhere in the CPE molecule.

While C-terminal CPE fragments such as CPE$_{171-319}$ bind well to receptors, they completely lack cytotoxicity (47, 48). This result indicates that sequences in the N-terminal half of CPE are necessary for cytotoxic activity and implies that CPE (like most bacterial toxins) segregates its binding and toxicity domains. Recent site-directed mutagenesis studies (65) have identified residues D48 and I51 in the N-terminal half of CPE as being important for cytotoxicity. Specifically, these two CPE residues were found to mediate formation of the ~155-kDa large complex that is responsible for initiating membrane permeability alterations.

Interestingly, the extreme N-terminal amino acids present in native CPE partially inhibit biological activity (47, 48). For example, limited trypsin or chymotrypsin digestion (which, respectively, removes the first 25 or 36 N-terminal amino acids from the toxin) increases CPE toxicity by two- to threefold. It has been suggested that these intestinal proteases might similarly activate CPE in the intestine during infection. Since trypsin-activated CPE exhibits the same binding properties as native CPE, removing N-terminal CPE sequences must facilitate some postbinding step in CPE action. A deletion mutagenesis study (39) identified the nature of this effect by showing that two recombinant N-terminal CPE-deletion fragments (CPE$_{37-319}$ and CPE$_{45-319}$) exhibit the same two- to threefold activation of cytotoxic activity as trypsin- or chymotrypsin-activated native CPE and possess binding properties similar to those of native CPE, yet form two to three times more ~155-kDa complex than does the native CPE. That is, removing up to the first 45 N-terminal amino acid residues from native CPE apparently enhances toxicity by promoting formation of the ~155-kDa large complex required for CPE cytotoxicity.

CPE Genetics and Expression

CPE is expressed only during sporulation, where it often accounts for 15% or more of the total protein inside the sporu-

lating *C. perfringens* cell (47, 48). This abundant CPE expression does not result from a gene dosage effect, since each *cpe*-positive *C. perfringens* cell apparently carries only a single copy of the *cpe* gene. Northern blot analyses and studies with *cpe* reporter constructs (48) indicate that CPE expression is regulated at the transcriptional level, with *cpe* transcription starting soon after the onset of sporulation. Three *cpe* promoter elements that can drive sporulation-associated CPE expression have been identified upstream of the *cpe* gene (71). Several observations (48, 71) suggest that the *cpe* gene is transcribed as a monocistronic message, as follows. (i) Northern blot analyses detected a single 1.2-kb *cpe* message in sporulating cultures of *cpe*-positive isolates. (ii) All three *cpe* promoters lie immediately upstream of the *cpe* open reading frame (ORF). (iii) A putative loop structure has been identified 36 bp downstream of the 3′ end of the *cpe* ORF; since this downstream loop is followed by an oligo(dT) tract, it may function as a Rho-independent transcription terminator.

The DNA sequences located immediately upstream of the putative transcriptional start sites of the three *cpe* promoters exhibit significant homology to *Bacillus subtilis* consensus sequences for RNA polymerase containing SigE or SigK (71). Since genes encoding SigE and SigK homologs are present in the *C. perfringens* genome (62), SigE and SigK may play a role in initiating CPE synthesis during sporulation. Recent studies (31) demonstrated that the SpoOA regulator, which initiates sporulation when phosphorylated, is necessary for CPE expression. Collectively, these findings might indicate that SpoOA regulates SigE and SigK production, which then specifies RNA polymerase binding to *cpe* promoters to initiate *cpe* transcription. Another recent study (68) found that a second regulatory protein, CcpA, is also needed for both *C. perfringens* sporulation and CPE expression. Interestingly, CcpA may (through unknown pathways) also repress *cpe* transcription during exponential growth.

As mentioned, type A isolates can carry the *cpe* gene on either their chromosome or on a large plasmid in type A isolates, although no single type A isolate has yet been found to carry both a chromosomal and plasmid-borne *cpe* gene (18–20). Regardless of its location, the *cpe* ORF is remarkably conserved in *cpe*-positive type A isolates. Not one base pair difference was detected (18) in the *cpe* ORFs from eight different type A *C. perfringens* isolates, which originated from diverse sources, including food poisoning, veterinary diarrhea, and non-food-borne human GI diseases. Some type C and D isolates also express serologically identical CPEs (47, 48), and recent observations (S. Sayeed et al., unpublished data) indicate that the *cpe* ORF present in those type C and D isolates is identical to the *cpe* ORF found in type A isolates. Interestingly, most (or all) *C. perfringens* type E isolates carry a plasmid with *cpe* sequences that are 90% homologous to the type A *cpe* ORF yet are silent due to (i) lack of the *cpe* initiation codon, promoters, and ribosome-binding site; and (ii) the presence of nine nonsense and two frameshift mutations (5). Those silent *cpe* sequences are always found adjacent to the iota-toxin genes and are highly conserved among type E isolates, despite the fact that most type E isolates themselves do not share any apparent clonal relationship. Collectively, these results suggest that the progenitor type E isolate originated when a DNA element containing iota-toxin genes was acquired from another bacterium, such as *Clostridium spiroforme* or *C. difficile*, both of which are known to carry iota toxin-like genes (see below). After this interspecies transfer, a recombinational or insertional event presumably occurred between the foreign DNA carrying the iota-toxin genes and a cpe-containing plasmid already present in the recipient C. perfringens cell. This insertion or recombination event resulted in loss of the cpe promoter; once CPE expression was disrupted, other mutations then accumulated in the cpe ORF. The highly conserved nature of the cpe sequences in a number of apparently unrelated type E isolates strongly suggests that the plasmid carrying both silent cpe sequences and iota-toxin genes has recently mobilized from the progenitor type E isolate to several different cpe-negative C. perfringens type A isolates, thereby converting those recipients to type E isolates. Although transfer of the type E plasmid carrying silent cpe genes has not yet been demonstrated, this possibility is supported by the documented (14) conjugative transfer of the cpe plasmid of F4969, a type A non-food-borne human GI disease isolate.

The 3,031,430-bp genome sequence of cpe-negative C. perfringens type A isolate strain 13 has an overall G+C content of 28.6% and carries 2,660 potential ORFs (62). When this chapter was written, The Institute for Genomic Research had also completed but not yet published or made available for public blast searches the genome sequence of SM101, a transformable derivative of cpe-positive C. perfringens food poisoning isolate NCTC8798. However, earlier genome mapping studies (47, 48) indicated that the chromosomal cpe gene is located in a highly variable region of the NCTC8798 chromosome. That finding is consistent with proposals (13) that the chromosomal cpe gene is present on Tn5565, a putative 6.3-kb mobile genetic element with flanking IS1470 sequences that has inserted into the chromosome of NCTC8798. This insertion occurred between two housekeeping genes, nadC, which potentially encodes a quinolate phosphoribosyltransferase, and uapC, which encodes a putative purine permease. Recent studies (51) suggest that the cpe locus is similarly arranged in most (or all) other type A isolates carrying a chromosomal cpe gene.

In contrast, both the upstream and downstream IS1470 sequences of the chromosomal cpe locus are absent from the plasmid cpe locus of type A isolates (51). Instead, type A plasmid cpe isolates carry either IS1151 or defective IS1470-like sequences downstream of their plasmid cpe gene. Interestingly, an IS1469 sequence is present immediately upstream of both the plasmid and chromosomal cpe genes of type A isolates, suggesting a common origin for all the cpe genes found in type A isolates.

Immunology

Protective immune responses do not contribute substantially to the rapid resolution of C. perfringens type A food poisoning (47, 48). Instead, symptoms of this illness are thought to quickly abate in large part because diarrhea flushes the GI tract, thus removing unbound CPE and CPE-producing C. perfringens cells from the intestines. Most Americans have detectable levels of CPE antibodies in their serum, but there is no evidence that prior exposure to C. perfringens type A food poisoning provides any significant long-term protection against future acquisition of this illness.

Treatment and Prevention

Treatment of C. perfringens type A food poisoning is symptomatic. Because antimicrobials are not used to treat this food-borne illness, there are few reliable data available regarding the antibiotic resistance patterns of C. perfringens food poisoning isolates.

No vaccine is currently licensed for the prevention of C. perfringens type A food poisoning. However, epitope-mapping studies (47, 48) have demonstrated the presence of a neutralizing linear epitope in the 30 C-terminal amino acids of CPE. Since C-terminal CPE fragments are not cytotoxic, these C-terminal CPE fragments hold promise as potential candidates for the development of a CPE vaccine. This hypothesis receives support from studies (47, 48) showing that mice developed high titers of CPE-neutralizing serum antibodies when they were immunized with a conjugate containing the 30 C-terminal CPE amino acids coupled to a thyroglobulin carrier. To obtain a strong CPE-neutralizing immunoglobulin A response in the intestines, further studies are needed to develop antigen delivery systems for these C-terminal CPE fragments. If successful, a CPE vaccine might potentially be used in such high-risk populations as nursing home residents or perhaps in veterinary medicine, to protect domestic animals from CPE-associated disease.

CPE-Associated Non-Food-Borne Human GI Disease

In the past 20 years, CPE-producing C. perfringens type A isolates have become established as a cause of non-food-borne human GI diseases, with some surveys (15) suggesting that these bacteria may be responsible for up to ~10 to 15% of all cases of AAD, a frequency approaching that of C. difficile-induced AAD and up to 5 to 20% of all cases of sporadic diarrhea. Some intriguing evidence (15) also suggests that a single AAD patient can become sick from a coinfection involving both C. difficile and CPE-producing C. perfringens.

The antibiotic sensitivity of CPE-positive C. perfringens AAD isolates remains unclear, but some preliminary information (9, 15) suggests that treatment with several different antibiotics (e.g., penicillins) can predispose patients to C. perfringens AAD. Preliminary epidemiological data also suggest (9, 15) that, like C. difficile-induced AAD, C. perfringens AAD can be acquired from the environment; i.e., this illness is a true infection. Presumably, AAD then develops when the acquired CPE-positive C. perfringens isolates proliferate in the gut of patients whose GI microflora have been disturbed by antibiotics or other factors. Most individuals suffering from C. perfringens AAD are compromised (e.g., transplant patients, burn patients, and AIDS patients), but it is unclear whether these compromising conditions predispose patients to this GI disease or if this association simply reflects higher rates of antibiotic usage in these compromised patient populations.

The pathogenesis of sporadic diarrhea from CPE-positive C. perfringens is also poorly understood at present. The only predisposing factor identified to date for this illness is age; i.e., sporadic C. perfringens-associated diarrhea is more common in the elderly (15).

As mentioned earlier, C. perfringens isolates associated with non-food-borne human GI disease consistently carry a plasmid-borne cpe gene, which distinguishes them from food poisoning isolates carrying a chromosomal cpe gene (19, 48). These genotypic differences between CPE-associated non-food-borne human GI disease isolates and food poisoning isolates are very interesting, since the symptoms of CPE-associated non-food-borne human GI disease are typically more severe and long lasting (non-food-borne GI illnesses can last for several weeks) than the symptoms of *C. perfringens* type A food poisoning. In part, this may be attributable (14) to the initial infecting isolates conjugatively transferring their *cpe* plasmid to the *cpe*-negative *C. perfringens* isolates found in normal GI flora. Acquisition of the *cpe* plasmid should convert those normal GI flora isolates to virulence, which may be pathogenically significant, since those normal flora isolates presumably have been under selection for their ability to persist in the GI tract.

C. DIFFICILE

C. difficile is an opportunistic pathogen that causes nosocomial diarrhea and colitis after the normal GI flora has been altered, most typically by antibiotics (3, 4). This organism causes a spectrum of GI diseases of the colon ranging from milder AAD to fulminating pseudomembranous colitis, which can be lethal if not treated. *C. difficile* causes essentially all cases of pseudomembranous colitis but only 25% or fewer of AAD cases; perhaps up to 15% of AAD cases are due to CPE-positive *C. perfringens* type A, with the remaining cases of AAD being currently undiagnosed. These cases could be due to infection by other clostridia, *Salmonella* sp., *Escherichia coli*, *Candida albicans*, or *Staphylococcus aureus* or could result from unconjugated bile acids reaching the colon.

C. difficile-mediated disease develops from the production of two toxins, toxin A and toxin B, which in some papers are referred to as the enterotoxin and cytotoxin, respectively. *Clostridium sordellii*, which is primarily a wound pathogen, produces toxin hemorrhagic toxin (HT) and LT, which immunologically cross-react with these *C. difficile* toxins; toxin A and HT are both enterotoxic as well as cross-reactive, while toxin B and LT are cross-reactive. These four toxins, along with alpha toxin from *Clostridium novyi*, comprise a new class of large clostridial cytotoxins whose cytotoxic, enterotoxic, and lethal activity is due to their effects on the cytoskeleton, causing cell rounding and eventually cell death. The level of cytotoxic activity varies among these toxins, with toxin B the most lethal; it is 100 to 1,000 times more lethal than toxin A against most cell lines. Toxin A, HT, and LT are comparable in their cytotoxic activity. These four toxins are all lethal when injected into animals but vary greatly in their enterotoxic activity. Toxin A binds to specific carbohydrates on the mucosa of the colon and causes extensive damage, whereas toxin B binds to cells that are not surrounded by a carbohydrate layer and thus is toxic for cells exposed by the action of toxin A. There is some evidence that toxins A and B act synergistically during the onset and development of disease (40). Some strains of *C. difficile* produce slight variants of these toxins that are more like *C. sordellii* toxins. In addition, some *C. difficile* strains produce toxin B but not toxin A (67); toxin B from these strains may have more enterotoxic activity than toxin B from A- and B-positive (A$^+$/B$^+$) strains.

The Sanger Institute has recently completed sequencing the genome of *C. difficile* strain 630 (epidemiologic type X). Although still considered preliminary, a database containing this genome sequence is available at the Sanger Center webpage (http://www.sanger.ac.uk/Projects/Microbes/) for blast searches. These preliminary genomic data indicate strain 630 carries a circular chromosome of 4,290,252 bp with a G+C content of ~29%, along with a 7.9-kb plasmid.

Bacteriology and Diagnosis

The ability of *C. difficile* to produce subterminal spores makes this organism difficult to eradicate in hospitals. Unlike the expression of CPE, there is no direct correlation between sporulation and toxin production in *C. difficile*. Sporulation of *C. difficile* can be stimulated by sodium cholate and sodium taurocholate. This property can be used to improve recovery of the organism from clinical specimens.

C. difficile fluoresces yellow when grown on solid media, especially media containing blood. It also produces isocaproic acid, a volatile fatty acid produced during fermentation, and *p*-cresol, which is produced during tyrosine metabolism (6). Although isocaproic acid production is unusual, it cannot be used for definitive identification, since this fatty acid is also produced by other clostridia, including *Clostridium bifermentans*, *C. sordellii*, and *Clostridium sporogenes*. In the clinical laboratory, isolates are often identified by their characteristic horse dung odor.

Laboratory tests used as diagnostic aids for *C. difficile* disease include bacterial culture, latex agglutination, tissue culture, and ELISAs (12, 37, 43). Bacterial culture is difficult and inconsistent, owing to variations in anaerobiosis, specimen collection, and media. Cycloserine-cefoxitin-fructose agar is the selective medium most often used for isolation of *C. difficile*. This medium consists of an egg yolk agar base containing cycloserine (500 μg/ml) and cefoxitin (16 μg/ml), along with 5% egg yolk suspension (25). Morphologically, the organism grows on cycloserine-cefoxitin-fructose agar as flat colonies that have a yellow fluorescence. These colonies are best viewed after a 72-h incubation. Broth media can be made selective for *C. difficile* by the addition of cycloserine and cefoxitin.

Isolation of *C. difficile* is not sufficient for diagnosis because ~20% of *C. difficile* isolates do not produce toxins and, therefore, are not pathogenic. Additional testing by tissue culture or ELISA must be performed to demonstrate that the isolates can produce toxin. Latex agglutination, membrane tests, and ELISAs are available that detect glutamate dehydrogenase (GDH), which is a nontoxic metabolic enzyme produced by all strains of *C. difficile* (42). Some other anaerobes, including *C. sporogenes*, certain types of *C. botulinum*, and *Peptostreptococcus anaerobius*, produce GDH enzymes that cross-react with *C. difficile* GDH and thus produce false-positive reactions in some of the older tests. As with bacterial culture, additional testing must be done to confirm the presence of *C. difficile* toxin in specimens testing positive for GDH.

The presence of *C. difficile* toxin in stools can be detected by tissue culture assay, ELISA, or rapid membrane tests. The tissue culture test is the more sensitive assay when performed properly, although some of the newer ELISAs are approaching similar levels of sensitivity. The detection of toxin by tissue culture is less controlled than that with ELISAs, and its accuracy varies greatly between hospital laboratories. Tissue culture also requires 24 to 48 h to obtain results, and the toxin must remain active in the stool for detection. ELISAs are a more controlled test, with a rapid turnaround time (<3 h), and can detect inactive toxin that remains serologically active. Recently, several membrane-based tests have become available commercially.

A common misconception about *C. difficile*-induced diarrhea is that *C. difficile* strains are resistant to the inciting antibiotic. In fact, most *C. difficile* strains are susceptible to a wide variety of antibiotics associated with the onset of disease, including penicillin, ampicillin, erythromycin, tetracycline, chloramphenicol, and clindamycin. *C. difficile*-induced GI disease often starts only after the antibiotic has been discontinued and antibiotic levels have dropped in the intestine. Some *C. difficile* strains have been identified that are relatively resistant to chloramphenicol, rifampin, erythromycin, clindamycin, and tetracycline; resistance to some of these antibiotics is conferred by chromosomally carried determinants, and resistance to tetracycline may result from a resistance determinant carried by a conjugative transposon. However, since antibiotic levels in the colon are often much higher than those in the bloodstream, this resistance must be interpreted cautiously. For practical purposes, all human *C. difficile* strains are susceptible to vancomycin and metronidazole, the antibiotics most commonly used to treat *C. difficile*-induced GI disease; however, there have been reports of metronidazole-resistant *C. difficile* in horses (44).

Epidemiology

C. difficile is the major cause of nosocomial diarrhea in industrialized countries (41). Outbreaks of this disease are increasingly common in hospitals and health care facilities, particularly Veterans Administration hospitals and large medical centers with large numbers of older susceptible patients. The increased incidence of *C. difficile*-induced GI disease results, in part, from increased recognition of this illness, with most health care facilities now testing for the presence of *C. difficile* toxins or antigen (GDH) in stools. Of utmost importance to the health care professional is knowledge of how this organism is transmitted and the role that health care workers play in this transmission. In a study of the acquisition and spread of *C. difficile* that was carried out over an 11-month period in an endemic setting (50), it was shown that (i) patient-to-patient transmission is common, (ii) not all people who become infected develop clinical symptoms, and (iii) health care workers often transmit the organism. The number of asymptomatic carriers in these settings may exceed the number of persons who develop disease; these asymptomatic carriers are an epidemiologic concern, since they may spread *C. difficile* to more-susceptible patients.

The persistence of spores in the hospital makes this disease especially difficult to control. Spores are present throughout the rooms of patients, and their persistence is directly associated with increased risk of infection and disease. Spores can also be isolated from the hands, clothes, and shoes of health care workers, and typing studies have identified isolates that are the same as those from patients. Exposure to patients is not considered a threat to health care workers, although there is one reported incident in which healthy workers developed diarrhea following exposure to a patient with *C. difficile* disease (21).

Up to 50% of asymptomatic infants (up to the age of several months) carry toxigenic *C. difficile* in their normal flora (43). The mechanism responsible for their protection from disease is not known, but some evidence indicates that infants tend to be colonized with serogroups that are not typically associated with adult disease and that often produce lower levels of toxin in vitro, suggesting that infants tend to be colonized with weakly toxigenic isolates. However, there are some instances in which asymptomatic infants have as much toxin in their colon as adults suffering from diarrhea and pseudomembranous colitis. Many infants also carry serogroup F strains, which behave phenotypically, at least in vitro, as A⁻/B⁺ isolates. Additional studies are needed to examine the toxin genes of these isolates at the molecular level.

The role of infants as a reservoir for the organism is still unclear, and currently no precautionary measures are used to limit the potential spread of the organism from nurseries to other hospital areas. The resistance of infants or further

identification of host specificity may represent a key to designing new treatments for *C. difficile* disease. The resistance afforded by age is not limited to infants, since most affected patients with disease are over the age of 50; younger patients are much more likely to have toxin in their colons but remain asymptomatic.

Patients can harbor multiple *C. difficile* strains at one time and can have both toxigenic and nontoxigenic isolates simultaneously present in their colon (43). Relapses are a common problem, especially in those patients remaining in hospitals for extended periods. In most instances, relapse results from the initial inciting strain, not from infection with a new strain. Most relapses probably occur from incomplete eradication of the organism following treatment with vancomycin or metronidazole or from reacquisition of the strain from spores present in the hospital room (41).

As mentioned earlier, toxin A⁻/B⁺ isolates are clinically important because they can cause *C. difficile* disease. The incidence of infection with these atypical isolates is still being established, but it is not as common as disease caused by typical isolates that produce both toxins. Even so, atypical isolates have been associated with outbreaks and will be missed by in vitro diagnostic tests that detect only toxin A. However, they will be detected by ELISAs that detect both toxins or by tissue culture. In addition, they will be detected by tests that detect antigen (e.g., GDH).

Pathogenesis

C. difficile cannot compete with the normal flora of adult humans or other animals and only grows to high numbers when the flora have been eliminated or drastically altered by antibiotics or chemotherapy. When the antibiotic declines below levels inhibitory to *C. difficile*, the ingested spores germinate and vegetative cells grow to high density in the colon. Whether or not the organism attaches to the colonic wall is unknown.

As the organism grows, it releases the toxins due to autolysis; thus, like CPE, neither toxin A nor toxin B appears to be secreted. The toxins then enter the host cell via receptor-mediated endocytosis and generalized pinocytosis and exert their molecular action. The initial diarrhea induced by *C. difficile* is caused by the toxins directly injuring the intestinal mucosa; however, the action of these toxins alone cannot explain the massive tissue destruction that occurs in pseudomembranous colitis. Much of this damage results from extensive inflammation due to infiltrating leukocytes. The role of inflammation in this process is substantiated by the presence of necrotic tissue and dead leukocytes in the pseudomembranes that develop in untreated disease. There are also elevated levels of prostaglandin E₂ and leukotriene B₄, both of which are associated with inflammation. Toxin A also has chemotactic properties, which most likely exacerbates the disease (56).

The hamster has been used as the animal model of choice (54) for studying the pathogenic mechanisms of *C. difficile* because hamsters are more susceptible than rats or mice and because treatment with antibiotics predisposes the hamster, like humans, to *C. difficile* disease. For example, when hamsters are given an ~3-mg dose of clindamycin intragastrically, they become infected with spores from contaminated bedding and then proceed to develop diarrhea and lethargy, with death following within several days. Furthermore, strains isolated from human cases of *C. difficile*-induced GI disease are as pathogenic for hamsters as are hamster isolates; as with humans, disease in hamsters is caused only by toxin-producing strains (3, 7, 43).

Virulence Factors

Immunization of hamsters against toxins A and B is completely protective (36, 41), indicating that the symptoms of *C. difficile* disease are mediated by these two toxins. Immunization of hamsters with toxin A alone results in some protection against diarrhea, and mortality is reduced. Interestingly, *C. difficile* strains that produce the same amount of toxin can vary in virulence for hamsters, suggesting that other, as-yet-unrecognized virulence factors also contribute to disease, perhaps by facilitating *C. difficile* growth in vivo. Some *C. difficile* isolates produce a thin capsule (40 to 80 nm in thickness) whose synthesis appears to be regulated by the level of glucose in the environment (6, 7). However, no association has been demonstrated between this capsule and either virulence or toxin production. Some isolates also produce fimbriae (8), but little is known about their production; again, no correlation has been demonstrated between these fimbriae and virulence. Additionally, a low percentage of isolates produce a binary toxin (see below), but the role of this toxin in *C. difficile*-mediated disease is not known.

Toxins A and B have M_rs of 308,000 and 269,000, respectively, which make them the largest known bacterial toxins (22). Both toxins are produced as single polypeptides and then released following autolysis of the bacterial cell. The toxin structural genes, *tcdA* and *tcdB*, have extensive homology; at the amino acid level, they are 49% identical and 63% similar when conservative substitutions are considered. Both toxins have a complex contiguous series of repeating units at the C terminus. These units comprise one-third of each toxin and consist of small repeating units within larger units. In toxin A, the repeating units bind to galactose-containing residues; the repeating units of toxin B may also be involved in binding, but the receptor has not been identified. The repeating subunits of toxins A and B have extensive similarity to the glucosyltransferases of *Streptococcus mutans* and *Streptococcus sobrinus* (GtfB, GtfC, and GtfI), which bind to carbohydrates (69). Most of the antibody elicited by immunization with native toxin A reacts with the repeating units of this toxin; two sites (starting at amino acids 2097 and 2355) in this repeating region appear to be especially immunodominant, and these regions react with monoclonal antibodies that neutralize the enterotoxic activity of toxin A and inhibit toxin A binding to carbohydrates.

Upstream of the repeating units are several conserved features, including four cysteine residues located in almost identical positions, a central hydrophobic region, and a putative nucleotide-binding region (Fig. 2B). The N-terminal portions of toxins A and B, which include all of these regions, exhibit about 50% homology. The hydrophobic region, which is composed of a stretch of ~50 amino acids, may represent a membrane-spanning region that is involved in toxin uptake and intracellular processing. Altering the conserved cysteines, deleting the internal hydrophobic region, or removing the repeating units of toxin B by site-directed or deletion mutagenesis results in a 90% loss of cytotoxic activity. Modifying the histidine residue located in the potential nucleotide-binding site to glutamine causes a 99% loss in activity, suggesting a key role for this region in the toxic activity of toxin B (2).

The genes encoding toxins A and B are located in a 19.6-kb pathogenicity locus (PaLoc) that is chromosomally located and contains five distinct genes, designated *tcdR*, *tcdB*, *tcdE*, *tcdA*, and *tcdC* (Fig. 2A). The *tcdR*, *tcdB*, *tcdE*, and *tcdA* genes are transcribed in the same direction, but *tcdC* is read in the opposite direction. The *tcdB* gene is located ~1 kb upstream of the *tcdA* gene. The proteins encoded by *tcdR*, *tcdE*, and *tcdC* are not directly involved in the toxicity of toxins A or B, as evident from the fact that the *tcdA* and *tcdB* genes, when cloned independently, encode fully active toxins A and B (22, 32).

The boundaries of the PaLoc have been defined using a series of strains that vary by 5 orders of magnitude in their in vitro toxigenicity. Interestingly, the boundaries in both weak and strong toxin-producing strains are highly conserved, and there are no obvious discrepancies in the sizes of the inserts that could explain the broad range of toxin expression among *C. difficile* isolates. In nontoxigenic strains, a small chromosomal region, which was found to be 127 bp in one study (28) but 115 bp in a second study (11), occupies the same chromosomal site as the PaLoc. There are no insertion sequences flanking the PaLoc that might provide a mechanism for genetic exchange. However, there are inverted repeat regions within the small insertion region of the nontoxigenic strains (11).

Most toxigenic strains of *C. difficile* produce both toxin A and B; nontoxigenic strains, which are also nonpathogenic, do not carry the toxin genes. Studies of the PaLoc have resulted in the development of a toxinotyping scheme (58). The toxinotypes are categorized by their changes within PaLoc and their pulsed-field gel electrophoretic patterns. At least 20 toxinotypes have been identified, some of which have been associated with specific serogroups. This toxinotyping scheme is particularly valuable as an epidemiologic tool for identifying and following the transmission of specific isolates. For example, toxinotypes have been identified as endogenous strains based upon their association with outbreaks.

Toxin production occurs during the stationary phase, under conditions (such as growth in dialysis tubing) that limit the growth of the organism. Toxin production is repressed by glucose (23); when the organism is grown in rich medium in free-standing culture, very little toxin is produced.

There is remarkable DNA sequence conservation in the PaLoc of both weakly and highly toxigenic *C. difficile* strains. Furthermore, the level of toxin-specific mRNA present in a *C. difficile* isolate directly correlates with the amount of toxin produced by that isolate, implying that the wide variation in toxin expression involves transcriptional regulation. The individual transcripts for *tcdR*, *tcdB*, *tcdE*, and *tcdA* are produced during stationary phase, whereas *tcdC* is transcribed during exponential growth. In addition to these smaller transcripts, a much larger (17.5-kb) transcript has been observed that apparently carries the information from *tcdR*, *tcdB*, *tcdE*, and *tcdA*. This large transcript is present at very low levels and probably results from incomplete transcriptional termination during high-level expression. Most transcription of *tcdA* and *tcdB* probably initiates from the individual promoters for these two genes; both promoters are very strong when the *tcdR* gene product is present. The *tcdR* gene encodes a small (22-kDa) protein with sequence similarity to several other clostridial regulatory proteins, including UviA of *C. perfringens*, BotR of *C. botulinum*, and TetR of *Clostridium tetani* (23, 45, 52). Like those other clostridial regulators, TcdR positively regulates toxin expression (23, 45, 52). Experiments have shown that *E. coli* supplied with TcdR in *trans* increased its expression of reporter sequences fused to either the *tcdA* or the *tcdB* promoter. TcdR appears to be a sigma factor that turns on

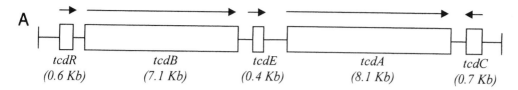

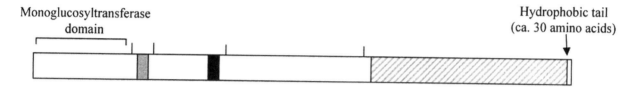

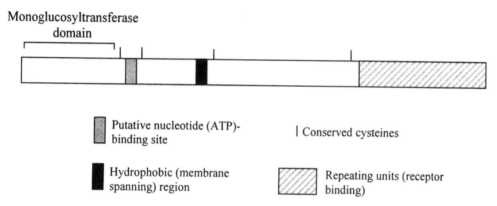

FIGURE 2 (A) PaLoc carrying the *tcdA* and *tcdB* genes of *C. difficile*. The DNA region comprising the PaLoc is approximately 19.6 kb. (B) Structural features conserved between toxins A and B.

production in response to environmental conditions, as well as stimulating its own synthesis (23, 45, 52). The level of transcription increases as the level of *tcdR* product increases. This mechanism is supported by results showing that the positive regulator encoded by *tcdR* affects an RNA polymerase, resulting in recognition of the *tcdA* and *tcdB* promoters. During slow growth, *tcdR* repression decreases, resulting in stimulation of *tcdR* and increased transcription of *tcdA* and *tcdB*. Alternatively, during rapid growth, very little toxin A or B is produced because *tcdR* is repressed.

Toxins A and B are monoglucosyltransferases (1) that cleave UDP-glucose and transfer glucose to Rho proteins, which are members of a subfamily of small GTP-binding proteins involved in the regulation of the cellular cytoskeleton and other cell functions, such as cell adhesion, microfilament organization, and nuclear signaling. Toxins A and B modify several Rho subtypes (RhoA, RhoB, and RhoC), as well as the functionally related Rac and Cdc42 proteins. The acceptor amino acid for enzymatic action by toxin A or B is residue Thr-37 in Rho. The monoglucosylation of Thr-37 in Rho represents a novel mechanism that offers molecular biologists new tools for studying the cytoskeletal machinery of mammalian cells. The enzymatic region of

toxins A and B resides in their N-terminal portion, since a recombinant fragment comprising the 63-kDa N-terminal region of these toxins retains enzymatic activity (30). More-detailed analysis (1) suggests that the region centered at residues 516 to 542 of these toxins is especially critical for activity. The effect of these toxins on the cytoskeletal system through Rho modification may cause detachment of the epithelial cells lining the mucosa, resulting in leakage of serous fluid into the intestines, which is then followed by diarrhea and the onset of colitis.

Toxin A binds to Galα1-3Galβ1-4GlcNAc and other galactose-containing residues (43). Toxin A is a more potent cytotoxin for tissue culture lines such as F9 teratocarcinoma mouse cells and HT-29 cells, both of which have large amounts of this trisaccharide on their surface. Humans do not produce this trisaccharide, but toxin A also binds to the carbohydrate antigens designated I, X, and Y (the latter two are also referred to as Lewis X or Le^x and Lewis Y or Le^y), which are produced in abundance by human intestinal epithelial cells. One theory for the resistance of infants to *C. difficile* disease is that infants lack toxin receptors; however, toxin A binds to human infant intestinal mucosa, which is known to contain the carbohydrate antigens that bind toxin A. In addition to uptake by receptor-mediated mechanisms, both toxins probably enter cells via generalized pinocytosis.

A low percentage of C. difficile strains produce CDT, a binary ADP-ribosylating toxin that modifies actin (10, 26, 27, 55, 57). The toxin is composed of two unlinked proteins, one an enzymatic component (48 kDa) and the other a binding component (94 kDa). CDT is related to the iota toxins produced by C. spiroforme and C. perfringens. The components from each of these ADP-ribosylating toxins cross-react immunologically. CDT has only been associated with a few toxinotypes, and its role as a virulence factor is unknown. However, the fact that the iota-like toxins of C. spiroforme and C. perfringens have been established as virulence factors suggests that CDT and its role in disease need to be studied further.

Immunology

The important contribution of inflammation to C. difficile disease has already been described. With regard to protective immune effects, little is known about the specific immunity that develops to C. difficile disease, although there is some evidence that convalescent patients develop mucosal and systemic antibodies against toxin A during infection (33). Relapses, which in the United States occur in nearly 20% of patients (41), are a major problem and represent a serious and debilitating form of the disease. Multiple relapses can occur, probably because the patient's intestinal tract is more susceptible after the initial damage and because C. difficile spores often remain in the patient's environment.

Therapy and Prevention

Since the onset of C. difficile disease in hospitalized patients can be triggered by any agent that disrupts the normal intestinal microflora, the disease can often be treated simply by stopping the inciting agent. If illness continues, vancomycin or metronidazole can be administered orally (4, 56). Vancomycin is poorly absorbed, so high GI concentrations can be achieved; however, there are concerns that overuse of vancomycin for treating non-life-threatening cases of C. difficile disease might contribute to the development of vancomycin resistance among commensal and pathogenic GI flora (e.g., vancomycin-resistant enterococci). Although metronidazole is readily absorbed, it is therapeutically useful for treating C. difficile disease, particularly for the milder AAD cases. Metronidazole is also less expensive than vancomycin.

Alternative therapeutic approaches now being evaluated include the use of probiotics. For example, in a 23-month study, the incidence of AAD decreased from 31% in patients receiving a placebo to 9% in patients receiving Saccharomyces boulardii, a nonpathogenic yeast that remains viable in the human intestinal tract (49). Another probiotic, a human isolate of Lactobacillus sp. named strain GG, has also been given to humans; patients suffering relapses of C. difficile disease responded favorably following daily doses of 10^{10} Lactobacillus cells over a period of 7 to 10 days. Patients who received fecal enemas (consisting of freshly passed stool from the patient's spouse or child) were cured within 24 h. Administration of bacterial mixtures containing Streptococcus faecalis, Clostridium innocuum, Clostridium ramosum, Bacteroides ovatus, Bacteroides vulgatus, Bacteroides thetaiotaomicron, E. coli, C. bifermentans, and Peptostreptococcus productus also reportedly caused a return to normal bowel habits within 24 h (43). Although probiotic therapy holds promise, it should be emphasized that these approaches have only been evaluated with small numbers of patients, so it remains unclear which specific approach should be used. However, probiotic therapy is appealing, since this approach is aimed at restoration of the normal intestinal flora, which offers the most effective means of inhibiting the growth of C. difficile.

Antibodies against toxins A and B can neutralize all of their toxin activities, and immunization of laboratory animals against these toxins is completely protective against disease (43). These observations suggest that passive immunotherapy, such as oral administration of bovine milk antibodies against toxins A and B or giving human immunoglobulin, holds promise for treating C. difficile disease. Furthermore, passive immunotherapy may be valuable for prevention, as well as therapy, since this approach protects against disease in laboratory animals (43). Passive immunotherapy should be of potential benefit to patients at risk, especially those undergoing prophylactic antibiotic therapy and/or those situated in endemic settings such as nursing homes or Veterans Administration hospitals. The administration of antibodies also shows promise as an alternative to vancomycin or metronidazole therapy for patients suffering milder forms of disease and as adjunctive therapy in combination with other treatments.

Certain basic precautions should be taken to help control outbreaks of C. difficile disease. In endemic settings, disposable vinyl gloves should be used, and proper washing procedures need to be implemented (41, 43). Proper disinfection should be performed and equipment should be cleaned. Patient isolation may be appropriate, and cleaning patients' rooms with agents such as phosphate-buffered hypochlorite (1,600 ppm) may reduce transmission. In most instances, the incidence of disease can be reduced simply by educating health care workers about the disease and how it is spread. Antibiotic usage patterns should also be monitored and physicians should be alerted when endemics are first detected.

Because of page limits, this chapter has cited published reviews for older studies. We thank James Smedley III for preparing Fig. 1.

Some of the research cited for C. perfringens enterotoxin was supported by Public Health Service grant R37 AI19844 from the National Institute of Allergy and Infectious Diseases and by National Research Initiative Competitive Grant 2005 352d 15387 from the USDA Cooperative State Research, Education and Extension Service (to B.A.M.). Some of the research cited for the molecular biology of C. difficile toxins A and B was supported by U.S. Public Health Service grant AI15749 from the National Institute of Allergy and Infectious Diseases (to T.D.W. and D.M.L.).

REFERENCES

1. **Aktories, K., and I. Just.** 1995. Monoglucosylation of low-molecular-mass GTP-binding Rho proteins by clostridial cytotoxins. *Trends Cell Biol.* **5:**441–443.
2. **Barroso, L. A., J. S. Moncrief, D. M. Lyerly, and T. D. Wilkins.** 1994. Mutagenesis of the *Clostridium difficile* toxin B genes and effect on cytotoxic activity. *Microb. Pathog.* **16:**297–303.
3. **Bartlett, J. G.** 1994. *Clostridium difficile*: history of its role as an enteric pathogen and the current state of knowledge about the organism. *Clin. Infect. Dis.* **18:**S265–S272.
4. **Bartlett, J. G.** 1995. Antibiotic-associated diarrhea, p. 893–904. *In* M. J. Blaser, P. D. Smith, J. I. Ravdin, H. B. Greenberg, and R. L. Guerrant (ed.), *Infections of the Gastrointestinal Tract.* Raven Press, New York, N.Y.
5. **Billington, S. J., E. U. Wieckowski, M. R. Sarker, D. Bueschel, J. G. Songer, and B. A. McClane.** 1998. *Clos-*

tridium perfringens type E animal isolates with highly conserved, silent enterotoxin gene sequences. *Infect. Immun.* 66:4531–4536.

6. **Bongaerts, G. P. A., and D. M. Lyerly.** 1997. Role of bacterial metabolism and physiology in the pathogenesis of *Clostridium difficile* disease. *Microb. Pathog.* 22:253–256.

7. **Borriello, S. P.** 1990. Pathogenesis of *Clostridium difficile* of the gut. *J. Med. Microbiol.* 33:207–215.

8. **Borriello, S. P., H. A. Davies, and F. E. Barclay.** 1988. Detection of fimbriae amongst strains of *Clostridium difficile.* *FEMS Microbiol. Lett.* 49:65–67.

9. **Borriello, S. P., H. E. Lawson, F. E. Barclay, and A. R. Welch.** 1987. *Clostridium perfringens* enterotoxin-associated diarrhoea, p. 33–42. *In* S. P. Borriello (ed.), *Recent Advances in Anaerobic Microbiology.* Martinus Nijhoff, Boston, Mass.

10. **Braun, M., C. Herholz, R. Straub, B. Choisat, J. Frey, J. Nicolet, and P. Kuhnert.** 2000. Detection of the ADP-ribosyltransferase toxin gene (*cdtA*) and its activity in *Clostridium difficile* isolates from Equidae. *FEMS Microbiol. Lett.* 184:29–33.

11. **Braun, V., T. Hundsberger, P. Leukel, M. Sauerborn, and C. von Eichel-Streiber.** 1996. Definition of the single integration site of the pathogenicity locus in *Clostridium difficile.* *Gene* 181:29–38.

12. **Brazier, J. S.** 1998. The diagnosis of *Clostridium difficile*-associated disease. *J. Antimicrob. Chemother.* 41:29–40.

13. **Brynestad, S., and P. E. Granum.** 1999. Evidence that Tn*5565*, which includes the enterotoxin gene in *Clostridium perfringens,* can have a circular form which may be a transposition intermediate. *FEMS Microbiol. Lett.* 170: 281–286.

14. **Brynestad, S., M. R. Sarker, B. A. McClane, P. E. Granum, and J. I. Rood.** 2001. Conjugative transfer of the plasmid carrying the *Clostridium perfringens* enterotoxin gene. *Infect. Immun.* 69:3483–3487.

15. **Carman, R. J.** 1997. *Clostridium perfringens* in spontaneous and antibiotic-associated diarrhoea of man and other animals. *Rev. Med. Microbiol.* 8(Suppl. 1):S43–S46.

16. **Chakrabarti, G., and B. A. McClane.** 2005. The importance of calcium influx, calpain, and calmodulin for the activation of CaCo-2 cell death pathways by *Clostridium perfringens* enterotoxin. *Cell. Microbiol.* 7:129–146.

17. **Chakrabarti, G., X. Zhou, and B. A. McClane.** 2003. Death pathways activated in CaCo-2 cells by *Clostridium perfringens* enterotoxin. *Infect. Immun.* 71:4260–4270.

18. **Collie, R. E., J. F. Kokai-Kun, and B. A. McClane.** 1998. Phenotypic characterization of enterotoxigenic *Clostridium perfringens* isolates from non-foodborne human gastrointestinal diseases. *Anaerobe* 4:69–79.

19. **Collie, R. E., and B. A. McClane.** 1998. Evidence that the enterotoxin gene can be episomal in *Clostridium perfringens* isolates associated with non-food-borne human gastrointestinal diseases. *J. Clin. Microbiol.* 36:30–36.

20. **Cornillot, E., B. Saint-Joanis, G. Daube, S. Katayama, P. E. Granum, B. Carnard, and S. T. Cole.** 1995. The enterotoxin gene (*cpe*) of *Clostridium perfringens* can be chromosomal or plasmid-borne. *Mol. Microbiol.* 15:639–647.

21. **Delmee, M.** 1989. *Clostridium difficile* infection in healthcare workers. *Lancet* ii:1095.

22. **Dove, C. H., S. Z. Wang, S. B. Price, C. J. Phelps, D. M. Lyerly, T. D. Wilkins, and J. L. Johnson.** 1990. Molecular characterization of the *Clostridium difficile* toxin A gene. *Infect. Immun.* 58:480–488.

23. **Dupuy, B., and A. L. Sonenshein.** 1998. Regulated transcription of *Clostridium difficile* toxin genes. *Mol. Microbiol.* 27:107–120.

24. **Fujita, K., J. Katahira, Y. Horiguchi, N. Sonoda, M. Furuse, and S. Tsukita.** 2000. *Clostridium perfringens* enterotoxin binds to the second extracellular loop of claudin-3, a tight junction integral membrane protein. *FEBS Lett.* 476:258–261.

25. **George, W. L., V. L. Sutter, D. Citron, and S. M. Finegold.** 1979. Selective and differential medium for isolation of *Clostridium difficile.* *J. Clin. Microbiol.* 9:214–219.

26. **Goncalves, C., D. Decre, F. Barbut, B. Burghoffer, and J. C. Petit.** 2004. Prevalence and characterization of a binary toxin (actin-specific ADP-ribosyltransferase) from *Clostridium difficile.* *J. Clin. Microbiol.* 42:1933–1939.

27. **Gulke, I., G. Pfeifer, J. Liese, M. Fritz, F. Hofmann, K. Aktories, and H. Barth.** 2001. Characterization of the enzymatic component of the ADP-ribosyltransferase toxin CDTa from *Clostridium difficile.* *Infect. Immun.* 69:6004–6011.

28. **Hammond, G. A., and J. L. Johnson.** 1995. The toxigenic element of *Clostridium difficile* strain VPI 10463. *Microb. Pathog.* 19:203–213.

29. **Hardy, S. P., C. Ritchie, M. C. Allen, R. H. Ashley, and P. E. Granum.** 2001. *Clostridium perfringens* type A enterotoxin forms mepacrine-sensitive pores in pure phospholipid bilayers in the absence of putative receptor proteins. *Biochim. Biophys. Acta* 1515:38–43.

30. **Hofmann, F., C. Busch, U. Prepens, I. Just, and K. Aktories.** 1997. Localization of the glucosyltransferase activity of *Clostridium difficile* toxin B to the N-terminal part of the holotoxin. *J. Biol. Chem.* 272:11074–11078.

31. **Huang, I. H., M. Waters, R. R. Grau, and M. R. Sarker.** 2004. Disruption of the gene (spoOA) encoding sporulation transcription factor blocks endospore formation and enterotoxin production in enterotoxigenic *Clostridium perfringens* type A. *FEMS Microbiol. Lett.* 233:233–240.

32. **Johnson, J. L., C. Phelps, L. Barroso, M. D. Roberts, D. M. Lyerly, and T. D. Wilkins.** 1990. Cloning and expression of the toxin B gene of *Clostridium difficile.* *Curr. Microbiol.* 20:397–401.

33. **Johnson, S., W. D. Sypura, D. N. Gerding, S. L. Ewing, and E. N. Janoff.** 1995. Selective neutralization of a bacterial enterotoxin by serum immunoglobulin A in response to mucosal disease. *Infect. Immun.* 63:3166–3173.

34. **Katahira, J., N. Inoue, Y. Horiguchi, M. Matsuda, and N. Sugimoto.** 1997. Molecular cloning and functional characterization of the receptor for *Clostridium perfringens* enterotoxin. *J. Cell Biol.* 136:1239–1247.

35. **Katahira, J., H. Sugiyama, N. Inoue, Y. Horiguchi, M. Matsuda, and N. Sugimoto.** 1997. *Clostridium perfringens* enterotoxin utilizes two structurally related membrane proteins as functional receptors in vivo. *J. Biol. Chem.* 272:26652–26658.

36. **Kim, P. H., J. P. Iaconis, and R. D. Rolfe.** 1987. Immunization of adult hamsters against *Clostridium difficile*-associated ileocecitis and transfer of protection to infant hamsters. *Infect. Immun.* 55:2984–2992.

37. **Knoop, F. C., M. Owen, and I. C. Crocker.** 1993. *Clostridium difficile*: clinical disease and diagnosis. *Clin. Microbiol. Rev.* 6:251–265.

38. **Kokai-Kun, J. F., and B. A. McClane.** 1996. Evidence that a region(s) of the *Clostridium perfringens* enterotoxin molecule remains exposed on the external surface of the mammalian plasma membrane when the toxin is sequestered in small or large complexes. *Infect. Immun.* 64:1020–1025.

39. **Kokai-Kun, J. F., and B. A. McClane.** 1997. Deletion analysis of the *Clostridium perfringens* enterotoxin. *Infect. Immun.* 65:1014–1022.

40. **Libby, J. M., B. S. Jortner, and T. D. Wilkins.** 1982. Effects of the two toxins of *Clostridium difficile* in antibiotic-associated cecitis in hamsters. *Infect. Immun.* **36:**822–829.

41. **Lyerly, D. M.** 1993. Epidemiology of *Clostridium difficile* disease. *Clin. Microbiol. Newsl.* **15:**49–52.

42. **Lyerly, D. M., L. A. Barroso, and T. D. Wilkins.** 1991. Identification of the latex-reactive protein of *Clostridium difficile* as glutamate dehydrogenase. *J. Clin. Microbiol.* **29:** 2639–2642.

43. **Lyerly, D. M., and T. D. Wilkins.** 1995. *Clostridium difficile*, p. 867–891. *In* M. J. Blaser, P. D. Smith, J. I. Ravdin, H. B. Greenberg, and R. L. Guerrant (ed.), *Infections of the Gastrointestinal Tract.* Raven Press, New York, N.Y.

44. **Magdesian, K. G., J. E. Madigan, D. C. Hirsh, S. S. Jang, Y. J. Tang, T. E. Carpenter, L. M. Hansen, and J. Silva, Jr.** 1997. *Clostridium difficile* and horses: a review. *Rev. Med. Microbiol.* **8:**S46–S48.

45. **Mani, N., D. Lyras, L. Barroso, P. Howarth, T. Wilkins, J. I. Rood, A. L. Sonenshein, and B. Dupuy.** 2002. Environmental response and autoregulation of *Clostridium difficile* txeR, a sigma factor for toxin gene expression. *J. Bacteriol.* **184:**5971–5978.

46. **McClane, B. A.** 2000. *Clostridium perfringens* enterotoxin and intestinal tight junctions. *Trends Microbiol.* **8:**145–146.

47. **McClane, B. A.** 2000. The action, genetics and synthesis of *Clostridium perfringens* enterotoxin, p. 247–272. *In* J. W. Cary, M. A. Stein, and D. Bhatnagar (ed.), *Microbial Foodborne Diseases: Mechanisms of Pathogenesis and Toxin Synthesis.* Technomic Press, Lancaster, Pa.

48. **McClane, B. A.** 2001. *Clostridium perfringens*, p. 351–372. *In* M. P. Doyle, L. R. Beuchat, and T. J. Montville (ed.), *Food Microbiology: Fundamentals and Frontiers*, 2nd ed. ASM Press, Washington, D.C.

49. **McFarland, L. V., and G. W. Elmer.** 1995. Biotherapeutic agents: past, present and future. *Microecol. Ther.* **23:** 46–73.

50. **McFarland, L. V., C. M. Surawicz, and W. E. Stamm.** 1990. Risk factors for *Clostridium difficile* carriage and C. *difficile*-associated diarrhea in a cohort of hospitalized patients. *J. Infect. Dis.* **162:**678–684.

51. **Miyamoto, K., G. Chakrabarti, Y. Morino, and B. A. McClane.** 2002. Organization of the plasmid *cpe* locus of *Clostridium perfringens* type A isolates. *Infect. Immun.* **70:** 4261–4272.

52. **Moncrief, J. S., L. A. Barroso, and T. D. Wilkins.** 1997. Positive regulation of *Clostridium difficile* toxins. *Infect. Immun.* **65:**1105–1108.

53. **Moncrief, J. S., D. M. Lyerly, and T. D. Wilkins.** 2000. Molecular biology of large clostridial toxins, p. 333–359. *In* K. Aktories and I. Just (ed.), *Handbook of Experimental Pharmacology.* Springer-Verlag, Berlin, Germany.

54. **Onderdonk, A. B.** 1988. Role of the hamster model of antibiotic-associated colitis in defining the etiology of the disease, p. 115–125. *In* R. D. Rolfe and S. M. Finegold (ed.), *Clostridium difficile: Its Role in Intestinal Disease.* Academic Press, Inc., New York, N.Y.

55. **Perelle, S., M. Gibert, P. Bourlioux, G. Corthier, and M. R. Popoff.** 1997. Production of a complete binary toxin (actin-specific ADP-ribosyltransferase) by *Clostridium difficile* CD196. *Infect. Immun.* **65:**1402–1407.

56. **Pothoulakis, C., I. Castagliuolo, C. P. Kelly, and J. T. Lamont.** 1993. *Clostridium difficile*-associated diarrhea and colitis: pathogenesis and therapy. *Int. J. Antimicrob. Agents* **3:**17–32.

57. **Rupnik, M.** 2002. Binary toxin-producing *Clostridium difficile.* *Anaerobe* **8:**164–165.

58. **Rupnik, M., V. Avesani, M. Janc, C. von Eichel-Streiber, and M. Delmee.** 1998. A novel toxinotyping scheme and correlation of toxinotypes with serogroups of *Clostridium difficile* isolates. *J. Clin. Microbiol.* **36:**2240–2247.

59. **Sarker, M. R., R. J. Carman, and B. A. McClane.** 1999. Inactivation of the gene (*cpe*) encoding *Clostridium perfringens* enterotoxin eliminates the ability of two *cpe*-positive C. *perfringens* type A human gastrointestinal disease isolates to affect rabbit ileal loops. *Mol. Microbiol.* **33:**946–958.

60. **Sarker, M. R., R. P. Shivers, S. G. Sparks, V. K. Juneja, and B. A. McClane.** 2000. Comparative experiments to examine the effects of heating on vegetative cells and spores of *Clostridium perfringens* isolates carrying plasmid versus chromosomal enterotoxin genes. *Appl. Environ. Microbiol.* **66:**3234–3240.

61. **Shih, N. J., and R. G. Labbe.** 1996. Sporulation-promoting ability of *Clostridium perfringens* culture fluids. *Appl. Environ. Microbiol.* **62:**1441–1443.

62. **Shimizu, T., K. Ohtani, H. Hirakawa, K. Ohshima, A. Yamashita, T. Shiba, N. Ogasawara, M. Hattori, S. Kuhara, and H. Hayashi.** 2002. Complete genome sequence of *Clostridium perfringens*, an anaerobic flesh-eater. *Proc. Natl. Acad. Sci. USA* **99:**996–1001.

63. **Singh, U., C. M. Van Italie, L. L. Mitic, J. M. Anderson, and B. A. McClane.** 2000. CaCo-2 cells treated with *Clostridium perfringens* enterotoxin form multiple large complex species, one of which contains the tight junction protein occludin. *J. Biol. Chem.* **275:**18407–18417.

64. **Singh, U., L. L. Mitic, E. U. Wieckowski, J. M. Anderson, and B. A. McClane.** 2001. Comparative biochemical and immunocytochemical studies reveal differences in the effects of *Clostridium perfringens* enterotoxin on polarized CaCo-2 cells versus Vero cells. *J. Biol. Chem.* **276:**33402–33412.

65. **Smedley, J. G., III, and B. A. McClane.** 2004. Fine mapping of the N-terminal cytotoxicity region of *Clostridium perfringens* enterotoxin by site-directed mutagenesis. *Infect. Immun.* **72:**6914–6923.

66. **Songer, J. G.** 1996. Clostridial enteric diseases of domestic animals. *Clin. Microbiol. Rev.* **9:**216–234.

67. **Torres, J. F.** 1991. Purification and characterization of toxin B from a strain of *Clostridium difficile* that does not produce toxin A. *J. Med. Microbiol.* **35:**40–44.

68. **Varga, J., V. L. Stirewalt, and S. B. Melville.** 2004. The CcpA protein is necessary for efficient sporulation and enterotoxin gene (*cpe*) regulation in *Clostridium perfringens.* *J. Bacteriol.* **186:**5221–5229.

69. **von Eichel-Streiber, C., M. Sauerborn, and H. K. Kuramitsu.** 1992. Evidence for a modular structure of the homologous repetitive C-terminal carbohydrate-binding sites of *Clostridium difficile* toxins and *Streptococcus mutans* glucosyltransferases. *J. Bacteriol.* **174:**6707–6710.

70. **Wieckowski, E. U., J. F. Kokai-Kun, and B. A. McClane.** 1998. Characterization of membrane-associated *Clostridium perfringens* enterotoxin following pronase treatment. *Infect. Immun.* **66:**5897–5905.

71. **Zhao, Y., and S. B. Melville.** 1998. Identification and characterization of sporulation-dependent promoters upstream of the enterotoxin gene (*cpe*) of *Clostridium perfringens.* *J. Bacteriol.* **180:**136–142.

Histotoxic Clostridia

DENNIS L. STEVENS AND JULIAN I. ROOD

58

Histotoxic clostridial infection is a general term coined over a century ago that referred to gas gangrene and malignant edema in humans and blackleg in cattle (Table 1). In the last half of the 20th century, novel histotoxic infections have been described, such as necrotic enteritis, neutropenic enterocolitis, and spontaneous gas gangrene—all of which occur exclusively in humans—and abomasal ulceration in cattle (Table 1). These infections are rapidly progressive, are associated with gas in tissue, and manifest impressive tissue destruction, shock, and frequently death (for a review, see reference 65).

EMERGING CLOSTRIDIAL INFECTIONS

During the last 2 years, a variety of histotoxic or necrotizing soft tissue infections caused by pathogenic clostridia have been described in published reports from throughout the world. Cases have been described of fatal gas gangrene caused by *Clostridium sordellii* and *Clostridium perfringens* among recipients of allografts of tendons and other connective tissue, having a frequency of 0.12% (39). Severe gas gangrene requiring enucleation has also been described in a patient receiving a cadaveric corneal transplant (21); spontaneous endophthalmitis caused by *Clostridium septicum* (54) and *C. sordellii* (84) has been recently reported. *C. perfringens* has caused necrotizing pancreatitis (77), necrotizing infections in infants (26), omental abscess (51), and spontaneous gas gangrene in neutropenic children undergoing cancer chemotherapy (78). *C. septicum*, in addition to causing traumatic gas gangrene and spontaneous gas gangrene in patients with either gastrointestinal malignancy or neutropenia, has also been recently implicated in cases of endophthalmitis and myocotic aneurysm in humans. Gas gangrene caused by *C. sordellii* and/or *C. perfringens* has been reported following intracutaneous injection of black tar heroin (3, 42). Finally, *Clostridium fallax* has recently been implicated as a cause of spontaneous myonecrosis (33).

GENERAL CHARACTERISTICS OF HISTOTOXIC CLOSTRIDIA

Although the histotoxic clostridia are classified as gram-positive, spore-forming, anaerobic bacilli, not all of them are definitely so. Specifically, these bacteria readily lose the crystal violet stain in vitro and appear as gram-negative rods after overnight culture. Similarly, in vivo smears made of gangrenous lesions invariably demonstrate predominantly gram-negative rods. Further, not all of the histotoxic clostridia are strict anaerobes. For example, *Clostridium histolyticum* is aerotolerant and will form colonies on freshly prepared blood agar medium incubated aerobically. *C. septicum* is less aerotolerant but will grow slowly at ambient oxygen tensions as well. In contrast, *C. perfringens* and *Clostridium novyi* require anaerobic conditions, the latter being inhibited by oxygen tensions of <0.05% (63).

MICROBIOLOGICAL NICHE OF HISTOTOXIC CLOSTRIDIA

The main habitats of all of the histotoxic clostridia are soils and the intestinal contents of humans and animals. *C. perfringens* is the most widespread of the histotoxic clostridia, with the quantity of organisms in soil being proportional to the degree and duration of animal husbandry in the region. For example, *C. perfringens* can reach as high as 5×10^5 per g of soil in the fertile valleys of Europe. The number of *C. perfringens* organisms carried by healthy humans has varied from 100 per g of stool among Swedish subjects to as high as 10^9 per g of stool among some Japanese subjects (49). These concepts are important, since the source of bacteria in histotoxic infections of humans is invariably related to either endogenous infection from a bowel source or contamination of deep wounds by soil. In contrast to all the other histotoxic clostridia in nature, *Clostridium chauvoei* resides largely in the gut of ruminants; hence, blackleg in cattle is invariably an infection caused by endogenous flora.

Gram-Positive Pathogens, 2nd edition, edited by Vincent A. Fischetti et al.
© 2006 ASM Press, Washington, D.C.

TABLE 1 Histotoxic clostridial infections

Type of infection	Species infected	Organism
Gas gangrene		
Traumatic	Humans	*Clostridium perfringens* type A
		Clostridium septicum
		Clostridium histolyticum
		Clostridium sordellii
		Clostridium novyi
		Clostridium fallax
Spontaneous	Humans	*Clostridium septicum*
Malignant edema	Humans	*Clostridium septicum*
Blackleg	Cattle	*Clostridium chauvoei*
Necrotic enteritis	Humans	*Clostridium perfringens* type C
Neutropenic enterocolitis	Humans	*Clostridium septicum*
Abomasal ulceration	Cattle	*Clostridium perfringens* type A

VIRULENCE FACTORS OF THE HISTOTOXIC CLOSTRIDIA

Our present understanding of the potent toxins produced by histotoxic clostridia is based upon studies done between World Wars I and II when gas gangrene was a major complication of battlefield injuries. Investigators of this period labeled the major lethal toxins of these bacteria with Greek letters; the letter alpha was always used to designate the most potent toxin. A review of these data can be found in the monograph by Smith (62). Over the next 50 years, modern technology has provided a greater understanding of mechanisms of action of some of these factors (Table 2).

STRUCTURE/FUNCTION OF THE MAJOR HISTOTOXIC CLOSTRIDIAL EXOTOXINS

The major *C. perfringens* extracellular toxins implicated in gas gangrene are alpha-toxin and theta-toxin. Alpha-toxin is a lethal, hemolytic toxin that has both phospholipase C and sphingomyelinase activities. The *C. perfringens* alpha-toxin has an N-terminal domain with sequence similarity to phospholipase C enzymes from other bacteria and a smaller β-sandwich C-terminal domain that is responsible for calcium-dependent membrane binding and hemolysis. Both domains are required for toxicity. Analysis of the alpha-toxin structure by X-ray crystallography has confirmed the two-domain structure, with the larger α-helical N-terminal domain containing the active enzymatic site. The C-terminal C2-like domain has strong structural analogy to eukaryotic phospholipid- and/or calcium-binding C2 domains that are found in intracellular second-messenger proteins and human arachidonate 5′-lipoxygenase (79). The tyrosine residues of the alpha-toxin C2-like domain interact with the calcium/phosphatidylcholine complex of eukaryotic cell membranes, leading to binding of the toxin to the membrane surface (30) and the generation of intracellular messengers such as diacylglycerol and ceramide (20).

The theta-toxin (perfringolysin O) from *C. perfringens* is a member of the cholesterol-dependent cytolysin family of pore-forming toxins. These toxins contain a conserved ECTGLAWEWWR motif in their C-terminal domain. Modification of the invariant cysteine residue leads to loss of hemolytic activity; however, this residue can be replaced by alanine without loss of function. The crystal structure of theta-toxin has also been determined (53). The protein has an unusual elongated shape with 40% β-sheet structure and four discontinuous domains. The conserved C-terminal tryptophan-rich motif described above is located in an elongated loop near the tip of domain 4. This motif appears to be close to the putative cholesterol-binding domain. Based on the crystal structure, a model for the insertion of theta-toxin into the membrane was proposed (53). This model involved the oligomerization of the theta-toxin monomers upon contact with the cell membrane, followed by insertion of the oligomer into the membrane via the formation of a hydrophobic structure involving the binding of cholesterol to the protruding domain 4, which was proposed to span the cell membrane. However, more recent fluorescence spectroscopy and cysteine-scanning mutagenesis studies have indicated that this membrane insertion model is incorrect (56, 57). Membrane insertion involves the formation of a membrane-spanning dual β-hairpin structure by conformational changes in α-helices in domain 3 that form two extended amphipathic antiparallel β-sheets (56). Domain 4 does not penetrate the membrane; instead, upon contact, vertical collapse of the elongated theta-toxin structure brings the newly formed domain 3 β-hairpins close enough to traverse the membrane (25).

Insertion of the oligomers into the membrane results in the formation of a membrane pore and leads to cell lysis. Although theta-toxin can lyse mammalian cells, genetic studies have shown that theta-toxin by itself is not essential in causing mortality, since an insertionally inactivated chromosomal structural gene (*pfoA*) mutant is still lethal in the mouse myonecrosis model (4). However, these same genetic studies as well as passive immunization studies in mice (14) suggest that theta-toxin does contribute to pathogenesis by enhancing the morbidity associated with gas gangrene, probably by its ability to modulate the inflammatory response to infection (see below) (14, 76). Other genetic data suggest that in vivo theta-toxin works synergistically with alpha-toxin (6, 27).

The other major toxin that has been extensively studied is the alpha-toxin from *C. septicum*. While its importance in pathogenesis has been demonstrated (8), its effects on host systems have not been determined. The toxin is secreted as an inactive prototoxin that is cleaved near the C terminus by eukaryotic proteases such as trypsin or membrane-bound furin to form the active toxin (29). The toxin then oligomerizes on the membrane and inserts, forming a membrane pore and resulting in colloid-osmotic lysis (55). Crystallographic analysis of the related toxin, aerolysin (9), indicates that the proteolytic cleavage site is located in an exposed loop that also appears to be present in alpha-toxin.

GENETIC REGULATION OF TOXIN PRODUCTION IN *C. PERFRINGENS*

Global Regulation of Toxin Production

One of the most common mechanisms for regulating the expression of bacterial virulence genes involves two-component signal transduction. In this process, an environmental or chemical signal, which may be specific for the human host, interacts with a sensor histidine kinase located in the bacterial cell membrane. The resultant autophosphorylation of the kinase enables it to act as a phosphodonor and leads to the phosphorylation of a cytoplasmic protein,

TABLE 2 Major virulence factors of the histotoxic clostridia

Clinical infection	Organism	Virulence factor	Mechanism of action
Traumatic gas gangrene	*Clostridium perfringens* type A	Alpha-toxin	Phospholipase C
		Theta-toxin	Cytolysin
		Kappa-toxin	Collagenase
		Mu-toxin	Hyaluronidase
		Nu antigen	Deoxyribonuclease
Traumatic gas gangrene	*Clostridium histolyticum*	Alpha-toxin	Hemolytic, cytotoxic, lethal, antigenically related to alpha-toxin of *Clostridium septicum*
		Beta-toxin	Collagenase
		Gamma-toxin	Thiol-activated protease
		Delta-toxin	Elastase
		Epsilon-toxin	Thiol-activated cytolysin
Traumatic gas gangrene	*Clostridium novyi*	Alpha-toxin	Dermonecrotic; causes gelatinous edema
		Gamma-toxin	Phospholipase C
		Delta-toxin	Thiol-activated cytolysin
Traumatic and spontaneous gas gangrene and neutropenic enterocolitis in humans	*Clostridium septicum*	Alpha-toxin	Cytotoxic, lethal, hemolytic, antigenically related to alpha-toxin of *Clostridium histolyticum*
		Delta-toxin	Thiol-activated cytolysin
Gas gangrene and malignant edema	*Clostridium sordellii*	Alpha-toxin	Phospholipase C
		Beta-toxin	"Edema-producing factor"
		Gamma-toxin	Protease
		?	Phospholipase A
		?	Lysolecithinase
		?	Thiol-activated cytolysin
Enteritis necroticans	*Clostridium perfringens* type C	Beta-toxin	Cytolytic for intestinal villi
Abomasal ulceration in cattle	*Clostridium perfringens* type A	Beta-toxin	
Blackleg in cattle	*Clostridium chauvoei*	Alpha-toxin	Lethal, necrotizing
		Beta-toxin	Deoxyribonuclease
		Gamma-toxin	Hyaluronidase
		Delta-toxin	Thiol-activated cytolysin

the cognate response regulator. The activated response regulator usually interacts with specific virulence genes to activate or repress their transcription (37).

Little is known about which *C. perfringens* genes are differentially expressed in vivo or in vitro, although it is clear that all of the major extracellular toxins are produced when the cells are grown in artificial medium. However, there is convincing experimental evidence that toxin production in *C. perfringens* is regulated by a two-component signal transduction system that consists of a sensor histidine kinase, VirS, and a response regulator, VirR (43, 58, 60). These genes are located in an operon (12, 43) on the *C. perfringens* chromosome, at a site quite distinct from the regions carrying the various toxin structural genes (24). Mutation of either *virS* or *virR* leads to the complete loss of the ability to produce theta-toxin and a reduction in the amount of alpha-toxin, collagenase (kappa-toxin), sialidase, and protease that is produced. Complementation of the mutants with the respective cloned wild-type *virS* or *virR* genes restores the wild-type toxin phenotype (43, 58). In addition, virulence studies carried out with the mouse myonecrosis model have shown that a Tn916-derived *virS* mutant has significantly reduced virulence, which is consistent with its reduced toxin production. These effects on virulence could be reversed by complementation with a functional *virS* gene (43).

Analysis of mRNA transcripts from the wild-type, mutant, and complemented derivatives has confirmed that regulation occurs at the level of transcription (12). These stud-

ies have also shown that the VirS/R system differentially activates the transcription of the various toxin genes. The theta-toxin structural gene *pfoA* is expressed from a major promoter whose expression is totally dependent upon the presence of a functional *virRS* operon (12). Subsequent studies have shown that the VirR protein binds independently to two directly repeated sequences, called VirR boxes, located immediately upstream of the *pfoA* promoter (22). Both VirR boxes are required for biological activity and the spacing between the VirR boxes and their distance from the −35 region of the promoter are critical for VirR-dependent transcriptional activation (23).

The N-terminal domain of VirR contains two regions that are conserved in the LytTR family of response regulators (45, 46, 48). Extensive mutagenesis studies have shown that these FxRxHrS and SKHR motifs are essential for biological activity and are involved in the binding of VirR to its target VirR boxes (45, 46).

The alpha-toxin structural gene *plc* is also expressed from a single promoter, but transcription from this promoter is only partially VirR dependent; there is a basal level of VirR-independent transcription. By contrast, transcription of the *colA* gene, which encodes the extracellular collagenase, is transcribed from two separate promoters, one of which is VirR dependent and one of which is VirR independent (12). These data explain why theta-toxin production is totally dependent on the VirS/R system, whereas mutation of either *virS* or *virR* only partially reduces the production of alpha-toxin or collagenase. VirR-dependent regulation of the *plc* and *colA* genes and of several other chromosomal and plasmid genes known to be regulated indirectly by the VirS/R system (10, 50) has been shown to be mediated by a regulatory RNA molecule, VR-RNA, whose expression is VirR dependent (61).

Initial studies suggested that the production of theta-toxin may be controlled not only by the VirR/S regulon but by another gene, *pfoR*, located 591 bp upstream of *pfoA* (59). However, more recent experiments have shown that the *pfoR* gene is not involved in the regulation of *pfoA* in *C. perfringens* (5).

Although some knowledge has been gained regarding the effect of the VirS/R system on its target structural genes, almost nothing is known about the signal that triggers the regulatory cascade. It is possible that a small, intercellular signaling molecule produced by *C. perfringens*, termed substance A, may function as a quorum sensor and control the expression of the VirS/R regulon (for a review, see reference 60). This hypothesis suggests that once the concentration of *C. perfringens* cells reaches a certain density in host tissues, the level of substance A becomes sufficient to bind to VirS, activate the regulatory cascade, and increase toxin production. The net effect of the increased expression of the cytolysins, hydrolytic enzymes, and toxins such as theta-toxin, alpha-toxin, and collagenase would be increased tissue destruction, with the concomitant liberation of essential nutrients.

Role of Upstream Sequences in the Regulation of Alpha-Toxin Production

The production of alpha-toxin is also regulated by a temperature-dependent mechanism that involves repeated AT-rich regions located upstream of the *plc* gene, with higher levels of alpha-toxin being produced at lower temperatures (60). Immediately upstream of the -35 box of the *plc* promoter are three directly repeated $d(A)_{5-6}$ sequences, with a 10- to 11-bp periodicity, located within a 31-bp AT-rich (30 of 31 nucleotides) region that confers significant DNA bending, suggesting that DNA topology plays a role in the regulation of alpha-toxin production (80). Deletion of the poly(A) tracts progressively decreases the amount of DNA curvature that is observed in vitro, as well as decreasing the levels of *plc*-specific mRNA observed in *C. perfringens*. In addition, disruption of the periodicity of the repeats by the insertion of bases between the repeats has similar effects. These results indicate that the AT repeats are important for optimal *plc* transcription and alpha-toxin production (44). It was postulated that the poly(A) tracts, by conferring a temperature-dependent DNA bending on the region upstream of the *plc* promoter, increased *plc* transcription and alpha-toxin production under lower-temperature saprophytic conditions, such as those experienced when the organism encounters decaying animal material in the environment (44). It has now been shown that the alpha subunit of RNA polymerase preferentially binds to these phased poly(A) tracts at lower temperatures (41), explaining the mechanistic basis for their regulatory effects (41). Increasing the supply of nutrients to the bacterial population again appears to be the major force driving the regulatory process.

Other studies have involved the comparison of a *C. perfringens* type A strain and a type C strain that produces 30 times less alpha-toxin (40). Gel mobility shift assays carried out with crude extracts from both strains suggested that the type A strain but not the type C strain contained a protein

that was able to bind a 376-bp fragment internal to the *plc* gene. Although it was suggested that this putative protein was involved in the regulation of alpha-toxin production, it has not yet been purified or identified. Comparison of other type A and B strains of *C. perfringens* also indicates that strain-dependent regulatory systems are important in the control of alpha-toxin production (19).

PATHOGENESIS OF HISTOTOXIC INFECTIONS

The events leading to full-blown histotoxic infection in humans or animals occur in three distinct stages. Stage 1 can be initiated by either bacteria or spores, although the elemental composition and oxidative reduction potential of the tissue is a critical factor in either event. In traumatic gas gangrene, this is most critical for *C. perfringens* and *C. novyi*. It is important to recognize that for these two organisms, trauma must be extensive. Indeed, 40 to 60% of cases of gas gangrene caused by these organisms are associated with trauma sufficient to interrupt the blood supply to large muscle groups. Interruption of blood flow rapidly results in a drop in oxidation reduction potential, from $+170$ to $+50$ mV. At neutral pH, the latter value is sufficient for growth of most anaerobic bacteria, including the clostridia. Anaerobic conversion of muscle glycogen into lactic acid by endogenous muscle enzymes increases the hydrogen ion concentration, allowing clostridia to proliferate at oxidation reduction potentials higher than $+50$ mV (62). Hypoxic muscle tissue releases endogenous lysosomal enzymes, converting muscle protein into peptides and amino acids that are also vital to clostridial growth and toxin production. Release of these factors is particularly critical for *C. perfringens*, which requires over 20 amino acids and vitamins for growth (63). In contrast, the more aerotolerant *C. septicum* can infect muscle and tissue with normal oxygen tensions; hence, it can cause either nontraumatic or spontaneous gas gangrene. In addition, the nutritional requirements for *C. septicum* are less demanding in terms of amino acids, although vitamins such as thiamine, riboflavin, pyridoxine, niacin, biotin, and cobalamin are required.

In either case, once the organism begins to proliferate (stage 2), the pH and redox potential decline further, providing ideal conditions for growth and toxin production. Factors known to increase production of alpha-toxin of *C. perfringens* include (i) peptides containing glycine and branched-chain amino acids such as valine- or tyrosine-containing peptides of low molecular weight; (ii) a carbohydrate source of starch, dextran or fructose, but not glucose; (iii) a pH below 7.0; and (iv) a total electrolyte concentration of 0.1 to 0.15 M, with a ratio of sodium to potassium of 2:1. Interestingly, injection of recombinant alpha-toxin results in a major drop in local pH as measured by nuclear magnetic resonance imaging (66).

Stage 3 is the consequence of the elaboration of potent extracellular toxins that cause local and regional tissue destruction and, upon reaching the systemic circulation, cause organ failure, circulatory collapse, and death. Two independent studies have shown that the alpha-toxin is an essential toxin in the disease process. First, vaccination with a purified recombinant protein consisting of the C-terminal alpha-toxin domain (amino acids 247 to 370) has been shown to protect mice from experimental *C. perfringens* infection (83). Second, an alpha-toxin (*plc*) mutant constructed by allelic exchange has been shown to be avirulent in a mouse myonecrosis model (4). Complementation of the chromosomal mutation with a recombinant plasmid carrying a wild-type *plc* gene restores the ability to cause disease, providing clear genetic evidence for the essential role of alpha-toxin in the disease process.

HISTOTOXIC CLOSTRIDIA AND THE HUMAN IMMUNE SYSTEM: THE INITIAL ENCOUNTER

Toxin production has been documented in vivo by demonstrating the progressive appearance of both theta- and alpha-toxins at the site of the experimental infection by 4 h (49). Still, continued toxin production and extension of infection are dependent upon the outcome of the initial interaction of locally proliferating bacteria with the immune system. It is clear that *C. perfringens* is capable of activation of the complement cascade with the generation of serum-derived chemotactic factors and opsonins via the alternative complement cascade (73). This suggests that *C. perfringens* is opsonized and ingested by human polymorphonuclear leukocytes (PMN) in the presence of nonimmune serum. Migration of PMN to the site of clostridial proliferation is driven by the generation of bacteria-induced serum-derived chemotactic factors (C3a and C5a) (73). This response is likely to be amplified in vivo by the generation of the potent chemokine interleukin-8 (IL-8), which is induced by bacterial exotoxins such as alpha-toxin (17). This initial and rapid response by the host arrests bacterial proliferation in the vast majority of patients with wounds contaminated with *C. perfringens*, since progression of infection to full-blown clinical gas gangrene develops in only 0.03 to 5.2% of traumatic open wounds, despite the fact that 3.8 to 39% of these wounds are contaminated with histotoxic clostridial species such as *C. perfringens* (1). Thus, clinical evidence suggests that host factors dominate in the initial encounter and overcome the constitutive and inducible arsenal of toxins produced by these microbes. Clinical factors within the wound that tip the balance in favor of bacterial proliferation include the number of bacteria, the presence of foreign material such as soil, and, importantly, interruption of arterial blood supply.

Clostridial exotoxin production in situ dramatically affects the next stage of the host response. As shown by Bryant et al. (14) with experimental models, PMN influx (chemotaxis and diapedesis) is intact and luxurious where killed *C. perfringens* organisms are injected into thigh muscles of mice (Table 3). In wild-type infection, both experimentally (14) and in human cases (64), a conspicuous absence of leukocytes at the site of *C. perfringens* proliferation is the hallmark of gas gangrene. The anti-inflammatory effect (suppression of PMN influx) by both theta-toxin and alpha-toxin in vivo has been demonstrated in studies utilizing either purified or recombinant toxins (Table 3) (14, 76) or isogenic mutants lacking theta-toxin or alpha-toxin (4, 76). The mechanism of attenuation of this host response can be inferred from in vitro studies demonstrating direct concentration-dependent effects of theta-toxin upon PMN viability (73), directed and random migration (49), ability to generate superoxide anions (14, 73), reduced ability to ingest opsonized zymosan (14) or killed *C. perfringens* (66), and heterologous desensitization of chemotactic factor receptors (14, 73). The effects of alpha-toxin on phagocyte function are no less remarkable. Alpha-toxin primes resting PMN respiratory burst in response to opsonized zymosan, with a resultant maximal generation of oxygen radical production (18, 73). Morphologically, PMN treated with alpha-toxin assume a cytokineplast form, with multiple daughter cells connected by a cytoplasmic bridge (18, 69). Premature priming of cells derails anaerobic glycolysis in favor of hexose monophosphate shunt activity, subverting generation of ATP necessary for the cytoskeletal rearrangements involved in chemotaxis and ingestion.

PROPAGATION OF TISSUE HYPOXIA BEYOND THE SITE OF INITIAL TRAUMA

Recent experimental studies by Bryant, Stevens, and colleagues (14, 75), as well as the classical clinical observations by McNee and Dunn (47), demonstrate not only an absence of PMN at the site of active infection (see above), but also accumulation or entrapment of PMN within the postcapillary venules. In vitro studies suggest that theta-toxin and alpha-toxin up-regulate endothelial adherence molecules such as P-selectin (20), E-selectin, and intracellular adhesion molecule 1 (17). Additionally, theta-toxin up-regulates the important neutrophil adherence molecule CD11b/CD18 (14). Theta- and alpha-toxins are also capable of rapidly inducing endothelial synthesis of the lipid autocoid platelet-activating factor (PAF) but by different mechanisms. PAF synthesis induced by theta-toxin is likely the consequence of increased cytosolic calcium concentrations, due to increased influx or mobilization of intracellular stores, with rapid activation of endogenous phospholipase A_2 leading to hydrolysis of phosphatidylcholine with release of arachidonic acid and incorporation of acetyl-coenzyme A into the Sn-2 position of lysophosphatidylcholine, yielding PAF (82). Alpha-toxin activates endothelial cells by direct formation of the intracellular messengers ceramide and diacylglycerol, the latter a known regulator of protein kinase C

TABLE 3 Anti-inflammatory effects of *C. perfringens* alpha- and theta-toxins[a]

Material injected	PMN influx		Tissue destruction (myonecrosis)	
	Early	Late	Early	Late
Viable, wild-type *C. perfringens*	−	−	+	++++
Killed, washed *C. perfringens*	+++	++++	−	−
Killed *C. perfringens* + alpha-toxin	−	−	++	+++
Killed *C. perfringens* + theta-toxin	−	+	++	+++

[a]Adapted from reference 76.

(20). In either case, PAF, a potent endothelial receptor for platelets and PMN, is produced. Coincubation of PMN and endothelial cells in the presence of theta-toxin demonstrates tight adherence of PMN to endothelium via a CD11b/CD18 mechanism (14). Such adherence reaches sufficient magnitude to attenuate blood flow while further damaging endothelial cells by neutrophil-dependent mechanisms such as degranulation (release of hydrolytic enzymes) and oxygen radical production (76).

In the absence of a host response sufficient to limit bacterial growth, clostridial proliferation is rapid, leading to the local accumulation of toxin. Higher in situ concentrations of *C. perfringens* toxins further inhibit PMN influx and reach concentrations sufficient to cause cytotoxicity to emigrated cells. Absorption of toxin in postcapillary venules most likely activates the endothelium lining these vessels, resulting in propagation of the previously described endothelial cell-PMN interactions to more proximal tissue.

In vivo alpha-toxin induced rapid and irreversible decline in tissue perfusion, as measured by laser Doppler blood flow analysis (15). Interestingly, blood flow dropped by 50% within 10 min of toxin injection in a rat model of tissue perfusion and was associated with the appearance by 11 min of intravascular aggregates of platelets, neutrophils, and fibrin. Using flow cytometry, Bryant et al. demonstrated that these aggregates were the consequence of alpha-toxin-induced activation of gpIIbIIIa receptors on platelets (13). Neutralization of alpha-toxin (13) or addition of competitive inhibitors of gpIIbIIIa prevented these aggregates in vitro. In vivo, alpha-toxin induced platelet-neutrophil complexes did not form in neutropenic animals and could be prevented in animals treated with heparin (16). Taken together, these results demonstrate important new targets for treatment, which could prevent or attenuate vascular compromise in patients with gas gangrene caused by *C. perfringens*.

PATHOGENESIS OF SHOCK AND ORGAN FAILURE

Cardiovascular collapse and end organ failure occur late in the course of gas gangrene caused by histotoxic clostridia such as *C. perfringens* and *C. septicum*. "Tachycardia with feebleness of the pulse has followed the onset of pain and characteristically, has been out of proportion to the degree of elevation of the temperature" (1). This description offers a striking contrast to the early manifestations of septic shock caused by gram-negative bacteria such as *Escherichia coli*, in which a hyperdynamic picture is most common even in the face of low blood pressure. For example, a rapid heart rate is generally associated with high cardiac output and a bounding pulse. In gas gangrene, at the onset of tachycardia, the blood pressure is normal but then drops precipitously. Although these descriptions from the 1970s are useful, the subsequent clinical literature has not offered more modern measurements of cardiovascular parameters in humans with clostridial gas gangrene. Experimental studies of animals do provide some useful insights into the dynamics of cardiovascular dysfunction induced by *C. perfringens* toxins. Initial studies demonstrated that purified as well as recombinant theta- and alpha-toxins were lethal to mice by intravenous injection, with alpha-toxin having the greatest potency.

In an awake rabbit model, recombinant theta-toxin, but not recombinant alpha-toxin, caused a significant increase in heart rate that was apparent within 60 min of toxin infusion (2). Interestingly, theta-toxin did not cause a signifi-

cant drop in mean arterial pressure (Fig. 1). In contrast, crude toxin preparations induced a dramatic drop in mean arterial blood pressure within 2 h after the infusion of toxin (Fig. 1). Alpha-toxin also caused a significant reduction in blood pressure (Fig. 1), but the onset of alpha-toxin-induced hypotension was 1 h later (i.e., at 180 min posttoxin infusion) than that caused by the crude toxin (120 min posttoxin infusion). These data suggest additive or even synergistic interactions of toxins such as alpha- and theta-toxins. Both recombinant alpha-toxin and crude toxin caused significant drops in cardiac index as early as 90 min after infusion of the respective toxins (Fig. 2) (2).

MOLECULAR MECHANISMS OF TOXIN-INDUCED SHOCK

As reflected by increased mortality in rabbits receiving recombinant alpha-toxin or crude toxin, a greater and more rapid reduction in cardiac index was also measured in these groups than in those receiving recombinant theta-toxin or normal saline (Fig. 2) (2). These data suggest that one mechanism of cardiovascular dysfunction induced by alpha-toxin is related to direct myocardial toxicity. Indeed, a direct and dose-dependent reduction in myocardial contractility (*df/dt*) occurred in isolated atrial strips bathed with recombinant alpha-toxin (2). Alpha-toxin has also been shown to affect the inotropic cardiac response in isolated embryonic chick heart preparations (52). However, alpha-toxin may also contribute indirectly to shock by stimulating production of endogenous mediators such as tumor necrosis factor (TNF) (Fig. 3 and 4) (67) and PAF (20).

Theta-toxin reduces systemic vascular resistance and markedly increases cardiac output (2, 75). This afterload reduction occurs, undoubtedly, through induction of endogenous mediators that cause relaxation of blood vessel wall tension such as PAF and prostaglandin I_2 (82). Reduced vascular tone develops rapidly and, to maintain adequate tissue perfusion, a compensatory host response is required to either increase cardiac output or rapidly expand the intravascular blood volume. In contrast, patients with gram-negative sepsis compensate for hypotension by markedly increasing cardiac output; however, this adaptive mechanism may not be possible in *C. perfringens*-induced shock, owing to direct suppression of myocardial contractility by alpha-toxin (75). Theta-toxin, like other cholesterol-dependent cytolysins, likely contributes to septic shock through additional indirect routes, including augmented release of TNF, IL-1, and IL-6 (31, 38). The roles of TNF, IL-1, and IL-6, as well as of the potent endogenous vasodilators bradykinin and nitric oxide, in shock associated with *C. perfringens* gas gangrene have not been elucidated.

TREATMENT OF *C. PERFRINGENS* GAS GANGRENE

Aggressive debridement of devitalized tissue, as well as rapid repair of compromised vascular supply, and prophylactic antibiotics greatly reduce the frequency of gas gangrene in contaminated deep wounds (11, 28, 64, 65). Intramuscular epinephrine, prolonged application of tourniquets, and surgical closure of traumatic wounds should be avoided. Patients presenting with gas gangrene of an extremity have a better prognosis than those with truncal or intra-abdominal gas gangrene, largely because it is difficult to adequately debride such lesions (11, 28, 32). In addition, patients with

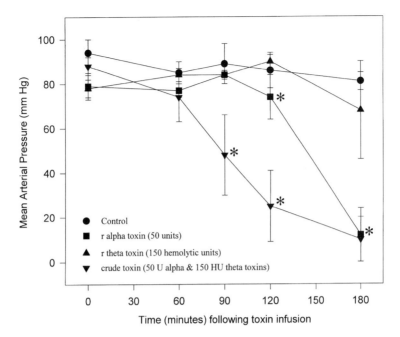

FIGURE 1 Effects of clostridial exotoxins on mean arterial pressure. Rabbits with stable vital signs were given intravenous infusions of normal saline, crude toxin preparation, or recombinant alpha-toxin or theta-toxin. Mean arterial pressures were measured continuously via a catheter placed in the carotid arterty. Each data point represents the mean arterial pressure (± standard error) determined by using six animals, with triplicate determinations for each time point. Asterisks indicate values significantly different from control values (P < 0.05) using Student's test. Circles, control (normal saline); squares, recombinant alpha-toxin; triangles, recombinant theta-toxin; inverted triangles, crude toxin. (Reprinted, with permission, from reference 2.)

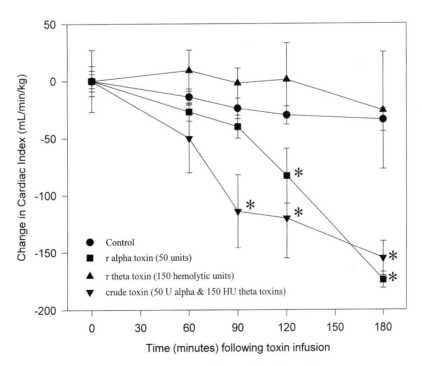

FIGURE 2 Effects of clostridial exotoxins on cardiac index. Rabbits with stable vital signs were given an intravenous infusion of normal saline, crude toxin preparation, or recombinant alpha-toxin or theta-toxin. Cardiac index was measured over a 3-h period using a thermodilution technique. Each data point represents the mean cardiac index (± standard error) determined by using six animals, with triplicate determinations for each time point. Asterisks indicate values significantly different from control values (P < 0.05). Circles, control (normal saline); squares, recombinant alpha-toxin; triangles, recombinant theta-toxin; inverted triangles, crude toxin. (Reprinted, with permission, from reference 2.)

associated bacteremia and intravascular hemolysis have the greatest likelihood of progressing to shock and death. Patients who are in shock at the time diagnosis is made have the highest rates of mortality (32).

SUPPRESSION OF TOXIN SYNTHESIS BY SPECIFIC CLASSES OF ANTIBIOTICS CORRELATES WITH IN VIVO EFFICACY

Based strictly on in vitro susceptibility data, most textbooks state that penicillin is the drug of choice (11, 28). However, experimental studies in mice suggest that clindamycin has the greatest efficacy and penicillin has the least (70, 71). Other agents with greater efficacy than penicillin include erythromycin, rifampin, tetracycline, chloramphenicol, and metronidazole (70, 71). Slightly higher rates of survival have been observed with animals receiving both clindamycin and penicillin; in contrast, antagonism was observed with penicillin plus metronidazole (70). Because some strains (2 to 5%) are resistant to clindamycin, a combination of penicillin and clindamycin is warranted. Based on his experimental studies and his vast clinical experience with gas gangrene, William Altemeier also recommended a protein synthesis inhibitor, tetracycline, plus penicillin (1). Thus, given an absence of efficacy data from a clinical trial with humans, the best treatment would appear to be a combination of antibiotics such as clindamycin and penicillin.

The failure of penicillin in experimental clostridial myonecrosis may be related to continued toxin production by filamentous forms of the organism induced by this cell wall-active agent (72). In contrast, the efficacy of clindamycin and tetracycline may be related to their ability to rapidly inhibit toxin synthesis (72).

The use of hyperbaric oxygen (HBO) is controversial, although some nonrandomized studies have reported excellent results with HBO therapy when combined with antibiotics and surgical debridement (7, 32, 34). Some experimental

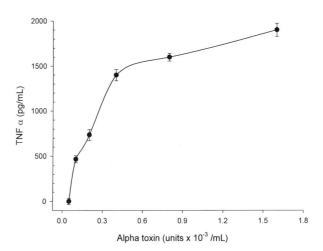

FIGURE 4 Alpha-toxin-induced TNF-α production by human mononuclear cells. TNF-α was measured in supernatant from 10^6 human mononuclear cells stimulated with recombinant alpha-toxin. Each data point represents the mean TNF-α production (± standard error) on samples collected at 24 h and assayed in duplicate using a commercial enzyme-linked immunosorbent assay. (Reprinted, with permission, from reference 43.)

studies demonstrated that HBO alone was an effective treatment if the inoculum was small and treatment was begun immediately (36). In contrast, other studies have demonstrated that HBO was only of slight benefit when combined with penicillin (68). In these studies, however, survival was greater with clindamycin alone than with either HBO alone, penicillin alone, or HBO plus penicillin (68). The benefit of HBO, at least theoretically, is to inhibit bacterial growth (35), preserve marginally perfused tissue, and inhibit toxin production (81). Interestingly, Altemeier and Fullen did not use HBO in patients and were able to realize a mortality rate of <15% by using only surgical debridement and antibiotics (tetracycline plus penicillin) (1).

Therapeutic strategies directed against toxin expression in vivo, such as neutralization with specific antitoxin antibody or inhibition of toxin synthesis, may be valuable adjuncts to traditional antimicrobial regimens. Currently, antitoxin is no longer available. Future strategies may target endogenous proadhesive molecules, such that toxin-induced vascular leukostasis and resultant tissue injury are attenuated.

PREVENTION

Recent studies have demonstrated that immunization of mice with the C-domain of alpha-toxin provided 80 to 90% protection against challenge with 3.75×10^9 or 3.75×10^8 CFU of log-phase *C. perfringens* cells (74). In contrast, none of the sham-immunized animals so challenged survived. Furthermore, immunization with the C-domain localized the infection and prevented ischemia of the feet. Histopathologic findings demonstrated limited muscle necrosis, reduced microvascular thrombosis, and enhanced granulocytic influx in C-domain immunized mice. Thus, immunization with the C-domain of alpha-toxin is a viable strategy for the prevention of gas gangrene caused by *C. perfringens*.

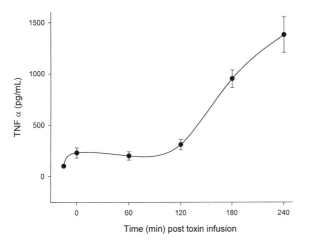

FIGURE 3 In vivo TNF-α production is induced by *C. perfringens* alpha-toxin. Serum samples were obtained from rabbits infused with recombinant alpha-toxin. TNF-α was measured by enzyme-linked immunosorbent assay. Each data point represents the mean TNF-α production (± standard error) of duplicate serum samples run in triplicate. (Reprinted, with permission, from reference 43.)

REFERENCES

1. **Altemeier, W. A., and W. D. Fullen.** 1971. Prevention and treatment of gas gangrene. *JAMA* **217:**806–813.

2. **Asmuth, D. A., R. D. Olson, S. P. Hackett, A. E. Bryant, R. K. Tweten, J. Y. Tso, T. Zollman, and D. L. Stevens.** 1995. Effects of *Clostridium perfringens* recombinant and crude phospholipase C and theta toxins on rabbit hemodynamic parameters. *J. Infect. Dis.* **172:**1317–1323.

3. **Assadian, O., A. Assadian, C. Senekowitsch, A. Makristathis, and G. Hagmuller.** 2004. Gas gangrene due to Clostridium perfringens in two injecting drug users in Vienna, Austria. *Wien. Klin. Wochenschr.* **116:**264–267.

4. **Awad, M. M., A. E. Bryant, D. L. Stevens, and J. I. Rood.** 1995. Virulence studies on chromosomal α-toxin and θ-toxin mutants constructed by allelic exchange provide genetic evidence for the essential role of α-toxin in *Clostridium perfringens*-mediated gas gangrene. *Mol. Microbiol.* **15:**191–202.

5. **Awad, M. M., and J. I. Rood.** 2002. Perfringolysin O expression in *Clostridium perfringens* is independent of the upstream *pfoR* gene. *J. Bacteriol.* **184:**2034–2038.

6. **Awad, M. M., D. M. Ellemor, R. L. Boyd, J. J. Emmins, and J. I. Rood.** 2001. Synergistic effects of alpha-toxin and perfringolysin O in *Clostridium perfringens*-mediated gas gangrene. *Infect. Immun.* **69:**7904–7910.

7. **Bakker, D. J.** 1988. Clostridial myonecrosis, p. 153–172. *In* J. C. Davis and T. K. Hunt (ed.), *Problem Wounds: the Role of Oxygen.* Elsevier Science Publishing, Inc., New York, N.Y.

8. **Ballard, J., A. Bryant, D. Stevens, and R. K. Tweten.** 1992. Purification and characterization of the lethal toxin (alpha-toxin) of *Clostridium septicum. Infect. Immun.* **60:**784–790.

9. **Ballard, J., B. A. Crabtree, J. Roe, and R. K. Tweten.** 1995. The primary structure of *Clostridium septicum* alphatoxin exhibits similarity with that of *Aeromonas hydrophila* aerolysin. *Infect. Immun.* **63:**340–344.

10. **Banu, S., K. Ohtani, H. Yaguchi, T. Swe, S. T. Cole, H. Hayashi, and T. Shimizu.** 2000. Identification of novel VirR/VirS-regulated genes in *Clostridium perfringens. Mol. Microbiol.* **35:**854–864.

11. **Bartlett, J. G.** 1990. Gas gangrene (other clostridium-associated diseases), p. 1851–1860. *In* G. L. Mandell, R. G. Douglas, and J. E. Bennett (ed.), *Principles and Practice of Infectious Diseases.* Churchill Livingstone, Ltd., New York, N.Y.

12. **Ba-Thein, W., M. Lyristis, K. Ohtani, I. T. Nisbet, H. Hayashi, J. I. Rood, and T. Shimizu.** 1996. The *virR/virS* locus regulates the transcription of genes encoding extracellular toxin production in *Clostridium perfringens. J. Bacteriol.* **178:**2514–2520.

13. **Bryant, A. E., C. R. Bayer, S. M. Hayes-Schroer, and D. L. Stevens.** 2003. Activation of platelet gpIIbIIIa by phospholipase C from *Clostridium perfringens* involves store-operated calcium entry. *J. Infect. Dis.* **187:**408–417.

14. **Bryant, A. E., R. Bergstrom, G. A. Zimmerman, J. L. Salyer, H. R. Hill, R. K. Tweten, H. Sato, and D. L. Stevens.** 1993. *Clostridium perfringens* invasiveness is enhanced by effects of theta toxin upon PMNL structure and function: the roles of leukocytotoxicity and expression of CD11/CD18 adherence glycoprotein. *FEMS Immunol. Med. Microbiol.* **7:**321–336.

15. **Bryant, A. E., R. Y. Chen, Y. Nagata, Y. Wang, C. H. Lee, S. Finegold, P. H. Guth, and D. L. Stevens.** 2000. Clostridial gas gangrene I: cellular and molecular mechanisms of microvascular dysfunction induced by exotoxins of *C. perfringens. J. Infect. Dis.* **182:**799–807.

16. **Bryant, A. E., R. Y. Chen, Y. Nagata, Y. Wang, C. H. Lee, S. Finegold, P. H. Guth, and D. L. Stevens.** 2000. Clostridial gas gangrene II: phospholipase C-induced activation of platelet gpIIb/IIIa mediates vascular occlusion and myonecrosis in *C. perfringens* gas gangrene. *J. Infect. Dis.* **182:**808–815.

17. **Bryant, A. E., and D. L. Stevens.** 1996. Phospholipase C and perfringolysin O from *Clostridium perfringens* upregulate endothelial cell-leukocyte adherence molecule 1 and intercellular leukocyte adherence molecule 1 expression and induce interleukin-8 synthesis in cultured human umbilical vein endothelial cells. *Infect. Immun.* **64:**358–362.

18. **Bryant, A., D. Stevens, and J. Tso.** 1991. Effects of alpha toxin from *Clostridium perfringens* (Cp) on PMNL, abstr. B-311, p. 77. *Abstr. 91st Gen. Meet. Am. Soc. Microbiol. 1991.* American Society for Microbiology, Washington, D.C.

19. **Bullifent, H. L., A. Moir, M. M. Awad, P. T. Scott, and R. W. Titball.** 1996. The level of expression of α-toxin by different strains of *Clostridium perfringens* is dependent upon differences in promoter structure and genetic background. *Anaerobe* **2:**365–371.

20. **Bunting, M., D. E. Lorant, A. E. Bryant, G. A. Zimmerman, T. M. McIntyre, D. L. Stevens, and S. M. Prescott.** 1997. Alpha toxin from *Clostridium perfringens* induces proinflammatory changes in endothelial cells. *J. Clin. Investig.* **100:**565–574.

21. **Centers for Disease Control and Prevention.** 2003. Clostridial endophthalmitis after cornea transplantation—Florida,. *Morb. Mortal. Wkly. Rep.* **52:**1176–1179.

22. **Cheung, J. K., and J. I. Rood.** 2000. The VirR response regulator from *Clostridium perfringens* binds independently to two imperfect direct repeats located upstream of the *pfoA* promoter. *J. Bacteriol.* **182:**57–66.

23. **Cheung, J. K., B. Dupuy, D. S. Deveson, and J. I. Rood.** 2004. The spatial organization of the VirR boxes is critical for VirR-mediated expression of the perfringolysin O gene, *pfoA*, from *Clostridium perfringens. J. Bacteriol.* **186:**3321–3330.

24. **Cole, S. T., and B. Canard.** 1997. Structure, organization and evolution of the genome of *Clostridium perfringens*, p. 49–63. *In* J. I. Rood, B. A. McClane, J. G. Songer, and R. W. Titball (ed.), *The Clostridia: Molecular Biology and Pathogenesis.* Academic Press, Inc., London, United Kingdom.

25. **Czajkowsky, D. M., E. M. Hotze, Z. Shao, and R. K. Tweten.** 2004. Vertical collapse of a cytolysin prepore moves its transmembrane beta-hairpins to the membrane. *EMBO J.* **23:**3206–3215.

26. **de la Cochetiere, M. F., H. Piloquet, C. des Robert, D. Darmaun, J. P. Galmiche, and J. C. Roze.** 2004. Early intestinal bacterial colonization and necrotizing enterocolitis in premature infants: the putative role of clostridium. *Pediatr. Res.* **56:**366–370.

27. **Ellemor, D. M., R. N. Baird, M. M. Awad, R. L. Boyd, J. I. Rood, and J. J. Emmins.** 1999. Use of genetically manipulated strains of *Clostridium perfringens* reveals that both alpha-toxin and theta-toxin are required for vascular leukostasis to occur in experimental gas gangrene. *Infect. Immun.* **67:**4902–4907.

28. **Gorbach, S. L.** 1992. *Clostridium perfringens* and other clostridia, p. 1587–1596. *In* S. L. Gorbach, J. G. Bartlett, and N. R. Blacklow (ed.), *Infectious Diseases.* The W. B. Saunders Co., Philadelphia, Pa.

29. Gordon, V., R. Benz, K. Fujii, S. Leppla, and R. Tweten. 1997. *Clostridium septicum* alpha-toxin is proteolytically activated by furin. *Infect. Immun.* 65:4130–4134.

30. Guillouard, I., P. M. Alzari, B. Saliou, and S. T. Cole. 1997. The carboxy-terminal C2-like domain of the α-toxin from *Clostridium perfringens* mediates calcium-dependent membrane recognition. *Mol. Microbiol.* 26:867–876.

31. Hackett, S. P., and D. L. Stevens. 1992. Streptococcal toxic shock syndrome: synthesis of tumor necrosis factor and interleukin-1 by monocytes stimulated with pyrogenic exotoxin A and streptolysin O. *J. Infect. Dis.* 165:879–885.

32. Hart, G. B., R. C. Lamb, and M. B. Strauss. 1983. Gas gangrene. I. A collective review. *J. Trauma* 23:991–1000.

33. Hausmann, R., F. Albert, W. Geissdorfer, and P. Betz. 2004. *Clostridium fallax* associated with sudden death in a 16-year-old boy. *J. Med. Microbiol.* 53:581–583.

34. Heimbach, R. D., I. Boerema, W. H. Brummelkamp, and W. G. Wolfe. 1977. Current therapy of gas gangrene, p. 153–176. *In* J. C. Davis and T. K. Hunt (ed.), *Hyperbaric Oxygen Therapy.* Undersea Medical Society, Bethesda, Md.

35. Hill, G. B., and S. Osterhout. 1972. Experimental effects of hyperbaric oxygen on selected clostridial species. I. In vitro studies. *J. Infect. Dis.* 125:17–25.

36. Hill, G. B., and S. Osterhout. 1972. Experimental effects of hyperbaric oxygen on selected clostridial species. II. In vivo studies on mice. *J. Infect. Dis.* 125:26–35.

37. Hoch, J. A., and T. J. Silhavy (ed.). 1995. *Two-Component Signal Transduction.* American Society for Microbiology, Washington, D.C.

38. Houldsworth, S., P. W. Andrew, and T. J. Mitchell. 1994. Pneumolysin stimulates production of tumor necrosis factor alpha and interleukin-1β by human mononuclear phagocytes. *Infect. Immun.* 62:1501–1503.

39. Kainer, M. A., J. V. Linden, D. N. Whaley, H. T. Holmes, W. R. Jarvis, D. B. Jernigan, and L. K. Archibald. 2004. Clostridium infections associated with musculoskeletal-tissue allografts. *N. Engl. J. Med.* 350:2564–2571.

40. Katayama, S. I., O. Matsushita, J. Minami, S. Mizobuchi, and A. Okabe. 1993. Comparison of the alpha-toxin genes of *Clostridium perfringens* type A and C strains: evidence for extragenic regulation of transcription. *Infect. Immun.* 61:457–463.

41. Katayama, S., O. Matsushita, E. Tamai, S. Miyata, and A. Okabe. 2001. Phased A-tracts bind to the alpha subunit of RNA polymerase with increased affinity at low temperature. *FEBS Lett.* 509:235–238.

42. Kimura, A. C., J. I. Higa, R. M. Levin, G. Simpson, Y. Vargas, and D. J. Vugia. 2004. Outbreak of necrotizing fasciitis due to Clostridium sordellii among black-tar heroin users. *Clin. Infect. Dis.* 38:e87–e91.

43. Lyristis, M., A. E. Bryant, J. Sloan, M. M. Awad, I. T. Nisbet, D. L. Stevens, and J. I. Rood. 1994. Identification and molecular analysis of a locus that regulates extracellular toxin production in *Clostridium perfringens. Mol. Microbiol.* 12:761–777.

44. Matsushita, C., O. Matsushita, S. Katayama, J. Minami, K. Takai, and A. Okabe. 1996. An upstream activating sequence containing curved DNA involved in activation of the *Clostridium perfringens plc* promoter. *Microbiology* 142:2561–2566.

45. McGowan, S., I. S. Lucet, J. K. Cheung, M. M. Awad, J. C. Whisstock, and J. I. Rood. 2002. The FxRxHrS Motif: a conserved region essential for DNA binding of the VirR response regulator from *Clostridium perfringens. J. Mol. Biol.* 322:997–1011.

46. McGowan, S., J. R. O'Connor, J. K. Cheung, and J. I. Rood. 2003. The SKHR motif is required for biological function of the VirR response regulator from *Clostridium perfringens. J. Bacteriol.* 185:6205–6208.

47. McNee, J. W., and J. S. Dunn. 1917. The method of spread of gas gangrene into living muscle. *Br. Med. J.* 1:727–729.

48. Nikolskaya, A. N., and M. Y. Galperin. 2002. A novel type of conserved DNA-binding domain in the transcriptional regulators of the AlgR/AgrA/LytR family. *Nucleic Acids Res.* 30:2453–2459.

49. Noyes, H. E., W. L. Pritchard, F. B. Brinkley, and J. A. Mendelson. 1964. Analyses of wound exudates for clostridial toxins. *J. Bacteriol.* 87:623–629.

50. Ohtani, K., H. I. Kawsar, K. Okumura, H. Hayashi, and T. Shimizu. 2003. The VirR/VirS regulatory cascade affects transcription of plasmid-encoded putative virulence genes in *Clostridium perfringens* strain 13. *FEMS Microbiol. Lett.* 222:137–141.

51. Otagiri, N., J. Soeda, T. Yoshino, H. Chisuwa, H. Aruga, H. Kasai, M. Komatsu, T. Ohmori, K. Tauchi, and H. Koike. 2004. Primary abscess of the omentum: report of a case. *Surg. Today* 34:261–264.

52. Regal, J. F., and F. E. Shigeman. 1980. The effect of phospholipase C on the responsiveness of cardiac receptors. I. Inhibition of the adrenergic inotropic response. *J. Pharmacol. Exp. Ther.* 214:282–290.

53. Rossjohn, J., S. C. Feil, W. J. McKinstry, R. K. Tweten, and M. W. Parker. 1997. Structure of a cholesterol-binding, thiol-activated cytolysin and a model of its membrane form. *Cell* 88:685–692.

54. Schickner, D. C., A. Yarkoni, P. Langer, L. Frohman, X. Chen, R. Folberg, and L. V. Del Priore. 2004. Panophthalmitis due to clostridium septicum. *Am. J. Ophthalmol.* 137:942–944.

55. Sellman, B. R., B. L. Kagan, and R. K. Tweten. 1997. Generation of a membrane-bound, oligomerized pre-pore complex is necessary for pore formation by *Clostridium septicum* alpha toxin. *Mol. Microbiol.* 23:551–558.

56. Shatursky, O., A. Heuck, L. Shepard, J. Rossjohn, M. Parker, A. Johnson, and R. Tweten. 1999. The mechanism of membrane insertion of a cholesterol-dependent cytolysin: a novel paradigm for pore-foring toxins. *Cell* 99:293–299.

57. Shepard, L. A., A. P. Heuck, B. D. Hamman, J. Rossjohn, M. W. Parker, K. R. Ryan, A. E. Johnson, and R. K. Tweten. 1998. Identification of a membrane-spanning domain of the thiol-activated pore-forming toxin *Clostridium perfringens* perfingolysin O: an alpha-helical to beta-sheet transition identified by fluorescence spectroscopy. *Biochemistry* 37:14563–14574.

58. Shimizu, T., W. Ba-Thein, M. Tamaki, and H. Hayashi. 1994. The *virR* gene, a member of a class of two-component response regulators, regulates the production of perfringolysin O, collagenase, and hemagglutinin in *Clostridium perfringens. J. Bacteriol.* 176:1616–1623.

59. Shimizu, T., A. Okabe, J. Minami, and H. Hayashi. 1991. An upstream regulatory sequence stimulates expression of the perfringolysin O gene of *Clostridium perfringens. Infect. Immun.* 59:137–142.

60. Shimizu, T., A. Okabe, and J. I. Rood. 1997. Regulation of toxin production in *C. perfringens*, p. 451–470. *In* J. I.

Rood, B. A. McClane, J. G. Songer, and R. W. Titball (ed.), *The Clostridia: Molecular Biology and Pathogenesis*. Academic Press, Inc., London, United Kingdom.

61. **Shimizu, T., H. Yaguchi, K. Ohtani, S. Banu, and H. Hayashi.** 2002. Clostridial VirR/VirS regulon involves a regulatory RNA molecule for expression of toxins. *Mol. Microbiol.* **43:**257–265.

62. **Smith, L. D. S.** 1975. Clostridial wound infections, p. 321–324. *In* L. D. S. Smith (ed.), *The Pathogenic Anaerobic Bacteria*. Charles C. Thomas, Springfield, Ill.

63. **Smith, L. D. S.** 1975. Clostridium, p. 109–114. *In* L. D. S. Smith (ed.), *The Pathogenic Anaerobic Bacteria*. Charles C. Thomas, Springfield, Ill.

64. **Stevens, D. L.** 1995. Clostridial infections, p. 13.1–13.9. *In* D. L. Stevens and G. L. Mandell (ed.), *Atlas of Infectious Diseases*. Churchill Livingstone, Ltd., Philadelphia, Pa.

65. **Stevens, D. L.** 1996. Clostridial myonecrosis and other clostridial diseases, p. 2090–2093. *In* J. C. Bennett and F. Plum (ed.), *Cecil Textbook of Medicine*. The W. B. Saunders Co., Philadelphia, Pa.

66. **Stevens, D. L.** Unpublished data.

67. **Stevens, D. L., and A. E. Bryant.** 1997. Pathogenesis of *Clostridium perfringens* infection: mechanisms and mediators of shock. *Clin. Infect. Dis.* **25:**S160–S164.

68. **Stevens, D. L., A. E. Bryant, K. Adams, and J. T. Mader.** 1993. Evaluation of hyperbaric oxygen therapy for treatment of experimental *Clostridium perfringens* infection. *Clin. Infect. Dis.* **17:**231–237.

69. **Stevens, D. L., A. E. Gibbons, and R. A. Bergstrom.** 1989. Ultrastructural changes in human granulocytes induced by purified exotoxins from *Clostridium perfringens*, abstr. J-17, p. 244. Abstr. 89th Annu. Meet. Am. Soc. Microbiol. 1989. American Society for Microbiology, Washington, D.C.

70. **Stevens, D. L., B. M. Laine, and J. E. Mitten.** 1987. Comparison of single and combination antimicrobial agents for prevention of experimental gas gangrene caused by *Clostridium perfringens*. *Antimicrob. Agents Chemother.* **31:**312–316.

71. **Stevens, D. L., K. A. Maier, B. M. Laine, and J. E. Mitten.** 1987. Comparison of clindamycin, rifampin, tetracycline, metronidazole, and penicillin for efficacy in prevention of experimental gas gangrene due to *Clostridium perfringens*. *J. Infect. Dis.* **155:**220–228.

72. **Stevens, D. L., K. A. Maier, and J. E. Mitten.** 1987. Effect of antibiotics on toxin production and viability of *Clostridium perfringens*. *Antimicrob. Agents Chemother.* **31:**213–218.

73. **Stevens, D. L., J. Mitten, and C. Henry.** 1987. Effects of alpha and theta toxins from *Clostridium perfringens* on human polymorphonuclear leukocytes. *J. Infect. Dis.* **156:**324–333.

74. **Stevens, D. L., R. W. Titball, M. Jepson, C. R. Bayer, S. M. Hayes-Schroer, and A. E. Bryant.** 2004. Immunization with the C-domain of alpha-toxin prevents lethal infection, localizes tissue injury, and promotes host response to challenge with Clostridium perfringens. *J. Infect. Dis.* **190:**767–773.

75. **Stevens, D. L., B. E. Troyer, D. T. Merrick, J. E. Mitten, and R. D. Olson.** 1988. Lethal effects and cardiovascular effects of purified alpha- and theta-toxins from *Clostridium perfringens*. *J. Infect. Dis.* **157:**272–279.

76. **Stevens, D. L., R. K. Tweten, M. M. Awad, J. I. Rood, and A. E. Bryant.** 1997. Clostridial gas gangrene: evidence that alpha and theta toxins differentially modulate the immune response and induce acute tissue necrosis. *J. Infect. Dis.* **176:**189–195.

77. **Stockinger, Z. T., and R. L. Corsetti.** 2004. Pneumoperitoneum from gas gangrene of the pancreas: three unusual findings in a single case. *J. Gastrointest. Surg.* **8:**489–492.

78. **Temple, A. M., and N. J. Thomas.** 2004. Gas gangrene secondary to Clostridium perfringens in pediatric oncology patients. *Pediatr. Emerg. Care* **20:**457–459.

79. **Titball, R. W., D. L. Leslie, S. Harvey, and D. Kelly.** 1991. Hemolytic and sphingomyelinase activities of *Clostridium perfringens* alpha-toxin are dependent on a domain homologous to that of an enzyme from the human arachidonic acid pathway. *Infect. Immun.* **59:**1872–1874.

80. **Toyonaga, T., O. Matsushita, S.-I. Katayama, J. Minami, and A. Okabe.** 1992. Role of the upstream regulon containing an intrinsic DNA curvature in the negative regulation of the phospholipase C gene of *Clostridium perfringens*. *Microbiol. Immunol.* **36:**603–613.

81. **van Unnik, A. J. M.** 1965. Inhibition of toxin production in *Clostridium perfringens* in vitro by hyperbaric oxygen. *Antonie Leeuwenhoek* **31:**181–186.

82. **Whatley, R. E., G. A. Zimmerman, D. L. Stevens, C. J. Parker, T. M. McIntyre, and S. M. Prescott.** 1989. The regulation of platelet activating factor production in endothelial cells—the role of calcium and protein kinase C. *J. Biol. Chem.* **264:**6325–6333.

83. **Williamson, E. D., and R. W. Titball.** 1993. A genetically engineered vaccine against alpha-toxin of *Clostridium perfringens* protects against experimental gas gangrene. *Vaccine* **11:**1253–1258.

84. **Zink, J. M., R. Singh-Parikshak, A. Sugar, and M. W. Johnson.** 2004. *Clostridium sordellii* endophthalmitis after suture removal from a corneal transplant. *Cornea* **23:**522–523.

Corynebacterium diphtheriae: Iron-Mediated Activation of DtxR and Regulation of Diphtheria Toxin Expression

JOHN F. LOVE AND JOHN R. MURPHY

59

During the 1990s, the Newly Independent States (NIS) of the former Soviet Union experienced a sweeping epidemic of diphtheria. The principal factor contributing to this outbreak was the decreased proportion of the public vaccinated against diphtheria toxin, the primary virulence factor produced by toxigenic strains of *Corynebacterium diphtheriae*. Although the diphtheria toxoid vaccine has been available since the 1920s and is widely used in most industrialized countries, in 1990 only 68% of Russian children had received appropriate vaccination (47). In contrast, in the United States the childhood vaccination rates against diphtheria during the same period were 90 to 95% (3). The result, possibly exacerbated by falling living standards and increased stress on the medical infrastructure, was an outbreak of epidemic diphtheria which disproportionately affected the Russian Federation. The incidence of clinical diphtheria in the NIS escalated each year, peaking in 1995 with >50,000 cases, yielding an incidence of 16.9 cases per 100,000 people (103). Since 1990, >150,000 cases and 4,000 deaths have been reported (103). Although vaccination programs have slowed the advance of the epidemic, diphtheria remains a serious health issue for many countries in the region.

The interplay between *C. diphtheriae* and its environment, governing toxin production and human disease, is well documented (63). Since the mechanism of classical pathogenesis by *C. diphtheriae* is relatively uncomplicated, diphtheria lends itself well as a model and/or paradigm for other toxin-mediated diseases. Before a given strain of *C. diphtheriae* can cause clinical diphtheria, it first must be infected and lysogenized by a strain of corynebacteriophage that carries the structural gene for diphtheria toxin, *tox*. While diphtheria toxin is the primary virulence factor expressed by *C. diphtheriae*, the control of *tox* expression is regulated by an iron-activated regulatory element, the diphtheria toxin repressor (DtxR), which is encoded on the genome of the bacterial host. The DtxR-mediated regulation of diphtheria *tox* expression has been well studied and has served as a paradigm for the control of iron-sensitive virulence gene expression.

C. diphtheriae is a gram-positive, nonmotile, club-shaped bacillus. Three distinct colony types (*mitis*, *intermedius*, and *gravis*) have been identified. Most strains of *C. diphtheriae* require the addition of nicotinic and pantothenic acid to culture medium for growth. Corynebacteria have been shown to have cell walls that contain arabinose, galactose, mannose, and a toxic 6,6'-diester of trehalose that is composed of corynemycolic and corynemycolenic acids.

Clinical diphtheria is most commonly seen as an infection of the upper respiratory tract; however, cutaneous diphtheria has also been reported. In each instance, the site of infection is characterized by a pseudomembrane, a fibrin membrane at the base of an ulcerative lesion that forms as a result of the combined effects of bacterial colonization and growth, toxin production, necrosis of the underlying tissue, and the immune response of the host. The symptoms of clinical nasopharyngeal diphtheria range from mild pharyngitis with fever to potentially fatal airway obstruction, due to the combined effects of the pseudomembrane and profound supraglottic edema (bull neck diphtheria). Since diphtheria toxin is delivered from the site of the primary lesion(s) to all organ systems by the circulation, life-threatening systemic complications may also be associated with loss of motor function, and congestive heart failure may develop as a result of the toxin's action on peripheral neurons and myocardium.

While the pathogenesis of diphtheria is based upon the ability of a given strain to both colonize a given host and produce diphtheria toxin, to date little is known about colonization factors that may be associated with virulence. In contrast, there is now a detailed understanding of the molecular biology and genetics of both the regulation of diphtheria toxin expression and the structure-function relationships and mode of action of the toxin. The existence of virulence determinants in *C. diphtheriae* beyond those associated with diphtheria toxin is demonstrated by the recently reported outbreak of invasive disease caused by a clonal focus of nontoxigenic *C. diphtheriae* among intravenous drug users in Switzerland (25). In this instance, multiple individuals presented with *C. diphtheriae* bacteremia, and several

Gram-Positive Pathogens, 2nd edition, edited by Vincent A. Fischetti et al.

otherwise-healthy young people developed endocarditis and subsequently died from infection with this nontoxigenic strain. At present, the function and relative significance of additional virulence factors that may be associated with this strain remain unclear.

IRON-SENSITIVE EXPRESSION OF VIRULENCE FACTORS IN *C. DIPHTHERIAE*

The diphtheria toxin structural gene *tox* has long been known to be expressed by *C. diphtheriae* under conditions of iron starvation; work by Pappenheimer and Johnson as early as 1936 (62) demonstrated the inhibitory effect of iron on toxin production. Like the expression of siderophores and heme oxygenase (84, 86), diphtheria toxin is expressed at maximal rates only when iron becomes the growth rate-limiting substrate in in vitro culture medium (63, 64). The production of siderophores and heme oxygenase by *C. diphtheriae* is an adoptive response to low iron and results in the acquisition of this essential nutrient, while the derepression of the diphtheria *tox* gene appears to act toward this end indirectly by the induction of tissue damage. Moreover, recent studies have suggested that adherence and cell surface carbohydrates are altered by *C. diphtheriae* in response to environmental iron concentrations (50), although the relevance of this observation to virulence remains unproven. While the structural genes encoding siderophores and heme oxygenase are carried on the *C. diphtheriae* genome, the structural gene for diphtheria toxin is carried on the genomes of a family of closely related corynebacteriophages (9, 102). Interestingly, the regulation of expression of the *tox* gene, as well as the genes involved in iron acquisition and utilization, is under the control of the *C. diphtheriae*-encoded iron-activated repressor DtxR.

THE DIPHTHERIA TOXIN REPRESSOR, DtxR

The discovery that the diphtheria *tox* gene was carried by corynebacteriophage β raised an interesting and important question: was the regulation of *tox* mediated by a viral or corynebacterial determinant? Work by Murphy et al. (53) first suggested that the corynephage *tox* gene was regulated

by a corynebacterial protein and that Fe^{2+} by itself does not directly control the expression of diphtheria toxin. Further work by Kanei et al. (35) and Murphy et al. (54) with various *C. diphtheriae* and corynephage β mutants led Murphy and Bacha in 1979 (52) to propose a model for the regulation of *tox* expression in which a *cis*-acting viral element and a *trans*-acting *C. diphtheriae* product together control the expression of *tox* in an iron-dependent manner.

As shown in Fig. 1, the *cis*-acting viral element was found to be the *tox* operator, a 27-bp interrupted palindromic sequence (23, 35, 78) that overlaps the −10 region of the *tox* promoter (5). This region upstream from the diphtheria toxin structural gene has been shown to be the *tox* promoter-operator *toxPO*. In 1989, Fourel et al. (19) used DNase protection assays to show that an element or elements found in crude *C. diphtheriae* extracts bind specifically to the *tox* operator sequence. Mutant strains of *C. diphtheriae* that were *cis* dominant for toxin production (106) were later found to carry mutations in the *tox* operator alone (37).

The *trans*-acting factor hypothesized by Murphy and Bacha in 1979 (52) was cloned by Boyd et al. (6, 7) by screening genomic libraries of nontoxigenic, nonlysogenic *C. diphtheriae* in recombinant *Escherichia coli* that carried a transcriptional fusion between *toxPO* and *lacZ*. The factor found, named diphtheria toxin regulatory protein, or DtxR, was a 226-amino-acid protein. DtxR was shown to repress the expression of β-galactosidase from a *toxPO-lacZ* fusion in an iron-dependent fashion. The functional activity of DtxR in *C. diphtheriae* was subsequently demonstrated by studies performed by Schmitt and Holmes (83): expression of *dtxR* in C7hm723(β*tox+*), a strain of *C. diphtheriae* in which toxin expression is iron insensitive, resulted in conversion to the wild-type iron-sensitive phenotype.

DtxR: BIOCHEMICAL AND GENETIC ANALYSIS

In 1992, Tao et al. (93) used gel mobility shift assays with purified DtxR and ^{32}P-labeled *toxPO* as a probe to explore the transition metal ion activation of the repressor and binding to the *tox* operator. This work demonstrated that

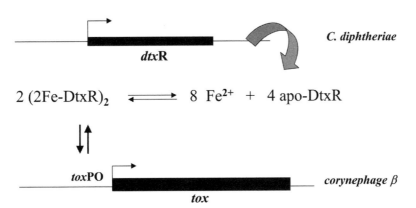

FIGURE 1 Expression of the diphtheria toxin structural gene, *tox*, is regulated by the Fe^{2+}-activated repressor DtxR. The *dtxR* gene is carried on the genome of *C. diphtheriae*, while *tox* is carried by a family of closely related corynebacteriophages. In the presence of Fe^{2+}, apo-DtxR forms active dimers [(2Fe-DtxR)₂], and the two dimers bind to the diphtheria *tox* operator and repress expression. Under iron-limiting conditions, the ternary complex, the 2(2Fe-DtxR)₂-*tox* operator, dissociates and diphtheria toxin is expressed.

activation of apo-DtxR and subsequent binding to the *tox* operator required a divalent transition metal ion. Since binding to the ^{32}P-labeled probe was blocked by the addition of either excess nonradioactive probe, anti-DtxR antisera, or the cation chelator 2,2′-dipyridyl, binding was clearly specific and dependent upon metal ion activation. Tao and Murphy (94) and Schmitt and Holmes (84) both showed that DtxR protects *toxPO* from DNase digestion, following activation by one of a number of divalent cations (e.g., Co^{2+}, Fe^{2+}, Ni^{2+}, Cd^{2+}, or Mn^{2+}). Zinc (Zn^{2+}) was a weak activator of DtxR, and Cu^{2+} failed to activate repressor activity. This in vitro analysis of metal ion activation of apo-DtxR confirmed and extended earlier observations on the inhibitory effect of transition metal ions on the expression of diphtheria toxin by toxigenic strains of *C. diphtheriae* (24).

In 1994, Tao and Murphy (96) demonstrated by in vitro affinity enrichment and gel mobility shift selection that the minimal nucleotide sequence necessary for DtxR binding is a 9-bp palindromic sequence separated by a single C or G nucleotide as follows: 5′-TA/TAGGTTAGG/CCTAACC-TA/TA-3′. Since a family of related target sequences that bound DtxR with the same apparent affinity as the 27-bp *tox* operator was isolated, it was apparent that DtxR was able to function as a global regulatory element in the regulation of iron-sensitive genes in *C. diphtheriae*, as earlier postulated (83). Recently, a number of genes with upstream DtxR-binding sites were isolated and characterized (39, 73, 83, 84, 91). Several studies employed site-directed mutagenesis of DtxR to elucidate structure-function relationships within the repressor, especially the metal ion-binding and DNA-binding domains of DtxR. Since DtxR contains only a single cysteine residue and disulfide bond-linked dimers of DtxR are inactive, Tao and Murphy (95) used saturation site-directed mutagenesis of Cys-102 to demonstrate that this residue plays a role in the coordination of Fe^{2+}. The replacement of Cys by all 20 amino acids, except Asp, resulted in the complete loss of repressor activity. DtxR(C102D) retained an iron-dependent active phenotype, albeit not to wild-type levels. Wang et al. (105) utilized bisulfite mutagenesis of *dtxR* to analyze the role of particular residues in DtxR function. These investigators also postulated that the ion-binding domain included Cys-102 and further that the DNA-binding domain (residues 1 to 52) formed a helix-turn-helix motif similar to the DNA-binding domains of many other regulatory proteins.

CRYSTALLOGRAPHIC ANALYSES OF DtxR

DtxR was the first transition metal ion dependent repressor for which an X-ray crystal structure was solved. In 1995, both Qiu et al. (74) and Schiering et al. (82) published structures at similar resolution. Initially, both groups were able to assign coordinates to amino acids 1 through 136; however, they were unable to assign coordinates to the C-terminal 90 amino acid residues, owing to the high thermal factors of this portion of the repressor. More recently, Qiu et al. (75) reported the DtxR coordinates at 2.0-Å resolution, at which level the C-terminal domain was found to fold into an SH3-like structure. White et al. (107) solved the structure of the Ni^{2+}-activated DtxR(C102D) *toxPO* complex. With a few exceptions, additional X-ray crystal structures have provided minimal further insights (11, 14, 15, 67–70).

DtxR contains a total of eight α-helices, six of which are contained within the N-terminal two-thirds of the protein.

Helices B and C and the three-amino-acid connecting loop between them (residues 27 to 50) constitute the helix-turn-helix motif and form the DNA-binding site. Surprisingly, activated DtxR(C102D) binds to *toxO* as two pairs of dimers (107). Each DtxR dimer binds to one of the palindromic sequences on almost opposite faces of the operator (Fig. 2). Knowledge of the structure of the ternary complex confirms and extends earlier observations that the DtxR footprint encompassed a region of 30 bp (94).

While the overall mechanism of target DNA recognition by DtxR is similar to that of other prokaryotic repressors, there are some noteworthy interactions. The C α-helix of DtxR is responsible for most interactions with DNA and inserts into the major groove of the double helix. Each helix-turn-helix motif makes a total of nine interactions with backbone phosphate groups: three from α-helix B, two from the turn, and four from α-helix C. The guanidinium group of Arg-60 binds in the minor groove of DNA, thereby making a bridge to additional phosphates. Owing to a structural rearrangement of DtxR upon binding to the *tox* operator, Thr-7 in α-helix A also interacts with a backbone phosphate. Originally, only a single amino acid, Gln-43, appeared to associate directly with a DNA base. Additional data, however, demonstrated that Ser-37 and Pro-39 specifically recognized the methyl groups of particular thymine bases within the *toxO* sequence through van der Waals interactions (11), providing an additional level of target specificity. Again, observations made by crystallographic analysis confirmed and extended earlier biochemical genetic analysis of the repressor (84, 105).

Metal Ion Binding and Activation of Repressor Activity

Early observations made by saturation and equilibrium dialysis suggested that DtxR carried a single class of metal ion-binding site with an apparent K_d of 2×10^{-6} to 9×10^{-7} M (97, 105). Paradoxically, X-ray crystallographic analysis of metal ion-DtxR complexes clearly revealed two metal ions bound per monomer (74, 82). Using site-directed mutagenesis of each of the coordinating amino acid residues, Ding et al. (15) demonstrated that the primary metal ion-binding site involved in activation of DNA binding is composed of amino acids that form an octahedral coordination center. The primary metal ion-binding site required for DtxR(C102D) and DtxR activation is composed of the side chains of Met-10, Asp-102, Cys-102, Glu-105, His-106, and a water molecule. This structural water molecule is also held in place by the main chain carbonyl oxygen of Leu-4.

For many years, the function, if any, of the ancillary site was unclear, but recent studies have begun to delineate its role in repressor activity. Early crystallographic data suggested the ancillary site is comprised of only three residues, His-79, His-98, and Glu-83 (74, 82). A later structure raised the possibility that two residues from the C-terminal domain of DtxR (Glu-170 and Gln-173) are also ligands for metal ions at this site (69). Recent in vivo analysis confirms that these residues contribute to the ancillary metal ion-binding site and that the ancillary site facilitates the activation of the repressor (42). These studies were possible only after the development of a more sensitive in vivo reporter system (41).

The structural changes of apo-DtxR upon binding metal ions were originally described using X-ray crystal structures, and a helix-to-coil transition of residues 1 through 4 and a pincer-like movement of α-helix C were proposed (68).

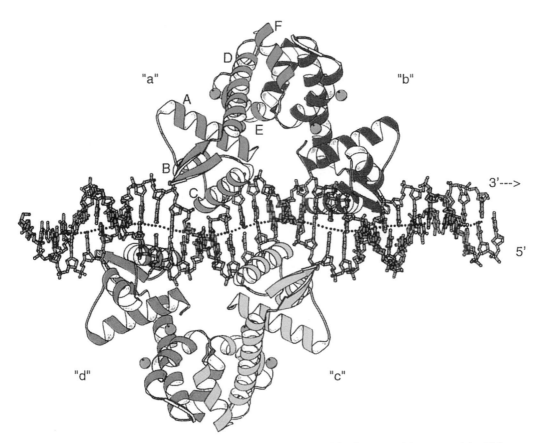

FIGURE 2 *(See the separate color insert for the color version of this illustration.)* Structure of the 2[Ni DtxR(C102D)]$_2$-*tox* operator complex. Residues 3 to 120 in each DtxR(C102D) monomer are designated "a" to "d." Ribbons and arrows are used to indicate α-helices and β-strands in each monomer. The 33-bp DNA segment carries the 27-bp interrupted palindromic *tox* operator sequence. (Adapted from White et al. [107].)

These conclusions, however, were limited by the solid phase necessary for crystallographic studies. More recently, nuclear magnetic resonance solution structures combined with other biophysical data have suggested that apo-DtxR exists in a partially unstructured molten globule form (101). Coordination of divalent transition metal ions results in a conversion to a discrete, ordered tertiary structure, capable of dimerization and binding *toxO*. This transition represents another level of regulation of DtxR's action by limiting the repressor activity of apo-DtxR. Exquisite in vitro studies have confirmed that DtxR has two functional metal ion-binding sites and that coordination of metal ions lowers the K_d for DtxR dimerization (90).

The activating metal ion does not form a bridge between DtxR and target operators. In contrast, the *trp* repressor, for which tryptophan is a corepressor, also binds to its DNA recognition site as two pairs of dimers. However, the *trp* repressor dimers interact with each other (38), and the corepressor tryptophan is bound at the DNA-protein interface (110). Although both DtxR and the *trp* repressor mediate repression through corepressors and bind to their respective operators as two dimers, the molecular mechanisms involved in repression are in fact very different.

DtxR Protein-Protein Interactions

The protein-protein interactions stabilizing DtxR dimers arise mostly from hydrophobic associations. Several nonpolar amino acids contained within α-helices D, E, and F and their intervening loops compose this hydrophobic network (82). This region contains amino acids 85 through 116 of DtxR. One internal hydrogen bond exists between Glu-100 and Trp-104; mutation of either of these residues drastically reduces repressor activity (105). This network of hydrophobic interactions is likely responsible for the weak equilibrium between monomeric and dimeric apo-DtxR that was demonstrated by Tao et al. (97).

C-Terminal Domain of DtxR

The X-ray structure of the C-terminal domain of DtxR shows that this region of the repressor is composed of five antiparallel β-sheets and two short α-helices (75). Twigg et al. (100) and Wang et al. (104) recently expressed the C-terminal SH3-like domain of DtxR, DtxRΔ(1-129), and determined its solution structure by heteronuclear nuclear magnetic resonance. These studies have confirmed the observations of Qiu et al. (75) that this region of DtxR folds into an SH3-like structure. Interestingly, Wang et al. (104) also report that, like eukaryotic SH3 domains of Hck, Src, and Itk, the C-terminal domain of DtxR may associate with a proline-rich peptide derived from the linker region between the N- and C-terminal domains of the repressor. The interactions between the C-terminal domain and the proline-rich linker region of DtxR (i.e., amino acids 125 to 139) appear to stabilize the monomeric form of the repressor. A hyperactive mutant of DtxR, DtxR(E175K), has been isolated, which possesses a single amino acid change in the C-terminal domain (92). This mutant is active at lower metal ion concentrations than wild-type DtxR (43). Further studies of this mutant have confirmed an important role of the C-terminal domain and its interaction with the proline-rich linker region of DtxR as a modulator of the repressor's activation (43).

Based upon a series of biophysical and genetic analyses, Love et al. (43) have proposed a three-step model for the transition metal ion activation of apo-DtxR to its fully functional conformation. Once an activating transition metal ion binds to a metal ion-binding site, the N-terminal domain of the repressor undergoes a partial structural change that results in the formation of the other metal ion-binding site. At present, the order of metal ion binding to the ancillary and primary sites remains unclear. Finally, transition metal ion binding to the second site gives rise to a further structural organization, leading to the formation of the fully active form of the repressor.

DtxR HOMOLOGS IN OTHER PROKARYOTES

Since the cloning of DtxR from C. diphtheriae in 1990, homologous proteins have been identified in a number of other microorganisms. IdeR (for iron-dependent regulator) has been found to be the DtxR homolog in several species of Mycobacterium: Mycobacterium tuberculosis H37Rv, Mycobacterium bovis BCG, Mycobacterium leprae, Mycobacterium avium, and Mycobacterium smegmatis (16, 85). IdeR and DtxR are 78% identical and 90% homologous over their N-terminal 140 amino acids (16). DtxR homologs in numerous bacterial species have now been identified (Fig. 3), including Enterococcus faecalis (44), Streptococcus mutans (36), Staphylococcus aureus (31), Streptomyces pilosus (27), and Rhodococcus equi (4). Some bacterial species, such as M. tuberculosis (13) and C. diphtheriae itself (87), have been shown to express multiple DtxR-like proteins, likely responsive to different transition metal ions. Recently, the gram-negative E. coli has been shown to possess a DtxR homolog (66).

Remarkably, in many cases these homologous iron-dependent regulatory proteins have been shown to be functional homologs of DtxR and to be capable of recognizing and regulating gene expression through the diphtheria tox operator (26, 61, 85). Expression of the hyperactive mutant DtxR(E175K) in merodiploid strains of both M. tuberculosis (46) and S. aureus (1) markedly attenuates virulence. These results suggest that there is a high degree of similarity in function and targets among DtxR homologs.

Two microbes express truncated DtxR homologs, encoding a protein corresponding to only the N-terminal domain of DtxR. The genome of Treponema pallidum, the spirochete that causes syphilis, carries the troR gene, in which only the N-terminal 153 amino acids are present (28, 72). Although the function of this homolog is largely unknown, the conservation of amino acids essential for DtxR function suggests that they have a similar role. In Bacillus subtilis, the single-domain homolog MntR has been extensively studied for its role in manganese homeostasis (22, 40, 76).

DIPHTHERIA TOXIN: BIOCHEMISTRY AND MECHANISM OF ACTION

Diphtheria toxin is the primary virulence factor expressed by toxigenic strains of C. diphtheriae. As discussed above, the structural gene encoding diphtheria toxin is carried by a family of closely related corynebacteriophages, the best studied of which is corynephage β (9). The diphtheria tox mRNA has been shown to be associated with membrane-bound polysomes, and the nascent toxin is cotranslationally secreted in precursor form (89). Following cleavage of the 19-amino-acid signal sequence, the mature form of the toxin is released into the surrounding medium.

In mature form, diphtheria toxin is a 535-amino-acid single-chain protein (molecular weight, 58,342) with a 50% lethal dose in sensitive mammalian species of approximately 100 ng/kg of body weight (20, 23, 34, 78, 108). Following mild digestion with trypsin or other serine proteases, the toxin may be separated under denaturing conditions into two polypeptide chains: the N-terminal 193-amino-acid fragment A and the C-terminal 342-amino-acid fragment B (17, 21). Biochemical and genetic analyses have shown that the A fragment, or catalytic (C) domain, of diphtheria toxin is an enzyme that catalyzes the NAD^+-dependent ADP ribosylation of eukaryotic elongation factor 2 (EF2) according to the following equation:

$$NAD^+ + EF2 \xrightarrow{\text{Fragment A}} \begin{array}{c} ADPR\text{-}EF2 \\ + \text{ nicotinamide } + H^+ \end{array}$$

The B fragment of diphtheria toxin has been shown to facilitate the delivery of fragment A across the eukaryotic cell membrane to the cytosol and to carry the native receptor-binding (R) domain. Choe et al. (12) first described the X-ray crystal structure of diphtheria toxin, which was subsequently corrected by Bennett et al. (2). The X-ray crystal structure analysis of diphtheria toxin clearly demonstrated that the toxin is a three-domain protein: it has a C domain (fragment A), the fragment B-associated transmembrane (T) domain, and an R domain (Fig. 4).

The diphtheria toxin-mediated intoxication of sensitive eukaryotic cells has been shown to be a sequential process in which each structural domain plays an essential role. The intoxication process is initiated by the interaction of diphtheria toxin with its cell surface receptor. Naglich et al. (57) have shown that the diphtheria toxin receptor is the heparin-binding epidermal growth factor-like precursor (HB-EGF-LP). In addition, DAP-27 (the 27,000-molecular-weight monkey homolog of human CD9) has been shown to modulate the binding of diphtheria toxin to the cell surface and to enhance the sensitivity of given cells toward the toxin (8, 33, 49). Since an association between HB-EGF-LP and DAP-27 has not yet been detected, it is likely that the role of DAP-27 is to maintain large numbers of HB-EGF-LP on the cell surface.

Once bound to its cell surface receptor, diphtheria toxin is internalized by receptor-mediated endocytosis in coated pits (51). Either on the cell surface or in the early endocytic vesicle, the α-carbon backbone of diphtheria toxin is cleaved by the cellular protease furin after Arg-193, which is positioned in a 14-amino-acid protease-sensitive loop that is subtended by a disulfide bond between Cys-184 and Cys-201 that forms the boundary between fragments A and B of the toxin (80).

The diphtheria toxin T domain, which is composed of nine α-helices and their connecting loops, appears to play at least two important roles in the intoxication process (32). The first helical layer (i.e., helices 1 to 3) is amphipathic and appears to stabilize the toxin on the luminal surface of the endocytic vesicle membrane. As the lumen of the early endosome is acidified through the action of the

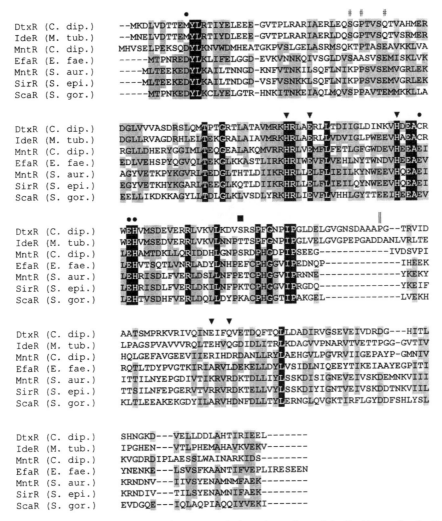

FIGURE 3 Amino acid sequence alignments of selected members of the DtxR superfamily. DtxR is compared to the two-domain homologs IdeR (*M. tuberculosis*), MntR (*C. diphtheriae*), EfaR (*E. faecalis*), MntR (*S. aureus*), SirR (*Staphylococcus epidermidis*), and ScaR (*Streptococcus gordonii*). Residues identical in all proteins are shown as white letters with black background, while conservative differences are highlighted in grey. Circles (●) and triangles (▼) mark the residues of DtxR's primary and ancillary metal ion-binding sites, respectively. The solid square (■) above S124 of DtxR marks the end of the N-terminal domain, while two vertical lines (‖) above P148 mark the beginning of the C-terminal domain, with the intervening sequence constituting the tether region. A number sign (#) marks the residues of DtxR involved in interactions with specific base pairs of target operators (S37, P39, and Q43).

vesicular ATPase, the decrease in pH causes a partial denaturation of the T domain and spontaneous insertion of helices 5 to 9 into the plane of the vesicle membrane, thereby forming a pore or channel through the membrane. This insertion event appears to begin with the third helix layer, channel-forming helixes 8 and 9, which is then stabilized by the insertion of the second helix layer, composed of helixes 5, 6, and 7. As the channel is forming, an extended denatured C domain appears to be inserted into the nascent pore by a mechanism that may involve positioning of the carboxy-terminal end of the C domain by T-domain helix 1.

Using an in vitro assay based upon purified early endosomes that have been preloaded with the diphtheria toxin-related fusion protein toxin DAB$_{389}$IL-2 (interleukin-2), Ratts et al. (79) recently demonstrated that C-domain translocation from the lumen of the endosome to the external mileu requires a cytosolic translocation factor complex. Mass spectrometry sequencing and inhibitor analysis have further demonstrated that the chaperonin heat shock protein Hsp 90 and thioredoxin reductase were required for C-domain translocation from the lumen of purified early endosomes to the external milieu in vitro. While studies of the molecular steps of C-domain translocation from the endosomal lumen are in their infancy, this work has begun to shed light on this step of the intoxication process. It is tempting to speculate that following furin-mediated nicking of the toxin, the carboxy-terminal end of the C domain and the amino-terminal end of the T domain, which are linked by the disulfide bond between Cys-186 and Cys-201, are threaded through the channel formed by the T domain.

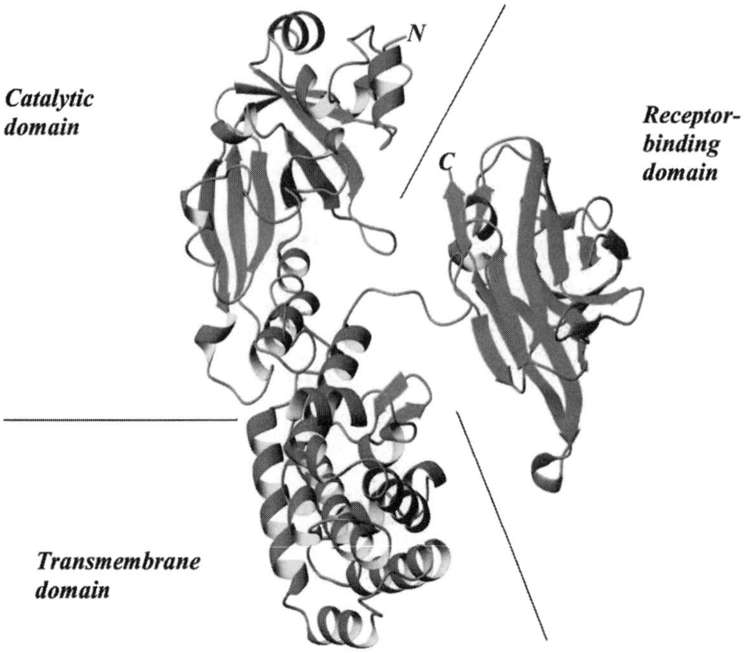

FIGURE 4   Ribbon diagram of the X-ray crystal structure of native diphtheria toxin. The relative portions of the C, T, and R domains are indicated. N, amino terminus; C, carboxy terminus.

The finding that thioredoxin reductase is an essential component of the cytosolic translocation factor complex suggests that the disulfide bond between the C and T domains is reduced in the cytosol by this enzyme. Further, once the first few hydrophobic residues of a completely extended C domain reach the cytosol, it is possible that cellular chaperonins, including Hsp 90, facilitate the entry process by sequentially binding and refolding the denatured polypeptide into an active conformation as it emerges from the lumen of the early endosome, in a process analogous to that of protein import into mitochondria (45, 108, 109).

TURNING THE TABLES ON DIPHTHERIA TOXIN: FUSION PROTEIN TOXINS AND THE DEVELOPMENT OF A NEW CYTOTOXIC AGENT FOR THE TREATMENT OF CUTANEOUS T-CELL LYMPHOMA (CTCL)

Since the first step in the intoxication of eukaryotic cells is receptor binding, the use of protein engineering methods to redesign native diphtheria toxin by R domain substitution has been used to construct fusion proteins that display highly selective target cell specificity. In 1986, Murphy et al. "envisioned that these chimeric molecules might serve as . . . targeted toxins for the treatment of human malignancies" (55).

In the ensuing period, a family of fusion protein toxins, each with a different targeting ligand, has been constructed and characterized (56). One such fusion protein toxin, composed of the C and T domains of diphtheria toxin to which human IL-2 has been genetically fused (DAB$_{389}$IL-2, or denileukin diftitox; brand name, Ontak), has been extensively evaluated in human clinical trials for the treatment of IL-2 receptor-positive leukemias and lymphomas (18).

In February 1999, the U.S. Food and Drug Administration approved DAB$_{389}$IL-2 for the treatment of CTCL refractory to other therapies, making it the first recombinant receptor-targeted biologic to achieve approval. Unlike many other non-Hodgkin's lymphomas, CTCL is a clearly devastating malignancy in which patients suffer a substantial and often disfiguring disability during the course of their illness. Phase III trials enrolled patients with CTCL who had failed other therapies; these studies demonstrated that 30% of the total number of patients who met inclusion criteria had a 50% or greater reduction in their tumor burden for at least 6 weeks following treatment with DAB$_{389}$IL-2. In this refractory patient population, 10% of the total number of patients who were evaluable had either a complete response (in the absence of histologic analysis) or a complete clinical response (documented by histology). Consistent with earlier clinical studies, the most common adverse events experienced by this patient population were chills, fever, malaise, nausea, and vomiting. Less-frequent adverse events included hypotension, edema, rash, and capillary leak syndrome. Currently, clinical trials are exploring the ability of DAB$_{389}$IL-2 to treat other non-Hodgkin's lymphomas, B-cell chronic lymphocytic leukemia, and psoriasis.

Ontak has been recently shown to be remarkably effective in the treatment of steroid resistant graft-versus-host disease. Ho et al. (30) reported that 71% of patients responded following treatment, with 50% achieving complete resolution and the remaining 21% achieving a partial resolution of their steroid-resistant graft-versus-host disease.

C. DIPHTHERIAE IN THE AGE OF THE GENOME

In 1995, *Haemophilus influenzae* became the first bacterium for which a complete genome sequence was published. As of June 2004, The Institute for Genomic Research listed over 140 complete bacterial genomes in its Comprehensive Microbial Resource (http://www.tigr.org/tdb/). The availability of genomic data has revolutionized molecular microbiology by allowing comparisons between different strains and species, often leading to greater understanding of virulence determinants, environmental adaptations, and the process of speciation.

The complete genome sequence of *C. diphtheriae* NCTC13129 (*tox*$^+$) was recently reported (10). This strain was isolated from the pharynx of a British woman who had contracted diphtheria while on a Baltic cruise in 1997. The single circular chromosome has 2.48 Mbp and has, on average, a 53.48% G+C content, consistent with the well-documented G+C-rich nature of the corynebacteria. No plasmid or other extrachromosomal DNA was isolated from *C. diphtheriae* NCTC13129 (*tox*$^+$).

Surprisingly, the G+C content of the *C. diphtheriae* genome is not homogeneous. Ceredeño-Tárraga et al. (10) found that a stretch of approximately 740 kb at the replication terminus had a significantly lower G+C content (averaging 49.99%) than the rest of the genome (at 54.96%). This change was absent in the genomes of *Corynebacterium glutamicum* and *Corynebacterium efficiens*, although an alignment of the translated open reading frames of each genome demonstrated strict linear conservation of genes. Interestingly, this G+C drop is most profound in the third nucleotide position of codons, a position often inconsequential to protein sequence due to codon wobble. In this way, the *C. diphtheriae* genome appears to retain function despite its lowered G+C content.

Although the *C. diphtheriae* genome sequence has been complete for <1 year, insights into the pathophysiology of diphtheria have already been gleaned. A number of putative pathogenicity islands were identified as part of the genome sequence, and genes potentially involved in iron acquisition, fimbria production, and cell wall biosynthesis have been proposed (10). The existence of sortase-dependent pili on the surface of *C. diphtheriae* NCTC13129 has been identified through gene sequence similarity to pili and sortase genes found in other gram-positive microbes (98). What role, if any, these pili play in diphtheria pathogenesis remains unstudied.

The *C. diphtheriae* genome sequence has also permitted examination of microbial speciation and evolution. One study compared the genomes of several members of the taxonomic order *Actinomycetales* (58). These authors noted that the genomes of *C. diphtheriae*, *C. efficiens*, and *C. glutamicum* are nearly superimposable in their linear sequences of homologous genes, meaning that little if any shuffling in genome order has occurred. In contrast, the genome of *M. tuberculosis* has a markedly altered arrangement from the three corynebacterium species or even *M. leprae*, despite high conservation of orthologous genes. These authors note that *Corynebacterium* species and *M. leprae* lack the *recBCD* operon; in *E. coli*, the Rec BCD pathway plays a role in homologous recombinations and chromosomal inversions. They propose that the absence of *recBCD* has yielded a relatively static genomic structure in these species.

EPIDEMIOLOGY OF DIPHTHERIA

Outbreaks of clinical diphtheria almost always occur in individuals who have not become immunized and who have been exposed to a carrier (i.e., an individual who carries toxigenic *C. diphtheriae* in his or her nasopharyngeal flora). The infection is spread by droplet from person to person. While colonization of a susceptible individual with a toxi-

genic strain certainly plays a role in pathogenesis, Pappenheimer and Murphy (65) have also demonstrated that transmission of toxigenicity may occur by in situ lysogenic conversion of an autochthonous nontoxigenic strain of *C. diphtheriae* to toxigenicity.

In 1923, Ramon (77) demonstrated that the treatment of diphtheria toxin with formaldehyde resulted in the formation of a "toxoid" that could be used as an immunogen for vaccination against diphtheria. Remarkably, immunization with diphtheria toxoid results in the production of neutralizing antibodies that not only block native diphtheria toxin from binding its cell surface receptor but also protect against clinical disease (63, 111). Mass immunizations of populations with diphtheria toxoid have led to a dramatic decrease in the incidence of diphtheria. One of the best-chronicled studies of the effects of mass immunization comes from Romania. Before a large-scale immunization program began in 1958, only 60% of the population was immune to diphtheria toxin, as measured by Schick test reactivity; by 1979, this proportion had jumped to 97% (64, 81). Over the same period, diphtheria morbidity dropped from approximately 600 per 100,000 in 1958 to only 1 per 100,000 in 1972 (64, 81).

Saragea et al. (81) also observed a concomitant reduction in the relative proportion of toxigenic strains of *C. diphtheriae* that were isolated during that same period in Romania. Between 1955 and 1966, 86% of *C. diphtheriae* clinical isolates were toxigenic; that number fell to 5% by 1977. Importantly, the overall prevalence of *C. diphtheriae* isolated from the general population did not change over this period. The results are clear: immunization with diphtheria toxoid eliminated a selective advantage for the spread of toxigenic *C. diphtheriae* isolates among the population. However, nontoxigenic strains of *C. diphtheriae* continued to fill the ecologic niche.

Data from the recent diphtheria epidemic in Russia and the NIS teach a similar lesson (29). The relative number of Russian children vaccinated with diphtheria toxoid decreased from >80% in 1980 to 68% in 1990 (47), a value that falls well below the 95% minimum that the World Health Organization suggests is necessary to prevent an epidemic. The decrease in the relative number of individuals who were fully vaccinated or in whom titers of antidiphtheria toxin antibodies achieved protective levels has allowed toxigenic strains of *C. diphtheriae* to spread through the population.

The NIS epidemic cannot be explained exclusively by a fall in the proportion of vaccinated children, especially as adults accounted for the majority of diphtheria cases. A recent study of American adults found that a decreasing portion of the population has protective levels of diphtheria toxin-neutralizing antibodies (48). The protected proportion drops with age: 90% of 6- to 11-year-olds and 65% of 30- to 39-year-olds, reaching a nadir of 30% of the studied population over the age of 70 (48). Yet no epidemic of diphtheria has been reported in the United States, despite confirmation of the continued presence of toxigenic strains (71).

In reviewing the NIS epidemic, it has been proposed that a number of additional factors may have contributed to this outbreak (103). These investigators cite changes in immunization schedules, contraindications, and lowered antigen content, leading to increased childhood susceptibility. They point to the large portion of the Soviet population which was militarized, requiring mixing of people from across the former Soviet Union, frequently in crowded conditions with poor hygiene. The civilian Soviet public was highly urbanized, with dense urban populations and a dropping standard of living. There was increased travel within and without the Soviet Union, including mass population movements of refugees. At the microbial level, a change in the predominant *C. diphtheriae* biotype was noted in the late 1980s in the Soviet Union. More recent ribotype analyses have suggested that the epidemic strain of *C. diphtheriae* was present in several geographically distant regions of Russia from 1985 through 1987, a full 5 years before the epidemic began (88), placing even greater emphasis on the human factors contributing to the outbreak and suggesting important lessons about the reemergence of pandemic diseases.

Molecular epidemiologic analysis of toxigenic strains of *C. diphtheriae* isolated from this epidemic has provided further insight into the virulence of this pathogen. The *tox* and *dtxR* genes from 72 strains of *C. diphtheriae* that were isolated from Russia and Ukraine have been PCR amplified and sequenced. The *tox* gene was highly conserved: <6% of the strains carried a mutation in *tox*; importantly, none of those mutations encoded an amino acid change. This extraordinary conservation of amino acid sequence again suggests the importance of the structure-function relationships within diphtheria toxin. Sequencing of *dtxR* genes provided somewhat different results: only 20% of the strains lacked a mutation in the diphtheria toxin repressor. While most mutations in *dtxR* did not give rise to amino acid substitutions, nine substitutions were found. Importantly, all nine residue substitutions were found in the C-terminal domain of DtxR (i.e., distal to residue 147) (59, 60), which appears to subtly affect DtxR activity (41, 43).

The extreme conservation of the diphtheria toxin primary sequence suggests that the evolution of new *tox* alleles that would encode a toxin sufficiently different to avoid neutralization by anti-diphtheria toxoid antibodies is highly unlikely. This is good news and argues against fears that an evolved "superdiphtheria" could make vaccination with diphtheria toxoid obsolete.

The high conservation of the N-terminal half of DtxR underscores the importance of this iron-dependent repressor in the control of virulence gene expression. Mutations in DtxR that give rise to a loss of repressor activity would allow constitutive expression of not just *tox* but all other iron-sensitive genes as well. Constitutive expression of siderophores could conceivably result in iron loading that might lead to DNA damage (99).

We thank Timothy Logan, Department of Chemistry and Biophysics, Florida State University, for the ribbon diagram of native diphtheria toxin.

J.R.M. is partially supported by U.S. Public Health Service grant AI21628 from the National Institute of Allergy and Infectious Diseases.

REFERENCES

1. **Ando, M., Y. C. Manabe, P. J. Converse, E. Miyazaki, R. Harrison, J. R. Murphy, and W. R. Bishai.** 2003. Characterization of the role of the divalent metal ion-dependent transcriptional repressor MntR in the virulence of *Staphylococcus aureus*. *Infect. Immun.* **71**:2584–2590.
2. **Bennett, M. J., S. Choe, and D. Eisenberg.** 1994. Domain swapping: entangling alliances between proteins. *Proc. Natl. Acad. Sci. USA* **91**:3127–3131.
3. **Bisgard, K. M., I. R. B. Hardy, T. Popovic, P. M. Strebel, M. Wharton, R. T. Chen, and S. C. Hadler.** 1998. Respiratory diphtheria in the United States, 1980 through 1995. *Am. J. Public Health* **88**:787–791.
4. **Boland, C. A., and W. G. Meijer.** 2000. The iron dependent regulatory protein IdeR (DtxR) of *Rhodococcus equi*. *FEMS Microbiol. Lett.* **191**:1–5.
5. **Boyd, J., and J. R. Murphy.** 1988. Analysis of the diphtheria *tox* promoter by site-directed mutagenesis. *J. Bacteriol.* **170**:5949–5952.
6. **Boyd, J., M. Oza, and J. R. Murphy.** 1990. Molecular cloning and DNA sequence analysis of an iron dependent diphtheria *tox* regulatory element (*dtxR*) from *Corynebacterium diphtheriae*. *Proc. Natl. Acad. Sci. USA* **87**:5968–5972.
7. **Boyd, J., K. Hall, and J. R. Murphy.** 1992. Characterization of *dtxR* alleles from *Corynebacterium diphtheriae* strains PW8, 1030, and C7hm723. *J. Bacteriol.* **174**:1268–1272.
8. **Brown, J. G., B. D. Almond, J. G. Naglich, and L. Eidels.** 1993. Hypersensitivity to diphtheria toxin by mouse cell expressing both diphtheria toxin receptor and CD9 antigen. *Proc. Natl. Acad. Sci. USA* **90**:8184–8188.
9. **Buck, G. A., R. E. Cross, T. P. Wong, J. Lorea, and N. Groman.** 1985. DNA relationships among some *tox*-bearing corynebacteriophages. *Infect. Immun.* **49**:679–684.
10. **Ceredeño-Tárraga, A. M., A. Efstratiou, L. G. Dover, M. T. G. Holden, M. Phallen, S. D. Bentley, G. S. Besra, C. Churcher, K. D. James, A. De Zoysa, T. Chillingsworth, A. Cronin, L. Dowd, T. Feltwell, N. Hamlin, S. Holroyd, K. Jagels, S. Moule, M. A. Quail, E. Rabbinowitsch, K. M. Rutherford, N. R. Thomson, L. Unwin, S. Whitehead, B. G. Barrell, and J. Parkhill.** 2003. The complete genome sequence and analysis of *Corynebacterium diphtheriae* NCTC13129. *Nucleic Acids Res.* **31**: 6516–6523.
11. **Chen, C. S., A. White, J. Love, J. R. Murphy, and D. Ringe.** 2000. Methyl groups of thymine bases are important for nucleic acid recognition by DtxR. *Biochemistry* **39**:10397–10407.
12. **Choe, S., M. J. Bennett, G. Fugii, P. M. G. Curmi, K. A. Kantardjieff, R. J. Collier, and D. Eisenberg.** 1992. The crystal structure of diphtheria toxin. *Nature* **357**:216–222.
13. **Cole, S. T., R. Brosch, J. Parkhill, T. Garnier, C. Churcher, D. Harris, S. V. Gordon, K. Eiglmeier, S. Gas, C. E. Barry III, F. Tekaia, K. Badcock, D. Basham, D. Brown, T. Chillingworth, R. Connor, R. Davies, K. Devlin, T. Feltwell, S. Gentles, N. Hamlin, S. Holroyd, T. Hornsby, K. Jagels, and B. G. Barrell.** 1998. Deciphering the biology of *Mycobacterium tuberculosis* from the complete genome sequence. *Nature* **393**:537–544.
14. **D'Aquino, J. A., and D. Ringe.** 2003. Determinants of the Src homology domain 3-like fold. *J. Bacteriol.* **185**:4081–4086.
15. **Ding, X., H. Zeng, N. Schiering, D. Ringe, and J. R. Murphy.** 1966. Identification of the primary metal ion-activation sites of the diphtheria tox repressor by X-ray crystallography and site-directed mutagenesis. *Nat. Struct. Biol.* **3**:382–387.
16. **Doukhan, L., M. Predich, G. Nair, O. Dussurget, I. Manic-Mulec, S. T. Cole, D. R. Smith, and I. Smith.** 1995. Genomic organization of the mycobacterial sigma gene cluster. *Gene* **165**:67–70.
17. **Drazin, R., J. Kandel, and R. J. Collier.** 1971. Structure and activity of diphtheria toxin. II. Attack by trypsin at a specific site within the intact toxin molecule. *J. Biol. Chem.* **246**:1504–1510.
18. **Foss, F. M., M. N. Saleh, J. G. Krueger, J. C. Nichols, and J. R. Murphy.** 1997. Diphtheria toxin fusion proteins. *Curr. Top. Microbiol. Immunol.* **234**:63–81.
19. **Fourel, G., A. Phalipon, and M. Kaczorek.** 1989. Evidence for direct regulation of diphtheria toxin gene tran-

scription by an Fe^{2+}-dependent DNA-binding repressor, DtoxR, in *Corynebacterium diphtheriae. Infect. Immun.* **57:**3221–3225.

20. **Gill, D. M.** 1982. Bacterial toxins: a table of lethal amounts. *Microbiol. Rev.* **46:**86–94.

21. **Gill, D. M., and A. M. Pappenheimer, Jr.** 1971. Structure-activity relationships in diphtheria toxin. *J. Biol. Chem.* **246:**1492–1495.

22. **Glasfeld, A., E. Guedon, J. D. Helmann, and R. G. Brennan.** 2003. Structure of the manganese-bound manganese transport regulator of *Bacillus subtilis. Nat. Struct. Biol.* **10:**652–657.

23. **Greenfield, L., M. J. Bjorn, G. Horn, D. Fong, G. A. Buck, R. J. Collier, and D. A. Kaplan.** 1983. Nucleotide sequence of the structural gene for diphtheria toxin carried by corynebacteriophage β. *Proc. Natl. Acad. Sci. USA* **80:**6853–6857.

24. **Groman, N. B., and K. Judge.** 1979. Effects of metal ions on diphtheria toxin production. *Infect. Immun.* **26:**1065–1070.

25. **Gubler, J., C. Huber-Schneider, E. Gruner, and M. Altwegg.** 1998. An outbreak of nontoxigenic *Corynebacterium diphtheriae* infection: single bacterial clone causing invasive infection among Swiss drug users. *Clin. Infect. Dis.* **27:**1295–1298.

26. **Günter, K., C. Toupet, and T. Schupp.** 1993. Characterization of an iron-regulated promoter involved in desferrioxamine B synthesis in *Streptomyces pilosus:* repressor-binding site and homology to the diphtheria toxin gene promoter. *J. Bacteriol.* **175:**3295–3302.

27. **Günter-Seeboth, K., and T. Schupp.** 1995. Cloning and sequence analysis of the *Corynebacterium diphtheriae dtxR* homologue from *Streptomyces lividans* and *S. pilosus* encoding a putative iron repressor. *Gene* **166:**117–119.

28. **Hardham, J. M., L. V. Stamm, S. F. Porcella, J. G. Frye, N. Y. Barnes, J. K. Howell, S. L. Mueller, J. D. Radolf, G. M. Weinstock, and S. J. Norris.** 1997. Identification and transcriptional analysis of a *Treponema pallidum* operon encoding a putative ABC transport system, an iron-activated repressor protein homolog, and a glycolytic pathway enzyme homolog. *Gene* **197:**47–64.

29. **Hardy, I. R. B., S. Dittman, and R. W. Sutter.** 1996. Current situation and control strategies for resurgence of diphtheria in newly independent states of the former Soviet Union. *Lancet* **347:**1739–1744.

30. **Ho, V. T., D. Zahrieh, E. Hochberg, E. Micale, J. Levin, C. Reynolds, S. Steckel, C. Cutler, D. C. Fisher, S. J. Lee, E. P. Alyea, J. Ritz, R. J. Soiffer, and J. H. Antin.** 2004. Safety and efficacy of denileukin diftitox in patients with steroid-refractory acute graft- versus-host disease after allogeneic hematopoietic stem cell transplantation. *Blood* **104:**1224–1226.

31. **Horsburgh, M. J., S. J. Wharton, A. G. Cox, E. Ingham, S. Peacock, and S. J. Foster.** 2002. MntR modulates expression of the PerR regulon and superoxide resistance in *Staphylococcus aureus* through control of manganese uptake. *Mol. Microbiol.* **44:**1269–1286.

32. **Hu, H.-Y., P. D. Hunth, J. R. Murphy, and J. C. vanderSpek.** 1998. The effects of helix breaking mutations in the diphtheria toxin transmembrane domain helix layers of the fusion toxin DAB$_{389}$IL-2. *Protein Eng.* **11:**101–107.

33. **Iwamoto, R., H. Senoh, Y. Okada, and E. Mekada.** 1991. An antibody that inhibits the binding of diphtheria toxin to cells revealed the association of a 27-kDa membrane protein with the diphtheria toxin receptor. *J. Biol. Chem.* **266:**20463–20469.

34. **Kaczorek, M., F. Delpyyroux, N. Chenciner, R. E. Streek, J. R. Murphy, P. Boquet, and P. Tiollais.** 1983. Nucleotide sequence and expression in *Escherichia coli* of the CRM228 diphtheria toxin gene. *Science* **221:**855–858.

35. **Kanei, C., T. Uchida, and M. Yoneda.** 1977. Isolation from *Corynebacterium diphtheriae* C7(β) of bacterial mutants that produce toxin in medium containing excess iron. *Infect. Immun.* **18:**203–209.

36. **Kitten, T., C. L. Munro, S. M. Michalek, and F. L. Macrina.** 2000. Genetic characterization of a *Streptococcus mutans* LraI family operon and role in virulence. *Infect. Immun.* **68:**4441–4451.

37. **Krafft, A. E., S. P. Tai, C. Coker, and R. K. Holmes.** 1992. Transcription analysis and nucleotide sequence of tox promoter/operator mutants of corynebacteriophage beta. *Microb. Pathog.* **13:**85–92.

38. **Lawson, C. L., and J. Carey.** 1993. Tandem binding in crystals of a trp repressor/operator half site complex. *Nature* **366:**178–182.

39. **Lee, J. H., T. Wang, K. Ault, J. Liu, M. P. Schmitt, and R. K. Holmes.** 1997. Identification and characterization of three new promoter/operators from *Corynebacterium diphtheriae* that are regulated by the diphtheria toxin repressor (DtxR) and iron. *Infect. Immun.* **65:**4273–4280.

40. **Lieser, S. A., T. C. Davis, J. D. Helmann, and S. M. Cohen.** 2003. DNA-binding and oligomerization studies of the manganese(II) metalloregulatory protein MntR from *Bacillus subtilis. Biochemistry* **42:**12634–12642.

41. **Love, J. F., and J. R. Murphy.** 2002. Design and development of a novel genetic probe for the analysis of repressor-operator interactions. *J. Microbiol. Methods* **51:**63–72.

42. **Love, J. F., J. C. vanderSpek, and J. R. Murphy.** 2003. The *src* homology 3-like domain of the diphtheria toxin repressor (DtxR) modulates repressor activation through interaction with the ancillary metal ion-binding site. *J. Bacteriol.* **185:**2251–2258.

43. **Love, J. F., J. C. vanderSpek, V. Marin, L. Guerrero, T. M. Logan, and J. R. Murphy.** 2004. Genetic and biophysical studies of diphtheria toxin repressor and the hyperactive mutant DtxR(E175K) support a multistep model for activation. *Proc. Natl. Acad. Sci. USA* **101:**2506–2511.

44. **Low, Y. L., N. S. Jakubovics, J. C. Flatman, H. F. Jenkinson, and A. W. Smith.** 2003. Manganese-dependent regulation of the endocarditis-associated virulence factor EfaA of *Enterococcus faecalis. J. Med. Microbiol.* **52:**113–119.

45. **Lund, R. A.** 1995. The role of molecular chaparones in vivo. *Essays Biochem.* **29:**113–129.

46. **Manabe, Y. C., B. J. Saviola, L. Sun, J. R. Murphy, and W. R. Bishai.** 1999. Attenuation of virulence in *Mycobacterium tuberculosis* expressing a constitutively active iron repressor. *Proc. Natl. Acad. Sci. USA* **96:**12844–12848.

47. **Maurice, J.** 1995. Russian chaos breeds diphtheria outbreak. *Science* **167:**1416–1417.

48. **McQuillan, G. M., D. Kruszon-Moran, A. Deforest, S. Y. Chu, and M. Wharton.** 2002. Serologic immunity to diphtheria and tetanus in the United States. *Ann. Intern. Med.* **136:**660–666.

49. **Mitamura, T., R. Iwamoto, T. Umata, T. Yomo, I. Urabe, M. Tsuneoka, and E. Mekada.** 1992. The 27-kD diphtheria toxin receptor-associated protein (DRAP27) from Vero cells is the monkey homologue of human CD9 antigen: expression of DRAP27 elevates the number of diphtheria toxin receptors on toxin-sensitive cells. *J. Cell Biol.* **118:**1389–1399.

50. Moreira, L. O., A. F. B. Andrade, M. D. Vale, S. M. S. Souza, R. Hirata, Jr., L. M. O. B. Asad, N. R. Asad, L. H. Monteiro-Leal, J. O. Previato, and A. L. Mattos-Guaraldi. 2003. Effects of iron limitation on adherence and cell surface carbohydrates of *Corynebacterium diphtheriae* strains. *Appl. Environ. Microbiol.* **69:**5907–5913.

51. Moya, M., A. Dautry-Versat, B. Goud, D. Louvard, and P. Boquet. 1985. Inhibition of coated-pit formation in Hep2 cells blocks the cytotoxicity of diphtheria toxin but not ricin toxin. *J. Cell Biol.* **101:**548–559.

52. Murphy, J. R., and P. Bacha. 1979. Studies of the regulation of diphtheria toxin production, p. 181–186. *In* D. Schlessinger (ed.), *Microbiology—1979*. American Society for Microbiology, Washington, D.C.

53. Murphy, J. R., A. M. Pappenheimer, Jr., and S. Tayart de Borms. 1974. Synthesis of diphtheria *tox* gene products in *Escherichia coli* extracts. *Proc. Natl. Acad. Sci. USA* **71:**11–15.

54. Murphy, J. R., J. Skiver, and G. McBride. 1976. Isolation and partial characterization of a corynebacteriophage *tox* operator constitutive-like mutant lysogen of *Corynebacterium diphtheriae*. *J. Virol.* **18:**235–244.

55. Murphy, J. R., W. Bishai, M. Borowski, A. Miyanohara, J. Boyd, and S. Nagle. 1986. Genetic construction, expression, and melanoma selective cytotoxicity of a diphtheria toxin α-melanocyte stimulating hormone fusion protein. *Proc. Natl. Acad. Sci. USA* **83:**8258–8262.

56. Murphy, J. R., and J. C. vanderSpek. 1995. Targeting diphtheria toxin to growth factor receptors. *Semin. Cancer Biol.* **6:**259–267.

57. Naglich, J. G., J. E. Matherall, D. W. Russell, and L. Eidels. 1992. Expression cloning of a diphtheria toxin receptor: identity with a heparin-binding EGF-like growth factor precursor. *Cell* **69:**1051–1061.

58. Nakamura, Y., Y. Nishio, K. Ikeo, and T. Gojobori. 2003. The genome stability in *Corynebacterium* species due to lack of the recombinational repair system. *Gene* **317:**149–155.

59. Nakao, H., I. K. Mazurova, T. Glushkevich, and T. Popovic. 1997. Analysis of heterogeneity of *Corynebacterium diphtheriae* toxin gene, *tox*, and its regulatory element, *dtxR*, by direct sequencing. *Res. Microbiol.* **148:**45–54.

60. Nakao, H., J. M. Pruckler, I. K. Mazurova, O. V. Narvskaia, T. Glushkevich, V. F. Marijevski, A. N. Kravetz, B. I. Fields, I. K. Wachsmuth, and T. Popovic. 1996. Heterogeneity of diphtheria toxin gene, *tox*, and its regulatory element, *dtxR*, in *Corynebacterium diphtheriae* strains causing epidemic diphtheria in Russia and Ukraine. *J. Clin. Microbiol.* **34:**1711–1716.

61. Oguiza, J. A., X. Tao, A. T. Marcos, J. F. Martin, and J. R. Murphy. 1995. Molecular cloning and characterization of the *Corynebacterium diphtheriae dtxR* homolog from *Brevibacterium lactofermentum*. *J. Bacteriol.* **177:**465–467.

62. Pappenheimer, A. M., Jr., and S. J. Johnson. 1936. Studies in diphtheria toxin production. I: The effects of iron and copper. *Br. J. Exp. Pathol.* **17:**335–341.

63. Pappenheimer, A. M., Jr. 1977. Diphtheria toxin. *Annu. Rev. Biochem.* **46:**69–94.

64. Pappenheimer, A. M., Jr. 1980. Diphtheria: studies on the biology of an infectious disease, p. 45–73. *In The Harvey Lectures*, series 76. Academic Press, Inc., New York, N.Y.

65. Pappenheimer, A. M., Jr., and J. R. Murphy. 1983. Studies on the molecular epidemiology of diphtheria. *Lancet* **ii:**923–926.

66. Patzer, S. I., and K. Hantke. 2001. Dual repression by Fe^{2+}-Fur and Mn^{2+}-MntR of the *mntH* gene, encoding an NRAMP-like Mn^{2+} transporter in *Escherichia coli*. *J. Bacteriol.* **183:**4806–4813.

67. Pohl, E., X. Qiu, L. M. Must, R. K. Holmes, and W. G. J. Hol. 1997. Comparison of high-resolution structures of the diphtheria toxin repressor in complex with cobalt and zinc at the cation-anion binding site. *Protein Sci.* **6:**1114–1118.

68. Pohl, E., R. K. Holmes, and W. G. J. Hol. 1998. Motion of the DNA-binding domain with respect to the core of the diphtheria toxin repressor (DtxR) revealed in the crystal structures of apo- and holo-DtxR. *J. Biol. Chem.* **273:**22420–22427.

69. Pohl, E., R. K. Holmes, and W. G. J. Hol. 1999. Crystal structure of a cobalt-activated diphtheria toxin repressor-DNA complex reveals a metal-binding SH3-like domain. *J. Mol. Biol.* **292:**653–667.

70. Pohl, E., J. Goranson-Siekierke, M. K. Choi, T. Roosild, R. K. Holmes, and W. G. J. Hol. 2001. Structures of three diphtheria toxin repressor (DxtR) variants with decreased repressor activity. *Acta Crystallogr. D Biol. Crystallogr.* **57:**619–627.

71. Popovic, T., C. Kim, J. Reiss, M. Reeves, H. Nakao, and A. Golaz. 1999. Use of molecular subtyping to document long-term persistence of *Corynebacterium diphtheriae* in South Dakota. *J. Clin. Microbiol.* **37:**1092–1099.

72. Posy, J. E., J. M. Hardham, S. J. Norris, and F. C. Gherardini. 1999. Characterization of a manganese-dependent regulatory protein, TroR, from *Treponema pallidum*. *Proc. Natl. Acad. Sci. USA* **96:**10887–10892.

73. Qian, Y., J. H. Lee, and R. K. Holmes. 2002. Identification of a DtxR-regulated operon that is essential for siderophore-dependent iron uptake in *Corynebacterium diphtheriae*. *J. Bacteriol.* **184:**4846–4865.

74. Qiu, X., C. L. Verlinde, S. Zhang, M. P. Schmitt, R. K. Holmes, and W. G. J. Hol. 1995. Three-dimensional structure of the diphtheria toxin repressor in complex with divalent cation co-repressors. *Structure* **3:**87–100.

75. Qiu, X., E. Pohl, R. K. Holmes, and W. G. J. Hol. 1996. High-resolution structure of the diphtheria toxin repressor complexed with cobalt and manganese reveals an SH3-like third domain and suggests a possible role of phosphate as co-repressor. *Biochemistry* **35:**12292–12302.

76. Que, Q., and J. D. Helmann. 2000. Manganese homeostasis in *Bacillus subtilis* is regulated by MntR, a bifunctional regulator related to the diphtheria toxin repressor family of proteins. *Mol. Microbiol.* **35:**1454–1468.

77. Ramon, G. 1923. Sur la concentration du serum antidiphtherique et l'isolement de la antitoxine. *C. R. Seances Soc. Biol. Fil.* **88:**167–168.

78. Ratti, G., R. Rappuoli, and G. Giannini. 1983. The complete nucleotide sequence of the gene coding for diphtheria toxin in the corynephage omega (*tox+*) genome. *Nucleic Acids Res.* **11:**6589–6595.

79. Ratts, R., H. Zeng, E. A. Berg, C. Blue, M. E. McComb, C. E. Costello, J. C. vanderSpek, and J. R. Murphy. 2003. The cytosolic entry of diphtheria toxin catalytic domain requires a host cell cytosolic translocation factor complex. *J. Cell Biol.* **160:**1139–1150.

80. Ryser, H.-J., R. Mandel, and F. Ghani. 1991. Cell surface sulfhydryls are required for the cytotoxicity of diphtheria toxin but not ricin toxin in Chinese hamster ovary cells. *J. Biol. Chem.* **266:**18439–18442.

81. Saragea, A., P. Maximescu, and E. Meitert. 1979. *Corynebacterium diphtheriae*: microbiological methods

used in clinical and epidemiological investigations. *Methods Microbiol.* **13**:61–176.

82. **Schiering, N., X. Tao, H. Zeng, J. R. Murphy, G. A. Petsko, and D. Ringe.** 1995. Structures of the apo- and the metal ion-activated forms of the diphtheria tox repressor from *Corynebacterium diphtheriae*. *Proc. Natl. Acad. Sci. USA* **92**:9843–9850.

83. **Schmitt, M. P., and R. K. Holmes.** 1991. Characterization of a defective diphtheria toxin repressor (*dtxR*) allele and analysis of *dtxR* transcription in wild-type and mutant strains of *Corynebacterium diphtheriae*. *Infect. Immun.* **59**:3903–3908.

84. **Schmitt, M. P., and R. K. Holmes.** 1994. Cloning, sequence, and footprint analysis of two promoter/operators from *Corynebacterium diphtheriae* that are regulated by the diphtheria toxin repressor (DtxR) and iron. *J. Bacteriol.* **176**:1141–1149.

85. **Schmitt, M. P., M. Predich, L. Doukhan, I. Smith, and R. K. Holmes.** 1995. Characterization of an iron-dependent regulatory protein (IdeR) of *Mycobacterium tuberculosis* as a functional homolog of the diphtheria toxin repressor (DtxR) from *Corynebacterium diphtheriae*. *Infect. Immun.* **63**:4284–4289.

86. **Schmitt, M. P.** 1997. Transcription of the *Corynebacterium diphtheriae hmuO* gene is regulated by iron and heme. *Infect. Immun.* **65**:4634–4641.

87. **Schmitt, M. P.** 2002. Analysis of a DtxR-like metalloregulatory protein, MntR, from *Corynebacterium diphtheriae* that controls expression of an ABC metal transporter by an Mn^{2+}-dependent mechanism. *J. Bacteriol.* **184**:6882–6892.

88. **Skogen, V., V. V. Cherkasova, N. Maksimova, C. K. Marston, H. Sjursen, M. W. Reeves, O. Olsvik, and T. Popovic.** 2002. Molecular characterization of *Corynebacterium diphtheriae* isolates, Russia, 1957–1987. *Emerg. Infect. Dis.* **8**:516–518.

89. **Smith, W. P., P. C. Tai, J. R. Murphy, and B. D. Davis.** 1980. A precursor in the cotranslational secretion of diphtheria toxin. *J. Bacteriol.* **141**:184–189.

90. **Spiering, M. M., D. Ringe, J. R. Murphy, and M. A. Marletta.** 2003. Metal ion stoichiometry and functional studies of the diphtheria toxin repressor. *Proc. Natl. Acad. Sci. USA* **100**:3808–3813.

91. **Sun, L., and J. R. Murphy.** Unpublished data.

92. **Sun, L., J. C. vanderSpek, and J. R. Murphy.** 1998. Isolation and characterization of iron-independent positive dominant mutants of the diphtheria toxin repressor DtxR. *Proc. Natl. Acad. Sci. USA* **95**:14985–14990.

93. **Tao, X., J. Boyd, and J. R. Murphy.** 1992. Specific binding of the diphtheria *tox* regulatory element DtxR to the *tox* operator requires divalent cations and a 9-base-pair interrupted palindromic sequence. *Proc. Natl. Acad. Sci. USA* **89**:5897–5901.

94. **Tao, X., and J. R. Murphy.** 1992. Binding of the metalloregulatory protein DtxR to the diphtheria *tox* operator requires a divalent heavy metal ion and protects the palindromic sequence from DNase I digestion. *J. Biol. Chem.* **267**:21761–21764.

95. **Tao, X., and J. R. Murphy.** 1993. Cysteine-102 is positioned in the metal binding activation site of the

Corynebacterium diphtheriae regulatory element DtxR. *Proc. Natl. Acad. Sci. USA* **90**:8524–8528.

96. **Tao, X., and J. R. Murphy.** 1994. Determination of the DtxR consensus binding site by in vitro affinity selection. *Proc. Natl. Acad. Sci. USA* **91**:9646–9650.

97. **Tao, X., H. Zeng, and J. R. Murphy.** 1995. Heavy metal ion activation of the diphtheria *tox* repressor (DtxR) results in the formation of stable homodimers. *Proc. Natl. Acad. Sci. USA* **92**:6803–6807.

98. **Ton-Than, H., and O. Schneewind.** 2003. Assembly of pili on the surface of *Corynebacterium diphtheriae*. *Mol. Microbiol.* **50**:1429–1438.

99. **Touate, D., M. Jacques, B. Tardat, L. Bouchard, and S. Despied.** 1995. Lethal oxidative damage and mutagenesis are generated by iron in Δ*fur* mutants of *Escherichia coli*: protective role of superoxide dismutase. *J. Bacteriol.* **177**:2305–2314.

100. **Twigg, P. D., G. P. Wylie, G. Wang, D. L. D. Caspar, J. R. Murphy, and T. M. Logan.** 1999. Expression and assignment of the 1H, 15N, and 13C resonances of the C-terminal domain of the diphtheria toxin repressor. *J. Biomol. NMR* **13**:197–198.

101. **Twigg, P. D., G. Parthasarathy, L. Guerrero, T. M. Logan, and D. L. Caspar.** 2001. Disordered to ordered folding in the regulation of diphtheria toxin repressor activity. *Proc. Natl. Acad. Sci. USA* **98**:11259–11264.

102. **Uchida, T., D. M. Gill, and A. M. Pappenheimer, Jr.** 1971. Mutation in the structural gene for diphtheria toxin carried by temperate phage β. *Nature New Biol.* **233**:8–11.

103. **Vitek, C. R., and M. Wharton.** 1998. Diphtheria in the former Soviet Union: reemergence of a pandemic disease. *Emerg. Infect. Dis.* **4**:539–550.

104. **Wang, G., G. P. Wylie, P. D. Twigg, D. L. D. Caspar, J. R. Murphy, and T. M. Logan.** 1999. Solution structure and peptide binding studies of the C-terminal Src homology 3-like domain of the diphtheria toxin repressor protein. *Proc. Natl. Acad. Sci. USA* **96**:6119–6124.

105. **Wang, Z., M. P. Schmitt, and R. K. Holmes.** 1994. Characterization of mutations that inactivate the diphtheria toxin repressor. *Infect. Immun.* **62**:1600–1608.

106. **Welkos, S. L., and R. K. Holmes.** 1981. Regulation of toxinogenesis in *Corynebacterium diphtheriae*. I. Mutations in bacteriophage β that alter the effect of iron on toxin production. *J. Virol.* **37**:936–945.

107. **White, A., X. Ding, J. R. Murphy, and D. Ringe.** 1998. Structure of metal ion-activated diphtheria toxin repressor/*tox* operator complex. *Nature* **394**:502–506.

108. **Yamaizumi, M., E. Mekada, T. Uchida, and Y. Okada.** 1978. One molecule of diphtheria toxin fragment A introduced into a cell can kill the cell. *Cell* **15**:245–250.

109. **Zeng, H., and J. R. Murphy.** Unpublished data.

110. **Zhang, H., D. Zhao, M. Revington, W. Lee, X. Jia, C. Arrowsmith, and O. Jardetzky.** 1994. The solution structures of the *trp* repressor-operator DNA complex. *J. Mol. Biol.* **238**:592–614.

111. **Zucker, D. R., and J. R. Murphy.** 1984. Monoclonal antibody analysis of diphtheria toxin. I. Localization of epitopes and neutralization of cytotoxicity. *Mol. Immunol.* **21**:785–793.

Actinomyces and *Arcanobacterium* spp.: Host-Microbe Interactions

B. HELEN JOST AND STEPHEN J. BILLINGTON

60

The organisms informally known as the actinomycetes are gram-positive rods, which vary considerably in their size and morphology. Recently, the taxonomy of this genus has undergone dramatic changes. With the advent of DNA hybridization and 16S rRNA sequencing, there has been substantial phylogenetic reorganization, including the addition of many new species and the reclassification of *Actinomyces pyogenes* into the genus *Arcanobacterium*. This chapter describes our current knowledge of the mechanisms of *Actinomyces* and *Arcanobacterium* spp. host-microbe interactions, with an emphasis on recent advances in the molecular analyses of toxins and factors involved in bacterial adherence.

ACTINOMYCES SPP.

The genus *Actinomyces* now contains over 35 species, and these organisms are generally found in association with mammals. *Actinomyces* spp. have been isolated either as commensals or from disease processes in humans, rodents, and bovine, canine, feline, and porcine species. Human, commensal *Actinomyces* spp. are predominantly found in the oral cavity or the urogenital tract. Members of the genus *Actinomyces* are nonmotile and non-spore-forming and although facultative anaerobes, most species grow best under anaerobic conditions. The cellular morphology of these organisms may be diphtheroidal or filamentous, with or without branching; they contain genomes with a high G+C content of 65 to 69%.

Precise delineation of the genus *Actinomyces* has been difficult, as there is considerable phenotypic variation between species. Hence, differentiating isolates on the basis of conventional biochemical tests is problematic. The use of molecular techniques such as DNA hybridization for the taxonomic analysis of the oral actinomycetes has resulted in significant changes in the taxonomy of this genus (25). *Actinomyces naeslundii* serotype I is now considered *A. naeslundii* genospecies 1. *A. naeslundii* serotypes II, III, and NV plus *Actinomyces viscosus* serotype II have been grouped together as *A. naeslundii* genospecies 2. Nonhuman *A. viscosus* maintains its designation. *Actinomyces israelii* serotype II

has also retained its nomenclature, but *A. israelii* serotype I is now reclassified as *Actinomyces gerencseriae*. In addition, new *Actinomyces* species are being identified using 16S rRNA sequencing, resulting in a substantial number of new species in this genus over the past several years.

ORAL *ACTINOMYCES* SPP.

Along with the streptococci, *Actinomyces* spp. are the major members of the >500 bacterial species present in the oral cavity, and these organisms persist from infancy to adulthood (45). Most of the bacteria present in the oral cavity are in the form of biofilms, also known as dental plaque. Oral biofilms form on distinctly different surfaces: hard surfaces, such as the enamel and cementum associated with the tooth, and the soft surfaces of the oral mucosa, which include the buccal, lingual, and gingival epithelia. Biofilms on teeth are at least several bacterial layers thick; if left undisturbed by teeth cleaning, they can increase to a thickness of several hundred micrometers. In contrast, mucosal biofilms usually consist of only bacterial monolayers.

Bacterial adherence is an important first step in colonization and biofilm formation. Bacteria which colonize the oral cavity have evolved a number of mechanisms which allow them to adhere to and persist in this microenvironment. Saliva contains a number of factors important to oral microbial ecology, dental biofilm formation, and health. For example, in this unique habitat, several salivary glycoproteins, such as agglutinins or mucins, and the normal flushing action of saliva flow promote bacterial clearance. The formation of a biofilm, such as dental plaque, on tooth and mucosal surfaces follows a definite sequence characterized by the initial presence of facultatively anaerobic gram-positive bacteria such as *Actinomyces* spp., to a flora rich in anaerobic gram-negative bacteria in the later stages (18).

A. naeslundii genospecies 1 and 2 and *Actinomyces odontolyticus* are some of the first bacterial species to colonize the oral cavity after birth (18). *A. naeslundii* is one of the pioneer organisms found on the cleaned tooth surface; as a primary colonizer, this organism plays an important role in modulation of the microbial flora of the mouth. Adherence

Gram-Positive Pathogens, 2nd edition, edited by Vincent A. Fischetti et al.
© 2006 ASM Press, Washington, D.C.

of *A. naeslundii* to the tooth and mucosal surfaces provides a substrate for the adherence of other bacteria, such as the oral streptococci and gram-negative anaerobes (18). *A. naeslundii* also participates in coaggregation with other oral bacteria such as the streptococci, and the action of its enzymes, such as neuraminidase, modifies host receptors to allow the adhesion of yet other bacterial species. *Actinomyces meyeri, Actinomyces georgiae,* and *A. gerencseriae,* which have been isolated from peridontically normal gingival crevices, and *A. odontolyticus,* which has been found on the tongue surface but not the teeth or the buccal epithelium, are not implicated in periodontal disease.

Oral *Actinomyces* spp. coaggregate with streptococci and some gram-negative bacteria which are putative periodontal pathogens, and at least six different adherence mechanisms have been characterized which mediate coaggregation among *Actinomyces* spp. and the oral streptococci (33). Many of these interactions involve protein-carbohydrate interactions, which are inhibitable by the addition of mono- or disaccharides. Of the >80% of oral streptococci that coaggregate with *A. naeslundii* genospecies 1 and 2, over 85% of these interactions are reversible in the presence of lactose (33). However, coaggregation of *A. naeslundii* and some oral bacteria is not inhibited by lactose, indicating the presence of multiple adhesins. Cocultures of *A. naeslundii* genospecies 2 and *Streptococcus oralis* produce a luxuriant biofilm, in contrast to monoculture of these organisms, which results in only poor biofilm formation (56). As coaggregation among plaque bacteria results in growth and accumulation of the biofilm, the ability to form bacterial aggregates may actually enhance the pathogenic properties of these bacteria. For example, bacterial aggregates consisting of *Actinomyces* spp. and oral streptococci are resistant to phagocytosis and killing by polymorphonuclear leukocytes (PMNs) in vitro and in vivo (54). Therefore, it is clear that these interactions are beneficial for the participating organisms.

Numerous in vivo and in vitro studies have demonstrated the significance of *Actinomyces* spp. in the initiation and progression of plaque development. High numbers of these bacteria are recovered within minutes from the cleanest tooth surfaces. An increasing number of these organisms are recovered from enamel chips placed in the oral cavity of human subjects during the initial 24 h of plaque development (53). These early colonizers facilitate further bacterial colonization and maintain biofilm integrity, as well as protect the host against colonization by more pathogenic bacterial species; the numbers of *A. naeslundii* genospecies 1 and 2 in the oral biofilm correlate with health and disease. Studies with several animal models have established that gram-negative periodontal pathogens fail to adhere and colonize the oral cavity in the absence of the primary colonizing *Actinomyces* spp. As many putative periodontal pathogens are known to express proteases and other enzymes associated with cytokine induction and/or host tissue inflammation and destruction (66), these early colonizers are crucial to the maintenance of health and disease in the oral environment.

Streptococcus mutans and *A. naeslundii* are the major pathogenic bacteria found associated with surface root caries. Early studies of the microbiology of cariogenesis emphasized the role of *A. naeslundii* genospecies 1 and 2. As well as being present in lesions, these organisms were also shown to produce caries in experimental animal models. However, there is little direct evidence to support a definitive role in caries formation, gingivitis, and/or periodontitis, due to the polymicrobial nature of many of these infections. Despite their pathogenic potential, the oral *Actinomyces*

spp. are now generally considered commensals, although it is clear that their key role in colonization of the oral cavity, including teeth and the gingival crevices, modulates subsequent periodontal disease. Traditional factors such as sugar intake and toothbrushing often correlate poorly with the development of dental caries. However, the association of large numbers of *S. mutans* and *A. naeslundii* with high- and low-caries levels of development, respectively, suggests an inverse relationship of adhesion of cariogenic and commensal bacteria with this disease (68). Therefore, the relative susceptibility or resistance of an individual to dental caries and/periodontal disease is significantly influenced by the oral microbial flora present, which is ultimately a function of the adhesins expressed by these bacteria.

Actinomyces Fimbriae

Fimbriae (pili) are filamentous, proteinaceous surface appendages that are key components in the cell-to-surface or cell-to-cell adherence of numerous bacterial pathogens. While fimbrial production is widespread in gram-negative bacteria, to date, only a few gram-positive organisms have been shown to produce fimbriae. They include *Actinomyces* spp. (11), some oral streptococci (75), and several *Corynebacterium* spp. and *Arcanobacterium pyogenes* (72, 76). Fimbriae expressed by gram-positive bacteria are morphologically similar to those of gram-negative bacteria, at 200 to 700 nm in length and 2.5 to 4.5 nm in width (11, 72, 76).

Fimbrial production is a common characteristic of the oral *Actinomyces* spp., including *A. naeslundii* genospecies 1 and 2, *A. viscosus, A. odontolyticus,* and *A. israelii* (22, 78); much of our knowledge of the gram-positive fimbriae comes from the study of *A. naeslundii* genospecies 2 strain T14V, formerly *A. viscosus.* These fimbriae were first isolated by Cisar and Vatter (11), and subsequent molecular characterization led to the identification of two fimbrial types, designated 1 and 2, which are involved in adherence to different surfaces in the oral cavity. Type 1 fimbriae mediate adherence to the hard tooth surfaces by interaction with salivary acidic proline-rich proteins (APRPs) and/or statherins that coat the tooth enamel (36). In contrast, type 2 fimbriae promote adherence to mucosal epithelial cells, and this adherence is enhanced by neuraminidase treatment of host cells (7, 63). Type 2 fimbriae bind to β-linked galactose- and N-acetylgalactosamine-containing glycoconjugates of host cells, as well as polysaccharides produced by some oral streptococci; hence, they are involved in coaggregation of these species. The *A. naeslundii* type 1 and type 2 fimbrial subunits share 34% amino acid identity, but there is no significant global homology to the fimbrial subunits of gram-negative bacteria (79). *Actinomyces* spp. may express either or both of the fimbrial types, but type 1 fimbriae are more commonly found on *A. naeslundii* genospecies 2 (22).

Type 1 Fimbriae

Type 1 fimbriae preferentially bind to APRPs and/or statherins, which are polymorphic, multifunctional, salivary proteins. These fimbriae bind avidly to APRPs adsorbed onto tooth surfaces, but they do not interact with APRPs in solution. Adsorbed APRPs undergo conformational changes, and the type 1 fimbriae recognize cryptic domains, which are only exposed in adsorbed molecules. This provides the bacteria with a mechanism for efficiently attaching to teeth while suspended in saliva. Humans and animals produce structurally distinct APRPs and statherins, and the fimbrial specificity of an actinomycete can result in host and/or tissue tropism. *A. naeslundii* fimbriae with preferential binding to APRPs are isolated from the human oral cavity, whereas those with a preference for statherin are isolated from the rodent mouth or human soft tissue infections (36). Recognition of the different substrates correlates with the amino acid sequence of the major structural subunit, FimP (36, 37). However, binding has not been unequivocally ascribed to the subunit protein; accessory proteins such as the minor subunits and/or tip proteins may be involved in binding specificity, as monoclonal antibodies which bind to FimP do not inhibit adhesion of fimbriae to APRPs (50).

The type 1 fimbriae of *A. naeslundii* genospecies 2 strain T14V preferentially bind to proline-149 and glutamine-150 in the C terminus of APRPs (36). The type 1 subunit gene, *fimP*, is found clustered with five other genes, *orf1* to -6, required for fimbrial biogenesis (80, 83). The FimP subunit contains a signal peptide and a cell wall sorting domain, containing the motif LPxTG, which are required for secretion of FimP through the cell membrane and subsequent covalent linkage of the protein to the cell wall, via the enzyme sortase. Sortases catalyze cleavage of proteins between the threonine and glycine residues of the cell wall sorting domain and covalently link proteins to the peptidoglycan, subsequently resulting in cell surface localization (42). The FimP protein also contains a pilin motif and an E box, sequences which are required for the assembly of gram-positive fimbriae (73). Upstream of *fimP* are *orf3*, *orf2*, and *orf1* (Fig. 1; Table 1). Neither Orf2 nor Orf3 has any similarity to any known protein, although they contain N-terminal signal peptides, suggesting that they are secreted proteins. *orf1* encodes a protein with amino acid similarity to the FimP subunit (83). In addition, Orf1 has a signal peptide, a cell wall sorting domain, and an E box; this protein is thought to be incorporated into the fimbrial strand. Knockout mutants with deletions in *orf1* or *orf2* do not express FimP, resulting in a concomitant loss of type 1 fimbriae (83). In contrast, an *orf3* knockout mutant expressed Orf1 and FimP, but the proteins were not assembled into fimbriae. Downstream of *fimP* are *orf4*, *orf5*, and *orf6* (Fig. 1; Table 1). Orf4 is a homolog of sortase, the enzyme involved in anchoring proteins to the gram-positive cell wall via their cell wall sorting domain. A knockout mutant in this gene had the same phenotype as an *orf3* mutant, in that FimP and Orf1 accumulated in the cell but were not assembled (83). *orf5* encodes a protein with a putative membrane-spanning domain, but no signal peptide or cell wall sorting domain, and it displays no significant similarity to any known protein (37). Orf6 displays similarity to a *Pseudomonas putida* type 4 pre-pilin peptidase (37). The relevance of this similarity is unknown, as FimP does not possess a type 4-like signal sequence. However, it is possible that other proteins involved in fimbrial biogenesis require cleavage by this peptidase. Attempts to construct knockout mutants in *orf5* and *orf6* were unsuccessful (83). Human *A. naeslundii* genospecies 2, which preferentially binds APRPs, has conserved *fimP* genes with >98% nucleotide identity and >97% amino acid identity (36).

Molecular characterization of statherin-specific type 1 fimbriae expressed by *A. naeslundii* genospecies 2 strain ATCC 19246 revealed a related but unique fimbrial gene operon with 81.3% nucleotide identity to that of strain T14V (Fig. 1; Table 1). Statherin-specific fimbriae bind to the C-terminal threonine-42 and phenylalanine-43 in statherins (36). The FimP protein, which displays 80.6% amino acid identity to FimP of strain T14V, also contains a signal peptide and cell wall sorting domain, as well as a pilin motif and an E box. In addition to *fimP*, the operon contains three other genes, *orfA*, *orfB*, and *orfC* (37). *orfA*, upstream of *fimP*, has 79.7% nucleotide identity to *orf3*, *orf2*, and *orf1* of strain T14V; the C terminus of OrfA displays 40% amino acid similarity to the FimP subunit (37). OrfA contains a signal peptide, a cell wall sorting domain, and the E box, found associated with gram-positive fimbrial subunits. Like Orf1 from strain T14V, OrfA does not have a pilin motif. OrfA shows 35 to 40% amino acid similarity to usher-like proteins, autotransporters, and K88 pilin subunit proteins (37). Downstream of *fimP* is *orfB* with 80.8% nucleotide identity to *orf4* and the first two-thirds of *orf5* from strain T14V (37); like Orf4, OrfB is a sortase homolog. *orfC*, downstream of *orfB*, has 87.7% nucleotide identity with *orf6* from strain T14V and encodes a protein with 40% amino acid similarity to the type 4 pre-pilin peptidase from *P. putida* (37).

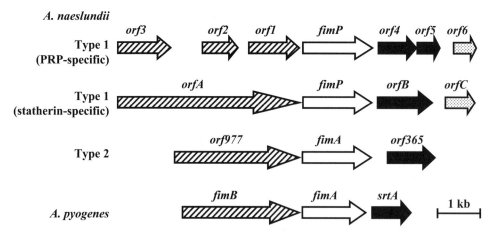

FIGURE 1 Organization of the fimbrial operons from *Actinomyces* spp. and *A. pyogenes*. The size and direction of transcription of each gene are shown by the arrows. Genes with similar functions are shaded as follows: hatched lines, adhesin-minor subunit-accessory protein; white, major structural subunit; black, sortase; stippled, pre-pilin peptidase-like protein. Bar, 1 kb (lower right).

TABLE 1 *A. naeslundii* fimbrial proteins

Protein	Function	Molecular mass (kDa)	Motif(s)	Amino acid similarity
Type 1 fimbriae (APRP specific)				
FimP	Structural subunit	56.9	Pilin motif, E box, cell wall sorting domain	*Actinomyces* spp. and *C. diphtheriae* fimbrial subunits; Orf1
Orf1	Putative minor structural protein	39.3	E box, cell wall sorting domain	*Actinomyces* spp. and *C. diphtheriae* fimbrial subunits; OrfA
Orf2	Unknown; possible accessory protein	44.0	Signal peptide	None
Orf3	Unknown; possible accessory protein	28.1	Signal peptide	None
Orf4	Sortase	30.9		Sortases
Orf5	Unknown; possible accessory protein	17.8		None
Orf6	Putative pre-pilin peptidase	16.7	Signal peptide	*P. putida* pre-pilin peptidase; OrfC
Type 1 fimbriae (statherin specific)				
FimP	Structural subunit	56.4	Pilin motif, E box, cell wall sorting domain	*Actinomyces* spp. and *C. diphtheriae* fimbrial subunits; OrfA
OrfA	Putative minor structural protein	149.1	E box, cell wall sorting domain	*Actinomyces* spp. and *C. diphtheriae* fimbrial subunits; Orf1
OrfB	Sortase	41.9		Sortases; Orf4 and Orf5
OrfC	Putative pre-pilin peptidase	27.3	Signal peptide	*P. putida* pre-pilin peptidase; Orf6
Type 2 fimbriae				
FimA	Structural subunit	56.0	Pilin motif, E box, cell wall sorting domain	*A. pyogenes* and *C. diphtheriae* fimbrial subunits
Orf977	Putative structural protein and fibronectin-binding adhesin	103.4	E box, fibronectin-binding domain, cell wall sorting domain	FimA and FimP; *A. pyogenes* FimB, *S. gordonii* CshA; streptococcal SfbII
Orf365	Sortase	39.4		Sortases

DNA hybridization experiments have indicated that *orf1* to *orf6* and *fimP* are highly conserved in *A. naeslundii* genospecies 2 (37). Similarly, these genes are highly conserved in *A. naeslundii* genospecies 1 with APRP-binding specificity, with the exception of *orf5*, which is either significantly divergent or absent in these bacteria (37). Genes from strains of *A. viscosus* with statherin-binding specificity show less conservation, with very divergent or absent *orf3* and *orf5* (37). In *Actinomyces* spp., *orf4* and *orf6*, encoding a sortase and a putative pre-pilin peptidase, respectively, are the most highly conserved genes, being found in all type 1 fimbriate strains (37), possibly indicating that the pathway for fimbrial biogenesis is highly conserved in these organisms.

Type 2 Fimbriae

The 6.2-kb *A. naeslundii* genospecies 2 strain T14V type 2 fimbrial operon is surrounded by a gene encoding elongation factor TU and the S10 ribosomal protein operon (24, 82). Type 2 fimbriae are assembled from FimA subunits (17, 81), which are 56-kDa proteins with a signal peptide, cell wall sorting domain, pilin motif, and E box. In addition to *fimA*, the operon contains two other genes involved in fimbrial biogenesis, *orf977* and *orf365* (Fig. 1; Table 1). Immediately upstream of *fimA* is *orf977*, which encodes a 103.4-

kDa protein containing a signal peptide, a cell wall sorting domain, an E box, and a C-terminal fibronectin-binding domain. The C terminus of Orf977 displays amino acid similarity to *A. naeslundii* fimbrial subunits FimA and FimP, and the presence of the E box suggests that this protein may be incorporated into the fimbrial strand. Hoflack and Yeung hypothesized that Orf977 may act as a fimbrial adhesin (24), based on amino acid similarity with a *Streptococcus gordonii* surface-expressed, fibronectin-binding adhesin, CshA (43). Klier et al. identified a 95-kDa protein associated with type 2 fimbriae in *Actinomyces* serotype WVA963; this protein could act as adhesin in the absence of expression of the fimbrial subunit (32). While this protein was not further characterized, it may be similar to that encoded by *orf977*, given their similar molecular masses. *orf365*, downstream of *fimA*, displays significant amino acid similarity to bacterial sortases. Mutants in *orf365* express and secrete FimA, but the protein is not assembled into functional fimbriae (82). Unlike type 1 fimbriae, a gene encoding a putative pre-pilin peptidase was not associated with the type 2 fimbrial operon.

Fimbrial Biogenesis

Gram-positive fimbriae are significantly more chemically and physically resistant than fimbriae from gram-negative bacteria, suggesting the presence of covalent linkages between subunits, and hence a novel system of polymerization and assembly. The *Corynebacterium diphtheriae* genome contains homologs of the fimbrial gene operons of *A. naeslundii* types 1 and 2; the mechanism for fimbrial biogenesis in this organism was recently described (74). The best studied of the *C. diphtheriae* fimbrial operons consists of a major subunit, SpaA; two minor subunits, SpaB, spaced at regular intervals along the shaft; and SpaC, located at the tip of the fimbriae (74). SpaA and SpaC show amino acid similarity with FimP and Orf1/OrfA, respectively, but SpaB displays no similarity to any known protein. Based on the location of SpaC in the *C. diphtheriae* fimbrial strand, it is tempting to speculate that Orf1 and/or OrfA of *A. naeslundii* type 1 fimbriae also functions as a tip protein.

Each of the *C. diphtheriae* subunits is synthesized as a precursor with an N-terminal signal peptide and C-terminal cell wall sorting domain, and these proteins contain additional conserved domains. A pilin motif with a conserved lysine (WxxxVxVYP**K**) (in boldface type) is found in all *Actinomyces*, *Arcanobacterium*, and *Corynebacterium* spp. major fimbrial subunits sequenced to date. The pilin motif is not found in any of the minor subunit or accessory proteins encoded by the fimbrial operons of these organisms. Site-directed mutagenesis experiments indicated that both the pilin motif and the cell wall sorting domain were required for subunit polymerization and fimbrial assembly (74). An additional element found in all fimbrial subunits and putative minor subunits (e.g., Orf1, OrfA, and Orf977 of *Actinomyces* spp. and FimB of *A. pyogenes*) is the E box (YxLxET xAPxGY), so named because of its conserved glutamic acid (in boldface type) (72). This motif is found at the C terminus of the protein, between the pilin motif and the cell wall sorting domain. Site-directed mutagenesis has determined that the E box is not absolutely required for polymerization, but it appears to be important for incorporation of the minor subunits and/or tip proteins into the fimbrial strand (72).

SrtA, a sortase homolog also encoded by the fimbrial operon, catalyzes the polymerization or ordered cross-linking of subunits into the fimbrial strand. Sortase first cleaves the fimbrial subunit between threonine and glycine in the C-terminal cell wall sorting domain (LPxTG) and generates subsequent covalent linkages between individual subunits using the side chain amino acid groups of the pilin motif (72). Deletion of *srtA*, but not any of the other five sortases encoded in the *C. diphtheriae* genome, abolished fimbrial assembly, indicating that SrtA is a fimbriae-specific sortase (74). It is not yet clear how the additional subunits and/or tip proteins are assembled into the fimbriae in the absence of a pilin motif.

While the evidence for this method of fimbrial biogenesis in *Actinomyces* spp. is still indirect, Ton-That and Schneewind have demonstrated that the *A. naeslundii* genospecies 2 strain T14V FimA also requires the pilin motif and cell wall sorting domain for polymerization into a fimbrial strand (74). Furthermore, when expressed in *C. diphtheriae*, *A. naeslundii* FimA polymerization was also sortase dependent, although in this case it was dependent on SrtD, the sortase encoded in the *C. diphtheriae* fimbrial operon with the most similarity to the *A. naeslundii* type 2 fimbrial operon (74).

Role of Fimbriae in Host Colonization and Potential Virulence

Type 1 fimbriae are thought to participate in early plaque development by adhesion to the newly cleaned tooth sur-

face. With respect to development of microbial communities such as dental biofilms, APRPs appear to promote adhesion of commensal organisms, such as *A. naeslundii* and the oral streptococci, but statherins promote adhesion of potential pathogens such as *Porphyromonas* spp. and *Candida albicans*. Similarly, non-oral *A. naeslundii* genospecies 1 expresses fimbriae that display weaker binding to statherin and none to APRPs, in contrast to the fimbriae of oral *A. naeslundii* genospecies 1, which bind avidly to APRPs and statherins (37). Type 2 fimbriae mediate late plaque development by coaggregation with other plaque bacteria such as the oral streptococci, and these fimbriae also mediate colonization of the oral mucosa. Therefore, expression of the different fimbrial types by the oral *Actinomyces* spp. results in tropism for specific host species and/or an ecological niche, e.g., the oral cavity or skin and soft tissue infections. Furthermore, in the case of a commensal organism, the ability to shift ecological niches could be sufficient to drive the organism to a more pathogenic lifestyle.

Neuraminidase

Neuraminidases (or sialidases) cleave terminal sialic acid residues from carbohydrates or glycoproteins. Enzyme specificity can depend on the type of O- or N-linked substitutions or the nature of the alpha-glycosidic linkages coupling the sialic acid to its underlying molecule. Neuraminidases are expressed by a number of bacteria, most notably those associated with the host mucosal surface, including bacteria found in the oral cavity. These enzymes can play a number of roles in virulence. Neuraminidases impair the host immune response as desialylation increases the susceptibility of mucosal immunoglobulin A (IgA) to bacterial proteases. *A. naeslundii* and *A. odontolyticus* neuraminidases can desialylate mucosal IgA (20), which may positively influence bacterial survival in the oral cavity. Furthermore, neuraminidases enhance bacterial adhesion and hence colonization. Neuraminidases unmask cryptic receptors on host cells, promoting binding by other bacterial adhesins. In addition, the action of neuraminidase can decrease mucous viscosity, enhancing the colonization of the underlying tissues. Given these activities, it should not be unexpected that neuraminidase has a significant effect on the distribution of bacterial species in the oral cavity.

Neuraminidase activity is expressed by most oral *Actinomyces* spp., with this enzyme detected in 79% of clinical isolates from five species and in 85 and 100% of *A. naeslundii* genospecies 1 and 2, respectively (44). Adhesion of *Actinomyces* spp. to host epithelial (63), phagocytic (61, 62), and red blood (hemagglutination) (13) cells requires the action of neuraminidase to expose cryptic host cell receptors. In all cases, adhesion was mediated by type 2, but not type 1, fimbriae. The host receptor on buccal epithelial cells was identified as a 160-kDa glycoprotein (6). Neuraminidase was required to promote nonopsonic phagocytosis of *A. naeslundii* genospecies 1 and 2 by PMNs; in its absence, bacterial internalization and subsequent killing did not occur (62). The major receptor on host PMNs was identified as CD43 (leukosialin) (61). CD45 (leukocyte common antigen), although present in lower abundance on the PMN surface, also acted as a receptor for type 2 fimbriae (61).

The *nanH* gene encoding neuraminidase was cloned and sequenced from *A. naeslundii* genospecies 2 strain T14V. The gene encoded a 92.9-kDa protein which displayed significant amino acid similarity to the catalytic domain and Asp box motifs of other bacterial neuraminidases (77). Furthermore, this protein contained a cell wall sorting domain (77), and cell wall anchoring is consistent with the location of the majority of neuraminidase activity associated with the cell (13). DNA hybridization experiments also indicated that genes with substantial DNA identity to *nanH* were present in 18 strains from five *Actinomyces* spp. (77). Neuraminidase expression levels varied within strains of a single species (13), but the Southern blotting experiments described by Yeung indicate that the *nanH* gene is present in only one copy (77). Two tandem promoters, identified upstream of the *A. naeslundii* genospecies 2 *nanH* gene, regulate gene expression during log- and stationary-phase growth, and these promoters may modulate gene expression under different environmental conditions (77).

Adherence to ECM Components

The host extracellular matrix (ECM) is composed of a number of proteins and glycoproteins which are responsible for providing and maintaining the form of the body. In the oral cavity specifically, teeth with caries have exposed type I collagen on the cemental and tooth root surfaces, and the surface of the tooth itself is exposed to other ECM components, such as fibronectin and fibrinogen from plasma-derived crevicular fluid. Many commensal and pathogenic bacteria utilize these ubiquitous host ECM molecules as substrates for adhesion.

A. *naeslundii* genospecies 1 and 2, A. *israelii*, A. *gerencseriae*, and A. *odontolyticus* bound to human type I collagen, although not all strains within a species did so (23, 39, 40). Mutants of A. *naeslundii* genospecies 2 that possessed type 1, but not type 2, fimbriae bound to collagen, suggesting that adhesion was associated with type 1 fimbriae (39). However, at least one collagen-binding strain, A. *naeslundii* genospecies 1 strain ATCC 12104, does not possess type 1 fimbriae (40), so it appears that collagen binding may be multifactorial. As well as to collagen, A. *naeslundii* genospecies 1 and 2 also bound to fibrinogen and fibronectin (9, 23, 40), and this activity was associated with the bacterial cell surface (23). In A. *naeslundii* genospecies 1 strain ATCC 12104, the presence of free fibrinogen reduced bacterial binding to collagen-coated beads (23), suggesting that the same adhesin may be responsible for collagen and fibrinogen binding. Interestingly, portions of APRP molecules are structurally related to collagen, which may explain the tropism of type 1 fimbriae from some *Actinomyces* spp. for both APRPs and collagen.

Therefore, the presence of specific ECM-binding adhesins is not ubiquitous in all strains of a given species, and adherence to a specific ECM component may occur by different mechanisms in different *Actinomyces* spp. While some of these ECM-binding activities are specific to type 1 fimbriae, the identity of the other adhesins has not been elucidated. It is tempting to speculate that some ECM binding may be due to the presence of cell surface-exposed MSCRAMM (microbial surface components recognizing adhesive matrix molecules)-like proteins similar to those found in other gram-positive, mucosal pathogens (58). The ability to adhere to type I collagen could contribute to the ability of these bacteria to penetrate into dental tissue during cariogenesis. Bacterial adhesion may also interfere with the ability of the tooth matrix to remineralize after the acid challenge has abated; hence, collagen binding may contribute to the cariogenic potential of the oral *Actinomyces* spp.

Urease

Ureases are multisubunit, nickel-containing enzymes that catalyze the hydrolysis of urea to carbon dioxide and ammonia, and strains of A. *naeslundii* genospecies 1, but not 2, are usually urease positive upon isolation. An A. *naeslundii*

genospecies 1 urease gene cluster was shown to be composed of seven contiguously arranged open reading frames with significant similarity to the *ureABCEFGD* genes from other organisms (47). Three genes, *ureABC*, encode the structural enzyme, while the remaining genes are required for the synthesis of catalytically active urease. Transcription of the operon was from a single promoter upstream of *ureA* (47). Interestingly, expression of the A. *naeslundii* urease operon was up-regulated under nitrogen-limiting conditions (46), which often exist when carbohydrates are present in excess.

A carbohydrate-rich diet promotes the overgrowth of acidogenic streptococci, resulting in decreased plaque pH, which is directly responsible for the development of dental caries. In the oral cavity, ureolysis of salivary urea raises the plaque pH, protecting the less acid-tolerant A. *naeslundii* genospecies 1 from the lethal effects of acidification. These organisms subsequently utilize the ammonia liberated during ureolysis as a nitrogen source, which gives them an additional growth advantage, as ammonia is not readily utilized by the acidogenic streptococci (47). As the balance between the more- and less-acidogenic organisms is one of the most important determinants of caries susceptibility, ureolysis by A. *naeslundii* could ultimately inhibit caries formation by preventing the shift to a very acidogenic flora (8). The effects of A. *naeslundii* urease in the oral ecology could be even more significant during the early stages of plaque formation, as this organism is one of the pioneers in colonization of the oral cavity. However, ureolysis may also promote calculus (tartar) formation, and exposure to elevated levels of ammonia and its derivatives enhances the inflammatory processes and tissue loss associated with periodontal disease (8).

Levansucrase and Levanse Activity

It has been long hypothesized that the ability of *Actinomyces* spp. to hydrolyze dietary sucrose is a virulence characteristic that contributes to the initiation and progression of dental caries and periodontal disease. The levansucrases (fructosyltransferases) from *Actinomyces* spp. catabolize sucrose to form β2,6-linked (levans or fructans) or β2,1-linked (inulin) homopolymers of fructose. In *Actinomyces* spp., levansucrase activity is both extracellular and cell wall associated. Sequence analysis of the A. *naeslundii* genospecies 2 *levJ* gene revealed that it encoded a levansucrase with a cell wall sorting domain, consistent with the location of the majority of enzyme activity associated with the cell (51). The *ftf* gene from A. *naeslundii* genospecies 1, also encoding a levansucrase, displayed more similarity to gram-negative than gram-positive homologs, including *levJ* from A. *naeslundii* genospecies 2, suggesting that this gene may have been acquired by horizontal transfer (2). Furthermore, transcriptional analysis indicated that expression of the *ftf* gene was constitutive and was not affected by carbon source or growth phase, which is similar to that seen with levansucrase genes of gram-negative origin (2).

Levans are thought to accumulate in dental plaque to serve as storage polysaccharides, which are subsequently metabolized when other, more readily utilized carbohydrates are exhausted. A secreted levanase (fructanase) subsequently degrades levan for use as a carbon source (for a review, see reference 79). Levan metabolism increases plaque acidification and hence promotes dental caries. Production of levans may directly contribute to periodontal disease, as the levans themselves can trigger host inflammation and B-cell activation. Furthermore, a recent study has indicated that levans may play a role in the adhesion and colonization of oral streptococci and *Actinomyces* spp. Adhesion of A. *naeslundii* genospecies 2 to saliva-treated hydroxyapatite, which models the tooth surface, was levan dependent, as pretreatment with levanase significantly reduced the ability of this organism to adhere (59). In contrast, adhesion of oral *Streptococcus* spp. was positively influenced by levanase treatment (59).

Other Infections Caused by *Actinomyces* spp.

In addition to dental caries and periodontal disease, *Actinomyces* spp. can cause a variety of opportunistic infections in people with compromised immune systems, such as the elderly or those undergoing chemotherapy. *A. naeslundii*, *A. israelii*, and *A. meyeri* are associated with a variety of infections, including those of the central nervous system, vis- ceral organs, and joints. Three recently described *Actinomyces* spp., *Actinomyces turicenesis*, *Actinomyces radingae*, and *Actinomyces europaeus*, are also associated with skin, genital, urinary, and reproductive tract infections. In many cases, these bacteria are coisolated with other potential pathogens, so the actual role of the actinomycetes in these polymicrobial infections is unclear.

Actinomyces bovis causes a type of osteomyelitis in cattle and exotic ungulates. Infection with *A. bovis* results in the formation of protuberances on the upper and lower jawbones, hence the aptly descriptive common name lumpy jaw. DNA hybridization experiments indicated that *A. bovis* possessed a *fimP* gene (78); hence, this organism may express type 1 fimbriae. Interestingly, the ability to bind collagen, which has been ascribed to *Actinomyces* type 1 fimbriae (39), has been shown to be a virulence factor in osteomyelitis caused by *Staphylococcus aureus*. However, little or no other molecular characterization of the non-oral *Actinomyces* spp. has been undertaken, and nothing is known of the virulence factors or the pathogenic mechanisms associated with these organisms.

ARCANOBACTERIUM SPP.

The genus *Arcanobacterium* is closely related to that of *Actinomyces*, so much so in fact that until recently, members of these genera were intermingled with each other phylogenetically. The two predominant species within the arcanobacteria are the ubiquitous animal commensal and opportunistic pathogen *A. pyogenes* and the human pathogen *Arcanobacterium haemolyticum*. While arcanobacterial species are, in general, not well characterized, recent developments in the molecular characterization of *A. pyogenes* have provided insights into how this organism adheres to mucosal surfaces and causes disease, and several of these mechanisms overlap with those identified for the *Actinomyces* spp. The genus *Arcanobacterium* contains four additional species. *Arcanobacterium bernardiae* has been found in a variety of human clinical samples and is considered an opportunistic pathogen. *Arcanobacterium hippocoleae*, *Arcanobacterium phocae*, and *Arcanobacterium pluranimalium* have been isolated from horses, seals, and deer and porpoises, respectively, but nothing is known about the virulence factors involved in infection with these organisms.

A. haemolyticum

A. haemolyticum was first described in 1946 as the pathogenic agent causing pharyngitis and cutaneous infections among United States servicemen and South Pacific islanders (41). The organism was initially described as *Corynebacterium haemolyticum*, but was reclassified as the first member of the genus *Arcanobacterium* in 1982 (12). *A. haemolyticum* is a pleomorphic, nonmotile, non-spore-forming bacterium. As suggested by its name, its most defining characteristic is the presence of beta-hemolysis on blood agar after 48 h of incubation. The organism is a facultative anaerobe, growing best in the presence of increased CO_2. Most *A. haemolyticum* isolates produce a black, opaque dot at the center of each colony, which remains when the colony is scraped away (41). Smooth and rough biotypes have been described, with the rough biotype predominantly isolated from the respiratory tract (10).

Disease

The incidence of *A. haemolyticum* pharyngitis is 0.5 to 2.5% of all pharyngitis cases (38). Ninety percent of cases occur in teenagers and young adults in their 20s and 30s, and the prevalence of *A. haemolyticum* pharyngitis in this age group is approximately 2% (1). However, underreporting probably occurs, as *A. haemolyticum* is not readily detected by standard diagnostic culture, due to its slow-growing nature and specific lack of hemolysis on sheep blood agar. Humans are the main environmental reservoir of *A. haemolyticum*, but it does not appear to be part of the normal human flora, as carriage in the upper respiratory tract occurs in only 0.2 to 1.3% of healthy individuals. However, carriage on the skin has not been investigated. Epidemiologic contact studies suggest that the organism is disseminated via droplet spread from infected individuals.

Clinically, the pharyngitis resembles that from *Streptococcus pyogenes*, and sore throat and pharyngeal edema are always present. Fever, cervical or submandibular lymphadenopathy, tonsillar exudates, and a nonproductive cough also occur in approximately half of all cases. A unique feature of *A. haemolyticum* pharyngitis is the presence of a skin rash in about one-third of patients (38). The rash, which is usually described as maculopapular or erythematous, develops 1 to 4 days after the pharyngitis; occasionally, it is the initial manifestation of the infection. The rash begins on the extremities, spreading to the neck and the trunk, and the pathophysiology of the rash is unknown. However, the hypothesis that the rash is caused by an *A. haemolyticum* exotoxin is reasonable. In addition, *A. haemolyticum* can cause a variety of diseases such as wound infections, meningitis, endocarditis, and osteomyelitis, although these infections are uncommon. Interestingly, the skin rash is only found in patients with pharyngitis and is never associated with the other types of *A. haemolyticum* infections. Infection in animals has only been rarely reported.

Virulence Factors and Pathogenesis

Initial virulence studies were performed by intracutaneous injection of bacteria into guinea pigs and rabbits, resulting in elevated abscesses with necrosis and a pronounced PMN infiltration after 24 to 48 h (41). Phospholipase D is the best-characterized *A. haemolyticum* virulence factor, with 64% similarity to the phospholipase D of *Corynebacterium pseudotuberculosis* (15). Despite the fact that this is the only virulence factor cloned and sequenced to date, little is known about its role in pathogenesis, although the dermonecrosis associated with *A. haemolyticum* abscesses is attributed to this toxin (67). Human volunteers were given intracutaneous injections of bacteria, which resulted in a papular lesion surrounded by erythema and edema (41). However, an attempt to induce pharyngitis by introduction of bacteria onto the pharynx of seven adult volunteers was unsuccessful (41).

A. *haemolyticum* isolates are hemolytic, but the identity of the hemolysin is unknown. However, its activity is most pronounced with erythrocytes of human or rabbit origin, with zones of hemolysis significantly larger on rabbit blood agar than on sheep or bovine blood agar (21). DNA hybridization studies suggest that A. *haemolyticum* carries a gene related to the erythrogenic toxin of *S. pyogenes* (14). A. *haemolyticum* also expresses DNase, lipase, and neuraminidase activities (12, 48, 65), the latter of which may be involved in adherence to the host, similar to that seen with the neuraminidases of A. *naeslundii* (63) and A. *pyogenes* (30). A. *haemolyticum* also binds several human plasma proteins, including fibrinogen (34), which also may play a role in host cell adhesion. Furthermore, A. *haemolyticum* is internalized and survives within HEp-2 cells for at least 72 h (55), possibly contributing to antibiotic failure with drugs such as ampicillin that do not penetrate well into host cells. Infection with A. *haemolyticum* results in serum antibodies against cell wall components of 80, 60, and 30 kDa (52). Despite its potential as a respiratory pathogen and a prevalence of 5 to 13% of streptococcal pharyngitis (21), this organism has been somewhat neglected, and it is clear there is need for molecular studies to elucidate the pathogenesis of infections by A. *haemolyticum*.

A. *pyogenes*

A. *pyogenes* is a nonmotile, non-spore-forming, short, rod-shaped bacterium. This organism was reclassified from the genus *Actinomyces* on the basis of 16S rRNA sequences (57) and is the most prevalent pathogen within the *Arcanobacterium* genus. A. *pyogenes* grows under aerobic and strictly anaerobic conditions, but optimal growth is obtained in a CO_2-enriched atmosphere. All A. *pyogenes* isolates exhibit beta-hemolysis on agar media containing bovine or ovine blood, although isolates of porcine origin are generally more hemolytic (26, 71).

Disease

A. *pyogenes* is a common commensal of the upper respiratory, urogenital, and gastrointestinal tracts of economically important animal species (27, 49, 71). The source of infecting A. *pyogenes* is probably autogenous, and this organism is capable of acting as a primary pathogen. However, infection most often follows a physical or microbial trauma to the mucous membrane, allowing dissemination of the organism. A. *pyogenes* is one of the most common opportunistic pathogens in domestic ruminants and pigs, causing a variety of suppurative infections involving the skin, joints, and visceral organs, including liver abscessation in feedlot cattle, mastitis in dairy cattle, and osteomyelitis and polyarthritis in pigs (71). Abortion and infertility are also sequelae to A. *pyogenes* uterine infections. The disease is not limited to domestic animals and frequently occurs in various species of wildlife and birds, but it is uncommon in humans or companion animals. Like the oral *Actinomyces* spp., A. *pyogenes* infections are often polymicrobial, involving interactions with other bacteria. These infections may be facilitated by the ability of A. *pyogenes* to form biofilms, but this property has not been examined in detail.

Virulence Factors and Pathogenesis

A. *pyogenes* expresses several known and putative virulence factors, which are summarized in Table 2. These factors are required for adherence and subsequent colonization and to cause the host tissue damage associated with A. *pyogenes* infection.

PLO

The best-defined virulence factor of A. *pyogenes* is the secreted toxin pyolysin (PLO). This toxin is the most divergent member of a family of pore-forming toxins called the cholesterol-dependent cytolysins (CDCs), which includes listeriolysin O, pneumolysin, and perfringolysin O (3). CDCs

TABLE 2 Known and putative virulence factors of A. *pyogenes*

Factor	Activities in the host
Pyolysin, cholesterol-dependent cytolysin	Forms pores in cholesterol-containing cell membranes, resulting in hemolysis and cytolysis of immune cells; alters host cytokine expression
Extracellular matrix-binding proteins	
CbpA collagen-binding protein	Binds to collagen, promoting adhesion to collagen-rich tissues
Fibrinogen-binding protein	Binds to fibrinogen; promotes phagocytosis by PMNs
Fibronectin-binding protein	Binds to fibronectin; may promote adhesion
Exoenzymes	
DNase	Degrades host nucleic acids, making nucleotides available as nutrients
NanH and NanP neuraminidases	Cleaves sialic acids from host molecules; exposes host receptors for increased adhesion; reduces mucus viscosity; makes IgA more susceptible to proteolytic attack
Proteases	Degrades host proteins; releases amino acids as nutrients; degrades host defense proteins, such as IgA
Other	
Type 2 fimbriae	Aids in adhesion
Ability to invade epithelial cells	Hides in a protected niche; escapes host defenses and certain antimicrobial drugs
Ability to survive within macrophages	Evades macrophage killing
Ability to form biofilms	May aid in bacterial persistence; may play a role in virulence

have a C-terminal undecapeptide sequence implicated in the initial tethering of these toxins to host membranes, which with some notable exceptions is almost invariant among members of the family. However, PLO possesses a variant undecapeptide sequence, which site-directed mutagenesis experiments have determined is absolutely required for its cytolytic activity (5). PLO differs from the better-known members of the CDC family in that its activity is not subject to oxygen inactivation, a characteristic which can be attributed to a substitution of a conserved cysteine residue in the undecapeptide sequence with alanine (5). Despite the fact that all *A. pyogenes* isolates express PLO, the *plo* gene shows characteristics of horizontal inheritance. *plo* is located on a genomic islet between two essential genes, *smc* and *ftsY*, which are usually adjacent in other genomes (60). In addition, *plo* contains a significantly reduced percent G+C content compared to the surrounding genes, suggesting that this gene was acquired by horizontal transfer (60).

PLO plays a prominent role in the infectious process, as a PLO knockout mutant is significantly attenuated for virulence in a mouse model of infection, and both passive and active immunization of mice with chemical and genetic PLO toxoids protects against experimental challenge (3, 28, 31). However, the role of PLO in infection has not been precisely defined. PLO is cytolytic for a number of cell types including erythrocytes, epithelial cells, macrophages, and PMNs (16, 28), and its action against host immune cells appears likely to play a role in the dissemination of *A. pyogenes* cells during infection. Furthermore, PLO up-regulates tumor necrosis factor alpha expression by J744 macrophages (26) and so may modulate host immune response through the alteration of expression of various cytokines and other inflammatory substances, as is seen with other CDCs (4).

Host Cell Invasion and Intracellular Survival

A. pyogenes is able to invade and survive within epithelial cells (26), which may allow the organism to hide in a protected niche. Furthermore, *A. pyogenes* is taken up by macrophages and PMNs in a nonopsonic manner. The organisms survive within macrophages for up to 96 h but do not appear to multiply intracellularly, and this property is independent of PLO expression (26).

Neuraminidase

As the first step in many pathogenic processes is adherence, *A. pyogenes* would neither be able to colonize as a commensal without this ability nor progress to the pathogenic state. *A. pyogenes* adheres to a number of epithelial and fibroblast cell lines, and this adhesion is the result of the interaction of surface-exposed or -secreted bacterial proteins with host cells and molecules. *A. pyogenes* expresses two cell wall-bound neuraminidases, NanH and NanP (29, 30). NanH is present in all isolates examined, while NanP is found in approximately two-thirds of *A. pyogenes* isolates and is preferentially associated with isolates of porcine origin, suggesting possible tissue tropism for this enzyme (30). A neuraminidase-deficient knockout mutant of *A. pyogenes* had a 53% decrease in its ability to adhere to epithelial cells compared to the wild-type strain (30), and it is proposed that removal of sialic acid moieties from host molecules by *A. pyogenes* neuraminidases exposes cryptic receptors for bacterial attachment, similar to that seen with *Actinomyces* spp. (13, 62, 63).

Adherence to ECM Components

A. pyogenes binds to several ECM components, including collagen (19), fibrinogen (35), and fibronectin (26), which may also play a role in host cell adhesion. The best characterized ECM-binding protein is CbpA, a collagen-binding MSCRAMM-like protein expressed on the cell surface of ~50% of *A. pyogenes* isolates (19). A *cbpA* knockout mutant had significantly reduced binding to epithelial and fibroblast cell lines. Therefore, it is likely that the ability to bind to collagen promotes adherence and subsequent colonization in collagen-rich tissue. Collagen-binding proteins may act as disease-specific virulence factors by promoting adhesion to collagen-rich tissues, such as cartilage, bone, and dentin. While not as well characterized, the ability of *A. pyogenes* to bind fibronectin and fibrinogen may also promote host cell adhesion. Furthermore, the binding of fibrinogen to *A. pyogenes* significantly increases the phagocytic capacity of bovine PMNs (35), which at first may seem a self-defeating property, but with the ability of *A. pyogenes* to survive inside host phagocytes, may provide the organism with a protective niche in which to hide from other innate and specific host defenses.

Fimbriae

Yanagawa and Honda first identified fimbriae on the surface of a number of corynebacterial species including *A. pyogenes* (76). We recently identified a fimbrial gene operon in *A. pyogenes* strain BBR1 (26). The *A. pyogenes* fimbrial genes are similar in genetic organization to those from the *A. naeslundii* type 2 strain T14V fimbrial gene operon and consist of *fimB*, *fimA*, and *srtA* (Fig. 1). *fimA* encodes a protein with 27% identity and 41% similarity to the *A. naeslundii* type 2 fimbrial subunit, FimA (82); the 45.7-kDa FimA protein contains a signal peptide, cell wall sorting domain, pilin motif, and E box, which are required for gram-positive fimbrial biogenesis (74). The C terminus of the 90.5-kDa FimB has 42% amino acid similarity to Orf977, the putative adhesin from *A. naeslundii* type 2 fimbriae (24); like that protein, it contains a signal peptide, cell wall sorting domain, fibronectin-binding domain, and E box. *srtA* encodes a sortase of 32.5 kDa with 68% amino acid similarity to Orf365 from *A. naeslundii* (82). Given the conservation of the operons, it is hypothesized that *A. pyogenes* fimbriae are assembled in a manner similar to that proposed for *Actinomyces* spp. and *C. diphtheriae* (74).

Protease and DNase Activities

A. pyogenes produces four distinct Ca^{2+}-dependent serine proteases with caseinase and/or gelatinase activity (64, 70). While there is currently no compelling evidence to support a role for *A. pyogenes* proteases in pathogenesis, proteases are expressed during infection. A total of 35% of randomly selected pigs had antibodies to proteases, while 93% of pigs with *A. pyogenes* abscesses were antibody positive (69), suggesting that protease expression by *A. pyogenes* plays a role in pathogenesis. In addition, all *A. pyogenes* isolates secrete a DNase, which may aid in the depolymerization of highly viscous DNA released from disintegrating PMNs in inflammatory lesions, making nucleotides available as nutrients.

Immunity and Vaccines

Previous vaccination experiments using whole cells or culture supernatant were largely unsuccessful in protecting domestic animals from *A. pyogenes* infections. However, PLO shows promise as a subunit vaccine. The serum samples from mice immunized with formalin-inactivated or genetically toxoided, recombinant PLO exhibited high neutralizing titers to PLO hemolytic activity and were protected from challenge with *A. pyogenes* (28, 31). It is uncertain how these results will translate to economically important livestock in which *A. pyogenes* is a pathogen. However, in dairy and feedlot cattle vaccinated with recombinant PLO, good levels of neutralizing antibody titers have been obtained (26).

CONCLUSIONS

Initially, the study of bacterial pathogenesis was targeted at the individual organism and its arsenal of virulence factors. More recently, research has focused on host-microbe interactions, as the host ultimately influences the expression of bacterial virulence determinants. However, especially in the case of commensal, opportunistic pathogens, there is another dimension to the pathogenesis equation, that of intra- and interspecies bacterial interactions. Communication, cooperation, and/or antagonism between resident bacteria will also influence the expression of virulence factors and hence disease progression. This communication also allows proliferation of one or more components of the normal flora or suppresses overgrowth by other pathogens, potentially altering the host niches available to the commensal bacteria.

Furthermore, modulation of virulence expression may be triggered by bacterial adhesion to a previously unoccupied niche exposed through the action of trauma or other microbial damage. Alternately, host factors may influence expression of specific adhesins, resulting in a concomitant up-regulation of other virulence determinants. The better-studied *Actinomyces* and *Arcanobacterium* spp. share a number of adhesion mechanisms, and further research may determine that these methods are common to all members of the two genera. However, the virulence factors possessed by the organisms appear to be significantly different, with *A. pyogenes* at least well-equipped to function as a pathogen. Its array of virulence factors, including a potent CDC, is probably a major reason why this organism is such a versatile pathogen, causing disease in a number of organs in a wide variety of host species.

So in the case of opportunistic pathogens, opportunity is the key word. As adhesion is often the first step in the disease process, these commensal opportunists prepare for their pathogenic potential by adherence to the host. Further research investigating the mechanisms and regulation of the adhesins of *Actinomyces* and *Arcanobacterium* spp., as well as their interactions with other bacteria in polymicrobial infections, will be necessary to complete our understanding of disease pathogenesis.

Unpublished research presented here was supported, in part, by USDA/NRICGP awards 97-35204-4750 and 99-35204-7818.

We thank Hien T. Trinh, Jeremy W. Coombs, and Dawn M. Bueschel for their excellent technical assistance. In addition, we recognize the contributions of the late Maria K. Yeung to the molecular biology of virulence factors of Actinomyces spp.

REFERENCES

1. **Banck, G., and M. Nyman.** 1986. Tonsillitis and rash associated with *Corynebacterium haemolyticum. J. Infect. Dis.* **154:**1037–1040.

2. **Bergeron, L. J., E. Morou-Bermudez, and R. A. Burne.** 2000. Characterization of the fructosyltransferase gene of *Actinomyces naeslundii* WVU45. *J. Bacteriol.* **182:**3649–3654.

3. **Billington, S. J., B. H. Jost, W. A. Cuevas, K. R. Bright, and J. G. Songer.** 1997. The *Arcanobacterium* (*Actinomyces*) *pyogenes* hemolysin, pyolysin, is a novel member of the thiol-activated cytolysin family. *J. Bacteriol.* **179:**6100–6106.

4. **Billington, S. J., B. H. Jost, and J. G. Songer.** 2000. Thiol-activated cytolysins: structure, function and role in pathogenesis. *FEMS Microbiol. Lett.* **182:**197–205.

5. **Billington, S. J., J. G. Songer, and B. H. Jost.** 2002. The variant undecapeptide sequence of the *Arcanobacterium pyogenes* haemolysin, pyolysin, is required for full cytolytic activity. *Microbiology* **148:**3947–3954.

6. **Brennan, M. J., J. O. Cisar, and A. L. Sandberg.** 1986. A 160-kilodalton epithelial cell surface glycoprotein recognized by plant lectins that inhibit the adherence of *Actinomyces naeslundii. Infect. Immun.* **52:**840–845.

7. **Brennan, M. J., J. O. Cisar, A. E. Vatter, and A. L. Sandberg.** 1984. Lectin-dependent attachment of *Actinomyces naeslundii* to receptors on epithelial cells. *Infect. Immun.* **46:**459–464.

8. **Burne, R. A., and Y. Y. Chen.** 2000. Bacterial ureases in infectious diseases. *Microbes Infect.* **2:**533–542.

9. **Carlén, A., S. G. Rüdiger, I. Loggner, and J. Olsson.** 2003. Bacteria-binding plasma proteins in pellicles formed on hydroxyapatite in vitro and on teeth in vivo. *Oral Microbiol. Immunol.* **18:**203–207.

10. **Carlson, P., K. Lounatmaa, and S. Kontiainen.** 1974. Biotypes of *Arcanobacterium haemolyticum. J. Clin. Microbiol.* **32:**1654–1657.

11. **Cisar, J. O., and A. E. Vatter.** 1979. Surface fibrils (fimbriae) of *Actinomyces viscosus* T14V. *Infect. Immun.* **24:**523–531.

12. **Collins, M. D., D. Jones, and G. M. Schofield.** 1982. Reclassification of 'Corynebacterium haemolyticum' (MacLean, Liebow & Rosenberg) in the genus *Arcanobacterium* gen. nov. as *Arcanobacterium haemolyticum* nom. rev., comb. nov. *J. Gen. Microbiol.* **128:**1279–1281.

13. **Costello, A. H., J. O. Cisar, P. E. Kolenbrander, and O. Gabriel.** 1979. Neuraminidase-dependent hemagglutination of human erythrocytes by human strains of *Actinomyces viscosus* and *Actinomyces naeslundii. Infect. Immun.* **26:**563–572.

14. **Coyle, M. B., and B. A. Lipsky.** 1990. Coryneform bacteria in infectious diseases: clinical and laboratory aspects. *Clin. Microbiol. Rev.* **3:**227–246.

15. **Cuevas, W. A., and J. G. Songer.** 1993. *Arcanobacterium haemolyticum* phospholipase D is genetically and functionally similar to *Corynebacterium pseudotuberculosis* phospholipase D. *Infect. Immun.* **61:**4310–4316.

16. **Ding, H., and C. Lämmler.** 1996. Purification and further characterization of a haemolysin of *Actinomyces pyogenes. J. Vet. Med. B* **43:**179–188.

17. **Donkersloot, J. A., J. O. Cisar, M. E. Wax, R. J. Harr, and B. M. Chassy.** 1985. Expression of *Actinomyces viscosus* antigens in *Escherichia coli*: cloning of a structural gene (*fimA*) for type 2 fimbriae. *J. Bacteriol.* **162:**1075–1078.

18. **Ellen, R. P.** 1982. Oral colonization by gram-positive bacteria significant to periodontal disease, p. 98–111. *In* R. J. Genco and S. E. Mergenhagen (ed.), *Host-Parasite Interactions in Periodontal Disease.* American Society for Microbiology, Washington, D.C.

19. **Esmay, P. A., S. J. Billington, M. A. Link, J. G. Songer, and B. H. Jost.** 2003. The *Arcanobacterium pyogenes* collagen binding protein, CbpA, promotes adhesion to host cells. *Infect. Immun.* **71:**4368–4374.

20. **Frandsen, E. V. G.** 1994. Carbohydrate depletion of immunoglobulin A1 by oral species of gram-positive rods. *Oral Microbiol. Immunol.* **9:**352–358.

21. **Funke, G., A. von Graevenitz, J. E. Clarridge III, and K. A. Bernard.** 1997. Clinical microbiology of coryneform bacteria. *Clin. Microbiol. Rev.* **10:**125–159.

22. Hallberg, K., C. Holm, U. Ohman, and N. Strömberg. 1998. *Actinomyces naeslundii* displays variant *fimP* and *fimA* fimbrial subunit genes corresponding to different types of acidic proline-rich protein and β-linked galactosamine binding specificity. *Infect. Immun.* **66**:4403–4410.

23. Hawkins, B. W., R. D. Cannon, and H. F. Jenkinson. 1993. Interactions of *Actinomyces naeslundii* strains T14V and ATCC 12104 with saliva, collagen and fibrinogen. *Arch. Oral Biol.* **38**:533–535.

24. Hoflack, L., and M. K. Yeung. 2001. *Actinomyces naeslundii* fimbrial protein Orf977 shows similarity to a streptococcal adhesin. *Oral Microbiol. Immunol.* **16**:319–320.

25. Johnson, J. L., L. V. Moore, B. Kaneko, and W. E. Moore. 1990. *Actinomyces georgiae* sp. nov., *Actinomyces gerencseriae* sp. nov., designation of two genospecies of *Actinomyces naeslundii*, and inclusion of *A. naeslundii* serotypes II and III and *Actinomyces viscosus* serotype II in *A. naeslundii* genospecies 2. *Int. J. Syst. Bacteriol.* **40**:273–286.

26. Jost, B. H., and S. J. Billington. Unpublished data.

27. Jost, B. H., K. W. Post, J. G. Songer, and S. J. Billington. 2002. Isolation of *Arcanobacterium pyogenes* from the porcine gastric mucosa. *Vet. Res. Comm.* **26**:419–425.

28. Jost, B. H., J. G. Songer, and S. J. Billington. 1999. An *Arcanobacterium (Actinomyces) pyogenes* mutant deficient in production of the pore-forming cytolysin pyolysin has reduced virulence. *Infect. Immun.* **67**:1723–1728.

29. Jost, B. H., J. G. Songer, and S. J. Billington. 2001. Cloning, expression and characterization of a neuraminidase gene from *Arcanobacterium pyogenes*. *Infect. Immun.* **69**:4430–4437.

30. Jost, B. H., J. G. Songer, and S. J. Billington. 2002. Identification of a second *Arcanobacterium pyogenes* neuraminidase, and involvement of neuraminidase activity in host cell adhesion. *Infect. Immun.* **70**:1106–1112.

31. Jost, B. H., H. T. Trinh, J. G. Songer, and S. J. Billington. 2003. Immunization with genetic toxoids of the *Arcanobacterium pyogenes* cholesterol-dependent cytolysin, pyolysin, protects mice against infection. *Infect. Immun.* **71**:2966–2969.

32. Klier, C. M., A. G. Roble, and P. E. Kolenbrander. 1998. *Actinomyces* serovar WVA963 coaggregation-defective mutant strain PK2407 secretes lactose-sensitive adhesin that binds to coaggregation partner *Streptococcus oralis* 34. *Oral Microbiol. Immunol.* **13**:337–340.

33. Kolenbrander, P. E. 1991. Coaggregation: adherence in the human oral microbial ecosystem, p. 309–329. *In* M. Dworkin (ed.), *Microbial Cell-Cell Interactions*. American Society for Microbiology, Washington, D.C.

34. Lämmler, C. 1994. Studies on biochemical and serological characteristics and binding properties of *Arcanobacterium haemolyticum* for human plasma proteins. *Med. Microbiol. Lett.* **3**:66–71.

35. Lämmler, C., and H. Ding. 1994. Characterization of fibrinogen-binding properties of *Actinomyces pyogenes*. *J. Vet. Med. B* **41**:588–596.

36. Li, T., I. Johansson, D. I. Hay, and N. Strömberg. 1999. Strains of *Actinomyces naeslundii* and *Actinomyces viscosus* exhibit structurally variant fimbrial subunit proteins and bind to different peptide motifs in salivary proteins. *Infect. Immun.* **67**:2053–2059.

37. Li, T., M. K. Khah, S. Slavnic, I. Johansson, and N. Strömberg. 2001. Different type 1 fimbrial genes and tropisms of commensal and potentially pathogenic *Actinomyces* spp. with different salivary acidic proline-rich protein and statherin ligand specificities. *Infect. Immun.* **69**:7224–7233.

38. Linder, R. 1997. *Rhodococcus equi* and *Arcanobacterium haemolyticum*: two "coryneform" bacteria increasingly recognized as agents of human infection. *Emerg. Infect. Dis.* **3**:145–153.

39. Liu, T., R. J. Gibbons, D. I. Hay, and Z. Skobe. 1991. Binding of *Actinomyces viscosus* to collagen: association with the type 1 fimbrial adhesin. *Oral Microbiol. Immunol.* **6**:1–5.

40. Loo, C. Y., M. D. P. Willcox, and K. W. Knox. 1994. Surface-associated properties of *Actinomyces* strains and their potential relation to pathogenesis. *Oral Microbiol. Immunol.* **9**:12–18.

41. MacLean, P. D., A. A. Liebow, and A. A. Rosenberg. 1946. A haemolytic bacterium resembling *Corynebacterium ovis* and *Corynebacterium pyogenes* in man. *J. Infect. Dis.* **79**:69–90.

42. Mazmanian, S. K., H. Ton-That, and O. Schneewind. 2001. Sortase-catalysed anchoring of surface proteins to the cell wall of *Staphylococcus aureus*. *Mol. Microbiol.* **40**:1049–1057.

43. McNab, R., H. Forbes, P. S. Handley, D. M. Loach, G. W. Tannock, and H. F. Jenkinson. 1999. Cell wall-anchored CshA polypeptide α259 kilodaltons) in *Streptococcus gordonii* forms surface fibrils that confer hydrophobic and adhesive properties. *J. Bacteriol.* **181**:3087–3095.

44. Moncla, B. J., and P. Braham. 1989. Detection of sialidase (neuraminidase) activity in *Actinomyces* species by using $2'$-(4-methylumbelliferyl)α-D-N-acetylneuraminic acid in a filter paper spot test. *J. Clin. Microbiol.* **27**:182–184.

45. Moore, W. E. C., and L. V. H. Moore. 1994. The bacteria of periodontal diseases. *Periodontol. 2000* **5**:66–77.

46. Morou-Bermudez, E., and R. A. Burne. 2000. Analysis of urease expression in *Actinomyces naeslundii* WVU45. *Infect. Immun.* **68**:6670–6676.

47. Morou-Bermudez, E., and R. A. Burne. 1999. Genetic and physiologic characterization of urease of *Actinomyces naeslundii*. *Infect. Immun.* **67**:504–512.

48. Mueller, H. E. 1973. The occurrence of neuraminidase and acylneuraminate pyruvate-lyase in *Corynebacterium haemolyticum* and *Corynebacterium pyogenes*. *Zentralbl. Bakteriol. Orig. A* **225**:59–65.

49. Narayanan, S., T. G. Nagaraja, N. Wallace, J. Staats, M. M. Chengappa, and R. D. Oberst. 1998. Biochemical and ribotypic comparison of *Actinomyces pyogenes* and *A. pyogenes*-like organisms from liver abscesses, ruminal wall, and ruminal contents of cattle. *Am. J. Vet. Res.* **59**:271–276.

50. Nesbitt, W. E., J. E. Beem, K. P. Leung, R. Stroup, R. Swift, W. P. McArthur, and W. B. Clark. 1996. Inhibition of adherence of *Actinomyces naeslundii* (*Actinomyces viscosus*) T14V-J1 to saliva-treated hydroxyapatite by a monoclonal antibody to type 1 fimbriae. *Oral Microbiol. Immunol.* **11**:51–58.

51. Norman, J. M., K. L. Bunny, and P. M. Giffard. 1995. Characterization of *levJ*, a sucrase/fructanase-encoding gene from *Actinomyces naeslundii* T14V, and comparison of its product with other sucrose-cleaving enzymes. *Gene* **152**:93–98.

52. Nyman, M., K. R. Alugupalli, S. Stromberg, and A. Forsgren. 1997. Antibody response to *Arcanobacterium haemolyticum* infection in humans. *J. Infect. Dis.* **175**:1515–1518.

53. Nyvad, B., and M. Kilian. 1987. Microbiology of the early colonization of human enamel and root surfaces in vivo. *Scand. J. Dent. Res.* **95**:369–380.

54. **Ochiai, K., T. Kurita-Ochiai, Y. Kamino, and T. Ikeda.** 1993. Effect of co-aggregation on the pathogenicity of oral bacteria. *J. Med. Microbiol.* **39:**183–190.

55. **Österlund, A.** 1995. Are penicillin treatment failures in *Arcanobacterium haemolyticum* pharyngotonsillitis caused by intracellularly residing bacteria? *Scand. J. Infect. Dis.* **27:**131–134.

56. **Palmer, R. J., Jr., K. Kazmerzak, M. C. Hansen, and P. E. Kolenbrander.** 2001. Mutualism versus independence: strategies of mixed-species oral biofilms in vitro using saliva as the sole nutrient source. *Infect. Immun.* **69:**5794–5804.

57. **Pascual Ramos, C., G. Foster, and M. D. Collins.** 1997. Phylogenetic analysis of the genus *Actinomyces* based on 16S rRNA gene sequences: description of *Arcanobacterium phocae* sp. nov., *Arcanobacterium bernardiae* comb. nov., and *Arcanobacterium pyogenes* comb. nov. *Int. J. Syst. Bacteriol.* **47:**46–53.

58. **Patti, J. M., B. L. Allen, M. J. McGavin, and M. Höök.** 1994. MSCRAMM-mediated adherence of microorganisms to host tissues. *Annu. Rev. Microbiol.* **48:**585–617.

59. **Rozen, R., G. Bachrach, M. Bronshteyn, I. Gedalia, and D. Steinberg.** 2001. The role of fructans on dental biofilm formation by *Streptococcus sobrinus, Streptococcus mutans, Streptococcus gordonii* and *Actinomyces viscosus. FEMS Microbiol. Lett.* **195:**205–210.

60. **Rudnick, S. T., B. H. Jost, J. G. Songer, and S. J. Billington.** 2003. The gene encoding pyolysin, the pore-forming toxin of *Arcanobacterium pyogenes*, resides within a genomic islet flanked by essential genes. *FEMS Microbiol. Lett.* **225:**241–247.

61. **Ruhl, S., J. O. Cisar, and A. L. Sandberg.** 2000. Identification of polymorphonuclear leukocyte and HL-60 cell receptors for adhesins of *Streptococcus gordonii* and *Actinomyces naeslundii. Infect. Immun.* **68:**6346–6354.

62. **Sandberg, A. L., L. L. Mudrick, J. O. Cisar, M. J. Brennan, S. E. Mergenhagen, and A. E. Vatter.** 1986. Type 2 fimbrial lectin-mediated phagocytosis of oral *Actinomyces* spp. by polymorphonuclear leukocytes. *Infect. Immun.* **54:**472–476.

63. **Saunders, J. M., and C. H. Miller.** 1983. Neuraminidase-activated attachment of *Actinomyces naeslundii* ATCC 12104 to human buccal epithelial cells. *J. Dent. Res.* **62:**1038–1040.

64. **Schaufuss, P., R. Sting, and C. Lämmler.** 1989. Isolation and characterization of an extracellular protease of *Actinomyces pyogenes. Zentralbl. Bakteriol.* **271:**452–459.

65. **Sneath, P. H. A., N. S. Mair, M. E. Sharpe, and J. G. Holt.** 1986. *Bergey's Manual of Systematic Bacteriology*, vol. 2. Williams and Wilkins, Baltimore, Md.

66. **Socransky, S. S., and A. D. Haffajee.** 1991. Microbial mechanisms in the pathogenesis of destructive periodontal diseases: a critical assessment. *J. Periodont. Res.* **26:**195–212.

67. **Soucek, A., and A. Souckova.** 1974. Toxicity of bacterial sphingomyelinases D. *J. Hyg. Epidemiol. Microbiol. Immunol.* **18:**327–335.

68. **Stenudd, C., A. Nordlund, M. Ryberg, I. Johansson, C. Kallestal, and N. Stromberg.** 2001. The association of bacterial adhesion with dental caries. *J. Dent. Res.* **80:**2005–2010.

69. **Takeuchi, S., R. Azuma, Y. Nakajima, and T. Suto.** 1979. Diagnosis of *Corynebacterium pyogenes* in pigs by immunodiffusion test with protease antigen. *Natl. Inst. Anim. Health Q.* **19:**77–82.

70. **Takeuchi, S., T. Kaidoh, and R. Azuma.** 1995. Assay of proteases from *Actinomyces pyogenes* isolated from pigs and cows by zymography. *J. Vet. Med. Sci.* **57:**977–979.

71. **Timoney, J. F., J. H. Gillespie, F. W. Scott, and J. E. Barlough.** 1988. *Hagan and Bruner's Microbiology and Infectious Diseases of Domestic Animals*, 8th ed. Cornell University Press, Ithaca, N.Y.

72. **Ton-That, H., L. A. Marraffini, and O. Schneewind.** 2004. Sortases and pilin elements involved in pilus assembly of *Corynebacterium diphtheriae. Mol. Microbiol.* **53:**251–261.

73. **Ton-That, H., and O. Schneewind.** 2004. Assembly of pili in gram-positive bacteria. *Trends Microbiol.* **12:**228–234.

74. **Ton-That, H., and O. Schneewind.** 2003. Assembly of pili on the surface of *Corynebacterium diphtheriae. Mol. Microbiol.* **50:**1429–1438.

75. **Wu, H., and P. M. Fives-Taylor.** 2001. Molecular strategies for fimbrial expression and assembly. *Crit. Rev. Oral Biol. Med.* **12:**101–115.

76. **Yanagawa, R., and E. Honda.** 1976. Presence of pili in species of human and animal parasites and pathogens of the genus *Corynebacterium. Infect. Immun.* **13:**1293–1295.

77. **Yeung, M.** 1993. Complete nucleotide sequencing of the *Actinomyces viscosus* T14V sialidase gene: presence of a conserved repeating sequence among strains of *Actinomyces* spp. *Infect. Immun.* **61:**109–116.

78. **Yeung, M. K.** 1992. Conservation of an *Actinomyces viscosus* T14V type 1 fimbrial subunit homolog among divergent groups of *Actinomyces* spp. *Infect. Immun.* **60:**1047–1054.

79. **Yeung, M. K.** 1999. Molecular and genetic analyses of *Actinomyces* spp. *Crit. Rev. Oral Biol. Med.* **10:**120–138.

80. **Yeung, M. K., B. M. Chassy, and J. O. Cisar.** 1987. Cloning and expression of a type 1 fimbrial subunit of *Actinomyces viscosus* T14V. *J. Bacteriol.* **169:**1678–1683.

81. **Yeung, M. K., and J. O. Cisar.** 1988. Cloning and nucleotide sequence of a gene for *Actinomyces naeslundii* WVU45 type 2 fimbriae. *J. Bacteriol.* **170:**3803–3809.

82. **Yeung, M. K., J. A. Donkersloot, J. O. Cisar, and P. A. Ragsdale.** 1998. Identification of a gene involved in assembly of *Actinomyces naeslundii* T14V type 2 fimbriae. *Infect. Immun.* **66:**1482–1491.

83. **Yeung, M. K., and P. A. Ragsdale.** 1997. Synthesis and function of *Actinomyces naeslundii* T14V type 1 fimbriae require the expression of additional fimbria-associated genes. *Infect. Immun.* **65:**2629–2639.

The Pathogenesis of *Nocardia*

BLAINE L. BEAMAN

61

Nocardiae are gram-positive, partially acid-fast, filamentous bacteria that grow by apical extension, forming elongated cells with lateral branching (Fig. 1A). These filaments divide by fragmentation into shorter rods and coccoid cells. Therefore, all of the nocardial cells are branching filaments at the log phase of growth, whereas stationary-phase nocardiae are short pleomorphic rods, coccobacilli, and cocci (6). The nocardiae are relatively slow-growing (doubling time, >2 h), and they tend to form variable, hard, tenacious colonies with aerial filamentation (Fig. 1B). The nocardiae belong to the aerobic actinomycetes group of bacteria (Fig. 1); nevertheless, they are phylogenetically related to the corynebacteria, mycobacteria, rhodococci, and other mycolic acid-containing organisms. Most species of *Nocardia* have been recovered from soil, plant material, and water in most regions of the world. Therefore, it is generally believed that diseases caused by the nocardiae result from either a respiratory or a traumatic exposure to these sources. However, soil isolates of pathogenic species of *Nocardia* often exhibit either no or low virulence toward laboratory animals. On the other hand, clinical isolates of the same species are usually moderately to highly virulent (5). These observations suggest that some intermediate host may be involved in upregulating and maintaining nocardial virulence in the environment. Protozoa such as amoebae and small worms (e.g., nematodes) have been suggested as possible intermediate hosts that could maintain an evolutionary pressure on the preservation of invasion and virulence genes within many environmental pathogens (30, 64). This is an engaging hypothesis that may be supported by observations that *Nocardia* spp. reside naturally in dinoflagellates (69). Furthermore, additional pressure to maintain nocardial virulence in nature may occur as the result of survival and growth within the gut of various insects, as was reported for cockroaches (71).

HUMAN NOCARDIOSIS

Diseases in humans caused by nocardiae may be divided into at least six general categories based on the route of infection, site of disease, and subsequent pathological responses (5, 9).

There are no pathognomonic clinical manifestations for nocardiosis, and diseases caused by these organisms can mimic a wide variety of other conditions (9, 47, 49, 58). Furthermore, the nocardiae in clinical samples may be quite variable in Gram-stained preparations, where they may appear as beaded filamentous cells, rods, cocci, diphtheroids, variably gram-positive bacteria, and variably gram-negative organisms (Fig. 1C). The nocardiae are also variably acid fast, so some strains may be mistaken for mycobacteria. Exposure of most humans to nocardiae probably occurs frequently; based on experimental studies of animals, some of these exposures almost certainly lead to infection (34). Therefore, it is reasonable to assume that subclinical nocardial infections in humans may be relatively common. However, progression of nocardial infections to clinical disease is not recognized with high frequency, suggesting that both innate and acquired host defenses against nocardiae are quite effective (9). Numerous factors contribute to the progression from an unapparent infection to overt disease (9, 47, 49).

The incidence of human disease caused by nocardiae is not known; however, nocardiosis has been reported in most regions of the world (5, 9, 20, 45, 47, 58). Many of these investigators suggest that infections caused by *Nocardia* spp. are on the rise, and various studies suggest an incidence rate of disease varying from 0.3% to as much as 4.2% of all individuals with pulmonary diseases. On a global scale, these observations suggest that >1 million people develop clinical signs of nocardiosis each year. Unfortunately, most of these cases are misdiagnosed, and the correct diagnosis occurs either serendipitously or retrospectively (9, 19, 45, 47). The incidence of subclinical infection by the nocardiae is even less well documented. Hubble et al. (34), using serological analyses with antigens produced and secreted only by nocardiae during growth, reported that as many as 50% of the elderly people in Kansas were subclinically infected by nocardiae. At the same time, it was shown that approximately 20% of university students on a California campus had antibody to the secreted, diagnostic nocardial antigens. These studies suggest a very high incidence of exposure with possible subclinical infection by nocardiae (9, 34). Nocardial infections in humans may have a regional preva-

Gram-Positive Pathogens, 2nd edition, edited by Vincent A. Fischetti et al.
© 2006 ASM Press, Washington, D.C.

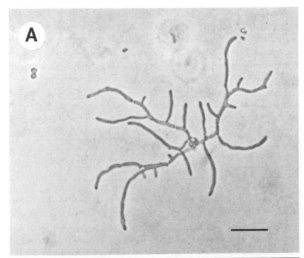

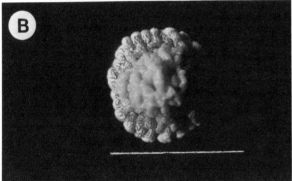

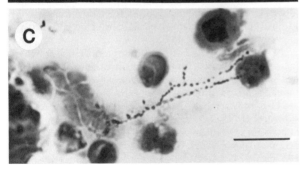

FIGURE 1 General characteristics of nocardiae. (A) Phase-contrast micrograph of *Nocardia* spp. grown on tryptone agar for 12 h. Note the typical branching, filamentous growth characterized as nocardioform morphology. Bar, 10 μm. (Reprinted from reference 12 with permission from the publisher.) (B) Typical colonial morphology of *N. asteroides* grown on glucose yeast extract agar at 37°C for 14 days. Bar, 1 cm. (C) Gram stain of *N. asteroides* in a smear from an abscess. Note the typical beaded appearance of the branching filaments. Bar, 15 μm.

lence, since the southwestern desert area of the United States seems to have a high incidence of diagnosed cases (60). For example, Saubolle and Sussland report an average of about 100 cases per year in the Phoenix, Ariz., region alone (60).

Pulmonary Nocardiosis

It is generally thought that the nocardiae residing on dust particles become airborne and are then inhaled (9, 47, 49).

Depending on the size of the particle, nocardiae can be deposited on the epithelial surface of either the upper or lower airways. The subsequent events occurring at this location depend upon the numbers of bacteria, the growth state, and the virulence properties of individual nocardial cells present on the particle. If a virulent nocardial cell is present, then infection may ensue. Log-phase cells of nocardiae are typically much more virulent and invasive for animals and tissue culture cells than are stationary-phase organisms from the same culture (5, 9). Indeed, log-phase cells of the most virulent strains of *Nocardia asteroides* behave as invasive, primary pathogens, whereas stationary-phase cells act more like opportunistic pathogens requiring some host deficit for infection to occur (5, 9). It is most likely that nocardiae aerosolized with dust particles would be dehydrated, starved, mixed with other microorganisms, and not in the log phase of growth. Therefore, this encounter in an immunologically intact host should be abortive, but in an immunocompromised individual the nocardiae may survive and grow. Pulmonary encounters with healthy nocardiae in the log phase of growth would be predicted to occur infrequently. These findings might provide some explanation for the observations that pulmonary nocardiosis is recognized most frequently in compromised hosts but that it can occasionally present as an aggressive disease in previously healthy individuals with no identifiable deficits in host defenses (9, 47, 49).

In the lung, nocardiae grow as facultative intracellular pathogens that induce predominantly a polymorphonuclear neutrophil (PMN) response with abscess formation, with or without calcification, although granulomas have been reported (9, 58). Either the organisms may remain localized within the bronchioles and alveoli, or they may invade aggressively through the lung parenchyma. The lesions may present as a localized, expanding infiltration; mild, self-limited infiltration; diffuse pneumonia; multinodular pulmonary infiltrates; lobar consolidation; interstitial infiltration; necrotizing pneumonia; indolent, progressive fibrosis; or empyema with pleural effusions (5, 9, 17, 72). The clinical presentation of pulmonary nocardiosis is extremely variable, and it may be confused with many other pulmonary diseases. Misdiagnosis of pulmonary nocardiosis is often reported (9, 17, 58, 72).

Extrapulmonary Nocardiosis

Primary nocardiosis outside of the lungs is probably the result of direct implantation from a contaminated source. The most frequently recognized site is in the eye, usually following some form of trauma (5, 9, 63). Osteomyelitis, bone infections, and arthritis have been reported after various injuries (9), and endocarditis and pericarditis have been reported subsequent to surgical procedures (5, 9).

Systemic Nocardiosis

Systemic nocardiosis is defined as a disease in which two or more locations in the body become infected by the same nocardial strain. Regardless of the site of the original primary infection, dissemination of the nocardiae to other sites may occur (9, 47). The skin, central nervous system (CNS), retinas of the eyes, and kidneys present frequent secondary foci; however, any region of the body can become involved following dissemination (5, 9, 47).

CNS Nocardiosis

Approximately one-fourth of the cases of nocardiosis reported in the literature (excluding mycetomas) involve the CNS, especially the brain (5, 9). Contrary to the general belief of many clinicians, brain infections caused by nocardiae may be primary, with no identifiable lesions elsewhere in previously healthy individuals (5, 9). Indeed, >40% of these CNS infections occur in previously healthy individuals with no identifiable predisposing conditions (9, 53). Therefore, the perception that CNS infection by nocardia occurs only in immunocompromised patients is not accurate (5, 53). Animal models show that intravenous (i.v.) inoculation with log-phase cells of most pathogenic *Nocardia* spp. results in preferential invasion of the brain (9). Taken together, these observations indicate that nocardiae may be primary brain pathogens in humans and other mammals (9).

Nocardiae may induce a variable response in the brain depending upon the patient, the strain of nocardia, and the location of the infection (9, 53). Nocardial infections in the brain may be silent, with diagnosis made by accident (9); they may be insidious and difficult to diagnose (5, 9); they may be acute with rapid progression (9); or they may be chronic with gradual progression, resulting in neurological deficits lasting from months to years (5). The types of lesions produced in the brain may be just as variable. Abscess formation is common, but there can be diffuse infiltration with no focal lesions, as well as organized granulomata with giant cells (5, 9, 53).

Cutaneous, Subcutaneous, and Lymphocutaneous Nocardiosis

Many nonimmunocompromised individuals probably develop cutaneous and subcutaneous infections caused by all of the pathogenic species of *Nocardia* following traumatic exposure to contaminated soil (e.g., a rose thorn puncture). In most instances, these infections are self-limited and induce transitory cellulitis, pustules, or abscesses at the site of injury (38, 39, 60, 62). However, in some individuals, these persist and progress to form a chronic, expanding cutaneous or subcutaneous lesion that requires medical attention (5, 38, 39, 60). Occasionally, the infection extends into the lymphatics, affecting the regional lymph nodes. This form of lymphocutaneous nocardiosis may resemble sporotrichosis, which is usually caused by the fungus *Sporothrix schenckii*, and it is called sporotrichoid nocardiosis. Most cases of lymphocutaneous and sporotrichoid nocardiosis are caused by *Nocardia brasiliensis* (5, 9, 60, 62).

Mycetoma

A mycetoma is a chronic, progressive, pyogranulomatous disease that usually develops at the site of a localized injury such as a thorn prick (1, 9, 60). Mycetomas are divided into two types, based on the specific etiology. Those caused by a variety of fungi are called eumycetomas; those caused by aerobic actinomycetes are referred to as actinomycetomas (1, 5). Three species of actinomycetes (*Actinomadura madurae*, *N. brasiliensis*, and *Streptomyces somaliensis*) account for most cases of actinomycetomas recognized throughout the tropical and subtropical regions of the world. Occasionally, other aerobic actinomycetes, including all species of nocardia, cause actinomycetomas (1, 60).

SPECIES OF *NOCARDIA* PATHOGENIC FOR ANIMALS

The nocardiae were first recognized as pathogens in animals such as cows and dogs (58, 60); however, many species of *Nocardia* cause diseases in a wide variety of animals, including birds, fish, invertebrates, and mammals (5, 9). The first disease attributed to nocardia is bovine farcy, described by

Nocard in 1888 (5, 7). At about the same time that Nocard recognized actinomycetes in bovine farcy, similar organisms were found to cause disease in dogs and in humans (5, 7, 9). Indeed, in 1889 Trevisan named Nocard's bovine isolate *Nocardia farcinica*, and in 1896 Blanchard listed Eppinger's human isolate as *N. asteroides* (5, 7, 9).

It is well established that the genus *Nocardia* is composed of a heterogeneous group of incompletely characterized aerobic, gram-positive bacteria. The type species for the genus is currently *N. asteroides* (37), although at one time *N. farcinica* was considered the type species for this genus. This is also a heterogeneous taxon that probably consists of several different species. With the advent of new methods of taxonomy, including molecular taxonomy utilizing 16S ribosomal RNA sequencing, the *N. asteroides* complex has been divided into many subtypes and different species. To help provide a stable anchor to this taxonomic confusion, it was recommended that *N. asteroides* ATCC 19247 be the working type strain to establish *N. asteroides* sensu stricto. All other nocardiae would then be compared to this strain (60). Many members of the *N. asteroides* complex have been renamed and they include *N. cyriacigeorgica*, *N. farcinica*, *N. nova*, *N. transvalensis*, as well as possibly *N. puris*, *N. abscessus*, *N. veterana*, *N. africana*, *N. cerradoensis*, *N. tenerifensis*, *N. inohanensis*, *N. yamanashiensis*, *N. niigatensis*, *N. neocaledoniensis*, *N. asiatica*, *N. testacea*, and *N. senatus* (57, 59). The other characterized species include *N. brasiliensis*, *N. pseudobrasiliensis*, *N. otitidiscaviarum*, *N. seriolae*, *N. salmonicida*, *N. vaccinii*, *N. crassostreae*, *N. brevicatena*, *N. carnea*, *N. lactamdurans*, *N. paucivorans*, *N. flavorosea*, *N. pseudovaccinii*, *N. uniformis*, *N. fusca*, *N. pseudosporangifera*, *N. violaceofuscai*, *N. cummidelens*, *N. fluminea*, *N. soli*, *N. corynebacterioides*, *N. beijingensis*, and *N. arthritidis* (59). Unfortunately, because of this complex heterogeneity, the taxonomy of these organisms remains in flux and newly described species are being added to the above list regularly (22, 48).

N. asteroides ATCC 19247 is not pathogenic for experimental hosts, while in contrast, many other isolates identified as *N. asteroides* are either moderately or highly virulent for experimental animals. One of these stains, *N. asteroides* GUH-2, belongs to *N. asteroides* Wallace subgroup VI (this subgroup of *N. asteroides* is most closely related to *N. cyriacigeorgica*). *N. asteroides* GUH-2 is quite virulent for mice and monkeys (9, 10). These two strains of *N. asteroides* (ATCC 19247 and GUH-2) represent opposite ends of the taxonomic continuum for this species. If one accepts that nocardiae identical to *N. asteroides* ATCC 19247 are the only legitimate *N. asteroides* sensu stricto, then *N. asteroides* is probably not a human pathogen per se because most clinical isolates from humans and animals are more closely related to *N. cyriacigeorgica* (57), and organisms identical to ATCC 19247 are only rarely isolated from clinical material (59, 60). Since most clinical isolates including many of those from the historical past preceding *N. cyriacigeorgica* appear to belong to *N. asteroides* subgroup VI, the argument can be made that *N. asteroides* subgroup VI (*N. cyriacigeorgica*) should be reassigned to the status of *N. asteroides* sensu stricto. (The type strain for this species might more correctly be *N. asteroides* ATCC 14759.) This would mean that *N. cyriacigeorgica* should probably become either redefined or invalidated as a separate species (48, 57, 59, 60). Thus, *N. asteroides* ATCC 19247 should be redefined as a unique species, perhaps either as *N. bowmanii* or *N. gordonii* in honor of those individuals who either first isolated the organism or published the original taxonomic description (31, 57, 59).

Nocardial infections are well described in the veterinary literature. In general, the same species of *Nocardia* that cause disease in humans (*N. asteroides* and relatives *N. brasiliensis*, *N. farcinica*, *N. nova*, *N. otitidiscaviarum*, *N. pseudobrasiliensis*, and *N. transvalensis*) induce the equivalent disease in most other mammals, with *N. asteroides* Wallace type VI representing the dominant pathogen (5, 9, 58). *N. seriolae* induces abscesses in fish (5, 29), *N. crassostreae* produces lesions in oysters (29), and *N. vaccinii* causes galls in blueberries (5). These last three species of *Nocardia* have not been identified from human infections. *N. brevicatena* and *N. carnea* have been recovered occasionally from clinical material, but their roles in disease are not established, and such organisms as *N. flavorosea* have not been associated with disease (22).

N. asteroides, *N. farcinica*, and *N. otitidiscaviarum* have caused significant outbreaks worldwide in dairy cattle, usually in the form of mastitis (5, 16). In one published series from California, >1,600 cows at five dairies were involved. Many of these animals either died or had to be euthanized, thus representing a large economic loss (5). Interestingly, *N. brasiliensis* has not been reported to cause mastitis in dairy animals, even though it was found to cause mastitis in a human (52). None of the other nocardial species have been reported to cause infections in cows (5).

Since the initial reports on bovine farcy in the late 19th century, there have been hundreds of cases of nocardiosis described in animals. These include both domestic animals such as dogs, cats, sheep, pigs, goats, and horses (5, 7) and wild animals such as raccoons, antelope, gazelles, whales, dolphins, porpoises, seals, deer, monkeys and other primates, armadillos, fox, mongoose (5, 7), water buffalo (56), a variety of birds (5, 7), crayfish (5, 7), various fish (5, 7, 29), and oysters (29).

ANIMAL MODELS TO STUDY NOCARDIOSIS, PATHOGENESIS, AND HOST RESPONSES

A variety of animal species were utilized as models to investigate nocardiosis. The first reported use of rabbits and guinea pigs as experimental models to study human disease caused by *N. asteroides* was in 1891 by Eppinger (5, 7, 9). In 1902, MacCallum (5, 7, 9) demonstrated that healthy dogs were quite susceptible to systemic infection by i.v. inoculation with *N. asteroides*. He suggested that dogs were ideal experimental models for studying human nocardiosis (5, 7). Over the next 70 years, many investigators attempted to establish mice and rats as the preferred experimental models for investigating nocardial pathogenesis. The results of these investigations were variable, often contradictory; every conceivable outcome was reported (5, 7, 58). These vacillating reports were the result of early misconceptions regarding the nature of nocardiae and the general lack of standardization of methods. Many other animals, such as monkeys, were explored as models for investigating nocardial pathogenesis (5, 7). These studies showed that a wide variety of animals were susceptible to nocardial infection (5, 7). Experimentally, the nocardiae induced similar types of diseases following inoculation into most mammals, including mice, which indicated a general lack of unique host responses (5, 7). Therefore, the responses observed with murine models probably had their counterparts in most other mammalian species, including humans. Regardless of the earlier erratic responses reported by different investigators utilizing mice as models for nocardiosis, the mouse remained the best, most versatile general model for investigating mechanisms of host-nocardia interactions (5, 7, 9, 58).

TISSUE CULTURES TO INVESTIGATE NOCARDIA-HOST INTERACTIONS

Nocardiae are facultatively intracellular pathogens that resist the microbicidal activities of PMNs. They grow within mononuclear phagocytes from mice, rabbits, guinea pigs, and humans (5, 9). Once inside macrophages, virulent strains of *N. asteroides* inhibit phagosome-lysosome fusion (Fig. 2A), block phagosomal acidification, and grow (5, 9, 24, 58). In contrast, phagosome-lysosome fusion is facilitated in activated macrophages in the presence of *Nocardia*-specific antibody and T lymphocytes from preimmunized animals (Fig. 2B); the intracellular bacteria are either inhibited or killed (24). Furthermore, virulent strains of *N. asteroides* actively invade a variety of host cells both in vitro (Fig. 3) and in vivo (Fig. 4). The invasion of these host cells appears to be both specific and multifactorial. The differential adherence to and specificity for invasion of cells by *N. asteroides* is growth stage dependent. In addition, different strains of nocardiae possess distinct surface-associated ligands that bind to either epithelial cells in the lung (Fig. 4C and D) or endothelial cells in the brain (Fig. 4A and B). A model strain, *N. asteroides* GUH-2, has all the attributes listed above, and it is neuroinvasive. Therefore, the mechanisms of nocardia-host interactions have been studied most extensively using this strain of *N. asteroides*. By comparing and contrasting a variety of human, veterinary, and environmental isolates of different species of *Nocardia* with *N. asteroides* GUH-2, a spectrum of virulence patterns and a progression of host interactions have been recognized (4, 9, 12, 13, 58).

The in vitro interactions of nocardiae with monocytes, macrophages, and PMNs were studied extensively. Many of these investigations focused on determining the mechanisms for the ability of *N. asteroides* to resist intracellular killing and then grow within these phagocytes (9, 58). It was shown that virulent strains of *N. asteroides* secreted a superoxide dismutase (SOD) that became surface associated. This SOD, in combination with high levels of intracytoplasmic catalase, protected the nocardiae from the oxidative killing mechanisms of PMN, monocytes, and macrophages (5, 9, 58). Once phagocytized, these organisms remained within phagosomes wherein they inhibited phagosome-lysosome fusion (Fig. 2). In addition, the nocardiae blocked or completely neutralized the acidification of the phagosome so that the intraphagosomal pH remained above pH 7 (5). At the same time, virulent strains of *N. asteroides* modulated lysosomal enzyme content, especially within macrophages (5, 9, 58). There was selective reduction in acid phosphatase activity, while lysozyme and esterase-neutral protease activities either increased or remained unchanged (5). The strains of nocardiae that dramatically reduced acid phosphatase activity in macrophages also utilized this enzyme as a carbon source (5). Furthermore, when added to culture medium containing glutamate as the carbon source, acid phosphatase synergistically enhanced nocardial growth. These effects were not observed with strains of *N. asteroides* that do not alter acid phosphatase activity in macrophages (5, 9, 58).

In addition to macrophages, monocytes, and PMNs, the interactions of *N. asteroides* with a wide variety of both primary and established tissue culture cells were reported (Fig. 3A through F). Primary cultures of astrocytes and microglia from newborn mice demonstrated differential interactions with *N. asteroides* GUH-2 (12). Stationary-phase cells had specific longitudinal adherence to the surface of cells that had a stellate morphology consistent with type II astroglia (Fig. 3F). In contrast, there was little or no adherence of

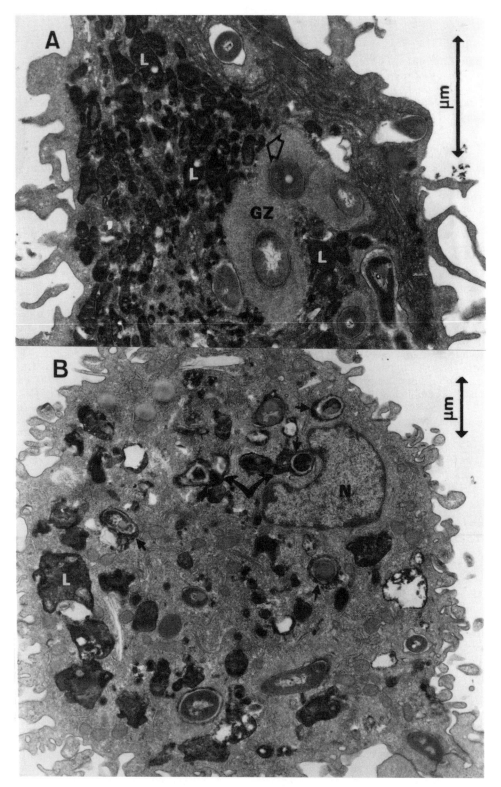

FIGURE 2 Electron micrographs of phagosome-lysosome interactions in activated rabbit alveolar macrophages infected with *N. asteroides* GUH-2. The lysosomes were prelabeled with horseradish peroxidase for the purpose of ultrastructural and histochemical localization. (A) Section showing inhibition of phagosome-lysosome fusion and the intact appearance of the nocardiae. Many of the bacteria appear to be surrounded by a large granular zone (GZ) that prevents contact between the lysosome (L) and the phagosome (open arrow). The nocardiae were preincubated with 20% normal rabbit serum. Bar, 1.0 μm. (Reprinted from reference 19 with permission from the publisher.) (B) Phagosome-lysosome fusion and bacterial cellular damage are prominent (arrows) in the same preparation of macrophages shown in panel A, except that these phagocytes were incubated with specifically primed lymph node lymphocytes; the bacteria were preincubated with sera and pulmonary lavage fluid from immunized rabbits. Note that the extensive bacterial damage and enhanced phagosome-lysosome fusion presented in this figure did not occur if any one of the components (primed lymphocytes, antibody, or pulmonary lavage fluid from immunized rabbits) were deleted. Bar, 1.0 μm. (Reprinted from reference 24 with permission from the publisher.)

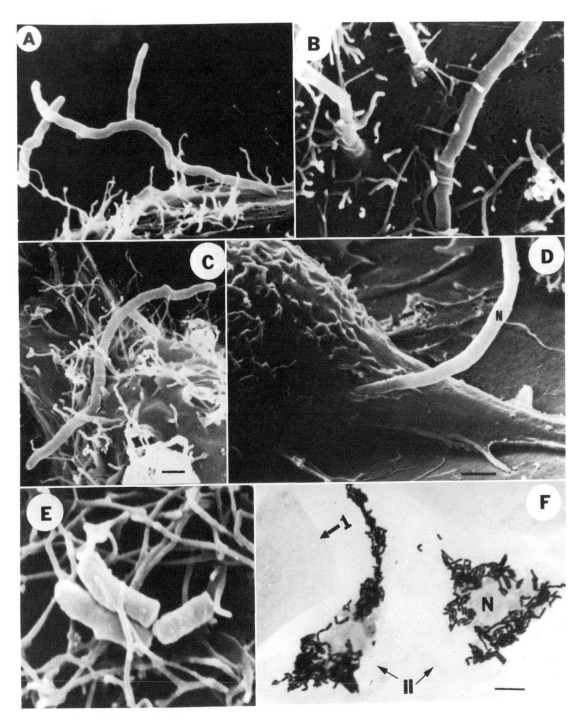

FIGURE 3   The comparative interactions of *N. asteroides* with different types of host cells grown in tissue culture. (A) Apical attachment of a log-phase nocardial filament to the surface of a HeLa cell. Note that this tip-associated adherence precedes penetration and invasion. (Reprinted from reference 8 with permission from the publisher.) (B) Apical penetration of the HeLa cell by three nocardial filaments (arrows). (Reprinted from reference 8 with permission from the publisher.) (C) Longitudinal adherence of a log-phase cell of *N. asteroides* to the surface of a HeLa cell. Note that only the bacterial filaments that attach by way of the filament tip appear to penetrate into the host cell. Bar, 1.0 μm. (Reprinted from reference 8 with permission from the publisher.) (D) Nocardial filament (N) tip invading through the surface (arrow) of an astrocytoma cell (CCF-STTG1) even after pretreatment of the tissue culture with cytochalasin. Bar, 1.0 μm. (Reprinted from reference 7 with permission from the publisher.) (E) Longitudinal association of stationary-phase bacilli to the microvilli of the HeLa cell. Bacteria attached in this manner do not invade the host cell. (Reprinted from reference 8 with permission from the publisher.) (F) Light micrograph of stationary-phase cells of *N. asteroides* GUH-2 adherent to the surface of type II astroglia (II). Bar, 10.0 μm. Note the arrangement of bacteria clustered around the nuclear region (N). Note, in contrast, the adjacent cuboidal type I astroglia (1) with a total absence of adherent nocardiae.

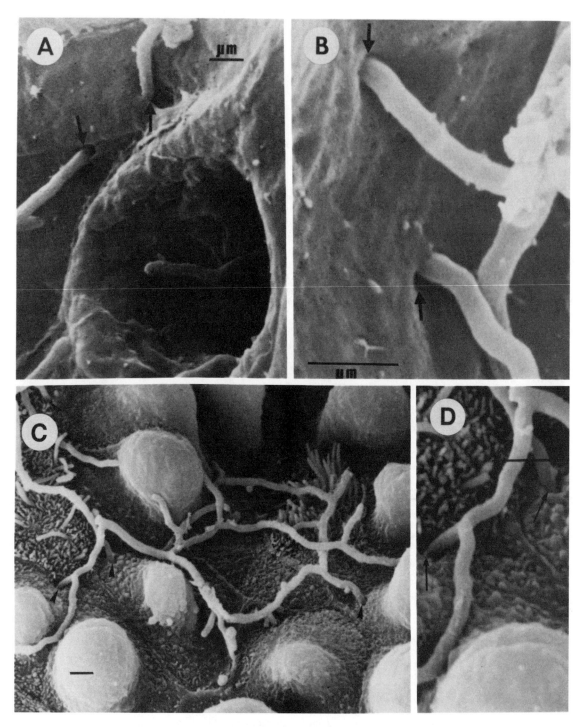

FIGURE 4   Scanning electron micrographs of differential attachment to and penetration of cells within the brain and lungs of mice by log-phase cells of *N. asteroides* GUH-2. (A) Apical penetration of capillary endothelial cells within an arteriole in the thalamus by nocardial filaments (arrows). Bar, 1.0 μm. (Reprinted from reference 11 with permission from the publisher.) (B) High-magnification view of two nocardial filaments penetrating into the endothelium of an arteriole in the region of the hypothalamus (arrows). Bar, 1.0 μm. (Reprinted from reference 11 with permission from the publisher.) (C) Nocardial interactions in the bronchiole of a C57BL/6 mouse 3 h after intranasal administration of log-phase cells of *N. asteroides* GUH-2. Note the association with nonciliated epithelial cells and apical penetration of Clara cells (arrowheads). Bar, 1.0 μm. (D) High-magnification view of nocardial penetration into bronchiolar epithelial cells as shown in panel C. Bar, 1.0 μm.

nocardiae to the surface of the more prevalent astroglia that had the type I polygonal morphology (Fig. 3F) (12). In these preparations, the microglia rapidly phagocytosed both filamentous and coccoid cells of nocardiae. However, only filamentous bacteria exhibiting apical adherence penetrated into the astroglia. Once inside, the nocardiae grew in astroglia, whereas growth in microglia was inhibited (12).

Comparative and differential interactions of log-phase and stationary-phase cells of *N. asteroides* GUH-2 were investigated with monocyte-macrophage cell lines J774A.1 and P388D1, bovine pulmonary artery endothelial cell line CPAE, rat glial cell line C6, and human astrocytoma cell lines CCF-STTG1 and U-373 MG (13). The interactions of log-phase and stationary-phase cells of *N. asteroides* GUH-2 in these six cell lines were different. There was a significant difference in uptake between log- and stationary-phase nocardiae in the phagocytic cell lines J774A.1 and P388D1. In cell line J774A.1, significantly more (greater than fivefold) log-phase cells of GUH-2 were internalized than stationary-phase organisms, and cytochalasin inhibited this uptake (>95% reduction) (13). Nevertheless, filaments of log-phase GUH-2 cells still adhered to the surface and penetrated into the cell in the absence of phagocytosis in cytochalasin-treated cells of J774A.2 (13). In contrast, human astrocytoma U-373 MG cells demonstrated little difference in effectiveness of internalization between log-phase and stationary-phase nocardiae, and cytochalasin did not block uptake (<10% reduction) (13). In all cell lines, only log-phase cells of *N. asteroides* GUH-2 exhibited a distinct filament tip-associated invasion process that appeared to be quite different from typical or classical phagocytosis (13). This filament tip-associated penetration did not appear to be affected by either cytochalasin or colchicine. On the other hand, stationary-phase cells of *N. asteroides* GUH-2 adhered longitudinally to the host cell surface with no evidence of apical penetration. These data suggest that there are multiple growth-stage-dependent surface components that are involved in the adherence to and internalization of *N. asteroides* (13).

Several investigations on the interactions of *N. asteroides* GUH-2 with PC12 cell cultures have been reported (19, 44, 68). This cell line is derived from a rat pheochromocytoma. As a consequence of being a dopaminergic cell line, PC12 cells are utilized extensively as models for investigating dopamine metabolism, serving as an in vitro model for investigating various components of Parkinson's disease (PD) (19, 68). As with other cell lines, log-phase cells of *N. asteroides* GUH-2 adhered apically to PC12 cells, and GUH-2 induced apoptotic cellular death (68). Concentrated culture filtrates from GUH-2 grown in a chemically defined medium also induced apoptosis, as well as dopamine depletion (19). A low-molecular-weight, nonproteinaceous compound secreted by the nocardiae was responsible for this dopamine depletion and might be involved in the induction of apoptosis (19, 44). However, cell death per se was not a prerequisite for dopamine depletion in PC12 cells by *N. asteroides* GUH-2 (44).

DIFFERENTIAL ADHERENCE TO AND INVASION OF HOST CELLS IN VIVO AND IN VITRO

Host-pathogen interactions studied in tissue culture systems are artificial by nature, and these types of investigations pose serious concerns about the validity of the conclusions drawn from the resultant data. Therefore, data obtained by

in vitro studies must be validated in a relevant animal model system in vivo. The mouse may be utilized as an ideal model for analyzing nocardia-host interactions, because responses to nocardiae in mice, rats, guinea pigs, rabbits, monkeys, dogs, and humans appear to be the same (9, 58). Furthermore, data on nocardial interactions in mice indicate that nocardiae behave the same in host cells maintained in vitro (9, 58). For example, nocardial interactions on the pulmonary epithelium in mice appear to be the same as those with pulmonary epithelial cells grown in tissue culture. Therefore, nocardial interactions with specific types of host cells grown in vitro should be relevant for studying pathogenic mechanisms (5, 8, 9, 12, 13, 58).

Cells of *N. asteroides* GUH-2 in the log phase of growth adhered by way of the filament tip to the surface of both pulmonary epithelial cells (Clara cells) (Fig. 4C and D) and brain capillary endothelial cells in mice (Fig. 4A and B) (4, 8, 11). In contrast, stationary-phase organisms from the same culture did not adhere apically either to these Clara cells in the bronchioles or to capillary endothelial cells in the brain (4, 8, 11). Unlike log-phase cells, stationary-phase organisms of *N. asteroides* GUH-2 adhered longitudinally to host cells, and in vitro they exhibited a high specificity for primary type II astroglial cells (12). As noted above, only nocardial filaments attached apically to the surface of the host cells (Fig. 3A), and then the bacterium rapidly penetrated this surface to become internalized (Fig. 3B and D) (4, 12, 13). The same pattern of adherence to and invasion of epithelial, endothelial, and astroglial cell lines occurs in vitro (Fig. 3A through D) as observed in vivo (Fig. 4), following incubation with log-phase cells of *N. asteroides* GUH-2 (4, 8, 11–13).

The differential and selective adherence displayed by nocardiae both in vitro and in vivo suggested distinct multiple ligands for host cells on the nocardial surface (4, 8, 11–13). Furthermore, the expression of these components on the nocardial surface appeared to be growth stage dependent (4, 8, 11–13). Utilizing different antibodies to block nocardial attachment supported the hypothesis that nocardiae possessed growth-stage-dependent ligands that recognized surface receptors on different types of host cells (8). Some of these nocardial components appeared to be specific proteins expressed only at the filament tip of nocardiae during active growth (43- and 36-kDa proteins), whereas others appeared to be glycolipids (trehalose dimycolates) incorporated into the entire cell envelope of the nocardiae during all stages of growth (8, 9). A 43-kDa protein on the filament tip of log-phase cells of *N. asteroides* GUH-2 was involved in apical attachment to and invasion of epithelial cells in the murine lung and HeLa cells in vitro (8). At the same time, a 36-kDa filament tip protein seemed to be involved in apical adherence to and penetration of capillary endothelial cells in the murine brain. Antibody against this 36-kDa protein blocked attachment and invasion in the brain but not in the lung. Conversely, antibody specific for the 43-kDa tip protein blocked interactions in the lung but had no apparent effect on the interactions in the brain (8). At the same time, monoclonal antibody prepared against trehalose dimycolates from *N. asteroides* GUH-2 decreased longitudinal adherence of both log- and stationary-phase nocardiae to many types of host cells (8, 9, 58).

NOCARDIAL INVASION OF THE BRAIN

N. asteroides, *N. farcinica*, *N. nova*, *N. pseudobrasiliensis*, *N. transvalensis*, *N. brasiliensis*, and *N. otitidiscaviarum* cause

CNS infections in humans (5, 9, 47, 73) and animals (7). These infections are most frequently recognized as brain abscesses (5).

The interactions of *N. asteroides* in experimental CNS infections in animals have been studied. Certain strains, exemplified by *N. asteroides* GUH-2, were selectively neuroinvasive (3–5, 9, 11). These neuroinvasive nocardiae adhered to capillary endothelial cells within specific regions of the brain, wherein they invaded through the endothelial cell into the brain parenchyma (Fig. 4A and B). During this invasion process, there frequently was no ensuing inflammatory response. Furthermore, the integrity of the blood-brain barrier often remained intact (3–5, 9, 11). We demonstrated that both mice and monkeys infected i.v. with neuroinvasive nocardiae developed either overt or cryptic CNS infections (3–5, 9, 10).

Even though it is well documented that nocardiae induce lesions in the human brain, the nature of the early interactions is not known. Based on animal models, the nocardiae probably reach the brain initially by way of the bloodstream, either by dissemination from a distant site (e.g., the lungs) or by direct inoculation following traumatic exposure (e.g., a puncture wound by a contaminated thorn or splinter). In addition to invasion of the brain, many neuroinvasive strains of *N. asteroides* also enter bronchiolar epithelial cells (Fig. 4C and D). In mice, log-phase cells of neuroinvasive nocardiae injected i.v. enter neurons and reside intracellularly, surrounded by numerous layers of membrane of unknown origin (Fig. 5A) (3). The innermost layer of these membranes always appears to be tightly associated with the outermost layer of the nocardial cell wall (Fig. 5A, arrow). Exactly the same process is observed in the brain of monkeys, following i.v. injection with the same nocardial strain (Fig. 5B, arrow) (10). Since this process occurs identically in mice and monkeys, it seems reasonable to suggest that the same events take place when nocardiae become blood-borne in humans. It is not known whether cryptic disease ensues following this invasion in humans.

Overt invasion by nocardiae results in lesions that vary pathologically from a diffuse cellular infiltration of PMNs to either an abscess or a granuloma in mammals such as mice, monkeys, rabbits, and humans (9). On the other hand, cryptic invasion induces few or no inflammatory infiltrates but instead induces subtle cellular changes, including neurodegeneration and axonal alterations, especially along the myelin sheath (3, 10). In both mice and monkeys, the nocardiae invade neurons in the basal ganglia. Furthermore, intraneuronal and intraparenchymal inclusions are produced in the brain after a prolonged incubation period of several months (21, 41). Indeed, cryptic invasion of the brain in mice following i.v. inoculation of log-phase cells of *N. asteroides* GUH-2 results in dopaminergic neurodegeneration in the substantia nigra, leading to a levodopa-responsive movement disorder that shares many features with parkinsonism (41). Thus, sublethal i.v. inoculation of mice with log-phase cells of *N. asteroides* GUH-2 results in an infectious disease model of parkinsonism (41). The relationships, if any, of this process to the human disease are unknown.

In the 1980s, B. L. Beaman and S. Scates (unpublished data) noticed that a significant number of mice that survived for >2 weeks following an i.v. injection of *N. asteroides* GUH-2 developed abnormal behavior, with vertical head-shaking, rigidity, and dyskinesia being most prominent. In 1991, Kohbata and Beaman published obervations that many of the mice that exhibited abnormal movement responded temporarily to treatment with a combination of L-dopa and carbidopa (41); thus, these mice had an L-dopa-responsive movement disorder. Furthermore, the substantia nigra region of the brain was affected most in these head-shake mice, exhibiting loss of Nissl staining, a decrease in tyrosine hydroxylase immunoreactivity, and the occasional presence of hyaline-like inclusion bodies (41).

The cardinal symptoms of PD are tremor, rigidity, dyskinesia or akinesia, and postural instability (18, 43). Possessing any two of these features at the same time is sufficient to make a tentative diagnosis of PD (9, 18, 21, 43, 58). In humans, many of these behaviors are temporarily ameliorated by treatment with both carbidopa and L-dopa; thus, PD may be defined as an L-dopa-responsive movement disorder (41). Some of the mice experimentally infected with *N. asteroides* GUH-2 exhibited all of these signs within the same animal, suggesting that they had a form of parkinsonism (9, 41). Although postencephalic parkinsonism (encephalitis lethargica) was known to follow some viral infections, idiopathic PD, the most common form, is not thought to have a foundation in infectious diseases (9, 21, 41, 58). It is generally thought that PD is a result of exposure to some unidentified toxic substance from the environment, combined with normal aging. Currently, no bacterial or viral agent is believed to be involved as an etiology of PD. There are many cogent arguments against infectious agents as etiologies of PD and a significant amount of data supporting the role of toxic substances from the environment (9, 21, 41, 58), but the real etiology (or etiologies) of PD remains unknown. Nevertheless, the observations of Kohbata and Beaman (41) legitimately raise the question. Is it possible that nocardiae may be an etiology of some cases of PD?

Numerous observations concerning nocardial effects on the brain that might lead to parkinsonism have been published. For example, there are at least three murine models of L-dopa-responsive movement disorders caused by nocardioform *Actinomycetes* reported in the literature (25, 41, 50). In summary, one of these models has focused on log-phase cells of *N. asteroides* GUH-2 (41). Strain GUH-2 exhibits specificity for adherence to the substantia nigra pars compacta (SN) in both mice and monkeys. This region of the brain is the focal point of study for such neurodegenerative diseases in humans as PD (10, 11). Furthermore, electron microscopy shows that these adherent nocardiae actively invade through the endothelium and basal lamina and grow within neurons, astroglia, and the neuropil in the SN in both mice and monkeys (3, 10). Not only do microdissected regions of monkey and murine brains show that there are site-specific adherence, invasion, and growth of *N. asteroides* GUH-2 within the SN (3, 5, 10, 11, 58), but histological and electron microscopic observations of the SN after invasion by nocardiae fail to reveal any host-cell infiltrates at the sites of nocardial invasion in the SN (3, 9–11, 58). Thus, it appears that log-phase cells of *N. asteroides* GUH-2 induce several changes in neuronal structure typical for neurodegeneration characteristic for parkinsonism in both mice and monkeys (10, 11). For instance, *N. asteroides* GUH-2 induces selective reduction of striatal dopamine and altered neurochemistry similar to that observed with other agents that induce parkinsonism in mice (35). One of the suggested mechanisms for PD is the selective dopaminergic neuronal dropout in the substantia nigra by apoptosis. *N. asteroides* GUH-2 induces apoptosis in dopaminergic neurons in the substantia nigra in mice (Fig. 6) and dopaminergic PC12 cells in vitro (68). Also, culture filtrates of *N. asteroides* GUH-2 cause depletion of dopamine in PC 12 cells in vitro, similar

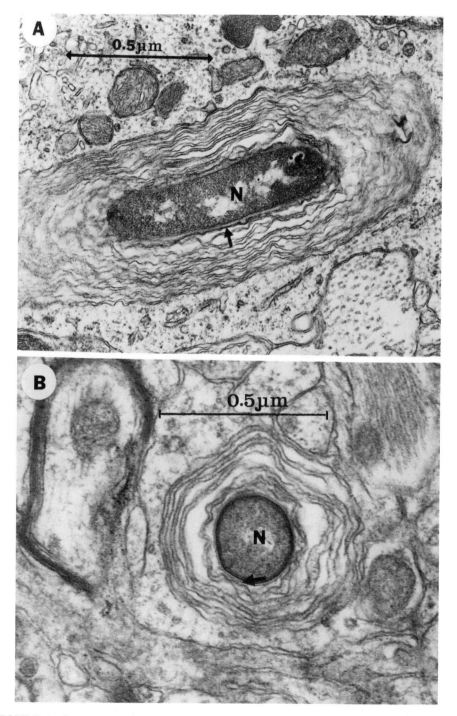

FIGURE 5 Comparative ultrastructure of *N. asteroides* GUH-2 growing within cells of the brain of a mouse and a monkey. Note that there is no inflammatory infiltration at the site of nocardial invasion in either animal. (A) Cell of *N. asteroides* GUH-2 growing within a neuron in the midbrain of a mouse 24 h after tail vein injection of a suspension of log-phase nocardiae. Note the numerous layers of membrane surrounding the ultrastructurally intact bacterium, with the innermost layer of membrane tightly adherent to the bacterial surface (arrow). N, nocardial filament. Bar, 0.5 μm. (Reprinted from reference 3 with permission from the publisher.) (B) A cross section of a nocardial cell growing within the brain of a monkey 48 h after i.v. injection (leg vein) of a suspension of log-phase cells of *N. asteroides* GUH-2 as in panel A. Note the numerous layers of membrane, with the innermost layer adherent to the bacterial surface (arrow). Bar, 0.5 μm.

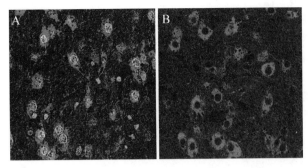

FIGURE 6 *(See the separate color insert for the color version of this illustration.)* Confocal micrographs of nocardia-induced apoptosis. (A) Apoptosis of dopaminergic neurons in the substantia nigra in a head-shake mouse 14 days after infection. The red stain localizes dopaminergic neurons. Free 3'-OH ends of DNA from apoptotic nuclei were labeled with nucleotides conjugated to fluorescein isothiocyanate (green). (B) Uninfected control. Dopaminergic neurons in the substantia nigra in a healthy uninfected mouse. The red stain localizes dopaminergic neurons. Note that there is no apoptosis. (Reproduced from reference 68.)

to agents such as 1,2,3,6-methylphenyltetrahydropyridine that represent a chemical model for PD (19, 44). One of the major pathological components of PD is the presence of intracellular inclusion bodies (Lewy bodies) composed of both ubiquitin and alpha synuclein. Monkeys develop similar inclusion bodies after infection with *N. asteroides* GUH-2 (21). Finally, in humans, many of the inclusion bodies as well as some cells of the SN (including patients diagnosed with idiopathic PD) label positively for a nocardia-specific 16S rRNA probe (21).

As noted above, nocardiae are aerobic members of the order *Actinomycetes*, related to *Streptomyces* (5, 9, 37, 47). It is well established that many of these organisms including nocardiae secrete secondary metabolites such as antibiotics (5, 9, 47); several *Streptomyces* species secrete a variety of proteasome inhibitors (46). It was discovered that an unidentified aerobic actinomycete secreted epoximicin, which is a potent proteasome inhibiter. Proteasomes play a central role in the directed degradation of proteins within many cells. These proteins are first labeled with ubiquitin, and then they are degraded by the proteasome. Inhibition of this function will result in the abnormal accumulation of intracellular ubiquinated proteins (46). It is thought that within neurons in the brain, ubiquitin becomes bound to alpha synuclein; altered function of the proteasome results in the accumulation of this ubiquinated alpha synuclein complex to form the intraneuronal inclusion body known as a Lewy body (46). Proteasome inhibitors such as epoximicin can have this function (46). Indeed, epoximicin has been reported to induce progressive parkinsonism in animals with key features almost identical to those described for the nocardial model of parkinsonism (7, 9, 35, 41, 46, 58, 68). McNaught et al. suggested that "environmental proteasome inhibitors are candidate PD-inducing toxins" (46). Analysis of the genome of *N. farcinica* revealed a large number of genes that encoded a variety of secondary metabolites similar to *Streptomyces* spp., making nocardiae metabolically very diverse (37).

Contrary to current belief, these published observations justify the idea that infectious agents such as nocardiae may play a role in the development of neurodegenerative dis-

eases (i.e., PD and Lewy body dementia) (18, 21, 43). It is tempting to speculate that opportunistic organisms of relatively low virulence like nocardiae may invade regions of the brain and induce localized and progressive damage. Because these organisms are not highly virulent, their infectivity in an immunologically intact host is only transient, they can be controlled by local innate responses, and they do not evoke a strong inflammatory response. Nevertheless, their passing becomes evident in the brain by placing into motion a series of events that lead to neurodegeneration. In contrast, more-virulent organisms may establish a progressive infection leading to damage that requires a significant host response to control it. Several species of *Nocardia* including *N. asteroides* GUH-2 reside on the borderline between being opportunistic pathogens of low virulence and true pathogens that are aggressively invasive (9, 58).

THE GLYCOLIPIDS OF NOCARDIAE

Mycobacterium, Corynebacterium, Nocardia, Rhodococcus, Tsukamurella, and *Gordonia* are bacterial genera that possess mycolic acids in their cell envelopes (5, 9, 27). All of these genera contain species that may be pathogenic for humans; therefore, this group of organisms is collectively named mycolic acid-containing pathogens (Beaman, unpublished). Mycolic acids are unique α-branched, β-hydroxylated fatty acids that represent a major component of the cell wall, and they define many of the basic biological properties of mycolic acid-containing pathogens (5, 9, 74). Relatively small amounts of mycolic acid reside freely within the envelope. Instead, most of these long-chained fatty acids are covalently linked to a sugar moiety to form a glycolipid. Indeed, the bulk of the mycolic acids are bound to the arabinose portion of the arabinogalactan polymer, which, in turn, is covalently linked to the peptidoglycan of the cell wall (9, 32). In addition, there are significant amounts of freely associated glycolipids consisting of mycolic acids covalently bound to other sugars, such as glucose and trehalose. The biological activities of the trehalose mycolates have received most of the research attention, and the literature on these compounds is extensive. Trehalose dimycolate may be the most biologically active of these glycolipids (2, 5, 9, 32, 66).

In 1950, Bloch (32) isolated a compound from *Mycobacterium tuberculosis* that was believed to be responsible for both virulence and a characteristic serpentine pattern of growth. Because the virulence of *M. tuberculosis* for guinea pigs corresponded to the presence of the cordlike growth and this growth was associated further with a substance that was soluble in organic solvents, this substance was named cord factor (5, 32). Bloch demonstrated that petroleum ether dissolved this substance from *M. tuberculosis* H37RV, and it imparted important biological activities to unrelated bacilli that were coated with this extract. Later studies demonstrated that this compound was composed of two mycolic acid molecules linked to trehalose to form trehalose 6,6'-dimycolate (TDM). Thus, TDM is synonymous with cord factor, and these two terms are currently used interchangeably by numerous investigators. Data suggest that TDM may actually impart serpentine cord formation in growing strains of virulent *M. tuberculosis*, even though earlier investigators expressed skepticism about this relationship (15). It is beyond the scope of this chapter to discuss the extensive literature on the structure and biology of TDMs (5, 32).

TDM has been shown to possess a wide variety of biological activities, depending upon the source of the TDM (and thus different chemical structures) and the manner of preparation and presentation within various bioassays. The published effects of TDM on biological systems include the following. (i) TDM from *N. asteroides* in oil is lethal for mice (5, 9). (ii) TDM targets and disrupts mitochondrial function (5, 32). (iii) TDM uncouples oxidative phosphorylation (5, 32). (iv) TDM induces apoptotic death in cells (55). (v) TDM inhibits phagosome-lysosome fusion in macrophages (74). (vi) TDM prevents calcium-dependent fusion in liposomal membranes (23). (vii) TDM is membrane interactive, altering hydration, rigidity, fluidity, and permeability in lipid bilayers (23). (viii) TDM induces granulomas in mice (5, 9, 51). (ix) TDMs are potent immunoadjuvants (42). (x) TDMs stimulate γδ T cells and modulate cytokine production from lymphocytes (33). (xi) TDM has potent anticancer potential (5, 32, 52). (xii) TDM interferes with the coagulation system in blood (5, 32). (xiii) TDM is antigenic (40). (xiv) TDM can be used as a diagnostic antigen (27). (xv) TDM activates protein kinase C and induces tumor necrosis factor (65). (xvi) TDM can induce serpentine cord formation (15). (xvii) TDM coupled to a protein carrier is an effective vaccine against virulent *M. tuberculosis* in mice (5, 32). Not all TDMs are equally active in the above-listed properties; indeed, the TDMs in the reports described above were obtained from different organisms grown under different culture conditions. Nevertheless, strains of nocardia that possess TDMs are more virulent than those that do not (36), and TDMs of nocardiae possess most of the same properties described above (5, 36). These observations support the hypothesis that the structures of the mycolic acid moieties are important for biological activities of the different TDMs. It is difficult to assess the role of chemical structure on a specific biological activity, because TDMs purified from most mycolic acid-containing bacteria are mixtures that have different sizes of mycolic acid side chains linked to trehalose (32). It is possible to synthesize artificial TDMs in the laboratory, but the biological relationships of these TDMs to the naturally occurring mixtures appear not to be the same.

It has been established clearly that TDM is a potent immunomodulator and immunoadjuvant, even though the mechanisms for these activities are not understood. Mycobacterial TDMs stimulate an increase of γδ T cells in human cord blood T lymphocytes (70), and TDM from *Mycobacterium bovis* preferentially increases the numbers of γδ T cells in the lungs of mice (61). Oswald et al. (54) reported that TDMs from *M. tuberculosis* induced interleukin-12 to mediate macrophage priming through induction of gamma interferon (54), and Beckman et al. (14) demonstrated that mycolic acids from the TDMs of *M. tuberculosis* are CD1b-restricted antigens (14). Also, mycobacterial TDMs cause an increase in natural killer, intermediate CD3, and γδ T cells during granuloma formation in murine lungs (67).

The biological properties of some of these glycolipids in nocardiae and their possible relationship to nocardial pathogenesis have received a modicum of attention. It was first suggested in the early 1980s that the amount of trehalose dimycolate present in the cell wall of *N. asteroides* correlated directly with virulence (5). Thus, strains with large amounts of TDM were highly virulent, whereas strains with no detectable TDM were either avirulent or had very low virulence for mice (5, 9, 36). In addition, small amounts of TDM from *N. asteroides* were lethal for mice, following intraperitoneal injection of a mineral oil emulsion. Strains of *N. asteroides* that possessed TDM inhibited phagosome-lysosome fusion in macrophages, while strains

that did not have TDM did not inhibit fusion (5, 36). Furthermore, TDMs purified from *N. asteroides* and coated onto yeast cells prevented phagosome-lysosome fusion when phagocytized by human monocytes. In contrast, these same yeast cells without TDM induced strong phagosome-lysosome fusion in monocytes (5, 23). Similar results were reported using TDMs from *M. tuberculosis*. Based on the above observations, it is probable that TDM plays a critical role in nocardial pathogenesis and host responsiveness to nocardial infection (5, 9). Investigators have focused on the roles of other glycolipids in virulence of mycobacteria. Many of these studies have focused on lipoarabinomannans, but these have not been reported for nocardiae (9, 58; B. L. Beaman and P. Pujic, unpublished data).

NOCARDIAL FACTORS IMPORTANT FOR PATHOGENESIS

The mechanisms of nocardial pathogenesis are not known, since the nocardial genome has been sequenced only recently (37). Our knowledge about the molecular mechanisms and genes involved in pathogenesis lags far behind what is understood with other bacterial pathogens (37). Nevertheless, some of the factors contributing to pathogenesis and virulence of nocardiae have been determined (5, 9, 58). Both the conditions and phase of growth affect nocardial virulence (58), and nocardial strains capable of survival within macrophages are most virulent (5, 9, 58). The inhibition of phagosome-lysosome fusion is important for the intracellular survival and growth of nocardiae (5). This inhibition appears to be the result of surface glycolipids such as TDM (5, 23). Furthermore, there is a relationship between TDM content, inhibition of phagosome-lysosome fusion, and nocardial virulence for mice (5, 9, 58). Therefore, as described above, TDMs (cord factors) appear to be virulence factors (5, 36). Other cell wall components, as well as the structure of individual mycolic acids in the envelope, correlate with nocardial virulence (58). The cell walls of nocardiae contain powerful mitogens for both B and T cells. At least one of these is a T-cell-independent B-cell mitogen (5). The most virulent strains of nocardiae secrete SOD, and they have increased levels of intracytoplasmic catalase (5, 9, 58). Thus, both SOD and catalase are important virulence factors for nocardiae (5, 9, 58), since the level of nocardial resistance to the oxidative killing mechanisms of PMN and macrophages is important as a virulence determinant (5, 9, 58). Virulent strains of *N. asteroides* block or neutralize phagosomal acidification in macrophages (5, 9, 58), and they decrease intralysosomal acid phosphatase and modulate lysosomal enzyme content in macrophages. These same strains utilize acid phosphatase as a sole carbon source. All of these activities are concordant with increased nocardial virulence, and they may be involved in pathogenesis (5, 9, 58). Cell-associated and -secreted hemolysins appear to be associated with nocardial virulence, and they may be important for nocardial invasion of host cells (5). The nocardiae are closely related phylogenetically to mycobacteria, including *M. tuberculosis*; therefore, pathogenic mechanisms identified for these bacteria may be similar in many of the nocardiae (9, 58).

Cell wall-defective nocardial cells (L forms) are involved in nocardial persistence, chronic infection, and cryptic invasion. L forms of certain strains of nocardiae are involved in pathogenesis (5, 9, 58). Some strains of *N. asteroides* are neuroinvasive, whereas others are not. The strains of *N. asteroides* that invade both the lungs and brain of mice equally well are significantly more virulent than those strains that invade only the lungs (4). The mechanisms are not known, but nocardial L forms may be involved in cryptic invasion and persistence in the CNS (5, 9, 58).

Many of the neuroinvasive strains of *N. asteroides* selectively invade the basal ganglia, with specificity for the substantia nigra (3–5, 11, 41). Cryptic invasion of the basal ganglia results in neurodegeneration and selective destruction of dopaminergic neurons in the substantia nigra of mice, resulting in a parkinsonian movement disorder (41). Furthermore, this selective adherence to and invasion of host tissues by log-phase cells of *N. asteroides* are mediated by filament tip-associated proteins. For example, a 43-kDa protein on the surface of the tip of log-phase cells of *N. asteroides* GUH-2 is involved in adherence to epithelial cells in the lung. At the same time, a 36-kDa tip-associated protein appears to be involved with adherence to capillary endothelial cells in the brain (8).

THE FIRST GENOMIC SEQUENCE OF A *NOCARDIA* SPECIES: WHAT DOES IT SAY ABOUT NOCARDIAL PATHOGENESIS?

Ishikawa et al. (37) sequenced the entire genome of *N. farcinica* strain IFM 10152, which was originally isolated from a human infection. This strain was selected because both the taxonomy and antimicrobial resistances among members of the *N. asteroides* complex are notoriously variable. However, *N. farcinica* IFM 10152 (a member of the *N. asteroides* complex) is most characteristic for this species, which in Japan is taxonomically quite homogeneous (37). Strain IFM 10152 has a single circular chromosome containing 6,021,225 bp with an average G+C content of 70.8%. Unlike many nocardial strains, it also contains two circular plasmids (37) with mostly unknown functions. The genomic sequence of IFM 10152 was compared to other bacterial genomes with the BLASTp program. Based on this analysis, *N. farcinica* IFM 10152 is most closely related to *M. tuberculosis*, *Corynebacterium glutamicum*, and (to a lesser extent) *Streptomyces avermitilis* (37). The deduced proteins encoded by the genome were then clustered, and paralogous families of these clusters were identified. It was deduced that 552 paralogous families consisting of between 2 and 86 proteins were encoded by 36% of the IFM 10152 genome (37). Paralog analysis suggested that *N. farcinica* evolved to survive in both the soil (similar to *Streptomyces*) and the mammalian host (similar to *M. tuberculosis*) and that all of these organisms probably originated from a common ancestor (37).

The *M. tuberculosis* genome has at least four mammalian cell entry (*mce*) operons that appear to be virulence factors involved in entry into and survival within macrophages. By way of comparison, the genome of IFM 10152 encodes six *mce* operons, whereas nonpathogenic *Streptomyces* species only have one *mce* operon (37). Also, *M. tuberculosis* possesses the antigen 85 family of proteins, which are fibronectin-binding proteins that are important for mycobacterial pathogenesis. *M. tuberculosis* encodes 4 proteins in this family, while in contrast, the genome of *N. farcinica* IFM 10152 encodes 12 different proteins in this family (37).

In addition to the above, Ishikawa et al. (37) reported that there were many other genes found that were probably associated with nocardial virulence and pathogenesis. These included components responsible for adherence to and invasion of host cells such as Nfa34810, which has high homology to mycobacterial invasions (37). They also found genes encoding for four catalases, two SODs, an alkylhydroperoxidase, two nitrate reductases, gene clusters for iron acquisition (i.e., siderophores related to mycobactin), and mycolic acid biosynthesis such as KasA (37). *N. farcinica* IFM 10152 did not have genes for PE/PPE/PGRS, a family of genes thought to be essential for pathogenicity of *M. tuberculosis* (37).

HOST FACTORS IMPORTANT FOR PROTECTION AGAINST NOCARDIAE

Many components of both the innate and acquired host responses are involved in protection from pathogenic species of *Nocardia* (5, 9, 58). First, PMN and macrophages are central to host resistance to these bacteria (5, 9, 58). However, PMN and macrophages alone do not kill the most virulent strains of *N. asteroides*. These professional phagocytes require the complete, intact machinery of the immune system to kill virulent nocardiae (5, 9, 24, 58). In addition, an immunologically specific subtype of T lymphocyte from immunized animals binds to the surface of log-phase cells of *N. asteroides* and induces lysis of these bacteria. This T-cell-mediated killing of nocardiae is not associated with natural killer cell activity and appears to be independent of antibody and complement (5, 9, 58). In contrast, B lymphocytes appear not to be required for host resistance to nocardiae; their role in protection of the host against nocardial infection is not clear (5, 9, 58). Immunosuppressive treatment with compounds such as cyclophosphamide, prednisone, or other substances that decrease cell-mediated responses significantly enhances host susceptibility to progressive, disseminated disease caused by even the least virulent strains of *N. asteroides* (5, 9, 58). Thus, T lymphocytes, activated macrophages, and cell-mediated immune responses appear to be most important for host resistance to nocardiae (5, 9, 58). However, such factors as the normal microflora of the animal as well as the γδ subset of T cells are important in the early stages of innate host resistance to nocardial infection (28, 58).

The route of exposure to nocardiae has a profound effect on host susceptibility, suggesting that host resistance toward nocardiae is compartmentalized (5, 9, 58). For example, following i.v. inoculation, certain neuroinvasive strains of *N. asteroides* invade the substantia nigra and cryptically induce dopaminergic neuronal dropout and neurodegenerative changes in mice (41, 68), leading to parkinsonian alterations in the basal ganglia. Often, there is no recognizable inflammatory response, and no PMNs accumulate at this site of invasion. During this invasion of the basal ganglia, the bacteria appear either to disappear completely through elimination by microglia or to be induced into cryptic persistence (perhaps as L forms). The host response during this process appears to be progressively degenerative and not inflammatory (3–5, 11, 41). In contrast, the host responses to the same organism introduced intranasally into the lungs of the same animal are very different, with a strong cellular infiltration by PMN becoming prominent (9, 58).

Even though the mechanisms of host resistance to nocardial infection are complex, multifactorial, and not completely understood, a more comprehensive picture is emerging (9, 58). It is clear that a timed and coordinated sequence of events is necessary for host protection against these organisms. The first host cell to respond to nocardia-induced damage during nocardial invasion is the neutrophil (PMNs). As stated earlier, this host cell is not able to eliminate the invading nocardia, but it does prevent further growth of the pathogen (9, 26, 58). Thus, PMNs become the first line of defense against nocardial invasion. In response to the nocardiae, PMNs secrete a variety of cytokines, including gamma interferon, which initiate a cascade of host responses that include summoning subpopulations of lymphocytes. In the lungs, it appears that γδ T lymphocytes replace PMNs as the prominent host cell. These γδ T cells are involved in the earliest innate host response to nocardiae, and cross talk between resident subtypes of γδ T cells and PMNs appears to occur (26). The result is that the nocardiae are removed. This process is followed by what appears to be a second wave of different γδ T-cell subpopulations (26). These newly arriving γδ T cells seem to be associated with repairing the damage caused by the invading nocardiae (26, 58). In the absence of a strong PMN and γδ T-cell response in the lung, other factors that may involve macrophages take control to eliminate the nocardiae and repair the damage (58). In the lungs of mice infected with *N. asteroides*, there appears to be no evidence for the involvement of B cells, αβ T cells (either CD4$^+$ or CD8$^+$ cells), or NK cells (58). These observations may partly explain why individuals with AIDS, who have a severely depleted CD4$^+$ T-cell population, may not be greatly predisposed to nocardial infections (58).

CONCLUSIONS

The nocardiae are gram-positive, mycolic acid-containing bacteria belonging to the aerobic actinomycetes. They are relatively common environmental organisms that are important causes of acute and chronic diseases in a wide variety of hosts, including humans. The mechanisms of pathogenesis and host resistance to these facultatively intracellular pathogens are incompletely understood (5). The most frequently recognized species causing disease in the United States is *N. asteroides* (5). This species is taxonomically heterogeneous, and many investigators suggest that this taxon consists of several undefined species (5, 36). Therefore, it is not surprising that the host responses, pathogenic potential, and pathogenesis of individual isolates are quite diverse. Defining relative nocardial virulence by using murine models, following i.v. inoculation, results in a continual spectrum from nonpathogenic (i.e., >10^9 CFU to cause disease) to highly virulent (i.e., <10^5 CFU to cause the death of a healthy mouse). Some understanding of the mechanisms for pathogenesis and host responses has emerged from comparing and contrasting strains of nocardiae that are nonpathogenic with those that are virulent. In general, clinical isolates of nocardiae are more virulent for mice than soil isolates (5, 9, 58). Furthermore, the observation that the virulence of most strains of *N. asteroides* is growth stage dependent has provided additional insight. It appears that with most clinical isolates of *N. asteroides*, organisms in the stationary phase of growth behave in the host as opportunistic pathogens, whereas log-phase cells from the same culture are more like primary pathogens in healthy hosts. It should be noted that most studies of nocardial pathogenesis have focused on *N. asteroides* sensu stricto. There are few investigations of the mechanisms of pathogenesis of the other pathogenic species of *Nocardia* (5, 9, 58). Whether the mechanisms of host resistance and pathogenesis for species such as *N. brasiliensis*, *N. farcinica*, and *N. otitidiscaviarum* are the same as those for *N. asteroides* is not known. A few reports suggest significant differences among these species. Therefore, understanding and defining nocardial pathogenesis are in the earliest stages (9, 58). There have been very few studies of the genetics of nocardiae, and no virulence genes have been recognized. There remains a long way to go to unravel the contents in the mysterious "black box" of nocardial pathogenesis.

Portions of this work were supported by Public Health Service grants RO1-AI42144 from the National Institute of Allergy and Infectious Diseases and RO1-HL69426 from the National Heart, Lung, and Blood Institute.

REFERENCES

1. **Arenas, R.** 1997. Mycetoma vs. nocardiosis. *J. Dermatol.* **24:**68.
2. **Baba, T., Y. Natsuhara, K. Kaneda, and I. Yano.** 1997. Granuloma formation activity and mycolic acid composition of mycobacterial cord factor. *Cell. Mol. Life Sci.* **53:**227–232.
3. **Beaman, B. L.** 1993. Ultrastructural analysis of growth of *Nocardia asteroides* during invasion of the murine brain. *Infect. Immun.* **61:**274–283.
4. **Beaman, B. L.** 1996. Differential binding of *Nocardia asteroides* in the murine lung and brain suggests multiple ligands on the nocardial surface. *Infect. Immun.* **64:**4859–4862.
5. **Beaman, B. L., and L. Beaman.** 1994. *Nocardia* species: host-parasite relationships. *Clin. Microbiol. Rev.* **7:**213–264.
6. **Beaman, B. L., and D. M. Shankel.** 1969. Ultrastructure of *Nocardia* cell growth and development on defined and complex agar media. *J. Bacteriol.* **99:**876–884.
7. **Beaman, B. L., and A. M. Sugar.** 1983. *Nocardia* in naturally acquired and experimental infections in animals. *J. Hyg.* **91:**393–419.
8. **Beaman, B. L., and L. Beaman.** 1998. Filament tip-associated antigens (FTAAs) involved in adherence to and invasion of pulmonary epithelia in vivo and HeLa cells in vitro by *Nocardia asteroides*. *Infect. Immun.* **66:**4676–4689.
9. **Beaman, B. L., and L. V. Beaman.** 2000. *Nocardia asteroides* as an invasive, intracellular pathogen of the brain and lungs, p. 167–198. *In* J. Hacker and T. Oelschlaeger (ed.), *Bacterial Invasion into Eukaryotic Cells: Subcellular Biochemistry.* Springer-Verlag, Berlin, Germany.
10. **Beaman, B. L., D. Canfield, J. Anderson, B. Pate, and D. Calne.** 2000. Site specific invasion of the basal ganglia by *Nocardia asteroides* GUH-2. *Med. Microbiol. Immunol.* **188:**161–168.
11. **Beaman, B. L., and S. A. Ogata.** 1993. Ultrastructural analysis of attachment to and penetration of capillaries in the murine pons, midbrain, thalamus and hypothalamus by *Nocardia asteroides*. *Infect. Immun.* **61:**955–965.
12. **Beaman, L., and B. L. Beaman.** 1993. Interaction of *Nocardia asteroides* with murine glia cells in culture. *Infect. Immun.* **61:**343–347.
13. **Beaman, L., and B. L. Beaman.** 1994. Differences in the interactions of *Nocardia asteroides* with macrophage, endothelial, and astrocytoma cell lines. *Infect. Immun.* **62:**1787–1798.
14. **Beckman, E. M., S. A. Porcelli, C. T. Morita, S. M. Behar, S. T. Furlong, and M. B. Brenner.** 1994. Recognition of a lipid antigen by CD1-restricted αβ$^+$ T-cells. *Nature* **372:**691–694.
15. **Behling, C. A., B. Bennett, K. Takayama, and R. L. Hunter.** 1993. Development of a trehalose 6,6'-dimycolate model which explains cord formation by *Mycobacterium tuberculosis*. *Infect. Immun.* **61:**2296–2303.

16. **Brown, J. M., K. N. Pham, M. M. McNeil, and B. A. Lasker.** 2004. Rapid identification of *Nocardia farcinica* clinical isolates by a PCR assay targeting a 314-base-pair species-specific DNA fragment. *J. Clin. Microbiol.* **42:** 3655–3660.

17. **Buckley, J. A., A. R. Padhani, and J. E. Kuhlman.** 1995. CT features of pulmonary nocardiosis. *J. Comput. Assist. Tomogr.* **19:**726–732.

18. **Camicioli, R.** 2002. Identification of parkinsonism and Parkinson's disease. *Drugs Today* (Barcelona) **38:**677–686.

19. **Camp, D. M., D. A. Loeffler, B. A. Razoky, S. Tam, B. L. Beaman, and P. A. LeWitt.** 2003. *Nocardia asteroides* culture filtrates cause dopamine depletion and cytotoxicity in PC12 cells. *Neurochem. Res.* **28:**1359–1367.

20. **Castro, L. G., W. Belda, Jr., A. Salebian, and L. C. Cuce.** 1993. Mycetoma: a retrospective study of 41 cases seen in Sao Paulo, Brazil, from 1978 to 1989. *Mycoses* **36:** 89–95.

21. **Chapman, G., B. L. Beaman, D. A. Loeffler, D. M. Camp, E. F. Domino, D. W. Dickson, W. G. Ellis, I. Chen, S. E. Bachus, and P. A. LeWitt.** 2003. In situ hybridization for detection of nocardial 16S rRNA: reactivity within intracellular inclusions in experimentally infected cynomolgus monkeys—and in Lewy body-containing human brain specimens. *Exp. Neurol.* **184:**715–725.

22. **Chun, J., C. N. Seong, K. S. Bae, K. J. Lee, S. O. Kang, M. Goodfellow, and Y. C. Hah.** 1998. *Nocardia flavorosea* sp. nov. *Int. J. Syst. Bacteriol.* **48:**901–905.

23. **Crowe, L. M., B. L. Spargo, T. Ioneda, B. L. Beaman, and J. H. Crowe.** 1994. Interaction of cord factor (alpha, alpha'-trehalose-6,6'-dimycolate) with phospholipids. *Biochim. Biophys. Acta* **1194:**53–60.

24. **Davis-Scibienski, C., and B. L. Beaman.** 1980. Interaction of alveolar macrophages with *Nocardia asteroides*: immunological enhancement of phagocytosis, phagosome lysosome fusion, and microbicidal activity. *Infect. Immun.* **30:**578–587.

25. **Diaz-Corrales, F. J., C. Colasante, Q. Contreras, M. Puig, J. A. Serrano, L. Hernandez, and B. L. Beaman.** 2004. *Nocardia otitidiscaviarum* (GAM-5) induces Parkinsonian-like alterations in the mouse. *Braz. J. Med. Biol. Res.* **37:**539–548.

26. **Ellis, T. N., and B. L. Beaman.** 2002. Polymorphonuclear neutrophils are activated to produce interferon-γ in response to pulmonary infection with *Nocardia asteroides*. *J. Leukoc. Biol.* **72:**373–381.

27. **Enomoto, K., S. Oka, N. Fujiwara, T. Okamoto, Y. Okuda, R. Maekura, T. Kuroki, and I. Yano.** 1998. Rapid serodiagnosis of *Mycobacterium avium-intracellulare* complex infection by ELISA with cord factor (trehalose 6,6'-dimycolate), and serotyping using the glycopeptidolipid antigen. *Microbiol. Immunol.* **42:**689–696.

28. **Ferrick, D. A., R. K. Braun, H. D. Lepper, and M. D. Schrenzel.** 1996. Gamma delta T cells in bacterial infections. *Res. Immunol.* **147:**532–541.

29. **Friedman, C. S., B. L. Beaman, J. Chun, M. Goodfellow, A. Gee, and R. P. Hedrick.** 1998. *Nocardia crassostreae* sp. nov., the causal agent of nocardiosis in Pacific oysters. *Int. J. Syst. Bacteriol.* **48:**237–246.

30. **Gao, L. Y., O. S. Harb, and Y. Abu Kwaik.** 1997. Utilization of similar mechanisms by *Legionella pneumophila* to parasitize two evolutionarily distant host cells, mammalian macrophages and protozoa. *Infect. Immun.* **65:**4738–4746.

31. **Gordon, R. E., and J. M. Mihm.** 1959. A comparison of *Nocardia asteroides* and *Nocardia brasiliensis*. *J. Gen. Microbiol.* **20:**129–135.

32. **Goren, M. B.** 1990. Mycobacterial fatty acid esters of sugars and sulfosugars, p. 363–461. *In* M. Kaitz (ed.), *Handbook of Lipid Research*, vol. 6. *Glycolipids, Phospholipids, and Sulfoglycolipids*. Plenum Press, New York, N.Y.

33. **Hoq, M. M., T. Suzutani, T. Toyoda, G. Horiike, I. Yoshida, and M. Azuma.** 1997. Role of gamma delta TCR+ lymphocytes in the augmented resistance of trehalose 6,6'-dimycolate-treated mice to influenza virus infection. *J. Gen. Virol.* **78:**1597–1603.

34. **Hubble, J. P., T. Cao, J. A. Kjelstrom, W. C. Koller, and B. L. Beaman.** 1995. *Nocardia* species as an etiologic agent in Parkinson's disease: serological testing in a case-control study. *J. Clin. Microbiol.* **33:**2768–2769.

35. **Hyland, K., B. L. Beaman, P. LeWitt, and A. DeMaggio.** 2000. Monoamine changes in the brain of BALB/c mice following sub-lethal infection with *Nocardia asteroides* (GUH-2). *Neurochem. Res.* **25:**443–448.

36. **Ioneda, T., B. L. Beaman, L. Viscaya, and E. T. deAlmeida.** 1993. Composition and toxicity of diethyl ether soluble lipids from *Nocardia asteroides* GUH-2 and *Nocardia asteroides* 10905. *Chem. Phys. Lipids* **65:**171–178.

37. **Ishikawa J., A. Yamashita, Y. Mikami, Y. Hoshino, H. Kurita, K. Hotta, T. Shiba, and M. Hattori.** 2004. The complete genomic sequence of *Nocardia farcinica* IFM 10152. *Proc. Natl. Acad. Sci. USA* **101:**14925–14930.

38. **Kannon, G. A., M. K. Kuechle, and A. B. Garrett.** 1996. Superficial cutaneous *Nocardia asteroides* infection in an immunocompetent pregnant woman. *J. Am. Acad. Dermatol.* **35:**1000–1002.

39. **Karakayali, G., A. Karaarslan, F. Artz, N. Alli, and A. Tekeli.** 1998. Primary cutaneous *Nocardia asteroides*. *Br. J. Dermatol.* **139:**919–920.

40. **Kashima, K., S. Oka, A. Tabata, K. Yasuda, A. Kitano, K. Kobayashi, and I. Yano.** 1995. Detection of anti-cord factor antibodies in intestinal tuberculosis for its differential diagnosis from Crohn's disease and ulcerative colitis. *Dig. Dis. Sci.* **40:**2630–2634.

41. **Kohbata, S., and B. L. Beaman.** 1991. L-Dopa-responsive movement disorder caused by *Nocardia asteroides* localized in the brains of mice. *Infect. Immun.* **59:**181–191.

42. **Koike, Y., Y. C. Yoo, M. Mitobe, T. Oka, K. Okuma, S. Tono-oka, and I. Azuma.** 1998. Enhancing activity of mycobacterial cell-derived adjuvants on immunogenicity of recombinant human hepatitis B virus vaccine. *Vaccine* **16:**1982–1989.

43. **Lisitchkina, H. V., and H. P. Ludin.** 2002. Associated symptoms and relevant associated illnesses in idiopathic Parkinson syndrome. *Schweiz. Rundsch. Med. Prax.* **91:** 395–401.

44. **Loeffler, D. A., D. M. Camp, S. Qu, B. L. Beaman, and P. A. LeWitt.** 2004. Characterization of dopamine-depleting activity of *Nocardia asteroides* strain GUH-2 culture filtrate on PC 12 cells. *Microb. Pathog.* **37:**73–85.

45. **Lopez Martinez, R., L. J. Mendez Tovar, P. Lavalle, O. Welsh, A. Saul, and E. Macotela Ruiz.** 1992. Epidemiology of mycetoma in Mexico: study of 2105 cases. *Gac. Med. Mex.* **128:**477–481.

46. **McNaught, K. S., D. P. Perl, A. L. Brownell, and C.W. Olanow.** 2004. Systemic exposure to proteasome inhibitors causes a progressive model of Parkinson's disease. *Ann. Neurol.* **56:**149–162.

47. **McNeil, M. M., and J. M. Brown.** 1994. The medically important aerobic actinomycetes: epidemiology and microbiology. *Clin. Microbiol. Rev.* **7:**357–417.

48. **Mellmann, A., J. L. Cloud, S. Andrees, K. Blackwood, K. C. Carroll, A. Kabani, A. Roth, and D. Harmsen.**

2003. Evaluation of RIDOM, MicroSeq, and Genbank services in the molecular identification of *Nocardia* species. *Int. J. Med. Microbiol.* **293:**359–370.

49. **Menendez, R., P. J. Cordero, M. Santos, M. Gobernado, and V. Marco.** 1997. Pulmonary infection with *Nocardia* species: a report of 10 cases and review. *Eur. Respir. J.* **10:**1542–1546.

50. **Min, Y., M. Asano, M. Kohanawa, and T. Minagawa.** 1999. Movement disorders in encephalitis induced by *Rhodococcus aurantiacus* infection relieved by the administration of L-dopa and anti-T-cell antibodies. *Immunology* **96:**10–15.

51. **Natsuhara, Y., S. Oka, K. Kaneda, Y. Kato, and I. Yano.** 1990. Parallel antitumor, granuloma-forming and tumor-necrosis-factor-priming activities of mycoloyl glycolipids from Nocardia rubra that differ in carbohydrate moiety: structure-activity relationships. *Cancer Immunol. Immunother.* **31:**99–106.

52. **Navarro, V., and M. Salavert.** 1997. Mastitis caused by *Nocardia brasiliensis* in an immunocompetent patient. *Enferm. Infec. Microbiol. Clin.* **15:**339–340. (In Spanish.)

53. **Ng, C. S., and W. C. Hellinger.** 1998. Superficial cutaneous abscess and multiple brain abscesses from *Nocardia asteroides* in an immunocompetent patient. *J. Am. Acad. Dermatol.* **39:**793–794.

54. **Oswald, I. P., C. M. Dozois, J. F. Petit, and G. Lemoire.** 1997. Interleukin-12 synthesis is a required step in trehalose dimycolate-induced activation of mouse peritoneal macrophages. *Infect. Immun.* **65:**1364–1369.

55. **Ozeki, Y., K. Kaneda, N. Fujiwara, M. Morimoto, S. Oka, and I. Yano.** 1997. In vivo induction of apoptosis in the thymus by administration of mycobacterial cord factor (trehalose 6,6'-dimycolate). *Infect. Immun.* **65:**1793–1799.

56. **Pal, M.** 1997. *Nocardia asteroides* as a cause of pneumonia in a buffalo calf. *Rev. Sci. Tech.* **16:**881–884.

57. **Patel, J. B., R. J. Wallace, Jr., B. A. Brown-Elliott, T. Taylor, C. Imperatrice, D. G. Leonard, R. W. Wilson, L. Mann, K. C. Jost, and I. Nachamkin.** 2004. Sequence-based identification of aerobic actinomycetes. *J. Clin. Microbiol.* **42:**2530–2540.

58. **Pujic, P., and B. L. Beaman.** 2001. *Actinomyces* and *Nocardia*, p. 937–960. *In* M. Sussman (ed.), *Molecular Medical Microbiology*. Academic Press, New York, N.Y.

59. **Roth, A., S. Andrees, R. M. Kroppenstedt, D. Harmsen, and H. Mauch.** 2003. Phylogeny of the genus *Nocardia* based on reassessed 16S rRNA gene sequences reveals underspeciation and division of strains classified as *Nocardia asteroides* into three established species and two unnamed taxons. *J. Clin. Microbiol.* **41:**851–856.

60. **Saubolle, M. A., and D. Sussland.** 2003. Nocardiosis: review of clinical and laboratory experience. *J. Clin. Microbiol.* **41:**4497–4501.

61. **Sazaki, K., I. Yoshida, and M. Azuma.** 1992. Mechanisms of augmented resistance of cyclosporin A-treated mice to influenza virus infection by trehalose-6,6'-dimycolate. *Microbiol. Immunol.* **36:**1061–1075.

62. **Seddon, M., D. Parr, and R. B. Ellis-Pegler.** 1995. Lymphocutaneous *Nocardia brasiliensis* infection: a case report and review. *N. Z. Med. J.* **108:**385–386.

63. **Sridhar, M. S., S. Sharma, M. K. Reddy, P. Mruthyunjay, and G. N. Rao.** 1998. Clinicomicrobiological review of Nocardia keratitis. *Cornea* **17:**17–22.

64. **Steinert, M., K. Birkness, E. White, B. Fields, and F. Quinn.** 1998. *Mycobacterium avium* bacilli grow saprozoically in coculture with *Acanthamoeba polyphaga* and survive within cyst walls. *Appl. Environ. Microbiol.* **64:**2256–2261.

65. **Sueoka, E., S. Nishiwaki, S. Okabe, N. Iida, M. Suganuma, I. Yano, K. Aoki, and H. Fujiki.** 1995. Activation of protein kinase C by mycobacterial cord factor, trehalose 6-monomycolate, resulting in tumor necrosis factor-alpha release in mouse lung tissues. *Jpn. J. Cancer Res.* **86:**749–755.

66. **Syed, S. S., and R. L. Hunter, Jr.** 1997. Studies on the toxic effects of quartz and a mycobacterial glycolipid, trehalose 6,6'-dimycolate. *Ann. Clin. Lab. Sci.* **27:**375–383.

67. **Tabata, A., K. Kaneda, H. Watanabe, T. Abo, and I. Yano.** 1996. Kinetics of organ-associated natural killer cells and intermediate CD3 cells during pulmonary and hepatic granulomatous inflammation induced by mycobacterial cord factor. *Microbiol. Immunol.* **40:**651–658.

68. **Tam, S., D. Barry, L. Beaman, and B. L. Beaman.** 2002. Neuroinvasive *Nocardia asteroides* GUH-2 induces apoptosis in the substantia nigra in vivo and dopaminergic cells in vitro. *Exp. Neurol.* **177:**453–460.

69. **Tosteson, T. R., D. L. Ballantine, C. G. Tosteson, V. Hensley, and A. T. Bardales.** 1989. Associated bacterial flora, growth, and toxicity of cultured benthic dinoflagellates *Ostreopsis lenticularis* and *Gambierdiscus toxicus*. *Appl. Environ. Microbiol.* **55:**137–141.

70. **Tsuyuguchi, I., H. Kawasumi, C. Ueta, I. Yano, and S. Kishimoto.** 1991. Increase of T-cell receptor γ/δ-bearing T cells in cord blood of newborn babies obtained by in vitro stimulation with mycobacterial cord factor. *Infect. Immun.* **59:**3053–3059.

71. **Umunnabuike, A. C., and E. A. Irokanulo.** 1986. Isolation of *Campylobacter* subsp. *jejuni* from Oriental and American cockroaches caught in kitchens and poultry houses in Vom, Nigeria. *Int. J. Zoonoses* **13:**180–186.

72. **Uttamchandani, R. B., G. L. Daikos, R. R. Reyes, M. A. Fischl, G. M. Dickinson, E. Yamaguchi, and M. R. Kramer.** 1994. Nocardiosis in 30 patients with advanced human immunodeficiency virus infection: clinical features and outcome. *Clin. Infect. Dis.* **18:**348–353.

73. **Wallace, R. J., Jr., B. A. Brown, Z. Blacklock, R. Ulrich, K. Jost, J. M. Brown, M. M. McNeil, G. Onyi, V. A. Steingrube, and J. Gibson.** 1995. New *Nocardia* taxon among isolates of *Nocardia brasiliensis* associated with invasive disease. *J. Clin. Microbiol.* **33:**1528–1533.

74. **Yano, I.** 1998. Biochemistry and bioactivities of mycobacterial components. *Nippon Rinsho* **56:**3008–3016.

ANTIBIOTIC RESISTANCE MECHANISMS

SECTION EDITOR: Richard P. Novick

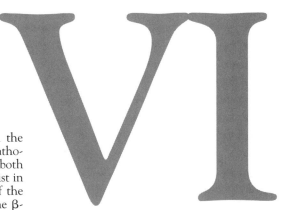

GRAM-POSITIVE PATHOGENS have acquired or developed much the same set of antibiotic resistances as have the gram-negative pathogens; in many cases, the same mechanisms are employed by both groups, and homologous genes are involved. Significant exceptions exist in relation to antibiotics that do not penetrate the outer membrane of the gram-negative pathogens, such as glycopeptides, macrolides, and some β-lactams, to which the gram-negative pathogens are intrinsically resistant. Recent alarming developments have occurred in resistance to just these particular groups of antibiotics. Mutational resistance to vancomycin has begun to occur among patients receiving long-term treatment; more disturbing, the mobile vancomycin resistance systems of enterococci have made their way to *Staphylococcus aureus*, threatening to eliminate the last available remedy for multiresistant infections. Also alarming is the appearance of methicillin and/or oxacillin resistance among community-acquired staphylococci, carried by variants of the mobile staphylococcal chromosome cassette *mec* elements that are not closely related to those seen with nosocomial strains. As has always been the case, the introduction of new antibiotics is soon followed by the development of resistance, and so the contest between resistant bacteria and the efforts to control them continues to escalate, with the former now regrettably having the upper hand.

This section covers resistance to β-lactams, glycopeptides, tetracyclines, and quinolones—areas in which there have been important developments since the first edition of this volume. Resistances that are more important in particular groups of organisms are discussed within the respective sections.

Mechanisms of Resistance to β-Lactam Antibiotics

DOUGLAS S. KERNODLE

62

The introduction of penicillin into clinical use in 1941 had a profound impact on the treatment of diseases caused by gram-positive pathogens. Life-threatening infections became curable, including meningitis caused by *Streptococcus pneumoniae* and endocarditis caused by streptococci, staphylococci, and enterococci. In the preantibiotic era, these infections would almost certainly have caused the patient's death.

Within a short period of time, however, resistant bacteria that failed to respond to penicillin therapy were identified. In 1944, Kirby reported the production of a penicillin-inactivating enzyme by *Staphylococcus aureus* (45), and analysis of stored isolates from earlier decades established that occasional resistant strains had existed prior to the clinical use of penicillin (17, 18). Over the next decade, antibiotic-resistant clones became more abundant as the selective pressure provided by the clinical use of antimicrobial agents gave these clones a survival advantage over antibiotic-susceptible strains. β-Lactamase-producing isolates of *S. aureus* became endemic in many hospitals, as well as in the community, rendering penicillin ineffective in the treatment of most staphylococcal infections and prompting the development of penicillinase-resistant penicillins.

Shortly after methicillin was introduced in 1960, *S. aureus* isolates exhibiting resistance to methicillin (MRSA) were reported, apparently by an intrinsic mechanism not involving drug inactivation (7, 39). Although MRSA isolates were initially concentrated within hospitals and remained primarily nosocomial pathogens for several decades, genetically distinct community-associated MRSA (CA-MRSA) isolates have recently emerged worldwide and become important pathogens among previously healthy persons, including children (74). Similarly, although early in the antibiotic era *S. pneumoniae* was reliably and exquisitely susceptible to penicillin, isolates with intermediate resistance were identified by 1967 (35). Highly resistant pneumococcal isolates appeared in South Africa by 1977 (2), again by a mechanism other than penicillin degradation. It has subsequently been shown that the mechanism of resistance underlying these examples involves changes in membrane proteins involved in peptidoglycan biosynthesis that are the normal targets of penicillin.

Antibiotic degradation by β-lactamase and alterations in penicillin-binding membrane proteins remain the major mechanisms by which gram-positive pathogens express resistance to β-lactam antibiotics. Antibiotic tolerance is a phenomenon wherein the concentration of a β-lactam required to kill an isolate greatly exceeds the concentration needed to inhibit it (minimum bactericidal concentration [MBC]/MIC ≥ 32) and might also be of clinical importance among some gram-positive species. Diminished β-lactam permeability through the outer cellular structures, which is a major mechanism of resistance to β-lactams among gram-negative species, possibly contributes to the resistance phenotype in a few gram-positive pathogens such as nocardia and mycobacteria but does not appear to be a major mechanism of resistance among most gram-positive species.

Although β-lactam resistance mechanisms have had a profound impact upon the management of infections caused by staphylococci, pneumococci, and enterococci, other important gram-positive pathogens including group A streptococci have remained highly susceptible to penicillin. It has been suggested that the continued high penicillin susceptibility of group A streptococci might be related to the poor expression or toxicity of β-lactamase and low-affinity penicillin-binding proteins (PBPs) (38). Furthermore, β-lactamase production has not become established in all gram-positive species, e.g., pneumococci. Such examples suggest either that the circumstances favorable for the exchange of resistance determinants occur at very low frequency between certain species or that there are as-yet-unclear mechanisms of interference that restrict the host range of certain resistance genes.

MECHANISM OF ACTION OF β-LACTAMS

Gram-positive pathogens are characterized by a thick peptidoglycan layer, which is a relatively inelastic structure that confers shape on the organism and protects it against damage from differences in the osmotic pressure of its cytoplasm and its external environment. Peptidoglycan consists of alternating subunits of two amino sugars, N-acetylglucosamine and N-acetylmuramic acid. The N-acetylmuramic acid units

Gram-Positive Pathogens, 2nd edition, edited by Vincent A. Fischetti et al.
© 2006 ASM Press, Washington, D.C.

are linked to peptide chains, many of which are interlinked to each other, providing cell wall stability.

Peptidoglycan biosynthesis is a dynamic process that requires the breaking of covalent bonds by autolytic enzymes and the insertion of newly synthesized amino sugar cell wall subunits. Bacterial autolytic enzymes include murein hydrolases with glycosidase, glucosaminidase, and amidase activities. Synthetic enzymes have glycosyltransferase, transpeptidase, and carboxypeptidase activities.

The major theory involving the mechanism of action of β-lactams concerns their structural similarity to the D-alanyl-D-alanine carboxy-terminal region of peptides involved in the cross-linking of peptidoglycan. Penicillins, cephalosporins, and other β-lactams acylate the active-site serine of cell transpeptidases, forming a stable acylenzyme that lacks synthetic activity (24, 27). Thus, an imbalance between synthetic and autolytic capacity develops which leads to cell lysis. There is evidence in some species that this imbalance is heightened by the induction of autolysins in the presence of β-lactams—e.g., in *S. pneumoniae*, a secreted peptide triggers the production of autolytic enzymes via binding to a two-component signal transduction system (59). β-Lactams might also exert an antibacterial effect via other mechanisms, as cell lysis is not reliably seen in all species. For example, benzylpenicillin induces a dose-dependent, rapid loss of total cellular RNA in group A streptococci without overt cell lysis, and the loss of RNA correlates with the loss of cell viability (50). Autolysis-dependent and autolysis-independent mechanisms of killing in pneumococci have been previously described (52).

The penicillin-interactive enzymes involved in cell wall biosynthesis are specialized acyl serine transferases localized on the outer face of the cytoplasmic membrane. They catalyze the rupture of the β-lactam bond of penicillin to form a serine ester-linked penicilloyl enzyme (Fig. 1). Thus, protein bands corresponding to these enzymes can be visualized on fluorograms following sodium dodecyl sulfate-polyacrylamide gel electrophoresis of bacterial membranes using radioactive penicillin for labeling, hence the name PBPs (28, 47).

The strong antibacterial efficacy of β-lactams, combined with their low toxicity for eukaryotic cells, has helped to make them the most highly developed class of antibacterial agents in clinical use. Historically, the most important β-lactam classes have been the penicillins and the cephalosporins, which were introduced into clinical practice in the 1940s and the 1960s, respectively. Over the past 2 decades, additional antibacterial agents based on other modifications of the basic β-lactam structure have been introduced into clinical practice including carbapenems, monobactams, and carbacephems.

β-LACTAMASE

β-Lactamases catalyze the inactivation of penicillin and other β-lactams by covalently binding to the β-lactam ring. This is the same acylation reaction that occurs when β-lactams bind to the active site of PBPs. Yet whereas the PBPs are inactivated because the reaction is not appreciably reversible, with β-lactamases there is a subsequent deacylation step whereby the covalent bond is hydrolyzed by a molecule of water (Fig. 1). As a result, the antibiotic is inactivated by cleavage of its β-lactam ring, and there is regeneration of the active β-lactamase.

β-Lactamases can be divided into four major groups based on their active site, molecular mass, and primary structure (24, 47). Class A enzymes have a serine-active

site, a molecular mass of about 30 kDa, and usually greater penicillinase activity levels than cephalosporinase activity. Class B contains the Zn^{2+} active-site metallo-β-lactamases, which usually exhibit a broad spectrum of activity including the hydrolysis of carbapenems. They are found in *Bacillus*

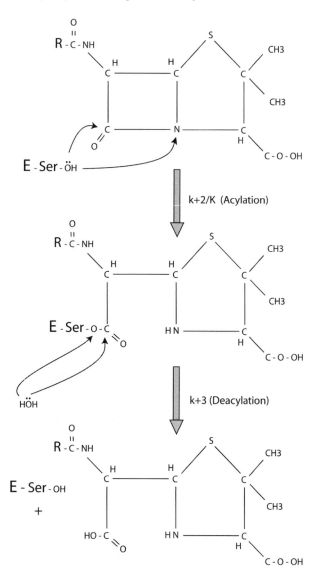

FIGURE 1 Interaction between the β-lactam ring and the active-site serine of penicillin-interactive enzymes. The chemical reaction for binding of penicillin and other β-lactams to PBPs and β-lactamases is represented by the equation

$$E + S \overset{K}{\rightleftharpoons} E{\cdot}S \overset{k+2}{\longrightarrow} E - S^* \underset{+H_2O}{\overset{k+3}{\nearrow}} E + P$$

where E = enzyme, S = β-lactam substrate, $E{\cdot}S$ = Michaelis complex, $E - S^*$ = acylenzyme, P = product, K = dissociation constant, and $k+2$ and $k+3$ are first-order rate constants, respectively, for the acylation step and for hydrolysis of the acylenzyme (deacylation). With PBPs, $k+3$ is low, such that the acylenzyme is inert or hydrolyzed very slowly. Thus, the acylenzyme form can be identified after sodium dodecyl sulfate-polyacrylamide gel electrophoresis of bacterial membranes using radioactive penicillin for labeling. With β-lactamases, $k+3$ is large, and the antibiotic is rapidly hydrolyzed.

and *Bacteroides* spp. and in some gram-negative aerobic species. Class C enzymes also have a serine-active site, but they are larger than class A β-lactamases, with a molecular mass of about 40 kDa, and exhibit predominantly cephalosporinase activity. Class D enzymes have a serine-active site and exhibit oxacillinase activity. β-Lactamase classification schemes based on functional characteristics including whether the enzyme exhibits predominant penicillinase or cephalosporinase activity have also been described and provide a useful and more detailed means for cataloging multiple enzymes (6). Excepting the class B β-lactamases produced by *Bacillus* spp. and a recently described class D β-lactamase in *Bacillus subtilis* (10), all of the known β-lactamases produced by gram-positive species are class A enzymes.

The serine-active site β-lactamases strongly resemble PBPs with several conserved primary structural elements that comprise the active site, including the SxxK tetrad, SDN triad, and KTG triad, as well as similarities in the three-dimensional positioning of α-helices and β-strands (Fig. 2) (24, 40). Some PBPs have detectable β-lactamase activity including PBP4 of *S. aureus* and peptide fragments of the pneumococcal PBPs (20, 26). The clinical relevance of this activity is unclear and it does not appear to contribute much to resistance.

The ability of β-lactamase to hydrolyze a β-lactam can be expressed in simple kinetic terms: REH = k_{cat}/K_m, where REH is the relative efficiency of hydrolysis, k_{cat} is the turnover number (i.e., the number of molecules of substrate degraded per second per molecule of enzyme), and K_m is the Michaelis constant, which reflects the affinity of the enzyme for its substrate. From this equation, two basic approaches for re-

TABLE 1 β-Lactam hydrolysis by *S. aureus* β-lactamase[a]

Antibiotic	Value			
	k_{cat} (s^{-1})	K_m (μM)	REH[b]	Stability[c]
Benzylpenicillin	171	50	3.4	1
Methicillin	17	10,000	0.0017	2,000
Cephalothin	0.015	6.8	0.0022	1,545
Cefazolin	1.01	18.4	0.054	63

[a]Assays were performed using purified type A *S. aureus* β-lactamase (42, 80).
[b]Relative efficacy of hydrolysis (REH) = k_{cat}/K_m.
[c]Stability is expressed as a ratio of the stability of benzylpenicillin, which was set at 1.

ducing REH become evident: (i) to increase the K_m and thus decrease the binding affinity of the drug for the enzyme such that a higher concentration of the antibiotic is required to achieve half-maximal binding and (ii) to reduce the k_{cat} and thus make the β-lactam intrinsically more resistant to enzymatic hydrolysis. A third strategy is to inhibit β-lactamase activity by the addition of a competing substance.

Each of these strategies has been used in the development of β-lactams active against β-lactamase-producing *S. aureus*. For example, via the addition of bulky side chains, the K_m values of the antistaphylococcal penicillins such as methicillin and oxacillin are much higher than the K_m of benzylpenicillin (Table 1). Because of their high K_m values, at clinically obtainable peak serum antibiotic levels of about 60 μg/ml there is minimal binding of methicillin or the isoxazolyl penicillins to *S. aureus* β-lactamase. Without binding, acylenzyme formation and hydrolysis cannot proceed, and thus the antibiotic is not inactivated. In contrast, some of the cephalosporins (e.g., cephalothin) have much lower k_{cat} values while having K_m values that are comparable to or even lower than that of benzylpenicillin. Although such agents form a Michaelis complex with β-lactamase at clinically relevant concentrations, drug inactivation proceeds slowly. Finally, clavulanic acid, sulbactam, and tazobactam bind covalently to the active site of staphylococcal β-lactamase and inactivate the enzyme. These agents have been combined with labile penicillins, e.g., ampicillin and amoxicillin, and the combinations exhibit activity against β-lactamase-producing staphylococci. The similar structure of the penicillin-interactive regions of β-lactamases and PBPs (40, 47) complicates attempts to develop new β-lactams, as it is necessary to make structural modifications that result in diminished binding and/or hydrolysis by β-lactamase while simultaneously retaining the capacity for binding to the target PBPs.

The resistance phenotype in β-lactamase-producing gram-positive bacteria differs from that observed with gram-negative species and is associated with an inoculum effect in which the MIC depends upon the number of bacteria tested (63). That is, a higher concentration of antibiotic is needed to inhibit the growth of a large amount of inoculum of β-lactamase-producing *S. aureus* than to inhibit a small amount of inoculum. This effect is related to the fact that with a large amount of inoculum there is more preformed β-lactamase, as well as more bacteria, that can be induced to make β-lactamase than with the small amount of inoculum. Whereas a large amount of inoculum is better able to clear the medium of the β-lactam and grow, a small amount of inoculum lacks sufficient enzyme to clear the medium of the β-lactam and thus is killed by the same antibiotic concentration. Accordingly, MICs are higher with a large amount

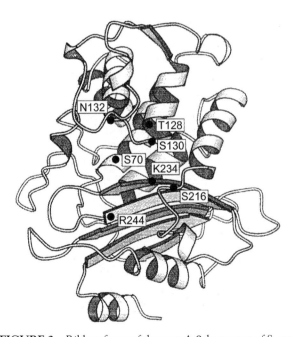

FIGURE 2 Ribbon figure of the type A β-lactamase of *S. aureus*, illustrating key motifs of penicillin-interactive enzymes. S70 indicates the active-site serine in the SxxK motif, S130 and N132 are in the SDN loop, and K234 begins the KTG triad. R244 is a highly conserved residue in class A β-lactamases that is important in catalysis. Differences in amino acids positioned near the active-site cleft at residues 128 and 216 that are responsible for the kinetic differences observed among the wild-type variants of *S. aureus* β-lactamase are summarized in Table 2.

of inoculum than with a small amount. With gram-negative bacteria, an inoculum effect is not generally observed, as the presence of an outer membrane and the sequestration of β-lactamase within the periplasmic space enable even single bacterial cells to exhibit high-level resistance.

EXAMPLES OF β-LACTAMASE-MEDIATED RESISTANCE IN GRAM-POSITIVE PATHOGENS

S. aureus

Around 95% of S. aureus isolates recovered from clinical specimens produce β-lactamase. The structural gene for β-lactamase, blaZ, and two regulatory genes, blaI and blaR1, usually reside on a plasmid, although a chromosomal location has been identified for some strains (17, 18). Some β-lactamase plasmids contain genes for resistance to other antibiotics, in particular, gentamicin and erythromycin.

Four variants of S. aureus β-lactamase have been described, initially by using serologic techniques and subsequently by detecting differences in the kinetics of hydrolysis of selected β-lactam substrates (17, 18, 44). All are class A β-lactamases based on their active site, molecular mass, and primary structure (see above); yet based on a nomenclature developed in the 1960s that antedates the class-based designation of β-lactamases, the variants are called types A, B, C, and D (i.e., not to be confused with classes A, B, C, and D). The enzymes are >95% identical at the amino acid level, and kinetic differences derive from amino acid substitutions close to the active site, specifically, differences at residue 128, which is close to the SDN loop (amino acids 130 to 132 in the numbering scheme of Ambler) (1) and residue 216, which is close to the β3 strand containing the KTG triad) (Fig. 2; Table 2). Differences at residue 216 influence the K_m values for various substrates, while differences at residue 128 are reflected by more subtle differences in k_{cat} values (75, 76). Other differences in amino acid sequence are present but appear to be kinetically silent. The type B enzyme exhibits the greatest divergence in amino acid sequence, differing from each of the type A, type C, and type D enzymes by 15 or 16 amino acid residues; it is encoded by a chromosomal gene in some phage group II isolates (75).

The type A β-lactamase is the most common type found among clinical isolates of S. aureus. Among 809 clinical and reference isolates that underwent β-lactamase typing by substrate profile analysis in my laboratory between 1988 and 1998, 432 (53%) produced type A β-lactamase. Another 283 isolates produced either type B or C β-lactamase, both of which have similar substrate profiles and can be difficult to distinguish unless the enzyme is purified or the isolate produces a large amount of enzyme (42). Only six isolates

produced type D β-lactamase, and all of these originated from Europe. Seventy-four isolates were either β-lactamase negative or exhibited activity too weak to enable typing. Only 14 isolates were found that exhibited substrate profiles distinct from isolates producing type A, B, C, or D S. aureus β-lactamases—these could potentially reflect either novel β-lactamases or the production of more than one β-lactamase by a single isolate.

Most isolates produce both extracellular and cell membrane-bound forms of β-lactamase, with the exoenzyme usually comprising 30 to 60% of total β-lactamase activity. Following cleavage of the 24-residue leader peptide, the exoenzyme contains 257 amino acids with a high proportion of basic residues. Its highly basic nature (pI values of 9.7 to 10.1) (80) may facilitate adherence to anionic cell wall structures via electrostatic interactions and thus keep the enzyme in the vicinity of the bacteria and thereby enhance their protection from β-lactams. The cell-associated form is six amino acids longer than the extracellular enzyme and is attached to fatty acids in the cell membrane via a glyceride-thioether bond (57). Some strains make only the cell-associated form and cannot make extracellular β-lactamase, a property that appears to be attributable to amino acid substitutions at the normal proteolytic cleavage site for separation of the exoenzyme from the leader peptide (19).

The regulatory system for the induction of β-lactamase production in S. aureus involves derepression (Fig. 3) (17, 18). Under basal (uninduced) conditions, a repressor binds to the promoter of blaZ but with exposure to β-lactam antibiotics (induced) derepression occurs, typically producing a 30- to 100-fold increase in penicillinase activity. Two verified regulatory loci (blaI and blaR1) are located directly upstream of blaZ. blaI encodes the repressor. It has been cloned and expressed, and DNA footprinting studies show that BlaI dimers bind specifically to the blaZ operator region (17, 30). blaR1 encodes a high-molecular-weight (HMW) class C PBP of 585 amino acids that exhibits sensor-transducer properties and includes penicillin-binding and prometallo-protease components. After β-lactam binding to the sensor region, BlaR1 undergoes autocatalytic cleavage, which yields the mature metalloprotease that then either directly cleaves or participates with other factors to convert BlaI from a 14-kDa active repressor to inactive 11-kDa and 3-kDa fragments containing the DNA-binding and dimerization motifs, respectively (79). In contrast to typical bacterial two-component signaling systems that modify proteins by phosphorylating histidine or other amino acids, the β-lactamase induction pathway is unusual in that it depends on a series of cleavage and self-activation steps. The existence of another β-lactamase regulatory locus (blaR2) on the chromosome has been proposed but remains unverified—this proposal was based on studies showing that in some isolates, the amount of β-lactamase a strain produces appears to depend on the nature of the host isolate rather than plasmid-borne blaZ, blaR1, and blaI genomic elements (9, 17).

Various isolates produce different amounts of β-lactamase and have been classified as to the type of mutation in the induction pathway (18). Magnoconstitutive mutants produce large amounts of β-lactamase even in the absence of an inducer. Mutations in blaI that lead to impaired repression of the blaZ operator can produce this phenotype. The magnoconstitutive phenotype appears to be linked to the chromosome in some mutants (9, 17). Similarly, mesoinducible mutants have reduced induction capability; this may involve a chromosomal defect, as wild-type plasmids assume the mesoinducible phenotype when transduced into a

TABLE 2 Effect of amino acid differences upon kinetic profiles of S. aureus β-lactamase

β-Lactamase type	Amino acid at position:	
	128	216
A	Threonine	Serine
B	Lysine	Asparagine
C	Threonine	Asparagine
D	Alanine	Serine

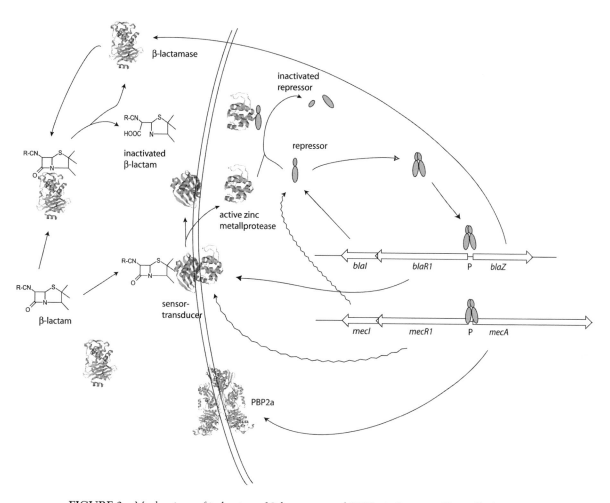

FIGURE 3 Mechanisms of induction of β-lactamase and PBP2a in *S. aureus*. Normally the penicillin-interactive proteins are produced in only small amounts because of the binding of a *blaI*-encoded repressor to the promoters (P) of *blaZ* and *blaR1*. Induction of β-lactamase is initiated by the binding of a β-lactam to the sensor-transducer encoded by *blaR1*. Binding induces autocatalytic cleavage within *blaR1*, which unmasks a metalloprotease domain that cleaves the *blaI* repressor into two fragments, either directly or via interaction with another factor. Thus, the repressor can no longer bind to P, and β-lactamase and *blaR1* sensor-transducer production are induced via derepression. The induced β-lactamase is excreted to the extracellular space, where it inactivates the β-lactam. The induction of the low-affinity PBP2a, encoded by *mecA*, is regulated in a fashion similar to that regulating β-lactamase, and PBP2a mediates resistance by cross-linking peptidoglycan in the presence of β-lactams. There is homology between *mecI* and *blaI*, *mecR1* and *blaR1*, and the promoter and N-terminal portions of *blaZ* and *mecA*. The zigzag arrows denoting the products of *mecI* and *mecR1* indicate that these genes encode homologs of the *blaI* and *blaR1* products, respectively, with similar functions in induction signaling. The homology is strong enough that a plasmid-derived *blaI-blaR1-blaZ* genomic element can restore the normal inducible phenotype to MRSA clonotypes that produce large amounts of PBP2a constitutively because of deletions in the *mecI-mecR1* region.

mesoinducible host (17). Microconstitutive mutants produce a small amount of β-lactamase and are little affected by inducers (17). Mutations in the *blaR1* membrane sensor-transducing protein are believed to account for this phenotype.

The production of large amounts of β-lactamase in isolates possessing the normal penicillin-sensitive PBPs has been associated with borderline susceptibility to the antistaphylococcal penicillins. The term borderline-susceptible (or borderline-resistant) *S. aureus* (BSSA) became popular

after a 1986 report of *S. aureus* isolates exhibiting a distinctive phenotype which included all of the following: (i) borderline MICs to oxacillin, (ii) lowering of oxacillin MICs into the clearly susceptible range by β-lactamase inhibitors, (iii) rapid hydrolysis of the chromogenic cephalosporin nitrocefin, and (iv) high MICs of benzylpenicillin (49). Most of the isolates exhibiting all of these traits belong to phage group 94/96 and possess a 17.2-kb plasmid (51). They make large amounts of type A *S. aureus* β-lactamase and apparently represent the same group of isolates reported in the

1970s to hydrolyze cefazolin relatively efficiently (22, 63). When cured of β-lactamase production, they still exhibit an MIC slightly above that of other non-β-lactamase-producing strains (3). Furthermore, non-phage group 94/96 *S. aureus* isolates transformed with the 17.2-kb plasmid from a BSSA strain do not exhibit the BSSA phenotype and make less type A β-lactamase than a phage group 94/96 host strain. This suggests that the phage group 94/96 host background is critical to the full expression of the BSSA phenotype, although the mechanism is unclear.

The distinctive phenotypic, phage, and plasmid characteristics of BSSA strains suggest that they comprise a widespread and clinically important clone of relatively resistant *S. aureus* isolates. BSSA isolates have been associated with *S. aureus* wound infections in surgery that occur following prophylaxis with cefazolin, a cephalosporin that is hydrolyzed relatively rapidly by type A β-lactamase (43). Epidemiologic monitoring of phage types associated with bacteremic infection over recent decades by the Statens Serum Institut in Copenhagen, Denmark, shows that phage 94/96 strains were first detected in the early 1970s, rose rapidly to comprise about 20% of all isolates in the late 1970s through the mid-1980s, and subsequently fell to about 5% of all isolates by 2000 (68). There are still sporadic reports of cefazolin treatment failure associated with *S. aureus* isolates that produce large amounts of type A β-lactamase (55), but these isolates appear to be less of a clinical problem now than in the 1970s and 1980s.

Terms including borderline-resistant, low-level resistant, and borderline-susceptible have also been applied to other isolates of *S. aureus* that do not exhibit high-level β-lactamase production. Most of such strains either (i) have alterations in their normal PBPs such that they have reduced binding affinity for β-lactams (71) or (ii) contain *mecA*, the gene encoding PBP2a, but exhibit MICs around the breakpoint for designation as methicillin susceptible or methicillin resistant rather than the high MICs exhibited by most MRSA strains (8).

Coagulase-Negative Staphylococci

β-Lactamase is produced by most clinical isolates of coagulase-negative staphylococci. There is some evidence for β-lactamases with different substrate specificities, at least two of which are similar to the *S. aureus* enzymes (70).

Enterococci

β-Lactamase production appears to be a recently acquired trait in enterococci. It was first recognized in 1981 when an *Enterococcus faecalis* isolate was serendipitously observed to exhibit an inoculum effect to penicillin; it has subsequently been identified in strains from several countries (54). Most of the described isolates make the type A variant of *S. aureus* β-lactamase, although a second β-lactamase with several amino acid substitutions has also been identified (72). Although some of the β-lactamase-producing strains contain intact β-lactamase-regulatory genes, i.e., *blaI* and *blaR1*, enterococci lack inducibility and constitutively produce small amounts of β-lactamase (72). In most β-lactamase-producing strains of enterococci, *blaZ* has been reported to reside on plasmid DNA, although chromosomal integration has occasionally been observed.

Bacillus Species

Bacillus species including *Bacillus cereus*, *Bacillus licheniformis*, *B. subtilis*, *Bacillus mycoides*, and *Bacillus anthracis* make class A β-lactamases (6, 47). Some *Bacillus* spp. produce a class B zinc-ion-dependent metallo-β-lactamase (24, 48), and a low-activity class D β-lactamase in *B. subtilis* has also recently been described (10). The class A enzymes exhibit about 40% amino sequence homology with *S. aureus* β-lactamase, suggesting a common ancestral origin (18). Furthermore, class A β-lactamase induction in *B. licheniformis* involves a regulatory mechanism similar to that described for *S. aureus*, and their *blaR1*-encoded proteins share strong homology (17). Although penicillin G has historically been an effective antibiotic in treating inhalation and cutaneous forms of anthrax, about 15% of clinical and environmental isolates exhibit resistance to penicillin. Recent investigations show that *B. anthracis* isolates possess genes for class A and class B β-lactamases (48), which exhibit predominant penicillinase and cephalosporinase activity, respectively, but appear to be poorly expressed in penicillin-susceptible isolates.

Other Gram-Positive Pathogens

Nocardia spp. and some isolates of *Rhodococcus equi* make class A β-lactamases (21, 77). β-Lactamase production has not been observed with *Listeria monocytogenes*, *S. pneumoniae*, the viridans streptococci, or other streptococcal species, including group A streptococci. The absence of β-lactamase production from the various streptococcal species is remarkable, as they cohabit the oropharynx, tonsils, and nasopharynx with *S. aureus* and other β-lactamase-producing organisms.

PBPs

Most bacterial species contain four to eight PBPs with molecular masses of 35 to 120 kDa (25, 32). By convention, the PBPs for a particular species are numbered in order of diminishing molecular mass, with the largest designated PBP1. Once the number of PBPs for a species is assigned, additional enzymes are numbered as derivatives of the previously established ones (25). In general, PBPs are species specific (e.g., *S. pneumoniae* PBPs are different from *S. aureus* PBPs). However, the PBPs of *S. pneumoniae* and viridans streptococcal species including *Streptococcus mitis*, *Streptococcus oralis*, and *Streptococcus sanguis* contain identical or closely related DNA sequences (32), and there is a high degree of homology in PBPs from different enterococcal species (46).

The understanding of the relationships among and between various PBPs and other acyltransferases is evolving and includes categorization into HMW and low-molecular-weight (LMW) PBPs (47) and identification of multimodular or multidomain PBPs that have PBP and non-PBP (nPBP) domains. A recent classification scheme for the SxxK acyltransferase superfamily places peptidoglycan biosynthetic enzymes that cross-link the carboxy-terminal D-alanine at position 4 of a stem tetrapeptide to the lateral amino group at position 3 of another tetrapeptide (i.e., a typical 4 → 3 peptidoglycan synthesis) into group I SxxK acyltransferases (29). Although marked heterogeneity exists in the amino acid sequences of group I enzymes, they all possess the SxxK, SxN, and KTG motifs characteristic of penicillin-interactive enzymes.

Group I SxxK acyltransferases include class A and class B PBPs (Table 3), which are also referred to as HMW or multimodular PBPs and possess a two-domain structure (27, 28, 47). In class A PBPs, a C-terminal PBP domain is fused with an N-terminal glycosyltransferase (nPBP) module. These bifunctional enzymes catalyze transglycosylation, as

TABLE 3 Classification of penicillin-binding proteins and examples from prominent gram-positive organisms

Group and class[a]	Staphylococcus aureus	Streptococcus pneumoniae	Enterococcus	Bacillus subtilis
Group I				
Class A				
A3	PBP2	PBP1a	PonA	PBP1
A4		PBP2a	PbpF	PBP2c
A5		PBP1b	PbpZ	PBP4
A6				PBP2d
Class B				
B1	PBP2a (in MRSA)		Pbp5	PBP3
B3				PBPVD
B4	PBP1	PBP2x	PbpB	PBP2b
B5	PBP3	PBP2b	PbpA	PBP2a
Nonmodular LMW	PBP4	PBP3	Pbp6	PBP5, DacC
Group II, class C (sensor-transducer)	BlaR1	CiaR-CiaH		BlaR1

[a]Classification hierarchy as outlined by Goffin and Ghuysen (28).

well as transpeptidation, and thereby polymerize the disaccharide pentapeptide subunits borne by the lipid carrier molecule into nascent $(4 \rightarrow 3)$ peptidoglycan. Class A HMW PBPs are generally accepted as the lethal targets for β-lactam antibiotics; in general, there is a direct relationship between the binding affinity of a β-lactam with these PBPs and the inhibition of bacterial growth, unless other low-affinity PBPs can complement their activity.

In class B PBPs, the C-terminal PBP domain is fused with an N-terminal nPBP domain that mediates protein-protein interactions needed for peptidoglycan assembly in a cell cycle-dependent manner. These enzymes control the shape of the cell wall and direct septum formation and cell division. The class B PBPs include several low-affinity PBPs that have been implicated in β-lactam resistance in gram-positive species and are discussed in more detail below.

The third subgroup of the group I enzymes are nonmodular or freestanding PBPs. This subgroup includes LMW PBPs with single-domain structures that exhibit carboxypeptidase activity (47) and help to control the balance between various peptidoglycan precursors and thereby influence cell morphogenesis.

Group II SxxK acyltransferases are either not involved or only indirectly involved in peptidoglycan biosynthesis. Some serve to protect the group I enzymes from the effects of penicillin. Some, such as BlaR1 of S. aureus, were previously called HMW class C PBPs, are multimodular, and mediate the signal transduction associated with the induction of β-lactamase and PBP2a production in S. aureus (47, 79). Others, such as the freestanding PBPs of Streptomyces spp., appear to function as decoy high-affinity PBPs that bind β-lactams and thus interfere with the binding of β-lactam to group I enzymes (29). Such freestanding PBPs are believed to have been the precursors of the serine active-site β-lactamases and thus were the intermediate step toward evolution to an even more efficient means for protecting the critical group I enzymes.

As different gram-positive species are heterogeneous in their PBPs, there is a wide range in the natural susceptibility of different species to penicillin and other β-lactams.

Historically, group A streptococci and isolates of pneumococci from before 1967 have been exquisitely susceptible to benzylpenicillin, with MICs of about 0.01 μg/ml. Similarly, non-β-lactamase-producing strains of S. aureus generally exhibit MICs of benzylpenicillin of about 0.03 μg/ml. At the other extreme, penicillin-susceptible enterococci are intrinsically 60- to 200-fold less sensitive to penicillin, with MICs of about 2 μg/ml. Although L. monocytogenes is susceptible to benzylpenicillin and the aminopenicillins, cephalosporins have such low affinity for its essential PBP (PBP3) that this species is considered to be intrinsically resistant to the cephalosporins. For a similar reason, enterococci are uniformly resistant to the cephalosporins.

The terms susceptible, intermediately resistant, and resistant are defined differently for different gram-positive species (Table 4). For example, with the pneumococci even a small elevation in the MIC of ampicillin to the 0.12 to 1.0 μg/ml range indicates intermediate resistance; isolates for which MICs are ≥2.0 μg/ml are considered resistant. In contrast, an enterococcus for which an MIC is 8 μg/ml is still considered to be susceptible to ampicillin (56).

TABLE 4 MIC interpretive standards for gram-positive pathogens against ampicillin[a]

Pathogen(s)	Ampicillin MIC (μg/ml) required for pathogen to be classified as:		
	Susceptible	Intermediate	Resistant
Staphylococci	≤0.12		≥0.25
Enterococci	≤8		≥16
S. pneumoniae	≤0.06	0.12–1	≥2
Other streptococci	≤0.12	0.25–2	≥4

[a]MIC interpretive standards as recommended by CLSI (formerly NCCLS) (56).

EXAMPLES OF PBP-MEDIATED RESISTANCE IN GRAM-POSITIVE PATHOGENS

There are several PBP-related mechanisms by which gram-positive pathogens exhibit increased resistance to β-lactams. These include (i) acquisition of a gene encoding a new PBP from an entirely exogenous source or via recombination with a host PBP or β-lactamase gene, (ii) overexpression of a normal PBP that has low affinity for β-lactams compared to the other normal PBPs, and (iii) point mutations in normal PBPs that reduce the binding affinity of β-lactams.

S. aureus

S. aureus exhibits PBP-mediated resistance to β-lactams by each of the mechanisms described above, but from a clinical and epidemiologic perspective the most important mechanism involves the production of a low-affinity PBP called PBP2a. Typical methicillin-susceptible S. aureus isolates produce four major PBPs and a sensor-transducer, BlaR1 (Table 3). MRSA also produces an additional PBP of 78 kDa, called PBP2a and encoded by a unique gene called mecA. PBP2a is a multimodular class B1 acyltransferase that can replace the essential peptidoglycan transpeptidase activity but not the transglycolase activity of PBP2 when PBP2 has been inactivated by β-lactams (60). The origin of mecA is unclear; however, it exhibits 91% amino acid identity and 96% similarity with the transpeptidase domain of an intrinsic PBP from a colonizing species of squirrels, Staphylococcus sciuri (7, 36). Most S. sciuri isolates do not exhibit a methicillin-resistant phenotype, yet this appears to be related to low expression of the mecA homolog. More recently, isolates with higher expression exhibiting methicillin resistance have been described (11). The promoter and N-terminal portions (i.e., the nPBP domain) of PBP2a bear strong homology to S. aureus β-lactamase, and it has been suggested that mec might have originated from the fusion of genes encoding β-lactamase and a PBP (67). The atypical codon usage patterns and skewed G+C content at third codon positions within mec suggest possible acquisition from another species more recently than most other genes in the S. aureus chromosome (36).

Classic hospital-acquired MRSA strains throughout the world derive from at least five distinct MRSA clonotypes that represent an independent introduction of the mecA gene into S. aureus, followed by spread of the resistant clones (36). The clonotypes derive from three distinct polymorphisms of mec accompanied by 21 to 60 kb of surrounding DNA that are called staphylococcal cassette chromosome mec (SCCmec) types I, II, and III. The SCCmec elements have integrated into three distinct methicillin-susceptible chromosomal backgrounds, termed A, B, and C, to yield the five clonotypes. All three SCCmec types are found within the ribotype A chromosomal background, and the type II SCCmec has also integrated into type B and C chromosomal backgrounds.

A more recent development is the discovery of SCCmec type IV (13), which has been associated with MRSA isolates recovered from infections in previously healthy persons in a community setting without a clear epidemiologic link to the hospital environment. CA-MRSA infections represent a new and ominous development, as these isolates have potent virulence characteristics and have quickly been associated with infections worldwide (74). Genetic analysis of multiple type IV SCCmec isolates from three continents suggests that most of the isolates have an agr (for accessory gene regulator) type 3 background. However, rather than dissemination of a single type IV SCCmec CA-MRSA clone, there appear to be multiple CA-MRSA subclones that have coevolved independently as the result of repeated introduction of type IV SCCmec into the chromosomes of different but related isolates.

As with β-lactamase, the production of PBP2a in most MRSA isolates is inducible. Furthermore, the genetic regulatory systems for the induction of β-lactamase and PBP2a production are related, with similar sensor-transducer and repressor proteins. The regulatory genes for PBP2a production, mecI and mecR1, are 61 and 34% homologous, respectively, with blaI and blaR1 (5), including their positioning upstream of the promoter for the respective structural genes, mecA and blaZ (Fig. 3).

Most MRSA isolates exhibit heterogeneity in the expression of resistance, with a small fraction of the total population growing at a much higher concentration of antibiotic than the majority population. External factors such as temperature, osmolality, and light influence the proportion of the bacterial cell population that exhibits resistance (7). Transposon mutagenesis studies indicate that the expression of methicillin resistance is also influenced by chromosomal genes involved in complex regulatory functions (sigmaB) and cell wall biosynthesis (murE, pbp2, glmM, glnR, llm, etc.), including the fem (factors essential for the expression of methicillin resistance) genes as well as genes affecting autolysis (4, 15). This suggests that the optimal expression of resistance involves the cooperative functioning of a number of genes in cell wall metabolism, as well as the stress response. In contrast to the situation with β-lactamase and the level of penicillin resistance, the amount of PBP2a a strain produces often does not correlate very well with the MIC of methicillin among different MRSA isolates. This observation may reflect the effect of these other genes.

Other PBP-mediated mechanisms of resistance to β-lactams have been reported for S. aureus but do not appear to be important clinically, despite the clinical use of β-lactams for 6 decades. For example, isolates of S. aureus that overexpress PBP4 exhibit small but reproducible increases in β-lactam resistance (4). Also, point mutations in the penicillin-binding domains of PBP2 render low-level resistance to methicillin (7, 71). It has recently been shown that amino acid substitutions within PBP2a can render high-level resistance to new β-lactams being developed for clinical use against MRSA isolates (41), showing once again the resiliency of the staphylococcus and its capacity to rapidly develop resistance.

Coagulase-Negative Staphylococci

Although less well studied, in general the mechanism of methicillin resistance in the coagulase-negative staphylococci generally involves a PBP2a that is nearly identical to that seen with S. aureus. A recent report of SCCmec typing performed upon 44 bloodstream S. epidermidis isolates recovered between 1973 and 1983 showed that 1 (2%) harbored SCCmec type I, 15 (34%) harbored type II, 12 (28%) harbored type III, and 16 (36%) harbored type IV (78). Thus, not only do the coagulase-negative staphylococci possess the same SCCmec types as S. aureus, they have harbored SCCmec type IV for decades and may be the reservoir from which SCCmec type IV is being introduced into S. aureus. Finally, a few methicillin-resistant strains which lack mecA have been previously described (69).

Enterococci

Enterococcal isolates produce at least five and occasionally as many as nine PBPs (32). The major mechanism of enterococcal resistance to penicillins involves increased production of PBP5. Similar to the staphylococcal PBP2a with which it shares about 33% amino acid identity, enterococcal PBP5 is a group I, class B1 PBP (Table 3) that exhibits low affinity for penicillin (23). Studies involving a swine-derived isolate of *Enterococcus hirae* (ATCC 9790) demonstrate a direct correlation between how much PBP5 an isolate produces and its penicillin MIC, which corresponds to the concentration of penicillin that saturates PBP5. Conversely, PBP5-deficient mutants are highly susceptible to benzylpenicillin, for which MICs are produced that are similar to those for group A and other streptococcal species that do not produce a low-affinity PBP. Antiserum raised against PBP5 of *E. hirae* is cross-reactive with PBPs of similar sizes in *Enterococcus faecium* and *E. faecalis* (46). PBP5 is normally repressed by the product of the PBP5 synthesis repressor gene (*psr*), which is located just upstream of the gene encoding PBP5. PBP5 overproduction is generally attributable to deletion of (or defects in) *psr*. Recent data suggest that enterococci synthesize 3 → 3 peptidoglycan involving L,D-amino acid substrates, in addition to 4 → 3 peptidoglycan; however, the role of PBP5 in this reaction is unclear (29). A second PBP-related mechanism of β-lactam resistance has been described in which amino acid substitutions close to the SDN triad within PBP5 result in even lower affinity for penicillin (62).

S. pneumoniae

Penicillin-susceptible strains of *S. pneumoniae* possess five HMW PBPs (PBP1a, -1b, -2x, -2a, and -2b) and one LMW PBP (PBP3) (32), with different strains exhibiting considerable antigenic variation. PBP2x and -2b are the primary targets for penicillins. Alterations in PBP2x and -2b confer low-level resistance to penicillin and can be derived in the laboratory by selection on antibiotic-containing agar.

Most penicillin-resistant clinical isolates exhibit a PBP pattern more complex than just a combination of point mutations. They exhibit a wide range of PBP profiles, and the large number of nucleotide changes suggests that the resistant PBPs originated by lateral transfer of PBP genes or portions thereof from other streptococcal species. Homologs of the pneumococcal PBP2x gene in *S. oralis* and *S. mitis* have been identified (66). These genes contain regions of heterologous DNA with similar or identical junctions flanking them that mediate recombination to produce mosaic PBPs (Fig. 4). The development of low-affinity PBPs in *S. pneumoniae* appears to involve a combination of the accumulation of point mutations in the PBP genes of commensal streptococcal species, followed by lateral transfer and recombination of these PBP genes with their homologs in *S. pneumoniae* (16, 33). High-level resistance to penicillins involves mosaic genes in at least three of the five HMW PBPs: PBP1a, -2x, and -2b. Resistance to the expanded-spectrum cephalosporins can be seen involving alterations in just PBP1a and PBP2x, as PBP2b is not a natural target for the cephalosporins (32). PBP2x and PBP1a are closely linked on the chromosome; high-level resistance to cefotaxime and intermediate resistance to penicillin have been transferred to a susceptible *S. pneumoniae* via a single round of transformation (53).

Selected capsular serotypes typically observed with pneumococcal infections in children are overrepresented among penicillin-resistant strains, including serotypes 6, 9, 14, 19, and 23 (14). Resistant strains within a region frequently appear to be clonal, with the same serotype and PBP banding profile. However, strains from different regions indicate the presence of multiple clones and support the hy-

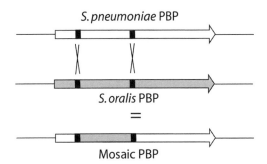

FIGURE 4 Model for the formation of mosaic PBPs in *S. pneumoniae*. An isolate of *S. pneumoniae* in the nasopharynx or oropharynx of a colonized person becomes transformed with PBP-encoding DNA from *S. oralis*, *S. mitis*, or another commensal streptococcal species that has accumulated point mutations related to exposure to β-lactams. Recombination occurs between regions of identity or high homology between species (black) such that heterologous DNA from the donor (gray) is inserted into the PBP (white) of *S. pneumoniae*, thereby producing a mosaic PBP that contains regions of both PBPs and exhibits reduced β-lactam-binding affinity.

pothesis that resistance has emerged in *S. pneumoniae* independently on multiple occasions with extensive dissemination of the highly resistant clones.

Viridans Streptococci

Historically, the viridans streptococci have been highly susceptible to β-lactam antibiotics, and most clinical isolates have remained susceptible. However, cefotaxime- and benzylpenicillin-resistant isolates of *S. mitis* and *S. oralis* with low-affinity PBPs have been identified (61) and typically have alterations in multiple PBPs. Interspecies and intraspecies gene transfer and recombination appear to be the major mechanism underlying β-lactam resistance for viridans streptococci, as well as for *S. pneumoniae*.

Other Gram-Positive Pathogens

Clinical isolates of group A streptococci have remained highly susceptible to benzylpenicillin (MIC < 0.01 μg/ml); however, intermediate resistance (MIC, 0.2 μg/ml) mediated by changes in PBP2 and PBP3 can be selected by antibiotic exposure in the laboratory (31). Group B streptococci are intrinsically about 10-fold less susceptible than group A streptococci to penicillin. Most clinical isolates have remained penicillin susceptible, although intermediate resistance (MICs, 0.25 to 2.0 μg/ml) has been described. The PBPs of *B. subtilis* have been extensively studied and include a low-affinity class B1 HMW PBP (Table 3); however, less is known about PBPs of other *Bacillus* spp. Some *Bacillus* PBPs play a role in sporulation. Altered PBP patterns have been reported in imipenem-resistant isolates of *R. equi* (58).

TOLERANCE

The term tolerance has been applied to bacteria that are inhibited but not killed by β-lactam antibiotics. It was first used in 1970 to describe laboratory strains of *S. pneumoniae* with a reduced rate of autolysis in response to penicillin inhibition. As a result, the strains lost viability at much lower rates than usual, following exposure to penicillin at concentrations above the MIC (73).

Tolerance differs from other types of resistance in that it does not involve an increase in the MIC of the drug compared to other isolates of the same species (73). In many clinical laboratories, it has been defined as an MBC/MIC ratio of ≥32; however, it has been noted that by using this arbitrary breakpoint, the original pneumococcal isolate in which tolerance was reported would not have been considered tolerant (73). The use of different definitions of tolerance along with numerous technical factors which affect MBC test results have led to inconsistencies and confusion about the clinical relevance of tolerance in the medical literature (34, 65).

The genetic basis of tolerance suggests several mechanisms (34). Among tolerant mutants isolated in the laboratory, tolerant *S. pneumoniae* isolates produce subnormal amounts of *N*-acetylmuramic acid–L-alanine amidase, whereas tolerant *E. faecium* isolates have diminished muramidase production. In some strains of *B. subtilis*, tolerance involves the inactivation of autolysin by a protease. In group A streptococci, tolerance has been associated with changes in PBP expression. Multiple other defects that might lead to reduced killing of bacteria without affecting the MIC have been proposed (34). Furthermore, tolerance attributable to reduced autolytic activity is not unique to β-lactams but is also observed with other cell wall-active antimicrobial agents. For example, reduced autolysis in the presence of glycopeptides in *S. aureus* is exhibited by *agr* null mutants of *agr* group 2 isolates; this may be a factor in the development of intermediate-level glycopeptide heteroresistance (64).

In clinical isolates, tolerance appears to be more prevalent among gram-positive than gram-negative species (34). Group A, B, C, and G streptococci exhibiting tolerance to penicillin have been recovered from clinical specimens (12, 31). In viridans streptococci, tolerance has been reported to adversely affect the response to antibiotic therapy in animal models of endocarditis (73). Clinical isolates of enterococci are now almost always tolerant to penicillin (65), in marked contrast to the so-called penicillin virgin isolates of *E. faecalis* recovered from persons in the Solomon Islands in the 1960s (37). However, the latter group can be easily induced to develop tolerance when exposed to pulsed doses of benzylpenicillin in vitro, and it has been suggested that the frequent use of β-lactams in clinical practice might have contributed to the widespread prevalence of tolerant enterococci among clinical isolates (37). The prevalence of tolerance in *S. aureus* has been debated and it has been more difficult to demonstrate the clinical relevance of tolerance with *S. aureus* than with other gram-positive species (65, 73).

Phenotypic tolerance refers to the ability of all bacteria to avoid the bactericidal activity of antibiotics at times when they are dormant or only growing slowly. It may be a more important clinical problem than genotypic tolerance and likely contributes to difficulties in eradicating gram-positive pathogens from infections involving prosthetic devices.

SPREAD OF RESISTANCE AND IMPLICATIONS FOR THE FUTURE

After >6 decades of use, the effectiveness of β-lactams in infections caused by gram-positive pathogens has been seriously eroded by the development and spread of β-lactam resistance. Plasmid-encoded β-lactamase production is now widespread in *S. aureus* and has been transferred to enterococci. Low-affinity PBPs are widespread in staphylococci, enterococci, and pneumococci and have reduced the usefulness of β-lactams as agents to treat infections caused by

such pathogens. These changes bear strong evidence of the power of selective antibiotic pressure in the clinical setting and the genetic versatility of gram-positive bacteria. Perhaps the most important lesson to be learned from the antibiotic era is that bacteria quickly evolve to become resistant to things that hurt them. Thus, success in containing the dissemination of β-lactam-resistant gram-positive pathogens must include strategies to control antibiotic utilization.

REFERENCES

1. **Ambler, R. P., A. F. Coulson, J. M. Frere, J. M. Ghuysen, B. Joris, M. Forsman, R. C. Levesque, G. Tiraby, and S. G. Waley.** 1991. A standard numbering scheme for the class A beta-lactamases. *Biochem. J.* **276:**269–270.

2. **Appelbaum, P. C., A. Bhamjee, J. N. Scragg, A. F. Hallett, A. J. Bowen, and R. C. Cooper.** 1977. *Streptococcus pneumoniae* resistant to penicillin and chloramphenicol. *Lancet* **ii:**995–997.

3. **Barg, N., H. Chambers, and D. Kernodle.** 1991. Borderline susceptibility to antistaphylococcal penicillins is not conferred exclusively by the hyperproduction of β-lactamase. *Antimicrob. Agents Chemother.* **35:**1975–1979.

4. **Berger-bächi, B.** 1997. Resistance not mediated by β-lactamase (methicillin resistance), p. 158–174. *In* K. B. Crossley and G. L. Archer (ed.), *The Staphylococci in Human Disease*. Churchill Livingstone, Ltd., New York, N.Y.

5. **Brakstad, O. G., and J. A. Maeland.** 1997. Mechanisms of methicillin resistance in staphylococci. *APMIS* **105:**264–276.

6. **Bush, K., G. A. Jacoby, and A. A. Medeiros.** 1995. A functional classification scheme for β-lactamases and its correlation with molecular structure. *Antimicrob. Agents Chemother.* **39:**1211–1233.

7. **Chambers, H. F.** 1997. Methicillin resistance in staphylococci: molecular and biochemical basis and clinical implications. *Clin. Microbiol. Rev.* **10:**781–791.

8. **Chambers, H. F., and M. Sachdeva.** 1990. Binding of β-lactam antibiotics to penicillin-binding proteins in methicillin-resistant *Staphylococcus aureus*. *J. Infect. Dis.* **161:**1170–1176.

9. **Cohen, S., and H. M. Sweeney.** 1968. Constitutive penicillinase formation in *Staphylococcus aureus* owing to a mutation unlinked to the penicillinase plasmid. *J. Bacteriol.* **95:**1368–1374.

10. **Colombo, M. L., S. Hanique, S. L. Baurin, C. Bauvois, K. De Vriendt, J. J. Van Beeumen, J. M. Frere, and B. Joris.** 2004. The *ybxI* gene of *Bacillus subtilis* 168 encodes a class D β-lactamase of low activity. *Antimicrob. Agents Chemother.* **48:**484–490.

11. **Couto, I., S. W. Wu, A. Tomasz, and H. de Lencastre.** 2003. Development of methicillin resistance in clinical isolates of *Staphylococcus sciuri* by transcriptional activation of the *mecA* homologue native to the species. *J. Bacteriol.* **185:**645–653.

12. **Dagan, R., M. Ferne, M. Sheinis, M. Alkan, and E. Katzenelson.** 1987. An epidemic of penicillin–tolerant group A streptococcal pharyngitis in children living in a closed community: mass treatment with erythromycin. *J. Infect. Dis.* **156:**514–516.

13. **Daum, R. S., T. Ito, K. Hiramatsu, F. Hussain, K. Mongkolrattanothai, M. Jamklang, and S. Boyle-Vavra.** 2002. A novel methicillin-resistance cassette in community-acquired methicillin-resistant *Staphylococcus aureus* iso-

lates of diverse genetic backgrounds. *J. Infect. Dis.* **186:** 1344–1347.

14. **de Lencastre, H., and A. Tomasz.** 2002. From ecological reservoir to disease: the nasopharynx, day-care centres and drug-resistant clones of *Streptococcus pneumoniae*. *J. Antimicrob. Chemother.* **50**(Suppl. S2):75–81.

15. **de Lencastre, H., S. W. Wu, M. G. Pinho, A. M. Ludovice, S. Filipe, S. Gardete, R. Sobral, S. Gill, M. Chung, and A. Tomasz.** 1999. Antibiotic resistance as a stress response: complete sequencing of a large number of chromosomal loci in *Staphylococcus aureus* strain COL that impact on the expression of resistance to methicillin. *Microb. Drug Resist.* **5:**163–175.

16. **Dowson, C. G., T. J. Coffey, C. Kell, and R. A. Whiley.** 1993. Evolution of penicillin resistance in *Streptococcus pneumoniae*; the role of *Streptococcus mitis* in the formation of a low affinity PBP2B in S. *pneumoniae. Mol. Microbiol.* **9:** 635–643.

17. **Dyke, K., and P. Gregory.** 1997. Resistance to β-lactam antibiotics: resistance mediated by β-lactamases, p. 139–157. *In* K. B. Crossley and G. L. Archer (ed.), *The Staphylococci in Human Disease*. Churchill Livingstone, Ltd., New York, N.Y.

18. **Dyke, K. G. H.** 1979. β-Lactamases of *Staphylococcus aureus*, p. 291–310. *In* J. M. T. Hamilton-Miller and J. T. Smith (ed.), *Beta-Lactamases*. Academic Press, Inc., New York, N.Y.

19. **East, A. K., S. P. Curnock, and K. G. Dyke.** 1990. Change of a single amino acid in the leader peptide of a staphylococcal β-lactamase prevents the appearance of the enzyme in the medium. *FEMS Microbiol. Lett.* **57:**249–254.

20. **Ellerbrok, H., and R. Hakenbeck.** 1988. Penicillin-degrading activities of peptides from pneumococcal penicillin-binding proteins. *Eur. J. Biochem.* **171:**219–224.

21. **Fierer, J., P. Wolf, L. Seed, T. Gay, K. Noonan, and P. Haghighi.** 1987. Non-pulmonary *Rhodococcus equi* infections in patients with acquired immune deficiency syndrome (AIDS). *J. Clin. Pathol.* **40:**556–558.

22. **Fong, I. W., E. R. Engelking, and W. M. Kirby.** 1976. Relative inactivation by *Staphylococcus aureus* of eight cephalosporin antibiotics. *Antimicrob. Agents Chemother.* **9:**939–944.

23. **Fontana, R., M. Aldegheri, M. Ligozzi, H. Lopez, A. Sucari, and G. Satta.** 1994. Overproduction of a low-affinity penicillin-binding protein and high-level ampicillin resistance in *Enterococcus faecium. Antimicrob. Agents Chemother.* **38:**1980–1983.

24. **Frere, J. M.** 1995. β-Lactamases and bacterial resistance to antibiotics. *Mol. Microbiol.* **16:**385–395.

25. **Georgopapadakou, N. H.** 1993. Penicillin-binding proteins and bacterial resistance to β-lactams. *Antimicrob. Agents Chemother.* **37:**2045–2053.

26. **Georgopapadakou, N. H., and F. Y. Liu.** 1980. Binding of β-lactam antibiotics to penicillin-binding proteins of *Staphylococcus aureus* and *Streptococcus faecalis*: relation to antibacterial activity. *Antimicrob. Agents Chemother.* **18:** 834–836.

27. **Ghuysen, J. M.** 1991. Serine β-lactamases and penicillin-binding proteins. *Annu. Rev. Microbiol.* **45:**37–67.

28. **Goffin, C., and J. M. Ghuysen.** 1998. Multimodular penicillin-binding proteins: an enigmatic family of orthologs and paralogs. *Microbiol. Mol. Biol. Rev.* **62:**1079–1093.

29. **Goffin, C., and J. M. Ghuysen.** 2002. Biochemistry and comparative genomics of SxxK superfamily acyltransferases offer a clue to the mycobacterial paradox: presence of penicillin-susceptible target proteins versus lack of effi-

ciency of penicillin as therapeutic agent. *Microbiol. Mol. Biol. Rev.* **66:**702–738.

30. **Gregory, P. D., R. A. Lewis, S. P. Curnock, and K. G. Dyke.** 1997. Studies of the repressor (BlaI) of β-lactamase synthesis in *Staphylococcus aureus. Mol. Microbiol.* **24:** 1025–1037.

31. **Gutmann, L., and A. Tomasz.** 1982. Penicillin-resistant and penicillin-tolerant mutants of group A streptococci. *Antimicrob. Agents Chemother.* **22:**128–136.

32. **Hakenbeck, R., and J. Coyette.** 1998. Resistant penicillin-binding proteins. *Cell Mol. Life Sci.* **54:**332–340.

33. **Hakenbeck, R., A. Konig, I. Kern, M. van der Linden, W. Keck, D. Billot-Klein, R. Legrand, B. Schoot, and L. Gutmann.** 1998. Acquisition of five high-M$_r$ penicillin-binding protein variants during transfer of high-level β-lactam resistance from *Streptococcus mitis* to *Streptococcus pneumoniae. J. Bacteriol.* **180:**1831–1840.

34. **Handwerger, S., and A. Tomasz.** 1985. Antibiotic tolerance among clinical isolates of bacteria. *Rev. Infect. Dis.* **7:** 368–386.

35. **Hansman, D., H. Glasgow, J. Sturt, L. Devitt, and R. Douglas.** 1971. Increased resistance to penicillin of pneumococci isolated from man. *N. Engl. J. Med.* **284:** 175–177.

36. **Hiramatsu, K., L. Cui, M. Kuroda, and T. Ito.** 2001. The emergence and evolution of methicillin-resistant *Staphylococcus aureus. Trends Microbiol.* **9:**486–493.

37. **Hodges, T. L., S. Zighelboim-Daum, G. M. Eliopoulos, C. Wennersten, and R. C. Moellering, Jr.** 1992. Antimicrobial susceptibility changes in *Enterococcus faecalis* following various penicillin exposure regimens. *Antimicrob. Agents Chemother.* **36:**121–125.

38. **Horn, D. L., J. B. Zabriskie, R. Austrian, P. P. Cleary, J. J. Ferretti, V. A. Fischetti, E. Gotschlich, E. L. Kaplan, M. McCarty, S. M. Opal, R. B. Roberts, A. Tomasz, and Y. Wachtfogel.** 1998. Why have group A streptococci remained susceptible to penicillin? Report on a symposium. *Clin. Infect. Dis.* **26:**1341–1345.

39. **Jevons, M. P.** 1961. "Celebenin"-resistant staphylococci. *Brit. Med. J.* **1:**124–125.

40. **Joris, B., P. Ledent, O. Dideberg, E. Fonze, J. Lamotte-Brasseur, J. A. Kelly, J. M. Ghuysen, and J. M. Frere.** 1991. Comparison of the sequences of class A β-lactamases and of the secondary structure elements of penicillin-recognizing proteins. *Antimicrob. Agents Chemother.* **35:**2294–2301.

41. **Katayama, Y., H. Z. Zhang, and H. F. Chambers.** 2004. PBP 2a mutations producing very-high-level resistance to β-lactams. *Antimicrob. Agents Chemother.* **48:**453–459.

42. **Kernodle, D. S.** Unpublished data.

43. **Kernodle, D. S., D. C. Classen, C. W. Stratton, and A. B. Kaiser.** 1998. Association of borderline oxacillin-susceptible strains of *Staphylococcus aureus* with surgical wound infections. *J. Clin. Microbiol.* **36:**219–222.

44. **Kernodle, D. S., C. W. Stratton, L. W. McMurray, J. R. Chipley, and P. A. McGraw.** 1989. Differentiation of β-lactamase variants of *Staphylococcus aureus* by substrate hydrolysis profiles. *J. Infect. Dis.* **159:**103–108.

45. **Kirby, W. M. M.** 1944. Extraction of a highly potent penicillin inactivator from penicillin resistant staphylococci. *Science* **99:**452–453.

46. **Ligozzi, M., M. Aldegheri, S. C. Predari, and R. Fontana.** 1991. Detection of penicillin-binding proteins immunologically related to penicillin-binding protein 5 of *Enterococcus hirae* ATCC 9790 in *Enterococcus faecium* and *Enterococcus faecalis. FEMS Microbiol. Lett.* **67:**335–339.

47. Massova, I., and S. Mobashery. 1998. Kinship and diversification of bacterial penicillin-binding proteins and β-lactamases. *Antimicrob. Agents Chemother.* **42**:1–17.

48. Materon, I. C., A. M. Queenan, T. M. Koehler, K. Bush, and T. Palzkill. 2003. Biochemical characterization of β-lactamases Bla1 and Bla2 from *Bacillus anthracis*. *Antimicrob. Agents Chemother.* **47**:2040–2042.

49. McDougal, L. K., and C. Thornsberry. 1986. The role of β-lactamase in staphylococcal resistance to penicillinase-resistant penicillins and cephalosporins. *J. Clin. Microbiol.* **23**:832–839.

50. McDowell, T. D., and K. E. Reed. 1989. Mechanism of penicillin killing in the absence of bacterial lysis. *Antimicrob. Agents Chemother.* **33**:1680–1685.

51. McMurray, L. W., D. S. Kernodle, and N. L. Barg. 1990. Characterization of a widespread strain of methicillin-susceptible *Staphylococcus aureus* associated with nosocomial infections. *J. Infect. Dis.* **162**:759–762.

52. Moreillon, P., Z. Markiewicz, S. Nachman, and A. Tomasz. 1990. Two bactericidal targets for penicillin in pneumococci: autolysis-dependent and autolysis-independent killing mechanisms. *Antimicrob. Agents Chemother.* **34**:33–39.

53. Munoz, R., C. G. Dowson, M. Daniels, T. J. Coffey, C. Martin, R. Hakenbeck, and B. G. Spratt. 1992. Genetics of resistance to third-generation cephalosporins in clinical isolates of *Streptococcus pneumoniae*. *Mol. Microbiol.* **6**: 2461–2465.

54. Murray, B. E. 1992. β-Lactamase-producing enterococci. *Antimicrob. Agents Chemother.* **36**:2355–2359.

55. Nannini, E. C., K. V. Singh, and B. E. Murray. 2003. Relapse of type A β-lactamase-producing *Staphylococcus aureus* native valve endocarditis during cefazolin therapy: revisiting the issue. *Clin. Infect. Dis.* **37**:1194–1198.

56. National Committee for Clinical Laboratory Standards. 1998. *Performance Standards for Antimicrobial Susceptibility Testing*. Eighth informational supplement. NCCLS approved standard M100-S8. National Committee for Clinical Laboratory Standards, Wayne, Pa.

57. Nielsen, J. B., and J. O. Lampen. 1982. Membrane-bound penicillinases in gram-positive bacteria. *J. Biol. Chem.* **257**:4490–4495.

58. Nordmann, P., M. H. Nicolas, and L. Gutmann. 1993. Penicillin-binding proteins of *Rhodococcus equi*: potential role in resistance to imipenem. *Antimicrob. Agents Chemother.* **37**:1406–1409.

59. Novak, R., E. Charpentier, J. S. Braun, and E. Tuomanen. 2000. Signal transduction by a death signal peptide: uncovering the mechanism of bacterial killing by penicillin. *Mol. Cell* **5**:49–57.

60. Pinho, M. G., H. de Lencastre, and A. Tomasz. 2001. An acquired and a native penicillin-binding protein cooperate in building the cell wall of drug-resistant staphylococci. *Proc. Natl. Acad. Sci. USA* **98**:10886–10891.

61. Reichmann, P., A. Konig, J. Linares, F. Alcaide, F. C. Tenover, L. McDougal, S. Swidsinski, and R. Hakenbeck. 1997. A global gene pool for high-level cephalosporin resistance in commensal *Streptococcus* species and *Streptococcus pneumoniae*. *J. Infect. Dis.* **176**:1001–1012.

62. Rybkine, T., J. L. Mainardi, W. Sougakoff, E. Collatz, and L. Gutmann. 1998. Penicillin-binding protein 5 sequence alterations in clinical isolates of *Enterococcus faecium* with different levels of β–lactam resistance. *J. Infect. Dis.* **178**:159–163.

63. Sabath, L. D., C. Garner, C. Wilcox, and M. Finland. 1975. Effect of inoculum and of beta-lactamase on the anti-staphylococcal activity of thirteen penicillins and cephalosporins. *Antimicrob. Agents Chemother.* **8**:344–349.

64. Sakoulas, G., G. M. Eliopoulos, R. C. Moellering, Jr., R. P. Novick, L. Venkataraman, C. Wennersten, P. C. DeGirolami, M. J. Schwaber, and H. S. Gold. 2003. *Staphylococcus aureus* accessory gene regulator (*agr*) group II: is there a relationship to the development of intermediate-level glycopeptide resistance? *J. Infect. Dis.* **187**:929–938.

65. Sherris, J. C. 1986. Problems in in vitro determination of antibiotic tolerance in clinical isolates. *Antimicrob. Agents Chemother.* **30**:633–637.

66. Sibold, C., J. Henrichsen, A. Konig, C. Martin, L. Chalkley, and R. Hakenbeck. 1994. Mosaic *pbpX* genes of major clones of penicillin-resistant *Streptococcus pneumoniae* have evolved from *pbpX* genes of a penicillin-sensitive *Streptococcus oralis*. *Mol. Microbiol.* **12**:1013–1023.

67. Song, M. D., M. Wachi, M. Doi, F. Ishino, and M. Matsuhashi. 1987. Evolution of an inducible penicillin-target protein in methicillin-resistant *Staphylococcus aureus* by gene fusion. *FEBS Lett.* **221**:167–171.

68. Statens Serum Institut. 2003. *Annual Report on Staphylococcus aureus Bacteraemia Cases in Denmark, 2001*. Staphylococcus Laboratory, National Center For Antimicrobials and Infection Control. Statens Serum Institut, Copenhagen, Denmark. [Online.] http://www.ssi.dk/graphics/dk/overvagning/ sygdomsovervaagning/Annual01.pdf.

69. Suzuki, E., K. Hiramatsu, and T. Yokota. 1992. Survey of methicillin-resistant clinical strains of coagulase-negative staphylococci for *mecA* gene distribution. *Antimicrob. Agents Chemother.* **36**:429–434.

70. Thore, M. 1992. β-Lactamase substrate profiles of coagulase-negative skin staphylococci from orthopaedic inpatients and staff members. *J. Hosp. Infect.* **22**:229–240.

71. Tomasz, A., H. B. Drugeon, H. M. de Lencastre, D. Jabes, L. McDougall, and J. Bille. 1989. New mechanism for methicillin resistance in *Staphylococcus aureus*: clinical isolates that lack the PBP 2a gene and contain normal penicillin-binding proteins with modified penicillin-binding capacity. *Antimicrob. Agents Chemother.* **33**:1869–1874.

72. Tomayko, J. F., K. K. Zscheck, K. V. Singh, and B. E. Murray. 1996. Comparison of the β-lactamase gene cluster in clonally distinct strains of *Enterococcus faecalis*. *Antimicrob. Agents Chemother.* **40**:1170–1174.

73. Tuomanen, E., D. T. Durack, and A. Tomasz. 1986. Antibiotic tolerance among clinical isolates of bacteria. *Antimicrob. Agents Chemother.* **30**:521–527.

74. Vandenesch, F., T. Naimi, M. C. Enright, G. Lina, G. R. Nimmo, H. Heffernan, N. Liassine, M. Bes, T. Greenland, M. E. Reverdy, and J. Etienne. 2003. Community-acquired methicillin-resistant *Staphylococcus aureus* carrying Panton-Valentine leukocidin genes: worldwide emergence. *Emerg. Infect. Dis.* **9**:978–984.

75. Voladri, R. K., and D. S. Kernodle. 1998. Characterization of a chromosomal gene encoding type B β-lactamase in phage group II isolates of *Staphylococcus aureus*. *Antimicrob. Agents Chemother.* **42**:3163–3168.

76. Voladri, R. K., M. K. Tummuru, and D. S. Kernodle. 1996. Structure-function relationships among wild-type variants of *Staphylococcus aureus* β-lactamase: importance of amino acids 128 and 216. *J. Bacteriol.* **178**:7248–7253.

77. Wallace, R. J., Jr., P. Vance, A. Weissfeld, and R. R. Martin. 1978. β-Lactamase production and resistance to β-lactam antibiotics in *Nocardia*. *Antimicrob. Agents Chemother.* **14**:704–709.

78. Wisplinghoff, H., A. E. Rosato, M. C. Enright, M. Noto, W. Craig, and G. L. Archer. 2003. Related clones con-

taining *SCCmec* type IV predominate among clinically significant *Staphylococcus epidermidis* isolates. *Antimicrob. Agents Chemother.* **47:**3574–3579.

79. **Zhang, H. Z., C. J. Hackbarth, K. M. Chansky, and H. F. Chambers.** 2001. A proteolytic transmembrane sig-

naling pathway and resistance to β-lactams in staphylococci. *Science* **291:**1962–1965.

80. **Zygmunt, D. J., C. W. Stratton, and D. S. Kernodle.** 1992. Characterization of four β-lactamases produced by *Staphylococcus aureus. Antimicrob. Agents Chemother.* **36:**440–445.

Resistance to Glycopeptides in Gram-Positive Pathogens

HENRY S. FRAIMOW AND PATRICE COURVALIN

63

Glycopeptide antibiotics including vancomycin and teicoplanin have been critical components of our therapeutic arsenal for treating infections caused by gram-positive pathogens for the past 40 years. Initially introduced into clinical practice in the 1960s, usage of the glycopeptide vancomycin skyrocketed during the 1980s as one of the few antimicrobials reliably effective against methicillin-resistant staphylococci (Fig. 1) (43). Intrinsic resistance to vancomycin was observed in several generally nonpathogenic or rarely seen gram-positive bacterial genera such as *Leuconostoc*, *Pediococcus*, *Lactobacillus*, and *Erysipelothrix*, but such resistance excited little interest among either clinicians or microbiologists. The emergence of acquired glycopeptide resistance in clinically significant pathogens was not anticipated; in fact, some even suggested that such resistance would be unlikely ever to occur. Glycopeptide-resistant enterococci (GRE) were first isolated in Europe in 1986 (44, 66), appeared shortly thereafter in the United States, and have since become a worldwide problem (13, 48). Glycopeptide resistance in staphylococci was initially observed with coagulase-negative staphylococci (CNS), especially in *Staphylococcus haemolyticus* and less commonly *Staphylococcus epidermidis*, but beginning in 1996 there have been multiple reports of clinical isolates of *Staphylococcus aureus* with reduced susceptibility to vancomycin associated with clinical treatment failure (18, 39, 40). In 2002, the medical community's greatest fears about glycopeptide resistance were finally realized with the first appearance of strains of methicillin-resistant *S. aureus* (MRSA) with acquired high-level vancomycin resistance (19–21). As glycopeptides have remained the antibiotics of last resort for gram-positive infections, the broader dissemination of glycopeptide resistance in this virulent pathogen is potentially catastrophic. Coincident with the emergence of glycopeptide resistance in enterococci and staphylococci, there has been a tremendous explosion of information on the epidemiology, mechanisms, and genetics of glycopeptide resistance in both pathogenic and nonpathogenic gram-positive bacteria.

STRUCTURE AND MECHANISM OF ACTION OF GLYCOPEPTIDES

Glycopeptide antimicrobials are natural products that are produced by various soil-dwelling species of the order *Actinomycetales* such as *Amycolatopsis orientalis* and *Amycolatopsis coloradensis* (55, 73). Thus far, over 200 different glycopeptides have been isolated. Two of these, vancomycin and teicoplanin, have been successfully developed and approved for use in humans. Vancomycin is widely used throughout the United States, Europe, Japan, and many other parts of the world. Teicoplanin is available in Europe but not in the United States. Vancomycin and teicoplanin are primarily used as parenteral agents due to their poor oral bioavailability, although oral formulations are available for treatment of specific enteric infections such as *Clostridium difficile*. Another agent, avoparcin, has been widely used in some parts of Europe in animal farming and in veterinary medicine (70). With the recent emergence of glycopeptide-resistant enterococci and staphylococci, there has been renewed interest by the pharmaceutical industry in the development of modified semisynthetic glycopeptides with enhanced activity against resistant gram-positive pathogens including glycopeptide-resistant organisms. Several agents, including the vancomycin derivatives oritavancin and telavancin and the teicoplanin derivative dalbovancin, are currently in late stages of clinical development in the United States and Europe (68).

Glycopeptide antibiotics are inhibitors of peptidoglycan synthesis. Unlike other cell wall synthesis inhibitors such as the beta-lactams, glycopeptides do not bind directly to cell wall-biosynthetic enzymes but instead interfere with peptidoglycan synthesis by complexing with carboxy-terminal alanine residues on peptidoglycan precursors (55). All glycopeptides are large, complex heptapeptide structures of molecular weights of 1,200 to 1,500, consisting of an aglycone core of six peptide residues linked to a triphenyl ether, with a variety of complex sugar groups attached to the heptapeptide core (68, 73). Due to their large, rigid, hydrophobic structures, glycopeptides do not penetrate bacterial cellular

Gram-Positive Pathogens, 2nd edition, edited by Vincent A. Fischetti et al.
© 2006 ASM Press, Washington, D.C.

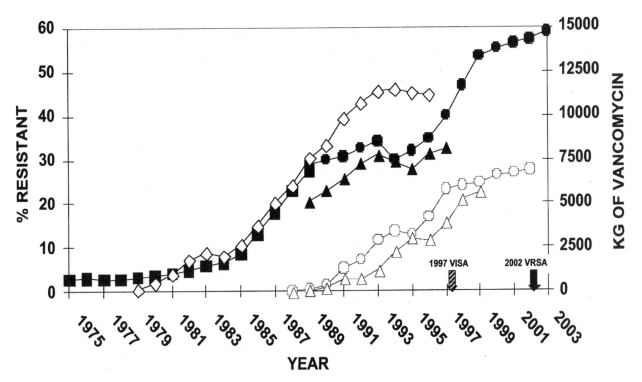

FIGURE 1 Increases in vancomycin usage (in kilograms) worldwide (primarily in the United States) from 1979 to 1983 and in the United States from 1984 to 1996 (◇—◇) that parallels the emergence of MRSA as a percentage of S. aureus isolates from large teaching hospitals in the United States from 1975 to 1989 (■ ■) and as a percentage of nosocomial S. aureus isolates from ICU (■—■) and non-ICU (▲—▲) settings from 1989 to 2003 (33, 43, 50). Rates of VRE in ICU (▢—▢) and non-ICU (△—△) settings begin to rise a decade after the dramatic increase in vancomycin usage (48). Also shown is the appearance of clinical VISA isolates in the United States in 1997 and VRSA in 2002, though these remain a very small percentage of total S. aureus strains.

membranes. They can only interact with gram-positive cell wall precursors after translocation of these precursors on the undecaprenol lipid carrier onto the outer surface of the cytoplasmic membrane. The terminal D-alanyl–D-alanine (D-Ala-D-Ala) moiety of the UDP-*N*-acetyl muramyl pentapeptide precursors binds tightly within the groove in the aglycone portion of the glycopeptide molecule and blocks progression to the subsequent transglycosylation steps in peptidoglycan synthesis (Fig. 2 and 3). Complexing of glycopeptide to the D-Ala-D-Ala residues also interferes with the subsequent reactions catalyzed by D,D transpeptidases and D,D-carboxypeptidases necessary for the anchoring of the peptidoglycan complex. Blockage of transglycosylation also leads to trapping of lipid carrier and accumulation of peptidoglycan precursors in the cytoplasm. Some glycopeptides, including teicoplanin and the newer semisynthetic glycopeptides oritavancin, dalbavancin, and telavancin have lipophilic side chains facilitating binding to the cell membrane, which may enhance the interaction of these glycopeptides with peptidoglycan precursors. The semisynthetic glycopeptides differ from vancomycin by demonstrating rapid, concentration-dependent bactericidal activity, the mechanism of which remains incompletely understood (68). Both the enhanced membrane anchoring of these glycopeptides and the ability of these glycopeptides to form dimers may be important in their relative potency against even some GRE. The glycopeptide target structures, pentapeptides terminating in D-Ala-D-Ala, are ubiquitous among

most species of bacteria. The spectrum of activity of glycopeptides encompasses nearly all clinically important gram-positive cocci including staphylococci, streptococci, enterococci, gram-positive rods including *Bacillus* species, and most *Corynebacterium* and other diphtheroids. Glycopeptides also have activity against many gram-positive anaerobes, including *C. difficile*.

OVERVIEW OF POTENTIAL MECHANISMS OF RESISTANCE TO GLYCOPEPTIDES

Due to glycopeptides' unique mechanism of action and ability to interfere with multiple critical reactions in peptidoglycan synthesis, acquired resistance to glycopeptides was deemed unlikely. Resistance to glycopeptides could theoretically occur by a variety of mechanisms including degradation of the antibiotic, exclusion of the antibiotic from the target site, or modification or elimination of the D-Ala-D-Ala target. Thus far, no naturally occurring glycopeptide-modifying or -degrading enzymes have been characterized from resistant organisms or from glycopeptide-producing strains. Gram-negative bacteria are generally resistant to glycopeptides, due to the presence of the outer membrane, which effectively prevents drug access by the large glycopeptide molecules to the developing cell wall. Certain outer-membrane-deficient *Escherichia coli* mutants have increased susceptibility to vancomycin.

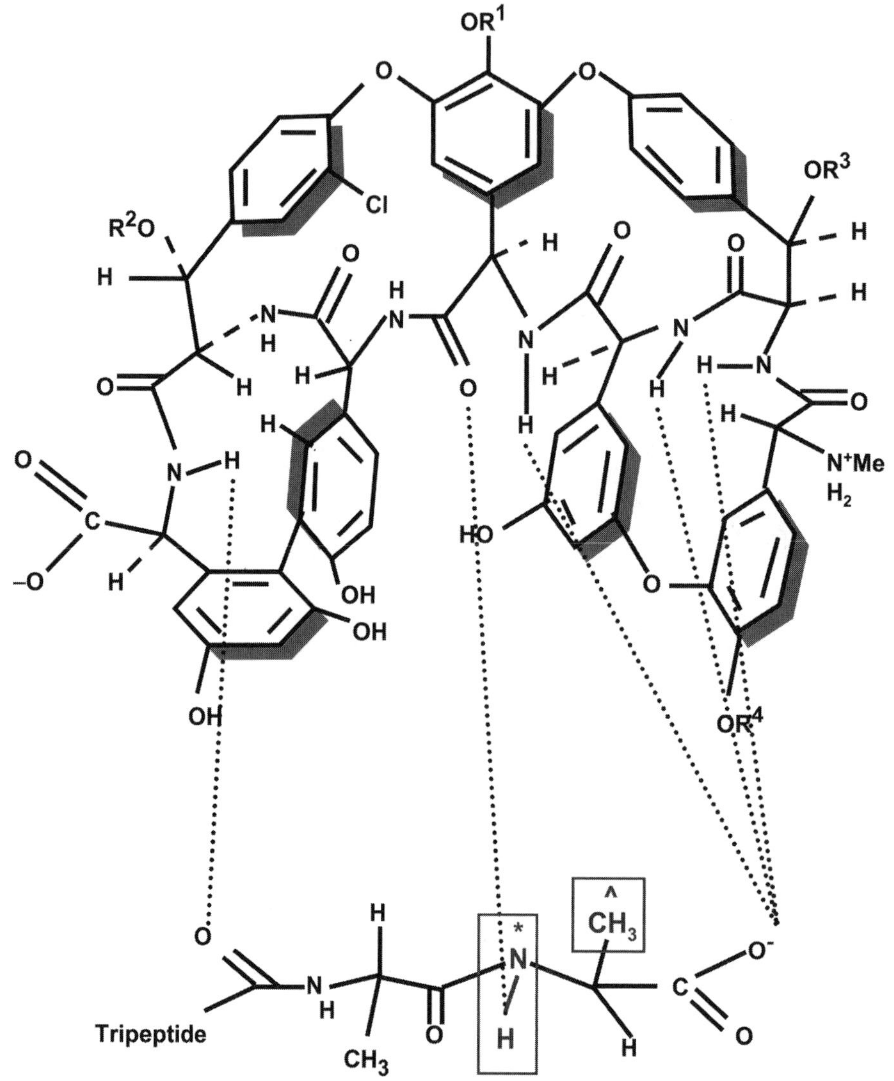

FIGURE 2 Structure of the glycopeptide; the peptidyl-D-Ala-D-Ala complex. Binding of the glycopeptide to the D-Ala-D-Ala residue on a peptidoglycan precursor involves five hydrogen bond interactions (indicated by the dashed lines). In D-Ala-D-Lac-terminating precursors, the NH of the amide group (*) is replaced by an oxygen in an ester linkage, eliminating the central hydrogen bond. In D-Ala-D-Ser-terminating precursors, the methyl side chain of the carboxy-terminal D-Ala (^) is replaced by a hydroxymethyl (CH$_2$OH) side chain.

Several gram-positive bacterial species associated with human colonization and infection are intrinsically highly resistant to vancomycin, including all species of *Leuconostoc*, *Pediococcus*, and *Erysipelothrix* and most species of *Lactobacillus*. Intrinsic resistance in these species is due to the presence of alternative cell wall synthesis intermediates, most commonly a depsipeptide that terminates in D-Ala-D-Lac, rather than the D-Ala-D-Ala-terminating pentapeptide (12). The conformational change created by the ester rather than the amide bond in the depsipeptide significantly alters its interaction within the groove in the aglycone structure of the glycopeptide and eliminates one hydrogen bond, resulting in binding affinity that is >1,000-fold less than that of the pentapeptide terminating in D-Ala-D-Ala (15, 69). Intrinsic low-level resistance can also be seen with alternative pentapeptides terminating in D-Ala-D-Ser, which are found in the uncommon motile enterococcal

species *Enterococcus gallinarum*, *Enterococcus casseliflavus*, and *Enterococcus flavescens* and in *Clostridium innocuum* (3, 24, 57). The terminal residue of the pentapeptide or depsipeptide is cleaved during subsequent cross-linking reactions; thus, there should be little consequence of employing this altered precursor structure on the ultimate composition and stability of the mature peptidoglycan, provided that the modified precursors can be efficiently processed by other enzymes in the chromosomal cell wall synthesis pathways. Some intrinsically glycopeptide-resistant species, including the intrinsically resistant enterococcal species above, possess multiple D,D-type ligases, allowing for the synthesis of either D-Ala-D-Ala or alternative precursors. The alternate ligase is typically under the control of a two-component regulatory element. Thus, although resistance may be intrinsic to these species, it is not necessarily constitutively expressed. The rationale for maintaining diverse cell wall

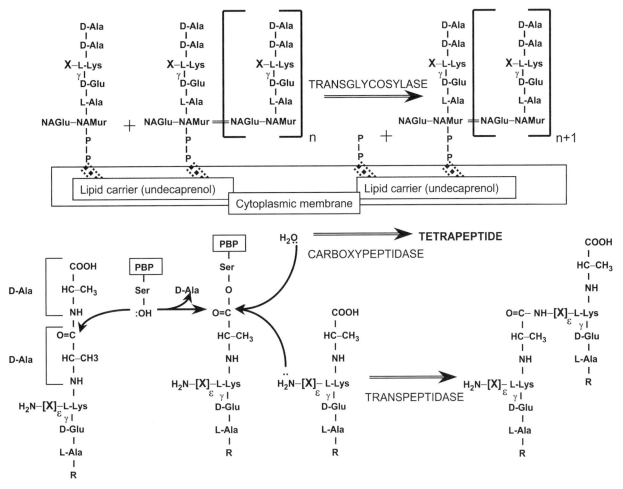

FIGURE 3 Transglycosylation and subsequent reactions in the cell wall synthesis pathway catalyzed by PBPs. Binding of glycopeptide to the terminal D-Ala-D-Ala stem inhibits both the transglycosylation reaction and the subsequent carboxypeptidase and transpeptidase steps. In the initial step of the carboxypeptidase or transpeptidase reactions, the terminal D-Ala is cleaved during the initial binding of the active serine site of the PBPs to form the acyl-enzyme complex. The same acyl-enzyme intermediate would be produced if the terminal D-Ala residue were replaced by either D-Lac or D-Ser. (Adapted from reference 9.)

synthesis pathways under natural conditions in the absence of selective antimicrobial pressure is uncertain. The specific signals required to induce expression of these alternate ligases are also unknown.

Acquired resistance to glycopeptides is defined as the development of resistance in strains of a genus previously considered to be uniformly susceptible; strains can acquire one or even multiple different resistance mechanisms. Clinically important acquired resistance occurs predominantly in enterococci and staphylococci, although the resistance elements found in *Enterococcus faecalis* and *Enterococcus faecium* have also been found in clinical isolates of other organisms. The mechanisms of acquired resistance to glycopeptides in enterococci have been well characterized and reflect changes in target analogous to those in intrinsically resistant organisms (9, 69). Acquired glycopeptide resistance in enterococci can result in decreased susceptibility to vancomycin alone or cross-resistance to both vancomycin and teicoplanin; resistant strains are thus generally referred to as GRE. GRE possess one of several usually inducible and potentially transferable gene clusters that not only are responsible for

synthesis of alternative cell wall precursors terminating in D-Ala-D-Lac or D-Ala-D-Ser but concurrently eliminate precursors terminating in D-Ala-D-Ala. The specifics of the various enterococcal resistance elements are described in detail below. Glycopeptide resistance in staphylococci can also occur by acquisition of the enterococcal *vanA* resistance cluster. However, the more commonly observed mechanism of decreased glycopeptide susceptibility in staphylococcal occurs without the introduction of new genetic material and is selectable in vitro (23, 61). Such resistance does not appear to be mediated by alteration in the D-Ala-D-Ala glycopeptide-binding site but instead reflects a variety of more general alterations in cell wall composition that limit access of the glycopeptide to nascent cell wall precursors.

The newer, semisynthetic glycopeptides oritavancin and telavancin have enhanced activity against some strains of enterococci with acquired glycopeptide resistance (68). However, resistance to oritavancin can also occur in both *vanA* and *vanB* strains in a single step through a variety of different mechanisms (6).

GLYCOPEPTIDE RESISTANCE IN ENTEROCOCCI: HISTORICAL AND EPIDEMIOLOGICAL FEATURES

The first well-characterized isolates of *E. faecalis* and *E. faecium* resistant to vancomycin were described in Europe and in the United States in the mid-1980s (44, 59). Subsequently, a large-scale outbreak in a renal failure unit at the Dulwich Infirmary in England in 1986 was reported (66). This prolonged outbreak consisted of numerous patients colonized or infected with isolates of *E. faecalis* and *E. faecium* for which vancomycin MICs were >64 μg/ml. Since these initial reports, GRE have emerged as major nosocomial pathogens throughout Europe and the United States, although with significant variation in the rates of isolation and pattern of spread in different geographic regions. Data from the Centers for Disease Control and Prevention's National Nosocomial Infections Surveillance Project have charted the emergence of GRE in the United States (Fig. 1) (33, 48). Vancomycin resistance in United States hospitals has steadily increased from 0.3% of nosocomial isolates in 1989 to 27.5% of intensive care unit (ICU) isolates in 2002, and the prevalence of nosocomial non-ICU isolates has also steadily increased. GRE have also been reported from throughout all of Europe. The prevalence rates in hospitalized European patients vary widely but in general are lower than those found in the United States (13). GRE have also been reported with various levels of prevalence from South America, Australia, Japan, China, and other parts of Asia.

Numerous studies have specifically addressed the epidemiology of GRE colonization and infection in the hospital setting (13, 48). The major reservoir for enterococci in humans is the lower gastrointestinal tract. Factors associated with spread of GRE are those permitting the establishment of resistant enterococci in the gastrointestinal tract of hospitalized patients, as well as those increasing the enterococcal burden and facilitating dissemination of resistant organisms through person-to-person or environmental contact. The predominant risk factor for acquisition of GRE is exposure to antimicrobial agents. Enterococci are intrinsically resistant to antimicrobials routinely used to treat hospitalized patients including cephalosporins, clindamycin, and other antianaerobic agents. Prior to the emergence of GRE, cephalosporin exposure was demonstrated to be a risk factor for the increased density of gastrointestinal colonization by enterococci and was implicated in the overall increasing incidence of enterococcal infections in hospitalized patients. In various studies, exposure to both parenteral and oral vancomycin, cephalosporins, metronidazole, and other antimicrobial agents, as well as overall intensity of antimicrobial exposure, has been linked to the acquisition of GRE (48). Other risk factors include general debility, longer durations of hospital stay, and exposure to particular hospital units such as ICUs, hematology-oncology units, liver transplant units, and renal failure units. These all reflect both increased risk of antibiotic exposure, as well as increased risk of exposure to GRE from sicker patients who are more heavily colonized and actively shedding organisms. Differences in the rates of colonization and infection in Europe and the United States may reflect differences in hospital antimicrobial usage patterns, especially glycopeptide usage (13).

Once colonization is established, GRE may be shed from the gastrointestinal tract for extended periods, even in the absence of continued antimicrobial pressure. Persistently colonized patients remain an ongoing reservoir for dissemination of GRE when transferred to other hospitals, intermediate-care facilities, or nursing homes. Prevalence of colonization may be up to 10 times higher than infection in the midst of an outbreak. Diarrhea and fecal incontinence markedly increase the risk of skin colonization and potential spread of organisms from persistently colonized patients (13, 48). Enterococci are durable and can survive up to a week on inanimate surfaces in the hospital environment; they have been recovered from countertops, rectal thermometer probes, bed rails, telephones, television remote controls, and other surfaces in patient rooms, as well as on the hands and stethoscopes of health care personnel.

VANCOMYCIN RESISTANCE PHENOTYPES AND GENOTYPES

The initial clinical isolates of GRE were classified by specific phenotypic characteristics into VanA, VanB, and VanC phenotypes, and these phenotypic classifications are still widely used (9). VanA strains predominantly include *E. faecium* and *E. faecalis*, but occasionally other enterococcal species such as *Enterococcus avium*, *Enterococcus durans*, *Enterococcus hirae*, and *Enterococcus raffinosus* and demonstrate inducible high-level resistance to both vancomycin and teicoplanin, with vancomycin MICs of ≥256 μg/ml. VanB strains are almost exclusively *E. faecium* and *E. faecalis*, demonstrate a range of vancomycin resistance levels from as low as 4 μg/ml to >1,000 μg/ml, and are generally susceptible to teicoplanin. VanC strains include the low-virulence, motile enterococcal species *E. casseliflavus*, *E. gallinarum*, and *E. flavescens*, and are characterized by intrinsic low levels of glycopeptide resistance (MICs, 2 to 32 μg/ml). The genetic and molecular bases for these different patterns of resistance are discussed below. However, optimal characterization of strains is now based on genotypic strategies such as DNA hybridization or gene amplification by PCR or similar methods for detecting specific glycopeptide resistance genes (28). Occasional isolates may demonstrate discordance between phenotypic and genotypic profiles, such as *vanB* genotype strains that are resistant to teicoplanin (26, 37). Genotypic methods have also aided in localization of resistance elements to the host chromosome or to plasmids and have facilitated the recognition of novel genotypes including *vanD*, *vanE*, and *vanG* (2, 25, 27). The current genotypic classification of vancomycin-resistant enterococci (VRE) is summarized in Table 1. New enterococcal resistance clusters are designated by their alternate ligase, e.g., *vanA*, *vanB*, *vanC*, and *vanD*; additional genes in the cluster are designated in relation to their first described homologs in the other enterococcal clusters, e.g., *vanH*, *vanH*$_B$, and *vanH*$_D$. There are also closely related variants within each genotype, e.g., *vanC1*, *vanC2*, and *vanC3*. The various proteins involved in glycopeptide resistance are shown in Table 2. As the genetic elements of the various vancomycin resistance clusters in enterococci and their homologs in other species have been characterized, the designations for newly identified resistance genes and their protein products have become increasingly confusing and will require more rigorous standardization.

NONHUMAN SOURCES FOR GRE

There remains considerable speculation on the initial sources of GRE isolates found in humans. Although GRE were initially isolated at similar times in Europe and the United States, studies demonstrate very different patterns of spread in the community setting (13, 48). Despite the rapid dissemination of both *vanA* and *vanB* genotypes of GRE in

TABLE 1 Genotypic and phenotypic characterization of GRE

Resistance genotype	Predominant phenotype[a]	Mode of expression	Predominant location	Transferable element(s)	Alternate precursor	Found in the following species:
vanA	Vanco ≥256 Teico ≥32	Inducible	Plasmid (chromosome[b])	Tn1546 and related	D-ALA-D-Lac	E. faecium, E. faecalis, E. hirae, E. avium, E. durans, and other species; S. aureus; B. circulans, O. turbata, A. haemolyticum
vanB	Vanco 4–1,000 Teico ≤1	Inducible	Chromosome (plasmid[b])	Tn1547, Tn1549, Tn5382	D-ALA-D-Lac	E. faecium, E. faecalis, S. bovis, S. agalactiae
vanC1, vanC2, vanC3	Vanco 2–32 Teico ≤1	Constitutive or inducible	Chromosome	?	D-ALA-D-Ser	E. gallinarum (vanC-1), E. casseliflavus (vanC-2), E. flavescens (vanC-3)
vanD	Vanco 32–256[c] Teico 4–32[c]	Constitutive	Chromosome	?	D-ALA-D-Lac	E. faecium, E. faecalis
vanE	Vanco = 16 Teico = 0.5	Inducible	Chromosome	?	D-ALA-D-Ser	E. faecalis
vanG	Vanco = 16 Teico = 0.5	Inducible	Chromosome	Yes	D-ALA-D-Ser	E. faecalis

[a]Expressed as MIC of vancomycin (Vanco) or teicoplanin (Teico) as micrograms per milliliter.
[b]Rare isolates described.
[c]MICs lower for E. faecalis strains.

United States hospitals, presumed to be fueled by high levels of antimicrobial usage, the rates of fecal colonization in healthy nonhospital volunteers have remained low. In Europe, rates of colonization in healthy volunteers are much higher, and the wide clonal diversity of these predominantly *vanA* genotype community isolates is also mirrored by the broad diversity of hospital isolates. In comparison, the overall clonal diversity of isolates from United States hospitals is lower than that in Europe. In Belgium, fecal carriage rates in the community of up to 28% were reported from healthy volunteers given an oral glycopeptide challenge, although most European studies demonstrate lower, but still significant, colonization rates of 2 to 5% (13). GRE were recovered from sewage in several European cities at times when hospital infections were still uncommon, and GRE have been recovered from other environmental sources including animal feces, especially those from poultry and pig farms. GRE have also been recovered from chicken carcasses and ground meat, suggesting a route of spread from these agricultural sources into the human food chain. GRE with indistinguishable pulsed-field gel electrophoresis patterns have been isolated from farm animals and humans, further supporting a role for these animal reservoirs as a source of human infection (1). The widespread use of the glycopeptide avoparcin as a food additive in European countries beginning in the 1970s that predated the marked increases in vancomycin usage in humans has been strongly implicated in the emergence of GRE in Europe. Avoparcin has also been widely used in other developed and developing countries, but it was never approved for use in the United States. Rates of GRE may be higher in countries with more-intensive avoparcin usage; individual farms where avoparcin was used have shown much higher rates of recovery of GRE from an-

imal feces (1). Other nonglycopeptide antimicrobials used in animal feeds such as tylosin and avilamycin may provide additional selective pressure for persistence of GRE, once established in animal populations (70). These observations together support a model of widespread multiclonal emergence and dissemination of predominantly *vanA* genotype enterococcal strains in animals, with subsequent colonization of humans.

LABORATORY DETECTION OF GRE

The unexpected arrival of glycopeptide resistance in enterococci quickly demonstrated the inadequacy of the then-standard testing methods for the detection of GRE (63). VanA-type strains generally are readily detected by all commercial testing methods, including disk diffusion, Etest, and automated broth microdilution systems. However, low- to intermediately resistant VanB-type strains were not well detected by either broth microdilution or diffusion methods. In 1992, CLSI (formerly NCCLS) in the United States revised the criteria for vancomycin testing of enterococci by disk diffusion, changing the requirement for a definition of fully susceptible to a zone size of ≥17 mm on Mueller-Hinton agar. There have also been improvements made in the newer versions of automated test panels to include longer incubation times to facilitate expression of resistance and enhance its detection. A commercial plate-based screening method using brain heart infusion agar with 6 μg of vancomycin/ml spotted with 10^6 CFU of inoculum is recommended as an adjunct method for confirming resistance in equivocal or critical isolates. A variety of vancomycin-containing media are available for screening for GRE colonization from complex laboratory specimens such as stool

TABLE 2 Van alphabet

Prototype enterococcal protein	Resistance operon	Protein function	Enterococcal homologs[a]	Closest nonenterococcal homolog(s)
Van A	VanA	D-Ala-D-Lac ligase	VanB, VanD	VanF$_{P.\ popilliae}$, VanA$_{S.\ coelicolor}$, DdlM$_{S.\ toyocaensis}$, DdlN$_{A\ orientalis}$, other glycopeptide producers
VanH	VanA	Lactate dehydrogenase	VanH$_B$, VanH$_D$	VanH$_{P.\ popilliae}$, VanH$_{S.\ coelicolor}$, VanH$_{S.\ toyocaensis}$, VanH$_{A\ orientalis}$, other glycopeptide producers
VanX	VanA	D,D-Dipeptidase	VanX$_B$, VanX$_D$	VanX$_{P.\ popilliae}$, VanX$_{S.\ coelicolor}$, VanX$_{S.\ toyocaensis}$, VanX$_{A\ orientalis}$, other glycopeptide producers
VanY	VanA	D,D-Carboxypeptidase	VanY$_B$, VanY$_D$, VanY$_G$	VanY$_{P.\ popilliae}$, carboxypeptidases of many *Bacillus* spp.
VanZ	VanA	Unknown (?) teicoplanin resistance	None	VanZ$_{P.\ popilliae}$
VanR	VanA	Response regulator	VanR$_B$, VanR$_C$, VanR$_D$, VanR$_E$, VanR$_G$	VanZ$_{P.\ popilliae}$, VanR$_{P.\ popilliae}$, VanR$_{S.\ coelicolor}$, VanR$_{S.\ toyocaensis}$, many other two-component response regulators
VanS	VanA	Glycopeptide sensor histidine kinase	VanS$_B$, VanS$_C$, VanS$_D$, VanS$_E$, VanS$_G$	VanS$_{P.\ popilliae}$, VanS$_{S.\ coelicolor}$, VanS$_{S.\ toyocaensis}$, many other two-component sensor histidine kinases
VanXY$_C$	VanC	Bifunctional dipeptidase-carboxypeptidase	VanXY$_E$, VanXY$_G$	None
VanT	VanC	Serine racemase	VanT$_E$, VanT$_G$	*C. innocuum* serine racemase
VanU	VanG	Component of VanG; regulator (?); transcriptional activator	None	*Clostridium thermocellum* protein of unknown function
VanW	VanB	Unknown	VanW$_G$	None

[a] In enterococcal glycopeptide resistance operons.

samples and rectal swabs. A number of genotypic methods for detection of *vanA*, *vanB*, and *vanC* genes have been described; several are now in commercial development, including PCR, hybridization probes, and cycled probe reactions. Recently, a multiplex PCR technique allowing detection of infrequent genotypes such as *vanD*, *vanE*, and *vanG* was described (28). Rapid laboratory identification of resistant strains is a critical component of public health recommendations for control of GRE such as those proposed by the Hospital Infection Control Practices Advisory Committee (17).

OVERVIEW OF GENETIC AND MOLECULAR BASIS OF ENTEROCOCCAL GLYCOPEPTIDE RESISTANCE

The basis of both intrinsic and acquired glycopeptide resistance in enterococci involves alteration of the composition of the terminal dipeptide in muramyl pentapeptide cell wall precursors, resulting in a structure with decreased binding affinity for glycopeptides (11, 12). This is achieved by replacing the terminal D-Ala by either D-Lac or D-Ser. Replacement of the D-Ala-D-Ala pentapeptide terminus by a D-Ala-D-Lac ester linkage changes the conformation of the ligand, eliminates the central hydrogen bond critical to glycopeptide binding, and results in >1,000-fold-lower binding affinity (69). Replacement of D-Ala-D-Ala by D-Ala-D-Ser reduces binding affinity sevenfold (11). The terminal D-Lac or D-Ser residue should be cleaved during subsequent cross-linking reactions and should not alter the final composition of mature peptidoglycan. The bacterial combining enzyme that ligates D-Ala-D-Ala to tripeptide to form pentapeptide precursors readily incorporates D-Ala-D-X structures other than D-Ala-D-Ala. However, depsipeptide precursors terminating in D-Lac may be processed differently than pentapeptide precursors by penicillin-binding proteins (PBPs) during subsequent transpeptidation reactions.

The central enzyme required for glycopeptide resistance in enterococci is thus a cellular ligase of altered specificity preferentially synthesizing either D-Ala-D-Lac (VanA, VanB, and VanD) or D-Ala-D-Ser (VanC, VanE, and VanG) rather than D-Ala-D-Ala. Resistance also requires a mechanism for synthesis of the D-X (D-Lac or D-Ser) residue; for D-Lac, this is a dehydrogenase such as VanH that generates D-Lac from pyruvate; for D-Ser, this is a D-serine racemase. Synthesis of exclusively D-Lac-terminating depsipeptide should yield a high-level glycopeptide-resistant phenotype analogous to the intrinsically highly glycopeptide-resistant organisms. However, the enterococcal resistance pathway also requires an additional mechanism for elimination of precursors terminating in D-Ala-D-Ala, since coproduction of both D-Ala- and D-Lac-terminating precursors will not yield a resistant phenotype (5). If sufficient D-Ala-terminating precursors generated by the constitutively expressed D-Ala-D-Ala ligase housekeeping gene are present, glycopeptides can still complex to these precursors while they are bound to the lipid carrier, sequestering the lipid carrier and blocking transport of additional precursors across the cellular membrane. The other essential function required for expression of resistance is an enzyme with D,D-dipeptidase activity such as VanX or VanXY$_C$, which hydrolyzes D-Ala-D-Ala, creating a futile cycle of synthesis and subsequent hydrolysis of D-Ala-D-Ala-dipeptide and markedly depleting the pool of D-Ala-terminating precursors (9, 58). Mature pentapeptide target for vancomycin can also be eliminated by D,D-carboxypeptidases (VanY) that cleave the terminal D-alanine from pentapeptide, although this activity is not essential. The requirements for a dehydrogenase or related activity, ligase activity, and D,D-dipeptidase activity are universal features of enterococcal glycopeptide resistance clusters, as well as glycopeptide resistance clusters of glycopeptide-producing organisms (9, 53).

RESISTANCE MEDIATED THROUGH PRECURSORS TERMINATING IN D-ALA-D-LAC

The *vanA* Resistance Cluster

The first enterococcal resistance cluster to be characterized in detail was the *vanA* operon, conveying high-level resistance to both vancomycin and teicoplanin. The *vanA* cluster was initially found as part of Tn*1546*, a nonconjugative transposon of the Tn3 superfamily of transposons (8). In most strains that have been studied in detail, the *vanA* gene cluster is incorporated in Tn*1546* or other highly conserved, related elements and is usually found on self-transferable conjugative plasmids, accounting for the appar-

ent ease of transfer of *vanA* resistance. The *vanA* element has also been found on a plasmid in association with an enterococcal pheromone response gene, providing another mechanism to facilitate horizontal dissemination. The *vanA*-containing element or portions of the *vanA* gene cluster can also be found on the bacterial chromosome and can be associated with other complex transposable elements. Tn*1546* (Fig. 4) is composed of nine open reading frames (ORFs) flanked by 38-bp imperfect inverted repeats. ORF1 and ORF2 encode a transposase and a resolvase required for mobilization of the transposon. The remaining seven ORFs encode proteins specifically related to expression of glycopeptide resistance. The structure and function of these genes, their protein products, and the resulting cell wall precursor structures have been elegantly studied in vitro in functional assays, as well as in laboratory enterococcal constructs carrying partial complements of the *vanA* gene cluster in *cis* or in *trans* (Fig. 5). The *vanHAX* cluster comprises a single transcriptional element with a single promoter and includes the three genes essential for expression of glycopeptide resistance. The VanH protein is a dehydrogenase catalyzing production of D-Lac from pyruvate and is structurally similar to other cellular dehydrogenases (15). The product of *vanA* is a D,D ligase that preferentially synthesizes D-Ala-D-Lac, although small amounts of D-Ala-D-Ala can also be produced. The VanA ligase has approximately 30 to 35% amino acid identity with other bacterial D-Ala-D-Ala (Ddl) ligases as well as with the D-Ala-D-Lac ligases of the intrinsically resistant organisms *Leuconostoc*, *Pediococcus*, and *Lactobacillus*, but *vanA* is phylogenetically distinct from both of these groups and more highly related to other glycopeptide resistance D-Ala-D-Lac ligases and the ligases of glycopeptide-producing members of the order *Actinomycetales* (30, 47). The *vanX* gene encodes a metallodipeptidase that preferentially hydrolyzes D-Ala-D-Ala.

Transcriptional activation of *vanHAX* is regulated by the *vanRS* two-component regulatory system. The *vanR* and *vanS* genes encode the signal sensor molecule VanS and the response regulator VanR, respectively (4, 7). VanS responds to subinhibitory concentrations of glycopeptides in the cellular medium by autophosphorylation, followed by phosphorylation of VanR and resulting in transcription of the glycopeptide resistance genes. The N-terminal portion of the VanS protein consists of a membrane-associated domain that appears to provide specificity for signal detection and demonstrates low homology to other known proteins. The C-terminal portion of the *vanS* gene encodes histidine kinase, dimerization-phosphoacceptor, and ATP-binding domains that are highly conserved in many histidine kinase sensor proteins and are necessary for transfer of a phosphate group to the response regulator VanR. In the phosphorylated state, VanR functions as a transcriptional activator at the

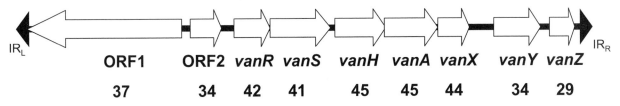

FIGURE 4 Structure of Tn*1546* (10,851 bp) carrying the *vanA* vancomycin resistance gene cluster. The nine ORFs are delineated by 38-bp imperfect inverted repeats and include genes with resolvase and transposase activity (ORF1 and -2), as well as genes involved in regulation (*vanR* and *vanS*), synthesis of D-Ala-D-Lac (*vanH* and *vanA*), and hydrolysis of D-Ala-D-Ala precursors (*vanX* and *vanY*). The percent G+C content of each gene is also shown.

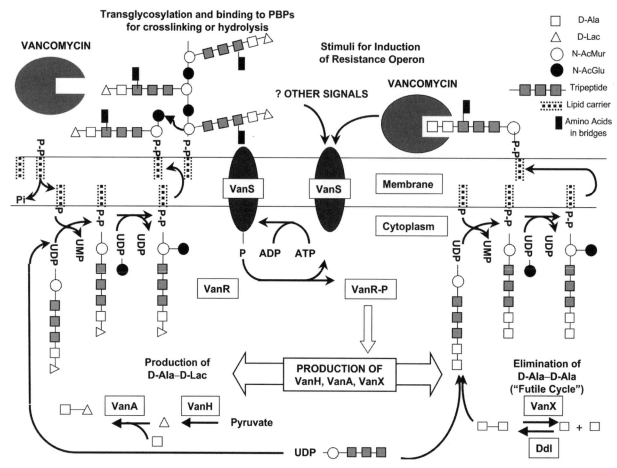

FIGURE 5 General scheme of peptidoglycan synthesis in GRE. The enterococcal Ddl ligases produce D-Ala-D-Ala residues that are incorporated into peptidoglycan precursors. D-Ala-D-Ala-terminating precursors interact with glycopeptides, preventing transglycosylation and leading to the accumulation of cytoplasmic precursors. The VanRS signal-transducing system responds directly or indirectly (via accumulation of cytoplasmic precursors) to the presence of glycopeptides leading to induction of transcription of the *vanHAX* operon in the cytoplasm. Induction of these genes results in production of D-Lac-containing precursors, as well as elimination of the pool of D-Ala-D-Ala precursors. (Adapted from reference 9.)

P_R and P_H promoters located upstream of *vanRS* and *vanHAX*, respectively (4). The specific molecular signals for induction of the VanA resistance cluster are unknown. Other transglycosylation inhibitors with sites of action distinct from glycopeptides including moenomycin also cause induction of VanA, suggesting that the stimulus for VanS may be inhibition of transglycosylation and accumulation of cell wall precursors, rather than the presence of the glycopeptide itself. Other structurally and mechanistically unrelated compounds may also induce expression of the *vanA* system. When inserted into *E. coli* strains, cross talk can occur between the *vanRS* system and other two-component systems like *phoA*, and there is evidence that similar cross talk may occur between *vanRS* and other enterococcal response regulators in vivo (62).

The remainder of the *vanA* complex includes two additional genes, *vanY* and *vanZ*. Unlike the *vanHAX* cluster, neither of these is obligatory for the expression of glycopeptide resistance. The VanY protein is a D,D-carboxypeptidase that cleaves terminal D-Ala from pentapeptide residues, and it can increase the level of glycopeptide resistance by further eliminating binding targets. Similar D,D-carboxypeptidases

linked to two-component regulators but not associated with other glycopeptide resistance genes have been found in the genomes of several glycopeptide-susceptible *Bacillus* species. The function of *vanZ* is poorly understood, but the product appears to be important in mediating increased resistance to teicoplanin (8). The *vanZ* gene also has homologs in some *Bacillus* species.

The general structural features of *vanA*-containing elements are highly conserved, but with the more widespread dissemination of *vanA* over the past decade numerous strains have been found that contain additional insertional elements such as IS*1216V* or IS*1251* incorporated within the *vanA* cluster (72). Most commonly, these insertions occur in intergenic regions, but occasional insertions are found that disrupt the function of the noncritical *vanY* and *vanZ* genes. Insertions into essential components of the *vanA* cluster result in a susceptible phenotype.

The G+C content of the glycopeptide resistance genes *vanRS* and *vanHAXYZ* ranges from 29 to 51% and differs from that of the transposase and resolvase genes (37 and 34%), as well as from that of the enterococcal chromosome (35%) (8). This suggests that these genes were initially acquired from one or multiple nonenterococcal sources and subsequently integrated into transposable elements prior to their acquisition by enterococci.

Transmission of the *vanA* Cluster

The *vanA* resistance cluster is capable of horizontal transfer to both enterococci and other species. Conjugative transfer of VanA-type plasmids between enterococcal strains of many species occurs readily in the laboratory, and there are numerous examples of clinical outbreaks of GRE involving multiple different strains and species of enterococci all harboring the same VanA-type resistance plasmids. This may also explain the clonal heterogeneity of VanA-type strains found among environmental isolates and population-based surveys of healthy outpatients. In early experiments using *vanA* strain BM4165, vancomycin resistance was transferable in vitro to *Streptococcus lactis*, *Streptococcus sanguinis*, *Streptococcus pyogenes*, and *Listeria monocytogenes* with various levels of expression in the recipients (9, 44). The *vanA* operon was also successfully transferred to *S. aureus* in vitro by an indirect selection method, with expression of high-level resistance in the recipient strain and no apparent deleterious effects on growth (49). Recently, three unique vancomycin-resistant *S. aureus* strains containing *vanA* have been described (19–21). Such strains are postulated to have evolved in vivo from the acquisition of *vanA* by *S. aureus* from GRE concurrently colonizing these patients. The *vanA* operon has also been found in other gram-positive species isolated from clinical material, including rare strains of *Oerskovia turbata*, *Arcanobacterium haemolyticum*, and *Bacillus circulans* (45). The structure of the *vanA*-containing elements and sequences of the *vanA* genes in the *Oerskovia* and *Arcanobacterium* strains are remarkably similar to those found in Tn1546, and their appearance following outbreaks of vancomycin resistance in enterococci suggests that these were acquired from enterococci rather than the converse. In the *B. circulans* isolate, the structure of the vancomycin resistance genes was highly similar to that found in Tn1546, but resistance was carried on the chromosome in a distinct genetic element (45).

Structure and Function of the *vanB* Resistance Cluster

vanB genotype strains are characterized by variable levels of resistance to vancomycin and, with rare exceptions, susceptibility to teicoplanin. *vanB* is found almost exclusively in *E. faecium* and *E. faecalis* but has also been reported in a *Streptococcus bovis* strain and a *Streptococcus agalactiae* strain. The structure of the *vanB* resistance cluster from the initial United States isolate, V583, is shown in Fig. 6 (31). The organization of the *vanB* cluster is similar to that of the *vanA* operon; however, there appears to be much more heterogeneity in both specific gene sequences within the cluster and in the overall organization of the resistance elements. The VanR$_B$ and VanS$_B$ regulatory elements of the *vanB* cluster have only low-level amino acid identity with VanR and VanS of the *vanA* cluster (34 and 23%) but share similar motifs that distinguish them as a two-component regulatory system homologous to VanRS. There is minimal relatedness between the N-terminal sensing portions of VanS and VanS$_B$, suggesting that they may respond to different signals. There are also functional differences between VanS and VanS$_B$. VanS$_B$ does not respond to the presence of teicoplanin, although teicoplanin-inducible derivatives can be selected with defined mutations in both the N-terminal sensing and the C-terminal kinase regions of VanS$_B$ (10). Analogous to the *vanA* system, VanR$_B$S$_B$ controls transcrip-

tional activation of the downstream *vanH$_B$*, *vanB*, and *vanX$_B$* genes required for expression of vancomycin resistance. These gene products are closely related to the D,D-dehydrogenase, D-Ala-D-Lac ligase, and D,D-dipeptidase of the *vanA* cluster, with amino acid identities of 67, 76, and 71%, respectively. Sequencing of *vanB* genes from numerous strains reveals heterogeneity of up to 5% between different isolates, which may have implications for sequence-based detection methods for *vanB* strains. In addition to the *vanH$_B$BX$_b$* element, the *vanB* cluster also contains the *vanY$_B$* gene encoding a D,D-carboxypeptidase, VanY$_B$, with low-level homology to VanY, as well as a novel gene *vanW* that is unique to the *vanB* cluster and of unknown function. There is no homolog of *vanZ* in the *vanB* cluster.

Higher levels of D,D-dipeptidase activity and the resulting higher ratio of D-Lac-containing depsipeptide precursors to pentapeptide precursors correlate with increased resistance to vancomycin in *vanB* strains, suggesting that *vanX$_B$* expression and the efficiency of elimination of D-Ala-D-Ala are an important variable in the phenotypic diversity of VanB-type strains (10). Resistance to teicoplanin can be selected in laboratory and is also found in clinical isolates via constitutive expression of the *vanH$_B$BX$_B$* complex, bypassing the need for a specific induction signal (37). In constitutive *vanB* strains, discrete mutations are found in the N-terminal kinase portion of the *vanS$_B$* gene, resulting in either altered or truncated VanS$_B$ (4, 10). Teicoplanin resistance can also occur through alterations in the induction specificity of the VanS$_B$ sensor, as described above.

The VanB cluster is found most commonly in the chromosome and can be transferred in absence of plasmids by conjugation in association with transfer of large blocks of DNA in sizes of 90 to 250 kb. The first *vanB*-transposable element to be described was Tn1547, found in *E. faecalis* strain V583. This large composite transposon contains the *vanB* gene cluster flanked by insertion sequences related to the IS256 family (54). Recently, VanB-type resistance in multiple *E. faecium* isolates from northeastern Ohio was found to be carried in a 27-kb element of a transposon-like structure, designated Tn5382, with proposed excisase and integrase genes and terminal regions with high similarity to the broad-host-range conjugative transposon Tn916 (16). Tn5382 is also capable of transfer between enterococci as part of an even larger genetic element that also includes an *E. faecium* PBP5 gene, which confers high-level resistance to ampicillin, perhaps accounting for the simultaneous appearance of both high-level ampicillin resistance and vancomycin resistance among many *E. faecium* strains. The *vanB* cluster has also occasionally been found in plasmid-borne transposons, such as Tn1549 (36).

The *vanD* Operon

VanD-type resistance in a small number of strains from Europe, Australia, and the Americas has recently been described (27). These strains are predominantly *E. faecium* and demonstrate constitutively expressed, moderate-level resistance to vancomycin for which MICs were 64 to 128 μg/ml as well as low-level resistance to teicoplanin for which MICs were 4 μg/ml. The few VanD *E. faecalis* strains reported show low-level vancomycin resistance (MIC, 16 μg/ml) but are susceptible to teicoplanin (27). The *vanD* clusters from several strains that have been analyzed in detail are structurally similar to the *vanA* and *vanB* clusters (27). The *vanD* operon includes the *vanD* gene encoding a D-Ala-D-Lac ligase with approximately 65 to 70% homology to VanA and VanB. The cluster also includes VanH$_D$, a homolog of VanH; VanX$_D$, a dipeptidase similar to VanX;

Enterococcal D-Ala—D-Lac Glycopeptide Resistance Operons

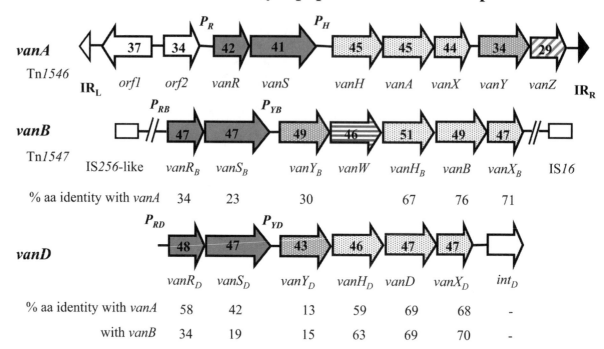

% aa identity with *vanA*	34	23	30		67	76	71	
% aa identity with *vanA*	58	42	13	59	69	68	-
with *vanB*	34	19	15	63	69	70	-

Enterococcal D-Ala—D-Ser Glycopeptide Resistance Operons

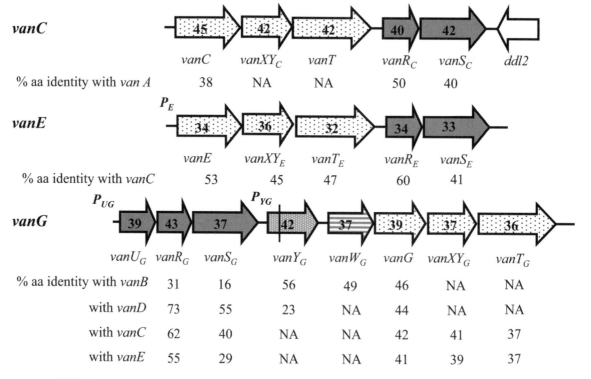

| % aa identity with *van A* | 38 | NA | NA | 50 | 42 | |

| % aa identity with *vanC* | 53 | 45 | 47 | 60 | 41 |

% aa identity with *vanB*	31	16	56	49	46	NA	NA
with *vanD*	73	55	23	NA	44	NA	NA
with *vanC*	62	40	NA	NA	42	41	37
with *vanE*	55	29	NA	NA	41	39	37

FIGURE 6 Comparison of the organization of the enterococcal D-Ala-D-Lac glycopeptide resistance operons *vanA*, *vanB*, and *vanD* and the enterococcal D-Ala-D-Ser glycopeptide resistance operons *vanC*, *vanE*, and *vanG*. The G+C content of the respective genes is also shown. The percent amino acid identities of the various genes to their homologs in the other enterococcal resistance clusters are listed underneath each cluster (Table 2).

and VanY$_D$, a carboxypeptidase of the PBP family that is only distantly related to the penicillin-insensitive, Zn^{2+}-dependent VanY and VanY$_B$. Upstream of these are genes encoding the VanR$_D$S$_D$ two-component response regulator analogous to VanRS. *vanD* has been identified exclusively in the chromosome thus far and has not been transferred, although integrase genes have been found within some *vanD* clusters. The constitutive expression of *vanD* requires mutations inactivating the host Ddl ligase gene, as well as additional mutations in the two-component regulatory element (27). At least eight different variants of the VanD ligase have been described.

RESISTANCE MEDIATED THROUGH PRECURSORS TERMINATING IN D-ALA-D-SER

Enterococcal *vanC* Genotypes

VanC-type glycopeptide resistance is confined to several species of enterococci that are occasionally isolated from clinical material, especially fecal specimens, and are only rarely associated with human infection. These include *E. gallinarum*, *E. casseliflavus*, and *E. flavescens*. Most strains of *E. gallinarum* and *E. casseliflavus* are motile, and some *E. casseliflavus* strains produce a yellow pigment. *E. flavescens* is genetically very closely related, if not identical, to *E. casseliflavus*. These organisms demonstrate low-level vancomycin resistance for which MICs were 2 to 32 μg/ml (57). VanC-type strains with very high vancomycin MICs have been described; however, such strains carried both *vanC* and *vanA* resistance genes. Resistance is an intrinsic feature of all these organisms but may be either constitutively or inducibly expressed. The mechanism of VanC-type resistance is due to production of pentapeptide terminating in D-Ser rather than D-Ala or D-Lac; vancomycin binds less avidly to D-Ala-D-Ser residues, but binding is not decreased to the extent occurring with precursors terminating in D-Lac (11). VanC-type strains remain susceptible to teicoplanin. The genes encoding the D-Ala-D-Ser ligases from *E. gallinarum* (*vanC-1*), *E. casseliflavus* (*vanC-2*), and *E. flavescens* (*vanC-3*) have been sequenced and functionally characterized in vitro (29, 57). The VanC proteins constitute a family of closely related D-Ala-D-Ser ligases with high identity to each other but low-level similarity to either D-Ala-D-Ala or D-Ala-D-Lac ligases (30). Additional D-Ala-D-Ala-specific ligases have also been identified in *E. gallinarum* and *E. casseliflavus*; the presence of such host ligases would be required for the inducible expression of vancomycin resistance in VanC-type strains. The complete *vanC-1* operon from *E. gallinarum* has been sequenced and characterized (3). In addition to the VanC-1 D-Ala-D-Ser ligase, this operon includes the *vanT* and *vanXY*$_C$ genes. *vanT* encodes a membrane-bound serine racemase that catalyzes conversion of L-Ser to D-Ser. The product of *vanXY*$_C$ is a unique bifunctional D,D-dipeptidase–D,D-carboxypeptidase that provides activities similar to both VanX and VanY from the *vanA* cluster by eliminating the pool of both D-Ala-D-Ala-dipeptide and -pentapeptide precursors (56). When cloned into susceptible *E. faecalis*, *vanC*, *vanXY*$_C$, and *vanT* are all required for expression of low-level vancomycin resistance. Located downstream from the clustered *vanC-1*, *vanT*, and *vanXY*$_C$ genes are reading frames whose predicted products have 50 and 42% amino acid identity with VanR and VanS and which constitute the presumed glycopeptide sensor and response regulator VanR$_C$S$_C$ for this resistance system (3).

The organization of the *vanC-2* cluster of *E. casseliflavus* is structurally very similar to that of *E. gallinarum*, and the corresponding genes have a high level of identity.

The *vanE* and *vanG* Resistance Operons

Two additional enterococcal resistance genotypes, *vanE* and *vanG*, have also recently been described (2, 25). Both are characterized by low-level vancomycin resistance mediated through the synthesis of precursors terminating in D-Ala-D-Ser analogous to the mechanism of VanC-type resistance (57). The chromosomally located five-gene *vanE* resistance cluster from *E. faecalis* BM4405 is organized very similarly to the *vanC* operon and includes genes encoding the VanE ligase, the VanXY$_E$ D,D-dipeptidase–D,D-carboxypeptidase, and the VanT$_E$ serine racemase, as well as the VanR$_E$S$_E$ two-component regulator. The chromosomal *vanG* resistance cluster, although also mediated through synthesis of D-Ala-D-Ser-containing precursors, has a more complex structure than the *vanC* or *vanE* clusters (25). The 3′ end of the chromosomally located *vanG* cluster encodes VanG, VanXY$_G$, and VanT$_G$, similar to VanC, VanXY$_C$, and VanT. The region upstream of this also encodes VanW$_G$ of unknown function and VanY$_G$, a D,D-carboxypeptidase. The 5′ end of the cluster includes a putative, constitutively expressed three-component regulatory element VanU$_G$R$_G$S$_G$. The entire chromosomal *vanG* cluster could be transferred to susceptible *E. faecalis* by conjugation as part of a much larger genetic element.

GLYCOPEPTIDE-DEPENDENT ENTEROCOCCI

An unusual vancomycin resistance phenotype described in several clinical GRE isolates is vancomycin dependence (32, 67). These strains grow poorly or not at all in the absence of vancomycin but grow well in vancomycin-supplemented media. Most vancomycin-dependent enterococci have been isolated serendipitously from clinical specimens. Initial growth of organisms can occur in blood culture bottles or plates due to carryover of low concentrations of vancomycin from clinical material, but organisms will then fail to grow on subculture material without vancomycin supplementation. Most strains have been isolated from patients receiving long-term vancomycin therapy and are of the *vanB* genotype, but a few vancomycin-dependent *vanA* strains have also been isolated. Such isolates remain more a laboratory curiosity than a major clinical problem. The mechanism of vancomycin dependence is due to loss of function of the host constitutive enterococcal Ddl ligase by mutations in *ddl* genes, resulting in either truncated protein or alteration of essential amino acids (10, 67). In a Ddl-inactivated glycopeptide-dependent strain, cell wall synthesis is dependent on activity of the D-Ala-D-Lac ligase and can occur only after induction and expression of the glycopeptide resistance cluster. Loss of the dependent phenotype can evolve either through further mutations leading to constitutive expression of the resistance operon or by mutations restoring D-Ala-D-Ala ligase activity (67). Mutations in Ddl similar to those found in glycopeptide-dependent strains have also been found in constitutively expressing *vanD* strains.

ORIGINS OF GLYCOPEPTIDE RESISTANCE GENES

The lack of any documented high-level glycopeptide resistance in enterococci prior to 1986 and the significant differences in G+C content of the *vanA* and *vanB* glycopeptide resistance clusters from other enterococcal housekeeping genes strongly imply the relatively recent acquisition of these genes by enterococci (9). However, the natural reservoir for these genes remains obscure. *Leuconostoc, Lactobacillus, Pediococcus,* and other intrinsically resistant organisms do contain D-Ala-D-Lac ligases functionally analogous to VanA and VanB, but their ligase genes are phylogenetically distinct from *vanA, vanB,* and the other acquired enterococcal resistance ligase genes (30). *Leuconostoc* and *Lactobacillus* strains also contain homologs of the VanH and VanH$_B$ lactate dehydrogenases, but these gene products are only distantly related (~30% amino acid homology) to the enterococcal dehydrogenases and the genes are not found clustered with the D-Ala-D-Lac ligase genes.

The genomes of the glycopeptide-producing organisms *Streptomyces toyocaenis, A. orientalis,* and other glycopeptide-producing members of the order *Actinomycetales* contain gene clusters with close homology to the dehydrogenase–ligase–D,D-dipeptidase complexes of the *vanA, vanB,* and *vanD* clusters (47, 53). The *ddlM* gene product of *S. toyocaenis* and *ddlN* gene product of *A. orientalis* demonstrate >60% amino acid sequence similarity with the deduced sequences of VanA, VanB, and VanD, but much lower homology with other D-Ala-D-Lac or D-Ala-D-Ala ligases. Similarly, the VanH homologs in these strains have 51 to 55% amino acid identity with the VanH and VanH$_B$ gene products, and the VanX homologs are 61 to 63% identical with the enterococcal VanX proteins. The organization of the *vanH-ddl-vanX* clusters in the glycopeptide producers mirrors that of the *vanA* and *vanB* clusters, with these three genes closely linked and under control of a single promoter (53). The *vanH-ddlM-vanX* element of *S. toyocaenis* is located within a large gene cluster encoding a glycopeptide synthesis pathway that also includes a postulated two-component regulator homologous to *vanRS*. No genes homologous to *vanY, vanW,* or *vanZ* have been found linked to the dehydrogenase–ligase–D,D-dipeptidase elements of these glycopeptide-producing organisms. It is unlikely that the glycopeptide-producing *Actinomycetes* are the immediate source of the resistance genes found in the genetically very different enterococci. The G+C content of the clusters from glycopeptide-producing organisms is much higher than that of the *vanA* or *vanB* clusters (>60% versus 40 to 44%), reflecting the higher G+C content of the *Actinomycetales* in general. However, these organisms are potential reservoirs for a large family of glycopeptide resistance genes that may have been incorporated by other soil organisms and ultimately found their way into enterococci.

Recently, a postulated complete glycopeptide resistance cluster has been found in the genome of the nonglycopeptide-producing *Actinomycetales* species *Streptomyces coelicolor* (41). This seven-gene cluster includes a transcript for proteins with >60% similarity to the products of *vanHAX,* as well as a two-component regulator with low-level homology (25 to 30%) to VanRS. The cluster also contains two other transcriptional elements encoding the novel proteins VanJ and VanK. VanK is a member of the Fem family and is required for expression of resistance (41). All the genes of the *S. coelicolor* cluster have G+C content of 60 to 70%, similar to that of other *Actinomycetales* genes but much higher than that of the acquired enterococcal resistance elements. Another homolog of the *vanA* and *vanB* cluster has been found in nearly all strains of the soil-dwelling biopesticide *Paenibacillus popilliae* (51). The organization and the G+C content of the inducible glycopeptide resistance cluster in *P. popilliae* parallel that of the *vanA* operon and are linked to a two-component regulator similar to that of *vanRS*. However, the *P. popilliae vanRS*-like element is also very similar to a family of two-component regulators linked to VanY-like carboxypeptidases found in several glycopeptide-susceptible *Bacillus* species. Thus, soil-dwelling *Bacillus*-like organisms such as *P. popilliae* may provide a unique evolutionary milieu where genes from glycopeptide-producing *Actinomycetales* can combine with the two-component regulators and associated carboxypeptidase genes of glycopeptide-susceptible *Bacillus* species as essential steps in the evolution of the enterococcal acquired glycopeptide resistance clusters.

GLYCOPEPTIDE RESISTANCE IN *S. AUREUS*

Recent reports of clinical isolates of *S. aureus* with reduced susceptibility to glycopeptides from patients failing glycopeptide therapy have brought renewed attention to several observations regarding glycopeptide therapy for staphylococcal infections (38, 64). Investigators have described in vitro selection of *S. aureus* strains with reduced vancomycin susceptibility after serial passage on vancomycin (61). Vancomycin MICs for laboratory-selected resistant derivatives could be as high as 100 μg/ml and could be generated in vitro from both methicillin-susceptible *S. aureus* (MSSA) and MRSA strains. The mechanisms of increased resistance of these mutants were at first poorly understood, and their clinical relevance was not widely appreciated, especially in the apparent absence of any documented clinical resistance during the first 30 years of vancomycin usage (Fig. 1). Clinicians have also noted that patients with *S. aureus* infections caused by apparently susceptible isolates have failed glycopeptide therapy. Since resistance may be only one of multiple potential explanations for treatment failure, such treatment failures were generally attributed to causes other than drug resistance, although many clinicians have considered vancomycin to be only a mediocre antistaphylococcal drug. In retrospect, some of these failures may have resulted from acquired glycopeptide resistance similar to that documented in recent cases.

Terminology and Laboratory Detection

In discussing glycopeptide resistance in *S. aureus,* it is important to clarify the various terminologies used to define these strains (23, 64). In the United States, where vancomycin is the only glycopeptide currently available, strains for which vancomycin MICs were >4 and <32 μg/ml are considered intermediately resistant to vancomycin or are referred to as VISA strains; strains for which the MICs were ≥32 μg/ml are vancomycin-resistant *S. aureus* (VRSA). The terms glycopeptide intermediately resistant *S. aureus* (GISA) and glycopeptide-resistant *S. aureus* are used to include isolates that are intermediately resistant or resistant to either teicoplanin or vancomycin. This term is more widely used in European countries, where both vancomycin and teicoplanin are available. Although strains with increased resistance to vancomycin demonstrate increased resistance to teicoplanin, the converse may not be true. To add to the confusion, there are also differences in the MIC breakpoints used to define susceptibility, intermediate-level resistance, and complete resistance that have been established by the various national laboratory standards committees (Table 3) (23). Another term used for describing decreased vancomycin susceptibility is vancomycin heteroresistance, in which only a small subpopulation of cells of a given strain demonstrates decreased susceptibility to vancomycin (40, 46). Vancomycin-heteroresistant *S. aureus* (VHRSA) strains

TABLE 3 Characteristics and methods used for detection of *Staphylococcus aureus* clinical isolates with reduced susceptibility to glycopeptides

Resistance phenotype	Acronym(s) and designation(s)[a]	Methods(s) of detection described	Vancomycin or teicoplanin MIC	Apparent prevalence (representative studies)	Postulated risk factor(s)
Heterogeneously teicoplanin resistant	h-GISA	Teicoplanin screen agar (various), high-inoculum Etest, population analysis profile	Teicoplanin, ≤4 μg/ml; subpopulations grow on >4 μg/ml	2–3% of MRSA, 1–2% of MSSA (France, Netherlands, Belgium)	? Glycopeptide exposure
Teicoplanin resistant	GISA (Europe)	Teicoplanin MIC (with/without prior screen as per h-GISA)	Teicoplanin, ≥8 μg/ml	2.2% (France)	Glycopeptide exposure
Heterogeneously vancomycin resistant	Hetero-VRSA, VHRSA, h-VISA	Teicoplanin or vancomycin screen agar, high-inoculum Etest, population analysis profile	≤4 μg/ml; subpopulations (1 in 10⁴–10⁵) grow at ≥8 μg/ml[b]	1.3% of MRSA (Japan), 0.5% of MRSA (France), 13% of patients failing vancomycin therapy (Australia)	Vancomycin exposure; ? failure of vancomycin therapy, high-inoculum infection
Intermediate resistance to vancomycin	VISA (United States), GISA (France), VRSA[a] (Britain, Sweden, Japan)	Broth dilution vancomycin MIC, ≥24-h incubation (with/without prior screening as per VHRSA)	Vancomycin MIC, 8–16 μg/ml[a]	~20 strains total (United States); multiple isolates or clusters reported from Asia, Europe, South America but <0.1–1% in several large surveys	Prolonged vancomycin exposure, hemodialysis, indwelling foreign material, or intravenous device
Vancomycin resistant	VRSA	Broth dilution vancomycin MIC with ≥24-h incubation, multiplex PCR	Vancomycin MIC, ≥32 μg/ml	Only 3 strains isolated to date, all from the United States	Colonization or infection with both *vanA* VRE and MRSA

[a]Strains for which MICs of 8 to 16 μg/ml are defined as intermediately resistant by the NCCLS and the Comité de l' Antibiogramme de la Société Française de Microbiologie but are defined as resistant by the British Society for Antimicrobial Chemotherapy and the Swedish Reference Group for Antibiotics.

may be much more common than intermediately resistant strains, but the optimal detection and even the significance of VHRSA remain in dispute. The current "gold standard" for detection of VHRSA is population analysis profiling, but this technique is not routinely performed, as it is technically difficult and time-consuming.

The method of testing, media used, and incubation time are also relevant for defining strains with decreased glycopeptide susceptibility (46, 64). Disk diffusion methods and rapid automated susceptibility panels will fail to detect many intermediately resistant strains and even fully vancomycin-resistant staphylococci. Optimal testing methods that have recently been proposed by the Centers for Disease Control and Prevention incorporate screening plates with brain heart infusion agar plus 6 μg of vancomycin/ml in addition to disk diffusion and automated MIC or use of a nonautomated MIC method, such as reference broth microdilution, agar dilution, or Etest that allows at least a full 24-h incubation time. The Société Française de Microbiologie has proposed the use of Mueller-Hinton screen agar with 5 μg of teicoplanin/ml as a screening method for GISA strains, and glycopeptide Etest with a large (2.0 McFarland standard) inoculum has also been used in some studies. The inoculum size is also important for detection of VHRSA strains. In practice, any *S. aureus* isolate for which a vancomycin MIC is ≥4 μg/ml by any method should have confirmatory testing done by a reference laboratory. Other screening media and methods are currently being evaluated.

EMERGENCE OF VISA CLINICAL ISOLATES

Teicoplanin-resistant *S. aureus* strains were shown to emerge in patients being treated with teicoplanin, but these observations initially caused little concern, as these teicoplanin-resistant strains remained susceptible to vancomycin (42). The first well-documented clinical VISA strain was isolated from Japan in 1996 (39). This strain, designated Mu50, was an MRSA, isolated from an immunocompromised child with a history of prior vancomycin exposure who was failing vancomycin therapy, for which the vancomycin MIC was 8 μg/ml. This MIC remained stable during serial passage. Subsequent epidemiological studies from Japan have described large numbers of VHRSA strains (40). These strains have reproducibly selectable subpopulations that will grow on brain heart infusion agar containing vancomycin at concentrations of 8 to 16 μg/ml. VHRSA strains made up to 20% of MRSA in the hospital where Mu50 was isolated and 1.3% of MRSA from throughout Japan, although no other true resistant strains similar to Mu50 were found. The selected derivatives of VHRSA strains with higher-level vancomycin resistance of ≥8 μg/ml behaved like strain Mu50 in vitro. The Japanese VISA and VHRSA strains were of the same clonal type. Subsequently, there were several well-documented reports of unrelated MRSA strains for which vancomycin MICs were 8 μg/ml isolated in the United States during 1997 and 1998 from Michigan, New Jersey, and New York (64). All were associated with clinical failure of vancomycin therapy and occurred in patients with extensive prior vancomycin exposure for treatment of MRSA; in addition, all patients were receiving dialysis. Extensive epidemiological investigations revealed no secondary cases of infection or colonization in association with these. A series of 19 cases of *S. aureus* with reduced vancomycin susceptibility (MIC ≥4 μg/ml), including four true VISA isolates from the United States, was recently reported; the only apparent risk factors were prior MRSA infection and antecedent vancomycin use (34). There have been multiple reports of similar strains in Europe, South Africa, South America, Korea, and Australia, including reports of possible clusters of strains. There have now been over 20 well-documented cases of VISA infection (MIC ≥8 μg/ml) worldwide, and a much higher number of infections due to VHRSA strains. However, several other large surveys of clinical *S. aureus* isolates from a number of countries have failed to reveal any VISA or VHRSA strains, including subsequent studies from Japan. Worldwide, VISA and VHRSA strains comprise multiple MLST types, suggesting that resistance has emerged concurrently in a variety of MRSA lineages. Most strains are *agr* type I or II.

MECHANISMS OF RESISTANCE IN GISA STRAINS

The mechanisms of intermediate resistance to glycopeptides in *S. aureus* are distinct from those elaborated by enterococci. Phenotypic features have been described that are common to many of the clinical and laboratory GISA strains. These include slower growth rates and heterogeneous morphology when plated on vancomycin-containing medium, a markedly thickened cell wall, altered susceptibility to lysostaphin, and in some (but not all) instances a paradoxically increased susceptibility to beta-lactams (14, 23). Cell wall thickening in particular in some studies correlates well with the level of vancomycin susceptibility (22). In Mu50, there is evidence of increased activation of cell wall synthesis including increased levels of PBP2 and PBP2' and increased incorporation of *N*-acetyl glucosamine, consistent with the observed thickening of the cell wall. Other observations include an increase in the synthesis of non-amidated muropeptides and reduced peptidoglycan cross-linking within the thickened cell wall (Table 4). Other clinical and laboratory isolates have demonstrated increased amounts of PBP2 or other PBPs (14). Glycopeptide-resistant strains also appear to have an enhanced ability to absorb vancomycin from antibiotic-containing medium, perhaps due to binding of glycopeptide to a surplus of terminal D-Ala-D-Ala residues in the loosely adherent, thickened cell wall. One proposed mechanism for resistance that incorporates many of these observations is that increased sponging of vancomycin into the thickened cell wall effectively excludes vancomycin from accessing the actively metabolizing peptidoglycan adjacent to the membrane (23). The increased levels of PBPs may also contribute to resistance by competing more aggressively with vancomycin for binding to D-Ala-D-Ala-terminal residues. Increased levels of PBP2' may be more important for increased resistance to teicoplanin than to vancomycin. Despite the intriguing phenomenological observations, the current understanding of specific mechanisms or genetic events responsible for glycopeptide resistance remains incomplete. However, based on observations on resistant strains that have been generated by serial passage in vitro, it is likely that resistance is related to a series of mutational and regulatory events rather than to a single genetic event (14, 23, 61). Strains with higher mutational frequencies may develop resistance more rapidly. In addition, although glycopeptide resistance can be selected in vitro in both MSSA and MRSA strains, nearly all clinical glycopeptide-resistant isolates thus far have been MRSA, suggesting that MRSA strains are more readily primed for evolution toward glycopeptide resistance.

TABLE 4 Mechanisms of resistance and clinical significance of VISA, VHRSA, and VRSA strains

Glycopeptide resistance phenotype	Proposed mechanisms of resistance or associations with resistance	Clinical relevance
VISA	Thickened cell wall, increased glycopeptide trapping	Failure of glycopeptide therapy, ? clonal dissemination of VISA strains
	Other phenotypic observations[a]: alteration in cell wall composition, increased precursors and cell wall turnover, decreased muropeptide amidation, decreased cross-linking, alterations in autolysis, hypersensitivity to beta-lactams, increased PBP2 and PBP2', decreased PBP4	
	Other genetic associations[a]: alterations in *sigB*, *ddh*, *tca*, *vraR*; increased purine biosynthesis; alterations in *agr*	
VHRSA	Same as for VISA strains	? Precursor for VISA, ? failure of glycopeptide therapy
VRSA	Acquisition of *vanA* cluster from *vanA* VRE	Failure of glycopeptide therapy, potential for horizontal transmission of resistance

[a]Described for some clinical or laboratory strains.

VANCOMYCIN RESISTANCE IN *S. AUREUS* MEDIATED BY THE *vanA* OPERON

The first fully vancomycin-resistant clinical *S. aureus* isolate was identified in the United States from a patient in Michigan in 2002 (19). Since then, there have been two additional VRSA isolates, one from Pennsylvania in 2002 and one from New York in 2004 (20, 65). MICs for these three strains ranged from 32 to >256 μg/ml when tested by broth microdilution reference methods, although initial reported MICs from clinical laboratories for all three strains were much lower, raising concern of whether there have been similar VRSA strains that went undetected. Intensive epidemiologic investigations found no evidence of secondary transmission of any of the VRSA strains. In all cases, the mechanism of resistance was acquisition of a plasmid-borne Tn*1546* element containing the *vanA* resistance cluster by an MRSA strain (52, 65, 71). All three VRSA isolates are clinically and genetically unrelated to each other and are presumed to have arisen from independent events. In all cases, the patients were colonized or infected with both MRSA and GRE prior to isolation of VRSA. The presumed vancomycin-resistant donor strains and the recipient MRSA strain from the Michigan isolate have been analyzed in detail. The Tn*1546* transposon on the broad-host-range conjugative plasmid pAM830 appears to have been the source for transposition of Tn*1546* into the staphylococcal conjugative plasmid pAM829. The *vanA*-containing plasmid in the highly resistant Michigan VRSA isolate remained fully stable in vitro, but the analogous *vanA*-containing plasmid in the Pennsylvania VRSA isolate was lost at high frequency, possibly accounting for the lower level of resistance in this strain (52). Although thus far transfer of *vanA* to MRSA has been a rare event, there will be increasing potential opportunities for this to occur as more and more hospitalized patients are simultaneously colonized with both GRE and MRSA. Guidelines for management of patients with suspected *S. aureus* isolates with reduced susceptibility to vancomycin have been established (18). The VRSA strains identified to date remain susceptible to the novel glycopeptides telavancin and oritavancin.

GLYCOPEPTIDE RESISTANCE IN CNS

Glycopeptide resistance is more common in CNS than in *S. aureus*, although these organisms do not present the same threat posed by glycopeptide-resistant *S. aureus*. Clinical isolates of glycopeptide-resistant CNS were first described in 1987, and there have been numerous reports of similar isolates since that time (60). However, over the past decade, the rate of CNS with decreased susceptibility to glycopeptides has remained quite low, at <1% in several recent surveys. It is not clear if this reflects a true lack of emergence of resistance or inadequacy of testing methods, as reliable methods for detection of intermediately resistant and fully resistant strains in CNS are even less well defined than for *S. aureus*. Most vancomycin resistance has been described for *S. haemolyticus* and *S. epidermidis*, with the highest prevalence in strains of *S. haemolyticus* (35). Vancomycin MICs for some *S. haemolyticus* strains have been as high as 128 μg/ml. Most resistant isolates are recovered from patients receiving prolonged or repeated courses of glycopeptide therapy. Such strains have been particularly problematic in renal failure and oncology units and in neonatal ICUs. Stepwise selection of vancomycin-resistant mutants of *S. haemolyticus* and *S. epidermidis* is readily ac-

complished in vitro. As with *S. aureus*, resistance to teicoplanin appears to develop more readily than resistance to vancomycin.

None of the glycopeptide-resistant CNS strains studied to date has contained a VanA-type or VanB-type gene cluster or other acquired glycopeptide resistance clusters. Despite the occasionally much higher levels of resistance found, resistance mechanisms in CNS appear to be generally similar to those in VISA strains. Some of the laboratory-derived and clinical glycopeptide-resistant CNS share phenotypic characteristics of intermediately resistant *S. aureus* strains, including thicker cell walls, smaller colony size with slower growth rates, increased levels of PBP2, and increased susceptibility to beta-lactams. Novel membrane proteins of various sizes, generally from 35 to 40 kDa, have been detected in some isolates, but the significance of this is unclear. Analysis of cell wall precursor pools in one glycopeptide-resistant *S. haemolyticus* isolate revealed small amounts of peptide terminating in D-Lac, although the low concentration of lactate-containing precursors seemed inadequate to account for the resistance observed in this strain. Alterations in the composition of the peptidoglycan peptide cross bridges have also been described for some resistant strains.

REFERENCES

1. Aarestrup, F. M., P. Ahrens, M. Madsen, L. V. Pallesen, R. L. Poulsen, and H. Westh. 1996. Glycopeptide susceptibility among Danish *Enterococcus faecium* and *Enterococcus faecalis* isolates of animal and human origin and PCR identification of genes within the VanA cluster. *Antimicrob. Agents Chemother.* **40:**1938–1940.
2. Abadia-Patino, L., P. Courvalin and B. Perichon. 2002. *vanE* gene cluster of vancomycin-resistant *Enterococcus faecalis* BM4405. *J. Bacteriol.* **184:**6457–6464.
3. Arias, C. A., P. Courvalin, and P. E. Reynolds. 2000. VanC cluster of vancomycin-resistant *Enterococcus gallinarum* BM4174. *Antimicrob. Agents Chemother.* **44:**1660–1666.
4. Arthur, M., F. Depardieu, and P. Courvalin. 1999. Regulated interactions between partner and non-partner sensors and response regulators that control glycopeptide resistance gene expression in enterococci. *Microbiology* **145:**1849–1858.
5. Arthur, M., F. Depardieu, P. Reynolds, and P. Courvalin. 1996. Quantitative analysis of the metabolism of soluble cytoplasmic peptidoglycan precursors of glycopeptide-resistant enterococci. *Mol. Microbiol.* **21:**33–44.
6. Arthur, M., F. Depardieu, P. Reynolds, and P. Courvalin. 1999. Moderate-level resistance to glycopeptide LY333328 mediated by genes of the *vanA* and *vanB* clusters in enterococci. *Antimicrob. Agents Chemother.* **43:**1875–1880.
7. Arthur, M., C. Molinas, and P. Courvalin. 1992. The VanS-VanR two-component regulatory system controls synthesis of depsipeptide peptidoglycan precursors in *Enterococcus faecium* BM4147. *J. Bacteriol.* **174:**2582–2591.
8. Arthur, M., C. Molinas, F. Depardieu, and P. Courvalin. 1993. Characterization of Tn*1546*, a Tn3-related transposon conferring glycopeptide resistance by synthesis of depsipeptide peptidoglycan precursors in *Enterococcus faecium* BM4147. *J. Bacteriol.* **175:**117–127.
9. Arthur, M., P. Reynolds, and P. Courvalin. 1996. Glycopeptide resistance in enterococci. *Trends Microbiol.* **4:**401–407.

10. **Baptista, M., F. Depardieu, P. E. Reynolds, P. Courvalin, and M. Arthur.** 1997. Mutations leading to increased levels of resistance to glycopeptide antibiotics in VanB-type enterococci. *Mol. Microbiol.* **25:**93–105.

11. **Billot-Klein, D., D. Blanot, L. Gutmann, and J. van Heijenoort.** 1994. Association constants for the binding of vancomycin and teicoplanin to N-acetyl-D-alanyl-D-alanine and N-acetyl-D-alanyl-D-serine. *Biochem. J.* **304:**1021–1022.

12. **Billot-Klein, D., L. Gutmann, S. Sable, E. Guittet, and J. van Heijenoort.** 1994. Modification of peptidoglycan precursors is a common feature of the low-level vancomycin-resistant VANB-type *Enterococcus* D366 and of the naturally glycopeptide-resistant species *Lactobacillus casei, Pediococcus pentosaceus, Leuconostoc mesenteroides,* and *Enterococcus gallinarum. J. Bacteriol.* **176:**2398–2405.

13. **Bonten, M. J., R. Willems, and R. A. Weinstein.** 2001. Vancomycin-resistant enterococci: why are they here, and where do they come from? *Lancet Infect. Dis.* **1:**314–325.

14. **Boyle-Vavra, S., H. Labischinski, C. C. Ebert, K. Ehlert, and R. S. Daum.** 2001. A spectrum of changes occurs in peptidoglycan composition of glycopeptide-intermediate clinical *Staphylococcus aureus* isolates. *Antimicrob. Agents Chemother.* **45:**280–287.

15. **Bugg, T. D. H., G. D. Wright, S. Dutka-Malen, M. Arthur, P. Courvalin, and C. T. Walsh.** 1991. Molecular basis for vancomycin resistance in *Enterococcus faecium* BM4147: biosynthesis of a depsipeptide peptidoglycan precursor by vancomycin resistance proteins VanH and VanA. *Biochemistry* **30:**10408–10415.

16. **Carias, L. L., S. D. Rudin, C. J. Donskey, and L. B. Rice.** 1998. Genetic linkage and co-transfer of a novel, *vanB*-containing transposon (Tn5382) and a low-affinity penicillin-binding protein 5 gene in a clinical vancomycin-resistant *Enterococcus faecium* isolate. *J. Bacteriol.* **180:**4426–4434.

17. **Centers for Disease Control and Prevention.** 1995. Recommendations for preventing the spread of vancomycin resistance: recommendations of the Hospital Infection Control Practices Advisory Committee (HICPAC). *Morb. Mortal. Wkly. Rep. Recomm. Rep.* **44**(RR-12):1–20.

18. **Centers for Disease Control and Prevention.** 1997. Interim guidelines for prevention and control of staphylococcal infection associated with reduced susceptibility to vancomycin. *Morb. Mortal. Wkly. Rep.* **46:**626–628, 635–636.

19. **Centers for Disease Control and Prevention.** 2002. *Staphylococcus aureus* resistant to vancomycin—United States, 2002. *Morb. Mortal. Wkly. Rep.* **51:**565–567.

20. **Centers for Disease Control and Prevention.** 2004. Vancomycin-resistant *Staphylococcus aureus*—New York, 2004. *Morb. Mortal. Wkly. Rep.* **53:**322–323.

21. **Chang, S., D. M. Sievert, J. C. Hageman, M. L. Boulton, F. C. Tenover, F. P. Downes, S. Shah, J. T. Rudrik, G. R. Pupp, W. J. Brown, D. Cardo, S. K. Fridkin, and Vancomycin-Resistant *Staphylococcus aureus* Investigative Team.** 2003. Infection with vancomycin-resistant *Staphylococcus aureus* containing the *vanA* resistance gene. *N. Engl. J. Med.* **348:**1342–1347.

22. **Cui, L., X. Ma, K. Sato, K., K. Okuma, F. C. Tenover, E. M. Mamizuka, C. G. Gemmell, M.-N. Kim, M.-C. Ploy, N. El Solh, V. Ferraz, and K. Hiramatsu.** 2003. Cell wall thickening is a common feature of vancomycin resistance in *Staphylococcus aureus. J. Clin. Microbiol.* **41:**5–14.

23. **Cui, L., and K. Hiramatsu.** 2003. Vancomycin-resistant *Staphylococcus aureus*, p. 187–212. *In* A. C. Fluit and F.-J. Schmitz (ed.), *MRSA: Current Perspectives*. Caister Academic Press, Norfolk, England.

24. **David, V., B. Bozdogan, J. L. Mainardi, R. Legrand, L. Gutmann, and R. Leclercq.** 2004. Mechanism of intrinsic resistance to vancomycin in *Clostridium innocuum* NCIB 10674. *J. Bacteriol.* **186:**3415–3422.

25. **Depardieu, F., M. G. Bonora, P. E. Reynolds, and P. Courvalin.** 2003. The *vanG* glycopeptide resistance operon from *Enterococcus faecalis* revisited. *Mol. Microbiol.* **50:**931–948.

26. **Depardieu, F., P. Courvalin, and T. Msadek.** 2003. A six amino acid deletion, partially overlapping the VanS$_B$ G2 ATP-binding motif, leads to constitutive glycopeptide resistance in VanB-type *Enterococcus faecium. Mol. Microbiol.* **50:**1069–1083.

27. **Depardieu, F., M. Kolbert, H. Pruul, J. Bell, and P. Courvalin.** 2004. VanD-type vancomycin-resistant *Enterococcus faecium* and *Enterococcus faecalis. Antimicrob. Agents Chemother.* **48:**3892–3904.

28. **Depardieu, F., B. Perichon, and P. Courvalin.** 2004. Detection of the *van* alphabet and identification of enterococci and staphylococci at the species level by multiplex PCR. *J. Clin. Microbiol.* **42:**5857–5860.

29. **Dutka-Malen, S., C. Molinas, M. Arthur, and P. Courvalin.** 1992. Sequence of the *vanC* gene of *Enterococcus gallinarum* BM4174 encoding a D-alanine:D-alanine ligase-related protein necessary for vancomycin resistance. *Gene* **112:**53–58.

30. **Evers, S., B. Casadewall, M. Charles, S. Dutka-Malen, M. Galimand, and P. Courvalin.** 1996. Evolution of structure and substrate specificity in D-alanine:D-alanine ligases and related enzymes. *J. Mol. Evol.* **42:**706–712.

31. **Evers, S., and P. Courvalin.** 1996. Regulation of VanB-type vancomycin resistance gene expression by the VanS$_B$-VanR$_B$ two-component regulatory system in *Enterococcus faecalis* V583. *J. Bacteriol.* **178:**1302–1309.

32. **Fraimow, H. S., D. L. Jungkind, D. W. Lander, D. R. Delso, and J. L. Dean.** 1994. Urinary tract infection with an *Enterococcus faecalis* isolate that requires vancomycin for growth. *Ann. Intern. Med.* **121:**22–26.

33. **Fridkin, S. K., and R. P. Gaynes.** 1999. Antimicrobial resistance in intensive care units. *Clin. Chest Med.* **20:**303–316.

34. **Fridkin, S. K., J. Hageman, L. K. McDougal, J. Mohammed, W. R. Jarvis, T. M. Perl, F. C. Tenover, and the Vancomycin-Intermediate *Staphylococcus aureus* Epidemiology Study Group.** 2003. Epidemiological and microbiological characterization of infections caused by *Staphylococcus aureus* with reduced susceptibility to vancomycin, United States, 1997–2001. *Clin. Infect. Dis.* **36:**429–439.

35. **Froggatt, J. W., J. L. Johnston, and G. L. Archer.** 1989. Antimicrobial resistance in nosocomial isolates of *Staphylococcus haemolyticus. Antimicrob. Agents Chemother.* **33:**460–466.

36. **Garnier, F., S. Taourit, P. Glaser, P. Courvalin, and M. Galimand.** 2000. Characterization of transposon Tn1549 conferring VanB type resistance in *Enterococcus* spp. *Microbiology* **146:**1481–1489.

37. **Hayden, M. K., G. M. Trenholme, J. E. Schultz, and D. F. Sahm.** 1993. In vivo development of teicoplanin resistance in a VanB *Enterococcus faecium* isolate. *J. Infect. Dis.* **167:**1224–1227.

38. **Hiramatsu, K.** 1998. The emergence of *Staphylococcus aureus* with reduced susceptibility to vancomycin in Japan. *Am. J. Med.* **104:**7S–10S.

39. **Hiramatsu, K., H. Hanaki, T. Ino, K. Yabuta, T. Oguri, and F. C. Tenover.** 1997. Methicillin-resistant *Staph-*

ylococcus aureus clinical strain with reduced vancomycin susceptibility. *J. Antimicrob. Chemother.* **40:**135–136.

40. **Hiramatsu, K., N. Aritaka, H. Hanaki, S. Kawasaki, Y. Hosoda, S. Hori, Y. Fuckuchi, and I. Kobayashi.** 1997. Vancomycin-resistant *Staphylococcus aureus*: dissemination of heterogeneously resistant strains in a Japanese hospital. *Lancet* **350:**1670–1673.

41. **Hong, H. E., M. I. Hutchings, J. M. Neu, G. D. Wright, M. S. B. Paget, and M. Buttner.** 1994. Characterization of an inducible vancomycin resistance system in *Streptomyces coelicolor* reveals a novel gene (*vanK*) required for drug resistance. *Mol. Microbiol.* **52:**1107–1121.

42. **Kaatz, G. W., S. M. Seo, N. J. Dorman, and S. A. Lerner.** 1990. Emergence of teicoplanin resistance during therapy of *Staphylococcus aureus* endocarditis. *J. Infect. Dis.* **162:**103–108.

43. **Kirst, H. A., D. G. Thompson, and T. I. Nicas.** 1998. Historical yearly usage of vancomycin. *Antimicrob. Agents Chemother.* **42:**1303–1304.

44. **Leclercq, R., E. Derlot, J. Duval, and P. Courvalin.** 1988. Plasmid-mediated vancomycin and teicoplanin resistance in *Enterococcus faecium. N. Engl. J. Med.* **319:**157–161.

45. **Ligozzi, M., G. Lo Cascio, and R. Fontana.** 1998. *vanA* gene cluster in a vancomycin resistant clinical isolate of *Bacillus circulans. Antimicrob. Agents Chemother.* **42:**2055-2059.

46. **Liu, C., and H. F. Chambers.** 2003. *Staphylococcus aureus* with heterogeneous resistance to vancomycin: epidemiology, clinical significance and critical assessment of diagnostic methods. *Antimicrob. Agents Chemother.* **47:**3040–3045.

47. **Marshall, C. G., G. Broadhead, B. K. Leskiw, and G. D. Wright.** 1997. D-Ala-D-Ala ligases from glycopeptide antibiotic-producing organisms are highly homologous to the enterococcal vancomycin-resistance ligases VanA and VanB. *Proc. Natl. Acad. Sci. USA* **94:**6480–6483.

48. **Martone, W. J.** 1998. Spread of vancomycin-resistant enterococci: Why did it happen in the United States? *Infect. Control Hosp. Epidemiol.* **19:**539–545.

49. **Noble, W. C., Z. Virani, and R. Cree.** 1992. Co-transfer of vancomycin and other resistance genes from *Enterococcus faecalis* NCTC 12201 to *Staphylococcus aureus.* FEMS *Microbiol. Lett.* **93:**195–198.

50. **Panlilio, A. L., D. H. Culver, R. P. Gaynes, S. Banerjee, T. S. Henderson, J. S. Tolson, and W. J. Martone.** 1992. Methicillin-resistant *Staphylococcus aureus* in U.S. hospitals, 1975–1991. *Infect. Control Hosp. Epidemiol.* **13:**582–586.

51. **Patel, R., K. Piper, F. R. Cockerill III, J. M. Steckelberg, and A. A. Yousten.** 2000. The biopesticide *Paenibacillus popilliae* has a vancomycin resistance gene cluster homologous to the enterococcal VanA vancomycin resistance gene cluster. *Antimicrob. Agents Chemother.* **44:**705–709.

52. **Perichon, B., and P. Courvalin.** 2004. Heterologous expression of the enterococcal *vanA* operon in methicillin-resistant *Staphylococcus aureus. Antimicrob. Agents Chemother.* **48:**4281–4285.

53. **Pootoolal, J., M. G. Thomas, C. G. Marshall, J. M. Neu, B. K. Hubbard, C. T. Walsh, and G. D. Wright.** 2002. Assembling the glycopeptide antibiotic scaffold: the biosynthesis of A47934 from *Streptomyces toyocaensis* NRRL15009. *Proc. Natl. Acad. Sci. USA* **99:**8962–8967.

54. **Quintiliani, R. J., and P. Courvalin.** 1996. Characterization of Tn*1547*, a composite transposon flanked by the

IS*16* and IS*256*-like elements that confers vancomycin resistance in *Enterococcus faecalis* BM428. *Gene* **172:**1–8.

55. **Reynolds, P. E.** 1989. Structure, biochemistry, and mechanism of action of glycopeptide antibiotics. *Eur. J. Clin. Microbiol. Infect. Dis.* **8:**943–950.

56. **Reynolds, P. E., C. A. Arias, and P. Courvalin.** 1999. Gene *vanXY*$_C$ encodes D,D-dipeptidase (VanX) and D,D-carboxypeptidase (VanY) activities in vancomycin-resistant *Enterococcus gallinarum* BM4174. *Mol. Microbiol.* **34:**341–349.

57. **Reynolds, P. E., and P. Courvalin.** 2005. Vancomycin resistance in enterococci due to synthesis of precursors terminating in D-alanyl-D-serine. *Antimicrob. Agents Chemother.* **49:**21–25.

58. **Reynolds, P. E., F. Depardieu, S. Dutka-Malen, M. Arthur, and P. Courvalin.** 1994. Glycopeptide resistance mediated by enterococcal transposon Tn*1546* requires production of VanX for hydrolysis of D-alanyl-D-alanine. *Mol. Microbiol.* **13:**1065–1070.

59. **Sahm, D., J. Kissinger, M. S. Gilmore, P. R. Murray, R. Mulder, J. Solliday, and B. Clarke.** 1989. In vitro susceptibility studies of vancomycin-resistant *Enterococcus faecalis. Antimicrob. Agents Chemother.* **33:**1588–1591.

60. **Schwabe, R. S., J. T. Stapleton, and P. H. Gilligan.** 1987. Emergence of vancomycin resistance in coagulase-negative staphylococci. *N. Engl. J. Med.* **316:**927–931.

61. **Sieradzki, K., and A. Tomasz.** 1997. Inhibition of cell wall turnover and autolysis by vancomycin in a highly vancomycin-resistant mutant of *Staphylococcus aureus. J. Bacteriol.* **179:**2557–2566.

62. **Silva, J. C., A. Haldimann, M. K. Prahalad, C. T. Walsh, and B. L. Wanner.** 1998. In vivo characterization of the type A and B vancomycin resistant enterococci (VRE) VanRS two-component systems in *Escherichia coli*: a nonpathogenic model for studying the VRE signal transduction pathways. *Proc. Natl. Acad. Sci. USA* **95:**11951–11956.

63. **Swenson, J. M., N. C. Clark, M. J. Ferraro, D. F. Sahm, G. Doern, M. A. Pfaller, L. B. Reller, M. P. Weinstein, R. J. Zabransky, and F. C. Tenover.** 1994. Development of a standardized screening method for detection of vancomycin-resistant enterococci. *J. Clin. Microbiol.* **32:**1700–1704.

64. **Tenover, F. C., M. V. Lancaster, B. C. Hill, C. D. Steward, S. A. Stocker, G. A. Hancock, C. M. O'Hara, S. K. McAllister, N. C. Clark, and K. Hiramatsu.** 1998. Characterization of staphylococci with reduced susceptibilities to vancomycin and other glycopeptides. *J. Clin. Microbiol.* **36:**1020–1027.

65. **Tenover, F. C., L. M. Weigel, P. C. Appelbaum, L. K. McDougal, J. Chaitram, S. McAllister, N. Clark, G. Killgore, C. M. O'Hara, L. Jevitt, J. B. Patel, and B. Bozdogan.** 2004. Vancomycin-resistant *Staphylococcus aureus* isolate from a patient in Pennsylvania. *Antimicrob. Agents Chemother.* **48:**275–280.

66. **Uttley, A. H., C. H. Collins, J. Naidoo, and R. C. George.** 1988. Vancomycin-resistant enterococci. *Lancet* **i:**57–58.

67. **Van Bambeke, F., M. Chauvel, P. E. Reynolds, H. S. Fraimow, and P. Courvalin.** 1999. Vancomycin-dependent *Enterococcus faecalis* clinical isolates and revertant mutants. *Antimicrob. Agents Chemother.* **43:**41–47.

68. **Van Bambeke, F., Y. Van Laethem, P. Courvalin, and P. M. Tulkens.** 2004. Glycopeptide antibiotics: from conventional molecules to new derivatives. *Drugs* **64:**913–936.

69. Walsh, C. T., S. L. Fisher, L.-S. Park, M. Prahalad, and Z. Wu. 1996. Bacterial resistance to vancomycin: five genes and one missing hydrogen bond tell the story. *Chem. Biol.* **3:**21–28.

70. Wegener, H. C. 2003. Antibiotics in animal feed and their role in resistance development. *Curr. Opin. Microbiol.* **6:**439–445.

71. Weigel, L. M., D. B. Clewell, S. R. Gill, N. C. Clark, L. K. McDougal, S. E. Flannagan, J. F. Kolonay, J. Shetty, G. E. Killgore, and F. C. Tenover. 2003. Genetic analysis of a high-level vancomycin-resistant isolate of *Staphylococcus aureus*. *Science* **302:**1569–1571.

72. Woodford, N. 2001. Epidemiology of the genetic elements responsible for acquired glycopeptide resistance in enterococci. *Microb. Drug Resist.* **7:**229–236.

73. Yao, R. C., and L. W. Crandall. 1994. Glycopeptides: classification, occurrence and discovery, p. 1–27. *In* R. Nagarajan (ed.), *Glycopeptide Antibiotics*. Marcel Dekker, New York, N.Y.

Tetracycline Resistance Determinants in Gram-Positive Bacteria

LAURA M. McMURRY AND STUART B. LEVY

64

Tetracyclines are natural products of *Streptomyces* soil organisms and were discovered in the late 1940s. After the discovery of penicillin, the tetracyclines and their semisynthetic derivatives formed a second wave of antibiotics because of their broad spectrum of activity and relatively low toxicity and cost. By 1971, tetracyclines represented 30% of all antibiotics consumed for human use (106). The emergence of tetracycline resistance in the mid-1950s, initially in gram-negative bacteria and then in gram-positive bacteria, resulted in the declining usefulness of tetracyclines (83). Today, the tetracyclines are primarily used to treat Lyme disease, acne, rickettsia, chlamydia, and periodontal disease. They can sometimes be used against other infective strains whose susceptibility has been verified.

Because of the worldwide problem with resistance even to newer antibiotics, there has been recent interest, and some success, in developing new derivatives of the tetracyclines that are effective against resistant strains. These include the glycylcyclines (168) and the aminomethylcyclines (12). Earlier reviews of tetracyclines and tetracycline resistance mechanisms in general include references 14, 18, 30, 31, 62, 85, 106, 133, 145, and 153.

(Note that in this chapter, we sometimes cite a reference which is not itself a primary reference but in which a primary reference is discussed and cited; in these cases, the words "see reference" are used.)

CHEMICAL PROPERTIES OF TETRACYCLINES

Tetracycline and most of its analogs have an aromatic D ring and three ionizable groups with pK_a values as shown in Fig. 1 (94). In an aqueous environment at pH 7.4, most of the molecules have approximately one negative and one positive charge on the A ring and an average partial negative charge on the oxygens near atoms 11 and 12. This is the form in which the drug is crystallized from water (156). When it is crystallized under conditions in which water is carefully excluded, the molecule has no charge at any loca-

tion but instead has intramolecular hydrogen bonding (156). This may be the form that crosses biological membranes. The uncharged form has been calculated from the microscopic ionization constants for each group to be a surprising 7% of the total drug at pH 7.4 in water (118).

Groups at positions 5, 6, 7, 8, and 9 (Fig. 1) can be modified with some retention of antibacterial activity, while others generally cannot be modified (14, 18, 106). The exception is that removal of the group at position 4 lessens activity against gram-negative organisms but has no effect against gram-positive ones (14). The phenol-diketone region (positions 10, 11, and 12) is responsible for chelation of divalent cations (18). Tetracycline chelates cations such as Fe^{3+}, Co^{2+}, Fe^{2+}, Mn^{2+}, Mg^{2+}, and Ca^{2+} (in order of decreasing affinity) (94), and chelation plays an important role in tetracycline function. Activity against intact cells also requires the tricarbonylmethane group (positions 1, 2, and 3) of the A ring (18).

ENTRY OF TETRACYCLINES INTO MICROBIAL CELLS

Gram-positive organisms are generally more susceptible to tetracyclines than are gram-negative ones, traditionally attributed to a more rapid entry of the drug in gram-positive organisms, which lack an outer membrane. An energy-dependent tetracycline accumulation by bacterial cells derives from simple equilibration of the multiprotonated acid-base tetracycline with the transmembrane pH gradient portion of the proton motive force across the cytoplasmic membrane, as has been shown both in gram-positive *Enterococcus faecalis* (109) and in gram-negative *Escherichia coli* (see reference 118); no protein transporter is required. In a cell in which the internal pH is 7.8 and the external pH is 6.1, it has been calculated that tetracycline should be concentrated two-fold by the proton motive force (118), although the energy-dependent concentration measured in gram-negative cells appears to be greater (97, 178; see also reference 169).

Gram-Positive Pathogens, 2nd edition, edited by Vincent A. Fischetti et al.
© 2006 ASM Press, Washington, D.C.

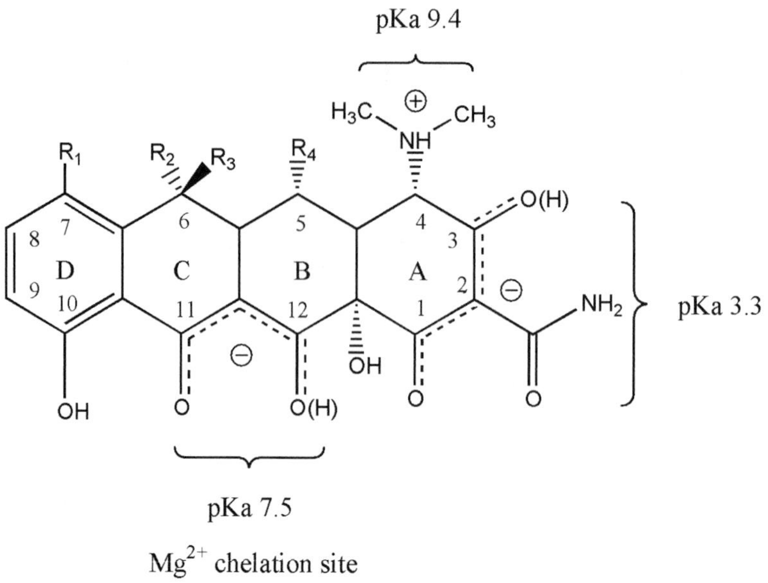

pKa 9.4

Mg^{2+} chelation site

	R^1	R^2	R^3	R^4
chlortetracycline	Cl	CH$_3$	OH	H
doxycycline	H	CH$_3$	H	OH
minocycline	NH(CH$_3$)$_2$	H	H	H
oxytetracycline	H	CH$_3$	OH	OH
tetracycline	H	CH$_3$	OH	H

FIGURE 1 Structure of tetracycline and some of its clinically used analogs.

INHIBITION OF PROTEIN SYNTHESIS BY TETRACYCLINES

Tetracyclines are bacteriostatic antibiotics that act by inhibiting protein synthesis. These drugs inhibit the binding of aminoacyl-tRNA to the ribosome (158) in the acceptor (A) site during elongation (52). Older studies have been summarized (30, 126). Most of the work involving protein synthesis has been done with *E. coli*, but the same picture was seen with the gram-positive *Bacillus megaterium* (35). The elongation process in protein synthesis has been previously reviewed (88, 182).

Tetracycline binds to ribosomes at a single high-affinity site at a K_d value of 1 to 20 μM (37, 51, 170) and at numerous sites with lower affinity (see reference 145). UV light-stimulated covalent photoincorporation of [^{3}H]tetracycline into ribosomes occurs mostly at protein S7 from the 30S subunit when careful correction for artifactual binding of tetracycline photolysis products is made (51). That the incorporation is physiologically meaningful seems likely, since the labeling is both stereospecific and saturable, with a K_d value similar to that of the high-affinity binding (51). 16S rRNA is also photolabeled by tetracycline (121). An en-

hancement of the chemical reactivity of certain residues by tetracycline suggests a conformational change in the rRNA (see reference 145). Ribosomes reconstituted so as to lack the neighboring (15, 19) S7 or S14 proteins lose the tight binding of [^{3}H]tetracycline, as assayed by filtration and sucrose density centrifugation (19).

More recently, two independent crystal structures of the *Thermus thermophilus* 30S ribosomal subunit complexed with tetracycline-Mg^{2+} showed the major binding site to be on the 16S rRNA at helix 34 and the stem-loop of helix 31 (16, 127). These regions, along with protein S7, are adjacent to the ribosomal A site. Secondary binding sites include one at helix 29. These locations were also found in *E. coli* ribosomes by using tetracycline-Fe^{2+}-mediated cleavage of rRNA (9). The cross-linking in *E. coli* ribosomes described in reference 121 is also consistent with these findings (9).

Tetracycline-resistant clinical strains have been found with 16S rRNA mutations, in helix 34 of *Propionibacterium acnes* (135), and in the stem-loop of helix 31 in *Helicobacter pylori* (see "Tetracycline Resistance by Mutated rRNA," below) (171). Plasmid-encoded, laboratory-mutated *E. coli* rRNA genes expressed in an *E. coli* strain deleted for all seven copies of the rRNA genes confirmed that a mutation of *E. coli* 16S RNA in helix 34 or in the helix 31 loop increases the resistance to tetracycline severalfold (9).

In summary, tetracycline (likely as a tetracycline-divalent cation complex) probably sterically blocks proper accommodation of the amino-acyl tRNA complex at the A site of the prokaryotic ribosome via binding in a pocket formed by helix 34 and the stem-loop of helix 31 of the 16S rRNA, thereby inhibiting protein synthesis and so bacterial growth.

TETRACYCLINE RESISTANCE MECHANISMS AND RESISTANCE DETERMINANTS: OVERVIEW

There are three different tetracycline resistance mechanisms of clinical importance found among gram-positive organisms. They are active efflux, ribosomal protection, and mutated rRNA (133, 135, 145, 153, 171). Degradation of the drug appears to be relatively unimportant.

We define a tetracycline resistance determinant as a contiguous genetic unit encoding all the involved genes and specifying only resistance to tetracyclines. Mutations in genes encoding rRNA are not counted as resistance determinants. Two determinants were historically classified as different if they showed no DNA hybridization with each other at high stringency (105). With DNA sequencing now routine, a >20% difference in structural protein sequence identity has typically defined a new determinant (87). Because some determinants have more than one gene, it was decided to make a distinction between the name of a gene and the class name assigned to the determinant (87). For example, the class P determinant is called Tet P; it has two genes, *tetA(P)* and *tetB(P)*, encoding the proteins TetA(P) (efflux) and TetB(P) (ribosomal protection). The class M determinant is called Tet M; it has a single gene, *tet*(M), encoding a single protein, Tet(M). Since all letters of the alphabet have now been used for the labeling of determinants, the nomenclature for new determinants has turned to numerals (86). Note that when searching databases, to retrieve intact a term containing parentheses, such as TetA(P), or a space, such as Tet M, it is often necessary to type the term within quotation marks. Recently, some proteins of the ribosomal protection group have been found to be hybrids of two different determinants (103, 147).

Multidrug transporters are able to efflux substrates that are chemically diverse (43, 124), and some from gram-negative bacteria include tetracycline in their repertoire, for example, AcrAB (*E. coli*) and MexAB/OprM (*Pseudomonas* sp.) (117). The only multidrug transporter from gram-positive bacteria known to transport tetracycline is the TetAB transporter from *Corynebacterium* (see below).

The emphasis in this chapter is on proteins found in gram-positive organisms. However, we have chosen to include selected information about certain related proteins from gram-negative organisms in cases where those proteins are better studied and offer insights.

TETRACYCLINE RESISTANCE BY ACTIVE EFFLUX

To date, there are about 20 known tetracycline resistance determinants that encode the active efflux of tetracycline as a resistance mechanism. Of these, nine are found in gram-positive organisms (Table 1). The 20 determinants fall into six groups, based on the degree of identity of amino acid sequence of the efflux proteins (Table 1). From one group to another, there is little identity. A phylogenetic tree of these determinants can be found in reference 58. Group 1, found mostly in gram-negative bacteria, comprises classes A to E, G, H, J, Y, Z, 30, 31, 33, and probably I (86). The proteins in group 1 share 41 to 78% identity (86, 145, 164). The only members of group 1 to be found in a gram-positive organism are classes Z and 33, both on large plasmids from *Corynebacterium glutamicum* (162, 164). Group 1 proteins have 12 predicted transmembrane α-helices, with a relatively long, central, nonconserved cytoplasmic loop connecting helices 6 and 7 (Fig. 2), and are regulated by a class-specific transcriptional repressor protein, TetR. Group 2 includes classes K and L, 58 to 59% identical, predominating in gram-positive species but also found in *Veillonella* and *Haemophilus* spp. (133). Group 2 proteins have 14 predicted transmembrane helices (Fig. 2). Class L has two subclasses that are 81% identical, one from the *Bacillus subtilis* chromosome and one from a variety of plasmids. The newly described Tet(38) protein has been placed in this group because of its (low) identity with Tet(K) and Tet(L) (63). Group 3 contains the chromosomally encoded OtrB and Tcr3 proteins (56% identical) with a topology like that of group 2 but with longer amino termini and longer loops 5 to 6 and 13 to 14 (and, for OtrB, a longer carboxy terminus) (Fig. 2). Members of this group, from the tetracycline-producing organisms *Streptomyces* spp., do not follow the standard nomenclature. The only member of group 4 is the TetA(P) protein from *Clostridium perfringens*, with 12 putative transmembrane helices but no central loop (Fig. 2) (5, 149). The class P determinant was found on a conjugative plasmid and is unusual in having two overlapping genes (149). Group 5 also has but a single member protein, Tet(V), from the *Mycobacterium smegmatis* genome, with 10 to 11 helices (38). Group 6 has the *otrC* tetracycline resistance gene from *Streptomyces* spp. Its recent sequence suggests an ABC transporter (see reference 43) comprising a membrane protein of six transmembrane helices and an ATP-binding protein; the only resistance it is known to specify is that to tetracyclines (132).

tetAB from a *Corynebacterium striatum* plasmid encodes two related putative ABC transport proteins, TetA and TetB, both of which are required for resistance and each of which has five to six putative transmembrane helices and a carboxy-terminal ATP-binding cassette (163). Since this determinant also encodes oxacillin resistance, however, the proteins actually constitute a multidrug exporter. TetAB is not given a group number (Table 1).

A low level of sequence identity is seen among all proton-coupled transporters, of which there are many (124). The first and second halves of group 1 Tet proteins and the first six helices of group 2 Tet proteins appear to have a common ancestor (139; see references 83 and 124). However, the different groups of tetracycline efflux proteins are more related to other transporters than to each other, and it is therefore likely that tetracycline efflux has evolved more than once (125, 148). The topology of these transporters can be predicted by computer-assisted hydropathy analysis, which locates stretches of hydrophobic residues sufficiently long to span a lipid bilayer as an α-helix. The experimentally determined topology for Tet efflux proteins has been consistent with such theoretical analyses, as shown for TetA(B) (75, 160; see reference 145), TetA(C) (4), and Tet(K) (49, 60). In addition, Tcr3 is 25% identical to the lipophilic organic cation efflux protein QacA, which has a topology

TABLE 1 Tetracycline resistance genes from gram-positive organisms: efflux[a]

Group	Gene(s)	AA[b]	Host of sequenced gene	GenBank nucleotide sequence accession no. (source)	% Identity of protein to that of pTHT15	Regulation	Reference(s)	Locations of class members[d]
1	tetA(B)	401	Gram-negative bacteria	J01830, citation 2; Tn10	Prototype protein from gram-negative bacteria	Inducible (repressor)	116	C, P**
1	tetA(Z)	~397	Corynebacterium glutamicum	AF121000 (pAG1)	Related to tet(A-H) of gram-negative bacteria	Inducible (repressor)	164	P**
1	tet(33)	407	Corynebacterium glutamicum	AJ420072 (pTET3)	Related to TetA(Z)	Inducible (repressor)	162	P**
2	tet(K)[c]	459	Staphylococcus aureus	M16217 (pNS1), S67449 (pT181)	59	Inducible	54, 120	C, P**
2	tet(L)	458	Bacillus stearothermophilus	M11036 (pTHT15)	100	Inducible[e]	64	P**
2	tet(L) (chromosomal)	458	Bacillus subtilis	X08034 (strain GSY908 called tetBS908; identical in B. subtilis 168)	81	Inducible	141	C
2	tet(38)	450	Staphylococcus aureus	See AP003129 and BAB41352	25	Unknown	63	C
3	otrB	563	Streptomyces rimosus	AF079900 (chromosomal DNA)	20	Unknown	99	C
3	tcr3 (tcrC)	512	Streptomyces lividans	D38215 (chromosomal DNA)	23	Unknown	36	C
4	tetA(P)	420	Clostridium perfringens	L20800 (pCW3)	20	Inducible	149	P**
5	tet(V)	419	Mycobacterium smegmatis mc^2	AF030344 (chromosomal DNA)	19	Unknown	38	C
6	otrC	281, 352	Streptomyces rimosus	AY509111	Putative two-part ABC transporter	Unknown	132	C
See text	tetAB	513, 528	Corynebacterium striatum M82B	U21300 (pTP10)	Putative two-part multidrug ABC transporter	Unknown	163	P**

[a]A specific sequenced gene was chosen to represent each class. Information in the first eight columns applies to this gene. The last column indicates the locations of typical class members. Groups coincide well with the data of Guillaume et al. (58), except that (the gram-negative) Tet Y has been moved from group 6 into group 1.

[b]Number of amino acid residues in protein. For class Z, the N-terminal residue has been located 13 residues upstream of that originally annotated, to coincide with the putative translational start of Tet(33) and to agree with the topology of other group 1 proteins.

[c]C, chromosome; P, plasmid; P**, conjugative plasmid; T, conjugative transposon.

[d]The two tet(K) sequences, one from 1986 and one from 1993, are identical; in intervening years, inaccurate sequences were reported (54, 84).

[e]Some plasmid-mediated Tet L determinants are constitutive (see text).

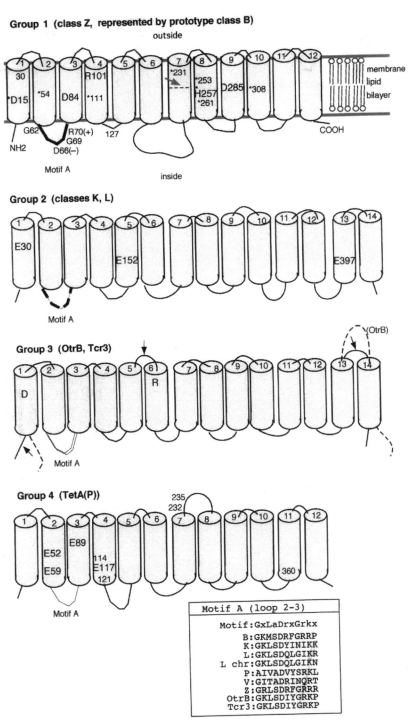

FIGURE 2 Predicted or verified (see text) transmembrane topology of tetracycline efflux proteins from groups 1 to 4. The inset at the bottom shows the consensus motif A sequence in loop 2 to 3, followed by the motif A sequences for Tet proteins of groups 1 to 4. L chr, the protein from the chromosomal Tet L determinant. The topologies of groups 5 and 6 are less certain (see text) and are not included. In group 1, the better-studied gram-negative TetA(B) protein replaces the gram-positive TetA(Z) and TetA(33) proteins. Residues marked in TetA(B) are identical in TetA(Z) and TetA(33), except for H257 and residue 253. Transmembrane α-helices (presumed) are shown in gray and are numbered. Charged residues predicted or known to be within transmembrane helices are shown with a large font and lettering. In motif A of TetA(B), a plus or minus sign indicates the necessity of that charge at that site. Gly-62 and, to a lesser extent, Gly-69 probably cause a β-turn (see references 121 and 145). Other residues mentioned in the text are shown in small fonts without letters. An asterisk indicates residues possibly involved in tetracycline binding in TetA(B). A large arrow indicates the Fe^{2+}-mediated cleavage site within helix 7, and the dotted line there shows the membrane impermeability barrier, which is not marked in other helices. Small arrows show regions in Tcr3 that are longer than similar regions in Tet(K) and Tet(L). A dotted line in group 3 similarly shows where OtrB has regions longer than those in Tcr3.

consistent with 14 transmembrane α-helices (see reference 124). The high α-helical content expected of TetA(B) has been seen in the purified protein by circular dichroism analysis (3).

With the presumed exception of the putative ABC transporters mentioned above, all of the tetracycline efflux systems use a proton gradient across the cell membrane (91) as the energy source to drive tetracycline out of the cell against a concentration gradient. Calculations and experiments have shown that the efflux achieves an appropriately low cytoplasmic concentration of tetracycline (169). For a heuristic model, we can imagine that binding of tetracycline to a region of a Tet protein on the inside surface of the cell, when (and only when) coupled with binding of a proton to the protein on the opposite surface, results in a conformational change of the protein and allows movement of tetracycline across the membrane. Then, release of tetracycline to the outside and a proton to the inside permits return of the protein to its pretransport conformation. Some questions of interest include the following. (i) Does the protein transport tetracycline in the unchelated or in the chelated form? (ii) Which regions of the protein bind tetracycline and which bind the proton? (iii) What is the attendant conformational change(s) and how does it result in the transfer of tetracycline from one side of the membrane to the other?

The properties of efflux systems are often conveniently studied with cell-free everted (inside-out) membrane vesicles, in which an everted membrane proton gradient is generated by electron transport (from the addition of electron transport substrates such as lactate) or by the membrane F_0F_1 ATPase (with the addition of ATP). That group 1 tetracycline resistance determinants encode an energy-dependent (active) efflux was initially demonstrated from the active uptake of [3H]tetracycline by everted vesicles from cells bearing the gram-negative determinants of classes A to D (100). The principles behind these better-studied efflux systems can aid in understanding the gram-positive systems.

A tetracycline-divalent cation complex is the species transported by those Tet efflux proteins examined to date. This point was originally suggested by the proportionality between the effectiveness of a given divalent cation in stimulating efflux and its affinity for tetracycline (190) and was proved by the cotransport of tetracycline and radiolabeled Co^{2+} for TetA(B) (190), Tet(K) (188), and chromosomal Tet(L) (57). A single divalent cation atom is transported per tetracycline molecule for TetA(B) (190) and for Tet(L) (29). Therefore, +1 is the net charge of the actual substrate complex (tetracycline, −1, plus divalent cation, +2). An energy-dependent efflux catalyzed by the TetA(P) protein can be seen in *E. coli* cells (149), but everted vesicles have yielded no activity. The K_m for TetA(B), Tet(K), and Tet(L) in membrane vesicles is 10 to 50 μM, and the V_{max} is 1 to 30 nmol mg^{-1} min^{-1} (57, 100, 188, 190), which for TetA(B) corresponds to a k_{cat} of approximately 0.03 to 1 s^{-1}. Purified TetA(B) (150) or Tet(L) (29) reconstituted as proteoliposomes (vesicles having a lipid bilayer, made in vitro from purified lipids plus purified protein) can perform tetracycline-proton antiport, showing that only the Tet proteins are necessary for this activity. Reconstituted Tet(L) was reported to have a k_{cat} for tetracycline as great as 180 s^{-1} (29).

The number of protons used per substrate complex transported was found to be 1 for the gram-negative TetA(B) (185), meaning that the antiport exchange is electroneu-

tral, that is, +1 (proton) comes into the cell while +1 (tetracycline-divalent cation) goes out. In the case of the chromosomally encoded Tet(L) from *B. subtilis*, however, transport of tetracycline is electrogenic; that is, it is energized by the electrical potential (it is also energized by the chemical potential) (29, 56, 57), indicating that more than one proton is exchanged per tetracycline-metal complex. This mechanism is relevant to the additional role of this protein in pH homeostasis under alkaline conditions (see the following section). Tet(K) is also a tetracycline-proton antiporter (61) and Tet(K)-mediated tetracycline transport is also electrogenic (56).

Defining the protein residues involved in binding the tetracycline-cation complex awaits a cocrystal structure. An illustration of a tetracycline-Mg^{2+} complex bound specifically to another protein (the crystallized TetR cytoplasmic repressor protein) can be seen in reference 59. In this case, the net charge on the tetracycline moiety is −1. Since the tetracycline-divalent cation complex has a net +1 charge, negatively charged acidic residues within transmembrane helices of TetA are candidates for such residues. Indeed, all tetracycline transporters, at least in the better-characterized groups 1 to 4 (Fig. 2), have transmembrane acidic residues, and mutagenesis studies have shown their importance in all determinants studied in detail. The conserved Asp-285 (transmembrane helix 9) is essential for the gram-negative TetA(B) protein (184) and likely for the gram-positive TetA(Z) and TetA(33) proteins. Photoaffinity labeling of TetA(B) protein with [3H]tetracycline in membranes occurs at a very low level and disappears if Asp-285 is mutated to Ala (77), showing that Asp-285 is involved, directly or indirectly, in substrate binding. In both Tet(K) and Tet(L), a negative charge is required at each of three conserved residues, Glu-30 (transmembrane helix 1), Glu-152 (transmembrane helix 5), and Glu-397 (transmembrane helix 13) (45, 69). The gram-positive class P efflux protein TetA(P) has four acidic residues within putative transmembrane helices (5) (Fig. 2). Glu-52 and Glu-59, both in helix 2, are essential (71). A negative charge must be present at Glu-89 (transmembrane helix 3) (71) and at Glu-117 (transmembrane helix 4) (5).

Positively charged residues within transmembrane helices have also been studied. In TetA(B), Arg-101, located just within transmembrane helix 4, is essential (74). On the other hand, His-257, in helix 8, can be mutated to other residues with retention of some activity (187) and is not even present in Tet Z, Tet 30, or Tet 33. In cytoplasmic loops 4 to 5 of TetA(B), there must be a positive charge at Arg-127 (74).

Other residues which may be involved in tetracycline binding are those which influence affinity or specificity. Most of these studies have involved gram-negative proteins. For TetA(B), mutations near the centers of transmembrane helix 4 (Gly-111), helix 8 (nonconserved Leu-253), and helix 10 (Leu-308) and toward the periplasmic end of helix 7 (Trp-231) enhance resistance to the normally inhibitory analog glycylcycline while decreasing resistance to tetracycline (Fig. 2) (173; see reference 145). Asp-15–Asn (helix 1) (101), Gln-54–Ala (helix 2), and Gln-261–Ala (helix 8) mutations raise the K_m value for the transport of tetracycline (see reference 145). An His-257–Asp mutation raises the K_m 4.5-fold (183), as does an Asp-285–Asn/Ala-220–Asp double mutation (186). Cleavage of the TetA(B) backbone by a Fe^{2+}-tetracycline complex shows that Gln225 in helix 7 is within 12 Å of the Fe^{2+} (96). These results together show that residues involved in substrate binding likely occur in transmembrane regions in both halves of the protein. Of note, involvement of the cytoplasmic interdomain loop in specificity was seen for the class A protein (172).

Tet proteins have several sequence motifs in common with other proton-dependent transporters (124). Particularly striking is motif A in the cytoplasmic loop connecting transmembrane helices 2 and 3 (loop 2 to 3); this motif is found in almost all such transporters (124, 125), including all of the proton-dependent tetracycline efflux proteins (Fig. 2). Such a broadly conserved motif is probably not substrate specific but rather may be involved in interactions of the coupling proton with the transporters. The effects of mutations in this loop in TetA(B) have been previously reviewed (124, 145). In this loop, Asp-66 is essential. A Ser-65–Cys mutant protein is active and can be inactivated by the sulfhydral reagent N-ethylmaleimide (NEM) (these experiments use an active TetA in which the single Cys at residue 377 has been replaced by Ala). Reactivity of the Cys of the Ser-65–Cys mutant protein with NEM increases in the presence of tetracycline-divalent cation (73), perhaps due to a conformational change in motif A resulting from substrate binding. Interestingly, this increase is prevented by the inactivating mutation Asp-66–Ala, but not by the inactivating mutation Arg-70–Ala (73). Therefore, Arg-70–Ala mutation appears to inactivate without preventing substrate binding. The Asp-66–Ala mutation may directly or indirectly prevent tetracycline binding, or it may prevent a conformational change in motif A caused by binding elsewhere. Conformational changes (as judged by NEM reactivity of the Cys substitutions at various positions), including those accompanying binding of tetracycline, are transmitted between cytoplasmic loop 2 to 3 and other regions of the TetA(B) protein located both within the membrane and even on its periplasmic side (76, 151). In the case of the gram-positive Tet(L) and Tet(K) proteins, substitutions at Asp-74, corresponding to the essential Asp-66 in TetA(B), only partly decrease tetracycline resistance, although efflux activity is more severely affected (46, 69). Therefore, this residue may not be essential. Replacement of Gly-70 by Arg, but not by Ser, eliminated resistance in Tet(L) (69). In Tet(K), Asp-318 in a secondary motif A sequence (cytoplasmic loop 10 to 11) is essential and cannot be replaced by Asn or Glu (46). In TetA(P), Pro-61 and Arg-71 of motif A are essential for activity (5); mutations at Asp-67 [= Aps-66 of TetA(B)] destabilized the protein (71).

Random mutagenesis in gram-negative Tet(C) revealed the importance of some of the same residues mentioned above, as well as others (102). For class P, random mutagenesis identified certain mutations that severely lower resistance without affecting the amount of transporter: Gly-114–Asp and Ala-121–Glu (transmembrane helix 5), Glu-232–Lys and Asp-235–Asn (external loops 7 to 8), and Ser-360–Phe (transmembrane helix 11) (6; topology as given in reference 5).

The topology and geometry of TetA(B) have been demonstrated by cysteine scanning mutagenesis of the entire protein and reactivity of the Cys residues with the membrane-permeable reagent NEM, which, like all sulfhydrals, requires water for the reaction. In this way, helical regions facing water were distinguished from those in a hydrophobic environment. Use of membrane-impermeant sulfhydral reagents allowed mapping of the inward- and outward-facing surfaces of the protein. The location of suppressor mutations that restored activity to mutant proteins helped identify

neighboring helices (143, 144), as did cross-linking studies. The results of many publications on these topics are integrated into a single model for the TetA(B) protein (160, 161). In this picture, 8 of the 12 transmembrane helices of TetA(B) appear to have part or all of one surface facing an aqueous channel. Within the protein, about halfway across the membrane and parallel to the membrane surface, lies a barrier to the passage of hydrophilic molecules (Fig. 1, dotted line in helix 7). This proposed structure of TetA(B) has been compared to the more precise structures recently solved by X-ray crystallography for two distantly related transporters, LacY (1) and GlpT (65) and concluded to be rather different (174).

Mutations in the first half of TetA(B) can complement those in the second half on a second polypeptide, suggesting possible dimerization (see references 84 and 145). These two halves do not, however, have independent functions, since a hybrid protein that has its first half from class B and its second half from class C, or vice versa, is inactive (137). An A/C hybrid, from more closely related classes, is active (137). Surprisingly, single mutations in either half of the inactive C/B hybrid can restore some tetracycline resistance (143). As is true for TetA(B), a mutation in the second half of TetA(C) can suppress a mutation in the first half on the same polypeptide, and vice versa (see reference 145). These results show that the two halves of the protein interact functionally.

Both Tet(B) and Tet(K) can function if individually inactive fragments are present together in the same cell. This type of complementation is seen with the first and second halves of TetA(B) (138, 189). In the case of Tet(K), the second half of the protein, together with an inactive protein having 62 consecutive amino acids deleted near the carboxy terminus, restores about 60% of the wild-type resistance (107). Therefore, certain domains of the proteins can be held together by strong noncovalent bonds.

Tet(L) in the membrane is a dimer which is retained upon solubilization, although no genetic complementation for tetracycline transport was seen between three pairs of mutant proteins in which one protein had a mutation in the N-terminal half while the other was mutated in the C-terminal half (140). Therefore, it may be that the protein functions as a monomer and that the dimer provides a regulatory function; alternatively, the pairs of mutations tested may have been in noncomplementary domains. The intracistronic genetic complementation between two different polypeptides of TetA(B) mentioned above suggested the formation of multimers, as did biochemical studies (98), but no such complementation was seen for TetA(C) (102).

The mechanisms of the other gram-positive efflux determinants have been less well characterized. The gene from *Streptomyces rimosus*, identical or similar to *otrB* (99), results in a decreased accumulation of tetracycline in intact cells (122), although no further biochemical or genetic studies have been reported. *otrB* (formerly known as *tet347* when its protein product was thought to have 347 residues) was resequenced, showing a predicted membrane protein of 563 residues and 14 transmembrane helices (99). The newly described group 2 determinant Tet 38, isolated from the *Staphylococcus aureus* chromosome, encodes a protein with 14 putative transmembrane helices and only 25% identity to Tet(K) and Tet(L) (63). The *otrC* tetracycline resistance gene from *Streptomyces* sp. appears from its sequence to encode an ABC transporter. The chromosomal *tet*(V) from *Mycobacterium* sp. causes active efflux from cells when cloned in multicopy (38).

TRANSPORT OF SODIUM, POTASSIUM, AND OTHER SUBSTRATES BY TETRACYCLINE EFFLUX PROTEINS

Some tetracycline efflux proteins from both gram-negative and gram-positive organisms can transport monocationic substrates other than tetracycline and protons. In gram-negative bacteria, a cellular defect for K^+ uptake was twice unexpectedly discovered to be alleviated by TetA(C) but not by TetA(B) (41, 111). Strikingly, the first 97 amino acid residues (three helices) are apparently sufficient for this activity (53, 111). An N-terminal region of Tet(K) is also sufficient (55). However, removal of a large carboxy-terminal region from Tet(L) stops K^+ transport (56). Mutations in TetA(C) that eliminate tetracycline resistance still allow the potassium transport function (102).

In the gram-positive B. subtilis, screening for genetic loci whose disruption by Tn917 made cells more sensitive to sodium or high pH pointed to the chromosomally located tet(L) gene (27). The effect is most obvious when the complete gene is deleted (28), which causes an enhanced susceptibility to tetracycline. Protection against high pH is completely restored by a cloned tet(L) gene and partially restored by a cloned tet(K) gene (28). The explanation for these results is that these Tet proteins can perform electrogenic Na^+/H^+ antiport and K^+/H^+ antiport (seen using radiolabeled Rb^+ representing K^+), so that more than one H^+ enters per exiting Na^+ or K^+, thereby maintaining the internal pH at about 7.7 even when the external pH is 8.5 (29, 56, 57). tet(L) appears to be the chromosomal gene most responsible for pH homeostasis mediated by $Na^+(K^+)/H^+$ antiport in B. subtilis (28). Purified Tet(L) reconstituted in liposomes catalyzes the $Na^+(K^+)/H^+$ exchange (29).

K^+ can take the place of the proton in the $Na^+(K^+)/H^+$ antiport function of Tet(L) and even more so in that of Tet(K), resulting in the electrogenic uptake of more than one K^+ ion per single K^+ or Na^+ ion exported (56). That is, both Tet(K) and chromosomal Tet(L) catalyze the net uptake of potassium. In fact, Na^+, K^+, or a tetracycline-metal complex can all be exported from the cell in exchange for an incoming K^+ by both Tet(K) and Tet(L), allowing efflux at high pH if K^+ is present (67).

Substitutions at Glu-397 in Tet(L) reveal that the negative charge required here for tetracycline transport (see above) is also critical for transport of Na^+ and K^+ (69), suggesting a pathway in common for all three positively charged substrates. This idea is supported by competition among the substrates (see reference 67). However, different substrates also require different residues. For example, alteration of certain regions of Tet(L) diminishes only tetracycline efflux without affecting transport of Na^+ or K^+: substitutions at Gly-70 and Asp-74 (motif A, cytoplasmic loop 2 to 3) (69), deletion of transmembrane helix 7 and transmembrane helix 8 (68), or changing Arg-110 to Cys (69).

Motif A (see above) and motif C are found in many transporters (124). Asp-318, in the secondary motif A in cytoplasmic loop 10 to 11, is critical in Tet(K) (45) but not in Tet(L) (69), showing a difference in behavior of these two proteins. The identity of residue 157 in motif C in transmembrane helix 5 is related to the greater strength of K^+ transport in Tet(K) than in Tet(L) (67). Antiport of tetracycline with a proton was reduced by mutations at Pro-156 and Gly-159 in motif C of Tet(K) (50).

The functioning of Tet proteins as monovalent cation antiporters may provide a selective advantage, e.g., with regard to pH homeostasis or potassium uptake, something to be considered in understanding the persistence of tetracycline resistance (56). Krulwich and coworkers also mention that these Tet proteins may eliminate most of the pH gradient component of the proton motive force, thereby lowering the pH gradient-dependent uptake of tetracycline in cells (26). Cells lacking chromosomal tet(L) grow slowly even at neutral pH, suggesting that Tet(L) may also transport an unknown metabolite (28).

TETRACYCLINE RESISTANCE BY RIBOSOMAL PROTECTION

Ribosomal protection is the most widespread of the tetracycline resistance mechanisms. It is mediated by a cytoplasmic protein that reduces the susceptibility of ribosomes to tetracycline. Tet(M), Tet(O), and OtrA proteins have been shown to function as ribosomal protection proteins, while Tet(S), Tet(T), Tet(Q), TetB(P), Tet(W), Tet(32), Tet(36), and "Tet" are presumed to do so because of their related sequences (Table 2). "Tet" and OtrA, both from Streptomyces sp., are 65% identical (40). A dendrogram showing the relationships among most of these determinants can be seen in reference 180. Four different (but unnamed) presumed ribosomal protection classes from streptococci have also been cloned using degenerate PCR primers based on known ribosomal protection genes; they do not hybridize with each other or with class M, O, P, Q, S, or T (32). Mosaic proteins formed from Tet(W) and Tet(O) sequences have also been recovered (154). What was previously described as "Tet(32)" (104) has been found upon resequencing to be a mosaic comprising both Tet(O) and Tet(32), better termed Tet(O32O) (147).

Ribosomal protection was first described with S. rimosus, an organism that produces oxytetracycline. The chromosomal gene, now called otrA, encodes a ribosome-associated factor that causes resistance to tetracycline of protein synthesis in vitro and dissociates in high salt (122). More-detailed studies have subsequently been done of Tet(M) and Tet(O). The 639-amino-acid Tet(M) and other ribosomal protection proteins are homologous to translational elongation factors EF-G and EF-Tu. The closest identity is to EF-G, particularly with the GTP-binding domain at the N-terminal region (20, 92, 165). The three-dimensional structure of EF-G is known (2).

Protein synthesis in crude extracts from cells bearing the Tet M determinant from Tn916 is resistant to tetracycline (21). The Tet(M) protein was overproduced, purified, and shown to confer resistance upon ribosomes from nonresistant cells (20). Like EF-G and OtrA, Tet(M) is released from resistant ribosomes washed with high levels of salt; like EF-G, it has ribosome-dependent GTPase activity (20). However, Tet(M) is not an alternative EF-G, since it cannot complement a temperature-sensitive EF-G in cells (20), nor can it replace EF-G (or EF-Tu) during in vitro protein synthesis (22). The related Tet(O) protein is also a ribosome-dependent GTPase, and it binds both GDP and GTP (166). Site-directed mutations in Tet(O) at Asn-128, conserved in six different ribosomal protection proteins and known to correspond to important residue 142 in the GTP-binding site of EF-G, cause a decrease in Tet(O) activity, bolstering the notion that GTP binding is important to Tet(O) function (see reference 165).

Further studies showed that Tet(M) permits aminoacyl-tRNA to bind to the A site of the ribosome in the presence of a normally inhibitory 50 μM tetracycline (22). That

TABLE 2 Tetracycline resistance genes from gram-positive organisms: ribosomal protection[a]

Gene	AA[b]	Host of sequenced gene	GenBank nucleotide sequence accession no. (source)	% Identity of protein to Tet(M) protein from Tn916	Regulation	Reference(s)	Locations of class members[c]
tet(M)	639	Enterococcus faecalis	M85225 (Tn916)	100	Inducible	157	C, P, T
tet(O)	639	Streptococcus mutans DL5	M20925 (chromosomal DNA)	76–77	Constitutive	80	C
	639	Campylobacter jejuni[d]	M18896 (pUA466) (M74450 upstream)			93, 177	P**
tetB(P)	652	Clostridium perfringens	L20800 (pCW3)	38	Inducible	149	P**
tet(Q)[e]	641 or 657	Bacteroides thetaiotaomicron (gram negative)	X58717 (conjugative transposon)	40	Constitutive	119	C, P, T
tet(S)	641	Listeria monocytogenes	L09756 (pIP811)	79	Unknown	25	C, P, T
tet(T)	651	Streptococcus pyogenes A498	L42544 (chromosomal DNA)	44	Unknown	32	C
tet(W)	639	Butyrivibrio fibrisolvens[f]	AJ222769 (chromosomal DNA)	68	Inducible	8	C, T
tet(O32O)	639	Clostridium related	AJ295238 (resequence shows hybrid; see text)	71	Unknown	104, 147	C, T?
tet(36)[e]	640	Bacteroides sp. (gram negative)	AJ514254	43	Unknown	180	?
"tet"	639	Streptomyces lividans 1326	M74049 (chromosomal DNA)	34	Unknown	40	C
otrA	663	Streptomyces rimosus	X53401 (chromosomal DNA)	33	Unknown	42	C

[a] A specific sequenced gene was chosen to represent each class. Information in the first eight columns applies to this gene. The last column indicates the locations of typical class members.
[b] AA, number of amino acid residues in the protein.
[c] C, chromosome; P, plasmid; P**, conjugative plasmid; T, conjugative transposon.
[d] A tet(O) sequence taken from the gram-negative Campylobacter is included because of the availability of additional upstream sequences of interest.
[e] First discovered in a gram-negative bacterium, but also found in gram-positive organisms (see Table 3).
[f] Butyrivibrio has recently been reclassified as a gram-positive-like organism (181).

Tet(M) brings about the actual release of the antibiotic, rather than just helping the ribosomes cope with bound tetracycline, was shown by the GTP-dependent loss of tetracycline from ribosomes within seconds of the addition of Tet(M) (22). Tet(M) does not itself bind tetracycline (22). Similarly, Tet(O) reduces binding of tetracycline to ribosomes in the presence of GTP (and not GDP), and binding of radiolabeled GTP to Tet(O) is enhanced by ribosomes (170). A Tet(O)-[^{35}S]GTPγS-labeled ribosome complex was identified by gel filtration (170), as was a [^{3}H]Tet(M)-GTP-ribosome complex (37). A nonhydrolyzable GTP analog catalyzed Tet(M) binding to ribosomes (37). Whether the energy from GTP hydrolysis serves to cause the conformational change which releases tetracycline from the ribosome or to dissociate the ribosomal protection protein from the ribosome to allow its reuse has been investigated using the nonhydrolyzable GTP analog. In the case of Tet(M), the data were interpreted in favor of the first possibility (22) and for Tet(O) in favor of the second (170), although the difference may reflect the details of the experiments rather than the different proteins themselves.

The binding of Tet(M) to the ribosome is not affected by tetracycline but is inhibited by thiostrepton, which interferes with EF-G interaction with the ribosomal GTPase center. Moreover, EF-G and Tet(M) compete for binding, with Tet(M) having a greater affinity (37). Cryoelectron microscope studies confirmed binding of these two proteins to similar regions on the ribosome (152; see reference 33).

How Tet(O) recognizes ribosomes which contain tetracycline may involve the preference of Tet(O) for ribosomes with an empty A site (34). The Tet(O)-mediated ribosomal conformational change that leads to release of tetracycline may persist for a while after release of Tet(O), slowing rebinding by free tetracycline (33).

Chromosomal mutations at two loci, *miaA* and *rpsL*, reduce the effectiveness of ribosomal protection proteins (23, 167). *miaA* encodes an enzyme that catalyzes the first step in modification of A37 on those tRNAs that read codons beginning with U (13). A37 is just 3′ to the anticodon, and its undermodification leads to a decrease in the rate of elongation, an increase in reading errors at the first position of the codon, and a decrease in errors at the third position (13). Resistance due to Tet(M) but not to Tet(O) was reduced severalfold by mutations in *miaA* (23, 167). Mutations in *rpsL*, encoding protein S12 of the S30 ribosomal subunit, cause streptomycin resistance or dependence and reduce the ability of Tet(M) and Tet(O) to cause tetracycline resistance in cells, in one case by as much as 32-fold (167). How the decrease in the rate of elongation and change in translational accuracy caused by mutations in *miaA* and *rpsL* might be related to their effects on the activity of ribosomal protection proteins is not known.

We can conclude that ribosomal protection proteins bind to ribosomes and somehow alter the ribosomal conformation to eliminate tetracycline binding without harming protein synthesis. The hydrolysis of GTP to GDP may provide energy either for the conformational change or for the departure of the protection protein from the ribosome to get out of the way of EF-G, which shares the Tet(M)-binding site and whose action is required for the translocation step. Tetracycline and the ribosomal protection protein may cycle on and off the ribosome in synchrony with peptide bond formation. The effect of the *miaA* and *rpsL* mutations may involve the kinetics of sampling of aminoacyl-tRNAs by the decoding site. A model of the mechanism of action of the protection proteins has been proposed (33, 170).

TETRACYCLINE RESISTANCE BY MUTATED rRNA

The first ribosomal mutation giving rise to clinical tetracycline resistance was described in 1998 for isolates of *P. acnes* (135). A change of G to C at a position cognate with nucleotide (nt) 1058 of *E. coli* 16S rRNA was seen in 15 of 21 resistant isolates (64-fold increase in mean MIC of tetracycline-doxycycline) and was not seen in any susceptible strains. When this mutation was recreated in *rrnB*, the *E. coli* gene for 16S rRNA, and cloned on a multicopy plasmid, *E. coli* strains bearing the plasmid were more resistant to tetracycline (and had a longer lag when grown without the drug, perhaps reflecting a slight loss of ribosome function). The mutation is in helix 34 of the 16S rRNA, near the tetracycline-binding site (see above) (135). Curiously, only the clinical strains isolated after 1988 had this mutation, while tetracycline-resistant mutants created in the laboratory did not (135). Mutations in *H. pylori* have also been isolated clinically in helix 31 (47, 171), which is also at the binding site (see above).

TETRACYCLINE RESISTANCE BY DEGRADATION OR UNKNOWN MECHANISM

Inactivation of tetracyclines has been observed in the gram-negative bacteria *Bacteroides* (Tet X) (191), *Pseudomonas* (110), and total oral flora (Tet 37) (39) and involves NADPH-dependent oxidoreductases which modify the drug. *tet*(U), a plasmid-borne gene reported to give low-level resistance to both tetracycline and minocycline in *Enterococcus faecium*, encodes a predicted protein of only 102 amino acids with little sequence identity to any other tetracycline resistance protein (130); no further work has been reported.

REGULATION OF EXPRESSION OF GROUP 1 EFFLUX (Tet Z AND Tet 33)

The group 1 determinants each have two genes, *tetA* (transporter) and *tetR* (repressor), divergently transcribed from a common, complex regulatory region containing the promoters and two palindromic operators, O1 and O2. The prototype of group 1 is Tet B, carried by Tn*10* (59). In the absence of tetracycline, the repressor typically binds to both operators in such a way that transcription of *tetA* is inhibited more severely, while enough *tetR* expression occurs to maintain repression. In the presence of tetracycline at levels far below those inhibitory to protein synthesis, a repressor–tetracycline-divalent cation complex forms at a very high affinity (on the order of $K_d = 3 \times 10^{-4}$ μM), causing dissociation of the repressor from DNA with attendant transcription and translation of *tetA* and *tetR* (59). The three-dimensional structure of dimeric (gram-negative) TetR(D) protein complexed with tetracycline-Mg^{2+} (see reference 59) or in the absence of inducer (123) has been determined, permitting understanding of the switch mechanism by which induction occurs.

REGULATION OF EXPRESSION OF GROUP 2 EFFLUX (Tet K AND Tet L)

Tet(K)-mediated resistance is inducible by tetracycline, and the presence of sequences upstream from *tet*(K) suggested that it might be regulated by translational attenuation (72), as are two other systems involving antibiotics acting on the ribosome, the *cat* chloramphenicol acetyltransferase and *erm* methylase genes (90). In translational attenuation, the product of translation of the first few codons of a leader peptide initiated using ribosome-binding site 1 (RBS1) inhibits the peptidyl transferase activity of the ribosome, causing ribosomal pausing on mRNA. If an inducer that blocks protein synthesis is present, the ribosome becomes permanently stalled, resulting in unfolding of a downstream mRNA stem-loop to unmask an RBS2, which is then used by other ribosomes to initiate translation of the protein itself (90). Such a mechanism has not been proven experimentally for Tet K, although the fold of Tet(K) protein induction is about four times that of its mRNA (107). An induction of mRNA that is also seen may represent stabilization (107; see reference 89).

Induction of the chromosomal *tet*(L)-mediated tetracycline resistance also occurs at the translational level, but by a mechanism termed translational reinitiation (10). This mechanism also involves a leader peptide but apparently not an uncovering of a downstream RBS2. Instead, the stalling of ribosomes within the leader peptide by tetracycline allows formation of an mRNA stem-loop which extends from the downstream half of the leader peptide sequence to just upstream of the Tet(L) RBS2. This stem-loop then physically guides the stalled ribosomes to the RBS2 for the Tet(L) protein. Since the RBS1 for the leader peptide is about 55-fold-more efficient than RBS2 for the Tet(L) protein, induction occurs. A minor tetracycline induction at the transcriptional level of unknown mechanism also occurs (10, 28, 155).

Whether the plasmid-mediated Tet L determinant has the same mechanism of regulation as that determined experimentally for the chromosomal determinant is unclear, since the complete set of upstream mRNA secondary structures predicted for the two kinds of determinants is not the same (Genetics Computer Group programs StemLoop and Mfold). Translational attenuation was proposed to regulate the plasmid *tet*(L) (64, 146), although no experiments on the mechanism of induction in plasmid-mediated Tet L have been reported. Naturally occurring constitutive plasmid *tet*(L) genes have truncated leader peptides, while inducible genes have intact ones. Specifically, a single-nucleotide difference between the sequences of the *tet*(L) region of the inducible pTHT15 and that of the constitutive pLS1 (derived from pMV158; GenBank accession number M29725) (79) and the constitutive pJH1 (bearing an identical determinant; GenBank accession number U17153) (128) results in leader termination after four residues in the constitutive determinants, while a 7-nt insertion in the constitutive pAMα1 results in leader termination after five residues (GenBank accession number D26045) (66).

As discussed earlier, the chromosomal *tet*(L) of *B. subtilis* is involved not only in tetracycline resistance but also in response of the cell to elevated pH and the presence of Na⁺ and K⁺. The Tet L determinant appears to be regulated by those factors differently than by tetracycline. When a sequence beginning 150 nt upstream from the transcriptional start site and extending to codon 381 of Tet(L) is fused in frame to LacZ, enzyme activity is up-regulated somewhat by elevated pH, Na⁺, and K⁺, but most dramatically by tetracycline (28). However, a promoterless *lacZ* gene inserted between the −35 and −10 elements of the *tet*(L) promoter is regulated only by pH, Na⁺, and K⁺. These findings suggest that tetracycline regulation may be mostly at the translational level, while regulation by pH, Na⁺, and K⁺ may be largely transcriptional, involving a second promoter 5′ to the *tet*(L) promoter. Because Tet(K) has many of the non-tetracycline-related functions of Tet(L), it may also be regulated by factors other than tetracycline, although no studies are available.

REGULATION OF EXPRESSION OF RIBOSOMAL PROTECTION (EXCEPT IN *STREPTOMYCES* SPP.)

Resistance mediated by Tet M (21), but not Tet O, is inducible by tetracycline (see reference 177). Transcription attenuation was proposed as a regulatory mechanism for Tet M in *E. faecalis* in view of several potential overlapping mRNA stem-loop structures and an overlapping putative leader peptide of 28 residues upstream of the apparent initiation codon for the Tet(M) protein (157). The downstream-most stem-loop is predicted to be a rho-independent transcriptional terminator. As a mechanism, transcriptional attenuation differs from translational attenuation (discussed above) in that ribosomal stalling on the leader peptide in cells induced, e.g., by tetracycline would lead to formation of an alternative antiterminator stem-loop accompanied by unfolding of a transcriptional terminator, so that transcription proceeds. Studies of *tet*(M) mRNA revealed three transcripts (157). One transcript, 0.25 kb, hybridized with a probe corresponding to the upstream stem-loop region but not with one representing the terminus of the *tet*(M) gene. It was postulated to be the attenuated transcript and was seen (as expected) in the absence of tetracycline. Two transcripts, 2.5 and 3.2 kb [the latter caused by readthrough past *tet*(M)], both of which presumably represent the antitermination products, hybridized with both probes and were induced by tetracycline. Unexpectedly, the amount of the 0.25-kb transcript did not decrease upon induction, and the total transcript amount appeared to increase (157). This may be an artifact reflecting the mRNA stabilization seen in the presence of inhibitors of protein synthesis (89) or may represent an added transcriptional layer of regulation not based on attenuation. On the other hand, disappearance of the 0.25-kb transcript has been seen for *tet*(M) from *Staphylococcus* sp. (78). The mRNA findings in general were supported by studies on mobilization of Tn*916* (see below) (24).

Tet O, Tet S, Tet T, and Tet W each have an upstream rho-independent terminator sequence similarly positioned to that of Tet M, but they do not all have the clear putative leader peptide or, probably, appropriate overlapping stem-loops as seen for Tet M. Tet O is not inducible (see below). Three upstream inverted repeat pairs proposed for Tet T do not overlap, nor is there a leader peptide (32); in any case, whether Tet T (or Tet S) is inducible by tetracycline is not known. Tet W is inducible when cloned into the heterologous host *E. coli* (7), yet it does not have a leader peptide (8). The mechanism of regulation in such a case is not known.

Tet O is noninducible both in the original host, *Campylobacter coli*, and in *E. coli*. In the latter host, it uses two transcription initiation sites controlled by promoters P2 and P1 (respectively, 273 nt and 42 nt upstream of the initiation codon). Promoter P2 corresponds to that used by Tet M in *E. faecalis* (157). In *C. coli*, only promoter P1 is used (177); this promoter is downstream from both the rho-independent terminator and a putative leader peptide of only 10 residues, perhaps accounting for the lack of inducibility of Tet O in *C. coli*. A deletion of *tet*(O) sequences encompassing P2 and a region 5′ thereto, leaving the rho-independent terminator and P1 intact, causes a reduction in the level of resistance in *E. coli*. Replacement of the deletion with a *lac* promoter does not restore resistance unless the terminator between the *lac* promoter and the *tet*(O) gene is deleted

(177). Therefore, P2 and/or upstream regions may assist in readthrough of the terminator or otherwise enhance the level of class O-mediated resistance in *E. coli*. In the native *C. coli*, P1 is probably sufficient for resistance, although plasmid instability precluded determinations of MIC values (177). Tetracycline resistance by one *tet*(Q) gene is not inducible, and its sequence in the upstream region is quite different from that of *tet*(M) (82). However, a *tet*(Q) gene from a different *Bacteroides* species appears to be regulated by tetracycline by translational attenuation (176). Resistance encoded by the *tetAB*(P) genes on the naturally occurring plasmid pCW3 is inducible at the transcriptional level, but inducibility is lost upon cloning (149) and requires a host factor (70).

REGULATION OF EXPRESSION OF TETRACYCLINE RESISTANCE IN *STREPTOMYCES* SPP.

The ribosomes of those *Streptomyces* spp. that synthesize tetracyclines are sensitive to the antibiotics (122). In these *Streptomyces* spp., access of tetracyclines to ribosomes is prevented by both efflux (OtrB and Tcr3) and ribosomal protection (OtrA and "Tet") mechanisms. Synthesis of the resistance proteins is turned on concurrently with the production of the antibiotics. In the oxytetracycline producer *S. rimosus*, tetracycline resistance mediated by "*tetA*" (likely the same gene as *otrA*) (42) is inducible by tetracycline (122), but this is in the vegetative stage rather than during production of tetracycline (95). During starvation for phosphate, when the antibiotics are made, an OmpR-like mechanism turns on the synthetic genes and also, by readthrough, the *otrA* resistance gene (42, 95). *tcr3* mRNA is highly expressed in a high-level *Streptomyces aureofaciens* producer of chlortetracycline (36). The transcriptional starts of "*tet*" have been mapped to 109 to 110 nt upstream from the translational start, and an inverted repeat of unknown significance is seen between the −35 and −10 regions of the promoter. Highly resistant mutant strains do not show gene amplification, nor do they have mutations in the 300 nt immediately upstream of the protein start (40).

DISTRIBUTION OF TETRACYCLINE RESISTANCE DETERMINANTS AND MODES OF TRANSFER

Tetracycline resistance determinants are widely spread among different gram-positive genera (Table 3). Unlike in gram-negative bacteria, it is common for more than one type of determinant to be found in a single gram-positive organism (131). Tetracycline resistance spreads because the determinants are often located on conjugative elements, either plasmids or transposons. Tet L and Tet K determinants in gram-positive organisms are usually specified by small transmissible plasmids and occasionally by the chromosome (Table 1). Tet M is often, but not always (133), found on a conjugative transposon that is generally chromosomal, for example, Tn916 (18.5 kb) and Tn5253 (60 kb) (142). Such conjugative transposons can also provide a conjugation pore for the transmission of mobilizable plasmids (142). Tet O has been found on self-transmissible plasmids and on the chromosome (131, 133). Tet Q has been best characterized in a gram-negative *Bacteroides* sp. and its relatives, where it occurs on several different very large conjugative transposons. A translational attenuation mecha-

nism appears responsible for the enhancement of both transposon excision and synthesis of the *tet*(Q) operon that is mediated by tetracycline and certain other inhibitors of protein synthesis, although the mechanism is not clear (176). Nonconjugative Tet Q has also been seen in gram-negative (142) and gram-positive (81) organisms. Tet S occurs on a self-transmissible plasmid in a *Listeria* sp. and as part of an uncharacterized 40-kb unit on the chromosome in an *Enterococcus* sp.; in the latter case, a resident plasmid causes conjugative transfer of the chromosome, thereby transferring Tet S (44). Tet W, located on the chromosome, is transferred by conjugation in the rumen *Butyrivibrio* sp. via a 40- to 50-kb transposon, Tn*B1230* (8, 103); Tet 32 is also conjugative, probably by a transposon (104).

Conjugation of Tn916 (bearing Tet M) is induced by tetracycline after excision and circularization of the transposon in the donor strain (24, 108). The mechanism involves induction of *tet*(M) by transcriptional attenuation, as described above, followed by readthrough into downstream genes, resulting in up-regulation of a downstream promoter, P*orf7*. Since circularization of the excised transposon has placed the transfer genes downstream of this promoter, transfer is induced (24).

DEVELOPMENT OF DRUGS TO COMBAT TETRACYCLINE RESISTANCE

Because the tetracyclines have many desirable properties, efforts have been made to devise analogs not affected by the resistance mechanisms. Glycylcyclines are functional tetracyclines having an *N*,*N*-dimethylglycylamido group [-NH-CO-CH$_2$-N(CH$_3$)$_2$] at position 9 and H at positions 5 and 6 (168), against which neither the tetracycline efflux proteins (group 1 or 2) nor the ribosomal protection proteins are effective (129, 168). They have been shown to be effective against infections in mice. They are not recognized as substrates by the TetA(B) efflux protein (see reference 145). They bind five times more strongly to ribosomes than does tetracycline and are correspondingly more inhibitory; however, this only partially explains their effectiveness in the presence of ribosomal protection proteins (11). Glycylcyclines also work against tetracycline-resistant, rapidly growing mycobacteria (17). Interestingly, when a dimethylamino group replaces the *N*,*N*-dimethylglycylamido group at position 9, the analog becomes a substrate for the TetA(B) efflux protein (115). The aminomethylcyclines also circumvent both efflux and ribosomal protection resistances, particularly in gram-positive organisms (12).

Several analogs of tetracycline having a rather large hydrophobic group at position 6 interfere specifically with the ability of the TetA(B) protein to efflux tetracycline (114, 115). One such analog can synergistically enhance the effectiveness of doxycycline against cells bearing efflux proteins from classes A, B, K, and L (115). In one case examined in more detail, the analog competes with tetracycline for binding to the TetA(B) protein yet is transported poorly itself (as assayed via H$^+$ antiport) (113). Finally, one of these analogs was about eight times as effective as doxycycline against resistance mediated by Tet M (113). The Tet(K) pump is inhibited by reserpine (48) and epigallocatechin gallate (159). Dactylocycline is a naturally occurring tetracycline analog glycosylated at position 6 active against tetracycline-resistant gram-positive organisms of uncharacterized resistance mechanisms (179). Siderophores (Fe^{3+} chelators of biological origin) in the uncomplexed state

TABLE 3 Distribution of tetracycline resistance determinants in gram-positive and certain other bacteria[a]

Determinant class or gene	Resistance mechanism	Genera in which determinant is found
Tet K	Efflux	*Bacillus, Bacterionema, Bifidobacterium, Clostridium, Corynebacterium, Enterococcus, Eubacterium, Gardnerella, Gemella, Listeria, Mycobacterium, Mycoplasma, Nocardia, Peptostreptococcus, Staphylococcus, Streptococcus, Streptomyces, Ureaplasma*
Tet L	Efflux	*Actinomyces, Bacillus, Clostridium, Enterococcus, Listeria, Mycobacterium, Nocardia, Peptostreptococcus, Staphylococcus, Streptococcus, Streptomyces*
Tet M	Ribosomal protection	*Abiotrophia,[b] Afipia,[b] Actinomyces, Aerococcus, Bacillus, Bacterionema, Bifidobacterium, Clostridium, Corynebacterium, Enterococcus, Erysipelothrix,[b] Eubacterium, Gardnerella, Gemella, Lactobacillus,[b] Listeria, Microbacterium,[b] Mycobacterium,[b] Mycoplasma, Nocardia, Peptostreptococcus, Staphylococcus, Streptococcus, Streptomyces,[b] Ureaplasma*
Tet O	Ribosomal protection	*Aerococcus, Butyrivibrio, Enterococcus, Lactobacillus, Megasphaera, Mobiluncus, Peptostreptococcus, Staphylococcus, Streptococcus*
Tet O32O	Ribosomal protection (hybrid)	*Clostridium* related
Tet P [*tetA*(P)]	Efflux	*Clostridium*
Tet P [*tetB*(P)]	Ribosomal protection	*Clostridium*
Tet Q	Ribosomal protection	*Clostridium,[b] Eubacterium, Gardnerella,[b] Lactobacillus, Mitsuokella, Mobiluncus, Peptostreptococcus, Streptococcus*
Tet S	Ribosomal protection	*Enterococcus, Lactobacillus,[b] Lactococcus, Listeria*
Tet T	Ribosomal protection	*Streptococcus*
Tet U	Unknown	*Enterococcus*
Tet V	Efflux	*Mycobacterium*
Tet W	Ribosomal protection	*Actinomyces, Arcanobacterium,[b] Bacillus, Bifidobacterium, Butyrivibrio, Clostridiaceae, Lactobacillus, Megasphaera, Mitsuokella, Roseburia, Selenomonas, Staphylococcus, Streptococcus, Streptomyces*
"tet"	Ribosomal protection	*Streptomyces*
tcr3	Efflux	*Streptomyces*
Tet Z	Efflux	*Corynebacterium*
Tet 32	(See Tet O32O)	
Tet 33	Efflux	*Corynebacterium*
Tet 36	Ribosomal protection	*Butyrivibrio, Clostridium, Eubacterium, Lactobacillus*
Tet 38	Efflux	*Staphylococcus aureus*
otrA	Ribosomal protection	*Streptomyces, Mycobacterium*
otrB	Efflux	*Streptomyces, Mycobacterium*
otrC	Efflux	*Streptomyces, Mycobacterium*

[a]Includes *Actinomyces, Mycobacterium, Mycoplasma, Nocardia, Streptomyces,* and *Ureaplasma* spp. Based on reference 131, with additions from references 8, 32, 63, 81, 103, 130, 132 (for *otrC*), 154, 162–164, 175, and 180. *Butyrivibrio* sp. has been reclassified as a gram-positive-like organism (181); we have therefore included the related genera *Mitsuokella* and *Selenomonas.*

[b]In addition to the references listed in footnote *a*, reference 134 was also used.

inhibit both Tet K- and Tet B-mediated resistance (136), but the reason is not known.

SUMMARY

Tetracycline resistance is widespread across the bacterial kingdom. Tetracycline affects all prokaryotes by binding to the 16S rRNA in the ribosome, altering the A site and inhibiting protein synthesis. Three clinically important resistance mechanisms have emerged in gram-positive bacteria: active efflux, ribosomal protection, and rRNA mutation. In the case of the efflux genes tetA(Z), tetA(33), tet(K), tet(L), tet(38), tetA(P), otrB, and tcr3, a membrane protein of 12 to 14 (putative) transmembrane helices uses the proton gradient across the cytoplasmic membrane to pump out a tetracycline-divalent cation complex. The putative ABC efflux transporters encoded by the tetAB and otrC genes presumably use ATP to energize efflux. Ribosomal protection proteins [genes tet(M), tet(O), and probably tetB(P), tet(Q), tet(S), tet(T), tet(W), tet(32), otrA, and "tet"] are GTPases that bind to the ribosome, thereby preventing binding of tetracycline. Finally, mutations in the 16S rRNA of the ribosome itself lead to reduced tetracycline effectiveness.

Tetracycline resistance is often inducible by tetracycline. For classes Z and 33, studies of regulation in closely related determinants suggest that, when the repressor protein TetR binds tetracycline, the protein loses its hold on the operator regulating the efflux protein. For chromosomal class L, induction involves translational reinitiation. For classes K and plasmid class L, the induction mechanism may involve translational attenuation. For class M, induction probably occurs by transcriptional attenuation. In the *Streptomyces* spp. that synthesize tetracyclines, induction of tetracycline resistance determinants may coordinate with drug production. The class K and L tetracycline efflux proteins also have a semi-independent $H^+(K^+)/K^+$, Na^+ transport function and play a role in pH homeostasis. Regulation of these functions may differ from regulation of tetracycline transport.

Tetracycline resistance determinants are often found on conjugative elements, both plasmid and chromosomal, allowing their ready transfer among widely different microorganisms. New tetracycline analogs and other drugs effective against efflux and ribosomal protection resistance mechanisms are being developed.

Work in our laboratory is supported in part by grants from the National Institutes of Health. This article was completed in December 2004.

REFERENCES

1. **Abramson, J., I. Smirnova, V. Kasho, G. Verner, H. R. Kaback, and S. Iwata.** 2003. Structure and mechanism of the lactose permease of *Escherichia coli*. *Science* **301:**610–615.

2. **Aevarsson, A., E. Brazhnikov, M. Garber, J. Zheltonosova, Y. Chirgadze, A. Al-Karadaghi, L. A. Svensson, and A. Liljas.** 1994. Three-dimensional structure of the ribosomal translocase: elongation factor G from *Thermus thermophilus*. *EMBO J.* **13:**3669–3677.

3. **Aldema, M. L., L. M. McMurry, A. R. Walmsley, and S. B. Levy.** 1996. Purification of the Tn*10*-specified tetracycline efflux antiporter TetA in a native state as a polyhistidine fusion protein. *Mol. Microbiol.* **19:**187–195.

4. **Allard, J. D., and K. P. Bertrand.** 1992. Membrane topology of the pBR322 tetracycline resistance protein: TetA-PhoA gene fusions and implications for the mechanism of

5. **Bannam, T. L., P. A. Johanesen, C. L. Salvado, S. J. Pidot, K. A. Farrow, and J. I. Rood.** 2004. The *Clostridium perfringens* TetA(P) efflux protein contains a functional variant of the Motif A region found in major facilitator superfamily transport proteins. *Microbiology* **150:**127–134.

6. **Bannam, T. L., and J. I. Rood.** 1999. Identification of structural and functional domains of the tetracycline efflux protein TetA(P) from *Clostridium perfringens*. *Microbiology* **145:**2947–2955.

7. **Barbosa, T. M.** Personal communication.

8. **Barbosa, T. M., K. P. Scott, and H. J. Flint.** 1999. Evidence for recent intergeneric transfer of a new tetracycline resistance gene, *tet*(W), isolated from *Butyrivibrio fibrisolvens*, and the occurrence of *tet*(O), in ruminal bacteria. *Environ. Microbiol.* **1:**53–64.

9. **Bauer, G., C. Berens, S. J. Projan, and W. Hillen.** 2004. Comparison of tetracycline and tigecycline binding to ribosomes mapped by dimethylsulphate and drug-directed Fe^{2+} cleavage of 16S rRNA. *J. Antimicrob. Chemother.* **53:**592–599.

10. **Bechhofer, D. H., and S. J. Stasinopoulos.** 1998. *tet*A(L) mutants of a tetracycline-sensitive strain of *Bacillus subtilis* with the polynucleotide phosphorylase gene deleted. *J. Bacteriol.* **180:**3470–3473.

11. **Bergeron, J., M. Ammirati, D. Danley, L. James, M. Norcia, J. Retsema, C. A. Strick, W.-G. Su, J. Sutcliffe, and L. Wondrack.** 1996. Glycylcyclines bind to the high-affinity tetracycline ribosomal binding site and evade Tet(M)- and Tet(O)-mediated ribosomal protection. *Antimicrob. Agents Chemother.* **40:**2226–2228.

12. **Bhatia, B., T. Bowser, J. Chen, M. Ismail, L. Mcintyre, R. Mechiche, M. Nelson, K. Ohemeng, and A. Verma.** 2003. PTK0796 and other novel tetracycline derivatives exhibiting potent in vitro and in vivo activities against antibiotic resistant gram-positive bacteria. Progr. Abstr. 43rd Intersci. Conf. Antimicrob. Agents Chemother., abstr. 2420. American Society for Microbiology, Washington, D.C.

13. **Bjork, G.** 1996. Stable RNA modification, p. 861–886. *In* F. C. Neidhardt, R. Curtiss III, J. L. Ingraham, E. C. C. Lin, K. B. Low, B. Magasanik, W. S. Reznikoff, M. Riley, M. Schaechter, and H. E. Umbarger (ed.), *Escherichia coli and Salmonella: Cellular and Molecular Biology*, 2nd ed. ASM Press, Washington, D.C.

14. **Blackwood, R. K., and A. R. English.** 1970. Structure-activity relationships in the tetracycline series, p. 237–266. *In* D. Perlman (ed.), *Advances in Applied Microbiology*. Academic Press, New York, N.Y.

15. **Brimacombe, R., B. Greuer, P. Mitchell, M. Osswald, J. Rinke-Appel, D. Schueler, and K. Stade.** 1990. Three-dimensional structure and function of *Escherichia coli* 16S and 23S rRNA as studied by cross-linking techniques, p. 93–106. *In* W. E. Hill (ed.), *The Ribosome: Structure, Function, and Evolution*. American Society for Microbiology, Washington, D.C.

16. **Brodersen, D. E., W. M. Clemons, Jr., A. P. Carter, R. J. Morgan-Warren, B. T. Wimberly, and V. Ramakrishnan.** 2000. The structural basis for the action of the antibiotics tetracycline, pactamycin, and hygromycin B on the 30S ribosomal subunit. *Cell* **103:**1143–1154.

17. **Brown, B. A., R. J. Wallace, Jr., and G. Onyi.** 1996. Activities of the glycylcyclines *N,N*-dimethylglycylamido-minocycline and *N,N*-dimethylglycylamido-6-demethyl-6-deoxytetracycline against *Nocardia* spp. and tetracycline-

resistant isolates of rapidly growing mycobacteria. *Antimicrob. Agents Chemother.* **40:**874–878.

18. **Brown, J. R., and D. S. Ireland.** 1978. Structural requirements for tetracycline activity. *Adv. Pharmacol. Chemother.* **15:**161–202.

19. **Buck, M. A., and B. S. Cooperman.** 1990. Single protein omission reconstitution studies of tetracycline binding to the 30S subunit of *Escherichia coli* ribosomes. *Biochemistry* **29:**5374–5379.

20. **Burdett, V.** 1991. Purification and characterization of Tet(M), a protein that renders ribosomes resistant to tetracycline. *J. Biol. Chem.* **266:**2872–2877.

21. **Burdett, V.** 1986. Streptococcal tetracycline resistance mediated at the level of protein biosynthesis. *J. Bacteriol.* **165:**564–569.

22. **Burdett, V.** 1996. Tet(M)-promoted release of tetracycline from ribosomes is GTP dependent. *J. Bacteriol.* **178:**3246–3251.

23. **Burdett, V.** 1993. tRNA modification activity is necessary for Tet(M)-mediated tetracycline resistance. *J. Bacteriol.* **175:**7209–7215.

24. **Celli, J., and P. Trieu-Cuot.** 1998. Circularization of Tn*916* is required for expression of the transposon-encoded transfer functions: characterization of long tetracycline-inducible transcripts reading through the attachment site. *Mol. Microbiol.* **28:**103–117.

25. **Charpentier, E., G. Gerbaud, and P. Courvalin.** 1993. Characterization of a new class of tetracycline-resistance gene tet(S) in *Listeria monocytogenes* BM4210. *Gene* **131:**27–34.

26. **Cheng, J., K. Baldwin, A. A. Guffanti, and T. A. Krulwich.** 1996. Na$^+$/H$^+$ antiport activity conferred by *Bacillus subtilis* tetA(L), a 5′ truncation product of tetA(L), and related plasmid genes upon *Escherichia coli*. *Antimicrob. Agents Chemother.* **40:**852–857.

27. **Cheng, J., A. A. Guffanti, and T. A. Krulwich.** 1994. The chromosomal tetracycline resistance locus of *Bacillus subtilis* encodes a Na$^+$/H$^+$ antiporter that is physiologically important at elevated pH. *J. Biol. Chem.* **269:**27365–27371.

28. **Cheng, J., A. A. Guffanti, W. Wang, T. A. Krulwich, and D. H. Bechhofer.** 1996. Chromosomal tetA(L) gene of *Bacillus subtilis*: regulation of expression and physiology of a tetA(L) deletion strain. *J. Bacteriol.* **178:**2853–2860.

29. **Cheng, J., D. B. Hicks, and T. A. Krulwich.** 1996. The purified *Bacillus subtilis* tetracycline efflux protein TetA(L) reconstitutes both tetracycline-cobalt/H$^+$ and Na$^+$(K$^+$)/H$^+$ exchange. *Proc. Natl. Acad. Sci. USA* **93:**14446–14451.

30. **Chopra, I.** 1985. Mode of action of the tetracyclines and the nature of bacterial resistance to them, p. 317–392. *In* J. J. Hlavka and J. H. Boothe (ed.), *The Tetracyclines.* Springer-Verlag, Berlin, Germany.

31. **Chopra, I., P. M. Hawkey, and M. Hinton.** 1992. Tetracyclines, molecular and clinical aspects. *J. Antimicrob. Chemother.* **29:**245–277.

32. **Clermont, D., O. Chesneau, G. de Cespedes, and T. Horaud.** 1997. New tetracycline resistance determinants coding for ribosomal protection in streptococci and nucleotide sequence of tet(T) isolated from *Streptococcus pyogenes* A498. *Antimicrob. Agents Chemother.* **41:**112–116.

33. **Connell, S. R., D. M. Tracz, K. H. Nierhaus, and D. E. Taylor.** 2003. Ribosomal protection proteins and their mechanism of tetracycline resistance. *Antimicrob. Agents Chemother.* **47:**3675–3681.

34. **Connell, S. R., C. A. Trieber, G. P. Dinos, E. Einfeldt, D. E. Taylor, and K. H. Nierhaus.** 2003. Mechanism of

Tet(O)-mediated tetracycline resistance. *EMBO J.* **22:**945–953.

35. **Cundliffe, E., and K. McQuillen.** 1967. Bacterial protein synthesis: the effects of antibiotics. *J. Mol. Biol.* **30:**137–146.

36. **Dairi, T., K. Aisaka, R. Katsumata, and M. Hasegawa.** 1995. A self-defense gene homologous to tetracycline effluxing gene essential for antibiotic production in *Streptomyces aureofaciens.* *Biosci. Biotechnol. Biochem.* **59:**1835–1841.

37. **Dantley, K. A., H. K. Dannelly, and V. Burdett.** 1998. Binding interaction between Tet(M) and the ribosome: requirements for binding. *J. Bacteriol.* **180:**4089–4092.

38. **De Rossi, E., M. C. Blokpoel, R. Cantoni, M. Branzoni, G. Riccardi, D. B. Young, K. A. De Smet, and O. Ciferri.** 1998. Molecular cloning and functional analysis of a novel tetracycline resistance determinant, tet(V), from *Mycobacterium smegmatis.* *Antimicrob. Agents Chemother.* **42:**1931–1937.

39. **Diaz-Torres, M. L., R. McNab, D. A. Spratt, A. Villedieu, N. Hunt, M. Wilson, and P. Mullany.** 2003. Novel tetracycline resistance determinant from the oral metagenome. *Antimicrob. Agents Chemother.* **47:**1430–1432.

40. **Dittrich, W., and H. Schrempf.** 1992. The unstable tetracycline resistance gene of *Streptomyces lividans* 1326 encodes a putative protein with similarities to translational elongation factors and Tet(M) and Tet(O) proteins. *Antimicrob. Agents Chemother.* **36:**1119–1124.

41. **Dosch, D. C., F. F. Salvacion, and W. Epstein.** 1984. Tetracycline resistance element of pBR322 mediates potassium transport. *J. Bacteriol.* **160:**1188–1190.

42. **Doyle, D., K. J. McDowall, M. J. Butler, and I. S. Hunter.** 1991. Characterization of an oxytetracycline-resistance gene, otrA, of *Streptomyces rimosus.* *Mol. Microbiol.* **5:**2923–2933.

43. **Fath, M. J., and R. Kolter.** 1993. ABC transporters: bacterial exporters. *Microbiol. Rev.* **57:**995–1017.

44. **Francois, B., M. Charles, and P. Courvalin.** 1997. Conjugative transfer of tet(S) between strains of *Enterococcus faecalis* is associated with the exchange of large fragments of chromosomal DNA. *Microbiology* **143:**2145–2154.

45. **Fujihira, E., T. Kimura, Y. Shiina, and A. Yamaguchi.** 1996. Transmembrane glutamic acid residues play essential roles in the metal-tetracycline/H$^+$ antiporter of *Staphylococcus aureus.* *FEBS Lett.* **391:**243–246.

46. **Fujihira, E., T. Kimura, and A. Yamaguchi.** 1997. Roles of acidic residues in the hydrophilic loop regions of metal-tetracycline/H$^+$ antiporter Tet(K) of *Staphylococcus aureus.* *FEBS Lett.* **419:**211–214.

47. **Gerrits, M. M., M. R. de Zoete, N. L. Arents, E. J. Kuipers, and J. G. Kusters.** 2002. 16S rRNA mutation-mediated tetracycline resistance in *Helicobacter pylori.* *Antimicrob. Agents Chemother.* **46:**2996–3000.

48. **Gibbons, S., and E. E. Udo.** 2000. The effect of reserpine, a modulator of multidrug efflux pumps, on the in vitro activity of tetracycline against clinical isolates of methicillin resistant *Staphylococcus aureus* (MRSA) possessing the tet(K) determinant. *Phytother. Res.* **14:**139–140.

49. **Ginn, S. L., M. H. Brown, and R. A. Skurray.** 1997. Membrane topology of the metal-tetracycline/H$^+$ antiporter TetA(K) from *Staphylococcus aureus.* *J. Bacteriol.* **179:**3786–3789.

50. **Ginn, S. L., M. H. Brown, and R. A. Skurray.** 2000. The TetA(K) tetracycline/H$^+$ antiporter from *Staphylococcus aureus*: mutagenesis and functional analysis of motif C. *J. Bacteriol.* **182:**1492–1498.

51. **Goldman, R. A., T. Hasan, C. C. Hall, W. A. Strycharz, and B. S. Cooperman.** 1983. Photoincorporation of tetracycline into *Escherichia coli* ribosomes. Identification of the major proteins photolabeled by native tetracycline and tetracycline photoproducts and implications for the inhibitory action of tetracycline on protein synthesis. *Biochemistry* **22:**359–368.

52. **Gottesman, M. E.** 1967. Reaction of ribosome-bound peptidyl transfer ribonucleic acid with aminoacyl transfer ribonucleic acid or puromycin. *J. Biol. Chem.* **242:**5564–5571.

53. **Griffith, J. K., T. Kogoma, D. L. Corvo, W. L. Anderson, and A. L. Kazim.** 1988. An N-terminal domain of the tetracycline resistance protein increases susceptibility to aminoglycosides and complements potassium uptake defects in *Escherichia coli. J. Bacteriol.* **170:**598–604.

54. **Guay, G. G., S. A. Khan, and D. M. Rothstein.** 1993. The tet(K) gene of plasmid pT181 of Staphylococcus aureus encodes an efflux protein that contains 14 transmembrane helices. *Plasmid* **30:**163–166.

55. **Guay, G. G., M. Tuckman, P. McNicholas, and D. M. Rothstein.** 1993. The *tet*(K) gene from *Staphylococcus aureus* mediates the transport of potassium in *Escherichia coli. J. Bacteriol.* **175:**4927–4929.

56. **Guffanti, A. A., J. Cheng, and T. A. Krulwich.** 1998. Electrogenic antiport activities of the gram-positive Tet proteins include a $Na^+(K^+)/K^+$ mode that mediates net K^+ uptake. *J. Biol. Chem.* **273:**26447–26454.

57. **Guffanti, A. A., and T. A. Krulwich.** 1995. Tetracycline/H^+ antiport and Na^+/H^+ antiport catalyzed by the *Bacillus subtilis* TetA(L) transporter expressed in *Escherichia coli. J. Bacteriol.* **177:**4557–4561.

58. **Guillaume, G., V. Ledent, W. Moens, and J. M. Collard.** 2004. Phylogeny of efflux-mediated tetracycline resistance genes and related proteins revisited. *Microb. Drug Resist.* **10:**11–26.

59. **Hillen, W., and C. Berens.** 1994. Mechanism underlying expression of Tn10 encoded tetracycline resistance. *Annu. Rev. Microbiol.* **48:**345–369.

60. **Hirata, T., E. Fujihira, T. Kimura-Someya, and A. Yamaguchi.** 1998. Membrane topology of the staphylococcal tetracycline efflux protein Tet(K) determined by antibacterial resistance gene fusion. *J. Biochem.* (Tokyo) **124:**1206–1211.

61. **Hirata, T., R. Wakatabe, J. Nielsen, Y. Someya, E. Fujihira, T. Kimura, and A. Yamaguchi.** 1997. A novel compound, 1,1-dimethyl-5-(1-hydroxypropyl)-4,6,7-trimethylindan, is an effective inhibitor of the *tet*(K) gene-encoded metal-tetracycline/H^+ antiporter of *Staphylococcus aureus. FEBS Lett.* **412:**337–340.

62. **Hlavka, J. J., and J. H. Boothe (ed.).** 1985. *The Tetracyclines.* Springer-Verlag, Berlin, Germany.

63. **Hooper, D. C.** 2004. Personal communication.

64. **Hoshino, T., T. Ikeda, N. Tomizuka, and K. Furukawa.** 1985. Nucleotide sequence of the tetracycline resistance gene of pTHT15, a thermophilic *Bacillus* plasmid: comparison with staphylococcal Tc^R controls. *Gene* **37:**131–138.

65. **Huang, Y., M. J. Lemieux, J. Song, M. Auer, and D. N. Wang.** 2003. Structure and mechanism of the glycerol-3-phosphate transporter from *Escherichia coli. Science* **301:**616–620.

66. **Ishiwa, H., and H. Shibahara.** 1985. New shuttle vectors for *Escherichia coli* and *Bacillus subtilis.* III. Nucleotide sequence analysis of tetracycline resistance gene of pAMa1 and *ori*-177. *Jpn. J. Genet.* **60:**485–498.

67. **Jin, J., A. A. Guffanti, D. H. Bechhofer, and T. A. Krulwich.** 2002. Tet(L) and Tet(K) tetracycline-divalent metal/H^+ antiporters: characterization of multiple catalytic modes and a mutagenesis approach to differences in their efflux substrate and coupling ion preferences. *J. Bacteriol.* **184:**4722–4732.

68. **Jin, J., A. A. Guffanti, C. Beck, and T. A. Krulwich.** 2001. Twelve-transmembrane-segment (TMS) version (ΔTMS VII-VIII) of the 14-TMS Tet(L) antibiotic resistance protein retains monovalent cation transport modes but lacks tetracycline efflux capacity. *J. Bacteriol.* **183:**2667–2671.

69. **Jin, J., and T. A. Krulwich.** 2002. Site-directed mutagenesis studies of selected motif and charged residues and of cysteines of the multifunctional tetracycline efflux protein Tet(L). *J. Bacteriol.* **184:**1796–1800.

70. **Johanesen, P. A., D. Lyras, and J. I. Rood.** 2001. Induction of pCW3-encoded tetracycline resistance in *Clostridium perfringens* involves a host-encoded factor. *Plasmid* **46:**229–232.

71. **Kennan, R. M., L. M. McMurry, S. B. Levy, and J. I. Rood.** 1997. Glutamate residues located within putative transmembrane helices are essential for TetA(P)-mediated tetracycline efflux. *J. Bacteriol.* **179:**7011–7015.

72. **Khan, S. A., and R. P. Novick.** 1983. Complete nucleotide sequence of pT181, a tetracycline-resistance plasmid from *Staphylococcus aureus. Plasmid* **10:**251–259.

73. **Kimura, T., Y. Inagaki, T. Sawai, and A. Yamaguchi.** 1995. Substrate-induced acceleration of N-ethylmaleimide reaction with the Cys-65 mutant of the transposon Tn10-encoded metal-tetracycline/H^+ antiporter depends on the interaction of Asp-66 with the substrate. *FEBS Lett.* **362:**47–49.

74. **Kimura, T., M. Nakatani, T. Kawabe, and A. Yamaguchi.** 1998. Roles of conserved arginine residues in the metal-tetracycline/H^+ antiporter of *Escherichia coli. Biochemistry* **37:**5475–5480.

75. **Kimura, T., M. Ohnuma, T. Sawai, and A. Yamaguchi.** 1997. Membrane topology of the transposon 10-encoded metal-tetracycline/H^+ antiporter as studied by site-directed chemical labeling. *J. Biol. Chem.* **272:**580–585.

76. **Kimura, T., T. Sawai, and A. Yamaguchi.** 1997. Remote conformational effects of the Gly-62→Leu mutation of the Tn10-encoded metal-tetracycline/H^+ antiporter of *Escherichia coli* and its second-site suppressor mutation. *Biochemistry* **36:**6941–6946.

77. **Kimura, T., and A. Yamaguchi.** 1996. Asp-285 of the metal-tetracycline/H^+ antiporter of *Escherichia coli* is essential for substrate binding. *FEBS Lett.* **388:**50–52.

78. **Kornblum, J., and R. Novick.** Personal communication.

79. **Lacks, S. A., P. Lopez, B. Greenberg, and M. Espinosa.** 1986. Identification and analysis of genes for tetracycline resistance and replication functions in the broad-host-range plasmid pSL1. *J. Mol. Biol.* **192:**753–765.

80. **LeBlanc, D. J., L. N. Lee, B. M. Titman, C. J. Smith, and F. C. Tenover.** 1988. Nucleotide sequence analysis of tetracycline resistance gene *tetO* from *Streptococcus mutans* DL5. *J. Bacteriol.* **170:**3618–3626.

81. **Leng, Z., D. E. Riley, R. E. Berger, J. N. Krieger, and M. C. Roberts.** 1997. Distribution and mobility of the tetracycline resistance determinant *tetQ. J. Antimicrob. Chemother.* **40:**551–559.

82. **Lepine, G., J.-M. Lacroix, C. B. Walker, and A. Progulske-Fox.** 1993. Sequencing of a *tet*(Q) gene isolated from *Bacteroides fragilis* 1126. *Antimicrob. Agents Chemother.* **37:**2037–2041.

83. **Levy, S. B.** 1992. Active efflux mechanisms for antimicrobial resistance. *Antimicrob. Agents Chemother.* **36:**695–703.

84. **Levy, S. B.** 1992. *The Antibiotic Paradox. How Miracle Drugs Are Destroying the Miracle.* Plenum Publishing, New York, N.Y.

85. **Levy, S. B.** 1984. Resistance to the tetracyclines, p. 191–240. *In* L. E. Bryan (ed.), *Antimicrobial Drug Resistance.* Academic Press, Inc., New York, N.Y.

86. **Levy, S. B., L. M. McMurry, T. M. Barbosa, V. Burdett, P. Courvalin, W. Hillen, M. C. Roberts, J. I. Rood, and D. E. Taylor.** 1999. Nomenclature for new tetracycline resistance determinants. *Antimicrob. Agents Chemother.* **43:**1523–1524.

87. **Levy, S. B., L. M. McMurry, V. Burdett, P. Courvalin, W. Hillen, M. C. Roberts, and D. E. Taylor.** 1989. Nomenclature for tetracycline resistance determinants. *Antimicrob. Agents Chemother.* **33:**1373–1374.

88. **Lewin, B.** 1997. *Genes VI.* Oxford University Press, Oxford, United Kingdom.

89. **Lopez, P. J., I. Marchand, O. Yarchuk, and M. Dreyfus.** 1998. Translation inhibitors stabilize *Escherichia coli* mRNAs independently of ribosome protection. *Proc. Natl. Acad. Sci. USA* **95:**6067–6072.

90. **Lovett, P. S., and E. J. Rogers.** 1996. Ribosome regulation by the nascent peptide. *Microbiol. Rev.* **60:**366-385.

91. **Maloney, P. C., and T. H. Wilson.** 1996. Ion-coupled transport and transporters, p. 1130–1148. *In* F. C. Neidhardt, R. Curtiss III, J. L. Ingraham, E. C. C. Lin, K. B. Low, B. Magasanik, W. S. Reznikoff, M. Riley, M. Schaechter, and H. E. Umbarger (ed.), Escherichia coli *and* Salmonella: *Cellular and Molecular Biology*, 2nd ed. ASM Press, Washington, D.C.

92. **Manavathu, E. K., C. L. Fernandez, B. S. Cooperman, and D. E. Taylor.** 1990. Molecular studies on the mechanism of tetracycline resistance mediated by Tet(O). *Antimicrob. Agents Chemother.* **34:**71–77.

93. **Manavathu, E. K., K. Hiratsuka, and D. E. Taylor.** 1988. Nucleotide sequence analysis and expression of a tetracycline-resistance gene from *Campylobacter jejuni*. *Gene* **62:**17–26.

94. **Martin, R. B.** 1985. Tetracyclines and daunorubicin, p. 19–52. *In* H. Sigel (ed.), *Metal Ions in Biological Systems*. Marcel Dekker, Inc., New York, N.Y.

95. **McDowall, K. J., A. Thamchaipenet, and I. S. Hunter.** 1999. Phosphate control of oxytetracycline production by *Streptomyces rimosus* is at the level of transcription from promoters overlapped by tandem repeats similar to those of the DNA-binding sites of the OmpR family. *J. Bacteriol.* **181:**3025–3032.

96. **McMurry, L. M., M. L. Aldema-Ramos, and S. B. Levy.** 2002. Fe^{2+}-tetracycline-mediated cleavage of the Tn*10* tetracycline efflux protein TetA reveals a substrate binding site near glutamine 225 in transmembrane helix 7. *J. Bacteriol.* **184:**5113–5120.

97. **McMurry, L. M., J. C. Cullinane, and S. B. Levy.** 1982. Transport of the lipophilic analog minocycline differs from that of tetracycline in susceptible and resistant *Escherichia coli* strains. *Antimicrob. Agents Chemother.* **22:**791–799.

98. **McMurry, L. M., and S. B. Levy.** 1995. The NH_2-terminal half of the tetracycline efflux protein from Tn*10* contains a functional dimerization domain. *J. Biol. Chem.* **270:**22752–22757.

99. **McMurry, L. M., and S. B. Levy.** 1998. Revised sequence of OtrB (Tet347) tetracycline efflux protein from *Streptomyces rimosus*. *Antimicrob. Agents Chemother.* **42:**3050.

100. **McMurry, L. M., R. E. Petrucci, Jr., and S. B. Levy.** 1980. Active efflux of tetracycline encoded by four genetically different tetracycline resistance determinants in *Escherichia coli*. *Proc. Natl. Acad. Sci. USA* **77:**3974–3977.

101. **McMurry, L. M., M. Stephan, and S. B. Levy.** 1992. Decreased function of the class B tetracycline efflux protein Tet with mutations at aspartate 15, a putative intramembrane residue. *J. Bacteriol.* **174:**6294–6297.

102. **McNicholas, P., I. Chopra, and D. M. Rothstein.** 1992. Genetic analysis of the *tet*A(C) gene on plasmid pBR322. *J. Bacteriol.* **174:**7926–7933.

103. **Melville, C. M., R. Brunel, H. J. Flint, and K. P. Scott.** 2004. The *Butyrivibrio fibrisolvens tet*(W) gene is carried on the novel conjugative transposon Tn*B1230*, which contains duplicated nitroreductase coding sequences. *J. Bacteriol.* **186:**3656–3659.

104. **Melville, C. M., K. P. Scott, D. K. Mercer, and H. J. Flint.** 2001. Novel tetracycline resistance gene, *tet*(32), in the *Clostridium*-related human colonic anaerobe K10 and its transmission in vitro to the rumen anaerobe *Butyrivibrio fibrisolvens*. *Antimicrob. Agents Chemother.* **45:**3246–3249.

105. **Mendez, B., C. Tachibana, and S. B. Levy.** 1980. Heterogeneity of tetracycline resistance determinants. *Plasmid* **3:**99–108.

106. **Mitscher, L. A.** 1978. *The Chemistry of the Tetracycline Antibiotics*. Marcel Dekker, Inc., New York, N.Y.

107. **Mojumdar, M., and S. A. Khan.** 1988. Characterization of the tetracycline resistance gene of plasmid pT181 of *Staphylococcus aureus*. *J. Bacteriol.* **170:**5522–5528.

108. **Mullany, P., A. P. Roberts, and H. Wang.** 2002. Mechanism of integration and excision in conjugative transposons. *Cell. Mol. Life Sci.* **59:**2017–2022.

109. **Munske, G. R., E. V. Lindley, and J. A. Magnuson.** 1984. *Streptococcus faecalis* proton gradients and tetracycline transport. *J. Bacteriol.* **158:**49–54.

110. **Nakamura, A., M. Nakagawa, H. Yoshikoshi, S. Shoutou, K. O'Hara, and T. Sawai.** 2002. Novel enzymatically determined minocycline-resistance in a *Pseudomonas aeruginosa* clinical isolate. Progr. Abstr. 42st Intersci. Conf. Antimicrob. Agents Chemother., abstr. C1-1603. American Society for Microbiology, Washington, D.C.

111. **Nakamura, T., Y. Matsuba, A. Ishihara, T. Kitagawa, F. Suzuki, and T. Unemoto.** 1995. N-terminal quarter part of tetracycline transporter from pACYC184 complements K^+ uptake activity in K^+ uptake-deficient mutants of *Escherichia coli* and *Vibrio alginolyticus*. *Biol. Pharm. Bull.* **18:**1189–1193.

112. Reference omitted.

113. **Nelson, M. L., and S. B. Levy.** 1999. Reversal of tetracycline resistance mediated by different bacterial tetracycline resistance determinants by an inhibitor of the Tet(B) antiport protein. *Antimicrob. Agents Chemother.* **43:**1719–1724.

114. **Nelson, M. L., B. H. Park, J. S. Andrews, V. A. Georgian, R. C. Thomas, and S. B. Levy.** 1993. Inhibition of the tetracycline efflux antiport protein by 13-thio-substituted 5-hydroxy-6-deoxytetracyclines. *J. Med. Chem.* **36:**370–377.

115. **Nelson, M. L., B. H. Park, and S. B. Levy.** 1994. Molecular requirements for the inhibition of the tetracycline antiport protein and the effect of potent inhibitors on the growth of tetracycline-resistant bacteria. *J. Med. Chem.* **37:**1355–1361.

116. **Nguyen, T. T., K. Postle, and K. P. Bertrand.** 1983. Sequence homology between the tetracycline-resistance determinants of Tn*10* and pBR322. *Gene* **25:**83–92.

117. **Nikaido, H.** 1996. Multidrug efflux pumps of gram-negative bacteria. *J. Bacteriol.* **178:**5853–5859.

118. **Nikaido, H., and D. G. Thanassi.** 1993. Penetration of lipophilic agents with multiple protonation sites into bacterial cells: tetracyclines and fluoroquinolones as examples. *Antimicrob. Agents Chemother.* **37:**1393–1399.

119. **Nikolich, M. P., N. B. Shoemaker, and A. A. Salyers.** 1992. A *Bacteroides* tetracycline resistance gene represents a new class of ribosome protection tetracycline resistance. *Antimicrob. Agents Chemother.* **36:**1005–1012.

120. **Noguchi, N., T. Aoki, M. Sasatsu, M. Kono, K. Shishido, and T. Ando.** 1986. Determination of the complete nucleotide sequence of pNS1, a staphylococcal tetracycline-resistance plasmid propagated in *Bacillus subtilis*. *FEMS Microbiol. Lett.* **37:**283–288.

121. **Oehler, R., N. Polacek, G. Steiner, and A. Darta.** 1997. Interaction of tetracycline with RNA: photoincorporation into ribosomal RNA of *Escherichia coli*. *Nucleic Acids Res.* **25:**1219–1224.

122. **Ohnuki, T., T. Katch, T. Imanaka, and S. Aiba.** 1985. Molecular cloning of tetracycline resistance genes from *Streptomyces rimosus* in *Streptomyces griseus* and characterization of the cloned genes. *J. Bacteriol.* **161:**1010–1016.

123. **Orth, P., F. Cordes, D. Schnappinger, W. Hillen, W. Saenger, and W. Hinrichs.** 1998. Conformational changes of the Tet repressor induced by tetracycline trapping. *J. Mol. Biol.* **279:**439–447.

124. **Paulsen, I. T., M. H. Brown, and R. A. Skurray.** 1996. Proton-dependent multidrug efflux systems. *Microbiol. Rev.* **60:**575–608.

125. **Paulsen, I. T., and R. A. Skurray.** 1993. Topology, structure and evolution of two families of proteins involved in antibiotic and antiseptic resistance in eukaryotes and prokaryotes—an analysis. *Gene* **124:**1–11.

126. **Peska, S.** 1971. Inhibitors of ribosome function. *Annu. Rev. Microbiol.* **25:**487–562.

127. **Pioletti, M., F. Schlunzen, J. Harms, R. Zarivach, M. Gluhmann, H. Avila, A. Bashan, H. Bartels, T. Auerbach, C. Jacobi, T. Hartsch, A. Yonath, and F. Franceschi.** 2001. Crystal structures of complexes of the small ribosomal subunit with tetracycline, edeine and IF3. *EMBO J.* **20:**1829–1839.

128. **Platteeuw, C., F. Michiels, H. Joos, J. Seurinck, and W. M. de Vos.** 1995. Characterization and heterologous expression of the *tet*L gene and identification of *iso*-ISS*1* elements from *Enterococcus faecalis* plasmid pJH1. *Gene* **160:**89–93.

129. **Rasmussen, B. A., Y. Gluzman, and F. P. Tally.** 1994. Inhibition of protein synthesis occurring on tetracycline-resistant TetM-protected ribosomes by a novel class of tetracyclines, the glycylcyclines. *Antimicrob. Agents Chemother.* **38:**1658–1660.

130. **Ridenhour, M. B., H. M. Fletcher, J. E. Mortensen, and L. Daneo-Moore.** 1996. A novel tetracycline-resistant determinant, tet(U), is encoded on the plasmid pKQ10 in *Enterococcus faecium*. *Plasmid* **35:**71–80.

131. **Roberts, M. C.** 1997. Genetic mobility and distribution of tetracycline resistance determinants, p. 206–218. *In* D. J. Chadwick (ed.), *Antibiotic Resistance: Origins, Evolution, Selection, and Spread.* John Wiley and Sons, Chichester, United Kingdom.

132. **Roberts, M. C.** 2004. Personal communication.

133. **Roberts, M. C.** 1996. Tetracycline resistance determinants: mechanisms of action, regulation of expression, genetic mobility, and distribution. *FEMS Microbiol. Rev.* **19:**1–24.

134. **Roberts, M. C.** 2004. Distribution of tetracycline resistance genes among gram-positive bacteria, *Mycobacterium*, *Mycoplasma*, *Nocardia*, *Streptomyces* and *Ureaplasma*. [Online.] http://faculty.washington.edu/marilynr/tetweb3.pdf.

135. **Ross, J. I., E. A. Eady, J. H. Cove, and W. J. Cunliffe.** 1998. 16S rRNA mutation associated with tetracycline resistance in a gram-positive bacterium. *Antimicrob. Agents Chemother.* **42:**1702–1705.

136. **Rothstein, D. M., M. McGlynn, V. Bernan, J. McGahren, J. Zaccardi, N. Cekleniak, and K. P. Bertrand.** 1993. Detection of tetracyclines and efflux pump inhibitors. *Antimicrob. Agents Chemother.* **37:**1624–1629.

137. **Rubin, R. A., and S. B. Levy.** 1990. Interdomain hybrid tetracycline proteins confer tetracycline resistance only when they are derived from closely related members of the *tet* gene family. *J. Bacteriol.* **172:**2303–2312.

138. **Rubin, R. A., and S. B. Levy.** 1991. Tet protein domains interact productively to mediate tetracycline resistance when present on separate polypeptides. *J. Bacteriol.* **173:**4503–4509.

139. **Rubin, R. A., S. B. Levy, R. L. Heinrikson, and F. J. Kézdy.** 1990. Gene duplication in the evolution of the two complementing domains of gram-negative bacterial tetracycline efflux proteins. *Gene* **87:**7–13.

140. **Safferling, M., H. Griffith, J. Jin, J. Sharp, M. De Jesus, C. Ng, T. A. Krulwich, and D. N. Wang.** 2003. TetL tetracycline efflux protein from *Bacillus subtilis* is a dimer in the membrane and in detergent solution. *Biochemistry* **42:**13969–13976.

141. **Sakaguchi, R., H. Amano, and K. Shishido.** 1988. Nucleotide sequence homology of the tetracycline-resistance determinant naturally maintained in *Bacillus subtilis* Marburg 168 chromosome and the tetracycline-resistance gene of *B. subtilis* plasmid pNS1981. *Biochim. Biophys. Acta* **950:**441–444.

142. **Salyers, A. A., N. B. Shoemaker, A. M. Stevens, and L.-Y. Li.** 1995. Conjugative transposons: an unusual and diverse set of integrated gene transfer elements. *Microbiol. Rev.* **59:**579–590.

143. **Saraceni-Richards, C. A., and S. B. Levy.** 2000. Evidence for interactions between helices 5 and 8 and a role for the interdomain loop in tetracycline resistance mediated by hybrid Tet proteins. *J. Biol. Chem.* **275:**6101–6106.

144. **Saraceni-Richards, C. A., and S. B. Levy.** 2000. Second-site suppressor mutations of inactivating substitutions at Gly247 of the tetracycline efflux protein, Tet(B). *J. Bacteriol.* **182:**6514–6516.

145. **Schnappinger, D., and W. Hillen.** 1996. Tetracyclines: antiobiotic action, uptake, and resistance mechanisms. *Arch. Microbiol.* **165:**359–369.

146. **Schwarz, S., M. Cardoso, and H. C. Wegener.** 1992. Nucleotide sequence and phylogeny of the *tet*(L) tetracycline resistance determinant encoded by plasmid pSTET1 from *Staphylococcus hyicus*. *Antimicrob. Agents Chemother.* **36:**580–588.

147. **Scott, K. P.** 2004. Personal communication.

148. **Sheridan, R. P., and I. Chopra.** 1991. Origin of tetracycline efflux proteins: conclusions from nucleotide sequence analysis. *Mol. Microbiol.* **5:**895–900.

149. **Sloan, J., L. M. McMurry, D. Lyras, S. B. Levy, and J. I. Rood.** 1994. The *Clostridium perfringens* Tet P deter-

minant comprises two overlapping genes: tetA(P), which mediates active tetracycline efflux, and tetB(P), which is related to the ribosomal protection family of tetracycline-resistance determinants. *Mol. Microbiol.* **11**:403–415.

150. **Someya, Y., Y. Moriyama, M. Futai, T. Sawai, and A. Yamaguchi.** 1996. Reconstitution of the metal-tetracycline/H$^+$ antiporter of *Escherichia coli* in proteoliposomes including F$_0$F$_1$-ATPase. *FEBS Lett.* **374**:72–76.

151. **Someya, Y., and A. Yamaguchi.** 1997. Second-site suppressor mutations for the Arg70 substitution mutants of the Tn*10*-encoded metal-tetracycline/H$^+$ antiporter of *Escherichia coli*. *Biochim. Biophys. Acta* **1322**:230–236.

152. **Spahn, C. M., G. Blaha, R. K. Agrawal, P. Penczek, R. A. Grassucci, C. A. Trieber, S. R. Connell, D. E. Taylor, K. H. Nierhaus, and J. Frank.** 2001. Localization of the ribosomal protection protein Tet(O) on the ribosome and the mechanism of tetracycline resistance. *Mol. Cell* **7**:1037–1045.

153. **Speer, B. S., N. B. Shoemaker, and A. A. Salyers.** 1992. Bacterial resistance to tetracycline: mechanisms, transfer, and clinical significance. *Clin. Microbiol. Rev.* **5**:387–399.

154. **Stanton, T. B., J. S. McDowall, and M. A. Rasmussen.** 2004. Diverse tetracycline resistance genotypes of *Megasphaera elsdenii* strains selectively cultured from swine feces. *Appl. Environ. Microbiol.* **70**:3754–3757.

155. **Stasinopoulos, S. J., G. A. Farr, and D. H. Bechhofer.** 1998. *Bacillus subtilis* tetA(L) gene expression: evidence for regulation by translational reinitiation. *Mol. Microbiol.* **30**:923–932.

156. **Stezowski, J. J.** 1976. Chemical-structural properties of tetracycline derivatives. 1. Molecular structure and conformation of the free base derivatives. *J. Am. Chem. Soc.* **98**:6012–6018.

157. **Su, Y. A., H. Ping, and D. B. Clewell.** 1992. Characterization of the tet(M) determinant of Tn*916*: evidence for regulation by transcription attenuation. *Antimicrob. Agents Chemother.* **36**:769–778.

158. **Suarez, G., and D. Nathans.** 1965. Inhibition of aminoacyl-sRNA binding to ribosomes by tetracycline. *Biochem. Biophys. Res. Commun.* **18**:743–750.

159. **Sudano Roccaro, A., A. R. Blanco, F. Giuliano, D. Rusciano, and V. Enea.** 2004. Epigallocatechin-gallate enhances the activity of tetracycline in staphylococci by inhibiting its efflux from bacterial cells. *Antimicrob. Agents Chemother.* **48**:1968–1973.

160. **Tamura, N., S. Konishi, S. Iwaki, T. Kimura-Someya, S. Nada, and A. Yamaguchi.** 2001. Complete cysteine-scanning mutagenesis and site-directed chemical modification of the Tn*10*-encoded metal-tetracycline/H$^+$ antiporter. *J. Biol. Chem.* **276**:20330–20339.

161. **Tamura, N., S. Konishi, and A. Yamaguchi.** 2003. Mechanisms of drug/H$^+$ antiport: complete cysteine-scanning mutagenesis and the protein engineering approach. *Curr. Opin. Chem. Biol.* **7**:570–579.

162. **Tauch, A., S. Gotker, A. Puhler, J. Kalinowski, and G. Thierbach.** 2002. The 27.8-kb R-plasmid pTET3 from *Corynebacterium glutamicum* encodes the aminoglycoside adenyltransferase gene cassette aadA9 and the regulated tetracycline efflux system Tet 33 flanked by active copies of the widespread insertion sequence IS6100. *Plasmid* **48**:117–129.

163. **Tauch, A., S. Krieft, A. Puhler, and J. Kalinowski.** 1999. The tetAB genes of the *Corynebacterium striatum* R-plasmid pTP10 encode an ABC transporter and confer tetracycline, oxytetracycline and oxacillin resistance in

Corynebacterium glutamicum. *FEMS Microbiol. Lett.* **173**:203–209.

164. **Tauch, A., A. Puhler, J. Kalinowski, and G. Thierbach.** 2000. TetZ, a new tetracycline resistance determinant discovered in gram-positive bacteria, shows high homology to gram-negative regulated efflux systems. *Plasmid* **44**:285–291.

165. **Taylor, D. E., and A. Chau.** 1996. Tetracycline resistance mediated by ribosomal protection. *Antimicrob. Agents Chemother.* **40**:1–5.

166. **Taylor, D. E., L. J. Jerome, J. Grewal, and N. Chang.** 1995. Tet(O), a protein that mediates ribosomal protection to tetracycline, binds, and hydrolyses GTP. *Can. J. Microbiol.* **41**:965–978.

167. **Taylor, D. E., C. A. Trieber, G. Trescher, and M. Bekkering.** 1998. Host mutations (miaA and rpsL) reduce tetracycline resistance mediated by Tet(O) and Tet(M). *Antimicrob. Agents Chemother.* **42**:59–64.

168. **Testa, R. T., P. J. Petersen, N. V. Jacobus, P.-E. Sum, V. J. Lee, and F. P. Tally.** 1993. In vitro and in vivo antibacterial activities of the glycylcyclines, a new class of semisynthetic tetracyclines. *Antimicrob. Agents Chemother.* **37**:2270–2277.

169. **Thanassi, D. G., G. S. Suh, and H. Nikaido.** 1995. Role of outer membrane barrier in efflux-mediated tetracycline resistance of *Escherichia coli*. *J. Bacteriol.* **177**:998–1007.

170. **Trieber, C. A., N. Burkhardt, K. H. Nierhaus, and D. E. Taylor.** 1998. Ribosomal protection from tetracycline mediated by Tet(O): Tet(O) interaction with ribosomes is GTP-dependent. *Biol. Chem.* **379**:847–855.

171. **Trieber, C. A., and D. E. Taylor.** 2002. Mutations in the 16S rRNA genes of *Helicobacter pylori* mediate resistance to tetracycline. *J. Bacteriol.* **184**:2131–2140.

172. **Tuckman, M., P. J. Petersen, and S. J. Projan.** 2000. Mutations in the interdomain loop region of the tetA(A) tetracycline resistance gene increase efflux of minocycline and glycylcyclines. *Microb. Drug Resist.* **6**:277–282.

173. **Tuckman, M., and S. Projan.** 1998. Characterization of TetA(B) glycylglycine-resistant mutants. Progr. Abstr. 38th Intersci. Conf. Antimicrob. Agents Chemother., abstr. C-97, p. 96. American Society for Microbiology, Washington, D.C.

174. **Vardy, E., I. T. Arkin, K. E. Gottschalk, H. R. Kaback, and S. Schuldiner.** 2004. Structural conservation in the major facilitator superfamily as revealed by comparative modeling. *Protein Sci.* **13**:1832–1840.

175. **Villedieu, A., M. L. Diaz-Torres, N. Hunt, R. McNab, D. A. Spratt, M. Wilson, and P. Mullany.** 2003. Prevalence of tetracycline resistance genes in oral bacteria. *Antimicrob. Agents Chemother.* **47**:878–882.

176. **Wang, Y., N. B. Shoemaker, and A. A. Salyers.** 2004. Regulation of a *Bacteroides* operon that controls excision and transfer of the conjugative transposon CTnDOT. *J. Bacteriol.* **186**:2548–2557.

177. **Wang, Y., and D. E. Taylor.** 1991. A DNA sequence upstream of the tet(O) gene is required for full expression of tetracycline resistance. *Antimicrob. Agents Chemother.* **35**:2020–2025.

178. **Weckesser, J., and J. A. Magnuson.** 1979. Light-induced, carrier-mediated transport of tetracycline by *Rhodopseudomonas sphaeroides*. *J. Bacteriol.* **138**:678–683.

179. **Wells, S. J., J. O'Sullivan, C. Aklonis, H. A. Ax, A. A. Tymiak, D. R. Kirsch, W. H. Trejo, and P. Principe.** 1992. Dactylocyclines: novel tetracycline derivatives produced by a *Dactylosporangium* sp. *J. Antibiot.* **45**:1892–1898.

180. **Whittle, G., T. R. Whitehead, N. Hamburger, N. B. Shoemaker, M. A. Cotta, and A. A. Salyers.** 2003. Identification of a new ribosomal protection type of tetracycline resistance gene, *tet*(36), from swine manure pits. *Appl. Environ. Microbiol.* **69:**4151–4158.

181. **Willems, A., M. Amat-Marco, and M. D. Collins.** 1996. Phylogenetic analysis of *Butyrivibrio* strains reveals three distinct groups of species within the *Clostridium* subphylum of the gram-positive bacteria. *Int. J. Syst. Bacteriol.* **46:**195–199.

182. **Wilson, K. S., and H. F. Noller.** 1998. Molecular movement inside the translational engine. *Cell* **92:**337–349.

183. **Yamaguchi, A., K. Adachi, T. Akasaka, N. Ono, and T. Sawai.** 1991. Metal-tetracycline/H$^+$ antiporter of *Escherichia coli* encoded by a transposon Tn*10*: histidine 257 plays an essential role in H$^+$ translocation. *J. Biol. Chem.* **266:**6045–6051.

184. **Yamaguchi, A., T. Akasaka, N. Ono, Y. Someya, M. Nakatani, and T. Sawai.** 1992. Metal-tetracycline/H$^+$ antiporter of *Escherichia coli* encoded by transposon Tn*10*. Roles of the aspartyl residues located in the putative transmembrane helices. *J. Biol. Chem.* **267:**7490–7498.

185. **Yamaguchi, A., Y. Iwasaki-Ohba, N. Ono, M. Kaneko-Ohdera, and T. Sawai.** 1991. Stoichiometry of metal-tetracycline /H$^+$ antiport mediated by transposon Tn*10*-encoded tetracycline resistance protein in *Escherichia coli*. *FEBS Lett.* **282:**415–418.

186. **Yamaguchi, A., R. O'yauchi, Y. Someya, T. Akasaka, and T. Sawai.** 1993. Second-site mutation of Ala-220 to Glu or Asp suppresses the mutation of Asp-285 to Asn in the transposon Tn*10*-encoded metal-tetracycline/H$^+$ antiporter of *Escherichia coli*. *J. Biol. Chem.* **268:**26990–26995.

187. **Yamaguchi, A., T. Samejima, T. Kimura, and T. Sawai.** 1996. His257 is a uniquely important histidine residue for tetracycline/H$^+$ antiport function but not mandatory for full activity of the transposon Tn*10*-encoded metal-tetracycline/H$^+$ antiporter. *Biochemistry* **35:**4359–4364.

188. **Yamaguchi, A., Y. Shiina, E. Fujihira, T. Sawai, N. Noguchi, and M. Sasatsu.** 1995. The tetracycline efflux protein encoded by the *tet*(K) gene from *Staphylococcus aureus* is a metal-tetracycline/H$^+$ antiporter. *FEBS Lett.* **365:**193–197.

189. **Yamaguchi, A., Y. Someya, and T. Sawai.** 1993. The in vivo assembly and function of the N- and C-terminal halves of the Tn*10*-encoded TetA protein in *Escherichia coli*. *FEBS Lett.* **324:**131–135.

190. **Yamaguchi, A., T. Udagawa, and T. Sawai.** 1990. Transport of divalent cations with tetracycline as mediated by the transposon Tn*10*-encoded tetracycline resistance protein. *J. Biol. Chem.* **265:**4809–4813.

191. **Yang, W., I. F. Moore, K. P. Koteva, D. C. Bareich, D. W. Hughes, and G. D. Wright.** 2004. TetX is a flavin-dependent monooxygenase conferring resistance to tetracycline antibiotics. *J. Biol. Chem.* **279:**52346–52352.

Mechanisms of Quinolone Resistance

DAVID C. HOOPER

65

Quinolones are antimicrobials that are widely used in clinical medicine. Initial members of this class were used largely for treatment of infections caused by gram-negative bacteria, but over time quinolones with increasing potency against gram-positive bacteria have been developed. With the increasing use of quinolones for the treatment of gram-positive bacterial infections, an understanding of the mechanisms of quinolone resistance in gram-positive bacteria is of considerable importance. In this chapter, I summarize current understanding of established mechanisms of resistance to this class of antimicrobial agents in gram-positive bacteria. Much of what has been found in gram-positive bacteria has parallels in gram-negative bacteria, from which much of the original data was obtained. There are, however, important differences between gram-positive and gram-negative bacteria both in target enzyme sensitivity and in the means by which efflux resistance mechanisms operate that are of clinical and fundamental importance.

ACTIVITY AND MECHANISM OF ACTION OF QUINOLONES AGAINST GRAM-POSITIVE PATHOGENS

Structure-Activity Relationships

The earliest-developed quinolones had their greatest potency against gram-negative pathogens with less activity against gram-positive pathogens, but extensive studies of the structure-activity relationships of this class of compounds identified features that enhanced potency against gram-positive pathogens (19, 20, 66). Some of these structural features have recently been incorporated into newly released congeners and additional compounds under development. Selected quinolone structures are shown in Fig. 1. All are based on a dual-ring structure with adjacent carbonyl and carboxyl groups at positions 3 and 4 of one ring, which are essential for activity. Additional substituents at other ring positions have been varied with effects on potency. The term fluoroquinolone derives from the presence of a fluorine at position 6, which is a feature of virtually all drugs of this

class developed since the 1980s. Substituents at positions 5 and 7, in particular, have been noted to have effects on activity against gram-positive bacteria. At position 5, addition of an amino group (sparfloxacin) or, to a lesser extent, a methyl group (grepafloxacin) improves activity against gram-positive bacteria relative to a hydrogen, which was commonly present in the initially developed fluoroquinolones (norfloxacin, enoxacin, ciprofloxacin, ofloxacin, and lomefloxacin). A piperazinyl ring was a common substituent at position 7, because in combination with a fluorine at position 6, it substantially improved potency against gram-negative bacteria. Addition of one (grepafloxacin, ofloxacin, or levofloxacin) or two (sparfloxacin) methyl groups to the piperazinyl ring enhanced activity against gram-positive bacteria, but an amino-substituted pyrrolidinyl ring at the 7 position of the quinolone rings (clinafloxacin) further enhanced activity against gram-positive bacteria relative to piperazinyl substituents. Derivatives of a pyrrolidinyl ring with an attached second ring (trovafloxacin, moxifloxacin, and sitafloxacin) have also been exploited for enhancement of activity against gram-positive bacteria. Most recently, quinolones lacking a fluorine at position 6 have been developed and shown to have potent activity against gram-positive bacteria (91, 92).

Type 2 Topoisomerases

Quinolones interact with both of the two type 2 topoisomerases in eubacteria, DNA gyrase and topoisomerase IV, which are essential for bacterial DNA replication. Type 2 topoisomerases act by breaking both strands of duplex DNA, passing another DNA strand through the break, and resealing the initial broken strands (32, 108, 109). This activity occurs at the expense of ATP, with ATP hydrolysis serving to reset the enzyme for another cycle of strand passage (8). DNA gyrase, which is composed of two GyrA and two GyrB subunits encoded by the *gyrA* and *gyrB* genes, respectively, is the only enzyme that introduces negative superhelical twists into and removes positive superhelical twists from bacterial DNA. Negative superhelical twists are important for initiation of DNA replication, and they facilitate binding of initiation proteins. Positive superhelical twists accumulate ahead of the

Gram-Positive Pathogens, 2nd edition, edited by Vincent A. Fischetti et al.
© 2006 ASM Press, Washington, D.C.

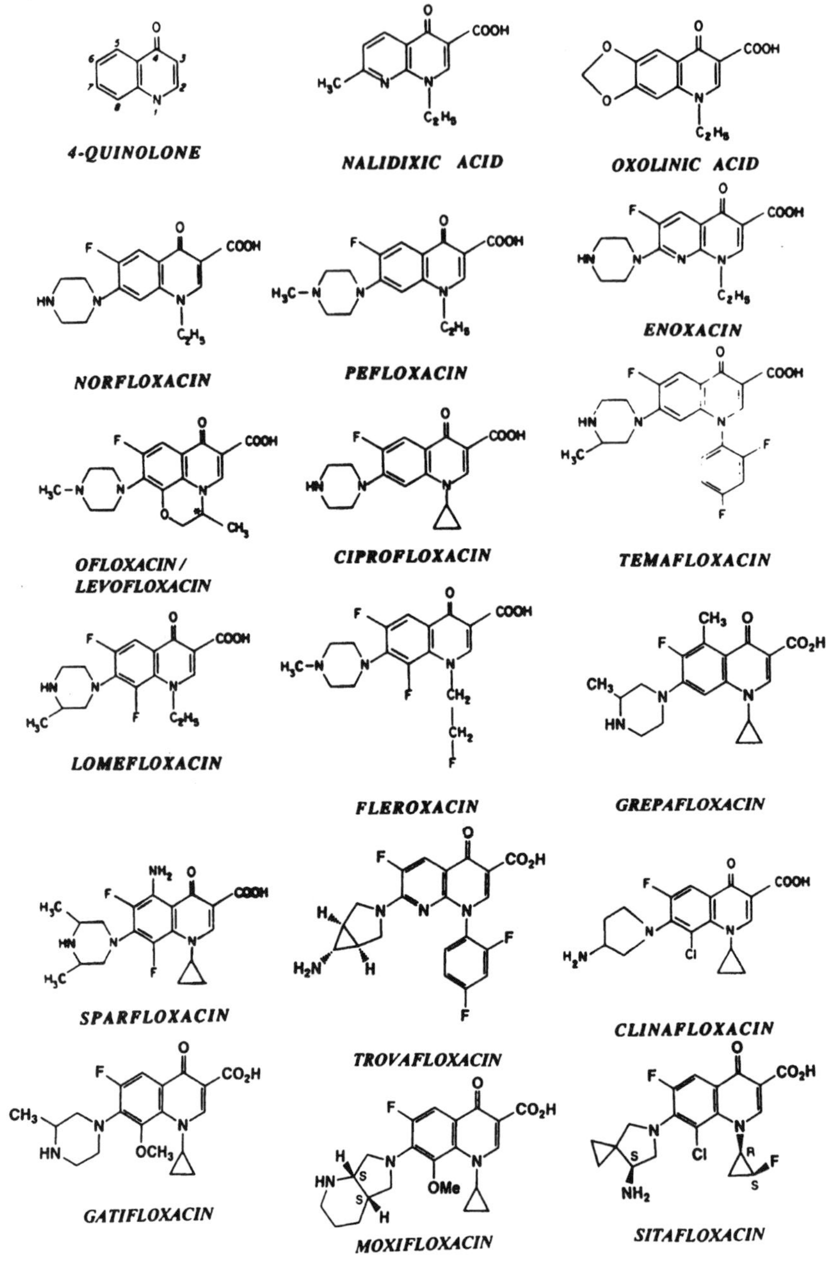

FIGURE 1 Structures of old and newer quinolones.

progressing bacterial replication fork and must be removed to prevent stalling of the fork.

The two subunits of topoisomerase IV are homologous to those of DNA gyrase. ParC (also termed GrlA in *Staphylococcus aureus*) encoded by the *parC* gene is homologous to GyrA, and ParE (also termed GrlB in *S. aureus*) encoded by the *parE* gene is homologous to GyrB (24, 57, 58). Topoisomerase IV is the principal enzyme that removes the interlinking of daughter chromosomes at the completion of a round of DNA replication and allows their segregation into daughter cells (120). It is also able to remove DNA supercoils, removing positive supercoils more efficiently than negative supercoils (17). Although these physiologic roles for topoisomerase IV and DNA gyrase have been determined largely from studies of *Escherichia coli*, it is presumed that they are fundamentally similar in gram-positive bacte-

ria. The catalytic properties of DNA gyrase and topoisomerase IV purified from *E. coli* and *S. aureus* appear to be similar for these two species, although the *S. aureus* enzymes function optimally in the presence of high concentrations of potassium glutamate (10), reflecting adaptation to the high intracellular concentrations of both of these ions in *S. aureus* (114).

Differences in Mechanism of Fluoroquinolone Action in Gram-Negative and Gram-Positive Bacteria

DNA gyrase was demonstrated to be a primary quinolone target in *E. coli*, based on genetic studies in which mutations in either the GyrA or GyrB subunit of this enzyme conferred increments in drug resistance (33, 39, 97). In contrast with *S. aureus*, initial genetic studies showed that common first-step drug resistance mutations were not in the subunits of DNA gyrase but in a distinct genetic locus, *flqA* (104). *flqA* mutants were shown subsequently to have mutations in either the ParC (GrlA) (74) or ParE (GrlB) (27) subunit of topoisomerase IV. Whereas first-step mutants have mutations in topoisomerase IV, second-step mutants have mutations in DNA gyrase (23, 41). In addition, mutations in DNA gyrase subunits of *S. aureus* when present in the absence of mutations in a subunit of topoisomerase IV cause no change in susceptibility to most quinolones, except nalidixic acid (27). Thus, topoisomerase IV in *S. aureus* is a primary target of action of many quinolones (100). Recent studies, however, suggest that sparfloxacin, nadifloxacin, and garenoxacin (BMS-284756), a novel desfluoroquinolone with potent gram-positive activity, have DNA gyrase as their primary target in *S. aureus* (18, 100), but this property is not consistent for all desfluoroquinolones (92).

These differences between *S. aureus* and *E. coli* appear to relate to the differences in the relative quinolone sensitivities of DNA gyrase and topoisomerase IV in the two species (10). For many quinolones, *E. coli* DNA gyrase is more sensitive than *E. coli* topoisomerase IV (42). In contrast, for at least some of this same group of quinolones, *S. aureus* topoisomerase IV is more sensitive than *S. aureus* DNA gyrase (10). Thus, the more sensitive of the two enzymes within a given species appears to determine the primary drug target indicated by genetic tests.

Although data are less complete, a similar pattern is apparent for some quinolones in two other gram-positive species studied. Ciprofloxacin-selected, first-step mutants of *Streptococcus pneumoniae* have mutations in the ParC or ParE subunits of topoisomerase IV, similar to those reported for *S. aureus* (80, 85). In addition in *Enterococcus faecalis*, second- but not first-step resistant mutants selected with ciprofloxacin have mutations in GyrA (59). Furthermore, in clinical resistant isolates, ParC mutations were found in low-level resistant isolates without a mutation in GyrA, and high-level resistant isolates had mutations in both ParC and GyrA, suggesting that first-step mutants may have mutations in topoisomerase IV (15, 21, 56). Thus, ciprofloxacin and other quinolones may have as their primary target topoisomerase IV rather than DNA gyrase in *S. pneumoniae* and *E. faecium*, as well as in *S. aureus*.

This pattern may vary, however, with some quinolones in some species. Sparfloxacin-selected, first-step mutants of *S. pneumoniae* have mutations in DNA gyrase, and the mutations in topoisomerase IV that affected ciprofloxacin activity have no effect on the activity of sparfloxacin or clinafloxacin (81, 82). Similar data in which first-step resistant mutants of *S. pneumoniae* have mutations in *gyrA* or *gyrB* have been reported for gatifloxacin (30) and moxifloxacin (70). These data vary with species as well, since first-step resistant mutants of *S. aureus* selected with gatifloxacin or moxifloxacin have mutations in *parC* but not *gyrA* (44, 45). Furthermore, in *S. pneumoniae*, removal of the 8-methoxy group of gatifloxacin results in a compound, AM-1121, which selects for first-step *parC* mutants (31). Data on the primary target of moxifloxacin in the gram-positive anaerobe, *Clostridium difficile*, are not yet conclusive, since although moxifloxacin-resistant clinical isolates of *C. difficile* had mutations in GyrA, first-step mutants selected directly with moxifloxacin in the laboratory did not, and mutations in ParC and ParE were not evaluated (1). Thus, the patterns of primary and secondary targets of quinolones depend on drug structure and cannot be generalized among different gram-positive species.

Although the model in which the principal target of quinolones is determined by the more sensitive of the two target enzymes is conceptually appealing for quinolones, which act as DNA poisons trapping an enzyme-DNA complex and thereby blocking progression of the DNA replication complex, there are inconsistencies that have not been fully explained. In particular, in many cases with quinolones that target DNA gyrase of *S. pneumoniae* in vivo (based on selection of first-step resistant mutants), purified topoisomerase IV is nevertheless substantially more sensitive than gyrase to these same quinolones (40, 68, 81). The approximately threefold excess of cellular levels of the subunits of DNA gyrase relative to those of topoisomerase IV were thought to be insufficient to compensate for the 20- to 40-fold differences in drug sensitivity found with the purified enzymes (83). Furthermore, in some cases first-step selections with the same quinolone may generate diverse resistant mutants, some with alterations in topoisomerase IV and others with alterations in DNA gyrase (107). In addition for *S. aureus*, single mutations in ParC or GyrA both produce increments in resistance for some quinolones (garenoxacin and gemifloxacin) that exhibit similar inhibitory activities against purified gyrase and topoisomerase IV (48, 49). Thus, differences between results in vitro and in vivo and primary and secondary target interactions suggest that other factors, as yet unidentified, may contribute to the interactions of quinolones with their target enzymes and to the phenotypic effects of resistance mutations in these targets in vivo.

RESISTANCE DUE TO ALTERED DRUG TARGET ENZYMES

Topoisomerase IV

Among antimicrobials in common clinical use for treatment of bacterial infections, quinolones and rifampin are distinctive for having predominant resistance mechanisms, due to de novo selection of alterations in drug targets in contrast to acquisition of preexisting resistance determinants.

The largest body of information concerning quinolone resistance in gram-positive bacteria comes from studies of *S. aureus*. Mutations resulting in single amino acid changes in either the ParC or ParE subunits of topoisomerase IV cause four- to eightfold increments in resistance to ciprofloxacin and many other quinolones (23, 74). ParC mutations appear more common than ParE mutations among clinical resistant strains and have clustered in the amino terminus, with position 80 being most commonly altered from serine (Ser) to either phenylalanine (Phe) or tyrosine (Tyr) (Table 1) (23, 25, 74, 93, 99, 101, 111, 117). Homologous mutations have been found in *parC* of strains of *E. coli* that are highly resistant and also have mutations in *gyrA*. Also commonly altered is position 84 from glutamic acid (Glu) to lysine (Lys) or leucine (Leu). Alterations in position 116 from alanine (Ala) to Glu or proline (Pro) are less common but also cause quinolone resistance similar to that due to mutations at positions 80 or 84. In addition, Ala116Glu/Pro mutations confer slight increases in susceptibility to novobiocin and coumermycin (28), coumarins which act as competitive inhibitors of the ATPase activity of topoisomerase IV and DNA gyrase. Mutations at position 116 are in proximity to the active site Tyr122.

In *S. pneumoniae*, similar *parC* mutations have been reported at positions 79 (Ser changed to either Tyr or Phe)

TABLE 1 Mutations in the GyrA subunit of DNA gyrase and the ParC subunit of topoisomerase IV associated with quinolone resistance[a]

Species	GyrA			ParC		
	Amino acid position	Wild-type amino acid	Mutant amino acid	Amino acid position	Wild-type amino acid	Mutant amino acid
S. aureus				23	Lys	Asn[b]
				41	Val	Gly
				43	Arg	Cys[b]
				45	Ile	Met
				48	Ala	Thr
				52	Ser	Arg
				69	Asp	Tyr[b]
				78	Gly	Cys[b]
	84	Ser	Leu,[b] Ala, Val, Lys	80	Ser	Phe,[b] Tyr[b]
	85	Ser	Pro[b]	81	Ser	Pro
	86	Glu	Lys,[b] Gly	84	Glu	Lys,[b] Leu,[b] Val, Ala, Gly, Tyr
	88	Glu	Val, Lys	103	His	Tyr
	106	Gly	Asp	116	Ala	Glu,[b] Pro[b]
				157	Pro	Leu[b]
				176	Ala	Thr,[b] Gly
				327	Asn	Lys[b]
Coagulase-negative staphylococci	84	Ser	Leu, Phe, Ala	80	Ser	Leu, Phe, Tyr
				84	Asp	Tyr, Gly, Asn
S. pneumoniae				63	Ala	Tyr[b]
				78	Asp	Asn
	81	Ser	Phe,[b] Tyr,[b] Cys	79	Ser	Phe,[b] Tyr,[b] Ala
	84	Ser	Phe,[b] Tyr[b]			
	85	Glu	Lys,[b] Gln,[b] Gly	83	Asp	Asn,[b] Gly,[b] Val
				85	Asp	Gly
				93	Lys	Glu[b]
				95	Arg	Cys
				102	His	Tyr
				115	Ala	Pro, Val
				129	Tyr	Ser
				137	Lys	Asn
S. oralis, S. mitis, Streptococcus sanguinis, Streptococcus anginosus	81	Ser	Phe,[b] Tyr[b]	79	Ser	Phe,[b] Ile, Leu,[b] Tyr, Arg
	85	Glu	Gln,[b] Lys,[b] Gly	83	Asp	Asn, His, Tyr
Streptococcus pyogenes	81	Ser	Phe	79	Ser	Tyr
	99	Met	Leu			
E. faecalis	83	Ser	Arg, Ile, Asn	80	Ser	Arg, Ile
	87	Glu	Lys, Gly	84	Glu	Ala
E. faecium	83	Ser	Arg, Leu, Ile, Tyr	80	Ser	Ile, Arg
	87	Glu	Leu, Gly, Lys	84	Glu	Lys, Thr
				97	Ser	Asn
C. difficile	83	Thr	Val, Ile			

[a]Rows reflect alignments of homologous amino acids.
[b]Amino acids for which genetic data support a role for the mutation in causing resistance. Other mutant amino acids have been associated with resistance in clinical isolates.

and 83 (aspartic acid [Asp] changed to asparagine [Asn] or glycine [Gly]) in resistant mutants selected in the laboratory and clinical resistant isolates (Table 1) (36, 53, 69, 79, 80, 82, 102). Clinical resistant isolates of *E. faecalis* have also been recently shown to have similar mutations in ParC (Ser80 changed to arginine [Arg] or isoleucine [Ile] and Glu84Ala). In one strain with low-level resistance, a mutation in ParC without a mutation in GyrA was found, and no GyrA mutants were found in the absence of a ParC mutation (56). High-level quinolone-resistant strains of *Enterococcus faecium* have been shown to contain mutations in both ParC and GyrA (21). A similar pattern has been seen in resistant clinical isolates of viridans streptococci in which a Ser79Phe was found in a low-level resistant isolate in the absence of mutations in ParE, GyrA, and GyrB, and isolates with high-level resistance had an additional mutation in GyrA or GyrB (35). Resistant isolates with Ser79Ile and Ser79Tyr ParC mutations in addition to GyrA mutations were also found, and DNA from a Ser79Tyr mutant was able to transform low-level resistance to a susceptible recipient. Although genetic studies of *E. faecalis* have not yet proved the role of ParC mutations in resistance, the analogy to the patterns found in *S. aureus* and *S. pneumoniae*, for which genetic data are available, strongly suggests that the enterococcal ParC mutations contribute to resistance.

Although mutations in ParC are most commonly found in the quinolone resistance-determining region (QRDR) of the enzyme subunit, a broader array of mutations outside this region have been found in mutants of *S. aureus* selected with new highly potent quinolones (45, 47–50). In a number of cases, mutations (Lys23Asn, Arg43Cys, Asp69Tyr, Pro157Leu, Ala176Thr, and Asn327Lys) have been proved by allelic exchange to contribute to the resistance phenotype.

Mutations in ParE of *S. aureus* (27) and *S. pneumoniae* (85) have also been found in quinolone-resistant isolates. Changes at positions 470 (*S. aureus*, from Asn to Asp) and 435 (*S. pneumoniae*, from Asp to Asn) have been shown to cause quinolone resistance; other changes at positions 422 (Glu to Asp), 432 (Asp to Gly), and 451 (Pro to Ser) have been reported in resistant clinical isolates of *S. aureus* but not yet shown to contribute to the resistance phenotype (Table 1). No ParE mutations have yet been reported in *E. faecalis*. The Asn470Asp mutation of ParE, like the ParC Ala116Glu mutation, causes hypersusceptibility to coumarins (27). These mutations are in the same region also reported for resistance mutations in ParE of *E. coli* (11).

DNA Gyrase

DNA gyrase of *S. aureus* appears substantially less sensitive to several fluoroquinolones than is *E. coli* DNA gyrase (10). Resistance mutations in GyrA in *S. aureus* and *S. pneumoniae* are generally not found in first-step mutants except those in *S. pneumoniae* selected with sparfloxacin, as noted above. Second-step mutations in either GyrA or GyrB, however, cause increased resistance in mutants with first-step mutations in ParC or ParE. Initial studies of resistant clinical isolates of *S. aureus* prior to the recognition of the role of topoisomerase IV in resistance commonly showed GyrA and (less often) GyrB mutations (52, 96), but when ParC and ParE were also evaluated, GyrA or GyrB quinolone resistance mutations were not found in the absence of mutations in topoisomerase IV (24, 94). As noted above, *gyrA* single mutants constructed in the laboratory have no fluoroquinolone resistance but interestingly exhibit a four-

fold increase in resistance to nalidixic acid, an early nonfluorinated quinolone with low potency against gram-positive bacteria (28), suggesting that nalidixic acid has limited or no activity against *S. aureus* topoisomerase IV.

Mutations in GyrA have been clustered in the QRDR, in the amino terminus, that is homologous to similar regions in *E. coli* GyrA and ParC in which resistance mutations have been clustered. Similarly, resistance mutations in GyrB of *S. aureus* are in regions similar to those involved in resistance in *E. coli* GyrB and ParE (Table 2). Mutations in GyrA are found more commonly than mutations in GyrB and most often occur at positions 83 (*S. pneumoniae*) and 84 (*S. aureus*) with changes from Ser to Leu (*S. aureus*) or to Tyr or Phe (*S. pneumoniae*). Changes at positions 86 (*S. aureus*) and 87 (*S. pneumoniae*) from Glu to Lys are also commonly reported among GyrA mutants. All of the above-mentioned GyrA mutations and the less common Ser85Pro mutation of *S. aureus* have been shown to contribute incrementally to ciprofloxacin resistance in the presence of mutant topoisomerase IV. The Ser83Phe or Tyr mutation of *S. pneumoniae* also confers resistance to sparfloxacin in the absence of mutant resistant topoisomerase IV (82), but the similar Ser84Leu GyrA mutation of *S. aureus* does not confer resistance to this or other quinolones studied in the absence of mutant topoisomerase IV (27). Other mutations as listed in Table 2 have been associated with resistance in clinical isolates but not yet been shown directly to contribute to resistance. Noteworthy is the similarity of the mutations in GyrA found in resistant clinical isolates of *E. faecalis* (Ser83Arg [or Ile or Asn] and Glu87Lys [or Gly]) to those found in *S. aureus* and *S. pneumoniae* (56, 59, 102). In the one study that evaluated both ParC and GyrA, no GyrA mutations were found in the absence of a ParC mutation (56). A GyrA subunit purified from another highly resistant but genetically uncharacterized clinical strain of *E. faecalis* when combined with a wild-type GyrB subunit was also shown to confer quinolone resistance on enzymatic activity (71). Thus, it is likely that resistance caused by GyrA mutation in *E. faecalis* is highly similar to that in *S. pneumoniae* and *S. aureus*.

GyrB mutations are found least often in laboratory and clinical resistant isolates and have been found at only two positions, 437 (*S. aureus*) and 435 (*S. pneumoniae*) with a change from Asp to Asn and position 458 (*S. aureus*) with a change from Arg to Glu. No GyrB resistance mutations in *E. faecalis* have yet been reported. Data indicate that each of these mutations contributes to resistance only in the presence of mutant topoisomerase IV.

Models for How Altered Drug Targets Cause Resistance

Direct data on quinolone interactions with topoisomerases come from studies of *E. coli* gyrase, for which it has been shown that quinolones bind to a complex of DNA and gyrase rather than gyrase alone (95). The Ser83Trp mutation in GyrA was also associated with reduced binding of norfloxacin to the gyrase-DNA complex (115). Thus, reductions in the affinity of drug for the enzyme-DNA target may mediate resistance for some classes of mutants, particularly those with mutations in and around position 83 (position 84 for *S. aureus* GyrA and position 80 for *S. aureus* ParC) in the QRDR. The X-ray crystallographic structure of a fragment of *E. coli* GyrA localizes the QRDR to a positively charged surface along which DNA has been modeled to bind (67). This region is adjacent to the two tyrosines (Tyr

TABLE 2 Mutations in the GyrB subunit of DNA gyrase and the ParE subunit of topoisomerase IV associated with quinolone resistance[a]

Species	GyrB			ParE		
	Amino acid position	Wild-type amino acid	Mutant amino acid	Amino acid position	Wild-type amino acid	Mutant amino acid
S. aureus				25	Pro	His[b]
				410	Ser	Pro
				422	Glu	Asp
	437	Asp	Asn[b]	432	Asp	Asn,[b] Gly, Val
				451	Pro	Gln,[b] Ser
	458	Arg	Gln[b]			
				470	Asn	Asp,[b] Ile[b]
	477	Glu	Ala	472	Glu	Val, Lys
				478	His	Tyr
S. pneumoniae				103	His	Tyr[b]
	406	Gly	Ser, Asp			
				431	Leu	Ile
	435	Asp	Asn,[b] Glu,[b] Val	435	Asp	Asn,[b] Val
	440	Ser	Tyr			
	445	Arg	Ser			
				447	Arg	Cys, Ser
	454	Pro	Ser	454	Pro	Ser[b]
				473	Asn	Ile
	474	Glu	Lys[b]	474	Glu	Lys
	475	Glu	Val			
				476	Ile	Phe
S. mitis, S. oralis				424	Pro	Gln
				474	Glu	Lys
	494	Ser	Thr			

[a]Rows reflect alignments of homologous amino acids.
[b]Amino acids for which genetic data support a role for the mutation in causing resistance. Other mutant amino acids have been associated with resistance in clinical isolates.

122, one from each GyrA subunit) that are linked to DNA during strand breakage and might be considered as a candidate for the site of quinolone binding. Thus, in one model, amino acid changes in the QRDR of GyrA (and, by homology, ParC) alter the structure of a quinolone-binding site near the interface of the enzyme and DNA, and resistance is effected by reduced drug affinity for the modified enzyme-DNA complex. As yet, no topoisomerase crystal structures that include a quinolone molecule have been reported to enable direct assessment of this model.

Resistance mediated by mutations in ParE or GyrB might act by a different mechanism based on structural data. The crystal structure of a fragment of topoisomerase II of yeast, which has homology with both GyrA (carboxy terminus) and GyrB (amino terminus), suggested that the region of quinolone resistance mutations in ParE of S. aureus and S. pneumoniae is distant from the QRDR region of GyrA or ParC (9), but in other conformations these regions appear to be in proximity (22), suggesting that a quinolone-binding site involving residues of both GyrA (or ParC) and GyrB (or ParE) could form at one stage of the enzyme catalytic cycle.

Because quinolones act as poisons with a quinolone-enzyme-DNA complex serving to block movement of the DNA replication fork, mutants or conditions that reduce the number of drug target complexes of enzyme and DNA could cause drug resistance. This mechanism of resistance has been seen for amsacrine, which targets yeast topoiso-

merase II in a manner similar to the action of quinolones on bacterial topoisomerases. Some amsacrine-resistant yeast mutants have mutations encoding topoisomerase II with reduced catalytic function (112). The coumarin hypersusceptibility of S. aureus ParE mutants suggests the possibility that impairment of catalytic function may contribute to quinolone resistance in these mutants (27). Because ParC116 mutants also exhibit a coumarin hypersusceptibility phenotype and are in proximity to the Tyr122 active site, they too might be postulated to have such a mechanism (28). Topoisomerase IV reconstituted with this and other novel ParC subunits with resistance mutations outside the QRDR exhibit reduced catalytic function (X. Zhang and D. Hooper, unpublished data) relative to the wild-type enzyme or topoisomerase IV containing ParC with the common QRDR mutation (Phe80). Most recently for S. aureus a mutation upstream of parEC has been shown to cause reduced levels of ParE and resistance to a range of quinolones, establishing reduced amounts of target enzyme as an additional mechanism of quinolone resistance in S. aureus (46), as it is for amsacrine resistance in yeast. This mutant also demonstrated a growth defect that could be compensated for by overexpression of topB (encoding topoisomerase III) and gyrBA, the gene products of which have DNA decatenating activity that may complement the reduced amounts of decatenation activity resulting from reduced topoisomerase IV.

Interactions of Resistance Mutations Affecting Dual Targets

Stepwise incremental quinolone resistance often occurs by a series of mutations, first in the primary enzyme target and followed by mutations in the secondary enzyme target. Second and third mutations in primary and secondary targets may follow with further increments in resistance. The level of resistance conferred by an individual mutation is predicted to be determined by at least two factors, the extent to which the mutation alters the sensitivity of the mutant enzyme and the level of sensitivity of the other target enzyme. Thus, after mutation, whichever of the two topoisomerases (the mutant primary target topoisomerase and the wild-type secondary target topoisomerase) is the more sensitive determines the ceiling on the increment in resistance.

This sequence of resistance mutations implies that when there are differences in the sensitivity of the two target enzymes to a particular quinolone, a single spontaneous mutation may cause some increment in resistance and could be selected at a frequency typical of spontaneous mutations. In contrast, for a quinolone congener with equal activity against both topoisomerase IV and DNA gyrase, two mutations, one in each target, would need to be present simultaneously for the first step in resistance due to an altered target to occur, and thus resistance would be substantially less frequent. This association of a low frequency of selection of resistant mutants ($<10^{-10}$) and similar drug activities against both gyrase and topoisomerase IV has been seen with clinafloxacin in S. pneumoniae (81) and with garenoxacin and gemifloxacin in S. aureus (48, 49).

RESISTANCE DUE TO ALTERED ACCESS OF DRUG TO TARGET ENZYMES

Quinolones must traverse the cell wall and cytoplasmic membrane of gram-positive bacteria to reach DNA gyrase and topoisomerase IV present in the cytoplasm. The cell wall itself is thought to provide little or no barrier to diffusion of small molecules such as quinolones, which have molecular masses of 300 to 400 Da. Accumulation of quinolones by whole cells of S. aureus is nonsaturable under usual experimental conditions and likely occurs by simple diffusion across the cytoplasmic membrane. All active quinolones have a negatively charged carboxyl group at position 3, and most quinolones developed since 1985 have an additional positively charged group at position 7 (piperazinyl or pyrrolidinyl ring derivatives) and thus are zwitterionic. The proportions of positively charged, negatively charged, dually charged, and uncharged species of a given quinolone vary with pH, and it is presumed that it is the uncharged species that diffuses freely across the membrane and reaches equilibrium with the cytoplasm (76). Differences in the pH between the medium and the cytoplasm thus may affect partitioning of drug by altering the proportions of charged species that are trapped in the cytoplasmic compartment. These factors are likely responsible for the reductions in activity of zwitterionic quinolone congeners that occur when the pH of the environment falls below 7. Little is known about the nonspecific binding of quinolones to bacterial cytoplasmic proteins or any compartmentalization of quinolones within the bacterial cell.

Active transport of fluoroquinolones has been shown to contribute to quinolone resistance in a number of gram-positive bacteria as will be discussed below. In all cases identified thus far, the transporters have had broad substrate pro-files; quinolones, which are synthetic compounds, appear to be incidental substrates for these pumps. No specific quinolone transporters have been reported. Regulation of expression of these multidrug resistance (MDR) pumps by changes in physiologic conditions or mutation modulates the level of drug resistance, and characteristically each bacterium has multiple pumps with differing substrate and regulatory profiles.

NorA of S. aureus

Acquired quinolone resistance that correlates with reduced quinolone accumulation in gram-positive bacteria has been identified for S. aureus and is due to active efflux of drug across the cytoplasmic membrane. Resistance of this type has been shown to be due to increased levels of expression of norA, a chromosomal gene that encodes a protein with 12 predicted membrane-spanning domains that is a member of the major facilitator superfamily of transporters (75, 118). MDR transporters of this type have broad substrate profiles, and NorA mediates pleiotropic resistance that includes ethidium bromide, rhodamine 6G, tetraphenylphosphonium (TPP), chloramphenicol, and hydrophilic quinolones such as norfloxacin, ciprofloxacin, and ofloxacin (73, 73, 118). The activities of hydrophobic quinolones such as sparfloxacin, moxifloxacin, and trovafloxacin are less affected by NorA. Cloned norA mediates a similar resistance phenotype in E. coli. In everted membrane vesicles prepared from E. coli cells containing cloned norA but not cells containing the vector plasmid alone, uptake of labeled norfloxacin (which represents drug efflux because of the reversed membrane orientation in everted vesicles) is saturable and dependent on the energy generated by the proton gradient across the membrane (75). Norfloxacin uptake in everted vesicles and the resistance phenotype associated with norA expression are inhibited by reserpine and verapamil (63, 75), which also inhibit other MDR transporters. Purified NorA has been reconstituted into liposomes and shown to transport the dye Hoechst 33342 out of the liposomal membrane in response to an artificial proton gradient (119). Transport of Hoechst 33342 by NorA in everted membranes is competitively inhibited by the hydrophilic quinolones norfloxacin and ciprofloxacin and noncompetitively inhibited by the hydrophobic quinolone sparfloxacin, indicating that the hydrophilic quinolones and the dye share a transport-binding site or pathway and that the NorA-binding site of sparfloxacin differs from that of norfloxacin and ciprofloxacin. The specific quinolone-binding sites on NorA are not yet known.

The level of expression of norA is regulated. Increased steady-state levels of norA mRNA and pleiotropic resistance are associated with single nucleotide changes upstream of norA (55, 55, 75). A change from T to G at a position four nucleotides downstream of the transcriptional start site in the 5' untranslated region of the gene has been associated with increased stability of norA transcripts (29). Mutation in the two-component regulatory system arlRS produces low-level quinolone resistance and a two- to threefold increase in norA expression (26). It is not yet known if the effect of arlRS on norA expression is direct or indirect, but another global regulator, MgrA (also initially termed NorR), was identified by its binding to the norA promoter-operator region. Overexpression of mgrA from a plasmid increases expression of norA and an increase in resistance to NorA substrates (105). MgrA is member of the MarR family of transcriptional regulators and is predicted to have a winged helix structure similar to that found for MarR (4). An mgrA::cat knockout mutant was also found unexpectedly to have an MDR phenotype without a change in the levels of *norA* transcripts, suggesting that MgrA also serves as a negative regulator of other MDR pumps (105). Included in the resistance profile of the *mgrA::cat* mutant was increased resistance to sparfloxacin and moxifloxacin, quinolones that are not transported by NorA. Another chromosomally encoded efflux pump termed NorB, which is related to NorA, is overexpressed in the *mgrA* mutant and contributes to its resistance phenotype. Cloned *norB* confers resistance to sparfloxacin, moxifloxacin, and other drugs found in the *mgrA* mutant phenotype (104a). Thus, MgrA appears to have a central role in the regulation of efflux pumps in *S. aureus*, positively regulating NorA and negatively regulating others, such as NorB. MgrA is also involved in the regulation of capsule synthesis and autolysin production (51, 60), further emphasizing its central regulatory role in cellular physiology.

A mutant of *S. aureus* in which *norA* expression appeared to be induced by norfloxacin has also been described, but the nature of the mutation was not defined (55). Resistance has also been associated with a single mutation in the *norA* structural gene (78), but little is known about structure-activity relationships of NorA itself.

Also undefined are the normal functions of NorA in the cell. Although *norA* appears to be a nonessential gene (43), based on the isolation of viable knockout mutants, it appears to be present commonly if not universally on the chromosomes of clinical isolates of *S. aureus* and coagulase-negative staphylococci (55). The extent to which mutations that increase *norA* expression or enhance its function contribute to resistance in clinical isolates of *S. aureus* is not yet certain. Blocking of NorA function with reserpine, however, increases quinolone susceptibility (75) and reduces the frequency of selection of resistant mutants with norfloxacin (64), suggesting a role for NorA in determining quinolone susceptibility even in wild-type staphylococci.

Bmr, Blt, and Bmr3 of *Bacillus subtilis*

The Bmr protein of *B. subtilis* has 44% amino acid identity to NorA of *S. aureus* and also belongs to the major facilitator superfamily (72, 73). The expression of the chromosomal *bmr* gene results in a resistance phenotype similar to that due to expression of *norA*. Expression of *bmr* is induced by rhodamine 6G and TPP, two Bmr substrates, and *bmrR*, which is upstream of *bmr*, has homology to other transcriptional activator proteins (2). BmrR has been shown to bind the *bmr* promoter with increased affinity in the presence of these inducers, and the purified carboxy-terminal half of the Bmr protein has also been shown to bind rhodamine and TPP (63). The crystal structure of BmrR bound with TPP alone (122) and together with the *bmr* promoter (121), as well as the targeted structural modifications of many pumps (90), has suggested the structural basis for the binding of multiple structurally diverse molecules to BmrR that may be relevant to the mechanism by which the pump itself recognizes such a wide variety of structural substrates, many of which are either amphiphilic with a positive or neutral charge or hydrophobic. This general model includes a pocket with hydrophobic and nonpolar residues and a buried negatively charged amino acid, such as Glu, which is exposed on drug binding, stabilizing the drug-protein interaction. Glu residues located in a hydrophobic domain have been found in many secondary transporters and are often essential for function. Kinetic experiments also suggest, however, that transporters can have different functional sites of binding for different substrates (88, 90, 119).

A related transporter with an apparently identical substrate profile in *B. subtilis*, Blt, is not expressed in wild-type cells (3). Its expression is regulated by the product of the upstream gene, *bltR*, which is not induced by rhodamine. There is also an additional downstream gene, *bltD*, which is cotranscribed with *blt* and appears to encode a specific polyamine acetyltransferase (116). Blt also appears to efflux the polyamine spermidine into the medium. Thus, *bltD* and *blt* appear to have related, specific physiologic functions in the cell.

The most recently identified MDR transporter of *B. subtilis*, Bmr3, has 14 membrane-spanning segments; when overexpressed, it confers resistance to puromycin and to the quinolones norfloxacin and tosufloxacin, but not levofloxacin (77). Disruption of *bmr3* does not affect intrinsic susceptibility, indicating that the gene is not essential and is little expressed in wild-type cells. The level of *bmr3* transcripts appears to be lower in the late log phase of growth than early log phase.

PmrA of *S. pneumoniae*

Quinolone-resistant clinical and laboratory strains of *S. pneumoniae* have been shown to have reduced accumulation of quinolones that is reversible with reserpine, suggesting the involvement of an efflux system(s) in quinolone resistance in this organism as well (5, 13). In addition, a *norA* homolog, *pmrA*, has been identified in *S. pneumoniae*, and its inactivation by introduction of a *pmrA::cat* insertion in the chromosome caused increased quinolone susceptibility and decreased whole-cell efflux of ethidium bromide (34). Reserpine was also able to reduce the selection of ciprofloxacin-resistant mutants of *S. pneumoniae* (62), as it was in *S. aureus* (64). As many as one-third of resistant isolates studied appear to have at least a partially reserpine-reversible resistance phenotype, suggesting a common but not exclusive role for drug efflux, since target enzyme mutations were commonly present as well (7, 14, 16). A similar magnitude of a reserpine effect (two- to fourfold) can, however, be seen in susceptible isolates as well; the *pmrA* transcript levels correlated poorly with the level of quinolone resistance and MDR phenotype (87). Thus, efflux pump expression occurs in wild-type bacteria, and the extent to which mutational quinolone resistance in *S. pneumoniae* is due to overexpression of PmrA or other pumps is uncertain. A reserpine-reversible MDR phenotype that includes quinolones can also be selected in *pmrA::cat* mutants, suggesting that *S. pneumoniae* has other efflux pumps in addition to PmrA that can affect quinolone susceptibility (12, 86). The nature of mutations in *S. pneumoniae* that can lead to increased expression of PmrA or other efflux pumps is unknown.

Quinolone-resistant clinical isolates of viridans streptococci have also been shown to have an efflux phenotype defined as lower MICs of quinolones in the presence of reserpine. DNA from such strains of *Streptococcus mitis* and *Streptococcus oralis* was able to transform *S. pneumoniae* to this phenotype in the laboratory. Transformation of resistance determinants from viridans streptococci to pneumococci, however, appears to be uncommon among ciprofloxacin-resistant clinical isolates of *S. pneumoniae* (6).

Other Efflux Transporters in Gram-Positive Bacteria

Other MDR-type transporters have been identified in gram-positive bacteria. LmrA of *Lactococcus lactis* confers resistance to quinolones and multiple other compounds (90) and is distinctive because it is a member of the ABC transporter family, which utilizes ATP hydrolysis, rather than the membrane proton gradient, as a source of energy (61, 106). The other MDR transporter of *L. lactis*, LmrP, a secondary transporter the expression of which is increased when LmrA is inactivated, does not include quinolones in its substrate profile (89).

Some highly quinolone-resistant clinical isolates of *E. faecalis* also appear to accumulate smaller amounts of norfloxacin than susceptible strains, suggesting a possible role for efflux in resistance in this organism as well (71). Recently, a homolog of *norA*, named *emeA*, was identified in *E. faecalis* and shown to contribute to the level of susceptibility to norfloxacin and ethidium bromide in a wild-type strain and to the reserpine-reversible efflux of ethidium bromide from intact cells (54). The extent to which quinolone resistance in *E. faecalis* is caused by overexpression of or structural changes in EmeA remains to be determined.

OTHER POSSIBLE MECHANISMS OF RESISTANCE

No specific quinolone-degrading enzymes have been reported as a mechanism of resistance, but fungi that are capable of degrading quinolones through metabolic pathways have been reported (113). Plasmid-mediated resistance to quinolones in gram-negative bacteria has been identified (65, 110). Resistance is due to plasmid-encoded Qnr, a member of the pentapeptide repeat family of proteins that protects gyrase from quinolone action (103). No such plasmid-mediated quinolone resistance has been reported for natural isolates of gram-positive bacteria. One possible form of plasmid-mediated quinolone resistance in gram-positive bacteria, however, is due to MDR pumps. The QacA/B (84) and Smr (37) MDR pumps, which can cause resistance to other antimicrobials but not quinolones, are encoded on plasmids in *S. aureus*, and the *norA* gene cloned on a shuttle plasmid confers quinolone resistance in *S. aureus*, presumably due to increased gene dosage (118). In addition, mutant resistant alleles of *parC* and *parE* of topoisomerase IV present on shuttle plasmids exhibit codominance over their chromosomal wild-type alleles and confer resistance when introduced into wild-type *S. aureus* (27, 28). Overexpression of *norA* and genes for topoisomerases from plasmids are known, however, to have toxic effects on the cell that may limit the fitness of resistant bacteria containing them (38, 98). Thus, at present quinolone resistance in gram-positive bacteria is attributable exclusively to chromosomal mutations that affect quinolone targets or quinolone permeation to these targets.

REFERENCES

1. **Ackermann, G., Y. J. Tang, R. Kueper, P. Heisig, A. C. Rodloff, J. Silva, Jr., and S. H. Cohen.** 2001. Resistance to moxifloxacin in toxigenic *Clostridium difficile* isolates is associated with mutations in *gyrA. Antimicrob. Agents Chemother.* 45:2348–2353.

2. **Ahmed, M., C. M. Borsch, S. S. Taylor, N. Vazquez-Laslop, and A. A. Neyfakh.** 1994. A protein that activates expression of a multidrug efflux transporter upon binding the transporter substrates. *J. Biol. Chem.* 269:28506–28513.

3. **Ahmed, M., L. Lyass, P. N. Markham, S. S. Taylor, N. Vazquez-Laslop, and A. A. Neyfakh.** 1995. Two highly similar multidrug transporters of *Bacillus subtilis* whose expression is differentially regulated. *J. Bacteriol.* 177:3904–3910.

4. **Alekshun, M. N., S. B. Levy, T. R. Mealy, B. A. Seaton, and J. F. Head.** 2001. The crystal structure of MarR, a regulator of multiple antibiotic resistance, at 2.3 Å resolution. *Nat. Struct. Biol.* 8:710–714.

5. **Baranova, N. N., and A. A. Neyfakh.** 1997. Apparent involvement of a multidrug transporter in the fluoroquinolone resistance of *Streptococcus pneumoniae. Antimicrob. Agents Chemother.* 41:1396–1398.

6. **Bast, D. J., J. C. S. De Azavedo, T. Y. Tam, L. Kilburn, C. Duncan, L. A. Mandell, R. J. Davidson, and D. E. Low.** 2001. Interspecies recombination contributes minimally to fluoroquinolone resistance in *Streptococcus pneumoniae. Antimicrob. Agents Chemother.* 45:2631–2634.

7. **Bast, D. J., D. E. Low, C. L. Duncan, L. Kilburn, L. A. Mandell, R. J. Davidson, and J. C. S. De Azavedo.** 2000. Fluoroquinolone resistance in clinical isolates of *Streptococcus pneumoniae*: contributions of type II topoisomerase mutations and efflux to levels of resistance. *Antimicrob. Agents Chemother.* 44:3049–3054.

8. **Bates, A. D., M. H. O'Dea, and M. Gellert.** 1996. Energy coupling in *Escherichia coli* DNA gyrase: the relationship between nucleotide binding, strand passage, and DNA supercoiling. *Biochemistry* 35:1408–1416.

9. **Berger, J. M., S. J. Gamblin, S. C. Harrison, and J. C. Wang.** 1996. Structure and mechanism of DNA topoisomerase II. *Nature* 379:225–232.

10. **Blanche, F., B. Cameron, F. X. Bernard, L. Maton, B. Manse, L. Ferrero, N. Ratet, C. Lecoq, A. Goniot, D. Bisch, and J. Crouzet.** 1996. Differential behaviors of *Staphylococcus aureus* and *Escherichia coli* type II DNA topoisomerases. *Antimicrob. Agents Chemother.* 40:2714–2720.

11. **Breines, D. M., S. Ouabdesselam, E. Y. Ng, J. Tankovic, S. Shah, C. J. Soussy, and D. C. Hooper.** 1997. Quinolone resistance locus *nfxD* of *Escherichia coli* is a mutant allele of *parE* gene encoding a subunit of topoisomerase IV. *Antimicrob. Agents Chemother.* 41:175–179.

12. **Brenwald, N. P., P. Appelbaum, T. Davies, and M. J. Gill.** 2003. Evidence for efflux pumps, other than PmrA, associated with fluoroquinolone resistance in *Streptococcus pneumoniae. Clin. Microbiol. Infect.* 9:140–143.

13. **Brenwald, N. P., M. J. Gill, and R. Wise.** 1997. The effect of reserpine, an inhibitor of multidrug efflux pumps, on the in-vitro susceptibilities of fluoroquinolone-resistant strains of *Streptococcus pneumoniae* to norfloxacin. *J. Antimicrob. Chemother.* 40:458–460.

14. **Brenwald, N. P., M. J. Gill, and R. Wise.** 1998. Prevalence of a putative efflux mechanism among fluoroquinolone-resistant clinical isolates of *Streptococcus pneumoniae. Antimicrob. Agents Chemother.* 42:2032–2035.

15. **Brisse, S., A. C. Fluit, U. Wagner, P. Heisig, D. Milatovic, J. Verhoef, S. Scheuring, K. Köhrer, and F. J. Schmitz.** 1999. Association of alterations in ParC and GyrA proteins with resistance of clinical isolates of *Enterococcus faecium* to nine different fluoroquinolones. *Antimicrob. Agents Chemother.* 43:2513–2516.

16. **Broskey, J., K. Coleman, M. N. Gwynn, L. McCloskey, C. Traini, L. Voelker, and R. Warren.** 2000. Efflux and target mutations as quinolone resistance mechanisms in clinical isolates of *Streptococcus pneumoniae. J. Antimicrob. Chemother.* 45:95–99.

17. **Crisona, N. J., T. R. Strick, D. Bensimon, V. Croquette, and N. R. Cozzarelli.** 2000. Preferential relaxation of positively supercoiled DNA by *E. coli* topoisomerase IV in single-molecule and ensemble measurements. *Genes Dev.* 14:2881–2892.

18. **Discotto, L. F., L. E. Lawrence, K. L. Denbleyker, and J. F. Barrett.** 2001. *Staphylococcus aureus* mutants selected

by BMS-284756. *Antimicrob. Agents Chemother.* **45**:3273–3275.

19. **Domagala, J. M.** 1994. Structure-activity and structure-side-effect relationships for the quinolone antibacterials. *J. Antimicrob. Chemother.* **33**:685–706.

20. **Domagala, J. M., and S. E. Hagen.** 2003. Structure-activity relationships of the quinolone antibacterials in the new millennium: some things change and some do not, p. 3–18. *In* D. C. Hooper and E. Rubinstein (ed.), *Quinolone Antimicrobial Agents.* ASM Press, Washington, D.C.

21. **El Amin, N., S. Jalal, and B. Wretlind.** 1999. Alterations in GyrA and ParC associated with fluoroquinolone resistance in *Enterococcus faecium. Antimicrob. Agents Chemother.* **43**:947–949.

22. **Fass, D., C. E. Bogden, and J. M. Berger.** 1999. Quaternary changes in topoisomerase II may direct orthogonal movement of two DNA strands. *Nature Struct. Biol.* **6**:322–326.

23. **Ferrero, L., B. Cameron, and J. Crouzet.** 1995. Analysis of *gyrA* and *grlA* mutations in stepwise-selected ciprofloxacin-resistant mutants of *Staphylococcus aureus. Antimicrob. Agents Chemother.* **39**:1554–1558.

24. **Ferrero, L., B. Cameron, B. Manse, D. Lagneaux, J. Crouzet, A. Famechon, and F. Blanche.** 1994. Cloning and primary structure of *Staphylococcus aureus* DNA topoisomerase IV: a primary target of fluoroquinolones. *Mol. Microbiol.* **13**:641–653.

25. **Fitzgibbon, J. E., J. F. John, J. L. Delucia, and D. T. Dubin.** 1998. Topoisomerase mutations in trovafloxacin-resistant *Staphylococcus aureus. Antimicrob. Agents Chemother.* **42**:2122–2124.

26. **Fournier, B., R. Aras, and D. C. Hooper.** 2000. Expression of the multidrug resistance transporter NorA from *Staphylococcus aureus* is modified by a two-component regulatory system. *J. Bacteriol.* **182**:664–671.

27. **Fournier, B., and D. C. Hooper.** 1998. Mutations in topoisomerase IV and DNA gyrase of *Staphylococcus aureus*: novel pleiotropic effects on quinolone and coumarin activity. *Antimicrob. Agents Chemother.* **42**:121–128.

28. **Fournier, B., and D. C. Hooper.** 1998. Effects of mutations in GrlA of topoisomerase IV from *Staphylococcus aureus* on quinolone and coumarin activity. *Antimicrob. Agents Chemother.* **42**:2109–2112.

29. **Fournier, B., Q. C. Truong-Bolduc, X. Zhang, and D. C. Hooper.** 2001. A mutation in the 5′ untranslated region increases stability of *norA* mRNA, encoding a multidrug resistance transporter of *Staphylococcus aureus. J. Bacteriol.* **183**:2367–2371.

30. **Fukuda, H., and K. Hiramatsu.** 1999. Primary targets of fluoroquinolones in *Streptococcus pneumoniae. Antimicrob. Agents Chemother.* **43**:410–412.

31. **Fukuda, H., R. Kishii, M. Takei, and M. Hosaka.** 2001. Contributions of the 8-methoxy group of gatifloxacin to resistance selectivity, target preference, and antibacterial activity against *Streptococcus pneumoniae. Antimicrob. Agents Chemother.* **45**:1649–1653.

32. **Gellert, M.** 1981. DNA topoisomerases. *Annu. Rev. Biochem.* **50**:879–910.

33. **Gellert, M., K. Mizuuchi, M. H. O'Dea, T. Itoh, and J. I. Tomizawa.** 1977. Nalidixic acid resistance: a second genetic character involved in DNA gyrase activity. *Proc. Natl. Acad. Sci. USA* **74**:4772–4776.

34. **Gill, M. J., N. P. Brenwald, and R. Wise.** 1999. Identification of an efflux pump gene, *pmrA*, associated with fluoroquinolone resistance in *Streptococcus pneumoniae. Antimicrob. Agents Chemother.* **43**:187–189.

35. **González, I., M. Georgiou, F. Alcaide, D. Balas, and A. G. de la Campa.** 1998. Fluoroquinolone resistance mutations in the *parC*, *parE*, and *gyrA* genes of clinical isolates of viridans group streptococci. *Antimicrob. Agents Chemother.* **42**:2792–2798.

36. **Gootz, T. D., R. Zaniewski, S. Haskell, B. Schmieder, J. Tankovic, D. Girard, P. Courvalin, and R. J. Polzer.** 1996. Activity of the new fluoroquinolone trovafloxacin (CP-99,219) against DNA gyrase and topoisomerase IV mutants of *Streptococcus pneumoniae* selected in vitro. *Antimicrob. Agents Chemother.* **40**:2691–2697.

37. **Grinius, L. L., and E. B. Goldberg.** 1994. Bacterial multidrug resistance is due to a single membrane protein which functions as a drug pump. *J. Biol. Chem.* **269**:29998–30004.

38. **Grkovic, S., M. H. Brown, and R. A. Skurray.** 2002. Regulation of bacterial drug export systems. *Microbiol. Mol. Biol. Rev.* **66**:671–701.

39. **Hane, M. W., and T. H. Wood.** 1969. *Escherichia coli* K-12 mutants resistant to nalidixic acid: genetic mapping and dominance studies. *J. Bacteriol.* **99**:238–241.

40. **Heaton, V. J., J. E. Ambler, and L. M. Fisher.** 2000. Potent antipneumococcal activity of gemifloxacin is associated with dual targeting of gyrase and topoisomerase IV, an in vivo target preference for gyrase, and enhanced stabilization of cleavable complexes in vitro. *Antimicrob. Agents Chemother.* **44**:3112–3117.

41. **Hori, S., Y. Ohshita, Y. Utsui, and K. Hiramatsu.** 1993. Sequential acquisition of norfloxacin and ofloxacin resistance by methicillin-resistant and -susceptible *Staphylococcus aureus. Antimicrob. Agents Chemother.* **37**:2278–2284.

42. **Hoshino, K., A. Kitamura, I. Morrissey, K. Sato, J. Kato, and H. Ikeda.** 1994. Comparison of inhibition of *Escherichia coli* topoisomerase IV by quinolones with DNA gyrase inhibition. *Antimicrob. Agents Chemother.* **38**:2623–2627.

43. **Hsieh, P. C., S. A. Siegel, B. Rogers, D. Davis, and K. Lewis.** 1998. Bacteria lacking a multidrug pump: a sensitive tool for drug discovery. *Proc. Natl. Acad. Sci. USA* **95**:6602–6606.

44. **Ince, D., R. Aras, and D. C. Hooper.** 1999. Mechanisms and frequency of resistance to moxifloxacin in comparison with ciprofloxacin in *Staphylococcus aureus. Drugs* **58**:132–133.

45. **Ince, D., and D. C. Hooper.** 2001. Mechanisms and frequency of resistance to gatifloxacin in comparison to AM-1121 and ciprofloxacin in *Staphylococcus aureus. Antimicrob. Agents Chemother.* **45**:2755–2764.

46. **Ince, D., and D. C. Hooper.** 2003. Quinolone resistance due to reduced target enzyme expression. *J. Bacteriol.* **185**:6883-6892.

47. **Ince, D., and D. C. Hooper.** 2000. Mechanisms and frequency of resistance to premafloxacin in *Staphylococcus aureus*: novel mutations suggest novel drug-target interactions. *Antimicrob. Agents Chemother.* **44**:3344–3350.

48. **Ince, D., X. Zhang, L. C. Silver, and D. C. Hooper.** 2003. Topoisomerase targeting with and resistance to gemifloxacin in *Staphylococcus aureus. Antimicrob. Agents Chemother.* **47**:274–282.

49. **Ince, D., X. Zhang, L. C. Silver, and D. C. Hooper.** 2002. Dual targeting of DNA gyrase and topoisomerase IV: target interactions of garenoxacin (BMS-284756, T3811ME), a new desfluoroquinolone. *Antimicrob. Agents Chemother.* **46**:3370–3380.

50. **Ince, D., X. M. Zhang, and D. C. Hooper.** 2003. Activity of and resistance to moxifloxacin in *Staphylococcus aureus. Antimicrob. Agents Chemother.* **47**:1410–1415.

51. **Ingavale, S. S., W. Van Wamel, and A. L. Cheung.** 2003. Characterization of RAT, an autolysis regulator in *Staphylococcus aureus*. *Mol. Microbiol.* **48:**1451–1466.

52. **Ito, H., H. Yoshida, M. Bogaki-Shonai, T. Niga, H. Hattori, and S. Nakamura.** 1994. Quinolone resistance mutations in the DNA gyrase *gyrA* and *gyrB* genes of *Staphylococcus aureus*. *Antimicrob. Agents Chemother.* **38:**2014–2023.

53. **Janoir, C., V. Zeller, M. D. Kitzis, N. J. Moreau, and L. Gutmann.** 1996. High-level fluoroquinolone resistance in *Streptococcus pneumoniae* requires mutations in *parC* and *gyrA*. *Antimicrob. Agents Chemother.* **40:**2760–2764.

54. **Jonas, B. M., B. E. Murray, and G. M. Weinstock.** 2001. Characterization of *emeA*, a *norA* homolog and multidrug resistance efflux pump, in *Enterococcus faecalis*. *Antimicrob. Agents Chemother.* **45:**3574–3579.

55. **Kaatz, G. W., and S. M. Seo.** 1995. Inducible NorA-mediated multidrug resistance in *Staphylococcus aureus*. *Antimicrob. Agents Chemother.* **39:**2650–2655.

56. **Kanematsu, E., T. Deguchi, M. Yasuda, T. Kawamura, Y. Nishino, and Y. Kawada.** 1998. Alterations in the GyrA subunit of DNA gyrase and the ParC subunit of DNA topoisomerase IV associated with quinolone resistance in *Enterococcus faecalis*. *Antimicrob. Agents Chemother.* **42:**433–435.

57. **Kato, J., Y. Nishimura, R. Imamura, H. Niki, S. Hiraga, and H. Suzuki.** 1990. New topoisomerase essential for chromosome segregation in *E. coli*. *Cell* **63:**393–404.

58. **Kato, J., H. Suzuki, and H. Ikeda.** 1992. Purification and characterization of DNA topoisomerase IV in *Escherichia coli*. *J. Biol. Chem.* **267:**25676–25684.

59. **Korten, V., W. M. Huang, and B. E. Murray.** 1994. Analysis by PCR and direct DNA sequencing of *gyrA* mutations associated with fluoroquinolone resistance in *Enterococcus faecalis*. *Antimicrob. Agents Chemother.* **38:**2091–2094.

60. **Luong, T. T., S. W. Newell, and C. Y. Lee.** 2003. *mgr*, a novel global regulator in *Staphylococcus aureus*. *J. Bacteriol.* **185:**3703–3710.

61. **Margolles, A., M. Putman, H. W. Van Veen, and W. N. Konings.** 1999. The purified and functionally reconstituted multidrug transporter LmrA of *Lactococcus lactis* mediates the transbilayer movement of specific fluorescent phospholipids. *Biochemistry* **38:**16298–16306.

62. **Markham, P. N.** 1999. Inhibition of the emergence of ciprofloxacin resistance in *Streptococcus pneumoniae* by the multidrug efflux inhibitor reserpine. *Antimicrob. Agents Chemother.* **43:**988–989.

63. **Markham, P. N., M. Ahmed, and A. A. Neyfakh.** 1996. The drug-binding activity of the multidrug-responding transcriptional regulator BmrR resides in its C-terminal domain. *J. Bacteriol.* **178:**1473–1475.

64. **Markham, P. N., and A. A. Neyfakh.** 1996. Inhibition of the multidrug transporter NorA prevents emergence of norfloxacin resistance in *Staphylococcus aureus*. *Antimicrob. Agents Chemother.* **40:**2673–2674.

65. **Martínez-Martínez, L., A. Pascual, and G. A. Jacoby.** 1998. Quinolone resistance from a transferable plasmid. *Lancet* **351:**797–799.

66. **Mitscher, L. A., P. Devasthale, and R. Zavod.** 1993. Structure-activity relationships, p. 3–51. *In* D. C. Hooper and J. S. Wolfson (ed.), *Quinolone Antimicrobial Agents*. American Society for Microbiology, Washington, D.C.

67. **Morais Cabral, J. H., A. P. Jackson, C. V. Smith, N. Shikotra, A. Maxwell, and R. C. Liddington.** 1997. Crystal structure of the breakage-reunion domain of DNA gyrase. *Nature* **388:**903–906.

68. **Morrissey, I., and J. T. George.** 2000. Purification of pneumococcal type II topoisomerases and inhibition by gemifloxacin and other quinolones. *J. Antimicrob. Chemother.* **45:**101–106.

69. **Muñoz, R., and A. G. de la Campa.** 1996. ParC subunit of DNA topoisomerase IV of *Streptococcus pneumoniae* is a primary target of fluoroquinolones and cooperates with DNA gyrase A subunit in forming resistance phenotype. *Antimicrob. Agents Chemother.* **40:**2252–2257.

70. **Nagai, K., T. A. Davies, B. E. Dewasse, M. R. Jacobs, and P. C. Appelbaum.** 2001. Single- and multi-step resistance selection study of gemifloxacin compared with trovafloxacin, ciprofloxacin, gatifloxacin and moxifloxacin in *Streptococcus pneumoniae*. *J. Antimicrob. Chemother.* **48:**365–374.

71. **Nakanishi, N., S. Yoshida, H. Wakebe, M. Inoue, and S. Mitsuhashi.** 1991. Mechanisms of clinical resistance to fluoroquinolones in *Enterococcus faecalis*. *Antimicrob. Agents Chemother.* **35:**1053–1059.

72. **Neyfakh, A. A.** 1992. The multidrug efflux transporter of *Bacillus subtilis* is a structural and functional homolog of the *Staphylococcus* NorA protein. *Antimicrob. Agents Chemother.* **36:**484–485.

73. **Neyfakh, A. A., C. M. Borsch, and G. W. Kaatz.** 1993. Fluoroquinolone resistance protein NorA of *Staphylococcus aureus* is a multidrug efflux transporter. *Antimicrob. Agents Chemother.* **37:**128–129.

74. **Ng, E. Y., M. Trucksis, and D. C. Hooper.** 1996. Quinolone resistance mutations in topoisomerase IV: relationship of the *flqA* locus and genetic evidence that topoisomerase IV is the primary target and DNA gyrase the secondary target of fluoroquinolones in *Staphylococcus aureus*. *Antimicrob. Agents Chemother.* **40:**1881–1888.

75. **Ng, E. Y., M. Trucksis, and D. C. Hooper.** 1994. Quinolone resistance mediated by *norA*: physiologic characterization and relationship to *flqB*, a quinolone resistance locus on the *Staphylococcus aureus* chromosome. *Antimicrob. Agents Chemother.* **38:**1345–1355.

76. **Nikaido, H., and D. G. Thanassi.** 1993. Penetration of lipophilic agents with multiple protonation sites into bacterial cells: tetracyclines and fluoroquinolones as examples. *Antimicrob. Agents Chemother.* **37:**1393–1399.

77. **Ohki, R., and M. Murata.** 1997. *bmr3*, a third multidrug transporter gene of *Bacillus subtilis*. *J. Bacteriol.* **179:**1423–1427.

78. **Ohshita, Y., K. Hiramatsu, and T. Yokota.** 1990. A point mutation in *norA* gene is responsible for quinolone resistance in *Staphylococcus aureus*. *Biochem. Biophys. Res. Commun.* **172:**1028–1034.

79. **Pan, X. S., J. Ambler, S. Mehtar, and L. M. Fisher.** 1996. Involvement of topoisomerase IV and DNA gyrase as ciprofloxacin targets in *Streptococcus pneumoniae*. *Antimicrob. Agents Chemother.* **40:**2321–2326.

80. **Pan, X. S., and L. M. Fisher.** 1996. Cloning and characterization of the *parC* and *parE* genes of *Streptococcus pneumoniae* encoding DNA topoisomerase IV. Role in fluoroquinolone resistance. *J. Bacteriol.* **178:**4060–4069.

81. **Pan, X. S., and L. M. Fisher.** 1998. DNA gyrase and topoisomerase IV are dual targets of clinafloxacin action in *Streptococcus pneumoniae*. *Antimicrob. Agents Chemother.* **42:**2810–2816.

82. **Pan, X. S., and L. M. Fisher.** 1997. Targeting of DNA gyrase in *Streptococcus pneumoniae* by sparfloxacin: selective targeting of gyrase or topoisomerase IV by quinolones. *Antimicrob. Agents Chemother.* **41:**471–474.

83. **Pan, X. S., G. Yague, and L. M. Fisher.** 2001. Quinolone resistance mutations in *Streptococcus pneumoniae* GyrA

and ParC proteins: mechanistic insights into quinolone action from enzymatic analysis, intracellular levels, and phenotypes of wild-type and mutant proteins. *Antimicrob. Agents Chemother.* **45**:3140–3147.

84. **Paulsen, I. T., M. H. Brown, T. G. Littlejohn, B. A. Mitchell, and R. A. Skurray.** 1996. Multidrug resistance proteins QacA and QacB from *Staphylococcus aureus*: membrane topology and identification of residues involved in substrate specificity. *Proc. Natl. Acad. Sci. USA* **93**: 3630–3635.

85. **Perichon, B., J. Tankovic, and P. Courvalin.** 1997. Characterization of a mutation in the *parE* gene that confers fluoroquinolone resistance in *Streptococcus pneumoniae*. *Antimicrob. Agents Chemother.* **41**:1166–1167.

86. **Pestova, E., J. J. Millichap, F. Siddiqui, G. A. Noskin, and L. R. Peterson.** 2002. Non-PmrA-mediated multidrug resistance in *Streptococcus pneumoniae*. *J. Antimicrob. Chemother.* **49**:553–556.

87. **Piddock, L. J., M. M. Johnson, S. Simjee, and L. Pumbwe.** 2002. Expression of efflux pump gene *pmrA* in fluoroquinolone-resistant and -susceptible clinical isolates of *Streptococcus pneumoniae*. *Antimicrob. Agents Chemother.* **46**:808–812.

88. **Putman, M., L. A. Koole, H. W. Van Veen, and W. N. Konings.** 1999. The secondary multidrug transporter LmrP contains multiple drug interaction sites. *Biochemistry* **38**: 13900–13905.

89. **Putman, M., H. W. Van Veen, J. E. Degener, and W. N. Konings.** 2001. The lactococcal secondary multidrug transporter LmrP confers resistance to lincosamides, macrolides, streptogramins and tetracyclines. *Microbiology* **147**:2873–2880.

90. **Putman, M., H. W. Van Veen, and W. N. Konings.** 2000. Molecular properties of bacterial multidrug transporters. *Microbiol. Mol. Biol. Rev.* **64**:672–693.

91. **Rolston, K. V., S. Frisbee-Hume, B. M. LeBlanc, H. Streeter, and D. H. Ho.** 2002. Antimicrobial activity of a novel des-fluoro (6) quinolone, garenoxacin (BMS-284756), compared to other quinolones, against clinical isolates from cancer patients. *Diagn. Microbiol. Infect. Dis.* **44**: 187–194.

92. **Roychoudhury, S., C. E. Catrenich, E. J. McIntosh, H. D. McKeever, K. M. Makin, P. M. Koenigs, and B. Ledoussal.** 2001. Quinolone resistance in staphylococci: activities of new nonfluorinated quinolones against molecular targets in whole cells and clinical isolates. *Antimicrob. Agents Chemother.* **45**:1115–1120.

93. **Schmitz, F. J., B. Hofmann, B. Hansen, S. Scheuring, M. Lückefahr, M. Klootwijk, J. Verhoef, A. Fluit, H. P. Heinz, K. Köhrer, and M. E. Jones.** 1998. Relationship between ciprofloxacin, ofloxacin, levofloxacin, sparfloxacin and moxifloxacin (BAY 12-8039) MICs and mutations in *grlA, grlB, gyrA* and *gyrB* in 116 unrelated clinical isolates of *Staphylococcus aureus*. *J. Antimicrob. Chemother.* **41**:481–484.

94. **Schmitz, F. J., M. E. Jones, B. Hofmann, B. Hansen, S. Scheuring, M. F. A. Lückefahr, J. Verhoef, U. Hadding, H. P. Heinz, and K. Köhrer.** 1998. Characterization of *grlA, grlB, gyrA*, and *gyrB* mutations in 116 unrelated isolates of *Staphylococcus aureus* and effects of mutations on ciprofloxacin MIC. *Antimicrob. Agents Chemother.* **42**: 1249–1252.

95. **Shen, L. L., W. E. Kohlbrenner, D. Weigl, and J. Baranowski.** 1989. Mechanism of quinolone inhibition of DNA gyrase. Appearance of unique norfloxacin binding sites in enzyme-DNA complexes. *J. Biol. Chem.* **264**: 2973–2978.

96. **Sreedharan, S., M. Oram, B. Jensen, L. R. Peterson, and L. M. Fisher.** 1990. DNA gyrase *gyrA* mutations in ciprofloxacin-resistant strains of *Staphylococcus aureus*: close similarity with quinolone resistance mutations in *Escherichia coli*. *J. Bacteriol.* **172**:7260–7262.

97. **Sugino, A., C. L. Peebles, K. N. Kreuzer, and N. R. Cozzarelli.** 1977. Mechanism of action of nalidixic acid: purification of *Escherichia coli nalA* gene product and its relationship to DNA gyrase and a novel nicking-closing enzyme. *Proc. Natl. Acad. Sci. USA* **74**:4767–4771.

98. **Sun, L., S. Sreedharan, K. Plummer, and L. M. Fisher.** 1996. NorA plasmid resistance to fluoroquinolones: role of copy number and *norA* frameshift mutations. *Antimicrob. Agents Chemother.* **40**:1665–1669.

99. **Takahata, M., M. Yonezawa, S. Kurose, N. Futakuchi, N. Matsubara, Y. Watanabe, and H. Narita.** 1996. Mutations in the *gyrA* and *grlA* genes of quinolone-resistant clinical isolates of methicillin-resistant *Staphylococcus aureus*. *J. Antimicrob. Chemother.* **38**:543–546.

100. **Takei, M., H. Fukuda, R. Kishii, and M. Hosaka.** 2001. Target preference of 15 quinolones against *Staphylococcus aureus*, based on antibacterial activities and target inhibition. *Antimicrob. Agents Chemother.* **45**:3544–3547.

101. **Takenouchi, T., C. Ishii, M. Sugawara, Y. Tokue, and S. Ohya.** 1995. Incidence of various *gyrA* mutants in 451 *Staphylococcus aureus* strains isolated in Japan and their susceptibilities to 10 fluoroquinolones. *Antimicrob. Agents Chemother.* **39**:1414–1418.

102. **Tankovic, J., F. Mahjoubi, P. Courvalin, J. Duval, and R. Leclercq.** 1996. Development of fluoroquinolone resistance in *Enterococcus faecalis* and role of mutations in the DNA gyrase *gyrA* gene. *Antimicrob. Agents Chemother.* **40**:2558–2561.

103. **Tran, J. H., and G. A. Jacoby.** 2002. Mechanism of plasmid-mediated quinolone resistance. *Proc. Natl. Acad. Sci. USA* **99**:5638–5642.

104. **Trucksis, M., J. S. Wolfson, and D. C. Hooper.** 1991. A novel locus conferring fluoroquinolone resistance in *Staphylococcus aureus*. *J. Bacteriol.* **173**:5854–5860.

104a. **Truong-Bolduc, Q. C., P. M. Dunman, J. Strahilevitz, S. J. Projan, and D. C. Hooper.** 2005. MgrA is a multiple regulator of two new efflux pumps in *Staphylococcus aureus*. *J. Bacteriol.* **187**:2395–2405.

105. **Truong-Bolduc, Q. C., X. Zhang, and D. C. Hooper.** 2003. Characterization of NorR protein, a multifunctional regulator of *norA* expression in *Staphylococcus aureus*. *J. Bacteriol.* **185**:3127–3138.

106. **Van Veen, H. W., K. Venema, H. Bolhuis, I. Oussenko, J. Kok, B. Poolman, A. J. M. Driessen, and W. N. Konings.** 1996. Multidrug resistance mediated by a bacterial homolog of the human multidrug transporter MDR1. *Proc. Natl. Acad. Sci. USA* **93**:10668–10672.

107. **Varon, E., C. Janoir, M. D. Kitzis, and L. Gutmann.** 1999. ParC and GyrA may be interchangeable initial targets of some fluoroquinolones in *Streptococcus pneumoniae*. *Antimicrob. Agents Chemother.* **43**:302–306.

108. **Wang, J. C.** 2002. Cellular roles of DNA topoisomerases: a molecular perspective. *Nat. Rev. Mol. Cell Biol.* **3**:430–440.

109. **Wang, J. C.** 1996. DNA topoisomerases. *Annu. Rev. Biochem.* **65**:635–692.

110. **Wang, M. G., J. H. Tran, G. A. Jacoby, Y. Y. Zhang, F. Wang, and D. C. Hooper.** 2003. Plasmid-mediated quinolone resistance in clinical isolates of *Escherichia coli* from Shanghai, China. *Antimicrob. Agents Chemother.* **47**: 2242–2248.

111. **Wang, T., M. Tanaka, and K. Sato.** 1998. Detection of *grlA* and *gyrA* mutations in 344 *Staphylococcus aureus* strains. *Antimicrob. Agents Chemother.* **42:**236–240.

112. **Wasserman, R. A., and J. C. Wang.** 1994. Mechanistic studies of amsacrine-resistant derivatives of DNA topoisomerase. II. Implications in resistance to multiple antitumor drugs targeting the enzyme. *J. Biol. Chem.* **269:** 20943–20951.

113. **Wetzstein, H. G., N. Schmeer, and W. Karl.** 1997. Degradation of the fluoroquinolone enrofloxacin by the brown rot fungus *Gloeophyllum striatum*: identification of metabolites. *Appl. Environ. Microbiol.* **63:**4272–4281.

114. **Wilkinson, B. J.** 1997. Biology, p. 1–38. *In* K. B. Crossley and G. L. Archer (ed.), *The Staphylococci in Human Disease.* Churchill Livingstone, New York, N.Y.

115. **Willmott, C. J., and A. Maxwell.** 1993. A single point mutation in the DNA gyrase A protein greatly reduces binding of fluoroquinolones to the gyrase-DNA complex. *Antimicrob. Agents Chemother.* **37:**126–127.

116. **Woolridge, D. P., N. Vazquez-Laslop, P. N. Markham, M. S. Chevalier, E. W. Gerner, and A. A. Neyfakh.** 1997. Efflux of the natural polyamine spermidine facilitated by the *Bacillus subtilis* multidrug transporter Blt. *J. Biol. Chem.* **272:**8864–8866.

117. **Yamagishi, J. I., T. Kojima, Y. Oyamada, K. Fujimoto, H. Hattori, S. Nakamura, and M. Inoue.** 1996. Alterations in the DNA topoisomerase IV *grlA* gene responsible for quinolone resistance in *Staphylococcus aureus*. *Antimicrob. Agents Chemother.* **40:**1157–1163.

118. **Yoshida, H., M. Bogaki, S. Nakamura, K. Ubukata, and M. Konno.** 1990. Nucleotide sequence and characterization of the *Staphylococcus aureus* norA gene, which confers resistance to quinolones. *J. Bacteriol.* **172:**6942–6949.

119. **Yu, J. L., L. Grinius, and D. C. Hooper.** 2002. NorA functions as a multidrug efflux protein in both cytoplasmic membrane vesicles and reconstituted proteoliposomes. *J. Bacteriol.* **184:**1370–1377.

120. **Zechiedrich, E. L., and N. R. Cozzarelli.** 1995. Roles of topoisomerase IV and DNA gyrase in DNA unlinking during replication in *Escherichia coli*. *Genes Dev.* **9:** 2859–2869.

121. **Zheleznova, E. E., and R. G. Brennan.** 2000. Crystal structure of the transcription activator BmrR bound to DNA and a drug. *Nature* **409:**378–382.

122. **Zheleznova, E. E., P. N. Markham, A. A. Neyfakh, and R. G. Brennan.** 1999. Structural basis of multidrug recognition by BmrR, a transcription activator of a multidrug transporter. *Cell* **96:**353–362.

Index

Abscess(es)
  dental, 336
  kidney, animal models, 538
  staphylococcal, *color illustration for Chapter 41, Figure 1*
N-Acetylglucosamine, 120
ActA protein, *L. monocytogenes*, 650, 652–653
Actinomyces, 738–749
  extracellular matrix adherence, 743
  fimbriae, 739–742
  levansucrase activity, 743–744
  neuraminidase, 742–743
  oral, 738–744
Actinomyces pyogenes, 745–747
Adhesins, 89
  streptococci, 101–104, 350–352
Adhesion molecule, staphylococcal invasion, 520–521
Adhesive glycoproteins, 91
AFM, 7
agr system
  affecting virulon, *color illustration for Chapter 41, Figure 3*
  staphylococci, 501, 566–567
    specificity groups, 503
    translational regulation, 502–503
Albumin
  binding to group C streptococci, 216
  binding to group G streptococci, 216
Allelic replacement
  linear DNA, 61–62
Alpha C protein, group B streptococci, 175
Alpha-toxin, *S. aureus*, 464–465
Alveoli, *S. pneumoniae* interactions, 256
Aminoglycosides, 591
  resistance, 591
Amniotic fluid, group B streptococcal infection, 153
Aneurinibacillus thermoaerophilus, 9
Animal models
  staphylococcal disease, 535–543
  vaccination strategies for staphylococcal disease, 539
Anthrax, 659–671
  action of toxin, *color illustration for Chapter 54, Figure 2*
  and systemic diseases, 659–660
  EdTx and, 662
  genome, 660–661
  -host interactions, 662–663
  LeTx and, 661–662
  proteins, structures, *color illustration for Chapter 54, Figure 1*
Antibiotic resistance
  enterococci, 315–317

S. aureus, 439
  staphylococci, 529, 587–597
Antibiotic susceptibility
  S. caprae, 577
  S. haemolyticus, 574–575
  S. lugdunensis, 572
  S. saprophyticus, 574
  S. schleiferi, 575–579
  S. simulans, 575–579
Antibiotics
  histotoxic clostridial infection, 722
  in group A streptococcal infections, 147–148
  staphylococcal infections, 539–540
Antimyosin antibodies, in rheumatic fever, 78, 82
Antiphagocytic factors
  group C streptococci, 215–216
  group G streptococci, 215–216
Antistaphylococcal drugs, development, 593
Antistaphylococcal therapies, animal models, 539
Antistreptococcal/antimyosin MAb 3.B6
  reaction with endothelium and myocardium, *color illustration for Chapter 7, Figure 4*
Antistreptococcal/antimyosin monoclonal antibodies, 75, 76, 77
Apoptosis
  host cells, 106–107
  nocardia-induced, confocal micrographs, *color illustration for Chapter 61, Figure 6*
  staphylococci and, 521
Arcanobacterium haemolyticum, 744–745
arlAB system, affecting virulon, *color illustration for Chapter 41, Figure 3*
Arthritis
  group B streptococci, 187
  group C streptococci, 224
  group G streptococci, 224
  staphylococcal
    animal models, 537–538
Asymptomatic carriage, group A streptococci, 144, 148
Autolysin, *L. monocytogenes*, 648–650

β-Lactam resistance, 589
  Bacillus, 774
  coagulase-negative staphylococci, 776
  enterococci, 774
  penicillin-binding protein-mediated, 774–775
  S. aureus, 772–774, 776
  S. pneumoniae, 777

β-Lactam resistance *(continued)*
  spread, 778
  staphylococci, 589
  viridans streptococci, 777
β-Lactamase, 770–772
β-Lactam(s)
  antibiotic resistance, 769–781
  mechanism of action, 769–770
  tolerance, 777–778
Bacillus anthracis, 659–671
  capsule, 664
  exosporium, 664
  genetic tools, 666–667
  -macrophage interactions, 663–664
  physiology, morphology, and taxonomy, 660
  S-layer, 664
  Sterne strain, 667
  surface structures, 664–665
  toxins, 661–662
  vaccines, 667
  virulence gene expression, 665–666
Bacillus licheniformis, 4
Bacillus subtilis, 67, 312
  cell wall, 4, 6, 8
  peptidoglycan, 7
  quinolone resistance, 828
Bacillus thuringiensis, 7, 8
Bacteremia
  enterococci, 302–303
  group B streptococci, 186
  group C streptococci, 225
  group G streptococci, 225
  staphylococci, 526–527
Bacterial reactive oxidants, staphylococcal, 551
Bacteriophages
  attachment sites, genome distribution, 125–128
  group A streptococci, 50, 123–142
    bacteriophage A25, 123–124
    bacteriophage CS112, 125
    bacteriophage SF370, 125, 128, 135, 136, 137
    bacteriophage SP24, 134
    bacteriophage T12, 125
    diversity, modular exchange, 128–131
    early studies, 123
    late phage genes, 131
    lytic phages, 123–124
    temperate phages, 124–138
    virulence and, 124, 132–138
  S. aureus, 421
  S. aureus NCTC 8325, 382–410
  S. pneumoniae, 281–282
Beta C protein, group B streptococci, 176
Beta-toxin, *S. aureus*, 465–466
Biofilms
  formation by enterococci, 302–304
  formation by *S. epidermidis*, 561
    accumulation process, 563–564
    extracellular matrix proteins in, 562–563
    initial attachment, 561–562
  S. sanguinis, 352
  staphylococcal virulon expression, 508
Blood-brain barrier, *S. pneumoniae* interactions, 257–259
Bloodstream, *S. pneumoniae* access and survival, 256–257
Bloodstream infection, staphylococci, 528
Body fluids, antistaphylococcal action, 549
Bone infections, streptococci, 224
Botulinum neurotoxin
  detection, 692
  gene arrangement, 693–695
  inhibition of neurotransmission, 696–698
  safety in working with, 699
  structure, 695–696

Botulinum toxin, biological warfare agent, 690
Botulism. *See also Clostridium botulinum*
  clinical aspects, 688–691
  treatment, 690
  wound, 688
BPS protein, 177
Brain, nocardial invasion, 757–760

c protein, group B streptococci, 159
C5a peptidase
  group B streptococci, 177
  group C streptococci, 201
  group G streptococci, 201
  streptococci, 34, 119
CAMP phenomenon, group B streptococci, 156–157, 159–160
Capsular polysaccharide
  group A streptococci, 37–46
    attachment to epithelial cells and, 42
    capsule and host immune defenses, 40–41
    capsule as virulence factor, 39
    capsule production, 39
    entry into epithelial cells and, 43
    experimental infection models, 41–42
    mucosal colonization and, 42
  group B streptococci, 157–158, 169–174
    vaccine based on, 177–178
  S. pneumoniae, 241–252
    background, 241
    biosynthesis genes, 242–243
    biosynthesis in different pneumococcal serotypes, 243
    capsular transformation in vivo, 246–247
    cps loci in serotypes, compared, 243–244, 245
    regulation of production, 247–249
    type 3 and, 37
    type 3 CPS biosynthesis genes, 242–243
    vaccine based on, 242
    virulence and, 241–242
Carbohydrate
  -containing structures, *S. pneumoniae*, 269–270
  peptides and, immunological cross-reactivity, 82
Carbohydrate antigens, group B streptococci, 169–174
  capsular polysaccharide, 169–174
  genetics, 173–174
  group B carbohydrate, 169–170
  sialic acid, 171–173
Carbohydrate metabolism, staphylococci, 427–433
Carbon catabolite repression, staphylococci, 430–431
Carbon metabolism, lactococci, 358
Cardiolipin, 38
Cassette chromosome, staphylococci, 376–377
CbpA/polymeric immunoglobulin receptor, 255–256
CD14, 104
CD46, 32, 101
CD44-mediated tissue invasion, capsule and, 42–43
Cell division, staphylococci, 44
  murein hydrolases in, 451–452
Cell surface-associated proteins, 197–201
Cell wall
  chemistry
    peptidoglycan, 7
    secondary polymers, 8
  electron microscopy, 8
  gram-positive bacteria, 3–11
  mycobacteria, 9–10
  S. aureus, 4, 5, 8, 443–455
  S. epidermidis, 3, 4
  S. pneumoniae, 230–240
    anatomy, 230
    choline residue functions, 237–238
    composition and autolysis, 232
    covalent modifications, 237

growth zone and cell wall segregation, 231–232
host-related activities of wall components, 237
peptidoglycan, 232–236
teichoic and lipoteichoic acids, 236–237
S-layered, 10
scanning electron microscopy, 3–4
staphylococci, 443–455
changing image of cell walls, 443
complex functions, 451–452
historical overview, 443
methicillin-conditional mutants, 449–450
structure, 12, 13
techniques to study
AFM, 7
conventional embedding, 5
freeze-etching, 4
freeze-substitution, 5
hydrated cryosections, 5–7
negative staining, 5
shadow-casting, 3
transmission electron microscopy, 3, 4
turnover, 8–9
ultrastructure, 3–10
Cellulitis
group A streptococci, 59
group B streptococci, 186
Chimeric reporter protein, 69
Chloramphenicol, 592
Chloramphenicol acetyltransferase, 39
Chloramphenicol resistance, 592, 675, 682
Choline
pneumococcal surface, immunohistochemical/schematic
depiction, *color illustration for Chapter 21, Figure 1*
S. pneumoniae, 237, 270
Choline-binding protein, *S. pneumoniae*, 237–238
Choline-independent strains, 238
Choline phosphorylcholine, biology in *S. pneumoniae*, 271–272
Chromatography, high-performance liquid, 232
Chromosomal virulence genes, 499
Chromosome cassettes, staphylococcal, 376–377, 421–422
Clostridia
cell wall, 8
enterotoxic, 703–714
genetics, 672–687
histotoxic, 715–725
characteristics, 715
emerging infections, 715
exotoxins, 716–718
immune response, 719
microbiological niche, 715
pathogenesis of infections, 718–719
prevention, 722
shock and organ failure, 720
tissue hypoxia beyond site of initial trauma, 719–720
treatment of gas gangrene, 720–722
virulence factors, 716
neurotoxigenic, 688–702
microbiology, 691–692
phylogenetics, 672
Clostridium botulinum, 672–674
genetic manipulation, 674–675
genetics, 672–674
genomics, 698–699
Clostridium difficile, 703–714
bacteriology, 708–709
chloramphenicol resistance, 682
disease
pathogenesis, 709
prevention, 712
therapy, 712
erythromycin resistance, 682
genetic manipulation, 683–684

genetics, 682–684
S-layer, 9, 10
tetracycline, 682
toxin, epidemiology, 709
virulence factors, 710
Clostridium perfringens, 691, 703–714
antibiotic resistance, 675–677
food poisoning, 703–707
genetic manipulation, 681–682
genetics, 675–682
microbiology, 691–692
non-food-borne disease, 707–708
Clostridium tetani, 690
genetic manipulation, 675
genetics, 674–675
genomics, 698–699
Coagulase, staphylococci, 478–479
Coagulase-negative staphylococci, 782, 797
Cofilin, 652
Collagens, 89–91
binding to gram-positive pathogens, 92–93
binding to staphylococci, 489
Competence, development in streptococci, 348
Competence pheromones, mitis group streptococci, 348
Complement, streptococcal inhibitor, 53
Conjugation, *S. pneumoniae*, 281
Conjugative transposon, enterococci, 321–323
Corynebacterium diphtheriae, 726–737
genome sequence, 732–733
structure of 2[Ni DtrR(C102D)]$_2$-tox operator complex, *color
illustration for Chapter 59, Figure 2*
virulence factors, iron-sensitive expression, 727
cps genes, *S. pneumoniae*, 242–245
Cross-reactive antigens, group A streptococci, 74–88
historical perspective, 74–75
monoclonal antibodies to identify, 77–82
novel protein antigens, 82
recognition of myosin by monoclonal antibodies, 75–77
Crp protein, *L. monocytogenes*, 637
csr genes, group A streptococci, 39–40
C-terminal-anchored surface proteins, 12–20
C-terminal anchor region, 15, 18
M protein, 12–13
microscopy, 15
multifunctional nature, 20
seven-residue periodicity, 18–19
size variation, 19–20
surface-exposed region, 17, 18, 19
wall-associated region, 17
C-terminal end, M protein sequence, *color illustration for
Chapter 2, Figure 5*
Cysteine proteases, *S. aureus*, 480–481
Cytokines, response to *L. monocytogenes*, 520–521
Cytolytic toxin, group A streptococci, 52
Cytotoxicity, *S. pneumoniae*, 262

Dactylocycline, 812
Daptomycin, 589–590
Delta-toxin, *S. aureus*, 466–467
Dental abscess, 336
Dental caries, oral streptococci, 335–336
Dental plaque, metabolism, 335
Dermatitis, *S. aureus*, infiltrations of inflammatory
cells, *color illustration for Chapter 44, Figure 1*
Diabetes, group B streptococcal infections, 187
Diagnostics, staphylococci, 371–372
Diarrhea, *C. difficile*, 709
Diphtheria, 726–737
epidemiology, 733–734
Diphtheria toxin, 730–733
fusion protein toxins, 732

Diphtheria toxin repressor
    activation, 728
    crystallographic analyses, 728–730
    fusion protein toxin, 727–730
Directed mutagenesis, group A streptococci, 61–65
    allelic replacement using linear DNA, 61–62
    directed insertional inactivation, 62–63
    in-frame deletion, 63–65
DNA, linear, allelic replacement, 61–62
DNA gyrase, 825
DNA repair, *S. pneumoniae*, 280
Dpn restriction system, *S. pneumoniae*, 282–283

Ear, middle, *S. pneumoniae* invasion, 255
E-cadherin, 650
Ecology, group A streptococci, 143–144
    genotypic markers related to, 146
Elastin
    binding to gram-positive pathogens, 93
    structure and function, 91
Electron microscopy
    cell wall, 8
    C-terminal-anchored surface proteins, 15
Electron transport
    lactococci, 359
    S. aureus, 8
Embedding, study of cell wall, 5
emm genes, group A streptococci, 144–146
Endocarditis
    enterococci, 303–304
    group B streptococci, 187
    group G streptococci, 214, 223, 225–226
    L. monocytogenes, 604
    S. epidermidis, 560
    S. sanguinis, 352
    staphylococci, 426–427
        animal models, 536
Endophthalmitis, enterococci, 304
Endothelial cells
    interaction with staphylococci, 517–525
        clinicopathologic correlations, 521
        consequences of invasion, 520–521
        intracellular survival of bacteria, 519
        invasion, 518–519
        staphylococcal adherence, 517
    invasion by group B streptococci, 153–155
Endothelium, MAb 3.B6, antistreptococcal/antimyosin
        reaction with, *color illustration for Chapter 7,*
        Figure 4
α-Enolase, 20
Enolase, surface, 177
Enterococci
    β-lactam resistance, 774, 776–777
    biofilm formation, 302–304
    biology, 300
    colonization and translocation, 300–302
    conjugative transposons, 321–323
    environmental persistence, 300
    epidemiology, 300
    future perspectives, 306
    gene expression regulation, 313–315
    genetics, 312–331
        replication, inheritance, and repair, 312–313
    glycopeptide dependence, 793
    glycopeptide resistance, 786–793
    immune evasion, 304
    integrons, 323, 324
    lipoteichoic acid, 301, 304–305
    non-pheromone-responsive plasmids and, 321
    pathogenicity, 299–311
    pheromone-responsive plasmids, 317–321
    surface proteins, 14

    tissue damage, 304–305
    toxic metabolites, 306
    vancomycin resistance, 315–316, 786
    virulence factors, 299, 300
Enterococcus faecalis, 67, 69, 70, 299, 301, 302, 306
Enterococcus faecium, 299, 301, 306. *See also*
        Enterococci
Environment
    effect on virulence of *S. aureus*, 500–501
    lactococcal defense against, 361–363
    S. epidermidis, 566
Epidemiology
    enterococci, 300
    group B streptococci, 186–195
    group C streptococci, 222
    group G streptococci, 222
Epithelial cells
    adherence of group B streptococci, 152–153
    attachment of group A streptococci, 42
    entry of group A streptococci, 43
    entry of group B streptococci, 153–155
    interactions with pneumococci, 253–255
    staphylococcal adherence, 518
Erythromycin, 226
Erythromycin resistance, 675, 682
Escherichia coli, 61, 156, 177, 232
Evolution
    emm genes of group A streptococci, 144–146
    group A streptococci, 146–147
Exfoliative toxins, *S. aureus*, 471–472
Exosporium, *B. anthracis*, 664–445
Exotoxin, *S. aureus*, 664–445
Extracellular enzymes
    S. epidermidis, 564
    staphylococci, 478–485
Extracellular matrix, 89–92
    degradation by gram-positive pathogens, 95
    interaction with gram-positive pathogens, 92–95
    structure and function of matrix molecules, 89–92
Eye infection, staphylococci, animal models, 537

FAME
    staphylococci, 482
Fatty acid-modifying enzyme
    staphylococci, 482
FbpA, *L. monocytogenes*, 650
Fc receptors, endothelial cells invaded by staphylococci, 520
Fibrinogen
    binding to gram-positive pathogens, 94–95
    binding to staphylococci, 489
    structure and function, 92
Fibronectin
    binding to gram-positive pathogens, 93–94
    binding to group B streptococci, 152
    binding to group C streptococci, 199–200
    binding to group G streptococci, 199–200
    binding to staphylococci, 488
    structure and function, 91
Fluorescent microscopy
    L. monocytogenes infection of Vero cells, *color illustration for*
        Chapter 53, Figure 5
    surface proteins, 15, 17
Fluoroquinolone resistance, 591
Fluoroquinolones, 590–591, 822–823
Fn-binding proteins, *S. pyogenes*, 31–32
Focal adhesion kinase, phosphorylated, 33
Food-grade strains, lactococci, 364
Food poisoning
    C. botulinum, 688–689
    C. perfringens type A, 703–707
    L. monocytogenes, 602
    staphylococci, 468–469

Freeze-etching, cell wall, 4
Freeze-substitution, cell wall, 5
Fur-like protein, *S. epidermidis*, 567
Furuncle, cutaneous staphylococcal, *color illustration for Chapter 41, Figure 1*
Fusidic acid resistance, staphylococci, 593

Galactose, utilization in staphylococci, 429–430
Gamma-toxin, *S. aureus*, 468–469
Gas gangrene, 720–722
Gastrointestinal disease, *L. monocytogenes*, 605
Gastrointestinal tract, enterococci, 300–302
Gene expression
    group A streptococci, 69–71
        heterologous expression, 71
    L. monocytogenes, 621–627
Gene knockout mouse models, staphylococcal disease, 538–539
Gene regions, group IV prophages, *color illustration for Chapter 11, Figure 8*
Gene transfer, group A streptococci, 146
Genetic exchange
    group A streptococci, 59–60
    staphylococci, 413
Genetic map, *S. aureus*, 381–410
Genetic markers, *S. aureus*, 409–410
Genetics
    antibiotic resistance in staphylococci, 588
    enterococci, 312–331
    group A streptococci, 59–73
    group C streptococci, 196–212
    group G streptococci, 196–212
    lactococci, 356–368
    S. aureus, 413–426
    S. pneumoniae, 275–288
        capsular polysaccharide, 242
    S. sanguinis, 347–355
    staphylococcal capsule, 456–457
    staphylococci, 449–451
Genome distribution, bacteriophage attachment sites, 125–126
Genome NCTC 8325
    circular map, *color illustration for Chapter 32, Figure 3*
    role category assignments for ORFs for, *color illustration for Chapter 32, Figure 4*
Genome sequences
    group B streptococci, 161, 177
    S. pyogenes, 125–128
Genomes
    S. pneumoniae, 262, 275–276
    whole, compared using bitsum model, *color illustration for Chapter 32, Figure 2*
Gentamicin, 226
gfp gene, *L. monocytogenes*, 629
Glomerulonephritis, acute, 59, 74, 81, 223–224
Glucose, utilization in staphylococci, 429
β-Glucuronidase, *L. monocytogenes*, 629
Glyceraldehyde-3-phosphate dehydrogenase, 358, 359
Glycerol ester hydrolase, staphylococci, 481–482
Glycolipids, *Nocardia*, 760–761
Glycolytic enzymes/pathways
    staphylococci, 427
    surface proteins, 20
Glycopeptide dependence, enterococci, 793
Glycopeptide resistance, 590, 782–800
    coagulase-negative staphylococci, 797
    enterococci, 786
        genetic and molecular basis, 788–789
        laboratory detection, 787–788
        origin of resistance genes, 793–794
        van genes, 789–793
    mechanisms, 783–785
    S. aureus, 794–795, 797
    vancomycin, 795–797

Glycopeptides
    structure and mechanism of action, 782–783
Glycoproteins, adhesive, 91
Gram-positive bacteria
    collagen binding to, 92–93
    elastin binding to, 93
    fibrinogen binding to, 94–95
    laminin binding to, 93
    surface proteins on, 12–25
        mucin interaction, 101
    thrombospondin binding to, 95
    vitronectin binding to, 94
Gram-positive pathogens
    extracellular matrix degradation by, 95
    extracellular matrix interactions, 92–95
    fibronectin and, 93–94
Green fluorescent protein, 69
Group A streptococcal disease, 47–48, 59, 143–144
Group A streptococcal pharyngitis, 144
    vaccine, 113–122
Group A streptococci
    asymptomatic carriage, 144, 148
    attachment to epithelial cells, 42
    bacteriophages, 123–142
        bacteriophage A25, 123–124
        bacteriophage CS112, 125
        bacteriophage SF370, 125, 128, 135, 136, 137
        bacteriophage SP24, 134
        bacteriophage T12, 125
        diversity, modular exchange, 128–131
        early studies, 123
        late phage genes, 131
        lytic phages, 123–124
        temperate phages, 124–138
        virulence and, 124, 132–138
    capsular polysaccharide, 37–46, 82
        capsule and host immune defenses, 40–41
        capsule as virulence factor, 39
        capsule production, 39
        experimental infection models, 41–42
    cross-reactive antigens, 74–88
        historical perspective, 74–75
        monoclonal antibodies to identify, 77–82
        novel protein antigens, 82
        recognition of myosin by monoclonal antibodies, 75–77
    cytolytic toxins, 52
    directed mutagenesis, 61–65
        allelic replacement using linear DNA, 61–62
        directed insertional inactivation, 62–63
        in-frame deletion, 63–65
    ecology, 143–144
        genotypic markers related to, 146
    emm genes, 144–146
    entry into epithelial cells, 43
    evolution, 146–147
    flow of genetic information in natural populations, 138–139
    gene expression
        analysis, 69–71
        heterologous expression, 71
    gene transfer, 146
    genetic exchange, 59–60
    genetics, 59–73
    hyaluronidase, 124
    integrins, 104–105
    mucosal colonization, 42
    plasmid technology, 60–61
    pyrogenic exotoxins, 51–52. *See also* Streptococcal pyrogenic exotoxins
    resistance to phagocytosis, 40–41
    serological markers, 143
    streptokinase, 204
    superantigens, 47–58
    surface proteins, 100

Group A streptococci (*continued*)
  systemic disease, 143–144
  toxins, 47–58
  transduction, 59–60, 138–139
  transformation, 60
  transmission dynamics, 147–148
  transposon mutagenesis, 65–69
    analysis of mutants, 68–69
    Tn916, 66–67
    Tn917, 67
    Tn4001, 67–68
  vaccine, 113–122
  virulence factors, 41–42, 124
Group B alphalike protein, 176
Group B streptococcal disease
  future aspects, 192
  in pregnancy, 188–190
    duration of colonization, 188
    risk factors for maternal carriage, 188–189
    sites of maternal colonization, 188
  neonate, 189–191
    colonization, 189
    early-onset prevention, 191–192
    infections, 189
    prevention, 190–191
    risk factors, 190
  nonpregnant adults, 186–188
    incidence, 186
    risk factors, 187–188
    syndromes, 186–187
  nosocomial, 188
  sepsis syndrome, 160–161
Group B streptococci
  adherence to epithelial surfaces, 152–153
  avoidance of immune clearance, 157–160
  beta C protein, 176
  c protein, 159
  CAMP phenomenon, 156–157, 159–160
  capsular polysaccharide, 157–158, 169–174
  carbohydrate antigens, 169–174
  disease incidence by age, 186, 187
  epidemiology, 186–195
  glutamine synthetase, 177
  group B alphalike protein, 175–176
  hemolysins, 155
  hyaluronate lyase, 156
  injury to host cellular barriers, 155–157
  invasion of epithelial and endothelial cells, 153–155
  pathogenic mechanisms, 152–168
  protein antigens, 173–177
  R proteins, 176
  Rib protein, 176
  sialic acid, 157
  surface structures, 169–185
  syndromes caused by, 186–187
  tandem repeat-containing proteins, 174–176
  transposon insertion sites, 158
  vaccines, 177–181
    capsular polysaccharide, 177
    clinical trials with Ia-TT and Ib-TT vaccines, 180
    clinical trials with II-TT vaccine, 180
    clinical trials with III-TT vaccine, 180
    clinical trials with V-TT and V-CRM197 vaccines, 180–181
    polysaccharide-protein conjugate, 178–181
  virulence factors, 152–168
  X protein, 177
Group C streptococcal disease
  animal infections, 223
  bacteremia, 225
  deep tissue infections, 225
  human infections, 223–226
  in pregnancy, 224–225
  invasive disease, 224

neonate, 224–225
  pharyngitis, 223–224
  skin and soft tissue infections, 224
  treatment, 226
Group C streptococci
  adherence mechanisms, 214–215
  albumin-binding proteins, 216
  antiphagocytic factors, 215–216
  bacteriophage C1, 124
  C5a peptidase, 216
  classification, 196–197
  collagen-binding proteins, 215
  cytoplasmic membrane-associated enzymes, 201–202
  enzymes, 216–217
  epidemiology, 222
  extracellular proteins, 202–204
  fibronectin-binding proteins, 199–200
  genetics, 196–212
  hyaluronan synthase, 201
  immunoglobulin-binding protein, 198–199, 216
  lipoprotein acid phosphatase, 201–202
  M and M-like proteins, 197–198, 216
  morphologic groups, 213
  pathogenicity factors, 196–212, 213–221
  plasmin(ogen)-binding proteins, 200–201
  protein G, 216
  streptodornase, 204
  streptokinase, 202–204, 217
  streptolysin, 204, 217
  surface proteins, 197–201
  taxonomy, 213–214, 222
  toxins, 216–217
  vitronectin-binding proteins, 215
Group F streptococci, pyrogenic toxin, 52
Group G streptococcal disease
  animal infections, 223
  arthritis, 224
  bacteremia, 225
  endocarditis, 214, 223, 225–226
  human infections, 223–226
  invasive disease, 224
  neonate, 224–225
  pharyngitis, 214, 223–224
  skin and soft tissue infections, 224
  treatment, 226
Group G streptococci
  adherence mechanisms, 214–215
  albumin-binding proteins, 216
  antiphagocytic factors, 215–216
  C5a peptidase, 216
  classification, 196–197
  collagen-binding proteins, 215
  cytoplasmic membrane-associated enzymes, 201–202
  enzymes, 216–217
  epidemiology, 222
  extracellular proteins, 202–204
  fibronectin-binding proteins, 199–200
  genetics, 196–212
  hyaluronan synthase, 201
  immunoglobulin-binding protein, 198–199
  lipoprotein acid phosphatase, 201–202
  M and M-like proteins, 197–198, 216
  α_2-macroglobulin-binding proteins, 216
  pathogenicity factors, 196–212, 213–221
  plasmin(ogen)-binding proteins, 200–201
  protein G, 216
  streptodornase, 204
  streptokinase, 202–204, 217
  streptolysin, 204, 217
  surface proteins, 197–201
  taxonomy, 213–214, 222
  toxins, 216–217
  vitronectin, 216–217

Haemophilus influenzae, 247, 255
has genes, group A streptococci, 37–38
Health care-associated infections, staphylococcal, 528–529
Heme iron transport, sortase-anchored proteins and, 489
Heme source, lactococci, 360
Hemin auxotrophs, *S. aureus*, 438–439
Hemolysins, group B streptococci, 155
Hepatitis, *L. monocytogenes*, 604
Heterologous gene expression, group A streptococci, 71
Hex mismatch repair system, *S. pneumoniae*, 280
Hexose-uptake system, *L. monocytogenes*, 652
Host cell barriers, group B streptococci, 155
Host cell receptors, 100
    binding of streptococcal adhesions, 101–103
    S. pyogenes, 30
Host cell signaling, streptococcus-mediated, *color illustration for Chapter 9, Figure 1*, 100–112
    integrins in, 104
    membrane receptor-originated, 105–106
    targeting the nucleus, 105–106
Host defense
    against *S. epidermitis*, 564–565
Hyaluronan synthase
    group A streptococci, 37, 38
    group C streptococci, 201
    group G streptococci, 201
Hyaluronate lyase
    group B streptococci, 156
    staphylococci, 482–483
Hyaluronic acid
    biosynthesis, 37–38
        regulation, 38–40
    immunogenicity, 40
Hyaluronic acid synthase, 37
Hyaluronidase, group A streptococci, 124

Immune clearance, avoidance of group B streptococci, 157–160
Immunization, active, conserved-region peptides, 116–117
Immunoglobulin
    binding to group C streptococci, 198–199, 216
    binding to group G streptococci, 198–199, 216
Impetigo, 144
In-frame deletion, group A streptococci, 63–65
In utero infection, group B streptococci, 153
Infection
    primary, group A streptococci, 143–144
    skin. *See* Skin infections
    soft tissue. *See* Soft tissue infections
    staphylococcal
        animal models, 535–543
        community-acquired, 526–528
        epidemiology, 526–534
        health care-associated, 528–529
        in health care setting, 530
    streptococcal, non-M-protein and, 119–120
    wound, staphylococcal, animal models, 536–537
Inflammatory cells, *S. aureus* dermatitis, *color illustration for Chapter 44, Figure 1*
Inflammatory response
    S. epidermitis, 564–565
    S. pneumoniae disease, 259–262
    staphylococcal disease, 549–550
Insertion-duplication mutagenesis, *S. pneumoniae*, 283–284
Insertion sequence(s)
    C. perfringens, 678–681
    S. pneumoniae, 283
Integrin-linked kinase, 34
Integrins
    group A streptococci, 104
    in host cell signaling, 104–105
Integrons, enterococcal, 323, 324

Internalins (In1A)
    L. monocytogenes, 646–647
    receptor, 650
Invasins, *S. pyogenes*, 31–32
Iron
    acquisition by *S. epidermitis* inside host, 565

Kidney abscess, staphylococci, animal models, 538

Lactococci
    antigenicity, 364–365
    carbon metabolism, 358
    cell lysis system, 364
    conjugation, 363
    electron transport chain, 359
    expression strain, 364
    fermentation, 358–359
    food-grade strains, 364
    gene replacement, 363
    genetics, 356–368
    metabolic options, 357–361
    nucleotide metabolism, 361
    phylogenetic tree, 356–368
    promoters, 364
    protein export reporter, 364
    respiration, 359–361
        prototype, 360–361
    secretion and anchoring signals, 364
    site-specific single-copy integration, 363
    stress resistance, 361–363
    sugar metabolism, 358
    survival, respiration metabolism, 360
    transposition, 363
Lactococcus lactis, 356–358. *See also* Lactococci
Lactose, utilization in staphylococci, 429
Laminins
    binding to gram-positive pathogens, 93
    structure and function, 91
Lantibiotics, 567–568
Leukocidin, staphylococci, 527
LIM kinase, *L. monocytogenes*, 652
Lipoprotein acid phosphatase
    group C streptococci, 201–202
    group G streptococci, 201–202
Lipoproteins, surface proteins, 21
Lipoteichoic acid, 152
    enterococci, 301, 304–305
    S. pneumoniae, 236–237, 293
Listeria, 34
Listeria ivanovii, 634–635
Listeria monocytogenes
    ActA protein, 650
    bacteriophage, 629
    chromosomes, 620–621
    epidemiology, 601–608
    genetic tools, 620–633
    heterologous expression systems, 629
    host cell immune system, 653
    invasion and intracellular growth, 646–656
    lipid rafts, 651
    P35, 629–630
    plasmid vectors, 621
    regulatory proteins, 634
    reporter genes, 627–629
    U153, 630
    Vero cell infection, fluorescence microscopy, *color illustration for Chapter 53, Figure 5*
    virulence genes, 634–645
Listeria monocytogenes disease, 601–608
    clinical disease, 603
    cutaneous listeriosis, 604

Listeria monocytogenes disease (*continued*)
 diagnosis, 605
 endocarditis, 604
 entry into nonphagocytic cells, 646
 food-borne outbreaks, 602
 gastroenteritis, 605
 hepatitis, 604
 host cell interactions, 653
 immune response, 611–615
 meningoencephalitis, 603–604
 mouse model, 609–619
 musculoskeletal infection, 605
 peritonitis, 604
 sepsis, 604
 treatment, 605
Listeriosis. *See Listeria monocytogenes* disease
Liver abscess, *L. monocytogenes,* 604, 609
LLO, *L. monocytogenes,* 650, 652
LPXTG motif, 17
LPXTGase, 17
Lymphoma, cutaneous T-cell, 732
Lytic bacteriophages, group A streptococci, 123–124

M-like proteins
 group C streptococci, 197–198
 group G streptococci, 197–198
M protein sequence, C-terminal end, *color illustration for
  Chapter 2, Figure 5*
M proteins, 12–15, 37
 acute glomerulonephritis and, 81–82
 cross-reactive T cells and, 80–81
 group A streptococci, 143
 group C streptococci, 197–198, 216
 group G streptococci, 197–198, 216
 rheumatic fever and, 78–80
 S. pyogenes, 34
 structure and function, 113–114
MAb, mouse antistreptococcal, reaction with myocardial
  tissue section, *color illustration for Chapter 7, Figure 1*
MAb 3.B6, antistreptococcal/antimyosin
 reaction with endothelium and myocardium, *color illustration for
  Chapter 7, Figure 4*
Macrolide-lincosamide-streptogramin, 592
Macrolide-lincosamide-streptogramin resistance, 592
Major histocompatibility complex, 47, 49
Maltose, utilization in staphylococci, 430
Mannitol, utilization in staphylococci, 430
Mastitis, staphylococci, animal models, 536
Membrane-associated enzymes, cytoplasmic, 201–202
Meningitis, group B streptococci, 155
Meningoencephalitis, *L. monocytogenes,* 603–604
Metalloprotease, staphylococci, 481
Methicillin-conditional mutants, staphylococci, 449–450
Micrococcaceae, 371, 372
Mitogenic factor, 52
Modular exchange, phage diversity, 128–131
Molecular mimicry, 74
Monoclonal antibodies
 cross-reactive with group A streptococci, 74–77
 identification of streptococcal cross-reactive antigens, 77–82
Moraxella catarrhalis, 255
Mouse models, gene knockout models of staphylococcal
  disease, 538–539
MSCRAMMs, 488–489
Mucin, interactions with surface proteins, 101
Mucosal colonization, group A streptococci, 42
Mucosal vaccine, group A streptococcal pharyngitis, 115–116
Multifunctional surface proteins, 20
Multilocus enzyme electrophoresis, typing staphylococci, 374
Multilocus sequence typing, staphylococci, 375–376
Mupirocin resistance, staphylococci, 593
murMN operon, 232–235
Muropeptide, staphylococci, 446

Musculoskeletal infection, *L. monocytogenes,* 605
Mutagenesis
 directed. *See* Directed mutagenesis
 S. pneumoniae, 283–284
 transposon. *See* Transposon mutagenesis
Mycetoma, 752
Mycobacteria, cell walls, 9–10
Mycobacterium kansasii, 9
Mycobacterium tuberculosis, 9
Myocardium, MAb 3.B6, antistreptococcal/antimyosin
  reaction with, *color illustration for Chapter 7, Figure 4*
Myocardium tissue section, mouse antistreptococcal MAb
  reaction with, *color illustration for Chapter 7, Figure 1*
Myosin, antimyosin antibodies in streptococcal infection, 78

N-terminal-anchored surface lipoproteins, 21
Nasopharynx, epithelial cells, pneumococcal interactions, 253–
  255
NCTC 8325 genome
 circular map, *color illustration for Chapter 32, Figure 3*
 role category assignments for ORFs for, *color illustration for
  Chapter 32, Figure 4*
Necrotizing fasciitis, 59, 186, 214
Negative staining, cell wall, 5
Neonate
 group B streptococcal disease
  colonization, 189
  infections, 189
  prevention, 190–191
  risk factors, 189–190
 group C streptococcal disease, 224–225
 group G streptococcal disease, 224–225
Nephritis, staphylococci, animal models, 538
Newborn. *See* Neonate
Nocardia
 adherence and invasion of host cells, 757
 animal pathogens, 752–753
 apoptosis induced by, confocal micrographs, *color illustration for
  Chapter 61, Figure 6*
 genomic sequence, 762
 glycolipids, 760–761
 host factor protection, 762–763
 pathogenesis, 750–765
 tissue culture to study nocardia-host interactions, 753–757
Nocardia asteroides, 753–757, 759
Nocardiosis
 human, 750–752
NorA protein, *S. aureus,* 827–828
Nosocomial infection
 enterococci, 299–300
 group B streptococci, 186–187
 S. epidermitis, 560
Nuclease, staphylococci, 483
Nucleotide metabolism, lactococci, 361

Obsessive-compulsive disorder, 59
Oligosaccharide/oligonucleotide binding, 48
Opacity factor, 143
Oral streptococci
 acquisition, 334
 adherence to oral surfaces, 334–335
 as pathogens, 335
 colonization, 334–335
 dental abscesses, 336
 dental caries, 335–336
 names, 334
 pathogenesis, 332–339
 periodontal disease, 336
 systemic infections, 336–337
 taxonomy, 332–334
 virulence factors, 337

Osteomyelitis
staphylococci, 527
animal models, 536
Otitis media, *S. pneumoniae*, 253, 255
Oxazolidinones, 593

Pathogenicity factors
enterococci, 299–311
group B streptococci, 152–168
group C streptococci, 196–212, 213–221
group G streptococci, 196–212, 213–221
staphylococci, 332–339, 496–516
Pathogenicity islands, *S. aureus*, 409–410, 422–423
PCR typing, staphylococcal, 374–375
Penicillin
group A streptococcal infections, 147
in group C streptococcal infections, 226
in group G streptococcal infections, 226
prevention of early-onset group B streptococcal disease, 192
Penicillin-binding proteins, 230, 450–451, 774–775
Peptides
and carbohydrate, immunological cross-reactivity, 82
conserved-region, acute immunization, 116–117
Peptidoglycan, 12
chemistry, 7
electron microscopy, 6
S. pneumoniae
composition, 232–236
structure, 232
staphylococci
effect of antibiotics, 447
effect of growth phase and medium composition, 447–448
high-resolution analysis, 445–446
muropeptide composition, 446
S. aureus, 445–446
teicoplanin resistance and, 448
vancomycin resistance and, 448
variations in composition, 447–449
Peptidoglycan *N*-acetylglucosamine deacetylase, 237
Peptostreptococcus magnus, 14
Periodontal disease, oral streptococci, 336
Periplasm, gram-positive, 10
Phages. *See* Bacteriophages
Phagocytes, antistaphylococcal action, 544–547
oxygen-dependent, 545, 550–554
oxygen-independent, 546–547, 551–554
respiratory burst oxidase, 550–551
Phagocytosis
resistance in group A streptococci, 40–41
zipper phagocytosis of streptococci, 30–31
Pharyngitis
group A streptococci, 144
vaccine, 113–122
group C streptococci, 223–224
group G streptococci, 214, 223–224
Phase variation, *S. pneumoniae*, 268–274
characteristics of phase variants, 268–269, 271
contribution to host clearance mechanisms, 272
correlation between opacity variation and infection, 269
opacity variation and carbohydrate-containing structures, 269–270
variation in surface proteins, 270–271
Phenotypic variation, 268. *See also* Phase variation
Pheromone response, regulatory circuitry, *color illustration for Chapter 26, Figure 2*
Pheromone-responsive plasmids
enterococci, 317–321
Phosphocholine, teichoic acids, 293
Phosphoglycerate kinase, 20
Phosphoglycerate mutase, 20
Phosphoinositide, *L. monocytogenes*, 651–652
Phosphoryl choline residues, enzymatic removal, 237
Phosphorylcholine, *S. pneumoniae*, 237

Phosphotransferase system, staphylococci, 427–429
Phylogenetic relations, lactococci, 356, 357
PI-phospholipase C, 482
Placental membrane rupture, induced by group B streptococci, 156
Plasmid(s)
C. perfringens, 677–678
enterococci, 317–321
group A streptococci, 60–61
L. monocytogenes, 621–627
S. pneumoniae, 280, 281–283
staphylococci, 413–417, 499
conjugative plasmids, 416–417, 418
multiresistance plasmids, 415–416, 417
pSK639 family, 414–415
small rolling-circle plasmids, 413–414
Plasmin(ogen)
binding to group C streptococci, 200–201
binding to group G streptococci, 200–201
Platelet(s)
antistaphylococcal action, 547–549
interactions, staphylococcal, 518
Platelet kinocidins, 552–553
Platelet microbicidal proteins, 552–553
Pneumococcal C-polysaccharide and F-antigen, 235–236
Pneumococcal capsular polysaccharide vaccine, 242
Pneumococcal surface protein A, 230, 231
Pneumococci. *See Streptococcus pneumoniae*
Pneumolysin, 291–292
Pneumonia
group B streptococci, 186–187
staphylococci, 528
Polymers, secondary, 8
Polysaccharide-protein conjugate vaccine
group B streptococci, 177–180
S. pneumoniae, 290
Porphyromonas gingivalis, 20
Pregnancy
group B streptococcal disease, 187
risk factors for maternal carriage, 187
site of maternal colonization, 187
Premature delivery, response to group B streptococci, 156
PrfA protein, *L. monocytogenes*, 635–637, 639
Proinflammatory response, host cells after interaction with streptococci, 106
Promoters, *S. pneumoniae*, 283
Prophage genomes, decay, 136–138
Prophages
group IV, late gene regions, *color illustration for Chapter 11, Figure 8*
S. pyogenes, 125–128, 129–131
Proteases, staphylococci, 479–481
cysteine, 480–481
metalloprotease, 481
serine protease, 480
Protein(s)
ActA, *L. monocytogenes*, 650, 652–653
anthrax toxin, ribbon diagrams, *color illustration for Chapter 54, Figure 1*
cell surface-associated, 197–201
extracellular, 202–204
penicillin-binding, 230, 450–451, 774–775
secretion by group B streptococci, 156
surface. *See* Surface proteins
X, group B streptococci, 177
Protein A
S. aureus, 12, 486
staphylococci, 486
Protein vaccine, *S. pneumoniae*, 291
Pseudomonas aeruginosa, 101
Pseudomonas putida, 312
Pseudotransduction, *S. pneumoniae*, 281
Psoriasis, 59
PspA protein, 20–21, 291

PspC protein, 291
Pullulanase, 101
Pulmonary abscess, staphylococcal, *color illustration for Chapter 41, Figure 1*
Pyrogenic exotoxins, streptococci. *See* Streptococcal pyrogenic exotoxins

Quinolone resistance, 821–833

R proteins, group B streptococci, 176
Rel$_{Seq}$1-385 structure, *color illustration for Chapter 16, Figure 5*
Reporter genes
    group A streptococci, 69, 70
    L. monocytogenes, 627–629
Respiration
    L. lactis, 359–361
    S. aureus, 434–435, 439
    staphylococcal, 434–435
Respiratory tract
    S. pneumoniae in, 272–273
    S. pneumoniae invasion, 255–256
Restriction fragment length polymorphism, typing staphylococci, 374
Rheumatic fever, 59, 74
    M proteins and, 78–80
Rheumatic heart disease, 75, 81, 82
Ribosomal protection, tetracycline resistance, 811–812
Ribosomal RNA, mutation to tetracycline resistance, 810
Rifampin resistance, staphylococci, 590
RNA, ribosomal, mutation to tetracycline resistance, 810
RNA I-RNA II interaction, *color illustration for Chapter 26, Figure 4*

S-layer, 10
Sak, *S. aureus*, 479
sar locus, staphylococci, 505, 507
Scalded skin syndrome, staphylococcal, 471
Scanning electron microscopy, cell wall, 3–4
Scarlet fever, 59, 125
SCC*mec* elements in MRSA, *color illustration for Chapter 31, Figure 5*
SCPA, 34
Sdr protein, staphylococci, 489
Sepsis
    group B streptococci, 160–161
    S. aureus, animal models, 535–536
    S. pyogenes, 29
Septicemia, *S. epidermidis*, 560
Serine protease, staphylococci, 480
Serological markers, group A streptococci, 143
Shadow-casting, cell wall, 3
Sialic acid, group B streptococci, 157, 171–173
Signal transduction, membrane receptor-originated, 105–106
Signaling, host cell. *See* Host cell signaling
Sip protein, 176–177
Skin infections
    group C streptococci, 224
    group G streptococci, 224
    staphylococci, 526
        animal models, 536–537
Small-colony variants, *S. aureus*, 434–442
    antibiotic resistance, 439
    clinical studies, 438
    history, 435
    instability of phenotype, 439
    phenotypic changes and interrupted electron transport, 435–437
Soft tissue infections
    group C streptococci, 224
    group G streptococci, 224
    S. pyogenes, 224
    staphylococci, 526

Sortase, 17
Sortase A, staphylococcal, 486–488
Sortase A-anchored surface proteins, 488–489
Sortase A-dependent protein attachments, 237
Sortase-anchored proteins, and heme iron transport, 489
Sortase B, 489
Spleen, *L. monocytogenes*, 611
Sporosarcina, 8
srhRS system, affecting virulon, *color illustration for Chapter 41, Figure 3*
Staphylococcal abscesses, *color illustration for Chapter 41, Figure 1*
Staphylococcal disease
    animal models, 535–543
    arthritis
        animal models, 535–543
    bacteremia, 526–527
    bloodstream infections, 528
    coagulase-negative staphylococci, 782, 797
    community-acquired, 526–528
    endocarditis, 526–528
        animal models, 536
    epidemiology, 526–534
    eye infection, animal models, 537
    food-borne, 468–469
    gene knockout mouse models, 538–539
    host defense, 544–559
        antistaphylococcal action of body fluids, 549
        antistaphylococcal action of phagocytes, 549
        antistaphylococcal action of platelets, 547–549
        cathelicidins, 554
        cellular and extracellular defenses, 544–559
        defensins, 553–554
        group IIA PLA2, 551–552
        molecular, 550–554
        platelet microbicidal proteins, 552–553
    infection control, 530
    inflammatory response, 549–550
    kidney abscess, animal models, 558
    mastitis, animal models, 536
    nephritis, animal models, 538
    nosocomial infection, 526
    osteomyelitis, 527
        animal models, 536
    pathogenesis, 496–497
    pneumonia, 528
    S. aureus-eukaryotic cell interactions, 517–525
    scalded skin syndrome, 471
    sepsis, 535–536
    skin infections, 526
        animal models, 536–537
    soft tissue infections, 526
    surgical site infection, 528–529
    toxic shock syndrome, 527, 535
    toxin-mediated, 527
    transmission, 530
    treatment, 530–531, 539–540
    wound infection, animal models, 536–537
Staphylococci
    accessory gene regulatory organization, 508–511
    agr system, 501
        translational regulation, 502–503
    antibiotic resistance, 529, 587–597
        consequences, 587
        source of infection, 587–588
    apoptosis in mammary epithelial cells, 521
    capsular polysaccharide, 456–457
    capsule, 456–463
        biosynthesis, 458–460
        genetics, 456–457
        in virulence, 460
        regulation of capsule expression, 457–458
    carbohydrate catabolism, 427–433
    carbon catabolite repression, 430–431

cassette chromosome, 376–377, 421–422
cell division, 444
    murein hydrolases, 451–452
cell wall, 443–455
    changing image of cell walls, 443
    complex functions, 451–452
    historical overview, 443
    methicillin-conditional mutants, 449–450
coagulase, 478–479
coagulase-negative, β-lactam resistance, 776
collagen-binding proteins, 489
diagnostics, 371–372
extracellular enzymes, 478–485
FAME, 482
fatty acid-modifying enzyme, 482
fibrinogen-binding proteins, 489
fibronectin-binding proteins, 488
galactose utilization, 429–430
genetic exchange, 413
genetics, 413–426
glucose utilization, 429
glycerol ester hydrolase, 481–482
glycolytic pathways, 427
hyaluronate lyase, 482–483
infections
    community-acquired, 526–528
    epidemiology, 526–534
    health care-associated, 528–529
    in health care setting, 530
insertion sequences, 417–421
interaction with endothelial cells, 517–525
    adherence, 517–518
    clinicopathologic correlations, 521
    consequences of invasion, 520–521
    intracellular survival, 519
    invasion, 520–521
lactose utilization, 429–430
lipase, 481–582
maltose utilization, 430
mannitol utilization, 430
mupirocin resistance, 593
nuclease, 483
pathogenicity factors, 422–423, 496–516, 572–586
penicillin-binding proteins, 450–451
peptidoglycan
    effect of antibiotics, 447
    effect of growth phase and medium composition, 447–448
    high-resolution analysis, 445–446
    muropeptide composition, 446
    teicoplanin resistance and, 448
    vancomycin resistance and, 448
    variations in composition, 447–449
phosphotransferase system, 427–429
PI-phospholipase, 482
plasmids, 413–417, 499
    conjugative plasmids, 416–417, 418
    multiresistance plasmids, 415–416, 417
    pSK639 family, 414–415
    small rolling-circle plasmids, 413–414
protein A, 486
regulatory genes, 501
respiration, 434–435
streptogramin combination resistance, 592–593
sucrose utilization, 430
sulfonamide-trimethoprim resistance, 593
surface proteins, 486–495
    anchoring to cell wall, 486
    cell wall-associated, 489–491
    sortase A-anchored, 488–489
TCS modules, 501–505
transposons, 417–421, 499
typing, 372–378
    amplified fragment length polymorphism, 375

arbitrarily primed PCR, 374–375
combined use of molecular methods, 377–378
general aspects, 372–374
genotypic methods, 373–374, 378
multilocus enzyme electrophoresis, 374–375
PCR, 374–375
PCR for DNA flanked by insertion elements, 375
PCR of repetitive sequences, 375
PCR of rRNA gene spacer sequences, 375
phenotypic methods for S. aureus, 373
plasmid profile analysis, 374
requirements for typing systems, 372–373
restriction fragment length polymorphism, 374
sequence-based, 375–376
spa-sequence typing, 376, 377
vaccines, 460–461
virulon
    genetics, 497–499
    regulation, 499–511
xylose utilization, 430
Staphylococcus aureus
alpha-toxin, 464–465
antibiotic resistance, 439
β-lactam resistance, 772–774, 776
β-lactam tolerance, 777–778
beta-toxin, 465–466
bicomponent toxins, 467–468
carrier state, 530–531
cell wall, 4, 5, 8, 443–455
community-associated pneumonia, 528
delta-toxin, 466–467
dermatitis, infiltrations of inflammatory cells, *color illustration for Chapter 44, Figure 1*
electron transport, 437, 439
endothelial cell adherence, 517–518
endothelial cell invasion, 518–519
eukaryotic cell interactions, 517–525
exfoliative toxins, 471–472
exotoxins, 464–477
extracellular enzymes, 478–485
gamma-toxin, 467–468
genetic map, 381–410
genetic markers, 409–410
genotyping systems, 378
glycopeptide resistance, 794–795
hemin- or menaquinone-deficient strains, 437–438
leukocidin, 467–468
NCTC 8325 genome, 381–412
NorA, 827–828
pathogenicity islands, 409–410
peptidoglycan, 445–446
phenotyping methods, 373
protein A, 12, 486
respiration, 434–435, 439
small-colony variants, 434–442
    antibiotic resistance and, 439
    clinical studies, 438
    history, 435
    instability phenotype, 439
    phenotypic changes and interrupted electron transport, 435–437
staphylococcal pyrogenic toxin superantigen, 468–471
surface proteins, 12, 486–495
tissue site adherence, 518
toxins
    production, 439
    with superantigen activity, 468–471
vancomycin resistance, 448–449, 794–795, 797
virulon, regulation, 499–511
VISA isolates, 448
Staphylococcus capitis, 579–580
Staphylococcus caprae, 577
    antibiotic susceptibility, 577

Staphylococcus caprae (continued)
  diseases, 577
  genomic diversity and natural habitat, 577
  pathogenicity, 577
Staphylococcus carnosus, 582
Staphylococcus chromogenes, 582
Staphylococcus cohnii, 581
Staphylococcus epidermidis disease, 560–565
  endocarditis, 560
  host defense, 564–565
  inflammatory reaction, 564–565
  nosocomial infections, 560
  septicemia, 560
  spectrum, 560–561
Staphylococcus epidermitis
  agr system, 566–567
  biofilms, 561–564
    accumulation process, 563–564
    extracellular matrix proteins and, 562–563
    formation, 561
    initial attachment, 561–562
  cell wall, 3, 4
  DtxR homolog, 567
  environmental factors, 566
  extracellular enzymes, 564
  Fur-like protein, 567–568
  future aspects, 568
  iron acquisition inside host, 565
  lantibiotics, 567–568
  phase variation, 567
  sar locus, 567
  sigB operon, 566
  stress, 566
  toxins, 564
  virulence factors, 564–567
Staphylococcus gallinarum, 582
Staphylococcus haemolyticus, 574–575
  antibiotic susceptibility, 574
  diseases, 574
  genomic diversity and natural habitat, 574
  pathogenicity, 574–575
Staphylococcus hominis, 580
Staphylococcus hyicus, 582
Staphylococcus intermedius, 580–581
  diseases and habitat, 580
  genomic diversity, 581
  pathogenicity, 581
Staphylococcus lugdunensis, 572–574
  antibiotic susceptibility, 512
  diseases, 572
  genomic diversity and natural habitat, 572
  pathogenicity, 572–574
Staphylococcus pasteuri, 579
Staphylococcus saprophyticus, 574
  antibiotic susceptibility, 574
  diseases, 574
  genomic diversity and natural habitat, 574
  pathogenicity, 574
Staphylococcus schleiferi, 575–579
  antibiotic susceptibility, 575
  disease, 575
  genomic diversity and natural habitat, 575
  pathogenicity, 577–579
Staphylococcus sciuri, 581–582
Staphylococcus simulans, 579
  antibiotic susceptibility, 579
  diseases, 579
  genomic diversity and natural habitat, 579
  pathogenicity, 579
Staphylococcus warneri, 577
  diseases, 577
  pathogenicity, 577
Staphylococcus xylosus, 580

Sterne strain, *B. anthracis*, 667
Streptococcal genomes, 53, 347–348, 349
Streptococcal inhibitor of complement, 53
Streptococcal pyrogenic exotoxins, 51–52
  SPE A toxin, 50
  SPE B toxin, 51–52
  SPE C toxin, 52
  SPE F toxin, 52
  SSA toxin, 50
Streptococcal superantigen, 50, 53
Streptococcal surface dehydrogenase, 101–104
Streptococcal toxic shock syndrome, 47–48, 53
Streptococci
  focal adhesion and signal transduction, 33–34
  group A. *See* Group A streptococci
  group B. *See* Group B streptococci
  group C. *See* Group C streptococci
  group G. *See* Group G streptococci
  mediation of host signaling, *color illustration for Chapter 9,*
    Figure 1
  intracellular, 34–35
  mitis group
    competence pheromones, 348
    genomes, 347–348
  oral. *See* Oral streptococci
  salivarius group, 147
  sanguis group
    development, 348–349
    transformation, 349–350
  superantigens, 47–58
    T-cell activation complexes for, *color illustration for Chapter 5,*
    Figure 1
  tetracycline resistance, 808–810
  viridans
    β-lactam resistance, 777
    β-lactam tolerance, 777–778
  zipper phagocytosis, 30–31
Streptococcus agalactiae, 34, 186
Streptococcus anginosus, 213, 214, 222
Streptococcus canis, 222
Streptococcus constellatus, 214, 222
Streptococcus dysgalactiae, 14, 94, 205, 222, 225, 226
Streptococcus equi, 213, 222
Streptococcus equisimilis, 205, 213, 222, 226
Streptococcus gordonii, 352–353
  genome, 347–348
  streptococcal vaccine, 118–119
Streptococcus intermedius, 222
Streptococcus milleri, 214, 222
Streptococcus mitis, 214, 347, 348
Streptococcus mutans
  biofilm formation, 341–342
  dental caries reduction, 343–344
  genetic analysis, 340–341
  genome sequencing, 342–343
  interactions with other plaque bacteria, 342
  signaling pathways, 341
  surface proteins, 14
  tetracycline resistance, 810
  virulence factors, issues defining, 343
  virulence properties, 340–355
    identification, 340
Streptococcus oralis, 347
Streptococcus pneumoniae, 347
  β-lactam resistance, 777
  bacteriophage, 281–282
  blood-brain barrier interactions, 257–259
  bloodstream, access and survival, 256–257
  capsular polysaccharide, 241–252
    background, 241
    biosynthesis genes, 242–243
    biosynthesis in different pneumococcal serotypes, 243
    capsular transformation in vivo, 246–247

cps loci in serotypes, compared, 243, 248
  genetics, 242
  regulation of CPS production, 247–249
  types 3 and 37 CPS biosynthesis genes, 242–243
  vaccine based on, 242
  virulence and, 241–242
cell wall, 104, 230–240
  anatomy, 230
  choline residue functions, 237–238
  composition, 232
  covalent modifications, 237
  growth zone and cell wall segregation, 231–232
  host-related activities of wall components, 237
  peptidoglycan, 232–236
choline, *color illustration for Chapter 21, Figure 1*
choline-binding protein, 237–238
choline phosphorylcholine, biology, 271–272
conjugation, 281
cytotoxicity, 262
disease, 253
Dpn restriction system, 282–283
fluoroquinolone resistance, 591
gene transfer, 276–281
generation and analysis of mutants, 283–284
genetics, 275–288
genomes
  diversity and, 275–276
  pathogenesis, 262
inflammation, 259–262
invasion, 253–267
lipoteichoic acid, 236–237
middle ear invasion, 255–256
peptidoglycan, 232–236
phase variation, 268–274
  characteristics of phase variants, 268–269, 271
  contribution to host clearance mechanisms, 272
  correlation between opacity variation and infection, 269
  opacity variation and carbohydrate-containing structures, 269–270
  relation to other respiratory pathogens, 272–273
  variation in surface proteins, 270–271
promotors, 283
pseudotransduction, 281
quinolone resistance, 828
surface proteins, 20
teichoic acid, 230, 236–237, 293
transformation, 276–281
  artificial, 280–281
  chromosomal integration, 279–280
  DNA release, binding, and uptake, 279
  DNA repair, 280
  Hex mismatch repair system, 280
  induction of competence, 277–279
  mismatch repair, 280
  plasmid, 280
upper respiratory tract invasion, 255–256
vaccines, 289–298
  capsular polysaccharide vaccine, 242
  multivalent protein vaccine, 293
  noncapsular polysaccharide vaccine, 291–293
  pneumococcal capsular surface protein, 289–290
  pneumolysin, 291–292
  polysaccharide-protein conjugate, 290
  protein, 291
virulence factors, 241–242, 262
Streptococcus pyogenes, 12, 47, 60, 71, 94, 105, 113, 125, 215
  disease, 29
  fate of intracellular streptococci, 34–35
  Fn-binding proteins as invasins, 29
  host cell interactions leading to adherence and invasion, 30
  host receptors, 30

intracellular invasion, 29–30
  invasins, 30
  M proteins, 34
  surface proteins, 12
  tonsils as reservoir for, 30
Streptococcus salivarius, 147
Streptococcus sanguinis
  adhesion, 350–352
    multiple adhesive interactions, 350
  biofilms, 352
  competence pheromones, 348
  genetics, 347–355
  heterologous gene expression, 352–353
  pathogenesis of oral streptococci, 347
  surface proteins, 350–352
  transformation, 349–350
Streptococcus sobrinus, 14
Streptococcus suis, 14
Streptococcus zooepidemicus, 213, 222, 226
Streptodornase
  group C streptococci, 204
  group G streptococci, 204
Streptogramins, combinations, resistance in staphylococci, 592–593
Streptokinase
  group A streptococci, 204
  group C streptococci, 202–204
  group G streptococci, 202–204
Streptolysin
  group C streptococci, 204
  group G streptococci, 204
Streptolysin O, 52
Streptomyces, tetracycline resistance, 811–812
Stress resistance
  lactococci, 361–363
Sucrose, utilization in staphylococci, 430
Sugar metabolism, lactococci, 358
Sulfonamide-trimethoprim, 593
Superantigens
  group A streptococci, 47–58
  S. aureus toxins, 468–470
  streptococci, 47–58
    T-cell activation complexes for, *color illustration for Chapter 5, Figure 1*
Surface proteins
  anchored by charge and/or hydrophobic interactions, 20–21
  C-terminal-anchored proteins, 12–20
  glycolytic enzymes, 20
  group A streptococci, 100
  group B streptococci, 174–176, 177
  group C streptococci, 197–201
  group G streptococci, 197–201
  mucin interactions, 101
  multifunctional, 20
  on gram-positive bacteria, 12–25
  S. aureus, 12, 486–495
  S. pneumoniae, 20
  S. pyogenes, 12
  S. sanguinis, 350–352
  staphylococci, 486–495
    anchoring to cell wall, 486
    cell wall-associated, 489–491
    sortase A-anchored, 488–489
Surface structures, group B streptococci, 169–185
Sydenham chorea, 82–83
Systemic disease
  group A streptococci, 143–144
  oral streptococci, 336–337

T-cell activation complexes
  streptococcal superantigens, *color illustration for Chapter 5, Figure 1*

T-cell antigen receptor, 47
T-cell lymphoma, treatment, 732
T-cell responses, to *L. monocytogenes*, 613–615
T cells, cross-reactive, M proteins and, 80–81
T lymphocytes
    adhesion and extravasation into ARF valve in valvulitis, *color illustration for Chapter 7, Figure 5*
    cross-reactive, M proteins and, 80–81
Taxonomy
    group C streptococci, 213–214, 222
    group G streptococci, 213–214, 222
    oral streptococci, 332–334
TCS modules, affecting virulon, *color illustration for Chapter 41, Figure 3*
TCS modules, staphylococci, 501–505
Teichoic acid, 4, 5
    S. pneumoniae, 230, 236–237, 293
        structure, 236–237
Teichuronic acid, 5
Teicoplanin resistance, staphylococci, 448
Temperate bacteriophages, group A streptococci, 124–138
Tetanus, clinical aspects, 690–691
Tetanus neurotoxin
    detection, 692
    gene arrangement, 693–695
    inhibition of neurotransmission, 696–698
    regulation of synthesis, 693–695
    safety in working with, 699
    structure, 695–696
Tetracycline(s), 591
    chemical properties, 801
    entry into microbial cells, 801
    inhibition of protein synthesis, 802–803
Tetracycline resistance, 591–592
    by active efflux, 803–811
    by ribosomal protection, 808–810
    C. difficile, 682
    C. perfringens, 676–678
    drugs to combat, 812–814
    gram-positive bacteria, 801–820
    mutated rRNA, 810
    streptococci, 808–810
    Streptomyces, 812
Thrombospondin
    binding to gram-positive pathogens, 95
    structure and function, 92
Thromboxane A$_2$, 160
Tissue damage, enterococci, 304–305
Tn916
    genetic organization, *color illustration for Chapter 26, Figure 5*
    transposon mutagenesis, 66–67
Tonsils, reservoir for *S. pyogenes*, 30
Topoisomerase IV, 823–825
Topoisomerases, type 2, 821–822
Toxic shock syndrome, 214, 217
    animal models, 535–536
    group A streptococci, 47–48, 53
    staphylococci, 527
Toxins
    B. anthracis, 661–662
    C. botulinum, 672–674
    C. difficile, 709
    C. tetani, 674
    coagulase-negative staphylococci, 782
    group A streptococci, 47–58
    group C streptococci, 216–217
    group G streptococci, 216–217
    S. aureus, 467–468
    S. epidermidis, 564
Transduction
    B. anthracis, 666
    group A streptococci, 59–60, 138–139
    L. monocytogenes, 629

Transformation
    C. botulinum, 674
    C. perfringens, 681–682
    C. tetani, 675
    group A streptococci, 60
    S. pneumoniae, 276–281
        artificial transformation, 280–281
        chromosomal integration, 279–280
        DNA release, binding, and uptake, 279
        DNA repair, 280
        Hex mismatch repair system, 280
        induction of competence, 277–279
        mismatch repair, 280
        plasmid transformation, 280
    S. sanguinis, 349–350
Transmission
    group A streptococci, 147–148
    staphylococcal disease, 530
Transmission electron microscopy, cell wall, 3, 4, 5
Transposon(s)
    C. botulinum, 674
    C. difficile, 682–683
    C. perfringens, 678–681
    C. tetani, 675
    enterococci, 312, 321–323
    insertion sites
        group B streptococci, 158
        L. monocytogenes, 621
    staphylococci, 417–421, 499
Transposon mutagenesis
    C. perfringens, 682
    group A streptococci, 65–69
        analysis of mutants, 68–69
        Tn916, 66–67
        Tn917, 67
        Tn4001, 67–68
Treponema denticola, 214
Triosephosphate isomerase, 20
Tumor necrosis factor, response to streptococci, 106
Typing, staphylococci, 372–378
    amplified fragment length polymorphism, 375
    arbitrarily primed PCR, 374–375
    combined use of molecular methods, 377–378
    general aspects, 372–374
    genotypic methods, 373–374, 378
    multilocus enzyme electrophoresis, 374
    PCR, 374–375
    PCR for DNA flanked by insertion elements, 375
    PCR for repetitive sequences, 375
    PCR for rRNA gene spacer sequences, 375
    phenotypic methods for *S. aureus*, 373
    plasmid profile analysis, 374
    requirement for typing systems, 372–373
    restriction fragment length polymorphism, 374
    sequence-based, 375–376
    spa-sequence typing, 376, 377

Urinary tract infection, enterococci, 303

Vaccine(s)
    B. anthracis, 667
    delivery, *L. monocytogenes*, 615
    diphtheria, 726, 733
    group A streptococcal pharyngitis, 113–122
        gram-positive commensals as vectors, 117
        immunization with conserved region peptides, 116–117
        mucosal vaccine, 115–116
        new generation vaccine, 118–119
        non-M-protein approaches, 119–120
        passive protection, 116
        type-specific protection, 114–115
        vaccinia virus as vector, 117

group B streptococci, 177–181
    capsular polysaccharide vaccine, 177–178
    clinical trials with Ia-TT and Ib-TT vaccines, 180
    clinical trials with II-TT vaccine, 180
    clinical trials with III-TT vaccine, 180
    clinical trials with V-TT and V-CRM197 vaccines, 180–181
    polysaccharide-protein conjugate vaccine, 177–180
  S. pneumoniae, 289–298
    capsular polysaccharide vaccine, 242
    multivalent protein vaccine, 293
    pneumococcal capsular surface protein, 289–290
    pneumolysin, 291–292
    polysaccharide-protein conjugate vaccines, 290
    protein vaccines, 291
  staphylococci, 460–461
    strategies using animal models, 539
Vaccinia virus, vector for group A streptococcal vaccine, 117
Valvulitis
  adhesion and extravasation of T lymphocytes into ARF
    valve in, *color illustration for Chapter 7, Figure 5*
van genes, enterococci, 789–793
Vancomycin, 786
Vancomycin resistance, 786–787
  enterococci, 315–316
  staphylococci, 448–449
  VanA and VanB systems, *color illustration for Chapter 26,*
    Figure 1
Variable genetic elements, *S. aureus*, 410
Virulence factors
  C. difficile, 710
  C. diphtheriae, 727
  enterococcal, 299, 300
  group A streptococci, 53, 124, 134–136
  group B streptococci, 152–168

  oral streptococci, 337
  S. epidermidis, 564–567
  S. mutans, 343
  S. pneumoniae, 241–242, 262
  staphylococcal capsule, 460
Virulence genes
  B. anthracis, 665–666
  fixation, prophage genomes, 136–138
  L. ivanovii, 634–635
  L. monocytogenes, 634–645
  staphylococcal, 497–499
Virulon
  genetics, *S. aureus*, 497–499
  regulation, *S. aureus*, 499–511
  TCS modules affecting, *color illustration for Chapter 41, Figure 3*
Vitronectin
  binding to gram-positive pathogens, 94
  binding to group C streptococci, 215
  binding to group G streptococci, 215
  structure and function, 91–92

Wound botulism, 688
Wound infection, staphylococci, animal models, 536–537

X protein, group B streptococci, 177
Xylose, utilization in staphylococci, 430

Yersinia pseudotuberculosis, 30

Zipper phagocytosis, streptococci, 30–31